AF726499

Commonly Used Notations

Symbol	Meaning
α	Fine structure constant
a_0	Bohr radius (atomic unit)
a_s	Scattering length
A_e	Electron affinity
$\boldsymbol{B}$	Magnetic field vector
c	Velocity of light
c.c.	Complex conjugate
c.m.	Center of mass
CTMC	Classical trajectory Monte Carlo
Δ	Optical detuning $\omega_0 - \omega$
$\hat{\boldsymbol{\epsilon}}$	Polarization vector
e	Elementary charge
E_h	Hartree energy (atomic unit)
$\boldsymbol{E}$	Electric field vector
${}_2F_1(a, b; c; z)$	Hypergeometric function
h	Planck's constant ($\hbar = h/2\pi$)
H	Hamiltonian
IR	Infrared
I_P	Ionization potential
Im	Imaginary part
$\boldsymbol{k}$	Wave vector (magnitude $k = \|\boldsymbol{k}\|$)
k_B	Boltzmann constant
k_R	Rutherford coefficient $2\pi\alpha^4 a_0^2 m_e c^2 Z^2$
λ	Wavelength
λ_{dB}	de Broglie wavelength
λ_{mfp}	Mean free path
$\bar{\lambda}$	$\lambda/2\pi$
l, ℓ	Angular momentum quantum number
$\boldsymbol{L}$	Angular momentum vector
$\boldsymbol{\mu}$	Magnetic moment
μ_B	Bohr magneton
m_e	Electron mass
n	Principal quantum number
n_e	Electron density
N	Heavy particle density
ν	Frequency ($\omega = 2\pi\nu$)
ω_p	Plasma frequency
$\wp$	Principal value
$\boldsymbol{p}, \boldsymbol{P}$	Momentum vector
q	Generic charge
$\boldsymbol{r}$	Position vector
r_0	Classical electron radius
R_∞	Rydberg constant (infinite mass)
R_M	Rydberg constant (finite mass)
Re	Real part
T	Temperature
$\mathcal{T}$	Kinetic energy

Commonly Used Notations

Symbol	Meaning
TDHF	Time dependent Hartree–Fock
u	Atomic mass unit
UV	Ultraviolet
v	Velocity
v_B	Bohr velocity (atomic unit)
V	Potential energy
Z	Nuclear charge

Physical Quantities in Atomic Units

Quantity	Unit	Value
Length	a_0	$0.529\,177\,2108\,(18) \times 10^{-10}$ m
Mass	m_e	$0.910\,938\,26\,(16) \times 10^{-30}$ kg
Time	$\hbar/E_h$	$2.418\,884\,326\,505\,(16) \times 10^{-17}$ s
Velocity	$v_B \equiv \alpha c$	$2.187\,691\,2633\,(73) \times 10^6$ m s^{-1}
Energy	E_h	$4.359\,744\,17\,(75) \times 10^{-18}$ J
Action	$\hbar$	$1.054\,571\,68\,(18) \times 10^{-34}$ J s
Force	E_h/a_0	$0.823\,872\,25\,(14) \times 10^{-7}$ N
Power	$E_h^2/\hbar$	$0.180\,237\,811\,(31)$ W
Intensity	$\dfrac{E_h^2}{\hbar a_0^2}$	$64.364\,091\,(11) \times 10^{18}$ W m^{-2}
Charge	e	$1.602\,176\,53\,(14) \times 10^{-19}$ C
Electric potential	E_h/e	$27.211\,3845\,(23)$ V
Electric field	$\dfrac{E_h}{ea_0} = \dfrac{\alpha\hbar c}{ea_0^2}$	$0.514\,220\,642\,(44) \times 10^{12}$ V m^{-1}
Magnetic flux density	$\dfrac{E_h}{ea_0\alpha c}$	$2.350\,517\,42\,(20) \times 10^5$ T

Atomic Unit Definitions

Length:	$a_0 = \dfrac{4\pi\epsilon_0\hbar^2}{m_e e^2} = \dfrac{\hbar}{\alpha m_e c}$,
Velocity:	$v_B = \dfrac{e^2}{4\pi\epsilon_0\hbar} = \alpha c$,
Time:	$\tau_0 = \dfrac{16\pi^2\epsilon_0^2\hbar^3}{m_e e^4} = \dfrac{\hbar}{\alpha^2 m_e c^2}$,
Energy:	$E_h = \dfrac{e^2}{4\pi\epsilon_0 a_0} = \alpha^2 m_e c^2$,

Energy Conversion Factors

1 eV $= 1.602\,176\,53\,(14) \times 10^{-19}$ J
$= 2.417\,989\,40\,(21) \times 10^{14}$ Hz $\times h$
$= 8065.544\,45\,(69)$ cm$^{-1} \times hc$
$= 3.674\,932\,45\,(31) \times 10^{-2}\, E_h$
$= 1.160\,4505\,(20) \times 10^4$ K $\times k_B$
$= 96.485\,3383\,(83)$ kJ mol^{-1}

Springer Handbook of Lasers and Optics

Springer Handbooks provide a concise compilation of approved key information on methods of research, general principles, and functional relationships in physical sciences and engineering. The world's leading experts in the fields of physics and engineering will be assigned by one or several renowned editors to write the chapters comprising each volume. The content is selected by these experts from Springer sources (books, journals, online content) and other systematic and approved recent publications of physical and technical information.

The volumes are designed to be useful as readable desk reference books to give a fast and comprehensive overview and easy retrieval of essential reliable key information, including tables, graphs, and bibliographies. References to extensive sources are provided.

Springer Handbook of Lasers and Optics

Frank Träger (Ed.)

With CD-ROM, 978 Figures and 136 Tables

Springer

Editor:

Prof. Dr. Frank Träger
University of Kassel
Department of Physics
Heinrich-Plett-Str. 40
34132 Kassel
Germany

Library of Congress Control Number: 2007920818

ISBN-10: 0-387-95579-8 e-ISBN: 0-387-30420-7
ISBN-13: 978-0-387-95579-7 Printed on acid free paper

Production and typesetting: LE-TeX GbR, Leipzig
Handbook Manager: Dr. W. Skolaut, Heidelberg
Typography and layout: schreiberVIS, Seeheim
Illustrations:
schreiberVIS, Seeheim & Hippmann GbR, Schwarzenbruck
Cover design: eStudio Calamar Steinen, Barcelona
Cover production: WMXDesign GmbH, Heidelberg
Printing and binding: Stürtz AG, Würzburg

SPIN 10892328 100/3100/YL 5 4 3 2 1 0

Foreword

After working for more than four decades in the field of laser science, I am delighted that Springer-Verlag has devoted one of the first volumes of the new Springer Handbook series to lasers and optics. The exhilarating pace of technological advances in our field is still accelerating, and lasers and optical techniques are becoming ever more indispensible as enabling tools in almost any field of science or technology. Since no single physicist, engineer, or graduate student can be an expert in all the important subfields of optical science, a concise, balanced, and timely compilation of basic principles, key applications, and recent advances, written by leading experts, will make a most valuable desk reference. The chosen readable style and attractive, well-illustrated layout is even inviting to casual studying and browsing. I know that I will keep my Springer Handbook of Lasers and Optics close at hand, despite the infinite amount of information (and misinformation) that is readily accessible via the Internet.

Munich, February 2007 Theodor W. Hänsch

Prof. Theodor W. Hänsch
Nobel Laureate in Physics 2005
Department of Physics
Ludwig-Maximilians-Universität (LMU)
Munich, Germany

Preface

It is often said that the 21st century is the century of the photon. In fact, optical methods, materials, and components have reached an advanced state of sophistication hitherto unknown. Optical techniques, particularly those based on lasers, not only find applications in the classical fields of physics and engineering but have expanded into many other disciplines such as medicine, the life sciences, chemistry and environmental research, to mention only a few examples. Nevertheless, progress in optics, photonic materials and coherent light sources continues at a rapid pace: new laser materials are being developed; novel concepts such as optics far beyond the diffraction limit, or nanooptics, are being explored; and coherent light sources generate wavelengths in ranges not previously accessible.

In view of the pronounced interdisciplinary nature of optics, the *Springer Handbook of Lasers and Optics* is designed as a readable desk reference book to provide fast, up-to-date, comprehensive, and authoritative coverage of the field. The handbook chapters are grouped into four parts covering basic principles and materials; fabrication and properties of optical components; coherent and incoherent light sources; and, finally, selected applications and special fields such as terahertz photonics, X-ray optics and holography.

I hope that all readers will find this Springer Handbook useful and will enjoy using it.

Kassel, February 2007 Frank Träger

Prof. Frank Träger
Universität Kassel
Experimentalphysik I
Germany

List of Authors

Andreas Assion
Femtolasers Produktions GmbH
Fernkorngasse 10
Vienna, 1100, Austria
e-mail: *andreas.assion@femtolasers.com*

Thomas E. Bauer
JENOPTIK Polymer Systems GmbH
Coating Department
Am Sandberg 2
07819 Triptis, Germany
e-mail: *thomas.bauer@jenoptik-ps.de*

Thomas Baumert
Universität Kassel
Institut für Physik
Heinrich-Plett-Str. 40
34132 Kassel, Germany
e-mail: *baumert@physik.uni-kassel.de*

Dietrich Bertram
Philips Lighting
Philipsstr. 8
52068 Aachen, Germany
e-mail: *dietrich.bertram@philips.com*

Klaus Bonrad
Schott Spezialglas AG
Division Luminescence Technology
Hattenbergstr. 10
55014 Mainz, Germany
e-mail: *klaus.bonrad@schott.com*

Matthias Born
Philips Research Laboratories Aachen
Weisshausstr. 2
52066 Aachen, Germany
e-mail: *matthias.born@philips.com*

Annette Borsutzky
Technische Universität Kaiserslautern
Fachbereich Physik
Erwin-Schrödinger-Str. 49
67663 Kaiserslautern, Germany
e-mail: *borsu@uni-kl.de*

Hans Brand
Friedrich-Alexander-University of
Erlangen-Nürnberg LHFT
Department of Electrical, Electronic
and Communication Engineering
Cauerstr. 9
91058 Erlangen, Germany
e-mail: *hans@lhft.eei.uni-erlangen.de*

Robert P. Breault
Breault Research Organization, Inc.
6400 East Grant Road, Suite 350
Tucson, AZ 85715, USA
e-mail: *bbreault@breault.com*

Matthias Brinkmann
University of Applied Sciences Darmstadt
Mathematics and Natural Sciences
Schoefferstr. 3
64295 Darmstadt, Germany
e-mail: *brinkmann@fh-darmstadt.de*

Uwe Brinkmann
Bergfried 16
37120 Bovenden, Germany
e-mail: *ubri@arcor.de*

Robert Brunner
Carl Zeiss AG
Central Research and Technology
Carl-Zeiss-Promenade 10
07745 Jena, Germany
e-mail: *r.brunner@zeiss.de*

Geoffrey W. Burr
IBM Almaden Research Center
650 Harry Road
San Jose, CA 95120
e-mail: *burr@almaden.ibm.com*

Karsten Buse
University of Bonn
Institute of Physics
Wegelerstr. 8
53115 Bonn, Germany
e-mail: *kbuse@uni-bonn.de*

Carol Click
Schott North America
Regional Research and Development
400 York Avenue
Duryea, PA 18642, USA
e-mail: *carol.click@us.schott.com*

Hans Coufal[†]
IBM Research Division
San Jose, CA, USA

Mark J. Davis
Schott North America
Regional Research and Development
400 York Avenue
Duryea, PA 18642, USA
e-mail: *mark.davis@us.schott.com*

Wolfgang Demtröder
TU Kaiserslautern
Department of Physics
Am Harzhübel 80
67663 Kaiserslautern, Germany
e-mail: *demtroed@physik.uni-kl.de*

Henrik Ehlers
Laser Zentrum Hannover e.V.
Department of Thin Film Technology
Hollerithallee 8
30419 Hannover, Germany
e-mail: *h.ehlers@lzh.de*

Rainer Engelbrecht
Friedrich-Alexander-University of Erlangen-Nürnberg
Department of Electrical, Electronic and Communication Engineering
Cauerstr. 9
91058 Erlangen, Germany
e-mail: *rainer@lhft.eei.uni-erlangen.de*

Martin Fally
University of Vienna
Faculty of Physics, Department for Experimental Physics
Boltzmanngasse 5
Vienna, 1090, Austria
e-mail: *martin.fally@univie.ac.at*

Yun-Hsing Fan
University of Central Florida
College of Optics and Photonics
4000 Central Florida Blvd.
Orlando, 32816, USA
e-mail: *yfan@mail.ucf.edu*

Enrico Geißler
Carl Zeiss AG
Central Research and Technology
Carl Zeiss Promenade 10
07745 Jena, Germany
e-mail: *e.geissler@zeiss.de*

Ajoy Ghatak
Indian Institute of Technology Delhi
Physics Department
Hauz Khas
New Delhi, 110016, India
e-mail: *ajoykghatak@yahoo.com*

Alexander Goushcha
SEMICOA
333 McCormick Avenue
Costa Mesa, CA 92626, USA
e-mail: *ogoushcha@semicoa.com*

Daniel Grischkowsky
Oklahoma State University
Electrical and Computer Engineering
Engineering South 202
Stillwater, OK 74078, USA
e-mail: *grischd@ceat.okstate.edu*

Richard Haglund
Vanderbilt University
Department of Physics and Astronomy
6301 Stevenson Center Lane
Nashville, TN 37235-1807, USA
e-mail: *richard.haglund@vanderbilt.edu*

Stefan Hansmann
AI Technologies GmbH
Deutsche-Telekom-Allee 3
64295 Darmstadt, Germany
e-mail: *shansmann@al-technologies.de*

Joseph Hayden
Schott North America
Regional Research and Development
400 York Avenue
Duryea, PA 18642, USA
e-mail: *joe.hayden@us.schott.com*

Joachim Hein
Friedrich-Schiller University Jena
Institute for Optics and Quantum Electronics
Max-Wien-Platz 1
07743 Jena, Germany
e-mail: *jhein@ioq.uni-jena.de*

Stefan W. Hell
Max Planck Institute for Biophysical Chemistry
Am Fassberg 11
37077 Göttingen, Germany
e-mail: *hell@nanoscopy.de*

Jürgen Helmcke
Physikalisch-Technische Bundesanstalt (PTB)
Braunschweig
Former Head Quantum Optics and Length Unit
(retired)
Bundesallee 100
38116 Braunschweig, Germany
e-mail: *juergen.helmcke@ptb.de*

Hartmut Hillmer
University of Kassel
Institute of Nanostructure Technologies and
Analytics (INA)
Heinrich-Plett-Str. 40
34128 Kassel, Germany
e-mail: *hillmer@ina.uni-kassel.de*

Günter Huber
Universität Hamburg
Institut für Laser-Physik
Department Physik
Luruper Chaussee 149
22761 Hamburg, Germany
e-mail: *huber@physnet.uni-hamburg.de*

Mirco Imlau
University of Osnabrück
Department of Physics
Barbarastr. 7
49069 Osnabrück, Germany
e-mail: *mimlau@uos.de*

Kuon Inoue
Chitose Institute of Science and Technology
Department of Photonics
758-65 Bibi
066-8655 Chitose, Japan
e-mail: *inoue@guppy.chitose.ac.jp*

Thomas Jüstel
University of Applied Sciences Münster
Stegerwaldstr. 39
48565 Steinfurt, Germany
e-mail: *tj@fh-muenster.de*

Jeffrey L. Kaiser
Spectra-Physics
Division of Newport Corporation
1335 Terra Bella Avenue
Mountain View, CA 94043, USA
e-mail: *jeff.kaiser@spectra-physics.com*

Ferenc Krausz
Max Planck Institut für Quantenoptik
Hans-Kopfermann-Str. 1
85748 Garching, Germany
e-mail: *ferenc.krausz@mpq.mpg.de*

Eckhard Krätzig
University of Osnabrück
Physics Department
Barbarastr. 7
49069 Osnabrück, Germany
e-mail: *eckhard.kraetzig@uos.de*

Stefan Kück
Physikalisch-Technische Bundesanstalt
Optics Division
Bundesallee 100
38116 Braunschweig, Germany
e-mail: *stefan.kueck@ptb.de*

Anne L'Huillier
University of Lund
Department of Physics
P.O. Box 118
22100 Lund, Sweden
e-mail: *anne.lhuillier@fysik.lth.se*

Bruno Lengeler
Aachen University (RWTH)
II. Physikalisches Institut
Templergraben 55
52056 Aachen, Germany
e-mail: *lengeler@physik.rwth-aachen.de*

Martin Letz
Schott Glas
Materials Science, Central Research
Hattenbergstr. 1
55014 Mainz, Germany
e-mail: *martin.letz@schott.com*

Gerd Leuchs
University of Erlangen-Nuremberg
Institute of Optics, Information and Photonics
Guenther-Scharowsky-Str. 1
91058 Erlangen, Germany
e-mail: *leuchs@physik.uni-erlangen.de*

Norbert Lindlein
Friedrich-Alexander University of
Erlangen-Nürnberg
Max-Planck Research Group
Institute of Optics Information and Photonics
Staudtstr. 7/B2
91058 Erlangen, Germany
e-mail:
norbert.lindlein@optik.physik.uni-erlangen.de

Dennis Lo†
The Chinese University of Hong Kong
Hong Kong, P. R. China

Stefano Longhi
University of Politecnico di Milano
Department of Physics
Piazza Leonardo da Vinci 32
20133 Milano, Italy
e-mail: *longhi@fisi.polimi.it*

Ralf Malz
LASOS Lasertechnik GmbH
Research and Development
Carl-Zeiss-Promenade 10
07745 Jena, Germany
e-mail: *ralf.malz@lasos.com*

Wolfgang Mannstadt
Schott AG
Research and Technology Development
Hattenbergstr. 10
55122 Mainz, Germany
e-mail: *wolfgang.mannstadt@schott.com*

Gerd Marowsky
Laser-Laboratorium Göttingen e.V.
Hans-Adolf-Krebs-Weg 1
37077 Göttingen, Germany
e-mail: *gmarows@gwdg.de*

Dietrich Martin
Carl Zeiss AG
Corporate Research and Technology
Microstructured Optics Research
Carl-Zeiss-Promenade 10
07745 Jena, Germany
e-mail: *d.martin@zeiss.de*

Bernhard Messerschmidt
GRINTECH GmbH
Research and Development, Management
Schillerstr. 1
07745 Jena, Germany
e-mail: *messerschmidt@grintech.de*

Katsumi Midorikawa
RIKEN
Laser Technology Laboratory
Hirosawa 2-1, Wako
351-0198 Saitama, Japan
e-mail: *kmidori@riken.jp*

Gerard J. Milburn
The University of Queensland
Center for Quantum Computer Technology
School of Physical Sciences
St. Lucia, QLD 4072, Australia
e-mail: *milburn@physics.uq.edu.au*

Kazuo Ohtaka
Chiba University
Center for Frontier Science
Photonic Crystals
1-33 Yayoi
263-8522 Chiba, Japan
e-mail: *ohtaka@cfs.chiba-u.ac.jp*

Motoichi Ohtsu
Department of Electronics Engineering
The University of Tokyo
2-11-16 Yayoi, Bunkyo-ku
113-8656 Tokyo, Japan
e-mail: *ohtsu@ee.t.u-tokyo.ac.jp*

Roger A. Paquin
Advanced Materials Consultant
1842 E. Pole Star Place
Oro Valley, AZ 85737, USA
e-mail: *materials.man@att.net*

Alan B. Peterson
Spectra-Physics
Division of Newport Corporation
1335 Terra Bella Avenue
Mountain View, CA 94043, USA
e-mail: *alan.peterson@spectra-physics.com*

Klaus Pfeilsticker
Universität Heidelberg
Institut für Umweltphysik
Fakultät für Physik und Astronomie
Im Neuenheimer Feld 366
69121 Heidelberg, Germany
e-mail: *klaus.pfeilsticker@iup.uni-heidelberg.de*

Ulrich Platt
Universität Heidelberg
Institut für Umweltphysik
Fakultät für Physik und Astronomie
Im Neuenheimer Feld 366
69121 Heidelberg, Germany
e-mail: *ulrich.platt@iup.uni-heidelberg.de*

Markus Pollnau
University of Twente
MESA+ Institute for Nanotechnology
P.O. Box 217
7500 AE, Enschede, The Netherlands
e-mail: *m.pollnau@ewi.utwente.nl*

Steffen Reichel
SCHOTT Glas
Service Division Research
and Technology Development
Hattenbergstr. 10
55014 Mainz, Germany
e-mail: *steffen.reichel@schott.com*

Hans-Dieter Reidenbach
University of Applied Sciences Cologne
Institute of Communications Engineering
Institute of Applied Optics and Electronics
Betzdorfer Str. 2
50679 Cologne, Germany
e-mail: *hans.reidenbach@fh-koeln.de*

Hongwen Ren
University of Central Florida
College of Optics and Photonics
Central Florida Blvd. 162700
Orlando, FL 32816, USA
e-mail: *hren@mail.ucf.edu*

Detlev Ristau
Laser Zentrum Hannover e.V.
Department of Thin Film Technology
Hollerithallee 8
30419 Hannover, Germany
e-mail: *d.ristau@lzh.de*

Simone Ritter
Schott AG
Division Research and Technology Development
Material Development
Hattenbergstr. 10
55122 Mainz, Germany
e-mail: *simone.ritter@schott.com*

Evgeny Saldin
Deutsches Elektronen Synchrotron (DESY)
Notkestr. 85
22607 Hamburg, Germany
e-mail: *evgueni.saldin@desy.de*

Roland Sauerbrey
Forschungszentrum
Dresden-Rossendorf e.V.
Bautzner Landstr. 128
01328 Dresden, Germany
e-mail: *r.sauerbrey@fzd.de*

Evgeny Schneidmiller
Deutsches Elektronen Synchrotron (DESY)
Notkestr. 85
22607 Hamburg, Germany
e-mail: *evgeny.schneidmiller@desy.de*

Bianca Schreder
Schott Glas
Division Research and Technology Development
Material Development
Hattenbergstr. 10
55122 Mainz, Germany
e-mail: *bianca.schreder@schott.com*

Christian G. Schroer
Dresden University of Technology
Institute of Structural Physics
Zellescher Weg 16
01062 Dresden, Germany
e-mail: *schroer@physik.tu-dresden.de*

Markus W. Sigrist
ETH Zurich, Institute of Quantum Electronics
Department of Physics
Schafmattstr. 16
8093 Zurich, Switzerland
e-mail: *sigrist@iqe.phys.ethz.ch*

Glenn T. Sincerbox
University of Arizona
Optical Sciences
1630 East University Boulevard
Tucson, AZ 85721, USA
e-mail: *sinbox@cox.net*

Elisabeth Soergel
University of Bonn
Institute of Physics
Wegelerstr. 8
53115 Bonn, Germany
e-mail: *soergel@uni-bonn.de*

Steffen Steinberg
LASOS Lasertechnik GmbH
Carl-Zeiss-Promenade 10
07745 Jena, Germany
e-mail: *steffen.steinberg@lasos.com*

Sune Svanberg
Lund University
Division of Atomic Physics
P.O. Box 118
22100 Lund, Sweden
e-mail: *sune.svanberg@fysik.lth.se*

Orazio Svelto
Politecnico di Milano
Department of Physics
Piazza Leonardo da Vinci 32
20133 Milan, Italy
e-mail: *orazio.svelto@fisi.polimi.it*

Bernd Tabbert
Semicoa
Engineering Department
333 McCormick Avenue
Costa Mesa, CA 92626, USA
e-mail: *btabbert@semicoa.com*

K. Thyagarajan
Indian Institute of Technology Delhi
Physics Department
Hauz Khas
New Delhi, 110016, India
e-mail: *ktrajan@physics.iitd.ernet.in*

Mary G. Turner
Engineering Synthesis Design, Inc.
310 S. Williams Blvd.
Tucson, AZ 85711, USA
e-mail: *mary.t@engsynthesis.com*

Giuseppe Della Valle
Polytechnic Institute of Milan
Department of Physics
Piazza Leonardo da Vinci 32
20133 Milan, Italy
e-mail: *giuseppe.dellavalle@polimi.it*

Michael Vollmer
University of Applied Sciences Brandenburg
Department of Physics
Magdeburger Str. 50
14770 Brandenburg, Germany
e-mail: *vollmer@fh-brandenburg.de*

Silke Wolff
SCHOTT Spezialglas AG
Department of Research & Technology
Development, Material Development
Optical Glasses
Hattenbergstr. 10
55122 Mainz, Germany
e-mail: *silke.wolff@schott.com*

Matthias Wollenhaupt
Universität Kassel
Institut für Physik
Heinrich-Plett-Str. 40
34132 Kassel, Germany
e-mail: *wollenhaupt@physik.uni-kassel.de*

Shin-Tson Wu
University of Central Florida
College of Optics and Photonics
Central Florida Blvd. 162700
Orlando, FL 32816, USA
e-mail: *swu@mail.ucf.edu*

Helen Wächter
ETH Zurich, Institute of Quantum Electronics
Department of Physics
Schafmattstr. 16
8093 Zurich, Switzerland
e-mail: *waechter@phys.ethz.ch*

Mikhail Yurkov
Deutsches Elektronen Synchrotron (DESY)
Notkestr. 85
22607 Hamburg, Germany
e-mail: *mikhail.yurkov@desy.de*

Aleksei Zheltikov
M.V. Lomonosov Moscow State University
Physics Department
Vorobyevy gory
Moscow, 119992, Russia
e-mail: *zheltikov@phys.msu.ru*

Contents

FP	Fabry–Pérot
FR	Faraday rotator
FROG	frequency-resolved optical gating
FTS	Fourier-transform spectroscopy
FWHM	full width at half-maximum
FWM	four-wave mixing
FZP	Fresnel zone plate
FoM	figure of merit

G

GAC	grating assisted coupler
GASMAS	gas in scattering media absorption spectroscopy
GC	gain-coupled
GCF	geometrical configuration factor
GDD	group delay dispersion
GLAS	geoscience laser altimeter system
GLS	sulfide glasses GaLaS
GRIN	gradient index
GSA	ground-state absorption
GTI	Gires–Tournois interferometer
GVD	group velocity dispersion

H

HDPE	high-density polyethylene
HDSS	holographic data storage system
HHG	high-order-harmonic generation
HMO	heavy metal oxide
HOMO	highest occupied molecular orbital
HR	highly reflecting
HVPE	hydride-vapor-phase epitaxy

I

IAD	ion-assisted deposition
IBAD	ion-beam-assisted deposition
IBS	ion-beam sputtering
ICLAS	International Coordination Group for Laser Atmospheric Studies
ICP	inductively coupled plasma
IL	interference lithography
ILRC	International Laser Radar Conferences
IR	infrared
ITO	indium–tin oxide

K

KB	Kirkpatrick–Baez

L

lcp	left-circularly polarized light
LC–SLM	liquid-crystal spatial light modulator
LCVD	laser(-induced) chemical vapor deposition
LCoS	liquid crystal on silicon
LD	laser diode
LEAF	large effective area
LH	left-handed
LIBS	laser-induced breakdown spectroscopy
LIDAR	light detecting and ranging
LIDT	laser-induced damage threshold
LIF	laser-induced fluorescence
LMJ	laser megajoule
LOQC	linear optics quantum computing
LPCVD	low-pressure CVD
LPE	liquid-phase epitaxy
LSHB	longitudinal spatial hole burning
LSO	laser safety officer
LT	low-temperature
LT-GaAs	low-temperature GaAs
LTG-GaAs	low-temperature-grown GaAs
LUMO	lowest unoccupied molecular orbital
LWFA	laser wakefield acceleration

M

MCP	microchannel plate
MCVD	modified chemical vapor deposition
MFD	multilayer fluorescent disk
MI	modulation instability
MIS	metal–insulator–semiconductor
MMA	methyl methacrylate
MOCVD	metalorganic chemical vapour epitaxy
MOPA	master-oscillator power-amplifier
MOS	metal–oxide–semiconductor
MOT	magnetooptical trap
MPC	metallic photonic crystal
MPE	maximum permissible exposure
MPMMA	modified poly(methyl methacrylate)
MQW	multiquantum well
MSR	magnetic super-resolution
MTF	modulation transfer function

N

NA	numerical aperture
NCPM	noncritical phase matching
NEP	noise equivalent power
NFL	nanofocusing lenses
NGL	next-generation lithography
NIF	National Ignition Facility
NIM	nearly index-matched
NLSE	nonlinear Schrödinger equation
NLSG	nonlinear signal generator
NLTL	nonlinear transmission line
NOHD	nominal ocular hazard distance
NRI	nonresonant intrinsic
NSIC	National Storage Industry Consortium

O

OCT	optical coherence tomography
OFA	optical fibre amplifier
OFHC	oxygen-free high conductivity
OFI	optical-field ionization
OLED	organic light-emitting device
OP	oriented-patterned
OPA	optical parametric amplifier
OPCPA	optical parametric chirped pulse amplification
OPD	optical path difference
OPG	optical parametric generation
OPL	optical path length
OPO	optical parametric oscillator
OPS	optically pumped semiconductor laser
ORMOSIL	organically modified silicates
OSNR	optical signal-to-noise ratio
OTF	optical transfer function
OVD	outside vapor deposition

P

PA	photon avalanche
PB	photonic band
PBG	photonic band gap
PBS	photonic band structure
PBS	polarizing beam splitter
PC	photonic crystal
PCB	printed circuit board
PCF	photonic-crystal fibers
PD	photodetector
PDH	Pound-Drever-Hall technique
PDLC	polymer-dispersed liquid crystal
PDMS	polydimethylsiloxane
PECVD	plasma-enhanced chemical vapor deposition
PEDT/PSS	polyethylenedioxythiophene/ polystyrylsulfonat
PESRO	pump-enhanced SRO
PG	polarization gate
PIC	particle-in-cell
PICVD	plasma impulse CVD
PL	photoluminescence
PLD	pulsed-laser deposition
PMD	polarization mode dispersion
PMMA	polymethylmethacrylate
PMT	photomultiplier tube
POLLIWOG	polarization-labeled interference versus wavelength for only a glint
PPE	personal protective equipment
PPKTP	periodically poled potassium titanyl phosphate
PPLN	periodically poled lithium niobate
PPV	poly-para-phenylenevinylene
PQ	phenanthraquinone
PS	polystyrene
PSA	projected solid angle
PSF	point spread function
PTV	peak-to-valley
PVD	physical vapor deposition
PWM	pulse width modulator
PZT	piezoelectric transducer

Q

QC	quasicrystals
QCL	quantum cascade laser
QD	quantum dot
QDIP	quantum-dot infrared photodetector
QND	quantum nondemolition
QPM	quasi-phase matching
QW	quantum well
QWIP	quantum well infrared photodetector
QWOT	quarter-wave optical thickness
QWP	quarter-wave plate

R

RAM	residual amplitude modulation
rcp	right-circularly polarized light
RCWA	rigorous coupled wave analysis
RDE	rotating disc electrode
RDS	relative dispersion slope
RE	rare-earth
RESOLFT	reversible saturable optical fluorescence transition
RF	radio frequency
RFA	Raman fiber amplifier
RGB	red, green and blue
RIE	reactive-ion etching
RIKE	Raman-induced Kerr effect
RLVIP	reactive low-voltage ion plating
RMS	root-mean-square
RPE	retinal pigment epithelium
R/W	rewritable

S

SAR	synthetic-aperture radar
SASE	self-amplified spontaneous emission
SBS	stimulated Brillouin scattering
SC	supercontinuum
SCP	stretcher–compressor pair
SEM	scanning electron microscope
SFG	sum-frequency generation
SG	sampled grating
SHG	second-harmonic generation
si	semiinsulating
SI	Système International
SIL	solid-immersion lens
SLAR	side-looking airborne radar

SLM	spatial light modulator
SM-LWFA	self-modulated laser wakefield acceleration
SMF	single-mode fiber
SMSR	side-mode suppression ratio
SNR	signal-to-noise ratio
SOA	semiconductor optical amplifier
SOI	silicon-on-insulator
SOS	silicon-on-sapphire
SPDC	spontaneous parametric down conversion
SPIDER	spectral phase interferometry for direct electric field reconstruction
SPM	self-phase modulation
SRO	singly resonant OPO
SRS	stimulated Raman scattering
SS	stainless-steel
SSDL	solid-state dye laser
SSFS	soliton self-frequency shift
SSG	superstructure grating
SSI	spatial–spectral interference
STED	stimulated emission depletion
STP	standard temperature and pressure
STPA	sequential two-photon absorption
STRUT	spectrally temporally resolved upconversion technique
SVEA	slowly varying envelope approximation

T

TADPOLE	temporal analysis by dispersing a pair of light E-fields
TCE	transient collisional excitation
TDSE	time-dependent Schrödinger equation
TEF	trap enhanced field
TEM	transverse electric magnetic
TF	thin film
THG	third-harmonic generation
THz-TDS	THz time-domain spectroscopy
TIP	truncated inverted pyramid
TIR	total internal reflection
TNSA	target normal sheath acceleration
TOD	third-order dispersion
TRO	triply resonant OPO
TS	total scattering
TTF	TESLA test facility
TTG	tunable twin guide
TV	television

U

UV	ultraviolet

V

VC	vertical cavity
VCSEL	vertical-cavity surface-emitting laser
VLSI	very large scale integration
VPE	vapor-phase epitaxy

W

WDM	wavelength division multiplexing
WG	waveguide
WGP	wire-grid polarizer
WORM	write-once, read-many times

X

XFEL	X-ray FELS
XPM	cross-phase modulation
XUV	extreme ultraviolet (soft X-ray)

Y

YAG	yttrium aluminium garnet
YAP	yttrium aluminium perovskite
YLF	yttrium lithium fluoride
YSZ	yttria-stabilized zirconia
YVO	yttrium vanadate

Z

ZAP	zero additional phase
ZAP-SPIDER	zero-additional-phase SPIDER

Part A

Basic Prin

Part A Basic Principles and Materials

1. The Properties of Light

The mystery of light has formed the core of creation stories in every culture, and attracted the earnest attentions of philosophers since at least the fifth century BCE. Their questions have ranged from how and what we see, to the interaction of light with material bodies, and finally to the nature of light itself. This chapter begins with a brief intellectual history of light from ancient Greece to the end of the 19th century. After introducing the physical parameterization of light in terms of standard units, three concepts of light are introduced: light as a wave, light as a quantum particle, and light as a quantum field. After highlighting the distinctive characteristics of light beams from various sources – thermal radiation, luminescence from atoms and molecules, and synchrotron light sources – the distinctive physical characteristics of light beams are examined in some detail. The chapter concludes with a survey of the statistical and quantum-mechanical properties of light beams. In the appropriate limits, this treatment not only recovers the classical description of light waves and the semiclassical view of light as a stream of quanta, but also forms a consistent description of quantum phenomena – such as interference phenomena generated by single photons – that have no classical analogs.

that light has a new property, which came to be called polarization, that does not fit within the corpuscular picture at all. The earliest systematic studies of polarization phenomena associated with the propagation of light waves are due to Étienne Malus in 1808, in response to a prize competition offered by the Paris Academy for a mathematical description of the phenomenon of double refraction in Iceland spar (calcite). Malus's discoveries led to the recognition that light is a transverse electromagnetic wave, in which the electric and magnetic fields are perpendicular to each other and to the direction of propagation. Malus, using his ingenious refractometer, demonstrated in 1807 that the phenomenon of double refraction could be explained mathematically by Huygens' construction. Fresnel, a dozen years later, was to win the prize competition for his theory of diffraction, even anticipating the objection of Poisson that light diffracted around a tiny opaque object would produce a bright spot in the middle of the geometrical shadow – to be known afterwards as *Poisson's spot* [1.4]. Moreover, increasingly powerful mathematical descriptions [1.14] were applied to the phenomena of interference and diffraction studied experimentally by Thomas Young, the London polymath, and Fresnel. It was at last becoming clear that light constituted a qualitatively new kind of wave in which the vibrations were transverse to the direction of propagation of the light [1.15]. Indeed, the transverse character of the vibrations was first suggested by Young in a letter to Arago in 1812, thus hinting that Young was already reinterpreting his interference experiments in a way that differed sharply from previous thinking based on analogies with acoustic waves [1.16].

At virtually the same time, Biot and Savart, Ampère and Faraday were generating the experimental underpinnings for the eventual unification of optics and electromagnetism. Galvani's experiments on the stimulus of what was then called *animal electricity* had shifted attention from electrostatics, the major preoccupation of the eighteenth century, to time-dependent phenomena associated with electricity. However, it was Alessandro Volta who successfully showed that this phenomenon was not due to some *vital* magnetic force, but that it was no different from ordinary magnetism. While electrophysiology continued to be of major interest to biologists and students of medicine, it was thereafter studied by physicists primarily in relation to other electromagnetic phenomena. Oersted, by showing the deflection of a compass placed next to a current-carrying wire, demonstrated the interconnection of electrical and magnetic phenomena. And Faraday, in 1845, showed that the polarization of light could be rotated by applying a strong magnetic field to a medium through which the light was propagating.

Thus the stage was set for the grand synthesis of classical electromagnetic theory. The first step was the publication of James Clerk Maxwell's theory of electromagnetism in 1869. Maxwell's theoretical prediction of electromagnetic radiation was verified experimentally by Heinrich Hertz in 1888 with the discovery of *Hertzian waves* in what now would be called the radio-frequency range of the spectrum. The classical theory of the electron developed by H. A. Lorentz would be the next step in the creation of a 19th-century *theory of everything*. The only clouds on the horizon were the unsolved problems of black-body radiation and the photoelectric effect, problems whose solutions would lead to the development of quantum physics and the evolution of a new view of light based on its dual character as wave and particle, and later of its accommodation into a fully quantum-mechanical field theory.

1.2 Parameterization of Light

The properties of light are parameterized in similar ways in both the classical (wave) and semiclassical (photon) pictures of light. The fundamental physical properties of an electromagnetic wave are its wavelength λ, frequency ν and polarization state; alternatively, the first two of these properties may be stated in the form of a *wave number* $k = 2\pi\lambda$ and *angular frequency* $\omega = 2\pi\nu$. The photon model associates with individual light quanta a particle-like *photon energy* $E_{\text{photon}} = \hbar\omega$ and momentum $p_{\text{photon}} = \hbar k$, where $h = 2\pi\hbar$ is Planck's constant. Photons are also associated with a helicity (photon spin) of ± 1 that can be related to wave polarization.

The properties of light have been defined by international commissions in four kinds of units now in general use, depending on what properties are to be emphasized: radiometric units, based on the physical units, such as energy, and power, are used to describe the properties of electromagnetic waves or photons; photometric units, which refer to the properties of light as discerned by the human eye; photon units analogous to radiometric units that are normalized to photon energy; and spectral units that parameterize light in terms of its properties at specific frequencies or wavelength.

1.2.1 Spectral Regions and Their Classification

The electromagnetic spectrum extends over an enormous range of frequencies and wavelengths, from low-frequency radio wavelength vibrations to extremely high-energy, short-wavelength nuclear gamma radiation. Figure 1.1 shows a typical classification scheme, relating wavelengths, frequencies, wave numbers and photon energies to the common designations of spectral regions of interest in optics, extending from the vacuum ultraviolet through the far-infrared. Some of the units employed match the Système International (SI) convention, others are habitually used in specialized science or technology communities.

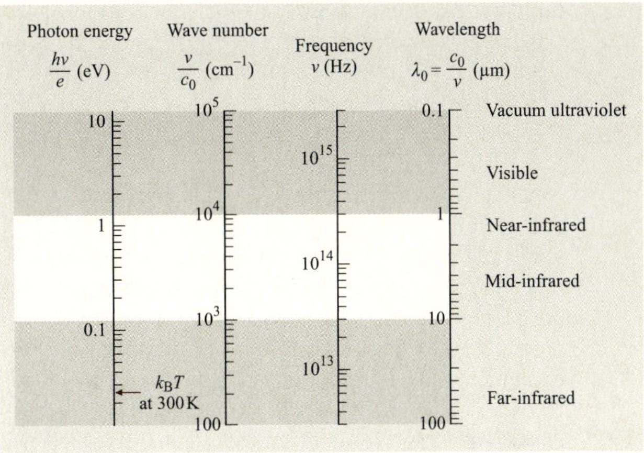

Fig. 1.1 Chart showing the wavelengths, frequencies, wave numbers and photon energies of electromagnetic radiation of interest in optics. (After [1.17])

1.2.2 Radiometric Units

Radiometric units measure the properties of light in terms of physical units of energy and power, without reference to wavelength, and are therefore the most fundamental of the parameters used to describe light [1.19, 20]. The fundamental radiometric units are: radiance, a vector $\boldsymbol{L}$ whose magnitude is the power passing through a surface of unit area into a unit solid angle about the normal to the surface; irradiance, again a vector $\boldsymbol{E}$, defined as the total power per unit spectral interval passing through a surface of unit area. As shown in Fig. 1.2, the magnitude of the radiance and irradiance depends on the shape of the surface over which one integrates, that is, over the projected area $A_\perp$ as well as the solid angle $\mathrm{d}\Omega$ into which light is emitted and the perpendicular area of the detector. The definition of spectral interval is not uniform; depending on the resolution or the parameterization desired, it might be given in Å, nm, cm^{-1} (not the same as 1/cm), or Hz, as here. To convert any radiometric unit X to the corresponding spectral radiometric unit X_ν, recall that $X = X_\nu \,\mathrm{d}\nu$ for values of the frequency lying between ν and $\nu + \mathrm{d}\nu$.

1.2.3 Photometric Units

Photometry refers to the measurement of light as it is perceived by the human eye; thus these units pertain principally to light with wavelengths of 380–760 nm. In astronomy, photometry also refers to the measurement of apparent magnitudes of celestial objects. Since these quantities depend on the spectral amplitude of light, it is not possible to convert photometric values directly into energy values. The photometric units use the same terminology and symbols as the radiometric units, but with a subscript V for *visual*.

The four fundamental photometric quantities, listed in Table 1.2, are: *luminous intensity*, the amount of light emitted by a source; *luminous flux*, the quantity of light transmitted in a given direction; *illuminance*, the measure of light falling on a surface; and luminance, which measures the brightness of a surface

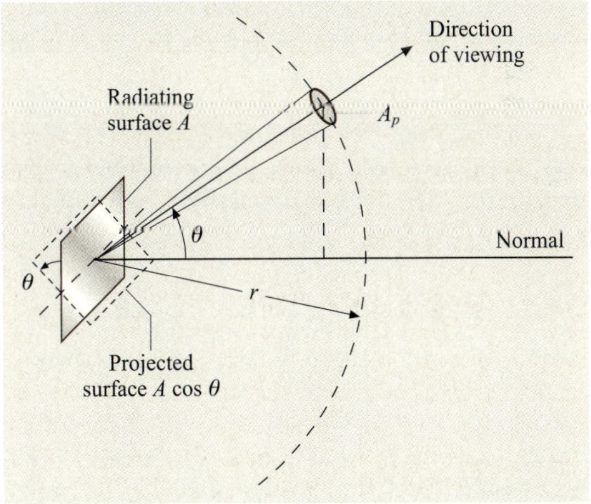

Fig. 1.2 Geometry used to define radiometric units of radiance and irradiance in terms of emitting area, detecting area and solid angle of emission. The projected surface area in a given angular direction Θ is $A_\perp = A\cos\Theta$, while the solid angle in radiometric units is determined by the projected detector area A_p perpendicular to the viewing direction, $\mathrm{d}\Omega = A_\mathrm{p}/r^2$. (After [1.18])

Table 1.1 Radiometric units

	Symbol	SI unit	Definition
Radiant energy	Q_e	J = W s	–
Radiant energy density	w_e	J/m^{-3}	$w_e = \langle dQ_e/dV \rangle$
Radiant flux (power)	Φ_e	W	$\Phi_e = \langle dQ_e/dt \rangle$
Radiant exitance	M_e	$W\,m^{-2}$	$M_e = \langle d\Phi_e/dA \rangle$
Irradiance	E_e	$W\,m^{-2}$	$E_e = \langle dQ_e/dt \rangle$
Radiant intensity	I_e	$W\,sr^{-1}$	$I_e = \langle d\Phi_e/d\Omega \rangle$
Radiance	L_e	$W\,m^{-2}\,sr^{-1}$	$L_e = I_e/\Delta A \equiv \langle d^2\Phi_e/d\Omega \cdot dA \rangle$

Table 1.2 Photometric units

	Symbol	SI unit	Photometric unit	Definition
Luminous energy	Q_V	J=W s	lm s (talbot)	—
Luminous energy density	W_V	J/m^3	$lm\,s/m^3$	$w_V = \overline{dQ_V/dV}$
Luminous intensity	I_V	$W\,sr^{-1}$	$lm\,sr^{-1}$ = candela (cd)	$I_V = \overline{d\Phi_V/d\Omega}$
Luminous power	Φ_V	W	lm (lumen)	$\Phi_V = \overline{dQ_V/dt}$
Luminous exitance	M_V	$W\,m^{-2}$	$lm\,m^{-2}$	$M_V = d\Phi_V/dA$
Illuminance	E_V	$W sr^{-1}$	lux (lx) = $lm\,m^{-2}$	$E_V = d\Phi_V/dA$
Luminance (Apostilb)	L_V	$W\,m^{-2}\,sr^{-1}$	asb = $1/\pi$ cd/m^2	$L_V = \overline{d^2\Phi_V/dA \cdot d\Omega}$

considered as a light source. The standard source, or international standard candle, is defined as the intensity of a black-body radiator with an area of $1/60\,cm^2$ heated to the melting point of platinum. Two auxiliary quantities, *luminous energy* and *luminous energy density*, correspond to the analogous radiometric units. The photometric units carry a subscript V for *visual*, to distinguish them from their radiometric counterparts; the overbar in the table below signifies an averaged quantity.

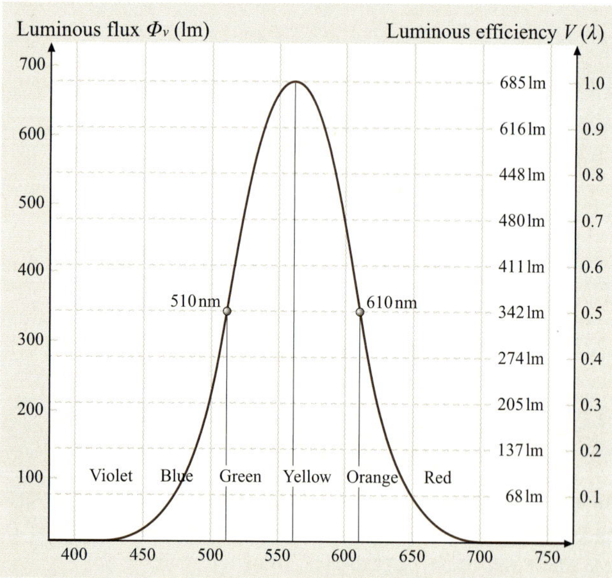

Fig. 1.3 The standard CIE luminous efficacy curve for the human eye, used as the basis for converting between photometric and radiometric units

The Commission Internationale de l'Eclairage (CIE) has developed a standard luminous efficacy curve for the human eye, with respect to which the photometric units are referred (Fig. 1.3). The lumen is defined such that the peak of the photopic (light-adapted) vision spectrum of an average eye has a luminous efficacy of 683 lm/W.

1.2.4 Photon and Spectral Units

In the photon picture, there is a different set of descriptive quantities normalized to photon energy or photon number, as shown in Table 1.3. The overbarred quantities denote an average over photon wavelengths as well as over area and solid angle.

In some cases – for example, when discussing the spectral brightness of laser or synchrotron sources – it is useful to distinguish physical quantities by their frequency ν. For example, in most cases involving spectroscopy or materials processing with lasers, the

Table 1.3 Photon units

	Symbol	SI unit	Definition
Photon number	$\bar{n}$	Number	$\bar{n} = Q_\text{v}/\hbar\bar{\omega}$
Photon density	w_n	Number m^{-3}	$w_\text{n} = \text{d}\bar{n}/\text{d}V$
Photon flux (power)	Φ_n	Number s^{-1}	$\Phi_\text{n} = \text{d}\bar{n}/\text{d}t$
Photon irradiance	E_n	Number s^{-1} m^{-2}	$E_\text{n} = \text{d}\Phi_\text{n}/\text{d}A$
Photon intensity	I_n	Number s^{-1} sr^{-1}	$I_\text{n} = \text{d}\Phi_\text{n}/\text{d}\Omega$
Photon radiance	L_n	Number s^{-1} m^{-2} sr^{-1}	$L_\text{n} = \text{d}^2\Phi_\text{n}/\text{d}A \cdot \text{d}\Omega$

Table 1.4 Spectral radiometric units

	Symbol	SI unit	Definition
Spectral radiant energy	Q_ν	J Hz^{-1} = W s Hz^{-1}	–
Spectral radiant energy density	w_ν	J Hz^{-1} m^{-3}	$w_\nu = \overline{\text{d}Q_\nu/\text{d}V}$
Spectral radiant flux (power)	Φ_ν	W Hz^{-1}	$\Phi_\nu = \overline{\text{d}Q_\nu/\text{d}t}$
Spectral radiant exitance	M_ν	W m^{-2} Hz^{-1}	$M_\nu = \overline{\text{d}\Phi_\nu \text{d}A}$
Spectral irradiance	E_ν	W m^{-2} Hz^{-1}	$E_\nu = \overline{\text{d}Q_\nu/\text{d}t}$
Spectral radiant intensity	I_ν	W sr^{-1} Hz^{-1}	$I_\nu = \overline{\text{d}\Phi_\nu/\text{d}\Omega}$
Spectral radiance	L_ν	W m^{-2} sr^{-1} Hz^{-1}	$L_\nu = \overline{\text{d}^2\Phi_\nu/\text{d}\Omega \cdot \text{d}A}$

quantity of interest is not simply the intensity or radiance, but the intensity or radiance available within a certain spectral bandwidth that defines the effective region of laser-materials interaction, in other words, the *spectral brilliance*. In Table 1.4, the overbarred quantities indicate averages over time, space or solid angle, but *not* over frequency. To convert any radiometric unit X to the corresponding spectral radiometric unit X_ν, recall that $X = X_\nu \text{d}\nu$ for values of the frequency lying between ν and $\nu + \text{d}\nu$.

1.3 Physical Models of Light

By the end of the 19th century, it was generally accepted that the battle between the corpuscular (or emission) and wave theories of light had been resolved in favor of the latter. However, experimental developments – and the inability of classical electrodynamics and statistical mechanics to account for either the photoelectric effect or the shape of the black-body spectrum – drove Planck, Einstein and de Broglie to develop a semiclassical theory of the light quantum (eventually christened the "photon" by *G. N. Lewis* [1.21]). The picture of light as a quantum field emerged in the second half of the 20th century as it became experimentally possible to investigate phenomena associated both with small numbers of photons, statistically distinct ensembles of photons, and correlations between photons emitted in atomic cascades. We now consider the basic concepts underlying the classical, semiclassical and quantum models of light.

1.3.1 The Electromagnetic Wave Picture

Maxwell's classical theory of electromagnetism [1.22] is based on: Gauss's law, a mathematical relationship governing the spatial properties of vector fields with or without sources; Faraday's law of magnetic induction; and Ampère's law linking currents with spatial variations in magnetic field. Maxwell's recognition of the displacement current – the induced current due to time-varying electromagnetic fields – made it possible finally to establish the static and dynamical relationships between the electric field $\boldsymbol{E}$ (the force per unit charge) and magnetic induction $\boldsymbol{B}$ (the force per unit current). Together with the constitutive relations for the electric displacement and the magnetic field, Maxwell's four equations, given below in differential and integral form, govern all classical electromagnetic phenomena including electromagnetic waves.

$$\begin{aligned}
&\nabla \cdot \boldsymbol{D} \equiv \nabla \cdot (\varepsilon_0 \boldsymbol{E} + 4\pi\varepsilon_0 \boldsymbol{P}) = \rho\,,\\
&\int\limits_{\text{volume}} \nabla \cdot \boldsymbol{D} \mathrm{d}V = \frac{Q}{\varepsilon_0}\,,\\
&\nabla \cdot \boldsymbol{B} = 0\,,\\
&\int\limits_{\text{volume}} \nabla \cdot \boldsymbol{B} \mathrm{d}V = 0\,,\\
&\nabla \times \boldsymbol{E} = -\frac{\partial \boldsymbol{B}}{\partial t}\,,\\
&\oint\limits_{\text{open surface}} (\nabla \times \boldsymbol{E}) \cdot \mathrm{d}\boldsymbol{S} = -\frac{\partial}{\partial t} \oint\limits_{\text{loop}} \boldsymbol{B} \cdot \mathrm{d}\boldsymbol{s}\,,\\
&\nabla \times \boldsymbol{H} = \left(\boldsymbol{J} + \frac{\partial \boldsymbol{D}}{\partial t}\right),\\
&\oint\limits_{\text{surface}} (\nabla \times \boldsymbol{H}) \cdot \mathrm{d}\boldsymbol{S} = \oint\limits_{\text{loop}} \left(\boldsymbol{J} + \frac{\partial \boldsymbol{D}}{\partial t}\right) \cdot \mathrm{d}\boldsymbol{s}
\end{aligned} \tag{1.1}$$

where ρ is the charge density and Q the enclosed charge. The constitutive relations electric displacement vector $\boldsymbol{D}$ and for magnetic field $\boldsymbol{H}$ are

$$\begin{aligned}
&D_\alpha = \varepsilon_0 E_\alpha + \left(P_\alpha - \sum_\beta \frac{\partial}{\partial x_\beta} Q^{(2)}_{\alpha\beta} + \ldots\right),\\
&H_\alpha = \frac{1}{\mu_0} B_\alpha - (M_\alpha + \ldots)\,,
\end{aligned} \tag{1.2}$$

where the terms in parentheses represent the contributions of dielectric and magnetic materials to the electric field (e.g., linear polarization P_α, quadrupole contributions $Q^{(2)}_{\alpha\beta}$, and so on) and the magnetic induction (e.g., the bulk magnetization M_α). The dielectric function and the magnetic susceptibilities are properties of materials, and vanish in vacuum. Further information on the constitutive relations, including their relativistic forms, may be found in textbooks [1.23, 24]

The second term in Ampère's law – the last of the four Maxwell equations – was deduced by James Clerk Maxwell from the asymmetry between the electric and magnetic fields that would exist if this *displacement current* were omitted. The last two of these equations can be combined using the vector identity $\nabla \times (\nabla \times \boldsymbol{V}) = \nabla(\nabla \cdot \boldsymbol{V}) - \nabla^2 \boldsymbol{V}$ to yield a wave equation that describes the propagation of transverse electromagnetic waves in a medium with a dielectric function $\varepsilon = n^2$ and magnetic susceptibility μ_0

$$\begin{aligned}
&\nabla^2 \boldsymbol{E} - \frac{n^2}{c^2} \frac{\partial^2 \boldsymbol{E}}{\partial t^2} = 0\,,\\
&\nabla^2 \boldsymbol{B} - \frac{n^2}{c^2} \frac{\partial^2 \boldsymbol{B}}{\partial t^2} = 0\,,\\
&c^2 = (\varepsilon_0 \mu_0)^{-1}\,.
\end{aligned} \tag{1.3}$$

The solutions to this equation include near-field or *static-field* terms that decrease as $1/r^3$, an intermediate-range oscillatory field that exists in the so-called *induction zone*, and the propagating-wave terms whose amplitude decreases as $1/r$.

A time-dependent scalar field satisfying the wave equation in a source-free region may be decomposed into Fourier components at angular frequencies ω_i, $i = 0, 1, 2, 3, \ldots$ that satisfy a Helmholtz wave equation, as follows:

$$\begin{aligned}
&E_\alpha(\boldsymbol{r}, t) = \int_{-\infty}^{\infty} E_\alpha(\boldsymbol{r}, \omega_i)\, \mathrm{e}^{-\mathrm{i}\omega_i t} \mathrm{d}\omega_i\,,\\
&\Rightarrow \left(\nabla^2 + k_i^2\right) E_\alpha(\boldsymbol{r}, \omega_i) = 0\,,\\
&k_i^2 = \frac{\omega_i^2}{c^2}\,.
\end{aligned} \tag{1.4}$$

The Maxwell equations are generally presented in terms of the *force fields* $\boldsymbol{E}$ and $\boldsymbol{B}$. However, it is often more convenient to represent electromagnetic waves in terms of vector and scalar potentials; this is particularly true for making the transition between the classical and quantum field pictures. Given the two homogeneous Maxwell equations, we can define a vector potential that is related to the electric field, the electric potential and the magnetic induction. Furthermore, because the magnetic induction remains unchanged by the addition of the gradient of a scalar function, we have an additional degree of freedom, the choice of gauge, which, in this case, is conveniently chosen to be the so-called *Lorentz gauge*:

$$\begin{aligned}
&\boldsymbol{B} = \nabla \times \boldsymbol{A}\,,\\
&\boldsymbol{E} + \frac{\partial \boldsymbol{A}}{\partial t} = -\nabla \Phi\,,\\
&\nabla \cdot \boldsymbol{A} + \frac{1}{c^2} \frac{\partial \Phi}{\partial t} = 0\,.
\end{aligned} \tag{1.5}$$

With this choice of gauge, the inhomogeneous Maxwell equations can be decoupled to form a pair of inhomogeneous wave equations in the potentials, in which the sources of the wave fields are the charge and current densities [1.24]:

$$\nabla^2 \Phi - \frac{1}{c^2}\frac{\partial^2 \Phi}{\partial t^2} = -\frac{\rho}{\varepsilon_0}$$
$$\nabla^2 \boldsymbol{A} - \frac{1}{c^2}\frac{\partial^2 \boldsymbol{A}}{\partial t^2} = -\mu_0 \boldsymbol{J} \,. \tag{1.6}$$

The other common gauge is the Coulomb or transverse gauge, which is so named because the source term is only the transverse component of the current density $\boldsymbol{J}$. The radiation fields are determined by the vector potential alone; the Coulomb potential contributes only to the near field. However, although it will not be discussed further here, the optical near field has assumed significant importance as a probe of nanoscale phenomena now that there are optical techniques for sampling it and coupling it to far-field radiation that can be transported and observed [1.25, 26].

The ratio of electric to magnetic field amplitudes derivable from the Maxwell equations for a plane wave is $|E(\boldsymbol{r},t)| = c|B(\boldsymbol{r},t)|$; in fact, it turns out that this is a general property of electromagnetic waves. This means that for almost all practical purposes in describing the properties of electromagnetic waves, it is safe to focus on the electric field alone.

For most purposes, three traveling-wave solutions of the scalar wave equation are sufficient to encompass the most common phenomena of wave optics: the spherical wave typical of light emanating from a point source; the plane wave that is the asymptotic form of the wavefront of a spherical wave at large distances from the source; and the Gaussian beam that describes light emitted by a laser source that is constrained in two spatial dimensions by the laser resonator.

Spherical Wave

The scalar solution for electromagnetic radiation emitted from a point source, in spherical coordinates, takes the form of a spherical wave

$$E(\boldsymbol{r},\omega) = \sum_{l,m}\left[A^{(1)}_{lm} h^{(1)}_l(kr) + A^{(2)}_{lm} h^{(2)}_l(kr)\right] \times Y_{lm}(\vartheta,\varphi)\,\mathrm{e}^{\mathrm{i}\omega t} \,, \tag{1.7}$$

where $h^{(i)}_l(kr)$, $i = 1, 2$ are the Hankel functions of the first and second kind, $Y_{lm}(\vartheta,\varphi)$ are the spherical harmonics, and the coefficients $A^{(i)}_{lm}$, $i = 1, 2$ are determined by the boundary conditions. Since the electric field in general is a vector, the general solution is significantly more complex but systematic presentations may be readily found elsewhere [1.24].

Plane Wave

The simplest possible solution to Maxwell's equations is the *plane wave*, in which the phase fronts are infinite planar surfaces perpendicular to the direction of propagation. In Cartesian coordinates, assuming that the wave propagates in the z-direction, the plane wave for the Fourier component with frequency ω is described by the equation

$$\boldsymbol{E}(\boldsymbol{r},t) = \boldsymbol{E}_0(x,y)\exp[\mathrm{i}(k_z z - \omega t)] \,. \tag{1.8}$$

The plane wave is a convenient approximation to a small segment of an electromagnetic wave far from a point source where the spherical wavefronts have a very large radius of curvature.

Gaussian Beam

The beam of light emitted by a laser has properties that are determined by the geometry of the optical resonator for the laser, and for a cylindrically symmetric beam propagating in the z-direction is described by a function of the form [1.23]:

$$E(r) = E_0(r,z)\exp\left(-\frac{r^2}{w^2(z)}\right),$$
$$w^2(z) = w_0^2\left[1 + \left(\frac{\lambda z}{\pi w_0^2}\right)^2\right],$$
$$E_0(r,z) = \frac{1}{w(z)}\left[\frac{\sqrt{2}r}{w(z)}\right]^m L^m_l\left(\frac{2r^2}{w^2(z)}\right) \times \mathrm{e}^{\mathrm{i}m\varphi}\exp\left[\mathrm{i}\left(\Phi(z) + \frac{kr^2}{2R(z)}\right)\right], \tag{1.9}$$

$$\Phi(z) = -(2l+m+1)\tan^{-1}\left(\frac{z \quad z_0}{z_R}\right),$$
$$R(z) = z\left(1 + \frac{z_R^2}{z^2}\right),$$
$$z_R = \frac{\pi w_0^2}{\lambda} \,. \tag{1.10}$$

Here the functions $L^m_l(r,z)$ are the generalized Laguerre polynomials, a complete orthogonal set of functions. The indices $\{l, m\}$ represent different spatial modes of the Gaussian beam; for most laser experiments, one tries very hard to have a beam in the *fundamental Gaussian mode*, in which $\{l = 0, m = 0\}$. The quantity w_0 is frequently called the *beam waist*, and represents the minimum focal spot size that will be reached at some point along the propagation axis. The intensity falls off exponentially from the central maximum, and the rate at which that happens is governed by the beam waist at the given z coordinate. At that point, the wavefront is planar and perpendicular to the axis of propagation. The behavior of a Gaussian beam is often specified by the *Rayleigh range* z_R (or the *confocal beam parameter* $b = 2z_R$) and the *divergence angle* Θ, shown schematically in Fig. 1.4; these are given respectively by

$$b = \frac{2\pi w_0^2}{\lambda} ,$$
$$\Theta = \frac{2\lambda}{\pi w_0} \tag{1.11}$$

Wave Packets

Although all of these solutions to the Maxwell equations are monochromatic waves, in fact light beams often are mixtures of waves of differing frequencies. This comes about because of the spread in wavelengths of white-light sources, the natural line width of even spectrally pure atomic sources, the broadening mechanisms (e.g., Doppler broadening) typical of many atomic or molecular light sources, and because of mode hopping from one frequency to another in laser sources. Mathematically, of course, this fact causes no difficulties because of the superposition principle for harmonic functions. The term *wave packet* refers to such a superposition of waves of many different modes. This should not be confused with the so-called wave-packet model of the photon found in many elementary texts; this model, usually presented as a way of thinking about wave–particle duality, presents a number of philosophical and pedagogical problems [1.27].

For coherent sources, there is an additional constraint on pulse duration and bandwidth, rather like the indeterminacy principle in quantum mechanics and sometimes derived from it. This constraint says that, for a source with duration $\Delta\tau$, the spectral bandwidth satisfies the condition $\Delta\tau \cdot \Delta\nu \geq 1$, where ν is the frequency. Hence, light that comes from continuous-wave coherent sources can have extremely narrow bandwidths (i. e., high spectral purity). The best achieved so far is a bandwidth of order 1 kHz. On the other hand, the shortest laser pulses made to date, with durations of 4 fs, have spectral bandwidths of 200 nm or more.

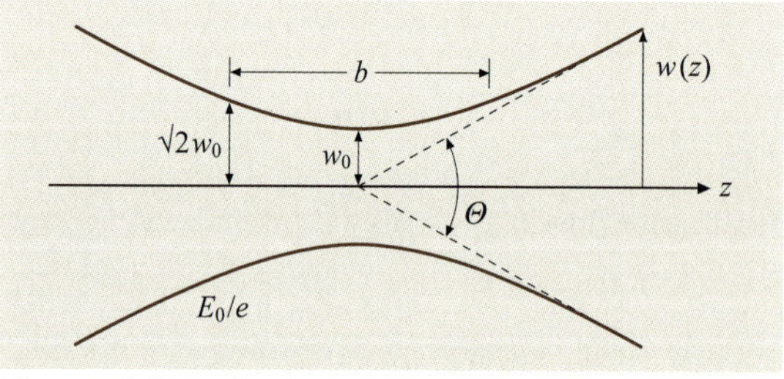

Fig. 1.4 Schematic of the spatial profile of a Gaussian beam near a beam waist, showing the Rayleigh range or confocal beam parameter and the divergence angle Θ. The envelope of the beam represents the point at which the field has decreased to $1/e$ of its maximum value on the beam axis

1.3.2 The Semiclassical Picture: Light Quanta

The theory of light quanta developed by Planck and Einstein explained the salient characteristics of blackbody radiation spectra in a semiclassical picture that quantized the allowable frequencies of radiation, but otherwise treated electromagnetic radiation in terms of classical fields [1.28]. This picture was incorporated into the quantum theory developed by Schrödinger, Heisenberg and especially Dirac; importantly, it avoided the troublesome mathematical divergences that occurred in a quantum-mechanical treatment of electromagnetic radiation. It eventually turned out, as noted by Heitler in a classic text, that these difficulties were primarily of a formal mathematical character [1.29] that was resolved in the modern version of quantum electrodynamics by Dyson, Feynman, Schwinger and Tomonaga [1.30]

The spectrum of radiation from a body at thermal equilibrium at a temperature T, when modeled on the basis of classical theory, produced a catastrophic divergence of the predicted radiated intensity at ultraviolet wavelengths. Planck solved this problem by treating the radiation field as a collection of simple harmonic oscillators in the cavity (Hohlraumstrahlung), and then taking a new approach to calculating the density of modes based on Boltzmann statistics [1.31]. Since the radiation field must satisfy Maxwell's wave equation, the three components of the electric field must satisfy the equations

$$\begin{aligned}
E_x(\boldsymbol{r}, t) &= E_x(t)\cos(k_x x)\sin(k_y y)\sin(k_z z) ,\\
k_x &= \pi\nu_x L , \quad \nu_x = 0, 1, 2, 3, \ldots\\
E_y(\boldsymbol{r}, t) &= E_x(t)\sin(k_x x)\cos(k_y y)\sin(k_z z) ,\\
k_y &= \pi\nu_y L , \quad \nu_y = 0, 1, 2, 3, \ldots\\
E_z(\boldsymbol{r}, t) &= E_x(t)\sin(k_x x)\sin(k_y y)\cos(k_z z) ,\\
k_z &= \pi\nu_z L , \quad \nu_z = 0, 1, 2, 3, \ldots .
\end{aligned} \tag{1.12}$$

Any possible radiation field must be expressible as a sum of these cavity modes; the *spatial quantization* here is the consequence of the boundary conditions at the cavity walls applied to the Maxwell equations. The density of modes is readily calculated to be

$$\begin{aligned}\rho_k \, \mathrm{d}k &= \frac{k^2 \, \mathrm{d}k}{\pi^2} \\ \Rightarrow \rho_\omega \, \mathrm{d}\omega &= \frac{\omega^2 \, \mathrm{d}\omega}{\pi^2 c^3} \,.\end{aligned} \tag{1.13}$$

For a system of harmonic oscillators in thermal equilibrium, the probability that an oscillator is excited into the nth mode is given by the Boltzmann probability

$$\begin{aligned}P_n &= \frac{\exp(-E_n/k_\mathrm{B}T)}{\sum\limits_m \exp(-E_m/k_\mathrm{B}T)} \\ &= \frac{\exp[-(n+1/2)\hbar\omega/k_\mathrm{B}T]}{\sum\limits_{m=0}^{m} \exp[-(m+1/2)\hbar\omega/k_\mathrm{B}T]} \\ &= \frac{\exp(-n\hbar\omega/k_\mathrm{B}T)}{\sum\limits_{m=0}^{m} \exp(-m\hbar\omega/k_\mathrm{B}T)} \,,\end{aligned} \tag{1.14}$$

where $\hbar$ is Planck's constant. From this, with the substitution $U \equiv \exp(-\hbar\omega/k_\mathrm{B}T)$ the average occupation number of a cavity radiation mode for one polarization direction is calculated to be

$$\begin{aligned}\langle n \rangle &= \sum_m m P_m = (1-U) \sum_m m U^m \\ &= (1-U)\, U \frac{\partial}{\partial U} \sum_m U^m \\ &= \frac{1}{\exp\,(\hbar\omega/k_\mathrm{B}T)} \,.\end{aligned} \tag{1.15}$$

This distribution function correctly reproduced the extant spectral measurements on black-body radiators, as well as the empirically derived Wien displacement law, which gave the shift in the measured maximum intensity as a function of wavelength.

The energy density per unit angular frequency in a Hohlraum or cavity radiator is directly proportional to the irradiance (or intensity) of the radiation, and is found by multiplying the density of radiation modes (1.13) by the energy per quantum times the average occupation number:

$$\begin{aligned}U(\omega)\,\mathrm{d}\omega &= \hbar\omega\rho_\omega \,\langle n \rangle \,\mathrm{d}\omega \\ &= \frac{\hbar\omega^3}{\pi^2 c^3} \frac{1}{\exp\,(\hbar\omega/k_\mathrm{B}T) - 1} \mathrm{d}\omega \,.\end{aligned} \tag{1.16}$$

1.3.3 Light as a Quantum Field

Neither Planck's explanation of black-body radiation nor Einstein's theory of the photoelectric effect required one to think of light as a quantum object. In their semiclassical approach, light was treated as a classical electromagnetic wave and only the interaction with matter (e.g., in absorption or emission) was described in terms of quanta. However, working from experimental data of *Kocher* and *Commins*, [1.32], *Clauser* showed that polarization correlations of photons emitted in cascaded atomic transitions could not be accounted for by a semiclassical theory [1.33]. This turned out to be only one instance of what now is understood to be generally true: higher-order photon correlations, single-photon experiments, photon entanglement and photon squeezing cannot be properly described without a fully quantum-mechanical theory of the radiation field.

Such a theory requires, first of all, a prescription for converting the field variables $\boldsymbol{E}$ and $\boldsymbol{B}$ of classical radiation theory into a quantum mechanical operators [1.34]. This is accomplished by introducing the vector potential $\boldsymbol{A}$ in the Coulomb gauge, satisfying the conditions

$$\begin{aligned}&\boldsymbol{B} = \nabla \times \boldsymbol{A} \,, \\ &\boldsymbol{E} = -\nabla\phi \,, \\ &\nabla \cdot \boldsymbol{A} = 0 \,.\end{aligned} \tag{1.17}$$

In free space, the vector potential satisfies the same wave equation that is satisfied by the electric and magnetic fields. For traveling waves subject to periodic boundary conditions, the vector potential can be expanded in a Fourier series. The vector potentials associated with the kth frequency mode of the radiation field can be expressed in terms of generalized position and momentum variables, Q_k and P_k, associated with the kth oscillatory mode, as follows

$$\begin{aligned}\boldsymbol{A}_k &= \frac{1}{4\pi\varepsilon_0\omega_k^2} \frac{1}{V} \left(\omega_k Q_k + \mathrm{i} P_k\right) \varepsilon_k \\ \boldsymbol{A}_k^* &= \frac{1}{4\pi\varepsilon_0\omega_k^2} \frac{1}{V} \left(\omega_k Q_k - \mathrm{i} P_k\right) \varepsilon_k \quad \Rightarrow \\ \bar{E}_k &= \frac{1}{2} \int\limits_{\text{cavity}} \left(\varepsilon_0 \overline{\boldsymbol{E}_k^2} + \mu_0^{-1} \overline{\boldsymbol{B}_k^2}\right) \mathrm{d}V \\ &= \frac{1}{2} \left(P_k^2 + \omega_k^2 Q_k^2\right) \,,\end{aligned} \tag{1.18}$$

where the overbar indicates spatial and temporal averages. Thus the quantum-mechanical description of light associates with each mode of the radiation field an oscillator with a frequency ω_k, and likewise associates the quantum canonical coordinates with the electric and magnetic fields.

The quantized radiation field is obtained by converting the dynamical variables $\{Q_k, P_k\}$ into corresponding operators $\{\hat{q}_k, \hat{p}_k\}$, and inserting these operators into the appropriate Hamiltonian for the field. The Hamiltonian operator for a harmonic oscillator of mass $m=1$ (in some natural units) is $\hat{H} = (\hat{p}^2 + \omega^2\hat{q}^2)$; the operators obey the commutation relation $[\hat{q}, \hat{p}] = \mathrm{i}\hbar$.

It turns out to be useful to work not with these artificial coordinates and momenta, but with the so-called creation and annihilation operators for the oscillators, which are defined, together with their commutator, as follows:

$$\begin{aligned}\hat{a} &= \left(\frac{1}{2\omega\hbar}\right)^{1/2} (\omega\hat{q} + \mathrm{i}\hat{p}) \,, \\ \hat{a}^\dagger &= \left(\frac{1}{2\omega\hbar}\right)^{1/2} (\omega\hat{q} - \mathrm{i}\hat{p}) \,, \\ \left[\hat{a}, \hat{a}^\dagger\right] &= 1 \,. \end{aligned} \tag{1.19}$$

With these definitions, the harmonic-oscillator Hamiltonian for the radiation field turns out to be

$$\begin{aligned}\hat{H} &= \frac{1}{2}\left(\hat{p}^2 + \omega^2\hat{q}^2\right) \\ &= \frac{1}{2}\left[\mathrm{i}^2\left(\frac{\hbar\omega}{2}\right)\left(\hat{a} - \hat{a}^\dagger\right)^2 \right. \\ &\quad \left. + \omega^2\left(\frac{\hbar}{2\omega}\right)\left(\hat{a} - \hat{a}^\dagger\right)^2\right] \\ &= \hbar\omega\left(\hat{a}^\dagger\hat{a} + 1/2\right) \,. \end{aligned} \tag{1.20}$$

A particularly important combination of these operators is the *number operator* given by $\hat{n} = \hat{a}^\dagger\hat{a}$. Rewriting the Hamiltonian and the corresponding eigenvalue equation for the radiation field in terms of the annihilation and creation operators, we have

$$\begin{aligned}\hat{H} &= \hbar\omega\left(\hat{a}^\dagger\hat{a} + \frac{1}{2}\right) \\ &\equiv \hbar\omega\left(\hat{n} + \frac{1}{2}\right) \Rightarrow \end{aligned}$$

$$\hat{H}\,|n\rangle = E_n\,|n\rangle = \hbar\omega\left(n + \frac{1}{2}\right)|n\rangle \,. \tag{1.21}$$

As expected, the vacuum – the state with no quanta ($n = 0$) – still has an associated zero-point energy. The field operators corresponding to the number states are found by substituting the annihilation and creation operators into the expressions for the vector potentials and the electromagnetic fields; the operator for the electric field, for example, is found to be [1.34]

$$\begin{aligned}\hat{E}_k &= \mathrm{i}\left(\frac{\hbar\omega_k}{2\varepsilon_0 V}\right)^{1/2} \\ &\quad \times \hat{\varepsilon}_k\left[\hat{a}_k \exp\left(-\mathrm{i}\omega_k t + \mathrm{i}\boldsymbol{k}\cdot\boldsymbol{r}\right)\right. \\ &\quad \left. - \hat{a}^\dagger \exp\left(\mathrm{i}\omega_k t - \mathrm{i}\boldsymbol{k}\cdot\boldsymbol{r}\right)\right] \,, \end{aligned} \tag{1.22}$$

where $\hat{\varepsilon}_k$ is a unit polarization vector corresponding to the wave-vector mode $\boldsymbol{k}$ and the distance vector $\boldsymbol{r}$ has its usual meaning. This provides the required correspondence between the classical electromagnetic vector fields and the field operators needed for quantum field theory.

In this quantum-mechanical model, the photon is not treated as a classical particle with definite energy, momentum and helicity, but as a quantum excitation associated with the normal modes of the electromagnetic field, specified by a wavevector $\boldsymbol{k}$ and the polarization $\boldsymbol{\varepsilon}$. Localization of the photon – in essence, defining a wave packet comprising photons – is achieved by introducing a linear superposition of one-photon Fock states $|1, \boldsymbol{k}, \boldsymbol{\varepsilon}\rangle$ (states with a well-defined number of photons), rather like any other quantum-mechanical superposition of states

$$\begin{aligned}|\psi\rangle &= C\sum_{\boldsymbol{k}} \exp\left[-\left(\boldsymbol{k} - \boldsymbol{k}_0\right)^2/2\sigma^2\right] \\ &\quad \times \exp\left[-\mathrm{i}\boldsymbol{k}\cdot r_0\right]|1, \boldsymbol{k}, \boldsymbol{\varepsilon}\rangle \,, \end{aligned} \tag{1.23}$$

where C is a normalization constant. This wave function may be regarded as the quantum-mechanical analog of the classical wave packet described earlier.

1.4 Thermal and Nonthermal Light Sources

The early papers on light quanta by Planck and Einstein implicitly assumed that it was possible to treat light as a classical electromagnetic wave, while treating matter quantum mechanically. This approach was successful in providing a theoretical understanding of many important phenomena, including the spectrum of black-body radiation, spontaneous emission, stimulated absorption and stimulated emission, resonance fluorescence, the photo-

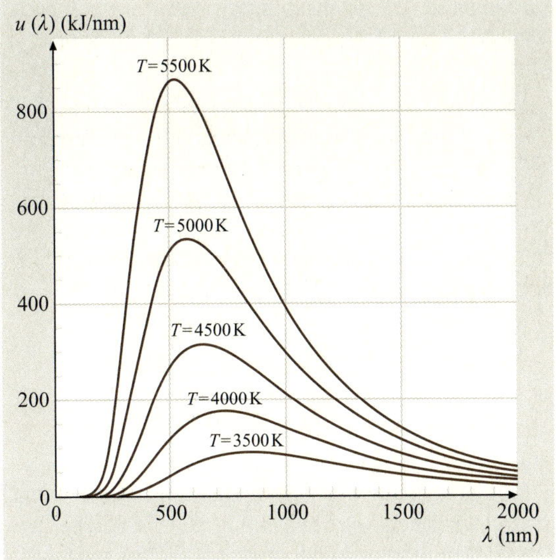

Fig. 1.5 Calculated spectrum of black-body radiators at several different temperatures based on (1.16)

electric effect, the Lamb shift and vacuum polarization. In this context, it is useful to note the different characteristics of light from thermal, nonthermal (or luminescent) and particle-beam sources.

Both thermal and nonthermal light originate from atomic or molecular transitions from higher- to lower-energy states, conventionally labeled as a transition from state 2 to state 1. In the semiclassical picture, these transitions are the outcome of a quantum process that may either be spontaneous or stimulated; the transition rates are described by the Einstein A and B coefficients:

$$A_{2\to1} = \frac{1}{\tau} ,$$
$$B_{2\to1} = \frac{\pi^2 c^3}{\hbar\omega^3} A_{2\to1} = \frac{\pi^2 c^3}{\hbar\omega^3 \tau} , \qquad (1.24)$$

where τ is the mean lifetime of a given species. Because of the quantum-mechanical indeterminacy in the mean lifetime at which an atom or molecule will decay, there is a corresponding uncertainty in the distribution of frequencies that is described by the Lorentzian line-shape function of (1.45)

$$|E(\omega)|^2 = \frac{E_0^2}{(\omega-\omega_0)^2+\gamma^2} , \gamma \equiv 1/\tau . \qquad (1.25)$$

The Lorentzian line shape is also predicted for a driven classical harmonic oscillator with natural frequency ω_0. However, in this case the lifetime is not determined by the properties of individual atoms, as it would be in quantum theory; instead, it is given by a classical radiative lifetime function defined by

$$\tau = \left(6\pi\varepsilon_0 m_e c^2\right) / e^2\omega^2 , \qquad (1.26)$$

where m_e is the mass of the electron. This classical lifetime turns out to be of the order of nanoseconds, typical for the lifetimes of many atomic transitions in the visible as calculated from quantum theory and measured using modern spectroscopic techniques.

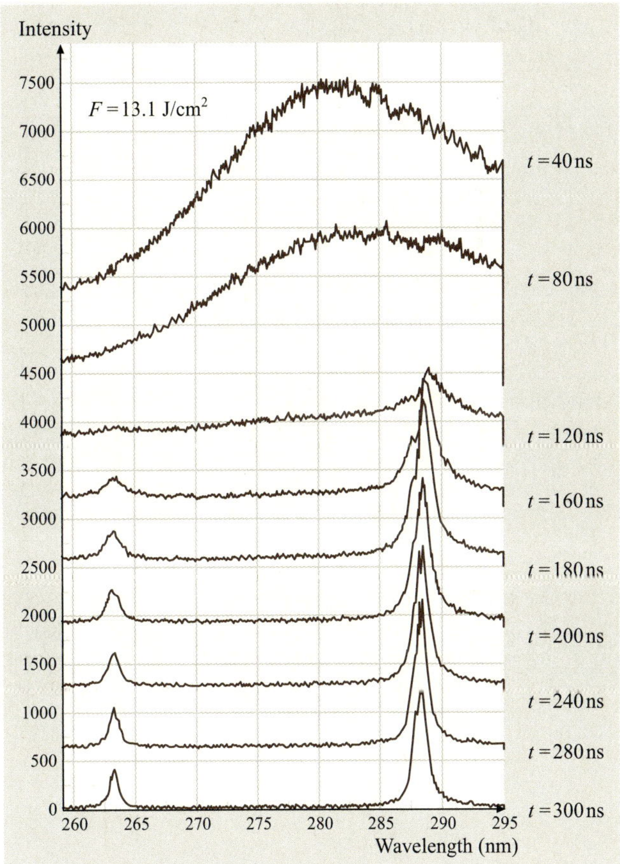

Fig. 1.6 Time-lapse spectra of a laser-produced plasma generated by Nd : YAG laser irradiation of silicon at a fluence of 13.1 J/cm^2. At early times, one observes only the black-body radiation of the large volume of material ablated from the surface; note that the peak of the black-body radiation curve shifts to the red with time, indicating a decrease in temperature. At later times, the spectra exhibit the luminescence lines from atomic transitions in silicon superimposed on a gradually disappearing black-body background. (After [1.35])

1.4.1 Thermal Light

Thermal or chaotic light sources are represented by the idealized black-body radiator, a large ensemble of emitters in thermal equilibrium with each other and with their surroundings. The intensity spectrum of a black body calculated from (1.16) is shown in Fig. 1.5 for several different temperature. The Planck distribution formula in (1.16) yields two important empirical relations between the spectral and thermodynamic properties of black-body radiation laws. The peak wavelength of black-body spectra is described by Wien's displacement law, which shows a shift in the wavelength λ_{max} at which the distribution function is a maximum proportional to the inverse temperature

$$\lambda_{max} = \frac{2.8978\times10^{-6}}{T}\,, \tag{1.27}$$

where T is the absolute temperature in Kelvin. This law can be derived by substituting in (1.16) and finding its extreme value in the usual way.

The areal power density radiated through a small aperture in an enclosure containing radiation in thermodynamic equilibrium at a temperature T is given by the Stefan-Boltzmann law. This law was discovered empirically by Josef Stefan in 1879, and postulated on the basis of thermodynamics by Ludwig Boltzmann in 1884. This law states that the radiated power per unit area is

$$\frac{P(T)}{A} \equiv E(T) = \sigma T^4\,,$$
$$\sigma = \frac{\pi^2}{60}\frac{k_B^4}{h^3c^2} = 5.67\times10^{-8}\,\mathrm{W/(m^2K^4)}\,. \tag{1.28}$$

This relationship requires that one integrate the radiated power for an ensemble of atoms over all wavelengths, again beginning with the Planck distribution. This integral can be carried out with the aid of (1.25) by a change of variables $x = \hbar\omega/k_BT$ to yield the required result:

$$E(T) = \frac{c}{4}\int_0^\infty \frac{\hbar\omega^3}{\pi^2c^3}\frac{d\omega}{\exp(\hbar\omega/k_BT)-1}$$
$$= \frac{\hbar}{4\pi^2c^2}\left(\frac{k_BT}{\hbar}\right)^4\int_0^\infty\frac{x^4\,dx}{e^x-1} \propto T^4\,. \tag{1.29}$$

1.4.2 Luminescence Light

Electromagnetic radiation emitted by excited atoms and molecules not in thermal equilibrium with their surroundings typically exhibits much narrower spectra than black-body radiation, and is called luminescence. The excitation could come from any number of energy sources: energetic electrons (cathodoluminescence), light (photoluminescence), applied electric

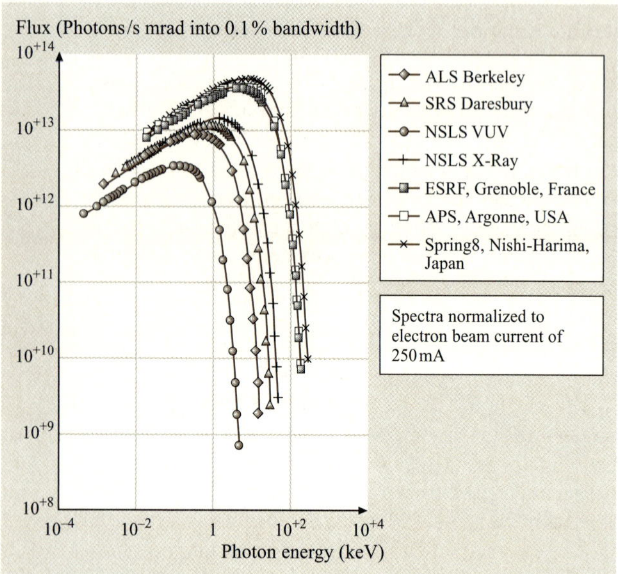

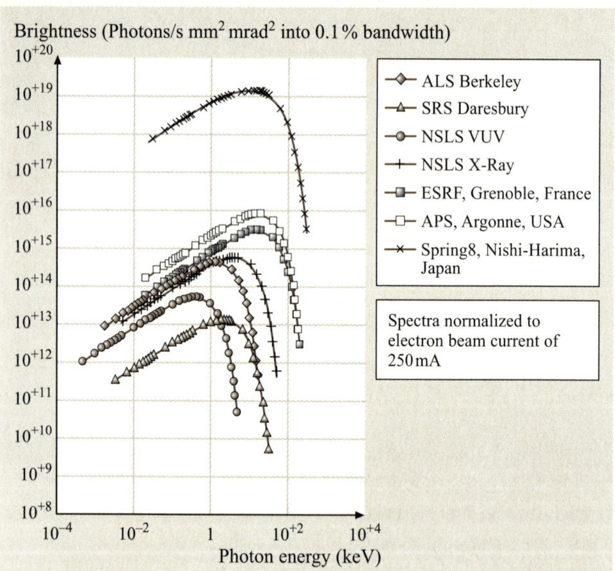

Fig. 1.7a,b Spectral distribution of light from various synchrotron light sources identified in the caption. **(a)** Measured photon flux from the synchrotron light sources. **(b)** Measured spectral brilliance of several synchrotron sources. (After [1.36])

fields (electroluminescence), sound waves (sonoluminescence), or chemical reactions (chemiluminescence). Light produced in this way exhibits a spectrum that is indicative both of its natural frequency and line width, and also characteristic of its environment. For example, impurity atoms in a solid, or excited atoms in a gas-discharge tube at moderate pressure, will exhibit line widths that convolute the natural (Lorentzian) line width with the Gaussian profile due to homogeneous or inhomogeneous broadening by the local electronic environment to produce the Voigt profile of (1.52).

It is also possible for light sources to produce thermal and nonthermal light simultaneously from coupled equilibrium and nonequilibrium processes. Light produced by high-intensity laser irradiation of materials, for example, often has this characteristic. Figure 1.6 shows the spectrum of light emitted by a laser-produced plasma that at later times shows narrow luminescence lines riding on a broader, gradually decreasing black-body background. In this case, the luminescence has the characteristics of radiation from individual atoms (with a line shape that convolutes the natural line width, Doppler and pressure broadening) and black-body radiation (due to hot ablated material in local thermodynamic equilibrium).

1.4.3 Light from Synchrotron Radiation

In recent years, accelerator-driven light sources have assumed an important place in both science and technology. Sources such as synchrotron storage rings [1.36,37] and free-electron lasers [1.38, 39] generate light by bending relativistic electron beams either with dipole magnets or in insertion devices called undulators or wigglers. From such sources, extremely broad spectra can be generated that range from X-rays to the infrared region of the spectrum, with spectral brightnesses far exceeding those of any thermal source. The temporal and spatial characteristics of such radiation are not related to atomic or molecular properties, but instead to the particular characteristics of the associated electron accelerators or bending magnets. Bending magnets produce relatively narrow-band but incoherent radiation, and wavelength selectivity is achieved by the use of monochromators. Undulators and wigglers produce coherent beams of extremely high spectral brightness. In free-electron lasers, a quasi-monochromatic spectrum is produced by placing the wiggler inside an optical resonator.

The spectral profile of light from synchrotron sources follows a universal curve derived from the classical theory of radiation by electrons moving on the arc of a circle. This distribution is quite unlike that of black-body radiation. This is illustrated in Fig. 1.7a, where the measured photon flux from a number of synchrotron sources is compared for constant electron current. The spectral brilliance of some of the same sources is shown in Fig. 1.7b, illustrating the point that this quantity can be highly variable even for sources that have similar photon flux. Besides its broad spectral distribution, synchrotron light produced by wigglers and undulators (*insertion devices*) can have a high degree of spectral coherence.

1.5 Physical Properties of Light

This section outlines the measurable physical properties of light associated with both the wave and photon pictures. These include intensity or irradiance, velocity of propagation, polarization, energy, power and momentum transport. These properties underlie the radiometric, photometric and spectral characterizations of light presented in Sect. 1.2.

1.5.1 Intensity

Almost all measurements of the effects of light interacting with matter depend either on energy incident per unit volume, energy incident per unit area (fluence), power per unit area (intensity or irradiance), or a combination of these quantities. The time-averaged intensity or irradiance of an harmonic electromagnetic wave described by an electric field vector $\boldsymbol{E}(\boldsymbol{r},t) \equiv \boldsymbol{E}_0(\boldsymbol{r})\mathrm{e}^{\pm\mathrm{i}\omega t}$ is

$$\begin{aligned} I(\boldsymbol{r}) &= \frac{1}{T} c\varepsilon_0 \int_0^T |E(\boldsymbol{r},t)|^2 \,\mathrm{d}t \\ &= c\varepsilon_0 \left\langle |E(\boldsymbol{r},t)|^2 \right\rangle = \frac{c\varepsilon_0}{2} |E_0(\boldsymbol{r})|^2 \,. \end{aligned} \tag{1.30}$$

In the photon picture, the analog of the intensity is the areal number density of photons per unit time, or photon flux; it is related to the energy density by $I(\omega) = cU(\omega)/4$, thus

$$\begin{aligned} I(\omega) &= \frac{cU(\omega)}{4} \\ &= \frac{\hbar\omega^3}{4\pi^2 c^2} \frac{1}{\exp(\hbar\omega/k_\mathrm{B}T) - 1} \,. \end{aligned} \tag{1.31}$$

The fluence is the integral of the intensity over a suitable time interval; in the case of laser beams, this interval is often simply the duration of the laser pulse.

1.5.2 Velocity of Propagation

Two different concepts of velocity are associated with the idea of electromagnetic radiation. Phase velocity refers to the propagation of points with the same phase on a wavefront, while group velocity generally refers to the propagation velocity of energy or information in a wave packet. The phase velocity for a wave of frequency ν and vacuum wavelength λ in a medium with an index of refraction n is

$$v_{\mathrm{p}} = \frac{c}{n} = \frac{\lambda}{n}\nu = \frac{\omega}{kn} \Rightarrow \omega = v_{\mathrm{p}} kn \,. \tag{1.32}$$

While this velocity represents the movement of the phase fronts, it is not the speed at which energy or information propagates; that requires some interaction with the oscillations that make up the wave. The group velocity v_g is conventionally defined in terms of these quantities and the index of refraction of the medium in which the wave propagates as

$$\begin{aligned} \frac{\mathrm{d}\omega}{\mathrm{d}k} &= \frac{c}{n}\left(1 - \frac{k}{n}\frac{\mathrm{d}n}{\mathrm{d}k}\right) \\ &= v_{\mathrm{p}}\left(1 - \frac{k}{n}\frac{\mathrm{d}n}{\mathrm{d}k}\right) . \end{aligned} \tag{1.33}$$

Evidently in a medium with constant index of refraction, the phase velocity and group velocities are equal, while in a medium which exhibits normal dispersion ($\mathrm{d}n/\mathrm{d}k > 0$) the group velocity is at most equal to the phase velocity. However, in the case of anomalous dispersion ($\mathrm{d}n/\mathrm{d}k < 0$), the group velocity may actually exceed the speed of light in vacuum c; in such cases, it is always possible to demonstrate that the group velocity is not the speed at which information or energy propagates. This is the case, for example, in hollow waveguides with a cut-off frequency ω_0, in which the dispersion is $k = \sqrt{\omega^2 - \omega_0^2}/c$, and also on the blue side, of many atomic resonances.

1.5.3 Polarization

The polarization of a light wave is widely used as a marker and diagnostic in various spectroscopic techniques, [1.40] especially in laser spectroscopy, [1.41] and is conventionally described by three formalisms: the Mueller calculus, the Jones calculus, and the Poincaré sphere.

The Mueller calculus is based on a scheme first given by G. G. Stokes for measuring the polarization observables of a light beam; it uses four-component vectors to describe the measured intensities (all real numbers), and 4×4 matrices to describe the interaction of the light with various polarizing elements or materials. The Jones calculus is somewhat simpler computationally because it uses only two-component vectors and 2×2 matrices to calculate the polarization in a light beam; however, the parameters in the Jones vectors are complex. The Stokes vectors do not, however, constitute the basis of a linear vector space, while the Jones vectors do. The Poincaré sphere method is based, not on a numerical parameterization of polarized light, but on a mathematical way of relating different polarization forms to points on a sphere; transformations from one polarization form to another are embodied by rotations of the sphere. It is a useful qualitative construct, particularly in making experimental judgments where phase retardation effects rather than changes of intensity are involved. Various states of polarization are illustrated schematically in Fig. 1.8. The matrices that represent the operation of material polarizers such as wave-plates and phase retarders on Jones or Mueller vectors can be found in most optics textbooks.

The Jones Calculus

Consider a light wave propagating in the z-direction, so that the electric field has components only in the x–y-plane. In the previous notation, the electric field is then

$$\begin{aligned} \boldsymbol{E} = {} & E_{0x} \exp\left[\mathrm{i}\left(kz - \omega t + \phi_x\right)\right] \hat{\boldsymbol{x}} \\ & + E_{0y} \exp\left[\mathrm{i}\left(kz - \omega t + \phi_y\right)\right] \hat{\boldsymbol{y}} \,. \end{aligned} \tag{1.34}$$

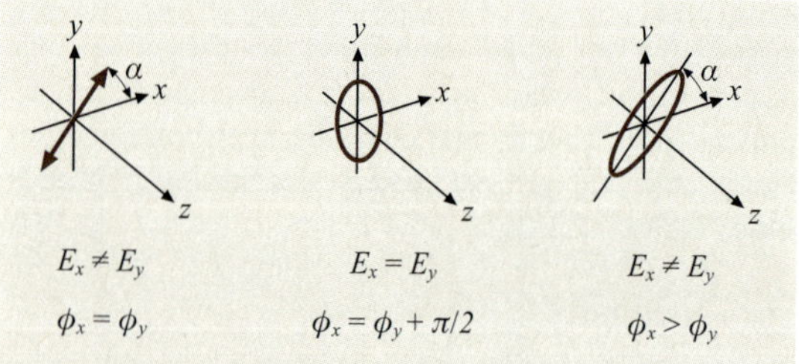

Fig. 1.8 Schematic representation of linear, right-hand circular and elliptical states of polarization. The propagation direction is taken to be the z-axis of a Cartesian coordinate system. The various polarization states are characterized by differing relative values of electric-field amplitudes E_x and E_y and of the phase shifts Φ_x and Φ_y

Factoring out the time and z dependence, and writing the x- and y-components of the electric field as a column vector, we have

$$\begin{aligned}\boldsymbol{E} &= \exp\left[\mathrm{i}\,(kz-\omega t)\right]\begin{bmatrix}E_{0x}\,\mathrm{e}^{\mathrm{i}\phi_x}\\ E_{0y}\,\mathrm{e}^{\mathrm{i}\phi_y}\end{bmatrix}\\ &\equiv \exp\left[\mathrm{i}\,(kz-\omega t)\right]\boldsymbol{J}\,. \end{aligned} \tag{1.35}$$

The column vector, omitting the z and t dependence is called the *full Jones vector* $\boldsymbol{J}$ (or in some circles, the *Maxwell column*) of the light wave. The use of the same symbol as the electric current density rarely causes problems in optics, where electrical currents are usually negligible.

Because the Jones vectors are the bases of a vector space, it is easy to calculate the action of a polarizing device on light having a given polarization state by simple matrix multiplication. For example, consider linearly polarized light with the polarization making equal 45° angles with the x- and y-axes of a quarter-wave plate (QWP). The action of the QWP, which introduces a quarter-wave phase delay between x- and y-components of the electric field, is represented by the matrix $(M)_{\mathrm{QWP}}$. As shown below, the outgoing light beam is right-circularly polarized.

$$\begin{aligned}\boldsymbol{E}_{\mathrm{out}} &= (M)_{\mathrm{QWP}}\,(\boldsymbol{J})_{\mathrm{in}}\\ &= \exp\left[\mathrm{i}\,(kz-\omega t)\right]\begin{pmatrix}1 & 0\\ 0 & i\end{pmatrix}\begin{pmatrix}E_{0x}\\ E_{0y}\end{pmatrix}\\ &\equiv \exp\left[\mathrm{i}\,(kz-\omega t)\right]\begin{pmatrix}E_{0x}\\ \mathrm{e}^{\mathrm{i}\pi/2}E_{0y}\end{pmatrix} \end{aligned} \tag{1.36}$$

Mueller Matrices

The Mueller calculus for describing the polarization of a light beam is based on a measurement scheme for polarization developed by *Stokes* [1.42] The column-vector representation of the Stokes parameterization is

$$\begin{aligned}\boldsymbol{S} = \begin{pmatrix}I\\ Q\\ U\\ V\end{pmatrix} &= \begin{pmatrix}I\\ I_{0^\circ}-I_{90^\circ}\\ I_{+45^\circ}-I_{-45^\circ}\\ I_{\mathrm{rcp}}-I_{\mathrm{lcp}}\end{pmatrix}\\ &= \begin{pmatrix}E_{0x}^2+E_{0y}^2\\ E_{0x}^2-E_{0y}^2\\ 2E_{0x}E_{0y}\cos\Delta\\ 2E_{0x}E_{0y}\sin\Delta\end{pmatrix}. \end{aligned} \tag{1.37}$$

The Stokes parameter I is the total intensity; the quantity I_ϕ is the intensity measured at the angle ϕ; the phase differential Δ satisfies $-\pi \le \Delta = \phi_y - \phi_x \le \pi$; and rcp/lcp refers to right-/left-circularly polarized light. The derivation of the relationships in the second column vector above is complicated, but may be found in a variety of sources [1.40]. An important advantage of the Mueller calculus is the fact that this formalism is able to describe unpolarized or partially polarized light, by taking time averages in time over the Stokes parameters [1.43]. For unpolarized light, only I is nonvanishing, and for partially polarized light, the remaining Stokes parameters must satisfy $0 < (Q^2+U^2+V^2) < I$. On the other hand, for a completely polarized beam, $(Q^2+U^2+V^2) = I$, so that the points (Q,U,V) lying on a unit sphere represent well-defined polarization states.

The Poincaré Sphere

In general, the polarization of a light wave is described by an ellipse, which can be characterized by the angle α describing the inclination of the ellipse with respect to a polar axis and by its ellipticity ε. These quantities can be related to the Stokes parameters by

$$\begin{aligned} S_1 &\equiv Q/I = \cos(2\varepsilon)\cos(2\alpha)\ ,\\ S_2 &\equiv U/I = \cos(2\varepsilon)\sin(2\alpha)\ ,\\ S_3 &\equiv V/I = \sin(2\varepsilon)\ . \end{aligned} \tag{1.38}$$

With these definitions, any desired polarization state can be represented as a point or collection of points on a sphere, as shown in Fig. 1.9 [1.44]. States of constant ellipticity are represented by the locus of points

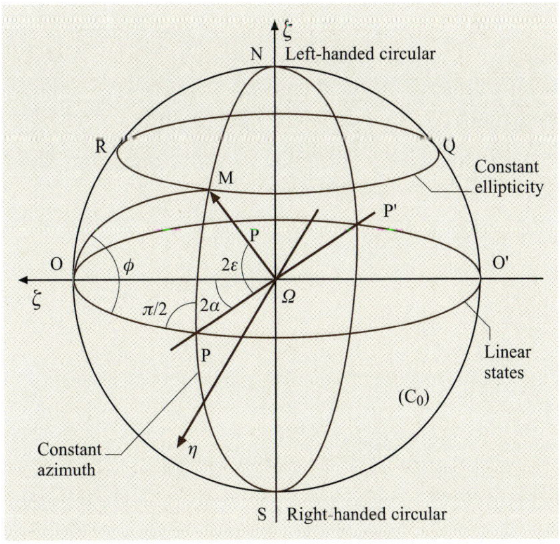

Fig. 1.9 Schematic representation of polarization states on the surface of the Poincaré sphere

on circles parallel to the equatorial plane; the equatorial plane itself is the locus of points representing linear polarization. States of circularly polarized light are represented by the poles. Unpolarized light is represented by the points constituting the surface of the sphere, while partially polarized light would be represented by a subset of points on the surface clustered around a particular point representing the highest degree of polarization.

Photon Spin and Polarization

In classical electromagnetic theory, the polarization of a light wave is determined by the transverse character of vibrations in the electromagnetic field. In the quantum theory of the photon, it becomes necessary to endow the photon with a property of angular momentum just as one does for the electron, but it is also necessary that this concept should be consistent with the wave description of polarization. The photon is a boson and has unit spin; since the photon is massless, it does not have orbital angular momentum, but only its intrinsic spin $s = 1$. A spin-1 particle can have possible spin projections $+1$, 0 and -1 with respect to some given axis; however, since light waves are transverse, the 0-component is ruled out.

From quantum electrodynamics, it is known that photons with helicities of ± 1 are related to left- and right-handed circular polarization states. The basis helicity states of the photon can be represented as column vectors assuming that the quantization axis is the direction of propagation:

$$|s=1\rangle \equiv |+\rangle = \begin{pmatrix}1\\0\end{pmatrix}, \; |s=-1\rangle \equiv |-\rangle = \begin{pmatrix}0\\1\end{pmatrix}. \tag{1.39}$$

Taking the quantization axis to be the z-axis, as well as the direction of light-wave propagation, we construct linearly polarized light in the x- and y-directions from these basis states as follows:

$$\begin{aligned}|e_x\rangle &= \frac{1}{\sqrt{2}}(|+\rangle + |-\rangle)\,,\\ |e_y\rangle &= \frac{1}{\sqrt{2}}(|+\rangle - |-\rangle)\,.\end{aligned} \tag{1.40}$$

Once this correspondence is established, the overall polarization of an ensemble of photons, as in a light beam, can be treated in the same way that other ensembles of quantum particles are treated in the density-matrix formalism [1.45]. Suppose we have a beam of photons that is a mixture of two beams, prepared independently in polarization states characterized by the state vectors and $\boldsymbol{e}_a$, and $\boldsymbol{e}_b$ with intensities I_a and I_b, respectively. If we use the vectors of (1.39) as the basis states, then the state vectors characterizing the polarization are, respectively,

$$\begin{aligned}|e_a\rangle &= c_+^{(a)}|+\rangle + c_-^{(a)}|-\rangle\,,\\ |e_b\rangle &= c_+^{(b)}|+\rangle + c_-^{(b)}|-\rangle\,.\end{aligned} \tag{1.41}$$

This seemingly quite abstract expression corresponds directly to the Stokes parameters defined for a classical light beam, using the standard projection-operator formalism of quantum mechanics. The normalized density matrix for this system of two beams in two possible states is

$$\begin{aligned}\rho &= I_a|e_a\rangle\langle e_a| + I_b|e_b\rangle\langle e_b|\\ &= \begin{pmatrix}\langle +1|+1\rangle & \langle +1|-1\rangle\\ \langle -1|+1\rangle & \langle -1|-1\rangle\end{pmatrix}\\ &\equiv \begin{pmatrix}\rho_{+1,+1} & \rho_{+1,-1}\\ \rho_{-1,+1} & \rho_{-1,-1}\end{pmatrix}.\end{aligned} \tag{1.42}$$

Now consider, for example, the Stokes parameter V, which is defined to be the difference of the right- and left-circular polarized intensities $I_{rcp} - I_{lcp}$. From the definition of the photon helicity, it is clear that V and the density matrix elements are related by $IV = -(\rho_{1,-1} + \rho_{-1,1})$. Following the same approach for the other Stokes parameters, it is found that

$$\rho = \frac{I}{2}\begin{pmatrix}1+U & -V+\mathrm{i}Q\\ -V-\mathrm{i}Q & 1-U\end{pmatrix}, \tag{1.43}$$

thus clearly exhibiting the link between the classical and quantum descriptions of the polarization of the light wave and/or photons. Not surprisingly, it is the Stokes parameters – the classical observables – that also turn up in the quantum-mechanical treatment.

1.5.4 Energy and Power Transport

In general, the energy density in a region of free space occupied by electric and magnetic fields E and B is given by

$$\begin{aligned}u &= \frac{1}{2}\left(\frac{B^2}{\mu_0} + \varepsilon_0 E^2\right)\\ &= \frac{\varepsilon_0}{2}\left(\frac{B^2}{\varepsilon_0\mu_0} + E^2\right)\\ &= \frac{\varepsilon_0}{2}\left(c^2B^2 + E^2\right).\end{aligned} \tag{1.44}$$

In material media, this energy density would be modified by replacing the vacuum values of the dielectric permittivity and magnetic permeability by their values in the material, namely, ε and μ. If we now consider a plane electromagnetic wave traveling in the z-direction for the sake of definiteness, the electric and magnetic fields are given respectively by

$$\begin{aligned} E_y &= E_{0y}\cos(kz-\omega t+\phi)\,, \\ B_x &= -B_{0x}\cos(kz-\omega t+\phi)\,. \end{aligned} \tag{1.45}$$

The time-dependent volumetric energy density associated with this wave is therefore given by

$$u = \frac{\varepsilon_0}{2}\left(c^2 B_{0x}^2 + E_{0y}^2\right)\cos^2(kz-\omega t+\phi)\,. \tag{1.46}$$

This equation exhibits explicitly the wave-like transport of energy associated with the existence of the electromagnetic wave.

The total energy and power associated with the radiation field can be calculated by introducing the hypothesis that the energy in the field is related to the classical average field energy by

$$\begin{aligned} \langle U\rangle_{\text{field}} &= \frac{1}{2}\int\limits_{\text{cavity}} \varepsilon_0\,|\boldsymbol{E}_\omega(\boldsymbol{r},t)|^2\,\mathrm{d}V \\ &= (n+1/2)\,\hbar\omega\,. \end{aligned} \tag{1.47}$$

1.5.5 Momentum Transport: The Poynting Theorem and Light Pressure

The momentum and radiation pressure exerted by a light beam can be developed from the definition of energy contained in the beam. The total energy U delivered by a light beam in a time Δt is related to the change in radiative momentum transferred and the radiation pressure p_{rad} by:

$$\begin{aligned} U &= uAc\Delta t \\ \Rightarrow \Delta p_{\text{rad}} &= \frac{U}{c} = uA\Delta t \\ \Rightarrow F_{\text{rad}} &= \frac{\Delta p_{\text{rad}}}{\Delta t} = uA \\ \Rightarrow P_{\text{rad}} &= \frac{F_{\text{rad}}}{A} = u\,. \end{aligned} \tag{1.48}$$

The flow of energy and momentum carried by electromagnetic waves is described by Poynting's theorem for a vector $\boldsymbol{S} = \boldsymbol{E}\times\boldsymbol{B}/\mu_0$ with units of energy/(time × area). With this definition, the energy density of the electromagnetic field U and the momentum density $\boldsymbol{g} = \varepsilon_0\boldsymbol{E}\times\boldsymbol{B}$ satisfy

$$\begin{aligned} \frac{\partial U}{\partial t} &= -\nabla\cdot\boldsymbol{S} \\ \frac{\partial \boldsymbol{g}}{\partial t} &= -\rho\boldsymbol{E} - \overleftrightarrow{\mathbf{T}}^{(\mathrm{M})}\cdot\hat{\boldsymbol{n}} \end{aligned} \tag{1.49}$$

where $\rho\boldsymbol{E}$ is the Lorentz force per unit volume, $\overleftrightarrow{T}^{(\mathrm{M})}$ is the Maxwell stress tensor and the unit vector $\hat{\boldsymbol{n}}$ is directed along the *outward* normal to the surface bounding the relevant volume. In vacuum, the energy and momentum densities are equal apart from a factor $1/c$; in material media the situation is much more complicated [1.23].

Planck's quantum hypothesis asserts that the energy of a light quantum is related to the angular frequency of an electromagnetic wave ω by $E=\hbar\omega$, where $\hbar\equiv h/2\pi$, with h being Planck's constant 6.67×10^{-34} J s. Similarly, the momentum of a light quantum can be expressed as $p=h/\lambda=\hbar k$. A similar relationship for the momentum of a material particle, of course, was famously postulated by de Broglie and has been verified in experiments with electrons and other quantum particles.

1.5.6 Spectral Line Shape

One of the most important physical characteristics of a stream of light is its spectral line shape, which gives the probability density of intensity or irradiance as a function of wavelength or frequency. The line shape is also a unique signature of the source from which the light emanates.

The line shape of a thermal or chaotic source is that of the black-body spectrum described by the Planck irradiance function, (1.16). The position of the peak in that spectrum is characteristic of the temperature, while its overall form is governed by the Bose–Einstein distribution, the second factor in (1.16). The shape of this function is illustrated in Fig. 1.7a for several temperatures. The shifting position of the peak wavelength at each temperature follows Wien's displacement law (Sect. 1.4.1).

Light from electronic transitions between energy levels E_2 and E_1 in an ensemble of non-interacting atoms (as, for example, in a dilute gas) has a resonance or Lorentzian line shape having the normalized form

$$L(\omega;\omega_0,\gamma) = \frac{\gamma}{\pi\left[(\omega-\omega_0)^2+\gamma^2\right]}\,, \tag{1.50}$$

where F_0 is an appropriately normalized field amplitude, ω is the angular frequency of the light, $\omega_0=(E_2-E_1)/\hbar$ is the frequency at the centroid of the line, and γ is called the natural line width. The natural line width γ is related to the intrinsic lifetime of the atom τ_0 by $\gamma = 1/2\tau_0$.

Should those same atoms be in an environment in which the spectrum of light from the atoms is broadened because of, say, collisions or the Doppler effect, the line shape typically has a Gaussian intensity distribution. Its

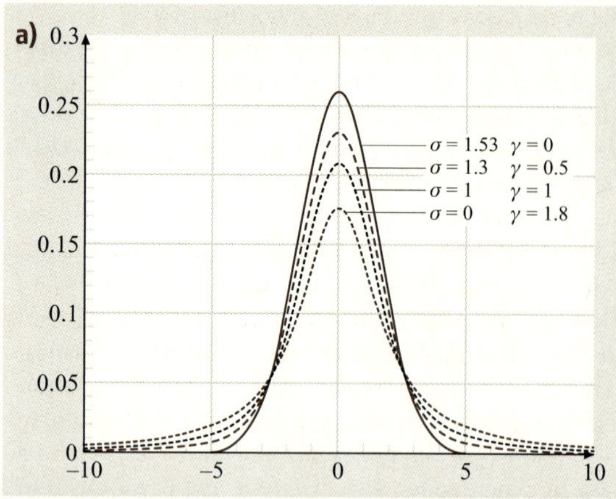

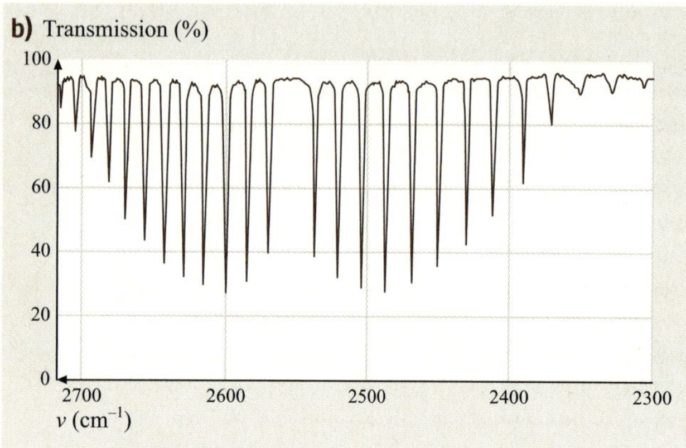

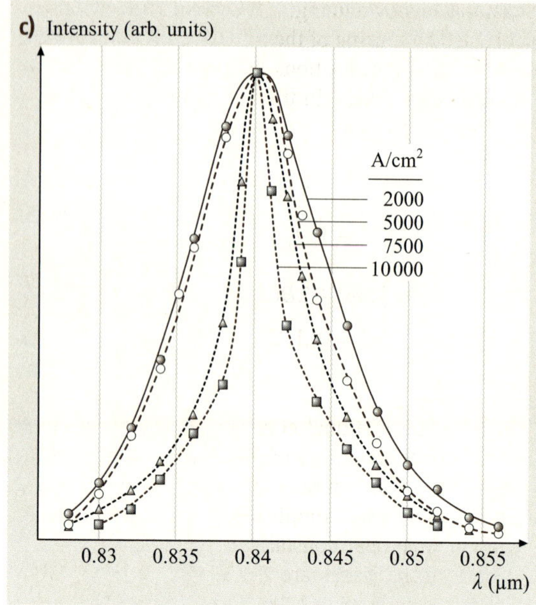

Fig. 1.10a–c Spectral characteristics of light from various sources. **(a)** Comparison of spectral line shapes from isolated atoms or molecules (Lorentzian), collisionally broadened sources (Gaussian) and from sources exhibiting both characteristics (Voigt profile). **(b)** Vibrational-rotational spectrum of HBr molecules, illustrating the band-like structure of radiation from even quite simple molecules. (After [1.46]). **(c)** Spectrum of light from a GaAs semiconductor laser as a function of driving current, showing the narrowing of the spectrum as the lasing threshold is exceeded. (After [1.47])

normalized form is

$$G(\omega;\omega_0,\sigma)=\frac{1}{\sigma\sqrt{\pi}}\exp\left(-\frac{(\omega-\omega_0)^2}{2\sigma^2}\right), \quad (1.51)$$

where the angular frequencies have the same meaning as before and σ is the variance of the line-shape. For Doppler broadening of an atom with mass m_0 at a temperature T, the variance is $\sigma=\omega_0\sqrt{k_B T/m_0 c^2}$, where k_B is Boltzmann's constant. When the atomic emitters all experience the same broadening mechanism, the line shape is said to be homogeneously broadened; when the broadening mechanism differs within the ensemble, such as when the emitting atoms occupy different kinds of lattice sites in a solid, the emission line is said to be inhomogeneously broadened.

In most realistic situations, of course, the broadening of atomic or molecular lines arises from a combination of Lorentzian and Gaussian mechanisms. In this case, the line shape assumes the so-called Voigt profile, which is the convolution of the Gaussian and Lorentzian functions:

$$\begin{aligned}V(\omega,\omega_0;\sigma,\gamma)&=\int_{-\infty}^{\infty}G(\omega',\omega_0;\sigma)L(\omega',\omega_0;\gamma)\,\mathrm{d}\omega'\\&=\frac{1}{\sigma\sqrt{2\pi}}\operatorname{Re}\left[\operatorname{erf}\left(\frac{\omega-\omega_0+\mathrm{i}\gamma}{\sigma\sqrt{2}}\right)\right],\end{aligned} \quad (1.52)$$

where erf is the complex error function. The contrasting line shapes are illustrated in Fig. 1.10a; notice the greater amplitude of the Gaussian function far from the line center frequency ω_0.

Light emission from molecules may arise from electronic, vibrational and/or rotational transitions, and this produces complex spectra in which the individual spectral lines are similar to those of atoms, but which appear in groups or bands of lines with rather regular spacings.

An example is shown in Fig. 1.10b for the diatomic molecule HBr. The spacing of the lines is determined by the selection rules for transitions between rotational or vibrational energy levels. In this example, the spacing is approximately given by

$$\Delta E\,(v, J) = \begin{cases} \hbar\omega_\text{v} - 2BJ\,, \\ J = 1, 2, 3, \ldots (P \text{ branch}) \\ \hbar\omega_\text{v} + 2B(J+1)\,, \\ J = 0, 1, 2, \ldots (R \text{ branch}) \end{cases}, \tag{1.53}$$

where B is the rotational constant of the molecule, J is the rotational quantum number, and ω_v is the angular frequency of the vibrational transition. In practice, molecular spectra are complicated by anharmonicities, changes in rotational constants induced by stretching and so on; these issues are treated in a multitude of specialized monographs [1.48].

Coherent light from laser sources exhibits another characteristic line-shape phenomenon, the narrowing of the emission line as laser action begins. This line narrowing, illustrated in Fig. 1.10c, is due to the higher emission probability – and hence preferential amplification – of wavelengths near the line center. However, depending on the mechanism of line broadening, the narrowing effect may actually be reversed as amplification enters the saturation regime.

1.5.7 Optical Coherence

The coherence phenomena associated with fundamental properties of light can be divided into spatial and temporal coherence [1.49]. The former refers to the effects on electromagnetic radiation emitted by sources of finite size, the latter to radiation emitted by sources with finite bandwidth.

A beam of light is said to be spatially coherent when the phase difference between points on the wavefront remains constant in time, even if the phase fluctuates randomly at any given point. Thus an extended source comprising an ensemble of randomly fluctuating point sources can produce spatially coherent light if the interference fringes from nearby point sources accidentally overlap. For example, starlight of wavelength λ is coherent over a spot of diameter d_coh given by

$$d_\text{coh} = 0.16\frac{\lambda R}{\rho} = 0.16\frac{\lambda}{\vartheta}\,, \tag{1.54}$$

where the stellar diameter is ρ, the distance to the point of observation is R, and ϑ is the angle that the star subtends at the point of observation. The coherence area associated with this diameter is $A_\text{coh} = \pi(d_\text{coh}/2)^2$.

A light beam in which phase differences between points on the wavefront remain constant in time is said to be temporally coherent. Given the Fourier-transform relationship between time and frequency, this also implies that a temporally coherent beam has a high degree of spectral purity. If a light source emits a beam with frequencies ranging from ν to $\nu + \delta\nu$, hence with a spectral bandwidth of $\delta\nu$, the extreme frequencies in the beam will lose temporal synchronization in a comparatively short time, called the coherence time, given by

$$\tau_\text{coh} = \frac{1}{2\pi\delta\nu}\,. \tag{1.55}$$

The coherence length associated with this coherence time is simply $l_\text{coh} = c\tau_\text{coh}$. The coherence volume for a source with given coherence properties is then $V_\text{coh} = A_\text{coh} l_\text{coh}$.

The physical process giving rise to light emission largely determines the coherence of a source. In a thermal source, such as a gas-discharge lamp or an incandescent bulb, light is produced by microscopic or even atomic sources emitting spontaneously, hence at random times relative to one another. Wave packets from different emission events are essentially uncorrelated, and the degree of temporal and spatial coherence is low, though not zero [1.50]. In a laser source, on the other hand, the light is produced by stimulated emission and the degree of coherence is high; however, since temporally and spatially independent transverse and longitudinal modes can coexist simultaneously, only single-mode lasers can achieve the highest degree of temporal coherence.

The coherence of a light source can be quantitatively characterized by interferometry [1.51]. For example, in a two-slit interference experiment (*Young's experiment*), slits separated by a distance d and illuminated by monochromatic light of wavelength λ will produce intensity maxima and minima on a screen at a distance L, separated by a distance $\Delta y = \lambda L/d$. The contrast between maxima and minima can be characterized by the fringe visibility defined in terms of the measured intensities:

$$V = \frac{I_\text{max} - I_\text{min}}{I_\text{max} + I_\text{min}}\,. \tag{1.56}$$

An important development in the last half century has been the recognition that coherence and polarization are inextricably intertwined [1.52]. This convergence has made it possible to develop deep and detailed connections between the classical and quantum theories of light. It has also resulted in the discovery that single-point descriptions of polarization – such as those provided by the Jones or Mueller calculus – are inadequate to explain experimental observations of changes in polarization that occur during propagation, even in empty space [1.53, 54]. These discoveries have led to the development of a generalized coherence matrix in a vector electromagnetic theory that correctly describes these interrelated coherence and polarization phenomena. The form of this hermitian coherence matrix is similar to the matrices found in the Jones calculus:

$$J = \begin{bmatrix} \langle E_x(\boldsymbol{r},t) E_x^*(\boldsymbol{r},t)\rangle & \langle E_x(\boldsymbol{r},t) E_y^*(\boldsymbol{r},t)\rangle \\ \langle E_x^*(\boldsymbol{r},t) E(\boldsymbol{r},t)\rangle & \langle E_y(\boldsymbol{r},t) E_y^*(\boldsymbol{r},t)\rangle \end{bmatrix},$$
$$\mathrm{tr(J)} = \bar{I}. \tag{1.57}$$

However, in this case, the elements of the coherence matrix are not simply complex scalars, as in the Jones matrices, but components of the generalized coherence tensors.

1.6 Statistical Properties of Light

The foregoing discussions are based on models either involving light rays or single photons with well-defined properties of wavelength and frequency, polarization, momentum and energy. However, all real light sources fluctuate in frequency and polarization, and a full treatment of the properties of light requires an accounting for this statistical character through probability density functions, correlation functions and such standard statistical measures as the variance.

The next paragraphs present a statistical characterization of thermal (sometimes called chaotic) or black-body sources, coherent sources (such as lasers) and nonclassical light sources. More detailed treatments are available in many recent textbooks and review articles.

1.6.1 Probability Density as a Function of Intensity

Consider a beam of unpolarized light from a thermal source, in a coordinate system defined such that the beam propagates in the z-direction. Such a beam comprises an equal mixture of x- and y-polarized components, with random amplitude and phase, each of which can be shown to obey a Rayleigh probability distribution given by [1.55].

$$p(I_x) = \frac{2}{\bar{I}} \exp\left(-2\frac{I_x}{\bar{I}}\right),$$
$$p(I_y) = \frac{2}{\bar{I}} \exp\left(-2\frac{I_y}{\bar{I}}\right), \tag{1.58}$$

where the average intensity $\bar{I}$ is related to the standard deviation σ_{I} by $\sigma_{\mathrm{I}} = \bar{I}/\sqrt{2}$. The joint density function of the two independent random variables I_x and I_y is found from a standard theorem of probability theory to be equal to

$$p(I) = \frac{2}{\bar{I}} \int_0^I \exp\left(-2\frac{\xi}{\bar{I}}\right) \exp\left(-2\frac{I-\xi}{\bar{I}}\right) \mathrm{d}\xi$$
$$= \left(\frac{2}{\bar{I}}\right)^2 I \exp\left(-2\frac{I}{\bar{I}}\right). \tag{1.59}$$

A plot of this function is shown in Fig. 1.11; the physical interpretation of this is that the probability has a maximum value of 0.5, while the integral under the curves, as with all probability functions has the value 1. Applying the same general analysis, the probability distribution for light with varying degrees of polarization P from a thermal source is given by:

$$p_{\mathrm{I}}(I) = \frac{1}{P\bar{I}} \left[\exp - \left(\frac{2I}{(1+P)\bar{I}}\right) - \exp\left(\frac{2I}{(1-P)\bar{I}}\right)\right]. \tag{1.60}$$

It is instructive to compare this distribution function to that of light from a laser source. The laser is proverbially a source of *highly organized* light, but its statistical character changes dramatically depending on whether the laser is operated below, near or well above the threshold for laser oscillation [1.56]. For a laser in the steady state, Mandel and Wolf show that the probability density $p(I)$ as a function of intensity can be written as [1.57]

$$p(I) = C \exp\left[-\frac{1}{4}(I-a)^2\right], \tag{1.61}$$

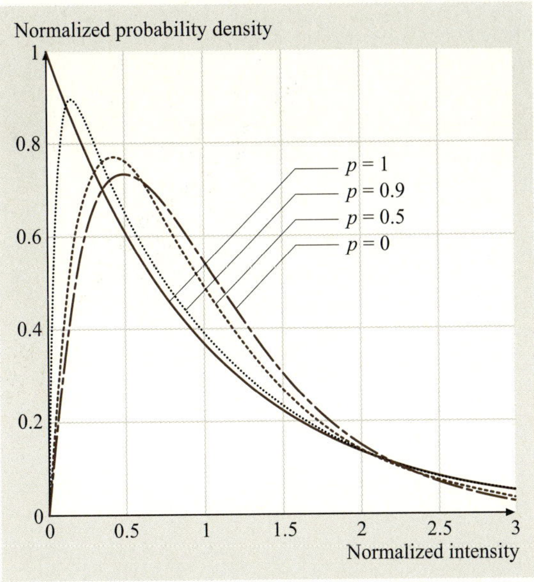

Fig. 1.11 Probability density $p(I)$ for light with varying degrees of polarization from a thermal or chaotic source, calculated from (1.58) and (1.59)

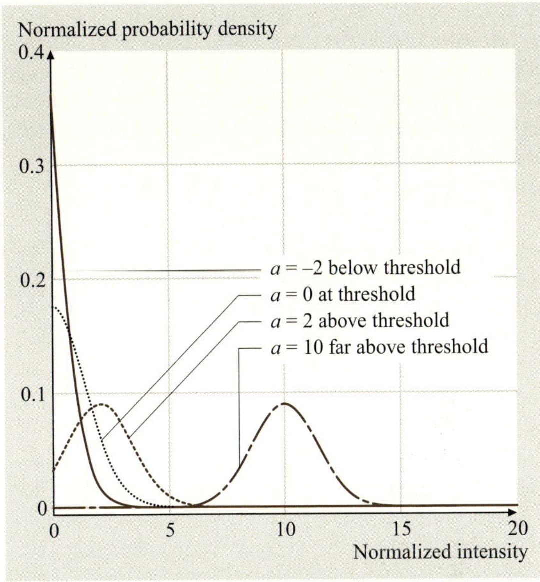

Fig. 1.12 Probability densities $p(I)$ for a laser as a function of various pump parameters, as described above

where a is a pump parameter and C is a normalization constant. Below threshold, $a < 0$ and the thermal or chaotic output of the laser follows Bose–Einstein statistics; the probability density falls off exponentially. At threshold, $a = 0$ and the probability density has the form of a half-Gaussian cut off at $I = 0$. Above threshold, $a > 0$, and the probability density more nearly resembles the Gaussian distribution characteristic of coherent oscillators; far above threshold it is the closest we can make to a completely coherent source. The probability densities for these several cases are illustrated in Fig. 1.12.

1.6.2 Statistical Correlation Functions

A particularly useful set of statistical measures are the correlation functions between either classical field variables or quantum field operators. The first-order classical and quantum-mechanical correlation functions are identical in form, and are given by

$$g^{(1)}(\boldsymbol{r}_1t_1, \boldsymbol{r}_2t_2) \equiv \frac{\langle E^*(\boldsymbol{r}_1t_1)\, E(\boldsymbol{r}_2t_2)\rangle}{\left[\langle |E(\boldsymbol{r}_1t_1)|^2\, |E(\boldsymbol{r}_2t_2)|^2\rangle\right]^{1/2}}, \tag{1.62}$$

where $\{\boldsymbol{r}_1t_1, \boldsymbol{r}_2t_2\}$ are the space–time variables that define the two electromagnetic fields. The quantum-mechanical expression is constructed analogously, with the field operators defined in (1.62) replacing the classical field variables. The significance of the function $g^{(1)}(\boldsymbol{r}_1t_1, \boldsymbol{r}_2t_2)$ is that it falls in the interval between 0 and 1, with 1 representing coherent light, 0 incoherent or chaotic light, and intermediate values representing varying degrees of partially coherent light. The results that follow from the first-order coherence properties of light, such as the results of Young's experiment, are identical regardless of whether one chooses the classical or quantum-mechanical description. This follows from the fact that $0 \le |g^{(1)}| \le 1$. Experimentally this is tantamount to saying that the same kinds of interference phenomena are observed classically as well as quantum mechanically.

On the other hand, the second-order coherence function leads to quite different and unanticipated differences between the quantum and classical cases. The second-order coherence for quantum fields is defined by analogy with the classical quantities, and is

$$\begin{aligned} &g^{(2)}(\boldsymbol{r}_1t_1, \boldsymbol{r}_2t_2; \boldsymbol{r}_2t_2, \boldsymbol{r}_1t_1) \\ &\equiv \frac{\langle E^-(\boldsymbol{r}_1t_1)\, E^-(\boldsymbol{r}_2t_2)\, E^+(\boldsymbol{r}_2t_2)\, E^+(\boldsymbol{r}_1t_1)\rangle}{\left[\langle E^-(\boldsymbol{r}_1t_1)\, E^+(\boldsymbol{r}_2t_2)\rangle\langle E^-(\boldsymbol{r}_2t_2)\, E^+(\boldsymbol{r}_1t_1)\rangle\right]}. \end{aligned} \tag{1.63}$$

The classical version of this equation is generated by replacing operators with the corresponding classical fields, whereupon one finds that the classical second-order correlation function satisfies $1 \leq g^{(2)}(0) \leq \infty$, $g^{(2)}(\tau) \leq g^{(2)}(0)$. To see what this implies, consider a simple model for a thermal source such as an atomic discharge lamp, with Doppler-broadened lines and a Gaussian line shape. Its second-order correlation function turns out to be

$$g^{(2)}(\tau) = 1 + \exp\left[-(\tau/\tau_{\text{coll}})^2\right] , \quad (1.64)$$

where τ_{coll} is the mean time between collisions. For perfectly coherent light, it is also easy to show that (1.54) implies $g^{(2)}(\tau) = 1$ for all values of τ. Figure 1.13 illustrates the differences between the two correlation functions; an experimental verification of this comparison for a discharge lamp and a laser was first published in 1966 [1.58].

In addition to these rather straightforward examples, however, there is also a range of values for which $0 \leq g^{(2)}(0) < 1$, $g^{(2)}(0) < g^{(2)}(\tau)$ that corresponds to quantum-mechanical second-order coherence phenomena that have no equivalent in the classical realm, most commonly for photon streams with small, well-defined photon number [1.59].

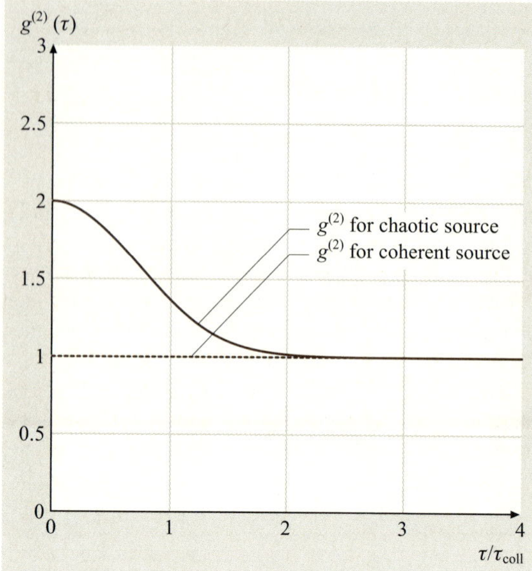

Fig. 1.13 Second-order correlation functions $g^{(2)}(\tau/\tau_{\text{coll}})$ for perfectly coherent light and chaotic light from a Doppler-broadened atomic discharge lamp

1.6.3 Number Distribution Functions of Light Sources

Another way of classifying the statistical properties of light beams is by comparing the number distribution functions of photons impinging on a detector, and comparing the variance in those distributions for different light sources.

The statistical distribution of photon numbers arriving at a detector depends on the source of the light. We can generally distinguish thermal light, in which the process producing the photons is random, and coherent light, when it is not. A coherent light field can be viewed as one which is the closest thing to a *perfectly classical* state attainable in the quantum realm; such states are readily achieved in lasers, as seen in the previous section [1.60]. If the arrival and detection of photons in a stream are independent events, the probability distribution $p(n)$ for n photons is given by the Poissonian distribution, with mean $\bar{n}$ and standard deviation σ_n^2 is given by

$$p(n) = \frac{\bar{n}^n \, e^{-\bar{n}}}{n!} , \sigma_n^2 = \bar{n} . \quad (1.65)$$

In other words, for a Poisson distribution, the mean photon number equals the variance.

Chaotic light, on the other hand, has a distribution function that can be described as *super-Poissonian*, which is derivable from the Bose–Einstein or Planck distribution of photons in thermal equilibrium with their surroundings. That distribution function is readily shown

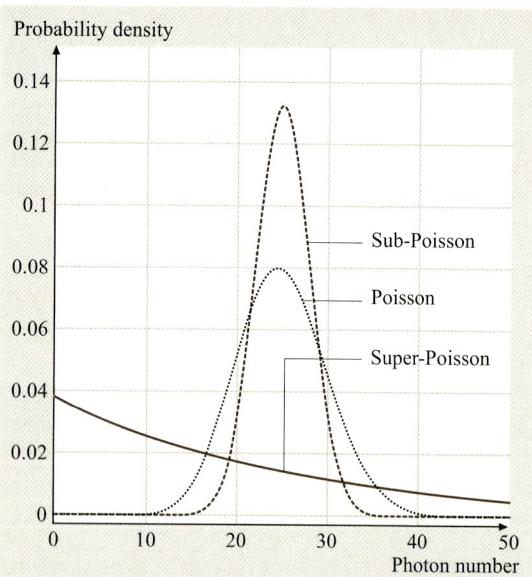

Fig. 1.14 Probability density as a function of mean photon number $\bar{n}$ for Poissonian (coherent), sub-Poissonian (nonclassical) and super-Poissonian (chaotic) light sources

to be [1.17]

$$p(n) = \frac{1}{\bar{n}+1}\left(\frac{\bar{n}}{\bar{n}+1}\right)^n \propto \left(\mathrm{e}^{-\hbar\omega/k_\mathrm{B}T}\right)^n ,$$
$$\bar{n} = \frac{1}{\mathrm{e}^{\hbar\omega/k_\mathrm{B}T}-1} . \tag{1.66}$$

The variance for the Bose–Einstein distribution is $\sigma_n^2 = \bar{n}+\bar{n}^2$, which is always greater than it is for a Poissonian distribution. Interestingly, another characteristic of this thermal light is that the signal-to-noise ratio is always less than 1, so that thermal light cannot be used to transmit digital information. The difference between the two cases is illustrated in Fig. 1.14.

In the case of nonclassical light, on the other hand, there are states in which $\sigma_n^2 < \bar{n}$ as shown in Figure 1.14; squeezed states of light (discussed in more detail below) are an example.

A single-parameter way of characterizing the statistical characteristics of the various light sources is the Mandel Q-parameter, defined by

$$Q \equiv \frac{\left\langle(\hat{n})^2\right\rangle - \langle\hat{n}\rangle}{\langle\hat{n}\rangle} = \frac{\left\langle(\hat{n})^2\right\rangle}{\langle\hat{n}\rangle} - 1 . \tag{1.67}$$

Evidently, the case of chaotic or thermal light corresponds to $Q > 0$, coherent light to $Q = 0$, and sub-Poissonian light when $0 > Q > -1$. Nonclassical light can be created in a number of different ways; some examples are fluorescence emitted by a single ion in an electromagnetic trap, [1.61] fluorescence from trapped atoms [1.62] and semiconductor quantum dots, [1.63, 64] and essentially arbitrary photon streams generated through cavity quantum electrodynamics [1.65–67].

1.7 Characteristics and Applications of Nonclassical Light

Although the application of classical light – whether viewed in the electromagnetic wave or photon (semiclassical) models – are now ubiquitous, and the stuff of undergraduate textbooks, the study of nonclassical light, in the quantum-field model, is still an intensely active field of research. In this section, we describe in greater detail some of the characteristics of nonclassical light and consider one of its many interesting applications.

1.7.1 Bunched Light

A particularly interesting statistical property of light is the arrival-time distribution of photons coming from the various sources described previously [1.68, 69]. It was experimentally observed a long time ago that light coming from thermal sources tends to be *bunched*, that is, the probability of a second hit on a detector immediately after one photon is detected is enhanced over the chances of a random detection event [1.70]. In contrast, the arrival-time distribution of photons from a coherent source is random, in keeping with the Poissonian nature of the emission process from a laser. Finally, however, there are light sources that generate arrival-time distributions exhibiting *anti-bunching*, with photons being detected at regular intervals. This anti-bunching phenomenon was first detected in fluorescence from sodium atoms [1.71], and has since been observed in quantum dots [1.72], atom [1.62] and ion traps, and even in atom lasers [1.73]. The transition between bunching and anti-bunching behavior has also been observed [1.74, 75]. However, there is apparently no connection between sub-Poissonian statistics and anti-bunching, [1.76] so that the underlying physics of the two statistical properties may well likewise be different.

1.7.2 Squeezed Light

In a classical wave, the phase and amplitude are independent variables. In quantum mechanics, however, these quantities are coupled through the uncertainty principle applied to the operators that define the quantum fields. Squeezed states are states of minimum uncertainty in which one or the other variable – number density or phase – is reduced below the Poissonian level [1.77–79] The origin of these states can be understood quantitatively by rewriting (1.22) as follows:

$$\hat{E}_k = \mathrm{i}\left(\frac{\hbar\omega_k}{2\varepsilon_0 V}\right)^{1/2} \times \hat{\varepsilon}_k\left[\hat{q}_k \cos(\boldsymbol{k}\cdot\boldsymbol{r}-\omega_k t) - \hat{p}_k \sin(\boldsymbol{k}\cdot\boldsymbol{r}-\omega_k t)\right] , \tag{1.68}$$

where the operators or canonical variables $\{\hat{q}_k, \hat{p}_k\}$ are the quadrature amplitudes into which the electromagnetic field operator can be decomposed. These operators satisfy the standard indeterminacy relationship for the canonical variables in quantum mechanics. It follows, therefore, that while the product of their fluctuations

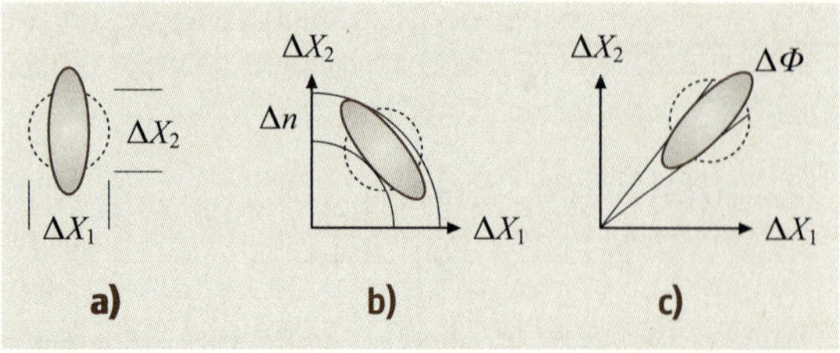

Fig. 1.15a–c Relationships between the various quadrature states for light that is **(a)** quadrature squeezed in the vacuum state; **(b)** number-squeezed; and **(c)** phase-squeezed

must satisfy $\langle(\Delta\hat{q}_k)^2\rangle\langle(\Delta\hat{p}_k)^2\rangle \geq 1$, it is possible to generate coherent light in which the fluctuations in one variable are decreased at the expense of the other.

It is conventional to redefine the complex electric field operator in (1.60) to make so-called dimensionless field quadratures X_1 and X_2, as follows:

$$\begin{aligned}\hat{X}_1(t) &\equiv \left(\frac{\varepsilon_0 V}{4\hbar\omega}\right)^{1/2} \hat{E}_{k0}\sin(\omega t)\\ &= \left(\frac{\hbar}{2\omega}\right)^{1/2} \hat{q}_k(t)\ ,\\ \hat{X}_2(t) &\equiv \left(\frac{\varepsilon_0 V}{4\hbar\omega}\right)^{1/2} \hat{E}_{k0}\cos(\omega t)\\ &= \left(\frac{\hbar}{2\omega}\right)^{1/2} \hat{p}_k(t)\ .\end{aligned} \tag{1.69}$$

In this formalism, it is evident that the field quadratures obey the same indeterminacy relations as the quantum harmonic oscillator. Both the vacuum state and coherent states have minimum uncertainties, and moreover, the uncertainties in the two field quadratures can in principle be equal

$$\Delta X_1 = \Delta X_2 = \frac{1}{2}\ . \tag{1.70}$$

However, the indeterminacy relation requires only that the product of the uncertainties be a minimum. Hence, it is possible to imagine and to construct minimum-uncertainty states in which the uncertainty in one field quadrature is reduced at the cost of increasing the uncertainty in the other quadrature [1.81]. Coherent light with such properties is said to be quadrature squeezed; the properties of such states are explored in detail in [1.57].

Another kind of coherent light is formed from the so-called number states of the electromagnetic field. These states, first used to describe optical correlations by *Glauber* [1.82–84] and *Sudarshan* [1.85], can in fact exhibit the properties of light that is number- or amplitude-squeezed. Additionally, of course, it is possible to squeeze the conjugate phase variable of the number states. Figure 1.15 illustrates the different relationships of the quadratures in the various kinds of squeezed light.

1.7.3 Entangled Light

In 1935, Einstein, Podolsky and Rosen (EPR) published their observation that for *entangled states* (Schrödinger's terminology [1.86]), quantum mechanics predicted that measurements on objects far apart could show a correlation. How then, argued Einstein, was it possible for quantum mechanics to be a complete, causal description of reality if the local results of a measurement could be influenced by quantum correlations at a large distance? Three decades later, J. S. Bell transformed this philosophical question into an experimentally testable issue by demonstrating that the quantum-mechanical predictions for a pair of interacting spin-1/2 particles were not consistent with the predictions of EPR for a system based on the idea of strict local causality [1.87, 88]

In the event, it turned out to be easier to test these predictions using spin-1 photons instead of spin-1/2 particles, following procedures suggested by *Clauser* and *Shimony* [1.89]. Experiments by *Aspect* [1.90] and later by *Zeilinger* et al. [1.80] showed conclusively that photons emitted from a single source violated Bell's inequality, and were thus entangled by correlations at a distance of precisely the kind that EPR argued were impossible.

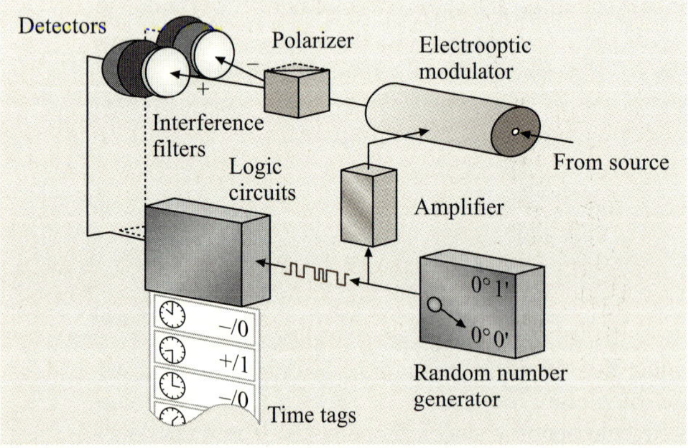

Fig. 1.16 Schematic of an experiment demonstrating photon entanglement by correlations at a distance ([1.80]). The details of the experiment are described in the text

The schematic of the experiment is shown in Fig. 1.16. Two photons are emitted from a common source, and pass through polarization-measuring devices that measure the photons emitted with orientations a and b perpendicular to the z-axis. At each terminus of the experiment, a computer monitors the orientation of the polarization as measured and the precise timing signal provided by an atomic clock. In the Zeilinger experiments, the polarization measurements are separated by 400 m, a distance that precludes semiclassical correlations by virtue of the use of ultrashort light pulses and random switching of the preferred orientation of the polarization analyzers.

The results appear to clearly demonstrate that quantum correlations at large distance are one of the novel realities of quantum physics. They also have applications in quantum cryptography [1.91, 92] and quantum teleportation [1.93, 94].

1.8 Summary

The properties of light at this point in the history of optical science are as varied as the multiple ways in which light can now be generated and detected. The kinds of sources that can generate light range from classical thermal (chaotic) sources and coherent sources such as lasers, wigglers and undulators – and most recently to nonclassical sources generated by exquisite control of light-emission processes at the single-quantum level. While the statistical properties of light governed by first-order correlations are virtually indistinguishable for both classical and quantum light sources, the second-order correlations reveal distinctive differences, particularly in the statistical properties of the light generated by these different sources. The wide-ranging impact of these studies, underlined by the recent award of a Nobel prize for the theory of quantum coherence, suggests not only that the field has reached the stature of a major part of physics, but that it has grown measurably in importance and is likely to continue to do so in the future.

From a review written for the World Year of Physics in 2005 [1.95], one finds that two of the three celebrated papers in Einstein's *annus mirabilis* had deep intellectual connections with questions about the properties of light. The paper on special relativity was derived from a profoundly insightful critique of the classical theory of electromagnetic radiation [1.96]. In the second paper, Einstein literally solidified the concept of the light quantum (later christened the photon), which Planck had regarded as a mere heuristic, a mathematical device for saving appearances [1.97]. Einstein did so by showing from the Planck distribution that the entropy of cavity radiation has the same volumetric dependence as the entropy of an ideal gas – and therefore, he inferred, it was appropriate to treat light quanta as particles. But Einstein was also profoundly critical of the quantum physics of his time, even though, or perhaps because, his own interpretation of quantum phenomena – as in the EPR paradox – turned out not to be commensurate with experimental reality [1.98]. Nevertheless, and notwithstanding his objections, perhaps Einstein would also find it amusing that contemporary studies of both classical and quantum optics continue to generate informed criticism and new, exciting interpretations of the quantum physics whose foundations he helped to lay.

References

1.1 V. Ronchi: *The Nature of Light: An Historical Survey* (Heinemann, London 1970)

1.2 Plato: *Timaeus* (Bobbs-Merrill, Indianapolis 1959) p. 117

1.3 D. C. Lindberg: *The Beginnings of Western Science: The European Scientific Tradition in Philosphical, Religious, and Institutional Context, 600 B.C. to A.D. 1450* (Univ. Chicago Press, Chicago 1992) p. 455

1.4 D. Park: *The Fire Within the Eye: A Historical Essay on the Nature and Meaning of Light* (Princeton Univ. Press, Princeton 1997)

1.5 D. C. Lindberg: *Studies in the History of Medieval Optics* (Variorum Reprints, London 1983)

1.6 A. R. Hall: *The Revolution in Science 1500–1750*, 2 edn. (Longman, London 1984)

1.7 A. I. Sabra: *Theories of Light from Descartes to Newton* (Cambridge Univ. Press, Cambridge 1981)

1.8 L. Davidovich: Sub-Poissonian processes in quantum optics, Rev. Mod. Phys. **68**(1), 127–173 (1996)

1.9 E. Hecht: *Optics*, 3 edn. (Addison-Wesley, Reading 1998)

1.10 X. Chen: *Instrumental Traditions and Theories of Light: The Uses of Instruments in the Optical Revolution* (Kluwer, Dordrecht 2000)
1.11 I. Newton: A new theory about light and colors, Am. J. Phys. **61**(2), 108–112 (1993)
1.12 G. N. Cantor: *Optics after Newton: Theories of Light in Britain and Ireland, 1704-1840* (Manchester Univ. Press, Manchester 1983)
1.13 C. Hakfoort: *Optics in the Age of Euler: Conceptions of the Nature of Light, 1700–1795* ((Cambride Univ. Press, Cambridge 1995)
1.14 J. Z. Buchwald: *The Rise of the Wave Theory of Light: Optical Theory and Experiment in the Early Nineteenth Century* (Univ. Chicago Press, Chicago 1989)
1.15 E. Frankel: Corpuscular optics and wave theory of light – Science and politics of a revolution in physics, Social Studies Sci. **6**(2), 141–184 (1976)
1.16 E. T. Whittaker: *A History of the Theories of Aether and Electricity* (Thomas Nelson, London 1951)
1.17 B. E. A. Saleh, M. C. Teich: *Wiley Ser. Pure and Appl. Optics: Fundamentals of Photonics* (Wiley, New York 1991)
1.18 L. S. Pedrotti, F. L. Pedrotti: *Optics and Vision* (Prentice–Hall, Upper Saddle River 1998)
1.19 L. S. Pedrotti, F. L. Pedrotti: *Optics and Vision* (Prentice Hall, Saddle River 1998) p. 395
1.20 F. E. Nicodemus: Optical resource letter on radiometry, Am. J. Phys. **38**(1), 43 (1970)
1.21 G. N. Lewis: The conservation of photons, Nature **118**, 874–875 (1926)
1.22 W. T. Scott: Resource letter FC-1 on the evolution of the electromagnetic field concept, Am. J. Phys. **31**(11), 819–826 (1963)
1.23 C. A. Brau: *Modern Problems in Classical Electrodynamics* (Oxford Univ. Press, Oxford 2004)
1.24 J. D. Jackson: *Classical Electrodynamics*, 3 edn. (Wiley, New York 1999)
1.25 M. A. Paesler, P. J. Moyer: *Near-Field Optics: Theory, Instrumentation and Applications* (Wiley, New York 1996) p. 355
1.26 M. Ohtsu, H. Hori: *Near-Field Nano-Optics: From Basic Principles to Nano-Fabrication and Nano-Photonics* (Kluwer, New York 1999) p. 386
1.27 R. Kidd, J. Ardini, A. Anton: Evolution of the modern photon, Am. J. Phys. **57**(1), 27–35 (1989)
1.28 M. Planck: Deduktion der Strahlungs-Entropie aus dem zweiten Hauptsatz der Thermodynamik, Verh. Phys. Ges. **2**, 37–43 (1900)
1.29 W. Heitler: *The Quantum Theory of Radiation* (Oxford Univ. Press, Oxford 1954)
1.30 S. S. Schweber: *QED and the Men who Made It* (Princeton Univ. Press, Princeton 1994) p. 732
1.31 T. S. Kuhn: *Blackbody Theory and the Quantum Discontinuity, 1894-1912* (Oxford Univ. Press, Oxford 1978) p. 378
1.32 C. A. Kocher, E. D. Commins: Polarization correlation of photons emitted in an atomic cascade, Phys. Rev. Lett. **18**(15), 575 (1967)
1.33 J. F. Clauser: Experimental limitations to validity of semiclassical radiation theories, Phys. Rev. A **6**(1), 49 (1972)
1.34 R. Loudon: *The Quantum Theory of Light*, 2 edn. (Oxford Univ. Press, Oxford 1983)
1.35 X. L. Mao, A. J. Fernandez, R. E. Russo: "Behavior of laser induced emission intensity versus laser power density during breakdown of optical materials, Proc. SPIE **2428**, 271–280 (1995)
1.36 P. J. Duke: *Synchrotron Radiation: Production and Properties* (Oxford Univ. Press, Oxford 2000) p. 251
1.37 G. Margaritondo: *Introduction to Synchrotron Radiation* (Oxford Univ. Press, New York 1988) p. 280
1.38 C. A. Brau: *Free-Electron Lasers* (Academic, Boston 1990) p. 420
1.39 E. L. Saldin, E. V. Scheidmiller, M. V. Yurkov: *The Physics of Free Electron Lasers* (Springer, Berlin, Heidelberg 2000)
1.40 D. S. Kliger, J. W. Lewis, C. E. Randall: *Polarized Light in Optics and Spectroscopy* (Academic, San Diego 1990)
1.41 W. Demtröder: *Laser Spectroscopy*, 3 edn. (Springer, Berlin, Heidelberg 2002)
1.42 G. G. Stokes: On the composition and resolution of streams of polarized light from different sources, Trans. Cambridge Philos. Soc. **9**, 399 (1852)
1.43 W. S. Bickel, W. M. Bailey: Stokes vectors, Mueller matrices, and polarized scattered light, Am. J. Phys. **53**(5), 468–478 (1985)
1.44 S. Huard: *Polarization of Light* (Wiley, Chicester 1997)
1.45 K. Blum: *Density Matrix Theory and Applications* (Plenum, New York 1981)
1.46 U. Fano, L. Fano: *Physics of Atoms and Molecules: An Introduction to the Structure of Matter* (Univ. Chicago Press, Chicago 1972)
1.47 A. Yariv, R. C. C. Leite: Superradiant narrowing in fluorescence radiation of inverted populations, J. Appl. Phys. **34**(11), 3410–3411 (1963)
1.48 P. Atkins, R. Friedman: *Molecular Quantum Mechanics*, 4 edn. (Oxford Univ. Press, Oxford 2005)
1.49 M. Born, E. Wolf: *Principles of Optics*, 7 edn. (Cambridge Univ. Press, Cambridge 2005)
1.50 G. S. Agarwal, G. Gbur, E. Wolf: Coherence properties of sunlight, Opt. Lett. **29**(5), 459–461 (2004)
1.51 H. F. Schouten, T. D. Visser, E. Wolf: New effects in Young's interference experiment with partially coherent light, Opt. Lett. **28**(14), 1182–1184 (2003)
1.52 L. Mandel, E. Wolf: Coherence properties of optical fields, Rev. Mod. Phys. **37**(2), 231 (1965)
1.53 E. Wolf: Unified theory of coherence and polarization of random electromagnetic beams, Phys. Lett. A **312**(5–6), 263–267 (2003)
1.54 E. Wolf: Correlation-induced changes in the degree of polarization, the degree of coherence, and the spectrum of random electromagnetic beams on propagation, Opt. Lett. **28**(13), 1078–1080 (2003)
1.55 J. W. Goodman: *Statistical Optics* (Wiley, New York 2000)

1.56 H. Risken: Statistical Properties of Laser Light, Prog. Opt. **8**, 241–296 (1970)

1.57 L. Mandel, E. Wolf: *Optical Coherence and Quantum Optics* (Cambridge Univ. Press, Cambridge 1995) p. 1166

1.58 F. T. Arecchi, E. Gatti, A. Sona: Time distribution of photons from coherent and gaussian sources, Phys. Lett. **20**(1), 27 (1966)

1.59 R. Loudon: Non-classical effects in the statistical properties of light, Rep. Prog. Phys. **43**(7), 913–949 (1980)

1.60 W. H. Zurek, S. Habib, J. P. Paz: Coherent states via decoherence, Phys. Rev. Lett. **70**(9), 1187–1190 (1993)

1.61 F. Diedrich, H. Walther: Nonclassical radiation of a single stored ion, Phys. Rev. Lett. **58**(3), 203–206 (1987)

1.62 B. Darquie, M. P. A. Jones, J. J. Dingjan, J. Beugnon, S. Bergamini, Y. Sortais, G. Messin, A. Browaeys, P. Grangier: Controlled single-photon emission from a single trapped two-level atom, Science **309**(5733), 454–456 (5733)

1.63 G. Messin, J. P. Hermier, E. Giacobino, P. Desbiolles, M. Dahan: Bunching and antibunching in the fluorescence of semiconductor nanocrystals, Opt. Lett. **26**(23), 1891–1893 (2001)

1.64 E. Moreau, I. Robert, L. Manin, V. Thierry-Mieg, J. M. Gerard, I. Abram: Quantum cascade of photons in semiconductor quantum dots, Phys. Rev. Lett. **8718**, 18 (2001)

1.65 H. Walther: Generation of photons on demand using cavity quantum electrodynamics, Ann. Phys. **14**(1-3), 7–19 (2005)

1.66 A. S. Parkins, P. Marte, P. Zoller, H. J. Kimble: Synthesis of arbitrary quantum states via adiabatic transfer of zeeman coherence, Phys. Rev. Lett. **71**(19), 3095–3098 (1993)

1.67 B. T. H. Varcoe, S. Brattke, M. Weidinger, H. Walther: Preparing pure photon number states of the radiation field, Nature **403**(6771), 743–746 (6771)

1.68 M. Fox: *Quantum Optics: An Introduction* (Oxford Univ. Press, Oxford 2006)

1.69 M. C. Teich, B. E. A. Saleh: *Photon Bunching and Antibunching, Prog. Opt.* (Elsevier, Amsterdam 1988) pp. 1–106

1.70 B. L. Morgan, L. Mandel: Measurement of photon bunching in a thermal light beam, Phys. Rev. Lett. **16**(22), 1012 (1966)

1.71 H. J. Kimble, M. Dagenais, L. Mandel: Photon antibunching in resonance fluorescence, Phys. Rev. Lett. **39**(11), 691–695 (1977)

1.72 Z. L. Yuan, B. E. Kardynal, R. M. Stevenson, A. J. Shields, C. J. Lobo, K. Cooper, N. S. Beattie, D. A. Ritchie, M. Pepper: Electrically driven single-photon source, Science **295**(5552), 102–105 (5552)

1.73 J. McKeever, A. Boca, A. D. Boozer, J. R. Buck, H. J. Kimble: Experimental realization of a one-atom laser in the regime of strong coupling, Nature **425**(6955), 268–271 (6955)

1.74 M. Hennrich, A. Kuhn, G. Rempe: Transition from antibunching to bunching in cavity QED, Phys. Rev. Lett. **94**(5), 053604 (2005)

1.75 A. Beige, G. C. Hegerfeldt: Transition from antibunching to bunching for two dipole-interacting atoms, Phys. Rev. A **58**(5), 4133–4139 (1998)

1.76 X. T. Zou, L. Mandel: Photon-antibunching and subpoissonian photon statistics, Phys. Rev. A **41**(1), 475–476 (1990)

1.77 D. F. Walls: Squeezed states of light, Nature **306**(5939), 141–146 (5939)

1.78 R. W. Henry, S. C. Glotzer: A squeezed-state primer, Am. J. Phys. **56**(4), 318–328 (1988)

1.79 M. C. Teich, B. E. A. Saleh: Squeezed and Antibunched Light, Phys. Today **43**(6), 26–34 (1990)

1.80 G. Weihs, T. Jennewein, C. Simon, H. Weinfurter, A. Zeilinger: Violation of Bell's inequality under strict Einstein locality conditions, Phys. Rev. Lett. **81**(23), 5039–5043 (1998)

1.81 R. E. Slusher, L. W. Hollberg, B. Yurke, J. C. Mertz, J. F. Valley: Observation of squeezed states generated by 4-wave mixing in an optical cavity, Phys. Rev. Lett. **55**(22), 2409–2412 (1985)

1.82 R. J. Glauber: Photon correlations, Phys. Rev. Lett. **10**(3), 84–86 (1963)

1.83 R. J. Glauber: Coherent and incoherent states of the radiation field, Phys. Rev. **131**(6), 2766–2788 (1963)

1.84 R. J. Glauber: The quantum theory of optical coherence, Phys. Rev. **130**(6), 2529–2539 (1963)

1.85 E. C. G. Sudarshan: Equivalence of semiclassical and quantum mechanical descriptions of statistical light beams, Phys. Rev. Lett. **10**(7), 277–279 (1963)

1.86 E. Schrödinger: Die gegenwärtige Situation in der Quantenmechanik, Naturwiss. **48**, 807–812 (1935)

1.87 J. S. Bell: On the Einstein–Podolsky–Rosen paradox, Physics **1**, 195–200 (1964)

1.88 J. S. Bell: *Speakable and Unspeakable in Quantum mechanics* (Cambridge Univ. Press, Cambridge 1993)

1.89 J. F. Clauser, A. Shimony: Bells theorem – experimental tests and implications, Rep. Prog. Phys. **41**(12), 1881–1927 (1978)

1.90 A. Aspect, J. Dalibard, G. Roger: Experimental test of bell inequalities using time-varying analyzers, Phys. Rev. Lett. **49**(25), 1804–1807 (1982)

1.91 E. Knill, R. Laflamme, G. J. Milburn: A scheme for efficient quantum computation with linear optics, Nature **409**(6816), 46–52 (6816)

1.92 T. Jennewein, C. Simon, G. Weihs, H. Weinfurter, A. Zeilinger: Quantum cryptography with entangled photons, Phys. Rev. Lett. **84**(20), 4729–4732 (2000)

1.93 A. Kitagawa, K. Yamamoto: Teleportation-based number-state manipulation with number-sum measurement, Phys. Rev. A **68**(4), 042324 (2003)

1.94 D. Bouwmeester, J. W. Pan, K. Mattle, M. Eibl, H. Weinfurter, A. Zeilinger: Experimental quantum teleportation, Nature **390**(6660), 575–579 (1997)

1.95 A. Zeilinger, G. Weihs, T. Jennewein, M. Aspelmeyer: Happy centenary, photon, **433**(7023), 230–238 (2005)

1.96 A. Einstein: Zur Elektrodynamik bewegter Körper, Ann. Phys. **17**, 891–921 (1905)
1.97 A. Einstein: Über einen die Erzeugung und Verwandlung des Lichtes betreffenden heutistischen Gesichtspunkt, Ann. Phys. **17**, 132–148 (1905)
1.98 A. Pais: Einstein and the quantum theory, Rev. Mod. Phys. **51**, 863–914 (1979)

2. Geometrical Optics

This chapter shall discuss the basics and the applications of geometrical optical methods in modern optics. Geometrical optics has a long tradition and some ideas are many centuries old. Nevertheless, the invention of modern personal computers which can perform several million floating-point operations in a second also revolutionized the methods of geometrical optics and so several analytical methods lost importance whereas numerical methods such as ray tracing became very important. Therefore, the emphasis in this chapter is also on modern numerical methods such as ray tracing and some other systematic methods such as the paraxial matrix theory.

We will start with a section showing the transition from wave optics to geometrical optics and the resulting limitations of the validity of geometrical optics. Then, the paraxial matrix theory will be used to introduce the traditional parameters such as the focal length and the principal points of an imaging optical system. Also, an extension of the paraxial matrix theory to optical systems with non-centered elements will be briefly discussed. After a section about stops and pupils the next section will treat ray tracing and several extensions to analyze imaging and non-imaging optical systems. A section about aberrations of optical systems will give a more vivid insight into this matter than a systematic treatment. At the end a section about the most important optical instruments generally described with geometrical optics will be given. These are the achromatic lens, the camera, the human eye, the telescope, and the microscope.

For more information about the basics of geometrical optics we refer to text books such as [2.1–8].

2.1 The Basics and Limitations of Geometrical Optics

2.1.1 The Eikonal Equation

Geometrical optics is normally defined to be the limiting case of wave optics for very small wavelengths $\lambda \to 0$. In fact it is well-known that electromagnetic waves with a large wavelength λ such as radio waves cannot generally be treated with geometrical optical methods. X-rays and gamma radiation on the other hand propagate nearly like rays. They can generally be described quite well with geometrical optical methods provided the size of the optical elements (especially stops) is at least several hundred wavelengths. The accuracy of a geometrical optical calculation increases if the size of the optical element increases compared to the wavelength of the light used.

The basic equations of geometrical optics [2.1, 7] are derived directly from the Maxwell equations. The restriction here is that only linear and isotropic materials are considered. Additionally, the electric charge density ρ is assumed to be zero.

In this case the four Maxwell equations are (see Sects. 1.3, 3.1.1):

$$\nabla \times \boldsymbol{E}(\boldsymbol{r}, t) = -\frac{\partial \boldsymbol{B}(\boldsymbol{r}, t)}{\partial t}, \quad (2.1)$$

$$\nabla \times \boldsymbol{H}(\boldsymbol{r}, t) = \frac{\partial \boldsymbol{D}(\boldsymbol{r}, t)}{\partial t} + \boldsymbol{j}(\boldsymbol{r}, t), \quad (2.2)$$

$$\nabla \cdot \boldsymbol{B}(\boldsymbol{r}, t) = 0, \quad (2.3)$$

$$\nabla \cdot \boldsymbol{D}(\boldsymbol{r}, t) = 0, \quad (2.4)$$

where the following quantities of the electromagnetic field are involved: electric vector $\boldsymbol{E}$, magnetic vector $\boldsymbol{H}$, electric displacement $\boldsymbol{D}$, magnetic induction $\boldsymbol{B}$ and electric current density $\boldsymbol{j}$. The arguments illustrate that all quantities are in the general case functions of the spatial coordinates x, y, z with position vector $\boldsymbol{r} = (x, y, z)$ and of the time t. The so called nabla operator

$$\nabla = \begin{pmatrix} \frac{\partial}{\partial x} \\ \frac{\partial}{\partial y} \\ \frac{\partial}{\partial z} \end{pmatrix} \quad (2.5)$$

is used and the symbol "×" indicates the vector product of two vectors whereas "·" indicates the scalar product of two vectors.

The material equations in the case of linear and isotropic materials link the electromagnetic quantities with each other:

$$\boldsymbol{D}(\boldsymbol{r}, t) = \epsilon(\boldsymbol{r})\epsilon_0 \boldsymbol{E}(\boldsymbol{r}, t), \quad (2.6)$$

$$\boldsymbol{B}(\boldsymbol{r}, t) = \mu(\boldsymbol{r})\mu_0 \boldsymbol{H}(\boldsymbol{r}, t), \quad (2.7)$$

$$\boldsymbol{j}(\boldsymbol{r}, t) = \sigma(\boldsymbol{r}) \boldsymbol{E}(\boldsymbol{r}, t). \quad (2.8)$$

The function ϵ is the dielectric function, μ is the magnetic permeability and σ is the specific conductivity. The constants ϵ_0 and μ_0 are the dielectric constant of vacuum and the magnetic permeability of vacuum, respectively. A quite general approach for stationary monochromatic waves is used to describe the electric and the magnetic field

$$\boldsymbol{E}(\boldsymbol{r}, t) = \boldsymbol{e}(\boldsymbol{r}) e^{ik_0 L(\boldsymbol{r})} e^{-i\omega t}, \quad (2.9)$$

$$\boldsymbol{H}(\boldsymbol{r}, t) = \boldsymbol{h}(\boldsymbol{r}) e^{ik_0 L(\boldsymbol{r})} e^{-i\omega t}. \quad (2.10)$$

The real function L is the optical path length and the vectors $\boldsymbol{e}$ and $\boldsymbol{h}$ are in the general case complex-valued to be able to represent all polarization states. The surfaces with constant optical path length L are the wave fronts, and the term $\Phi(\boldsymbol{r}) = k_0 L(\boldsymbol{r})$ is the phase of the wave; $\boldsymbol{e}$ and $\boldsymbol{h}$ are slowly varying functions of the position $\boldsymbol{r}$ whereas the term $\exp(ik_0 L)$ varies rapidly because the constant k_0 is defined as $k_0 = 2\pi/\lambda$ with the vacuum wavelength λ. The angular frequency ω of the wave is linked to λ by $\omega = 2\pi c/\lambda = ck_0$, where c is the speed of light in vacuum.

By applying these equations to the Maxwell equations the so called time-independent Maxwell equations result:

$$\nabla \times \left[\boldsymbol{e}(\boldsymbol{r}) e^{ik_0 L(\boldsymbol{r})}\right] = i\omega\mu(\boldsymbol{r})\mu_0 \boldsymbol{h}(\boldsymbol{r}) \cdot e^{ik_0 L(\boldsymbol{r})}, \quad (2.11)$$

$$\nabla \times \left[\boldsymbol{h}(\boldsymbol{r}) e^{ik_0 L(\boldsymbol{r})}\right] = [-i\omega\epsilon(\boldsymbol{r})\epsilon_0 + \sigma(\boldsymbol{r})] \cdot \boldsymbol{e}(\boldsymbol{r}) e^{ik_0 L(\boldsymbol{r})}, \quad (2.12)$$

$$\nabla\cdot\left[\mu(\boldsymbol{r})\mu_0\boldsymbol{h}(\boldsymbol{r})\mathrm{e}^{\mathrm{i}k_0L(\boldsymbol{r})}\right]=0\,, \tag{2.13}$$

$$\nabla\cdot\left[\epsilon(\boldsymbol{r})\epsilon_0\boldsymbol{e}(\boldsymbol{r})\mathrm{e}^{\mathrm{i}k_0L(\boldsymbol{r})}\right]=0\,. \tag{2.14}$$

Equation (2.13) is not independent of (2.11) because it is well known that the quantity $\nabla\cdot(\nabla\times\boldsymbol{f})$ of an arbitrary vector function $\boldsymbol{f}$ is always zero [2.9]. Therefore, if (2.11) is fulfilled, (2.13) will also be fulfilled. In the case of nonconducting materials, i. e. $\sigma=0$, the same is valid for the relation between (2.12) and (2.14). In the more general case $\sigma\neq 0$, (2.12) and (2.14) require that

$$\nabla\cdot\left[\sigma(\boldsymbol{r})\boldsymbol{e}(\boldsymbol{r})\mathrm{e}^{\mathrm{i}k_0L(\boldsymbol{r})}\right]=0\,. \tag{2.15}$$

Using the rules of the nabla calculus the left-hand sides of (2.11) and (2.12) can be transformed to

$$\begin{aligned}&\nabla\times\left[\boldsymbol{e}(\boldsymbol{r})\mathrm{e}^{\mathrm{i}k_0L(\boldsymbol{r})}\right]\\&=\nabla\left(\mathrm{e}^{\mathrm{i}k_0L(\boldsymbol{r})}\right)\times\boldsymbol{e}(\boldsymbol{r})+\mathrm{e}^{\mathrm{i}k_0L(\boldsymbol{r})}\nabla\times\boldsymbol{e}(\boldsymbol{r})\\&=\left[\mathrm{i}k_0\nabla L(\boldsymbol{r})\times\boldsymbol{e}(\boldsymbol{r})+\nabla\times\boldsymbol{e}(\boldsymbol{r})\right]\mathrm{e}^{\mathrm{i}k_0L(\boldsymbol{r})}\,,\end{aligned} \tag{2.16}$$

$$\begin{aligned}&\nabla\times\left[\boldsymbol{h}(\boldsymbol{r})\mathrm{e}^{\mathrm{i}k_0L(\boldsymbol{r})}\right]\\&=\left[\mathrm{i}k_0\nabla L(\boldsymbol{r})\times\boldsymbol{h}(\boldsymbol{r})+\nabla\times\boldsymbol{h}(\boldsymbol{r})\right]\mathrm{e}^{\mathrm{i}k_0L(\boldsymbol{r})}\,.\end{aligned} \tag{2.17}$$

So, (2.11) and (2.12) give:

$$\begin{aligned}&\nabla L(\boldsymbol{r})\times\boldsymbol{e}(\boldsymbol{r})-c\mu(\boldsymbol{r})\mu_0\boldsymbol{h}(\boldsymbol{r})\\&=\frac{\mathrm{i}}{k_0}\nabla\times\boldsymbol{e}(\boldsymbol{r})\,,\end{aligned} \tag{2.18}$$

$$\begin{aligned}&\nabla L(\boldsymbol{r})\times\boldsymbol{h}(\boldsymbol{r})+c\epsilon(\boldsymbol{r})\epsilon_0\boldsymbol{e}(\boldsymbol{r})\\&=\frac{\mathrm{i}}{k_0}\left[\nabla\times\boldsymbol{h}(\boldsymbol{r})-\sigma(\boldsymbol{r})\boldsymbol{e}(\boldsymbol{r})\right]\,.\end{aligned} \tag{2.19}$$

For the limiting case $\lambda\to 0$, i. e. $k_0\to\infty$, the right sides of both equations become zero

$$\nabla L(\boldsymbol{r})\times\boldsymbol{e}(\boldsymbol{r})-c\mu(\boldsymbol{r})\mu_0\boldsymbol{h}(\boldsymbol{r})=0\,, \tag{2.20}$$

$$\nabla L(\boldsymbol{r})\times\boldsymbol{h}(\boldsymbol{r})+c\epsilon(\boldsymbol{r})\epsilon_0\boldsymbol{e}(\boldsymbol{r})=0\,. \tag{2.21}$$

Now, (2.20) is inserted into (2.21) and the calculus for a double vector product is applied

$$\begin{aligned}&\frac{1}{c\mu(\boldsymbol{r})\mu_0}\nabla L(\boldsymbol{r})\times[\nabla L(\boldsymbol{r})\times\boldsymbol{e}(\boldsymbol{r})]+c\epsilon(\boldsymbol{r})\epsilon_0\boldsymbol{e}(\boldsymbol{r})=0\\&\Rightarrow\quad[\nabla L(\boldsymbol{r})\cdot\boldsymbol{e}(\boldsymbol{r})]\,\nabla L(\boldsymbol{r})\\&\qquad-[\nabla L(\boldsymbol{r})]^2\,\boldsymbol{e}(\boldsymbol{r})+n^2(\boldsymbol{r})\boldsymbol{e}(\boldsymbol{r})=0\,.\end{aligned} \tag{2.22}$$

Here, $\mu_0\epsilon_0=1/c^2$ and $\mu\epsilon=n^2$ are used, where n is the refractive index of the material.

Equation (2.21) shows that the scalar product $\nabla L\cdot\boldsymbol{e}$ is zero and the final result is the well-known eikonal equation

$$[\nabla L(\boldsymbol{r})]^2=n^2(\boldsymbol{r})\,. \tag{2.23}$$

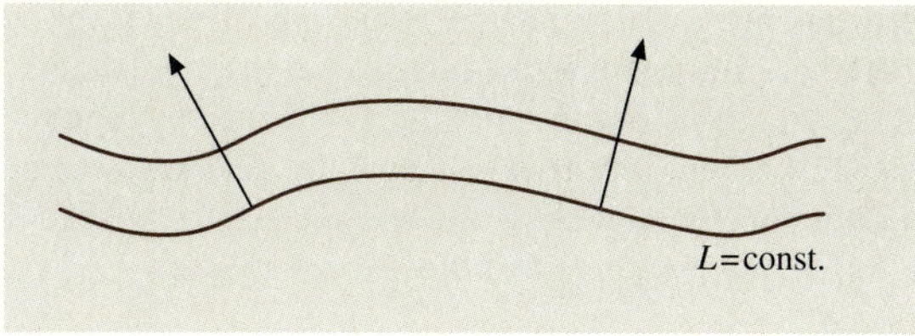

Fig. 2.1 Optical ray: the trajectory which is perpendicular to the surfaces of equal optical path length L

This is the basic equation of geometrical optics which provides e.g. the basis for the concept of optical rays. A ray is defined as that trajectory which is always perpendicular to the wave fronts, which are the surfaces of equal optical path length L (Fig. 2.1). Therefore, a ray points in the direction of ∇L. Equation (2.23) has the name eikonal equation because the optical path length L is for historical reasons sometimes called the eikonal [2.1].

2.1.2 The Orthogonality Condition of Geometrical Optics

Equations (2.20) and (2.21) can be solved for $\boldsymbol{e}$ and $\boldsymbol{h}$:

$$\boldsymbol{h}(\boldsymbol{r})=\frac{1}{c\mu(\boldsymbol{r})\mu_0}\nabla L(\boldsymbol{r})\times\boldsymbol{e}(\boldsymbol{r})\,, \tag{2.24}$$

$$\boldsymbol{e}(\boldsymbol{r})=-\frac{1}{c\epsilon(\boldsymbol{r})\epsilon_0}\nabla L(\boldsymbol{r})\times\boldsymbol{h}(\boldsymbol{r})\,. \tag{2.25}$$

This shows on the one hand that $\boldsymbol{h}$ is perpendicular to $\boldsymbol{e}$ as well as ∇L and on the other hand that $\boldsymbol{e}$ is perpendicular to $\boldsymbol{h}$ as well as ∇L. Therefore, in the limiting case $\lambda\to 0$ ∇L, $\boldsymbol{e}$ and $\boldsymbol{h}$ have to form an orthogonal triad of vectors. This confirms the well-known fact that electromagnetic waves are transversal waves.

At the end of the last section a light ray has been defined as being parallel to ∇L and in Sect. 2.4 the important method of ray tracing will be explained. An extended method of ray tracing is polarization ray tracing where the polarization state of a ray which locally represents a wave is transported along with each ray [2.10, 11]. Using the results of this section it is clear that the vector $\boldsymbol{e}$, which indicates the polarization (and amplitude) of the ray, has to be perpendicular to the ray direction ∇L.

2.1.3 The Ray Equation

A surface with constant values L is a surface of equal optical path length. Now, a ray is defined as that trajectory starting from a certain point in space which is perpendicular to the surfaces of equal optical path length. Therefore, ∇L points in the direction of the ray. We use the arc length s along the curve that is defined by the ray (Fig. 2.2). Then, if $\boldsymbol{r}$ describes now the position vector of a point on the ray, $d\boldsymbol{r}/ds$ is a unit vector which is tangential to the ray curve and the eikonal equation (2.23) delivers:

$$\nabla L = n\frac{d\boldsymbol{r}}{ds} \tag{2.26}$$

Here and in the following L and n are not explicitly indicated as functions of the position to tighten the notation. From (2.26) a differential equation for the ray can be derived by again using (2.23) and the definition of $d\nabla L/ds$ as being the directional derivative of ∇L along $\mathrm{d}\boldsymbol{r}/\mathrm{d}s$:

$$\begin{aligned} \frac{\mathrm{d}}{\mathrm{d}s}\left(n\frac{\mathrm{d}\boldsymbol{r}}{\mathrm{d}s}\right) &= \frac{\mathrm{d}}{\mathrm{d}s}\nabla L = \frac{\mathrm{d}\boldsymbol{r}}{\mathrm{d}s}\cdot\nabla(\nabla L) \\ &= \frac{1}{n}\nabla L\cdot\nabla(\nabla L) = \frac{1}{2n}\nabla(\nabla L)^2 \\ &= \frac{1}{2n}\nabla n^2 = \nabla n\,, \\ \Rightarrow \quad \frac{\mathrm{d}}{\mathrm{d}s}\left(n\frac{\mathrm{d}\boldsymbol{r}}{\mathrm{d}s}\right) &= \nabla n\,. \end{aligned} \tag{2.27}$$

This is the differential equation for a ray in a general inhomogeneous isotropic and linear substance. Such materials where the refractive index is a function of the position are often called graded index (GRIN) materials. In this case the solution of the differential equation may be a quite complex curve.

However, the most important case is that n is independent of the position, i.e. the ray propagates in a homogeneous material. Then, a simple differential equation for the ray is obtained:

$$\frac{\mathrm{d}^2\boldsymbol{r}}{\mathrm{d}s^2} = 0\,. \tag{2.28}$$

The solution of this equation is a straight line. So, the ray equation in homogeneous materials is

$$\boldsymbol{r} = s\boldsymbol{a} + \boldsymbol{p} \tag{2.29}$$

with constant vectors $\boldsymbol{a}$ and $\boldsymbol{p}$. This means that light rays propagate rectilinearly in a homogeneous and isotropic material if the eikonal equation is valid. Moreover, $\boldsymbol{a}$ has to be a unit vector, i.e. $|\boldsymbol{a}| = 1$, because s is the geometrical path length along the ray. So, for the distance between two points P_1 and P_2 with position vectors $\boldsymbol{r}_1$ and $\boldsymbol{r}_2$ we have ($s_2 > s_1$):

$$|\boldsymbol{r}_2 - \boldsymbol{r}_1| = (s_2 - s_1)\,|\boldsymbol{a}| = s_2 - s_1 \;\Rightarrow\; |\boldsymbol{a}| = 1\,.$$

The limitations of the validity of the eikonal equation will be investigated in the next section.

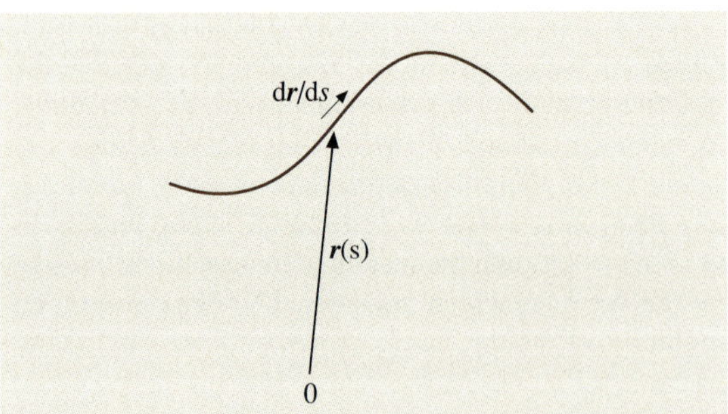

Fig. 2.2 A curved light ray in a general (inhomogeneous) material. $\boldsymbol{r}(s)$ is the position vector of a point on the ray, where s is the arc length along the curve and O is the origin of the coordinate system. Then, the vector $\mathrm{d}\boldsymbol{r}/\mathrm{d}s$ is a unit vector tangential to the ray

2.1.4 Limitations of the Eikonal Equation

Besides using directly the Maxwell equations the eikonal equation can also be derived from the wave equation and, in the case of a monochromatic wave, from the Helmholtz equation. This will be done in the following for a homogeneous, isotropic and linear dielectric material, i.e. where n is constant and $\sigma = 0$. Moreover, it is assumed that the scalar case is valid, i.e. that polarization effects can be neglected and only one component $u(\boldsymbol{r})$ of the electric or magnetic vector has to be considered. In this limiting case it is easier to start directly with the scalar Helmholtz equation (Sect. 3.1.5) than to start as in Sect. 2.1.1 with the Maxwell equations and then to make the transition to the scalar case.

The scalar Helmholtz equation is:

$$\left[\nabla\cdot\nabla + (nk_0)^2\right]u(\boldsymbol{r}) = 0\,. \tag{2.30}$$

Analogously to (2.9) or (2.10) the following approach for u is used

$$u(\boldsymbol{r}) = A(\boldsymbol{r})\,\mathrm{e}^{\mathrm{i}k_0 L(\boldsymbol{r})}\,, \tag{2.31}$$

where the amplitude A and the optical path length L are both real functions of the position and A varies only slowly with the position.

Then, by omitting the arguments of the functions we can write

$$\begin{aligned}\nabla u &= \nabla\left(A\,\mathrm{e}^{\mathrm{i}k_0 L}\right) = \mathrm{e}^{\mathrm{i}k_0 L}\nabla A + \mathrm{i}k_0 A\,\mathrm{e}^{\mathrm{i}k_0 L}\nabla L\\ &= \left(\frac{\nabla A}{A} + \mathrm{i}k_0 \nabla L\right) u\,,\\ \Delta u &= \nabla\cdot\left[\left(\frac{\nabla A}{A} + \mathrm{i}k_0\nabla L\right)u\right]\\ &= \left(\frac{\nabla A}{A} + \mathrm{i}k_0\nabla L\right)^2 u\\ &\quad + \left(\frac{\Delta A}{A} - \frac{(\nabla A)^2}{A^2} + \mathrm{i}k_0\Delta L\right)u\\ &= \left[\frac{\Delta A}{A} - k_0^2(\nabla L)^2\right.\\ &\quad \left. + 2\mathrm{i}k_0\frac{\nabla A\cdot\nabla L}{A} + \mathrm{i}k_0\Delta L\right]u\,.\end{aligned}$$

Here, $\Delta := \nabla\cdot\nabla$ is the Laplace operator or Laplacian. So, by inserting the expression for Δu into the Helmholtz equation and dividing it by u the result is

$$\begin{aligned}&\frac{\Delta A}{A} - k_0^2(\nabla L)^2 + n^2k_0^2 + 2\mathrm{i}k_0\frac{\nabla A\cdot\nabla L}{A}\\ &\quad + \mathrm{i}k_0\Delta L = 0\,. \end{aligned} \tag{2.32}$$

Since A, L, k_0 and n are all real quantities the real and the imaginary part of (2.32) can be simply separated and both have to be zero.

To obtain the eikonal equation only the real part is considered:

$$\begin{aligned}&\frac{\Delta A}{A} - k_0^2(\nabla L)^2 + n^2k_0^2 = 0\,,\\ \Rightarrow\quad &(\nabla L)^2 = n^2 + \underbrace{\frac{1}{k_0^2}\frac{\Delta A}{A}}_{=:\gamma}\,.\end{aligned} \tag{2.33}$$

In the limiting case $\lambda \to 0 \Rightarrow k_0 \to \infty$ the term γ can be neglected and again the eikonal equation (2.23) is obtained

$$(\nabla L)^2 = n^2\,.$$

But (2.33) shows that the eikonal equation can also be fulfilled for a finite value of λ with good approximation as long as the term γ is much smaller than 1, because the order of magnitude of n^2 is typically between 1 (vacuum) and 12 (silicon for infrared light). Therefore, the condition is:

$$\gamma \ll 1 \Rightarrow \frac{\lambda^2}{4\pi^2}\frac{\Delta A}{A} \ll 1\,. \tag{2.34}$$

This is fulfilled with good approximation if A is a slowly varying function of the position, i. e. if the relative curvature of A over the distance of a wavelength is very small. If the term γ is not very small the right-hand side of (2.33) depends on the position (because A depends generally on $\boldsymbol{r}$) even though n is constant. Formally this is equivalent to an eikonal equation with position-dependent refractive index n so that light rays would formally be bent in regions of a rapidly changing amplitude, as e.g. in the focus. Therefore, the results of ray-tracing calculations (Sect. 2.4), which assume rectilinear rays in a homogeneous material, are not correct in the neighborhood of the focus where the amplitude changes rapidly. If aberrations are present the variation of the amplitude in the focal region will be less severe and the accuracy of geometrical optical calculations improves with increasing aberrations. In practice, a rule of thumb is that the focal region of an aberrated wave calculated with ray tracing approximates the actual focus very well if the result of the ray-tracing calculation gives a focus that has several times the size of the corresponding diffraction limited focus (Airy disc), which can easily be estimated (Sect. 3.4, PSF (point spread function)).

There are also scalar waves which fulfill exactly the eikonal equation so that the term γ is exactly zero. One example is a plane wave with $u(\boldsymbol{r}) = u_0\exp(\mathrm{i}nk_0\boldsymbol{a}\cdot\boldsymbol{r})$. $\boldsymbol{a}$ is a constant unit vector in the direction of propagation and u_0 is also a constant. So, we have

$$\begin{aligned}&A = u_0 \Rightarrow \Delta A = 0 \Rightarrow \gamma = 0\,,\\ &L = n\boldsymbol{a}\cdot\boldsymbol{r} \Rightarrow \nabla L = n\boldsymbol{a} \Rightarrow (\nabla L)^2 = n^2\,.\end{aligned}$$

Of course, a plane wave is also a solution of the Maxwell equations.

A second example is a spherical wave, which is a solution of the scalar Helmholtz equation but not of the Maxwell equations themselves because the orthogonality conditions (2.24) and (2.25) cannot be fulfilled for a spherical wave in the whole of space. Nevertheless, a spherical wave $u(r) = u_0\exp(\mathrm{i}nk_0 r)/r$ with $r = |\boldsymbol{r}|$ is a very important approximation in many cases and a dipole radiation behaves in the far field in a plane perpendicular to the dipole axis like a spherical wave. For the spherical wave we obtain:

$$\begin{aligned}A &= \frac{u_0}{r} \Rightarrow \Delta A = u_0\nabla\cdot\left(-\frac{\boldsymbol{r}}{r^3}\right)\\ &= -\frac{3u_0}{r^3} + \frac{3u_0\boldsymbol{r}\cdot\boldsymbol{r}}{r^5} = 0 \Rightarrow \gamma = 0\,,\\ L &= nr \Rightarrow \nabla L = n\frac{\boldsymbol{r}}{r} \Rightarrow (\nabla L)^2 = n^2\,.\end{aligned}$$

Here, the coordinate system has been chosen in such a way that the center of curvature of the spherical wave is at the origin. Of course, it is quite straightforward to formulate the spherical wave with an arbitrary center of curvature $\boldsymbol{r}_0$ by replacing r with $|\boldsymbol{r}-\boldsymbol{r}_0|$.

So, plane waves and spherical waves, which are very important in geometrical optics, fulfill both the eikonal equation (2.23) not only in the limiting case $\lambda \to 0$ but also for finite wavelengths λ.

2.1.5 Energy Conservation in Geometrical Optics

The imaginary part of (2.32) gives information about the intensity of the amount of transported light:

$$\Delta L + 2\nabla L \cdot \frac{\nabla A}{A} = 0 . \tag{2.35}$$

Since the intensity I of a light wave is proportional to the square of the amplitude A^2 the following equality holds:

$$\frac{\nabla I}{I} = \frac{\nabla\left(A^2\right)}{A^2} = \frac{2A\nabla A}{A^2} = 2\frac{\nabla A}{A} .$$

Therefore, (2.35) delivers

$$I\Delta L + \nabla L \cdot \nabla I = 0$$

or

$$\nabla \cdot (I\nabla L) = 0 . \tag{2.36}$$

Now, the integral theorem of Gauss can be applied

$$\int_V \nabla \cdot (I\nabla L)\, \mathrm{d}V = \oint_S I\nabla L \cdot \mathrm{d}\boldsymbol{S} = 0 , \tag{2.37}$$

where the left integral symbolizes a volume integral over a volume V and the right integral a surface integral over the closed surface S which confines the volume V.

A light tube (Fig. 2.3) is a tube-like entity (simple forms are e.g. a cylinder or a cone) where light rays form the mantle surface. Therefore, on the mantle surface the vectors ∇L (ray direction) and $\mathrm{d}\boldsymbol{S}$ (surface normal) are perpendicular to each other and therefore $\nabla L \cdot \mathrm{d}\boldsymbol{S} = 0$. At the two face surfaces of the light tube (refractive index n) with surface vectors $\mathrm{d}\boldsymbol{S}_1$ and $\mathrm{d}\boldsymbol{S}_2$, which are assumed to have an infinitesimally small size, the electromagnetic power flux P_1 and P_2 are given by

$$P_j = \frac{I_j}{n}|\nabla L_j \cdot \mathrm{d}\boldsymbol{S}_j| ; \quad j \in \{1, 2\} .$$

Using (2.37) the result is

$$\oint_S I\nabla L \cdot \mathrm{d}\boldsymbol{S} = 0 = I_1\nabla L_1 \cdot \mathrm{d}\boldsymbol{S}_1 + I_2\nabla L_2 \cdot \mathrm{d}\boldsymbol{S}_2 . \tag{2.38}$$

Hereby, the surface normals $\mathrm{d}\boldsymbol{S}_1$ or $\mathrm{d}\boldsymbol{S}_2$ always point out of the closed surface S and therefore $\nabla L_1 \cdot \mathrm{d}\boldsymbol{S}_1$ and $\nabla L_2 \cdot \mathrm{d}\boldsymbol{S}_2$ have opposite algebraic signs. In total, the power flux $P_1 = -I_1\nabla L_1 \cdot \mathrm{d}\boldsymbol{S}_1/n$ which enters the light tube at the left is equal to the power flux $P_2 = I_2\nabla L_2 \cdot \mathrm{d}\boldsymbol{S}_2/n$ which leaves the light tube at the right: $P_1 = P_2$. This means that the energy is conserved and we can formulate the following lemma: In the scope of geometrical optics the electromagnetic power (energy) is transported along the light rays and the total light power is conserved in a light tube if no light is absorbed.

2.1.6 Law of Refraction

Let us consider the interface between two materials with refractive index n_1 on one side and n_2 on the other. This interface is assumed to be replaced by a very thin layer in which the refractive index varies quite rapidly but continuously from n_1 to n_2. An infinitesimally small rectangular closed loop C is then constructed at the interface in such a way that two of the edges of the loop are parallel to the interface and the other two edges are parallel to the surface normal $\boldsymbol{N}$ ($|\boldsymbol{N}| = 1$) of the interface

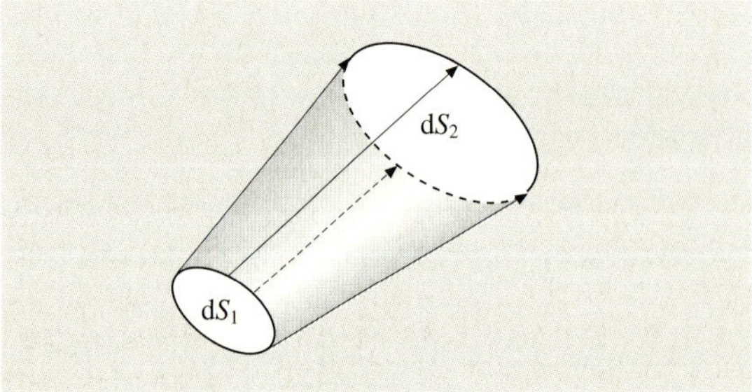

Fig. 2.3 Scheme of a light tube which is bounded by a bundle of rays

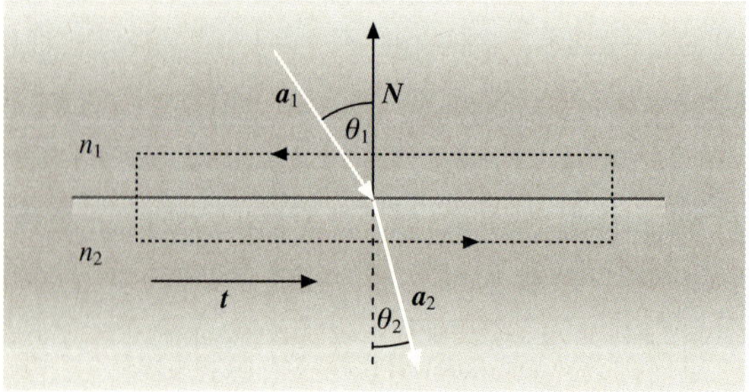

Fig. 2.4 The parameters used in deriving Snell's law

where N points from material 2 to material 1 (Fig. 2.4). Since the direction vectors a of the light rays can be expressed as the gradient of a scalar function (2.26) the following identity is valid:

$$\nabla\times\left(n\frac{\mathrm{d}r}{\mathrm{d}s}\right)=\nabla\times\nabla L=0\,. \qquad (2.39)$$

The ray direction vector is written in the following as $a=\mathrm{d}r/\mathrm{d}s$.

Using the integral theorem of Stokes equation (2.39) delivers

$$\int_S \nabla\times(na)\cdot\mathrm{d}S=\oint_C na\cdot\mathrm{d}r=0\,,$$

where the left integral is a surface integral over the infinitesimally small rectangular surface S which is bounded by the closed loop C. The right integral is a line integral over the closed loop C.

If now the length of the side lines of the loop C parallel to N tends to zero the line integral is

$$0=lt\cdot(n_2a_2-n_1a_1) \qquad (2.40)$$

with l being the length of a side line of the loop parallel to the interface and t being a unit vector parallel to the interface. Another unit vector b is defined as being perpendicular to both N and t and therefore also perpendicular to the surface S. This means that N, t and b form an orthogonal triad of unit vectors with $t=b\times N$ and therefore it holds that

$$(b\times N)\cdot(n_2a_2-n_1a_1)=0\,,$$
$$\Rightarrow\quad [N\times(n_2a_2-n_1a_1)]\cdot b=0\,.$$

But the rectangular integration area can be rotated about N serving as axis. Therefore, the direction of b can be chosen arbitrarily as long as it is perpendicular to N. By fulfilling the upper equation for an arbitrary vector b we obtain the following equation, which is the vectorial formulation of the law of refraction,

$$N\times(n_2a_2-n_1a_1)=0\,. \qquad (2.41)$$

This means that $n_2a_2-n_1a_1$ is parallel to N (or $n_2a_2-n_1a_1=0$ what is only possible for the trivial case $n_1=n_2$) and therefore all three vectors a_1, a_2 and N have to lie in the same plane. This means particularly that the refracted ray with direction vector a_2 lies in the plane of incidence formed by N and a_1.

By defining the acute angles θ_j between the rays a_j ($j\in\{1,2\}$) and the surface normal N (Fig. 2.4) the modulus of (2.41) results in:

$$n_1\sin\theta_1=n_2\sin\theta_2\,. \qquad (2.42)$$

This is the well-known Snell's law.

If n_2 is bigger than n_1 there is always a solution θ_2 for a given angle θ_1. But, if n_2 is smaller than n_1 there is the so-called critical angle of total internal reflection $\theta_{1,\text{critical}}$ for which the refracted ray grazes parallel to the interface, i. e. $\theta_2=\pi/2$:

$$n_1\sin\theta_{1,\text{critical}}=n_2\sin\theta_2=n_2$$
$$\Rightarrow\ \theta_{1,\text{critical}}=\arcsin\frac{n_2}{n_1}\,. \qquad (2.43)$$

For angles $\theta_1>\theta_{1,\text{critical}}$ there exists no refracted ray because the sine function $\sin\theta_2$ cannot be larger than 1. Then, all light is reflected at the interface and only a reflected ray exists.

2.1.7 Law of Reflection

If a plane wave enters the interface between two materials there is, as well as a refracted wave, a reflected wave and in the case of total internal reflection there is only a reflected wave. A ray represents locally a plane wave and the law of reflection is formally obtained from (2.41) by setting $n_2=n_1$. Of course, the algebraic signs of the scalar products $N\cdot a_1$ and $N\cdot a_2$ have to be different in order to obtain a reflected ray whereas they have to be identical to obtain a refracted ray. This will be discussed later in Sect. 2.4 by finding explicit solutions of (2.41).

After having now discussed the basics of geometrical optics the next section will treat the paraxial ray tracing through an optical system by using a matrix theory [2.12–15].

2.2 Paraxial Geometrical Optics

2.2.1 Paraxial Rays in Homogeneous Materials

Some Basic Definitions

In a homogeneous material the refractive index n is constant and therefore, according to the ray equation (2.29), a ray propagates rectilinear. This means that a ray is definitely described by the position vector p of one point on the ray and the ray direction vector a. So, six scalar parameters (each vector with three components) are necessary. In principle one component of a is redundant (apart from the algebraic sign of this component) because $\boldsymbol{a}$ is a unit vector ($a_x^2+a_y^2+a_z^2=1$). Another component can be saved if a reference plane is defined, e.g. the x–y plane at $z=0$. Then, the x- and y-components of the point of intersection of the ray with this plane are sufficient. However, in the case of non-paraxial ray tracing (Sect. 2.4) all components of the two vectors $\boldsymbol{p}$ and $\boldsymbol{a}$ are stored and used since not all rays start in the same plane and the algebraic sign of each component of $\boldsymbol{a}$ is needed. Moreover, it is often more efficient to store a redundant parameter instead of calculating it from other parameters.

Most convenient optical systems consist of a sequence of rotationally symmetric centered refractive and reflective components. The rotation axis is called the optical axis of the system. For a simple lens with two spherical surfaces the optical axis is defined by the two centers of curvature C_1 and C_2 of the spherical surfaces (Fig. 2.5). Using (2.41) it has been shown that a refracted ray (and also a reflected ray) remains in the plane of incidence. Therefore, it is useful to define the meridional plane, which is a plane containing an object point P and the optical axis (Fig. 2.6). All rays which come from the object point P and lie in the meridional plane are called meridional rays. A plane perpendicular to the meridional plane which contains a special reference ray, mostly the chief ray (Sect. 2.3), is called the sagittal plane and rays lying in it are called sagittal rays. In this section only meridional rays are discussed and moreover only so-called paraxial rays are considered. Paraxial rays are rays which fulfill the following conditions:

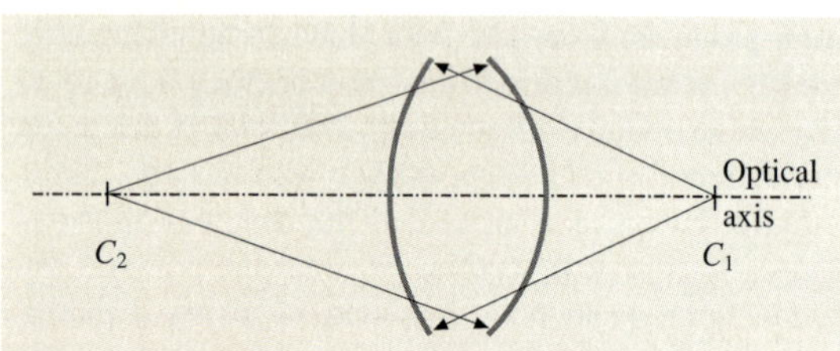

Fig. 2.5 The optical axis of a lens

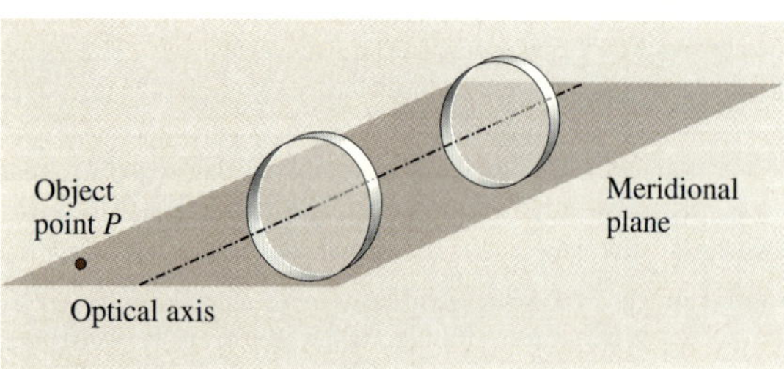

Fig. 2.6 The meridional plane of an optical system

- The distance x of the ray from the optical axis is small compared to the focal length of each optical element of the system.
- The angle φ between the optical axis and the ray is small, i.e. $\varphi \ll 1$. The same has to be valid for other angles, e.g. for the refraction angles at a lens.

For the angles this means that the following approximations have to be valid:

$$\sin\varphi \approx \tan\varphi \approx \varphi\,, \quad \cos\varphi \approx 1\,.$$

The most important optical systems consist of optical elements (refractive, reflective or diffractive elements) which are embedded into piecewise homogeneous materials. Therefore, ray tracing (in the paraxial as well as non-paraxial case) through an optical system consists of the alternating sequence of propagation in a homogeneous material and refraction (or reflection or diffraction) at an element.

Optical Imaging

At this point some words on the term optical imaging have to be said. An object point which is either illuminated by external light or self-illuminating emits a ray fan, i.e. in geometrical optics an object point is the source of a ray fan. On the other side an image point is the drain of a ray fan and in the ideal case all rays of the fan should intersect each other in the image point (Fig. 2.7a). Therefore, the image point can be defined in the ideal case by the point of intersection of only two rays. However, this is only useful in the case of paraxial ray tracing where all aberrations of the optical system are neglected. If the aberrations of an optical system also have to be taken into account, non-paraxial ray tracing (Sect. 2.4), also simply called ray tracing, has to be used. Then, there are several definitions of an image point because there is in general no longer a single point of intersection of all rays of the ray fan coming from the object point (Fig. 2.7b). The lateral deviation of the actual point of intersection of a ray with the image plane from the ideal image point is called ray aberration.

A more advanced definition of optical imaging has of course to take into account interference effects between the different rays coming from the object point since the image point is a multiple-beam interference phenomenon. A typical example where the simple ray-based model fails would be an ideal spherical wave where a half-wave plate is introduced in half of the aperture (Fig. 2.7c). Then, the ray directions are unchanged and an ideal point of intersection of all rays exists, i.e. there are no ray aberrations. But, the image point would be

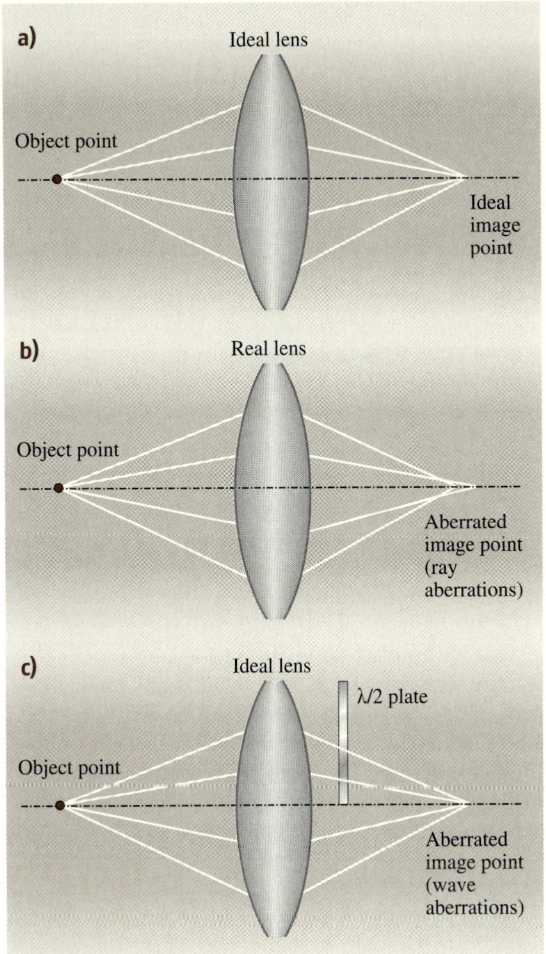

Fig. 2.7a–c Schematic display of three different situations in optical imaging: **(a)** ideal imaging, **(b)** image point showing ray aberrations (and of course also wave aberrations), **(c)** image point showing no ray aberrations but nevertheless wave aberrations

massively disturbed because there is destructive interference in the center of the image point due to the different optical path lengths of the rays. Therefore, a more advanced ray-based model additionally calculates the optical path length along each ray. The deviation in the optical path length of a ray from the ideal optical path length is called wave aberration.

However, in this section we will treat the very simple model of paraxial ray tracing, which takes into account neither ray aberrations nor wave aberrations. Aberrations will be taken into account in Sect. 2.4 when we consider non-paraxial ray tracing.

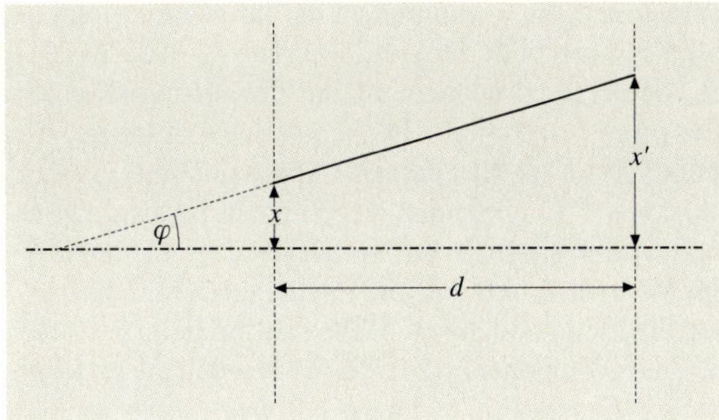

Fig. 2.8 Scheme showing the parameters of paraxial ray tracing for the transfer between two parallel planes with distance d

A Note on the Validity of the Paraxial Approximation

The approximation of $\sin\varphi$ by φ means that the next term of the Taylor series $-\varphi^3/6$ and all higher order terms are neglected. In the case of $\tan\varphi$ the next term of the Taylor series which is neglected is $+\varphi^3/3$. So, the equivalence of $\sin\varphi$ and $\tan\varphi$ is only valid if the difference of both third-order terms $\varphi^3/2$ is so small that it can be neglected. This is the case if the alternation of the optical path length from the object point to the image point by neglecting this term is smaller than the Rayleigh criterion of $\lambda/4$, where λ is again the wavelength. In the case of two rays with an optical path difference of $\lambda/4$ the phase difference is $\Delta\Phi = \pi/2$, i. e. the rays are in phase quadrature and the intensities have to be added because the interference term $\cos(\Delta\Phi)$ is then zero. If the optical path difference is $\lambda/2$ the phase difference is $\Delta\Phi = \pi$ and the amplitudes of both rays cancel each other (if the amplitudes have equal modulus). Then, the image point is strongly aberrated. So, the validity of the Rayleigh criterion is useful to define the limitations of the paraxial approximation.

In practice, the paraxial theory is quite important because it allows the definition of such important parameters as the focus, the focal length or the principal points of a lens or optical system. An optical designer [2.16–19] will always first design an optical system by using the paraxial matrix theory (or another paraxial method) so that the paraxial parameters are right. Afterwards he will try to optimize the non-paraxial parameters using ray tracing in order to correct aberrations of the system.

Definition of a Paraxial Ray

In the paraxial theory only rays in the meridional plane, which is here defined as the x–z-plane, are regarded. Then, the y-component of the ray direction vector $\boldsymbol{a}$ and the y-component of the starting point $\boldsymbol{p}$ of the ray are both zero: $a_y = 0$ and $p_y = 0$. We define for the x-component of the ray direction vector $a_x = \sin\varphi \approx \tan\varphi \approx \varphi$. The z-component of the ray direction vector in the paraxial approximation is then $a_z = \cos\varphi \approx 1$. Therefore, a meridional paraxial ray at a certain z-position z can be described by the angle φ with the optical axis and the ray height x, which is indeed the x-component p_x of the starting point $\boldsymbol{p}$ of the ray. The z-component p_z of a ray is noted in the paraxial matrix theory externally because in many cases several rays starting at the same z-position $z = p_z$ but having different values x and φ are considered.

So, in total a paraxial ray is described by x and φ. Since matrix methods play an important role in optics [2.12, 13] these two parameters are noted as the components of a vector

$$\begin{pmatrix} x \\ \varphi \end{pmatrix}$$

so that the optical operations which we will discuss now can be represented as 2×2 matrices.

Transfer Equation

The paraxial ray tracing between two planes with a separation d and which are perpendicular to the optical axis is one of the basic operations. Here, only the lines of intersection of these two planes with the meridional plane are regarded (Fig. 2.8), even though later we will use the term planes slightly incorrectly. The ray parameters in the first plane shall be x and φ and those in the second plane x' and φ'. Then, the transfer from the first plane to the second plane with distance d is done by (Fig. 2.8):

$$\begin{pmatrix} x' \\ \varphi' \end{pmatrix} = \begin{pmatrix} x + \varphi d \\ \varphi \end{pmatrix} . \tag{2.44}$$

This means that ray directions are not changed during the propagation of paraxial rays in a homogeneous material. Equation (2.44) can also be written by using a two times two matrix [2.12–15]:

$$\begin{pmatrix} x' \\ \varphi' \end{pmatrix} = \begin{pmatrix} 1 & d \\ 0 & 1 \end{pmatrix} \begin{pmatrix} x \\ \varphi \end{pmatrix} = M_{\mathrm{T}} \begin{pmatrix} x \\ \varphi \end{pmatrix} . \tag{2.45}$$

The matrix M_{T} is called the paraxial transfer matrix in a homogeneous material.

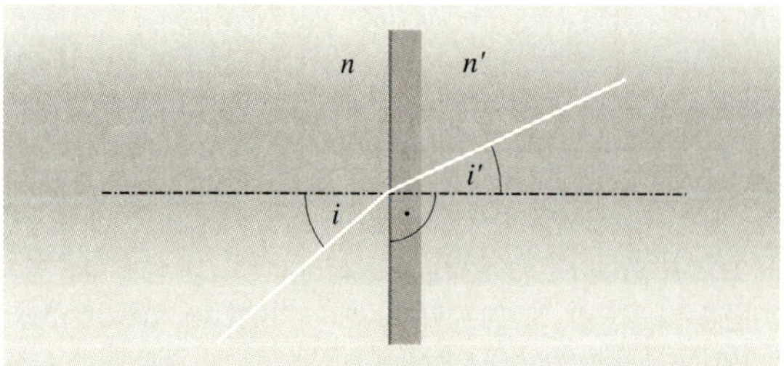

Fig. 2.9 Paraxial refraction at a (locally) plane surface

2.2.2 Refraction in the Paraxial Case

Paraxial Law of Refraction

The law of refraction connects the angle i between the incident ray and the surface normal with the angle i' between the refracted ray and the surface normal (Fig. 2.9). The law of refraction (2.42) in the paraxial formulation is

$$ni = n'i' \tag{2.46}$$

where n and n' are the refractive indices of the two homogeneous materials in front of the surface and behind the surface, respectively.

Refraction at a Plane Surface

A paraxial ray with parameters x and φ hits a plane surface which is perpendicular to the optical axis (Fig. 2.9). The refractive index is n in front of the surface and n' behind the surface. Then, the ray height x remains unchanged and only the ray parameter φ changes according to the paraxial law of refraction (2.46):

$$\begin{pmatrix} x' \\ \varphi' \end{pmatrix} = \begin{pmatrix} 1 & 0 \\ 0 & \frac{n}{n'} \end{pmatrix} \begin{pmatrix} x \\ \varphi \end{pmatrix} = M_{\mathrm{R}} \begin{pmatrix} x \\ \varphi \end{pmatrix} \tag{2.47}$$

The matrix M_{R} is the paraxial matrix for refraction at a plane surface.

Refraction at a Plane Parallel Plate

The plane parallel plate is the simplest case for a sequence of several surfaces and can be used to demonstrate the principle of tracing paraxial rays through an optical system by using the paraxial matrix theory. It is well known that the order of two matrices A and B is very important if two matrices have to be multiplied, i. e.

$$AB \neq BA .$$

Therefore, the matrix for the first operation has to be positioned immediately to the left of the vector (x, φ) of the paraxial ray that has to be traced through the system. The matrix of the next operation then has to be multiplied from the left side and so on for all other matrices. Using the notations of Fig. 2.10 the parameters of the paraxial ray on the right side of a plane parallel plate with thickness d and refractive index n_P are:

$$\begin{aligned}\begin{pmatrix}x'\\ \varphi'\end{pmatrix} &= \begin{pmatrix}1 & 0\\ 0 & \frac{n_P}{n'}\end{pmatrix}\begin{pmatrix}1 & d\\ 0 & 1\end{pmatrix}\begin{pmatrix}1 & 0\\ 0 & \frac{n}{n_P}\end{pmatrix}\begin{pmatrix}x\\ \varphi\end{pmatrix}\\ &= \begin{pmatrix}1 & d\\ 0 & \frac{n_P}{n'}\end{pmatrix}\begin{pmatrix}1 & 0\\ 0 & \frac{n}{n_P}\end{pmatrix}\begin{pmatrix}x\\ \varphi\end{pmatrix}\\ &= \begin{pmatrix}1 & d\frac{n}{n_P}\\ 0 & \frac{n}{n'}\end{pmatrix}\begin{pmatrix}x\\ \varphi\end{pmatrix}.\end{aligned}$$

Here, n is the refractive index left of the plane parallel plate and n' is the refractive index right to the plane parallel plate.

In total the parameters x' and φ' of a paraxial ray immediately behind the plane parallel plate are obtained from the parameters x and φ of the incident ray immediately in front of the plane parallel plate by multiplying them by the matrix M_P of a plane parallel plate

$$\begin{pmatrix}x'\\ \varphi'\end{pmatrix} = \begin{pmatrix}1 & d\frac{n}{n_P}\\ 0 & \frac{n}{n'}\end{pmatrix}\begin{pmatrix}x\\ \varphi\end{pmatrix} = M_P\begin{pmatrix}x\\ \varphi\end{pmatrix}. \tag{2.48}$$

The most important practical case is a plane parallel plate in air ($n = n' = 1$). Then it holds that

$$\begin{pmatrix}x'\\ \varphi'\end{pmatrix} = \begin{pmatrix}1 & \frac{d}{n_P}\\ 0 & 1\end{pmatrix}\begin{pmatrix}x\\ \varphi\end{pmatrix}. \tag{2.49}$$

This means that the matrix of a plane parallel plate is identical to the transfer matrix in a homogeneous material on substitution of the transfer distance d in the

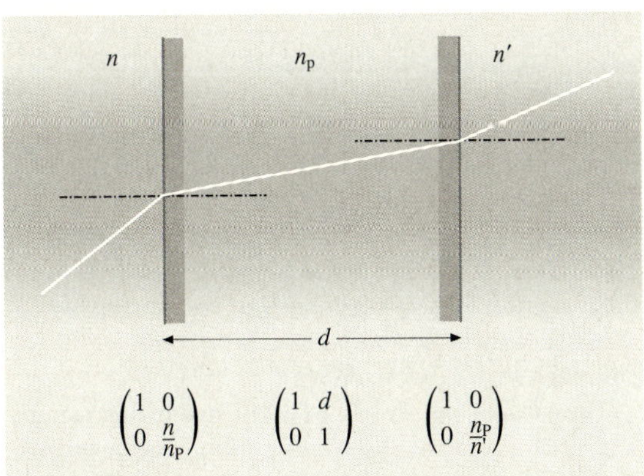

Fig. 2.10 Application of the paraxial matrix theory at a plane-parallel plate. The order of the multiplication of the matrices has to be reverse since the first matrix has to be directly left of the paraxial ray vector and the second matrix left of the first matrix, and so on

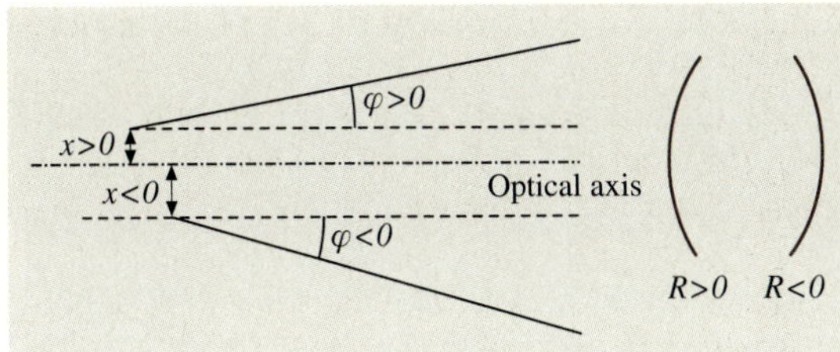

Fig. 2.11 Sign conventions in the paraxial matrix theory

homogeneous material by the term d/n_P. A lateral ray shift $\Delta x = x_{\text{without plate}} - x_{\text{with plate}}$ results at a plane parallel plate with thickness d compared to the propagation in air by a distance d

$$\Delta x = x + \varphi d - \left(x + \varphi\frac{d}{n_P}\right) = \varphi d\frac{n_P - 1}{n_P}.$$

For normal glass with $n_P \approx 1.5$ the lateral ray shift is $\Delta x = \varphi d/3$. This effect is used in optical systems to introduce a lateral shift where the size of the shift increases with the ray angle φ. So, in practice the plane parallel plate is tilted by an angle φ with respect to the optical axis of the system to introduce such a lateral shift. However, a plane parallel plate can also introduce aberrations if the incident wave is not a plane wave. Therefore, the introduction of a lateral shift by using a plane parallel plate has to be used with care.

Some Notes on Sign Conventions

Up to now no sign conventions have been made for the paraxial matrix theory. This will be included now and is illustrated graphically in Fig. 2.11.

- Ray angles φ are positive if the acute angle between the optical axis and the ray is mathematically positive.
- Refraction angles are all treated as positive angles.
- All angles are acute angles.
- Light rays always travel from left to right for positive propagation distances d. A negative propagation distance d means that the light is traveling from right to left and is only used for virtual rays.
- Ray heights x are upwards positive.
- Radii of curvature R are positive if the center of curvature is right of the vertex of the surface.

Refraction at a Spherical Surface

A spherical surface with radius of curvature R and refractive indices n in front of and n' behind the surface is hit by a paraxial ray with parameters x and φ. In the paraxial approximation the ray height x at the point of

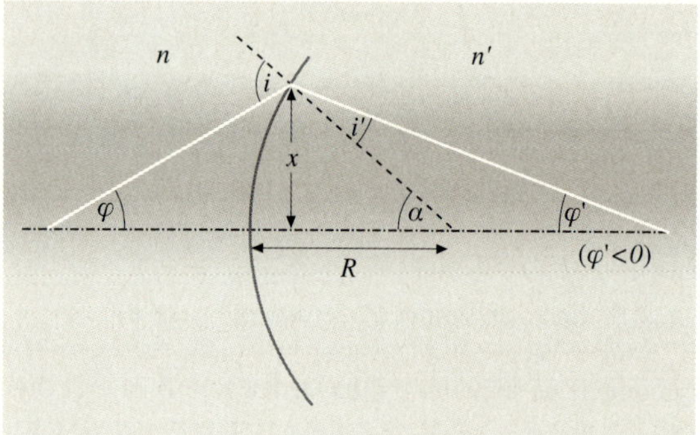

Fig. 2.12 Parameters for refraction at a spherical surface

intersection of the ray with the spherical surface is the same as in the vertex plane since the radius of curvature R is assumed to be large compared to x. According to Fig. 2.12 and the paraxial law of refraction (2.46) the following relations are valid:

$$\varphi' + \alpha = i' ,$$
$$\varphi + \alpha = i \quad \Rightarrow \quad \varphi' = i' - \alpha = \frac{n}{n'}(\varphi + \alpha) - \alpha ,$$
$$n'i' = ni .$$

Additionally, the angle α between the optical axis and the line connecting the center of curvature of the spherical surface and the point of intersection of the ray with the spherical surface is, in the paraxial approximation, defined as

$$\sin\alpha = \frac{x}{R} \Rightarrow \alpha = \frac{x}{R} .$$

Together, these results allow one to express the ray angle φ' of the refracted paraxial ray as a function of the parameters of the incident ray and the spherical surface

$$\varphi' = \frac{n}{n'}\varphi - \frac{n'-n}{n'}\frac{x}{R} . \tag{2.50}$$

The ray height x itself remains constant in the case of refraction. Therefore, the matrix M_S for refraction at a spherical surface is defined as

$$\begin{pmatrix} x' \\ \varphi' \end{pmatrix} = \begin{pmatrix} 1 & 0 \\ -\frac{n'-n}{n'R} & \frac{n}{n'} \end{pmatrix} \begin{pmatrix} x \\ \varphi \end{pmatrix} = M_S \begin{pmatrix} x \\ \varphi \end{pmatrix} . \tag{2.51}$$

The validity of the sign conventions can be shown by regarding some concrete cases:

- For $\varphi = 0$, $n' > n > 1$ and $R > 0$ (convex surface) a positive ray height x of the incident ray results in a negative ray angle φ' of the refracted ray. This means that the convex spherical surface with lower refractive index on the left side has a positive refractive power and focuses a plane wave.
- For $\varphi = 0$, $n' > n > 1$ and $R < 0$ (concave surface) the angle φ' of the refracted ray is positive if the ray height x of the incident ray is also positive. This means that two rays would only intersect virtually in front of the lens. Therefore, a concave spherical surface with a lower refractive index on the left-hand side has a negative focal power.

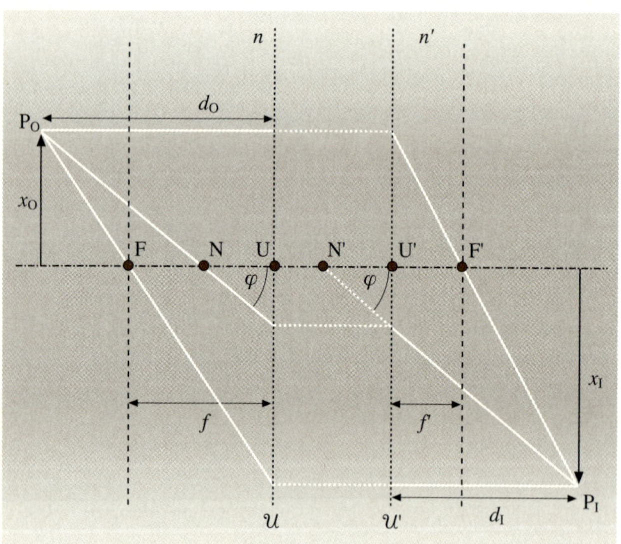

Fig. 2.13 Cardinal points of an optical system: F and F′ are the foci in the object space and the image space, respectively. N and N′ are the nodal points in the object and image space and U and U′ are the unit or principal points in the object and image space

2.2.3 The Cardinal Points of an Optical System

An optical imaging system has several cardinal points [2.1, 3], and by knowing these values the paraxial properties of the optical system are determined definitely. The cardinal points are the principal points, the nodal points and the focal points and they are all situated on the optical axis. In order to define them some additional definitions have to be made.

The cardinal points will be calculated in this section for a general optical system using the paraxial matrix theory [2.6]. At the end of this section the cardinal points of the simplest optical system, a refracting spherical surface, will be calculated explicitly to demonstrate the method.

Assume a general optical imaging system such as that symbolized in Fig. 2.13. An object point P_O with a lateral distance x_O from the optical axis, which is called the object height, is imaged by the optical system to an image point P_I with lateral distance x_I, called the image height. The refractive indices are n in the object space and n' in the image space.

The lateral magnification β of an imaging system is defined as the ratio of the image height x_I to the object height x_O:

$$\beta := \frac{x_I}{x_O} . \qquad (2.52)$$

According to our sign convention the lateral magnification in Fig. 2.13 is negative since x_O is positive and x_I negative.

The Principal Points

The principal plane $\mathcal{U}$ or unit plane in the object space is the plane perpendicular to the optical axis that has the property that an object point in this principal plane is imaged to a point in the principal plane $\mathcal{U}'$ of the image space with a lateral magnification $\beta = +1$. The points of intersection of the principal planes in the object and image space with the optical axis are called the principal or unit points U and U′, respectively. So, U′ is the image of U.

An important practical property of the principal planes following from this definition is that a ray which intersects the principal plane $\mathcal{U}$ in the object space at a height x is transferred to the principal plane $\mathcal{U}'$ of the image space at the same height (Fig. 2.13). This property is e.g. used to construct the path of a paraxial ray graphically.

The Nodal Points

The second cardinal points of an optical system are the nodal points N (in the object space) and N′ (in the image space). A ray in the object space which intersects the optical axis in the nodal point N at an angle φ intersects the optical axis in the image space in the nodal point N′ at the same angle $\varphi' = \varphi$. Therefore, the angular magnification γ defined as

$$\gamma := \frac{\varphi'}{\varphi} \qquad (2.53)$$

is $\gamma = 1$ for rays going through the nodal points. Additionally, since this has to be valid for arbitrary angles φ, the nodal point N′ is the image of the nodal point N.

The Focal Points

The focal points F (in the object space) and F′ (in the image space), also called principal foci or foci for short, have the following properties. A ray starting from the focus F in the object space is transformed into a ray parallel to the optical axis in the image space. Vice versa, a ray which is parallel to the optical axis in the object space intersects the focus F′ in the image space. The planes perpendicular to the optical axis which intersect the optical axis in the focal points are called the focal planes. The distance between the principal point U and the focus F is called the focal length f in the object space and the distance between the principal point U′ and the focal point F′ is called the focal length f' in the image space. In geometrical optics the sign convention for the focal length is usually that it is positive if the focus is right of the principal point. In Fig. 2.13 this means e.g. that the focal length f in the object space is negative whereas the focal length f' in the image space is positive.

A more general property of the focal planes is that rays starting from a point with object height x_O in the focal plane of the object space form in the image space a bundle of parallel rays making an angle $\varphi' = -x_O/f'$ with the optical axis. The relation for φ' can be easily understood by the fact that a ray starting from the object point parallel to the optical axis is transferred at the principal planes from $\mathcal{U}$ to $\mathcal{U}'$ with the same height x_O and then passes, after a distance f', the focal point F′ in the image space. The negative sign has to be taken due to the sign convention.

Calculation of the Cardinal Points of a General Optical System

Assume that we have a general optical system that is formed by an arbitrary combination of refracting spherical and plane surfaces which are all situated on a common optical axis (Fig. 2.14). Then, the system can be described by a 2×2 matrix M which is the product of a sequence of matrices M_T, M_R and M_S (or further matrices for other optical elements). The matrix

$$M = \begin{pmatrix} A & B \\ C & D \end{pmatrix} = M_{S,m} \cdot M_{T,m-1} \cdot M_{S,m-1} \cdot \ldots \cdot M_{T,2} \cdot M_{S,2} \cdot M_{T,1} \cdot M_{S,1} \qquad (2.54)$$

describes the propagation of a ray from a plane immediately in front of the vertex of the first surface (surface 1) to a plane immediately behind the vertex of the last surface (surface m). Here, only matrices M_S of refractive spherical surfaces are taken because a plane surface with matrix M_R can be represented as a spherical surface with radius of curvature $R = \infty$. Additionally, behind each surface (apart from the last surface) the transfer to the

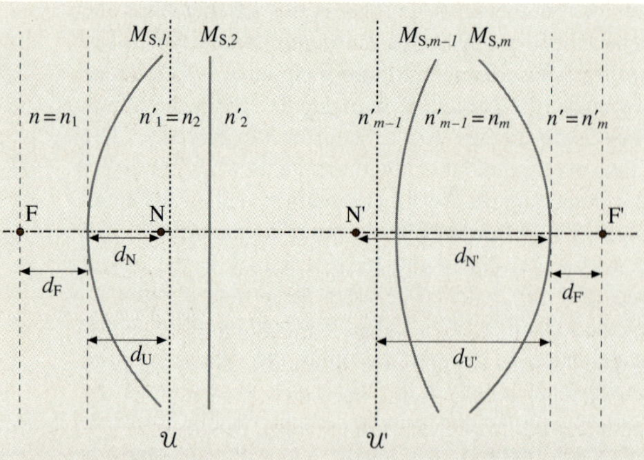

Fig. 2.14 Distances between the cardinal points in the object space and the vertex of the first surface (quantities without apostrophe) and the vertex of the last surface and the cardinal points in the image space (quantities marked by an apostrophe) of a general optical system consisting of refractive surfaces. d_U and $d'_{U'}$ are negative in the scheme, d_F and $d'_{F'}$ are positive and d_N and $d'_{N'}$ are again negative. n and n' are the refractive indices in front of and behind the whole system, whereas n_i and n'_i are the refractive indices in front of and behind the single refracting surface number i

next surface is described by using a matrix $M_{T,i}$. In the special case of a thin lens (which does not exist in reality but which is an important idealization in geometrical optics) the propagation distance can just be set to zero so that the transfer matrix is identical to the unit matrix.

The restriction to spherical surfaces is not stringent because in paraxial optics an aspheric surface is identical to a spherical surface if the radius of curvature of the aspheric surface at the vertex is identical to the radius of curvature of the spherical surface. From a mathematical point of view the determination of the radius of curvature in the paraxial regime just means that in both cases the parabolic terms are taken. Moreover, also cylindrical surfaces can be calculated with this method if the radius of curvature in the selected x–z-plane is taken. For a plane which contains the cylinder axis the cylindrical surface behaves like a plane surface whereas the cylindrical surface behaves like a spherical surface if the cylinder axis is perpendicular to the considered x–z-plane.

A ray starts in front of the optical system in a material with refractive index $n := n_1$ and ends behind the system in a material with refractive index $n' := n'_m$ (Fig. 2.14). n_i and n'_i with $i \in \{1, 2, \dots, m\}$ are the refractive indices in front of and behind each refracting surface, which is described by the matrix $M_{S,i}$. Of course, there is the relation

$$n_i = n'_{i-1} \quad \text{for} \quad i \in \{2, 3, \dots, m\} . \qquad (2.55)$$

Now, a matrix M' is calculated which describes the propagation of a ray from a plane $\mathcal{P}$ through the optical system to a plane $\mathcal{P}'$. The plane $\mathcal{P}$ is a distance d from the vertex of the first surface of the optical system, whereas the vertex of the last surface is a distance d' from the plane $\mathcal{P}'$. Using the paraxial sign conventions d is positive if $\mathcal{P}$ is in front of (i.e. left of) the first surface. Similar d' is positive if $\mathcal{P}'$ is behind (i.e. right of) the last surface. It is very important to remember that d is measured from the plane $\mathcal{P}$ to the vertex of the first surface whereas d' is measured from the vertex of the last surface to the plane $\mathcal{P}'$. For these quantities the usual sign conventions are valid, i.e. they are positive if the propagation is from left to right and negative if the propagation is in the opposite direction.

By using equation (2.45) the matrix M' is:

$$\begin{aligned} M' &= \begin{pmatrix} A' & B' \\ C' & D' \end{pmatrix} = M_{T'} M M_T \\ &= \begin{pmatrix} 1 & d' \\ 0 & 1 \end{pmatrix} \begin{pmatrix} A & B \\ C & D \end{pmatrix} \begin{pmatrix} 1 & d \\ 0 & 1 \end{pmatrix} \\ &= \begin{pmatrix} A + Cd' & Ad + B + Cdd' + Dd' \\ C & Cd + D \end{pmatrix} . \end{aligned} \qquad (2.56)$$

Principal Points. To calculate the principal planes $\mathcal{U}$ and $\mathcal{U}'$ of the system the definition is used. If $\mathcal{P}$ is identical to the principal plane $\mathcal{U}$ and $\mathcal{P}'$ is identical to $\mathcal{U}'$ an object point in $\mathcal{P}$ has to be imaged to $\mathcal{P}'$ with lateral magnification $\beta = +1$. Imaging means that all rays with arbitrary ray angles φ starting from the object point with height x have the same height x' in $\mathcal{P}'$, independent of φ. Since the relation

$$\begin{pmatrix} x' \\ \varphi' \end{pmatrix} = \begin{pmatrix} A' & B' \\ C' & D' \end{pmatrix} \begin{pmatrix} x \\ \varphi \end{pmatrix} = \begin{pmatrix} A'x + B'\varphi \\ C'x + D'\varphi \end{pmatrix} \qquad (2.57)$$

holds, this means that the matrix element B' has to be zero in order to have imaging between the planes $\mathcal{P}$ and $\mathcal{P}'$. Therefore, we have as a first condition

$$B' = Ad + B + Cdd' + Dd' = 0 . \qquad (2.58)$$

The second condition $\beta = 1$ means, by using $B' = 0$,

$$\begin{aligned} & x' = A'x + B'\varphi = (A + Cd')\, x = x , \\ & \Rightarrow A + Cd' = 1 \Rightarrow d'_{U'} = \frac{1 - A}{C} . \end{aligned} \qquad (2.59)$$

Here, we use the name $d'_{U'} := d'$ (Fig. 2.14) to indicate that it is the distance from the vertex of the last surface of the optical system to the principal point U′. From the first condition we then obtain the distance $d_U := d$ between the principal point U and the vertex of the first surface

$$B' = Ad + B + (1-A)d + D\frac{1-A}{C} = 0\,,$$
$$\Rightarrow d_U = \frac{D}{C}(A-1) - B\,. \tag{2.60}$$

It has to be mentioned that, in the case of optical imaging, the coefficient A' of the matrix M' has a concrete meaning. It is:

$$x' = A'x \Rightarrow \beta = \frac{x'}{x} = A' \tag{2.61}$$

Therefore, the coefficient A' is identical to the lateral magnification β defined by (2.52).

Nodal Points. If $\mathcal{P}$ contains the nodal point N and $\mathcal{P}'$ contains the nodal point N′ the conditions for the ray parameters are $x = x' = 0$ and $\varphi' = \varphi$. Using (2.57) this gives:

$$0 = x' = A'x + B'\varphi = B'\varphi$$
$$\Rightarrow B' = Ad + B + Cdd' + Dd' = 0\,,$$
$$\varphi' = C'x + D'\varphi = D'\varphi = \varphi$$
$$\Rightarrow D' = Cd + D = 1\,. \tag{2.62}$$

Then, the distances $d_N := d$ between the nodal point N and the vertex of the first surface of the optical system on the one hand and $d'_{N'} := d'$ between the vertex of the last surface of the optical system and the nodal point N′ on the other hand are:

$$d_N = \frac{1-D}{C}\,, \tag{2.63}$$
$$d'_{N'} = \frac{A}{C}(D-1) - B\,. \tag{2.64}$$

Focal Points and Focal Lengths. For the calculation of the focus F in the object space it is assumed that F is in the plane $\mathcal{P}$. Then, all rays starting from the height $x = 0$ have to be in the image space rays parallel to the optical axis, i. e. $\varphi' = 0$. Since this has to be valid in all planes in the image space the distance d' in (2.56) is set to zero. So, the condition for the distance $d_F := d$ between the focus F and the vertex of the first surface of the optical system is

$$\varphi' = C'x + D'\varphi = D'\varphi = 0$$
$$\Rightarrow D' = Cd + D = 0 \Rightarrow d_F = -\frac{D}{C}\,. \tag{2.65}$$

The focal length f is defined as the distance between the principal point U and the focus F, where the sign convention in geometrical optics is that f is positive if F is right of U. Therefore, by using the sign conventions for d_U and d_F it is:

$$f = d_U - d_F = \frac{D}{C}(A-1) - B + \frac{D}{C} = \frac{AD}{C} - B\,.$$
$$= \frac{AD - BC}{C} \tag{2.66}$$

The focus F′ in the image space can be calculated analogously. There, rays parallel to the optical axis (i. e. $\varphi = 0$) in front of the optical system in an arbitrary plane $\mathcal{P}$, e.g. at $d = 0$, have to focus in the image space in the focus F′ at $x' = 0$. If F′ is in the plane $\mathcal{P}'$, the distance $d'_{F'} := d'$ between the vertex of the last surface of the optical system and the focus F′ is, by using (2.57),

$$x' = A'x + B'\varphi = A'x = 0$$
$$\Rightarrow A' = A + Cd' = 0 \Rightarrow d'_{F'} = -\frac{A}{C}\,. \tag{2.67}$$

Analogously, the focal length f', which is positive if F′ is to the right of U′, can be calculated by

$$f' = d'_{F'} - d'_{U'} = -\frac{A}{C} - \frac{1-A}{C} = -\frac{1}{C} \Rightarrow C = -\frac{1}{f'}\,. \tag{2.68}$$

Now, the concrete meaning of the matrix coefficient C as the negative reciprocal value of the focal length f' in the image space becomes clear.

By summarizing (2.59), (2.60), (2.63–2.65) and (2.67) the distances d_U, d_N and d_F between the cardinal points in the object space and the vertex of the first surface of the optical system as well as the distances $d'_{U'}$, $d'_{N'}$ and $d'_{F'}$ between the vertex of the last surface of the optical system and the cardinal points in the image space are:

$$d_U = \frac{D}{C}(A-1) - B\,,$$
$$d_N = \frac{1-D}{C}\,,$$
$$d_F = -\frac{D}{C}\,,$$
$$d'_{U'} = \frac{1-A}{C}\,,$$
$$d'_{N'} = \frac{A}{C}(D-1) - B\,,$$
$$d'_{F'} = -\frac{A}{C}\,. \tag{2.69}$$

Also, the focal lengths can now be expressed as functions of the coefficients A, B, C and D of the matrix M by summarizing (2.66) and (2.68):

$$f = \frac{AD - BC}{C} = \frac{\det(M)}{C}$$
$$f' = -\frac{1}{C} \tag{2.70}$$

Relation Between the Focal Lengths in the Object and Image Spaces

There is a very interesting relation between the focal length f in the object space and the focal length f' in the image space. To derive it the ratio f'/f is calculated by using (2.66) and (2.68):

$$\frac{f'}{f} = \frac{-1/C}{(AD - BC)/C} = -\frac{1}{AD - BC} = -\frac{1}{\det(M)} \tag{2.71}$$

Here, the determinant $\det(M)$ of the matrix M, defined by (2.54), has been used.

According to the calculus of linear algebra the determinant of the product of several matrices is equal to the product of the determinants of these matrices. Therefore, it holds that

$$\det(M) = \det(M_{S,m}) \cdot \det(M_{T,m-1}) \cdot \det(M_{S,m-1}) \cdot \ldots \cdot \det(M_{T,1}) \cdot \det(M_{S,1}) . \tag{2.72}$$

So, we have first to calculate the determinants of the two elementary matrices of (2.45) and (2.51):

$$M_{T,i} = \begin{pmatrix} 1 & d_i \\ 0 & 1 \end{pmatrix} \Rightarrow \det(M_{T,i}) = 1 , \tag{2.73}$$

$$M_{S,i} = \begin{pmatrix} 1 & 0 \\ -\frac{n'_i - n_i}{n'_i R_i} & \frac{n_i}{n'_i} \end{pmatrix} \Rightarrow \det(M_{S,i}) = \frac{n_i}{n'_i} . \tag{2.74}$$

Again, n_i and n'_i are the refractive indices in front of and behind the respective surface; d_i is the distance between surface i and $i+1$ ($i \in \{1, 2, \ldots, m-1\}$); and R_i is the radius of curvature of surface i.

Now, we define again the refractive index in front of the first surface as $n := n_1$ and the refractive index behind the last surface of the optical system as $n' := n'_m$. Since the determinants of the transfer matrices $M_{T,i}$ are one the determinant of M is:

$$\begin{aligned} \det(M) &= \prod_{i=1}^{m} \det(M_{S,m+1-i}) = \prod_{i=1}^{m} \frac{n_{m+1-i}}{n'_{m+1-i}} \\ &= \frac{n_m}{n'_m} \cdot \frac{n_{m-1}}{n'_{m-1}} \cdot \ldots \cdot \frac{n_2}{n'_2} \cdot \frac{n_1}{n'_1} \\ &= \frac{n'_{m-1}}{n'} \cdot \frac{n'_{m-2}}{n'_{m-1}} \cdot \ldots \cdot \frac{n'_1}{n'_2} \cdot \frac{n}{n'_1} = \frac{n}{n'} . \end{aligned} \tag{2.75}$$

Here, relation (2.55) for the refractive indices of neighboring surfaces has been used.

Therefore, the ratio of the focal length f' and f is, according to (2.71),

$$\frac{f'}{f} = -\frac{1}{\det(M)} = -\frac{n'}{n} \quad \text{or} \quad \frac{f'}{n'} = -\frac{f}{n} . \tag{2.76}$$

The Cardinal Points of an Optical System with Identical Surrounding Refractive Indices

An interesting special case is when the refractive indices n in front of the first surface of the optical system and n' behind the last surface of the optical system are identical: $n = n'$. Then the determinant of the matrix M is, according to equation (2.75), $\det(M) = 1$. Therefore, the focal lengths in the object and image spaces have, due to (2.76), equal absolute value but different signs (due to the sign conventions of geometrical optics):

$$f' = -f . \tag{2.77}$$

A second quite interesting property of an optical system with identical refractive indices in front of the first surface and behind the last surface is that the principal points and the nodal points coincide. This can easily be derived from (2.59), (2.60), (2.63) and (2.64) by using $\det(M) = AD - BC = 1$:

$$\begin{aligned} d_U &= \frac{D}{C}(A-1) - B = \frac{AD - D - BC}{C} \\ &= \frac{1-D}{C} = d_N \end{aligned} \tag{2.78}$$

and

$$\begin{aligned} d'_{N'} &= \frac{A}{C}(D-1) - B = \frac{AD - A - BC}{C} \\ &= \frac{1-A}{C} = d'_{U'} . \end{aligned} \tag{2.79}$$

The Cardinal Points of a Spherical Refracting Surface

The simplest optical imaging system is a single spherical refracting surface. As an application of (2.69) and (2.70) the cardinal points of a spherical refracting surface shall be determined.

In this special case the matrix M is, according to (2.51),

$$M = \begin{pmatrix} A & B \\ C & D \end{pmatrix} := M_S = \begin{pmatrix} 1 & 0 \\ -\frac{n'-n}{n'R} & \frac{n}{n'} \end{pmatrix} . \qquad (2.80)$$

Then, according to (2.69), the result is:

$$\begin{aligned} d_U &= \frac{D}{C}(A-1) - B = 0 \, , \\ d_N &= \frac{1-D}{C} = -R \, , \\ d_F &= -\frac{D}{C} = \frac{nR}{n'-n} \, , \\ d'_{U'} &= \frac{1-A}{C} = 0 \, , \\ d'_{N'} &= \frac{A}{C}(D-1) - B = R \, , \\ d'_{F'} &= -\frac{A}{C} = \frac{n'R}{n'-n} \, . \end{aligned} \qquad (2.81)$$

This means (Fig. 2.15):

1. Both principal points coincide with the vertex of the spherical surface ($d_U = d'_{U'} = 0$).
2. Both nodal points coincide with the center of curvature of the spherical surface ($-d_N = d'_{N'} = R$). To understand this, the sign conventions have to be noticed: d_N is positive if the vertex of the surface is right to the nodal point N, but $d'_{N'}$ is positive if the vertex of the surface is left to the nodal point N′.
3. For a convex surface ($R > 0$) and $n' > n$ the surface has a positive optical power and the focus F is in front of the surface and F′ behind the surface. For a concave surface ($R < 0$) but still $n' > n$ the surface has a negative optical power and the foci change their positions, i. e. F is right of the vertex of the surface and F′ is left of it.

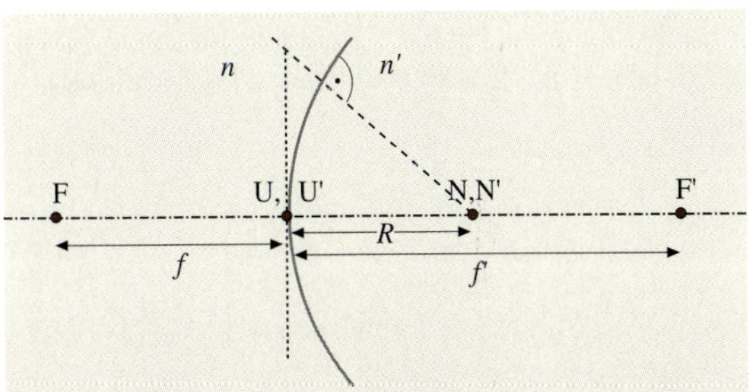

Fig. 2.15 Cardinal points and parameters of a spherical refracting surface. (The position of F and F′ is drawn for the example $n = 1$ and $n' = 1.8$)

Similar the focal lengths are calculated using (2.70):

$$\begin{aligned} f &= \frac{AD-BC}{C} = \frac{\det(M)}{C} = -\frac{nR}{n'-n} \\ f' &= -\frac{1}{C} = \frac{n'R}{n'-n} \, . \end{aligned} \qquad (2.82)$$

Since the principal points coincide with the vertex of the surface, the focal length f is of course identical to $-d_F$ and the focal length f' is $f' = d'_{F'}$. The general equation (2.76) $f'/n' = -f/n$ is of course also valid.

2.2.4 The Imaging Equations of Geometrical Optics

The Lens Equation

On page 46 it has already been shown what imaging means. A point P_O lying in the plane $\mathcal{P}$ a distance d in front of the vertex of the first surface of an optical system with matrix M (2.54) is imaged onto a point P_I in a plane $\mathcal{P}'$ a distance d' from the vertex of the last surface of the system. This is only the case if the matrix

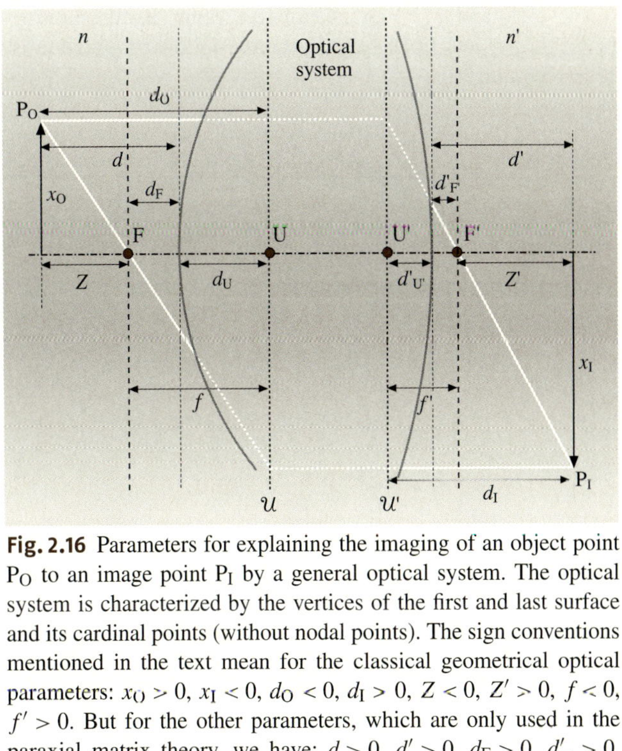

Fig. 2.16 Parameters for explaining the imaging of an object point P_O to an image point P_I by a general optical system. The optical system is characterized by the vertices of the first and last surface and its cardinal points (without nodal points). The sign conventions mentioned in the text mean for the classical geometrical optical parameters: $x_O > 0$, $x_I < 0$, $d_O < 0$, $d_I > 0$, $Z < 0$, $Z' > 0$, $f < 0$, $f' > 0$. But for the other parameters, which are only used in the paraxial matrix theory, we have: $d > 0$, $d' > 0$, $d_F > 0$, $d'_{F'} > 0$, $d_U < 0$, $d'_{U'} < 0$

element B' of the matrix M' (2.56), which describes the complete ray propagation from $\mathcal{P}$ to $\mathcal{P}'$, is zero:

$$B' = Ad + B + Cdd' + Dd' = 0 . \tag{2.83}$$

Then, all rays starting from the object point P_O intersect in the image point P_I. The distance from the object point to the principal plane $\mathcal{U}$ in the object space will be named d_O and the distance from the principal plane $\mathcal{U}'$ in the image space to the image point P_I will be d_I (Fig. 2.16). According to the sign conventions of geometrical optics d_O is positive if the object point is right of $\mathcal{U}$ (i. e. d_O is negative in Fig. 2.16) and d_I is positive if the image point is also right of $\mathcal{U}'$ (i. e. d_I is positive in Fig. 2.16).

Then, the relations between d, d_O and d_U (the distance between $\mathcal{U}$ and the vertex of the first surface) on the one hand and d', d_I and $d'_{U'}$ (the distance between the vertex of the last surface and $\mathcal{U}'$) on the other hand are:

$$d_O = d_U - d , \tag{2.84}$$

$$d_I = d' - d'_{U'} . \tag{2.85}$$

Here, the different sign conventions for d_O (d_I) and d_U ($d'_{U'}$) and d (d') are taken into account. By substituting (2.84) and (2.85) into (2.83) the following equation is obtained:

$$\begin{aligned} &A\,(d_U - d_O) + B + C\,(d_U - d_O)\left(d_I + d'_{U'}\right) \\ &\qquad + D\left(d_I + d'_{U'}\right) = 0 . \end{aligned} \tag{2.86}$$

By using (2.59) and (2.60) to express d_U and $d'_{U'}$ as functions of the matrix elements A, B, C, D of M and some calculations the result is:

$$\begin{aligned} &A\left[\frac{D}{C}(A-1) - B - d_O\right] + B \\ &\quad + \left\{C\left[\frac{D}{C}(A-1) - B - d_O\right] + D\right\} \\ &\quad \times\left(d_I + \frac{1-A}{C}\right) \\ &= \frac{AD}{C}(A-1) - AB - Ad_O + B \\ &\quad + (AD - BC - Cd_O)\left(d_I + \frac{1-A}{C}\right) \\ &= -Ad_O + B + ADd_I - BCd_I - Cd_Od_I \\ &\quad - B - (1-A)d_O \\ &= (AD - BC)\,d_I - Cd_Od_I - d_O \\ &= \det(M)d_I - Cd_Od_I - d_O = 0 . \end{aligned} \tag{2.87}$$

The determinant of M is, according to (2.75), $\det(M) = n/n'$, where n is the refractive index in the object space and n' is the refractive index in the image space. Additionally, according to (2.68) it is $C = -1/f'$, where f' is the focal length in the image space. So, the final result is:

$$d_O - \frac{n}{n'}d_I = \frac{d_Od_I}{f'} \quad \text{or} \quad \frac{d_O}{n} - \frac{d_I}{n'} = \frac{d_Od_I}{nf'} . \tag{2.88}$$

An equivalent formulation of this equation is the well-known imaging equation of geometrical optics, which is often called the lens equation although it is valid for quite complex optical imaging systems:

$$\frac{n'}{d_I} - \frac{n}{d_O} = \frac{n'}{f'} = -\frac{n}{f} . \tag{2.89}$$

At the right side of equation (2.89), (2.76) has been used.

If the refractive indices n and n' are identical the equation is

$$\frac{1}{d_I} - \frac{1}{d_O} = \frac{1}{f'} . \tag{2.90}$$

As defined above, the object distance d_O and the image distance d_I are measured in the lens equation relative to the principal planes.

Newton Equation

Another formulation of the imaging equation is the Newton equation where the object distance and the image distance are measured relative to the focal points. Therefore, we define the distance between the focal point F in the object space and the object point P_O as Z. Analogous, Z' is the distance from the focal point F′ in the image space to the image point P_I. Both quantities are again positive if the object/image point is right of the focus F/F′. In Fig. 2.16 Z is negative and Z' positive. From this figure, using the sign conventions and (2.84) and (2.85), it is clear that the following relations are valid:

$$\begin{aligned} &Z = d_F - d = d_F + d_O - d_U \\ &\Rightarrow d_O = Z + d_U - d_F = Z + f , \end{aligned} \tag{2.91}$$

$$\begin{aligned} &Z' = d' - d'_{F'} = d_I + d'_{U'} - d'_{F'} \\ &\Rightarrow d_I = Z' - d'_{U'} + d'_{F'} = Z' + f' , \end{aligned} \tag{2.92}$$

where (2.66) and (2.68) have been used.

Substituting these equations into the lens equation (2.88) and using (2.76) gives

$$\frac{Z+f}{n}-\frac{Z'+f'}{n'}=\frac{(Z+f)(Z'+f')}{nf'} \Rightarrow$$

$$\frac{Z}{n}+\frac{f}{n}-\frac{Z'}{n'}-\frac{f'}{n'}=\frac{ZZ'}{nf'}+\frac{Z}{n}+\frac{Z'f}{nf'}+\frac{f}{n} \Rightarrow$$

$$-\frac{f'}{n'}=\frac{ZZ'}{nf'}$$

and finally

$$ZZ'=ff' . \tag{2.93}$$

This is the well-known Newton equation for the imaging of an object point into an image point. The advantage of the Newton equation is its quite simple and symmetric form, which does not explicitly depend on n and n'. Of course, the dependence on the refractive indices in the object and image spaces is hidden in the ratio of f and f'.

Relation Between Lateral and Longitudinal Magnification

The Newton equation can also be easily explained by looking at Fig. 2.17. Due to the similar triangles the following relations are valid in the object space and the image space where the signs have to be noticed:

$$\frac{x_O}{-Z}=\frac{x_I}{f} \Rightarrow \beta=\frac{x_I}{x_O}=-\frac{f}{Z}$$

$$\frac{-x_I}{Z'}=\frac{x_O}{f'} \Rightarrow \beta=\frac{x_I}{x_O}=-\frac{Z'}{f'} \Rightarrow ZZ'=ff' . \tag{2.94}$$

Here, the lateral magnification β defined by (2.52) has been used.

The longitudinal magnification is defined as $\mathrm{d}Z'/\mathrm{d}Z$, i.e. the ratio of the axial (longitudinal) shift of the image plane to an axial shift of the object plane. According to the Newton equation (2.93) and the relation (2.76) between f and f' it holds that

$$Z'=\frac{ff'}{Z} \Rightarrow \frac{\mathrm{d}Z'}{\mathrm{d}Z}=-\frac{ff'}{Z^2}=\frac{n'}{n}\left(\frac{f}{Z}\right)^2=\frac{n'}{n}\beta^2 . \tag{2.95}$$

This means that the longitudinal magnification is proportional to the square of the lateral magnification.

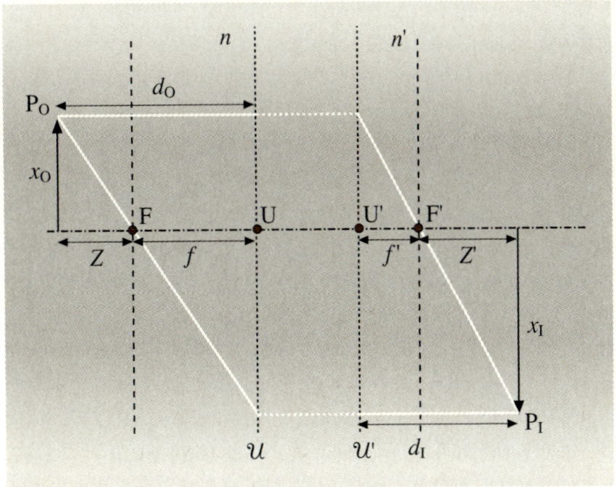

Fig. 2.17 Graphical explanation of the Newton equation. The distances drawn have the following signs due to our sign conventions: $d_O<0$, $d_I>0$, $Z<0$, $Z'>0$, $f<0$, $f'>0$, $x_O>0$, $x_I<0$

2.2.5 The Thin Lens

A quite important element in the paraxial theory is a so-called thin lens [2.2–4, 20]. This means that the transfer from the first to the second surface is neglected (the thickness d_1 of the lens is assumed to be zero) and a paraxial ray that intersects the first surface at a ray height of x also has, immediately behind the second surface, the same ray height x':

$$x'=x .$$

The refractive indices are $n=n_1$ in front of the first surface, $n_L:=n'_1=n_2$ between the two surfaces and $n'=n'_2$ behind the second surface. The radii of curvature of the spherical surfaces are R_1 for the first surface and R_2 for the second surface (see Fig. 2.18). The thin lens, as defined here, does of course not exist in reality but

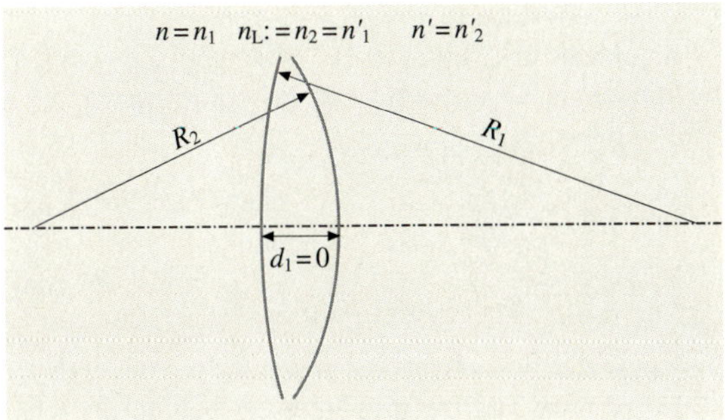

Fig. 2.18 Parameters of a thin lens

it is a good approximation for lenses which are thin compared to their focal length.

The matrix M_{L0} of a thin lens is obtained from (2.54) using $m = 2$ and $d_1 = 0$ ($\Rightarrow M_{T,1}$ is the unit matrix) by multiplying the two matrices $M_{S,1}$ and $M_{S,2}$ for refraction at the two spherical surfaces in the correct order

$$M_{L0} = M_{S,2}M_{S,1} = \begin{pmatrix} 1 & 0 \\ -\frac{n'-n_L}{n'R_2} & \frac{n_L}{n'} \end{pmatrix} \begin{pmatrix} 1 & 0 \\ -\frac{n_L-n}{n_L R_1} & \frac{n}{n_L} \end{pmatrix}$$
$$= \begin{pmatrix} 1 & 0 \\ -\frac{n'-n_L}{n'R_2} - \frac{n_L-n}{n'R_1} & \frac{n}{n'} \end{pmatrix} . \tag{2.96}$$

In the following the important case when the external materials of the lens are identical, i. e. $n' = n$, is considered. Then, the matrix is

$$M_{L0} = \begin{pmatrix} 1 & 0 \\ -\frac{n-n_L}{nR_2} - \frac{n_L-n}{nR_1} & 1 \end{pmatrix}$$
$$= \begin{pmatrix} 1 & 0 \\ -\frac{n_L-n}{n}\left(\frac{1}{R_1} - \frac{1}{R_2}\right) & 1 \end{pmatrix}$$
$$\Rightarrow M_{L0} = \begin{pmatrix} 1 & 0 \\ -\frac{1}{f'} & 1 \end{pmatrix} \tag{2.97}$$

with

$$\frac{1}{f'} = \frac{n_L - n}{n}\left(\frac{1}{R_1} - \frac{1}{R_2}\right) . \tag{2.98}$$

Here, the focal length f' of a thin lens in the image space has been defined according to (2.70) and the focal length f in the object space is of course $f = -f'$. By using (2.69) ($d_U = d'_{U'} = 0$) it can be seen that the principal points U and U' of the thin lens coincide with the vertices of the two surfaces, which themselves coincide. Of course, the nodal points also coincide with the vertices because the nodal points coincide with the principal points due to $n = n'$.

In total the ray parameters x', φ' immediately behind the thin lens are connected to the ray parameters x, φ in front of the lens by

$$x' = x , \tag{2.99}$$

$$\varphi' = \varphi - \frac{x}{f'} . \tag{2.100}$$

For a lens with a positive focal power the focal length f' is also positive and the rays intersect behind the lens in a real focus. For a lens with a negative focal power

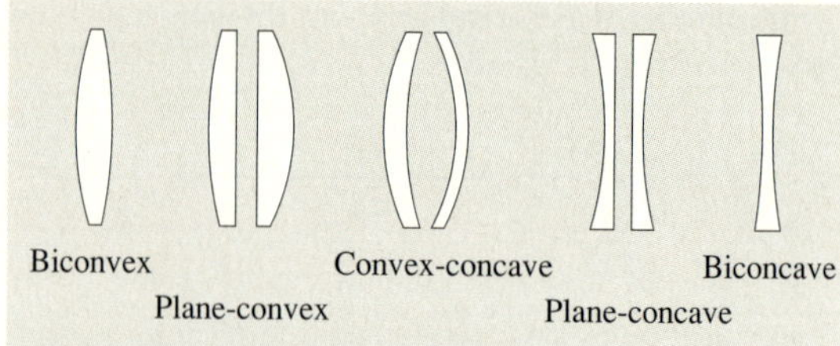

Fig. 2.19 Different types of lenses

the focal length f' is negative and this means that rays that were originally parallel to the optical axis would intersect in a so-called virtual focus in front of the lens. Of course, a virtual focus has its name because there is in reality no focus at this position in front of the lens but the rays behind the lens seem to come from this virtual focus.

There are several different types of lenses depending on their radii of curvature (see Fig. 2.19):

- biconvex: $R_1 > 0$ and $R_2 < 0$,
- plane–convex: $R_1 > 0$ and $R_2 = \infty$ (or $R_1 = \infty$, $R_2 < 0$)
- convex–concave (meniscus lens): $R_1 > 0$ and $R_2 > 0$ (or both negative)
- plane–concave: $R_1 < 0$ and $R_2 = \infty$ (or $R_1 = \infty$, $R_2 > 0$)
- biconcave: $R_1 < 0$ and $R_2 > 0$.

These lenses have different focal powers. For the case $n_L > n$ (e.g. for a lens made of glass which is used in air) biconvex and plane–convex lenses have generally positive focal lengths, i. e. they are positive lenses. On the other side, biconcave and plane–concave lenses have negative focal lengths, i. e. they are negative lenses. Meniscus lenses can be either positive (if the convex surface has the smaller radius of curvature) or negative (if the convex surface has the larger radius of curvature). Pay attention to the fact that in the case $n_L < n$ (which can be realized e.g. by a hollow lens made of thin plastic which is filled with air and used in water) the properties of the different types of lenses are reversed. In this case a biconvex lens has e.g. a negative focal length.

2.2.6 The Thick Lens

In the case of a thick lens the ray transfer through the thickness $d := d_1$ between the two spherical surfaces is taken into account (see Fig. 2.20). Of course, the radii of curvature of the two spherical surfaces are still assumed to be so large that the point of intersection of a paraxial ray with the surface is in the same plane as the vertex of

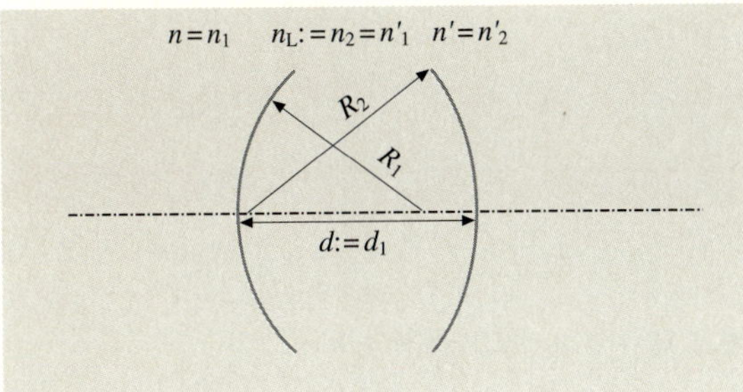

Fig. 2.20 Parameters of a thick lens

the surface. The matrix M_{Ld} of a thick lens is the product of three single matrices: matrix $M_{\mathrm{S},1}$ for refraction at the first spherical surface with radius of curvature R_1, matrix $M_{\mathrm{T},1}$ for the transfer between the two surfaces by the distance d and matrix $M_{\mathrm{S},2}$ for the refraction at the second spherical surface with the radius of curvature R_2. The refractive indices in front of, in, and behind the lens are $n = n_1, n_{\mathrm{L}} := n'_1 = n_2$ and $n' = n'_2$, respectively. Then the matrix M_{Ld} of a thick lens is:

$$
\begin{aligned}
&M_{\mathrm{Ld}}\\
&=\begin{pmatrix} 1 & 0 \\ -\frac{n'-n_{\mathrm{L}}}{n'R_2} & \frac{n_{\mathrm{L}}}{n'} \end{pmatrix}\begin{pmatrix} 1 & d \\ 0 & 1 \end{pmatrix}\begin{pmatrix} 1 & 0 \\ -\frac{n_{\mathrm{L}}-n}{n_{\mathrm{L}}R_1} & \frac{n}{n_{\mathrm{L}}} \end{pmatrix}\\
&=\begin{pmatrix} 1 & 0 \\ -\frac{n'-n_{\mathrm{L}}}{n'R_2} & \frac{n_{\mathrm{L}}}{n'} \end{pmatrix}\begin{pmatrix} 1-\frac{n_{\mathrm{L}}-n}{n_{\mathrm{L}}R_1}d & \frac{n}{n_{\mathrm{L}}}d \\ -\frac{n_{\mathrm{L}}-n}{n_{\mathrm{L}}R_1} & \frac{n}{n_{\mathrm{L}}} \end{pmatrix}\\
&=\begin{pmatrix} 1-\frac{n_{\mathrm{L}}-n}{n_{\mathrm{L}}R_1}d & \frac{n}{n_{\mathrm{L}}}d \\ -\frac{n_{\mathrm{L}}-n}{n'R_1}-\frac{n'-n_{\mathrm{L}}}{n'R_2}+\frac{(n'-n_{\mathrm{L}})(n_{\mathrm{L}}-n)}{n_{\mathrm{L}}n'R_1R_2}d & \frac{n}{n'} \quad \frac{n(n'-n_{\mathrm{L}})}{n_{\mathrm{L}}n'R_2}d \end{pmatrix}
\end{aligned}
\tag{2.101}
$$

In the most important case of identical external materials $n' = n$, (2.101) reduces to

$$
\begin{aligned}
&M_{\mathrm{Ld}}\\
&=\begin{pmatrix} 1-\frac{n_{\mathrm{L}}-n}{n_{\mathrm{L}}R_1}d & \frac{n}{n_{\mathrm{L}}}d \\ \underbrace{-\frac{n_{\mathrm{L}}-n}{n}\left[\left(\frac{1}{R_1}-\frac{1}{R_2}\right)+\frac{n_{\mathrm{L}}-n}{n_{\mathrm{L}}}\frac{d}{R_1R_2}\right]}_{=-1/f'} & 1+\frac{n_{\mathrm{L}}-n}{n_{\mathrm{L}}R_2}d \end{pmatrix}
\end{aligned}
\tag{2.102}
$$

The matrix element C in the first column of the second row is, according to (2.70), defined as $-1/f'$, where f' is the focal length of the thick lens in the image space

$$
f' = \frac{nn_{\mathrm{L}}R_1R_2}{(n_{\mathrm{L}}-n)\left[n_{\mathrm{L}}(R_2-R_1)+(n_{\mathrm{L}}-n)d\right]}\,. \tag{2.103}
$$

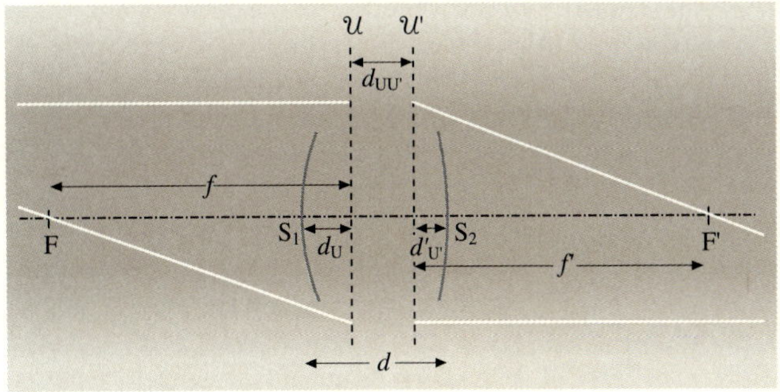

Fig. 2.21 The principal planes of a thick lens. Notice that d_{U} and $d'_{\mathrm{U'}}$ are both negative in the figure due to our sign convention

Because $n = n'$, the focal length f in the object space is $f = -f'$ and the nodal points and the principal points coincide. So, it is now necessary to calculate the positions of the principal points U and U' (Fig. 2.21). By using (2.69) one obtains

$$
\begin{aligned}
d_{\mathrm{U}} &= \frac{D}{C}(A-1)-B = \frac{AD-D-BC}{C} = \frac{1-D}{C}\\
&= \frac{-\frac{n_{\mathrm{L}}-n}{n_{\mathrm{L}}R_2}d}{-\frac{n_{\mathrm{L}}-n}{n}\left(\frac{1}{R_1}-\frac{1}{R_2}+\frac{n_{\mathrm{L}}-n}{n_{\mathrm{L}}}\frac{d}{R_1R_2}\right)}\\
&= \frac{ndR_1}{n_{\mathrm{L}}(R_2-R_1)+(n_{\mathrm{L}}-n)\,d}\,,\\
d'_{\mathrm{U'}} &= \frac{1-A}{C} = \frac{\frac{n_{\mathrm{L}}-n}{n_{\mathrm{L}}R_1}d}{-\frac{n_{\mathrm{L}}-n}{n}\left(\frac{1}{R_1}-\frac{1}{R_2}+\frac{n_{\mathrm{L}}-n}{n_{\mathrm{L}}}\frac{d}{R_1R_2}\right)}\\
&= -\frac{ndR_2}{n_{\mathrm{L}}(R_2-R_1)+(n_{\mathrm{L}}-n)\,d}\,.
\end{aligned}
\tag{2.104}
$$

The distance $d_{\mathrm{UU'}}$ between the two principal planes, which is positive if $\mathcal{U}'$ is right of $\mathcal{U}$, is:

$$
\begin{aligned}
d_{\mathrm{UU'}} &= d + d_{\mathrm{U}} + d'_{\mathrm{U'}}\\
&= d\left(1-\frac{n(R_2-R_1)}{n_{\mathrm{L}}(R_2-R_1)+(n_{\mathrm{L}}-n)d}\,.\right)
\end{aligned}
\tag{2.105}
$$

A Thick Lens in Air

Since the special case of a thick lens in air ($n = 1$) is the most important in practice, (2.102) for $1/f'$, (2.104) for d_{U} and $d'_{\mathrm{U'}}$ and (2.105) for $d_{\mathrm{UU'}}$ shall be repeated for this case:

$$\frac{1}{f'} = (n_L - 1)\left[\left(\frac{1}{R_1} - \frac{1}{R_2}\right) + \frac{n_L - 1}{n_L}\frac{d}{R_1 R_2}\right], \tag{2.106}$$

$$d_U = \frac{dR_1}{n_L(R_2 - R_1) + (n_L - 1)d}, \tag{2.107}$$

$$d'_{U'} = -\frac{dR_2}{n_L(R_2 - R_1) + (n_L - 1)d}, \tag{2.108}$$

$$d_{UU'} = d\left(1 - \frac{R_2 - R_1}{n_L(R_2 - R_1) + (n_L - 1)d}\right). \tag{2.109}$$

In the following, three important cases of thick lenses in air will be described to illustrate the optical parameters of lenses.

Special Cases of Thick Lenses in Air

Ball Lens. For a ball lens with radius of curvature $R > 0$ and refractive index n_L the lens parameters are (Fig. 2.22)

$$R_1 = R, \quad R_2 = -R, \quad d = 2R.$$

This means, according to (2.106–2.109), for the parameters in air

$$\begin{aligned} \frac{1}{f'} &= \frac{2(n_L - 1)}{Rn_L} \Rightarrow f' = \frac{n_L R}{2(n_L - 1)}, \\ d_U &= -R, \\ d'_{U'} &= -R, \\ d_{UU'} &= 0. \end{aligned} \tag{2.110}$$

This means that the principal points coincide and are at the center of curvature of the ball lens. For the special case $n_L = 2$ the focal length would be equal to the radius of curvature $f' = R$ so that the focus in the image space would be on the backside of the sphere. For $n_L < 2$ (e.g. nearly all glasses) the focus is outside of the sphere, whereas for $n_L > 2$ (e.g. a silicon ball lens illuminated with infrared light) the focus would be inside the sphere.

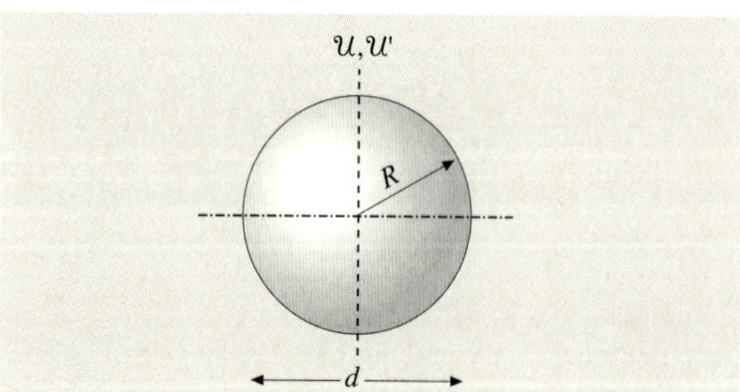

Fig. 2.22 Parameters of a ball lens

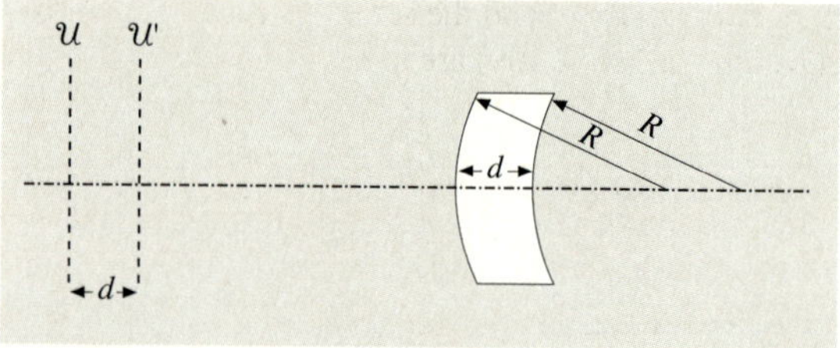

Fig. 2.23 The meniscus of Hoegh

The Meniscus Lens of Hoegh. For the meniscus lens of Hoegh (Fig. 2.23) with refractive index n_L and thickness d the two radii of curvature are identical, i. e. $R_1 = R_2 = R$. Then, (2.106–2.109) deliver:

$$\begin{aligned} \frac{1}{f'} &= \frac{(n_L - 1)^2 d}{n_L R^2}, \\ d_U &= \frac{R}{n_L - 1}, \\ d'_{U'} &= -\frac{R}{n_L - 1}, \\ d_{UU'} &= d. \end{aligned} \tag{2.111}$$

A thin meniscus with identical radii of curvature would have no optical effect. Contrary to this the thick meniscus of Hoegh has a positive optical power. At least one of the principal points is always outside of the lens and their separation is identical to the thickness of the lens (Fig. 2.23).

Plane–Convex or Plane–Concave Lenses. We assume now that the first surface of the thick lens with refractive index n_L and thickness d is curved (either convex, i. e.

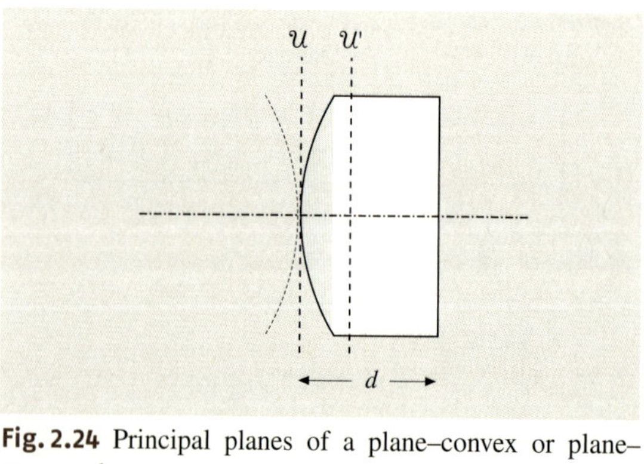

Fig. 2.24 Principal planes of a plane–convex or plane–concave lens

$R_1 > 0$, or concave) and the second is plane ($R_2 = \infty$). Equations (2.106–2.109) are in this case (Fig. 2.24):

$$\frac{1}{f'} = \frac{n_L - 1}{R_1},$$
$$d_U = 0,$$
$$d'_{U'} = -\frac{d}{n_L},$$
$$d_{UU'} = d\left(1 - \frac{1}{n_L}\right) = \frac{(n_L - 1)d}{n_L}. \tag{2.112}$$

This means that the first principal point coincides with the vertex of the curved surface. Moreover, the focal length of a lens with one plane surface is calculated like the focal length of a thin lens. This is not astonishing since the plane–convex/plane–concave lens can be interpreted as a combination of a thin lens with focal length f' and a plane parallel plate with thickness d and refractive index n_L. This can easily be shown by calculating the matrix $M = M_P M_{L0}$ and comparing it with M_{Ld} of (2.102) for $R_2 = \infty$.

2.2.7 Reflecting Optical Surfaces

Up to now only refracting surfaces have been treated, which form lenses and complete objectives. But there are of course also reflecting surfaces which are e.g. very important in astronomical telescopes [2.21] or which will be very important in the near future for optical lithography systems [2.22] in the extreme ultraviolet (EUV) at a wavelength of 13 nm. However, a reflecting surface can easily be included in a paraxial design by calculating its paraxial 2×2 matrix and including it instead of the surface of a refracting surface in (2.54). We will see that the determinant of the matrix of a reflecting surface is one so that our general discussions concerning the relation between the focal lengths f and f' are valid.

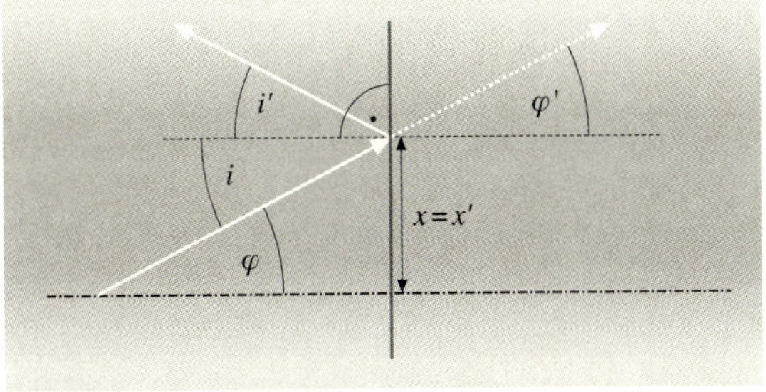

Fig. 2.25 Reflection at a plane surface

A Plane Reflecting Surface

The reflection at a plane surface, which is perpendicular to the optical axis, is shown in Fig. 2.25. The law of reflection means that the angle i' of the reflected ray with the surface normal is identical to the angle i of the incident ray, i.e. $i = i'$. In the paraxial theory it is common practice not to take the reflected ray since then the light would travel from right to left. Instead, the unfolded ray path is taken, which is obtained by mirroring the reflected ray at the reflecting surface. By doing this the dashed ray in Fig. 2.25 is obtained and there is no change of the paraxial ray parameters x and φ. So, the paraxial ray matrix M_{RP} of a reflecting plane surface is just the unit matrix:

$$M_{RP} = \begin{pmatrix} 1 & 0 \\ 0 & 1 \end{pmatrix}. \tag{2.113}$$

Its determinant is of course one.

A Spherical Reflecting Surface

The reflection at a spherical surface is treated analogously to the case of a plane surface and is shown for a convex mirror in Fig. 2.26. The ray that is reflected at the spherical surface is mirrored at a plane which goes through the vertex of the surface and is perpendicular to the optical axis. So, the dashed ray in Fig. 2.26 results. All angles in Fig. 2.26 are positive so that the following

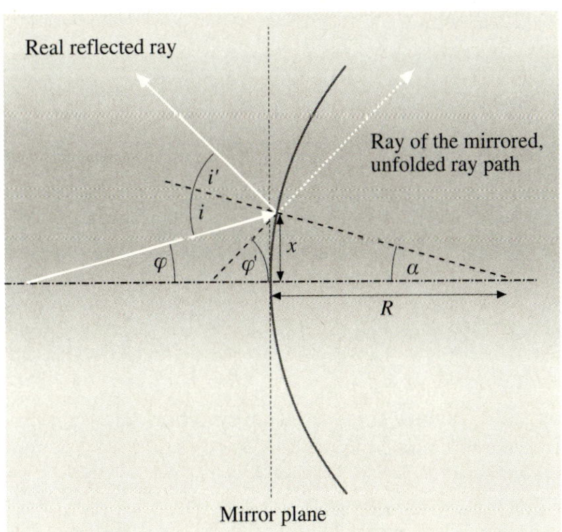

Fig. 2.26 Reflection at a spherical surface

relations are valid:

$$\left.\begin{aligned} i &= \varphi+\alpha \\ \alpha+i' &= \varphi' \\ i &= i' \\ \alpha &= x/R \end{aligned}\right\} \Rightarrow \varphi' = \varphi+2\alpha = \varphi+2\frac{x}{R} . \tag{2.114}$$

Since the ray height x remains constant during reflection the paraxial ray matrix M_{RS} is

$$M_{RS} = \begin{pmatrix} 1 & 0 \\ \frac{2}{R} & 1 \end{pmatrix} . \tag{2.115}$$

Again the determinant is one.

The matrix (2.115) is also valid for a concave mirror. There, the radius of curvature R is negative so that the angle φ' is smaller than the angle φ for a positive ray height x. This is just the effect of a concave mirror with a positive optical power.

As an exercise the cardinal points of a spherical mirror shall be calculated by using (2.69) and (2.70):

$$\begin{aligned} d_U &= \frac{D}{C}(A-1)-B=0 , \\ d_N &= \frac{1-D}{C}=0 , \\ d_F &= -\frac{D}{C}=-\frac{R}{2} , \\ d'_{U'} &= \frac{1-A}{C}=0 , \\ d'_{N'} &= \frac{A}{C}(D-1)-B=0 , \\ d'_{F'} &= -\frac{A}{C}=-\frac{R}{2} , \\ f &= \frac{AD-BC}{C}=\frac{\det(M)}{C}=\frac{R}{2} , \\ f' &= -\frac{1}{C}=-\frac{R}{2} . \end{aligned} \tag{2.116}$$

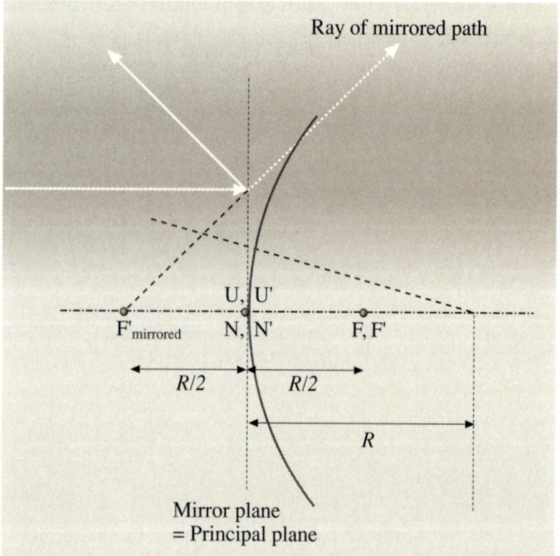

Fig. 2.27 Cardinal points of a convex mirror. A ray coming from the left parallel to the optical axis has to go into the image space (virtually) through the focus $F'_{mirrored}$. $F'_{mirrored}$ is the focus of the unfolded ray path, which is mirrored at the vertex plane. The real reflected ray would virtually go through the focus F' which then coincides with F

So, the principal points U, U′ and the nodal points N, N′ all coincide with the vertex of the spherical mirror (Fig. 2.27). The focus F in the object space is at half the distance between the center of curvature of the spherical surface and the vertex. On the other side, the focus F′ in the image space would coincide with F for the real reflected ray. But since the unfolded ray path with the mirrored ray is taken the focus $F'_{mirrored}$ is also mirrored at the principal plane through the vertex that is perpendicular to the optical axis. The focal length is of course half the radius of curvature and a convex mirror has a negative optical power whereas a concave mirror has a positive optical power.

If an optical system containing refractive and reflective surfaces has to be analyzed so that the same lens is e.g. passed twice or more it is necessary, on the way back, to change the order of surfaces and refractive indices and also the signs of the radii of curvature.

2.2.8 Extension of the Paraxial Matrix Theory to 3×3 Matrices

The paraxial 2×2 matrix theory can only be used as long as all elements are centered around the optical axis and symmetric with respect to the optical axis. A tilted refractive plane surface or a diffraction grating, which both introduce a global tilt of all rays, can e.g. not be included in the 2×2 matrix theory. But there is an extension of this method by using 3×3 matrices [2.15]. This will be described in the following.

Paraxial Ray Tracing at a Diffraction Grating

A ray representing a plane wave with wavelength λ which hits a diffraction grating with a period Λ is diffracted according to the well-known grating equation [2.1] (Fig. 2.28):

$$\sin\varphi' = \sin\varphi + m\frac{\lambda}{\Lambda} . \tag{2.117}$$

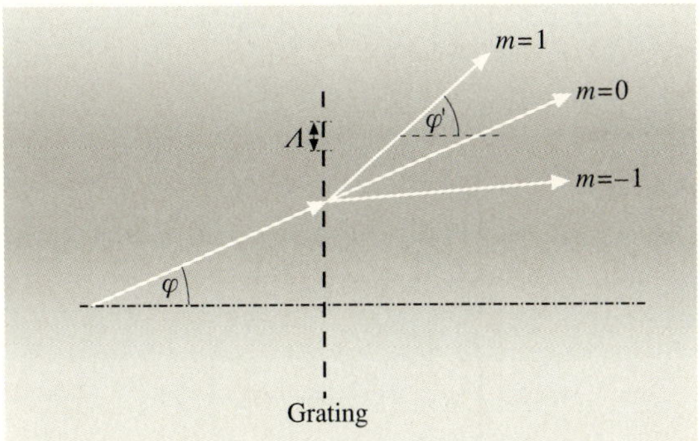

Fig. 2.28 The diffraction at a grating symbolized here as an amplitude grating with period Λ. Three different diffraction orders m are shown

Here, the integer m is the diffraction order of the grating and, depending on the type of the grating, there may be only one efficient order (e.g. for blazed gratings or volume holograms) or many orders with non-vanishing efficiency (e.g. for binary phase elements or amplitude gratings) [2.23–25]. In the case of many diffraction orders each order has to be calculated separately. The angles φ and φ' are the angles of the incident and diffracted ray, respectively.

In the paraxial approximation the sine of the angles is replaced by the angle itself so that the grating equation is

$$\varphi' = \varphi + m\frac{\lambda}{\Lambda} . \tag{2.118}$$

Together with the equation for the ray height x ($x' = x$), which does not change in the case of diffraction at a grating, there are two equations relating the ray parameters before and after diffraction at the grating. However, it is no longer possible to write these two equations in a pure 2×2 matrix notation since it would be

$$\begin{pmatrix} x' \\ \varphi' \end{pmatrix} = \begin{pmatrix} 1 & 0 \\ 0 & 1 \end{pmatrix} \begin{pmatrix} x \\ \varphi \end{pmatrix} + \begin{pmatrix} 0 \\ m\lambda/\Lambda \end{pmatrix} . \tag{2.119}$$

So, a constant additive vector at the end would be necessary and the calculation of one 2×2 matrix for a complete optical system containing one or more diffraction gratings and several other optical elements would be impossible. However, there is a possibility to change this by using 3×3 matrices instead of 2×2 matrices and a paraxial ray vector with three components instead of two, where the third component is always 1. The 3×3 matrices and the paraxial ray vectors are of the form

$$M_{3\times3} = \begin{pmatrix} A & B & \Delta x \\ C & D & \Delta\varphi \\ 0 & 0 & 1 \end{pmatrix} \Rightarrow \begin{pmatrix} x' \\ \varphi' \\ 1 \end{pmatrix} = M_{3\times3} \begin{pmatrix} x \\ \varphi \\ 1 \end{pmatrix}$$
$$= \begin{pmatrix} M \begin{pmatrix} x \\ \varphi \end{pmatrix} + \begin{pmatrix} \Delta x \\ \Delta\varphi \end{pmatrix} \\ 1 \end{pmatrix} , \tag{2.120}$$

where M is the normal paraxial 2×2 matrix with coefficients A, B, C and D. The coefficients Δx and $\Delta\varphi$ are constant values which symbolize a lateral shift or a tilt which is exerted on the incident paraxial ray by the element. To obtain the 3×3 matrix appendant to a pure paraxial 2×2 matrix the coefficients Δx and $\Delta\varphi$ just have to be set to zero.

The solution of our original example to define the paraxial 3×3 matrix $M_{G,3\times3}$ of a (non-tilted) diffraction grating is now quite easy:

$$M_{G,3\times3} = \begin{pmatrix} 1 & 0 & 0 \\ 0 & 1 & m\frac{\lambda}{\Lambda} \\ 0 & 0 & 1 \end{pmatrix} . \tag{2.121}$$

Tilted Refractive Plane Surface

A refractive plane surface shall have a normal vector that is tilted by a small angle α with respect to the optical axis. The surface with refractive indices n in front of and n' behind the surface is hit by a paraxial ray with ray parameters x and φ (Fig. 2.29). Since, the tilt angle α has to be small and the ray heights x also, the variation of the z-coordinates at the points of intersection of the tilted surface and rays with different heights x can be

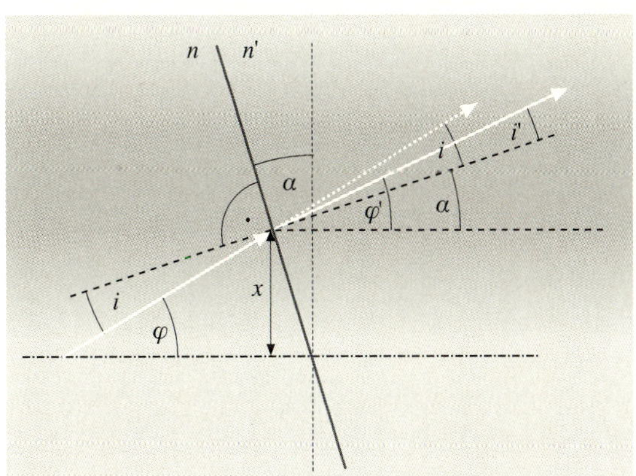

Fig. 2.29 The refraction at a tilted plane surface with tilt angle α

neglected:

$$\Delta z = x \tan\alpha \approx x\alpha \approx 0\,, \tag{2.122}$$

i. e. it is of second order and only first-order terms are taken into account in the paraxial approximation.

Additionally, the ray height x remains constant for refraction. The ray angles, tilt angles and refraction angles depend on each other through the following equations:

$$\left.\begin{aligned}\varphi' &= i' + \alpha\\ \varphi &= i + \alpha\\ ni &= n'i'\end{aligned}\right\} \Rightarrow \varphi' = \frac{n}{n'}i + \alpha = \frac{n}{n'}(\varphi - \alpha) + \alpha$$

$$= \frac{n}{n'}\varphi + \frac{n'-n}{n'}\alpha\,. \tag{2.123}$$

So, the 3×3 matrix $M_{\mathrm{R},\alpha,3\times3}$ for refraction at a tilted plane surface is

$$M_{\mathrm{R},\alpha,3\times3} = \begin{pmatrix}1 & 0 & 0\\ 0 & \frac{n}{n'} & \frac{n'-n}{n'}\alpha\\ 0 & 0 & 1\end{pmatrix}\,. \tag{2.124}$$

As an application and to see how the matrix of a complete system is determined the matrix of a thin prism will be calculated in the next paragraph.

Thin Prism

A thin prism consists of two tilted refractive surfaces and we assume that the prism is made of a material with refractive index n' and the refractive index outside of the prism is n at both sides. Since the prism is assumed to be thin the propagation between the two refractive surfaces is neglected and the total matrix $M_{\mathrm{Prism},3\times3}$ of the prism is obtained by just multiplying the 3×3 matrices of the single surfaces. The tilt angles of the two surfaces are α_1 and α_2 so that we have

$$M_{\mathrm{Prism},3\times3} = M_{\mathrm{R},\alpha_2,3\times3} M_{\mathrm{R},\alpha_1,3\times3}$$

$$= \begin{pmatrix}1 & 0 & 0\\ 0 & \frac{n'}{n} & \frac{n-n'}{n}\alpha_2\\ 0 & 0 & 1\end{pmatrix}\begin{pmatrix}1 & 0 & 0\\ 0 & \frac{n}{n'} & \frac{n'-n}{n'}\alpha_1\\ 0 & 0 & 1\end{pmatrix}$$

$$= \begin{pmatrix}1 & 0 & 0\\ 0 & 1 & \frac{n'-n}{n}(\alpha_1 - \alpha_2)\\ 0 & 0 & 1\end{pmatrix}\,. \tag{2.125}$$

By defining the prism angle $\gamma := \alpha_1 - \alpha_2$ the total deflection angle δ of a thin prism with prism angle γ is $\delta = (n'-n)\gamma/n$. For the most important case of a thin prism in air ($n = 1$) it is $\delta = (n'-1)\gamma$.

The Transformation Matrices

The matrix of a tilted plane surface or other tilted and also laterally shifted surfaces can be calculated quite formally by introducing the paraxial transformation matrix between two coordinate systems. The first coordinate system with axes x and z will be named the global coordinate system. The second coordinate system with axes x' and z' is called the local coordinate system because, in this coordinate system, the surface will have a simple

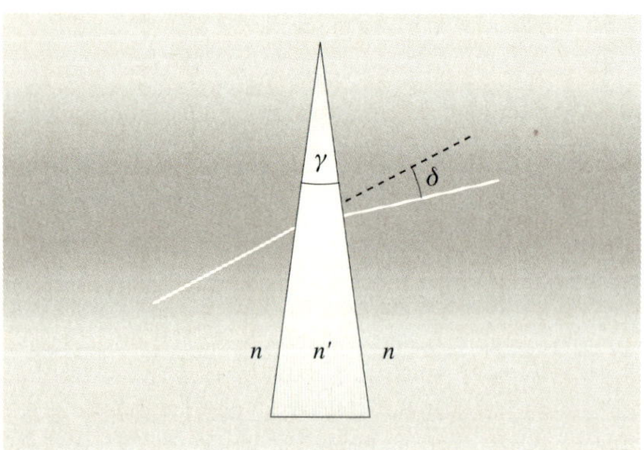

Fig. 2.30 The refraction at a thin prism with prism angle γ. The incident ray is deflected by an angle δ

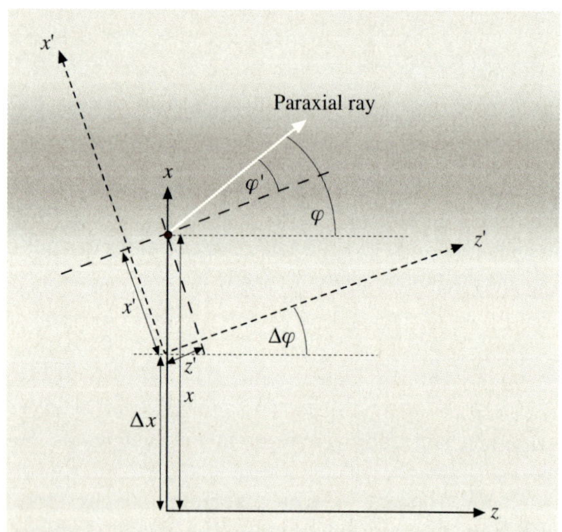

Fig. 2.31 Transformation between two relatively tilted and shifted coordinate systems. The local x'–z'-coordinate system is laterally shifted with respect to the global x–z-coordinate system by a distance Δx and tilted by an angle $\Delta\varphi$

form, i. e. it is non-tilted and non-shifted in this local coordinate system. The local coordinate system is obtained from the global one by shifting a copy laterally in the x-direction by the small distance Δx and rotating it by an angle $\Delta\varphi$ (Fig. 2.31). So, a paraxial ray with ray parameters (x, φ) in the global coordinate system has ray parameters (x', φ') in the local coordinate system and the following relations are valid

$$x' = (x - \Delta x)\cos(\Delta\varphi) \approx x - \Delta x\,, \tag{2.126}$$

$$z' = (x - \Delta x)\sin(\Delta\varphi) \approx (x - \Delta x)\,\Delta\varphi \approx 0\,, \tag{2.127}$$

$$\varphi' = \varphi - \Delta\varphi\,. \tag{2.128}$$

Here, the paraxial approximations are used and since Δx, $\Delta\varphi$, x and φ are all paraxial (i. e. small) quantities, only terms of first order are taken into account whereas terms of second order such as $(x - \Delta x)\Delta\varphi$ are set to zero. So, the z'-coordinate remains zero if the ray has a global coordinate $z = 0$, what is always the case by choosing the global coordinate system accordingly.

Therefore, the matrix $M_{G\to L,3\times3}$ for the transformation of a paraxial ray from the global coordinate system to a local coordinate system is

$$M_{G\to L,3\times3} = \begin{pmatrix} 1 & 0 & -\Delta x \\ 0 & 1 & -\Delta\varphi \\ 0 & 0 & 1 \end{pmatrix}\,. \tag{2.129}$$

Vice versa, the matrix $M_{L\to G,3\times3}$ for the transformation of a paraxial ray from the local coordinate system to the global coordinate system is the inverse matrix to $M_{G\to L,3\times3}$:

$$M_{L\to G,3\times3} = M^{-1}_{G\to L,3\times3} = \begin{pmatrix} 1 & 0 & \Delta x \\ 0 & 1 & \Delta\varphi \\ 0 & 0 & 1 \end{pmatrix}\,. \tag{2.130}$$

It is also important to notice that in the paraxial approximation with small shifts Δx and small angles $\Delta\varphi$ the order of shifting and tilting is arbitrary whereas this is not the case for finite quantities. Mathematically, this can be proved by calculating that the two matrices for a pure shift (i. e. $\Delta\varphi = 0$) and for a pure tilt (i. e. $\Delta x = 0$) permute:

$$\begin{aligned} \begin{pmatrix} 1 & 0 & \Delta x \\ 0 & 1 & 0 \\ 0 & 0 & 1 \end{pmatrix} \begin{pmatrix} 1 & 0 & 0 \\ 0 & 1 & \Delta\varphi \\ 0 & 0 & 1 \end{pmatrix} &= \begin{pmatrix} 1 & 0 & \Delta x \\ 0 & 1 & \Delta\varphi \\ 0 & 0 & 1 \end{pmatrix} \\ &= \begin{pmatrix} 1 & 0 & 0 \\ 0 & 1 & \Delta\varphi \\ 0 & 0 & 1 \end{pmatrix} \begin{pmatrix} 1 & 0 & \Delta x \\ 0 & 1 & 0 \\ 0 & 0 & 1 \end{pmatrix}\,. \end{aligned} \tag{2.131}$$

This means, that in the paraxial approximation it is identical if the coordinate system is first tilted and afterwards shifted or if it is first shifted and then tilted. So, we can use one matrix for the whole transformation without taking care of the order of the single transformations.

As an application of the transformation matrices the 3×3 matrix for refraction at a tilted and laterally shifted spherical surface with a radius of curvature R shall be calculated. The refractive indices are again n in front of the surface and n' behind it. The vertex of the spherical surface is laterally shifted by a distance Δx with respect to the optical axis (global coordinate system) and the surface is rotated around an axis perpendicular to the meridional plane by an angle $\Delta\varphi$. The local coordinate system is of course that system in which the surface is neither tilted nor rotated. Then, a ray in the local coordinate system can be calculated by multiplying the incident ray (in the global coordinate system) by the transformation matrix $M_{G\to L,3\times3}$. In the local coordinate system, the ray is multiplied by the matrix of a normal non-tilted and non-shifted spherical surface $M_{S,3\times3}$. Afterwards, the ray in the local coordinate system is transformed back into the global system by multiplying it with $M_{L\to G,3\times3}$. So, the matrix $M_{S,\Delta x,\Delta\varphi,3\times3}$ for refraction at a tilted and shifted spherical surface in the global coordinate system is just the product of the three matrices:

$$\begin{aligned} M_{S,\Delta x,\Delta\varphi,3\times3} &= M_{L\to G,3\times3} M_{S,3\times3} M_{G\to L,3\times3} \\ &= \begin{pmatrix} 1 & 0 & \Delta x \\ 0 & 1 & \Delta\varphi \\ 0 & 0 & 1 \end{pmatrix} \begin{pmatrix} 1 & 0 & 0 \\ -\frac{n'-n}{n'R} & \frac{n}{n'} & 0 \\ 0 & 0 & 1 \end{pmatrix} \begin{pmatrix} 1 & 0 & -\Delta x \\ 0 & 1 & -\Delta\varphi \\ 0 & 0 & 1 \end{pmatrix} \\ &= \begin{pmatrix} 1 & 0 & \Delta x \\ 0 & 1 & \Delta\varphi \\ 0 & 0 & 1 \end{pmatrix} \begin{pmatrix} 1 & 0 & -\Delta x \\ -\frac{n'-n}{n'R} & \frac{n}{n'} & \frac{n'-n}{n'R}\Delta x - \frac{n}{n'}\Delta\varphi \\ 0 & 0 & 1 \end{pmatrix} \\ &= \begin{pmatrix} 1 & 0 & 0 \\ -\frac{n'-n}{n'R} & \frac{n}{n'} & \frac{n'-n}{n'R}\Delta x + \frac{n'-n}{n'}\Delta\varphi \\ 0 & 0 & 1 \end{pmatrix}\,. \end{aligned} \tag{2.132}$$

The result shows that the ray height x remains, as expected, unchanged by refraction at the surface ($x' = x$) and that there is for the ray angle φ', besides the usual term of a spherical surface, an additional term which does not depend on the angle of incidence but on the shift Δx and the tilt $\Delta\varphi$. It can also be seen that this additional term is zero if the condition $\Delta x/R = -\Delta\varphi$ is fulfilled. This is the well-known fact that a lateral shift of a spherical surface can be compensated by tilting it.

A special case is $R \rightarrow \infty$ so that the spherical surface becomes a plane surface. In this case the matrix of (2.132) becomes

$$M_{\text{S},\Delta x,\Delta\varphi,3\times3} \overset{R\rightarrow\infty}{\Rightarrow} \begin{pmatrix} 1 & 0 & 0 \\ 0 & \frac{n}{n'} & \frac{n'-n}{n'}\Delta\varphi \\ 0 & 0 & 1 \end{pmatrix} = M_{\text{R},\Delta\varphi,3\times3} \,. \quad (2.133)$$

This is of course the same result for the matrix $M_{\text{R},\alpha,3\times3}$ for refraction at a tilted plane surface with $\Delta\varphi = \alpha$ as that obtained in (2.124), which we derived directly from Fig. 2.29.

2.3 Stops and Pupils

In the preceding section about paraxial optics only rays and object points in the neighborhood of the optical axis were considered. So, in the paraxial calculations stops have no influence. But this changes dramatically for the case of non-paraxial optics. There, stops are quite important optical elements which determine the light-gathering power of an optical system, its resolution, the amount of aberrations, its field and so on. In the following only some elementary definitions about stops and pupils can be given. For more information we refer to the literature [2.1, 3, 8, 20]. There are two especially important stops, the aperture stop and the field stop.

2.3.1 The Aperture Stop

Assume first of all a light-emitting object point which radiates in all directions. Then, the aperture stop is that physical stop which limits the cross-section of the image-forming pencil of rays. To determine the aperture stop the size and position of the images of all stops (e.g. lens apertures or real stops) in the system by that part of the system which precedes the respective stop have to be calculated. To do this the paraxial matrix theory of the last section can, for example, be used. If the distance of the image of stop i from the object point is l_i and the diameter of the stop image is d_i, then the aperture angle φ_i which can pass that stop is

$$\tan\varphi_i = \frac{d_i}{2l_i} \,. \quad (2.134)$$

The aperture stop is now that stop number i that provides the minimum value φ_O of φ_i. The image of the aperture stop made by that part of the optical system which precedes the aperture stop is called the entrance pupil and the image of the aperture stop made by that part of the optical system which follows the aperture stop is called the exit pupil. The full aperture angle $2\varphi_\text{O}$ is called the angular aperture on the object side and the corresponding quantity $2\varphi_\text{I}$ on the image side is called the angular aperture on the image side. φ_I can be determined by calculating the diameter d_I of the exit pupil and the distance l_I between the exit pupil and the image point and using again an equation like (2.134), replacing d_i with d_I and l_i with l_I.

If the aperture stop is in front of the optical system the aperture stop and the entrance pupil will be identical. On the contrary, if the aperture stop is behind the whole optical system the aperture stop and the exit pupil are identical. In the general case, where the aperture stop is somewhere in the optical system the entrance pupil and the exit pupil can also be somewhere and they can be real or virtual images of the aperture stop. If an optical system consists of only one (thin) single lens the aperture stop, entrance pupil and exit pupil are of course all identical to the aperture of the lens itself. Another interesting case is e.g. an optical system where the aperture stop is in the back focal plane of the preceding part of the optical system. Then the entrance pupil is at infinity and the system is called telecentric on the object side. In this case all chief rays on the object side (see later in this section) are parallel to the optical axis. Similarly, if the aperture stop is in the front focal plane of the part of the optical system which follows the aperture stop the exit pupil will be at infinity and the system is called telecentric on the image side. Optical systems which are telecentric on both sides are quite important in optical metrology because in this case object points in different object planes have the same lateral magnification because the chief rays in the object and image spaces are both parallel to the optical axis. Therefore, the measured size of the object will be correct in a given image plane even though the object may be out of the object plane that is imaged sharply.

A quite important quantity to characterize an optical system is the numerical aperture NA. The numerical aperture NA_O on the object side is defined as

$$NA_O = n_O \sin\varphi_O \tag{2.135}$$

and the numerical aperture NA_I on the image side is

$$NA_I = n_I \sin\varphi_I\,, \tag{2.136}$$

where n_O and n_I are the refractive indices in the object and image space, respectively. It is an elementary property of optical imaging systems that NA_O and NA_I are connected by the lateral magnification β (2.52) of the optical system if the sine condition is fulfilled

$$NA_I = \frac{NA_O}{\beta}\,. \tag{2.137}$$

In fact, by replacing β by the ratio x_I/x_O of the image size and the object size this equation can be written as

$$x_I n_I \sin\varphi_I = x_O n_O \sin\varphi_O \tag{2.138}$$

which is the usual formulation of the sine condition [2.1]. For the paraxial case this invariant reduces to the Smith–Helmholtz invariant:

$$x_I n_I \varphi_I = x_O n_O \varphi_O \tag{2.139}$$

The numerical aperture determines how much light the optical system can gather from the object. It also determines (in the case of no aberrations) the resolution of the system due to diffraction. We will see in Sect. 2.5.3 that many aberrations depend on the numerical aperture. The position of the aperture stop in an optical system also influences the aberrations [2.26].

Another quite important definition of geometrical optics is the so called chief ray or principal ray. This is that ray coming from the object point (which can of course be off-axis) which passes through the center of the aperture stop. Since the entrance pupil and the exit pupil are both images of the aperture stop the chief ray also passes through the centers of the entrance pupil and exit pupil. If there are strong aberrations in the system this may not be exactly the case for object points which are strongly off-axis.

2.3.2 The Field Stop

The second quite important stop is the field stop, which limits the diameter of the object field which can be imaged by an optical system. To find the field stop we calculate again the images of all stops by that part of the optical system which precedes the respective stop. Let us assume that the image of stop number i has again a diameter d_i and that the distance between the image of the stop and the entrance pupil of the system is L_i. The field stop is then that stop which has the smallest value ϕ_O of all values ϕ_i with

$$\tan\phi_i = \frac{d_i}{2L_i} \tag{2.140}$$

The value $2\phi_O$ is called the field angle. The image of the field stop by that part of the optical system which precedes the field stop is called the entrance window and the image by that part of the optical system which follows the field stop is called the exit window.

If the line connecting an (off-axis) object point and the center of the entrance pupil is blocked by the entrance window the chief ray cannot pass the field stop and so this object point cannot be imaged in most cases. However, there are cases where other rays coming from the object point can pass anyway and then there is no sharp border of the object field but the outer parts of the object field are imaged with lower intensity. This effect is called vignetting.

2.4 Ray Tracing

It has been shown in Sect. 2.1.3 that light can be described by rays as long as the approximation of geometrical optics is valid. The propagation of such rays through an optical system is a very important tool to develop optical systems and to calculate their expected quality. The propagation of light rays through an optical system is called ray tracing [2.27, 28] and it is the basic tool of optical design, i. e. the design and optimization of optical systems concerning their imaging quality or other properties (e.g. tolerance of a system against misalignments or fabrication errors of components). In this section the principle of ray tracing and some applications will be described. There is of course no room to discuss the basics of optical design itself. For this we refer to the literature [2.16–19].

2.4.1 Principle

According to (2.29) a light ray propagates rectilinearly in a homogeneous and isotropic material. At an interface to another material the ray is partially refracted and partially reflected depending on the properties of the in-

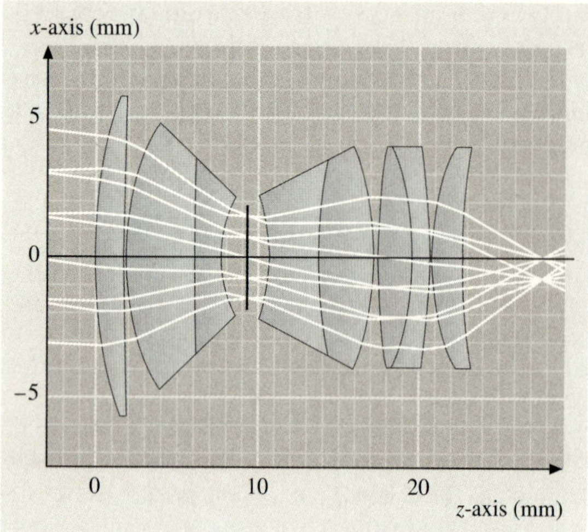

Fig. 2.32 Propagation of some light rays in a typical microscopic objective (NA = 0.4, magnification 20×, focal length $f' = 11.5$ mm) calculated with our internal software RAYTRACE. In this case the microscopic objective is used in the reverse order, i. e. to focus light

terface. If a material is inhomogeneous (e.g. in GRIN lenses or in air films with different temperatures) the light ray is curved during the propagation and the path of the ray has to be calculated by solving (2.27), in most cases numerically [2.29, 30]. However, in this section it is assumed that the optical system consists of different homogeneous materials which are separated by refracting or reflecting interfaces.

Ray tracing means that the path of a bundle of rays, which are e.g. emitted by an object point or form a plane wave (i. e. object point with infinite distance), is determined in an optical system (see Fig. 2.32 for tracing rays through an microscopic objective). In the approximation of geometrical optics the calculation is exact and no other approximations, such as e.g. paraxial approximations, are made. Since ray tracing can be easily automated with the help of computers it is nowadays the most important tool for designing lenses, telescopes and complete optical systems [2.16–19]. For complex optical systems it is even today with the help of modern computers not possible to replace ray tracing by pure wave-optical methods. Moreover, for most macroscopic optical systems, ray tracing in combination with wave-optical evaluation methods like the calculation of the point spread function [2.18], assuming that only the exit pupil of the system introduces diffraction, is a sufficiently accurate method to analyze imaging systems. Another quite modern application of ray tracing is the analysis of illumination systems with incoherent light. This will be discussed in Sect. 2.4.8 about non-sequential ray tracing.

A precondition for ray tracing is that the optical system is known very well. It is not sufficient to know some paraxial parameters but it is necessary to know the following data of the surfaces as well as the materials:

- Type of the surface, e.g. plane, spherical, parabolic, cylindrical, toric or other aspheric surface.
- Characteristic data of the surface itself, e.g. the radius of curvature in the case of a spherical surface or the aspheric coefficients in the case of an aspheric surface.
- Shape and size of the boundary of the surface, e.g. circular with a certain radius, rectangular with two side lengths or annular with an interior and an outer radius.
- Position and orientation of the surface in all three directions of space.
- Refractive indices of all materials and their dependence on wavelength.

The tracing of a given ray through an optical system has the following structure:

1. Determine the point of intersection of the ray with the following optical surface. If there is no point of intersection or if the hit surface is absorbing mark the ray as invalid and finish the tracing of this ray. Depending on the type of ray tracing it may also be necessary in this case to leave the ray unchanged and to go to step 4. If there is a point of intersection go to step 2.
2. Calculate the surface normal in the point of intersection.
3. Apply the law of refraction or reflection (or another law e.g. in the case of diffractive optical elements [2.31, 32]). Then, the new direction of the deflected ray is known and the point of intersection with the surface is the new starting point of the ray.
4. If there is another surface in the optical system go back to step 1 or, if not, then finish the tracing of this ray.

In the case of step 1 the following surface can either be the physically next surface of the optical system which will really be hit by the ray (i. e. non-sequential ray tracing) or just the next surface in the computer list of surfaces where the order of the surfaces has been determined by the user (i. e. sequential ray tracing).

In the next sections the mathematical realization of the different steps of ray tracing will be described.

2.4.2 Mathematical Description of a Ray

A light ray (in a homogeneous material) can be described mathematically as a straight line with a starting point $\boldsymbol{p}$ and a direction vector $\boldsymbol{a}$ parallel to the ray (see Fig. 2.33). Here, $\boldsymbol{a}$ is a unit vector, i. e. $|\boldsymbol{a}| = 1$. According to (2.29) an arbitrary point on the ray with position vector $\boldsymbol{r}$ is described by

$$\boldsymbol{r} = \boldsymbol{p} + s\boldsymbol{a} . \tag{2.141}$$

The scalar parameter s is the arc length of the ray, i. e. in this case of rectilinear rays it is just the distance between $\boldsymbol{r}$ and $\boldsymbol{p}$. The virtual part of the ray is described by $s < 0$ whereas that part where there is really light has $s \geq 0$. In practice, there is also a maximum value s_{max} if the ray hits a surface where it is deflected.

2.4.3 Determination of the Point of Intersection with a Surface

The determination of the point of intersection of a light ray described by (2.141) with a surface requires of course a mathematical description of a surface. It is well known from mathematics that a surface can be described in an implicit form with a function F fulfilling

$$F(\boldsymbol{r}) = 0 . \tag{2.142}$$

Concrete examples will be given later. By combining (2.141) and (2.142) the determination of the point of intersection is mathematically equivalent to the determination of the roots of a function G with the variable s

$$G(s_0) := F(\boldsymbol{p} + s_0\boldsymbol{a}) = 0 . \tag{2.143}$$

After having determined the value s_0 at the root of G the position vector $\boldsymbol{r}_0$ of the point of intersection itself is obtained by applying s_0 to (2.141).

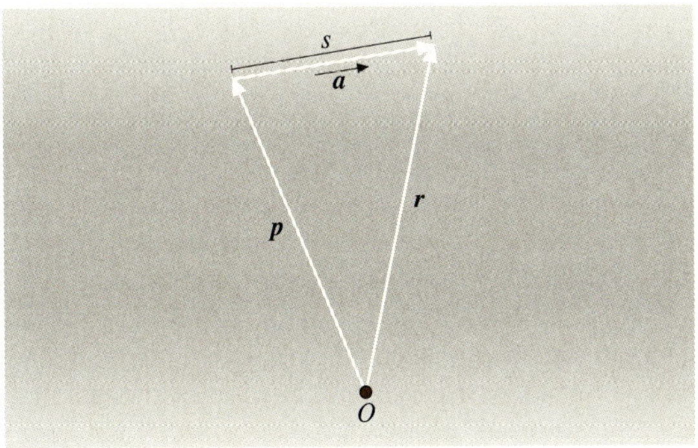

Fig. 2.33 Mathematical representation of a light ray as a straight line. O symbolizes the origin of the coordinate system

In many cases there can be several roots of G and it is also necessary to check whether the point of intersection is in the valid part of the surface, which is in practice limited by a boundary. Then, that root with the smallest positive value of s_0 lying in the valid part of the surface has to be taken. These queries can be quite complex in a computer program.

For general aspheric surfaces the solution of (2.143) will only be possible numerically, but for some simple cases the analytic solutions will be given in the following.

Plane Surface

A plane surface can be described by the position vector $\boldsymbol{C}$ of a point on the surface (typically this point is in the center of the plane surface) and by the surface normal $\boldsymbol{n}_z$ (see Fig. 2.34). Then, each point $\boldsymbol{r}$ of the surface fulfills

$$F(\boldsymbol{r}) = (\boldsymbol{r} - \boldsymbol{C}) \cdot \boldsymbol{n}_z = 0 . \tag{2.144}$$

The solution of (2.143) in this case is

$$(\boldsymbol{p} - \boldsymbol{C}) \cdot \boldsymbol{n}_z + s_0\boldsymbol{a} \cdot \boldsymbol{n}_z = 0$$
$$\Rightarrow \quad s_0 = \frac{(\boldsymbol{C} - \boldsymbol{p}) \cdot \boldsymbol{n}_z}{\boldsymbol{a} \cdot \boldsymbol{n}_z} . \tag{2.145}$$

In the case $\boldsymbol{a} \cdot \boldsymbol{n}_z = 0$ there is no definite point of intersection with the surface.

Of course, (2.144) describes an unlimited surface whereas the surfaces of an optical system are limited. Therefore, it has to be checked whether the point of intersection is in the valid area of the surface.

For a circular surface with radius R and center $\boldsymbol{C}$ this means e.g. that the point of intersection $\boldsymbol{r}_0$ has to fulfill the condition $|\boldsymbol{r}_0 - \boldsymbol{C}| \leq R$.

For a rectangular surface a second vector $\boldsymbol{n}_x$ (with $\boldsymbol{n}_x \cdot \boldsymbol{n}_z = 0$ and $|\boldsymbol{n}_x| = 1$) along one of the sides (the vector $\boldsymbol{n}_y$ along the second side is then just

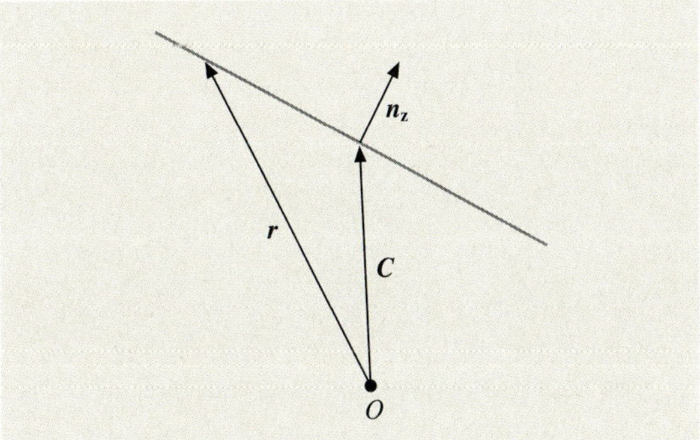

Fig. 2.34 Mathematical description of a plane surface

$\boldsymbol{n}_y := \boldsymbol{n}_z \times \boldsymbol{n}_x$) and the side lengths l_x and l_y of the rectangle have to be defined, additionally. Then, it has to be checked whether the conditions $|(\boldsymbol{r}_0 - \boldsymbol{C}) \cdot \boldsymbol{n}_x| \leq l_x/2$ and $|(\boldsymbol{r}_0 - \boldsymbol{C}) \cdot \boldsymbol{n}_y| \leq l_y/2$ are fulfilled.

Spherical Surface

A sphere whose center of curvature has position vector $\boldsymbol{C}$ and whose radius is R is described by

$$F(\boldsymbol{r}) = |\boldsymbol{r} - \boldsymbol{C}|^2 - R^2 = 0\,. \tag{2.146}$$

Therefore, (2.143) results in a quadratic equation for s_0:

$$s_0^2 + 2s_0\,(\boldsymbol{p} - \boldsymbol{C}) \cdot \boldsymbol{a} + |\boldsymbol{p} - \boldsymbol{C}|^2 - R^2 = 0\,.$$

The two solutions are

$$s_0^{1,2} = (\boldsymbol{C} - \boldsymbol{p}) \cdot \boldsymbol{a} \pm \sqrt{[(\boldsymbol{C} - \boldsymbol{p}) \cdot \boldsymbol{a}]^2 - |\boldsymbol{C} - \boldsymbol{p}|^2 + R^2}\,, \tag{2.147}$$

where the superscript 1, 2 is an index marking the two solutions. Depending on the argument of the square root there exist no (if the argument is negative), one (if the argument is zero) or two (if the argument is positive) solutions.

After having determined the points of intersection with the full sphere it has to be checked whether the points of intersection are in the valid part of the spherical surface. To do this an additional vector $\boldsymbol{n}_z$ ($|\boldsymbol{n}_z| = 1$) along the local optical axis and the lateral diameter D of the surface have to be defined (Fig. 2.35). The radius of curvature R is positive if the vector $\boldsymbol{n}_z$ points from the vertex to the center of curvature. In Fig. 2.35 R is, for example, positive. By using some trigonometric relations

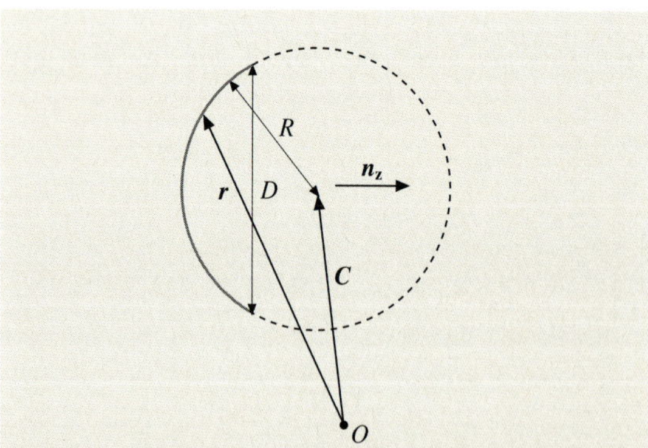

Fig. 2.35 Mathematical representation of a spherical surface (*solid line*) which is part of a full sphere (*dashed plus solid line*). O symbolizes the origin of the coordinate system

it is easy to see that the condition

$$\frac{(\boldsymbol{C} - \boldsymbol{r}_0) \cdot \boldsymbol{n}_z}{R} \geq \sqrt{1 - \frac{D^2}{4R^2}}$$

has to be fulfilled by the point of intersection $\boldsymbol{r}_0$ if it lies on the valid spherical surface.

General Surface $z = f(x,y)$

There are many important surfaces in optics, e.g. aspheric surfaces, which are described by a function f and the equation $z = f(x, y)$. The implicit formulation with the function F is then

$$F(\boldsymbol{r}) = z - f(x, y) = 0\,, \tag{2.148}$$

with $\boldsymbol{r} = (x, y, z)$.

For a general function f the points of intersection of such a surface with a ray cannot be calculated analytically. But there are numerical methods such as Newton's method combined with bracketing [2.33] to determine the roots of (2.143) where F of (2.148) is used.

An important case is e.g. the description of rotationally symmetric aspheric surfaces with their axis of rotation along z by using the function [2.18]:

$$z = f(x, y) = f(h) = \frac{ch^2}{1 + \sqrt{1 - (K+1)c^2h^2}} + \sum_{i=1}^{i_{\max}} a_i h^i \tag{2.149}$$

with $h = \sqrt{x^2 + y^2}$; $c = 1/R$ is the curvature of the conical part of the surface with the conic constant K ($K < -1$ for a hyperboloid, $K = -1$ for a paraboloid and $K > -1$ for ellipsoids, with the special case $K = 0$ for a sphere). a_i are aspheric coefficients describing a polynomial of h. In most cases, only coefficients with even integers $i \geq 4$ are used and $i_{\max}$ is in most cases less than or equal to ten. But in modern aspheric surfaces there may also be odd terms of i and $i_{\max} > 10$.

Coordinate Transformation

In many cases there is a quite simple description of a surface in a local coordinate system [e.g. the description of a rotationally symmetric aspheric surface by using (2.149)] and it would not be useful to find the implicit function F in the global coordinate system if the surface is e.g. tilted. In these cases it is more useful to transform the ray parameters $\boldsymbol{p}$ and $\boldsymbol{a}$ from the global coordinate system to the local system. Then, finding the point of intersection with the surface and the refraction or reflection (or "diffraction" if the element is a diffractive

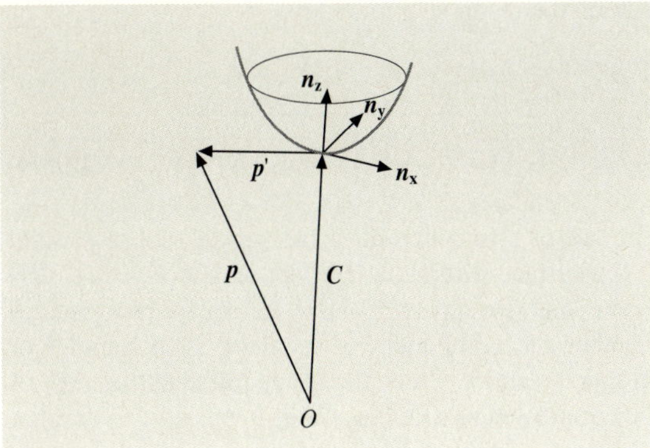

Fig. 2.36 Parameters for transforming a vector $\boldsymbol{p}$ defined in a global coordinate system to a vector $\boldsymbol{p}'$ defined in a local coordinate system. O symbolizes the origin of the global coordinate system

optical element) are done in the local coordinate system. Afterwards, the new ray is transformed back into the global coordinate system.

Assume that the origin of the local coordinate system has the position vector $\boldsymbol{C}$ in the global coordinate system and the three unit vectors along the coordinate axes of the local system are $\boldsymbol{n}_x$, $\boldsymbol{n}_y$ and $\boldsymbol{n}_z$ in the global system (Fig. 2.36). To transform between the position vector $\boldsymbol{p} = (p_x, p_y, p_z)$ in the global system and $\boldsymbol{p}' = (p'_x, p'_y, p'_z)$ in the local system one has the equations:

$$\boldsymbol{p} = \boldsymbol{C} + p'_x \boldsymbol{n}_x + p'_y \boldsymbol{n}_y + p'_z \boldsymbol{n}_z \tag{2.150}$$

and

$$\begin{aligned} p'_x &= (\boldsymbol{p} - \boldsymbol{C}) \cdot \boldsymbol{n}_x , \\ p'_y &= (\boldsymbol{p} - \boldsymbol{C}) \cdot \boldsymbol{n}_y , \\ p'_z &= (\boldsymbol{p} - \boldsymbol{C}) \cdot \boldsymbol{n}_z . \end{aligned} \tag{2.151}$$

For the ray direction vector with coordinates $\boldsymbol{a}' = (a'_x, a'_y, a'_z)$ in the local system and $\boldsymbol{a} = (a_x, a_y, a_z)$ in the global system the analogous equations are valid (but with $\boldsymbol{C} = 0$ because direction vectors are measured from the origin of the respective coordinate system and can be shifted arbitrarily):

$$\boldsymbol{a} = a'_x \boldsymbol{n}_x + a'_y \boldsymbol{n}_y + a'_z \boldsymbol{n}_z \tag{2.152}$$

and

$$\begin{aligned} a'_x &= \boldsymbol{a} \cdot \boldsymbol{n}_x , \\ a'_y &= \boldsymbol{a} \cdot \boldsymbol{n}_y , \\ a'_z &= \boldsymbol{a} \cdot \boldsymbol{n}_z . \end{aligned} \tag{2.153}$$

Of course it would also be possible to write the coordinate transformation using 3×3 matrices with the vectors $\boldsymbol{n}_x$, $\boldsymbol{n}_y$ and $\boldsymbol{n}_z$ as column or row vectors, but we prefer the vector notation here.

2.4.4 Calculation of the Optical Path Length

The optical path length L along a ray at the point of intersection with the next surface is calculated by adding to the original optical path length L_0 at the starting point $\boldsymbol{p}$ of the ray the distance s_0 between the starting point of the ray and the point of intersection with the next surface multiplied by the refractive index n of the material in which the ray propagates. Therefore, the optical path length is

$$L = L_0 + n s_0 . \tag{2.154}$$

If the optical path length on another point $\boldsymbol{r} = \boldsymbol{p} + s\boldsymbol{a}$ on the ray has to be calculated this is done by just replacing s_0 in equation (2.154) by s.

2.4.5 Determination of the Surface Normal

If the function F of the implicit representation of the surface is known the surface normal $\boldsymbol{N}$ at the point of intersection is defined as the normalized gradient of F at the point of intersection $\boldsymbol{r}_0$

$$\boldsymbol{N} = \frac{\nabla F}{|\nabla F|} . \tag{2.155}$$

Some examples of surface normals are given in the following.

Plane Surface

According to (2.144), one has

$$F(\boldsymbol{r}) = (\boldsymbol{r} - \boldsymbol{C}) \cdot \boldsymbol{n}_z = 0 \Rightarrow \boldsymbol{N} = \boldsymbol{n}_z . \tag{2.156}$$

Spherical Surface

A spherical surface is described by (2.146):

$$\begin{aligned} F(\boldsymbol{r}) &= |\boldsymbol{r} - \boldsymbol{C}|^2 - R^2 = 0 , \\ \Rightarrow \boldsymbol{N} &= \frac{\boldsymbol{r}_0 - \boldsymbol{C}}{|\boldsymbol{r}_0 - \boldsymbol{C}|} . \end{aligned} \tag{2.157}$$

General Surface $z = f(x,y)$

$$\begin{aligned} F(\boldsymbol{r}) &= z - f(x, y) = 0 , \\ \Rightarrow \boldsymbol{N} &= \frac{(-f_x, -f_y, 1)}{\sqrt{1 + f_x^2 + f_y^2}} , \end{aligned} \tag{2.158}$$

Part A | 2.4

where $f_x := \partial f/\partial x$ and $f_y := \partial f/\partial y$ are the partial derivatives of f at the point of intersection $\boldsymbol{r}_0$ with the surface.

2.4.6 Law of Refraction

For ray tracing a vectorial formulation of the law of refraction is necessary. In (2.41) an implicit formulation of the law of refraction (and also of the law of reflection) was given

$$\boldsymbol{N} \times (n_2 \boldsymbol{a}_2 - n_1 \boldsymbol{a}_1) = 0 ,$$

where n_1 and n_2 are the refractive indices of the two materials and $\boldsymbol{a}_1$ and $\boldsymbol{a}_2$ are the unit direction vectors of the incident and refracted ray, respectively (Fig. 2.37). $\boldsymbol{N}$ is the local surface normal at the point of intersection of the incident ray with the surface.

A solution of this equation can be found by the following steps:

$$\left(\boldsymbol{a}_2 - \frac{n_1}{n_2}\boldsymbol{a}_1\right) \times \boldsymbol{N} = 0 .$$

This means that the term in round brackets has to be parallel to $\boldsymbol{N}$ or itself zero. The later case is only possible for $n_1 = n_2$ so that, for $n_1 \neq n_2$, we have

$$\boldsymbol{a}_2 = \frac{n_1}{n_2}\boldsymbol{a}_1 + \gamma \boldsymbol{N}$$

with a real value γ. Taking the square of both sides yields (as $\boldsymbol{a}_1$, $\boldsymbol{a}_2$ and $\boldsymbol{N}$ are all unit vectors, i. e. $|\boldsymbol{a}_1| = |\boldsymbol{a}_2| = |\boldsymbol{N}| = 1$)

$$1 = \left(\frac{n_1}{n_2}\right)^2 + \gamma^2 + 2\gamma \frac{n_1}{n_2}\boldsymbol{a}_1 \cdot \boldsymbol{N}$$

and therefore

$$\gamma_{1,2} = -\frac{n_1}{n_2}\boldsymbol{a}_1 \cdot \boldsymbol{N} \pm \sqrt{1 - \left(\frac{n_1}{n_2}\right)^2 \left[1 - (\boldsymbol{a}_1 \cdot \boldsymbol{N})^2\right]} .$$

In total the result is:

$$\boldsymbol{a}_2 = \frac{n_1}{n_2}\boldsymbol{a}_1 - \frac{n_1}{n_2}(\boldsymbol{a}_1 \cdot \boldsymbol{N})\,\boldsymbol{N} \pm \sqrt{1 - \left(\frac{n_1}{n_2}\right)^2 \left[1 - (\boldsymbol{a}_1 \cdot \boldsymbol{N})^2\right]}\boldsymbol{N} . \quad (2.159)$$

The vector term in front of the square root is parallel to the surface (the scalar product with $\boldsymbol{N}$ is zero). This means that the sign in front of the square root decides whether the component of $\boldsymbol{a}_2$ along $\boldsymbol{N}$ is parallel or antiparallel to $\boldsymbol{N}$. Since the ray is refracted the sign of the component of $\boldsymbol{a}_1$ along $\boldsymbol{N}$ has to be equal to the sign of the component of $\boldsymbol{a}_2$ along $\boldsymbol{N}$:

$$\operatorname{sign}(\boldsymbol{a}_1 \cdot \boldsymbol{N}) = \operatorname{sign}(\boldsymbol{a}_2 \cdot \boldsymbol{N}) , \quad (2.160)$$

where the sign function is $+1$ for a positive argument and -1 for a negative argument.

Therefore, (2.159) can be written independently of the relative direction of $\boldsymbol{N}$ with respect to $\boldsymbol{a}_1$ as

$$\boldsymbol{a}_2 = \frac{n_1}{n_2}\boldsymbol{a}_1 - \frac{n_1}{n_2}(\boldsymbol{a}_1 \cdot \boldsymbol{N})\,\boldsymbol{N} + \operatorname{sign}(\boldsymbol{a}_1 \cdot \boldsymbol{N}) \times \sqrt{1 - \left(\frac{n_1}{n_2}\right)^2 \left[1 - (\boldsymbol{a}_1 \cdot \boldsymbol{N})^2\right]}\boldsymbol{N} \quad (2.161)$$

So, this equation allows the calculation of the direction vector $\boldsymbol{a}_2$ of the refracted ray if the incident ray (direction vector $\boldsymbol{a}_1$), the local surface normal $\boldsymbol{N}$ and the two refractive indices n_1 and n_2 are known.

2.4.7 Law of Reflection

The law of reflection is also formally described by (2.41) and therefore also by (2.159). But, in the case of reflection, first of all the refractive indices are identical for the incident and the reflected ray, i. e. $n_1 = n_2$, and sec-

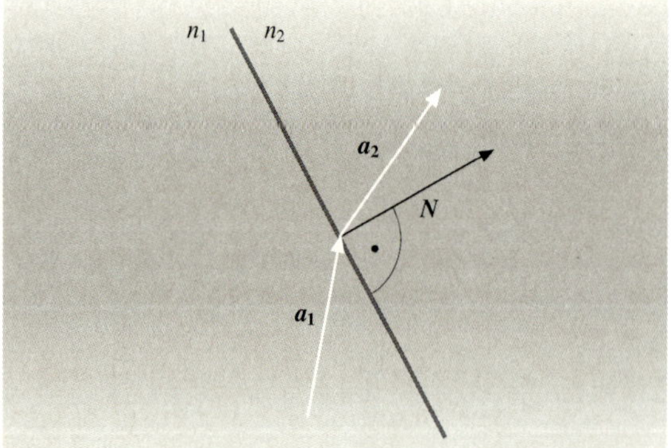

Fig. 2.37 Parameters for refraction of a ray at a surface

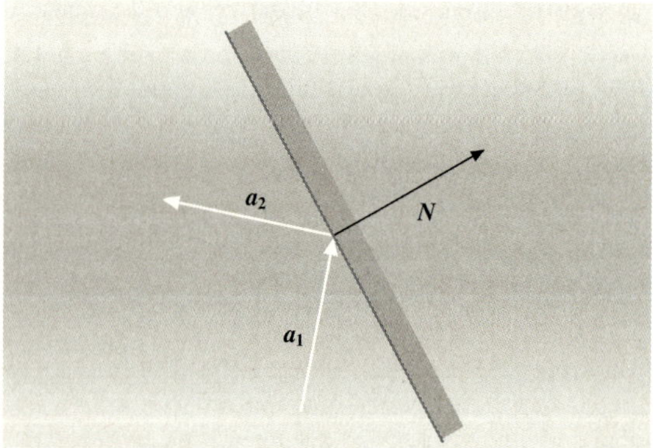

Fig. 2.38 Parameters for the reflection of a ray at a surface

ond the component of $\boldsymbol{a}_2$ along $\boldsymbol{N}$ has the opposite sign to the component of $\boldsymbol{a}_1$ along $\boldsymbol{N}$ (Fig. 2.38). This means that the other sign in front of the square root has to be taken and (2.159) results in

$$\boldsymbol{a}_2 = \boldsymbol{a}_1 - (\boldsymbol{a}_1 \cdot \boldsymbol{N})\,\boldsymbol{N} - \operatorname{sign}(\boldsymbol{a}_1 \cdot \boldsymbol{N}) \times \sqrt{(\boldsymbol{a}_1 \cdot \boldsymbol{N})^2}\boldsymbol{N} = \boldsymbol{a}_1 - 2\,(\boldsymbol{a}_1 \cdot \boldsymbol{N})\,\boldsymbol{N}\,. \quad (2.162)$$

It is easy to prove that this equation correctly describes the reflection of a ray at a surface because all three vectors lie in a common plane (linearly dependent vectors) and the angle of incidence is equal to the angle of the reflected ray. The later can be seen by calculating the modulus of the cross product of (2.162) with $\boldsymbol{N}$. Third, $\boldsymbol{a}_2$ really describes a reflected ray since twice the component of $\boldsymbol{a}_1$ along $\boldsymbol{N}$ is subtracted from $\boldsymbol{a}_1$ to obtain $\boldsymbol{a}_2$.

Besides refraction and reflection there is also a third quite important law for deflecting a ray at a surface, the vectorial local grating equation, which is used for ray tracing on holographic and more general diffractive optical elements. But for this equation and its solution we refer to the literature [2.31, 32, 34, 35].

2.4.8 Non-Sequential Ray Tracing and Other Types of Ray Tracing

The normal mode in most ray-tracing computer programs is so-called sequential ray tracing, i. e. the user defines the order in which the different surfaces of the optical system are passed by a ray. But this method is e.g. not useful for the analysis of illumination systems where the path of a ray and the order of surfaces can be different for each ray. The stability analysis of laser resonators [2.14, 15] is also quite exhausting with the sequential mode because the user knows the order of the surfaces but not how many times they will be hit by a ray. Of course, a stable resonator will be crossed by a light ray with an infinite number of cycles. But for unstable resonators there is a finite number of cycles before the ray leaves the resonator.

Therefore, non-sequential ray tracing is used in these cases. In this, the computer automatically calculates the next surface that is physically hit by each ray. This is done, for example, by calculating the points of intersection of the ray with all surfaces and taking the surface with the smallest positive distance s_0. If there is no point of intersection with $s_0 > 0$ the ray does not hit any surface of the system. Of course, non-sequential ray tracing is quite expensive in terms of computing time and therefore is normally only used if really necessary.

Another speciality of non-sequential ray tracing is that a ray can be split at a surface into a refracted and a reflected ray (and in the case of diffractive optical elements also into more than two rays representing the different diffraction orders). Each ray is then recursively traced through the optical system.

Some interesting modern optical systems such as Shack–Hartmann wavefront sensors [2.36] or beam homogenizers [2.37] use microlens arrays in combination with macroscopic optics. These array systems can also be analyzed with sequential or non-sequential ray tracing to obtain a first insight [2.38, 39]. Of course, one has to be careful in these cases because diffraction and interference effects (for coherent or partially coherent illumination) cannot be neglected in several cases [2.40].

Sophisticated modern computer programs for sequential or non-sequential ray tracing implement in addition polarization ray tracing [2.10, 11]. In this, the local polarization state of each ray is taken into account and, for example, the split of the local power transported by each ray to the refracted and reflected ray for refraction/reflection at a surface is calculated according to the Fresnel equations [2.1].

A third type of ray tracing is so called differential ray tracing or generalized ray tracing [2.28, 35, 41]. In this case, each ray is assumed to represent a local wave front with two principal curvatures and two principal directions. These parameters are then traced in addition to the normal ray parameters for each ray during propagation through the optical system. This allows, for example, the calculation of the local astigmatism of the wave front belonging to the ray by just tracing one ray. It also allows the calculation of the change of local intensity of the wave during the propagation.

2.5 Aberrations

Whereas in the paraxial case the imaging quality of an optical system is ideal there are in practice aberrations of an optical system which deteriorate its imaging quality [2.1, 6, 8, 26, 42]. To explain the nature of aberrations, see Fig. 2.39. At the exit pupil of an optical system there is a real wave front (solid line), i. e. the surface of equal optical path length, which intersects the exit pupil on the optical axis and which has its paraxial focus at the

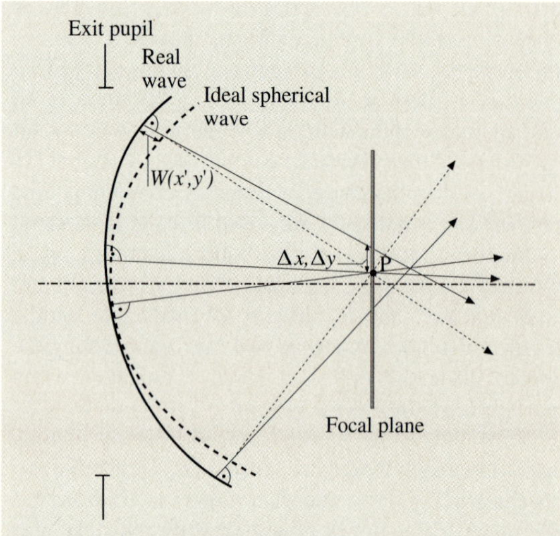

Fig. 2.39 Explanation of the wave aberrations W and the ray aberrations Δx, Δy. The solid curve is the real wave front and the dashed curve the ideal spherical wave front. The *solid rays* are rays starting from the real wave front whereas the *dashed rays* are rays starting from the ideal spherical wave front. P is the (paraxial) focus of the wave front

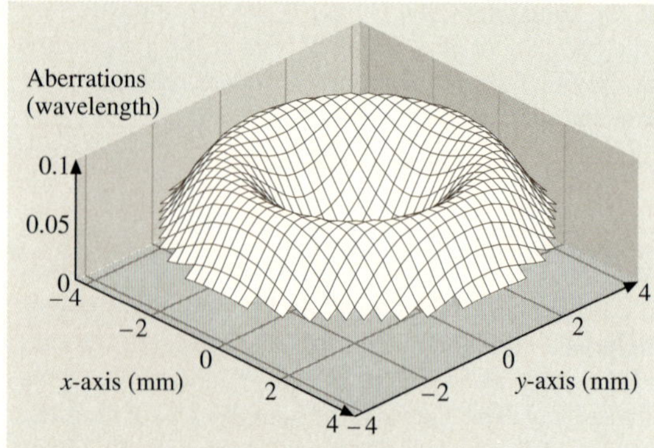

Fig. 2.40 Wave aberrations for the on-axis point of the microscopic objective of Fig. 2.32 (NA = 0.4, magnification 20×, focal length $f' = 11.5$ mm). The reference sphere is around the best focus of the wave aberrations. The focused spot is diffraction-limited since the peak-to-valley value is just 0.1 wavelengths (for a wavelength of 587.6 nm)

point P, which lies in the focal plane of the system. However, there are deviations between an ideal spherical wave front (dashed line) with its center of curvature at P and the real wave front in the non-paraxial region. So, a ray starting at the point (x', y') of the exit pupil has an optical path length difference between the real wave front and the ideal spherical wave front, which is called the wave aberration $W(x', y')$. Additionally, a ray with aberrations does not intersect the focal plane at the focus P but at a point with lateral displacements Δx and Δy in the x- and y-directions. These lateral deviations from the paraxial focus P are called the ray aberrations. Of course, the wave aberrations and ray aberrations are not independent of each other [2.43] and with good approximation the ray aberrations are proportional to the partial derivatives of the wave aberrations with respect to x' and y'.

2.5.1 Calculation of the Wave Aberrations

The wave aberrations can be calculated by ray tracing. To do this a sphere which intersects the exit pupil on the optical axis and which has the (paraxial) focus P as its center of curvature is defined. Then, the optical path lengths $L(x', y')$ of the points of intersection of this reference sphere with rays starting at the exit pupil at the points (x', y') are calculated by using (2.147) and (2.154). The optical path length $L(0, 0)$ of the chief ray is subtracted from the optical path length values of all other rays, resulting in the wave aberrations $W(x', y')$:

$$W(x', y') = L(x', y') - L(0, 0) . \qquad (2.163)$$

So, the wave aberrations are known for a grid of rays, i.e. the points (x', y') in the exit pupil.

In some cases it is useful not to take the paraxial focus for P but the so-called best focus. This is the point where either the wave aberrations or the ray aberrations have their smallest mean value (Fig. 2.40). So, in fact there are two different definitions of the best focus. If there is, for example, field curvature the best focus will not be in the focal plane but on a sphere which intersects the optical axis in the focal plane.

2.5.2 Ray Aberrations and the Spot Diagram

Ray aberrations can also be calculated by ray tracing. These aberrations are just the lateral deviations Δx and Δy between the focus P itself (which can be the paraxial focus or the best focus) and the points of intersection of the rays with a plane through the focus P. The surface normal of this plane is assumed to be $\boldsymbol{n}_z$ (in most cases $\boldsymbol{n}_z$ will be parallel to the optical axis) and the focus P has the position vector $\boldsymbol{P}$. Additionally, the two unit vectors $\boldsymbol{n}_x$ and $\boldsymbol{n}_y$ lying in this plane and defining the local x- and y-axis are known ($\boldsymbol{n}_x$, $\boldsymbol{n}_y$ and $\boldsymbol{n}_z$ form an orthogonal triad

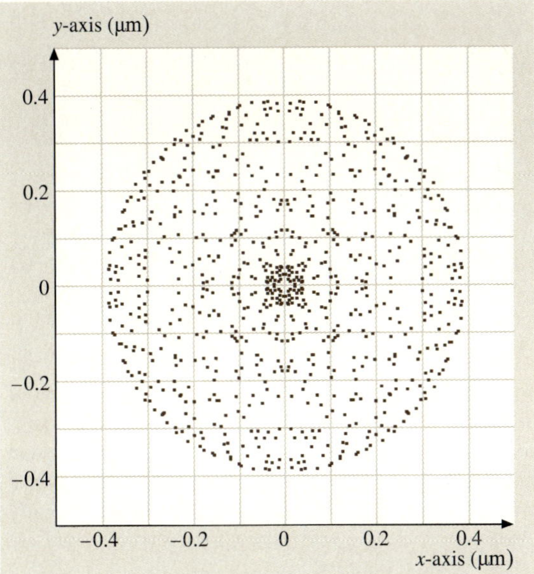

Fig. 2.41 Spot diagram for the on-axis point of the microscopic objective of Fig. 2.32 with NA = 0.4, magnification 20×, focal length $f' = 11.5$ mm. Since the numerical aperture of the lens used to focus the light is NA = 0.4 on the image side and the wavelength is $\lambda = 587.6$ nm the diffraction-limited Airy disc would have a diameter of $1.22\lambda/\text{NA} = 1.8\,\mu\text{m}$, i. e. larger than the ray aberrations. So, as in Fig. 2.40, it can also be seen from the ray aberrations that the on-axis spot of this lens is diffraction-limited

of unit vectors). Then, a ray number i with starting point $\boldsymbol{p}_i$ and direction vector $\boldsymbol{a}_i$ has its point of intersection $\boldsymbol{r}_i$ with the plane according to (2.141) and (2.145) at

$$\boldsymbol{r}_i = \boldsymbol{p}_i + \frac{(\boldsymbol{P} - \boldsymbol{p}_i) \cdot \boldsymbol{n}_z}{\boldsymbol{a}_i \cdot \boldsymbol{n}_z} \boldsymbol{a}_i \,. \tag{2.164}$$

The ray aberrations are then defined as:

$$\Delta x = (\boldsymbol{r}_i - \boldsymbol{P}) \cdot \boldsymbol{n}_x \,, \tag{2.165}$$

$$\Delta y = (\boldsymbol{r}_i - \boldsymbol{P}) \cdot \boldsymbol{n}_y \,. \tag{2.166}$$

A quite demonstrative representation of the ray aberrations is a spot diagram. There, the points of intersection of the rays with a plane are graphically displayed by just drawing them as points (Fig. 2.41). This means that the spot diagram is a graphical representation of the ray aberrations (Δx, Δy). Sometimes it is useful to determine the spot diagram not only in a plane through the focus but also in other planes to track the focusing of the rays.

2.5.3 The Seidel Terms and the Zernike Polynomials

In classical aberration theory [2.1,3] the primary aberration terms of Seidel (fourth-order wave aberration terms or third-order ray aberration terms) play an important role. The different terms are: spherical aberration, coma, astigmatism, curvature of field and distortion. Whereas the first three terms are point aberrations, i. e. aberrations that generate a blurred image point, the last two terms just cause a shift of the image point relative to the ideal paraxial image point but the image point itself would be sharp. There is not time in this chapter to go into details and to give a mathematical derivation so that only some facts will be stated for the different aberration terms. The distance of the object point from the optical axis will be called the object height r_O in the following whereas the distance of a ray from the optical axis in the exit pupil will be called r_A (from aperture height). For lenses with a small numerical aperture the maximum value r_A is proportional to the numerical aperture NA of the image forming pencil of rays. Therefore, in the following the numerical aperture NA and the object height r_O will be used to describe the functionality of the different Seidel terms.

Spherical Aberration

Spherical aberration is the only classical aberration that also occurs for object points on the optical axis of a rotationally symmetric optical system, i. e. for $r_\text{O} = 0$. Spherical aberration of a normal single lens causes rays with a large height r_A in the exit pupil of the lens to be refracted more strongly, so that they intersect the optical axis in front of the paraxial focus. In general optical systems the off axis rays can also intersect the optical axis behind the paraxial focus. A typical property of spherical aberration is that it increases with the fourth power of the numerical aperture NA of the ray pencil forming the image point:

$$\text{spherical aberration} \propto (\text{NA})^4 \tag{2.167}$$

As already mentioned above the spherical aberration is independent of the object height r_O.

Coma

Coma is an aberration which occurs only for off-axis points (of a rotationally symmetric optical system), i. e. $r_\text{O} \neq 0$. The name coma is caused by the deformation of the image point, which looks like the coma of a comet. The coma depends on the third power of the numerical aperture and linearly on the image height

$$\text{coma} \propto r_{\mathrm{O}}(\mathrm{NA})^3 \,. \tag{2.168}$$

This is the reason why coma especially occurs for large numerical apertures whereas astigmatism dominates for small numerical apertures and large object heights (see the next paragraph).

Coma can, for example, be generated in the microscopic objective of Fig. 2.32 by a lateral shift of the first lens. A shift of 0.1 mm results in the aberrations shown in Fig. 2.42, which are dominated by coma although the spherical aberration of the original lens is still present.

Astigmatism

Astigmatism means that rays of the meridional plane and of the sagittal plane focus in different planes perpendicular to the optical axis. So, the geometrical shape of the image point is in general an ellipse. In two special planes, called the meridional and sagittal focal planes, the ellipses degenerate into two focal lines. The focal lines are perpendicular to each other. Between the meridional and the sagittal focal plane there is another plane where the shape of the image point is a circle, but of course this circle is extended whereas an ideal image point in geometrical optics would be a mathematical point. The astigmatism of an optical system is proportional to the square of the numerical aperture and the square of the object height:

$$\text{astigmatism} \propto r_{\mathrm{O}}^2(\mathrm{NA})^2 \,. \tag{2.169}$$

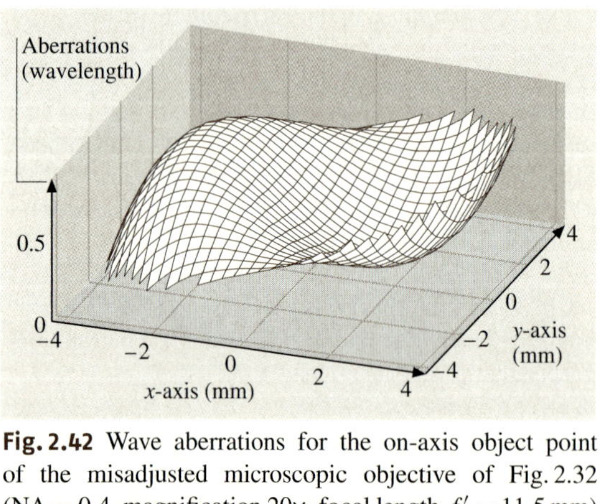

Fig. 2.42 Wave aberrations for the on-axis object point of the misadjusted microscopic objective of Fig. 2.32 (NA = 0.4, magnification 20×, focal length $f' = 11.5$ mm) by shifting the first lens laterally by 0.1 mm. The image point which is then no more on-axis shows mainly coma but of course mixed with the spherical aberration of Fig. 2.40

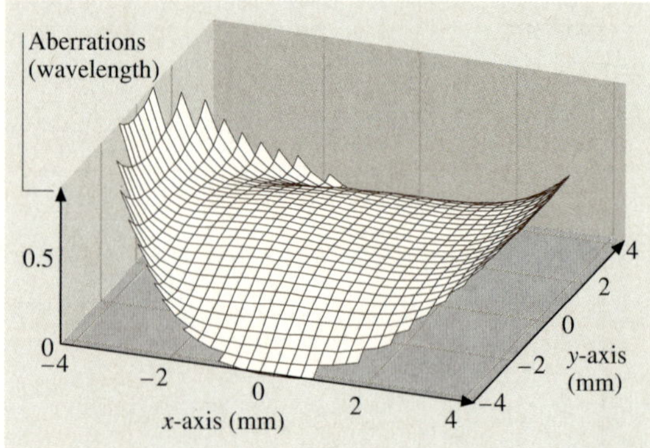

Fig. 2.43 Wave aberrations for an off-axis object point of the microscopic objective of Fig. 2.32 (NA = 0.4, magnification 20×, focal length $f' = 11.5$ mm). The image point shows in this case mainly astigmatism but of course mixed with the spherical aberration of Fig. 2.40. There is nearly no coma because the well-adjusted microscopic objective fulfills the sine condition

As mentioned above, this functionality is the reason that astigmatism also occurs for quite narrow pencils of rays. If there are cylindrical or toric surfaces in an optical system, astigmatism also occurs on the optical axis, whereas in the usual case of rotationally symmetric optical systems astigmatism occurs only for off-axis points.

If we again take the microscopic objective of Fig. 2.32 but now with an off-axis object point (object height 15 mm, resulting image height 0.74 mm because of curvature of the field) the resulting aberrations are mainly astigmatism, showing the typical saddle shape. Of course, the spherical aberration which is present on-axis remains, so that in fact the resulting aberrations represented in Fig. 2.43 are a mixture of astigmatism with a peak-to-valley value of about one wavelength and spherical aberration of 0.1 wavelength. Nearly no coma appears for off-axis points because a microscopic objective fulfills the sine condition (2.138), which guarantees that object points in the neighborhood of the optical axis are imaged without coma.

Curvature of Field

As mentioned above the curvature of field is not a point aberration but a field aberration, i. e. the image point can be sharp but the position of the image point is shifted relative to the ideal paraxial value. In the case of the curvature of the field the image points are situated on a spherical surface and, in connection with astigmatism, there are even two different spheres

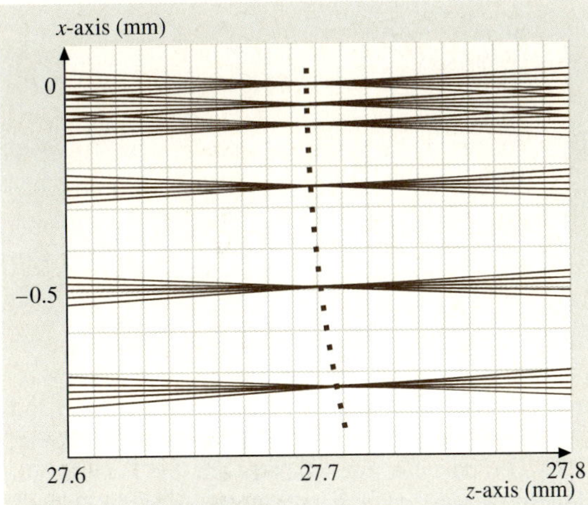

Fig. 2.44 Curvature of field: ray-tracing picture of the image points of different field points of the microscopic objective of Fig. 2.32 (NA = 0.4, magnification 20×, focal length $f' = 11.5$ mm) which is used in the reverse order, i.e. to build an image with reduced size. It can be seen that the image points are situated on a curve (*dotted curve*) which is in fact part of a sphere. Note that the scaling is quite different for the x- and z-axes to see the effect of the curvature of the field

for rays in the meridional plane and in the sagittal plane.

Figure 2.44 shows the curvature of the field in the image plane of the microscopic objective of Fig. 2.32. For the off-axis points the best focus of the image points is behind the focal plane (light is coming, as usual, from the left). Of course, the off-axis points also show astigmatism so that the image points are blurred.

Distortion

The last Seidel term is distortion, which is also a field aberration and not a point aberration. Distortion means that the lateral magnification for imaging is not a constant for all off-axis points but depends to some extent on the object height r_O. The result is that each straight line in the object plane that does not pass through the optical axis is curved in the image plane. A regular grid like that in Fig. 2.45b is either pincushion-distorted (Fig. 2.45a) or barrel-distorted (Fig. 2.45c).

The Zernike Polynomials

A quite important method to calculate the different terms of the wave aberrations of an optical system is to fit the so-called Zernike polynomials [2.1, 44, 45] to it. The wave aberration data for this procedure can either be theoretically determined, e.g. by ray tracing, or experimentally determined, e.g. by interferometry.

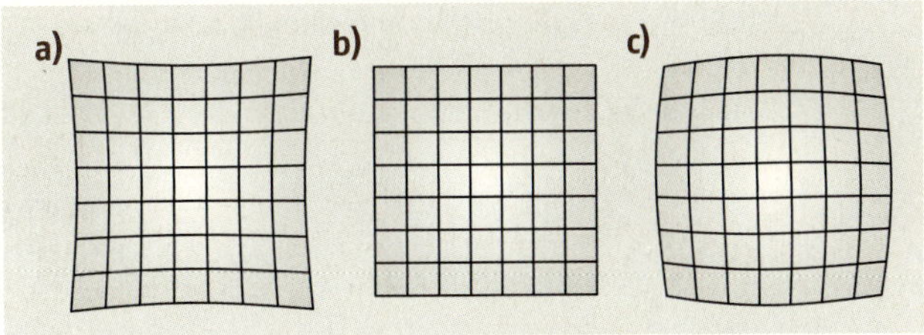

Fig. 2.45a–c Effect of distortion. The regular grid of (**b**) in the object plane is either distorted in the image plane to a pincushion shape (**a**) or a barrel shape (**c**). In (**a**) the lateral magnification increases with increasing object height whereas it decreases in (**c**)

The condition for using Zernike polynomials is that the aperture of the optical system is circular because the Zernike polynomials are only orthogonal on the unit circle. There, they build a complete set of orthogonal polynomials and some of the terms correspond to the classical Seidel terms for the point aberrations, i.e. spherical aberration, coma and astigmatism. Besides this there are other terms such as trefoil or tetrafoil that result, for example, if the optical elements are stressed by fixing them at three or four points. It has to be emphasized that there are no terms corresponding to the Seidel terms of curvature of field and distortion since the Zernike polynomials can only represent point aberrations and not field aberrations.

2.5.4 Chromatic Aberrations

Up to now it was implicitly assumed that only light of one wavelength is considered and the presented aberrations were all monochromatic aberrations. Besides this, there are so-called chromatic aberrations, which are a result of the dispersion of a material, i.e. the dependence of the refractive index of a material on the wavelength (or, if there are diffractive optical elements in the system, the dispersion results from the strong dependence of the grating equation on the wavelength). Dispersion changes the paraxial parameters such as the focal length of a lens. For a thin lens with refractive index n in air we have for example (2.98):

$$\frac{1}{f'} = (n-1)\left(\frac{1}{R_1} - \frac{1}{R_2}\right) . \qquad (2.170)$$

If now n depends on the wavelength λ we have

$$\frac{\mathrm{d}}{\mathrm{d}\lambda}\left(\frac{1}{f'}\right) = -\frac{\mathrm{d}f'/\mathrm{d}\lambda}{f'^2} = \frac{\mathrm{d}n}{\mathrm{d}\lambda}\left(\frac{1}{R_1}-\frac{1}{R_2}\right) = \frac{\mathrm{d}n/\mathrm{d}\lambda}{n-1}\frac{1}{f'} . \tag{2.171}$$

By replacing the differentials by finite differences we can write, to a good approximation,

$$\frac{\Delta f'}{f'} = -\frac{\Delta n}{n-1} . \tag{2.172}$$

To characterize the dispersion of a material the so-called Abbe number V_d is used; it is defined as

$$V_\mathrm{d} = \frac{n_\mathrm{d}-1}{n_\mathrm{F}-n_\mathrm{C}} = \frac{n(\lambda_\mathrm{d}=587.6\,\mathrm{nm})}{n(\lambda_\mathrm{F}=486.1\,\mathrm{nm})-n(\lambda_\mathrm{C}=656.3\,\mathrm{nm})} . \tag{2.173}$$

So, we have to a good approximation:

$$\frac{f'(\lambda_\mathrm{C}=656.3\,\mathrm{nm})-f'(\lambda_\mathrm{F}=486.1\,\mathrm{nm})}{f'(\lambda_\mathrm{d}=587.6\,\mathrm{nm})} = \frac{1}{V_\mathrm{d}} . \tag{2.174}$$

For glasses with normal dispersion the Abbe number is a positive constant that has a small value for materials with high dispersion (e.g. materials like SF10) and a large value for materials with small dispersion (like e.g. BK7). The positive sign indicates that the focal length of a lens increases with increasing wavelength.

2.6 Some Important Optical Instruments

In this section some important optical instruments such as the achromatic lens, the camera, the human eye, the telescope and the microscope will be discussed. However, this will be done in some cases quite briefly because there are many text books on geometrical optics which treat these subjects quite amply [2.3, 5, 8].

2.6.1 The Achromatic Lens

In paragraph 2.5.4 the chromatic aberrations of a single lens, i. e. the dependence of the focal length on the wavelength, were treated. An achromatic lens should have, in the ideal case, no chromatic aberrations. However, in practice the most important achromatic lens is an achromatic doublet consisting of two cemented lenses, for which the focal length can be identical for only two different wavelengths. So, in technical optics the term achromatic lens normally means a lens where the focal length is identical for two different wavelengths. For applications in the visible range these two wavelengths are commonly $\lambda_\mathrm{F} = 486.1$ nm (blue line of atomic hydrogen) and $\lambda_\mathrm{C} = 656.3$ nm (red line of atomic hydrogen), which are near the edge of the visible range. A lens where the focal length is identical for three wavelengths is called an apochromatic lens.

To understand the principle of the achromatic correction of a lens doublet the paraxial matrix M of a combination of two thin lenses (paraxial matrices M_1 and M_2) situated in air with a zero separation is calculated. Of course, this is a simplification because in practice no thin lens really exists and, if the principal points of thick lenses are taken as reference elements, the distance between two lenses will normally be different from zero. But nevertheless, the calculation with two zero-distant thin lenses explains the principle and, according to equation (2.97), the result is

$$M = M_2 M_1 = \begin{pmatrix} 1 & 0 \\ -\frac{1}{f_2'} & 1 \end{pmatrix}\begin{pmatrix} 1 & 0 \\ -\frac{1}{f_1'} & 1 \end{pmatrix} = \begin{pmatrix} 1 & 0 \\ -\frac{1}{f_1'}-\frac{1}{f_2'} & 1 \end{pmatrix} , \tag{2.175}$$

where the focal lengths of the two thin lenses are f_1' and f_2'. Therefore, the focal length f' of the combination of these two lenses is

$$\frac{1}{f'} = \frac{1}{f_1'} + \frac{1}{f_2'} ; \tag{2.176}$$

that is, the optical powers of the single lenses are just added.

The refractive index of the first lens is n_1 and that of the second lens is n_2, and both are situated in air. The optical powers $1/f_i'$ ($i \in \{1,2\}$) of refractive thin lenses are then, according to (2.98),

$$\frac{1}{f_i'(\lambda)} = [n_i(\lambda)-1]\left(\frac{1}{R_{i,1}}-\frac{1}{R_{i,2}}\right) =: [n_i(\lambda)-1]\,C_i , \tag{2.177}$$

where the term C_i depends only on the two radii of curvature $R_{i,1}$ and $R_{i,2}$ of the thin lenses and is independent of the wavelength λ, whereas the refractive index n_i depends on the wavelength.

For an achromatic lens the optical powers at the two wavelengths λ_F and λ_C (or for two other wavelengths depending on the application) have to be identical. By using (2.176) and (2.177) this means:

$$\begin{aligned}&\frac{1}{f_1'(\lambda_F)}+\frac{1}{f_2'(\lambda_F)}=\frac{1}{f_1'(\lambda_C)}+\frac{1}{f_2'(\lambda_C)}\\ &\Rightarrow [n_1(\lambda_F)-1]\,C_1+[n_2(\lambda_F)-1]\,C_2\\ &=[n_1(\lambda_C)-1]\,C_1+[n_2(\lambda_C)-1]\,C_2\,,\\ &\Rightarrow [n_1(\lambda_F)-n_1(\lambda_C)]\,C_1\\ &=-[n_2(\lambda_F)-n_2(\lambda_C)]\,C_2\,.\end{aligned}$$

By using (2.177) again the terms C_1 and C_2 can be expressed using the refractive indices at an intermediate wavelength between λ_F and λ_C, in our case $\lambda_d = 587.6$ nm (the yellow line of helium), and the focal lengths at this wavelength, and the result is:

$$\begin{aligned}&\frac{n_1(\lambda_F)-n_1(\lambda_C)}{[n_1(\lambda_d)-1]\,f_1'(\lambda_d)}=-\frac{n_2(\lambda_F)-n_2(\lambda_C)}{[n_2(\lambda_d)-1]\,f_2'(\lambda_d)}\\ &\Rightarrow V_{1,d}\,f_1'(\lambda_d)=-V_{2,d}\,f_2'(\lambda_d)\,. \qquad (2.178)\end{aligned}$$

Here, the Abbe numbers $V_{i,d}$ ($i \in \{1,2\}$) of the materials with refractive indices n_i are defined by (2.173).

Since the Abbe number of a refractive material is always positive one of the refractive thin lenses has to be a negative lens and one a positive lens to fulfill (2.178). However, if one of the two thin lenses is not a refractive but a diffractive lens it formally has a constant negative Abbe number $V_d = -3.453$ ([2.23], Chap. 10). So, in the case of a so-called hybrid achromatic lens, which consists of a refractive and a diffractive lens, both lenses will have the same sign of optical powers, and so a positive hybrid achromatic lens consists of two positive single lenses: a refractive lens with a high optical power and a high Abbe number and a diffractive lens with a small optical power and a negative Abbe number with small modulus.

However, a positive purely refractive achromatic lens consists of a positive lens with high optical power and high Abbe number (made of a crown glass such as BK7) and a negative lens with smaller optical power and smaller Abbe number (made of a highly dispersive flint glass such as SF10), so that in total a positive optical power results.

In our mathematical description only the case of two thin lenses with zero separation is treated. But it is no problem to use the paraxial matrix theory to calculate the matrix of a real achromatic doublet consisting of two cemented lenses, i. e. three refractive spherical surfaces with finite distances embedding two different materials. However, in this case not only the focal length will depend on the wavelength but also to some degree the position of the principal planes. So, the position of the focus itself can vary a little, although the focal length is identical for the two selected wavelengths λ_F and λ_C.

In practice, refractive achromatic doublets that can be bought do not only correct the chromatic errors but also fulfill the sine condition (2.138). This is possible because there are three surfaces with different radii of curvature, whereas to fulfill the paraxial properties only two of these three parameters are determined.

Examples of Designing Achromatic Doublets

In this paragraph the paraxial properties of different achromatic doublets will be calculated and compared with those of single refractive lenses. It will be assumed as above that the two lenses of the achromatic doublet are thin lenses with zero distance between the two lenses. This is of course a simplification, but nevertheless it is a good approximation for most cases.

Due to (2.178) the focal lengths $f_1'(\lambda_d)$ and $f_2'(\lambda_d)$ of the two lenses of the achromatic doublet at the wavelength $\lambda_d = 587.6$ nm have to fulfill the condition:

$$\begin{aligned}&V_{1,d}\,f_1'(\lambda_d)=-V_{2,d}\,f_2'(\lambda_d)\,,\\ \Rightarrow\ &f_1'(\lambda_d)=-\tfrac{V_{2,d}}{V_{1,d}}\,f_2'(\lambda_d)\quad\text{or}\\ &f_2'(\lambda_d)=-\tfrac{V_{1,d}}{V_{2,d}}\,f_1'(\lambda_d)\,.\end{aligned}$$

Here, $V_{1,d}$ and $V_{2,d}$ are the Abbe numbers of the materials of the two lenses. Additionally, the focal length f' of the achromatic doublet can be calculated according to (2.176) by

$$\frac{1}{f'}=\frac{1}{f_1'}+\frac{1}{f_2'}\,.$$

By combining both equations the focal lengths of the two single lenses can be expressed as functions of the focal length of the achromatic doublet:

$$\begin{aligned}f_1'(\lambda_d)&=\frac{V_{1,d}-V_{2,d}}{V_{1,d}}\,f'(\lambda_d)\,,\\ f_2'(\lambda_d)&=\frac{V_{2,d}-V_{1,d}}{V_{2,d}}\,f'(\lambda_d)\,. \qquad (2.179)\end{aligned}$$

A refractive achromatic doublet made of BK7 and SF10 has, for example, Abbe numbers of $V_{1,d} = 64.17$ (BK7) and $V_{2,d} = 28.41$ (SF10). Therefore, the focal lengths of the two single lenses are in this case, due to (2.179),

$$\text{Lens made of BK7}: \quad f_1'(\lambda_d) = 0.557 f'(\lambda_d)\,,$$
$$\text{Lens made of SF10}: \quad f_2'(\lambda_d) = -1.259 f'(\lambda_d)\,.$$

So, the second lens, made of the highly dispersive material SF10, is a negative lens if the achromatic doublet itself is a positive lens.

As mentioned previously a diffractive lens (DOE) can be described by a negative and material-independent Abbe number $V_d = -3.453$. Therefore, in the following the focal lengths of the two single lenses of a hybrid achromatic doublet made of one refractive lens and one DOE shall be considered. First, the refractive lens with focal length f_1' is made of BK7 and the second lens with focal length f_2' is a DOE. According to equations (2.179) the focal lengths are:

$$\text{Lens made of BK7}: \quad f_1'(\lambda_d) = 1.054 f'(\lambda_d)\,,$$
$$\text{DOE}: \quad f_2'(\lambda_d) = 19.588 f'(\lambda_d)\,.$$

So, as mentioned previously, both lenses are positive lenses if the achromatic doublet has a positive optical power. Of course, most of the optical power is delivered by the refractive lens.

A second hybrid achromatic doublet can be made, for example, by taking a refractive lens made of SF10 and a DOE:

$$\text{Lens made of SF10}: \quad f_1'(\lambda_d) = 1.122 f'(\lambda_d)\,,$$
$$\text{DOE}: \quad f_2'(\lambda_d) = 9.230 f'(\lambda_d)\,.$$

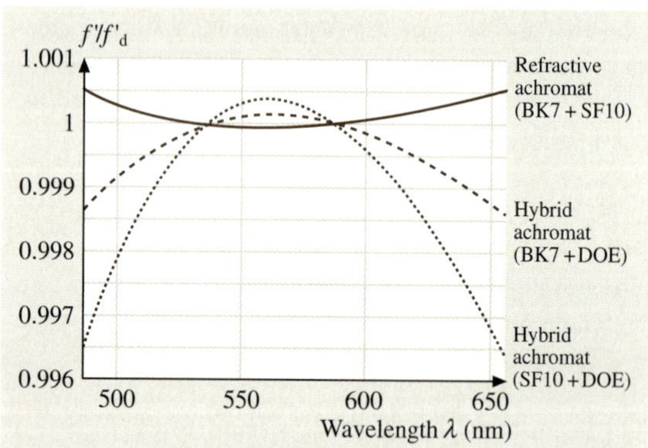

Fig. 2.46 The focal length f' of different achromatic doublets (a refractive achromatic doublet made of BK7 and SF10, a hybrid achromatic doublet made of BK7 and a diffractive lens (DOE) and a hybrid achromatic doublet made of SF10 and a DOE) as a function of the wavelength λ normalized by the focal length f_d' at $\lambda_d = 587.6$ nm

The remaining chromatic aberrations of an achromatic doublet, i. e. the variation of the focal length with the wavelength of the illuminating light, can be calculated by using (2.176). In this equation the optical power of a refractive lens as function of the wavelength is calculated by (2.177) and the optical power of a diffractive lens as a function of the wavelength is [2.23]

$$\frac{1}{f_{\text{DOE}}'(\lambda)} = \frac{\lambda}{\lambda_d f_{\text{DOE}}'(\lambda_d)} =: C\lambda\,.$$

Here, $C = 1/[\lambda_d f_{\text{DOE}}'(\lambda_d)]$ is a constant value which depends on the focal length $f_{\text{DOE}}'(\lambda_d)$ of the DOE at the wavelength $\lambda_d = 587.6$ nm. So, the optical power of the DOE increases linearly with wavelength. This is easy to explain because, due to the paraxial grating equation (2.118), the angle of the diffracted light also increases linearly with the wavelength.

The chromatic aberrations for the different types of achromatic doublets are shown in Fig. 2.46. The chromatic aberrations of a single refractive lens (made of either BK7 or SF10) compared to that of a refractive achromatic doublet (made of BK7 and SF10) are shown in Fig. 2.47. The result is that the best correction of the chromatic aberrations is made by the refractive achromatic doublet. However, the hybrid achromatic doublet made of a BK7 lens and a DOE also has quite small chromatic aberrations. Nevertheless, all types of achromatic doublets (purely refractive or hybrid) have lower chromatic aberrations than a single refractive lens and of course much lower chromatic aberrations than a single DOE, which is not shown in the figures.

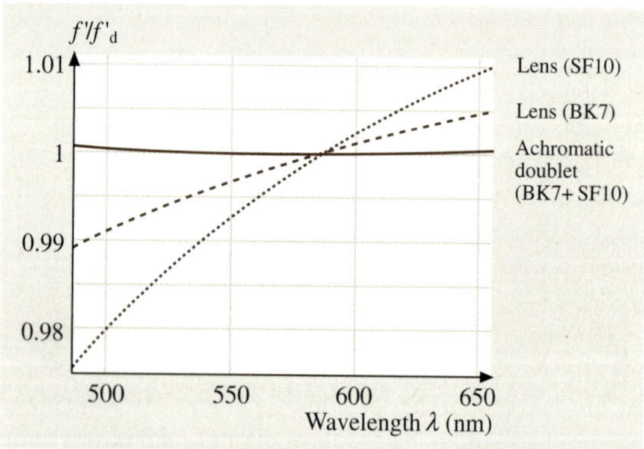

Fig. 2.47 The focal length f' of two single lenses (one made of BK7 and the other made of SF10) and a refractive achromatic doublet (made of BK7 and SF10) as a function of the wavelength λ normalized by the focal length f_d' at $\lambda_d - 587.6$ nm

2.6.2 The Camera

One of the simplest optical instruments is the camera [2.1, 8]. Of course, modern cameras are highly sophisticated technical instruments with complex wide-angle or zoom lenses. However, the basic principle of each camera (see Fig. 2.48) is that a lens forms a real inverted picture of an object on a photosensitive surface, which can be a photographic film or an electronic detector such as a charge-coupled device (CCD) chip. Additionally, each camera has a diaphragm near the lens.

The standard camera objective of a miniature camera has a focal length of $f' = 50$ mm so that each object at a distance of several meters can be assumed to be at an infinite distance and the object distance d_O in the lens equation (2.89) can be assumed to be $d_O \rightarrow -\infty$. Then, the image is practically formed in the focal plane of the lens, i. e. $d_I \approx f'$. So, the size x of the image of an object is determined by the angular extension φ of the object by

$$x \approx \varphi f' . \tag{2.180}$$

The moon has, for example, an angular extension of about half a degree if observed from the earth so that its image on a standard camera would be just $x = 0.44$ mm. This is the reason why the moon on a photo made with a miniature camera with a film size of 24 mm × 36 mm is really small and details cannot be detected. However, this can be changed by using a telescope (Sect. 2.6.4) in front of the camera, which changes the angular extension φ of the object. In astronomical cameras the eyepiece of the telescope is commonly omitted and the detector is positioned directly in the focal plane of the objective lens or mirror, which has a large focal length f' and which serves as the camera lens. Nevertheless, such a device is still called an astronomical telescope.

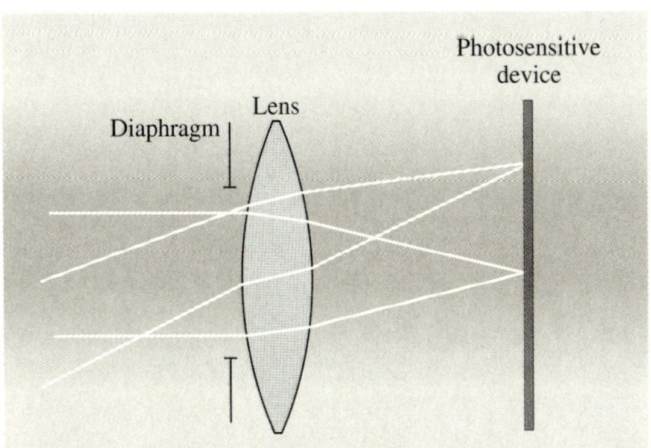

Fig. 2.48 Principle of a camera. The distance from the object to the lens of the camera compared to the focal length of the lens is in the presented case so large that the object can be assumed to be at infinite distance

The Depth of Field

In geometrical optics an ideal camera lens (without aberrations) images one object plane very sharply onto the photosensitive image plane. But in reality each image point is imperfect for two reasons: firstly, due to the wave nature of light, it is not an ideal mathematical point but an airy disc, and secondly the resolution of the detector is in many cases smaller than the maximum possible resolution given by the wave nature of light. Object points in other planes than the ideal object plane are imaged to planes in front of or behind the detector plane (Fig. 2.49). Therefore, in the detector plane they will form small image discs, and if the diameter of the image discs is smaller than the pixel size p of the detector these other planes will also be imaged without loss of resolution onto the detector, which limits the resolution.

The ideal object plane, which is imaged very sharply onto the detector, has object distance d_O while the detector plane has image distance d_I (where $d_O < 0$ and $d_I > 0$ for a real image in a camera). An object plane that is nearer the camera lens than $|d_O|$ and where the light rays of the object points form small discs in the detector plane with a diameter of exactly p is the nearest object plane which is imaged onto the detector with the maximum resolution, given by the pixel distance p. Its object distance is called $d_{O,N}$ (index "N" for "near") and its image distance is called $d_{I,N}$ (Fig. 2.49). Similarly, that object plane a larger distance than $|d_O|$ from the lens where the rays coming from the object points also form discs in the detector plane with a diameter p is the farthest object plane which is imaged with the maximum resolution given by the detector. It has object

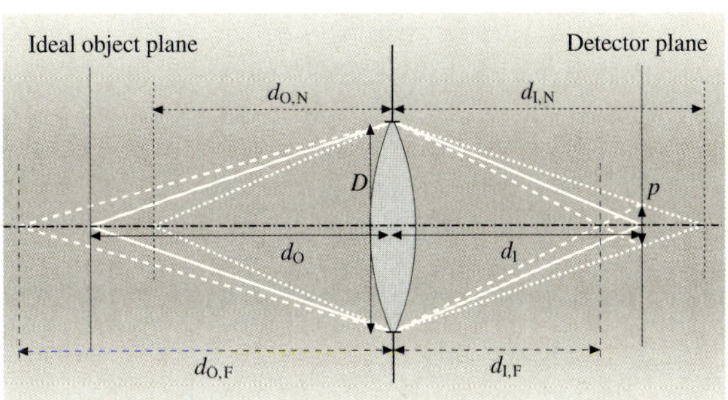

Fig. 2.49 Calculation of the depth of field for the case of a camera with a thin lens

distance $d_{O,F}$ and corresponding image distance $d_{I,F}$ (the index F indicates "far").

The depth of field is now defined as the axial extension of the object space between the near object plane and the far object plane, which are both just imaged with the maximum resolution of the detector. The depth of field depends of course on the diameter D of the aperture stop and on the resolution of the detector, i.e. the pixel distance p. We assume in the following that we have a thin ideal lens with a focal length f' on the image side and that the aperture stop is directly in the plane of the lens. Then, the aperture stop is also the entrance pupil and the exit pupil. An important quantity is the so called F number, $f\#$, of the lens which is defined as the ratio of the focal length f' of the lens and the diameter D of the entrance pupil

$$f\# = \frac{f'}{D} . \tag{2.181}$$

If the image is formed in the focal plane, as is nearly the case for a camera imaging a distant object, and if the diameter D is small compared to the focal length f' the F number and the numerical aperture on the image side defined by (2.136) are connected to a good approximation by:

$$\mathrm{NA_I} = n_I \sin\varphi_I \approx n_I \frac{D}{2f'} = n_I \frac{1}{2f\#} . \tag{2.182}$$

Here, φ_I is the half aperture angle of the light cone on the image side and n_I is the refractive index on the image side. In most cases there will be air on the image side, i.e. $n_I = 1$. But for some camera-like systems such as the human eye n_I will be different from 1 (Sect. 2.6.3).

According to the lens equation (2.89), where n_O and n_I are the refractive indices on the object and image side, respectively, we have three equations for the different object and image distances:

$$\frac{n_I}{d_I} - \frac{n_O}{d_O} = \frac{n_I}{f'} \Rightarrow d_I = \frac{n_I f' d_O}{n_O f' + n_I d_O} , \tag{2.183}$$

$$\frac{n_I}{d_{I,N}} - \frac{n_O}{d_{O,N}} = \frac{n_I}{f'} \Rightarrow d_{I,N} = \frac{n_I f' d_{O,N}}{n_O f' + n_I d_{O,N}} , \tag{2.184}$$

$$\frac{n_I}{d_{I,F}} - \frac{n_O}{d_{O,F}} = \frac{n_I}{f'} \Rightarrow d_{I,F} = \frac{n_I f' d_{O,F}}{n_O f' + n_I d_{O,F}} . \tag{2.185}$$

Additionally, according to the theorem on intersecting lines, we have two additional equations (Fig. 2.49):

$$\frac{D}{d_{I,N}} = \frac{p}{d_{I,N} - d_I} \Rightarrow d_{I,N} - d_I = \frac{p}{D} d_{I,N} , \tag{2.186}$$

$$\frac{D}{d_{I,F}} = \frac{p}{d_I - d_{I,F}} \Rightarrow d_I - d_{I,F} = \frac{p}{D} d_{I,F} . \tag{2.187}$$

By putting (2.183) and (2.184) in (2.186) and solving for $d_{O,N}$ the result is

$$d_{O,N} = \frac{n_O f' d_O}{n_O f' - \frac{p}{D}(n_O f' + n_I d_O)} = \frac{d_O}{1 - \frac{p}{D}\left(1 + \frac{n_I d_O}{n_O f'}\right)} . \tag{2.188}$$

In the same way by combining (2.183), (2.185) and (2.187) the result for $d_{O,F}$ is

$$d_{O,F} = \frac{n_O f' d_O}{n_O f' + \frac{p}{D}(n_O f' + n_I d_O)} = \frac{d_O}{1 + \frac{p}{D}\left(1 + \frac{n_I d_O}{n_O f'}\right)} . \tag{2.189}$$

It is common practice in photography to use the lateral magnification β, which was defined by (2.52) as the ratio of the image height x_I and the object height x_O. For a lens which fulfills the sine condition (2.138) the principal planes are in reality principal spheres which are centered around the object and the image point, respectively. The same is valid for the entrance and exit pupil [2.3]. Then, using (2.138), the lateral magnification can be expressed as

$$\beta = \frac{x_I}{x_O} = \frac{n_O \sin\varphi_O}{n_I \sin\varphi_I} = \frac{n_O D/(2d_O)}{n_I D/(2d_I)} = \frac{n_O}{n_I} \frac{d_I}{d_O} . \tag{2.190}$$

By multiplying (2.183) with d_O/n_O it holds:

$$\frac{n_I d_O}{n_O d_I} - 1 = \frac{n_I d_O}{n_O f'} \Rightarrow \frac{1}{\beta} = 1 + \frac{n_I d_O}{n_O f'} . \tag{2.191}$$

So, (2.188) and (2.189) can be expressed by

$$d_{O,N} = \frac{d_O}{1 - \frac{p}{D\beta}} = \frac{d_O}{1 - \frac{p f\#}{f'\beta}} , \tag{2.192}$$

$$d_{O,F} = \frac{d_O}{1 + \frac{p}{D\beta}} = \frac{d_O}{1 + \frac{p f\#}{f'\beta}} . \tag{2.193}$$

In the last step, the F number, $f\#$, defined by (2.181) is used.

In the case of a camera the focal length f' is positive and the lateral magnification β is always negative since a real image is formed, i. e. $\beta < 0$. So, there is the interesting special case that the denominator of (2.193) can be zero:

$$1+\frac{p}{D\beta}=0 \Rightarrow \beta=-\frac{p}{D}$$

$$\Rightarrow d_{\mathrm{O,C}}=-\frac{n_{\mathrm{O}}f'}{n_{\mathrm{I}}}\left(1+\frac{D}{p}\right)$$

$$=-\frac{n_{\mathrm{O}}f'}{n_{\mathrm{I}}}\left(1+\frac{f'}{pf\#}\right). \tag{2.194}$$

On the right-hand side, (2.191) was used and solved for d_{O}. So, if the camera is focused to the critical object distance $d_{\mathrm{O,C}}$ given by (2.194) it holds that $|d_{\mathrm{O,F}}| \to \infty$ and all objects which are farther from the camera lens than $|d_{\mathrm{O,N}}| = |d_{\mathrm{O,C}}|/2$ (this follows from (2.192)) will be imaged onto the detector with the maximum resolution, i. e. the image will look sharp. Of course, if the modulus $|d_{\mathrm{O}}|$ of the actual object distance is larger than the modulus of the critical value $|d_{\mathrm{O,C}}|$ given by (2.194), $d_{\mathrm{O,F}}$ will formally be positive. This means that a virtual object behind the lens with distance $d_{\mathrm{O,F}}$, which can be produced by some auxiliary optics, can also be imaged sharp onto the detector. In fact, this still means that all real objects farther from the lens than $|d_{\mathrm{O,N}}|$ will be imaged sharp onto the detector.

If we have for example a camera with $f' = 50\,\mathrm{mm}$, a minimum F number $f\# = 2.8$, $n_{\mathrm{O}} = n_{\mathrm{I}} = 1$, and a pixel distance $p = 11\,\mu\mathrm{m}$ (typical for a CCD chip) the critical object distance $d_{\mathrm{O,C}}$ of (2.194) is $d_{\mathrm{O,C}} = -81.2\,\mathrm{m}$. Therefore, all objects at a distance of more than $|d_{\mathrm{O,N}}| = |d_{\mathrm{O,C}}|/2 = 40.6\,\mathrm{m}$ from the camera will be imaged sharply if the camera is focused to $d_{\mathrm{O,C}}$. If the F number is $f\# = 16$ all objects with a distance of more than 7.1 m will be imaged sharp for a focusing distance of $|d_{\mathrm{O,C}}| = 14.3\,\mathrm{m}$. However, for larger F numbers the wave nature of light begins to limit the resolution because the radius r_{diff} of a diffraction limited spot will be $r_{\mathrm{diff}} = 0.61\lambda/\mathrm{NA} \approx 1.22\lambda f\# = 10.7\,\mu\mathrm{m} \approx p$ for a wavelength $\lambda = 550\,\mathrm{nm}$ and $f\# = 16$.

Of course, a larger F number means that the light intensity on the detector decreases because the light intensity on the detector is proportional to the effective area $\pi D^2/4$ of the light gathering lens and therefore proportional to $1/f\#^2 = D^2/f'^2$. So, a larger F number means that the exposure time has to be increased proportional to $f\#^2$. All these facts are well-known from photography.

If $d_{\mathrm{O,F}}$ has a finite value, e.g. if the camera is focused on a near object, it is useful to calculate the axial extension $\Delta d = d_{\mathrm{O,N}} - d_{\mathrm{O,F}}$ of the sharply imaged object space. By using (2.192) and (2.193) the result is

$$\Delta d = d_{\mathrm{O,N}} - d_{\mathrm{O,F}} = 2d_{\mathrm{O}}\frac{\frac{pf\#}{f'\beta}}{1-\left(\frac{pf\#}{f'\beta}\right)^2}$$

$$=2\frac{n_{\mathrm{O}}}{n_{\mathrm{I}}}pf\#\frac{\left(\frac{1}{\beta}-1\right)\frac{1}{\beta}}{1-\left(\frac{pf\#}{f'\beta}\right)^2}. \tag{2.195}$$

In the last step (2.191) has been used to express the object distance d_{O} by the lateral magnification β because these two quantities are of course coupled to each other.

Again, we see in (2.195) the limiting case that the denominator can approach zero [if (2.194) is fulfilled] and that therefore the depth of field has an infinite range. But, for near objects (for example $|d_{\mathrm{O}}| \le 1\,\mathrm{m}$) we normally have the case that $f'|\beta| \gg pf\#$. Then, Δd has first of all a finite positive value and second there is a quite good approximation which is often used for the photography of near objects [2.8]

$$\Delta d = 2\frac{n_{\mathrm{O}}}{n_{\mathrm{I}}}pf\#\frac{\left(\frac{1}{\beta}-1\right)\frac{1}{\beta}}{1-\left(\frac{pf\#}{f'\beta}\right)^2}$$

$$\approx 2\frac{n_{\mathrm{O}}}{n_{\mathrm{I}}}pf\#\left(\frac{1}{\beta}-1\right)\frac{1}{\beta} = 2\frac{n_{\mathrm{O}}}{n_{\mathrm{I}}}pf\#\frac{1-\beta}{\beta^2}. \tag{2.196}$$

As an example we take again a common electronic camera with $f' = 50\,\mathrm{mm}$, $p = 11\,\mu\mathrm{m}$ and $n_{\mathrm{O}} = n_{\mathrm{I}} = 1$. The F number is assumed to be $f\# = 10$ and the object is at $d_{\mathrm{O}} = -1\,\mathrm{m}$. Then, the lateral magnification is, according to (2.191), $\beta = -0.05263$. The extension of the depth of field Δd is according to the exact (2.195) $\Delta d = 83.75\,\mathrm{mm}$ and according to the approximate (2.196) $\Delta d = 83.60\,\mathrm{mm}$. So, the error of the approximate equation is about 0.2% and the depth of field has an extension of about 8.4 cm, i. e. objects with an axial extension in this range (for a medium object distance $|d_{\mathrm{O}}| = 1\,\mathrm{m}$) will be imaged without loss of resolution onto the detector.

2.6.3 The Human Eye

The human eye is based on the principle of a camera that builds an inverted real image of the surroundings on the retina [2.1,8]. However, the actual structure and performance of the human eye is quite complex [2.46–48] so that only the most important features of the normal emmetropic eye can be discussed in this section.

The optical power of the eye is delivered by the cornea and the deformable crystalline lens (Fig. 2.50). The main part of the optical power is delivered by the cornea with about 43 diopters because at the first surface the difference between the refractive indices of air and the cornea (1.376) is quite high. The crystalline lens, with a refractive index between 1.386 in the outer parts and 1.406 in the core, is immersed on one side in the aqueous humour and on the other side in the vitreous body, both of which have a refractive index of 1.336. Therefore, the lens has about 19 diopters in the case of distant vision. The resulting total optical power of the eye is, due to the finite distance between the cornea and the crystalline lens, about 59 diopters for distant vision. The accommodation of the eye for near objects, which is performed by the crystalline lens, can vary between about 14 diopters in young age and nearly 0 diopters above 50 years of age because the crystalline lens loses its flexibility with increasing age. Since the normal distance for reading is about 25–30 cm an accommodation of less than 3–4 diopters has to be corrected by wearing eyeglasses for reading.

The photosensitive surface of the eye is the curved retina and the diaphragm of the eye is the iris, which can change its diameter between about 2 mm and 8 mm to control the irradiance on the retina depending on the intensity of the illuminating light. The effective focal length of the eye, which, as mentioned above, is an immersion system, is $f'/n' \approx 1/(59\,\text{diopters}) \approx 17\,\text{mm}$ ($n' = 1.336$ is the refractive index of the vitreous body between the eye lens and the retina). The so-called least distance of distinct vision of a normal adult eye is about 25 cm, which requires an accommodation of 4 diopters.

The angular resolution $\Delta\varphi$ of a normal eye is about 1' (one arc min) and, under optimal conditions, it can achieve 30". The later corresponds to a distance of $\Delta x = \Delta\varphi f'/n' = 2.5\,\mu\text{m}$ on the retina. So, the light-sensitive cells (cones) in the fovea (about 200 μm in diameter), which is the central part of the retina, have to be about 2.5 μm or less in diameter and distance. In the fovea there are mainly color-sensitive cones whereas in the outer parts the rods, which are more sensitive to light but which cannot distinct between different colors, dominate.

It is interesting to note that a human eye with normal vision is a nearly diffraction-limited optical system for a diameter of the pupil of up to 3 mm (the diameter for sharpest vision). This can be seen because in this case the radius r of the airy disc which limits the resolution according to the Rayleigh criterion (see Sect. 3.4.2) is about $r = 0.61\lambda/\text{NA} = 3.8\,\mu\text{m}$ for a wavelength of $\lambda = 0.55\,\mu\text{m}$ and a numerical aperture

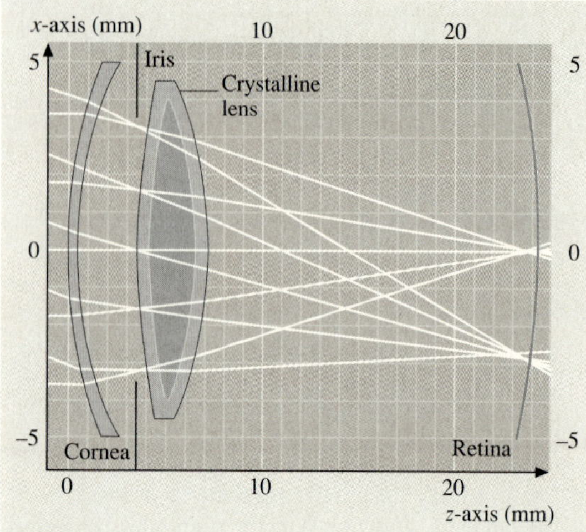

Fig. 2.50 Ray tracing of an eye model as used in ophthalmology. Here, the crystalline lens consists of a nucleus with higher refractive index (1.406) and a cortex with lower refractive index (1.386). The first surface of the cornea and all surfaces of the crystalline lens are aspheric surfaces. It can be seen that the foci show aberrations for this large pupil (iris) diameter. The drawn off-axis point with quite high aberrations is of course far out of the fovea so that it is only used in the eye to detect motions and not to image an object

$\text{NA} \approx 1.5\,\text{mm}/17\,\text{mm} \approx 0.088$. So, the aforementioned value $\Delta x = 2.5\,\mu\text{m}$ for the smallest resolvable distance on the retina is even slightly smaller than the distance r given by the Rayleigh criterion due to diffraction, which assumes that a drop of 26 percent in irradiance can be detected. The reason is that the Rayleigh criterion is slightly arbitrary and, under optimal conditions, the eye can also detect smaller drops in irradiance between two adjacent points. For larger diameters of the pupil than 3–4 mm spherical aberration and chromatic aberrations of the eye reduce the resolution. Therefore, at night or in badly illuminated rooms the resolution of the eye is reduced and all tasks which need a high resolution, for example reading, are more difficult or impossible if the irradiance on the retina is too low even for the largest pupil diameter.

2.6.4 The Telescope

One of the most important optical instruments is the telescope [2.1, 3, 8]. It has well-known applications in terrestrial and astronomical observations [2.21]. But there are at least as important applications in optics to expand or compress a collimated (laser) beam, to filter spatial frequencies in an optical system, to transport an intermediate optical image to another plane, and so on.

A telescope consists in principle of two lenses or two other focusing optical elements such as spherical or aspheric mirrors. Here, to demonstrate the principle we assume that it consists of two lenses with focal lengths f_1' and f_2' and a distance d between the two lenses. In order to have a telescope the image-sided focus F_1' of the first lens and the object-sided focus F_2 of the second lens have to coincide (Fig. 2.51). Additionally, we assume that the two lenses are situated in air, so that we have $f_2' = -f_2$ for the image-and object-sided focal length. So, by taking into account the sign conventions for the focal lengths the condition for the distance between the two lenses of a telescope is

$$d = f_1' - f_2 = f_1' + f_2' \,. \tag{2.197}$$

The paraxial matrix M of the telescope from the object-sided principal plane $\mathcal{U}_1$ of the first lens to the image-sided principal plane $\mathcal{U}_2'$ of the second lens is

$$\begin{aligned} M &= \begin{pmatrix} 1 & 0 \\ -\frac{1}{f_2'} & 1 \end{pmatrix} \begin{pmatrix} 1 & d \\ 0 & 1 \end{pmatrix} \begin{pmatrix} 1 & 0 \\ -\frac{1}{f_1'} & 1 \end{pmatrix} \\ &= \begin{pmatrix} 1-\frac{d}{f_1'} & d \\ -\frac{1}{f_1'} - \frac{1}{f_2'} + \frac{d}{f_1' f_2'} & 1-\frac{d}{f_2'} \end{pmatrix} \\ &= \begin{pmatrix} -\frac{f_2'}{f_1'} & f_1' + f_2' \\ 0 & -\frac{f_1'}{f_2'} \end{pmatrix} . \end{aligned} \tag{2.198}$$

So, the coefficient C of the ABCD-matrix M, which is according to (2.68) the negative value of the optical power, is zero and therefore the focal length of the telescope is infinity. Such a system with zero optical power is called an afocal system. So, a telescope can also be defined to be an afocal optical system where the trivial case that all lenses themselves have zero optical power, i. e. $1/f_1' = 1/f_2' = 0$, is excluded.

Telescope as a Beam Expander and Imaging System for Far-Distant Objects

An important property of an afocal system is that it transforms a collimated bundle of rays into another collimated bundle of rays. The application as a beam expander for a collimated beam or as an imaging system for far-distant objects can easily be seen from (2.198) by taking two parallel rays with paraxial ray parameters (x_1, φ_1) and (x_2, φ_2) $(\varphi_2 = \varphi_1)$ in front of the telescope. The paraxial ray parameters (x_1', φ_1') and (x_2', φ_2') of the

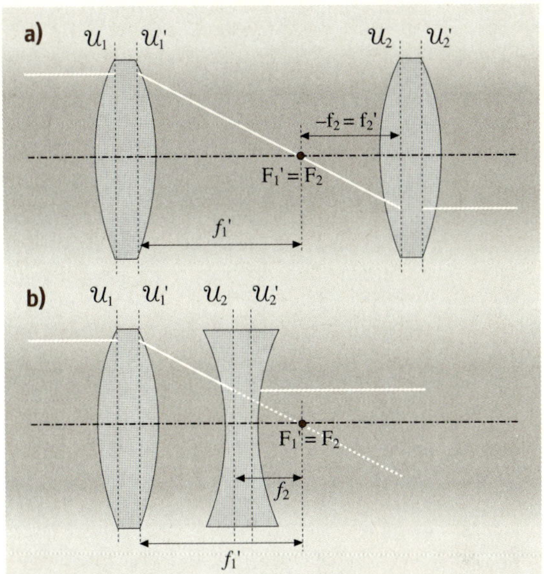

Fig. 2.51a,b Principle set ups of an astronomical telescope (**a**) and a Galilean telescope (**b**)

rays behind the telescope are then:

$$\begin{aligned} \begin{pmatrix} x_i' \\ \varphi_i' \end{pmatrix} &= \begin{pmatrix} -\frac{f_2'}{f_1'} & f_1' + f_2' \\ 0 & -\frac{f_1'}{f_2'} \end{pmatrix} \begin{pmatrix} x_i \\ \varphi_i \end{pmatrix} \\ &= \begin{pmatrix} -\frac{f_2'}{f_1'} x_i + (f_1' + f_2')\varphi_i \\ -\frac{f_1'}{f_2'} \varphi_i \end{pmatrix} \end{aligned} \tag{2.199}$$

with $i \in \{1, 2\}$.

The angular magnification γ, defined by equation (2.53), i. e. the ratio of the angle $\varphi' := \varphi_1' = \varphi_2'$ between the bundle of rays and the optical axis behind the afocal system to the angle $\varphi_i := \varphi_1 = \varphi_2$ in front of the system, in the paraxial case is

$$\gamma = \frac{\varphi'}{\varphi} = -\frac{f_1'}{f_2'} \,. \tag{2.200}$$

So, the angular magnification, which determines the size of the image of a far-distant object, only depends on the ratio of the focal lengths of the two lenses.

The beam-expanding property can be seen by calculating the distance Δx between two parallel rays $(\varphi_2 = \varphi_1)$ in front of the telescope and the distance $\Delta x'$ behind the telescope:

$$\Delta x' = x_2' - x_1' = -\frac{f_2'}{f_1'}(x_2 - x_1) = -\frac{f_2'}{f_1'}\Delta x \tag{2.201}$$

So, the beam expanding ratio $\Delta x'/\Delta x$ is the reciprocal of the angular magnification.

Imaging Property of a Telescope for Finite Distant Objects

Although a telescope has zero optical power it images an object from one plane to another plane. This can be seen by calculating the paraxial matrix M' from an object plane with a distance d_1 to the principal plane $\mathcal{U}_1$ of the first lens (keep in mind that, contrary to the normal sign conventions of paraxial optics, in the paraxial matrix theory d_1 is positive if the object plane is in front of $\mathcal{U}_1$ and negative if it is behind $\mathcal{U}_1$) to an image plane at a distance d_2 behind the principal plane $\mathcal{U}_2'$ of the second lens (d_2 is positive if the image plane is real and behind $\mathcal{U}_2'$ and negative if it is a virtual image plane in front of $\mathcal{U}_2'$). Figure 2.52 shows the parameters to calculate M':

$$M' = \begin{pmatrix} 1 & d_2 \\ 0 & 1 \end{pmatrix} \begin{pmatrix} -\frac{f_2'}{f_1'} & f_1'+f_2' \\ 0 & -\frac{f_1'}{f_2'} \end{pmatrix} \begin{pmatrix} 1 & d_1 \\ 0 & 1 \end{pmatrix}$$

$$= \begin{pmatrix} -\frac{f_2'}{f_1'} & f_1'+f_2'-d_1\frac{f_2'}{f_1'}-d_2\frac{f_1'}{f_2'} \\ 0 & -\frac{f_1'}{f_2'} \end{pmatrix} . \qquad (2.202)$$

In the case of imaging, the parameter B of the matrix has to be zero. So, the condition for the distances d_1 and d_2 is

$$f_1'+f_2'-d_1\frac{f_2'}{f_1'}-d_2\frac{f_1'}{f_2'}=0\,,$$

$$\Rightarrow d_2 = f_2' + \frac{(f_2')^2}{f_1'} - d_1\frac{(f_2')^2}{(f_1')^2} . \qquad (2.203)$$

As mentioned before, the image is real if d_2 is positive and it is virtual for a negative value d_2. So, a real image of a real object point (i. e. $d_1 \geq 0$) means

$$d_2 \geq 0 \Rightarrow f_1' + \frac{f_1'^2}{f_2'} \geq d_1 \geq 0 \Rightarrow \frac{1}{f_1'} + \frac{1}{f_2'} \geq 0 . \qquad (2.204)$$

It can easily be seen that the Galilean telescope (see next paragraph) cannot deliver a real image of a real object point whereas the astronomical telescope delivers real images as long as $0 \leq d_1 \leq f_1' + f_1'^2/f_2'$.

A quite interesting property of a telescopic imaging system is the lateral magnification β [see (2.52)]. It is, according to equation (2.202) for the imaging case, i. e. matrix element $B = 0$, equal to the matrix element A:

$$\beta = \frac{x'}{x} = -\frac{f_2'}{f_1'} . \qquad (2.205)$$

So, the lateral magnification of the telescopic system depends only on the focal lengths of the two lenses and is independent of the axial position of the object point. If we additionally place the aperture stop into the focal plane of the first lens (only for the astronomical telescope) the telescopic imaging system is telecentric (see also page 60).

A very important system is the so-called $4f$-system with $f := f_1' = f_2' > 0$. Then we have, with the help of (2.203),

$$d_2 = f + f - d_1 \quad \Rightarrow \quad d_1 + d_2 = 2f . \qquad (2.206)$$

This means that the sum of the two distances d_1 and d_2 is always $2f$ and in total the distance from the object plane to the image plane is $4f$ (in the case of thin lenses where the thickness of the lenses can be neglected compared to $4f$) because the length of the telescope has to be added. This also means that, for a $4f$-system, the shift of the image plane is equal to the shift of the object plane and, therefore, the telescope itself can, for example, be shifted relative to the object and image plane without changing the imaging situation. Of course, in the non-paraxial realm aberrations will change the imaging quality if the telescope of a $4f$-system is moved because the aberrations depend on the actual position of the telescope relative to the object and image plane.

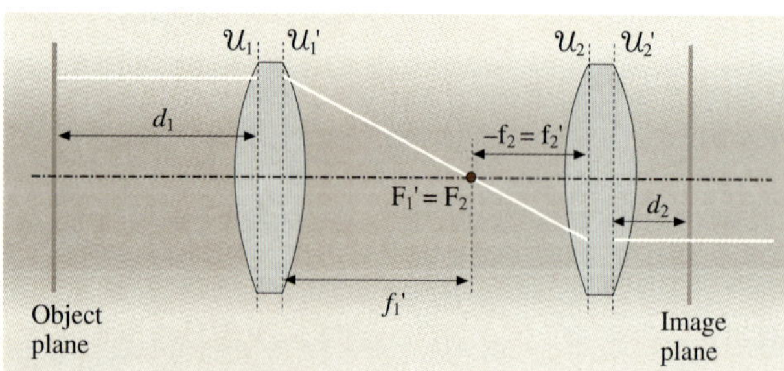

Fig. 2.52 Parameters to calculate the paraxial matrix M' for imaging of an object point to an image point with the help of a telescope

The Astronomical and the Galilean Telescope

There are two different types of telescopes (Fig. 2.51). The astronomical telescope (also called a Kepler telescope) and the Galilean telescope (also called a Dutch telescope).

Astronomical Telescope. The astronomical telescope (Fig. 2.51a or Fig. 2.53) consists of two positive lenses so that the first lens (called the objective) forms a real image of a far-distant object near the focal plane (or exactly in the focal plane for an object with infinite distance). Then, the second lens (called the eyepiece) also forms an infinite-distant image but with an increased or decreased angular magnification. Since the focal lengths f_1' and f_2' are both positive the angular magnification γ is, according to (2.200), $\gamma = -f_1'/f_2' < 0$. Therefore, the image is upside down so that an astronomical telescope without additional optics to reverse the image is not practical for terrestrial inspections. However, for astronomical purposes or for image transfers in optical systems this is no disadvantage. Additionally, the advantage of the astronomical telescope is that the entrance pupil coincides with the objective in the case of imaging infinite-distant objects. This means that the exit pupil, which is the image of the objective formed by the eyepiece, is typically near the focal plane of the eyepiece if $f_1' \gg f_2'$, which is necessary to obtain an image with angular magnification $|\gamma| \gg 1$. Therefore, the pupil of the eye can be positioned at the exit pupil of the telescope and all light rays with the same off-axis angle (i. e. from the same infinite-distant object point) entering the telescope contribute to the image on the retina of the eye. Another advantage of the astronomical telescope is, as mentioned previously, that it can deliver a telecentric real image of objects with a finite distance.

It is quite interesting to think a little bit more about the position of the aperture stop and the field stop for the two cases of imaging infinite-distant objects or finite-distant objects (Fig. 2.53). As mentioned, the aperture stop for the imaging of infinite-distant objects (Fig. 2.53a) is the aperture of the first lens. The field stop lies in this case in the back focal plane of the first lens. For the case of imaging of finite-distant objects (Fig. 2.53b) the situation is different and it is useful to put the aperture stop in the back focal plane of the first lens to have a well-defined numerical aperture for all object points which are not too far away from the optical axis. Then, the aperture of the first lens can act as the field stop. Of course, in this case the field stop has no sharp rim because parts of the light cone of points with a similar distance from the optical axis than the radius of the first lens can pass the system if the aperture stop is large enough. In this case a kind of vignetting occurs. So, an additional stop directly in the object or image plane serving as a field stop would be desirable.

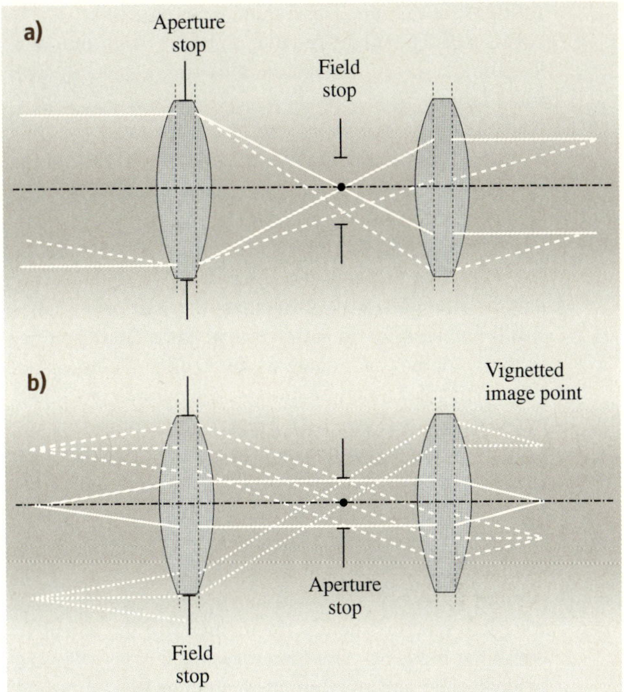

Fig. 2.53a,b The aperture stop and the field stop for the imaging of **(a)** infinite-distant objects or **(b)** finite-distant objects (here shown for the case that the object plane is in the front focal plane of the first lens) with the help of an astronomical telescope

For astronomical observations most modern telescopes use mirrors as focusing elements instead of lenses [2.21]. There are telescopes with a primary mirror with a diameter of $D = 8$ m. From wave optics it is well known that two (infinite-distant) object points can just be resolved by an aberration-free telescope with a diameter D of the primary mirror and a wavelength λ of the observed light if their angular separation is larger or equal to $\Delta\varphi$ with

$$\Delta\varphi = k\frac{\lambda}{D}\,, \qquad (2.207)$$

where k is a constant of approximately $k = 1$ (for a full circular aperture it is $k = 1.22$). The exact value of k depends on the actual design of the instrument because a reflective telescope has in many cases an annular or more complicated aperture because the secondary mirror and the mounting shadow the central and other parts of the primary mirror. So, a telescope with a large-diameter primary mirror has of course an increased light-gathering power and an increased angular resolution.

Galilean Telescope. The Galilean telescope (Fig. 2.51b) consists of a positive lens (the objective) with the focal length $f_1' > 0$ and a negative lens (the eyepiece) with focal length $f_2' < 0$ or $f_2 = -f_2' > 0$ and $|f_2'| < |f_1'|$. Of course, the telescope can also be rotated by 180° so that it reduces the angular magnification. But, in the following we assume that $f_1' > 0$ and $f_2' < 0$. The total length of the Galilean telescope is only $|f_1'| - |f_2'|$ (for thin lenses) compared to $|f_1'| + |f_2'|$ for the astronomical telescope (we use here the absolute values of the focal lengths although f_1' is always positive and only f_2' has a different sign for an astronomical and a Galilean telescope). Another advantage of the Galilean telescope is that the angular magnification γ is, according to (2.200) positive, $\gamma = -f_1'/f_2' > 0$. Therefore, the image is upright and can be directly used for terrestrial inspections. However, a disadvantage of the Galilean telescope is that the image of the objective made by the second lens is between the two lenses. Therefore, the exit pupil of the Galilean telescope is not accessible for the eye and the pupil of the eye works itself as the aperture stop of the complete system whereas the diameter of the objective limits the field. So, Galilean telescopes have a limited field of view and only small magnifications of two to five are useful. Another disadvantage is that the Galilean telescope cannot deliver a real image of a real object. So, a Galilean telescope cannot be used to transport a real intermediate image to another plane in an optical system.

However, the compact overall length and the positive angular magnification provide applications for a Galilean telescope as a beam expander or terrestrial telescope like a lorgnette.

2.6.5 The Microscope

The last important optical instrument that will be discussed here is the microscope [2.1, 3, 8]. Whereas, the telescope, especially an astronomical telescope, is used to achieve an angular magnification of distant objects, a microscope is used to obtain a magnified image of a very small near object.

The Magnifier

If somebody wants to see details of a small object he brings the object as close to the eye as possible since then the image of the object on the retina of the eye is as large as possible. However, a typical human eye can only form a sharp image of an object at a smallest distance of about $d_S = 25$ cm, which is the standard distance for distinct vision. So, it is obvious that a positive lens, called a magnifier, directly in front of the eye can be used to obtain at the distance $|d_I| = d_S$ a magnified virtual image of an object that has itself a smaller separation $|d_O|$ from the eye than the standard distance d_S. The image at a distance d_I from the image-sided principal plane of the lens with focal length f' has to fulfill the imaging equation (2.89) whereby the refractive index on the image side has to be $n' = 1$ because the human eye is normally used in air and only in this case delivers a sharp image. Then, the imaging equation is

$$\frac{1}{d_I} - \frac{n}{d_O} = \frac{1}{f'} . \qquad (2.208)$$

Here, n is the refractive index on the object side, which is often 1 (object in air) but sometimes also larger than 1 if the object is in immersion (for example in water or oil). Due to the sign conventions of geometrical optics d_O is negative since the object is in front of the lens. The image distance d_I is also negative for a virtual image. Then, the lateral magnification β of the image is, according to (2.52), (2.208) and Fig. 2.54,

$$\beta = \frac{x_I}{x_O} = \frac{\varphi_I d_I}{\varphi_O d_O} = n\frac{d_I}{d_O} = 1 - \frac{d_I}{f'} = 1 + \frac{d_S}{f'} . \qquad (2.209)$$

Here, we use that, in the paraxial case, the angles φ_I and φ_O have to fulfill the condition $n'\varphi_I = n\varphi_O$, where $n' = 1$ is valid in our case. Additionally, it is used that the virtual image is at the standard distance for distinct

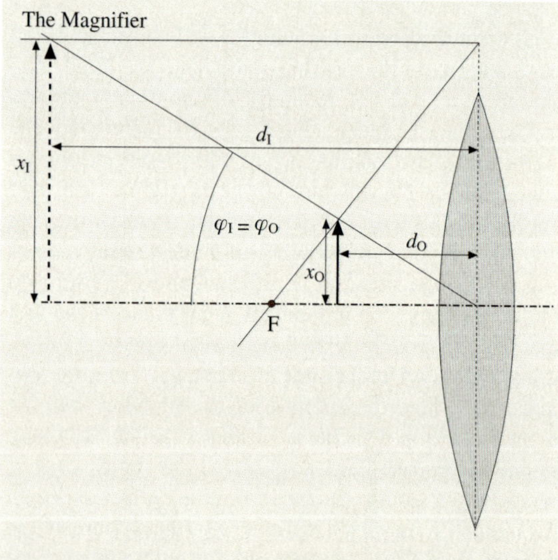

Fig. 2.54 The principle of a magnifier. A thin lens is used and it is assumed that the refractive indices in the object and image space are both equal so that $\varphi_I = \varphi_O$

vision so that the image distance d_{I}, which is negative, is replaced by $-d_{\mathrm{S}}$, where d_{S} is the absolute value of the standard distance for distinct vision.

If the lens has, for example, a focal length of $f' = 5\,\mathrm{cm}$ a lateral magnification of $\beta = 1 + 25/5 = 6$ is obtained. In order to have a large field of view without aberrations and especially without chromatic aberrations the magnifier itself is in practice not a single lens but an achromatic combination of different single lenses.

The Two-Stage Microscope

There is of course a limitation to the lateral magnification achievable by using a magnifier because the object has to be very close to the magnifier and therefore also to the eye to achieve large lateral magnifications. Therefore, microscopes have been invented that make a magnification of the object in two stages (Fig. 2.55). First, a magnified real image of the object with magnification $\beta_{\text{objective}}$ is formed by using a lens with a small focal length, called the objective. This real image is of course inverted. Then, a magnifier, called the eyepiece, with a (mostly) larger focal length is used to form a magnified virtual image of the intermediate real image, where the virtual image is at the standard distance of distinct vision of the eye. The lateral magnification for this second operation is β_{eyepiece}. This means, that the lateral magnifications of both operations are multiplied and the total lateral magnification of the microscope $\beta_{\text{microscope}}$ is

$$\beta_{\text{microscope}} = \beta_{\text{objective}} \beta_{\text{eyepiece}} \,. \qquad (2.210)$$

In practice, the objective of a microscope is a quite complex lens consisting of many single lenses to correct the aberrations (especially spherical aberration, coma and chromatic aberrations) of the objective and to guarantee a large field of view [2.8]. Moreover, modern microscope objectives are corrected for infinity. This means that their aberrations are only corrected if the object is exactly in the object-sided focal plane. Therefore, the image would be at infinite distance and an additional lens (called the tubus lens) with a fixed focal length (the so-called tubus length, which is often 160 mm) must be used to get the real image with the magnification imprinted on the objective. For biological investigations, where the object is often covered by a thin coverslip, the spherical aberrations, which result from a high-NA spherical wave passing through a plane-parallel plate, also have to be corrected.

Another very important parameter of the objective is its numerical aperture NA (2.135). It determines on the one hand the light-gathering power of the objective and on the other the resolution which is possible. From wave optics we know that the smallest distance Δx between two points which can be resolved by a microscope is

$$\Delta x = k \frac{\lambda}{\mathrm{NA}} \,, \qquad (2.211)$$

where λ is the wavelength of the used light and k is a constant (typically about 0.5) that depends on the illumination conditions (coherence) and the exact aperture shape of the objective (mostly circular).

If the image of a microscope has to be on a camera chip (for example a CCD chip) a real image has to be made on the camera chip. Therefore, the eyepiece, which produces a virtual image, cannot be used and indeed it is sufficient just to bring the CCD chip to the position of the real image of the objective (plus the tubus lens). A typical magnification of a high-NA objective in air with, for example, $|\beta| = 50$ is sufficient if a CCD chip with a typical pixel size of $11\,\mu\mathrm{m}$ is used. This would mean that a structure size of $0.22\,\mu\mathrm{m}$ on the object is magnified to the size of a pixel of the CCD chip. However, due to (2.211), $0.22\,\mu\mathrm{m}$ is approximately the resolution of an objective with NA< 1 and a wavelength in the visible spectral range. By bringing an immersion oil between the object and the objective, which has to be a special immersion objective, the NA can be increased up to about 1.4. So, the resolution can be increased accordingly. Another possibility is of course to reduce the wavelength. Modern microscopes for the inspection of integrated circuits use ultraviolet light with a wavelength of 248 nm.

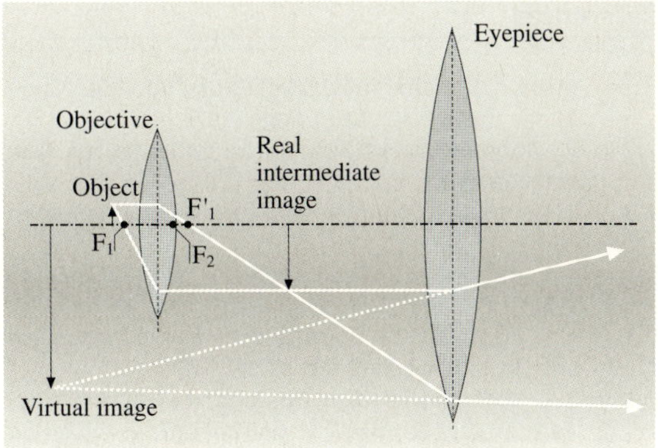

Fig. 2.55 The principle of a microscope illustrated by using thin lenses. The objective forms a real magnified intermediate image of the object, which is then transformed by the eyepiece into a virtual further-magnified image. The distance of this virtual image from the eyepiece and the eye, which is directly behind the eyepiece, has to be the standard distance of distinct vision

References

2.1 M. Born, E. Wolf: *Principles of Optics*, 6th edn. (Cambridge Univ. Press, Cambridge 1997)

2.2 R. Ditteon: *Modern Geometrical Optics* (Wiley, New York 1998)

2.3 H. Haferkorn: *Optik*, 4th edn. (Wiley–VCH, Weinheim 2003)

2.4 E. Hecht: *Optics*, 3rd edn. (Addison–Wesley, Reading 1998)

2.5 R. S. Longhurst: *Geometrical and Physical Optics*, 3rd edn. (Longman, New York 1973)

2.6 V. N. Mahajan: *Optical Imaging and Aberrations, Part I: Ray Geometrical Optics* (SPIE, Bellingham 1998)

2.7 D. Marcuse: *Light Transmission Optics*, 2nd edn. (Van Nostrand, New York 1982)

2.8 H. Naumann, G. Schröder: *Bauelemente der Optik*, 5th edn. (Hanser, München 1987)

2.9 I. N. Bronstein, K. A. Semendjajew: *Taschenbuch der Mathematik*, 23rd edn. (Thun, Frankfurt/Main 1987)

2.10 R. A. Chipman: Mechanics of polarization ray tracing, Opt. Eng. **34**, 1636–1645 (1995)

2.11 E. Waluschka: Polarization ray trace, Opt. Eng. **28**, 86–89 (1989)

2.12 W. Brouwer: *Matrix Methods in Optical Instrument Design* (Benjamin, New York 1964)

2.13 A. Gerrard, J. M. Burch: *Introduction to Matrix Methods in Optics* (Wiley, London 1975)

2.14 H. Kogelnik, T. Li: Laser beams and resonators, Appl. Opt. **5**, 1550–1567 (1966)

2.15 A. E. Siegman: *Lasers* (Univ. Science Books, Mill Valley 1986)

2.16 R. Kingslake: *Lens Design Fundamentals* (Academic, San Diego 1978)

2.17 R. Kingslake: *Optical System Design* (Academic, New York 1983)

2.18 D. Malacara, Z. Malacara: *Handbook of Lens Design* (Dekker, New York 1994)

2.19 D. C. O'Shea: *Elements of Modern Optical Design* (Wiley, New York 1985)

2.20 M. V. Klein, Th. E. Furtak: *Optics*, 2nd edn. (Wiley, New York 1986)

2.21 R. Riekher: *Fernrohre und ihre Meister*, 2nd edn. (Verlag Technik, Berlin 1990)

2.22 H. J. Levinson: *Principles of Lithography* (SPIE, Bellingham 2001)

2.23 H. P. Herzig: *Micro-Optics* (Taylor Francis, London 1997)

2.24 B. Kress, P. Meyrueis: *Digital Diffractive Optics* (Wiley, Chicester 2000)

2.25 S. Sinzinger, J. Jahns: *Microoptics* (Wiley–VCH, Weinheim 1999)

2.26 B. Dörband: Abbildungsfehler und optische Systeme. In: *Technische Optik in der Praxis*, ed. by G. Litfin (Springer, Berlin Heidelberg New York 1997) pp. 73–101

2.27 G. H. Spencer, M. V. R. K. Murty: General ray tracing procedure, J. Opt. Soc. Am. **52**, 672–678 (1962)

2.28 O. N. Stavroudis: *The Optics of Rays, Wavefronts, and Caustics* (Academic, New York 1972)

2.29 A. Sharma, D.V. Kumar, A. K. Ghatak: Tracing rays through graded-index media: a new method, Appl. Opt. **21**, 984–987 (1982)

2.30 A. Sharma: Computing optical path length in gradient-index media: a fast and accurate method, Appl. Opt. **24**, 4367–4370 (1985)

2.31 R. W. Smith: A note on practical formulae for finite ray tracing through holograms and diffractive optical elements, Opt. Commun. **55**, 11–12 (1985)

2.32 W. T. Welford: A vector raytracing equation for hologram lenses of arbitrary shape, Opt. Commun. **14**, 322–323 (1975)

2.33 W. H. Press, B. P. Flannery, S. A. Teukolsky, W. T. Vetterling: *Root Finding and Nonlinear Sets of Equations, Numerical Recipes in C* (Cambridge Univ. Press, Cambridge 1988) pp. 255–289

2.34 J. N. Latta: Computer-based analysis of holography using ray tracing, Appl. Opt. **10**, 2698–2710 (1971)

2.35 N. Lindlein, J. Schwider: Local wave fronts at diffractive elements, J. Opt. Soc. Am. A **10**, 2563–2572 (1993)

2.36 I. Ghozeil: Hartmann and other screen tests. In: *Optical Shop Testing*, ed. by D. Malacara (Wiley, New York 1978) pp. 323–349

2.37 Z. Chen, W. Yu, R. Ma, X. Deng, X. Liang: Uniform illumination of large targets using a lens array, Appl. Opt. **25**, 377 (1986)

2.38 N. Lindlein, F. Simon, J. Schwider: Simulation of micro-optical array systems with RAYTRACE, Opt. Eng. **37**, 1809–1816 (1998)

2.39 N. Lindlein: Simulation of micro-optical systems including microlens arrays, J. Opt. A: Pure Appl. Opt. **4**, S1–S9 (2002)

2.40 A. Büttner, U. D. Zeitner: Wave optical analysis of light-emitting diode beam shaping using microlens array, Opt. Eng. **41**, 2393–2401 (2002)

2.41 J. A. Kneisly: Local curvature of wavefronts in an optical system, J. Opt. Soc. Am. **54**, 229–235 (1964)

2.42 W. T. Welford: *Aberrations of Optical Systems* (Hilger, Bristol 1986)

2.43 J. L. Rayces: Exact relation between wave aberration and ray aberration, Opt. Acta **11**, 85–88 (1964)

2.44 C.-J. Kim, R. R. Shannon: *Catalog of Zernike Polynomials, Applied Optics and Optical Engineering*, Vol. X, ed. by R. R. Shannon, J. C. Wyant (Academic, San Diego 1987) pp. 193–221

2.45 F. M. Küchel, Th. Schmieder, H. J. Tiziani: Beitrag zur Verwendung von Zernike-Polynomen bei der

automatischen Interferenzstreifenauswertung, Optik **65**, 123–142 (1983)
2.46 W. N. Charman: *Optics of the Eye*, *Handbook of Optics*, Vol. I, 2nd edn., ed. by M. Bass (McGraw-Hill, New York 1995) pp. 24.3–24.54
2.47 W. S. Geisler, M. S. Banks: *Visual Performance*, *Handbook of Optics*, Vol. I, 2nd edn., ed. by M. Bass (McGraw-Hill, New York 1995) pp. 25.1–25.55
2.48 P. L. Kaufman, A. Alm: *Adler's Physiology of the Eye*, 10th edn. (Mosby, St. Louis 2002)

3. Wave Optics

The quest to understand the nature of light is centuries old and today there can be at least three answers to the single question of what light is depending on the experiment used to investigate light's nature: (i) light consists of rays that propagate, e.g., rectilinear in homogeneous media, (ii) light is an electromagnetic wave, (iii) light consists of small portions of energy, or so-called photons. The first property will be treated in the chapter about geometrical optics, which can be interpreted as a special case of wave optics for very small wavelengths. On the other hand, the interpretation as photons is unexplainable with wave optics and, above all, contradictory to wave optics. Only the theory of quantum mechanics and quantum field theory can simultaneously explain light as photons and electromagnetic waves. The field of optics treating this subject is generally called quantum optics.

In this chapter about wave optics the electromagnetic property of light is treated and the basic equations describing all relevant electromagnetic phenomena are Maxwell's equations. Starting with the Maxwell equations, the wave equation and the Helmholtz equation will be derived. Here, we will try to make a tradeoff between theoretical exactness and a practical approach. For an exact analysis see, e.g., [3.1]. After this, some basic properties of light waves like polarization, interference, and diffraction will be described. The propagation of coherent scalar waves is especially important in optics. Therefore, the section about diffraction will treat several propagation methods like the method of the angular spectrum of plane waves, which can be easily implemented on a computer, or the well-known diffraction integrals of Fresnel–Kirchhoff as well as Fresnel and Fraunhofer. In modern physics and engineering, lasers are very important and therefore the propagation of a coherent laser beam is of special interest. A good approximation for a laser beam is a Hermite–Gaussian mode and the propagation of a fundamental Gaussian beam can be performed very easily if some approximations of paraxial optics are valid. The formulae for this are treated in the last section of this chapter.

3.1 Maxwell's Equations and the Wave Equation

3.1.1 The Maxwell Equations

The well known *Maxwell* equations regarding electrodynamics [3.1] are the basis for our considerations and will be presented here in international SI units. The following physical quantities are used:

- $\boldsymbol{E}$: electric (field) vector, unit $[E] = 1\,\mathrm{V/m}$
- $\boldsymbol{D}$: electric displacement, unit $[D] = 1\,\mathrm{A\,s/m^2}$
- $\boldsymbol{H}$: magnetic (field) vector, unit $[H] = 1\,\mathrm{A/m}$
- $\boldsymbol{B}$: magnetic induction, unit $[B] = 1\,\mathrm{V\,s/m^2} = 1\,\mathrm{T}$
- $\boldsymbol{j}$: electric current density, unit $[j] = 1\,\mathrm{A/m^2}$
- ρ: electric charge density, unit $[\rho] = 1\,\mathrm{A\,s/m^3}$

All quantities can be functions of the spatial coordinates with position vector $\boldsymbol{r} = (x, y, z)$ and time t. In the following, this explicit functionality is mostly omitted in the equations if it is clear from the context.

The Maxwell equations are formulated in differential form by using the so-called nabla operator

$$\nabla := \begin{pmatrix} \frac{\partial}{\partial x} \\ \frac{\partial}{\partial y} \\ \frac{\partial}{\partial z} \end{pmatrix} . \tag{3.1}$$

The four Maxwell equations and the physical interpretation are

$$\nabla \times \boldsymbol{E}(\boldsymbol{r}, t) = -\frac{\partial \boldsymbol{B}(\boldsymbol{r}, t)}{\partial t} . \tag{3.2}$$

The vortices of the electric field $\boldsymbol{E}$ are caused by temporal variations of the magnetic induction $\boldsymbol{B}$.

$$\nabla \times \boldsymbol{H}(\boldsymbol{r}, t) = \frac{\partial \boldsymbol{D}(\boldsymbol{r}, t)}{\partial t} + \boldsymbol{j}(\boldsymbol{r}, t) . \tag{3.3}$$

The vortices of the magnetic field $\boldsymbol{H}$ are either caused by an electric current with density $\boldsymbol{j}$ or by temporal variations of the electric displacement $\boldsymbol{D}$. The quantity $\partial \boldsymbol{D}/\partial t$ is called the electric displacement current.

$$\nabla \cdot \boldsymbol{D}(\boldsymbol{r}, t) = \rho(\boldsymbol{r}, t) \tag{3.4}$$

The sources of the electric displacement $\boldsymbol{D}$ are the electric charges with density ρ.

$$\nabla \cdot \boldsymbol{B}(\boldsymbol{r}, t) = 0 \tag{3.5}$$

The magnetic field (induction) is solenoidal, i. e., there exist no "magnetic charges".

The Continuity Equation

From (3.3) and (3.4) the conservation of the electric charge can be obtained by using the mathematical identity $\nabla \cdot (\nabla \times \boldsymbol{H}) = 0$,

$$\frac{\partial \rho}{\partial t} + \nabla \cdot \boldsymbol{j} = 0 . \tag{3.6}$$

This equation is called the continuity equation of electrodynamics because it is analogous to the continuity equation of hydrodynamics. By integrating over a volume V with a closed surface A and applying Gauss' theorem the following equation is obtained:

$$\int_V \frac{\partial \rho}{\partial t}\,\mathrm{d}V = -\int_V \nabla \cdot \boldsymbol{j}\,\mathrm{d}V = -\oint_A \boldsymbol{j} \cdot \mathrm{d}\boldsymbol{A} . \tag{3.7}$$

Note that the integral $\int_V f\,\mathrm{d}V$ always indicates a volume integral of a scalar function f over the volume V, whereas the symbol $\oint_A \boldsymbol{a} \cdot \mathrm{d}\boldsymbol{A}$ always indicates a surface integral of the vector function $\boldsymbol{a}$ over the closed surface A, which borders the volume V. The vector $\mathrm{d}\boldsymbol{A}$ always points outwards from the closed surface.

Therefore, the left side of (3.7) is the temporal variation of the total electric charge Q in the volume V and the right-hand side of (3.7) is the net electric current I_{net} (i. e., the current of positive charges flowing out of the surface plus the current of negative charges flowing in the surface minus the current of positive charges flowing in the surface minus the current of negative charges flowing out of the surface), which flows through the closed surface A

$$\frac{\partial Q}{\partial t} = -I_{\mathrm{net}} . \tag{3.8}$$

If the net current I_{net} is positive the total charge in the volume decreases during time, i. e., it becomes "more negative."

Energy Conservation in Electrodynamics

From (3.2) and (3.3) a law of energy conservation of electrodynamics can be deduced by calculating the scalar product of $\boldsymbol{E}$ with (3.3) minus the scalar product of $\boldsymbol{H}$ with (3.2):

$$\begin{aligned}&\boldsymbol{E}\cdot(\nabla\times\boldsymbol{H})-\boldsymbol{H}\cdot(\nabla\times\boldsymbol{E})\\&=\boldsymbol{E}\cdot\frac{\partial\boldsymbol{D}}{\partial t}+\boldsymbol{E}\cdot\boldsymbol{j}+\boldsymbol{H}\cdot\frac{\partial\boldsymbol{B}}{\partial t}\,.\end{aligned}$$

According to the rules of calculating using a nabla operator, the following equation is obtained:

$$\begin{aligned}\nabla\cdot\boldsymbol{S}&=\nabla\cdot(\boldsymbol{E}\times\boldsymbol{H})\\&=-\left[\boldsymbol{E}\cdot(\nabla\times\boldsymbol{H})-\boldsymbol{H}\cdot(\nabla\times\boldsymbol{E})\right]\,.\end{aligned}$$

The quantity $\boldsymbol{S}$

$$\boldsymbol{S}=\boldsymbol{E}\times\boldsymbol{H} \tag{3.9}$$

is called the Poynting vector and has the physical unit of an intensity: $[S]=1\,\mathrm{V\,A/m^2}=1\,\mathrm{W/m^2}$, i. e., power per surface area. The Poynting vector arises due to the property of a cross product of two vectors perpendicular to both the electric and magnetic vector. Its absolute value describes the flow of energy per unit area and unit time through a surface perpendicular to the Poynting vector. It therefore describes the energy transport of the electromagnetic field.

The sources of $\boldsymbol{S}$ are connected with temporal variations of the electric displacement, or of the magnetic induction, or with explicit electric currents.

$$\nabla\cdot\boldsymbol{S}=-\left(\boldsymbol{E}\cdot\frac{\partial\boldsymbol{D}}{\partial t}+\boldsymbol{E}\cdot\boldsymbol{j}+\boldsymbol{H}\cdot\frac{\partial\boldsymbol{B}}{\partial t}\right)\,. \tag{3.10}$$

In the next section it will be shown for the special case of an isotropic dielectric material that this equation can be interpreted as an equation of energy conservation.

Energy Conservation in the Special Case of Isotropic Dielectric Materials

In Sect. 3.1.3 we will see that isotropic dielectric materials are described with the following equations. The charge density and the electric currents are both zero

$$\rho=0,\quad \boldsymbol{j}=0\,. \tag{3.11}$$

Additionally, there are the following linear interrelations between the electric and magnetic quantities

$$\begin{aligned}\boldsymbol{D}(\boldsymbol{r},t)&=\varepsilon_0\varepsilon(\boldsymbol{r})\,\boldsymbol{E}(\boldsymbol{r},t)\,,\\\boldsymbol{B}(\boldsymbol{r},t)&=\mu_0\mu(\boldsymbol{r})\,\boldsymbol{H}(\boldsymbol{r},t)\,,\end{aligned} \tag{3.12}$$

where ε is the dielectric function of the material and μ the magnetic permeability. Both are functions of the position $\boldsymbol{r}$. The dielectric constant of the vacuum $\varepsilon_0=8.8542\times10^{-12}\,\mathrm{A\,s\,V^{-1}\,m^{-1}}$ and the magnetic permeability of the vacuum $\mu_0=4\pi\times10^{-7}\,\mathrm{V\,s\,A^{-1}\,m^{-1}}$ are related to the speed of light in a vacuum c by:

$$c=\frac{1}{\sqrt{\varepsilon_0\mu_0}} \tag{3.13}$$

with $c=2.99792458\times10^8\,\mathrm{m/s}$. In fact, the light speed in a vacuum is defined in the SI system as a fundamental constant of nature to exactly this value so that the basic unit of length (1 m) can be connected with the basic unit of time (1 s). The magnetic permeability of the vacuum is also defined in order to connect the basic SI unit of the electric current (1 A) with the mechanical basic SI units of mass (1 kg), length (1 m) and time (1 s). So, only the dielectric constant of the vacuum has to be determined by experiment, whereas c and μ_0 are defined constants in the SI system.

In dielectrics, by using (3.11) and (3.12), (3.10) is reduced to the following equation

$$\begin{aligned}\nabla\cdot\boldsymbol{S}&=-\left(\varepsilon_0\varepsilon\boldsymbol{E}\cdot\frac{\partial\boldsymbol{E}}{\partial t}+\mu_0\mu\boldsymbol{H}\cdot\frac{\partial\boldsymbol{H}}{\partial t}\right)\\&=-\frac{1}{2}\frac{\partial}{\partial t}\left(\varepsilon_0\varepsilon\boldsymbol{E}\cdot\boldsymbol{E}+\mu_0\mu\boldsymbol{H}\cdot\boldsymbol{H}\right)\,.\end{aligned} \tag{3.14}$$

By integrating both sides of (3.14) over a volume V, which is bounded by a closed surface A (see Fig. 3.1),

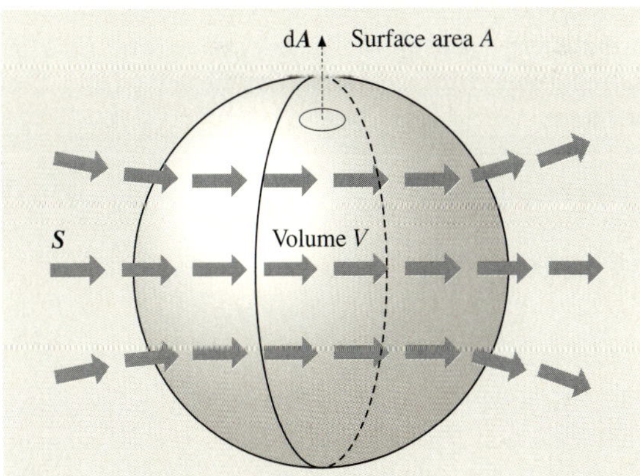

Fig. 3.1 Illustration of the quantities used for applying Gauss' theorem. The surface A need not be the surface of a sphere with volume V, but can be an arbitrary closed surface. The *small dotted* vector symbolizes the infinitesimal surface vector d$\boldsymbol{A}$, whereas the other vectors represent the vector field of the local Poynting vectors $\boldsymbol{S}$ at a fixed time

Gauss' theorem can be applied:

$$\begin{aligned} P_{\text{net}} &:= \oint_A \boldsymbol{S}\cdot \mathrm{d}\boldsymbol{A} = \int_V \nabla\cdot \boldsymbol{S}\mathrm{d}V \\ &= \int_V \left[-\frac{1}{2}\frac{\partial}{\partial t}(\varepsilon_0\varepsilon \boldsymbol{E}\cdot\boldsymbol{E}+\mu_0\mu\boldsymbol{H}\cdot\boldsymbol{H})\right]\mathrm{d}V \\ &= -\frac{\partial}{\partial t}\int_V \left(\frac{1}{2}\varepsilon_0\varepsilon \boldsymbol{E}\cdot\boldsymbol{E}+\frac{1}{2}\mu_0\mu\boldsymbol{H}\cdot\boldsymbol{H}\right)\mathrm{d}V \\ &= -\frac{\partial}{\partial t}\int_V w\,\mathrm{d}V\,. \end{aligned} \tag{3.15}$$

The integral $\int_V \nabla\cdot\boldsymbol{S}\,\mathrm{d}V$ symbolizes the volume integral of $\nabla\cdot\boldsymbol{S}$ over the volume V, whereas the integral $\oint_A \boldsymbol{S}\cdot\mathrm{d}\boldsymbol{A}$ symbolizes the surface integral of the Poynting vector over the closed surface A. The vector $\mathrm{d}\boldsymbol{A}$ in the integral always points outwards in the case of a closed surface. Therefore, P_{net} is equal to the net amount of the electromagnetic power (difference between the power flowing out of the closed surface and the power flowing in the closed surface), which flows through the closed surface. So, a positive value of P_{net} indicates that more energy flows out of the surface than into the system. Since the right-hand side of (3.15) must therefore also have the physical unit of power (unit $1\,\mathrm{W} = 1\,\mathrm{J/s}$), it is clear that the quantity

$$w := \frac{1}{2}(\varepsilon_0\varepsilon\boldsymbol{E}\cdot\boldsymbol{E}+\mu_0\mu\boldsymbol{H}\cdot\boldsymbol{H}) = w_{\text{e}}+w_{\text{m}} \tag{3.16}$$

is the energy density of the electromagnetic field in isotropic dielectric materials having the unit $1\,\mathrm{J/m^3}$. The first term,

$$w_{\text{e}} = \frac{1}{2}\varepsilon_0\varepsilon\boldsymbol{E}\cdot\boldsymbol{E} \tag{3.17}$$

is the electric energy density and the second term,

$$w_{\text{m}} = \frac{1}{2}\mu_0\mu\boldsymbol{H}\cdot\boldsymbol{H} \tag{3.18}$$

is the magnetic energy density. The negative sign on the right-hand side of (3.15) just indicates that the amount of energy in the volume decreases over time if the net amount of power P_{net} flowing through the surface is positive, because this means that in total energy flows out of the closed surface. If the net electromagnetic power flow through the surface is zero, i. e., $P_{\text{net}} = 0$, the total amount of electromagnetic energy $\int_V w\,\mathrm{d}V$ in the volume is constant. This again shows that it is useful to interpret w as an energy density.

The Wave Equation in Homogeneous Dielectrics

In this section the behavior of light in homogeneous dielectric materials will be discussed. In homogeneous materials the dielectric function ε and the magnetic permeability μ are both constants. A special case is the vacuum, where both constants are equal to 1 ($\varepsilon = 1$, $\mu = 1$). The conclusion that electromagnetic waves can also exist in a vacuum without any matter was one of the most important discoveries in physics in the 19th century.

In homogeneous dielectrics the Maxwell equations (3.2–3.5) can be simplified by using (3.11) and (3.12) with ε and μ being constant

$$\nabla\times\boldsymbol{E}(\boldsymbol{r},t) = -\mu_0\mu\frac{\partial\boldsymbol{H}(\boldsymbol{r},t)}{\partial t}\,, \tag{3.19}$$

$$\nabla\times\boldsymbol{H}(\boldsymbol{r},t) = \varepsilon_0\varepsilon\frac{\partial\boldsymbol{E}(\boldsymbol{r},t)}{\partial t}\,, \tag{3.20}$$

$$\nabla\cdot\boldsymbol{E}(\boldsymbol{r},t) = 0\,, \tag{3.21}$$

$$\nabla\cdot\boldsymbol{H}(\boldsymbol{r},t) = 0\,. \tag{3.22}$$

These equations are completely symmetrical to a simultaneous replacement of $\boldsymbol{E}$ with $\boldsymbol{H}$ and $\varepsilon_0\varepsilon$ with $-\mu_0\mu$.

In the following, the vector identity

$$\begin{aligned} \nabla\times(\nabla\times\boldsymbol{E}) &= \nabla(\nabla\cdot\boldsymbol{E}) - (\nabla\cdot\nabla)\boldsymbol{E} \\ &= \nabla(\nabla\cdot\boldsymbol{E}) - \Delta\boldsymbol{E} \end{aligned} \tag{3.23}$$

must be used. Thereby, the Laplacian operator $\Delta = \partial^2/\partial x^2 + \partial^2/\partial y^2 + \partial^2/\partial z^2$ has to be applied to each component of $\boldsymbol{E}$!

Equation (3.19) together with (3.21) then results in

$$\begin{aligned} \nabla\times(\nabla\times\boldsymbol{E}) &= -\Delta\boldsymbol{E} = -\mu_0\mu\nabla\times\frac{\partial\boldsymbol{H}}{\partial t} \\ &= -\mu_0\mu\frac{\partial}{\partial t}(\nabla\times\boldsymbol{H})\,. \end{aligned} \tag{3.24}$$

Using (3.20) the wave equation for the electric vector in a homogeneous dielectric is obtained

$$\begin{aligned} -\Delta\boldsymbol{E} &= -\mu_0\mu\frac{\partial}{\partial t}\left(\varepsilon_0\varepsilon\frac{\partial\boldsymbol{E}}{\partial t}\right) \\ &\Rightarrow \Delta\boldsymbol{E} - \varepsilon_0\mu_0\varepsilon\mu\frac{\partial^2\boldsymbol{E}}{\partial t^2} = 0\,. \end{aligned} \tag{3.25}$$

By using (3.13) this is usually written as

$$\Delta\boldsymbol{E} - \frac{n^2}{c^2}\frac{\partial^2\boldsymbol{E}}{\partial t^2} = 0\,. \tag{3.26}$$

The refractive index n of a homogeneous dielectric is defined as

$$n = \sqrt{\varepsilon\mu}\,. \tag{3.27}$$

Because of the symmetry in $\boldsymbol{E}$ and $\boldsymbol{H}$ of the Maxwell equations (3.19–3.22) in homogeneous dielectrics, the same equation also holds for the magnetic vector

$$\Delta \boldsymbol{H} - \frac{n^2}{c^2}\frac{\partial^2 \boldsymbol{H}}{\partial t^2} = 0 . \tag{3.28}$$

Plane Waves in Homogeneous Dielectrics

By defining the so-called phase velocity

$$v = \frac{c}{n} \tag{3.29}$$

a solution of (3.26) or (3.28) is

$$\boldsymbol{E}(\boldsymbol{r}, t) = \boldsymbol{f}(\boldsymbol{e}\cdot\boldsymbol{r} \mp vt) , \tag{3.30}$$

$$\boldsymbol{H}(\boldsymbol{r}, t) = \boldsymbol{g}(\boldsymbol{e}\cdot\boldsymbol{r} \mp vt) . \tag{3.31}$$

This can be seen by using

$$u := \boldsymbol{e}\cdot\boldsymbol{r} \mp vt = e_x x + e_y y + e_z z \mp vt \quad \text{with} \quad e_x^2 + e_y^2 + e_z^2 = 1 \tag{3.32}$$

so that it holds

$$\Delta \boldsymbol{E} = \left(\frac{\partial^2}{\partial x^2} + \frac{\partial^2}{\partial y^2} + \frac{\partial^2}{\partial z^2}\right)\begin{pmatrix} E_x \\ E_y \\ E_z \end{pmatrix} = \left(e_x^2 + e_y^2 + e_z^2\right)\frac{\partial^2 \boldsymbol{f}(u)}{\partial u^2} = \frac{\partial^2 \boldsymbol{f}(u)}{\partial u^2} ,$$

$$\frac{\partial^2 \boldsymbol{E}}{\partial t^2} = v^2 \frac{\partial^2 \boldsymbol{f}(u)}{\partial u^2} ,$$

and thus the same is valid for $\boldsymbol{H}$ in (3.31). The quantity nu has the physical unit of a path and the first term $n\boldsymbol{e}\cdot\boldsymbol{r}$ is called the optical path difference OPDbecause it is the product of the geometrical path and the refractive index n.

A solution of type (3.30) or (3.31) is called a plane wave because for the following reason. The value of $\boldsymbol{E}$ remains constant for a constant argument $u = u_0$ defined by (3.32). The same is valid for $\boldsymbol{H}$. Now, if we consider e.g., $u = u_0 = 0$ and take the negative sign in (3.32) we obtain:

$$u = 0 \quad \Rightarrow \quad \boldsymbol{e}\cdot\boldsymbol{r} = vt . \tag{3.33}$$

The geometrical path at time $t = 0$ must then also be zero and this means that the plane then passes through the origin of the coordinate system. For a fixed value of t_0, (3.33) describes a plane surface in space (see Fig. 3.2) at a distance vt_0 from the origin. At the time $t_0 + \Delta t$ it again describes a plane surface parallel to the plane at $t = 0$ but with a distance $v(t_0 + \Delta t)$ from the origin. The

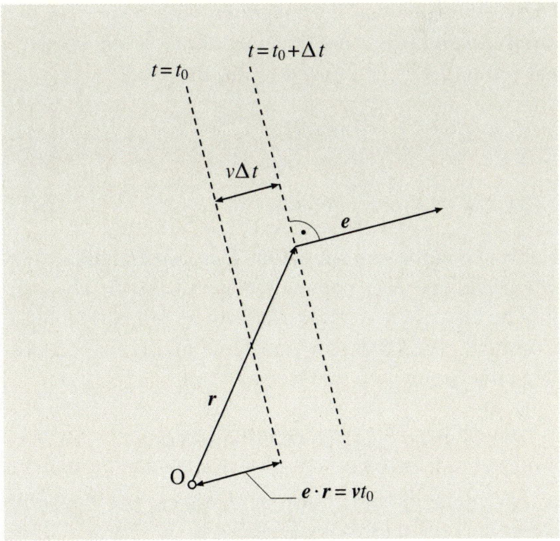

Fig. 3.2 The plane surfaces of constant optical path in the case of a plane wave. O is the origin of the coordinate system and the *dashed lines* indicate the planes at two different times for the fixed value $u = 0$

unit vector $\boldsymbol{e}$ is perpendicular to the planes and points in the direction of propagation if the negative sign is used in (3.32) (as done here and in the following) and points in the opposite direction if the positive sign is used.

The Orthogonality Condition for Plane Waves in Homogeneous Dielectrics

The Maxwell equations (3.19–3.22) do not allow all orientations of the electric and magnetic vector relative to the propagation direction $\boldsymbol{e}$ of a plane wave. Equations (3.30) and (3.32) deliver

$$\frac{\partial \boldsymbol{E}}{\partial t} = \mp v \frac{\partial \boldsymbol{f}}{\partial u}$$

and

$$(\nabla \times \boldsymbol{E})_x = \frac{\partial E_z}{\partial y} - \frac{\partial E_y}{\partial z} = \frac{\partial f_z}{\partial u}e_y - \frac{\partial f_y}{\partial u}e_z = \left(\boldsymbol{e}\times\frac{\partial \boldsymbol{f}}{\partial u}\right)_x \quad \Rightarrow \quad \nabla\times\boldsymbol{E} = \boldsymbol{e}\times\frac{\partial \boldsymbol{f}}{\partial u} .$$

An analogous expression is valid for $\boldsymbol{H}$.

Therefore, the Maxwell equations (3.19) and (3.20) deliver

$$\boldsymbol{e}\times\frac{\partial \boldsymbol{f}}{\partial u} = \pm\mu_0\mu v \frac{\partial \boldsymbol{g}}{\partial u} , \tag{3.34}$$

$$\boldsymbol{e}\times\frac{\partial \boldsymbol{g}}{\partial u} = \mp\varepsilon_0\varepsilon v \frac{\partial \boldsymbol{f}}{\partial u} . \tag{3.35}$$

These equations can be integrated with respect to the variable u and by setting the integration constant to zero and using (3.13), (3.27) and (3.29) the result is

$$E = f = \mp\sqrt{\frac{\mu_0\mu}{\varepsilon_0\varepsilon}}\, e \times H\,, \tag{3.36}$$

$$H = g = \pm\sqrt{\frac{\varepsilon_0\varepsilon}{\mu_0\mu}}\, e \times E\,. \tag{3.37}$$

These two equations show that E is perpendicular to e and H and that H is perpendicular to e and E. This can only be the case if e, E and H form an orthogonal triad of vectors. Therefore, a plane wave in a homogeneous dielectric is always a transversal wave.

The Poynting Vector of a Plane Wave

In this section, the physical interpretation of the Poynting vector will be illustrated for plane waves. The Poynting vector defined in (3.9) is parallel to e

$$S = E \times H = \left(\mp\sqrt{\frac{\mu_0\mu}{\varepsilon_0\varepsilon}}\, e \times H\right) \times \left(\pm\sqrt{\frac{\varepsilon_0\varepsilon}{\mu_0\mu}}\, e \times E\right)$$
$$= -\,[(e \times H)\cdot E]\, e + [(e \times H)\cdot e]\, E\,. \tag{3.38}$$

The second scalar product is zero so that only the first term remains. By using (3.36) this finally results in

$$S = -\,[(e \times H)\cdot E]\, e = \pm\sqrt{\frac{\varepsilon_0\varepsilon}{\mu_0\mu}}\,(E \cdot E)\, e\,. \tag{3.39}$$

This means that the energy transport is along the direction of propagation of the plane wave and that the absolute value of the Poynting vector, i.e., the intensity of the light wave, is proportional to $|E|^2$.

By using the vector identity $(a \times b)\cdot c = -\,(a \times c)\cdot b$ and (3.37) we can also write

$$S = -\,[(e \times H)\cdot E]\, e = [(e \times E)\cdot H]\, e$$
$$= \pm\sqrt{\frac{\mu_0\mu}{\varepsilon_0\varepsilon}}\,(H \cdot H)\, e\,. \tag{3.40}$$

This means that the absolute value of the Poynting vector is also proportional to $|H|^2$ and that the following equality holds:

$$\sqrt{\frac{\mu_0\mu}{\varepsilon_0\varepsilon}}\,|H|^2 = \sqrt{\frac{\varepsilon_0\varepsilon}{\mu_0\mu}}\,|E|^2$$
$$\Rightarrow \quad \mu_0\mu\,|H|^2 = \varepsilon_0\varepsilon\,|E|^2\,. \tag{3.41}$$

Comparing this with (3.16), (3.17) and (3.18) for the energy density of the electromagnetic field we have for a plane wave in a homogeneous dielectric

$$w_e = w_m = \frac{1}{2}w$$
$$\Rightarrow \quad w = \varepsilon_0\varepsilon\,|E|^2 = \mu_0\mu\,|H|^2\,. \tag{3.42}$$

By again using (3.13), (3.27) and (3.29) the equations (3.39) and (3.40) can be transformed to

$$S = \pm v\mu_0\mu\,|H|^2\, e = \pm v\varepsilon_0\varepsilon\,|E|^2\, e = \pm v w e\,. \tag{3.43}$$

This means that the absolute value of the Poynting vector is, in a homogeneous dielectric, the product of the energy density (energy per volume) of the electromagnetic field and the phase velocity of light. This confirms the interpretation of the Poynting vector as being the vector of the electromagnetic wave transporting the energy of the electromagnetic field with the phase velocity of light. Figure 3.3 illustrates this. In the infinitesimal time interval dt the light covers the infinitesimal distance $dz = v\,dt$. We assume that the distance dz is so small that the local energy density w of the electromagnetic field is constant in the volume $dV = A\,dz$, whereby A is the area of a small surface perpendicular to the Poynting vector. Therefore, all the energy $dW = w\,dV$ that is contained in the infinitesimal volume dV passes the surface area A in the time dt and we have, for the intensity I (electromagnetic power per area),

$$I = \frac{dW}{A\,dt} = \frac{w\,dV}{A\,dt} = \frac{wA\,v\,dt}{A\,dt} = wv = |S|\,. \tag{3.44}$$

This is exactly the absolute value of the Poynting vector S. The light intensity on a surface, which is not perpendicular to the direction of the Poynting vector is calculated by the equation

$$I = S \cdot N\,, \tag{3.45}$$

where N is a local unit vector perpendicular to the surface.

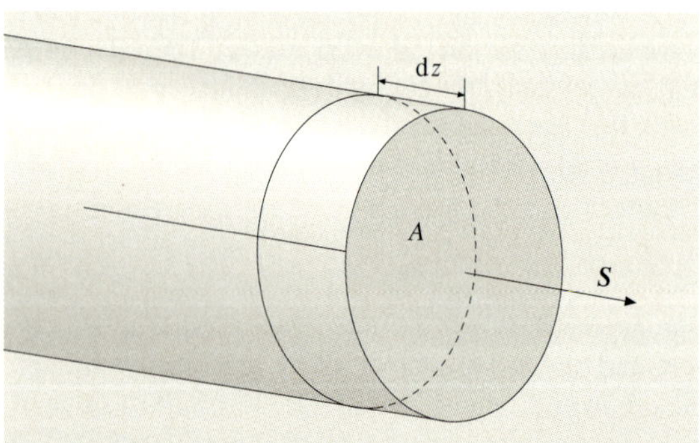

Fig. 3.3 Illustration of the Poynting vector S as transporting the energy of the electromagnetic field. A is the area of the circular surface

A Time-Harmonic Plane Wave

Up to now a plane wave was defined with (3.30) and (3.31) to be $\boldsymbol{E}(\boldsymbol{r},t)=\boldsymbol{f}(u)$ and $\boldsymbol{H}(\boldsymbol{r},t)=\boldsymbol{g}(u)$. The argument u is defined by (3.32) to be $u=\boldsymbol{e}\cdot\boldsymbol{r}-vt$. This means that all points with position vector $\boldsymbol{r}$ at a fixed point t in time lie on a plane surface for a constant value u. Additionally, we saw that $\boldsymbol{e}$, $\boldsymbol{E}$ and $\boldsymbol{H}$ have to form an orthogonal triad of vectors (3.36, 3.37). But the concrete form of the functions $\boldsymbol{f}$ and $\boldsymbol{g}$ can be quite arbitrary to fulfill these conditions. A wave which is very important in optics because of its simple form is a time-harmonic wave. Furthermore, it should be linearly polarized, i. e., the direction of the electric and magnetic vector are both constant. A linearly polarized time-harmonic plane wave is represented by the equations:

$$\begin{aligned}\boldsymbol{E}(\boldsymbol{r},t)&=\boldsymbol{E}_0\cos\left(\frac{2\pi n}{\lambda}u\right)\\&=\boldsymbol{E}_0\cos\left[\frac{2\pi n}{\lambda}(\boldsymbol{e}\cdot\boldsymbol{r}-vt)\right],\end{aligned}\tag{3.46}$$

$$\begin{aligned}\boldsymbol{H}(\boldsymbol{r},t)&=\boldsymbol{H}_0\cos\left(\frac{2\pi n}{\lambda}u\right)\\&=\boldsymbol{H}_0\cos\left[\frac{2\pi n}{\lambda}(\boldsymbol{e}\cdot\boldsymbol{r}-vt)\right].\end{aligned}\tag{3.47}$$

Here, we have introduced the value λ, which has the physical unit of a length so that the argument of the cosine function has no physical unit. Its meaning will soon become clear.

The characteristic property of a time harmonic wave is that it has, for a fixed point $\boldsymbol{r}$, periodically the same value after a certain time interval. The smallest time interval for which this is the case is called period T:

$$\begin{aligned}\boldsymbol{E}(\boldsymbol{r},t+T)&=\boldsymbol{E}_0\cos\left\{\frac{2\pi n}{\lambda}[\boldsymbol{e}\cdot\boldsymbol{r}-v(t+T)]\right\}\\&=\boldsymbol{E}_0\cos\left[\frac{2\pi n}{\lambda}(\boldsymbol{e}\cdot\boldsymbol{r}-vt)\right]=\boldsymbol{E}(\boldsymbol{r},t)\\\Rightarrow\quad&\frac{2\pi n}{\lambda}vT=2\pi\\\Rightarrow\quad&vT=\frac{\lambda}{n}\quad\text{or}\quad cT=\lambda\,.\end{aligned}\tag{3.48}$$

Therefore, λ/n is the distance the light covers in the material in the period T and is called the wavelength of the harmonic wave in the material. The quantity λ itself is the wavelength in a vacuum. The reciprocal of T is called the frequency ν of the wave and the term $2\pi\nu=2\pi/T$ is called the angular frequency ω of the wave. Therefore the two following equations are valid

$$cT=\lambda\quad\Rightarrow\quad c=\lambda\nu\,,\tag{3.49}$$

$$\frac{2\pi}{\lambda}c=\frac{2\pi}{T}=2\pi\nu=\omega\,.\tag{3.50}$$

Additionally, we introduce the wave vector $\boldsymbol{k}$, which is defined by

$$\boldsymbol{k}=\frac{2\pi n}{\lambda}\boldsymbol{e}\,.\tag{3.51}$$

Then (3.46) and (3.47) for $\boldsymbol{E}$ and $\boldsymbol{H}$ can be written as

$$\boldsymbol{E}(\boldsymbol{r},t)=\boldsymbol{E}_0\cos(\boldsymbol{k}\cdot\boldsymbol{r}-\omega t)\,,\tag{3.52}$$

$$\boldsymbol{H}(\boldsymbol{r},t)=\boldsymbol{H}_0\cos(\boldsymbol{k}\cdot\boldsymbol{r}-\omega t)\,.\tag{3.53}$$

Because of the orthogonality condition $\boldsymbol{k}$ (or $\boldsymbol{e}$, which is parallel to $\boldsymbol{k}$), $\boldsymbol{E}_0$ and $\boldsymbol{H}_0$ have to form an orthogonal triad. This can be explicitly seen in this case by using Maxwell's first equation (3.19) in a homogeneous dielectric and the mathematical rules for the nabla operator

$$\begin{aligned}\nabla\times\boldsymbol{E}&=\nabla\times[\boldsymbol{E}_0\cos(\boldsymbol{k}\cdot\boldsymbol{r}-\omega t)]\\&=[\nabla\cos(\boldsymbol{k}\cdot\boldsymbol{r}-\omega t)]\times\boldsymbol{E}_0\\&=-\boldsymbol{k}\times\boldsymbol{E}_0\sin(\boldsymbol{k}\cdot\boldsymbol{r}-\omega t)\,,\\-\mu_0\mu\frac{\partial\boldsymbol{H}}{\partial t}&=-\omega\mu_0\mu\boldsymbol{H}_0\sin(\boldsymbol{k}\cdot\boldsymbol{r}-\omega t)\,;\\\Rightarrow\quad&\omega\mu_0\mu\boldsymbol{H}_0=\boldsymbol{k}\times\boldsymbol{E}_0\\\Rightarrow\quad&\boldsymbol{H}_0=\frac{1}{\omega\mu_0\mu}\boldsymbol{k}\times\boldsymbol{E}_0=\frac{\lambda}{2\pi c\mu_0\mu}\boldsymbol{k}\times\boldsymbol{E}_0\\&=\sqrt{\frac{\varepsilon_0\varepsilon}{\mu_0\mu}}\boldsymbol{e}\times\boldsymbol{E}_0\,.\end{aligned}\tag{3.54}$$

In the last step (3.50), (3.13) and (3.27) are used and the geometrical interpretation of the result is that $\boldsymbol{H}_0$ is perpendicular to both $\boldsymbol{e}$ and $\boldsymbol{E}_0$. The third Maxwell equation (3.21) delivers

$$\begin{aligned}\nabla\cdot\boldsymbol{E}&=\nabla\cdot[\boldsymbol{E}_0\cos(\boldsymbol{k}\cdot\boldsymbol{r}-\omega t)]\\&=\boldsymbol{E}_0\cdot[\nabla\cos(\boldsymbol{k}\cdot\boldsymbol{r}-\omega t)]\\&=-\boldsymbol{E}_0\cdot\boldsymbol{k}\sin(\boldsymbol{k}\cdot\boldsymbol{r}-\omega t)=0\,;\\\Rightarrow\quad&\boldsymbol{E}_0\cdot\boldsymbol{k}=0\,.\end{aligned}\tag{3.55}$$

This also means that $\boldsymbol{k}$ (or $\boldsymbol{e}$) and $\boldsymbol{E}_0$ are perpendicular to each other. The two other Maxwell equations (3.20) and (3.22) are automatically fulfilled because of the symmetry in $\boldsymbol{E}$ and $\boldsymbol{H}$.

3.1.2 The Complex Representation of Time-Harmonic Waves

In paragraph "A Time harmonic Plane Wave" of Sect. 3.1.1 a linearly polarized time-harmonic plane wave is expressed with real cosine functions for the electric and the magnetic vector. Because $\boldsymbol{E}$ and $\boldsymbol{H}$ are observable physical quantities, they must, of course, be expressed by real functions. The fact that usual detectors are not fast enough to detect the electric and magnetic vector of light waves directly does not matter here. Nevertheless, the calculation with a complex exponential function is more convenient than the calculation with real cosine or sine functions. Now, the Maxwell equations (3.2–3.5) are linear. Therefore, if the functions $\boldsymbol{E}_1$, $\boldsymbol{D}_1$, $\boldsymbol{H}_1$, $\boldsymbol{B}_1$, $\boldsymbol{j}_1$ and ρ_1 on the one hand and $\boldsymbol{E}_2$, $\boldsymbol{D}_2$, $\boldsymbol{H}_2$, $\boldsymbol{B}_2$, $\boldsymbol{j}_2$ and ρ_2 on the other are both solutions of the Maxwell equations, then a linear combination of these functions is also a solution:

$$
\begin{aligned}
&\nabla\times\boldsymbol{E}_1 = -\frac{\partial \boldsymbol{B}_1}{\partial t}\,,\\
&\nabla\times\boldsymbol{H}_1 = \frac{\partial \boldsymbol{D}_1}{\partial t}+\boldsymbol{j}_1\,,\\
&\nabla\cdot\boldsymbol{D}_1 = \rho_1\,,\\
&\nabla\cdot\boldsymbol{B}_1 = 0\,,\\
&\nabla\times\boldsymbol{E}_2 = -\frac{\partial \boldsymbol{B}_2}{\partial t}\,,\\
&\nabla\times\boldsymbol{H}_2 = \frac{\partial \boldsymbol{D}_2}{\partial t}+\boldsymbol{j}_2\,,\\
&\nabla\cdot\boldsymbol{D}_2 = \rho_2\,,\\
&\nabla\cdot\boldsymbol{B}_2 = 0\,;\\
&\Rightarrow
\end{aligned}
$$

$$\nabla\times(\alpha\boldsymbol{E}_1+\beta\boldsymbol{E}_2) = -\frac{\partial(\alpha\boldsymbol{B}_1+\beta\boldsymbol{B}_2)}{\partial t}\,, \quad (3.56)$$

$$\nabla\times(\alpha\boldsymbol{H}_1+\beta\boldsymbol{H}_2) = \frac{\partial(\alpha\boldsymbol{D}_1+\beta\boldsymbol{D}_2)}{\partial t}+\alpha\boldsymbol{j}_1+\beta\boldsymbol{j}_2\,, \quad (3.57)$$

$$\nabla\cdot(\alpha\boldsymbol{D}_1+\beta\boldsymbol{D}_2) = \alpha\rho_1+\beta\rho_2\,, \quad (3.58)$$

$$\nabla\cdot(\alpha\boldsymbol{B}_1+\beta\boldsymbol{B}_2) = 0\,, \quad (3.59)$$

where α and β are arbitrary real or complex constants. The Euler equation delivers:

$$\mathrm{e}^{\mathrm{i}x} = \cos x + \mathrm{i}\sin x \quad (3.60)$$

or

$$\cos x = \frac{\mathrm{e}^{\mathrm{i}x}+\mathrm{e}^{-\mathrm{i}x}}{2} = \frac{\mathrm{e}^{\mathrm{i}x}+\left(\mathrm{e}^{\mathrm{i}x}\right)^*}{2}\,. \quad (3.61)$$

Due to the linearity of the Maxwell equations it is obvious that if a function containing a cosine function is a solution of the Maxwell equations the replacement of the cosine function by a complex exponential function will then also be a solution. Therefore, it is quite normal that waves are expressed by using a complex function although only the real part of this function represents the real physical quantity. The addition, subtraction, integration and differentiation of such a complex function is also a linear operation, so that we can in the end build the real part and have the real solution:

$$
\begin{aligned}
&z_1(x) = a_1(x)+\mathrm{i}b_1(x)\\
&z_2(x) = a_2(x)+\mathrm{i}b_2(x)\\
&\qquad \mathrm{Re}\,(z_1+z_2) = \mathrm{Re}\,(z_1)+\mathrm{Re}\,(z_2)\\
&\qquad \mathrm{Re}\,(z_1-z_2) = \mathrm{Re}\,(z_1)-\mathrm{Re}\,(z_2)\\
&\Rightarrow\quad \mathrm{Re}\left(\frac{\mathrm{d}z_1}{\mathrm{d}x}\right) = \frac{\mathrm{d}}{\mathrm{d}x}\mathrm{Re}\,(z_1)\\
&\qquad \mathrm{Re}\left(\textstyle\int z_1\,\mathrm{d}x\right) = \textstyle\int \mathrm{Re}\,(z_1)\,\mathrm{d}x\\
&\qquad \mathrm{Re}\,(fz_1) = f\mathrm{Re}\,(z_1)
\end{aligned}
\quad (3.62)
$$

Here, f is an arbitrary real function or constant. Only if two complex functions have to be multiplied or divided, or the absolute value has to be built, do we have to be careful:

$$
\begin{aligned}
&\mathrm{Re}\,(z_1z_2) = a_1a_2-b_1b_2 \neq a_1a_2 = \mathrm{Re}\,(z_1)\,\mathrm{Re}\,(z_2)\,,\\
&\mathrm{Re}\left(\frac{z_1}{z_2}\right) = \frac{a_1a_2+b_1b_2}{a_2^2+b_2^2} \neq \frac{a_1}{a_2} = \frac{\mathrm{Re}\,(z_1)}{\mathrm{Re}\,(z_2)}\,,\\
&\mathrm{Re}\,(z_1z_1) = a_1^2-b_1^2 \neq a_1^2+b_1^2 = |z_1|^2\,.
\end{aligned}
\quad (3.63)
$$

So, if the Poynting vector or products of the electric or magnetic vectors have to be calculated it is not allowed to just take the complex functions. Nevertheless, there are some useful applications of the complex notation. As we mentioned before, the frequency of a light wave is so high that no usual detector can directly measure the vibrations. For a typical wavelength of visible light of 500 nm the frequency of a wave in vacuum is determined according to (3.49)

$$\nu = \frac{c}{\lambda} = \frac{2.998\times10^8\,\mathrm{m\,s^{-1}}}{5.0\times10^{-7}\,\mathrm{m}} = 5.996\times10^{14}\,\mathrm{s^{-1}}\,,$$

$$\Rightarrow\quad T = \frac{1}{\nu} = 1.668\times10^{-15}\,\mathrm{s} = 1.668\,\mathrm{fs}\,.$$

So, the period is just a little bit more than a femtosecond. This means that in most cases only the time average of the light intensity over many periods is measured.

A general time-harmonic wave with the angular frequency ω has the representation

$$\boldsymbol{E}(\boldsymbol{r},t)=\begin{pmatrix}A_x(\boldsymbol{r})\cos[\Phi_x(\boldsymbol{r})-\omega t]\\A_y(\boldsymbol{r})\cos[\Phi_y(\boldsymbol{r})-\omega t]\\A_z(\boldsymbol{r})\cos[\Phi_z(\boldsymbol{r})-\omega t]\end{pmatrix}$$

$$=\mathrm{Re}\left\{\begin{pmatrix}A_x(\boldsymbol{r})\,\mathrm{e}^{\mathrm{i}\Phi_x(\boldsymbol{r})-\mathrm{i}\omega t}\\A_y(\boldsymbol{r})\,\mathrm{e}^{\mathrm{i}\Phi_y(\boldsymbol{r})-\mathrm{i}\omega t}\\A_z(\boldsymbol{r})\,\mathrm{e}^{\mathrm{i}\Phi_z(\boldsymbol{r})-\mathrm{i}\omega t}\end{pmatrix}\right\}$$

$$=\mathrm{Re}\left\{\mathrm{e}^{-\mathrm{i}\omega t}\begin{pmatrix}A_x(\boldsymbol{r})\,\mathrm{e}^{\mathrm{i}\Phi_x(\boldsymbol{r})}\\A_y(\boldsymbol{r})\,\mathrm{e}^{\mathrm{i}\Phi_y(\boldsymbol{r})}\\A_z(\boldsymbol{r})\,\mathrm{e}^{\mathrm{i}\Phi_z(\boldsymbol{r})}\end{pmatrix}\right\}$$

$$=:\mathrm{Re}\left\{\mathrm{e}^{-\mathrm{i}\omega t}\hat{\boldsymbol{E}}(\boldsymbol{r})\right\}. \qquad (3.64)$$

A_x, A_y, A_z, Φ_x, Φ_y and Φ_z are all real functions that depend only on the position $\boldsymbol{r}$. Additionally, A_x, A_y and A_z, which are called the components of the amplitude, are slowly varying functions of the position. On the other hand, the exponential terms with the components of the phase Φ_x, Φ_y and Φ_z are rapidly varying functions of the position. The components of the complex vector $\hat{\boldsymbol{E}}(\boldsymbol{r})$ are often called the complex amplitudes of the electric vector of the wave.

Equation (3.39) gives the relation between the Poynting vector and the electric vector in a homogeneous dielectric

$$\boldsymbol{S}=\sqrt{\frac{\varepsilon_0\varepsilon}{\mu_0\mu}}(\boldsymbol{E}\cdot\boldsymbol{E})\,\boldsymbol{e}.$$

Now, the time average $\bar{S}$ of the absolute value of the Poynting vector, i. e., the intensity, which is really measured by a common detector, will be calculated for the general time-harmonic wave. Therefore, we have to integrate the absolute value S of the Poynting vector over one period T and divide it by T

$$\bar{S}(\boldsymbol{r}):=\frac{1}{T}\int_0^T|\boldsymbol{S}(\boldsymbol{r},t)|\,\mathrm{d}t$$

$$=\sqrt{\frac{\varepsilon_0\varepsilon}{\mu_0\mu}}\frac{1}{T}\int_0^T\boldsymbol{E}(\boldsymbol{r},t)\cdot\boldsymbol{E}(\boldsymbol{r},t)\,\mathrm{d}t. \qquad (3.65)$$

Using (3.64) for a general time-harmonic wave and (3.13) we obtain

$$\bar{S}=\frac{\varepsilon_0 c}{T}\sqrt{\frac{\varepsilon}{\mu}}\int_0^T\left[A_x^2\cos^2(\Phi_x-\omega t)+A_y^2\cos^2(\Phi_y-\omega t)+A_z^2\cos^2(\Phi_z-\omega t)\right]\mathrm{d}t$$

$$=\sqrt{\frac{\varepsilon}{\mu}}\frac{\varepsilon_0 c}{2}\left(A_x^2+A_y^2+A_z^2\right). \qquad (3.66)$$

But, if we calculate directly the square of the absolute value of the time-independent vector $\hat{\boldsymbol{E}}$ we also obtain

$$\left|\hat{\boldsymbol{E}}\right|^2=\hat{\boldsymbol{E}}\cdot\hat{\boldsymbol{E}}^*=A_x^2+A_y^2+A_z^2. \qquad (3.67)$$

By combining equations (3.66) and (3.67) we finally obtain

$$\bar{S}(\boldsymbol{r})=\sqrt{\frac{\varepsilon}{\mu}}\frac{\varepsilon_0 c}{2}\left|\hat{\boldsymbol{E}}(\boldsymbol{r})\right|^2=\sqrt{\frac{\varepsilon}{\mu}}\frac{\varepsilon_0 c}{2}\hat{\boldsymbol{E}}(\boldsymbol{r})\cdot\left[\hat{\boldsymbol{E}}(\boldsymbol{r})\right]^*. \qquad (3.68)$$

Therefore, the complex representation of time-harmonic waves allows a fast calculation of the time average of the Poynting vector, i. e., of the intensity of the light wave.

3.1.3 Material Equations

In the last two sections we often concentrated on the electromagnetic field in an isotropic and homogeneous dielectric material where the Maxwell equations are simplified to (3.19–3.22). In other materials the general Maxwell equations (3.2–3.5) have to be used and more complex interrelations between the electric displacement and the electric vector on one hand and the magnetic induction and the magnetic vector on the other have to be found. Since the atomic distances are small compared to the wavelength of light, a macroscopic description with smooth functions is possible. To calculate the influence of the material, first of all the interrelations between $\boldsymbol{D}$ and $\boldsymbol{E}$ as well as $\boldsymbol{B}$ and $\boldsymbol{H}$ are considered in vacuum. These equations in a vacuum are obtained from (3.12) for the case $\mu=\varepsilon=1$. Then, additional terms are added to the equations in a vacuum. The electric polarization $\boldsymbol{P}$ and the magnetization $\boldsymbol{M}$ are introduced by:

$$\boldsymbol{D}(\boldsymbol{r},t)=\varepsilon_0\boldsymbol{E}(\boldsymbol{r},t)+\boldsymbol{P}(\boldsymbol{r},t), \qquad (3.69)$$

$$\boldsymbol{B}(\boldsymbol{r},t)=\mu_0\boldsymbol{H}(\boldsymbol{r},t)+\boldsymbol{M}(\boldsymbol{r},t). \qquad (3.70)$$

The atomic theory goes far beyond our scope. But in a macroscopic theory the effect of the atoms (i. e., mainly

where $\varepsilon(\boldsymbol{r})$, $\mu(\boldsymbol{r})$ means that the material functions will in general depend on the position. An explicit dependence on time is most often not the case so it is omitted here.

Additionally, we assume that the charge density is zero and

(3.73) is valid

$$\rho = 0\,, \tag{3.82}$$

$$\boldsymbol{j}(\boldsymbol{r},t) = \sigma(\boldsymbol{r})\,\boldsymbol{E}(\boldsymbol{r},t)\ . \tag{3.83}$$

If ε, μ and σ are constant, i.e., independent of the position, the material is called homogeneous.

Due to the dispersion theory, which will not be treated here, the material functions will, in general, depend on the frequency of the stimulating electric or magnetic fields. Therefore, the Fourier components of the electric and magnetic field have to be calculated and treated separately. The electric vector and the electric displacement are written as Fourier integrals, i.e., as a superposition of time-harmonic waves with the angular frequency ω:

$$\boldsymbol{E}(\boldsymbol{r},t) = \frac{1}{\sqrt{2\pi}}\int_{-\infty}^{+\infty} \tilde{\boldsymbol{E}}(\boldsymbol{r},\omega)\,\mathrm{e}^{-\mathrm{i}\omega t}\,\mathrm{d}\omega\,,$$

$$\boldsymbol{D}(\boldsymbol{r},t) = \frac{1}{\sqrt{2\pi}}\int_{-\infty}^{+\infty} \tilde{\boldsymbol{D}}(\boldsymbol{r},\omega)\,\mathrm{e}^{-\mathrm{i}\omega t}\,\mathrm{d}\omega\,. \tag{3.84}$$

The magnetic vector and the magnetic induction are treated in the same way so that we can omit this. If the function $\boldsymbol{E}$ is given $\tilde{\boldsymbol{E}}$ is calculated by

$$\tilde{\boldsymbol{E}}(\boldsymbol{r},\omega) = \frac{1}{\sqrt{2\pi}}\int_{-\infty}^{+\infty} \boldsymbol{E}(\boldsymbol{r},t)\,\mathrm{e}^{\mathrm{i}\omega t}\,\mathrm{d}t\,. \tag{3.85}$$

Since $\boldsymbol{E}$ is a real function the complex Fourier components have to fulfill the condition

$$\tilde{\boldsymbol{E}}(\boldsymbol{r},-\omega) = \tilde{\boldsymbol{E}}^*(\boldsymbol{r},\omega)\ . \tag{3.86}$$

The same is valid for the electric displacement, the magnetic vector and the magnetic induction.

In total, the material equations can be written for isotropic and linear materials with the Fourier components of the four electromagnetic vector quantities:

$$\tilde{\boldsymbol{D}}(\boldsymbol{r},\omega) = \varepsilon_0\varepsilon(\boldsymbol{r},\omega)\,\tilde{\boldsymbol{E}}(\boldsymbol{r},\omega)$$

$$\tilde{\boldsymbol{B}}(\boldsymbol{r},\omega) = \mu_0\mu(\boldsymbol{r},\omega)\,\tilde{\boldsymbol{H}}(\boldsymbol{r},\omega)\ . \tag{3.87}$$

In the following the tilde on the different quantities will mostly be omitted to simplify the notation. This is equivalent to just deal with time-harmonic waves of a certain angular frequency ω, where the quantities ε and μ are functions of ω.

3.1.4 The Wave Equations

The Maxwell equations (3.2–3.5) contain the five vector fields $\boldsymbol{E}$, $\boldsymbol{D}$, $\boldsymbol{H}$, $\boldsymbol{B}$ and $\boldsymbol{j}$ and the scalar field ρ. These quantities are related to each other by the material equations. Here, only the case of isotropic, linear and uncharged ($\rho = 0$) materials will be treated. Additionally, all material parameters like ε, μ and σ will be independent of the time t, but functions of the position $\boldsymbol{r}$ (and the frequency or wavelength of the light). In the following the explicit dependence of the functions on $\boldsymbol{r}$ and t will be omitted, but there are the following functionalities: $\boldsymbol{E}(\boldsymbol{r},t)$, $\boldsymbol{H}(\boldsymbol{r},t)$, $\mu(\boldsymbol{r})$, $\varepsilon(\boldsymbol{r})$, $\sigma(\boldsymbol{r})$.

Using (3.81) and (3.83) for this case results in the following specialized Maxwell equations

$$\nabla\times\boldsymbol{E} = -\mu_0\mu\frac{\partial\boldsymbol{H}}{\partial t}\,, \tag{3.88}$$

$$\nabla\times\boldsymbol{H} = \varepsilon_0\varepsilon\frac{\partial\boldsymbol{E}}{\partial t} + \sigma\boldsymbol{E}\,, \tag{3.89}$$

$$\nabla\cdot(\varepsilon\boldsymbol{E}) = 0\,, \tag{3.90}$$

$$\nabla\cdot(\mu\boldsymbol{H}) = 0\,. \tag{3.91}$$

These equations contain, for given material functions ε, μ and σ, only the electric and the magnetic vector. To eliminate the magnetic vector the cross product of the nabla operator with (3.88) is taken

$$\begin{aligned}\nabla\times(\nabla\times\boldsymbol{E}) &= -\mu_0\nabla\times\left(\mu\frac{\partial\boldsymbol{H}}{\partial t}\right)\\ &= -\mu_0\mu\frac{\partial}{\partial t}(\nabla\times\boldsymbol{H}) - \mu_0(\nabla\mu)\times\frac{\partial\boldsymbol{H}}{\partial t}\,.\end{aligned} \tag{3.92}$$

By using (3.88), (3.89) and the nabla operator calculus for a double cross product, this results in

$$\begin{aligned}\nabla(\nabla\cdot\boldsymbol{E}) - \Delta\boldsymbol{E} = &-\varepsilon_0\mu_0\varepsilon\mu\frac{\partial^2\boldsymbol{E}}{\partial t^2} - \mu_0\mu\sigma\frac{\partial\boldsymbol{E}}{\partial t}\\ &+[\nabla(\ln\mu)]\times(\nabla\times\boldsymbol{E})\ .\end{aligned} \tag{3.93}$$

Here, $\Delta = \nabla\cdot\nabla$ is the Laplacian operator, which has to be applied on each component of $\boldsymbol{E}$. Equation (3.90) can be used to eliminate the term $\nabla\cdot\boldsymbol{E}$ from (3.93)

$$\nabla\cdot(\varepsilon\boldsymbol{E}) = \boldsymbol{E}\cdot\nabla\varepsilon + \varepsilon\nabla\cdot\boldsymbol{E} = 0\,;$$

$$\Rightarrow\quad \nabla\cdot\boldsymbol{E} = -\frac{1}{\varepsilon}\boldsymbol{E}\cdot\nabla\varepsilon - -\boldsymbol{E}\cdot\nabla(\ln\varepsilon)\ .$$

So, (3.93) gives the final so-called wave equation for the electric vector $\boldsymbol{E}$

$$\Delta \boldsymbol{E}+\nabla\left[\boldsymbol{E}\cdot\nabla(\ln\varepsilon)\right]-\frac{\varepsilon\mu}{c^2}\frac{\partial^2\boldsymbol{E}}{\partial t^2}-\mu_0\mu\sigma\frac{\partial\boldsymbol{E}}{\partial t}+\left[\nabla(\ln\mu)\right]\times(\nabla\times\boldsymbol{E})=0\,. \tag{3.94}$$

Additionally, (3.13) was used to replace $\varepsilon_0\mu_0$ by $1/c^2$.

An analogue equation for the magnetic vector can be found using (3.88), (3.89) and (3.91):

$$\begin{aligned}\nabla\times(\nabla\times\boldsymbol{H}) &= \nabla\underbrace{(\nabla\cdot\boldsymbol{H})}_{=-\boldsymbol{H}\cdot\nabla(\ln\mu)}-\Delta\boldsymbol{H}\\ &= -\nabla\left[\boldsymbol{H}\cdot\nabla(\ln\mu)\right]-\Delta\boldsymbol{H}\\ &= \varepsilon_0\nabla\times\left(\varepsilon\frac{\partial\boldsymbol{E}}{\partial t}\right)+\nabla\times(\sigma\boldsymbol{E})\\ &= \varepsilon_0\varepsilon\nabla\times\frac{\partial\boldsymbol{E}}{\partial t}+\varepsilon_0(\nabla\varepsilon)\times\frac{\partial\boldsymbol{E}}{\partial t}\\ &\quad+\sigma\nabla\times\boldsymbol{E}+(\nabla\sigma)\times\boldsymbol{E}\\ &= -\varepsilon_0\mu_0\varepsilon\mu\frac{\partial^2\boldsymbol{H}}{\partial t^2}+\varepsilon_0(\nabla\varepsilon)\\ &\quad\times\frac{\partial\boldsymbol{E}}{\partial t}-\mu_0\mu\sigma\frac{\partial\boldsymbol{H}}{\partial t}+(\nabla\sigma)\times\boldsymbol{E}\,.\end{aligned} \tag{3.95}$$

Using (3.89) again, this equation can be resolved with respect to $\partial\boldsymbol{E}/\partial t$:

$$\begin{aligned}&\nabla\times\boldsymbol{H}=\varepsilon_0\varepsilon\frac{\partial\boldsymbol{E}}{\partial t}+\sigma\boldsymbol{E}\\ \Rightarrow\quad &\frac{\partial\boldsymbol{E}}{\partial t}=\frac{1}{\varepsilon_0\varepsilon}(\nabla\times\boldsymbol{H}-\sigma\boldsymbol{E})\,.\end{aligned}$$

Then, $\partial\boldsymbol{E}/\partial t$ can be eliminated in (3.95) resulting in:

$$\begin{aligned}&\Delta\boldsymbol{H}+\nabla\left[\boldsymbol{H}\cdot\nabla(\ln\mu)\right]-\frac{\varepsilon\mu}{c^2}\frac{\partial^2\boldsymbol{H}}{\partial t^2}\\ &-\mu_0\mu\sigma\frac{\partial\boldsymbol{H}}{\partial t}+\left[\nabla(\ln\varepsilon)\right]\times(\nabla\times\boldsymbol{H})\\ &+\left[\nabla\sigma-\sigma\nabla(\ln\varepsilon)\right]\times\boldsymbol{E}=0\,.\end{aligned} \tag{3.96}$$

Unfortunately, it is not possible to completely eliminate the electric vector from this wave equation of the magnetic field.

Equations (3.94) and (3.96) are nearly symmetrical for a replacement of $\boldsymbol{E}$ with $\boldsymbol{H}$ and ε with μ. Only the terms containing the conductivity σ are not symmetrical. Nevertheless, there are two important special cases that provide symmetries of the wave equations of the electric and magnetic vector.

Wave Equations for Pure Dielectrics

If the material is a pure dielectric the conductivity σ, which is responsible for absorption is zero. Then (3.94) and (3.96) reduce to:

$$\begin{aligned}&\Delta\boldsymbol{E}+\nabla\left[\boldsymbol{E}\cdot\nabla(\ln\varepsilon)\right]-\frac{\varepsilon\mu}{c^2}\frac{\partial^2\boldsymbol{E}}{\partial t^2}\\ &\quad+\left[\nabla(\ln\mu)\right]\times(\nabla\times\boldsymbol{E})=0\,,\end{aligned} \tag{3.97}$$

$$\begin{aligned}&\Delta\boldsymbol{H}+\nabla\left[\boldsymbol{H}\cdot\nabla(\ln\mu)\right]-\frac{\varepsilon\mu}{c^2}\frac{\partial^2\boldsymbol{H}}{\partial t^2}\\ &\quad+\left[\nabla(\ln\varepsilon)\right]\times(\nabla\times\boldsymbol{H})=0\,.\end{aligned} \tag{3.98}$$

Here, the equations are really symmetrical for a replacement of $\boldsymbol{E}$ with $\boldsymbol{H}$ and ε with μ. In practice, there are, of course, no materials that are completely transparent to light. But in the visible or infrared region most glasses can be assumed with a good approximation to be dielectrics.

Wave Equations for Homogeneous Materials

The second interesting special case is for homogeneous materials. Here, ε, μ and σ are constants that do not depend on $\boldsymbol{r}$ and their gradients are zero. In this case (3.94) and (3.96) reduce to

$$\Delta\boldsymbol{E}-\frac{\varepsilon\mu}{c^2}\frac{\partial^2\boldsymbol{E}}{\partial t^2}-\mu_0\mu\sigma\frac{\partial\boldsymbol{E}}{\partial t}=0\,, \tag{3.99}$$

$$\Delta\boldsymbol{H}-\frac{\varepsilon\mu}{c^2}\frac{\partial^2\boldsymbol{H}}{\partial t^2}-\mu_0\mu\sigma\frac{\partial\boldsymbol{H}}{\partial t}=0\,. \tag{3.100}$$

These equations are symmetrical to a replacement of $\boldsymbol{E}$ with $\boldsymbol{H}$. In practice, homogeneous materials are the most important case because all conventional lenses (with the exception of graded index lenses, or GRIN lenses) are made of homogeneous glasses or at least of glasses with a very small inhomogeneity.

3.1.5 The Helmholtz Equations

Assume a time-harmonic wave with angular frequency ω represented in its complex notation (3.64):

$$\boldsymbol{E}(\boldsymbol{r},t)=\hat{\boldsymbol{E}}(\boldsymbol{r})\,\mathrm{e}^{-\mathrm{i}\omega t}\,, \tag{3.101}$$

$$\boldsymbol{H}(\boldsymbol{r},t)=\hat{\boldsymbol{H}}(\boldsymbol{r})\,\mathrm{e}^{-\mathrm{i}\omega t}\,. \tag{3.102}$$

As long as linear operations are made using this complex representation is permitted. The partial derivatives with respect to t can be calculated directly where again the functionalities are omitted in our notation:

$$\frac{\partial \boldsymbol{E}}{\partial t} = -\mathrm{i}\omega \hat{\boldsymbol{E}} \mathrm{e}^{-\mathrm{i}\omega t} ,$$
$$\frac{\partial \boldsymbol{H}}{\partial t} = -\mathrm{i}\omega \hat{\boldsymbol{H}} \mathrm{e}^{-\mathrm{i}\omega t} ,$$
$$\frac{\partial^2 \boldsymbol{E}}{\partial t^2} = -\omega^2 \hat{\boldsymbol{E}} \mathrm{e}^{-\mathrm{i}\omega t} ,$$
$$\frac{\partial^2 \boldsymbol{H}}{\partial t^2} = -\omega^2 \hat{\boldsymbol{H}} \mathrm{e}^{-\mathrm{i}\omega t} .$$

These equations can be used in the wave equations (3.94) and (3.96) for linear and isotropic materials. The result is

$$\Delta \hat{\boldsymbol{E}} + \nabla \left[\hat{\boldsymbol{E}} \cdot \nabla (\ln \varepsilon) \right] + \omega^2 \frac{\varepsilon\mu}{c^2} \hat{\boldsymbol{E}} + \mathrm{i}\omega\mu_0\mu\sigma \hat{\boldsymbol{E}} + [\nabla (\ln \mu)] \times \left(\nabla \times \hat{\boldsymbol{E}} \right) = 0 , \tag{3.103}$$

$$\Delta \hat{\boldsymbol{H}} + \nabla \left[\hat{\boldsymbol{H}} \cdot \nabla (\ln \mu) \right] + \omega^2 \frac{\varepsilon\mu}{c^2} \hat{\boldsymbol{H}} + \mathrm{i}\omega\mu_0\mu\sigma \hat{\boldsymbol{H}} + [\nabla (\ln \varepsilon)] \times \left(\nabla \times \hat{\boldsymbol{H}} \right) + [\nabla\sigma - \sigma\nabla (\ln \varepsilon)] \times \hat{\boldsymbol{E}} = 0 . \tag{3.104}$$

These two time-independent equations are called the Helmholtz equations for the electric and the magnetic vector. Since only the position-dependent complex electric and magnetic vectors $\hat{\boldsymbol{E}}$ and $\hat{\boldsymbol{H}}$ are used, only time-averaged quantities can be calculated using the Helmholtz equations. Again two special cases are of interest.

Helmholtz Equations for Pure Dielectrics

For pure dielectric materials the conductivity is zero ($\sigma = 0$). In this case (3.103) and (3.104) result in

$$\Delta \hat{\boldsymbol{E}} + \nabla \left[\hat{\boldsymbol{E}} \cdot \nabla (\ln \varepsilon) \right] + \omega^2 \frac{\varepsilon\mu}{c^2} \hat{\boldsymbol{E}} + [\nabla (\ln \mu)] \times \left(\nabla \times \hat{\boldsymbol{E}} \right) = 0 , \tag{3.105}$$

$$\Delta \hat{\boldsymbol{H}} + \nabla \left[\hat{\boldsymbol{H}} \cdot \nabla (\ln \mu) \right] + \omega^2 \frac{\varepsilon\mu}{c^2} \hat{\boldsymbol{H}} + [\nabla (\ln \varepsilon)] \times \left(\nabla \times \hat{\boldsymbol{H}} \right) = 0 . \tag{3.106}$$

Again, both equations are symmetric with regards to a replacement of $\hat{\boldsymbol{E}}$ with $\hat{\boldsymbol{H}}$ and ε with μ.

Helmholtz Equations for Homogeneous Materials

For homogeneous materials the gradients of ε, μ and σ are zero. In this case (3.103) and (3.104) obtain a quite simple form

$$\Delta \hat{\boldsymbol{E}} + \omega^2 \frac{\varepsilon\mu}{c^2} \hat{\boldsymbol{E}} + \mathrm{i}\omega\mu_0\mu\sigma \hat{\boldsymbol{E}} = 0 , \tag{3.107}$$

$$\Delta \hat{\boldsymbol{H}} + \omega^2 \frac{\varepsilon\mu}{c^2} \hat{\boldsymbol{H}} + \mathrm{i}\omega\mu_0\mu\sigma \hat{\boldsymbol{H}} = 0 . \tag{3.108}$$

Both equations are completely symmetric in $\hat{\boldsymbol{E}}$ and $\hat{\boldsymbol{H}}$. The angular frequency ω is defined as $2\pi\nu$ and the frequency ν and the wavelength in vacuum λ are connected by (3.49): $\nu\lambda = c$. Therefore, (3.107) and (3.108) can be written as

$$\left(\Delta + \hat{k}^2 \right) \hat{\boldsymbol{E}} = 0 , \tag{3.109}$$

$$\left(\Delta + \hat{k}^2 \right) \hat{\boldsymbol{H}} = 0 , \tag{3.110}$$

with

$$\hat{k}^2 = \omega^2 \frac{\varepsilon\mu}{c^2} + \mathrm{i}\omega\mu_0\mu\sigma = \left(\frac{2\pi}{\lambda} \right)^2 \left(\varepsilon\mu + \mathrm{i}\frac{\lambda}{2\pi c\varepsilon_0}\mu\sigma \right) = \left(\frac{2\pi\hat{n}}{\lambda} \right)^2 . \tag{3.111}$$

Here, the refractive index $\hat{n}$ is defined as

$$\hat{n}^2 = \varepsilon\mu + \mathrm{i}\frac{\lambda}{2\pi c\varepsilon_0}\mu\sigma . \tag{3.112}$$

This means that $\hat{n}$ is a complex number if the conductivity σ is not zero. Therefore, the real part n and the imaginary part n_I of $\hat{n}$ can be calculated:

$$\hat{n} = n + \mathrm{i}n_\mathrm{I} \quad \Rightarrow \quad \hat{n}^2 = n^2 - n_\mathrm{I}^2 + 2\mathrm{i}nn_\mathrm{I} ; \tag{3.113}$$

$$\Rightarrow \quad n^2 - n_\mathrm{I}^2 = \varepsilon\mu \quad \text{and} \quad 2nn_\mathrm{I} = \frac{\lambda}{2\pi c\varepsilon_0}\mu\sigma ,$$

$$\Rightarrow \quad n_\mathrm{I}^2 = \left(\frac{\lambda}{4\pi c\varepsilon_0 n}\mu\sigma \right)^2 ,$$

$$\Rightarrow \quad n^4 - \varepsilon\mu n^2 - \left(\frac{\lambda}{4\pi c\varepsilon_0}\mu\sigma \right)^2 = 0 .$$

$$\Rightarrow \quad n = \sqrt{\frac{\varepsilon\mu + \sqrt{\varepsilon^2\mu^2 + \frac{\lambda^2\mu^2\sigma^2}{4\pi^2c^2\varepsilon_0^2}}}{2}} ; \tag{3.114}$$

$$n_\mathrm{I} = \frac{\lambda}{4\pi c\varepsilon_0}\mu\sigma \sqrt{\frac{2}{\varepsilon\mu + \sqrt{\varepsilon^2\mu^2 + \frac{\lambda^2\mu^2\sigma^2}{4\pi^2c^2\varepsilon_0^2}}}} . \tag{3.115}$$

Only the positive solution of the two solutions of the quadratic equation with the variable n^2 is taken since n should be a real number and additionally only the positive square root of n^2 is taken since n should be a positive real number. For a pure dielectric ($\sigma = 0$) the imaginary part vanishes and the refractive index is a real number like it was defined in (3.27):

$$\sigma = 0 \quad \Rightarrow \quad \hat{n} = n = \sqrt{\varepsilon\mu} \tag{3.116}$$

A Simple Solution of the Helmholtz Equation in a Homogeneous Material

A simple solution of (3.109) is, e.g., a linearly polarized plane wave propagating in the z direction

$$\hat{\boldsymbol{E}} = \hat{\boldsymbol{E}}_0 \mathrm{e}^{\mathrm{i}\hat{k}z} = \hat{\boldsymbol{E}}_0 \mathrm{e}^{\mathrm{i}2\pi\hat{n}z/\lambda} \,. \tag{3.117}$$

Here, $\hat{\boldsymbol{E}}_0$ is a constant vector and its modulus represents the amplitude of the electric vector at $z = 0$. If $\hat{n}$ is complex the effective position-dependent amplitude decreases exponentially

$$\hat{n} = n + \mathrm{i}n_\mathrm{I} \quad \Rightarrow \quad \hat{\boldsymbol{E}} = \hat{\boldsymbol{E}}_0 \mathrm{e}^{-2\pi n_\mathrm{I} z/\lambda} \mathrm{e}^{\mathrm{i}2\pi n z/\lambda} \,. \tag{3.118}$$

So, the extinction of a wave can be formally included in the notation of a wave using complex exponential terms by just assuming a complex refractive index. The real part of this complex refractive index is responsible for the "normal" refractive properties and the imaginary part is responsible for absorption. For metals, n_I can be larger than one so that the wave can enter the metal for only a fraction of a wavelength before the electric (and magnetic) vector vanishes.

Instead of using the imaginary part n_I of the refractive index the so-called absorption coefficient α is often used. It is defined by

$$\alpha := \frac{4\pi}{\lambda} n_\mathrm{I} \quad \Rightarrow \quad \hat{\boldsymbol{E}} = \hat{\boldsymbol{E}}_0 \mathrm{e}^{-\alpha z/2} \mathrm{e}^{\mathrm{i}2\pi n z/\lambda} \,. \tag{3.119}$$

After having propagated a distance $z = 1/\alpha$ the electric energy density of the wave, which is proportional to $|\hat{\boldsymbol{E}}|^2$, decreases to $1/\mathrm{e}$ of its starting value.

Whereas α is, according to our definition, always positive in lossy materials there are also active gain media, e.g., in lasers, which have a negative coefficient α. Then α is not an absorption but an amplification or gain coefficient.

Inhomogeneous Plane Waves

The solution of the Helmholtz equation (3.109) in a homogeneous but lossy material defined by (3.118) is the simplest form of a so-called inhomogeneous plane wave [3.1, 4]. The general inhomogeneous plane wave is obtained from (3.109) by the approach

$$\hat{\boldsymbol{E}} = \hat{\boldsymbol{E}}_0 \mathrm{e}^{\mathrm{i}\hat{\boldsymbol{k}}\cdot\boldsymbol{r}} = \hat{\boldsymbol{E}}_0 \mathrm{e}^{-\boldsymbol{g}\cdot\boldsymbol{r}} \mathrm{e}^{\mathrm{i}\boldsymbol{k}\cdot\boldsymbol{r}}$$
$$\Rightarrow \quad -\hat{\boldsymbol{k}}\cdot\hat{\boldsymbol{k}} + \hat{k}^2 = 0 \,, \tag{3.120}$$

where $\hat{\boldsymbol{k}} = \boldsymbol{k} + \mathrm{i}\boldsymbol{g}$ is a constant but complex wave vector with the real part $\boldsymbol{k}$ and the imaginary part $\boldsymbol{g}$. In the general case, $\hat{\boldsymbol{E}}_0$ is also a constant but complex electric vector so that all polarization states can be represented (Sect. 3.2.3).

By using the (3.111), (3.113) and (3.119) the complex quantity $\hat{k}$ is defined as

$$\hat{k} = \frac{2\pi}{\lambda}\hat{n} = \frac{2\pi}{\lambda}(n + \mathrm{i}n_\mathrm{I}) = \frac{2\pi n}{\lambda} + \mathrm{i}\frac{\alpha}{2} \tag{3.121}$$
$$\Rightarrow \quad \hat{k}^2 = \frac{4\pi^2 n^2}{\lambda^2} - \frac{\alpha^2}{4} + \mathrm{i}\frac{2\pi n\alpha}{\lambda} \,.$$

This means that the two vectors $\boldsymbol{k}$ and $\boldsymbol{g}$ have to fulfill the conditions

$$\begin{aligned}\hat{\boldsymbol{k}}\cdot\hat{\boldsymbol{k}} &= (\boldsymbol{k} + \mathrm{i}\boldsymbol{g})\cdot(\boldsymbol{k} + \mathrm{i}\boldsymbol{g}) \\ &= |\boldsymbol{k}|^2 - |\boldsymbol{g}|^2 + 2\mathrm{i}\boldsymbol{g}\cdot\boldsymbol{k} \\ &= \hat{k}^2 \\ &= \frac{4\pi^2 n^2}{\lambda^2} - \frac{\alpha^2}{4} + \mathrm{i}\frac{2\pi n\alpha}{\lambda} \,.\end{aligned} \tag{3.122}$$

A separation of the real and imaginary part gives

$$|\boldsymbol{k}|^2 - |\boldsymbol{g}|^2 = \frac{4\pi^2 n^2}{\lambda^2} - \frac{\alpha^2}{4} \,, \tag{3.123}$$

$$\boldsymbol{g}\cdot\boldsymbol{k} = \frac{\pi n\alpha}{\lambda} \,. \tag{3.124}$$

So, the projection of the vector $\boldsymbol{g}$ onto the vector $\boldsymbol{k}$ has to fulfill the second equation. An important and interesting case is that of a lossless material, i.e., $\alpha = 0$. Then $\boldsymbol{g}$ and $\boldsymbol{k}$ have to be perpendicular to each other meaning that the planes of constant phase, which are perpendicular to $\boldsymbol{k}$, and the planes of constant amplitude, which are perpendicular to $\boldsymbol{g}$, are also perpendicular to each other.

Inhomogeneous plane waves do not exist in the whole space because the amplitude decreases exponentially along the direction of $\boldsymbol{g}$, but on the other side it increases exponentially for the direction antiparallel to $\boldsymbol{g}$ and would tend to infinity. Therefore, only the half-space with the exponentially decreasing part can exist in the real world, whereas in the other direction there has to be a limit. An example of an inhomogeneous plane wave is an evanescent wave in the case of total internal reflection at an interface between two dielectric materials with different refractive indices. There, a plane wave propagating in the material with higher refractive index with an angle of incidence at the interface of more than the critical angle of total internal reflection is reflected. Besides the reflected wave, however, there exists an evanescent wave in the material with the lower refractive index. Its vector $\boldsymbol{k}$ is parallel to the interface between the two materials, while its amplitude decreases exponentially with increasing distance from the interface. The evanescent wave transports no energy into the material with the lower refractive index, but the total energy is in the reflected wave.

In the following sections some basic properties of light waves will be described. For more information see textbooks on optics such as e.g., [3.1, 5–11].

3.2 Polarization

In Sect. 3.1.1 it is shown that a so-called linearly polarized plane wave is a solution of Maxwell's equations. There, the electric vector has a well-defined direction which remains constant during the propagation of the wave. There are other solutions of Maxwell's equations where the direction of the electric vector does not remain constant during the propagation, but nevertheless, it has at a certain point and at a certain time a well-defined direction. All these solutions are called polarized light.

Contrary to this, light that is emitted by an electric bulb is unpolarized. This means that there are many light waves with stochastically distributed phase relations to each other, i. e., incoherent light, and where the polarization varies in time. So, these light waves are added incoherently and there is no preferred direction of the electric vector. In practice, light is often partially polarized, i. e., some of the light is unpolarized and the other is polarized. Natural sun light on the earth is, e.g., partially polarized because of the influence of the atmosphere to the originally unpolarized light of the sun.

Here, only the case of a fully polarized plane wave in a homogeneous dielectric material will be investigated. In the paragraph "The Orthogonality Condition for Plane Waves in Homogeneous Dielectrics" it is shown that the electric vector $\boldsymbol{E}$ and the magnetic vector $\boldsymbol{H}$ of a plane wave in a homogeneous and isotropic linear material are always perpendicular to each other and both are perpendicular to the direction of propagation $\boldsymbol{e}$ of the plane wave. Therefore, for a given direction of propagation $\boldsymbol{e}$ it is sufficient to consider only the electric vector. The magnetic vector is then automatically defined by (3.37)

$$\boldsymbol{H} = \pm\sqrt{\frac{\varepsilon_0\varepsilon}{\mu_0\mu}}\,\boldsymbol{e}\times\boldsymbol{E}\,.$$

The electric vector $\boldsymbol{E}$ has to fulfill the wave equation (3.99) with $\sigma = 0$

$$\Delta\boldsymbol{E} - \frac{n^2}{c^2}\frac{\partial^2\boldsymbol{E}}{\partial t^2} = 0\,.$$

Without loss of generality, the direction of propagation will be parallel to the z-axis, i. e., $\boldsymbol{e} = (0, 0, 1)$. Because of the orthogonality relation $\boldsymbol{E}$ can then only have a x and a y component. A quite general plane wave solution of the wave equation is in this case

$$\boldsymbol{E}(z,t) = \begin{pmatrix} E_x(z,t) \\ E_y(z,t) \\ E_z(z,t) \end{pmatrix} = \begin{pmatrix} A_x\cos(kz-\omega t+\delta_x) \\ A_y\cos\left(kz-\omega t+\delta_y\right) \\ 0 \end{pmatrix}\,. \tag{3.125}$$

Here, $k = 2\pi n/\lambda = \omega n/c$ (3.50) is the modulus of the wave vector $\boldsymbol{k} = 2\pi n\boldsymbol{e}/\lambda$. It holds $A_x \geq 0$ and $A_y \geq 0$. By applying a trigonometric theorem, introducing the parameter $\alpha := kz - \omega t + \delta_x$ and the phase difference $\delta := \delta_y - \delta_x$ this equation can be written as

$$\begin{pmatrix} E_x(\alpha) \\ E_y(\alpha) \end{pmatrix} = \begin{pmatrix} A_x\cos\alpha \\ A_y\cos(\alpha+\delta) \end{pmatrix} = \begin{pmatrix} A_x\cos\alpha \\ A_y\cos\alpha\cos\delta - A_y\sin\alpha\sin\delta \end{pmatrix}\,. \tag{3.126}$$

This equation is the parametric representation of an ellipse, which is formed by the apex of the two-dimensional vector (E_x, E_y) in the xy-plane for different values of the parameter α. Unfortunately, in general, the principal axes of this ellipse will be rotated with respect to the x-axis and y-axis. Therefore, a transformation has to be done to calculate the principal axes of this ellipse with lengths $2a$ and $2b$. To do this, the following quantity is calculated where the argument α of E_x and E_y is omitted in the notation

$$\begin{aligned}
&\left(\frac{E_x}{A_x}\right)^2+\left(\frac{E_y}{A_y}\right)^2-2\frac{E_xE_y}{A_xA_y}\cos\delta\\
&=\cos^2\alpha+(\cos\alpha\cos\delta-\sin\alpha\sin\delta)^2\\
&\quad-2\cos\alpha\cos\delta\,(\cos\alpha\cos\delta-\sin\alpha\sin\delta)\\
&=\cos^2\alpha+(\cos\alpha\cos\delta-\sin\alpha\sin\delta)\\
&\quad\times(-\cos\alpha\cos\delta-\sin\alpha\sin\delta)\\
&=\cos^2\alpha\left(1-\cos^2\delta\right)+\sin^2\alpha\sin^2\delta=\sin^2\delta;\\
&\Rightarrow\quad\left(\frac{E_x}{A_x\sin\delta}\right)^2+\left(\frac{E_y}{A_y\sin\delta}\right)^2\\
&\quad-2\frac{E_xE_y}{A_xA_y\sin^2\delta}\cos\delta=1\,. \qquad (3.127)
\end{aligned}$$

This is the implicit representation of an ellipse that is rotated with respect to the x- and y-axis (see Fig. 3.4). On the other side an ellipse with the half-axes a and b parallel to the coordinate axes x' and y' and the coordinates of the electric vector E'_x and E'_y are written as

$$\left(\frac{E'_x}{a}\right)^2+\left(\frac{E'_y}{b}\right)^2=1\,. \qquad (3.128)$$

This equation is transformed into a coordinate system with axes x and y whereby the system (x', y') is rotated by an angle φ relative to the system (x, y) by applying

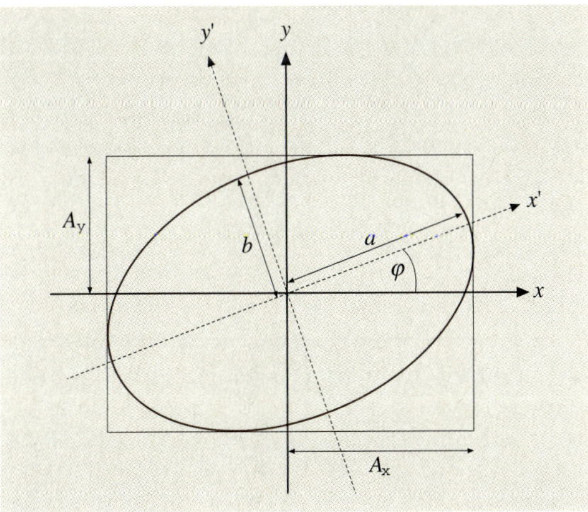

Fig. 3.4 The polarization ellipse on which the apex of the electric vector moves if the time t or the coordinate z changes

the well-known rotation matrix to the coordinates

$$\begin{aligned}
\begin{pmatrix}E'_x\\E'_y\end{pmatrix}&=\begin{pmatrix}\cos\varphi & \sin\varphi\\ -\sin\varphi & \cos\varphi\end{pmatrix}\begin{pmatrix}E_x\\E_y\end{pmatrix}\\
&=\begin{pmatrix}E_x\cos\varphi+E_y\sin\varphi\\ -E_x\sin\varphi+E_y\cos\varphi\end{pmatrix}.
\end{aligned}$$

So, the equation of the rotated ellipse in the coordinate system (x, y) indexed by terms in E_x^2, E_y^2 and E_xE_y is

$$\begin{aligned}
&E_x^2\left(\frac{\cos^2\varphi}{a^2}+\frac{\sin^2\varphi}{b^2}\right)\\
&+E_y^2\left(\frac{\sin^2\varphi}{a^2}+\frac{\cos^2\varphi}{b^2}\right)\\
&+E_xE_y\sin(2\varphi)\left(\frac{1}{a^2}-\frac{1}{b^2}\right)=1\,.
\end{aligned}$$

In the last step the trigonometric equality $\sin(2\varphi)=2\sin\varphi\cos\varphi$ is used.

By comparing this equation with (3.127) the coefficients of the terms in E_x^2, E_y^2 and E_xE_y have to be equal to calculate the rotation angle φ and the principal axes. This results in three equations

$$\begin{aligned}
\frac{\cos^2\varphi}{a^2}+\frac{\sin^2\varphi}{b^2}&=\frac{1}{A_x^2\sin^2\delta},\\
\frac{\sin^2\varphi}{a^2}+\frac{\cos^2\varphi}{b^2}&=\frac{1}{A_y^2\sin^2\delta},\\
\sin(2\varphi)\left(\frac{1}{a^2}-\frac{1}{b^2}\right)&=-2\frac{\cos\delta}{A_xA_y\sin^2\delta}\,.
\end{aligned}$$

By (1) adding the second equation to the first equation and (2) subtracting the second equation from the first equation and applying the trigonometric equation $\cos(2\varphi)=\cos^2\varphi-\sin^2\varphi$ two new equations are obtained and written together with the old third equation

$$\begin{aligned}
\frac{1}{a^2}+\frac{1}{b^2}&=\frac{1}{\sin^2\delta}\left(\frac{1}{A_x^2}+\frac{1}{A_y^2}\right),\\
\cos(2\varphi)\left(\frac{1}{a^2}-\frac{1}{b^2}\right)&=\frac{1}{\sin^2\delta}\left(\frac{1}{A_x^2}-\frac{1}{A_y^2}\right),\\
\sin(2\varphi)\left(\frac{1}{a^2}-\frac{1}{b^2}\right)&=-2\frac{\cos\delta}{A_xA_y\sin^2\delta}\,.
\end{aligned}$$

Division of the third equation by the second equation gives the tangent of twice the rotation angle φ of the ellipse

$$\tan(2\varphi) = \frac{-2\cos\delta}{A_x A_y\left(\frac{1}{A_x^2} - \frac{1}{A_y^2}\right)} = \frac{-2A_x A_y \cos\delta}{A_y^2 - A_x^2}\,. \tag{3.129}$$

So, the rotation angle φ can be calculated from the known variables A_x, A_y and δ. There is a null of the denominator for the case $A_x = A_y$. Then the argument of the tangent is $2\varphi = \pm\pi/2$, i.e., $\varphi = \pm\pi/4$. Only if the numerator is also zero, i.e., $\cos\delta = 0$ and therefore $\delta = (2n+1)\pi/2$, the angle φ is not defined. In this case, the ellipse is degenerated into a circle as we will see later and then the rotation angle is, of course, not defined. It should also be mentioned that the rotation angle φ is only defined between $-\pi/4$ and $\pi/4$ and therefore 2φ between $-\pi/2$ and $\pi/2$ so that the arctangent function is well defined. This is sufficient because first, a rotation of an ellipse by π (= 180 degrees) does not change anything. And second, the principal axes of the ellipse can be chosen in such a way that either a is the large axis or b. So, the rotation angle φ has to be defined only in an interval of the length $\pi/2$, i.e., $[-\pi/4; \pi/4]$.

After having calculated φ, the half-axes a and b of the polarization ellipse can also be calculated using the first two of the above equations

$$(i)\quad \frac{1}{a^2} + \frac{1}{b^2} = \frac{1}{\sin^2\delta}\left(\frac{1}{A_x^2} + \frac{1}{A_y^2}\right),$$

$$(ii)\quad \frac{1}{a^2} - \frac{1}{b^2} = \frac{1}{\cos(2\varphi)\sin^2\delta}\left(\frac{1}{A_x^2} - \frac{1}{A_y^2}\right);$$

$$(i)+(ii) \Rightarrow \frac{1}{a^2} = \frac{1}{2\sin^2\delta}\left[\frac{1}{A_x^2} + \frac{1}{A_y^2} + \frac{1}{\cos(2\varphi)}\left(\frac{1}{A_x^2} - \frac{1}{A_y^2}\right)\right], \tag{3.130}$$

$$(i)-(ii) \Rightarrow \frac{1}{b^2} = \frac{1}{2\sin^2\delta}\left[\frac{1}{A_x^2} + \frac{1}{A_y^2} - \frac{1}{\cos(2\varphi)}\left(\frac{1}{A_x^2} - \frac{1}{A_y^2}\right)\right]. \tag{3.131}$$

Using a trigonometric relation and (3.129) it holds that

$$\frac{1}{\cos(2\varphi)} = \sqrt{1+\tan^2(2\varphi)} = \sqrt{1+4\frac{A_x^2A_y^2\cos^2\delta}{\left(A_y^2 - A_x^2\right)^2}} = \frac{\sqrt{\left(A_y^2 - A_x^2\right)^2 + 4A_x^2A_y^2\cos^2\delta}}{\left|A_y^2 - A_x^2\right|}\,.$$

Remember that 2φ is only defined between $-\pi/2$ and $\pi/2$ so that $\cos(2\varphi) \geq 0$. Defining the value s by

$$s = \begin{cases} +1 \text{ if } A_y^2 - A_x^2 \geq 0 \\ -1 \text{ if } A_y^2 - A_x^2 < 0 \end{cases}, \tag{3.132}$$

the reciprocal values of the squares of the half-axes can be explicitly written as

$$\frac{1}{a^2} = \frac{1}{2A_x^2A_y^2\sin^2\delta} \times \left[A_y^2 + A_x^2 + s\sqrt{\left(A_y^2 - A_x^2\right)^2 + 4A_x^2A_y^2\cos^2\delta}\right], \tag{3.133}$$

$$\frac{1}{b^2} = \frac{1}{2A_x^2A_y^2\sin^2\delta} \times \left[A_y^2 + A_x^2 - s\sqrt{\left(A_y^2 - A_x^2\right)^2 + 4A_x^2A_y^2\cos^2\delta}\right]. \tag{3.134}$$

Now it is easy to calculate the ratio of the squares of the half-axes, whereby the equality $\cos^2\delta - 1 = -\sin^2\delta$ is used

$$\frac{b^2}{a^2} = \frac{A_y^2 + A_x^2 + s\sqrt{\left(A_y^2 + A_x^2\right)^2 - 4A_x^2A_y^2\sin^2\delta}}{A_y^2 + A_x^2 - s\sqrt{\left(A_y^2 + A_x^2\right)^2 - 4A_x^2A_y^2\sin^2\delta}} = \frac{1 + s\sqrt{1 - 4\sin^2\delta\frac{A_x^2A_y^2}{\left(A_y^2+A_x^2\right)^2}}}{1 - s\sqrt{1 - 4\sin^2\delta\frac{A_x^2A_y^2}{\left(A_y^2+A_x^2\right)^2}}}\,. \tag{3.135}$$

So, for given parameters A_x, A_y and δ the half-axes a and b of the ellipse can be calculated by (3.133) and (3.134). The rotation angle can be calculated by (3.129) and the ratio of the half-axes by (3.135). There are several interesting special cases of polarization states and these will be discussed in the following.

3.2.1 Different States of Polarization

Linear Polarization

An important and quite simple polarization state is the case of linear polarization. In this case the polarization ellipse degenerates to a line and the apex of the electric vector just oscillates on a line. This is the case if either the numerator or the denominator of (3.135) is zero so that a or b is zero. This means

$$1-\sqrt{1-4\sin^2\delta\frac{A_x^2A_y^2}{\left(A_y^2+A_x^2\right)^2}}=0 \Rightarrow \sin\delta=0$$

$$\text{or } A_x=0 \text{ or } A_y=0\,. \tag{3.136}$$

The two cases $A_x=0$ or $A_y=0$ are obvious because in this case there is only an x or a y component of the electric vector. If both components are different from zero there is nevertheless linear polarization if the phase difference δ between the two components of the electric vector is $\delta=0$ or $\delta=\pi$.

Circular Polarization

If the apex of the electric vector moves on a circle, the polarization state is called circular polarization. This means that both half-axes have to be equal: $a=b$. Using (3.135) this requires

$$\frac{b^2}{a^2}=1 \Rightarrow 2\sin\delta\frac{A_xA_y}{A_y^2+A_x^2}=\pm1\,.$$

Since $|\sin\delta|\le 1$ this requires (A_x and A_y are both positive)

$$\frac{A_xA_y}{A_y^2+A_x^2}\ge\frac{1}{2} \Rightarrow 2A_xA_y\ge A_x^2+A_y^2$$

$$\Rightarrow 0\ge\left(A_x-A_y\right)^2\,.$$

This condition can only be fulfilled for the case $A_x=A_y$ and then there is the additional condition $\sin\delta=\pm1$. Finally, the conditions for circular polarization are

$$A_x=A_y \wedge \delta=\frac{\pm\pi}{2}\,. \tag{3.137}$$

Using the original (3.125) this means for the electric vector

$$\begin{pmatrix} E_x(z,t) \\ E_y(z,t)\end{pmatrix}=\begin{pmatrix} A_x\cos(kz-\omega t+\delta_x) \\ A_x\cos\left(kz-\omega t+\delta_x\pm\frac{\pi}{2}\right)\end{pmatrix} = \begin{pmatrix} A_x\cos(kz-\omega t+\delta_x) \\ \mp A_x\sin(kz-\omega t+\delta_x)\end{pmatrix}. \tag{3.138}$$

So, the two different signs of the phase difference δ correspond to different directions of rotation of the apex of the electric vector. These two cases are called right-handed circular polarization ($\delta=\pi/2$) and left-handed circular polarization ($\delta=-\pi/2$). The definition of right-handed and left-handed is not always identical in textbooks so that we use the definition of [3.1].

Elliptic Polarization

The general polarization state is, of course, the so-called elliptic polarization. In this case the apex of the electric vector moves on an ellipse if the time t or the position z is changed. This state is the case if neither $\delta=0$ or $\delta=\pi$ nor $\delta=\pm\pi/2$. Also if $\delta=\pm\pi/2$ the light is elliptically polarized if $A_x\ne A_y$. There, we have to again distinguish between right-handed and left-handed elliptic polarization.

3.2.2 The Poincaré Sphere

A method to visualize the different states of polarization is the so-called Poincaré sphere, which was introduced by H. Poincaré in 1892. By using (3.126) as the definition of the electric field, the so-called Stokes parameters of a plane monochromatic wave can be defined [3.1]

$$\begin{aligned} s_0 &:= A_x^2+A_y^2\,, \\ s_1 &:= A_x^2-A_y^2\,, \\ s_2 &:= 2A_xA_y\cos\delta\,, \\ s_3 &:= 2A_xA_y\sin\delta\,. \end{aligned} \tag{3.139}$$

It is obvious that the four quantities are connected by the relation

$$\begin{aligned} s_1^2+s_2^2+s_3^3 &= A_x^4+A_y^4-2A_x^2A_y^2+4A_x^2A_y^2 \\ &= A_x^4+A_y^4+2A_x^2A_y^2=s_0^2\,. \end{aligned} \tag{3.140}$$

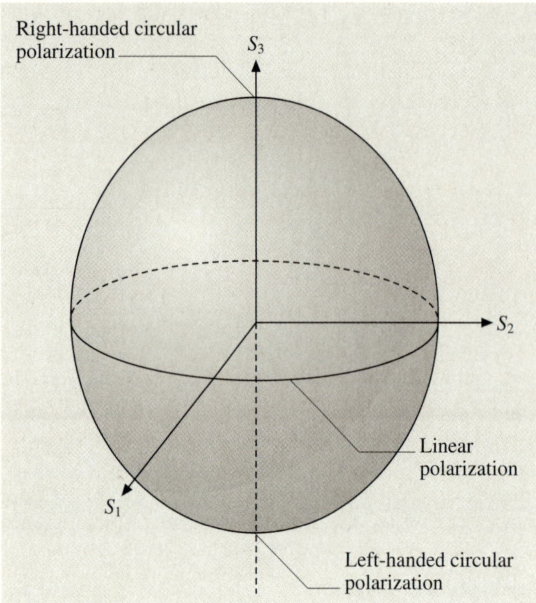

Fig. 3.5 The Poincaré sphere and the visualization of the different states of polarization

So, only three of the parameters are independent and the parameter s_0 is proportional to the intensity of the wave. If s_1, s_2 and s_3 are now used as the cartesian coordinates of a point in space, all allowed combinations will be situated according to (3.140) on a sphere with radius s_0. The radius s_0 is proportional to the intensity of the wave and the sphere is called the Poincaré sphere (Fig. 3.5).

The different polarization states correspond to different positions on the Poincaré sphere. For linearly polarized light it is, e.g., either $A_x = 0$ or $A_y = 0$ or $\delta = 0$ or $\delta = \pi$. In all four cases the parameter s_3 will be zero. This means that points lying in the equatorial plane of the Poincaré sphere represent linear polarization. Another interesting case is circular polarization. In this case the conditions are according to (3.137): $A_x = A_y$ and $\delta = \pm\pi/2$. Therefore, s_1 and s_2 are both zero and circular polarization corresponds to the poles of the Poincaré sphere. At the north pole ($s_1 = s_2 = 0$ and $s_3 = s_0$) the light is right-handed circularly polarized ($\delta = \pi/2$). At the south pole ($s_1 = s_2 = 0$ and $s_3 = -s_0$) the light is left-handed circularly polarized ($\delta = -\pi/2$). All other states of polarization (elliptic polarization) correspond to points somewhere else on the Poincaré sphere. In the upper hemisphere of the Poincaré sphere ($s_3 > 0$) the light is always right-handed polarized and in the lower hemisphere ($s_3 < 0$) the light is left-handed polarized.

3.2.3 Complex Representation of a Polarized Wave

In (3.125) the electric vector is expressed as a real quantity. As we have seen in other sections it is in many cases useful to take a complex notation, where only the real part has a physical meaning

$$\begin{aligned} \boldsymbol{E}(z,t) &= \begin{pmatrix} E_x(z,t) \\ E_y(z,t) \\ E_z(z,t) \end{pmatrix} \\ &= \begin{pmatrix} A_x \cos(kz - \omega t + \delta_x) \\ A_y \cos\left(kz - \omega t + \delta_y\right) \\ 0 \end{pmatrix} \\ &= \mathrm{Re}\left[\begin{pmatrix} A_x e^{i\delta_x} \\ A_y e^{i\delta_y} \\ 0 \end{pmatrix} e^{ikz} e^{-i\omega t}\right] \\ &= \mathrm{Re}\left(\hat{\boldsymbol{A}} e^{ikz} e^{-i\omega t}\right). \end{aligned} \tag{3.141}$$

In this case the vector $\hat{\boldsymbol{A}}$ is a constant but complex vector in order to represent all possible states of polarization. With a real vector only a linear polarization state could be represented. The complex notation has the advantage that it is quite easy to calculate the time average of the intensity. According to (3.68) the time average of the intensity I is proportional to the square of the absolute value of the time-independent complex electric vector $\hat{\boldsymbol{E}}$:

$$\hat{\boldsymbol{E}}(z) = \hat{\boldsymbol{A}} e^{ikz} \Rightarrow I \propto \hat{\boldsymbol{E}} \cdot \hat{\boldsymbol{E}}^* = A_x^2 + A_y^2 . \tag{3.142}$$

Therefore, it is clear that a detector sensitive only to the intensity of a light wave cannot distinguish between different polarization states.

3.2.4 Simple Polarizing Optical Elements and the Jones Calculus

There are, of course, optical elements that influence the polarization state like polarizers, quarter-wave plates, half-wave plates and many others. Here, only the basic idea of their effects can be discussed. For more information about the treatment of polarizing optical elements see, e.g., [3.12–14]. In this section only fully polarized light is treated. This can be produced from natural unpolarized light with the help of a polarizer. In the following the word *polarizer* is always used for a polarization filter whereas other polarizing elements are simply called *polarizing elements* or *polarizing optical elements*. The case of partially polarized light will not be treated.

A quite useful algorithm for the treatment of fully polarized light is the so-called Jones calculus, which was invented by *Jones* [3.15]. If we again have a plane wave propagating in the z direction the polarization state can be described by a two-dimensional vector $\boldsymbol{J}$ in the x and y direction containing the x and y components of the vector $\hat{\boldsymbol{A}}$ of (3.141):

$$\boldsymbol{J}=\begin{pmatrix} J_x \\ J_y \end{pmatrix}=\begin{pmatrix} A_x e^{i\delta_x} \\ A_y e^{i\delta_y} \end{pmatrix}. \tag{3.143}$$

This vector is called the Jones vector.

Now each polarizing optical element can be represented by a 2×2 matrix, the Jones matrix. The resulting Jones vector of the light that has passed such a polarizing optical element is calculated by multiplying the Jones vector of the incident light with the Jones matrix. Several polarizing optical elements can be passed by simply multiplying the Jones matrices of these elements.

Polarizer

A polarizer is, e.g., a device that produces linearly polarized light from an arbitrary polarization state. If the polarizer passes only light polarized in x direction its Jones matrix is

$$\mathbf{P_x}=\begin{pmatrix} 1 & 0 \\ 0 & 0 \end{pmatrix}. \tag{3.144}$$

Similarly Jones matrices of polarizers in other directions (y direction, 45° or −45°) can be represented by

$$\mathbf{P_y}=\begin{pmatrix} 0 & 0 \\ 0 & 1 \end{pmatrix};\ \mathbf{P}_{45°}=\frac{1}{2}\begin{pmatrix} 1 & 1 \\ 1 & 1 \end{pmatrix};$$

$$\mathbf{P}_{-45°}=\frac{1}{2}\begin{pmatrix} 1 & -1 \\ -1 & 1 \end{pmatrix}. \tag{3.145}$$

As an example consider a plane wave linearly polarized in the x direction, which first passes a polarizer in the x direction and then in the direction 45°. The resulting vector is:

$$\begin{aligned}\boldsymbol{J}_{\text{final}}&=\frac{1}{2}\begin{pmatrix} 1 & 1 \\ 1 & 1 \end{pmatrix}\begin{pmatrix} 1 & 0 \\ 0 & 0 \end{pmatrix}\begin{pmatrix} E_0 \\ 0 \end{pmatrix}\\&=\frac{1}{2}\begin{pmatrix} 1 & 1 \\ 1 & 1 \end{pmatrix}\begin{pmatrix} E_0 \\ 0 \end{pmatrix}=\frac{1}{2}\begin{pmatrix} E_0 \\ E_0 \end{pmatrix}.\end{aligned}$$

Of course, the first polarizer in the x direction has no effect and the second polarizer selects the component of the electric vector in the 45° direction. The resulting intensity of the wave is then proportional to $|\boldsymbol{J}_{\text{final}}|^2=E_0^2/2$, i. e., half of the intensity is absorbed. Another well-known effect is the combination of two crossed polarizers (e.g., in the x and y directions). Their matrix is, of course,

$$\mathbf{P}_{\text{crossed}}=\mathbf{P_y}\mathbf{P_x}=\begin{pmatrix} 0 & 0 \\ 0 & 1 \end{pmatrix}\begin{pmatrix} 1 & 0 \\ 0 & 0 \end{pmatrix}=\begin{pmatrix} 0 & 0 \\ 0 & 0 \end{pmatrix}.$$

This means that no light passes this combination. Only if other polarizing elements are between the two crossed polarizers can light pass. If, e.g., a polarizer with direction 45° is inserted between the two crossed polarizers light can pass this combination if it originally has a component in the x direction. The resulting light will, of course, only have a y component:

$$\begin{aligned}\mathbf{P}_{\text{crossed}+45°}&=\mathbf{P_y}\mathbf{P}_{45°}\mathbf{P_x}\\&=\begin{pmatrix} 0 & 0 \\ 0 & 1 \end{pmatrix}\frac{1}{2}\begin{pmatrix} 1 & 1 \\ 1 & 1 \end{pmatrix}\begin{pmatrix} 1 & 0 \\ 0 & 0 \end{pmatrix}\\&=\frac{1}{2}\begin{pmatrix} 0 & 0 \\ 1 & 0 \end{pmatrix}.\end{aligned}$$

Though this is a well-known example it is nevertheless a typical example that polarizing optical elements can produce quite astonishing results by inserting additional elements.

Quarter-Wave Plate

Another elementary polarizing optical element is a quarter-wave plate ($\lambda/4$-plate), which consists of a birefringent material. If the axes of the material are correctly oriented the refractive index for light polarized in x direction is, e.g., different from the refractive index for polarized light in y direction. The resulting effect is a phase difference between the two components of the Jones vector of $\pi/2$. The Jones matrix for a higher phase velocity in the y direction is, e.g.,

$$\mathbf{P}_{\lambda/4}=e^{i\pi/4}\begin{pmatrix} 1 & 0 \\ 0 & i \end{pmatrix}. \tag{3.146}$$

This means that linearly polarized light with components in only the x or y direction remains linearly polarized

$$= A_{\parallel,1}^2 + A_{\parallel,2}^2 + 2A_{\parallel,1}A_{\parallel,2}\,(\boldsymbol{e}_1 \cdot \boldsymbol{e}_2) \times \cos\left[(\boldsymbol{k}_1 - \boldsymbol{k}_2)\cdot \boldsymbol{r} + \delta_{\parallel}\right] + A_{\perp,1}^2 + A_{\perp,2}^2 + 2A_{\perp,1}A_{\perp,2} \times \cos\left[(\boldsymbol{k}_1 - \boldsymbol{k}_2)\cdot \boldsymbol{r} + \delta_{\perp}\right] . \tag{3.156}$$

The two terms that depend on the parameters of both waves and are functions of the position are called interference terms. These interference terms distinguish the superposition of coherent waves and incoherent waves. In the case of incoherent waves the interference terms vanish and the resulting intensity is just the sum of the single intensities of both waves.

It can be seen that the interference pattern resolves into terms that depend only on TM components and those that depend only on TE components. Since both are perpendicular to each other, the intensities of both waves can also be divided into the sum of a "TM intensity" and a "TE intensity"

$$I_1 = a\left(A_{\parallel,1}^2 + A_{\perp,1}^2\right) = I_{\parallel,1} + I_{\perp,1} ,$$
$$I_2 = a\left(A_{\parallel,2}^2 + A_{\perp,2}^2\right) = I_{\parallel,2} + I_{\perp,2} . \tag{3.157}$$

Here, the constant of proportionality a is used, which was explained above. Then, the intensity of the interference pattern is

$$I_{1+2} = I_{\parallel,1} + I_{\parallel,2} + 2\sqrt{I_{\parallel,1}I_{\parallel,2}}\,(\boldsymbol{e}_1 \cdot \boldsymbol{e}_2)\cos\left[(\boldsymbol{k}_1 - \boldsymbol{k}_2)\cdot \boldsymbol{r} + \delta_{\parallel}\right] + I_{\perp,1} + I_{\perp,2} + 2\sqrt{I_{\perp,1}I_{\perp,2}}\cos\left[(\boldsymbol{k}_1 - \boldsymbol{k}_2)\cdot \boldsymbol{r} + \delta_{\perp}\right] . \tag{3.158}$$

The Grating Period and the Fringe Period

Equation (3.158) shows that the surfaces of constant intensity are plane surfaces with

$$(\boldsymbol{k}_1 - \boldsymbol{k}_2)\cdot \boldsymbol{r} = \text{constant} . \tag{3.159}$$

The planes are perpendicular to the so-called grating vector $\boldsymbol{G}$ (Fig. 3.6) with

$$\boldsymbol{G} = \boldsymbol{k}_1 - \boldsymbol{k}_2 . \tag{3.160}$$

Since the cosine function is periodic the distance between the two neighboring planes of equal intensity is called the grating period Λ of the interference pattern. It can be calculated by taking a point $\boldsymbol{r}_1$ on the first plane and a point $\boldsymbol{r}_2$ on the neighboring second plane, so that the vector $\Delta\boldsymbol{r} := \boldsymbol{r}_2 - \boldsymbol{r}_1$ is parallel to $\boldsymbol{G}$ and simultaneously perpendicular to the planes. Its modulus is the grating period

$$\cos\left[(\boldsymbol{k}_1 - \boldsymbol{k}_2)\cdot \boldsymbol{r}_2\right] = \cos\left[(\boldsymbol{k}_1 - \boldsymbol{k}_2)\cdot \boldsymbol{r}_1\right];$$
$$\Rightarrow (\boldsymbol{k}_1 - \boldsymbol{k}_2)\cdot \boldsymbol{r}_2 = (\boldsymbol{k}_1 - \boldsymbol{k}_2)\cdot \boldsymbol{r}_1 + 2\pi$$
$$\Rightarrow \boldsymbol{G}\cdot \Delta\boldsymbol{r} = |\boldsymbol{G}|\,|\Delta\boldsymbol{r}| = 2\pi$$
$$\Rightarrow \Lambda = |\Delta\boldsymbol{r}| = \frac{2\pi}{|\boldsymbol{G}|} = \frac{2\pi}{\frac{2\pi n}{\lambda}\,|\boldsymbol{e}_1 - \boldsymbol{e}_2|} = \frac{\lambda}{n\,|\boldsymbol{e}_1 - \boldsymbol{e}_2|} . \tag{3.161}$$

Here, n is the refractive index of the material in which the waves propagate and $\lambda = 2\pi c/\omega$ would be the wavelength in vacuum. So, the grating period is infinity if both waves are propagating parallel, i. e., $\boldsymbol{e}_1 = \boldsymbol{e}_2$, and the smallest grating period can be obtained if both waves are propagating antiparallel, i. e., $\boldsymbol{e}_1 = -\boldsymbol{e}_2$. Then the grating period $\Lambda_{\min}$ is

$$\Lambda_{\min} = \frac{\lambda}{2n} .$$

In the general case the grating period can be expressed by using the angle φ between the two wave vectors (Fig. 3.6)

$$|\boldsymbol{e}_1 - \boldsymbol{e}_2| = \sqrt{(\boldsymbol{e}_1 - \boldsymbol{e}_2)\cdot(\boldsymbol{e}_1 - \boldsymbol{e}_2)} = \sqrt{2 - 2\boldsymbol{e}_1 \cdot \boldsymbol{e}_2}$$
$$\Rightarrow \Lambda = \frac{\lambda}{n\sqrt{2\,(1 - \cos\varphi)}} . \tag{3.162}$$

This can also be seen in Fig. 3.6 using the trigonometric identity $\sqrt{2(1 - \cos\varphi)} = 2\sin(\varphi/2)$.

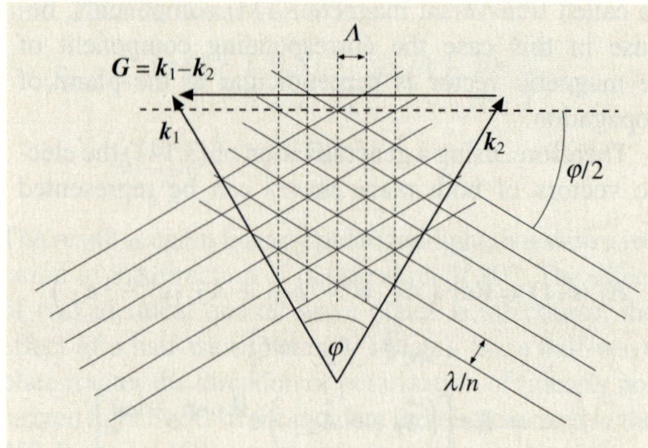

Fig. 3.6 Interference of two plane waves. The *solid lines* indicate the planes of constant phase of the two plane waves at a fixed time having the distance λ/n, respectively, whereas the *dashed lines* indicate the interference planes with constant intensity having a distance Λ. The planes themselves are perpendicular to the drawing plane

What is really observed are the lines of intersection of the planes of constant intensity with a detector plane. The resulting lines are called the interference fringes. In the case of the interference of two plane waves the interference fringes are straight, parallel and equidistant lines with the distance p, called the period of the fringes. Only in the case when the grating vector $\boldsymbol{G}$ is parallel to the detector plane is the fringe period p equal to the grating period Λ. In the general case, only the component of the grating vector parallel to the detector plane has to be used to calculate the fringe period. This relation can be easily seen by taking the plane with $z=0$ as detector plane. Then the fringes in the xy-plane are described in analogy to (3.159) by

$$\left(k_{x,1}-k_{x,2}\right)x+\left(k_{y,1}-k_{y,2}\right)y=\boldsymbol{G}_{\parallel}\cdot\boldsymbol{r}=\text{constant}\,. \tag{3.163}$$

In analogy to (3.161) the result is

$$p=\frac{2\pi}{|\boldsymbol{G}_{\parallel}|}=\frac{2\pi}{|\boldsymbol{G}-(\boldsymbol{G}\cdot\boldsymbol{N})\,\boldsymbol{N}|}\,, \tag{3.164}$$

where $\boldsymbol{N}$ is a unit vector perpendicular to the detector plane. By defining the two angles of incidence β_1 and β_2 of the plane waves onto the detector plane as

$$\cos\beta_1:=\boldsymbol{e}_1\cdot\boldsymbol{N};\quad \cos\beta_2:=\boldsymbol{e}_2\cdot\boldsymbol{N};$$

the fringe period p can be written as:

$$\begin{aligned}p&=\frac{2\pi}{\sqrt{[\boldsymbol{G}-(\boldsymbol{G}\cdot\boldsymbol{N})\,\boldsymbol{N}]^2}}=\frac{2\pi}{\sqrt{|\boldsymbol{G}|^2-(\boldsymbol{G}\cdot\boldsymbol{N})^2}}\\&=\frac{2\pi}{\sqrt{|\boldsymbol{k}_1-\boldsymbol{k}_2|^2-[(\boldsymbol{k}_1-\boldsymbol{k}_2)\cdot\boldsymbol{N}]^2}}\\&=\frac{\lambda}{n\sqrt{2\,(1-\cos\varphi)-(\cos\beta_1-\cos\beta_2)^2}}\end{aligned}$$

or

$$\begin{aligned}p&=\frac{2\pi}{\sqrt{|\boldsymbol{G}|^2-(\boldsymbol{G}\cdot\boldsymbol{N})^2}}=\frac{\Lambda}{\sqrt{1-\cos^2\alpha}}\\&=\frac{\Lambda}{\sin\alpha}=\frac{\lambda}{n\sqrt{2\,(1-\cos\varphi)}\sin\alpha}\,.\end{aligned} \tag{3.165}$$

In the last equation the angle α between the grating vector $\boldsymbol{G}$ and the surface normal $\boldsymbol{N}$, i. e., the angle α between the grating planes and the detector plane, is defined as $\cos\alpha=(\boldsymbol{G}\cdot\boldsymbol{N})/|\boldsymbol{G}|$. From (3.165) it can be seen that the fringe period is infinity if the grating planes are parallel to the detector plane and that the fringe period is minimal if the grating planes are perpendicular to the detector plane, i. e., if $\boldsymbol{G}$ is parallel to the detector plane. Then the fringe period p is equal to the grating period Λ.

3.3.2 Interference Effects for Plane Waves with Different Polarization States

In the calculation of the grating period and the fringe period we assumed that the interference terms in (3.158) are different from zero so that interference occurs. But, this is not always the case as we will now discuss.

A quite interesting quantity in interferometry is the visibility V of the interference fringes. It is defined as

$$V:=\frac{I_{\max}-I_{\min}}{I_{\max}+I_{\min}}\,, \tag{3.166}$$

where $I_{\max}$ is the maximum intensity and $I_{\min}$ the minimum intensity at a point in the interference pattern when the phase (i. e., the argument of the cosine function) of the interference terms is varied in a range of 2π. For the interference of plane waves the maximum and minimum intensity can also be taken at different points because the intensity of the two single waves is then independent of the position. The visibility can vary between 0 for $I_{\min}=I_{\max}$, i. e., no interference occurs, and 1 for $I_{\min}=0$.

Linearly Polarized Plane Waves

For the case of linearly polarized plane waves the phase constants of each wave $\delta_{\parallel,1}$ and $\delta_{\perp,1}$ on the one hand and $\delta_{\parallel,2}$ and $\delta_{\perp,2}$ on the other are equal or differ only by π. There are, in fact, effectively two different cases:

$$\left.\begin{aligned}\delta_{\parallel,1}=\delta_{\perp,1}\ \wedge\ \delta_{\parallel,2}=\delta_{\perp,2}\\\delta_{\parallel,1}=\delta_{\perp,1}+\pi\ \wedge\ \delta_{\parallel,2}=\delta_{\perp,2}+\pi\end{aligned}\right\}$$
$$\begin{aligned}&\delta_{\perp}=\delta_{\perp,1}-\delta_{\perp,2}\\\Rightarrow&=\delta_{\parallel,1}-\delta_{\parallel,2}=\delta_{\parallel}\\&=:\delta\text{ and }s:=+1\end{aligned} \tag{3.167}$$

$$\left.\begin{aligned}\delta_{\parallel,1}=\delta_{\perp,1}+\pi\ \wedge\ \delta_{\parallel,2}=\delta_{\perp,2}\\\delta_{\parallel,1}=\delta_{\perp,1}\ \wedge\ \delta_{\parallel,2}=\delta_{\perp,2}+\pi\end{aligned}\right\}$$
$$\begin{aligned}&\delta_{\perp}=\delta_{\perp,1}-\delta_{\perp,2}\\\Rightarrow\ &\delta_{\parallel,1}-\delta_{\parallel,2}\mp\pi=\delta_{\parallel}\mp\pi\\&=:\delta\mp\pi\text{ and }s:=-1\end{aligned} \tag{3.168}$$

The parameter s characterizes the different cases and is either $+1$ or -1. Then the intensity of the interference

Part A | 3.3

pattern (see (3.158)) can be expressed as

$$\begin{aligned} I_{1+2} &= I_{\parallel,1} + I_{\parallel,2} \\ &\quad + 2\sqrt{I_{\parallel,1} I_{\parallel,2}} \cos\varphi \cos\left[(\boldsymbol{k}_1 - \boldsymbol{k}_2)\cdot\boldsymbol{r} + \delta\right] \\ &\quad + I_{\perp,1} + I_{\perp,2} \\ &\quad + 2s\sqrt{I_{\perp,1} I_{\perp,2}} \cos\left[(\boldsymbol{k}_1 - \boldsymbol{k}_2)\cdot\boldsymbol{r} + \delta\right] . \end{aligned} \tag{3.169}$$

Here, the angle φ between the directions of propagation of the two waves is used. So, there are several interesting special cases of (3.169):

- Both waves have only TE components, i. e., $I_{\parallel,1} = I_{\parallel,2} = 0$, $I_{\perp,1} \neq 0$ and $I_{\perp,2} \neq 0$. Then, we obtain

$$\begin{aligned} I_{\text{TE,TE}} &= I_{\perp,1} + I_{\perp,2} \\ &\quad + 2s\sqrt{I_{\perp,1} I_{\perp,2}} \cos\left[(\boldsymbol{k}_1 - \boldsymbol{k}_2)\cdot\boldsymbol{r} + \delta\right] . \end{aligned} \tag{3.170}$$

This is the well-known interference equation that is also used for scalar waves where an arbitrary component of the electric vector, but for a small angle φ between the two wave vectors of the waves, is regarded. In this case the visibility $V_{\text{TE,TE}}$ (defined with (3.166)) of the interference pattern is

$$V_{\text{TE,TE}} = \frac{2\sqrt{I_{\perp,1} I_{\perp,2}}}{I_{\perp,1} + I_{\perp,2}} . \tag{3.171}$$

If the intensities of both waves are equal the visibility is 1.

- One wave has only a TE component, i. e., $I_{\parallel,1} = 0$ and $I_{\perp,1} \neq 0$, and the other wave has only a TM component, i. e., $I_{\parallel,2} \neq 0$ and $I_{\perp,2} = 0$. Then the interference terms vanish and the intensity is constant

$$I_{\text{TE,TM}} = I_{\perp,1} + I_{\parallel,2} = \text{constant} . \tag{3.172}$$

This just means that orthogonally polarized waves cannot interfere.

- Both waves have only TM components, i. e., $I_{\perp,1} = I_{\perp,2} = 0$, $I_{\parallel,1} \neq 0$ and $I_{\parallel,2} \neq 0$. Then, we obtain

$$\begin{aligned} I_{\text{TM,TM}} &= I_{\parallel,1} + I_{\parallel,2} \\ &\quad + 2\sqrt{I_{\parallel,1} I_{\parallel,2}} \cos\varphi \cos\left[(\boldsymbol{k}_1 - \boldsymbol{k}_2)\cdot\boldsymbol{r} + \delta\right] . \end{aligned} \tag{3.173}$$

The visibility $V_{\text{TM,TM}}$ is in this case:

$$V_{\text{TM,TM}} = \frac{2\sqrt{I_{\parallel,1} I_{\parallel,2}} \cos\varphi}{I_{\parallel,1} + I_{\parallel,2}} . \tag{3.174}$$

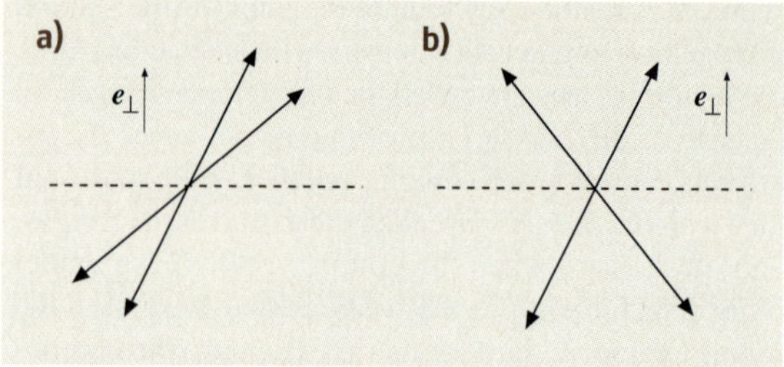

Fig. 3.7a,b The different possibilities for the interference of linearly polarized waves. **(a)** $s = +1$, i. e., the two electric vectors are oscillating in the same quadrant; **(b)** $s = -1$, i. e. the two electric vectors are oscillating in different quadrants. The *dashed line* indicates the plane in which the wave vectors of both waves are situated

So, the visibility is always smaller than 1 and the interference term vanishes if both waves are propagating perpendicular to each other, i. e., $\boldsymbol{e}_1 \cdot \boldsymbol{e}_2 = \cos\varphi = 0$. For the case $I_{\parallel,1} = I_{\parallel,2}$, where the visibility is 1 in the TE polarized case, the visibility in the TM polarized case is $V_{\text{TM,TM}} = \cos\varphi$. Only for small angles between the directions of propagation of the two plane waves the visibility is high. Of course, for small angles there is in fact no real difference between TE- and TM-polarized light and for $\varphi = 0$ or $\varphi = \pi$ the difference between TE and TM exists no longer.

- TE and TM components of both waves are present. Then the value of the constant s is important. For $s = +1$ the interference terms have identical signs and add. But for $s = -1$ the interference terms have different signs and cancel each other out if

$$\sqrt{I_{\parallel,1} I_{\parallel,2}} \cos\varphi = \sqrt{I_{\perp,1} I_{\perp,2}} . \tag{3.175}$$

The meaning of this is that for $s = -1$ the electric vectors of both waves are oscillating in different quadrants. If the above equation is fulfilled they are again perpendicular to each other and cannot interfere. This is explained in Fig. 3.7.

Circularly Polarized Plane Waves

Circularly polarized plane waves exist according to (3.137) only if

$$\begin{aligned} I_{\parallel,1} &= I_{\perp,1} = \frac{1}{2} I_1; \quad \delta_{\parallel,1} - \delta_{\perp,1} = \pm\frac{\pi}{2} , \\ I_{\parallel,2} &= I_{\perp,2} = \frac{1}{2} I_2; \quad \delta_{\parallel,2} - \delta_{\perp,2} = \pm\frac{\pi}{2} . \end{aligned}$$

Then, the intensity of the interference pattern is determined according to (3.158)

$$I_{1+2} = \frac{1}{2}\Big\{I_1 + I_2 + 2\sqrt{I_1 I_2} \times \cos\varphi \cos\left[(\boldsymbol{k}_1 - \boldsymbol{k}_2)\cdot\boldsymbol{r} + \delta_\parallel\right]\Big\} + \frac{1}{2}\Big\{I_1 + I_2 + 2\sqrt{I_1 I_2} \times \cos\left[(\boldsymbol{k}_1 - \boldsymbol{k}_2)\cdot\boldsymbol{r} + \delta_\perp\right]\Big\}$$
$$= I_1 + I_2 + \sqrt{I_1 I_2} \times \Big\{\cos\varphi\cos\left[(\boldsymbol{k}_1 - \boldsymbol{k}_2)\cdot\boldsymbol{r} + \delta_\parallel\right] + \cos\left[(\boldsymbol{k}_1 - \boldsymbol{k}_2)\cdot\boldsymbol{r} + \delta_\perp\right]\Big\}. \quad (3.176)$$

Now, we have to differentiate between several cases:

- Both waves have the same chirality, i.e., either $\delta_{\parallel,1} - \delta_{\perp,1} = \pi/2$ and $\delta_{\parallel,2} - \delta_{\perp,2} = \pi/2$ or $\delta_{\parallel,1} - \delta_{\perp,1} = -\pi/2$ and $\delta_{\parallel,2} - \delta_{\perp,2} = -\pi/2$. Then, the phase differences are

$$\delta_\parallel = \delta_{\parallel,1} - \delta_{\parallel,2}; \quad \delta_\perp = \delta_{\perp,1} - \delta_{\perp,2} = \delta_\parallel =: \delta$$

and (3.176) reduces to

$$I_{\uparrow\uparrow} = I_1 + I_2 + \sqrt{I_1 I_2} \times (\cos\varphi + 1)\cos\left[(\boldsymbol{k}_1 - \boldsymbol{k}_2)\cdot\boldsymbol{r} + \delta\right]. \quad (3.177)$$

The visibility (3.166) is

$$V_{\uparrow\uparrow} = \frac{\sqrt{I_1 I_2}\,(\cos\varphi + 1)}{I_1 + I_2}. \quad (3.178)$$

For small angles φ between the directions of propagation of the two waves the intensity is identical to the interference pattern of two linearly TE-polarized waves and the visibility can reach 1 for equal intensities in both waves. If both waves are propagating perpendicular to each other the interference term has only half the size and the visibility is for equal intensities only $1/2$. If the angle φ is larger than $\pi/2$ and approaches π the interference term and the visibility vanish. This means that waves propagating antiparallel and with the same chirality cannot interfere.

- Both waves have different chirality, i.e., either $\delta_{\parallel,1} - \delta_{\perp,1} = \pi/2$ and $\delta_{\parallel,2} - \delta_{\perp,2} = -\pi/2$ or $\delta_{\parallel,1} - \delta_{\perp,1} = -\pi/2$ and $\delta_{\parallel,2} - \delta_{\perp,2} = \pi/2$. Then, the phase differences are

$$\delta_\parallel = \delta_{\parallel,1} - \delta_{\parallel,2} =: \delta;$$
$$\delta_\perp = \delta_{\perp,1} - \delta_{\perp,2} = \delta_\parallel \pm \pi = \delta \pm \pi$$

and (3.176) reduces to

$$I_{\uparrow\downarrow} = I_1 + I_2 + \sqrt{I_1 I_2} \times (\cos\varphi - 1)\cos\left[(\boldsymbol{k}_1 - \boldsymbol{k}_2)\cdot\boldsymbol{r} + \delta\right]. \quad (3.179)$$

The visibility (3.166) is

$$V_{\uparrow\downarrow} = \frac{\sqrt{I_1 I_2}\,(1 - \cos\varphi)}{I_1 + I_2}. \quad (3.180)$$

So, for different chirality the behavior is reverse to that of equal chirality. Waves that are propagating parallel ($\varphi = 0$) and have different chirality cannot interfere, whereas waves that are propagating antiparallel ($\varphi = \pi$) and have different chirality interfere very well.

The Application of Two-Beam Interference for an Electron Accelerator

A quite interesting modern application of the interference of two waves is the laser-driven electron accelerator [3.21],[3.22]. This accelerator can in principle also be used for any other charged particle whereby the efficiency increases if the velocity of the particle is nearly identical to the speed of light. So, heavy particles need a lot of kinetic energy (at best a multiple of their rest energy m_0c^2) if they shall be accelerated by laser light with a good efficiency. Here, only the basic principle can be discussed.

Figure 3.8 shows two interfering waves both with linear TM polarization, equal amplitudes and a phase

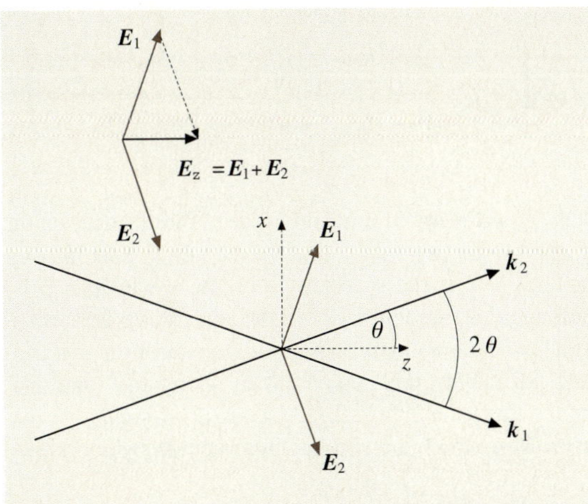

Fig. 3.8 The interference of two linearly polarized waves with TM polarization and π phase difference results in a longitudinal component $\boldsymbol{E}_z$ of the electric field

difference of π on points along the z-axis. First of all, we assume that the waves are plane. The z-axis bisects the angle 2θ between the wave vectors $\boldsymbol{k}_1$ and $\boldsymbol{k}_2$ of both waves so that θ is the angle between the z-axis and one of the wave vectors. Then, the electric vectors $\boldsymbol{E}_1$ and $\boldsymbol{E}_2$ are oriented as indicated and the components perpendicular to the z-axis, i. e., parallel to the x-axis, cancel each other. The magnetic vectors of both waves, which are perpendicular to the drawing plane, i. e., parallel to the y-axis, also cancel each other out on points along the z-axis because they are antiparallel. However, due to the configuration of the two interfering waves there is a resulting longitudinal component $\boldsymbol{E}_z$ of the electric vector for points on the z-axis. From a mathematical point of view, we have for the electric vectors using the coordinate system of Fig. 3.8

$$\begin{aligned}
&\boldsymbol{E}_1\,(x,z,t)\\
&=\begin{pmatrix} E_0\cos\theta\cos\left[\frac{2\pi}{\lambda}\,(-x\sin\theta+z\cos\theta)-\omega t\right]\\ 0\\ E_0\sin\theta\cos\left[\frac{2\pi}{\lambda}\,(-x\sin\theta+z\cos\theta)-\omega t\right]\end{pmatrix},\\
&\boldsymbol{E}_2\,(x,z,t)\\
&=\begin{pmatrix} -E_0\cos\theta\cos\left[\frac{2\pi}{\lambda}\,(x\sin\theta+z\cos\theta)-\omega t\right]\\ 0\\ E_0\sin\theta\cos\left[\frac{2\pi}{\lambda}\,(x\sin\theta+z\cos\theta)-\omega t\right]\end{pmatrix};\\
&\Rightarrow \boldsymbol{E}_z\,(x=0,z,t)\\
&=\boldsymbol{E}_1\,(x=0,z,t)+\boldsymbol{E}_2\,(x=0,z,t)\\
&=\begin{pmatrix}0\\0\\2E_0\sin\theta\cos\left(\frac{2\pi}{\lambda}z\cos\theta-\omega t\right)=:E_z\,(0,z,t)\end{pmatrix}.
\end{aligned} \tag{3.181}$$

Here, E_0 is the maximum amplitude of the electric vector of one of the two interfering waves ($E_0=\max|\boldsymbol{E}_1|=\max|\boldsymbol{E}_2|$), λ is the wavelength of the waves and ω is their angular frequency. So, we see that along the z-axis with $x=0$ there exists only a component of the electric field parallel to the z-axis and by using the relations $\lambda\nu=c$ and $\omega=2\pi\nu=2\pi c/\lambda$ (ν is the frequency) for our waves, which are situated in a vacuum, we obtain

$$\begin{aligned}
&E_z\,(x=0,z,t)\\
&=2E_0\sin\theta\cos\left[2\pi\frac{c}{\lambda}\left(z\frac{\cos\theta}{c}-t\right)\right].
\end{aligned} \tag{3.182}$$

Now, a relativistic electron, i. e., the speed v of the electron should be nearly the speed of light c, travels along the z-axis from left to right. It should pass the regarded point at that time where $E_z=-2E_0\sin\theta$. Then, the electron will be maximally accelerated along the z-axis due to its negative electric charge. But, the interference maximum with $E_z=-2E_0\sin\theta$ seems to propagate along the z-axis with the phase velocity $c/\cos\theta$, which is faster than the speed of light for $\theta>0$. Of course, this is in no contradiction to special relativity because no information nor energy is transported with this speed. So, for a small angle θ the relativistic electron with $v\approx c$ (but nevertheless $v<c$) travels a certain distance nearly in phase with the accelerating electric field before it comes out of phase because of $c/\cos\theta>c>v$. The distance after which the electron is out of phase can also be calculated very easily. The electron travels in the laboratory framework in the time interval t, a distance $z=vt$, i. e., $t=z/v$. The velocity v can be assumed to be constant during the acceleration process because it should be nearly the speed of light and therefore it does not change considerably although the electron may gain a lot of kinetic energy. By introducing this into (3.182) the argument Φ of the cosine function as a function of the position z on the z-axis is:

$$\Phi(z)=2\pi\frac{c}{\lambda}\left(z\frac{\cos\theta}{c}-t\right)=2\pi\frac{c}{\lambda}z\left(\frac{\cos\theta}{c}-\frac{1}{v}\right). \tag{3.183}$$

If the phase Φ of the electric field, which affects the electron changes by $\pm\pi$, the electric field that first accelerated the electron will now slow it down. So, the distance Δz on the z-axis between being accelerated and decelerated by the electric field is

$$\begin{aligned}
\Delta\Phi&=2\pi\frac{c}{\lambda}\Delta z\left(\frac{\cos\theta}{c}-\frac{1}{v}\right)\\
&=\pm\pi\ \Rightarrow\ \Delta z=\pm\frac{\lambda}{2\,(\cos\theta-c/v)}.
\end{aligned} \tag{3.184}$$

For $v\approx c$ the longest distance for being in phase would be achieved by $\theta=0$. But then, the accelerating electric field itself would be zero because of the factor $\sin\theta$ in (3.182). So, in practice, a tradeoff has to be found between a big value $\sin\theta$ and another big value $\cos\theta$. Additionally, the velocity of the electron (or other charged particle) should be nearly the speed of light.

However, we see that for the interference of two infinitely extended plane waves the electron would be accelerated and slowed down periodically. But, wc can replace the plane waves by focused laser beams, i. e., Gaussian beams (Sect. 3.5), so that the region with a high electric field has a quite limited length smaller than Δz. The beam waist of each laser beam should be at the crossing point of the two Gaussian beams on the z-axis to achieve a high amplitude of the electric vector. Then, the electric field amplitude E_0 is not constant along the z-axis but decreases outside of the beam waist like a Gaussian curve. So, it is possible to achieve a net acceleration of the electron if it is harmonized with the phase of the electric field of the laser beams when it crosses the beam waist. If the electron is not harmonized with the phase of the laser beams it can also be slowed down. The concrete calculation in the case of Gaussian beams is, of course, a little bit more complex than with plane waves because the wave vector and therefore the direction of the electric vector changes locally in the case of a Gaussian beam. By using ultrashort and focused laser pulses the resulting electric field in the beam waist, which accelerates the electron can be as high as 1 GV/m or higher. Of course, the acceleration distance is only as long as the beam waist, i. e., several μm for a strongly focused laser beam. So, some keV of kinetic energy can be gained by the electron. But, by repeating many acceleration devices in series (and, of course, harmonized in phase) the effective acceleration distance can be increased so that the laser-driven electron accelerator may become an alternative to conventional particle accelerators in the future.

3.3.3 Interference of Arbitrary Scalar Waves

The interference phenomena are, of course, not restricted to plane waves but can occur for arbitrary waves. Since for arbitrary waves the polarization can change locally we will neglect the polarization in this section and concentrate on so-called scalar waves.

Some Notes on Scalar Waves

In the case of scalar waves only one Cartesian component of the electric (or magnetic) vector is regarded and the complete polarization state is neglected. Nevertheless, for two linearly polarized interfering waves, which are both TE polarized, the result of the scalar calculation is identical to the exact result. Since the orthogonality condition is neglected for scalar waves the result of the scalar wave equation is not automatically a solution of the Maxwell equations.

A scalar wave, which is often used in optics, is a spherical wave with its center of curvature at the point $\boldsymbol{r}_0$ with the complex amplitude

$$u(\boldsymbol{r}) = a\frac{\mathrm{e}^{\mathrm{i}k|\boldsymbol{r}-\boldsymbol{r}_0|}}{|\boldsymbol{r}-\boldsymbol{r}_0|} \tag{3.185}$$

and the modulus of the wave vector $k = 2\pi n/\lambda$, which is also called a wave number. a is a constant. It should be mentioned that a spherical wave is a solution of the scalar Helmholtz equation of homogeneous materials (Sect. 3.1.5), where only one component of the electric or magnetic vector is regarded. But, a spherical wave is not a solution of the Maxwell equations itself because this would violate the orthogonality conditions of an electromagnetic wave. But, in the far field and in the plane perpendicular to its dipole axis, a dipole radiation is a good approximation for a spherical wave.

Here, the scalar complex amplitude u was introduced, which can stand, apart from a constant of proportionality, for one component of the electric or magnetic vector. Again, the intensity of this scalar wave is at least proportional to the square of the modulus of u and in the following we just define for scalar waves

$$I := uu^* . \tag{3.186}$$

A general scalar wave can be described by

$$u(\boldsymbol{r}) = A(\boldsymbol{r})\,\mathrm{e}^{\mathrm{i}\Phi(\boldsymbol{r})} . \tag{3.187}$$

In this case, A is a real function, which changes only slowly with the position $\boldsymbol{r}$, where Φ is also a real function but the complex exponential factor $\exp(\mathrm{i}\Phi)$ varies rapidly with the position $\boldsymbol{r}$.

The Interference Equation for Scalar Waves

By using two general scalar waves (3.187)

$$u_1(\boldsymbol{r}) = A_1(\boldsymbol{r})\,\mathrm{e}^{\mathrm{i}\Phi_1(\boldsymbol{r})}, \quad u_2(\boldsymbol{r}) = A_2(\boldsymbol{r})\,\mathrm{e}^{\mathrm{i}\Phi_2(\boldsymbol{r})}$$

instead of using plane waves in (3.158) or (3.170) the interference equation of scalar waves is obtained by

$$I_{1+2} = I_1 + I_2 + 2\sqrt{I_1 I_2}\cos\Phi \tag{3.188}$$

with $I_1 = A_1^2$, $I_2 = A_2^2$ and $\Phi = \Phi_1 - \Phi_2$. In some cases it is more convenient to write this equation as

$$I_{1+2} = I_0\,(1 + V\cos\Phi) . \tag{3.189}$$

Here, $I_0 = I_1 + I_2$ is defined as the resulting intensity for incoherent light, where only the intensities of the single waves have to be added. The visibility V is defined in (3.166) and is here

$$V = \frac{I_{max} - I_{min}}{I_{max} + I_{min}} = \frac{2\sqrt{I_1 I_2}}{I_1 + I_2} = \frac{2\sqrt{I_1 I_2}}{I_0} . \tag{3.190}$$

For general scalar waves the fringe period in the detector plane with coordinates (x, y) is not constant but will vary. But at a point (x, y) in the neighborhood of a fixed point (x_0, y_0) the phase function Φ can be written as a Taylor expansion neglecting all terms of second and higher order:

$$\begin{aligned}\Phi(x, y) &\approx \Phi(x_0, y_0) + \begin{pmatrix} \frac{\partial\Phi(x_0,y_0)}{\partial x} \\ \frac{\partial\Phi(x_0,y_0)}{\partial y} \end{pmatrix} \cdot \begin{pmatrix} x - x_0 \\ y - y_0 \end{pmatrix} \\ &= \Phi(x_0, y_0) + \nabla_\perp \Phi(x_0, y_0) \cdot \Delta\boldsymbol{r} \,. \end{aligned} \tag{3.191}$$

Here, the two-dimensional nabla operator $\nabla_\perp$ is introduced. The local fringe period p is defined as the distance between two fringes, i. e., the distance where the phase function increases or decreases by 2π taken along a path parallel to the local phase gradient. Therefore, we have for the vector $\Delta\boldsymbol{r}$ pointing from one fringe to the next neighboring fringe at the position (x_0, y_0)

$$\begin{aligned}&\Delta\boldsymbol{r} = p\frac{\nabla_\perp\Phi}{|\nabla_\perp\Phi|} \Rightarrow \nabla_\perp\Phi \cdot \Delta\boldsymbol{r} = p|\nabla_\perp\Phi| = 2\pi ; \\ &\Rightarrow p = \frac{2\pi}{|\nabla_\perp\Phi|} = \frac{2\pi}{\sqrt{\left(\frac{\partial\Phi}{\partial x}\right)^2 + \left(\frac{\partial\Phi}{\partial y}\right)^2}} \,. \end{aligned} \tag{3.192}$$

All quantities have to be calculated at the point (x_0, y_0). By comparing this equation with (3.164) it is clear that the component $\boldsymbol{G}_\parallel$ of the grating vector in the xy-plane is defined by

$$\boldsymbol{G}_\parallel = \begin{pmatrix} \frac{\partial\Phi}{\partial x} \\ \frac{\partial\Phi}{\partial y} \\ 0 \end{pmatrix} . \tag{3.193}$$

The grating vector $\boldsymbol{G}$ itself is defined as

$$\boldsymbol{G} = \nabla\Phi = \begin{pmatrix} \frac{\partial\Phi}{\partial x} \\ \frac{\partial\Phi}{\partial y} \\ \frac{\partial\Phi}{\partial z} \end{pmatrix} , \tag{3.194}$$

with Φ defined as a function of (x, y, z).

The local fringe frequency ν is defined as the reciprocal of the local fringe period

$$\nu = \frac{1}{p} = \frac{\sqrt{\left(\frac{\partial\Phi}{\partial x}\right)^2 + \left(\frac{\partial\Phi}{\partial y}\right)^2}}{2\pi} \tag{3.195}$$

and describes the number of fringes per length unit. If the fringe frequency is too high the interference pattern cannot be resolved in practice since common detector arrays like a CCD camera only have a limited number of pixels per length unit and integrate the light intensity over the area of one pixel.

Interference of Scalar Spherical and Plane Waves

The interference of two plane waves is investigated in detail in the last section for general polarization states. Simple examples of the interference of two scalar waves are the interference of a spherical wave and a plane wave or the interference of two spherical waves. In principle, the general statements to the effects of different polarization states also hold for spherical waves as long as the numerical aperture is not too high. So, the investigation of scalar waves is not really a restriction.

Two spherical waves with the wavelength λ and the corresponding wave number $k = 2\pi n/\lambda$ having their center of curvature at the points $\boldsymbol{r}_1 = (x_1, y_1, z_1)$ and $\boldsymbol{r}_2 = (x_2, y_2, z_2)$ have the complex amplitude functions

$$\begin{aligned} u_1(\boldsymbol{r}) &= a_1 \frac{e^{ik|\boldsymbol{r}-\boldsymbol{r}_1|}}{|\boldsymbol{r}-\boldsymbol{r}_1|} , \\ u_2(\boldsymbol{r}) &= a_2 \frac{e^{ik|\boldsymbol{r}-\boldsymbol{r}_2|}}{|\boldsymbol{r}-\boldsymbol{r}_2|} . \end{aligned}$$

In the following, only the interference pattern in the xy-plane at $z = 0$ in an area centered around the origin of the coordinate system at $x = y = z = 0$ is evaluated. Additionally, the distances $|\boldsymbol{r}_i|$ $(i = 1, 2)$ of the centers of curvature of both spherical waves from the origin of the coordinate system shall be large compared to the maximum distance $|\boldsymbol{r}| = \sqrt{x^2 + y^2}$ of the origin of the coordinate system from a point lying in the evaluated aperture of the interference pattern. Then the amplitude of the spherical waves can be assumed to be constant because of

$$\begin{aligned} &|\boldsymbol{r}_i| \gg |\boldsymbol{r}| \Rightarrow \\ &\quad |\boldsymbol{r}-\boldsymbol{r}_i| = \sqrt{(\boldsymbol{r}-\boldsymbol{r}_i)\cdot(\boldsymbol{r}-\boldsymbol{r}_i)} \\ &\quad = \sqrt{|\boldsymbol{r}_i|^2 + |\boldsymbol{r}|^2 - 2|\boldsymbol{r}_i||\boldsymbol{r}|\cos\alpha} \approx |\boldsymbol{r}_i| \,. \end{aligned}$$

Here, α is the angle between the two vectors $\boldsymbol{r}$ and $\boldsymbol{r}_i$. Therefore, the two spherical waves are written as

$$\begin{aligned} u_1(\boldsymbol{r}) &= A_1 e^{ik|\boldsymbol{r}-\boldsymbol{r}_1|} , \\ u_2(\boldsymbol{r}) &= A_2 e^{ik|\boldsymbol{r}-\boldsymbol{r}_2|} \end{aligned}$$

with constant amplitudes $A_1 = a_1/|\boldsymbol{r}_1|$ and $A_2 = a_2/|\boldsymbol{r}_2|$. The arguments of the complex exponential functions can, of course, not be replaced by the constant terms $|\boldsymbol{r}_i|$ because these are very fast oscillating functions. Then, the intensity I_{1+2} of the interference pattern is, according to (3.188),

$$I_{1+2}(x, y) = I_1 + I_2 + 2\sqrt{I_1 I_2}\cos\Phi(x, y)$$
$$\text{with } I_1 = A_1^2;\; I_2 = A_2^2$$

and

$$\begin{aligned}\Phi(x, y) &= k\left(|\boldsymbol{r}-\boldsymbol{r}_1| - |\boldsymbol{r}-\boldsymbol{r}_2|\right)\\ &= k\left(|\boldsymbol{r}_1|\sqrt{1+\frac{x^2+y^2-2xx_1-2yy_1}{|\boldsymbol{r}_1|^2}}\right.\\ &\quad\left.-|\boldsymbol{r}_2|\sqrt{1+\frac{x^2+y^2-2xx_2-2yy_2}{|\boldsymbol{r}_2|^2}}\right).\end{aligned} \tag{3.196}$$

The square roots can be developed into a Taylor series according to $\sqrt{1+x} \approx 1 + x/2 - x^2/8$ for $x \ll 1$. Since $|\boldsymbol{r}_i| \gg |\boldsymbol{r}|$, the most important terms are

$$\begin{aligned}&|\boldsymbol{r}_i|\sqrt{1+\frac{x^2+y^2-2xx_i-2yy_i}{|\boldsymbol{r}_i|^2}}\\ &\approx |\boldsymbol{r}_i| + \frac{x^2+y^2}{2|\boldsymbol{r}_i|} - \frac{xx_i+yy_i}{|\boldsymbol{r}_i|}\\ &\quad - \frac{\left(x^2+y^2-2xx_i-2yy_i\right)^2}{8|\boldsymbol{r}_i|^3}.\end{aligned} \tag{3.197}$$

The last term can be neglected for

$$\begin{aligned}&k\frac{\left(x^2+y^2-2xx_i-2yy_i\right)^2}{8|\boldsymbol{r}_i|^3} \ll 1\\ &\rightarrow \frac{\left(x^2+y^2-2xx_i-2yy_i\right)^2}{8|\boldsymbol{r}_i|^3} \ll \frac{\lambda}{2\pi n}.\end{aligned} \tag{3.198}$$

By using spherical coordinates $r_i, \vartheta_i, \varphi_i$ for $\boldsymbol{r}_i$, where r_i is the distance from the origin, ϑ_i is the polar angle and φ_i is the azimuthal angle,

$$x_i = r_i \cos\varphi_i \sin\vartheta_i\,, \tag{3.199}$$
$$y_i = r_i \sin\varphi_i \sin\vartheta_i\,, \tag{3.200}$$
$$z_i = r_i \cos\vartheta_i\,, \tag{3.201}$$

the condition for neglecting the last term is

$$\begin{aligned}&\frac{\left(x^2+y^2\right)^2}{4r_i^3} - \frac{\left(x^2+y^2\right)(x\cos\varphi_i + y\sin\varphi_i)\sin\vartheta_i}{r_i^2}\\ &+\frac{(x\cos\varphi_i + y\sin\varphi_i)^2\sin^2\vartheta_i}{r_i} \ll \frac{\lambda}{\pi n}.\end{aligned} \tag{3.202}$$

If this condition is fulfilled for both spherical waves the phase function of the interference pattern can be written as

$$\begin{aligned}\Phi(x, y) &\approx \delta + \frac{\pi n}{\lambda}\left(\frac{1}{r_1}-\frac{1}{r_2}\right)\left(x^2+y^2\right)\\ &\quad - \frac{2\pi n}{\lambda}(\cos\varphi_1\sin\vartheta_1 - \cos\varphi_2\sin\vartheta_2)\,x\\ &\quad - \frac{2\pi n}{\lambda}(\sin\varphi_1\sin\vartheta_1 - \sin\varphi_2\sin\vartheta_2)\,y\,.\end{aligned} \tag{3.203}$$

The phase constant δ is defined as $\delta = 2\pi n(r_1 - r_2)/\lambda$.

The term dependent on x^2+y^2 is called defocus and is proportional to the difference of the curvatures of both spherical waves. The linear terms in x and y are called tilts and are only present if the centers of curvature of the two spherical waves and the origin do not lie on a common line. In the interferometric testing of spherical surfaces or in the measurement of the wave aberrations of lenses there often appears defocus and tilts due to an axial ($\Rightarrow$ defocus) or lateral ($\Rightarrow$ tilts) misalignment of the test object. Then the coefficients of these terms are determined by a least-squares fit of the function

$$\Phi_{\text{misalign}} = a + bx + cy + d\left(x^2+y^2\right) \tag{3.204}$$

to the measured phase function Φ_{measured}. Afterwards, the phase function Φ_{reduced}, which is freed from misalignment aberrations is calculated by

$$\Phi_{\text{reduced}} = \Phi_{\text{measured}} - \Phi_{\text{misalign}} \tag{3.205}$$

using the fitted coefficients a, b, c and d. The phase function Φ_{reduced} then only contains the desired wave aberrations or the desired surface deviations from the ideal surface plus systematic errors of the experimental setup.

A special case of (3.203) is that one of the waves is a plane wave. Without loss of generality the second wave shall be plane. This means that the parameter r_2 is infinity. Then (3.203) reduces to:

$$\begin{aligned}\Phi(x, y) &\approx \delta + \frac{\pi n}{\lambda r_1}\left(x^2+y^2\right)\\ &\quad - \frac{2\pi n}{\lambda}\left(\cos\varphi_1\sin\vartheta_1 - e_{x,2}\right)x\\ &\quad - \frac{2\pi n}{\lambda}\left(\sin\varphi_1\sin\vartheta_1 - e_{y,2}\right)y\,,\end{aligned} \tag{3.206}$$

where $e_{x,2} := \cos\varphi_2\sin\vartheta_2$ and $e_{y,2} := \sin\varphi_2\sin\vartheta_2$ are the x and y components of the unit vector $\boldsymbol{e}_2 = \boldsymbol{k}_2/|\boldsymbol{k}_2|$ parallel to the wave vector $\boldsymbol{k}_2$ of the plane wave. Of course, the phase constant δ, which is only defined modulus 2π, is therefore not infinity, but has a certain value that depends on r_1 and the phase offset of the plane wave at $(x=0,\ y=0,\ z=0)$. In the case of the interference of a spherical and a plane wave the defocus term is directly proportional to the curvature of the spherical wave. The tilt terms again depend on both waves but for the case that either the spherical wave has its center of curvature at $x=y=0$ ($\Rightarrow \sin\vartheta_1 = 0$) or the plane wave is perpendicular to the xy-plane ($\Rightarrow e_{x,2}=e_{y,2}=0$) the tilt terms depend only on the parameters of one wave.

Two Examples of Interference Patterns

Assume that we have two interfering monochromatic waves with a wavelength $\lambda = 0.5\,\mu\text{m}$. Additionally, we know that the first of the waves is a plane wave that propagates parallel to the optical axis, which is defined to be parallel to the surface normal of the detector and to intersect the detector in its center. So, the parameters of the first wave in (3.203) are $r_1 \to \infty$, $\vartheta_1 = 0$ and $\varphi_1 = 0$

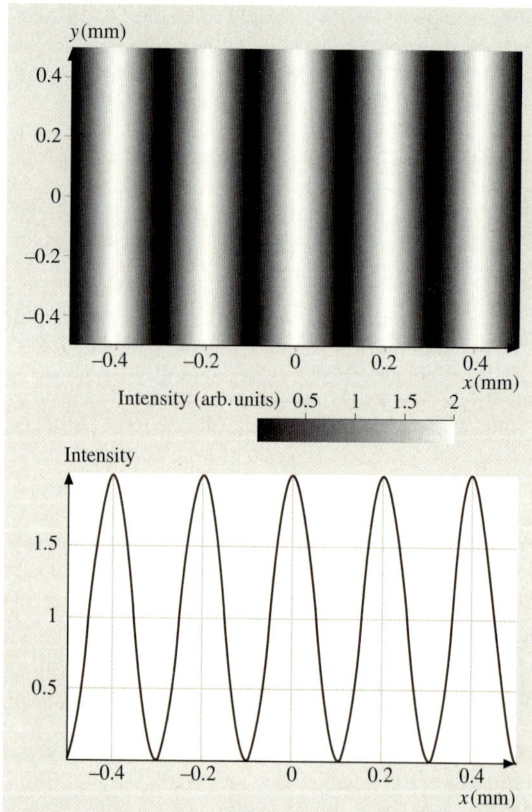

Fig. 3.9 Example of an interference pattern (interferogram) with straight, parallel and equidistant fringes. *Top:* (simulated) camera picture as also seen directly with the eye. *Bottom:* section through the intensity function

and the phase function Φ of the interference pattern depends only on the parameters of the second wave

$$\Phi(x,y) \approx \delta - \frac{\pi}{\lambda r_2}\left(x^2+y^2\right) + \frac{2\pi}{\lambda}\cos\varphi_2 \sin\vartheta_2\, x + \frac{2\pi}{\lambda}\sin\varphi_2 \sin\vartheta_2\, y\,. \tag{3.207}$$

The refractive index n has been set to 1 because the measurements should be made in air.

Let us now assume that we detect the interference pattern, which is displayed in Fig. 3.9. Such an interference pattern is often called an interferogram. In this case it has straight, parallel and equidistant fringes parallel to the y-axis. Therefore, the second wave also has to be a plane wave ($r_2 \to \infty$), which can be described by the two angles φ_2 and ϑ_2. The interference pattern only changes along the x-axis and there the period p is 0.2 mm. Therefore, the phase function Φ has to be of the form $\Phi(x,y) = ax$ with $a = 2\pi/p = 10\pi/\text{mm}$. Comparing this with (3.207) results in

$$\begin{aligned}
&\delta + \frac{2\pi}{\lambda}\cos\varphi_2\sin\vartheta_2\, x + \frac{2\pi}{\lambda}\sin\varphi_2\sin\vartheta_2\, y = ax;\\
&\Rightarrow \quad \delta = 0 \wedge \varphi_2 = 0 \wedge \sin\vartheta_2 = \frac{a\lambda}{2\pi} = 0.0025\,.
\end{aligned} \tag{3.208}$$

By looking at the intensity pattern it can also be seen that the visibility has the maximum value of $V = (I_{\max} - I_{\min})/(I_{\max} + I_{\min}) = (2-0)/(2+0) = 1$. Therefore, the second plane wave has the same intensity as the first wave.

In a second (simulated) measurement the interferogram of Fig. 3.10 is obtained. It can be estimated that the local fringe frequency increases linearly with the distance from the center and that we therefore have a defocus term with a quadratic phase function. No linear phase function is present and therefore the tilt angle ϑ_2 of the second wave has to be zero. A more detailed evaluation of the intensity pattern using the interference equation (3.188) confirms the estimation that the phase function is $\Phi(x,y) = b(x^2+y^2)$ with $b = 20\pi/\text{mm}^2$. Additionally, it can be seen that the visibility is $V = (1.28-0.72)/(1.28+0.72) = 0.28$.

Using (3.207) to calculate the radius of curvature r_2 of the second wave results in

$$\begin{aligned}
&\Phi(x,y) = -\frac{\pi}{\lambda r_2}\left(x^2+y^2\right) = b\left(x^2+y^2\right)\\
&\Rightarrow |r_2| = \frac{\pi}{b\lambda} = 100\,\text{mm}\,.
\end{aligned} \tag{3.209}$$

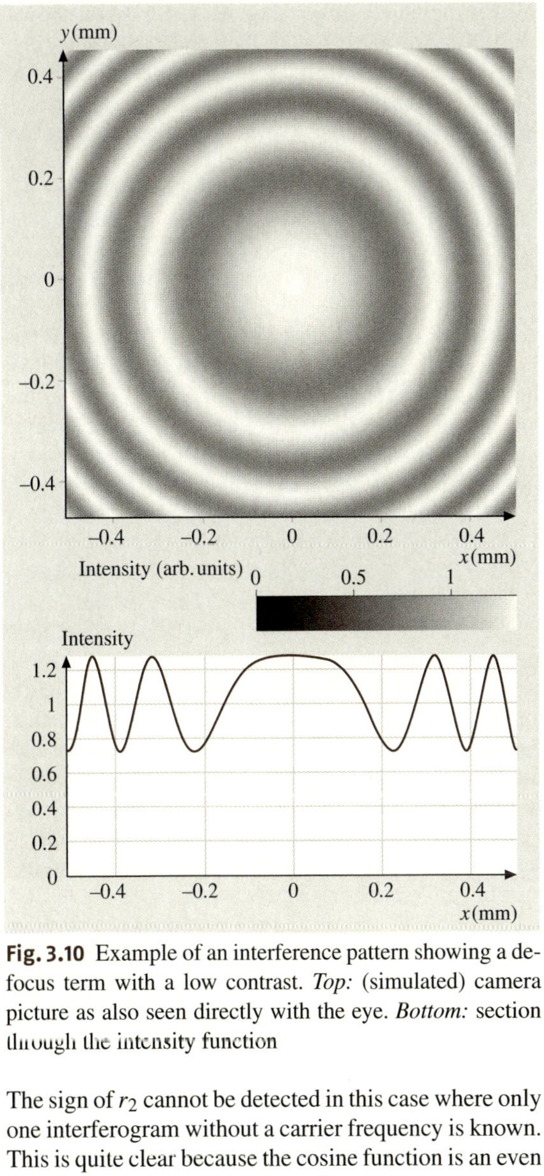

Fig. 3.10 Example of an interference pattern showing a defocus term with a low contrast. *Top:* (simulated) camera picture as also seen directly with the eye. *Bottom:* section through the intensity function

The sign of r_2 cannot be detected in this case where only one interferogram without a carrier frequency is known. This is quite clear because the cosine function is an even function and so $\cos\Phi = \cos(-\Phi)$.

By using (3.190) and the approach $I_2 = \alpha I_1$ the coefficient α can be calculated using the visibility V

$$V = \frac{2\sqrt{\alpha}}{1+\alpha}$$

$$\Rightarrow \alpha = \frac{2}{V^2} - 1 \pm \sqrt{\left(\frac{2}{V^2} - 1\right)^2 - 1} = \begin{cases} 49 \\ 0.02 \end{cases} . \tag{3.210}$$

This means that the intensity of the second wave is either about 50 times higher than that of the first wave or about 50 times smaller. Which one of these two values is valid can be determined by measuring first the intensity I_1 of the first wave alone and then the intensity of the interference pattern and therefore also $I_1 + I_2$. The example also shows that the visibility decreases quite slowly if the intensity difference factor α increases. This is the reason why, e.g., a scattered spherical wave of a dust particle, which has quite a small intensity compared to the intensity of the illuminating coherent wave, produces in many cases quite high-contrast fringes that disturb the measurement.

These two examples are, of course, quite simple and can be evaluated manually. However, already in the second example it can be seen that it is not so easy to decide whether it is really a pure defocus term or mixed with some other terms. Therefore, in practice, an automated evaluation of the interference pattern has to be made [3.20]. One step to this is the phase shifting technique which will be discussed briefly on page 121.

3.3.4 Some Basic Ideas of Interferometry

The basic principle of an interferometer is that an incident wave is divided into two waves, which can then interfere with each other. In most interferometers, like, e.g., a Michelson or a Mach–Zehnder interferometer, there exists a so-called reference arm and an object arm. The object arm often contains an object to be tested that changes the object wave. Together with the unchanged wave of the reference arm the interference pattern is formed and carries information about the test object. Nevertheless, there are also interferometers, like, e.g., shearing interferometers, which do not have an object and a reference arm but two copies of an object wave that interfere.

The applications of interferometers are the measurement of surface deviations or aberrations of optical elements and wavefront characterization. Another application is the high-precision length measurement. In the following, the principles of the most important two-beam interferometers with monochromatic light are described. Other types of interferometers, which are not treated here, are interferometers with two or more wavelengths and so-called white-light interferometers with a broad spectrum of wavelengths. Additionally, there are multiple beam interferometers where three or more light beams interfere, e.g., a Fabry–Perot interferometer. For more information about interferometry we refer to [3.17–20, 23–28].

Michelson Interferometer

One of the simplest interferometers is the Michelson interferometer (Fig. 3.11). A plane wave is divided by a beam splitter into two plane waves. One of these plane waves hits the reference mirror and the other hits the object mirror. Both waves are then reflected back and each wave is again divided into two plane waves. Therefore, the Michelson interferometer has two exits, where one is identical to the entrance so that only the other can really be used. If the beam splitter is exactly oriented at 45° relative to the incoming plane wave and both mirrors are exactly perpendicular to the plane waves the wave vectors $\boldsymbol{k}_1$ and $\boldsymbol{k}_2$ of the two plane waves at the exit are parallel, i. e., $\boldsymbol{k}_1 = \boldsymbol{k}_2$. Then, according to (3.170), which is also valid for scalar waves, the intensity in the whole space behind the exit, e.g., in the detector plane, which shall be perpendicular to the z-axis, depends only on the phase difference δ, which is constant over the whole exit pupil:

$$\begin{aligned} I_{1+2} &= I_1 + I_2 + 2\sqrt{I_1 I_2}\cos\left[(\boldsymbol{k}_1 - \boldsymbol{k}_2)\cdot\boldsymbol{r} + \delta\right] \\ &= I_1 + I_2 + 2\sqrt{I_1 I_2}\cos\delta\,. \end{aligned} \tag{3.211}$$

In this case one speaks of fluffed out fringes because no interference fringes are present. Depending on the optical path difference δ between object and reference arm the intensity can have a maximum or a minimum. If the object mirror is axially shifted the intensity on the detector changes periodically and one period corresponds to an axial shift Δz of half a wavelength in the material with refractive index n, in which the light propagates (normally air), because of the double pass arrangement

$$\Delta z = \frac{\lambda}{2n}\,. \tag{3.212}$$

If the beam splitter or one of the mirrors are tilted, there are interference fringes on the detector. An axial shift of the object mirror then causes a lateral movement of the fringes. Again, the fringes move by one period if the axial shift is $\lambda/(2n)$. A typical application of a Michelson interferometer is the length measurement whereby the relative shift of the object mirror can be measured.

There are several variations of the Michelson interferometer. One is a Twyman–Green interferometer, where a lens is placed in the object arm and instead of using a plane object mirror a spherical mirror is used. In the case of the ideal adjustment of the interferometer the focus of the lens and the center of curvature of the spherical mirror have to coincide. If the quality of either the lens or the spherical mirror (and of all other components of the interferometer) is known, the resulting interference pattern can be used to determine the errors of the other component.

Mach–Zehnder Interferometer

Another very important interferometer is a Mach–Zehnder interferometer (Fig. 3.12). There the light of an incoming plane wave is again divided by a beam splitter into two waves. Then the transmitted plane wave is reflected at the upper mirror and passes the second beam splitter or is reflected at it. The plane wave reflected at the first beam splitter is reflected at the lower mirror and can pass an optional transmissive object to be tested. At the second beam splitter this wave can be transmitted or reflected. So, the Mach–Zehnder interferometer has two exits, which can both be used. Similarly to the case of the Michelson interferometer there are fluffed out fringes if all mirrors and beam splitters are oriented by exactly 45° relative to the incoming plane wave. If one of the mirrors or beam splitters is tilted there are fringes on the detector.

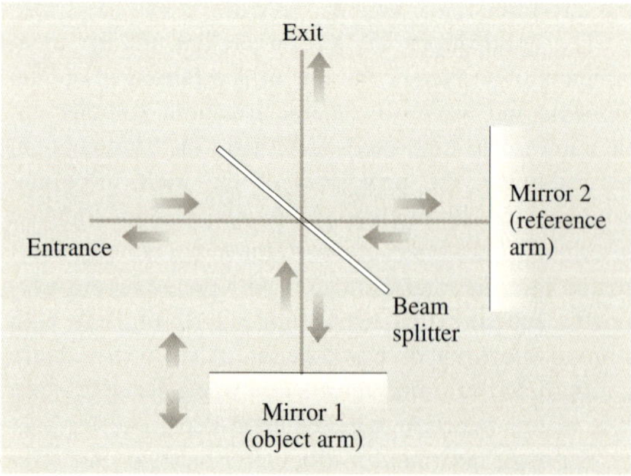

Fig. 3.11 Basic principle of a Michelson interferometer

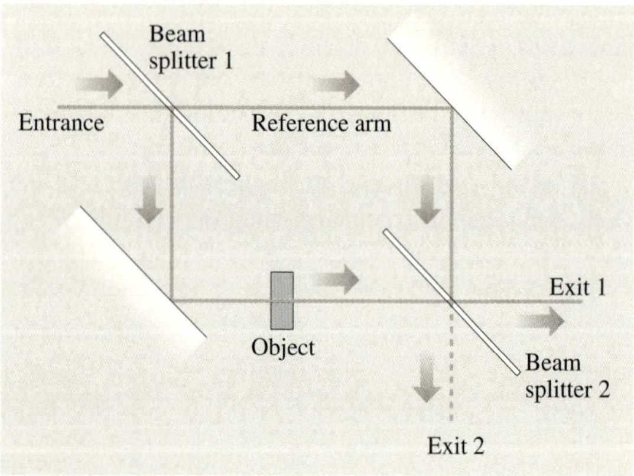

Fig. 3.12 Basic principle of a Mach–Zehnder interferometer

A Mach–Zehnder interferometer can be used to detect inhomogeneities of an optional object in the object arm. A special type is, e.g., an interferometer where the object is a tube filled with gas. By changing the pressure or the temperature in the tube the refractive index n is changed. Then the optical path difference between the object and the reference arm also changes and the fringes move or, in the case of fluffed out fringes, the overall intensity changes. So, the dependence of the refractive index of the gas on the pressure and temperature can be measured. In another type of Mach–Zehnder interferometer the object is a combination of a well-known lens and a lens to be tested, which together form a telescope. Using phase shifting interferometry (see later) by shifting one of the mirrors the errors of the lens to be tested can be determined. Of course, in practice there are always so-called adjustment errors that have to be eliminated from the measurement results and there are also systematic errors of the other components in the setup. Additionally, a very important fact is that the object to be tested has to be imaged onto the detector using auxiliary optics. Then, it is possible to say that the measured errors correspond to errors of the object at a certain point. This is, of course, also valid for the Twyman–Green interferometer and all other interferometers used for the measurement of an optical object with a limited depth. In an interferometer for the measurement of refractive index changes where the object is very long, it is not possible to image the complete object sharply onto the detector.

Shearing Interferometer

An interesting interferometer that needs no external reference arm is a shearing interferometer. There, by some means, a copy of the object wave is generated, which is either laterally or radially sheared [3.19, 29]. Here, only the case of lateral shearing will be discussed. The coordinate system is chosen in such a way that the shearing is along the x-axis by a distance Δx. Then, the phases of the two copies are $\Phi_1(x,y) = \Phi_o(x+\Delta x, y)$ and $\Phi_2(x,y) = \Phi_o(x,y)$, whereby Φ_o is the phase of the object wave itself. The phase difference Φ, which appears in the interference term of the interference (3.188), is then

$$\begin{aligned}\Phi(x,y) &= \Phi_1(x,y) - \Phi_2(x,y)\\ &= \Phi_o(x+\Delta x, y) - \Phi_o(x,y)\\ &\approx \Delta x \frac{\partial \Phi_o(x,y)}{\partial x} .\end{aligned} \tag{3.213}$$

The approximation of taking the first partial derivative at the point (x, y) is valid for small shearing distances Δx. The shearing interferometer can be similarly evaluated as other interferometers with phase shifting techniques and phase unwrapping, whereby a continuous function for $\Phi \approx \Delta x \partial \Phi_o / \partial x$ results. To obtain the phase Φ_o of the wavefront itself a kind of integration has to be made [3.19], [3.30]. To obtain an unambiguous wavefront both partial derivatives of Φ_o in the x and y directions have to be determined before the integration can be performed.

Phase-Shifting Interferometry

A typical technique to extract the object phase from the measured intensity values is phase-shifting interferometry [3.25, 31]. There, the reference mirror (or the object) is axially shifted by a well-known small distance and at least three different intensity distributions with different but well-known reference phases have to be observed, whereby there are also other possibilities to shift the phase [3.32]. By shifting the reference phase by $\delta\phi$, which is a well-defined integer multiple m of $\pi/2$ the intensity of the interference pattern $I_{1+2}^{(m)}$ changes according to (3.188) as

$$\begin{aligned}I_{1+2}^{(m)} &= I_1 + I_2 + 2\sqrt{I_1 I_2} \cos(\Phi + \delta\phi)\\ &= I_1 + I_2 + 2\sqrt{I_1 I_2} \cos\left(\Phi + m\frac{\pi}{2}\right) .\end{aligned} \tag{3.214}$$

Three measurements with different reference phases are in principle enough because (3.214) contains the three unknowns I_1, I_2 and the desired phase Φ. It is quite simple to combine the three measured intensity distributions $I_{1+2}^{(0)}$, $I_{1+2}^{(1)}$ and $I_{1+2}^{(2)}$ with different values m in order to calculate Φ:

$$\begin{aligned}I_{1+2}^{(0)} &= I_1 + I_2 + 2\sqrt{I_1 I_2} \cos\Phi ,\\ I_{1+2}^{(1)} &= I_1 + I_2 + 2\sqrt{I_1 I_2} \cos\left(\Phi + \frac{\pi}{2}\right)\\ &= I_1 + I_2 - 2\sqrt{I_1 I_2} \sin\Phi ,\\ I_{1+2}^{(2)} &= I_1 + I_2 + 2\sqrt{I_1 I_2} \cos(\Phi + \pi)\\ &= I_1 + I_2 - 2\sqrt{I_1 I_2} \cos\Phi ;\\ \Rightarrow\ & I_{1+2}^{(0)} + I_{1+2}^{(2)} = 2(I_1 + I_2) ,\\ & I_{1+2}^{(0)} - I_{1+2}^{(2)} = 4\sqrt{I_1 I_2} \cos\Phi ,\\ & I_{1+2}^{(0)} + I_{1+2}^{(2)} - 2I_{1+2}^{(1)} = 4\sqrt{I_1 I_2} \sin\Phi ,\\ \Rightarrow\ & \tan\Phi = \frac{I_{1+2}^{(0)} + I_{1+2}^{(2)} - 2I_{1+2}^{(1)}}{I_{1+2}^{(0)} - I_{1+2}^{(2)}} .\end{aligned} \tag{3.215}$$

However, this simple phase shifting algorithm with only three measurements is quite sensitive to phase shifting errors, i. e., if the phase shifts are not integer multiples of $\pi/2$. If more measurements are made, the correction of phase shifting errors is possible [3.20, 26, 31].

A principal problem of two beam interferometry is that the phase values obtained with a phase shifting algorithm, like, e.g., with (3.215), are only defined modulo 2π. (Note: The arctangent function itself is only unambiguously defined between $-\pi/2$ and $+\pi/2$. But by evaluating the signs of the numerator and denominator of (3.215) the phase Φ can be unambiguously calculated between $-\pi$ and $+\pi$.) Therefore, so-called phase unwrapping algorithms [3.20] have to be used in order to obtain a continuous phase profile of, e.g., the surface deviations or wave aberrations of a lens.

Some Ideas on the Energy Conservation in Interferometers

Here, some principal ideas to the conservation of energy in an interferometer shall be given because the laws of energy conservation have to be fulfilled everywhere in optics.

Let us consider, e.g., the Mach–Zehnder interferometer of Fig. 3.12. We assume that each beam splitter has a splitting ratio of 1:1, i. e., half of the light power is transmitted and half is reflected. So, if the intensity of the incoming plane wave is I_0 at the entrance, the intensities of the transmitted and reflected plane waves are each $I_0/2$. The two mirrors are assumed to reflect all light without losses and diffraction effects at apertures are also neglected because we use plane waves. So, at the second beam splitter each of the plane waves is again divided into two waves with equal intensity, i. e., each of the four waves now has the intensity $I_0/4$. At exit 1, two of these waves interfere and we assume that the phase difference Φ between these two waves shall be zero or an integer number times 2π. Then, the resulting intensity I_{1+2} is, according to (3.188),

$$\begin{aligned} I_{1+2} &= I_1 + I_2 + 2\sqrt{I_1 I_2}\cos\Phi \\ &= \frac{I_0}{4} + \frac{I_0}{4} + 2\sqrt{\frac{I_0^2}{16}} = I_0 \,. \end{aligned} \tag{3.216}$$

This, however, means that all of the incoming light power has to be at exit 1. Therefore, the intensity I'_{1+2} at exit 2 has to automatically be zero. Since the intensities of the two single waves, which interfere at exit 2 are also $I_0/4$, this is only possible if the phase difference Φ' is an odd-numbered multiple of π

$$\begin{aligned} I'_{1+2} &= \frac{I_0}{4} + \frac{I_0}{4} + 2\sqrt{\frac{I_0^2}{16}}\cos\Phi' \\ &= 0 \Rightarrow \Phi' = (2m+1)\pi \end{aligned} \tag{3.217}$$

with an integer number m. So, if we take the basic solution $\Phi' = \pi$ this requires that the phase shift between the transmitted and the reflected wave at each beam splitter is half of this value, i. e., $\pi/2$. Then, the law of energy conservation is fulfilled. To explain this further consider Fig. 3.12 regarded.

The phase differences between the two waves interfering at exits 1 and 2, respectively, due to the geometrical path are identical for both exits. The same is valid for reflections at the mirrors because each wave is exactly reflected once at a mirror. So, there has to be a phase shift between a reflected and a transmitted wave at a beam splitter in order to fulfill the law of energy conservation. The two waves that interfere at exit 1 (symmetrical exit) are each reflected one time a a beam splitter and transmitted one time by a beam splitter. So, the assumed phase shifts of $\pi/2$ due to a reflection at a beam splitter cancel out each other because the phase difference between both interfering waves is taken. But, at the exit 2 (antisymmetric exit) the first wave is transmitted by both beam splitters and the other wave is reflected at both beam splitters. Therefore, the phase difference between the two interfering waves is in this case

$$\Phi' = \Phi + 2\frac{\pi}{2} = \Phi + \pi \,. \tag{3.218}$$

This guarantees that the sum of the intensities $I_{1+2} + I'_{1+2}$ at both exits is equal to the intensity at the entrance:

$$\begin{aligned} I_{1+2} + I'_{1+2} &= \left(\frac{I_0}{4} + \frac{I_0}{4} + 2\sqrt{\frac{I_0^2}{16}}\cos\Phi\right) \\ &\quad + \left(\frac{I_0}{4} + \frac{I_0}{4} + 2\sqrt{\frac{I_0^2}{16}}\cos\Phi'\right) \\ &= \frac{I_0}{2}(1+\cos\Phi) + \frac{I_0}{2}(1-\cos\Phi) = I_0 \,. \end{aligned} \tag{3.219}$$

So, the energy is conserved if the phase shift of a wave, which is reflected at a beam splitter, is $\pi/2$ compared to the phase of the transmitted wave.

3.4 Diffraction

Up to now we have mostly investigated the propagation of plane waves and other waves that are not affected by any limiting apertures. A plane wave has, e.g., an infinite spatial extension and therefore it does not exist in the real world. Nevertheless, if the diameter of the limiting aperture is very large compared to the wavelength of the light, a plane wave can be a quite good approximation if the propagation distance is not very large. But also in this case there are disturbances at the rim of the wave which are called diffraction effects. In this section diffraction theory will mostly be treated for scalar waves and only at the end of this section will the influence of polarization effects on the electric energy density in the focal region of a lens be treated [3.33,34]. In contrast to most textbooks on optics like [3.1,35], we will not start historically with the Huygens–Fresnel principle or with the integral theorem of Helmholtz and Kirchhoff, but with the angular spectrum of plane waves. Only Kirchhoff's boundary conditions will be used, i. e., a wave that is incident on an absorbing screen with a hole will be undisturbed in the area of the hole and completely absorbed in the other parts of the screen. Starting from the angular spectrum of plane waves the Fresnel–Kirchhoff diffraction integral will be derived and it will be shown that both formulations are nearly equivalent [3.36–38]. The approximations of Fresnel diffraction and Fraunhofer diffraction will be discussed afterwards. A quite interesting application of Fraunhofer diffraction is, e.g., the calculation of the intensity distribution in the focal region of a lens [3.1, 39]. Afterwards, some ideas to the numerical implementation of scalar diffraction formula are given [3.38]. At the end of this section we will briefly reflect on the combination of polarization and diffraction by using the superposition of plane waves taking into account their polarization states. This is used to calculate the influence of polarization effects to the electric energy density in the focal region of a lens.

There are many modern applications of diffraction and interference effects in optical holography [3.40–43] and computer-generated diffractive optics [3.44–50]. But, there is no space to cover these fields in this chapter and we refer therefore to the literature.

3.4.1 The Angular Spectrum of Plane Waves

The knowledge of the angular spectrum of plane waves allows the exact propagation of a complex amplitude function u from one plane (which is chosen perpendicular to the z-axis) to another parallel plane in the distance z_0 in a homogeneous and isotropic material with the refractive index n. The only approximation here is that u is assumed to be a scalar function. But, since a plane wave can easily be defined by taking into account polarization (Sect. 3.2) an extension of this formalism is possible, but will not be treated here.

According to (3.52) a scalar plane wave

$$u(\boldsymbol{r}) = u_0 \mathrm{e}^{\mathrm{i}\boldsymbol{k}\cdot\boldsymbol{r}} \tag{3.220}$$

fulfills the Helmholtz equation (3.109), which is written for scalar waves as

$$\left(\nabla^2 + k^2\right) u\,(x, y, z) = 0\,. \tag{3.221}$$

The condition for the modulus $|\boldsymbol{k}| = k$ of the wave vector is

$$|\boldsymbol{k}| = \sqrt{k_x^2 + k_y^2 + k_z^2} = \frac{2\pi n}{\lambda}\,. \tag{3.222}$$

According to the linearity of the Helmholtz equation a sum of plane waves with different directions of propagation is also a solution of the Helmholtz equation and in the limit a continuous spectrum of plane waves is a solution (Fig. 3.13). There, the integration has to be done over two angles or more exactly over two components of the wave vector. The third component is then automatically defined by (3.222) as long as only plane waves propagating in the positive z direction are taken into account, which will be the case here. Since the complex amplitude is always regarded in a plane perpendicular to the z-axis, the two components of the wave

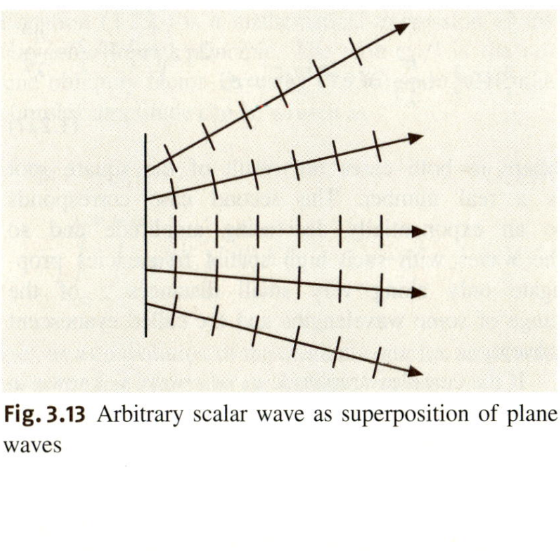

Fig. 3.13 Arbitrary scalar wave as superposition of plane waves

Part A | 3.4

$1/r$ is valid and the impulse response of (3.238) can be written as

$$h(x,y,z_0) = -\frac{1}{2\pi}\left(\mathrm{i}k - \frac{1}{r}\right)\frac{z_0}{r}\frac{\mathrm{e}^{\mathrm{i}kr}}{r}$$
$$\approx -\mathrm{i}\frac{n}{\lambda}\frac{z_0}{r}\frac{\mathrm{e}^{\mathrm{i}kr}}{r}\,. \tag{3.240}$$

Then (3.236) results in the more familiar but less general Rayleigh–Sommerfeld diffraction formula

$$u(x,y,z_0)$$
$$\approx -\mathrm{i}\frac{n}{\lambda}\int\limits_{-\infty}^{+\infty}\int\limits_{-\infty}^{+\infty} u_0(x',y',0)\,\frac{z_0}{l}\frac{\mathrm{e}^{\mathrm{i}kl}}{l}\,\mathrm{d}x'\mathrm{d}y'\,. \tag{3.241}$$

In the case that the complex amplitude u_0 is different from zero inside an aperture A and zero outside of the aperture A (3.241) is also known as the Fresnel–Kirchhoff diffraction integral [3.1]. Then, the effective integration is not carried out from minus infinity to plus infinity but on the area of the aperture A.

It has to be mentioned that (3.236) can also be written in a similar form that does not explicitly assume that the original complex amplitude, which is now named $u_0(\boldsymbol{r}')$ has to be known in a plane and that the complex amplitude $u(\boldsymbol{r})$, which has to be calculated, is also defined in a plane. The generalization is

$$u(\boldsymbol{r}) = -\frac{1}{2\pi}\iint\limits_A u_0(\boldsymbol{r}')\left(\mathrm{i}k - \frac{1}{l}\right)$$
$$\times \frac{\boldsymbol{N}\cdot(\boldsymbol{r}-\boldsymbol{r}')}{l}\frac{\mathrm{e}^{\mathrm{i}kl}}{l}\,\mathrm{d}S\,. \tag{3.242}$$

Here, $\boldsymbol{r}' = (x',y',z')$ defines an arbitrary point on a curved surface (the aperture A) where the original complex amplitude u_0 is defined and $\boldsymbol{r} = (x,y,z)$ is an arbitrary point on the second curved surface on which the complex amplitude u has to be calculated. Additionally, $\boldsymbol{N}$ is a unit vector perpendicular to the aperture A at the point $\boldsymbol{r}'$. The integration is done over the aperture A and the integration element $\mathrm{d}S$ indicates a two-dimensional integration. Additionally, the distance l is then defined by:

$$l := \sqrt{|\boldsymbol{r}-\boldsymbol{r}'|^2}\,. \tag{3.243}$$

Nevertheless, in the following we will always assume that both complex amplitudes u_0 and u are defined in parallel planes. This allows, as we will see in the next section, that approximate integrals like the Fresnel diffraction integral can be numerically calculated using the efficient fast Fourier transformation.

3.4.3 The Fresnel and the Fraunhofer Diffraction Integral

In the following, we always assume that the plane aperture A on which the integration of the diffraction integral is carried out is limited and has a maximum diameter of D. There is no other restriction on the form of the aperture, which can be circular, rectangular or irregularly formed. The parameter D is therefore the diameter of a circle, which contains the aperture and which is centered around the z-axis. Again, we have the complex amplitude $u_0(x',y',0)$ in a first plane at $z=0$, which is zero outside of the aperture. Additionally, the distance z_0 of the second plane to the first parallel plane is much larger than the diameter D of the aperture A, i.e., $D \ll z_0$.

Then, the distance l (see (3.237)) of a point $P' = (x',y',0)$ in the first plane and a point $P=(x,y,z_0)$ in the second plane can be written as (Fig. 3.14)

$$l = \sqrt{(x-x')^2+(y-y')^2+z_0^2}$$
$$= \sqrt{x^2+y^2+z_0^2+x'^2+y'^2-2xx'-2yy'}$$
$$= \sqrt{x^2+y^2+z_0^2}\sqrt{1+\frac{x'^2+y'^2-2xx'-2yy'}{x^2+y^2+z_0^2}}\,. \tag{3.244}$$

We define the term l_0 as

$$l_0 := \sqrt{x^2+y^2+z_0^2} \gg D \Rightarrow \frac{D}{l_0} \ll 1\,. \tag{3.245}$$

Because of $x' \leq D/2$, $y' \leq D/2$ and (3.245) all terms on the order of x'^3/l_0^3, y'^3/l_0^3 and higher can be neglected. Then l can be approximated by the first terms of its

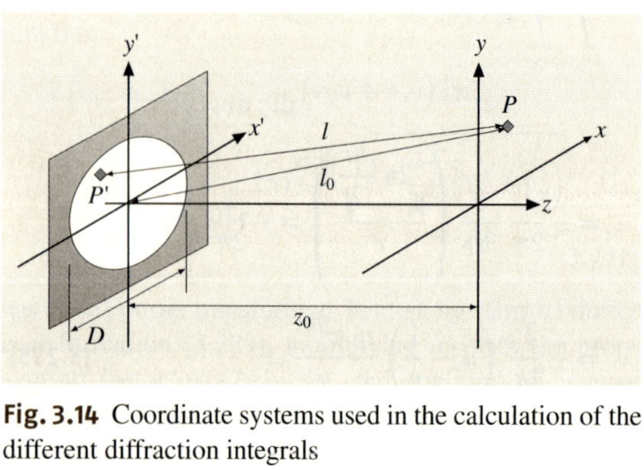

Fig. 3.14 Coordinate systems used in the calculation of the different diffraction integrals

Taylor expansion:

$$l = l_0\sqrt{1+\frac{x'^2+y'^2-2xx'-2yy'}{l_0^2}}$$

$$\approx l_0\left(1+\frac{x'^2+y'^2}{2l_0^2}-\frac{xx'+yy'}{l_0^2}\right.$$

$$\left.-\frac{\left(x'^2+y'^2-2xx'-2yy'\right)^2}{8l_0^4}\right)$$

$$= l_0+\frac{x'^2+y'^2}{2l_0}-\frac{xx'+yy'}{l_0}-\frac{(xx'+yy')^2}{2l_0^3}$$

$$+\frac{\left(x'^2+y'^2\right)(xx'+yy')}{2l_0^3}-\frac{\left(x'^2+y'^2\right)^2}{8l_0^3}\,. \tag{3.246}$$

The last two terms are on the order of x'^3/l_0^3, y'^3/l_0^3 or higher. Therefore, they can be neglected and the result is

$$l \approx l_0+\frac{x'^2+y'^2}{2l_0}-\frac{xx'+yy'}{l_0}-\frac{(xx'+yy')^2}{2l_0^3}\,, \tag{3.247}$$

where l in the denominator of the Huygens wavelet of (3.241) can then be replaced with a good approximation by only the first term of the Taylor expansion, i. e., l_0, and the cosine obliquity factor z_0/l can be replaced by the term z_0/l_0. This means that the Fresnel–Kirchhoff diffraction integral (3.241) can be approximated by

$$u(x, y, z_0)$$

$$\approx -\mathrm{i}\frac{n}{\lambda l_0}\frac{z_0}{l_0}\int\limits_{-\infty}^{+\infty}\int\limits_{-\infty}^{+\infty} u_0\left(x', y', 0\right)\mathrm{e}^{\mathrm{i}kl}\,\mathrm{d}x'\,\mathrm{d}y'\,, \tag{3.248}$$

where l is defined via (3.247). This equation is valid as long as the last two terms of (3.246) can be neglected in the exponential factor $\exp(\mathrm{i}kl)$. Therefore, two conditions have to be valid:

$$k\frac{\left(x'^2+y'^2\right)(xx'+yy')}{2l_0^3} \ll \pi$$

$$\Rightarrow n\frac{\left(x'^2+y'^2\right)(xx'+yy')}{\lambda l_0^3} \ll 1\,, \tag{3.249}$$

$$k\frac{\left(x'^2+y'^2\right)^2}{8l_0^3} \ll \pi \Rightarrow n\frac{\left(x'^2+y'^2\right)^2}{4\lambda l_0^3} \ll 1\,. \tag{3.250}$$

There are two especially interesting approximations of the Fresnel–Kirchhoff diffraction integral. The first is the Fresnel diffraction integral where only points P in the neighborhood of the axis, i. e., in the paraxial regime, are considered and the second is the Fraunhofer diffraction integral where only points in the far field or in the focal plane of a lens are considered.

The Fresnel Diffraction Integral

In the paraxial regime, i. e., $x^2+y^2 \le D^2$, we can write

$$x^2+y^2 \le D^2 \ll z_0^2$$

$$\Rightarrow l_0 = z_0\sqrt{1+\frac{x^2+y^2}{z_0^2}} \approx z_0+\frac{x^2+y^2}{2z_0}\,. \tag{3.251}$$

The second term is only interesting in the rapidly oscillating exponential factor $\exp(\mathrm{i}kl_0)$. In the other cases we can write $l_0 \approx z_0$ and $z_0/l_0 \approx 1$. Additionally, we have to approximate $1/l_0$ in some terms in the exponential factor

$$\frac{1}{l_0} = \frac{1}{z_0}\left(1+\frac{x^2+y^2}{z_0^2}\right)^{-1/2} \approx \frac{1}{z_0}-\frac{x^2+y^2}{2z_0^3}\,. \tag{3.252}$$

Since the term $1/l_0$ appears only in terms that are themselves on the order of x'^2, y'^2 or xx', yy', the second term $(x^2+y^2)/(2z_0^3)$ can be neglected because it would lead to terms of higher order. In the paraxial regime the last term in (3.247) can also be neglected. Finally, we obtain from (3.248):

$$u(x, y, z_0)$$

$$= -\mathrm{i}\frac{n}{\lambda z_0}\mathrm{e}^{\mathrm{i}\frac{2\pi n z_0}{\lambda}}\mathrm{e}^{\mathrm{i}\pi n\frac{x^2+y^2}{\lambda z_0}}$$

$$\times\iint\limits_A u_0\left(x', y', 0\right)\mathrm{e}^{\mathrm{i}\pi n\frac{x'^2+y'^2}{\lambda z_0}}$$

$$\times\mathrm{e}^{-2\pi\mathrm{i}n\frac{xx'+yy'}{\lambda z_0}}\,\mathrm{d}x'\,\mathrm{d}y'\,. \tag{3.253}$$

This is the Fresnel diffraction integral, where the integration is made over the aperture A. The condition for the validity of the Fresnel diffraction integral is, according to (3.249) and (3.250),

$$n\frac{\left(x'^2+y'^2\right)\left(xx'+yy'\right)}{\lambda l_0^3} \ll 1$$

and $$n\frac{\left(x'^2+y'^2\right)^2}{4\lambda l_0^3} \ll 1$$

$$\Rightarrow Q_{\text{Fresnel}} := \frac{n(D/2)^4}{\lambda z_0^3} = \frac{nD^4}{16\lambda z_0^3} \ll 1\,. \tag{3.254}$$

As an example, we take $n=1$, $\lambda = 0.5\,\mu\text{m}$, $D = 10\,\text{mm}$ and $z_0 = 1\,\text{m}$. Then the term Q_{Fresnel} has the value $Q_{\text{Fresnel}} = 0.00125$ and the condition for the validity of the Fresnel diffraction integral is very well fulfilled. If the distance z_0 is only 0.1 m the term Q_{Fresnel} is 1.25 and the Fresnel approximation is at the limit of its validity. This shows that the Fresnel diffraction integral is a good approximation in a distance between the near and the far field. In the near field (which ranges from $z_0 = 0$ up to a distance z_0 of several times D) the Fresnel–Kirchhoff diffraction integral or the angular spectrum of plane waves has to be used. In the far field there is another more simple approximation, the Fraunhofer diffraction formula, which will be discussed in the next section. But before doing this the Fresnel diffraction integral of (3.253) will be discussed a little further.

Equation (3.253) shows that the integral itself is formally the Fourier transformation of the function

$$f\left(x',y'\right) = \begin{cases} u_0\left(x',y',0\right)\exp\left(\mathrm{i}\pi n\frac{x'^2+y'^2}{\lambda z_0}\right) & \text{if } (x',y') \in A \\ 0 & \text{if } (x',y') \notin A \end{cases}\,. \tag{3.255}$$

This gives the quite efficient possibility of calculating the integral numerically by using a fast Fourier transformation (FFT) [3.52]. But (3.253) can also be written in a different form

$$u(x,y,z_0) = -\mathrm{i}\frac{n}{\lambda z_0}\mathrm{e}^{\mathrm{i}\frac{2\pi n z_0}{\lambda}} \times \iint_A u_0\left(x',y',0\right)\mathrm{e}^{\mathrm{i}\pi n\frac{\left(x-x'\right)^2+\left(y-y'\right)^2}{\lambda z_0}}\,\mathrm{d}x'\,\mathrm{d}y'. \tag{3.256}$$

So, this form shows the Fresnel diffraction integral as a convolution of the functions u_0 and $\exp[\mathrm{i}\pi n(x'^2 + y'^2)/(\lambda z_0)]$. According to the convolution theorem the Fresnel diffraction integral can then formally be written as

$$u(x,y,z_0) = -\mathrm{i}\frac{n}{\lambda z_0}\mathrm{e}^{\mathrm{i}\frac{2\pi n z_0}{\lambda}}\,\mathrm{FT}^{-1} \times \left\{\mathrm{FT}\left\{u_0(x',y',0)\right\}\cdot\mathrm{FT}\left\{\mathrm{e}^{\mathrm{i}\pi n\frac{x'^2+y'^2}{\lambda z_0}}\right\}\right\}. \tag{3.257}$$

Now, according to (3.229) we have

$$\tilde{u}_0\left(\nu_x,\nu_y,0\right) = \int_{-\infty}^{+\infty}\int_{-\infty}^{+\infty} u_0(x',y',0)\mathrm{e}^{-2\pi\mathrm{i}(\nu_x x'+\nu_y y')}\,\mathrm{d}x'\,\mathrm{d}y' = \mathrm{FT}\left\{u_0(x',y',0)\right\}\,. \tag{3.258}$$

The second Fourier pair is

$$\mathrm{FT}\left\{\mathrm{e}^{\mathrm{i}\pi n\frac{x'^2+y'^2}{\lambda z_0}}\right\} = \int_{-\infty}^{+\infty}\int_{-\infty}^{+\infty}\mathrm{e}^{\mathrm{i}\pi n\frac{x'^2+y'^2}{\lambda z_0}}\,\mathrm{e}^{-2\pi\mathrm{i}(\nu_x x'+\nu_y y')}\,\mathrm{d}x'\,\mathrm{d}y' = \mathrm{i}\frac{\lambda z_0}{n}\mathrm{e}^{-\mathrm{i}\pi\frac{\lambda z_0}{n}\left(\nu_x^2+\nu_y^2\right)}\,. \tag{3.259}$$

In total, (3.257) results in

$$u(x,y,z_0) = \mathrm{e}^{\mathrm{i}\frac{2\pi n z_0}{\lambda}}\int_{-\infty}^{+\infty}\int_{-\infty}^{+\infty}\tilde{u}_0\left(\nu_x,\nu_y,0\right) \times \mathrm{e}^{-\mathrm{i}\pi\frac{\lambda z_0}{n}\left(\nu_x^2+\nu_y^2\right)} \times \mathrm{e}^{2\pi\mathrm{i}(\nu_x x+\nu_y y)}\,\mathrm{d}\nu_x\,\mathrm{d}\nu_y\,. \tag{3.260}$$

This is the Fresnel diffraction integral expressed in the Fourier domain. The same equation can also be obtained from the angular spectrum of plane waves (3.231) by expanding the square root of the transfer function of free space in a Taylor series

$$\sqrt{1-\frac{\lambda^2}{n^2}\left(\nu_x^2+\nu_y^2\right)} \approx 1-\frac{\lambda^2}{2n^2}\left(\nu_x^2+\nu_y^2\right);$$
$$\Rightarrow \mathrm{e}^{2\pi\mathrm{i}\frac{nz_0}{\lambda}\sqrt{1-\frac{\lambda^2}{n^2}\left(\nu_x^2+\nu_y^2\right)}}$$
$$\approx \mathrm{e}^{2\pi\mathrm{i}\frac{nz_0}{\lambda}}\,\mathrm{e}^{-\mathrm{i}\pi\frac{\lambda z_0}{n}\left(\nu_x^2+\nu_y^2\right)}\,. \tag{3.261}$$

This approximation is valid as long as the higher-order terms do not contribute to a considerable variation of the exponential factor. The condition for this is

$$2\pi\frac{nz_0}{\lambda}\frac{\lambda^4}{8n^4}\left(\nu_x^2+\nu_y^2\right)^2 = \pi\frac{\lambda^3 z_0}{4n^3}\left(\nu_x^2+\nu_y^2\right)^2 \ll \pi;$$
$$\Rightarrow Q_{\text{Fresnel, Fourier}} := \frac{\lambda^3 z_0}{4n^3}\left(\nu_x^2+\nu_y^2\right)^2 \ll 1\,. \tag{3.262}$$

To give an estimation of this term a spherical wave with the half-aperture angle φ is regarded. Then, the maximum spatial frequency of the spherical wave will be $n\sin\varphi/\lambda$. The function $\tilde{u}_0$ will be considerably different from zero only for spatial frequencies with $\nu_x^2+\nu_y^2 < n^2\sin^2\varphi/\lambda^2$. Therefore, the condition (3.262) can be transformed to

$$Q_{\text{Fresnel, Fourier}} = \frac{nz_0}{4\lambda}\sin^4\varphi \ll 1\,. \tag{3.263}$$

It is obvious that the error term $Q_{\text{Fresnel, Fourier}}$ increases with increasing distance z_0 whereas the error term Q_{Fresnel} of (3.254) decreases with increasing z_0.

So, the formulation of the Fresnel diffraction integral in the Fourier domain (3.260) is better used for the near field whereas the Fresnel diffraction integral of (3.253) is better used for the medium or far field. So, both formulations are contrary. The numerical evaluation of both diffraction integrals can be done using FFT's whereas only one FFT is necessary in the formulation of (3.253) and two FFT's (one for calculating $\tilde{u}_0$ and one for the integral itself) in the formulation of (3.260).

The Fraunhofer Diffraction Formula

An approximation of the Fresnel–Kirchhoff diffraction integral for the far field can be obtained from (3.247) and (3.248). First, we define the direction cosines α and β as

$$\alpha := \frac{x}{l_0}\,; \quad \beta := \frac{y}{l_0} \tag{3.264}$$

so that (3.247) can be written as

$$l \approx l_0 + \frac{x'^2+y'^2}{2l_0} - \left(\alpha x'+\beta y'\right) - \frac{\left(\alpha x'+\beta y'\right)^2}{2l_0}\,. \tag{3.265}$$

With increasing l_0 the second and fourth term decrease more and more and only the first and the third term remain. The condition that only these two terms have to be considered is that the contribution of the other terms does not remarkably vary the exponential factor in (3.248). This is fulfilled if

$$\pi n\frac{x'^2+y'^2}{\lambda l_0} \ll \pi \Rightarrow Q_{\text{Fraunhofer}} := \frac{nD^2}{4\lambda z_0} \ll 1\,. \tag{3.266}$$

In the last step it is required that all points in the aperture A fulfill the condition $x'^2+y'^2 \le (D/2)^2$, where D is again the maximum diameter of the aperture. We also used that according to (3.245) $l_0 \ge z_0$.

Then, (3.248) can be written as

$$u(\alpha,\beta,z_0)$$
$$= -\mathrm{i}\frac{n}{\lambda l_0}\frac{z_0}{l_0}\mathrm{e}^{\mathrm{i}kl_0}$$
$$\times \iint_A u_0\left(x',y',0\right)\mathrm{e}^{-2\pi\mathrm{i}\frac{n}{\lambda}\left(\alpha x'+\beta y'\right)}\,\mathrm{d}x'\,\mathrm{d}y'\,. \tag{3.267}$$

This is the well known Fraunhofer diffraction integral. It means that the complex amplitude in the far field is the Fourier transform of the complex amplitude at $z=0$.

The importance of (3.267) would be quite marginal if it is only valid for the far field. This can be seen by the following example. We assume $n=1$, $\lambda = 0.5\,\mu\text{m}$, $D = 10\,\text{mm}$. Then (3.266) would require

$$z_0 \gg \frac{nD^2}{4\lambda} = 50\,\text{m}\,. \tag{3.268}$$

But, there is another quite important case: the complex amplitude in the focal plane of a lens.

The Complex Amplitude in the Focal Plane of a Lens

We assume a complex amplitude u_0 in the starting plane that is defined as different from zero in the aperture A and zero outside of the aperture. The influence of an ideal thin lens which is positioned in the starting plane would be that u_0 has to be multiplied by the transmission function $t_{\text{lens, ideal}}$ of the lens, which is an exponential phase factor of the form

$$t_{\text{lens, ideal}}\left(x', y'\right) = \mathrm{e}^{-\mathrm{i}k\left(f\sqrt{1+\frac{x'^2+y'^2}{f^2}}-f\right)} =: \mathrm{e}^{-\mathrm{i}kl_{\text{lens}}} . \quad (3.269)$$

Here, f is the focal length of the lens and a positive value f corresponds to a positive lens, whereas in our case we will only have a positive lens. Of course, an ideal lens does not exist in reality and a more adapted transmission function is

$$t_{\text{lens}}\left(x', y'\right) = \mathrm{e}^{-\mathrm{i}kl_{\text{lens}}+\mathrm{i}W\left(x', y'\right)} , \quad (3.270)$$

where W are the wave aberrations of the lens. However, in reality the wave aberrations of a lens will depend on the incident wavefront and a wave-optical simulation of a lens including aberrations is not so easy. But here, we assume that W is known for a given complex amplitude u_0. For an ideal lens we just have to put W equal to zero.

So, the new complex amplitude u_0' behind the lens is defined by

$$u_0'\left(x', y', 0\right) = u_0\left(x', y', 0\right) t_{\text{lens}}\left(x', y'\right) \quad (3.271)$$

and the complex amplitude in a parallel plane at the distance z_0 is, according to the Fresnel–Kirchhoff diffraction integral (3.241),

$$u(x, y, z_0) = -\mathrm{i}\frac{n}{\lambda}\iint_A u_0'\left(x', y', 0\right) \frac{z_0}{l}\frac{\mathrm{e}^{\mathrm{i}kl}}{l}\,\mathrm{d}x'\mathrm{d}y' , \quad (3.272)$$

with l defined by (3.237) as

$$l = \sqrt{(x-x')^2+(y-y')^2+z_0^2} .$$

Now, we are only interested in points in the neighborhood of the Gaussian focus of the lens at $(0, 0, f)$ and since the radius $D/2$ of the aperture shall be several times smaller than the focal length f of the lens we have the following conditions and approximations

$$f = z_0(1+\varepsilon) \quad \text{with}$$

$$|\varepsilon| \ll 1, \quad f \gg \frac{D}{2} \gg \frac{\lambda}{n};$$

$$x'^2+y'^2 \le \frac{D^2}{4} \ll z_0^2,$$

$$\frac{x}{z_0} \ll \frac{D/2}{z_0} \ll 1 \quad \text{and} \quad \frac{y}{z_0} \ll \frac{D/2}{z_0} \ll 1 .$$

Then, similar to the case of the Fresnel diffraction integral the cosine obliquity factor z_0/l is one and the distance l in the denominator of (3.272) can be replaced by z_0. Only l in the exponential phase factor of (3.272) has to be considered carefully since the phase factor will rapidly oscillate if l varies by more than one wavelength λ/n. So, using (3.270), (3.272) and the approximations the intermediate result is

$$u(x, y, z_0) = -\mathrm{i}\frac{n}{\lambda z_0}\iint_A u_0\left(x', y', 0\right)\mathrm{e}^{\mathrm{i}W\left(x', y'\right)} \times \mathrm{e}^{\mathrm{i}k\left(l-l_{\text{lens}}\right)}\,\mathrm{d}x'\mathrm{d}y' . \quad (3.273)$$

The term $l-l_{\text{lens}}$ has to be evaluated:

$$\begin{aligned} l-l_{\text{lens}} &= z_0\sqrt{1+\frac{x^2+y^2+x'^2+y'^2-2\left(xx'+yy'\right)}{z_0^2}} - f\sqrt{1+\frac{x'^2+y'^2}{f^2}}+f \\ &\approx z_0+\frac{x^2+y^2}{2z_0}+\frac{x'^2+y'^2}{2z_0}-\frac{xx'+yy'}{z_0}-\frac{x'^2+y'^2}{2f} \\ &= z_0+\frac{x^2+y^2}{2z_0}+\frac{\Delta z}{fz_0}\frac{x'^2+y'^2}{2}-\frac{xx'+yy'}{z_0} \end{aligned} \quad (3.274)$$

with

$$\Delta z := f-z_0 .$$

As in the case of the Fresnel diffraction integral the terms of higher than second order in x, y, x' or y' have been neglected because of our conditions. Nevertheless, it should be pointed out that the restrictions on the sine of the half-aperture angle φ of the lens ($\sin\varphi \approx D/(2f)$) are not so severe as in the case of the Fresnel diffraction integral since only points in the neighborhood of the Gaussian focus, i. e., when x, y and Δz are small, are interesting. To make this clear the higher-order terms have to be estimated whereby only one section along x and x' is considered. But this is no restriction if the aperture is circular or if the section along x, x' has a larger diameter than along y, y'. The maximum value of x' is $D/2$ and the maximum interesting value of x is just some wavelengths if we are near the focal plane of the lens. Additionally, we can replace the term $D/(2z_0)$ with a good approximation by $\sin\varphi$ because of $f \approx z_0$. For the same reason we also have $1/z_0^3 - 1/f^3 \approx 3\Delta z/f^4$. The fourth-order terms of $k(l - l_{\text{lens}})$ are therefore of the following form

$$k\left(\frac{\left(x^2+x'^2-2xx'\right)^2}{8z_0^3}-\frac{x'^4}{8f^3}\right)$$
$$=2\pi n\left[\frac{x^4}{8\lambda z_0^3}+\left(\frac{1}{8\lambda z_0^3}-\frac{1}{8\lambda f^3}\right)x'^4 + \frac{3x^2x'^2}{4\lambda z_0^3}-\frac{x^3x'}{2\lambda z_0^3}-\frac{xx'^3}{2\lambda z_0^3}\right]$$
$$\le \pi n\left(\frac{x_{\max}^4}{4\lambda f^3}+\frac{3\Delta z}{4\lambda}\sin^4\varphi+\frac{3x_{\max}^2\sin^2\varphi}{2\lambda f} - \frac{x_{\max}^3\sin\varphi}{\lambda f^2}-\frac{x_{\max}\sin^3\varphi}{\lambda}\right). \quad (3.275)$$

Again, these terms have to be much smaller than π so that they can be neglected. A numerical example can illustrate this: A lens with a focal length of $f = 100$ mm and $D/2 = 30$ mm, i. e., $\sin\varphi = 0.29$, is illuminated with light of the wavelength $\lambda = 0.5\,\mu$m. The refractive index for the light propagation is $n = 1$. The propagation distance behind the lens is $z_0 = 99.9$ mm and the maximum value of x, which is interesting for us, is $x_{\max} = 10\,\mu$m. Later, in (3.294), we will see that the radius of the diffraction-limited Airy disc of a lens with $\sin\varphi = 0.29$ and $\lambda = 0.5\,\mu$m is $\rho_0 = 0.61\,\lambda/\text{NA} = 0.61\lambda/(n\sin\varphi) \approx 1\,\mu$m. Therefore, an area with radius $x_{\max} = 10\,\mu$m contains all interesting structures of the intensity distribution of the focus. Using these values, the higher-order terms are

$$\pi n\frac{x_{\max}^4}{4\lambda f^3} = 5\times10^{-12}\pi,$$
$$\pi n\frac{3\Delta z}{4\lambda}\sin^4\varphi = 1.1\,\pi,$$
$$\pi n\frac{3x_{\max}^2\sin^2\varphi}{2\lambda f} = 2.5\times10^{-4}\pi,$$
$$\pi n\frac{x_{\max}^3\sin\varphi}{\lambda f^2} = 6\times10^{-8}\pi,$$
$$\pi n\frac{x_{\max}\sin^3\varphi}{\lambda} = 0.49\pi.$$

We see that the "defocus" term of higher order is $1.1\,\pi$ and cannot be neglected. But if we go directly in the focal plane, i. e., $z_0 = f$, this term will completely vanish. The second term that cannot be neglected is the last term ($0.49\,\pi$), which is proportional to $x_{\max}/\lambda$ and the third power of the $\sin\varphi$. But in the direct neighborhood of the Airy disc, which has in the diffraction-limited case a radius of about 1 μm this term will be a factor of 10 smaller. This means that for $\sin\varphi$ of 0.3 (and the given other parameters) the higher order terms can only be neglected in the direct neighborhood of the focus and the complex amplitude calculated outside may have some errors. But by decreasing the numerical aperture of the lens the accuracy of the calculation increases.

So, finally we have the following result for the complex amplitude in the neighborhood of the Gaussian focus by neglecting higher-order terms

$$u(x,y,z_0) = -\mathrm{i}\frac{n}{\lambda z_0}\mathrm{e}^{ikz_0}\mathrm{e}^{ik\frac{x^2+y^2}{2z_0}} \times \iint_A u_0\left(x',y',0\right)\mathrm{e}^{\mathrm{i}W\left(x',y'\right)}\mathrm{e}^{ik\frac{\Delta z}{z_0 f}\frac{x'^2+y'^2}{2}} \times \mathrm{e}^{-ik\frac{xx'+yy'}{z_0}}\,\mathrm{d}x'\,\mathrm{d}y'. \quad (3.276)$$

This integral is similar to the Debye integral. In [3.1] a slightly different but in fact nearly identical form of this integral is evaluated for an ideal lens, i. e., $W = 0$, using the Lommel functions (E. Lommel invented these functions when he was a professor of physics at the University of Erlangen in 1868–1886). For us it is especially interesting that (3.276) expresses the complex amplitude in the neighborhood of the Gaussian focus as a Fourier transformation of the pupil function G

$$G\left(x',y'\right) = \begin{cases} u_0(x',y',0)\exp\left[\mathrm{i}W(x',y')\right]\exp\left(ik\frac{\Delta z}{fz_0}\frac{x'^2+y'^2}{2}\right) & \text{if } (x',y')\in A \\ 0 & \text{if } (x',y')\notin A\,. \end{cases} \quad (3.277)$$

Here, u_0 is the complex amplitude of the incident wave, the term $\exp(\mathrm{i}W)$ describes the influence of the wave aberrations of the lens and the third term is a defocus term. In the focal plane itself the defocus term vanishes since then $z_0 = f$. So, in fact we see that in the focal plane of an ideal lens (i. e., $W = 0$ and $\Delta z = 0$) we again have a kind of Fraunhofer diffraction and the complex amplitude u is calculated by a Fourier transformation of u_0.

The intensity distribution I in the focal plane for an incident plane on-axis wave, i. e., $u_0(x',y',0) =$

$a=$ constant with $I_0=a^2$, is according to (3.276)

$$I(x,y,z_0=f)=I_0\frac{n^2}{\lambda^2 f^2} \times\left|\iint\limits_A \mathrm{e}^{\mathrm{i}W(x',y')}\mathrm{e}^{-2\pi \mathrm{i}n\frac{xx'+yy'}{\lambda f}}\,\mathrm{d}x'\,\mathrm{d}y'\right|^2, \quad (3.278)$$

where I_0 is the intensity of the incident plane wave. The intensity distribution in the focal plane is also called the point spread function (PSF) of the lens. It is often usual to normalize the PSF by dividing it by the intensity I_F of an nonaberrated lens of the same type at the Gaussian focus. Using (3.278) I_F is obtained by setting $W=0$ and $(x,y)=(0,0)$:

$$I_F(0,0,f)=I_0\frac{n^2}{\lambda^2 f^2}\left|\iint\limits_A \mathrm{d}x'\,\mathrm{d}y'\right|^2=I_0\frac{n^2}{\lambda^2 f^2}S^2. \quad (3.279)$$

Here, $S=\int\int_A \mathrm{d}x'\,\mathrm{d}y'$ is the surface area of the aperture A. Then, the normalized point spread function PSF is

$$\mathrm{PSF}(x,y)=\frac{I(x,y,f)}{I_F(0,0,f)} \quad (3.280)$$

$$=\frac{1}{S^2}\left|\iint\limits_A \mathrm{e}^{\mathrm{i}W(x',y')}\mathrm{e}^{-2\pi \mathrm{i}n\frac{xx'+yy'}{\lambda f}}\,\mathrm{d}x'\,\mathrm{d}y'\right|^2.$$

The dimensionless number $\sigma=\mathrm{PSF}(0,0)$ of the aberrated lens at the Gaussian focus is called the Strehl ratio of the lens and is defined by

$$\sigma=\mathrm{PSF}(0,0)=\frac{1}{S^2}\left|\iint\limits_A \mathrm{e}^{\mathrm{i}W(x',y')}\,\mathrm{d}x'\,\mathrm{d}y'\right|^2. \quad (3.281)$$

In this section we calculated the PSF and the Strehl ratio only for a thin lens with the aperture in the plane of the lens. For general optical systems the same concept and the same equations are used but then the aperture A is the exit pupil of the optical system and f is the distance of the Gaussian focus from the exit pupil of the system.

Two Examples of Fraunhofer Diffraction

Two simple examples that can be solved analytically are the Fraunhofer diffraction at a rectangular aperture or at a circular aperture.

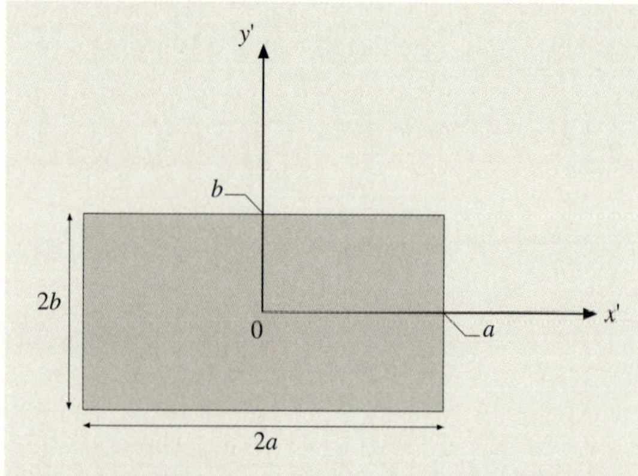

Fig. 3.15 Parameters of a rectangular transparent aperture in an opaque screen

Fraunhofer Diffraction at a Rectangular Aperture. The intensity distribution in the focal plane of an ideal lens (focal length f) with a rectangular aperture of the diameter $2a$ in x direction and $2b$ in y direction has to be calculated (Fig. 3.15). The wavelength of the light is λ and the refractive index of the material in which the wave propagates is n. The lens itself is illuminated with a uniform plane on-axis wave with the intensity I_0. Then, according to (3.278) the intensity in the focal plane of the ideal lens (i. e., $W=0$) is

$$\begin{aligned} I(x,y,z_0=f) &= I_0\frac{n^2}{\lambda^2 f^2}\left|\iint\limits_A \mathrm{e}^{-2\pi \mathrm{i}n\frac{xx'+yy'}{\lambda f}}\,\mathrm{d}x'\,\mathrm{d}y'\right|^2 \\ &= I_0\frac{n^2}{\lambda^2 f^2}\left|\int_{-a}^{a} \mathrm{e}^{-2\pi \mathrm{i}n\frac{xx'}{\lambda f}}\,\mathrm{d}x'\right|^2 \\ &\quad\times\left|\int_{-b}^{b} \mathrm{e}^{-2\pi \mathrm{i}n\frac{yy'}{\lambda f}}\,\mathrm{d}y'\right|^2. \end{aligned} \quad (3.282)$$

The first integral is

$$\begin{aligned} \int_{-a}^{a} \mathrm{e}^{-2\pi \mathrm{i}n\frac{xx'}{\lambda f}}\,\mathrm{d}x' &= \left[\frac{-\lambda f}{2\pi \mathrm{i}nx}\mathrm{e}^{-2\pi \mathrm{i}n\frac{xx'}{\lambda f}}\right]_{x'=-a}^{x'=a} \\ &= \frac{2\lambda f}{2\pi nx}\sin\left(2\pi n\frac{xa}{\lambda f}\right). \end{aligned} \quad (3.283)$$

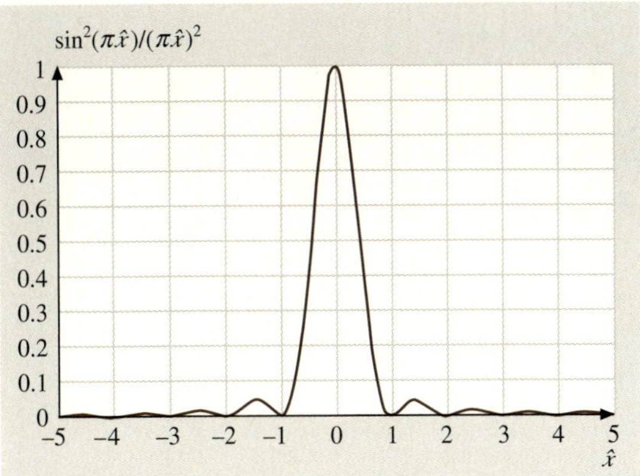

Fig. 3.16 Normalized intensity distribution along the x-axis in the focal plane of a lens for a rectangular aperture. Shown is the function $\left[\sin(\pi\hat{x})/(\pi\hat{x})\right]^2$. The function $\left[\sin(\pi\hat{y})/(\pi\hat{y})\right]^2$ along the y-axis is identical

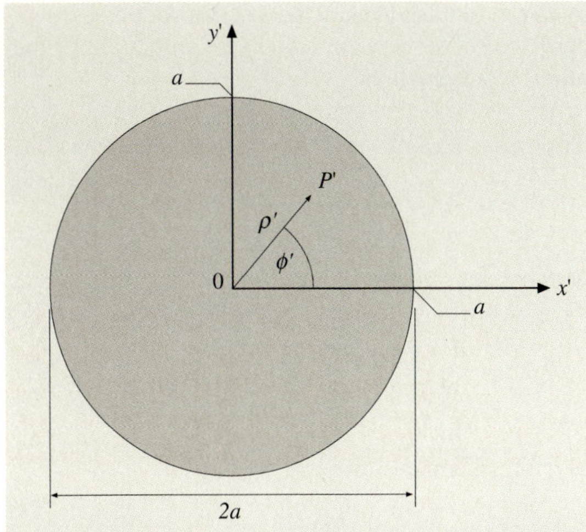

Fig. 3.17 Parameters of a circular transparent aperture in an opaque screen

The second integral is analogous and so the intensity in the focal plane is

$$I(x, y, z_0 = f) = I_0 \left(\frac{n\,2a\,2b}{\lambda f}\right)^2 \left(\frac{\sin\left(\pi\frac{nx\,2a}{\lambda f}\right)}{\pi\frac{nx\,2a}{\lambda f}}\right)^2 \times \left(\frac{\sin\left(\pi\frac{ny\,2b}{\lambda f}\right)}{\pi\frac{ny\,2b}{\lambda f}}\right)^2 . \tag{3.284}$$

By introducing the variables $\hat{x} = nx\,2a/(\lambda f)$ and $\hat{y} = ny\,2b/(\lambda f)$, which are pure numbers without a physical unit, the normalized intensity distribution along one of the axes (along the x- or y-axis) can be easily calculated and is shown in Fig. 3.16. The minima of the intensity distribution along the x-axis are at

$$\hat{x} = m \text{ with } m = 1, 2, 3 \ldots \Rightarrow x = m\frac{\lambda f}{2na} \approx m\frac{\lambda}{2\mathrm{NA}_x} . \tag{3.285}$$

Here, the numerical aperture $\mathrm{NA}_x := n\sin\varphi \approx na/f$ with the half-aperture angle $\varphi \ll 1$ of the lens in the x direction has been used.

Fraunhofer Diffraction at a Circular Aperture. The intensity distribution in the focal plane of an ideal lens (focal length f) with a circular aperture of the radius a can also be calculated using (3.278) (Fig. 3.17). The wavelength of the light is again λ and the refractive index of the material in which the wave propagates is n. The lens itself is again illuminated with a uniform plane on-axis wave with intensity I_0. Because of the circular symmetry it is useful to introduce polar coordinates

$$\left.\begin{array}{l} x' = \rho'\cos\phi';\quad x = \rho\cos\phi \\ y' = \rho'\sin\phi';\quad y = \rho\sin\phi \end{array}\right\} \Rightarrow xx' + yy' = \rho\rho'\cos\left(\phi' - \phi\right) . \tag{3.286}$$

Then, according to (3.278) the intensity in the focal plane of the ideal lens (i.e., $W = 0$) is written in polar coordinates

$$\begin{aligned} I(\rho, \phi, z_0 = f) &= I_0 \frac{n^2}{\lambda^2 f^2} \left| \iint_A \mathrm{e}^{-2\pi \mathrm{i} n \frac{xx' + yy'}{\lambda f}}\, \mathrm{d}x'\,\mathrm{d}y' \right|^2 \\ &= I_0 \frac{n^2}{\lambda^2 f^2} \left| \int_0^a \int_0^{2\pi} \mathrm{e}^{-2\pi \mathrm{i} n \frac{\rho\rho'\cos\left(\phi' - \phi\right)}{\lambda f}}\, \rho'\,\mathrm{d}\rho'\mathrm{d}\phi' \right|^2 \\ &= I_0 \frac{n^2}{\lambda^2 f^2} \left| \int_0^a \int_0^{2\pi} \mathrm{e}^{-2\pi \mathrm{i} n \frac{\rho\rho'\cos\phi'}{\lambda f}}\, \rho'\,\mathrm{d}\rho'\,\mathrm{d}\phi' \right|^2 . \end{aligned} \tag{3.287}$$

To solve the double integral the well-known Bessel functions [3.51] of the first kind $J_m(x)$ are introduced by the integral representation

$$J_m(x) = \frac{i^{-m}}{2\pi} \int_0^{2\pi} e^{ix\cos\alpha} e^{im\alpha}\, d\alpha\,. \tag{3.288}$$

For $m = 0$ we obtain

$$J_0(x) = \frac{1}{2\pi} \int_0^{2\pi} e^{ix\cos\alpha}\, d\alpha\,. \tag{3.289}$$

Therefore, the intensity is

$$I(\rho, \phi, z_0 = f) = I_0 \left(\frac{2\pi n}{\lambda f}\right)^2 \left| \int_0^a J_0\left(-2\pi n \frac{\rho\rho'}{\lambda f}\right) \rho'\, d\rho' \right|^2 . \tag{3.290}$$

There is another integral relation which connects the two Bessel functions J_0 and J_1:

$$x J_1(x) = \int_0^x x' J_0(x') dx'\,. \tag{3.291}$$

By substituting $x' = -2\pi n\rho\rho'/(\lambda f)$ and $dx' = -2\pi n\rho/(\lambda f)\, d\rho'$ we obtain

$$\begin{aligned} I(\rho, \phi, z_0 = f) &= I_0 \left(\frac{2\pi n}{\lambda f}\right)^2 \left| \frac{\lambda^2 f^2}{4\pi^2 n^2 \rho^2} \int_0^{-2\pi n \frac{\rho a}{\lambda f}} x' J_0(x')\, dx' \right|^2 \\ &= I_0 \left(\frac{2\pi n}{\lambda f}\right)^2 \left[-\frac{\lambda f a}{2\pi n\rho} J_1\left(-2\pi n \frac{\rho a}{\lambda f}\right) \right]^2 \\ &= I_0 \left(\frac{2\pi n a^2}{\lambda f}\right)^2 \left(\frac{J_1\left(2\pi n \frac{\rho a}{\lambda f}\right)}{2\pi n \frac{\rho a}{\lambda f}} \right)^2 . \end{aligned} \tag{3.292}$$

Here, the symmetry of the Bessel function $J_1(-x) = J_1(x)$ has been used. By defining the variable $\hat{\rho} := 2n\rho a/(\lambda f)$, which is a pure number without a physical unit, the intensity in the focal plane of the ideal lens

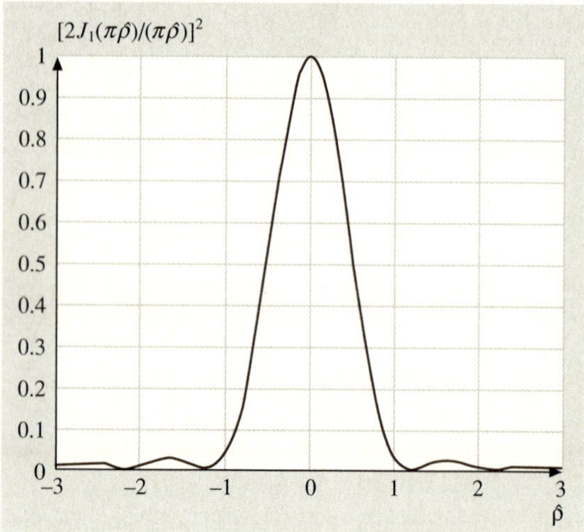

Fig. 3.18 Normalized intensity distribution in the focal plane of a lens. Shown is the function $\left[2J_1(\pi\hat{\rho})/(\pi\hat{\rho})\right]^2$

can be written as

$$I(\hat{\rho}, z_0 = f) = I_0 \left(\frac{n\pi a^2}{\lambda f}\right)^2 \left(2\frac{J_1(\pi\hat{\rho})}{\pi\hat{\rho}}\right)^2 . \tag{3.293}$$

The function $\left[2J_1(\pi\hat{\rho})/(\pi\hat{\rho})\right]^2$ is shown in Fig. 3.18. The first minimum is at the value $\hat{\rho}_0 = 1.22$, i. e., at the radius ρ_0 with

$$\rho_0 = 1.22 \frac{\lambda f}{2na} = 0.61 \frac{\lambda f}{na} = 0.61 \frac{\lambda}{\mathrm{NA}}\,. \tag{3.294}$$

Here, again the numerical aperture of the lens $\mathrm{NA} \approx na/f$ is used. The area inside the first minimum of the diffraction limited focus is called the Airy disc. It is also interesting to compare the maximum intensity $I(\hat{\rho} = 0, z_0 = f)$ in the central peak of the focus with the intensity I_0 of the incident plane wave. The ratio is

$$\frac{I(\hat{\rho} = 0, z_0 = f)}{I_0} = \left(\frac{n\pi a^2}{\lambda f}\right)^2 . \tag{3.295}$$

For a lens with a focal lens of $f = 10$ mm, an aperture radius of $a = 1$ mm and a wavelength of $\lambda = 0.5$ μm and $n = 1$ we obtain, e.g., $I(\hat{\rho} = 0, z_0 = f)/I_0 = (200\pi)^2 \approx 4\times10^5$. The quantity $na^2/(\lambda f)$ is also known as the Fresnel number F of a lens, which is the number of Fresnel zones of the lens in the paraxial case. This can be easily seen by calculating thc optical path difference OPDbetween a ray from the center of the lens to the focus and a ray from the rim of the lens to the focus. It is

$$\mathrm{OPD} = n\left(\sqrt{a^2+f^2}-f\right) = n\left(f\sqrt{1+\frac{a^2}{f^2}}-f\right)$$
$$\approx \frac{na^2}{2f} = F\frac{\lambda}{2};$$
$$\Rightarrow F = \frac{na^2}{\lambda f}. \quad (3.296)$$

This means that the intensity in the central peak of the focus of an ideal lens is proportional to the square of the Fresnel number of this lens.

3.4.4 Numerical Implementation of the Different Diffraction Methods

Many of the proposed diffraction integrals can be solved by performing one or two Fourier transformations (Table 3.1). For a numerical implementation, a discrete Fourier transformation is necessary and to increase the speed of the calculation it makes sense to take a fast Fourier transformation (FFT) [3.52]. Of course, it has to be noted that the sampling theorem is fulfilled and that the size of the field is large enough. In practice, large field sizes of, e.g., more than 2048 × 2048 samples need a lot of computer memory and computing time. To use a FFT the field of the complex amplitude u_0 is in each spatial direction x and y uniformly sampled at $N_x \times N_y$ points, where N_x and N_y are powers of two. The diameters of the field in the spatial domain are called D_x in the x and D_y in the y direction. In most cases the field will be quadratic and sampled with equal number of points, i. e., $N_x = N_y$ and $D_x = D_y$. Nevertheless, there are cases (e.g., systems with cylindrical or toric optical elements) where it is useful to have different sampling rate and number of sampling points along x and y. Then, the sampling interval Δx and Δy in the x and y direction between two neighboring sampling points in the spatial domain are (Fig. 3.19)

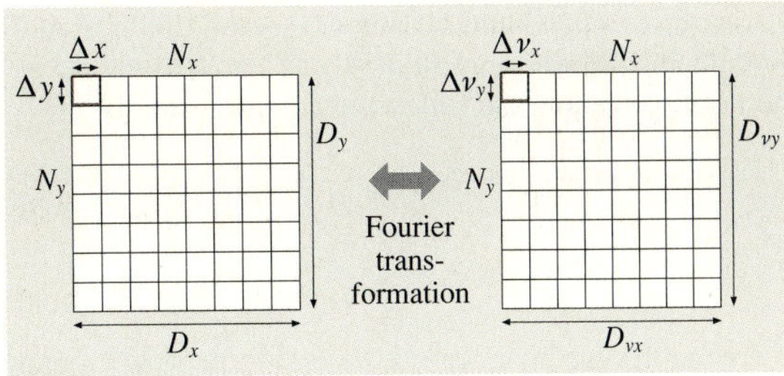

Fig. 3.19 Discrete fields for solving diffraction integrals using a FFT

$$\Delta x = \frac{D_x}{N_x},$$
$$\Delta y = \frac{D_y}{N_y}. \quad (3.297)$$

We call the conjugated variables in the Fourier domain ν_x and ν_y, where these can be really the spatial frequencies defined in (3.223) or other variables like in the Fresnel diffraction integral, which just have the physical dimension of a spatial frequency. Table 3.1 shows the conjugated variables for the different diffraction integrals. The diameters of the fields in the Fourier domain are called $D_{\nu x}$ and $D_{\nu y}$ along the ν_x- and ν_y direction. The associated sampling intervals between two neighboring sampling points in the Fourier domain are

$$\Delta \nu_x = \frac{D_{\nu x}}{N_x},$$
$$\Delta \nu_y = \frac{D_{\nu y}}{N_y}. \quad (3.298)$$

For a FFT the product of the respective diameters in the spatial domain and in the Fourier domain is equal to the number of sampling points. Therefore, the two relations are valid

$$D_x D_{\nu x} = N_x,$$
$$D_y D_{\nu y} = N_y. \quad (3.299)$$

Table 3.1 Conjugated variables and number of FFTs for calculating the different diffraction integrals

Diffraction method	Spectrum of plane waves	Fresnel (Fourier domain)	Fresnel (convolution)	Fraunhofer, Debye integral
Equation	(3.231)	(3.260)	(3.253)	(3.267), (3.276)
Conjugated variables	$(x,y) \leftrightarrow (\nu_x,\nu_y)$	$(x,y) \leftrightarrow (\nu_x,\nu_y)$	$(x',y') \leftrightarrow \left(\frac{nx}{\lambda z_0},\frac{ny}{\lambda z_0}\right)$	$(x',y') \leftrightarrow \left(\frac{n\alpha}{\lambda},\frac{n\beta}{\lambda}\right)$, $(x',y') \leftrightarrow \left(\frac{nx}{\lambda z_0},\frac{ny}{\lambda z_0}\right)$
Number of FFTs	2	2	1	1

By using (3.297) and (3.298) it is clear that for the product of the sampling intervals in the spatial and in the Fourier domain the following equations apply

$$\Delta x \Delta \nu_x = \frac{1}{N_x} \Rightarrow \Delta \nu_x = \frac{1}{D_x}; \ \Delta x = \frac{1}{D_{\nu x}},$$
$$\Delta y \Delta \nu_y = \frac{1}{N_y} \Rightarrow \Delta \nu_y = \frac{1}{D_y}; \ \Delta y = \frac{1}{D_{\nu y}} \quad (3.300)$$

The variables are in our case always symmetrical around the origin of the coordinate system. Therefore, the minimum and maximum values of the respective variables are

$$x_{\min} = -\frac{D_x}{2}, \qquad x_{\max} = \frac{D_x}{2};$$
$$y_{\min} = -\frac{D_y}{2}, \qquad y_{\max} = \frac{D_y}{2};$$
$$\nu_{x,\min} = -\frac{N_x}{2}\Delta\nu_x = -\frac{N_x}{2D_x},$$
$$\nu_{x,\max} = +\frac{N_x}{2}\Delta\nu_x = +\frac{N_x}{2D_x};$$
$$\nu_{y,\min} = -\frac{N_y}{2}\Delta\nu_y = -\frac{N_y}{2D_y},$$
$$\nu_{y,\max} = +\frac{N_y}{2}\Delta\nu_y = +\frac{N_y}{2D_y}. \quad (3.301)$$

Since a discrete Fourier transformation has periodic boundary conditions the function values at the left boundary will be equal to those at the right boundary, i. e., $u(x_{\min}, y) = u(x_{\max}, y)$, $u(x, y_{\min}) = u(x, y_{\max})$, and so on. This fact is important because it generates aliasing effects if the field size and the sampling are not correct. In the following some special aspects of the different diffraction integrals, which can be calculated using one or two FFTs are presented.

Numerical Implementation of the Angular Spectrum of Plane Waves or the Fresnel Diffraction in the Fourier Domain

To solve the diffraction integrals (3.231) and (3.260) two FFTs are necessary. The first is to transform the complex amplitude u_0 into the Fourier domain with the spatial frequencies ν_x and ν_y. In order to represent all propagating waves the maximum spatial frequencies have to fulfill the conditions

$$\nu_{x,\max} \geq \frac{n}{\lambda} \Rightarrow N_x \geq \frac{2nD_x}{\lambda},$$
$$\nu_{y,\max} \geq \frac{n}{\lambda} \Rightarrow N_y \geq \frac{2nD_y}{\lambda}. \quad (3.302)$$

Here, (3.223) and (3.301) have been used. However, it is in general not necessary that all spatial frequencies belonging to propagating waves can be represented. Especially, in the case of the Fresnel diffraction integral (3.260) formulated in the Fourier domain high spatial frequencies will in most cases not be allowed because the equation is, depending on the propagation distance z_0, only valid for small spatial frequencies (3.262). Let us assume that the angular spectrum of plane waves has a maximum tilt angle φ for which the Fourier transform $\tilde{u}_0$ of u_0 has a function value that is noticeably different from zero. Then, it is sufficient that instead of (3.302) the following conditions are used

$$\nu_{x,\max} \geq \frac{n}{\lambda}\sin\varphi \Rightarrow N_x \geq \frac{2nD_x}{\lambda}\sin\varphi,$$
$$\nu_{y,\max} \geq \frac{n}{\lambda}\sin\varphi \Rightarrow N_y \geq \frac{2nD_y}{\lambda}\sin\varphi. \quad (3.303)$$

If these conditions are not fulfilled aliasing effects occur and the numerical result is wrong. It is clear that only for microoptical elements the condition (3.302), i. e., $\sin\varphi$ is allowed to be 1, can be fulfilled. For a wavelength of $\lambda = 0.5\,\mu m$, refractive index $n = 1$ and a field diameter in the x direction of $D_x = 1$ mm, we obtain, e.g., $N_x \geq 2nD_x/\lambda = 4000$. For a two-dimensional FFT a field size of 4096×4096 samples is at the upper limit of modern personal computers.

Also in the case of the second (inverse) FFT, which calculates u using (3.231) or (3.260) aliasing effects can occur. Graphically, this means that parts of a diverging propagating wave leave the border of the field. Because of the periodic boundary conditions these parts of the waves will then enter the field at the opposite boundary. Figure 3.20 illustrates this effect for a diverging spherical wave that has a diameter D_0 at the starting plane and would have a diameter $D_{z0} > D_x$ in the distance z_0. If the half-aperture angle of the spherical wave is φ, where we assume that φ is so small that $\sin\varphi \approx \tan\varphi$,

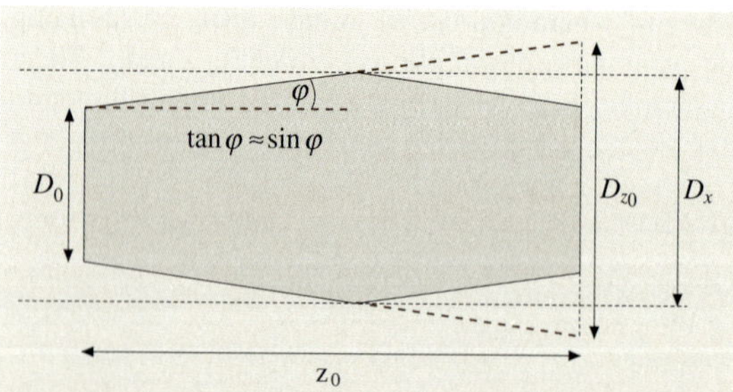

Fig. 3.20 Illustration of aliasing effects. The parts of the wave that leave the field at a boundary enter the field at the opposite boundary and interfere with the other parts of the wave

the diameter D_{z_0} will be

$$D_{z_0} \approx D_0 + 2z_0 \sin\varphi\,. \tag{3.304}$$

If $D_{z_0} > D_x$ aliasing effects will appear. This effect can be useful to numerically simulate the self-imaging Talbot effect for infinitely extended periodic structures [3.53]. However, in most cases aliasing effects are disturbing and have to be avoided. For converging waves, aliasing effects will not appear as long as the propagation distance is not so large that the wave passes the focus and becomes diverging. In practice, the limitations on the number of samples will limit the application of the propagation of a wave using the angular spectrum of plane waves. It is also very interesting to note that the field diameters D_x and D_y normally do not change between the two planes at $z = 0$ and $z = z_0$ in the case of the propagation using the angular spectrum of plane waves or the Fresnel integral in the Fourier domain formulation. Only, if the field size in the Fourier domain is manipulated will the field size in the spatial domain change. This is only possible by taking each second sampling point and embedding the new field with so many zeros that the total number of sampling points remains unchanged. Then the effective diameters in the Fourier domain $D_{\nu x}$ and $D_{\nu y}$ are doubled and therefore, according to (3.299) the diameters D_x and D_y in the spatial domain will be halved. But this manipulation reduces the sampling density in the Fourier domain and is therefore a kind of high-pass filtering operation suppressing the long-periodic spatial structures.

Numerical Implementation of the Fresnel (Convolution Formulation) and the Fraunhofer Diffraction

Using the method of the angular spectrum of plane waves guarantees that the field sizes D_x and D_y do not change during the propagation as long as no manipulations are made in the Fourier domain. The reason for this is that two FFTs are used, one "normal" and one inverse FFT. If the Fresnel diffraction integral of (3.253) or the Fraunhofer diffraction integrals (3.267) or (3.276) are used, only one FFT is made. Therefore, the field size changes because the conjugated variables are now, according to Table 3.1, (x', y') and $[nx/(\lambda z_0), ny/(\lambda z_0)]$ or $[(n\alpha/\lambda, n\beta/\lambda)$, where $\alpha = x/z_0$ and $\beta = y/z_0]$. This means that according to (3.299) the following relations are valid if we introduce the field diameters $D_{x,0} = D_x$ and $D_{y,0} = D_y$ in the first plane and the field diameters D_{x,z_0} and D_{y,z_0} in the second plane, where $D_{\nu x} = nD_{x,z_0}/(\lambda z_0)$ and $D_{\nu y} = nD_{y,z_0}/(\lambda z_0)$

$$D_x D_{\nu x} = N_x \Rightarrow D_{x,z_0} = N_x \frac{\lambda z_0}{nD_{x,0}}\,,$$

$$D_y D_{\nu y} = N_y \Rightarrow D_{y,z_0} = N_y \frac{\lambda z_0}{nD_{y,0}}\,. \tag{3.305}$$

Of course, the sampling densities Δx_{z_0} and Δy_{z_0} in the second plane are then

$$\Delta x_{z_0} = \frac{\lambda z_0}{nD_{x,0}}\,,$$

$$\Delta y_{z_0} = \frac{\lambda z_0}{nD_{y,0}}\,. \tag{3.306}$$

Let us calculate the intensity distribution in the focal plane of a lens with focal length f ($z_0 = f$) using (3.276). The aperture of the lens shall be quadratic with diameters $2a = 2b$. Then, it is enough to just consider one dimension, e.g., the x direction. If the field size $D_{x,0}$ would now only be $2a$ the sampling interval would be

$$\Delta x_{z_0} = \frac{\lambda f}{n2a} \approx \frac{\lambda}{2\mathrm{NA}}\,, \tag{3.307}$$

where, the numerical aperture of the lens $\mathrm{NA} := n\sin\varphi \approx na/f$ has been used. But by comparing this result with (3.285) it is clear that the sampling density is so low that the secondary maxima are not observed because the sampling is only in the minima. Therefore, it is necessary that the aperture of the lens is embedded into a field of zeros so that the effective field diameter $D_{x,0}$ is at least doubled $D_{x,0} \geq 4a$. So, by embedding the aperture of the lens by more and more zeros the effective sampling density in the focal plane is increased. Increasing the field size by a factor m and filling the new area with zeros reduces the sampling interval by a factor m to $\Delta x_{z_0} = \lambda/(2m\mathrm{NA})$. In other words we can say: the total field size D_{x,z_0} in the focal plane is proportional to the number N_x/m of samples with which the lens aperture is sampled whereas the sampling density is proportional to the factor m of the zero embedding.

Of course, there is also another reason for embedding the lens aperture with zeros: the aliasing effects. A quadratic lens aperture with $D_{x,0} = 2a$ would, due to the periodic boundary conditions, mean that the aperture is repeated periodically without spaces in between and fills the whole space. Therefore, no diffraction at all would occur and the focus would be a delta peak as in geometrical optics.

3.4.5 The Influence of Polarization Effects to the Intensity Distribution Near the Focus

In the previous cases of diffraction we considered only the scalar case. But, in this section we want to discuss the influence of polarization effects to the intensity distribution in the focal region of a lens. However, we will only discuss a simple numerical simulation method, which is in fact identical to the semi-analytical vectorial Debye integral formulation discussed, e.g., in [3.54] or [3.34]. *Richards* and *Wolf* [3.34] were one of the first who calculated that the light distribution in the focus of a lens, which is illuminated by a linearly polarized plane wave, is asymmetrical for a high numerical aperture of the lens.

Some Elementary Qualitative Explanations

To illustrate the influence of polarization effects to the focus refer to Fig. 3.21. Assume a plane wave with linear polarization (electric vector in the y direction), which is focused by a lens. Then, there is one plane (xz-plane) where the electric vectors are perpendicular to the plane of refraction of the rays (Fig. 3.21a). There, the electric vectors in the focus add like scalars and a quite large transversal component is obtained. But, there is also the yz-plane, where the electric vectors change their direction after being refracted by the lens. Then, they add in the focus like real vectors and a smaller transversal component than in the xz-plane is obtained (Fig. 3.21b). Especially, for very steep rays corresponding to high numerical aperture rays the electric vectors nearly cancel each other in the focus. Due to the broken symmetry of

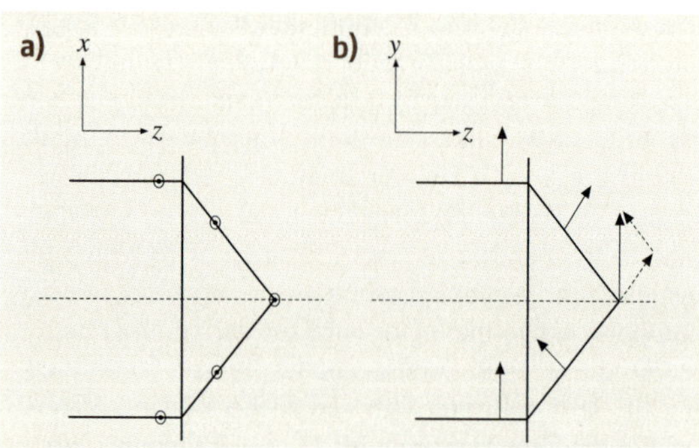

Fig. 3.21a,b Addition of the electric vectors in the focus of a lens for linearly polarized light. **(a)** The electric vectors are perpendicular to the xz-plane, i.e., they add arithmetically like scalars; **(b)** the electric vectors lie in the yz-plane and add vectorially to a transversal component

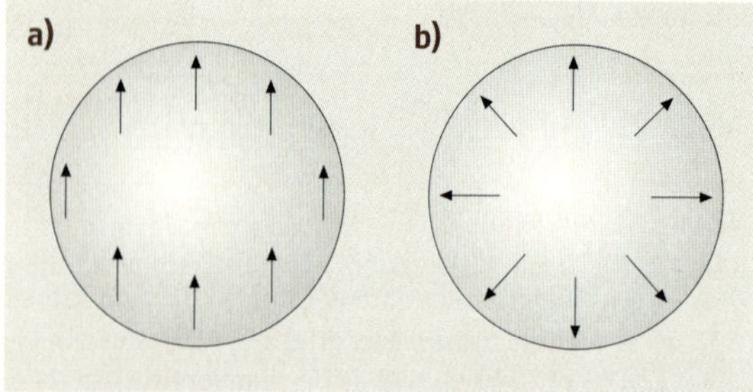

Fig. 3.22a,b Local polarization vectors in the aperture in front of the lens for **(a)** linearly polarized light and **(b)** a radially polarized doughnut mode

this problem the intensity distribution in the focus will also be nonrotationally symmetric. Of course, this effect is only visible for very high numerical apertures, because otherwise the vector character of the electric field is not so obvious. It is especially visible for annular apertures and a high numerical aperture lens.

But there is a rotationally symmetric polarization pattern, called the doughnut mode [3.33], where the direction of the electric vector varies locally, so that it always points away from the optical axis in radial direction. Figure 3.22 shows in (a) the electric vectors in the aperture of the lens at a certain time for linearly polarized light, i.e., they all point in the same direction. In (b) the electric vectors for the radially polarized doughnut mode are drawn at a certain time. They all point radially away from the optical axis. Of course, in the case of the doughnut mode there has to be the intensity zero on the optical axis due to the symmetry of the problem. So, the intensity distribution in the aperture of the lens is not homogeneous in the case of the radially symmetric doughnut mode but in fact it is the superposition of a Hermite–Gaussian TEM_{10} and a TEM_{01} mode (Fig. 3.32) with relative phase difference zero. Then, the time-independent complex-valued electric vector $\boldsymbol{E}_{\mathrm{rad}}$ before the lens has the value

$$\boldsymbol{E}_{\mathrm{rad}}(x, y, z=0) = E_0 \begin{pmatrix} x \\ y \\ 0 \end{pmatrix} \mathrm{e}^{-\left(x^2+y^2\right)/w_0^2} . \tag{3.308}$$

Here, w_0 is the beam waist of the Gaussian function and E_0 is a constant. The maximum of the amplitude of the electric vector is at $\sqrt{x^2+y^2} = w_0/\sqrt{2}$ as can be easily calculated.

But now, for the radially polarized doughnut mode in all planes containing the optical axis, the electric vectors are oriented as in Fig. 3.23. Then, the electric vectors add

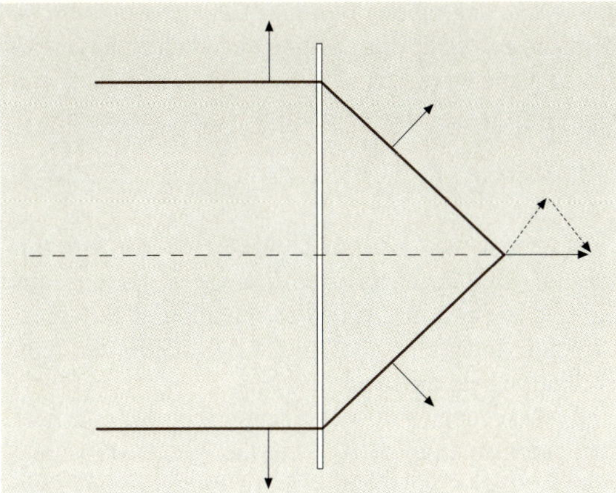

Fig. 3.23 Orientation of the electric vectors in front of and behind a lens for the radially symmetric doughnut mode. The electric vectors add to a longitudinal component parallel to the optical axis

in the focus to a longitudinal component and this is the case for all planes containing the optical axis. Therefore, the focus is completely rotationally symmetric.

Numerical Calculation Method

To calculate the electric energy density in the focal region of a lens the vectorial Debye integral of [3.34] or the method of [3.55] can be used. Both say that the electric vector in the focal region can be written as a superposition of plane waves, which propagate along the direction of the rays running from the exit pupil of the lens to the geometrical focus. In this model, diffraction effects at the rim of the aperture are neglected, as is also done in the scalar formulation of the Debye integral. But this is a quite good approximation as long as the diameter of the aperture of the lens is large compared to the wavelength ($2r_{\text{aperture}} \gg \lambda$) and as long as the numerical aperture of the lens is sufficiently high (which is always the case here because polarization effects are only interesting for high numerical apertures).

For the numerical calculation we make a uniform sampling of rays in the planar entrance pupil of the lens, e.g., $N \times N$ rays in an orthogonal pattern, where the rays outside of the aperture (e.g., circular or annular aperture) have zero amplitude. The optical axis shall be the z-axis. Each ray number (i, j) at the coordinate (x_i, y_j, z_0) in the entrance pupil has a wave vector $\boldsymbol{k}_{i,j} = (0, 0, k) = (0, 0, 2\pi/\lambda)$ and a generally complex polarization vector $\boldsymbol{P}_{i,j} = (P_{x,i,j}, P_{y,i,j}, 0)$, which is in fact proportional to the electric vector $\boldsymbol{E}$ at this point if the sampling is uniform. Of course, $\boldsymbol{P}_{i,j}$ is orthogonal to $\boldsymbol{k}_{i,j}$ because of the orthogonality condition of plane waves (Sect. 3.1.1) (see the paragraph on page 91). Examples of different polarization states are (see Sect. 3.2.4):

- A linearly polarized homogeneous plane wave (polarization in the y direction) with the polarization vector $\boldsymbol{P}_{i,j} = (0, P_0, 0)$ with a constant real value P_0.
- A circularly polarized homogeneous plane wave with the polarization vector $\boldsymbol{P}_{i,j} = (P_0/\sqrt{2}, \mathrm{i}P_0/\sqrt{2}, 0)$, where the factor $1/\sqrt{2}$ is introduced to have $|\boldsymbol{P}_{i,j}|^2 = P_0^2$.
- The radially polarized doughnut mode with the polarization vectors according to (3.308).

The lens itself fulfills the sine condition so that we have for the refracted ray behind the lens

$$h = \sqrt{x^2 + y^2} = f \sin\vartheta \quad \Rightarrow \quad \sin\vartheta = \frac{\sqrt{x^2+y^2}}{f}\,, \tag{3.309}$$

where, f is the focal length of the lens and ϑ the polar angle (Fig. 3.24). By defining the azimuthal angle φ we have the following relations for the rays in front of the lens and behind the lens

$$\begin{aligned} x &= h\cos\varphi \\ y &= h\sin\varphi \end{aligned} \quad \Rightarrow \quad \varphi = \arctan\frac{y}{x} \tag{3.310}$$

and

$$\boldsymbol{k}' = k \begin{pmatrix} -\cos\varphi\sin\vartheta \\ -\sin\varphi\sin\vartheta \\ \cos\vartheta \end{pmatrix}. \tag{3.311}$$

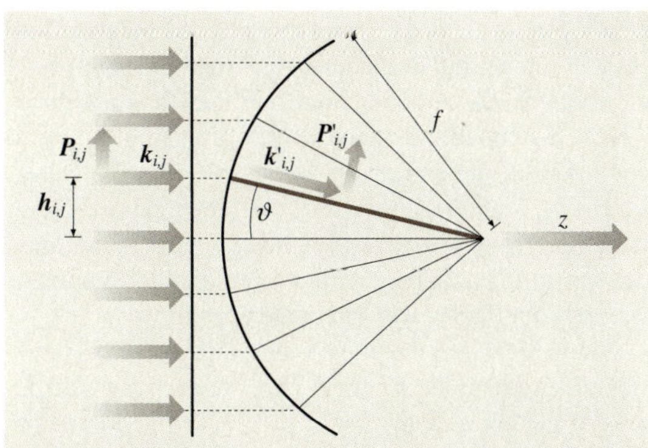

Fig. 3.24 Principal scheme of the distribution of rays used to calculate the electric energy density in the focus. The lens has to fulfill the sine condition so that the ray height $h = \sqrt{x^2+y^2}$ is $h = f\sin\vartheta$

The polarization vector $\boldsymbol{P}$ of each ray has to be separated into a component lying in the plane of refraction and a component perpendicular to it. The unit vector $\boldsymbol{e}_{\parallel}$ along the component in the plane of refraction in front of the lens is

$$\boldsymbol{e}_{\parallel}=\begin{pmatrix}\cos\varphi\\ \sin\varphi\\ 0\end{pmatrix}. \tag{3.312}$$

Behind the lens the new unit vector $\boldsymbol{e}'_{\parallel}$ along the component in the plane of refraction is

$$\boldsymbol{e}'_{\parallel}=\begin{pmatrix}\cos\varphi\cos\vartheta\\ \sin\varphi\cos\vartheta\\ \sin\vartheta\end{pmatrix}. \tag{3.313}$$

So, the polarization vector $\boldsymbol{P}'$ behind the lens can be calculated by keeping in mind that the component perpendicular to the plane of refraction remains unchanged and that the component in the plane of refraction keeps its amplitude but is now parallel to $\boldsymbol{e}'_{\parallel}$. In total this means for $\boldsymbol{P}'$

$$\boldsymbol{P}'=\frac{1}{\sqrt{\cos\vartheta}}\left[\boldsymbol{P}-(\boldsymbol{P}\cdot\boldsymbol{e}_{\parallel})\,\boldsymbol{e}_{\parallel}+(\boldsymbol{P}\cdot\boldsymbol{e}_{\parallel})\,\boldsymbol{e}'_{\parallel}\right]$$

$$=\frac{\boldsymbol{P}-(P_x\cos\varphi+P_y\sin\varphi)\begin{pmatrix}\cos\varphi\,(1-\cos\vartheta)\\ \sin\varphi\,(1-\cos\vartheta)\\ -\sin\vartheta\end{pmatrix}}{\sqrt{\cos\vartheta}}. \tag{3.314}$$

Here, the factor $1/\sqrt{\cos\vartheta}$ is necessary in order to conserve the energy of the tilted plane wave [3.56]. In other words, behind the lens the density of the numerical sampling by the polarization vectors $\boldsymbol{P}'_{i,j}$ depends on the angle ϑ whereas in front of the lens the sampling of $\boldsymbol{P}_{i,j}$ is selected to be uniform. So, the electric vector $\boldsymbol{E}$ behind the lens is no longer proportional to the polarization vector $\boldsymbol{P}'$ alone, but to the ratio of $\boldsymbol{P}'$ and the small surface element dA perpendicular to $\boldsymbol{P}'$ which is assigned to each ray by the distance of the ray sampling, and which is itself depending on the angle ϑ. So, the factor $1/\sqrt{\cos\vartheta}$ corrects the non uniform numerical sampling of the polarization vectors behind the lens. If the lens does not fulfill the sine condition but another equation the correction factor will of course change.

For an ideal lens the plane waves along the rays have to be all in phase at the focus. So, we just can set the focus to the coordinate $\boldsymbol{r}'=(x'=0,\,y'=0,\,z'=0)$. To calculate the electric vector $\boldsymbol{E}$ at a point $\boldsymbol{r}'$ near the focus the plane waves have to be summarized according to

$$\boldsymbol{E}(\boldsymbol{r}')\propto\sum_{i,j}\boldsymbol{P}'_{i,j}\mathrm{e}^{\mathrm{i}\boldsymbol{k}'_{i,j}\cdot\boldsymbol{r}'}. \tag{3.315}$$

Small wave aberrations $W(x,y)$ of the lens can also be taken into account by just adding them to the phase term of each plane wave, i. e., $\boldsymbol{k}'_{i,j}\cdot\boldsymbol{r}'\rightarrow\boldsymbol{k}'_{i,j}\cdot\boldsymbol{r}'+W_{i,j}$ with $W_{i,j}:=W(x_i,y_j)$. Then we have:

$$\boldsymbol{E}(\boldsymbol{r}')\propto\sum_{i,j}\boldsymbol{P}'_{i,j}\mathrm{e}^{\mathrm{i}\boldsymbol{k}'_{i,j}\cdot\boldsymbol{r}'+\mathrm{i}W_{i,j}}. \tag{3.316}$$

It can be seen that in the focal plane $z'=0$ and for linearly polarized light with a small numerical aperture, this sum reduces to a discretized formulation of the scalar diffraction integral of (3.276) for calculating the light distribution near the focus of a lens.

By just changing the polarization vectors of the field distribution in front of the lens the effects of a quite arbitrary state of polarization to the focus can be investigated. The squares of the x, y, or z components of the electric field $\boldsymbol{E}$ can also be calculated separately. Of course, the electric energy density w_{e}, which is the physical quantity normally detected by a light detector, can then be calculated by

$$w_{\mathrm{e}}(\boldsymbol{r}')\propto|\boldsymbol{E}|^2=\boldsymbol{E}\cdot\boldsymbol{E}^*. \tag{3.317}$$

The proportionality factors are omitted because we are only interested in relative energy distributions and not in absolute values.

Some Simulation Results

Figures 3.25, 3.26 and 3.27 show the results of some simulations for the electric energy density in the focal plane (xy-plane) of an ideal lens. In all cases the numerical aperture of the lens is assumed to be 1.0 and the wavelength of the illuminating light is $\lambda=632.8$ nm. The typical number of plane waves used to sample the aperture of the lens is $N\times N=100\times100$ (or 200×200), whereby in the case of a circular aperture (normal case) and especially in the case of an annular aperture (see later) the effective number of plane waves is smaller (factor $\pi/4$ smaller for a circular aperture) because the sampling is made in an orthogonal and uniformly sampled xy-pattern and plane waves outside of the aperture just have zero amplitude. The number of sampling points in the focal plane is 200×200 and so a simulation on a modern 1 GHz Pentium PC takes less than two minutes.

In Fig. 3.25 the lens is illuminated by a linearly polarized plane wave (polarization in y direction). The squares of the different components of the electric vector in the focal plane are displayed. The x component in Fig. 3.25a is quite small and vanishes in the center of the geometrical focus (pay attention to the scale). The biggest component is the y component in Fig. 3.25b, i. e., along the direction in which the light is polarized

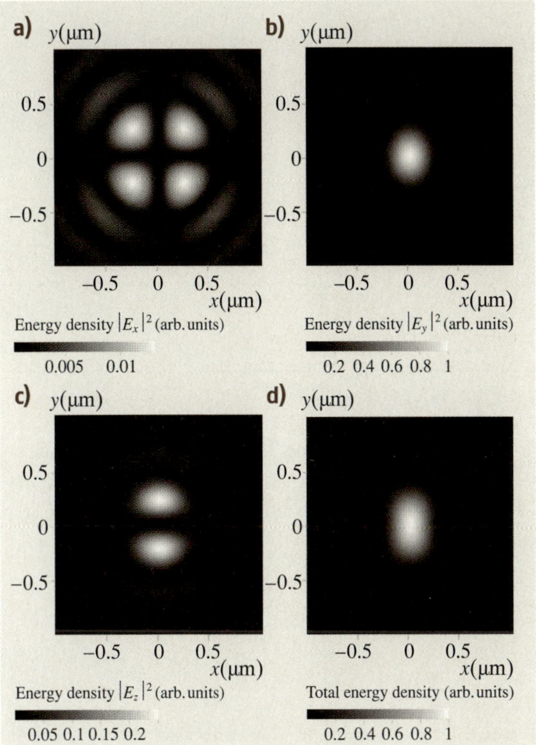

Fig. 3.25a–d Squares of the components of the electric vector in the focal plane of an ideal lens with NA = 1.0, which is illuminated by a plane linearly polarized (in the y direction) wave with $\lambda = 632.8$ nm. **(a)** x component, **(b)** y component, **(c)** z component and **(d)** sum of all components, i. e., total electric energy density

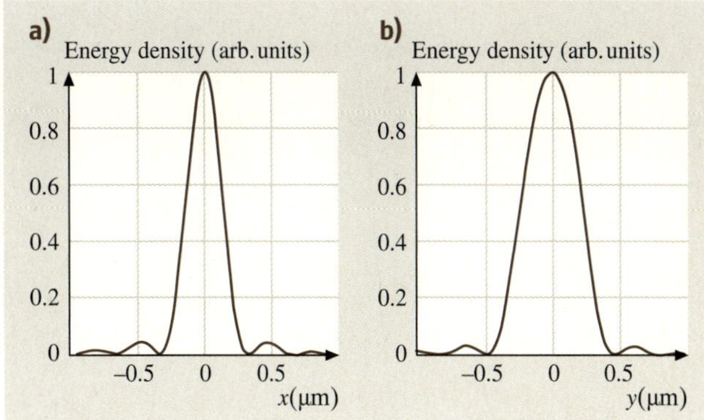

Fig. 3.26a,b Sections along **(a)** the x-axis and **(b)** the y-axis of the total electric energy density in the focal plane of an ideal lens with NA = 1.0, which is illuminated by a plane linearly polarized (in the y direction) wave with $\lambda = 632.8$ nm (see also Fig. 3.25d)

in front of the lens. But, there is also a z component (Fig. 3.25c) for points besides the geometrical focus. This component is mainly responsible for the asymmetric shape of the total electric energy density, which is displayed in Fig. 3.25d. Figure 3.26 shows sections along the x- and y-axis of the total electric energy density. It can be seen that the diameter of the central maximum along the y-axis is increased whereas the diameter along the x-axis has nearly the value of the scalar calculation $d_{\text{focus}} = 1.22\lambda/\text{NA} = 0.77\,\mu\text{m}$. All quantities are normalized in such a way that the total energy density has a maximum value of 1.

In Fig. 3.27 the lens is illuminated by a radially polarized doughnut mode, whereby the beam waist w_0 is at 95% of the lens aperture radius. Then, there are radial, i. e., transversal components of the electric field in Fig. 3.27a, which are of course rotationally symmetric due to the symmetry of the field. But, the strongest component is the longitudinal z component in Fig. 3.27b, which is also rotationally symmetric and has a central maximum with a diameter slightly smaller than the value of the scalar calculation. The total electric energy density in Fig. 3.27c is also rotationally symmetric but the diameter of the central maximum is increased due to the transversal components of the electric field. However, the surface area S, which is covered in the focus by a total electric energy density of more than half the maximum value is $S = 0.29\lambda^2$ in the linearly polarized case and only $S = 0.22\lambda^2$ for the radially polarized doughnut mode. So, if the light spot is used to write into a nonlinear material, which is only sensitive to a total electric energy density beyond a certain threshold a tighter spot can be obtained using the radially polarized doughnut mode.

Figure 3.28 shows for the same parameters as in Figs. 3.26 and 3.27 the total electric energy density near the focus in planes containing the optical axis (z-axis) and one of the lateral coordinates. In Figs. 3.28a,b the xz-plane and the yz-plane are shown for the case of an illumination of the lens with linearly polarized homogeneous light (polarization in the y direction). Figure 3.28c shows the same for the case of the radially polarized doughnut mode, where here the focal region is completely rotationally symmetric so that only one section has to be shown. It can again be seen that in the case of linear polarization the lateral diameter of the focus is broader in the yz-plane than in the xz-plane. The focus of the radially polarized doughnut mode has a lateral diameter nearly that of the small axis of the linearly polarized case focus.

From the scalar theory it is well-known that the diameter of the central maximum of the focus is decreased if an annular aperture is used instead of a circular one. It is

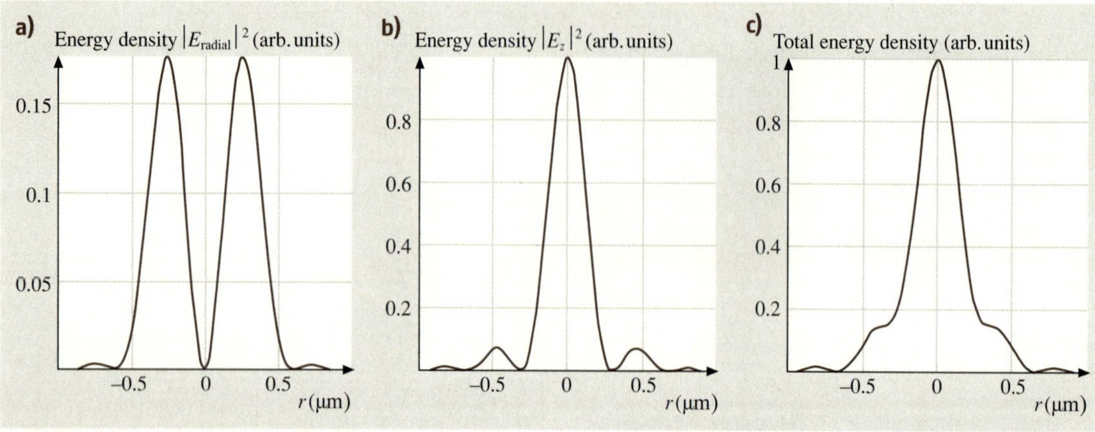

Fig. 3.27a–c Squares of the components of the electric vector in the focal plane of an ideal lens with NA = 1.0, which is illuminated by a radially polarized doughnut mode with $\lambda = 632.8$ nm. The beam waist w_0 of the Gaussian function is $w_0 = 0.95 r_{aperture}$, whereby $r_{aperture}$ is the illuminated aperture radius of the lens. **(a)** Radial components, i. e., $|E_x|^2 + |E_y|^2$, **(b)** z component, **(c)** sum of all components, i. e., total electric energy density

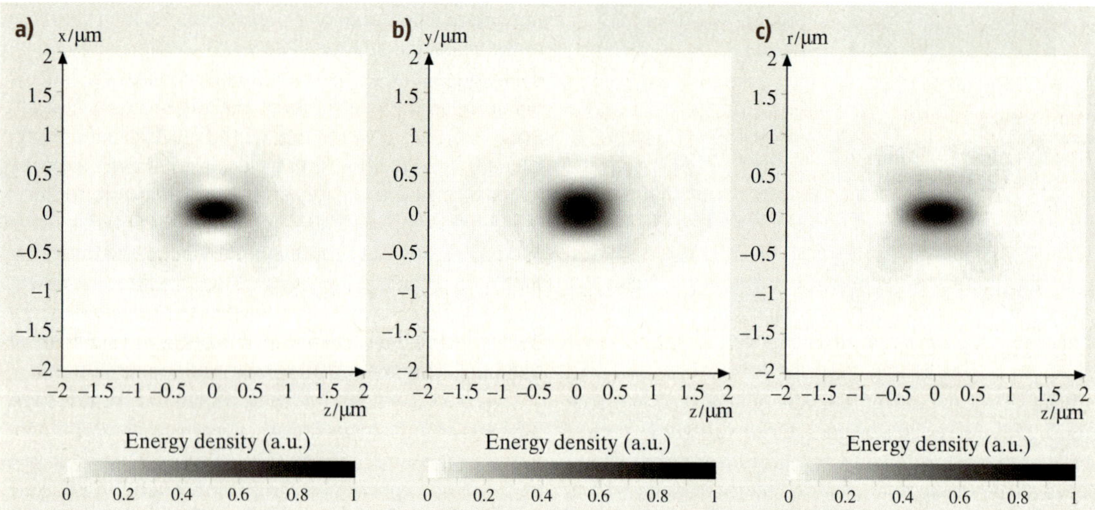

Fig. 3.28a–c Simulation of the total electric energy density in the focal region by using a circular aperture (NA = 1.0, $\lambda = 632.8$ nm). **(a)** Linearly polarized homogeneous light (xz-plane), **(b)** linearly polarized homogeneous light (yz-plane), **(c)** radially polarized doughnut mode with $w_0 = 0.95 r_{aperture}$ (xz-plane or yz-plane because of rotational symmetry)

also well-known from the scalar theory that in this case, the secondary maxima increases compared to a circular aperture. So, by using an annular aperture a smaller focus should be reached and this should especially be the case for the radially polarized doughnut mode. The reason is that in this case the transversal components in the focus of the doughnut mode decrease so that the total energy density is more and more dominated by the longitudinal z component, which has a small diameter. So, by using a radially symmetric doughnut mode, an annular aperture (e.g., $r_{annulus} = 0.9 r_{aperture}$), and a high numerical aperture lens (NA > 0.9) a very tight rotationally symmetric spot with a small surface area can be obtained. This effect can be used to achieve a higher resolution in optical data storage or in optical lithography where a mask is written spot by spot by a so-called laser pattern

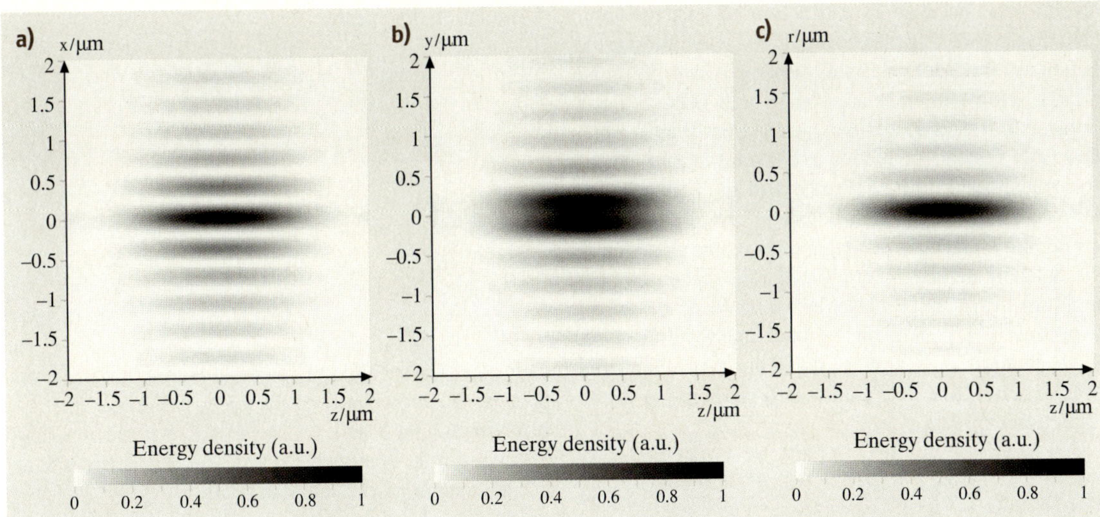

Fig. 3.29a–c Simulation of the total electric energy density in the focal region by using an annular aperture with an inner radius of 90% of the full aperture (NA = 1.0, $\lambda = 632.8$ nm). **(a)** linearly polarized homogeneous light (*xz*-plane), **(b)** linearly polarized homogeneous light (*yz*-plane), **(c)** radially polarized doughnut mode with $w_0 = 0.95r_{\text{aperture}}$ (*xz*-plane or *yz*-plane because of rotational symmetry)

generator. Figure 3.29 shows the simulation results of the electric energy density in the focal region for an annular aperture with $r_{\text{annulus}} = 0.9r_{\text{aperture}}$. As in the former cases, the numerical aperture itself is NA = 1.0 and the wavelength is $\lambda = 632.8$ nm. Figures 3.29a and b show the *xz*-plane and *yz*-plane, respectively, for the case of linear polarization and Fig. 3.29c shows the same for the doughnut mode ($w_0 = 0.95r_{\text{aperture}}$), where the result is rotationally symmetric so that only one section is shown. It can be seen that in the case of an annular aperture the lateral diameter of the focus is decreased but on the other side the depth of focus along the optical axis is increased compared to the full aperture case. Besides the central maximum there are also, as expected, some secondary maxima with increased height. Nevertheless, a comparison of the cases of linear polarization and the doughnut mode shows that the lateral diameter of the focus is decreased for the doughnut mode, especially in the *yz*-plane. By calculating the surface area S again, which is covered in the focus by a total electric energy density of more than half the maximum value, the result is $S = 0.29\lambda^2$ for the linearly polarized light and only $S = 0.12\lambda^2$ for the radially polarized doughnut mode. Moreover, the height of the secondary maxima is not as high for the doughnut mode as for the linear polarization. In some applications these secondary maxima can be disturbing, but in applications where a certain threshold value of the energy density has to be reached in order to obtain an effect, these secondary maxima have no influence. So, the annular aperture, especially combined with the radially polarized doughnut mode, allows a high lateral resolution. Experiments verify the simulation results for linearly polarized light [3.57] and for the radially polarized doughnut mode [3.33, 54].

3.5 Gaussian Beams

Gaussian beams are a paraxial solution of the scalar Helmholtz equation and are suitable to describe the propagation of coherent laser beams [3.10, 11, 58]. However, the influence of apertures on the laser beam is not considered in this description because apertures would generally disturb the Gaussian beam. The transformation of laser beams at a lens is, of course, also only treated in a paraxial sense and aberrations of the lens are not taken into account.

3.5.1 Derivation of the Basic Equations

The typical property of a collimated laser beam is that it propagates straight on in a homogeneous material.

Nevertheless, because of diffraction effects, the laser beam diverges during the propagation. Depending on the diameter of the laser beam this effect will be quite small or large. However, the laser beam will behave along the direction of propagation (z-axis) nearly as a plane wave. But instead of having a constant amplitude, as for a plane wave, the amplitude will be a function of the transversal coordinates (x and y) and also a slowly varying function of the propagation distance along the z-axis. Mathematically, this means that the scalar complex amplitude $u(x, y, z)$ of a laser beam can be described by the product of a (generally) complex function $\Psi(x, y, z)$, which changes only slowly along the z-axis, and the complex amplitude of a plane wave propagating in the z direction $\exp(\mathrm{i}kz)$

$$u(x, y, z) = \Psi(x, y, z)\mathrm{e}^{\mathrm{i}kz} . \tag{3.318}$$

The constant k is again defined as $2\pi n/\lambda$. n is the refractive index of the homogeneous material in which the Gaussian beam propagates and λ is the wavelength in a vacuum. To simplify the notation the wavelength $\lambda_n = \lambda/n$ in the material is used in the following.

By using the scalar Helmholtz equation (3.221)

$$\left(\nabla^2 + k^2\right) u(x, y, z) = 0 \tag{3.319}$$

the following equation for u is obtained

$$\begin{aligned}
\frac{\partial u}{\partial x} &= \frac{\partial \Psi}{\partial x}\mathrm{e}^{\mathrm{i}kz} , \\
\frac{\partial^2 u}{\partial x^2} &= \frac{\partial^2 \Psi}{\partial x^2}\mathrm{e}^{\mathrm{i}kz} , \\
\frac{\partial u}{\partial z} &= \frac{\partial \Psi}{\partial z}\mathrm{e}^{\mathrm{i}kz} + \mathrm{i}k\Psi\mathrm{e}^{\mathrm{i}kz} , \\
\frac{\partial^2 u}{\partial z^2} &= \frac{\partial^2 \Psi}{\partial z^2}\mathrm{e}^{\mathrm{i}kz} + 2\mathrm{i}k\frac{\partial \Psi}{\partial z}\mathrm{e}^{\mathrm{i}kz} - k^2\Psi\mathrm{e}^{\mathrm{i}kz} ; \\
\Rightarrow \left(\nabla^2 + k^2\right) u &= \left(\frac{\partial^2 \Psi}{\partial x^2} + \frac{\partial^2 \Psi}{\partial y^2}\right)\mathrm{e}^{\mathrm{i}kz} \\
&\quad + \frac{\partial^2 \Psi}{\partial z^2}\mathrm{e}^{\mathrm{i}kz} + 2\mathrm{i}k\frac{\partial \Psi}{\partial z}\mathrm{e}^{\mathrm{i}kz} = 0 .
\end{aligned} \tag{3.320}$$

According to our assumption that Ψ changes only slowly along the z direction the term $\partial^2\Psi/\partial z^2$ is assumed to be so small that it can be neglected. This is the case if the relative variation of $\partial\Psi/\partial z$ during the propagation by one wavelength is much smaller than one. In a mathematical formulation this means

$$\begin{aligned}
&\left|\frac{\partial^2 \Psi}{\partial z^2}\right| \ll \left|2k\frac{\partial \Psi}{\partial z}\right| = \frac{4\pi}{\lambda_n}\left|\frac{\partial \Psi}{\partial z}\right| \\
&\Rightarrow \left.\frac{|\Delta(\partial\Psi/\partial z)|}{|\partial\Psi/\partial z|}\right|_{\Delta z = \lambda_n} \ll 4\pi .
\end{aligned} \tag{3.321}$$

Using this simplification the following equation for Ψ is obtained

$$\frac{\partial^2 \Psi}{\partial x^2} + \frac{\partial^2 \Psi}{\partial y^2} + 2\mathrm{i}k\frac{\partial \Psi}{\partial z} = 0 . \tag{3.322}$$

This equation is called the paraxial Helmholtz equation because it corresponds to the case of Fresnel diffraction (Sects. 3.4.3 and 3.5.2). To solve it we first use a quite simple approach for Ψ, which corresponds to a fundamental mode Gaussian beam

$$\Psi(x, y, z) = \Psi_0 \exp\left\{\mathrm{i}\left[P(z) + \frac{k\left(x^2 + y^2\right)}{2q(z)}\right]\right\} \tag{3.323}$$

with the two complex functions P and q, which are both functions of z. Ψ_0 is a constant that depends on the amplitude of the Gaussian beam and is determined by the boundary conditions. Using the notations $P' := \mathrm{d}P/\mathrm{d}z$ and $q' := \mathrm{d}q/\mathrm{d}z$ our approach gives

$$\begin{aligned}
\frac{\partial \Psi}{\partial z} &= \left(\mathrm{i}P' - \frac{\mathrm{i}k\left(x^2 + y^2\right)}{2q^2}q'\right)\Psi , \\
\frac{\partial \Psi}{\partial x} &= \mathrm{i}\frac{kx}{q}\Psi , \\
\frac{\partial^2 \Psi}{\partial x^2} &= \frac{\mathrm{i}k}{q}\Psi - \frac{k^2x^2}{q^2}\Psi .
\end{aligned}$$

Inserting these equations in (3.322) results in the following conditions for P and q

$$\frac{2\mathrm{i}k}{q} - \frac{k^2\left(x^2 + y^2\right)}{q^2} - 2kP' + \frac{k^2\left(x^2 + y^2\right)}{q^2}q' = 0 . \tag{3.324}$$

This equation has to be fulfilled for arbitrary values of x and y. Therefore, the equation finally gives two equations

$$P' = \frac{\mathrm{i}}{q} \quad \text{and} \quad q' = 1 . \tag{3.325}$$

By integration we obtain

$$q(z) = q_0 + z , \tag{3.326}$$

$$P(z) = \mathrm{i}\ln\left(1 + \frac{z}{q_0}\right) . \tag{3.327}$$

The integration constant of P has been put to zero because it would just introduce a constant phase factor in Ψ.

Equation (3.323) has a similar form like a paraxial spherical wave, i. e., a parabolic wave, if q is interpreted as a kind of complex radius of curvature. Therefore, it is useful to split $1/q$ into a real and an imaginary part

$$\frac{1}{q(z)} = \frac{1}{R(z)} + \mathrm{i}\frac{\lambda_n}{\pi w^2(z)} . \qquad (3.328)$$

The real part is the curvature of the wave and R is the real radius of curvature. The selection of the imaginary part of $1/q$ becomes obvious by inserting (3.328) into (3.323). It shows that the real function w describes the distance $\sqrt{x^2+y^2}$ from the z-axis at which the amplitude decreases to $1/\mathrm{e}$ of the maximum value. Therefore, w is called the beam radius; w and R are both real functions of z.

A further simplification can be made by choosing $q_0 = q(0)$ as an imaginary number. This means that the radius of curvature R is infinity at $z = 0$, i. e., the curvature of the wave is zero at $z = 0$

$$\frac{1}{q_0} = \mathrm{i}\frac{\lambda_n}{\pi w_0^2} \quad \Rightarrow \quad q_0 = -\mathrm{i}\frac{\pi w_0^2}{\lambda_n} . \qquad (3.329)$$

The propagation constant w_0, which corresponds to the curvature $1/R_0 = 0$, is called the beam waist. Later it will be shown (3.337) that the beam waist is the smallest beam radius of a Gaussian beam during its propagation.

In summary by using (3.323), (3.326), (3.327), (3.328), and (3.329) the function Ψ of a fundamental mode Gaussian beam can be written as:

$$\Psi(x, y, z) = \Psi_0 \frac{1}{1+\mathrm{i}\frac{\lambda_n z}{\pi w_0^2}} \mathrm{e}^{-\frac{x^2+y^2}{w^2(z)}} \mathrm{e}^{\mathrm{i}\frac{k(x^2+y^2)}{2R(z)}} . \qquad (3.330)$$

3.5.2 The Fresnel Diffraction Integral and the Paraxial Helmholtz Equation

In Sect. 3.4.3 the Fresnel diffraction integral is derived as a paraxial solution of the Fresnel–Kirchhoff diffraction formula. Here, it will be shown that the Fresnel diffraction integral describes the propagation of waves with complex amplitudes that fulfill the paraxial Helmholtz equation (3.322). So, the name "paraxial Helmholtz equation" is appropriate.

The Fresnel diffraction integral is determined according to (3.256), where the argument z_0 is substituted by z, because this equation is not only valid in a plane, and k is defined as usual as $k = 2\pi n/\lambda$

$$\begin{aligned} u(x, y, z) &= -\frac{\mathrm{i}k}{2\pi z}\mathrm{e}^{\mathrm{i}kz} \\ &\quad \times \iint_A u_0(x', y', 0)\, \mathrm{e}^{\mathrm{i}k\frac{(x-x')^2+(y-y')^2}{2z}} \mathrm{d}x'\mathrm{d}y' \\ &= \Psi(x, y, z)\mathrm{e}^{\mathrm{i}kz}; \end{aligned} \qquad (3.331)$$

$$\begin{aligned} \Rightarrow \Psi(x, y, z) &= -\frac{\mathrm{i}k}{2\pi z} \\ &\quad \times \iint_A u_0(x', y', 0)\, \mathrm{e}^{\mathrm{i}k\frac{(x-x')^2+(y-y')^2}{2z}} \mathrm{d}x'\mathrm{d}y' . \end{aligned}$$

So, the function Ψ is defined in accordance with (3.318) and it has to be shown that this function Ψ is a solution of the paraxial Helmholtz equation (3.322). We have the following equations for the partial derivatives of Ψ

$$\begin{aligned} \frac{\partial\Psi}{\partial x} &= \frac{k^2}{2\pi z^2}\iint_A u_0(x', y', 0)\,(x-x') \\ &\quad \times \mathrm{e}^{\mathrm{i}k\frac{(x-x')^2+(y-y')^2}{2z}} \mathrm{d}x'\mathrm{d}y' , \end{aligned} \qquad (3.332)$$

$$\begin{aligned} \frac{\partial^2\Psi}{\partial x^2} &= \mathrm{i}\frac{k^3}{2\pi z^3}\iint_A u_0(x', y', 0)\,(x-x')^2 \\ &\quad \times \mathrm{e}^{\mathrm{i}k\frac{(x-x')^2+(y-y')^2}{2z}} \mathrm{d}x'\mathrm{d}y' \\ &\quad + \frac{k^2}{2\pi z^2}\iint_A u_0(x', y', 0) \\ &\quad \times \mathrm{e}^{\mathrm{i}k\frac{(x-x')^2+(y-y')^2}{2z}} \mathrm{d}x'\mathrm{d}y' , \end{aligned}$$

$$\begin{aligned} \frac{\partial^2\Psi}{\partial y^2} &= \mathrm{i}\frac{k^3}{2\pi z^3}\iint_A u_0(x', y', 0)\,(y-y')^2 \\ &\quad \times \mathrm{e}^{\mathrm{i}k\frac{(x-x')^2+(y-y')^2}{2z}} \mathrm{d}x'\mathrm{d}y' \\ &\quad + \frac{k^2}{2\pi z^2}\iint_A u_0(x', y', 0) \\ &\quad \times \mathrm{e}^{\mathrm{i}k\frac{(x-x')^2+(y-y')^2}{2z}} \mathrm{d}x'\mathrm{d}y' , \end{aligned}$$

Part A | 3.5

$$\frac{\partial \Psi}{\partial z} = \frac{ik}{2\pi z^2} \iint_A u_0(x', y', 0) \times e^{ik\frac{(x-x')^2+(y-y')^2}{2z}} dx' dy' - \frac{k^2}{4\pi z^3} \iint_A u_0(x', y', 0) \times \left[(x-x')^2+(y-y')^2\right] \times e^{ik\frac{(x-x')^2+(y-y')^2}{2z}} dx' dy' .$$

So, it is clear that the function Ψ of (3.331) fulfills the paraxial Helmholtz equation

$$\frac{\partial^2 \Psi}{\partial x^2} + \frac{\partial^2 \Psi}{\partial y^2} + 2ik\frac{\partial \Psi}{\partial z} = 0 .$$

Therefore, the Fresnel diffraction integral and the paraxial Helmholtz equation correspond to each other. The propagation of a Gaussian beam can be made either by calculating the Fresnel diffraction integral if the complex amplitude u_0 is given in a plane, or the propagation rules can be directly derived from the paraxial Helmholtz equation as has been done to obtain (3.326) and (3.327) [3.10].

3.5.3 Propagation of a Gaussian Beam

The parameters w and R of a Gaussian beam change during the propagation of the beam along the z-axis. An explicit representation of w and R can be obtained by combining (3.326), (3.328), and (3.329)

$$\frac{1}{R} + i\frac{\lambda_n}{\pi w^2} = \frac{1}{z - i\frac{\pi w_0^2}{\lambda_n}} = \frac{z + i\frac{\pi w_0^2}{\lambda_n}}{z^2 + \left(\frac{\pi w_0^2}{\lambda_n}\right)^2} .$$

To simplify the notation the so-called Rayleigh length is defined as

$$z_R := \frac{\pi w_0^2}{\lambda_n} . \tag{3.333}$$

So, by separating the real and the imaginary part, two equations are obtained

$$\frac{1}{R} = \frac{z}{z^2+z_R^2} , \tag{3.334}$$

$$\frac{\lambda_n}{\pi w^2} = \frac{z_R}{z^2+z_R^2} ; \tag{3.335}$$

$$\Rightarrow \quad R(z) = z + \frac{z_R^2}{z} = z + \frac{\pi^2 w_0^4}{\lambda_n^2 z} , \tag{3.336}$$

$$w^2(z) = \frac{\lambda_n}{\pi}\frac{z^2+z_R^2}{z_R} = w_0^2 + \frac{\lambda_n^2 z^2}{\pi^2 w_0^2} . \tag{3.337}$$

The last equation shows that the beam waist w_0 is indeed the smallest value of the beam radius w and that it is obtained at $z = 0$. Simultaneously, the radius of curvature R is infinity at $z = 0$. The equation also shows that the beam radius of the Gaussian beam has the value $w = \sqrt{2}w_0$ at the distance $z = z_R$ (Rayleigh length) from the beam waist.

Another interesting limiting case is the far field, i. e., $z \to \pm\infty$. Then we have

$$R(z) = z , \tag{3.338}$$

$$w(z) = \frac{\lambda_n |z|}{\pi w_0} . \tag{3.339}$$

The far field angle θ of a Gaussian beam is

$$\theta \approx \tan\theta = \frac{w(z)}{|z|} = \frac{\lambda_n}{\pi w_0} . \tag{3.340}$$

So, by measuring the far field angle θ and the wavelength λ_n of a laser diode its beam waist w_0 can be calculated if we assume that the fundamental Gaussian beam is a good description for the wavefront of a laser diode.

By using (3.337) the function Ψ (3.330) can be written in a more illustrating way

$$\frac{1}{1+i\frac{\lambda_n z}{\pi w_0^2}} = \frac{w_0}{w_0 + i\frac{\lambda_n z}{\pi w_0}} = \frac{w_0\left(w_0 - i\frac{\lambda_n z}{\pi w_0}\right)}{w_0^2 + \frac{\lambda_n^2 z^2}{\pi^2 w_0^2}} .$$

The term in parentheses of the numerator can be expressed as

$$w_0 - i\frac{\lambda_n z}{\pi w_0} = A\,e^{i\Phi} = A\cos\Phi + iA\sin\Phi$$

with

$$A = \sqrt{w_0^2 + \frac{\lambda_n^2 z^2}{\pi^2 w_0^2}} = w(z)$$

and

$$\cos\Phi = \frac{w_0}{w(z)} ,$$

$$\sin\Phi = -\frac{\lambda_n z}{\pi w_0 w(z)} ;$$

$$\Rightarrow \quad \tan\Phi = -\frac{\lambda_n z}{\pi w_0^2} .$$

In summary we have

$$\frac{1}{1+\mathrm{i}\frac{\lambda_n z}{\pi w_0^2}} = \frac{w_0}{w(z)}\mathrm{e}^{\mathrm{i}\Phi(z)} \quad \text{with} \quad \tan\Phi(z) = -\frac{\lambda_n z}{\pi w_0^2}\,. \tag{3.341}$$

The complex amplitude u of a Gaussian beam can then be expressed using (3.318), (3.330), and (3.341)

$$u(x,y,z) = \Psi_0 \frac{w_0}{w(z)} \mathrm{e}^{-\frac{x^2+y^2}{w^2(z)}} \mathrm{e}^{\mathrm{i}\Phi(z)} \mathrm{e}^{\mathrm{i}\frac{k(x^2+y^2)}{2R(z)}} \mathrm{e}^{\mathrm{i}kz}\,. \tag{3.342}$$

This means that a Gaussian beam has a Gaussian profile for a constant value z (Fig. 3.30). The term $w_0/w(z)$ ensures that the total power P_G of the beam is conserved during the propagation along the z direction:

$$\begin{aligned} P_G(z) &= \int_{-\infty}^{+\infty}\int_{-\infty}^{+\infty} |u(x,y,z)|^2\,\mathrm{d}x\,\mathrm{d}y \\ &= \int_{-\infty}^{+\infty}\int_{-\infty}^{+\infty} \Psi_0^2 \frac{w_0^2}{w^2(z)} \mathrm{e}^{-2\frac{x^2+y^2}{w^2(z)}}\,\mathrm{d}x\,\mathrm{d}y \\ &= \Psi_0^2 \frac{w_0^2}{w^2(z)} \frac{\pi w^2(z)}{2} = \Psi_0^2 \frac{\pi w_0^2}{2} = \text{constant}\,. \end{aligned} \tag{3.343}$$

By interpreting the beam radius w as lateral extension of the Gaussian beam it can be graphically symbolized as in Fig. 3.31. At the beam waist the local curvature of the Gaussian beam is zero. In the far field, the radius of curvature R increases proportional to z like the radius of curvature of a spherical wave.

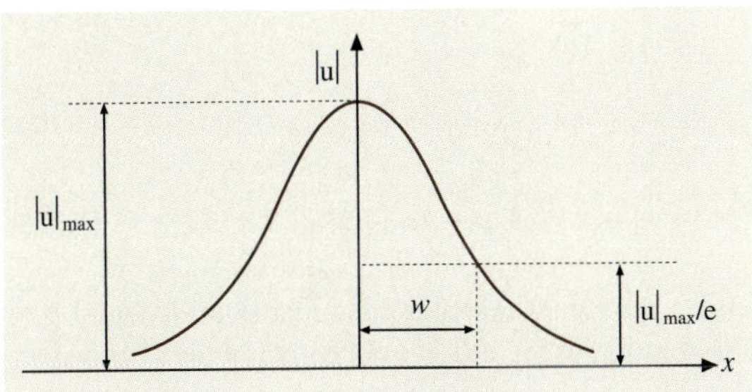

Fig. 3.30 Amplitude of a Gaussian beam at a constant value z

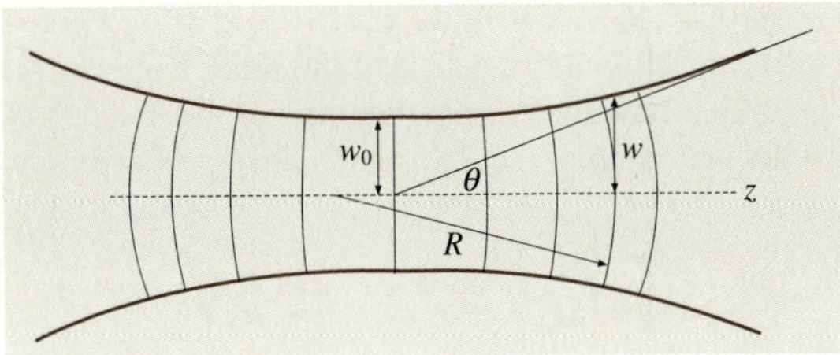

Fig. 3.31 Scheme showing the propagation of a Gaussian beam along the z-axis. The Gaussian beam is laterally limited by the beam radius w and its wavefront has the local radius of curvature R

3.5.4 Higher-Order Modes of Gaussian Beams

In (3.323) a quite simple approach has been selected for the function Ψ, which mainly describes the lateral variation of the Gaussian beam. This approach is the fundamental mode in the case of rotational symmetry. In the following a more general approach is made. The beam can now have two different principal curvatures along the local x and y direction and also higher order modes are taken into account. The different principal curvatures are useful to describe the radiation of laser diodes (e.g., edge emitter), which often have different beam radii and radii of curvature in the x and y directions. Therefore, the following approach is taken for Ψ [3.58]:

$$\Psi(x,y,z) = g\left(\sqrt{2}\frac{x}{w_x(z)}\right) h\left(\sqrt{2}\frac{y}{w_y(z)}\right) \times \mathrm{e}^{\mathrm{i}\left[P(z) + \frac{kx^2}{2q_x(z)} + \frac{ky^2}{2q_y(z)}\right]}\,. \tag{3.344}$$

The functions g and h have to describe the lateral variations of the amplitude of the different modes and therefore it is useful to take the normalized quantities x/w_x and y/w_y, which are pure numbers without a physical unit. Here, w_x and w_y are again the beam radii in the x and y directions, respectively. The factor $\sqrt{2}$ in the arguments of g and h seems to be quite arbitrary at the moment. But it will be shown in the following that this leads to a well-known differential equation for g and h. This approach for Ψ has to fulfill (3.322). The functions are written without arguments in the following to simplify the notation. Additionally, the first or second derivative of a function $f(\eta)$ with respect to its argument η is written as f' or f'', respectively. However, it should be kept in mind that, e.g., g is in fact a function of x and z (since w_x is a function of z) and therefore the derivative $g'(\eta) := \mathrm{d}g(\eta)/\mathrm{d}\eta$ with $\eta := \sqrt{2}x/w_x(z)$ remains a function of x and z. On the other hand, q_x is, e.g., only a function of z so that the derivative $q_x'(z)$ is just defined as $\mathrm{d}q_x(z)/\mathrm{d}z$. So, using these notations, we have:

$$\begin{aligned}\frac{\partial\Psi}{\partial z} &= \left[gh\left(\mathrm{i}P' - \frac{\mathrm{i}kx^2}{2q_x^2}q_x' - \frac{\mathrm{i}ky^2}{2q_y^2}q_y'\right)\right.\\ &\quad\left.+\left(-gh'\sqrt{2}\frac{y}{w_y^2}w_y' - g'h\sqrt{2}\frac{x}{w_x^2}w_x'\right)\right]\\ &\quad\times \mathrm{e}^{\mathrm{i}\left(P+\frac{kx^2}{2q_x}+\frac{ky^2}{2q_y}\right)},\end{aligned}$$

$$\frac{\partial\Psi}{\partial x} = \left(gh\frac{\mathrm{i}kx}{q_x} + \sqrt{2}\frac{g'}{w_x}h\right)\mathrm{e}^{\mathrm{i}\left(P+\frac{kx^2}{2q_x}+\frac{ky^2}{2q_y}\right)},$$

$$\begin{aligned}\frac{\partial^2\Psi}{\partial x^2} &= \left(\frac{\mathrm{i}k}{q_x}gh - \frac{k^2x^2}{q_x^2}gh + \sqrt{2}\frac{\mathrm{i}kx}{q_xw_x}g'h\right.\\ &\quad\left.+2\frac{g''}{w_x^2}h + \sqrt{2}\frac{\mathrm{i}kx}{q_xw_x}g'h\right)\\ &\quad\times \mathrm{e}^{\mathrm{i}\left(P+\frac{kx^2}{2q_x}+\frac{ky^2}{2q_y}\right)},\end{aligned}$$

$$\begin{aligned}\frac{\partial^2\Psi}{\partial y^2} &= \left(\frac{\mathrm{i}k}{q_y}gh - \frac{k^2y^2}{q_y^2}gh + \sqrt{2}\frac{\mathrm{i}ky}{q_yw_y}gh'\right.\\ &\quad\left.+2\frac{h''}{w_y^2}g + \sqrt{2}\frac{\mathrm{i}ky}{q_yw_y}gh'\right)\\ &\quad\times \mathrm{e}^{\mathrm{i}\left(P+\frac{kx^2}{2q_x}+\frac{ky^2}{2q_y}\right)}.\end{aligned}$$

By inserting these functions in (3.322) and dividing by gh the following equation is obtained

$$\begin{aligned}&\frac{\mathrm{i}k}{q_x} + \frac{\mathrm{i}k}{q_y} - 2kP' + \frac{k^2x^2}{q_x^2}(q_x'-1) + \frac{k^2y^2}{q_y^2}(q_y'-1)\\ &+2\frac{g''}{gw_x^2} - 2\sqrt{2}\frac{\mathrm{i}kx}{w_x^2}\frac{g'}{g}\left(w_x' - \frac{w_x}{q_x}\right)\\ &+2\frac{h''}{hw_y^2} - 2\sqrt{2}\frac{\mathrm{i}ky}{w_y^2}\frac{h'}{h}\left(w_y' - \frac{w_y}{q_y}\right) = 0\,. \qquad (3.345)\end{aligned}$$

This equation also has to be fulfilled for $x \to \infty$ and $y \to \infty$. Then, the terms proportional to x^2 and y^2 are very large compared to the other terms and, similarly to the case of the fundamental mode, the following two conditions have to be fulfilled:

$$q_x' = 1 \quad \Rightarrow \quad q_x = q_{x,0} + z$$

and

$$q_y' = 1 \quad \Rightarrow \quad q_y = q_{y,0} + z\,. \qquad (3.346)$$

Additionally, q_x and q_y are analogous to the fundamental mode split up into real and imaginary parts

$$\frac{1}{q_x} = \frac{1}{R_x} + \mathrm{i}\frac{\lambda_n}{\pi w_x^2} \quad \text{and} \quad \frac{1}{q_y} = \frac{1}{R_y} + \mathrm{i}\frac{\lambda_n}{\pi w_y^2}\,. \qquad (3.347)$$

Calculating the derivative with respect to z delivers for the first equation

$$-\frac{q_x'}{q_x^2} = -\frac{1}{q_x^2} = -\frac{R_x'}{R_x^2} - 2\mathrm{i}\frac{\lambda_n w_x'}{\pi w_x^3}\,. \qquad (3.348)$$

This equation is added to the square of (3.347). Then, the real and the imaginary part are split resulting in two equations

$$R_x' = 1 - \frac{\lambda_n^2 R_x^2}{\pi^2 w_x^4} \quad \text{and} \quad w_x' = \frac{w_x}{R_x}\,. \qquad (3.349)$$

Analogous results are obtained for R_y' and w_y'. Inserting these results into (3.345) finally delivers

$$\begin{aligned}&\frac{\mathrm{i}k}{q_x} + \frac{\mathrm{i}k}{q_y} - 2kP'\\ &+2\frac{g''}{gw_x^2} - 4\sqrt{2}\frac{x}{w_x^3}\frac{g'}{g}\\ &+2\frac{h''}{hw_y^2} - 4\sqrt{2}\frac{y}{w_y^3}\frac{h'}{h} = 0\,. \qquad (3.350)\end{aligned}$$

Now, the terms in the first row depend only on z, whereas the terms in the second row depend on x and z and the terms in the third row on y and z. Therefore, a separation approach has to be made

$$\frac{\mathrm{i}k}{q_x} + \frac{\mathrm{i}k}{q_y} - 2kP' = -f_x(z) - f_y(z)\,, \qquad (3.351)$$

$$2\frac{g''}{gw_x^2} - 4\sqrt{2}\frac{x}{w_x^3}\frac{g'}{g} = f_x(z)\,, \qquad (3.352)$$

$$2\frac{h''}{hw_y^2} - 4\sqrt{2}\frac{y}{w_y^3}\frac{h'}{h} = f_y(z)\,, \qquad (3.353)$$

where f_x and f_y are functions that only depend on z. The solution of the differential equation for g (and analogously for h) shall be described briefly because it is a quite general solution scheme, which is often applied in optics and physics. First, the differential equation is written by using $\eta = \sqrt{2}x/w_x$ and the abbreviation $\alpha := f_x w_x^2$ as

$$\frac{\mathrm{d}^2 g(\eta)}{\mathrm{d}\eta^2} - 2\eta \frac{\mathrm{d}g(\eta)}{\mathrm{d}\eta} - \frac{1}{2}\alpha g(\eta) = 0\,. \tag{3.354}$$

The usual approach to solve such a differential equation is to write g as a polynomial

$$\begin{aligned} g(\eta) &= \sum_{m=0}^{\infty} a_m \eta^m\,, \\ \frac{\mathrm{d}g(\eta)}{\mathrm{d}\eta} &= \sum_{m=1}^{\infty} m a_m \eta^{m-1}\,, \\ \frac{\mathrm{d}^2 g(\eta)}{\mathrm{d}\eta^2} &= \sum_{m=2}^{\infty} m(m-1) a_m \eta^{m-2}\,. \end{aligned} \tag{3.355}$$

Inserting of this approach into the differential (3.354) and arranging for equal powers of η gives

$$\begin{aligned} &\sum_{m=2}^{\infty} m(m-1) a_m \eta^{m-2} \\ &\quad - 2\sum_{m=1}^{\infty} m a_m \eta^m - \frac{1}{2}\alpha \sum_{m=0}^{\infty} a_m \eta^m \\ &= \sum_{m=0}^{\infty} \Bigg[(m+2)(m+1) a_{m+2} \\ &\quad - \left(2m + \frac{1}{2}\alpha\right) a_m \Bigg] \eta^m = 0\,. \end{aligned} \tag{3.356}$$

This equation can only be fulfilled for all possible values of η if each coefficient in front of η^m is zero, i. e.,

$$\begin{aligned} &(m+2)(m+1) a_{m+2} - \left(2m + \frac{1}{2}\alpha\right) a_m = 0 \\ &\Rightarrow \quad a_{m+2} = \frac{2m + \frac{1}{2}\alpha}{(m+2)(m+1)} a_m\,. \end{aligned} \tag{3.357}$$

Now, if there would be no stop criterion for the progression of coefficients a_m this equation would tend towards very large values of m to

$$\lim_{m\to\infty} a_{m+2} = \frac{2}{m} a_m \tag{3.358}$$

because α has a finite value. But this is the same progression of coefficients as for $\exp(\eta^2)$

$$\begin{aligned} \mathrm{e}^{\eta^2} &= \sum_{m=0}^{\infty} \frac{\left(\eta^2\right)^m}{m!} = \sum_{m=0}^{\infty} \frac{\eta^{2m}}{m!} \\ &= \sum_{m=0,2,4,\ldots}^{\infty} \frac{1}{\left(\frac{m}{2}\right)!} \eta^m = \sum_{m=0,2,4,\ldots}^{\infty} b_m \eta^m\,. \end{aligned} \tag{3.359}$$

So, the progression of coefficients will in this case be:

$$\begin{aligned} b_m &= \frac{1}{\left(\frac{m}{2}\right)!} \\ \Rightarrow \quad b_{m+2} &= \frac{1}{\left(\frac{m+2}{2}\right)!} = \frac{2}{m+2} b_m\,. \end{aligned} \tag{3.360}$$

Therefore, for very large values m the progression of coefficients b_m will have the same behavior as the coefficients a_m and the amplitude $|\Psi|$ of the higher-order mode Gaussian beam would tend to infinity for large values η because the compensating term [(3.344), (3.347)] only has the form $\exp(-x^2/w_x^2) = \exp(-\eta^2/2)$. But, for physical reasons, $|\Psi|$ has to tend to zero for large values of η. Therefore, there has to be a stop criterion for the progression of coefficients, which just means that the variable α has to fulfill the following equation

$$\begin{aligned} &\alpha = f_x w_x^2 = -4j; \quad j = 0, 1, 2, \ldots \\ &\Rightarrow \quad f_x = -\frac{4j}{w_x^2}\,. \end{aligned} \tag{3.361}$$

By inserting this into (3.354) the well-known differential equation for the Hermite polynomials H_j is obtained

$$\frac{\mathrm{d}^2 g(\eta)}{\mathrm{d}\eta^2} - 2\eta \frac{\mathrm{d}g(\eta)}{\mathrm{d}\eta} + 2jg(\eta) = 0\,. \tag{3.362}$$

The progression of coefficients of the Hermite polynomials fulfill, according to (3.357) and (3.361), the condition

$$\begin{aligned} a_{m+2} &= \frac{2m + \frac{1}{2}\alpha}{(m+2)(m+1)} a_m \\ &= \frac{2(m-j)}{(m+2)(m+1)} a_m\,. \end{aligned} \tag{3.363}$$

But, we have two progressions of coefficient, one for odd numbers m and one for even numbers m. So, if j is odd, only the odd coefficient progression will stop and vice versa with even values of j. Therefore, one of the coefficients a_0 or a_1, which are the two integration constants of our second-order differential equation, must additionally be zero. So, we now have the possibility to calculate the Hermite polynomials H_j, which are the solutions of g and h. The Hermite polynomials are, in most textbooks, normalized but by using (3.363) it is only possible to calculate the unnormalized Hermite polynomials. But this is no problem since we do not need the normalized polynomials.

If we take $a_0 \neq 0$ and $a_1 = 0$ for the even Hermite polynomials H_j and $a_0 = 0$ and $a_1 \neq 0$ for the odd Hermite polynomials we obtain up to the third order apart from the normalization constant

$$\begin{aligned} H_0(\eta) &= 1, \\ H_1(\eta) &= \eta, \\ H_2(\eta) &= -2\eta^2 + 1, \\ H_3(\eta) &= -\frac{2}{3}\eta^3 + \eta. \end{aligned} \tag{3.364}$$

Equation (3.351) for P' results, together with (3.361) (taking m instead of j) and the analogous equation for f_y (taking n instead of j), in

$$\begin{aligned} \frac{\mathrm{d}P}{\mathrm{d}z} &= \frac{\mathrm{i}}{2}\left(\frac{1}{q_x} + \frac{1}{q_y}\right) - m\frac{\lambda_n}{\pi w_x^2} - n\frac{\lambda_n}{\pi w_y^2} \\ &= \frac{\mathrm{i}}{2}\left(\frac{1}{q_x} + \frac{1}{q_y}\right) - m\,\mathrm{Im}\left(\frac{1}{q_x}\right) - n\,\mathrm{Im}\left(\frac{1}{q_y}\right). \end{aligned} \tag{3.365}$$

Using (3.346) and

$$\begin{aligned} \frac{1}{q_x} &= \frac{1}{q_{x,0} + z} = \frac{1}{\mathrm{Re}(q_{x,0}) + \mathrm{i}\,\mathrm{Im}(q_{x,0}) + z} \\ &= \frac{z + \mathrm{Re}(q_{x,0}) - \mathrm{i}\,\mathrm{Im}(q_{x,0})}{z^2 + 2z\,\mathrm{Re}(q_{x,0}) + |q_{x,0}|^2} \end{aligned} \tag{3.366}$$

finally gives

$$\begin{aligned} P(z) &= \mathrm{i}\ln\left(\sqrt{1 + \frac{z}{q_{x,0}}}\sqrt{1 + \frac{z}{q_{y,0}}}\right) \\ &\quad - m\arctan\left(\frac{z + \mathrm{Re}(q_{x,0})}{-\mathrm{Im}(q_{x,0})}\right) \\ &\quad - n\arctan\left(\frac{z + \mathrm{Re}(q_{y,0})}{-\mathrm{Im}(q_{y,0})}\right). \end{aligned} \tag{3.367}$$

Note that $-\mathrm{Im}(q_{x,0})$ and $-\mathrm{Im}(q_{y,0})$ are used because these quantities are positive as will be seen in (3.369). In summary, the function Ψ of the higher-order mode Gaussian beams (Hermite–Gaussian modes) in the case of a cartesian coordinate system can be written by using (3.344)–(3.367)

$$\begin{aligned} \Psi(x, y, z) &= H_m\left(\sqrt{2}\frac{x}{w_x(z)}\right) H_n\left(\sqrt{2}\frac{y}{w_y(z)}\right) \\ &\quad \times \frac{1}{\sqrt{\left(1 + \frac{z}{q_{x,0}}\right)\left(1 + \frac{z}{q_{y,0}}\right)}} \\ &\quad \times \mathrm{e}^{-\mathrm{i}\left[m\arctan\left(\frac{z + \mathrm{Re}(q_{x,0})}{-\mathrm{Im}(q_{x,0})}\right) + n\arctan\left(\frac{z + \mathrm{Re}(q_{y,0})}{-\mathrm{Im}(q_{y,0})}\right)\right]} \\ &\quad \times \mathrm{e}^{\mathrm{i}\pi\left(\frac{x^2}{\lambda_n R_x(z)} + \frac{y^2}{\lambda_n R_y(z)}\right)} \\ &\quad \times \mathrm{e}^{-\left(\frac{x^2}{w_x^2(z)} + \frac{y^2}{w_y^2(z)}\right)}. \end{aligned} \tag{3.368}$$

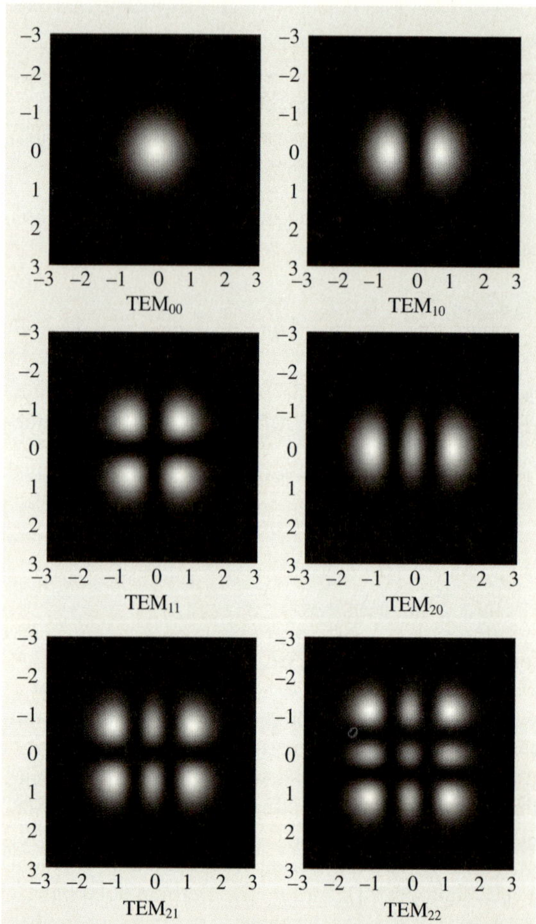

Fig. 3.32 Simulation of the intensity distributions of some Hermite–Gaussian modes using the normalized coordinates x/w_x and y/w_y

The functions w_x and R_x are obtained by comparing the real and imaginary parts of (3.347) and (3.366)

$$w_x^2(z) = \frac{\lambda_n}{\pi} \frac{z^2 + 2z\mathrm{Re}(q_{x,0}) + |q_{x,0}|^2}{-\mathrm{Im}(q_{x,0})}, \quad (3.369)$$

$$R_x(z) = \frac{z^2 + 2z\mathrm{Re}(q_{x,0}) + |q_{x,0}|^2}{z + \mathrm{Re}(q_{x,0})}. \quad (3.370)$$

Analogous equations are, of course, valid for w_y and R_y, which are obtained by substituting the index x for y.

Figure 3.32 shows the typical intensity distribution $|\Psi|^2$ of some lower-order Hermite–Gaussian modes with

$$|\Psi(x,y,z)|^2 = \left[H_m\left(\sqrt{2}\frac{x}{w_x(z)}\right) H_n\left(\sqrt{2}\frac{y}{w_y(z)}\right)\right]^2 \times \left|\frac{1}{\sqrt{\left(1+\frac{z}{q_{x,0}}\right)\left(1+\frac{z}{q_{y,0}}\right)}}\right|^2 \times \mathrm{e}^{-2\left(\frac{x^2}{w_x^2(z)}+\frac{y^2}{w_y^2(z)}\right)}. \quad (3.371)$$

They are named TEM_{mn}, where m is the index of the Hermite polynomial H_m with the argument $\sqrt{2}x/w_x$ and n is the index of the Hermite polynomial H_n with the argument $\sqrt{2}y/w_y$. The number of zeros is equal to the mode number and the area covered by the modes increases with the mode number.

So, the complete behavior of the higher order Gaussian beam is well defined if the complex quantities $q_{x,0}$ and $q_{y,0}$ at the plane $z = 0$ are known. This is the case if the beam radii $w_{x,0}$ and $w_{y,0}$ and the radii of curvature $R_{x,0}$ and $R_{y,0}$ of the wavefront at the plane $z = 0$ are known. Hereby, the beam waists in x and y direction can be in different planes. If both beam waists are in the same plane the coordinate system can be chosen such that the beam waists are in the plane $z = 0$ and $q_{x,0} = -\mathrm{i}\pi w_{x,0}^2/\lambda_n$ with the beam waist $w_{x,0}$ in the x direction. Then a simplification similar to the case of the fundamental mode of a Gaussian beam can be made and (3.369) and (3.370) reduce to (3.336) and (3.337). Also, (3.368) can then be simplified.

3.5.5 Transformation of a Fundamental Gaussian Beam at a Lens

The transformation of a fundamental Gaussian beam at a (thin) lens is performed using a paraxial approximation. This means that it is assumed that the beam radius immediately in front of the lens is identical to the beam radius immediately behind the lens. Additionally, the radius of curvature of the Gaussian beam changes in the same way as that of a spherical wave. The sign convention is that a positive lens has a positive focal length $f > 0$ and that a divergent spherical wave coming from the negative z direction (i. e., from "left" using the optical agreement) has a positive radius of curvature $R > 0$. R_1 is the radius of curvature immediately in front of the lens and R_2 is the radius of curvature immediately behind the lens. Then, a lens with focal length f transforms the radii of curvature according to the paraxial imaging equation of geometrical optics (Fig. 3.33)

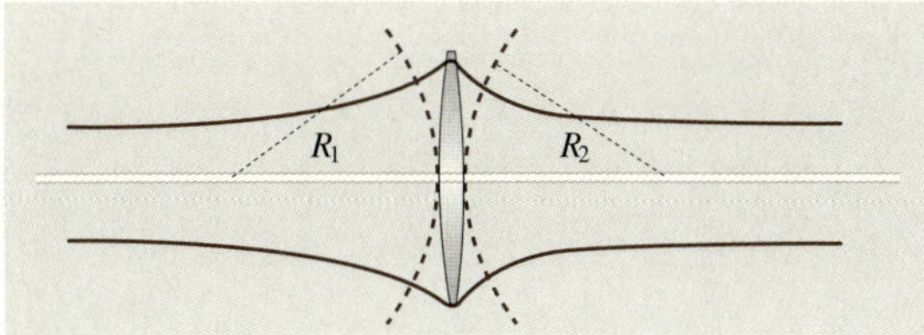

Fig. 3.33 Transformation of a Gaussian beam at a thin ideal lens

$$\frac{1}{R_2} = \frac{1}{R_1} - \frac{1}{f}. \quad (3.372)$$

Since the beam radius remains constant, the complex beam parameters q_1 immediately in front of the lens and q_2 immediately behind the lens transform also with

$$\frac{1}{q_2} = \frac{1}{q_1} - \frac{1}{f}. \quad (3.373)$$

In the case of a thick lens or a lens system the two principal planes of the lens system have to be taken as reference planes for q_1 and q_2 according to the laws of paraxial geometrical optics. If the q parameters are different in x and y direction as in (3.347) both sets of parameters just have to be treated separately using (3.373).

To calculate the relation between the q parameter q_1 in the distance d_1 in front of a lens with the focal length f and q_2 in the distance d_2 behind the lens, (3.326) and (3.373) have to be combined. We call q_L the Gaussian beam parameter immediately in front of the lens and q_R the beam parameter immediately behind the lens, where

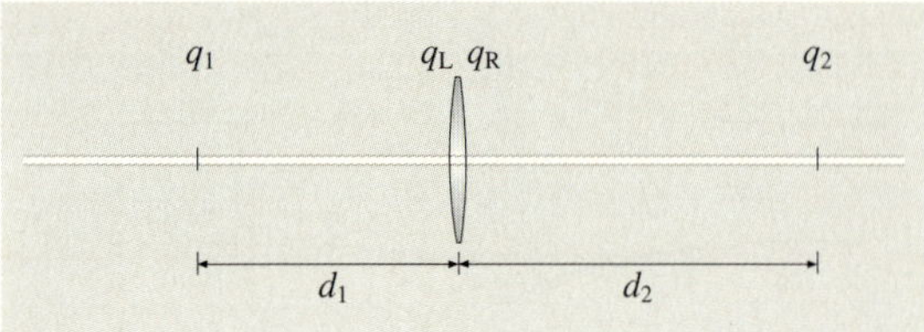

Fig. 3.34 Scheme showing the complex beam parameters for the transformation of a Gaussian beam from a plane in the distance d_1 in front of a lens (with focal length f) to a plane in the distance d_2 behind the lens

the parameters are illustrated in Fig. 3.34. Then we have

$$q_L = q_1 + d_1 ,$$

$$\frac{1}{q_R} = \frac{1}{q_L} - \frac{1}{f} = \frac{1}{q_1+d_1} - \frac{1}{f} ,$$

$$\Rightarrow q_R = \frac{f(q_1+d_1)}{f-q_1-d_1} ;$$

$$q_2 = q_R + d_2 = \frac{fq_1 + fd_1 + fd_2 - d_2 q_1 - d_1 d_2}{f-q_1-d_1} ,$$

$$\Rightarrow q_2 = \frac{q_1\left(1-\frac{d_2}{f}\right) + \left(d_1+d_2-\frac{d_1 d_2}{f}\right)}{-\frac{q_1}{f} + \left(1-\frac{d_1}{f}\right)} . \quad (3.374)$$

3.5.6 ABCD Matrix Law for Gaussian Beams

The propagation through an optical system can be described in the paraxial geometrical optics by an ABCD matrix (see the chapter about geometrical optics or [3.11,58–60]). We compare the terms of (3.374) with the paraxial ABCD matrix for the propagation from a plane with the distance d_1 in front of a lens with focal length f to a plane with distance d_2 behind the lens

$$\begin{pmatrix} A & B \\ C & D \end{pmatrix} = \begin{pmatrix} 1-\frac{d_2}{f} & d_1+d_2-\frac{d_1 d_2}{f} \\ -\frac{1}{f} & 1-\frac{d_1}{f} \end{pmatrix} . \quad (3.375)$$

We see that the Gaussian-beam parameter transforms as

$$q_2 = \frac{Aq_1+B}{Cq_1+D} . \quad (3.376)$$

It can be shown that this ABCD matrix law is valid quite generally as long as the paraxial approximation holds. In the following it will be shown for a sequence of (thin) lenses and free space propagation.

Free Space Propagation

The free space propagation in a homogeneous material with refractive index n is described by (3.326): $q_2 = q_1 + z$. On the other hand the paraxial matrix for free space propagation between two planes with a distance z is

$$\begin{pmatrix} A & B \\ C & D \end{pmatrix} = \begin{pmatrix} 1 & z \\ 0 & 1 \end{pmatrix}$$

$$\Rightarrow q_2 = q_1 + z = \frac{1 \cdot q_1 + z}{0 \times q_1 + 1} = \frac{Aq_1+B}{Cq_1+D} . \quad (3.377)$$

So, the free space propagation fulfills the ABCD matrix law of Gaussian beams (3.376).

Thin Lens

For the transformation of a Gaussian beam at a thin lens, (3.373) is valid. The paraxial matrix of a thin lens with focal length f is

$$\begin{pmatrix} A & B \\ C & D \end{pmatrix} = \begin{pmatrix} 1 & 0 \\ -\frac{1}{f} & 1 \end{pmatrix}$$

$$\Rightarrow q_2 = \frac{fq_1}{f-q_1} = \frac{1 \times q_1 + 0}{-\frac{q_1}{f}+1} = \frac{Aq_1+B}{Cq_1+D} . \quad (3.378)$$

So, the transformation of a Gaussian beam at a thin lens also fulfills the ABCD matrix law of (3.376).

A Sequence of Lenses and Free Space Propagation

We assume that M_1 and M_2 are the paraxial matrices for two subsequent operations like free space propagation or transformation at a thin lens. The Gaussian beam has the q parameters q_0 before the first operation, q_1 after the first operation and q_2 after the second operation. Both (3.377) and (3.378) fulfill (3.376). Therefore, we have the relations

$$q_1 = \frac{A_1 q_0 + B_1}{C_1 q_0 + D_1} \quad \text{and} \quad q_2 = \frac{A_2 q_1 + B_2}{C_2 q_1 + D_2} . \quad (3.379)$$

Substitution of q_1 into q_2 gives

$$q_2 = \frac{A_2 \frac{A_1 q_0 + B_1}{C_1 q_0 + D_1} + B_2}{C_2 \frac{A_1 q_0 + B_1}{C_1 q_0 + D_1} + D_2} \quad (3.380)$$

$$= \frac{(A_2 A_1 + B_2 C_1) q_0 + (A_2 B_1 + B_2 D_1)}{(C_2 A_1 + D_2 C_1) q_0 + (C_2 B_1 + D_2 D_1)} .$$

However, the paraxial matrix M of both operations is

$$M=\begin{pmatrix}A & B\\ C & D\end{pmatrix}=M_2M_1$$
$$=\begin{pmatrix}A_2 & B_2\\ C_2 & D_2\end{pmatrix}\cdot\begin{pmatrix}A_1 & B_1\\ C_1 & D_1\end{pmatrix}$$
$$=\begin{pmatrix}A_2A_1+B_2C_1 & A_2B_1+B_2D_1\\ C_2A_1+D_2C_1 & C_2B_1+D_2D_1\end{pmatrix}. \quad (3.381)$$

Summarizing, the relation between q_2 and q_0 is

$$q_2=\frac{Aq_0+B}{Cq_0+D}. \quad (3.382)$$

This shows that the ABCD matrix law is valid for two subsequent operations of free space propagation or transformation at a thin lens. Therefore, it also has to be valid for an arbitrary number of those operations. Geometrical optics shows that a thick lens can be replaced by a thin lens and free space propagation. Therefore, the ABCD matrix law can also be applied for thick lenses or a system consisting of many lenses. We assume that the paraxial ABCD matrix of such a system is known for describing the propagation between two planes with the optical system in between. The transformation of the Gaussian beam parameter q_1 at the first plane to the parameter q_2 at the second plane is then described by (3.376). Of course, it is always assumed that no apertures are in the system and that the paraxial approximation is valid, i. e., the optical system is ideal and does not introduce any aberrations.

3.5.7 Some Examples of the Propagation of Gaussian Beams

Transformation in the Case of Geometrical Imaging

A Gaussian beam with beam parameter q_1 at the distance d_1 in front of the first principal plane of a lens (or a lens system) with focal length f is examined at the distance d_2 behind the second principal plane of the lens. There, the Gaussian beam has the beam parameter q_2. Additionally, it is assumed that the distances d_1 and d_2 and the focal length of the lens fulfill the imaging equation of paraxial geometrical optics

$$\frac{1}{d_1}+\frac{1}{d_2}=\frac{1}{f}\Rightarrow \quad (3.383)$$
$$1-\frac{d_2}{f}=-\frac{d_2}{d_1}=\beta\,, \quad (3.384)$$
$$d_1+d_2-\frac{d_1d_2}{f}=0\,, \quad (3.385)$$
$$1-\frac{d_1}{f}=-\frac{d_1}{d_2}=\frac{1}{\beta}. \quad (3.386)$$

By using (3.374) we obtain

$$q_2=\frac{\beta q_1}{-\frac{q_1}{f}+\frac{1}{\beta}}\Rightarrow\frac{1}{q_2}=-\frac{1}{\beta f}+\frac{1}{\beta^2 q_1}. \quad (3.387)$$

Using (3.328) to split the complex q parameters into their real variables delivers

$$\frac{1}{R_2}+\mathrm{i}\frac{\lambda_n}{\pi w_2^2}=-\frac{1}{\beta f}+\frac{1}{\beta^2R_1}+\mathrm{i}\frac{\lambda_n}{\beta^2\pi w_1^2};$$
$$\Rightarrow\frac{1}{R_2}=\frac{1}{\beta^2R_1}-\frac{1}{\beta f}\quad\text{and}\quad w_2=|\beta|w_1\,. \quad (3.388)$$

The result is that the beam radius transforms from one plane to another plane with the lateral magnification β if the imaging equation is fulfilled for these two planes, i. e., if the first plane is imaged by the lens onto the second plane.

Position and Size of the Beam Waist Behind a Lens

It is assumed that a Gaussian beam has its beam waist w_0 at the distance d_1 in front of a lens with focal length f so that the beam parameter in the first plane is, according to (3.329), $q_1=-\mathrm{i}\pi w_0^2/\lambda_n$. The position and size of the beam waist behind the lens has to be calculated. The desired parameters are the size w_2 of the beam waist and its distance d_2 from the lens (Fig. 3.35).

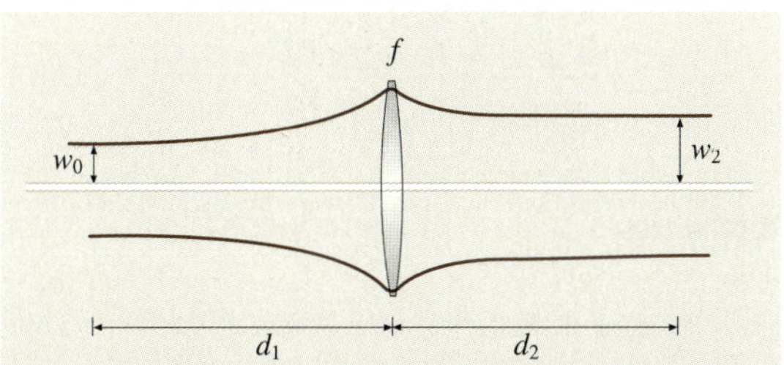

Fig. 3.35 Scheme showing the parameters to calculate the position and the size of the beam waist of a Gaussian beam behind a lens

Part A | 3.5

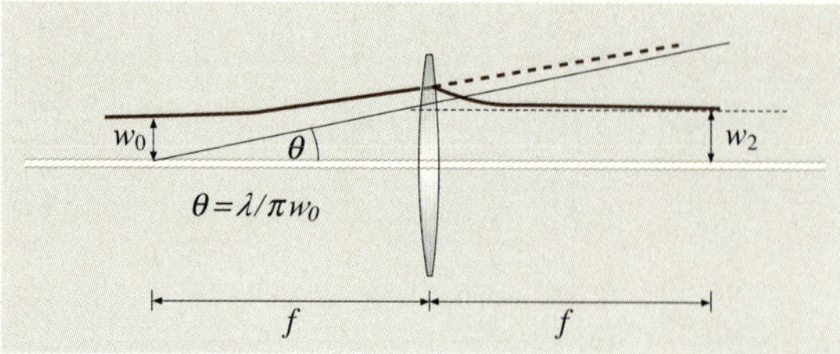

Fig. 3.36 Special case of the transformation of a Gaussian beam at a lens where the beam waist lies in the front focal plane of the lens

Using (3.328) and (3.374) results in

$$\frac{1}{q_2} = \frac{1}{R_2} + \mathrm{i}\frac{\lambda_n}{\pi w_2^2}$$

$$= \frac{\mathrm{i}\frac{\pi w_0^2}{\lambda_n f} + \left(1 - \frac{d_1}{f}\right)}{-\mathrm{i}\left(1 - \frac{d_2}{f}\right)\frac{\pi w_0^2}{\lambda_n} + \left(d_1 + d_2 - \frac{d_1 d_2}{f}\right)} ;$$

$$\Rightarrow \frac{1}{R_2} + \mathrm{i}\frac{\lambda_n}{\pi w_2^2}$$

$$= \frac{\left\{\begin{array}{l} \mathrm{i}\frac{\pi w_0^2}{\lambda_n} - \left(1 - \frac{d_2}{f}\right)\frac{\pi^2 w_0^4}{\lambda_n^2 f} + \\ \left(1 - \frac{d_1}{f}\right)\left(d_1 + d_2 - \frac{d_1 d_2}{f}\right) \end{array}\right\}}{\left(1 - \frac{d_2}{f}\right)^2 \frac{\pi^2 w_0^4}{\lambda_n^2} + \left(d_1 + d_2 - \frac{d_1 d_2}{f}\right)^2} . \qquad (3.389)$$

At the position of the beam waist the real part $1/R_2$ of this equation has to vanish. This gives a condition for calculating d_2:

$$-\left(1 - \frac{d_2}{f}\right)\frac{\pi^2 w_0^4}{\lambda_n^2 f} + \left(1 - \frac{d_1}{f}\right)\left(d_1 + d_2 - \frac{d_1 d_2}{f}\right) = 0 ;$$

$$\Rightarrow d_2 = \frac{\frac{\pi^2 w_0^4}{\lambda_n^2 f} - d_1\left(1 - \frac{d_1}{f}\right)}{\frac{\pi^2 w_0^4}{\lambda_n^2 f^2} + \left(1 - \frac{d_1}{f}\right)^2} . \qquad (3.390)$$

In the limiting case of geometrical optics, i. e., $w_0 \to 0$, this equation is equal to the paraxial imaging equation.

The beam waist w_2 can be determined using the imaginary part of (3.389) and replacing then the term $d_1 + d_2 - d_1 d_2/f$ with the help of the first part of (3.390)

$$\frac{\lambda_n}{\pi w_2^2} = \frac{\frac{\pi w_0^2}{\lambda_n}}{\left(1 - \frac{d_2}{f}\right)^2 \frac{\pi^2 w_0^4}{\lambda_n^2}\left(1 + \frac{\frac{\pi^2 w_0^4}{\lambda_n^2 f^2}}{\left(1 - \frac{d_1}{f}\right)^2}\right)} ; \qquad (3.391)$$

$$\Rightarrow w_2^2 = w_0^2\left(1 - \frac{d_2}{f}\right)^2\left(1 + \frac{\frac{\pi^2 w_0^4}{\lambda_n^2 f^2}}{\left(1 - \frac{d_1}{f}\right)^2}\right) .$$

Equation (3.390) delivers

$$1 - \frac{d_2}{f} = \frac{1 - \frac{d_1}{f}}{\frac{\pi^2 w_0^4}{\lambda_n^2 f^2} + \left(1 - \frac{d_1}{f}\right)^2} \qquad (3.392)$$

so that the final result for w_2 is

$$w_2^2 = \frac{w_0^2}{\frac{\pi^2 w_0^4}{\lambda_n^2 f^2} + \left(1 - \frac{d_1}{f}\right)^2} . \qquad (3.393)$$

Of special interest is the case that the beam waist w_0 in front of the lens lies in the front focal plane of the lens, i. e., $d_1 = f$. Equations (3.390) and (3.393) reduce in this special case to

$$d_2 = f \quad \text{and} \quad w_2 = \frac{\lambda_n f}{\pi w_0} . \qquad (3.394)$$

So, if the beam waist of the incident Gaussian beam lies in the front focal plane of the lens, the beam waist of the transformed Gaussian beam lies in the back focal plane of the lens (Fig. 3.36). Additionally, its size w_2 will be the product of the focal length f of the lens and the far field angle (3.340) of the incident Gaussian beam. This result shows that the transformation of Gaussian beams should not be confused with the transformation of paraxial spherical waves of geometrical optics.

References

3.1 M. Born, E. Wolf: *Principles of Optics*, 6th edn. (Cambridge University Press, Cambridge 1997)

3.2 R.W. Boyd: *Nonlinear Optics* (Academic, San Diego 2003)

3.3 A. Yariv, P. Yeh: *Optical Waves in Crystals* (Wiley, New York 1984)

3.4 A. E. Siegman: Propagating modes in gain-guided optical fibers, J. Opt. Soc. Am. A **20**, 1617–1628 (2003)

3.5 J. W. Goodman: *Introduction to Fourier Optics*, 2nd edn. (McGraw-Hill, New York 1996)

3.6 H. Haferkorn: *Optik*, 4th edn. (Wiley-VCH, Weinheim 2003)

3.7 E. Hecht: *Optics*, 3rd edn. (Addison-Wesley, Reading 1998)

3.8 M. V. Klein, Th. E. Furtak: *Optics*, 2nd edn. (Wiley, New York 1986)

3.9 V. N. Mahajan: *Optical Imaging and Aberrations, Part II: Wave Diffraction Optics* (SPIE Press, Bellingham 2001)

3.10 D. Marcuse: *Light Transmission Optics*, 2nd edn. (Van Nostrand, New York 1982)

3.11 A. E. Siegman: *Lasers* (University Science Books, Mill Valley 1986)

3.12 R. A. Chipman: Polarization analysis of optical systems, Opt. Eng. **28**, 90–99 (1989)

3.13 R. A. Chipman: Mechanics of polarization ray tracing, Opt. Eng. **34**, 1636–1645 (1995)

3.14 E. Waluschka: Polarization ray trace, Opt. Eng. **28**, 86–89 (1989)

3.15 R. C. Jones: A new calculus for the treatment of optical systems, J. Opt. Soc. Am. **31**, 488–503 (1941)

3.16 J. W. Goodman: *Statistical Optics* (Wiley, New York 1985)

3.17 M. Françon: *Optical Interferometry* (Academic, New York 1984)

3.18 P. Hariharan: *Optical Interferometry* (Academic, Sydney 1985)

3.19 D. Malacara (Ed.): *Optical Shop Testing*, 2nd edn. (Wiley, New York 1991)

3.20 D. W. Robinson, G. T. Reid (Eds.): *Interferogram Analysis* (IOP Publ., Bristol 1993)

3.21 C. M. Haaland: Laser electron acceleration in vacuum, Opt. Commun. **114**, 280–284 (1995)

3.22 Y. C. Huang, R. L. Byer: A proposed high-gradient laser-driven electron accelerator using crossed cylindrical laser focusing, Appl. Phys. Lett. **69**, 2175–2177 (1996)

3.23 R. Dändliker: Two-Reference-Beam Holographic Interferometry. In: *Holographic Interferometry*, ed. by P. K. Rastogi (Springer, Berlin, Heidelberg 1994) pp. 75–108

3.24 R. Dändliker: *Heterodyne Holographic Interferometry*, Prog. Opt., Vol. 17, ed. by E. Wolf (Elsevier, New York 1980) pp. 1–84

3.25 G. Schulz, J. Schwider: *Interferometric Testing of Smooth Surfaces*, Prog. Opt., Vol. 13, ed. by E. Wolf (Elsevier, New York 1976) pp. 93–167

3.26 J. Schwider: *Advanced Evaluation Techniques in Interferometry*, Prog. Opt., Vol. 28, ed. by E. Wolf (Elsevier, New York 1990) pp. 271–359

3.27 W. H. Steel: *Two-Beam Interferometry*, Prog. Opt., Vol. 5, ed. by E. Wolf (Elsevier, New York 1966) pp. 145–197

3.28 H. J. Tiziani: Optical metrology of engineering surfaces – scope and trends. In: *Optical Measurement Techniques and Applications*, ed. by P. K. Rastogi (Artech House, Norwood 1997) pp. 15–50

3.29 O. Bryngdahl: *Applications of Shearing Interferometry*, Prog. Opt., Vol. 4, ed. by E. Wolf (Elsevier, New York 1965) pp. 37–83

3.30 H. Sickinger, O. Falkenstörfer, N. Lindlein, J. Schwider: Characterization of microlenses using a phase-shifting shearing interferometer, Opt. Eng. **33**, 2680–2686 (1994)

3.31 K. Creath: *Phase-Mesurement Interferometry Techniques*, Prog. Opt., Vol. 26, ed. by E. Wolf (Elsevier, New York 1988) pp. 349–393

3.32 A. Hettwer, J. Kranz, J. Schwider: Three channel phase-shifting interferometer using polarization-optics and a diffraction grating, Opt. Eng. **39**, 960–966 (2000)

3.33 S. Quabis, R. Dorn, M. Eberler, O. Glöckl, G. Leuchs: Focusing light to a tighter spot, Opt. Commun. **179**, 1–7 (2000)

3.34 B. Richards, E. Wolf: Electromagnetic diffraction in optical systems II. Structure of the image field in an aplanatic system, Proc. R. Soc. A **253**, 358–379 (1959)

3.35 M. Françon: *Diffraction* (Pergamon, Oxford 1966)

3.36 J. E. Harvey: Fourier treatment of near-field scalar diffraction theory, Am. J. Phys. **47**, 974–980 (1979)

3.37 E. Lalor: Conditions for the validity of the angular spectrum of plane waves, J. Opt. Soc. Am. **58**, 1235–1237 (1968)

3.38 U. Vokinger: Propagation, modification, analysis of partially coherent light fields. Dissertation (University of Neuchatel (UFO), Allensbach 2000)

3.39 J. J. Stamnes: *Waves in Focal Regions* (Hilger, Bristol 1986)

3.40 H. J. Caulfield: *Handbook of Optical Holography* (Academic, New York 1979)

3.41 P. Hariharan: *Basics of Holography* (Cambridge University Press, Cambridge 2002)

3.42 E. N. Leith, J. Upatnieks: *Recent Advances in Holography*, Prog. Opt., Vol. 6, ed. by E. Wolf (Elsevier, New York 1967) pp. 1–52

3.43 G. Saxby: *Practical Holography* (Prentice Hall, New York 1988)

3.44 B. R. Brown, A. W. Lohmann: Computer generated binary holograms, IBM J. **13**, 160–168 (1969)

3.45 O. Bryngdahl, F. Wyrowski: *Digital holography – Computer-Generated Holograms*, Prog. Opt., Vol. 28, ed. by E. Wolf (Elsevier, New York 1990) pp. 1–86

3.46 H. P. Herzig: *Micro-Optics* (Taylor & Francis, London 1997)

3.47 B. Kress, P. Meyrueis: *Digital Diffractive Optics* (Wiley, Chichester 2000)

3.48 W.-H. Lee: *Computer-Generated Holograms: Techniques, Applications*, Prog. Opt., Vol. 16, ed. by E. Wolf (Elsevier, New York 1978) pp. 119–232

3.49 D. Maystre: *Rigorous Vector Theories of Diffraction Gratings*, Prog. Opt., Vol. 11, ed. by E. Wolf (Elsevier, New York 1984) pp. 1–67

3.50 S. Sinzinger, J. Jahns: *Microoptics* (Wiley-VCH, Weinheim 1999)

3.51 I. N. Bronstein, K. A. Semendjajew: *Taschenbuch der Mathematik*, 23rd edn. (Thun, Frankfurt 1987)

3.52 W. H. Press, B. P. Flannery, S. A. Teukolsky, W. T. Vetterling: *Numerical Recipes in C* (Cambridge University Press, Cambridge 1991) pp. 398–470

3.53 B. Besold, N. Lindlein: Fractional Talbot effect for periodic microlens arrays, Opt. Eng. **36**, 1099–1105 (1997)

3.54 S. Quabis, R. Dorn, M. Eberler, O. Glöckl, G. Leuchs: The focus of light-theoretical calculation and experimental tomographic reconstruction, Appl. Phys. B **72**, 109–113 (2001)

3.55 M. Mansuripur: Distribution of light at and near the focus of high-numerical-aperture objectives, J. Opt. Soc. Am. A **3**, 2086–2093 (1986)

3.56 M. Mansuripur: Distribution of light at and near the focus of high-numerical-aperture objectives: Erratum, J. Opt. Soc. Am. A **10**, 382–383 (1993)

3.57 R. Dorn, S. Quabis, G. Leuchs: The focus of light – linear polarization breaks the rotational symmetry of the focal spot, J. Mod. Opt. **50**, 1917–1926 (2003)

3.58 H. Kogelnik, T. Li: Laser beams and resonators, Appl. Opt. **5**, 1550–1567 (1966)

3.59 W. Brouwer: *Matrix Methods in Optical Instrument Design* (Benjamin, New York 1964)

3.60 A. Gerrard, J. M. Burch: *Introduction to Matrix Methods in Optics* (Wiley, London 1975)

4. Nonlinear Optics

This chapter provides a brief introduction into the basic nonlinear-optical phenomena and discusses some of the most significant recent advances and breakthroughs in nonlinear optics, as well as novel applications of nonlinear-optical processes and devices.

Nonlinear optics is the area of optics that studies the interaction of light with matter in the regime where the response of the material system to the applied electromagnetic field is nonlinear in the amplitude of this field. At low light intensities, typical of non-laser sources, the properties of materials remain independent of the intensity of illumination. The superposition principle holds true in this regime, and light waves can pass through materials or be reflected from boundaries and interfaces without interacting with each other. Laser sources, on the other hand, can provide sufficiently high light intensities to modify the optical properties of materials. Light waves can then interact with each other, exchanging momentum and energy, and the superposition principle is no longer valid. This interaction of light waves can result in the generation of optical fields at new frequencies, including optical harmonics of incident radiation or sum- or difference-frequency signals.

Although the observation of most nonlinear-optical phenomena requires laser radiation, some classes of nonlinear-optical effects were known long before the invention of the laser. The most prominent examples of such phenomena include Pockels and Kerr electrooptic effects [4.1], as well as light-induced resonant absorption saturation, described by *Vavilov* [4.2, 3]. It was, however, only with the advent of lasers that systematic studies of optical nonlinearities and the observation of a vast catalog of spectacular nonlinear-optical phenomena became possible.

In the first nonlinear-optical experiment of the laser era, performed by *Franken* et al. in 1961 [4.4], a ruby-laser radiation with a wavelength of 694.2 nm was used to generate the second harmonic in a quartz crystal at the wavelength of 347.1 nm. This seminal work was followed by the discovery of a rich diversity of nonlinear-optical effects, including sum-frequency generation, stimulated Raman scattering, self-focusing, optical rectification, four-wave mixing, and many others. While in the pioneering work by Franken the efficiency of second-harmonic generation (SHG) was on the order of 10^{-8}, optical frequency doublers created by early 1963 provided 20%–30% efficiency of frequency conversion [4.5, 6]. The early phases of the development and the basic principles of nonlinear optics have been reviewed in the most illuminating way in the classical books by *Bloembergen* [4.7] and *Akhmanov* and *Khokhlov* [4.8], published in the mid 1960s.

Over the following four decades, the field of nonlinear optics has witnessed an enormous growth, leading to the observation of new physical phenomena and giving rise to novel concepts and applications. A systematic introduction into these effects along with a comprehensive overview of nonlinear-optical concepts and devices can be found in excellent textbooks by *Shen* [4.9], *Boyd* [4.1], *Butcher* and *Cotter* [4.10], *Reintjes* [4.11] and others. One of the most recent up-to-date reviews of the field of nonlinear optics with an in-depth discussion of the fundamental physics underlying nonlinear-optical interactions was provided by *Flytzanis* [4.12]. This chapter provides a brief introduction into the main nonlinear-optical phenomena and discusses some of the most significant recent advances in nonlinear optics, as well as novel applications of nonlinear-optical processes and devices.

4.1 Nonlinear Polarization and Nonlinear Susceptibilities

Nonlinear-optical effects belong to a broader class of electromagnetic phenomena described within the general framework of macroscopic Maxwell equations. The Maxwell equations not only serve to identify and classify nonlinear phenomena in terms of the relevant nonlinear-optical susceptibilities or, more generally, nonlinear terms in the induced polarization, but also govern the nonlinear-optical propagation effects. We assume the absence of extraneous charges and currents and write the set of Maxwell equations for the electric, $\boldsymbol{E}(\boldsymbol{r},t)$, and magnetic, $\boldsymbol{H}(\boldsymbol{r},t)$, fields in the form

$$\nabla\times\boldsymbol{E} = -\frac{1}{c}\frac{\partial\boldsymbol{B}}{\partial t}\,, \tag{4.1}$$

$$\nabla\times\boldsymbol{B} = \frac{1}{c}\frac{\partial\boldsymbol{D}}{\partial t}\,, \tag{4.2}$$

$$\nabla\cdot\boldsymbol{D} = 0\,, \tag{4.3}$$

$$\nabla\cdot\boldsymbol{B} = 0\,. \tag{4.4}$$

Here, $\boldsymbol{B}=\boldsymbol{H}+4\pi\boldsymbol{M}$, where $\boldsymbol{M}$ is the magnetic dipole polarization, c is the speed of light, and

$$\boldsymbol{D} = \boldsymbol{E}+4\pi\int_{-\infty}^{t}\boldsymbol{J}(\zeta)\,\mathrm{d}\zeta\,, \tag{4.5}$$

where $\boldsymbol{J}$ is the induced current density. Generally, the equation of motion for charges driven by the electromagnetic field has to be solved to define the relation between the induced current $\boldsymbol{J}$ and the electric and magnetic fields. For quantum systems, this task can be fulfilled by solving the Schrödinger equation. In Sect. 4.5 of this chapter, we provide an example of such a self-consistent analysis of nonlinear-optical phenomena in a model two-level system. Very often a phenomenological approach based on the introduction of field-independent or local-field-corrected nonlinear-optical susceptibilities can provide an adequate description of nonlinear-optical processes.

Formally, the current density $\boldsymbol{J}$ can be represented as a series expansion in multipoles:

$$\boldsymbol{J} = \frac{\partial}{\partial t}(\boldsymbol{P}-\nabla\cdot\boldsymbol{Q})+c\,(\nabla\times\boldsymbol{M})\,, \tag{4.6}$$

where $\boldsymbol{P}$ and $\boldsymbol{Q}$ are the electric dipole and electric quadrupole polarizations, respectively. In the electric dipole approximation, we keep only the first term on the right-hand side of (4.6). In view of (4.5), this gives the following relation between the $\boldsymbol{D}$, $\boldsymbol{E}$, and $\boldsymbol{P}$ vectors:

$$\boldsymbol{D} = \boldsymbol{E}+4\pi\boldsymbol{P}. \tag{4.7}$$

We now represent the polarization $\boldsymbol{P}$ as a sum

$$\boldsymbol{P} = \boldsymbol{P}_{\mathrm{L}}+\boldsymbol{P}_{\mathrm{nl}}\,, \tag{4.8}$$

where $\boldsymbol{P}_{\mathrm{L}}$ is the part of the electric dipole polarization linear in the field amplitude and $\boldsymbol{P}_{\mathrm{nl}}$ is the nonlinear part of this polarization.

The linear polarization governs linear-optical phenomena, i. e., it corresponds to the regime where the optical properties of a medium are independent of the field intensity. The relation between $\boldsymbol{P}_{\mathrm{L}}$ and the electric field $\boldsymbol{E}$ is given by the standard formula of linear optics:

$$\boldsymbol{P}_{\mathrm{L}} = \int\chi^{(1)}(t-t')\boldsymbol{E}(t')\,\mathrm{d}t'\,, \tag{4.9}$$

where $\chi^{(1)}(t)$ is the time-domain linear susceptibility tensor. Representing the field $\boldsymbol{E}$ and polarization $\boldsymbol{P}_{\mathrm{L}}$ in the form of elementary monochromatic plane waves,

$$\boldsymbol{E} = \boldsymbol{E}(\omega)\exp(\mathrm{i}\boldsymbol{kr}-\omega t)+\mathrm{c.c.} \tag{4.10}$$

and

$$\boldsymbol{P}_{\mathrm{L}} = \boldsymbol{P}_{\mathrm{L}}(\omega)\exp\left(\mathrm{i}\boldsymbol{kr}-\omega t\right)+\mathrm{c.c.}\,, \tag{4.11}$$

we take the Fourier transform of (4.9) to find

$$\boldsymbol{P}_{\mathrm{L}}(\omega) = \chi^{(1)}(\omega)\boldsymbol{E}(\omega)\,, \tag{4.12}$$

where

$$\chi^{(1)}(\omega) = \int\chi^{(1)}(t)\exp(\mathrm{i}\omega t)\,\mathrm{d}t\,. \tag{4.13}$$

In the regime of weak fields, the nonlinear part of the polarization $\boldsymbol{P}_{\mathrm{nl}}$ can be represented as a power-series expansion in the field $\boldsymbol{E}$:

$$\begin{aligned}\boldsymbol{P}_{\mathrm{nl}} &= \iint\chi^{(2)}(t-t_1,t-t_2):\boldsymbol{E}(t_1)\boldsymbol{E}(t_2)\,\mathrm{d}t_1\,\mathrm{d}t_2\\ &\quad+\iiint\chi^{(3)}(t-t_1,t-t_2,t-t_3)\\ &\quad\vdots\boldsymbol{E}(t_1)\boldsymbol{E}(t_2)\boldsymbol{E}(t_3)\,\mathrm{d}t_1\,\mathrm{d}t_2\,\mathrm{d}t_3+\ldots\,,\end{aligned} \tag{4.14}$$

where $\chi^{(2)}$ and $\chi^{(3)}$ are the second- and third-order nonlinear susceptibilities.

Representing the electric field in the form of a sum of plane monochromatic waves,

$$\boldsymbol{E} = \sum_i \boldsymbol{E}_i(\omega_i)\exp(\mathrm{i}\boldsymbol{k}_i\boldsymbol{r}-\omega_i t)+\mathrm{c.c.}\,, \tag{4.15}$$

we take the Fourier transform of (4.14) to arrive at

$$\boldsymbol{P}_{\mathrm{nl}}(\omega) = \boldsymbol{P}^{(2)}(\omega)+\boldsymbol{P}^{(3)}(\omega)+\ldots\,, \tag{4.16}$$

where

$$P^{(2)}(\omega) = \chi^{(2)}(\omega; \omega_i, \omega_j) : E(\omega_i)E(\omega_j)\,, \quad (4.17)$$

$$P^{(3)}(\omega) = \chi^{(3)}(\omega; \omega_i, \omega_j, \omega_k) \vdots E(\omega_i)E(\omega_j)E(\omega_k)\,, \quad (4.18)$$

$$\chi^{(2)}(\omega; \omega_i, \omega_j) = \chi^{(2)}(\omega = \omega_i + \omega_j) = \iint \chi^{(2)}(t_1, t_2) \exp[\mathrm{i}(\omega_i t_1 + \omega_j t_2)]\,\mathrm{d}t_1\,\mathrm{d}t_2 \quad (4.19)$$

is the second-order nonlinear-optical susceptibility and

$$\chi^{(3)}(\omega; \omega_i, \omega_j, \omega_k) = \chi^{(3)}(\omega = \omega_i + \omega_j + \omega_k) = \iint \chi^{(3)}(t_1, t_2, t_3) \exp[\mathrm{i}(\omega_i t_1 + \omega_j t_2 + \omega_k t_3)]\,\mathrm{d}t_1\,\mathrm{d}t_2\,\mathrm{d}t_3 \quad (4.20)$$

is the third-order nonlinear-optical susceptibility.

The second-order nonlinear polarization defined by (4.17) gives rise to three-wave mixing processes, optical rectification and linear electrooptic effect. In particular, setting $\omega_i = \omega_j = \omega_0$ in (4.17) and (4.19), we arrive at $\omega = 2\omega_0$, which corresponds to second-harmonic generation, controlled by the nonlinear susceptibility $\chi^{(2)}_{\mathrm{SHG}} = \chi^{(2)}(2\omega_0; \omega_0, \omega_0)$. In a more general case of three-wave mixing process with $\omega_i = \omega_1 \neq \omega_j = \omega_2$, the second-order polarization defined by (4.17) can describe sum-frequency generation (SFG) $\omega_{\mathrm{SF}} = \omega_1 + \omega_2$ Fig. 4.1 or difference-frequency generation (DFG) $\omega_{\mathrm{DF}} = \omega_1 - \omega_2$, governed by the nonlinear susceptibilities $\chi^{(2)}_{\mathrm{SFG}} = \chi^{(2)}(\omega_{\mathrm{SF}}; \omega_1, \omega_2)$ and $\chi^{(2)}_{\mathrm{DFG}} = \chi^{(2)}(\omega_{\mathrm{DF}}; \omega_1, -\omega_2)$, respectively.

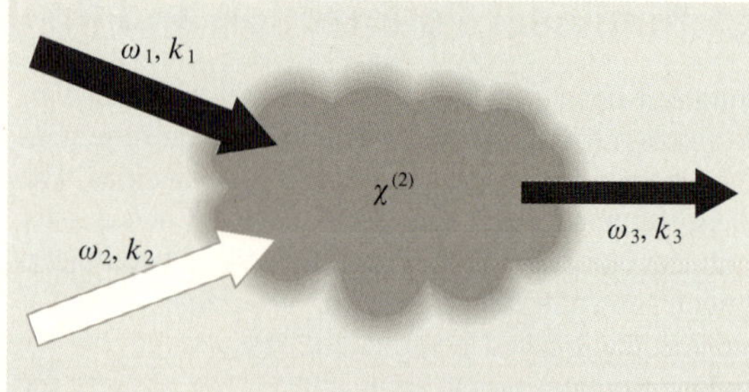

Fig. 4.1 Sum-frequency generation $\omega_1 + \omega_2 = \omega_3$ in a medium with a quadratic nonlinearity. The case of $\omega_1 = \omega_2$ corresponds to second-harmonic generation

The third-order nonlinear polarization defined by (4.18) is responsible for four-wave mixing (FWM), stimulated Raman scattering, two-photon absorption, and Kerr-effect-related phenomena, including self-phase modulation (SPM) and self-focusing. For the particular case of third-harmonic generation, we set $\omega_i = \omega_j = \omega_k = \omega_0$ in (4.18) and (4.20) to obtain $\omega = 3\omega_0$. This type of nonlinear-optical interaction, in accordance with (4.18) and (4.20), is controlled by the cubic susceptibility $\chi^{(3)}_{\mathrm{THG}} = \chi^{(3)}(3\omega_0; \omega_0, \omega_0, \omega_0)$. A more general, frequency-nondegenerate case can correspond to a general type of an FWM process. These and other basic nonlinear-optical processes will be considered in greater details in the following sections.

4.2 Wave Aspects of Nonlinear Optics

In the electric dipole approximation, the Maxwell equations (4.1–4.4) yield the following equation governing the propagation of light waves in a weakly nonlinear medium:

$$\nabla \times (\nabla \times E) - \frac{1}{c^2}\frac{\partial^2 E}{\partial t^2} - \frac{4\pi}{c^2}\frac{\partial^2 P_{\mathrm{L}}}{\partial t^2} = \frac{4\pi}{c^2}\frac{\partial^2 P_{\mathrm{nl}}}{\partial t^2}\,. \quad (4.21)$$

The nonlinear polarization, appearing on the right-hand side of (4.21), plays the role of a driving source, inducing an electromagnetic wave with the same frequency ω as the nonlinear polarization wave $P_{\mathrm{nl}}(r, t)$. Dynamics of a nonlinear wave process can be then thought as a result of the interference of induced and driving (pump) waves, controlled by the dispersion of the medium.

Assuming that the fields have the form of quasi-monochromatic plane waves propagating along the z-axis, we represent the field E in (4.21) by

$$E(r, t) = \mathrm{Re}\left[e A(z, t) \exp(\mathrm{i}kz - \omega t)\right] \quad (4.22)$$

and write the nonlinear polarization as

$$P_{\mathrm{nl}}(r, t) = \mathrm{Re}\left[e_{\mathrm{p}} P_{\mathrm{nl}}(z, t) \exp\left(\mathrm{i}k_{\mathrm{p}}z - \omega t\right)\right], \quad (4.23)$$

where k and $A(z, t)$ are the wave vector and the envelope of the electric field, k_p and $P_{\mathrm{nl}}(z, t)$ are the wave vector and the envelope of the polarization wave.

If the envelope $A(z, t)$ is a slowly varying function over the wavelength, $|\partial^2 A/\partial z^2| \ll |k \partial A/\partial z|$, and $\partial^2 P_{\mathrm{nl}}/\partial t^2 \approx -\omega^2 P_{\mathrm{nl}}$, (4.21) is reduced to [4.9]

$$\frac{\partial A}{\partial z} + \frac{1}{u}\frac{\partial A}{\partial t} = \frac{2\pi \mathrm{i}\omega^2}{kc^2} P_{\mathrm{nl}} \exp(\mathrm{i}\Delta kz)\,, \quad (4.24)$$

where $u = (\partial k/\partial\omega)^{-1}$ is the group velocity and $\Delta k = k_p - k$ is the wave-vector mismatch.

In the following sections, this generic equation of slowly varying envelope approximation (SVEA) will be employed to analyze the wave aspects of the basic second- and third-order nonlinear-optical phenomena.

4.3 Second-Order Nonlinear Processes

4.3.1 Second-Harmonic Generation

In second-harmonic generation, a pump wave with a frequency of ω generates a signal at the frequency 2ω as it propagates through a medium with a quadratic nonlinearity (Fig. 4.1). Since all even-order nonlinear susceptibilities $\chi^{(n)}$ vanish in centrosymmetric media, SHG can occur only in media with no inversion symmetry.

Assuming that diffraction and second-order dispersion effects are negligible, we use (4.24) for a quadratically nonlinear medium with a nonlinear SHG susceptibility $\chi^{(2)}_{\mathrm{SHG}} = \chi^{(2)}(2\omega; \omega, \omega)$ to write a pair of coupled equations for the slowly varying envelopes of the pump and second-harmonic fields $A_1 = A_1(z,t)$ and $A_2 = A_2(z,t)$:

$$\frac{\partial A_1}{\partial z} + \frac{1}{u_1}\frac{\partial A_1}{\partial t} = \mathrm{i}\gamma_1 A_1^* A_2 \exp(\mathrm{i}\Delta kz)\,, \tag{4.25}$$

$$\frac{\partial A_2}{\partial z} + \frac{1}{u_2}\frac{\partial A_2}{\partial t} = \mathrm{i}\gamma_2 A_1^2 \exp(-\mathrm{i}\Delta kz)\,, \tag{4.26}$$

where

$$\gamma_1 = \frac{2\pi\omega_1^2}{k_1 c^2}\chi^{(2)}(\omega; 2\omega, -\omega) \tag{4.27}$$

and

$$\gamma_2 = \frac{4\pi\omega_1^2}{k_2 c^2}\chi^{(2)}_{\mathrm{SHG}} \tag{4.28}$$

are the nonlinear coefficients, u_1 and u_2 are the group velocities of the pump and second-harmonic pulses, respectively, and $\Delta k = 2k_1 - k_2$ is the wave-vector mismatch for the SHG process.

If the difference between the group velocities of the pump and second-harmonic pulses can be neglected for a nonlinear medium with a given length and if the intensity of the pump field in the process of SHG remains much higher than the intensity of the second-harmonic field, we set $u_1 = u_2 = u$ and $|A_1|^2 = |A_{10}|^2$ = const. in (4.25) and (4.26) to derive in the retarded frame of reference with $z' = z$ and $\eta = t - z/u$

$$A_2(L) = \mathrm{i}\gamma_2 A_{10}^2 \frac{\sin\left(\frac{\Delta kL}{2}\right)}{\frac{\Delta kL}{2}} L \exp\left(\frac{\mathrm{i}\Delta kL}{2}\right), \tag{4.29}$$

where L is the length of the nonlinear medium.

The intensity of the second-harmonic field is then given by

$$I_2(L) \propto \gamma_2^2 I_{10}^2 \left(\frac{\sin\left(\frac{\Delta kL}{2}\right)}{\frac{\Delta kL}{2}}\right)^2 L^2, \tag{4.30}$$

where I_{10} is the intensity of the pump field.

Second-harmonic intensity I_2, as can be seen from (4.30) oscillates as a function of L Fig. 4.2 with a period $L_c = \pi/|\Delta k| = \lambda_1(4|n_1 - n_2|)^{-1}$, where λ_1 is the pump wavelength and n_1 and n_2 are the values of the refractive index at the frequencies of the pump field and its second harmonic, respectively. The parameter L_c, defining the length of the nonlinear medium providing the maximum SHG efficiency, is referred to as the coherence length.

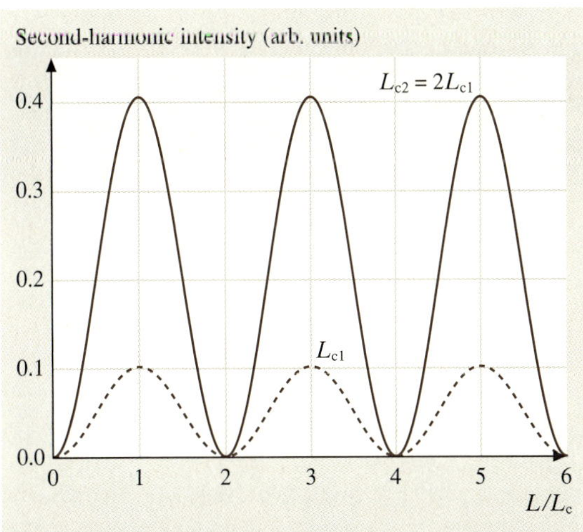

Fig. 4.2 Second-harmonic intensity as a function of the length L of the nonlinear medium normalized to the coherence length L_c for two values of L_c: (*dashed line*) L_{c1} and (*solid line*) $L_{c2} = 2L_{c1}$

Although the solution (4.29) describes the simplest regime of SHG, it is very instructive as it visualizes the significance of reducing the wave-vector mismatch Δk for efficient SHG. Since the wave vectors k_1 and k_2 are associated with the momenta of the pump and second-harmonic fields, $p_1 = \hbar k_1$ and $p_2 = \hbar k_2$, with $\hbar$ being the Planck constant, the condition $\Delta k = 0$, known as the phase-matching condition in nonlinear optics, in fact, represents momentum conservation for the SHG process, where two photons of the pump field are requited to generate a single photon of the second harmonic.

Several strategies have been developed to solve the phase-matching problem for SHG. The most practically significant solutions include the use of birefringent nonlinear crystals [4.13, 14], quasi-phase-matching in periodically poled nonlinear materials [4.15, 16] and waveguide regimes of nonlinear interactions with the phase mismatch related to the material dispersion compensated for by waveguide dispersion [4.7]. Harmonic generation in the gas phase, as demonstrated by *Miles* and *Harris* [4.17], can be phase-matched through an optimization of the content of the gas mixture. Figure 4.3 illustrates phase matching in a birefringent crystal. The circle represents the cross section of the refractive-index sphere $n_0(\omega)$ for an ordinary wave at the pump frequency ω. The ellipse is the cross section of the refractive-index ellipsoid $n_e(2\omega)$ for an extraordinary wave at the frequency of the second harmonic 2ω. Phase matching is achieved in the direction where $n_0\omega = n_e(2\omega)$, corresponding to an angle θ_{pm} with respect to the optical axis c of the crystal in Fig. 4.3.

When the phase-matching condition $\Delta k = 0$ is satisfied, (4.29) and (4.30) predict a quadratic growth of the

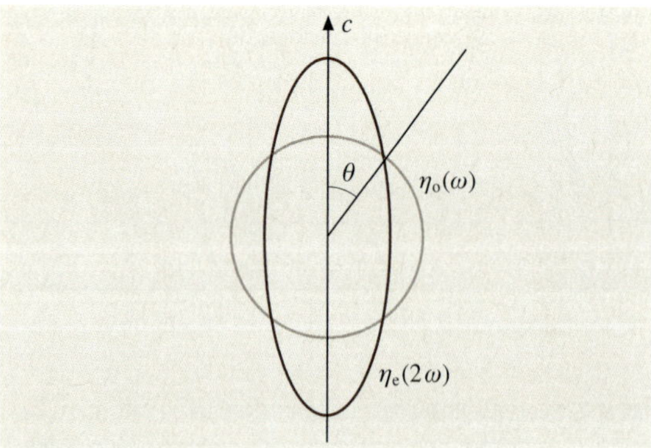

Fig. 4.3 Phase-matching second-harmonic generation in a birefringent crystal

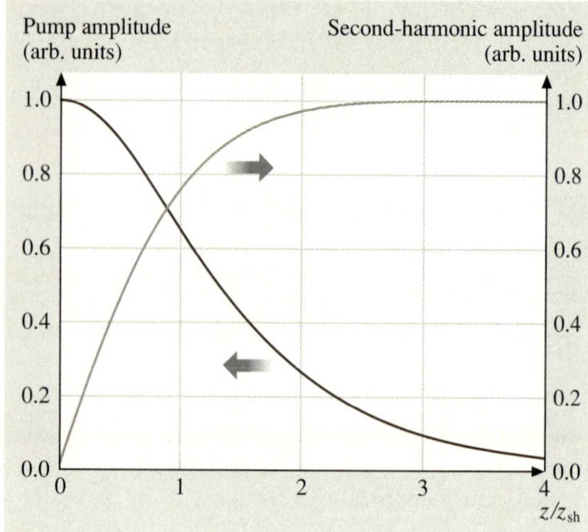

Fig. 4.4 The amplitudes of the pump and second-harmonic fields as functions of the normalized propagation distance z/z_{sh} with $z_{sh} = [\gamma\rho_{10}(0)]^{-1}$

second-harmonic intensity as a function of the length L of the nonlinear medium. This scaling law holds true, however, only as long as the second-harmonic intensity remains much less than the pump intensity. As $|A_2|$ becomes comparable with $|A_1|$, depletion of the pump field has to be taken into consideration. To this end, we introduce the real amplitudes ρ_j and phases φ_j of the pump and second-harmonic fields, $A_j = \rho_j \exp(\mathrm{i}\varphi_j)$, with $j = 1, 2$. Then, assuming that $u_1 = u_2 = u$ and $\gamma_1 = \gamma_2 = \gamma$, we derive from (4.25) and (4.26)

$$\rho_1(\eta, z) = \rho_{10}(\eta)\,\mathrm{sech}\,[\gamma\rho_{10}(\eta)z]\ , \qquad (4.31)$$

$$\rho_2(\eta, z) = \rho_{10}(\eta)\tanh[\gamma\rho_{10}(\eta)z]\ . \qquad (4.32)$$

The solutions (4.31) and (4.32) show that the entire energy of the pump field in the phase-matching regime can be transferred to the second harmonic. As the pump field becomes depleted Fig. 4.4, the growth of the second-harmonic field saturates.

Effects related to the group-velocity mismatch become significant when the length of the nonlinear medium L exceeds the length $L_g = \tau_1/|u_2^{-1} - u_1^{-1}|$, where τ_1 is the pulse width of the pump field. The length L_g characterizes the walk-off between the pump and second-harmonic pulses caused by the group-velocity mismatch. In this nonstationary regime of SHG, the amplitude of the second harmonic in the constant-pump-field approximation is given by

$$A_2(z,t) = \mathrm{i}\gamma_2 \int_0^z A_{10}^2\left[t - z/u_2 + \xi\left(u_2^{-1} - u_1^{-1}\right)\right] \times \exp(-\mathrm{i}\Delta k\xi)\,\mathrm{d}\xi\,. \quad (4.33)$$

Group-velocity mismatch may lead to a considerable increase in the pulse width of the second harmonic τ_2. For $L \gg L_g$, the second harmonic pulse width, $\tau_2 \approx |u_2^{-1} - u_1^{-1}|z$, scales linearly with the length of the nonlinear medium and is independent of the pump pulse width.

4.3.2 Sum- and Difference-Frequency Generation and Parametric Amplification

In sum-frequency generation Fig. 4.1, two laser fields with frequencies ω_1 and ω_2 generate a nonlinear signal at the frequency $\omega_3 = \omega_1 + \omega_2$ in a quadratically nonlinear medium with a nonlinear susceptibility $\chi_{\text{SFG}}^{(2)} = \chi^{(2)}(\omega_3; \omega_1, \omega_2)$. In the first order of dispersion theory, the coupled equations for slowly varying envelopes of the laser fields $A_1 = A_1(z,t)$ and $A_2 = A_2(z,t)$ and the nonlinear signal $A_3 = A_3(z,t)$ are written as

$$\frac{\partial A_1}{\partial z} + \frac{1}{u_1}\frac{\partial A_1}{\partial t} = \mathrm{i}\gamma_1 A_3 A_2^* \exp(\mathrm{i}\Delta kz)\,, \quad (4.34)$$

$$\frac{\partial A_2}{\partial z} + \frac{1}{u_2}\frac{\partial A_2}{\partial t} = \mathrm{i}\gamma_2 A_3 A_1^* \exp(\mathrm{i}\Delta kz)\,, \quad (4.35)$$

$$\frac{\partial A_3}{\partial z} + \frac{1}{u_3}\frac{\partial A_3}{\partial t} = \mathrm{i}\gamma_3 A_1 A_2 \exp(-\mathrm{i}\Delta kz)\,, \quad (4.36)$$

where

$$\gamma_1 = \frac{2\pi\omega_1^2}{k_1 c^2}\chi^{(2)}(\omega_1; \omega_3, -\omega_2)\,, \quad (4.37)$$

$$\gamma_2 = \frac{2\pi\omega_2^2}{k_2 c^2}\chi^{(2)}(\omega_2; \omega_3, -\omega_1)\,, \quad (4.38)$$

$$\gamma_3 = \frac{2\pi\omega_3^2}{k_3 c^2}\chi_{\text{SFG}}^{(2)} \quad (4.39)$$

are the nonlinear coefficients, u_1, u_2, and u_3 and k_1, k_2, and k_3 are the group velocities and the wave vectors of the fields with frequencies ω_1, ω_2, and ω_3, respectively, and $\Delta k = k_1 + k_2 - k_3$ is the wave-vector mismatch for the SFG process.

As long as the intensity of the sum-frequency field remains much less than the intensities of the laser fields, the amplitudes of the laser fields can be can be assumed to be given functions of t, $A_1(z,t) = A_{10}(t)$ and $A_2(z,t) = A_{20}(t)$, and the solution of (4.36) yields

$$A_3(z,t) = \mathrm{i}\gamma_3 \int_0^z A_{10}\left[t - z/u_3 + \xi\left(u_3^{-1} - u_1^{-1}\right)\right] \times A_{20}\left[t - z/u_3 + \xi\left(u_3^{-1} - u_2^{-1}\right)\right] \times \exp(-\mathrm{i}\Delta k\xi)\,\mathrm{d}\xi\,. \quad (4.40)$$

The efficiency of frequency conversion, as can be seen from (4.40) is controlled by the group delays $\Delta_{21} \approx |u_2^{-1} - u_1^{-1}|L$, $\Delta_{31} \approx |u_3^{-1} - u_1^{-1}|L$, and $\Delta_{32} \approx |u_3^{-1} - u_2^{-1}|L$ between the pulses involved in the SFG process. In particular, the laser fields cease to interact with each other when the group delay Δ_{21} starts to exceed the pulse width of the faster laser pulse.

In difference-frequency generation (DFG), two input fields with frequencies ω_1 and ω_2 generate a nonlinear signal at the frequency $\omega_3 = \omega_1 - \omega_2$. This process is of considerable practical significance as it can give rise to intense coherent radiation in the infrared range. In the limiting case of $\omega_1 \approx \omega_2$, this type of nonlinear-optical interaction corresponds to optical rectification, which has been intensely used over the past two decades for the generation of terahertz radiation.

If the field at the frequency ω_1 is strong and remains undepleted in the process of nonlinear-optical interaction, $A_1(z,t) = A_{10}(t)$, the set of coupled equations governing the amplitudes of the remaining two fields in the stationary regime is written as

$$\frac{\partial A_2}{\partial z} + \frac{1}{u_2}\frac{\partial A_2}{\partial t} = \mathrm{i}\gamma_2 A_1 A_3^* \exp(\mathrm{i}\Delta kz)\,, \quad (4.41)$$

$$\frac{\partial A_3}{\partial z} + \frac{1}{u_3}\frac{\partial A_3}{\partial t} = \mathrm{i}\gamma_3 A_1 A_2^* \exp(-\mathrm{i}\Delta kz)\,, \quad (4.42)$$

where,

$$\gamma_2 = \frac{2\pi\omega_2^2}{k_2 c^2}\chi^{(2)}(\omega_2; \omega_1, -\omega_3)\,, \quad (4.43)$$

$$\gamma_3 = \frac{2\pi\omega_3^2}{k_3 c^2}\chi^{(2)}(\omega_3; \omega_1, -\omega_2) \quad (4.44)$$

are the nonlinear coefficientsm and $\Delta k = k_1 - k_2 - k_3$ is the wave-vector mismatch for the DFG process.

With no signal at ω_3 applied at the input of the nonlinear medium, $A_3(0,t) = 0$, the solution to (4.41) and (4.42) in the stationary regime is given by [4.12]

$$A_2(z) = A_2(0)\left[\cosh(\kappa z) + \mathrm{i}\frac{\Delta k}{2\kappa}\sinh(\kappa z)\right]\,, \quad (4.45)$$

$$A_3(z) = \mathrm{i}A_2(0)\sinh(\kappa z)\,, \quad (4.46)$$

Part A | 4.3

where

$$\kappa^2 = 4\gamma_2\gamma_3^* |A_1|^2 - (\Delta k)^2 \, . \tag{4.47}$$

Away from the phase-matching condition, the amplification of a weak signal is achieved only when the intensity of the pump field exceeds a threshold,

$$I_1 > I_{\text{th}} = \frac{n_1 n_2 n_3 c^3 (\Delta k)^2}{32\pi^3 \left|\chi^{(2)}_{\text{DFG}}\right|^2 \omega_2\omega_3} \, , \tag{4.48}$$

where we took

$$\chi^{(2)}(\omega_2; \omega_1, -\omega_3) \approx \chi^{(2)}(\omega_3; \omega_1, -\omega_2) = \chi^{(2)}_{\text{DFG}} \, .$$

Above, this threshold, the growth in the intensity I_2 of a weak input signal is governed by

$$I_2(z) = I_2(0)\left[\frac{\gamma_2\gamma_3^* |A_{10}|^2}{\kappa^2}\sin^2(\kappa z) + 1\right] . \tag{4.49}$$

This type of three-wave mixing is often referred to as optical parametric amplification. A weak input field, referred to as the signal field (the field with the amplitude A_2 in our case), becomes amplified in this type of process through a nonlinear interaction with a powerful pump field (the undepleted field with the amplitude A_1 in the case considered here). In such a scheme of optical parametric amplification, the third field (the field with the amplitude A_3) is called the idler field.

Part A | 4.4

We now consider the regime of optical parametric amplification $\omega_1 = \omega_2 + \omega_3$ where the pump, signal and idler pulses are matched in their wave vectors and group velocities. Introducing the real amplitudes ρ_j and phases φ_j of the pump, signal, and idler fields, $A_j = \rho_j \exp(i\varphi_j)$, where $j = 1, 2, 3$, assuming that $\gamma_2 = \gamma_3 = \gamma$ in (4.35) and (4.36), $A_1(z, t) = A_{10}(t)$ and $A_3(0, t) = 0$, we write the solution for the amplitude of the signal field as [4.18]

$$A_2(\eta, z) = A_{20}(\eta)\cosh[\gamma\rho_{10}(\eta) z] \, . \tag{4.50}$$

The idler field then builds up in accordance with

$$A_3(\eta, z) = A_{20}^*(\eta)\exp[i\varphi_{10}(\eta)]\sinh[\gamma\rho_{10}(\eta) z] \, . \tag{4.51}$$

As can be seen from (4.50), optical parametric amplification preserves the phase of the signal pulse. This property of optical parametric amplification lies at the heart of the principle of optical parametric chirped-pulse amplification [4.19], allowing ultrashort laser pulses to be amplified to relativistic intensities. It also suggests a method of efficient frequency conversion of few-cycle field waveforms without changing the phase offset between their carrier frequency and temporal envelope, making few-cycle laser pulses a powerful tool for the investigation of ultrafast electron dynamics in atomic and molecular systems.

In the nonstationary regime of optical parametric amplification, when the pump, signal, and idler fields propagate with different group velocities, useful and important qualitative insights into the phase relations between the pump, signal, and idler pulses can be gained from energy and momentum conservation, $\omega_1 = \omega_2 + \omega_3$ and $k_1 = k_2 + k_3$. These equalities dictate the following relations between the frequency deviations $\delta\omega_j$ in the pump, signal, and idler fields ($j = 1, 2, 3$):

$$\delta\omega_1 = \delta\omega_2 + \delta\omega_3 \tag{4.52}$$

and

$$\delta\omega_1/u_1 = \delta\omega_2/u_2 + \delta\omega_3/u_3 \, . \tag{4.53}$$

In view of (4.52) and (4.53), we find

$$\delta\omega_2 = q_2\delta\omega_1 \tag{4.54}$$

and

$$\delta\omega_3 = q_3\delta\omega_1 \, , \tag{4.55}$$

where $q_2 = (u_1^{-1} - u_3^{-1})/(u_2^{-1} - u_3^{-1})$, $q_3 = 1 - q_2$.

In the case of a linearly chirped pump, $\varphi_1(t) = \alpha_1 t^2/2$, the phases of the signal and idler pulses are given by $\varphi_m(t) = \alpha_m t^2/2$, where $\alpha_m = q_m\alpha_1$, $m = 2, 3$. With $q_m \gg 1$, the chirp of the signal and idler pulses can thus considerably exceed the chirp of the pump field.

4.4 Third-Order Nonlinear Processes

Optical nonlinearity of the third order is a universal property, found in any material regardless of its spatial symmetry. This nonlinearity is the lowest-order nonvanishing nonlinearity for a broad class of centrosymmetric materials, where all the even-order nonlinear susceptibilities are identically equal to zero for symmetry reasons. Third-order nonlinear processes include a vast variety of four-wave mixing processes, which are extensively used for frequency conversion of laser radiation and as powerful methods of nonlinear spectroscopy. Frequency-degenerate, Kerr-effect-type phenomena constitute another important class of third-order nonlinear processes. Such effects lie at the heart of optical compressors, mode-locked femtosecond lasers, and numerous photonic devices, where one laser pulse is used to switch, modulate, or gate another laser pulse. In this section, we provide a brief overview of the main third-order nonlinear-optical phenomena and discuss some of their practical applications.

4.4.1 Self-Phase Modulation

The third-order nonlinearity gives rise to an intensity-dependent additive to the refractive index:

$$n = n_0 + n_2 I(t) , \tag{4.56}$$

where n_0 is the refractive index of the medium in the absence of light field, $n_2 = (2\pi/n_0)^2 \chi^{(3)}(\omega; \omega, \omega, -\omega)$ is the nonlinear refractive index, $\chi^{(3)}(\omega; \omega, \omega, -\omega)$ is the third-order nonlinear-optical susceptibility, referred to as the Kerr-type nonlinear susceptibility, and $I(t)$ is the intensity of laser radiation. Then, the nonlinear (intensity-dependent) phase shift of a pulse at a distance L is given by

$$\Phi(t) = \frac{\omega}{c} n_2 I(t) L . \tag{4.57}$$

Due to the time dependence of the radiation intensity within the light pulse, the nonlinear phase shift is also time-dependent, giving rise to a generally time-dependent frequency deviation:

$$\Delta\omega(t) = \frac{\omega}{c} n_2 L \frac{\partial I}{\partial t} . \tag{4.58}$$

The resulting spectral broadening of the pulse can be estimated in the following way:

$$\Delta\omega = \frac{\omega}{c} n_2 L \frac{I_0}{\tau} , \tag{4.59}$$

where I_0 is the peak intensity of the light pulse and τ is the pulse duration.

The first-order dispersion-theory equation for the slowly varying envelope $A(t, z)$ of a laser pulse propagating in a medium with a Kerr-type nonlinearity is written as [4.9]

$$\frac{\partial A}{\partial z} + \frac{1}{u}\frac{\partial A}{\partial t} = \mathrm{i}\tilde{\gamma} |A|^2 A , \tag{4.60}$$

where u is the group velocity of the laser pulse and

$$\tilde{\gamma} = \frac{3\pi\omega}{2n_0^2 c} \chi^{(3)}(\omega; \omega, -\omega, \omega) . \tag{4.61}$$

is the nonlinear coefficient.

In the retarded frame of reference, $z' = z$ and $\eta = t - z/u$, the solution to (4.60) is written as

$$A(\eta, z) = A_0(\eta) \exp\left[\mathrm{i}\tilde{\gamma} |A_0(\eta)|^2 z\right] , \tag{4.62}$$

where $A_0(\eta)$ is the initial field envelope.

Since the group-velocity dispersion was not included in (4.60), the shape of the pulse envelope remains unchanged as the pulse propagates through the nonlinear medium. The intensity-dependent change in the refractive index gives rise to a nonlinear phase shift

$$\varphi_{\mathrm{nl}}(\eta, z) = \gamma_{\mathrm{SPM}} I_0(\eta) z , \tag{4.63}$$

where $\gamma_{\mathrm{SPM}} = 2\pi n_2/\lambda$ and $I_0(\eta)$ is the initial intensity envelope.

The deviation of the instantaneous frequency of the pulse is given by

$$\delta\omega(\eta, z) = -\frac{\partial \varphi_{\mathrm{nl}}(\eta, z)}{\partial t} = -\gamma_{\mathrm{SPM}} \frac{\partial I_0(\eta)}{\partial \eta} z . \tag{4.64}$$

A quadratic approximation of the pulse envelope,

$$I_0(\eta) \approx I_0(0)\left(1 - \frac{\eta^2}{\tau_0^2}\right) , \tag{4.65}$$

where τ_0 is the pulse width, which is valid around the maximum of the laser pulse, gives a linear chirp of the

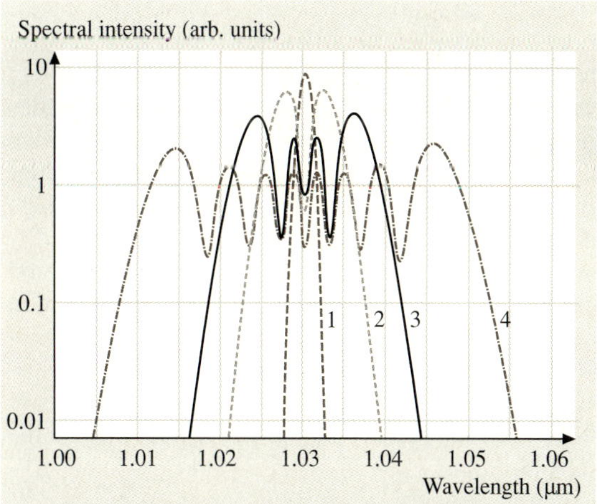

Fig. 4.5 Self-phase-modulation-induced spectral broadening of a laser pulse with a central wavelength of 1.03 μm (an input spectrum is shown by *curve 1*) in a fused-silica optical fiber with $n_2 = 3.2 \times 10^{-16}\,\mathrm{cm^2/W}$: $\gamma_{\mathrm{SPM}} I_0(0) z = 1.25$ (*curve 2*), 2.50 (*curve 3*), and 6.25 (*curve 4*)

pulse

$$\delta\omega(\eta, z) \approx -2\gamma_{\mathrm{SPM}} \frac{I_0(0)}{\tau_0^2} \eta z . \tag{4.66}$$

The spectrum of a self-phase-modulated pulse is given by

$$S(\omega) = \left| \int_0^\infty I(\eta) \exp\left[\mathrm{i}\omega\eta + \mathrm{i}\varphi_{\mathrm{nl}}(\eta)\right] \mathrm{d}\eta \right|^2 . \tag{4.67}$$

Figure 4.5 illustrates SPM-induced spectral broadening of a short laser pulse with a central wavelength of 1.03 μm, typical of ytterbium fiber lasers, in a fused-silica optical fiber with nonlinear refractive index $n_2 = 3.2\times10^{-16}\,\mathrm{cm^2/W}$.

Thus, self-phase modulation results in spectral broadening of a light pulse propagating through a hollow fiber. This effect allows compression of light pulses through the compensation of the phase shift acquired by the pulse in a hollow fiber. Compensation of a linear chirp, corresponding to a linear time dependence of the instantaneous frequency, is straightforward from the technical point of view. Such a chirp arises around the maximum of a light pulse, where the temporal pulse envelope can be approximated with a quadratic function of time [see (4.65, 4.66)].

It is physically instructive to consider the compression of chirped light pulses in the time domain. Since the frequency of a chirped pulse changes from its leading edge to its trailing edge, dispersion of our compressor should be designed in such a way as to slower down the leading edge of the pulse with respect to the trailing edge of the pulse. In other words, the group velocities for the frequencies propagating with the leading edge of the pulse should be lower than the group velocities for the frequencies propagating with the trailing edge of the pulse. This can be achieved by designing a dispersive element with the required sign of dispersion and appropriate dispersion relation. Systems of diffraction gratings and, recently, multilayer chirped mirrors [4.20] are now widely used for the purposes of pulse compression. In certain regimes of pulse propagation, self-phase modulation and pulse compression may take place in the same medium.

4.4.2 Temporal Solitons

The nonlinear phase shift acquired by a laser pulse propagating through a medium with a Kerr nonlinearity can be balanced by group-velocity dispersion, giving rise to pulses propagating through the nonlinear dispersive medium with an invariant or periodically varying shape: optical solitons.

Optical solitons is a special class of solutions to the nonlinear Schrödinger equation (NLSE)

$$\mathrm{i}\frac{\partial q}{\partial \xi} + \frac{1}{2}\frac{\partial^2 q}{\partial \tau^2} + |q|^2 q = 0 . \tag{4.68}$$

The NLSE can describe the evolution of optical wave packets including the dispersion $\beta(\omega)$ of optical waves in a bulk material or in a waveguide structure through the power series expansion

$$\beta(\omega) \approx \beta(\omega_0) + \frac{1}{u}(\omega - \omega_0) + \frac{1}{2}\beta_2(\omega-\omega_0)^2 + \ldots , \tag{4.69}$$

where ω_0 is the central frequency of the wave packet, $u = (\partial\beta/\partial\omega|_{\omega=\omega_0})^{-1}$ is the group velocity, and $\beta_2 = \partial^2\beta/\partial\omega^2|_{\omega=\omega_0}$. Thus, with the NLSE (4.68) projected on laser pulses propagating in a nonlinear medium, q is understood as the normalized pulse envelope, $q = A/(P_0)^{1/2}$, with ξ being the normalized propagation coordinate, $\xi = z/L_\mathrm{d}$, $L_\mathrm{d} = \tau_0^2/|\beta_2|$ being the dispersion length, P_0 and τ_0 defined as the pulse width and the pulse peak power, respectively, and $\tau = (t - z/u)/\tau_0$.

The canonical form of the fundamental soliton solution to (4.68) is [20]

$$q(\xi, \tau) = \operatorname{sech}(\tau) \exp\left(\mathrm{i}\frac{\xi}{2}\right) . \tag{4.70}$$

The radiation peak power required to support such a soliton is given by

$$P_0 = |\beta_2| / (\gamma\tau_0^2) . \tag{4.71}$$

Solitons retain their stable shape as long as their spectrum lies away from the spectrum of dispersive waves that can propagate in the medium. High-order dispersion perturbs solitons, inducing Cherenkov-type wave-matching resonances between solitons and dispersive waves [4.21, 22]. Under these conditions solitons tend to lose a part of their energy in the form of blue-shifted dispersive-wave emission. For low pump-field powers, the generic wave-matching condition for such soliton–dispersive wave resonances is written [4.22] $\Omega = 1/2\varepsilon$, where Ω is the frequency difference between the soliton and the resonant dispersive wave and ε is the parameter controlling the smallness of perturbation of the nonlinear Schrödinger equation, which can be represented as $\varepsilon = |\beta_3/6\beta_2|$ for photonic-crystal fibers

(PCFs) with second-order dispersion $\beta_2 = \partial^2\beta/\partial\omega^2$ and third-order dispersion $\beta_3 = \partial^3\beta/\partial\omega^3$. This dispersive-wave emission of solitons is an important part of supercontinuum generation in nonlinear optical fibers, including photonic-crystal fibers.

4.4.3 Cross-Phase Modulation

Cross-phase modulation (XPM) is a result of nonlinear-optical interaction of at least two physically distinguishable light pulses (i. e., pulses with different frequencies, polarizations, mode structures, etc.) related to the phase modulation of one of the pulses (a probe pulse) due to the change in the refractive index of the medium induced by another pulse (a pump pulse).

The cross-action of a pump pulse with a frequency ω_1 on a probe pulse with a frequency ω_2 gives rise to a phase shift of the probe pulse, which can be written as [4.23].

$$\Phi_{\mathrm{XPM}}(\eta, z) = \frac{3\pi\omega_2^2}{c^2 k_2}\chi^{(3)}(\omega_\mathrm{s}; \omega_\mathrm{s}, \omega_\mathrm{p}, -\omega_\mathrm{p}) \times \int_0^z \left|A_\mathrm{p}\left(\eta - \frac{\zeta}{\sigma}, 0\right)\right|^2 \mathrm{d}\zeta \,, \tag{4.72}$$

where $\chi^{(3)}(\omega_\mathrm{s}; \omega_\mathrm{s}, \omega_\mathrm{p}, -\omega_\mathrm{p})$ is the third-order nonlinear-optical susceptibility of the medium; $1/\sigma = 11/u_1 - 1/u_2$; u_1 and u_2 are the group velocities of the pump and probe pulses, respectively; and k_2 is the wave number of the pump pulse. Taking the time derivative of the nonlinear phase shift, we arrive at the following expression for the frequency deviation of the probe pulse

$$\delta\omega_{\mathrm{XPM}}(\eta, z) = -\frac{3\pi\omega_2^2}{c^2 k_2}\chi^{(3)}(\omega_\mathrm{s}; \omega_\mathrm{s}, \omega_\mathrm{p}, -\omega_\mathrm{p}) \times\sigma\left[|A_\mathrm{p}(\eta, 0)|^2 - \left|A_\mathrm{p}\left(\eta - \frac{z}{\sigma}, 0\right)\right|^2\right] . \tag{4.73}$$

Similarly to self-phase modulation, cross-phase modulation can be employed for pulse compression. The dependence of the chirp of the probe pulse on the pump pulse intensity can be used to control the parameters of ultrashort pulses [4.24]. Cross-phase modulation also opens the ways to study the dynamics of ultrafast nonlinear processes, including multiphoton ionization, and to characterize ultrashort light pulses through phase measurements on a short probe pulse [4.25].

4.4.4 Self-Focusing

Self-focusing is a spatial counterpart of self-phase modulation. While SPM originates from the time-dependent change in the refractive index induced by a laser pulse with an intensity envelope $I(t)$ varying in time, self-focusing is related to a nonlinear lens induced by a laser beam with a spatially nonuniform intensity distribution $I(r)$. Given a transverse intensity profile $I(r)$, the nonlinear additive to the refractive index is written as

$$n(r) = n_0 + n_2 I(r) \,. \tag{4.74}$$

If the field intensity peaks at the center of the beam at $r = 0$, the nonlinear change in the refractive index also reaches its maximum at $r = 0$, yielding a focusing or defocusing lens, depending on the sign of n_2.

The stationary regime of self-focusing is governed by [4.9]

$$2\mathrm{i}k\frac{\partial A}{\partial z} + \Delta_\perp A = -2k^2\frac{\Delta n}{n_0}A \,, \tag{4.75}$$

where $\Delta n = n_2 I = \tilde{n}_2|E|^2$, $\Delta_\perp$ is the transverse part of the Laplacian.

We consider a Gaussian beam and assume that this beam retains its profile as it propagates through the nonlinear medium,

$$A(r, z) = \frac{A_0}{f(z)}\exp\left[-\frac{r^2}{2a_0^2 f^2(z)} + \mathrm{i}\psi(z)\right] , \tag{4.76}$$

where a_0 is the initial beam size, $f(z)$ characterizes the evolution of the beam size along the propagation coordinate z [$f(0) = 1$], and the function $\psi(z)$ describes the spatial phase modulation of the field.

In the paraxial approximation, $r \ll a_0 f(z)$, (4.75) and (4.76) give [4.18]

$$\frac{\mathrm{d}^2 f}{\mathrm{d}z^2} = \frac{\left(L_\mathrm{df}^{-2} - L_\mathrm{nl}^{-2}\right)}{f^3(z)} \,, \tag{4.77}$$

where $L_\mathrm{df} = 2\pi a_0^2/\lambda$ and $L_\mathrm{nl} = a_0[2n_0/(n_2|A|^2)]^{1/2}$ are the characteristic diffraction and nonlinear lengths, respectively.

Solving (4.77), we arrive at

$$f^2(z) = 1 + \left(\frac{z}{L_\mathrm{df}}\right)^2\left(1 - \frac{P_0}{P_\mathrm{cr}}\right) , \tag{4.78}$$

where P_0 is the total power of the laser beam and

$$P_\mathrm{cr} = \frac{c\lambda^2}{16\pi^2 n_2} \tag{4.79}$$

is the critical power of self-focusing. The focal length of the nonlinear lens is given by

$$L_\mathrm{sf} = \frac{L_\mathrm{df}}{\left(\frac{P_0}{P_\mathrm{cr}} - 1\right)^{1/2}} \,. \tag{4.80}$$

With $P_0 > P_{cr}$, the nonlinear lens leads to a beam collapse. In reality, beam collapse can be arrested by the saturation of optical nonlinearity occurring at high field intensities.

Beyond the paraxial approximation, the scenario of self-focusing is much more complicated. The beam does not collapse as a whole, as the focal length of the nonlinear lens for peripheral beams differs from the one for paraxial beams. In the quasistationary regime, i. e., when the pulse duration τ_0 is much larger than the characteristic response time of optical nonlinearity τ_{nl}, the length of self-focusing is a function of time, giving rise to moving foci [4.26]. In the nonstationary regime, i. e., on time scales less than τ_{nl}, the leading edge of the pulse experiences no focusing, but induces a nonlinear lens that focuses the trailing edge of the pulse. As a result, the beam becomes distorted, evolving to a hornlike pattern [4.27].

The equation of self-focusing (4.75) allows a waveguide solution [4.28], which corresponds to the regime where the nonlinear lens exactly compensates for the diffraction of the laser beam. This solution is, however, unstable with respect to infinitely small fluctuations, which either give rise to a diffraction divergence or lead to a beam collapse. Such nonlinear waveguides can be stabilized, as shown by *Fibich* and *Gaeta* [4.29], by reflections of light from guiding boundaries in optical waveguides. To illustrate this regime of nonlinear beam dynamics, we consider a cylindrical gas-filled hollow waveguide and define nondimensional cylindrical coordinates r and z as $r = R/R_0$ (R is the dimensional radial coordinate and R_0 is the inner radius of the hollow fiber) and $z = Z/L_{df}$ (Z is the dimensional longitudinal coordinate). The radial profiles $Q_\beta(r)$ of light intensity distribution in the waveguide solutions $\psi \propto \exp(i\beta z)Q_\beta(r)$ to the NLSE governing self-focusing (i. e., waveguides induced through the Kerr nonlinearity along the z-axis on a bounded domain with a circular symmetry) are described [28, 29] by solutions to the ordinary differential equation $\Delta_\perp Q_\beta - \beta Q_\beta + Q_\beta^3 = 0$ $[\Delta_\perp = \partial^2/\partial r^2 + (1/r)\partial/\partial r]$, subject to the boundary conditions $\mathrm{d}Q_\beta(0)/\mathrm{d}r = 0$, $Q_\beta(1) = 0$. Although this model neglects the fields outside the fiber core (e.g., radiation modes), it provides useful physical insights into the spatial self-action in hollow PCFs. This differential equation has an infinite number of solutions $Q_\beta^{(n)}$, $n = 0, 1, 2, \ldots$. The Hamiltonians for all the guided modes $Q_\beta^{(n)}$ are positive, preventing the blowup of the profiles $Q_\beta^{(n)}$ in the presence of small fluctuations, thus stabilizing the nonlinear waveguides in a hollow fiber. The ground-state solution $Q_\beta^{(0)}$ is a monotonically decreasing function of r, tending to a zeroth-order Bessel function in the case of low field amplitudes $Q_\beta^{(0)} \propto \varepsilon J_0(2.4r)$, where ε is a small parameter controlled by the field intensity and the nonlinearity of the gas filling the fiber.

Although the modes $Q_\beta^{(n)}$ are stable with respect to small perturbations, these solutions are centers, rather than attractors, in a conservative system [4.30]. However, mode solutions corresponding to Kerr-nonlinearity-induced waveguides may become attractors in systems with dissipation, e.g., in hollow waveguides

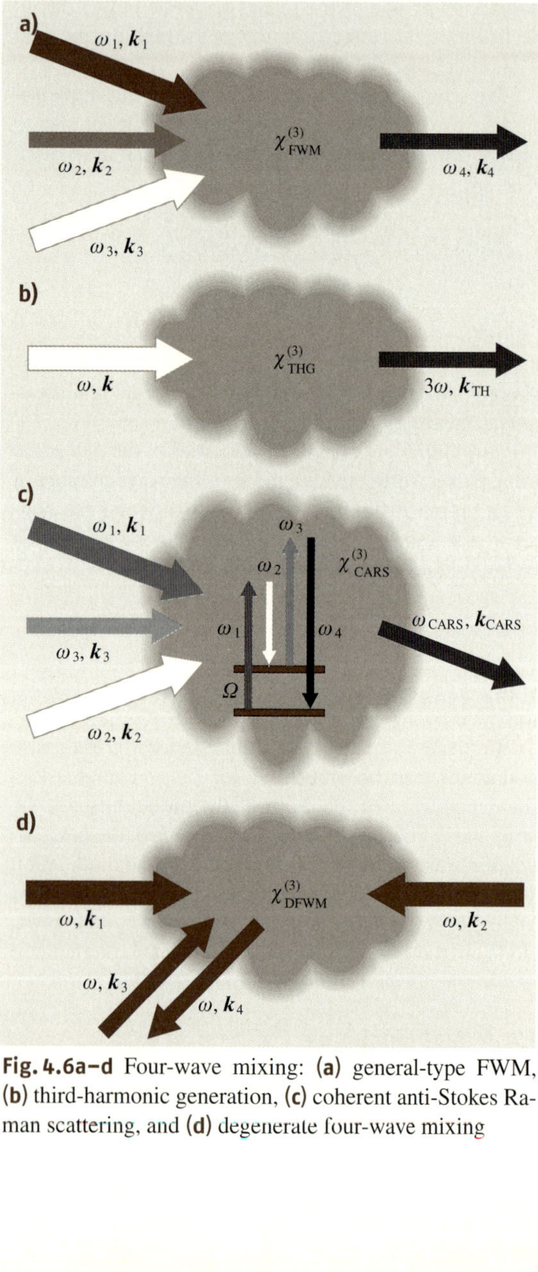

Fig. 4.6a–d Four-wave mixing: **(a)** general-type FWM, **(b)** third-harmonic generation, **(c)** coherent anti-Stokes Raman scattering, and **(d)** degenerate four-wave mixing

with losses. The circularly symmetric field distribution is then formed at the output of the hollow waveguide regardless of the initial beam profile [30]. As demonstrated by *Moll* et al. [4.31], collapsing light beams tend to form universal, Townesian profiles [4.28] while undergoing self-focusing on an infinite domain in bulk materials. Unlike Townesian beam profiles, which are known to be unstable in free space, the ground-state waveguide modes observed in hollow photonic-crystal fibers [4.32] remain stable with respect to small fluctuations, in agreement with the theory of self-focusing on bounded domains [4.29, 30], resulting in no blowup until the critical power of beam collapse is reached.

4.4.5 Four-Wave Mixing

In general-type four-wave mixing (Fig. 4.6a), three laser fields with frequencies ω_1, ω_2, and ω_3 generate the fourth field with a frequency $\omega_{\text{FWM}} = \omega_1 \pm \omega_2 \pm \omega_3$. In the case when all three laser fields have the same frequency (for example, when all the three pump photons are taken from the same laser field), $\omega_{\text{FWM}} = 3\omega$ (Fig. 4.6b), and we deal with third-harmonic generation (THG), which is considered in greater detail for short-pulse interactions in Sect. 4.4.9. If the frequency difference $\omega_1 - \omega_2$ of two of the laser fields is tuned to a resonance with a Raman-active mode of the nonlinear medium (Fig. 4.6c), the FWM process $\omega_{\text{FWM}} = \omega_1 - \omega_2 + \omega_3 = \omega_{\text{CARS}}$ is refereed to as coherent anti-Stokes Raman scattering (CARS). An FWM process involving four fields of the same frequency (Fig. 4.6d) with $\omega_{\text{DFWM}} = \omega = \omega - \omega + \omega$ corresponds to degenerate four-wave mixing (DFWM). The FWM field in this nonlinear-optical process is the phase-conjugate of one of the laser fields, giving rise to another name of this type of FWM – optical phase conjugation.

For the general-type FWM $\omega_{\text{FWM}} = \omega_1 + \omega_2 + \omega_3$, we represent the pump fields as

$$\boldsymbol{E}_j = \boldsymbol{A}_j \exp\left[\mathrm{i}\left(\boldsymbol{k}_j \mathrm{r} - \omega_j t\right)\right] + \text{c.c.}\,, \tag{4.81}$$

where $j = 1, 2, 3$ and $\boldsymbol{k}_j = \boldsymbol{k}'_j + \mathrm{i}\alpha_j$ are the complex wave vectors of the pump fields.

The FWM signal is written as

$$\boldsymbol{E}_{\text{FWM}} = \boldsymbol{A}_{\text{FWM}} \exp\left[\mathrm{i}\left(\boldsymbol{k}_{\text{FWM}} \mathrm{r} - \omega_{\text{FWM}} t\right)\right] + \text{c.c.} \tag{4.82}$$

where $\boldsymbol{k}_{\text{FWM}} = \boldsymbol{k}'_{\text{FWM}} + \mathrm{i}\alpha_{\text{FWM}}$ is the complex wave vector of the FWM fields.

The third-order nonlinear polarization responsible for the considered FWM process is

$$\boldsymbol{P}^{(3)}_{\text{FWM}} = \chi^{(3)}\left(\omega_{\text{FWM}}; \omega_1, \omega_2, \omega_3\right) \vdots \boldsymbol{E}(\omega_1)\, \boldsymbol{E}(\omega_2)\, \boldsymbol{E}(\omega_3)\,. \tag{4.83}$$

With no depletion of the pump fields, the SVEA equations give the following expression for the envelope of the i-th Cartesian component of the FWM field [4.9]:

$$\begin{aligned}[A_{\text{FWM}}(z)]_i = &-\frac{2\pi\omega^2_{\text{FWM}}}{k_{\text{FWM}} c^2} \chi^{(3)}\left(\omega_{\text{FWM}}; \omega_1, \omega_2, \omega_3\right) \\ &\times A_{1j} A_{2k} A_{3l} \exp\left(\frac{\mathrm{i}\Delta k z}{2}\right) \\ &\times \exp\left(-\alpha_{\text{FWM}} z\right) \frac{\sin\left(\frac{\Delta k z}{2}\right)}{\frac{\Delta k z}{2}} z\,,\end{aligned} \tag{4.84}$$

where

$$\begin{aligned}\Delta k = &\, k'_1 + k'_2 + k'_3 - k'_{\text{FWM}} \\ &+ \mathrm{i}\left(\alpha_1 + \alpha_2 + \alpha_3 - \alpha_{\text{FWM}}\right)\end{aligned} \tag{4.85}$$

is the z-component of the wave-vector mismatch.

Phase matching, as can be seen from (4.84), is the key requirement for high efficiency of frequency conversion in FWM. The fields involved in FWM can be phase matched by choosing appropriate angles between the wave vectors of the laser fields and the wave vector of the nonlinear signal or using the waveguide regime with the phase mismatch related to the material dispersion compensated by the waveguide dispersion component.

4.4.6 Optical Phase Conjugation

Optical phase conjugation is generally understood as the generation of an optical field with a time-reversed wave front, or with a conjugate phase. This effect can be used to correct aberrations in certain types of optical problems and systems [4.33]. Suppose that a light beam with an initially plane wave front propagates through an aberrating medium, such as, for example a turbulent atmosphere or a material with inhomogeneities of the refractive index. The wave front of the light beam transmitted through such a medium is distorted. We now use an optical phase-conjugate process to generate the field with a wave front time-reversed with respect to the wave front of the beam transmitted through the aberrating system. As the phase-conjugate beam now propagates through the aberrating medium in the backward direction, its wave front becomes restored.

A phase-conjugate field can be produced through the degenerate four-wave mixing of light fields

$$E_j(\boldsymbol{r},t)=A_j(\boldsymbol{r},t)\exp\left[\mathrm{i}\left(\boldsymbol{k}_j\boldsymbol{r}-\omega t\right)\right]+\text{c.c.}\,, \tag{4.86}$$

where $j=1,2,3,4$.

The phase-conjugate geometry of DFWM is shown in Fig. 6d. In this scheme, two strong counterpropagating pump fields $\boldsymbol{E}_1$ and $\boldsymbol{E}_2$ with the same frequency ω and wave vectors $\boldsymbol{k}_1$ and $\boldsymbol{k}_2=-\boldsymbol{k}_1$ illuminate a medium with a cubic nonlinearity $\chi^{(3)}_{\text{DFWM}}=\chi^{(3)}(\omega;\omega,-\omega,\omega)$. The DFWM interaction of these two pump fields with a weak signal of the same frequency ω and an arbitrary wave vector $\boldsymbol{k}_3$ gives rise to a field with a frequency ω that propagates in the opposite direction to the signal beam and that is phase-conjugate of the signal field. The phase-conjugate field generated through DFWM can be instructively thought of as a result of scattering of the forward pump field off the grating induced by the backward pump and the signal field or as a result of scattering of the backward pump from the grating induced by the forward pump and the signal field.

The nonlinear polarization responsible for phase conjugation in DFWM is

$$\boldsymbol{P}^{(3)}_{\text{DFWM}}=6\chi^{(3)}_{\text{DFWM}}\vdots\boldsymbol{E}_1\boldsymbol{E}_2\boldsymbol{E}_3^*\,. \tag{4.87}$$

If the depletion of the pump fields is negligible, the nonlinear propagation equation (4.24) for the considered DFWM process is reduced to the following two equations for the amplitude of the signal field and its phase-conjugate [4.1]

$$\frac{\mathrm{d}A_3}{\mathrm{d}z}=\mathrm{i}\kappa_3A_3+\mathrm{i}\kappa_4A_4^*, \tag{4.88}$$

$$\frac{\mathrm{d}A_4}{\mathrm{d}z}=-\mathrm{i}\kappa_3A_4-\mathrm{i}\kappa_4A_3^*, \tag{4.89}$$

where

$$\kappa_3=\frac{12\pi\omega}{cn}\chi^{(3)}_{\text{DFWM}}\left(|A_1|^2+|A_2|^2\right)\,, \tag{4.90}$$

$$\kappa_4=\frac{12\pi\omega}{cn}\chi^{(3)}_{\text{DFWM}}A_1A_2\,. \tag{4.91}$$

By introducing

$$A_3=B_3\exp\left(\mathrm{i}\kappa_3z\right)\,, \tag{4.92}$$

$$A_4=B_4\exp\left(-\mathrm{i}\kappa_3z\right)\,, \tag{4.93}$$

we reduce the set of equations (4.88) and (4.89) to

$$\frac{\mathrm{d}B_3}{\mathrm{d}z}=\mathrm{i}\kappa_4B_4^*\,, \tag{4.94}$$

$$\frac{\mathrm{d}B_4}{\mathrm{d}z}=-\mathrm{i}\kappa_4B_3^*\,. \tag{4.95}$$

The solution can now be written as

$$B_3^*(z)=-\frac{\mathrm{i}\,|\kappa_4|}{\kappa_4}\frac{\sin\left(|\kappa_4|\,z\right)}{\cos\left(|\kappa_4|\,L\right)}B_4(L)+\frac{\cos\left[|\kappa_4|\,(z-L)\right]}{\cos\left(|\kappa_4|\,L\right)}B_3^*(0)\,, \tag{4.96}$$

$$B_4(z)=\frac{\cos\left(|\kappa_4|\,z\right)}{\cos\left(|\kappa_4|\,L\right)}B_4(L)-\frac{\mathrm{i}\kappa_4}{|\kappa_4|}\frac{\sin\left[|\kappa_4|\,(z-L)\right]}{\cos\left(|\kappa_4|\,L\right)}B_3^*(0)\,, \tag{4.97}$$

where $B_3(0)$ and $B_4(L)$ are the boundary conditions for the signal and DFWM fields. With $B_4(L)=0$, (4.97) gives

$$B_4(0)=\frac{\mathrm{i}\kappa_4}{|\kappa_4|}\tan\left(|\kappa_4|\,L\right)B_3^*(0)\,. \tag{4.98}$$

This expression visualizes the structure of the phase-conjugate field generated through DFWM. The intensity reflection coefficient of the DFWM medium serving as a phase-conjugate mirror is given by

$$R_{\text{DFWM}}=\tan^2\left(|\kappa_4|\,L\right)\,. \tag{4.99}$$

As can be seen from (4.99), the reflectivity of the DFWM-based phase-conjugate mirror can exceed 100%. This becomes possible due to the energy supplied by the strong pump fields.

4.4.7 Optical Bistability and Switching

In optical bistability or multistability, an optical system has two or more than two stable states, most often represented by the intensity at the output of the system as a function of the input intensity. With the level of the output intensity determined by a certain operation on a light beam, a bi- or multistable system makes a decision in which state it will operate, acting as a switch for optical communications or optical data processing.

As an example of an optical bistable system, we consider a Fabry–Perot cavity with a Kerr-nonlinear medium inside. We assume that the cavity mirrors are identical and have an amplitude reflectance r and transmittance t. Let A_1, A_2 and A_3 be the amplitudes of the incident, reflected, and transmitted fields, respectively. The amplitudes of forward and backward waves inside the cavity will be denoted as B_1 and B_2, respectively. The amplitudes of the fields inside and outside the cavity are related by the expressions [4.1]

$$B_2=rB_1\exp\left(2\mathrm{i}kL-\alpha L\right)\,, \tag{4.100}$$

$$B_1=tA_1+rB_2\,, \tag{4.101}$$

where k is the wave number and α is the intensity absorption coefficient.

Solving (4.100) and (4.101), we arrive at the relation

$$B_1 = \frac{tA_1}{1 - r^2 \exp(2ikL - \alpha L)} , \tag{4.102}$$

which is known as the Airy equation. When k or α is a strongly nonlinear function of the light intensity, (4.102) leads to a bistable behavior of the system in the transmitted field.

Let us assume that the absorption is negligible and rewrite (4.102) as

$$B_1 = \frac{tA_1}{1 - R\exp(i\delta)} , \tag{4.103}$$

where $R = \rho^2 \exp(-i\varphi)$ is the intensity reflectance of the cavity mirrors and the phase shift δ, corresponding to a full round trip around the cavity, is given by

$$\delta = \delta_0 + \delta_{nl} , \tag{4.104}$$

where

$$\delta_0 = \varphi = 2n_0 \frac{\omega}{c} L \tag{4.105}$$

is the linear phase shift and

$$\delta_{nl} = 2n_2 I \frac{\omega}{c} L \tag{4.106}$$

is the nonlinear phase shift, with $I = [cn/(2\pi)](|B_1|^2 + |B_2|^2) \approx [cn/(\pi)]|B_1|^2$

The ratio of the field intensity inside the cavity $I_2 = [cn/(2\pi)]|B_1|^2$, to the intensity of the incident field $I_1 = [cn/(2\pi)]|A_1|^2$ is now given by

$$\frac{I_2}{I_1} = F(I_2) , \tag{4.107}$$

where

$$F(I_2) = \frac{1}{T + \frac{4R}{T}\sin^2\left(\frac{\delta}{2}\right)} , \tag{4.108}$$

where T is the intensity transmittance of the cavity mirrors, and

$$\delta = \delta_0 + \frac{4n_2\omega L}{c} I_2 . \tag{4.109}$$

In Fig. 4.7, we present the function $F(I_2)$ and the ratio I_2/I_1, i. e., the left- and right-hand sides of (4.107), plotted as functions of the intensity I_2 for different input field intensities I_1. The points where the straight lines representing the left-hand side of (4.107) cross the plot of the Airy-type function $F(I_2)$ define the operation points of the nonlinear Fabry–Perot cavity. The circles in the Airy-function curve show the range of input intensities where the operation is unstable. As the input intensity I_1 is increased, the system displays a bistable behavior and hysteresis. More hysteresis loops and a multistable behavior are observed, as can be seen from Fig. 4.7, with a further increase in I_1.

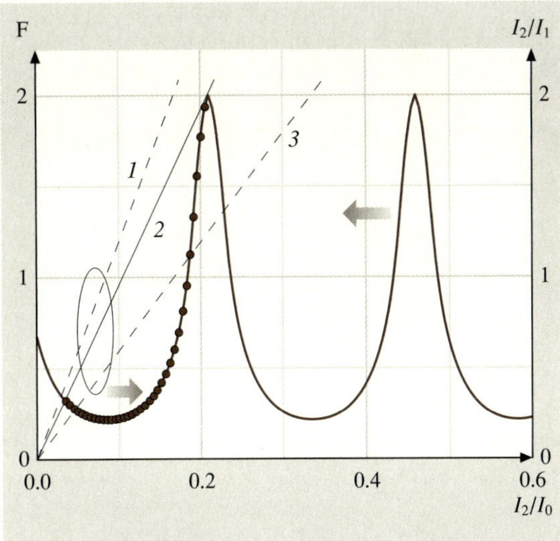

Fig. 4.7 The function $F(I_2)$ and the ratio I_2/I_1, appearing on the right- and left-hand sides of (4.107), plotted as functions of the intensity I_2 normalized to $I_0 = \lambda/(n_2 L)$ for different input field intensities I_1: (1) $I_1 = I'$, (2) $I_1 = I''$, and (3) $I_1 = I'''$, $I''' > I'' > I'$. The *circles* in the Airy-function curve show the range of input intensities where the operation of the Fabry–Perot cavity is unstable

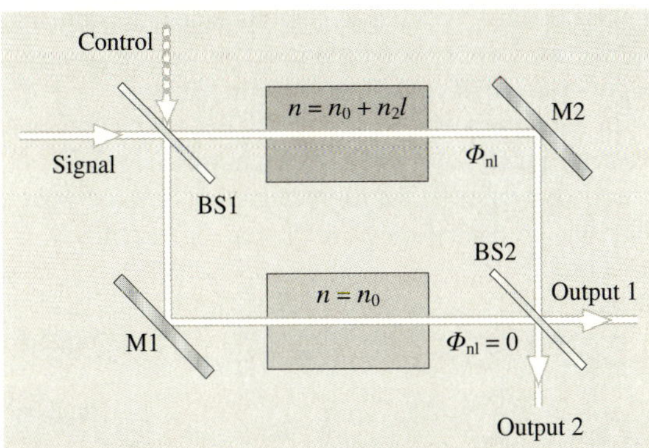

Fig. 4.8 Mach–Zehnder interferometer with a nonlinear medium in one of its arms, functioning as an all-optical switching device: BS1, BS2, beam splitters; M1, M2, mirrors

Optical switching can now be implemented [4.1] by including a nonlinear medium into one of the arms of a Mach–Zehnder interferometer Fig. 4.8. The intensities at the output ports 1 and 2 are determined by the interference of light fields transmitted through the arms of the interferometer. If there is only one input beam applied to the system with symmetric beam splitters BS1 and BS2, output intensities display an oscillating behavior as functions of the nonlinear phase shift Φ_{nl} acquired by the light field in one of the arms of the interferometer. When $\Phi_{nl} = 0$, the intensity at output port 1 is at its maximum, while the intensity at output port 2 is at its minimum. An opposite relation between the output intensities is achieved with $\Phi_{nl} = \pi$. The requirement $\Phi_{nl} = \pi$ is typical of a broad class of all-optical switching devices.

4.4.8 Stimulated Raman Scattering

Vibrations or rotations of molecules, electronic motions in atoms or collective excitations of matter can interact with light, shifting its frequency through an inelastic scattering process by the frequency Ω of Raman-active motions (as shown in Fig. 4.6c). This phenomenon was discovered by *Raman* and *Krishnan* [4.34] and almost simultaneously by *Mandelstam* and *Landsberg* [4.35] in 1928. In an intense laser field, pump laser photons and frequency-shifted photons act coherently to resonantly drive molecular motions, leading to the amplification of the Raman-shifted signal. This effect is called stimulated Raman scattering (SRS). In the SRS regime, Raman-active modes of a material function as optical modulators, forcing the driving laser field to oscillate at new frequencies. An intense laser field under these conditions not only gives rise to photons at new frequencies through interaction with Raman-active modes, but also amplifies the light consisting of those photons.

In the continuous-wave regime, the interaction between the pump field and the frequency-shifted (Stokes) signal is governed by the following set of coupled equations for the intensities of the pump I_p and Stokes I_s fields [4.23]:

$$\frac{dI_s}{dz} = g_R I_p I_s - \alpha_s I_s \,, \tag{4.110}$$

$$\frac{dI_p}{dz} = -\frac{\omega_p}{\omega_s} g_R I_p I_s - \alpha_p I_p \,. \tag{4.111}$$

Here, α_p and α_s are the losses at the frequencies of the pump and Stokes fields, ω_p and ω_s are the frequencies of the pump and Stokes fields, and g_R is the Raman gain, which is related to the imaginary part of the relevant third-order susceptibility [4.9],

$$g_R \propto \mathrm{Im}\left[\chi^{(3)}\left(\omega_s; \omega_p, -\omega_p, \omega_s\right)\right] . \tag{4.112}$$

Neglecting the action of the Stokes field on the pump, i. e., omitting the first term on the right-hand side of (4.111), we arrive at the following solution for the intensity of the Stokes field at the output of a Raman-active medium with a length L:

$$I_s(L) = I_s(0) \exp\left(g_R I_0 L_{eff} - \alpha_s L\right) , \tag{4.113}$$

where I_0 is the incident pump intensity at $z = 0$ and

$$L_{eff} = \frac{1}{\alpha_p}\left[1 - \exp\left(-\alpha_p L\right)\right] . \tag{4.114}$$

Thus, pump absorption restricts the nonlinear interaction length to L_{eff}.

In the absence of the Stokes field at the input of a Raman-active medium, $I_s(0) = 0$, the Stokes field builds up from spontaneous Raman scattering inside the medium. The power of the Stokes signal at the output of a medium with a length L is then given by

$$P_s(L) = \bar{P}_0 \exp\left[g_R(\omega_s) I_0 L_{eff} - \alpha_s L\right] , \tag{4.115}$$

where

$$\bar{P}_0 = \hbar\omega_s B_{eff}$$

$$B_{eff} = (2\pi)^{1/2}\left[|g_2(\omega_s)| I_0 L_{eff}\right]^{-1/2} ,$$

and

$$g_2(\omega_s) = (\partial^2 g_R/\partial\omega^2)|_{\omega=\omega_s} .$$

With an assumption of a Lorentzian Raman gain band, the critical pump power corresponding to the threshold of SRS effect is given by an approximate formula

$$P_{cr} \approx \frac{16 S_{eff}}{g_R L_{eff}} , \tag{4.116}$$

where S_{eff} is the effective mode area of the pump field.

In Sect. 4.6.3, we will consider in greater detail the influence of SRS on the propagation of ultrashort laser pulses in a nonlinear medium. It will be shown, in particular, that the Raman effect in fibers gives rise to a continuous frequency downshift of solitons propagating in optical fibers.

4.4.9 Third-Harmonic Generation by Ultrashort Laser Pulses

Third-harmonic generation (THG) is one of the basic nonlinear-optical processes, which has been intensely studied and employed for numerous applications since the early days of nonlinear optics [4.7–9]. The seminal work by *Miles* and *Harris* [4.17] has demonstrated a tremendous potential of direct THG related to the cubic optical nonlinearity $\chi^{(3)}$ of gases for efficient frequency conversion of laser radiation and for the diagnostics of the gas phase. Solid-state strategies of frequency conversion, on the other hand, mainly rely on the quadratic nonlinearity $\chi^{(2)}$ of non-centrosymmetric crystals, with frequency tripling conventionally implemented through cascaded second-order nonlinear-optical processes, phase matched by crystal anisotropy [4.13, 14] or periodic poling of nonlinear materials [4.16].

Photonic-crystal fibers (PCFs) [4.36, 37] – optical fibers of a new type where the solid continuous cladding of standard fibers is replaced by a microstructure with an array of air holes running along the fiber parallel to its core Fig. 4.9 – have opened a new phase in nonlinear optics [4.38, 39]. Controlled dispersion of guided modes [4.40] and large interaction lengths provided by these fibers for light fields strongly confined in a small fiber core [4.41] result in a radical enhancement of nonlinear-optical frequency conversion and spectral transformation of laser radiation through self- and cross-phase modulation (SPM and XPM) [4.38], supercontinuum generation [4.42–44], four-wave mixing (FWM) [4.45], third-harmonic generation [4.42–52], modulation instabilities [4.53], and soliton frequency shifting [4.54]. Third-order nonlinear-optical processes enhanced in PCFs now offer a useful alternative to frequency-conversion schemes using $\chi^{(2)}$ nonlinear crystals.

Highly efficient THG has recently been observed in fused silica [4.46–50] and multicomponent glass PCFs [4.51], as well as in tapered fibers [4.52]. These

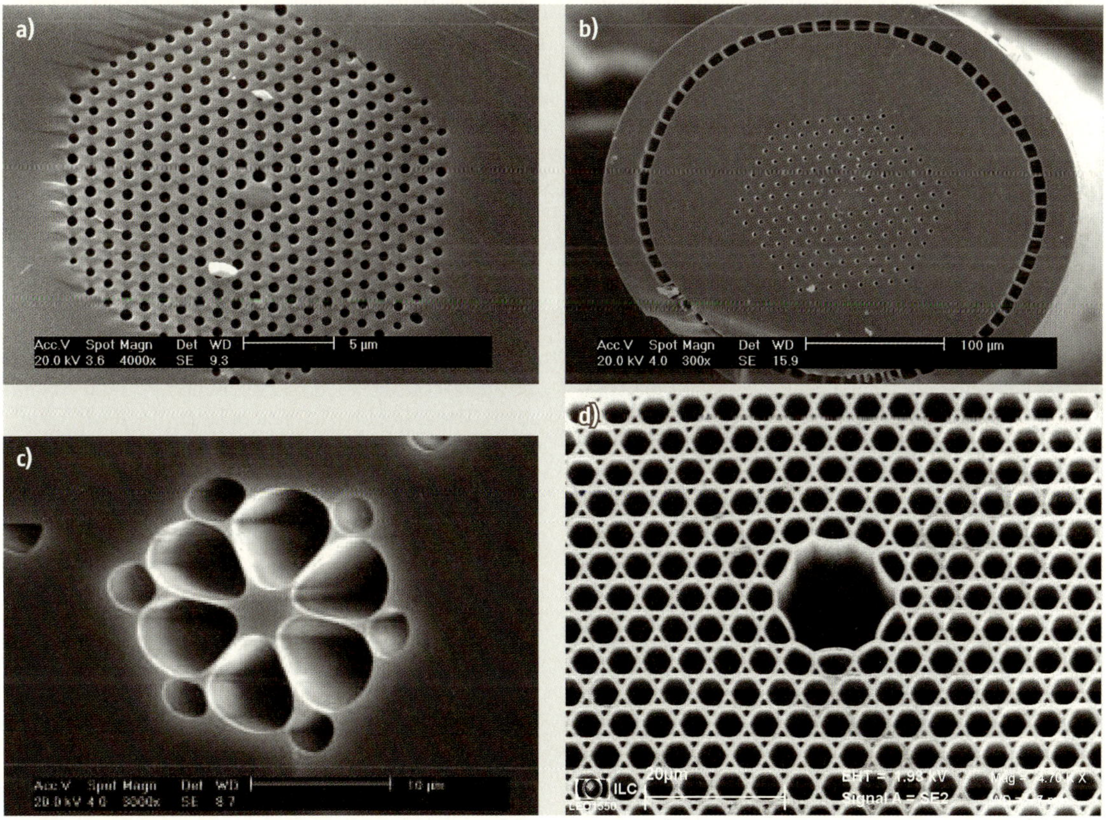

Fig. 4.9a–d SEM images of photonic-crystal fibers: **(a)** periodic- and **(b)** double-cladding fused-silica PCFs, **(c)** high-index-step birefringent PCF, and **(d)** hollow-core PCF

experiments not only demonstrated the significance of THG for efficient, guided-wave frequency tripling of femtosecond laser pulses, but also revealed several new interesting nonlinear-optical phenomena. The third-harmonic signal has been shown to display asymmetric spectral broadening [4.51, 52] or even a substantial frequency shift. We will demonstrate here that such a behavior is a universal intrinsic feature of multimode guided-wave THG. Based on the arguments of the slowly varying envelope approximation (SVEA), we will show that the sign and the absolute value of the third-harmonic frequency shift, observed in many PCF experiments, is controlled by the phase- and group-index mismatch for the interacting pair of pump and third-harmonic modes. The possibility to tune the frequency of the main spectral peak in the spectrum of the third harmonic by varying the group-velocity mismatch is a unique property of THG-type processes, which is not typical of standard parametric FWM processes, where the first-order dispersion terms cancel out of the balance of the field momenta. New regimes of THG will be identified with no signal produced at the central frequency of the third harmonic $3\omega_0$ and with the pump energy efficiently converted to spectrally isolated narrowband frequency components, which can be tuned within a spectral range of several tens of terahertz from the $3\omega_0$ frequency.

We start with qualitative arguments illustrating phase matching for third-harmonic generation generalized to include the phase and group-velocity mismatch of the pump and third-harmonic fields, as well as the Kerr effect, broadening the pump spectrum through SPM. We represent the wave numbers (or the propagation constants of guided modes in a waveguide regime) k_p and k_h at the frequencies of the pump field and the third harmonic as

$$k_\mathrm{p}(\omega) \approx k(\omega_0) + v_\mathrm{p}^{-1}\Omega/3 + \kappa_\mathrm{SPM}P \,, \qquad (4.117)$$

$$k_\mathrm{h}(3\omega) \approx k(3\omega_0) + v_\mathrm{h}^{-1}\Omega + 2\kappa_\mathrm{XPM}P \,, \qquad (4.118)$$

where ω_0 is the central frequency of the pump field; $v_{\mathrm{p,h}} = (\partial k/\partial\omega)^{-1}_{\omega_0,3\omega_0}$ are the group velocities of the pump and its third harmonic; $\Omega = 3\omega - 3\omega_0$; $\kappa_\mathrm{SPM} = \omega_0 n_2/cS_\mathrm{eff}$ and $\kappa_\mathrm{XPM} = 3\omega_0\bar{n}_2/cS_\mathrm{eff}$ are the SPM and XPM nonlinear coefficients (S_eff is the effective beam, or mode, area and n_2 and $\bar{n}_2$ are the nonlinear refractive indices at ω_0 and $3\omega_0$, respectively); and P is the power of the pump field. In writing (4.117) and (4.118), we neglect group-velocity dispersion and higher-order dispersion effects, as well as the SPM of the third-harmonic field. With $n_2 \approx \bar{n}_2$, the phase mismatch is then given by

$$\Delta k = k_\mathrm{h} - 3k_\mathrm{p} \approx \Delta k_0 + \xi\Omega + 3\kappa_\mathrm{SPM}P \,, \qquad (4.119)$$

where $\Delta k_0 = k(3\omega_0) - 3k(\omega_0)$ is the phase mismatch of the pump and third-harmonic wave numbers at the central frequencies of these fields and $\xi = v_\mathrm{h}^{-1} - v_\mathrm{p}^{-1}$ is the group-velocity mismatch.

As can be seen from (4.119), the group delay of the pump and third-harmonic pulses is an important factor in THG momentum conservation. In this respect, the balance of momenta for THG radically differs from standard phase-matching conditions for parametric FWM processes [4.23], where the first-order dispersion order terms cancel out, reducing the FWM momentum balance to group-velocity dispersion (GVD)-related issues.

The phase-matching condition (4.119) suggests the possibility of substantially frequency shifting the maximum in the spectrum of the third harmonic. However, the amplitude of an Ω-shifted spectral component in the spectrum of the third harmonic and, hence, the efficiency of Ω-shifted peak generation is determined by the spectrum of the pump field. To specify this dependence, we proceed with an SVEA analysis of THG in the field of SPM-broadened pump field by writing SVEA coupled equations for the envelopes of the pump and third-harmonic fields, $A(t, z)$ and $B(t, z)$:

$$\left(\frac{\partial}{\partial t} + \frac{1}{v_\mathrm{p}}\frac{\partial}{\partial z}\right) A = \mathrm{i}\gamma_1 A\,|A|^2 \,, \qquad (4.120)$$

$$\left(\frac{\partial}{\partial t} + \frac{1}{v_\mathrm{h}}\frac{\partial}{\partial z}\right) B = \mathrm{i}\beta\,(A)^3 \exp(-\mathrm{i}\Delta k_0 z) + 2\mathrm{i}\gamma_2 B\,|A|^2 \,, \qquad (4.121)$$

where v_p and v_h are the group velocities of the pump and third-harmonic pulses, respectively, and γ_1, γ_2 and β are the nonlinear coefficients responsible for SPM, XPM, and THG, respectively; and $\Delta k = k_\mathrm{h} - 3k_\mathrm{p}$ is the phase mismatch (or the difference of propagation constants in the guided-wave regime) in the absence of the nonlinear phase shifts of the pump and third-harmonic fields.

Solution of (4.120) and (4.121) yields [4.24, 55]

$$A\left(t_\mathrm{p}, z\right) = A_0\left(t_\mathrm{p}\right) \exp\left[\mathrm{i}\varphi_\mathrm{SPM}\left(t_\mathrm{p}, z\right)\right] \,, \qquad (4.122)$$

$$B\left(t_\mathrm{h}, z\right) = \mathrm{i}\sigma \int_0^z \mathrm{d}z' A_0^3\left(t_\mathrm{h} + \xi z'\right) \times \exp\left[-\mathrm{i}\Delta\beta_0 z' + 3\mathrm{i}\varphi_\mathrm{SPM}\left(t_h + \xi z', z'\right) + \mathrm{i}\varphi_\mathrm{XPM}\left(t_\mathrm{h}, z', z'\right)\right] \,, \qquad (4.123)$$

where $t_l = (t - z/v_l)$ with $l = \mathrm{p,h}$ for the pump and the field, respectively; $A_0(t)$ is the initial-condition envelope of the pump pulse;

$$\varphi_{\mathrm{SPM}}\left(t_{\mathrm{p}}, z\right)=\gamma_1\left|A_0\left(t_{\mathrm{p}}\right)\right|^2 z \tag{4.124}$$

is the SPM-induced phase shift of the pump field; and

$$\varphi_{\mathrm{XPM}}\left(t_{\mathrm{h}}, z', z\right)=2\gamma_2\int_{z'}^{z}\left|A_0\left(t_{\mathrm{h}}+\xi z'\right)\right|^2 \mathrm{d}z' \tag{4.125}$$

is the XPM-induced phase shift of the third-harmonic field.

In the regime where the nonlinear phase shifts, given by (4.124) and (4.125), are small, the Fourier transform of (4.123) yields the following expression for the spectrum of the third-harmonic intensity:

$$I(\Omega, z) \propto \beta^2 \frac{\sin^2\left[(\Delta k_0+\Omega\xi)\frac{z}{2}\right]}{(\Delta k_0+\Omega\xi)^2} \times\left|\int_{-\infty}^{\infty}\int_{-\infty}^{\infty} A\left(\Omega-\Omega'\right) A\left(\Omega'-\Omega''\right) \times A\left(\Omega''\right) \mathrm{d}\Omega' \mathrm{d}\Omega''\right|^2, \tag{4.126}$$

where $A(\Omega)$ is the spectrum of the input pump field. Analysis of the regime of small nonlinear phase shifts is thus very instructive as it allows phase-matching effects to be decoupled from the influence of the spectrum of the pump field. While the phase matching is represented by the argument in the exponential in the first factor on the right-hand side of (4.126), the significance of the pump spectrum is clear from the convolution integral appearing in this expression. Depending on the signs of the phase and group-velocity mismatch, Δk_0 and ξ, the peak in the spectrum of the third harmonic can be either red- or blue-shifted with respect to the frequency $3\omega_0$. The spectral width of this peak, as can be seen from (4.126), is given by $\delta \approx 2\pi(|\xi|z)^{-1}$, decreasing as z^{-1} with the growth in the propagation coordinate z (see inset in Fig. 4.10).

With low pump powers, the generalized phase-matching condition (4.119), as can be seen also from the exponential factor in (4.126), defines the central frequency $3\omega_0+\Omega_{\max}$, $\Omega_{\max}=-\Delta k_0/\xi$, of the peak in the spectrum of the third harmonic. The amplitude of this peak, as shown by (4.126), is determined by the amplitudes of the pump field components with frequencies $\omega_1=\omega_0+\Omega_{\max}-\Omega'$, $\omega_2=\omega_0+\Omega'-\Omega''$ and $\omega_3=\omega_0+\Omega''$, which can add up to transfer the energy to the $3\omega_0+\Omega_{\max}$ component in the spectrum of the

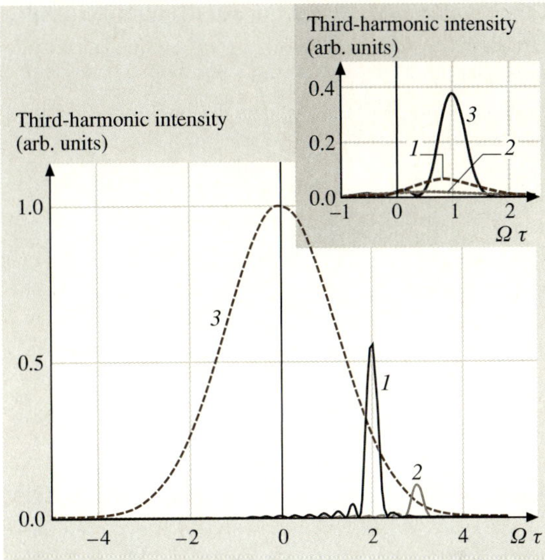

Fig. 4.10 Spectra of the third harmonic generated in the regime of weak self-phase modulation of the pump field with $\Delta k_0\tau/\xi=-2$ (*curve 1*), −3 (2) and 0 (3). The *inset* illustrates narrowing of the main peak in the spectrum of the third harmonic with $\Delta k_0\tau/\xi=-1$ and $\xi L/\tau=2$ (1), 4 (2), and 10 (3)

third harmonic through the $\omega_1+\omega_2+\omega_3=3\omega_0+\Omega_{\max}$ process. The spectrum of the pump field should therefore be broad enough to provide a high amplitude of this peak. In the regime when the SPM-induced broadening of the pump spectrum is small, the tunability range of the third harmonic (i. e., the range of frequency offsets Ω) is mainly limited by the bandwidth of the input pump field. This regime of THG is illustrated in Fig. 4.10. As the ratio $\Delta k_0\tau/\xi$ changes from −2 (curve 1) to −3 (curve 2), the peak in the spectrum of the third harmonic is bound to shift from $\Omega_{\max}=2/\tau$ to $3/\tau$. A similar spectral shift limited by a field-unperturbed pump spectrum has been earlier predicted by *Akhmanov* et al. for second-harmonic generation [4.56]. For large $|\Delta k_0/\xi|$, however, the pump power density at the wings of the spectrum is too low to produce a noticeable peak with $\Omega_{\max}=-\Delta k_0/\xi$ in the spectrum of the third harmonic (cf. curves 2 and 3 in Fig. 4.10).

In the general case of non-negligible nonlinear phase shifts, phase-matching effects cannot be decoupled from the influence of the pump spectrum, and we resort to approximate integration methods to identify the main features of the THG process. For a pump pulse with a Gaussian envelope, $A_0(t_p)=\tilde{A}\exp[-t_{\mathrm{p}}^2/(2\tau^2)]$, where $\tilde{A}$ and τ are the amplitude and the initial duration of the pump pulse, the spectrum of the third harmonic, in view of (4.123), is given by

$$B(\Omega,z)=\mathrm{i}\beta\tilde{A}^3\int_{-\infty}^{\infty}\mathrm{d}t_\mathrm{h}\int_0^z\mathrm{d}z'$$
$$\times\exp\left[-\frac{3\left(t_\mathrm{h}+\xi z'\right)^2}{2\tau^2}-\mathrm{i}\Delta k_0 z'+3\mathrm{i}\gamma_1\left|\tilde{A}\right|^2\right.$$
$$\left.\times\exp\left(-\frac{\left(t_\mathrm{h}+\xi z'\right)^2}{\tau^2}\right)z'\right]$$
$$\times\exp\left[2\mathrm{i}\gamma_2\left|\tilde{A}\right|^2\int_{z'}^{z}\mathrm{d}z''\right.$$
$$\left.\times\exp\left(-\frac{\left(t_\mathrm{h}+\zeta z''\right)^2}{\tau^2}\right)+\mathrm{i}\Omega t_\mathrm{h}\right] \qquad (4.127)$$

Changing the order of integration in (4.127), we apply the saddle-point method to estimate the integral in $\mathrm{d}t_h$:

$$B(\Omega,z)\propto\mathrm{i}\beta\tilde{A}^3\tau\int_0^z\mathrm{d}z'$$
$$\times\exp\left\{-\mathrm{i}\left(\Delta k_0+\Omega\xi-3\gamma_1|\tilde{A}|^2\right)z'\right.$$
$$\left.+2\mathrm{i}\gamma_2|\tilde{A}|^2\frac{\tau}{\xi}\Phi\left[(z-z')\frac{\xi}{\tau}\right]\right\}, \qquad (4.128)$$

where $\Phi(x)=\int_0^x\exp(-x^2)\mathrm{d}x$.

With $\zeta z/\tau\gg 1$ and $\Phi[(z-z')\zeta]$ constant, the phase matching is controlled by the factor

$$F(\Omega,z)=\frac{\sin^2\left[\left(\Delta k+\Omega\xi+3\gamma_1\left|\tilde{A}\right|^2\right)\frac{z}{2}\right]}{\left(\Delta k+\Omega\xi+3\gamma_1\left|\tilde{A}\right|^2\right)^2}. \qquad (4.129)$$

Thus, the saddle-point estimate of the SVEA integral for the third-harmonic field recovers the generalized phase-matching condition in the form of (4.119). The frequency offset Ω_max providing the maximum efficiency of THG is now determined by the dispersion of the material, its nonlinear properties, and the intensity of the pump field.

Spectral broadening of the pump field due to SPM radically expands the tunability range of the third harmonic (Fig. 4.11). With $\gamma I_0 L=1$ (L is the interaction length and I_0 is the pump-field peak intensity), the spectrum of the pump field is broad enough to produce a high-amplitude peak at $\Omega_\mathrm{max}=-6/\tau$. The peak becomes narrower as the length of the nonlinear medium increases (cf. curves 2 and 3 in Fig. 4.11), suggesting a convenient way for an efficient generation of short-wavelength radiation with a well-controlled spectrum for spectroscopic and metrological applications. In the time domain, the third harmonic tends to break up into two pulses, as shown in the inset to Fig. 4.11 for $\gamma I_0 L=1$ (curve 1) and 2 (curve 2) with $\xi L\tau=20$. The first peak at $t_\mathrm{h}\tau=-20$ represents a third-harmonic pulse that propagates with the pump pulse and that is phase-matched to the pump field in the sense of (4.119). The second pulse propagates with the group velocity of the third harmonic and is group-delayed in our case with respect to the pump field. Due to the phase matching, the ratio of the amplitudes of the first and second peaks increases with the growth in $\gamma I_0 L$ (cf. curves 1 and 2 in the inset to Fig. 4.11).

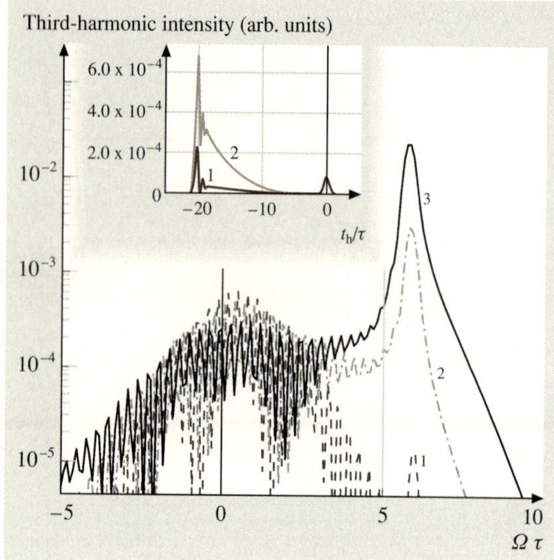

Fig. 4.11 Spectra of the third harmonic generated with no SPM of the pump (*curve 1*) and with an SPM-broadened pump (*curves 2, 3*) with $\gamma I_0 L=1$ (1) and 2 (2); $\Delta k_0\tau/\xi=-6$, $\xi L=20$. The insets show the temporal structure of the third harmonic for $\gamma I_0 L=1$ (1) and 2 (2) with $\xi L/\tau=20$

Examples of frequency-shifted THG can be found in the recent literature on nonlinear optics of PCFs and tapered fibers [4.48–52]. Interesting collinear and Cherenkov-type intermode phase-matching options [4.50, 51] have been highlighted. The nature of the frequency shift has been identified in [4.57, 58]. Generation of an asymmetrically broadened and spectrally shifted third harmonic is the most frequently encountered situation, observed in nonlinear-optical ex-

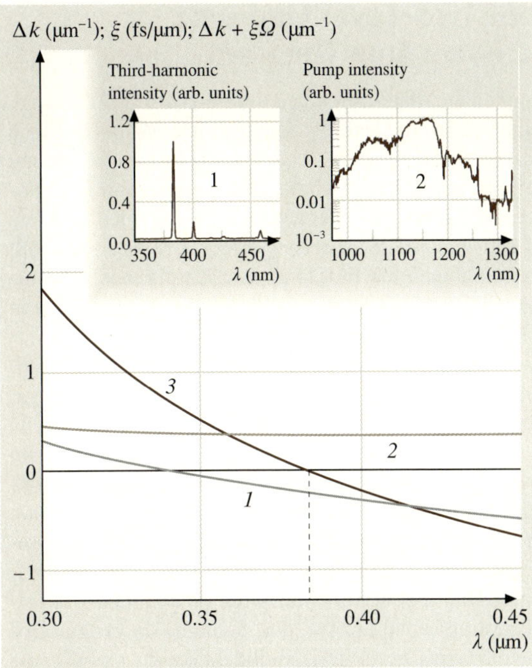

Fig. 4.12 The phase mismatch Δk_0(1), group-velocity mismatch ξ (2), and the effective phase mismatch $\Delta k_0 + \xi\Omega$ (3) for third-harmonic generation in an air-clad fused-silica fiber with a core radius of 0.9 μm. The pump wavelength is 1.25 μm. The vertical dashed line shows the wavelength $\lambda = 383$ nm, where $\Delta k_0 + \xi\Omega = 0$. The insets show a typical experimental spectrum of blue visible emission (1) and the spectrum of the Cr:forsterite laser pump field (2) at the output of a photonic-crystal fiber with a length of 10 cm (after [4.57])

periments with PCFs [4.49–51] and tapered fibers [4.52]. The two most striking recent experimental findings include the observation of a 260 nm spectral component generated by Ti:sapphire laser pulses in high-delta PCFs [4.49] and the generation of a spectrally isolated frequency component at 380 nm by femtosecond pulses of 1.25 μm Cr:forsterite laser radiation [4.59]. In the former case, the central wavelength of the ultraviolet (UV) component remained stable [4.49] as the central pump wavelength was tuned from 770 to 830 nm [which seems to agree well with the phase-matching condition of (4.119)]. In the latter case, generation of the blue-shifted third harmonic allowed a highly efficient frequency conversion of pump radiation over the frequency interval exceeding 540 THz [4.59]. The spectral peak of visible blue emission was shifted from the frequency $3\omega_0$ in these experiments by 34 nm (a typical spectrum is presented in the inset to Fig. 4.12), with no signal produced at $3\omega_0$.

To identify the main features of the dramatic frequency shift in the spectrum shown in the inset to Fig. 4.12, we consider the dispersion of guided modes in an air-clad fused-silica thread as a generic example of PCF dispersion. In Fig. 4.12, we plot the phase and group-velocity mismatch, Δk_0 and ξ, as well as the effective phase mismatch $\Delta k_{\text{eff}} = \Delta k_0 + \xi\Omega$ for intermodal THG in an air-clad fused silica fiber with a core radius of 0.9 μm. We assume that the pump field is coupled into the fundamental mode of the fiber, while the third harmonic is generated in one of high-order HE_{13}-type modes. The effective phase mismatch Δk_{eff} passes through zero at the wavelength of 383 nm, corresponding to a frequency shift of about 63 THz relative to $3\omega_0$, which agrees well with the typical spectrum of blue emission from a PCF [4.59] presented in inset 1 to Fig. 4.12. The main features and dominant tendencies of the third-harmonic spectrum are adequately described by (4.119) and (4.127). Comparison of the pump (inset 2 to Fig. 4.12) and third-harmonic spectra presented in Fig. 4.11 also suggests that a blue shifting of the main peak in the spectrum of the third harmonic by more than 60 THz becomes possible due to a strong broadening of the pump field, which is, of course, not only due to SPM in realistic conditions of PCF experiments.

We have shown that third-harmonic generation in the field of spectrally broadened short pump pulses can display interesting and practically significant new features. A short-pulse pump field broadened due to self-phase modulation can generate its third harmonic within a broad spectral range. However, the phase-matching condition generalized to include the phase and group-velocity mismatch of the pump and third-harmonic fields, as well as the Kerr nonlinearity induced spectral broadening of the pump field, tends to select a narrow spectral region of efficient THG. The possibility to tune the frequency of the main spectral peak in the spectrum of the third harmonic by varying the group-velocity mismatch is a unique property of THG-type processes, which is not typical of standard parametric FWM processes, where the first-order dispersion terms cancel out of the balance of the field momenta. For pump fields with large nonlinear frequency deviations, this spectral region may lie several tens of terahertz away from the central frequency of the third harmonic $3\omega_0$. This non-3ω THG process, leading to no signal at $3\omega_0$, is shown to result in interesting and practically significant spectral-transformation phenomena in photonic-crystal fibers.

4.5 Ultrashort Light Pulses in a Resonant Two-Level Medium: Self-Induced Transparency and the Pulse Area Theorem

4.5.1 Interaction of Light with Two-Level Media

The phenomenological approach based on nonlinear-optical susceptibilities does not include nonstationary phenomena related to a dynamic modification of the nonlinear medium by the laser field. Interaction of a resonant laser field with a two-level system is a physically interesting and methodologically important regime where the equations governing the evolution of the laser field in a nonlinear medium can be solved self-consistently with the equations of motion for the quantum system interacting with the field. Such a self-consistent analysis reveals the existence of a remarkable regime of nonlinear-optical interactions. A resonance laser pulse whose amplitude and pulse width are carefully matched to the two-level system can propagate in a two-level medium with no absorption-induced attenuation of the pulse amplitude, a phenomenon known as self-induced transparency.

The interaction of laser radiation with a two-level system is a classic problem of laser physics [4.60]. It has been extensively studied with the use of various approximate analytical approaches and numerical procedures over four decades, revealing important aspects of the interaction of laser radiation with a two-level system and, generally, laser–matter interactions. The standard approach based on the slowly varying envelope approximation (SVEA) and rotating-wave approximation gives Maxwell–Bloch equations in the case of a two-level atom [4.60, 61]. These equations provide an adequate description of laser radiation propagating in resonant media within a broad range of parameters and give the key to understanding such fundamental resonant optical phenomena as the formation of 2π solitons and self-induced transparency [4.60–63].

As shown in the classic works by *McCall* and *Hahn* [4.62, 63], light pulses with pulse areas multiple of π can propagate in a two-level medium with no changes in their shape, while pulses with other pulse areas tend to change their areas during the propagation in a two-level medium evolving to pulses with areas multiple of p. Much analytical and numerical work has been done over the past decade to extend this SVEA result to ultrashort pulses. *Eberly* [4.64] has rederived the area theorem for the case of short light pulses, modifying this theorem to include pulse chirping and homogeneous damping. *Ziolkowski* et al. [4.65] applied the finite-difference time-domain (FDTD) technique [4.66] to solve the semiclassical Maxwell–Bloch equations numerically. This approach revealed several important features of short-pulse propagation in a two-level medium and allowed a more detailed analysis of self-induced transparency effects. *Hughes* [4.67, 68] has employed the FDTD approach to demonstrate the possibility of generating sub-femtosecond transients in a two-level medium.

Tarasishin et al. [4.69] have applied FDTD technique to integrate jointly the Maxwell and Schrödinger equations for an ultrashort light pulse propagating in a two-level medium. Below in this section, we explain the FDTD-based algorithm [4.69] solving the Maxwell and Schrödinger equations to model the interaction of ultrashort laser pulses with an ensemble of two-level atoms. FDTD simulations presented below reveal interesting regimes of short-pulse propagation and amplification in two-level media, including the evolution of the pulse to a 2π soliton, amplification of a single-cycle pulse in a medium with a spatially modulated distribution of dipole moments of resonant transitions, and amplification of chirped light pulses.

4.5.2 The Maxwell and Schrödinger Equations for a Two-Level Medium

We shall start with the extension of the standard FDTD procedure to the case of short pulses propagating in a two-level medium, when the Maxwell equation for the fields and the Schrödinger equation for the wave functions should be solved without any assumptions that are usually employed in the SVEA approach. In the one-dimensional case, the FDTD algorithm involves step-by-step integration of two Maxwell curl equations

$$\frac{\partial D_z(x,t)}{\partial t} = \frac{\partial H_y(x,t)}{\partial x}, \tag{4.130}$$

$$\frac{\partial H_y(x,t)}{\partial t} = \frac{\partial E_z(x,t)}{\partial x}. \tag{4.131}$$

To perform this integration, we have to define the relation between the components of the electromagnetic induction and the electromagnetic field. This can be done through the equation for the polarization of the medium. In our case of a two-level medium, this involves the solution of the Schrödinger equation for the wave functions of the energy levels.

We will consider an ensemble of noninteracting two-level atoms or molecules whose wave functions can be represented as superpositions of two basis states 1 and 2:

$$\psi(t) = a(x,t)\psi_1 + b(x,t)\psi_2, \quad (4.132)$$

where ψ_1 and ψ_2 are the eigenfunctions of an unperturbed system corresponding to the states with energies E_1 and E_2 (we assume for definiteness that $E_1 > E_2$), respectively, and $a(x,t)$ and $b(x,t)$ are complex coefficients. Then, the Schrödinger equation for the wave function yields the following set of differential equations:

$$\mathrm{i}\hbar \frac{\mathrm{d}a(x,t)}{\mathrm{d}t} = E_1 a(x,t) - \mu E_z(x,t) b(x,t), \quad (4.133)$$

$$\mathrm{i}\hbar \frac{\mathrm{d}b(x,t)}{\mathrm{d}t} = E_2 b(x,t) - \mu E_z(x,t) a(x,t), \quad (4.134)$$

where μ is the dipole moment of transition between the levels 1 and 2. Following *Feynman* et al. [4.68], we introduce real combinations of the complex quantities $a(x,t)$ and $b(x,t)$:

$$r_1(x,t) = a(x,t)b^*(x,t) + a^*(x,t)b(x,t), \quad (4.135)$$

$$r_2(x,t) = \mathrm{i}[a(x,t)b^*(x,t) - a^*(x,t)b(x,t)], \quad (4.136)$$

$$r_3(x,t) = a(x,t)a^*(x,t) - b(x,t)b^*(x,t), \quad (4.137)$$

$$r_4(x,t) = a(x,t)a^*(x,t) + b(x,t)b^*(x,t). \quad (4.138)$$

The physical content of the parameters defined by (4.135–4.138) is well known from classic textbooks on coherent optics [4.60, 61]. It can be easily verified using the set (4.133) and (4.134) that $r_4(x,t)$ is independent of time and can be interpreted as the probability to find the system in either state 1 or state 2. The quantities $r_1(x,t)$ and $r_3(x,t)$ play an especially important role. Depending on the sign of $r_3(x,t)$, resonant electromagnetic radiation can be either amplified or absorbed by a two-level system. The quantity $r_1(x,t)$ controls the polarization of the medium in the case of a linearly polarized light field:

$$P_z(x,t) = 4\pi\mu r_1(x,t)N, \quad (4.139)$$

where N is the volume density of two-level atoms or molecules. Thus, the quantity $r_1(x,t)$ defines the sought relation between the components of the electromagnetic induction and the electromagnetic field.

Using (4.133) and (4.134), we arrive at the following set of equations for the quantities defined by (4.135–4.137):

$$\frac{\mathrm{d}r_1(x,t)}{\mathrm{d}t} = -\omega_0 r_2(x,t), \quad (4.140)$$

$$\frac{\mathrm{d}r_2(x,t)}{\mathrm{d}t} = \omega_0 r_1(x,t) + 2(\mu/\hbar) E_z(x,t) r_3(x,t), \quad (4.141)$$

$$\frac{\mathrm{d}r_3(x,t)}{\mathrm{d}t} = -2(\mu/\hbar) E_z(x,t) r_2(x,t), \quad (4.142)$$

where $\omega_0 = (E_1 - E_2)/\hbar$. Differentiating (4.141), using (4.140) for $\mathrm{d}r_1(x,t)/\mathrm{d}t$, and taking into consideration that $r_1(x,t) = P_z(x,t)/(\mu N) = [D_z(x,t) - E_z(x,t)]/(4\pi\mu N)$, we find that

$$\frac{\mathrm{d}^2 D_z(x,t)}{\mathrm{d}t^2} - \frac{\mathrm{d}^2 E_z(x,t)}{\mathrm{d}t^2} + \omega_0^2 [D_z(x,t) - E_z(x,t)] + 2\frac{4\pi\mu^2\omega_0 N}{\hbar} E_z(x,t) r_3(x,t) = 0, \quad (4.143)$$

$$\frac{\mathrm{d}r_3(x,t)}{\mathrm{d}t} = \frac{2}{4\pi\hbar\omega_0 N} E_z(x,t) \times \left(\frac{\mathrm{d}D_z(x,t)}{\mathrm{d}t} - \frac{\mathrm{d}E_z(x,t)}{\mathrm{d}t}\right). \quad (4.144)$$

The FDTD approach involves difference approximation [4.66] of time and spatial derivatives involved in (4.130) and (4.131):

$$D_{z,i}^{n+1} = D_{z,i}^{n} + \frac{\Delta t}{\Delta x}\left(H_{y,i+1/2}^{n+1/2} - H_{y,i-1/2}^{n+1/2}\right), \quad (4.145)$$

$$H_{y,i+1/2}^{n+3/2} = H_{y,i+1/2}^{n+1/2} + \frac{\Delta t}{\Delta x}\left(E_{z,i+1}^{n+1} - E_{z,i}^{n+1}\right), \quad (4.146)$$

where i and n indicate the values of discrete spatial and temporal variables, respectively, $x = i\Delta x$, $t = n\Delta t$, Δx and Δt are the steps of discretization in spatial and temporal variables. This approach yields the following set of difference equations:

$$E_{z,i}^{n+1} = D_{z,i}^{n+1} + D_{z,i}^{n-1} - E_{z,i}^{n-1} + \frac{\left(4E_{z,i}^{n} - 4D_{z,i}^{n} + 4\Delta t^2 \frac{4\pi\mu^2\omega_0 N}{\hbar} E_{z,i}^{n-1} r_{3,i}^{n-1}\right)}{2 + \Delta t^2 \omega_0^2}, \quad (4.147)$$

$$r_{3,i}^{n+1} = r_{3,i}^{n-1} + 0.5\frac{2}{4\pi\hbar\omega_0 N}\left(E_{z,i}^{n+1} + E_{z,i}^{n}\right) \times \left(D_{z,i}^{n+1} - D_{z,i}^{n-1} - E_{z,i}^{n+1} + E_{z,i}^{n-1}\right). \quad (4.148)$$

Thus, we arrive at the following closed algorithm of numerical solution:

1. substitute $D_{z,i}^{n+1}$ determined from (4.145) for the current value of the discrete-time variable into the set

of (4.147) and (4.148) and determine the values of $r_{3,i}^{n+1}$ and $E_{z,i}^{n+1}$,

2. substitute the quantities thus determined into (4.146) and determine the values of the magnetic field $H_{y,i}^{n+3/2}$,
3. substitute these values of the magnetic field into (4.145) and repeat the procedure for the next value of the discrete-time variable.

4.5.3 Pulse Area Theorem

To test the FDTD-based procedure of simulations described in Sect. 4.5.2, we model the propagation of light pulses

$$E(x,t) = A(x,t)\mathrm{e}^{\mathrm{i}\phi+\mathrm{i}kz-\mathrm{i}\omega t} + \text{c.c.}\,, \tag{4.149}$$

where $A(x,t)$ is the pulse envelope, in a two-level medium and compare the results of these simulations with the predictions of the McCall–Hahn area theorem [4.62, 63]. This plan can be accomplished by keeping track of the pulse area

$$\theta(x) = \int_{-\infty}^{\infty} \Omega(x,\tau)\mathrm{d}\tau\,, \tag{4.150}$$

where $\Omega(x,t) = \frac{2\mu}{\hbar}A(x,t)$ is the real Rabi frequency.

The SVEA equation for the pulse envelope $A(x,t)$ can be represented as

$$\frac{\partial A(x,t)}{\partial x} + \frac{n}{c}\frac{\partial A(x,t)}{\partial t} = \mathrm{i}\frac{2\pi\omega}{nc}P(x,t)\,, \tag{4.151}$$

where n is the refractive index of the medium, $P(x,t)$ is the slowly varying amplitude of the polarization induced in the medium. As shown by *McCall* and *Hahn* [4.62, 63], on exact resonance, $P(x,t) = \mathrm{i}\mu N\sin[\theta(x,t)]$, with $\theta(x,t) = \int_{-\infty}^{t}\Omega(x,\tau)\mathrm{d}\tau$. Hence, (4.151) yields

$$\frac{\partial A(x,t)}{\partial x} + \frac{n}{c}\frac{\partial A(x,t)}{\partial t} = -\frac{2\pi\omega\mu N}{nc}\sin[\theta(x,t)]\,. \tag{4.152}$$

The celebrated solution to (4.152) is a pulse with a hyperbolic-secant shape, which propagates in a resonant two-level medium with no changes in its envelope:

$$A(x,t) = \frac{\hbar}{\mu\tau}\operatorname{sech}\left(\frac{t-x/V}{\tau}\right), \tag{4.153}$$

where V is the pulse velocity in the medium,

$$V = \left(\frac{4\pi\mu^2\omega N\tau^2}{\hbar nc} + \frac{n}{c}\right)^{-1}. \tag{4.154}$$

In the regime of weak absorption (when the pulse area only slightly deviates from its stable values), the evolution of the pulse area is governed by the equation

$$\frac{\partial\theta(x)}{\partial x} = -\frac{2\pi\omega\mu N}{ncE_0}\theta(x)\sin\theta(x), \tag{4.155}$$

where E_0 is the initial pulse amplitude. With $\theta \approx 2\pi$, we arrive at

$$\frac{\partial\theta(x)}{\partial x} = -\frac{\alpha}{2}\sin\theta(x), \tag{4.156}$$

where $\alpha = \frac{8\pi^2\omega\mu N}{ncE_0}$.

According to (4.156), pulses whose areas $\theta(x_0)$ are multiples of π propagate in a two-level medium with no changes in their envelopes (the soliton propagation regime). However, solitons with pulse areas equal to π, $3\pi, 5\pi, \ldots$ are unstable. Thus, a pulse with an arbitrary initial area changes its waveform propagating in a two-level medium until its pulse area becomes multiple of 2π. The characteristic spatial scale of this process is estimated as α^{-1}.

Numerical simulations [4.69] have been performed for hyperbolic-secant pulses:

$$E(x_0,t) = E_0\frac{2\cos[\omega(t-t_0)]}{\exp[-(t-t_0)/T] + \exp[(t-t_0)/T]}\,. \tag{4.157}$$

The parameters of the medium and the pulse were chosen as follows: $\Omega = 2\mu E_0/2\hbar = 0.0565\omega$, $4\pi\mu N r_3(0) = -0.12E_0$, and $\omega = \omega_0$ (exact resonance).

In accordance with the area theorem, a pulse with a duration $T = 5.631\,2\pi/\omega$ should propagate in the two-level medium with these specified parameters values without any changes in its envelope. FDTD simulations [4.69] reveal no changes in the waveform and the amplitude of such a pulse within a distance of 100 λ with an accuracy better than 0.1%.

Now, we examine how an arbitrary light pulse evolves to a 2π pulse in a two-level medium. FDTD simulations were performed for the propagation of a pulse (4.157) with a duration $T = 7.0392\pi/\omega$ in a two-level medium with these specified parameters values. While the amplitude of such a pulse increases by 15% as it propagates through the medium, the duration of the pulse decreases by a factor of about two. The pulse areas FDTD-simulated for these values of the propagation coordinate are equal to 2.5π, 2.33π, 2.24π, and 2.1π, while the pulse areas for the same values of x calculated from (4.156) are equal to 2.5π, 2.35π, 2.22π, and 2.09π, respectively. Thus, the results of FDTD simulations agree very well with the predictions of the area theorem, indicating the adequacy of the numerical approach.

4.5.4 Amplification of Ultrashort Light Pulses in a Two-Level Medium

Since the numerical algorithm based on the FDTD technique is intended to simulate the evolution of very short pulses, it enables one to explore many important aspects of short-pulse amplification in a two-level medium, providing a deeper understanding of the problems arising in the amplification of short pulses and the ways that can be employed to resolve these problems.

Propagation of π pulses seems to provide optimal conditions for amplification in a two-level system, since such pulses transfer atoms (or molecules) in an initially inverted medium to the lower state. However, the Rabi frequency increases as a light pulse propagates through the medium and its amplitude increases due to the amplification. Because of this change in the Rabi frequency, the pulse now cannot transfer inverted atoms or molecules to the lower state, but leaves some excitation in a medium, which reduces the gain and leads to pulse lengthening due to the amplification of the pulse trailing edge in a medium with residual population inversion. Thus, some precautions have to be taken to keep the area of the pulse constant in the process of pulse amplification in the medium. Below, we use the FDTD technique to explore two possible ways to solve this problem: (i) modulation of the spatial distribution of the dipole moments of resonant transitions in a two-level system and (ii) amplification of a frequency-detuned chirped pulse.

Modulation of the Spatial Distribution of the Dipole Moment

First, let us explain why we expect that the modulation of the spatial distribution of the dipole moment may improve the efficiency of amplification of an ultrashort light pulse in a two-level medium. To do this, we multiply (4.152) by $A(x,t)$ and integrate the resulting expression in time. Then, introducing the pulse energy, $\Phi(x)=\int_{-\infty}^{\infty} A^2(x,t)\,\mathrm{d}t$, and taking into consideration that the pulse energy remains constant if $\theta(x_0)=0, 2\pi, \ldots$, we arrive at the following equation relating the pulse energy and area:

$$\frac{\partial \Phi(x)}{\partial x}=\frac{2\pi\omega N\hbar}{nc}\left[\cos\theta(x)-1\right] . \tag{4.158}$$

Thus, if the pulse area is kept equal to π, as the pulse propagates through an inverted two-level medium, then the energy of such a pulse would grow linearly as a function of the distance:

$$\Phi(x)=(1+\beta x)\,\Phi(0) , \tag{4.159}$$

where $\beta=\frac{4\pi\omega N\hbar}{\Phi(0)nc}$ and $\Phi(0)$ is the initial energy of the pulse.

By looking at (4.150) and (4.159) and taking into consideration that $\Phi(0)=2\tau E_0^2=2E_0\hbar/\mu(0)$ for a hyperbolic-secant pulse (4.157), we find that the pulse area can be kept equal to π by modulating the spatial distribution of the dipole moment in accordance with

$$\mu(x)=\frac{\mu_0}{\sqrt{1+\beta x}} . \tag{4.160}$$

Such a modulation of the spatial distribution of dipole moments can be achieved, for example, by preliminarily orienting molecules in the medium.

In accordance with the area theorem, the evolution of the amplitude of a π pulse propagating in an inverted two-level medium with a spatial profile of the dipole moment described by (4.160) is governed by

$$E(x)=E_0\sqrt{1+\beta x} . \tag{4.161}$$

This growth of the pulse amplitude is exactly compensated by the decrease in the dipole moment of transitions in the two-level medium. The net effect is that the pulse area remains constant and equal to π.

Figures 4.13a–d present the results of FDTD simulations for the amplification of a light pulse with a duration $T=2\pi/\omega$ corresponding to a single optical cycle in an inverted two-level medium with a uniform distribution of the dipole moment (Figs. 4.13a,b) and in an inverted two-level medium where the spatial distribution of dipole moments of resonant transitions is modulated in accordance with (4.160) (Figs. 4.13c,d). Simulations were performed for the case when $\Omega=2\mu_0E_0/2\hbar=0.159\omega$, $4\pi\mu Nr_3(0)=0.12E_0$, and $\omega=\omega_0$. The time in these plots is measured from the moment when the center of the pulse passes through the entrance boundary of the medium.

Figures 4.13a,b display the evolution of a π pulse in an inverted uniform two-level medium and the population difference in this medium at the distances 0, $3\beta^{-1}$ and $6\beta^{-1}$. These plots show that a two-level medium with a uniform distribution of the dipole moment of transitions cannot ensure an efficient amplification of a transform-limited resonant π pulse. Figures 4.13c,d present the results of FDTD simulation for a π pulse propagating through a medium with a spatial evolution of the dipole moment described by (4.160). These simulations illustrate the possibility of efficient amplification of a single-cycle pulses in such a medium. Note that the pulse amplitude in Figs. 4.13c and 4.13d increases linearly with the propagation coordinate x (the deviation from the linear dependence does not exceed 10^{-4}). This result agrees very well with the predictions of the area theorem and can also be considered as another test of the reliability of the developed algorithm.

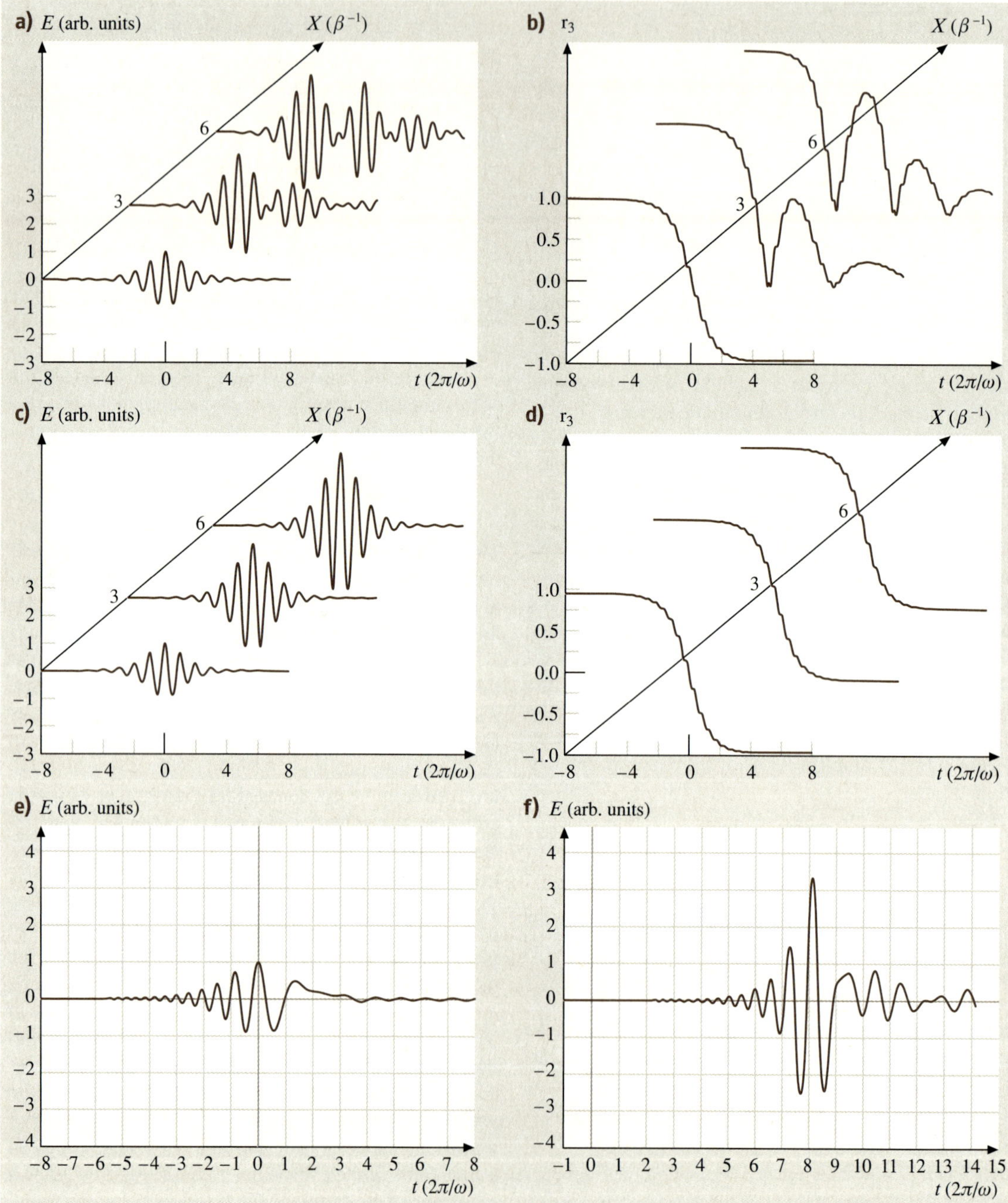

Frequency-Detuned Chirped Pulses

The second idea of keeping the area of the amplified pulse unchanged is to use chirped pulses whose frequency is detuned from the exact resonance. Such a pulse experiences compression as it propagates in the considered two-level medium. With an appropriate choice of pulse parameters, this decrease in pulse duration may serve to preserve the area of the light pulse undergoing amplification.

The Rabi frequency in this case is written

$$\Omega = \frac{1}{2}\sqrt{\Delta^2 + \left(\frac{2\mu_0 E_0}{\hbar}\right)^2}\,, \tag{4.162}$$

where $\Delta = \omega - \omega_0$ is the detuning of the central frequency of the pulse from the resonance.

Figures 4.13e,f present the results of FDTD simulations for a single-cycle pulse with a quadratic initial chirp

$$E(0,t) = E_0 \frac{2\cos\left(\omega t + kt^2/T^2\right)}{\exp(-t/T) + \exp(-t/T)}\,, \tag{4.163}$$

with $k = 1.4$, propagating in a uniform two-level medium with the same parameters as in Figs. 4.13a,b. The distance covered by the pulse in the medium is $6\beta^{-1}$, and the frequency detuning is $\Delta = \omega - \omega_0 = -0.02\omega_0$. As can be seen from Figs. 4.13e,f, although higher-order dispersion effects noticeably distort the waveform of the pulse, the peak intensity of an initially chirped pulse at the output of an amplifying uniform medium is nearly twice as high as the output energy of a transform-limited pulse. Furthermore, comparison of Figs. 4.13a,b,e,f shows that, chirping a light pulse, one can avoid undesirable pulse-lengthening effects.

Fig. 4.13 (a)–(d) The evolution of **(a)**, **(c)** the pulse waveform and **(b)**, **(d)** the population difference for a single-cycle light pulse propagating in an inverted two-level medium with **(a)**, **(b)** a uniform distribution of the dipole moment and **(c)**, **(d)** a spatial distribution of dipole moments of resonant transitions modulated in accordance with (4.160) simulated with the use of the FDTD technique [4.69] for $\Omega = 2\mu_0 E_0/2\hbar = 0.159\omega$, $4\pi\mu N r_3(0) = 0.12 E_0$, and $\omega = \omega_0$. The time in these plots is measured from the moment when the center of the pulse passes through the input boundary of the medium. **(e)**, **(f)** Amplification of a single-cycle chirped laser pulse (4.163) in a two-level medium: **(e)** the input pulse and **(f)** the output pulse simulated with the use of the FDTD technique for $\Omega = 2\mu_0 E_0/2\hbar = 0.159\omega$, $4\pi\mu N r_3(0) = 0.12 \times 10^0$, and $\omega = \omega_0$ ◀

4.5.5 Few-Cycle Light Pulses in a Two-Level Medium

The results of FDTD simulations, as shown by *Tarasishin* et al. [4.70], typically agree very well with the predictions of *McCall* and *Hahn* theory [4.62, 63] for light pulses propagating in a two-level medium until the pulse duration T becomes less than the duration T_0 of a single optical cycle. In agreement with the general predictions of McCall and Hahn, a 2.9π pulse with $T = T_0$, $2\mu E_0 \pi T/\hbar = 2.9\pi$, $\mu N/E_0 = 0.00116$, for example, is transformed until its area becomes equal to 2π. The peak amplitude of the pulse increases by a factor of 1.31 under these conditions, while its duration decreases to $0.55T_0$.

Noticeable deviations from the McCall–Hahn regime were observed for pulses with durations shorter than the duration of a single field cycle. In particular, half-cycle 2π pulses become asymmetric as they propagate through a two-level medium (Fig. 4.14a), and the characteristic length corresponding to the phase shift (4.9) equal to π estimated on the basis of FDTD simulations for such pulses was equal to 9.4 λ, which appreciably differs from the SVEA estimate for the characteristic length corresponding to the π phase shift, $L = (\hbar nc)/(4\omega^2\mu^2 T^2 N) = (E_0^2 nc)/(4\omega^2 N\hbar)$. Quarter-cycle 2π pulses display noticeable distortions and lengthening in the process of propagation through a two-level medium (Fig. 4.14b).

Deviations observed in the behavior of very short 2π pulses from the McCall–Hahn scenario are due to the fact that, although, formally, such pulses have an area of 2π, the cycle of interaction between light and a two-level system remains incomplete in this case, as the pulses do not even contain a full cycle of the field (Figs. 4.14a,b). As a result, such pulses leave some excitation in a two-level medium (Fig. 4.14d) instead of switching excited-state population back to the ground state, as in the case of longer 2π pulses (Fig. 4.14c).

The amplitude of the leading edge of the pulse becomes higher than the amplitude of its trailing edge, and the pulse waveform becomes noticeably asymmetric (Fig. 4.14a). The group velocity of such very short pulses increases due to this incompleteness of the light-two-level-system interaction cycle, leading to a discrepancy between the SVEA estimate and the FDTD result for the characteristic length corresponding to the π phase shift. The residual population in the medium and the asymmetry of the pulse waveform increase with pulse shortening.

The results of FDTD simulations presented in this section thus show that the general predictions of McCall and Hahn for the evolution of the amplitude and the

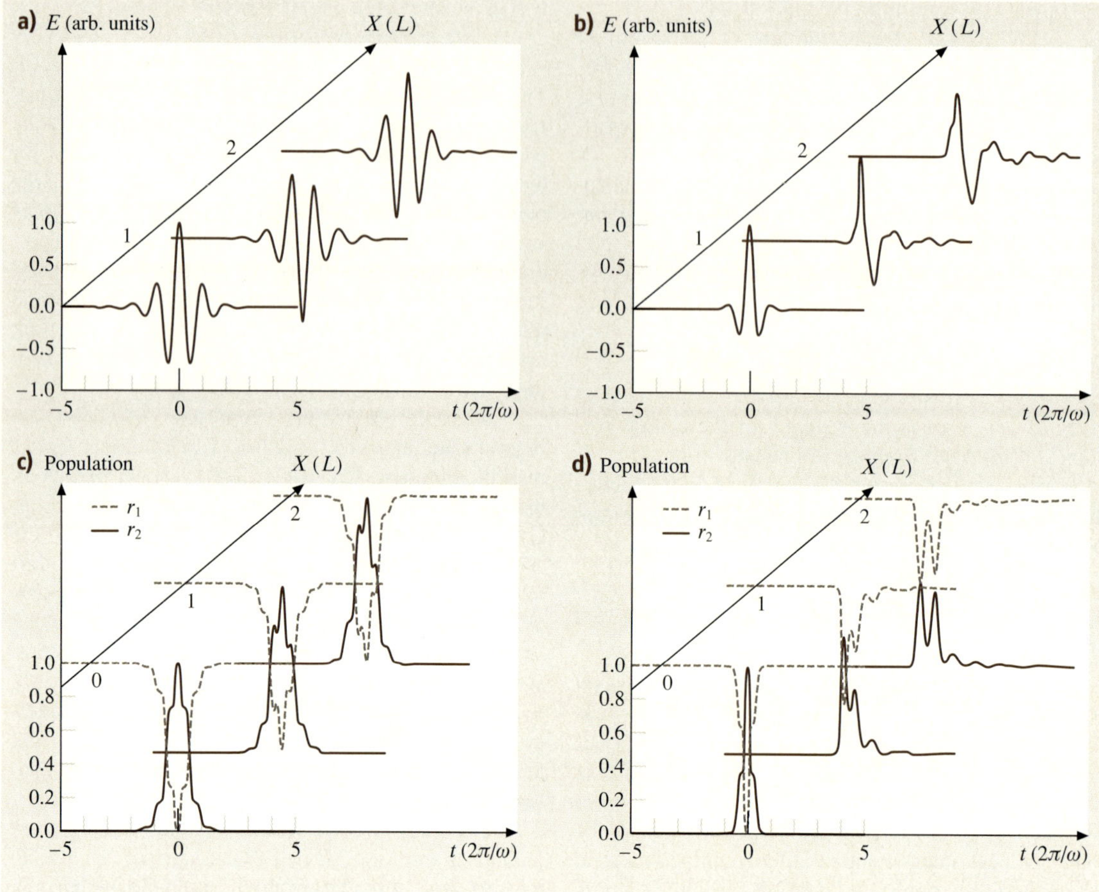

Fig. 4.14 **(a)** Evolution of a half-cycle 2π pulse in a two-level medium: $T = 0.5T_0$, $2\mu E_0 \pi T/\hbar = \pi$, and $\mu N/E_0 = 0.0016$, **(b)** Evolution of a quarter-cycle 2π pulse in a two-level medium: $T = 0.25T_0$, $2\mu E_0 \pi T/\hbar = 2\pi$, and $\mu N/E_0 = 0.0032$. **(c)**, **(d)** Evolution of the excited- and ground-state populations in a two-level medium under the action of **(c)** a half-cycle 2π pulse and **(d)** a quarter-cycle 2π pulse: (*dashed line*) ground-state population r_1 and (*solid line*) excited-state population r_2. Simulations were performed with the use of the FDTD technique [4.70]

phase of short pulses in a two-level medium generally agree reasonably well with the results of numerical simulations until the pulse duration becomes less than the duration of a single optical cycle. Numerical analysis reveals several interesting physical features in the formation of 2π solitons produced as a result of splitting of single-cycle pulses propagating in a two-level medium. In particular, the resulting pulses may have different amplitudes, durations, and group velocities, allowing the formation of sub-femtosecond pulses and slowing down of the light in two-level media. Noticeable deviations from the McCall–Hahn regime can be observed for pulses with durations shorter than the duration of a single field cycle. Half-cycle 2π pulses become asymmetric as they propagate through a two-level medium, while quarter-cycle 2π pulse display considerable distortions and lengthening in the process of propagation through a two-level medium. Deviations observed in the behavior of very short 2π pulses from the McCall–Hahn scenario are due to the fact that the cycle of interaction between light and a two-level system remains incomplete in this case, and light pulses leave some excitation in a two-level medium instead of switching excited-state population back to the ground state.

4.6 Let There be White Light: Supercontinuum Generation

Supercontinuum (SC) generation – a physical phenomenon leading to a dramatic spectral broadening of laser pulses propagating through a nonlinear medium – was first demonstrated in the early 1970s [4.71, 72] (see [4.73] for an overview of early experiments on supercontinuum generation). Presently, more than three decades after its discovery, supercontinuum generation is still one of the most exciting topics in laser physics and nonlinear optics [4.44], the area where high-field science meets the physics of low-energy unamplified ultrashort pulses in the most amazing way. The advent of photonic-crystal fibers [4.36, 37], capable of generating supercontinuum emission with unamplified, nano- and even sub-nanojoule femtosecond pulses, has resulted in revolutionary changes in frequency metrology [4.74–77] opened new horizons in ultrafast science [4.78, 79] and allowed the creation of novel wavelength-tunable and broadband fiber-optic sources for spectroscopic [4.80] and biomedical [4.81] applications. The rainbow of colors produced by a laser beam Fig. 4.15 has become an optical instrument and a practical tool.

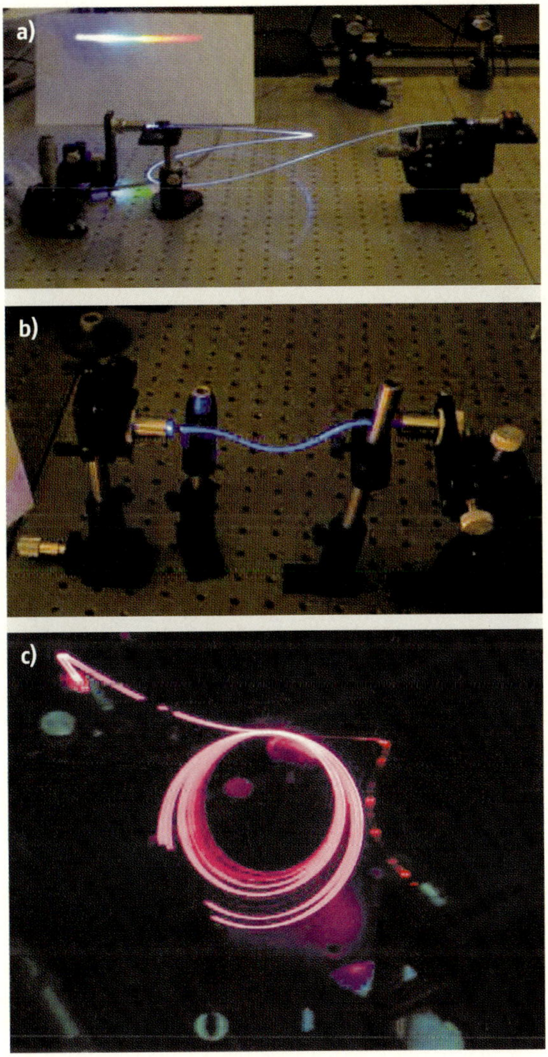

Fig. 4.15a–c Spectral transformations of ultrashort pulses in photonic-crystal fibers: **(a)** supercontinuum generation, **(b)** third-harmonic generation, and **(c)** frequency shifting

As a physical phenomenon, supercontinuum generation involves the whole catalog of classical nonlinear-optical effects, such as self- and cross-phase modulation, four-wave mixing, stimulated Raman scattering, solitonic phenomena and many others, which add up to produce emission with an extremely broad spectra, sometimes spanning over a couple of octaves. Below, we discuss the basic physical processes contributing to supercontinuum generation in greater detail, with special emphasis made on self-phase modulation, four-wave mixing, and modulation instabilities (Sect. 4.6.1), cross-phase modulation (Sect. 4.6.2), as well as the solitonic phenomena and stimulated Raman scattering (Sect. 4.6.3).

4.6.1 Self-Phase Modulation, Four-Wave Mixing, and Modulation Instabilities in Supercontinuum-Generating Photonic-Crystal Fibers

Propagation of laser pulses in PCFs is always accompanied by SPM-induced spectral broadening. The basic features of SPM are discussed in Sect. 4.4.1. For very short laser pulses and broadband field waveforms, SPM can be thought of as a four-wave mixing $\omega_{p1} + \omega_{p2} = \omega_3 + \omega_4$ with frequency components ω_{p1} and ω_{p2} from the spectrum of the laser field serving as pump photons generating new frequency components ω_3 and ω_4. In the case of a frequency-degenerate pump, $\omega_{p1} = \omega_{p2} = \omega_p$, the new frequency components generated through FWM appear as Stokes and anti-Stokes sidebands at frequencies ω_s and ω_a in the spectrum of the output field. Such FWM processes become especially efficient, as emphasized in Sect. 4.4.4, when phase matching is achieved for the fields involved in the nonlinear-optical interaction. Under certain conditions, the pump fields modify the effective refractive indices of the guided modes involved in the FWM process, inducing the phase matching for the FWM and leading to a rapid growth of spectral sidebands representing the Stokes and anti-Stokes fields. This regime of four-wave mixing, referred to as modulation instability (MI), plays an especially important role in nonlinear-optical spectral transformations of ultrashort pulses in PCFs.

In the simplest, scalar regime of modulation instability, two photons of the pump field with the frequency ω_p generate Stokes and anti-Stokes sidebands with frequencies $\omega_s = \omega_p - \Omega$ and $\omega_a = \omega_p + \Omega$. To illustrate this regime of MI, we represent the propagation constants of the Stokes and anti-Stokes sidebands as Taylor-series expansions around ω_p,

$$\beta(\omega_p+\Omega) \approx \beta_0(\omega_p) + \frac{1}{u_p}\Omega + \frac{1}{2}\beta_2(\omega_p)\Omega^2 + 2\gamma P\,, \tag{4.164}$$

$$\beta(\omega_p-\Omega) \approx \beta_0(\omega_p) - \frac{1}{u_p}\Omega + \frac{1}{2}\beta_2(\omega_p)\Omega^2 + 2\gamma P\,, \tag{4.165}$$

where P is the peak power of the pump field, $\beta_0(\omega_p)$ is the Kerr-effect-free propagation constant of the pump field mode (i. e., the propagation constant of the pump field corresponding to the regime with $P = 0$), $u_p = (\partial\beta/\partial\omega|_{\omega=\omega_p})^{-1}$ is the group velocity of the pump pulse, $\beta_2(\omega_p) = \partial^2\beta/\partial\omega^2|_\omega = \omega_p$, $\gamma = (n_2\omega_p)/(cS_{\text{eff}})$ is the nonlinear coefficient, and $S_{\text{eff}} = [\int_{-\infty}^{\infty}\int_{-\infty}^{\infty}|F(x,y)|^2\,\mathrm{d}x\,\mathrm{d}y]^2 / \int_{-\infty}^{\infty}\int_{-\infty}^{\infty}|F(x,y)|^4\,\mathrm{d}x\,\mathrm{d}y$ is the effective area of a guided mode with the transverse field profile $F(x, y)$.

With the propagation constant of the pump field written as

$$\beta(\omega_p) = \beta_0(\omega_p) + \gamma P\,, \tag{4.166}$$

the mismatch of the propagation constants of the fields involved in the FWM process is given by

$$\Delta\beta_{\text{FWM}} = \beta(\omega_p+\Omega) + \beta(\omega_p+\Omega) - 2\beta(\omega_p) \approx \beta_2(\omega_p)\Omega^2 + 2\gamma P\,. \tag{4.167}$$

The phase matching can thus be achieved for this type of FWM at

$$\Omega = \pm\left(\frac{2\gamma P}{|\beta_2(\omega_p)|}\right)^{1/2} \tag{4.168}$$

only when the central frequency of the pump field lies in the range of anomalous group-velocity dispersion, where $\beta_2(\omega_p) < 0$.

Figure 4.16 illustrates typical features of the scalar MI in PCFs observed by *Fedotov* et al. [4.82]. In those experiments, unamplified 50 fs pulses of 790–810 nm Ti:sapphire laser radiation with a repetition rate of 10 MHz and an energy of 0.1–1.4 nJ were coupled into micro-waveguide channels off the central core of the PCF (the inset in Fig. 4.16). The zero group-velocity dispersion (GVD) wavelength for the micro-waveguide channel used to observe MI is estimated as $\lambda_0 \approx 720$ nm, providing an anomalous GVD for the pump field. With a special choice of the pump pulse power, it is then possible to use the SPM phase shift to induce phase matching for efficient four-wave mixing (Fig. 4.16). This process can be understood as SPM-induced modulation instability, resulting in an exponential growth of spectral sidebands phase-matched with the pump field. In the output spectrum presented in Fig. 4.16, the 795 nm pump field generates sidebands centered at 700 and 920 nm. To understand this result, we use the standard result of MI theory (4.168) for the frequency shift corresponding to the maximum MI gain. With $\gamma \approx 50\,\text{W}^{-1}\text{km}^{-1}$, $D \approx 30$ ps/(nm · km), pulse energy $E \approx 0.5$ nJ, initial pulse duration $\tau \approx 50$ fs, we find $\Omega_{\max}/2\pi \approx 50$ THz, which agrees well with the frequency shifts of the sidebands observed in the output spectra presented

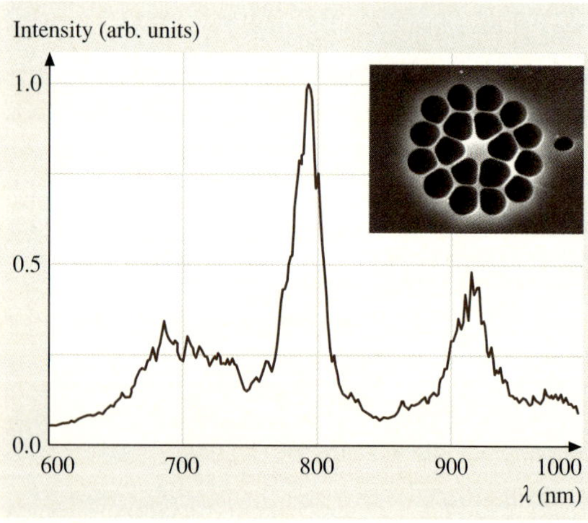

Fig. 4.16 Sideband generation through modulation instability in the spectrum of an ultrashort pulse transmitted through a photonic-crystal fiber (shown in the *inset*). The input energy of laser pulses is 0.5 nJ. A scanning electron microscopy (SEM) image of the photonic-crystal fiber is shown in the *inset*

in Fig. 4.16. SPM-induced broadening sufficient to seed the considered MI-type FWM process is achieved within a fiber length $z \approx (2L_\mathrm{d}L_\mathrm{nl})^{1/2}$, where $L_\mathrm{d} = \tau^2/|\beta_2|$ and $L_\mathrm{nl} = (\gamma P)^{-1}$ are the dispersion and nonlinear lengths, respectively. For the above-specified parameters of laser pulses and a PCF, we have $L_\mathrm{d} \approx 25$ cm and $L_\mathrm{nl} \approx 0.2$ cm. SPM-induced broadening can thus provide seeding for sideband generation at Ω_max with PCF lengths exceeding 3.2 cm. This condition was satisfied in our experiments, where the PCF length was equal to 8 cm. Our experimental results thus demonstrate efficient regimes of SPM-induced MI in microchannel waveguides of PCFs, which offer much promise for parametric frequency conversion and photon-pair generation.

4.6.2 Cross-Phase-Modulation-Induced Instabilities

Self-induced MI in PCFs has been shown [4.53, 82] to provide a convenient means for broadband parametric amplification, permitting the creation of compact and convenient all-fiber optical parametric oscillators. Cross-phase-modulation-induced MI, on the other hand, does not require anomalous dispersion [4.23], suggesting convenient and practical control knobs for the frequency conversion of ultrashort pulses in PCFs, allowing the frequency shifts and amplitudes of sidebands in output spectra of the probe field to be tuned by varying the amplitude of the pump field.

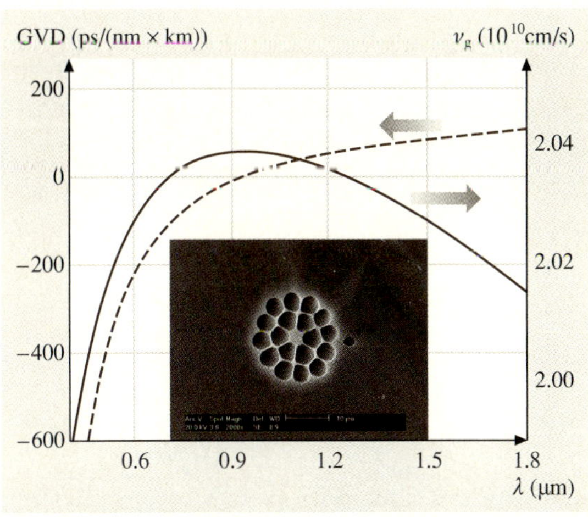

Fig. 4.17 The group velocity (the *solid curve* and the *right-hand axis*) and the group-velocity dispersion (the *dashed curve* and the *left-hand axis*) calculated as functions of the radiation wavelength λ for the fundamental mode of the PCF with a cross-sectional structure shown in the inset

Experiments [4.83] were performed with high-index-step fused-silica PCFs having a cross-sectional geometry shown in the inset to Fig. 4.17 with a core diameter of 4.3 μm. Figure 4.17 displays the group velocity and GVD calculated for this PCF using the polynomial expansion technique [4.84]. The standard theory of XPM-induced MI, as presented by *Agrawal* [4.85], was used to analyze the main features of this phenomenon for fundamental-wavelength and second-harmonic femtosecond pulses of a Cr:forsterite laser co-propagating in a PCF with the above-specified structure. This theory predicts that stationary solutions to slowly varying envelope approximation equations for the pump and probe fields including dispersion up to the second order become unstable with respect to a small harmonic perturbation with the wave vector K and the frequency Ω if K has a nonzero imaginary part. The domains of this instability can be found by analyzing the dispersion relation

$$\left[(K-\Omega\delta g/2)^2-h_1\right]\left[(K+\Omega\delta g/2)^2-h_2\right]=C^2\,, \quad (4.169)$$

where

$$h_j=\beta_{2j}^2\Omega^2\left(\Omega^2+4\gamma_j P_j g/\beta_{2j}\right)/4\,, \quad (4.170)$$

$$C=2\Omega^2\sqrt{\beta_{21}\beta_{22}\gamma_1\gamma_2 P_1 P_2}\,, \quad (4.171)$$

$\gamma_j = n_2\omega_j/(cS_j)$ are the nonlinear coefficients, $\delta = (v_{g2})^{-1} - (v_{g1})^{-1}$, $\beta_{2j} = (\mathrm{d}^2\beta_j/\mathrm{d}\omega^2)_{\omega=\omega_j}$, P_j, ω_j, v_{gj} and β_j are the peak powers, the central frequencies, the group velocities, and the propagation constants of the pump ($j = 1$) and probe ($j = 2$) fields, n_2 is the nonlinear refractive index, and S_j are the effective mode areas for the pump and probe fields. The gain of instabilities with a wave number K is given by $G(\Omega) = 2\,\mathrm{Im}(K)$.

Analysis of the dispersion properties of the PCFs employed in our experiments (Fig. 4.17) yields $\beta_{21} \approx -500\,\mathrm{fs}^2/\mathrm{cm}$, $\beta_{22} \approx 400\,\mathrm{fs}^2/\mathrm{cm}$, and $\delta = 150$ fs/cm. The dimensionless frequency shift of the probe field, $f = \Omega/\Omega_\mathrm{c}$ [where $\Omega_\mathrm{c} = (4\gamma_2 P_2/|\beta_{22}|)^{1/2}$] changes from approximately 3.3 up to 3.8 as the $\gamma_1 P_1/\gamma_2 P_2$ ratio is varied from 0.3 to 2.5. As highlighted by *Agrawal* [4.85], such a weak dependence of the frequency shift of the probe field on the pump power is typical of XPM-induced MI in the regime of pump–probe group-velocity mismatch. With $\gamma_2 P_2 \approx 1.5\,\mathrm{cm}^{-1}$, the frequency shift $f \approx 3.8$ gives sidebands shifted by $\Omega/2\pi \approx 74$ THz with respect to the central frequency ω_2 of the second harmonic (which corresponds to approximately 90 nm on the wavelength scale). As will be shown below, this prediction agrees well with our experimental results.

The laser system used in experiments [4.83] consisted of a Cr^{4+}:forsterite master oscillator, a stretcher, an optical isolator, a regenerative amplifier, and a compressor. The master oscillator, pumped with a fiber ytterbium laser, generated 30–60 fs light pulses of radiation with a wavelength of 1.23–1.25 μm at a repetition rate of 120 MHz. These pulses were then transmitted through a stretcher and an isolator, to be amplified in a Nd:YLF-laser-pumped amplifier and recompressed to the 170 fs pulse duration with the maximum laser pulse energy up to 40 μJ at 1 kHz. A 1 mm-thick β barium borate (BBO) crystal was used to generate the second harmonic of amplified Cr:forsterite laser radiation. Fundamental-wavelength, 1235 nm radiation of a femtosecond Cr:forsterite laser and its second harmonic were used as pump and probe fields, respectively. As can be seen from Fig. 4.16, the pump wavelength falls within the area of anomalous dispersion for the fundamental mode of the PCF, while the second-harmonic probe lies in the range of normal dispersion. The faster pump pulse Fig. 4.17 was delayed in our experiments with respect to the slower probe pulse at the input of the PCF by a variable delay time of τ.

Figure 4.18 presents the results of experimental measurements performed with 170 fs pump pulses (the fundamental radiation of the Cr:forsterite laser) with an energy ranging from 0.2 up to 20 nJ and 3 nJ, 180 fs probe pulses (the second-harmonic output of the Cr:forsterite laser) transmitted through a 5 cm PCF with the cross-sectional structure shown in the inset to Fig. 4.17. For delay times τ around zero, the slower probe pulse sees only the trailing edge of the faster moving pump pulse. In such a situation, XPM predominantly induces a blue shift of the probe field. For $\tau \approx \delta L \approx 750$ fs, where $L = 5$ cm is the PCF length, the leading edge of the pump pulse catches up with the probe field closer to the output end of the fiber, which results in a predominant red shift of the probe. To symmetrize the interaction between the pump and probe fields with respect to the XPM-induced frequency shift, we choose the delay time $\tau = \delta L/2 \approx 375$ fs. In the regime of low peak pump powers (less than 3 kW), the output spectrum of the probe field displays only slight broadening due to self-phase modulation (Fig. 4.18a). Pump pulses with higher peak powers lead to radical changes in the output spectra of the probe field, splitting the central spectral component of the probe field and giving rise to intense symmetric sidebands around the central frequency ω_2 (Figs. 4.18b–d).

The general tendencies in the behavior of the output spectrum of the probe field as a function of the pump power agree well with the prediction of the standard theory of XPM-induced MI. In view of the splitting and slight blue-shifting of the central spectral component of the probe field (Figs. 4.18b–d), we define the effective central wavelength of the pump-broadened probe spectrum as 605 nm. As the pump power changes from 5 kW up to 42 kW in our experiments, the shift of the short-wavelength sideband in the output spectrum of the second harmonic increases from 80 nm up to approximately 90 nm. The theory predicts the wavelength shifts of 76 nm and 90 nm, respectively, indicating the predominant role of XPM-induced MI in the observed spectral transformations of the probe field. The ampli-

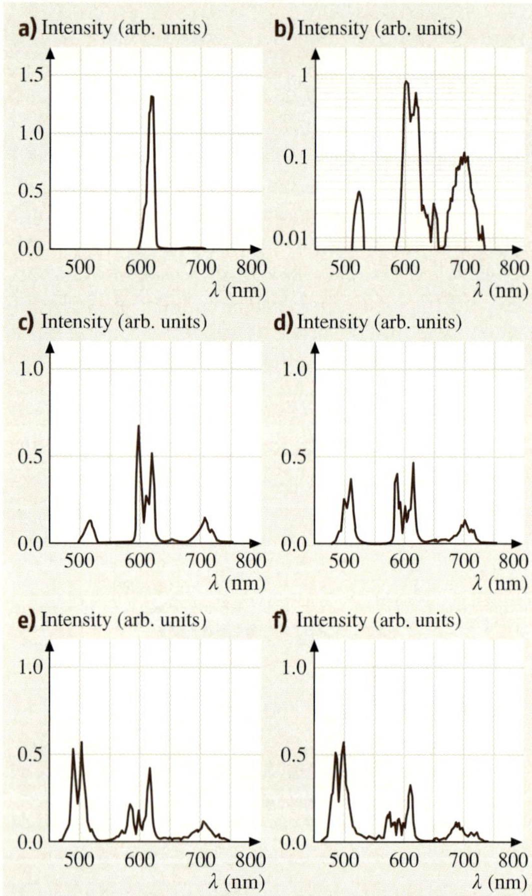

Fig. 4.18a–f Output spectra of the second-harmonic field transmitted through a 5 cm PCF. The power of the pump pulses is **(a)** 3, **(b)** 7, **(c)** 30, **(d)** 42, **(e)** 70, and **(f)** 100 kW. The power of the probe pulse is 8 kW

tudes of sidebands generated by pump pulses with a peak power of about 40 kW, as can be seen from Fig. 4.18d, become comparable or may even exceed the amplitude of the spectral components at the central part of the probe spectrum. The maximum frequency shift of the probe-field sidebands achieved in our experiments with 45 kW pump pulses is estimated as 80 THz, which is substantially larger than typical frequency shifts resulting from XPM-induced MI in conventional fibers [4.23]. With pump powers higher than 50 kW, both the central spectral components of the probe field and its sidebands featured a considerable broadening (Figs. 4.18e,f) and tended to merge together, apparently due to the cross-phase modulation induced by the pump field.

XPM-induced instabilities thus open an efficient channel of parametric FWM frequency conversion in photonic-crystal fibers. Fundamental-wavelength femtosecond pulses of a Cr:forsterite laser were used in our experiments as a pump field to generate intense sidebands around the central frequency of co-propagating second-harmonic pulses of the same laser through XPM-induced MI in a PCF. This effect leads to efficient pump-field-controlled sideband generation in output spectra of the second-harmonic probe field.

4.6.3 Solitonic Phenomena in Media with Retarded Optical Nonlinearity

Optical solitons propagating in media with noninstantaneous nonlinear response experience reshaping and continuous frequency down-shifting due to the Raman effect phenomenon, called soliton self-frequency shift (SSFS) [4.86, 87]. Photonic-crystal fibers substantially enhance this nonlinear-optical process due to strong field confinement in a small-size fiber core and the possibility to tailor dispersion of guided modes by varying the fiber structure. *Liu* et al. [4.54] have shown that 200 fs input pulses of 1.3 μm laser radiation can generate sub-100 fs soliton pulses with a central wavelength tunable down to 1.65 μm through the SSFS in a tapered PCF. Photonic-crystal fibers with the wavelength of zero group-velocity dispersion (GVD) shifted to shorter wavelengths have been used for the soliton frequency downshifting of 800–1050 nm laser pulses [4.88, 89]. *Abedin* and *Kubota* [4.90] have employed a PCF to demonstrate a 120 nm SSFS for 10 GHz-repetition-rate picosecond pulses. In recent experiments [4.91, 92], PCFs with a special dispersion profile have been shown to provide an efficient spectral transformation of chirped sub-6 fs Ti:sapphire laser pulses through SSFS, leading to the generation of a well-resolved solitonic spectral component centered at 1.06 μm. Red-shifted soliton signals formed by sub-6 fs laser pulses in PCFs have been demonstrated to allow to a synchronized seeding of a picosecond Nd:YAG pump laser, permitting a considerable simplification of a few-cycle optical parametric chirped-pulse amplification (OPCPA) scheme [4.79].

With many of the key tendencies in the evolution of ultrashort pulses in PCFs analyzed in the extensive literature, we focus here on the possibility of using the SSFS phenomenon for widely tunable frequency shifting of few-cycle laser pulses. Our theoretical analysis is based on the numerical solution of the generalized nonlinear Schrödinger equation [4.93]

$$\frac{\partial A}{\partial z} = \mathrm{i}\sum_{k=2}^{6} \frac{(\mathrm{i})^k}{k!} \beta^{(k)} \frac{\partial^k A}{\partial \tau^k} + \mathrm{i}\gamma \left(1 + \frac{\mathrm{i}}{\omega_0}\frac{\partial}{\partial \tau}\right) \times \left[A(z,\tau) \int_{-\infty}^{\infty} R(\eta)\,|A(z,\tau-\eta)|^2\, \mathrm{d}\eta \right], \tag{4.172}$$

where A is the field amplitude, $\beta^{(k)} = \partial^k \beta / \partial \omega^k$ are the coefficients in the Taylor-series expansion of the propagation constant β, ω_0 is the carrier frequency, τ is the retarded time, $\gamma = (n_2\omega_0)/(cS_{\mathrm{eff}})$ is the nonlinear coefficient, n_2 is the nonlinear refractive index of the PCF material,

$$S_{\mathrm{eff}} = \frac{\left[\int_{-\infty}^{\infty}\int_{-\infty}^{\infty} |F(x,y)|^2\, \mathrm{d}x\,\mathrm{d}y \right]^2}{\int_{-\infty}^{\infty}\int_{-\infty}^{\infty} |F(x,y)|^4\, \mathrm{d}x\,\mathrm{d}y} \tag{4.173}$$

is the effective mode area [$F(x, y)$ is the transverse field profile in the PCF mode], and $R(t)$ is the retarded nonlinear response function. For fused silica, we take $n_2 \approx 3.2\times10^{-16}\,\mathrm{cm^2/W}$, and the $R(t)$ function is represented in a standard form [4.93, 94]:

$$R(t) = (1 - f_{\mathrm{R}})\delta(t) + f_{\mathrm{R}}\Theta(t)\frac{\tau_1^2 + \tau_2^2}{\tau_1\tau_2^2} \mathrm{e}^{-\frac{t}{\tau_2}} \sin\left(\frac{t}{\tau_1}\right), \tag{4.174}$$

where $f_R = 0.18$ is the fractional contribution of the Raman response; $\delta(t)$ and $\Theta(t)$ are the delta and the Heaviside step functions, respectively; and $\tau_1 = 12.5$ fs and $\tau_2 = 32$ fs are the characteristic times of the Raman response of fused silica.

We now apply (4.172) and (4.174) to compute the evolution of ultrashort pulses in two types of PCFs (Figs.

Part A | 4.6

4.19, 4.20). PCFs of the first type consist of a fused silica core with a diameter of $1.6\,\mu m$, surrounded with two cycles of air holes (inset in Fig. 4.20a). To find the parameters $\beta^{(k)}$ for these fibers, we numerically solved the Maxwell equations for the transverse components of the electric field in the cross section of a PCF using a modification of the method of polynomial expansion in localized functions [4.84]. Polynomial approximation of the frequency dependence of the propagation constant β for the fundamental mode of the PCF computed with the use of this numerical procedure with an accuracy better than 0.1% within the range of wavelengths 580–1220 nm yields the following $\beta^{(k)}$ coefficients for the central wavelength of 800 nm: $\beta^{(2)} \approx -0.0293\,\mathrm{ps^2/m}$, $\beta^{(3)} \approx 9.316\times10^{-5}\,\mathrm{ps^3/m}$, $\beta^{(4)} \approx -9.666\times10^{-8}\,\mathrm{ps^4/m}$, $\beta^{(5)} \approx 1.63\times10^{-10}\,\mathrm{ps^5/m}$, $\beta^{(6)} \approx -3.07\times10^{-13}\,\mathrm{ps^6/m}$. For the fundamental mode of such PCFs, the GVD, defined as $D = -2\pi c\lambda^{-2}\beta^{(2)}$, vanishes at $\lambda_z \approx 690\,\mathrm{nm}$. Fibers of the second type are commercial NL-PM-750 PCFs (from Crystal Fibre). The core diameter for these PCFs was equal to $1.8\,\mu m$. The parameters $\beta^{(k)}$ for these PCFs were defined as polynomial expansion coefficients for the dispersion profile of the fundamental mode of these fibers provided by the manufacturer. The group-velocity dispersion for PCFs of this type vanishes at $\lambda_z \approx 750\,\mathrm{nm}$.

In the case studied here, the laser field at the input of a PCF has the form of a few-cycle pulse (the upper panel in Fig. 4.19a) with a broad spectrum (the upper panel in Fig. 4.19b) and a complicated chirp [4.79, 92]. For both types of PCFs, the short-wavelength part of the spectrum lies in the range of normal dispersion, while the wavelengths above λ_z experience anomalous dispersion. A typical scenario of spectral and temporal evolution of a few-cycle laser pulse in PCFs of the considered types is illustrated by Figs. 4.19a,b. The initial stage of nonlinear-optical transformation of a few-cycle pulse involves self-phase modulation, which can be viewed as four-wave mixing of different frequency components belonging to the broad spectrum of radiation propagating through the fiber. Frequency components lying near the zero-GVD wavelength of the PCF then serve, as shown in the classical texts on nonlinear fiber optics [4.23], as a pump for phase-matched FWM. Such phase-matched FWM processes, which involve both frequency-degenerate and frequency-nondegenerate pump photons, deplete the spectrum of radiation around the zero-GVD wavelength and transfer the radiation energy to the region of anomalous dispersion (spectral components around 920 nm for $z = 2\,\mathrm{cm}$ in Fig. 4.19b). A part of this frequency-downconverted radiation then couples into a soliton, which undergoes continuous frequency downshifting due to the Raman effect (Fig. 4.19b), known as soliton self-frequency shift [4.23, 85, 86]. In the time domain, the red-shifted solitonic part of the radiation field becomes

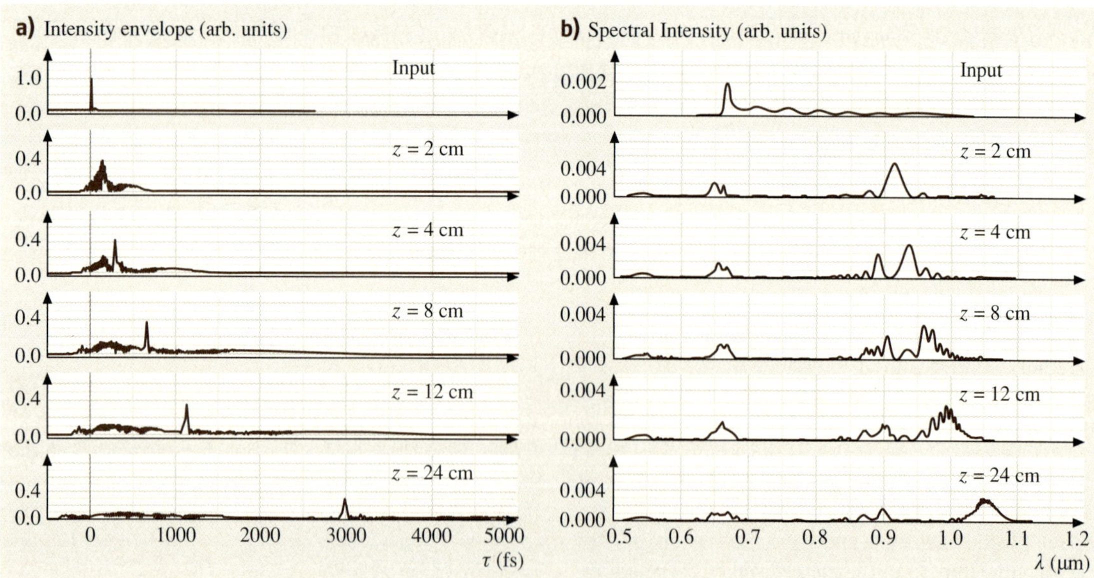

Fig. 4.19 Temporal **(a)** and spectral **(b)** evolution of a laser pulse with an initial energy of 0.25 nJ and an input temporal envelope and chirp shown in Fig. 4.1b propagating through the second-type PCF (shown in the inset)

delayed with respect to the rest of this field (Fig. 4.19a) because of the anomalous GVD of the fiber. As a result of these processes, the red-shifted soliton becomes increasingly isolated from the rest of the light field in both the time and frequency domains, which reduces, in particular, the interference between the solitonic and nonsolitonic part of radiation, seen in Fig. 4.19b.

High-order fiber dispersion induces soliton instabilities, leading to Cherenkov-type emission of dispersive waves [4.21, 22] phase-matched with the soliton, as discussed in the extensive literature (see, e.g., [4.95, 96]). This resonant dispersive-wave emission gives rise to a spectral band centered around 540 nm in Fig. 4.19b. As a result of the above-described nonlinear-optical transformations, the spectrum of the radiation field for a PCF with a characteristic length of 20 cm typically features four isolated bands, representing the remainder of the FWM-converted pump field (the bands centered at 670 and 900 nm in Fig. 4.19b), the red-shifted solitonic part (reaching 1.06 μm for $z = 24$ cm in Fig. 4.19b), and the blue-shifted band related to the Cherenkov emission of dispersive waves in the visible. In the time domain, as

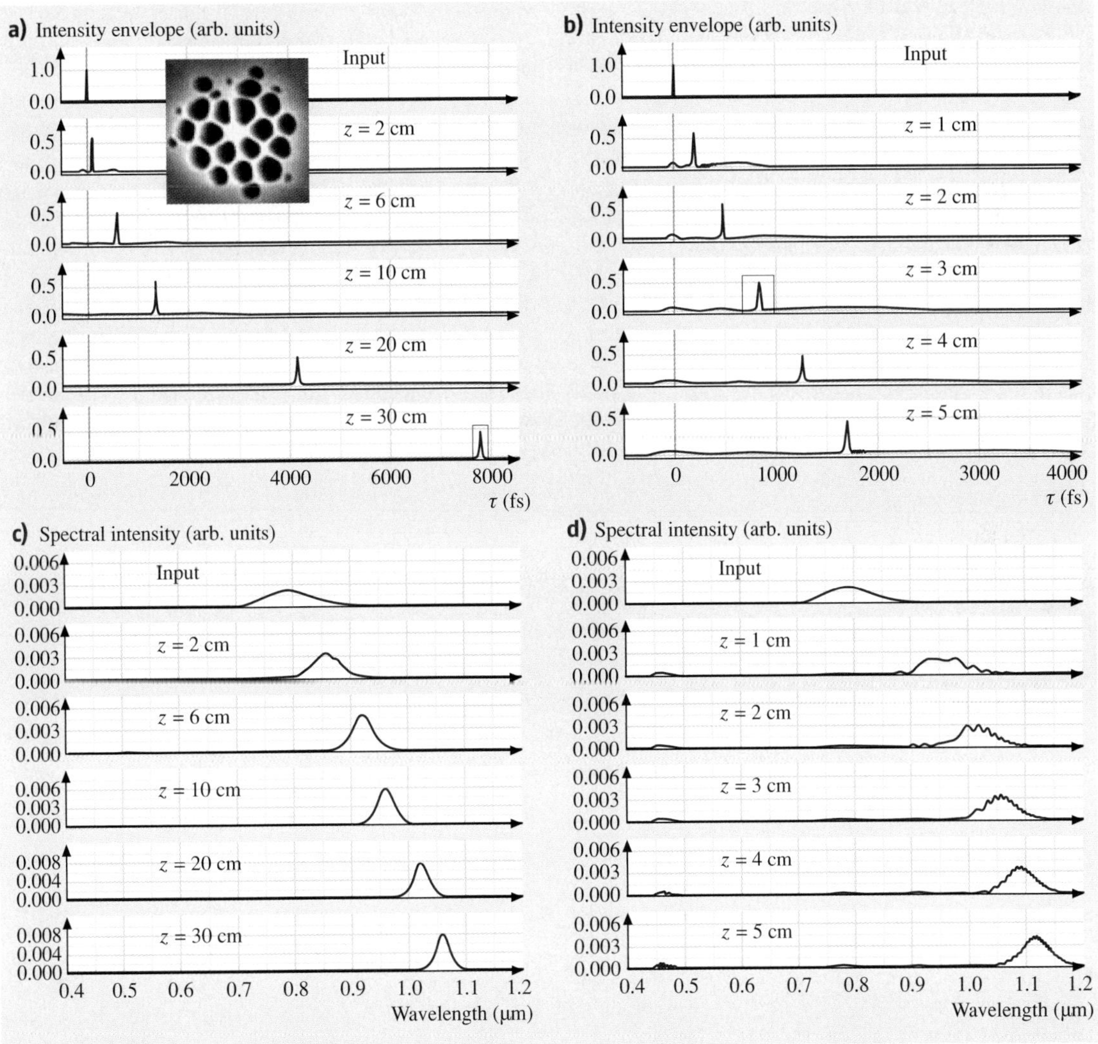

Fig. 4.20 Temporal **(a)**, **(b)** and spectral **(c)**, **(d)** evolution of laser pulses with an initial energy of **(a)**, **(c)** 0.15 nJ and **(b)**, **(d)** 0.5 nJ and an initial pulse width of 6 fs in the first-type PCF (shown in the *inset*). The input pulses are assumed to be transform limited

can be seen from Fig. 4.19a, only the solitonic part of the radiation field remains well localized in the form of a short light pulse, the remaining part of the field spreading out over a few picoseconds.

In Figs. 4.20a–d, we illustrate tunable frequency shifting of few-cycle laser pulses through SSFS in PCFs by presenting the results of simulations performed for an idealistic input pulse with an initial pulse width of 6 fs and a Gaussian pulse shape. For the first-type PCF (shown in the inset to Fig. 4.20a), almost the entire spectrum of the input pulse falls within the range of anomalous dispersion, and the pulse tends to form solitons, observed as well-resolved prominent spikes in the time domain (Figs. 4.20a,b). In the frequency domain, the Raman effect leads to a continuous frequency downshifting of the soliton (Figs. 4.20c,d). The rate of this frequency shift $d\nu/dz$, where ν is the carrier frequency and z is the propagation coordinate, rapidly grows with a decrease in the pulse duration τ_0. With a linear approximation of the Raman gain as a function of the frequency, the integration of the nonlinear Schrödinger equation, as shown by *Gordon* [4.97], yields $d\nu/dz \propto \tau_0^{-4}$. Although high-order dispersion and deviations of the Raman gain curve from the linear function generally make the relation between $d\nu/dz$ and τ in soliton dynamics much more complicated [4.98], the soliton pulse width remains one of the key parameters controlling the soliton frequency shift for a given Raman gain profile. In the case considered here, the short duration of 6 fs input pulses provides a high rate of soliton-frequency shifting at the initial stage of pulse propagation through the PCF. As the spectrum of the soliton is shifted toward the spectral range with larger values of GVD, the pulse width increases, which slows down the frequency shift.

It is instructive to illustrate the main tendencies in the spectral and temporal evolution of few-cycle laser pulses in PCFs using the results of analysis of ideal solitons, i. e., solitons governed by the nonlinear Schrödinger equation [(4.68) in Sect. 4.4.2]. The NLSE (4.68) is recovered from (4.172) by setting $\beta^{(k)} = 0$ for $k \geq 3$, taking $f_R = 0$, and keeping only the term representing the Kerr effect, i. e., the term proportional to $i\gamma A\,|A|^2$, in the nonlinear part of the equation. In normalized, soliton units, the energy carried by a soliton j is [4.98] $E_j = 4\xi_j$, where $\xi_j = W - j + 0.5$ is the soliton eigenvalue, controlled by the input pulse energy $2W^2$. The soliton pulse width is given by $\tau_j = \tau_0/2\xi_j$, where τ_0 is the input pulse width. The soliton pulse width can thus be reduced, leading to higher SSFS rates, by increasing the energy of the input pulse.

In the case of solitary waves evolving in fibers with high-order dispersion and retarded nonlinearity, the results of NLSE analysis for the soliton energy and the soliton pulse width are no longer valid. In particular, as the soliton spectrum is shifted toward larger values of GVD, the soliton pulse width is bound to increase, while the soliton amplitude decreases (Figs. 4.20a,b). These changes in the soliton pulse width and amplitude are dictated by the balance between the dispersion and the nonlinearity, necessary for the existence of the soliton. On the qualitative level, however, being applied to short sections of a fiber, these simple relations provide important clues for the physical understanding of the evolution of Raman-shifted solitons in a PCF. Indeed, as can be seen from the comparison of the results of simulations performed for input pulses with the same initial pulse width (6 fs), but different energies, the SSFS rate in the case of higher energy pulses can substantially exceed the frequency-shift rate of solitons produced by lower energy pulses. A pulse with an input energy of 0.15 nJ, as can be seen from Fig. 4.20c, is coupled into a soliton, which undergoes a permanent red-shift as it propagates through the fiber. At $z = 30$ cm, the spectrum of this soliton peaks at 1.06 μm. A similar input pulse that has an initial energy of 0.5 nJ forms a soliton that exhibits a much faster frequency downshift. The central wavelength of this soliton reaches 1.12 μm already at $z = 5$ cm.

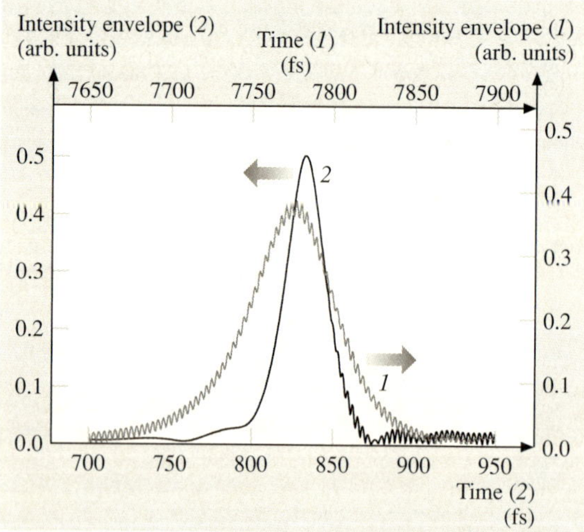

Fig. 4.21 Temporal envelopes of red-shifted solitons (close-up of the peaks labeled with boxes in Figs. 4.20a,b) generated by laser pulses with an initial pulse width of 6 fs and an initial energy of (1) 0.15 nJ and (2) 0.5 nJ in a PCF at $z = 30$ cm (1) and 3 cm (2)

To provide illustrative physical insights into the observed behavior of red-shifted solitons in PCFs as a function of the input pulse energy, we plot in Fig. 4.21 the snapshots of temporal envelopes of the solitonic part of the field corresponding to the input energies of 0.15 nJ (curve 1) and 0.5 nJ (curve 2). We take these snapshots of solitons, representing close-up views of intensity envelope sections labeled with boxes in Figs. 4.20a,b, for two different values of the propagation coordinate, $z = 30$ cm in the case of a 0.15 nJ input pulse and $z = 3$ cm for the 0.5 nJ input energy. With the spectra of red-shifted solitons centered around 1.06 μm in both cases (Figs. 4.20c,d), these values of the propagation coordinate allow a fair comparison of SSFS dynamics in terms of the dependence of the frequency shift rate on the soliton pulse width. As is seen from Fig. 4.21, the SSFS rate correlates well with the soliton pulse width. While the soliton produced by a pulse with an initial energy of 0.15 nJ has a pulse width of about 50 fs (curve 1 in Fig. 4.21), the pulse width of the soliton emerging from the 0.5 nJ laser pulse is about 20 fs. In qualitative agreement with predictions of *Gordon* [4.97] and *Lucek* and *Blow* [4.98], this shorter soliton in Figs. 4.20b,d displays a faster downshifting as compared with the longer soliton in Figs. 4.20a,c. In the following section, this dependence of the SSFS rate on the energy of the pulse launched into the fiber will be used for the experimental demonstration of widely tunable soliton frequency shift of 6 fs pulses produced by a Ti:sapphire oscillator.

At higher input powers, the spectral features originating from FWM, SSFS, dispersive-wave emission of solitons experience broadening due to SPM and XPM effects, merging together and giving rise to a broadband white-light emission (Figs. 4.15, 4.22). This supercontinuum radiation generated in PCFs has been intensely employed through the past few years to measure and control the offset between the carrier and envelope phases of ultrashort laser pulses, as well as for the creation of novel broadband sources for nonlinear spectroscopy, biomedical applications, and photochemistry. Examples of applications of PCF light sources based on enhanced nonlinear-optical interactions of guided modes will be given in Sect. 4.9.

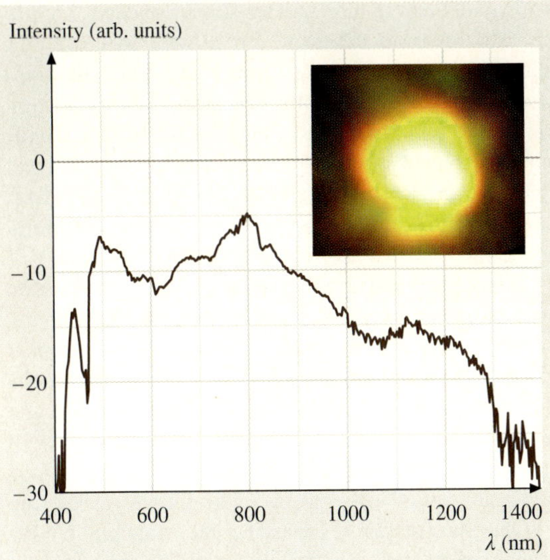

Fig. 4.22 The spectrum of supercontinuum emission produced by 820 nm pump pulses with an initial duration of 35 fs and an input power of 320 mW in a microstructure fiber with a length of 30 cm and the cross-section structure shown in the inset to Fig. 4.20a.

4.7 Nonlinear Raman Spectroscopy

Nonlinear Raman spectroscopy is one of the most powerful techniques of nonlinear spectroscopy, which has found numerous applications in condensed- and gas-phase analysis, plasma diagnostics, investigation of molecular relaxation processes, temperature and concentration measurements, condensed-phase studies, and femtochemistry. While many of modifications of nonlinear Raman spectroscopy have become a routine tool of modern optical experiments, giving rise to many successful engineering applications, some of nonlinear Raman experiments carried out in the last decade have shown that the potential of this technique for many topical and sometimes interdisciplinary problems of modern physics, chemistry, and biology is far from being completely realized. Similar to frequency-tunable sources of coherent radiation, which revolutionized nonlinear optics in its early days, allowing many delicate spectroscopic experiments, including nonlinear spectroscopic studies, to be performed, the impressive progress of femtosecond lasers in the 1990s has resulted in the breakthrough of the nonlinear Raman spectroscopy to new unexplored areas, giving rise to several elegant new ideas and approaches, permitting more-complicated systems and problems to be attacked, and leading to the measurements of fundamental importance. This new phase of nonlinear Raman spectroscopy also promoted the development of new spectroscopic concepts, including time-resolved schemes, broadband spectroscopy, polarization measurements, and CARS generalizations based on higher-order nonlinear processes. This conceptual and technical progress achieved in the last decade shows us some very important features of what nonlinear Raman spectroscopy is going to be in the nearest future, encouraging the application of new ideas, techniques, and methods in this area of spectroscopy.

This section provides a brief introduction to the main principles of nonlinear Raman spectroscopy, giving a general idea of how the measurements are performed and the spectroscopic data are extracted from the results of these measurements. Following this plan, we will first give a brief introduction to the basic concepts of nonlinear Raman spectroscopy. Then, we will consider various modifications of coherent Raman four-wave mixing (FWM) spectroscopy, including the standard CARS scheme, stimulated Raman scattering, Raman-induced Kerr effect, degenerate four-wave mixing (DFWM), and coherent hyper-Raman scattering. We will also briefly describe polarization techniques for nonlinear Raman spectrometry and coherent ellipsometry, allowing selective investigation of multicomponent molecular and atomic systems and permitting the sensitivity of nonlinear Raman spectrometry to be radically improved. Finally, in the context of the growing interest in the applications of short-pulse spectroscopy for the investigation of ultrafast processes, we will provide an introduction to time-resolved nonlinear Raman spectroscopy.

4.7.1 The Basic Principles

Nonlinear Raman spectroscopy is based on nonlinear-optical interactions in Raman-active media. The nonlinear character of light interaction with a medium implies that molecular, atomic, or ionic vibrations in a medium are no longer independent of the light field. Instead, pump light waves with frequencies ω_1 and ω_2 modulate Raman-active vibrations in a medium at the frequency $\Omega \approx \omega_1 - \omega_2$, which can be then probed with another light beam, generally having a frequency ω_3 (Fig. 4.6c). This wave-mixing process involving the inelastic scattering of the probe wave by molecular vibrations gives rise to coherent Stokes- and anti-Stokes-frequency-shifted signals, whose amplitude I, polarization (the ellipticity χ and the tilt angle ψ of the principal axis of the polarization ellipse), and phase φ carry the spectroscopic information concerning the medium under study. This is the general idea of nonlinear Raman spectrometry, illustrated by Fig. 4.6c.

In practical terms, to undertake a simple three-color CARS experiment, one generally needs three laser sources generating radiation with the frequencies meeting the requirements specified above. In the most popular scheme of two-color CARS, where the anti-Stokes signal is generated through the frequency-mixing scheme $\omega_a = 2\omega_1 - \omega_2$, the number of lasers required is reduced to two. The light beams have to be brought into coincidence in space to excite Raman-active transitions in a medium and to generate the anti-Stokes signal. Introducing some delay time between pumping and probing pulses, one can also perform time-resolved CARS measurements to keep track of the temporal dynamics of excitation in the system under investigation.

While spontaneous Raman scattering often suffers from the low quantum yield, which eventually results in the loss of sensitivity, coherent Raman scattering allows much more intense signals to be generated, thus allowing very high sensitivities to be achieved. Coherent Raman scattering also offers several other very important advantages that stem from the coherent character of the signal, which is thus well collimated and generated in a precisely known direction.

Generally, the analysis of the amplitude, phase, and polarization of the signal resulting from a nonlinear wave-mixing process in a Raman-resonant medium involves the calculation of the nonlinear response of the medium in terms of the relevant nonlinear susceptibilities and the solution of Maxwell equations for the field of the signal. Below, we restrict our brief introduction to the theory of nonlinear Raman processes to the description of the basic notions and terminology, which will be employed later to explain the main concepts of various nonlinear Raman methods, including frequency- and time-domain CARS, coherent ellipsometry, and the Raman-induced Kerr effect (RIKE).

Excitation of Raman Modes

Within the framework of a simple, but physically instructive, semiclassical model of a nonlinear medium consisting of noninteracting molecules with a Raman-active vibration with frequency Ω, the interaction of light with molecules (atoms or ions) can be described in terms of the electronic polarizability of molecules α depending on the generalized normal coordinate Q (e.g., defined as the distance between the nuclei in a molecule) [4.99]:

$$\alpha(Q) = \alpha_0 + \left(\frac{\partial \alpha}{\partial Q}\right)_0 Q + \dots , \tag{4.175}$$

where α_0 is the equilibrium polarizability of a molecule, $(\partial\alpha/\partial Q)_0$ is the derivative of the electronic polarizability in the normal coordinate taken for the equilibrium position of nuclei, and we restrict our consideration to terms in this expansion linear in Q. The term $(\partial\alpha/\partial Q)_0 Q$ in (4.175) is responsible for the modulation of light by molecular vibrations, as it gives rise to new frequency components in the induced polarization of the system, whose frequency shift is determined by the frequency of molecular vibrations. This can be seen from the expression for the polarization of a medium:

$$P = Np\,, \tag{4.176}$$

where N is the number density of Raman-active molecules,

$$p = \alpha(Q)E = \alpha_0 E + \left(\frac{\partial\alpha}{\partial Q}\right)_0 QE + \ldots \tag{4.177}$$

is the dipole moment of a molecule and E is the electric field strength. The energy of a molecule in a light field is written as

$$H = -pE = -\alpha\,(Q)\,E^2\,. \tag{4.178}$$

Thus, the light-induced force driving molecular vibrations is given by

$$F = -\frac{\partial H}{\partial Q} = \frac{\partial\alpha}{\partial Q}E^2\,. \tag{4.179}$$

As can be seen from (4.179) the force acting on a molecule in a light field may result in a resonant excitation of Raman-active vibrations with a frequency Ω if the field involves frequency components ω_1 and ω_2 with $\omega_1 - \omega_2 \approx \Omega$.

Wave Mixing in Raman-Active Media

The propagation of light waves in a nonlinear medium is governed by the wave equation

$$\Delta \boldsymbol{E} - \frac{n^2}{c^2}\frac{\partial^2 \boldsymbol{E}}{\partial t^2} = \frac{4\pi}{c^2}\frac{\partial^2 \boldsymbol{P}_{\text{nl}}}{\partial t^2}\,, \tag{4.180}$$

where $\boldsymbol{E}$ is the field in the light wave, n is the refractive index, c is the speed of light, and $\boldsymbol{P}_{\text{nl}}$ is the nonlinear polarization of the medium.

Consider for example the CARS process, as an example of a nonlinear Raman process, assuming that it involves plane and monochromatic waves,

$$\begin{aligned}\boldsymbol{E}\,(\boldsymbol{r},t) &= \boldsymbol{E}_1 \exp\left(-\mathrm{i}\omega_1 t + \mathrm{i}\boldsymbol{k}_1\boldsymbol{r}\right)\\ &+ \boldsymbol{E}_2 \exp\left(-\mathrm{i}\omega_2 t + \mathrm{i}\boldsymbol{k}_2\boldsymbol{r}\right)\\ &+ \boldsymbol{E} \exp\left(-\mathrm{i}\omega t + \mathrm{i}\boldsymbol{k}\boldsymbol{r}\right)\\ &+ \boldsymbol{E}_\text{a} \exp\left(-\mathrm{i}\omega a t + \mathrm{i}\boldsymbol{k}_\text{a}\boldsymbol{r}\right) + \text{c.c.}\,,\end{aligned} \tag{4.181}$$

where ω_1 and ω_2 are the frequencies of pump waves, ω is the frequency of the probe wave, $\omega_\text{a} = \omega + \omega_1 - \omega_2$, the field envelopes $\boldsymbol{E}$, $\boldsymbol{E}_1$, $\boldsymbol{E}_2$, and $\boldsymbol{E}_\text{a}$ are slowly varying functions of coordinates $\boldsymbol{r}$ and time t, $\boldsymbol{k}$, $\boldsymbol{k}_1$, $\boldsymbol{k}_2$, and $\boldsymbol{k}_\text{a}$ are the wave vectors of light waves with frequencies ω, ω_1, ω_2, and ω_a, respectively, and c.c. stands for a complex conjugate. Then, representing the nonlinear polarization in the medium as a superposition of plane waves, we can write the equation for the amplitude of the anti-Stokes wave as

$$\frac{n_\text{a}}{c}\frac{\partial \boldsymbol{E}_\text{a}}{\partial t} + \frac{\partial \boldsymbol{E}_\text{a}}{\partial z} = \mathrm{i}\frac{2\pi\omega_\text{a}}{cn_\text{a}}\boldsymbol{P}_{\text{nl}}\,(\omega_\text{a})\exp\left(\mathrm{i}k_\text{a}\,z\right)\,, \tag{4.182}$$

where $\boldsymbol{P}_{\text{nl}}(\omega_\text{a})$, n_a, and k_a are the amplitude of the nonlinear polarization, the refractive index, and the z-component of the wave vector at the frequency ω_a. The pump wave amplitudes will be assumed to be constant. Now, the nonlinear polarization of a medium has to be found with the use of some model of the nonlinear medium.

Raman-Resonant Nonlinear Polarization and Nonlinear Susceptibilities

Within the framework of our model of a nonlinear medium, the third-order nonlinear polarization of the medium at the anti-Stokes frequency ω_a is given by [4.99]

$$P^{(3)}\,(\omega_\text{a}) = \frac{N}{4MD\,(\Omega,\omega_1-\omega_2)}\left(\frac{\partial\alpha}{\partial Q}\right)_0^2 EE_1E_2^*\,, \tag{4.183}$$

where

$$\begin{aligned}D\,(\Omega,\omega_1-\omega_2) &= \Omega^2 - (\omega_1-\omega_2)^2\\ &- 2\mathrm{i}\Gamma\,(\omega_1-\omega_2)\,,\end{aligned} \tag{4.184}$$

derivatives are taken in the equilibrium position, Γ is the phenomenologically introduced damping constant, M is the reduced mass of a molecule, N is the number density of molecules, an asterisk indicates a complex conjugate, Q is the amplitude of the Raman-active molecular vibration, which can be expressed in terms of the density matrix ρ of an ensemble of molecules,

$$Q = \text{Sp}\,(\rho q) = \rho_{ab}q_{ba}\,, \tag{4.185}$$

where Sp is the trace operator and q is the operator of the vibrational coordinate.

Introducing the third-order nonlinear-optical susceptibility of the medium,

$$\chi^{(3)R} = \frac{N}{24MD\,(\Omega,\omega_1-\omega_2)}\left(\frac{\partial\alpha}{\partial Q}\right)_0^2\,, \tag{4.186}$$

and solving (4.182), we arrive at the following expression for the intensity of the CARS signal I_a:

$$I_a \propto \left|\chi^{(3)}(\omega_a; \omega, \omega_1, -\omega_2)\right|^2 \times I I_1 I_2 l^2 \left(\frac{\sin\left(\frac{\Delta kl}{2}\right)}{\frac{\Delta kl}{2}}\right)^2 , \quad (4.187)$$

where I, I_1, I_2 are the intensities of the pump and probe beams, l is the length of the nonlinear medium, and $\Delta k = |\Delta \boldsymbol{k}| = |\boldsymbol{k}_a - \boldsymbol{k} - \boldsymbol{k}_1 + \boldsymbol{k}_2|$ is the wave-vector mismatch. As can be seen from (4.187), the anti-Stokes wave is generated especially efficiently in the direction of phase matching, where $\Delta k = 0$.

The cubic nonlinear-optical susceptibility of a medium $\chi^{(3)}_{ijkl}$ is a fourth-rank tensor. The knowledge of the form of this tensor is important, in particular, for understanding the polarization properties of the signal of nonlinear Raman scattering. The form of the $\chi^{(3)}_{ijkl}$ tensor is determined by the symmetry properties of a medium. For an isotropic medium, only 21 of 81 tensor components of $\chi^{(3)}_{ijkl}$ are nonvanishing. Only three of these components are independent of each other, as the relations

$$\chi^{(3)}_{1111} = \chi^{(3)}_{2222} = \chi^{(3)}_{3333} , \quad (4.188)$$

$$\chi^{(3)}_{1122} = \chi^{(3)}_{1133} = \chi^{(3)}_{2211} = \chi^{(3)}_{2233} = \chi^{(3)}_{3311} = \chi^{(3)}_{3322} , \quad (4.189)$$

$$\chi^{(3)}_{1212} = \chi^{(3)}_{2121} = \chi^{(3)}_{1313} = \chi^{(3)}_{3131} = \chi^{(3)}_{2323} = \chi^{(3)}_{3232} , \quad (4.190)$$

$$\chi^{(3)}_{1221} = \chi^{(3)}_{2112} = \chi^{(3)}_{1331} = \chi^{(3)}_{3113} = \chi^{(3)}_{2332} = \chi^{(3)}_{3223} , \quad (4.191)$$

$$\chi^{(3)}_{1111} = \chi^{(3)}_{1122} = \chi^{(3)}_{1212} = \chi^{(3)}_{1221} \quad (4.192)$$

hold true for an isotropic medium [4.9, 99]. The form of the $\chi^{(3)}_{ijkl}$ tensor for all the crystallographic classes can be found in many textbooks on nonlinear optics [4.9, 10].

Phase Matching of Focused Beams

Generally, in the case of focused beams, phase-matching effects in nonlinear Raman scattering are taken into account through the phase-matching integral. In particular, for Gaussian beams, the expression for the overall power of the two-color CARS signal occurring in accordance with the scheme $\omega_a - 2\omega_1 - \omega_2$ is written as [4.11, 100, 101]

$$P_a = \left(\frac{3\pi^2\omega_a}{c^2 n_a}\right)^2 \left|\chi^{(3)}\right|^2 \times P_1^2 P_2 \frac{32}{\pi^3} \left(\frac{b}{w_0^3}\right)^2 \int_0^\infty 2\pi r\, |J|^2\, \mathrm{d}r . \quad (4.193)$$

Here, P_1 and P_2 are the powers of the pump beams with frequencies ω_1 and ω_2, respectively, w_0 is the waist size of the focused pump beams, b is the confocal parameter,

$$J = \int_{-C_1}^{C_2} \frac{\exp\left(-\frac{r^2}{bH}\right)}{(1 + \mathrm{i}\xi')(k'' - \mathrm{i}k'\xi')\, H}\, \mathrm{d}\xi' , \quad (4.194)$$

where $-C_1$ and C_2 are the coordinates of the boundaries of the nonlinear medium,

$$H = \frac{1 + \xi'^2}{k'' - \mathrm{i}k'\xi'} - \mathrm{i}\frac{\xi' - \zeta}{k'} , \quad (4.195)$$

$k'' = 2k_1 + k_2$, $k' = 2k_1 - k_2$, $\zeta = 2(z - f)/b$ is the normalized coordinate along the z-axis (f is the coordinate of the beam waist along the z-axis).

Equations (4.193–4.195) show that the information on the nonlinear cubic susceptibility in the CARS signal may be distorted by phase-matching effects. This problem has been analyzed both theoretically and experimentally for different modifications of nonlinear Raman spectroscopy [4.102]. The influence of absorption at the wavelength of pump and probe waves and at the frequency of the FWM signal can be taken into account by including the imaginary parts of the relevant wave vectors. The integral in (4.194) can be calculated in an analytical form for several particular cases, giving a clear physical understanding of the role of phase-matching and absorption effects in coherent FWM spectroscopy and imaging [4.103].

4.7.2 Methods of Nonlinear Raman Spectroscopy

In this section, we will briefly consider standard and widely used schemes for nonlinear Raman spectroscopy (Fig. 4.2), including coherent Raman scattering, stimulated Raman scattering, and the Raman-induced Kerr effect and provide a brief introduction into the vast area of DFWM.

Stimulated Raman Scattering

The idea of using stimulated Raman scattering (SRS) as a spectroscopic technique is based on the measurement of the frequency dependence of the SRS small-signal gain, which is proportional to the imaginary part of the nonlinear cubic susceptibility of a Raman-active medium [4.9, 99]. The power of the pump wave in such measurements has to be chosen in such a way as to avoid uncontrollable instabilities and to obtain noticeable SRS gain. The SRS-based approach was successfully employed, in particular, for high-resolution spectroscopy of Raman transitions [4.104].

Limitations of SRS as a spectroscopic technique are due to instabilities arising for light intensities exceeding the threshold SRS intensity. These instabilities build up under conditions when several nonlinear processes, including self-focusing and self-phase modulation, compete with each other, often rendering the SRS method impractical for spectroscopic applications.

Coherent Anti-Stokes Raman Scattering

Instead of measuring the gain of one of two waves, as is done in SRS, Maker and Terhune [4.105] have demonstrated a spectroscopic technique based on measuring the frequency dependence of the intensity of a new wave generated at the anti-Stokes frequency $\omega_a = 2\omega_1 - \omega_2$ in the presence of two light waves with frequencies ω_1 and ω_2 chosen in such a way as to meet the condition of a Raman resonance with a Raman-active transition in a medium: $\omega_1 - \omega_2 \approx \Omega$ (Fig. 4.6c). This approach, called coherent anti-Stokes Raman scattering, has become one of the most widespread nonlinear Raman methods, allowing many urgent spectroscopic problems to be successfully solved and stimulating numerous engineering applications of nonlinear laser spectroscopy (so-called three-color CARS with $\omega_a = \omega_1 - \omega_2 + \omega_3$ is shown in Fig. 4.6c). Similar to the SRS process described in the previous section, CARS involves the stimulated scattering of light in a Raman-active medium. However, in contrast to the standard SRS scheme, where a Stokes wave is generated or amplified, the CARS process gives rise to the appearance of a new frequency component, suggesting a spectroscopic approach that is free of instabilities typically arising in SRS due to the competition of different nonlinear processes.

Due to its high spatial, temporal, and spectral resolution, the possibilities of studying highly luminous objects, and a rich variety of polarization methods, the CARS technique has gained a wide acceptance for temperature and concentration measurements in excited gases, combustion, and flames [4.99, 106–108], gas-phase analysis [4.99, 109, 110], high-resolution molecular spectroscopy [4.111, 112]. Short-pulse CARS gives an access to ultrafast processes and wave-packet dynamics in molecular systems [4.113]. The nonlinear nature and the spectral selectivity of CARS make this method an ideal tool for nonlinear spectroscopy [4.114, 115]. The most recent advances in nonlinear Raman techniques include coherence-controlled CARS [4.116], enhancing the potential of CARS microscopy [4.117], and CARS in photonic-crystal fibers [4.118, 119].

The widely employed geometry of nonlinear Raman measurements implies the use of collinear focused laser beams. While focusing allows the intensity sufficient to ensure a reliable detection of the nonlinear Raman signal to be achieved, the collinear geometry of wave mixing increases the length of nonlinear interaction. However, such an approach is reasonable as long as the spatial resolution along the propagation coordinate is not important. The scheme of nonlinear Raman spectroscopy becomes resolvable along the propagation coordinate as soon as collinear beams are replaced by a noncollinear one (Fig. 4.6c). In CARS, this technique is called the boxcars geometry [4.120]. The interaction area is confined in this case to the region where the beams intersect each other, allowing a high spatial resolution to be achieved.

In the broad-beam CARS geometry Fig. 4.23, focused laser beams are replaced with broad or sheetlike beams. This approach allows the nonlinear Raman signal to image the whole areas of a nonlinear medium

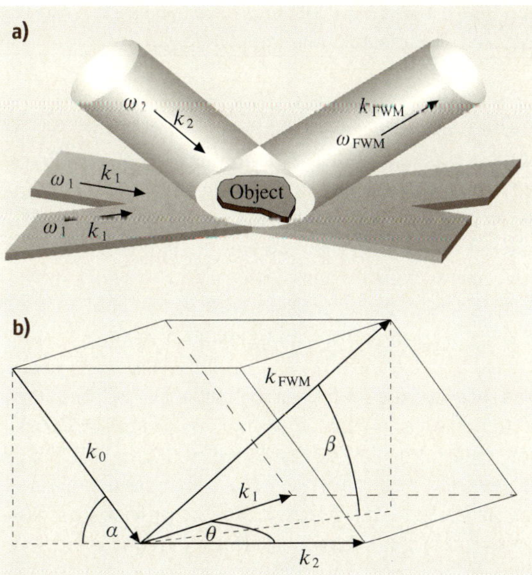

Fig. 4.23a,b Broad-beam folded coherent-anti-Stokes Raman scattering: **(a)** beam arrangements and **(b)** wave-vector diagram

on a charge-coupled device (CCD) camera. The idea of broad-beam CARS imaging, which was discussed by *Regnier* and *Taran* [4.121] in 1973, has later proved to be efficient for the solution of many problems of gas-phase and plasma diagnostics [4.102, 122, 123]. Significant progress in extracting the data concerning parameters of a gas medium was achieved, allowing CARS signals from molecules of different types to be simultaneously detected, with the development of a dual broadband CARS scheme [4.124, 125] and angularly resolved CARS [4.126].

Figure 4.23 illustrates the application of the broad-beam folded CARS geometry for the investigation of excited and ionized gases [4.122, 123]. In this scheme, a pair of cylindrically focused coplanar broad light beams with frequencies ω_1 and ω_2 and wave vectors $\boldsymbol{k}_1$ and $\boldsymbol{k}_2$, forming a small angle θ, irradiate a thin plasma layer in a plane parallel to the plane of the target (Fig. 4.23a). A cylindrically focused or a collimated nonfocused laser beam with frequency ω_3 and wave vector $\boldsymbol{k}_3$, which makes an angle α with the plane of the $\boldsymbol{k}_1$ and $\boldsymbol{k}_2$ vectors (Fig. 4.23b), irradiates the laser-produced spark from above. The FWM signal is generated in the direction $\boldsymbol{k}_{\text{FWM}}$ determined by phase-matching conditions, forming an angle β with the plane of the target (Figs. 4.23a,b). Imaging the one-dimensional FWM signal onto a CCD array, we were able to map the spatial distribution of resonant particles in the plasma line by line. The use of a collimated unfocused beam ω_3 enables slice-by-slice plasma imaging [4.102].

Raman-Induced Kerr Effect

The Raman-induced Kerr effect (RIKE) [4.9] is understood as an optical birefringence induced in an initially isotropic medium due to an anisotropic Raman-resonant third-order polarization of a medium. In this scheme, a nonlinear medium is irradiated with a pair of light beams with frequencies ω_1 and ω_2 whose difference is tuned, in accordance with the general idea of probing Raman-active vibrations, to a resonance with Raman-active transitions in a medium. Then, the polarization of a probe wave at the frequency ω_1 becomes perturbed due to the anisotropic nonlinear polarization induced in the Raman-active medium, which can be detected with the use of a polarization analyzer and a detector. The RIKE technique provides us with a convenient method for measuring the frequency dependence of the cubic susceptibility $\chi^{(3)}_{ijkl}$ around a Raman resonance. The transmission coefficient T of a polarization analyzer aligned in such a way as to block the probe beam in the absence of the pump beam is given by [4.99]

$$T_{\text{l}}(\omega_1-\omega_2) \propto \sin^2 2\gamma \left| \chi^{(3)}_{1122}(\omega_1;\omega_2,\omega_1,-\omega_2) + \chi^{(3)}_{1221}(\omega_1;\omega_2,\omega_1,-\omega_2) \right|^2 I_2^2 \tag{4.196}$$

in the case of a linearly polarized pump (I_2 is the intensity of the pump beam and γ is the angle between the polarization vectors of the pump and probe waves) and

$$T_{\text{c}}(\omega_1-\omega_2) \propto \left| \chi^{(3)R}_{1122}(\omega_1;\omega_2,\omega_1,-\omega_2) - \chi^{(3)R}_{1221}(\omega_1;\omega_2,\omega_1,-\omega_2) \right|^2 I_2^2 \tag{4.197}$$

in the case of a circularly polarized pump. As can be seen from (4.197), the coherent background is completely suppressed in the case of a circularly polarized pump.

Importantly, RIKE is one of the four-wave mixing processes where the phase-matching condition is satisfied automatically regardless of the arrangement of the wave vectors of pump and probe waves, since the phase-matching condition $\boldsymbol{k}_1 = \boldsymbol{k}_2 + \boldsymbol{k}_1 - \boldsymbol{k}_2$ becomes an identity in this case.

Degenerate Four-Wave Mixing

Although, rigorously speaking, degenerate four-wave mixing does not employ Raman transitions and the models used to describe DFWM may sometimes differ from the standard ways of CARS description [4.9, 10], it is reasonable to briefly introduce DFWM as a nonlinear technique here, as it very frequently offers a useful alternative to CARS, allowing valuable data on a medium to be obtained in a convenient and physically clear way. DFWM is closely related to CARS as both processes are associated with the third-order nonlinearity of a medium. The main difference between these methods is that CARS implies the use of a two-photon Raman-type resonance (Fig. 4.6c), while DFWM is a frequency-degenerate process (Fig. 4.6d), involving either four one-photon resonances or a pair of two-photon resonances. With modern lasers capable of generating very short pulses, having large spectral widths, the DFWM signal can be detected simultaneously with CSRS and CARS in the same experimental geometry with the same molecular system by simply tuning the detection wavelength [4.127]. The combination of these nonlinear-optical approaches allows a more elaborate study of molecular relaxation and photochemistry processes, providing a much deeper insight into the ultrafast molecular and wave-packet dynamics [4.127, 128].

The main advantages of DFWM as a spectroscopic technique are associated with the technical simplicity of this approach, which requires only one laser source and allows phase-matching conditions to be automatically satisfied regardless of the dispersion of the medium under study. Broadband DFWM [4.129] makes it possible to measure the temperature of excited gases, including atomic gases [4.130], with a single laser pulse. Folded broad-beam DFWM schemes are employed in several convenient and elegant methods for two-dimensional imaging of spatial distributions of gas parameters [4.131, 132].

4.7.3 Polarization Nonlinear Raman Techniques

Methods of polarization-sensitive four-photon spectroscopy provide an efficient tool for the solution of many problems arising in the investigation of Raman resonances. In particular, the polarization technique is a standard method to suppress the coherent background in CARS measurements [4.99, 106], which makes it possible to considerably improve the sensitivity of spectroscopic measurements [4.99] and improves the contrast in CARS microscopy [4.115]. Polarization techniques in FWM spectroscopy can separately measure the real and imaginary parts of the relevant third-order nonlinear-optical susceptibility [4.133, 134], resolve closely spaced lines in FWM spectra of molecules [4.99, 135] and atoms [4.102], and improve the contrast of cubic-susceptibility dispersion curves near Raman resonances [4.136, 137]. Polarization methods in nonlinear Raman spectroscopy [4.138–140] help to analyze the interference of vibrational Raman resonances with one- and two-photon electronic resonances in CARS spectra [4.141], and can be used to determine invariants of atomic and molecular Raman and hyper-Raman scattering tensors [4.102] and to perform conformational analysis for complex organic molecules [4.141]. A comprehensive review of polarization techniques employed for molecular spectroscopy was provided by *Akhmanov* and *Koroteev* [4.99].

Polarization Properties of the Coherent FWM Signal

When analyzing polarization properties of the FWM signal, one has to take into account the interference of resonant components of FWM related to various (molecular or atomic) transitions in the medium and the nonresonant coherent background. In particular, it is the interference of the resonant FWM component with the nonresonant coherent background that ensures the possibility to record complete spectral information concerning the resonance under study, including the data on the phase of resonant FWM.

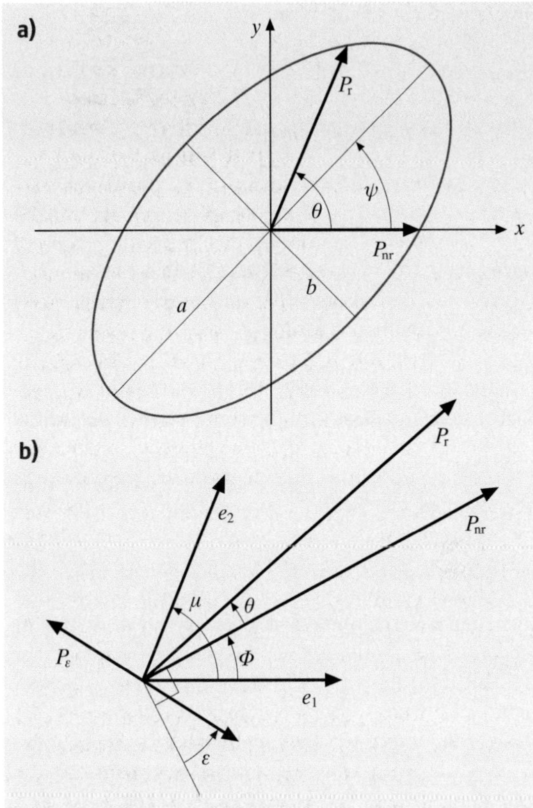

Fig. 4.24a,b Polarization technique of nonlinear Raman spectrometry: **(a)** coherent ellipsometry and **(b)** polarization suppression of the nonresonant background in CARS spectroscopy

The polarization ellipse of a Raman-resonant FWM signal is characterized by its ellipticity χ (which is defined as $\chi = \pm\text{atan}(b/a)$, where atan stands for the arctangent function, b and a are the small and principal semiaxes of the polarization ellipse, respectively) and the tilt angle ψ of its principal axis (Fig. 4.24a). These parameters are related to the Cartesian components of the third-order polarization of a medium P_x and P_y by the following expressions [4.99]:

$$\tan(2\psi) = \tan(2\beta)\cos(\delta) , \quad (4.198)$$

$$\sin(2\chi) = \sin(2\beta)\sin(\delta) , \quad (4.199)$$

where β and δ are defined as

$$\tan(\beta) = \left|\frac{P_y}{P_x}\right| , \quad (4.200)$$

$$\delta = \arg(P_y) - \arg(P_x) . \quad (4.201)$$

The frequency dependencies of the FWM polarization ellipse parameters, as can be seen from (4.198) and (4.199), provide information concerning the phase of the resonant FWM component, allowing a broad class of phase measurements to be performed by means of nonlinear Raman spectroscopy.

Suppressing the Nonresonant Background

Polarization suppression of the nonresonant background in CARS is one of the most useful, practical, and widely employed polarization techniques in nonlinear Raman spectroscopy. Physically, the possibility of suppressing the nonresonant background in coherent Raman spectroscopy is due to the fact that the resonant and nonresonant components of the nonlinear polarization induced in a Raman-active medium are generally polarized in different ways. Let us illustrate this technique for the CARS process $\omega_a = 2\omega_1 - \omega_2$, where ω_a is the frequency of the anti-Stokes signal and ω_1 and ω_2 are the frequencies of pump waves, in the case of an isotropic medium. The third-order polarization responsible for the generation of the signal with the frequency ω_a is then written as

$$\boldsymbol{P}^{(3)} = [\boldsymbol{P}_r + \boldsymbol{P}_{nr}] E_1^2 E_2^* , \quad (4.202)$$

where E_1 and E_2 are the amplitudes of the light fields, $\boldsymbol{P}_r$ and $\boldsymbol{P}_{nr}$ are the resonant and nonresonant components of the third-order polarization induced in the Raman-active medium.

In the case of an isotropic medium, relations (4.188–4.192) are satisfied for both resonant and nonresonant components of the nonlinear-optical susceptibility. However, only the nonresonant part of the cubic susceptibility satisfies the Kleinman relations [4.9, 99],

$$\chi_{1111}^{(3)nr} = 3\chi_{1122}^{(3)nr} = 3\chi_{1221}^{(3)nr} = 3\chi_{1212}^{(3)nr} , \quad (4.203)$$

while the resonant part of the cubic susceptibility is usually characterized by a considerable dispersion near a Raman resonance, which implies that the resonant cubic susceptibility tensor components are not invariant with respect to the permutation of their frequency arguments. Taking into account relations (4.203) for the nonresonant part of the cubic susceptibility, we arrive at the following expression for the nonresonant and resonant components of the third-order polarization of a Raman-active medium:

$$\boldsymbol{P}_{nr} = \chi_{1111}^{(3)nr} \left[2\boldsymbol{e}_1 \left(\boldsymbol{e}_1\boldsymbol{e}_2^*\right) + \boldsymbol{e}_2^* \left(\boldsymbol{e}_1\boldsymbol{e}_1\right)\right] , \quad (4.204)$$

$$\boldsymbol{P}_r = 3\chi_{1111}^{(3)r} \left[\left(1-\bar{\rho}\right) \boldsymbol{e}_1 \left(\boldsymbol{e}_1\boldsymbol{e}_2^*\right) + \bar{\rho}\boldsymbol{e}_2^* \left(\boldsymbol{e}_1\boldsymbol{e}_1\right)\right] , \quad (4.205)$$

where $\bar{\rho} = \chi_{1221}^{(3)r} / \chi_{1111}^{(3)r}$ and $\boldsymbol{e}_1$ and $\boldsymbol{e}_2$ are the unit polarization vectors of the light fields with frequencies ω_1 and ω_2, respectively (Fig. 4.24b).

Suppose that the pump fields with frequencies ω_1 and ω_2 are linearly polarized and their polarization vectors $\boldsymbol{e}_1$ and $\boldsymbol{e}_2$ are oriented at an angle μ with respect to each other, as shown in Fig. 4.24b. The vectors $\boldsymbol{P}_{nr}$ and $\boldsymbol{P}_r$, in accordance with (4.204) and (4.205), generally have different orientations in space making a nonzero angle θ with each other. Therefore, by orienting the polarization analyzer in such a way as to suppress the nonresonant component of the CARS signal (by setting the angle ε measured from the direction perpendicular to the vector $\boldsymbol{P}_{nr}$ equal to zero, see Fig. 4.24b), one can analyze background-free CARS spectra.

In many situations, nonlinear Raman study of Raman-active media would be simply impossible without this technique. This is the case, for example, when the CARS signal from a resonant gas under investigation is too weak to be reliably detected against the nonresonant CARS signal from the windows of a gas cell. Another example is the CARS spectroscopy on low concentrations of complex biological molecules, when the coherent background due to solvent molecules may be so strong that it leaves no way to detect the CARS signal from the molecules being studied without polarization suppression of the nonresonant nonlinear Raman signal. Even small deviations of the orientation of the polarization analyzer from the background-suppression position may be crucial, leading to dramatic changes in the signal-to-noise ratio. Another important conclusion that can be made from (4.204) and (4.205) is that, measuring the ratio of the CARS signals for different polarization arrangements is a convenient way of determining the properties of the nonlinear susceptibility tensor and, thus, characterizing the symmetry properties of molecular transitions under investigation.

Coherent Ellipsometry

Coherent ellipsometry, i. e., the measurement of the parameters of the polarization ellipse corresponding to the FWM signal, is one of widely used modifications of polarization sensitive four-photon spectroscopy. Below, we consider the main physical principles and discuss the main ideas of coherent ellipsometry.

The possibility of reconstructing the real and imaginary parts of the nonlinear-optical susceptibility of a medium as functions of frequency and time is due to the interference of the resonant FWM component with the nonresonant coherent background, which ensures the recording of the phase information for the resonance being studied. For a broad class of problems, the nonlinear polarization of a medium responsible for coherent FWM processes can be represented as a sum of the nonresonant and resonant components described by a real vector $\boldsymbol{P}_{nr}$ and a complex vector $\boldsymbol{P}_r$, respectively. Choosing the x-axis along the vector $\boldsymbol{P}_{nr}$ (Fig. 4.24b), we can write the Cartesian components of the total polarization of a medium cubic in the external field as

$$P_x = P_{nr} + P_r e^{i\varphi}\cos(\theta), \tag{4.206}$$

$$P_y = P_r e^{i\varphi}\sin(\theta), \tag{4.207}$$

where φ is the phase of the resonant component of the nonlinear polarization and θ is the angle between the resonant and nonresonant components.

Knowing the parameters of the polarization ellipse from ellipsometric measurements, we can reconstruct, with the use of (4.198), (4.199), (4.206), (4.207), the real and imaginary parts of the resonant nonlinear polarization as functions of frequency and time from the experimental data of coherent ellipsometry.

In the important particular case when the resonant component of the FWM signal can be considered as a small correction to the nonresonant component, the general procedure of separating the real and imaginary parts of the nonlinear polarization of a medium becomes especially simple. One can easily verify that the relations [4.99]

$$\psi = \beta\cos(\varphi) \propto \mathrm{Re}(P) \tag{4.208}$$

$$\chi = \beta\sin(\varphi) \propto \mathrm{Im}(P) \tag{4.209}$$

are satisfied in this case, showing that the spectral or temporal dependencies of the parameters ψ and χ of the FWM polarization ellipse respectively reproduce the spectral or temporal dependencies of the real and imaginary parts of the resonant component of the nonlinear polarization of a medium.

Thus, the data obtained by means of coherent ellipsometry enable one to extract complete information concerning the resonant component of the nonlinear polarization of a medium, including information on its phase [4.134]. Note that no assumptions regarding the shape of the line observed in a nonlinear Raman spectrum was made in our consideration, which means that this approach can be applied to a broad class of spectral lines. This procedure can be also extended to the time domain, allowing not only the spectra but also the time dependencies of the real and imaginary parts of the nonlinear polarization of a medium to be reconstructed. Finally, close molecular and atomic lines unresolvable in amplitude nonlinear Raman spectra can be also resolved in certain cases with the use of the phase information stored in coherent nonlinear spectra.

4.7.4 Time-Resolved Coherent Anti-Stokes Raman Scattering

The method of time-resolved FWM spectroscopy implies that information on the parameters of atomic or molecular systems is extracted from an impulse response of a coherently excited system rather than from the frequency dispersion of nonlinear susceptibilities, as is done in frequency-domain FWM spectroscopy. The original idea of time-resolved CARS is that a light pulse with a duration shorter than the characteristic transverse relaxation time T_2 induces coherent molecular vibrations with amplitude $Q(t)$, and the decay kinetics of these vibrations is analyzed with the use of another, probe light pulse, which is delayed in time with respect to the pump pulses. The complete set of equations governing the processes related to time-domain CARS includes the SVEA wave equation (4.182) and the equations for the amplitude of coherent molecular vibrations $Q(t)$, defined in accordance with (4.185), and the normalized population difference between the levels involved in the Raman resonance, $n = \rho_{aa} - \rho_{aa}$:

$$\frac{\partial^2 Q}{\partial t^2} + \frac{2}{T_2}\frac{\partial Q}{\partial t} + \Omega^2 Q = \frac{1}{2M}\frac{\partial\alpha}{\partial Q} n E^2 , \tag{4.210}$$

$$\frac{\partial n}{\partial t} + \frac{n-1}{T_1} = \frac{1}{2\hbar\Omega}\frac{\partial\alpha}{\partial Q} E^2 \frac{\partial Q}{\partial t} , \tag{4.211}$$

where T_1 is the population relaxation time.

In many cases, (4.182), (4.210), and (4.211) can be simplified with the use of the slowly varying envelope approximation. In this approximation, the energy of the CARS signal as a function of the delay time τ of the probe pulse in the scheme of time-resolved CARS with short light pulses and $\Delta k = 0$ is given by [4.99]

$$W_a(\tau) \propto \int_{-\infty}^{\infty} |Q(t) A(t-\tau)|^2 \, dt . \tag{4.212}$$

Part A | 4.7

Since the intensity of the CARS signal in such a scheme is determined by (4.212), the use of sufficiently short probe pulses makes it possible to measure the kinetics of $Q(t)$. Experiments by *Alfano* and *Shapiro* [4.142] and *von der Linde* et al. [4.143] have demonstrated the possibility of using the time-domain CARS technique to directly measure the T_2 time for Raman-active modes in crystals and organic liquids.

Formulas (4.185–4.187) and (4.212) show the relation between the information that can be obtained by frequency- and time-domain CARS spectroscopy. In fact, this information is essentially the same, and frequency- and time-domain CARS methods can successfully complement each other in studies of complex inhomogeneously broadened spectral bands. For example, in frequency-domain CARS spectroscopy, the phase information on molecular resonances can be extracted through polarization measurements and coherent ellipsometry (see the discussion above), and the level of coherent nonresonant background can be suppressed by means of the relevant polarization technique [see (4.204), (4.205) and Fig. 4.24b]. In time-domain CARS, on the other hand, the nonresonant background appears only at zero delay time between the pump and probe pulses, having no influence on the transient signal, while the impulse-response measurements may directly provide the information not only on the amplitude, but also on the phase of a molecular or atomic resonance. The development of femtosecond laser systems resulted in the impressive technical and conceptual progress of time-domain FWM spectroscopy, allowing photochemistry processes and molecular dynamics to be monitored in real time (see [4.109, 110, 113] for a review).

4.8 Waveguide Coherent Anti-Stokes Raman Scattering

4.8.1 Enhancement of Waveguide CARS in Hollow Photonic-Crystal Fibers

The general idea of waveguide CARS [4.144–149] is to improve the efficiency of four-wave mixing by increasing the interaction length and increasing the intensity of pump waves with given pump powers by reducing the transverse size of the wave-guiding layer in planar waveguides or the core diameter in optical fibers. CARS spectroscopy of the gas phase leaves no alternative to hollow waveguides and fibers. Since the refractive index of the core in such waveguides is lower than the refractive index of the cladding, the modes guided in the hollow core are always characterized by nonzero losses. The magnitude of these losses scales [4.150, 151] as λ^2/a^3 preventing fibers with small inner diameters to be used in nonlinear-optical experiments, which limits the waveguide CARS enhancement factors attainable with such fibers. In hollow-core photonic-crystal fibers [4.36, 152, 153], which guide light due to photonic band gaps, optical losses can often be kept low even in the case of small core diameters. Hollow PCFs with an inner diameter of $\approx 15\,\mu$m demonstrated by *Benabid* et al. [4.154] have the magnitude of optical losses on the order of 1–3 dB/m. We will show below that this property of hollow-core microstructure fibers makes them very attractive for waveguide CARS spectroscopy.

We start with the expression [4.11] for the power of the three-color CARS signal generated at the frequency $\omega_s = \omega_0 + \omega_1 - \omega_2$ by pump fields with frequencies ω_0, ω_1 and ω_2:

$$P_{CARS} = 1.755 \times 10^{-5} \frac{\omega_s^4 k_0 k_1 k_2}{c^4 k_s^2 k'} D^2 \times \left|\chi_{eff}^{(3)}\right|^2 P_0 P_1 P_2 F_2, \quad (4.213)$$

where k_0, k_1, k_2, k_s are the wave numbers of light fields with frequencies ω_0, ω_1, ω_2, ω_s, respectively; P_0, P_1, P_2 are the powers of the fields with frequencies ω_0, ω_1, ω_2, respectively; $\chi_{eff}^{(3)}$ is the effective combination of cubic nonlinear-optical susceptibility tensor components corresponding to the chosen set of polarization vectors of pump and signal fields; D is the frequency degeneracy factor of the four-wave mixing process defined after *Maker* and *Terhune* [4.105];

$$F_2 = \frac{2k'}{\pi b} \int_0^\infty 2\pi R \,\mathrm{d}R \times \left| \int_{-\zeta}^{\xi} \mathrm{d}\xi' \frac{\exp\left(\frac{\mathrm{i}b\Delta k \xi'}{2}\right)}{(1+\mathrm{i}\xi')\,(k'' - \mathrm{i}k'\xi')\,H} \exp\left(-\frac{R^2}{bH}\right) \right|^2 \quad (4.214)$$

is the phase-matching integral, $\Delta k = k_s - (k_0 + k_1 - k_2)$, $k' = k_0 + k_1 - k_2$, $k'' = k_0 + k_1 + k_2$, $\xi = 2(z-f)/b$, $\zeta = 2f/b$, $b = n_j \omega_j w_0^2/c$ is the confocal parameter, w_0 is the beam waist diameter,

$$H=\frac{(1+\xi')^2}{(k''-\mathrm{i}k'\xi')}-\mathrm{i}\frac{\xi'-\xi}{k'}. \tag{4.215}$$

In the limiting case of tight focusing, when the confocal parameter b is much less than the length of the nonlinear medium l, $b \ll l$, no increase in the CARS power can be achieved by reducing the pump-beam waist radius because of the simultaneous decrease in the interaction length. Mathematically, this well-known result is a consequence of the tight-focusing limit existing for the phase-matching integral (4.214). For small phase mismatches, $\Delta kl \ll \pi$, the phase-matching integral can be written in this limiting case as

$$F_2=\frac{4\pi^2}{\left(1+\frac{k''}{k'}\right)^2}. \tag{4.216}$$

In the opposite limiting case of loosely focused pump beams, $b \gg l$, weak absorption and negligible phase mismatches, the phase-matching integral is reduced to

$$F_2=\frac{k'}{k''}\frac{4l^2}{b^2}. \tag{4.217}$$

Since the latter regime is exactly the case of waveguide CARS, we can use (4.216) and (4.217) to estimate the enhancement of waveguide CARS with respect to the regime of tight focusing. Phase mismatches in waveguide CARS should be understood as difference of propagation constants of waveguide modes involved in the wave-mixing process, and the mode-overlapping integral should generally be included to allow for the contribution of waveguide effects, in particular, the influence of higher-order waveguide modes.

Assuming that the beam waist radius of focused pump beams is matched to the inner radius of a hollow fiber, a, we find from (4.216) and (4.217) that the waveguide CARS enhancement factor scales as $\lambda^2 l^2/a^4$. The length l can be made very large in the case of fibers, but the fundamental limitation of waveguide CARS in hollow fibers comes from optical losses, whose magnitude scales as λ^2/a^3. The influence of optical losses and phase-mismatch effects on the CARS process in the loose-focusing regime can be included through the factor

$$M \propto \exp[-(\Delta\alpha+\alpha_4)l] \times \left(\frac{\sinh^2\left(\frac{\Delta\alpha l}{2}\right)+\sin^2\left(\frac{\Delta kl}{2}\right)}{\left(\frac{\Delta\alpha l}{2}\right)^2+\left(\frac{\Delta kl}{2}\right)^2}\right) l^2, \tag{4.218}$$

where $\Delta\alpha=(\alpha_1+\alpha_2+\alpha_3-\alpha_4)/2$, α_1, α_2, α_3, α_4 are the magnitudes of optical losses at frequencies ω_0, ω_1, ω_2, ω_s, respectively.

It is straightforward to see from (4.218) that the amplitude of the CARS signal in a lossy waveguide reaches its maximum at some optimal length l_{opt}^{CARS}, which is given by

$$l_{opt}^{CARS}=\frac{1}{\Delta\alpha}\ln\left(\frac{\alpha_1+\alpha_2+\alpha_3}{\alpha_4}\right). \tag{4.219}$$

With $\alpha_1 \approx \alpha_2 \approx \alpha_3 \approx \alpha_4 = \alpha$, (4.219) yields

$$l_{opt}^{CARS}=\frac{\ln 3}{\alpha}. \tag{4.220}$$

Then, setting $\Delta k = 0$ for phase matching and $w_0 = 0.73a$ for the best matching of input beams with the fiber mode radius, assuming that the refractive index of the gas filling the fiber core is approximately equal to unity, and taking into consideration that $M=(3^{1/2}-3^{-1/2})^2/(3\ln 3)^2 \approx 0.123$ for $\Delta k=0$ and $l=l_{opt}^{CARS}=\ln 3/\alpha$, we arrive at the following expression for the waveguide CARS enhancement factor:

$$\mu = 1.3\times 10^{-3}\frac{(k'+k'')^2}{k'k''}\frac{\lambda^2}{\alpha^2 a^4}. \tag{4.221}$$

We can now see from (4.221), that the waveguide CARS enhancement factor scales as $\lambda^2/\alpha^2 a^4$ and is limited by fiber losses. We will show in the next section that, due to the physically different mechanism behind light guiding, hollow microstructure fibers allow CARS enhancement factors to be substantially increased with respect to standard, solid-cladding hollow fibers. We will examine also the CARS enhancement factors as functions of the core radius for the fibers of both types and investigate the influence of the phase mismatch.

We start with the case of standard, solid-cladding hollow fibers. The magnitude of optical losses for EH_{mn} modes in such fibers is given by [4.150]

$$\alpha=\left(\frac{u_{mn}}{2\pi}\right)^2\frac{\lambda^2}{a^3}\frac{n^2+1}{\sqrt{n^2-1}}, \tag{4.222}$$

where u_{mn} is the eigenvalue of the characteristic equation for the relevant hollow-fiber mode (the mode parameter), n is the refractive index of the fiber cladding, and the refractive index of the gas filling the fiber core is set equal to unity.

Plugging optical losses into the CARS enhancement factor by substituting (4.222) into (4.221) with $u_n = 2.4$ for the limiting eigenvalue of the EH_{11} mode of a hollow fiber, we derive the following expression for the factor of CARS enhancement in a solid-cladding hollow fiber relative to the tight-focusing regime in the case of exact phase matching:

$$\rho = 6.1\times 10^{-2}\frac{(k'+k'')^2}{k'k''}\left(\frac{a}{\lambda}\right)^2\frac{n^2-1}{\left(n^2+1\right)^2}. \quad (4.223)$$

Optical losses, which grow with decreasing inner radius a, limit the CARS enhancement, with the factor ρ rapidly lowering with decreasing a for small values of the fiber inner radius. The situation radically changes in the case of a microstructure fiber. The magnitude of optical losses for such fibers, as mentioned above, may be on the order of 1–3 dB/m in the case of fibers with a hollow core diameter of about 15 μm [4.154]. In the case of small inner radii, microstructure fibers provide much higher CARS enhancement factors than solid-core hollow fibers. The CARS enhancement factor in hollow microstructure fibers with the magnitude of optical losses equal to 0.1 and 0.01 cm^{-1} starts to exceed the CARS enhancement factor in a solid-cladding hollow fiber for core radii less than 20 and 45 μm, respectively. For hollow fibers with small core radii, the factor μ may be several orders of magnitude higher than the enhancement factor ρ.

An additional source of radiation losses in hollow fibers is related to radiation energy transfer to higher-order waveguide modes. The efficiency of this nonlinear-optical mode cross-talk process depends on radiation intensity and the mismatch Δk_c of propagation constants of waveguide modes involved in energy exchange. Starting with the standard expression for the propagation constants of EH_{mn} modes in a hollow fiber, we arrive at the following formula for the coherence length $l_c = \pi(2|\Delta k_c|)^{-1}$ of the mode cross-talk process:

$$l_c = \frac{2\pi^2}{\lambda}\frac{a^2}{\left|u_2^2-u_1^2\right|}, \quad (4.224)$$

where u_2 and u_1 are the parameters of cross-talking fiber modes. The coherence length l_c, as can be seen from (4.224), becomes very small for guided modes of high orders, making the efficiencies of energy transfer from the fundamental to very high order modes negligible. For the cross-talk between the lowest order EH_{11} and EH_{12} modes, with $u_1 \approx 2.4$ and $u_2 \approx 5.5$, the coherence length can be estimated as $l_c \approx 0.8a^2/\lambda$. The coherence length of such a cross-talk process is typically much smaller than the optimal length for the wave-mixing process (4.220). However, for high-intensity pump beams, the efficiency of this cross-talk process increases [4.155], and the energy lost from the fundamental mode due to the excitation of higher-order modes may become comparable with radiation energy leakage with the characteristic length governed by (4.222).

Importantly, the scaling law of the waveguide CARS enhancement factor as a function of the magnitude of optical losses, fiber inner radius, and radiation wavelength differs from a similar scaling law of the waveguide SRS enhancement factor [4.23], $\eta = \lambda/\alpha a^2$. Physically, this difference stems from differences in scattering mechanisms involved in SRS and CARS, with SRS and CARS signals building up in different fashions as functions of the interaction length and pump field amplitudes. The difference in waveguide enhancement factors for SRS and CARS suggests different strategies for optimizing fibers designed to enhance these processes.

Phase mismatch, resulting from the difference in propagation constants of guided modes involved in the CARS process, is another important factor limiting the efficiency of CARS in a hollow fiber. In the case of nonzero phase mismatch Δk, the optimal length for the CARS process can be found from a transcendental equation that immediately follows from (4.218):

$$\begin{aligned}&\Delta\alpha \sinh\left(\Delta\alpha l_{\mathrm{opt}}^{\mathrm{CARS}}\right)+\Delta k \sin\left(\Delta k l_{\mathrm{opt}}^{\mathrm{CARS}}\right)\\&+(\Delta\alpha+\alpha_4)\\&\times\left[\cos\left(\Delta k l_{\mathrm{opt}}^{\mathrm{CARS}}\right)-\cosh\left(\Delta\alpha l_{\mathrm{opt}}^{\mathrm{CARS}}\right)\right]=0\,.\end{aligned} \quad (4.225)$$

Phase mismatch reduces the maximum waveguide CARS enhancement attainable with a hollow microstructure fiber, with the power of the CARS signal becoming an oscillating function of the fiber length. The characteristic period of these oscillations is determined by the coherence length. Oscillations become less pronounced and then completely flatten out as optical losses build up. No oscillations is observed when the attenuation length becomes less than the coherence length. An important option offered by hollow microstructure fibers is the possibility to compensate for the phase mismatch related to the gas dispersion with an appropriate choice of waveguide parameters due to the waveguide dispersion component, scaling as a^{-2} in the case of a hollow fiber.

We have shown in this section that hollow microstructure fibers offer a unique opportunity of implementing nonlinear-optical interactions of waveguide modes with transverse sizes of several microns in a gas medium, opening the ways to improve the efficiency of nonlinear-optical processes, including four-wave mixing and coherent anti-Stokes Raman scattering, and suggesting the principle for the creation of highly sensitive gas-phase sensors based on nonlinear spectroscopic techniques. Hollow-core microstructure fibers have been demonstrated to allow the waveguide CARS efficiency to be substantially increased as compared to standard, solid-cladding hollow fibers. The theorem predicting an l^2/a^4 enhancement for a waveguide CARS process in a hollow fiber with an inner radius a and length l has been extended to include new solutions offered by microstructure fibers. The maximum CARS enhancement in a hollow microstructure fiber was shown to scale as $\lambda^2/\alpha^2 a^4$ with radiation wavelength λ, radiation losses α, and the inner fiber radius, allowing CARS efficiency to be substantially improved in such a fiber. This $\lambda^2/\alpha^2 a^4$ CARS enhancement factor differs from the $\lambda/\alpha a^2$ ratio, characterizing waveguide SRS enhancement in a hollow microstructure fiber, which is related to the difference in the physical nature of SRS and CARS signals and suggests different strategies for optimizing fibers designed to enhance CARS and SRS processes.

4.8.2 Four-Wave Mixing and CARS in Hollow-Core Photonic-Crystal Fibers

Hollow-core photonic-crystal fibers (PCFs) [4.36, 152, 153] offer new interesting options for high-field physics and nonlinear optics. Waveguide losses can be radically reduced in such fibers relative to standard, solid-cladding hollow fibers due to the high reflectivity of a periodically structured fiber cladding within photonic band gaps (PBGs) [4.152, 153, 156], allowing transmission of high-intensity laser pulses through a hollow fiber core in isolated guided modes with typical transverse sizes of 10–20 μm. Due to this unique property, hollow PCFs can substantially enhance nonlinear-optical processes [4.157], including stimulated Raman scattering [4.154, 158, 159], four-wave mixing (FWM) [4.160], coherent anti-Stokes Raman scattering (CARS) [4.118], and self-phase modulation [4.161]. Air-guided modes in hollow PCFs can support high-power optical solitons [4.162, 163], allow transportation of high-energy laser pulses for technological [4.68, 164] and biomedical [4.165] applications.

In this section, we discuss phase-matched FWM of millijoule nanosecond pulses in hollow PCFs with a period of the photonic-crystal cladding of about 5 μm and a core diameter of approximately 50 μm. We will show that Raman-resonant FWM in large-core hollow PCFs enhances the potential of waveguide CARS in hollow fibers, providing a convenient sensing tool for condensed-phase species adsorbed on the inner fiber walls and trace-gas detection.

Large-core-area hollow PCFs employed in experiments [4.119, 166] were fabricated using a standard procedure, which involves stacking glass capillaries into a periodic array and drawing this preform at a fiber-drawing tower. Several capillaries have been omitted from the central part of the stack, to produce a hollow core of the fiber. While in standard hollow PCFs, the number of omitted capillaries is seven, PCFs used in our experiments had a hollow core in the form of a regular hexagon with each side corresponding to five cane diameters. The inset in Fig. 4.25 shows an image of a hollow PCF with a period of the cladding of approximately 5 mm and a core diameter of about 50 μm. The baking of capillaries forming the photonic-crystal structure, as shown in the image, allows a hollow waveguide with a nearly ideal 50 μm-diameter hexagonal core to be fabricated. It is still to be explored whether this technique can be scaled up to the fabrication of hollow PCFs with even larger core diameters. Transmission spectra

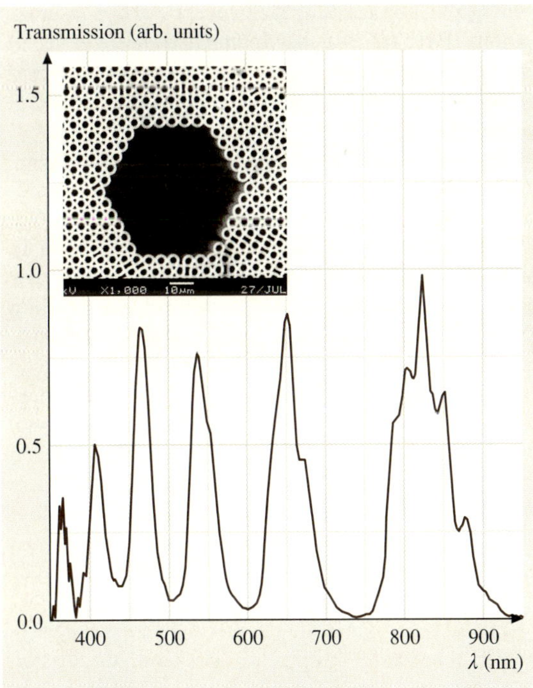

Fig. 4.25 Transmission spectrum of the hollow-core PCF. The *inset* shows the cross-section view of the PCF with a period of the cladding structure of about 5 μm

of hollow PCFs employed in our experiments display well-pronounced passbands (Fig. 4.25), indicating the PBG guidance of radiation in air modes of the fiber.

The laser system used in experiments [4.119, 166] consisted of a Q-switched Nd:YAG master oscillator, Nd:YAG amplifiers, frequency-doubling crystals, a dye laser, as well as a set of totally reflecting and dichroic mirrors and lenses adapted for the purposes of CARS experiments. The Q-switched Nd:YAG master oscillator generated 15 ns pulses of 1.064 μm radiation, which were then amplified up to about 30 mJ by Nd:YAG amplifiers. A potassium dihydrogen phosphate (KDP) crystal was used for the frequency doubling of the fundamental radiation. This second-harmonic radiation served as a pump for the dye laser, generating frequency-tunable radiation within the wavelengths ranges 540–560 and 630–670 nm, depending on the type of dye used as the active medium for this laser. All the three outputs of the laser system, viz., the fundamental radiation, the second harmonic, and frequency-tunable dye-laser radiation, were employed as pump fields in FWM, as described below. The frequency dependencies of the anti-Stokes signals produced through different FWM processes were measured point by point by scanning the frequency of dye-laser radiation. The energies of these pump fields were varied in our experiments from 0.5 up to 10 mJ at the fundamental wavelength, from 0.5 to 8 mJ in the second harmonic, and from 0.05 to 0.7 mJ for dye-laser radiation. To couple the laser fields into the fundamental mode of the PCF, we focused laser beams into spots with a diameter of 35 μm at the input end of the fiber. The PCF could withstand the energy of fundamental radiation up to 10 mJ, corresponding to a laser fluence of approximately 630 J/cm^2, without an irreversible degradation of fiber performance because of optical breakdown. Laser-induced breakdown on PCF walls was judged by a dramatic irreversible reduction in fiber transmission and intense sideward scattering of laser radiation, visible through the fiber cladding. While the achieved level of input energies was sufficient to produce reliably detectable FWM signals in our experiments, a further increase in the laser radiation energy coupled into the PCF is possible through a more careful optimization of the coupling geometry.

FWM processes with the CARS-type frequency-mixing scheme $\omega_a = 2\omega_1 - \omega_2$ (ω_1 and ω_2 are the frequencies of the pump fields and ω_a is the frequency of the anti-Stokes signal produced through FWM) were studied in our experiments for two different sets of pump and signal frequencies. In the first FWM process, used in our experiments to test phase matching and assess the influence of waveguide losses, two waves with the wavelength $\lambda_1 = 2\pi c/\omega_1$ ranging from 630 to 665 nm, provided by the dye laser, are mixed with the fixed-frequency field of the fundamental radiation at $\lambda_2 =$ 1064 nm, to generate an anti-Stokes signal within the range of wavelengths λ_a from 445 to 485 nm. The second FWM process, designed to demonstrate the potential of CARS spectroscopy with hollow PCFs, is a standard Nd:YAG-laser CARS arrangement with $\lambda_1 = 532$ nm and λ_2 ranging from 645 to 670 nm.

To assess the influence of phase matching and radiation losses on the intensity of the FWM signal generated in a hollow PCF, we use (4.213), (4.218), and (4.225) to write the power of the anti-Stokes signal as $P_a \propto |\chi^{(3)}_{\text{eff}}|^2 P_1 P_2^2 M$, where P_1 and P_2 are the powers of the fields with frequencies ω_1 and ω_2, respectively; $\chi^{(3)}_{\text{eff}}$ is the effective combination of cubic nonlinear-optical susceptibility tensor components; and the factor M includes optical losses and phase-mismatch effects: $M(\Delta\alpha l, \alpha_a l, \delta\beta l) = \exp[(\Delta\alpha + \alpha_a)l][\sinh^2(\Delta\alpha l/2) + \sin^2(\delta\beta l/2)][(\Delta\alpha l/2)^2 + (\delta\beta l/\delta\beta l/2)^2]^{-1} l^2$, where $\Delta\alpha = (2\alpha_1 + \alpha_2 - \alpha_a)/2$, α_1, α_2, and α_a are the magnitudes of optical losses at frequencies ω_1, ω_2, and ω_a, respectively, and $\delta\beta$ is the mismatch of the propagation constants of waveguide modes involved in the FWM process. In order to provide an order of magnitude estimate on typical coherence lengths $l_c = \pi/(2|\delta\beta|)$ for FWM processes in hollow PCFs and to choose PCF lengths L meeting the phase-matching requirement $L \leq l_c$ for our experiments, we substitute the dispersion of a standard hollow fiber with a solid cladding for the dispersion of PCF modes in these calculations. As shown by earlier work on PBG waveguides [4.167], such an approximation can provide a reasonable accuracy for mode dispersion within the central part of PBGs, but fails closer to the passband edges. For the waveguide FWM process involving the fundamental modes of the pump fields with $\lambda_1 = 532$ nm and $\lambda_2 = 660$ nm, generating the fundamental mode of the anti-Stokes field in a hollow fiber with a core radius of 25 μm, the coherence length is estimated as $l_c \approx$ 10 cm. Based on this estimate, we choose a fiber length of 8 cm for our FWM experiments. With such a choice of the PCF length, effects related to the phase mismatch can be neglected as compared with the influence of radiation losses.

Phase matching for waveguide CARS in the PCF was experimentally tested by scanning the laser frequency difference $\omega_1 - \omega_2$ off all the Raman resonances (with λ_1 ranging from 630 to 665 nm and $\lambda_2 = 1064$ nm)

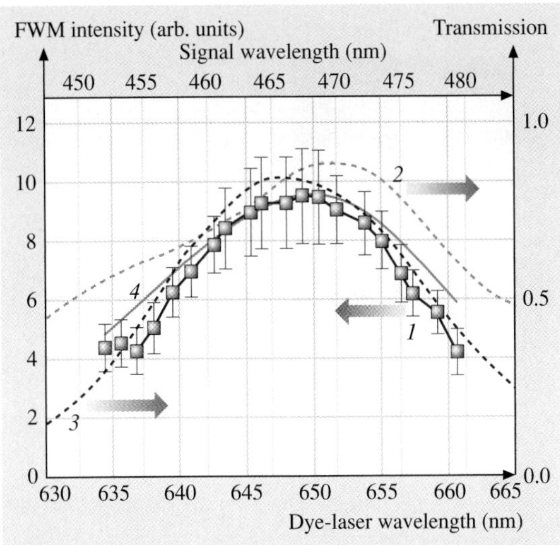

Fig. 4.26 Intensity of the $\omega_a = 2\omega_1 - \omega_2$ four-wave mixing signal from the hollow PCF with the length of 8 cm versus the wavelength of dye-laser radiation with λ_1 ranging from 630 to 665 nm and $\lambda_2 = 1064$ nm: (1) the measured spectrum of the FWM signal, (2) fiber transmission for dye-laser radiation, (3) fiber transmission for the FWM signal, and (4) the spectral profile of the factor M

and using the above expression for $M(\Delta\alpha l, \alpha_a\, l, \delta\beta l)$ with $\delta\beta L \approx 0$ to fit the frequency dependence of the FWM signal. Dots with error bars (line 1) in Fig. 4.26 present the intensity of the anti-Stokes signal from hollow PCFs measured as a function of the frequency of the dye laser. Dashed lines 2 and 3 in this figure display the transmission of the PCF for dye-laser radiation and the anti-Stokes signal, respectively. Solid line 4 presents the calculated spectral profile of the factor $M(\Delta\alpha L, \alpha_a\, L, 0)$. Experimental frequency dependencies of the FWM signals, as can be seen from the comparison of lines 1 and 4 in Fig. 4.26, are fully controlled by the spectral contours of PCF passbands (lines 2 and 3), indicating that phase-mismatch effects are much less significant for the chosen PCF lengths than variations in radiation losses.

The second series of experiments was intended to demonstrate the potential of waveguide CARS in a hollow PCF for the sensing of Raman-active species. For this purpose, the frequency difference of the second-harmonic and dye-laser pump fields was scanned through the Raman resonance, $\omega_1 - \omega_2 = 2\pi c\Omega$, with O–H stretching vibrations of water molecules, adsorbed on the inner PCF walls. The frequencies Ω of O–H stretching vibrations of water molecules typically fall within a broad frequency band of 3200–3700 cm^{-1}. The frequency dependence of the FWM signal from the PCF substantially deviates from the spectral profile of the factor $M(\Delta\alpha L, \alpha_a\, L, 0)$ (cf. lines 1 and 4 in Fig. 4.27), clearly indicating the contribution of Raman-active species to the FWM signal. To discriminate between the CARS signal related to water molecules adsorbed on the PCF walls against the OH contamination of the PCF cladding, we measured the spectrum of the CARS signal from a PCF heated above a burner. Heating by 30 K during 30 min reduced the amplitude of the Raman resonance in the spectrum of the CARS signal by a factor of about seven. The high level of the CARS signal was then recovered within several days. This spectrum of the CARS signal from the dry PCF was subtracted from the CARS spectrum recorded at the output of the hollow PCF under normal conditions. The difference spectrum was normalized to the spectral profile of the factor $M(\Delta\alpha L, \alpha_a\, L, 0)$. The result of this normalization is shown by line 5 in Fig. 4.27.

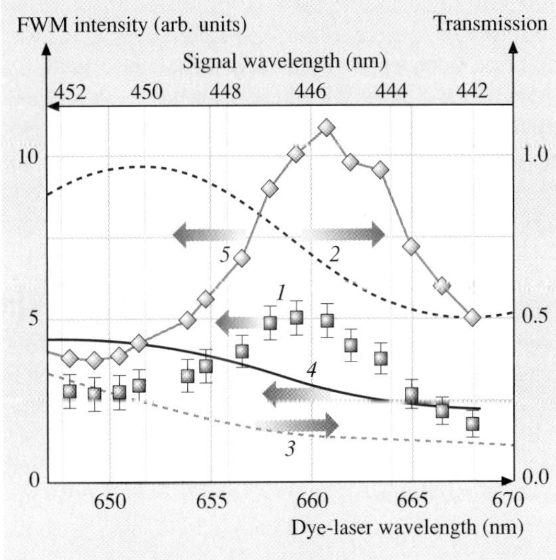

Fig. 4.27 Intensity of the $\omega_a = 2\omega_1 - \omega_2$ four-wave mixing signal from the hollow PCF with the length of 8 cm versus the wavelength of dye-laser radiation with $\lambda_1 = 532$ nm and λ_2 ranging from 645 to 670 nm: (1) the measured spectrum of the FWM signal, (2) fiber transmission for dye-laser radiation, (3) fiber transmission for the FWM signal, (4) the spectral profile of the factor M and (5) the spectrum of the FWM signal corrected for the factor M upon the subtraction of the spectrum of the CARS signal from the heated hollow PCF

Notably, the contrast of the experimental wavelength dependence of the FWM intensity (squares with error bars) in Fig. 4.26 is higher than the contrast of a similar dependence for the CARS signal in Fig. 4.27. This variation in the ratio of the maximum amplitude of the nonlinear signal correlates well with the behavior of transmission for dye-laser radiation and the nonlinear signal, shown by curves 2 and 3 in both figures. With the dye-laser radiation wavelength set around 650 nm, both the pump and nonlinear signal wavelengths λ_1 and λ_a in Fig. 4.26 are close to the respective maxima of PCF transmission. The CARS signal, on the other hand, is detected away from the maximum transmission for the dye-laser radiation and the nonlinear signal (Fig. 4.27). It is therefore important to normalize the measured CARS spectrum to the wavelength dependence of the *M* factor, taking into account wavelength-dependent losses introduced by the PCF. This normalization procedure considerably improves the contrast of the CARS spectrum, as shown by curve 5 in Fig. 4.27.

Experiments presented above demonstrate the potential of waveguide CARS in PCFs to detect trace concentrations of Raman-active species, suggesting PCF CARS as a convenient diagnostic technique. However, CARS signals detected in these experiments do not allow the origin of Raman-active species in the fiber to be reliably identified, as it is not always possible to discriminate between the contributions to the CARS signal provided by the hollow core and PCF walls. As an example of a more easily quantifiable Raman medium, permitting the CARS signal from the PCF core to be separated from the signal from PCF walls, we chose gas-phase molecular nitrogen from atmospheric-pressure air filling the hollow core of a PCF. A two-color Raman-resonant pump field used in these experiments consisted of 15 ns second-harmonic pulses of Nd:YAG laser radiation with a wavelength of 532 nm (ω_1) and dye-laser radiation (ω_2) with a wavelength of 607 nm. The dye-laser frequency was chosen in such a way as to satisfy the condition of Raman resonance $\omega_1 - \omega_2 = \Omega$ with a *Q*-branch Raman-active transition of N_2 at the central frequency $\Omega = 2331\,\mathrm{cm}^{-1}$. Coherently excited *Q*-branch vibrations of N_2 then scatter off the second-harmonic probe field, giving rise to a CARS signal at the frequency $\omega_{\mathrm{CARS}} = 2\omega_1 - \omega_2$ (corresponding to a wavelength of 473 nm). The hollow PCF (shown in inset 1 to Fig. 4.28) was designed to simultaneously provide high transmission for the air-guided modes of the second harmonic, dye-laser radiation, and the CARS signal. With an appropriate fiber structure, as can be seen from inset 2 to Fig. 4.28, PCF transmission peaks can be centered around the carrier wavelengths of the input light fields and the CARS signal (shown by vertical lines in inset 2 to Fig. 4.28). Phase matching for CARS with the chosen set of wavelengths has been confirmed [4.168] (inset 3 in Fig. 4.28) by a numerical analysis of PCF dispersion based on a modification of the field-expansion technique developed by *Poladian* et al. [4.169].

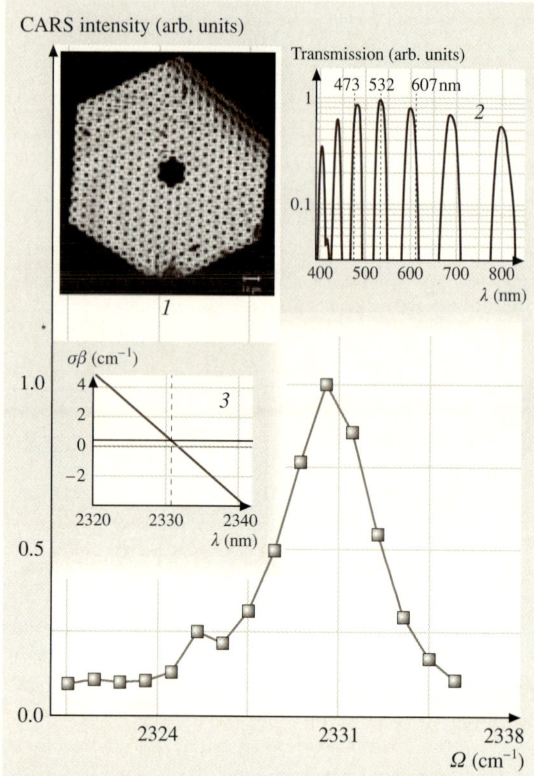

Fig. 4.28 CARS spectrum of *Q*-branch Raman-active vibrations of N_2 molecules in the atmospheric-pressure air filling the hollow core of the PCF. The *insets* show: (1) an image of the PCF cross section; (2) the transmission spectrum of the hollow PCF designed to simultaneously support air-guided modes of the pump, probe, and CARS signal fields (their wavelengths are shown by *vertical lines*) in the hollow core of the fiber; and (4.3) the mismatch of the propagation constants $\delta\beta = 2\beta_1 - \beta_2 - \beta_a$ calculated for the $\omega_a = 2\omega_1 - \omega_a 2$ CARS process in fundamental air-guided modes of the hollow PCF with β_1, β_2, and β_a being the propagation constants of the Nd:YAG-laser second-harmonic (ω_1) and dye-laser (ω_2) pump fields and the CARS signal (ω_a) in the hollow PCF

The resonant CARS signal related to Q-branch vibrations of N_2 in these experiments can be reliably separated from the nonresonant part of the CARS signal originating from the PCF walls. The spectra of the CARS signal at the output of the PCF Fig. 4.28 are identical to the N_2 Q-branch CARS spectrum of the atmospheric air [4.99] measured in the tight-focusing regime. In view of this finding, the CARS signal can be completely attributed to the coherent Raman scattering in the gas filling the fiber core with no noticeable contribution from the nonlinearity of PCF walls.

Results presented here show that large-core-area hollow PCFs bridge the gap between standard, solid-cladding hollow fibers and hollow PCFs in terms of effective guided-mode areas, allowing energy fluence scaling of phase-matched waveguide four-wave mixing of laser pulses. We used hollow PCFs with a core diameter of about 50 μm to demonstrate phase-matched FWM for millijoule nanosecond laser pulses. Intense CARS signal has been observed from stretching vibrations of water molecules inside the hollow fiber core, suggesting CARS in hollow PCFs as a convenient sensing technique for pollution monitoring and trace gas detection. Hollow PCFs have been shown to offer much promise as fiber-optic probes for biomedical Raman applications, suggesting the way to substantially reduce the background related to Raman scattering in the core of standard biomedical fiber probes [4.170].

4.9 Nonlinear Spectroscopy with Photonic-Crystal-Fiber Sources

4.9.1 Wavelength-Tunable Sources and Progress in Nonlinear Spectroscopy

The progress in wavelength-tunable light sources through the past decades has been giving a powerful momentum to the development of nonlinear laser spectroscopy. Nonlinear Raman spectroscopy, in particular, has benefited tremendously more than 30 years ago from the application of tunable laser sources, as optical parametric oscillators [4.171] and dye lasers [4.172] were demonstrated to greatly simplify measurements based on coherent anti-Stokes Raman scattering (CARS), making this technique much more informative, efficient, and convenient. Broadband laser sources later contributed to the technical and conceptual progress in nonlinear Raman spectroscopy [4.99, 106–110], allowing single-shot CARS measurements. In the era of femtosecond lasers, several parallel trends have been observed in the development of laser sources for nonlinear Raman spectroscopy [4.113]. One of these tendencies was to adapt broadband femtosecond pulses for spectroscopic purposes [4.113, 173, 174] and to use different spatial phase-matching geometries to simultaneously generate coherent Stokes and anti-Stokes, as well as degenerate four-wave mixing signals [4.113, 127, 128]. The rapid progress in nonlinear materials, on the other hand, resulted in the renaissance of optical parametric oscillators and amplifiers (OPOs and OPAs) for nonlinear spectroscopy [4.175]. Chirped pulses were used [4.176, 177] to probe broad spectral regions and large ranges of delay times, suggesting efficient single-shot nonlinear spectroscopic approaches [4.178–180].

In this section, we focus on the potential of photonic-crystal fibers [4.36,37] as novel efficient sources for nonlinear spectroscopy. PCFs are unique waveguide structures allowing dispersion [4.40] and spatial field [4.181] profiles to be engineered by modifying the design of the fiber structure. Nonlinear-optical PCF components and novel PCF-based light sources have been intensely used through the past few years in frequency metrology [4.74, 77], biomedical optics [4.81], ultrafast photonics [4.78, 79], and photochemistry [4.182].

Coherent nonlinear spectroscopy and microscopy open up a vast area for applications of PCF light sources and frequency shifters. Efficient frequency conversion and supercontinuum generation in PCFs have been shown to enhance the capabilities of chirped-pulse CARS [4.183] and coherent inverse Raman spectroscopy [4.184]. Cross-correlation frequency-resolved optically gated CARS (XFROG CARS) has recently been demonstrated [4.80] using specially designed PCF frequency converters for ultrashort laser pulses. Novel light sources based on frequency shifting in PCFs provide a useful tool for the measurement of second-order optical nonlinearities in organic materials [4.185] and offer interesting new options in CARS microscopy [4.186]. Efficient spectral broadening of ultrashort pulses in PCFs with carefully engineered dispersion profiles [4.42–44] makes these fibers ideal light sources for pump–supercontinuum probe time- and frequency-resolved nonlinear-optical measurements [4.187].

Below in this section, we demonstrate applications of PCF light sources for chirped-pulse CARS and nonlinear absorption spectroscopy. We will show that PCFs can provide efficient nonlinear-optical transformations of femtosecond Cr:forsterite laser pulses, delivering linearly chirped frequency-shifted broadband light pulses with central wavelengths ranging from 400 to 900 nm. These pulses were cross-correlated in our experiments with the femtosecond second-harmonic output of the Cr:forsterite laser in toluene solution, used as a test object, in boxcars geometry to measure CARS spectra of toluene molecules (XFROG CARS). The blue-shifted chirped-pulse output of a photonic-crystal fiber with a spectrum stretching from 530 to 680 nm is shown to be ideally suited for the nonlinear absorption spectroscopy of one- and two-exciton bands of thiacarbocyanine J aggregates in a polymer film excited by femtosecond second-harmonic pulses of the Cr:forsterite laser.

4.9.2 Photonic-Crystal Fiber Frequency Shifters

Spectroscopic measurements were performed with multicomponent-glass PCFs [4.51, 188], fabricated with the use of the standard PCF technology [4.36, 37]. In PCFs used in our experiments (Fig. 4.29a), the solid fiber core is surrounded by a single ring of thin-wall capillaries whose outer diameters are equal to the diameter of the PCF core. The outer part of the microstructure cladding consists of 11 rings of capillaries with outer diameters approximately three times larger than the diameter of the PCF core and a high air-filling fraction. Dispersion and nonlinearity of the PCFs were managed by scaling the geometric sizes of the PCF structure. Technologically, this was realized by using the same preform to fabricate PCFs with the same type of the structure, but with different magnifying factors. This procedure allowed to scale the sizes of PCF structure without changing its geometry.

The laser system used in experiments consisted of a Cr^{4+}:forsterite master oscillator, a stretcher, an optical isolator, a regenerative amplifier, and a compressor [4.189]. The master oscillator, pumped with a fiber ytterbium laser, generated 30–60 fs light pulses of radiation with a wavelength of 1.23–1.25 μm at a repetition rate of 120 MHz. These pulses were then transmitted through a stretcher and an isolator, to be amplified in a Nd:YLF-laser-pumped amplifier and recompressed to the 200 fs pulse duration with the maximum laser pulse energy up to 20 μJ at 1 kHz. The energy of laser pulses used in experiments presented in this paper ranged from 0.5 up to 200 nJ.

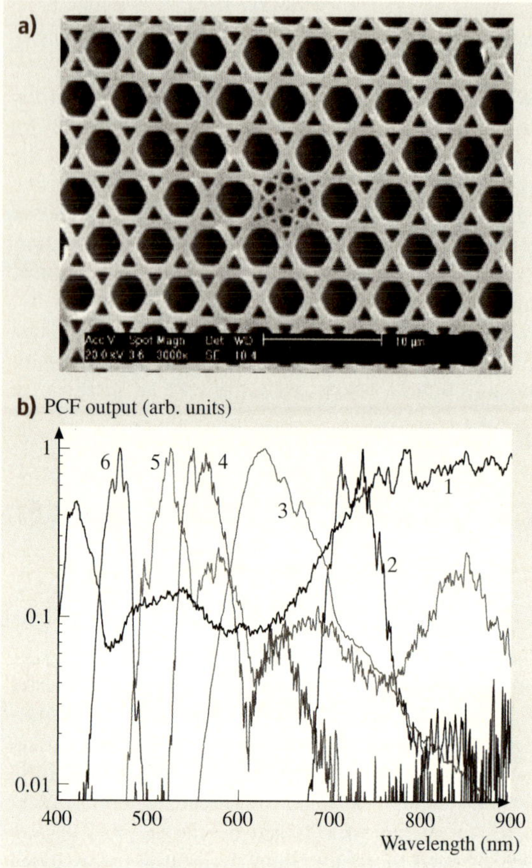

Fig. 4.29 **(a)** An SEM image of soft-glass photonic-crystal fibers. **(b)** Intensity spectra of the dispersion-managed blue-shifted output of soft-glass PCFs. The wavelength offset δ is 50 nm (1), 110 nm (2), 150 nm (3), 190 nm (4), 220 nm (5), and 300 nm (6)

For the PCFs used in experiments, the central wavelength of Cr:forsterite laser pulses falls within the range of anomalous dispersion. Such pulses can therefore form solitons in the fiber. High-order dispersion induces wave-matching resonances between solitons and dispersive waves [4.21, 22], giving rise to intense blue-shifted emission, observed as prominent features in the PCF output spectra (Fig. 4.29b). The central wavelength of this blue-shifted emission correlates well with the wavelength where dispersive waves guided in the fundamental mode of the PCF are phase-matched with the soliton produced by Cr:forsterite laser pulses. The phase matching between the soliton produced by the input laser pulse and the dispersive wave, which defines the frequency of the dominant peaks in output spectra of PCFs, is controlled by the dispersion of the fiber. The central frequency of the blue-shifted signal in the output of PCFs can thus be tuned by modifying the dispersion profile of the fiber. The GVD profiles of waveguide modes were modified in our experiments by scaling the geometric sizes of the fiber without changing the type of the structure shown in Fig. 4.29a. The core diameter of PCFs fabricated for these experiments was varied from 0.9 to 3.8 μm. Figure 4.29b shows the spectra of the blue-shifted output of the fiber, tuned by changing the offset $\delta = \lambda_0 - \lambda_z$ between the central wavelength of the input laser field, λ_0, and the zero-GVD wavelength λ_z. With the PCF length remaining unchanged (10 cm), larger blue shifts are achieved by increasing the offset δ (cf. curves 1–6 in Fig. 4.29b). The power of the input laser field needs to be increased for larger δ in these experiments to keep the amplitude of the blue-shifted signal constant. In these experiments, dispersion-managed soft-glass PCFs are shown to serve as frequency shifters of femtosecond Cr:forsterite laser pulses, providing an anti-Stokes output tunable across the range of wavelengths from 400 to 900 nm.

Experiments presented in this section show that the structural dispersion and nonlinearity management of multicomponent-glass photonic-crystal fibers allows a wavelength-tunable frequency shifting and white-light spectral transformation of femtosecond Cr:forsterite laser pulses. We have explored the ways toward optimizing non-silica PCFs for frequency shifting and white-light spectral superbroadening of femtosecond Cr:forsterite laser pulses and identified important advantages of multicomponent-glass PCFs over silica microstructure fibers for the spectral transformation of laser pulses in the 1.2–1.3 μm spectral range. By coupling 200 fs pulses of 1.24 μm Cr:forsterite laser radiation into different types of multicomponent-glass PCFs, where the zero-GVD wavelength is tuned by scaling the sizes of the fiber structure, we have demonstrated spectrally tailored supercontinuum generation and frequency upshifting providing a blue-shifted output tunable across the range of wavelengths from 400 to 900 nm.

4.9.3 Coherent Anti-Stokes Raman Scattering Spectroscopy with PCF Sources

In this section, we show that PCF frequency shifters can serve as convenient sources of chirped wavelength-tunable pulses for CARS spectroscopy. In CARS

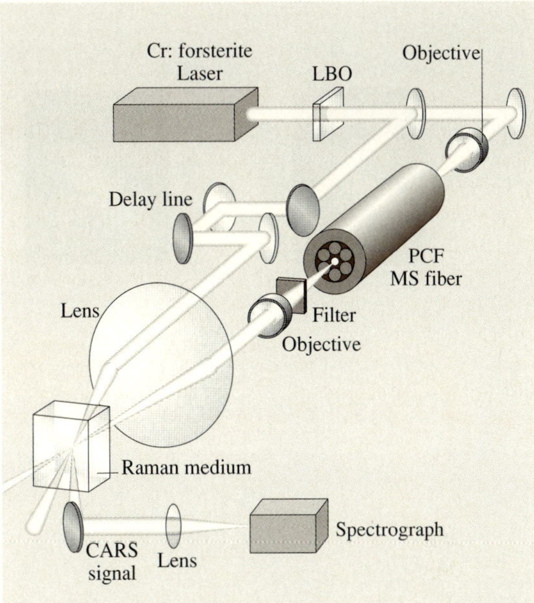

Fig. 4.30 Diagram of femtosecond CARS spectroscopy with the use of ultrashort pulses frequency-upconverted and chirped in a photonic-crystal fiber. The *inset* shows the cross-sectional view of the central part of the photonic-crystal fiber

experiments, sub-microjoule Cr:forsterite-laser pulses with an initial duration of about 90 fs are launched into the central core of the PCF (Fig. 4.30), resulting in the efficient generation of a blue-shifted signal (Fig. 4.31), with a central wavelength dictated by phase matching for dispersive-wave emission and controlled by fiber dispersion. The wavelength of the blue-shifted signal can be finely tuned by changing the intensity of the pump pulse (Fig. 4.31) due to the nonlinear change in the refractive index of the fiber core and the spectral broadening of the pump pulse.

Cross-correlation frequency-resolved optical gating (XFROG) [4.190, 191] was used to characterize the blue-shifted output of the PCF. An XFROG signal was generated by mixing the blue-shifted signal from the fiber E_a with the 620 nm 90 fs second-harmonic output of the Cr:forsterite laser E_{SH} in a BBO crystal. A two-dimensional XFROG spectrogram, $S(\omega, \tau) \propto |\int_{-\infty}^{\infty} E_a(t)\, E_{SH}(t-\tau) \exp(-i\omega t)\, dt|^2$, was then plotted by measuring the XFROG signal as a function of the delay time τ between the second-harmonic pulses and the blue-shifted output of the PCF and spectrally dispersing the XFROG signal. The XFROG spectrogram shown in inset 1 to Fig. 4.31 visualizes

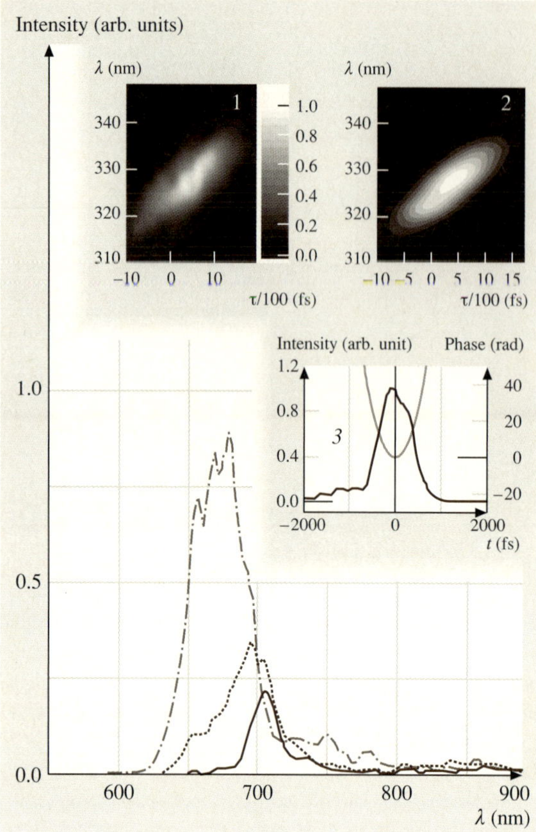

Fig. 4.31 The spectrum of the blue-shifted output of a photonic-crystal fiber pumped by 1.24 μm 90 fs Cr:forsterite-laser pulses with an input energy of (*solid line*) 170 nJ, (*dashed line*) 220 nJ, and (*dash–dotted line*) 270 nJ. The insets show: (1) the intensity of the sum-frequency signal generated in a BBO crystal by the second-harmonic pulse from the Cr:forsterite laser and the blue-shifted PCF output as a function of the wavelength and the delay time τ between the second-harmonic and anti-Stokes pulses, (2) theoretical fit of the XFROG trace, and (3) the pulse envelope and the phase of the anti-Stokes pulse providing the best fit

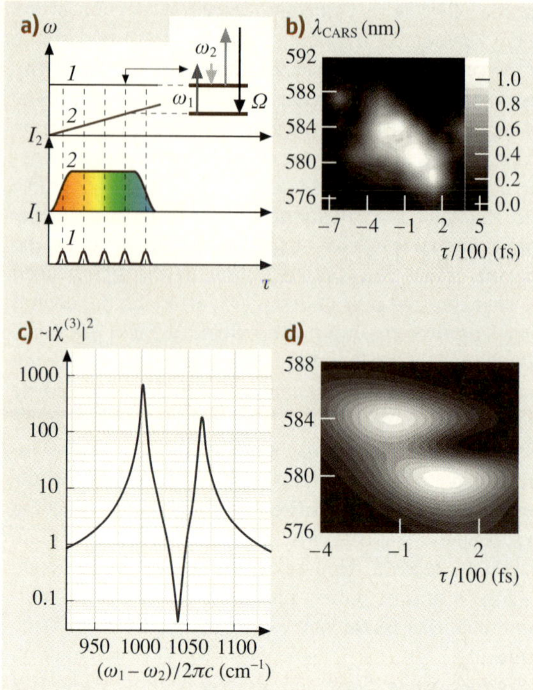

Fig. 4.32 **(a)** A diagram of femtosecond CARS spectroscopy with chirped pulses is shown on the *left*. The first pulse (frequency ω_1) is transform limited. The second pulse (frequency ω_2) is linearly chirped. The linear chirp of the second pulse maps the delay time between the pulses on the frequency, allowing the frequency difference $\omega_1 - \omega_2$ to be scanned through Raman resonances Ω by tuning the delay time τ. **(b)** The intensity of the CARS signal generated in a noncoplanar boxcars geometry from toluene solution as a function of the wavelength and the delay time τ between the second-harmonic and the blue-shifted PCF output pulses used as a biharmonic pump. **(c)** A model of the Raman spectrum, including a doublet of Lorentzian lines and a frequency-independent nonresonant background used in the theoretical fit. **(d)** Theoretical fit of the XFROG CARS spectrogram

the temporal envelope, the spectrum, and the chirp of the blue-shifted PCF output. A reasonable fit of the experimental XFROG trace was achieved (inset 2 in Fig. 4.31) with a blue-shifted pulse having a duration of about 1 ps and a linear positive chirp corresponding to the phase $\varphi(t) = \alpha t^2$ with $\alpha = 110\,\text{ps}^{-2}$. The spectrum and the phase φ of the blue-shifted pulse reconstructed with this procedure are shown in inset 3 in Fig. 4.31.

The linear chirp defines a simple linear mapping between the instantaneous frequency of the blue-shifted PCF output and the delay time τ, allowing spectral measurements to be performed by varying the delay time between the pump pulses. Experiments were performed with 90 fs second-harmonic pulses of the Cr:forsterite laser (at the frequency ω_1) and the linearly chirped pulses from the PCF (the frequency ω_2) as a biharmonic pump for the CARS spectroscopy of toluene solution. The frequency difference $\omega_1 - \omega_2$ was scanned through the frequencies of Raman-active modes of toluene molecules by tuning the delay time between the pump pulses (Fig. 4.32a). The second-harmonic pulse also served as a probe in our CARS scheme, generating the CARS signal at the frequency $\omega_{\text{CARS}} = 2\omega_1 - \omega_2$ through the scattering from Raman-active vibrations coherently excited by the pump fields. The light beams with frequencies ω_1 and ω_2 were focused into a cell with toluene solution at a small angle with respect to each other (Fig. 4.30). The CARS signal generated in the area of beam interaction in this non-coplanar boxcars geometry had a form of a sharply directed light beam with a low, phase-matching-controlled angular divergence spatially separated (Fig. 4.30) from the pump beams. Figure 4.32b presents the map of CARS spectra from the toluene solution measured for different delay times τ between the biharmonic pump pulses. This procedure of measurements, in fact, implements the XFROG technique [4.190, 191]. However, while FROG-based techniques [4.192] are usually employed to characterize ultrashort pulses, our goal here is to probe Raman-active modes of toluene molecules, used as a test object, by means of CARS spectroscopy.

In the case of a positively chirped pulse from the PCF (insets 2 and 3 in Fig. 4.31), small delay times τ correspond to the excitation of low-frequency Raman-active modes ($\tau \approx -200$ fs in Fig. 4.32b). In particular, the $1004\,\text{cm}^{-1}$ Raman mode of toluene is well resolved in the presented XFROG CARS spectrogram. This mode is excited with the second harmonic of Cr:forsterite-laser radiation and the spectral slice around the wavelength of $\lambda_2 \approx 661$ nm picked with an appropriate τ (Fig. 4.32a) from the positively chirped blue-shifted PCF output, giving rise to a CARS signal with the wavelength $\lambda_{\text{CARS}} \approx 584$ nm. Raman modes with higher frequencies are probed at larger delay times ($\tau \approx 100$–200 fs in Fig. 4.32b).

As a simple model of the Raman spectrum of toluene molecules in the studied spectral range, we employ a doublet of Lorentzian lines interfering with the coherent nonresonant background (Fig. 4.32c). The frequencies of the peaks in these spectra are used as fitting parameters. The ratio of the peak value of the resonant part of the CARS susceptibility, $\bar{\chi}^{(3)}$, to the nonresonant susceptibility, $\chi^{(3)\text{nr}}$, was estimated by measuring the intensities of the CARS signal on and off the Raman resonances, yielding $|\chi^{(3)\text{nr}}|/|\bar{\chi}^{(3)}| \approx 0.05$. The best fit (Fig. 4.32d) is achieved with the Raman peaks centered at 1004 and $1102\,\text{cm}^{-1}$, which agrees well with earlier CARS studies of toluene [4.99].

Although the slope of the XFROG CARS trace and positions of Raman peaks can be adequately described with the use of this simple model, some of the spectroscopic features of the experimental XFROG CARS trace deviate from the theoretical fit. These deviations may originate from variations in the coherent background as a function of the frequency. For a quantitative spectroscopic analysis, these inhomogeneities in the frequency dependence of the coherent background should be carefully measured and included in the fit. On the other hand, the nonresonant contribution to CARS spectral profiles can be efficiently suppressed [4.99, 106] by using three input pulses in the CARS arrangement instead of two and by introducing the delay time between the third pulse (probe) and the two-color pump, tuned to a Raman resonance under study.

4.9.4 Pump-Probe Nonlinear Absorption Spectroscopy using Chirped Frequency-Shifted Light Pulses from a Photonic-Crystal Fiber

Time-resolved nonlinear-optical spectroscopy of molecular dynamics and fast excitation transfer processes typically involves specifically designed sequences of pump and probe pulses with a variable delay time and a smoothly tunable frequency of the probe field. In a laboratory experiment, such pulse sequences can be generated by femtosecond optical parametric amplifiers (OPAs). Femtosecond OPAs, however, inevitably increase the cost of laser experiments and make the laser system more complicated, unwieldy, and difficult to align. An interesting alternative strategy of pump–probe spectroscopy employs a pump field in the form of a broadband radiation (supercontinuum) with a precisely characterized chirp. A time–frequency map, defined by the chirp of such a supercontinuum pulse, allows time- and frequency-resolved measurements to be performed by tuning the delay time between the transform-limited pump pulse and chirped supercontinuum pulse [4.176,193]. The supercontinuum probe pulse for pump–probe experiments is most often generated by focusing amplified femtosecond pulses into a silica or sapphire plate. The chirp of the probe pulse in such an arrangement is dictated by the regime of nonlinear-optical spectral transformation and dispersion of the nonlinear material, leaving little space for the phase tailoring of the probe field.

In this section, we demonstrate the potential of PCFs as compact and cost-efficient fiber-optic sources of probe pulses with a tunable frequency and tailored phase for a time-resolved nonlinear spectroscopy of molecular aggregates. We will show that PCFs can provide efficient nonlinear-optical transformations of femtosecond Cr:forsterite laser pulses, delivering linearly chirped frequency-shifted broadband light pulses, optimized for pump–probe nonlinear absorption spectroscopy of molecular aggregates. The blue-shifted output of a photonic-crystal fiber with a spectrum stretching from 530 to 680 nm will be used to probe one- and two-exciton bands of thiacarbocyanine J aggregates in a polymer film excited by femtosecond second-harmonic pulses of the Cr:forsterite laser.

Molecular aggregates are interesting and practically significant objects encountered in many physical, chemical and biological systems [4.194]. Interactions between the molecules forming aggregates give rise to collective electronic states, which can be delocalized over large chains of molecules, modifying the optical response of the system [4.195]. Specific types of molecular aggregation, known as J and H aggregation, give rise to a pronounced spectral shift and dramatic narrowing of absorption bands, indicating the cooperative nature of the optical response of molecular aggregates [4.196]. In natural systems, molecular aggregates are involved in the processes and functions of vital importance as they play the key role in light harvesting and primary charge separation in photosynthesis [4.197]. On optical resonances, the nonlinear susceptibility of molecular aggregates displays a collective enhancement [4.195], scaling as N^2 with the number of particles N forming an aggregate. Dramatic enhancement of optical nonlinearity and the available pathways for ultrafast relaxation of excited states [4.194] suggest a variety of interesting applications of molecular aggregates, such as terahertz demultiplexing of optical signals [4.198], spectral sensitization in optical data storage and photography [4.199], energy transfer from light-harvesting antennas and complexes in artificial photosynthesis [4.197], mode locking in laser cavities, and creation of novel devices for ultrafast photonics.

Collective electronic eigenstates in aggregates of strongly coupled molecules are grouped into excitonic bands [4.194–196]. In the ground state of this band structure, all N molecules of the aggregate reside in the ground state. In the lowest excited band, the molecules coupled into an aggregate share one excitation. The eigenstates of this first excited band (one-excitons, or Frenkel excitons) are represented by linear combinations of basis states corresponding to one excited and $N-1$ ground-state molecules. Eigenstates that can be reached from the ground state via two optical transitions, making the molecules of an aggregate share two excitations, form a two-exciton band. The third-order susceptibility $\chi^{(3)}$, responsible for four-wave mixing (FWM) processes, can involve only one- and two-excitons. Although three-exciton states can be reached

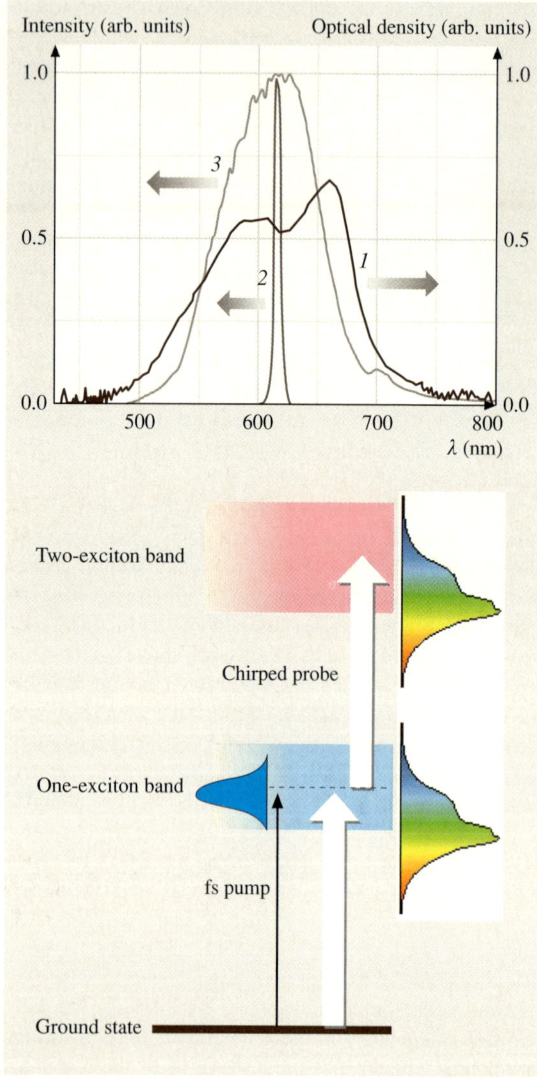

Fig. 4.33 (a) Absorption spectrum of the thiacarbocyanine film (1), spectrum of the second-harmonic output of the Cr:forsterite laser (2), and spectrum of the blue-shifted PCF output (3). **(b)** Diagram of the pump–probe nonlinear-absorption spectroscopy of exciton bands in molecular aggregates using a femtosecond (fs) pump pulse and a broadband chirped probe field

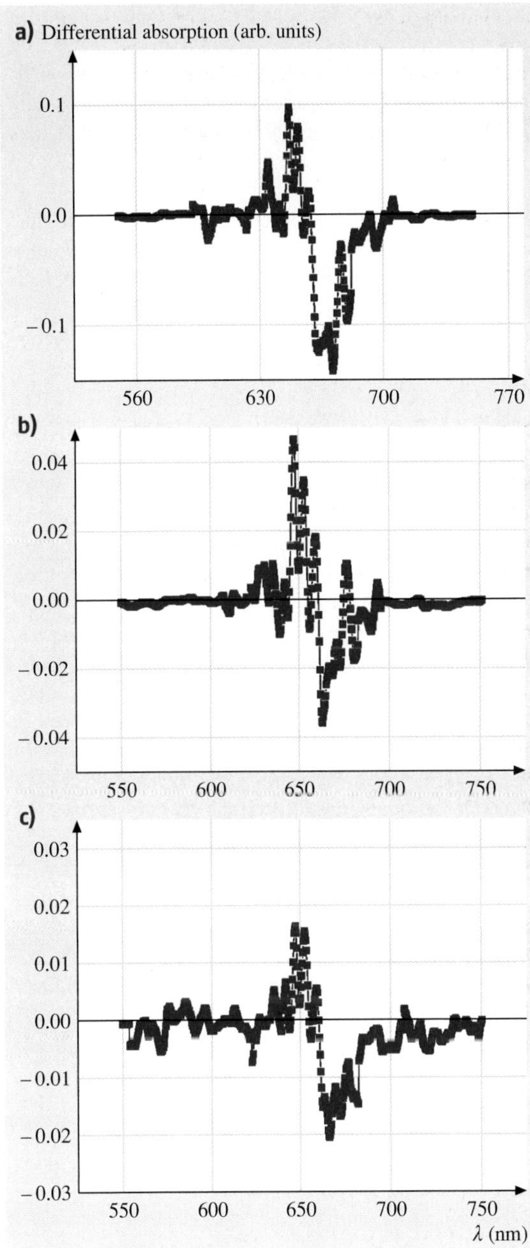

Fig. 4.34a–c Differential spectra of nonlinear absorption measured by the pump–probe technique for thiacarbocyanine J aggregates in a polymer film with a delay time between the probe and pump pulses equal to (**a**) 100 fs, (**b**) 500 fs, and (**c**) 1100 fs

by three transitions, they do not contribute to the third-order polarization as they do not have a transition dipole that would couple them to the ground state. Time-resolved nonlinear spectroscopic methods based on $\chi^{(3)}$ processes, such as spectroscopy of nonlinear absorption, degenerate four-wave mixing, and third-harmonic generation, have proven to be convenient and informative techniques for the characterization of one- and two-exciton states of aggregates, providing the data on the strength of dipole–dipole interaction between molecules in aggregates, as well as on the disorder and typical relaxation and exciton annihilation times in aggregates [4.194].

Experiments [4.200] were performed with thin-film samples of J aggregates of thiacarbocyanine dye. Film samples of J aggregates were prepared by spin-coating the solution of thiacarbocyanine dye on a thin substrate. The concentration of thiacarbocyanine dye was 5×10^{-3} mol/l. The spinning speeds ranged from 1000 up to 3000 revolutions per minute. A mixture of acetonitrile, dichloroethane, and chloroform with a volume ratio of 2:2:1 was used as a solvent. The thickness of the dye layer applied to a substrate was estimated as 30 nm, and the total thickness of the sample was about 1 μm.

Absorption spectra of J-aggregate films (curve 1 in Fig. 4.33a) display two pronounced peaks. The broader peak centered at 595 nm corresponds to thiacarbocyanine monomers, while the narrower peak at 660 nm represents the excitonic absorption of J aggregates. To provide a rough estimate of the delocalization length N_d of excitons in aggregates, we apply the formula [4.201] $N_\mathrm{d}^W \approx (3\pi^2|J|3\pi^2|J|W)^{1/2} - 1$, which expresses N_d through the half-width at half-maximum W of the aggregate peak in the absorption spectrum and the energy J of dipole–dipole interaction between nearest-neighbor molecules in the aggregate ($J < 0$ for J aggregates). With the J parameter estimated as $J \approx 900\,\mathrm{cm}^{-1}$ from the aggregation-induced red shift of the absorption peak in Fig. 4.33a, we find $N_\mathrm{d} \approx 6$. Experimental results presented below in this paper show that, with such a delocalization length of excitons in molecular aggregates, spectra of nonlinear absorption display well-pronounced nonoverlapping features indicating [4.194–196] bleaching through transitions between the ground and one-exciton states and induced absorption via transitions between one- and two-exciton states of molecular aggregates.

The laser system used in experiments [4.200] was based on the Cr^{4+}:forsterite laser source with regenerative amplification, as described in Sect. 4.9.2. A 1 mm-thick BBO crystal was used to generate the second harmonic of amplified Cr:forsterite laser radiation. The spectrum of the second-harmonic output of the Cr:forsterite laser was centered at 618 nm (curve 2 in Fig. 4.33a). Second-harmonic pulses with a pulse width of about 120 fs and the energy ranging from 10 to 80 nJ were used as a pump field in our experiments on the nonlinear spectroscopy of molecular aggregates.

Frequency-tunable upconversion of fundamental-wavelength Cr:forsterite-laser pulses was performed through the nonlinear-optical spectral transformation of these pulses in soft-glass PCFs with the cross-section structure shown in Fig. 4.29a. The properties of such fibers and the methods of frequency conversion of Cr:forsterite laser pulses in these PCFs have been discussed in Sect. 4.9.2 Intensity spectrum of the frequency-shifted output of the PCF best suited as a probe field for time-resolved nonlinear-absorption spectroscopy of J aggregates is presented by curve 3 in Fig. 4.33a. At the level of 20% of its maximum, the intensity spectrum of the blue-shifted PCF output stretches from 530 to 680 nm. Dispersion of the PCF frequency shifter used in our experiments provided a linear chirp of the output pulse Sect. 4.9.3 with the pulse chirp rate controlled by the fiber length.

The spectrum of 120 fs second-harmonic pulses of the Cr:forsterite laser partially overlaps the absorption spectrum of molecular aggregates (Fig. 4.33a). These pulses were used in our experiments to excite the aggregates through the transitions from the ground state to the one-exciton band (Fig. 4.33b). The spectra of absorption modified by the pump field were measured by chirped blue-shifted pulses delivered by the PCF (Fig. 4.33b). Figures 4.34a–c present the results of experimental measurements upon the subtraction of absorption spectra measured in the absence of the pump pulse and normalization to the spectrum of the probe field. Nonlinear absorption spectra shown in Figs. 4.34a–c display well-pronounced minima at 665 nm and blue-shifted peaks at 640 nm. Such features are typical of nonlinear absorption spectra of J aggregates measured by the pump–probe technique (see [4.194] for a review). The negative feature is indicative of bleaching through pump-induced transitions between the ground state and the one-exciton band, while the blue-shifted peak originates from induced absorption due to transitions between one- and two-exciton bands of molecular aggregates (Fig. 4.34a–c).

For highly ordered aggregates, the spectrum of nonlinear absorption is dominated by transitions between the ground state and lowest one- and two-exciton states [4.195, 196]. The exciton delocalization length can be then estimated from the spectral shift Δ of the induced-absorption peak relative to the bleaching minimum using the following formula [4.202]: $N_{\mathrm{d}}^{\Delta} \approx (3\pi^2|J|/\Delta)^{1/2} - 1$. With the spectral shift estimated as $\Delta \approx 470\,\mathrm{cm}^{-1}$, we find that $N_{\mathrm{d}}^{\Delta} \approx 6$, in perfect agreement with the value of N_{d}^{W} obtained from aggregate absorption spectra.

The amplitudes of both positive and negative features in nonlinear absorption spectra decay on a sub-picosecond time scale with increasing delay time between the pump and probe pulses (Fig. 4.34a–c), indicating a sub-picosecond relaxation rate of the one-exciton state of molecular aggregates in our experiments. This finding suggests, in agreement with earlier studies of ultrafast excitation energy transfer processes in molecular aggregates, that the relaxation dynamics of aggregates in our experimental conditions is mainly controlled by the quenching of excited states of aggregates through exciton–exciton annihilation [4.194].

We have thus shown that photonic-crystal fibers with a specially designed dispersion offer the ways to create efficient sources of ultrashort pulses for coherent nonlinear spectroscopy. These fibers provide a high efficiency of frequency upconversion of femtosecond laser pulses, permitting the generation of sub-picosecond linearly chirped anti-Stokes pulses ideally suited for femtosecond coherent anti-Stokes Raman scattering spectroscopy. Experimental studies demonstrate that PCFs can deliver linearly chirped frequency-shifted broadband light pulses, optimized for pump–probe nonlinear absorption spectroscopy.

4.10 Surface Nonlinear Optics, Spectroscopy, and Imaging

In this section, we will dwell upon the potential of nonlinear-optical methods for the investigation of surfaces and interfaces. The ability of second-order nonlinear-optical processes, such as SHG and SFG (Sect. 4.3), to probe surfaces and interfaces is most clearly seen in the case of a centrosymmetric material. In this situation, the electric-dipole SHG and SFG response from the bulk of the material vanishes. At a surface or an interface, the bulk symmetry is broken, and electric-dipole second-order nonlinear-optical effects are allowed. Such surface-specific SHG and SFG processes enable a highly sensitive nondestructive local optical diagnostics of surfaces and interfaces (Fig. 4.35a–b). Illuminating and physically insightful discussion of this technique can be found in the classical texts on nonlinear optics [4.9, 203].

The $\chi^{(2)}$ signal from a surface or an interface is, however, not entirely background-free, as the second-order nonlinear optical processes are not strictly forbidden even in a centrosymmetric medium. Beyond the electric-dipole approximation, the second-order nonlinear signal, as can be seen from (4.6–4.8) and (4.14), can be generated through the electric-quadrupole and magnetic-dipole effects. It is generally very difficult, often impossible, to completely distinguish between the surface and bulk contributions to the nonlinear signal. Luckily enough, the electric-quadrupole and magnetic-dipole components in the nonlinear-optical response are typically ka times less significant than the dipole-allowed part [4.9], with $k = 2\pi/\lambda$ and a being a typical size of an atom or a unit cell in a crystal. The ratio of the surface dipole-allowed susceptibility $\chi_s^{(2)}$ to the bulk susceptibility $\gamma_b^{(2)}$ can be then estimated as $|\chi_s^{(2)}|/|\gamma_b^{(2)}| \sim d/(ka)$, where d is the thickness of the surface layer. In reflection SHG (Fig. 4.35a), the bulk contribution is typically generated in a subsurface layer with a thickness of $d \sim \lambda/(2\pi)$. The ratio of the surface part of the total reflected SHG signal to the bulk contributions in this case is on the order of d^2/a^2, which can be easily made much larger than unity. This ratio can be substantially enhanced on frequency resonances or with an appropriate choice of polarization arrangement.

A combination of the high spatial and temporal resolution with a spectral selectivity makes $\chi^{(2)}$ techniques a powerful tool for time-resolved species-selective studies of surfaces and buried interfaces (see, e.g., [4.204] for a comprehensive review of recent results), detection and size and shape analysis of adsorbed species, nanoparticles, and clusters on surfaces [4.205, 206], as well as imaging and microscopy of biological species [4.207]. The sensitivity and selectivity of the $\chi^{(2)}$ technique are enhanced when the frequency of one of the laser fields (ω_1 in the inset to Fig. 4.35b) is tuned to a resonance with a frequency of one of the vibrations typical of species on a surface or an interface (Fig. 4.35b). This method of surface analysis is referred to as sum-frequency surface vibrational spectroscopy. The capabilities of this technique have been impressively

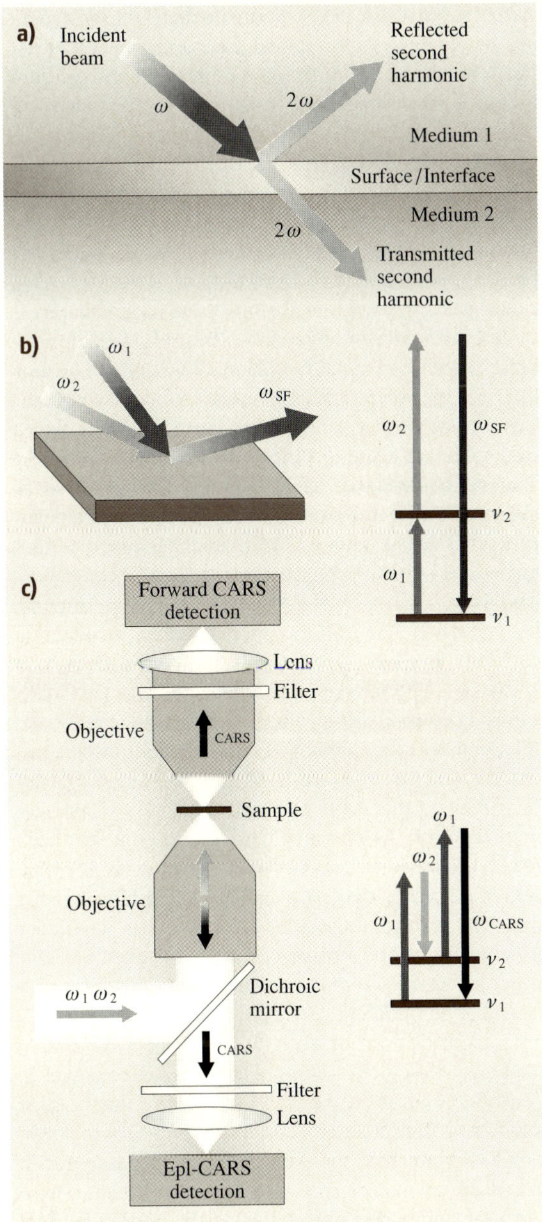

Fig. 4.35 Nonlinear optics, spectroscopy, and imaging of surfaces and buried structures. **(a)** Nonlinear-optical probing of surfaces and interfaces using second-harmonic generation. **(b)** Surface vibrational spectroscopy using sum-frequency generation. Diagram of vibrational transitions probed by SFG is shown on the *right*. **(c)** Nonlinear microscopy based on coherent anti-Stokes Raman scattering. Diagram of Raman-active transitions selectively addressed through CARS is shown on the *left*

demonstrated for vapor–liquid and liquid–solid interfaces [4.205].

In sum-frequency surface vibrational spectroscopy [4.207] an infrared laser pulse $\boldsymbol{E}_1$ with a frequency ω_1 overlaps on the surface of a sample with the second laser pulse $\boldsymbol{E}_2$, which typically has a frequency ω_2 in the visible, to induce a second-order polarization at the sum frequency $\omega_{\mathrm{SF}} = \omega_1 + \omega_2$:

$$\boldsymbol{P}_{\mathrm{SF}}^{(2)}(\omega_{\mathrm{SF}}) = \widehat{\chi}^{(2)}(\omega_{\mathrm{SF}}; \omega_1, \omega_2) : \boldsymbol{E}_1 \boldsymbol{E}_2 . \quad (4.226)$$

The intensity of the optical signal at the sum frequency is given by

$$I_{\mathrm{SF}} \propto \left|\chi_{\mathrm{eff}}^{(2)}\right|^2 I_1 I_2 . \quad (4.227)$$

Here, I_1 and I_2 are the intensities of the laser beams and

$$\chi_{\mathrm{eff}}^{(2)} = \left(\widehat{L}_{\mathrm{SF}} \cdot \boldsymbol{e}_{\mathrm{SF}}\right) \cdot \widehat{\chi}_{\mathrm{s}}^{(2)} : \left(\widehat{L}_2 \cdot \boldsymbol{e}_2\right) \left(\widehat{L}_1 \cdot \boldsymbol{e}_1\right) , \quad (4.228)$$

where $\widehat{L}_1$, $\widehat{L}_2$ and $\widehat{L}_3$ are the Fresnel factors at the frequencies ω_1, ω_2, and ω_{SF}, respectively, $\boldsymbol{e}_1$, $\boldsymbol{e}_2$, and $\boldsymbol{e}_{\mathrm{SF}}$ are the unit polarization vectors of the laser and sum-frequency fields, and the surface quadratic susceptibility $\widehat{\chi}_{\mathrm{s}}^{(2)}$ is written as

$$\widehat{\chi}_{\mathrm{s}}^{(2)} = \widehat{\chi}^{(2)\mathrm{nr}} + \sum_q \frac{\widehat{a}_q}{\omega_1 - \omega_q + \mathrm{i}\Gamma_q} , \quad (4.229)$$

with $\widehat{\chi}^{(2)\mathrm{nr}}$ being the nonresonant quadratic susceptibility and $\widehat{a}_q$, ω_q, and Γ_q being the strength, the frequency, and the damping constant for the q-th vibrational mode, respectively.

When the infrared field is scanned over the frequency of the q-th vibrational mode, the SFG signal is resonantly enhance, and its spectrum gives the spectrum of the vibrational mode. Tunable dye lasers [4.204], optical parametric oscillators and amplifiers [4.205], or PCF frequency shifters [4.185, 208] are employed as sources of frequency tunable radiation, allowing selective probing of vibrational (as well as electronic and excitonic) transitions in molecules and molecular aggregates.

Through the past few years, nonlinear-optical methods of surface spectroscopy have been extensively involved in the rapid growth of nonlinear microscopic techniques based on $\chi^{(2)}$ and $\chi^{(3)}$ processes. In particular, SHG and THG processes have proven to be convenient techniques for high-resolution three-dimensional microscopy of biological objects [4.207, 209], as well as laser-produced plasmas and micro-explosions [4.210]. In early experimental demonstrations of SHG microscopy, grain structures and defects on the surface of thin films were visualized using SHG in transmission [4.211] and surface monolayers were imaged by reflection-geometry SHG microscopy [4.212]. In recent years, progress in laser technologies and the advent of new-generation imaging and scanning systems made it possible to extend nonlinear microscopy techniques to three-dimensional structures, buried objects, and biological tissues [4.207, 209]. In this modification of nonlinear microscopy, the nonlinear signal is generated in the focal region of the laser beam in the bulk of a sample, originating from optical micro-inhomogeneities, which break the point-group symmetry of the medium or change phase-matching conditions. In two transverse dimensions, the high spatial resolution of nonlinear microscopy techniques is controlled by the nonlinear nature of the process, tightly confining the area where the signal is generated to the focal region. Resolution in the direction of probing is achieved either due to symmetry breaking, similar to nonlinear-optical surface-imaging techniques, or through phase-matching modification.

In CARS microscopy [4.207, 209, 213, 214], the nonlinear signal is resonantly enhanced when a frequency difference between two laser fields is tuned to a Raman-active mode of molecules under study, as described in Sect. 4.4.8 (see also Fig. 4.35c). This makes microscopy also species-selective as Raman resonances serve as fingerprints of a certain type of molecules or molecular aggregates. Forward CARS and backward CARS (also called epi-CARS) geometries have been developed (Fig. 4.35c). In the forward-CARS microscopy, the mismatch $|\Delta \boldsymbol{k}|$ of the wave vectors involved in wave mixing (Sect. 4.7) is typically much smaller than $2\pi/\lambda_{\mathrm{CARS}}$, where λ_{CARS} is the wavelength of the CARS signal. For epi-CARS, $|\Delta \boldsymbol{k}|$ is larger than $2\pi/\lambda_{\mathrm{CARS}}$. The intensity of the epi-CARS signal can thus be comparable with the forward-CARS signal intensity only for a thin sample with a sample thickness d meeting the inequality $|\Delta \boldsymbol{k}| d < \pi$. CARS microscopy has been one of the most successful recent developments in the field of nonlinear-optical microscopy. In addition to the high spatial resolution, this technique has a number of other important advantages over, for example, microscopy based on spontaneous Raman scattering. In particular, in CARS microscopy, laser beams with moderate intensities can be used, which reduces the risk of damaging biological tissues. The anti-Stokes signal in CARS microscopy is spectrally separated from the laser beams and from fluorescence, as the anti-Stokes wavelength is shorter than the wavelengths of the laser beams. In transparent materials, CARS microscopy can be used to image tiny buried objects inside the sample in three dimensions. Due to the coherent nature of CARS, the capabilities of CARS microscopy can be enhanced by means of coherence control, as recently demonstrated by *Dudovich* et al. [4.117].

4.11 High-Order Harmonic Generation

4.11.1 Historical Background

The invention of the laser in 1960 opened many new fields of research. One of them is *nonlinear optics* triggered by *Franken* [4.215] with the first demonstration of frequency doubling in a crystal (1961) and pioneered by Bloombergen and coworkers. Third-harmonic generation in gases was observed for the first time by *New* and *Ward* in 1967 [4.216], fifth and higher-order harmonic generation a few years later by *Reintjes* and others [4.217]. The main objectives of this research were to increase the conversion efficiency, to cover a large (continuous) spectral range with, for example, frequency mixing processes and to reach very short wavelengths. The most natural route for the latter goal was to use fundamental fields with the shortest possible wavelength, to produce short-wavelength radiation through a low-order nonlinear-optical process.

Another research area, which started in the late sixties, is the study of *atoms in strong laser fields*. The objective of this fundamental research was simply to understand the behavior of atoms and molecules exposed to intense electromagnetic fields. This field evolved in parallel with the development of pulsed lasers towards increasing peak powers, increasing repetition rates and decreasing pulse durations. The character of the laser–atom interaction also evolved from being essentially perturbative for laser intensities below $10^{13}\,\mathrm{W/cm^2}$ to strongly nonperturbative for higher intensities. For many years, this regime of nonlinear optics was studied only by looking at ionization processes. The number and charge of the produced ions, as well as the energy and angular emission of the electrons, were experimentally detected and compared with theoretical predictions.

At the end of the 1980s, it was realized that looking at the emitted photons would bring complementary information on the physical processes taking place. Indeed, efficient photon emission in the extreme ultraviolet (XUV) range, in the form of high-order harmonics of the fundamental laser field, was observed for the first time in 1987, in Saclay [4.218] (33rd harmonic of a Nd:YAG laser) and in Chicago [4.219] (17th harmonic of a KrF laser). The harmonic spectra were characterized by a decrease in efficiency for the first harmonics, followed by a broad *plateau* of nearly constant conversion efficiency, ending up by an abrupt cutoff. The existence of such a plateau was clearly a nonperturbative signature of the laser-atom interaction. Most of the early work concentrated on the extension of the plateau, i. e. the generation of harmonics of higher frequency and shorter wavelength going progressively from $\approx 20\,\mathrm{nm}$ at the end of the 1980s to $\approx 7\,\mathrm{nm}$ by the middle of the 1990s [4.220–224]. It was soon realized that, in contrast to the ideas promoted in the nonlinear-optics community, the shortest wavelengths were obtained with long-wavelength lasers. Today, harmonic spectra produced with short and intense laser pulses extend to the water window (below the carbon K-edge at 4.4 nm) [4.225, 226].

A breakthrough in the theoretical understanding of high-order harmonic generation processes was initiated by *Krause* and coworkers [4.227] who showed that the cutoff position in the harmonic spectrum follows the universal law $I_\mathrm{p}+3U_\mathrm{p}$, where I_p is the ionization potential, whereas $U_\mathrm{p}=e^2E^2/4m\omega^2$, is the ponderomotive potential, i. e. the mean kinetic energy acquired by an electron oscillating in the laser field. Here e is the electron charge, m is its mass, and E and ω are the laser electric field and its frequency, respectively. An explanation of this universal fact in the framework of a simple semiclassical theory was found shortly afterwards [4.228, 229], and confirmed by quantum-mechanical calculations [4.230, 231].

Progress in experimental techniques and theoretical understanding stimulated numerous studies of harmonic generation. The influence of the laser polarization was investigated in great detail [4.232–239] (see also [4.240–242] for the theory). The nonlinear conversion process was optimized with respect to the laser parameters [4.243, 244], and to the generating medium [4.245–250]. Finally, the spatial [4.251–256] and temporal [4.257–263] properties of the radiation were characterized and optimized. The specifications of the harmonic emission (ultrashort pulse duration, high brightness, good coherence) make it a unique source of XUV radiation, used in a growing number of applications ranging from atomic [4.264, 265] and molecular [4.266–268] spectroscopy to solid-state [4.269–

271] and plasma [4.272–274] physics. Finally, it has recently been demonstrated that the low-order harmonics are intense enough to induce nonlinear optical processes in the XUV range [4.275–277].

Almost immediately after the first observation of the harmonic plateau, it was realized that, if the harmonics were emitted in phase, i. e. phase-locked, the temporal structure of the radiation emitted from the medium would consist of a "train" of attosecond pulses separated by half the laser period [4.278, 279]. There is a clear analogy here with mode-locked lasers, where axial modes oscillating in a laser cavity are locked in phase, leading to the production of trains of short pulses. Attosecond pulses have remained, however, a theoretical prediction until recently [4.280–282]. A first possible indication of harmonic radiation containing an attosecond sub-structure was in a high-order autocorrelation of the driving laser pulse [4.283]. This was followed by a beautiful experiment showing evidence for phase-locking between five consecutive harmonics generated in argon, thus indicating that trains of 250 as pulses were formed [4.284]. In a series of experiments performed in Vienna, single pulses of duration of a few hundred attoseconds were demonstrated by using ultrashort (5 fs) laser pulses and spectrally filtering a few harmonics in the cutoff spectral region [4.285, 286]. These experimental results are the beginning of a new field of research *attophysics*, where processes in atoms and molecules can be studied at an unprecedented time scale.

The purpose of this section is to present to the nonexpert reader a simple description of high-order-harmonic generation, and its application to attosecond metrology. In Sect. 4.11.2, we describe the most important aspects of high-order-harmonic generation processes. We begin with a short description of the experimental set up needed to obtain high-order harmonics. Then we discuss the microscopic and macroscopic physics underlying the generation of high-order harmonics. We focus on the physics of importance for the generation of attosecond pulse trains and single attosecond pulses. We refer the reader to several review articles for a more complete overview of this research topic [4.245, 287, 288]. In Sect. 4.12, the emerging field of attosecond science is reviewed. The different measurement techniques are described and the first application of attosecond pulses is presented.

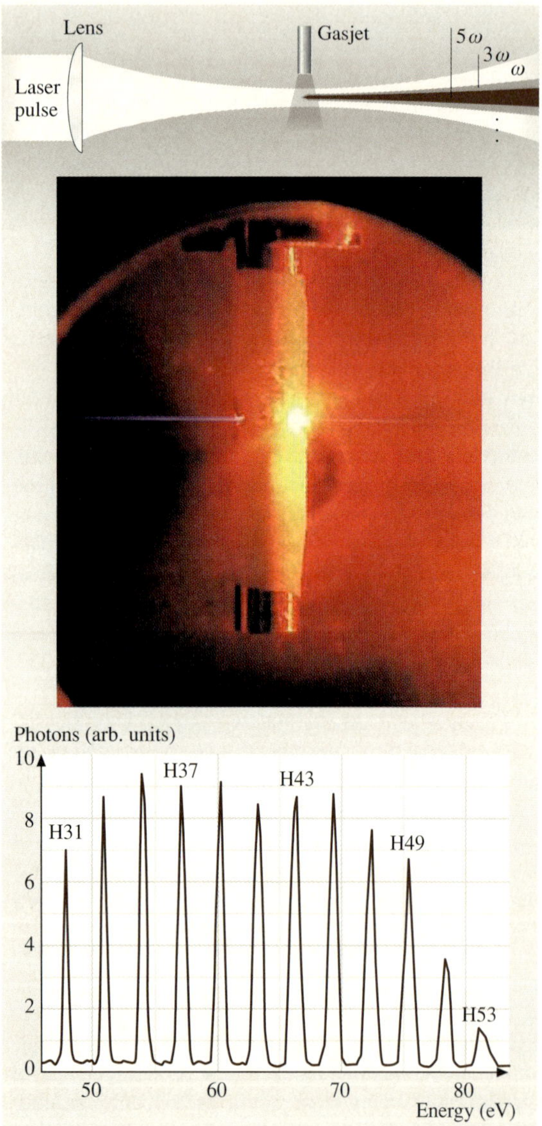

Fig. 4.36 Schematic representation of a typical experimental setup for high-order-harmonic generation. An intense short pulse laser is focused into a vacuum chamber containing a gas medium. Harmonics are emitted along the laser propagation axis. A photograph of a gas target is shown *below*. A typical spectrum obtained in neon is shown at the *bottom*. It shows a plateau ending with a sharp cutoff

4.11.2 High-Order-Harmonic Generation in Gases

Experimental Method

Generating high-order harmonics is experimentally simple. A typical set up is shown in Fig. 4.36. A laser pulse

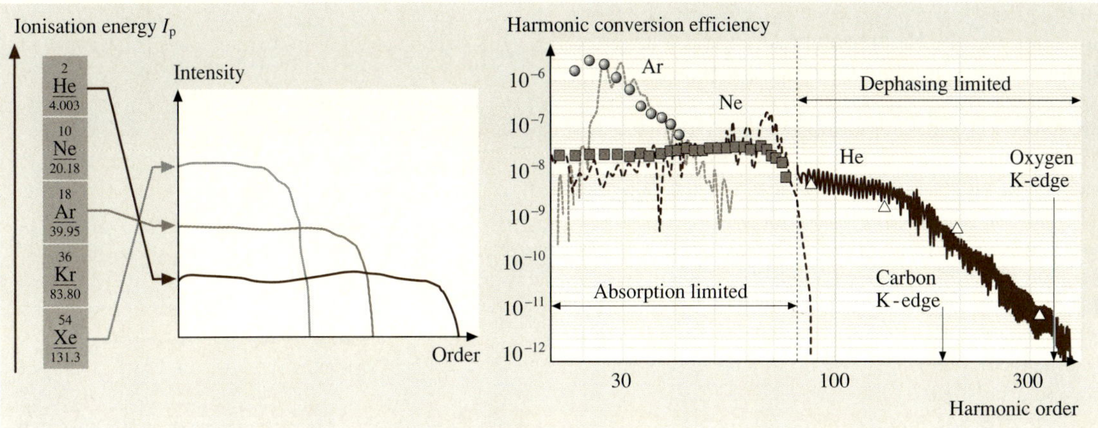

Fig. 4.37 Illustration of the influence of the gas species on the generation efficiency (*left*). The result presented at the *right* has been obtained with 5 fs 800 nm laser pulses (after [4.302])

is focused into a vacuum chamber containing a rather small gas target with an atomic pressure of at least a few mbar. The harmonics (only of odd order, owing to inversion symmetry) are emitted along the laser propagation axis. Many types of lasers have been used, excimers, Nd:YAG, Nd:glass, Ti:sapphire, dye lasers, etc. [4.218,219,222,223,289,290]. In addition, the second harmonic of these lasers [4.291–293] as well as the radiation from sum- or difference-frequency mixing processes [4.294, 295] for example to get into the mid-infrared range [4.296], have also been employed. Typical energies are between a fraction to a few tens of mJ, typical pulse durations are from a few femtoseconds to a few tens of picoseconds. In the last five years, the favorite tool has become the Ti:sapphire laser, providing very short pulse lengths, high laser intensities at high repetition rates. The shortest laser pulses used today to study atoms in strong laser fields are about 5 fs (two cycles) long [4.285]. The advantage of using short pulses is that atoms get exposed to higher laser intensity before they ionize, leading to higher-order harmonics. The different parameters of the laser pulses, such as the polarization, the focusing characteristics, the spatial and temporal profiles are often varied and optimized in the experiments. Recent studies do not simply vary a given parameter, but attempt to shape a laser pulse (by varying, for example, its phase [4.244], its degree of ellipticity [4.297–299], or its spatial properties [4.300, 301]) to tailor the harmonics for different applications.

The gas medium is provided by a gas jet, hollow fiber [4.246–248] or a (small) gas cell [4.249, 250, 303]. Figure 4.36 shows a photograph of such a gas target containing argon atoms irradiated by an intense laser pulse. Rare gases are the favored species, for obvious technical reasons. In addition, some work has been done with alkali atoms [4.296], ions [4.220, 304], molecules [4.305–308] and clusters [4.309–311]. Photons are separated in energy and detected by an XUV spectrometer, including a grating, sometimes a refocusing mirror, and a detector (electron multiplier, microchannel plates, etc.). A typical experimental spectrum is shown in Fig. 4.36. This result has been obtained in a gas of neon using a 100 fs 800 nm Ti:sapphire laser [4.312]. It shows odd harmonic peaks up to the 53rd order, with a rapid decrease beyond the 49th harmonic, characteristic of the cutoff region. The spectral range of the harmonic emission as well as the conversion efficiency depends strongly on the gas medium. As schematically illustrated in Fig. 4.37 (left), the efficiency is highest in the heavy atoms Ar, Xe, Kr, but the highest photon energies are obtained in He and Ne [4.222, 245, 302]. Figure 4.37 (right) presents

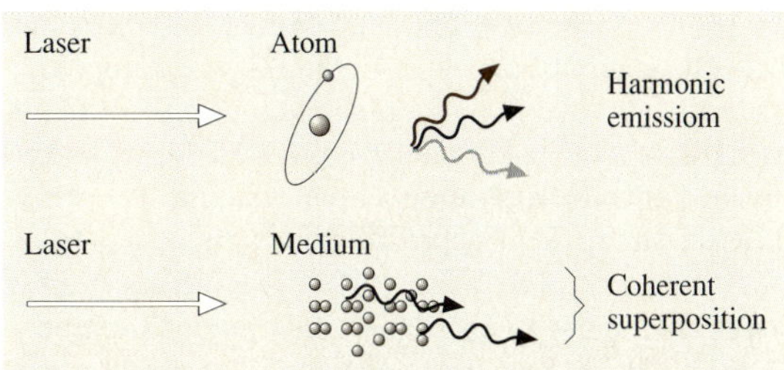

Fig. 4.38 Illustration of the two aspects of high-order-harmonic generation. (*Top*) harmonic emission from a single atom. (*Bottom*) phase matching in the nonlinear medium

Part A | 4.11

a comparison between the generation efficiency for three rare gases Ar, Ne, He, when ultrashort laser pulses (of duration 5 fs) are used for the excitation [4.245, 302].

As illustrated in Fig. 4.38, two conditions are necessary to observe efficient harmonic emission. First, each individual atom must generate light at these frequencies, requiring a highly nonlinear response to the radiation field. Second, the harmonic field results from the coherent superposition of all the emitting atoms in the medium. Harmonic generation will be efficient only if phase matching is achieved, requiring the generated field to be in phase with the nonlinear polarization creating it over the medium's length. We discuss these two aspects in more details in the following sections.

4.11.3 Microscopic Physics

Electrons in an atom in the presence of a time-dependent radiation field oscillate. This is described by a dipole moment $d(t) = \langle \Phi(t)|er|\Phi(t)\rangle$, where $|\Phi(t)\rangle$ is the time-dependent electronic wavefunction, the solution of the Schrödinger equation. When the radiation field is weak, there is mainly one oscillation frequency, that of the field. In a strong radiation field, the oscillatory motion becomes distorted and the dipole moment now includes a series of higher-order frequencies, odd harmonics of the fundamental one. The harmonic emission from a single atom can thus be calculated by taking the Fourier transform of the dipole moment. Theorists often use a single-active-electron approximation, assuming that the interaction with the field involves essentially one active electron, to describe the response of the atom to a strong laser field. A number of methods have been proposed to solve this problem and it is beyond the scope of this paper to review all of these approaches. The most realistic approach is probably the numerical solution of the time-dependent Schrödinger equation, pioneered by *Kulander* at the end of the 1980s [4.313]. Many important results, such as the determination of the cutoff law for high-order harmonics in 1992 [4.227] or proposals for single attosecond pulse generation using few-cycle laser pulses at the end of the 1990s [4.245, 314, 315] were obtained directly from numerical calculations. The semiclassical strong field approximation, originating from a seminal paper of *Keldysh* in 1964 [4.316], applied by *Lewenstein* in the 1990s to high-order-harmonic generation [4.230, 231] allows to explore a larger parameter space as well as to gain intuitive insight. This model and a related one approximating the atomic potential by a $\delta(r)$ potential [4.240] have been extensively used to interpret experimental results.

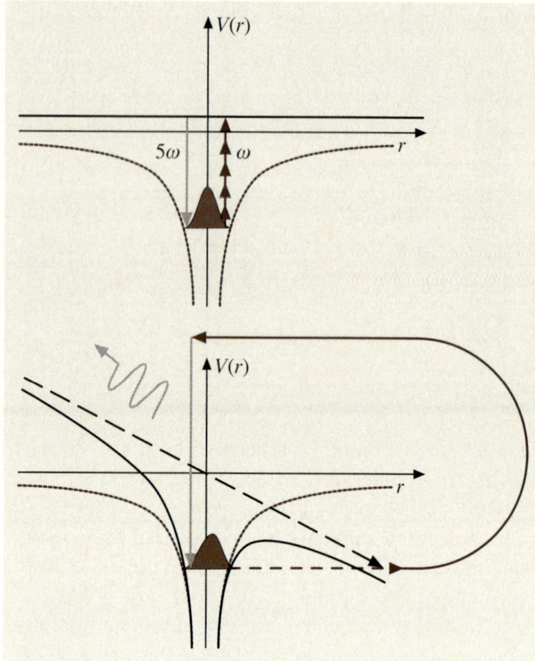

Fig. 4.39 Schematic picture of different regimes for harmonic generation. The multiphoton, perturbative regime is illustrated on the *top*, while the semiclassical model is presented on the *bottom*

Many insights in the physical understanding of the interaction between atoms and strong laser fields have been provided by a simple semiclassical model, proposed first by *Van der Linden*, *van der Heuvell*, and *Muller* in the context of above-threshold ionization and extended by *Corkum* and others [4.228, 229] to multiple ionization and high-order-harmonic generation. According to this model, illustrated in Fig. 4.39 (bottom), the electron tunnels through the Coulomb energy barrier modified by the presence of the relatively slowly varying linearly polarized electric field of the laser. It then undergoes (classical) oscillations in the field, during which the influence of the Coulomb force from the nucleus is practically negligible. If the electron comes back to the vicinity of the nucleus, it may be rescattered one or several times by the nucleus, thus acquiring a high kinetic energy, and in some cases, kicking out a second or third electron. It may also recombine back to the ground state, thus producing a photon with energy I_p, the ionization potential, plus the kinetic energy acquired during the oscillatory motion. We also show in Fig. 4.39 (top) for comparison the more traditional harmonic-generation process based on multiphoton absorption in the (barely perturbed) atomic potential.

An intuitive understanding of some of the properties of harmonic generation can be gained by elementary classical calculations of the electron motion outside the binding potential. Assuming the electron to have zero velocity immediately after it has tunneled through the potential barrier at time $t = t_0$, and the laser field to be simply described by $E = E_0 \sin(\omega t)$, we obtain:

$$v = -v_0 \cos(\omega t) + v_0 \cos(\omega t_0)\,, \tag{4.230}$$

$$x = -\frac{v_0}{\omega}\sin(\omega t) + \frac{v_0}{\omega}\sin(\omega t_0) + (t - t_0) v_0 \cos(\omega t_0)\,, \tag{4.231}$$

where $v_0 = qE_0/m\omega$. Depending on the time at which the electron is released into the continuum (t_0), it will follow different trajectories, as illustrated in Fig. 4.40. Only those electrons released between $T/4$ and $T/2$ (where T is the laser period) are of interest for harmonic generation. When the laser field changes its sign, they come back towards the core (at $x = 0$) with a certain kinetic energy. This energy, which determines the emitted harmonic order, is proportional to the square of the slope of the trajectory as it crosses the time axis [open circles in Fig. 4.40 (top)]. Except for the trajectory starting at approximately $0.3T$, giving the highest kinetic energy (cutoff), there are two (main) trajectories leading the same kinetic energy. This is illustrated in Fig. 4.40 (bottom), showing the kinetic energy when the electron returns to the core (solid line), as well as the time spent in the continuum (dot-dashed line), as a function of release time. As shown in the figure, for each energy (dashed line), and hence

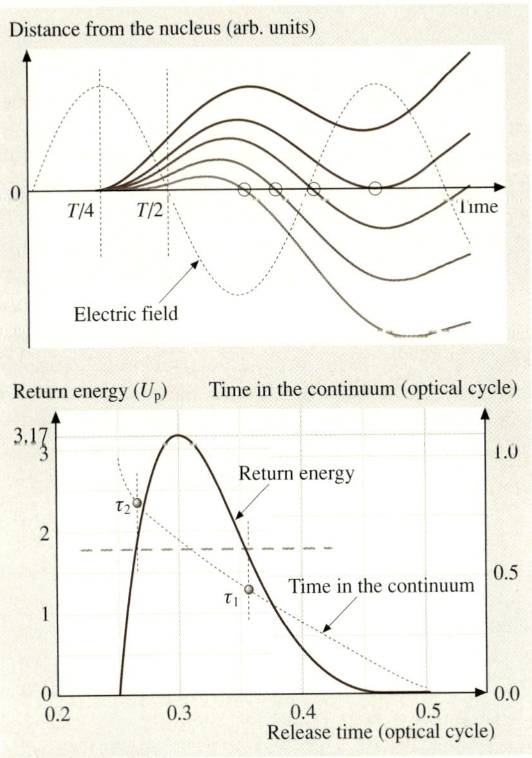

Fig. 4.40 (*Top*) Electron trajectories in the continuum corresponding to different release times. The laser electric field is represented in *dotted line*. (*Bottom*) kinetic energy (*solid line*) and time in the continuum (*dot-dashed line*)as a function of the release time

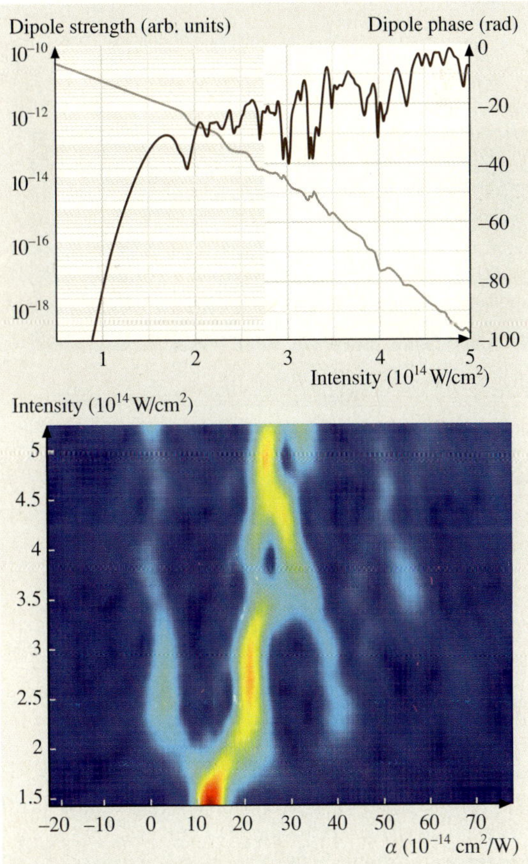

Fig. 4.41 (*Top*) Single atom response within the strong-field approximation. Intensity (*dark line*) and phase (*light line*) of the 35-th harmonic in neon as a function of the laser intensity. (*Bottom*) Quantum path analysis of the same harmonic

for each harmonic order, there are mainly two trajectories, a short and a long one, contributing to the harmonic emission. The periodicity (for a pulse several cycles long) of the process implies that the light emission is not continuous but at discrete (odd-harmonic) frequencies.

The influence of the complex electron dynamics inherent to the harmonic-generation process is clearly visible on the intensity dependence of the harmonic components of the quantum-mechanical dipole moment. Figure 4.41 shows for example the 35-th harmonic generated in neon, calculated within the strong field approximation. The intensity and phase are represented respectively in solid and dashed line as a function of the laser intensity. The sharp intensity and phase variation at low intensity corresponds to the cutoff region. In the plateau, oscillations are clearly visible in the variation of both the intensity and the phase. They originate from interference effects between the contributions from the multiple trajectories. This fascinating conclusion stimulated the development of analysis techniques to extract the contributing electron trajectories (or rather the relevant quantum paths) from quantum-mechanical approaches [4.317, 318]. The result of such an analysis for the 35-th harmonic in neon is shown at the bottom in Fig. 4.41. The contributions to the dipole phase varying linearly with the intensity $\phi_{dip}(I) = \alpha_j I$, corresponding to a quantum path (j), are identified as vertical lines in Fig. 4.41. In this case, the dominant trajectory is the "second" one, with a coefficient $\alpha_2 = 24 \times 10^{-14}\,cm^2\,W$. At low intensity, in the cutoff region, there is only one quantum path.

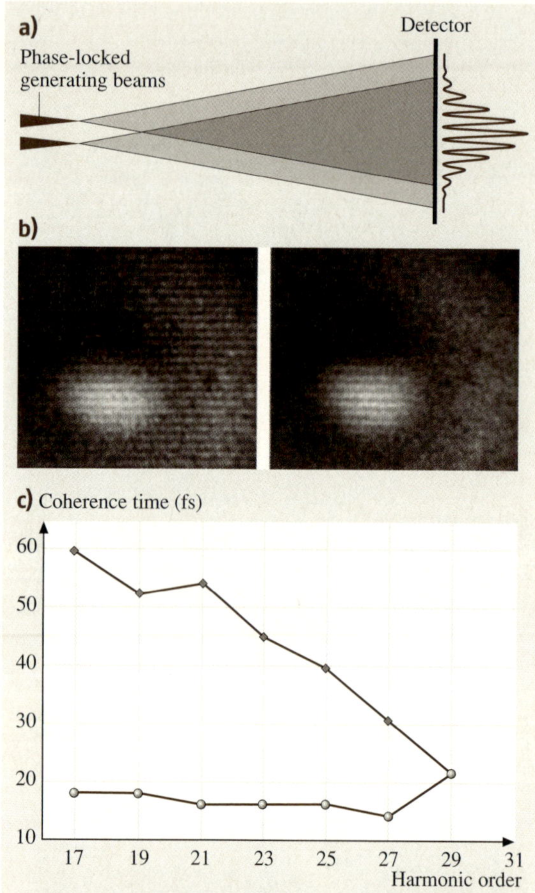

Fig. 4.42a–c Spatially-resolved temporal coherence measurements of high-order harmonics. The principle of the measurement is shown in **(a)**. Two images obtained on the 13-th harmonic in xenon obtained for two different time delays (0 and 25 fs) between the two pulses are shown in **(b)**. A summary of the measured coherence times in the two spatial regions for high harmonics in xenon is presented in **(c)**

The microscopic physics (the quantum orbits) behind high-order-harmonic generation was shown in a series of experiments investigating the temporal coherence of the harmonics [4.319–321]. We present in Fig. 4.42 the principle as well as some results obtained in xenon [4.320]. Two phase-locked spatially separated harmonic sources are created by splitting the laser into two replicas in a Michelson interferometer and by slightly misaligning one of the arms. A variable time delay can be introduced between the two pulses. The generated harmonics are separated by a grating and interfere in the far field. The variation of the contrast of the fringes as a function of time delay gives the coherence time. Figure 4.42b shows images obtained for the 13-th harmonic in xenon. (For experimental reasons, the images are not symmetrical.) These images present two spatial regions with different coherence times. The central region exhibits a long coherence time, whereas the outer region a much shorter one. Figure 4.42c summarizes measurements on the harmonics generated in xenon.

These results can be interpreted in a simple way by recalling that the harmonic field is a sum of the contributions from each quantum path j

$$E_q(\boldsymbol{r}, t) = \sum_j A_j(\boldsymbol{r}, t)\,e^{-[iq\omega t + \alpha_j I(\boldsymbol{r}, t)]} \qquad (4.232)$$

where A_j is an envelope function, representing the strength of each path. The temporal variation of the laser intensity $I(t)$ will produce a chirp in the emitted field, and consequently a spectral broadening (or a reduced coherence time). The radial variation of the laser intensity $I(r)$ will affect the curvature of the phase front of the harmonics, and therefore their divergence. The contribution from the quantum paths with a long excursion time in the continuum (Fig. 4.40), resulting in a large α_j, will have a short coherence time and a pronounced curvature in the far field, whereas that from the quantum paths with a short excursion time will have a longer coherence time and be more collimated [4.322]. Similar evidence for the existence of (at least) two quantum paths has been obtained by analyzing the harmonic spectra in combination with measuring the harmonic pulse duration [4.321]. Contributions of different quantum paths can be selected macroscopically, either by spatial or spectral filtering.

4.11.4 Macroscopic Physics

We now turn to the second aspect of harmonic generation, namely the response of the whole medium. To achieve phase matching, i. e. to ensure efficient conversion, the wave vector difference (traditionally called the phase mismatch) between the harmonic wave and the polarization must be minimized, so that the phase difference varies as little as possible over the medium's length. For an incident Gaussian beam, this phase difference on the propagation axis (z) is given by

$$\begin{aligned}\delta\phi_q &= \phi_q - \phi_q^{\text{pol}}\\ &\approx k_q z - qk_1 z + q\arctan(2z/b) - \phi_{\text{dip}}\,. \end{aligned} \tag{4.233}$$

In this equation, the first two terms characterize the difference in dispersion at the fundamental and q-th harmonic frequencies, mainly due to the effect of the free electrons in the medium. The third one is a geometrical term originating from the Gouy phase shift of a Gaussian beam with confocal parameter b across the focus. The fourth term is the dipole phase described above. According to the previous discussion, it is more physically correct to consider separately the contributions from the different quantum paths, before making a (coherent) sum. The phase difference to minimize depends on the trajectory and is given by

$$\begin{aligned}\delta\phi_{q,j} &= \phi_q - \phi_q^{\text{pol}}\\ &\approx k_q z - qk_1 z + q\arctan(2z/b) - \alpha_j I(r,z,t)\,. \end{aligned} \tag{4.234}$$

Figure 4.43 illustrates the variation of $\delta\phi_q$ along the propagation axis (z). Most contributions are monotonic. Focusing as well as the free electron dispersion (not shown in Fig. 4.43) lead to a contribution to $\delta\phi_q$ increasing with z, while the (normal) dispersion leads to

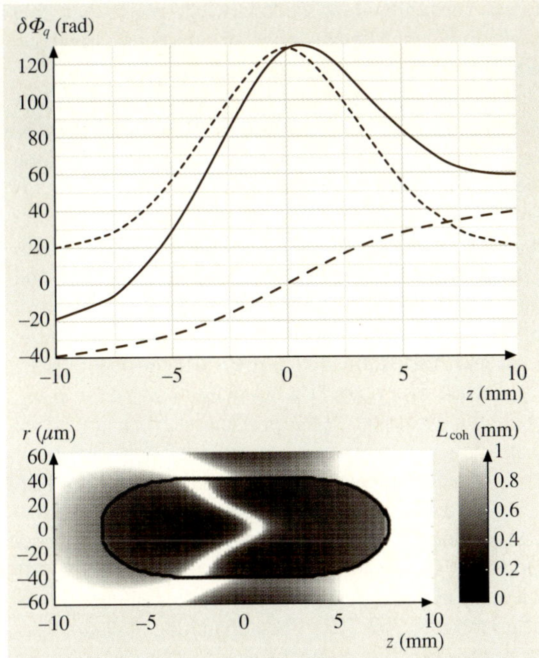

Fig. 4.43 (*Top*) phase difference between the harmonic wave and the polarization for the 35-th harmonic generated in neon on the propagation axis. In *dashed line*, we show the term coming from the geometrical phase shift of a Gaussian beam and in *dotted line*, the term coming from the dipole phase. The laser intensity is $6\times10^{14}\,\text{W/cm}^2$. The confocal parameter is chosen to be $b = 1$ cm. (*Bottom*) corresponding phase matching map (after [4.317])

a term decreasing with z, usually quite small. In contrast, the contribution of the dipole phase, here due mainly to the second trajectory, with a transition to the cutoff for $|z| \geq 7.5$ mm, first increases, then decreases with z. In the particular case shown in Fig. 4.43, phase matching on the propagation axis is best realized for $z \geq 7$ mm, requiring the medium to be located after the focus. In general, the situation can be rather complicated. Methods have been developed to visualize where (in the nonlinear medium) and when (during the laser pulse) phase matching was best realized [4.317], using in particular three-dimensional maps representing the local coherence length, defined as

$$L_q = \pi\left(\frac{\partial\delta\phi_q}{\partial z}\right)^{-1}\,. \tag{4.235}$$

An example of such a map for the situation corresponding to Fig. 4.43 (top) is shown at the bottom of the same figure. The white areas indicate where phase-matching is best realized. The two parts of the figure correspond to the transition between cutoff and plateau.

The brief analysis presented above can easily be generalized to include propagation in waveguides [4.246–248], which modifies somewhat the phase-matching conditions, since the geometrical and dipole phase effects are much reduced (owing to a constant intensity in the waveguide). Absorption in the nonlinear medium starts to play a role for high conversion efficiencies and long medium lengths and/or high pressures. The optimization of phase matching of high-order-harmonic generation has stimulated a great deal of efforts during the last five years. The so-called *absorption limit*, where the limitation on the conversion efficiency is due to absorption, and not to the coherence length, has been reached in different wavelength ranges [4.249,250,302]. Recently, the use of extremely long focusing geometries (which minimize both the effect of the dipole phase and that of the Gouy phase shift) has led to conversion efficiencies as high as a few times 10^{-5} and energies in the μJ range [4.303,323]. Phase matching of high-order-harmonic generation is by no means a solved issue, since it is a complicated three-dimensional problem, involving a number of parameters (laser focusing, pressure, medium length and geometry, laser intensity). In addition, the problem is quite different for the low-order harmonics with energy around a few tens of eV, where most of the work has been done, and for the high-order harmonics at 100 eV or more. An interesting idea, similar to quasi-phase-matching has recently been investigated [4.324]. A modulated waveguide, used for the generation of high-order harmonics, induces a periodic variation of the degree of ionization leading to enhanced conversion efficiency, especially for the high harmonics.

Finally, a conclusion of importance for the generation of attosecond pulse trains is that the different quantum paths contributing to harmonic generation discussed in the previous section are not phase matched in the same conditions (4.234). The axial variation of the laser intensity $I(z)$ leads to different phase matching conditions for the contributions from the different quantum paths. In other words, depending on the geometry, ionization, pressure conditions, phase matching will enhance one of the contributing trajectories to the detriment of the others. This conclusion is extremely important for the generation of attosecond pulse trains. As illustrated in Fig. 4.44 (top), the electron trajectories contributing to harmonic generation in the single-atom response lead to bursts of light at different times during the laser half-cycle. Phase locking between consecutive harmonics is not realized. In some conditions, however, phase matching results in efficient generation of only one of these contributions [Fig. 4.44 (bottom)] [4.325, 326], leading to a train of attosecond pulses. Another possibility, that makes use of the different spatial properties of the contributing trajectories, is to select the shortest trajectory by spatially filtering the harmonic beam with an aperture. Finally, a spectral filter can be used to select only the harmonics in the cutoff region, where the electron dynamics is much simpler, with only one electron trajectory in a single-atom response.

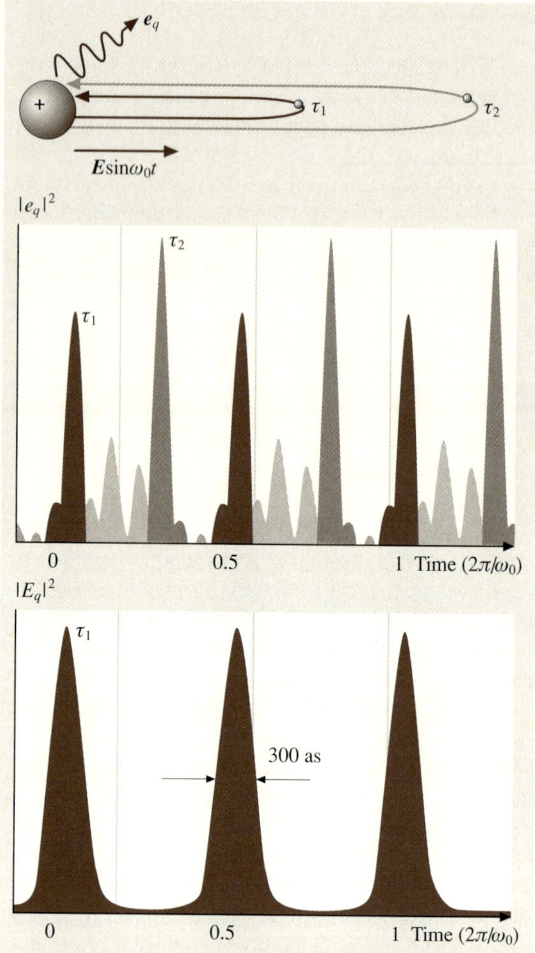

Fig. 4.44 Illustration of the temporal structure of a few harmonics in the single atom response (*top, middle*). In some conditions, phase matching selects only one burst of light per half cycle, leading to phase locking (*bottom*)

4.12 Attosecond Pulses: Measurement and Application

4.12.1 Attosecond Pulse Trains and Single Attosecond Pulses

The first experimental indication of harmonic radiation containing a substructure was observed in a high-order autocorrelation of the driving laser pulse [4.283]. The experimental set up was similar to that described in Fig. 4.42 (top), except that the two harmonics sources were spatially overlapped and no spectral selection with a grating was done. The signal was studied as a function of time delay, and showed substructures.

More-recent experiments [4.284] showed conclusive evidence for a group of harmonics beating together to form a sequence of XUV bursts. In perfect phase-locking conditions, a combination of $N = 5$ harmonics generated by a 800 nm laser can form a train of pulses separated by half the laser period $T_0/2 = 1.35$ fs, with a duration as short as $T_0/2N$. In the experiment described in [4.284], the signal used to study the phase locking of the harmonics were the sidebands in the photoelectron spectra, which are due to the coherent superposition of two processes, combined absorption of a harmonic and infrared (IR) laser photon, and absorption of the next harmonic and emission of a laser photon [Fig. 4.45 (top)]. The sideband intensity depends on the phase difference between the two harmonics involved, as well as on the time delay between the XUV and IR pulses. Monitoring the intensity as a function of time delay for several sideband peaks allows one to study the variation of the phase difference between consecutive harmonics [Fig. 4.45 (bottom)]. The relative phases of the harmonics were found to be fairly constant, yielding a train of 250 as pulses spaced by half the oscillation period of the laser field [Fig. 4.45 (middle)].

The high repetition rate (twice the laser frequency) of this train of pulses may pose problems when it comes to spectroscopic applications, because the interpretation of pump–probe experiments tends to be ambiguous. Using such a train for either triggering or probing a dynamic process that does not completely come to a standstill within $T_0/2(\approx 1.3$ fs for 800 nm driver laser light from a Ti:sapphire laser) gives rise to multiple pumping or probing, complicating the interpretation of measured data. This recognition led to the proposal that a single pulse could possibly be selected from the attosecond train by a driving laser pulse with a polarization state rapidly varying in time to confine pure linear polarization (which is a prerequisite for harmonic emission) to a single laser cycle at the pulse peak [4.280, 281, 327]. This *polarization gating* technique has been experimentally implemented using different methods to achieve the modulation of the degree of ellipticity of the fundamental laser pulse. A significant confinement of the harmonic emission (however not in the attosecond regime) has been shown using both spectral and temporal diagnostics [4.297,298,328,329]. This technique could be a way to isolate single as pulses in the plateau energy region.

A more straightforward approach to confining high-harmonic emission to a single laser cycle became feasible with the availability of intense few-cycle (sub-10 fs) laser pulses. When driven with a laser pulse consisting of only a few oscillation cycles, the harmonic emission at the highest photon energies (near cutoff region) can be confined to one half of the oscillation period near the pulse peak. By selecting a well-defined photon energy band near the cutoff region, which requires the highest driving field intensity, it is expected to be possible to isolate one single sub-femtosecond burst, as illustrated in Fig. 4.46. This diagram also shows the sensitivity of isolated pulse generation to the absolute (or carrier-envelope) phase φ of the fundamental laser pulse

$$E_L(t) = E_a(t)\cos(\omega_L t + \varphi). \quad (4.236)$$

Only pulses with $\varphi \approx 0$ have the potential for producing a single burst. For such a fundamental pulse with a "cosine" waveform, the peak intensity of the laser pulse can be adjusted to give rise to a single burst within a preselected frequency band (beige band in Fig. 4.46). Recently, carrier-envelope-phase-controlled few-cycle pulses opened the door to the reproducible generation of single sub-femtosecond XUV pulses [4.330].

The feasibility of a single sub-femtosecond pulse produced by few-cycle-driven high-order harmonic generation was corroborated by numerical studies based on a computer code [4.331] solving Maxwell's wave equations in three dimensions and calculating the radiation of the strongly driven atomic dipoles using the quantum theory of *Lewenstein* et al. [4.231]. These numerical calculations predict the feasibility of isolated near-bandwidth-limited XUV pulse generation with durations down to the 100 as range.

4.12.2 Basic Concepts for XUV Pulse Measurement

The most direct information about the duration of short flashes of electromagnetic radiation can be gained from time-domain measurements. To this end, the burst to

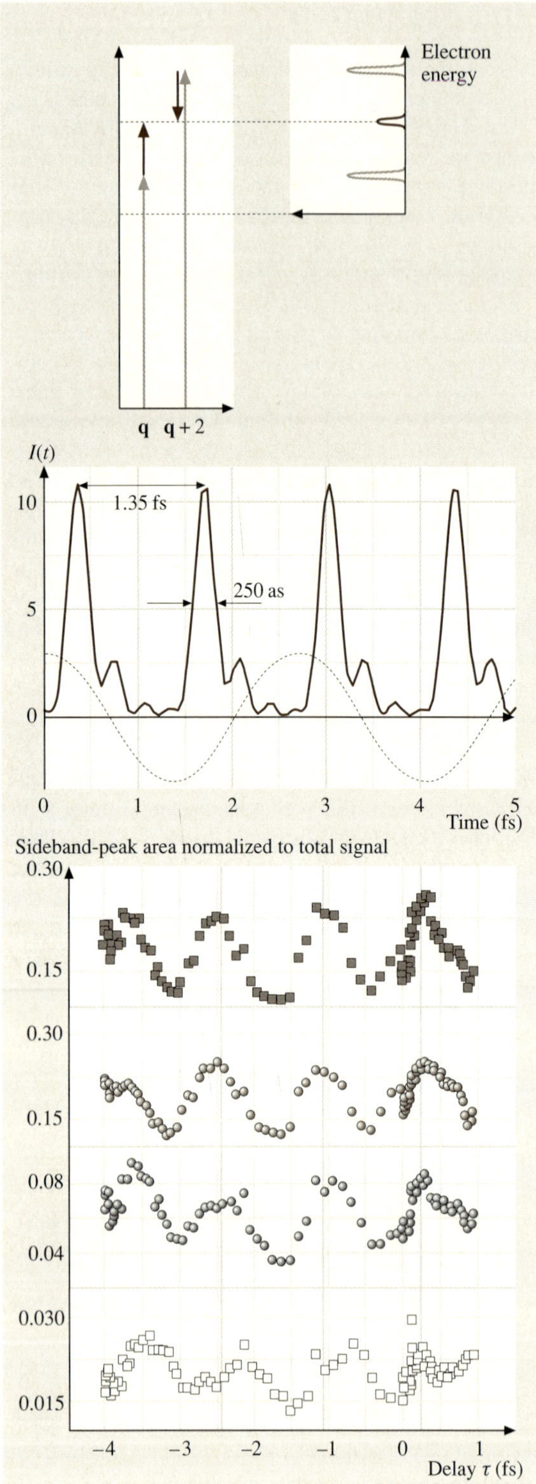

Fig. 4.45 Experimental evidence for phase-locking of high-order harmonics; principle of measurement (*top*), experimental results (*bottom*) and reconstructed temporal intensity profile (*middle*). Reprinted from [4.284] with permission from the authors

be characterized must be probed by a sampling pulse that is comparable to or shorter than the duration of the burst. In the optical regime this concept has been implemented by nonlinear autocorrelation techniques, where the burst to characterize is probed by itself. The extension of these techniques to the XUV spectral range is not trivial, because of the unfavorable scaling of the nonlinear polarizability with wavelength and because of the low power of the XUV pulse to be measured. Only a few results have been obtained so far for relatively low-order harmonics. The nonlinear process used for the autocorrelation is two-photon ionization of a gas target. Experiments have demonstrated two-photon ionization in Xe, Kr [4.277, 332] by low-order harmonics and recently two-photon ionization in He with a combination of harmonics from the 7-th to the 13-th [4.333]. In the first autocorrelation experiments [4.275, 334], harmonics generated by a 1 TW 30 fs laser beam were used to ionize He atoms. The number of He^+ ions was measured as a function of the delay between two identical ninth-harmonic pulses (14.1 eV) generated by two replicas of the fundamental pulse. A pulse width of 27 fs was obtained by assuming a $sech^2$ pulse shape (Fig. 4.47). In a more recent experiment a two-photon-absorption-based auto-correlation was implemented for a frequency-resolved optical gating (FROG) measurement [4.335]. Here the two replicas of the fifth harmonic of a 7 mJ Ti:sapphire laser pulse were focused into a Xe gas jet. From the recorded FROG trace the intensity profile and chirp of a high-harmonic pulse could be retrieved. In another recent experiment [4.336], an autocorrelation of a bunch of harmonics (from the 7-th to the 11-th) gives evidence for attosecond bunching.

The extension of auto-correlation techniques to higher-order harmonics, with energies in the XUV/soft-X-ray region, remains, however, a formidable challenge. The atomic cross section of the absorption process σ scales with λ^6, implying a dramatic decrease of the two-photon transition probability with decreasing wavelength. As a consequence auto-correlation schemes require photon fluences orders of magnitude higher than those available from existing harmonic sources in the XUV regime. In order to circumvent the difficulties resulting from low two-photon transition cross sections

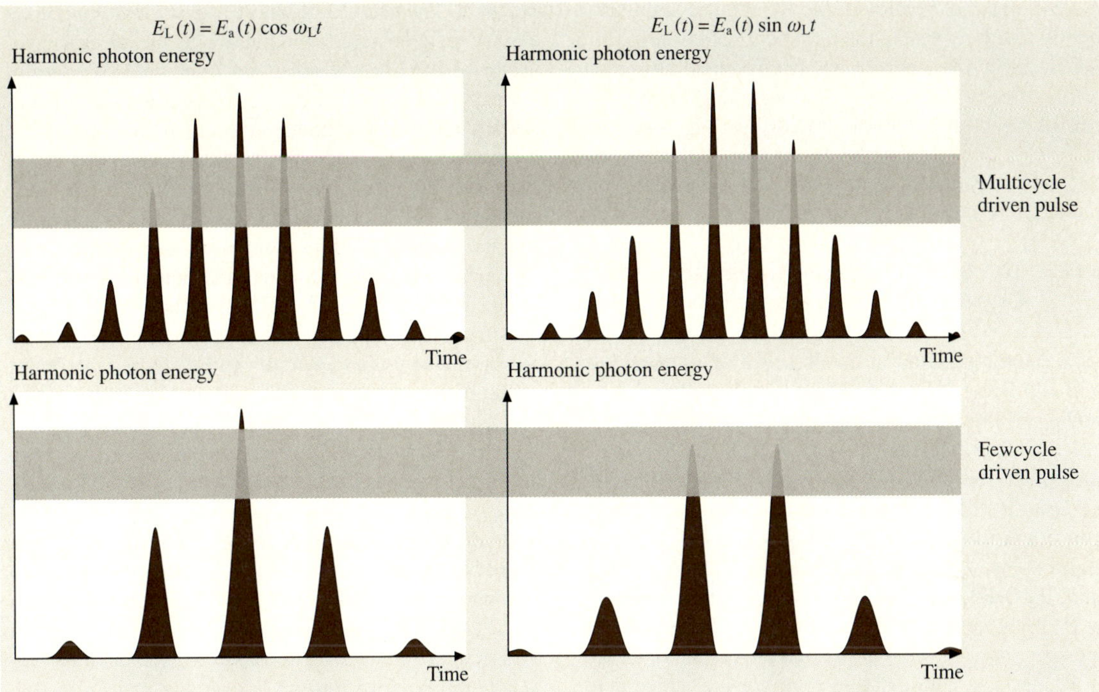

Fig. 4.46 Comparison of the temporal structure of harmonics of multi-cycle or few-cycle pulses for different values of the carrier-envelope phase. The peaks sketch the temporal structure (*horizontal axis*) of emitted XUV radiation as a function of XUV photon energy (*vertical axis*)

at short wavelengths, the XUV pulse can be correlated with the strong laser pulse previously used for its generation. The cross-correlation can be implemented by XUV photoionization in the presence of the laser pulse. XUV pulses have been cross-correlated with the femtosecond infrared (IR) laser pulses in a number of experiments over the past decade [4.257–260,262,337,338]. All these experiments had in common that the temporal intensity envelope of the laser pulse served as a sampling function for measuring the XUV pulse shape and duration, which limits the temporal resolution to several femtoseconds or more.

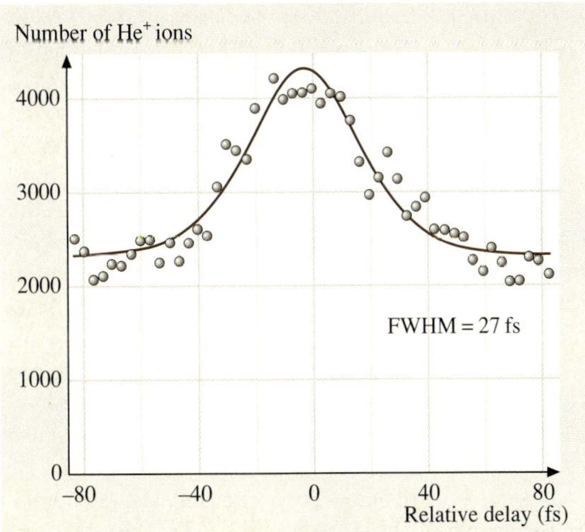

Fig. 4.47 Auto-correlation trace of the ninth harmonic by two-photon ionization of He (after [4.334])

Here we review a novel implementation of XUV/IR cross-correlation technique that correlates the XUV pulse *envelope* with the oscillating laser *field* rather than its envelope, providing a sub-$T_0/4$ probe (T_0 is the laser cycle) for measuring the XUV pulse duration. The method exploits the dependence of the final kinetic energy of the XUV-pulse-generated photoelectron on the phase (and strength) of the laser field at the instant of photoionization. The width and position of the resultant photoelectron spectra are measured as a function of the relative delay between XUV and laser pulse and yield a convolution of the oscillating laser light field with the XUV pulse envelope. Careful deconvolution allows one to determine the duration of the XUV pulse and its timing jitter with respect to the light field with as resolution [4.261, 285]. The concept has been termed attosecond streak camera because it is the streaking of the electron's kinetic energy by the rapidly varying laser field that yields direct time-domain information of the time structure of emitted electron wave packet, which mimics that of the XUV pulse [4.339–341].

4.12.3 The Optical-Field-Driven XUV Streak Camera Technique

XUV photoionization in the presence of a strong laser field can be accounted for by a simple quasi-classical model, similar to that presented above, treating the interaction as a two-step process [4.261]. According to this model, the photoelectron is first ejected by a short XUV pulse with a distribution of initial momenta known from conventional photoionization studies [4.342]. Subsequently, it is accelerated (or decelerated) by the light field. For XUV pulse durations τ_x very short compared to $T_0/2$, the model predicts that, depending on the oscillation phase of the light field at the instant of "birth" of the electron, a momentum component is added to the initial momentum of the electron.

The concept of laser-field-assisted XUV photoionization can be implemented in various detection geometries. The two most important ones are shown in

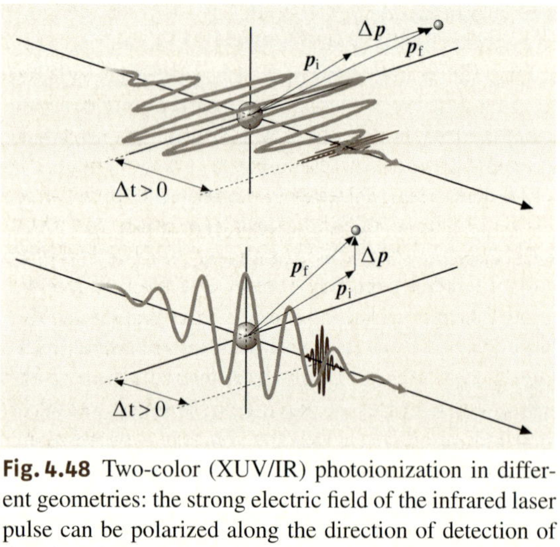

Fig. 4.48 Two-color (XUV/IR) photoionization in different geometries: the strong electric field of the infrared laser pulse can be polarized along the direction of detection of the photoelectrons or orthogonally to it, giving rise to substantial differences in the photoelectron's motion after its liberation. The polarization of the XUV pulse has little influence on the electron motion

Fig. 4.48. The photoelectrons are detected within a cone aligned parallel and orthogonally to the electric field vector of the linearly polarized laser field (henceforth referred to as the parallel and orthogonal detection geometry, respectively). Here we focus on the orthogonal geometry because the first sub-femtosecond pulse measurement used this configuration [4.285]. In the next section we shall demonstrate that the parallel geometry can also be employed for efficient as diagnostics.

Figure 4.49 depicts the final momentum distribution of the photoelectron for different instants of release in the strong laser field. In general, a laser-induced momentum component is added to the initial momentum vector of the photoelectron, resulting in a shift of the photoelectron angular distribution along the laser polarization in momentum space. The momentum transferred from the IR light field is largest if the electric field of the light wave is zero, i. e. its slope is maximum, at the instant of birth of the photoelectron. The width of the XUV photoelectron energy spectrum ΔW, which is equal to the bandwidth of the XUV pulse spectrum in the absence of the light field, increases with increasing momentum shift, within a finite detection cone aligned orthogonally to the laser polarization.

Scanning the instant of birth of the photoelectron through the light field oscillations by changing the relative delay t_d between the light pulse and XUV pulse results in a modulation of the center of gravity as well as the width $\Delta W(t_d)$ of the XUV photoelectron spectrum with a period equal to one half of the light oscillation period T_0. In the first time-resolved sub-femtosecond XUV pulse measurement [4.285] the modulation of the field-induced spectral broadening $\Delta W(t_d)$ was recorded. The detection of such a modulation relies on an XUV pulse short compared to half the laser period, i. e. $\tau_x \ll T_0/2$, implying also a fast (attosecond) ionization of the bound electrons by the XUV photons, and a timing firmly fixed (with as precision) of the XUV pulse with respect to the laser field. A finite XUV pulse duration or a finite timing jitter of the XUV pulse (relative to the phase of the light field) of any origin results in a reduced contrast of the predicted variation of $\Delta W(t_d)$ versus time delay (illustrated in Fig. 4.49). An XUV pulse duration or a timing jitter exceeding $T_0/2$ ($\approx$ 1.25 fs in the case of near-infrared laser light from the Ti:sapphire laser) would smear out the modulation completely. Hence, the modulation depth of $\Delta W(t_d)$ provides a sub-femtosecond probe for setting a reliable upper limit on the XUV pulse duration and timing jitter.

Figure 4.50 displays the experimental set up for the generation of sub-femtosecond XUV pulses by har-

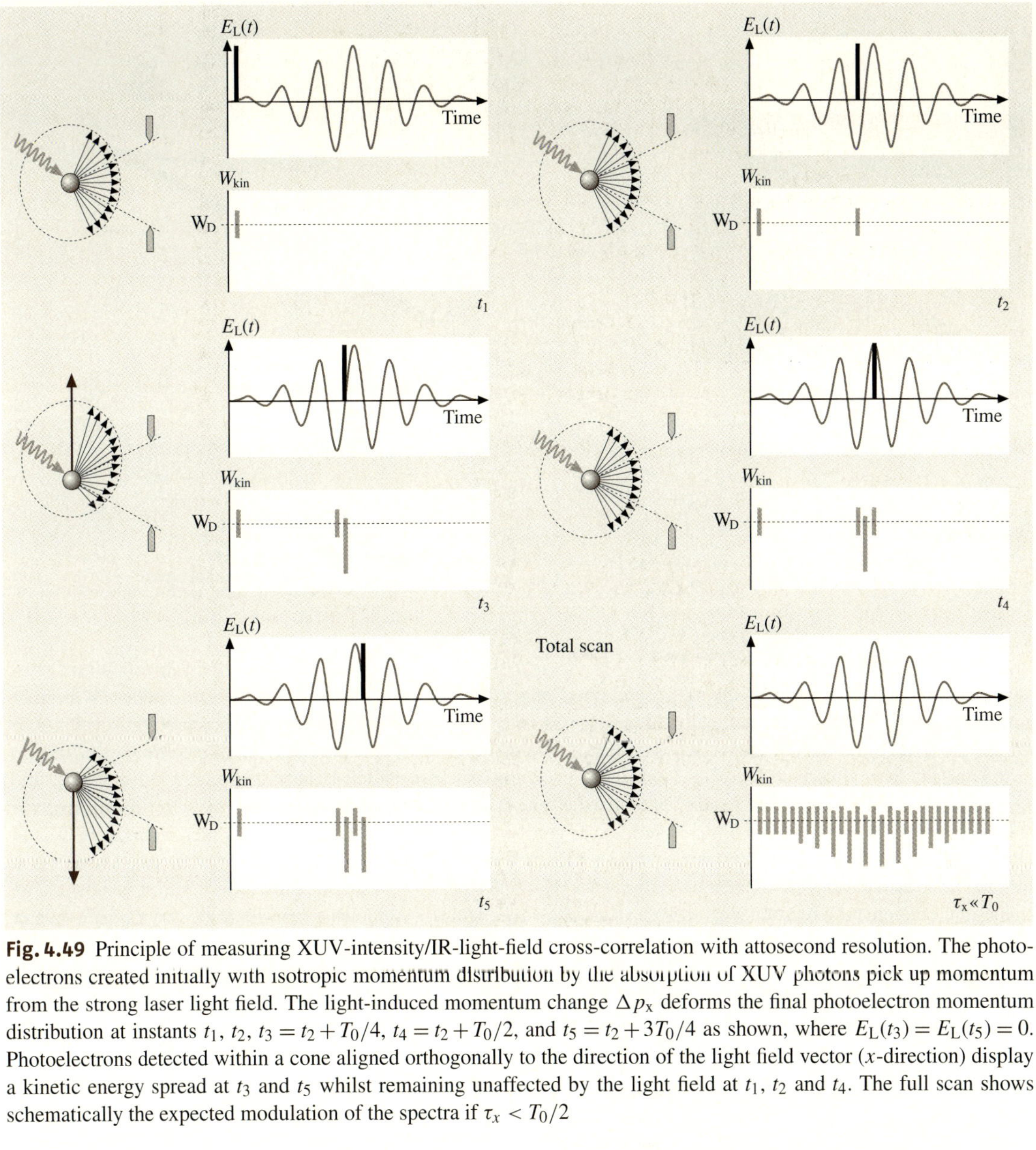

Fig. 4.49 Principle of measuring XUV-intensity/IR-light-field cross-correlation with attosecond resolution. The photoelectrons created initially with isotropic momentum distribution by the absorption of XUV photons pick up momentum from the strong laser light field. The light-induced momentum change Δp_x deforms the final photoelectron momentum distribution at instants t_1, t_2, $t_3 = t_2 + T_0/4$, $t_4 = t_2 + T_0/2$, and $t_5 = t_2 + 3T_0/4$ as shown, where $E_L(t_3) = E_L(t_5) = 0$. Photoelectrons detected within a cone aligned orthogonally to the direction of the light field vector (x-direction) display a kinetic energy spread at t_3 and t_5 whilst remaining unaffected by the light field at t_1, t_2 and t_4. The full scan shows schematically the expected modulation of the spectra if $\tau_x < T_0/2$

monic generation and their time-resolved measurement by the attosecond streak-camera technique described in the previous section. A 200 nm-thick zirconium foil with an aperture of 2 mm is placed 150 cm downstream from the XUV pulse source. The aperture is matched to the beam diameter of the harmonic beam, which has a low divergence of ≈ 0.7 mrad. This filter is placed in the beam to block the laser and low-order harmonics across the XUV beam, transmitting only photons with an energy of higher than 70 eV. It is mounted on a nitrocellulose membrane of 5 μm thickness to cover a 2 mm-diameter hole in the membrane. This device is virtually dispersion-free and produces an annular laser beam confining the XUV beam in its center. The energy in the laser beam is adjusted by a motorized iris and measured by a photodiode. Both the laser and the collinearly propagating XUV beam are focused with a two-component focusing mirror directing the XUV

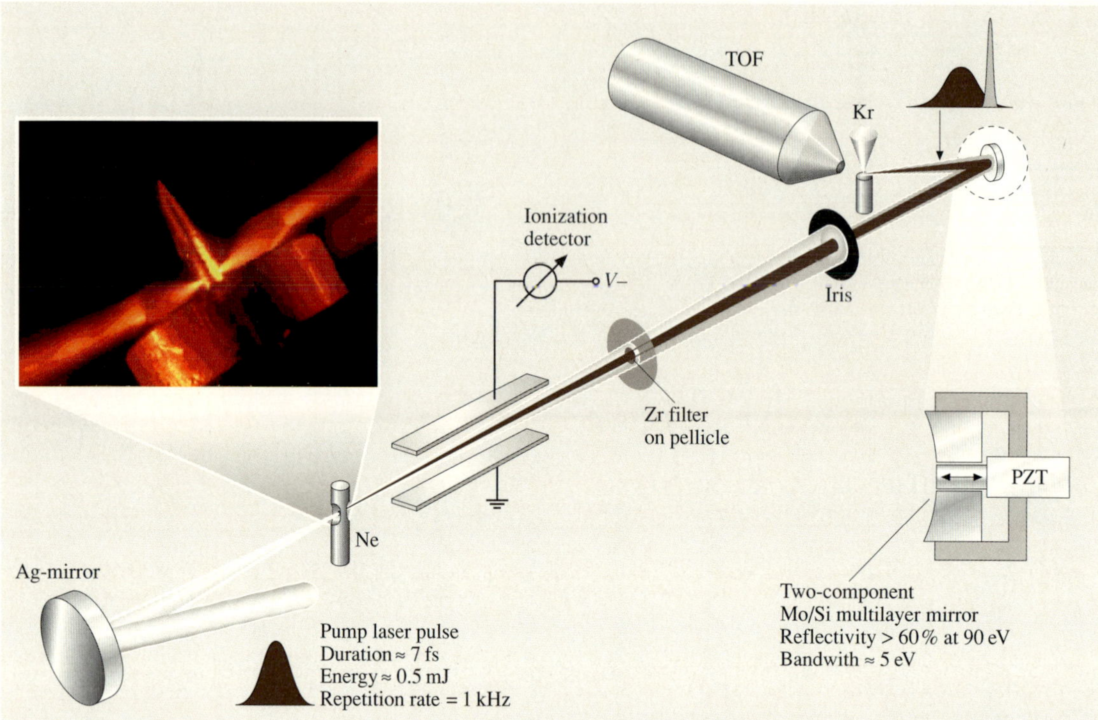

Fig. 4.50 Experimental set up for the attosecond pulse measurement. The focused 7 fs laser beam interacts with neon atoms to produce high-harmonic radiation. The laser and the highly collimated XUV beam co-propagate collinearly through a 2 m beam line towards the measurement. In the beaml ine the laser and harmonic beams pass through a 200 nm-thick 3 mm-diameter zirconium foil placed on a 5 μm-thick nitrocellulose pellicle to cover a hole of 2 mm diameter. The energy transported by the resulting annular beam can be adjusted with a motorized iris between a fraction of a microjoule and a few tens of microjoules. The Mo/Si multilayer consists of an annular part having an outer diameter of 10 mm with a concentric hole of 3 mm diameter hosting a miniature mirror of slightly smaller diameter. Both parts originate from the same substrate, ensuring identical radii of curvature ($R = 70$ mm). The miniature central mirror is mounted on a wide-range nanometer-precision piezo-driven stage, allowing alignment and translation with respect to the external part

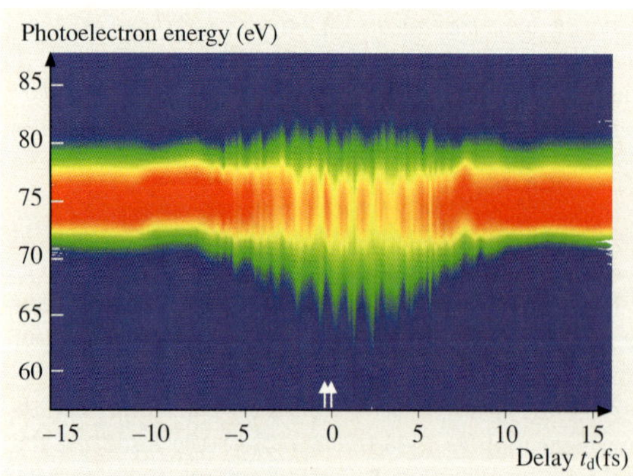

Fig. 4.51 Energy distributions of the Kr-4p photoelectrons (binding energy ≈15 eV) as a function of the delay t_d between the ionizing sub-femtosecond 90 eV XUV pulse and the *dressing* few-cycle laser wave over some tens of femtoseconds

and IR beam with an adjustable time delay into a common focus of the two focusing mirrors, where a nozzle supplying the target atoms is placed. The focusing mirror is coated with a Mo/Si multilayer stack designed to reflect photons with 90 eV energy within a 5 eV bandwidth. This bandwidth is large enough to support XUV pulses as short as 0.4 fs.

The contour plot in Fig. 4.51 depicts the variation of the energy distribution of photoelectrons knocked out from the 4 p shell of krypton by 90 eV XUV pulses (y-axis) as a function of delay t_d (x-axis). The XUV pulses were filtered out from the high-harmonic emission spectrum of Ne atoms driven by few-cycle (7 fs, 750 nm) laser pulses by the Mo/Si mirror within a 5 eV bandwidth at 90 eV near the cutoff of the emission spectrum. The XUV pulse were delayed by 150 as steps with respect to the few-cycle laser pulse to record the laser-affected XUV photoelectron spectra versus delay. The data clearly bring to light a quasi-periodic evolution of the photoelectron energy spectrum with a period of $\approx T_0/2$. From the depth of this modulation an XUV pulse duration of $\tau_{XUV} = 650\,\text{as} \pm 150\,\text{as}$ was evaluated [4.285].

Figure 4.52 shows two representative spectra sliced out from the contour plot in Fig. 4.51 at delays of $t_d = -450$ as and $t_d = 0$, as indicated by the arrows. The marked difference between the two spectra recorded at delays just 450 as apart provides a clear evidence for the photoelectron emission time (and hence the XUV pulse) being temporally confined to a fraction of a femtosecond.

In the analysis presented in the previous subsections, the laser field was used to probe the XUV pulse duration. With its duration known, the sub-femtosecond XUV pulse can take over the role of the probe as it is scanned across the laser pulse (in time) and measure the frequency of light for the first time in a time-resolved experiment. A possible sweep of instantaneous frequency ν_{inst} (or wavelength λ_{inst}) of the few-cycle laser pulse

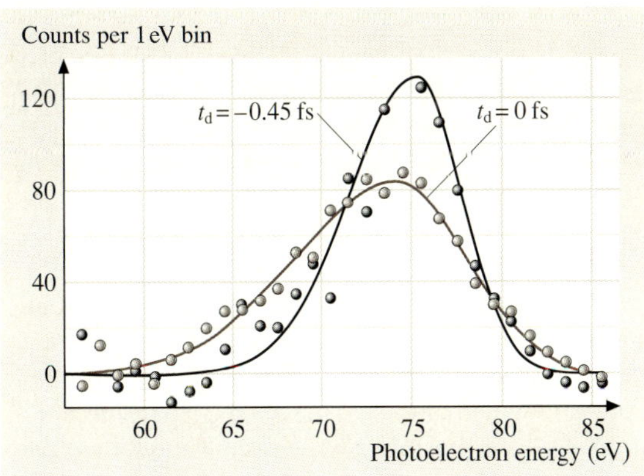

Fig. 4.52 Kr-4p photoelectron spectra extracted from the contour plot in Fig. 4.51 at delays indicated by the *arrows*. The *dots* represent spectra normalized to the same number of electron counts in the time-of-flight electron spectrometer. The *lines* show asymmetric Gaussian fits to the data

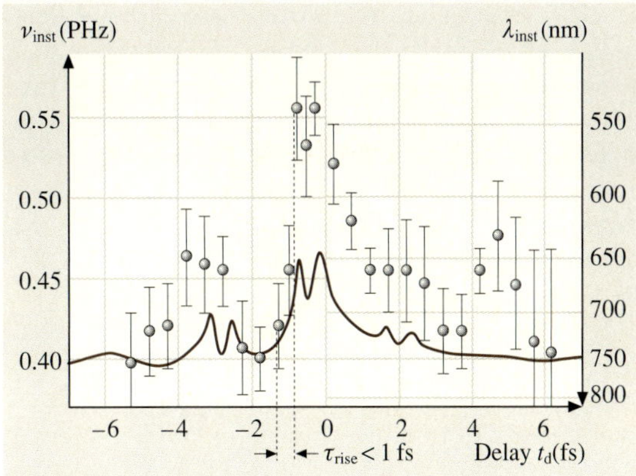

Fig. 4.53 Calculated (*line*) and measured blue shift of the fundamental pulse, probed by the XUV pulse. The steep rising edge indicates the presence of single XUV pulses without significant satellites

can be evaluated from the modulation in $\Delta W(t_d)$ by fitting a sinusoidal half-oscillation of adjustable period to $\Delta W(t_d)$ at different values of t_d. The sampling interval was scanned through the range of $-8\,\text{fs} \le t_d \le 8\,\text{fs}$. The dots in Fig. 4.53 show the carrier-frequency sweep evaluated in this manner, revealing a dynamic blue shift from a carrier wavelength of $\approx$750 to $\approx$550 nm.

This strong dynamic frequency shift carried by the few-cycle laser pulse has been found to originate from the interaction of the pulse with the ionizing gas source of the XUV pulses. In fact, the line in Fig. 4.53 is obtained from propagating a bandwidth-limited 7 fs laser pulse through the volume of neon gas emitting the high harmonics in a numerical experiment [4.331]. The measured $\approx$30% dynamic frequency blue shift at the center of the pulse is larger than predicted by the simulations but reflects qualitatively the predicted behavior. The qualitative agreement suggests that the observed blue shift can be attributed to ionization-induced self-phase modulation in the high-harmonic-generation process. This direct probing of the field oscillations of a light wave can be regarded as the first application of sub-femtosecond XUV pulses.

The observed sub-femtosecond rise time of the ionization-induced dynamic blue shift provides conclusive evidence for the isolated nature of the sub-femtosecond XUV burst. Any satellite of substantial energy spaced by some $T_0/2 \approx 1.30$ fs from the main burst would broaden this rise time to more than 1 fs. The agreement of the measured XUV pulse duration and that obtained from simulations [4.285] within the experimental error suggests that the timing jitter of the XUV pulse with respect to the phase of the IR light wave must be small compared to 1 fs, indicating that the sub-femtosecond XUV pulse is locked to the carrier wave of its generating few-cycle light pulse with as precision. The attosecond timing stability of the sub-femtosecond XUV pulse to a few-cycle IR light wave makes these pulses a unique tool for investigating the dynamic behavior of matter on an attosecond time scale. The light-field-controlled photoemission experiment already demonstrates this capability. In the measured energy distribution $\Delta W(t_d)$ it is implicit, that in the investigated spectral range near 90 eV bound–free electronic transitions from the 4p state in krypton respond to XUV excitation within less than 500 as, constituting

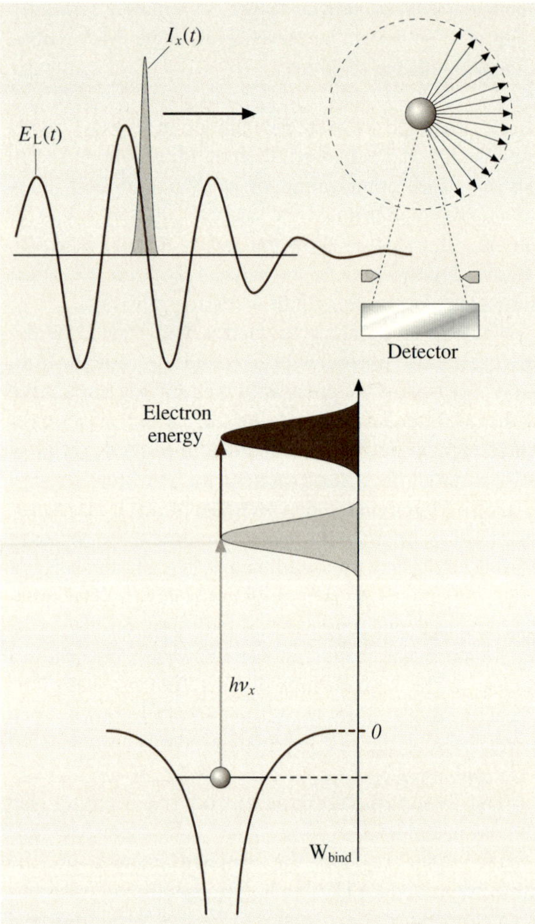

Fig. 4.54 Energy shift of XUV photoelectrons ejected nearly parallel to the polarization direction of a strong laser field near the instant of zero transition of the laser field by a sub-$T_0/2$-duration XUV pulse

the first time-resolved measurement on an atomic time scale.

Another implementation of the attosecond streak-camera technique is shown in Fig. 4.54 [4.286]. Here the photoelectrons are detected in the parallel geometry as sketched in the top of Fig. 4.48. If the photoelectron is ejected near the zero crossing of the laser electric field, its energy spectrum may get up- or downshifted by many electronvolts. This is in strong contrast with the orthogonal geometry used before, where the primary effect of the laser field is a broadening rather than a shift of the photoelectron energy spectrum.

The predicted spectral shift relies not only on a precise timing of the XUV pulse to the zero transition of the laser field. This shift without substantial broadening happens only if the generated XUV burst (and therefore the generated photoelectron wave packet) is very short compared to the laser field half-oscillation period $T_0/2$. If the XUV pulse and therefore the electron wave packet approaches or exceeds $T_0/2$, different portions of the wave packet experience different momentum transfer, resulting in a large spread of energies. A recent experiment using XUV pulses of 8 eV bandwidth verified this prediction and allowed to set an upper limit of 500 as and 200 as on the duration of the XUV pulse and its timing jitter with respect to its generating laser wave, respectively [4.286].

4.12.4 Applications of Sub-femtosecond XUV Pulses: Time-Resolved Spectroscopy of Atomic Processes

Finally, we show, how pump–probe spectroscopy of atomic excitation and relaxation processes – such as e.g. optical-field ionization and inner-shell relaxation processes – can be traced directly in the time domain by drawing on low-energy isolated X-ray pulses in combination with synchronized strong few-cycle laser pulses, i. e. on tools that are available now.

Pump–probe techniques offer the most direct experimental approach to tracing microscopic dynamics. The extension of time-resolved spectroscopy to ultrafast electronic processes taking place deep inside atoms has so far been frustrated by the simultaneous requirements of short wavelengths (i. e. high photon energy) and sub-femtosecond pulse duration. There is the additional difficulty that a straightforward interpretation of spectroscopic data requires isolated (single) pulses. With these pulses now available from few-cycle-driven high-order harmonic generation, extension of time-resolved spectroscopy into the attosecond domain can now be

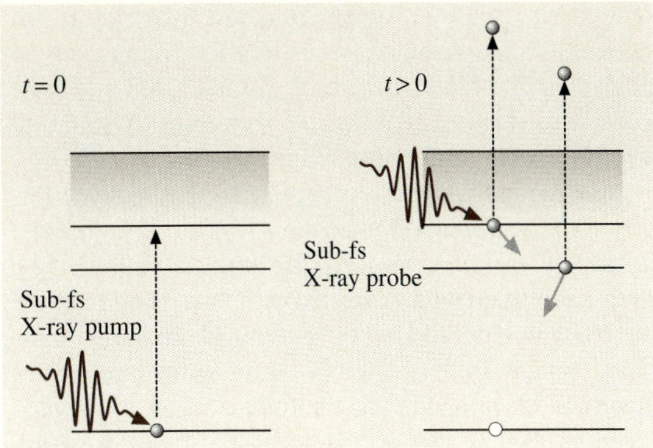

Fig. 4.55 Tracing inner-shell relaxation processes by XUV–pump/XUV–probe spectroscopy. The kinetic energy of photoelectrons detached by the probe pulse from the atom following excitation by an XUV pump pulse is analyzed as a function of the delay between the pump and the probe pulse

tackled. Figure 4.55 shows the principle of an XUV-pump/XUV-probe experiment. In this – conceptually most straightforward – implementation of as spectroscopy, the sub-femtosecond XUV pulses are used for both triggering and probing bound–bound or bound–free transitions in atoms or molecules.

However, the sub-femtosecond XUV pulses currently available do not have sufficient flux for XUV–pump/XUV–probe spectroscopy yet. This is because in this kind of experiments the physical quantity measured as a function of delay between pump and probe pulse relies on a two-XUV-photon transition, the probability of which scales with $\sigma I_{pump} I_{probe}$, where σ is the atomic cross section of the absorption process and I_{pump} and I_{probe} are the intensities of the pump and probe XUV bursts, respectively. Both I_{pump} and I_{probe} are many orders of magnitude less than what can be achieved in the optical regime. Moreover σ is also reduced by many orders of magnitude because it scales with λ^6.

Extension of the concept of laser-field-assisted XUV photoelectron emission to sampling the emission of secondary (Auger) electrons offers an alternative solution to time-resolving inner-shell atomic processes with as accuracy. The underlying principle is illustrated in Fig. 4.56. The sub-femtosecond XUV pulse excites a core electron and produces thereby a short-lived inner-shell vacancy. This is rapidly filled by an electron from a higher energy level (outer shell). The energy lost by the electron undergoing this transition is carried away either by an energetic (XUV/X-ray) photon or by a secondary (Auger) electron. The emission

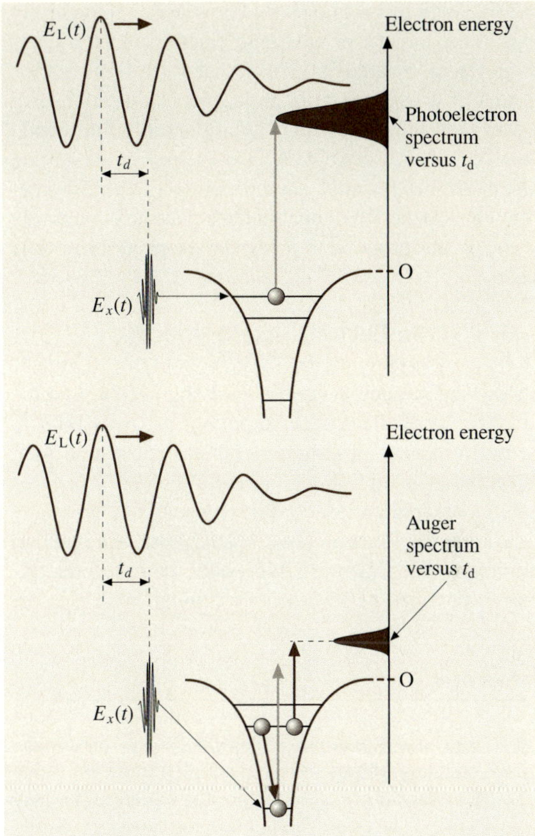

Fig. 4.56 Schematic representation of the Auger process and its temporal evolution displaying the decay of an inner-shell vacancy

time of this Auger electron corresponds exactly to the lifetime of the inner-shell vacancy. Hence, sampling the Auger electron emission in the same way as the primary photoelectron emission with the oscillating laser field enables researchers to gain direct time-domain access to inner-shell atomic processes with as resolution. The first proof-of-principle time-resolved inner-shell spectroscopic experiment was recently demonstrated [4.343]. Although the decay time measured in this experiment can also be inferred from energy-domain measurements, it served as a benchmark process for testing the feasibility of the methods described above for attosecond time-resolved spectroscopy.

The frontiers of both time-resolved spectroscopy and the control of microscopic dynamics are about to be radically extended due to the new technical capability that allows scientists to synthesize intense pulses containing a few cycles evolving in a precisely determined way

[4.330]. With this tool it should be possible to precisely control the motion of energetic electron wave packets around atoms on attosecond timescales just as the motion of nuclear wave packets in molecules can be controlled within a few femtoseconds. The single sub-femtosecond electron bunches and (XUV/X-ray) photon bursts that arise from the recently gained ability to control electron wave packets will enable the scientific community to excite and probe atomic dynamics on atomics time scales.

4.12.5 Some Recent Developments

Attosecond science is rapidly evolving [4.344] and the last three years have seen important progress both in the performances of femtosecond and attosecond light pulses based on high-order harmonic generation in gases and in their applications in different scientific areas.

Harmonic sources now reach pulse energies in the microjoule range [4.345] and their spectra extend to energies of several keV (though with a lower throughput) [4.346]. Applications are flourishing, going from the determination of vibration frequencies in molecules [4.347] to microscopy [4.348, 349] and even recently to seeding of X-ray laser plasmas [4.350] and possibly, in the future, free-electron lasers.

Attosecond pulses have been studied in more details with different techniques [4.351, 352] and the method to isolate a single pulse refined [4.353, 354]. Their time-frequency characteristics have been mapped out [4.355–357] and ways to control both the individual pulses [4.358] and the train structure [4.359] have been developed. The shortest isolated pulse produced to date is 130 as, using the polarization gating technique for the temporal confinement [4.360, 361]. Applications now include characterization of electromagnetic fields [4.362], tomography of molecular orbitals [4.363], molecular dynamics studies of simple molecules [4.364, 365], dynamical studies and interferometric measurements of electron wavepackets [4.366, 367], and time-resolved inner-shell spectroscopy in atoms and solids [4.368].

References

4.1 R. W. Boyd: *Nonlinear Optics*, 2nd edn. (Academic, San Diego 2003)

4.2 S. I. Wawilov, W. L. Lewschin: Die Beziehungen zwischen Fluoreszenz und Phosphoreszenz in festen und flüsssigen Medien, Z. Phys. **35**, 920–936 (1926)

4.3 S. I. Vavilov: *Microstructure of Light* (USSR Acad. Sci., Moscow 1950)

4.4 P. A. Franken, A. E. Hill, C. W. Peters, G. Weinrich: Generation of optical harmonics, Phys. Rev. Lett. **7**, 118 (1961)

4.5 R. Terhune P. Maker, C. Savage: Observation of saturation effects in optical harmonic generation, Phys. Rev. Lett. **2**, 54 (1963)

4.6 A. I. Kovrigin, A. S. Piskarskas, R. V. Khokhlov: On the generation of UV radiation by cascaded frequency conversion, Pis'ma Zh. Eksp. Teor. Fiz. **2**, 223 (1965)

4.7 N. Bloembergen: *Nonlinear Optics* (Benjamin, New York 1964)

4.8 S. A. Akhmanov, R. V. Khokhlov: *Problems of Nonlinear Optics* (VINITI, Moscow 1964) Engl. transl. New York, Gordon & Breach, 1972.

4.9 Y. R. Shen: *The Principles of Nonlinear Optics* (Wiley, New York 1984)

4.10 D. Cotter, P. N. Butcher: *The Elements of Nonlinear Optics* (Cambridge University Press, Cambridge 1990)

4.11 J. F. Reintjes: *Nonlinear Optical Parametric Processes in Liquids and Gases* (Academic, Orlando 1984)

4.12 Th. G. Brown, K. Creath, H. Kogelnik, M. A. Kriss, J. Schmit, M. J. Weber: *The Optics Encyclopedia*, ed. by Th. G. Brown, K. Creath, H. Kogelnik, M. A. Kriss, J. Schmit, M. J. Weber (Wiley-VCH, Weinheim 2004) Chap. Nonlinear Optics, p. 1617

4.13 R. W. Terhune, M. Nisenoff, C. M. Savage: Mixing of dispersion and focusing on the production of optical harmonics, Phys. Rev. Lett. **8**, 21 (1962)

4.14 J. A. Giordmaine: Mixing of light beams in crystals, Phys. Rev. Lett. **8**, 19 (1962)

4.15 N. Bloembergen, J. Ducuing, P. S. Pershan: Interactions between light waves in a nonlinear dielectric, Phys. Rev. **1217**, 1918 (1962)

4.16 M. M. Feier, G. A. Magel, D. H. Jundt, R. L. Byer: Quasi-phase-matched second-harmonic generation: tuning and tolerances, IEEE J. Quantum Electron. **28**, 2631 (1992)

4.17 R. B. Miles, S. E. Harris: Optical third-harmonic generation in alkali metal vapors, IEEE J. Quantum Electron. **9**, 470 (1973)

4.18 S. A. Akhmanov, V. A. Vysloukh, A. S. Chirkin: *Optics of Femtosecond Laser Pulses* (Nauka, Moscow 1988)

4.19 A. Dubietis, G. Jonušaskas, A. Piskarskas: Powerful femtosecond pulse generation by chirped and stretched ulse parametric amplification in BBO crystal, Opt. Commun. **88**, 437 (1992)

4.20 T. Brabec, F. Krausz: Intense few-cycle laser fields: Frontiers of nonlinear optics, Rev. Mod. Phys. **72**, 545 (2000)

4.21 H. H. Chen, Y. C. Lee: Radiations by solitons at the zero group-dispersion wavelength of single-mode optical fibers, Phys. Rev. A **41**, 426–439 (1990)

4.22 N. Akhmediev, M. Karlsson: Cherenkov radiation emitted by solitons in optical fibers, Phys. Rev. A **51**, 2602–2607 (1995)

4.23 G. P. Agrawal: *Nonlinear Fiber Optics* (Academic, Boston 1989)

4.24 N. I. Koroteev, A. M. Zheltikov: Chirp control in third-harmonic generation due to cross-phase modulatio, Appl. Phys. B **67**, 53–57 (1998)

4.25 S. P. Le Blanc, R. Sauerbrey: Spectral, temporal, and spatial characteristics of plasma-induced spectral blue shifting and its application to femtosecond pulse measurement, J. Opt. Soc. Am. B **13**, 72 (1996)

4.26 M. M. T. Loy, Y. R. Shen: Theoretical interpretation of small-scale filaments of light originating from moving focal spots, Phys. Rev. A **3**, 2099 (1971)

4.27 S. A. Akhmanov, A. P. Sukhorukov, R. V. Khokhlov: Self-focusing and diffraction of light in a nonlinear medium, Sov. Phys. Usp. **93**, 19 (1967)

4.28 R. Y. Chiao, E. Garmire, C. H. Townes: Self-trapping of optical beams, Phys. Rev. Lett. **13**, 479 (1964)

4.29 G. Fibich, A. L. Gaeta: Critical power for self-focusing in bulk media and in hollow waveguides, Opt. Lett. **25**, 335 (2000)

4.30 G. Fibich, F. Merle: Self-focusing on bounded domains, Physica D **155**, 132 (2001)

4.31 K. D. Moll, A. L. Gaeta, G. Fibich: Self-similar optical wave collapse: observation of the townes profile, Phys. Rev. Lett. **90**, 203902 (2003)

4.32 S. O. Konorov, A. M. Zheltikov, A. P. Tarasevitch, Ping Zhou, D. von der Linde: Self-channeling of subgigawatt femtosecond laser pulses in a ground-state waveguide induced in the hollow core of a photonic crystal fiber, Opt. Lett. **29**, 1521 (2004)

4.33 R. A. Fisher: *Optical Phase Conjugation* (Academic, New York 1983)

4.34 C. V. Raman, K. S. Krishnan: A new type of secondary radiation, Nature **121**, 501 (1928)

4.35 G. Landsberg, L. Mandelstam: Eine neue Erscheinung bei der Lichtzertreuung, Naturwiss. **16**, 557 (1928)

4.36 P. St. J. Russell: Photonic crystal fibers, Science **299**, 358–362 (2003)

4.37 J. C. Knight: Photonic crystal fibers, Nature **424**, 847–851 (2003)

4.38 C. M. Bowden, A. M. Zheltikov: Nonlinear optics of photonic crystals, J. Opt. Soc. Am. B **19**, 9 (2002), feature issue

4.39 A. M. Zheltikov: Nonlinear optics of microstructure fibers, Phys. Uspekhi **47**, 69–98 (2004)

4.40 W. H. Reeves, D. V. Skryabin, F. Biancalana, J. C. Knight, P.St.J. Russell, F. G. Omenetto, A. Efimov, A. J. Taylor: Transformation and control of ultra-short pulses in dispersion-engineered photonic crystal fibres, Nature **424**, 511–515 (2003)

4.41 D. A. Akimov, E. E. Serebryannikov, A. M. Zheltikov, M. Schmitt, R. Maksimenka, W. Kiefer, K. V. Dukel'skii, V. S. Shevandin, Yu.N. Kondrat'ev: Efficient anti-Stokes generation through phase-matched four wave mixing in higher-order modes of a microstructure fiber, Opt. Lett. **28**, 1948–1950 (2003)

4.42 J. K. Ranka, R. S. Windeler, A. J. Stentz: Visible continuum generation in air-silica microstructure optical fibers with anomalous dispersion at 800 nm, Opt. Lett. **25**, 25–27 (2000)

4.43 W. J. Wadsworth, A. Ortigosa-Blanch, J. C. Knight, T. A. Birks, T. P. M. Mann, P. St. J. Russell: Supercontinuum generation in photonic crystal fibers and optical fiber taperss: a novel light source, J. Opt. Soc. Am. B **19**, 2148–2155 (2002)

4.44 A. M. Zheltikov: Supercontinuum generation, Appl. Phys. B **77**, 2 (2003), special issue ed. by A. M. Zheltikov

4.45 S. Coen, A. H. L. Chau, R. Leonhardt, J. D. Harvey, J. C. Knight, W. J. Wadsworth, P. St.J. Russell: Supercontinuum generation by stimulated Raman scattering and parametric four-wave mixing in photonic crystal fibers, J. Opt. Soc. Am. B **19**, 753–764 (2002)

4.46 J. K. Ranka, R. S. Windeler, A. J. Stentz: Optical properties of high-delta air-silica microstructure optical fiber, Opt. Lett. **25**, 796–798 (2000)

4.47 F. G. Omenetto, A. J. Taylor, M. D. Moores, J. Arriaga, J. C. Knight, W. J. Wadsworth, P. St. J. Russell: Simultaneous generation of spectrally distinct third harmonics in a photonic crystal fiber, Opt. Lett. **26**, 1158–1160 (2001)

4.48 F. G. Omenetto, A. Efimov, A. J. Taylor, J. C. Knight, W. J. Wadsworth, P. St. J. Russell: Polarization dependent harmonic generation in microstructured fibers, Opt. Express **11**, 61–67 (2003)

4.49 A. Efimov, A. J. Taylor, F. G. Omenetto, J. C. Knight, W. J. Wadsworth, P. St. J. Russell: Nonlinear generation of very high-order UV modes in microstructured fibers, Opt. Express **11**, 910–918 (2003)

4.50 A. Efimov, A. J. Taylor, F. G. Omenetto, J. C. Knight, W. J. Wadsworth, P. St. J. Russell: Phase-matched third harmonic generation in microstructured fibers, Opt. Express **11**, 2567–2576 (2003)

4.51 A. N. Naumov, A. B. Fedotov, A. M. Zheltikov, V. V. Yakovlev, L. A. Mel'nikov, V. I. Beloglazov, N. B. Skibina, A. V. Shcherbakov: Enhanced $\chi^{(3)}$ interactions of unamplified femtosecond Cr: forsterite laser pulses in photonic-crystal fibers, **19**, 2183–2191 (2002)

4.52 D. A. Akimov, A. A. Ivanov, A. N. Naumov, O. A. Kolevatova, M. V. Alfimov, T. A. Birks, W. J. Wadsworth, P. St. J. Russell, A. A. Podshivalov, A. M. Zheltikov: Generation of a spectrally asymmetric third harmonic with unamplified 30-fs Cr:forsterite laser pulses in a tapered fiber, Appl. Phys. B **76**, 515–519 (2003)

4.53 J. D. Harvey, R. Leonhardt, S. Coen, G. K. L. Wong, J. C. Knight, W. J. Wadsworth, P. St. J. Russell: Scalar modulation instability in the normal dispersion regime by use of a photonic crystal fiber, Opt. Lett. **28**, 2225–2227 (2003)
4.54 X. Liu, C. Xu, W. H. Knox, J. K. Chandalia, B. J. Eggleton, S. G. Kosinski, R. S. Windeler: Soliton self-frequency shift in a short tapered air-silica microstructure fiber, Opt. Lett. **26**, 358–360 (2001)
4.55 A. N. Naumov, A. M. Zheltikov: Asymmetric spectral broadening and temporal evolution of cross-phase-modulated third harmonic pulses, Opt. Express **10**, 122–127 (2002)
4.56 S. A. Akhmanov, A. P. Sukhorukov, A. S. Chirkin: Second-harmonic generation in nonlinear crystals by group-delayed ultrashort pulses, **55**, 1430–1435 (1968)
4.57 A. M. Zheltikov: Third-harmonic generation with no signal at 3ω, Phys. Rev. A. **72**, 043812 (2005)
4.58 A. M. Zheltikov: Multimode guided-wave non-3ω third-harmonic generation by ultrashort laser pulses, J. Opt. Soc. Am. B **22**, 2263 (2005)
4.59 S. O. Konorov, A. A. Ivanov, M. V. Alfimov, A. M. Zheltikov: Polarization-sensitive non-3ω third-harmonic generation by femtosecond Cr: forsterite laser pulses in birefringent microchannel waveguides of photonic-crystal fibers, **81**, 219–223 (2005)
4.60 L. Allen, J. H. Eberly: *Optical Resonance and Two-Level Atoms* (Wiley, New York 1975)
4.61 M. Sargent III, M. O. Scully, W. E. Lamb Jr.: *Laser Physics* (Addison-Wesley, London 1974)
4.62 S. L. McCall, E. L. Hahn: Self-induced transparency by pulsed coherent light, Phys. Rev. Lett. **18**, 908–911 (1967)
4.63 S. L. McCall, E. L. Hahn: Self-induced transparency, Phys. Rev. **183**, 457–485 (1969)
4.64 J. H. Eberly: Area Theorem rederived, Opt. Express **2**, 173–176 (1998)
4.65 R. W. Ziolkowski, J. M. Arnold, D. M. Gogny: Ultrafast pulse interactions with two-level atoms, Phys. Rev. A **52**, 3082–3094 (1995)
4.66 A. Taflove: *Computational Electrodynamics: The Finite-Difference Time-Domain Method* (Artech House, Norwood 1995)
4.67 S. Hughes: Subfemtosecond soft-x-ray generation from a two-level atom: Extreme carrier-wave Rabi flopping, Phys. Rev. A **62**, 055401–055405 (2000)
4.68 S. Hughes: Breakdown of the area theorem: carrier-wave Rabi flopping of femtosecond optical pulses, Phys. Rev. Lett **81**, 3363–3366 (1998)
4.69 A. V. Tarasishin, S. A. Magnitskii, A. M. Zheltikov: Propagation and amplification of ultrashort light pulses in a resonant two-level medium: finite-difference time-domain analysis, Opt. Commun. **193**, 187 (2001)
4.70 A. V. Tarasishin, S. A. Magnitskii, V. A. Shuvaev, A. M. Zheltikov: Evolution of ultrashort light pulses in a two-level medium visualized with the finite-difference time domain technique, Opt. Express **8**, 452–457 (2001)
4.71 R. R. Alfano, S. L. Shapiro: Emission in the region 4000 to 7000 Å via four-photon coupling in glass, Phys. Rev. Lett. **24**, 584 (1970)
4.72 R. R. Alfano, S. L. Shapiro: Observation of self-phase modulation and small-scale filaments in crystals and glasses, Phys. Rev. Lett. **24**, 592 (1970)
4.73 R. Alfano: *The Supercontinuum Laser Source*, ed. by R. Alfano (Springer, Berlin 1989)
4.74 D. J. Jones, S. A. Diddams, J. K. Ranka, A. Stentz, R. S. Windeler, J. L. Hall, S. T. Cundiff: Carrier-envelope phase control of femtosecond mode-locked lasers and direct optical frequency synthesis, Science **288**, 635–639 (2000)
4.75 R. Holzwarth, T. Udem, T. W. Hänsch, J. C. Knight, W. J. Wadsworth, P.St.J. Russell: Optical frequency synthesizer for precision spectroscopy, Phys. Rev. Lett. **85**, 2264–2267 (2000)
4.76 S. A. Diddams, D. J. Jones, S. T. Cundiff. Jun Ye, J. L. Hall, J. K. Ranka, R. S. Windeler, R. Holzwarth, T. Udem, T. W. Hänsch: Direct link between microwave and optical frequencies with a 300 THz femtosecond laser comb, Phys. Rev. Lett. **84**, 5102 (2000)
4.77 Th. Udem, R. Holzwarth, T. W. Hänsch: Optical frequency metrology, Nature **416**, 233 (2002)
4.78 A. Baltuska, T. Fuji, T. Kobayashi: Self-referencing of the carrier-envelope slip in a 6-fs visible parametric amplifier, Opt. Lett. **27**, 1241–1243 (2002)
4.79 C. Y. Teisset, N. Ishii, T. Fuji, T. Metzger, S. Köhler, R. Holzwarth, A. M. Zheltikov, F. Krausz: Soliton-based pump-seed synchronization for few-cycle OPCPA, Opt. Express **13**, 6550–6557 (2005)
4.80 S. O. Konorov, D. A. Akimov, E. E. Serebryannikov, A. A. Ivanov, M. V. Alfimov, A. M. Zheltikov: Cross-correlation FROG CARS with frequency-converting photonic-crystal fibers, Phys. Rev. E **70**, 057601 (2004)
4.81 I. Hartl, X. D. Li, C. Chudoba, R. K. Rhanta, T. H. Ko, J. G. Fujimoto, J. K. Ranka, R. S. Windeler: Ultrahigh-resolution optical coherence tomography using continuum generation in an air-silica microstructure optical fiber, Opt. Lett. **26**, 608–610 (2001)
4.82 A. B. Fedotov, S. O. Konorov, E. E. Serebryannikov, D. A. Sidorov-Biryukov, V. P. Mitrokhin, K. V. Dukel'skii, A. V. Khokhlov, V. S. Shevandin, Y. N. Kondrat'ev, M. Scalora, A. M. Zheltikov: Assorted nonlinear optics in microchannel waveguides of photonic-crystal fibers, Opt. Commun. **255**, 218–224 (2005)
4.83 E. E. Serebryannikov, S. O. Konorov, A. A. Ivanov, M. V. Alfimov, M. Scalora, A. M. Zheltikov: Cross-phase-modulation-induced instability in photonic-crystal fibers, Phys. Rev. E **72**, 027601 (2005)
4.84 T. M. Monro, D. J. Richardson, N. G. R. Broderick, P. J. Bennet: Holey optical fibers: An efficient modal method, J. Lightwave Technol. **17**, 1093–1102 (1999)

4.85 G. P. Agrawal: Modulation instability induced by cross-phase modulation, Phys. Rev. Lett. **59**, 880 (1987)

4.86 F. M. Mitschke, L. F. Mollenauer: Discovery of the soliton self-frequency shift, Opt. Lett. **11**, 659–661 (1986)

4.87 E. M. Dianovand A. Y. Karasik, P. V. Mamyshev, A. M. Prokhorov, V. N. Serkin, M. F. Stel'makh, A. A. Fomichev: Stimulated-Raman conversion of multisoliton pulses in quartz optical fibers, JETP Lett. **41**, 294 (1985)

4.88 B. R. Washburn, S. E. Ralph, P. A. Lacourt, J. M. Dudley, W. T. Rhodes, R. S. Windeler, S. Coen: Tunable near-infrared femtosecond soliton generation in photonic crystal fibers, Electron. Lett. **37**, 1510–1512 (2001)

4.89 J. H. Price, K. Kurosawa, T. M. Monro, L. Lefort, D. J. Richardson: Tunable, femtosecond pulse source operating in the range 1061.33 μm based on an Yb3+-doped holey fiber amplifier, J. Opt. Soc. Am. B **19**, 1286–1294 (2002)

4.90 K. S. Abedin, F. Kubota: Widely tunable femtosecond soliton pulse generation at a 10 GHz repetition rate by use of the soliton self-frequency shift in photonic crystal fiber, Opt. Lett. **28**, 1760–1762 (2003)

4.91 E. E. Serebryannikov, A. M. Zheltikov, N. Ishii, C. Y. Teisset, S. Köhler, T. Fuji, T. Metzger, F. Krausz: Soliton self-frequency shift of 6-fs pulses in photonic-crystal fibers, Appl. Phys. B **81**, 585 (2005)

4.92 E. E. Serebryannikov, A. M. Zheltikov, N. Ishii, C. Y. Teisset, S. Köhler, T. Fuji, T. Metzger, F. Krausz, A. Baltuška: Nonlinear-optical spectral transformatin of few-cycle laser pulses in photonic-crystal fibers, Phys. Rev. E **72**, 056603 (2005)

4.93 K. J. Blow, D. Wood: Theoretical description of transient stimulated Raman scattering in optical fibers, IEEE J. Quantum Electron. **25**, 2665–2673 (1989)

4.94 R. H. Stolen, J. P. Gordon, W. H. Tomlinson, H. A. Haus: Raman response function of silica-core fibers, J. Opt. Soc. Am. B **6**, 1159–1169 (1989)

4.95 J. Herrmann, U. Griebner, N. Zhavoronkov, A. Husakou, D. Nickel, J. C. Knight, W. J. Wadsworth, P. St. J. Russell, G. Korn: Experimental evidence for supercontinuum generation by fission of higher-order solitons in photonic fibers, Phys. Rev. Lett. **88**, 173901 (2002)

4.96 D. V. Skryabin, F. Luan, J. C. Knight, P. St. J. Russell: Soliton self-frequency shift cancellation in photonic crystal fibers, Science **301**, 1705–1708 (2003)

4.97 P. Gordon J. Theory of the soliton self-frequency shift, Opt. Lett. **11**, 662–664 (1986)

4.98 J. K. Lucek, K. J. Blow: Soliton self-frequency shift in telecommunications fiber, Phys. Rev. A **45**, 6666 (1992)

4.99 S.A. Akhmanov, N. I. Koroteev: *Methods of Nonlinear Optics in Light Scattering Spectroscopy* (Nauka, Moscow 1981)

4.100 J. F. Ward, G. H. C. New: Optical third harmonic generation in gases by a focused laser beam, Phys. Rev **185**, 57 (1969)

4.101 G. C. Bjorklund: Effects of focusing on third-order nonlinear process in isotropic media, IEEE J. Quantum Electron. **QE-11**, 287 (1975)

4.102 A. M. Zheltikov, N. I. Koroteev: Coherent four-wave mixing in excited and ionized gas media, Phys. Uspekhi **42**, 321 (1999)

4.103 A. M. Zheltikov, N. I. Koroteev, A. N. Naumov: Phase matching for hyper-Raman-resonant coherent four-wave mixing, Quantum Electron. **24**, 1102 (1994)

4.104 A. Owyoung, E. D. Jones: Stimulated Raman spectroscopy using low-power cw lasers, Opt. Lett. **1**, 152–154 (1977)

4.105 P. D. Maker, R. W. Terhune: Study of optical effects due to an induced polarization third order in the electric field strength, Phys. Rev. A **137**, 801 (1965)

4.106 G. L. Eesley: *Coherent Raman Spectroscopy* (Pergamon, Oxford 1981)

4.107 A. C. Eckbreth: *Laser Diagnostics for Combustion Temperature and Species* (Abacus, Cambridge 1988)

4.108 S. A. J. Druet, J.-P. E. Taran: CARS spectroscopy, Progress Quantum Electron. **7**, 1 (1981)

4.109 P. Radi, A. M. Zheltikov: Nonlinear Raman spectroscopy, J. Raman Spect. **33**, 11/12 (2002), special Issue

4.110 P. Radi, A. M. Zheltikov: Nonlinear Raman spectroscopy II, J. Raman Spect. **34**, 12 (2003), special Issue

4.111 D. N. Kozlov, A. M. Prokhorov, V. V. Smirnov: The methane $\nu_1(a_1)$ vibrational state rotational structure obtained from high-resolution CARS-spectra of the *Q*-branch, J. Molec. Spectrosc. **77**, 21 (1979)

4.112 W. Kiefer: Active Raman spectroscopy: High resolution molecular spectroscopical methods, J. Molec. Struct. **59**, 305 (1980)

4.113 W. Kiefer: Femtosecond coherent Raman spectroscopy, J. Raman Spect. **1/2.**, 31 (2000)Special Issue

4.114 A. Zumbusch, G. R. Holtom, X. Sunney Xie: Three-dimensional vibrational imaging by coherent anti-Stokes Raman scattering, Phys. Rev. Lett. **82**, 4142 (1999)

4.115 E. O. Potma, X. Sunney Xie: CARS microscopy for biology and medicine, Opt. Photonics News **15**, 40 (2004)

4.116 D. Oron, N. Dudovich, D. Yelin, Y. Silberberg: Quantum control of coherent anti-Stokes Raman processes, Phys. Rev. A **65**, 043408 (2002)

4.117 N. Dudovich, D. Oron, Y. Silberberg: Single-pulse coherently controlled nonlinear Raman spectroscopy and microscopy, Nature **418**, 512 (2002)

4.118 A. B. Fedotov, S. O. Konorov, V. P. Mitrokhin, E. E. Serebryannikov, A. M. Zheltikov: Coherent anti-Stokes Raman scattering in isolated air-guided modes of a hollow-core photonic-crystal fiber, Phys. Rev. A **70**, 045802 (2004)

4.119 S.O. Konorov, A.B. Fedotov, A.M. Zheltikov, R.B. Miles: Phase-matched four-wave mixing and sensing of water molecules by coherent anti-Stokes Raman scattering in large-core-area hollow photonic-crystal fibers, J. Opt. Soc. Am. B **22**, 2049 (2005)

4.120 A.C. Eckbreth: BOXCARS: Crossed-beam phase-matched CARS generation in gases, Appl. Phys. Lett. **32**, 421 (1978)

4.121 P.R. Regnier, J.P.-E. Taran: On the possibility of measuring gas concentrations by stimulated anti-Stokes scattering, Appl. Phys. Lett. **23**, 240 (1973)

4.122 D.A. Akimov, A.B. Fedotov, N.I. Koroteev, A.N. Naumov, D.A. Sidorov-Biryukov, A.M. Zheltikov: Application of coherent four-wave mixing for two-dimensional mapping of spatial distribution of excited atoms in a laser-produced plasma, Opt. Commun. **140**, 259 (1997)

4.123 D.A. Akimov, A.B. Fedotov, N.I. Koroteev, A.N. Naumov, D.A. Sidorov-Biryukov, A.M. Zheltikov, R.B. Miles: One-dimensional coherent four-wave mixing as a way to image the spatial distribution of atoms in a laser-produced plasma, Opt. Lett **24**, 478 (1999)

4.124 A.C. Eckbreth, T.J. Anderson: Dual broadband CARS for simultaneous, multiple species measurements, Appl. Opt. **24**, 2731 (1985)

4.125 M. Alden, P.-E. Bengtsson, H. Edner: Rotational CARS generation through a multiple four-color interaction, Appl. Opt. **25**, 4493 (1986)

4.126 G. Laufer, R.B. Miles: Angularly resolved coherent Raman spectroscopy (ARCS), Opt. Commun. **28**, 250 (1979)

4.127 M. Schmitt, G. Knopp, A. Materny, W. Kiefer: Femtosecond time-resolved four-wave mixing spectroscopy in iodine vapour, Chem. Phys. Lett. **280**, 339 (1997)

4.128 T. Chen, V. Engel, M. Heid, W. Kiefer, G. Knopp, A. Materny, S. Meyer, R. Pausch, M. Schmitt, H. Schwoerer, T. Siebert: Femtosecond pump-probe abd four-wave mixing spectroscopies applied to simple molecules, Vibr. Spectrosc. **19**, 23 (1999)

4.129 D.R. Meacher, A. Charlton, P. Ewart, J. Cooper, G. Alber: Phys. Rev. A, Degenerate four-wave mixing with broad-bandwidth pulsed lasers **42**, 3018 (1990)

4.130 P. Ewart, P. Snowdon: Multiplex degenerate four-wave mixing in a flame, Opt. Lett. **15**, 1403 (1990)

4.131 P. Ewart, P.G.R. Smith, R.B. Williams: Appl. Opt., Imaging of trace species distributions by degenerate four-wave mixing: diffraction effects, spatial resolution, and image referencing **36**, 5959 (1997)

4.132 D.J. Rakestraw, R.L. Farrow, T. Dreier: Opt. Lett., Two-dimensional imaging of OH in flames by degenerate four-wave mixing **15**, 709 (1990)

4.133 J.-L. Oudar, R.W. Smith, Y.R. Shen: Polarization-sensitive coherent anti-Stokes Raman spectroscopy, Appl. Phys. Lett. **34**, 758 (1979)

4.134 N.I. Koroteev: Interference phenomena in coherent active spectroscopy of light scattering and absorption: holographic multidimensional spectroscopy, Sov. Phys. Uspekhi **30**, 628 (1987)

4.135 N.I. Koroteev, M. Endemann, R.L. Byer: Resolved structure within the broad-band vibrational Raman line of liquid H_2O from polarization coherent anti-Stokes Raman spectroscopy, Phys. Rev. Lett. **43**, 398 (1979)

4.136 H. Lotem, R.T. Lynch, N. Bloembergen: Phys. Rev. B, Interference between Raman resonances in four-wave difference mixing **14**, 1748 (1976)

4.137 L.S. Aslanyan, A.F. Bunkin, N.I. Koroteev: Coherent-ellipsometry determination of the complex third-order nonlinear polarizabilities of dye molecules, Sov. Tech. Phys. Lett. **4**, 473 (1978)

4.138 M.D. Levenson, N. Bloembergen: Phys. Rev. B, Dispersion of the nonlinear optical susceptibility tensor in centrosymmetric media **10**, 4447 (1974)

4.139 M.D. Levenson, N. Bloembergen: J. Chem. Phys, Dispersion of the nonlinear optical susceptibilities of organic liquids and solutions **60**, 1323 (1974)

4.140 A.F. Bunkin, S.G. Ivanov, N.I. Koroteev: Coherent polarization spectroscopy of Raman scattering of light, Sov. Phys. Dokl. **22**, 146 (1977)

4.141 A.F. Bunkin, S.G. Ivanov, N.I. Koroteev: Observation of resonant interference of nonlinear optical susceptibilities of molecules in a solution, JETP Lett. **24**, 429 (1976)

4.142 R.R. Alfano, S.L. Shapiro: Optical phonon lifetime measured directly with picosecond pulses, Phys. Rev. Lett. **26**, 1247 (1971)

4.143 D. von der Linde, A. Laubereau, W. Kaiser: Molecular vibrations in liquids: direct measurement of the molecular dephasing time; determination of the shape of picosecond light pulses, Phys. Rev. Lett. **26**, 854 (1971)

4.144 R.B. Miles, G. Laufer, G.C. Bjorklund: Coherent anti-Stokes Raman scattering in a hollow dielectric waveguide, Appl. Phys. Lett. **30**, 417 (1977)

4.145 G.I. Stegeman, R. Fortenberry, C. Karaguleff, R. Moshrefzadeh, W.M. II.I. Hetherington, N.E. Van Wyck, J.E. Sipe: Coherent anti-Stokes Raman scattering in thin-film dielectric waveguides, Opt. Lett. **8**, 295 (1983)

4.146 W.P. de Boeij, J.S. Kanger, G.W. Lucassen, C. Otto, J. Greve: Waveguide CARS spectroscopy: A new method for background suppression, using dielectric layers as a model, Appl. Spectrosc. **47**, 723 (1993)

4.147 J.S. Kanger, C. Otto, J. Greve: Waveguide CARS: A method to determine the third-order polarizability to thin layers applied to a sioxny waveguide, Appl. Spectrosc **49**, 1326 (1995)

4.148 A.B. Fedotov, F. Giammanco, A.N. Naumov, P. Marsili, A. Ruffini, D.A. Sidorov-Biryukov, A.M. Zheltikov: Four-wave mixing of picosecond pulses in hollow fibers: expanding the possibilities of gas-phase analysis, Appl. Phys. B **72**, 575 (2001)

4.149 S.O. Konorov, D.A. Akimov, A.N. Naumov, A.B. Fedotov, R.B. Miles, J.W. Haus, A.M. Zheltikov: Bragg resonance-enhanced coherent anti-Stokes Raman scattering in a planar photonic band-gap waveguide, J. Raman Spectrosc. **33**, 955 (2002)

4.150 E.A.J. Marcatili, R.A. Schmeltzer: Hollow metallic and dielectric waveguides for long distance optical transmission and lasers, Bell Syst. Tech. J **43**, 1783 (1964)

4.151 M.J. Adams: *An Introduction to Optical Waveguides* (Wiley, New York 1981)

4.152 R.F. Cregan, B.J. Mangan, J.C. Knight, T.A. Birks, P.St.J. Russell, D. Allen, P.J. Roberts: Single-mode photonic bandgap guidance of light in air, **285**, 1537–1539 (1999)

4.153 S.O. Konorov, A.B. Fedotov, O.A. Kolevatova, V.I. Beloglazov, N.B. Skibina, A.V. Shcherbakov, A.M. Zheltikov: Guided modes of hollow photonic-crystal fibers, JETP Lett. **76**, 341 (2002)

4.154 F. Benabid, J.C. Knight, G. Antonopoulos, P. St. J. Russell: Stimulated Raman scattering in hydrogen-filled hollow-core photonic crystal fiber, Science **298**, 399–402 (2000)

4.155 O.A. Kolevatova, A. N. Naumov, A.M. Zheltikov: Guiding high-intensity laser pulses through hollow fibers: self-phase modulation and cross-talk of guided modes, Opt. Commun., 217 **217**, 169 (2003)

4.156 C.M. Smith, N. Venkataraman, M.T. Gallagher, D. Muller, J.A. West, N.F. Borrelli, D.C. Allan, K. Koch: Low-loss hollow-core silica/air photonic band-gap fibre, Nature **424**, 657 (2003)

4.157 A.M. Zheltikov: Isolated waveguide modes of high-intensity light fields, Physics-Uspekhi **47**(12), 1205–1220 (2004)

4.158 F. Benabid, F. Couny, J.C. Knight, T.A. Birks, P.St.J. Russell: Compact, stable and efficient all-fibre gas cells using hollow-core photonic crystal fibres, Nature **434**(7032), 488–491 (2005)

4.159 A.M. Zheltikov: The friendly gas phase, Nature Mater. **4**(4), 267–268 (2005)

4.160 S.O. Konorov, A.B. Fedotov, A.M. Zheltikov: Enhanced four-wave mixing in a hollow-core photonic-crystal fiber, Opt. Lett. **28**, 1448–1450 (2003)

4.161 S.O. Konorov, D.A. Sidorov-Biryukov, I. Bugar, D. Chorvat Jr., D. Chorvat, E.E. Serebryannikov, M.J. Bloemer, M. Scalora, R.B. Miles, A.M. Zheltikov: Limiting of microjoule femtosecond pulses in air-guided modes of a hollow photonic-crystal fiber, Phys. Rev. A **70**, 023807 (2004)

4.162 D.G. Ouzounov, F.R. Ahmad, D. Muller, N. Venkataraman, M.T. Gallagher, M.G. Thomas, J. Silcox, K.W. Koch, A.L. Gaeta: Generation of megawatt optical solitons in hollow-core photonic band-gap fibers, Science **301**, 1702–1704 (2003)

4.163 F. Luan, J.C. Knight, P.S.J. Russell, S. Campbell, D. Xiao, D.T. Reid, B.J. Mangan, D.P. Williams, P.J. Roberts: Femtosecond soliton pulse delivery at 800nm wavelength in hollow-core photonic bandgap fibers, Opt. Express **12**, 835–840 (2004)

4.164 S.O. Konorov A.B. Fedotov, O.A. Kolevatova, V.I. Beloglazov, N.B. Skibina, A.V. Shcherbakov, E. Wintner, A.M. Zheltikov: Laser breakdown with millijoule trains of picosecond pulses transmitted through a hollow-core photonic-crystal fibre, **36**, 1375–1381 (2003)

4.165 S.O. Konorov, A.B. Fedotov, V.P. Mitrokhin, V.I. Beloglazov, N.B. Skibina, A.V. Shcherbakov, E. Wintner, M. Scalora, A.M. Zheltikov: Laser ablation of dental tissues with picosecond pulses of 1.06 μm radiation transmitted through a hollow-core photonic-crystal fiber, Appl. Opt. **43**, 2251–2256 (2004)

4.166 S.O. Konorov E.E. Serebryannikov, A.B. Fedotov, R.B. Miles, A.M. Zheltikov: Phase-matched waveguide four-wave mixing scaled to higher peak powers with large-core-area hollow photonic-crystal fibers, Phys. Rev. E **71**, 057603 (2005)

4.167 S.G. Johnson, M. Ibanescu, M. Skorobogatiy, O. Weisberg, T.D. Engeness, M. Soljacic, S.A. Jacobs, J.D. Joannopoulos, Y. Fink: Low-loss asymptotically single-mode propagation in large-core OmniGuide fibers, Opt. Express **9**, 748–779 (2001)

4.168 S.O. Konorov, A.B. Fedotov, E.E. Serebryannikov, V.P. Mitrokhin, D.A. Sidorov-Biryukov, A.M. Zheltikov: Phase-matched coherent anti-Stokes Raman scattering in isolated air-guided modes of hollow photonic-crystal fibers, J. Raman Spectrosc. **36**, 129 (2005)

4.169 L. Poladian, N.A. Issa, T.M. Monro: Fourier decomposition algorithm for leaky modes of fibres with arbitrary geometry, Opt. Express **10**, 449–454 (2002)

4.170 J.T. Motz, M. Hunter, L.H. Galindo, J.A. Gardecki, J.R. Kramer, R.R. Dasari, M.S. Feld: Optical fiber probe for biomedical Raman spectroscopy, Appl. Opt. **43**, 542 (2004)

4.171 S.A. Akhmanov, V.G. Dmitriev, A.I. Kovrigin, N.I. Koroteev, V.G. Tunkin, A.I. Kholodnykh: Active spectroscopy of coherent anti-Stokes Raman scattering using an optical parametric oscillator, JETP Lett. **15**, 425 (1972)

4.172 M.D. Levenson, C. Flytzanis, N. Bloembergen: Interference of resonant and nonresonant three-wave mixing in diamond, Phys. Rev. B **6**, 3962 (1972)

4.173 R. Leonhardt, W. Holzapfel, W. Zinth, W. Kaiser: Terahertz quantum beats in molecular liquids, Chem. Phys. Lett. **133**, 373 (1987)

4.174 M. Motzkus, S. Pedersen, A.H. Zewail: Femtosecond real-time probing of reactions: Nonlinear (DFWM) techniques for probing transition states of uni- and bi-molecular reactions, J. Phys. Chem. **100**, 5620 (1996)

4.175 D. Brüggemann, J. Hertzberg, B. Wies, Y. Waschke, R. Noll, K.-F. Knoche, G. Herziger: Test of an optical parametric oscillator (OPO) as a compact and fast tunable stokes source in coherent anti-stokes

Raman spectroscopy (CARS), Appl. Phys. B **55**, 378 (1992)

4.176 E.T.J. Nibbering, D.A. Wiersma, K. Duppen: Ultrafast nonlinear spectroscopy with chirped optical pulses, Phys. Rev. Lett. **68**, 514 (1992)

4.177 T. Lang, M. Motzkus: Single-shot femtosecond coherent anti-Stokes Raman-scattering thermometry, J. Opt. Soc. Am. B **19**, 340 (2002)

4.178 A.M. Zheltikov, A.N. Naumov: High-resolution four-photon spectroscopy with chirped pulses, Quantum Electron. **30**, 606 (2000)

4.179 A.N. Naumov, A.M. Zheltikov: Frequency-time and time-space mappings for single-shot coherent four-wave mixing with chirped pulses and broad beams, J. Raman Spectrosc. **32**, 960–970 (2000)

4.180 A.N. Naumov, A.M. Zheltikov: Frequency-time and time-space mappings with broadband and supercontinuum chirped pulses in coherent wave mixing and pump-probe techniques, Appl. Phys. B **77**, 369–376 (2003)

4.181 A.N. Naumov, A.M. Zheltikov, A.P. Tarasevitch, D. von der Linde: Enhanced spectral broadening of short laser pulses in high-numerical-aperture holey fibers, Appl. Phys. B **73**, 181 (2001)

4.182 S.O. Konorov, A.M. Zheltikov: Frequency conversion of subnanojoule femtosecond laser pulses in a microstructure fiber for photochromism initiation, Opt. Express **11**, 2440–2445 (2003)

4.183 S.O. Konorov, D.A. Akimov, A.A. Ivanov, M.V. Alfimov, A.M. Zheltikov: Microstructure fibers as frequency-tunable sources of ultrashort chirped pulses for coherent nonlinear spectroscopy, **78**, 565–567 (2004)

4.184 H. Kano, H. Hamaguchi: Characterization of a supercontinuum generated from a photonic crystal fiber and its application to coherent Raman spectroscopy, Opt. Lett. **28**, 2360–2362 (2003)

4.185 S.O. Konorov, D.A. Akimov, A.A. Ivanov, M.V. Alfimov, A.V. Yakimanskii, A.M. Zheltikov: Probing resonant nonlinearities in organic materials using photonic-crystal fiber frequency converters, Chem. Phys. Lett. **405**, 310–313 (2005)

4.186 H.N. Paulsen, K.M. Hilligsøe, J. Thøgersen, S.R. Keiding, J.J. Larsen: Coherent anti-Stokes Raman scattering microscopy with a photonic crystal fiber based light source, Opt. Lett. **28**, 1123–1125 (2003)

4.187 A.M. Zheltikov: Limiting temporal and spectral resolution in spectroscopy and microscopy of coherent Raman scattering with chirped ultrashort laser pulses, JETP **100**, 833–843 (2005)

4.188 S. Konorov, A. Ivanov, D. Ivanov, M. Alfimov, A. Zheltikov: Ultrafast photonic-crystal fiber light flash for streak-camera fluorescence measurements, Opt. Express **13**, 5682–5688 (2005)

4.189 A.A. Ivanov, M.V. Alfimov, A.M. Zheltikov: Femtosecond pulses in nanophotonics, Phys. Uspekhi **47**, 687 (2004)

4.190 S. Linden, J. Kuhl, H. Giessen: Amplitude and phase characterization of weak blue ultrashort pulses by downconversion, **24**, 569–571 (1999)

4.191 X. Gu, L. Xu, M. Kimmel, E. Zeek, P. O'Shea, A.P. Shreenath, R. Trebino, R.S. Windeler: Frequencyresolved optical gating and single-shot spectral measurements reveal fine structure in microstructure fiber continuum, Opt. Lett. **27**, 1174–1176 (2002)

4.192 R. Trebino: *Frequency-Resolved Optical Gating: The Measurement of Ultrashort Laser Pulses* (Kluwer Academic, Boston 2002)

4.193 S.A. Kovalenko, A.L. Dobryakov, J. Ruthmann, N.P. Ernsting: Femtosecond spectroscopy of condensed phases with chirped supercontinuum probing, Phys. Rev. A **59**, 2369–2384 (1999)

4.194 T. Kobayashi: *J-Aggregates*, ed. by T. Kobayashi (World Scientific, Singapore 1996)

4.195 F.C. Spano, S. Mukamel: Nonlinear susceptibilities of molecular aggregates: enhancement of $\chi^{(3)}$ by size, Phys. Rev. A **40**, 5783–5801 (1989)

4.196 J. Knoester: Nonlinear-optical susceptibilities of disordered aggregates: a comparison of schemes to account for intermolecular interactions, Phys. Rev. A 47 **47**, 2083–2098 (1993)

4.197 O. Kuhn, V. Sundstrom: Pump-probe spectroscopy of dissipative energy transfer dynamics in photosynthetic antenna complexes: A density matrix approach, J. Chem. Phys. **104**, 4154–4164 (1997)

4.198 M. Furuki, M. Tian, Y. Sato, L.S. Pu, S. Tatsuura, O. Wada: Terahertz demultiplexing by a single-shot time-to-space conversion using a film of squarylium dye *J* aggregates, Appl. Phys. Lett. **77**, 472–474 (2000)

4.199 T. Tani: *J-Aggregates*, ed. by T. Kobayashi (World Scientific, Singapore 1996)

4.200 A.A. Ivanov, D.A. Akimov, P.V. Mezentsev, A.I. Plekhanov, M.V. Alfimov, A.M. Zheltikov: Pump-probe nonlinear absorption spectroscopy of molecular aggregates using chirped frequency-shifted light pulses from a photonic-crystal fiber, Laser Phys. **16**, 6 (2006)

4.201 L.D. Bakalis, J. Knoester: Linear absorption as a tool to measure the exciton delocalization length in molecular assemblies, J. Luminescence **87-89**, 66–70 (2000)

4.202 L.D. Bakalis, J. Knoester: Can the exciton delocalization length in molecular aggregates be determined by pump-probe spectroscopy?, J. Luminescence **83-84**, 115–119 (1999)

4.203 G.I. Stegeman, H.-E. Ponath: *Nonlinear Surface Electromagnetic Phenomena*, ed. by G.I. Stegeman, H.-E. Ponath (North-Holland, Amsterdam 1991)

4.204 J.F. McGilp: A review of optical second-harmonic and sum-frequency generation at surfaces and interfaces, J. Phys. D: Appl. Phys. **29**, 1812–1821 (1996)

4.205 Y.R. Shen: Surface studies by optical second harmonic generation: An overview, J. Vac. Sci. Technol. B **3**, 1464 (1985)

4.206 T. Götz, F. Träger, M. Buck, C. Dressler, F. Eisert: Optical second harmonic generation of supported metal clusters: size and shape effects, Appl. Phys. **607**, 60 (1995)

4.207 F.L. Labarthet, Y.R. Shen: *Nonlinear Optical Microscopyin Optical Imaging and Spectroscopy: Techniques and Advanced Systems*, ed. by P. Török, Fu-Jen Kao (Springer, Series in Optical Sciences, Berlin 2003) pp. 169–196

4.208 A.M. Zheltikov: Nanoscale nonlinear optics in photonic-crystal fibres, J. Opt. A: Pure Appl. Opt. **8**, 1–26 (2006)

4.209 M. Müller, G.J. Brakenhoff: *Parametric Nonlinear Optical Techniques in Microscopyin Optical Imaging and Spectroscopy: Techniques and Advanced Systems*, ed. by P. Török, Fu-Jen Kao (Springer, Series in Optical Sciences, Berlin 2003) pp. 197–218

4.210 A.N. Naumov, D.A. Sidorov-Biryukov, A.B. Fedotov, A.M. Zheltikov: Third-harmonic generation in focused beams as a method of three-dimensional microscopy of laser-produced plasma, Opt. Spectrosc. **90**, 863 (2001)

4.211 R. Hellwarth, P. Christensen: Rapid communications: Nonlinear optical microscope using second harmonic generation, Appl. Opt. **14**, 247 (1975)

4.212 G.T. Boyd, Y.R. Shen, T.W. Hänsch: Continuous-wave second-harmonic generation as a surface microprobe, Opt. Lett. **11**, 97–99 (1986)

4.213 M.D. Duncan, J. Reintjes, T.J. Manuccia: Scanning coherent anti-Stokes Raman microscope, Opt. Lett. **7**, 350–352 (1982)

4.214 A. Zumbusch, G.R. Holtom, X.S. Xie: Vibrational microscopy using coherent anti-Stokes Raman scattering, Phys. Rev. Lett. **82**, 4142 (1999)

4.215 P.A. Franken, A.E. Hill, C.W. Peters, G. Weinreich: Generation of optical harmonics, Phys. Rev. Lett. **7**, 118 (1961)

4.216 G.H.C. New, J.F. Ward: Optical third-harmonic generation in gases, Phys. Rev. Lett. **19**, 556 (1967)

4.217 J.F. Reintjes: *Nonlinear Optical Parametric Processes in Liquids and Gases in Quantum Electronics, Principles and Applications* (Academic, New York 1984)

4.218 M. Ferray, A. L'Huillier, X.F. Li, L.A. Lompré, G. Mainfray, C. Manus: Multiple-harmonic conversion of 1064 nm radiation in rare gases, J. Phys. B **21**, 31 (1988)

4.219 A. McPherson, G. Gibson, H. Jara, U. Johann, T.S. Luk, I. McIntyre, K. Boyer, C.K. Rhodes: Studies of multiphoton production of vacuum-ultraviolet radiation in the rare gases, J. Opt. Soc. Am. B **4**, 595 (1987)

4.220 S.G. Preston, A. Sanpera, M. Zepf, J.W. Blyth, C.G. Smith, J.S. Wark, M.H. Key, K. Burnett, M. Nakai, D. Neely, A.A. Offenberger: High-order harmonics of 248.6 nm KrF laser from helium and neon ions, Phys. Rev. **53**, 31 (1996)

4.221 J.J. Macklin, J.D. Kmetec, C.L.II.I. Gordon: High-order harmonic generation using intense femtosecond pulses, Phys. Rev. Lett. **70**, 766 (1993)

4.222 A. L'Huillier, Ph. Balcou: High-order harmonic generation in rare gases with an intense short-pulse laser, Phys. Rev. Lett. **70**, 774 (1993)

4.223 C.G. Wahlström, J. Larsson, A. Persson, T. Starczewski, S. Svanberg, P. Salieres: High-order harmonic generation in rare gases with an intense short-pulse laser, Phys. Rev. A. **48**, 4709 (1993)

4.224 J.K. Crane, M.D. Perry: High-field harmonic generation in helium, Opt. Lett. **917**, 1256 (1992)

4.225 Z. Chang, A. Rundquist, H. Wang, M.M. Murnane, H.C. Kapteyn: Generation of coherent soft x rays at 2.7 nm using high harmonics, Phys. Rev. Lett. **79**, 2967 (1997)

4.226 C. Spielmann, N. Burnett, S. Sartania, R. Koppitsch, M. Schnurer, C. Kan, M. Lenzner, P. Wobrauschek, F. Krausz: Generation of coherent X-rays in the water window using 5-femtosecond laser pulses, Science **278**, 661 (1997)

4.227 J.L. Krause, K.J. Schafer, K.C. Kulander: High-order harmonic generation from atoms and ions in the high intensity regime, Phys. Rev. Lett. **68**, 3535 (1992)

4.228 K.J. Schafer, B. Yang, L.F. DiMauro, K.C. Kulander: Above threshold ionization beyond the high harmonic cutoff, Phys. Rev. Lett. **70**, 1599 (1993)

4.229 P.B. Corkum: Plasma perspective on strong-field multiphoton ionization, Phys. Rev. Lett. **71**, 1994 (1993)

4.230 A. L'Huillier, M. Lewenstein, P. Salières, Ph. Balcou, M.Yu. Ivanov, J. Larsson, C.G. Wahlström: High-order harmonic generation cutoff, Phys. Rev. A **48**, R3433 (1993)

4.231 M. Lewenstein, Ph. Balcou, M.Yu. Ivanov, A. L'Huillier, P. Corkum: Theory of high-order harmonic generation by low-frequency laser fields, Phys. Rev. A **49**, 2117 (1994)

4.232 K.S. Budil, P. Salières, A. L'Huillier, T. Ditmire, M.D. Perry: Influence of ellipticity on harmonic generation, Phys. Rev. A **48**, 3437 (1993)

4.233 P. Dietrich, N.H. Burnett, M.Y. Ivanov, P.B. Corkum: High-harmonic generation and correlated two-electron multiphoton ionization with elliptically polarized light, Phys. Rev. A **50**, R3585 (1994)

4.234 Y. Liang, M.V. Ammosov, S.L. Chin: High-order harmonic generation in argon by elliptically polarized picosecond dye laser pulses, J. Phys. B **27**, 1296 (1994)

4.235 N.H. Burnett, C. Kan, P.B. Corkum: Ellipticity and polarization effects in harmonic generation in ionizing neon, Phys. Rev. A **51**, R3418 (1995)

4.236 F.A. Weihe, S.K. Dutta, G. Korn, D. Du, P.H. Bucksbaum, P.L. Shkolnikov: Polarization of high-intensity high-harmonic generation, Phys. Rev. A **51**, R3433 (1995)

4.237 Ph. Antoine, B. Carré, A. L'Huillier, M. Lewenstein: Polarization of high-order harmonics, Phys. Rev. A **55**, 1314 (1997)

4.238 F.A. Weihe, P.H. Bucksbaum: Measurement of the polarization state of high harmonics generated in gases, J. Opt. Soc. Am. B **13**, 157 (1996)

4.239 D. Schulze, M. Dörr, G. Sommerer, J. Ludwig, P.V. Nickles, T. Schlegel, W. Sandner, M. Drescher, U. Kleineberg, U. Heinzmann: Polarization of the 61st harmonic from 1053-nm laser radiation in neon, Phys. Rev. A **57**, 3003–3007 (1998)

4.240 W. Becker, S. Long, J.K. McIver: Higher-harmonic production in a model atom with short-range potential, Phys. Rev. A **41**, 4112 (1990)

4.241 Ph. Antoine A. L'Huillier, M. Lewenstein, P. Salières, B. Carré: Theory of high-order harmonic generation by an elliptically polarized laser field, Phys. Rev. A **53**, 1725 (1996)

4.242 W. Becker, A. Lohr, M. Kleber, M. Lewenstein: A unified theory of high-harmonic generation: Application to polarization properties of the harmonics, Phys. Rev. A **56**, 645 (1997)

4.243 Ph. Balcou, C. Cornaggia, A.S.L. Gomes, L.-A. Lompré, A. L'Huillier: Optimizing high-order harmonic generation in strong fields, J. Phys. B **35**, 4467 (1992)

4.244 R. Bartels, S. Backus, E. Zeek, L. Misoguti, G. Vdovin, I.P. Christov, M.M. Murnane, H.C. Kapteyn: Shaped-pulse optimization of coherent emission of high-harmonic soft X-rays, Nature **406**, 164 (2000)

4.245 T. Brabec, F. Krausz: Intense few-cycle laser fields: frontiers of nonlinear optics, Rev. Mod. Phys. **72**, 545 (2000)

4.246 A. Rundquist, C.II.I. Durfee, Z. Chang, C. Herne, S. Backus, M. Murnane, H.C. Kapteyn: Phase matched generation of coherent soft X-rays, Science **280**, 1412 (1998)

4.247 C.II.I. Durfee andA. Rundquist, S. Backus, C. Herne, M. Murnane, H.C. Kapteyn: Phase matching of high-order harmonics in hollow waveguides, Phys. Rev. Lett. **83**, 2187 (1999)

4.248 E. Constant D. Garzella, P. Breger, E. M'evel, Ch. Dorrer, C. Le Blanc, F. Salin, P. Agostini: Optimizing high harmonic generation in absorbing gases: model and experiment, Phys. Rev. Lett. **82**, 1668 (1999)

4.249 Y. Tamaki, Y. Nagata, Y. Nagata, M. Obara, K. Midorikawa: Phase-matched high-order harmonic generation in a gas-filled hollow fibre, Phys. Rev. A **59**, 4041 (1999)

4.250 Y. Tamaki, J. Itatini, Y. Nagata, M. Obara, K. Midorikawa: Phase-matched high-order harmonic generation in a gas-filled hollow fibre, Phys. Rev. Lett. **82**, 1422 (1999)

4.251 J.W.G. Tisch A.R. Smith, J.E. Muffet, M. Ciarocca: Angularly-resolved high-order harmonic generation in helium, Phys. Rev. A. **49**, R28 (1994)

4.252 J. Peatross, D.D. Meyerhofer: Angular distribution of high-order harmonics emitted from rare gases at low density, Phys. Rev. A **51**, R906 (1995)

4.253 P. Salières, T. Ditmire, M.D. Perry, A. L'Huillier, M. Lewenstein: Angular distributions of high-order harmonics generated by a femtosecond laser, J. Phys. B **29**, 4771 (1996)

4.254 M. Nisoli, E. Priori, G. Sansone, S. Stagira, G. Cerullo, S. De Silvestri, C. Altucci, R. Bruzzese, C. de Lisio, P. Villoresi, L. Poletto, M. Pascolini, G. Tondello: High-brightness high-order harmonic generation by truncated bessel beams in the sub-10-fs regime, Phys. Rev. Lett. **88**, 033902 (2002)

4.255 T. Ditmire, E.T. Grumbell, R.A. Smith, J.W.G. Tisch, D.D. Meyerhofer, M.H.R. Hutchinson: Spatial coherence measurement of soft X-ray radiation produced by high order harmonic generation, Phys. Rev. Lett **77**, 4756 (1996)

4.256 L. Le Déroff, P. Salières, B. Carré, D. Joyeux, D. Phalippou: Measurement of the degree of spatial coherence of high-order harmonics using a Fresnel-mirror interferometer, Phys. Rev. A **61**, 043802 (2000)

4.257 T.E. Glover, R.W. Schoenlein, A.H. Chin, C.V. Shank: Observation of laser assisted photoelectric effect and femtosecond high order harmonic radiation, Phys. Rev. Lett. **76**, 2468 (1996)

4.258 J.M. Schins, P. Breger, P. Agostini, R.C. Constantinescu, H.G. Muller, A. Bouhal, G. Grillon, A. Antonetti, A. Mysyrowicz: Cross-Correlation Measurements of Femtosecond Extreme-Ultraviolet High-Order Harmonics, J. Opt. Soc. Am. B. **13**, 197 (1996)

4.259 A. Bouhal, R. Evans, G. Grillon, A. Mysyrowicz, P. Breger, P. Agostini, R.C. Constantinescu, H.G. Muller, D. von-der-Linde: Cross-correlation measurement of femtosecond noncollinear high-order harmonics, J. Opt. Soc. Am. B. **14**, 950 (1997)

4.260 E.S. Toma, H.G. Muller, P.M. Paul, P. Breger, M. Cheret, P. Agostini, C. Le Blanc, G. Mullot, G. Cheriaux: Ponderomotive streaking of the ionization potential as a method for measuring pulse durations in the XUV domain with fs resolution, Phys. Rev. A **62**, R061801 (2000)

4.261 M. Drescher, M. Hentschel, R. Kienberger, G. Tempea, Ch. Spielmann, G. Reider, P.B. Corkum, F. Krausz: X-ray pulses approaching the attosecond frontier, Science **291**, 1923 (2001)

4.262 J. Norin, J. Mauritsson, A. Johansson, M.K. Raarup, S. Buil, A. Persson, O. Dühr, M.B. Gaarde, K.J. Schafer, U. Keller, C.G. Wahlström, A. L'Huillier: Time- frequency characterization of femtosecond XUV pulses, Phys. Rev. Lett. **88**, 193901 (2002)

4.263 T. Sekikawa, T. Katsura, S. Miura, S. Watanabe: Measurement of the intensity-dependent atomic dipole phase of a high harmonic by frequency-resolved optical gating, Phys. Rev. Lett. **88**, 193902 (2002)

4.264 J. Larsson, E. Mevel, R. Zerne, A. L'Huillier, C.G. Wahlström, S. Svanberg: Two-colour time-resolved spectroscopy of helium using high-order harmonics, J. Phys. B **28**, L53 (1995)

4.265 M. Gisselbrecht, D. Descamps, C. Lynga, A. L'Huillier, C.G. Wahlström, M. Meyer: Absolute photoionization cross sections of excited He states in the near-threshold region, Phys. Rev. Lett. **82**, 4607 (1999)
4.266 S.L. Sorensen, O. Bjorneholm, I. Hjelte, T. Kihlgren, G. Ohrwall, S. Sundin, S. Svensson, S. Buil, D. Descamps, A. L'Huillier: Femtosecond pump-probe photoelectron spectroscopy of predissociative states in acetylen, J. Chem. Phys. **112**, 8038 (2000)
4.267 M. Bauer, C. Lei, K. Read, R. Tobey, J. Gland, M.M. Murnane, H.C. Kapteyn: Direct Observation of Surface Chemistry Using Ultrafast Soft-X-Ray Pulses, Phys. Rev. Lett. **87**, 025501 (2001)
4.268 L. Nugent-Glandorf, M. Scheer, D.A. Samuels, A.M. Mulhisen, E.R. Grant, X. Yang, V.M. Bierbaum, S.R. Leone: Ultrafast time-resolved soft X-ray photoelectron spectroscopy of dissociating Br_2, Phys. Rev. Lett. **87**, 193002 (2001)
4.269 R. Haight, D.R. Peale: Antibonding state on the Ge(111):As surface: spectroscopy and dynamics, Phys. Rev. Lett. **70**, 3979 (1993)
4.270 F. Quéré, S. Guizard, G. Petite, Ph. Martin, H. Merdji, B. Carré, J.-F. Hergott: Hot-electron relaxation in quartz using high-order harmonics, Phys. Rev. B **61**, 9883 (2000)
4.271 T. Sekikawa, T. Ohno, Y. Nabekawa, S. Watanabe: Auger-free luminescence excited by high-order harmonics of a femtosecond Ti:sapphire laser, J. Lumin. **87**, 827 (2000)
4.272 W. Theobald, R. Hässner, C. Wülker, R. Sauerbrey: Temporally resolved measurement of electron densities ($>10^{23}$ cm^{-3}) with high harmonics, Phys. Rev. Lett. **77**, 298 (1996)
4.273 P. Saliéres, L. Le Déroff, T. Auguste, P. Monot, P. d'Oliveira, D. Campo, J.-F. Hergott, H. Merdji, B. Carré: Frequency domain interferometry in the XUV with high-order harmonics, Phys. Rev. Lett. **83**, 5483 (1999)
4.274 D. Descamps, C. Lynga, J. Norin, A. L'Huillier, C.G. Wahlström, J.F. Hergott, H. Merdji, P. Saliéres, M. Bellini, T.W. Hänsch: Extreme ultraviolet interferometry measurements with high-order harmonics, Opt. Lett. **25**, 135 (2000)
4.275 Y. Kobayashi, T. Sekikawa, Y. Nabekawa, S. Watanabe: 27-fs extreme ultraviolet pulse generation by high-order harmonics, Opt. Lett. **23**, 64 (1998)
4.276 T. Sekikawa, T. Ohna, T. Yamazaki, Y. Nabekawa, S. Watanabe: Pulse compression of a high-order harmonic by compensating the atomic dipole phase, Phys. Rev. Lett. **83**, 2564 (1999)
4.277 D. Descamps, L. Roos, C. Delfin, A. L'Huillier, C.G. Wahlström: Two and three-photon ionization of rare gases using femtosecond harmonic pulses generated in a gas medium, Phys. Rev. A **64**, R031401 (2001)
4.278 G. Farkas, C. Toth: Proposal for attosecond light pulse generation using laser induced multiple-harmonic conversion processes in rare gases, Phys. Lett. A. **168**, 447 (1992)
4.279 S.E. Harris, J.J. Macklin, T.W. Hänsch: Atomic scale temporal structure inherent to high-order harmonic generation, Opt. Commun. **100**, 487 (1993)
4.280 P. B. Corkum, N.H. Burnett, M.Y. Ivanov: Subfemtosecond pulses, Opt. Lett. **19**, 1870 (1994)
4.281 M. Ivanov, P.B. Corkum: Routes to control of intense-field atomic polarizability, Phys. Rev. Lett. **74**, 2933–2936 (1995)
4.282 P. Antoine: Attosecond pulse trains using high-order harmonics, Phys. Rev. Lett. **77**, 1234 (1996)
4.283 N.A. Papadogiannis, B. Witzel, C. Kalpouzos, D. Charalambidis: Observation of attosecond light localization in higher order harmonic generation, Phys. Rev. Lett. **83**, 4289 (1999)
4.284 P. M. Paul, E.S. Toma, P. Breger, G. Mullot, F. Augé, Ph. Balcou: Observation of a train of attosecond pulses from high harmonic generation, Science **292**, 1689–1692 (2001)
4.285 M. Hentschel, R. Kienberger, Ch. Spielmann, G.A. Reider, N. Milosevic, T. Brabec, P.B. Corkum, U. Heinzmann: Attosecond metrology, Nature **414**, 509–513 (2001)
4.286 R. Kienberger M. Hentschel, M. Uiberacker, Ch. Spielmann, M. Kitzler, A. Scrinzi, M. Wieland, Th. Westerwalbesloh, U. Kleineberg, U. Heinzmann: Steering attosecond electron wave packets with light, Science **297**, 1144 (2002)
4.287 P. Salières, A. L'Huillier, Ph. Antoine, M. Lewenstein: Study of the spatial and temporal coherence of high-order harmonics, Adv. Atom. Mol. Opt. Phys **41**, 83 (1999)
4.288 P. Salières, M. Lewenstein: Generation of ultrashort coherent XUV pulses by harmonic conversion of intense laser pulses in gases: towards attosecond pulses, Measur. Sci. Tech. **12**(11), 1818 (2001)
4.289 N. Sarukura, K. Hata, T. Adachi, R. Nodomi, M. Watanabe: Coherent soft X-ray generation by the harmonics of an ultrahigh-power KrF laser, Phys. Rev. A. **43**, 1669 (1991)
4.290 K. Myazaki: High-order harmonic generation in the tunneling regime, Phys. Rev. A. **52**, 3007 (1995)
4.291 M.D. Perry: High-order harmonic emission from mixed fields, Phys. Rev. A **48**, R4051 (1993)
4.292 S. Watanabe, K. Kondo, Y. Nabekawa, A. Sagisaka, Y. Kobayashi: Two-color phase control in tunneling ionization and harmonic generation by a strong laser field and its third harmonic, Phys. Rev. Lett. **73**, 2692 (1994)
4.293 H. Eichmann, A. Egbert, S. Nolte, C. Momma, B. Wellegehausen, W. Becker, S. Long, J.K. McIver: Polarization-dependent high-order two-color mixing, Phys. Rev. A **51**, R3414 (1995)
4.294 H. Eichmann, S. Meyer, K. Riepl, C. Momma, B. Wellegehausen: Generation of short-pulse tunable XUV radiation by high-order frequency mixing, Phys. Rev. A **50**, R2834 (1994)

4.295 M. B. Gaarde, P. Antoine, A. Persson, B. Carré, A. L'Huillier, C. G. Wahlström: High-order tunable sum- and difference frequency mixing in the XUV region, J. Phys. B **29**, L163 (1996)

4.296 B. Sheehy, J. D. D. Martin, L. F. DiMauro, P. Agostini, K. J. Schafer, M. B. Gaarde, K. C. Kulander: High harmonic generation at long wavelengths, Phys. Rev. Lett. **83**, 5270 (1999)

4.297 C. Altucci, Ch. Delfin, L. Roos, M. B. Gaarde, A. L'Huillier, I. Mercer, T. Starczewski, C. G. Wahlström: Frequency time-resolved high-order harmonics, Phys. Rev. A. **58**, 3934 (1998)

4.298 M. Kovacev, Y. Mariesse, E. Priori, H. Merdji, O. Tcherbakoff, P. Monchieourt, P. Breger, E. Mevel, E. Constant, P. Salieres, B. Carre, P. Agostini: Temporal confinement of harmonic emission by polarization gating, European Journal of Physics D **26**, 79 (2003)

4.299 I. Mercer, E. Mével, R. Zerne, A. L'Huillier, Ph. Antoine, C. G. Wahlström: Spatial mode control of high-order harmonics, Phys. Rev. Lett. **77**, 1731 (1996)

4.300 C. Altucci, R. Bruzzese, D. D. Amtuoni, C. de Lisio, S. Solimeno: Harmonic generation in gases by use of Bessel- Gauss laser beams, J. Opt. Soc. Am. **17**, 34 (2000)

4.301 L. Roos, E. Constant, E. Mevel, Ph. Balcou, D. Descamps, M. B. Gaarde, A. Valette, R. Haroutounian: Controlling phase-matching of high-order harmonic generation by manipulating the fundamental field, Phys. Rev. A **60**, 5010 (1999)

4.302 M. Schnürer, Z. Cheng, M. Hentschel, G. Tempea, P. Kálmán, T. Brabec, F. Krausz: Absorption-limited generation of coherent ultrashort soft-x-ray pulses, Phys. Rev. Lett. **83**, 722 (1999)

4.303 S. Kazamias, D. Douillet, F. Weihe, C. Valentin, A. Rousse, S. Sebban, G. Grillon, F. Augé, D. Hulin, Ph. Balcou: Global optimization of high harmonic generation, Phys. Rev. Lett. **90**, 193901 (2003)

4.304 C. G. Wahlström, S. Borgström, J. Larsson, S. G. Pettersson: High-order harmonic generation in laser-produced ions using a near-infrared laser, Phys. Rev. A **51**, 585 (1995)

4.305 M. Y. Ivanov, P. B. Corkum: Generation of high-order harmonics from inertially confined molecular ions, Phys. Rev. A. **48**, 580 (1993)

4.306 Y. Liang, S. August, S. L. Chin: High harmonic generation in atomic and diatomic molecular gases using intense picosecond laser pulses-a comparison, Phys. Rev. B. **27**, 5119 (1994)

4.307 C. Lyngå, A. L'Huillier, C. G. Wahlström: High-harmonic generation in molecular gases, J. Phys. B **29**, 3293 (1996)

4.308 N. Hay, R. de Nalda, T. Halfmann, K. J. Mendham, M. B. Mason, M. Castillejo, J. P. Marangos: High-order harmonic generation from organic molecules in ultra-short pulses, Eur. Phys. J. D **14**, 231 (2001)

4.309 T. D. Donnelly, T. Ditmire, K. Neumann, M. D. Perry, R. W. Falcone: High-Order Harmonic Generation in Atom Clusters, Phys. Rev. Lett. **76**, 2472 (1996)

4.310 J. W. G. Tisch, T. Ditmire, D. J. Fraser, N. Hay, M. B. Mason, E. Springate, J. P. Marangos, M. H. R. Hutchinson: Investigation of high harmonic generation from xenon atom clusters, J. Phys. B **30**, L709 (1997)

4.311 S. X. Hu, Z. Z. Xu: Enhanced harmonic emission from ionized clusters in intense laser pulses, Appl. Phys. Lett. **71**, 2605 (1997)

4.312 L. Roos: *Optimisation and Application of Intense High-Order Harmonic Pulses* (Lund University, Lund 2001) PhD thesis, ISBN 91-631.1370-8.

4.313 K. C. Kulander, B. W. Shore: Calculations of multiple-harmonic conversion of 1064-nm radiation in Xe, Phys. Rev. Lett. **62**, 524 (1989)

4.314 K. J. Schafer, K. C. Kulander: High harmonic generation from ultrafast pump lasers, Phys. Rev. Lett. **78**, 638 (1997)

4.315 I. P. Christov, M. M. Murnane, H. C. Kapteyn: High-harmonic generation of attosecond pulses in the single cycle regime, Phys. Rev. Lett. **78**, 1251 (1997)

4.316 L. V. Keldysh: Ionization in the field of a strong electromagnetic wave, Sov. Phys. JETP **20**, 1307 (1965)

4.317 Ph. Balcou, A. S. Dederichs, M. B. Gaarde, A. L'Huillier: Quantum-path analysis and phase-matching of high-order harmonic generation and high-order frequency mixing processes in strong laser fields, J. Phys. B **32**, 2973 (1999)

4.318 M. B. Gaarde, K. J. Schafer: Quantum path distributions for high-order harmonics in rare gas atoms, Phys. Rev. A **65**, R031406 (2002)

4.319 M. Bellini, A. Tozzi, M. B. Gaarde, C. Delfin, T. W. Hänsch, A. L'Huillier, C. G. Wahlström: Temporal coherence of ultrashort high-order harmonic pulses, Phys. Rev. Lett. **81**, 297 (1998)

4.320 C. Lyngå, M. B. Gaarde, C. Delfin, M. Bellini, T. W. Hänsch, A. L'Huillier, C. G. Wahlström: Studies of the temporal coherence of high-order harmonics, Phys. Rev. A **60**, 4823 (1999)

4.321 P. Saliéres, B. Carré, L. Le Déroff, F. Grasbon, G. G. Paulus, H. Walther, R. Kopold, W. Becker, A. Sanpera, M. Lewenstein: Feynman's path-integral approach for intense-laser-atom interactions, Science **292**, 902 (2001)

4.322 M. Gaarde, F. Salin, E. Constant, Ph. Balcou, K. J. Schafer, K. C. Kulander, A. L'Huillier: Spatiotemporal separation of high harmonic radiation into two quantum path components, Phys. Rev. A **59**, 1367 (1999)

4.323 J.-F. Hergott, M. Kovacev, H. Merdji, C. Hubert, Y. Mairesse, E. Jean, P. Breger, P. Agostini, B. Carré, P. Salières: Extreme-ultraviolet high-order harmonic pulses in the microjoule range, Phys. Rev. A **66**, 021801 (2002)

4.324 A. Paul, R. A. Bartels, R. Tobey, H. Green, S. Weiman, I. P. Chrsitov, M. M. Murnane, H. C. Kapteyn,

S. Backus: Quasi-phase-matched generation of coherent extreme-ultraviolet light, Nature **421**, 51 (2003)

4.325 Ph. Antoine, D.B. Milosevic, A. L'Huillier, M.B. Gaarde, P. Salières, M. Lewenstein: Generation of attosecond pulses in macroscopic media, Phys. Rev. A **56**, 4960 (1997)

4.326 M.B. Gaarde, K.J. Schafer: Space-time considerations for the phase locking of high-order harmonics, Phys. Rev. Lett. **89**, 213901 (2002)

4.327 V.T. Platonenko, V.V. Strelkov: Single attosecond soft-x-ray pulse generated with a limited laser beam, J. Opt. Soc. Am. B. **16**, 435 (1999)

4.328 O. Tcherbakoff, E. Mével, D. Descamps, J. Plumridge, F. Salin, E. Constant: Time gated high order harmonic generation, Phys. Rev. A **68**, 043804 (2003)

4.329 R. López-Martens, J. Mauritsson, P. Johnsson, A. L'Huillier, O. Tcherbakoff, A. Zaïr, E. Mevel, E. Constant: Time-resolved temporal confinement of high harmonic emission, Phys. Rev. A **69**, 053811 (2004)

4.330 A. Baltuska, Th. Udem, M. Uiberacker, M. Hentschel, E. Goulielmakis, Ch. Gohle, R. Holzwarth, V.S. Yakovlev, A. Scrinzi, F. Krausz: Attosecond control of electronic processes by intense light field, Nature **421**, 611 (2003)

4.331 N. Milosevic, A. Scrinzi: Ab initio numerical calculation of attosecond pulse generation, Phys. Rev. Lett. **88**, 093905 (1996)

4.332 D. Xenakis, O. Faucher, D. Charalambidis, C. Fotakis: Observation of two-XUV-photon ionization using harmonic generation from a short, intense laser pulse, J. Phys. B **29**, L457 (1996)

4.333 N.A. Papadogiannis, L.A.A. Nikolopoulos, D. Charalambidis, P. Tzallas, G. Tsakiris, K. Witte: Two XUV-photon ionization of He through a superposition of higher harmonics, Phys. Rev. Lett. **90**, 133902 (2003)

4.334 Y. Kobayashi, T. Ohno, T. Sekikawa, Y. Nabekawa, S. Watanabe: Pulse width measurement of high-order harmonics by autocorrelation, Appl. Phys. B. **70**, 389–394 (2000)

4.335 T. Sekikawa, T. Katsura, S. Miura, S. Watanabe: Meas-urement of the intensity-dependent atomic dipole phase of a high harmonic by frequency-resolved optical gating, Phy. Rev. Lett. **88**, 193902 (2002)

4.336 P. Tzallas, D. Charalambidis, N.A. Papadogiannis, K. Witte, G.D. Tsakiris: Direct observation of attosecond light bunching, Nature **426**, 267 (2003)

4.337 J.M. Schins, P. Breger, P. Agostini, R.C. Constantinescu, H.G. Muller, G. Grillon, A. Antonetti, A. Mysyrowicz: Observation of laser-assisted auger decayin Argon, Phys. Rev. Lett. **73**, 2180 (1994)

4.338 Y. Kobayashi, O. Yoshihara, Y. Nabekawa, K. Kondo, S. Watanabe: Femtosecond measurement of high-order harmonic pulse width and electron recombination time by field ionization, Opt. Lett. **21**, 417–419 (1996)

4.339 E. Constant, V.D. Taranukhin, A. Stolow, P.B. Corkum: Methods for the measurement of the duration of high-harmonic pulses, Phys. Rev. A **56**, 3870 (1997)

4.340 J. Itatani, F. Quéré, G.L. Yudin, M. Yu. Ivanov: Attosecond streak camera, Phys. Rev. Lett. **88**, 173903 (2002)

4.341 M. Kitzler, N. Milosevic: Theory of attosecond streak camera, Phys. Rev. Lett. **88**, 173904 (2002)

4.342 U. Becker, D.A. Shirley: *VUV and Soft X-Ray Photoionization* (Plenum, New York 1996)

4.343 M. Drescher, M. Hentschel, R. Kienberger, M. Uiberacker, V. Yakovlev, A. Scrinzi, Th. Westerwalbesloh, U. Kleineberg, U. Heinzmann, F. Krausz: Time-resolved atomic inner-shell spectroscopy, Nature **419**, 803 (2002)

4.344 P. Agostini, L.F. DiMauro: The physics of attosecond light pulses, Rep. Prog. Phys. **67**, 813 (2004)

4.345 E. Takahashi, Y. Nabekawa, M. Nurhuda, K. Midorikawa: Generation of high-energy high-order harmonics by use of a long interaction medium, J. Opt. Soc. Am. A **20**, 158 (2003)

4.346 J. Seres, E. Seres, A.J. Verhoef, G. Tempea, C. Streli, P. Wobrauschek, V. Yakovlev, A. Scrinzi, C. Spielmann, F. Krausz: Source of coherent kiloelectronvolt X-rays, Nature **433**, 596 (2005)

4.347 N.L. Wagner, A. Wüest, I.P. Christov, T. Popmintchev, X. Zhou, M.M. Murnane, H.C. Kapteyn: Monitoring molecular dynamics using coherent electrons from high harmonic generation, Proc. Nat. Ac. Sc. **103**, 13279 (2006)

4.348 A.-S. Morlens, J. Gautier, G. Rey, P. Zeitoun, J.-P. Caumes, M. Kos-Rosset, H. Merdji, S. Kazamias, K. Cassou, M. Fajardo: Submicrometer digital in-line holographic microscopy at 32 nm with high-order harmonics, Opt. Lett. **31**, 3095 (2006)

4.349 M. Wieland, Ch. Speilmann, U. Kleineberg, Th. Westerwalbesloth, U. Heinzmann, T. Wilhein: Toward time-resolved soft X-ray microscopy using pulsed fs-high-harmonic radiation, Ultramicroscopy **102**, 93 (2005)

4.350 Ph. Zeitoun, G. Faivre, S. Sebban, T. Mocek, A. Hallou, M. Fajardo, D. Aubert, Ph. Balcou, F. Burgy, D. Douillet, S. Kazamias, G. de Lachèze-Murel, T. Lefrou, S. le Pape, P. Mercère, H. Merdji, A.S. Morlens, J.P. Rousseau, C. Valentin: A high-intensity highly coherent soft X-ray femtosecond laser seeded by a high harmonic beam, Nature **431**, 426 (2004)

4.351 S.A. Aseyev, Y. Ni, L.J. Frasinski, H.G. Muller, M.J.J. Vrakking: Attosecond angle resolved photoelectron spectroscopy, Phys. Rev. Lett. **91**, 223902 (2003)

4.352 Y.T. Nabekawa, Y.T. Shimizu, T. Okino, K. Furusawa, H. Hasegawa, K. Yamanouchi, K. Midorikawa: Conclusive evidence of an attosecond pulse train observed with the mode resolved autocorrelation technique, Phys. Rev. Lett. **96**, 083901 (2006)

4.353 A. Baltuska, T. Udem, M. Uiberacker, M. Hentschel, E. Goulielmakis, C. Gohle, R. Holzwarth, V. S. Yakovlev, A. Scrinzi, T. W. Hänsch, F. Krausz: Attosecond control of electronic processes by intense light fields, Nature **421**, 611 (2003)

4.354 R. Kienberger, E. Goulielmakis, M. Uiberacker, A. Baltuska, V. Yakovlev, F. Bammer, A. Scrinzi, Th. Westerwalbesloh, U. Kleineberg, U. Heinzmann, M. Drescher, F. Krausz: Atomic transient recorder, Nature **427**, 817 (2004)

4.355 Y. Mairesse, A. De Bohan, L. J. Frasinski, H. Merdji, L. C. Dinu, P. Monchicourt, P. Bréger, M. Kovacev, R. Taieb, B. Carré, H. G. Muller, P. Agostini, P. Salieres: Attosecond synchronization of high-harmonic soft X-rays, Science **302**, 1540 (2003)

4.356 Y. Mairesse, A. De Bohan, L. J. Frasinski, H. Merdji, L. C. Dinu, P. Monchicourt, P. Breger, M. Kovacev, T. Auguste, B. Carré, H. G. Muller, P. Agostini, P. Salieres: Optimization of attosecond pulses, Phys. Rev. Lett. **930**, 163901 (2004)

4.357 K. Varjú, Y. Mairesse, P. Agostini, P. Breger, B. Carré, L. J. Frasinski, E. Gustafsson, P. Johnsson, J. Mauritsson, H. Merdji, P. Monchicourt, A. L'Huillier, P. Salières: Reconstruction of Attosecond Trains using an adiabatic phase expansion, Phys. Rev. Lett. **95**, 243901 (2005)

4.358 R. López-Martens, K. Varjú, P. Johnsson, J. Mauritsson, Y. Mairesse, P. Salières, M. B. Gaarde, K. J. Schafer, A. Persson, S. Svanberg, C.-G. Wahlström, A. L'Huillier: Amplitude and phase control of attosecond light pulses, Phys. Rev. Lett. **94**, 033001 (2005)

4.359 J. Mauritsson, P. Johnsson, E. Gustafsson, A. L'Huillier, K. J. Schafer, M. B. Gaarde: Attosecond pulse trains generated using two-color laser fields, Phys. Rev. Lett. **97**, 013001 (2006)

4.360 I. J. Sola, E. Mével, L. Elouga, E. Constant, V. Strelkov, L. Poletto, P. Villoresi, E. Benedetti, J-P. Caumes, S. Stagira, C. Vozzi, G. Sansone, M. Nisoli: Controlling attosecond electron dynamics by phase-stabilized polarization gating, Nature Physics **2**, 319 (2006)

4.361 G. Sansone, E. Benedetti, F. Calegari, C. Vozzi, L. Avaldi, R. Flammini, L. Poletto, P. Villoresi, C. Altucci, R. Velotta, S. Stagira, S. De Silvestri, M. Nisoli: Isolated single-cycle attosecond pulses, Science **314**, 443 (2006)

4.362 E. Goulielmakis, M. Uiberacker, R. Kienberger, A. Baltuska, V. Yakovlev, A. Scrinzi, Th. Westerwalbesloh, U. Kleineberg, U. Heinzmann, M. Drescher, F. Krausz: Direct measurement of light waves, Science **305**, 1267 (2004)

4.363 J. Itatani, J. Levesque, D. Zeidler, H. Niikura, H. Pepin, J. C. Kieffer, P. B. Corkum, D. M. Villeneuve: Tomographic imaging of molecular orbitals, Nature **432**, 867 (2004)

4.364 T. Kanai, S. Minemoto, H. Sakai: Quantum interference during high-order harmonic generation from aligned molecules, Nature **435**, 470 (2005)

4.365 S. Baker, J. S. Robinson, C. A. Haworth, H. Teng, R. A. Smith, C. C. Chirila, M. Lein, J. W. G. Tisch, J. P. Marangos: Probing proton dynamics in molecules on an attosecond time scale, Science **312**, 424 (2006)

4.366 P. Johnsson, R. López-Martens, S. Kazamias, J. Mauritsson, C. Valentin, T. Remetter, K. Varjú, M. B. Gaarde, Y. Mairesse, H. Wabnitz, P. Salières, Ph. Balcou, K. J. Schafer, A. L'Huillier: Attosecond electron wave packet dynamics in strong laser fields, Phys. Rev. Lett. **95**, 013001 (2005)

4.367 T. Remetter, P. Johnsson, J. Mauritsson, K. Varjú, Y. Ni, F. Lépine, M. Kling, J. Khan, E. Gustafsson, R. López-Martens, K. J. Schafer, M. J. J. Vrakking, A. L'Huillier: Attosecond electron wavepacket interferometry, Nature Physics **2**, 323 (2006)

4.368 F. Krausz, M. Ivanov: Attosecond physics, Rev. Mod. Phys. (2007) to be published

5. Optical Materials and Their Properties

This chapter provides an extended overview on today's optical materials, which are commonly used for optical components and systems. In Sect. 5.1 the underlying physical background on light–matter interaction is presented, where the phenomena of refraction (linear and nonlinear), reflection, absorption, emission and scattering are introduced. Sections 5.2 through 5.8 focus on the detailed properties of the most common types of optical materials, such as glass, glass ceramics, crystals, and plastics. In addition, special materials displaying "unusual nonlinear" or "quasi-nonreversible" optical behavior such as photorefractive or photorecording solids are described in Sect. 5.9. The reader could use this chapter as either a comprehensive introduction to the field of optical materials or as a reference text for the most relevant material information.

5.1 Interaction of Light with Optical Materials

In this section the general physics of the interaction of light with matter is briefly presented. A detailed insight into theoretical electrodynamics cannot be given here. The interested reader might refer to standard textbooks on electrodynamics e.g. [5.1, 2].

5.1.1 Dielectric Function

The starting point for an analysis of any interaction between electromagnetic waves with matter is Maxwell's equations. The static interaction for the dielectric displacement and the magnetic induction is described by

$$\begin{aligned} \nabla \cdot \boldsymbol{D} &= \rho , \\ \nabla \cdot \boldsymbol{B} &= 0 , \end{aligned} \tag{5.1}$$

whereas the dynamic interaction of the electric and magnetic fields is given by

$$\begin{aligned} \nabla \times \boldsymbol{E} &= -\dot{\boldsymbol{B}} , \\ \nabla \times \boldsymbol{H} &= \boldsymbol{j} + \dot{\boldsymbol{D}} . \end{aligned} \tag{5.2}$$

$\boldsymbol{E}$ and $\boldsymbol{B}$ are the electric and magnetic fields; $\boldsymbol{D}$ and $\boldsymbol{H}$ are the electric displacement and the auxiliary magnetic fields; ρ and $\boldsymbol{j}$ are the charge and the current density.

Material equations are needed to close Maxwell's equations:

$$\begin{aligned} \boldsymbol{D} &= \varepsilon_0 \boldsymbol{E} + \boldsymbol{P} , \\ \boldsymbol{B} &= \mu_0 \boldsymbol{H} + \boldsymbol{M} , \end{aligned} \tag{5.3}$$

where $\boldsymbol{P}$ and $\boldsymbol{M}$ are the polarization and magnetization densities. The vacuum permittivity (in SI units) is $\varepsilon_0 = 8.854 \times 10^{-12}$ A s/V m and the vacuum permeability is $\mu_0 = 4\pi \times 10^{-7}$ V s/A m.

The complete optical properties for any spatial combination of matter are included in the solution of (5.1), (5.2), which are closed by using the material equations (5.3) and by using appropriate boundary conditions. For only a few special cases such a solution can be written down directly. In the following we give a few examples.

Wave Equation in Vacuum

If we want to solve (5.1), (5.2) in infinite vacuum we have the following boundary conditions and material equations:

$$\boldsymbol{P}(\boldsymbol{r}) = 0 , \quad \boldsymbol{M}(\boldsymbol{r}) = 0 , \quad \rho(\boldsymbol{r}) = 0 , \quad \boldsymbol{j}(\boldsymbol{r}) = 0 , \tag{5.4}$$

where $\boldsymbol{r} = (x, y, z)$ are the three spatial coordinates. With these simplest possible boundary conditions the material equations (5.3) read

$$\begin{aligned} \boldsymbol{D} &= \varepsilon_0 \boldsymbol{E} , \\ \boldsymbol{B} &= \mu_0 \boldsymbol{H} . \end{aligned} \tag{5.5}$$

After applying a few vector operations, one gets the wave equation for the electromagnetic field $\boldsymbol{E}$ in vacuum:

$$\Delta \boldsymbol{E} - \mu_0 \varepsilon_0 \ddot{\boldsymbol{E}} = 0 . \tag{5.6}$$

An identical wave equation can be derived for the magnetic field $\boldsymbol{B}$. (5.6) immediately defines the speed of light c (in vacuum).

$$c = \sqrt{\frac{1}{\mu_0 \varepsilon_0}} . \tag{5.7}$$

Equation (5.6) is generally solved by all fields which fulfill $\boldsymbol{E}(\boldsymbol{r}, t) = \boldsymbol{E}_0 \cdot f(\boldsymbol{kr} \pm \omega t)$ involving any arbitrary scalar function f. The most common systems of function f are plane waves:

$$\boldsymbol{E}_{\mathrm{s}}(\boldsymbol{r}, t) = \boldsymbol{E}_0 \,\mathrm{Re}\left(\mathrm{e}^{-\mathrm{i}(\boldsymbol{kr} - \omega t)}\right) . \tag{5.8}$$

These plane waves, with a time and spatial dependent phase $\theta = \boldsymbol{kr}$, are described by a wave vector $\boldsymbol{k}$, an angular frequency ω, and a corresponding wavelength $\lambda = 2\pi/k = 2\pi c/\omega$, where $k = |\boldsymbol{k}|$ is the absolute value of the wave vector. Describing an arbitrary field $\boldsymbol{E}$ in terms of plane waves is identical to decomposing this electrical field into its Fourier components.

Wave Propagation in an Ideal Transparent Medium

We can describe an ideal material by simply replacing the speed of light in vacuum by that of the medium.

$$c \to \frac{c}{n} , \tag{5.9}$$

where n is the (in this case only real) refractive index of the material. Wave propagation in a dispersing or weakly absorbing medium is considered at the end of the present section. In fact, most parts of an optical design can be done by treating optical glasses as such ideal transparent materials (see Sect. 5.1.2). Even though such an ideal material cannot exist in reality, optical glasses come very close to it (for electromagnetic radiation in the visible range). For such an ideal material the wave equation (5.6) reads:

$$\Delta \boldsymbol{E} - \frac{n^2}{c^2} \ddot{\boldsymbol{E}} = 0 . \tag{5.10}$$

It is solved again by plane transverse waves. Where the speed of light is now reduced to the speed of light in the transparent medium $c_{\text{med}} = c/n$ and the wavelength of the lightwave is reduced to $\lambda_{\text{med}} = \lambda/n$.

Refraction and Reflection

We now derive the laws of refraction and reflection for the ideal transparent medium just described. They are obtained by solving Maxwell's equations at the (infinite) boundary between two materials of different refractive indices n_1 and n_2 (see Fig. 5.1). As boundary conditions one obtains that the normal component of the electric displacement (and magnetic induction) and the tangential component of the electric (and magnetic) field have to be continuous at the interface:

$$D_1^{\text{n}} = D_2^{\text{n}}\,, \quad E_1^{\text{t}} = E_2^{\text{t}}\,. \tag{5.11}$$

Further, a phase shift of an incoming wave occurs upon reflection

$$\theta_{\text{r}} = \pi - \theta_{\text{i}}\,, \tag{5.12}$$

where $\theta_{\text{r,i}}$ are the phases of the reflected and incident wave, respectively. If we solve Maxwell's equations for an incoming plane wave (applying the boundary conditions stated above), Snell's law of refraction is obtained

$$n_1 \sin\alpha_1 = n_2 \sin\alpha_2 \tag{5.13}$$

together with that of reflection

$$\alpha_{\text{r}} = \alpha_1\,. \tag{5.14}$$

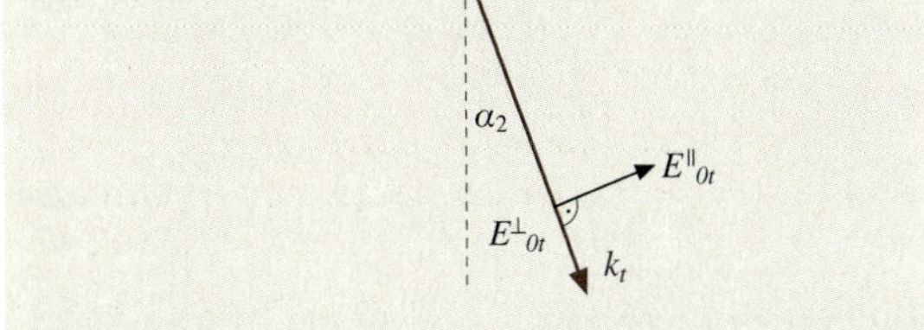

Fig. 5.1 The polarization directions of the E and B fields for reflection and refraction at an interface between two optical materials of different refractive indices are shown. A circle indicates that the vector is perpendicular to the plane shown

Now, the electric field $\boldsymbol{E}$ is decomposed into its components which are defined relative to the plane outlined by the three beams of incoming, transmitted and reflected light. This decomposition is shown in Fig. 5.1. The coefficients for reflection and transmission are defined as:

$$r_{\parallel} = \frac{E_{0\text{r}}^{\parallel}}{E_{0\text{i}}^{\parallel}}\,; \qquad t_{\parallel} = \frac{E_{0\text{t}}^{\parallel}}{E_{0\text{i}}^{\parallel}}\,,$$
$$r_{\perp} = \frac{E_{0\text{r}}^{\perp}}{E_{0\text{i}}^{\perp}}\,; \qquad t_{\perp} = \frac{E_{0\text{t}}^{\perp}}{E_{0\text{i}}^{\perp}}\,. \tag{5.15}$$

The Fresnel formula for these coefficients can be derived as

$$\begin{aligned}
r_{\perp} &= \frac{n_1\cos(\alpha_1) - n_2\cos\alpha_2}{n_1\cos(\alpha_1) + n_2\cos(\alpha_2)} = -\frac{\sin(\alpha_1-\alpha_2)}{\sin(\alpha_1+\alpha_2)}\,,\\
r_{\parallel} &= \frac{n_2\cos(\alpha_1) - n_1\cos\alpha_2}{n_1\cos(\alpha_2) + n_2\cos(\alpha_1)} = -\frac{\tan(\alpha_1-\alpha_2)}{\tan(\alpha_1+\alpha_2)}\,,\\
t_{\perp} &= \frac{2n_1\cos(\alpha_1)}{n_1\cos(\alpha_1) + n_2\cos(\alpha_2)}\\
&= -\frac{2\sin(\alpha_2)\cos(\alpha_1)}{\sin(\alpha_1+\alpha_2)}\,,\\
t_{\parallel} &= \frac{2n_1\cos(\alpha_1)}{n_1\cos(\alpha_2) + n_2\cos(\alpha_1)}\\
&= -\frac{2\sin(\alpha_2)\cos(\alpha_1)}{\sin(\alpha_1+\alpha_2)\cos(\alpha_1-\alpha_2)}\,.
\end{aligned} \tag{5.16}$$

Here the usual convention has been used that the coefficients of reflectivity obtain an additional minus sign in order to indicate backtraveling of light. The quantities that are measured in an experiment are intensities. The relationship between the intensities defines the reflectivity and transmissivity of a material

$$R_{\perp} := |r_{\perp}|^2\,; \qquad R_{\parallel} := |r_{\parallel}|^2\,,$$
$$T_{\perp} := |t_{\perp}|^2\,; \qquad T_{\parallel} := |t_{\parallel}|^2\,. \tag{5.17}$$

The angular-dependent coefficients of reflection from (5.16) are displayed in Fig. 5.2. In Fig. 5.2a the case of light propagating from an optically thin medium with refractive index n_1 to an optically thicker medium with refractive index $n_2 > n_1$ is plotted. At the so called Brewster angle α_{B} the reflected light is completely polarized; α_{B} is given by the condition $\alpha_1 + \alpha_2 = \pi/2$. Therefore, the Brewster angle α_{B} results as a solution of

$$\alpha_1 = \frac{\pi}{2} - \arccos\left(\frac{n_2}{n_1}\cos\alpha_1\right) \tag{5.18}$$

which gives $\alpha_{\text{B}} = \arctan\frac{n_2}{n_1}$. In Fig. 5.2b the case of light propagating from an optically thick to an opti-

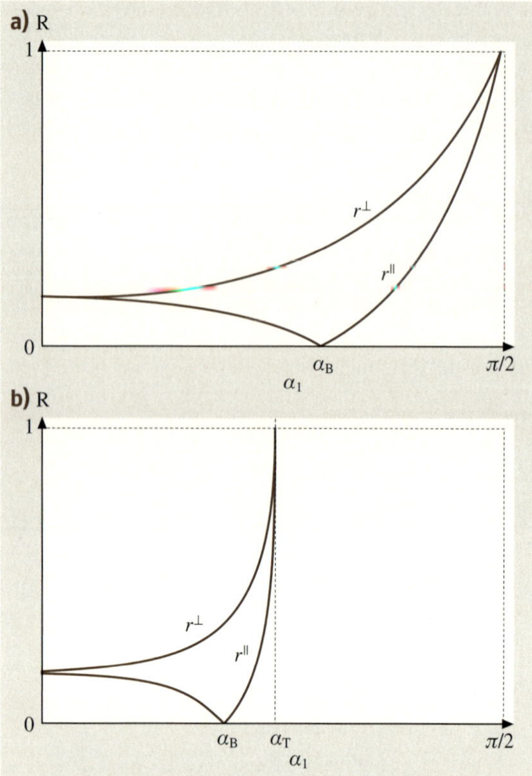

Fig. 5.2a,b The reflection coefficients are plotted as a function of incident scattering angle for light propagating from a medium of **(a)** smaller refractive index into a medium of larger index and **(b)** larger refractive index into a medium of smaller index. Here, total reflection occurs at an angle α_T and α_B is the Brewster angle

cally thin medium is plotted. Here an additional special angle occurs – the angle of total reflection, α_T. All light approaching the surface at an angle larger than α_T is totally reflected. At $\alpha_1 = \alpha_T$ the angle for refraction in the medium with refractive index n_2 is $\alpha_2 = \pi/2$. For α_T it follows that:

$$\alpha_T = \arcsin\left(\frac{n_2}{n_1}\right) . \tag{5.19}$$

Evaluating (5.16) for the special case of incident light as $\lim_{\alpha\to 0}$ allows one to calculate the reflectivity for normal incidence

$$R_{norm} = \left|\frac{n_1 - n_2}{n_1 + n_2}\right|^2 . \tag{5.20}$$

For practical applications it is important to note that the Fresnel equations remain valid in the case of weakly absorbing media discussed at the end of the present section. Here, only the refractive indices have to be replaced by the complex quantities of (5.36). It is further helpful to define a transmittivity and reflectivity for unpolarized light

$$R^{unpol} = \frac{E_{0r}^{\parallel 2} + E_{0r}^{\perp 2}}{E_{0i}^{\parallel 2} + E_{0i}^{\perp 2}} ; \quad T^{unpol} = \frac{E_{0t}^{\parallel 2} + E_{0t}^{\perp 2}}{E_{0i}^{\parallel 2} + E_{0i}^{\perp 2}} . \tag{5.21}$$

Inserting the expressions for the reflection coefficients we obtain e.g. the total reflectivity as a function of the incident and refracted angular

$$R_{all}^{unpol} = \frac{E_{0i}^{\parallel 2} \frac{\tan^2(\alpha_1-\alpha_2)}{\tan^2(\alpha_1+\alpha_2)} + E_{0i}^{\perp 2} \frac{\sin^2(\alpha_1-\alpha_2)}{\sin^2(\alpha_1+\alpha_2)}}{E_{0i}^{\parallel 2} + E_{0i}^{\perp 2}}$$

With the definitions from (5.21) the following sum rule must be fulfilled:

$$R^{unpol} + \frac{n_2}{n_1}\frac{\cos\alpha_2}{\cos\alpha_1} T^{unpol} = 1 . \tag{5.22}$$

The rule provides an easy check for transmitted and reflected total intensities, especially for normal incidence.

Wave Propagation in an Isotropic, Homogeneous Medium

We now consider wave propagation in an ideal optical material. This is a nonmagnetic, homogeneous, isotropic, perfectly insulating medium, which is further a perfectly linear optical material. Considering time dependence including retardation in the materials (5.2) leads to:

$$\boldsymbol{D}(\boldsymbol{r},t) = \varepsilon_0 \boldsymbol{E}(\boldsymbol{r},t) + \boldsymbol{P}(\boldsymbol{r},t) . \tag{5.23}$$

The polarizability is related to the electric field via the susceptibility χ. In the case of a homogeneous isotropic material χ is a scalar function. In Sect. 5.1.4 we will consider the case of optically anisotropic media, where χ becomes a second-rank tensor.

$$\boldsymbol{P}(\boldsymbol{r},t) = \int \mathrm{d}\boldsymbol{r}' \int_{-\infty}^{t} \mathrm{d}t' \chi(\boldsymbol{r}-\boldsymbol{r}', t-t') \boldsymbol{E}(\boldsymbol{r},t') . \tag{5.24}$$

Fourier transformation in time and space deconvolutes the integral and leads to:

$$\boldsymbol{P}(\boldsymbol{k},\omega) = \chi(\boldsymbol{k},\omega)\boldsymbol{E}(\boldsymbol{k},\omega) , \tag{5.25}$$

where $\chi(\boldsymbol{k},\omega)$ is, in general, a complex analytic function of the angular frequency ω. The complex function $\chi(\boldsymbol{k},\omega)$ unifies the two concepts of a low-frequency polarizability χ' and a low-frequency conductivity σ of mobile charges to a single complex quantity

$$\lim_{\omega\to 0}\chi(\omega)=\chi'(\omega)+4\pi\mathrm{i}\frac{\sigma(\omega)}{\omega}\,. \tag{5.26}$$

At larger frequencies the separation of the two concepts breaks down, since above the frequencies of optical phonon modes in the IR the bound charges are unable to follow the electric field, whereas below the phonon modes the charges can follow this motion (Sect. 5.1.3). The usual form in which the susceptibility enters the equations for optical purposes is via the dielectric function

$$\varepsilon(\boldsymbol{k},\omega)=1+\chi(\boldsymbol{k},\omega)\,. \tag{5.27}$$

Inserting the dielectric function into the material equation (5.23) gives:

$$\boldsymbol{D}(\boldsymbol{k},\omega)=\varepsilon_0\varepsilon(\boldsymbol{k},\omega)\boldsymbol{E}(\boldsymbol{k},\omega)\,. \tag{5.28}$$

Here, we restrict ourselves to ideal optically isotropic materials by neglecting the nature of the dielectric function as a second-rank tensor. In Sect. 5.1.4 we will extend our considerations to optically anisotropic materials. With the same steps as in Eqs. (5.5–5.6) a wave equation can be derived which has the following form in Fourier space:

$$\left[k^2-\varepsilon(\boldsymbol{k},\omega)\frac{\omega^2}{c^2}\right]\boldsymbol{E}_0=0\,, \tag{5.29}$$

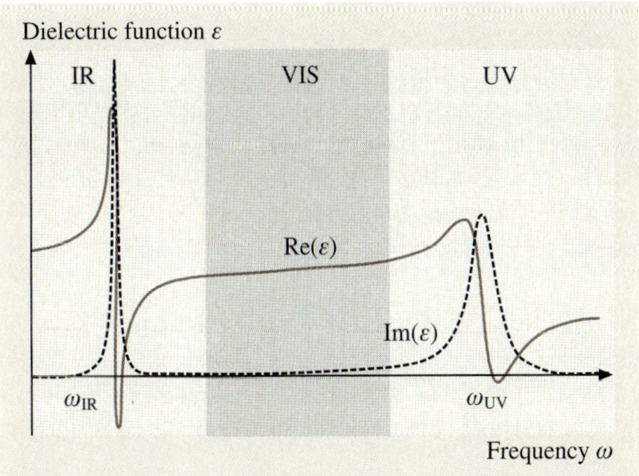

Fig. 5.3 Dielectric function $\varepsilon(\omega)$ for the "model optical solid" with one generic absorption in the infrared ω_{IR} and a second one in the ultraviolet ω_{UV}. The dielectric function is plotted on a logarithmic energy scale. The *solid line* is the real part and the *dashed line* is the imaginary part of $\varepsilon(\omega)$

where the relations $\partial^2/\partial t^2\boldsymbol{E}_\mathrm{s}(\boldsymbol{r},t)\to-\omega^2\boldsymbol{E}_\mathrm{s}(\boldsymbol{r},t)$ and $\Delta\boldsymbol{E}_\mathrm{s}(\boldsymbol{r},t)\to-k^2\boldsymbol{E}_\mathrm{s}(\boldsymbol{r},t)$ have been used. The expression in brackets in (5.29) defines the dispersion relation for an optically linear, homogeneous, isotropic material.

Poynting Vector and Energy Transport

The energy flux density of the electric field is obtained via the Poynting vector, given by

$$\boldsymbol{S}=\boldsymbol{E}\times\boldsymbol{H}\,. \tag{5.30}$$

It gives the rate at which electromagnetic energy crosses a unit area and has the unit $\mathrm{W/m^2}$. It points in the direction of energy propagation. The time average of the absolute value of the Poynting vector $\langle|\boldsymbol{S}|\rangle$ is called the intensity I of the electromagnetic wave

$$I=\langle|\boldsymbol{S}|\rangle=\frac{1}{2}|\boldsymbol{E}\times\boldsymbol{H}| \tag{5.31}$$

and is the energy flux density of the electromagnetic radiation. In thespecial case of propagation of transverse plane waves [as given in (5.8)], it simplifies to

$$I=\frac{1}{2}\frac{n}{c\mu_0}|\boldsymbol{E}_0|^2\,, \tag{5.32}$$

where in vacuum $n=1$ is valid.

General Form of the Dielectric Function

For most optical materials the dielectric function has a form in which a transparent frequency (or wavelength) window is bounded at the high energy site by electron–hole excitations (dominating the UV edge) and at the low energy site by IR absorptions given by optical phonon modes (lattice vibrations). The general form of the dielectric function is given by the Kramers–Heisenberg equation [5.3]

$$\varepsilon(\boldsymbol{k},\omega)=1+\sum_j\frac{\alpha_{\boldsymbol{k},j}}{\omega^2-\omega_{\boldsymbol{k},j}^2-\mathrm{i}\omega\eta_{\boldsymbol{k},j}}\,. \tag{5.33}$$

Here $\alpha_{\boldsymbol{k},j}$ is the amplitude, $\omega_{\boldsymbol{k},j}$ the frequency and $\eta_{\boldsymbol{k},j}$ the damping of the particular excitation j. A schematic view of the dielectric function is plotted in Fig. 5.3. Here we use a model for a transparent homogeneous, isotropic solid (such as glass) with one generic absorption at low energies (ω_{IR} in the infrared, IR) and another one at large phonon energies (ω_{UV} in the ultraviolet spectral range, UV). In the following this model solid is used to discuss optical material properties.

Dispersion Relation

Solving (5.29) gives two frequency-dependent solutions for the wave vector $\boldsymbol{k}$ as a function of ω since the left-hand side of (5.29) is quadratic in $\boldsymbol{k}$. Far away from

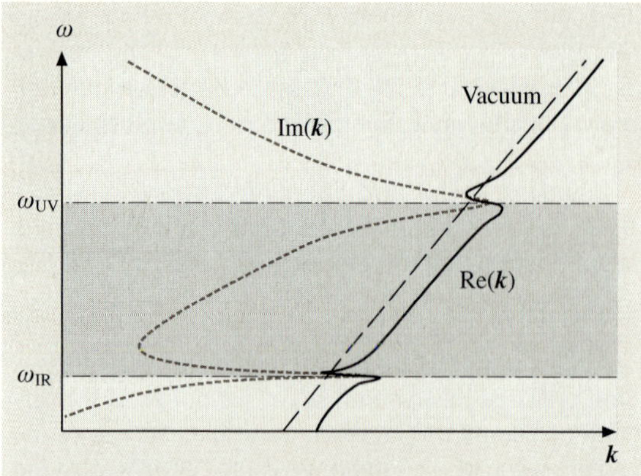

Fig. 5.4 Dispersion relation $\omega(k)$ for the "model optical solid" on a double logarithmic scale. The *solid line* is the real part of $\boldsymbol{k}$ and the *dashed line* the imaginary part of $\boldsymbol{k}$. For comparison the simply linear dispersion relation for light propagating in vacuum $\omega = ck$ is shown with a *long-dashed line*

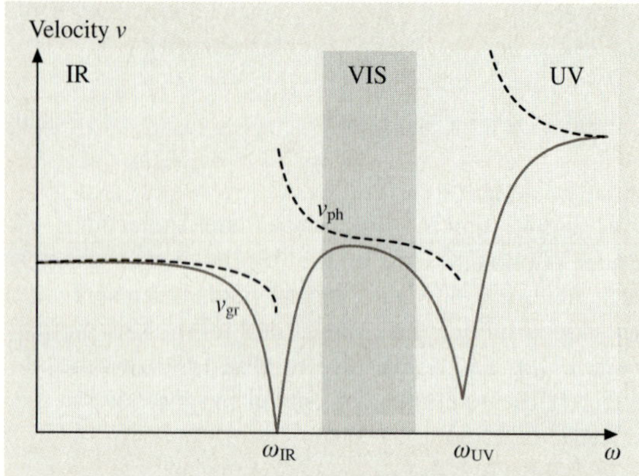

Fig. 5.5 The group velocity $v_{\text{gr}}(k) = \partial\omega/\partial k$ (*solid line*) and the phase velocity $v_{\text{ph}}(k) = \omega/k$ (*dashed line*) are plotted on a logarithmic frequency scale. Far away from absorptions both velocities approach each other while the group velocity is always smaller than the phase velocity

absorptions the dielectric function is real and positive. Here only one solution exists, which describes the wave propagating with the speed of light in the medium. Close to an absorption two solutions exist which are even more complicated. This means that near a resonance the dispersion of the light cannot be considered independently from the dispersion of the excitations in the material. They both form a composite "new" entity propagating in the medium. This is called the polariton [5.4]. For our model solid with two generic absorptions (ω_{IR} and ω_{UV}), the dispersion is plotted in Fig. 5.4.

Wave Propagation, Phase and Group Velocity

When an electromagnetic wave propagates through a medium one can define two velocities. The phase velocity is the speed with which a certain phase propagates. It is, for example, the velocity of the wavefront maxima moving through the medium. The phase velocity is given as

$$v_{\text{ph}} = \frac{\omega}{k} \,. \tag{5.34}$$

In Fig. 5.5 the phase velocity of the model solid is plotted on a logarithmic frequency scale. Close to the absorption edges of the material it loses its meaning because attenuation due to the absorption processes will dominate most processes. Far away from absorptions it reaches a nearly constant value. The second velocity is the group velocity

$$v_{\text{gr}}(k) = \frac{\partial\omega}{\partial k} \,. \tag{5.35}$$

This is the velocity at which a complete wave packet travels through the medium and is, hence, the speed with which information can travel through the system. It is plotted in Fig. 5.5 for the model solid. Reasonably far away from material absorptions the group and phase velocities approach each other. However, the group velocity is always smaller than the phase velocity.

Refractive Index

The refractive index n is the most widely used physical quantity in optical design. It is the square root of the dielectric function. The dynamic refractive index is generally a complex quantity

$$\tilde{n}(\omega) = n(\omega) + \mathrm{i}\kappa(\omega) \tag{5.36}$$

and must fulfill the Kramers–Kronig relations [5.3]. The refractive index for our generic model solid is plotted as a function of logarithmic frequency in Fig. 5.6. In practical use, the wavelength dependence is often exploited

$$\tilde{n}(\boldsymbol{k}, \lambda) = \sqrt{\varepsilon(\boldsymbol{k}, 2\pi c/\lambda)} \,. \tag{5.37}$$

With a few basic steps, (5.33) can be rewritten as a function of wavelength alone. If one further restricts to wavelengths which are far away from absorptions $\left(\omega^2 - \omega_{k,j}^2\right) \ll \omega\eta_{k,j}$, the Sellmeier formula (see

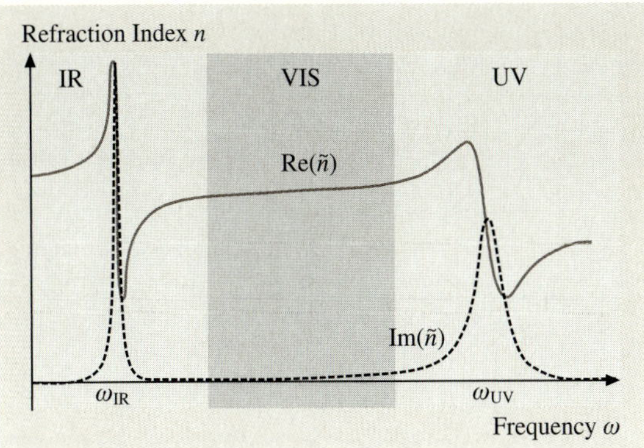

Fig. 5.6 The complex refractive index $n(\omega)=n^{*}(\omega)+\mathrm{i}\kappa(\omega)$ is plotted on a logarithmic frequency scale. The real part is plotted with a *solid line* and the imaginary part with a *dashed line*

Sect. 5.1.2), which is widely used for characterizing optical materials, is obtained

$$n(\lambda)^2 \approx 1+\sum_j \frac{B_j\lambda^2}{\lambda^2-\lambda_j^2}\,, \tag{5.38}$$

where $\lambda_i = 2\pi c/\omega_i$ with $i \in \{(\boldsymbol{k}, j)\}$ is used and $B_j = a_{k,j}\lambda_{k,j}/(2\pi c)^2$. Normally B_j and λ_j are just fitting constants to describe the dispersion of the refractive index over a certain wavelength range. They are, however, connected to the microscopic fundamental absorption behavior of the material. Sometimes also $n(\lambda)$ and not $n(\lambda)^2$ is approximated with a Sellmeier formula. Since as well $n(\lambda)$ as well as $n(\lambda)^2$ are complex differential (analytic) functions both formula give refractive indices and dispersions with the same accuracy. However care has to be taken, which quantity is expressed when using a Sellmeier formula.

Wave Propagation in Weakly Absorbing Medium

In this subsection the link between the attenuation of a wave and the imaginary part of the refractive index is given. A weakly absorbing medium is defined by the imaginary part of the refractive index (5.37) being much smaller than the real part

$$\kappa \ll n \tag{5.39}$$

(the coefficient κ/n is also called the attenuation index). In this case light propagates as transverse waves through the medium. We consider two points in our medium: P_1 and P_2. Between these points the light travels the distance l. In the absence of absorption, the electric and

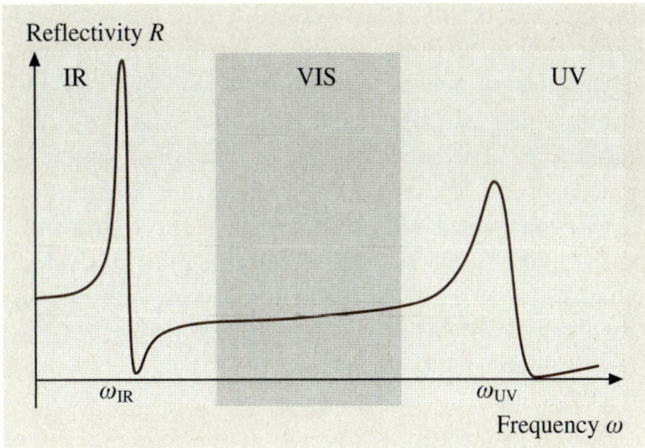

Fig. 5.7 Reflectivity R for normal incidence plotted on a logarithmic frequency scale

magnetic fields at point P_2 are given by

$$\boldsymbol{E}_2 = \boldsymbol{E}_1\,\mathrm{e}^{\mathrm{i}\frac{\omega}{c}\tilde{n}l}\,,\quad \boldsymbol{H}_2 = \boldsymbol{H}_1\,\mathrm{e}^{\mathrm{i}\frac{\omega}{c}\tilde{n}l}\,. \tag{5.40}$$

Using (5.31) we obtain for weak absorption the radiation intensity at point P_1. The radiation intensity at P_2 is:

$$\begin{aligned} I_2 &= \frac{1}{2}|\boldsymbol{E}_2\times\boldsymbol{H}_2|\,,\\ &= \frac{1}{2}|\boldsymbol{E}_1\times\boldsymbol{H}_1|\,\mathrm{e}^{-2\frac{\omega}{c}\kappa l}\,,\\ &= I_1\,\mathrm{e}^{-2\frac{\omega}{c}\kappa l} = I_1\,\mathrm{e}^{-\alpha l}\,. \end{aligned} \tag{5.41}$$

The absorption coefficient α is connected to the complex refractive index by

$$\alpha = 2\frac{\omega}{c}\kappa\,; \tag{5.42}$$

α can easily be measured and its importance for optical properties is discussed in Sect. 5.1.3.

It is also important to note that the Fresnel equations (5.16) remain valid in the case of a weakly absorbing medium if the complex refractive indices are used. As an example, we plot the reflectivity (at the interface air–model solid) near an absorption resulting from (5.20) for a complex refractive index $n_2 \to \tilde{n}_2$. In Fig. 5.7 the reflectivity is plotted on a logarithmic frequency scale. Note that the absorption seems to be shifted compared to the plots of the complex dielectric function or the complex refractive index. Measurement of the reflectivity is of importance for reflection spectroscopy.

5.1.2 Linear Refraction

As already introduced in Sect. 5.1.1, two phenomena occur when light impinges upon the surface of any optical material: reflection and refraction [5.5]. The reflected light bounces off the glass surface, while the refracted light travels through the material. The amount of light that is reflected depends on the refractive index of the sample, which also affects the refractive behavior of the sample [5.6]. The refractive index of optical materials turns out to be one of the most important factors that must be considered when designing systems to transmit and modulate light [5.7]. The refractive index is a complex material property that depends on temperature and wavelength [5.8]. The wavelength dependence of the refractive index is the dispersion [5.5].

Law of Refraction

When a light ray impinges upon a glass surface, a portion is reflected and the rest is either transmitted or absorbed. The material modulates the light upon transmission. The light travels at a different velocity as it is transmitted through the glass as compared to through vacuum. As introduced in Sect. 5.1.1, the index of refraction (n), is defined as the ratio of the speed of light in vacuum (c) to that in the material (c_m) [5.11]:

$$n = \frac{c}{c_m} . \quad (5.43)$$

Most commonly, the reported refractive indices are relative to the speed of light in air, rather than in vacuum, no matter which technique is used to measure the refractive index [5.12]. The index of refraction for vacuum, by definition, must be exactly 1. The index of refraction of air is 1.00029 at standard temperature (25 °C) and pressure (1 atm) (STP). Therefore, the index of refraction of optical matter (n_{rel}) relative to air (n_{air}), rather than vacuum is [5.11]:

$$n_{rel} = \frac{n_m}{n_{air}} . \quad (5.44)$$

Table 5.1 Indices of common materials at standard temperature and pressure at 587.56 nm (helium d line) [5.9]

Material	n_d	Material	n_d
Vacuum	1	Crown glass	1.52
Air	1.00029	Sodium chloride	1.54
Water	1.33	Polystyrene	1.55
Acetone	1.36	Carbon disulfide	1.63
Ethanol	1.36	Flint glass	1.65
Sugar solution (30 wt %)	1.38	Sapphire	1.77
Fused silica	1.46	Heavy flint glass	1.89
Sugar solution (80 wt %)	1.49	Diamond	2.42

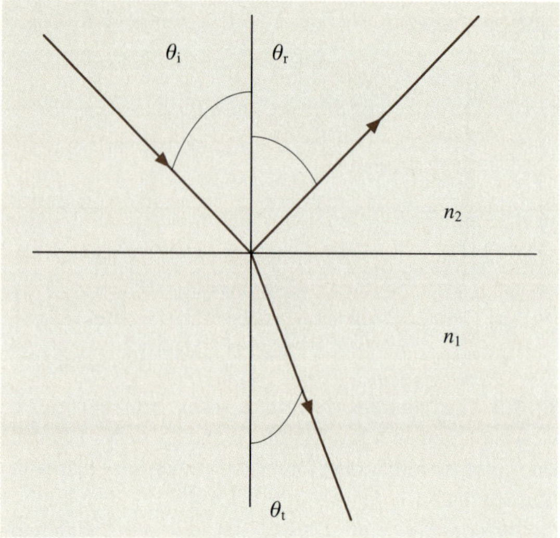

Fig. 5.8 Ray tracings of incident, reflected, and transmitted light from one medium to another representing the angles and indices necessary to apply Snell's law (after [5.10])

The STP indices of some common compounds, and classes of compounds, are shown in Table 5.1 [5.9].

As discussed in Sect. 5.1.1, when light hits a glass surface at an angle α_i, it is Fresnel-reflected back at an angle α_t. The angle of incidence is equal to the angle of reflection ($\alpha_i = \alpha_r$), as shown in Fig. 5.8 [5.10]. The percentage of light reflected for $\alpha_i = 0$ at each interface (R) relative to the incident intensity (Sect. 5.1.1) is dependent on the index of refraction of the two media the light is passing through, typically air (n_2) and glass (n_1) Fig. 5.8 and (5.20) [5.10]

$$R_{norm} = 100 \left(\frac{n_1 - n_2}{n_1 + n_2} \right)^2 . \quad (5.45)$$

Fresnel's formula (5.45) assumes smooth surfaces that produce only specular reflection. Diffuse reflection occurs when the surface is rough, so the incident light is reflected through a range of angles, thereby reducing the intensity of the reflected light at any given angle [5.10] (see Sect. 5.1.8). The specular reflection that is taken into consideration by Fresnel's relationship can be monitored and used to estimate the refractive index of samples in situ [5.13].

The angle of the light transmitted within the material (α_t), relative to the incident light transmitted through air, is dependent on the refractive indices of the air (n_{air}) and solid (n_m) and the incident light angle (α_i) [5.10]:

$$n_{air} \sin \alpha_i = n_m \sin \alpha_t . \quad (5.46)$$

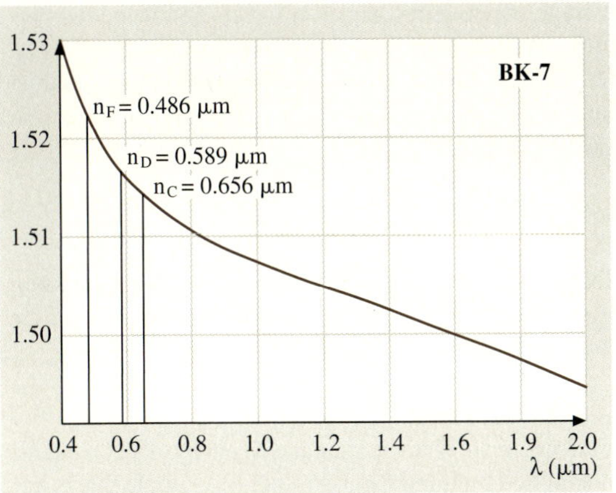

Fig. 5.9 Dispersion present in BK-7 optical glass. Common index of refraction measurement wavelengths are indicated

This is the general form of Snell's law of refraction to predict transmitted angles in media [5.14].

Dispersion Relationships in Glass

The refractive index of a medium is dependent upon the wavelength of the light being transmitted (see Sect. 5.1.1). This wavelength dependence is dispersion, which means that different wavelengths of light will be modulated differently by the same piece of matter [5.10]. Each wavelength of light will be subject to a different index of refraction. One ramification of dispersion is that white light can be separated into its principal visible components through a glass prism, or a simple raindrop. It is the dispersion of white light through raindrops that causes rainbows. The dispersion of light through optical materials results in the light being refracted at various angles because of Snell's law (5.46). The various components of white light experience different indices of refraction, which leads to different angles of exiting light. The difference in refractive index with wavelength is illustrated in Fig. 5.9 for BK-7 optical glass, which is a high dispersion material.

In normal dispersion, the index increases for shorter wavelengths of light [5.15]. Normal dispersion is valid only far away from absorption bands (see Fig. 5.6). Water has a normal dispersion response in the visible, so the red light is refracted by a lower index, and thus a greater angle, which is why red is on top in a rainbow. Discussion of rainbow formation is presented elsewhere in depth [5.10].

Anomalous dispersion is an increase in refractive index with an increase in wavelength. Anomalous dispersion typically occurs at frequencies that represent a crossover between the polarization mechanisms (dipolar, ionic, electronic etc.) that are responsible for absorption of electromagnetic radiation. This is further discussed in [5.11].

Due to dispersion, the index of refraction must be reported with the wavelength of measurement. The most common wavelengths at which n is measured are reported in Table 5.2. These wavelengths most often correspond to common sharp emission lines. The index can be determined most accurately $(\pm 1\times 10^{-6})$ by measuring the angle of minimum deviation of light in a prism [5.15]. However, a Pulfrich refractometer $(\pm 1\times 10^{-5})$ is most commonly used in industry. Details of measurement techniques are given in [5.16].

The index at various wavelengths is commonly referred to by the designations in Table 5.2: i. e. n_d is the refractive index measured at the yellow helium d line of 582.5618 nm. The dispersion is often given as a difference in n at two wavelengths. For instance, the primary dispersion is given by $n_F - n_C$ (hydrogen lines) and $n_{F'} - n_{C'}$ (cadmium lines). The most commonly reported measure of dispersion is the Abbe number (ν), which is

Table 5.2 Wavelength of spectral lines used for measuring index of refraction, with the common designation and spectral line source [5.11]

Wavelength (nm)	Designation	Spectral line
2325.4		Hg IR line
1970.1		Hg IR line
1529.6		Hg IR line
1013.98	t	H IR line
852.1101	s	Cs IR line
706.5188	r	He red line
656.2725	C	H red line
643.8469	C′	Cd red line
632.8		He–Ne laser line
589.2938	D	Na yellow line (center of doublet)
587.5618	d	He yellow line
546.074	e	Hg green line
486.1327	F	H blue line
479.9914	F′	Cd blue line
435.8343	g	Hg blue line
404.6561	h	Hg violet line
365.0146	i	Hg UV line

commonly given for two sets of conditions

$$\nu_d = \frac{n_d - 1}{n_F - n_C}, \quad \nu_e = \frac{n_e - 1}{n_{F'} - n_{C'}}. \tag{5.47}$$

The Abbe number is a measure of the ratio of the refractive power to the dispersion. In most optical materials catalogs, a six-digit number is assigned to the solid that is dependent upon the index and the Abbe number: $1000(n_d - 1) + 10\nu_d$. Using this property, e.g. optical glasses are divided into two general categories: crowns and flints (see Sect. 5.2). Crown glasses typically have a low index of refraction and a high Abbe number ($n_d < 1.60$ and $\nu_d > 55$), whereas flint glasses have high indices and low Abbe numbers ($n_d > 1.60$ and $\nu_d > 50$) [5.15]. The terms crown and flint have historical significance in that flint glasses typically had lead oxide added to them to increase the refractive index (see Sect. 5.2) and crown glasses were typically blown and had curvature – a crown. Typically, to make an achromatic optical system crown and flint lenses are combined in series. The n_d and ν_d are plotted for various types of optical glasses in Fig. 5.87. The n_d and ν_d are listed with glass type and manufacture in Table 5.37.

Often it is desired to have a mathematical representation of the index as a function of wavelength. A considerable number of models exist for just this purpose. Perhaps the best known, and most widely used, is the Sellmeier form [5.15]

$$n^2(\lambda) - 1 = \frac{B_1\lambda^2}{\lambda^2 - C_1} + \frac{B_2\lambda^2}{\lambda^2 - C_2} + \frac{B_3\lambda^2}{\lambda^2 - C_3}. \tag{5.48}$$

Most major optical material manufacturers supply the Sellmeier coefficients for their glasses on the product data sheets. With the six Sellmeier coefficients it is possible to estimate the index of refraction at any wavelength, given that it is not near a strong absorption. Numerous other dispersion models have been developed and are presented elsewhere [5.10, 15].

The index of refraction is also dependent upon temperature, and similar formulas and tables of constants have been developed for the differential change in n with temperature [5.15]

$$\frac{\mathrm{d}n_{abs}(\lambda, T)}{\mathrm{d}T} = \frac{[n^2(\lambda, T_0) - 1]}{2n(\lambda, T_0)} \times \left(D_0 + 2D_1\Delta T + 3D_2\Delta T^2 + \frac{E_0 + 2E_1\Delta T}{\lambda^2 - \lambda_{TK}^2}\right), \tag{5.49}$$

where T_0 is 20 °C, T is the temperature in °C, ΔT is $T - T_0$, λ is the wavelength in microns, and λ_{TK} is the average effective resonance wavelength for the temperature coefficients in microns. The constants E_0, E_1, D_0, D_1, and D_2 are provided on the manufacture product data sheets for each composition. The index of refraction must be measured.

5.1.3 Absorption

Based on the discussion in Sect. 5.1.1 the most common processes causing absorption in optical materials will be discussed here.

Introduction to Absorption

Absorption in glass is characterized by a decrease in transmitted light intensity through the sample that is not accounted for by reflection losses at the surface or scattering by inclusions [5.14]. As already introduced in Sect. 5.1.1 absorption is not uniform across all wavelengths of interest (UV–VIS–IR; ≈ 200–2000 nm) and can be characterized by absorption bands [5.17]. The absorption bands are due to both intrinsic and extrinsic effects [5.14, 18, 19].

The quantity used to discuss absorption as a function of wavelength in glass is the transmittance (T), which is the ratio of the transmitted light intensity (I) to the initial light intensity (I_0) after passing through a glass plate of thickness L [5.20]:

$$T = I/I_0. \tag{5.50}$$

The transmittance is related to the more common percent transmittance (%T) through:

$$\%T = 100\,T. \tag{5.51}$$

Sometimes it is more practical or convenient to look at the optical spectra in terms of absorbance rather than the transmittance. Absorbance (A) is defined as the $\log_{10}$ of $(1/T)$ [5.20]

$$A = \log_{10}\frac{I_0}{I} = \log_{10}\frac{1}{T} = 2 - \log_{10}\%T, \tag{5.52}$$

$$\%T = 100/10^A. \tag{5.53}$$

Absorbance is often referred to as optical density (OD). When the absorbance changes linearly with concentration of the absorbing species (C), then Beer's law, which relates absorbance with concentration, path length (l) and extinction coefficient (ε), is applicable [5.20]

$$A = \varepsilon Cl. \tag{5.54}$$

The typical units of ε are liters mole^{-1} cm^{-1} and are often omitted.

From Beer's law we see that the absorbance changes with path length. To account for this, the spectra (in A or $\%T$) must list the sample thickness. Absorbance can also be normalized to path length in terms of an absorption coefficient (α). The connection of an optical absorption to a microscopic property of an underlying absorption process is straightforward. A light beam propagating through a medium is exponentially damped if absorption is present. This follows directly from the fact that the attenuation inside the optical material is proportional to the present (wavelength-dependent) intensity $I(\lambda)$ [5.21]

$$\mathrm{d}I(\lambda) = N\sigma^{\text{abs}}(\lambda)I(\lambda)\,\mathrm{d}z\,. \tag{5.55}$$

Where $\mathrm{d}z$ is an infinitesimally small part of the propagation part (unit meter) and N is the density of absorbing centers (in units of m^{-3}). The quantity $\sigma^{\text{abs}}(\lambda)$ is the absorption cross section. Since both sides of (5.55) must have the same units, it follows that the absorption cross section is measured in m^2. It therefore has the meaning of an area and can be interpreted as the effective area perpendicular to the direction of an incoming beam that is seen by that beam and causes an absorption process. Figure 5.10 illustrates this. The solution of (5.55) for an optical plate of thickness L and initial intensity I_0 gives the intensity of the beam after traveling the distance L through the material [5.21]

$$I_{L,G}(\lambda) = I_{0,G}\,\mathrm{e}^{-\alpha(\lambda)L} \tag{5.56}$$

with $I_{0,G}$ and $I_{L,G}$ being taken at both surfaces, but inside the solid. They are related to the measurable values I_0 and I_L outside the material by $I_{0,G} = (1-R)I_0$ and $I_L = (1-R)I_{L,G}$ with R being the Fresnel losses as described in Sect. 5.1.1. Equation (5.56) is known as the law of Lambert–Beer, where the absorption coefficient $\alpha(\lambda)$ is defined as [5.21]

$$\alpha(\lambda) = N\sigma^{\text{abs}}(\lambda)\,. \tag{5.57}$$

In contrast to the absorbance A, the absorption coefficient takes into account the reduction in transmission at both surfaces due to Fresnel-type losses [5.15]

$$\alpha = \frac{-\ln\left[T/(1-R)^2\right]}{L}\,. \tag{5.58}$$

The absorption coefficient normalizes for path length, while the absorption cross section σ^{abs} normalizes for absorbing ion (AI) concentration [AI] (in ions/m^3) [5.15]

$$\sigma(\lambda) = \frac{\alpha(\lambda)}{[\text{AI}]}\,. \tag{5.59}$$

The absorption cross section allows for a more fundamental comparison of various concentrations of absorbing ions (as long as they obey Beer's law) within different matrices.

Summary of Absorption Mechanisms in Glass

Glasses, as typical candidates for the simple optical solid model described in Sect. 5.1.1, are transparent from the near UV to the near IR. The regions of transparency and semi-transparency in the visible and near IR are shown for a typical optical glass (BK-7) in Fig. 5.11. The intrinsic absorption is dominated by electronic transitions in the UV–VIS range, while in the IR it is dominated by molecular vibrations [5.20]. There are multiple extrinsic mechanisms, which occur with the inclusion of ionic

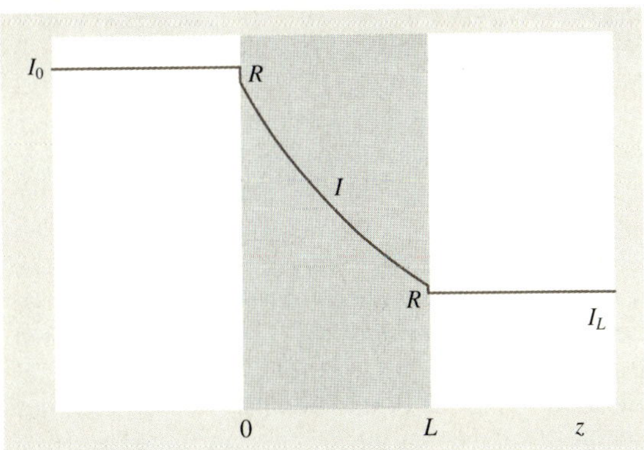

Fig. 5.10 Illustration of the Lambert–Beer law, displaying the bulk absorption inside the optical material and reflective Fresnel losses R at both surfaces

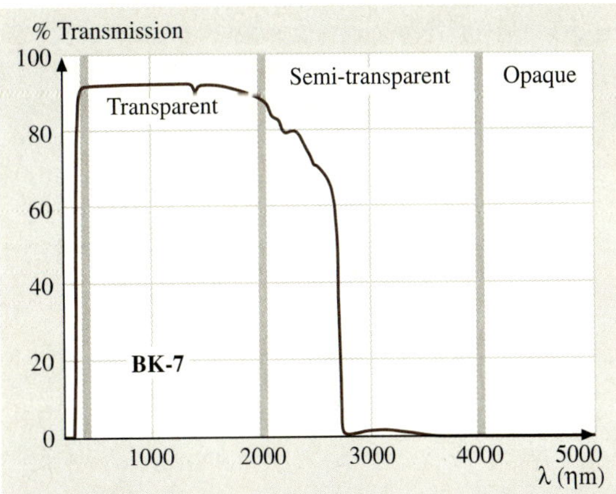

Fig. 5.11 Transmission through 10 mm thick BK-7 optical glass showing regions of transparency (visible), semi-transparency (near IR) and opacity (far IR)

complexes, insulators (crystals and phase separation), semiconductors, and conductors [5.19].

Intrinsic Absorption Mechanisms in Glass. The UV–VIS absorption is dominated by electronic transitions [5.22]. The electronic transitions that occur in the UV–VIS range involve a transition from the highest occupied molecular orbital (HOMO) to the lowest unoccupied molecular orbital (LUMO). The difference in energy (ΔE) between the LUMO (E_{LUMO}) and HOMO (E_{HOMO}) energy levels corresponds to the energy of the corresponding absorption maxima, which can be related to the wavelength, and frequency of the absorption through Plank's constant and the speed of light [5.22]

$$\Delta E = E_{\text{LUMO}} - E_{\text{HOMO}} = h\nu = hc/\lambda \,. \tag{5.60}$$

The bonds present dominate the energy gap between the HOMO and LUMO, and thus the energy of the transition [5.22]. For example, the UV absorption is very similar between crystalline and fused silica, indicating that the short-range structure dominates the electronic transitions [5.22]. It is in the long-range structure of silica that fused and crystalline differ due to the random connections of the SiO_4 tetrahedra in fused silica.

The absorption band present in Fig. 5.12 represents the intrinsic UV absorption in silica. The increase in peak width is due to the distribution of tetrahedra in the fused silica. The lowest energy peak is associated with a Wannier exciton, but the next three peaks are associated with electronic transitions between the oxygen 2p electrons with the silicon 3d electrons [5.15].

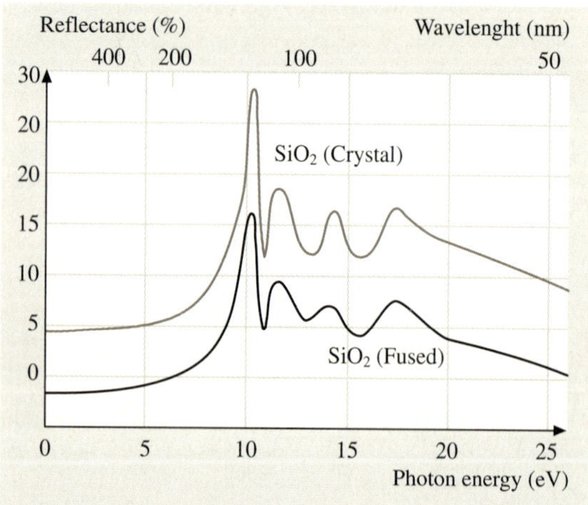

Fig. 5.12 Intrinsic absorption of glassy and crystalline SiO_2 (after [5.15]). The values for fused silica have been lowered by 5% for improved visibility

The intrinsic absorption of glass in the IR is due to molecular vibrations, which begin at the multiphonon edge [5.23]. These vibrations can be understood using a harmonic oscillator model. For a diatomic molecule, the frequency of vibration, ν_{Ph}, is dependent upon the reduced mass (M) and the force constant (f) [5.15]:

$$\nu_{\text{Ph}} = \frac{1}{2\pi}\sqrt{\frac{f}{M}} \,, \tag{5.61}$$

where the reduced mass between two atoms is

$$\frac{1}{M} = \frac{1}{m_1} + \frac{1}{m_2} \,. \tag{5.62}$$

However, in glasses we are not dealing with a simple diatomic system, which makes it really an *anharmonic* oscillator giving rise to overtone vibrations at higher energies (shorter wavelengths). However, the harmonic oscillator model can be used to gain a qualitative understanding of vibrational frequencies in various glasses. The most common extrinsic absorption centers in the IR are OH, CO_2 and iron [5.15]. The multiphonon edge is shifted to longer wavelengths (smaller frequencies) by heavier ions (greater M) and weaker bonds (smaller f), which is why chalcogenide glasses transmit further into the IR than more traditional oxide glasses [5.23].

In the visible wavelength region, glasses normally display no intrinsic absorption.

Ionic Absorption in Glass. Absorption in glass due to ionic impurities involves transition-metal (TM) and rare-earth (RE) ions that act as the central atom (CA) in ligand complexes absorbing light in different wavelength bands [5.20,22,24,25]. The absorption due to TM and RE ions results from electronic transitions within the d or f orbitals of the TM or RE elements [5.22]. The energy of the absorption band is dependent upon the coordination environment of the CA (coordination number and geometry), the chemical identity of the CA (Cr, Co, Cu ...) and the ligand (O versus F). The base composition of the glasses affects the nature of the ligand bonds available for the CA. This change affects the energy of the absorption bands and thus the resulting color of the glass [5.19]. This effect is taken advantage of when designing filter glass (see Sect. 5.3), and can be used to explain the differences in color of the same TM or RE in different glass matrices. The absorptions result from a splitting of the d orbitals for TM elements and the 4f orbitals for the RE elements in the presence of a ligand field. The ligand field theory that governs these splittings and interactions are covered in detail for TM elements in *Bersuker* [5.22] and in general in

Douglas et al. [5.24]. The absorption cross section for TM elements is approx. 100 times that of the absorption cross section for RE elements [5.15]. This results in distinct coloration at the ppm concentration level for TM elements.

Figure 5.13 shows the variation in absorption due to various TM elements in BK 7 glass at the 1 ppm dopant level [5.15]. The intensity of cobalt and nickel absorption is clearly shown.

These intraband transitions due to electronic transitions within the ionic species d or f orbitals are formally Laporte-forbidden transitions by selection rules, so the intensity is low [5.22, 24]. Charge-transfer transitions involve an electron transition between the CA and the coordinating ligand (interband). These interband transitions are formally allowed, and thus have approx. 100–1000 times greater intensity than the intraband electronic transitions [5.24].

Absorption by Semiconductor Particles. Semiconductor particles in glass are typically too small (1–10 nm) to scatter visible light, however they absorb light over a continuum of wavelengths corresponding to energies greater than the band gap (E_g) of the semiconductor particles [5.10]. The band gap of the semiconductor particles is controlled by the size of the particle and the chemical composition. Typically, semiconductor glasses are melted with Zn, Cd, S, Se and Te raw materials in the batch and upon casting the glasses cool colorless [5.15]. Secondary heat treatment (striking) results in the crystallization of various semiconductor crystal phases in the glass, or a mixture thereof: ZnS, ZnSe, ZnTe, CdS, CdSe, and CdTe. The size and distribution of the semiconductor particles can be controlled by the heat treatment and thus so can the optical properties of the resulting glasses. With the proper heat treatment, one glass can be "struck" into multiple glasses with absorptions leading to red, yellow and orange coloration [5.15]. When the band gap of the particles is large enough, the absorption edge is shifted into the UV and the glass appears colorless. The opposite can also occur when the band gap energy is so low that it absorbs all visible light, and the sample appears black [5.15].

Semiconductor-doped glasses are often used as low-pass filter glasses because of their sharp absorption cutoff. Semiconductor-doped glasses are covered in more detail in Sect. 5.3.

Absorption by Conducting Particles. The coloration of glasses due to small conducting particles in the glass is a combination of both absorption and scattering effects. For small metallic particles, the scattering is governed specifically by Mie's theory but can also be treated in general with the Rayleigh scattering theory [5.10]. Mie scattering is covered in depth in Sect. 5.1.7 and Rayleigh scattering in Sect. 5.1.8.

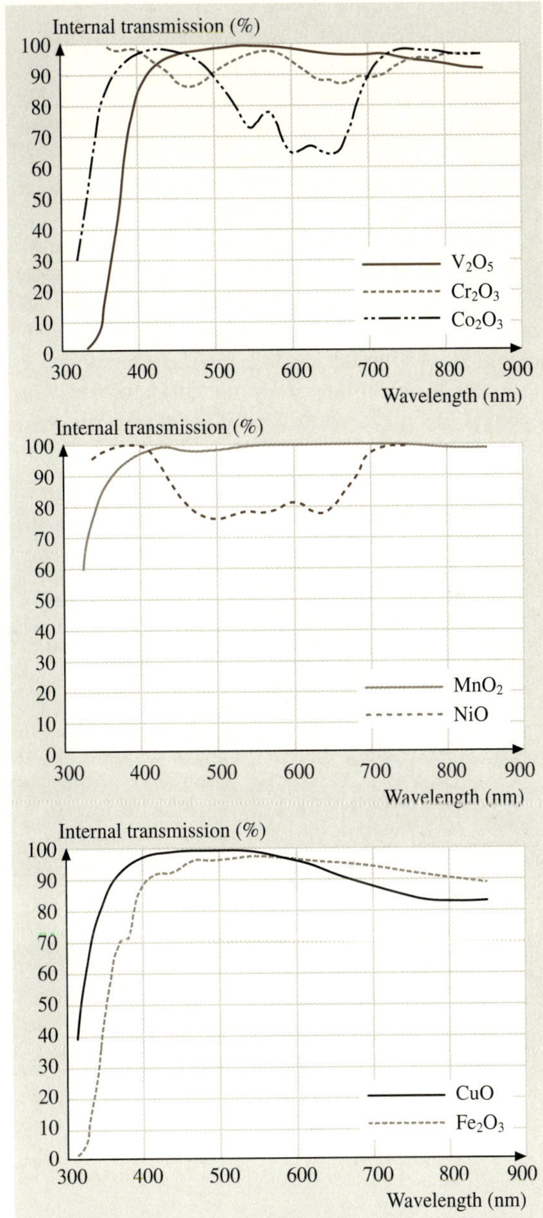

Fig. 5.13 Absorption due to 1 ppm of transition metal elements in BK 7 glass in 100 mm thick samples (after [5.15])

5.1.4 Optical Anisotropy

Maxwell's Equation in Dielectric Media

In the following section we will discuss the behavior of electromagnetic fields in media, especially in anisotropic materials. The basic interaction of light with matter consists of the interaction of the electric field with charges (electrons and ions) in a material (see Sects. 5.1–5.3). The propagation of light through matter is driven by the polarization of the material. Here we only consider linear effects in the material, i. e. the polarizability depends only linearly on the electric field strength. In a macroscopic formulation the material's response to the external field is given by the dielectric function (Sect. 5.1.1). From a microscopic point of view it is obvious that the polarization and thus the dielectric function depend on the atomic structure of the material. Looking at the various crystal structures [5.26] shown in Fig. 5.14 one can easily understand that along different crystal orientations the dielectric function has different values since the atomic distances have different values. Thus the dielectric function of a crystalline material will not just be a number, but will be represented by a matrix (a two-dimensional tensor). The well-known occurrence of birefringence in certain materials is a consequence of the orientation-dependent dielectric function.

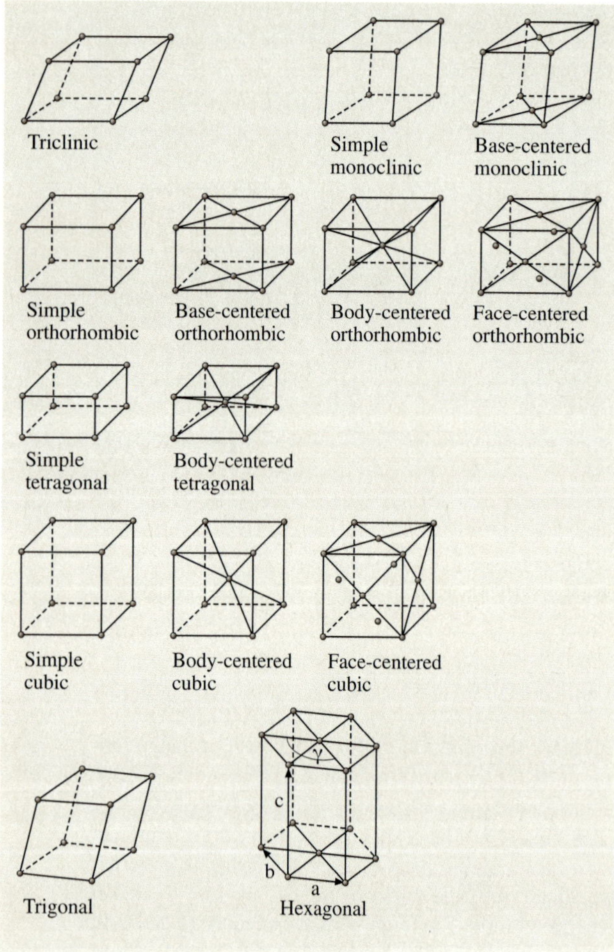

Fig. 5.14 The fourteen Bravais lattices [5.26]

We start again with the Maxwell equation [5.27] for the solution of a plane wave. This is appropriate since any solution can be described as the superposition of plane waves. Inserting a plane wave ansatz

$$\boldsymbol{E}=\boldsymbol{E}_0\,\mathrm{e}^{\mathrm{i}(\boldsymbol{k}\cdot\boldsymbol{r}-\omega t)}=\boldsymbol{E}_0\,\mathrm{e}^{\mathrm{i}\omega\left(\frac{k}{\omega}\boldsymbol{l}\cdot\boldsymbol{r}-t\right)}=\boldsymbol{E}_0\,\mathrm{e}^{\mathrm{i}\omega\left(\frac{1}{v}\boldsymbol{l}\cdot\boldsymbol{r}-t\right)} \tag{5.63}$$

(with the phase velocity $v=\omega/k$ and $\boldsymbol{l}=\boldsymbol{k}/k$) into Maxwell's equations we find the following relations for the field components

$$\boldsymbol{D}=-\frac{1}{\mu\mu_0\omega}\boldsymbol{k}\times(\boldsymbol{k}\times\boldsymbol{E}_0)\,\mathrm{e}^{\mathrm{i}(\boldsymbol{k}\cdot\boldsymbol{r}-\omega t)}\,, \tag{5.64}$$

$$\boldsymbol{B}=\frac{1}{\omega}(\boldsymbol{k}\times\boldsymbol{E}_0)\,\mathrm{e}^{\mathrm{i}(\boldsymbol{k}\cdot\boldsymbol{r}-\omega t)}\,, \tag{5.65}$$

and

$$\boldsymbol{H}=\frac{1}{\mu\mu_0\omega}(\boldsymbol{k}\times\boldsymbol{E}_0)\,\mathrm{e}^{\mathrm{i}(\boldsymbol{k}\cdot\boldsymbol{r}-\omega t)}\quad(\boldsymbol{B}=\mu\mu_0\boldsymbol{H})\,. \tag{5.66}$$

The relation of the linear response of the material to the external electric field together with the Maxwell equation will give conditions for the existence of the above ansatz of a plane wave. Not taking into account dispersion, i. e. the frequency dependence of the dielectric function, the material's polarization is given by [5.28]

$$D_i=\varepsilon_{ij}\varepsilon_0 E_j \tag{5.67}$$

(where i and j are indices for the spatial components and the Einstein sum convention is used). Since the dielectric tensor is symmetric one can always find a coordinate system, where the tensor is diagonal, the system of principle axes [5.28, 29]. In the following we will always work in this reference system.

$$\begin{pmatrix}\varepsilon_1 & 0 & 0\\ 0 & \varepsilon_2 & 0\\ 0 & 0 & \varepsilon_3\end{pmatrix}\,. \tag{5.68}$$

The relation between D and E now gives $D_i = \varepsilon_i \varepsilon_0 E_i$, (no sum over i). Inserting this into (5.64) and using the expansion of two cross products one finds

$$\varepsilon_i \varepsilon_0 \mu \mu_0 \omega^2 E_i = k^2 E_i - k_i(\boldsymbol{k} \cdot \boldsymbol{E}) . \tag{5.69}$$

This is the basic equation which determines the solution that can exist in the material. Introducing a unit vector in the direction of propagation $\boldsymbol{l}$ and defining a velocity v_i one gets:

$$\frac{\omega^2}{v_i} E_i = k^2 \left[E_i - l_i(\boldsymbol{l} \cdot \boldsymbol{E})\right] ,$$
$$\left(\boldsymbol{k} = k\boldsymbol{l} , \quad |\boldsymbol{l}| = 1 , \quad \frac{1}{v_i} = \sqrt{\varepsilon_i \varepsilon_0 \mu \mu_0} , \right.$$
$$\left. E_i = \frac{l_i(\boldsymbol{l} \cdot \boldsymbol{E})}{1 - v^2/v_i^2} , \quad v = \frac{\omega}{k} \right) . \tag{5.70}$$

Multiplying the last equation by l_i and summing over i results in a quadratic equation for v, the so called Fresnel equation of wave normals [5.28, 29]

$$\ell \cdot \boldsymbol{E} = \sum_i \frac{l_i^2}{1 - v^2/v_i^2} (\boldsymbol{l} \cdot \boldsymbol{E})$$
$$\Leftrightarrow 1 = \sum_i \frac{l_i^2}{1 - v^2/v_i^2} \left(1 = \sum_i l_i^2\right)$$
$$0 = \sum_i \frac{l_i^2}{1 - v^2/v_i^2} - l_i^2 = v^2 \sum_i \frac{l_i^2}{v^2 - v_i^2}$$
$$\Leftrightarrow \sum_i \frac{l_i^2}{v^2 - v_i^2} = 0 . \tag{5.71}$$

This is the equation usually discussed in crystal optics. Since it is a quadratic equation in the variable v, there are at most two solutions for the phase velocity. This is the origin of birefringence which we will discuss in detail in the section on uniaxial materials.

Classes of Optical Anisotropy Materials

From the representation of the dielectric tensor in the system of principle axes we see that there are at most three different values possible for the dielectric function. Before discussing in detail birefringence in matter, we discuss first the different classes of optical anisotropic materials resulting from the dielectric tensor [5.29].

Group I. Crystals in which all three values in the dielectric tensor in (5.68) are equal, $\varepsilon_1 = \varepsilon_2 = \varepsilon_3 = \varepsilon$, belongs to the so called cubic class. The material is said to be *optical isotropic*. The relation between D and E reduces to the well-known form $\boldsymbol{D} = \varepsilon \boldsymbol{E}$.

Group II. The next possible case is that in which two of the three axes are equivalent. The corresponding dielectric tensor has the property $\varepsilon_1 = \varepsilon_2 \neq \varepsilon_3$. Such a crystal is said to be *uniaxial*. This group of crystals belong to the trigonal, tetragonal and hexagonal system (Fig. 5.14).

Group III. The last group of crystals has three different values in the dielectric tensor $\varepsilon_1 \neq \varepsilon_2 \neq \varepsilon_3$. All three axes are different and the crystals belong to the orthorhombic, monoclinic and triclinic class (Fig. 5.14).

Uniaxial Materials: Ordinary and Extraordinary Rays

As discussed above a material is said to be uniaxial if two of the three values of the dielectric tensor are equal $\varepsilon_1 = \varepsilon_2 \neq \varepsilon_3$. As a result the velocities v_1 and v_2 are also equal ($v_1 = v_2$). The Fresnel equation of the wave normals (5.71) for this case have the following form

$$0 = \left(v^2 - v_1^2\right)\left(v^2 - v_3^2\right)\left(l_1^2 + l_2^2\right) + \left(v^2 - v_1^2\right)^2 l_3^2 \Leftrightarrow$$
$$0 = \left(v^2 - v_1^2\right)\left[\left(v^2 - v_3^2\right)\left(l_1^2 + l_2^2\right) + \left(v^2 - v_1^2\right) l_3^2\right] . \tag{5.72}$$

This equation has two solutions for the phase velocity

$$1: v = v_1$$
$$2: \left(v^2 - v_3^2\right)\left(l_1^2 + l_2^2\right) + \left(v^2 - v_1^2\right) l_3^2 = 0$$
$$\Rightarrow v^2 = v_3^2\left(l_1^2 + l_2^2\right) + v_1^2 l_3^2 . \tag{5.73}$$

Expressing the unit vector $\boldsymbol{l}$ in spherical coordinates, $l_3^2 = \cos^2(\vartheta)$, $l_1^2 + l_2^2 = \sin^2(\vartheta)$, one gets

$$v^2 = v_3^2 \sin^2(\vartheta) + v_1^2 \cos^2(\vartheta) . \tag{5.74}$$

Usually one uses the notation $v_o = v_1$ (index o: ordinary) and $v_e = v_3$ (index e: extraordinary) (Fig. 5.16). Finally we arrive at the two solutions for the phase velocity of a plane wave propagating through the medium

$$1. \; v = v_o ,$$
$$2. \; v = \sqrt{v_e^2 \sin^2(\vartheta) + v_o^2 \cos^2(\vartheta)} . \tag{5.75}$$

The z-axis is called the optical axis. For the refractive index one obtains

$$n = \frac{c}{v} \Rightarrow$$
$$n(\vartheta) = c \frac{1}{\sqrt{v_o^2 \cos^2(\vartheta) + v_e^2 \sin^2(\vartheta)}}$$
$$= \frac{n_o n_e}{\sqrt{n_o^2 + \left(n_e^2 - n_o^2\right) \cos^2(\vartheta)}} . \tag{5.76}$$

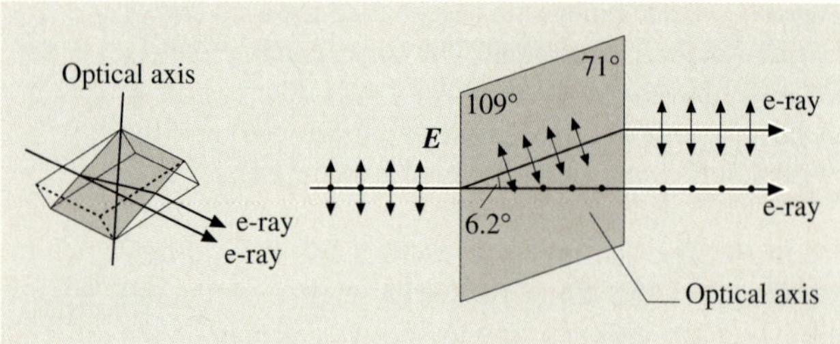

Fig. 5.15 Birefringence in calcite ($CaCO_3$), ordinary (o) and extraordinary ray (e) observed by perpendicular incidence to a surface that has an angle of approximately 45° to the optical axis [5.30]

The case $v_e > v_o$, corresponding to $n_o > n_e$, is called a positive crystal; the other case $v_e < v_o$ corresponding to $n_o < n_e$ is called a negative crystal.

Thus we have derived that in a uniaxial crystal there are always two solutions – the ordinary and extraordinary beam – with different phase velocities; one of them, the extraordinary beam, depends on the direction. To be able to understand the picture of these two rays traveling through for example a calcite crystal one must consider another quantity used in the theory of electro-magnetic radiation – namely the Poynting vector $\boldsymbol{S}$, which describes the energy flux of electromagnetic radiation (Sect. 5.1.1). For a plane wave with propagation vector $\boldsymbol{k}$ it follows from the Maxwell equation that the field vectors $\boldsymbol{E}, \boldsymbol{H}, \boldsymbol{S}$ and $\boldsymbol{D}, \boldsymbol{H}, \boldsymbol{k}$ form a set of orthogonal vectors:

$$\begin{aligned} \boldsymbol{S} &= \boldsymbol{E}\times\boldsymbol{H} \rightarrow \boldsymbol{S}\perp\boldsymbol{H} \text{ and } \boldsymbol{E}\,, \\ \boldsymbol{k}\cdot\boldsymbol{D} &= 0 \rightarrow \boldsymbol{k}\perp\boldsymbol{D}\,, \\ \boldsymbol{k}\cdot\boldsymbol{H} &= 0 \rightarrow \boldsymbol{k}\perp\boldsymbol{H} \\ \boldsymbol{D} &= -\frac{1}{\omega}(\boldsymbol{k}\times\boldsymbol{H})\,. \end{aligned} \tag{5.77}$$

The angle α between the vectors $\boldsymbol{k}$ and $\boldsymbol{S}$ is the same as that between $\boldsymbol{D}$ and $\boldsymbol{E}$ (Fig. 5.17). In an isotropic medium the angle α is zero, since D is proportional to E. As a consequence, $\boldsymbol{k}$ and $\boldsymbol{S}$ point in the same direction. In an anisotropic medium this is generally no longer true. The light traveling through a uniaxial material and in particular the two rays shown in the calcite example are in fact the $\boldsymbol{S}$ vectors of the ordinary and extraordinary rays [5.30].

The angle between the ordinary and extraordinary ray is given by α, since the ordinary ray travels along $\boldsymbol{k}$ (remember, $\boldsymbol{k}$ and $\boldsymbol{S}$ point in the same direction). Thus the angle αgives the splitting between the ordinary and extraordinary ray in a uniaxial crystal. In the following derivation we want to obtain a rather simple expression for the angle α determined by the refractive indices n_e and n_o. Since we are only interested in the angle between two vectors we can consider unit vectors. Taking $\boldsymbol{D}$ as a unit vector, $\boldsymbol{E}$ is given for the material equation. In the framework of the principle axis

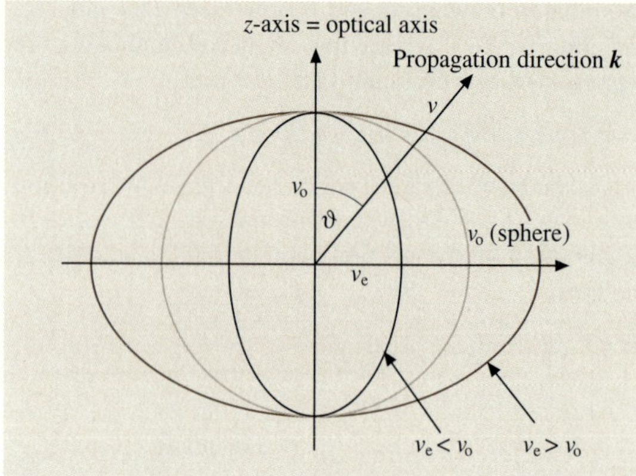

Fig. 5.16 Graphical representation of the solution to Fresnel's equation (Indikatrix)

$$\boldsymbol{k} = \begin{pmatrix} \sin(\vartheta)\cos(\varphi) \\ \sin(\vartheta)\sin(\varphi) \\ \cos(\vartheta) \end{pmatrix}, \quad \boldsymbol{D} = \begin{pmatrix} \sin(\vartheta')\cos(\varphi') \\ \sin(\vartheta')\sin(\varphi') \\ \cos(\vartheta') \end{pmatrix}, \tag{5.78}$$

since $\boldsymbol{k}\cdot\boldsymbol{D} = 0$, it follows that

$$\begin{aligned} &\cos(\vartheta)\cos(\vartheta') + \sin(\vartheta)\sin(\vartheta')\cos(\varphi-\varphi') = 0 \Leftrightarrow \\ &\cot(\vartheta') = -\tan(\vartheta)\cos(\varphi-\varphi')\,. \end{aligned}$$

For the angle-independent solution v_o it follows that

$$\begin{aligned} &\cos(\varphi-\varphi') = 0 \Rightarrow \varphi' = \varphi - \frac{\pi}{2} \\ &\text{and } \cot(\vartheta') = 0 \Rightarrow \vartheta' = \frac{\pi}{2}\,. \end{aligned} \tag{5.79}$$

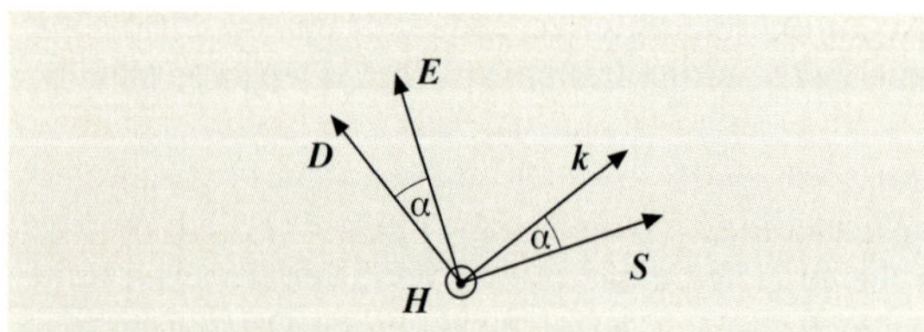

Fig. 5.17 Angle α between $\boldsymbol{k}$, $\boldsymbol{S}$, $\boldsymbol{D}$ and $\boldsymbol{E}$

For the second solution, which must be orthogonal to the first, it follows that: $\varphi' = \varphi$, $\vartheta' = \vartheta - \frac{\pi}{2}$.

Together with the relation between D and E one has

$$\boldsymbol{D}^{(1)} = \begin{pmatrix} -\cos(\vartheta)\cos(\varphi) \\ -\cos(\vartheta)\sin(\varphi) \\ \sin(\vartheta) \end{pmatrix}, \quad \boldsymbol{D}^{(2)} = \begin{pmatrix} \sin(\varphi) \\ -\cos(\varphi) \\ 0 \end{pmatrix}$$

$$\text{and } \boldsymbol{E}^{(1)} = \begin{pmatrix} -\cos(\vartheta)\cos(\varphi)\frac{1}{\varepsilon_1} \\ -\cos(\vartheta)\sin(\varphi)\frac{1}{\varepsilon_1} \\ \sin(\vartheta)\frac{1}{\varepsilon_3} \end{pmatrix}, \quad \boldsymbol{E}^{(2)} = \frac{1}{\varepsilon_1}\boldsymbol{D}^{(2)} . \tag{5.80}$$

In the second case $\boldsymbol{D} \propto \boldsymbol{E}$ and, hence, $\alpha = 0$, which is obviously the ordinary ray. For the first solution, the extraordinary ray, we calculate the angle α from the scalar product $\boldsymbol{D} \cdot \boldsymbol{E} = |\boldsymbol{D}|\,|\boldsymbol{E}|\cos(\alpha)$, substituting D and E it follows that

$$\begin{aligned}
&\frac{1}{\varepsilon_1}\cos^2(\vartheta)\cos^2(\varphi) \\
&\quad + \frac{1}{\varepsilon_1}\cos^2(\vartheta)\sin^2(\varphi) + \frac{1}{\varepsilon_3}\sin^2(\vartheta) = \\
&1\left(\frac{1}{\varepsilon_1^2}\cos^2(\vartheta)\cos^2(\varphi) + \frac{1}{\varepsilon_1^2}\cos^2(\vartheta)\sin^2(\varphi)\right. \\
&\quad \left. + \frac{1}{\varepsilon_3^2}\sin^2(\vartheta)\right)^{1/2} \cos(\alpha) \Leftrightarrow \\
&\frac{1}{\varepsilon_1}\cos^2(\vartheta) + \frac{1}{\varepsilon_3}\sin^2(\vartheta) = \\
&\left(\frac{1}{\varepsilon_1^2}\cos^2(\vartheta) + \frac{1}{\varepsilon_3^2}\sin^2(\vartheta)\right)^{1/2} \cos(\alpha) \Leftrightarrow \\
&\cos(\alpha) = \frac{\varepsilon_3\cos^2(\vartheta) + \varepsilon_1\sin^2(\vartheta)}{\left(\varepsilon_3^2\cos^2(\vartheta) + \varepsilon_1^2\sin^2(\vartheta)\right)^{1/2}} , \\
&\cos(\alpha) = \frac{n_e^2\cos^2(\vartheta) + n_o^2\sin^2(\vartheta)}{\left(n_e^4\cos^2(\vartheta) + n_o^4\sin^2(\vartheta)\right)^{1/2}} ,
\end{aligned} \tag{5.81}$$

with the relation $\varepsilon = n^2$ and the notation $n_3 = n_e$, $n_1 = n_o$. As an example let us consider the aforementioned calcite. The surface of the calcite that one gets by cleaving the crystal has an angle of approximately 45°. Light incident perpendicular to the surface thus gives $\cos^2(\vartheta) = \sin^2(\vartheta) = 0.5$. The refractive indices for the ordinary and extraordinary rays in calcite are: $n_e = 1.4864$, $n_o = 1.6584$. From these values one finds the angle between the ordinary and extraordinary ray as (Fig. 5.15):

$$\cos(\alpha) = \frac{1}{\sqrt{2}}\frac{n_e^2 + n_o^2}{\sqrt{n_e^4 + n_o^4}} \Rightarrow \alpha = 6.22° . \tag{5.82}$$

5.1.5 Nonlinear Optical Behavior and Optical Poling

Since the first observation of nonlinear (NL) effects, many materials have been investigated concerning their nonlinear optical properties [5.31,32]. Nonlinear materials can be found among gases and vapors, polymers, liquid crystals, organic solvents, or crystals – in short, in nearly every material system.

A linear dielectric medium is characterized by a linear relation between the polarization density $\boldsymbol{P}$ and the electric field $\boldsymbol{E}$, $\boldsymbol{P} = \varepsilon_0 \chi \boldsymbol{E}$ (5.25), where ε_0 is the vacuum permittivity and χ is the electric susceptibility of the medium (see Sect. 5.1.1). A nonlinear dielectric medium, on the other hand, is characterized by a nonlinear relation between $\boldsymbol{P}$ and $\boldsymbol{E}$, as illustrated in Fig. 5.18.

The nonlinearity may be of microscopic or macroscopic origin. The polarization density $\boldsymbol{P} = N\boldsymbol{p}$ is a product of the individual dipole moment $\boldsymbol{p}$ (which is induced by the applied electric field E) and the number density of dipole moments N. The nonlinear behavior may have its origin in either $\boldsymbol{p}$ or N.

The relation between $\boldsymbol{p}$ and $\boldsymbol{E}$ is linear when $\boldsymbol{E}$ is small, but becomes nonlinear as $\boldsymbol{E}$ acquires values comparable with interatomic electric fields (typically 10^5–10^8 V/m). This may be explained in terms of the simple Lorentz model in which the dipole moment is $\boldsymbol{p} = -\mathrm{e}\boldsymbol{x}$, where $\boldsymbol{x}$ is the displacement of a mass with charge $-\mathrm{e}$ to an applied electric force $-\mathrm{e}\boldsymbol{E}$. If the restraining elastic force is proportional to the displacement (i. e., if Hooke's law is satisfied), the equilibrium displacement $\boldsymbol{x}$ is proportional to $\boldsymbol{E}$. Hence, $\boldsymbol{P}$ is proportional to $\boldsymbol{E}$, and the medium is linear. However, if the restraining force is a nonlinear function of the

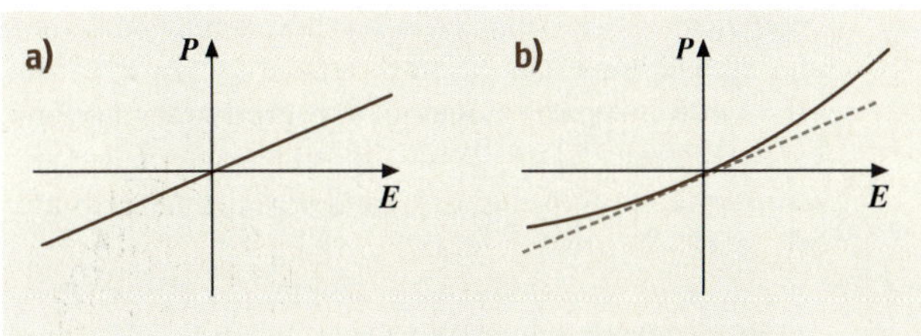

Fig. 5.18a,b The P–E relation for **(a)** a linear dielectric medium, and **(b)** a nonlinear medium

displacement, the equilibrium displacement $\boldsymbol{x}$ and the polarization density $\boldsymbol{P}$ are nonlinear functions of $\boldsymbol{E}$ and, consequently, the medium is nonlinear.

Another possible origin of the nonlinear response of an optical material to light is the dependence of the number density N on the optical field. An example is a laser medium for which the number of atoms occupying the energy levels involved in the absorption and emission of light is dependent upon the light intensity itself (see Sect. 5.1.7).

Since externally applied optical fields are typically small in comparison with characteristic interatomic or crystalline fields, even when focused laser light is used the nonlinearity is usually weak. The relation between $\boldsymbol{P}$ and $\boldsymbol{E}$ is then approximately linear for small $\boldsymbol{E}$, deviating only slightly from linearity as $\boldsymbol{E}$ increases (Fig. 5.18). Under these circumstances, it is possible to expand the function that relates $\boldsymbol{P}$ to $\boldsymbol{E}$ in a Taylor series about $\boldsymbol{E}=0$

$$\boldsymbol{P}(\boldsymbol{E})=\varepsilon_0\left(\chi^{(0)}\boldsymbol{E}+\chi^{(2)}\boldsymbol{E}^2+\chi^{(3)}\boldsymbol{E}^3+\ldots\right), \quad (5.83)$$

where $\chi^{(0)}$ is the linear susceptibility and $\chi^{(n>0)}$ are the higher-order coefficients.

For the nonlinear refractive index n, we get

$$n(\boldsymbol{E})=n_0+n_1\boldsymbol{E}+n_2\boldsymbol{E}^2+\ldots . \quad (5.84)$$

Equations (5.83) and (5.84) provide the basic description for a nonlinear optical medium. Anisotropy (Sect. 5.1.4), dispersion (Sect. 5.1.2), and inhomogeneities (Sect. 5.1.7) have been ignored both for simplicity and to enable us to focus here on the basic nonlinear effect.

In centrosymmetric media with inversion symmetry (like glass), in which the properties of the medium are not altered by the transformation $\boldsymbol{r}\rightarrow-\boldsymbol{r}$, the $\boldsymbol{P}(\boldsymbol{E})$ function must have odd symmetry so that the reversal of $\boldsymbol{E}$ results in the reversal of $\boldsymbol{P}$ without any other change. The second-order nonlinear coefficient $\chi^{(2)}$ must then vanish, and the lowest nonlinear term is of third order. Typical values of the second-order nonlinear coefficient $\chi^{(2)}$ for dielectric crystals, semiconductors, and organic materials lie in the range $\chi^{(2)}=10^{-24}$ to 10^{-21} m/V. Typical values of the third-order nonlinear coefficient $\chi^{(3)}$ for glasses, crystals, semiconductors, semiconductor-doped glasses, and organic materials are $\chi^{(3)}=10^{-34}$ to $10^{-29}\,(\mathrm{m/V})^2$.

Nonlinear Wave Equation

The propagation of light in a nonlinear medium is governed by the wave equation (5.29), which can be written in the form [5.34]:

$$\Delta\boldsymbol{E}-\frac{n^2}{c_0^2}\frac{\partial^2\boldsymbol{E}}{\partial t^2}=-\mu_0\frac{\partial^2\boldsymbol{P}_{\mathrm{NL}}}{\partial t^2} \quad (5.85)$$

where $\boldsymbol{P}_{\mathrm{NL}}$ is the nonlinear part of the polarization density:

$$\boldsymbol{P}_{\mathrm{NL}}=\varepsilon_0\left(\chi^{(2)}\boldsymbol{E}^2+\chi^{(3)}\boldsymbol{E}^3+\ldots\right). \quad (5.86)$$

It is useful to regard (5.85) as a wave equation in which the term $-\mu_0\partial^2\boldsymbol{P}_{\mathrm{NL}}/\partial t^2$ acts as a source radiating in a linear medium of refractive index n. Because $\boldsymbol{P}_{\mathrm{NL}}$ is a nonlinear function of $\boldsymbol{E}$, (5.85) is a nonlinear partial differential equation in $\boldsymbol{E}$. This is the basic equation that underlies the theory of nonlinear optics. There are two approximate approaches to solve the nonlinear wave equation: the Born approximation and the coupled wave theory.

Optical Classification of Nonlinear Effects

Generally, the term "nonlinear optics" describes that part of the interaction between light and matter in which the optical material properties of the light beam passing through a nonlinear medium are changed by the beam itself or another electromagnetic field. Table 5.3 gives a short overview over the various nonlinear effects and their order of nonlinearity.

Here, we will focus our discussion on one of the most prominent effects: second-harmonic generation.

Second-Harmonic Generation (SHG)

In the case of the main second-order application [5.31] – second-harmonic (SH) generation – which e.g. can

Table 5.3 Nonlinear processes [5.33]

Nonlinear optical effect	Order
nth harmonic generation (e.g. SHG, rectification)	n
n-wave mixing	$n-1$
nth photon absorption	n
Phase conjugation	3
Intensity-dependent refraction (e.g. Kerr effect, self phase modulation, self modulation)	2, 3
Optical parametric oscillation	2
Induced opacity	2
Induced reflectivity	2
Raman and Brillouin scattering	2

be used to convert IR laser light to Vis–UV light, the amount of second-harmonic light generated by the material depends on $\chi^{(2)}$.

When applying (5.85) to planar waves (5.8) the source term $-\mu_0 \partial^2 \boldsymbol{P}_{\mathrm{NL}}/\partial t^2$ in (5.85) has a complex amplitude $4(\omega/c_0)^2 \chi^{(2)} E^2(\omega)$ which radiates an optical field component at frequency 2ω. Thus the SH optical field has a component at the second harmonic of the incident optical field. Fig. 5.19 illustrates the SHG effect.

Since the amplitude of the emitted second-harmonic light is proportional to $\omega^2 \chi^{(2)} I$, its intensity is proportional to $\omega^4 (\chi^{(2)})^2 I^2$, where I is the intensity of the incident wave. The intensity of the second-harmonic wave is therefore proportional to $(\chi^{(2)})^2$, to $1/\lambda^4$, and to I^2. Consequently, the efficiency of second-harmonic generation is proportional to $I = P/A$, where P is the incident power and A is the cross-sectional area.

It is therefore essential that the incident wave have the largest possible power and be focused to the smallest possible area to produce strong second-harmonic radiation. Pulsed lasers are convenient in this respect since they deliver large peak powers. To enhance the efficiency of second-harmonic generation, the interaction region should also be as long as possible. Since diffraction effects limit the distances within which light remains confined, waveguide structures that confine light for relatively long distances offer a clear advantage.

Third-Order Nonlinear Effects in Glass

For materials which have a center of inversion $\boldsymbol{P}(\boldsymbol{E})$ is an odd function. This means that even terms are all zero and no even harmonics can be observed. This is the case for isotropic media like glass. Third-order effects are orders of magnitude smaller than second-order ones. Besides polarization, the effective refractive index n_{eff} of the material depends on the intensity of the radiation (5.84). The nonlinear refractive index n_2 is usually determined by the polarizability of the ionic components of the material. In conventional glass systems, n_2 is determined by the amount of nonbridging oxygens and thus on the concentration of glass modifiers (see Sect. 5.2).

Material-Based Classification of Nonlinear Optical Effects

When considering nonlinear effects [5.34], one must distinguish between resonant (where the optical frequency is resonant with an electronic transition in the material), nonresonant, and intrinsic and extrinsic effects [5.35–38].

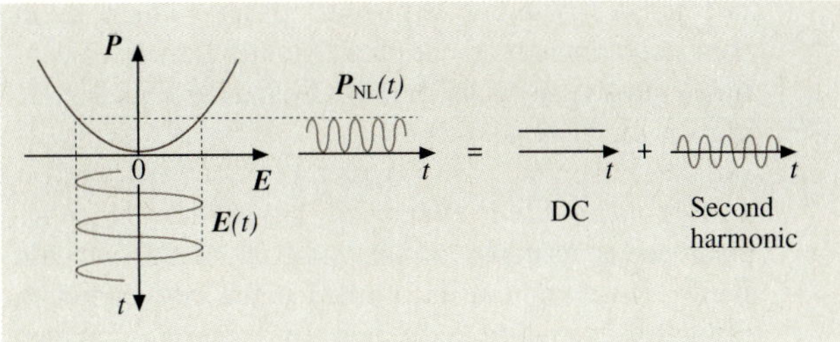

Fig. 5.19 A planar wave of frequency ω in a second-order nonlinear optical medium creates a polarization with component 2ω and a steady (dc) component (optical rectification)

Resonant Nonlinearities are found when the light wavelength is in the spectral range of a fundamental absorption due to electronic transitions or molecular vibrations. Resonant effects are usually rather large, but have slow relaxation times, depending on the relaxation time of the excited carriers. They show higher power dissipation and are accompanied by sample heating.

Nonresonant Nonlinearities are of purely electronic origin and arise from the distortion of the electron orbits (polarizability). The electronic effects range in a sub-ps time scale, with minimum heating of the device, but have only small nonlinearities and thus require high intensities [5.39].

Intrinsic effects are nonlinear effects where the material itself is responsible for the NL properties, for instance the Kerr effect. In crystallites as well as in glasses with intrinsic nonlinearity, three types of anion groups are believed to be mainly responsible for large nonlinearities [5.40]:

- distorted Met$-$O$_6$ octahedra: Tm^{3+}, Yb^{3+}, Sr^{2+};
- anion groups with a lone electron pair: e.g. Bi^{3+};
- anion groups with dislocated conjugated π-bond planar structure, like organic molecules or $[BO_3]^{3-}$-plane.

Extrinsic effects are properties of materials or dopants embedded in the material, such as rare-earth (RE) ions, microcrystallites, dyes or impurities, where the nonlinear effect is associated with the dopant, or are due to light-induced compositional changes, such as reversible or irreversible chemical reactions.

As a rule of thumb one can say that to get a NRI (nonresonant intrinsic) at least in the range of 10^{-11}, a linear refractive index $n_{\mathrm{D}} > 2$ and Abbe number $\nu < 20$ are required. Most of the oxide glasses have refractive indices below 2 but some glasses which fulfill these requirements are the chalcogenide glasses, chalcohalides, heavy metal oxide (HMO)- and RE-doped oxide glasses [5.37, 38, 41–43].

Coherent Nonlinear effects are characterized by an instantaneous response of the material to the incident fields. This condition is fulfilled if the corresponding electronic excitation in the dielectric material is created only virtually – using the language of coupled wave theory, see p. 266 – or, within the quantum mechanical treatment, if we treat only interactions of coherent polaritons [5.31].

Incoherent Nonlinear effects involve modifications of the optical properties by really or incoherently excited species, e.g. electron–hole pairs, excitons, or phonons. These species have finite lifetimes τ which can be on the order of ns to ms. Due to this finite lifetime, their density does not instantaneously follow the incident light field, but depends on the generation and depopulation rate, which are characterized by an instantaneous response of the material to the incident field. Table 5.4 lists some effects in optical materials and typical values.

Creation of Artificial Second-Order Nonlinearity: Optical Poling

As mentioned before, the SHG process is related to the second-order susceptibility and thus allowed only in non-centrosymmetric materials. On the other hand, third-order nonlinear materials such as glass offer advantages such as easier and cheaper fabrication with high homogeneity. Thus, attempts have been made either to increase the field strength (electric poling) or create artificial symmetry (thermal or optical poling [5.37, 38, 44, 45]), or combine the two methods.

There exist two types of electric field poling: electrode poling (in which a DC field is applied to a pair of electrodes) and corona discharge poling (where the surface of a film is charged under a strong potential). Annealing in vacuum is effective in enhancing the SHG intensity, whereas addition of hydrogen at room temperature increases the lifetime of the optical nonlinearity. Further methods are poling by mechanical stress or with a chemical dilutant [5.33, 40, 42, 46–48].

Higher SHG intensities than those obtained with thermal or electric poling can usually be derived by optical poling. This is due to the fact that an artificial phase matching (quasi phase matching) can be achieved by creating a stable $\chi^{(2)}$-grating in the glass (see Sect. 5.6). Thus, with optical poling phase matching over a sample length of several cm can be achieved, whereas the active region for thermal or electric poling is only between 5–15 μm.

Optical poling is achieved by intensive irradiation by, e.g. 1064 nm light and irradiation by additional light of the SH wavelength during the poling procedure (seeding light). After this procedure intensive SHG (e.g. in Pb–silica, Ge–silica, metal- or SC-doped, or $ZnO-TeO_2$ glasses) has been observed. Fundamental and SH cause a semipermanent DC electric field in the glass (by photoinduced current or structural orientation) which cancels the inversion symmetry and allows the periodic phase matching of the SHG (see also Table 5.22).

Optical poling can be also used for organic materials, e.g. dyes, which have asymmetric structure and exhibit polarization. Dispersed in a glass matrix, they have random orientation and thus are isotropic. Even growth of crystals in the glass does not result in the desired second-order nonlinearity because of the usually centrosymmetric (antiparallel) arrangement of the polar molecules. By irradiation with two laser beams or by applying an electric field, the random orientation can be changed to an ordered configuration.

In detail, poling is a process which can be found especially in the fabrication of photonic devices. Most

Table 5.4 Third-order susceptibility, response time τ, and magnitude of n_2 for various sources of n_2 (NRI) [5.43] (SC = semiconductor, MQW = multi quantum well)

Mechanism	Typical materials	$\chi^{(3)}$ (esu)	Response time τ (s)	n_2 (esu)	Type
Semiconductor bandfilling, exciton effects	GaAs, MQW, SC-doped glasses	$10^{-2}-10^{-8}$	$\approx 10^{-8}$		Resonant
Nonlinear electronic polarizability	PTS, glass	$10^{-11}-10^{-14}$	$10^{-14}-10^{-15}$	$10^{-8}-10^{-14}$	Non-resonant
Optical Kerr effect			$10^{-11}-10^{-12}$	$10^{-11}-10^{-12}$	
Electrostriction			$10^{-7}-10^{-8}$	$10^{-7}-10^{-8}$	
Photorefractive effect	$BaTiO_3$, CdTe	$10^{-4}-10^{-5}$	$10^{-3}-1$		Resonant
Molecular reorientation	CS_2	$10^{-12}-10^{-14}$	$\approx 10^{-12}$		Non-resonant
Thermal effects	ZnS, ZnSe	$10^{-4}-10^{-6}$	$10^{-3}-1$	$10^{-4}-10^{-5}$	Resonant

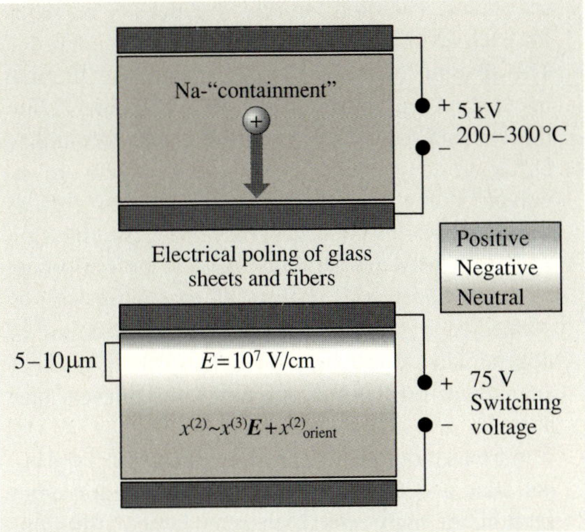

Fig. 5.20 Electrical poling of glass sheets

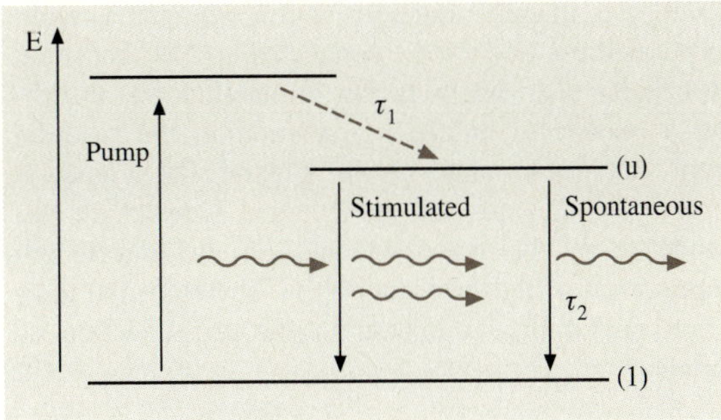

Fig. 5.21 Schematic plot of a three level laser system Via a pump level with a very short lifetime $\tau_1 \ll \tau_2$ the upper laser level (u) is populated. Depolulation can either occur via stimulated emission, where copies of photons are made or via sponateous emission where the probability is determined by the lifetime τ_2

of the optical poling experiments to date have been performed in fibers. But additional SHG in waveguide geometries is also possible. In the case of waveguides, the poling was performed near the glass transition point of a nonlinear core polymer in order to establish a permanent orientation of the nonlinear chromophor dipoles. In the case of e.g. glasses containing Na^+ and H^+ a charge field is established by wandering of these mobile ions in a strong electric field (thermo-electric poling).

5.1.6 Emission

For numerous applications for active optical materials the interrelation between specific absorption and emission processes [5.49] is of crucial importance. One example is a laser or an optical amplifier, where a specific absorption on a well-defined level in the optically active material is needed for pumping. This leads to the building of an inversion in a lower laying state. The inversion can either be emptied by spontaneous emission, which causes noise, or by stimulated emission which amplifies a laser (or signal) mode.

The microscopic processes between these absorption and emission levels can have different origins, for example several levels of 4f electron systems of a rare earth. These levels only interact weakly with lattice vibrations and therefore barely show thermal broadening (zero phonon lines). Examples for laser materials are Nd^{3+}-doped YAG (yttrium aluminate garnet) single crystals and Nd^{3+}-doped phosphate glasses. Examples of optical amplifier materials are Er^{3+}-doped fused silica and Er^{3+}/Yb^{3+}-doped multicomponent glasses (see Fig. 5.23). Here the optical emission of Er^{3+} in the wavelength range of optical telecommunication is used. Further laser systems are e.g. ruby lasers (Cr^{3+} + doped Al_2O_3); semiconductor lasers, where the pumping is done electronically; gas lasers such as the helium–neon laser, where the energy levels of He are used for pumping and the inversion is built up in the energy levels of Ne; or dye lasers where broad π-electron systems of organic molecules are used to provide the necessary tunable energy levels. For more details on laser physics see also [5.49]. The absorption and the absorption cross sections can be directly measured and are described in Sect. 5.1.3. As we will see, the approach to treating spectral emission of an excited material is more sophisticated.

Emission Cross Section

Although absorption has a well-defined meaning with respect to attenuation of light and can be measured quantitatively via the absorption coefficient α it has a quantummechanical origin. Especially for making absorption (and emission) quantitative the quantum-mechanical concept of transition probabilities is needed. It is also usually expressed via oscillator strengths of certain transitions. The fact that light is absorbed at certain frequencies ω (or energies $\hbar\omega$) means that light is portioned in packages (photons) of energy given by

$$\hbar\omega = \frac{hc}{\lambda}\,, \tag{5.87}$$

where h is Planck's quantum and $\hbar = h/2\pi$. The absorption cross section is connected to the transition probability R_{ij}, which is the probability for absorbing a photon to induce a transition in the medium from state i to state j. In general the transition probability depends on the detailed wavelength dependence of the transition and on the wavelength dependence of the incoming light. Therefore the transition probability is related to the cross section as follows:

$$R_{ij} = \int \sigma(\omega)\Phi(\omega)\mathrm{d}\omega = \int \frac{-2\pi c}{\lambda^2}\sigma(\lambda)\Phi(\lambda)\mathrm{d}\lambda\,, \tag{5.88}$$

where Φ is the photon flux measured in units of m^{-2}, which is related to the light intensity $I = \hbar\omega\Phi$.

Quantum Emissions

Furthermore, emission and emission cross sections have a quantum-mechanical nature. Historically the work of Einstein did not only bring its author a Nobel prize but also laid the foundation of laser physics. Einstein introduced the fundamental concepts of stimulated and spontaneous emission. The stimulated emission process makes copies of photons with identical energy and – more importantly – phase. This is the process behind laser activity and amplification. While stimulated emission is just the opposite process of absorption, spontaneous emission is caused by the finite lifetime of an excited state. It is responsible for noise in a laser or optical amplifier. The lifetime τ_2 of the upper, excited level with only one transition $2 \rightarrow 1$ is the inverse of the transition probability A_{21}

$$\tau_2 = \frac{1}{A_{21}}\,. \tag{5.89}$$

In the case of different decay channels the lifetime is the inverse of a sum over all transition probabilities

$$\tau_2 = \frac{1}{\sum_m A_{2m}}\,. \tag{5.90}$$

We now want to define an emission cross section. If we consider an ideal system with a single nondegenerate transition and no nonradiative decay channels, the emission and absorption cross sections are identical $\sigma^{\mathrm{abs}} = \sigma^{\mathrm{emis}}$.

Real systems deviate from this in the following ways:

- The upper as well as the lower laser level are composed of an ensemble of sublevels. Especially in disordered systems such as glasses an inhomogeneous broadening of the energy levels can become important. Therefore an effective density of states for each level has to be taken into account.
- The spacing of the sublevels is such that thermal occupation has to be considered, i. e. where $\Delta E = h\Delta\omega < k_\mathrm{B}T$ where k_B is the Boltzmann constant.
- Different degeneracies of the energy levels are involved. If the upper level has a degeneracy of g_2 and the lower level a degeneracy of g_1, the emission and absorption probabilities have to be multiplied by these degeneracy factors. If an energy state is connected with a magnetic moment, a canceling of degeneracies due to crystal field splitting can occur, which leads to an ensemble of sublevels (see above).
- There exist nonradiative processes that empty the upper state and decrease the radiative emission cross section. At high enough energies lattice vibrations (phonons) can empty an excited state in a nonradiative way.

For these reasons the effective emission cross section for a real system deviates from the absorption cross section in a nontrivial way

$$\sigma^{\mathrm{abs}}(\lambda) \neq \sigma^{\mathrm{emis}}(\lambda)\,. \tag{5.91}$$

An ideal two-level system and a real system are sketched schematically in Fig. 5.22a,b. The emission cross section is related to the transition probability by

$$\sigma^{\mathrm{emis}}(\lambda) = \frac{\lambda^2 A_{21}}{8\pi} S(c/\lambda)\,, \tag{5.92}$$

where $S(c/\lambda) = S(\nu)$ is the lineshape function. The lineshape function is normalized to unity $\int S(\nu)\mathrm{d}\nu = 1$. The

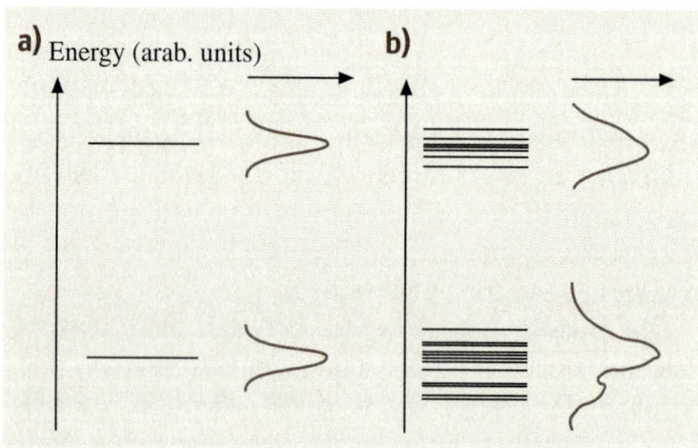

Fig. 5.22a,b In **(a)** the schematic terms for an ideal, only lifetime-broadened emision process from an excited state to the ground state is plotted. Many real systems sketched in **(b)** show a splitting into different sublevels, different degeneracies and nonradiative transitions

Part A | 5.1

question remains if and how it is possible to measure the emission cross section. While for the absorption cross section such a measurement is straightforward, it is far more complicated to complete this process for an emission cross section. The lineshape can be determined by saturating a transition with a light source, then switching off the light and measuring the resulting fluorescence, which is decaying exponentially in time, with a spectrometer. The wavelength dependent signal of the spectrometer is proportional to the emission cross section. If a sample of volume V is irradiated with light until a stationary state is reached, the inversion level $0 < N_2(\boldsymbol{x}) < 1$ is a measure of the excited optical centers. The probability for an emission process to occur in the wavelength range between λ and $\lambda + \Delta\lambda$ is

$$W(\lambda) = \frac{\sigma^{\text{emis}}(\lambda)\Delta\lambda}{\int \sigma^{\text{emis}}(\lambda)\,\mathrm{d}\lambda} . \qquad (5.93)$$

Therefore the light intensity emitted by an infinitesimal volume element $\mathrm{d}V(\boldsymbol{x})$ is given by

$$\mathrm{d}I(\lambda, \boldsymbol{x}) = NN_2\tau^{-1}\,\mathrm{d}V(\boldsymbol{x})h\nu W(\lambda) . \qquad (5.94)$$

Here again N is the concentration of optical centers and $h\nu = hc/\lambda$ is the energy of each emitted phonon with Planck's constant h. For a measurement of $W(\lambda)$, this intensity, which is emitted with equal probability in all spatial directions, has to be collected in a spectrometer which normally covers only a small part of the spatial angle in which the emission takes place. On the way to the spectrometer the light may also travel through other volume elements causing absorption or stimulated emission. Therefore the absolute signal measured in the spectrometer is related in a nontrivial way to the emission cross section. For measuring cross sections the following special conditions are thus used:

- Excitation and measurement are done close to the surface of a sample to avoid the emitted light traveling through large parts of the sample. This is done either in a backscattering geometry or at a corner of the material.
- The excitation is so strong (e.g., with a laser) that saturation is reached $N_2 \approx 1$. In this way absorption of emitted light traveling through the sample is suppressed.
- Standards are used for a comparison. These are e.g. rare-earth ions of well-defined concentration in single crystals, glasses with well-controlled compositions or fluid organic dyes.

Even if it is a nontrivial task to measure the emission cross sections, it is the most important microscopic quantity for the laser activity of a transition.

Example for Er Ions

As an example we show here the emission and absorption cross sections of Er^{3+} for the transition around 1550 nm. This transition is important for telecommunication applications since it takes place in the wavelength range where silica fibers exhibit minimum light attenuation. The states involved in this transition have the spectroscopic symbols $E_1 - \rightarrow {}^4\mathrm{I}_{15/2}$ and $E_2 \rightarrow {}^4\mathrm{I}_{13/2}$ which stand for spin, angular and total moment of the correlated eleven 4f electron system which forms the Er^{3+} system. In Fig. 5.23 we plot an example of this transition of Er^{3+}, which is of high technological importance. The value of the emission cross section is influenced by the fact that the ground state is eightfold degenerate whereas the degeneracy of the excited state is sevenfold. It is further influenced by nonradiative decay channels due to lattice vibrations. The spectral shape of the cross sections is formed by the level splitting in the local crystal field formed by the glass environment and the thermal occupation of the different sublevels. In addition there is inhomogeneous broadening due to the disordered structure of the glass.

5.1.7 Volume Scattering

This section deals with the interaction of light with particles inside a certain volume leading to light scattering. Before going into details the term scattering must be defined and explained. Afterwards scattering is subdivided

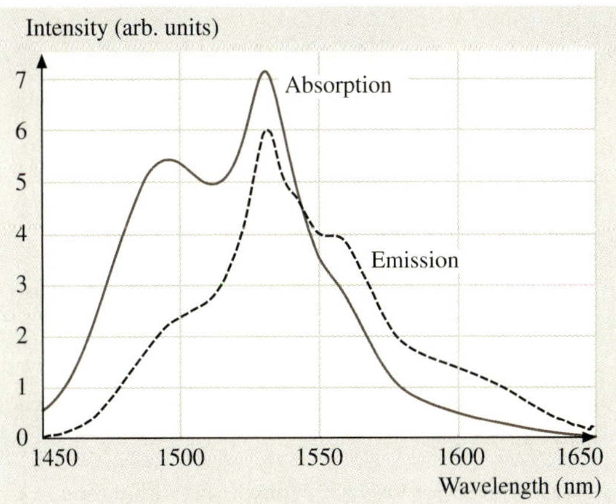

Fig. 5.23 Plot of the emission and absorption cross section for an Er^{3+}-doped glass

into single scattering, multiple scattering, and coherent scattering.

Definition and Basics

Scattering is defined here as energy absorption of incident light followed by re-emission of part of this light at the same frequency. Thus inelastic effects such as Raman scattering or Brillouin scattering are not discussed in this section.

The *origins* of light scattering are surface irregularities. From this point, volume scattering is scattering at the surface of particles in a certain volume. Surface scattering is discussed in Sect. 5.1.8.

Obviously scattering is strongly related to diffraction, as seen from the definition of scattering. Indeed, diffraction is scattering by a flat particle [5.50].

In the following, volume scattering is subdivided into (Fig. 5.24)

- single scattering,
- multiple scattering,
- coherent scattering.

Volume scattering can be regarded as the sum of single scattering events as long as the density of scattering particles is not too high. Mathematically this is expressed with the packing fraction η defined by

$$\eta := \frac{N V_{\text{scat}}}{V_{\text{vol}}} , \tag{5.95}$$

where N is the number (integer) of the single scatterer, V_{scat} is the volume of a single scatterer, and V_{vol} is the whole volume in which the N identical scatterers are located.

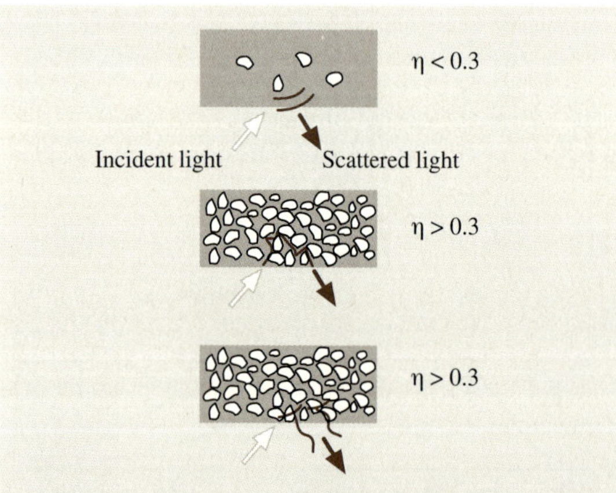

Fig. 5.24 Illustration of the subdivision of volume scattering: single scattering (*top*), multiple scattering (*middle*), and coherent scattering (*bottom*)

Before now discussing these three different subdivisions of volume scattering a few basic items must be described.

Scattering attenuates light, as does absorption by the medium itself. Both effects together are referred to as *extinction*, where: extinction = absorption + scattering [5.50].

The total (over all directions) scattered power P_{scat} (in units of W) can be calculated from the scattering cross section C_{scat} (in units of m^2) by

$$P_{\text{scat}} = C_{\text{scat}} I_{\text{in}} , \tag{5.96}$$

where I_{in} is the incident light intensity (in units of W/m^2). Typically one wants to know the power scattered in a certain direction (Fig. 5.25) and this is described by the differential scattering cross section $\text{d}C_{\text{scat}}/\text{d}\Omega$ (Ω: spatial angular)

$$\frac{\text{d}C_{\text{scat}}}{\text{d}\Omega} = R^2 \frac{I_{\text{scat}}}{I_{\text{in}}} \tag{5.97}$$

where R is the distance from scatterer to observer and I_{scat} is the scattering intensity. The scattering cross section can be obtained from the differential scattering cross section by:

$$C_{\text{scat}} = \int_{4\pi} \frac{\text{d}C_{\text{scat}}}{\text{d}\Omega} \text{d}\Omega . \tag{5.98}$$

Unfortunately the scattering cross section and the differential scattering cross section can only be calculated for a few geometries, such as spheres (Mie scattering).

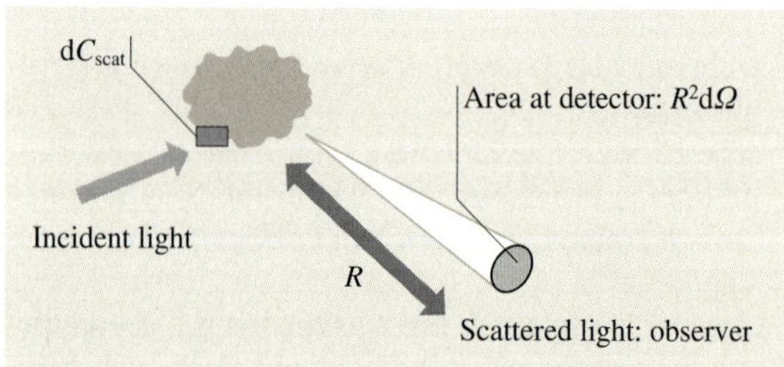

Fig. 5.25 Scattering in a certain direction described by the scattering cross section

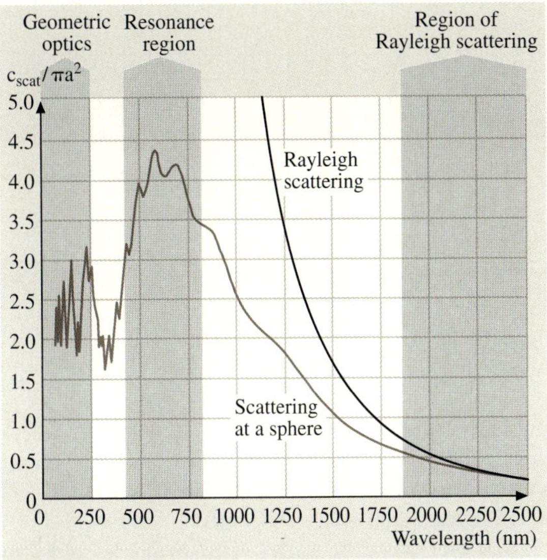

Fig. 5.26 Scattering cross section per sphere area (cross section) versus wavelength for a glass sphere ($n_{\text{sphere}} = 1.50$) of radius $a = 400$ nm in air ($n_{\text{medium}} = 1.0$)

Single Scattering (Mie Scattering)

A very important geometry of single scatterers is the sphere. In 1908 *Mie* [5.51] derived an analytic theory that completely describes the scattering from a (conducting or nonconducting) sphere embedded in a nonconducting medium. A compact survey of the formulas can be found in [5.52]. Many basic characteristics of scattered power can be obtained by studying scattering by a sphere. The calculated scattering cross section normalized to the geometric cross section of a glass sphere ($= \pi a^2$ where a is the sphere radius) can be seen in Fig. 5.26.

Figure 5.26 demonstrates the different wavelength dependence of the scattering cross section and thus of the scattered power according to (5.96). Depending on the geometric sphere size (characterized by the diameter $2a$) relative to the wavelength, three important regions can be identified:

- $\lambda \ll 2a/n_{\text{medium}}$, region of *geometric optics*. Surprisingly the term $C_{\text{scat}}/\pi a^2 = 2$. This effect is called the *extinction paradox* and the factor 2 is due to diffraction effects [5.53, 54].
- $\lambda \approx 2a/n_{\text{medium}}$, *resonance effect*, the term $C_{\text{scat}}/\pi a^2$ has its maximum. Therefore the highest scattered power can be measured if the geometric size of an object is of the order of the wavelength.
- $\lambda \gg 2a/n_{\text{medium}}$, *Rayleigh scattering*. The scattered power is proportional to $1/\lambda^4$. This effect is responsible for the blue sky, where small particles inside the atmosphere scatter blue light (≈ 400 nm) more than red light (≈ 800 nm).

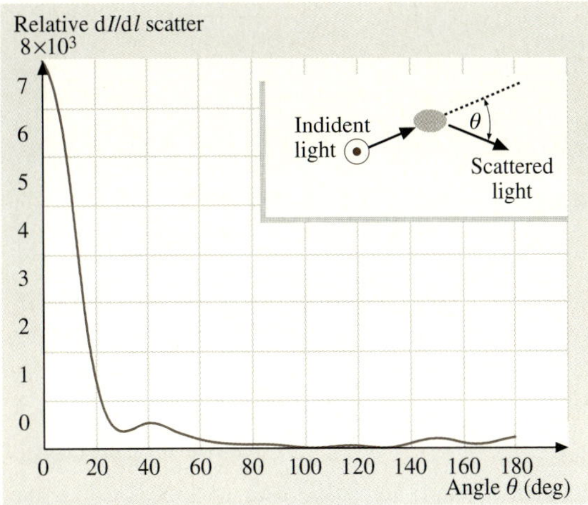

Fig. 5.27 Differential scattering cross section multiplied by 1.5×10^9 particles/mm^3 for a glass sphere ($n_{\text{sphere}} = 1.51$) of radius $a = 400$ nm embedded in a medium with $n_{\text{medium}} = 1.50$ at a wavelength of 550 nm

These basic characteristics of a single scatterer – here shown for the special case of a sphere – are typical for all single scatterers of arbitrary geometry. As discussed and shown in Fig. 5.26 different behavior of the scattered power can be observed based on the size of the scatterer relative to the wavelength.

Multiple Scattering

In the preceding section scattering at a single isolated particle has been described. Now, multiple scattering is discussed, which is the weighted superposition of many single scatterers without interference effects. Multiple scattering with interference effects is called coherent scattering and is discussed below.

Multiple scattering can only be described numerically and must be used whenever the packing fraction η is larger than about 0.3. Based on the knowledge of the scattering function (described by the differential scattering cross section) of a single scatterer, the overall scattering function is calculated by tracing rays throughout the volume.

For example, multiple scattering of 1.5×10^9 particles/mm^3 resulting from a packing fraction of $\eta = 0.8$ is calculated here (this can be the case for a glass ceramic, Sect. 5.5). The single scatterer is assumed to be a sphere of radius 400 nm with a refractive

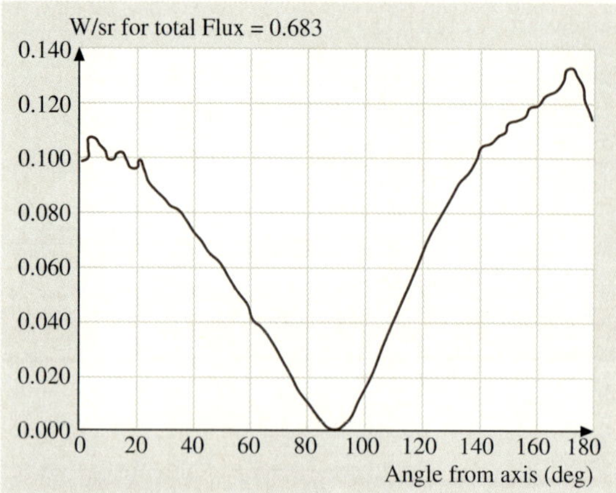

Fig. 5.28 Multiple scattering with $\eta = 0.8$ based on the data from Fig. 5.27. The differential scattering cross section due to multiple scattering inside the volume is totally different from that shown in Fig. 5.27 (the angle is defined in the inset of Fig. 5.27)

index of $n_{\text{sphere}} = 1.51$ embedded in a glass with refractive index $n_{\text{medium}} = 1.50$ at a wavelength of 550 nm. The differential scattering cross section of this single scatterer multiplied by 1.5×10^9 particles/mm^3 is shown in Fig. 5.27.

The differential scattering cross section of a single scatterer (as seen in Fig. 5.27) is used to calculate the differential scattering cross section of the whole volume with a packing fraction of $\eta = 0.8$. The result, shown in Fig. 5.28, has been obtained by calculating the multiple scattering numerically.

As shown in Fig. 5.28 most of the power will be scattered in the backward direction (180°). Less power will be scattered into the forward direction (0°) and no power will be scattered in the sideway direction (90°).

By comparing Fig. 5.27 with Fig. 5.28 it can be seen that the differential scattering cross sections are totally different. The reason for that is multiple scattering inside the volume.

Coherent Scattering

Coherent scattering – the phenomenon in which a series of independent single-particle scattering events is replaced by collective light manipulation from an ensemble of different scattering centers – is the focus of this section. When dealing with materials containing scatterers separated by distances greater than the coherence length (the distance necessary for propagating waves to lose their coherence), the scattering events can be treated as independent occurrences even under

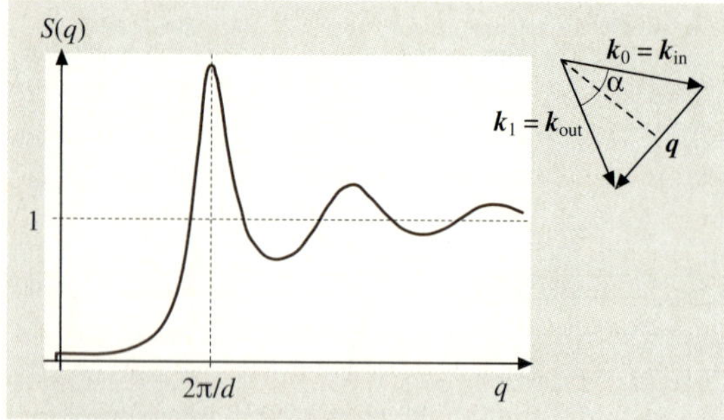

Fig. 5.29 (a) An example of the static structure factor, for the case of hard spheres. **(b)** The scattering vector $\boldsymbol{q}$

multiple scattering conditions. Such scattering is called incoherent, and the resulting intensity of such radiation is simply a summation of the intensity contributions of all the independent scattering centers.

However, when the distances between scatterers are on the order of or less than the coherence length, coherent scattering effects must be considered [5.55]. In this case, wave package interference takes place as the package, is capable of interacting with more than one scattering center at a time; several scatterers distort the photon package simultaneously. Thus, a relationship exists between the phases of the light signals arising from the different scatterers. The events are no longer separate; they are correlated, and the resulting intensity of transmitted light is no longer a simple sum. Thus, in coherent scattering, wave packages experience a combined interaction with several scattering centers that affects their transport through the medium. In photon group interference (a quantum effect), the wave package comes apart upon scattering but the different sinusoidal components meet again and interfere. This interference affects the intensity of light transmitted through the sample; thus, coherent scattering studies can yield valuable structural information.

As discussed, scattering is coherent when the phases of the light signals arising from different scattering centers are correlated and incoherent when the phases are not. Hence, the propagation of coherently scattered light depends strongly on the direction of the scattering vector $\boldsymbol{q}$ – the difference between the incident and outgoing wave vectors, $\boldsymbol{k}_{\text{in}} - \boldsymbol{k}_{\text{out}}$ (Fig. 5.29b) – whereas incoherently scattered light can propagate in any direction regardless of the phase relation between waves from different scattering centers [5.56]. When the average distance between scattering centers (d) is on the order of the coherence length or less, the interference effects are significant and quantitatively describable by the static structure factor $S(q)$, which gives correlations in the positions of particle centers (Fig. 5.29a). $S(q)$ is the link between the theoretical description of structural inhomogeneities and the actual experimental scattering measurements [5.57]. An extreme case of coherent scattering is a strictly periodic configuration of scattering centers. The scattering intensity shows maxima at Bragg peaks and the structure factor consists of sharp δ peaks. Note that the relation between incoherent and coherent scattering terms is usually expressed via the Debye-Waller factor.

The plot of the structure factor in real space is the pair correlation function, $g(r)$, which gives the probability of finding a pair of centers at a specific distance r apart in a sample. The relationship between $g(r)$ and $S(q)$ is

$$S(q) = 1 + 4\pi\rho \int_0^\infty r^2 [g(r) - 1] \frac{\sin(kr)}{kr} \, dr \,, \tag{5.99}$$

where ρ is the density of scattering centers. There is a mathematical connection between the cross section of the scatterers and the structure factor. For the general case of scattering from a correlated group of particles the scattering cross section is

$$C_{\text{scat}} = \frac{1}{k_0^2} \int_0^{2k_0 a} F(y, k_0 a) S(y) y \, dy \,, \tag{5.100}$$

where $y = qa$, $F(y)$ is the form factor, a is the radius of the scatterer, and k_0 is the incident wave vector [5.58].

Coherently scattered photons have a phase relationship, consequently exhibiting more wavelike behavior. In experiments in which coherent scattering is a prominent effect, the interference between scattering paths must be considered – and can be exploited. For example, light from a monochromatic coherent source scatters from a sample and exhibits a characteristic speckle pattern – an array of bright, non-overlapping spots due to interference effects – based on the composition and structure of the sample, as long as single scattering is the dominant effect [5.56]. Thus, such patterns contain structural information. An additional field is the analysis of backscattering cones. Another example of coherent scattering effects is evidenced by the unexpectedly high transparency of a glass ceramic – a phenomenon that cannot be explained by Rayleigh or Mie scattering theories (Sect. 5.5).

5.1.8 Surface Scattering

This section discusses in more detail light scattering due to surface irregularities. The Rayleigh criterion assesses when scattering is important and must be included. How roughness is related to the scattered light can be estimated by the so called total integrated scatter (TIS). TIS is an integrated (over all directions) value, so the angular dependence of scattering is not covered. The angular dependence of surface scattering can be estimated by assuming a sinusoidal surface roughness. A general (scalar) scattering theory is also briefly discussed.

More details on surface scattering can be found in [5.59, 60].

Basics and Rayleigh Criterion

Perfectly smooth surfaces only reflect light specularly. As stated in Sect. 5.1.7, due to surface irregularities light is also scattered in directions other than the specular one. This is illustrated in Fig. 5.30.

As can be seen from Fig. 5.30 in addition to the specular reflected light, scattered light is also generated due to the rough surface. This stray light is scattered over a wide angle around the specular direction. The rougher the surface the more light is scattered away from the specular direction.

But when is a surface smooth? This is answered by the so called *Rayleigh criterion* [5.60].

As can be seen from Fig. 5.31, a plane wave incident on a rough surface experiences a phase difference between the two outer "rays". If this phase difference Φ

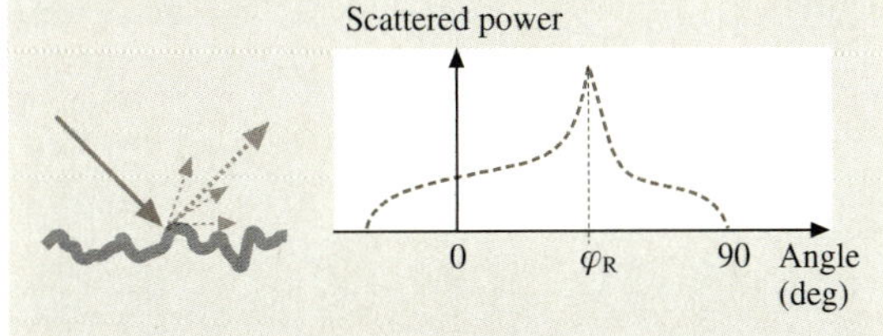

Fig. 5.30 Specular light reflection at a perfectly smooth surface (*top*) and light scattering at a rough surface (*bottom*). *Dotted line* = scattered (reflected) light, *solid line* = incident light

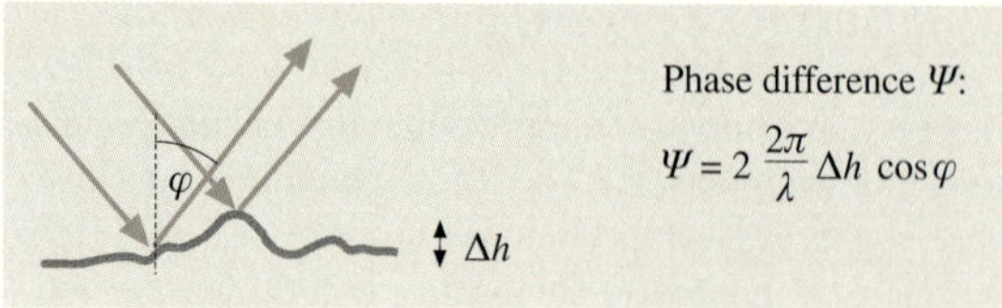

Fig. 5.31 A plane wave incident on a rough surface and its phase difference due to different optical paths

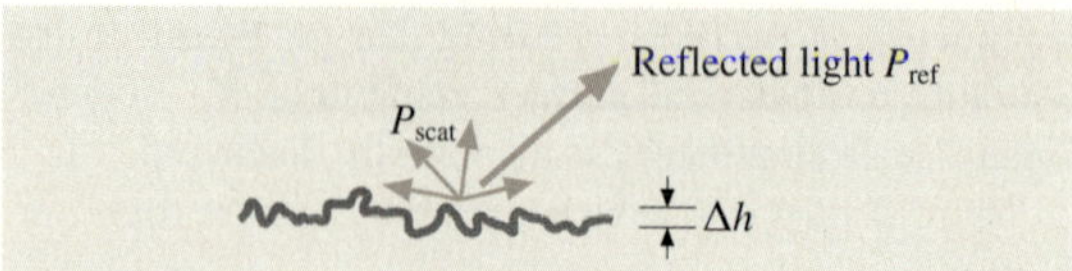

Fig. 5.32 Surface roughness and TIS

is smaller than $\pi/2$, the surface is called smooth (i. e. Rayleigh criterion)

$$\Delta h < \frac{\lambda}{8\cos\alpha}\,, \tag{5.101}$$

where Δh is the RMS roughness (standard deviation).

Scattering (TIS) and Surface Roughness

The way in which surface roughness is related to scattered light can be seen in the so called total integrated scatter (TIS). The TIS is the ratio of scattered power P_{scat} (without specular reflected light) in one hemisphere to the specularly reflected power P_{ref}. If the surface roughness is smaller than the wavelength, then TIS is given

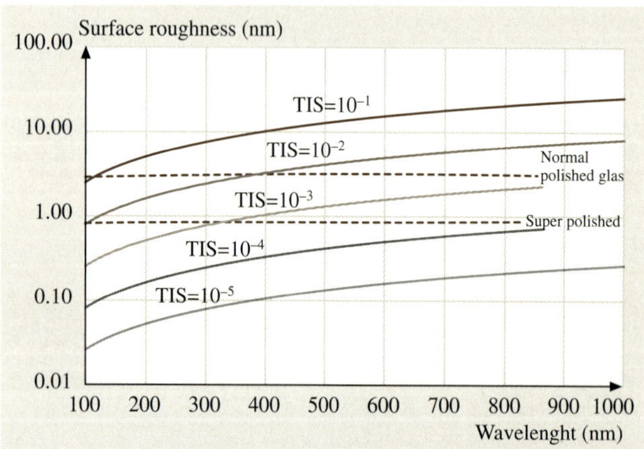

Fig. 5.33 Surface roughness plotted against wavelength for various TIS values. Typical roughness values for super and normal polished glass are also plotted. Note the logarithmic scale of the roughness axis

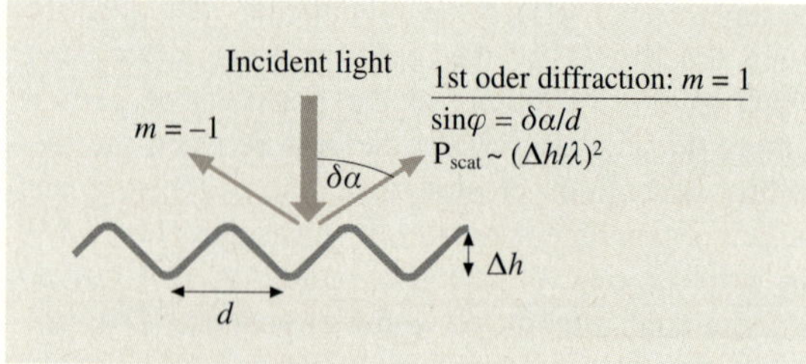

Fig. 5.34 Sinusoidal surface roughness and its parameters for the 1st diffraction order

by [5.59]:

$$TIS = \frac{P_{scat}}{P_{ref}} = \left[\frac{4\pi\cos\alpha_R}{\lambda}\right]^2 (\Delta h)^2\,, \tag{5.102}$$

where α_R is the angle of incidence (reflectance) as shown in Fig. 5.30 and Δh is the RMS roughness of the surface (Fig. 5.32).

From (5.102) it can be seen that the scattered power is proportional to the square of the surface roughness and depends as $1/\lambda^2$ on the wavelength. The surface roughness vs. wavelength for various TIS values is plotted in Fig. 5.33. Consider a specific glass surface and incident light at 500 nm. For this wavelength, only a thousandth part of the reflected light should be allowed to be scattered (TIS $= 10^{-3}$). For this wavelength and TIS value the RMS roughness is just 1.3 nm. This value is between normal and super-polished glass surface. Hence, normal polishing is *not* sufficient here. Therefore, a more expensive super-polishing must be used.

The TIS can also be used for estimating the surface roughness. By measuring the TIS at a determined wavelength, as described in [5.59, 61], the surface roughness can be calculated with the help of (5.102).

Angular Scattering from a Sinusoidal Surface Roughness

As already mentioned TIS is an integrated value, and thus angular dependencies of the scattered can not be described by TIS. But in many cases the direction of scattered light must be known, e.g. to determine the scattered contribution perpendicular (90°) to the incident light. To get a feeling for the angular dependence the easiest form of surface roughness – a sinusoidal surface – is considered (Fig. 5.34). An arbitrary surface roughness can be regarded as the superposition of sinusoidal rough surfaces. This approach is used in [5.60]. As already mentioned, scattering is strongly related to diffraction.

Such a sinusoidal grating diffracts (scatters) light into many orders. Light scattered into the m-th diffraction order is given by [5.59]

$$P^m_{\text{scat}} \propto \left(\frac{\Delta h}{\lambda}\right)^{2|m|}, \tag{5.103}$$

where the scattering angle belonging to this m-th order can be calculated from the grating equation

$$\sin \Delta\alpha_m = m\frac{\lambda}{d}. \tag{5.104}$$

$\Delta\alpha_m$ is the diffraction angle measured relatively to the direction of specular reflection. Here we will restrict the following considerations to the case of normal incidence, i. e. $\Delta\alpha = \alpha$. As an example, the case of a smooth surface is considered.

Smooth surface, i. e. $\Delta h/\lambda \ll 1$ and thus only the first-order diffraction need to be considered, as can be seen in Fig. 5.34. For that case the angle of the first diffraction order is only determined by the periodicity d and wavelength of the grating – *not* by the roughness Δh. The scattered power is proportional to $(\Delta h/\lambda)^2$, as it was obtained from TIS (which is valid only for smooth surfaces). The scattering angle $\Delta\alpha$ depends only on the periodicity d and the wavelength λ. Here three cases are of special importance:

- $\lambda/d \lll 1$, the periodicity is much larger than the wavelength and thus according to (5.104) with $m = 1$, the angle $\Delta(\alpha)$ equals 0. This means that the light will mainly scatter in the specular direction.
- $\lambda/d \approx 1$, the periodicity is of the order of the wavelength and so the scattering angle equals 90°.
- $\lambda/d < 1$, the periodicity is larger than the wavelength and light will be scattered between 0° and 90°.

All three cases for a smooth surface are illustrated in Fig. 5.35, which clearly shows the angular light scattering difference for different periodicities d (compared to the wavelength).

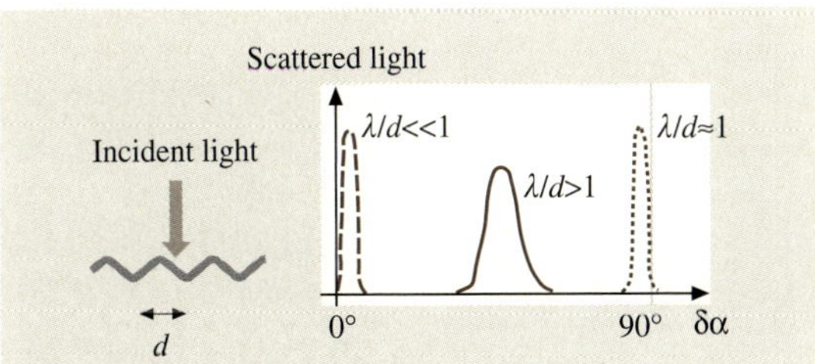

Fig. 5.35 Angular scattered light of a smooth surface for different periodicities d

Unfortunately, reality is much more complicated and a sinusoidal surface roughness cannot be found. But, as previously explained, a random surface roughness can be decomposed into a superposition of sinusoidal surfaces. This is done in much more detail in [5.60]. The angular scattering behavior of a surface is normally measured by the so called angular-resolved scatter (ARS). These measurements can be used to extract the angular scattering characteristics (the bidirectional scattering distribution function, BSDF) used in simulation tools for predicting scattering characteristics.

General Scalar Scattering Theory (First Born Approximation)

If the geometry of the roughness and the different refractive indices are known, then a general theory for calculating the scattered electric field can be developed. This theory is accurate for small refractive index changes (the so called first Born approximation), but after iterative use of the result stronger changes in the refractive index can also be calculated.

The total electric field $U(\boldsymbol{r}, \omega)$, i. e. the incident field U^{i} and the scattered field U^{s} ($U = U^{\text{i}} + U^{\text{s}}$), obeys the following differential equation [5.62] (in SI units):

$$\Delta U + \left(\frac{2\pi}{\lambda_0}\right)^2 U = -4\pi F(\boldsymbol{r}, \omega)\, U, \tag{5.105}$$

where λ_0 is the free space wavelength, $\boldsymbol{r}$ is the arbitrary point of observation, and the scattering potential F is

$$F(\boldsymbol{r}, \omega) = \frac{1}{4\pi}k^2\left[n^2(\boldsymbol{r}, \omega) - 1\right]. \tag{5.106}$$

The right-hand side of (5.105) describes the source of the scattering. Thus, as can be seen from (5.106), this source is only determined by the spatially and frequency-dependent refractive index n. Here it is clearly shown that the source of light scattering is the spatially dependent refractive index (rough surface).

As shown in [5.62] the solution of (5.105) for the scattered electric field is given by

$$U^{\text{s}}(\boldsymbol{r}, \omega) = \int_V F(\boldsymbol{x}, \omega) U(\boldsymbol{x}, \omega) \frac{e^{ik|\boldsymbol{r}-\boldsymbol{x}|}}{|\boldsymbol{r}-\boldsymbol{x}|}\, d^3x. \tag{5.107}$$

This form has the drawback that the total electric field inside the integral (5.107) must be known. But U also contains the scattered electric field. This problem can be overcome by using the so called first Born approximation (or Rayleigh–Gans theory). The first Born approximation assumes a smooth surface (weak change in the refractive index) and also a far field [5.62]:

$$U \approx U_1 = e^{i\boldsymbol{k}_i\boldsymbol{r}} + \frac{e^{ikr}}{r} \int_V F(\boldsymbol{x}) e^{i(\boldsymbol{k}_i - \boldsymbol{k})\boldsymbol{r}}\, d^3x\,, \quad (5.108)$$

where $\boldsymbol{k}_i$ is the wave vector of the incident wave and $\boldsymbol{k}$ is the wave vector of the origin and point of observation.

Equation (5.108) is the first Born approximation and the scattered electric field is given by

$$U^s = \frac{e^{ikr}}{r} \int_V F(\boldsymbol{x}) e^{i(\boldsymbol{k}_i - \boldsymbol{k})\vec{r}}\, d^3x\,. \quad (5.109)$$

The theory is limited to smooth surfaces, but iterative use of (5.109), as described in [5.54], allows its extension to rougher surfaces.

Unfortunately, the spatial distribution of the refractive index must be known. This is a problem and thus the elegant (5.109) has its limits. Therefore, TIS and ARS measurements are mainly used for describing the scattering of rough surfaces.

5.1.9 Other Effects

In addition to the previously discussed "classical" optical material properties:

- (non-)linear refractive index
- absorption coefficient
- emission cross section
- scattering cross section

there exist numerous effects which occur in optical materials when additional physical fields such as:

- quasi-DC electric field
- quasi-DC magnetic field
- mechanical stress

additionally influence the optical properties of the solid. From these phenomena, the resultant electro-optical and piezo-optical effects are described below.

For some optical materials, incoming light of certain intensity or energy causes (quasi-)permanent changes in optical properties. The unwanted effects of solarization and laser damage are described in Sect. 5.4. On the other hand, the desired permanent property changes one can encounter are:

- photochromism
- photorecordability
- photorefractivity.

Photochromism is briefly introduced here while photorecordability and photorefractivity are extensively discussed in Sect. 5.9.3.

The Electro-Optic Effect

The electro-optic effect [5.63] (Pockels effect) in special crystalline optical materials is the change in the indices of refraction of the ordinary and extraordinary rays (Sect. 5.1.4) that is caused by and is proportional to an applied electric field. The electro-optic effect was discovered by Roentgen and was extensively investigated by Pockels around the turn of the nineteenth century. In principle, the electro-optic effect is a special case of nonlinear optical behavior, where the perturbing electromagnetic field is of quasi-DC type.

The electro-optic effect provides a convenient means of controlling the intensity or phase of the propagating radiation. The linear effect can be characterized by a perturbative expansion of the dielectric tensor (see Sect. 5.1.4)

$$\varepsilon_{ij} = \varepsilon_{ij}^0 + \Delta\varepsilon_{ijk}^E E_k\,, \quad (5.110)$$

where ε_{ij}^0 describes the electric-field-independent contribution to the dielectric tensor and $\Delta\varepsilon_{ijk}^E$ is the first derivative of ε_{ij} with respect to each component k of the applied electric field E.

From symmetry arguments similar to the consideration in Sect. 5.1.4, the 27 different coefficients $\Delta\varepsilon_{ijk}^E$ collapse to only 18 parameters, because $\Delta\varepsilon_{ijk}^E = \Delta\varepsilon_{jik}^E$.

Table 5.5 Properties of most important electrooptical materials (after [5.63])

Material	Electrooptic constant r_{63} (pm/V)	Typical half-wave voltage at 546 nm (kV)	Approximate n_o
Ammonium dihydrogen phosphate (ADP)	8.5	9.2	1.526
Potassium dihydrogen phosphate (KDP)	10.5	7.5	1.51
Potassium dideuterium phosphate (KD*P)	26.4	2.6–3.4	1.52
Potassium dihydrogen arsenate (KDA)	10.9	6.4	1.57
Rubidium dihydrogen phosphate (RDP)	11.0	7.3	
Ammonium dihydrogen arsenate (ADA)	5.5	13	1.58

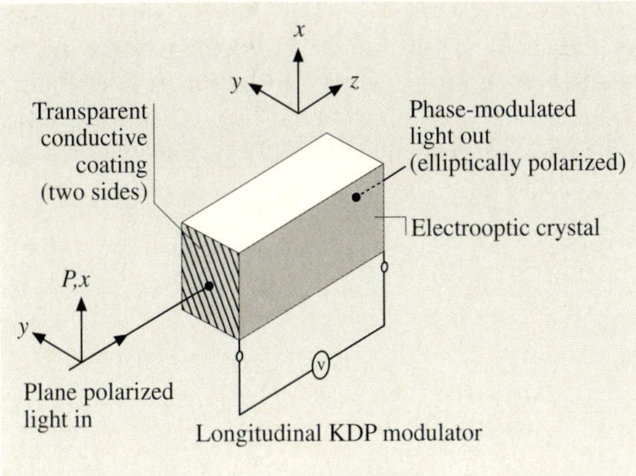

Fig. 5.36 Longitudinal electrooptical modulator (after [5.63])

These coefficients can be measured and provide the information necessary to compute the electrically induced birefringence. Some of the materials which exhibit the electro-optic effect and are used in electro-optic devices are listed in Table 5.5.

A potassium dihydrogen phosphate (KDP) longitudinal modulator is illustrated in Fig. 5.36. When no voltage is applied, the indices of refraction for the ordinary (o) and extraordinary (e) rays are identical along the optic axis. The incoming beam is plane-polarized with the plane of polarization aligned to the x- or y-axis. When a voltage is applied parallel to the light beam the crystal becomes birefringent. The extraordinary ray is retarded with respect to the ordinary ray and, upon emerging from the output face of the electro-optic crystal, the two polarized beams have acquired a relative phase shift due to the retardation of the light in the plane of the slow axis – resulting in an elliptically polarized beam. The retardation is given by

$$\Phi_{eo} = \frac{n_o^3 \Delta\varepsilon_{123}^{E} U}{\lambda} , \quad (5.111)$$

where Φ_{eo} is the number of wavelengths of retardation, n_o is the ordinary index of refraction, $\Delta\varepsilon_{123}^{E}$ is the electro-optic coefficient (μm/V), U is the applied voltage (V), and λ is the wavelength of incident light (μm). For the value $\Phi_{eo} = 1/2$, $U = U_{1/2}$ is defined as the half-voltage (see Table 5.5).

A transverse electro-optic effect is also exhibited by some crystals. In this case the birefringence changes are induced by a voltage applied transversely to the beam propagation.

Materials of prime importance concerning electro-optic properties are ADP, KDP, KD*P, KTN, $LiNbO_3$, $LiTaO_2$, $Ba_3NaNb_50_{15}$, $Sr_xBa_{1-x}Nb_20_6$, CuCl, ZnSe, hexamethylenetetramine, and Se. Some materials have a quadratic rather than linear electro-optic effect. That is, the retardation is proportional to the square of the applied voltage. Among these are strontium titanate, potassium tantalate, and KTN.

Piezo-Optic Effect

The Piezo-optic effect [5.63] is analogous to the electro-optic effect. When pressure is applied to a piezo-optic crystal, the birefringence of the material changes. Some of these piezo-optic crystals are originally isotropic and cubic, but when stressed become birefringent. If such a crystal is stressed in a direction perpendicular to a light beam, the phase retardation is proportional to the stress. This effect is applied in acoustooptic modulators. These devices function based on the fact that refraction and diffraction effects occur when light passes through a solid medium transversely to a high-frequency (ultrasonic) acoustic field propagating in the same medium. The interaction can affect the deflection, polarization, phase, frequency or amplitude of the optical beam. Ultrasonic waves range from 10^5 Hz (above the upper range, approximately 2×10^4 Hz, of the human ear) to 10^9 Hz. Light passing through such a crystal in a direction transverse to the acoustic-wave direction is modulated by the wave.

Table 5.6 lists some of the more important acoustooptic materials and their characteristics. There are many figures of merit used in comparing these materials, two that are commonly employed are

$$F_1 = \frac{n^6 (\Delta\varepsilon^{P})^2}{\varrho v^3} ,$$
$$F_2 = \frac{n^7 (\Delta\varepsilon^{P})^2}{\varrho v} , \quad (5.112)$$

where n is the optical index of refraction, $\Delta\varepsilon^{P}$ is the appropriate component of the photoelastic tensor (analogous to $\Delta\varepsilon^{E}$ in the electro-optic effect), ϱ is the mass density, and v is the acoustic phase velocity. Since $\Delta\varepsilon^{P}$, v, and ϱ exhibit no extreme variations in the various acoustooptic materials, the candidates of most interest are those with good optical and acoustic qualities with large n, although optical and acoustic parameters are not included in the figures of merit.

In Table 5.6, the longitudinal (L) wave is one where the displacement u is parallel (‖) to the acoustic wave propagation direction ($\boldsymbol{k}$ vector), whereas the transverse (T) wave is one where the displacement u is normal (⊥) to the acoustic-wave propagation direction ($\boldsymbol{k}$ vector).

The polarization of the optical wave is defined as parallel (∥) or perpendicular (⊥) to the plane formed by the acoustic and optic $\boldsymbol{k}$ vectors.

Photochromism

As discussed in the previous subsections, the interaction of electromagnetic waves with matter is generally

Table 5.6 Properties of most important piezooptical materials (after [5.63])

Material	λ (μm)	n	Polarization and direction [a]		Figure of merit	
			Acoustic wave	Optical wave [b]	$\frac{n^6p^2}{\rho v^3}\times 10^{-18}$	$\frac{n^7p^2}{\rho v}\times 10^{-7}$
Fused Quarz	0.63	1.46	L	⊥	1.51	7.89
			T	∥ or ⊥	0.467	0.963
GaP	0.63	3.31	L, [110]	∥	44.6	590
			T, [110]	∥ or ⊥, [010]	24.1	137
GaAs	1.15	3.37	L, [110]	∥	104	925
			T, [100]	∥ or ⊥, [010]	46.3	155
TiO_2	0.63	2.58	L, [11$\bar{2}$0]	⊥, [001]	3.93	62.5
$LiNbO_3$	0.63	2.20	L, [11$\bar{2}$0]		6.99	66.5
YAG	0.63	1.83	L, [100]	∥	0.012	0.16
			L, [110]	⊥	0.073	0.98
YIG	1.15	2.22	L, [100]	⊥	0.33	3.94
$LiTaO_3$	0.63	2.18	L, [001]	∥	1.37	11.4
As_2S_3	0.63	2.61	L	⊥	433	762
	1.15	2.46	L	∥	347	619
SF-4	0.63	1.616	L	⊥	4.51	1.83
β-ZnS	0.63	2.53	L, [110]	∥, [001]	3.41	24.3
			T, [110]	∥ or ⊥, [001]	0.57	10.6
α-Al_2O_3	0.63	1.76	L, [001]	∥, [11$\bar{2}$0]	0.34	7.32
CdS	0.63	2.44	L, [11$\bar{2}$0]	∥	12.1	51.8
ADP	0.63	1.58	L, [100]	∥, [010]	2.78	16.0
			T, [100]	∥ or ⊥, [001]	6.43	3.34
KDP	0.63	1.51	L, [100]	∥, [010]	1.91	8.72
			T, [100]	∥ or ⊥, [001]	3.83	1.57
H_2O	0.63	1.33	L		160	43.6
Te	10.6	4.8	L, [11$\bar{2}$0]	∥, [0001]	4400	10 200
α-HIO_3 [c]	0.63		L-a	a–c	48.2	
				b–c	20.8	
				c–b	46.0	
			L-b	a–c	41.6	
				b–c	58.9	
				c–a	32.8	
			L-c	a–b	83.5	
				b–a	77.5	
				c–a	63.0	
			Shear a–b	a–c	17.1	

[a] L = longitudinal, T = transverse (shear)

[b] Polarization is defined parallel (∥) or perpendicular (⊥) to the plane formed by the acoustic and optic propagation directions ($\boldsymbol{k}$-vectors)

[c] Lattice constants: $a = 5.888$ Å, $b = 7.733$ Å, $c = 5.538$ Å

characterized by terms such as dispersion, absorption, scattering, refraction etc. Here we deal with absorptive effects which are induced in optical solids by the electromagnetic waves themselves.

Electromagnetic waves can cause electronic transitions from one state to another, create new particles or quasi-particles, or excite vibrations of ions and ensembles of charged particles. These effects are well-known absorption mechanisms. Obviously, the absorption spectrum depends on the electronic and ionic configuration. Consequently, the transition from one electronic or ionic configuration into a different configuration necessarily changes the absorption spectrum of the system under consideration. The control of the absorbance spectrum of any material by photons is called photochromism [5.64], regardless of the spectral range in which the absorbance is changed and regardless of the energy of the photons that induce the photochromism. Following this definition, a rich variety of possible effects exists that can be exploited for photochromism: the famous transition of F-centers into Q- and F′-centers or, more generally, the transformation from one to another color centre by illumination, the photon-induced orientation of color centers (photodichromism), the spectral hole burning as well as photon-induced polymerization or decomposition of compounds and photon-induced bleaching of absorption bands for Q-switching of a laser, just to name a few.

The change in the absorption spectrum, however, is hard to see if it occurs in a spectral region with a large absorption constant. Therefore, one must focus on materials and spectral regions where both the absolute and relative change of the absorbance is large if one wants to exploit the corresponding effect of photochromism for technical applications. Challenging mass applications of photochromic effects are optical data storage and "smart" windows (see Sect. 5.8.3). However, several qualitative and quantitative requirements must be fulfilled; for instances the requirement of special spectral ranges optimized both for the inducing photons and for the induced absorbance is very important in many applications. It is equally important to discern whether the effect is reversible spontaneously or under irradiation, whether it shows fatigue, and how efficiently the absorption spectrum changes per incident photon. As a consequence, the photon-induced control of the absorbance – which is obviously very intriguing from a theoretical point of view – is reduced to only a few applications in practice.

The most prominent and widespread application of a photochromic effect seems to have occurred in photochromic eye-glasses (see Fig. 5.37 left). Although photochromic effects have long been known to occur in glasses, the basis of a material which is useful for photochromic sunglasses was developed by *Armistead* and *Stookey* [5.65] in the 1950s and 1960s. These scientists introduced silver halides into suitable glass melts and succeeded in forming small silver halide precipitates or particles in the glass. The basic effect, namely the photolytical decomposition of silver halides, is very well known in chemistry. In a test tube, the silver halide is decomposed into Ag clusters, which are responsible for the absorbance in the visible spectral region, and volatile halogens. The photolytical decomposition is not reversible in a test tube, however, if the halogens can escape. In a glassy matrix, on the other hand, the photolytic decomposition products remain close together and can recombine into silver halides. The chemical and phys-

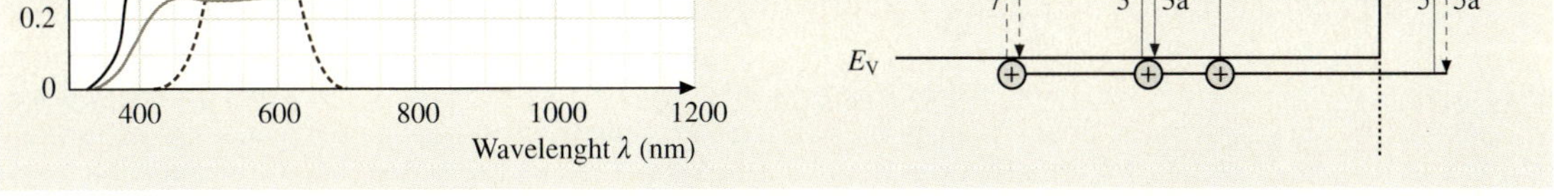

Fig. 5.37 *Left*: Transmittance of a commercially available photochromic glass as a function of wavelength before and under solar irradiation after about 15 min. *Right*: Energy band diagram of different photoelectronic processes which can occur in the silver halide particles at the interface between halide particle and vitreous glass matrix. (after [5.64])

ical details of the photochromism of glasses doped with silver halides (Fig. 5.37 right) have been discussed e.g. in [5.64].

Besides silver halides, many other different compounds can be decomposed due to light changing their absorption spectra under irradiation. To list a small arbitrary selection of inorganic systems, we mention the Pb-, Cu-, and Cd-halides, the transition metal oxides, and some other compounds. However, most systems are interesting only from a scientific point of view and have little technical potential – or other practical disadvantages reduce their applicability.

5.2 Optical Glass

Every discussion about optical glass, especially an intensive one such as that following, has to start with the definition of the term *optical glass* and its delimitation from other possible types of glasses, e.g. filter glass, technical glass and others.

(Classic) optical glass is the traditional generic term for all glass types which possess the ability to influence the path of transmitting electromagnetic rays in the region between 200 nm to 1500 nm, e.g. to guide light [5.66]. This optical effect, and therefore the optical glass type itself, is usually fixed by the two main optical values: the refractive index (n_d) and the Abbe number (ν_d, 1/dispersion) at the wavelength of the spectral d-line (587.6 nm, yellow helium line) [5.66, 67]. The relevant optical range was traditionally seen between $n_d = 1.38-2.20$ and $\nu_d = 18-100$. However, actual application trends lead to less extreme regions, so a reduction to $n_d = 1.40-2.00$ and $\nu_d = 20-90$ seems to be as suggestive as the prefix "classic". This prefix preserves the delimitation from other types, nowadays often declared as subgroups of optical glass, e.g. color and filter glass and active optical glass. Here this prefix will be neglected, the term optical glass always refers to the aforementioned traditional definition.

Accordingly, optical glass is mainly used for components (lenses, prisms) in light-guiding application fields where there are imaging optics (microscopy, photo objectives, etc.), (digital) projection (color management systems, beamers, cinema objectives), telecommunications (hosting for active elements, ball, and rod lenses, graded index lenses, lens arrays), optical communications engineering (light and image fiber guides, data pick-up systems, read–write systems) and laser technology (laser optics, hosting for active elements).

An overview of available optical glass types is usually given in optical-glass diagrams, wherein they are specified by their two main optical values n_d/ν_d only. Those diagrams, provided (currently even by internet) by all well-known glass manufacturers (Schott [5.66], Ohara, Hoya, Sumita, Pilkington, in former times Corning), itemize the glasses independently of possible applications and other optical and physico-chemical properties. Up to now they are divided into historically developed sections, see for example Fig. 5.87.

5.2.1 Chronological Development

The first known optics shops and the early optical industry (ophthalmic, astronomy and microscopy) suffered from the lack of well-defined, reproducible optical glass. To cut a long story short, they had to take and to work with what they obtained from more or less irreproducible pot melts which suffered from poor raw material purity and availability. In addition to the poor optical quality of materials produced in this way, there did not exist any assured knowledge about the relation between composition and properties. At most, raw materials and their effects could be classified by supplier [5.68, 69].

Increasing chemical knowledge allowed the identification of raw material additives and impurities. This enabled, in combination with increasing abilities in chemical separation, the first relations between composition/compounds and properties/defects to be developed. This in turn allowed for the defined adjustment of optical properties, whereas until that time attention was tuned only by the optical position, i. e. the explicit values of n_d/ν_d as mentioned above. Otto Schott, a German chemist from a family of glass makers, was the first to investigate the composition/property relation with scientific methods and achieved a fundamental knowledge about melting consistent predeterminable glasses. Ernst Abbe and Carl Zeiss, owners of a firm for high-quality optical instruments, recognized the advantages for their business and, by association with Schott, easily emerged to be the most successful in their profession [5.69].

Prior to 1880, only simple crown and flint glasses were known (approximately 30 types) [5.70]. Crown glasses, consisting of soda–lime silicates, exhibit low refractive indices and high Abbe numbers, although from a current point of view the Abbe numbers would be better judged as moderate. Flint glasses [5.71], with high refractive indices and low Abbe numbers, were represented solely by lead alkali silicates. The modern borderline between flint and crown is defined by the Abbe number, with flint glasses having $\nu_d < 50$ at $n_d > 1.60$ and $\nu_d < 55$ at $n_d < 1.60$. At that time, the distinction resulted naturally via composition.

From 1880, Otto Schott enlarged the glass map by establishing two new glass formers besides silicate: fluorine (very high ν_d and low n_d) and boron (moderate n_d/ν_d). Additionally he started to work with network modifiers by using BaO as a compound (moderate ν_d, high n_d). So, a variety of new glasses with a broader range of optical positions was generated, which led to the additional classifications: heavy (high refractive index) and light (low refractive index) besides crown and flint (the indication of dispersion/Abbe number) [5.72].

Around 1930 a new era in glass development started, when compounds such as rare earths (especially lanthanum), titania, zirconia and the additional network former phosphorous could be obtained and used in sufficient quantity and quality [5.73, 74]. These glasses led to further new optical positions, whose regions in the Abbe diagram were named by adding the new compound's chemical symbol to the traditional heavy/light crown/flint nomenclature, e.g. PSK (phosphorous heavy crown, the German *phosphor schwerkron*) or LaSF (lanthanum heavy flint, the German *Lanthanschwerflint*) [5.72].

In this context, it is easy to understand that a strong correlation between optical position (n_d/ν_d) and composition evolved, which led to the historical borders in the Abbe diagram and optically defined glass families. Hence, a glass's name, only defined by its optical position followed by a consecutive number, gave strong hints towards its composition.

Even though these boundaries have been valid until now, this strong correlation no longer exists because of the availability of highly pure (classical) raw materials and the accessibility of other, new raw materials obtained through new ways of mining, cleaning, purification etc.. Therefore, it is now possible to reach one optical position by several base-glass compositions, which leads to an increased number of compositional glass families with strong overlaps that cross the traditional boundaries. As a result, the information given by the glass type name (e.g. PSK 54) no longer provides reliable information about the components and glass family and thus about their properties. In this case, one cannot be sure about the existence of phosphorous, because, for example, PSK types can be obtained with and without phosphorous. And though both of these glass families occupy the same optical position (e.g. Schott's N-LaSF 31A, Hikari's E-LaSF 08 and Hoya's TaFD 30), their physico-chemical properties are quite different (chemical resistance, Knoop hardness, temperature–viscosity profile, etc.).

It remains to remark that mainly the Japanese glass manufacturers Ohara and Hoya recently diverged from this traditional nomenclature and implemented their own representative system of glass naming. Nevertheless they maintained the traditional borderlines in the Abbe diagram.

5.2.2 Compositions of Modern Optical Glass

Today we know of more than 300 different types of optical glasses that can be classified by their chemical composition Figs. 5.38 and 5.87, and Table 5.37). Because of better raw material availability, there exist more than several compositional methods to reach an optical position. Deviation from the traditional base systems and known compositions is often useful because of current enhanced glass-development specifications. Those specifications comprise customer- and process-referred profiles in which it is not unusual to find up to 15 properties, besides the well-known pure optical-position

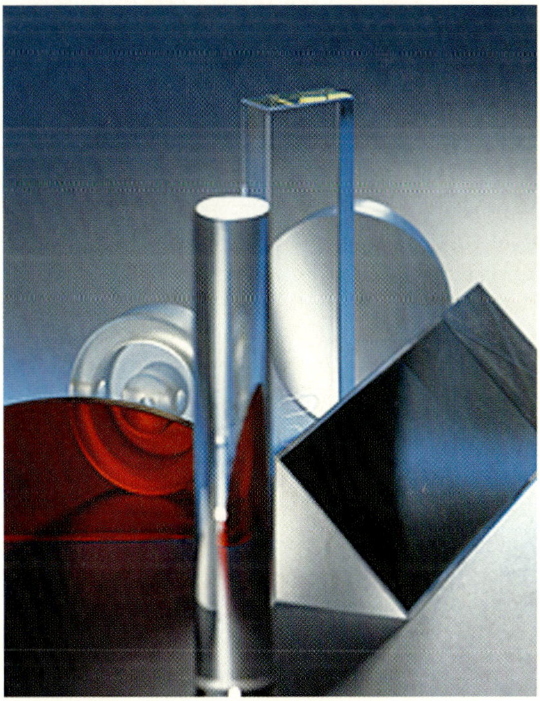

Fig. 5.38 Today, optical glasses with high quality exist in a broad range of shapes and products

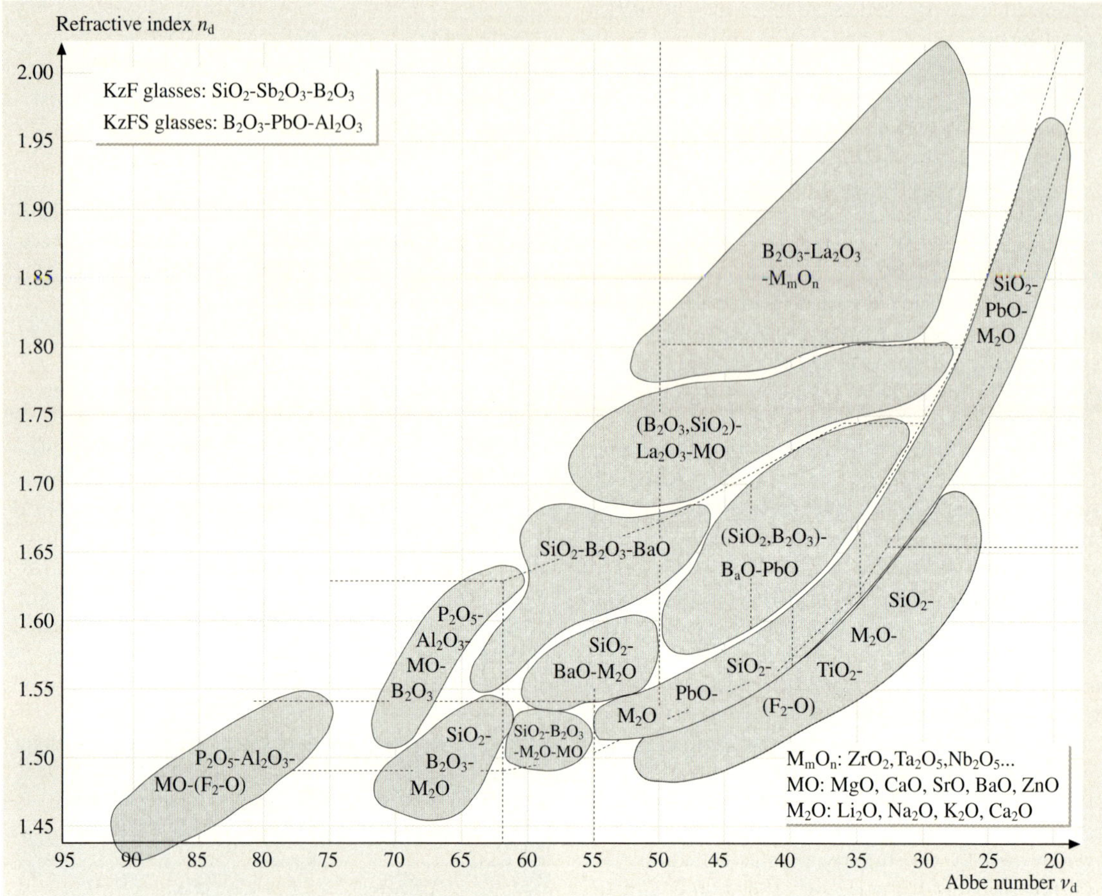

Fig. 5.39 Glass systems of optical glass in the n_d/ν_d plot

specification. Commonly specified properties belong to the following fields:

- *application*: T_g, CTE, specific density, spectral transmission, chemical resistance, solarization, compaction, birefringence coefficient, fracture behavior;
- *pollution control and operational safety*: chemical and thermal reactivity of batch and raw materials, environmental impact of raw materials (mining, cleaning, use, application);
- *melting and hot forming*: viscosity–temperature profile, CTE, adjustment to standard processes, crystallization, aggressiveness against fireproof material;
- *cold working*: CTE, stiffness, hardness, fracture behavior, abrasion rate, chemical resistance;
- *patent law* (very active international patent awareness): adaptation of compositions to relevant legal surrounding, including variations, admixtures, etc.

Increasingly a transition has occurred from developing a) as far as possible an optimized glass out of the traditional optical glass families to b) specially designed single glasses for specific application or process fields arranged on the Abbe diagram only for traditional reasons. Therefore only the most important base systems, as shown in Fig. 5.39, will be discussed in detail.

1. $SiO_2-B_2O_3-M_2O$

This system is formed by the network formers SiO_2 and B_2O_3. Alkaline oxides (M_2O) have been introduced as network modifiers. The presence of B_2O_3 and alkaline oxides is essential, as the glass-forming oxide SiO_2 has melting temperatures that are too high for conventional melting techniques. The addition of B_2O_3 and alkaline oxides facilitates the meltability of these glasses. The optical glass BK 7 is one of the main representative glasses of this system. It is the optical glass most frequently used and can be produced with extreme homogeneity.

2. $SiO_2-B_2O_3-BaO$ and 3. $SiO_2-BaO-M_2O$
If BaO is used as the main network modifier instead of the alkaline oxide or B_2O_3, this results in two large glass systems. In contrast to other earth-alkaline network modifiers, the introduction of BaO has various advantages. Besides PbO there is no other divalent oxide that increases the refractive index as strongly as BaO. Additionally, BaO neither decreases the Abbe number nor shifts the UV transmission edge significantly to greater wavelengths as PbO does. BaO-containing glasses normally show good abrasion hardness. In some glasses, BaO is partly replaced by ZnO.

4. $(SiO_2, B_2O_3)-M_2O-MO$
To obtain glasses of this glass system, divalent oxides (MO) are exchanged for B_2O_3. From the divalent oxides only CaO, ZnO, and PbO are used in this system. For the production of glasses with high chemical resistance and good crystallization stability, ZnO is particularly important since it increases durability with respect to water and acids much more than CaO, provided that the ZnO content exceeds approximately 10 wt %. It is also possible to reduce the melting temperatures by a simultaneous exchange of ZnO for SiO_2 and to increase the alkaline content without harming durability against weathering. In terms of these characteristics, ZnO is much more effective than B_2O_3. The use of CaO leads to the production of glasses with good mechanical properties and increased abrasion hardness. The chemical durability of these glasses increases with increasing CaO content.

5. $(SiO_2, B_2O_3)-BaO-PbO$
This glass system is a mixture of the glass systems $SiO_2-BaO-M_2O$ and $SiO_2-PbO-M_2O$. Thus these glasses represent a transition from one glass system to the other. In the system $SiO_2-BaO-M_2O$, the exchange of PbO for BaO leads to the production of glasses with a lower tendency to crystallize. Compared with similar BaO-containing glasses, PbO-containing glasses are characterized by lower softening points and higher expansion coefficients. In some glasses of this system, BaO is partly replaced by ZnO.

6. $SiO_2-PbO-M_2O$
Glasses of this system have been known for a very long time. They are widely used for optical glasses and crystal glasses. As a glass component, PbO plays a vital role because on the one hand it increases the refractive index and on the other it decreases the Abbe number very strongly. Furthermore, PbO significantly affects the partial dispersion of glasses since a high PbO content leads to a positive deviation from the "normal line". High amounts of PbO lead to a high glass density and a rather low chemical resistance, but effects an interesting birefringence coefficient near zero. PbO can be introduced into glasses at more than 80 wt %, which indicates that it does not only have network modifying properties. Normally, if a network modifier is introduced into a glass, e.g. Na_2O into a SiO_2 glass, bridging oxygen atoms are converted into non-bridging oxygen atoms:

$$\begin{aligned}&-\mathrm{Si-O-Si-}+\mathrm{Na_2O}\\&\quad\Rightarrow -\mathrm{Si-O-Na^+}+\mathrm{Na^+-O-Si-}\,.\end{aligned}$$

With an increasing content of network modifiers the network progressively weakens and the viscosity of the glass melt decreases. If the content of the network modifiers is increased further, more and more bridging oxygen atoms are converted into non-bridging atoms. For a ratio of network modifiers to network formers of about 1 : 1, a three-dimensional network can no longer be formed and glass formation becomes impossible. However, the glass-forming tendency of binary $PbO-SiO_2$ systems leads to a PbO content of more than 70 mol %. According to *Zachariasen* [5.75, 76] and *Warren* and *Loring* [5.77] these glasses would normally not exist if PbO was only a network modifier. *Fajans* and *Kreidl* [5.78] and *Stanworth* [5.79] explained how the introduction of such a high amount of PbO is possible. In small concentrations, PbO has the same function as any other network modifier. High concentrations of PbO, however, cause an increase in the formation of $[PbO_4]$ tetrahedra. These $[PbO_4]$ tetrahedra can act as network formers and are incorporated into the three-dimensional network. Glasses that contain large amounts of PbO are usually more or less yellow in color. Impurities, such as Fe_2O_3 or Cr_2O_3, have very strong effects on the color of PbO-containing glasses, even stronger than in soda–potash–lime glasses. It was further found that, first, besides the different structures of the PbO-containing glasses, there exists an equilibrium between Pb^{2+} and Pb^{4+} and, second, brown PbO_2 is formed if the melt is supersaturated with oxygen. Unlike the yellow color that is due to impurities, the yellow tint caused by the equilibrium between Pb^{2+} and Pb^{4+} can be reversed by a variation of the redox condition of the melt. For a long time, attempts have been made to increase the chemical stability of optical glasses with high PbO contents. Replacing parts of the PbO content with TiO_2 (in the range of some wt%) has been a common method to increase the chemical durability. A sufficient TiO_2 content not only improves the chemical durability but, unfortunately, also leads to a transmission loss in the UV range. A solution to this problem is reached by replacing parts of the PbO content with both TiO_2 and ZrO_2. The chemical durability is improved and the optical properties of the original glass (SiO_2−PbO−M_2O) are again obtained.

7. P_2O_5−MO−B_2O_3−Al_2O_3

Glasses in which P_2O_5 acts as a network former have been known for a long time. P_2O_5 forms glasses with rather low dispersions (high Abbe numbers) and high refractive indices, compared with the classic glasses based on B_2O_3 or SiO_2. Glasses with a high P_2O_5 content have long dispersion spectra in the blue range. These glasses are normally used to correct chromatic aberration. The second and very important property of glasses containing P_2O_5 is their low birefringence coefficient and excellent UV transparency. A disadvantage of phosphate glasses is that their chemical durability is poor compared with the durability of SiO_2-based glasses.

8. P_2O_5−Al_2O_3−MO−(F_2−O)

In order to obtain glasses with lower dispersions than those of phosphate glasses, one can introduce fluorine. Most of these glasses are based on a P_2O_5−Al_2O_3−MO glass system which contains high amounts of fluorine. This glass system shows a relatively high chemical durability and a relatively low devitrification tendency. Fluorine takes the places of oxygen atoms so that one oxygen is replaced by two fluorine atoms. The formation of the fluoro-phosphate glasses of this system is rather complex and can be described with amorphous fluoro-alumino-phosphates. As fluorine has a smaller ionic radius and a higher electronegativity than oxygen, the polarization of the cations in these glasses is increased so that the refractive index and the dispersion are decreased. The replacement of oxygen by fluorine leads to glasses that have distinct LgK characteristics, which leads to a long spectrum in the blue spectral region, a significantly positive $\Delta P_{g;F}$ (partial dispersion deviation coefficient) and hence the aptitude for correcting achromatic aberration in combination with a KzF(S) type (short flint) with a short spectrum in the blue spectral region (negative $\Delta P_{g;F}$). The term LgK was neglected for a long time in favor of the systematic terms FK (fluoro crown) and PK (phosphate crown) which both indicate the optical position of fluoro-phosphate types (see Abbe diagram).

9. (B_2O_3, SiO_2)−La_2O_3−MO

In connection with B_2O_3, glasses containing large amounts of La_2O_3 form a very extensive glass system. Glasses belonging to this system normally have a high refractive index with a relatively low dispersion. As their chemical resistance is rather poor, their durability can be increased by replacing SiO_2 with B_2O_3 and by the introduction of divalent oxides. In addition, the introduction of ZrO_2 and TiO_2 is another possibility for achieving better chemical properties. In the past, ThO_2 and CdO have also been used to achieve this aim, but for environmental protection these oxides have been eliminated from these glasses. The large disadvantage of these glasses is their poor crystallization stability, especially if SiO_2 is used.

10. B_2O_3−La_2O_3

These glasses are based on the B_2O_3−La_2O_3 glass system. They all have high refractive indices and high dispersion. In order to obtain high refractive indices, large numbers of oxides that are known to cause a high refractive index, such as Gd_2O_3, Y_2O_3, TiO_2, Nb_2O_5, Ta_2O_5, WO_3, and ZrO_2 are incorporated in the glasses of this system. The large number of network modifiers leads to glasses whose viscosities depend to a high degree on temperature (i. e., they are called short) and which show a very strong tendency to crystallize. On the other hand, these glasses have a distinct hardness and chemical resistance, which increase further with higher refractive indices. In this respect, the glasses differ from those of the system SiO_2−PbO−M_2O, which also have a high refractive index.

11. SiO_2−Sb_2O_3−B_2O_3

The replacement of PbO by Sb_2O_3 leads to an extensive glass system. Sb_2O_3 reduces the dispersion of the blue part of the visible spectrum. Consequently, short-flint glasses are obtained. The introduction of Sb_2O_3 into a glass gives results similar to those in the case of PbO, although Sb_2O_3 glasses seem to form long chains of −Sb−O−Sb−O− units [5.80].

12. B_2O_3−PbO−Al_2O_3

If SiO_2 is replaced by B_2O_3, glasses with primarily very poor chemical properties are obtained. Glasses of this system have an intensive short-flint characteristic, since the PbO content induces absorption in the blue range of the visible spectrum, and B_2O_3 causes strong absorption in the infrared range of the spectrum. Al_2O_3 is added to improve the chemical properties to an acceptable level.

13. $SiO_2-M_2O-TiO_2-(F_2-O)$
The PbO content in glasses of the system SiO_2-M_2O-PbO can be replaced by TiO_2. As TiO_2 causes an increase in the refractive index, glasses of the system $SiO_2-M_2O-TiO_2-(F_2-O)$ are characterized by high refractive indices. The dispersion of these glasses is increased to a higher degree than that of glasses containing the same amount of PbO. The introduction of TiO_2 into glasses is limited, because when the TiO_2 content is increased the glasses tend to crystallize strongly. This should be avoided by the introduction of alkaline oxides into the glass. In spite of the large content of alkaline oxides, these glasses are highly resistant to the influences of water and acids. Glasses with high TiO_2 contents show strong absorption in the UV range. This absorption is stronger than that of glasses containing PbO or Sb_2O_3. Both glass types are characterized by a positive deviation of the partial dispersion from the normal line.

The break up of traditional glass families into ambitious, multi-property specified designer glasses led to rethinking, and thereby to new methods, in glass development. The well-known stepwise successive method has (where suggestive) now been replaced by an intensive field-screening method supported by computer-assisted statistical experimental design (SED; multi-linear regression-driven analysis).

This breaking of traditional borders and innovation concerning methods and materials is the reason for drastic changes in the assortments of optical glasses during the last five to ten years. First, the environmentally driven conversion to lead- and arsenic-free optical glasses took place [5.81] (see next section). In combination, a simplification of the assortments has been carried out. Old glass types with small market volume have been eliminated and high-performance glasses which were offered by several glass manufacturers were slightly redesigned to be optical interoperable. This led to a remarkable shrinking of the available glass assortment. In contrast the development of new glass types with opportunity for less crystallizing types and better application- and production-fitted glasses expanded the list of available glasses. Presumably, this will proceed in cycles in the future.

5.2.3 Environmentally Friendly Glasses

New recipes for many traditional optical glasses have recently been developed by all the major glass suppliers to make them ecologically friendly. The primary elements of attention in the development of ecologically friendly optical glasses are arsenic and lead, although other toxic elements have also required attention. For example, in 1980, the use of ThO_2 was stopped completely and the use of CdO was restricted to colored optical glass.

In the late 1980s, when the public became increasingly sensitive to health problems caused by Pb, the whole glass-producing industry started developing PbO-free glasses as a first step towards the production of environmentally friendly glass. Apart from the health aspect, the reduction and replacement of PbO in the glass composition led to a considerable weight reduction, which is advantageous in several respects for outdoor applications. Since 1995 the use of another substance, As_2O_3, has been restricted; it is used as a refining agent in trace amounts only. With the intention of replacing PbO and As_2O_3, the glass manufacturers initiated a comprehensive program that has not yet been completed [5.81].

The main target has been to redesign all glass types so as to preserve n_d and ν_d while gaining a Pb- and As-free composition. Moreover, the partial-dispersion values $P_{g,F}$ and $\Delta P_{g,F}$ have been preserved for several special glasses. There are, however, some special glasses whose optical properties cannot be preserved without the addition of PbO. Of course, these glass types will be available in the future in their known quality and composition. They will be clearly marked as Pb-containing products. The bio-availability of the pure toxic substances in the products that are finally sold is negligibly low. For example, even high amounts of PbO in the composition of a glass would not cause any health problems. But although there is no objective reason, the customers' negative associations regarding Pb or As lead to a strong rejection of products containing these substances. There are, however, serious problems caused by hazardous compounds that make expensive and time-consuming development necessary. High amounts of fine dust are produced during acquisition of the raw material and during batch preparation. Of course, the dust does not contain a higher percentage of toxic substances than the base material, but it has a very much larger specific surface. This large surface in combination with the fact that dust tends to deposit in the lung tissue in large amounts causes a bio-availability of the toxic compounds that is in the latent-toxic range. High yields of grinding waste result from finishing processes, particularly from grinding and polishing. These must be disposed of at high costs on hazardous-waste disposal sites. Experience in this field shows that there is a high risk of incorrect handling and a tendency towards illegal disposal because of the high costs. Improving technical measures to protect directly affected persons from the unhealthy effects of dust and waste is theoretically possible, but hard to control. In particular in foreign production, for example in developing countries, safety could not be guaranteed [5.81].

For these reasons, there has been a tendency to dispense with these substances and to use TiO_2, Nb_2O_5, ZrO_2, WO_3 and others instead of PbO and As_2O_3 as a refining agent is replaced by Sb_2O_3. All glass compositions reformulated in this way are designated by a prefix to the original name, e.g. "N-" with Schott, "E-" with Hikari, "S-" with Ohara. For 50% of all glass compositions originally containing As_2O_3 but no PbO, replacing As_2O_3 by Sb_2O_3 does not change their properties.

For the compositions originally containing As_2O_3 and PbO (up to 80%), some physical and optical properties are significantly changed upon the replacement of both elements. Besides improvements, these changes also include some disadvantages. Improvements are: better chemical resistance; higher Knoop hardness; higher T_g point; lower density. Disadvantages are: higher tendency of crystallization; higher T_g, reduced transmission in the UV region; changed values for $P_{g,F}$ and $\Delta P_{g,F}$ (except for the N-KzFS series).

5.2.4 How to Choose Appropriate Optical Glasses

In this section practical guidelines for choosing appropriate glasses for optical applications are given. As stated above, optical glasses are ranked by refractive index and dispersion. However, during this selection process, not only the optical properties but also mechanical and chemical properties should be considered. We present each mechanical and chemical property individually along with a brief explanation of its importance.

Transmission and Absorption

Absorption in optical glass can be expressed in a number of ways and the reader is referred to Sect. 5.1.3 of this text for a more rigourous treatment. We simply state here that most optical glass manufacturers list for their glasses internal transmission values that neglect the losses from reflection at air–glass interfaces, and absorption values are thus conveniently calculated according to $\alpha = -\ln(\tau_1/\tau_2)/\Delta x$, where τ_1 and τ_2 are the internal transmission values for samples of thickness x_1 and x_2 ($x_1 > x_2$), Δx is the difference in thickness between the two samples, and α is the absorption coefficient (expressed in units of cm^{-1} when x is in cm).

Partial Dispersion

Since the refractive index varies with wavelength, not all colors are focused at the same point with a simple single-component lens. This effect is referred to as chromatic aberration. For the fabrication of optical systems, achromatic and apochromatic systems are often employed. In an achromatic system, chromatic aberration is minimized for two colors, e.g., red and blue. The other part of the visible spectrum remains uncorrected and is referred to as the "secondary spectrum". The chromatic aberration in an apochromatic system, however, is corrected through the whole visible region. Most optical glasses show normal dispersion where the partial dispersion of refractive index varies nearly linearly with the Abbe number. These glasses are not suitable for the construction of apochromatic systems. Therefore, glasses that deviate from this normal dispersion behavior are required for the construction of apochromatic systems. As a result, glasses with a high dispersion (short) in the blue region of the visible spectrum have been developed. Deviation of partial dispersion from the normal, or linear, behavior is described by a series of partial dispersion deviation coefficients, $\Delta P_{X,Y}$, where X and Y indicate the wavelengths of interest.

Photoelastic Constants

The refractive index of glasses is altered by the application of stress. The change is generally different for light linearly polarized in a plane parallel to the direction of stress than for light linearly polarized perpendicular to the stress. The magnitude of the change is given by

$$n_{\parallel} = n_0 + K_{\parallel} S \quad \text{and } n_{\perp} = n_0 + K_{\perp} S \tag{5.113}$$

where n_0 is the refractive index in the absence of applied stress, S, and $K_{\parallel}$ and $K_{\perp}$ are the photoelastic constants and $n_{\parallel}$and $n_{\perp}$ are the refractive indices for light linearly polarized parallel and perpendicular to the direction of stress, respectively. The commonly used stress optic coefficient (K) is the difference between the two photoelastic constants.

Thermal Expansion

Inorganic glasses expand upon heating and shrink upon cooling. The coefficient of linear thermal expansion α is defined as $\alpha = \Delta L / L_0 \Delta T$, where ΔL is the change in length of a material of length L_0 that experiences a temperature change of ΔT. The value of α is generally a function of temperature. The coefficient of thermal expansion values found in optical glass catalogs represent only a linear fit and care must be taken when comparing two glasses that the expansion information is taken over the same temperature range.

Thermal expansion is an important property since it is one indicator of the extent to which a particular glass will be prone to thermal fracture. Higher expansion values are generally an indicator that a glass is relatively weak against thermal shock.

Transformation Range

The thermal expansion curve for glasses begins to deviate significantly from a roughly linear shape over a temperature interval referred to as the transformation range. Other temperature-dependent glass properties also change rapidly with temperature over this same range. In addition, if cooling through this range takes place rapidly or unevenly, permanent residual stresses can be imparted to the glass due to unequal amounts of thermal contraction.

The transformation temperature, T_g, is found in glass catalogs and can be used as an indication of where the transformation range exists and what processing or environmental temperatures need to be avoided to avoid changes in surface figure or the creation of residual stresses that can reduce the quality of a transmitted optical wavefront through stress birefringence effects. In this regards, as a rule of thumb, parts should not be heated above a value of $T_g - 200\,°\mathrm{C}$.

Softening Point

Similar to the transformation temperature discussed above, another characteristic temperature for glass is the temperature at which the glass begins to deform under its own weight. The corresponding material viscosity is $10^{7.6}$dPa.

Thermal Properties

Additional thermal properties include heat capacity, c_P, and thermal conductivity, λ. The former provides information on the quantity of heat, $\mathrm{d}Q$, required to raise the material temperature by an amount $\mathrm{d}T$: $c_\mathrm{P} = \mathrm{d}Q/\mathrm{d}T$; and the latter on the heat flux, Q, that will pass through a slab of material of thickness L, in a time period t, when placed across a thermal gradient generated by a temperature difference ΔT; $\lambda = QLt/\Delta T$. Another thermal property found in some optical glass catalogs is the thermal diffusivity K, which is related to the thermal conductivity, heat capacity and glass density ϱ by $K = \lambda/\varrho c_\mathrm{P}$.

All three thermal properties discussed above have a weak dependence on temperature up to values just below the transformation range. As with thermal expansion information, comparison of glasses in critical situations requires that property information be stated at the same temperature.

Elastic Properties

When a glass bar is placed under tension it stretches. The amount of stress, S, and strain, ε, experienced by the bar are related by $S = E\varepsilon$ where E is the Young's modulus of the material (i. e. Hooke's law). The ratio of lateral to longitudinal stress in the bar is Poisson's ratio, μ. In the case where a compressive stress is applied, the proportionality constant between stress and strain is now the bulk modulus, K. Finally in the case of an applied shear stress, the proportionality constant is the shear modulus, G.

Young's modulus and Poisson's ratio values are typically provided in optical glass catalogs. These elastic constants are interrelated such that if any two are known the remaining pair can be readily calculated. We have for example, $K = E/[3(1-2\mu)]$ and $G = E/[2(1+\mu)]$.

Hardness

The classic hardness scale (Moh scale) is based on a series of minerals and their ability to scratch one another. Modern hardness values are collected based on indentation methods using points of prescribed geometry applied at specific load levels and times. Two common hardness measurement methods, with different indentor geometries and load condition, lead to values of Knoop and Vickers hardness. Because test conditions vary between different techniques it is important to make comparisons only between measurements employing the same standard.

Chemical Resistance

Selection of glasses based on chemical properties, com monly chemical resistance test results in optical glass catalogs, is difficult at best since a wide range of different test standards exist among the various suppliers. Consequently, only a brief introduction is provided here, based on some of the more common types of chemical durability tests with some examples of specific test recipes.

Please note that there are some optical glasses that show no staining or other evidence of chemical attack in one test type, but have a low chemical resistance in one or more other tests. Consequently it is important when evaluating the chemical behavior of optical glasses to consider all resistance test results.

Climate Resistance Test. Climate resistance is intended to give guidance concerning the sensitivity of a glass to water vapor in ambient air. Typically, but not always, a slow process, this form of chemical attack gradually leads to the formation of a cloudy surface film that cannot be wiped off. An accelerated procedure is thus typically used for testing the climatic resistance of glasses.

In one possible test technique, polished uncoated glass plates are exposed to a water-vapor-saturated atmosphere the temperatures of which are alternated between 40 °C and 50 °C on an hourly basis. Since the temperature increase in the glass plates follows that of the atmosphere, water condenses on the glasses during the warming phase. In the cooling phase the temperature of the atmosphere initially falls faster than that of the glass plates causing a drying of the glass surface. After an exposure time of, for example, 30 h the glasses are taken out of the climatic chamber. The difference, pH, between the transmission haze before and after testing is used as a measure of the resulting surface change. Classifications can then be done based on the transmission haze increase, pH, after the test period has been completed.

Resistance to Staining. The test procedure gives information on possible changes in the glass surface (stain formation) under the influence of lightly acidic water (for example perspiration, acidic condensates) without vaporization. The class of stain resistance is typically determined by placing a plane-polished glass sample in contact with a test solution (for example a standard acetate at $p\text{H} = 4.6$ or a sodium acetate buffer at $p\text{H} = 5.6$, depending on the relative durability of the glass) using a prescribed temperature and exposure time.

Interference color stains develop as a result of decomposition of the surface of the glass by the test solution. One possible measure for classifying glasses is the time which elapses before the first stain occurs. This change in color indicates a chemical change in the previously defined surface layer of 0.1mm thickness.

Resistance to Acids. When an acidic aqueous medium reacts with the surface of a glass stains can form, the glass can be decomposed, or both can occur. In contrast to the staining resistance tests discussed above, acid resistance testing provides information concerning dissolution since testing is done with larger quantities of solutions, often at lower pH.

ISO 8424 is a standard for evaluating acid resistance of glass. Here, the time required to dissolve a layer with a thickness of 0.1mm serves as a measure of the resistance to acids. Two aggressive solutions are used in determining the resistance to acids: a strong acid, nitric acid, pH 0.3 is used for the more resistant glass types and a weak acidic solution with a pH value of 4.6 (standard acetate) is used for glasses with lower acid resistance.

Resistance to Alkali. These tests concern resistance to alkalis and are provided because many fabrication processes occur in water-based media (for example, grinding and polishing compounds) that usually become increasingly alkaline through the chemical reactions of the water and the abraded glass particles over time. This particularly applies when such solutions are recycled.

ISO 10629 is a standard for evaluating alkali resistance of glass. This test also takes into consideration the fact that higher temperatures can occur as a result of the polishing process through abrasion. The alkali resistance classes indicate the resistance of glasses to alkaline solutions and therefore as a warning signal to finishers of glass of possible problematic materials. As with the acid resistance testing discussed above, alkali resistance is based on the time required to remove a layer thickness of glass of 0.1 mm in an alkaline solution (sodium hydroxide, $c = 0.01$ mol/l, $p\text{H} = 12$). In ISO 106529, the layer thickness is calculated from the weight loss per surface area and the density of the glass.

5.3 Colored Glasses

5.3.1 Basics

Color in glass arises from the selective attenuation or amplification of the incident light within special regions of wavelength in the visible part of the spectrum, i.e. in the range from about 380 nm to 780 nm [5.82]. Generally, the shape of the transmission curve versus the wavelength λ is described by (5.114) (see also Sect. 5.1.3)

$$\tau_\text{i}(\lambda) = \exp \sum_m \left(-\varepsilon_{\lambda,c,m} c_m d\right) , \tag{5.114}$$

$$\tau_\text{i}(\lambda) = \frac{(\Phi_{e\lambda})_\text{ex}}{(\Phi_{e\lambda})_\text{in}} , \tag{5.115}$$

$$\tau(\lambda) = \frac{(\Phi_{e\lambda})_\lambda}{\Phi_{e\lambda}} = P\tau_\text{i}(\lambda) , \tag{5.116}$$

where ε and c are the extinction coefficient and the concentration of the coloring species m, respectively, d is the thickness of the probe and τ_i the internal transmission. The internal spectral transmittance $\tau_i(\lambda)$ in (5.115) is the ratio of the emerging spectral radiant flux $(\Phi_{e\lambda})_{ex}$ to the penetrating radiant flux $(\Phi_{e\lambda})_{in}$. It describes the transmission of a homogeneous absorbing filter material without allowing for reflection losses; $\tau(\lambda)$ characterizes virtually every optical filter, regardless of its structure or mode of action. The extinction coefficient for a given base glass is a function of the wavelength, the coloring species m and, in some cases, the concentration of this species. Keep in mind that one coloring ion in two different oxidation states acts like two different species (e.g. Fe^{2+}, Fe^{3+}).

Unfortunately, there is no way to calculate ε from theoretical considerations so it must be taken from experiments for every coloring species. Some data have been compiled by *Bamford* [5.83], but not for all elements and combinations. Additionally, all data taken from the literature depend on the special base-glass composition. So, for a given new problem, the extinction coefficients must first be established experimentally.

In many cases (5.114) can be simplified. Very often, glasses are colored by only one species in one oxidation state and the extinction coefficient is not a function of concentration. This is especially true if the concentration of the species is low. Then (5.114) can be written as (Lambert–Beer law)

$$\tau_i(\lambda) = \exp(-\varepsilon_\lambda cd) . \qquad (5.117)$$

From this equation, the transmission plots for different concentrations of the coloring species and different probe thicknesses can easily be calculated from one experiment, since ε is then a constant for a given wavelength.

As already mentioned, the color of a glass can be taken from a plot of its internal transmission against wavelength. A translucent body showing an internal transmission of $\tau_i = 1$ at every wavelength in the visible range is called colorless (see Fig. 5.40); the usual optical glasses are examples of these. When the internal transmission at a range of wavelengths is attenuated by any of the mechanisms discussed in Sect. 5.1.3, a color arises. The color depends on the part of the spectrum missing (Table 5.7).

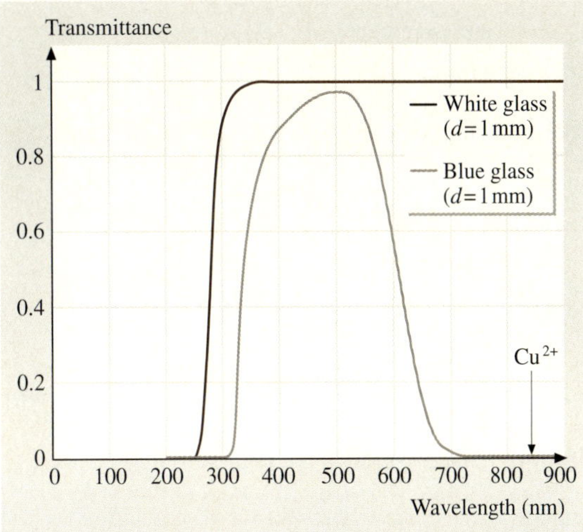

Fig. 5.40 Transmission spectra of white and blue glasses

Figure 5.40 shows the internal transmission plot of a white glass and a blue glass. The color of the blue glass is obtained by doping with Cu^{2+}. This kind of representation is chosen for scientific applications and construction of optical instruments because it advantageously shows the spectral distribution of transmitted light. Another way to describe the color of the glass is the use of color coordinates. When this method is used, the color is defined by two coordinates x and y, which are obtained by weighting the $\tau(\lambda)$ curve of the glass with the intensity distribution of the standardized light source for the colors red, green and blue. In the ideal case, the color portions of the light source add to clear white light. The procedure has been extensively described by *Bamford* [5.83].

Figure 5.41 shows the distribution coefficients for the red (P_x), green (P_y), and blue (P_z) part of a special standardized light source (illuminant D65) as has been set up by the International Commission on Light-

Table 5.7 Relation between color and absorbing wavelength range

$\lambda_{absorbed}$ (nm)	400	425	450	490	510	530	550	590	640	730
Color of $light_{absorbed}$	violet	indigo blue	blue	blue green	green	yellow green	yellow	orange	red	magenta
Color of $light_{resulted}$	green yellow	yellow	orange	red	magenta	violet	indigo blue	blue	blue green	green

Part A | 5.3

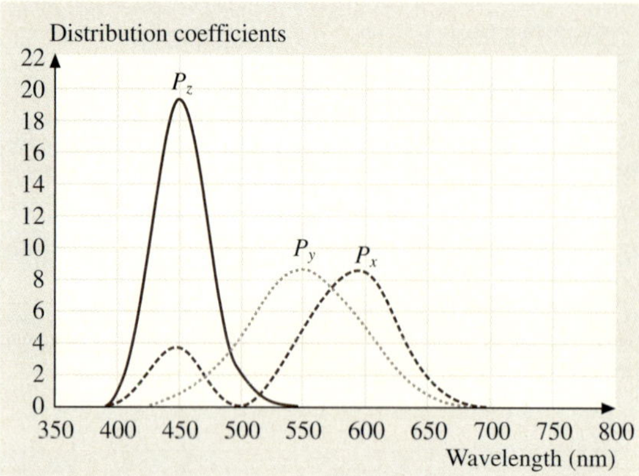

Fig. 5.41 Distribution coefficients for red (P_x), green (P_y), and blue (P_z) part of light of illuminant D65 (standardized light source)

ing (CIE). After multiplication of the transmission curve by the distribution coefficients, and normalization of the sum to unity, one obtains the color coordinates. The "z" coordinate has been omitted because it can easily be calculated from the condition that $x+y+z=1$.

This method is used when the color impression is needed; e.g. for traffic signs and glasses for artists.

From the chemical point of view, the base-glass composition of colored glasses ranges within the compositional region of normal optical glasses, as discussed in Sect. 5.2. Historically, colored glasses have been developed from optical glasses by the empirical addition of so called "coloring elements". There has been a lot of experimental work on this topic, of which a survey has been given by *Weyl* [5.84] and to some extent by *Vogel* [5.85]. The coloring elements additionally present in the base glasses, transition metals and lanthanides, contain partly filled d and f electronic shells. For some special applications, the elements of the sixth (S to Te) and seventh (Cl to I) main groups are used in combination with zinc and cadmium or copper and silver, respectively, in addition to the transition metals and lanthanides.

5.3.2 Color in Glass

The mechanisms of generating colors have been described in [5.86]. From these 15 mechanisms, only four are very important for practical use in inorganic oxidic bulk glasses: ligand field effects [5.82, 87–91], charge-transfer effects [5.82, 92], metal colloids [5.93–95] and semiconductor-doped glasses [5.85, 96–98]. These are in essence described in Sect. 5.1.3 and [5.82]. Therefore we will not explain this mechanism here.

The usual colors of some of the more important ions are listed in Table 5.8 [5.84], which can be described by ligand field effects.

The color of glass is determined by several parameters such as the kind of coloring species, its valency, the coordination of colored species, and the base glass.

The influence of the base glass on the spectra can be attributed to two items. The first is the intensity of the absorption band caused by the transition. This is influenced by the probability of electron transfer. Different base glasses show different coupling modes and therefore different extinction coefficients for a given coloring ion. Commonly, phosphate glasses show lower extinction and the color of a given ion is fainter than in silicate glasses [5.87]. The second reason is the difference of absorption peak maxima between silicate, phosphate and borate glasses. If it is only about 1–2% the arrangements of the coloring ion ligands stays unchanged. The differences between the peak maxima positions are much more pronounced. Some transitions show no correspondence at all. The explanation follows from changes in the coordination. The coordination, for instance, of Ni^{2+} in silicate glass is unexpectedly fourfold (tetrahedral), resulting in a purple color, whereas the dark yellow of Ni^{2+} in the phosphate glass results from the normal octahedral coordination [5.87].

Another coloring species of great importance is $Fe^{2+/3+}$. The interpretation of the spectra of iron is quite a complicated task. One not only has to consider two oxidation states with different peak positions but also that these two species each appear in tetrahedral as well as in octahedral positions. Adding further confusion, parts of these bands can occur at nearly the same wavelengths. Additionally, charge-transfer bands of Fe^{3+} with oxy-

Table 5.8 Usual colors of some ions in glasses, according to [5.84]

Element	Valency	Color
Fe	2+	Green, sometimes blue
Fe	3+	Brown
Cu	2+	Blue, turquoise
Cr	3+	Green
Ni	2+	Violet (tetrahedral)
Ni	2+	Yellow (octahedral)
Co	2+	Blue
Mn	2+	Pale yellow
Mn	3+	Violet
Pr	3+	Green
Nd	3+	Violet
Er	3+	Pale red

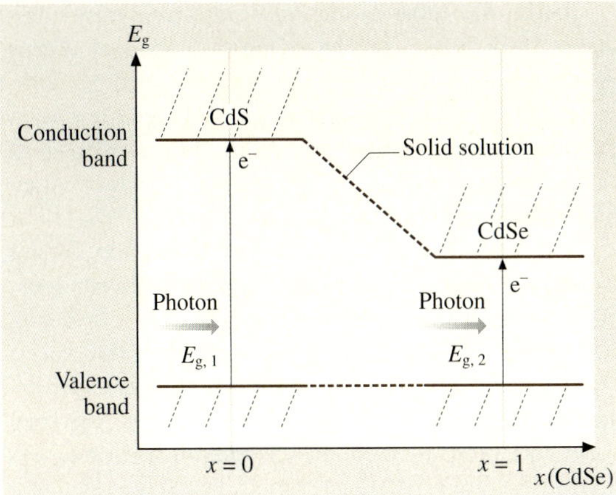

Fig. 5.42 Semiconductor-doped glasses: Band structure and absorption of light for II–VI semiconductors

gen occur. The color of the glass depends strongly on the redox conditions during melting. By the application of a low oxygen partial pressure the glass becomes very pale green to blue according to the transition at about 1100 nm. Under oxidizing conditions the glass turns brown because of the manifold bands of trivalent iron present all over the visible and UV range of the spectrum.

Beside the colored glasses described above, which are based on absorption bands, there are also long-pass filters, which are prepared by doping with semiconductors such as ZnS, ZnSe, ZnTe, CdS, CdSe, and CdTe followed by an additional annealing process to grow semiconductor crystals in the glass. These semiconductor-doped glasses show very different behavior with interesting spectral properties. The absorption mechanism is explained in Sect. 5.1.3.

Table 5.9 Band-gap energies for some semiconductor crystals (macrocrystals)

Compound	E_g(eV)/λ_c (nm)	E_g(eV)/λ_c (nm)
ZnS	3.7/335	3.65/340
ZnSe	2.7/459	–/–
ZnTe	2.2/563	–/–
CdS	2.4/517	2.42/512
CdSe	1.7/729	1.73/717
CdTe	1.5/827	–/–

The essential properties of the transmission spectrum are the wavelengths λ_c, λ_s and λ_p. The wavelength λ_c is defined as the middle of the maxima between the blocking and the pass area ($\tau_i = 0.5$). In the blocking part ($\leq \lambda_s$) all light will be absorbed and at the pass part ($\geq \lambda_p$) the incident light will be transmitted.

Commercially available filter glasses of this series have absorption edges from about 400 nm to 850 nm. As indicated in Fig. 5.42 the band gap, and therefore the absorption edge, depend on the chemical composition and size of the crystals. The transition from Zn to Cd and from S over Se to Te leads to an increasing red shift of λ_c. Decreasing the size of the crystals also moves the effective band gap to higher energies and creates a corresponding blue shift of λ_c. The exact energies of the band gap can be taken from Table 5.9.

By annealing [5.99, 100], influencing the diameter of the crystals, and by mixing appropriate amounts of the pure compounds, all values for λ_c between 350 nm and 850 nm can be achieved.

5.4 Laser Glass

5.4.1 Common Laser Glasses and Properties

Laser glass is a solid-state material with the ability to amplify light by stimulated emission of radiation. In its most common form it is a multicomponent oxide glass doped with a lasing ion such as neodymium. In this section we deal predominantly with multicomponent glasses in contrast to the rare-earth-doped fused silica such as that commonly employed in Er-doped fiber amplifiers.

Although the dominant commercial market for laser glass is in large laser systems for inertial confinement fusion research with application to fusion energy and weapons physics science [5.101], these materials have also found their way into a number of industrial and laboratory environments. For example, one leading application is in the field of laser shock peening [5.102].

Commercial laser glasses fall into three broad categories depending on the operating regime of the applied laser system. Laser glass composition development is a mature field and for each specific application glass compositions and/or processing details of these glasses have already been identified by suppliers that provide optimized properties for each situation.

There are, for example, laser glasses designed for high peak power. Here, laser properties have been optimized to offer the highest possible stored energy and extraction efficiency, resulting in high peak powers in single shots that are separated in time. Repetition rates of such systems are at most a few Hertz, and more commonly are one to several laser shots per day.

There are also laser glasses for high average power applications where repetition rates can be in the range of 1–20 Hz. Such laser systems are in general aggressively cooled to remove heat deposited into the glass during repeated optical pumping. In addition to good laser properties, these glasses also possess enhanced thermomechanical properties consistent with high thermal loading without part fracture.

Still another example are glasses developed for guided wave optics. Here, glass compositions are selected for high stability to devitrification during redrawing into fibers or for compatibility with structuring technologies, for example ion exchange, used in preparing planar waveguide structures. Figure 5.43 shows representative laser glass parts, the largest being a slab more than 1 m along on its diagonal axis for use in high peak power laser fusion research.

Another way laser glasses are categorized is by the glass type. The first laser glass to be identified was a glass based on silica (commonly called a silicate glass) [5.103]. Today, the multicomponent glasses of greatest commercial value are phosphate types and the compositions are essentially free of any silica. In addition, there are glasses based on other glass forming systems, as well as glasses that contain, in addition or in place of oxygen, other anions such as fluorine.

Fig. 5.43 Laser glasses

Today, the most commonly employed multicomponent laser glasses are based on phosphorous. These glasses are available in high optical quality from a number of commercial vendors with a wide range of dopants and doping levels, and in particular offer superior performance for resistance to laser damage when operated in high-fluence applications.

Although neodymium is the most common dopant and lasing ion, through selection of various dopant ions, either alone or in combination, a wide range of lasing wavelengths are possible. When more than one active ion is selected, usually one or more ions acts as a sensitizer for the primary lasing ion. The classical case is the codoping of laser glass containing erbium with ytterbium. Table 5.10 lists typical lasing wavelengths available in common glass hosts along with common sensitizing ions. The emission wavelength is, however, a function of glass composition so these values should only serve as a rough guide.

Specification of Laser Glass Doping Level

A common problem in identification of the optimal laser glass composition is the selection of the doping level of the lasing species or other sensitizing ions. The doping level is largely driven by two factors: the need to pump the entire laser glass volume uniformly and the avoidance of a phenomena called concentration quenching, in which two neighboring lasing ions in the glass can exchange energy by a radiationless process that steals available energy from the laser beam of interest. A rough guideline for dealing with the former case is provided in Table 5.11, which lists typical doping levels as a function of rod diameter for typical flashlamp pumped laser systems that provide nominally uniform excitation of the complete active volume.

Table 5.10 Laser wavelengths from selected active ions in glass

Active ion	Approximate emission wavelength (μm)	Sensitizing ion(s)
Nd^{3+}	0.93, 1.06, 1.35	Cr^{3+}, Mn^{2+}, Ce^{3+}, Eu^{3+}, Tb^{3+}, U^{3+}, Bi^{3+}
Er^{3+}	1.30, 1.54, 1.72, 2.75	Cr^{3+}, Yb^{3+}, Nd^{3+}
Yb^{3+}	1.03	Nd^{3+}, Cr^{3+}
Dy^{3+}	1.32	
Sm^{3+}	0.65	
Ho^{3+}	0.55, 1.38, 2.05	Er^{3+}, Yb^{3+}
Tm^{3+}	0.80, 1.47, 1.95, 2.25	Er^{3+}, Yb^{3+}
Tb^{3+}	0.54	Ce^{3+}, Cu^{+}
Pr^{3+}	0.89, 1.04, 1.34	

Table 5.11 Guidelines for neodymium dopant level for various rod diameters

Rod diameter (mm)	Nd concentration in phosphate laser glass (wt% Nd_2O_3)	Nd concentration in silicate laser glass (wt% Nd_2O_3)
≤ 5	8.0	3.5
5 to 7	6.0	3.0
6 to 10	4.0	3.0
9 to 13	3.0	3.0
12 to 16	2.5	3.0
15 to 20	2.0	2.0
20 to 26	1.5	2.0
≥ 27	1.0	2.0

Figure 5.44 shows the effect of concentration quenching for neodymium where the lifetime of the excited state used in the laser transition is shown to decrease with dopant level for some representative neodymium-doped laser glasses. The effect of concentration quenching is often estimated by the following equation

$$\tau = \tau_0 / \left[1 + (N/Q)^n\right] , \quad (5.118)$$

where τ is the lifetime at neodymium concentration level N (in units of 10^{20} ions per cm^3), τ_0 is the effective lifetime for a negligible doping level in the glass and Q is a concentration quenching factor that corresponds to the doping level at which the lifetime value falls to $\frac{1}{2}$ of τ_0. The value of n is generally equal to 1 or 2, and depends more on the amount of data available to determine the fit to (5.118) than on material considerations. The concentration quenching effect varies with glass type, and in

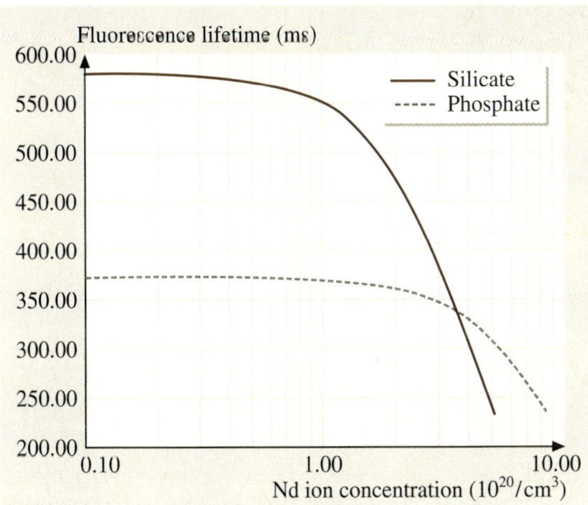

Fig. 5.44 Typical concentration quenching curves

particular is less pronounced in phosphate compositions than in silicate glasses. Table 5.11 also lists different optimal doping levels for both the silicate and phosphate glass types.

Rare-earth ion concentration is expressed in a number of ways, most commonly as either a weight percent of oxide in the glass or as the number of active ions per cubic centimeter. It is straightforward to convert between these two values if one knows the density of the glass. For the case of various lanthanide cations the following conversion formulae between lanthanide (Ln) ion density, in units of $10^{20}/cm^3$, and weight percent of the lanthanide oxide can be employed:

$$\text{Ln ion density} = \varrho F(\text{wt\%Ln}_2\text{O}_3) , \quad (5.119)$$

$$\text{wt\%Ln}_2\text{O}_3 = \text{Ln ion density}/(\varrho F) , \quad (5.120)$$

where ϱ is the glass density in g/cm^3 and the value of F for a number of common rare-earth elements is given in Table 5.12.

Glasses for High Peak Power

Glasses for high peak power are generally characterized by enhanced laser properties. This can perhaps be expressed as having a large value of the following laser figure of merit FOM_{laser},

$$FOM_{laser} = \sigma Q / n_2 , \quad (5.121)$$

where σ is the cross section for stimulated emission and Q is the concentration quenching factor (see p. 294), which together are a measure of the laser gain available in the glass, and n_2 is the nonlinear index, a measure of glass weakness to resist damage when transmitting a high-intensity pulse of laser light (see Sect. 5.4.2). These properties for some typical commercial high peak power laser glasses are given in Table 5.13.

Table 5.12 Rare earth active ion content conversion factors

Rare earth cation	F
Nd	0.358
Er	0.315
Yb	0.306
Tm	0.313
Ho	0.319
Pr	0.402

Table 5.13 Properties of some typical high peak power laser glasses

	Schott LG-770	Schott LG-750	Hoya LHG-80	Hoya LHG-8	Kigre Q-88
Cross section for stimulated emission σ (10^{-20} cm^2)	3.9	3.7	4.2	3.6	4.0
Concentration quenching factor Q (10^{20} cm^{-3}) [$n=2$ in (5.118)]	8.8	7.4	10.1	8.4	6.6
Nonlinear refractive index n_2 (10^{-13} esu)	1.01	1.08	1.24	1.12	1.14
Fracture toughness K_{IC} (MPa m$^{1/2}$)	0.48	0.48	0.46	–	–
Thermal conductivity K (W/m K)	0.57	0.60	0.59	0.58	0.84
Poisson's ratio ν	0.25	0.26	0.27	0.26	0.24
Thermal expansion α (10^{-6}/K)	13.4	13.2	13.0	12.7	10.4
Young's modulus E (GPa)	47.3	50.1	50.0	50.1	69.8
Index at 1060 nm n_{1060}	1.499	1.518	1.534	1.521	1.536
$dn/dT_{20-40\,°C}$ at 1060 nm	−4.7	−5.1	−3.8	−5.3	−0.5

Table 5.14 Properties of some typical high average power laser glasses

	Schott APG-1	Schott APG-2	Hoya HAP-4	Kigre QX-Nd
Cross section for stimulated emission σ (10^{-20} cm^2)	3.35	2.39	3.6	3.34
Concentration quenching factor Q (10^{20} cm^{-3}) [$n=1$ in (5.118)]	16.7	10.6	–	–
Nonlinear refractive index n_2 (10^{-13} esu)	1.13	1.02	1.21	1.17
Fracture toughness K_{IC} (MPa m$^{1/2}$)	0.60	0.67	0.83	–
Thermal conductivity K (W/m K)	0.83	0.86	1.02	0.85
Poisson's ratio ν	0.24	0.24	0.24	0.24
Thermal expansion α (10^{-6}/K)	7.6	6.4	7.2	8.4
Young's modulus E (GPa)	71	64	70	71
Index at 1060 nm n_{1060}	1.529	1.503	1.534	1.530
$dn/dT_{20-40\,°C}$ at 1060 nm	1.2	4.0	1.8	1.0

Glasses for High Average Power

Glasses for high average power are generally characterized by enhanced thermomechanical properties as well as good laser properties. This is often expressed as having a large value of the following thermomechanical figure of merit $\mathrm{FOM_{tm}}$,

$$\mathrm{FOM_{tm}} = K_{IC}K(1-\nu)/\alpha E\,, \qquad (5.122)$$

where K_{IC} is the fracture toughness, K is the thermal conductivity, ν is the Poisson's ratio, α is the thermal expansion and E is the Young's modulus of the glass. These properties for some typical high average power commercial laser glasses, along with the laser properties discussed earlier, are given in Table 5.14.

Planar and Fiber Waveguide Lasers and Amplifiers

The glasses described in the previous sections can also be drawn into fiber form for preparation of fiber lasers and amplifiers. In addition, a number of glasses are also available for preparing planar waveguide structures by techniques such as ion exchange. The ion-exchange structuring technique is normally performed at temperatures below the transformation point of the glass. Properties of some of these glasses specifically developed for planar waveguide applications are provided in Table 5.15. Since these glasses have been commonly employed in the preparation of active planar guided-wave devices based on the erbium emission in the 1540 nm telecommunications region, Table 5.15 contains some associated properties for this active ion.

Estimating Refractive Index

Glass manufacturers typically report refractive index and dispersion as measured at wavelengths near 587.6 nm. It is a common requirement to estimate the refractive index of laser glass at additional wavelengths, for example for purposes of coating design at a pump or lasing wavelength. In addition, refractive index is a function of doping identity and content, so that catalog index values and those listed here in Tables 5.13, 5.14, and 5.15 are only rough guides. For this, a useful relationship for the refractive index at a different wavelength $n(\lambda)$ is

$$n(\lambda) = n_D - [(n_D - 1)/V_D](1.5079 - 523\,640/\lambda^2) \tag{5.123}$$

where n_D is the refractive index at 589.3 nm, V_D is given by

$$V_D = (n_D - 1)/(n_F - n_C)\,, \quad \text{(see Sect. 5.1.2)} \tag{5.124}$$

and n_F is the index at 486.1 nm, n_C is the index at 656.3 nm, and λ is in nm. The accuracy of (5.123) is given as ± 0.001 in the cases where index measurements are accurate to ± 0.0001, but there is no applicable wavelength range provided [5.104].

Alternatively, a similar equation based on index data measured at the mercury d-line at 587.6 nm is the following:

$$n(\lambda) = n_d - [(n_d - 1)/V_d](1.5079 - 523\,640/\lambda^2)\,, \tag{5.125}$$

where n_d is the refractive index at 587.6 nm, V_d is given by

$$V_d = (n_d - 1)/(n_F - n_C)\,, \tag{5.126}$$

and n_F is the index at 486.1 nm, n_C is the index at 656.3 nm, and λ is again in nm. Equation (5.125) is expected to be accurate to ± 0.003 for IR wavelengths up to 2.3 μm in the cases where index measurements are accurate to ± 0.00002.

5.4.2 Laser Damage

Advances in high-power and high-energy laser systems have made power densities in excess of $10^9\,\text{W/cm}^2$ available. At such power levels a number of different damage mechanisms come into play that can lead to temporary or even permanent changes in the optical quality of optical materials, including laser glass. Such changes may occur at a surface or within the bulk material, and can both render an optic unusable and/or alter the transmitted beam profile in such a way as to place other downstream components in an optical system at increased risk of damage.

Transient Thermal Effects

Since the refractive index is generally a function of temperature, thermal gradients within laser glass parts can lead to optical distortion of propagating light through inhomogeneous changes in the optical pathlength. For some applications, it is possible to have laser glasses that behave essentially as an athermal component in which the thermal expansion of the glass offsets the change in optical pathlength due to the variation of index with temperature.

For a laser glass component within a cavity of an otherwise fixed length (for example as defined by a frame constructed from near zero expansion materials such as Zerodur or Invar), the variation in optical pathlength

Table 5.15 Properties of typical waveguide laser glasses

Glass type	Schott IOG-10 Silicate	Schott IOG-1 Phosphate	Hoya LHG-5 Phosphate	Kigre MM-2 Phosphate
n_d	1.530	1.523	1.541	1.54
V_d	56.6	67.5	63.5	–
Thermal expansion $\alpha_{20-40\,°C}$ $(10^{-6}/\text{K})$	6.8	9.3	8.4	7.3
Er peak emission wavemength (nm)	1536	1534	–	1535
Er FWHM fluorescence linewidth, $\Delta\lambda_{emm}$ (nm)	18.5	26.5	–	55
Er fluorescence lifetime (ms)	17.8	10.7	–	7.9
Glass transformation point T_g (°C)	569	474	455	506

with temperature, dS/dT, is given by

$$dS/dT = \alpha(n-1) + dn/dT \,, \tag{5.127}$$

where the glass is characterized by a thermal expansion value of α, refractive index of n, and temperature change of refractive index, relative to air, with temperature of dn/dT.

An alternative situation occurs if the laser glass part is coated in such a way as to define the length of the optical cavity, as in the case where the end mirrors of a laser oscillator cavity are directly applied to the ends of a laser rod. In this case the cavity length variation with temperature is given by

$$dS/dT = n\alpha + dn/dT \,. \tag{5.128}$$

Typical values for α, n, and dn/dT are included with other laser glass properties in Tables 5.13–5.15.

Surface Damage

Surface damage on active (and also passive) optical materials in high-intensity laser systems normally appears as circular elevations (pustules) or depressions (pits) on transmission surfaces. The number density and size of these features is somewhat determined by the laser energy density and pulse length, respectively. Laser glass surfaces are vulnerable to laser damage since they typically become contaminated through routine storage and handling as well as the gradual accumulation over time of airborne particles. Surface contamination serves as a nucleation point for damage. The initial preparation of the surface can potentially play a critical role in determining the likelihood of experiencing laser damage. After polishing of a surface is completed, there can still exist residual scratches, defects, and subsurface flaws that can retain contaminants of polishing compounds and cleaning materials. Such sites on optical material surfaces serve as additional nucleation points for laser damage.

Induced damage on beam exit surfaces is often correlated with absorption of laser energy at such surface defect sites, followed by formation of a plasma which in turn can enhance reflection of the laser light back on itself, further increasing the local electric field intensity associated with the laser beam. Finishing vendors for high-damage-threshold optics need to avoid leaving residue of polishing compounds that absorb strongly at the intended wavelength of exposure. Surface damage threshold is also clearly improved by proper cleaning of optical surfaces and subsequent protection from atmospheric contaminants, including water, which can alter surface properties through condensation and diffusion.

Self-Focusing Damage

Self-focusing occurs due to the dielectric breakdown of an optical material from localized focusing of a propagating laser beam by increase of the refractive index with the applied light intensity. This effect leads to one or more threadlike damage sites oriented along the main optical axis that are often referred to by a number of different terms including *tracks* or *angel hair*.

The damage event often begins as a localized hot spot within a laser beam that causes an increase in local refractive index within the glass. Since the optical pathlength is increased at such a hot spot, and drops towards the original value away from the high intensity region, this portion of the laser glass essentially acts as a positive lens. This artificial lens further converges the laser beam, compounding the effect. This convergence continues until the electric field associated with the laser increases to a point that atoms are ionized and a plasma is produced within the glass. An example of such self-focusing damage is shown in Fig. 5.45.

The phenomena of self-focusing is driven by the laser glass nonlinear refractive index n_2, the accumulated optical pathlength, and the presence/absence of localized hot spots in the laser beam. Internal optical quality of laser glass has improved over time to the point that meter-class sections are available that are essentially free of all bubbles and inclusions that can cause localized fluctuations in beam intensity through diffraction effects, so the nonlinear index plays a prominent role in determining if laser damage of this type will occur. It is for this reason that n_2 appears in (5.121) for FOM_{laser} in Sect. 5.4.1. In addition to selection of a laser glass with low n_2, laser cavity and beamline designs need to minimize the possibilities of high-fluence locations in

Fig. 5.45 Self-focusing damage in laser glass

beams passing through laser glass components, i. e. the laser energy should be spread out over as large an area as practically possible.

Multiphoton Induced Damage

Multiphoton transitions involve two or more photons whose energies are combined to a value sufficient to excite a real transition in a material. High-fluence laser sources provide high photon flux levels that make these low-probability multiphoton transitions in glass observable. Damage from multiphoton absorption can occur when electrons are excited into an optical material's effective conduction band, where they then migrate to stable traps in the material and induce structure and property changes. Damage takes two principle forms: as a localized discoloration and/or as a permanent change in the refractive index of the damaged region. This type of laser damage is almost always viewed as problematic and the observed absorption characteristics as a function of wavelength are generally similar to that observed with high-energy radiation damage, such as that found with UV, X-ray, or particle (proton, electron, etc.) exposure.

For a particular laser design, the extent of multiphoton damage is determined by the number of laser photons required to achieve a combined energy sufficient to lead to damage. As a rough guide, one can use an effective band gap of the laser glass, as calculated from the onset of absorption in the high-energy end of the electromagnetic spectrum, compared to the energy available from a single photon at the laser wavelength.

Point-Defect Laser Damage

Nearly all optical materials, including laser glass, contain some level of localized microscopic bulk defects including bubbles, dielectric inclusions, and precious metal particles. All such defects have some relationship to the original manufacturing process. Most inclusions and metal particles represent either unmelted compounds from the original glass melt or portions of the manufacturing equipment that were incorporated into the final laser glass component. Bubbles by themselves are not typically linked with the creation of localized damage sites, however they can cause diffraction of a propagating laser beam with associated small-spatial-scale high-intensity power spikes that can then damage other optical elements in a laser system through nonlinear effects (see for example the discussion in the subsection on surface damage).

Metal particles, chiefly platinum, remain in laser glass since for optical quality reasons these glasses are produced within pots and crucibles made of precious metals and their alloys. Subsequent particulate-related damage from these inclusions appears as isolated damage sites within a bulk optic. Figure 5.46 shows a magnified image of such a damage site that measures 0.25 mm in diameter. The damage mechanism is believed to be direct absorption of laser radiation by the dielectric or conducting-defect site, followed by heating of the particle to above its boiling point, with subsequent crack nucleation and growth from both the vaporization-related shock wave and localized thermal-induced stress occuring [5.105].

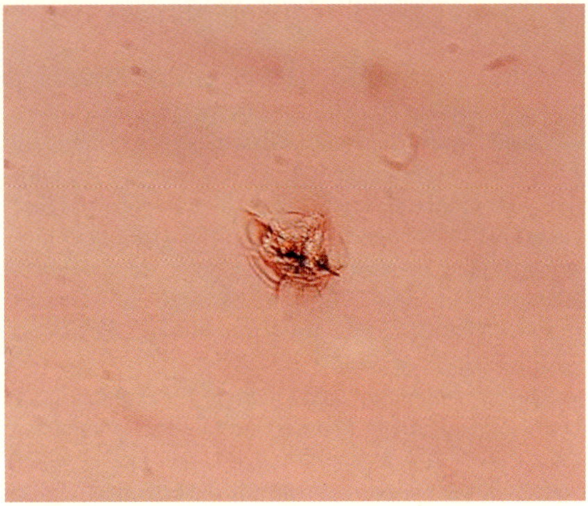

Fig. 5.46 Particulate damage in laser glass due to a platinum particle

Since the initial discovery of the laser in 1960, the largest single improvement in laser glass damage-threshold performance has been the development of manufacturing technologies capable of producing meter-class phosphate laser glass components completely free of all inclusions or internal defects capable of becoming particulate damage sites [5.106]. The specification of such platinum-particle-free glass should be made when laser fluence levels approach the damage threshold, a condition approximated for pulsed 1.064 μm laser exposure by

$$\text{Damage threshold in J/cm}^2 = 2.5\times(\text{pulse length in ns})^{0.3}\,. \tag{5.129}$$

Although laser damage is an extensively studied field, there is no comprehensive database of the enormous amount of accumulated experimental data on this subject that would make it possible to predict the laser damage level for a particular situation. Interested readers are referred to several textbooks that deal exclusively with this field of investigation [5.107, 108]

5.4.3 Storage and Handling of Laser Glass

Laser glasses are typically brittle materials that are easily chipped or broken upon thermal or mechanical loading. As a result, heating and cooling of glass parts should be at rates not exceeding 20–30 °C/h. Laser glasses with good laser properties are generally of compositions that also exhibit low chemical durability. The low durability can sometimes be used to an advantage, since deliberate chemical etching [5.109] or chemical ion-exchange processes [5.110] can be used to enhance mechanical strength.

There exists no universal method for evaluating all possible chemical attack paths and laser glass compositions. However, as a general rule, the presence of water is nearly always a prerequisite for the chemical attack of glass. As a consequence, long-term storage of sensitive glasses is best accomplished within a closed, and preferably evacuated, environment containing a desiccant material.

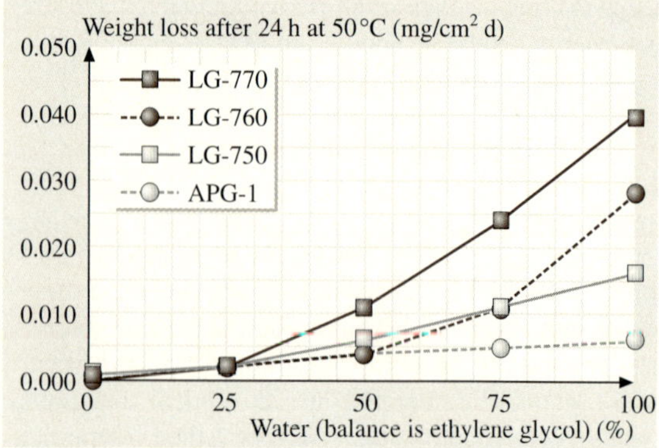

Fig. 5.47 Chemical durability in EtOH/water mixtures

Methods to Enhance Part Strength

Laser glasses can be strengthened by techniques such as acid etching and ion exchange. In the former, surface removal by etching is accompanied by blunting of crack tips left following fabrication or mechanical handling. Blunted crack tips require a higher energy level to propagate into cracks or fractures, effectively increasing the strength of the glass part to breakage. Ion exchange involves the substitution of a smaller cation within the glass structure by a larger cation, effectively placing the surface of the glass under compression. To experience fracture, the applied stress must now exceed not only the initial strength level of the glass but the compressive surface stress as well. Both processes involve some level of chemical attack to the glass surfaces and are not suitable for polished optical surfaces.

Liquid Cooling of Laser Glasses

Solid-state laser systems employing laser glass as the gain medium are typically cooled by a recirculating liquid coolant system. Chemical attack by the coolant solution is a variable to be considered, in particular when phosphate laser glasses are employed and the component is to be left in contact with coolant during periods of no operation. In this case, it has been shown that addition of ethylene glycol to water can provide improved protection to chemical attack, as seen in Fig. 5.47.

5.5 Glass–Ceramics for Optical Applications

5.5.1 Overview

Glass–ceramics – glasses that are crystallized via a deterministic and controlled process – combine the relative ease of manufacturing associated with glass with the added functionalities that exist only in crystalline states of matter [5.111, 112]. For example, the combination of a crystalline phase that exhibits negative thermal expansion, a relatively rare occurrence in nature, with the normal positive expansion for glass has led to the development of ultra-low thermal expansion glass–ceramics used for such things as telescope mirrors, IR-transparent cooker tops, and reflectors for digital projection [5.113]. Dental glass–ceramics combine high strength with the proper amount of controlled translucency to mimic natural teeth precisely [5.114]. And a wide variety of nonlinear phenomena have been demonstrated in glass–ceramics, including ferroelectric and pyroelectric behavior [5.115], second-harmonic generation [5.116], upconversion [5.117], and scintillation [5.118].

Contrary to one's first impression, the presence of crystals in a glass framework need not significantly inhibit transmission. The ability to control the crystallization process often allows one to keep crystal sizes small relative to the wavelength of the incident radiation, thereby minimizing scattering losses, particularly at longer wavelengths. In addition, the precipitation of very small, insoluble crystals in a soluble matrix allows one to form crystals in compositions that are otherwise difficult if not impossible to control. For example, *Beall* [5.119] developed a Cr-doped forsterite ($Cr{:}Mg_2SiO_4$) glass–ceramic; stoichiometric forsterite glass by itself is extremely unstable and only very special processing techniques are able to form it [5.120]. However, by relying on phase separation to create Cr-doped, forsterite-like sub-domains which subsequently crystallized, Beall was able to combine the ease of glass production with the formation of a desired crystal. This tactic is a familiar one in glass–ceramic development.

Glass–ceramic production begins with the formation of suitable glass castings using now-standard manufacturing techniques [5.124]. Detailed inspection techniques are then used to determine the internal quality of the part owing to the inherent high transparency of glass. After inspection, a controlled thermal cycling program, termed "ceramization", is used, whereby the glass part is partially or wholly transformed into a glass–ceramic. Typical thermal histories take advantage of the disparate temperature regimes in which crystal nucleation and subsequent growth take place [5.125]. At a minimum, three heating/cooling ramp rates and two pairs of temperature/time conditions need to be specified in a ceramization process (Fig. 5.48). This, coupled with the fact that both nucleation and growth are thermally activated processes with inherently strong nonlinear dependencies on temperature [5.126], can make ceramization a challenging process.

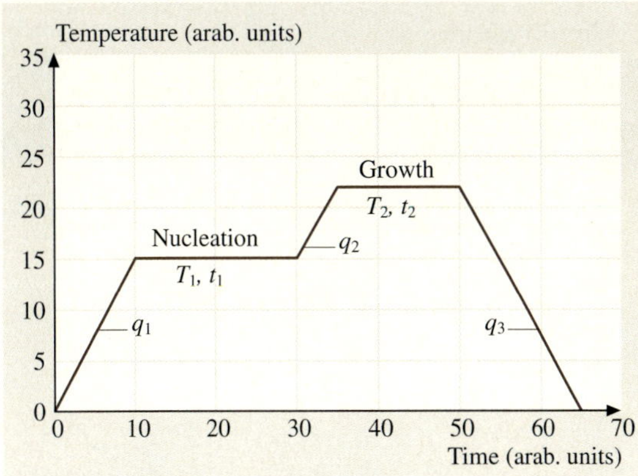

Fig. 5.48 Schematic temperature–time path used for the production of glass–ceramics, a process termed ceramization; q_i refers to heating and cooling rates, T_i and t_i refer to temperatures and time used during crystal nucleation and growth stages, respectively

5.5.2 Properties of Glass–Ceramics

Compared to their precursor glasses, glass–ceramics typically have increased strength and higher elastic constants. Thermal expansion, on the other hand, can be either higher or lower than the precursor glass. Glass–ceramics normally exhibit less dielectric loss than their glassy precursors, due in part to tying up the more mobile species in a more rigid, crystalline framework (though exceptions are known [5.111, 127]) whereas relative permittivities are generally higher in glass–ceramics. Finally, chemical durability is typically enhanced over glass, in some cases quite dramatically [5.112].

Thermomechanical Properties

Thermal Expansion. Coefficient of thermal expansion values (CTE) for glass–ceramics range from nearly -10 to $+25\times10^{-6}\,°C^{-1}$, a remarkably wide range few materials can approach (Table 5.16) [5.111]. Expansion matching to most metals (including Si) has been accomplished thereby allowing for the production of multi-material devices under stress-free or stress-controlled conditions (Fig. 5.49) [5.111,128,129]. A possible complexity with glass–ceramics arises from the often nontrivial expansion curves for the crystalline phases. For example, glass–ceramics that contain a silica polymorph typically show a pronounced change

Table 5.16 Comparison of optical and physical parameter values for a range of glass–ceramic materials

Material	n	T_i (%)	λ (nm)	Thickness (mm)	CTE (ppm/°C)	T range (°C)	E (GPa)
Zerodur[a]	1.52	99.6	1550	1	0±0.05	0 to 50	90
SA-O2[b]	1.56	≥ 98	1550	1	1.8	30 to 500	98
WMS-15[b]	1.52	99.9	1550	1	11.4	−30 to +70	96
NEX-C[b]		99.3	1550	10	−2	−40 to +80	94
Oxyfluoride[c]		99.99	1500	1			
Telluride[d]	2.1	≈ 80	800	1			
Pyroceram[e]					5.7	20 to 320	120

(Sources: [a] Zerodur product brochure, [b] Ohara product brochure, [c] [5.121], [d] [5.122], [e] [5.123]; note that Pyroceram, while opaque in the visible region, is transparent in its operative, radio-frequency region)

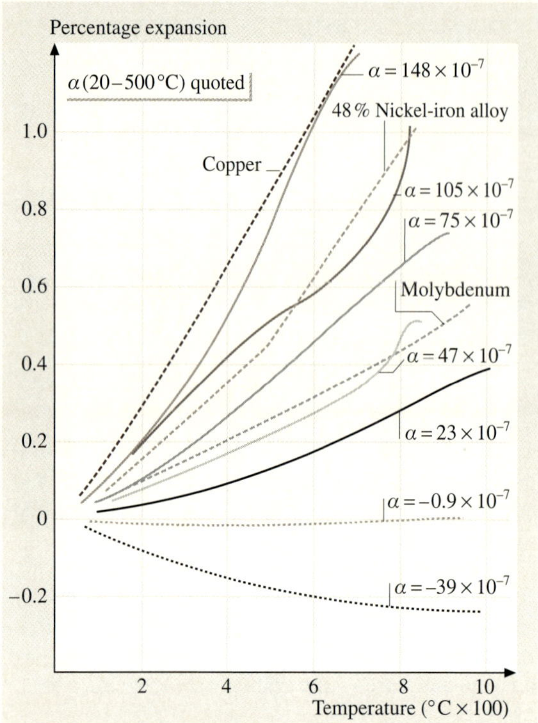

Fig. 5.49 Range of thermal expansion curves ($\Delta L/L$) attainable with glass–ceramics. Also plotted are expansion curves for typical metals (here, α is the CTE of the material). (After [5.111])

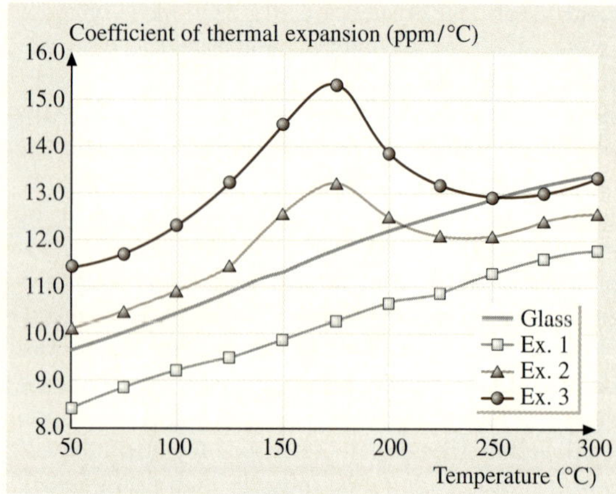

Fig. 5.50 Example of the range of CTEs that can be attained in glass–ceramics for a single composition. The curve labeled "glass" is for the unceramized precursor glass whereas those labeled "Ex. 1" and so on are for 3 different ceramization protocols. The large bump in CTE near 175 °C is due to the polymorphic transition of a cristobalite solid-solution phase

in slope in a plot of CTE versus temperature, associated with a polymorphic transition of the silica phase (Fig. 5.50). This can be dealt with, in principle, through a suitable choice of the particular crystalline phases that most modify the CTE relative to the glass precursor. Additionally, the value of CTE can be tuned over fairly substantial ranges for a single bulk composition through proper choice of heat treatment conditions (Fig. 5.50). To a fairly good degree, the CTE of a glass–ceramic depends linearly on the relative amounts of the various phases, though more complicated treatments exist [5.130]. Cryogenic thermal expansion data are known for only a few glass–ceramics [5.131].

Elastic Constants. Due to their lack of crystalline order, most glasses have elastic moduli and related parameters (e.g., Young's modulus E, sound velocities) considerably lower than those of their crystalline counterparts. Thus, glass–ceramics commonly have E values between those of crystals and glasses (e.g., 80 to 150 GPa), though in some cases nearly double that of their precursor glasses (Table 5.16) [5.111]. As for CTE, the Young's modulus also displays a nearly additive relationship with respect to the phase content [5.132].

Strength, Hardness, and Fracture Toughness. Enhanced strength is one of the hallmarks of glass–ceramics. Here, the moduli of rupture measured via four-point bend tests or associated techniques can be double or more those of the precursor glasses [5.133], though surface flaws still ultimately control the maximum attained strength values. Unlike thermal expansion and elastic properties, strength is not only controlled by the proportion of the various phases present, but also by the internal microstructure, particularly at or near the surface [5.112, 134]. The presence or absence of interlocking crystals, preferred crystalline orientation, or even the slip system associated with the crystalline phases present can all play a role in dictating glass–ceramic strength.

As a brittle material, the fracture behavior of glass–ceramics is commonly characterized by hardness and fracture toughness values. The former is normally quantified via diamond micro-indentations and the latter by the propagation of either a preexisting flaw of known dimensions or the development of indentation fractures [5.135, 136]. Here too crystal content and microstructure play dominant roles [5.137].

Electrodynamical Properties

Dielectric Properties. Dielectric property characterization – typically restricted to sub-optical frequencies and often grouped with physical properties – is analogous to refractive index and absorption characterization at optical frequencies. Dielectric properties are generally restricted to GHz and lower frequencies (microwaves and radio-frequency), whereas optical characterization is normally restricted to tens of THz and above (infrared, visible and UV primarily). Little is known about the response of glass–ceramics, or of many substances for that matter, in the so called terahertz regime (0.1 THz to 10 THz) [5.138].

The most important dielectric parameters in the GHz and below regime are the dielectric constant (relative permittivity) and dielectric loss (loss tangent). The power absorbed per unit volume P may be written:

$$P = \frac{1}{2} E_0^2 \omega \varepsilon_0 \varepsilon_r \tan\delta \,, \tag{5.130}$$

where E_0 is the magnitude of the electric field, ω is the frequency of the field, ε_0 is the permittivity of vacuum, ε_r is the *relative* permittivity of the material (not to be confused with the permittivity of the material, given by the product of ε_r and ε_0), and $\tan\delta$ is the loss tangent [5.139]. The relative permittivity of the dielectric material is related to its refractive index via

$$\varepsilon_r = n^2 \,. \tag{5.131}$$

Note that this latter expression is only strictly true if the frequency with which both permittivity and refractive index are made are equivalent, or, somewhat more loosely, if the polarization loss mechanisms are unchanged between the two measurements. Thus, ε_r is related to the speed of propagation within the dielectric whereas $\tan\delta$ is a measure of the absorption. Typical values for permittivity start at values similar to glass (≈ 5) but can exceed 1000 for glass–ceramics containing ferroelectric crystals [5.140–142]. Loss tangents also start close to glass values (e.g., 0.0002) and range up to nearly 1 (Table 5.17) [5.111]. The dielectric strength of glass–ceramics – the magnitude of the electric field just sufficient to initiate breakdown of the dielectric – is generally superior to that of ceramics and comparable to that of glasses (Table 5.17).

Refractive Index and Absorption. Optical properties of glass–ceramics are quite similar to those of their glass precursors and the crystalline phases contained therein. As introduced in Sect. 5.1.1, the two fundamental quantities, refractive index and absorption, can be described collectively via the complex refractive index

$$\tilde{n}(\omega) = n(\omega) + \mathrm{i}\kappa(\omega) \,, \tag{5.132}$$

where n, the refractive index, and κ, the bulk absorption, are both frequency dependent [5.143]; κ is in turn related to the more often used absorption per unit length α via

$$\alpha = \frac{2\omega\kappa}{c} \,, \tag{5.133}$$

(see Sect. 5.1.3) where c is the speed of light and α is used to characterize the total transmission T through a sample via

$$\frac{I_t}{I_0} = T = RT_i = R\exp(-\alpha z) \,, \tag{5.134}$$

(Fig. 5.10) where I_t is the transmitted intensity $(\mathrm{W/m^2})$, I_0 is the incident intensity, T_i is the internal transmission (not including surface reflection losses), R is a factor that accounts for reflection losses, and z is the sample thickness [5.144].

Refractive indices of most glass–ceramics are in the range 1.5 to 2.0 across much of the visible and infrared region, whereas transmission values are strongly dependent on the wavelength of interest (Table 5.16). In some applications, absorption across a wide range of

Table 5.17 Dielectric properties of various glass, glass–ceramic, and ceramic materials

Material	ε_r	$\tan\delta$	Test frequency	Dielectric strength (kV/mm)	Test frequency (Hz)
Lithium–zinc silicate glass-ceramic [a]	5.0	0.0023	1 MHz	47	50
Fused silica [b]	3.8	0.0004	1 MHz	36	100
Alumina (99.5%) [c]	9.7	0.0001	1 MHz	8.7	60
Y-stabilized zirconia [c]	9.0	0.001	1 MHz	9.0	60
Steatite [d]	6.1	0.0008	1 MHz	9.1	60
Pyroceram [e]	5.5	0.00033	8.5 GHz		

(Sources: [a] [5.111], [b] GE Quartz commercial literature, [c] Coors Ceramics Company commercial literature, [d] CeramTec commercial literature, [e] [5.123])

wavelengths is of interest. For example, in the characterization of dental glass–ceramics with controlled translucency, standardized tests exist that quantify a material's integrated absorption in the visible region when recorded against a white or black background [5.114] and may be quantified with respect to standardized color coordinates [5.145].

Scattering. Volume scattering (see Sect. 5.1.7) of radiation by crystals is one of the most interesting aspects of glass–ceramics. Glass–ceramics have in fact been used recently as test materials for wavelength-dependent scattering experiments and analysis in the UV–VIS portion of the spectrum [5.146]. Depending on the application, the wavelength of light may be greater or less than the average crystal size. This distinction, among other factors, leads to a variety of phenomena that dictate the ultimate utility of glass–ceramics for optical applications. The theoretical basis for scattering in condensed matter began over a hundred years ago with the work of Rayleigh and with important generalizations some twenty years later by Mie (Sect. 5.1.7). More recent efforts relevant to glass–ceramics have been devoted to tying the early, theoretical work to the particular set of conditions that glass–ceramics present.

Based on the theoretical arguments of Sect. 5.1.7, we consider here the most important type of scattering for optical glass–ceramics: pure Rayleigh scattering, in which the size of the particle is less than that of the wavelength of light and the scattering cross section is defined by [5.147–149]:

$$C_{\text{scat}} = \pi a^2 Q_{\text{scat}} , \tag{5.135}$$

where a is the particle radius and Q_{scat} is the scattering efficiency factor. The latter is in turn defined by

$$Q_{\text{scat}} = \frac{8}{3}\left|\frac{m^2-1}{m^2+2}\right|^2 x^4 , \tag{5.136}$$

where m is often approximated by the ratio of the refractive index of the particle to that of the matrix (but see [5.149]), and x is a size parameter defined by

$$x = \frac{2\pi a}{\lambda} , \tag{5.137}$$

where λ is the light wavelength. Taken together, we can write for the scattering cross section (cm^2):

$$C_{\text{scat}} = \frac{128}{3}\pi^5 \frac{a^6}{\lambda^4}\left|\frac{m^2-1}{m^2+2}\right|^2 . \tag{5.138}$$

The λ^{-4} dependence is a hallmark of Rayleigh scattering as is the strong dependence on particle size.

An alternative definition, based on a quasi-continuum model due to *Hopper* [5.150], is given by

$$C_{\text{scat}} = \left(6.3\times10^{-4}\right) k^4 \overline{w}^3 \left(\overline{n}\Delta n\right)^2 N_{\text{V}}^{-1} , \tag{5.139}$$

where k is the wavenumber $(2\pi/\lambda)$, $\overline{w}$ is a mean width, here taken as the particle radius (but see explanation in [5.150]), $\overline{n}$ is the mean refractive index, and N_{V} is the number of particles per unit volume. Here we have used Hopper's equation (70) and the definition of the absorption due solely to scattering, sometimes called the turbidity τ (cm^{-1})

$$C_{\text{scat}} = \frac{\tau}{N_{\text{V}}} . \tag{5.140}$$

The often cited subcase of Rayleigh–Gans scattering has a further stipulation that there is negligible reflection at the particle–matrix interface. In this case, (5.136) is modified [5.149]

$$Q_{\text{scat}} = |m-1|^2 x^4 \int_0^{\pi} G^2(u)\left(1+\cos^2\theta\right)\sin\theta\,\mathrm{d}\theta , \tag{5.141}$$

where

$$G(u) = \frac{2}{u^3}(\sin u - u\cos u) \tag{5.142}$$

and

$$u = 2x\sin\frac{\theta}{2} . \tag{5.143}$$

Another definition of the scattering cross section has no dependence on particle size, but is dependent on the particle number density [5.151]

$$C_{\text{scat}} = \frac{8}{3}\left(\frac{\pi\left(n^2-1\right)}{N_{\text{V}}\lambda^2}\right)^2 . \tag{5.144}$$

Finally, work by *Hendy* takes recent findings in structure factors in phase-separated media into account to arrive at a scattering cross section with quite different dependencies on both particle size and wavelength but which explicitly includes the volume fraction crystallized φ [5.152]

$$C_{\text{scat}} = \frac{14}{15\pi}\varphi(1-\varphi)k^8 a^7\left(\frac{\Delta n}{\overline{n}}\right)^2 N_v^{-1} . \tag{5.145}$$

To relate scattering cross sections to experimental measurables, we make use of the equationthat relates the normalized transmitted intensity to the path length z [5.151]

$$\frac{I_{\text{scat}}}{I_0} = \exp\left(-C_{\text{scat}} N_{\text{V}} z\right) , \tag{5.146}$$

where here (5.146) ignores losses due to absorption and reflection.

As an example of scattering in a glass–ceramic, we will analyze data from the nanocrystalline glass–ceramic Zerodur. This material, described more fully below, contains approximately 70 vol. % of 50 nm crystals. The number density for the microstructure is $\approx 10^{16}$ crystals/cm^3 [5.153]. Figure 5.51 plots uncorrected transmission data for both glassy and ceramized Zerodur. Note that the composition of these two samples is identical and so the difference between the two curves should be dominated by scattering losses. Figure 5.52 plots this difference, here normalized to remove the reflection losses which are virtually identical for both glassy and ceramized material. Rearranging (5.146) we have

$$-\ln\left(\frac{I_{\text{scat}}}{I_0}\right) = C_{\text{scat}} N_V z \,. \tag{5.147}$$

The measured quantity on the left-hand side (LHS) of this expression is directly proportional to the scattering cross section. Dividing the LHS of (5.147) by the path length (≈ 5 mm) and by the crystal number density for Zerodur, we arrive at the scattering cross section itself (small symbols in Fig. 5.53). Note the nearly linear appearance when plotted on this semilog scale, suggesting a power-law dependence on wavelength (the best-fit slope to data in the range 450 to 800 nm is -4.3, consistent with Rayleigh scattering). The reason for the small deviation from the λ^{-4} dependence at short wavelengths (< 425 nm) is not clear. Also plotted in Fig. 5.53 is the prediction of the Hopper cross section (5.139). We used an estimated ratio of indices (m) of $1.55/1.5 \approx 1.03$ for this example. For comparison, note that the observed cross section at 500 nm was 1.2×10^{-17} cm^2. Although both the Rayleigh and Hopper equations predict the observed λ^{-4} dependence of the cross section, the pure Rayleigh model is some three orders of magnitude too high (3.3×10^{-14} cm^2), whereas the Hopper model predicts 1.0×10^{-17} cm^2, very close to

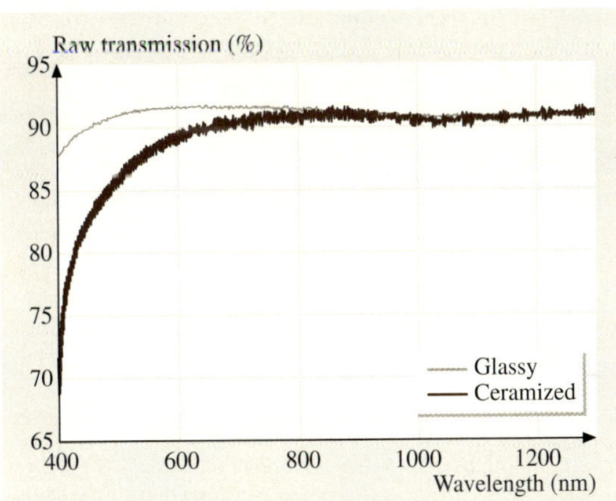

Fig. 5.51 Raw transmission data for glassy and ceramized Zerodur (thickness = 5.05 mm for both samples). Note the high transmission of both samples through the visible region (400 to 800 nm) and the nearly identical values for both ceramized and unceramized (glassy) samples at wavelengths greater than ≈ 1000 nm

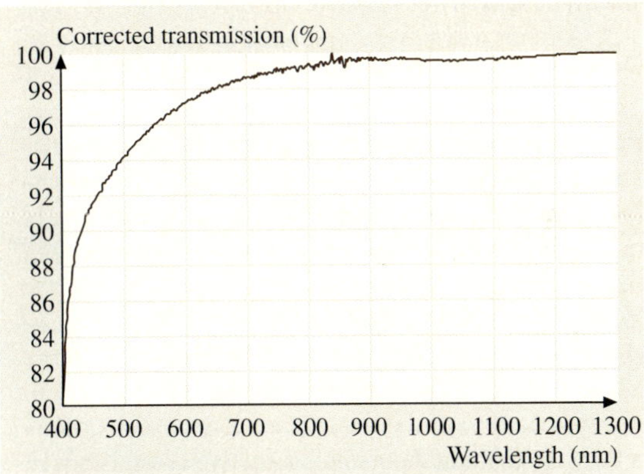

Fig. 5.52 Pure scattering effect on the transmitted intensity in Zerodur. The difference spectrum from the transmission curves in Fig. 5.51 were renormalized to remove reflection losses common to both samples. The subtraction process itself removes common absorption losses. What then remains, and is plotted here, is only the modification to the observed intensity due to scattering

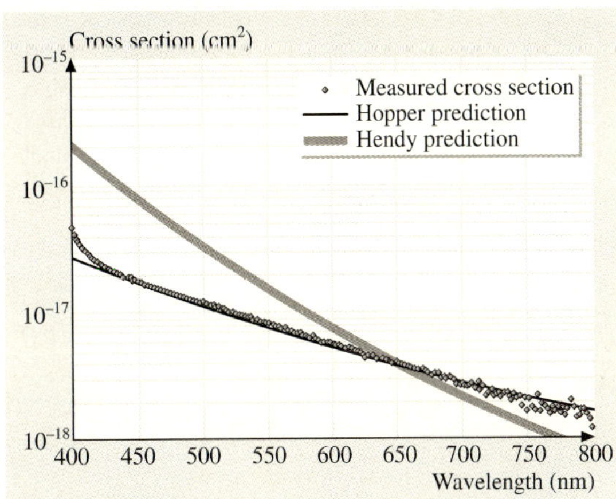

Fig. 5.53 Measured and calculated scattering cross sections in Zerodur

the observed value. The particle-size-independent model (5.144) gave comparable results to the pure Rayleigh model at 500 nm ($6.6\times10^{-14}\,cm^2$). Numerically integrating the Rayleigh–Gans equations (5.141–5.143) yielded $1.4\times10^{-14}\,cm^2$ at 500 nm, a fairly small correction from the pure Rayleigh value. The model of Hendy yielded $3.3\times10^{-17}\,cm^2$, quite close to the measured one near 500 nm. However, the Hendy model predicts a too steep, λ^{-8} dependence, refuted by the very close agreement between the data in Fig. 5.53 and those models that predict a λ^{-4} dependence.

Nonlinear Properties (Electro-Optic, Pyroelectric, SHG, etc.). Although nonlinear crystals have been in use for over 100 years [5.143], the use of glass–ceramics as a host for such crystals has only been known for some 50 years. *Herczog* [5.140] reported the crystallization of microcrystalline $BaTiO_3$ from suitable borosilicate glasses and the production of glass–ceramics with relative permittivities exceeding 1000, whereas *Borrelli* et al. [5.159] and *Borelli* [5.160] reported quadratic electro-optic behavior in $NaNbO_3$-containing glass–ceramics. Nanometer-sized crystals in related transparent glass–ceramics that previously failed to exhibit ferroelectricity were shown to exhibit pyroelectricity, thereby confirming their status as active crystals but revealing a size dependence on the glass–ceramic's measured functionality [5.115]. More recent work regarding transparency requirements in systems containing ferroelectric crystals has focused on crystal number densities, sizes and refractive index contrasts [5.161, 162].

A wide variety of compositional families that contain nonlinear crystals have been explored, including transparent titanate- and tantalate-bearing glass–ceramics [5.161, 163]. Telluride systems offer the advantage of a relatively small refractive index contrast between the high-index active crystals and the high-index residual glass matrix [5.122, 164–166]. Transparent bismuth–borate glass–ceramics have been developed to yield materials that exhibit second-harmonic generation [5.155, 167, 168] as have potassium–niobate [5.154] and potassium–titanium phosphate compositions [5.169]. Results on glass–ceramics containing superconducting $Bi_3Sr_2Ca_2Cu_3O_{12-x}$ crystals have been reported [5.170] as have glass–ceramics for X-ray scintillators [5.118, 171].

Glass–ceramics have been developed to serve as frequency upconverters, luminescent materials, laser hosts, Q-switches, and optical amplifiers (Table 5.18) [5.117, 172–177]. The typical design goal is to partition all elements that participate actively in the nonlinear process into the crystal phase, as in the *Beall* example noted earlier [5.119], thereby increasing the efficiency of the active process (e.g., narrowing a laser linewidth). Although early reports of Nd-doped $Li_2O-Al_2O_3-SiO_2$ glass–ceramics reported lasing [5.178, 179], subsequent work showed that Nd was not incorporated into the crystal [5.180], thereby limiting the advantage of the glass–ceramic route. Much recent work has focused on high-transparency rare-earth-doped fluoride- and oxyfluoride-based compositions for optical amplifiers [5.181–191]. Q-switching has been demonstrated in Co-doped $MgO-Al_2O_3-SiO_2$ glass–ceramics [5.177, 192, 193]. Long-lasting (> 12 h) phosphorescence has been reported for Mn^{2+}-doped $ZnO-B_2O_3-SiO_2$ glass–ceramics [5.194]. Cr-doped mullite-containing glass–ceramics have been developed for solar energy concentrators and laser hosts [5.195]. These have met the criteria for good absorption in the visible with luminescence in the near-IR suitable for conversion to electrical energy by Si photovoltaic strips, although the conversion efficiency is too low at present to make a viable alternative to existing solutions [5.112].

5.5.3 Applications

This section will highlight optical glass–ceramic applications that have actually made it to the commercialization stage, a stringent feat for any materials development program. The list includes glass–ceramics

Table 5.18 Examples of active functionalities attained in glass–ceramics

Material	Active feature	Characteristic feature
$K_2O-Nb_2O_5-4SiO_2$ [a]	Second-harmonic generation	Up to 600 times that of quartz
$Li_2B_4O_7-Bi_2WO_6$ [b]	Relative permittivity	Up to 10^6 at 673 K (100 Hz)
(Na, Sr)NbO_3 [c]	Kerr constant (electrooptic)	$\approx 2\times10^{-17}\,m^2/V^2$
$BaO-GeO_2-TiO_2$ [d]	Piezoelectric	$d_{33}\approx 6\,pC/N$
$Li_2O-SiO_2-B_2O_3$ [d]	Pyroelectric	$p_3\approx -8\,\mu C/m^2/K$
$BaO-B_2O_3$ [e]	Second-harmonic generation	0.12 times that of urea

(Sources: [a] [5.154], [b] [5.155], [c] [5.156], [d] [5.157], [e] [5.158])

used for both reflective (e.g., mirror) and transmissive (e.g., coating substrates) applications.

Reflective

One of the cornerstone applications for glass–ceramics is for mirror substrates. Typically this calls for ultra-low-expansion materials due to the exacting dimensional tolerances that must be attained to ensure satisfactory optical resolution in semi-controlled ambient environments such as an astronomical observatory. The use of negative-expansion high-quartz solid-solution crystals embedded in a positive expansion glass matrix to induce a nearly zero CTE has a long history [5.112, 196]. One of the most widely used low-expansion glass–ceramics for mirrors is Zerodur, developed by Schott Glas nearly 40 years ago [5.153]. In addition to control of the dependence of CTE on temperature [5.197], studies have reported precision polishing of Zerodur substrates using both conventional and ion-beam finishing [5.198, 199] although much knowledge on this topic remains proprietary to individual finishing houses. A common and nonproprietary technique to remove microcracks formed during grinding is the use of an HF−HCl acid etch to physically and chemically remove material [5.124]. Although HF is a well known and effective etchant of silicate glasses and glass–ceramics, the use of HCl is equally important to prohibit the precipitation of insoluble fluoride compounds on the surface of the material being etched. Recipes are also known for the removal of standard mirror coatings [5.124].

Although there are many examples of large Zerodur mirror blanks currently in service in astronomical observatories around the world, one is particularly noteworthy: the very-large telescope (VLT) array in Chile. There, four 8.2 m Zerodur-based telescopes use state-of-the-art adaptive optics and interferometric techniques to attain very high optical resolution with tremendous light-gathering capability [5.200, 201].

Transmissive

One of the earliest applications of glass–ceramics coincides with the emergence of glass–ceramics as a new material. *Stookey* [5.202] at Corning was able to fabricate a radio-frequency-transparent high-strength cordierite glass–ceramic called Pyroceram. This material is still used on the radar domes (radomes) of hypersonic missiles and has extremely low levels of rain erosion at these speeds; a chemical etch is used to increase the strength of the material [5.123].

Robax, a VIS-transparent, low-expansion glass–ceramic manufactured by Schott Glas, is widely used in oven and fireplace sight panels. The high transparency of the material (similar to Zerodur noted above), high thermal shock resistance and extremely high maximum service temperature (680 °C continuous, 750 °C for short times) has allowed Robax to replace borosilicate glass in many applications [5.124].

Telecommunication technologies have also driven glass–ceramic developments. Ohara has released IR-transparent high-Young's-modulus (E) high-CTE glass–ceramics called WMS-13 and WMS-15 for dense-wavelength-division multiplexing (DWDM). These materials also feature excellent chemical durability and the ability to achieve low RMS roughness values through superpolishing. The somewhat atypical combination of high E *and* high CTE while maintaining good chemical durability – required for the use of these materials as substrates for thin-film interference filters in field-based applications – is considerably more difficult to achieve in glass.

Thin film, transparent protective glass–ceramic coatings couple good transparency with extremely good wear resistance. *Rother* and *Mucha* [5.203] report using SiO_2 and Al_2O_3 powders and an electron beam evaporator to deposit a 3 μm coating on automobile wheel demonstrators. The mechanically stable coatings will undergo further testing before widespread commercialization.

5.6 Nonlinear Materials

5.6.1 Overview on Nonlinear Optical Materials

In Sect. 5.1.5, we have mentioned that, even among the second-order nonlinear (NL) materials [5.204–211], not many made their way into industrial application (because of disadvantages that NL materials are suffering due to other intensity dependent effects besides the desired one).

The materials with the highest nonlinearities can be found among the (photonic) crystals, such as the ferroelectrics ADP and KDP used in electrooptical and acoustooptical devices where the first crystals were applied for frequency conversion. Due to their lack of a center of symmetry, these crystals show rather high second-order susceptibilities, which are suitable for applications like frequency upconversion and others. Substitution of cations of an intrinsic nonlinear crystal (e.g.: Gd in Gd–Ca–Oxyborate) by optical active cations, like rare-earth cations (e.g. Nd^{3+}), results in self-frequency doubling crystallites, which combine the active ion with the nonlinear host. Photorefractive crystals can be used for many applications including phase conjugation, image amplification, information processing and optical computing. In the past, most of the experiments concerning optical nonlinearity have been performed on $BaTiO_3$ single crystals, which have excellent photorefractive properties, but are difficult to grow. One of the present leading crystals is K–Na–Sr–Ba–Niobate (KNSBN) because of its large electrooptic coefficients. Two further advantages of this material are that there is no room temperature phase transition and that there is a possibility of doping with Mn, which enables the formation of deep level centers for charged carrier recombination. On the other hand, single crystals, which are very expensive, are needed to avoid scattering effects [5.205, 212–216]. The high costs, and thus economic disadvantages, of crystalline materials arise due to the extreme difficulties to grow them. Nevertheless crystalline materials, especially the niobates will still be among the major materials for all-optical applications. But on the other hand, the limitations of Li-niobates will result in the research into materials like novel ceramic crystals, polymers and even fullerenes.

Other materials with high nonlinear refractive index and near resonance nonlinear effects are organic molecules, such as dyes, polymers (p-toluene-sulphonate), aromatic molecules, and molecules of the pyridine family, the stilbene family, and others which have extremely high $\chi^{(3)}$ values in the direction of the polymer chain, or chiral conjugated polymers and dendrimers (e.g. polydiacetylens) The disadvantages of these materials are their thermal instability, their softness, and the difficulty to get crystals of sufficient size. Furthermore, because of the mostly resonant nonlinearity, the response time for optical switching of these materials is rather low. A possible way to stabilize the organic molecules against degradation and to achieve larger sample sizes is to dope a glass host with these molecules. Glasses doped with organics will be further discussed Sect. 5.6.5 [5.205, 217, 218].

Currently it seems rather clear that the main improvements will come out of the field of organic crystals, as there exist the best possibilities to develop new high potential systems, whereas no real developments are expected to come from the field of inorganic crystals. Composites may also offer new possibilities to overcome the problems of the current materials.

An overview over many common NL materials is given in Table 5.19. Table 5.20 then lists some of the most common NL materials and their proposed (and sometimes already commercially realized) applications. Table 5.21 presents nonlinear susceptibilities for a wide range of exemplary inorganic and organic crystals.

According to *Hirao* [5.204, 219, 220] and others [5.221, 222], for the development of high efficient optical devices the following points have to be taken into account:

- Increase of the upconversion-efficiency and the transition probability of the radiating transition of RE-ion dopands by selection of glass hosts with low multiphonon emission rates.
- Processing speed ("turn-off" times in a time range $<$ ps)
- Phase matching
- Operating power and fraction of power, which can be switched: low switching energy and rapid switching times
- Nondispersive
- Heating stability
- Improvement of the crystallization stability
- Enhancement of IR-transparency and chemical resistance
- Avoiding release of toxic and aggressive substances
- Development of low cost fabrication methods
- Control and measurement of traces of contamination during the fabrication process

In contrast to the materials named above, materials which posses inversion symmetry (isotropic materials), like glass, have no second-order contribution of the optical susceptibility. Nonlinear glasses can be found among the oxide- halide- and chalcogenide glasses (intrinsic nonlinearity) and the semiconductor (SC), metal particle and organic dye-doped glasses (extrinsic nonlinearity) These materials have been intensively investigated in bulk structure as well as in waveguide and fiber structure and have shown many interesting properties: SC-micro-crystal doped films emit electroluminescence, dye-doped polymers exhibit SHG and Faraday effect glasses can be used for magneto-optical switches. Furthermore, methods for the fabrication of long period optical fiber gratings have been developed [5.204, 219, 220].

Table 5.19 The most important and other less important nonlinear optical materials [5.210]

Specialty **ceramics**: Li-, K-Niobate, K-Titanylphosphate (KTP, also very successful: used for doubling to Nd:YAG laser frequency to 532 nm), Li-Tantalate, Sr-Ba-Niobate, B-Ba-Borate, Li-Triborate (Other ceramic NLO materials: Ce-doped Sr-Ba-Niobate, Bi-silicon dioxide, Ba-Titanate)

Rare-earth elements doped in silica: Er and Pr in optical fiber amplifier (OFA)

Nanophase materials made from micrometer-sized powders: especially important, when used in composites

Polymers (including third-order NL-optical polymers, NL-dye functionalized thermoplastics)

Silicon-plastics: polysilicates and polysilynes

Liquid crystal polymers

Compound SC: transparent in the IR region

Ceramics

Semiconductors: excitation of carriers changes the optical properties, but difficult to find materials with low activation energies and large NLO properties, also difficult production.

Superlattices: excitation of carriers in the narrow bandgap part leads to migration of the carriers into the wide bandgap part, which creates a large charge separation. Possibility of tailoring the optical properties. But also difficult to produce.

Organic materials: disadvantages: thermal (operation temperatures about 80 °C) and chemical instabilities, but easy manipulation. Leading candidates are the Methacrylates (properties similar to $NiBO_3$, low cost production, but poor strength and easy evaporation), Poly- (di-) acetylenes (good NLO properties, but unstable, hard to dissolve), Nitroanilins, ...

Table 5.20 Applications and proposed nonlinear materials [5.222]

Application	Material
Communications	
Modulation and switching	$LiNbO_3$, KTP, II–V-SC, MQW's
Directional couplers	$LiNbO_3$
Lenses	$LiNbO_3$ with As_2S_3 glass
EO Grating modulator	Ti:$LiNbO_3$
Sensors	
Imaging/processing	KTP, $BaTiO_3$, $KNBO_3$
Single mode fiber optic sensors	$LiNbO_3$
Optical imiting	LCPs, (in-) organic materials
Instrumentation	
Fabry–Perot interferometers	$LiNbO_3$
Mach–Zahnder interferometers	$LiNbO_3$
Autocorrelators	KDP
Two photon spectroscopy	KTP, KDP
Holographic interferometry	$LiNbO_3$, KTP, $BaTiO_3$, ...
Computing	
Holographic gratings	$BaTiO_3$, KTP, BBO, ...
Storage media	$LiNbO_3$
Laser frequency doubling	$SrBaNbO_3$, Fe:$LiNbO_3$, Bi:SiO_2, III–V- and II–VI-SC

Table 5.21 Nonlinear susceptibilities of exemplary inorganic and organic crystals [5.211]

Material	Crystal symmetry neg.: $n_o > n_e$ pos: $n_e > n_o$	Point group	Transparency range (0.5 µm) from	to	Breakdown threshold (10^9 W/cm^2) λ = 1064 nm	τ (ns)	NL coefficient d_{36} (10^{-13} m/V) 1.06 µm	10.6 µm	Second harmonic generation: 1064 → 532 nm Active length (mm)	Conversion eff. (%)
KDP: KH_2PO_4	neg. uniaxial	$42m$	0.1765	1.7	23	0.2	4.35/6.3		14	82
DKDP: KD_2PO_4	neg. uniaxial	$42m$	0.2	2.0	6	0.25	4.02		40	70
ADP: $NH_4H_2PO_4$	neg. uniaxial	$42m$	0.184	1.5	6.4	15	5.28			
DADP: $ND_4D_2PO_4$	neg. uniaxial	$42m$	0.22	1.7			5.2			
RDP: RbH_2PO_4	neg. uniaxial	$42m$	0.22	1.5	>0.3	12	4.02			
CDA: CsH_2AsO_4	neg. uniaxial	$42m$	0.26	1.43	0.5	10	4.02		17.5	57
KDA: KH_2AsO_4	neg. uniaxial	$42m$	0.216	1.7			5.2			
$LiIO_3$	neg. uniaxial	6	0.3	6	3–8	0.12	$-55.3(d_{15})$/ $-72(d_{33})$		18	44
$LiNbO_3$	neg. uniaxial	$3m$	0.33	5.5	0.05	10	$-59.5(d_{31})$/ $27.6(d_{22})$		20	40
$K_3Li_2Nb_5O_{15}$	neg. uniaxial	$4mm$	0.4	5			$62(d_{15})$/ $127(d_{33})$			
BBO: β-BaB_2O_4	neg. uniaxial	$3m$	0.198	2.6	10	0.1	$22.2(d_{22})$/ $1.6(d_{31})$		6.8	70
Proustite: Ag_3AsS_3	neg. uniaxial	$3m$	0.6	13	0.02	20		$113(d_{15})$/ $180(d_{22})$		
Pyrargite: Ag_3SbS_3	neg. uniaxial	$3m$	0.7	14	0.02	17.5		$84(d_{15})$/ $92(d_{22})$		
$AgGaS_2$	neg. uniaxial	$42m$	0.5	13	0.025	35		134		
$AgGaSe_2$	neg. uniaxial	$42m$	0.71	18	0.03	35		330		
$ZnGeP_2$	pos. uniaxial	$42m$	0.74	12	0.003	30		754		
$CdGeAs_2$	pos. uniaxial	$42m$	2.4	18				2350		
GaSe	neg. uniaxial	$62m$	0.65	18	0.035	10		$544(d_{22})$		
CdSe	pos. uniaxial	$6mm$	0.75	20	>0.05	10		$180(d_{15})$		
HgS	pos. uniaxial	32	0.63	13.5	0.04	17		$502(d_{11})$		
Se	pos. uniaxial	32	0.7	21				$970(d_{11})$		
Te	pos. uniaxial	32	3.8	32				$6500(d_{11})$		
Quartz: α-SiO_2	pos. uniaxial	32	0.15	4.5			$3.64(d_{11})$			
Li-formate: $LiCOOH*H_2O$	neg. biaxial	$mm2$	0.23	1.2		cw	$1.0(d_{31})$/ $-1.16(d_{32})$/ $1.68(d_{33})$			
Ba-Formate: $Ba(COOH)_2$	pos. biaxial	222	0.245	2.2			$1.17(d_{36})$/ $1.13(d_{14})$/ $1.17(d_{25})$			
Na-formate: NaCOOH	neg. biaxial	$mm2$	0.26	1.28			$0.22(d_{31})$/ $-2.2(d_{32})$/ $3.3(d_{33})$			
LiB_3O_5	neg. biaxial	$mm2$	0.16	2.6	25	0.1	$10.9(d_{31})$/ $11.7(d_{32})$/ $0.65(d_{33})$			

Table 5.21 Nonlinear susceptibilities of exemplary inorganic and organic crystals [5.211], cont.

Material	Crystal symmetry neg.: $n_o > n_e$ pos: $n_e > n_o$	Point group	Transparency range (0.5 μm)		Breakdown threshold (10^9 W/cm^2)		NL coefficient d_{36} (10^{-13} m/V)		Second harmonic generation: 1064 → 532 nm	
			from	to	$\lambda = 1064$ nm	τ (ns)	1.06 μm	10.6 μm	Active length (mm)	Conversion eff. (%)
KTP: $KTiOPO_4$	pos. biaxial	$mm2$	0.35	4.5	15	1	65(d_{31})/ 50(d_{32})/ 137(d_{33})		4	60
$Ba_2NaNb_5O_{15}$	neg. biaxial	$mm2$	0.37	5	0.003	10	−132(d_{31})/ −182(d_{33})		3	20
$KNbO_3$	neg. biaxial	$mm2$	0.4	4.5	0.012	10	158(d_{31})/ −132(d_{32})/ −201(d_{33})			
$MgBaF_4$	neg. biaxial	$mm2$	0.185	10			0.248(d_{31})/ 0.37(d_{32})/ 0.1(d_{33})			
Urea: $CO(NH_2)_2$	pos. uniaxial	$42m$	0.2	1.8	5	10	13			
3-Methyl-4-nitro-pyridine-1-oxide	pos. biaxial	222	0.4	3	2	0.02	10			
MAP: $C_{10}H_{11}N_3O_6$	pos. biaxial	2	0.5	2	3	10	167(d_{21})/ 184(d_{22})/ 36.8(d_{23})/ −5.44			
COANP: $C_{13}H_{19}N_3O_2$	pos. biaxial	$mm2$	0.47	1.5			150(d_{31})/ 260(d_{32})/ 100(d_{33})			

The third-order susceptibility of glasses is generally much smaller than the second-order contributions of the crystalline material mentioned above. Additionally, the third-order nonlinearity makes a variety of processes possible, such as THG, FWM, DFWM (optical Kerr effect), as well as intensity dependent refractive index changes. But glass also shows some advantages compared to other second and third-order nonlinear materials [5.204, 206, 207, 222]:

- roducible with high optical quality and transparency/homogeneity even in large sizes
- easy fabrication of fiber and film waveguides with high quality
- high transparency over large spectral ranges
- high chemical, thermal and mechanical stability and durability, high optical damage threshold
- adjustable compositions
- quicker production process
- potential lower cost

A correlation between nonlinear properties and material structure would allow us to predict the nonlinear properties of new materials. A kind of quality coefficient can be used as a measure for the nonlinear properties of the materials. There are many suggestions for appropriate figures of merit for nonlinear materials, most of them reflect besides $\chi^{(3)}$, the absorption α and the response time τ of the material

$$F = \chi^3 / \tau\alpha . \tag{5.148}$$

For instance, the optical glass SF59 has a relative high figure of merit (FoM), not because of high nonlinearities (compared to typical second-order materials), but because of extremely low absorption and insensitivity to thermal effects [5.206, 207].

Due to the usually low third-order nonlinearities of glassy materials, nonlinear optical devices for signal processing or logic circuits for optical computing have to be produced in waveguide geometry to increase the power density. Optical waveguides can be produced via sputtering, CVD (Chemical vapor deposition), sol–gel coating, ion implantation and ion exchange. Besides the planar or channel waveguides, optical fibers are ideal for the study of nonlinear interactions as well as for the desired applications, because they combine strong beam confinement and low loss for propagation over long distances [5.206].

We now describe in more detail two representitive research areas for nonlinear materials, all optical switching and observation of second-harmonic generation in the isotropic material glass. Followng this is a discussion of the many glass systems, which have been intensely investigated for nonlinear effects over the last 20 years.

5.6.2 Application: All Optical Switching

All-optical switching [5.204–206, 219, 220, 223] is a technique based on the optical Kerr effect (linear electrooptical effect) and means the switching of a light beam by using a pump beam which changes the optical communication signal to be switched. The first ultra fast all-optical switch was demonstrated on a silica glass fiber. Silica is known to have only a small nonlinear susceptibility, but also a very small absorption coefficient at the operating wavelength. This allows all-optical switching between two waveguides embedded in the same fiber simply by controlling the pulse intensity. Generally, for optical switching at THz frequencies, there are several important material requirements besides $\chi^{(3)}$, interaction length, and light intensity that need to be present [5.205, 212]

- high damage threshold or thermal stability,
- low power switching threshold and minimal (two photon) absorption,
- wavelength tunability,
- sub-ps- nonlinear response times,
- compatibility with waveguide fabrication process.

The efficiency of a switching device is induced by confinement of the beam into the waveguide over a macroscopic distance. The "turn-on"-time of all nonlinearities is instantaneous, even for the thermal related effects. Thus, the most important parameter for optical switching is the relaxation time of the nonlinear index change. To avoid crosstalk, this relaxation time has to be significantly shorter than the time between two pulses. Two of the fastest switching speeds among the nonlinear materials, besides the organic para-toluene sulphonate, are found in the systems $Au{:}SiO_2$ and $Cu{:}SiO_2$. Generally, chalcogenide glasses, as well as metal and semiconductor doped glasses also are prospective materials for all-optical switching devices.

A figure of merit (FOM) for ultrafast photonic switching expresses the ratio of fast index change to the resultant thermal index change and is an indication for the number of switching operations that can be performed within the thermal lifetime [5.206, 207]:

$$F = n_2 c_p \varrho / \tau\alpha (\mathrm{d}n/\mathrm{d}T) . \qquad (5.149)$$

Due to this FOM, the ideal material for all-optical switching has high nonlinearities, fast response and minimal absorption. For instance, the lead silicate glass SF59 has a relative high FOM because of extremely low absorption and insensitivity to the thermal effects. SC-doped glasses show large and relatively fast NL-absorption changes, but the switching value of these SC-doped glasses is too low for practical application. Dye doped glasses have higher nonlinearity, but their response times are too slow for application. Another limitation of the resonant nonlinear materials for all-optical switching is two-photon-absorption (TPA).

Among the other nonresonant, nonlinear glasses, Bi_2O_3-glass with response times of ≈ 100 fs and a repetition rate of switching of about 1.6 THz is a very prospective candidate for all-optical switching. Other candidates are chalcogenide glass fibers (response time ≈ 200 fs) One disadvantage of these glasses may be a structural change occuring when irradiated with light near the bandgap wavelength. Additionally, photonic bandgap materials, like Au-doped, poled glasses are considered for optical switches [5.204, 219, 220].

Like the diffractive and the holographic optical elements, the switchable optical elements can be used in a great variety of advanced devices; from everyday consumer products (eyeglasses, cameras, CD-players, animated holograms and real time holographic display systems) to commercial instrumentation (tunable grating elements, fiber optic crossbar switch, tunable Bragg grating for DWDM) Many patents exist for applications from biochip reader to flexible flat-panel displays. Three existing kinds of optical switches are:

- **Directional coupler**: switching into one or the other channel; characteristics are controlled by the beam itself.
- **Mach–Zehnder interferometer**: switching is controlled by one or two separate control beams.
- **Nonlinear mode sorter**: a multi-mode beam is split into two channels. By varying the input power, the phase difference between two modes can be tuned. Thus, depending on the power of the signal, the

modes were separated and the output switched from one channel to another.

5.6.3 Second Harmonic Generation in Glass

Besides the Ge:SiO_2 fibers discussed in Sect. 5.1.5, other glasses which show photo-induced second-harmonic generation are Tm-doped alumosilicate fibers (resonant enhanced SHG) [5.224] and semiconductor microcrystallites-doped glasses [5.225], whereas Ce-doped lead germanate glasses show only low intensities (Table 5.22) [5.226].

SHG in HMO-Glass

Ionic bonds are believed to play a major role for encoded photoinduced SHG. Thus, it is not surprising that the SHG intensity in binary lead germanate glasses is found to be zero in contrast to binary lead silicates. Addition of Ce to the glass resulted in observation of saturable SHG. Low-Pb glass has a very low photoinduced SHG efficiency saturation level. On the other hand, for higher Pb-concentration the damage threshold of the glass decreases significantly. Further, the induced grating can be erased by an IR reading beam after some seconds. Comparison of lead silicate, Ce-Pb-germanate and Ge:silicate shows that high SH efficiencies can be achieved with Pb-containing glass at much lower preparation powers.

SHG in Rare Earth-Doped Glasses [5.227, 228]

Due to the absorption peaks near the SH wavelength, only a limited number of REs are suited for SHG with 1064 nm as fundamental wavelength. Furthermore, two or three photon absorption processes have been observed e.g. for Tm-doped glass, which lead to luminescence e.g. at 470 nm. Temporal and thermal stability is also a crucial factor in these glasses. As the formation of the $\chi^{(2)}$ grating is based on photooxidation, thermal and temporal stability of the gratings can be rather low, as the electron traps are usually rather shallow.

SHG in CdSSe-Doped Glasses [5.226]

SHG with pumping at 1064 nm was found to be more intensive for OG550 than for GG495. But GG495 is expected to be less sensitive to pumping and readout index changes. For instance, the SHG-intensity for OG550 ($7.7 \times 10^{-2}\,s^{-1}$ for 2 W) is found to rapidly decrease with time (optical erasure), whereas there is only a small decay for GG495 ($1.5 \times 10^{-2}\,s^{-1}$ for 2 W). Smaller decay effects due to heating are also found. Damage threshold of the glass is about $500\,W/\mu m^2$.

5.6.4 Glass Systems Investigated for Nonlinear Effects

Oxide Glasses Without Heavy Cations

The glasses of this group usually show only small nonlinearities, which are directly correlated with the nonbridging oxygen content [5.205, 206, 212, 229]. Measurements of various groups showed that the nonlinearity of nonbridging oxygen is significantly higher than that of bridging oxygens.

Despite the low nonlinearities of the oxide glass system, there have been recent experiments with silicon nanocrystals embedded in SiO_2 matrix. These samples show a $\chi^{(3)}$ of about 10^{-9} esu, dependent on the nanocrystal size. A much more interesting system, which is a kind of link to the heavy metal oxide glasses, is silica doped with various amounts of germania. The increase of the linear refractive index with increasing germania content consequently increases the nonlinear part of the refractive index. For about 50 wt % of GeO_2, a nonlinear refractive index of 1.1×10^{-13} esu has been found. Hydrogen-treatment even increases this value. Germania-doped silica fibers show photo-induced refractive index changes of up to 10^{-3} through exposure to UV light [5.230].

Heavy Metal Oxide Glasses (HMO)

The HMO-cations [5.231–233], like Pb, Bi and Ga posses large polarizabilities and therefore strong optical nonlinearities. Elements heavier than Bi might have larger NL-properties, but have disadvantages, such as their high toxicity, radioactivity and the sometimes poor optical quality of the resulting glasses, which make them useless for any practical application. Ba and La seem not to be responsible for nonlinearities. Tl^+ contributes strongly to the nonlinearity, for instance in the Tl-germanate, Tl-tellurite and Tl-antimonite glasses. It is the only metal which is a post-transition element and therefore forms bonds with large polarizability. In spite of this, Tl-containing glasses are not suitable for practical application, mainly due to the high toxicity.

Among the HMO-glasses, the Bi-oxide glasses and the Bi–Ga-oxide glasses are the most prospective for nonlinear devices. For example a Pb–Bi-Gallate glass was found to have the largest nonlinearity among the oxide glasses. In this glass system, Pb and Bi have an equivalent influence on the nonlinearity. Optical Kerr shutter experiments on Bi-glasses have shown that the response times of these glasses are in the range of 150 fs and THz optical switching can be performed. Thus, the conclusion was drawn, that these glasses are prospective materials for the realization of ultra fast response and high frequency switching (ultrafast switches, optical isolators and in advanced computing).

As a rule of thumb, one can say that the larger the linear refractive index, the higher the third-order susceptibility. Glasses with refractive indices higher than 2 have a $\chi^{(3)}$ larger than 10^{-13} esu. Typical ranges of $\chi^{(3)}$ for HMO-glasses are 10^{-12} to 10^{-11} esu, which are almost the same for all HMO-glasses. Due to the electronic nature of the nonlinearity, the response times of HMO-glasses lie in the sub-ps range. This makes the HMO-glasses good candidates for all-optical switching applications. For incident light with optical frequencies, possible present nuclear contributions would be mitigated, because of slower response times of nuclear motion.

Transition Metal-Doped Glasses

Oxide glasses show an increase in the NL-indices when they are doped with transition metal cations or index raising conditional network formers, like Ti, Ta and Nb, due to the high hyperpolarizability of these ions [5.206, 234–236]. Ions with an empty or unfilled d-shell (d^0 electron configuration), like Nb or Ti (i.e. Ti–O entities), contribute most strongly to the hyperpolarizabilities [5.231]. On the other side doping, for instance with rare earths, can lead to a strong increase of the absorption coefficient in some of the commercially interesting spectral regions (e.g. telecommunication windows) Thus, rare-earth- (RE) doped glasses were are prominent for their application as laser glasses (upconversion lasers) or optical amplifiers.

Compared to the heavy metal cations, the nonlinearity of this group seems to be caused by another mechanism. This is due to the fact that (e.g.) TiO_2-dopands are found to be responsible for nonresonant NL-effects in glass. However, the combined effect of Ti and HM-cations, like Bi and Tl, show NL-effects not higher than $10^{-18}\,m^2/W$. An explanation might be the fact, that glasses with high concentrations of Ta, Ti and Nb are found to have negative electrooptic coefficients, in contrast to Pb^{2+}. Therefore Ti^{4+} can be used to balance the positive electrooptic contribution of Pb^{2+}.

Studies with Nb-containing oxide glasses have shown that the d^0-transition metal is located in a $(Met-O)^6$-configuration, with the oxygen electrons strongly delocalized towards the transition metal. This delocalization results in a large hyperpolarizability and thus large nonlinearity, which is found to be directly proportional to the concentration of $(Met-O)^6$-entities [5.235].

Non-Oxide Glasses

Among the halide glasses, the fluoride glasses have the smallest NL, and the bromide and iodide glasses have the largest NL [5.206, 231, 237]. However, bromide and iodide glasses show very low glass-transition temperatures and poor chemical durability. Thus, they are not attractive for nonlinear applications, although the significant laser oscillation of the fluoride glasses at 1340 nm cannot be covered by oxide glasses.

Among the chalcogenide glasses, S glasses as well as Se and Te glasses show much higher nonlinearities than oxide glasses. As_2S_3 glass has shown the highest third-order susceptibility of all intrinsic nonresonant nonlinear glasses. Further examples of chalcogenide glasses with very high nonlinear susceptibilities are: $As_{40}S_{57}Se_3$, $As_{40}S_{60}$ and $As_{24}S_{58}Se_{38}$. Closer inspection of the origin of the nonlinearity of this glass exhibits a ratio $\chi^{(3)}_{2121}/\chi^{(3)}_{1111}$ of nearly zero, which is much smaller than the expected value for nonresonant NL of 1/3, the expected value for purely electronic contribution. The result of further investigations showed, that a kind of resonant contribution due to two-photon absorption is responsible for the high nonlinearity of this glass. Due to their very short response times, chalcogenide glasses are prospective materials for use as optical memory, switching devices, and (Bragg-) gratings in planar waveguides ($\Delta n \approx 10^{-4}$) Experiments with these glasses have shown that high quality, low loss waveguides for optical applications can be easily formed.

Tellurite and antimonate are both known as network formers, which contribute to $\chi^{(3)}$ via their stereochemical lonestanding electron pair in 5s (e.g. structural unit TeO_4) Both tellurite and antimonate glasses show a very good IR-transmission as well as very good chemical durability. Tellurite glasses have a nonlinear refractive index of 30–70 times that of silica. Pure tellurite containing glasses are difficult to prepare, thus, most of the investigated glasses have tellurite concentrations of about 70% [5.231].

5.6.5 NL-Effects in Doped Glasses

With the exception of the transition metal-doped glasses, all the glasses discussed above have intrinsic nonresonant nonlinearities [5.206]. Due to the fast response time of the usually electronic nonlinearity and the low absorption coefficient, the figures of merit of these materials can be rather high, despite the low $\chi^{(3)}$. In contrast to these glasses, the glasses discussed in the next part of the chapter have extrinsic resonant nonlinear optical properties, which are usually higher than the nonlinear ones at least in the spectral regions of fundamental absorption (Table 5.23).

Semiconductor-Doped Glasses

Among the studied nonlinear glasses, semiconductor-doped glasses [5.205,208,212,223,238,239], especially the Schott filter glasses based on $Zn_yCd_{1-y}S_xSe_{1-x}$, have been studied extensively. Other semiconductor dopants, besides CdSSe, are CuCl, CuBr, Cu_xSe, Ag(Cl, Br, I), CdTe, III–V compounds, Bi_2S_3, PbS, HgSe, In_2O_3, In_2Se_3, SnO_2 and AgI. Among all these glasses, the best materials for ultrafast switching, based on a FOM, are ZnTe, CdTe and GaAs. In some experiments, the optical switching is done with different pump and probe wavelengths. As a result, it has been found that one gets a larger $\chi^{(3)}$ for the small loss in the near resonance region.

Semiconductor-doped glasses exhibit large nonlinearities based on optical Kerr effect at photon energies near the bandgap. The NL-behavior of these materials is characterized by intensity dependent changes in absorption and related changes in the refractive index ($\Delta n \cong 10^{-4}$) with relaxation times in the range of ns or ps. A strong dispersion in the electronic nonlinearity is found near the two-photon absorption (TPA) edge. The nonlinear refractive index increases as the TPA-edge is approached and eventually becomes negative at higher energies. The nonlinear properties of CuCl-doped glasses are based on an enhancement of the nonlinearity in the resonance region of the CuCl Z_3-exciton [5.207]. The advantages of semiconductor-doped glasses (SDG) compared to the bulk semiconductor (SC) are [5.238, 239]:

- fast (ns–ps) response time,
- high threshold,
- weak absorption losses (for working wavelengths still near high absorption wavelengths!)
- possible usage at room temperature (but binding energies of these glasses are usually in the range of a view meV, thus, devices based on these materials better have to be operated at low temperatures).

On the other hand, the nonlinear optical effects in SDGs are smaller than those in bulk SC: χ^3 of the doped glasses is typically in the range of 3×10^{-9} to 10^{-4} esu. For example for CuCl-doped glasses, $\chi^{(3)}$ is in the range of 10^{-3} to 10^{-8} esu, depending on the crystallite size. Changing the crystallite size, doping with other semiconductors or metal colloids like Au or Ag, or changing the glass host composition may improve the nonlinear properties. Decreasing the size of the nanocrystallites, and thus enhancing the confinement, enhances the nonlinearity.

Metal-doped Glasses

Glasses containing very small metal particles exhibit an efficient, fast optical Kerr-effect [5.204, 206, 212, 219, 220]. The enhancement of the intrinsic $\chi^{(3)}$ arises from the surface mediated plasmon resonance, analogous to the SER-effect. The observed Kerr susceptibility or electronic nonlinearity of these materials includes contributions from intraband (transition into the conduction band), interband (saturation of interband transitions between d levels and the conduction band) and hot electron transitions (creation of photoexcited hot electrons due to the surface plasmon absorption) The maximum of the nonlinear susceptibility is found to be near the surface plasmon resonance.

Metal-doped glasses are expected to be good candidates for optical bistability and optical switching. Silver and gold particles dispersed in glass show NL effects, such as optical phase conjugation.

Gold-doped glasses: For gold spheres, $\chi^{(3)}$ is positive and found to be independent of particle size, with response times in the ps-regime. The nonlinear response of gold nanoparticle-doped glasses has been found to be as fast as 240 fs [5.240].

Silver-doped glasses: Glasses doped with silver nanoparticles exhibit an ultrafast nonlinear response of 360 fs and a $\chi^{(3)}$ of about 7.6 to 15×10^{-8} esu [5.204, 208, 241].

A nonlinearity of thermal origin has been reported for Mn doped glass. Due to a theoretical calculation, Al is suggested to have a larger nonlinearity than Ag and Au. By preparing core-shell systems, an enhancement of the nonlinearity by a factor of 10^8 might be possible [5.206]. Metal-doped glasses can be produced via sol–gel-technique, ion exchange, ion implantation or electrolytic coloration. Various metal-doped silica glasses, such as copper-, silver-, gold-, platinum-, lead-, tin-, iron-, germanium-, and phosphorous-doped glasses have been investigated. Of these glasses, the tin-implanted glasses have the highest nonlinear susceptibility. The size and the shape of the metal clusters can be modified by thermal treatment or by ion implantation with another element [5.206, 241].

Organic (Dye-) doped Glasses

Organic dyes, polymers and solvents, like 4-(N,N-diethylamino)-β-nitrostyrene (DEANST), acridine, fluorescein and rhodamine 590, have intense singlet–singlet-

Table 5.22 Poling of glass for second-harmonic generation [5.225, 227, 228, 242]

Glass	Poling	Parameters		SHG coefficient *d* (pm/V)
Fused silica, containing Na, OH	Thermal	275 °C	5 kV	1 (in a 10 μm layer)
GeO_2-SiO_2 , Herasil	Thermal + electric	280 °C	4.3 kV	1
Ge-silicate "half"-fiber	Thermal + electric	275 °C	−5 kV	0.3
Ge-silicate optical fibers	Electric, thermal electric and UV			0.002 to 6
GeO_2-SiO_2	Optical + electric field	ArF (193 nm)	Electric field	3.4–5.2
GeO_2-SiO_2 sputtered film	Optical + electric field	ArF 100 mJ/cm^2/pulse	Electric field	up to 12.5
15.7 GeO_2–84.2 SiO_2 (mol%)	Optical + electric field	ArF: 10–100 mJ/cm^2/pulse	3×10^5 V/cm	7×10^{-3}–2
Ge-silicate	fs-encoding	810 (960 mW)	405 nm (200 fs, 200 kHz)	
Er-, Tb- and Sm-doped Al-silicate fibers, Er-doped Ge:SiO_2 fiber	Optical	1065 nm (30–50 GW/cm^2)	Seed: 532 nm	Conversion effic.: 10^{-6} to 10^{-3}, 8×10^{-3} in Ge-silicate
Ce- and Eu-doped Al-silicate fibers	Optical	Nd:YAG: 1064 nm	Seed: 532 nm	Conversion effic.: 2×10^{-6} to 1.5% for Ce-doped fiber for 200 W pump
Tm-doped aluminosilica fiber	Optical	Nd:YAG: 1064 nm	Seed: 532 nm	SH output e.g. 4.6×10^{-2} W for 240 W pump at 1064 nm and 4.6×10^{-2} W for 55 W at 844 nm
Ce-doped lead germanate glass	Optical	1065 nm Nd:YAG	Seed: 532 nm KTP	
CdSSe-doped: GG495–RG630	Optical	1065 nm (3 W)	532 nm (1 W) pulsed laser	10^{-4}
Chalcogenide glasses: GeAsS	Optical	Nd:YAG 1064 nm	Seed: 532 nm	α(532 nm) = 1.99 cm^{-1} conversion effic.: 10^4x that of $TeO_2-Nb_2O_5$-glass
Fused silica fibers containing impurities: OH, Na, Fe	Gamma ray	^{60}Co source (10^6 rad at 400 rad/s)	1064 nm (Nd:YAG, 50 ns, 10 Hz, 1 kW) and 532 nm seed (300 W), 1 h	2×10^{-4} (natural SiO_2) 9×10^{-3} (Ge:SiO_2 + Ce) 2×10^{-2} (Ge:SiO_2 + Ce)

Table 5.23 Nonlinear susceptibilities of commercial glasses

Glass system	$\chi^{(3)}$ (10^{-14} esu)	Wavelength range
Heavy metal (Tl–Pb–Cd–Ga)	up to 56	High frequency
Pb-silicate	up to 16	
Ge–Ga–S	3000–27	600–1250 nm
Semiconductor doped (CdSSe)	up to 5 000 000	Near exciton absorption peak
Nb–Te-oxide	9.4	
Bi-oxide	930	
Chalcogenide	174	
Noble metal doped glass	up to 10 000 000	Near plasmon absorption
Dye doped glass	up to 1.4×10^{14}	Near absorption

transitions, which contribute to a high second-order susceptibility [5.204, 206]. The doping of glasses with dyes, polymers, and other organic materials also results in high nonlinearities, several orders of magnitude larger than that of inorganic glasses. Thus, these glasses are attractive candidates for various low power optical devices, such as optical power limiters, phase conjugated mirrors or optical logic gates. The large nonlinearities are caused by saturation of the near resonant intensity dependent ground state absorption. The excited dyes may return to their ground state either radiatively or nonradiatively or be transferred to the lowest triplet state by intersystem crossing. Triplet-singlet-transitions are forbidden, thus the triplet state has a lifetime many orders of magnitude longer than the singlet states. For sufficient intense excitation, the electrons accumulate in the triplet state and the ground state absorption is saturated. This produces a change in the refractive index via the Kramers–Kronig relation.

Dye-doped glasses are made by impregnation of porous silica or sol–gel-methods or require low melting glass hosts, such as lead–tin–fluorophosphate- (good optical quality and chemical stability) or boric acid-glass, because of the thermal degradation or decomposition of the dyes.

5.7 Plastic Optics

Injection moulded precision plastic optics in high volumes were first produced during the 1960's. After the development of sophisticated measuring and manufacturing methods in the late sixties also precise aspheric surfaces were as easy to make as spheric contours. Today plastic optics are a widely used low-cost option compared to glass with even more degrees of freedom for optical and component design. Polymer optical systems at present are used in sensor applications, visual systems, cameras (mobile phones, video-conferencing cameras), scanners, security systems and so on.

Of course the properties of glass materials are very different from that of plastic materials. But glass and plastic optics both offer unique advantages and help to solve various engineering problems.

Generally speaking, glass materials are harder and more durable than plastic materials. Glass materials are also more stable (temperature and humidity) than plastic. The variety of optical glass available from well-know suppliers comprises hundreds of different materials. Compared with this the choice for plastic materials is limited only to about 10 different materials (and even less optical parameter variations). The large selection of glass materials allows the designer to choose materials with desirable optical properties to gain better optical performance. However, plastic optics offers other design freedoms that are not achievable with glass optics.

The manufacturing technologies for glass and plastic optics are totally different. Glass lenses are made by a grinding and polishing process whereas precision plastic lenses are in the whole made by injection-moulding or compression moulding. Because of the materials characteristics and the manufacturing process plastic optics have some unique advantages.

- **High production numbers at low costs**
  Injection moulding is ideal for high volume production with low unit costs. Moderate raw material costs and multi cavity moulds (up to 32 cavities) allow large production volumes at a reasonable unit price. In spite of normally considerable tooling costs the cost break-even compared to glass design versions can be relatively low.
- **Lightweight and hardiness**
  For a given volume glass is much heavier than plastic (by a factor of 2.3 to 4.9). Plastic materials are relatively shatter and impact resistant. These features are important, e.g. for head-mounted systems.
- **Design potentials**
  Injection moulding makes it economical to produce sophisticated optical shapes such as aspheres, diffractive optical elements or even freeform surface structures. From the design point of view, the more sophisticated surface shapes reduce costs or obtain better performance or even performance unrealisable by optics in glass (Fresnel structures, lens arrays, diffractive optical elements).
- **Optical systems and component assembly**
  For typical optical systems, the optical components (mirrors, lenses, prisms etc.) must be fixed in a mounting. With plastic optics, it is possible to mould mounting elements, posts or alignment notches integrally with the optical component. This can reduce part and assembly costs considerably. Technologies adapted to plastic materials like ultrasonic and laser welding, gluing and integrated snap-in structures allow fast and cost efficient automated and manually operated assembling solutions. Two component injection moulding is the choice for integrating optical elements and mounts in one injection cycle.

5.7.1 Moulding Materials

As already mentioned, the choice for plastic optical material is limited only to about 10 different types

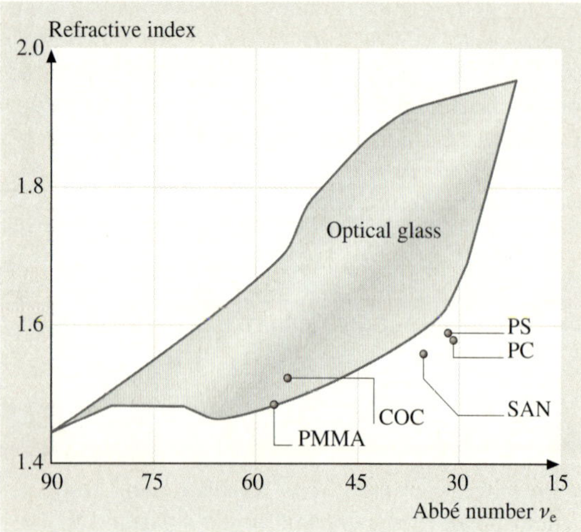

Fig. 5.54 Abbe diagram

of material. Optical properties (Abbé value, refractive index, Fig. 5.54) as well as mechanical, thermal and humidity boundary conditions are decisive for the material choice. Different to glass during plastic injection moulding the process affects not only the geometry but also the inner properties like refractive index and birefringence.

Optical Properties

Fundamental optical properties are defined by optical transmission, refractive index and dispersion. For plastic optics birefringence is an important parameter too. Although the total number of plastic materials has increased in recent years, the range of refractive index and dispersion characteristic is limited to almost two major groups – crown-like materials as acrylic (PMMA), polyolefin (COC, COP) and Flint-like materials as polystyrene, polycarbonate and SAN. This strongly limited variety of plastic optical materials restricts the optical design that can be made in plastic considerably.

Physical Properties

Important physical properties are weight, impact and abrasion resistance and thermal properties like temperature resistance and thermal expansion (Table 5.24).

Already during system design both mechanical and thermal properties have to be taken into consideration. The thermal expansion of optical plastics is approximately ten times higher than that of glass materials. In an optical system this effect has to be compensated by optical design and mounting.

Typically most of optical plastics can withstand temperatures up to 90 °C. The maximum service temperature of polycarbonate and polyolefine materials reaches 120 °C. The specific gravity of plastic optical materials ranges from 1 to 1.3. Polycarbonate has the highest impact resistance of all optical plastics and is used for windshields and crash helmets. Acrylic has the best abrasion resistance.

Most Common Materials

Some of the most common optical polymers are:

PMMA (Polymethyl methacrylate). Acrylic is the most commonly used optical plastic. It is moderately priced, easy to mould, scratch resistant and not very water absorptive. Its transparency is greater than that of most optical glasses. Additives to acrylic (as well as to several other plastics) considerably improve its ultraviolet transmittance and stability. So acrylic is used in almost 80% of all plastic optical applications.

(Poly-)Styrene. Polystyrene is a low-cost material with excellent moulding properties. Styrene has a higher index and a lower numerical dispersion value than other plastics. It is often used as the flint element in color-corrected plastic optical systems.

Compared with acrylic, styrene has lower transmission in the UV portion of the spectrum: It does not have the UV radiation resistance nor the scratch resistance of acrylic.

Because its surface is less durable, styrene is more typically used in nonexposed areas of a lens system.

NAS Copolymer. NAS is a copolymer of certain fractions of polystyrene and acrylic (typically 70/30). This allows to adjust refractive indexis a low-cost material with excellent moulding properties.

PC (Polycarbonate). Polycarbonate is very similar to styrene in terms of such optical properties as transmission, refractive index and dispersion. Polycarbonate, however, has a much broader operating temperature band of up to 120 °C. For this reason, it is used as the flint material for systems that have to withstand severe thermal conditions.

Another advantage is the high impact resistance of polycarbonate. Safety glasses and systems requiring durability often consist of polycarbonate.

Cyclic Olefin Polymer and Copolymer (COP/COC). Cyclic olefin (co-)polymer provides a high temperature alterna-

Table 5.24 Plastic material properties

Material / Properties	Characteristics	Acrylic (PMMA)	Poly-styrene (PS)	Poly-carbon-ate (PC)	Styrol–Acryl-nitril (SAN)	Cyclo-olefine (ZEONEX)	Polyether-sulfone (PES)	Acrylnitril–Butadien–Styrolcopoly-mere (ABS)	Optical glass (BK7)
Optical	Spectral passing band (nm)	390–1600	400–1600	360–1600	395–1600	300–1600			
	Refractive index at 587 nm and 20 °C	1.4918	1.5905	1.5855	1.5674	1.5261	1.6600	1.538	1.517
	Abbé value $(n_D - 1)/(n_F - n_C)$	57.2	30.7	30	34.8	56	19.4		64.4
	Transmittance (%) thickness 3.2 mm	92	88	90	88	92	80	85	
	Haze (%) thickness 3.2 mm	1.3	1.5	1.7	1.5	1.5			
Physical	Specific gravity (g/cm^3)	1.18	1.06	1.25	1.07	1.02	1.37	1.05	2.53
	Max. service temperature (°C)	80	90	120	95	125	200	90	400
	Linear expansion coefficient (1/K)	6.8×10^{-5}	8.0×10^{-5}	6.6×10^{-5}	7.0×10^{-5}	7.0×10^{-5}	5.5×10^{-5}	8.5×10^{-5}	1.1×10^{-5}
	Abrasion resistance (1–10)	10	4	2		6			
	Izod impact strength (kJ/m^2)	2.0	2.0			2.4	7.0	25.0	
Environ-mental	dn/dT (10^{-6})	−105							
	Sensitvity to humidity	high	low	low	medium	low	high	medium	
	Water absorbtion (weight %) 23 °C, ISO 61	0.60	0.10	0.15	0.30	0.01	0.70	0.45	0
Manufac-turability	Processability	Excellent	Good	Poor	Excellent	Good			
	Birefrigence	Low	High	High		Low			
Chemical	Resitance to alcohol	Limited	Good	Limited			Good		
Costs	Approx. material costs (EUR/kg)	3.3	2.5	4.4	4.4	27.1	21.0	3.5	25

tive to acrylic. Its refractive index and transmittance are similar but the heat distortion temperature is about 30 °C higher than for acrylic. Its water absorbtion capability is significant smaller.

5.7.2 Manufacturing Methods

The manufacturing processes for glass and plastic optics are totally different [5.243–247]. Glass lenses are made by grinding and polishing process. By contrast typical manufacturing methods for precise technical plastic optical parts are diamond turning, injection and compression moulding. Other methods like casting have their application in special fields like ophthalmics (eyeglasses). In any case, processing transparent plastic materials for optical purposes should take place in a clean environment.

Diamond Point Turning

Diamond point turning is an ultraprecision machining process carried out on special high performance, modular set up, multi-axis, linear motor driven, ultraprecision diamond CNC-machining systems.

In combination with vibration isolation systems, digital signal processor based machine control and integrated measuring systems an one nanometer programming resolution is possible.

This method can be expanded to full 3-D-milling systems. With such an arrangement the generation of free-form surface profiles is possible. Because of the long production time and the high machine costs, this technology is used for prototyping in plastic and nonferrous metals, mould inserts and series production plastic parts – which can not to be manufactured by injection moulding (because of size or precision demands).

Injection Moulding

Injection moulding produces several parts per shot (moulding cycle) in single or multicavity tools. This production method is used for most of the plastic optic parts. Other fabrication techniques are used only when moulding is inappropriate.

A plastic injection-moulding machine consists of a fixed platen, a moving platen, a clamping unit and an injection unit. Moulding of optical parts requires special machine configurations and auxiliary equipment.

Compression Moulding

Compression moulding is used for making of Fresnel lenses or other microoptical structures. The material is pressed between heated platens with accurately defined temperature cycling during pressing. Mould inserts are formed by electroplated copies, replicated from master structures. These masters can be, for instance, diamond turned structures or diffraction gratings made by holographic methods. The compression moulding process allows to realize small structures with high aspect ratios and tight angular and positioning tolerances.

5.7.3 Manufacturing Process

Optical and System Design

On the one hand optic design [5.243–249] for plastic optics in principle uses the same mathematical algorithms than optic design for glass. On the other hand, designing plastic optics requires a profound understanding of the material properties and the manufacturing processes [5.245, 248, 249]. Knowledge of production technologies, material characteristics and assembling methods together with design expertise are needed to fully exhaust what precision plastic optics can be.

Simply substituting the indices of refraction and reoptimising the design will not succeed. Expert design assistance is essential at this stage.

Designing plastic optics with modern design tools allows great design freedom. The advantage of combining integral mounting structures with the optical surfaces to create mounting flanges, alignment and snap features provides the ability to automated assembly. Aspheric, cylindric, or toroidal surfaces are as easy to realize as spherical ones. Microstructures such as diffractive optical elements can be integrated too.

Prototyping

After designing a plastic optical system, lens prototypes can be made by diamond point turning in various plastic materials. The best surface finishes can be obtained with PMMA. Materials such as polycarbonate, pleximid or Zeonex do not yield very smooth surface finishes.

A major problem is the availability of semifinished plastic blocks in various materials. For PMMA a wide variety of bars or plates is available. For other materials or colored options semifinished items have to be made by injection moulding.

Because of the high manufacturing costs diamond point turning is only recommended for making a limited number of prototypes to verify functionality of the optical design and to perform first tests.

In this stage optical systems are normally assembled from single elements. Housing parts often are made from aluminium. The resulting surface quality and system performance can not be a validation of the manufacturability by injection moulding. To get reliable knowledge about this, making a moulded prototype from a single-cavity prototype mould is recommended.

The Injection Mould

A high quality injection mould is obviously essential for precise plastic optic parts. The parts never can be better than the tool - but good tooling, however, does not guarantee good parts. A strong understanding of the whole manufacturing process is the key to producing precision plastic components.

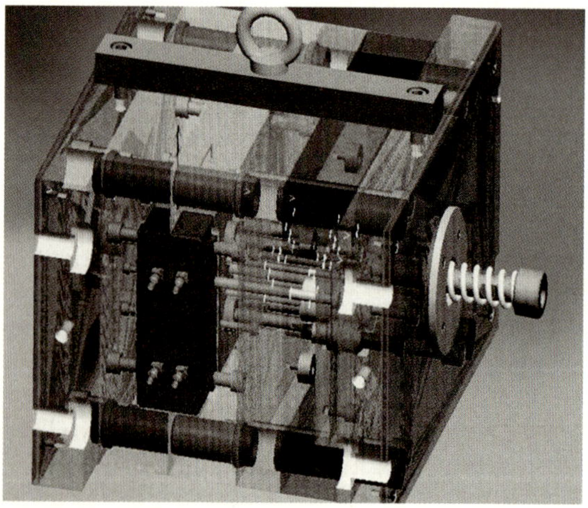

Fig. 5.55 Injection mould

Table 5.25 Tolerances of optical components

	Unit	Injection molding	High-end injection molding	Single-point diamond turning
Radius	% (+/-)	5	2	0.5
Refractive index	(+/-) (PMMA)	< 0.0005	< 0.0005	< 0.0005
EFL	% (+/-)	5	2	1
Thickness	mm (+/-)	0.13	0.05	0.02
Diameter	mm (+/-)	0.13	0.05	0.02
Surface figure	Fringes per inch of diameter	$< 10f$ (5λ)	$< 8f$ (4λ)	$< 2f$ (1λ)
Surface irregularity	Fringes per inch of diameter	$< 5f$ (2.5λ)	$< 3f$ (1.5λ)	$< 1f$ (0.5λ)
Surface rms error	nm	< 10	< 5	< 2
Surface S/D quality		80/50	60/40	40/20
Wedge (TIR)	mm	< 0.025	< 0.015	< 0.010
Radial displacement	mm	< 0.100	< 0.050	< 0.020
Diameter/thickness ratio		> 2 : 1	> 3 : 1	> 5 : 1
Repeatability	Focal length (part to part in one cavity)	< 2%	< 1%	< 0.5%
DOE depth	μm (+/-)	***	0.25	0.1
DOE min groove	mm	***	0.05	0.01

Any injection mould consists of three main parts. The upper half, which is affixed to the injection-side platen, the lower mould half, which is affixed to the ejector-side platen and the mould-ejection mechanism. Guide pins and taper locks ensure proper alignment of the mould halves.

Production volume and precision required from a tool influences the selection of mould materials and built-in maintanance features.

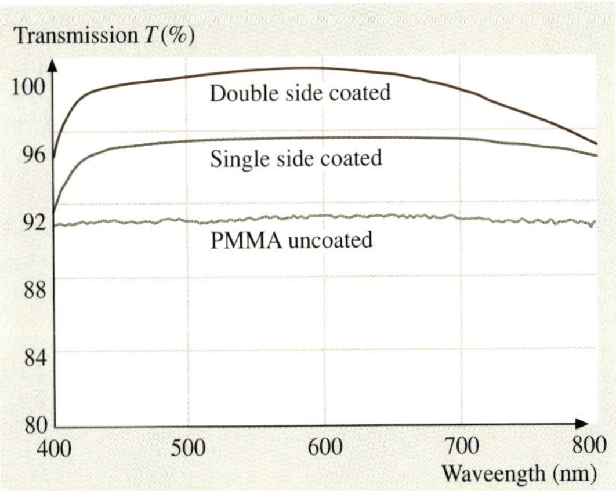

Fig. 5.56 Abbe values

Single-cavity and multi-cavity moulds from 2 to 8, 16, or even 32 cavities are used for plastic optic parts.

Thermoplastic material shrink during cooling down in the mould. This geometric effect has to be compensated for in the injection mould. Exact shrink rates can be calculated and the tool can be modified. For manufacturing optical plastic parts by injection moulding the optical surfaces (plano or aspherical shapes, diffractive, conical, lenticular and cylindrical surfaces) are generated as separate inserts in the tool.

Aspheric inserts are manufactured by two steps. First a best-fit curve is generated on a stainless steel substrate. The substrate is then subjected to a nickel plating process (electroless nickel) that deposits a thin layer of nickel (up to 500 μm). In the second step single-point diamond turning produces the final aspheric or diffractive curve in the nickel. Because the hardness of a nickel-plated insert is less than that of a steel spherical insert, it will be more susceptible to scratch defects.

Pre-Production

A pre-production stage is recommended to check the manufacturing process. Typically this is done by using a single-cavity prototype mould. This mould can be used to find optimal moulding conditions. Optical an mechanical design can be verified with real moulded components and design revision is possible to affordable conditions. Often the prototype moulds is used to start limited production since production tooling may take much more time.

Series Production

For series production of high volumes multi-cavity production moulds are required. Depending on quality, volume, throughput and cost, the production tool may have 2, 4, 8 up to 32 cavities. The production moulds function for at least several hundred of thousands of injection cycles.

5.7.4 Coating and Component Assembly

Coating

Because of their limited temperature and UV resistance plastic lenses must not be coated in an elevated temperature and radiation environment. During the deposition of thin films onto plastic, the coating chamber temperature is significantly lower than that for glass optics. This requires deposition techniques such as ion-assisted deposition to apply antireflective, conductive, mirror and beamsplitter coatings.

Today multi-layer dielectric coatings are routinely deposited on plastic components. Typical broad band anti-reflection coatings reduce reflection to about 0.5% per surface across the entire visible spectrum. Narrowband, multilayer antireflection coatings can achieve surface reflectances less than 0.2%.

Multi-layer dielectric coatings can be modified easily to scratch resistant designs for front lenses and windows.

Several front and back surface reflector coatings are available for plastic substrates. Standard coating metals include aluminium, silver and gold. Aluminium coatings provide surface reflectances greater than 88% across the visible spectrum, and gold coatings greater than 95% for the near infrared region.

Component Assembly

Many advantages of using plastic optics will come out in particular in manufacturing components. The integration of optical, mechanical and electronic elements allows to build low cost polymer optical components like small camera lenses or scanner heads f.i.

Plastic optical assemblies are usually done in a clean room environment by manual, semi or fully automated processes. The components have to be designed to ensure ease of assembly. Snap-on features, UV-cementing, heat-staking and ultrasonic and laser welding can also be employed with plastic materials.

Since most optical tolerances are additive, it is essential to establish quality check points in the manufacturing process to sort out non-conforming sub-assemblies. Automatic in-line optical performance monitoring such as MTF testing can be implemented. SPC techniques should be used here to ensure the process is not drifting out of the controllable range.

Component Tolerances

The high repeatability of the injection moulding process yields components with little dimensional variability. But the specific tolerances for an optical component strongly depend upon the part geometry and size, the plastic material, the mould design and the production process stability.

5.7.5 New Developments

In the last years several new developments in the field of plastic optics took place. The permanent improvement process in tooling technologies in combination with diamond turning, milling and grinding led to new design and tolerance levels. New materials like the cyclo-olefins led to better material performance and coating stability.

The fabrication process for diffractive elements (done primarily by single-point diamond turning) has improved considerably. Micro- and nanostructures like moth-eyed surfaces can be achieved by injection moulding and compression moulding and allow functionalised surfaces.

Even calibration structures for scanning probe microscopy applications can be replicated by injection moulding up to 25 nm structures.

Plastic optic technology permanently expands its traditional limits and fields of application. On the one hand plastic optic is still limited by material properties but on the other hand design and assembling freedoms allow new approaches.

5.8 Crystalline Optical Materials

In the following section we will discuss a few crystalline materials [5.250–252] that are used for optical applications. Among the oxides we want to focus on Sapphire, we mention the class of halides where CaF_2 is the most prominent example and we discuss a few semiconductors. The following subsections summarize material properties and fields of application. In each subsection, tables summarizes the optical and mechanical materials properties.

5.8.1 Halides, CaF_2

Halides are crystals whose anions are halogens, the elements of group seven in the periodic table, namely fluorine, chlorine, bromide and iodine. Halides tend to have rather simple ordered structures with a high degree of symmetry. The most famous halide is rock salt, NaCl. Halides can generally be synthesized as single crystals from molten salt by Czochralski process.

Table 5.26 Basic properties of CaF_2 crystals

Formula	CaF_2					
Max. sizes (diameter (mm))	180					
Transmission range, μm	0.15–9.0					
Density (g/cm^3)	3.18					
Melting point (°C)	1418					
Hardness (Mohs)	4					
Thermal expansion coefficient (10^{-6}/K)	16.2–19.4					
Thermal conductivity (W/m K)	9.17					
Specific heat capacity (J/kg K)	888					
Solubility in water (g/100 cm^3)	0.0016					
Wavelength (μm)	0.2	0.5	1.0	5.0	10.0	12.0
Refractive index	1.4951	1.4365	1.4289	1.3990	1.3002	1.2299
Absorption coefficient (cm^{-1})	0.10 at 0.2 μm; 0.01 at 0.4 μm; 0.03 at 2.7 μm					

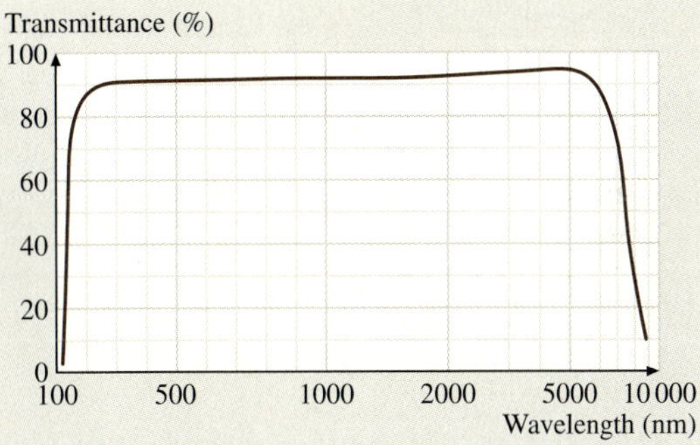

Table 5.27 Basic properties of MgF_2 crystals

Formula		MgF_2					
Maximum sizes (diameter)		80 mm					
Transmission range, μm		0.13–7.0					
Density (g/cm^3)		3.18					
Melting point (°C)		1255					
Hardness (Mohs)		6					
Specific heat capacity (J/kg K)		920					
Solubility in water (g/100 cm^3)		0.0076					
Thermal expansion coefficient (10^{-6}/K)		parallel c-axis: 10.86–14.54, vertical c-axis: 6.23–9.25					
Wavelength (μm)		0.2	0.5	1.0	3.0	5.0	7.0
Refractive index	n_o	1.4231	1.3797	1.3736	1.3618	1.3400	1.3044
	n_e	1.4367	1.3916	1.3852	1.3724	1.3487	1.3101
Absorption coefficient (cm^{-1})		0.07 at 0.2 μm; 0.02 at 5.0 μm					

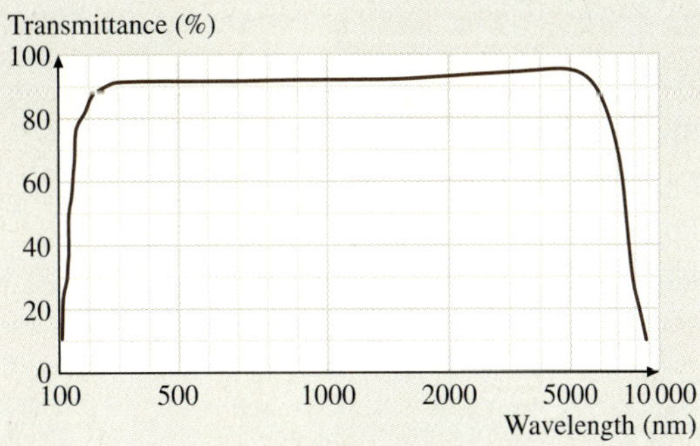

In optical applications two candidates are of particular importance: MgF_2 and CaF_2 (Table 5.26 and Table 5.27). These materials transmit well into the UV region and are used for most UV-optics. MgF_2 is an excellent material for eximer lasers. MgF_2 is slightly birefringent as can be seen from the data in Table 5.27.

CaF_2 is also an excellent material in laser optic field. It has a wide transmission range from 0.13–9.5 μm and an especially high transmission in the IR range. CaF_2 is widely used in laser, IR, and UV optics. It is slightly soluble in water and is susceptible to thermal shock.

Table 5.28 Main properties of Si

Chemical formula	Si
Molecular weight	28.09
Crystal class	Cubic
Lattice constant (Å)	5.43
Density at 293 K (g/cm^3)	2.329
Dielectric constant for 9.37×10^9 Hz	13
Melting point (K)	1690
Thermal conductivity (W/m K)	
at 125 K	598.6
at 313 K	163
at 400 K	105.1
Thermal expansion (1/K)	
at 75 K	-0.5×10^{-6}
at 293 K	2.6×10^{-6}
at 1400 K	4.6×10^{-6}
Specific heat (cal/g K)	
at 298 K	0.18
at 1800 K	0.253
Debye temperature (K)	640
Bandgap (eV)	1.1
Solubility in water	None
Knoop hardness (kg/mm^2)	1100
Mohs hardness	7
Young's modulus (GPa)	130.91
Shear modulus (GPa)	79.92
Bulk modulus (GPa)	101.97
Poisson's ratio	0.28
Wavelength (μm)	Refractive index
1.40	3.49
1.50	3.48
1.66	3.47
1.82	3.46
2.05	3.45
2.50	3.44
3.50–5.00	3.43
6.00–25.00	3.42

Table 5.29 Main properties of GaAs

Chemical formula	GaAs
Molecular weight	144.63
Density at 300 K (g/cm^3)	5.32
Absorption coefficient	
at 10.6 μm (1/cm)	< 0.02
Useful transmission range (μm)	1–11
Reflection losses	
for 2 surfaces at 12 μm (%)	45
Dielectric constant at 300 K	
high frequency	10.88
static	12.85
Melting point (K)	1511
Thermal conductivity	
at 300 K (W/m K)	55
Thermal expansion at 300 K (1/K)	5.7×10^{-6}
Specific heat at 273 K (cal/g K)	0.076
Debye temperature (K)	360
Knoop hardness (kg/mm^2)	731
Young's modulus (GPa)	82.68
Bulk modulus (GPa)	75.5
Poisson ratio	0.31
Bandgap (eV)	1.4
Solubility in water	None
Wavelength (μm)	Refractive index
8.0	3.34
10.0	3.13
11.0	3.04
13.0	2.97
13.7	2.89
14.5	2.82
15.0	2.73
17.0	2.59
19.0	2.41
21.9	2.12

CaF_2 and all halides are rather soft materials which are generally not very dense $(3.18\,\mathrm{g/cm^3})$. CaF_2 belongs to the cubic class and its cleavage plane is the 111-plane. In particular CaF_2 has been focused on as the material for future micro-lithography applications in the semiconductor industry because of its high transmission in the UV region. The band gap is larger than 11 eV. Further applications are laser windows and optical elements in the visible region. In addition CaF_2 has a low refractive index and its mechanical properties allow for rather easy polishing.

5.8.2 Semiconductors

Semiconductors have a wide range of applications in optical devices. In the following we will focus on Si (Table 5.28),Ge and CdTe as materials of interest in IR-transmitting elements. The crystal structure is cubic, and in the case of Si the diamond structure.

Silicon is commonly used as substrate material for infrared reflectors and windows in the 1.5–8 μm region. The strong absorption band at 9 μm makes it unsuitable for CO_2 laser transmission applications, but it is frequently used for laser mirrors because of its high thermal conductivity and low density. Silicon is also useful as a transmitter in the 20 μm range.

Germanium is used primarily in the 2–12 μm spectral region. Ge has a good thermal conductivity and excellent hardness and strength. It is the material of choice as a substrate for a variety of filters. One disadvantage of Ge over Si is its greater tendency to become opaque at higher temperatures due to its relatively low band-gap of 0.75 eV. This lower band-gap results in easier stimulation of electrons from the valence band to the conduction band. Cadmium telluride (CdTe) has an IR-transparent region from 1–25 μm. It is a very soft material and a low thermal conductivity. It is an excellent substrate material for filters in the 12–25 μm region in which many other materials have decreased transmittance due to the presence of absorption bands.

Gallium arsenide (GaAs) used for lenses and beam splitters provides an alternative to ZnSe in medium and high power CW CO_2 laser systems (Table 5.29). It is most useful in applications where toughness and durability are important. Its hardness and strength make it a good choice where dust or abrasive particles tend to build up on or bombard the optical surfaces. When frequent cleaning by wiping is required, GaAs is excellent. The material is nonhygroscopic, safe to use in laboratory and field conditions and is chemically stable except when in contact with strong acids.

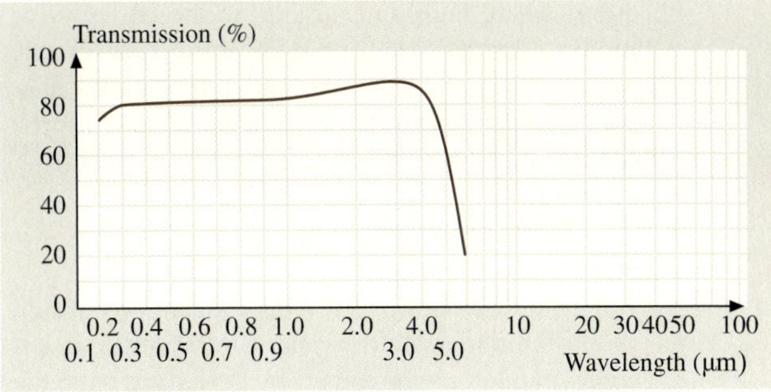

Fig. 5.57 Transmission of sapphire

Further areas of interest where semiconductor materials are used as optical materials are the LEDs and laser diodes. Depending on the spectral properties one is interested in the semiconductors of III–V type. Famous candidates are InAs or GaN as material for blue LEDs.

5.8.3 Sapphire

Sapphire is the leading material for highly sophisticated optical applications that require reliability, strength, and a wide range of light transmittance. Sapphire transmits light over a broad wavelength range spanning from 0.15 to 5 μm (Figs. 5.57, 5.58 and Table 5.30).

This unique ability to transmit over a broad range combined with its mechanical strength makes sapphire the material of choice for many applications. For imaging optics it is desirable for the refractive index of an optical material to have a low dependence on temperature. Because of its low $\mathrm{d}n/\mathrm{d}t$, a temperature gradient across a window will not cause image blur and foresight error. The United States National Bureau of Standards has extensively researched the index of refraction of sapphire. Recently a model was developed by *Thomas* et al. at the Advanced Physics Laboratory to predict $\mathrm{d}n/\mathrm{d}t$ from 0.7 to 5 μm.

Sapphire is a single crystal material from aluminium oxide, Al_2O_3. It has a hexagonal/rhombohedral structure. Optically it is a negative uniaxial crystal in the visible region with a small birefringence.

Sapphire has a band-gap of approximately 10 eV, which is one of the largest for an oxide crystal, permitting optical transmission in the above mentioned range. Like other crystals sapphire can be synthesized by a Czochralski process. Sapphire has a high melting point and is a very hard material which allows it to be used in applications of extrem temperature conditions. Sapphire is used in laser windows – in particular for ap-

Part A | 5.8

plications where high transmission in the IR-region is required. Its hardness and stiffness make it a suitable substrate material. In the semiconductor industry sapphire is used as a substrate to grow GaN on it for the production of blue LEDs.

5.8.4 Optic Anisotropy in Cubic Crystals

According to the Sect. 5.1.4 optical anisotropy would not be expected in a cubic crystal. Birefringence in cubic solids should only be caused by deviations from the cubic symmetry. There is, however, an effect that makes cubic solids optically anisotropic. In these crystals a symmetry breaking is provided by the electromagnetic radiation of the light wave. This is called induced spatial dispersion. Here the dependency of the dielectric function on the wavevector of the incident radiation $\boldsymbol{k}$ becomes important.

Table 5.30 Main Properties of sapphire

Property	Value
Chemical formula	Al_2O_3
Crystal class	Trigonal
Molecular weight	101.94
Density (20 °C) (g/cm^3)	3.98
Reflection loss for two surfaces	
at 4 μm (%)	12
Dielectric constant for 10^2–10^8 Hz	
at 298 K	
parallel	10.55
perpendicular	8.6
Dielectric strength (kV/mm)	17
Resistivity at 20 °C (Ω cm)	$> 10^{16}$
Melting temperature (K)	2300
Thermal conductivity at 300 K (W/m K)	
parallel	35.1
perpendicular	33.0
Thermal expansion at 293 K (1/K)	
parallel 5.6×10^{-6}	
perpendicular	5.0×10^{-6}
Specific heat at 298 K (cal/g K)	0.18
Bandgap (eV)	9.9
Solubility in water	None
Mohs hardness	9
Knoop hardness (kg/mm^2)	1370
Young's modulus (GPa)	335
Shear modulus (GPa)	148
Bulk modulus at 273 K (GPa)	240
Apparent elastic limit (MPa)	275
Poisson's Ratio	0.25

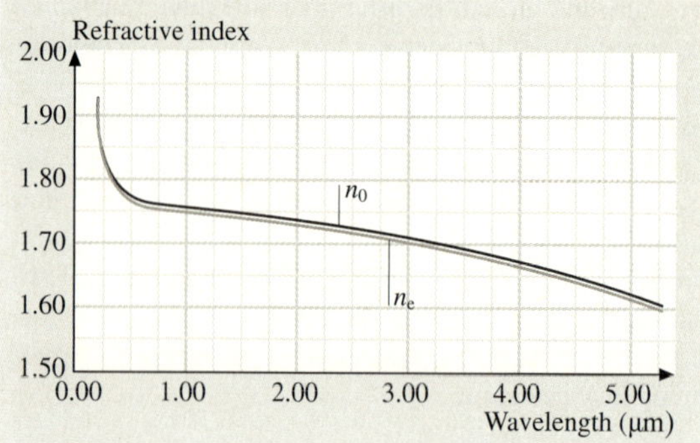

Fig. 5.58 Refractive index of the ordinary (n_o) and extraordinary (n_e) ray in sapphire

Let us try to estimate the order of magnitude of optical anisotropy which we can get due to the symmetry breaking of the light. Because of the extremely large velocity of the light, $c = 2.9979\times10^8$ m/s the corresponding wavevectors are very small

$$E_{\text{light}} = \hbar\omega = ck \,. \tag{5.150}$$

For visible light, e.g. at 600 nm, the photon energy is about 2.1 eV and this translates, using the dispersion relation (5.150), to $k = 2\pi/\lambda = 0.01\,\text{nm}^{-1}$. This has to be compared with the size and usual wavevectors occurring for electronic degrees of freedom in the solid. For a lattice constant of $a = 5.2\,\text{Å} = 0.52\,\text{nm}$ (CaF_2) the maximum wavevector at the edge of the Brillouin zone results to $k_{BZ} = 2\pi/a = 12.1\,\text{nm}^{-1}$. There are more than three orders of magnitude between the length scales on which the electromagnetic field is varying (λ) and the length at which the lattice is varying (a).

The spatial dispersion is the influence of the inhomogeneity of the electromagnetic field on the optical response of a system. This causes a non-locality. For perfectly homogeneous fields ($\lambda \to \infty$, $k \to 0$) the optical response of the cubic system is isotropic.

Let us assume now that our system allows for an expansion of the dielectric function around the zero wavevector [5.253]. Taylor expanding the (inverse) dielectric function with respect to the wavevector of the light gives an idea how the anisotropy enters into the optical response

$$\epsilon_{ij}^{-1}(\boldsymbol{k},\omega) = \epsilon_{0ij}^{-1}(\omega) + h_{ijkl}k_k k_l + O\left(k^4\right) \,. \tag{5.151}$$

The term linear in $\boldsymbol{k}$ vanishes due to the inversion symmetry of the cubic crystal. In order to obtain an upper boundary for the expected anisotropy one can do the following estimate [5.254]. Responsible for the anisotropy is the deviation of the electromagnetic field from uniformity along the elementary cell of the crystal. This is the quantity

$$\frac{a}{\lambda} = \frac{k}{k_{\mathrm{BZ}}} . \tag{5.152}$$

This means we get, using the k^2 dependency from (5.151) and the magnitude from (5.152) as an upper boundary for the magnitude of the expected anisotropy

$$\left|\epsilon_{ij}^{-1}(\boldsymbol{k}, \omega) - \epsilon_{0ij}^{-1}(\omega)\right| \ll \left(\frac{k}{k_{\mathrm{F}}}\right)^2 = \left(\frac{a}{\lambda}\right)^2 . \tag{5.153}$$

Since $k/k_{\mathrm{BZ}} = 8 \times 10^{-4}$, the anisotropy in ε is smaller than 10^{-6}. This effect can be only observed with extremly accurate high resolution interference measurements.

In the vicinity of a strong absorption in the material the physics changes. This was first described in 1958 by the 2003 nobel laureate *Ginzburg* [5.253]. The reason is that not the wavelength in vacuum as in (5.152) has to be compared with the length of the elementary cell but the wavelength in the medium $\lambda_{\mathrm{m}} = \lambda/n$. Since the refractive index n becomes large in the vicinity of a strong absorption (Sect. 5.1.1) the spatial dispersion induced optical anisotropy becomes better observable near an absorption in the material.

In semiconductor materials, the enhancing resonance can be provided by the vicinity of a band edge and has been shown in [5.255] for silicon and [5.256] for GaAs. It has also been measured in alkali halide crystals [5.257]. In strongly ionic crystals deep excitonic electron–hole bound states are present. For the strongly ionic crystal CuCl the exciton resonance was identified as the enhencing absorption [5.258]. Recently the effect of spatial dispersion induced birefringence became of special importance for the crystal $\mathrm{CaF_2}$ [5.259]. This material is used for UV lenses employed for microlithographic structuring of semiconductors. Here a direct connection between the exciton resonance and the spatial dispersion was made [5.260].

Finally, one has to note that – opposite to usual birefringence with according to Sect. 5.1.4 a maximum of two – the spatial dispersion induced birefringence in a cubic crystal has seven optical axes.

5.9 Special Optical Materials

5.9.1 Tunable Liquid Crystal Electronic Lens

The lens is a key element for use in optical systems. A conventional lens made by shaping a medium with a fixed refractive index brings light to focus. Glass, polymers and other transparent solid materials are commonly used for fabricating lenses. One major characteristic of the solid lens is that it has only one focus. To change the focal length, a zoom lens has been developed. A zoom lens system usually consists of groups of lenses. The separation distances between the lenses are mechanically adjusted. However, the adjusting processes are complicated, bulky and even inefficient. Therefore, developing a compact, lightweight, low-cost, and efficient tunable lens is highly desirable and urgently needed.

Due to high birefringence of nematic liquid crystals (LCs), which can be controlled by the applied voltage, an electrooptical effect based on the reorientation of LC molecules can be easily obtained. In addition to displays, adaptive optical components using LC materials have been demonstrated and application increased during the past decades. Among these photonic devices, tunable-focus LC lens is particularly attractive for optoelectronics, machine vision, imaging, and eyeglasses applications.

The use of nematic LCs for making tunable lenses was proposed as early as 1977 by *Bricot* [5.261]. Since then, various approaches have been developed, such as surface relief profile [5.262–265], hole-, line- or spherical patterned electrode [5.266–269], modal addressing [5.270–272], Fresnel zone lens [5.273–275], as well as LC/polymer UV pattern irradiation technique [5.276–278]. Each method has its own merits and demerits. In this chapter, we classify those LC lenses into three groups according to the underlying lens structure. In each group, we will choose a typical cell structure as an example for discussion. Especially, we put emphasis on the newly developed LC lens using the simple UV exposure technique. We describe the fabrication method and performance of this kind of lens in details.

Types of LC Lens

Similar to a conventional LC cell for displays, the LC lens is composed of LC material sandwiched between

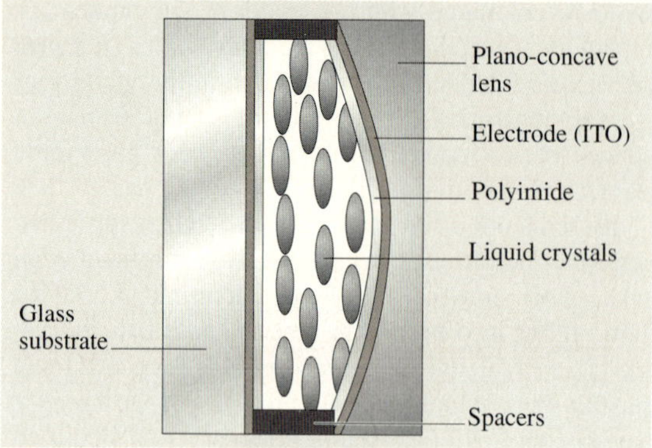

Fig. 5.59 A tunable-focus lens with an inhomogeneous LC layer

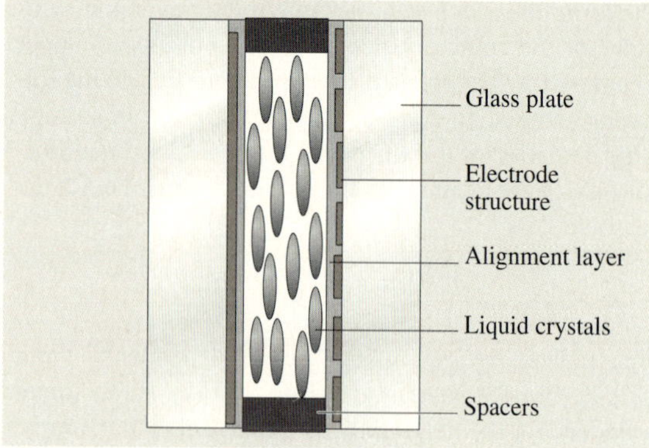

Fig. 5.60 A tunable-focus lens with striped electrodes and a homogeneous LC layer

two glass substrates. To obtain a switchable gradient profile within LC layer, several methods have been proposed. According to the device structure, we can classify LC lens into three different categories.

Curved Cell Gap. In some LC lenses, LC layers can have concave, convex or Fresnel lens structure [5.261–265]. Figure 5.59 shows such a structure. One glass substrate has plano-concave surface and the other is flat. The inner surfaces of the indium-tin-oxide (ITO) glass substrates are coated with thin polyimide layers and then rubbed in order to get a homogeneous alignment throughout the LC bulk. LC lens with such a structure can be operated in two modes.

In the voltage-off state, the incident unpolarized light can be separated into two polarization components: parallel and perpendicular to the LC director. The beam with polarization parallel to the LC directors is called extraordinary ray and the perpendicular one is called ordinary ray. The extraordinary and ordinary rays will be focused onto two different but fixed focal planes. Their focal lengths are given by $f_e = R/(n_e - n_g)$, and $f_o = R/(n_o - n_g)$, respectively, where R is the radius of the concave lens substrate, n_g the refractive index of the glass substrate, and n_e and n_o are the extraordinary and ordinary refractive indices of the LC material.

When a certain voltage is applied to the electrodes, the LC directors are reoriented along the electric field direction. Under such a circumstance, the extraordinary ray sees an effective refractive index (n_{eff}). In this case, the focal length is given by

$$f_{eff} = R/\left(n_{eff} - n_g\right) . \quad (5.154)$$

In (5.154), n_{eff} is equal to n_e at $V = 0$ and is reduced to n_o as $V \to \infty$. Therefore, the focal length can be tuned between f_e and f_o by the applied voltage. For the ordinary ray, it always sees the ordinary refractive index n_o regardless of the applied voltage. Thus, its focal length is not tunable by the applied voltage.

Because of the curved cell gap, liquid crystal in different positions will exhibit a different switching speed. The LC in the thicker area usually has a slower response time and presents light scattering due to the worse alignment. Another drawback of this kind of lens is the focal length can only be tuned between f_o and f_e. To widen the focal length, one has to use a high birefringence LC.

Planar Cell Gap. Several methods have been reported using a planar LC cell to realize the tunable-focus lens. To create a gradient refractive index profile within the LC layer, three different approaches have been explored. *Patterned Electrodes.* Several patterned electrodes, such as line-, hole-, spherical pattern electrode [5.266–269], and modal addressing have been consideredy [5.270–272]. A classical electrode structure is shown in Fig. 5.60. A homogeneous nematic LC layer is sandwiched between an ITO plate and a ground ITO plate. The ITO plate is a glass plate with a group of addressable discrete transparent ITO stripes. The ground plate is a glass plate with uniform ITO electrode covering the entire plate. Each electrode stripe on the glass substrate is individually addressed with voltages chosen so that the refractive index across the cell forms parabolic shape such that the incident light is brought to focus at a selected plane. From Fig. 5.60, one can see that only the light with its polarization axis parallel to the LC directors is affected. This kind of lens with linear electrode belongs to the cylindrical lens. The focal length has the following expression:

$$f=\frac{x_0^2}{2d(n_e-n_{eff})}\,, \tag{5.155}$$

where x_0 is the maximum aperture distance and n_{eff} is the effective refractive index.

Since $n_o \le n_{eff} \le n_e$, f can be tuned in the following range

$$\frac{x_0^2}{2d(n_e-n_o)} \le f \le \infty\,. \tag{5.156}$$

To get a true spherical lens, four cylindrical lenses are required to combine together. By designing the electrode with hole-, hybrid- or spherical pattern, or using modal addressing, a spherical LC lens is realized. The focal length of the spherical lens is expressed by

$$f=\frac{\pi r^2}{(\delta_c-\delta_e)\lambda}\,, \tag{5.157}$$

where r is the radius of the lens, λ is the wavelength, and $\delta_c-\delta_e$ is the phase difference at the lens center and edge.

The lens with a planar cell gap has several advantages. For instance, the focal length can be tuned within a wide range, be fabricated with large size, and the switching speed is uniform across the whole lens. The drawback of this type of LC lens is that the cell fabrication process is complicated and the addressing technique is difficult.

Spherical Electrode within a Planar Substrate. To fabricate an adaptive lens with a wide range tunable focal length, a new lens structure as shown in Fig. 5.61 is considered. The upper planar substrate actually consists of a convex and a matched concave spherical surface. One of the spherical surfaces is coated with indium-tin-oxide (ITO) as electrode. The lower substrate has a transparent electrode coated on its inner surface. Homogeneous LC layers are aligned between the substrates. In the null voltage state, no focusing effect occurs provided

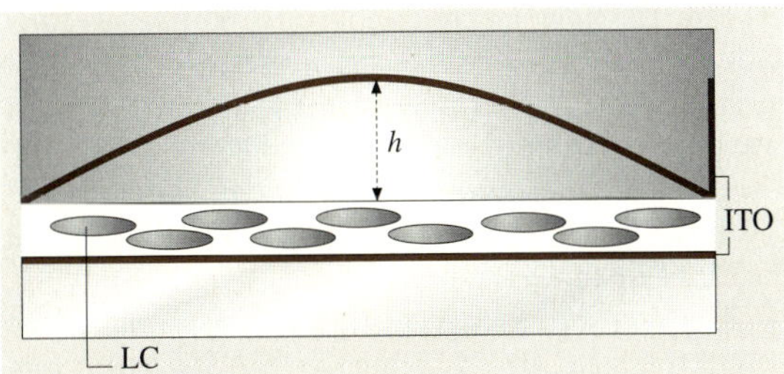

Fig. 5.61 A tunable-focus lens with a spherical electrode and a homogeneous LC layer

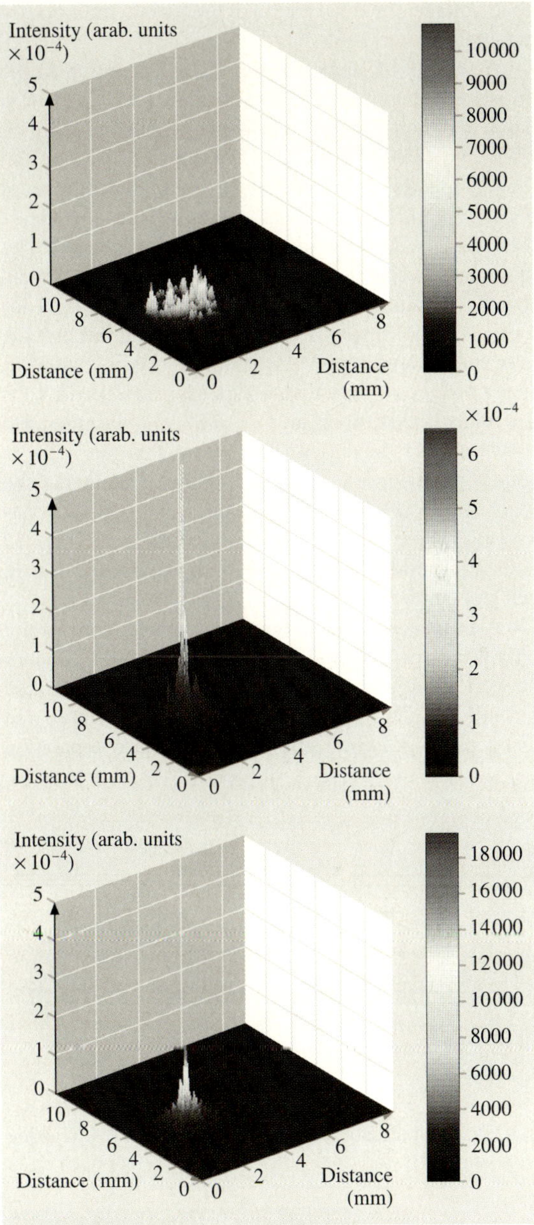

Fig. 5.62 Measured spot intensity profiles for the LC lens shown in Fig. 5.61. *Top*: $V = 0$; *middle*: $V = 18\,V_{rms}$; and *bottom*: $V = 35\,V_{rms}$. Lens diameter: 6 mm

that the refractive index of the upper convex and concave substrates is the same. When a voltage is applied across the LC layer, a centro-symmetrical gradient distribution of refractive index is generated. As a result, the LC layer causes light to focus. By controlling the applied voltage, the focal length of the lens is continuously tunable.

A sample LC lens was fabricated according to the design sketched in Fig. 5.61. The diameter of the lens is 6 mm and the LC cell gap is 40 μm which is filled with UCF-2 ($\Delta n = 0.4$) nematic LC mixture. The height from the top spherical electrode to the liquid crystal layer is $d = 1.03$ mm. The 3-D focal length intensity was measured using a CCD camera at different voltages. The distance between the LC lens and CCD camera is 78 cm. Results are shown in Fig. 5.62. At $V = 0$, the light intensity of a collimated He–Ne laser passing through the lens cell is very weak. As the voltage is increased to $V = 18\,V_{rms}$, a tightly focused spot appears. As the voltage is increased further, say to $V = 35\,V_{rms}$ the measured light intensity decreases. This means that the focus is somewhere between the LC cell and CCD camera. The focal length of this spherical lens can be tuned continuously from ∞ to 0.5 m by controlling the applied voltage. The response time is ≈ 200 ms for the 40 μm cell gap employed.

UV Exposure through Photomask. Tunable lenses using polymer dispersed liquid crystal (PDLC) and polymer network liquid crystal (PNLC) have been demonstrated recently [5.276, 277]. This kind of lenses is prepared using photo-polymerization induced phase separation technique. A unique feature of these polymer dispersed LCs is that the larger LC domain exhibits a lower threshold voltage. If LC presents an inhomogeneous centrosymmetric LC domain dispersed in a polymer matrix, then the applied homogeneous electric field would induce a gradient refractive index profile, which reveals a lens-like character.

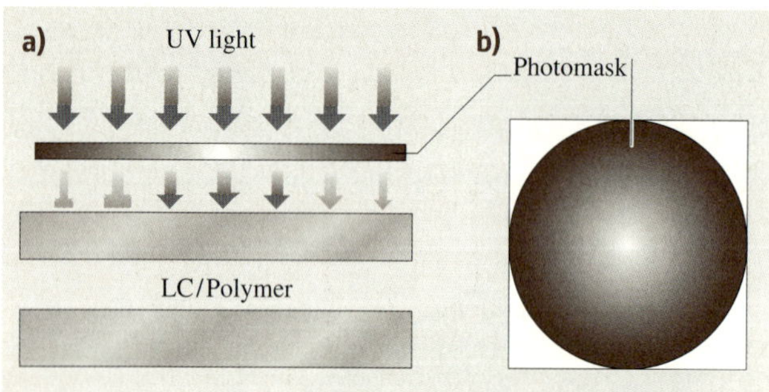

Fig. 5.63a,b Device fabrication of an inhomogeneous PNLC using a patterned photomask

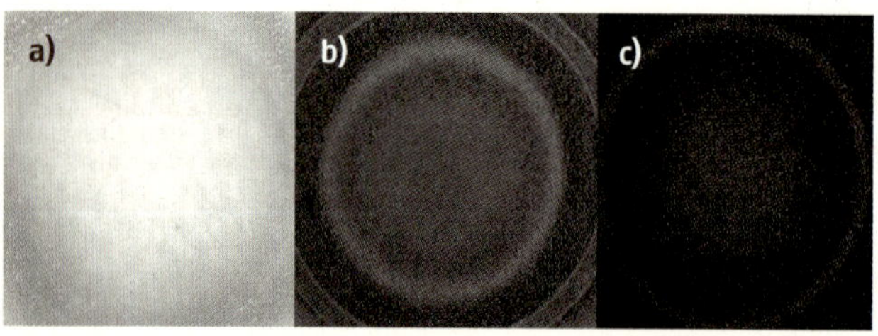

Fig. 5.64a–c Photographs of the PNLC sample at different operating voltages. **(a)** $V = 0$, **(b)** $V = 2\,V_{rms}$, and **(c)** $V = 35\,V_{rms}$

A PNLC lens using UV exposure fabrication technique has been proposed. Fig. 5.63a illustrates the fabrication method of a positive PNLC lens. The key element is the patterned photomask. To obtain a PNLC cell with an inhomogeneous centrosymmetric polymer network profile, a circular continuously variable optical density is used as the photomask, as shown in Fig. 5.63b. Its optical density increases outward radically from a minimum at the center to the maximum at the edges. When a uniform UV light passes through the photomask, the output intensity varies in parabolic profile across the LC sample. The area with a stronger UV light would accelerate the polymerization process and produce a higher polymer concentration. Conversely, the area with a weaker UV exposure would have a lower polymer network concentration. As a result, an inhomogeneous centro-symmetric polymer distribution is formed.

The PNLC cell with convex refractive index profile functions as a positive lens. When a uniform electric field is applied, the LC directors are reoriented to different degrees. The area with looser network packing (edges) has a lower threshold voltage than that of a denser zone (center). Thus, the convex refractive index is generated. As the applied voltage increases, the curvature of the gradient lens decreases. In the high voltage regime, nearly all the LC directors are aligned along the electric field direction. The gradient direction no longer exists and the lens effect vanishes.

To fabricate a positive PNLC lens, a mixture of 5 wt % UV curable monomer BAB-6 (containing 2% photoinitiator IRG-184) in a LC host (W-1331, $\Delta n = 0.229$) was injected to a homogeneous ITO coated LC cell. The monomer has a common rod-like structure with a reactive double bond at both sides. The inner surfaces of the LC cell were coated with thin polyimide layers. The cell gap of the homogeneous cell is 16.5 μm. The diameter of the photomask is 1.7 cm. The measured UV intensity after the photomask is 14 mW/cm^2, and the exposure time is 60 min.

The prepared sample is highly transparent in the voltage-off state. By observing the birefringence colors on a light table, one can check the gradient refractive index. Three photographs of the PNLC sample

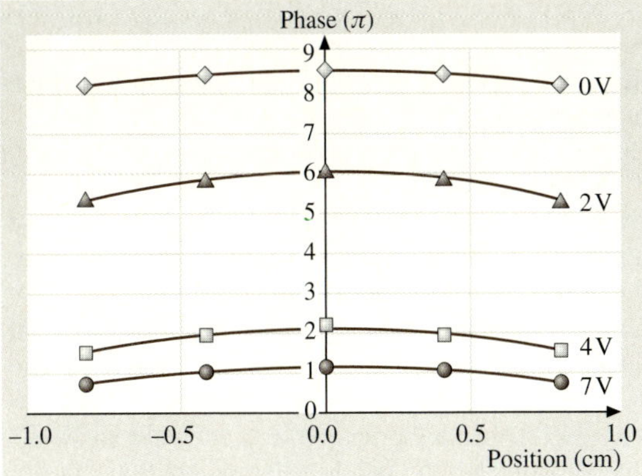

Fig. 5.65 Spatial phase profile of the positive lens at various voltages

cell at $V = 0$, 2 and 35 V_{rms} were taken, as shown in Fig. 5.64a–c, respectively. The rubbing direction of the cell was orientated at 45 °C with respect to the fast axis of the linear polarizer. The analyzer is crossed to the polarizer. At $V = 0$, the color is relatively uniform. As the applied voltage exceeds the threshold, color starts to change from the borders and gradually extends to the center, like a swallowing spot. A circular color ring was shown in Fig. 5.64b at $V = 2\,V_{rms}$. This implies that the central area has a higher threshold voltage than that of the edges. One the contrary, decreasing voltage changes the birefringence color from center to borders, like a vomiting spot. This cell acts as a positive lens.

The measured phase retardation of the cell is plotted in Fig. 5.65. At 2 V_{rms}, $\Delta\delta = 0.8\pi$ which corresponds to 120 meter focal length. As the voltage increases, $\Delta\delta$ decreases so that the focal length increases. Due to the long focus, this lens is suitable for satellite imaging and astronomy applications. To shorten the focal length, several methods can be chosen, such as improving the photomask pattern, using high birefringence LC material, increasing the cell gap, or reducing the diameter of the curved spot.

To fabricate a negative lens, one can choose a photomask with reversed optical density. By using patterned UV exposure method, both positive and negative lenses can be fabricated. Since the lens is a single pixel device, the fabrication process is simple and the lens size depends upon the photomask design.

Like the aforementioned LC lens, PNLC lens is polarization dependent too. To overcome this polarization dependence, two cells with orthogonal alignment directions should be considered. The PDLC with subwavelength droplet size is another technique for solving the polarization dependent problem for LC lens [5.276]. The drawbacks of the nano-PDLC lens are the high operating voltage and small refractive index change which results in a long focal length.

Besides the patterned photomask, exposure using a UV laser beam with Gaussian intensity distribution to the LC/monomer cell has been studied [5.278]. The cured spot presents a lens-like character under the action of a homogeneous electric field. The concern for this technology is that a high power laser is needed for scaling up the lens aperture.

Fresnel Zone Plate. In comparison with the above refractive lens, Fresnel zone plate lens belongs to the diffraction lens. It can be fabricated with large aperture size. Fresnel lenses are suitable for long distance optical communication, optical distance measurement and space navigation. In a binary-phase planar Fresnel lens, the phase difference between the adjacent zones can be tuned by the applied voltage. Maximum diffraction occurs when the phase difference is equal to an odd multiple of π while no diffraction occurs for even multiple of π. Through diffraction, the Fresnel zone plate works as a focusing element.

Due to higher-order Fourier components, Fresnel lens has multiple foci at f, $f/3$, $f/5$, etc. However, the majority of the incident light diffracts into the primary focus. The primary focal length can be expressed by $f = r_1^2/\lambda$, where r_1 is the innermost zone radius. The radius of the nth zone (r_n) is given by $r_n^2 = nr_1^2$, n is the zone number. Theoretically, the diffraction efficiency of the primary focus for the binary phase Fresnel lens is 41%.

A method of fabricating LC Fresnel lens was proposed by *Patel* and *Rastani* [5.273]. The idea to fabricate the lens is to make the neighboring zones with orthogonal LC directors on one substrate, and homeotropic alignment on the other. Another method of fabricating LC Fresnel lens is to etch the ITO electrode by photolithographic to form a zone plate [5.274]. However, the two fabrication methods are rather complicated.

Using UV pattern irradiation technique, one can simplify the device fabrication process [5.275]. A method of fabricating Fresnel lens using patterned photomask is similar to that as shown in Fig. 5.63, except that the photomask was produced with transparent odd zones and opaque even zones.

To fabricate a nanoscale PDLC Fresnel lens, 26 wt % LC E48 and 74 wt % UV curable prepolymer NOA81 were mixed and injected into a homogeneous empty

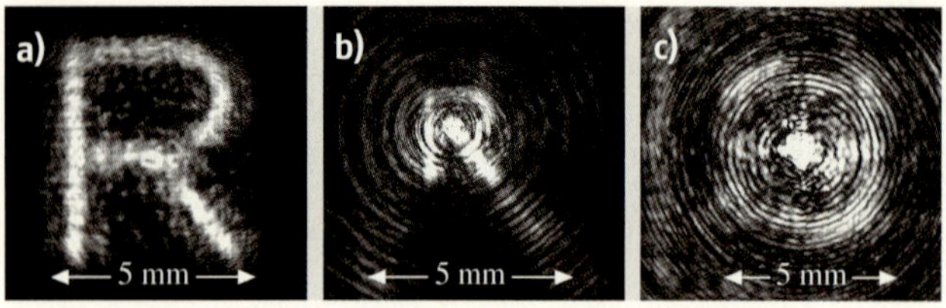

Fig. 5.66a–c Imaging properties of the PDLC Fresnel lens recorded by a CCD camera. **(a)** without sample, **(b)** with sample near the focal spot, and **(c)** with sample at focal point

cell. The cell gap is 15 μm. A photomask was produced by etching a chromium oxide layer using electron beam lithography. The radius of the innermost zone is 0.5 mm, and the zone plate consists of 80 zones in 1 cm aperture. The photomask is in proximity contact with the LC cell substrate during exposure to the UV light.

To evaluate the image quality of the PDLC Fresnel lens, a black cardboard with transparent alphabet **R** was placed before the sample. A collimated He–Ne laser has 1 cm diameter which fills the entire zone plate. A CCD camera was set at ≈ 25 cm behind the sample. Figure 5.66a and Fig. 5.66b show the photos without and with sample, respectively. When the sample is absent, no focusing effect occurs. Once the LC Fresnel lens is in position, a clear but smaller image is observed, although some circular noises exist due to diffraction. When the CCD camera was moved to the focal point, a tight spot appeared in the center, as shown in Fig. 5.66c. These results indicate that the sample indeed behaves like a lens.

The light efficiency of the PDLC Fresnel lens is about 32% at $V = 0$. As the applied voltage is increased, the diffraction efficiency is increased. At 180 V_{rms} (or 12 V/μm), the diffraction efficiency reaches nearly $\approx 39\%$. Further increasing the voltage, the diffraction efficiency will decrease. Because of the sub-wavelength PDLC droplets, the Fresnel lens is transparent in the voltage-off state, polarization independent, and its switching speed is in the sub-millisecond range. However, the operating voltage of this kind of lens is relatively high. To decrease the operating voltage, PNLC can be considered. The PNLC contains a lower polymer concentration than PDLC. As a result, its operating voltage is much lower except the PNLC lens is polarization dependent.

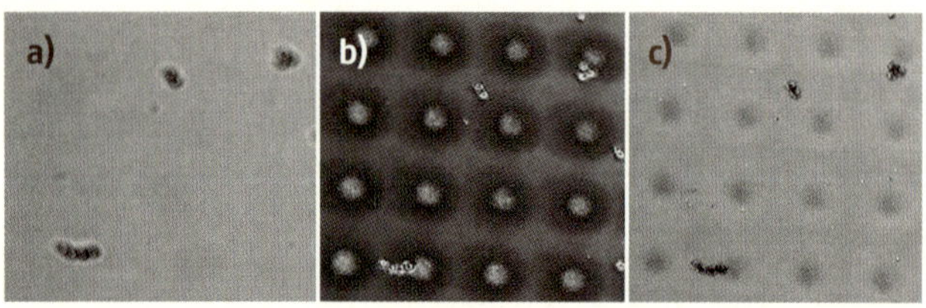

Fig. 5.67a–c Photographs of the PNLC sample cell at different operating voltages: **(a)** $V = 0$, **(b)** $V = 4\,V_{rms}$ and **(c)** $V = 30\,V_{rms}$

Microlens Arrays

Two-dimensional microlens array has a wide range of applications, including information processing, optoelectronics, integrated optics components and

optical communication systems. The hole-patterned electrode, hybrid-patterned electrode as well as surface relief profile can be used to demonstrate microlens arrays. To fabricate microlens arrays, we continue to use the UV exposure through a patterned photomask. According to diffraction theory, when UV light passes through a tiny hole, it will be diffracted and the diffracted intensity presents a parabolic-like profile. Therefore, when the UV curable monomer doped in LC host is exposed using the diffracted UV light, an inhomogeneous centro-symmetrical PNLC can be obtained. Similar to the results as shown in Fig. 5.64, the cured spot acts as a lens.

An example of fabricating LC lens array is demonstrated using PNLC technique. A mixture of 3 wt % monomer BAB6 (with a small amount of photoinitiator IRG-184) and 97 wt % LC E48 was injected to a homogeneous cell composing of ITO glass substrates. The inner surfaces of the cell were coated with polyimide layers and then buffed in the antiparallel direction. The LC cell gap and the substrate thickness are 15 μm and 1.1 mm, respectively. A chromium layer with circular aperture array deposited on a glass substrate is then placed on the top of the cell. The diameter of each aperture is 25 μm and the distance between the neighboring pixels is 85 μm from center to center. The cured UV intensity from the photomask side was 40 mW/cm^2 and the curing time was 45 min.

The cured sample was observed using a polarizing microscope. The rubbing direction of the cell was oriented at 45° with respect to the axis of a linear polarizer and the analyzer was crossed to the polarizer. Three photographs were taken at $V = 0$, 5 and 20 V_{rms} and results are shown in Fig. 5.67a–c, respectively.

At $V = 0$, the UV exposed array spots is very weak and cannot be observed clearly. The cured spot size is about 50 μm due to the diffraction effect. When the sample cell was rotated gradually, the array spots will disappear, this means that the PNLC cell is homogeneous and polarization dependent. When the applied electric field is increased slowly, a clear spot array appears as shown in Fig. 5.67b. In the mean time, a color change occurring starts from the spot borders, and gradually shrinks to its center. This implies that the spot center has a higher threshold voltage than that of its borders and the cured spot acts as a positive lens. As the applied voltage is sufficiently higher than the threshold, most of the bulk LC directors are orientated perpendicular to the substrates. Under such a circumstance, the spot profile becomes very weak, as shown in Fig. 5.67c.

To characterize the light focusing properties of the sample cell, A collimated He–Ne laser beam ($\lambda = 633$ nm) was normally incident on the sample with its polarization direction parallel to the cell rubbing direction. The laser beam was expanded to 5 mm diameter before illuminating the LC sample. A 10 times beam expander was positioned between the CCD and the LC sample in order to clearly resolve the output intensity. A CCD camera at $2cm$ behind the sample cell was used to record the light focusing ability. Figure 5.68a–c show the photos for $V = 0$, 5 and 20 V_{rms}, respectively. One can see in the voltage-off state, no focusing effect occurs at the exposed spots. At 5 V_{rms}, the light passing through the exposed spots was focused. The spots become bright, as shown in Fig. 5.68b. However, the focusing point is relatively large. When the voltage was increased to 20 V_{rms}, a tightly focused spot appeared in the center of each cured spot, as shown in Fig. 5.68c. These results indicate that each spot indeed behaves like a lens.

The voltage-dependent focal length of the microlens arrays was measured and results were shown in Fig. 5.69. As the applied voltage increases, the focal length increases too. This is because the lens profile becomes shallower with increasing voltage.

The focal length of the lens is expressed as $f = \pi r^2/\lambda\Delta\delta$, where r is the radius of the lens, λ the wavelength and $\Delta\delta$ is the phase difference between the lens center and border. Since $\Delta\delta$ can be adjusted by the applied voltage, the focal length is tunable. Theoretically, sufficiently high electric field would reorient the bulk LC directors perpendicular to the substrates so that $\Delta\delta \approx 0$ and $f \approx \infty$. However, in our microlens arrays the polymer networks exert the weakest stabilization on LC molecules at the border of the microlens, thus the radius of the microlens has the tendency to shrink at high voltage. In this condition, the focal length cannot be lengthened, it will gradually saturate as the voltage exceeds 10 V_{rms}, as shown in Fig. 5.69.

Comparing with other microlens technologies, the PNLC advantages are in uniform cell gap and using single electrode on both substrates. The tunable gradient profile is induced with the help of electric field. Like conventional PNLC cells, the PNLC lens has reasonably

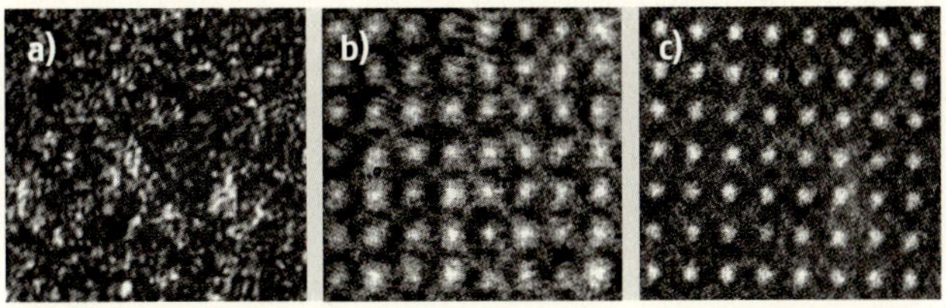

Fig. 5.68a–c Focusing properties of the PNLC lens arrays recorded by the CCD camera. **(a)** $V = 0$, **(b)** $V = 4\,V_{rms}$, and **(c)** $V = 6.7\,V_{rms}$

fast switching speed, wide range tunable focal length by optimizing the cell structure and UV curing conditions, and relatively low operating voltage. This kind of microlens is easily realized with a long focal length and, therefore, is suitable for optical fiber switches and incoherent correlator applications.

In summary, three basic mechanisms have been proposed to fabricate LC lens. They are: inhomogeneous electric field on inhomogeneous LC media, inhomogeneous electric field on homogeneous LC media, and homogeneous electric field in inhomogeneous LC media. Each technique has its own merits and demerits. In terms of fabrication process, the patterned UV exposure technique seems to be the simplest. Depending upon the specific photomask patterns, positive and negative spherical lens, elliptical lens, cylindrical lens, microlens array, prism, and prism grating can be fabricated. The nano-PDLC approach offers advantages in polarization independency and fast response time. However, its operating voltage exceeds 10 V/μm. On the other hand, polymer network has a lower polymer concentration. As a result, its operating voltage is reduced to 2 V/μm, except that the focusing effect is dependent on the light polarization.

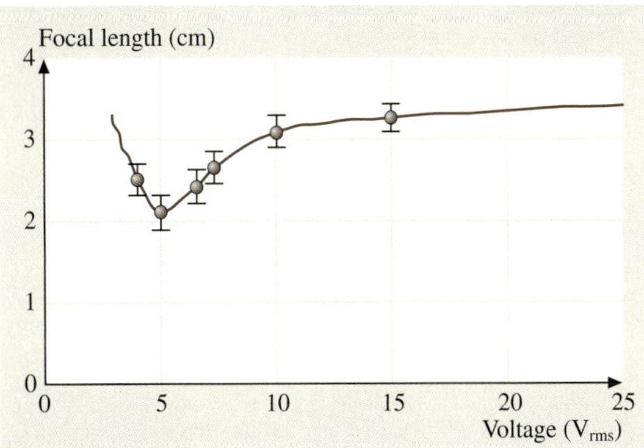

Fig. 5.69 Voltage dependent focal length of the PNLC microlens arrays. LC: E48, cell gap $d = 15\,\mu m$, and $\lambda = 633$ nm

Part A | 5.9

5.9.2 OLEDs

For over 100 years, people have been obtaining light from light bulbs. Unfortunately, the majority of the energy involved is converted into heat and only a tiny fraction actually produces light. In an attempt to overcome this inefficiency, very competitive technologies such as sodium pressure and fluorescent lamps and inorganic light-emitting diodes (LED) were developed. LEDs are highly efficient and available in all colors. Recently, a promising technology to convert electricity into light was developed by depositing thin layers of appropriate materials between two electrodes and applying an electric field (Fig. 5.70) [5.279–281].

The emissive material between the electrodes is an organic semiconducting material, and hence the entire device is now called an organic light-emitting diode (OLED). OLEDs combine the advantages of LEDs with those of organic materials, which include mechanical flexibility, full color availability and large lighting areas. To couple light out from the device one electrode must be transparent. Indium-tin-oxide (ITO) on glass or plastic substrates with a layer thickness of 100–200 nm is the most common transparent and conductive electrode.

This brief review cannot replace a physical textbook and not all elementary processes in an operating OLED can be examined herer intensively. However, the physics of a single layer OLED can be simplified. ITO works well, with its large work-function, as an anode and injects positive charge carriers (holes) into the organic semiconductor. On the other side of the organic layer, a metal electrode simultaneously injects electrons. After migration in the emissive layer, a hole and an electron combine to form an exciton. This neutral excited state is a high energy state far from thermal equilibrium. Unfortunately, this surplus energy cannot be released solely as radiation (light), but is also dissipated as heat. The emission color (wavelength, Table 5.31) is a material property and depends on the difference between the energy levels of the first electronic excited state and the ground state, i. e. the energy difference between the highest occupied molecular orbital (HOMO) and the lowest unoccupied molecular orbital (LUMO) of the molecule.

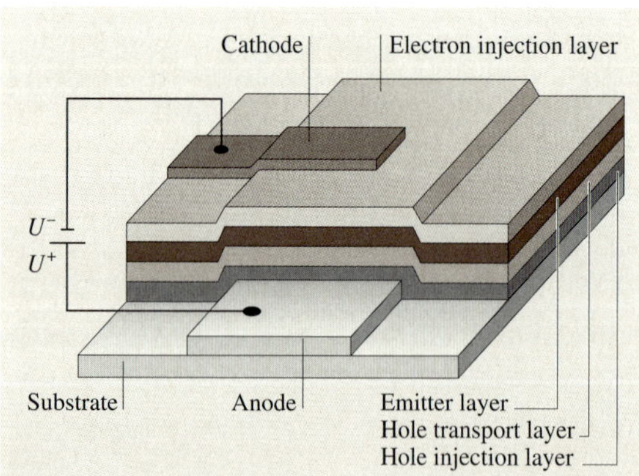

Fig. 5.70 Multi layer setup of an organic light-emitting diode

Table 5.31 Relation of energy gap of HOMO and LUMO, emission color, and wavelength of emitted light

Energy gap in emitters (eV)	Color	Wavelength (nm)
1.9–1.6	Red	650 780
2.3–1.9	Yellow	530–650
2.5–2.3	Green	490–530
3.1–2.5	Blue	420–490

An excited state can be generated by the absorption of light of suitable wavelength as well as by applying a voltage to an OLED device. Independent of the kind of excitation, the radiative decay of the excited state is physically identical and thus both emission spectra have the same shape. The processes are called photoluminescence (PL) and electroluminescence (EL), respectively.

Properties such as flatness and brightness, combined with the possibility of high resolution (pixelation) of the device, make OLEDs a very promising and competitive technology for liquid crystal (LC) displays. Extremely lightweight, full color and flexible displays are likely to be produced within a few years to penetrate the huge market dominated by LC, LEDs, and plasma technology. In recent years, enormous improvements upon OLED emitters allow the production of large bright areas, which will be interesting and useful for lighting applications. Developing a suitable coating technology is the key for making self-luminating large areas with low manufacturing costs. Although a few products equipped with OLED-technology are already available, several technical challenges still remain to be solved.

Emitter Materials

In the pioneering work by *Tang* and *van Slyke*, EL was first discovered in 1987 by using an aluminum quinolate (Alq_3) as an emitter material [5.282]. A second milestone in the field of organic EL occured when polymers were brought into the action in 1990, after workers at Cambridge University evidenced the generation of light from poly-para-phenylenevinylene (PPV) [5.283].

Part A | 5.9

The discovery of light emission from organic materials has led to tremendous research activity. As a consequence, the number of materials employed as the emitter material in OLEDs is vast, and a comprehensive list can not be published within the scope of this book. All organic materials used as emitters in OLEDs must combine several properties. First of all, they need to be of high purity and chemically and thermally stable. Secondly, they must exibit a high quantum yield. Finally, the materials need to be processible and have the ability to form thin films.

The most common classification for emitter materials used in OLEDs is "Small Molecules" (SM) versus "Polymers" (Fig. 5.71 and Fig. 5.72). The main difference between the two is the processing of the materials for fabrication of devices. SMs (compounds with a low molecular weight) are evaporated in a high vacuum process to deposit a film – ranging in thickness from a few layers to a few nanometers – on a surface. Due to the large molecular weight of polymer chains, a vacuum process cannot be used. Advantage can be taken of good solubility in various solvents to obtain homogeneous films by spin-coating or printing techniques. These solution coating processes offer enormous advantages concerning thin films on large areas. It is difficult to compare both classes so only advantages and drawbacks with regard to processability and OLED performance are considered.

Purification of the materials is an important issue since it is well known that tiny amounts of impurities can dramatically impair device performance. Deposition of pure low molecular weight compounds in high vacuum serves as an extra purification step during the OLED process itself. In contrast, polymers must be purified as well as possible before the coating process. For efficient devices, the number of positive and negative charge carriers in the emitter layer must be equal. This goal can be achieved by pursuing the following two strategies described. Using an evaporation process for SM, several materials can easily be deposited on top of each other. This process allows the manufacture of multilayer devices, in which each layer has its own function, such as hole injection or transport, light emission, electron transport or blocking, etc. Efficiency is optimized by material choice and order, layer thickness, co-evaporation, etc.

Polymeric multilayer devices are more difficult to produce due to polymer solubility in multiple solvents. Most two layer devices are made utilizing water soluble polyethylenedioxythiophene/polystyrylsulfonat (PEDT/PSS), which is rendered insoluble (even in water) after baking the film. In this case the emissive polymer can easily be deposited on top of PEDT/PSS by a solution coating process. Due to limitations on the number of polymeric layers that can be stacked, the ideal emitter should have a balanced number of holes and electrons during operation. Furthermore using appropriate electrodes gives nearly balanced electron and hole injection leading to efficient OLEDs.

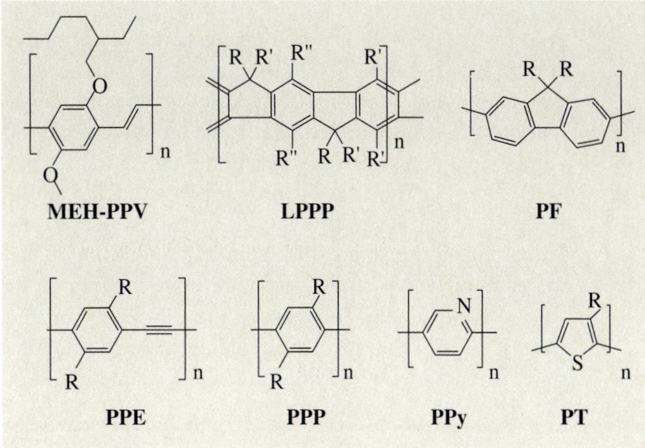

Fig. 5.71 Chemical structures of representative luminescent conjugated polymers

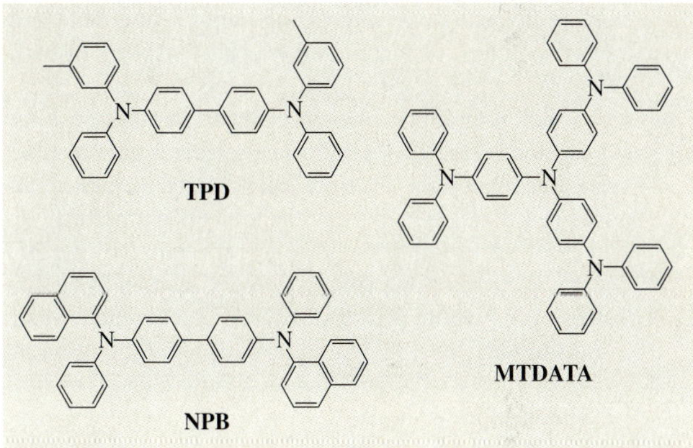

Fig. 5.72 Chemical structures of representative hole transport materials

Fluorescence and Phosphorescence

An important point to keep in mind is the fact that singlet and triplet excitons exist. Whereas during photoexcitation generally singlet excitons are formed, the spin statistics predict the ratio of singlet to triplet excitons created in an operating OLED device should be 1 : 3.

After the formation of singlet excitons by the combination of oppositely charged carriers, it is impossible to distinguish singlet excitons created by charge carriers from single excitons formed by photoexcitations. Without going into further details, the radiative decay of a singlet state is physically permitted and fast (lifetime $\approx 10^{-9}$ s). The transition from a triplet excited state to the ground state is strictly forbidden and only occurs under certain circumstances like spin–orbit coupling in the presence of heavy atoms, with lifetimes ranging from 10^{-6} s up to hours. To improve the efficiency of devices it is desirable to include radiative decay of both kinds of excitons by adding phosphorescence emitters like organic-heavy metal complexes.

Color

Yellow, red, and green emitters are already commercially available in large quantities. Recently, material suppliers are focusing on synthesizing long-lifetime blue and white emitters. In fact, it is the lack of stability in both blue and white emitters that prevents full color products with long lifetimes even under moderate operating conditions. Due to the high energy band gap of blue emitting materials, they are particularly unstable and side-reactions often damage the molecular structure.

Problems for full color displays are caused by the different lifetimes of the emissive colors, leading to a color shift over a short period of time. For example, white emitting devices – which are always a combination of several emission colors – show a slow change of hue during device operation.

Conjugated Polymers

Film formation, quantum yield, device efficiency, lifetime, and emission color can be tuned by modifying the chemical structure of the emitting materials. Since the discovery of EL in polymers many derivatives of poly-para-phenylenevinylene (PPV) as well as other classes of luminescent polymers have been synthesised and used successfully in OLEDs. PPV itself is an insoluble polymer. To make devices with PPV, a soluble non-conjugated precursor form must be coated on the device. The precursor is subsequently thermally transformed to PPV. PPV emits yellow-green light with a peak wavelength of 510 nm. For the sake of purification and processing a soluble polymer is highly desirable. Introducing side groups onto the polymer chain cannot only make the polymer soluble, but can also modify the electrical and optical emission characteristics. Poly[2-methoxy-5(2-ethylhexyloxy)-1,4-phenylenevinylene] (MEH-PPV), a material investigated and used in OLEDs in 1991 is one of the first examples of this approach. This molecule shows an emission in the orange-red region of the spectrum [5.284].

A disadvantage of PPV is the easy oxidation of the vinylene bridge between two phenylene-rings. This undesired side-reaction, which can occur during synthesis or device operation, breaks the conjugation and leads to a reduction in device efficiency. Again, two ways were identified to improve the oxidative stability: substitution at the vinylene carbon position or replacing the double bond by a triple bond ending with poly[2,5-bis(hexyloxy)-1,4-phenylene-(1-cyanovinylene)] (CN-PPV) [5.285] and poly(para-phenyleneethylene) (PPE), respectively [5.286].

Poly-para-phenylene (PPP) derivatives were synthesized to obtain stable blue emitters [5.287], however, conjugation in the polymer chain is reduced due to non-planar alignment of adjacent phenylene rings. Improved planarity is realised by bridging two phenyl rings with a methylene-group to form polyfluorene (PF). PF is a blue emitter which can be color tuned over the visible spectral range by adding different monomers during the polymerization of PF-precursors [5.288–290].

Researchers in Germany synthesised a PPP derivative where all adjacent rings are bridged, namely the ladder-PPP (LPPP)[5.291, 292]. The phenylene-rings are completely planarised and the stiff chains show high persistence lengths that are responsible for the relatively high mobility of charge carriers in this semiconducting polymer.

Another approach in polymer OLEDs is the application of polymer blends [5.293–295]. At least two polymers are mixed: a polymer for charge carrier transport and an emissive polymer. Recent research has been done on avoiding the problem of phase separation. For the polymer blend approach, either a copolymer possessing all properties of the blend itself is synthesized or the chains of the two polymers are crosslinked after processing as a thin film in the device.

The synthesis of polymer chains with defined chain ends has led to more stable derivatives in comparison to the polymer without defined chain ends. Such "end-capped" emitter polymers (for instance attached to a hole transport molecule at the chain end) show better device performance.

Along with conjugated polymers, other classes of organic materials are noteworthy. Heterocyclic polymers like polythiophene (PT) [5.296], polypyridine [5.297], and many others were synthesized and countless derivatives were investigated in OLEDs.

Small Molecules

Thermal stability is required in an evaporation process to deposit a molecule on a surface without decomposition. Furthermore, formation of stable and amorphous films of one or more functions, such as charge injection and transport, and emission must be combined within the material (Fig. 5.73) [5.298].

Promising compounds with reasonably efficient emission are organo-metallic complexes. In this case emission comes from either the ligands or the metal-center. The prototype for ligand emission is the aluminum triquinolate (Alq_3) [5.282] shown in Fig. 5.72. Many materials have been found to generate light of all colors. Rare-earth-complexes emit through electron transition in the metal atom [5.299]. These spectra are very narrow compared with polymer emission spectra, however short wavelength emission is hard to realise and quantum efficiency seems to be lower than in ligand-emission complexes, too.

In efficient SM devices multilayer systems are used. For a balanced number of oppositely charged carriers in the emissive layer and for decreasing the on-set voltage, injection and transport materials are necessary for holes and electrons in addition to device setup optimization.

Material improvements for SM-OLEDs can be shown exemplary at hole transport materials, where the most widely used compounds are aryl amines. To ensure the final materials are high temperature stable glassy films (instead of crystalline layers) many derivatives have been synthesised, including N,N'-diphenyl-N,N'-bis-(3-methylphenyl)-(1,1')-biphenyl-4-4'-diamine (TPD) [5.300] and N,N'-diphenyl-N,N'-denaphthyl-(1,1')-biphenyl-4,4'-diamine (NPB). An other large arylamine, the so-called starburst amine (MTDATA), shows a glass transition temperature (T_g) around 160 °C [5.301].

Two other approaches to increasing the T_g are raising the molecular weight or modifying the molecular structure of the material. The former approach is limited by the need to increase the evaporation temperature for higher molecular weights. In recent years *spiro*-compounds have been synthesized [5.302] to obtain materials with a high T_g. An advantage of this molecular element is the inhibition of crystallization to produce materials in a stable glassy state.

For generating light in an OLED, small fluorescent molecules are also used for doping both electron and hole transport materials. Due to the lower excited state energy of the dopants (as compared to the host), excitons formed in the binder are immediately transferred to a dopant which acts like an exciton trap. By optimizing the dopant/host ratio, emission only comes from the dopant. However, if the dopant concentrations are too high (> 2–3%), performance actually suffers due to concentration quenching of the luminescence. Examples of fluorescent dyes are perylene [5.303], 4-(dicyanomethylene)-2-methyl-6-(p-dimethylaminostyryl)-4H-pyran (DCM) [5.304, 305], and chinacridone [5.306].

Mechanism of Degradation and Device Failure

Depending on the desired properties of an OLED a long lifetime (ten thousands of hours) and/or a high brightness (1000 to $5000\,Cd/m^2$) have to be achieved (Fig. 5.74).

Two kinds of failure can be observed: a sudden death, where a short circuit prevents light emission, or a slow decrease, where the lighting area becomes inhomogeneous (black spots) and/or darker.

Immediate failure of the device occurs by connecting the ITO anode directly to the metal cathode during fabrication. In this case a large amount of current flows through this defect and the area becomes hot, damaging the metal and polymer alike. Interestingly, it was found that shorts can also arise during operation by a currently unknown mechanism. Non-emissive regions on the lighting area have to be avoided. These so-called black spots are very tiny in the beginning and grow larger with time. It was shown by different spectroscopic techniques that a large particle (in comparison to the film thickness) in the middle of such a spot can cause a pinhole in the cathode. These damages are supposed to be preferred channels for the penetration of reactive species like oxygen and water into the sensitive materials. This idea is also supported by the constant circular growth of these spots [5.307]. Clean room conditions during OLED preparation and manufacturing of a thin film of suitable polymer like PEDT/PSS to smooth the ITO-surface can prevent failure by shorts.

Lifetime is typically defined as the reduction of brightness to 50% of the initial brightness at constant current and depends strongly on the operation conditions. The brighter the device and the more charge carriers go through the layers the shorter the lifetime due to generating more light and heat in the device. Lifetime depends also on the emission color. Red, yellow and green emitters are stable for more than 20 000 h. Blue and white emitters are being improved intensively, but are still not appropriate for long life products yet.

A typical lifetime curve for a PPV-derivative such as that shown in Fig. 5.73 reveals two areas. During the first phase the brightness shows an "initial drop". After this short period the reduction of brightness be-

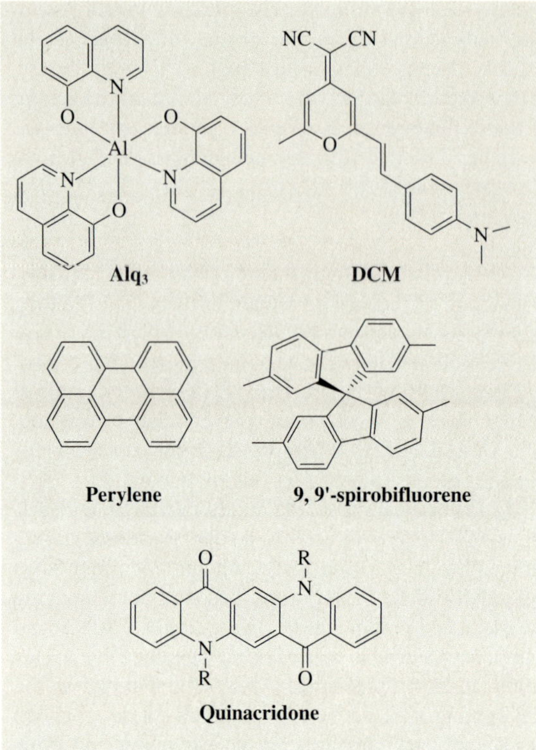

Fig. 5.73 Chemical structures of representative Small-Molecules luminescent dyes

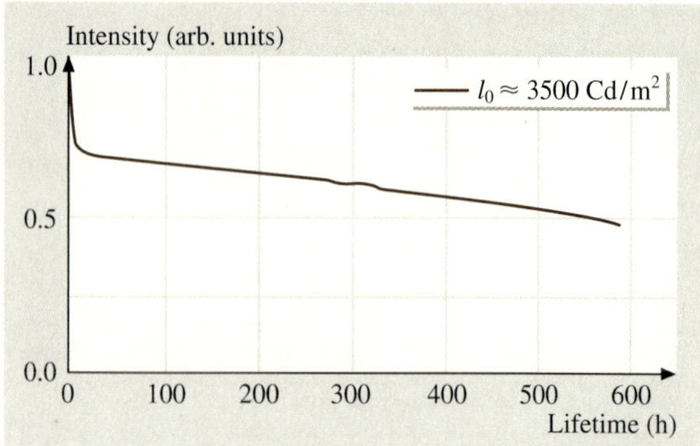

Fig. 5.74 Typical accelerated lifetime measurement of a PPV derivative at a high and constant current and high brightness (started at $3500\,\mathrm{Cd/m^2}$ and cancelled at $1750\,\mathrm{Cd/m^2}$ after several 100 h segments)

comes slower. Although a stable level is never reached and the brightness decreases continuously, lifetimes of PPV derivatives can still reach several 10 thousand hours under moderate conditions ($100\,\mathrm{Cd/m^2}$ at room temperature). The initial drop is often called burn-in and is not fully understood, whereas the slow reduction of brightness is attributed to slow degradation mechanisms.

OLEDs are generally very sensitive to water and oxygen; therefore the whole device must be hermetically encapsulated to minimize material damages. Thin metal cathodes are supposed to react with traces of water to decrease the ability of electron injection, particularly under operating conditions. Oxygen reacts with emitter polymers due to its potential for photooxidation of the conjugated polymer. Cleavages of polymer chains have been observed, consequently breaking the conjugated system [5.308–310]. But air is not the only oxygen source – ITO is another [5.311]. Again a dramatic improvement in lifetime was found by introducing a PEDT/PSS layer on top of ITO. A further advantage of this polymer coating is the prevention of indium migration in the device. Ion migration is also known to be a general problem in operating devices [5.312]. Several scenarios are described concerning this issue. First, the ion remains in his charge state and counteracts the applied electric field. Second, ions damage the materials by oxidizing and reducing. Third, the ions can react at the electrode and generate impurities which prevent the charge carrier injection. These descriptions of ion migration fit the observation that devices driven in a pulsed mode (forward/reverse bias) show slower aging than direct current driven devices.

A further drawback of organic materials is their sensitivity to UV-light. Formation of highly reactive radicals, cleavage of chemical bonds and polymer chain decomposition have negative effects on device performance. UV protection of OLEDs is necessary particularly for outdoor applications.

Summary

LEDs can be made of organic materials. These OLEDs possess a layer setup and are made of either Small Molecules or polymers, or a mixture of both. Thin amorphous films of Small Molecules are deposited chiefly by vacuum processes, whereas polymer films are obtained by wet coating techniques. All emission colors are available, including white. However, lifetimes of blue and white emitters (compared to red, yellow and green) are still too short for long life products and full color applications, respectively. A further issue is the color shift during device operation, which depends strongly on the device operation conditions. An encapsulation increases the device lifetime tremendously – however, hermetic protection against environmental impact is not available yet. Emitter materials, particularly polymers, have been improved astonishingly in the last decade. Detailed understanding of the device physics and the degradation mechanisms is necessary to furter develop material properties and device design.

For high-end products, some technical problems must still be solved. But visible progress has been made over the last decade in all the important fields like efficiency, color portfolio brightness, and lifetime to make OLED devices already competitive in some niches. Time will tell whether the replacement of spot lights (like light bulbs) by lighting areas made of diffuse OLEDs in applications will take place.

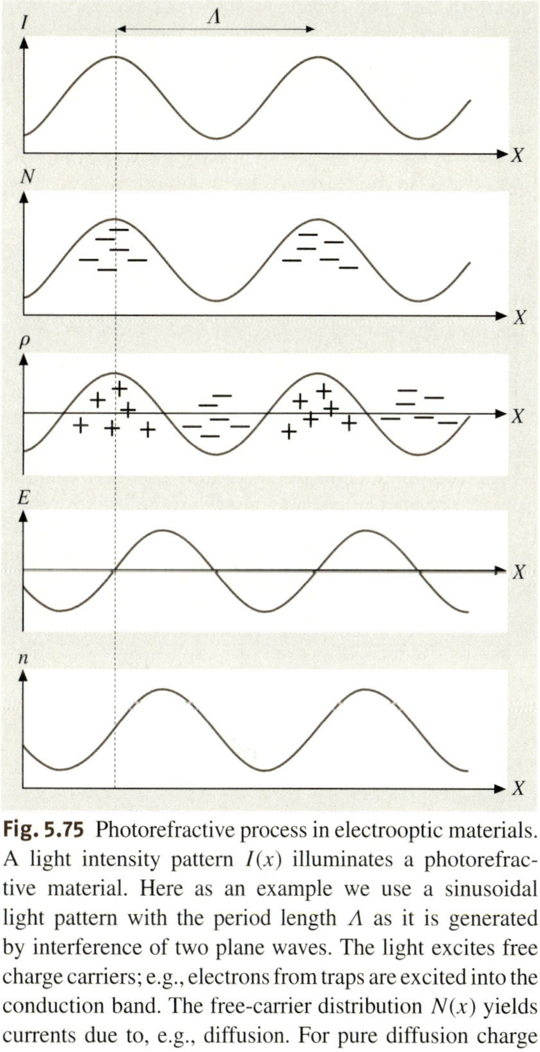

Fig. 5.75 Photorefractive process in electrooptic materials. A light intensity pattern $I(x)$ illuminates a photorefractive material. Here as an example we use a sinusoidal light pattern with the period length Λ as it is generated by interference of two plane waves. The light excites free charge carriers; e.g., electrons from traps are excited into the conduction band. The free-carrier distribution $N(x)$ yields currents due to, e.g., diffusion. For pure diffusion charge transport, the charge carriers move into the darker areas where they are captured by empty traps. An effective charge density distribution $\rho(x)$ builds up with missing electrons and effective positive charge in the bright and excess electrons and effective negative charge in the dark areas. This charge distribution yields a space–charge field $E(x)$ and, because of the linear electrooptic effect, also a refractive-index modulation $n(x)$

5.9.3 Photorefractive Crystals

Light-induced refractive index changes, so-called photorefractive effects, in inorganic electrooptic crystals were discovered by *Ashkin* et al. in 1966 [5.313]. Though in the beginning these effects seemed to be very undesirable ("optical damage"), *Chen* et al. recognized only two years later the significance for holographic storage [5.314]. In 1975, *Staebler* et al. reported the recording of 500 thermally fixed volume phase holograms in $LiNbO_3$: Fe, each hologram with more than 2.5% readout efficiency [5.315]. The method is based on the Bragg condition allowing the superposition of many volume ("thick") holograms at the same site under different angles.

During hologram recording a light pattern has to be transposed into a refractive index pattern. Interfering light beams generate bright and dark regions in an electrooptic crystal. When light of suitable wavelength is chosen, charge carriers – usually electrons [5.316] – are excited in the bright regions and become mobile. The charge carriers migrate in the crystal and are subsequently trapped at new sites. By this means electronic space–charge fields are set up that give rise to a modulation of refractive index via the electrooptic effect. Figure 5.75 illustrates the photorefractive process. Index changes up to 10^{-3} are obtained. The trapped charge can be released and the field pattern erased by uniform illumination or by heating. On the one hand, this reversibility is highly desired for, e.g., erasable holographic memories or adaptive optical components; on the other hand, the problem of destructive readout arises.

Photorefractive effects have been observed in many electrooptic crystals, among them $LiNbO_3$, $LiTaO_3$, the ferroelectric perovskites $BaTiO_3$, $Ba_{1-x}Ca_xTiO_3$, $KNbO_3$, and $KTa_{1-x}Nb_xO_3$ (KTN), the tungsten–bronze-type crystals $Ba_2NaNb_5O_{15}$ and $Sr_{1-x}Ba_xNb_2O_6$ (SBN), the nonferroelectric sillenites $Bi_{12}TiO_{20}$ (BTO), $Bi_{12}SiO_{20}$ (BSO), and $Bi_{12}GeO_{20}$ (BGO), the semiconductors GaAs and InP, and others. Very early crucial influences of dopants and thermal treatments were discovered [5.317, 318]. Photorefractive crystals are of particular interest for many unique devices, such as self-pumped phase-conjugating mirrors, parametric amplifiers and oscillators as well as static and dynamic optical filters [5.319, 320]. They are also considered for holographic data storage [5.321].

In the following we describe the charge transport processes that critically determine the performance of the material. Figures of merit for the hologram storage properties will be introduced, the limits will be evaluated, and the crystals mentioned above will be discussed. Finally, fixing techniques will be briefly reviewed.

Charge Transport

The charge driving forces are well known: diffusion, bulk photovoltaic currents [5.322], and drift of charge carriers in external or pyroelectric fields [5.323] generate the space–charge field. Compensating drift currents arise, and saturation is achieved if they are as large as the driving currents. Application of large external electric fields is of interest mostly for crystals with a relatively high conductivity, and large pyroelectric fields occur only in the case of high light intensities. Thus diffusion and the bulk photovoltaic effect are of major importance.

Extrinsic or intrinsic defects occurring in more than one valence state are sources and traps of the charge carriers. Figure 5.76 shows different charge transport situations in a band diagram. The energetic positions of the levels shown correspond to the thermal energy required to excite a charge carrier from the filled level. Photon energies necessary to release electrons are typically larger because of the Franck–Condon principle. The simplest and for many applications the desired mechanism is provided by a "one-center system" with monopolar conductivity (shown in Fig. 5.76 for electron conductivity). One impurity type occurs in two different valence states, and charge is redistributed via either conduction or valence band [5.324]. An example is iron-doped $LiNbO_3$ at usual continuous-wave laser intensities. Iron ions occur only in the valence states 2+ and 3+ [5.325]. Electrons are excited from Fe^{2+} to the conduction band and trapped by Fe^{3+} elsewhere.

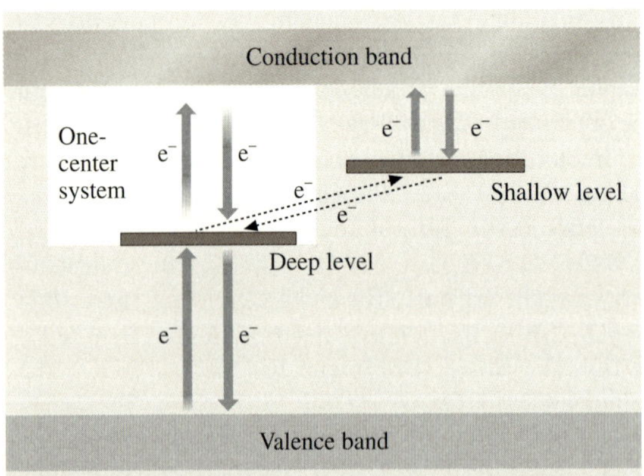

Fig. 5.76 Charge transport processes in a photorefractive crystal. *Arrows* indicate excitation and recombination of electrons (for details see text)

Electron–hole competition may complicates the situation [5.316, 326, 327]. Electrons and holes can be created at the same center, as shown in Fig. 5.76, but it is also possible that an additional independent center is present that generates holes. In any case, diffusion currents of electrons and holes compensate each other which is disadvantageous for hologram recording.

Many materials show more complicated transport mechanisms, because two or more photorefractive levels participate in the charge transport simultaneously. Additional independent centers [5.328, 329] or impurities occurring in three or more valence states [5.330] may be the origin of these additional levels. We have to distinguish between so-called deep and shallow levels. "Shallow" indicates here that charge can be released thermally from this level, which is practically impossible for a deep site. A system with the simultaneous presence of a deep and a shallow energy level (shown in Fig. 5.76 for electron conduction) has detrimental consequences for hologram recording. The shallow levels are empty in the dark because all charge carriers have been thermally excited and trapped by deep sites. Upon illumination, the shallow levels trap charge carriers from either the conduction or the valence band. As a consequence, the response time grows (smaller photoconductivity [5.331]) and, moreover, often the absorption increases, too, because the filled shallow traps absorb light effectively [5.329].

Deep and shallow levels exchange charge via conduction or valence band. Direct exchange of charge is possible, if the concentrations are large enough that different centers are located closely together. This may occur in the case of intrinsic defects [5.332]. For very high defect concentrations additional bands are created, because centers of the same type couple and exchange charge directly [5.333, 334]. More details about the charge transport may be found in a review [5.335].

The charge transport situation can be influenced by doping, thermal annealing, and selection of experimental parameters (light intensity and wavelength, crystal temperature). Thus it is possible to establish in many materials the desired one-center charge transport situation. We will concentrate in the following on a one-center system, because we are interested in the performance limits of different materials. Iron-doped lithium niobate ($LiNbO_3$: Fe) will be used as an example to describe in more detail the ongoing processes and the resulting dependences. However, most of the theoretical consideration is valid for any photorefractive crystal.

Quantitative description of the formation of the space–charge pattern under nonuniform illumination requires the solution of the Maxwell equations, together with the current and the rate equations. Generation and recombination of conduction band electrons are described by $(\mathrm{d}N_e)/(\mathrm{d}t) = +SIc_{\mathrm{Fe}^{2+}} - \gamma c_{\mathrm{Fe}^{3+}} N_e$, where N_e is the concentration of electrons in the conduction band, S is the photon absorption cross-section, I is the light intensity, $c_{\mathrm{Fe}^{2+}}$ and $c_{\mathrm{Fe}^{3+}}$ are the concentrations of Fe^{2+} and Fe^{3+} ions, and γ is the recombination coefficient. The general problem leads to a system of coupled nonlinear differential equations that cannot be solved analytically [5.324]. Reasonable approximations reveal that simple exponential evolutions govern build-up and decay of the space–charge field. Analysis yields the steady-state amplitude E_{sc} of the space–charge field [5.324]

$$E_{sc} = m\left(E_D^2 + E_{phv}^2\right)^{1/2}, \quad E_D = \frac{k_B T}{e} K,$$
$$E_{phv} = \frac{\beta\gamma}{e\mu S} c_{\mathrm{Fe}^{3+}}, \tag{5.158}$$

where m is the modulation degree ("visibility", "contrast") of the interference pattern, E_D is the diffusion field, E_{phv} is the bulk photovoltaic field, k_B is the Boltzmann constant, T is the temperature, e is the elementary charge, K is a typical spatial frequency of the hologram (2π/period length of the interference pattern), β is the bulk photovoltaic coefficient [bulk photovoltaic current density/$(c_{\mathrm{Fe}^{2+}} I)$], and μ is the charge carrier mobility. In the following we will assume a fully modulated light pattern, i. e., $m = 1$. The field E_{sc} might be reduced by space–charge limiting effects ($E_q < E_D$ or $E_q < E_{phv}$ with the space–charge limiting field $E_q = [e/(\varepsilon\varepsilon_0)](1/K)N_{eff}$, dielectric constant ε, permittivity of free space ε_0, and effective trap density N_{eff}).

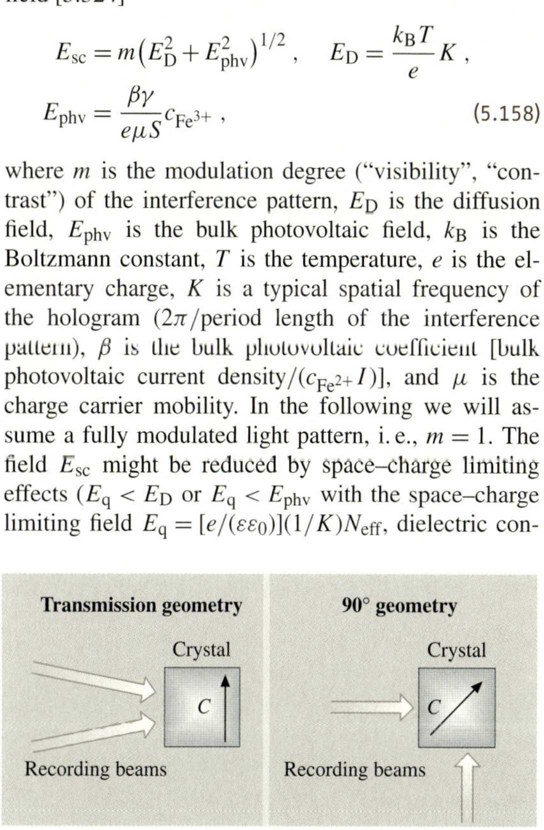

Fig. 5.77 Transmission and 90° recording geometries. The black *arrows* show the direction of the crystallographic *c*-axis

The bulk photovoltaic effect dominates in $LiNbO_3$ and $LiTaO_3$ crystals containing 0.01 wt% or more Fe, but diffusion is the major transport mechanism in many other materials. We have to distinguish between different geometries, e.g., transmission and 90° geometry (Fig. 5.77). Typical spatial frequencies in these geometries are $2\pi/0.4\,\mu\mathrm{m}^{-1}$ and $2\pi/0.15\,\mu\mathrm{m}^{-1}$, respectively. These values will be used in the following if we refer to transmission or 90° geometry. The corresponding diffusion fields at room temperature are 4 kV/cm and 10 kV/cm. Bulk photovoltaic fields in $LiNbO_3$ can reach values up to 100 kV/cm [5.322]. Larger fields cannot be created because electric breakdown takes place [5.336]. We will treat in the following the situation where the light is polarized perpendicularly to the plane of incidence (ordinary polarization), if nothing else is mentioned. In-plane polarization yields recording beams of crossed polarization in the case of the 90° geometry, and no intensity pattern is present. Extraordinary polarization can be used in the transmission geometry, and electrooptic coefficients, dynamic range and sensitivity might be higher, but extraordinarily polarized light tends to create strong holographic scattering [5.337], which is unacceptable for many applications.

The refractive index is modulated via the linear electrooptic effect [5.338], and the refractive index changes are

$$\Delta n = -\frac{1}{2} n^3 r E_{sc}, \tag{5.159}$$

where n is the refractive index and r is the effective electrooptic coefficient. Proper n and r values must be selected considering the material symmetry, the crystal cut, the polarization of light, and the direction of the space–charge field E_{sc}. For example, $LiNbO_3$ with ordinarily (o) or extraordinarily (e) polarized light and a field along the optical axis corresponds to $n = n_o$ and $r = r_{113}$, or $n = n_e$ and $r = r_{333}$, respectively.

Useful figures of merit are required in order to allow comparison and evaluation of the materials in terms of storage performance.

The *dark decay* of the refractive index modulation typically follows a monoexponential function $\Delta n(t) = \Delta n(t=0)\exp(-t/\tau_{dark})$. Large dark storage times τ_{dark} are usually desired.

Monoexponential growth of the space–charge fields and refractive index changes, $\Delta n(t) = \Delta n_s[1 - \exp(-t/\tau)]$, is characterized by the *response time* τ, which depends on light intensity according to $\tau = \varepsilon\varepsilon_0/\sigma_{ph}(I)$, where σ_{ph} is the photoconductivity.

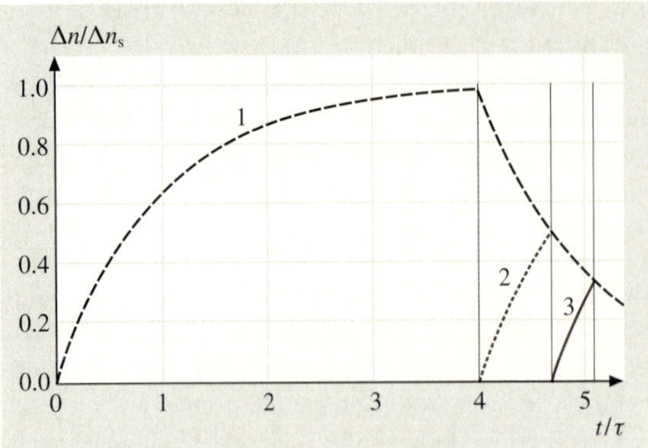

Fig. 5.78 Normalized refractive index modulation $\Delta n/\Delta n_s$ vs. normalized time t/τ for an exposure schedule that yields holograms of equal strength. Hologram Nr. 1 (*dashed curve*) is recorded into saturation. Recording of hologram Nr. 2 (*dotted curve*) erases part of hologram Nr. 1. Recording is stopped as soon as both holograms approach equal strength. At this point both holograms have both a Δn of $\Delta n_S/2$. Recording of hologram Nr. 3 (*full curve*) then erases the holograms Nr. 1 and 2 and so forth

Here Δn_S is the saturation value of the amplitude of a single hologram.

Subsequent superposition of several holograms requires a special exposure schedule, if holograms of equal efficiency are the target [5.352]. The diffraction efficiency (ratio between the intensities of diffracted and incident light) of an individual of M multiplexed holograms will be $\eta(M) = [(M\#)/M]^2$, if M is large enough to ensure that η is much smaller than 1. Figure 5.78 illustrates the required exposure schedule and derives graphically that for M holograms of equal strength the Δn per hologram is just $\Delta n_S/M$. Considering $\eta \propto \Delta n^2$ [5.353], we end up with the formula $\eta(M)$ given above.

The $M\#$ is a useful measure for the *storage density* and *capacity* of photorefractive crystals [5.354]. It is given by

$$M\# = \frac{\pi \Delta n_S d}{\lambda \cos \Theta} \exp(-1) \quad \text{for} \quad \alpha d = 2\,, \qquad (5.160)$$

if the time constants for recording and erasure are equal. Here d is the thickness of the crystal, λ is the vacuum wavelength of the light, Θ is half the angle between the recording beams in the medium, and α is the intensity absorption coefficient. Build-up of Δn requires the presence of excitable electrons, which cause absorption. Considering this trade-off, the largest $M\#$ is achieved for $\alpha d = 2$. The photon budget determines the minimum efficiency η necessary to read a hologram. Assuming, e.g., $\eta = 10^{-6}$, an $M\#$ of 10 indicates that 10 000 holograms can be multiplexed. However, the number of useful holograms can be reduced significantly owing to hologram cross-talk, light scattering, or homogeneous bulk photovoltaic fields.

There are various ways of defining the *sensitivity*. A useful measure is $W_{\eta=1\%}$, the exposure density (laser-light intensity multiplied by the recording time) required to record a hologram of 1% diffraction efficiency. Recording and erasure of the refractive index changes Δn are governed by monoexponential functions [5.324]. Using *Kogelnik's* equation [5.353] for small diffraction efficiencies $\left(\eta \approx [(\pi \Delta n d)/(\lambda \cos \Theta)]^2 \exp(-\alpha d)\right)$ we end up with

$$W_{\eta=1\%} = \frac{1}{10}\frac{\tau(I)I}{M\#} \quad \text{for} \quad \alpha d = 2\,. \qquad (5.161)$$

Table 5.32 Refractive index n_o for light polarized perpendicularly to the optical axis, electrooptic coefficient r, and dielectric coefficient ε for fields along the optical axis for different photorefractive crystals (light wavelength 514 nm; room temperature values)

Material	n_o			r (pm/V)			ε_{33}	
$LiNbO_3$	2.33		[5.339]	11	*1	[5.339]	28	[5.339]
$LiTaO_3$	2.21		[5.339]	8	*1	[5.339]	43	[5.339]
$BaTiO_3$	2.42		[5.340]	22	*	[5.341]	83	[5.341]
BCT [a]	2.47		[5.342]	36	*	[5.343]	230	[5.344]
KTN [b]	2.31		[5.345]	21	*	[5.346]	485	[5.345]
$KNbO_3$	2.33		[5.339]	40	*	[5.341]	35	[5.341]
SBN [c]	2.37		[5.347]	47	*	[5.348]	880	[5.339]
BTO [d]	2.65		[5.349]	5.75	+	[5.350]	52	[5.349]
GaAs	3.45	2	[5.351]	1.4	+2	[5.339]	12	[5.339]

[a] $Ba_{0.77}Ca_{0.23}TiO_3$, [b] $KTa_{0.52}Nb_{0.48}O_3$, [c] $Sr_{0.61}Ba_{0.39}Nb_2O_6$, [d] $Bi_{12}TiO_{20}$, * r_{113}, + r_{231}, [1] 633 nm, [2] 1150 nm

Small response times τ and large values of $M\#$ are required to achieve good sensitivities (small $W_{\eta=1\%}$ values). In addition to $W_{\eta=1\%}$ another measure for the sensitivity is also commonly used

$$S = \left.\frac{\sqrt{\eta}}{Idt}\right|_{t \ll \tau} , \qquad (5.162)$$

where η is the diffraction efficiency that the hologram reaches after the recording time t. For small recording times ($t \ll \tau$) we can write $\sqrt{\eta} = M/\# \times (t/\tau)$. From this the simple relation $S = (10dW_{\eta=1\%})^{-1}$ can be derived. For a sample of, e.g., $d = 1$ cm thickness and a $W_{\eta=1\%}$ of $10\,\mathrm{mJ/cm^2}$ we end up with $S = 10\,\mathrm{cm/J}$.

Performance Limits

There is no obvious theoretical limit to the *dark storage time*; very large values up to 10 y have been extrapolated [5.355]. For strongly doped crystals tunnelling of electrons between traps may reduce the dark storage time [5.333].

Crystals with thicknesses d up to 1 cm are available, and are supposed to have the highest *storage density*. Absorption is limited to $\alpha = 2/d = 2\,\mathrm{cm^{-1}}$, which determines the concentration $Fe^{2+} = 4.3 \times 10^{17}\,\mathrm{cm^{-3}}$ [5.325]. Thus the largest possible E_{sc} (all electrons are redistributed) is $E_{sc} = E_q = [e/(\varepsilon\varepsilon_0)](1/K)N_{eff} = 180\,\mathrm{kV/cm}$ and $66\,\mathrm{kV/cm}$ for transmission and 90° geometry, respectively (ε, Table 5.32). A field of 180 kV/cm is above the break-down threshold, but the $M\#$ describes the multiplexing properties, and the amplitude of each single hologram is well below the break-down field if many holograms are superimposed. We get $\Delta n_o = 13 \times 10^{-4}$ and $\Delta n_o = 5 \times 10^{-4}$ ((5.159), n_o and r_{113} from Table 5.32), which yields according to (5.160) the $M\#$ values 30 and 16 for transmission and 90° geometry respectively ($\lambda = 514$ nm, $\Theta = 45°$, $\alpha d = 2$). A similar estimate holds for $LiTaO_3$. However, in diffusion-controlled media the space–charge fields are lower, and limited by $E_{sc} = E_D$. The $M\#$ can be estimated using (5.158–5.160). $M\#$s range from 0.3 to 10 for different materials (Table 5.33).

Regarding *response time and sensitivity*, we have to consider the quantum limit, i.e., each incident photon moves in the optimum situation one electron to the desired position. Considering green light (photon energy 2.4 eV) and an energy difference of about 1 eV between the impurity level and the band, about 1.4 eV remains to move the electron against the space–charge field. Supposing a typical period length of 0.4 μm (transmission geometry), we have to move the electron about 0.2 μm against the field. Thus the energy of one photon is sufficient to move the electron against fields of up to 70 kV/cm. The fields are larger in the 90° geometry, but the transport lengths are smaller. From this perspective it is reasonable to treat the situation where each photon redistributes one electron. In this case we can replace $\tau(I)$ by $(\varepsilon\varepsilon_0/e)K(hc/\lambda)dE_{sc}(1/I)$, where h is the Planck constant and c is the vacuum speed of light, and get for $I = 1\,\mathrm{W/cm^2}$ and parameters of $LiNbO_3$ the result $\tau \approx 100$ ms (ε from Table 5.32, $K = 2\pi/0.4\,\mathrm{\mu m^{-1}}$, $\lambda = 514$ nm). This yields $W_{\eta=1\%} \approx 0.3\,\mathrm{mJ/cm^2}$.

Table 5.33 Various storage parameters for different photorefractive crystals (light wavelength 514 nm; room temperature values; Trans., transmission geometry; 90°, 90° geometry). The values for $M\#$ and $W_{\eta=1\%}$ are calculated from (5.158–5.161)

Material	Dark storage time		Response time (ms) @ 1 W/cm²		$M\#$ Storage capacity Trans.	90°	$W_{\eta=1\%}$ Sensitivity (mJ/cm²) Trans.	90°
$LiNbO_3$	1 a	[5.355]	3000	[5.356]	30	16	10	19
$LiTaO_3$	10 a	[5.355]	250 [1]	[5.355]	30	16	0.8	1.6
$BaTiO_3$	1 h – 1 a	[5.357]	500	[5.358]	1.5	5	33	10
BCT [a]	10 s	[5.359]	400	[5.359]	2.5	9	16	4
KTN [b]	0.6 a	[5.360]	200	[5.361]	1.2	4	16	5
$KNbO_3$	1 h – 30 d	[5.362, 363]	100	[5.364]	2.4	8	4	1.3
SBN [c]	1 h – 30 d	[5.365, 366]	100	[5.365]	3.0	10	3	1
BTO [d]	10 s	[5.367]	2 [2]	[5.368]	0.5	1.7	1.2	0.35
GaAs	1 ms	[5.369]	0.2 [3]	[5.370]	0.12	0.41	0.02	0.006

[a] $Ba_{0.77}Ca_{0.23}TiO_3$, [b] $KTa_{0.52}Nb_{0.48}O_3$, [c] $Sr_{0.61}Ba_{0.39}Nb_2O_6$, [d] $Bi_{12}TiO_{20}$, [1] 351 nm, [2] 633 nm, 3 W/cm² and $Bi_{12}SiO_{20}$, [3] 1160 nm, 0.13 W/cm²

Various Crystals

Table 5.32 summarizes important parameters, and Table 5.33 presents typical hologram recording characteristics of various inorganic photorefractive crystals.

Hologram recording performances of $LiNbO_3$ and $LiTaO_3$, of the perovskites $BaTiO_3$, $Ba_{0.77}Ca_{0.23}TiO_3$ (BCT), $KTa_{0.52}Nb_{0.48}O_3$ (KTN), and $KNbO_3$, of the tungsten–bronze structure $Sr_{0.61}Ba_{0.39}Nb_2O_6$ (SBN), of the sillenite $Bi_{12}TiO_{20}$ (BTO), and of the semiconductor GaAs are summarized in Table 5.33. *Dark storage* and *response time* may vary by orders of magnitude for the same material depending on doping (elements, concentrations), annealing (oxidation, reduction), and experimental conditions (light intensity, crystal temperature, etc.). Thus the values shown are only typical numbers. It is possible to improve one value at the expense of another. For example, the response time of $KNbO_3$ can be decreased to a few ms by electrochemical reduction [5.371], but simultaneously the dark storage time is decreased by orders of magnitude.

The largest possible *storage densities* are calculated [(5.158–5.160), Table 5.32] considering bulk photovoltaic charge transport for $LiNbO_3$ and $LiTaO_3$, and diffusion for the other materials. $LiNbO_3$ and $LiTaO_3$ have larger storage densities in the transmission geometry because more charge per interference fringe is available to build-up the space–charge field. However, the 90° geometry is advantageous for diffusion recording, because higher spatial frequencies yield larger diffusion fields. Using the response time τ and the storage capacity already determined, the *sensitivity* is estimated based on (5.161).

The number of holograms that can be multiplexed may be reduced significantly in many samples by several undesired effects. For example, light scattering reduces the visibility of the interference pattern, and therefore the fields and the dynamic range are also lowered. The required exposures $W_{\eta=1\%}$ [5.355, 372] for recording in $LiNbO_3$ and $LiTaO_3$ are one order of magnitude larger than the values presented in Table 5.33. It is also known that the measured refractive index changes in, e.g., $BaTiO_3$ and KTN crystals are often about 4 to 5 times below the calculated values [5.346, 373]. The numbers for storage capacity and sensitivity in Table 5.33 are optimum values which may be achieved in carefully selected and optimized samples.

Table 5.33 clearly shows which material is the best for which application. Crystals of the lithium-niobate family ($LiNbO_3$ and $LiTaO_3$) are the favorites for long-term high-capacity storage of holograms. The performance of the rarely used $LiTaO_3$ exceeds that of $LiNbO_3$, but the drawback in this case is that ultraviolet light is required. Good availability, excellent homogeneity, and high robustness are further advantages of $LiNbO_3$ and $LiTaO_3$. Furthermore, the processes occurring in $LiNbO_3$ are well understood, e.g., the proportionality between the space–charge field and the concentration of Fe^{3+} ions (5.158) is verified [5.356]. The crystals can be tailored for applications by optimization of the Fe concentration, by annealing and by additional doping. For example, doping with two deep independent levels which can be addressed with light of different wavelength allows optical fixing of the information [5.374]. Doping with large amounts of Mg increases the conductivity and improves the sensitivity [5.356]. More details about the specific properties of $LiNbO_3$ and other photorefractives can be found in a review [5.375].

The perovskites have shorter hologram lifetimes, but they are well suited for, e.g., holographic double-exposure interferometry [5.376]. The properties of all sillenite-type crystals $Bi_{12}TiO_{20}$ (BTO), $Bi_{12}SiO_{20}$ (BSO), and $Bi_{12}GeO_{20}$ (BGO) are very similar with one exception: The optical activity of BTO is much smaller than that of BSO and BGO [5.349], which is of importance especially for thick samples. Parameters of the semiconductor GaAs are shown for completeness. Small dark storage times and small storage capacities make GaAs unattractive for many photorefractive applications. However, perovskites, SBN crystals, crystals of the sillenite type, and photorefractive semiconductors can be of great interest for dynamic components because of the relatively large sensitivity that allows fast optical processing, fast adaption of optical components, and fast refreshment of storage systems.

The sensitivities of the materials are orders of magnitude away from the quantum performance limits, and the low sensitivity is indeed the major drawback of inorganic photorefractives. A detailed study of the properties of cerium-doped SBN revealed the reasons [5.377] which apply for many photorefractive crystals. Only a few percent of the photons release electrons; most of the light energy is immediately transferred to heat. Trapping of electrons is very efficient because of Coulomb attraction which limits the life-time and the transport lengths. It is unlikely that these problems can be completely overcome by a new inorganic photorefractive crystal.

Laser technology may solve the remaining problems of speed. The available output power of continuous-wave lasers increases rapidly. Pulsed systems become smaller and cheaper. Recording and reconstruction of

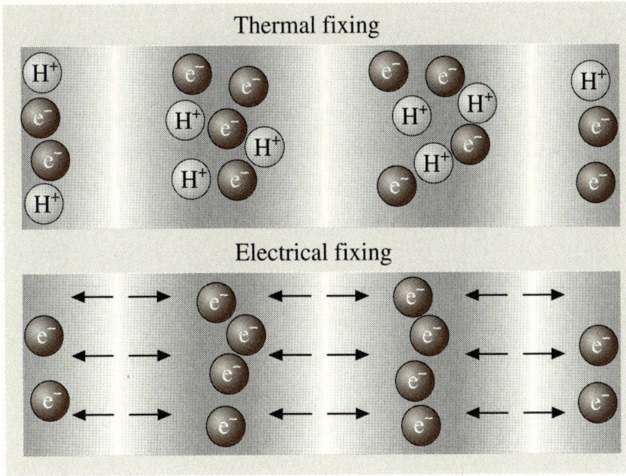

Fig. 5.79 *Thermal fixing*. During or after recording the crystal is heated. Ions become mobile and screen the electronic space–charge field. In the example shown, electrons are moved into the darker areas. This charge is compensated by H^+ ions. After cooling back to room temperature the ions are almost immobile. Now illumination moves part of the electrons back because of spatially inhomogeneous excitation, and a space–charge field appears that modulates the refractive-index through the electrooptic effect. Further erasure of the electronic pattern is impossible because drift of electrons in the space–charge field cancels all erasure currents. *Electrical fixing*. Similar to thermal fixing, but instead of heating and ions, external electrical fields and switched ferroelectric domains are involved. The *arrows* indicate the directions of the switched ferroelectric domains

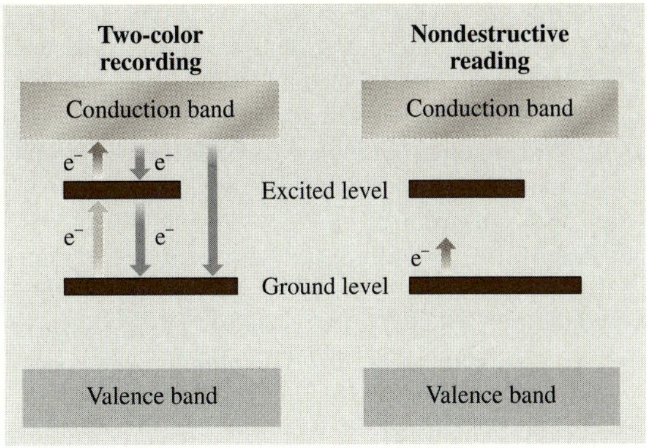

Fig. 5.80 Two-photon persistent holographic recording: The first photon excites an electron from a ground state into an excited state. This illumination can be, e.g., spatially homogeneous. Exposure with a holographic interference pattern of the recording wavelength excites in the illuminated areas electrons from the excited state into the conduction band where they move and recombine elsewhere with traps either through the excited state into the ground state or directly into the ground state. The wavelengths for the first and second transition must be carefully selected to make excitations through real levels possible. For reading, homogeneous light of the recording wavelength must be present. However, this light fails to excite the electrons from the ground state into the excited state because of insufficient photon energy. Thus no electrons are released, and the hologram is stable against further illumination

holograms in photorefractives by, e.g., femtosecond light pulses has been demonstrated [5.378]. Another approach to solve the problem of insufficient sensitivity is the usage of multiple-quantum wells [5.379, 380]. Electric space–charge fields yield large absorption changes which modulate the refractive index due to the Kramers–Kronig relations. Multiple-quantum wells are fast, and there seems to be a lot of room for further improvements.

Nondestructive Readout

The retrieval of information stored in electrooptic crystals requires homogeneous illumination and thus leads to erasure effects. In many cases asymmetries in the recording and read-out processes that are simply achieved by a reduction of the read-out light intensity are not sufficient. For this reason several methods have been developed to stabilize volume phase holograms versus the readout light.

Thermal fixing has already been discovered in $LiNbO_3$ in 1971 [5.381]. By heating, a hologram resulting from electronic and ionic charges is formed which is stable against homogeneous illumination. Protons (H^+) play the major role as mobile ions in the thermal fixing process [5.336, 382]. Electrical fixing has also been demonstrated [5.383, 384]. An external electric field slightly smaller than the coercive field has been applied to convert the space–charge pattern into a domain pattern. Figure 5.79 illustrates these fixing processes.

To preserve the possibility of desired optical erasure, the use of two-photon excitation has been proposed for hologram recording [5.385]. Then readout without erasure is possible using reduced light intensity [5.386]. Similar methods that avoid large recording intensities are based on the population of shallow levels [5.387] or of photochromic centers [5.374]. A general picture about two-photon persistent holographic recording is shown in Fig. 5.80.

There seems to exist a simple possibility for reading volume holograms nondestructively: the use of readout light of low photon energy, insufficient to excite charge carriers. However, information losses then arise because of the Bragg condition. A method has been proposed to improve the situation with the help of anisotropic diffraction [5.388]. Another solution to increase the bandwidth of the readout light has been developed [5.389, 390], based on a spectrum of spatially adapted wave vectors. Finally, the use of frequency-difference holograms has been suggested and demonstrated for nondestructive readout [5.391, 392]. The spatial frequency shift is achieved by nonlinear mixing of a hologram with a carrier frequency grating, but recording with two different wavelengths is required.

Conclusions

Good availability, excellent homogeneity, robustness, large dark storage times, large storage capacities, reversibility and the availability of efficient and convenient fixing techniques make $LiNbO_3$ and $LiTaO_3$ the favorites among the photorefractive crystals for applications that involve long-term storage of many strong holograms. This is the reason why many demonstrators are based on this material. Tungsten–bronze- and sillenite-type crystals are more sensitive, but small dark storage times and low storage capacities limit their applicability to short-term, high-resolution storage of images, e.g. for the purpose of holographic interferometry. Knowledge about the charge transport nowadays enables tailoring of the crystals considering system issues such as the desired Bragg selectivity, the light wavelength and the light intensity. Concentrations and valence states of impurities may be adjusted by doping or thermal annealing.

Holographic storage is only one of many possible applications of photorefractive crystals. Lithium niobate and lithium tantalate are, e.g., excellent to realize fixed Bragg wavelength filters for telecommunication. Materials like the perovskites barium titanate and potassium niobate, tungsten-bronze type and sillenite type photorefractive crystals as well as semiconductor photorefractive crystals can be of high interest for dynamic holography, optical signal processing, realization of adaptive optical components, for phase conjugation, and for many other advanced applications.

5.9.4 Metal Mirrors

Mirrors can be fabricated from any number of materials with an emphasis for high performance applications on near-zero expansion materials such as ULE™ [5.394] and Zerodur® [5.395]. However, metals and ceramics, primarily silicon carbide, are now being used in many applications. When designing a mirror, there are many metals from which to choose. After considering the properties with respect to the requirements, fabricability, compatibility with other components and dimensional stability must be considered. Metal and silicon carbide mirrors and support structures need to be as stable as their dielectric counterparts. After a brief discussion of dimensional stability – its causes and prevention, properties, selection criteria and recommended fabrication methods are given. More complete treatments have been given by *Paquin* [5.396, 397] and others [5.398].

Dimensional Stability

Dimensional stability of a component is determined by the amount of instability, recognized as dimensional change or distortion leading to reduction of system performance. Stability can only be achieved and controlled when the causes and sources are understood.

Instability Types. Basically there are two types of dimensional instability. First there are permanent changes such as temporal and hysteresis instabilities, and second, changes that may be permanent that happen when the environment is changed. Temporal instability is a dimensional change or distortion that occurs over time in a constant environment, while hysteresis is a change measured at constant conditions before and after changes in the environment such as temperature cycling or vibration (mechanical cycling). Thermal instability can be a change that is seen at a different temperature but returns to the original at the original temperature, or a change that depends on the rate of change of temperature and is not the same when returned to the original conditions.

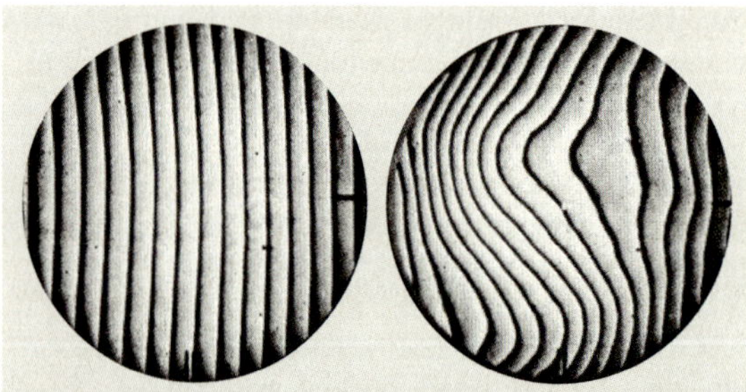

Fig. 5.81 Interferograms of an optical flat at 26 °C (*left*) and 85 °C showing thermal instability due to severe inhomogeneity of thermal expansion. The figure repeats no matter how many cycles are performed [5.393]

Instability Sources. The sources of these dimensional instabilities are the material, the fabrication process, the environment or a combination of them.

The Material

The material from which a component is fabricated is often the source of instability due to anisotropy and/or inhomogeneity. Anisotropy is direction dependent property variation and inhomogeneity is random spatial property variation looking like the material is "poorly mixed" as shown in Fig. 5.81 [5.393]. The result is thermal instability that is permanent in that the inhomogeneity and/or anisotropy is permanently in the material and cannot be changed, and repeatable in that the distortion at any temperature is repeatable time after time. To avoid this form of thermal distortion, the material selected must be homogeneous and isotropic. Single crystals are intrinsically homogeneous, but are uniformly anisotropic in some properties depending on crystal structure.

Metals can have microstructural changes from exposure to heat from any number of sources, such as processing operations or exposure in use. Amorphous metals such as as-deposited electroless nickel can crystallize and shrink, heat treatable metals such as 6061-T6 aluminum can overage and lose strength and metals with a low annealing temperature such as OFC copper can soften. Fabrication methods to eliminate or minimize these effects are discussed for each material in Sect. 5.9.4.

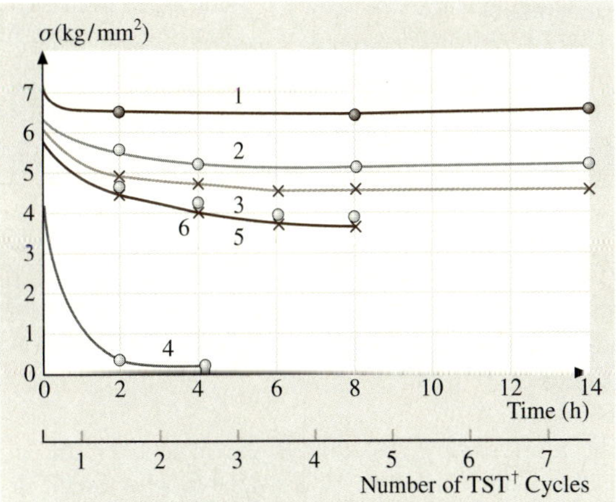

Curve	Thermal conditions
1	Isothermal; 100 °C
2	Isothermal; 190 °C
3	Isothermal; 400 °C
4	Isothermal; 600 °C
5	Cycled from −70 °C to +400 °C
6	Cycled from −196 °C to +400 °C

Fig. 5.82 Comparison of the effects of thermal cycling and isothermal exposure on stress relaxation in stressed pure beryllium. For Be, thermal cycling reduces stress more than isothermal treatment at the same upper temperature. (Thermocyclic stabilizing treatments) (After [5.399])

Another form of thermal instability is thermal mapping where the distortion is proportional to the intensity of impinging radiation and thermal properties of the material. This instability can be minimized by choosing materials that have low coefficient of linear thermal expansion (CTE or α), high thermal conductivity (k or λ) and high specific heat (c_p).

Fabrication

The most common instabilities are temporal and hysteresis. They are both caused by decrease of residual stress and associated strain. Machining and/or grinding operations induce residual stress, sub-surface damage and strain in the cut surface. This layer must be removed or the part will be unstable and subject to further surface damage from high energy laser impingement. Chemical etching can remove the damaged layer and heat treatment can reduce or eliminate the stress.

Temporal instability is the dimensional change associated with stress relaxation. The rate of decrease in stress (s) is proportional to the amount of stress present, that is,

$$(-\,\mathrm{d}s/\,\mathrm{d}t)\tau = s \tag{5.163}$$

and

$$s = s_\mathrm{o}\,\mathrm{e}^{(-t/\tau)}\,. \tag{5.164}$$

The relaxation time (τ) is the time (t) at which the stress has dropped to 0.37 of the original stress (s_o) and the decay is exponential. Similar equations can also be formulated for change in strain or surface figure. Stress relaxation is also a thermally sensitive process so that the rate at which the stress decreases, increases exponentially with increasing temperature,

$$1/\tau \propto \mathrm{e}^{(-E/kT)}\,. \tag{5.165}$$

These relationships are used in fabricating components to relieve stress, thereby increasing dimensional stability. *Lokshin* [5.399] has shown the effects of both thermal cycling and isothermal exposure on stress relaxation as shown in Fig. 5.82. Note that thermal cycling is more effective in reducing stress than isothermal treatment for the same elevated temperature.

Since thermal cycling effectively reduces residual stress it is obvious that hysteresis instabilities are caused by reduction of residual stress.

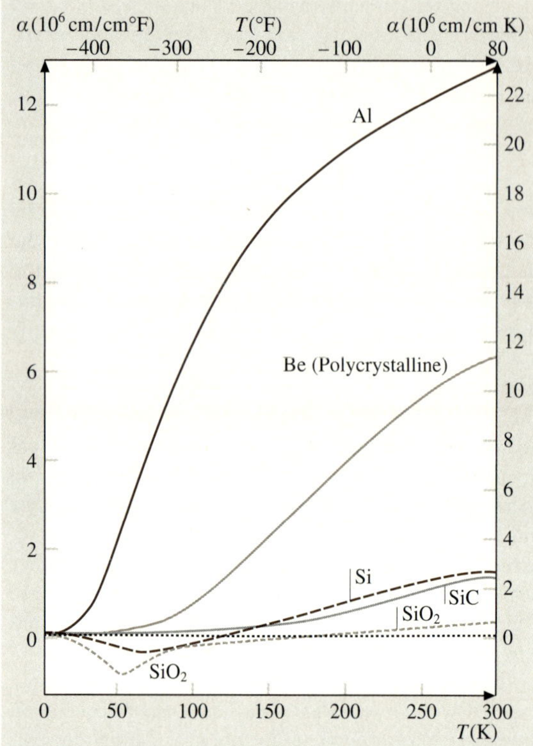

Fig. 5.83 Dependence of coefficient of linear thermal expansion on temperature for several mirror materials [5.400]

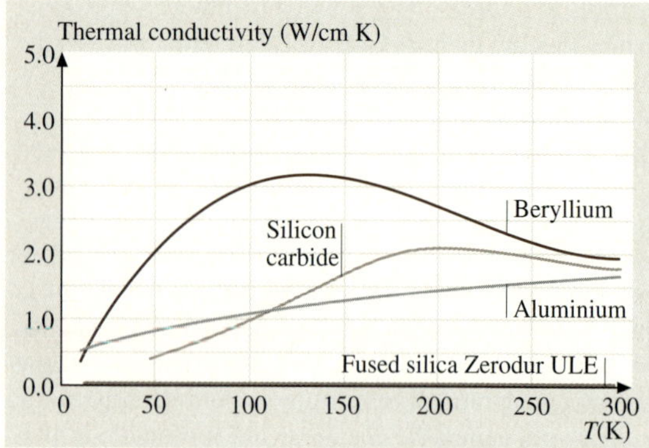

Fig. 5.84 Dependence of thermal conductivity on temperature for several mirror materials. (After [5.400])

Minimizing residual stress in fabrication of metal optics is best achieved in the fabrication process by using low stress machining/grinding techniques, chemical etching to remove damaged layers, annealing to eliminate stress, and thermal cycling to stabilize. Thermal cycling should be carried out between temperature extremes that are higher and lower than those the part will see in subsequent fabrication and assembly operations, shipping and storage, and in service.

Environment

A mirror's environment includes externally applied stresses incurred during fabrication and assembly, the attachment stresses, thermal stresses in storage and finally the use environment. To resist deformation due to these stresses, the material must be strong enough, both on a macro and micro scale and not be subject to temperatures that will cause any of the instabilities discussed above. Athermalization of systems to minimize thermal distortions is described elsewhere in this volume.

Mirror Materials. The room temperature properties of metal mirror materials are compared with those of other common mirror and structural materials in Table 5.34. The properties were obtained from many sources and are believed to be the most accurate at the time of preparation. Care should be taken in using these properties as they are typical and should not be used for design purposes. Properties vary with temperature, as shown in Figs. 5.83 and 5.84 [5.400], and the table values should not be assumed for other temperatures.

Some of the properties that may be unfamiliar are:

- Yield strength: stress at which plastic strain is 0.2% or 2 parts in 10^3
- Microyield strength: stress at which plastic strain is 1 ppm or 1 microstrain
- Steady state thermal distortion coefficient: distortion per unit of power in micrometers/watt
- Transient thermal distortion coefficient: related to the time in seconds for a distortion to reach steady state, per unit volume and unit temperature

Aluminum (Al)

The most readily available of the metals for use in mirrors and structures, Al is available as wrought or cast and heat treatable or not. Properties and heat treat temperatures are given in Table 5.35.

The most versatile alloy is the wrought, heat treatable 6061, the commonly used alloy for optical systems. The similar, but purer 6081 yields higher quality diamond turned (SPDT) surfaces. For either alloy the preferred fabrication sequence, as described in a study by NASA Goddard Space Flight Center [5.401], is: rough shape (forging preferred), solution treat, quench (in polyalkylene glycol/water solution) and age, followed by further machining, re-age, thermal cycle and then complete optical fabrication. The aircraft alloys 2024 and 7075 have

Table 5.34 Properties of mirror and structural materials

Preferred	Density ϱ (g/cm^3) Small	Young's modulus E (GPa) Large	Poisson's ratio ν Small	Yield strength σ_{YS} (MPa) Large	Microyield strength MYS (MPa) Large	Thermal expansion α $(10^{-6}/K)$ Small	Thermal conductivity k (W/m K) Large	Specific heat C_p (W s/kg K) Large	Thermal diffusivity D $(10^{-6}\,m^2/s)$ Large	Thermal distortion coefficients: Steady state α/k (μm/W) Small	Thermal distortion coefficients: Transient α/D $(s/m^2\,K)$ Small
Pyrex™7740	2.23	63	0.2	–	–	3.3	1.13	1050	0.65	2.92	5.08
Fused silica	2.19	72	0.17	–	–	0.5	1.4	750	0.85	0.36	0.59
ULE™fused silica	2.21	67	0.17	–	–	0.003	1.31	766	0.78	0.02	0.04
Zerodur®	2.53	91	0.24	–	–	0.05	1.64	821	0.77	0.03	0.07
Aluminum: 6061-T6	2.70	68	0.33	276	160.0	22.5	167.0	896	69.0	0.13	0.33
MMC: 30% SiC/Al	2.91	117	0.29	190	>150.0	12.4	123.0	870	57.0	0.10	0.22
AlBeMet®162	2.10	193	0.17	320	–	13.9	210	1506	62.3	0.06	0.22
Beryllium: O-30	1.85	303	0.043	296	35.0	11.38	216.0	1925	57.2	0.05	0.20
Beryllium: I-220-H	1.85	303	0.043	531	90.0	11.3	216.0	1925	57.2	0.05	0.20
Copper: OF	8.94	117	0.343	195	12.0	17.5	391.0	385	113.6	0.04	0.15
GlidCop®AL-15	8.75	131	0.35	352	>100.0	16.6	365.0	384	108.6	0.05	0.15
Invar 36	8.05	141	0.259	276	70.0	1.0	10.4	515	2.6	0.10	0.38
Super invar	8.13	148	0.26	230	75.0	0.3	10.5	515	2.5	0.03	0.12
Molybdenum	10.21	324	0.293	600	280.0	5.0	140.0	247	55.5	0.04	0.09
Silicon	2.33	131	0.42	–	–	2.6	156.0	710	94.3	0.02	0.03
SiC: sintered	3.16	420	0.14	–	–	2.0	190.0	700	85.9	0.02	0.04
SiC: CVD beta	3.21	465	0.21	–	–	2.4	198.0	733	84.2	0.01	0.03
SiC: RB-30% Si	2.89	330	0.24	–	–	2.6	155.0	670	80.0	0.02	0.03
Cesic®	2.65	197	0.25	–	–	2.0	125	700	71.5	0.02	0.03
SuperSiC®	2.55	214	–	–	–	–	143	–	–	–	–
Si infiltrated SuperSiC®	2.93	232	–	–	–	–	158	–	–	–	–
Stainless steel: 304	7.90	193	0.27	241	>300.0	14.7	16.2	500	4.2	0.91	3.63
Stainless steel: 416	7.70	215	0.283	950	>300	8.5	24.9	460	7.0	0.34	1.21
Stainless steel: 17-4PH	7.80	200	0.28	870	>300	10.4	22.2	460	6.2	0.47	1.68
Titanium: 6Al4V	4.43	114	0.31	830	>150.0	8.8	7.3	560	2.9	1.21	3.03
Electroless Ni: 12%	7.9	140	0.41	615	–	14.0	5.0	–	–	–	–
Electroplated Al	2.64	69	0.33	–	–	22.7	218	900	96.3	–	–

Pyrex™and ULE™are registered trademarks of Corning Glass Works, Corning, N.Y., USA; Zerodur®is a registered trademark of Schott Glaswerke, Mainz, Germany; GlidCop®is a registered trademark of OMG Americas Corp., Research Triangle Park, N.C., USA; Cesic®is a registered trademark of ECM Ingenieur-Unternehmen, Munich, Germany; SuperSiC®is a registered trademark of POCO Graphite, Inc., Decatur, TX, USA

Table 5.35 Properties of representative aluminum alloys

	Density ϱ (g/cm³)	Young's modulus E (GPa)	Yield strength σ_{YS} (MPa)	Thermal expansion[a] α (10^{-6}/K)	Thermal conductivity K (W/m K)	Anneal temperature (°C)	Solution treatment temperature (°C)	Aging temperature (4–5 h) (°C)
Wrought alloys								
1100-O	2.71	69	34	23.6	222	343	–	–
5056-O	2.64	71	152	24.1	117	415	–	–
2024-T6	2.77	72	393	23.2	151	385	500	190
6061-T6	2.70	68	276	23.6	167	415	530	170
7075-T6	2.81	72	462	23.6	130	415	470	120
Cast Alloys								
A201-T7	2.80	71	414	19.3	121	315	528	188
A356-T6	2.685	72	186	21.5	159	315	538	152
A357-T6	2.68	72	276	21.6	152	315	540	170
713	2.81	67	152	23.9	140	450[b]	N/A	120[c]
771-T7	2.823	71	372	24.7	138	–	588	140[d]

[a] 20–100 °C
[b] Stress relief temperature of 6 h and air cool yields stress free full strength component
[c] for 16 h; alternate treatment: room temperature for 21 d
[d] for 15 h

much higher strength, but are harder to stabilize and undergo dimensional changes during aging [5.402]. For large components with section thickness greater than 7.0 cm, the non-heat treatable alloys of the 5000 series are recommended, e.g., 5056 [5.403]. These alloys can be repeatedly annealed and still retain yield strengths of 150 MPa or more. For applications requiring minimal strength, 1100 is the preferred alloy for SPDT since it is 99% Al with no alloying that forms hard particles.

Cast alloys are generally more difficult to stabilize. The most commonly cast Al is A356, A357 being a cleaner version with tighter specificationss. A201 has lower alloying and when carefully cast can be SPDT finished directly. Alloys 713 (Tenzaloy) and 771 (Precedent 71A) are Al/Zn alloys that machine well and can be precision cast. Tenzaloy is usually not heat treated whereas Precedent 71A is. Precedent is the most stable of these alloys but requires special considerations in casting i. e., silicon free crucibles, chlorine purging and ceramic filters. Castings can be hot isostatic pressed (HIP'ed) to close porosity and increase properties, but is only cost effective in production.

Mirrors can be fabricated with bare surfaces finished by SPDT and/or polishing, but are usually plated with either electroless nickel (EN) that can be turned and/or polished, or electroplated Al for elimination of bimetallic thermal distortions [5.404]. Systems with Al mirrors and structures are a low cost approach for fully athermalized performance.

Beryllium (Be)
Beryllium is the lightest structural metal, with high elastic modulus, thermal conductivity and infrared reflectance, and is transparent to X-rays. The primary optical applications are space optics and structures, applications requiring low moment of inertia, and X-ray windows. All Be materials are made from powder, with HIP the preferred consolidation method [5.405]. The optical grade, O-30, is HIP'ed from spherical powder produced by gas atomization [5.406], is the most homogeneous and isotropic Be material, and is recommended for bare polished mirrors. Instrument grade I-220H, as used for the 1.1 m chopping secondary mirrors of the VLT telescopes [5.407], is a fine grained, high strength Be material made from HIP'ed impact ground powder that is often used for electroless nickel coated mirrors requiring higher strength than can be obtained with O-30 or other grades, e.g., S-200FH, S-65 or I-70H. Prior to the availability of O-30, I-70H, HIP'ed from impact ground powder, was the preferred mirror material as used for SIRTF [5.408], an all Be telescope launched in 2003.

The typical fabrication method for Be components requires machining thc shape from a HIP'ed billet,

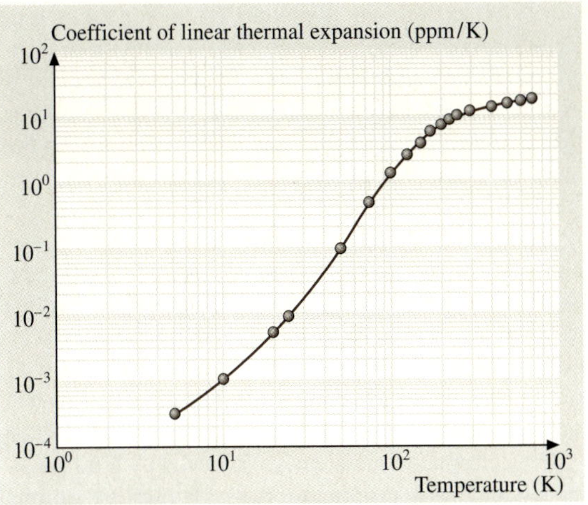

Fig. 5.85 Coefficient of linear thermal expansion of I-70H HIP'ed beryllium. Low-temperature data from *Swenson* [5.412]. High-temperature data from *Touloukian* et al. [5.413]

followed by chemical etch and anneal before finish machining. There are other near-net-shape fabrication techniques that can produce open or closed back mirrors or even integral cooling channels [5.405,406,409–411], but as yet, none of these methods has been extensively employed. Beryllium mirrors can be polished bare or with an EN coating that can be SPDT finished and/or polished.

Although Be has high CTE of 11.4 ppm/K at room temperature, it drops rapidly to 0.1 ppm/K at 50 K as shown in Fig. 5.85 [5.412,413]. Combined with its exceptionally high thermal conductivity, peaking at 150 K with 300 W/m K, as shown in Fig. 5.83, Be has exceptional cryogenic performance when bare polished and can be made very stable when careful fabrication methodology is strictly followed [5.414].

AlBeMet® 162 is an alloy of 62% Be, 38% Al, that has been used for space structural applications [5.415] but is now also being used as a mirror substrate material [5.411]. It is typically produced from atomized powder and extruded. It can be machined and welded much like Al and in mirror form must be plated to achieve an optical surface. Modifications of this alloy have been cast, but with mixed success due to porosity and segregation of the Be and Al phases.

Handling Be materials in solid form poses no health risk but inhalation of airborne Be may cause a serious lung disorder in susceptible individuals. The Brush Wellman website [5.416] contains extensive information on Be health and safety, specifications for all Be grades and a complete guide to designing with Be.

Copper
Copper has traditionally been used for high energy applications because of its high thermal conductivity. The two principal Cu alloys in use are commercially pure oxygen free copper (OFC) and dispersion strengthened GlidCop®. While OFC has the highest thermal conductivity, it loses most of its strength when brazed. It is this loss of strength that drove the selection of GlidCop®, a high strength Cu alloy with near OFC thermal conductivity, for brazed heat exchanger components [5.417,418]. The preferred grade is GlidCop® Al-15 LOX, dispersion strengthened with 0.3% Al_2O_3 in the form of 5–15 nm particles that retains > 70% of strength at temperatures up to 800 °C. Other grades have higher Al_2O_3 content but are more difficult to braze. Brazing is readily accomplished, but takes some care. Several brazing methods have been demonstrated [5.417,418] both without and with a pure Cu plated layer, and with either Ti/Cu/Ni or Au/Cu braze alloys. The former method seems simpler. Mirrors made from GlidCop® are usually EN plated and polished.

Ferrous Alloys
There are many uses for ferrous alloys in optical systems, but few mirrors are fabricated from these alloys. Among these are 17-4 PH precipitation hardening stainless and AISI 1010 mild steel [5.417]. The processing of 17-4 PH is simple when material is obtained in the solution treated and cold finished form. After machining to shape, a precipitation treatment of one hour at 480 °C with air cool to condition H900 provides a dimensionally stable substrate that can then be ground, thermal cycled, lapped and polished bare to very low roughness levels. Mild steel in the fully annealed condition has good strength and reasonably good thermal conductivity, but requires an EN coating for polishability.

Invar 36 and super invar have the lowest thermal expansion and therefore best thermal distortion performance of any metal in the temperature range of 0–80 °C. To achieve this low CTE and dimensional stability at the same time, the material must have < 0.02% carbon, low silicon and manganese and be given the three-step heat treatment developed by *Lement* et al. [5.419]. The recommended treatment is:

1. heat to 830 °C for 30 min and quench in polyalkylene glycol/water solution,
2. heat to 315 °C for 1 h and air cool,
3. heat to 95 °C for at least 48 h and air cool.

The fabrication/heat treat schedule then follows [5.417]. Shaping is be done before step 1) and finish machining after step 2), after which step 2) is repeated. Thermal cycling, not exceeding 300 °C, should also be performed in the final stages of fabrication. Step 3) should only be carried out after all fabrication steps are complete, and then that temperature, 95 °C, should never be exceeded.

Super invar has a somewhat lower expansion coefficient by virtue of a substitution of about 5% cobalt for some of the normal 36% Ni. It is fabricated and heat treated the same as regular invar 36, but undergoes a phase change at low temperatures that results in a sudden expansion and substantial increase in CTE. The phase change temperature is normally ≤ 80 °C, but if the material is to be used below about -40 °C, it should be dilatometrically tested.

Other Metals

There are many other metals that have been utilized in optical systems, chief among them being molybdenum (Mo). Many water cooled and uncooled Mo mirrors have been fabricated for use in high power laser optical systems. The preferred grade is low carbon vacuum arc cast, where fine grain material can be brazed and polished bare. Other grades (Mo 0.5% Ti and TZM) containing titanium (Ti) and/or zirconium (Zr) have fine particulates that make brazing more difficult and a fine polish impossible. Sputtered Mo coatings on Mo mirrors have produced the best finishes. For further information on Mo and its fabrication see the many papers in the proceedings of the Laser-Damage Conferences [5.420–424].

Titanium has also been used for some mirrors, primarily in high stress, low mass applications. In the 6Al4V alloy it has a very high strength-to-weight ratio. Very little has been published, but it is known to be difficult to polish well and also difficult to plate with Ni. Various schemes to improve plating adhesion include grit blasting the substrate, putting on a thin Ni or other electroplated metal strike followed by a diffusion anneal, and other proprietary methods used by various plating companies.

Polishable Coatings

Electroless nickel (EN) is the primary coating used to make metal mirror surfaces more polishable (including SPDT). It has been used on a variety of substrate materials including Al, Be, Cu, invar, steels, Ti and others. It is an alloy of Ni and phosphorous (P) and its properties depend on a number of factors including % P and post-plate bake time and temperature [5.425]. It deposits

Table 5.36 Major types of silicon carbide

SiC type	Structure/composition [a]	Density	Fabrication process	Remarks
Hot pressed	$\geq 98\%\ \alpha+$others [b]	> 98%	Powder pressed in heated dies	Simple shapes only
Hot isostatic pressed	> 98% SiC + others	> 99%	HIP encapsulated preform	Complex shapes can be formed; high equipment cost for large sizes
Chemically vapor deposited	100% β	$\geq$ 99.9%	Deposition on hot mandrel	Thin shell or plate forms and coatings
Sintered	$\geq 98\%\ \alpha+$others	> 98%	Preform with sintering aids vacuum fired	Preform machined from cold pressed billet; $\approx$ 15% shrinkage in firing; CVD SiC clad for mirrors
Reaction bonded/sintered	50–90% $\alpha+\approx 2\%\ \beta+$Si	> 99%	Prefired porous preform fired to infiltrate with Si	Complex shapes can be formed; properties depend on % Si, and SiC grain size; CVD Si or SiC clad for mirrors
Cesic®	50–60% α + 20–30% Si + 10–20% C	> 99%	Porous carbon/carbon preform fired to infiltrate with Si partly reacted to SiC	Complex shapes can be formed; properties depend on fibers and SiC:Si:C proportions; slurry clad for mirrors
SuperSiC®	100% β	$\approx$ 80%	Graphite preform converted to SiC	Complex shapes can be formed; CVD SiC clad to seal
Si-infiltrated SuperSiC®	82% β + Si	> 96%	Graphite preform converted to SiC	Complex shapes can be formed; CVD SiC clad for mirrors

Cesic® is a registered trademark of ECM Ingenieur-Unternehmen, Munich, Germany, for its short carbon fiber reinforced SiC

SuperSiC® is a registered trademark of POCO Graphite, Inc., Decatur, TX, USA for its converted graphite grades of SiC

[a] α is hexagonal SiC, β is face centered cubic SiC

[b] others are sintering aids such as Al_2O_3, Al, B, Be and C

as an amorphous coating at P contents of 6–15%, but with baking at moderate temperatures of 150–190 °C, the lower concentration alloys devitrify. Alloys above 10.5% P remain amorphous to greater than 300 °C and with a 2 h, 190 °C bake, they are ideal for SPDT. Lower concentration coatings are harder and are better for polishing. The CTE of EN varies inversely with % P from approximately 18 ppm/K at 6% P to 9 ppm/K at 13% P, all after baking.

All EN coatings have residual stresses [5.426] that come from three sources: intrinsic from deposition process, bimetallic from cooling from plating temperature ($\approx 90\,°C$) with CTE mismatch to substrate, and from shrinkage during post plate baking. The stresses can be measured [5.427] and controlled. For optical applications, great care must be taken to control the process parameters tightly to prevent variation in % P as plating progresses and to provide a cover on the bath to prevent contamination. Further details on processing are given by *Howells* and *Paquin* [5.417].

A more recent finishing method for virtual elimination of residual stress and bimetallic stress is electrolytic Al plating, a 99.9% pure Al coating called Alumiplate™ [5.428] that can be diamond turned. It was used on cryogenic Al mirrors for the Gemini Near Infrared Spectrometer with great success [5.429]. It has also been applied to Be mirrors.

Silicon Carbide

There are many types of silicon carbide (SiC) as can be seen from those shown in Table 5.34. The differences are substantial, both in the properties and in the fabrication methods as shown in Table 5.36.

Silicon carbide exists in two forms: alpha (α), the more common form with hexagonal crystal structure and anisotropic CTE and elastic properties, and beta (β), the face centered cubic form, isotropic in CTE but anisotropic in elastic properties. The α form is stable at all temperatures while β converts to α above about 1600 °C. Most properties of the two forms are within a few percent of each other. The fabrication methods described in Table 5.36 for the many SiC types yield combinations of α, β, Si and C that provide substantial differences in the usefulness and/or applicability for optical components.

Hot pressed SiC has been used for substrates in heat exchanger mirrors but while it is not normally used for optics, some applications have been reported [5.430]. HIP methods can provide small complex components, but have not been used because the extremely high temperatures and pressures needed require a large capital investment for an autoclave of practical size. Chemically vapor deposited (CVD) SiC [5.431] is extremely pure β, so it can be polished to very low roughness levels [5.432, 433], but can only be deposited as thin shells or plates. Many small mirrors have been fabricated, but attempts to produce large mirrors with lightweight back structure have been only partially successful. CVD SiC is used extensively to coat multiphase SiC types and has many other uses.

Sintered SiC has been used to fabricate a number of different optical forms and systems [5.434] including the 3.5 m diameter Herschel primary mirror of twelve brazed segments and the RocSat2 all SiC telescope with 600 mm primary. Fabrication consists of cold isostatic pressing a billet of α SiC with sintering aids, machining the part to near finish quality allowing for shrinkage ($\approx 15\%$, but known precisely) vacuum sintering and final grinding of critical surfaces. The finished surface contains approximately 2% voids of 2 μm average size and is typically coated with CVD SiC for final optical surface finishing.

Reaction bonded SiC, also known as reaction sintered or silconized SiC, is a two phase material consisting of interpenetrating networks of Si and α-SiC. Properties depend on the amount of Si that can be 10 to 50%. The process is flexible as practiced by numerous companies. The simplest form, siliconized SiC [5.435], uses a cast preform that is prefired, machined and final fired in the presence of Si to infiltrate to near full density. Higher SiC content provides improved properties, but requires more steps in the process. Carbonaceous material can be added at several points in the process, but these can lead to inhomogeneity. Preforms can be fabricated from reusable tooling to reduce cost of production parts. Meter-class mirrors have been fabricated [5.436]. Finishing can be performed on the bare two phase surface, but is usually performed on an amorphous PVD Si or CVD SiC surface [5.435, 437].

The multiphase short carbon fiber reinforced ceramic Cesic® [5.438] is an innovative approach to relatively rapid, low cost SiC optics and structures. Short fiber carbon felt is processed to graphitize it to a lightweight porous body that is readily machined. It is then infiltrated with Si that reacts with the carbon matrix and surfaces of the carbon fibers to form SiC. There is no noticeable shrinkage in the conversion process. Surfaces are not suitable for polishing, and components are slurry clad with material of the same composition but with a much finer microstructure allowing polishing to surface roughness of 2 nm rms. Mechanical properties, as shown in Table 5.34, are somewhat lower than other SiC types.

The SuperSiC® materials [5.439] are perhaps the simplest and quickest to get from raw material to finished optical component. A proprietary graphite body is machined to near finish dimensions and then converted to pure β SiC that has about 18% porosity. The porosity can be filled with Si. In either case, a coating of CVD SiC is applied to all surfaces. A complete list of properties is not yet available for this material.

5.10 Selected Data

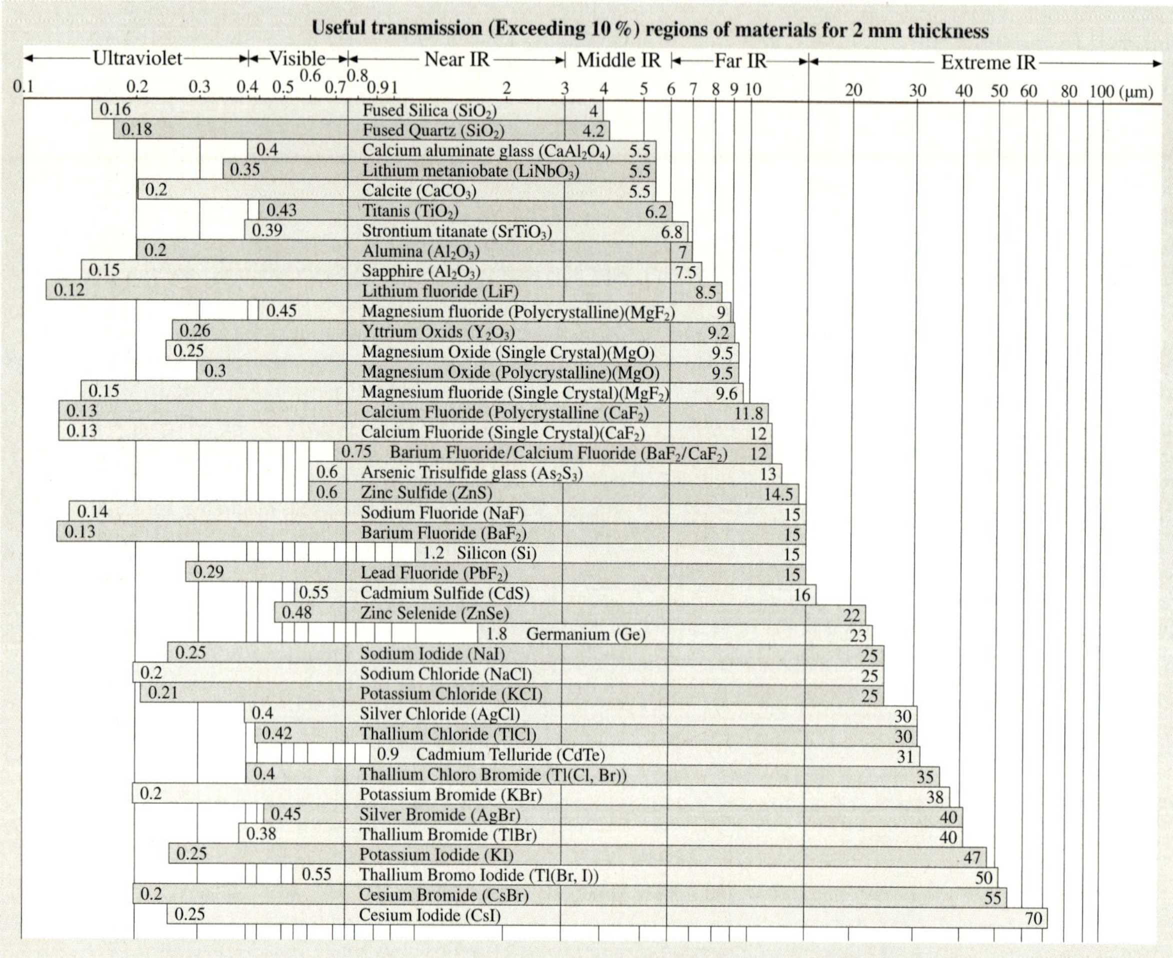

Fig. 5.86 Overview on optical material classes for different wavelength applications (After [5.440])

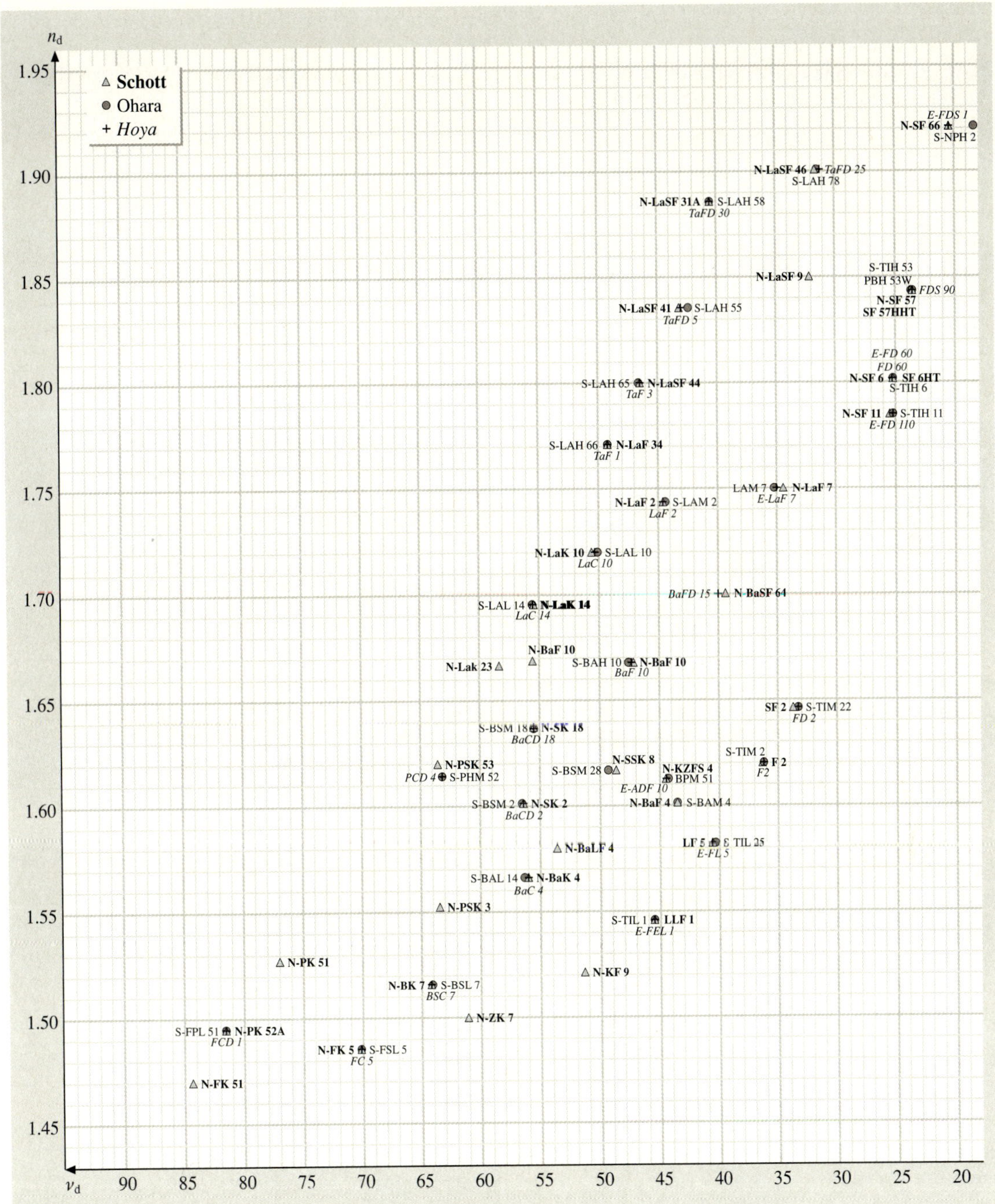

Fig. 5.87 Abbe diagram showing index of refraction versus the Abbe number for optical glasses

Part A | 5.10

Table 5.37 Commercially available optical glasses by code, glass type and manufacture (H = Hoya, O = Ohara, and S = SCHOTT) [5.441–443]

Code	Glass type	n_d	ν_d	Producer	Code	Glass type	n_d	ν_d	Producer
434950	N-FK 56	1.43430	95.0	S	571530	S-BAL 3	1.57135	53.0	O
439950	S-FPL 53	1.43875	95.0	O	573575	N-BAK 1	1.57250	57.5	S
456903	S-FPL 52	1.45600	90.3	O	573578	S-BAL 11	1.57250	57.8	O
457903	FCD 10	1.45650	90.3	H	575415	S-TIL 27	1.57501	41.5	O
487702	S-SL 5	1.48749	70.2	O	580539	N-BALF 4	1.57956	53.9	S
487704	FC 5	1.48749	70.4	H	581407	S-TIL 25	1.58144	40.7	O
487704	N-FK 5	1.48749	70.4	S	581408	N-LF 5	1.58144	40.8	S
487845	N-FK 51	1.48656	84.5	S	581409	E-FL 5	1.58144	40.9	H
497816	FCD 1	1.49700	81.6	H	581409	LF 5	1.58144	40.9	S
497816	S-FPL 51	1.49700	81.6	O	583464	BAM 3	1.58267	46.4	O
497816	N-PK 52	1.49700	81.6	S	583466	N-BAF 3	1.58272	46.6	S
498670	N-BK 10	1.49782	67.0	S	583594	S-BAL 42	1.58313	59.4	O
501564	K 10	1.50137	56.4	S	583595	BACD 12	1.58313	59.5	H
508612	N-ZK 7	1.50847	61.2	S	583595	M-BACD 12	1.58313	59.5	H
511604	K 7	1.51112	60.4	S	589612	S-BAL 35	1.58913	61.2	O
516641	S-BSL 7	1.51633	64.1	O	589613	BACD 5	1.58913	61.3	H
517522	E-CF6	1.51742	52.2	H	589613	M-BACD 5N	1.58913	61.3	H
517524	S-NSL 36	1.51742	52.4	O	589613	N-SK 5	1.58913	61.3	S
517642	BSC 7	1.51680	64.2	H	592684	N-PSK 57	1.59240	68.4	S
517642	N-BK 7	1.51680	61.2	S	593353	S-FTM 16	1.59270	35.3	O
518590	E-C 3	1.51823	59.0	H	593355	FF 5	1.59270	35.5	H
518590	S-NSL 3	1.51823	59.0	O	596392	E-F 8	1.59551	39.2	H
522595	N-K 5	1.52249	59.5	S	596392	S-TIM 8	1.59551	39.2	O
522598	S-NSL 5	1.52249	59.8	O	603380	E-F 5	1.60342	38.0	H
523515	N-KF 9	1.52346	51.5	S	603380	S-TIM 5	1.60342	38.0	O
529770	N-PK 51	1.52855	77.0	S	603380	F 5	1.60342	38.0	S
532488	E-FEL 6	1.53172	48.8	H	603606	N-SK 14	1.60311	60.6	S
532489	S-TIL 6	1.53172	48.9	O	603607	BACD 14	1.60311	60.7	H
532489	N-LLF 6	1.53169	48.9	S	603607	S-BSM 14	1.60311	60.7	O
540595	S-BAL 12	1.53996	59.5	O	603655	S-PHM 53	1.60300	65.5	O
540597	N-BAK 2	1.53996	59.7	S	606437	S-BAM 4	1.60562	43.7	O
541472	E-FEL 2	1.54072	47.2	H	606437	N-BAF 4	1.60568	43.7	S
541472	S-TIL 2	1.54072	47.2	O	606637	LBC 3N	1.60625	63.7	H
547536	N-BALF 5	1.54739	53.6	S	607567	BACD 2	1.60738	56.7	H
548458	E-FEL 1	1.54814	45.8	H	607567	N-SK 2	1.60738	56.7	S
548458	S-TIL 1	1.54814	45.8	O	607568	S-BSM 2	1.60738	56.8	O
548458	LLF 1	1.54814	45.8	S	609466	N-BAF 52	1.60863	46.6	S
548459	N-LLF 1	1.54814	45.9	S	613370	E-F 3	1.61293	37.0	H
551495	SBF 1	1.55115	49.5	H	613370	PBM 3	1.61293	37.0	O
552635	N-PSK 3	1.55232	63.5	S	613443	BPM 51	1.61340	44.3	O
558540	N-KZFS 2	1.55836	54.0	S	613443	KZFSN 4	1.61340	44.3	S
564607	S-BAL 41	1.56384	60.7	O	613444	E-ADF 10	1.61310	44.4	H
564608	BACD 11	1.56384	60.8	H	613445	N-KZFS 4	1.61336	44.5	S
564608	N-SK 11	1.56384	60.8	S	613586	BACD 4	1.61272	58.6	H
567428	E-FL 6	1.56732	42.8	H	613586	N-SK 4	1.61272	58.6	S

Table 5.37 Commercially available optical glasses by code, glass type and manufacture, cont.

Code	Glass type	n_d	ν_d	Producer	Code	Glass type	n_d	ν_d	Producer
567428	PBL 26	1.56732	42.8	O	613587	S-BSM 4	1.61272	58.7	O
569560	BAC 4	1.56883	59.0	H	614550	BSM 9	1.61405	55.0	O
569560	N-BAK 4	1.56883	56.0	S	617366	F 4	1.61659	36.6	S
569563	S-BAL 14	1.56883	56.3	O	618498	S-BSM 28	1.61772	49.8	O
569712	N-PSK 58	1.56907	71.2	S	618498	N-SSK 8	1.61773	49.8	S
571508	S-BAL 2	1.57099	50.8	O	618634	PCD 4	1.61800	63.4	H
618634	S-PHM 52	1.61800	63.4	O	667330	S-TIM 39	1.66680	33.0	O
620363	E-F 2	1.62004	36.3	H	667483	BAF 11	1.66672	48.3	H
620363	S-TIM 2	1.62004	36.3	O	667483	S-BAH 11	1.66672	48.3	O
620364	F 2	1.62004	36.4	S	670393	BAH 32	1.66998	39.3	O
620364	N-F 2	1.62005	36.4	S	670471	N-BAF 10	1.67003	47.1	S
620603	BACD 16	1.62041	60.3	H	670472	BAF 10	1.67003	47.2	H
620603	S-BSM 16	1.62041	60.3	O	670473	S-BAH 10	1.67003	47.3	O
620603	N-SK 16	1.62041	60.3	S	673321	S-TIM 25	1.67270	32.1	O
620622	ADC 1	1.62000	62.2	H	673322	E-FD 5	1.67270	32.2	H
620635	N-PSK 53	1.62014	63.5	S	673322	SF 5	1.67270	32.2	S
621603	SK 51	1.62090	60.3	S	673323	N-SF 5	1.67271	32.3	S
622532	BSM 22	1.62230	53.2	O	678506	LACL 9	1.67790	50.6	H
622533	N-SSK 2	1.62229	53.3	S	678507	S-LAL 56	1.67790	50.7	O
623569	E-BACD 10	1.62280	56.9	H	678549	LAKL 12	1.67790	54.9	S
623570	S-BSM 10	1.62280	57.0	O	678552	N-LAK 12	1.67790	55.2	S
623570	N-SK 10	1.62278	57.0	S	678553	S-LAL 12	1.67790	55.3	O
623580	N-SK 15	1.62296	58.0	S	678555	LAC 12	1.67790	55.5	H
623581	BACD 15	1.62299	58.1	H	689311	S-TIM 28	1.68893	31.1	O
623582	S-BSM 15	1.62299	58.2	O	689312	E-FD 8	1.68893	31.2	H
624471	E-BAF 8	1.62374	47.1	H	689313	N-SF 8	1.68894	31.3	S
626357	E-F 1	1.62588	35.7	H	691547	LAC 9	1.69100	54.7	H
626357	S-TIM 1	1.62588	35.7	O	691547	N-LAK 9	1.69100	54.7	S
638424	N-KZFSN 11	1.63775	42.4	S	691548	S-LAL 9	1.69100	54.8	O
639449	S-BAM 12	1.63930	44.9	O	694508	LACL 5	1.69350	50.8	H
639509	S-BSM 18	1.63854	55.4	O	694508	LAL 58	1.69350	50.8	O
639554	N-SK 18	1.63854	55.4	S	694532	M-LAC 130	1.69350	53.2	H
639555	BACD 18	1.63854	55.5	H	694532	S-LAL 13	1.69350	53.2	O
640345	S-TIM 27	1.63980	34.5	O	694533	LAC 13	1.69350	53.3	H
640346	E-FD 7	1.63980	34.6	H	694533	LAKN 13	1.69350	53.3	S
640601	S-BSM 81	1.64000	60.1	O	697485	LAFL 2	1.69700	48.5	H
640601	N-LAK 21	1.64049	60.1	S	697485	LAM 59	1.69700	48.5	O
640602	LACL 60	1.64000	60.2	H	697554	N-LAK 14	1.69680	55.4	S
648338	E-FD 2	1.64769	33.8	H	697555	LAC 14	1.69680	55.5	H
648338	S-TIM 22	1.64769	33.8	O	697555	S-LAL 14	1.69680	55.5	O
648338	SF 2	1.64769	33.8	S	699301	E-FD 15	1.69895	30.1	H
649530	E-BACED20	1.64850	53.0	H	699301	S-TIM 35	1.69895	30.1	O
649530	S-BSM 71	1.64850	53.0	O	699301	SF 15	1.69895	30.1	S
650557	LACL 2	1.65020	55.7	H	699302	N-SF 15	1.69892	30.2	S
651562	S-LAL 54	1.65100	56.2	O	700481	S-LAM 51	1.70000	48.1	O
651669	N-LAK 22	1.65113	55.9	S	702402	BAFD 15	1.70200	40.2	H

Table 5.37 Commercially available optical glasses by code, glass type and manufacture, cont.

Code	Glass type	n_d	ν_d	Producer	Code	Glass type	n_d	ν_d	Producer
652450	N-BAF 51	1.65224	45.0	S	702412	BAFD 7	1.70154	41.2	H
652584	LAC 7	1.65160	58.4	H	702412	S-BAH 27	1.70154	41.2	O
652585	S-LAL 7	1.65160	58.5	O	704394	N-BASF 64	1.70400	39.4	S
652585	N-LAK 7	1.65160	58.5	S	706302	N-SF 64	1.70591	30.2	S
654396	E-ADF 50	1.65412	39.6	H	713538	N-LAK 8	1.71300	53.8	S
654396	KZFSN 5	1.65412	39.6	S	713539	LAC 8	1.71300	53.9	H
654397	BPH 5	1.65412	39.7	O	713539	S-LAL 8	1.71300	53.9	O
658509	BACED 5	1.65844	50.9	H	717295	E-FD 1	1.71736	29.5	H
658509	S-BSM 25	1.65844	50.9	O	717295	PBH 1	1.71736	29.5	O
658509	N-SSK 5	1.65844	50.9	S	717295	SF 1	1.71736	29.5	S
664360	N-BASF 2	1.66446	36.0	S	717296	N-SF 1	1.71736	29.6	S
717479	S-LAM 3	1.71700	47.9	O	762401	S-LAM 55	1.76200	40.1	O
717480	LAF 3	1.71700	48.0	H	773496	TAF 1	1.77250	49.6	H
717480	N-LAF 3	1.71700	48.0	S	773496	S-LAH 66	1.77250	49.6	O
720347	BPH 8	1.72047	34.7	O	773496	N-LAF 34	1.77250	49.6	S
720420	LAM 58	1.72000	42.0	O	785257	FD 110	1.78472	25.7	H
720437	S-LAM 52	1.72000	43.7	O	785257	S-TIH 11	1.78472	25.7	O
720460	LAM 61	1.72000	46.0	O	785258	SF 11	1.78472	25.8	S
720502	S-LAL 10	1.72000	50.2	O	785261	FDS 30	1.78470	26.1	H
720503	LAC 10	1.72000	50.3	H	785261	N-SF 56	1.78470	26.1	S
720506	N-LAK 10	1.72003	50.6	S	785261	SF 56A	1.78470	26.1	S
722292	S-TIH 18	1.72151	29.2	O	785263	S-TIH 23	1.78470	26.3	O
723380	BAFD 8	1.72342	38.0	H	786439	NBFD 11	1.78590	43.9	H
723380	S-BAH 18	1.72342	38.0	O	786441	N-LAF 33	1.78582	44.1	S
724381	BASF 51	1.72373	38.1	S	786442	S-LAH 51	1.78590	44.2	O
728283	E-FD 10	1.72825	28.3	H	788474	S-LAH 64	1.78800	47.4	O
728284	SF 10	1.72825	28.4	S	788475	TAF 4	1.78800	47.5	H
728285	S-TIH 10	1.72825	28.5	O	788475	N-LAF 21	1.78800	47.5	S
728285	N-SF 10	1.72828	28.5	S	795454	TAF 2	1.79450	45.4	H
729547	TAC 8	1.72916	54.7	H	795455	N-LAF 32	1.79457	45.5	S
729547	S-LAL 18	1.72906	54.7	O	800422	S-LAH 52	1.79952	42.2	O
731405	M-LAF 81	1.73077	40.5	H	800423	NBFD 12	1.79950	42.3	H
734511	TAC 4	1.73400	51.1	H	801350	S-LAM 66	1.80100	35.0	O
734515	S-LAL 59	1.73400	51.5	O	801351	N-LASF 45	1.80100	35.1	S
740283	PBH 3	1.74000	28.3	O	804396	S-LAH 63	1.80440	39.6	O
741278	E-FD 13	1.74077	27.8	H	804465	TAF 3	1.80420	46.5	H
741278	S-TIH 13	1.74077	27.8	O	804465	N-LASF 44	1.80420	46.5	S
741526	TAC 2	1.74100	52.6	H	804466	S-LAH 65	1.80400	46.6	O
741527	S-LAL 61	1.74100	52.7	O	805254	S-TIH 6	1.80518	25.4	O
743492	NBF 1	1.74330	49.2	H	805254	N-SF 6	1.80518	25.4	S
743493	M-NBF 1	1.74330	49.3	H	805254	SF 6	1.80518	25.4	S
743493	S-LAM 60	1.74320	49.3	O	805255	FD 60	1.80518	25.5	H
744448	S-LAM 2	1.74400	44.8	O	805396	NBFD 3	1.80450	39.6	H
744449	LAF 2	1.74400	4.9	H	806333	NBFD 15	1.80610	33.3	H
744449	N-LAF 2	1.74397	44.9	S	806406	N-LASF 43	1.80610	40.6	S
749348	N-LAF 7	1.74950	34.8	S	806407	M-NBFD 130	1.80610	40.7	H

Table 5.37 Commercially available optical glasses by code, glass type and manufacture, cont.

Code	Glass type	n_d	ν_d	Producer	Code	Glass type	n_d	ν_d	Producer
750350	E-LAF 7	1.74950	35.0	H	806407	NBFD 13	1.80610	40.7	H
750350	LAFN 7	1.74950	35.0	S	806409	S-LAH 53	1.80610	40.9	O
750353	LAM 7	1.74950	35.3	O	808228	S-NPH 1	1.80809	22.8	O
752251	FF 8	1.75211	25.1	H	815370	M-NBFD 82	1.81474	37.0	H
754524	N-LAK 33	1.75398	52.4	S	816445	TAFD 10	1.81550	44.5	H
755274	N-SF 4	1.75513	27.4	S	816466	TAF 5	1.81600	46.6	H
755275	E-FD 4	1.75520	27.5	H	816466	S-LAH 59	1.81600	46.6	O
755275	S-TIH 4	1.75520	27.5	O	834372	S-LAH 60	1.83400	37.2	O
755276	SF 4	1.75520	27.6	S	834373	NBFD 10	1.83400	37.3	H
755523	TAC 6	1.75500	52.3	H	834373	N-LASF 40	1.83404	37.3	S
755523	S-YGH 51	1.75500	52.3	O	835427	S-LAH 55	1.83481	42.7	O
757477	NBF 2	1.75700	47.7	H	835430	TAFD 5	1.83500	43.0	H
757478	S-LAM 54	1.75700	47.8	O	835431	N-LASF 41	1.83501	43.1	S
762265	S-TIH 14	1.76182	26.5	O	847236	SFL 57	1.84666	23.6	S
762265	SF 14	1.76182	26.5	S	847238	FDS 90	1.84666	23.8	H
762266	FD 140	1.76182	26.6	H	847238	S-TIH 53	1.84666	23.8	O
847238	SF 57	1.84666	23.8	S	850322	LASFN 9	1.85025	32.2	S
881410	N-LASF 31	1.88067	41.0	S	883408	TAFD 30	1.88300	40.8	H
883408	S-LAH 58	1.88300	40.8	O	901315	LAH 78	1.90135	31.5	O
923209	E-FDS 1	1.92286	20.9	H	923209	SF 66	1.92286	20.9	S
923213	PBH 71	1.92286	21.3	O	1022291	N-LASF 35	2.02204	29.1	S

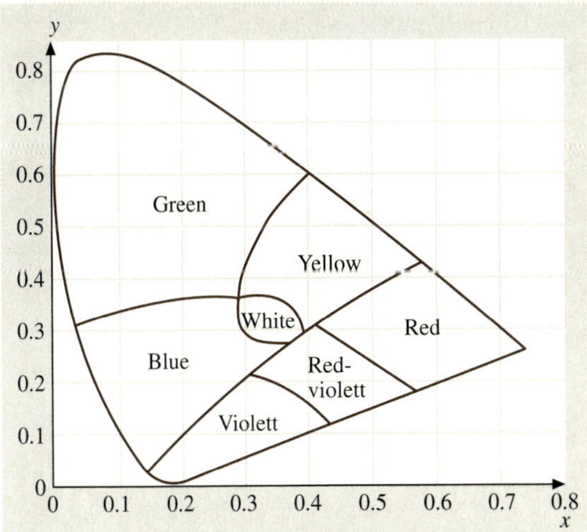

Fig. 5.88 Color chart (color coordinates) with some glass types and illuminant D65 (standardized light source)

References

5.1 J. Jackson: *Classical Electrodynamics* (Wiley, New York 1975)

5.2 L. D. Landau, E. M. Lifshitz: *The Classical Theory of Fields* (Addison Wesley, New York 1971)

5.3 Ch. Kittel: *Introduction to Solid State Physics* (Oldenbourg, Munich 1988)

5.4 H. Haken: *Quantum Field Theory of Solids* (Teubner, Stuttgart 1973)

5.5 A. Sommerfeld: *Optics: Lectures on Theoretical Physics*, Vol. IV (Academic, New York 1954)

5.6 A. C. Hardy, F. H. Perrin: *The Principles of Optics* (McGraw-Hill, New York 1932)

5.7 C. S. Williams, O. A. Becklund: *Optics: A Short Course for Engineers and Scientists* (Wiley, New York 1972)

5.8 W. G. Driscoll, W. Vaughan (Eds.): *Handbook of Optics* (McGraw-Hill, New York 1978)

5.9 D. Halliday, R. Resnick, J. Walker: *Fundamentals of Physics*, 4th edn. (Wiley, New York 1993)

5.10 J. H. Simmons, K. S. Potter: *Optical Materials* (Academic, New York 2000)

5.11 E. Hecht: *Optics*, 4th edn. (Addison Wesley, New York 2002)

5.12 S. Singh: Refractive index measurement and its applications, Phys. Scr. **65**(2), 167–180 (2002)

5.13 S. Tominaga, N. Tanaka: Refractive index estimation and color image rendering, Pattern Recognition Lett. **24**(11), 1703–1713 (2003)

5.14 J. E. Shelby: *Introduction to Glass Science and Technology* (The Royal Society of Chemistry, Cambridge 1997)

5.15 H. Bach, N. Neurorth (Eds.): *The Properties of Optical Glass* (Springer, Berlin, Heidelberg 1998)

5.16 J. V. Hughes: A new precision refractometer, J. Rev. Sci. Instrum. **18**, 234 (1941)

5.17 A. B. P. Lever: *Inorganic Electronic Spectroscopy* (Elsevier, New York 1968)

5.18 C. R. Bamford: *Colour Generation and Control in Glass* (Elsevier, New York 1977)

5.19 A. Paul: *Chemistry of Glass* (Chapman Hall, New York 1990)

5.20 D. C. Harris, M. D. Bertolucci: *Symmetry, Spectroscopy: An Introduction to Vibrational, Electronic Spectroscopy* (Dover, New York 1978)

5.21 M. Born, E. Wolf: *Principles of Optics* (Cambridge Univ. Press, Cambridge 1999) p. 218

5.22 I. B. Bersuker: *Electronic Structure and Properties of Transition Metal Compounds: Introduction to the Theory* (Wiley, New York 1996)

5.23 J. E. Shelby: *Introduction to Glass Science and Technology* (The Royal Society of Chemistry, Cambridge 1997)

5.24 B. Douglas, D. McDaniel, J. Alexander: *Concepts and Models of Inorganic Chemistry*, 3rd edn. (Wiley, New York 1994)

5.25 C. K. Jorgensen: Spectroscopy of transition-group complexes, Adv. Chem. Phys. **5**, 33–145 (1963)

5.26 N. W. Ashcroft, N. D. Mermin: *Solid State Physics* (Thomson, Stamford 2000)

5.27 J. D. Jackson: *Classical Electrodynamics* (Wiley, New York 1975)

5.28 J. F. Nye: *Physical Properties of Crystals* (Oxford Univ. Press, Oxford 1957)

5.29 M. Born, E. Wolf: *Principles of Optics* (Pergamon, Oxford 1986)

5.30 H. Vogel: *Gerthsen Physik* (Springer, Berlin, Heidelberg 1997)

5.31 B. E. A. Saleh, M. C. Teich: *Fundamentals of Photonics* (Wiley, New York 1991)

5.32 Y. R. Shen: *The Principles of Nonlinear Optics* (Wiley, New York 1984)

5.33 W. Nie: Optical nonlinearity: Phenomena, applications and materials, Adv. Mater. **5**, 520–545 (1993) and cited papers

5.34 C. F. Klingshirn: *Semiconductor Optics* (Springer, Berlin, Heidelberg 1997)

5.35 N. F. Borrelli, D. W. Hall: *Optical Properties of Glass*, ed. by Uhlmann, Kreidl (American Ceramic Society, Westerville 1991) pp. 87–124

5.36 P. Chakraborty: Metal nanoclusters in glasses as nonlinear photonic materials, J. Mater. Sci. **33**, 2235–2249 (1998)

5.37 E. M. Vogel, M. J. Weber, D. M. Krol: Nonlinear optical phenomena in glass, Phys. Chem. Glasses **32**, 231–250 (1991) and cited papers

5.38 E. M. Vogel: Glasses as nonlinear photonic materials, J. Am. Ceram. Soc. **72**, 719–724 (1989)

5.39 I. Kang, T. D. Krauss, F. W. Wise, B. G. Aitken, N. F. Borrelli: Femtosecond measurement of enhanced optical nonlinearities of sulfide glasses and heavy-metal-doped oxide glasses, J. Opt. Soc. Am. B **12**, 2053–2059 (1995)

5.40 H. Tanaka, K. Kashima, K. Hirao, N. Soga, A. Mito, H. Nasu: Second harmonic generation in poled tellurite glasses, Jpn. J. Appl. Phys **32**(2), 843 (1993)

5.41 K. Hirao, T. Mitsuyu, J. Si, J. Qiu: *Active Glasses for Photonic Devices* (Springer, Berlin, Heidelberg 2001)

5.42 K. Hirao: *Active Glass Project NEWS*, 99.8 Final Rep. No. 3 (Japan Science and Technology Agency, Kawaguchi City 1999)

5.43 H. Nasu, J. D. MacKenzie: Nonlinear optical properties of glass and glass or gel based compositions, Opt. Eng. **26**, 102–106 (1987)

5.44 J. S. Aitchinson, J. D. Prohaska, E. M. Vogel: The nonlinear optical properties of glass, Met. Mater. Proc. **8**, 277–290 (1996)

5.45 B. G. Potter, M. B. Sinclair: Photosensitive, rare earth doped ceramics for optical sensing, J. Electroceram. **2**, 295–308 (1998)

5.46 F. Ishh, T. Sawatari, A. Odajima: NMR study of chain orientation in drawn poly(vinylidene fluoride) films I. The effects of poling on the double orientation distribtion function, Jpn. J. Appl. Phys **27**, 1047–1053 (1988)

5.47 M. G. Kuzyk, K. D. Singer, H. E. Zahn, L. A. King: Second-order nonliniear-optical tensor properties of poled films under stress, J. Opt. Soc. Am. B **6**, 742 (1989)

5.48 S. Miyata, H. Sasabe: *Poled Polymers, their Applications to SHG and EO Devices* (Taylor Francis, New York 1997)

5.49 W. Koechner: *Solid-State Laser Engineering* (Springer, Berlin, Heidelberg 1976)

5.50 H. C. Van de Hulst: *Light Scattering by Small Particles* (Dover, New York 1981)

5.51 G. Mie: *Beiträge zur Optik trüber Medien, speziell kolloidaler Metallösungen*, Ann. Physik, Vol. 25, ed. by E. Grüneisen (Barth, Leipzig 1908) pp. 377–445

5.52 A. Ishimaru: *Wave Propagation and Scattering in Random Media* (IEEE, Piscataway 1997)

5.53 C. Bohren, D. Huffman: *Absorption, Scattering of Light by Small Particles* (Wiley, New York 1983)

5.54 M. Born, E. Wolf: *Principles of Optics*, 7th edn. (Cambridge Univ. Press, Cambridge 1999)

5.55 P. Debye, A. M. Bueche: J. Appl. Phys. **20**, 518 (1949)

5.56 C. S. Johnson, D. A. Gabriel: *Laser Light Scattering* (Dover, New York 1981)

5.57 A. Dogariu: Volume Scattering in Random Media. In: *Handbook of Optics, Volume III, 2nd edition, Part 1 Classical optics*, ed. by M. Bass (McGraw-Hill, New York 2001) Chap. 3

5.58 P. D. Kaplan, A. D. Dinsmore, A. G. Yodh: Diffuse-transmission spectroscopy: A structural probe of opaque colloidal mixtures, Phys. Rev. E **50**, 4827 (1994)

5.59 J. M. Elson, H. E. Bennett, J. M. Bennett: Scattering from optical surfaces. In: *Applied Optics and Optical Engineering*, Vol. 7, ed. by R. Shannon, J. Wyant (Academic, New York 1979) Chap. 7

5.60 A. Ishimaru: *Wave Propagation and Scattering in Random Media* (IEEE, Piscataway 1997)

5.61 A. Duparré: Surface characterization techniques for determining the root-mean-square roughness and power spectral densities of optical components, Appl. Opt. **41**, 154–171 (2002)

5.62 M. Born, E. Wolf: *Principles of Optics*, 7th edn. (Cambridge Univ. Press, Cambridge 1999)

5.63 S. Musikant: *Optical Materials* (Dekker, New York 1985)

5.64 H. Bach, N. Neuroth: *The Properties of Optical Glass*, Schott Series on Glass and Glass Ceramics (Springer, Berlin, Heidelberg 1995)

5.65 W. H. Armistead, S. D. Stookey: Photochromic silicate glasses sentized by silver halides, Science **144**, 150–158 (1964)

5.66 F.-T. Lentes: Refractive index and dispersion. In: *The Properties of Optical Glass*, ed. by H. Bach, N. Neuroth (Springer, Berlin, Heidelberg 1998) pp. 19–57

5.67 H. G. Pfänder: Optische Gläser und Brillengläser. In: *Schott Glaslexikon* (mvg, Landsberg 1997) pp. 129–142

5.68 B. Jaschke: Einfluß der Wissenschaften. In: *Glasherstellung* (Deutsches Museum, Munich 1997) pp. 79–86

5.69 H. G. Pfänder: Geschichte des Glases. In: *Schott Glaslexikon* (mvg, Landsberg 1997) pp. 13–23

5.70 N. J. Kreidl: Optical properties. In: *Handbook of Glass Manufacture*, ed. by F. V. Tooley (Books for Industry, New York 1974) pp. 957–997

5.71 B. Jaschke: Neuerungen in der Glasherstellung. In: *Glasherstellung* (Deutsches Museum, Munich 1997) pp. 65–78

5.72 M. K. Th. Clement: The chemical composition of optical glasses and its influence on the optical properties. In: *The Properties of Optical Glass*, ed. by H. Bach, N. Neuroth (Springer, Berlin, Heidelberg 1998) pp. 58–81

5.73 G. F. Brewster, N. J. Kreidl, T. G. Pett: Lanthanum and barium in glass-forming system, J. Soc. Glass Technol. **31**, 153–169 (1947)

5.74 W. Jahn: Mehrstoffsysteme zum Aufbau Optischer Gläser, Glastechn. Ber. **43**, 107–120 (1961)

5.75 W. H. Zachariasen: Die Struktur der Gläser, Glastechn. Ber. **11**, 120–123 (1933)

5.76 W. H. Zachariasen: The atomic arrangement in glass, J. Am. Chem. Soc. **54**, 3841–3851 (1932)

5.77 B. E. Warren, A. D. Loring: X-ray diffraction study of the structure of soda–silica glasses, J. Am. Ceram. Soc. **18**, 269–276 (1935)

5.78 K. Fajans, N. J. Kreidl: Stability of lead glasses, polarization of ions, J. Am. Ceram. Soc. **31**, 105–114 (1948)

5.79 J. E. Stanworth: On the structure of glass, J. Soc. Glass Technol. **32**, 154–172 (1948)

5.80 A. M. Bishay, P. Askalani: Properties of antimony glasses in relation to structure, C. R. VIIe Congrès International du Verre, Bruxelles 1965 (Maison d'Edition, Marcinelle 1965) Part 1, Nr. 24

5.81 S. Wolff, U. Kolberg: Environmental friendly optical glasses. In: *The Properties of Optical Glass*, ed. by H. Bach, N. Neuroth (Springer, Berlin, Heidelberg 1998) pp. 144–148

5.82 H. Bach, N. Neuroth: *The Properties of Optical Glass*, Schott Ser. Glass (Springer, Berlin, Heidelberg 1998)

5.83 C. R. Bamford: *Colour Generation and Control in Glass* (Elsevier, Amsterdam 1977)

5.84 W. A. Weyl: *Coloured Glasses* (Society of Glass Technology, Sheffield 1951)

5.85 W. Vogel: *Glaschemie* (Springer, Berlin, Heidelberg 1992) pp. 51–313

5.86 K. Nassau: The varied causes of colour in glass, Mat. Res. Soc. Symp. Proc. **61**, 427–439 (1986)

5.87 T. Bates: *Ligand field theory, absorption spectra of transition-metal ions in glasses*, Vol. 2, ed. by

J. D. Mackenzie (Butterworth, London 1962) pp. 195–254
5.88 L. E. Orgel: *An Introduction to Transition Metal Chemistry and Ligand Field Theory* (Methuen, London 1960)
5.89 H. L. Schläfer, G. Gliemann: *Einführung in die Ligandenfeldtheorie* (Akademische Verlagsgesellschaft, Frankfurt/Main 1967)
5.90 C. J. Ballhausen: *Introduction to Ligand Field Theory* (McGraw-Hill, New York 1962)
5.91 C. K. Jorgensen: *Absorption Spectra and Chemical Bonding in Complexes* (Pergamon, Oxford 1962)
5.92 A. Bishay, A. Kinawi: Absorption spectra of iron in phosphate glasses and ligand field theory, Phys. Non-Cryst. Solids **2**, 589–605 (1965)
5.93 C. F. Bohren: *Absorption and Scattering of Light by Small Particles* (Wiley, New York 1983)
5.94 M. Kerker: *The Scattering of Light and Other Electromagnetic Radiation* (Academic, New York 1969)
5.95 M. Born, E. Wolf: *Principles of Optics* (Pergamon, Oxford 1986)
5.96 H. P. Rooksby: The colour of selenium ruby glasses, J. Soc. Glass Technol. **16**, 171–181 (1932)
5.97 G. Schmidt: Optische Untersuchungen an Selenrubingläsern, Silikattechnik **14**, 12–18 (1963) (in German)
5.98 A. Rehfeld, R. Katzschmann: Farbbildung und Kinetik von Steilkanten-Anlaufgläsern, Silikattechnik **29**, 298–302 (1978) (in German)
5.99 G. Walter, R. Kranold, U. Lemke: Small angle X-ray scattering characterization of inorganic glasses, Makromol. Chem. Makromol. Symp. **15**, 361–372 (1988)
5.100 T. Yanagawa, Y. Sasaki, H. Nakano: Quantum size effects, observation of microcrystallites in coloured filter glasses, Appl. Phys. Lett. **54**, 1495–1497 (1989)
5.101 J. L. Emmett, W. F. Krupke, J. B. Trenholme: *The Future Development of High-Power Solid State Laser Systems* (Lawrence Livermore National Laboratory, Livermore 1982)
5.102 A. H. Clauer: New life for laser shock processing, Ind. Laser Rev., 7–9 (March 1996)
5.103 E. Snitzer: Optical laser action of Nd^{3+} in a barium crown glass, Phys. Rev. Lett. **7**, 444–446 (1961)
5.104 S. E. Stokowski, R. A. Saroyan, M. J. Weber: *Laser Glass Nd-Doped Glass Spectroscopic and Physical Properties* (Lawrence Livermore National Laboratory, Livermore 1981) pp. 1–9 M-95, Rev. 2
5.105 J. H. Pitts: Modeling laser damage caused by platinum inclusions in laser glass, laser induced damage in optical materials, Techn. Dig. Boulder Damage Symp., Boulder 1985 (NBS, Boulder 1986) 537–542
5.106 J. H. Campbell, E. P. Wallerstein, J. S. Hayden, D. L. Sapak, D. E. Warrington, A. J. Marker III, H. Toratani, H. Meissner, S. Nakajima, T. Izumitani: *Elimination of Platinum Inclusions in Phosphate Laser Glasses* (Lawrence Livermore National Laboratory, Livermore 1989)
5.107 R. Wood: *Laser Damage in Optical Materials* (IOP Publishing Limited, Bristol, Great Britain 1986)
5.108 D. C. Brown: *High-Peak-Power Nd:Glass Laser Systems* (Springer, Berlin, Heidelberg 1981) Chap. 6
5.109 D. H. Roach, A. R. Cooper: The effect of etch depth on strength of indented soda lime glass rods. In: *Strength of Inorganic Glass*, ed. by C. R. Kurkjian (Plenum, New York 1985) pp. 185–195
5.110 W. C. LaCourse: The strength of glass. In: *Introduction to Glass Science*, ed. by L. D. Pye, H. J. Stevensand W. C. LaCourse (Plenum, New York 1972) pp. 451–512
5.111 P. W. McMillan: *Glass-Ceramics* (Academic, London 1979) p. 285
5.112 W. Holand, G. H. Beall: *Glass-Ceramic Technology* (The American Ceramic Society, Westerville 2002) p. 372
5.113 W. Pannhorst: Low expansion glass ceramics – current developments, Proc. 7th Int. Otto Schott Coll., Jena, Germany 2002, ed. by C. Ruessel, G. Volksch (Verlag der Deutschen Glastechnischen Gesellschaft, Frankfurt 2002) 78
5.114 S. Cramer von Clausbruch, M. Schweiger, W. Hoeland, V. Rheinberger: Effect of ZnO on the crystallization, microstructure, properties of glass-ceramics in the $SiO_2-Li_2O-ZnO-K_2O-P_2O_5$ system, Glastech. Ber.-Glass Sci. Technol. **74**, 223 (2001)
5.115 M. M. Layton, J. W. Smith: Pyroelectric response in transparent ferroelectric glass, J. Am. Ceram. Soc. **58**, 435 (1975)
5.116 T. Komatsu, J. Onuma, H. G. Kim, J. R. Kim: Formation of Rb-doped crystalline phase with second harmonic generation in transparent $K_2O-Nb_2O_5-TeO_2$ glass ceramics, J. Mater. Sci. Lett. **15**, 2130 (1996)
5.117 F. C. Guinhos, P. C. Nobrega, P. A. Santa-Cruz: Compositional dependence of up-conversion process in $Tm^{3+}-Yb^{3+}$ codoped oxyfluoride glasses and glass-ceramics, J. Alloy. Compd. **323**, 358 (2001)
5.118 M. Secu, S. Schweizer, J. M. Spaeth, A. Edgar, G. V. M. Williams, U. Rieser: Photostimulated luminescence from a fluorobromozirconate glass-ceramic and the effect of crystallite size and phase, J. Phys. Cond. Matter **15**, 1097 (2003)
5.119 G. H. Beall: Glass-ceramics for photonic applications, Glass Sci. Technol.-Glastech. Ber. **73**, 3 (2000)
5.120 J. A. Tangeman, B. L. Phillips, A. Navrotsky, J. K. R. Weber, A. D. Hixson, T. S. Key: Vitreous forsterite (Mg_2SiO_4): Synthesis, structure, thermochemistry, Geophys. Res. Lett. **28**, 2517 (2001)
5.121 P. A. Tick, N. F. Borrelli, I. M. Reaney: The relationship between structure and transparency in glass-ceramic materials, Opt. Mater. **15**, 81 (2000)
5.122 K. Shioya, T. Komatsu, H. G. Kim, R. Sato, K. Matusita: Optical properties of transparent glass-ceramics in $K_2O-Nb_2O_5-TeO_2$ glasses, J. Non-Cryst. Solids **189**, 16 (1995)

5.123 H. A. Miska: Aerospace and military applications. In: *Ceramics and Glasses*, ed. by S. R. Lampman, M. S. Woods, T. B. Zorc (ASM Int., Materials Park 1991) p. 1016
5.124 H. Bach (Ed.): *Low Thermal Expansion Glass Ceramics* (Springer, Berlin, Heidelberg 1995) p. 223
5.125 P. F. James: Kinetics of crystal nucleation in lithium silicate glasses, Phys. Chem. Glasses **15**, 95 (1974)
5.126 A. C. Lasaga: *Kinetic Theory in the Earth Sciences* (Princeton Univ. Press, Princeton 1998) p. 811
5.127 P. W. McMillan, G. Partridge: Dielectric properties of certain $ZnO-Al_2O_3-SiO_2$ glass-ceramics, J. Mater. Sci. **7**, 847 (1972)
5.128 A. P. Tomsia, J. A. Pask, R. E. Loehman: Glass/metal and glass-ceramic/metal seals. In: *Ceramics and Glasses*, ed. by S. R. Lampman, M. S. Woods, T. B. Zorc (ASM Int., Materials Park 1991) p. 493
5.129 Ohara Corporation: *Glass-Ceramic Substrates for Planar Light Circuits (SA-02)*, Commercial Literature (Ohara Corporation, Japan 2003)
5.130 W. D. Kingery, H. K. Bowen, D. R. Uhlmann: *Introduction to Ceramics*, 2nd edn. (Wiley, New York 1976) p. 1032
5.131 R. B. Roberts, R. J. Tainsh, G. K. White: Thermal properties of Zerodur at low temperatures, Cryogenics **22**, 566 (1982)
5.132 S. W. Freiman, L. L. Hench: Effect of crystallization on mechanical properties of Li_2O-SiO_2 glass-ceramics, J. Am. Ceram. Soc. **55**, 86 (1972)
5.133 S. S. Bayya, J. S. Sanghera, I. D. Aggarwal, J. A. Wojcik: Infrared transparent germanate glass-ceramics, J. Am. Ceram. Soc. **85**, 3114 (2002)
5.134 Y. Tada, F. Kawano, M. Kon, N. Matsumoto, K. Asaoka: Influence of crystallization on strength and color of castable glass-ceramics containing two crystals, Biomed. Mater. Eng. **5**, 233 (1995)
5.135 B. R. Lawn, T. R. Wilshaw, T. I. Barry, R. Morrell: Hertzian fracture of glass ceramics, J. Mater. Sci. **10**, 179 (1975)
5.136 R. Morena, K. Niihara, D. P. H. Hasselman: Effect of crystallites on surface damage and fracture-behavior of a glass-ceramic, J. Am. Ceram. Soc. **66**, 673 (1983)
5.137 T. J. Hill, J. J. Melchosky, K. J. Anusavice: Fractal analysis of toughening behavior in $3BaO.5SiO_2$ glass-ceramics, J. Am. Ceram. Soc. **83**, 545(2000)
5.138 D. Mittleman (Ed.): *Sensing with Terahertz Radiation* (Springer, Berlin, Heidelberg 2003) p. 337
5.139 A. J. Moulson, J. M. Herbert: *Electroceramics* (Chapman Hall, London 1990) p. 464
5.140 A. Herczog: Microcrystalline $BaTiO_3$ by crystallization from glass, J. Am. Ceram. Soc. **47**, 107 (1964)
5.141 N. F. Borrelli: Electrooptic effect in transparent niobate glass-ceramic systems, J. Appl. Phys. **38**, 4243 (1967)
5.142 O. P. Thakur, D. Kumar, O. M. Parkash, L. Pandey: Crystallization, microstructure development and dielectric behaviour of glass ceramics in the system SrO center dot $TiO_2-2SiO_{(2)}$ center dot $B_2O_3-La_2O_3$, J. Mater. Sci. **37**, 2597 (2002)
5.143 F. Agullo-Lopez, J. M. Cabera, F. Agullo-Rueda: *Electrooptics: Phenomena, Materials and Applications* (Academic, London 1994) p. 345
5.144 H. Bach, N. Neuroth (Eds.): *The Properties of Optical Glass* (Springer, Berlin, Heidelberg 1995) p. 410
5.145 K. J. Anusavice, N.-Z. Zhang, J. E. Moorhead: Influence of P_2O_5, $AgNO_3$, $FeCl_3$ on color, translucency of lithia-based glass-ceramics, Dent. Mater. **10**, 230 (1994)
5.146 V. M. Khomenko, K. Langer, R. Wirth: On the influence of wavelength-dependent light scattering on the UV-VIS absorption spectra of oxygen-based minerals: A study on silicate glass ceramics as model substances, Phys. Chem. Min. **30**, 98 (2003)
5.147 G. H. Beall, D. A. Duke: Transparent glass ceramics, J. Mater. Sci. **4**, 340 (1969)
5.148 C. F. Bohren, D. R. Huffman: *Absorption and Scattering of Light by Small Particles* (Wiley, New York 1983) p. 530
5.149 F. M. Modest: *Radiative Heat Transfer* (McGraw-Hill, New York 1993) p. 832
5.150 R. W. Hopper: Stochastic-theory of scattering from idealized spinodal structures 2 scattering in general and for the basic late stage model, J. Non-Cryst. Solids **70**, 111 (1985)
5.151 R. Menzel: *Photonics: Linear and Nonlinear Interactions of Laser Light and Matter* (Springer, Berlin, Heidelberg 2001) p. 873
5.152 S. Hendy: Light scattering in transparent glass ceramics, Appl. Phys. Lett. **81**, 1171 (2002)
5.153 W. Pannhorst: Zerodur – a low thermal expansion glass ceramic for optical precision applications. In: *Low Thermal Expansion Glass Ceramics*, ed. by H. Bach (Springer, Berlin, Heidelberg 1995) p. 107
5.154 P. Pernice, A. Aronne, V. N. Sigaev, P. D. Sarkisov, V. I. Molev, S. Y. Stefanovich: Crystallization behavior of potassium niobium silicate glasses, J. Am. Ceram. Soc. **82**, 3447 (1999)
5.155 G. S. Murugan, K. B. R. Varma: Dielectric, linear and non-linear optical properties of lithium borate-bismuth tungstate glasses and glass-ceramics, J. Non-Cryst. Solids **279**, 1 (2001)
5.156 N. F. Borrelli, M. M. Layton: Dielectric and optical properties of transparent ferroelectric glass-ceramic systems, J. Non-Cryst. Solids **6**, 197 (1971)
5.157 A. Halliyal, A. S. Bhalla, R. E. Newnham, L. E. Cross: Glass-ceramics for piezoelectric and pyroelectric devices. In: *Glass and Glass Ceramics*, ed. by M. H. Lewis (Chapman Hall, London 1989) p. 273
5.158 Y.-H. Kao, Y. Hu, H. Zheng, J. D. Mackenzie, K. Perry, G. Bourhill, J. W. Perry: Second harmonic generation in transparent barium borate glass-ceramics, J. Non-Cryst. Solids **167**, 247 (1994)

5.159 N. F. Borrelli, A. Herczog, R. D. Maurer: Electro-optic effect of ferroelectric microcrystals in a glass matrix, Appl. Phys. Lett. **7**, 117 (1965)
5.160 N. F. Borrelli: Electrooptic effect in transparent niobate glass, J. Appl. Phys. **38**, 4243 (1967)
5.161 S. Ito, T. Kokubo, M. Tashiro: Transparency of $LiTaO_3-SiO_2-Al_2O_3$ glass-ceramics in relation to their microstructure, J. Mater. Sci. **13**, 930 (1978)
5.162 A. Herczog: Phase distribution and transparency in glass-ceramics based on a study of the sodium niobate–silica system, J. Am. Ceram. Soc. **73**, 2743 (1990)
5.163 T. Kokubo, M. Tashiro: Fabrication of transparent lead titanate(IV) glass, Bull. Inst. Chem. Res. Kyoto Univ. **54**, 301 (1976)
5.164 Y. Fujimoto, Y. Benino, T. Fujiwara, R. Sato, T. Komatsu: Transparent surface and bulk crystallized glasses with lanthanide tellurite nanocrystals, J. Ceram. Soc. Jpn. **109**, 466 (2001)
5.165 H. G. Kim, T. Komatsu, K. Shioya, K. Matusita, K. Tanaka, K. Hirao: Transparent tellurite-based glass-ceramics with second harmonic generation, J. Non-Cryst. Solids **208**, 303 (1996)
5.166 R. T. Hart, M. A. Anspach, B. J. Kraft, J. M. Zaleski, J. W. Zwanziger, P. J. DeSanto, B. Stein, J. Jacob, P. Thiyagarajan: Optical implications of crystallite symmetry and structure in potassium niobate tellurite glass ceramics, Chem. Mater. **14**, 4422 (2002)
5.167 N. S. Prasad, K. B. R. Varma: Nanocrystallization of $SrBi_2Nb_2O_9$ from glasses in the system $Li_2B_4O_7-SrO-Bi_2O_3-Nb_2O_5$, Mater. Sci. Eng. B **90**, 246 (2002)
5.168 G. S. Murugan, K. B. R. Varma: Characterization of lithium borate–bismuth tungstate glasses, glass-ceramics by impedance spectroscopy, Solid State Ionics **139**, 105 (2001)
5.169 V. N. Sigaev, P. Pernice, A. Aronne, O. V. Akimova, S. Y. Stefanovich, A. Scaglione: $KTiOPO_4$ precipitation from potassium titanium phosphate glasses, producing second harmonic generation, J. Non-Cryst. Solids **292**, 59 (2001)
5.170 Y. Balci, M. Ceylan, M. E. Yakinci: An investigation on the activation energy and the enthalpy of the primary crystallization of glass-ceramic Bi-rich BSCCOHTc superconductors, Mater. Sci. Eng. B **86**, 83 (2001)
5.171 A. Edgar, S. Schweizer, S. Assmann, J. M. Spaeth, P. J. Newman, D. R. MacFarlane: Photoluminescence and crystallization in europium-doped fluorobromozirconate glass-ceramics, J. Non-Cryst. Solids **284**, 237 (2001)
5.172 G. Muller, N. Neuroth: Glass ceramic as an active laser material, US Patent 3843551 (Jenaer Glaswerk Schott and Gen., USA 1974)
5.173 A. Lempicki, M. Edwards, G. H. Beall, D. Hall, L. J. Andrews: *Transparent Glass Ceramics* (Optical Society of America, Arlington, VA 1985) Laser Prospects, in Topical Meeting on Tunable Solid State Lasers
5.174 R. Reisfeld, C. K. Jorgensen: Excited-states of chromium(III) in translucent glass-ceramics as prospective laser materials, Struct. Bond. **69**, 63 (1988)
5.175 R. Reisfeld: Potential uses of chromium(III)-doped transparent glass ceramics in tunable lasers and luminescent solar concentrators, Mater. Sci. Engin. **71**, 375 (1985)
5.176 P. A. Tick, N. F. Borrelli, L. K. Cornelius, M. A. Newhouse: Transparent glass ceramics for 1300 nm, J. Appl. Phys. **78**, 6367 (1995)
5.177 A. M. Malyarevich, I. A. Denisov, K. V. Yumashev, O. S. Dymshits, A. A. Zhilin, U. Kang: Cobalt-doped transparent glass ceramic as a saturable absorber Q switch for Er: Glass lasers, Appl. Opt. **40**, 4322 (2001)
5.178 C. F. Rapp, J. Chrysochoos: Neodymium-doped glass-ceramic laser material, J. Mater. Sci. Lett. **7**, 1090 (1972)
5.179 G. Muller, N. Neuroth: Glass ceramic – A new laser host material, J. Appl. Phys. **44**, 2315 (1973)
5.180 C. F. Rapp, J. Chrysochoos: Fluorescence lifetimes of Neodymium-doped glasses and glass ceramics, J. Phys. Chem. **77**, 1016 (1973)
5.181 Y. H. Wang, J. Ohwaki: New transparent vitroceramics codoped with Er^{3+}, Yb^{3+} for efficient frequency up-conversion, Appl. Phys. Lett. **63**, 3268 (1993)
5.182 P. A. Tick, N. F. Borrelli, L. K. Cornelius, M. A. Newhouse: Transparent glass ceramics for 1300 nm amplifier applications, J. Appl. Phys. **78**, 6367 (1995)
5.183 M. J. Dejneka: The luminescence and structure of novel transparent oxyfluoride glass-ceramics, J. Non-Cryst. Solids **239**, 149 (1998)
5.184 Y. Kawamoto, R. Kanno, J. Qiu: Upconversion luminescence of Er^{3+} in transparent $SiO_2-PbF_2-ErF_3$ glass ceramics, J. Mater. Sci. **33**, 63 (1998)
5.185 M. Takahashi, M. Izuki, R. Kanno, Y. Kawamoto: Up-conversion characteristics of Er^{3+} in transparent oxyfluoride glass-ceramics, J. Appl. Phys. **83**, 3920 (1998)
5.186 M. Mortier, G. Patriarche: Structural characterisation of transparent oxyfluoride glass ceramics, J. Mater. Sci. **35**, 4849 (2000)
5.187 L. L. Kukkonen, I. M. Reaney, D. Furniss, M. G. Pellatt, A. B. Seddon: Nucleation and crystallisation of transparent, erbium III-doped, oxyfluoride glass-ceramics, J. Non-Cryst. Solids **290**, 25 (2001)
5.188 M. Mortier, A. Monteville, G. Patriarche, G. Maze, F. Auzel: New progresses in transparent rare-earth doped glass-ceramics, Opt. Mater. **16**, 255 (2001)
5.189 M. Mortier: Between glass and crystal: Glass-ceramics, a new way for optical materials, Philos. Mag. B **82**, 745 (2002)
5.190 V. K. Tikhomirov, D. Furniss, A. B. Seddon, I. M. Reaney, M. Beggiora, M. Ferrari, M. Montagna, R. Rolli: Fabrication and characterization of nanoscale,

Er^{3+}-doped, ultratransparent oxy-fluoride glass ceramics, Appl. Phys. Lett. **81**, 1937 (2002)

5.191 J. Mendez-Ramos, V. Lavin, I. R. Martin, U. R. Rodriguez-Mendoza, V. D. Rodriguez, A. D. Lozano-Gorrin, P. Nunez: Optical properties of rare earth doped transparent oxyfluoride glass ceramics, Radiat. Eff. Defects Solids **158**, 457 (2003)

5.192 A. M. Malyarevich, I. A. Denisov, K. V. Yumashev, O. S. Dymshits, A. A. Zhilin: Optical absorption, luminescence study of cobalt-doped magnesium aluminosilicate glass ceramics, J. Opt. Soc. Am. B **19**, 1815 (2002)

5.193 A. M. Malyarevich, I. A. Denisov, Y. V. Volk, K. V. Yumashev, O. S. Dymshits, A. A. Zhilin: Nanosized glass ceramics doped with transition metal ions: Nonlinear spectroscopy and possible laser applications, J. Alloy. Compd. **341**, 247 (2002)

5.194 C. Y. Li, Q. Su, S. B. Wang: Multi-color long-lasting phosphorescence in Mn^{2+}-doped $ZnO-B_2O_3-SiO_2$ glass ceramics, Mater. Res. Bull. **37**, 1443 (2002)

5.195 G. H. Beall: Glass-ceramics: Recent developments and applications. In: *Nucleation and Crystallization in Liquids and Glasses*, ed. by M. C. Weinberg (The American Ceramic Society, Westerville 1993)

5.196 G. H. Beall, L. R. Pinckney: Nanophase glass ceramics, J. Am. Ceram. Soc. **82**, 5 (1999)

5.197 I. Mitra, M. J. Davis, J. Alkemper, R. Mueller, H. Kohlmann, L. Aschke, E. Moersen, S. Ritter, H. Hack, W. Pannhorst: Thermal expansion behavior of proposed EUVL substrate materials, SPIE Proc. 4688–Emerging Lithographic Technologies VI, Santa Clara, CA 2002, ed. by R. L. Englestad (SPIE, Belham 2002) 462–468

5.198 N. Reisert: Application and machining of Zerodur for optical purposes, Proc. SPIE **1400**, 171 (1991)

5.199 L. N. Allen: Progress in ion figuring large optics, SPIE Proc. 2428–Laser-induced Damage in Optical Materials, Soileau 1995, ed. by H. E. Bennett, A. H. Guenther, M. R. Kozlowski, B. E. Newnam, M. J. Soileau (SPIE, Belham 1995) 237–247

5.200 L. Noethe: Active optics in modern large optical telescopes, Prog. Opt. **43**, 1 (2002)

5.201 C. A. Haniff: High angular resolution studies of stellar atmospheres, IAU Symp.: Galaxies and their Constituents at the Highest Angular Resolutions, (2001) 288–295

5.202 S. D. Stookey: Catalyzed crystallization of glass in theory and practice, Ind. Eng. Chem. **51**, 805 (1959)

5.203 B. Rother, A. Mucha: Transparent glass-ceramic coatings: Property distribution on 3D parts, Surf. Coat. Technol. **124**, 128 (2000)

5.204 J. Hirao, T. Mitsuyu, J. Si, J. Qiu: *Active Glasses for Photonic Devices* (Springer, Berlin, Heidelberg 2001)

5.205 W. Nie: Optical nonlinearity phenomena, applications, materials, Adv. Mater. **5**, 520–545 (1993) and cited papers

5.206 E. M. Vogel, M. J. Weber, D. M. Krol: Nonlinear optical phenomena in glass, Phys. Chem. Glasses **32**, 231–250 (1991) and cited papers

5.207 A. J. Hayden, A. J. Marker: Glass as a nonlinear optical material, SPIE Proc. **1327**, 132–144 (1990)

5.208 R. Shechter, E. Millul, Y. Amitai, A. A. Friesem, V. Weiss: Hybrid polymer-on-glass integrated optical diffractive structures for wavelength discrimination, Opt. Mater. **17**, 165–167 (2001)

5.209 Transparencies and oral communication J. Hayden: SCHOTT North America

5.210 Frost and Sullivan (study): Photonic Materials (2000)

5.211 V. G. Dmitriev, G. G. Gurzadya, D. N Nikogosyan: *Handbook of Nonlinear Optical Crystals* (Springer, Berlin, Heidelberg 1991)

5.212 P. Chakraborty: Metal nanoclusters in glasses as nonlinear photonic materials, J. Mater. Sci. **33**, 2235–2249 (1998)

5.213 H. R. Xia, J. H. Zou, H. C. Chen, D. L. Sun: Photorefractive properties of Co-doped potassium sodium strontium barium niobate crystals, Cryst. Res. Technol. **34**, 403–407 (1999)

5.214 H. R. Xia, C. J. Wang, H. C. Chen, X. L. Lu: Photorefractive properties of manganese-modified potassium sodium strontium barium niobate crystals, Phys. Rev. B **55**, 1292–1294 (1997)

5.215 S. Zhang, Z. Cheng, H. Chen: A new oxyborate crystal $GdCa_4O(BO_3)_3$: Defects and optical properties, Defect Diff. Forum **186–187**, 79–106 (2000)

5.216 V. Berger: Photonic crystals for nonlinear optical frequency conversion, Confined Photon Systems **531**, 366–392 (1999)Lect. Notes Phys.

5.217 H. Nasu, J. D. MacKenzie: Nonlinear optical properties of glas and glas or gel based compositions, Opt. Eng. **26**, 102–106 (1987)

5.218 F. Kajzar, J. Swalen: *Organic Thin Films for Waveguiding Nonlinear Optics*, Springer Ser. Adv. Nonlin. Opt., Vol. 3 (Springer, Berlin, Heidelberg 1996)

5.219 Hirao Active Glass Project NEWS, '99.8 Final Rep. No. 3 (1999)

5.220 Hirao Active Glass Project NEWS, '97.7 Final Rep. No. 2 (1997)

5.221 G. I. Stegeman, E. M. Wright, N. Finlayson, R. Zanoni, C. T. Seaton: Third order nonlinear integrated optics, J. Lightwave Technol. **6**, 953–967 (1988) and cited papers

5.222 R.W. Bryant: *Nonlinear Optical Materials: New Technologies, Applications, Markets* (Business and Coorporation Inc., Norwalk 1989)

5.223 Y. Kondo, Y. Kuroiwa, N. Sugimoto, T. Manabe, S. Ito, T. Tokizaki, A. Nakamura: Third-order optical nonlinearities of CuCl-doped glasses in a near resonance region, J. Non-Cryst. Solids **6**, 90–94 (1996)

5.224 D. M. Krol, D. J. DiGiovanni, W. Pleibel, R. H. Stolen: Observation of resonant enhancement of photoinduced second-harmonic generation in Tm-doped aluminosilicate glass fibers, Opt. Lett **18**, 1220 (1993)

5.225 N.M. Lawandy, R.L. MacDonald: Optically encoded phase-matched second-harmonic generation in semiconductor-microcrystallite-doped glasses, J. Opt. Soc. Am. B **8**, 1307 (1991)

5.226 E.M. Dianov, D.S. Starodubov, A.A. Izyneev: Photoinduced second-harmonic generation in fibers doped with rare-earth ions, Opt. Lett **19**, 936 (1994)

5.227 E.M. Dianov, L.S. Kornienko, V.I. Stupina, P.V. Chernov: Correlation of defect centers with photoinduced second-harmonic generation in Er- and Sm-doped aluminosilicate fibers, Opt. Lett. **20**, 1253–1255 (1995)

5.228 D.M. Krol, D.J. DiGiovanni, K.T. Nelson, W. Pleibel, R.H. Stolen: Observation of resonant enhancement of photoinduced second-harmonic generation in Tm-doped aluminosilicate glass fibers, Opt. Lett. **18**, 1220–1222 (1993)

5.229 I.S. Fogel, J.M. Bendickson, M.D. Tocci, M.J. Bloemer, M. Scalora, C.M. Bowden, J.P. Dowling: Spontaneous emission and nonlinear effects in photonic bandgap materials, Pure Appl. Opt. **7**, 393–407 (1998)

5.230 F. Oulette, K.O. Hill, D.C. Johnson: Enhancement of second-harmonic generation in optical fibres by hydrogen heat treatment, Appl. Phys. Lett **54**, 1086 (1989)

5.231 J.S. Aitchinson, J.D. Prohaska, E.M. Vogel: The nonlinear optical properties of glass, Met. Mater. Proc. **8**, 277–290 (1996)

5.232 I. Kang, T.D. Krauss, F.W. Wise, B.G. Aitken, N.F. Borrelli: Femtosecond measurement of enhanced optical nonlinearities of sulfide glasses and heavy-metal-doped oxide glasses, J. Opt. Soc. Am. B **12**, 2053–2059 (1995)

5.233 J. Fu, H. Yatsuda: New families of glasses based on Bi_2O_3, Phys. Chem. Glasses **36**, 211–215 (1995)

5.234 N.F. Borrelli, B.G. Aitken, M.A. Newhouse, D.W. Hall: Electric field induced birefringence properties of high-refractive-index glasses exhibiting large Kerr nonlinearities, J. Appl. Phys. **70**(5), 2774–2779 (1991)

5.235 S. Santran, L. Canioni, T. Cardinal, E. Fargin, G. Le Flem, C. Rouyer, L. Sarger: Precise and absolute measurements of the complex third-order optical susceptibility, Proc. SPIE **4106**, 349–359 (2000)

5.236 E. Fargin, A. Berthereau, T. Cardinal, G. Le Flem, L. Ducasse, L. Canioni, P. Segonds, L. Sarger, A. Ducasse: Optical nonlinearity in oxide glasses, J. Non-Cryst. Solids **203**, 96–101 (1996)

5.237 T. Cardinal, K.A. Richardson, H. Shim, A. Schulte, R. Beatty, K. Le Foulgoc, C. Meneghini, J.F. Viens, A. Villeneuve: Nonlinear optical properties of chalcogenide doped glasses in the system As–S–Se, J. Non-Cryst. Solids **256,257**, 353–360 (1999)

5.238 B. Speit, K.E. Remitz, N. Neuroth: Semiconductor doped glass as a nonlinear material, SPIE Proc. **1361**, 1128–1131 (1990)

5.239 B. Danielzik, K. Nattermann, D. von der Linde: Nanosecond optical pulse shaping in cadmium-sulfide–selenide glasses, Appl. Phys. B **38**, 31–36 (1985)

5.240 H. Inouye, K. Tanaka, I. Tanahashi, K. Hirao: Ultrafast dynamics of nonequilibrium electrons in a gold nanoparticle system, Phys. Rev. B **57**, 11334 (1998)

5.241 H. Inouye, K. Tanaka, I. Tanahashi, T. Hattori, H. Nakatsuka: Ultrafast optical switching in a silver nanoparticle system, Jpn. J. Appl. Phys. **39**, 5132–5133 (2000)

5.242 J. Khaled, T. Fujiwara, A.J. Ikushima: Optimization of second-order nonlinearity in UV-poled silica glass, Opt. Mater. **17**, 275–278 (2001)

5.243 B.G. Broome: *The Design of Plastic Optical Systems*, SPIE Short Course **384** (SPIE, San Diego 2001)

5.244 X. Ning, R.T. Hebert (Eds.): *Design, Fabrication and Application of Precision Plastic Optics*, SPIE Proc. **2600** (SPIE, Bellingham 1995)

5.245 Corning Precision Lens: *The Handbook of Plastic Optics*, 2nd edn. (Corning Precision Lens, OwensCorning 2000)

5.246 D.J. Butler: Plastic optics challenge glass, Photon. Spectra (May 2000) p. 168–171

5.247 Optical Coating Laboratory Inc. "An Introduction to the Design, Manufacture and Application of Plastic Optics" Technical Information, 2001

5.248 A. Ning: Plastic versus glass optics: Factors to consider, SPIE SC **384** (SPIE, San Diego 2001) short note

5.249 E. Bürkle, B. Klotz, P. Lichtinger: Durchblick im Spritzguss, KU Kunststoffe **11**, 54–60 (2001)

5.250 S. Musikant: *Optical Materials* (Dekker, New York 1990)

5.251 M.J. Weber: *Handbook of Optical Materials* (CRC, Boca Raton 2002)

5.252 K.S. Potter, J. Simmons: *Optical Materials* (Academic, New York 2000)

5.253 V.L. Ginzburg: Electromagnetic waves in isotropic and crystalline media characterized by dielectric permittivity with spatial dispersion, JETP **34**, 1096 (1958)

5.254 V.M. Agranovich, V.L. Ginzburg: *Crystal Optics with Spatial Dispersion and Excitons* (Springer, Berlin, Heidelberg 1984)

5.255 J. Pastrnak, K. Vedam: Optical anisotropy of silicon single crystals, Phys. Rev. B **3**, 2567 (1971)

5.256 P.Y. Yu, M. Cardona: Spatial dispersion in the dielectric constant of GaAs, Solid State Commun. **9**, 1421 (1971)

5.257 C. Zaldo, C. Lopez, F. Meseguer: Natural birefringence in alkali halide single crystals, Phys. Rev. B **33**, 4283 (1986)

5.258 E.G. Tsitsishvili: Optical anisotropy of cubic crystals induced by spatial dispersion, Sov. Phys. Semicond. **15**, 1152 (1981)

5.259 J.H. Burnett, Z.H. Levine, E.L. Shirley: Intrinsic birefringence in calcium fluoride and barium fluoride, Phys. Rev. B **64**, 241102R (2001)

5.260 M. Letz, L. Parthier, A. Gottwald, M. Richter: Spatial anisotropy of the exciton level in CaF_2 at 11.1 eV and its relation to the weak optical anisotropy at 157 nm, Phys. Rev. B **67**, 233101 (2003)

5.261 C. Bricot, M. Hareng, E. Spitz: Optical projection device and an optical reader incorporating this device, US Patent 4037929 (1977)

5.262 S. Sato: Liquid-crystal lens-cells with variable focal length, Jpn. J. Appl. Phys. **18**, 1679–1684 (1979)

5.263 S. Suyama, M. Date, H. Takada: Three-dimensional display system with dual-frequency liquid-crystal varifocal lens, Jpn J. Appl. Phys. **39**, 480–484 (2000)

5.264 L.G. Commander, S.E. Day, D.R. Selviah: Variable focal microlenses, Opt. Commun. **177**, 157–170 (2000)

5.265 Y. Choi, J.H. Park, J.H. Kim, S.D. Lee: Fabrication of a focal length variable microlens array based on a nematic liquid crystal, Opt. Mater. **21**, 643–646 (2002)

5.266 S.T. Kowel, D.S. Cleverly, P.G. Kornreich: Focusing by electrical modulation of refraction in aliquid crystal cell, Appl. Opt. **23**, 278–289 (1984)

5.267 T. Nose, S. Sato: A liquid crystal microlens with a nonuniform electric field, Liq. Cryst. **5**, 1425–1433 (1989)

5.268 W. Klaus, M. Ide, Y. Hayano, S. Morokawa, Y. Arimoto: Adaptive LC lens array and its application, SPIEProc. **3635**, 66–73 (1999)

5.269 B. Wang, M. Ye, M. Honma, T. Nose, S. Sato: Liquid crystal lens with spherical electrode, Jpn. J. Appl. Phys. **41**, L1232–1233 (2002)

5.270 A.F. Naumov, M.Yu. Loktev, I.R. Guealnik, G. Vdovin: Liquid crystal adaptive lenses with modal control, Opt. Lett. **23**, 992–994 (1998)

5.271 A.F. Naumov, G.D. Love, M.Yu. Loktev, F.L. Vladimirov: Control optimization of spherical modal liquid crystal lenses, Opt. Express **4**, 344–352 (1999)

5.272 G.D. Love, A.F. Naumov: Modal liquid crystal lenses, Liq. Cryst. Today **10**(1), 1–4 (2001)

5.273 J.S. Patel, K. Rastani: Electrically controlled polarization-independent liquid crystal Fresnel lens arrays, Opt. Lett. **16**, 532–534 (1991)

5.274 G. Williams, N.J. Powell, A. Purvis, M.G. Clark: Electrically controllable liquid crystal Fresnel lens, SPIE Proc. **1168**, 352–357 (1989)

5.275 H. Ren, Y.H. Fan, S.T. Wu: Tunable Fresnel lens using nanoscale polymer-dispersed liquid crystals, Appl. Phys. Lett. **83**, 1515–1517 (2003)

5.276 H. Ren, S.T. Wu: Inhomogeneous nanoscale polymer-dispersed liquid crystals with gradient refractive index, Appl. Phys. Lett. **81**, 3537–3539 (2002)

5.277 H. Ren, S.T. Wu: Tunable electronic lens using a gradient polymer network liquid crystal, Appl. Phys. Lett. **82**, 22–24 (2003)

5.278 V.V. Presnyakov, K.E. Asatryan, T.V. Galstian, A. Tork: Polymer-stabilized liquid crystal for tunable microlens applications, Opt. Express **10**, 865–870 (2002)

5.279 W. Helfrich, W.G. Schneider: Recombination radiation in anthracene crystals, Phys. Rev. Lett. **14**, 229 (1965)

5.280 M. Kawabe, K. Masuda, S. Nambu: Electroluminescence of green light region in doped anthracene, Jpn. J. Appl. Phys. **10**, 527 (1971)

5.281 C. Adachi, S. Tokito, S. Saito: Electroluminescence in organic films with three-layer structure, Jpn. J. Appl. Phys. **27**, L269 (1988)

5.282 C.W. Tang, S.A. van Slyke: Organic electroluminescent diodes, Appl. Phys. Lett. **51**, 913 (1987)

5.283 J.H. Burroughes, D.D.C. Bradley, A.R. Brown, R.N. Marks, K. Mackey, R.H. Friend, P.L. Burns, A.B. Holmes: Light-emitting-diodes based on conjugated polymers, Nature **347**, 539 (1990)

5.284 D. Braun, A.J. Heeger: Visible-light emission from semiconducting polymer diodes, Appl. Phys. Lett. **58**, 1982 (1991)

5.285 N.C. Greenham, S.C. Moratti, D.C.C. Bradley, R.H. Friend, A.B. Holmes: Efficient light-emitting-diodes based on polymers with high electron-affinities, Nature **365**, 628 (1993)

5.286 L.S. Swanson, J. Shinar, Y.W. Ding, T.J. Barton: Photoluminescence, electroluminescence and optically detected magnetic resonance study of 2,5-dialkoxy derivatives of poly(p-phenylene-acetylene) (PPA), PPA-based light emitting diodes, Synth. Met. **55-57**, 1–6 (1993)

5.287 M. Remmers, D. Neher, J. Grüner, R.H. Friend, G.H. Gelinck, J.M. Warman, C. Quattrocchi, D.A dos Santos, J.-L. Brédas: The optical, electronic, and electroluminescent properties of novel poly(p-phenylene)-related polymers, Macromol. **29**, 7432 (1996)

5.288 M. Kreyenschmidt, G. Klaerner, T. Fuhrer, J. Ashenhurst, S. Karg, W.D. Chen, V.Y. Lee, J.C. Schott, R.D. Miller: Thermally stable blue-light-emitting copolymers of poly(alkylfluorene), Macromol. **31**, 1099 (1998)

5.289 M.M. Grell, D.D.C. Bradley, M. Inbasekaran, E.P. Woo: A glass-forming conjugated main-chain liquid crystal polymer for polarized electroluminescence, Adv. Mater. **9**, 798 (1997)

5.290 Y. Ohmori, M. Uchida, K. Muro, K. Yoshino: Blue electroluminescent diodes utilizing poly(alkylfluorene), Jpn. J. Appl. Phys. **30**, 1941 (1991)

5.291 G. Grem, G. Leising: Electroluminescence of wide-bandgap chemically tunable cyclic conjugated polymers, Synth. Met. **55-57**, 4105 (1993)

5.292 U. Scherf, K. Müllen: Polyarylenes and poly(arylenevinylenes) a soluble ladder polymer via bridging of functionalized poly(para-phenylene)-precursors, Makromol. Chem. Rapid Commun. **12**, 489 (1991)

5.293 V. Cimrovà, D. Neher, M. Remmers: Blue light-emitting devices based on novel polymer blends, Adv. Mater. **10**, 676 (1998)

5.294 G. Yu, H. Nishino, A.J. Heeger, T.-A. Chen, R.D. Rieke: Enhanced electroluminescence from

5.361 S. Loheide, D. Sabbert, F. Mersch, H. Hesse, E. Krätzig: Influence of annealing treatments on photorefractive properties of $KTa_{1-x}Nb_xO_3$:Fe crystals, Ferroelectrics **166**, 99 (1995)

5.362 R.J. Reeves, M.G. Jani, B. Jassemnejad, R.C. Powell, G.J. Mizell, W. Fay: Photorefractive properties of $KNbO_3$, Phys. Rev. B **43**, 71 (1991)

5.363 M. Ewart, M. Ryf, C. Medrano, H. Wüest, M. Zgonik, P. Günter: High photorefractive sensitivity at 860 nm in reduced rhodium-doped $KNbO_3$, Opt. Lett. **22**, 781 (1997)

5.364 C. Medrano, M. Zgonik, N. Sonderer, S. Krucker, J. Seglins, H. Wüest, P. Günter: Photorefractive effect in Cu-, Ni-doped $KNbO_3$ in the visible and near infrared, J. Appl. Phys. **76**, 5640 (1994)

5.365 K. Buse, U. van Stevendaal, R. Pankrath, E. Krätzig: Light-induced charge transport properties of $Sr_{0.61}Ba_{0.39}Nb_2O_6$ crystals, J. Opt. Soc. Am. B **13**, 1461 (1996)

5.366 K. Megumi, H. Kozuka, M. Kobayashi, Y. Furuhata: High-sensitive holographic storage in Ce-doped SBN, Appl. Phys. Lett. **30**, 631 (1977)

5.367 P. Tayebati, D. Mahgerefteh: Theory of the photorefractive effect for $Bi_{12}SiO_{20}$ and $BaTiO_3$ with shallow traps, J. Opt. Soc. Am. B **8**, 1053 (1991)

5.368 M.P. Petrov, I.A. Sokolov, S.I. Stepanov, G.S. Trofimov: Non-steady-state photo-electromotive-force induced by dynamic gratings in partially compensated photoconductors, J. Appl. Phys. **68**, 2216 (1990)

5.369 Y. Fainman, J. Ma, S.H. Lee: Non-linear optical materials and applications, Mater. Sci. Rep. **9**, 53 (1993)

5.370 S. Bian, J. Frejlich: Photorefractive response time measurement in GaAs crystals by phase modulation in two-wave mixing, Opt. Lett. **19**, 1702 (1994)

5.371 D. Fluck, P. Amrhein, P. Günter: Photorefractive effect in crystals with a nonlinear recombination of charge carriers: Theory and observation in $KNbO_3$, J. Opt. Soc. Am. B **8**, 2196 (1991)

5.372 E. Krätzig, R.A. Rupp: Holographic storage properties of electrooptic crystals, SPIE Proc. **673**, 483 (1986)

5.373 U. van Stevendaal, K. Buse, H. Malz, H. Veenhuis, E. Krätzig: Reduction of light-induced refractive-index changes by decreased modulation of light patterns in photorefractive crystals, J. Opt. Soc. Am. B **15**, 2868 (1998)

5.374 K. Buse, A. Adibi, D. Psaltis: Non-volatile holographic storage in doubly doped lithium niobate crystals, Nature **393**, 665 (1998)

5.375 K. Buse: Light-induced charge transport processes in photorefractive crystals II: Materials, Appl. Phys. B **64**, 391 (1997)

5.376 D. Dirksen, F. Matthes, S. Riehemann, G. von Bally: Phase shifting holographic double exposure interferometry with fast photorefractive crystals, Opt. Commun. **134**, 310 (1997)

5.377 K. Buse, A. Gerwens, S. Wevering, E. Krätzig: Charge transport parameters of photorefractive strontium-barium niobate crystals doped with cerium, J. Opt. Soc. Am. B **15**, 1674 (1998)

5.378 L.H. Acioli, M. Ulman, E.P. Ippen, J.G. Fujimoto, H. Kong, B.S. Chen, M. Cronin-Golomb: Femtosecond temporal encoding in barium titanate, Opt. Lett. **16**, 1984 (1991)

5.379 Q.N. Wang, D.D. Nolte, M.R. Melloch: Two-wave mixing in photorefractive AlGaAs/GaAs quantum wells, Appl. Phys. Lett. **59**, 256 (1991)

5.380 Q.N. Wang, R.M. Brubaker, D.D. Nolte: Photorefractive phase-shift induced by hot-electron transport – multiple-quantum-well structures, J. Opt. Soc. Am. B **11**, 1773 (1994)

5.381 J.J. Amodei, D.L. Staebler: Holographic pattern fixing in electrooptic crystals, Appl. Phys. Lett. **18**, 540 (1971)

5.382 H. Vormann, G. Weber, S. Kapphan, E. Krätzig: Hydrogen as origin of thermal fixing in $LiNbO_3$:Fe, Solid State Commun. **40**, 543 (1981)

5.383 F. Micheron, G. Bismuth: Electrical control of fixation and erasure of holographic patterns in ferroelectric materials, Appl. Phys. Lett. **20**, 79 (1972)

5.384 J. Ma, T. Chang, J. Hong, R. Neurgaonkar: Electrical fixing of 1000 angle-multiplexed holograms in SBN:75, Opt. Lett. **22**, 1116 (1997)

5.385 D. von der Linde, A.M. Glass, K.F. Rodgers: Multiphoton photorefractive processes for optical storage in $LiNbO_3$, Appl. Phys. Lett. **25**, 155 (1974)

5.386 H. Vormann, E. Krätzig: Two step excitation in $LiTaO_3$:Fe for optical data storage, Solid State Commun. **49**, 843 (1984)

5.387 K. Buse, L. Holtmann, E. Krätzig: Activation of $BaTiO_3$ for infrared holographic recording, Opt. Commun. **85**, 183 (1991)

5.388 M.P. Petrov, S.I. Stepanov, A.A. Kamshilin: Holographic storage of information and peculiarities of light-diffraction in birefringent electrooptic crystals, Opt. Laser Technol. **11**, 149 (1979)

5.389 H.C. Külich: Reconstructing volume holograms without image field losses, Appl. Opt. **30**, 2850 (1991)

5.390 E. Chuang, D. Psaltis: Storage of 1000 holograms with use of a dual-wavelength method, Appl. Opt. **36**, 8445 (1997)

5.391 S. Fries, S. Bauschulte, E. Krätzig, K. Ringhofer, Y. Yacoby: Spatial frequency mixing in lithium niobate, Opt. Commun. **84**, 251 (1991)

5.392 S. Fries: Spatial frequency mixing in electrooptic crystals – application to nondestructive read-out of optically erasable volume holograms, Appl. Phys. A **55**, 104 (1992)

5.393 R.A. Paquin: Metal mirrors. In: *Handbook of Optomechanical Engineering*, ed. by A. Ahmad (CRC, Boca Raton 1996) p. 92

5.394 ULE^{TM} is a registered trademark of the Corning Glass Works, Corning, New York, USA for their ultra low expansion fused silica

5.395 Zerodur® is a registered trademark of Schott Glaswerke, Mainz, Germany, for their zero expansion glass ceramic
5.396 R. A. Paquin: Properties of metals. In: *Handbook of Optics, Devices, Measurements, and Properties*, Vol. II, 2nd edn., ed. by M. Bass (McGraw-Hill, New York 1994) pp. 35.1–35.78
5.397 R. A. Paquin: Materials for Optical Systems and Metal Mirrors. In: *Handbook of Optomechanical Engineering*, ed. by A. Ahmad (CRC, Boca Raton 1997) pp. 69–110
5.398 M. A. Ealey, R. A. Paquin, T. B. Parsonage (Eds.): *Advanced Materials for Optics and Precision Structures*, Crit. Rev. Opt. Sci. Technol., Vol. 67 (SPIE, Bellingham 1997)
5.399 I. Kh. Lokshin: Heat treatment to reduce internal stresses in beryllium, Metal. Sci. Heat Treat. **426**, 426 (1970)
5.400 R. A. Paquin: Advanced materials: An overview. In: *Advanced materials for optics and precision structures*, Crit. Rev. Opt. Sci. Technol., Vol. 67, ed. by M. A. Ealey, R. A. Paquin, T. B. Parsonage (SPIE, Bellingham 1997) p. 10
5.401 R. G. Ohl, M. P. Barthelmy, S. W. Zewari, R. W. Toland, J. C. McMann, D. F. Puckett, J. G. Hagopian, J. E. Hylan, J. E. Mentzell, R. G. Mink, L. M. Sparr, M. A. Greenhouse, J. W. Mac Kenty: Cryogenic optical systems and instruments IX, comparison of stress relief procedures for cryogenic aluminum mirrors, Proc. SPIE. **4822**, 51 (2002)
5.402 H. Y. Hunsicker: The metallurgy of heat treatment. In: *Aluminum 1:Properties, Physical Metallurgy and Phase Diagrams*, ed. by J. E. Hatch (American Society for Metals, Metals Park 1967)
5.403 J. B. C. Fuller Jr., P. Forney, C. M. Klug: Design and fabrication of aluminum mirrors for a large aperture precision collimator operating at cryogenic temperatures, Los Alamos Conference on Optics '81, Proc. SPIE **288**, 104 (1981)
5.404 D. Vukobratovich, K. Don, R. Sumner: Improved cryogenic aluminum mirrors, Cryogenic Optical Systems and Instruments VIII, Proc SPIE **3435**, 9–18 (1998)
5.405 R. A. Paquin: Hot isostatic pressed beryllium for large optics, Opt. Eng **25**, 1003 (1986)
5.406 D. Saxton, T. Parsonage: Advances in near net shape beryllium manufacturing technologies, Optical design, materials, fabrication, and maintenance, Proc. SPIE **4003**, 80 (2000)
5.407 M. Cayrel, R. A. Paquin, T. B. Parsonage, S. Stanghellini, K. H. Dost: Use of beryllium for the VLT secondary mirror, Advanced Materials for Optics and Precision Structures, Proc. SPIE **2857**, 86 (1996)
5.408 D. R. Coulter, S. A. Macenka, M. T. Stier, R. A. Paquin: ITTT: A state-of-the-art ultra-lightweight all-beryllium telescope. In: *Advanced Materials for Optics and Precision Structures*, Crit. Rev. Opt. Sci. Technol., Vol. 67, ed. by M. A. Ealey, R. A. Paquin, T. B. Parsonage (SPIE, Bellingham 1997) p. 277
5.409 G. Gould: Method and means for making a beryllium mirror, US Patent 4,492,669 (1985)
5.410 R. A. Paquin: New technology for beryllium mirror production, Current Developments in Optical Engineering and Commercial Optics, Proc. SPIE **1168**, 83 (1989)
5.411 T. Parsonage, J. Benoit: Advances in beryllium and AlBeMet® optical materials, Optomechanical Design And Engineering 2002, Proc. SPIE **4771**, 222 (2002)
5.412 C. A. Swenson: HIP beryllium: Thermal expansivity from 4 to 300 K and heat capacity from 1 to 108 K, J. Appl. Phys. **70**, 3046 (1991)
5.413 Y. S. Touloukian, R. K. Kirby, R. E. Taylor, P. D. Desai: *Thermal expansion, metallic elements and alloys*, Thermophys. Prop. Matter, Vol. 12 (IFI/Plenum, New York 1977) p. 23
5.414 R. A. Paquin, D. R. Coulter, D. D. Norris, A. C. Augason, M. T. Stier, M. Cayrel, T. Parsonage: New fabrication processes for dimensionally stable beryllium mirrors, Specification, Production, and Testing of Optical Components and Systems, Proc. SPIE **2775**, 480 (1996)
5.415 T. B. Parsonage: Development of aluminum beryllium for structural applications. In: *Advanced Materials for Optics and Precision Structures*, Crit. Rev. Opt. Sci. Technol., Vol. 67, ed. by T. B. Parsonage M. A. Ealey, R. A. Paquin (SPIE, Bellingham 1997) p. 236
5.416 www.brushwellman.com
5.417 M. R. Howells, R. A. Paquin: Optical substrate materials for synchrotron radiation beam lines. In: *Advanced Materials for Optics and Precision Structures*, Crit. Rev. Opt. Sci. Technol., Vol. 67, ed. by T. B. Parsonage M. A. Ealey, R. A. Paquin (SPIE, Bellingham 1997) p. 339
5.418 R. Valdiviez, D. Schrage, F. Martinez, W. Clark: The use of dispersion strengthened copper in accelerator designs, Proc. XX Int'l Linac Conf., Monterey 2000) 956
5.419 B. S. Lement, B. L. Averback, M. Cohen: The dimensional behavior of Invar, Trans. Am Soc. Met. **43**, 1072 (1951)
5.420 W. J. Spawr: Standard Industrial polishing of high energy laser optics, NBS Spec. Pub. **435**, 10–12 (1975)
5.421 P. A. Temple, D. K. Burge, J. M. Bennett: Optical properties of mirrors prepared by ultraclean dc sputter deposition, NBS Spec. Pub. **462**, 195–202 (1976)
5.422 G. E. Carver, B. O. Seraphin: CVD molybdenum films for high power laser mirrors. In: *Laser Induced Damage in Optical Materials* (NBS, Bolder 1979) pp. 287–292
5.423 H. Okamoto, M. Matsusue, K. Kitazima, K. Yoshida, Y. Ichikawa, M. Yamanaka, T. Yamanaka, Y. Tsunawaki: Laser-induced Mo mirror damage for high

power CO_2 laser. In: *Laser Induced Damage in Optical Materials* (NBS, Bolder 1985) pp. 248–260
5.424 M. Yamashita, S. Hara, H. Matsunaga: Ultrafine polishing of tungsten and molybdenum mirrors for CO_2 laser. In: *Laser-Damage in Optical Materials: Collected Papers, 1969–1998*, Vol. 08 (SPIE, Bellingham 1999)
5.425 D. L. Hibbard: Electroless nickel for optical applications. In: *Advanced Materials for Optics and Precision Structures*, Crit. Rev. Opt. Sci. Technol., Vol. 67, ed. by T. B. Parsonage M. A. Ealey, R. A. Paquin (SPIE, Bellingham 1997) p. 179
5.426 K. Parker, H. Shah: Residual stresses in electroless nickel plating, Plating **58**, 230 (1971)
5.427 K. Parker: Internal stress measurements of electroless nickel coatings by the rigid strip method. In: *Testing of Metallic and Inorganic Coatings*, Vol. STP 947, ed. by W. B. Hardig, G. D. di Ban (American Society for Testing and Materials, Philadelphia 1987) p. 111
5.428 Alumiplate™ is a registered trademark of Alumiplate, Inc., Coon Rapids, MN, USA for their aluminium plating process and coating
5.429 D. Vukobratovich, A. Gerzoff, M. K. Cho: Thermooptic analysis of bi-metallic mirrors, Optomechanical Design and Precision Instruments, Proc. SPIE **3132**, 12–23 (1997)
5.430 C. J. Shih, A. Ezis: Application of hot-pressed silicon carbide to large high-precision optical structures, Silicon Carbide Materials for Optics and Precision Structures, Proc. SPIE **2543**, 24 (1995)
5.431 J. S. Goela, M. A. Pickering: Optics applications of chemical vapor deposited β-SiC. In: *Advanced Materials for Optics and Precision Structures*, Crit. Rev. Opt. Sci. Technol., Vol. 67, ed. by T. B. Parsonage M. A. Ealey, R. A. Paquin (SPIE, Bellingham 1997) p. 71
5.432 V. Rehn, J. L. Stanford, A. D. Behr, V. O. Jones, W. J. Choyke: Total optical integrated scatter in the vacuum ultraviolet: Polished CVD SiC, Appl. Opt. **16**, 111 (1977)
5.433 V. Rehn, W. J. Choyke: Total optical integrated scatter in the vacuum ultraviolet: Polished CVD SiC, Nucl. Instrum. Methods **117**, 173 (1980)
5.434 E. Sein, F. Safa, D. Castel, P. Deny: Silicon carbide, a sound solution for space optics, Proc. 52Int'l Astronautical Congress Toulouse (2001)Paper IAF-01-I.1.06
5.435 M. A. Ealey, J. A. Wellman, G. Weaver: CERAFORM SiC: Roadmap to 2 meters and 2 kg/m^2 areal density. In: *Advanced Materials for Optics and Precision Structures*, Crit. Rev. Opt. Sci. Technol., Vol. 67, ed. by T. B. Parsonage M. A. Ealey, R. A. Paquin (SPIE, Bellingham 1997) p. 53
5.436 E. Tobin, M. Magida, S. Kishner, M. Krim: Design, fabrication, and Test of a meter-class reaction bonded sic mirror blank, Silicon Carbide Materials for Optics and Precision Structures, Proc. SPIE **2543**, 12 (1995)
5.437 R. A. Paquin, M. B. Magida: Low scatter surfaces on silicon carbide, Laser Induced Damage in Optical Materials: 1989, NIST Spec. Publ. **801**, 256 (1990)
5.438 M. Krödel, G. S. Kutter, M. Deyerler, N. Pailer: Short carbon-fiber reinforced ceramic – Cesic® – for optomechanical applications, Optomechanical Design And Engineering 2002, Proc. SPIE **4771**, 230 (2002)
5.439 R. Plummer, D. Bray: Guidelines for design of SuperSiC® silicon carbide mirror substrates, precision components, Optomechanical Design and Engineering 2002, Proc. SPIE **4771**, 265 (2002)
5.440 S. Musikant: *Optical Materials* (Dekker, New York 1985) p. 1985
5.441 Hoya Optical Glass Catalog
5.442 Ohara Optical Glass Catalog (Ohara, Brandsburg 1995)
5.443 Schott Optical Glass Catalog (Schott, Mainz 2001)

6. Thin Film Optical Coatings

Within the scientific conception of the modern world, thin film optical coatings can be interpreted as one-dimensional photonic crystals. In general, they are composed of a sequence of single layers which consist of different transparent dielectrics with a thickness in the nanometer scale according to the operation wavelength range. The major function of these photonic structures is to adapt the properties of an optical surface to the needs of specific applications. By application of optical thin film coatings with optimized designs, the spectral characteristics of a surface can be modified to practically any required transfer function for a certain wavelength range. For example, the Fresnel reflection of a lens or a laser window can be suppressed for a broad wavelength range by depositing an antireflective coating containing only a few single layers. On the basis of a layer stack with alternating high- and low-refracting materials, high reflectance values up to 99.999% can be achieved for a certain laser wavelength. In addition to these basic functions, optical coatings can realize a broad variety of spectral filter characteristics according to even extremely sophisticated demands in modern precision optics and laser technology. Moreover, recent developments in optical thin film technology provide the means to combine selected optical properties with other features concerning, for instance, the thermal, mechanical or chemical stability of a surface. The latest progress in ophthalmic coatings even includes the integration of self-cleaning, photoactive or anti-fogging functions in antireflective coatings on glass.

As a consequence of this enormous flexibility in adjusting the properties of functional surfaces, optical coatings can be found in nearly every product and development of modern optic today.

In order to keep pace with the rapid development of optical technology, innovations in the design, deposition processes and handling of optical coatings are some of the crucial factors. Also, high demands in respect to precision and reproducibility are imposed on the control of layer thickness during the production of the coating systems. For certain applications in fs lasers or optical measurement systems the individual layer thickness has to be controlled within the sub-nanometer scale, which can be only achieved on the basis of advanced in situ monitoring techniques of the growing layers. These skills have to be complemented by extended knowledge of characterization, because optimization and marketing of optical coatings can only be performed on the basis of reliable and standardized characterization techniques. The present chapter addresses these major aspects of optical coatings and concentrates on the essential topics of optical coatings in their theoretical modeling, production processes, and quality control.

6.1 Theory of Optical Coatings

Photonic structures were present in nature long before mankind. During their evolution, butterflies and other insects developed a variety of nanostructures on their wings and bodies for camouflage, deterrence or attraction [6.1]. The beginning of technical engineering of optical surfaces can be dated back to the Greek culture, when metal surfaces were polished to mirror quality. Tracing the history of optical coatings through the medieval times, the production of Venetian mirrors by covering glass with an amalgam during the 16th century can be considered as the first application of optical thin films. The first observation of an antireflective effect for transparent thin films may be assigned to *Fraunhofer* [6.2] who performed studies in aged glass surfaces and observed a reduction of reflectivity by the tarnish layer in the year 1817. In the same century, *Dennis Taylor* [6.3] investigated adapted etching techniques to achieve antireflective effects on different glass materials used in optical devices during these times, and *Augustin Jean Fresnel* (*1788; †1827) published his well-known equations describing the optical function of a single boundary. As a milestone in optical thin film technology, the conception of the Fabry–Perot theory [6.4] in 1899 opened the way towards the theoretical description of multilayer structures. This basic theory describes the interference of partial waves reflected from two parallel optical surfaces, which can be considered as the fundamental element of all coating systems (Fig. 6.1): layers of different dielectric materials are deposited in a defined sequence on the surface of the optical component. At least two layer materials with different refractive indices have to be selected to adjust the spectral transfer function of the layer stack to the specification defined by the application. Besides transparency in the operation wavelength range, major aspects for the choice of the coating materials include the contrast in the indices of refraction as well as their chemical and mechanical stability. The sequence of the materials and their corresponding thicknesses is often referred to as the design of the optical coating system and comprises all information for modeling of the spectral characteristics.

For the theoretical description of optical thin films, the application of the Fabry–Perot theory to a single layer (Fig. 6.2) can be considered as the starting point. In this approach a single layer of thickness d_1 and refractive index n_1 is formed by two boundaries between two semi-infinite media with indices of refraction n_0 and n_2, respectively. In the classical model, the light interacting with this structure is described by the function of a plane wave (point x, time t) with amplitude E_0 and wave number $k = 2\pi/\lambda$ incorporating the wavelength λ and frequency ω:

$$E(x,t) = E_0 \exp[-\mathrm{i}(kx - \omega t)]\,. \tag{6.1}$$

To calculate the spectral transfer function, this plane wave is followed on its way through the single layer, and the contributions of the individual partial waves are calculated and accumulated to form the total transmitted and reflected wave. At the first interface, a part of the incident plane wave is directly reflected with an amplitude of $A_0 = E_0 r_1$ (order zero), where the reflection coefficient r_1 can be calculated by the Fresnel formulae. The reminder of the wave is transmitted through the first

Fig. 6.1 Basic structure of a thin film system: transparent layers of at least two different materials (n_H, n_L) are deposited on a substrate with refractive index n_T

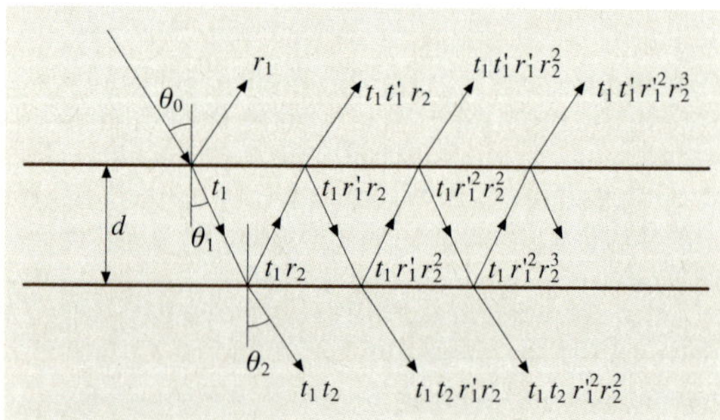

Fig. 6.2 Schematic path of a plane wave incident onto a single layer with refractive index n_1. The reflection and transmission coefficients for the amplitudes at the interfaces $m = 1, 2$ are denoted by t_m and r_m, respectively. A prime added to the coefficients indicates a transfer of the wave opposite to thc dircction of incidence

boundary (coefficient t_1), partially reflected back by the second boundary (coefficient r_1) and then reaches via the first boundary the incident medium with an amplitude $A_1 = E_0 t_1 r_2 t'_1$. On its double path through the layer the considered partial wave also undergoes a phase shift δ_1 which is dependent on the thickness d_1 and the refractive index n_1 of the layer:

$$\delta_1 = \frac{2\pi n_1 d_1 \cos\theta_1}{\lambda} . \tag{6.2}$$

In (6.2), the case of an arbitrary angle of incidence θ_1 in the layer is taken into account, which has to be calculated by Snellius' law. On the basis of the preceding considerations, the amplitude A_1 of the first partial wave emerging from the layer can be written as:

$$A_1 = t_1 t'_1 r_2 E_0 \exp(-\mathrm{i}2\delta_1) . \tag{6.3}$$

In this notation t'_1 indicates the transmission coefficient for waves passing the first boundary in the direction opposite to the incident wave. If this notation is applied in the same sense to the part of the wave reflected back into the layer by the first boundary (coefficient r'_1), the amplitude A_2 of this partial wave after back-reflection by the second boundary and passing the first boundary is given by:

$$A_2 = t_1 t'_1 r_2 E_0 \exp(-\mathrm{i}2\delta_1) r'_1 r_2 \exp(-\mathrm{i}2\delta_1) . \tag{6.4}$$

In addition to the expression for the first-order amplitude, the second-order partial wave includes a factor $r_2 r'_1 exp(-\mathrm{i}2\delta_1)$. Obviously, this additional factor appears for every further order of reflection, and therefore, the amplitudes A_k of a partial wave of the order k can be determined:

$$A_k = t_1 t'_1 r_2 E_0 \exp(-\mathrm{i}2\delta_1)[r'_1 r_2 \exp(-\mathrm{i}2\delta_1)]^{k-1} . \tag{6.5}$$

To calculate the reflectance of the layer, all partial waves emerging from the layer into the incident medium have to be added to determine the total amplitude A_{tot} of the reflected contributions. This sum can be expressed in closed form, because the partial waves follow the rule of a geometric expansion:

$$A_{\mathrm{tot}} = \left(r_1 + \frac{t_1 t'_1 r_2 \exp(-\mathrm{i}2\delta)}{1 - r'_1 r_2 \exp(-\mathrm{i}2\delta)} \right) E_0 . \tag{6.6}$$

Finally, the reflection coefficient r_S of the single layer is determined by the ratio between the total amplitude of the reflected wave and the amplitude of the incoming wave

$$r_s = \frac{A_{\mathrm{tot}}}{A_0} = \frac{r_1 + r_2 \exp(-\mathrm{i}2\delta)}{1 + r_1 r_2 \exp(-\mathrm{i}2\delta)} . \tag{6.7}$$

In this expression, the coefficients r'_1 and t'_1 have been replaced by using the following relations, which can be derived from the conditions of continuity valid for the amplitude at the boundaries:

$$t_1 t'_1 = (1 - r'_1)(1 - r_1) = (1 - r_1)(1 + r_1) = 1 - r_1^2 . \tag{6.8}$$

In view of the fact that only field intensities but not amplitudes can be measured, it is worthwhile to convert the expression for the amplitude reflection coefficient into the corresponding reflectivity value:

$$R_s = \frac{r_1^2 + 2r_1 r_2 \cos(2\delta_1) + r_2^2}{1 + 2r_1 r_2 \cos(2\delta_1) + r_1^2 r_2^2} . \tag{6.9}$$

In principle, (6.9) provides all the means for the description of the spectral transfer function for a single layer also including the cases of an arbitrary angle of incidence and an absorbing layer material. For this general case, the refractive index has to be expressed in its complex form with $n_1 = n_1 + ik_1$, where k_1 denotes the extinction coefficient of the layer material, and the Fresnel formulae as well as Snellius' law have to be applied in their complex form.

In order to get a deeper insight into the fundamental operating principle of coatings the discussion of a few special configurations is of particular interest. In optical thin film technology the layer thickness D_i is often expressed in units of quarter-wave optical thickness (QWOT) at the design wavelength λ_Z:

$$D_i = \frac{4n_i d_i}{\lambda_Z} . \tag{6.10}$$

If the layer thickness is an integer multiple of 1 QWOT, the phase shift of a wave with the design wavelength traveling perpendicularly through the layer corresponds to multiples of $\pi/2$. Evaluating (6.9) for a 2 QWOT of any dielectric material on the basis of the Fresnel formulae for normal incidence results in:

$$R_s = \left(\frac{r_1 + r_2}{1 + r_1 r_2} \right)^2 ,$$

with

$$r_1 = \frac{n_0 - n_1}{n_0 + n_1} , \quad r_2 = \frac{n_1 - n_2}{n_1 + n_2} ,$$

$$R_s = \left(\frac{n_0 - n_2}{n_0 + n_2} \right)^2 . \tag{6.11}$$

Apparently, the expression for the reflectance is reduced to the Fresnel formulae for an interface between the incident medium and the carrier medium, demonstrating that the single 2 QWOT layer has no influence on the spectral transmittance of the structure. For the case of a layer with a thickness of 1 QWOT at the design wavelength, (6.9) can be written in the form:

$$R_s = \left[\frac{n_2 n_0 - n_1^2}{n_2 n_0 + n_1^2}\right]^2 . \quad (6.12)$$

The reflectance of the single layer on the substrate can be totally suppressed if the condition $n_1^2 = n_0 n_2$ is fulfilled. For a further illustration of the effects typical for a single dielectric layer, corresponding reflectance spectra are depicted for different refractive indices in Fig. 6.3. Considering the reflectance values at the design wavelength λ_Z the cases $n_1^2 = n_2$ (zero reflectance), $n_1 < n_2$ (reflectance lower than the reflectance of the uncoated substrate), $n_1 = n_2$ (reflectance equal to the reflectance of the uncoated substrate), and $n_1 > n_2$ (reflectance higher than the reflectance of the uncoated substrate) can be distinguished. As a function of the wavelength, a cyclic behavior of the spectra with a period of one octave can be observed, which shows common points of tangency at the reflectance level of the bare substrate. These points can be attributed to the wavelength positions, where the layer thickness is according to an even multiple of QWOT resulting in the special condition described by (6.9). Thus, on one hand, a single 1 QWOT layer of a material matched to the refractive index of the substrate can be employed as an antireflective coating. On the other hand, a layer with a high index of refraction is suitable for an enhancement of the reflectance to a certain limit given by the availability of high-refracting materials. Actually, single-layer coatings can still be found nowadays on many optical components for laser applications, where only one wavelength has to be controlled.

In principle, the outlined theoretical approach for the single layer can also be transferred to the calculation of multilayer systems. However, considering the enormous number of partial waves, which increases exponentially with the number of layers in a stack, the resulting equations become extremely complicated and difficult to handle. For example, the expression for the reflectance of a three-layer system would already fill one printed page of this book. Therefore, the idea of a formalism assigning a matrix to each layer of the design was a substantial step forward to an understanding of thin film design. This so-called matrix formalism, which was born in the 1940s, can be deduced on the basis of the boundary conditions of the electric and magnetic field at the interfaces between the layers in a system [6.5–8]. Besides the clear representation of each layer by a single matrix $\boldsymbol{M}_i$, this approach offers a second major advantage which is related to a consecutive multiplication of these single-layer matrices to calculate the transfer function of a multilayer system. Thus, considering a stack, which is formed by a number K of single layers with layer 1 located at the first interface in respect to the incoming wave, the transfer matrix $\boldsymbol{M}_S$ of the entire stack can just be determined by the following multiplication:

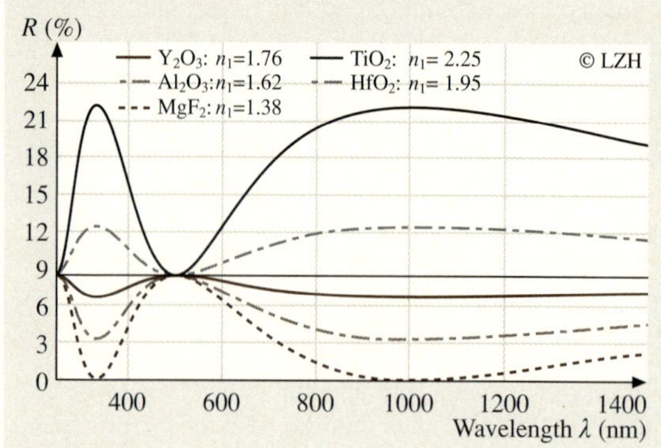

Fig. 6.3 Spectra calculated for a 1 QWOT single layer ($\lambda_Z = 1000$ nm) of different materials on a the surface of a Nd:YAG rod ($n_2 = 1.78$, ambient refractive index $n_0 = 1$). The refractive indices are typical values for e-beam deposition of the corresponding deposition materials: MgF_2: $n_1 = 1.38$, Al_2O_3: $n_1 = 1.62$, Y_2O_3: $n_1 = 1.76$, HfO_2: $n_1 = 1.95$, TiO_2: $n_1 = 2.25$. The line corresponds to the reflectivity of the uncoated substrate material

$$\boldsymbol{M}_S = \boldsymbol{M}_1 \boldsymbol{M}_2 \boldsymbol{M}_3 \dots \boldsymbol{M}_i \boldsymbol{M}_{i+1} \dots \boldsymbol{M}_K . \quad (6.13)$$

The elements of the single-layer matrix $\boldsymbol{M}_i$ can be derived from the boundary conditions in the layer structure relating the electric (E_{i-1}) and magnetic (H_{i-1}) field strength at the front face to the field strength values (E_i and H_i) at the rear face of the layer:

$$\begin{pmatrix} E_{i-1} \\ H_{i-1} \end{pmatrix} = \boldsymbol{M}_i \begin{pmatrix} E_i \\ H_i \end{pmatrix} = \begin{pmatrix} \cos\delta_i & \frac{\mathrm{i}}{n_i}\sin\delta_i \\ \mathrm{i}n_i \sin\delta_i & \cos\delta_i \end{pmatrix} \begin{pmatrix} E_i \\ H_i \end{pmatrix} . \quad (6.14)$$

In this formalism, the matrix elements $\boldsymbol{M}_{ij}$ contain the refractive index n_i of the layer and the phase shift δ_i, which are exclusively parameters of the layer number i. For a calculation of the reflection coefficient r_{SK} of the entire structure comprising also the substrate (index of refraction n_T) and the ambient medium (index of refraction n_0), again the ratios of the amplitudes have to be considered:

$$r_{SK} = \frac{n_0 M_{11} + i n_0 n_T M_{12} - i M_{21} - n_T M_{22}}{n_0 M_{11} + i n_0 n_T M_{12} + i M_{21} + n_T M_{22}} \quad (6.15)$$

In order to demonstrate the elegance of the matrix formalism, the reflectance of a high-reflecting dielectric mirror will be considered in the following. The standard design of such a mirror is a periodical 1 QWOT stack (see also Fig. 6.1) of two coating materials with a high n_H and a low index of refraction n_L. For an efficient description of the designs, a notation in capital letters, which indicate a 1 QWOT layer of a certain layer material, is often used in optical thin film technology. For example, a mirror stack with 11 layers in the described structure is represented by the sequence HLHLHLHLHLH (H: refractive index n_H L: refractive index n_L) or even more condensed by $(HL)^5H$. For the more general case of N layer pairs (design: $(HL)^N H$) the matrix M_{DM} of the dielectric mirror can be expressed by:

$$M_{DM} = \underbrace{\begin{pmatrix} 0 & i/n_H \\ i n_H & 0 \end{pmatrix} \begin{pmatrix} 0 & i/n_L \\ i n_L & 0 \end{pmatrix} \times \begin{pmatrix} 0 & i/n_H \\ i n_H & 0 \end{pmatrix} \dots \begin{pmatrix} 0 & i/n_H \\ i n_H & 0 \end{pmatrix}}_{(2N+1)\text{ matrices}} \quad (6.16)$$

Obviously, the single matrices for the 1 QWOT-layer reduce to simple expressions containing solely the refractive index of the materials for this special case of normal incidence. Executing the matrix multiplication, applying (6.15) and calculating the reflectance $R_{DM} = r_{DM} r_{DM}^*$ results in:

$$R_{DM} \approx 1 - 4 n_0 n_T \frac{n_L^{2N}}{n_H^{2(N+1)}} \,. \quad (6.17)$$

This expression is an approximation for a large number N of QWOT-layer pairs, which is typical for practical systems involving 10 to 40 layer pairs. An interpretation of this estimation (6.17) indicates that the contrast (n_L/n_H) between the indices of refraction governs the number of layer pairs necessary to achieve a defined reflectance value. A high contrast does not only reduce the number of layer pairs, but also results in a more extended reflection band in the spectrum of the QWOT stack (Fig. 6.4). In Fig. 6.5 the development of the reflectance spectrum is illustrated for the QWOT stack defined in Fig. 6.4 for different angles of incidence. The shift of the spectra towards shorter wavelength can be explained on the basis of a reduction of the phase difference between the partial waves in the layer structure. For arbitrary angles of incidence the spectra of p- and s-polarization have to be considered separately in the matrix formalism resulting in a broader reflectance band for the s-polarization compared to the p-polarization. These effects have to be taken into account if a dielectric mirror designed for normal incidence is operated under arbitrary angles of incidence in an application.

Besides the described high-reflecting stacks, antireflective coatings are often applied in laser technology to reduce reflection losses of windows, laser rods, or lenses. Compared to the mirror designs, antireflective (AR) coatings are regularly built only of a few layers, but often of more than two materials, especially to accomplish the demands of lowest residual reflectivity for a broad spectral range. For laser technology with its dominant single-wavelength applications and challenging specifications concerning the losses and the power-handling capability of the coatings, single layers or two-layer AR coatings are mainly employed. On the basis of two layers, a total suppression of the Fresnel reflection of most substrate materials can be accomplished for one wavelength. Such coatings, which are often called V-coatings according to the shape of their

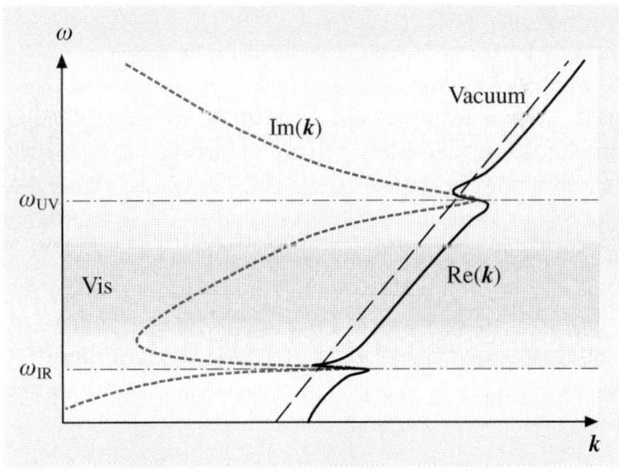

Fig. 6.4 Reflectance spectra calculated for 1 QWOT stacks ($\lambda_Z = 1064$ nm) of different high-refracting materials in conjunction with SiO_2 ($n_L = 1.46$) as the low-index material on a quartz surface ($n_T = 1.46$, refractive index of the ambient $n_0 = 1$). The refractive indices of the high-index layers are typical values for electron-beam deposition of the corresponding deposition materials: Al_2O_3: $n_H = 1.62$ (green), CeO_2: $n_H = 1.80$ (blue), HfO_2: $n_H = 1.95$ (magenta), TiO_2: $n_H = 2.25$ (black). The number of layer pairs is kept constant at $N = 6$ for all depicted spectra

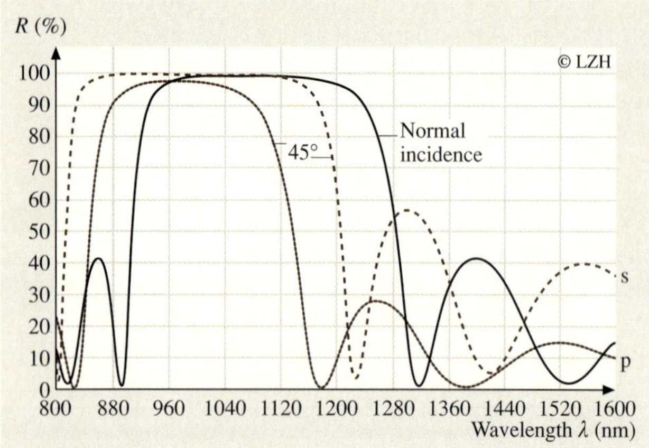

Fig. 6.5 Reflectance spectra calculated for 1 QWOT stack with parameters according to Fig. 6.4 and TiO_2 ($n_H = 2.25$) as the high-index material. Besides the spectrum for normal incidence, the reflectance is depicted for an angle of incidence of 45°. For arbitrary angles of incidence, s- and p-components have to be distinguished

spectra, offer the advantage of lowest total thickness providing minimum losses and highest laser-induced damage thresholds [6.9].

In regard to the important role of losses and stability of coating systems for laser technology, the designs are often also optimized with respect to the distribution of the electrical field strength in the structure. In view of the fact that the matrices associate the field strength values at different positions in the layer system, the calculation of the field strength distribution can be readily accomplished using the matrix method. Moreover, an implementation of the formalisms into modern computer algorithms, which allow for the calculation of all transfer parameters including many other important aspects in laser technology, such as absorbance, phase shift or group delay dispersion, is an uncomplicated task. In the course of the rapid development of computer technology, extensive software tools are available nowadays, which even allow for an inverse synthesis of multilayer designs. As input parameters of these modern optimization tools the desired spectral characteristics can be entered, and appropriate design solutions are developed on the basis of sophisticated algorithms. In summary, the challenges of optical thin film technology are no longer concentrated on the theoretical modeling and the calculation of designs. Present and future problems are coupled to the reproducible and precise production of coatings with high optical quality and environmental stability.

6.2 Production of Optical Coatings

Looking back, the production of optical coatings exhibits a long and successful technical history. Besides the already mentioned early aging and etching experiments of Fraunhofer and Taylor, in 1852 *Grove* described the sputtering effect, as it was later called, in a gas discharge apparatus [6.10]. In 1857, *Faraday* vaporized gold wires by high electrical currents and studied the optical properties of the deposited material, although he was mainly interested in the finest particles [6.11]. *Wright* reported in 1877 "on the production of transparent metallic films by the electrical discharge in exhausted tubes" [6.12]. In 1884, *Edison* filed an application for a patent, which described thermal evaporation within a vacuum vessel as well as arc vaporization to "plate one material with another" and was granted in 1894 (Fig. 6.6) [6.13]. In the following decades, a multitude of publications documented the development of industrial production methods for optical coatings, starting with single layers and resulting in increasingly complex multilayer systems [6.14, 15]. Within the scope of the present article, a short overview of the most common production techniques for optical thin film filters from the basics to some of the latest developments will be given. Further information about optical coating technology can be found in a variety of comprehensive technical books [6.9, 16–19].

Concerning the industrial production, there are fundamental demands on the deposition process, although their priorities depend on the specific applications. High reproducibility of the optical properties of the thin films is needed, as well as of the mechanical characteristics. Large areas have to be coated homogeneously, using high deposition rates with simultaneous realization of a precise layer thickness determination and termination. Furthermore, a high level of automation, a short process time, and inexpensive, nontoxic deposition materials are economically relevant. With regard to the thin film material choice, specific requirements also have to be fulfilled [6.20]. Dielectrics for use in multilayer coatings have to be transparent in the spectral region desired by the specific application. Therefore, besides an appropriate refractive index contrast, a low extinction coefficient has to be realized. In addition, an adequate

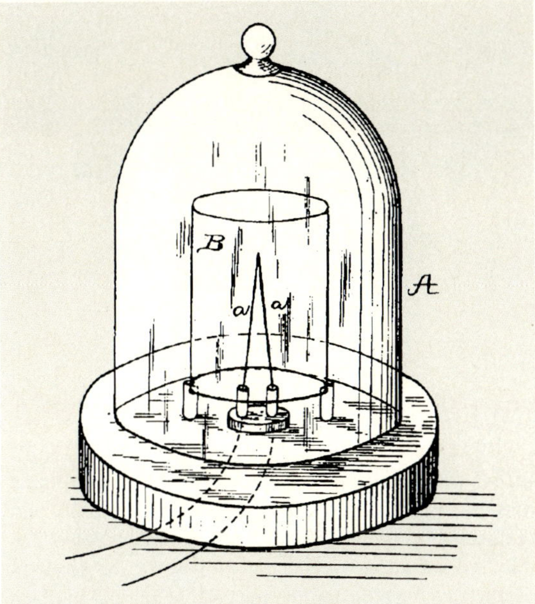

Fig. 6.6 Illustration from Edison's patent *Art of Plating One Material with Another*, granted in 1894 [6.13]. The conductive coating material is vaporized within a vacuum chamber by a continuous arc between the electrodes (*a*). For coating mirrors, Edison placed glass plates on the inner side of the apparatus. Furthermore, the patent describes thermal evaporation by resistive heating

mechanical and environmental stability must be ensured. Finally, the materials have to be technically controllable within the chosen deposition process, and have to exhibit the potential to form amorphous, compact thin film microstructures. In the visible and near-infrared spectral region, these requirements are covered best by the material class of oxides. Common examples are the low-refractive-index materials SiO_2 or Al_2O_3 and the high-refractive-index metal oxides Ta_2O_5, Nb_2O_5, or TiO_2. Going down to the ultraviolet and vacuum ultraviolet wavelength ranges, the oxides are complemented and successively replaced by fluorides, e.g., AlF_3, MgF_2, and LaF_3. In contrast, the mid- and far-infrared spectral region is the domain of materials such as Ge, Si, ZnSe, ZnS, and the radioactive ThF_4, which is still in use, mainly for high power CO_2 laser optics.

6.2.1 Thermal Evaporation

Within the most common deposition processes for optics, thin films are produced under vacuum conditions. In general, these processes are divided into two classification groups: on the one hand, the physical vapor deposition (PVD), comprising thermal vaporization as well as sputtering, and, on the other hand, chemical vapor deposition (CVD).

One of the basic ways to realize PVD is thermal evaporation of the thin film material by direct or indirect heating of the deposition material within a vacuum chamber. The evaporated material condenses on the substrates to be coated, which are usually located on a rotating spherical calotte above the evaporator. If the uniformity of the layer thickness distribution across the calotte does not fulfil the requirements on the basis of the given geometrical set up, additional masks are used. Under vacuum conditions, the evaporation rate exhibits a strong dependency on the temperature, and is mainly determined by the temperature-dependent equilibrium vapor pressure p^* of the evaporant. On the basis of the experimental work of *Hertz* [6.21] and *Knudsen* [6.22], a theoretical description of the evaporation rate became known as the Hertz–Knudsen equation:

$$\frac{\mathrm{d}N_\mathrm{e}}{A_\mathrm{e}\,\mathrm{d}t} = \alpha_\mathrm{e}\frac{p^* - p}{\sqrt{2\pi m k_\mathrm{B} T}}\,, \tag{6.18}$$

with

$\mathrm{d}N_\mathrm{e}/\,\mathrm{d}t$	number of evaporating atoms per time unit,
A_e	surface area of the evaporation source,
α_e	evaporation coefficient (sticking coefficient for vapor atoms onto the surface),
p^*	equilibrium vapor pressure of the evaporant,
p	hydrostatic pressure of the evaporant within the vacuum chamber,
m	atomic mass,
k_B	Boltzmann's constant,
T	temperature.

The strong temperature dependency of the evaporation rate results from the equilibrium vapor pressure as an exponential function of the temperature

$$p^* = p_0\,\mathrm{e}^{-\frac{L_0}{k_\mathrm{B}T}}\,, \tag{6.19}$$

where p_0 is a constant factor and L_0 is the latent heat of evaporation per atom or molecule.

With the focus on technical aspects, a variety of evaporation sources has been developed. In practice, unwanted contamination effects of the films with contact material from the evaporation source, as well as its significant corrosion, can occur due to chemical reactions at the required high temperatures. Furthermore, high evaporation temperatures between some hundred to several thousand Kelvin have to be reached within the evaporation source.

An elementary, but not often used, method is the sublimation of a conducting material by direct current heating. This technique is not applicable to many materials; potential examples are C, Cr, Fe, Ni, Rh, and Ti. An early developed and widely applied way to realize thermal evaporation is indirect heating of the coating material by a resistively heated reservoir. The characteristic shape of the reservoir gave this process variant its name: boat evaporation. Evaporation boats are made of conducting high-temperature resistant materials, e.g. W, Mo, Ta, or C, in addition partially equipped with a ceramic insert (liner). Obviously, the maximum evaporation temperature is limited by the melting point of the boat material. In a slightly modified alternative, the boat is reduced to a filament, wetted by molten material, which is kept in place by surface tension. However, due to the thermal inertia of the boat and the molten coating material, a nearly real-time rate control is difficult to realize. Furthermore, the melting can exhibit disadvantageous spatter behavior, leading to coating defects. This effect can be partially suppressed by equipping the boats with perforated caps or chimney constructions.

Another technical possibility to melt conductive materials is the application of induction heating. While inductive heaters are frequently used for large molten masses, e.g., in the field of crystal growth, this solution is rarely used for the deposition of optical coatings.

Today, the most common thermal evaporation method in the industrial production of optical coatings is the direct heating of the coating material within a crucible by an electron beam (Fig. 6.7). The characteristic parameters of the electron beam, which is directed by magnetic fields into the water-cooled crucibles, are an acceleration voltage of 6–10 kV and currents of 0.1–1.5 A. In order to achieve uniform evaporation, the electron beam can be swept across the coating material by deflection coils, optionally combined with a rotating crucible. In addition, multi-crucible electron-beam sources allow the sequential evaporation of different coating materials within one process; otherwise the deposition plants are equipped with two or more sources. Compared to boat evaporation, the electron-beam heating technique offers important advantages. Due to its high power densities, high-melting-point materials can be evaporated and, furthermore, the cooled crucible prevents contamination. With regard to rate control, electron-beam sources benefit from a low inertia in combination with the high reliability of today's technical solutions. However, it must be pointed out that high local temperatures may lead to decomposition, if chemical compounds are used as coating materials. In the case

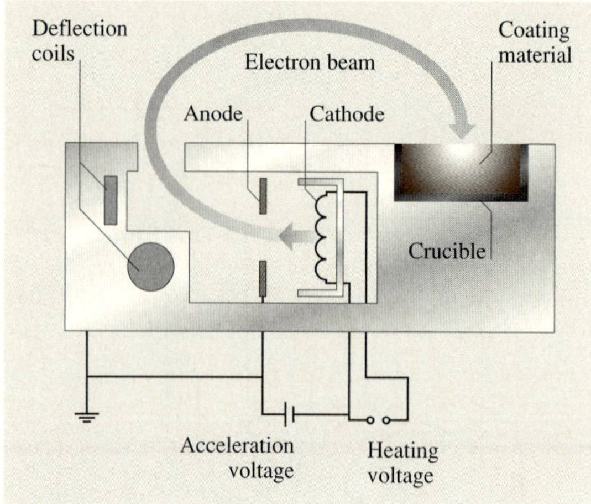

Fig. 6.7 Electron-beam evaporation source. The electrons are accelerated by a potential difference of several kV and are directed by a magnetic field into the water-cooled crucible. Additional deflection coils allow for the beam to write adapted patterns on the material to achieve uniform evaporation

of oxides, an additional oxygen inlet can overcome this problem. In common deposition systems, stoichiometric films can be attained applying a typical oxygen partial pressure range of 1 to 3×10^{-2} Pa. Thermal evaporation in the presence of reactive gases such as oxygen, nitrogen, or even fluorine is sometimes referred to as reactive evaporation.

Regarding the layer properties, thin films produced by thermal evaporation exhibit characteristic microstructures depending on the individual process parameters. The typical thermal energies below 0.3 eV of the condensing particles result in packing densities lower than the density values of the corresponding bulk materials. As a consequence of limited surface mobility of the condensing particles and shadowing effects, a columnar growth containing microstructural voids is most common for evaporated films [6.23, 24]. Derived from experimental studies, approved structure zone models have been developed that describe characteristic zones depending on the ratio between the substrate temperature and the melting point of the coating material [6.25]. Usually, the process parameters are optimized to achieve maximized packing densities to minimize voids, which results in the need for an additional substrate heating, typically in the range of 300 °C. Besides increasing mechanical stability, an enhanced packing density affects the optical layer properties, e.g., leading to higher refractive indices in most cases. A major disadvantage of the presence of microstructural voids in the thin films is the ingress of moisture from the environmental atmosphere. On the one hand, the adsorbed water and bonded OH groups result in optical absorption losses, especially in the near- and mid-infrared (MIR) wavelength range. On the other hand, water adsorption influences the effective refractive index of the layer. As a result of the varying water content, the spectral characteristic of a porous multilayer interference filter shows a strong dependency on environmental parameters such as temperature and humidity.

Furthermore, the microstructure of the growing layer is affected by the surface structure of the substrate in all deposition processes for optical filters. As a higher surface roughness of the substrate increases the roughness of the layer surface, the substrate should be chosen in view of the application to avoid unacceptably high optical scattering losses. In addition, cleanroom areas are indicated in optical thin film production, due to the fact that particle contamination results in serious layer defects.

6.2.2 Ion Plating and Ion-Assisted Deposition

As discussed above, thin films produced by thermal evaporation processes exhibit a distinct microstructure as a consequence of the low kinetic energies of the condensing molecules and atoms. Besides the thermal energy deposited into the growing layer by substrate heating, additional energy deposit by ion impact was quickly identified as a promising alternative. In the beginning, reactive evaporation processes were optimized by the application of ionized oxygen [6.26,27]. In 1967 a process concept was patented, in which a negative bias voltage accelerated positive ions onto the substrates [6.28]. Applied ion species included inert gas directly provided to maintain a discharge, as well as ionized evaporated coating material. In the course of further technological progress, this method was enhanced and became known as the ion plating process, which found its way into the optical thin film industry in different variations. An established concept is the reactive low-voltage ion plating (RLVIP) system, shown schematically in Fig. 6.8 [6.29]. In addition to the conventional thermal evaporation configuration, an argon discharge is directed from a hot cathode source mounted on a chamber wall into the crucible of the electron-beam evaporator, which acts as anode. As a consequence of the interaction with the argon plasma, the evaporated coating material is ionized and accelerated together with argon ions towards the substrates by the self-bias voltage of the electrically insulated substrate holder. This negative self-bias voltage usually ranges between −5 to −30 V with respect to the plasma potential and depends on the plasma parameters as well as the geometrical conditions. As a consequence of the additional energy input and the effective activation, thin films produced by reactive low-voltage ion plating, exhibit a compaction of the microstructure and optimized stoichiometry, respectively.

Employing ions to enhance the optical and mechanical thin film properties as well as the process stability, ion-assisted deposition (IAD), also referred to as ion-beam-assisted deposition (IBAD), represents a widely used process concept. As depicted in Fig. 6.9, the IAD process is characterized by a separate ion source, integrated in addition to the evaporation sources into the process chamber. The ion beam is superimposed upon the flux of condensing particles and leads to the desired densification of the growing layer. The dominant densification mechanism is attributed to momentum transfer, which was verified by calculations on the basis of collision cascade models in accordance with experimental results [6.30]. However, excessive ion energies

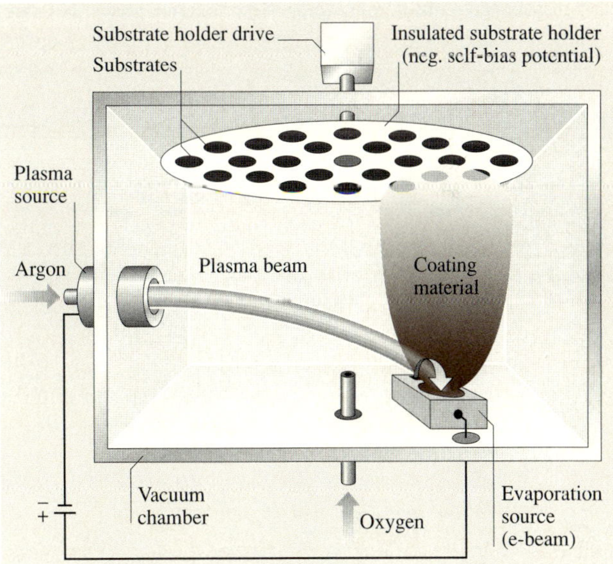

Fig. 6.8 Schematic diagram of a reactive low-voltage ion plating (RLVIP) system. An argon discharge is directed from a plasma source into the crucible of an electron-beam evaporator. The ionized coating material and a part of the argon ions are accelerated towards the substrates by the self-bias voltage of the electrically insulated substrate holder

may lead to ion-induced stoichiometry defects by the preferential sputtering effect, which results from different sputtering efficiencies of the single components of a compound [6.31]. In the case of the widely used metal oxides, the preferential sputtering of oxygen atoms results in an oxygen deficiency and consequently, in an increase of the optical absorption losses [6.32]. In practice, ion energies covering the range of several 10 eV to several 100 eV are applied within typical IAD processes. The usual ion-source operating media are inert or reactive gases, in the latter case, in particular, oxygen to assist oxide coatings. Consequently, besides the densification effect, the stoichiometry of oxides benefits from the reactive ions, resulting in homogeneous layers with reduced optical losses.

Concerning the IAD process environment, some central demands on the ion source can be stated. High ion currents at low ion energies are needed, and the ion current distribution across the calotte has to be extensive and homogeneous. The ion source must exhibit a high stability and reliability, even at long process times. An independent control of ion current and energy within a wide parameter field is also advantageous. Furthermore, contamination of the growing layers by material originating from the ion source has to be avoided. And finally, options for reactive gases that can

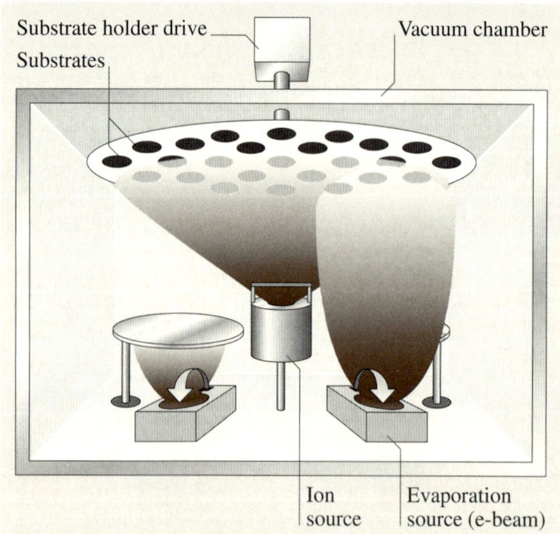

Fig. 6.9 Principle arrangement of an ion-assisted deposition (IAD) process. Besides the electron-beam evaporators, an ion source is implemented in the vacuum chamber. The substrates are exposed to the low-energy ion beam, which results in an additional energy transfer into the growing layer

be used, low levels of maintenance, and low operating costs are relevant. Nowadays, a variety of gridded and grid less ion sources, fulfilling nearly all these requirements for standard IAD applications, are commercially available [6.33, 34].

In the following, the positive effects of ion-induced layer densification are illustrated briefly. With regard to the microstructure, the aforementioned structure zone models for thermally evaporated films have been extended to integrate the dense structures resulting from ion-assisted deposition processes [6.35]. As the microstructural densification by the ions counteracts the void development of the columnar growth, the high optical losses attributed to adsorbed water can be prevented by applying IAD concepts. In Fig. 6.10, the optical losses of a SiO_2 single layer deposited with a thermal evaporation process are compared to those of a SiO_2 single layer produced within an ion-assisted deposition process. The plotted extinction coefficients are calculated from spectrophotometric transmittance and reflectance measurements and indicate optical losses in the range between 10% and 20% for the thermally deposited layer. Using spectrophotometric measurements, no water could be detected in the IAD coating. Besides the reduction of absorption losses, another positive effect of the dense and water-free microstructure is the drastic increase in thermal spectral stability of the coatings. Ion-assisted deposition can reduce the relative wavelength shift $\Delta\lambda/(\lambda °C)$ of the spectral characteristic from several hundred ppm/°C for a thermally deposited coating to a few ppm/°C [6.36].

Furthermore, ion-assisted deposition offers the possibility to tune the mechanical intrinsic stress of a coating from tensile to compressive stress, which can improve the layer adherence and the dimensional accuracy of critical optical components [6.37–39]. And finally, the implemented ion source enables options for substrate precleaning process steps.

6.2.3 Sputtering

Sputtering represents an established and versatile class of energetic vacuum deposition processes, applying condensing particles with significantly higher energies than those resulting from thermal evaporation methods. Its basic principle, derived from the early gas discharge experiments in the 19th century, is depicted in Fig. 6.11. Within a vacuum chamber at a pressure of about 1 Pa, a glow discharge is maintained by a direct-current (DC) voltage in the kV range between the anode, carrying the substrates to be coated, and the cathode. The cath-

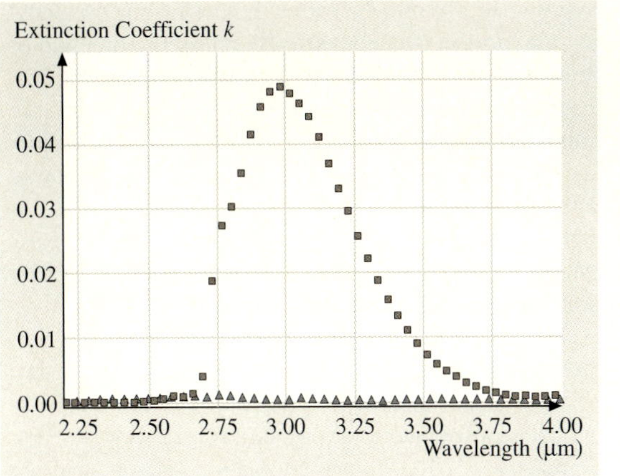

Fig. 6.10 Comparison of the optical losses of SiO_2 single layers: thermal evaporation versus ion-assisted deposition (absorption band of adsorbed water). The optical losses are calculated from measured spectrophotometric transmittance and reflectance data

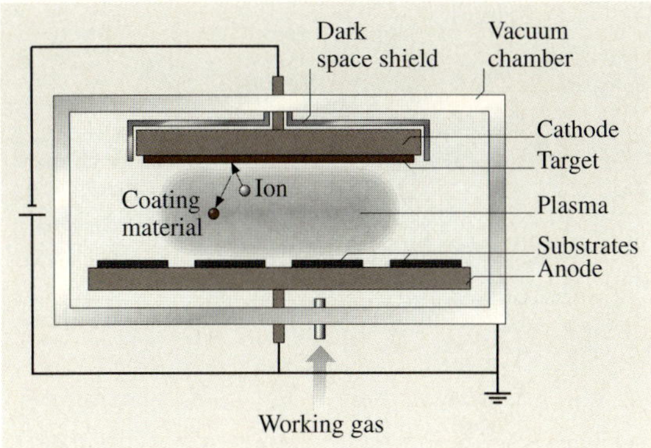

Fig. 6.11 Schematic diagram of a direct-current sputtering process. A glow discharge is driven by a DC voltage in the kV range. Coating material is sputtered from the target on the cathode and is deposited on the substrates positioned at the anode

ode consists of the coating material which has to be electrically conductive (*target*). In the discharge area, positive working gas ions, e.g., argon ions, are generated by impact ionization and accelerated towards the target. Ions impinging on a solid surface, besides other interactions, eject atoms, molecules, and clusters by collision cascades. This effect is called sputtering; its efficiency depends on the target material and microstructure as well as on the species, the energy, and the angle of incidence of the ions. Within the depicted DC discharge process, the particles sputtered from the target condense on the substrates and form a layer. As a consequence of the high kinetic energies of the adatoms in a typical range of several eV (maximum energy of several tens of eV), sputtered coatings exhibit a higher density in comparison to evaporated thin films.

A way to overcome the DC sputtering limitation to conductive targets was found in the technology of radio-frequency (RF) sputtering. In a common set up, the DC electrodes shown in Fig. 6.11 are converted to RF coupling electrodes. Choosing a geometrical arrangement in which the surface area of the substrate electrode exceeds the area of the target electrode, a DC potential with a negative target electrode results from the self-bias effect. Due to this DC potential, the sputtering occurs on the target, and not on the substrate electrode [6.40].

Today, magnetron sputtering is an important sputtering technology for the industrial production of optical coatings. The term magnetron is derived from a magnetic field, crossing the electric field of a DC discharge, and confining the plasma within the region in front of the target (Fig. 6.12). Resulting from the increased degree of ionization, a significantly higher deposition rate provides a reduction in process time. Furthermore, the layer quality benefits from the less intensive interaction with the plasma as well as from the reduced pressure, in the range of 10^{-1} Pa. However, the application of the DC magnetron source within a reactive process environment is linked to some limiting complications. The reactive gas interacts with the target material and the generated compounds form an insulating layer on the target surface, an effect called *target poisoning*. Due to the differences in sputtering efficiency of metals and, for example, oxides or nitrides, the deposition rate decreases strongly while the target is poisoned. Also a modification of the discharge conditions is attributed to the formation of an insulating layer, which can lead to arcs or even a termination of the discharge. Besides the poisoning of the target cathode, the whole chamber, acting as an anode, is covered successively by the insulating compound – the so-called disappearing anode. To overcome these limitations, the magnetron process can be driven in a transition mode between the metallic and the non-metallic mode. However, as this transition mode exhibits considerable instability, a complex active process control has to be implemented to stabilize the relevant parameters, e.g., on the basis of lambda probes for reactive oxygen [6.41].

A second technological approach is the sequential processing concept, in which the magnetron sources work in the metallic mode. Separated from the magnetrons, plasma sources provide a subsequent treatment of deposited sublayers, consisting of a few monolayers,

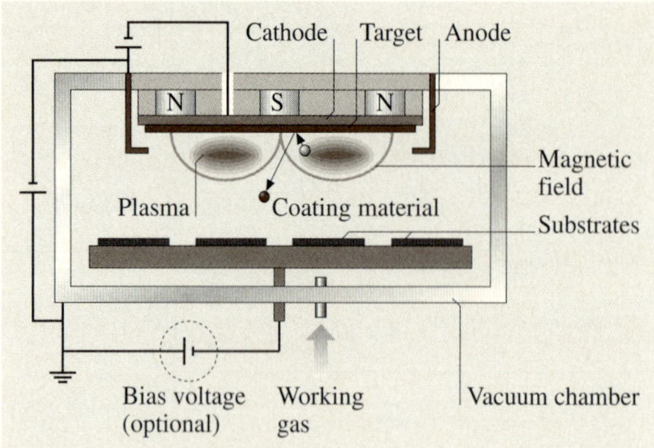

Fig. 6.12 Basic principle of the magnetron sputtering process. A magnetic field, crossing the electric field of a DC discharge, confines the plasma within a region in front of the target. An optional bias voltage applied to the substrate holder allows for an additional energy input into the growing layer

with activated reactive gases. In technical implementation, the magnetron and plasma sources are often located in a circular geometry with a fast rotating cylindrical substrate holder in its center [6.42, 43].

Important advances have been achieved on the basis of dual-magnetron (also called twin-magnetron) configurations, applying medium-frequency voltages in a typical range of 20–100 kHz [6.44]. In a dual-magnetron arrangement, both target electrodes working alternately as the cathode and anode, preventing electrostatic charging as well as the disappearing anode effect, mentioned above. Furthermore, improvements have been made by operating magnetrons in pulse mode, since the limiting thermal load on the target is decreased by this technique.

Nowadays, inline magnetron sputter systems are standard for the large-area deposition in the application fields of architectural glass, photovoltaic, and displays, whereas sequential processing concepts in combination with dual magnetrons are currently finding their way into precision optics.

6.2.4 Ion-Beam Sputtering

Ion-beam sputtering (IBS) represents its own class of sputtering deposition technology, applicable for the highest-quality optics. The basic principle of an ion-beam sputtering process is illustrated schematically in Fig. 6.13 [6.45]. Within a vacuum chamber, a separated ion source is directed to the target, which is consequently not in contact with the ion-generating plasma. The set up provides a rotating substrate holder for the condensing thin films in the geometrically preferred direction of the sputtered particles. Under typical process conditions, argon ions with kinetic energies in the rage of 1 kV are employed for sputtering, while an additional reactive-gas inlet provides an option for depositing compound layers from metallic targets. In some cases, a second ion source is directed to the substrates, and allows for assistance of the growing layer (compare IAD) as well as precleaning of the substrates.

In comparison to the described DC approach, alternating-current (AC) and magnetron sputtering processes, ion-beam sputtering offers a variety of advantages. A low working pressure (reactive $\approx 10^{-2}$ Pa, nonreactive $\approx 10^{-3}$ Pa) in combination with the absence of interactions between the substrates and the plasma result in high-quality thin films with minimum contamination and defects. However, the decisive factor for the excellent optical and mechanical film quality in ion-beam sputtering is the high energy (up to 100 eV) of the layer-forming particles. IBS coatings are dense, amorphous, and suitable for ultra-low-loss components, as total optical losses in the range of 1 ppm are achievable [6.46]. Furthermore, ion-beam sputtering is an extremely stable process, which allows for a high degree of automation. Also, energy and current density of the sputtering ions can be adjusted independently within a wide range.

Nevertheless, ion-beam sputtering is less often used in the industrial mass production of optical coatings. This niche position results from distinct economic disadvantages, namely a low deposition rate and technical difficulties in coating larger areas homogeneously. Compared to thermal evaporation or magnetron sputtering processes for precision optics, typical IBS deposition rates are 10 times lower (in the range 0.5 Å/s instead of 0.5 nm/s). Major current market segments for ion beam sputtering include high-end precision optics for special applications, mainly in the research and science sector, e.g. complex chirped mirrors for femtosecond lasers as well as next-generation lithography [6.47, 48].

6.2.5 Chemical Vapor Deposition (CVD)

Similar to the PVD processes described above, chemical vapor deposition offers a range of process variants. All CVD processes share the basic principle that the deposited layer is a product of a chemical reaction of gaseous reactants (precursors) [6.49]. This reaction is activated within various process types by different kinds of energy input, covering thermal, plasma, and radiation-

Fig. 6.13 Schematic illustration of the ion-beam sputtering process. A separate ion source is directed to a target, the sputtered particles condense on the rotating substrates. Automatically changeable targets allow for the production of multilayer coatings

induced CVD. Except for the coating material, all further reaction products have to be gaseous, as they have to be exhausted from the process chamber. Chemical vapor deposition is widely applied in hard coatings for wear-protection purposes, besides being the most widely used deposition method in the semiconductor industry. In the production of optical coatings, CVD processes still play a minor role in comparison with PVD technologies.

Chemical vapor deposition distinguishes two layer-forming mechanisms. In heterogeneous nucleation, the coating material is generated on the substrate surface. In homogeneous nucleation, the chemical reaction producing the coating material takes place above the substrate surface. Subsequently, the reaction product diffuses to the surface of the substrate, where the film is formed. Since coatings produced by homogeneous CVD processes in most cases are less dense and exhibit a lower layer quality than coatings by heterogeneous CVD processes, the latter variant is standard. However, some process concepts combine both types by applying a homogeneous gas-phase reaction to produce a reactant for the final heterogeneous layer-forming reaction. In all cases, complex gas dynamics have to be controlled by adaptation of the gas inlets, and a balanced flow and pressure control.

The classic way to perform chemical vapor deposition is by thermal activation of the reactions by heating of the substrates. Most of these reactions require high temperatures, typically exceeding 450 °C, which are incompatible with optical coatings on substrates for precision optics or on plastics. As an example, in the semiconductor industry the pyrolysis of silane is used to deposit silicon (g: gaseous, s: solid):

$$\mathrm{SiH_4(g)} \xrightarrow{\approx 600--700^\circ\mathrm{C}} \mathrm{Si(s)} + 2\mathrm{H_2(g)}\,. \tag{6.20}$$

Applying additional oxygen, silica coatings can be produced:

$$\mathrm{SiH_4(g)} + \mathrm{O_2(g)} \xrightarrow{\approx 450^\circ\mathrm{C}} \mathrm{SiO_2(s)} + 2\mathrm{H_2(g)}\,. \tag{6.21}$$

Thermally activated chemical vapor deposition processes are driven at atmospheric pressure (APCVD) or, more commonly today, in low-pressure reactors at a typical pressure range of 10^1-10^3 Pa (LPCVD). Atmospheric-pressure CVD offers higher deposition rates than LPCVD, which, in contrast, provides higher uniformity and a better covering of nonplanar geometrical shapes.

Another way of energy input is used in plasma-enhanced chemical vapor deposition (PECVD), a process variant rapidly gaining importance in many applications [6.50]. The PECVD technique offers the advantage of much lower process temperatures than those needed in thermally activated CVD. As a result of the applied DC, RF or microwave plasma excitation, the gaseous precursors can react while the substrate temperature stays below 300 °C. In Fig. 6.14, a basic RF-driven PECVD reactor is shown.

A special pulsed PECVD concept is based on microwave impulses and is therefore called plasma impulse CVD (PICVD) [6.51]. The PICVD process offers a feasible technical solution to apply coatings on complex-shaped surfaces, in particular on the inner surfaces of tubes, reflectors, or even bottles. Process pressures in the range of a few 10^{-2} Pa result in short pumping times. On the other hand, the pumping time is minimized by the smallest reactors, as, in the case of inner surfaces coatings, the process chamber is often geometrically defined by the substrate itself. Examples of this modularly scalable PICVD concept are cold light mirrors on halogen-lamp reflectors as well as barrier coatings on the inner walls of polyethylene (PET) bottles or of ampoules for pharmaceutical packaging. As every microwave pulse deposits a sublayer of a reproducible thickness in a few milliseconds, a precise thickness control can be realized by pulse counting. Consequently,

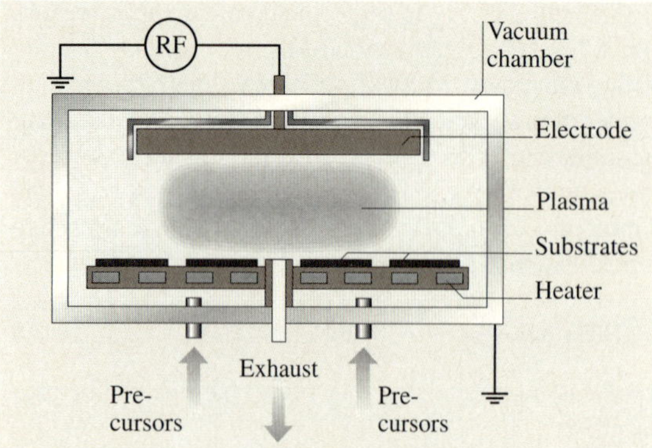

Fig. 6.14 Basic reactor set up for RF-excited plasma-enhanced chemical vapor deposition (PECVD). The plotted substrate heater is an option to assist the layer-forming mechanism. Without the RF electrode the CVD process can be driven as a non-PECVD process by thermal activation only

the deposition rate, which can exceed 10 nm/s for metal oxides, is determined by the pulse repetition rate and the conditions of the chemical reaction. For example, SiO_2 coatings are derived from a $SiCl_4$ precursor by the overall reaction:

$$SiCl_4(g) + O_2(g) \xrightarrow[\rightarrow \text{ plasma}]{\text{microwave pulses}} SiO_2(s) + 2Cl_2(g)\,. \quad (6.22)$$

In the process chamber, this reaction is divided into two parts. In a first homogeneous gas-phase reaction SiO is produced ($SiCl_4 + (1/2)O_2 \rightarrow SiO + 2Cl_2$), which diffuses to the surface and is oxidized in the final heterogeneous reaction ($SiO + (1/2)O_2 \rightarrow SiO_2$).

In an alternative approach for activation, the CVD reactions can be induced by laser radiation [laser CVD (LCVD)]. This process variant is based either on a pyrolytic, i. e., thermal, interaction or direct photolytic activation [6.52]. In contrast to the other mentioned CVD techniques, LCVD is normally not applied for optical coatings. Laser CVD is primarily suitable for micro-technical applications, as structures in the range of several 10 microns can be deposited selectively, e.g., for circuit or mask repair in the field of electronics and lithography, or for the creation of three-dimensional objects such as carbon fibres [6.53, 54].

Generally, chemical vapor deposition plays a minor role compared with physical vapor deposition in optical thin film technology. However, with regard to coatings on complex surface geometries, CVD processes are superior to PVD technologies, as diffusion is an undirected effect. Furthermore, the precursor chemistry offers the possibility to obtain intermediate refractive indices, e.g. for the production of rugate filters, by varying the mixture [6.55]. In this connection, it has to be mentioned that these chemicals are mostly toxic and often difficult to handle.

6.2.6 Other Methods

Besides the introduced common deposition techniques in the fields of precision, laser, and consumer optics, a variety of processes for special applications is available today. Analogous to the laser CVD mentioned above, a laser PVD process has been developed, which has become known as pulsed laser deposition (PLD). In PLD, target material is ablated by the laser pulses within a vacuum chamber and is subsequently deposited onto the substrates positioned in the plasma plume. The PLD process is mainly employed in laboratory-scale systems and offers superior reproduction of the target stoichiometry in the produced thin films of organic and inorganic compounds [6.56]. However, up-scaling of this technique to large coating areas and high deposition rates is problematic and still too cost-intensive.

A more common deposition method that is used in some application fields of optical coatings is the sol–gel process. Starting from a colloidal solution (sol), in most cases of silicon or metal alkoxides, a transition leads to a gel phase, which is the base material for the coating. In the case of dip-coating, the layer is formed by dipping the substrate into the solution. In the spin-coating process, the layer is produced by spreading the solution on the spinning substrate to achieve a uniform thickness distribution, which depends on material-specific parameters and the speed of rotation. Furthermore, the solution can also be applied by spraying, in a laminar-flow-coating process, using capillary forces, or by printing techniques. A densification of the initially porous sol–gel films is often done by baking at temperatures of up to 1000 °C. The domain of the sol–gel technology in optics is the production of large-area antireflection coatings, especially for high-power lasers [6.57].

6.2.7 Process Control and Layer Thickness Determination

With regard to the nanometer precision demanded in optical coating production, process control plays a key role in modern production environments. Recent trends to realize advanced process control strategies are often based on in situ monitoring techniques of important deposition parameters and coating properties. The central requirement for the production of optical interference filters is a precise thickness control of the growing layers.

In a straightforward approach, the layers can be terminated by time, if the deposition rate of the process is stable enough to fulfil the required thickness accuracy, e.g., in some sputtering processes. An exceptional position is taken by plasma impulse CVD, where the thickness can be determined by counting microwave pulses.

A widespread technical solution to measure the actual deposition rate and the layer thickness in the production of thin films is a quartz crystal monitoring system [6.58]. Since the resonance frequency of the oscillating quartz varies with the material deposited on its surface, the rate can be determined from the change in oscillation frequency on the basis of material-specific constants (e.g., specific weights). As a consequence of a position of the quartz crystal monitor differing from the substrate holder location, another calibration factor, called the tooling factor, has to be considered for an accurate determination of the deposition rate. In particular, these factors depend not only on coating materials and the process parameters, but often also on the layer thickness as a result of the developing microstructure in the growing thin films.

As indicated above, the performance of a multilayer filter depends on optical thicknesses, which are the products of the physical thicknesses and the refractive indices of the included single layers. Thus, monitoring methods that provide in situ analysis of the optical thickness of the growing layer are often superior to systems monitoring mass deposition, especially under the conditions of small variations in the optical properties. The natural way to determine optical thickness is a direct optical access in the form of a transmittance or reflectance measurement [6.59]. The theoretical analysis of a single layer reveals a cyclic behavior of the transmittance or reflectance at a fixed wavelength with increasing optical thickness. This effect can be employed to determine the actual optical and, moreover, the actual physical thickness of the deposited layer on the basis of the known optical constants. Especially, for desired single-layer thickness values of integer multiples of QWOT at the monitoring wavelength, the termination points are extremal values of the measured curves. Since certain wavelengths are distinguished, the monitoring wavelength has to be chosen in accordance with the coating design, and often multiwavelength monitoring systems are applied. In most cases, the measurements are performed on a centred test substrate, which remains stationary, in contrast to the products to be coated.

In the course of the rapid development of computer and spectrometer technology, advanced optical broadband monitoring systems have been implemented in deposition process environments [6.60]. These techniques allow for direct broadband in situ transmittance measurements on the moving substrates for each rotation of the calotte. For example, a spectral range of 350–1060 nm can be covered by a fibre-coupled charge-coupled device (CCD) spectrometer system in combination with a free-beam measurement set up inside the vacuum chamber (Fig. 6.15). Each in situ spectrum recorded can be calibrated, because three single measurements are typically performed during each revolution of the calotte: a dark measurement on an opaque part of the calotte, a reference measurement through an open position in the calotte, and the measurement through the substrate to be coated. Based on these three measurements, the absolute transmittance spectrum of the substrate is calculated. Furthermore, the online monitoring software calculates the current layer

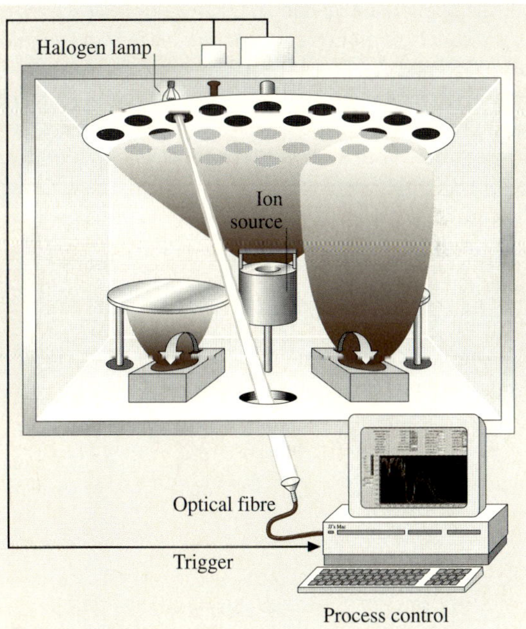

Fig. 6.15 Schematic view of the online monitoring system for optical broadband in situ transmittance measurements, realized within an IAD process environment. A halogen lamp mounted above the calotte inside of the process chamber is used as the light source, and the light is coupled into the detector fibre through a window in the chamber bottom

thickness on the basis of the known optical parameters of the coating material (dispersion values and extinction coefficients). Provided with a target multilayer design, the online system can control the deposition plant, determine the end of the layer currently being deposited, and start the next layer. Because of the large databases available, broadband monitoring systems are also useful as a supporting tool for process development and quality management in the production of modern optical coatings. In contrast to the test substrate arrangement in conventional single- and multiwavelength optical monitors, broadband measurements on the moving substrates make calibration factors and test runs redundant, if there are only the smallest deviations between the vacuum spectra and the spectra after venting (the *vacuum-to-air shift*). Therefore, an appropriate combination of an adequate deposition process and monitoring concept has to be chosen, e.g., an IAD process providing an extremely low vacuum-to-air shift, as shown above.

Besides a precise layer thickness determination, in the development and control of coating processes for optics, a variety of optimization criteria has to be taken into account. Large areas have to be coated homogenously in flexible and fast processes. Besides the optical quality, the coatings have to exhibit good adherence, low mechanical stress, high abrasion resistance, and adequate environmental stability. Parameters to be controlled are, for example, substrate temperatures, deposition rates, partial pressures of process gases, or ion current densities and energies. Finally, it has to be mentioned that the thin film processing concepts presented are not only applied for optical coatings, including precision, laser, and consumer optics, but also in the fields of electronics, semiconductors, displays, medical technology, tribology, and even decorative coatings.

6.3 Quality Parameters of Optical Coatings

For the application of optical coatings in laser technology, a variety of quality parameters have to be taken into account [Table 6.1, [6.61–68]]. The spectral characteristics are the fundamental properties of the coating system and have to be adjusted primarily for the optical functioning of the laser system or optical device [6.69]. But, as soon as high laser powers or requirements concerning the precision are involved, additional quality parameters describing the optical losses and the stability of the coatings have to be evaluated. For many applications of high-power lasers, the absorbance, which transforms a fraction of the impinging radiation energy into heat, inducing a temperature increase in the coating, is of major concern. As an effect of absorption a temperature profile is built up in the optical component, which may influence the transfer behavior or even lead to the thermal

Table 6.1 Selected quality parameters of optical coatings and surfaces in conjunction with corresponding ISO standards and measurement principles

Specification	Parameter	Unit	Standard	Measurement principle
Laser-induced damage threshold (LIDT)	cw-LIDT	W/cm	ISO 11254-1	Cw-laser irradiation
	1 on 1-LIDT	J/cm^2	ISO 11254-1	Irradiation with single pulses
	S on 1-LIDT	J/cm^2	ISO 11254-2	Repetitive irradiation with pulses
	Certification	J/cm^2	ISO 11254-3	Irradiation sequence
Optical losses	Absorptance	ppm	ISO 11551	Laser calorimetry
	Total scattering	ppm	ISO 13696	Integration of scattered radiation
Transfer function	Reflectance	%	ISO 13697	Precise laser ratiometric method
	Transmittance	%	ISO 15368	Spectrophotometry
Surface quality	Form tolerances Scratch/digs Roughness	λN	ISO 10110	13 parts containing different types of imperfections
Stability	Abrasion		ISO 9211	Different test methods
	Environmental Stability		ISO 9022	21 parts containing a variety of conditioning methods

destruction of the optical component [6.70]. Besides the technical problems directly related to thermal effects, absorption always implies a loss of expensive laser energy, demonstrating the economical dimension of optical losses in modern laser systems. Absorbance in optical coatings is mainly governed by defects, stoichiometric deficiencies, or contaminants generated in the coating material during the production steps. In particular, dielectrics often suffer from unbalanced stoichiometry with a slight excess of the metal component, which is caused by a decomposition of the coating materials during thermal evaporation or sputtering. Stoichiometric effects influence the absorbance behavior in the short-wavelength region whereas adsorbed water, which is the predominant contaminant in optical coatings, may lead to increased absorption values in the MIR spectral region, particularly at the wavelength 10.6 μm [6.71] of the CO_2 laser or the band of the Ho : YAG and Er : YAG lasers in the range 2–3 μm [6.72]. A standard procedure for the measurement of absorbance in laser components, which is based on the laser calorimetric principle, is outlined in ISO 11551 [6.62]. For calorimetric absorbance measurement, a temperature sensor is attached to the specimen, which is located in a thermally isolating chamber. According to the standard protocol, the sample is irradiated by a laser beam with known power for a heating time t_R after having reached thermal equilibrium with the environment. As a consequence of the heat flow coupled into the sample by the absorption, an exponential increase in temperature can be monitored by the sensor element. Subsequently, the laser is blocked and the sample temperature decreases in proportion to the heat dissipation into the environment. For the determination of the absorbtance, the recorded heating and cooling curve are evaluated according to the methods described in the standard. The laser-calorimetric measurement technique had been tested in various round-robin experiments [6.73, 74] and offers the advantage of an absolute and sensitive assessment of absorbance [6.75].

Scattering is the second loss channel in optical components and summarizes all effects that deflect the radiation from its specular direction [6.76, 77]. Besides the economic loss of radiation, scatter may induce a reduction of the imaging quality of optical systems and may even appear as a safety problem, if a significant fraction of laser power is diverted into the environment, endangering personal operating the laser device. Scattering losses of optical coating systems can mainly be attributed to microstructural imperfections and inclusions in the coatings as well as to the roughness of the surface and the interfaces between the individual layers [6.76]. Theoretical models on the basis of these effects reveal a scaling of the scattering with the wavelength which can be described on the basis of a $1/\lambda^x$ function with an exponent x of 1–4. In view of this fundamental relation, high scattering values dominating the losses of an optical component are expected especially for the vacuum ultraviolet (VUV)/UV spectral ranges. Following the wavelength scale towards longer wavelengths, optical scattering decreases and may be even neglected for the MIR spectral range, where absorption losses are more pronounced. During the last three decades, an extended scientific and technical background has been built up in the field of scattering measurements resulting in various standard measurement procedures for angle-resolved scattering (ARS) [6.78] and total scattering (TS) [6.63, 79]. Especially, for the determination of the TS value, which is defined by the total amount of radiation scattered into the 4π fullspace by an optical component, measurement set ups with an Ulbricht integrating sphere [6.80] or a Coblentz collecting sphere [6.81] are described in ISO 13696. The fundamental principle of the Ulbricht sphere is based on the integration of scattered radiation by a white highly diffuse reflecting coating on the inner wall of sphere and a subsequent monitoring of part of the integrated radiation, whereas the Coblentz sphere performs a direct collection and concentration of the scattered radiation onto the detector element. Also, the standard ISO 13696 for the measurement of TS values has been tested in various measurement campaigns [6.82] and has recently been qualified for the DUV/VUV spectral range [6.83, 84].

In particular, high-power lasers impose high demands on the power-handling capability of optical coatings, which is expressed in terms of the laser-induced damage threshold (LIDT). Fundamental parameters limiting the LIDT values of coatings are given by the inherent properties as the melting point, the thermal conductivity or the band-gap energy of the employed materials [6.85]. Besides these intrinsic properties, extrinsic effects related to defects and inclusions in the layer structure or special high-power mechanisms at the layer interfaces have to be taken into account [6.86]. Frequently, defect-induced damage mechanisms are observed in optical coatings (Fig. 6.16), which can be described on the basis of inclusions catastrophically heated under laser radiation [6.87]. In this theory, the generation of heat in the inclusion is modeled by the Mie absorption cross section, which is valid for particles with sizes in the range of the interacting wavelength. The diffusion of heat from the inclusion

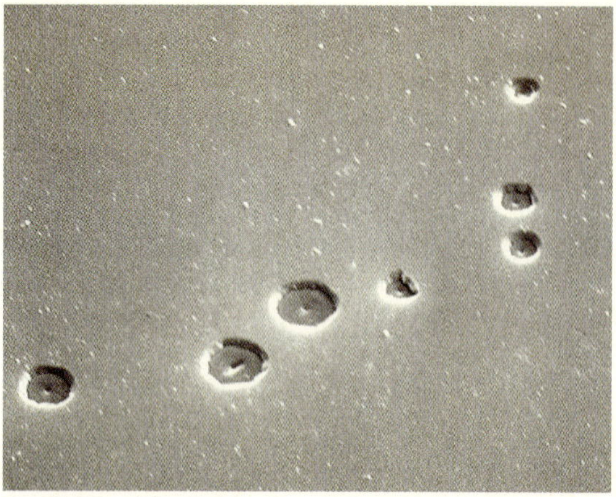

Fig. 6.16 Scanning electron microscope (SEM) picture of a damage site initiated by defect mechanisms in a single-layer coating of TiO_2 subjected to an energy density of 10.9 J/cm^2 in a Nd:YAG laser beam with a diameter of approximately 200 μm. The height of the picture corresponds to a scale of 20 μm

into the surrounding coating material can be expressed by a solution of the heat diffusion equation for this specific geometry with polar symmetry [6.88]. The point of damage is reached when the perimeter temperature of the inclusion attains the melting point of the surrounding coating material. During the last two decades, the defect model had been studied by a several working groups, resulting in a deep understanding of the underlying mechanisms and the corresponding properties of the coating defects [6.85, 89]. Besides inclusion-dominated breakdown other mechanisms based on absorption and on electronic effects are discussed for the development of high-power laser coatings. In the picture of absorption-induced damage, the energy is directly coupled into the layer structure by absorption in the interaction area with the laser beam and leads to homogeneous temperature increase until the damage temperature of the structure is reached. The damage temperature is specified by transitions in the crystalline structure, the crossing of a defined coating stress level (Fig. 6.18), or the melting temperature (Fig. 6.17) of the coating material [6.90]. For short laser pulses in the ps and fs time domain, typical diffusion lengths of thermal effects are very small compared with the thickness of the layer structure. As a consequence, thermal effects can be neglected for the modeling of ultrashort pulse damage, which is mainly governed by electronic processes. Present theories start from the assumption that catastrophic damage takes place at a critical electron density of 10^{21} cm^{-3} in the conduction band of the

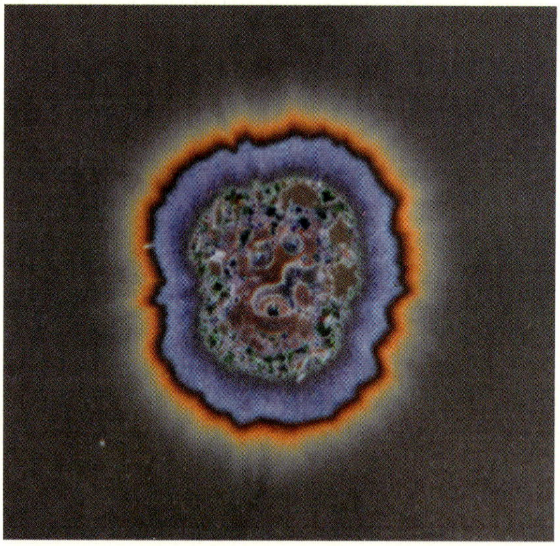

Fig. 6.17 Nomarski micrograph (color-enhanced presentation) of a damage site initiated by absorption-dominated mechanisms in a highly reflecting mirror of Ta_2O_5/SiO_2, objected to an energy density of 45.9 J/cm^2 in a Nd:YAG laser beam with a diameter of approximately 200 μm

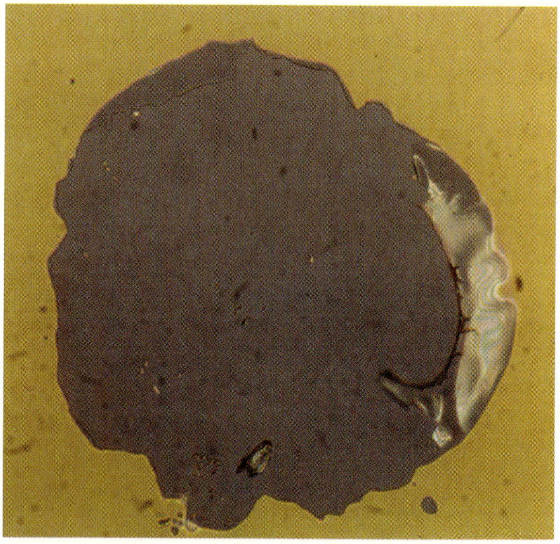

Fig. 6.18 Nomarski micrograph of a damage site initiated by absorption-dominated mechanisms in an antireflective coating of ZrO_2/MgF_2 on a quartz substrate, subjected to an energy density of 35 J/cm^2 in a Nd:YAG laser beam with a diameter of approximately 200 μm. In this case the catastrophic stress level was reached prior to the thermal deterioration of the coating material

coating material [6.91]. At this point laser energy is coupled efficiently into the electrons, resulting in ionization and disruption of the material. For the description of the development of the electron density in the conduction band before arriving at the critical density, rate equations involving multiphoton excitation and the avalanche effect are employed. The major outcome of these theories is a clear correlation of the damage threshold to the maximum internal field strength in the layer structure and to the band-gap energies of the layer materials involved. These theoretical predictions are supported by experiments on single-layer and coating systems tested with ultrashort-pulse lasers in a broad range of pulse durations [6.92]. Aside the numerous damage mechanisms in the depth of the layer structure, features on the surface of the component may also diminish the power-handling capability of the optical component. For example, the field strength of the incoming wave can be enhanced by more than a factor of two in the vicinity of grooves, cracks or other surface imperfection, leading to weak points on the component (Fig. 6.19) [6.93]. An additional cause of surface-initiated damage is the generation of plasmons, which is correlated to the surface roughness of the component [6.94]. In these models, the damage threshold values decrease with the increasing surface roughness of the layers. Conclusively, even though a variety of damage mechanisms have been identified and understood on the basis of adapted models, the power-handling capability of optical thin films is still a major and vivid research area which plays a key role in the development of high-power lasers and their commercialization.

In many technical applications, the optical component additionally has to withstand a variety of environmental influences, including mechanical abrasion, chemical corrosion or severe climatic conditions. For example, the most severe mechanical and chemical requirements are imposed onto coatings for ophthalmology or consumer optics, which are often cleaned with even abrasive and aggressive cleaning solvents and fabrics. In the course of the individual development of the different optical products a variety of standardized testing procedures has been cultivated in the different market sectors for the qualification of optical components. Some of these procedures originating from ophthalmology, defence and medical applications are compiled in Table 6.1, which illustrates the broad spectrum of test procedures available. In particular, in laser technology the requirements in respect to environmental stability are less demanding and in most cases certification according to the conditioning methods with a lower degree of severity described in the ISO standard series 9211 are sufficient. Typical conditioning methods cover abrasion resistance, which is tested by a series of strokes with a special cheese cloth or eraser applied to the optical surfaces, and climatic stability, which is assessed on the basis of defined conditioning cycles in a climatic chamber.

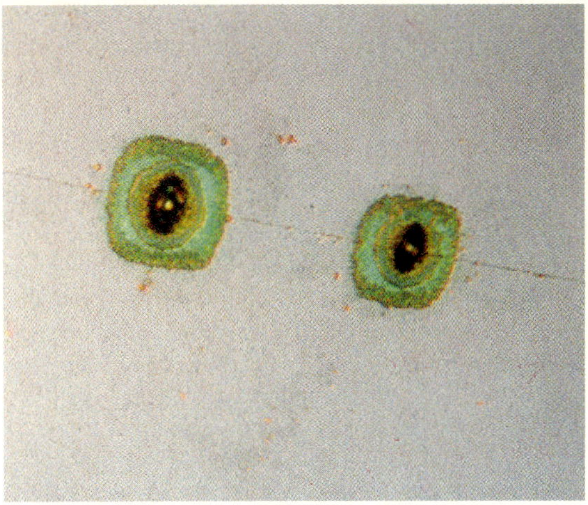

Fig. 6.19 Example of damage initiated by a surface imperfection in an antireflective coating of $ThF_4/ZnSe$ on a ZnSe substrate, subjected to an energy density of $12.7\,J/cm^2$ in a $TEA-CO_2$ laser beam. The width of the picture corresponds to a scale of 150 μm

6.4 Summary and Outlook

Even after more than 60 years of research and development, optical coatings are often still conceived as a technology with a large fraction of *magic art*, which can be mastered only by experience and always leaves a certain probability of unforeseen effects in production. Major challenges include the reliable and reproducible deposition on the basis of well-defined and stable process concepts as well as sensitive and qualified characterization of the coatings. In view of the rapid development of computer systems, online monitoring strategies of the deposition processes and the optimization of stable deposition processes, significant progress on the way towards precise and deterministic production of even complicated coating system are expected in the near future. Laser technology has always stimulated the development of coatings to a considerable quality level (Table 6.2) and will also be one of the major pace-setters for future innovations in the field.

Demanding future challenges will be imposed on optical coatings by the commercialization of new laser products and innovative applications, which are dependent on optical coatings with improved optical quality, higher stability and increased complexity in their functioning combining several surface properties. This new technology generation will also require improved flexibility and economy from optical factories, which can only be achieved on the basis of reproducible production processes and adapted characterization techniques. In conclusion, optical thin films will remain an enabling technology which will play a key role for many future applications and products.

Terms and Definitions

(in their order of appearance)

n_T	refractive index of the substrate
E_0	amplitude of a plane wave
k	wave number of a plane wave $k = 2\pi/\lambda$
λ	wavelength of a plane wave
ω	frequency of a plane wave
δ_i	phase shift of a plane wave in layer number i
d_i	thickness of the layer number i
n_i	refractive index of the layer number i
θ_i	angle of incidence in the layer number i
A_0	amplitude of the partial wave of order zero
A_1	amplitude of the partial wave, order one
A_2	amplitude of the partial wave, order two
t_1, t_1'	transmission coefficients of the ambient–layer interface
r_1, r_1'	reflection coefficients of the ambient–layer interface
r_2	reflection coefficient at the layer–substrate interface
A_k	amplitude of the partial wave of order k
r_S	reflection coefficient of a single layer
M_i	matrix of layer i
E_i, H_i	electric field strength, magnetic field strength at the rear plane interface of layer i
M_S	composite matrix of a layer stack of K single layers
r_{SK}	reflection coefficient of a layer stack of K single layers
n_0	index of refraction of the ambient medium
D_i	film thickness expressed in units of QWOT
QWOT	quarter-wave optical thickness; the unit for the thickness of the layer
λ_Z	design wavelength
n_H	refractive index of a high-index material
n_L	refractive index of a low-index material
R_S	reflectance of a QWOT stack

Table 6.2 Selected quality parameters of optical coating systems for laser applications (Types: HR: high-reflecting mirror, AR: antireflective coating, th: thermal evaporation, IBS: ion-beam sputtering)

Laser, wavelength	Type	Absorption ISO 11551	Total scattering ISO 13696	Laser-induced damage threshold, ISO 11254
157 nm, F_2-excimer	HR/th	1–4%	1–4%	
193 nm, ArF-excimer	AR/th	0.7–2.5%	0.2–0.5%	$1-2\,J/cm^2$ (1on1, 20 ns)
	HR/th	0.4–2.0%	0.2–2.5%	$2-4\,J/cm^2$ (1on1, 20 ns)
248 nm, KrF-excimer	AR/th		< 0.025%	$10\,J/cm^2$ (1on1, 30 ns)
	HR/th	< 500 ppm	< 0.2%	$> 20\,J/cm^2$ (1on1, 30 ns)
	HR/IBS		< 0.1%	$> 3\,J/cm^2$ (1on1, 30 ns)
633 nm, HeNe Laser	HR/th	< 30 ppm	< 30 ppm	-
	HR/IBS	< 5 ppm	< 5 ppm	-
1.064 μm, Nd:YAG	AR/th	< 20 ppm	< 100 ppm	$> 60\,J/cm^2$ (12 ns, 0.25 mm)
	HR/th	< 50 ppm	< 100 ppm	$> 100\,J/cm^2$ (12 ns, 0.25 mm)
	HR/IBS	< 1 ppm	< 1 ppm	$> 80\,J/cm^2$ (12 ns, 0.25 mm)
10.6 μm, CO_2-Laser	AR/th	< 0.16%	-	$> 20\,J/cm^2$ (100 ns, 1.4 mm) $> 2\,kJ/cm^2$ (1.2 ms, 250 μm) $> 3\,kW/mm$ (100 μm)
	HR/th	< 0.10%	-	$> 25\,J/cm^2$ (100 ns, 1.4 mm) $> 2\,kJ/cm^2$ (1.2 ms, 250 μm)

PVD	physical vapor deposition
CVD	chemical vapor deposition
$\mathrm{d}N_e/\mathrm{d}t$	number of evaporating atoms per unit time
A_e	surface area of the evaporation source
α_e	evaporation coefficient
p^*	equilibrium vapor pressure of the evaporant
p	hydrostatic pressure of the evaporant
m	atomic mass
k_B	Boltzmann's constant
T	temperature
p_0	constant factor (pressure)
L_0	latent heat of evaporation per atom or molecule
RLVIP	reactive low-voltage ion plating
I(B)AD	ion-(beam-)assisted deposition
IBS	ion-beam sputtering
APCVD	atmospheric pressure CVD
LPCVD	low-pressure CVD
PECVD	plasma-enhanced CVD
PICVD	plasma impulse CVD
LCVD	laser(-induced) CVD
PLD	pulsed laser deposition
ARS	angle-resolved scattering
TS	total scattering
LIDT	laser-induced damage threshold

References

6.1 P. Vukusic, J. R. Sambles: Photonic structures in biology, Nature **424**, 852–855 (2003)

6.2 J.v. Fraunhofer: *Versuche über die Ursachen des Anlaufens und Mattwerdens des Glases und die Mittel, denselben zuvorzukommen* (Verlag der Königlich Bayerischen Akademie der Wissenschaften, München 1888) p. 1817

6.3 H. D. Taylor: A Method of Increasing the Brilliancy of the Images Formed by Lenses, British patent No 29561 (1904)

6.4 C. Fabry, A. Perot: Théorie et applications d'une novelle méthode de spectroscopie interférentielle, Ann. Chim. Phys. **16**, 115–144 (1899)

6.5 M. Born, E. Wolf: *Principles of Optics*, 7 edn. (Cambridge Univ. Press, Cambridge 1999) p. 986

6.6 A. J. Thelen: *Design of Optical Interference Coatings* (McGraw-Hill, New York 1989) p. 255

6.7 H. M. Lidell, H. G. Jerrard: *Computer-Aided Techniques for the Design of Multilayer Filters* (Adam Hilger, Bristol 1981) p. 194

6.8 P. Baumeister: Computer software for optical coatings, Photon. Spectra **22**(9), 143–148 (1988)

6.9 A. Macleod: *Thin-Film Optical Filters*, 3 edn. (Inst, Publishing, London 2001) p. 641

6.10 W. R. Grove: On the electrochemical polarity of gases, Philos. Trans. R. Soc., London **142**, 87–101 (1852)

6.11 M. Faraday: The bakerian lecture: experimental relations of gold (and other metals) to light, Philos. Trans. R. Soc., London **147**, 145–181 (1857)

6.12 A. W. Wright: On the production of transparent metallic films by the electrical discharge in exhausted tubes, Am. J. Science Arts 3rd Ser. **13**(73), 49–55 (1877)

6.13 T. A. Edison: Art of Plating One Material with Another, US Patent 526,147 (1894)

6.14 A. Macleod: The early days of optical coatings, Journal of Optics A: Pure Appl. Opt. **1**, 779–783 (1999)

6.15 D. M. Mattox: *The Foundations of Vacuum Coating Technology* (Noyes Publications, Norwich 2003) p. 150

6.16 H. Bach, D. Krause: *Thin Films on Glass*, ed. by H. Bach, D. Krause (Springer, Berlin, Heidelberg 1997) p. 404

6.17 H. K. Pulker: *Coatings On Glass*, 2 edn. (Elsevier, Amsterdam 1999) p. 548

6.18 R. R. Willey: *Practical Design and Production of Optical Thin Films*, 2 edn. (Marcel Dekker, New York 2002) p. 547

6.19 P. W. Baumeister: *Optical Coating Technology* (SPIE Press, Bellingham 2004) p. 840

6.20 M. Friz, F. Waibel: Coatings Materials, Optical Interference Coatings, Springer Ser. Opt. Sci. **88**, 105–130 (2003)

6.21 H. Hertz: I. Über die Verdunstung der Flüssigkeiten, insbesondere des Quecksilbers, im luftleeren Raume, Ann. Phys. **17**, 177–200 (1882)

6.22 M. Knudsen: Die maximale Verdampfungsgeschwindigkeit des Quecksilbers, Ann. Phys. **47**, 697–708 (1915)

6.23 K. H. Guenther: Microstructure of vapor-deposited optical coatings **23**(21), 3806–3816 (1984)

6.24 I. Petrov, P. B. Barna, L. Hultman, J. E. Greene: Microstructural evolution during film growth, J. Vac. Sci. Technol. A **21**(5), 117–128 (2003)

6.25 B. A. Movchan, A. V. Demchishin: Study of the structure and properties of thick vacuum condesates of nickel, titanium, tungsten, aluminium oxide and zirconium dioxide, Phys. Met. Metallogr. **28**, 83 (1969)

6.26 M. Auwärter: Verfahren und Einrichtung zur Herstellung dünner Schichten aus Metallverbindungen enthaltenden Ausgangssubstanzen durch Aufdampfen im Vakuum und nach dem Verfahren erhaltene Aufdampfschicht, Swiss Patent No 322265 (1957)

6.27 M. Auwärter: Process for the Manufacture of Thin Films, US Patent 2,920,002 (1960)

6.28 D. M. Mattox: Apparatus for Coating a Cathodically Biased Substrate from Plasma of Ionized Coating Material, US Patent 3,329,601 (1967)

6.29 H. K. Pulker, W. Haag, M. Buhler, E. Moll: *Optical and mechanical properties of ion plated oxide films*, ed. by H. K. Pulker, W. Haag, M. Buhler, E. Moll (Proc. IPAT 85, 5th Int. Conf. on Ion and Plasma Assisted Techniques, Munich 1985) pp. 299–306

6.30 J. D. Targove, H. A. MacLeod: Verification of momentum transfer as the dominant densifying mechanism in ion-assisted deposition, Appl. Opt. **27, 18**, 3779–3781 (1988)

6.31 S. Berg, I. V. Katardjiev: Preferential sputtering effects in thin film processing, J. Vac. Sci. Technol. A **17**(4), 1916–1925 (1999)

6.32 J. B. Malherbe, S. Hofmann, J. M. Sanz: Preferential Sputtering of Oxides: A Comparison of Model Predictions with Experimental Data, Appl. Surf. Sci. **27**(3), 355–365 (1986)

6.33 W. Ensinger: Ion Sources for Ion Beam Assisted Thin-Film Deposition, Rev. Sci. Instrum. **63**(11), 5217–5233 (1992)

6.34 H. R. Kaufman, R. S. Robinson: *Operation of Broad-Beam Sources* (Commonwealth Scientific Corp., Alexandria 1987) p. 201

6.35 J. A. Thornton: Influence of apparatus geometry and deposition conditions on the structure and topography of thick coatings, J. Vac. Sci. Technol. **11**(4), 666–670 (1974)

6.36 H. Ehlers, M. Lappschies, D. Ristau: Ion assisted deposition processes for precision and laser optics, Proc. SPIE **5250**, 519–527 (2004)

6.37 C. R. Ottermann, K. Bange: Correlation between the density of TiO films and their properties, Thin Solid Films **286**(1-2), 32–34 (1996)

6.38 J. Y. Robic, H. Leplan, Y. Pauleau, B. Rafin: Residual stress in silicon dioxide thin films produced by ion-assisted deposition, Thin Solid Films **290-291**, 34–39 (1996)

6.39 J. E. Klemberg-Sapieha, J. Oberste-Berghaus, L. Martinu, R. Blacker, I. Stevenson, G. Sadkin, D. Morton, S. McEldowney, R. Klinger, P. J. Martin, N. Court, S. Dligatch, M. Gross, R. P. Netterfield: Mechanical characteristics of optical coatings prepared by various techniques: A comparative study, Appl. Opt. **43**(13), 2670–2679 (2004)

6.40 S. M. Rossnagel, J. J. Cuomo, W. D. Westwood: *Handbook of Plasma Processing Technology*, ed. by S. M. Rossnagel, J. J. Cuomo, W. D. Westwood (Noyes Publications, Park Ridge 1990) p. 523

6.41 I. Safi: Recent aspects concerning DC reactive magnetron sputtering of thin films: a review, Surface and Coatings Technology **127**(2-3), 203–218 (2000)

6.42 M. A. Scobey, R. I. Seddon, J. W. Seeser, R. R. Austin, P. M. Lefebvre, B. Manley: Magnetron Sputtering Apparatus and Process, US Patent 4,851,095 (1989)

6.43 R. I. Seddon, P. M. Lefebvre: MetaMode: a new method for high-rate MetaMode reactive sputtering, Proc. SPIE **1323**, 122–126 (1990)

6.44 J. Szczyrbowski, G. Braeuer, W. Dicken, M. Scherer, W. Maass, G. Teschner, A. Zmelty: Reactive sputtering of dielectric layers on large scale substrates using an AC twin magnetron cathode, Surface and Coatings Technology **93**(1), 14–20 (1997)

6.45 D. T. Wei, A. W. Louderback: Method for Fabricating Multi-Layer Optical Films, US Patent 4,142,958 (1978)

6.46 G. Rempe, R. J. Thompson, H. Kimble, R. Lalezari: Measurement of ultralow losses in an optical interferometer, Opt. Lett. **17**(5), 363–365 (1992)

6.47 D. H. Sutter, L. Gallmann, N. Matuschek, F. Morier-Genoud, V. Scheuer, G. Angelow, T. Tschudi, G. Steinmeyer, U. Keller: Sub-6-fs pulses from a SESAM-assisted Kerr-lens modelocked Ti: sapphire laser: at the frontiers of ultrashort pulse generation, Appl. Phys. B **70**, 5–12 (2000)

6.48 E. Quesnel, C. Teyssier, V. Muffato, J. Thibault: Study of ion-beam-sputtered Mo/Si mirrors for EUV lithography mask: influence of sputtering gas, Proc. SPIE **5250**, 88–98 (2004)

6.49 J. O. Carlsson: *Chemical Vapor Deposition, Handbook of Deposition Technologies for Films and Coatings*, 2 edn., ed. by R. F. Bunshah (Noyes Publications, Park Ridge 1994) p. 888

6.50 L. Martinu, D. Poitras: Plasma deposition of optical films and coatings: A review, Vacuum Science & Technology A **18**(6), 2619–2645 (2000)

6.51 D. Krause: *Plasma Impulse Chemical Vapour Deposition (PICVD) in Thin Films on Glass*, ed. by H. Bach, D. Krause (Springer, Berlin, Heidelberg 1997) Chap. 5.3, p. 404

6.52 C. Duty, D. Jean, W. J. Lackey: Laser chemical vapour deposition: materials, modelling, and process control, Int. Mat. Rev. **46**(6), 271–287 (2001)

6.53 S. Leppavuori, J. Remes, H. Moilanen: Laser chemical vapour deposition of Cu and Ni in integrated circuit repair, Proc. SPIE **2874**, 272–282 (1996)

6.54 C. Fauteux, R. Longtin, J. Pegna, M. Boman: Microstructure and growth mechanism of laser grown carbon microrods as a function of experimental parameters, J. Appl. Phys. **95**(5), 2737–2743 (2004)

6.55 P. L. Swart, B. M. Lacquet, A. A. Chtcherbakov, P. V. Bulkin: Automated electron cyclotron resonance plasma enhanced chemical vapor deposition system for the growth of rugate filters, J. Vac. Sci. & Tech. A **18**(1), 74–78 (2000)

6.56 P. R. Willmott: Deposition of complex multielemental thin films, Progress in Surface Science **76**, 163–217 (2004)

6.57 M. C. Ferrara, M. R. Perrone, M. L. Protopapa, J. Sancho-Parramon, S. Bosch, S. Mazzarelli: High mechanical-damage-resistant sol-gel coating for high-power lasers, Proc. SPIE **5250**, 537–545 (2004)

6.58 G. Sauerbrey: Use of quartz vibrator for weighing thin layers and as a micro-balance, Z. Phys. **155**(2), 206–222 (1959)

6.59 R. Richier, A. Fornier, E. Pelletier: *Optical Monitoring of Thin-Film Thickness, Thin Films for Optical*

Systems, ed. by R. Flory (Marcel Dekker, New York 1995) p. 585
6.60 D. Ristau: Characterisation and monitoring, Springer Ser. Opt. Sci. **88**, 181–205 (2003)
6.61 ISO: *ISO 11254: Test methods for laser induced damage threshold of optical surfaces. Part 1: 1 on 1-test, 2000, Part 2: S on 1 test, 2001, Part 3 (FDIS): Certification of power handling capability* (ISO, Geneva 2002)
6.62 ISO: *ISO 11551(1997): Test method for absorptance of optical laser components* (ISO, Geneva 1997)
6.63 ISO: *ISO 13696: Test method for radiation scattered by optical components* (ISO, Geneva 2002)
6.64 ISO: *ISO/DIS 15368: Measurement of reflectance of plane surfaces and transmittance of plane parallel elements* (ISO, Geneva 2000)
6.65 ISO: *ISO/DIS 13697: Optics and optical instruments. Lasers and laser related equipment. Test method for reflectance and transmittance of optical laser components* (ISO, Geneva 2004)
6.66 ISO: *ISO 9211: Optical coatings* (ISO, Geneva 1994–2004)
6.67 ISO: *ISO 9022: Optics and optical Instruments – Environmental test methods* (ISO, Geneva 2004)
6.68 ISO: *ISO 10110: Preparation of drawings for optical elements and systems* (ISO, Geneva 1997)
6.69 H. Czichos, T. Saito, L. Smith: *Springer Handbook of Material Measurement Methods*, Vol. 88, ed. by H. Czichos, T. Saito, L. Smith (Springer, Berlin, Heidelberg 2006)
6.70 ISO: *ISO/DTR 22588: Optics and Photonics – Lasers and laser-related equipment –Absorption induced effects in laser optical components* (ISO, Geneva 2004)
6.71 M. Rahe, D. Ristau, H. Schmidt: The effect of hydrogen concentration in conventional and IAD coatings on the absorption and laser induced damage at 10.6 μm, Proc. SPIE **1848**, 335–348 (1992)
6.72 T. Gross, F. Dreschau, D. Ristau, P. Fuhrberg, M. Adamik: Characterisation of laser components for high power Ho:YAG-laser, Proc. SPIE **3244**, 111–117 (1997)
6.73 D. Ristau, U. Willamowski, H. Welling: Evaluation of a round robin test on optical absorption at 10.6 μm, Proc. SPIE **2870**, 502–514 (1996)
6.74 D. Ristau, U. Willamowski, H. Welling: Measurement of optical absorptance according to ISO 11551: Parallel round-robin test at 10.6 μm, Proc. SPIE **3578**, 657–671 (1999)
6.75 U. Willamowski, D. Ristau, E. Welsch: Measuring the absolute absorptance of optical laser components, Appl. Opt. **37**(36), 8362–8370 (1998)
6.76 J. M. Bennett, L. Mattsson: *Introduction to Surface Roughness and Scattering*, 2 edn. (Opt. Soc. Am., Washington, D.C. 1999) p. 130
6.77 A. Duparré: *Light Scattering of Thin Dielectric Films in: Handbook of Optical Properties, Thin Films for Optical Coatings*, Vol. 1, ed. by R. E. Hummel, K. H. Guenther (CRC, Boca Raton 1995) pp. 273–303
6.78 ASTM: *E1392-90: Standard practice for angle resolved optical scatter measurements on specular or diffuse surfaces* (ASTM, Philadelphia 1990)
6.79 ASTM: *ASTM Doc. F1048-87: Standard test method for measuring the effective surface roughness of optical components by total integrated scattering* (ASTM, Philadelphia 1987)
6.80 R. Ulbricht: Die Bestimmung der mittleren räumlichen Lichtintensität durch nur eine Messung, Elektrotechnische Zeitschrift **29**, 595–597 (1900)
6.81 W. W. Coblentz: The diffuse reflecting power of various substances, Bull. Bur. Stand. **9**, 283–325 (1913)
6.82 P. Kadkhoda, A. Müller, D. Ristau, A. Duparré, S. Gliech, H. Lauth, U. N. Reng, M. R. Schuhmann, C. Amra, C. Deumie, C. Jolie, H. Kessler, T. Lindström, C. G. Ribbing, J. M. Bennett: International round-robin experiment to test the ISO total scattering draft standard, Appl. Opt. **39**(19), 3321–3332 (2000)
6.83 P. Kadkhoda, H. Welling, S. Günster, D. Ristau: Investigation on total scattering at 157 nm and 193 nm, Proc. SPIE **4099**, 65–74 (2000)
6.84 A. Hultacker, S. Gliech, N. Benkert, A. Duparré: VUV-Light scattering measurements of substrates and thin film coatings, Proc. SPIE **5188**, 115–122 (2003)
6.85 *Laser-induced damage in optical materials.* Proceedings of the Boulder Damage Symposium. NIST Spec. Publ. and SPIE-Publications, 1969 to 2006. NBS SP 372 (1972), NBS SP 387 (1973), NBS SP 414 (1974), NBS SP 435 (1975), NBS SP 462 (1976), NBS SP 509 (1977), NBS SP 541 (1978), Index of Papers 1969 - 1978 (1979), NBS SP 568 (1979), NBS SP 620 (1981), NBS SP 638 (1983), NBS SP 669 (1984), NBS SP 688 (1985), NBS SP 727 (1986), NBS SP 746 (1987), NBS SP 752 (1987), NBS SP 756 (1988), NBS SP 775 (1989), NBS SP 801, ASTM STP 1117, and SPIE Vol. 1438 (1989), ASTM STP 1141, and SPIE Vol. 1441 (1991), SPIE Vol. 1624 (1992), SPIE Vol. 1848 (1993), SPIE Vol. 2114 (1994), 25 Years Index: 1969 - 1993, SPIE Vol. 2162 (1994,) SPIE Vol. 2428 (1995), SPIE Vol. 2714 (1995), SPIE Vol. 2966 (1997), SPIE Vol. 3244 (1998), SPIE Vol. 3578 (1999), SPIE Vol. 3902 (2000), SPIE Vol. 4374 (2001), SPIE Vol. 4679 (2002), SPIE Vol. 4932 (2003), SPIE Vol. 5273 (2004), SPIE Vol. 5647 (2005), SPIE Vol. 5991 (2006) The publications of the first thirty years are available as CD-version: Laser Induced Damage in Optical Materials. Collected Papers published by the International Society of Optical Engineering, Washington 98227-0010
6.86 T. W. Walker, A. H. Guenther, P. Nielsen: Pulsed laser induced damage to thin film coatings, IEEE J. Quant. Elec. **QE17**(10), 2041–2065 (1981)
6.87 T. W. Walker, A. Vaidyanathan, A. H. Guenther, P. Nielsen: Impurity breakdown in thin films. In: *Proc.Symp. Laser Induced Damage Opt. Mat.*, ed. by T. W. Walker, A. Vaidyanathan, A. H. Guenther, P. Nielsen (NBS Special Publication, Washington 1979) pp. 479–495

6.88 H. Goldenberg, C. J. Tranter: Heat flow in an infinite medium heated by a sphere, Brit. J. Appl. Phys. **3**, 296–298 (1952)

6.89 D. Ristau: Laser damage in thin film coatings. In: *Encyclopedia of Modern Optics*, ed. by R. D. Guenther, D. G. Steel, L. Bayvel (Elsevier, Amsterdam 2004) pp. 339–349

6.90 D. Ristau, X. C. Dang, J. Ebert: Interface and bulk absorption of oxide layers and correlation to damage threshold at 1.064 μm, NBS Spec. Publ. **727**, 298–312 (1984)

6.91 J. Jasapara, A. V. V. Nampoothiri, W. Rudolph, D. Ristau, K. Starke: Femtosecond laser pulse induced breakdown in dielectric thin films, Phys. Rev. B **63**(4), 045117/1–5 (2001)

6.92 M. Mero, L. Jianhua, A. Sabbah, J. C. Jasapara, K. Starke, D. Ristau, J. K. McIver, W. G. Rudolph: Femtosecond pulse damage and pre-damage behavior of dielectric thin films, Proc. SPIE **4932**, 202–215 (2003)

6.93 N. Bloembergen: Role of cracks, pores and absorbing inclusions on laser damage threshold of transparent dielectrics, Appl. Opt. **12**(4), 661–664 (1973)

6.94 R. A. House: *The effects of surface structural properties on laser-induced damage at 1.06 micrometers* (Air Force Inst. of Tech., Wright-Patterson, OH 1975)

Part B

Fabricati

Part B Fabrication and Properties of Optical Components

7. Optical Design and Stray Light Concepts and Principles

To insure that an optical system performs to specifications, the optical engineer needs to fully consider several aspects of the design process. Each of these tasks can be aided with the use of software tools. The optical engineer needs to understand the strengths and limitations of the available software tools and how to best apply these programs to each design. There are several distinct steps in the implementation of an optical system: the first order optical layout, optimized design of the optical system, performing stray, scattered and ghost analysis, performing a tolerance analysis of the optical system and performing manufacturing analysis. Although each of these steps are often considered separately, and often require the use of several different software tools, it is imperative for the engineer to consider the entire process during each phase of system development so that issues arising from stray light or manufacturing tolerances do not force a redesign of the system. A thorough understanding of the optical design and analysis process as well as the proper use of available optical software tools are necessary to insure the optimum optical design for each specific application.

The design of an optical system is comprised of several distinct but interrelated steps. The initial phase in the optical design process is the design of the optical system. That is, establishing a set of optical parameters for each necessary optical component so that the design's performance level exceeds the design specifications. Another phase of the design process is to perform a tolerance analysis to verify that the system will still perform to specification when the limitations of material properties, manufacturing and assembly are considered. Another important phase is a separate analysis to consider the effects of stray light due to ghost reflections, scatter, and the design of proper baffling.

The optical designer is aided in each of these steps with computer programs specifically written for these tasks. The design and tolerancing of an optical system is generally performed using software referred to as sequential optical design programs or lens design codes. Stray light analysis is performed using a non-sequential, or unconstrained, optical analysis program. We will consider the capabilities and requirements of each of those optical programs.

7.1 The Design Process

The process of designing an optical system consists of establishing a set of optical components with optical properties, including materials (indices of refraction, dispersions), surface curvatures, element thicknesses and spacing, surface shape, etc., which in combination with system properties such as focal length, focal ratio, field of view, and others requirements, provide a design framework that will meet (or exceed) the required performance specification.

Although a lens design code is the most efficient method of determining the necessary parameter values, it is important for the designer to perform a first-order

Table 7.1 Basic optical system specifications and requirements

System parameters
Focal length
Focal ratio or numerical aperture
Aperture size
Wavelength (band, weighting)
Full field of view
Image size, shape
Magnification
Object, image locations
Performance
Transmission
Vignetting factors
Encircled energy
Field curvature
Distortion
Optical components
Spheres/aspheres
Materials (glass/plastic)
Number of elements
Coatings
Detector
Total size
Number of pixels (horizontal, vertical)
Pixel aspect ratio
Nyquist frequency
Package
Total track
Maximum length, width
Weight
Environment
Required temperature range
Humidity
Pressure
Shock/vibration

predesign before turning the process over to the computer. The predesign serves several purposes. First, it gives the designer another opportunity to consider all aspects of the design specifications closely to ensure that all necessary information and design specifications have been provided and to discuss any missing, incomplete or inconsistent requirements. Understanding the design requirements is a critical step in the design process. If any necessary specifications are uncertain, it is important to discuss any issues with the customer before the design begins. Any judgment call made by the optical designer can be overruled at any time, resulting in redesigns and costly delays. At this time the designer can also make a reasonable determination as to whether or not the desired system is physically realizable.

After determining that all necessary specification information is available, a first-order system layout can be determined (see Table 7.1 for a partial listing). As part of this analysis, it is important to consider any constraints placed on the design that may result in tolerance requirements that may exceed current fabrication limits. Although the full tolerance analysis will be performed after the design has been completed, it is at the designer's peril to ignore these aspects during the initial design phase. A similar warning applies to stray light issues as well.

Some of the system parameters that must be clearly understood before beginning the design include focal length, aperture, focal ratio and field of view (including a clear understanding of how the customer defines this parameter), wavelength band. Another necessary parameter is the actual performance requirement. In other words, how well must the system perform under a given set of operating conditions. The performance specification can be based on an encircled or ensquared energy size, modulation transfer function (MTF) values as well as consideration of some optical aberrations, in particular distortion and field curvature. Additional non-optical constraints may be provided including total footprint, volume and weight. Environmental issues such as humidity, pressure, salinity and temperature can degrade the performance of an optical system and can also limit the choices of optical materials as well as surface shape. As with any engineering task, allowable or measurable cost and time constraints will place limitations on the design. Parameters that can significantly impact the cost include aspheric surfaces, certain optical materials, and components with tight fabrication tolerances.

The stop surface is a critical surface in an optical design. The position and size of the stop significantly affect the ability of the design to perform as required. An optical system consists of a series of lenses and/or mirrors. The finite size of these surfaces limits the amount of light

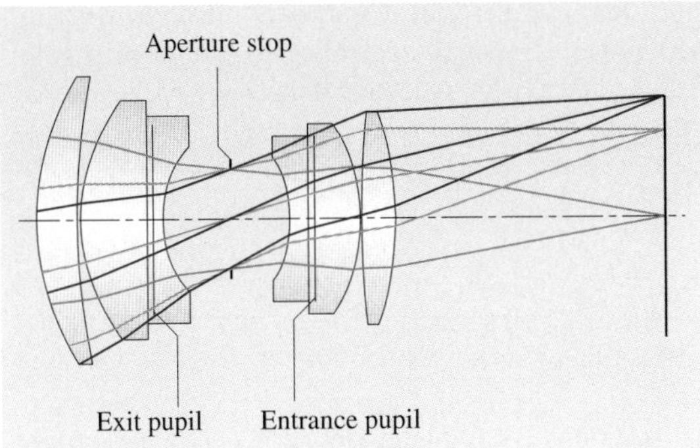

Fig. 7.1 Aperture stop and pupils

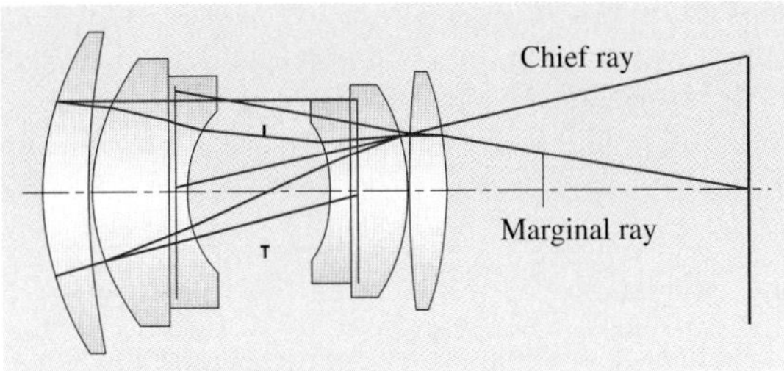

Fig. 7.2 System marginal and chief rays

that can pass through the optical system. There is one surface that will limit the angular extent of a light bundle entering the system from an on-axis point source. This surface is called the system's aperture stop. The stop can be formed directly by an optical surface in the system or by a separate mechanical surface, or iris. The stop can be located anywhere in the optical system. Two other related surfaces are the entrance pupil and the exit pupil. The entrance pupil is defined as the image of the aperture stop as seen from object space. If the aperture stop is at, or in front of, the first optical surface, the entrance pupil and aperture stop are coincident. The exit pupil is the image of the aperture stop as seen from image space. If the aperture stop is at, or behind, the last optical surface, the exit pupil and aperture stop are coincident.

In specifying an optical system in a lens design code, it is necessary for the designer to input information concerning the size and location of the system aperture. Often the entrance pupil size is used, as this is the parameter used to specify the system's focal ratio. When the entrance pupil is used to specify the aperture, the size and location of the actual stop surface can change during the design process. If the stop surface must be of a specific size, then the aperture is defined by the stop. In this case, the size of the entrance pupil will be determined by the magnification between the stop surface and the entrance pupil. Other aperture definitions may be available in the lens design program for use with other constraints. Figure 7.1 shows a typical optical system with an embedded stop and indicating the location of the entrance and exit pupils, which for this design are virtual surfaces.

The first-order optical design is used to provide a starting point for the actual design process. Information concerning the relative location of the system stop, the number of lenses or lens groups, and the effective power of each group is determined. First-order optics defines the properties of an aberration-free optical system. Many basic optical system parameters, such as focal length, focal ratio, magnification, and others, are derived from first-order, or paraxial, optics.

The first-order predesign locates the aperture stop, as well as the pupil and image locations in the optical design by tracing rays through the system. In particular, two specific rays, the marginal ray and the chief ray are used. The marginal ray is a ray that exits from an on-axis object point and travels through the top of the pupil or aperture stop before re-crossing the axis at the image. The marginal ray is used to determine the location of the image plane, the effective focal length and the focal ratio ($F/\#$). The chief ray is a ray that exits from the top of the object and crosses the optical axis at the aperture stop or pupil locations. The chief ray determines the location of the pupil planes, the field of view and the image height. Figure 7.2 shows the marginal and chief rays for the same optical system.

After performing the predesign to determine the general structure of the system, to include the number of elements, use of lenses and/or mirrors, relative stop location, and the basic system shape, the design of the actual system can be started. The synthesis of an optical system starts from the basic layout determined in the predesign. Using this as the basis, the designer uses the lens design program to determine the optical parameters necessary to meet the required performance specification. The lens design program will then take the information provided by the designer to find the combination of parameter values that best meet the performance goals. During this phase, the designer will need to determine whether the initial basic design had enough free parameters, called variables, to provide an acceptable solution. If not, the process begins anew.

The selection of the initial optical design form is a key role for the optical designer. The lens design program can only search for solutions within the solution space defined by the base design. Often a starting-point design is based on a prior known design used for a similar problem. In some instances an *off-the-shelf* solution, such as a camera lens or a microscope objective, might be the best option. Patent or literature searches can provide many useful ideas. These can provide information concerning the merits of reflective or refractive designs, the number of optical elements, the relative stop position and other important characteristics. Experience and intuition can also be employed in the predesign.

7.2 Design Parameters

The process of determining the optimal parameter values has become the domain of the lens design program. These programs, when used on a standard computer, can easily trace millions of rays through an optical system in seconds. Provided that the system has been properly defined and a proper set of constraints exist, the program can systematically adjust the parameter values until it reaches a solution that best matches the design goals. At this point, the designer needs to evaluate the information provided by the program to determine if the performance goals have actually been met. If not, a determination needs to be made as to whether the goals as defined to the program were sufficient and reasonable. If there were problems in the definition, then the goals need to be redefined before attempting further optimization. If the system and goals were properly defined and a solution was still not found, this indicates that there were an insufficient number of degrees of freedom in the design for a solution to be found. In this case the basic design needs to be reconsidered to allow for additional free parameters. This could also indicate that the design goals may need to be reevaluated.

In designing an optical system, the first and most frequent question asked should be: *does this make sense?* Design goals must be physically unrealizable. The optical designer needs to be able to recognize these issues and to offer alternative solutions. The designer must also be able to recognize when design goals compromise the performance of the system.

To use an optical design program, it is necessary to provide the program with sufficient information to understand the system under design. Lens design programs are called sequential ray traces because they trace geometrical rays through an optical system in a predefined sequence. These programs are based on a surface, rather than a component, model. Each surface defines a transition from one optical space to the next. Each surface has an object side and an image side. If we consider a simple optical system consisting of a singlet lens, there are four optical surfaces:

1. The object surface
2. The front lens surface
3. The rear lens surface
4. The image surface

Additionally, in this model, either the front or rear lens surface must be designated as the stop surface.

Each ray that is traced through the system starts at the object surface and is then traced sequentially through the front lens surface, the rear lens surface and on to the image surface. Rays cannot go through surfaces in any other order such as 1 to 3 to 2 to 4; this would require non-sequential ray tracing.

The optical design program simply applies Snell's Law to each ray on a sequential surface-by-surface basis. Some of the information the program will consider or provide includes the exact ray path, the effects of reflection and refraction, wavefront phase, aberrations and image quality. Some information concerning polarization effects may also be provided.

Equally important is the information that is ignored during sequential geometrical ray tracing, such as surface and volume scatter, if the system or any particular ray is physically realizable, or if there is any edge diffraction or other non-geometrical propagation of the wavefront.

The following information needs to be available to the optical design program before a system can be analyzed:

1. System aperture type and size
2. Surface information (sphere, asphere, others)
3. Number of surfaces
4. Which surface is the aperture stop
5. Wavelength band and weighting
6. Field of view

Additionally, optimization requires:

1. Variable parameters
2. Merit function

Optical design is an interaction between the designer and the computer program. The designer provides the program with information about the initial design form and the performance goals. An optical design program is constrained to work within the design space provided. The program can make significant changes to the values associated with any variable parameter, but it cannot add additional variables, or add new parameters, such as additional lenses or aspheric coefficients. The designer needs to use the analysis provided by the design program to determine if any changes need to be made to the design and what the most effective changes are.

As previously discussed, the system aperture is used to determine the size of the entrance pupil, which determines how large a light bundle enters the optical system.

Surface information includes the curvature, thickness or distance to the next surface, the optical material the ray is entering, and possibly other information including aspheric or other shape coefficients.

Wavelength information consists of one or more wavelengths. During optimization and analysis, rays are traced for all defined wavelengths. Wavelengths may be weighted to indicate importance. A common use of weighting would be to apply the photopic or scotopic curves for visual optical systems or an appropriate detector response curve. Rays are only traced at the defined wavelengths, not as a continuum. Additionally, one wavelength is designated as the primary wavelength. This wavelength is used to calculate wavelength-dependent system properties, such as the focal length.

The field of view is also defined as a collection of specific points. Field points can also be importance weighted. Field points serve as source locations for the rays to be traced. Rays from each of the defined wavelengths will be launched from each of the defined field points to perform analysis and optimization. The field of view can be defined in terms of angles or heights in either object or image space. The angle definitions are necessary for infinite conjugates; either angles or heights can be used for finite object distances. For rotationally symmetric optical system, it is only necessary to trace rays from half of the field.

Variables are any parameters in the system that can be adjusted during the optimization process. These include radii, thickness, refractive indices, Abbe numbers, conic constants, aspheric coefficients and others, depending on the type of surface. In some cases, wavelength and field of view can also be appropriate variables. During the optimization process, all of the variables are adjusted. It is important for the designer to limit the ability of the program to change the variables freely to insure a successful design. Common constraints to apply include a minimum and maximum lens thickness.

Although optical design codes can be used to analyze existing designs, the most significant reason to use a design code is optimization. This is the designing part of the process. As indicated earlier, for the program to perform optimization, the designer needs to have provided a set of variable parameters and to have defined a merit function. All lens design programs have multiple algorithms for optimization. The designer needs to determine which, if any, of the native optimization routines is appropriate for a particular design, and if one of these routines is not appropriate, the program needs to allow the designer to create a suitable merit function. Merit functions native to the design program are designed to maximize the image quality of the design. Typically they include procedures to minimize the spot size or wavefront error on either a root-mean-square (RMS) or peak-to-valley (PTV) basis. The spot size or wavefront error will be weighted over all the defined fields and wavelengths. Additionally, it may be necessary to select a point of reference on the image surface for the optimization; usually either the intercept coordinates of the primary-wavelength chief ray or the weighted center position over all the wavelengths, the centroid. The proper merit function depends on the required performance level of the system. For systems that will operate at or near the diffraction limit, wavefront-based optimizations are more appropriate. For designs that do not need to perform at the diffraction limit, the spot-size minimization should be used. In cases where a user-specified merit function is required, such as a merit function based on MTF constraints, it is often more efficient (faster) to begin with a default merit function until the design is nearly finished before switching to the necessary merit function for the final iterations.

The most common optimization algorithms use a damped least-squares process. Let the merit function be defined as:

$$\Phi = \phi_1^2 + \phi_2^2 + \phi_3^2 + \phi_4^2 + \cdots + \phi_m^2 \,. \tag{7.1}$$

In this case there are m items that are being considered or targeted. The contribution of each of these targets to the total merit function is determined by the difference between the actual value and the desired value:

$$\phi_i = v_i - t_i \,. \tag{7.2}$$

In the ideal case, the actual value exactly equals the targeted value, so the contribution of that item to the total merit function would be zero. The goal of the optimization is to determine the set of parameter values that will drive the total merit function value to zero.

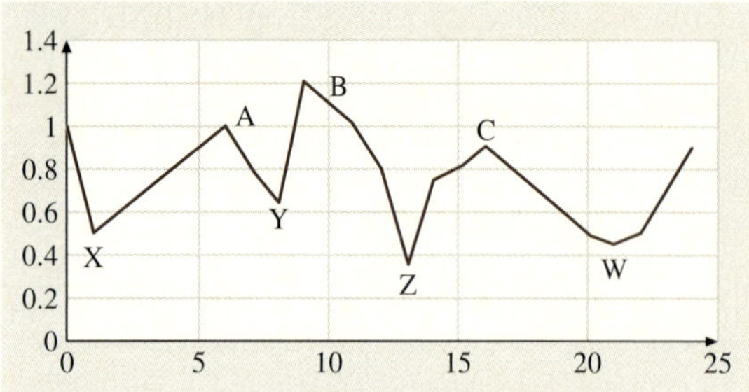

Fig. 7.3 Local and global minima

In these equations, each target or operand is given equal weight. In practice, applying weighting factors to each targeted value is necessary for optimal results. In this case the merit function is defined as:

$$\Phi^2 = \frac{\sum_i W_i(v_i - t_i)}{\sum_i W_i} \,. \tag{7.3}$$

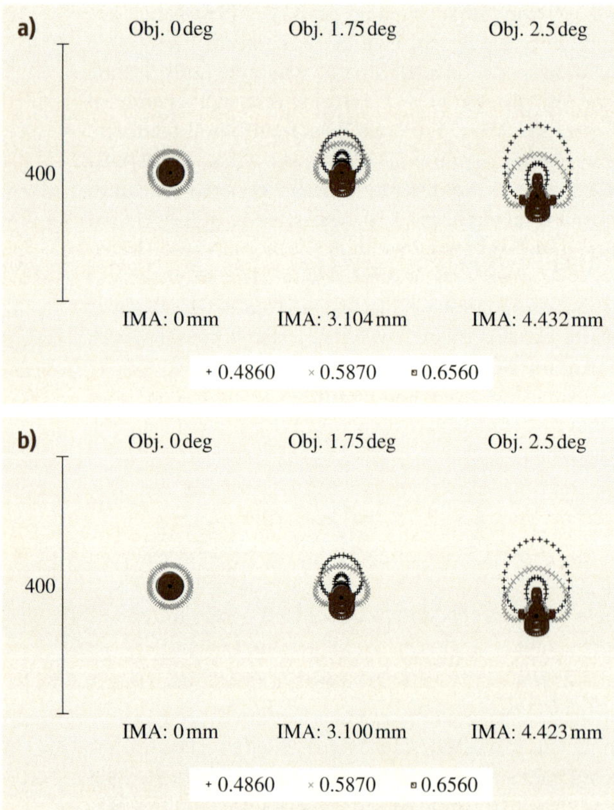

Fig. 7.4 **(a)** Spot diagram centered on the centroid **(b)** Spot diagram centered on the primary-wavelength chief ray

The values of each of the variables form an m-dimensional vector $\boldsymbol{X}$. The goal of the optimization is to find the value for $\boldsymbol{X}$ for which Φ is a minimum. Φ is minimized in a least-squares sense, where the possible movements in solution space are computed by determining the direction where the derivative matrix of $\boldsymbol{X}$ is minimized. Careful sampling of solution space by the design program is necessary to insure that the minimum can be isolated.

There are some inherent problems with damped least-squares optimizations algorithms. The first issue is that the optimization follows a downhill path. That is, the algorithm locates a direction (new location in solution space) where the total merit function is lower than the current merit function and changes the vector $\boldsymbol{X}$ to those coordinates. This process continues until movement in any direction around the final coordinate location results in an increase in the total merit function value. At this point the optimization stops. This value is a local minimum, not necessarily the global minimum. The algorithm cannot search further through the solution space for a better minimum. The solution found depends on the initial starting point. By changing some of the initial parameter values, the optimization could find another local minimum that may be better or worse than that previously found. Figure 7.3 shows a simple two-dimensional (2-D) model of solution space. A, B and C represent possible starting points, with W, X, Y and Z the local minimum values, Z being the global minimum in this space. An optimization starting from point B will most likely find the global minimum value. An optimization from starting point A will stop at the local minimum Y. The starting point C can finish at either W or Z, depending on the direction of the first step.

The second problem that can occur is called stagnation. This occurs when the derivatives of each of the targets with respect to all of the variables is so ill-conditioned that a suitable next step cannot be determined. In this case, the optimization can stop even before reaching a local minimum.

Most optical design programs have global optimization techniques available that can be effectively used to find alternate, better solutions. However, even with global optimization it can never be assumed that the solution is the *true* global minimum.

The key for the optical engineer is to understand that the solution found by the optimizer might not be the best solution. However, it is not necessary to find the best solution, only to find a suitable design form that meets the performance specifications. If not, then the optimization must continue either by changing the initial parametric values or by adding additional degrees of freedom in terms of additional variable parameters.

In constructing a merit function, the optical designer must perform several tasks. The first is the selection of the proper form of the merit function. As indicated earlier, RMS or PTV minimizations of the spot (blur) size or wavefront departure from spherical relative to either the weighted centroid or the primary-wavelength chief ray are the most common choices. Indeed, an RMS spot size relative to the centroid is often the most appropriate merit function for a design. Even for situations where the final design should be diffraction-limited, this form of optimization is often the best starting point. Figures 7.4a and 7.4b show the blurs for a double Gauss lens design. Although the physical distribution of the rays has not changed, the RMS and geometric values calculated differ for the off-axis field points.

The RMS spot size is a measure of image quality based on tracing geometrical rays through the optical system. Geometric rays propagate through the optical system according to Snell's law, ignoring the effects of diffraction from edges and apertures. The RMS spot size is calculated by tracing a selection of rays of each defined wavelength from each defined field point. The number of rays propagated is adjusted to consider any applied weighting factors. To calculate the RMS spot size relative to centroid, it is first necessary to locate the centroid position. This is the average image position determined by tracing a number of rays:

$$x_c = \sum_1^n \frac{x}{n} ,$$
$$y_c = \sum_1^n \frac{y}{n} . \tag{7.4}$$

The RMS spot size is then determined using

$$\text{RMS} = \left(\sum_1^n \frac{(x - x_c) + (y - y_c)}{n} \right)^{\frac{1}{2}} . \tag{7.5}$$

In attempting to optimize the design, the result will be a reduction in the aberration in the design usually without specifically targeting aberrations. It is also important for the designer to consider what is actually important in the design. The optimal result, where the total blur size is reduced, generally indicates optimal image quality. In using a native merit function, it is important for the engineer to understand what is and is not considered in the merit function. For example, distortion is not typically considered directly in optimization. If specific constrains are placed on distortion in the design specifications, it will be necessary for the designer to modify the merit function to include this constraint. Similar requirements may also apply to field curvature requirements.

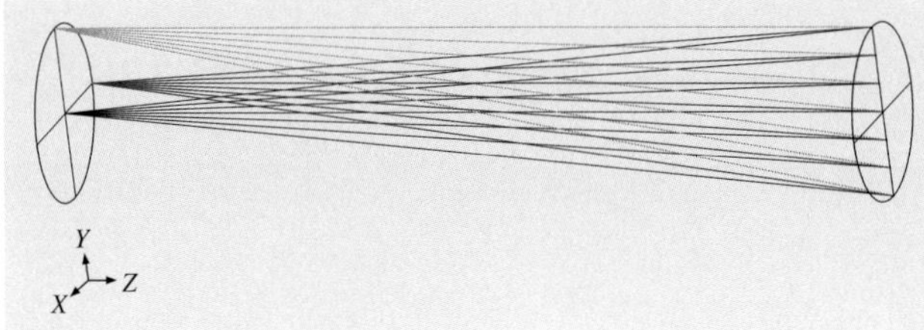

Fig. 7.5 Tangential ray fan

Additionally, the optical designer may need to add constraints to target specific system properties such as effective focal length, focal ratio or magnification. Boundary limitations also must be considered. Boundary constraints placed on lens components can assure that the lens is physically viable. The center thickness of the lens should be one tenth to one sixth of the lens diameter. Also for lenses with convex faces, the edge thickness needs to be sufficient to prevent chipping and to provide a solid surface for mounting. Often lenses will be oversized and *squared up* to prevent problems. However, if lenses are oversized during manufacture, it is necessary for the designer to insure that all light that could pass through the oversized lens is actually blocked by the lens housing, or the image quality may be degraded by excess light entering the system. Additional constraints that need to be considered and may be appropriate for the merit function include the total length of the system, the weight of any and all components, glass properties and grades, and transmission.

After the design program has reached a minimum solution, several graphical and numerical analytical tools are available to help evaluate the design. These tools provide information to help the designer understand the limitations or dominant aberrations present in the current design form. By understanding the system limitations, the designer can introduce the necessary degrees of freedom to correct for that limit. In terms of aberration correction, particular forms of variables may be effective for controlling some aberrations and be ineffective at correcting other aberrations. Introducing ineffective degrees of freedom can lead to stagnation of the optimization process.

Some of the tools include geometric-ray-based spot diagrams, encircled energy plots, ray fan and optical path difference (OPD) fan plots, wavefront maps, as well as plots of field curvature, distortion, longitudinal aberra-

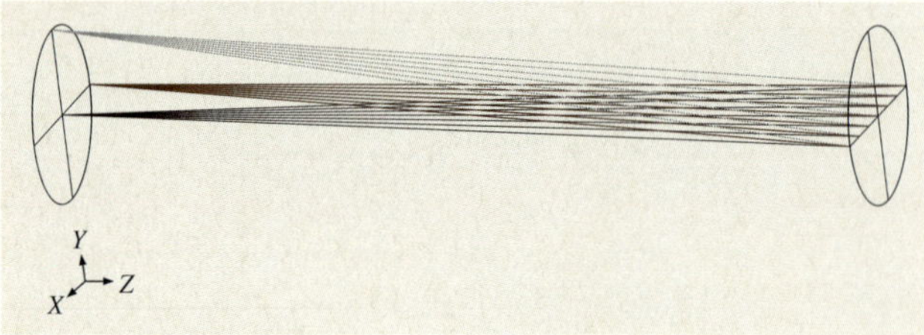

Fig. 7.6 Sagittal ray fan

tion and lateral color. Additional tools may be available that include edge diffraction effects. These tools should be used as the system's performance nears the diffraction limit. These tools include diffraction-based encircled energy plots, MTF plots and point spread function (PSF) plots.

Ray or OPD fan plots show the relative location of linear cross sections of rays through the entrance pupil. Fan plots are generated in the tangential and sagittal direction for each of the defined field points. Each wavelength is plotted separately. Figure 7.5 shows the tangential fan, which is directed along the y-axis of the pupil. Figure 7.6 shows the sagittal fan, which is directed along the x-axis of the pupil. The coordinate origin of the fan plots is the intercept of the primary-wavelength chief ray. Figure 7.7 shows the ray fan plot for a double Gauss design. The other points indicate how far each of the other rays landed from the chief-ray intercept in either the tangential or sagittal direction. The OPD plot (Fig. 7.8) indicates the difference in optical path traveled for each ray relative to the total optical path of the chief ray.

Although the fan plots are limited to showing information about a pair of linear ray distributions, each is very useful in determining the many first- and third-order aberrations present in the design. Each of these aberrations has a characteristic appearance on the fan plot. By recognizing these traits, the designer can determine which aberrations are limiting the design's performance.

As a system's performance level nears the diffraction limit, it becomes necessary to consider the effects of diffraction from apertures in the performance evaluation. Light has properties of both particles and waves. Rays are used to model particle-like behavior; diffraction and interference characterize wave-like behavior. At optical wavelengths light is a wave phenomenon, and the approximations used in geometric optics may not be sufficient to explain the images formed in such systems. To understand these effects, it is necessary to consider the effects of diffraction from edges and apertures. Although diffraction occurs in all optical systems, it is only

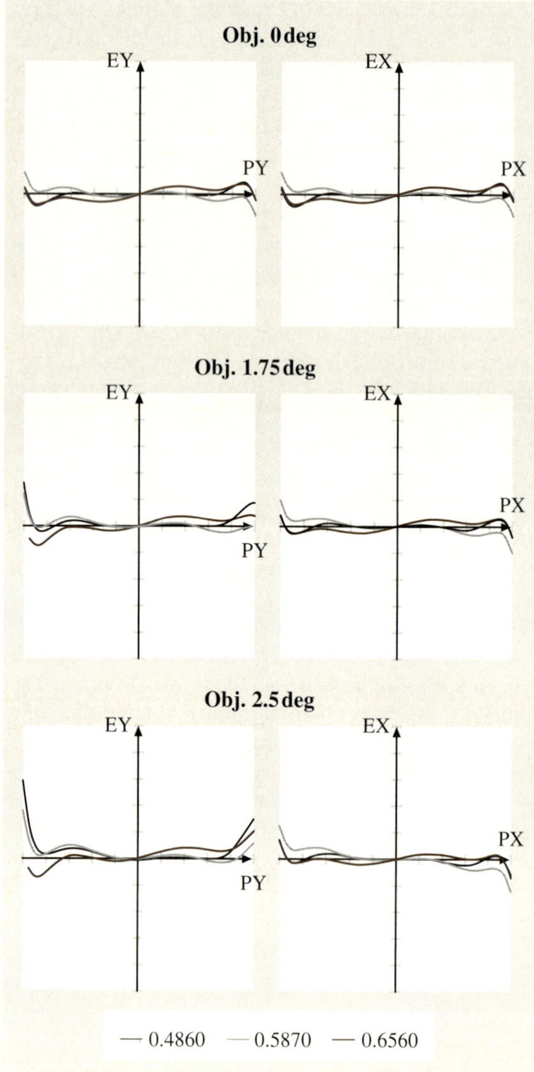

Fig. 7.7 Ray fan plot

necessary to consider diffraction effects when the scale of the image blur due to diffraction is on the same scale as the blur size due to geometric aberrations.

Diffraction theory says that the image formed by a converging wavefront is simply the Fourier transform of the complex wavefront in the exit pupil of the optical system. An important consequence of this is called *compact support*. This tells us that a signal band-limited in the spatial domain cannot be limited in its Fourier domain. As the wavefront passes through a physical aperture, some of the wavefront is clipped off. Because the wavefront is now limited in the spatial domain, it can-

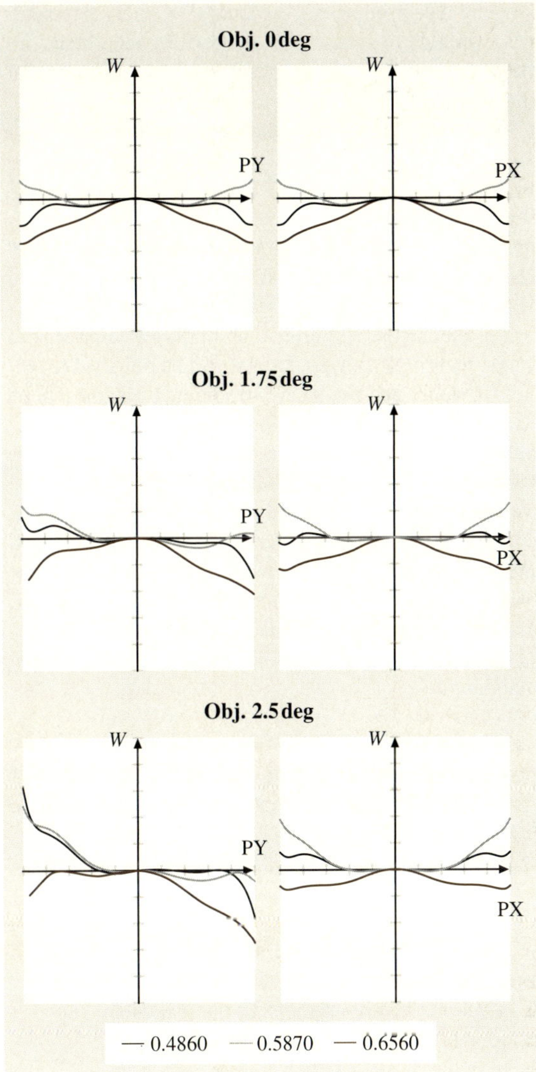

Fig. 7.8 Optical path difference fan plot

not be limited in the Fourier domain. Energy is spread out over all angular space. Due to this spreading of energy, point objects such as stars cannot form point images. The size of the image blur due to diffraction is a function of the focal ratio and wavelength of the system.

$$D = 1.22\lambda(F/\#)\,. \qquad (7.6)$$

Summing the plane waves incident on an optical system a long distance from the aperture gives the far-field diffraction pattern. In the common case of a planar wavefront incident on a circular aperture, the energy distribution is referred to as an Airy pattern. The irradiance

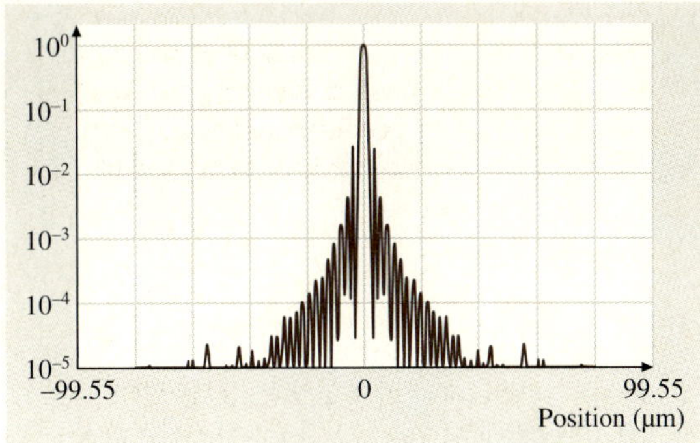

Fig. 7.9 Cross-sectional plot of the Airy function

distribution on a circular aperture is

$$A(\theta_x, \theta_y) = \frac{A_0}{\pi} \iint \mathrm{e}^{ik(x \sin\theta_x + y \sin\theta_y)}\, \mathrm{d}x\, \mathrm{d}y\,, \qquad (7.7)$$

where A_0 is the incident amplitude and π is an area normalization term. Integrating over the distribution yields

$$A(r) = 2A_0 \frac{J_1\left(\frac{kar}{z}\right)}{\frac{kar}{z}}\,, \qquad (7.8)$$

with k being $2\pi/\lambda$, r the aperture radial size and z the image distance and J_1 is a first-order Bessel function. The first root of $J_1(x)$ occurs at $1.22\lambda(F/\#)$. The Bessel function, or Airy disk pattern, describes a pattern of alternating bright and dark zones, with the location of the dark zones determined from the successive roots of the Bessel function. Figure 7.9 shows a cross section of the diffraction image. Figure 7.10 shows a 2-D image. Both figures are plotted on a logarithmic scale to allow further bands to be seen. The central core of the Airy pattern is useful in determining if a design is approaching the diffraction limit. Consider the spot diagrams shown in Fig. 7.11. The circle indicates the Airy disk. In the

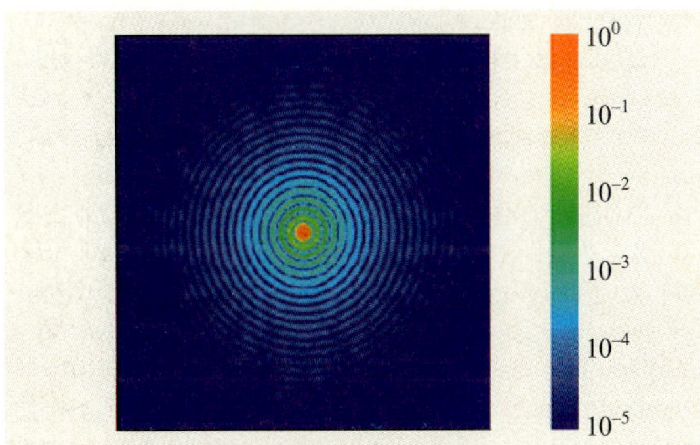

Fig. 7.10 2-D plot of the Airy function

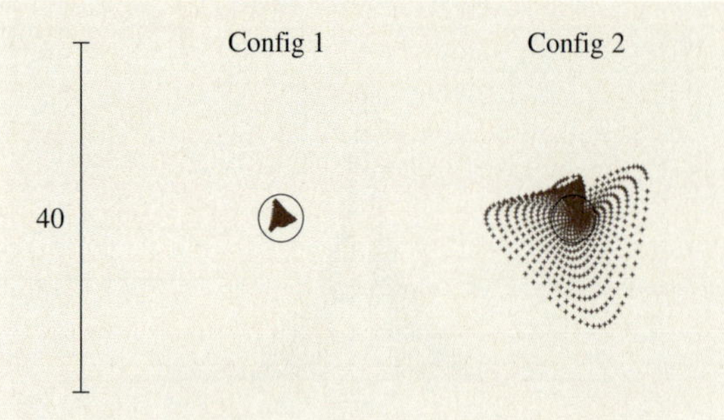

Fig. 7.11 Comparison of the blur size to the Airy disk

spot on the left, the geometric blur is contained within the Airy disk, and the effects of diffraction need to be considered; the size of the image blur will be larger than that calculated using only geometric analysis. For the system on the right, geometrical aberrations dominate the performance and the effects of diffraction can be ignored.

When performing diffraction-based calculations in a lens design code, it is important to understand the process being applied. Ray tracing programs propagate rays through the optical systems. Although each ray represents the normal to the wavefront, the actual wavefront is not propagated. Therefore some form of approximation is often used in diffractive calculations. The optical designer is responsible for determining the validity of the diffraction calculations for any given system.

Although there are situations where geometric or diffractive calculations must be applied, it is important to remember that the two can also be related: if geometric optics predicts a high level of performance for a design, than the system will perform close to its diffraction limit. If geometric optics predicts the blur size is smaller than the Airy disk, then the actual image blur will be near the size of the Airy disk. Also, if the geometric blur is smaller than the Airy disk, there is little value in attempting to reduce geometric aberrations further.

The design of an optical system requires the optical designer to develop a first-order design to insure that all necessary specifications have been provided and make sense. Then the basic system prescription including system parameters such as focal length, aperture, field of view and wavelength data, as well as the basic sequential optical layout with the desired number and types of surfaces and materials, needs to be defined in the lens design program. To perform optimization, variable parameters and an appropriate and well-constrained merit function must be defined. After allowing the optical design program to perform optimization, the designer needs to consider the appropriate geometrical or diffractive analysis available to insure that the performance goals are truly met. This ends the first and possibly easiest part of optical design.

The next step in designing the optical system is to perform a tolerance analysis. Tolerancing is one of the most complex aspects of optical design and engineering. Tolerance analysis is a statistical process during which changes or perturbations are introduced into the optical design to determine the performance level of an actual design manufactured and assembled to a set of manufacturing tolerances. To perform an optical tolerance is to accept reality. No optical surface will be polished to perfect curvature and figure. No mechanical mount will be perfectly machined. No component will be perfectly positioned. All these, as well as many other error sources will serve to degrade the performance of the assembled system. It is up to the optical engineer to define the fabrication limits properly for each optical and mechanical task. This needs to be done with a consideration for the system performance as well as an understanding of the cost impact of an overly constrained design.

There are several steps to performing a tolerance analysis.

1. Establish a tolerance budget,
2. Perform a test plate fitting,
3. Define a set of tolerance ranges that are easily within fabrication limits,
4. Define compensators that may be used to limit performance degradation,
5. Select an appropriate figure of merit for tolerancing,
6. Evaluate the tolerances to estimate the impact of each perturbation,
7. Generate random designs for statistical analysis,
8. Revise the tolerance ranges as necessary.

The error budget is the total system degradation that is available. If the basic design just met the performance requirement, then it is unlikely that a manufactured system would work to specification, as a system will not be made exactly to the prescription. The available error needs to be shared among all the factors that can impact the system including materials, manufacturing, assembly, environmental, and the residual design error. Table 7.2 lists some of the parameters that should be considered.

Test plate fitting refers to the process of matching the surfaces of all optical surfaces to a set of test plates available to the optician. The curvature of an optical surface is determined interferometrically by comparing the

Table 7.2 Possible tolerance defects

Fabrication errors
Incorrect radius of curvature
Incorrect center thickness
Surface shape incorrect
Curvature center offset from mechanical center
Incorrect/unwanted conic or aspheric terms
Material errors
Index accuracy
Index homogeneity
Index distribution
Dispersion accuracy
Assembly errors
Elements offset from mechanical axis (x, y)
Elements in wrong position along the optical axis (z)
Elements tilted relative to the optical axis
Elements reversed front to back
All the above for groups of lenses
Environmental errors
Thermal expansion/contraction of optical materials
Thermal expansion/contraction of mechanical materials
Thermal mismatch between optical and mechanical
Temperature dependent changes in refractive index
Pressure effect on refractive index
Alignment sensitivity due to shock or vibration
Mechanical stresses
Residual design error
System merit function value not zero

surface being polished to a surface of known curvature. If the design can be matched to the available test plates, the cost both in terms of fabrication as well as time will be significantly reduced. If matching all surfaces to test plates significantly degrades the performance, the optical engineer needs to determine if the cost of custom test plates is worth the cost and delay or if a system redesign should begin.

Setting the initial tolerance ranges is a balance: not so tight as to have a cost impact but not so loose that assembly is difficult. Understanding the capabilities of the optical fabrication shop as well as the mechanical fabrication shop is a good starting point for the parameter ranges. Start with readily achievable tolerances and then selectively tighten only those ranges that result in out-of-specification performance.

A compensator is a system parameter that can be adjusted to counteract the effect of an error in another parameter. The ability to use compensators can significantly loosen the parameter ranges, which can be useful in reducing costs. A common compensator is the image distance: the image location is set for each system based on the performance of that system. Compensators represent degrees of freedom, usually mechanical adjustments that need to be available and made during the assembly of the optical design. The cost impact of the compensator needs to be considered before including it in the analysis. A $500 kinematic mount cannot be included with each $7 optical system.

The tolerance figure of merit is usually similar to the final merit function used during the design phase. As each parameter is perturbed within the tolerance range, the compensator value is optimized to minimize the figure of merit. The difference between the original design merit function and the tolerance figure of merit describes the impact of each perturbation on the system performance.

During the tolerance process, each parameter is perturbed individually. This allows the optical engineer to determine which parameters are most sensitive. Parameters that perturb the design outside a portion of the error budget available for fabrication errors need to be more tightly constrained. Tightening constraints adds significantly to the manufacturing costs. Additionally there are some manufacturing limits that cannot be exceeded no matter what the cost. Designs that require such constraints should be deemed unbuildable and the engineer should return to the design process, with careful consideration as to why the design failed.

By generating a series of perturbed lenses, it is possible to determine what the performance level of any manufactured system would be. For these systems, each parameter is perturbed to a value within a specified statistical distribution of the nominal value. The accuracy of this analysis is directly related to the provision by the optical designer of the necessary information to the tolerance algorithm regarding the appropriate distribution. If such information is not available, a Gaussian distribution is normally assumed. This is not always the best choice. Consider the thickness of an optical window. It is unlikely that the window will be polished to the *thin* side. As soon as the component thickness is within the allowed thickness $+\Delta$, the polishing will stop. The fabricator will not continue to polish to get closer to the nominal value.

After analyzing the statistical information about the random designs, further tightening of the parameter ranges may be necessary to achieve the required pass rate. Again a careful balance needs to be achieved between the success rate (the percentage of the systems that will meet the required performance level) and the cost to increase it.

An optical design is not complete until a tolerance analysis has been carried out. The lens design needs to be manufacturable with some realistic range of parameter specifications to be useful to the customer. The best design is not necessarily the design that best matches the design specification. It is the design that can be built to best match the design specification.

Now that a finalized optical design has been established, it may be necessary to perform a stray light analysis to simulate the real working environment more completely and insure that the limitations imposed by sequential ray tracing have not hidden significant design flaws.

7.3 Stray Light Design Analysis

Years ago a stray light analysis was often an afterthought, if considered about at all. Today it is known that the stray light analysis should be considered along with the conception of the optical design in the very earliest stages of the design study. It might even play a major role in the selection of the starting optical design.

What is a stray light analysis? Stray light analysis is the calculation of how much unwanted radiation reaches the image plane/detector plane. The fine distinction being made here is that in some designs a detector is placed at the exit pupil plane where no traditional image is formed. In this chapter they will be used interchangeably. In a stray light analysis one must consider all the elements in the optical design. It must consider each surface as an independent element that will scatter directly to the image plane or that might create significant non-sequential stray light contributions due to ghost images or total internal reflections off any of its surfaces. In a lens system, all possible combinations of lens and mirror surfaces need to be considered. Each ground or polished edge of the lens elements are potential sources of stray light propagation, and each aperture edge of an optical element creates diffraction. All the mechanical structures that hold the elements in place are potential stray light propagators. Then there are often baffles, the main cylindrical tube-like structures that are used to either enclose the optical system or to baffle the stray light from reaching deep into the system. On these baffles, there are often vane structures to suppress the propagation of the unwanted light further, and those vanes usually have tapered edges, and as fine as they may get, they can sources of scattered stray light and diffracted stray light. These vanes and baffle surfaces are usually coated with paint or light-absorbing material that have a complicated bidirectional scattering function. There is no such thing as a Lambertian black coating at any wavelength. In the infrared there are also potential sources of thermally emitted stray light from the system itself.

In spite of all this complexity, significant stray light suppression design can be accomplished without doing any numerical calculations. Sooner or later one will eventually need a software program to quantify the performance and then a system performance measure to confirm the results of the analysis and fabrication. There are computer-aided design (CAD) tools available that greatly assist in inputting very complicated systems into stray light analysis codes.

Why do a stray light analysis? One needs to perform a stray light analysis and then a measurement when stray light might be a problem in systems that:

1. are in compromising environments that have one or more strong stray light sources
2. observe faint objects
3. make precise measurements
4. require high contrast
5. propagate high-power laser energy, when even small percentages of stray light can damage the system

One carries out the stray light analysis because in many systems it is necessary to assure that the system goals are achieved or even achievable. Stray light analysis usually improves system performance by a factor of 1000 and sometimes by a factor of 100 000 (e.g., in the Hubble telescope).

The adverse effects of stray light in a system are:

1. It can cause such severe problems that the design will not reach its desired optical performance;

Table 7.3 Basic optical system stray light specifications

System parameters
Optical design parameters as in Tables 7.1, 7.2
Are there possible locations for other stops
Aperture sizes
Wavelength and bandwidth
Image size, shape
Object, image locations.
Optical characteristics of intermediate images if any
Pupil aberrations
The purpose or mission of the sensor,
The environments in which it will be fabricated and function in
The optical design and optical performance requirements of the system
The mechanical design, size, and shape of the objects in the system
The thermal emittance characteristics for some systems
The scattering and reflectance characteristics of each surface for all input and output angles.
Performance
Stray light transmission
Vignetting factors
Central obscurations
Field positions of the stray light sources
Stray light propagation paths
Optical components
Each surface of all lenes
Alignment
Edge effects
Coatings
Mounting structures; especially sections with a "flat" profile
Detector
Total size
Number of pixels (horizontal, vertical)
Reflectivity
Package
Air movement
Maximum length, width
Weight
Environment
Clean room; dust, airborne volatile contaminants

2. It reduces the contrast on the image plane;
3. It obscures faint signals or creates false ones;
4. It produces false artifacts across the image plane that cause false alarms;
5. It causes magnitudinal errors in radiometric measurements;
6. It damages fragile optical components;
7. and stray light can burn out detectors.

Much like in the optical design, the design of stray light suppression in the system starts in the first-order optical design, and with initial specifications determined most often by the purpose, mission and environment of the instrument. The required basic thought analysis of a stray light designer must consider the items the designer considered in Tables 7.1 and 7.2 and some more given in Table 7.3. While stray light design is considered by some to be an esoteric art, it is not. Most good stray light design aspects of a system can be incorporated, in concept, without much precalculation. There are about eight basic principles in stray light design that simplify the process. They are explained in the text. They are:

1. Stray light analysis should be incorporated in the very earliest stages of a preliminary design study. It should play a role in the selection of the starting optical design.
2. Start from the detector.
3. Determine *critical objects*, i. e., objects seen from the image plane.
4. Move it or block it.
5. Trace rays out of the system from the image/detector plane.
6. Trace rays into the system from various field positions to determine the "illuminated" objects.
7. *Only* the aperture stop should not be *oversized*. All other elements have extra imaging surface from any given field point on the image surface. Be concerned about concessions made to oversizing elements for the convenience of manufacturability.

8. One wants and needs to know more than just how much stray light reaches the detector. One needs to know what path it took and the propagation method (scatter, total internal reflection (TIR), diffraction, etc.) in order to improve performance.

The concept is to think out the system before any stray light analysis is performed. It is encouraged to learn and appreciate the concepts so that, when one gets to the computer analysis, it will run faster and will probably produce a more accurate answer. Why? If one can limit the number of critical paths along which the unwanted stray light propagates there will be fewer interactions that can be miscalculated. Fewer interactions mean that the analyst will probably pay more attention to the fine details, and in the end the software will run faster because there are fewer calculations. Once you know the paths you do not have to make a Monte Carlo zoo of the analysis. After we look at the basic mathematical calculation that needs to be performed we will come back to this point.

7.4 The Basic Equation of Radiation Transfer

The fundamental equation relating differential power transferred from one section on an object, be it a baffle or an optical element, or even a diffracting edge, to another object is determined by

$$\mathrm{d}\Phi_\mathrm{c} = L_\mathrm{s}(\theta_\mathrm{i}, \phi_\mathrm{i}\, \theta_0, \phi_0)\,\mathrm{d}A_\mathrm{s}\frac{\cos(\theta_\mathrm{s})\,\mathrm{d}A_\mathrm{c}\cos(\theta_\mathrm{c})}{R_\mathrm{sc}^2}\,, \tag{7.9}$$

where $\mathrm{d}\Phi_\mathrm{c}$ is the differential power transferred, $L_\mathrm{s}(\theta_0, \phi_0)$ is the bidirectional radiance of the source section; $\mathrm{d}A_\mathrm{s}$ and $\mathrm{d}A_\mathrm{c}$ are the elemental areas of the source and collector; θ_s and θ_c are the angles that the line of sight from the source to the collector makes with their respective normals. This equation can be rewritten as three factors, which helps simplify the reduction of scattered radiation. With $E(\theta_\mathrm{i}, \phi_\mathrm{i})$ being the incident irradiance *incident on the scattering surface*, the three terms are:

$$\mathrm{d}\Phi_\mathrm{c} = \frac{L_\mathrm{s}(\theta_0, \phi_0)}{E(\theta_\mathrm{i}, \phi_\mathrm{i})}E(\theta_\mathrm{i}, \phi_\mathrm{i})\frac{\cos(\theta_\mathrm{s})\,\mathrm{d}A_\mathrm{c}\cos(\theta_\mathrm{c})}{R_\mathrm{sc}^2}\,. \tag{7.10}$$

The first term is the scatter function, known as the bidirectional reflectance distribution function (BRDF):

$$\mathrm{BRDF}(\theta_\mathrm{i}, \phi_\mathrm{i}; \theta_0, \phi_0) = \frac{L_\mathrm{s}(\theta_0, \phi_0)}{E(\theta_\mathrm{i}, \phi_\mathrm{i})}\,. \tag{7.11}$$

The second term is the power on the propagating surface section

$$\mathrm{d}\Phi_\mathrm{s} = E(\theta_\mathrm{i}, \phi_\mathrm{i})\,\mathrm{d}A_\mathrm{s}\,. \tag{7.12}$$

The final term has to do with the geometry. It represents the projected solid angle (PSA) of the collector as seen from the scattering surface, i.e., the source.

$$\mathrm{PSA}_\mathrm{sc} = \mathrm{d}\Omega_\mathrm{sc}\cos(\theta_\mathrm{s}) = \frac{\mathrm{d}A_\mathrm{c}\cos(\theta_\mathrm{c})}{R_\mathrm{sc}^2}\cos(\theta_\mathrm{s})\,. \tag{7.13}$$

$\mathrm{d}\Omega_\mathrm{sc}$ is the solid angle of the collector section as seen from the source. Another term used often is the geometrical configuration factor (GCF) introduced by thermal analysis engineers over a century ago. $\mathrm{GCF} = \mathrm{PSA}/\pi$ because the radiance of an emitting surface was assumed to be Lambertian (it includes the BRDF term).

$$\mathrm{d}\Phi_\mathrm{c} = \mathrm{BRDF}(\theta_\mathrm{i}, \phi_\mathrm{i}; \theta_0, \phi_0)\,\mathrm{d}\Phi_\mathrm{s}[\mathrm{d}\Omega_\mathrm{sc}\cos(\theta_\mathrm{s})]\,. \tag{7.14}$$

In words, the power propagated from one surface area section to another is equal to the power from the surface that is scattering the radiation times the solid angle the collector subtends as seen from the source times the scattering characteristics of the source. So stray light boils down to the repetitive use of the multiplication of just three numbers: the power on the scattering surface, the scatter value for the specific input and output of the scattering surface, and the projected solid angle of the collector. When conceptually designing a system, (7.14) is the equation to keep in mind.

All software programs implement some variation of this calculation, even the ray-based programs. In ray-based programs, the scattered ray is *weighted* by the power on the scattering surface, the BRDF for the directions of the incoming and outgoing ray direction and some form of weighted solid angle. The mathematics of a stray light analysis does not appear to be an overwhelming calculation; it is just the product of three numbers and the BRDF does not seem to be that hard to determine if one has measured data. All of which is correct. So why does stray light analysis seem so challenging? For one thing, in a typical analysis the equation is calculated 100 million times. No one has the time to do it by hand in detail.

7.4.1 Stray Radiation Paths

Since it is only the third term in (7.14) that can be reduced to zero, it should receive attention first

$$\mathrm{PSA}_{\mathrm{sc}} = \frac{\cos(\theta_{\mathrm{s}})\,\mathrm{d}A_{\mathrm{c}}\cos(\theta_{\mathrm{c}})}{R_{\mathrm{sc}}^2}\,. \quad (7.15)$$

How this factor can go to zero is hard to understand at first. The two cosine values can reduce the PSA to zero but seldom can this level of tilt be reached. Usually this is done on lens mounting structures, by tilting them completely out of the way. The diffraction effect remains, but is usually much lower than at the high angle of incidence of the forward-scattering path. For $\mathrm{d}A_{\mathrm{c}}$ the finite area of the collector is always present, so this seldom goes to zero. The PSA can be made to go to zero by moving the collector out of the scattering surface's field of view, in which case it is blocked. In some cases this can be done by a field or aperture stop. Alternatively, this can be achieved by placing vanes on a baffle surface so that the direct forward-scatter path is blocked and two scatterings are present in the path path (each absorbing maybe 99% of the energy with a very good black) and one additional PSA reduction by 90% along the path from the front surface of one vane to the back side of the preceding vane. This then results in a reduction by a factor of about 100 000 in the propagated energy along the path.

This is a crucial point in a stray light analysis. Most analysts make the mistake of working on the BRDF term first. They want the blackest black or the lowest-scatter optical surface without knowing if it will make any measurable difference. A stray light analysis will pay for itself in the long run.

7.4.2 Start from the Detector

Having explained the apparent possibilities for decreasing the $\mathrm{PSA}_{\mathrm{sc}}$ term, we now consider another concept, by considering the *start from the detector* and the *move it or block it* concepts. By placing baffles, stops, and apertures into a system in the correct places, many critical objects will be removed from the view of the detector. In other cases propagation paths from the directly illuminated surfaces to the critical objects can be blocked. The beautiful part of this is that one does not need to do the calculation. We all know what zero times something is zero. It is not unusual for an experienced stray light person to reduce the necessary calculations for a software program greatly without reducing the reliability of the result.

What is needed is a logical approach that first blocks off as many direct paths for unwanted energy to the detector as possible and then reduce the list of illuminated objects. Finally the $\mathrm{PSA}_{\mathrm{sc}}$ for the remaining paths from the illuminated objects to the critical objects is minimized.

Below several stray light paths are pointed out in detail. During the design phase the analyst designs out and reduces the number of critical objects that the detector can see. Then the direct stray light paths from the stray light sources to the illuminated objects are reduced using aperture stops, Lyot stops and intermediate field stops. Every effort is made to use these apertures to block the direct field of view from the image plane completely. Therefore the $\mathrm{PSA}_{\mathrm{sc}}$ in the above transfer goes to zero, maximizing stray light suppression for these paths.

The next step is to make sure that there is no section of an illuminated object that is both illuminated and seen (i. e., is also a critical object) by the detector. These are single-scatter paths from the stray light source to the detector and must be solved before proceeding. The good news is that, having done this, one more or less knows all the key stray light propagation paths without having done a calculation.

The analysis should not start from stray light sources as, even if some parts of the system are well lit, the emitted photons are not important if they are strongly attenuated before reaching the detector. Unfortunately many engineers first want to determine the destination of the stray light from each source, leading them to choose the best optics and the blackest blacks. This is not the right way to proceed, and the following two points should be borne in mind:

1. Only objects seen by the detector can contribute stray light;
2. The PSA is the only term that can go to zero.

Therefore the approach is first to determine the critical objects that can be seen from the image plane and list all objects that it can see either directly or through the various optics, be they reflective or refraction. This will be covered in detail shortly. The next step is to start moving the critical objects outside the field of view (FOV) or block them with baffles, apertures stops, field stops, or vanes. The PSA should be made to go to zero before a calculation is considered, with the number of critical objects that can be seen as low as possible.

The next step is to trace the propagation of stray light into the system and determine the illuminated objects. Having done the first step we now know where we do not want any of the direct energy to arrive, i.e., the critical objects, because it will create a single-scatter path to the detector. The significant stray light propagation paths will be defined as the paths:

- from the stray light sources of unwanted radiation to illuminated objects,
- then from these illuminated objects to critical objects,
- then from the critical objects to the detector.

This approach simplifies the analysis immensely and directs attention to the most productive solutions.

With computer software, one can then quantify the power propagated along these paths, which will then reveal which paths are the most significant. The analyst then gains the further advantage of working on the other two terms of (7.14). It is only at the end of the analysis that the analyst considers if the coating will make a difference. Its effect will usually be by less than a factor of five, maybe ten, for any single scatter between a great black and a mediocre black. Blacks are not the secret to good stray light design.

7.4.3 The Reverse Ray Trace

The purpose of a reverse ray trace is to determine what the detector can see. It is only from objects that the detector can see that there can be *any* direct contribution to stray light. An object in this sense can be a diffracting edge and any object seen in reflection or imaged by and through the optical system. This is not an easy calculation for most optical design programs as there is seldom a single aperture that defines the *reverse aperture stop*. Usually there are several apertures that define the limiting size of the ray bundle, so it can be difficult to use those codes.

Take for example a simple two-mirror Cassegrain reflective telescope. The on-axis incoming beam is limited by the aperture of the primary mirror, it is then reflected towards the secondary mirror where the circular beam footprint is then reflected back towards the primary, and often through a hole in the primary to an on-axis beam spot on the image plane behind the primary mirror.

From an off-axis position the incoming beam is again limited by the same aperture of the primary mirror, where it is again reflected towards the secondary mirror where the circular beam footprint is offset from the center of the secondary mirror and reflected back towards the primary to an off-axis beam spot on the image plane. So the secondary has to be bigger than the beam footprint from any given field position to account for the field of view (FOV) of the sensor. Other than diffraction and aberrational effects, this has little impact on the incoming beam. To the first order, it is geometrically similar, although it may be slightly elliptical.

However, looking out of the system it is dramatically different. All sorts of stray light concepts are involved. Consider the secondary as the aperture stop of the system in reverse; i. e., the full surface intercepts the ray bundle from the on-axis position that is reflected towards the primary mirror. However, when the ray bundle from an off-axis position on the detector reaches the plane of the rim of the primary mirror, it is bigger on one side than the diameter of the aperture of the primary. This is where the primary also becomes the limiting aperture and gives design programs a challenge. Nevertheless, it is because the instantaneous footprint of the *incoming* beam from an off-axis stray light source position needed a bigger secondary mirror aperture to accommodate the beam. The same extra mirror surface needed to accommodate the incoming beam allows the detector to see more in the reverse direction, which is how it sees the baffles, mirror mounts, struts, and (in other systems) the ground edges of lenses, etc. All of these are critical objects and are therefore more sources of scattered and diffracted light to the detector.

The amount of baffle seen increases as the point is moved to an off-axis position on the detector. If there were a baffle one would see much more of the baffle, which could be directly illuminated by a distant stray light source. A source that does not put any direct power on the primary mirror, but does to a black surface with a higher BRDF value is not good.

This concept applies to almost all optical systems. As briefly referred to above, the struts that support the secondary mirror in a typical two-mirror telescope are seen in double reflection off the secondary and them off the primary. As one moves off-axis there is a view to the edges of the struts if they are not tapered out of the field of view. It this case the forward scatter to the detector can be quite high.

These reverse ray traces are not easy in a conventional optical design package because, as the trace starts from the off-axis image plane position, there are often multiple apertures that limit the edge rays in the reverse direction. Once discipline is learned it is often easy, maybe even easier than using a program, to conceptual-

Table 7.4 Comparison of the advantages and disadvantages of a stray light system test and a computer-generated stray light analysis. It shows that every disadvantage of a stray light test is an advantage of a system analysis, and every disadvantage of a stray light analysis is an advantage of a stray light test

Stray light system test advantages	Stray light analysis disadvantages
1 The system is the "as built" system	1 There is the possibility that the system was input incorrectly; operator error. Not all the details are put in. If there are assembly errors they are not usually input into the analysis as they are unknown.
2 The BRDFs are the real coatings	2 The BRDFs are from representative measured data, not the real coatings performance.
Stray light system test disadvantages	**Stray light analysis advantages**
1 If the system does not pass, then what? A test is almost neveer designed to determine the paths of scatter or if the BRDF along a path is critical.	1 The analysis will easily point out the paths, the flaws, and the magnitudes of power propagated along the paths. You know what to do to fix it. You know what is not important.
2 If the system passes then it appears to have passed the ultimate test; but not in the real environment and the test chamber has been know to affect the results both better and worse than the actual system performance	2 An analysis can predetermine adverse effects resulting from the test chamber environment in comparison to the actual environment in which the system will need to perform in.
3 If the fully built system fails it has been an expensive and time-consuming experience.	3 The rejection of a design on paper is much easier to swallow than a completely built system. And it most often is much less expensive.
4 A stray light test can be relatively expensive and come after the system is built	4 A stray light analysis can be relatively inexpensive and can be performed ahead of time.

ize what would be seen from the various field positions on the detector.

The concept is this: all apertures that are not the aperture stop or conjugate to it will be *oversized* for any given field position and therefore are windows to other surfaces and surface areas; you see critical objects. Once the real hard stop is encountered the detector should not see any more surfaces beyond the ray bundle size. Note that central obscurations and strut-like objects in the field of view will be seen.

7.4.4 Field Stops and Lyot Stops

Field stops have not yet been discussed, but they too have an important role in stray light suppression. If one exists, it is placed at the position of an intermediate image position. Ray bundles traced from an out of for position into the system will be intercepted by this field stop; hence there should not be any directly illuminated objects passed by an intermediate field stop. Field stops limit the number of illuminated objects. Caution should be exercised in systems that have mirrors, as a beam focused out of field onto a field stop might be seen in the center of a reflection off a mirror, and would then be a bright point source in the field of view of the detector.

A Lyot stop is placed at the image of the aperture stop, often at the exit pupil. Lyot stops are often really *the* stop of the system as they are usually slightly undersized to account for aberrations and diffraction effects. Because they follow the defined aperture stop they are closer to the image plane in the series of optical elements and baffles that follow. Therefore the Lyot stop will limit the number of critical objects seen by the detector.

In detail the concept is more complicated and a software tool is needed to help speed up the calculation. For instance, the image at the intermediate field stop is often not of as high a quality as the final image. That is because the optical parameters are varied to control the final image quality. Therefore aberration will blur the image and increase the spot size, and some stray light might get through the aperture. The same goes for the aperture stop and its Lyot stop. An effect known as pupil aberration may also occur; if bad enough it will allow extra critical objects to be seen. Usually only a small portion of some new critical object will be seen. However, even if it is only a small piece that is directly illuminated, it could make orders of magnitude difference in the stray light background noise.

After the system is designed and analyzed a stray light system test is performed. One does not just believe the analysis because computers assume a perfect system, a perfectly clean system, a perfectly and completely well-coated system whereas the system will not be ideal. Then why do the stray light analysis at all? Why not just test it? If it does not work, then it can be fixed. However, this approach is not as easy as it first sounds, if you do not know the propagation paths then you do not know what to fix. A trial-and-error approach can be very time consuming and very expensive. It would be unfortunate if you found out that the system was not at fault but that the stray light test was the culprit.

Two comments from experience. In about 20 different stray light system tests on real built systems the test environment created a problem in 100% of the cases; no exceptions. In some cases the system to the pleasure of the designers exceeded performance goals but that was not correct, and several other times the system test indicated a failure when the failure was in the design of the chamber. It has gone both ways. Fortunately in most of the 20 cases the chamber was pre-evaluated first and a fix to the test procedure was made saving more time and money. In the other cases the program was halted until the fault in the chamber test was found via a stray light analysis. This happens often.

Table 7.4 gives a comparison between a stray light analysis and a system stray light test.

Every stray light analysis should be backed up with a stray light test. The chamber itself should have a performance analysis performed on it before it is used in any particular configuration. The two methods complement each other, the strengths of one cover the weaknesses of the other. Neither one alone should be considered sufficient.

7.5 Conclusion

During the process of the design and analysis of an optical system, it is necessary for the optical designer or engineer to consider many factors that will impact on the performance of the system in operation. These include the actual system design, usually defined using geometrical ray tracing, the effects of diffraction, manufacturing and material limitations, as well as the effects of stray and scattered light. Software tools are available to help perform these tasks, but it is up to the engineer to understand the strengths and limitations of these programs and to insure that a complete analysis of the system is performed before committing the design to hardware.

References

7.1 R. E. Fischer, B. Tadic-Galeb: *Optical System Design* (McGraw-Hill, New York 2000)

7.2 J. Geary: *Introduction to Lens Design with Practical ZEMAX Examples* (Willmann-Bell, Virginia 2002)

7.3 D. Goodman: *Introduction to Fourier Optics* (McGraw-Hill, New York 1969)

7.4 R. Kingslake: *Lens Design Fundamentals* (Academic, New York 1983)

7.5 D. O'Shea: *Elements of Modern Optical Design* (Wiley, New York 1985)

7.6 R. Shannon: *The Art and Science of Optical Design* (Cambridge Univ. Press, New York 1997)

7.7 G. Smith: *Practical Computer-Aided Lens Design* (Willmann-Bell, Virginia 1998)

7.8 W. Smith: *Modern Optical Engineering* (McGraw-Hill Professional, New York 2000)

7.9 Stray Light Problems in Optical Systems, SPIE Proceedings Vol. 107, April 18–21, 1977, (22 papers)

7.10 Radiation Scattering in Optical Systems, SPIE Proceedings Vol. 257, Sept. 30–Oct. 1, 1980, (28 papers)

7.11 Generation, Measurement, and Control of Stray Radiation III, SPIE Proceedings Vol. 384, Jan. 18–19, 1983, (15 papers)

7.12 Scattering in Optical Materials, SPIE Proceedings Vol. 362, Aug. 25–27, 1982, (28 papers)

7.13 Stray Radiation IV, SPIE Proceedings Vol. 511, Aug. 23, 1984, (14 papers)

7.14 Stray Radiation V, SPIE Proceedings Vol. 675, Aug. 18, 1986, (46 papers)

7.15 SPIE Proceedings Vol. 216, Feb. 4–5, 1980, Optics in Adverse Environments, (30 papers)

7.16 P. J. Peters: Stray Light Control, Evaluation, and Suppression, SPIE Proceedings Vol. 531, January 1985

7.17 W. Greynolds: Computer-Assisted Design of Well-Baffled Axially Symmetric Optical Systems, M.S. Thesis, Univ. of Az., 1981

7.18 D. A. Thomas: Light Scattering from Reflecting Optical Surfaces, Ph.D. Dissertation, Univ. of Az., 1980

7.19 R. P. Breault: Suppression of Scattered Light, Ph.D. Dissertation, Univ. of Az., 1979

7.20 W. G. Tifft, B. B. Fannin: Suppression of Scattered Light in Large Space Telescopes, NASA Contract NAS8-27804, Steward Observatory, Univ. of Az., 1973

7.21 Stray Light and Contamination in Optical Systems, SPIE Proceedings Vol. 967, Aug. 17–19, 1988, (33 papers)

8. Advanced Optical Components

This chapter describes a selection of advanced optical components including the underlying physical principles, production techniques and already existing or possible future applications.

Several of these optical elements, in particular variable lenses and photonic crystals, may replace conventional optical systems once their potential for applications has been fully explored. Other components such as high-quality optical fibres, though well established and used worldwide, still undergo a rapid further improvement and integration in communication systems.

Besides increased quality and versatility, a driving force and essential aspect in the development of optical components in general is low cost and mass production.

8.1 Diffractive Optical Elements

Today, the vast majority of optical instruments are based on refractive or reflective optical elements, most often restricted to a spherical shape. On the other hand, the demands on performance, volume and costs of optical imaging systems are increasing dramatically from one product generation to the next, so that the search for alternative design approaches is becoming increasingly important. Significant improvements of the optical performance of imaging systems can be achieved by the substitution of conventional refractive or reflective elements with diffractive optical lenses. Especially, the extremely different dispersive characteristics of diffractive optical elements (DOEs), compared to refractive elements, offer more degrees of freedom in the lens design and allow an advantageous combination of both to form hybrid systems with optimized performance and compactness.

The possible use of the hybrid concept covers a broad application spectrum: in eyepieces the combination of diffractive and refractive optical components allows one to overcome limitations in eye relief, exit pupil, and field of view [8.1–5]. Closely related to eyepieces are head-mounted displays (HMDs), a micro-display application near to the eye, with challenging issues in high-quality ergonomics combined with high optical performance in a minimal volume [8.6–8]. Further application examples of the hybrid concept comprise high-quality camera lenses working over the whole visible spectrum [8.9], optical inspection systems for the semiconductor industry [8.10, 11], and high-resolution microscopy with solid immersion lenses [8.12]. Extremely challenging projects concern space telescopes based on lightweight diffractive lenses with diameters larger than 25 m [8.13, 14] enabling a high resolution observation necessary e.g. for extrasolar planet detection.

8.1.1 The Fresnel Zone Plate Lens

A simple approach to describe the basic principles of the focusing properties of an imaging DOE is based on the analysis of the Fresnel zone plate (FZP) (Fig. 8.1). The FZP is characterized by an axially symmetrical and concentric zone system of alternating opaque and transparent rings. Consider a monochromatic plane wave which is incident onto the FZP. In accordance with the Huygens–Fresnel principle, a common focal point is guaranteed, when the optical path length from two adjacent transparent rings differ by an integer multiple of the wavelength λ. The radius r_j of the j-th transparent ring and the focal length f are related by:

$$r_j^2 = 2j\lambda f + (j\lambda)^2 . \tag{8.1}$$

The term $(j\lambda)^2$, which represents spherical aberration, can be ignored in the paraxial approximation for low numerical aperture (NA):

$$r_j \cong \sqrt{2j\lambda f} . \tag{8.2}$$

The focal length in this case is determined by:

$$f = \frac{r_1^2}{2\lambda} . \tag{8.3}$$

The last relation shows that the focal length scales inversely with the wavelength, which expresses a strong chromatic effect.

In many cases it is of practical interest to express the imaging properties of an FZP such as focal length, numerical aperture, and lens diameter D in terms of wavelength and the fabrication relevant quantities. These involves the total number of periods N and the minimum spatial period $\Delta r = r_N - r_{N-1}$. The derivation [8.15] starts with (8.2) for the outer region of the FZP ($j \to N$) and subtracts as

$$r_N^2 - r_{N-1}^2 = 2\lambda f . \tag{8.4}$$

On the other hand when using the definition for Δr with the approximation $2r_N \Delta r \gg (\Delta r)^2$ one can write:

$$r_N^2 - (r_N - \Delta r)^2 \cong 2r_N \Delta r . \tag{8.5}$$

The combination of the last two equations results in:

$$D\Delta r \cong 2\lambda f \tag{8.6}$$

with a total lens diameter $D = 2r_N$.

Finally, with the relation $f = r_N^2/(2\lambda N)$ from (8.2) the lens diameter and focal length become:

$$D \cong 4N\Delta r , \tag{8.7}$$

$$f \cong 2\frac{N\Delta r^2}{\lambda} \tag{8.8}$$

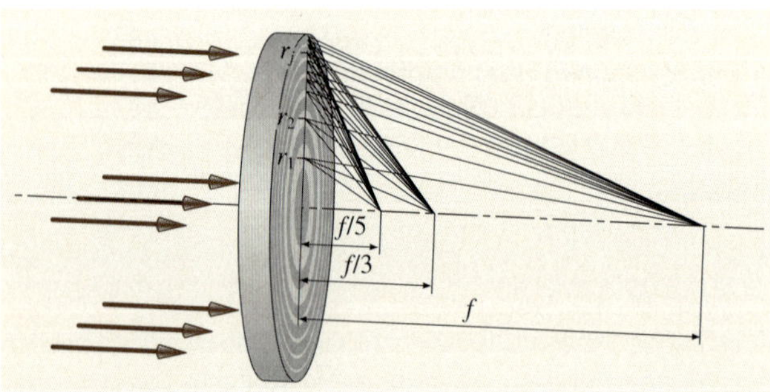

Fig. 8.1 Schematic view of the concentric ring system of a Fresnel zone-plate lens illuminated with a plane wave. The convergent first three positive diffraction orders are shown. The optical path length of the light deflected from adjacent zones towards a common focal point differ by an integral multiple of the design wavelength

These important relations show that the focal length of an FZP scales directly with the number of periods, with the square of the minimum period, and inversely with the wavelength. A further essential property of an imaging lens is the numerical aperture (NA), which is defined (in air) as NA $= \sin\theta$, where θ is half of the acceptance angle. From the geometry the NA of a zone plate is given by NA $= r_N/f = D/(2f)$ so that

$$\mathrm{NA} \cong \frac{\lambda}{\Delta r} \quad (8.9)$$

The numerical aperture of an imaging system plays an important role in the determination of both the lateral resolution Δd_{Rayl}, and the depth of focus or axial resolution, Δz. With the famous Rayleigh criterion for lateral resolution and the relation for the depth of focus one obtains:

$$\Delta d_{\mathrm{Rayl}} = 0.61\frac{\lambda}{\mathrm{NA}} \cong 0.61\,\Delta r\,, \quad (8.10)$$

$$\Delta z = \frac{1}{2}\frac{\lambda}{(\mathrm{NA})^2} \cong \frac{1}{2}\frac{\Delta r^2}{\lambda}\,, \quad (8.11)$$

which means the minimum feature size is the fundamental property that determines the resolution of the FZP.

The mathematical approach of adding a path length difference of exactly one wavelength for constructive interference caused by two adjacent transparent rings in first order can be extended to higher orders ($m = 2, 3, \ldots$) by adding a path length difference of $m\lambda$, leading to the corresponding radius

$$r_{j,m} \cong \sqrt{2jm\lambda f_m}\,. \quad (8.12)$$

This implies, in contrast to a conventional refractive lens, that an FZP causes multiple foci at distances of an integral fractions of the basic focal length

$$f_m = \frac{f}{m}\,. \quad (8.13)$$

The negative orders correspond to virtual foci of negative focal length. In the case of a transmission grating of equal sizes of lines and spaces only the odd orders occur ($f/3, f/5, \ldots$). If the alternating opaque and transparent structure is replaced by a phase-only diffractive element the light intensity obtained at the focus is increased dramatically. In this so-called Fresnel phase plate the phase difference between neighboring zones is π (180°), which is equivalent to a depth difference of $\lambda/2(n-1)$.

Dispersion Properties of a DOE

An essential difference between imaging DOEs and refractive lenses is their chromatic behavior. The performance of refractive lenses depends on material dispersion. Generally optical materials, e.g., inorganic glasses, crystals or polymers, show normal dispersion, i.e., the refraction index $n(\lambda)$ decreases with wavelength. This means for a refractive lens that the focus corresponding to the blue wavelength is located closer to the lens than the focus of the red wavelength (Fig. 8.2a). Quantitatively this can be expressed by the Abbe number $\nu_{\mathrm{e(ref)}}$, where the subscript defines the reference wavelength ($\lambda_{\mathrm{e}} = 546.1$ nm, $\lambda_{\mathrm{C}} = 656.3$ nm, $\lambda_{\mathrm{F}} = 486.1$ nm)

$$\nu_{\mathrm{e(ref)}} = \frac{n(\lambda_{\mathrm{e}}) - 1}{n(\lambda_{\mathrm{F}}) - n(\lambda_{\mathrm{C}})}\,. \quad (8.14)$$

The Abbe number of inorganic glasses ranges from 20 for very dense flint glasses up to 90 for fluor crown glasses. In contrast, the most prominent characteristic

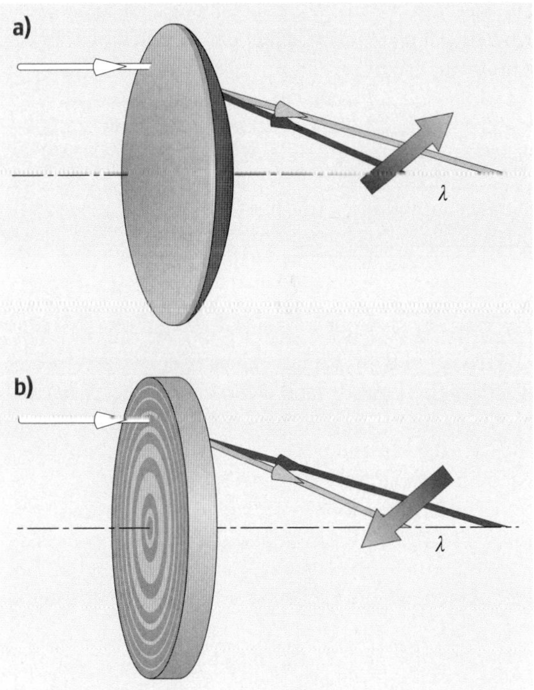

Fig. 8.2a,b Chromatic dependency of refractive and diffractive lenses. The refractive lens **(a)** is characterized by normal dispersion with an increasing focal length for increasing wavelength. The diffractive lens **(b)** shows the reverse behavior, that means decreasing focal length for increasing wavelength

of an imaging DOE is the strong and reversed dispersion, which results in a decreasing focal length with increasing wavelength (Fig. 8.2b). This characteristic can be expressed by an effective Abbe number $\nu_{e(diff)}$ for a diffractive lens, which is determined by the ratio of the wavelengths

$$\nu_{e(diff)} = \frac{\lambda_e}{\lambda_F - \lambda_C} = -3.21 . \tag{8.15}$$

Note, the Abbe number of an imaging DOE does not depend on the material or shape of the element but is only related to the application wavelength band.

For applications covering a broad optical spectrum chromatic aberrations are one of the most important factors limiting the optical performance. An elimination of the chromatic aberration for two selected wavelengths can be realized by combining lenses to an achromatic doublet with adapted dispersion properties and optical powers. The optical power of a lens ϕ is the reciprocal of the focal length f ($\phi = 1/f$). When the components of that achromat are in close contact, e.g., cemented or integrated elements, the relevant design equations are the achromatic condition

$$\frac{\phi_1}{\nu_1} + \frac{\phi_2}{\nu_2} = 0 \tag{8.16}$$

and the sum condition for the overall optical power ϕ of the final lens

$$\phi_1 + \phi_2 = \phi . \tag{8.17}$$

Here ϕ_1 and ϕ_2 are the values of the optical power of the single elements, and ν_1 and ν_2 are the associated Abbe numbers. The last two conditions allow one to calculate the optical power of the individual elements when the optical power of the final lens is given.

To fulfill the conditions (8.16) and (8.17) in a conventional all-refractive achromatic doublet it is necessary to combine lenses of both positive and negative optical power, which results in a partial compensation (Fig. 8.3a).

In contrast, due to the negative Abbe number of a DOE, a hybrid doublet combines diffractive and refractive components with optical power equal in sign (Fig. 8.3b). This implies that the optical power of each element is less than the total power, which allows a considerable reduction in the amount of material required and also a reduction of the weight and volume of the lens.

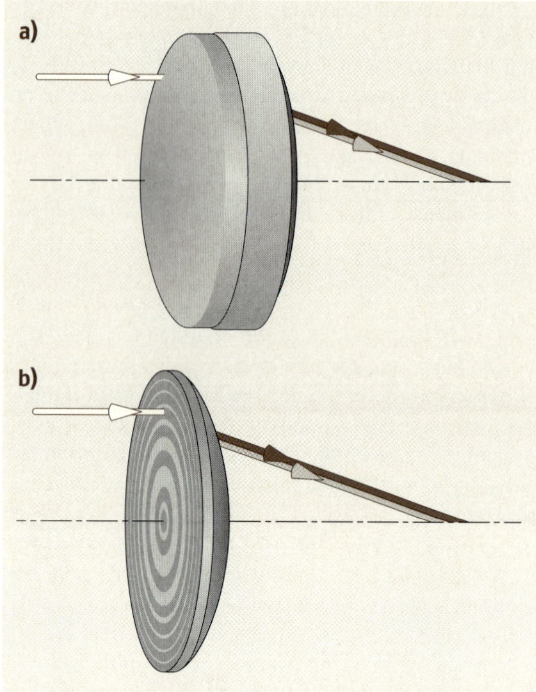

Fig. 8.3a,b Schematic view of achromatic lens elements. The refractive approach **(a)** combines positive and negative lens elements fabricated of different materials. In the hybrid diffractive/refractive solution **(b)** a monolithic design is possible

Achromatization with respect to two wavelengths does not allow a complete removal of the color error. The remaining chromatic aberration is known as the secondary spectrum, which tends to be more prominent for hybrid achromats than for all-refractive combinations [8.16]. A reduction of the secondary spectrum – apochromatic correction – could be achieved in the conventional approach by the introduction of a three-lens design, or alternatively in the hybrid approach by combing a single hybrid element with a refractive lens.

Diffraction Efficiency – Efficiency Achromatization

In a diffractive lens, the local grating period determines the direction of the deflected light for the various diffraction orders but it allows no prediction about the amount of light focused into the particular diffractive order. The latter is described by the diffraction efficiency η, which is strongly dependent on the nature of the periodic structure, e.g., geometry and material selection.

As a first example, the efficiency of a simple transmission binary phase grating with equal dimensions of lines and spaces can be calculated for the m-th diffraction order by:

$$\eta_{m,\text{binary,phase}} = \begin{cases} \left(\frac{2}{m\pi}\right)^2 & m \quad \text{odd}\,, \\ 0 & m \quad \text{even}\,. \end{cases} \tag{8.18}$$

In Fig. 8.4 the efficiencies for each of the diffraction orders are listed. The exact equivalence of lines and spaces guarantees the suppression of all nonzero even orders, and the phase depth of π results in destructive interference for the zeroth order.

To increase the diffraction efficiency for a selected diffraction order it is necessary to leave symmetric surface relief structures. Here, due to the advantageous realization possibilities, multilevel surface relief gratings are of enormous importance. These stepped gratings can be generated by projection photolithography using several binary masks and subsequent etching processes. When all steps of the N-level structure are of equal height then the number of processing steps can be reduced to $\log_2 N$. The first-order diffraction efficiency of these multilevel structures are given by

$$\eta_{1,N} = \text{sinc}^2\left(\frac{1}{N}\right)\,, \tag{8.19}$$

where

$$\text{sinc}\,(x) = \frac{\sin(\pi x)}{\pi x}\,. \tag{8.20}$$

Figure 8.5 shows the calculated efficiencies with the corresponding profile depths d_N for $N = 2$, 4, 8 and 16 phase levels. The strong increase of the diffraction efficiency with increasing phase levels is clearly observable. Note that the total profile depth d_N of the multilevel structure is also increasing with the number of phase levels N, which can be expressed with the relation

$$d_N = \frac{N-1}{N}\frac{\lambda}{n-1}\,. \tag{8.21}$$

It has to be mentioned that in practice fabrication errors such as etch depth errors and mask-to-mask misalignment result in reduced diffraction efficiencies.

With an increasing number of phase levels, the shape of the multilevel surface relief approaches the geometry of a sawtooth profile. Following the scalar theory, which is valid for grating periods much larger than the wavelength, the sawtooth profile of a blazed diffraction grating possesses the most favorable profile geometry.

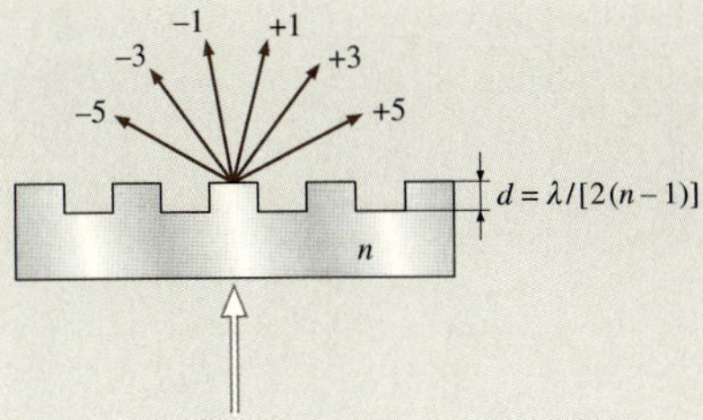

Orders	Efficiency (%)
0	0
+1, −1	40.5
+3, −3	4.5
+5, −5	1.6
>+7, −7	3.4

Fig. 8.4 Representation of a simple binary phase grating with the corresponding diffraction orders. In the table the efficiencies in dependence on the diffraction orders are listed. The fill factor of 50% guarantees the suppression of all the nonzero even orders, and the phase depth of π is responsible for zero diffraction efficiency into the zero order

The efficiency η_m in the desired order m for an ideal blazed profile is expressed by [8.17, 18]:

$$\eta_m(\lambda) = \text{sinc}^2(\alpha - m)\,, \tag{8.22}$$

where

$$\alpha = d[n(\lambda) - 1]\frac{1}{\lambda}\,. \tag{8.23}$$

These equations imply that 100% of the light ($\eta = 1$) will be diffracted into a specific diffraction order m if the denominator of the sinc-function is zero. For a given design wavelength λ_0 and design order $m = 1$

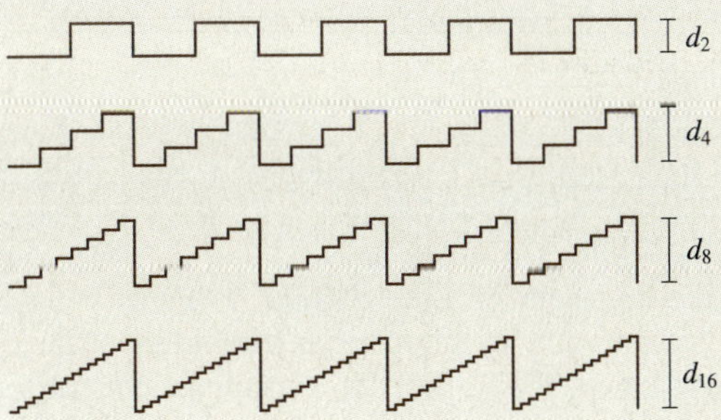

N	Efficiency (%)	$d_N = \frac{N-1}{N}\frac{\lambda}{n-1}$
2	40.5	$1/2\,\lambda/(n-1)$
4	81.1	$3/4\,\lambda/(n-1)$
8	94.9	$7/8\,\lambda/(n-1)$
16	98.7	$15/16\,\lambda/(n-1)$

Fig. 8.5 Multilevel gratings with 2, 4, 8, and 16 steps. With increasing step number the profile approximates the ideal blazed grating. The first-order diffraction efficiencies with the corresponding profile depths are given in the table

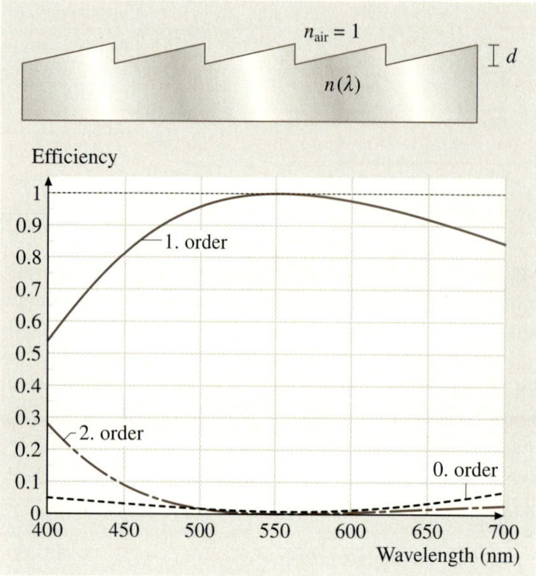

Fig. 8.6 Simple single-layer blazed grating with the interface to air. The diagram shows the calculated scalar diffraction efficiency for the diffracted orders $m = 0$, 1, and 2 and a design wavelength $\lambda_0 = 550$ nm (material: inorganic glass SSK3). The 100% maximum efficiency at the design wavelength slowly decays across the visible spectrum, where it is partly diffracted into the other orders

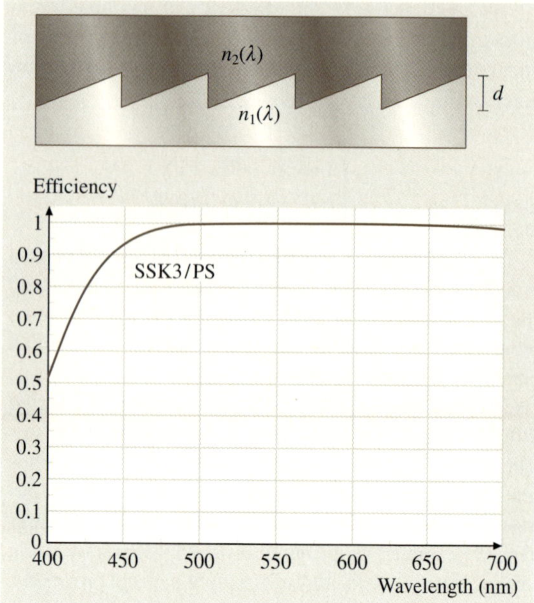

Fig. 8.7 Schematic view of an NIM DOE with two materials in direct contact at the diffractive interface. The diffraction efficiency was calculated for the combination of SSK3 and polystyrene. The design wavelength is 550 nm and the profile depth 24 μm. The efficiency is nearly 100% over a broad spectral range

the maximum efficiency needs an adjusted profile depth d of:

$$d = \frac{\lambda_0}{n(\lambda_0) - 1} . \tag{8.24}$$

Following (8.22), the calculated efficiencies as a function of the wavelength are shown in Fig. 8.6 for the diffracted orders $m = 0$, 1, and 2 and a design wavelength of $\lambda_0 = 550$ nm. Exemplary curves are calculated with the material parameters of the inorganic glass SSK3. The 100% maximum efficiency at the design wavelength slowly decays across the visible spectrum, where it is partly diffracted into the other orders. Nearly all of the energy within the visible spectral band is diffracted into the three orders shown.

This efficiency behavior has a dominant impact on the quality of diffractive or hybrid optical systems. On the one hand, due to the varying efficiency, the images are not generated in true colors and additionally the light transmitted into the zeroth order or into the higher non-design orders may add to the background illumination, reducing the image contrast. For that reason it becomes an attractive aim to find solutions for diffractive structures which allow a high efficiency (close to 100%) over a broad spectral bandwidth.

Nearly index-matched (NIM) elements [8.19] form one DOE type which show the capability of efficiency achromatization. These NIM-DOEs consist of a diffractive structure with large profile depth, separating two materials characterized by a small difference in their refractive index (Fig. 8.7). Compared to diffractive structures with an interface to air, the NIM concept has the advantage of reduced tolerances on profile depth and surface roughness.

The efficiency as a function of wavelength for this two-media approach shows a similar dependency compared to the previous situation with the interface to air. The important modification is the introduction of the refractive index of the second material

$$\eta_{NIM}(\lambda) = \mathrm{sinc}^2 \left\{ d_{NIM}[n_1(\lambda) - n_2(\lambda)] \frac{1}{\lambda} - 1 \right\} . \tag{8.25}$$

Also in this case the profile depth d_{NIM} has to be related to a design wavelength λ_0

$$d_{\mathrm{NIM}} = \frac{\lambda_0}{n_1(\lambda_0) - n_2(\lambda_0)} . \tag{8.26}$$

Due to the small difference in the refractive indices of both materials the adapted profile depth requires values in the range of some tens of microns.

In Fig. 8.7 the calculated functional form of the efficiency for this two-media system is shown exemplarily for the glass–polymer configuration of SSK3 and polystyrene (PS). The design wavelength was set to 550 nm, which corresponds to a profile depth of 24 μm. It is clearly observable that the efficiency is nearly 100% over a broad spectral range. A significant reduction occur only below 450 nm. Also remarkable is the asymmetric behavior over the visible spectral range.

From the equations above it follows that in the NIM configuration the dispersion of the two materials must be such that the change of the refractive index, Δn, at the interface remains proportional to λ or

$$\frac{\lambda}{\Delta n(\lambda)} = \text{const} . \tag{8.27}$$

This constraint limits the choice of possible material combinations. Furthermore, the material restriction becomes more significant with varying temperature, because $\mathrm{d}n/\mathrm{d}T$ often differs significantly for the two materials. The need for a large profile depth and a practicable aspect ratio in the fabrication process restrict the minimum zone width of the diffractive structure and therefore limits the diffractive optical power of the NIM-DOEs.

A second approach to realize DOEs with achromatized efficiency uses a multilayer combination of diffractive structures separated by an additional air gap [8.20, 21]. Due to the additional degrees of freedom, the above mentioned constraints on the dispersion properties of the materials are removed.

In one type of multilayer configuration, two single-layer DOE structures are arranged to face each other and are separated by an air gap of a few microns. The two single-layer DOEs are formed to have equal grating pitches but differ in grating heights (d_1, d_2) and in material parameters (n_1, n_2)(Fig. 8.8 top). In the scalar approximation, the efficiency takes the form

$$\eta_{\text{double layer}}(\lambda) = \operatorname{sinc}^2\left\{d_1[n_1(\lambda)-1]\frac{1}{\lambda} - d_2[n_2(\lambda)-1]\frac{1}{\lambda} - 1\right\} . \tag{8.28}$$

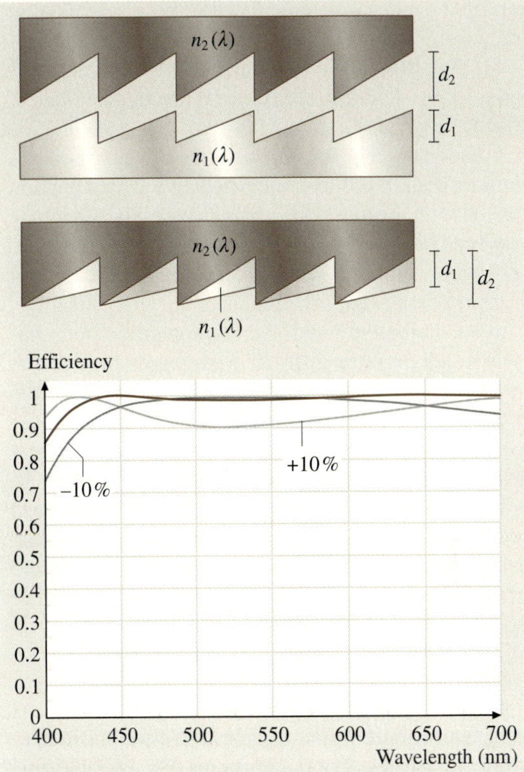

Fig. 8.8 Concepts for multilayer DOEs: *top* – two diffractive structures realized in different media with separating air gap; *center* – diffractive structures in direct contact, air gap moved to the exterior. *Bottom* – calculated diffraction efficiencies for optimum profile depths (*black curve*) and ± 10% deviation from ideal profile depths (*grey curves*)

This equation implies the condition for high diffraction efficiency:

$$d_1 \frac{n_1(\lambda)-1}{\lambda} - d_2 \frac{n_2(\lambda)-1}{\lambda} = 1 . \tag{8.29}$$

That means, a proper selection of layer materials and corresponding profile depths allows to keep this condition valid over a broad spectral range and therefore guarantees an efficiency achromatization.

Figure 8.8 (bottom) shows the efficiency curve (black) for a double layer system made of an inorganic glass (LLF6; n_{LLF6} (550 nm) $= 1.53404$, $\nu_{\mathrm{e,LLF6}} = 50$) and a transparent resin with $n_{\mathrm{res}}(550\,\mathrm{nm}) = 1.662$ and $\nu_{\mathrm{e,resin}} = 22$. The optical values of both materials are comparable to the ones discussed in [8.22]. From the design wavelengths $\lambda_{01} = 450$ nm and $\lambda_{02} = 650$ nm the profile depths were calculated to be $d_1 = 7.26\,\mu m$ and $d_2 = 4.97\,\mu m$.

All over the visible spectral range the efficiency is nearly 100%. A slight decay can be recognized below 430 nm.

Additionally Fig. 8.8 contains two efficiency curves assuming both, a positive and a negative 10% deviation from the ideal profile depths. In both cases the efficiency is above 90%. That means, the efficiency achromatization works well, also when the profile depths are not realized in the design optimum.

In an alternative multilayer configuration both materials which form the diffractive structures are in direct contact and the air gap is moved to the exterior of the structure [8.23] (Fig. 8.8 center). The choice of best layer configuration also depends upon polarization-dependent Fresnel losses and fabrication constraints.

The multilayer concept has recently been introduced in consumer camera lenses [8.24]. Compared to a classical dioptric lens design of the same specifications and performance, it was possible to reduce both the overall length by 27% and the weight by 31%.

Athermalization of Diffractive/Refractive Hybrid Lenses

Strong temperature variations may also effect the image-forming properties of optical systems. For example, this concerns high-volume consumer optics based on polymers, and also aerospace and military applications using infrared optical materials. For both application fields, the demands for operating temperature ranges covering 80 °C to 100 °C are not unusual. To describe the influence of temperature variations on the individual optical elements the optothermal expansion coefficient x_f was defined. This parameter relates the change in focal length f of a lens to a change in temperature.

For refractive lenses, the influence of temperature on the optical performance is dependent on two effects. Firstly, the thermal expansion of the lens material changes its geometry, and secondly the refractive index is a function of the environmental temperature [8.25]. For a thin refractive lens, the optothermal expansion coefficient $x_{f,\mathrm{ref}}$ is given by:

$$x_{f,\mathrm{ref}} = \frac{1}{f}\frac{\mathrm{d}f}{\mathrm{d}T} = \alpha_\mathrm{g} - \frac{1}{n-n_0}\left(\frac{\mathrm{d}n}{\mathrm{d}T} - n\frac{\mathrm{d}n_0}{\mathrm{d}T}\right), \tag{8.30}$$

where α_g is the linear thermal expansion coefficient, n is the refractive index of the lens material and n_0 is the refractive index of the image space. $\mathrm{d}n/\mathrm{d}T$ and $\mathrm{d}n_0/\mathrm{d}T$ express the change of refractive index with the temperature of the lens material and the image space, respectively. The change in focal length is given by

$$\Delta f = f x_{f,\mathrm{ref}} \Delta T\,. \tag{8.31}$$

In analogy to the previous, for diffractive lenses in the paraxial approximation, it is also possible to derive an optothermal expansion coefficient $x_{f,\mathrm{diff}}$. In this case, the characteristics of $x_{f,\mathrm{diff}}$ are attributed to the expansion or contraction of the zone spacing by temperature variation and also to the change of the refractive index of the image space with temperature [8.26–29]. This dependency is expressed by:

$$x_{f,\mathrm{diff}} = 2\alpha_\mathrm{g} + \frac{1}{n_0}\frac{\mathrm{d}n_0}{\mathrm{d}T}\,. \tag{8.32}$$

Here, it is of particular importance that there is no dependence on the refractive index change of the lens material. This is an essential difference between the thermal characteristics of refractive and diffractive lenses.

Investigations on the influence of temperature variations on the diffraction efficiency show an negligible effect [8.26]. Both variations of the profile height and changes in the refractive index of the material of the diffractive structure make no significant impact to the efficiency.

The comparison of refractive and diffractive elements for selected materials shows different characteristics and quantities in the optothermal expansion coefficients. In Table 8.1 some values for inorganic glasses, optical polymers and infrared materials are listed. The parameters for the inorganic glasses were taken from the Schott glass catalog with the reference wavelength of 546.1 nm (e-line). The change in index of the lens material relative to the change in index of air is related to the temperature range between +20–+40 °C. The linear thermal expansion coefficient is indicated for a temperature interval of −30–+70 °C. The parameters of the polymers polymethylmethacrylate (PMMA), polycarbonate (PC), polystyrene (PS) and cyclic olefin copolymer (COC) with the reference wavelength 589 nm were taken from [8.30]. The parameters for cyclic olefin polymer (COP) are related to [8.31]. Due to their very low water absorption in comparison to conventional polymers, both COC and COP became attractive materials for molded polymer optics. The quantities of the infrared materials silicon and germanium were calculated after [8.25].

The optothermal expansion coefficients for refractive lenses cover a wide range of values, with both positive and negative sign, depending on the magnitude

Table 8.1 List of refractive and diffractive optothermal expansion coefficients for different materials

Material	n_e	α (10^{-6}/K)	dn/dT (10^{-6}/K)	$x_{f,\text{ref}}$ (10^{-6}/K)	$x_{f,\text{diff}}$ (10^{-6}/K)
F2	1.624	8.20	4.4	1.15	16.4
N BK7	1.519	7.10	3.0	1.32	14.2
SF5	1.678	8.20	5.8	−0.35	16.4
SF11	1.792	6.10	12.9	−10.2	12.2
N-FK51A	1.488	12.7	−5.7	24.4	25.5
N-PK52A	1.498	13.0	−6.4	25.8	26.0
N-LASF40	1.839	6.9	9.3	−4.18	13.8
	n **(589 nm)**				
PC	1.586	70	−143	314	140
PS	1.59	70	−120	273	140
PMMA	1.491	65	−85	238	130
COP	1.525	60	−80	212	120
COC	1.533	60	−101	249	120
	n **(IR)**				
Ge	4.00	6.10	270	−83.9	12.2
Si	3.42	4.20	162	−62.7	8.4

of the thermal expansion coefficients and the variation of the refractive index with temperature. On the other hand, the optothermal expansion coefficients for diffractive lenses cover a smaller range and are dominated by the thermal expansion coefficients, which always assume positive quantities.

Comparable to the procedure to design achromatic lens doublets, it is also feasible to combine two optical elements to compensate the temperature dependencies of both. This method and the elements are called athermalization and athermats, respectively. In an all-refractive setup, it is necessary to use lens elements made of two different materials for athermalization. The alternative hybrid approach takes the advantage of the different values of optothermal coefficients for diffractive and refractive lenses made of the same material. This means that the realization of a hybrid athermat is possible, which offers advantages in reduced volume and weight compared to the all-refractive solution.

The total lens power of a hybrid athermat ϕ_{net} is the sum of the individual powers ϕ_{ref} and ϕ_{diff}

$$\phi_{\text{net}} = \phi_{\text{ref}} + \phi_{\text{diff}} \,. \tag{8.33}$$

The optothermal expansion coefficient for the doublet $x_{f,\text{net}}$ is given by

$$x_{f,\text{net}} = x_{f,\text{ref}} \frac{\phi_{\text{ref}}}{\phi_{\text{net}}} + x_{f,\text{diff}} \frac{\phi_{\text{diff}}}{\phi_{\text{net}}} \,. \tag{8.34}$$

These equations allow one to calculate the contributions of the diffractive and the refractive parts of a lens system when the total optical power is given. In order to realize a hybrid athermat with exclusively positive values for x_{ref} and x_{diff}, it is necessary to combine a positive and a negative lens element which partly compensate for each other in the optical power. Choosing a lens material with an opposite sign for x_{ref} and x_{diff} it is possible to combine lens elements with both positive or negative contributions so that their optical power is added and a more compact design can be realized.

8.1.2 Subwavelength Structured Elements

When light is incident on a surface-relief grating with period Λ the angles of the transmitted diffraction waves $\theta_{t,m}$ in the m-th diffraction order are given by the grating equation:

$$\sin\theta_{t,m} = \frac{m\lambda}{\Lambda n_2} + \frac{n_1}{n_2}\sin\theta_i \,. \tag{8.35}$$

Here n_1 and n_2 are the refraction indices of the incident and the transmitting medium, respectively, θ_i is the incidence angle as measured from the grating's surface normal, and λ is the incident free-space wavelength.

If the period of the diffraction grating becomes much smaller than the optical wavelength λ, we find that only the zeroth order is allowed to propagate and all the other orders are evanescent

$$\Lambda \ll \lambda \rightarrow m = 0 \,. \tag{8.36}$$

proximation are too small and therefore less suitable for accurate predictions.

Subwavelength structures which are periodic in two dimensions (2-D) show a polarization sensitivity that is dependent on the fill factor (L_x/Λ_x, L_y/Λ_y) and periodicity (Λ_x, Λ_y) in the orthogonal directions. A schematic view of such a 2-D structure is depicted in Fig. 8.11. It consists of an array of dielectric homogeneous pillars with a rectangular cross section periodically immersed in a second dielectric medium. In the totally symmetric case the periods and the widths of the pillars are identical in both dimensions which results in polarization insensitivity and a symmetry in the effective permittivity.

Instead of a corresponding EMT for the 2-D structures, several different approximations have been developed and compared to RCWA calculations [8.38, 39, 52]. The various approximations are based on different averaging procedures assuming that the effective permittivity of a 2-D structure is somewhere between the values for the orthogonal one-dimensional (1-D) structures. It has to be mentioned that a simple volume averaging is insufficient, but to get a more accurate result a complex consideration has to be carried out. In an exemplary extended averaging procedure the periodic binary rectangular structure is replaced by two different orthogonal one-dimensional structures. Here the permittivities of the two corresponding one-dimensional grating bars are set to be $\varepsilon_\perp$ and $\varepsilon_\parallel$, as calculated from the permittivity of the pillars [8.39].

Although 2-D effective-medium approximations show an acceptable validity especially in the long-wavelength limit, today increasingly rigorous design methods are applied. Here the further development in computing hardware and software reduces time and storage issues.

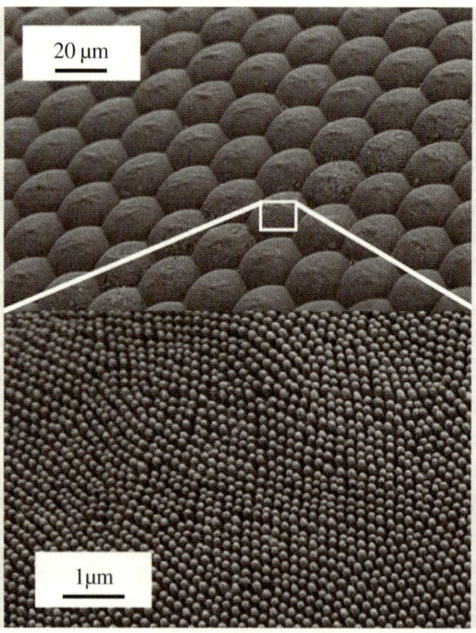

Fig. 8.12 Scanning electron microscopy images of a night-active insect. The upper part shows the hexagonal structure of the compound eye. With increased magnification (*bottom*) a subwavelength periodic surface corrugation becomes visible (courtesy of Prof. J. Spatz)

Antireflection Gratings

The reduction of Fresnel reflections at optical interfaces is an issue of enormous interest in a wide range of optical applications. Today antireflection (AR) coatings are most frequently based on multilayer interference structures with alternating high and low refractive indices. An alternative to the multilayer films are subwavelength structured antireflection surfaces, which provide a graded transition between the refractive indices of the two interfacing media. Such antireflective structures are also found in nature on the corneal surfaces of night-active insects [8.35, 53]. For example Fig. 8.12 shows scanning electron microscope images of a typical moth eye. In the upper image the hexagonal array of the facets of the compound eye is clearly visible. With increasing magnification (lower part) an area covered with conical and uniform protuberances is observable. The diameter of each protuberance is about 100 nm and the spatial periodicity is approximately 170 nm, well below the wavelength of the visible spectrum.

Different characteristics of subwavelength AR structures are advantageous compared to stacks of thin dielectric films, e.g., thin-film coatings may suffer from adhesion problems especially when the thermal expansion coefficients of substrate and layer material are different and the optical devices are used over a broad thermal range. Additionally, due to the restrictions in thin-film coating materials, the available number of discrete refractive indices is also limited. A further aspect concerns the sensitivity of the reflectance to incident angle or wavelength. Here, in general, a subwavelength AR structure tolerates more variations compared to simple layer configurations.

A binary pillar subwavelength structure (Fig. 8.9 top) shows an antireflection performance comparable to that achieved by an elementary single-layer dielectric coating. Following thin-film theory a single layer will exhibit minimum reflectivity at normal incidence when the reflectance from the air–film interface interferes destructively with the reflection at the film–substrate interface. This phase condition is fulfilled when the depth of the profile h satisfies

$$h = \frac{\lambda_0}{4 n_{\text{Layer}}} . \tag{8.45}$$

Additionally, complete destruction will only occur if the amplitudes of both reflecting waves are identical, which implies an amplitude condition for the refractive index of the layer n_{Layer}

$$n_{\text{Layer}} = \sqrt{n_1 \cdot n_2} . \tag{8.46}$$

Exemplary, to fulfill the last equation for the interface between air ($n_{\text{air}} = 1$) and a typical inorganic glass (assumption $n_g = 1.5$) the refractive index of the layer has to be $n_{\text{Layer}} \approx 1.22$ which is not available for thin-film materials. On the other hand, subwavelength structured effective media have the ability to control the effective index of refraction by simply adjusting the fill factor so that in principal each value is possible.

An extension to the single-layer equivalent binary subwavelength grating with improved optical performance is offered by the introduction of a multilevel stair-step surface relief (Fig. 8.9 center). A stair-step multilevel grating consisting of N steps shows similar optical characteristics compared to a traditional thin-film coating of N homogeneous layers. The optical properties of each individual layer are mainly dependent on the particular fill factor. In comparison to simple binary subwavelength gratings, the multilevel stair-step grating permits a large wavelength bandpass and is more tolerant against variations in the angle of incidence [8.54]. For triangular structured surface profiles (Fig. 8.9 bottom) the effective medium has gradient optical properties. By proper design of such surfaces extremely low reflectances can result [8.55].

Different techniques such as e-beam writing and mask-lithography [8.56] have been applied to realize master structures of subwavelength gratings. Among them, interference lithography (IL) has the advantage of allowing the realization of the complete array in a single exposure step over a large area. In the IL process, a photoresist-coated substrate is exposed with an interference fringe pattern generated by coherent beams [8.36, 57], where the grating period Λ is given by

$$\Lambda = \frac{\lambda}{2n \sin\theta} . \tag{8.47}$$

Here λ is the exposure wavelength, n is the refractive index, and θ is the half-angle between the interfering light waves under the condition of symmetric incidence. In air, the minimal obtainable grating period using counter-propagating waves ($\theta = 90°$) is therefore $\lambda/2$. To realize a periodic 2-D structure double- or multiple-exposure techniques are necessary with a defined rotation of the sample between the single exposure steps. During the recording process, a latent structure is created in the resist, which is transformed into a continuous surface phase profile in the subsequent development process. After the development process a stabilization of the pattern structure and an adjustment of the profile depth can be realized by selective reactive ion beam etching.

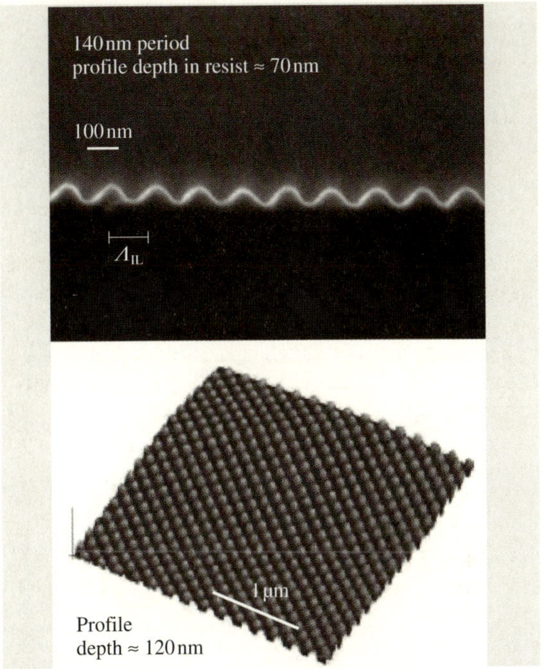

Fig. 8.13 *Moth eye* surfaces generated by interference lithography (IL)

At the top of Fig. 8.13 a scanning electron microscope image of a linear photoresist grating is presented in side view. The sinusoidal corrugation shows a period of 140 nm and a depth of 70 nm. The identical spatial period in a two-dimensional pattern is obtained by double exposure and rotation of the sample by 90° [atomic force microscope (AFM) scan at the bottom of Fig. 8.13].

The structures were realized by the use of a frequency-quadrupled Nd-YAG laser with 266 nm wavelength. Deep-UV interference lithography requires additional process steps including pre- or soft-bake and post-exposure bake. Soft bake is necessary for solvent removal, stress reduction and resist planarization. The

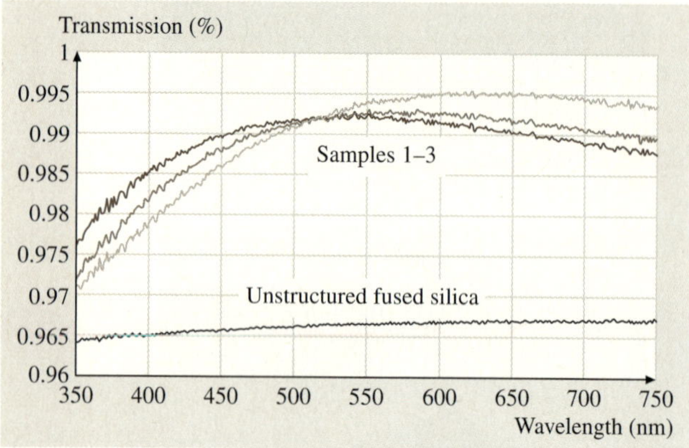

Fig. 8.14 Measured transmission in dependence on the wavelength for three samples of subwavelength crossed gratings in fused silica. The depth of the structure is 300 nm and the diameter of the pillars is about 85 nm. In comparison to an unstructured surface, the transmission is increased significantly across a broad spectral range in the VIS

process of post-exposure bake is an essential, critical and indispensable step to guarantee solubility of the optical exposed areas of these chemical amplified photoresists. These technological procedures have to be performed very carefully to ensure constant conditions and to obtain reproducible and defined profile parameters.

Figure 8.14 shows the measured transmission as a function of the wavelength of three samples of subwavelength crossed gratings. The profile was transferred into fused silica by a reactive-ion etching process. Here the depth of the etched structure is 300 nm and the diameter of the pillars is about 85 nm. In comparison to an unstructured surface, the transmission is increased significantly across a broad spectral range in the visible spectrum.

Using immersion configurations in the IL process allows a further reduction of feature sizes. In the accompanying exposure setup the photoresist-coated substrate is typically sandwiched between UV-transparent prisms using an immersion liquid with a matched refractive index. The minimal period is decreased by a factor given by the refractive index of the immersion medium. The scanning electron microscope images presented in Fig. 8.15 show patterns with a grating period of 93 nm (10 800 lines/mm) which were obtained by single (left) or double (right) exposure using 266 nm immersion lithography. This structures are developed to serve as AR coatings for deep-UV applications.

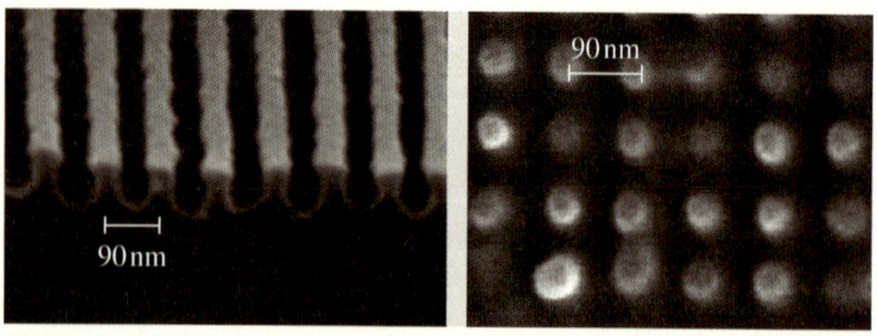

Fig. 8.15 Scanning electron microscope images of gratings with periods of 93 nm (10 800 l/mm). The gratings were obtained by single (*left*) and double (*right*) exposure technique using 266 nm immersion lithography

Polarization-Sensitive Elements

Historically, the wire-grid polarizer (WGP) was perhaps the first component to make use of the properties of subwavelength structured gratings as it was applied by Heinrich Hertz to investigate the properties of the newly discovered radio waves at the end of the 19-th century. Meanwhile, with constant progress in manufacturing technologies, the wire-grid polarizer is also available for the infrared and the optical wavelength region as well.

The classical WGP consists of a subwavelength periodic grid of parallel, self-supporting and perfectly conducting wires. This structure is characterized by a high reflectivity of transverse electric (TE) polarization and simultaneously a high transmittance of the transverse magnetic (TM) polarization of the incident electromagnetic radiation.

For applications in the visible spectral range WGPs are realized as subwavelength-period metal gratings deposited on a dielectric substrate. The introduction of additional dielectric layers sandwiched between the metal grid and the supporting substrate allows the adjustment of the spectral operation bandwidth and to relax the tolerances of the acceptance angle [8.58]. A further extension to control the polarization properties of the WGP is possible when the subwavelength-period grating is already formed in the dielectric medium and the metal stripes are supported on the dielectric bars.

Due to their specific optical characteristics and the compactness, WGPs became of particular interest in projection display applications especially when combined with liquid crystal on silicon (LCoS) imager systems [8.59, 60]. In these setups, the WGP acts as a beam splitter, separating the incoming illumination radiation and the back-reflected and polarization-rotated light coming from the LCoS panel.

Commonly the working principal of the WGP is explained by the dissimilar electron mobility parallel and perpendicular to the metal wires. If the incident wave is polarized along the wire direction, a corresponding elec-

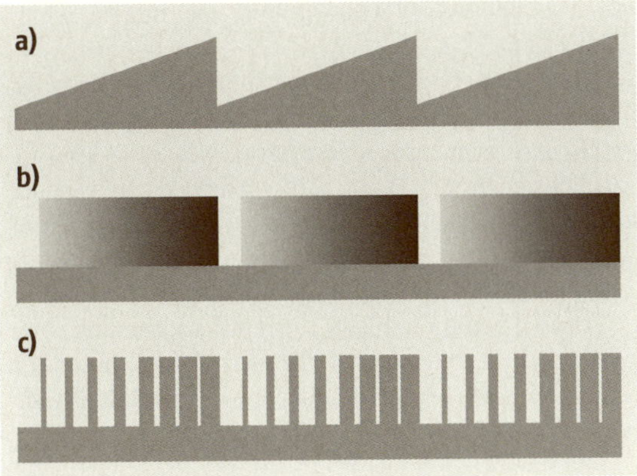

Fig. 8.16a–c Illustration of (**a**) blazed sawtooth profile, (**b**) blazed index and (**c**) blazed binary gratings

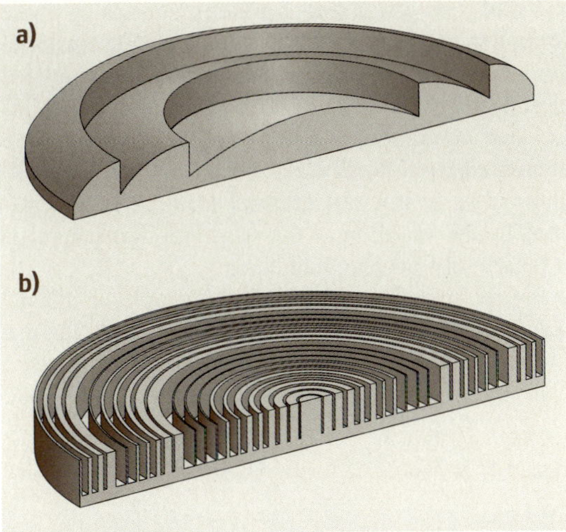

Fig. 8.17a,b Schematic comparison of diffractive lenses: (**a**) surface relief profile; (**b**) blazed binary approach

tron oscillation is induced. The reradiated wave cancels the incident wave in the forward direction, so that the wave is totally reflected. In contrast, if the incident wave is polarized perpendicular to the wire grid, and if the wire spacing is wider than the wavelength, then the oscillation of the electrons is restricted and the incident wave transmits the structure.

The approach of effective-medium theory (EMT) to describe the characteristics of subwavelength structured devices is also applicable to WGPs [8.61]. Hereby, the transmission and reflection properties of the wire grid polarizers can be estimated by inserting a complex value for the refractive index n_1 of the metal.

$$n_1 = n + \mathrm{i}\kappa \,. \tag{8.48}$$

For a highly conducting metal $\kappa \gg n$ and spacers made of perfect insulators (n_2) the refractive indices for both polarization directions become:

$$n_{\|} = \mathrm{i}\kappa\sqrt{f} \,, \tag{8.49}$$

$$n_{\perp} = n_2/\sqrt{1-f} \,. \tag{8.50}$$

In the case when the wire grids are made of perfect conductors and the dielectric spacers are made of perfect insulators, the incident radiation with a polarization parallel to the wire grids interacts with the medium equivalent to a metallic layer and is reflected. In contrast, for incident radiation polarized perpendicular to the wire grids, the element behaves as a perfect dielectric layer and the radiation is mainly transmitted.

In more-realistic scenarios the assumption of perfect conductivity is not suitable and therefore rigorous calculations become necessary to calculate the polarizing properties of the structures.

A different type of polarization-sensitive elements that provides a high polarization extinction ratio is based on a subwavelength structured multilayer system of alternating layers of dielectric materials with high and low refraction indices [8.40, 41]. In this approach the properties of anisotropic spectral reflectivity is realized by combining the effects of form birefringence of a high-spatial-frequency grating with the resonant reflectivity of a multilayer structure.

Subwavelength Blazed Gratings and Lenses

In the previous discussions on wave plates, antireflection coatings and polarizers based on subwavelength structures the properties of this artificial effective media were treated similarly to homogeneous thin films. Additionally, high-spatial-frequency structures can also be used to realize high-efficiency diffractive elements such as gratings and diffractive lenses.

The basic idea is a lateral tailoring of the local effective index of refraction by controlling the feature size of the subwavelength structures. Each individual period of a diffractive element which is larger than the wavelength is composed of a sublattice with variable width of lines and spaces forming a subwavelength binary surface profile.

The principal concept is illustrated in Fig. 8.16. A classic sawtooth blazed profile guarantees a continuous linear phase shift along each individual period by varying the depth of the structure in a material of constant refractive index and shows a phase step of 2π at the end of each period (Fig. 8.16 top). The analog phase distribution can be modeled by a gradient index material with constant height (Fig. 8.16 center). In this so-called blazed index grating, the refractive index is varying linearly and continuously from 1 to a maximum value within the period. In a blazed binary grating (Fig. 8.16 bottom), the continuous gradient material is approximated by the variation of the structural dimensions of the binary subwavelength features.

For the first time the physical principle of this artificial distributed-index media was verified by the use of water waves [8.45, 62]. The concept was transferred to the infrared spectral range at 10.6 μm [8.63] and was also applied to near-infrared and optical wavelengths [8.64, 65].

Blazed binary gratings allow a highly efficient coupling of light into the first diffraction order [8.46] which may even exceed the properties of a conventional blazed grating [8.48]. This enhanced performance has been interpreted as a pillar waveguiding effect responsible for a drastic reduction of the shadowing zone [8.66].

As an illustration, Fig. 8.17 shows a schematic comparison of a conventional diffractive lens based on a sawtooth surface profile and a subwavelength diffractive lens.

8.2 Electro-Optic Modulators

An electrooptic modulator (EOM) can be used to control the amplitude, phase, frequency, polarization or propagation direction of a light wave electrically. EOMs are typically used for external modulation of light sources, thus avoiding amplitude, frequency and linewidth stability issues associated with direct modulation.

EOMs in general consist of an electrooptic material embedded between a pair of plated electrodes. A voltage applied to these electrodes generates an electric field oriented mostly perpendicularly to the electrode surfaces. The field induces changes in the material structure or orientation and thus the refractive index or birefringence of the material is modulated by the electric field.

Two fundamental setups of EOMs can be differentiated: the *longitudinal-oriented modulation* where the light wave propagates parallel (Fig. 8.18a) to the electric field direction and *transverse-oriented modulation* with the light wave propagating perpendicularly (Fig. 8.18b) to the electric field. In the case of the longitudinal configuration the electrodes must be transparent to the light to be modulated or a small aperture in each electrode must allow the light to propagate through the modulator.

EOM devices can be constructed as bulk or as integrated optical devices (Fig. 8.18c). *Bulk modulators* are made out of discrete pieces of electrooptical materials such as lithium niobate $LiNbO_3$, and lithium tantalate $LiTaO_3$. These materials have high electrooptical coefficients and can be used over a wide range of wavelengths.

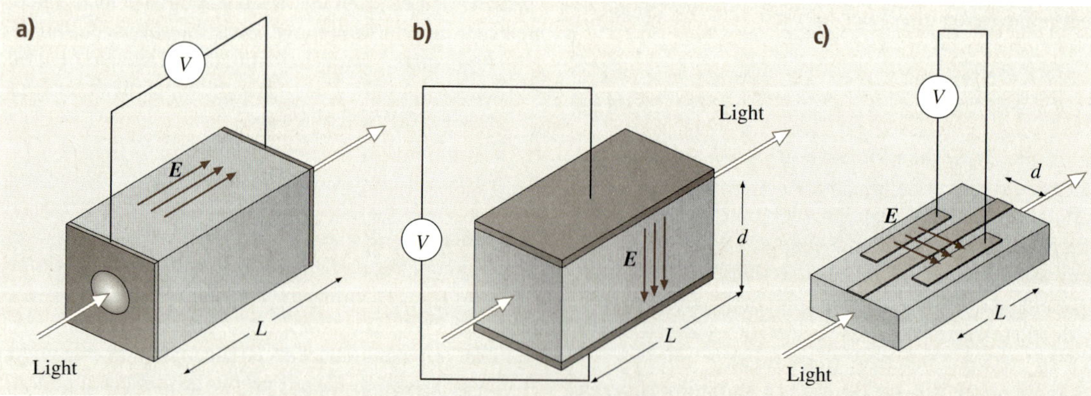

Fig. 8.18 **(a)** A longitudinal-oriented EOM has the ***E*** field applied parallel to the light-propagation direction, **(b)** whereas in the transverse oriented EOM the light wave is propagating perpendicularly to the ***E*** field. **(c)** An integrated optical modulator of transverse type

Integrated optical modulators are wavelength-specific due to the waveguide technology used. They are used in transverse modulation schemes that allow the width of the waveguide to be much smaller than its length. Thus, the half-wave voltage (8.56) is very small (typically a few volts) compared to the high voltage requirements of bulk modulators (up to tens of kV). In [8.67] a $BaTiO_3$-based modulator is reported with broadband modulation up to 40 GHz and a half-wave voltage of 3.9 V.

Some applications of electrooptic modulators are Q-switching, all optical switching, active mode locking and laser-disk recording.

Pockels and Kerr Effects

Electrooptic material change their optical properties when affected by an external electric field due to the altered position, orientation or shape of the material's structure. When an electric field E is applied the refractive index as a function of E (Fig. 8.19) can be expressed as

$$n(E) = n - \frac{1}{2} k_{\text{Pockels}} n^3 E - \frac{1}{2} k_{\text{Kerr}} n^3 E^2 \,. \tag{8.51}$$

If the change in refractive index is mostly linearly proportional to the applied field strength then the effect is known as the linear electrooptic effect or Pockels effect, with

$$n(E) = n - \frac{1}{2} r_{\text{Pockels}} n^3 E \,, \tag{8.52}$$

where r_{Pockels} is the linear electrooptic coefficient or Pockels coefficient.

The value of the electrooptic coefficient r depends on the directions of the light and the orientation of the applied electric field since the crystal is anisotropic.

In isotropic materials such as gases, liquids or centrosymmetric crystals the linear coefficient is zero and the change in the index of refraction depends on the square of the applied electric field. This is known as the quadratic electrooptic effect or Kerr effect, with

$$n(E) = n - \frac{1}{2} r_{\text{Kerr}} n^3 E^2 \,, \tag{8.53}$$

where r_{Kerr} is the quadratic electrooptic coefficient or Kerr coefficient.

Considering an average value for $r_{\text{Pockels}} = 10\,\text{pm/V}$ and a modulation voltage of $V = 10\,\text{kV}$ (which results in an electrical field strength of $E = 10^6\,\text{V/m}$ in a 1 cm-thick medium), the electrically induced change in refraction index is approximately 10^{-5}. The same

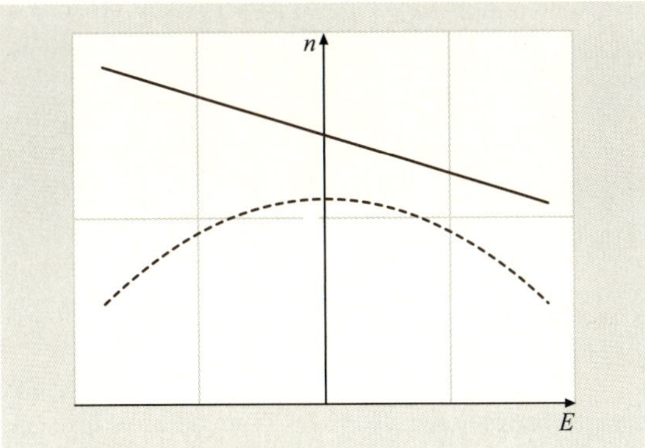

Fig. 8.19 The refraction index as a function of the applied electrical field strength E: (*solid line*) Pockels medium; (*dashed line*) Kerr medium

range is achieved when using the Kerr effect with an average value of $r_{\text{Kerr}} = 10\,\text{am}^2/\text{V}^2$. Although this change in refractive index is very small it will result in a significant phase shift for a wave propagation distance that is much greater than a wavelength of the light. For the given example the phase shift will be approximately $\pi/2$ for a wavelength of 600 nm.

An anisotropic electrooptical crystal with Pockels characteristic (with the electrooptic coefficient k_{Pockels}) will be considered in the following compilation of modulation techniques.

8.2.1 Phase Modulation

In analogy to the definition of the electric field strength between the electrodes of a parallel-plate capacitor, the field strength in a Pockels modulator can be expressed as

$$E = \frac{V}{d} \,, \tag{8.54}$$

where V is the applied quasistatic voltage on the electrodes of the modulator, and d is the distance between them (in the case of the longitudinal modulator it is equal to the length L).

In a *transversal Pockels modulator* (Fig. 8.18b) the phase shift is a function of the electric field strength E

$$\varphi = \varphi_0 - \pi \frac{n^3 L}{\lambda_0} k_{\text{Pockels}} E \,, \tag{8.55}$$

where $\varphi_0 = 2\pi\, n_{(V=0)} L/\lambda_0$ is the natural phase shift and λ_0 is the free-space wavelength.

By reducing the electrode distance d the voltage necessary for a given phase shift will be also reduced. This applies particularly in integrated optical devices.

The required voltage to achieve a phase shift $\varphi = \pi$ is called the *half-wave voltage* and can be expressed as

$$V_\pi = \frac{\lambda_0 d}{n^3 k_{\text{Pockels}} L} . \qquad (8.56)$$

It is convenient to express the phase shift by using V_π as

$$\varphi = \varphi_0 - \pi \frac{V}{V_\pi} . \qquad (8.57)$$

In a longitutional Pockels modulator (Fig. 8.18a) the distance between the electrodes is equal to the distance of light modulation $d = L$ and thus φ is independent of d and L:

$$\varphi = \varphi_0 - \pi \frac{n^3}{\lambda_0} k_{\text{Pockels}} V \qquad (8.58)$$

and the corresponding half-wave voltage is

$$V_\pi = \frac{\lambda_0}{n^3 k_{\text{Pockels}}} . \qquad (8.59)$$

Frequency modulation can be achieved by applying a sinusoidal phase modulation. This is due to the fundamental interrelation between these two kinds of modulation, where the frequency modulation can be seen as phase modulation with integrated modulation signal [8.68]. Thus sinusoidal phase modulation results in sinusoidal frequency modulation [8.69].

Another approach is to apply a circular electric modulation field of frequency ω_m to cause the principle axes to rotate with angular velocity $\omega_\text{m}/2$. A setup is used where two circular polarizers (inverted with respect to each other) at both ends of the modulator convert linearly polarized light to circularly polarized and vice versa.

The circularly polarized optical wave with frequency ω at the input will be shifted to $\omega \pm \omega_\text{m}$ at the output, depending on the direction of rotation and field strength of the applied electric field [8.70].

8.2.2 Polarization Modulation

A polarization state converter or dynamic wave plate allows the controlled change of an input polarization at the output of the modulator. This is achieved by coherent addition of two orthogonal waves (portions of the input wave) within the modulator.

The orientation of the electrooptic medium and the direction of the electric field are configured to have control of the fast and slow axis in the cross section normal to the light path. The corresponding refractive indices for this axis are

$$n_{x'} \approx n_x - \tfrac{1}{2} n_x^3 k_{x\,\text{Pockels}} E , \qquad (8.60)$$

$$n_{y'} \approx n_y - \tfrac{1}{2} n_y^3 k_{y\,\text{Pockels}} E , \qquad (8.61)$$

where n_x and n_y are the refractive indices in the absence of an electric field, and k_x and k_y are the appropriate electrooptic coefficients under the specific spatial and electric field conditions (for a detailed description see to [8.71] Chap. 18.2).

Light portions (the ordinary and extraordinary wave) propagate at different speeds through the system resulting in a relative phase difference Γ between them. This phase difference depends on the length of the system L and is also known as the retardation

$$\Gamma = \Gamma_0 + \frac{2\pi}{\lambda_0}(n_{x'} - n_{y'})L , \qquad (8.62)$$

where Γ_0 is the natural phase retardation with no voltage applied.

In the absence of natural birefringence ($n_x - n_y = 0$) the induced retardation in a transverse modulator is

$$\Gamma_i = \frac{\pi}{\lambda_0} \left(k_{\text{Pockels}\,y} n_y^3 - k_{\text{Pockels}\,x} n_x^3 \right) \frac{VL}{d} . \qquad (8.63)$$

The voltage required to achieve a retardation of $\Gamma_i = \pi$, which shifts a vertical polarized input wave to a horizontal polarized output wave, is the half-wave voltage V_π for the polarization modulator

$$V_\pi = \frac{\lambda_0}{k_{\text{Pockels}\,y} n_y^3 - k_{\text{Pockels}\,x} n_x^3} \frac{d}{L} . \qquad (8.64)$$

The corresponding formulas for a longitudional modulator are equal to (8.61) without the d/L term due to the modulator's independence of d and L.

8.2.3 Intensity Modulation

Intensity modulation is achieved by altering the transmission of light, which is the ratio of output to input intensity, $T = I_\text{o}/I_\text{i}$. As phase or retardation modulation does not affect the intensity of the output light an additional element or setup is needed.

One approach is to add a polarizing analyzer to a dynamic wave plate as described in [8.70] in such a way that the analyzer's axis is orthogonal to the input polarization (crossed polarizer setup). With no voltage applied the output intensity is zero (point A in Fig. 8.20b) and with $V = V_\pi$ (point C in Fig. 8.20b) the output intensity reaches the level of the input intensity. The transmittance T is a periodic function of V

$$T(V) = \sin^2\left(\frac{\Gamma_0}{2} - \frac{\pi}{2}\frac{V}{V_\pi}\right) . \tag{8.65}$$

By operating the modulation voltage near $V_\pi/2$ (point B in Fig. 8.20b) the output intensity is nearly linearly related to the voltage.

Figure 8.21a shows a typical setup. The light entering the modulator after passing the input polarizer is linearly polarized. Its polarization direction is tilted 45° with respect to the direction of the electric field. With a voltage applied the light will be split into two orthogonal components that will have a relative phase difference Γ_i between them after passing through the modulator. The superposition of these two components results in elliptically polarized light that will be analyzed by the output polarizer. With $V = V_\pi$ the superposition again results in linearly polarized light oriented at 90° with respect to the incoming light's polarization, and thus it will be fully transmitted by the analyzer.

It is also possible to use parallel polarizers, in which case the transmission behavior is inverted: full transmission at $V = 0$ and no transmission with $V = V_\pi$.

Another approach to modulate the intensity of light is to superpose the waves of the two branches of a Mach–Zehnder interferometer in which at least one branch contains a phase modulator (Fig. 8.22a). Assuming that the beam splitters divide the optical intensity equally, the intensity at the output I_o depends on the phase difference φ in the two branches:

$$I_o = \tfrac{1}{2}I_i + \tfrac{1}{2}I_i\cos\varphi = I_i\cos^2\frac{\varphi}{2} . \tag{8.66}$$

Considering the half-wave voltage expression for the phase shift (8.56), the transmittance $T = I_o/I_i$ is again

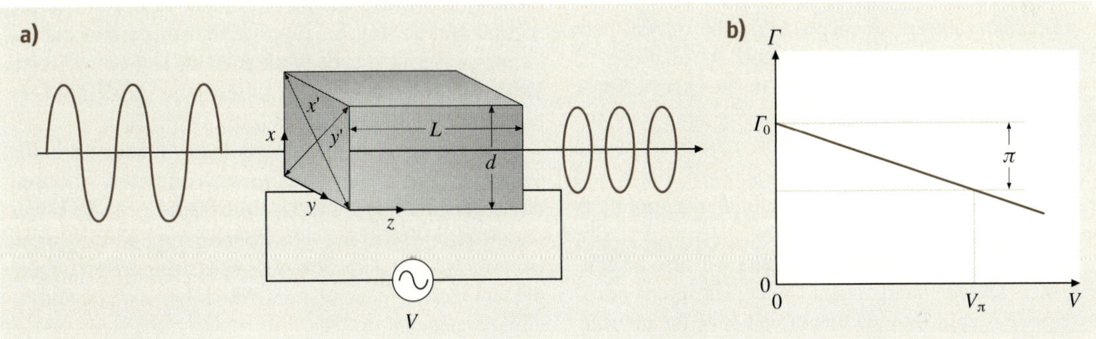

Fig. 8.20 **(a)** When linearly polarized light (parallel to the x-axis) enters the crystal, the output polarization state is in general elliptical. For $V = V_\pi$ the polarization is again linear, but turned 90° with respect to the input polarization. **(b)** The phase retardation is linearly related to the applied voltage. In the absence of an applied voltage only Γ_0 is present

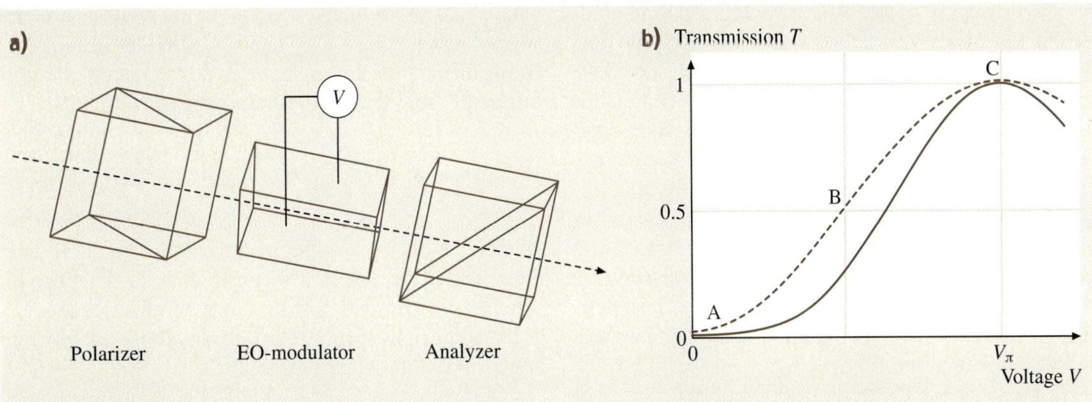

Fig. 8.21 **(a)** An intensity modulator setup implementing a dynamic wave plate between two crossed polarizers, placed at 45° with respect to the retarder's axes. **(b)** The transmittance T varies with the applied voltage V, following the *dashed* line for a Pockels characteristic modulator and the *solid* line for Kerr characteristic, both for $\Gamma_0 = 0$

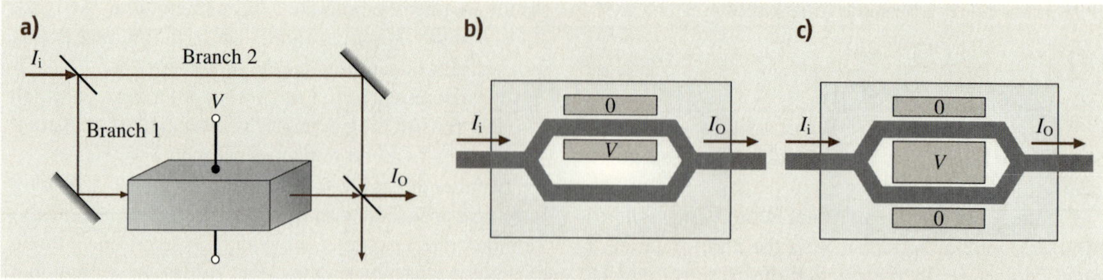

Fig. 8.22 **(a)** Mach–Zehnder interferometer where one branch contains a phase modulator. The transmittance of the interferometer T varies with the applied voltage V. The same setup can be used in an integrated optical device **(b)** where the input light is split into the two branches by a waveguide. With an additional electrode a push–pull configuration can be constructed **(c)** where the light in the two branches are modulated by the same voltage

a periodic function of V

$$T(V) = \cos^2\left(\frac{\varphi_0}{2} - \frac{\pi}{2}\frac{V}{V_\pi}\right) . \quad (8.67)$$

Here the value of φ_0 depends on the optical path difference when no voltage is applied. A value of $\pi/2$ is used to operate the modulator in the nearly linear region around $T = 1/2$ (point B in Fig. 8.21b). If the modulator is switched between full transmission and total extinction, φ_0 can be used to select whether $V = 0$ or $V = V_\pi$ corresponds to $T = 0$ and $T = 1$ (points A and C in Fig. 8.21b).

A Mach–Zehnder-based modulator can also be constructed as an integrated optical device. The input optical waveguide is split into the two branches of the interferometer and recombined at the output (Fig. 8.22b). In addition to the setup where a phase modulator is added to one of the branches there is a configuration called push–pull, which adds an additional modulator to the second branch (Fig. 8.22c). Full transmission occurs at 0° net phase difference while total extinction occurs at 180°. This configuration requires lower drive voltages to achieve a required phase shift.

Due to the small electrically induced changes in the refractive index, thermal influences and stress-induced double refraction as well as birefringence can be rather disturbing. Therefore, electrooptic modulators often contain two equal Pockels cells in a configuration where the temperature dependence of the relative phase shift is largely canceled [8.72].

8.3 Acoustooptic Modulator

Devices that apply an acoustic wave to a medium guiding optical waves are called acoustooptic modulators (AOM). They consist of at least one acoustic wave generator (usually a piezoelectric transducer) attached to an optical material such as tellurium dioxide, crystalline quartz or fused silica. The applied acoustic wave influences the spatial density distribution of the optical material (by compression and rarefaction) and thus a corresponding change in the index of refraction is generated. Due to the periodic structure of the acoustic wave the variation in the index of refraction is also periodic and acts as a diffraction grating for an optical wave present in the medium (Fig. 8.23).

The distance of the periodically repeated planes of equal density is given by

$$\Lambda = \frac{\nu_s}{f_s} , \quad (8.68)$$

where ν_s is the velocity of sound in the medium and f_s is the frequency of sound. In analogy to the formula for Bragg diffraction, the angles θ can be calculated for light diffracted from this periodic structure as (Fig. 8.20):

$$\sin\theta = \pm\frac{\lambda}{2\Lambda} , \quad (8.69)$$

where λ is the wavelength of light in the medium. For these angles the criteria for constructive interference is satisfied; θ represents both the angle of reflection as well as the angle of incidence. The intensity drops sharply for angles that differ slightly from θ.

Depending on the application there are many criteria for the choice of the material, including the transparency range, the optical damage threshold, and required size.

The characteristics of AOMs are given in Table 8.2.

Table 8.2 Characteristics of acoustooptic modulators [8.73]

Material	Bandwidth (MHz)	Rise time (ns)	Acoustic input power (W) saturation	Material wavelength range (μm)	Efficiency at maximum bandwidth (633 nm)
Glass	5–10	50	1–6	0.4–25	> 70%
$PbMoO_4$	> 80	4–6	0.5–1.5	0.42–5.5	> 70%
TeO_2	> 80	4–6	0.5–1.5	0.35–5	> 70%
Ge	10	70	20–30	2–20	> 50% at 10.6 μm
GaP	100–1500	0.7–10	0.02–2	0.6–1	> 80%
GaAs	50–200	5–20	0.02–2	1–1.6	> 30% at 1.5 μm

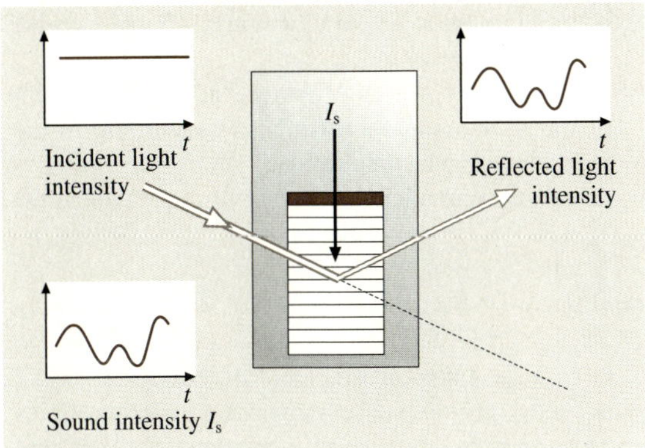

Fig. 8.23 AOM controlled by the sound intensity with a proportional modulated output intensity of the light

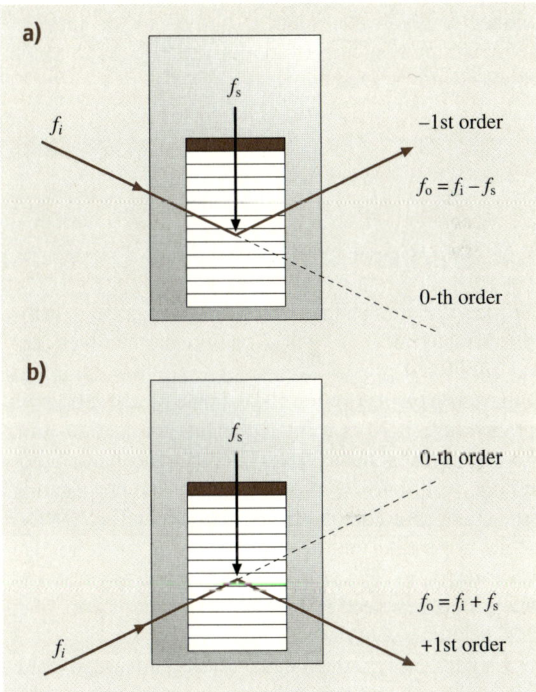

Fig. 8.24a,b Frequency shifter for (**a**) downshift and (**b**) upshift of the input light frequency

8.3.1 Intensity Modulator

The intensity of the reflected light is proportional to the intensity of the modulating sound wave for sufficiently weak sound intensity. At higher sound intensities nonlinear effects have a negative impact on the modulator.

The achievable bandwidth, B, is limited either by the size of the light beam waist (with the transit time $T = D/v_S$ of sound across the waist width D) to $B = 1/T$ (as shown in [8.71] Chap. 20.2) or by the transducer impedance mismatch, which can be overcome by inserting a quarter-wave plate.

8.3.2 Frequency Shifter

Partially reflected optical waves are Doppler-shifted by the refractive-index variations created by an acoustic plane wave that moves with a velocity v_S. While the angular frequency of the incident optical wave is $\omega = 2\pi f$ the reflected light is upshifted to an angular frequency $\omega + \Omega$ with $\Omega = 2\pi f_S$. This is achieved using a setup such as that shown in Fig. 8.24b where the acoustic wave propagates in the opposite direction to the incoming optical wave. In a geometry where light and sound waves travel in the same direction (Fig. 8.24a) the frequency of the reflected light is downshifted to $\omega - \Omega$.

Frequency shift over a range wider than the modulators range can be achieved by placing two modulators in series. When the first modulator applies an upshift and the second a downshift then the resulting output frequency will be shifted by the difference of the two frequencies. The sum of both frequencies will be achieved if both modulators apply an upshift. The overall efficiency is given by the combined diffraction efficiencies.

8.3.3 Deflector

By applying a wide acoustic frequency range with bandwidth Δf to an AOM the deflection (separation) angle 2θ can be made to vary proportionally to f_s. As θ is not only the reflected but also the required incidence angle a setup is needed where the incidence angle can be decoupled. This can be achieved by varying the direction of the acoustic wave corresponding to the applied acoustic frequency, for example by using multi-electrode transducers.

As the optical wave has a limited angular width the number of resolvable spots is limited to the ratio of the maximum scan angle variation $\Delta\theta = (\lambda/v_s)B$ to the angular width λ/D of the beam

$$N = \frac{D}{v_s} B . \tag{8.70}$$

Here, D/v_s is the transit time of sound through the light beam's width and B is the bandwidth of the sound wave.

Some applications of AOMs applying the modulation technics described above are:

- cavity dumping in a laser resonator (deflection),
- active mode locking,
- laser printing, where the AOM modulates the power of the laser beam,
- high-frequency (HF) spectrum analyzer, where the AOM deflects the length depending on the (to analyze) modulation signal to a charge-coupled device (CCD) line detector,
- laser projector, where AOMs modulate the intensity of the RGB components of light according to the video signal to be displayed,
- laser scanning microscopy.

8.4 Gradient Index Optical Components

The local variation of the refractive index $n(x, y, z)$ in the optical material of an optical component involves several additional possibilities to generate new functions of the component. In contrast to conventional refractive optics where light is refracted at the abrupt transition from one homogeneous medium with index n_1 to a second one with index n_2, the direction of the propagating light changes continuously in gradient index (GRIN) media, depending on the distribution of the refractive index $n(x, y, z)$. The propagation in terms of geometrical optics is governed by the rules of Fermat's principle. This is described in more detail in Sect. 8.4.1.

Usually, four gradient index types are distinguished. For axial gradients the index varies only in the direction of the optical axis z, while in planes perpendicular to the optical axis the index is constant. For radial gradients, nowadays the most important type, the index is solely a function of the radial coordinate r, which is perpendicular to the optical axis z. In one-dimensional or lateral GRIN media, the index depends only on one coordinate, y, which is perpendicular to the optical axis. Spherical gradients exhibit a variation of the index with a spherical radius R,

$$\begin{aligned} &n = n(z) && \text{axial gradient} \\ &n = n(r = x^2 + y^2) && \text{radial gradient} \\ &n = n(y) && \text{one-dimensional gradient} \\ &n = n(R = x^2 + y^2 + z^2) && \text{spherical gradient} \end{aligned} \tag{8.71}$$

For all the types of gradients described in (8.71), defined index profile functions exist that generate effects on incoming light beams similar to those of conventional optical components as, e.g., lenses. A radial (or one-dimensional) gradient with a nearly parabolic secant-hyperbolic index profile,

$$n(r) = n_0 \operatorname{sech}(gr) , \tag{8.72}$$

with a maximum index n_0 on the optical axis creates a focusing lens (Fig. 8.25) with a periodic sinusoidal ray trace [8.74, 75] for meridional rays. Period P solely depends on the gradient parameter g,

$$P = 2\pi/g . \tag{8.73}$$

A radial gradient with a parabolic index profile bending outwards

$$n^2(r) = n_0^2[1 + (gr)^2] \tag{8.74}$$

acts as a diverging lens [8.76] (Fig. 8.25). Both types (8.72) and (8.74) result in a lens function with plane optical surfaces, which is a tremendous advantage for the production and application of miniaturized gradient index lenses.

A unique spherical gradient is Maxwell's fisheye [8.77]

$$n(r) = \frac{n_0}{1 + (R/R_0)^2} , \tag{8.75}$$

which images all points on the surface ($R = R_0$) of a sphere in the surface on its opposite side; n_0 is the index at the center of the sphere. A modified version of the spherical or semi-spherical gradients is used in planar microlens arrays [8.78].

A so-called Luneburg lens [8.79] uses a spherical gradient

$$n(r) = n_0\sqrt{2-(R/R_0)^2} \quad (8.76)$$

to collimate light from a point source at the surface of the sphere. A generalized type of similar profiles is employed for producing fiber-based cylindrical collimating lenses with small focal lengths [8.80].

Axial gradients are most commonly used to correct the spherical aberration of lenses made with a homogeneous medium with spherical surfaces. An axial index profile that eliminates the spherical aberration of parallel light rays striking the spherical surface of a plano-convex lens on the gradient side (Fig. 8.25) was derived analytically by Murty [8.81],

$$n(z) = \frac{n_0}{\sqrt{1+2n_0(n_0-1)z/r_c}}, \quad (8.77)$$

where n_0 is the index at the vertex ($z = 0$) of the spherical surface with a radius of curvature r_c.

Most of these gradient examples represent analytical solutions of the ray equations (Sect. 8.4.1) but are difficult to fabricate in practice because of their exotic index value ranges. Gradient index materials used for the fabrication of optical components for the visible and near-infrared domain are most often silicate glasses and sometimes plastics, with indices between 1.4 and 1.9, allowing a limited index variation because of the chemistry, and mechanical and optical properties of the materials. Furthermore, applications often require the

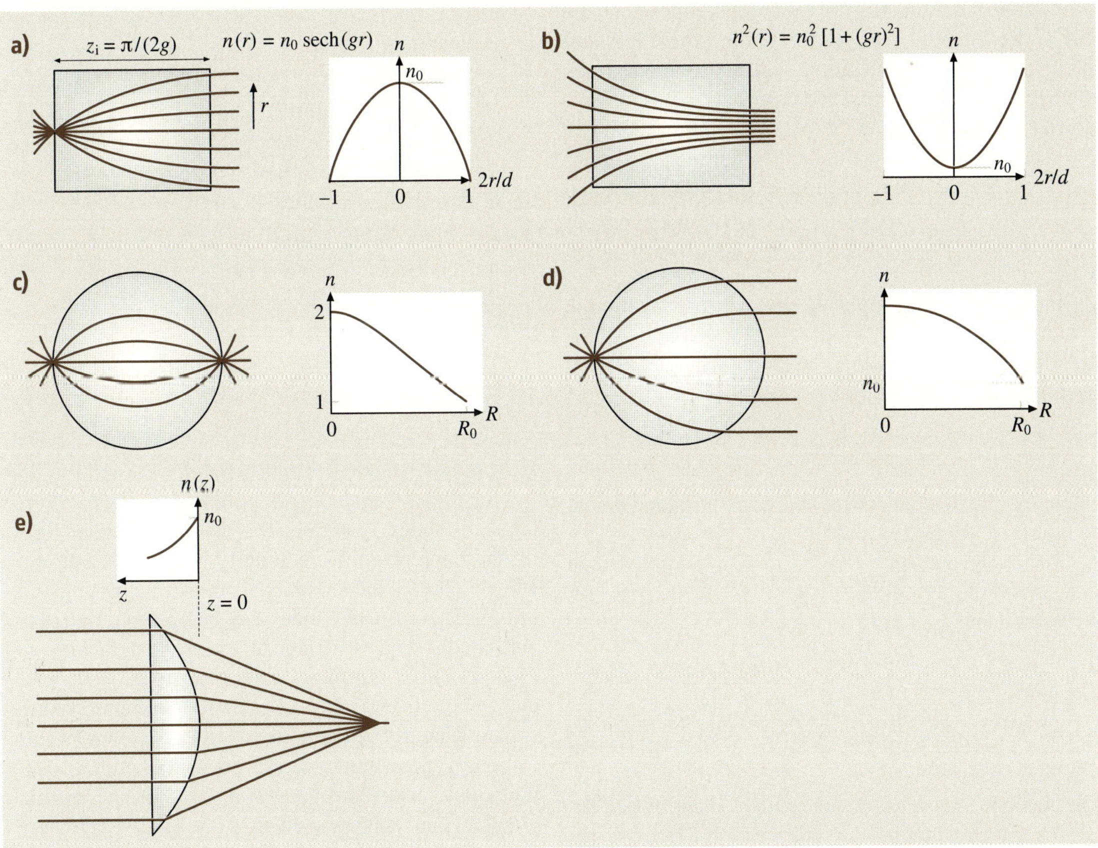

Fig. 8.25a–e Different types of gradient-index components with lens performance **(a)** radial focusing lens, **(b)** radial diverging lens, **(c)** spherical lens – Maxwell's fisheye ($n_0 = 2$) and **(d)** Luneburg lens, **(e)** plano–convex lens with axial gradient

integration of GRIN components in a configuration that deviates from the ideal and analytical examples presented in this chapter. Hence, the index profiles are formulated in model functions, normally as a polynomial expression, introducing an appropriate number of design parameters. Numerical ray tracing is then performed to determine optimum index profile parameters for a specific configuration or application, also considering the technological feasibility. Section 8.4.1 will outline some fundamentals of ray tracing in gradient index media. A more-detailed introduction to the fundamentals of gradient index optics is given in [8.74, 79, 82–84].

8.4.1 Ray Tracing in Gradient Index Media

Light propagation in gradient index media can be described by different physical models, i. e., by ray tracing in the sense of geometrical optics, by scalar wave propagation [8.85], or by solving the vector wave equations [8.84]. The suitability of the models depends on the size of the gradients with respect to the wavelength and the effects under investigation. Besides gradient index waveguide optics, where wave propagation is appropriate [8.86], ray tracing is the most common method to describe, simulate and optimize gradient index components, because its calculations are comparably fast and compatible with the principles of optical design software. Hence, this method will be introduced in more detail in this section.

In the geometrical optical sense, ray propagation in media with a nonuniform index distribution $n(x, y, z)$ is described with the vector ray equation, which is derived from Fermat's principle [8.87]

$$\frac{\mathrm{d}}{\mathrm{d}s}\left[n(x, y, z)\frac{\mathrm{d}\boldsymbol{r}}{\mathrm{d}s}\right] = \nabla n(x, y, z) , \tag{8.78}$$

where $\boldsymbol{r}$ is the position vector and $\mathrm{d}s$ is the path element along the ray trace (Fig. 8.26).

The vector equation (8.78) can also be represented as a system of differential equations of second order for the coordinates x, y and z,

$$\frac{\mathrm{d}^2x}{\mathrm{d}\tau^2} = \frac{\partial}{\partial x}\left(\frac{1}{2}n^2\right) ; \quad \frac{\mathrm{d}^2y}{\mathrm{d}\tau^2} = \frac{\partial}{\partial y}\left(\frac{1}{2}n^2\right) ;$$
$$\frac{\mathrm{d}^2z}{\mathrm{d}\tau^2} = \frac{\partial}{\partial z}\left(\frac{1}{2}n^2\right) . \tag{8.79}$$

The element $\mathrm{d}\tau$ results from the path element $\mathrm{d}s = n\,\mathrm{d}\tau$.

Analytical solutions of equation (8.79) exist for many index distributions, some well-known ones are presented before. Detailed descriptions are given in [8.74, 75, 79, 83].

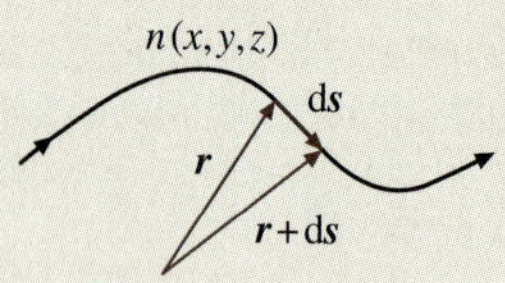

Fig. 8.26 Ray trace in a inhomogeneous medium

In routine cases, where the index distributions have to be adapted and optimized to a specific design configuration of an optical system and to profile shapes that can be fabricated by applying an appropriate technology, the ray equations (8.79) are solved numerically by the Runge–Kutta method [8.88]. This procedure is mostly used in commercially available optical design software. The index profiles are usually formulated as polynomial expansions in r, R, x, y, and z. The computation of wavefront aberrations and point-spread functions (PSF) considering the diffraction furthermore requires the calculation of the optical path length OPL of rays tracing in gradient index media,

$$\mathrm{OPL} = \int n(x, y, z)\,\mathrm{d}s ;$$
$$\mathrm{d}s = \sqrt{\mathrm{d}x^2 + \mathrm{d}y^2 + \mathrm{d}z^2} , \tag{8.80}$$

where $\mathrm{d}s$ is the geometrical path element. The described Runge–Kutta method uses $\mathrm{d}\tau$ as the numerical step parameter. Hence, the calculation of the optical path length (OPL) can easily be included in the numerical procedure [8.89] by

$$\mathrm{OPL} = \int n^2(x, y, z)\,\mathrm{d}\tau . \tag{8.81}$$

8.4.2 Fabrication Techniques

Gradient index materials for visible light and near-infrared applications are usually based on transparent silicate glasses and sometimes on organic polymer materials. The refractive index can be changed by locally varying the chemical composition and structure and related optical and mechanical properties. Early models [8.90] relate the optical properties of homogeneous silicate glasses to their composition. More-recent studies [8.91, 92] also involve aspects of the index gradient fabrication and chromatic properties of the gradient glasses. The appropriateness of a GRIN fabrication technology depends on the following optical aspects:

- the desired index change Δn
- the profile shape and symmetry

- the depth of the gradient, which determines the size of the GRIN element
- the chromatic variation of the index profile.

Although several techniques such as, the sol-gel [8.93] and the chemical vapor deposition (CVD) [8.94] processes are described in the literature, ion exchange or diffusion technology in glasses and plastics are of most practical importance for the production of gradient index lenses [8.95–98] and are described here in more detail. However, CVD processes are of great industrial relevance for the production of graded index fibers for optical data transmission as well as for the fabrication of Luneburg-type cylindrical lenses (Sect. 8.4.3). Sol-gel techniques have been investigated in more detail because of their potential to fabricate large radial-gradient lenses of diameters larger than 5 mm to open up new applications to camera lenses or eyepieces in optical instruments [8.99]. To produce even larger GRIN lens elements, organic or plastic materials have been investigated in detail [8.100]. A combination of the diffusion of different monomers with different refractive indices followed by copolymerization generates the final index gradient.

Ion Exchange and Diffusion Processes in Glass

Diffusion processes in glass are used to produce most of the commercially available GRIN lens types. Large axial gradients [8.101] and radial gradients [8.102] are fabricated by fusion–diffusion between different initially homogeneous glasses of distinguished refractive indices, which are combined into a plate stack for axial gradients or by inserting rods in tubes of different glasses. After combination, the compound is annealed at temperatures well above the transition temperature and the softening point of the glasses, causing the diffusion of many components of the glasses, with the initial step-like index distribution forming a continuous index

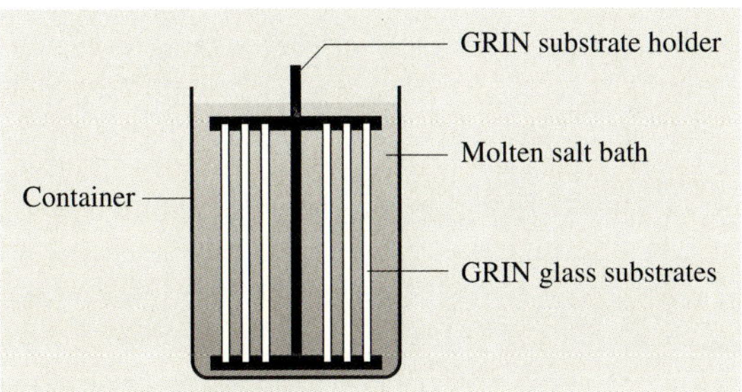

Fig. 8.27 Ion exchange scheme for borosilicate glass rods and slabs in molten salt baths

Table 8.3 Polarizability, ion radius and possible refractive index change of monovalent ions suitable for ion exchange

Ion	Electronic polarizability (A^3)	Ion radius (Å)	Δn
Na^+	0.41	0.95	–
Li^+	0.03	0.65	≤ 0.01
K^+	1.33	1.33	≤ 0.01
Ag^+	2.40	1.26	≤ 0.14
Cs^+	3.34	1.65	≤ 0.03
Tl^+	5.20	1.49	0.1 ... 0.2

gradient. In the case of one-dimensional axial gradient with index changes as large as 0.4, blanks are used to grind and polish spherical lenses with corrected spherical and chromatic aberration (Sect. 8.4.3). In the case of radial gradients, a larger preform with the appropriate index profile produced by the described diffusion is afterwards drawn out to the final lens diameter at lower temperatures.

For producing radial and lateral gradients in smaller diameters between 0.1 and 3.0 mm, the ion exchange of monovalent cations between molten salt baths and initially homogenous borosilicate glass rods or slabs is used [8.95, 96, 103] at temperatures between 300 and 600 °C. This is below or in the range of the transition temperature of the glass, where the network of the glass formers is still rigid, but monovalent ions have some mobility and the salts are liquid (Fig. 8.27). Numerous technological aspects affect the feasibility of producing a desired final refractive index profile. The ions to be exchanged determine the possible index change (Table 8.3) and the ion exchange depth after a certain exchange time, which results from different polarizabilities and ion radii. Table 8.3 compares the relevant properties of monovalent ions [8.104].

The fabrication time depends on the kinematic properties of the ion exchange, which is described as an interdiffusion process of the ions A and B in the glass and the salt melt,

$$A^+_{\text{Glass}} + B^+_{\text{Salt}} \leftrightarrow A^+_{\text{Salt}} + B^+_{\text{Glass}} \,. \tag{8.82}$$

It is modelled by Fick's diffusion equation

$$\frac{\partial c_{\text{A}}(\boldsymbol{r}, t)}{\partial t} = \nabla[D(c_{\text{A}}, T)\nabla c_{\text{A}}(\boldsymbol{r}, t)] \,, \tag{8.83}$$

where c_A is the normalized concentration of cation A, and D is the diffusion coefficient. It generally depends on the concentration and the temperature T and describes most of the process-relevant properties of the

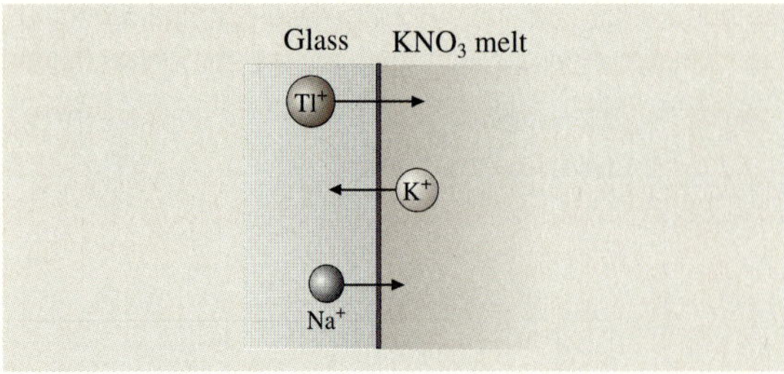

Fig. 8.28 Thallium/sodium–potassium ion exchange for producing high-NA radial GRIN lenses [8.95]

base glass composition and the kinematic behavior of the ions exchanged at different temperatures. Several empirical and physical models have been developed to enable a realistic simulation of the ion-exchange fabrication processes [8.104–107]. A process temperature has to be chosen where the diffusion coefficients are high enough to enable an economic production time, but still below or in the range of the softening point of the glass to avoid deformation of the glass blanks. Besides the temperature and the ion exchange time, the configuration of the initial and boundary conditions for the diffusion process, which can be controlled by varying the salt-bath composition, is an important tool to optimize the index profiles.

High-numerical-aperture lenses with radial gradients are obtained by large index changes of > 0.1. To produce GRIN lenses, in most cases they are generated by thallium/sodium–potassium ion exchange [8.95] (Fig. 8.28) or by the silver–sodium exchange [8.96, 103]. In contrast to silver, thallium is a very toxic material, larger portions of which, however, can initially be included in the base glass composition of the GRIN glass, because thallium oxide does not tend to reduce to metallic thallium during the melting process of the glass. Silver is almost nontoxic, but glasses with a high content of silver oxide larger than 15 mol % cannot be molten directly, because silver salts and oxides tend to reduce to metallic silver at typical melting temperatures above 1000 °C, which causes a dark color in the glass. The glass would then not be useful for GRIN-lens production. Hence, a two-step process is generally used for the production of focusing GRIN lenses [8.96, 103], where a sodium-containing base glass is initially immersed in a silver-containing salt melt. This step is carried out sufficiently long to form a nearly uniform refractive index increase Δn (ion stuffing) (Fig. 8.29). In a second *burying* step, silver ions in the glass are partially removed and replaced by index-lowering sodium ions, where the glass substrates are immersed in a sodium-containing salt melt, yielding the desired parabolic profile.

Many applications require longer ray periods P (8.73) connected with lower numerical apertures and minimized chromatic aberrations. Preferably, the lithium-sodium ion exchange in alumo-borosilicate glasses is used to produce these types of radial GRIN lenses with smaller index changes Δn [8.108–111].

It is not claimed that this chapter is a complete representation of all the available GRIN fabrication techniques; rather it reflects the author's experience.

8.4.3 Application

Radial and Lateral Gradient Lenses with Plane and Spherical Surfaces

Micro GRIN lenses with almost parabolic radial or lateral index profiles are an interesting alternative to conventional homogeneous glass lenses since the lens performance depends on a continuous change of the refractive index within the lens material. Instead of

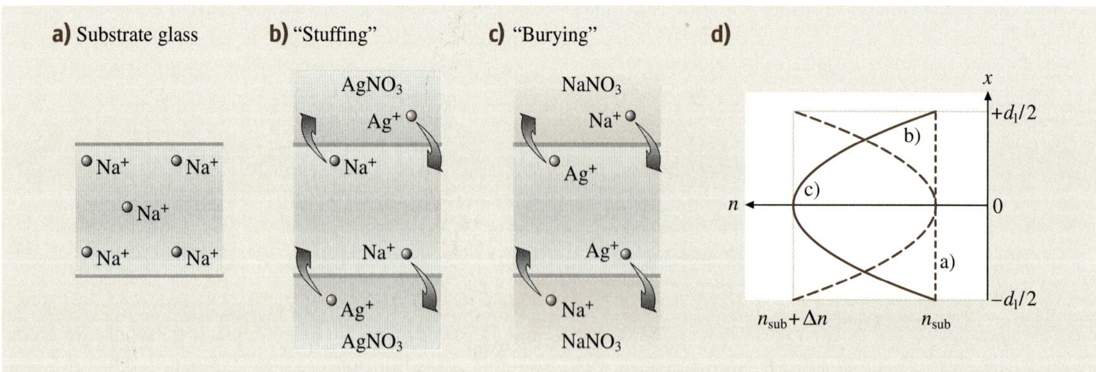

Fig. 8.29 Silver–sodium ion exchange for producing high-NA radial and one-dimensional GRIN lenses [8.103]

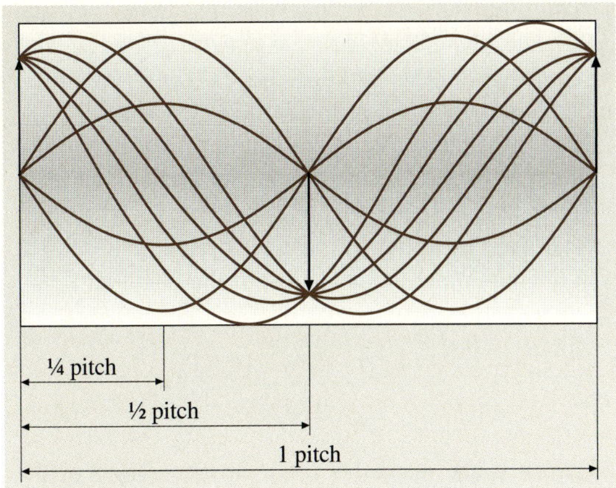

Fig. 8.30 Ray traces within a GRIN focusing lens of different pitch lengths

complicated surface shapes, plane optical surfaces are used. The light rays are continuously bent within the lens until, finally, they are focussed on a spot. Miniaturized lenses are fabricated with thickness or diameter of 3 mm down to 0.2 mm. The simple geometry allows their cost-effective production in larger volumes and essentially simplifies the assembly of rather complex systems, including optical fibers, prisms and beam splitters, because these plane optical components can be attached directly to the GRIN lenses. Varying the lens length yields an enormous flexibility to fit the lens parameters, e.g., the focal length and working distance, to a specific requirement. Preferred ranges of application are optical telecommunication components, laser diode-to-fiber couplers, miniaturized endoscopes, and a variety of optical sensors. New emerging fields are endomicroscopic probes with high resolution requirements and miniaturized heads for optical coherence tomography (OCT). Before these applications are presented in more detail, some fundamental aspects will be considered.

An almost parabolic radial focusing refractive-index profile (8.72) realizes a continuous cosine ray trace within a GRIN focussing lens, the period or pitch length P of which (8.73) does not depend on the entrance height and the entrance angle of the light ray (Fig. 8.30).

Various imaging configurations can be designed using the same index profile by choosing different lens lengths.

- The quarter-pitch lens images a point source on the entrance surface of the lens into infinity, or collimates it. This configuration is usually applied to the collimation of single-mode and multimode optical fibers and laser diodes, where the source can be attached directly to the lens. For high-power laser diodes, cylindrical GRIN lenses are used for fast-axis collimation. Together with other GRIN components they are easily integrated to form compact micro-optical systems.
- The half-pitch lens images an object on the entrance surface inverted on the exit surface of the lens (magnification $M = -1$).
- The 1 (2, 3, or more, respectively)-pitch lens images an object on the entrance surface of the lens identical to the exit surface (magnification $M = +1$). In endoscopes, those lenses are used as relay lenses, which transmit the image from the front part of the endoscope to the eyepiece (Fig. 8.31).
- Endoscope objective lenses are somewhat longer than a quarter-pitch lens. They image the object field to be viewed at a working distance on the exit surface of the lens on a reduced scale (Fig. 8.31).

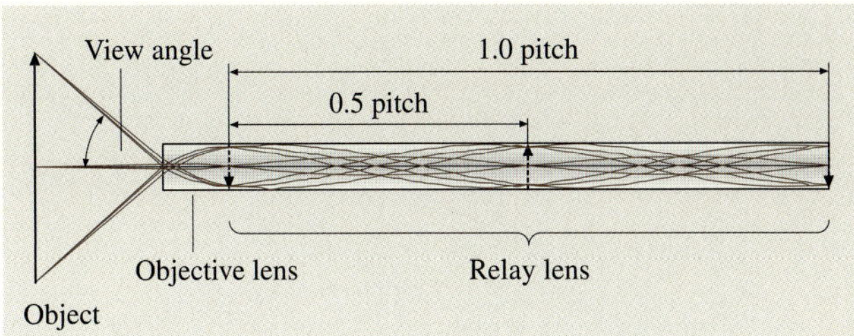

Fig. 8.31 GRIN endoscope without eyepiece function

The geometrical gradient constant g of the radial index profile (8.72) characterizes the steepness of the index gradient and, with the lens length z_1, determines the focal length, f, and the working distance, s, of the

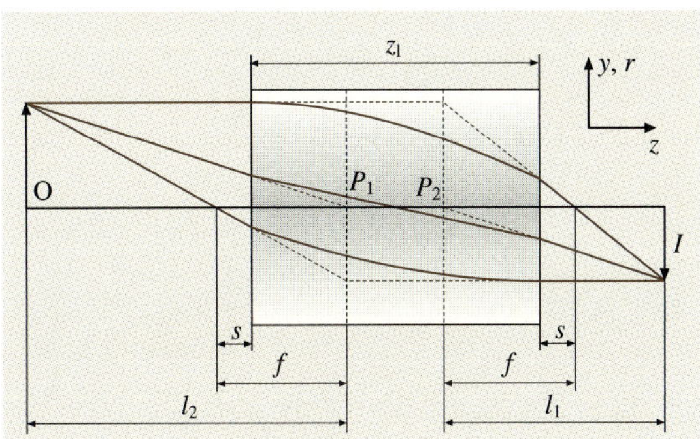

Fig. 8.32 Image formation by a GRIN focussing lens

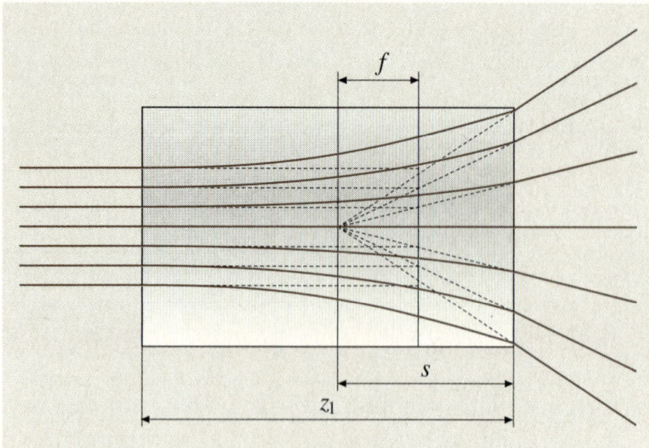

Fig. 8.33 Ray traces in a GRIN diverging lens

lens,

$$f = \frac{1}{n_0 g \sin(g z_1)}\,, \qquad s = \frac{1}{n_0 g \tan(g z_1)}\,. \tag{8.84}$$

Figure 8.32 illustrates the procedure of optically designing an imaging GRIN system using these parameters.

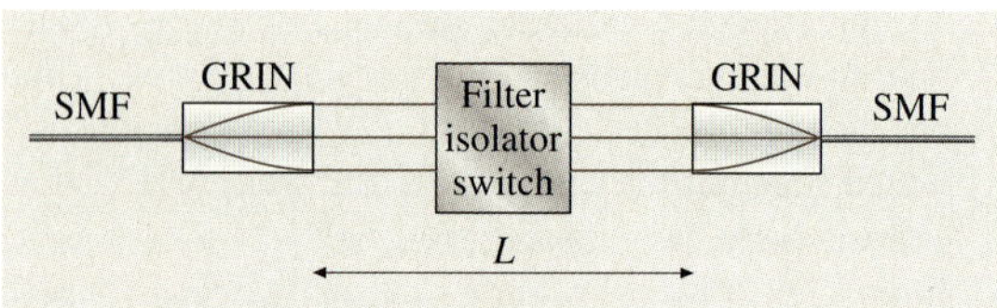

Fig. 8.34 Single-mode fiber coupling by a pair of GRIN lenses

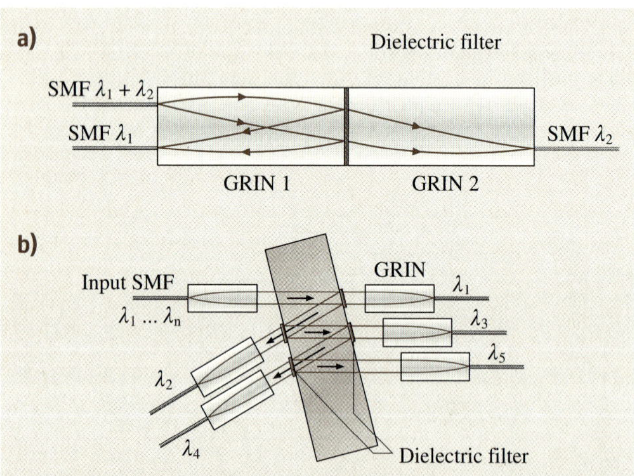

Fig. 8.35a,b Dense wavelength division multiplexing (DWDM) using GRIN lenses and narrow-band interference filters. **(a)** Two-channel multiplexer, **(b)** multichannel multiplexer [8.112]

The distance between the principal planes P_1 and P_2 indicates that GRIN lenses have to be treated as *thick* lenses. However, this fact does not influence the image quality and isoplanatic property of GRIN lenses.

The maximum acceptance angle of a GRIN collimation lens or the maximum viewing angle of a GRIN objective lens, respectively, ϑ, is determined by the numerical aperture (NA). As in fiber optics, it is derived from the maximum index change of the GRIN profile,

$$\begin{aligned} \sin(\vartheta) &= \mathrm{NA} = \sqrt{n_0^2 - n_R^2} \\ &= n_0\sqrt{1 - \operatorname{sech}^2(gd/2)}\,, \end{aligned} \tag{8.85}$$

where n_R is the refractive index at the margin of the profile, and d is the diameter or the thickness of the lens.

Diverging lenses with plane optical surfaces are obtained by parabolic-shaped refractive index profiles (8.74), with the minimum of the index, n_0, at the center of the profile.

A characteristic ray trace through a diverging lens is shown in Fig. 8.33. The comparably short focal length of the lens, f, is also determined by the lens length z_1,

$$f = -\frac{1}{n_0 g \sinh(g z_1)}\,, \qquad s = -\frac{1}{n_0 g \tanh(g z_1)}\,. \tag{8.86}$$

However, a periodic path of the rays is not obtained in this case. Those lenses are applied to the production of microoptical telescopes and scanners.

Optical communication systems use various fiber-optical components, where radial GRIN lenses with plane optical surfaces are successfully used on a large-volume scale, because of their advantageously small size, light weight, easy assembly and adjustment, and short focal length. Very well-established basic functions of GRIN lenses with diameters between 1.0 and 1.8 mm are the collimation of a single-mode fiber in the 1310 nm spectral band and in those between 1500 and 1600 nm and the low-loss coupling of a collimated beam again in a single-mode fiber (Fig. 8.34). The insertion loss can be kept very low (less than 0.5 dB) for distances L between a pair of GRIN lenses up to 25 mm and more, but in general, it is a function of this separation and other alignment factors [8.113, 114].

Optical elements, the function of which is sensitive to the angular orientation of the light, such as filters, optical crystals, gratings and moveable switching prisms, are placed between the GRIN lenses to build up a fiber-optical device. Most important are dense wavelength division multiplexing (DWDM) devices [8.112], whcrc

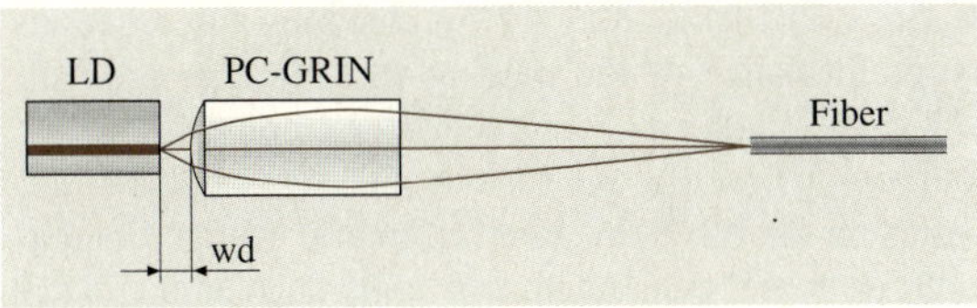

Fig. 8.36 Plano–convex radial GRIN lens with spherical surface for coupling laser diodes (LD) into fibers [8.115]

a multi-wavelength signal in an incoming single-mode fiber (SMF) needs to be separated into many wavelength channels (Fig. 8.35).

Fiber-optical attenuators and splitters can be built up similarly to Fig. 8.35 depending on the filter properties of the element between the GRIN lenses. Often, fibers and GRIN lenses with angled facets are used for optical interconnects to minimize back-reflections, which can disturb the performance of precisely wavelength-tuned devices and systems.

Radial GRIN lenses are also fabricated with spherical surfaces to enable the efficient coupling of laser diodes to optical fibers (Fig. 8.36). The spherical surface allows a working distance (WD) of a few hundred microns between the laser facet and the lens surface, while the NA of the lens is somewhat increased with respect to a similar lens with plane surfaces. The working distance is often necessary because of the can-type package of many laser diodes with a window. The lens matches the high NA of the laser diode to the lower NA of the fiber. The spherical aberration introduced by the spherical surface of the lens can be corrected by adapting the radial index profile [8.115].

A one-dimensional or cylindrical GRIN lens with plane surfaces is obtained by a parabolic-like one-dimensional index distribution, which is oriented perpendicular to and not varying along the optical axis (Fig. 8.37). It is produced by ion exchange in glass plates and can be applied to fast-axis collimation of laser diode bars [8.116], or to the anamorphic shaping of elliptically diverging and often astigmatic beams of laser diodes into the circular focussed beams that are necessary for efficient coupling into fibers, or for the pumping of solid-state lasers [8.117, 118].

In the field of conventional endoscopy, where an object in a barely accessible hollow space needs to be imaged with a large viewing angle through a small opening, the GRIN lens technology is important when the diameter of the optics has to be smaller than 2.0 mm, especially below 1.0 mm, as conventional lenses are difficult to manufacture and to assemble in long imaging systems. The complete imaging system of an endoscope can be built up with only two GRIN lenses: an objective lens of high numerical aperture (NA > 0.5), which generates a demagnified intermediate image of the object with a large viewing angle, and a relay lens of low numerical aperture (NA ≈ 0.1), which relays the intermediate image of the objective to the outside of the body (Fig. 8.31). Several ray periods or pitches of the GRIN relay with additional intermediate images may be necessary to generate the required length of the endoscope. It is also possible to integrate the eyepiece or ocular function within the same relay lens [8.118] to avoid additional optical elements. These GRIN systems do not lead to any internal Fresnel reflection losses because of the absence of internal optical surfaces within the imaging tube. The aberrations have been analyzed in comparison with those of conventional optical systems in endoscopes, and the possibilities of preventing and correcting the aberrations have been studied in detail [8.109, 110]. Flexible endoscopes use coherent imaging fiber bundles for image transmission over tens of centimeters. If the diameter of the optical channel is less than 1.0 mm, GRIN objective lenses, which can be attached directly to the fiber bundle, are preferably used as they yield good image quality. They are available in diameters down to 0.20 mm [8.119].

A novel field of application of GRIN lens systems is endomicroscopy, where small optical probes are needed to image biological tissue and corresponding specific functional features with subcellular resolution in the

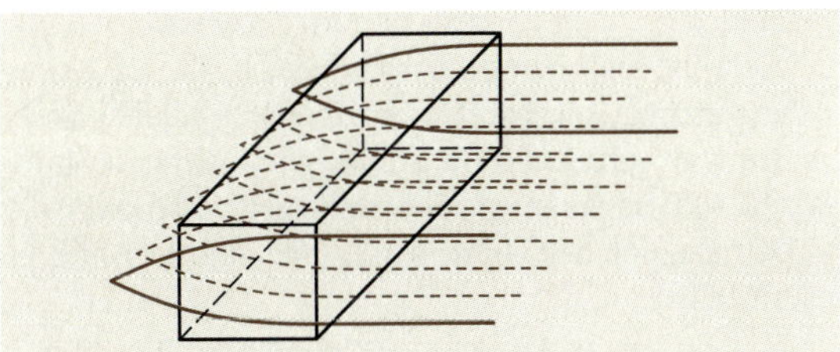

Fig. 8.37 Cylindrical GRIN lens with a one-dimensional lateral index profile

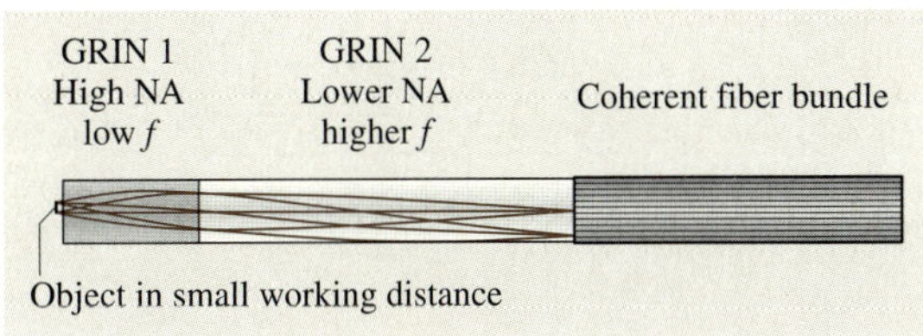

Fig. 8.38 Magnifying telecentric GRIN telescope connected to a coherent fiber bundle for endomicroscopic probes

micrometer range. GRIN lens systems have been developed which provide the required lateral and axial resolution below one micron and 10 microns, respectively. They generate magnified telecentric images of the object (Fig. 8.38), which can then be transferred by coherent fiber bundles [8.120], by rigid GRIN relay lenses [8.121], or by scanning an individual probe fiber [8.122]. These endomicroscopic probes are often operated in a confocal or laser scanning scheme, most recently also involving multiphoton excitation [8.122].

Optical coherence tomography (OCT) is another example of the preferred application of radial GRIN optics. The penetration depth of OCT is generally less than 2–3 mm and also requires miniaturized endoscopic probes for internal medical diagnostics [8.123]. GRIN lenses with plane optical surfaces are relatively easy to combine with the single-mode signal fiber and prisms for beam deflection to allow the scanning of the tissue by rotating the optical assembly or by some other scanning method (Fig. 8.39).

Many optical sensing applications are also based on GRIN focusing rod lenses with radial gradients. A variety of physical parameters such as displacement, pressure, temperature, or fluid flow have been measured by intensity-modulated fiber-optical sensors, where the coupling efficiency between two fibers is varied by lateral and axial displacements of the GRIN lens with respect to the fiber positions, by displacing mirror

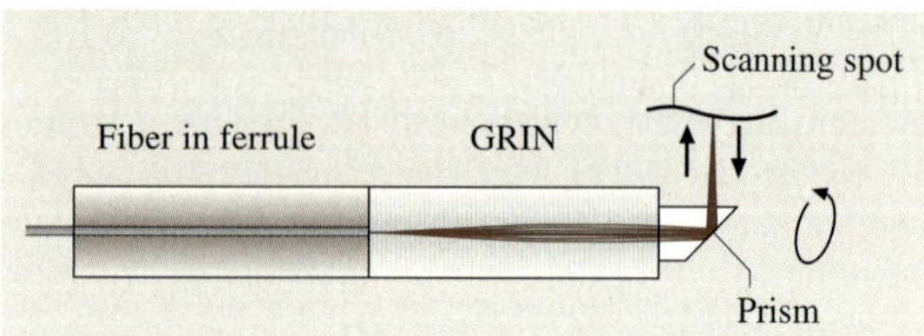

Fig. 8.39 OCT probe design with fiber, GRIN lens and scanning prism

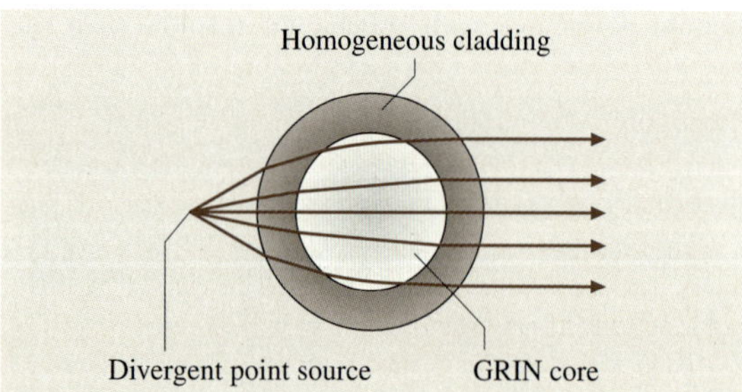

Fig. 8.40 Generalized non-full-aperture Luneburg lens [8.80]

coatings on membranes, or by changing the reflectivity performance of the coating [8.124]. The problem of a stable intensity reference is overcome by sensor solutions where the phase of coherent Gaussian light beams is analyzed by interferometric setups comparable to those used in optical coherence tomography (Fig. 8.39) [8.125].

Luneburg-Type Cylindrical Lenses

One type of radial index profiles in glass fibers is also successfully used for the fabrication of cylindrical lenses of a generalized non-full-aperture Luneburg type [8.80]. Earlier design studies investigated solutions of Luneburg-type lenses with index profiles where the required large index variation or the absolute index range was difficult to fabricate. However, it has been shown that radial index profiles with a peripheral cladding layer of a preform glass tube, which can be fabricated using chemical vapor deposition (CVD) technology, result in a cylindrical lens performance with well-corrected spherical aberration. Here, the rod is transversally radiated by a divergent line source of limited numerical aperture up to 0.5 (Fig. 8.40) [8.126].

By tuning the drawing process, diffraction-limited microlenses can be produced in the shape of fibers or rods with precise diameters and polished surface quality. The fused-silica cladding and the gradient index core of these lenses can withstand very high temperatures (the transition temperature is $\approx 1100\,°\mathrm{C}$). The lenses can be used in the visible and near-infrared spectrum. The rods or fibers are cut to the desired lengths. These lenses are used either as fast-axis collimators for single laser diodes or for building cylindrical-lens arrays for laser diode bars. In order to accommodate a small emitter pitch, the edge of the lens may be ground to a desired width [8.127]. Typical rod diameters are 0.06–3.0 mm with correspondingly scaled focal widths.

Large GRIN Lenses with Axial Gradients

In an axial-gradient lens, a spherically curved surface usually imparts the basic focussing power, whereas the axial index gradient controls or corrects aberrations, especially spherical aberration introduced by the refraction at the spherical surface. Commercially available axial gradients are produced by a bulk glass diffusion process [8.128] in large glass blanks with diameters up to 330 mm. For designing lenses, the index profile is described by

$$n(z, \lambda_{\mathrm{ref}}) = \sum_{k=0}^{11} n_k \left(\frac{z}{z_{\max}}\right)^k , \qquad (8.87)$$

where z is the axial distance from the blank surface and z_{max} is the blank thickness. In contrast to the simple analytical solution (8.77), which is however, difficult to realize, plano-convex, biconvex and meniscus-type lenses are fabricated from specially designed Gradium® glass blanks of standard diameters between 5 mm and 80 mm. The correction of spherical and chromatic aberration by adapting the index profiles is possible, which allows one to reduce the number of surfaces in optical systems and related reflection losses and mounting complexity. Binocular eyepieces and focus heads for high-power industrial Nd-YAG welding and cutting laser systems have been developed in addition to other applications [8.130].

GRIN Rod Lens Arrays

GRIN lenses, especially with plane optical surfaces, can easily be arranged into lens arrays because of their simple geometry and the possibility to grind and polish the optical surfaces of the individual lenses in the array simultaneously. A commercially very successful application is the SELFOC® lens array [8.129] which generates a non-inverted 1:1 image (Fig. 8.41). Usually, the arrays are composed of one or more rows of lenses. The images from adjacent lenses overlap and form a continuous erect image. The short total conjugate between the object plane and the image plane, typically 9–80 mm, allows the compact integration of the optical system, even if large plane objects of up to 60 cm width have to be imaged.

The generation of erect images by single lenses, which can overlap to a uniform large image, is ascribed to the unique property of radial focussing GRIN lenses to form intermediate images inside the lens rod according to the pitch length of the lenses. In general, the pitch length of rod lens arrays is between 0.5 and 1.0 pitches, resulting in a reduced intermediate image in the middle of the rod (Fig. 8.42)

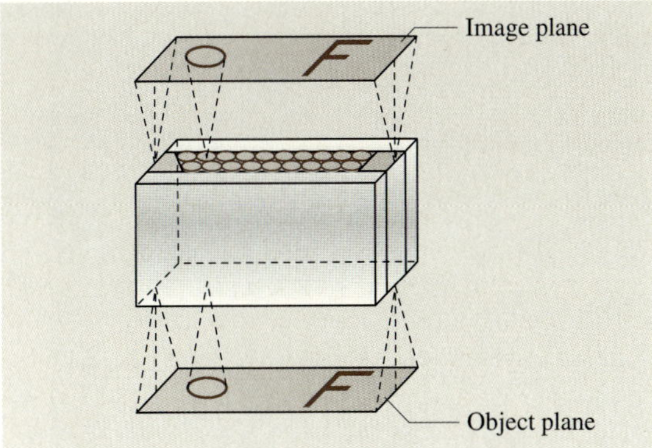

Fig. 8.41 Non-inverted 1:1 imaging by a GRIN rod lens array [8.129]

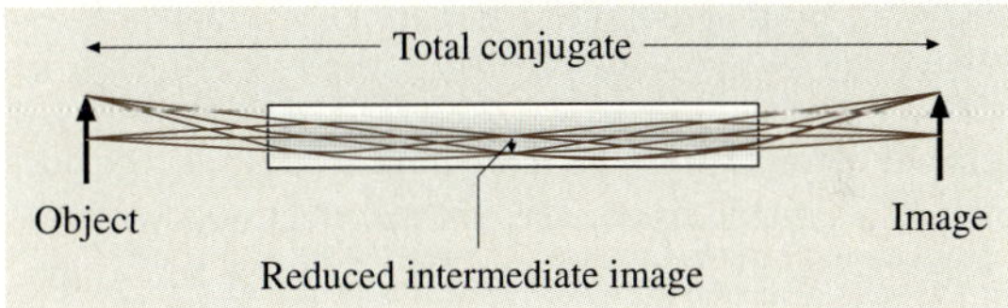

Fig. 8.42 Imaging ray traces of one array lens with reduced intermediate image

The resolution, transmission, and homogeneity of the image quality are determined by the rod lens diameter, by the numerical aperture of the lens and the related lens aperture angle (the image cone in Fig. 8.41), by the total conjugate and the working distance between the object and the lens surface, and finally also by adapting the index profile to the imaging conditions. The optimization of these parameters has been investigated in many studies [8.131–133]. Traditionally, these lens arrays have been widely applied to various photocopiers and scanning devices, to large printing machines, to medical imaging devices, and to many production lines requiring permanent control of surface quality and defect detection.

8.5 Variable Optical Components

Single-lens optical systems are only used for the least-demanding applications. They suffer, among other things, from severe chromatic and spherical aberration. To solve these problems, corrective lenses have to be inserted into the light path. If, additionally the focus needs to be changed, then the lens elements need to be moved relative to one another. This also requires additional lens elements. Therefore, high-quality optical systems with focusing ability need to have many elements, making them complex, heavy, and bulky.

Lenses are not the only optical components with the potential to change from fixed to variable optical properties. Other optical components that can do this include optical filters, polarizers, and attenuators.

Glass is a nearly perfect optical material, but it is static; after grinding and polishing it is fixed in

Fig. 8.43 Drop of water on CD acting as lens with the potential of a variable shape. The optical refractive power and the aberrations of it are obvious. The challenge here is the control of the surface shape of the lens and the encapsulation of such device to avoid influences from the environment such as evaporation or gravity effects

a particular shape, and a change in focal power is impossible. To make an optical zoom system with glass lenses, these fixed elements must be moved back and forth along the optical axis. Glass zoom lenses are big, heavy, and expensive, and suffer from a limited field of view. The whole zoom system also has to incorporate mechanical components, such as stepping motors or piezoelectric drivers and some kind of gearing mechanism.

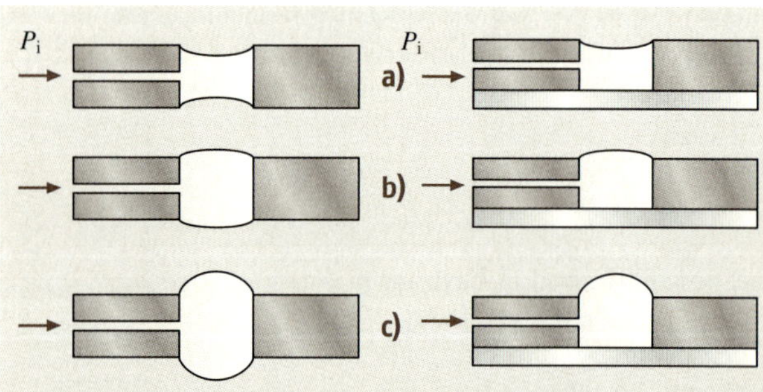

Fig. 8.44a–c Schematic cross section of bi- and plano–concave (**a**), convex (**b,c**) water lenses. The shape of the air–liquid interface depends on the inner pressure p_i and the capillary forces (after [8.134])

Variable optical components try to offer ways around these limitations. Different approaches, such as materials with internal microstructures that can be altered to change their shape and/or refractive index, have been studied. Micro-fluidic systems can produce such changes of optical interfaces by varying their internal pressure or actively changing liquids. The familiar technology of liquid crystals offers a way to modulate the refractive index. Liquid-crystal-based lenses and other devices are discussed in Sect. 5.9.1.

The goal of developing variable optical components has to be the integration of new (intrinsic) variable optical properties in such components, to make them cheaper, less prone to fail because of mechanical wear, and with a form factor as small as possible for a given purpose.

The focus of this section is on the setup, production, and function of such variable optical components and concepts discussed in research and academia in this field. In spite of the great potential of replacing more-complicated optical systems more work has to be done to develop the components to a higher level to make these components part of the standard catalogue of optical components.

8.5.1 Variable Lenses

The easiest device with the potential to work as a variable lens is a drop of optical transparent liquid, as shown in Fig. 8.43. There it is obvious that liquids can work as lenses.

A major challenge is the control of the shape of the drop and therefore its optical power as well as avoiding environmental influences such as gravity, vibration, evaporation, and centering the drop on the optical axis.

Water Lens

As early as the 17th century, the English scientist Stephen Gray built microscopes using water drops as lenses with a diameter of about 0.3 mm so that their curvature was not strongly influenced by gravity [8.135]. Due to the smooth surface of the drop, Gray found that the images these lenses created were quite good. The drops were kept from moving around by placing them in tiny holes drilled in a plate. Different hole diameters led to different lenses with different drop curvatures, and therefore different magnification factors.

Using the same idea of the refractive power of the interface between air and an optical liquid a variable-focus lens is easily realized. To control the shape of such a liquid interface, pressure can applied. For this purpose the liquid interface of water contained in small wells is used. A schematic sketch of the lens housing and possible lens types is shown in Fig. 8.44.

A capillary barrier develops at the exit of the well due to the abrupt area expansion. The shape of the air–liquid interface is described by the Young–Laplace equation of capillarity.

That such a lens is an open system is an obvious disadvantage. Environmental conditions will easily influence the shape of the air interface and the optical liquid used will most likely evaporate and therefore the lens will not have the same characteristic over a longer period of time.

A way to overcome a couple of the drawbacks mentioned above is the introduction of another liquid that is immiscible with the optical liquid. Ideally it should have the same density and a different refractive index. By using a second liquid, evaporation can be minimized since the whole system of both liquids can be encapsulated very tightly. Such an approach is shown in Fig. 8.45. An active optical surface develops at the interface between the water and silicon oil and changes its shape depending on the pressure of the water.

Using liquids with an equal density minimizes the influence of gravity and the optical interface between both liquids will therefore only be minimal disturbed. This approach is also used in electrowetting-based liquid lenses.

Electrowetting-Based Lenses

Electrowetting Phenomenon. Electrowetting describes the phenomenon of spreading liquid drops on solid surfaces due to an applied voltage across the liquid and a counter-electrode underneath the solid surface.

Lippman [8.136] started the study of this effect by observing the influence of an electrostatic charge distribution on the capillary forces and tension at an interface. The relation between the contact angle of a liquid drop on a solid surface and the interfacial tension at the three-phase point were formulated by *Young* [8.137]. Combining the results of both studies leads to the Young–Lippman equation, which describes the change of the contact angle of a liquid drop on top of a solid surface induced by electrostatic forces when a voltage is applied between the drop and the underlying counter-electrode. In earlier research on electrowetting phenomenon the electrolytic decomposition of water upon applying voltages beyond a few hundred millivolts was a problem.

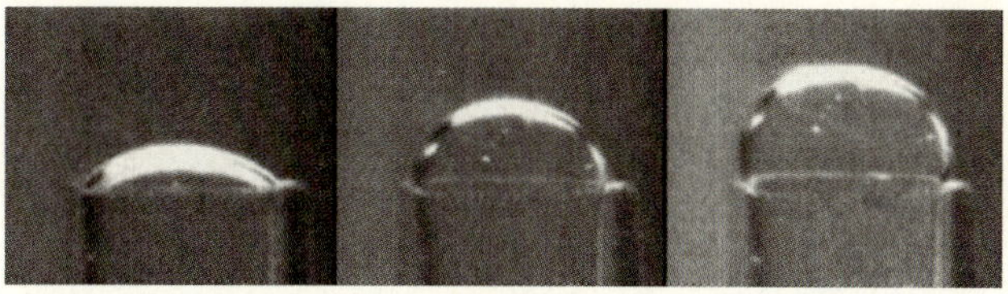

Fig. 8.45 Images of radius of curvature variation with pressure for liquid–liquid interface based lenses. The lens was made of water contained in a glass aperture (1 mm diameter) and immersed in low-viscosity silicon oil (after [8.134])

In the early 1990s *Berge* [8.138] introduced the idea of using a thin insulating layer to separate the conductive liquid from the metallic electrode in order to eliminate the problem of electrolysis. This new concept has become known as electrowetting on dielectrics (EWOD).

A detailed theoretical description of the electrowetting effect can be found in [8.139]. Only the basic idea is briefly discussed here. The phenomenon of electrowetting on dielectrics is shown in Fig. 8.46.

An electrically conducting fluid drop on an insulating surface usually forms a drop with a resulting contact angle θ_0 to the surface. The shape of the drop is determined by the minimization of the interfacial energies between the insulator and the droplet, the droplet and the surrounding gaseous phase, and the surrounding gaseous phase and the insulator.

If a voltage is applied between the droplet acting as one electrode and a (counter)electrode underneath the insulating film, the droplet spread over a wider area and the contact angle θ decreases. Additionally, the minimization of the interfacial energies leads to a change of the shape of the droplet. Depending on the applied voltage a wide variety of different shapes of the droplet are possible.

The basic concept behind such a variable lens becomes obvious, by imaging the droplet being an optical

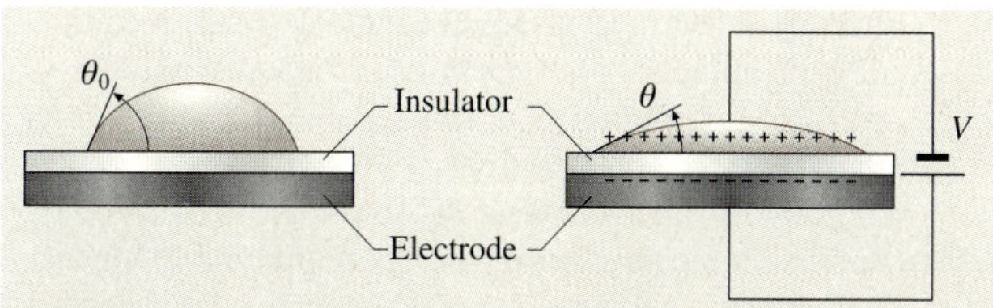

Fig. 8.46 Generic electrowetting setup. Conductive liquid droplet at zero voltage (*left*) and with applied voltage (*right*). See the text for details

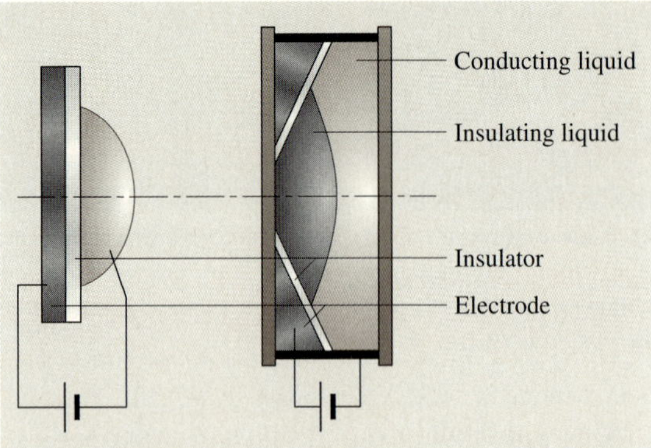

Fig. 8.47 Schematic description of the device evolution from electrowetting on dielectrics to competitive electrowetting. The shape of the refractive interface is formed between the conducting liquid and the insulating liquid. Density matching makes the device insensitive to gravity forces. A conducting liquid is used as the electrode

transparent and conductive liquid with a refractive index n_l sitting on a thin insulating layer on top a transparent counter electrode. The refracting interface is formed between the optical liquid and the air.

Optical devices based on electrowetting can usually be separated into two subgroups. The first uses the electrowetting on dielectrics effect by moving liquid drops macroscopically on electrodes. A concept for an optical switch based on electrowetting on dielectrics [8.141] is discussed in Sect. 8.5.2. The second group of optical devices employs so-called competitive electrowetting. This is especially used to make lenses.

Competitive Electrowetting. The electrowetting described above has the disadvantage that it usually works best for tiny droplets where the influence of gravity is negligible. An optimization is most likely possible for special mounting positions and temperatures. For applications requiring larger diameters and which are not limited to one position of use the influence of gravity on the droplet surface has to taken into account. Additionally the influence of the electrode structures in the beam path is very often a disadvantage due to diffraction or the occurrence of stray light.

One way to avoid these disadvantages is to introduce another liquid into the system. The requirements for such a liquid are as follows. It has to be an insulating liquid, immiscible with the conducting liquid and should have the same density as the conducting liquid.

A schematic cross section through a device is shown in Fig. 8.47. The drop of insulating liquid is covered with the conducting liquid. The device evolution is clearly visible. By applying a voltage between the conducting liquid and the counter-electrode the contact angles changes and therefore forces the liquid–liquid interface into different shapes. Due to the fact that the refractive indices for the two liquids are not equal a refracting surface is formed. This setup makes it possible to make a device insensitive to gravity forces due to the density match between the liquids. Because the packaging and the conducting liquid are used as electrodes, there will no electrodes disturb the optical path.

Components Based on Electrowetting.
Liquid Lenses. As mentioned competitive electrowetting allows the design of optical components with a refractive interface that can change its shape controlled by electrical voltage and that therefore have variable focal lengths. Such a lens was first described in [8.142]. The device is a closed cell filled with two immiscible liquids. A device with similar properties was presented in [8.143].

The first (insulating) liquid is a nonpolar oil while the other is a (conducting) aqueous salt solution. Both liquids were density-matched to within 10^{-3}. Due to this density matching of the two liquids the lens becomes insensitive to shocks and vibrations.

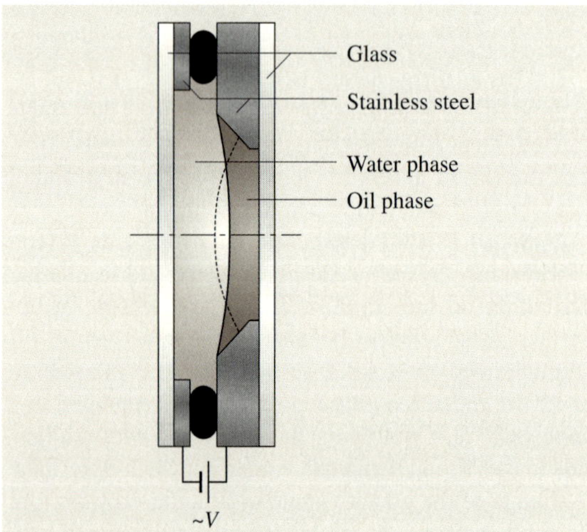

Fig. 8.48 Schematic representation of the liquid lens: two immiscible liquids (conducting salty liquid and insulating oily liquid) are trapped between two glass windows. The liquid–liquid interface moves from continuous ($V = 0$) to the *dashed line* when voltage is applied. The optical axis is shown as a *short line* made up of *long dashes* (after [8.140])

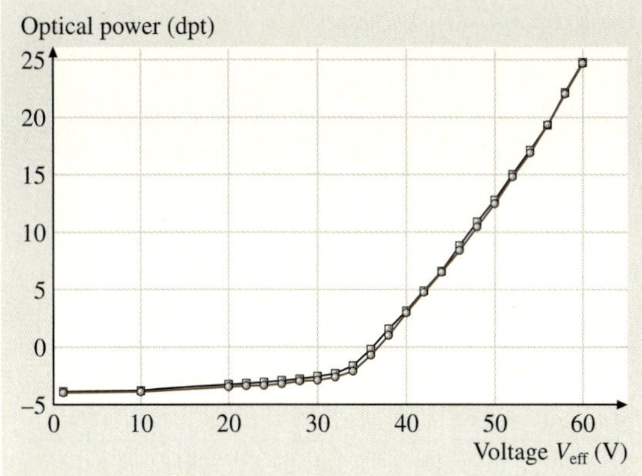

Fig. 8.49 Inverse focal length as a function of the applied voltage for an electrowetting-based lens. Two *curves* which correspond to increasing (*circle*) and decreasing (*square*) voltage are superimposed. The tiny hysteresis is most likely due to imperfect inner surfaces

The principal setup of such a liquid lens is shown in Fig. 8.48.

At low voltage, the water contact angle is high, i. e., the oil droplet has a low contact angle and vice versa. The optical power (or the inverse focal length) of a lens is shown as a function of the applied voltage in Fig. 8.49.

From 0 V to ≈ 36 V the lens has a concave and for higher voltages a convex shape. The focal length is therefore in the range −250 mm to infinity and from infinity to 40 mm.

Such electrowetting lenses seem to be very robust. It is reported that they can be switched more than 10^6 times without any sign of degradation. Switching speeds of the order of 10 ms have been achieved for a lens with a diameter of 2.5 mm [8.144]. To reach the maximal speed a precise adjustment of the liquids' viscosity is necessary, such that the drop motion is critically damped. Otherwise the droplet either oscillates several times or relaxes only slowly towards the new equilibrium shape.

Lenses with an aperture smaller 1 mm using competitive electrowetting can completely fabricated in MEMS technology [8.145]. To produce such lenses or lens arrays quadratic V-groove-defined holes are etched in a standard silicon wafer. These will later be used as cavities to hold the two liquids. Afterwards, all surfaces are passivated with an insulating layer consisting of silicon dioxide and a hydrophobic layer. For contacting the droplet, an ITO-structured Pyrex wafer is anodically bonded to the bottom of the wafer. The lens liquid is deposited in the middle of the V-groove, embedded in

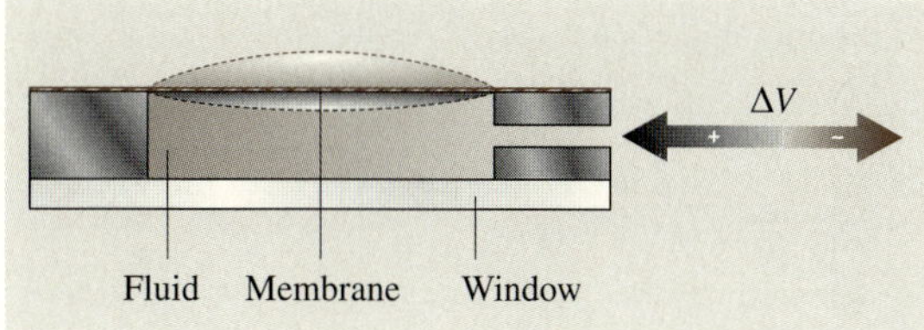

Fig. 8.50 Schematic cross section of a membrane lens. The fluid is confined in the lens body between the glass window and the flexible membrane. The enclosed volume is changed by means of an external actuator. It is possible to obtain a converging lens (convex membrane) as well as a diverging lens (concave membrane)

the surrounding liquid, and another glass cover closes the system on top. A water-based inorganic salt solution with a refractive index of 1.51 and a density of 2.1 g/cm^3 was used as lens liquid. As a surrounding liquid, a density-matched perfluorocarbon with a refractive index of 1.293 was used.

The lateral dimensions of prototype like system are 8 mm × 8 mm and the total thickness of the system is 1.525 mm. The lens aperture, which is equivalent to the size of the bottom of the V-groove, can be varied depending on application during lens production.

The back focal length of such a lens with an aperture of 300 μm can be varied between 2.3 mm and infinity using voltages between 0 V and 45 V.

Membrane Lenses

Variable lenses based on membranes have been discussed since the beginning of the 19-th century [8.146, 147]. More publications have been written and patents filed since then, but so far these devices have not made it to market, probably due to technical and material problems. They have mostly been discussed in the field of vision care in the form of spectacles with variable optical power.

Newer developments in the fields of material sciences, micromechanical technology, and optics make the successful development and production of such devices seem promising. The interest of companies and agencies looking for components for lightweight optical zoom systems [8.148] used, for example, in unmanned aerial vehicles, handheld optical devices (camera phones), and surveillance cameras is helpful in this area.

The general setup of such a lens is shown in Fig. 8.50. A cavity enclosing a transparent optical liquid with refractive index n_l has at least one variable and flexible side wall. For the optical liquids, water, salty solutions or index-matching liquids are used. The volume of the

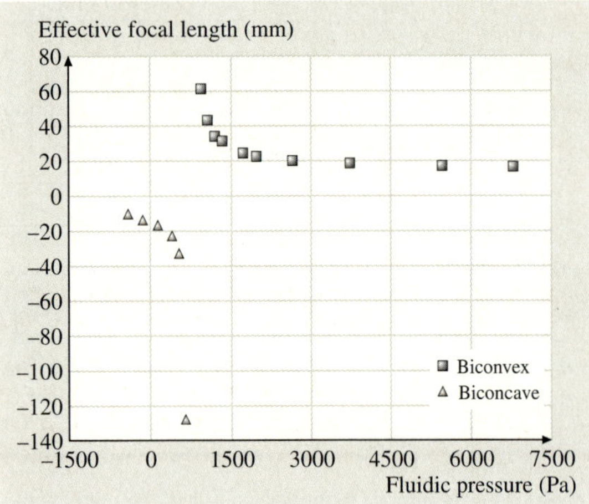

Fig. 8.51 Dependence of effective focal length on fluidic pressure of the fluidic adaptive lens during lens-type transformation between biconvex and biconcave lens (after [8.148])

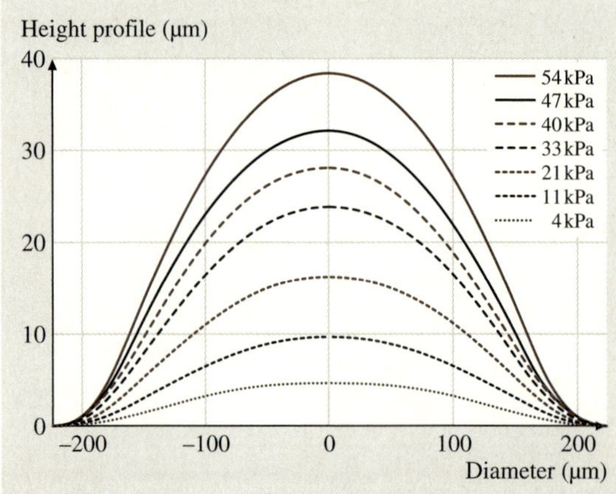

Fig. 8.52 Topography of a deflected membrane lens (aperture 400 μm) at different pressure values, measured with a surface profiler. The inflection points are seen on the outside of each graph (after [8.149])

liquid enclosed is varied by means of actuators and kept constant for a certain optical power of the optical device. It is easily seen that such a device can be used as a variable plano–convex/concave lens.

Lens diameters are 0.5–20 mm [8.149–151]. Microlens arrays can also be fabricated [8.149].

Polydimethylsiloxane (PDMS) is usually used as the membrane material [8.149, 151], although other materials such as polyvinylidene chloride, polyethylene, and polyvinyl chloride have also been discussed [8.152]; the membranes are usually 10–100 μm thick.

Setups with two flexible membranes on opposite sides of the device are als discussed in literature [8.153, 154] as well as devices with two separate cavities [8.155]. Such devices enlarge the possible choices of membrane-based variable lenses from plano–concave/convex to biconcave/convex and convex–concave and increase therefore the degrees of freedom in optical design setups.

Optical Performance of Membrane Lenses. The general optical performance of membrane lenses arises from the changes in the curvature of the membrane. Starting from a curvature with an infinite radius, where the lens acts as a parallel plate with no refractive power. With increasing pressure in the cavity, the liquid inside forms a convex lens, the lens radius decreases, and the refractive power becomes higher.

With decreasing cavity pressure the membrane forms a concave shape, and the radius decreases and its diverging refractive power increases. Due to the more-experimental character of these devices and the wide variety of different lens parameters, such as the aperture, membrane material, membrane thickness, and optical liquids used, it is impossible to present an exact characterization of the optical performance of such lenses.

In a more-general approach an example of the optical performance of a membrane lens with two membranes taken from [8.148] is shown in Fig. 8.51.

The fluidic lens consists of a 4 mm-thick cylinder with a diameter of 20 mm, covered with two 60 μm-thick PDMS membranes on both sides of the cylinder. As the optical liquid a 63% sodium chromate solution with a refractive index of $n_l = 1.5$ was used. The pressure inside the chamber was controlled by a battery-powered fluidic control consisting of a miniaturized pump, pressure sensor, valves and an electronic control unit.

The transformation of the membrane shape from concave to convex is clearly seen, as well as the increase in optical power with decreasing radius of curvature of the membrane.

Aberrations in Membrane Lenses. The focal length of these variable membrane lenses is not the only interesting parameter. Other interesting parameters include the degree of sphericality and the apertures of the lenses that can actually be used optically.

The results of measurements of the curvature as a function of the applied pressure of a membrane lens with an aperture of 200 μm and a PDMS membrane thickness of 50 μm [8.149], taken by using a mechanical profiler, are given in Fig. 8.52.

It is obvious that the lens profiles found are not ideally spherical. The first reason is the nonlinear deflection of the PDMS membrane. The nonlinearity of the elastic PDMS membrane, however, is an intrinsic effect of the material and is not alterable. Other, more-linear elastic materials have to be found. The other reason is the fixed edge of the membrane on a flat surface. This leads to an inflection line around the center of the lens at the lens edge. Outside this inflection line the lens curvature of a generally convex lens is concave. So the remaining convex usable aperture shrinks.

There are different ways to deal with these obstacles. The easiest is the development of such membrane lens for a special optical design, which means that the necessary change in focal length can be minimized and the mechanical setup, such as the aperture size, membrane thickness, and membrane fixtures of the lens can also be optimized.

The following countermeasures can be used in the special-purpose lens mentioned above, as well as in multipurpose lenses. To counteract the nonlinearity of the PDMS membrane a variation of thickness with the radius can be considered. This would probably increase the complexity of the manufacturing process and therefore raise the manufacturing costs. To counteract the observable inflection point a different fixture for the membrane on the lens edge may be possible. The fixture has to be formed in such a way that the membrane runs asymptotically smoothly onto the surrounding lens body. This change from a mostly 2-D approach to a more three-dimensional (3-D) approach unfortunately also probably means a change from a wafer-scale fabrication of multiple lens to a single-lens fabrication process. This leads to increased manufacturing costs for such devices.

The permeability of the membrane to gases and liquids is another issue that has to taken into account. A secondary experimental result observed while filling the lens mentioned in [8.149] is that any remaining bubbles quickly diffuse out of the system because the PDMS membrane is gas-permeable. On the other hand this means that gas is also able to permeate into the lens in the case of concave lenses, which makes this lens type more likely to fail during lifetime or makes the use of membrane lenses as concave lenses impossible.

For the thinnest membranes (thicknesses smaller than 10 μm) the membrane also becomes permeable to liquids. This leads to a loss of liquid over time and therefore to a slow change in the variable characteristic and a decay of the variable optical component in the long run.

More scientific research and development, i. e., the development of new membrane materials, the simulation of membranes in different fixtures, fabrication, optical, and environmental tests of membrane lenses are necessary to approach a perfect variable lens.

Production of Membrane Lenses. There are various ways to produce membrane lenses.

One method [8.151, 156] is it to produce the lens chamber and membrane separately using a soft lithography process. Afterwards the membrane and lens body are bonded together using an oxygen-plasma bonding technology. After the top of the lens has been formed, the bottom of the fluidic chamber is bonded to a thin glass slice for easy handling again using an oxygen-plasma-activated bonding technique.

The detailed fabrication process of a membrane lens array is illustrated in Fig. 8.53. At first a ca. 10 μm-thick layer of photoresist is patterned onto a silicon wafer to form the mold for the microfluidic network. A second ca.100 μm-thick layer of photoresist is patterned on top of the first layer to create circular chambers for the lenses. The PDMS prepolymer mixture is spin-cast over the mold and subsequently cured at 150 °C. The total thickness of the PDMS layer is approximately 140 μm. The thickness of the PDMS membrane in the lens area is ca. 40 μm. The cured PDMS is peeled off from the mold, treated with oxygen plasma and then irreversibly bonded to a glass substrate. Since the PDMS

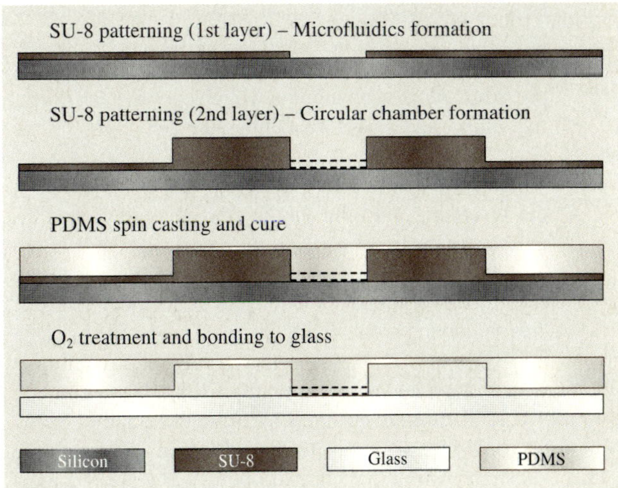

Fig. 8.53 See for details (after [8.156])

layer contains both the microfluidic network and the circular chambers, no alignment is required during the bonding process.

Another way of producing membrane lenses as well as membrane lens arrays is the use of standard micro-electromechanical-systems (MEMS) processing techniques [8.149]. This has the advantage that no transfer process is necessary and therefore all process steps can take place on the same mechanical basis, which ensures similar quality for all optical elements. Using clean-room facilities ensures particle-free optical surfaces. Another advantage of using wafer-scale technology is that otherwise critical alignment steps during the assembly of various parts of the lens can be avoided. This technology seems to make the low-cost mass fabrication of such lenses possible.

A summary of the process is shown schematically in Fig. 8.54. Processing starts with photolithography on a silicon wafer polished on both sides in which the lens openings and the microfluidic channels are defined. A masking layer is deposited in wet oxidation and plasma-enhanced chemical-vapor deposition oxidation steps. The first oxide mask defines the contours of the circular lens openings with diameters of 300–600 μm, the fluidic channels, and the reservoir. This layer is structured by photolithography and subsequently opened by reactive-ion etching (RIE). In a second photolithography step, the photoresist layer is patterned. This second masking step defines the contours of the lens chambers and the reservoirs. These are then etched through half of the silicon wafer by an inductively coupled plasma (ICP) RIE process. The resist layer is removed afterwards. After coating the substrate with a primer to improve the adhesion of the PDMS to the silicon or silicon oxide, the PDMS film is spin-coated on top of the wafer. The PDMS film is cured and forms the pressure-actuated membrane, which encloses the working liquid and therefore defines the curvature of the lens.

In a second ICP RIE etch step the lens chambers, the reservoirs, and the fluidic channels are etched simultaneously, passivated by the structured SiO_2 layer. The etching stops abruptly when the SiO_2 layer at the front side is reached due to the etch selectivity of SiO_2. The remaining front-side SiO_2 layer is removed in a final RIE step. This means that the PDMS membrane is now stretched freely over the silicon lens chamber. To cap the fluidic channels and chambers, a thin Pyrex wafer is being bonded to the back side of the silicon substrate with a UV-curable optical adhesive. To minimize scattered light and cross talk between the single lenses of an array, an aperture stop made of a 100 μm-thick structured chrome layer on the Pyrex wafer is required.

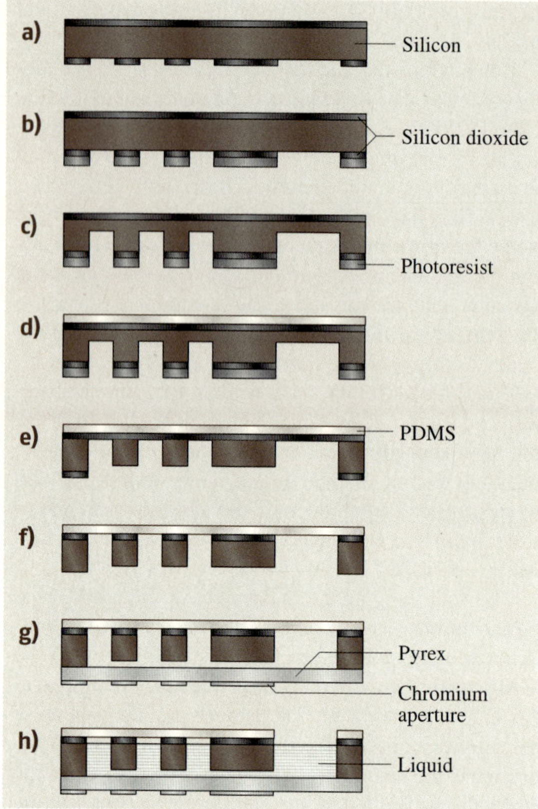

Fig. 8.54a–h Schematic process description: **(a)** Silicon wafer with both side SiO_2 layer and with back side opened by reactive-ion etching (RIE); **(b)** Photoresist, spin-coated and structured; **(c)** First inductively coupled plasma RIE etching; **(d)** Spin-coating of primer and PDMS; **(e)** Second ICP RIE etching; **(f)** SiO_2 removal; **(g)** Bonding of the Pyrex wafer and patterning of the back-side chromium layer; **(h)** Sawing and filling of the devices (after [8.149])

The wafer stack is then separated into individual chips by sawing. To prepare the fluidic lenses for operation, the cavity is filled with a liquid optical medium.

A similar process to fabricate membrane lenses with two flexible membrane sides is described in [8.154].

Actuation Mechanism. A rudimentary activation mechanism, mostly used in laboratory setups, is the use of a syringe pump to inject fluid into the lens chamber. Other actuation mechanisms also employ external pressure. To get a higher degree of control over the lens,

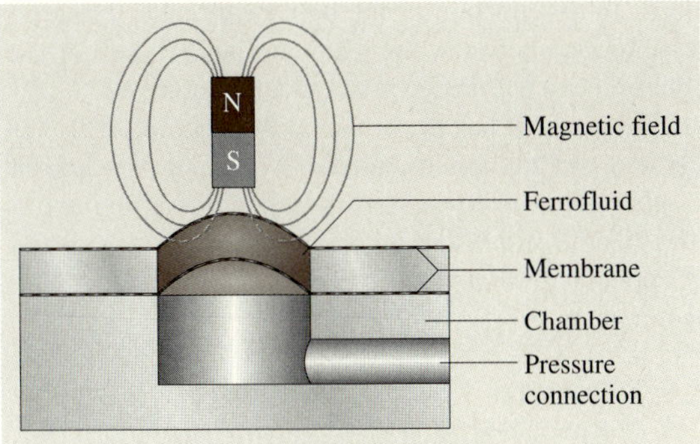

Fig. 8.55 Schematic view in a magnetic actuator. The ferrofluid is confined between two flexible membranes; the lower membrane separates the ferrofluid from the working fluid (pressure connection). The attractive force between the electromagnetic field and the ferrofluid leads to a decreasing pressure in the pressure connection and therefore a change in the optical active surface of the membrane lens connected to the actuator (after [8.158])

the pressure is built up by, for example, piezoelectric elements, MEMS-based miniature pumps or ferrofluids. For advanced variable optical lens systems it is necessary to employ automatic control including a pressure sensor to ensure uniformity of the optical power of the lens over a given amount of time. Highly integrated pressure sensors are made, for example, by MEMS technology [8.157]. Such controls are still in development and have to be integrated with the variable optical parts to give the new variable optical components a chance on the market.

External Pressure. Variable focusing in liquid-filled lenses has been demonstrated by deflecting the lens membrane using pressure-driven mechanisms such as external pumps [8.150] or piezoelectric stack actuators [8.160].

Piezoelectric stack actuators are suitable actuators for driving variable optical lenses even though their working range is usually too short ($\approx 10\,\mu m$) to actuate the lens surface directly. There are two ways to solve this problem; both mechanically amplify the stroke of the piezoelectric actuator stack. The first is a mechanical lever that transforms the small stroke into a larger one, the other way makes use of the incompressibility of the liquid. The cross-sectional area of the cylinder pressed on with the piezoelectric aactuator stack is several tens of times larger than that of the lens surface, and therefore the distortion of the lens surface is a couple of times that of the cylinder.

Magnetic Actuation. Ferrofluids provide another method to actuate a given variable membrane lens by magnetic forces. Ferrofluids are liquids with paramagnetic properties that combine rheological characteristics with strong force action. They consist of magnetite particles (Fe_3O_4 with a diameter of 10 nm dispersed in a liquid [8.161]. The possibility to fill any given or even changing geometry with ferrofluids and the easy integration into a system are the main advantages of this actuation method.

In Fig. 8.55 the actuator is shown as it generally functions [8.158]. A volume of ferrofluid is enclosed between two elastic membranes, which seal a chamber in a plastic housing. A magnetic field above the upper membrane attracts the ferrofluid and the upper membrane bulges upwards. The incompressible fluid causes the lower membrane to follow accordingly. Thereby, a negative pressure can be generated below the second membrane. This negative pressure can be used to pump a secondary fluid that fills the chamber below the double membrane and features a pressure connection to a membrane lens. The magnetic field can be produced by means of an electromagnet and thereby control the shape of the membrane lens by an electric voltage.

Due to the effect that ferrofluids can only be attracted by magnetic fields, this actuator is only able to decrease the optical power of a given membrane lens.

Stimuli-Responsive Hydrogels. Another way to change the optical power of a lens is the incorporation of stimuli-responsive hydrogels, as described in [8.159].

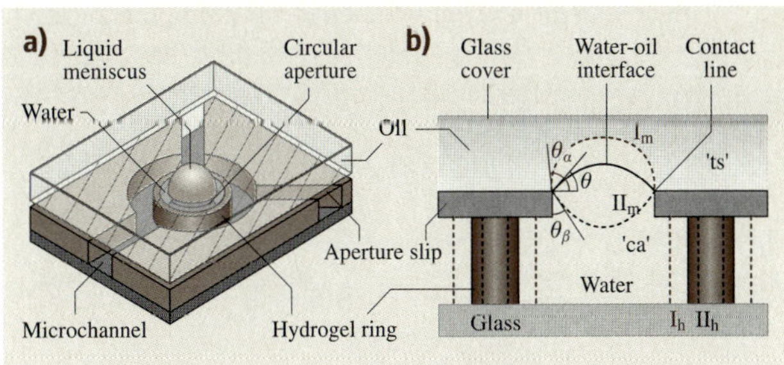

Fig. 8.56 Schematic view on a hydrogel lens setup. The interface between two immiscible optical liquids (water bottom, oil top) with different refractive indices forms an optical surface. The water is confined between the glass substrates, oil, and hydrogel ring. The outside of the hydrogel ring is accessible to various other liquids to change the volume of the hydrogel ring. Changes in the volume of the hydrogel ring lead to changes in the confined volume and therefore in the shape of the water–oil interface (after [8.159]).

The principal setup of such a lens system is shown in Fig. 8.56. There is a wide variety of hydrogels that react upon changing environmental conditions such as temperature or pH to incorporate or exclude water molecules in their structure, which results in a changing volume of the substances.

8.5.2 New Variable Optical Components

Electrooptical Switches Based on Electrowetting

A concept for a 1×2 optical switch based on total reflection and electrowetting [8.141] is outlined in Fig. 8.57.

The proposed sandwich structure consists of a top and bottom glass substrate each coated with an electrode structure and an insulating hydrophobic layer. One of the electrode structures is segmented and it is possible to switch the different electrodes in a timely manner to move the droplet in between the sandwiched structure using electrowetting forces.

A 45° prism is attached to both glass substrates to avoid total reflection at the air–glass surface and allows light to be coupled into and out of the device. The collimated light is guided through the prism into the electrowetting device. Placing the liquid droplet in the optical path, the gap between the two glass substrates is index-matched and the light is transmitted through the sandwich configuration to the output S2.

The liquid droplet can be driven out of the optical path by electrowetting forces due to the application of a voltage to the counter-electrodes in a timely manner. Now the gap between the top and bottom substrate is filled with air and not with index-matching liquids anymore. Total reflection of the light beam takes place on the interface between the glass and air. Therefore the light will exit the device at the second output S1.

The concept has the potential to become highly integrated and applied in a matrix switch. In comparison to other concepts for optical switches it is important to note that this setup does not need precise movement and all optical parts that require precise alignment have fixed positions.

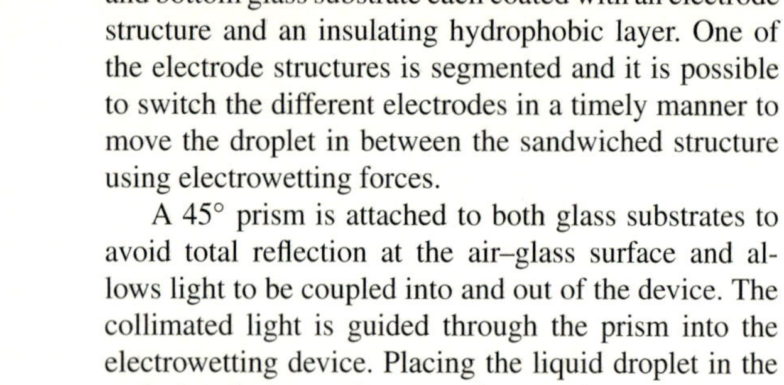

Fig. 8.57a,b Schematic drawing of the concept for an optical switch based on electrowetting. **(a)** The liquid droplet with a refractive index close to the refractive index of the hydrophobic layers and glass is positioned in the beam path of the light by electrowetting. The light is transmitted nearly without any losses to S2. **(b)** The droplet is moved out of the beam direction. Total reflection occurs at the interface between the dielectric layer and air (after [8.141])

8.5.3 Outlook on Variable Optical Components

Variable optical components have great potential to replace conventional optical systems in a wide variety of applications [8.154]. Applications in high-volume and low-cost markets, for example cell phones and web cameras, are particularly appropriate for early use of components with intrinsic variable properties. The advantages of integrating these properties into the components will probably result in smaller form factors, more freedom in the design of such devices, easier and quicker assembly, and finally lower manufacturing costs and higher profit margins. The liquid lens from Varioptics (Lyon, France), which is based on electrowetting, is the component closest to serious production. Electronic controls are also ready and will probably be seen in prototype mobile-phone cameras in the next months [8.162]. Membrane lenses are not at this stage yet. More work has to be done to develop new membrane materials and to integrate actuation and control into the variable component.

Other developments, such as variable filters or polarizers based on liquid crystals will easily find special niche applications in which to succeed.

All technologies for variable optical components face strong competition from existing technologies. They have to find there way into optical designs and prove there advantages and abilities to potential users. Since existing technologies are already approved and in the market there have to be clear advantages of using the new technologies. In the end this will require lower manufacturing costs to replace existing operational solutions. The key to success will probably be an application that is made possible only through such a new technology.

The development of existing technologies is also progressing and continuing to give easier and cheaper solutions to the problems, and some technologies may become extinct due to changes in other areas.

8.6 Periodically Poled Nonlinear Optical Components

Obtaining quasi-phase matching (QPM) by periodically alternating the sign of the second-order nonlinear coefficient $\chi^{(2)}$ was first proposed in 1962 in a landmark paper by *Armstrong* et al. [8.163]. It took, however, nearly 30 years until efficient generation of frequency-doubled laser light with QPM in a periodically poled nonlinear material was realized [8.164–166]. The final breakthrough of QPM occurred with the first tunable laser source in the infrared wavelength regime using a bulk periodically poled lithium niobate crystal (PPLN) in an optical parametric oscillator (OPO) setup [8.167, 168].

This section will briefly review the physics of quasi-phase matching (Sect. 8.6.1). Because it continues to be a challenge, a detailed description of the actually techniques used to fabricate periodically poled crystals will be given (Sect. 8.6.2) followed by the standard techniques for the visualization of the obtained structures (Sect. 8.6.3). Finally, the currently most prominent applications for periodically poled crystals are described (Sect. 8.6.4).

8.6.1 Fundamentals

Second-Harmonic Generation (SHG)

In second-harmonic generation (SHG, also called *frequency doubling*) photons interacting with a nonlinear medium are effectively combined to form photons with twice the energy, and therefore twice the frequency of the initial photons (Chap. 3).

For efficient SGH three features must mainly be fulfilled:

1. the pump intensity I must be sufficiently high over a certain propagation length since, at least for small conversion efficiencies, the SHG generation is proportional to I^2,
2. the nonlinear medium must have a high second-order nonlinear coefficient $\chi^{(2)}$, and
3. the light beams involved need to preserve their phase relationship over the propagation length.

Whereas (1) can be achieved by using high-power laser beams and for (2) a careful choice of the material is necessary, (3) requires additional measures since dispersion generally leads to a phase mismatch between the frequency of the pump laser beam (ω) and the frequency-doubled frequency (2ω).

This phase mismatch between ω and 2ω waves leads to destructive interference of the SHG light generated at

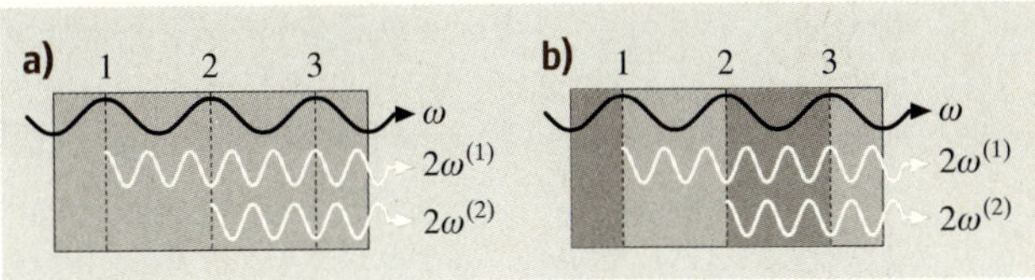

Fig. 8.58a,b Second-harmonic generation. **(a)** No phase matching: light generated at position 2 ($2\omega^{(2)}$) is out of phase with respect to the light generated at position 1 ($2\omega^{(1)}$). **(b)** Quasi-phase matching: by inversion of the sign of $\chi^{(2)}$, $2\omega^{(2)}$ is in phase with $2\omega^{(1)}$

the very beginning of the nonlinear crystal with that generated further inside the medium, thereby limiting the maximum output power (Fig. 8.58a). This problem can be overcome in birefringent crystals by choosing a particular orientation and keeping the crystal at a certain temperature such that both waves, ω and 2ω, propagate in the same direction and experience the same refractive index, so-called *noncritical phase matching* (NCPM). In general, these conditions are difficult to achieve, and because the value of $\chi^{(2)}$ also depends strongly on the orientation of the crystal, NCPM often results in an unsatisfying tradeoff.

There is, however, a very successful approach to preserving the phase relationship between ω and 2ω waves, so-called quasi-phase matching, which will be described in the following.

Quasi-Phase Matching (QPM)

For quasi-phase matching, the nonlinear susceptibility $\chi^{(2)}$ is modulated periodically to compensate for disper-

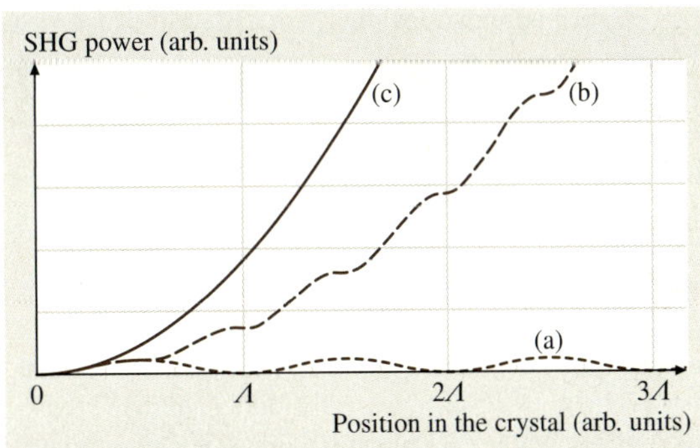

Fig. 8.59 (a) Efficiency of second-harmonic generation in a nonlinear crystal without phase matching, with quasi-phase matching **(b)**, and with perfect phase matching **(c)**. There Λ denotes the period length of the periodically poled nonlinear crystal

Table 8.4 Selected properties of frequently used nonlinear crystal materials utilized in periodically poled devices [8.169–171]. d_{im}: nonlinear optical coefficient; T: transparency range; opt. dam.: optical damage threshold for a given pulse duration

Crystal	d_{im} (pm/V)	T (μm)	Opt. dam. (GW/cm^2)
Lithium niobate ($LiNbO_3$)	$d_{33} = -27$	0.36–5.8	0.5 (10 ns)
Lithium tantalate ($LiTaO_3$)	$d_{33} = -21$	0.32–5.5	
Potassium titanyl phosphate (KTP, $KTiOPO_4$)	$d_{33} = 8.3$	0.34–3.2	4.4 (1.3 ns)
Beta barium borate (BBO, β-BaB_2O_4)	$d_{22} = 2.3$	0.2–2.6	9.9 (1.3 ns)
Cadmium selenide (CdSe)	$d_{33} = 36$	0.75–25	0.06 (10 ns)
Gallium arsenide (GaAs)	$d_{14} = 94$	0.9–17	

sion, i. e., in a nonlinear crystal, the orientation of the optical axis is periodically inverted. In ferroelectric materials this can be achieved by periodic inversion of the spontaneous polarization, i. e., of the ferroelectric domains (Fig. 8.58b). The period length Λ determines the wavelengths for which the specific nonlinear process is quasi-phase-matched. This Λ can be determined by an empirical relationship called the Sellmeier equation between the refractive indices n_i and wavelengths λ_i for a particular medium. Figure 8.59 schematically depicts the effect of QPM on the efficiency of second-harmonic generation in a nonlinear crystal.

The phase-matching condition for QPM and for the splitting of a pump wave into signal and idler waves is

$$\Delta k = k_p - k_s - k_i = 2\pi/\Lambda\,, \tag{8.88}$$

where k_p, k_s, and k_i are the wave vectors of the pump, signal, and idler waves, respectively, and Δk is the so-called phase mismatch, assuming that all wave vectors are collinear with the grating vector. Energy conservation requires $\omega_p = \omega_s + \omega_i$.

Frequently Used Materials

For practical applications plenty of different properties of nonlinear crystals are of importance, first of all of course the magnitude of the nonlinear coefficient $\chi^{(2)}$, particularly if the available optical intensities are low. However, the optical transparency for all the involved wavelengths is also of major importance, as well as the optical damage threshold. The latter is a measure of the maximum optical intensity for a given wavelength and pulse duration supported by the crystal without being destroyed. Although there has been substantial improvement in the past years, the optical damage resistance of nonlinear optical crystals is still a challenging research area [8.172–174]. Properties such as the material's potential to be periodically poled, the availability of crystals with consistently good quality, large size and a reasonable price and finally chemical durability may also influence the choice for a specific material. Table 8.4 lists the main properties of some relevant nonlinear crystals. A more-detailed list can be found in [8.169].

Very recently, a new type of materials, so called oriented-patterned (OP) materials have been developed, with gallium arsenide (GaAs) being the most promising example [8.175]. This is mainly due to its large nonlinear susceptibility, broad transparency range, low optical absorption, high thermal conductivity, high laser-damage threshold, and last but not least well-developed material technology.

8.6.2 Fabrication of Periodically Poled Structures

Depending on the application, periodically poled structures with period lengths of a few to some tens of microns are required. The necessary interaction length, i. e., the size of the periodically poled structure should reach several mm, over a thickness of some hundreds of microns to obtain reasonable output power. The latter specification can be circumvented by using waveguide techniques, thereby achieving a much higher interaction length for the confined light. Although advantageous because the required material length is shorter, and also because the periodically poled structures are necessary only at the surface layer, the crucial drawback of this approach lies in the optical damage threshold of the material, which is why higher output powers are generally achieved in bulk periodically poled crystals.

Electric Field Poling

Poling of domain patterns is usually achieved via ferroelectric domain engineering: applying electric fields of appropriate polarity and field strengths above the so-called coercive field E_c to cause domain reversal. The standard technique for the generation of periodic domain structures uses patterned electrodes generated using a photolithographic process (Fig. 8.60). This technique fails, however, for small poling periods ($\ll 5$ μm) in thick crystals (≥ 0.5 mm), because of an unavoidable spreading of the poled region with respect to the electrode size due amongst others to the enhanced electric fields at the electrode edges.

Corona Poling. To overcome these limitations, miniaturization of the electrodes is required. A possible solution relies on corona poling, thereby applying the required electric fields through an ultra-sharp tip (e.g., from a scanning force microscope) to the positions of the crystal to be poled [8.176]. A drawback of this technique is the sequential writing of domain patterns, which is therefore relatively slow. It is therefore preferred only for the generation of domain structures for storage applications [8.177].

E-Beam Poling. A similar technique uses the electron beam from an electron-beam lithography system to deposit electrons on top of the crystal, thereby generating the high electric fields required for poling. Gratings with periods as small as 1.6 μm across 500 μm-thick $LiNbO_3$ crystals have been obtained [8.178]. Even smaller structures of < 200 nm can be generated in thin (< 10 μm) liquid-phase epitaxy (LPE) $LiNbO_3$ films [8.179]. As for corona poling, this technique is slow because it is sequential and lacks a precise long-range periodicity due to stepping misalignment.

Light-Assisted Poling

To overcome these limitations on the depth of domain growth, a very promising new technique has been developed: light-induced poling. Controlled domain reversal is obtained by applying a homogeneous electric field slightly smaller than E_c in addition to a modulated UV-light intensity pattern (ideally generated by two interfering laser beams) [8.180]. Because in certain crystals (namely Mg-doped $LiNbO_3$) UV light reduces the coercive field [8.181], the light pattern is transferred into a domain pattern. In recent experiments, all-optical switching of ferroelectric domains has also been achieved, thereby avoiding the application of any electric field to the crystal.

Poled Growth

In some materials, namely $LiNbO_3$, poling of domain patterns during crystal growth has been successfully demonstrated with period lengths smaller than 4 μm [8.182, 183]. Although quite promising, the quality, namely the long-range periodicity of those periodically poled crystals, is not yet satisfactory, therefore this technique still needs improvement for wider application.

For the generation of orientation-patterned GaAs, all-epitaxial growth has been utilized. This process is based on photolithography and molecular-beam epitaxy to grow a thin-film template with the desired periodic crystal inversions. In a further step, a thick film is grown on this sample by hydride vapor-phase epitaxy to fabricate bulk orientation-patterned crystals [8.171].

8.6.3 Visualization of Ferroelectric Domain Structures

For the improvement of the quality of domain patterns a reliable, easy-to-use visualization technique is required. Freshly poled domain structures can be observed by means of optical microscopy due to stress-induced birefringence at the domain boundaries. The main drawbacks of this simple technique are its limited lateral resolution (a few microns) as well as the restriction to recently made domains. Among a multitude of visualization techniques [8.184], two standard methods are mostly applied, as discussed briefly below. Whereas the first is best suited for a large overview on the domain structures, the second enables an otherwise unreachable lateral resolution.

Domain-Selective Etching

This technique is based on the fact that the etch rate of a certain material depends on the orientation of the polar axis. As a consequence the domain configuration can be transferred into a topographical structure which can then be imaged by means of optical microscopy, or in the case of smaller domain patterns ($<$ micron-sized) as a topographical image by scanning force microscopy. Although it is destructive, this technique is still that most often used, mainly because of its ease of use [8.185].

Piezoresponse Force Microscopy (PFM)

This technique takes advantage of the fact that all crystals exhibiting ferroelectric domains are also piezoelectric. Furthermore, inverting the orientation of the spontaneous polarization also inverts the piezoelectric tensor element. This is why applying an electric field to a crystal reveals the domain structure as a topographical structure via the converse piezoelectric effect. In PFM, a standard scanning force microscope is modified to allow application of voltages to the tip, thereby generating an electric field inside the crystal at the position of the tip while it scans across the surface. Depending on the orientation of the ferroelectric domain, the crystal expands or contracts underneath the tip, thus allowing one to map the domain configuration with a lateral resolution of some 10 nm. The thickness changes to be measured are on the order of only some 10 pm and hence a frequency-selective detection scheme is used, applying an alternating voltage to the tip with subsequent lock-in detection [8.186].

8.6.4 Applications

In general periodically poled nonlinear materials are used to obtain quasi-phase-matching conditions for frequency conversion and therefore a higher output light power at the desired wavelength. Frequency conversion in nonlinear optical materials can be classified as follows:

- Second-harmonic generation (SHG): $\omega + \omega \rightarrow 2\omega$;
- Sum-frequency generation (SFG): $\omega_1 + \omega_2 \rightarrow \omega_3$;
- Difference-frequency generation (DFG): $\omega_3 - \omega_2 \rightarrow \omega_1$;
- Optical parametric generation (OPG): $\omega_3 \rightarrow \omega_1 + \omega_2$.

The current state of the art of the two main applications, frequency doubling via SHG and optical parametric generation as a source of infrared (IR), light will be presented in the following.

Second-Harmonic Generation (SHG)

Frequency doubling, thereby generating shorter wavelengths, is of special interest mainly because infrared light is relatively cheap, high-power IR laser diodes are available, howeven, laser light in the blue wavelength range is still difficult to obtain. For the experimental realization, there are several boundary conditions that need to be fulfilled. This is the reason why blue lasers based on frequency doubling in periodically poled nonlinear materials can only reach output powers of about 100 mW. Their main drawback currently lies in the optical damage threshold of the material. Note that the crystals need to have a high damage threshold for the pump as well as for the SHG light. Because the efficiency of SHG is proportional to I^2 of the IR pump light, a high power density inside the nonlinear medium is desired. From this point of view, waveguide techniques appear an ideal concept for efficient frequency doubling. However, optical damage of the crystals is still a crucial drawback. Currently, using MgO-doped periodically poled lithium niobate (PPLN), an output power of 25 mW (340 nm) out of 81 mW in the waveguide-based technique has been shown [8.187]. Typical values for an intracavity periodically poled potassium titanyl phosphate (PPKTP) frequency doubler, using a bulk crystal for frequency doubling, are 234 mW (461 nm) with 310 mW input power [8.188] or even 500 mW (473 nm) with 800 mW input power [8.189].

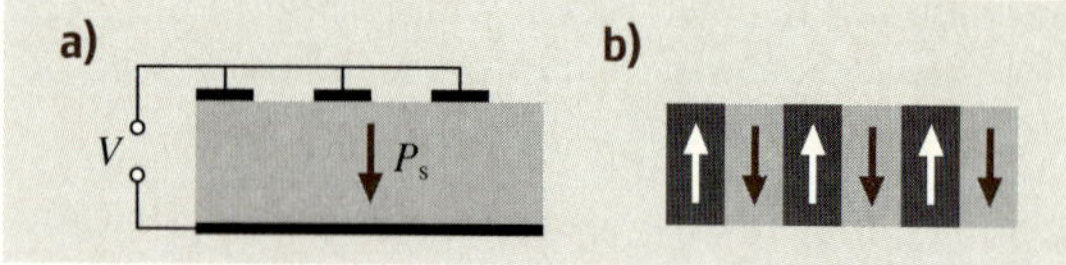

Fig. 8.60a,b Standard technique for the generation of periodically poled bulk nonlinear crystals. A single-domain crystal (spontaneous polarization P_s) is partially poled by applying a voltage through structured electrodes **(a)**. In **(b)** the resulting domain pattern is shown

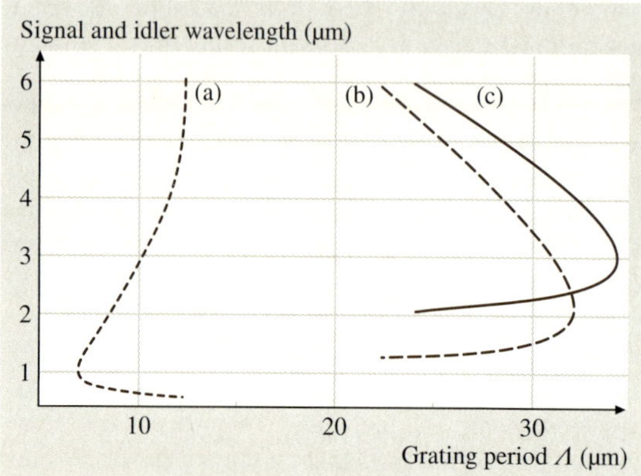

Fig. 8.61a–c Signal and idler output wavelengths of OPO operation for different grating periods Λ. The pump sources shown are **(a)**: Nd:YAG (2ω) laser – 532 nm; **(b)**: Nd:YAG laser – 1064 nm; **(c)**: Er-doped fiber laser – 1550 nm

Optical Parametric Generation

Many applications require laser sources in the mid-infrared wavelength between 2 μm and 20 μm, mainly for spectroscopic measurements of organic and inorganic molecules. A continuously tunable laser source with a narrow linewidth in this wavelength region is therefore highly desirable. A possible solution consists of using the OPG process, which allows the generation of the desired wavelength depending on the pump wavelength and the nonlinear material used. It can be tuned by choosing the phase-matching conditions, i. e., coarsely by the grating period Λ of the periodically poled nonlinear crystal and a fine adjustment by heating the crystal (Fig. 8.61). In order to make a powerful laser source, the nonlinear medium is brought into a cavity that is tuned on resonance for the non-desired wavelength. An overview of mid-IR frequency generation can be found in [8.169].

There are basically two different operation schemes for optical parametric oscillators (OPO). For continuous-wave operation a high pump-laser intensity is required to reach the threshold for frequency conversion. Another possibility consists of using an intracavity pumped setup, with the nonlinear medium placed inside the laser cavity.

Most OPOs, however, are pumped with nanosecond pulses from a Q-switched laser, thus easily reaching the required light intensities inside the nonlinear material. Typical emissions are nanosecond pulses in the near- or mid-infrared region with pulse energies up to a few millijoules.

8.7 Photonic Crystals

A photonic crystal (PC) is defined, in a narrow sense, as a crystal in which the dielectric constant (ε) periodically varies spatially in specific directions; for the reader's convenience we cite several textbooks and feature issues of journals for PCs [8.190–200]. In order to control the light field, the lattice constant (a) should be comparable to the wavelength λ of the relevant light. In the case where the variation is along one direction, it is called a 1-D PC. Similarly, we define 2-D and 3-D PCs, corresponding to the cases where ε varies along two and three independent directions, respectively.

Figure 8.62 shows schematics of representative examples. For a 1-D PC there is the well-known example of the multilayer dielectric film; in this review, however, there is little mention of 1-D PCs. In the case of 2-D PCs, shown in Fig. 8.62b, the intersection of air or dielectric rods with the plane perpendicular to the rod axis form a 2-D lattice; this plane is the 2-D PC plane. Let us consider the simple case of a 1-D PC. As is well known, states (or eigenmodes) of light in this PC with wave vector $\boldsymbol{k}$ parallel to this particular direction differ greatly from those in a uniform dielectric. Similarly, the respective eigenmodes of light with $\boldsymbol{k}$ in the 2-D plane for 2-D PC and with $\boldsymbol{k}$ in an arbitrary direction in the 3-D PC become unique compared to homogeneous 2-D and 3-D dielectrics. As for a 3-D PC, there are a variety of examples other than the two shown in Fig. 8.62c. An array of polystyrene spheres suspended in water, the first 3-D PC sample, was made as early as the 1960s and is the example shown on the right of Fig. 8.62c. Currently, a similar sample called an artificial opal crystal made of SiO_2 is also being intensively studied, together with so-called inverse opals.

The situation for the PC slabs shown in Fig. 8.63a also greatly differs from those of the corresponding 1-D

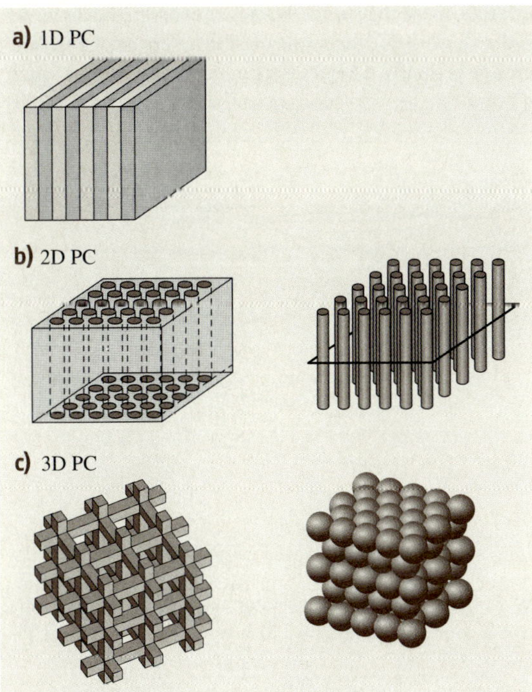

Fig. 8.62 Schematics of representative 1-D, 2-D and 3-D photonic crystals

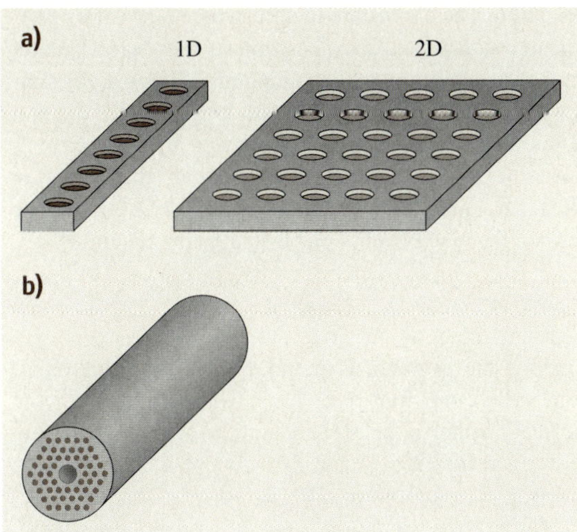

Fig. 8.63a,b Schematics of **(a)** photonic-crystal slabs and **(b)** photonic-crystal fiber

and 2-D PCs. Namely, if the thickness of the slab is of the same order as or smaller than the relevant wavelength of light, the situation is completely different from 1-D (2-D) PCs where the sample thickness is a priori assumed to be infinitely large. For the latter case, a plane wave propagating along one direction (1-D PC) or in the 2-D plane can be well defined, so that there is no need to treat a 2-D PC problem three-dimensionally. Figure 8.63b shows a schematic of a PC fiber, another important example of PCs where the plane perpendicular to the fiber forms a 2-D PC lattice except for the central portion along which light can propagate. There are two types of PC fibers: holey (index guiding) fibers, in which the central potion is made of a dielectric, and photonic band-gap fibers, in which it is filled with air.

Thus far we have classified PCs according to their dimensionality. However, there are related systems which are not covered by this classification based on the translational invariance. Quasi-crystal PCs having parallel rods arrayed on 2-D quasi-crystal lattice sites, and their relatives specified by well-controlled randomness of the rod sites are typical examples [8.201]. Since they are both characterized by the existence of a clear-cut photonic bandgap (PBG), we treat them as PCs in this article. Concerning fabrication methods of the representative samples in the optical region, see [8.200].

8.7.1 Photonic Band Structures

Light States in Photonic Crystal

In general the eigenstates (modes) of light are specified by a set $(\boldsymbol{k}, \hbar\omega, \boldsymbol{e})$, where $\boldsymbol{e}$ is the unit vector denoting polarization. The eigenstate in a PC where $\varepsilon(\boldsymbol{r})$ varies periodically with respect to $\boldsymbol{r}$ (position vector) differs greatly from that in a homogeneous medium. Suppose for simplicity that the dielectric (insulator) has the vacuum magnetic permeability μ_0. Assuming that the light is harmonic, i. e., $\boldsymbol{E}(\boldsymbol{r}, t) = \boldsymbol{E}(\boldsymbol{r})\exp(-\mathrm{i}\omega t)$ and $\boldsymbol{H}(\boldsymbol{r}, t) = \boldsymbol{H}(\boldsymbol{r})\exp(-\mathrm{i}\omega t)$, the equation, e.g., for $\boldsymbol{E}(\boldsymbol{r})$ can be derived from Maxwell's electromagnetic equations as

$$\mathrm{rot}\,\mathrm{rot}\,\boldsymbol{E}(\boldsymbol{r}) = [\omega^2\varepsilon(\boldsymbol{r})]\boldsymbol{E}(\boldsymbol{r})\,. \tag{8.89}$$

Namely, the problem is reduced to solving this equation, which was first done for a 3-D periodic array of dielectric spheres [8.202]. A similar equation can also be derived for $\boldsymbol{H}(\boldsymbol{r}, t)$, but $\boldsymbol{H}(\boldsymbol{r})$ is obtained from $\boldsymbol{E}(\boldsymbol{r})$ by using the relation, $\mathrm{rot}\,\boldsymbol{H}(\boldsymbol{r}) = -\mathrm{i}(\omega/c^2)\mu_0\varepsilon(\boldsymbol{r})\boldsymbol{E}(\boldsymbol{r})$.

The above equation can be solved by using, for example, the widely used plane-wave expansion method [8.197, 200]. Namely, the solution $\boldsymbol{E}_k(\boldsymbol{r})$ is expressed as a superposition of Bloch-type plane waves,

$$\boldsymbol{E}_k(\boldsymbol{r}) = \Sigma \boldsymbol{e}_k(\boldsymbol{G}_i)\exp[\mathrm{i}(\boldsymbol{k}+\boldsymbol{G}_i)\cdot\boldsymbol{r}]\,, \tag{8.90}$$

where $\boldsymbol{G}_i$ is the i-th reciprocal lattice vector, and the vector amplitude $\boldsymbol{e}_k(\boldsymbol{G}_i)$ is to be determined so as to satisfy the (8.89). Substituting (8.90) and the Fourier expansion version of $\varepsilon(\boldsymbol{r}) = \Sigma\varepsilon(\boldsymbol{G}_i)\exp(\mathrm{i}\boldsymbol{G}_i\cdot\boldsymbol{r})$ into (8.89) and setting the coefficients of each of the plane waves to zero, we get the eigenvalue equations for $\boldsymbol{e}_k(\boldsymbol{G}_i)$. From this equation, the secular equation for the eigenvalues ω^2 can be derived; needless to say, for each eigenstate characteristic spatial patterns of the electric $[\boldsymbol{E}_k(\boldsymbol{r})]$ and magnetic $[\boldsymbol{H}_k(\boldsymbol{r})]$ field are specified. We note that, in a 3-D PC with a sufficiently large variation of ε, i. e., $\varepsilon = 10$ against 1 (air), we often need to include more than 1000 plane waves to obtain very accurate values of ω^2; this is because the coupling between light and matter is much stronger in the optical (visible) region than in the X-ray region, because of the much stronger Bragg diffraction.

We plot the eigenfrequency (eigenenergy) $\omega(\boldsymbol{k})$ thus obtained as a function of $\boldsymbol{k}$ in the reduced Brillouin zone (BZ), which is called the *photonic band* (PB); The PB is split into many bands, which are numbered by the band index n in order of inverse energy. The photonic band structure (PBS) shown by $\omega_n(\boldsymbol{k})$ is unique and quite different from that in the homogeneous media in terms of the density of light states (DOS). A remarkable difference is that in a PBS a particular energy region often exists where no light states exist for any direction [8.203]

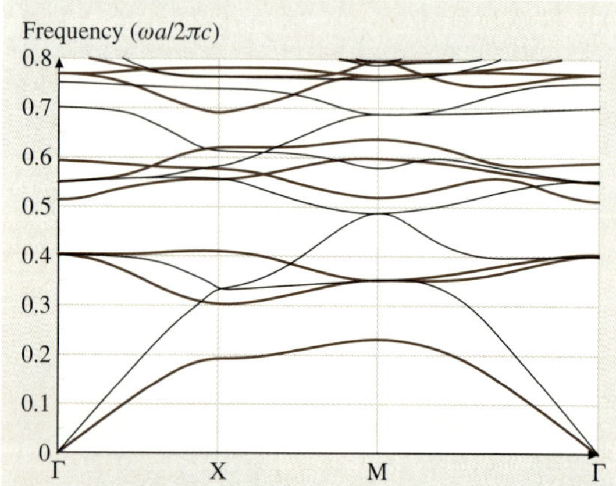

Fig. 8.64 Photonic band structure of a 2-D PC of cylinders arrayed in a square lattice: TM bands (*thick lines*) and TE bands (*thin lines*). The cylinders of $\varepsilon = 12$ are arrayed in a lattice in air. The ratio of the radius r of cylinders to a is 0.3. The frequency ω is shown in dimensionless units, i. e., by being normalized by $2\pi c/a$

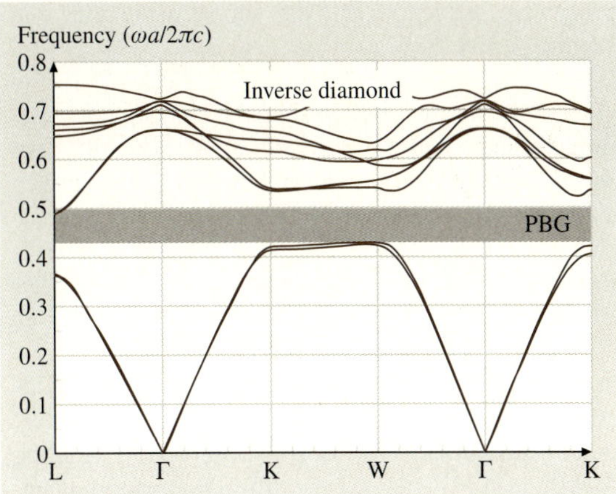

Fig. 8.65 3-D photonic band structure with the diamond lattice (a sample of inverse diamond structure; for details, see the text)

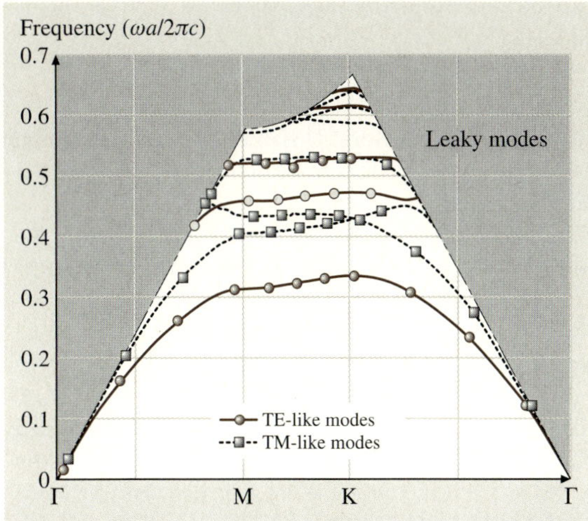

Fig. 8.66 Guided modes in a 2-D air-rod PC slab below the air-light line. *Filled* and *open circles* denote even and odd modes, respectively. The parameters are such that $\varepsilon = 11.9$, $r = 0.44a$ and $d = 2a/3$ where r and d refer to the hole radius and the slab thickness, respectively [8.204]

or for a range of directions often appears. We call the former a full (PBG) and the latter an incomplete photonic band gap (or stop band). Hereafter we call both of them simply as PBG when there is no need to emphasize the difference.

Here a question arises as to whether a PBS obtained for an infinitely large system makes sense in a realistic finite sample. This question is reasonable because the true bound states do not exist in the case of photons, in contrast to electrons which have definite bound states in the negative energy region. If the escape from a finite PC to the exterior region is large, the PB picture valid for an infinite system would lose its practical meaning. A pioneering work examined this question as early as the 1980s [8.205], revealing that the PBS in the ideal case can be used for a finite structure if the (lossless) eigenstate is regarded as leaky. Several important examples of PBS are presented below.

2-D Band Structure

Figure 8.64 shows an example of a 2-D PBS for a 2-D PC of dielectric cylinders ($\varepsilon = 12$) arrayed in a square lattice in air. The eigenmodes can be specified by their polarization, and are classified as TM or TE modes for $\boldsymbol{E}$ or $\boldsymbol{H}$ parallel to the cylinders, respectively; in this case, three 2-D PBGs are seen to open only for TM modes. Each band belongs to the specific irreducible representation of the relevant point group, which depends not only on the polarization, but also on the $\boldsymbol{k}$ direction. As a result, for example, if the group of a wave-vector $\boldsymbol{k}$ contains a mirror operation, then any state of that $\boldsymbol{k}$ is classified into even or odd symmetry according to whether $\boldsymbol{E}$ is symmetric or antisymmetric with respect to the mirror plane. This statement is important, for example, in coupling between external light and a PC state. Suppose an external light beam incident at normal incidence on a flat surface of a PC sample. Then, if the group of $\boldsymbol{k}$ with $\boldsymbol{k}$ the wave vector of incident light involves a mirror operation and the polarization is perpendicular to the mirror plane, the odd-parity (even-parity) modes cannot be coupled to even-parity (odd-parity) external light, and are called *uncoupled modes*. We note that the first and second bands are often called the dielectric and air band, respectively, since theelectric field is localized to a considerable extent around the dielectric cylinders and the air background, respectively [8.193].

3-D Band Structure

Figure 8.65 shows an example of a PBS for an inverse diamond lattice structure, which is composed of tetrabonds consisting of air rods (diameter r) in a dielectric of $\varepsilon = 7$; $r/a = 0.1$. The shape of the first BZ is the face-centered cubic (FCC) lattice. It is seen that a full PBG opens in all directions; the fact that a full PBG opens in a 3-D PC with a diamond structure has been experimentally evidenced in the optical region [8.199]. This PBG corresponds to the well-known BG of electrons in a semiconductor such as GaAs, which also has the same BZ structure. The photonic and electronic structures resemble each other in many respects, which arises from the resemblance of the two wave equations.

Band Structure of Slab-Type PCs

Air-hole semiconductor 2-D slab PCs have been studied intensively. In particular, two kinds of slab PCs are the most important: the air-bridge type, in which the core is suspended in air, and the clad type. A slab PC with the core made of AlGaAs or Si, and that of SiO_2/Si/SiO_2 are representative of the former and the latter, respectively. A typical PBS calculated for a triangular-lattice air-bridge PC is shown in Fig. 8.66.

The guided modes exist in the region below the so-called air light cone, where light can be confined due to total internal reflection [8.193]; light belonging to the modes inside this cone are leaky, i. e., light escapes from the slab region, just like the case for conventional slab waveguides (WGs). It is seen that for TE-like modes there is an energy region where no guided modes exist. Usually, in the case of slab PCs the term PBG is used for this situation. It is important to note that even in these 2-D PC slabs strong 3-D confinement of light is realized with a combined use of a 2-D PBG in the 2-D plane and total reflection in the direction perpendicular to the plane.

Band Structures of Other PCs

Metallic Photonic Crystal (MPC). In a MPC the plasmon modes of individual units can hop from one unit to another to form tight-binding plasmon bands. Due to the multipolar moments associated with the excited charge fluctuation, they can mix with light to become plasmon polariton bands. A large band gap, omnidirectional in favorable cases, originates from the large frequency difference between the different plasmon modes of individual units [8.206]. Thus, MPCs have potential applications in waveguiding and signal processing using their wide bandgaps. Furthermore, polariton bands in MPCs accompany a strong local field due to the charge fluctuation confined on the surfaces of the arrayed units. One can then obtain a strongly enhanced optical signal when a weak optical sign somehow excites plasmon polariton bands. Efficient biosensors or biomedical sensors can therefore be designed using MPCs. Another motivation for MPC research is provided by the strong transmission of light in an MPC. This is enabled by perforating a thick metal slab periodically with holes of subwavelength dimensions [8.207]. The use of a slab MPC with an array of holes has attracted renewed interest for the application of this extraordinary transmission to negative refraction and the superlensing effect in imaging [8.208, 209].

Quasi-PCs. Quasiperiodic systems, which are not translationally invariant in a strict sense, but still have long-range order, can have a PBG. Arrayed units stacked in a Fibonnaci sequence in 1-D or arranged by Penrose tiling in 2-D or 3-D space are examples of quasicrystals (QC). In QCs electrons or photons have a character intermediate between localization and delocalization. Therefore by developing a well-controlled quasi-PC, we can use well-localized states on one hand, which yields a very large quality factor when used, e.g., in lasing. At the same time, diffraction is observed by exciting a delocalized PB state, whose diffraction pattern is understood in terms of the reciprocal lattice of quasi-PCs. Using a delocalized state in lasing, the coexistence of the localization and delocalization characters has been demonstrated experimentally recently [8.210]. Besides Fibonnaci and Penrose lattices, several other relaated systems have been analyzed theoretically to demonstrate the presence of PBGs and a butterfly diagram in the BG [8.206].

How to Examine the Band Structure Experimentally

A simple and reliable method to investigate a sample's PBS experimentally is to observe either the transmittance (T) or reflectance (R_e) spectrum, or both in general, by varying the wavelength as well as the direction of incident light. To do this we need several samples with different surface normals. For wavelengths corresponding to a stop band (PBG), T drops greatly, whereas R should be unity, since there exist no photon modes inside in a particular direction for the external light to couple to. However, the reverse is not necessarily true. That is, observation that the external light is completely or totally reflected over a wavelength range does not necessarily indicate that the range corresponds to a stop band [8.200]. This is because complete reflection or drastic attenuation of T occurs for a portion of the uncoupled band; the extent to which external light couples to an individual coupled band depends on the character of the band itself.

8.7.2 Unique Characteristics

We summarize the unique and outstanding features that 2-D and 3-D PCs generally exhibit. PCs are well suited for controlling light, both the radiation field and light propagation characteristics. This important potential arises primarily from one or two of their features.

Photonic Band Gap

As seen in Fig. 8.64 there are three PBGs for TM modes, although a full 2-D PBG common to both polarized modes does not open in this case. It is known that a 2-D PBG common to TE and TM modes exists for a triangular lattice of air cylinders only if the ε difference between the background material and the air (cylinders or holes) is large enough [8.200]. For a 3-D PBG, the diamond structure is recognized to be the best for opening a full PBG with a large gap width, as shown in Fig. 8.65; an inverse diamond structure also belongs to the diamond structure. For other structures, it is rather difficult to realize a full 3-D PBG with a wide width. However, it is generally recognized that a variety of unique devices can be developed with a PC without using a 2-D or 3-D band gap, as will be shown below.

Defect Mode I. Classification

It is possible to introduce a defect or disorder at a particular lattice point or position, or thereby in an otherwise completely periodic PC lattice. Introducing this kind of defect creates a new eigenmode (eigenband) called a defect mode (defect band) in a PBG. An example of a zero-dimensional (0-D) or point defect in a 1-D or 2-D PC of air cylinders or dielectric pillars is to change the size (diameter) of the target cylinder/pillar compared to the host. One special case of this is to remove the cylinder or pillar completely. Another typical example of a 0-D defect in a 1-D PC is to alter the period at one lattice position, e.g. by changing the distance to the adjacent positions compared to that of the host. This kind of 0-D defect mode corresponds to an impurity state, analogous to a donor or acceptor state for electrons in a semiconductor; in the case of an air hole, increasing (decreasing) the hole size causes a decrease (increase) of the dielectric constant at that location, in good correspondence to the acceptor (donor) formation of electrons [8.193].

In addition to these 0-D defects, 1-D and 2-D types of defects, i.e., line and plane defects, can also be created and are important in controlling light. A 1-D defect in a 2-D PC of air cylinders or dielectric pillars can be created, for example, by leaving a single line of air cylinders imperforated, or removing a single line of pillars, i.e. introducing an empty (air) line, respectively. In a 3-D PC, e.g. made of air holes, a 1-D defect is such that a single line of holes are left unopened. A typical example of a 2-D defect is obtained, for example, by leaving one lattice plane structureless (homogeneous) in a 3-D PC.

Defect Mode II. Function

Let us discuss the physical properties of the defect modes. By exciting this mode, light can be localized around the defect, or the region with disorder.

(i) Microcavity. As the 0-D defect mode is localized in a small special region, the mode does not have a definite wave vector $\boldsymbol{k}$. This feature is similar to that for impurity states of electrons in the band gap. This kind of defect serves as an optical microcavity, exhibiting outstanding characteristics. Recently, various kinds of microcavities have been intensively explored using a semiconductor-based slab PC. As a result, a microcavity with a Q-value as large as, say, 3×10^5 but with an extremely small volume of the order of 3×10^{-14} cm^3 has even been developed [8.211]. Another type of superior microcavity or nanocavity has also been developed on the basis of the line-defect PC waveguide described below. Importantly, those extremely small microcavities provide strong coupling between light and matter, so that they have the potential to be used in physics and in application [8.212].

(ii) PC Waveguide. Unlike the point-defect case, the line-defect mode shows a particular dispersion in a PBG [8.193]. Let us consider an important application,

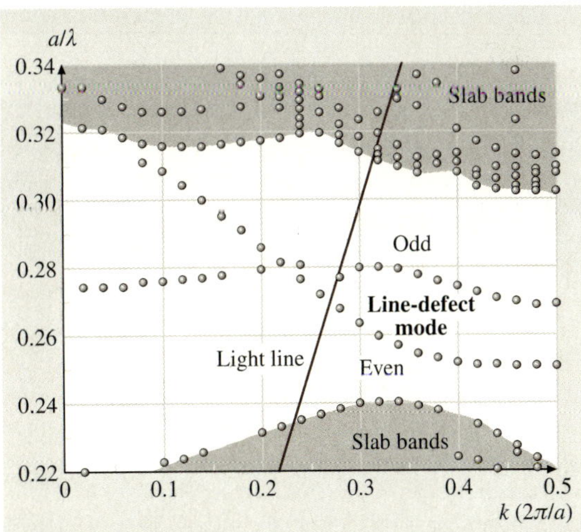

Fig. 8.67 Dispersion curves of a line-defect 2-D PC slab waveguide calculated by using the 3-D finite-difference time-domain (FDTD) method (for details, see the text): both the frequency ω and wavenumber k are shown in dimensionless units

i. e. a line defect in an air-bridge slab PC with air holes. The line defect is introduced by filling a single line of air holes along the $\Gamma - \text{K}$-direction (adopted as the x-axis) in the triangular lattice; the z-axis is normal to the 2-D plane formed by the x- and y-axes. Figure 8.67 shows the dispersion curves of the TE-polarized line-defect modes, where the frequency is plotted as a function of k_x only in the relevant region; remember that a large band gap opens for TE guided modes (Fig. 8.66). The hatched area, which are called the slab bands, are obtained by projecting all the band $\omega_n(k_x, k_y)$ in the first BZ onto the $\omega - k_x$ plane; one of the two modes is an *even mode*, characterized by its field profile symmetric with respect to the mirror plane (xz-plane) and the other mode having an antisymmetric profile is called an *odd mode*.

This line defect operates as a new type of waveguide (WG), known as a PC WG. It is easy to fabricate a bend WG in a 2-D plane by combining two line defects in directions that differ by 60° among six equivalent $\Gamma - \text{K}$ ones. Importantly, light belonging to the line-defect-guided modes propagates through a sharp corner without any loss [8.193]. Similarly, it is possible to design a stereoscopic PC WG using a combination of line defects with different directions in a 3-D PC.

Double-Heterostructured Microcavities. A line-defect slab PC WG can be designed as a heterostructure. This structure, e.g., in an air-hole PC WG, can easily be fabricated by introducing a somewhat larger lattice constant or slightly smaller air holes in a portion of the line defect. Double-heterostructured microcavities are such that the cavity position is sandwiched between two PC WGs. This serves as a microcavity because light with a frequency of $\omega_c^a > \omega > \omega_c^b$, where ω_c^a and ω_c^b are the cutoff frequency for the guided mode for the original and the sandwiched WGs, respectively, is localized in the particular portion [8.213].

Photonic-Crystal Fibers. We have already described two kinds of PC fibers, both of which were pioneered by *Russel* et al. [8.214, 215]. Those fibers have periodic structures in the cross-sectional plane and uniform structures along the long axis. The central portion of the holey fiber is a dielectric, while that of the PBG fiber is an air hole whose size is larger than the others. Therefore, the central portion is regarded as a kind of line defect introduced in the 2-D PC. The guided mode is confined there. The confinement is due to total reflection and due to the presence of a band gap for the two fibers, respectively. Note that in the holey fibers the average refractive index of the surrounding region is smaller than that of the core. Both fibers are known to exhibit superior characteristics.

Small Group Velocity

Both the group velocity $\boldsymbol{v}_\text{g}$ and the group-velocity dispersion (GVD) play an important role in light propagation. Theoretically, $\boldsymbol{v}_\text{g}$ is given by the slope $\nabla_{\boldsymbol{k}}\omega_n(\boldsymbol{k})$ of a band. Generally speaking, there are many portions or positions in a PC where the band slope is flat, which makes $\boldsymbol{v}_\text{g}$ anomalous. At the band edge $\boldsymbol{v}_\text{g}$ should always become extremely small. A position with such a flat energy dispersion is also encountered at anti-crossed (upper and lower) bands; a photon at those points is sometimes called a heavy photon in analogy with the heavy electron [8.216]. A relatively flat band with a small $\boldsymbol{v}_\text{g}$ is also frequently encountered in 2-D and 3-D PCs. Moreover, there are many bands where $\boldsymbol{v}_\text{g}$ becomes negative.

A small $\boldsymbol{v}_\text{g}$ indicates that the electric field strength becomes very large, or equivalently, that the DOS is very large. This fact is very important, since the effective interaction length between light and matter becomes longer due to the smallness of $\boldsymbol{v}_\text{g}$. More specifically, the enhanced light-matter interaction caused by the small $\boldsymbol{v}_\text{g}$ greatly intensifies, generally speaking, the radiation field involved in many optical processes in PCs. The examples are light absorption and emission, light scatterings such as Raman and Brillouin scattering, near-field phenomena near the PC surfaces, local field inducing an enhanced polarization and magnetization, and many nonlinear optical processes such as second-harmonic generation SHG). In all of them we can expect resonantly enhanced optical signals, often orders of magnitudes larger than those for a conventional sample of the same length. The small group velocity of light in PCs has indeed been one of the main stimulants of the PC research made until now. In the laser using PCs the threshold value of the electric current or fluence for pumping can be greatly reduced by this effect. However, the GVD is generally very large in a PC and this fact is disadvantageous for the propagation and processing of short light pulse.

Remarkable Polarization Dependence

In the case of 2-D PCs, and 2-D and 1-D PC slabs, the PBS is highly polarization dependent (see Figs. 8.64 and 8.66). Therefore, the propagation characteristics of light incident from outside differ remarkably for such a PC sample; in contrast, the polarization dependence is, generally speaking, less remarkable for 3-D PCs, in particular, with the diamond lattice. As a consequence, a unique polarizer can be developed by making use of a pair of specific bands for two different polarized light beams with the same frequency. A simple example is the case when there exists a stop band due to a PBG (or an uncoupled band) for one polarized beam, while there is a coupled band for the other.

Peculiar Band I. Anomalous Refraction

There are a variety of peculiar PBs or band points where anomalous features manifest themselves from the viewpoint of crystal optics; here we describe the cases of 2-D and 3-D PCs. However, we note that, even in a simple 1-D PC based on a periodic layered medium with each layer being isotropic, the whole structure is known to become anisotropic, i. e., birefringent [8.217]. Theoretically, the property of light in a PC as well as a homogeneous system is determined by the group of $\boldsymbol{k}$, i. e., the group of symmetry operations that leave the vector $\boldsymbol{k}$ in question at rest. The irreducible representations of the group of $\boldsymbol{k}$ thus entirely determine the anisotropic properties. From this group theoretical analog, PCs naturally embody all the optical properties exploited in crystal optics to date, often on a much larger scale. Furthermore, the band-folding effect operates in PCs in that the magnitude $|\boldsymbol{k}|$ also introduces remarkable optical diversity [8.216, 218].

In order to understand the propagation characteristics of light, we need to examine the equi-frequency (energy) surface (EFS) of frequency ω obtained by sweeping $\boldsymbol{k}$ over the entire BZ. The propagation characteristics of light at such a peculiar point are very complicated. The directions and magnitudes of the phase and group velocities are generally different from each other [8.216]. First of all, Snell's law of refraction does not hold in general in PCs. We list several other anomalies here.

1. Multi-refraction phenomena. This is because there exist several EFSs for a given ω in k space. As a result, if all of these are excited simultaneously by incident light of ω, refraction will be double, triple, quadruple, etc., according to the number of excited bands.
2. Conical refraction. There happens to be a singular point in $\boldsymbol{k}$ space where the group velocity, i. e., the vector normal to the EFS of the band, is not uniquely determined, e.g., at the intersections between two EFSs and at the point where an EFS shrinks to a point. In this phenomenon a light beam of that $\boldsymbol{k}$ propagates in the PC along the surface of a certain cone.
3. Superprism effects. Generally speaking, several bands participate simultaneously in constructing an EFS; there the direction of v_g, which is normal to the EFS, does not coincides with that of $\boldsymbol{k}$. A special case occurs such that the direction of v_g changes drastically when the direction of $\boldsymbol{k}$ is varied slightly. This is called the superprism effect [8.199].

Peculiar Band II. Negative Refraction and Left-Handed Properties

Left-handed (LH) material is usually defined to be the one having permeability μ and ε both negative [8.209]. When a LH system is homogeneous, a propagating plane-wave light of wave vector $\boldsymbol{k}$ has the LH property in the sense that the direction of the Poynting vector is reverse to that of $\boldsymbol{k}$ and, as a result, three vectors, i. e., electric field, magnetic field and $\boldsymbol{k}$, compose in this order the left-handed system. The LH system is well-defined in this way, when only a single $\boldsymbol{k}$ is involved. In a homogeneous LH system, a negative refraction takes place by necessity, in which an incident light beam is refracted at the entrance surface in the wrong way with respect to the surface normal. In the case of PCs, any PB state is composed of many plane waves whose wave vectors are mutually different. Therefore, both of the concepts of negative refraction and LH property have to be referred to the average of these plane waves. In this context we should note that the group velocity of the $(n, \boldsymbol{k})$ state, defined to be $\nabla_{\boldsymbol{k}}\omega_n(\boldsymbol{k})$, is just a quantity referring to this average. The direction of the group velocity is often quite different from that of $\boldsymbol{k}$ by the superprism effect mentioned above and the case of $\nabla_{\boldsymbol{k}}\omega_n(\boldsymbol{k})\cdot\boldsymbol{k} < 0$ is very frequently encountered. Therefore, the phenomenon of the negative refraction indeed occurs very often in PCs, in which the beam direction of an incident plane-wave light bends at the PC surface in the wrong way, the direction of the inside beam being an average of many plane waves composing an excited PB state. It is important to note that the negative refraction in PCs takes place in ordinary PCs, which have μ and ε both positive and is demonstrated by examining the inside beam direction using numerical simulation and experiment [8.219, 220]. Negative refraction is important in achieving superlensing effects in PCs [8.208]. The precise LH nature in PCs, on the other hand, is hard to define, because the direction of the inside beam is an average of many Poynting vectors. Some of them may compose a LH system with the wave vector $\boldsymbol{k}$, defined within the first Brillouin zone. But, as for the actual wave vector, taking an umklapp shift by a reciprocal lattice vector into account, that Poynting vector will certainly make up a normal right-handed system. In spite of this ambiguity, however, the negative refraction using PCs opens a way to pursue negative refraction and LH propagation in the optical region.

8.7.3 Applications

Light-Source Devices

To date, a variety of PC lasers have been developed with different lasing mechanisms and PC structures [8.221]. Most of them are based on PC slabs, because they are relatively easy to fabricate. These may be categorized into:

(1) point-defect lasers
(2) band-edge lasers

The concept of the lasers of category (1) was first put forward by *Yablonovitch* [8.203]. By using an ultrasmall point-defect optical cavity in a PBG, which enables ideal control over spontaneous emission, an ultra-low-threshold or threshold-less laser should be, in principle, possible. In fact, lasers of the literally point-defect type, lasers based on the double-heterostructured cavities described above, and stick resonators have already been developed. Lasers using quasi-crystal PCs may be considered to be a variation of category (1) laser.

The band-edge laser, i. e., category (2) laser, makes use of the small v_g and feedback mechanism arising from the Bragg diffractions contributing to the band edge; this feedback is 2-D, and therefore, this type of PC laser may be regarded as a 2-D laser, in contrast with the 1-D laser based on distributed feedback (DBF). In this case, there is essentially no need for a PC to have a PBG; in other words, even a weakly modulated PC suffices for this purpose. However, category (2) lasers utilizing a PBG-related band edge, in which the feedback is strong, has also been studied. A unique feature of this type PC laser is that coherent operation is possible over a wide 2-D area. In the case of the Γ point lasers using higher bands, the laser emission comes out in the vertical direction as a result of diffraction. This novel type of vertically emitting laser emits within a narrow solid angle and has the advantage that single-frequency operation is realized at high powers. Category (2) lasers were also developed based on line-defect PC WGs [8.222, 223]

As for the laser material, quantum wells (QWs, typically GaInAsP/InP), quantum dots (QDs, GaAs/InAs) and organic materials are employed by being embedded in a 2-D PC slab. In most cases, optical pumping is adopted, but in some cases, electric pumping is also employed.

It is now expected that 2-D PC slabs will be used to improve the extraction efficiency of light-emitting diodes (LEDs) that are embedded in other materials; the efficiency of light emitted to free space is usually poor (less than 10%), being restricted by, e.g., internal total reflection. Theoretically, a considerable improvement is expected by making use of leaky modes in a 2-D PC slab, but experimentally, for various reasons, the improvements achieved so far have been disappointing [8.221].

Other Optical and Communication Devices

Based on the unique characteristics of 2-D and 3-D PBSs described in Sect. 8.7.2, a variety of PC-based devices have been proposed; some of them are commercially available. Those include polarization beam splitters or polarization filters, wavelength filters, optical resonators, dispersion compensators, superprisms and superlenses. Some do not require a full PBG (3-D) or a 2-D PBG, while others, such as resonant wavelength filters using defect cavities, dispersion compensators based on coupled-cavity waveguides or polarization filters, do require a PBG. For example, a polarization-selective filter based on a 2-D PC allows vertically polarized light to be transmitted, while horizontally polarized light cannot pass because of the BG. A high performance with an extinction ratio of -50 dB can be attained. For these devices, the autocloning technology developed for the industrialization of PCs is highly suitable for the fabrication of large, homogeneous 2-D and 3-D PC samples [8.205].

Photonic-Crystal Fibers

PC fibers are commercially available. The holey fiber has unique and superior features [8.221]. The single-mode condition is maintained over a very wide frequency range from the visible to the near-infrared. Applying this condition to a small core size, strong confinement of light is realized. Making use of this feature, a variety of nonlinear optical phenomena such as Raman amplification, four-wave mixing, and the generation of supercontinuum radiation have been observed. Another feature is that a very large absolute value of positive and negative dispersion and a zero-dispersion wavelength can be arbitrarily achieved by designing the lattice and/or the air-hole diameter. These features are very attractive for novel types of optical fiber and devices, and PC fibers have already been used in place of conventional fibers. The PBG fiber also has unique features different from the holey fiber; their single-mode propagation and small bend radius possible with a large air-core size make these promising for high-power light transmission.

Planar Integrated Optical Circuits

Semiconductor-based PC slab WGs as well as PC slabs described in Sect. 8.7.1 and Sect. 8.7.2, respectively, are suited for the development of ultrasmall and ultrafast optical planar integrated circuits, in particular all-optical switches and optical buffer memory for future use in sub-terabit communication [8.200]. There are two reasons for this. One is that the PC slab WGs show superior light-propagation characteristics, particularly for an abruptly bent PC WG. The other reason is that good-quality samples can be fabricated by making use of state-of-the-art nanotechnology; besides, the fabrication process is rather simple. Currently research on this subject is therefore very active all over the world. It is remarkable that propagation loss can become as small as 0.5–0.7 dB/mm for straight PC WGs [8.224, 225]. To date, a variety of components including Y-splitters, WGs with intersections, directional couplers, dispersion compensators, and Mach–Zehnder-type switches have already been successfully developed [8.226].

Tunable Coherent Light Sources Based on the Smith–Purcel Effect

Electrons traveling along a straight trajectory are accompanied by a photon cloud, which decays exponentially away from the trajectory. The associated light can thus be observed only in the immediate neighborhood of the trajectory. If light scatterers are arrayed periodically within the decay range, the radiated photons suffers from umklapp scattering and become visible in the far field. The acquired momentum change shifts the phase-space point of the radiated photon from the outside to the inside of the light cone. This is called Smith–Purcell (SP) radiation. Usually SP radiation is observed using a metal grating. If a PC is used in place of a grating, the movement of the phase-space point is influenced by the presence of a band structure. In fact the SP spectrum, characterized by a series of sharp radiation signals, is observed by placing a PC near the high-energy electron beam and agrees satisfactorily with theoretical calculations [8.227, 228]. SP radiation is an interesting example of the use of PCs for electron–photon interactions.

8.7.4 Summary

The band structures of both bulk PCs and PCs with defects exhibit unique and marked features including the existence of PBGs, which are highly suitable for controlling the radiation field and light propagation characteristics. As a consequence, unique and novel devices are expected to be developed and currently, active research is being undertaken. This is particularly true for semiconductor-based PC slabs.

8.8 Optical Fibers

With the development of extremely low-loss optical fibers and their application to communication systems, a revolution has taken place during the last 35 years. In 2001, using glass fibers as the transmission medium and light waves as carrier waves, information was transmitted at a rate of more than 1 terabit per second (which is roughly equivalent to the transmission of about 15 million simultaneous telephone conversations) through one hair-thin optical fiber. Experimental demonstration of transmission at the rate of 14 terabits per second over a 160 km-long single fibers was demonstrated in 2006, which is equivalent to sending 140 digital high-definition movies in one second. This can be considered an extremely important technological achievement. In this section we will discuss the propagation characteristics of optical fibers with special applications to optical communication systems and also present some of their non-communication applications such as in sensing.

8.8.1 Historical Remarks

The idea of using light waves for communication can be traced as far back as 1880 when Alexander Graham Bell invented the photophone shortly after he invented the telephone in 1876. Actually according to reports (published in June 2002), an Italian immigrant, Antonio Meucci, was the inventor of the telephone; indeed Meucci demonstrated his *teletrfono* in New York in 1860. Alexander Graham Bell took out his patent 16 years later. This has apparently been recognized by the US Congress.

In the remarkable photophone experiment of Graham Bell, speech was transmitted by modulating a light beam, which traveled through air to the receiver. The transmitter consisted of a flexible reflecting diaphragm that could be activated by sound and was illuminated by sunlight. The reflected beam was received by a parabolic reflector placed at a distance. The parabolic reflector concentrated the light onto a photoconducting selenium cell, which formed part of a circuit with a battery and a receiving earphone. Sound waves present in the vicinity of the diaphragm vibrated the diaphragm, which caused a consequent variation of the light reflected by the diaphragm. The variation of the light falling on the selenium cell changed the electrical conductivity of the cell, which in turn changes the current in the electrical circuit. This changing current reproduces the sound on the earphone. This was the first experiment on optical communication. To quote from [8.229]: "*In 1880 he* (Graham Bell) *produced his "photophone" which to the end of his life, he insisted was " ... the greatest invention I have ever made, greater than the telephone ... ". Unlike the telephone it had no commercial value.*"

After this beautiful experiment by Alexander Graham Bell, not much work was carried out in the field of optical communications. This is because there was no suitable light source available that could reliably be used as the information carrier. The advent of lasers in 1960 immediately triggered a great deal of investigation aimed at examining the possibility of building optical analogues of conventional communication systems. The first such modern optical communication experiments involved laser-beam transmission through the atmosphere. However, it was soon realized that laser beams could not be sent in the open atmosphere for reasonably long distances to carry signals unlike, for example, microwave or radio systems operating at longer wavelengths. This is due to the fact that a light beam (of wavelength about 1 μm) is severely attenuated and distorted owing to scattering and absorption by the atmosphere. Thus for reliable long-distance light-wave communication in terrestrial environments it would be necessary to provide a transmission medium that could protect the signal-carrying light beam from the vagaries of the terrestrial atmosphere. In 1966, *Kao* and *Hockham* [8.230] made an extremely important suggestion; they noted that optical fibers based on silica glass could provide the necessary transmission medium if metallic and other impurities could be removed from the silica. In 1966, when the paper by Kao and Hockham was published, the most transparent glass available had extremely high losses (of more than about 1000 dB/km, implying a power loss by a factor of 100 in traversing only 20 m of the fiber); this high loss was primarily due to trace amounts of impurities present in the glass. Obviously this loss was too high even for short distances such as a few hundred meters. The 1966 paper of Kao and Hockham triggered the beginning of serious research into removing traces of impurities present in the glass, which resulted in the realization of low-loss optical fibers. In 1970, Kapron, Keck and Maurer (at Corning Glass, USA) were successful in producing silica fibers with a loss of about 17 dB/km at the helium–neon laser wavelength of 633 nm. Since then, technology has advanced with tremendous rapidity. By 1985 glass fibers were routinely being produced with extremely low losses: less than 0.25 dB/km, which corresponds to a transmission of more than 94% of the incident power after traversing 1 km of the optical fiber. Because of such low losses, the distance between two consecutive repeaters (used for amplifying and reshaping the attenuated signals) could be as large as 250 km.

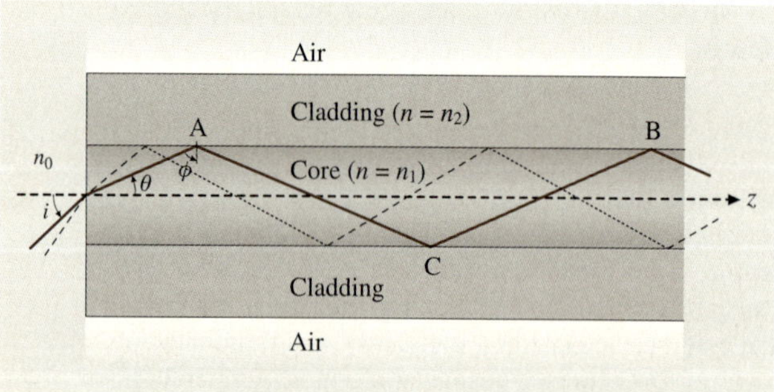

Fig. 8.68 Ray propagation through a step-index optical fiber

8.8.2 The Optical Fiber

At the heart of an optical communication system is the optical fiber that acts as the transmission channel carrying the light beam loaded with information. The light beam is guided through the optical fiber due to the phenomenon of total internal reflection (often abbreviated to TIR). Figure 8.68 shows an optical fiber, which consists of a (cylindrical) central dielectric core (of refractive index n_1) cladded by a material of slightly lower refractive index n_2 ($< n_1$). One usually defines a parameter $\Delta = (n_1 - n_2)/n_2$, which represents the fractional difference in the refractive index between the core and cladding. The necessity of a cladded fiber (Fig. 8.68) rather than a bare fiber, i. e., without cladding, was felt because of the fact that, for the transmission of light from one place to another, the fiber must be supported, and supporting structures may considerably distort the fiber, thereby affecting the guidance of the light wave. This can be avoided by choosing a sufficiently thick cladding.

When a light pulse propagates through an optical fiber it suffers from attenuation due to various mechanisms; also the pulse broadens in time, leading to what

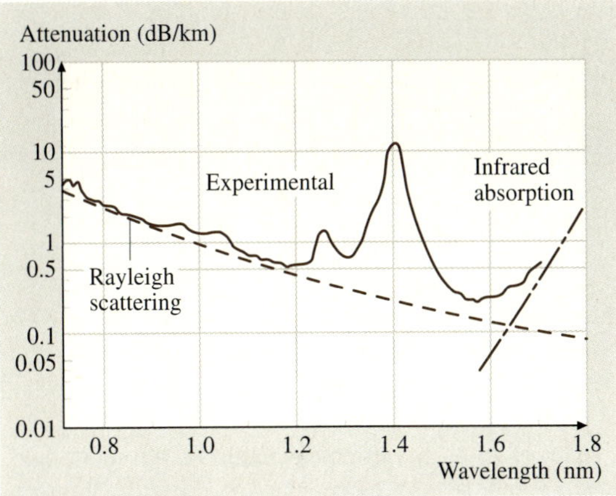

Fig. 8.69 Attenuation spectrum of a typical silica optical fiber (after [8.231])

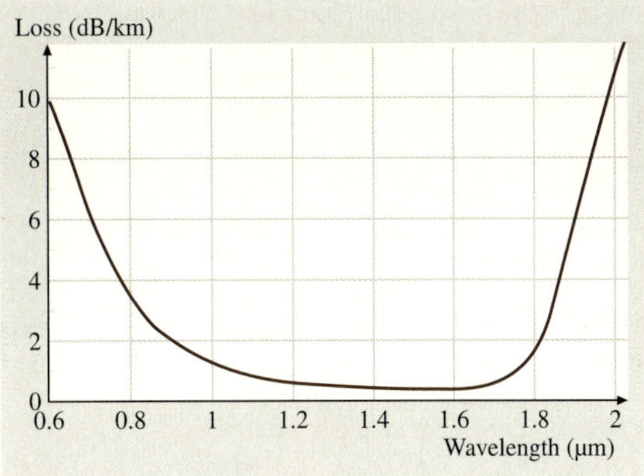

Fig. 8.70 Loss spectrum of an optical fiber with ultimately low water content. Note that the low-loss window extends from 1250 nm to 1650 nm ($\approx$ 50 THz) and such fibers are now available commercially (after [8.232])

is termed pulse dispersion. Apart from this, due to the high intensity of light present in the fiber, nonlinear optical effects come into play. Attenuation, pulse dispersion and nonlinear effects represent three of the most important characteristics that determine the information-transmission capacity of optical fibers. Obviously, the lower the attenuation (and similarly, the lower the dispersion and the smaller the nonlinear effects) the greater will be the required repeater spacing and therefore the higher the information-carrying capacity and the lower the cost of the communication system.

8.8.3 Attenuation in Optical Fibers

Losses in optical fiber are specified in terms of the unit dB/km, defined as

$$\alpha[\text{dB/km}] = \frac{10}{L[\text{km}]} \log\left(\frac{P_{\text{in}}}{P_{\text{out}}}\right) . \tag{8.91}$$

Here P_{in} and P_{out} are the input and output powers corresponding to an optical fiber of length L (km). Figure 8.69 shows a typical dependence of fiber attenuation coefficient α as a function of wavelength of a typical silica optical fiber [8.231]. It may be seen that the loss is about 0.25 dB/km at a wavelength of around 1550 nm. Various methods such as the modified chemical vapor deposition (MCVD) process or the outside vapor deposition (OVD) process allows one to fabricate such fibers with very low losses. The losses are caused due to various mechanisms such as Rayleigh scattering, absorption due to metallic impurities and water in the fiber, and due to intrinsic absorption by the silica molecule itself. The Rayleigh scattering loss varies as $1/\lambda_0^4$, i. e., shorter wavelengths scatter more than longer wavelengths. Here λ_0 represents the free-space wavelength. This is the reason why the loss coefficient decreases up to about 1550 nm. The two absorption peaks around 1240 nm and 1380 nm are primarily due to traces of OH^- ions and traces of metallic ions. For example, even 1 ppm (part per million) of iron can cause a loss of about 0.68 dB/km at 1100 nm. Similarly a concentration of 1 ppm of OH^- ions can cause a loss of 4 dB/km at 1380 nm. This shows the level of purity that is required to achieve low-loss optical fibers. If these impurities are completely removed, the two absorption peaks disappear (Fig. 8.70) and we will have very low losses over the entire range of wavelength starting from 1250 nm to 1650 nm [8.232]. In a typical commercially available fiber the loss is about 0.29 dB/km, 0.19 dB/km and 0.21 dB/km at $\lambda_0 = 1310$ nm, 1550 nm and 1625 nm respectively. Such fibers open up a bandwidth of more than 50 THz for communication. For $\lambda_0 > 1600$ nm the increase in the loss coefficient is due to the absorption of infrared light by the silica molecules. This is an intrinsic property of silica and no amount of purification can remove this infrared absorption tail.

As can be seen, there are two windows at which the loss attains its minimum value in silica fibers. The first window is around 1300 nm (with a typical loss coefficient of less than 1 dB/km) where fortunately (as we will see later) the material dispersion is negligible. However, the loss attains its absolute minimum value of about 0.2 dB/km around 1550 nm. The latter window has become extremely important in view of the availability of erbium-doped fiber amplifiers.

8.8.4 Modes of a Step-Index Fiber

The modal analysis of an optical fiber allows us to understand its propagation characteristics, which play an extremely important role in the design of a fiber optic communication system. The step-index fiber (Fig. 8.68) is characterized by the following refractive index distribution

$$\begin{aligned} n(r) &= n_1\,, && 0<r<a && \text{core}\,,\\ &= n_2\,, && r>a && \text{cladding}\,. \end{aligned} \quad (8.92)$$

In actual fibers $\Delta \ll 1$ and this allows the use of the so-called scalar wave approximation (also known as the weakly guiding approximation). In this approximation, the modal fields are assumed to be nearly transverse and can have an arbitrary state of polarization. Thus, the two independent sets of modes can be assumed to be x-polarized and y-polarized, and in the scalar approximation they have the same propagation constants. These linearly polarized modes are usually referred to as LP modes. In this approximation, the transverse component of the electric field (E_x or E_y) satisfies the scalar wave equation.

$$\nabla^2\Psi = \frac{n^2}{c^2}\frac{\partial^2\Psi}{\partial t^2}\,, \quad (8.93)$$

where c ($\approx 3\times10^8$ m/s) is the speed of light in free space. In most practical fibers n^2 depends only on the cylindrical coordinate r and therefore it is convenient to use the cylindrical system of coordinates and write the solution of (8.93) in the form

$$\Psi\,(r,\phi,z,t) = \psi\,(r,\phi)\;\mathrm{e}^{\mathrm{i}(\omega t-\beta z)}\,, \quad (8.94)$$

where ω is the angular frequency and β is known as the propagation constant. Equation (8.94) defines a propagating mode of the fiber. Since ψ depends only on r and ϕ, the modes represent transverse field configurations that do not change as they propagate through the optical fiber except for a phase change. From (8.94) we can define the phase velocity and group velocity of the mode as

$$v_\mathrm{p} = \frac{\omega}{\beta} = \frac{c}{n_\mathrm{eff}(\omega)} \quad \text{and} \quad v_\mathrm{g} = \left(\frac{\mathrm{d}\beta}{\mathrm{d}\omega}\right)^{-1}\,, \quad (8.95)$$

where

$$n_\mathrm{eff} = \frac{\beta}{k_0} \quad (8.96)$$

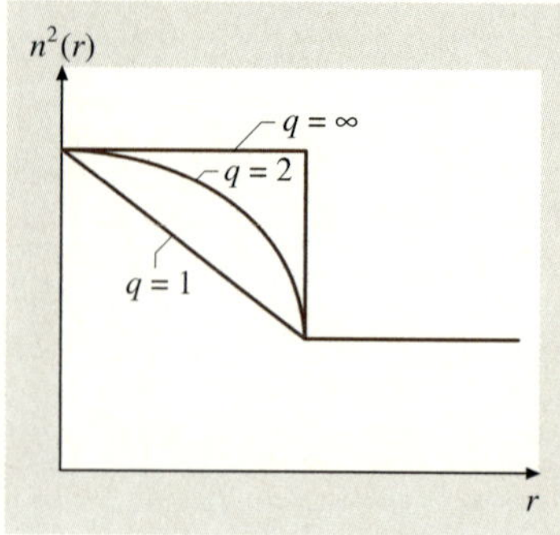

Fig. 8.71 Graded-index optical fiber with power-law profile

is referred to as the effective index of the mode and $k_0=\omega/c$ represents the free-space propagation constant.

Substituting for Ψ in (8.93) we obtain

$$\frac{\partial^2\psi}{\partial r^2} + \frac{1}{r}\frac{\partial\psi}{\partial r} + \frac{1}{r^2}\frac{\partial^2\psi}{\partial\phi^2} + \left[k_0^2 n^2(r) - \beta^2\right]\psi = 0\,. \quad (8.97)$$

Equation (8.97) can be solved by using the method of separation of variables and because the medium has cylindrical symmetry, the ϕ dependence will be of the form $\cos l\phi$ or $\sin l\phi$ and for the function to be single-valued [i. e., $\Phi\,(\phi+2\pi) = \Phi\,(\phi)$] we must have $l = 0, 1, 2, \ldots$ Thus the complete transverse modal field is given by

$$\Psi\,(r,\phi,z,t) = R(r)\,\mathrm{e}^{\mathrm{i}(\omega t-\beta z)} \begin{cases} \cos l\phi \\ \sin l\phi\,, \end{cases} \quad (8.98)$$

where $R(r)$ satisfied the radial part of the equation

$$r^2\frac{\mathrm{d}^2R}{\mathrm{d}r^2} + r\frac{\mathrm{d}R}{\mathrm{d}r} + \left\{\left[k_0^2n^2(r)-\beta^2\right]r^2 - l^2\right\}R = 0\,. \quad (8.99)$$

Before we write the solutions of this equation, we will make some general comments about the solutions of (8.99) for an arbitrary cylindrically symmetric profile having a refractive index that decreases monotonically from a value n_1 on the axis to a constant value n_2 beyond the core–cladding interface $r=a$ (Fig. 8.71). The solutions of (8.99) can be divided into two distinct classes: guided and radiation modes, which are described below.

Guided Modes

For β^2 satisfying

$$n_2^2 < \frac{\beta^2}{k_0^2} < n_1^2 \quad (8.100)$$

the fields $R(r)$ are oscillatory in the core and decay in the cladding and β^2 assumes only discrete values; these are known as the *guided modes* of the system. For a given value of l, there will be several guided modes, which are designated LP_{lm} modes ($m = 1, 2, 3, \ldots$); LP stands for linearly polarized. If one solves the vector wave equation, the modes are classified as HE_{lm}, EH_{lm}, TE_{0m}, and TM_{0m} modes, the correspondence is $\text{LP}_{0m} = \text{HE}_{lm}$; $\text{LP}_{1m} = \text{HE}_{2m}, \text{TM}_{0m}$, and $\text{LP}_{lm} = \text{HE}_{l+1,m}, \text{EH}_{l-1,m}$ ($l \geq 2$) [8.233]. Further, since the modes are solutions of the scalar wave equation, they can be assumed to satisfy the orthonormality condition

$$\int_{\sigma}^{\infty}\int_{0}^{2\pi} \psi_{lm}^{*}(r,\phi)\,\psi_{l'm'}(r,\phi)\,r\;\mathrm{d}r\;\mathrm{d}\phi = \delta_{ll'}\delta_{mm'}\;. \tag{8.101}$$

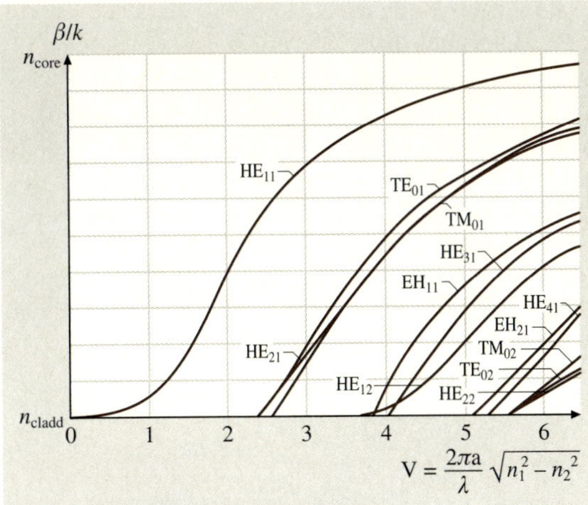

Fig. 8.72 Variation of effective index with V for a step-index fiber (after [8.234])

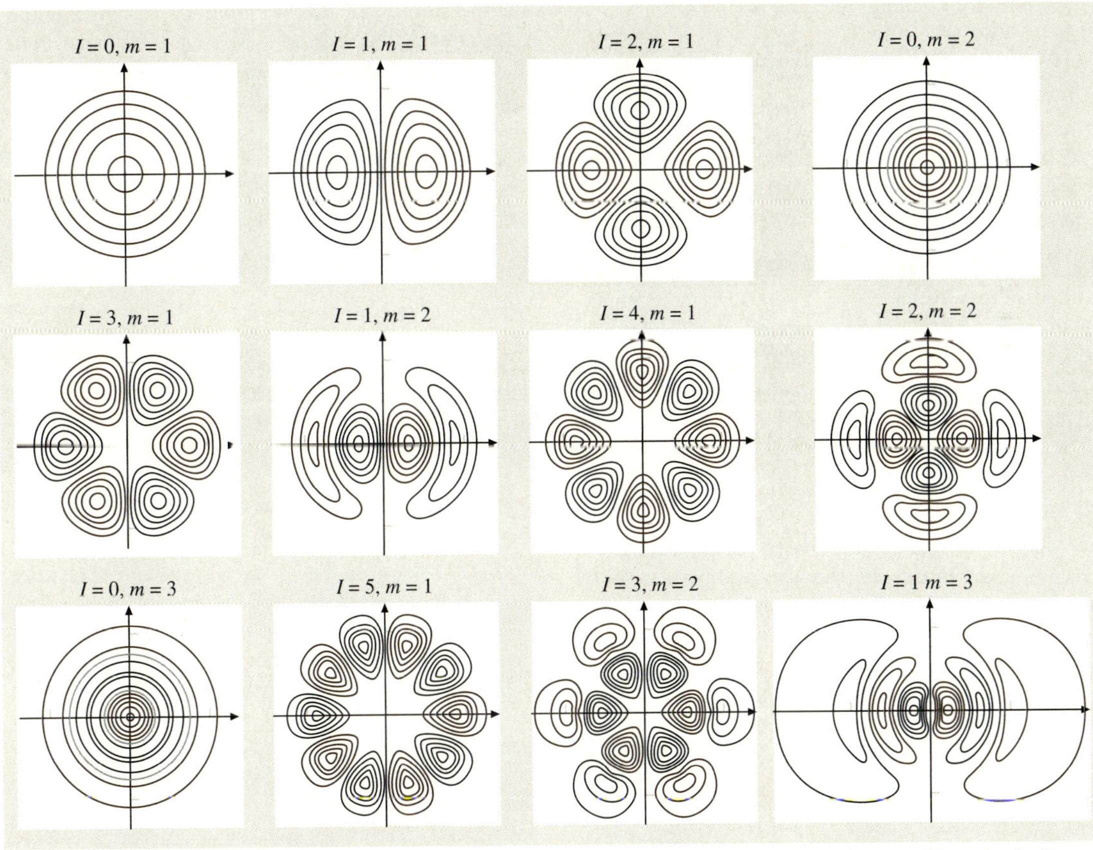

Fig. 8.73 Field patterns of some low-order guided modes (after http://www.rp-photonics.com/fibers.html, Rutger Paschotta, Date of last access: January 5, 2007)

Radiation Modes

For β values such that

$$0 < \frac{\beta^2}{k_0^2} < n_2^2 \qquad (8.102)$$

the fields are oscillatory even in the cladding and β can assume a continuum of values. These are known as the *radiation modes*.

Now, for a step-index profile, the well-behaved solution of (8.99) can be written in terms of the Bessel functions J_l and K_l:

$$\psi(r,\phi) = \begin{cases} \dfrac{A}{J_l(U)} J_l\left(\dfrac{Ur}{a}\right)\begin{pmatrix}\cos l\phi \\ \sin l\phi\end{pmatrix}, & r < a\,, \\ \dfrac{A}{K_l(W)} K_l\left(\dfrac{Wr}{a}\right)\begin{pmatrix}\cos l\phi \\ \sin l\phi\end{pmatrix}, & r > a\,, \end{cases} \qquad (8.103)$$

where A is a constant and we have assumed continuity of ψ at the core–cladding interface and

$$U \equiv a\sqrt{k_0^2 n_1^2 - \beta^2} \quad \text{and} \quad W \equiv a\sqrt{\beta^2 - k_0^2 n_2^2}\,. \qquad (8.104)$$

The normalized waveguide parameter V is defined by

$$V = \sqrt{U^2 + W^2} = k_0 a \sqrt{n_1^2 - n_2^2}\,. \qquad (8.105)$$

The waveguide parameter V (which also depends on the operating wavelength λ_0) is an extremely important quantity characterizing an optical fiber. For guided modes $n_2^2 k_0^2 < \beta^2 < n_1^2 k_0^2$ and therefore both U and W are real. It is convenient to define the normalized propagation constant

$$b = \frac{\beta^2/k_0^2 - n_2^2}{n_1^2 - n_2^2} = \frac{W^2}{V^2}\,. \qquad (8.106)$$

Thus $W = V\sqrt{b}$ and $U = V\sqrt{1-b}$. Using (8.100) we see that guided modes are characterized by $0 < b < 1$. Continuity of $\partial\psi/\partial r$ at $r = a$ and use of identities involving Bessel functions give us the following transcendental equations, which determine the allowed discrete values of the normalized propagation constant b of the guided LP_{lm} modes [8.235]:

$$\begin{aligned} &V(1-b)^{1/2} \frac{J_{l-1}\left[V(1-b)^{1/2}\right]}{J_l\left[V(1-b)^{1/2}\right]} \\ &= -Vb^{1/2} \frac{K_{l-1}\left(Vb^{1/2}\right)}{K_l\left(Vb^{1/2}\right)}\,, \quad l \geq 1 \end{aligned} \qquad (8.107)$$

and

$$\begin{aligned} &V(1-b)^{1/2} \frac{J_1\left[V(1-b)^{1/2}\right]}{J_0\left[V(1-b)^{1/2}\right]} \\ &= Vb^{1/2} \frac{K_1\left(Vb^{1/2}\right)}{K_0\left(Vb^{1/2}\right)}\,, \quad l = 0 \end{aligned} \qquad (8.108)$$

The solution of these equations will give us universal curves describing the dependence of b (and therefore U and W) on V. For a given value of l, there will be a finite number of solutions and the m-th solution ($m = 1, 2, 3, \ldots$) is referred to as the LP_{lm} mode. The variation of b with V forms a set of universal curves, which are plotted in Fig. 8.72 [8.234]. As can be seen, at a particular V value the fiber can support only a finite number of modes. Figure 8.73 shows the typical field patterns of some of the low-order LP_{lm} modes of a step-index fiber.

Guided and radiation modes form a complete set of solutions in the sense that any arbitrary field distribution in the optical fiber can be expressed as a linear combination of the discrete guided modes $\psi_j(x, y)$ and the continuum radiation modes $\psi(x, y, \beta)$:

$$\begin{aligned} \Psi(x, y, z) = &\sum a_j \psi_j(x, y) \mathrm{e}^{-\mathrm{i}\beta_j z} \\ &+ \int a(\beta)\psi(x, y, \beta)\mathrm{e}^{-\mathrm{i}\beta z}\,\mathrm{d}\beta \end{aligned} \qquad (8.109)$$

with $|a_j|^2$ proportional to the power carried by the j-th guided mode and $|a(\beta)|^2\,\mathrm{d}\beta$ proportional to the power carried by radiation modes with propagation constants lying between β and $\beta + \mathrm{d}\beta$. The constants a_j and $a(\beta)$ can be determined from the incident field distribution at $z = 0$.

8.8.5 Single-Mode Fiber (SMF)

As is obvious from Fig. 8.72, for a step-index fiber with $0 < V < 2.4048$ we will have only one guided mode – namely, the LP_{01} mode also referred to as the fundamental mode. Such a fiber is referred to as a single-mode fiber and is of tremendous importance in optical fiber communication systems. As an example, we consider a step-index fiber with $n_2 = 1.447$, $\Delta = 0.003$ and $a = 4.2\,\mu\mathrm{m}$, giving $V = 2.958/\lambda_0$, where λ_0 is measured in μm. Thus for $\lambda_0 > 1.23\,\mu\mathrm{m}$, the fiber will be single-mode. The wavelength for which $V = 2.4045$ is known as the *cut-off wavelength* and is denoted by λ_c. In this example, $\lambda_c = 1.23\,\mu\mathrm{m}$.

For a single-mode step-index fiber, a convenient empirical formula for $b(V)$ is given by

$$b(V) = \left(A - \frac{B}{V}\right)^2 , \quad 1.5 \lesssim V \lesssim 2.5 , \qquad (8.110)$$

where $A \approx 1.1428$ and $B \approx 0.996$.

Spot Size of the Fundamental Mode

The transverse field distribution associated with the fundamental mode of a single-mode fiber is an extremely important quantity and determines various important parameters such as splice loss at joints between fibers, launching efficiencies from sources, bending loss etc. For a step-index fiber one has analytical expression for the fundamental field distribution in terms of Bessel functions. For most single-mode fibers with a general transverse refractive-index profile, the fundamental mode field distributions can be well approximated by a Gaussian function, which may be written in the form

$$\psi(x, y) = A\,\mathrm{e}^{-\frac{x^2+y^2}{w^2}} = A\,\mathrm{e}^{-\frac{r^2}{w^2}} , \qquad (8.111)$$

where w is referred to as the spot size of the mode field pattern and $2w$ is called the mode field diameter (MFD). The MFD is a very important characteristic of a single-mode optical fiber. For a step-index fiber one has the following empirical expression for w [8.236]:

$$\frac{w}{a} \approx 0.65 + \frac{1.619}{V^{3/2}} + \frac{2.879}{V^6} , \quad 0.8 \leq V \leq 2.5 , \qquad (8.112)$$

where a is the core radius. As an example, for the step-index fiber considered earlier operating at 1300 nm we have $V \approx 2.28$ giving $w \approx 4.8\,\mu\mathrm{m}$. Note that the spot size is larger than the core radius of the fiber; this is due to the penetration of the modal field into the cladding of the fiber. The same fiber will have a V value of 1.908 at $\lambda_0 = 1550\,\mathrm{nm}$, giving a value of the spot size $\approx 5.5\,\mu\mathrm{m}$. *Thus, in general, the spot size increases with wavelength.* The standard single-mode fiber designated G.652 for operation at 1310 nm has an MFD of $(9.2 \pm 0.4)\,\mu\mathrm{m}$ and an MFD of $(10.4 \pm 0.8)\,\mu\mathrm{m}$ at 1550 nm.

For $V \geq 10$, the number of modes (for a step-index fiber) is approximately $1/2V^2$ and the fiber is said to be a multimoded fiber. Different modes (in a multimoded fiber) travel with different group velocities, leading to what is known as intermodal dispersion; in the language of ray optics, this is known as ray dispersion arising due to the fact that different rays take different amounts of time in propagating through the fiber. Indeed in a highly multimoded fiber, we can use ray optics to calculate pulse dispersion.

8.8.6 Pulse Dispersion in Optical Fibers

In digital communication systems, information to be sent is first coded in the form of pulses and then these pulses of light are transmitted from the transmitter to the receiver where the information is decoded. The larger the number of pulses that can be sent per unit time and still be resolvable at the receiver end, the larger will be the transmission capacity of the system. A pulse of light sent into a fiber broadens in time as it propagates through the fiber; this phenomenon is known as pulse dispersion and occurs primarily because of the following mechanisms.

1. In multimode fibers, dispersion is caused by different rays taking different times to propagate through a given length of the fiber. In the language of wave optics, this is known as *intermodal dispersion* because it arises due to different modes traveling with different group velocities.
2. Any given light source emits over a range of wavelengths and, because of the dependence of refractive index on wavelength, different wavelengths take different amounts of time to propagate along the same path. This is known as material dispersion and, obviously, it is present in both single-mode and multimode fibers.
3. In single-mode fibers, since there is only one mode, there is no intermodal dispersion. However, apart from material dispersion, we have what is known as waveguide dispersion. Physically, this arises due to the fact that the spot size (of the fundamental mode) depends explicitly on the wavelength.
4. A single-mode fiber can support two orthogonally polarized LP_{01} modes. In a perfectly circular core fiber laid along a perfectly straight path, the two polarizations propagate with the same velocity. However due to small random deviations from circularity of the core or due to random bends and twists present in the fiber, the orthogonal polarizations travel with slightly different velocities and get coupled randomly along the length of the fiber. This phenomenon leads to polarization mode dispersion (PMD) which becomes important for high-speed communication systems operating at 40 Gb/s and higher.

Obviously, waveguide dispersion and polarization mode dispersion are present in multimode fibers also – however, the effects are very small and can be neglected.

Dispersion in Multimode Fibers

A broad class of *multimoded* graded-index fibers can be described by the following refractive-index distribution:

$$\begin{aligned} n^2(r) &= n_1^2\left[1-2\Delta\left(\frac{r}{a}\right)^q\right], && 0<r<a,\\ &= n_2^2 = n_1^2(1-2\Delta), && r>a, \end{aligned} \tag{8.113}$$

where r corresponds to a cylindrical radial coordinate, n_1 represents the value of the refractive index on the axis (i.e., at $r=0$), and n_2 represents the refractive index of the cladding; $q=1, q=2$ and $q=\infty$ correspond to the linear, parabolic, and step-index profiles, respectively (Fig. 8.71). Equation (8.113) describes what is usually referred to as a power-law profile, which gives an accurate description of the refractive-index variation in most multimoded fibers. The total number of modes in a highly multimoded graded-index optical fiber characterized by (8.113) are approximately given by

$$N \approx \frac{q}{2(2+q)}V^2\,. \tag{8.114}$$

Thus, a parabolic-index ($q=2$) fiber with $V=10$ will support approximately 25 modes. Similarly, a step-index ($q=\infty$) fiber with $V=10$ will support approximately 50 modes. When the fiber supports such a large number of modes, then the continuum (ray) description gives very accurate results. For the power-law profile, it is possible to calculate the pulse broadening due to the fact that different rays take different amount of time in traversing a certain length of the fiber; details can be found in many text books; see for example [8.235]. The time taken to propagate through a length L of a multimode fiber described by a q-profile (8.113) is given by

$$\tau(L) = \left(A\tilde{\beta}+\frac{B}{\tilde{\beta}}\right)L\,, \tag{8.115}$$

where

$$A = \frac{2}{c(2+q)}\,,\qquad B = \frac{qn_1^2}{c(2+q)} \tag{8.116}$$

and for rays guided by the fiber $n_2<\tilde{\beta}<n_1$. Using (8.115) we can estimate the intermodal dispersion in fibers with different q values. Thus, for $n_1 \simeq 1.46$ and $\Delta \simeq 0.01$, the dispersion would be 50 ns/km for a step-index fiber ($q=\infty$), 0.25 ns/km for a parabolic index fiber ($q=2$) and for $q=2-2\Delta$ (referred to as the optimum profile exhibiting minimum dispersion) it will be 0.0625 ns/km.

Thus we find that for a parabolic-index fiber the intermodal (or ray) dispersion is reduced by a factor of about 200 in comparison to the step-index fiber and for the optimum profile there is a further reduction by a factor of four. It is because of this reason that first- and second-generation optical communication systems used near-parabolic-index fibers. In order to further decrease the pulse dispersion, it is necessary to use single-mode fibers because there will be no intermodal dispersion. However, in all fiber optic systems we will have material dispersion, which is a characteristic of the material itself and not of the waveguide; we will discuss this in the following section.

Material Dispersion

Above we have considered the broadening of an optical pulse due to different rays taking different amounts of time to propagate through a certain length of the fiber. However, every source of light has a certain wavelength spread, which is often referred to as the *spectral width of the source*. An LED would have a spectral width of about 25 nm and a typical laser diode (LD) operating at 1300 nm would have a spectral width of about 2 nm or less. The pulse broadening (due to wavelength dependence of the refractive index) is given in terms of the material dispersion coefficient D_m (which is measured in ps/(km nm)) and is defined by

$$D_m = -\frac{10^4}{3\lambda_0}\left(\lambda_0^2\frac{d^2n}{d\lambda_0^2}\right)\ \text{ps/(km nm)}\,. \tag{8.117}$$

λ_0 is measured in μm and the quantity inside the square brackets is dimensionless. Thus D_m represents the material dispersion in picoseconds per kilometer length of the fiber per nanometer spectral width of the source. At a particular wavelength, the value of D_m is a characteristic of the material and is (almost) the same for *all* silica fibers. When D_m is negative, it implies that the longer wavelengths travel faster; this is referred to as normal group velocity dispersion (GVD). Similarly, a positive value of D_m implies that shorter wavelengths travel faster; this is referred to as anomalous GVD.

The spectral width $\Delta\lambda_0$ of an LED operating around $\lambda_0 = 825$ nm is about 20 nm; at this wavelength for pure silica $D_m \approx 84.2$ ps/km nm. Thus a pulse will broaden by 1.7 ns per kilometer of fiber. It is interesting to note that, for operation around $\lambda_0 \approx 1300$ nm (where $D_m \approx 2.4$ ps/(km nm)), the resulting material dispersion is only 50 ps per kilometer of the fiber. The very small value of $\Delta\tau_m$ is due to the fact that the group velocity is approximately constant around $\lambda_0 = 1300$ nm. Indeed the wavelength $\lambda_0 \approx 1270$ nm is usually referred to as the zero-material-dispersion wavelength, and it is because of such low material dispersion that optical communication systems shifted their operation to around $\lambda_0 \approx 1300$ nm.

Optical communication systems in operation today use LDs with $\lambda_0 \approx 1550\,\mathrm{nm}$ having a spectral width of about 2 nm. At this wavelength the material dispersion coefficient is 21.5 ps/(km nm) and the material dispersion $\Delta\tau_m$ would be 43 ps/km.

Dispersion in Multimode Fibers

In multimode fibers the total dispersion consists of intermodal dispersion ($\Delta\tau_i$) and material dispersion ($\Delta\tau_m$) and is given by

$$\Delta\tau = \sqrt{(\Delta\tau_i)^2 + (\Delta\tau_m)^2} \quad (8.118)$$

In one type of extensively used coding referred to as non-return to zero (NRZ) the maximum permissible bit rate is approximately given by

$$B_{max} \approx \frac{0.7}{\Delta\tau} \quad (8.119)$$

Operation around 1310 nm minimizes $\Delta\tau_m$ and hence almost all multimode fiber systems operate in this wavelength region with optimum refractive-index profiles having small values of $\Delta\tau_i$.

Dispersion in Single-Mode Fibers

In the case of a single-mode optical fiber, the effective index $n_{eff} = \beta/k_0$ of the mode depends on the core and cladding refractive indices as well as the waveguide parameters (refractive-index profile shape and radii of various regions). Hence n_{eff} would vary with wavelength even if the core and cladding media were assumed to be dispersion-less (i. e., the refractive indices of core and cladding are assumed to be independent of wavelength). This dependence of effective index on wavelength is due to the wave-guiding mechanism and is referred to as *waveguide dispersion*. Waveguide dispersion can be understood from the fact that the effective index of the mode depends on the fraction of power in the core and the cladding at a particular wavelength. As the wavelength changes, this fraction also changes. Thus, even if the refractive indices of the core and the cladding are assumed to be independent of wavelength, the effective index will change with wavelength. It is this dependence of $n_{eff}(\lambda_0)$ that leads to waveguide dispersion.

Thus the total dispersion in the case of a single-mode optical fiber can be attributed to two types of dispersion, namely material dispersion and waveguide dispersion. Indeed it can be shown that the total dispersion coefficient D is given to a good accuracy by the sum of material (D_m) and waveguide (D_w) dispersion coefficients [8.237]. The material contribution is given by (8.117) while the waveguide contribution for a step-index fiber is given by

$$D_w = -\frac{n_2\Delta}{c\lambda_0}\left(V\frac{d^2(bV)}{dV^2}\right) \quad (8.120)$$

A simple empirical expression for waveguide dispersion for step-index fibers is

$$D_w = -\frac{n_2\Delta}{3\lambda_0}\times 10^7 \times\left[0.080 + 0.549(2.834 - V)^2\right]\mathrm{ps/(km\,nm)}\,, \quad (8.121)$$

where λ_0 is measured in nanometers.

In the single-mode regime, the quantity within the bracket in (8.120) is usually positive; hence the waveguide dispersion is negative. Since the sign of material

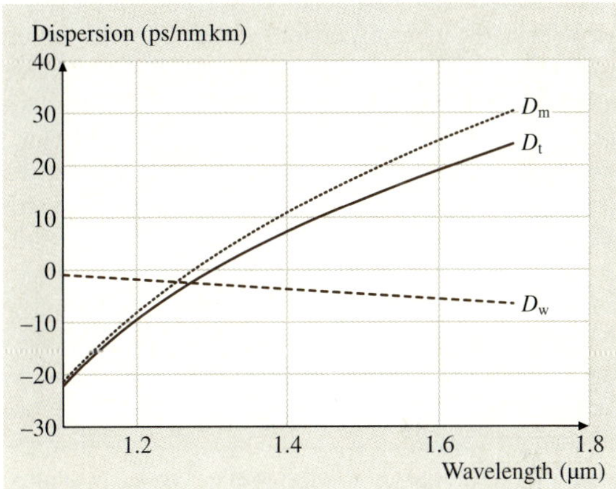

Fig. 8.74 Variation of material, waveguide and total dispersion with wavelength for a standard single-mode fiber

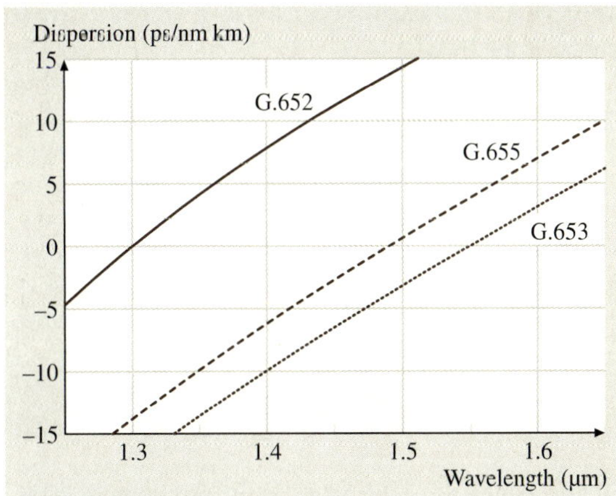

Fig. 8.75 Variation of total dispersion for three types of single-mode fibers

Table 8.5 Values of dispersion and dispersion slope for some standard fibers at 1550 nm (after [8.238])

Fiber type	D (ps/(km nm))	S (ps/(km nm^2))
Standard SMF (G.652)	17	0.058
LEAF (Corning)	4.2	0.085
Truewave-reduced slope (OFS)	4.5	0.045
TeraLight (Alcatel)	8.0	0.057

dispersion depends on the operating wavelength region, it is possible that the two effects namely, material and waveguide dispersions, cancel each other at a certain wavelength. Such a wavelength, which is a very important parameter of single-mode fibers, is referred to as the zero-dispersion wavelength (λ_{ZD}). For typical step-index fibers the zero-dispersion wavelength falls in the 1310 nm wavelength window. Since the lowest loss in an optical fiber occurs at a wavelength of 1550 nm and optical amplifiers are available in the 1550 nm window, fiber designs can be modified to shift the zero-dispersion wavelength to the 1550 nm wavelength window. Such fibers are referred to as dispersion-shifted fibers (with zero dispersion around 1550 nm) or nonzero-dispersion-shifted fibers (with finite but small dispersion around 1550 nm). With proper fiber refractive-index profile design it is also possible to have a flat dispersion spectrum, leading to dispersion flattened designs. Figure 8.74 gives the spectral variations of a standard SMF (with zero dispersion close to 1310 nm) while Fig. 8.75 shows the total dispersion in three standard types of fibers namely G.652, G.653 and G.655 fibers. The G.655 fibers have a small but finite dispersion around the 1550 nm wavelength. The small dispersion is required to avoid four-wave mixing.

It appears that when an optical fiber is operated at the zero-dispersion wavelength, the pulses will not suffer any dispersion at all. In fact zero dispersion only signifies that the second-order dispersive effects are absent. In this case the next-higher-order dispersion namely third-order dispersion characterized by $\mathrm{d}^3\beta/\mathrm{d}\omega^3$ will become the dominant term in determining the dispersion. Thus in the absence of second-order dispersion, we can write for dispersion suffered by a pulse as

$$\Delta\tau = \frac{L\,(\Delta\lambda_0)^2}{2}\frac{\mathrm{d}D}{\mathrm{d}\lambda_0}\,, \tag{8.122}$$

where $S = \mathrm{d}D/\mathrm{d}\lambda_0$ represents the dispersion slope at the zero-dispersion wavelength and is measured in units of ps/km nm^2. Third-order dispersion becomes important when operating close to the zero-dispersion wavelength. In the presence of only third-order dispersion, the pulse does not remain symmetric. Table 8.5 lists values of D and S for some standard fibers at 1550 nm.

Dispersion and Maximum Bit Rate in Single-Mode Fibers

In a digital communication system employing light pulses, pulse broadening would result in an overlap of adjacent pulses, resulting in inter-symbol interference leading to errors in detection. Apart from this, since the energy in the pulse gets reduced within the time slot, the corresponding signal-to-noise ratio (SNR) will decrease. One can offset this by increasing the power in the pulses. This additional power requirement is termed the *dispersion power penalty*. Increased dispersion would imply an increased power penalty.

In order to keep the interference between adjacent bits below a specified level, the root-mean-square width of the dispersed pulse needs to be kept below a certain fraction ε of the bit period. For a 2 dB power penalty, $\varepsilon \approx 0.491$ [8.239]. Using this condition we can estimate the maximum bit rate B for a given link length L and dispersion coefficient D operating at 1550 nm as

$$B^2 DL < 1.9\times10^5 \mathrm{Gb}^2\mathrm{ps/nm}\,, \tag{8.123}$$

where B is measured in Gbps, D in ps/(km nm) and L in km. Thus for a bit rate of 2.5 Gb/s the maximum allowed dispersion ($D\,L$) is approximately 30 400 ps/nm while for a bit rate of 10 Gb/s the maximum allowed dispersion is 1900 ps/nm.

Dispersion-Compensating Fibers

There already exist millions of kilometers of conventional single-mode fibers in underground ducts operating at 1310 nm and these fibers have very low dispersion at the operating wavelength. One could significantly increase the transmission capacity of these system by operating these fibers at 1550 nm (where the loss is extremely small) and we can have the added advantage of using erbium-doped fiber amplifiers (EDFAs) for optical amplification in this wavelength range. However, if we operate conventional single-mode fibers at 1550 nm, we will have a significant residual dispersion of about 17 ps/(km nm). Such a large dispersion would result in a significant decrease in the information-carrying capacity of the communication system. On the other hand, replacing the existing conventional single-mode fibers by lower-dispersion fibers would involve huge costs. As such, in recent years there has been considerable amount of work in upgrading of the installed 1310 nm optimized optical fiber links for operation at 1550 nm. This is achieved by developing fibers with very large negative dispersion coefficients, a few hundred meters to a kilometer of which can be used to compensate for dispersion accumulated over tens of kilometers of the fiber in the link.

By changing the refractive-index profile, one can alter the waveguide dispersion and hence the total dispersion. Indeed, it is possible to have specially designed fibers whose dispersion coefficient (D) is large and negative at 1550 nm. These types of fibers are known as dispersion-compensating fibers (DCFs). A short length of a DCF can be used in conjunction with the 1310 nm optimized fiber link so as to have small total dispersion value at the end of the link. There are a number of different fiber designs with optimized refractive-index profiles which have extremely large negative dispersion coefficient and a small length of which can compensate for the accumulated dispersion of a link. Some of the important ones include depressed clad designs, W-type fiber designs and dual-core coaxial designs [8.240–243].

If $D_T(\lambda_n)$ and L_T represent the dispersion coefficient and length of the transmission fiber and $D_C(\lambda_n)$ and L_C represent the corresponding quantities of the DCF, then to achieve zero net dispersion at a chosen wavelength λ_n we must have

$$D_T(\lambda_n)L_T + D_C(\lambda_n)L_C = 0 \qquad (8.124)$$

Hence for given values of $D_T(\lambda_n)$, L_T and $D_C(\lambda_n)$, the length of the DCF required is given by (8.124) showing that $D_C(\lambda_n)$ and $D_T(\lambda_n)$ should have opposite signs. Also the larger the value of $D_C(\lambda_n)$ the smaller the length of the required DCF. Since the wavelength dependence of dispersion of the link fiber and the dispersion-compensating fiber are in general different, in general the DCF would compensate for dispersion only at the design wavelength. However in a wavelength division multiplexed system it is necessary to compensate for the accumulated dispersion of all the wavelength channels simultaneously. For this to happen the relative dispersion slope (RDS), which is the ratio of the dispersion slope (S) and the dispersion coefficient (D) of the two fibers evaluated at the wavelength λ_n, must be equal. Typically the RDS of a G.652 fiber is about $0.0034\,nm^{-1}$ while that of large-effective-area (LEAF) fiber from Corning is about $0.0202\,nm^{-1}$. DCFs with similar RDS values are commercially available.

Polarization Mode Dispersion (PMD)

A perfectly circular single-mode fiber actually supports two orthogonally polarized modes with equal propagation constant. When such a fiber is laid along a perfect straight path, the two orthogonally polarized fundamental modes of the fiber have identical group velocities and any pulse launched into the fiber undergoes dispersion only due to material dispersion and waveguide dispersion effects. However in actual fibers, there is always a very small ellipticity of the fiber cores or there are nonsymmetric strains in the fiber or the fibers have bends and twists when laid in the link; this causes a difference of velocities between the two orthogonal polarized modes. This difference leads to the phenomenon of polarization mode dispersion. The effect of PMD is as if the fiber supported two modes with two different velocities; thus a pulse launched with some arbitrary polarization state would split into two pulses due to the two different velocities. Locally PMD is due to differential velocities of the two polarizations while globally it is combined with random polarization coupling along the fiber length. Since the fiber property as well as the external perturbation is random along the length of the fiber, this effect is a random phenomenon. Due to the random nature, the PMD effect increases as the square root of length rather than as length. PMD is usually measured in terms of the differential group delay (DGD) between the two polarizations. In order that PMD does not cause increased bit error rates, the differential group delay should be less than 10% of the bit period. Thus for a 2.5 Gb/s system, which consists of pulses of duration 400 ps, the max-

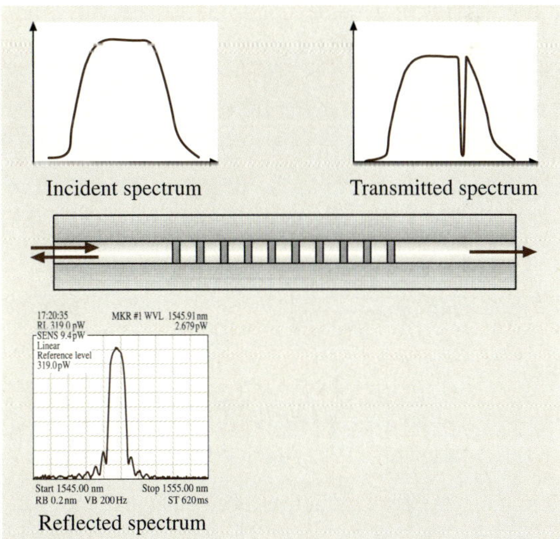

Fig. 8.76 The transmitted and reflected spectrum from an FBG. The reflected spectrum is an actual measured spectrum

imum allowed PMD is 40 ps and for a 40 Gb/s system the maximum allowed PMD is only 2.5 ps. If the PMD of the fiber is $0.5\,\text{ps/km}^{0.5}$, then for a 40 Gb/s link the maximum length due to PMD effects will be 25 km. Thus PMD effects become extremely important for systems operating at high bit rates, typically 40 Gb/s and higher. Commercially available single-mode fibers have typically PMD of less than $0.2\,\text{ps/km}^{0.5}$. In an actual system, a contribution to PMD also comes from other components used in the link, such as optical amplifiers, dispersion-compensating elements.

8.8.7 Fiber Bragg Gratings

When a germanium-doped silica-core fiber is exposed to ultraviolet radiation (with wavelength around 0.24 μm), the refractive index of the germanium-doped region increases; this is due to the phenomenon known as *photosensitivity* which was discovered by Kenneth Hill in 1974. The refractive index increase can be as large as 0.001 in the core of the fiber. If the fiber is exposed to a pair of interfering UV beams, then in regions of constructive interference, the refractive index increases while it does not change in regions of destructive interference. This results in a periodic variation of refractive index within the core of the fiber and this is referred to as a fiber Bragg grating (FBG). The period of the grating can be controlled by appropriately choosing the angle between the interfering beams. If we consider a polychromatic beam incident on the fiber, as shown in Fig. 8.76, then the reflection from the periodic structure will add up in phase when the following Bragg condition is satisfied:

$$\Lambda = \frac{\lambda_B}{2n_{\text{eff}}}\,, \tag{8.125}$$

where λ_B is referred to as the Bragg wavelength. The corresponding peak reflectivity R_p and bandwidth $\Delta\lambda$ are given by [8.233]

$$R_p = \tanh^2\left(\frac{\pi\,\Delta n\,LI}{\lambda_B}\right) \tag{8.126}$$

and

$$\frac{\Delta\lambda}{\lambda_B} \approx \frac{\lambda_B}{n_{\text{eff}}L}\left[1+\left(\frac{(\Delta n)LI}{\lambda_B}\right)^2\right]. \tag{8.127}$$

Here Δn is the peak refractive-index change in the grating and L is the grating length. The quantity I represents the transverse overlap integral of the modal distribution with the region where the grating is formed. It accounts for the fact that only a fraction of the mode interacts with the grating. If the modal distribution is assumed to be described by a Gaussian with a spot size of w_0 and if the grating is assumed to exist only within the core of radius a, then $I = (1 - \mathrm{e}^{-2a^2/w_0^2})$. For a typical single-mode fiber operating at 1550 nm, $n_{\text{eff}} \approx 1.4475$ and the required grating period is about 0.54 μm. Figure 8.76 also shows a typical measured reflection spectrum of an FBG. For a grating with $\Delta n = 4\times10^{-4}$ and $L = 2\,\text{mm}$, the peak reflectivity would be 0.86 and the bandwidth would be about 1 nm.

Some Applications of FBGs

Fiber Bragg gratings find many applications in telecommunication and sensing [8.244, 245]. These include applications in fiber grating sensors, add/drop multiplexers, to provide external feedback for laser-diode wavelength locking, and dispersion compensation. Here we discuss two applications.

FBG-Based Sensors. FBGs have a great potential for applications as sensors for sensing mechanical strain, temperature, acceleration etc. Since the Bragg wavelength depends on both the refractive index of the fiber as well as on the period of the grating, any external parameter that changes any of these would result in a change in the reflected wavelength. Thus by measuring the *changes in the reflected wavelength*, the external perturbations affecting the grating can be sensed. This is the basic principle of their application in sensing. The typical strain sensitivity of an FBG is about 1.3 pm/με at 1550 nm. Since the refractive index as well as the period of the grating change with change in temperature, the temperature change of an FBG would also result in a change in the peak reflected wavelength. The temperature sensitivity of FBGs is typically about 6 pm/°C. One of the important attributes of FBG sensors is that they can be multiplexed into a single fiber. FBGs with different Bragg wavelengths are fabricated at different points along the length of a single-mode fiber. Light from a broadband source is coupled into the fiber and light at different wavelengths get reflected from the individual gratings and are analyzed by the detection circuit. The wavelengths of the FBGs are so chosen that they do not overlap with each other and fall within the band of the source. By measuring the changes in the Bragg wavelength of individual FBGs, the strains or temperature changes at each of the location of the FBG can be independently measured. Among the various issues in connection with FBG sensors, is the problem of separating the changes brought about by temperature and strain. Fiber optic sensing technology is rapidly advancing and

FBGs are now being tested for structural monitoring. In the future it may be possible to multiplex perhaps 100 sensors on a single fiber, which gives enormous capability for structural monitoring.

Dispersion Compensation. Fiber Bragg gratings can also be used for dispersion compensation. In a uniform FBG, the period of the refractive-index modulation is constant along the length of the grating. If the period of the grating varies along its length, this is referred to as a *chirped fiber Bragg grating*. In such chirped FBGs since the period of the grating varies along the length, the Bragg wavelength also varies along the position in the grating (Fig. 8.77). When light propagates through such a grating, different wavelength components present in the incident wave will get reflected at different positions along the grating; this will lead to different wavelength components having different time delays to return to the input end. By using an appropriately chirped FBG one can indeed compensate for the differential delay of different wavelengths accumulated while propagating through an optical fiber link. Let us consider propagation of a pulse of light through an optical fiber operating at a wavelength longer than the zero-dispersion wavelength of the fiber. This would correspond to say operating a G.652 fiber (having a zero dispersion at 1310 nm) at a wavelength of 1550 nm. Thus if we consider three wavelength components such that $\lambda_1 > \lambda_2 > \lambda_3$ contained within the pulse, then, we can see that, due to dispersion in the fiber, wavelength λ_1 would suffer a larger delay than wavelength λ_2, which in turn will suffer a delay longer than λ_3 while propagating through the fiber. In order to compensate for this dispersion, we need to delay the wavelength component λ_3 more than the component λ_2, which in turn should suffer a delay more than at λ_1. In order to achieve this, the chirped grating is designed so that wavelength λ_1 reflects from the near end of the grating, λ_2 reflects from a portion farther away while λ_3 reflects from the far end so as to compensate for the differential delay between all the wavelength components and thus lead to dispersion compensation.

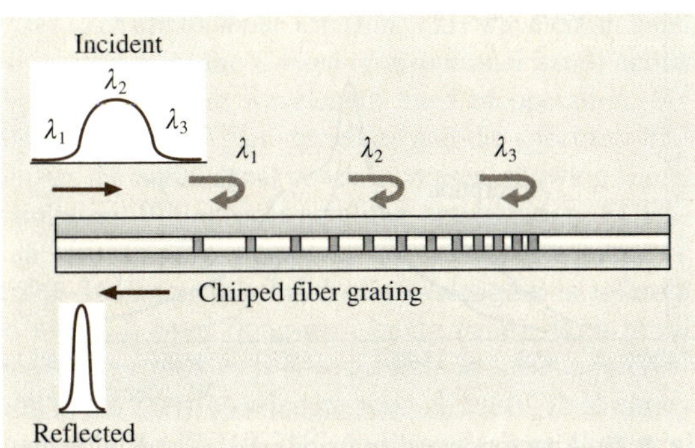

Fig. 8.77 Dispersion compensation using a chirped FBG

As an example a chirped grating of length 11 cm would have a dispersion of about 1380 ps/nm operating over a bandwidth of 0.61 nm. This grating can compensate for dispersion accumulated over a fiber with a dispersion coefficient of 17 ps/(km nm) and of length 81 km over a bandwidth of 0.61 nm, which approximately corresponds to 76 GHz of frequency bandwidth. It is interesting to note that the difference in period between the front and back end of the 11 cm-long grating is only about 0.25 nm while the average period of the grating is about 0.534 μm.

Chirped dispersion compensating gratings are commercially available for compensation of accumulated dispersion of up to 80 km of G.652 fiber for up to 32 wavelength channels. Unlike dispersion-compensating fibers, chirped FBGs provide the possibility of tweaking the required dispersion compensation, especially for 40 Gbps systems where the margin of dispersion available is rather small. Also by using nonlinearly chirped FBGs it has been shown that delay variation from −200 ps to −1200 ps is possible.

8.8.8 Erbium-Doped Fiber Amplifiers (EDFAs)

In traditional long-distance optical fiber communication systems, compensation of loss and dispersion is usually accomplished by using electronic regenerators in which the optical signals are first converted into electrical signals, then processed in the electrical domain and then reconverted into optical signals. Whenever the system limitation is due to insufficient optical power rather than dispersion, what is needed is just amplification of the signal and optical amplifiers can indeed perform this job. Optical amplifiers are devices that amplify the incoming optical signals in the optical domain itself without any conversion to the electrical domain, and have truly revolutionized long-distance fiber optic communications. Compared to electronic regenerators, optical amplifiers do not need any high-speed electronic circuitry, are transparent to bit rate and format, and most importantly can amplify multiple optical signals at different wavelengths simultaneously. Thus their development has ushered in the tremendous growth of communication capacity using wavelength division multiplexing (WDM) in which multiple wavelengths

placed within the amplifier between two gain units to optimize in terms of gain efficiency and noise characteristics. Filters based on fiber Bragg gratings, long-period gratings or thin-film filters are used. Typical gain flatness of better than 0.5 dB can be achieved and commercially available EDFAs are all gain-flattened. Other techniques involving modification of the refractive index profile to flatten the gain have also been demonstrated.

Noise in EDFA

In an EDFA population inversion between two energy levels of erbium ion leads to optical amplification by the process of stimulated emission. Erbium ions occupying the upper energy level can also make spontaneous transitions to the ground level and emit radiation. This radiation appears over the entire fluorescent band of emission of erbium ions and travels both in the forward as well as in the backward direction along the fiber. Part of the spontaneous emission generated at any point along the fiber gets coupled into the propagating mode of the fiber and can also get amplified just like the signal as it propagates through the population-inverted fiber. The resulting radiation is called amplified spontaneous emission (ASE). This ASE is the basic mechanism leading to noise in the optical amplifier [8.248]. ASE appearing in a wavelength region not coincident with the signal can be filtered using an optical filter. On the other hand, the ASE that appears in the signal wavelength region cannot be separated and constitutes the minimum added noise from the amplifier.

If P_{in} represents the signal input power (at frequency ν) into the amplifier and G represents the gain of the amplifier then the output signal power is given by GP_{in}. Along with this amplified signal, there is also ASE power which can be shown to be given by [8.248]

$$P_{ASE} = 2n_{sp}(G-1)h\nu B_o \,, \tag{8.130}$$

where B_o is the optical bandwidth over which the ASE power is being measured (which must be at least equal to the optical bandwidth of the signal), and $n_{sp} = N_2/(N_2 - N_1)$. Here N_2 and N_1 represent the population densities in the upper and lower amplifier energy levels of erbium in the fiber. The minimum value for n_{sp} corresponds to a completely inverted amplifier for which $N_1 = 0$ and thus $n_{sp} = 1$. For a gain of 20 dB, typical ASE powers within a band of 0.1 nm is about 0.6 μW(= −32 dB m) which corresponds to an ASE noise spectral density of −22 dBm/nm. We can define the optical signal-to-noise ratio (OSNR) as the ratio of the output optical signal power to the ASE power

$$\mathrm{OSNR} = \frac{P_{out}}{P_{ASE}} = \frac{GP_{in}}{2n_{sp}(G-1)h\nu B_o} \,, \tag{8.131}$$

where P_{in} is the average power input into the amplifier (which is about half of the peak power in the bit stream, assuming equal probability of ones and zeroes).

Each amplifier in a chain adds noise and thus in a fiber optic communication system consisting of multiple spans of optical fiber links with amplifiers, OSNR will keep falling and, at some point in the link when the OSNR falls below a certain value, the signal would need to be regenerated. Hence there is a maximum number of amplifiers that can be placed in a link beyond which the signal needs to be regenerated. For a link consisting of multiple spans of transmission fiber and EDFAs compensating the loss of each span, the OSNR is given by

$$\begin{aligned}\mathrm{OSNR(dB)} \approx\ & P_{out}(\mathrm{dBm}) - 10\log(n) + 58 \\ & - F(\mathrm{dB}) - 10\log(N+1) - L_{sp}(\mathrm{dB}) \,,\end{aligned} \tag{8.132}$$

where P_{out} is the total output power from the amplifier in dBm, n represents the number of wavelength channels in the link, F represents the noise figure of each EDFA (assumed to be the same), N represents the number of amplifiers, and L_{sp} the loss of each span. As a typical example we consider a link consisting of EDFAs with the following specifications: $P_{out} = 17$ dBm, $n = 32$, $F = 5$ dB, $L_{sp} = 20$ dB. If we require an OSNR of 22 dB at the end of the link, then using (8.132) the maximum number of amplifiers that can be used in the link comes out to be about 18. If more than this number of amplifiers are employed then the OSNR will fall below the required value of 22 dB. Thus for proceeding further along the length, the signal needs to be regenerated. It is also interesting to note from (8.132) that, to achieve the same OSNR at the output of the link, the number of amplifiers in the chain can be increased by reducing the noise figure of each amplifier or by increasing the output power of the amplifiers or by decreasing the span loss. Indeed by choosing smaller span loss, the number of amplifiers can be increased significantly so that the distance for regeneration can be made very large. Thus reducing each span loss by 3 dB would result in a doubling of the maximum allowed number of amplifiers (all other parameters being the same). Of course in this case, we would have to employ a larger number of amplifiers.

This discussion was based on the optical signal-to-noise ratio of the amplifier. When the amplified output is received by a detector, then the detector converts the optical signal into an electric current and the noise characteristics of the generated electrical signal are of importance. Apart from the optical signal, the amplified spontaneous emission within the bandwidth of the signal also falls on the photodetector. However, the ASE noise is completely random and contains no information. The photodetector would convert the total optical power received into electrical current and in the current there would be beating between the signal and noise fields and between noise fields at different frequencies. These lead respectively to signal–spontaneous beat noise and spontaneous–spontaneous beat noise. Under normal circumstances, the signal–spontaneous noise term and the signal shot-noise terms are the important noise terms and assuming the input to the amplifier to be shot noise limited we can calculate the output SNR from the noise terms. The amplifier noise figure defined as the ratio of input electrical signal-to-noise ratio to output electrical signal-to-noise ratio is given by

$$F = \frac{1 + 2n_{sp}(G-1)}{G} \,. \qquad (8.133)$$

Thus the noise figure depends on the inversion through n_{sp} and on the amplifier gain through G. For large gains $G \gg 1$, the noise figure is approximately given by $2n_{sp}$. Since the smallest value of n_{sp} is unity, the smallest noise figure is given by 2, or in decibel units as 3 dB.

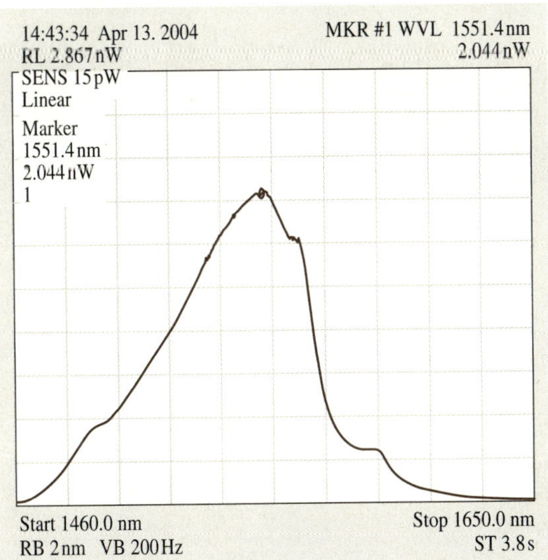

Fig. 8.81 Spontaneous Raman scattering from 25 km of a standard single-mode fiber

Noise figure is a very important characteristic of an amplifier and determines the overall performance of any amplified link. Noise figures of typical commercially available EDFAs are about 5 dB.

8.8.9 Raman Fiber Amplifier (RFA)

Another very important fiber based amplifier is the Raman fiber amplifier (RFA), which is based on the phenomenon of stimulated Raman scattering. The attractive feature of RFAs are that they can be made to work in any wavelength band by simply choosing appropriate pump wavelengths, they also have a large bandwidth and have lower noise figures compared to EDFAs. Apart from this, the link fiber can itself be used as the amplifier and thus the signal gets amplified as it covers the distance along the communication link itself; such amplifiers are referred to as distributed amplifiers.

When we send a strong light beam at a wavelength of 1450 nm through a long ($\approx$ 10 km) optical fiber, the light beam undergoes Raman scattering from the molecules of the glass fiber and this gives rise to scattered light at higher wavelengths. Figure 8.81 shows a typical spontaneous Raman spectrum from an optical fiber pumped by radiation at 1450 nm. As can be seen the scattered radiation occupies a large band and the peak of the scattered radiation lies at about 100 nm away from the pump wavelength. Indeed Raman scattering in silica leads to a Raman shift of between 13 and 14 THz, which corresponds to about 100 nm at the wavelength of 1550 nm.

If in addition to the strong pump light we also launch a weak light beam (referred to as the signal beam) with its wavelength lying within the band of spontaneous Raman scattering, then it leads to what is referred to as stimulated Raman scattering. In this case, the pump and signal wavelengths are coherently coupled by the Raman scattering process and the scattered radiation is coherent with the incident signal radiation much like

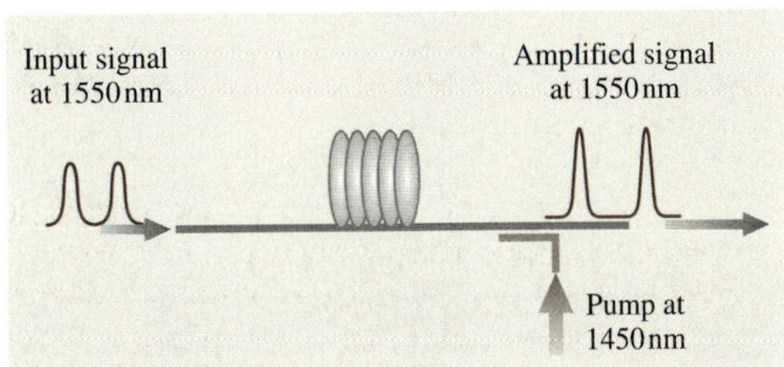

Fig. 8.82 Raman amplification by a counter-propagating pump

Part B | 8.8

stimulated emission that occurs in the case of a laser. It is this process that is used to build Raman fiber amplifiers (Fig. 8.82). Since the spontaneous Raman scattering spectrum is broad, the corresponding gain spectrum of the Raman amplifier is also very broad. The other interesting feature is that, no matter what the wavelength of the pump light is, the fiber can act like an amplifier in the wavelength range corresponding to the spontaneous Raman scattering spectrum.

The propagation equations describing the variation of signal power (P_s) and pump power (P_p) along the length of the fiber for a counter-pumped Raman amplifier are given by [8.251]

$$\frac{dP_s}{dz} = \gamma_R P_p P_s - \alpha_s P_s ;$$
$$\frac{dP_p}{dz} = \frac{\nu_p}{\nu_s} \gamma_R P_p P_s + \alpha_p P_p , \tag{8.134}$$

where ν_p and ν_s are the pump and signal frequencies, γ_R is the Raman gain efficiency defined by

$$\gamma_R = \frac{\iint g_R(\nu, r)\psi_p^2 \psi_s^2 r\, dr}{2\pi \iint \psi_p^2 r\, dr \iint \psi_s^2 r\, dr} \approx \frac{g_R(\nu)}{A_{eff}} , \tag{8.135}$$

where ψ_p and ψ_s are the transverse variation of pump and signal modal fields and g_R represents the material Raman gain coefficient and A_{eff} is the effective area defined by

$$A_{eff} = \frac{2\pi \iint \psi_p^2 r\, dr \iint \psi_s^2 r\, dr}{\iint \psi_p^2 \psi_s^2 r\, dr} . \tag{8.136}$$

Since the modal field profiles depend on the fiber refractive-index profile, the value of γ_R can be very different for various fibers. For example for standard SMFs, γ_R is $0.5–1\,W^{-1}km^{-1}$, for dispersion-compensating fibers it is $2.5–3\,W^{-1}km^{-1}$ and for highly nonlinear fibers it is about $6.5\,W^{-1}km^{-1}$. Photonic-crystal fibers and holey fibers can have extremely small mode effective areas and hence can provide extremely large Raman gains. The larger the value of γ_R, the larger is the stimulated Raman scattering and the larger the corresponding gains that are achievable. Figure 8.83 shows a typical length variation of pump and signal power of a counter pumped Raman amplifier.

Like any amplifier, in the case of Raman amplifiers also the amplified signal is accompanied by noise generated due to amplification of spontaneous Raman scattering that takes place within the fiber. Since amplification is taking place over a long length of the fiber, additional noise due to the phenomenon of double Rayleigh scattering (DRS) is also generated. Signals propagating in the forward direction suffer Rayleigh scattering and generate power in the backward direction. The backward propagating signals can undergo further Rayleigh scattering (double Rayleigh scattering, DRS) and generate power in the forward direction. These signals also use the same pump power to get amplified and constitute DRS noise. This noise becomes important for reasonably large pump powers and long interaction lengths when the gain becomes large. Apart from this, the amplified spontaneous Raman scattered light in the backward direction can get Rayleigh scattered to the forward direction which would also add to noise.

One can obtain approximate expressions for the Raman gain and spontaneously emitted noise power,

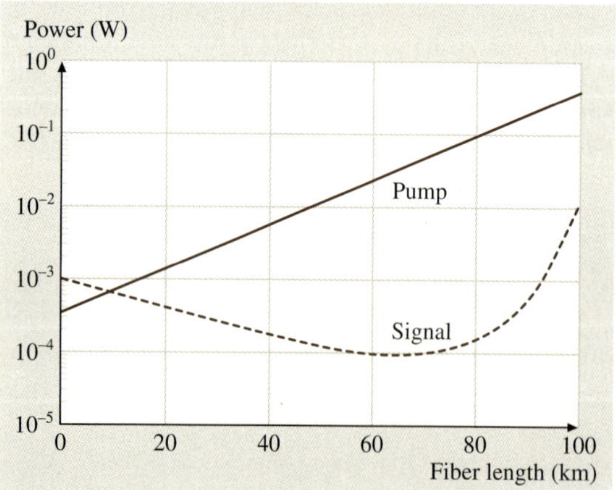

Fig. 8.83 Signal and pump power variation along the length of a Raman amplifier

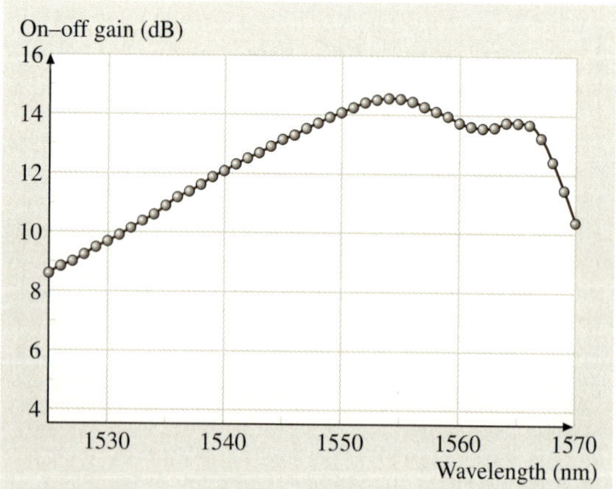

Fig. 8.84 Typical gain spectrum of a Raman amplifier

which are given by

$$G(\mathrm{dB}) = 10\log\left(\frac{P_s(L)}{P_s(0)\mathrm{e}^{-\alpha_s L}}\right) \approx 4.34\frac{\gamma_R}{\alpha_p}P_p(L)\,, \tag{8.137}$$

where the last approximation is valid for $\alpha_p L \gg 1$ and

$$P_{sp}(L) = 2h\nu\Delta\nu \times\left[\frac{4.34}{G}\exp\left(\frac{G}{4.34}\right) - \left(1 + \frac{4.34}{G}\right)\right]\,. \tag{8.138}$$

Here G defines the on–off gain, which is the ratio of output power with the Raman pump on to that of the output power when the Raman pump is switched off.

In Raman amplifiers the pump beam can propagate in the same direction as the signal or in the reverse direction. The former case is referred to as co-propagating (forward pumping) and the latter one as contra-propagating (backward pumping). Raman scattering phenomenon is an extremely fast process with time scales in the femtosecond (10^{-15} s) regime. This can lead to the transfer of power fluctuations from the pump to the signal. One way to avoid this is to have backward pumping (Fig. 8.82) wherein the pump fluctuations induced gain fluctuations get averaged out and thus the noise in the signal due to pump fluctuations is much lower.

Figure 8.84 shows a measured on–off gain spectrum of a backward-pumped Raman fiber amplifier corresponding to an input pump power of 750 mW at a wavelength of 1453 nm and an input signal power of 0.14 mW over a 25 km long span of single-mode fiber. On–off gains of greater than 12 dB are easily achievable.

Since the gain spectrum depends on the pump wavelength, it is indeed possible to achieve large flat gain using multiple pumps. Thus using 12 pumps with wavelengths lying between 1410 nm and 1510 nm, a total flat gain bandwidth of 100 nm from 1520 nm to 1620 nm (covering the C band and the L band) has been demonstrated. Since Raman fiber amplifiers can operate at any signal-wavelength region, they allow us to expand the operation region of fiber optic communication systems to other bands where EDFAs do not operate. Apart from this, Raman amplifiers can also be used to extend the operation of optical fiber communication systems in C-band. The repeater-less length can be increased by use of Raman amplifier at the terminal. Similarly when the bit rate of a communication channel is increased, then for the system to operate without signal degradation (within a certain bit error rate), the receiver would need more power. In this case, the additional Raman gain that is realized by propagating an appropriate pump with the signal can be used to increase the transmission bit rate for the same distance between the transmitter and receiver.

Since the gain coefficient depends on the effective area of the fiber, the Raman gain spectrum could be modified by proper fiber designs with appropriate spectral dependence of effective area. Thus, recently novel fiber designs have been proposed that can give flat Raman gain with just a single pump [8.252, 253]. For nice reviews of fiber Raman amplifiers, readers are referred to [8.251, 254, 255].

8.8.10 Nonlinear Effects in Optical Fibers

Consider a light beam having a power of 100 mW propagating through an optical fiber having an effective mode area of 50 μm^2. The corresponding optical intensity is 2×10^9 W/m^2. At such high intensities, the nonlinear effects in optical fibers start to influence the propagation of the light beam and can significantly influence the capacity of a WDM optical fiber communication system [8.256]. The most important nonlinear effects that affect optical fiber communication systems include self phase modulation (SPM), cross phase modulation (XPM) and four-wave mixing (FWM). Stimulated Raman scattering (SRS) and stimulated Brillouin scattering (SBS) are also important nonlinear phenomena and earlier we have seen how SRS can be used for optical amplification. In this section, we will discuss mainly SPM, XPM and FWM, which affect pulse propagation through optical fibers.

Self Phase Modulation (SPM)

The lowest-order nonlinearity present in an optical fiber is the third-order nonlinearity. Thus, in an optical fiber,

Table 8.6 Mode effective area of typical commercially available fibers

Fiber type	Effective area (μm^2)
Single-mode fiber (SMF) G652	≈ 85 μm^2
Dispersion-shifted fiber (DSF)	≈ 46 μm^2
Nonzero DSF (NZDSF)	≈ 52 μm^2 ($D > 0$), 56 μm^2 ($D < 0$) and 73 μm^2
Dispersion-compensating fiber (DCF)	≈ 23 μm^2 ($D < 0$)
Photonic-crystal fiber / holey fiber	≈ 3 μm^2

the polarization generated consists of a linear and a nonlinear term

$$P = \varepsilon_0 \chi E + \varepsilon_0 \chi^{(3)} E^3 , \tag{8.139}$$

where χ and $\chi^{(3)}$ represent the linear and third-order susceptibility of the medium (silica) and E represents the electric field of the propagating light wave/pulse. The nonlinearity due to $\chi^{(3)}$ results in an intensity-dependent refractive index given by

$$n = n_0 + n_2 I , \tag{8.140}$$

where $n_2 = 3\chi^{(3)}/4c\varepsilon_0 n_0$ and n_0 is the refractive index of the medium at low intensities. It is this intensity-dependent refractive index that gives rise to SPM.

Due to the intensity-dependent refractive index, the propagation constant of a mode becomes intensity dependent and can be written as

$$\beta_{NL} = \beta + \gamma P \tag{8.141}$$

where

$$\gamma = \frac{k_0 n_2}{\tilde{A}_{eff}} , \quad \tilde{A}_{eff} = 2\pi \frac{\left(\int \psi^2(r) r \, dr\right)^2}{\int \psi^4(r) r \, dr} \tag{8.142}$$

represent the nonlinear coefficient and the nonlinear mode effective area, respectively, and β is the propagation constant of the mode at low powers. If we assume the mode to be described by a Gaussian function, then $\tilde{A}_{eff} = \pi w_0^2$, where w_0 is the Gaussian-mode spot size. Note that the nonlinear coefficient γ of the fiber depends on the effective area of the mode; the larger the effective mode area, the smaller the nonlinear effects. Table 8.6 gives the mode effective area of some common fiber types.

If α represents the attenuation coefficient of the optical fiber, then the power propagating through the fiber decreases exponentially as $P(z) = P_0 \mathrm{e}^{-\alpha z}$ where P_0 is the input power. In α is the attenuation coefficient. In such a case, the phase shift suffered by an optical beam in propagating through a length L of the optical fiber is given by

$$\Phi = \int_0^L \beta_{NL} \, dz = \beta L + \gamma P_0 L_{eff} , \tag{8.143}$$

where

$$L_{eff} = \frac{\left(1 - \mathrm{e}^{-\alpha L}\right)}{\alpha} \tag{8.144}$$

is called the effective length of the fiber. If $\alpha L \gg 1$ then $L_{eff} \approx 1/\alpha$ and if $\alpha L \ll 1$ then $L_{eff} \approx L$. For single-mode fibers operating at 1550 nm, $\alpha \approx 0.25$ dB/km and

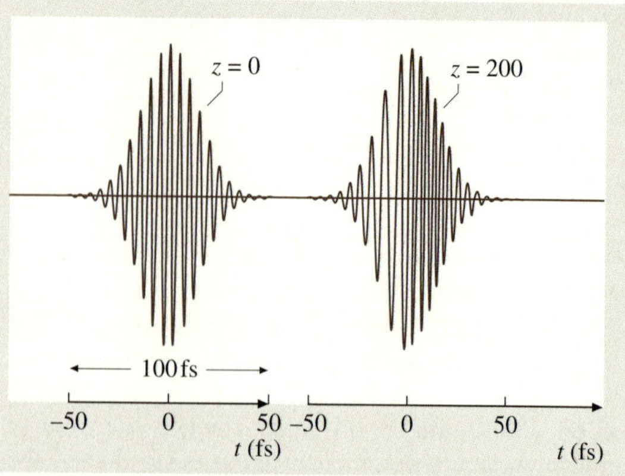

Fig. 8.85 An unchirped pulse at the input after propagating through an optical fiber gets chirped due to self phase modulation

thus $L_{eff} \approx L$ for $L \ll 20$ km and $L_{eff} \approx 20$ km for $L \gg 20$ km.

Since the propagation constant β_{NL} of the mode depends on the power carried by the mode, the phase Φ of the emergent wave depends on its power and hence this is referred to as *self phase modulation* (SPM).

For a light pulse P_0 in (8.145) becomes time dependent and this leads to an additional time dependent phase apart from $\omega_0 t$. Thus the output pulse is chirped and the instantaneous frequency of the output pulse is given by

$$\omega(t) = \frac{d}{dt}\left(\omega_0 t - \gamma P_0 L_{eff}\right) = \omega_0 - \gamma L_{eff} \frac{dP_0}{dt} . \tag{8.145}$$

The leading edge of the pulse corresponds to the positive values of dP_0/dt and the trailing edge to negative values of dP_0/dt. Thus in the presence of SPM, the leading edge gets downshifted in frequency while the trailing edge gets upshifted in frequency. The frequency at the center of the pulse remains unchanged from ω_0. Figure 8.85 shows an input unchirped and the output chirped pulse generated due to SPM. The output chirped pulse with the same temporal width has a larger frequency spectrum. These new frequencies have been generated by the nonlinear process.

The chirping due to nonlinearity without any corresponding increase in pulse width leads to increased spectral broadening of the pulse. This spectral broadening coupled with the dispersion in the fiber leads to modified dispersive propagation of the pulse in the presence of nonlinearity. In the normal-dispersion region the chirping due to dispersion is to downshift the leading edge and upshift the trailing edge of the pulse.

This is of the same sign as that due to SPM. Thus in the normal-dispersion regime (wavelength less than the zero-dispersion wavelength) the chirping due to dispersion and nonlinearity add. Thus at high powers, where the nonlinear effects are not negligible, the pulse will suffer additional dispersion as compared to the dispersion of the same pulse at low powers. On the other hand, in the anomalous-dispersion region (wavelength greater than the zero-dispersion wavelength), the chirping due to dispersion is opposite to that due to nonlinearity and thus in this wavelength region, nonlinearity and dispersion-induced chirpings can partially or even totally cancel each other. When total cancelation takes place, the pulse neither broadens in time nor in its spectrum and such a pulse is called a *soliton*. Such solitons can hence be used for dispersionless propagation of pulses to realize very high-bit-rate systems.

Assuming only second-order dispersion and $\chi^{(3)}$ nonlinearity, the amplitude $A(z, t)$ of the electric field of an incident pulse can be shown to satisfy the following equation [8.257]:

$$\frac{\partial A}{\partial z} = \mathrm{i}\frac{\beta_2}{2}\frac{\partial^2 A}{\partial T^2} - \mathrm{i}\gamma|A|^2 A\,, \tag{8.146}$$

where $\beta_2 = \mathrm{d}^2\beta/\mathrm{d}\omega^2$ and $T = t - z/v_g$. Equation (8.146) is referred to as the nonlinear Schroedinger equation and describes the propagation of a pulse in a medium possessing second-order dispersion and $\chi^{(3)}$ nonlinearity. The solution of the above equation give us solitons which are described mathematically by:

$$\tilde{A}(z, t) = \sqrt{P_0}\,\mathrm{sech}\left(\sqrt{\frac{P_0\gamma}{|\beta_2|}}T\right)\mathrm{e}^{-\mathrm{i}\gamma P_0 z/2}\,. \tag{8.147}$$

where $\tilde{A}(z, t)$ is the normalized electric field defined such that $|\tilde{A}(z, t)|^2$ gives the power carried by the pulse. Equation (8.147)) shows that the peak power required to form a soliton is related to the pulse width and dispersion coefficient D. As an example, for a soliton pulse at 1550 nm with full width at half maximum of $\tau_f = 10$ ps propagating in a fiber with $\gamma = 2.4\,\mathrm{W^{-1}km^{-1}}$ and $D = 2$ ps/(km nm) and the required peak power will be $P_0 = 33$ mW.

A heuristic derivation of the power required to form a soliton by cancelation of chirping due to dispersion and nonlinearity can be found in [8.235].

Even if the cancelation between dispersive and nonlinear chirping is not perfect, the nonlinear effects in an optical fiber lead to reduced pulse broadening in the anomalous-dispersion region. Thus the net dispersion suffered by the pulse decreases as the power increases. This fact needs to be kept in mind while designing dispersion-compensation schemes.

Cross Phase Modulation (XPM)

Consider the simultaneous launching of two or more different light beams of different wavelengths into an optical fiber. In such a case, each individual light wave will lead to a change in the refractive index of the fiber due to the intensity-dependent refractive index. This change in refractive index of the fiber then affects the phase of the other light beam(s); this results in what is referred to as cross phase modulation (XPM). Similar to the case of SPM, the instantaneous frequency of a signal at frequency ω_0 in the presence of XPM is given by

$$\omega(t) = \omega_0 - 2\gamma L_{\mathrm{eff}}\frac{\mathrm{d}P_2}{\mathrm{d}t}\,, \tag{8.148}$$

where P_2 is the power carried by the other wavelength. The part of the signal that is influenced by the leading edge of the pump will be downshifted in frequency (since in the leading edge $\mathrm{d}P_2/\mathrm{d}t > 0$) and the part overlapping with the trailing edge will be up shifted in frequency in frequency (since $\mathrm{d}P_2/\mathrm{d}t < 0$). This leads to a frequency chirping of the signal pulse just like in the case of SPM. Conventional detectors detect the intensity variation of the signal and hence are not affected by phase variations caused by XPM. However, these phase variations get converted to intensity variations due to dispersive effects in the fiber and these intensity variations can cause further bit errors in a fiber optic communication system.

If the two light waves are pulses, then XPM leads to chirping of the pulses. In the presence of finite dispersion (i. e., operation away from the zero-dispersion wavelength), the two pulses will move with different velocities and thus the pulses will walk off from each other. In case the pulses enter the fiber together, then due to walk off each pulse will see only the trailing or the leading edge of the other pulse which will lead to chirping. On the other hand, if the *collision* is complete, i. e., if the pulses start separately and walk through each other and again separate as they propagate through the fiber, then there would be no XPM-induced chirping since the pulses would have interacted with both the leading and the trailing edge of the other pulse. In an actual system, this cancelation will not be perfect since the pulses suffer attenuation and this leads to reduced nonlinear interaction as the pulses walk through each other.

Four-Wave Mixing (FWM)

Consider the incidence of three waves at three frequencies ω_2, ω_3 and ω_4 into an optical fiber. In the presence of the three waves, the third-order nonlinearity in the fiber leads to the generation of a nonlinear polarization at a frequency ω_1 given by

$$\omega_1 = \omega_3 + \omega_4 - \omega_2 \quad (8.149)$$

This nonlinear polarization can, under some circumstances, lead to the generation of electromagnetic waves at the new frequency ω_1. This phenomenon is referred to as four-wave mixing (FWM). In a dense wavelength division multiplexed (DWDM) system, FWM leads to cross talk among different wavelength channels of the system.

The nonlinear polarization at frequency ω_1 due to the presence of other waves is given by:

$$P_{\mathrm{NL}}^{\omega_1} = \frac{1}{2}\left(p_{\mathrm{nl}}\,\mathrm{e}^{\mathrm{i}(\omega_1 t - \beta_1 z)} + \mathrm{c.c.}\right) \quad (8.150)$$

where

$$p_{\mathrm{nl}} = \frac{3\varepsilon_0}{2}\chi^{(3)} A_2^* A_3 A_4 \psi_2 \psi_3 \psi_4 \,\mathrm{e}^{-\mathrm{i}\Delta\beta z}\,, \quad (8.151)$$

where $\Delta\beta = \beta_3 + \beta_4 - \beta_2 - \beta_1$ and A_i and ψ_i represent the amplitudes and modal field profiles corresponding to modes at frequency ω_i. From the expression for the nonlinear polarization it can be seen that, in general, the velocity of the nonlinear polarization is not equal to the velocity of the electromagnetic wave that it is generating. For efficient generation of the electromagnetic wave at the frequency ω_1, there must be phase matching, i. e., the velocity of the nonlinear polarization (which acts as the source of the radiation at frequency ω_1) and the electromagnetic wave (at frequency ω_1) must be equal. For this to happen we must have $\Delta\beta = 0$. Assuming the frequencies to be closely and equally spaced (i. e., $\omega_1 = \omega_2 - \Delta\omega$, $\omega_3 = \omega_2 - 2\Delta\omega$, $\omega_4 = \omega_2 + \Delta\omega$) and making a Taylor series expansion of all βs about a frequency ω_2, we get

$$\Delta\beta = -\frac{4\pi D\lambda^2}{c}(\Delta\nu)^2\,, \quad (8.152)$$

where $\Delta\omega = 2\pi\Delta\nu$ represents the frequency spacing between adjacent channels. Thus when the channels lie around the zero-dispersion wavelength of the fiber, $D = 0$ and we have phase matching and thus an efficient FWM. If one wishes to reduce FWM, then one must operate away from the zero-dispersion wavelength. This has led to the development of nonzero-dispersion-shifted fibers (NZ-DSF), wherein a finite but small dispersion (≈ 2 to $8\,\mathrm{ps/(km\,nm)}$) is introduced to reduce FWM effects in an actual fiber optic communication system. On the other hand, if a strong FWM is desired for an application such as all-optical signal processing or wavelength conversion, then the interacting wavelengths must lie close to the zero-dispersion wavelength.

Assuming that all frequencies have the same attenuation coefficient α and neglecting depletion due to conversion, the power generated at the frequency ω_1 due to mixing of the other three frequencies can be shown to be

$$P_1 = 4\gamma^2 P_2 P_3 P_4 L_{\mathrm{eff}}^2 \eta\, \mathrm{e}^{-\alpha L}\,, \quad (8.153)$$

where

$$\eta = \frac{\alpha^2}{\alpha^2 + \Delta\beta^2}\left(1 + \frac{4\mathrm{e}^{-\alpha L}\sin^2\frac{\Delta\beta L}{2}}{\left(1 - \mathrm{e}^{-\alpha L}\right)^2}\right). \quad (8.154)$$

As evident from (8.153), maximum conversion occurs when $\eta = 1$, i. e. $\Delta\beta = 0$. If we assume all wavelengths to carry the same power P_{in}, then under phase matching, the ratio of the power generated due to FWM to that exiting from the fiber is given by

$$R = \frac{P_\mathrm{g}}{P_{\mathrm{out}}} = \frac{P_1(L)}{P_{\mathrm{in}}\,\mathrm{e}^{-\alpha L}} = 4\gamma^2 P_{\mathrm{in}}^2 L_{\mathrm{eff}}^2\,. \quad (8.155)$$

Thus if in each channel we have an input power of 1 mW, then for $\gamma = 1.73 \times 10^{-3}\,\mathrm{m^{-1}W^{-1}}$, $L_{\mathrm{eff}} = 20\,\mathrm{km}$, the FWM generated output will be about 0.5% of the power exiting in the channel. This gives us the level of cross talk among the channels created due to FWM.

Figure 8.86 shows the output spectrum measured at the output of a 25 km-long dispersion-shifted fiber ($D = -0.2\,\mathrm{ps/(km\,nm)}$ at the central channel) when three 3 mW wavelengths are launched simultaneously.

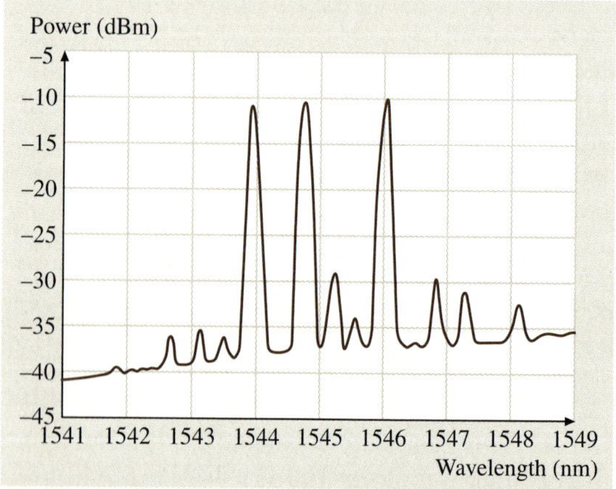

Fig. 8.86 New frequency components generated due to four wave mixing in an optical fiber operating close to the zero-dispersion wavelength. The input consists of the three large-amplitude frequencies (after [8.258])

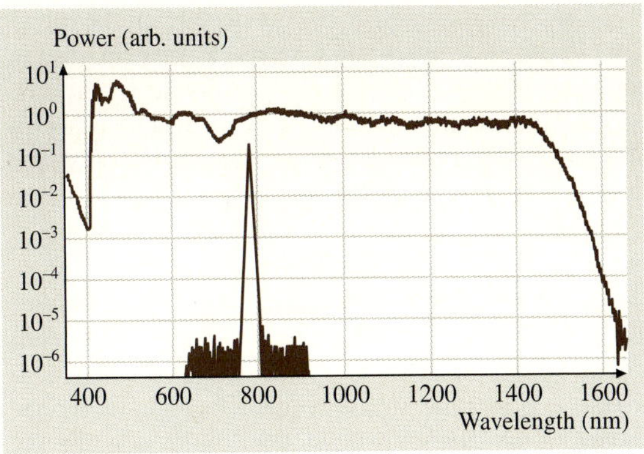

Fig. 8.87 Spectral broadening by supercontinuum generation (after [8.259])

Notice the generation of nine new frequencies with different amplitudes (with a maximum peak ratio of 1% to the input signals) due to FWM. Newly generated waves will interfere with power present in those channels and lead to cross talk. By choosing a nonzero value of dispersion, the four-wave mixing efficiency can be significantly reduced. The larger the dispersion coefficient, the smaller can be the channel spacing for the same cross talk.

Since dispersion leads to increased bit error rates in fiber optic communication systems, it is important to have low dispersion. On the other hand, lower dispersion leads to cross talk due to FWM. This problem can be resolved by noting that FWM depends on the local dispersion value in the fiber while the pulse spreading at the end of a link depends on the overall dispersion in the fiber link. If one chooses a link made up of positive and negative dispersion coefficients, then by an appropriate choice of the lengths of the positive and negative dispersion fibers, it would be possible to achieve a zero total link dispersion while at the same time maintaining a large local dispersion. This is referred to as dispersion management in fiber optic systems.

Although FWM leads to cross talk among different wavelength channels in an optical fiber communication system, it can be used for various optical processing functions such as wavelength conversion, high-speed time-division multiplexing, pulse compression, and optical amplification [8.260, 261]. For such applications, there is a concerted worldwide effort to develop highly nonlinear fibers with much smaller mode areas and higher nonlinear coefficients. Some of the very novel fibers that have been developed recently include holey fibers, photonic band-gap fibers or photonic crystal fibers which are very interesting since they posses extremely small mode effective areas ($\approx 2.5\,\mu m^2$ at 1550 nm) and can be designed to have zero dispersion even in the visible region of the spectrum [8.262, 263]. This is expected to revolutionize nonlinear fiber optics by providing new geometries to achieve highly efficient nonlinear optical processing at lower powers.

Supercontinuum Generation

Supercontinuum (SC) generation is the phenomenon in which a nearly continuous spectrally broadened output (bandwidth > 1000 nm) is produced through nonlinear effects on high-peak-power picosecond and femtosecond pulses. Such broadened spectra find applications in spectroscopy, optical coherence tomography, WDM sources for optical communication by slicing the spectrum etc. Supercontinuum generation in an optical fiber is a very convenient technique since the intensity levels can be maintained high over long interaction lengths by choosing small mode areas and the dispersion profile of the fiber can be appropriately designed by varying the transverse refractive-index profile of the fiber. The spectral broadening that takes place in the fiber is attributed to a combination of various third-order effects such as SPM, XPM, FWM, and Raman scattering. Since dispersion plays a significant role in the temporal evolution of the pulse, different dispersion profiles have been used in the literature to achieve broadband SC. Some studies have used dispersion-decreasing fibers, dispersion-flattened fibers, while others have used a constant anomalous-dispersion fiber followed by a normal-dispersion fiber. Figure 8.87 shows the input and output broadened spectra obtained by passing a light pulse through a photonic-crystal fiber [8.259].

8.8.11 Microstructured Fibers

A standard optical fiber guides light using the phenomenon of total internal reflection. Recently there

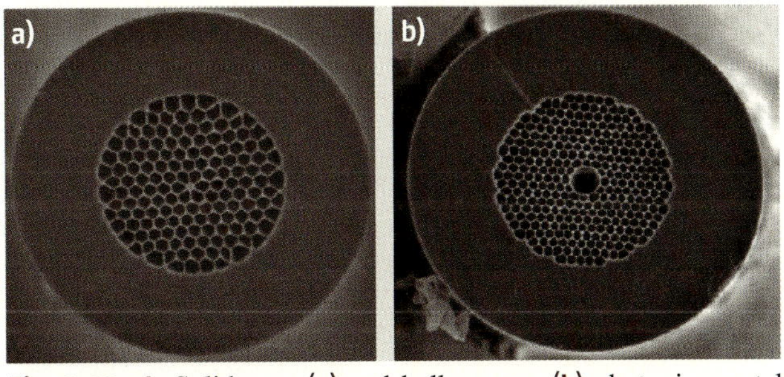

Fig. 8.88a,b Solid core **(a)** and hollow core **(b)** photonic-crystal fibers (Courtesy Blaze Photonics, Bath, UK)

has been intense activity in realization of optical fibers guiding light using Bragg reflections or photonic band-gap effects. Photonic crystals are periodic structures fabricated in an optical material with periodicities comparable to the optical wavelength. These can be one dimensional (like fiber Bragg gratings), two dimensional or three dimensional. Structures with appropriate symmetry and periodicities can exhibit photonic band gaps, which are regions of optical wavelengths where light cannot propagate through the structure, much like the band gaps for electrons in crystals. Thus such structures allow control of light propagation and find applications in semiconductor lasers, light modulators, integrated optical devices, nonlinear devices etc. Microstructured fibers have a periodic arrangement of holes in a silica background material running all along the length of the fiber and can lead to guidance of light using the band-gap effects. Figure 8.88 shows two types of such fibers, one having solid silica core and the other having a air hole at the center. Since confinement can be accomplished using the band-gap effect it is possible to realize optical fibers with air cores; such holey fibers are finding many applications. It is also possible to realize fibers with solid cores with mode effective areas in the region of $3\,\mu\mathrm{m}^2$. Since nonlinear effects depend on the intensity of light such fibers can give rise to very interesting nonlinear effects even at moderate power levels. Microstructured fibers exhibit the very interesting property of being single mode over a very large range of wavelengths. This can be understood from the fact that, due to the presence of holes in the cladding, as the wavelength changes the fraction of light power in the holes changes and this leads to an effective cladding refractive index which becomes highly wavelength dependent and increases with increasing wavelength, leading to an effective V number which results in single-mode operation. Such fibers are also referred to as endlessly single-mode fibers. The dispersion of microstructured fibers can also be controlled by appropriate design and it is possible to achieve zero dispersion in the visible range using silica which is otherwise not possible using conventional fiber designs. Supercontinuum generation which requires control of dispersion and nonlinearity is one such application and devices based on such fibers are now commercially available. With technological inputs, propagation loss in these fibers have been reduced significantly and photonic-crystal fibers with loss values of 0.3 dB/km have been realized recently [8.264].

References

8.1 M. D. Missig, G. M. Morris: Diffractive optics applied to eyepiece design, Appl. Opt. **34**, 2452–2461 (1995)

8.2 W. Knapp, G. Blough, K. Khajurival, R. Michaels, B. Tatian, B. Volk: Optical design comparison of 60° eyepieces: one with a diffractive surface and one with aspherics, Appl. Opt. **34**, 4756–4760 (1997)

8.3 Z.-Q. Wang, H.-J. Zhang, R.-L. Fu, G.-G. Mu, Z.-W. Lu, C. M. Cartwright, W. A. Gillespie: Hybrid diffractive refractive ultra-wide-angle eyepieces, Optik **113**, 159–162 (2002)

8.4 C. G. Blough, M. J. Hoppe, D. R. Hand, W. J. Peck: Achromatic eyepieces using acrylic diffractive lenses, Proc. SPIE **2600**, 93–99 (1995)

8.5 Z. Yun, Y. L. Lam, Y. Zhou, X. Yuan, L. Zhao, J. Liu: Eyepiece design with refractive-diffractive hybrid elements, Proc. SPIE **4093**, 474–480 (2000)

8.6 G. De Vos, G. Brandt: Use of holographic optical elements in HMDs, Proc. SPIE **1290**, 70–80 (1990)

8.7 J. A. Cox, T. A. Fritz, T. Werner: Application and demonstration of diffractive optics for head-mounted displays, Proc. SPIE **2218**, 32–40 (1994)

8.8 J. P. Rolland, M. W. Krueger, A. A. Goon: Dynamic focusing in head-mounted displays, Proc. SPIE **3639**, 463–470 (1999)

8.9 T. Nakai, H. Ogawa: Research on multi-layer diffractive optical elements and their application to camera lenses. In: *Diffractive Optics and Micro-Optics*, Techn. Dig. (Optical Society of America, Washington, DC 2002) pp. 5–7 (postconference edition)

8.10 R. Brunner, R. Steiner, K. Rudolf, H.-J. Dobschal: Diffractive-refractive hybrid microscope objective for 193 nm inspection systems, Proc. SPIE **5177**, 9–15 (2003)

8.11 R. Brunner, A. Menck, R. Steiner, G. Buchda, S. Weissenberg, U. Horn, A. M. Zibold: Immersion mask inspection with hybrid-microscopic systems at 193 nm, Proc. SPIE **5567**, 887–893 (2004)

8.12 R. Brunner, M. Burkhardt, A. Pesch, O. Sandfuchs, M. Ferstl, S. C. Hohng, J. O. White: Diffraction based solid immersion lens, J. Opt. Soc. Am. A **21**(7), 1186–1191 (2004)

8.13 R. A. Hyde: 1. Very large aperture diffractive telescopes, Appl. Opt. **38**(19), 4198–4212 (1999)

8.14 I. M. Barton, J. A. Britten, S. N. Dixit, L. J. Summers, I. M. Thomas, M. C. Rushford, K. Lu, R. A. Hyde, M. D. Perry: Fabrication of large-aperture lightweight diffractive lenses for use in space, Appl. Opt. **40**(4), 447–451 (2001)

8.15 D. Attwood: *Soft X-ray and Extreme Ultraviolet Radiation – Principles and Applications* (Cambridge Univ. Press, Cambridge 1999)
8.16 J. Nowak, J. Masajada: Hybrid apochromatic lens, Opt. Appl. **30**(2/3), 271–275 (2000)
8.17 D. A. Buralli: Optical performance of holographic kinoforms, Appl. Opt. **28**, 976–983 (1989)
8.18 C. Londono, P. P. Clark: Modeling diffraction efficiency effects when designing hybrid diffractive lens systems, Appl. Opt. **31**, 2248–2251 (1992)
8.19 S. M. Ebstein: Nearly index-matched optics for aspherical, diffractive, and achromatic-phase diffractive elements, Opt. Lett **21**, 1454–1456 (1996)
8.20 Y. Arieli, S. Noach, S. Ozeri, N. Eisenberg: Design of diffractive optical elements for multiple wavelengths, Appl. Opt. **37**, 6174–6177 (1998)
8.21 Y. Arieli, S. Ozeri, T. Eisenberg, S. Noach: Design of diffractive optical elements for wide spectral bandwidth, Opt. Lett. **23**, 823–824 (1998)
8.22 T. Nakai: Diffractive optical element and photographic optical system having the same, European Patent Application EP 1014150 A2 (1999)
8.23 H. P. Herzig, A. Schilling: Optical systems – design using microoptics. In: *Encyclopedia of Optical Engineering*, Vol. 2, ed. by R. G. Driggers (Marcel Dekker, New York 2003) pp. 1830–1842
8.24 T. Nakai, H. Ogawa: Research on multi-layer diffractive optical elementsand their applications to photographiclenses, 3rd Int. Conf. Optics–Photonics Design Fabrication, Tokyo 2002, ed. by T. Murakami (Optical Society of Japan, Tokio 2002) 61–62
8.25 T. H. Jamieson: Thermal effects in optical systems, Opt. Eng. **20**, 156–160 (1981)
8.26 G. P. Behrmann, J. P. Bowen: Influence of temperature on diffractive lens performance, Appl. Opt. **32**(14), 2483–2489 (1993)
8.27 J. Jahns, Y. H. Lee, C. A. Burrus, J. Jewell: Optical interconnects using top-surface-emitting microlasers and planar optics, Appl. Opt. **31**, 592–597 (1992)
8.28 C. Londono, W. T. Plummer, P. P. Clark: Athermalization of a single-component lens with diffractive optics, Appl. Opt. **32**, 2295–2302 (1993)
8.29 G. P. Behrmann, J. N. Mait: Hybrid (refractive/diffractive) optics. In: *Micro-optics: Elements, Systems and Application*, ed. by H. P. Herzig (Taylor Francis, London 1997) pp. 259–292
8.30 G. Khanarian: Optical properties of cyclic olefin copolymers, Opt. Eng. **40**(6), 1024–1029 (2001)
8.31 Nippon Zeon: *Zeonex™ brochure* (Nippon Zeon Co., Ltd, Tokyo 1998)
8.32 L. H. Cescato, E. Gluch, N. Streibl: Holographic quarterwave plates, Appl. Opt. **29**(22), 3286–3290 (1990)
8.33 I. Richter, P.-Ch. Sun, F. Xu, Y. Fainman: Design considerations of form birefringent microstructures, Appl. Opt. **34**(14), 2421–2429 (1995)
8.34 I. Richter, P.-Ch. Sun, F. Xu, Y. Fainman: Form birefringent microstructures: modeling and design, Proc. SPIE **2404**, 69–80 (1995)
8.35 P. B. Clapham, M. C. Hutley: Reduction of lens reflexion by the "moth eye" principle, Nature **244**(5414), 281–282 (1973)
8.36 S. J. Wilson, M. C. Hutley: The optical properties of 'moth eye' antireflection surfaces, Opt. Acta **29**(7), 993–1009 (1982)
8.37 T. K. Gaylord, W. E. Baird, M. G. Moharam: Zero-reflectivity high spatial-frequency rectangular-groove dielectric surface relief gratings, Appl. Opt. **25**(24), 4562–4567 (1986)
8.38 D. H. Raguin, G. M. Morris: Antireflection structured surfaces for the infrared spectral region, Appl. Opt. **32**(7), 1154–1167 (1993)
8.39 R. Bräuer, O. Bryngdahl: Design of antireflection gratings with approximate and rigorous methods, Appl. Opt. **33**(34), 7875–7882 (1994)
8.40 R.-Ch. Tyan, P.-Ch. Sun, Y. Fainman: Polarizing beam splitters constructed of form-birefringent multiplayer gratings, Proc. SPIE **2689**, 82–89 (1996)
8.41 R.-Ch. Tyan, A. A. Salvekar, H.-P. Chou, Ch.-Ch. Cheng, A. Scherer, P.-Ch. Sun, F. Xu, Y. Fainman: Design, fabrication, and characterization of form-birefringent multiplayer polarizing beam splitter, J. Opt. Soc. Am. A **14**(7), 1627–1636 (1997)
8.42 L. Pajewski, R. Borghi, G. Schettini, F. Frezza, M. Santarsiero: Design of a binary grating with subwavelength features that acts as a polarizing beam splitter, Appl. Opt. **40**(32), 5898–5905 (2001)
8.43 L. L. Soares, L. Cescato: Metallized photoresist grating as a polarizing beam splitter, Appl. Opt. **40**(32), 5906–5910 (2001)
8.44 P. Lalanne, J. Hazart, P. Chavel, E. Cambril, H. Launois: A transmission polarizing beam splitter grating, J. Opt. A **1**, 215–219 (1999)
8.45 H. Haidner, P. Kipfer, W. Stork, N. Streibl: Zero-order gratings used as an artificial distributed index medium, Optik **89**(3), 107–112 (1992)
8.46 M. W. Farn: Binary gratings with increased efficiency, Appl. Opt. **31**(22), 4453–4458 (1992)
8.47 M. Collischon, H. Haidner, P. Kipfer, A. Lang, J. T. Sheridan, J. Schwider, N. Streibl, J. Lindolf: Binary blazed reflection gratings, Appl. Opt. **33**(16), 3572–3577 (1994)
8.48 P. Lalanne, S. Astilean, P. Chavel, E. Cambril, H. Launois: Blazed binary subwavelength gratings with efficiencies larger than those of conventional echelette gratings, Opt. Lett. **23**(14), 1081–1083 (1998)
8.49 J. N. Mait, D. W. Prather, M. S. Mirotznik: Binary subwavelength diffractive lens design, Opt. Lett. **23**(17), 1343–1345 (1998)
8.50 M. Born, E. Wolf: *Principles of Optics*, 6 edn. (Pergamon, London 1980)
8.51 S. M. Rytov: Electromagnetic properties of a finely stratified medium, Soviet Phys. JETP **2**(3), 466–475 (1956)

8.52 E. B. Grann, M. G. Moharam, D. A. Pommet: Artificial uniaxial and biaxial dielectrics with use of two-dimensional subwavelength binary gratings, J. Opt. Soc. Am. A **11**(10), 2695–2703 (1994)

8.53 A. R. Parker: 515 million years of structural colors, J. Opt. A **2**, R15–R28 (2000)

8.54 E. N. Glytsis, T. K. Gaylord: High-spatial-frequency binary and multilevel stairstep gratings: polarization-selective mirrors and broadband antireflection surfaces, Appl. Opt. **31**(22), 4459–4470 (1992)

8.55 D. H. Raguin, G. M. Morris: Analysis of antireflection-structured surfaces with continuous one-dimensional surface profiles, Appl. Opt. **32**(14), 2582–2598 (1993)

8.56 M. E. Motamedi, W. H. Southwell, W. J. Gunning: Antireflection surfaces in silicon using binary optics technology, Appl. Opt. **31**(22), 4371–4376 (1993)

8.57 A. Gombert, K. Rose, A. Heinzel, W. Horbelt, Ch. Zanke, B. Bläsi, V. Wittwer: Antireflective submicrometer surface-relief gratings for solar applications, Solar Energy Mater. Solar Cell. **54**, 333–342 (1998)

8.58 R. T. Perkins, D. P. Hansen, E. W. Gardner, J. M. Thorne, A. A. Robbins: Broadband wire grid polarizer for the visible spectrum, US Patent 6122103 (2000)

8.59 M. Xu, H. P. Urbach, D. K. G. de Boer, H. J. Cornelissen: Wire-grid diffraction gratings used as polarizing beam splitter for visible light and applied in liquid crystal on silicon, Opt. Express **13**(7), 2303–2320 (2005)

8.60 T. Sergan, M. Lavrenzovich, J. Kelly, E. Gardner, D. Hansen: Measurement and modeling of optical performance of wire grids and liquid-crystal displays utilizing grid polarizers, J. Opt. Soc. Am. A **19**(9), 1872–1885 (2002)

8.61 P. Yeh: A new optical model for wire grid polarizers, Opt. Commun. **26**(3), 289–292 (1978)

8.62 W. Storck, N. Streibl, H. Haidner, P. Kipfer: Artificial distributed-index media fabricated by zero-order gratings, Opt. Lett. **16**(24), 1921–1923 (1991)

8.63 H. Haidner, P. Kipfer, T. Sheridan, J. Schwider, N. Streibl, M. Collischon, J. Hutfless, M. März: Diffraction grating with rectangular grooves exceeding 80% diffraction efficiency, Infrared Phys. **34**(5), 467–475 (1993)

8.64 M. E. Warren, R. E. Smith, G. A. Vawter, J. R. Wendt: High efficiency subwavelength diffractive optical element in GaAs for 975 nm, Opt. Lett. **20**(12), 1441–1443 (1995)

8.65 S. Astilean, P. Lalanne, P. Chavel, E. Cambril, H. Launois: High efficiency subwavelength diffractive element patterned in a high-refractive-index material for 633 nm, Opt. Lett. **23**(7), 552–554 (1998)

8.66 P. Lalanne: Waveguiding in blazed-binary diffractive elements, J. Opt. Soc. Am. A **16**(10), 2517–2520 (1999)

8.67 P. Tang, D. J. Towner, T. Hamano, A. L. Meier: Electrooptic modulation up to 40 GHz in a barium titanate thin film waveguide modulator, Opt. Express **12**(24), 5962–5967 (2004)

8.68 M. Hufschmid: *Winkelmodulation* (Fachhochschule, Basel 2002) (in German)

8.69 New Focus: *Practical Uses and Applications of Electrooptic Modulators* (Bockham Inc., San Jose 2001)

8.70 M. Bass: *Handbook of Optics*, Vol. 2 (McGraw-Hill, New York 1995)

8.71 B. E. A. Saleh, M. C. Teich: *Fundamentals of Photonics* (Wiley, New York 1991)

8.72 R. F. Enscoe, R. J. Kocka: *Systems and Applications Demands for Wider-Band Beam Modulation Challenge System Designers* (Conoptics Inc., Danbury 1981)

8.73 Laurin: *The Photonics Handbook* (Laurin, Pittsfield 2003) www.photonics.com/handbookHome.aspx

8.74 A. L. Mikaelian: Self-focusing media with variable index of refraction, Prog. Opt. **17**, 281–345 (1980)

8.75 K. Iga: Theory for gradient index imaging, Appl. Opt. **19**, 1039–1043 (1970)

8.76 H. Hovestädt: Cylindrical glass plates acting like diverging lenses. In: *Jena Glass and Its Scientific and Industrial Applications*, ed. by I. P. Everett, A. Everett (Macmillan, London 1902) Chap. 29, pp. 66–70

8.77 J. C. Maxwell: Solution of problems, Cambridge Dublin Math. J. **8**, 188 (1854)

8.78 K. Iga, S. Misawa: Distributed-index planar microlens and stacked planar optics: a review of progress, Appl. Opt. **25**, 3388–3396 (1986)

8.79 R. K. Luneburg: *Mathematical Theory of Optics* (Univ. California Press, Berkeley 1966) Chap. 27-30, pp. 164–195

8.80 S. Doric: Patent WO 96/14595 (1996)

8.81 M. V. R. K. Murty: Laminated lens, J. Opt. Soc. Am. **61**, 886–894 (1971)

8.82 K. Iga, Y. Kokubun, M. Gikawa: *Fundamentals of Microoptics* (Academic, New York 1984)

8.83 E. W. Marchand: *Gradient Index Optics* (Academic, New York 1978)

8.84 C. Gomez-Reino, M. V. Perez, C. Bao: *Gradient-Index Optics* (Springer, Berlin, Heidelberg 2002)

8.85 K. H. Brenner, W. Singer: Light propagation through microlenses: a new simulation method, Appl. Opt. **32**, 4984 (1993)

8.86 S. I. Najafi: *Introduction to Glass Integradted Optics* (Artech House, New York 1992)

8.87 M. Born, E. Wolf: *Principles of Optics* (Pergamon, Oxford 1980)

8.88 A. Sharma, D. V. Kumar, A. K. Ghatak: Tracing rays through graded-index media: a new method, Appl. Opt. **21**, 984–987 (1982)

8.89 A. Sharma: Computing optical path length in gradient-index media: a fast and accurate method, Appl. Opt. **24**, 4367–4370 (1985)

8.90 M. L. Huggins: The refractive index of silicate glasses as a function of composition, J. Opt. Soc. Am. **30**, 420 (1940)

8.91 S. D. Fantone: Refractive index and spectral models for gradient-index materials, Appl. Opt. **22**, 432–440 (1983)

8.92 D. P. Ryan-Howard, D. T. Moore: Model for the chromatic properties of gradient-index glass, Appl. Opt. **24**, 4356–4366 (1985)

8.93 K. Shingyouchi, S. Konishi: Gradient-index doped silica rod lenses produced by a solgel method, Appl. Opt. **29**, 4061–4063 (1990)

8.94 M. A. Pickering, R. L. Taylor, D. T. Moore: Gradient infrared optical material prepared by a chemical vapor deposition process, Appl. Opt. **25**, 3364–3372 (1986)

8.95 I. Kitano, K. Koizumi, H. Matsumura, T. Uchida, M. Furukawa: A light-focusing fiber guide prepared by ion-exchange techniques, J. Jpn. Soc. Appl. Phys. (Suppl.) **39**, 63–70 (1970)

8.96 S. Ohmi, H. Sakai, Y. Asahara, S. Nakayama, Y. Yoneda, T. Izumitani: Gradient-index rod lens made by a double ion-exchange process, Appl. Opt. **27**, 496 (1988)

8.97 R. H. Doremus: Ion exchange in glasses. In: *Ion Exchange*, ed. by J. Marinski (Marcel Decker, New York 1966)

8.98 Y. Ohtsuka, T. Sugano: Studies on the light-focusing plastic rod. 14: GRIN rod of CR-39-trifluoroethyl methacrylate copolymer by a vapor-phase transfer process, Appl. Opt. **22**, 413 (1983)

8.99 T. M. Che, J. B. Caldwell, R. M. Mininni: Sol-gel derived gradient index optical materials, Proc. SPIE **1328**, 145–159 (1990)

8.100 Y. Koike, H. Hidaka, Y. Ohtsuka: Plastic axial gradient-index lens, Appl. Opt. **24**, 4321 (1985)

8.101 P. K. Manhart, T. W. Stuhlinger, K. R. Castle, M. C. Ruda: Gradient refractive index lens elements, US Patent 5617252 (1997)

8.102 R. M. Ward, D. N. Pulsifer: Glass preform with deep radial gradient layer and method of manufacturing same, US Patent 5522003 (1996)

8.103 B. Messerschmidt, T. Possner, R. Goering: Colorless gradient-index cylindrical lenses with high numerical apertures produced by silver-ion exchange, Appl. Opt. **34**, 7825 (1995)

8.104 T. Findakly: Glass waveguides by ion exchange: a review, Opt. Eng. **24**, 244–252 (1985)

8.105 A. Tervonen, S. Honkanen: Model for waveguide fabrication in glass by two-step ion exchange with ionic masking, Opt. Lett. **13**, 71 (1988)

8.106 J. M. Inman, J. L. Bentley, S. N. Houde-Walter: Modeling ion-exchanged glass photonics: the modified quasi-chemical diffusion coefficient, J. Non-Cryst. Solids **191**, 209–215 (1995)

8.107 B. Messerschmidt, C. H. Hsieh, B. L. McIntyre, S. N. Houde-Walter: Ionic mobility in an ion exchanged silver–sodium boroaluminosilicate glass for micro-optics applications, J. Non-Cryst. Solids **217**, 264–271 (1997)

8.108 N. Haun, D. S. Kindred, D. T. Moore: Index profile control using Li+ for Na+ exchange in aluminosilicate glasses, Appl. Opt. **29**, 4056 (1990)

8.109 L. G. Atkinson, D. T. Moore, N. J. Sullo: Imaging capabilities of a long gradient-index rod, Appl. Opt. **21**, 1004 (1982)

8.110 D. C. Leiner, R. Prescott: Correction of chromatic aberrations in GRIN endoscopes, Appl. Opt. **22**, 383 (1983)

8.111 K. Fujii, S. Ogi, N. Akazawa: Gradient-index rod lens with a high acceptance angle for color use by Na+ for Li+ exchange, Appl. Opt. **33**, 8087 (1994)

8.112 Y. Mitsuhashi: SELFOC lenses: Applications in DWDM and optical data links, Proc. SPIE **3666**, 246–251 (1999)

8.113 S. Yuan, N. A. Riza: General formula for coupling-loss characterization of single-mode fiber collimators by use of gradient-index rod lenses, Appl. Opt. **38**, 3214–3222 (1999)

8.114 R. W. Gilsdorf, J. C. Palais: Single-mode fiber coupling efficiency with graded-index rod lenses, Appl. Opt. **33**, 3440 (1994)

8.115 I. Kitano, H. Ueno, M. Toyama: Gradient-index lens for low-loss coupling of a laser diode to single-mode fiber, Appl. Opt. **25**, 3336 (1986)

8.116 V. Blümel, B. Messerschmidt: Designs and applications of graded-index fast-axis-collimating lenses to high-power diode lasers, Proc. 10-th Microoptics Conference (MOC04), Jena 2004, ed. by Conventus Congress Management & Marketing GmbH (Elsevier, Amsterdam 2004) F-53

8.117 J. M. Stagaman, D. T. Moore: Laser diode to fiber coupling using anamorphic gradient-index lenses, Appl. Opt. **23**, 1730 (1984)

8.118 B. Messerschmidt, T. Possner, P. Schreiber: Gradient index optical systems for endoscope applications and beam shaping of laser diodes. In: *Diffractive Optics and Micro-Optics*, OSA Tech. Dig. (Optical Society of America, Washington, DC 2000) pp. 303–305

8.119 P. Rol, R. Jenny, D. Beck, F. Fankhauser, P. F. Niederer: Optical properties of miniaturized endoscopes for ophthalmic use, Opt. Eng. **34**, 2070–2076 (1995)

8.120 J. Knittel, L. G. Schnieder, G. Buess, B. Messerschmidt, T. Possner: Endoscope-compatible confocal microscope using a gradient-index lens system, Opt. Commun. **188**, 267–273 (2001)

8.121 J. C. Jung, M. J. Schnitzer: Multiphoton endoscopy, Opt. Lett. **28**, 902–904 (2003)

8.122 B. A. Flusberg, J. C. Jung, E. D. Cocker, E. P. Anderson, M. J. Schnitzer: In vivo brain imaging using a portable 3.9 gram two-photon fluorescence microendoscope, Opt. Lett. **30**, 2272–2274 (2005)

8.123 X. Li, C. Chudoba, T. Ko, C. Pitris, J. G. Fujimoto: Imaging needle for optical coherence tomography, Opt. Lett. **25**, 1520–1522 (2000)

8.124 J. M. Lopez-Higuera (Ed.): *Optical Sensors* (Cantabria Univ. Press, Santander 1998)
8.125 P. Drabarek: Modulation interferometer and fiberoptically divided measuring probe with light guided, US Patent WO 99/57506 (1999)
8.126 S. Doric: Generalized nonfull-aperture Luneburg lens: a new solution, Opt. Eng. **32**, 2118–2121 (1993)
8.127 Doric Lenses: www.doriclenses.com (Doric Lenses, Sainte-Foy 2006)
8.128 Lightpath Technologies: www.lightpath.com (Lightpath Technologies, Orlando 2006)
8.129 Nippon Sheet Glass: www.nsgamerica.com (NSG, Somerset 2006)
8.130 G. Goodman, B. Hunter: Photon. Spectra **9**, 132–138 (1999)
8.131 I. Kitano, K. Koizumi, H. Matsumura, K. Ikeda, T. Uchida: Image transmitter formed of a plurality of graded index fibers in bundled configuration, US Patent 3658407 (1972)
8.132 J. D. Rees: Non-Gaussian imaging properties of GRIN fiber lens arrays, Appl. Opt. **21**, 1009 (1982)
8.133 J. D. Rees, W. Lama: Some radiometric properties of gradient-index fiber lenses, Appl. Opt. **19**, 1065 (1980)
8.134 P. M. Moran, S. Dharmatilleke, A. H. Khaw, K. W. Tan, M. L. Chan, I. Rodriguez: Fluidic lenses with variable focal length, Appl. Phys. Lett. **88**, 041120-1–041120-3 (2006)
8.135 S. Gray: A letter from Mr. Stephen Gray, giving a further account of his water microscope, Phil. Trans. R. Soc. London **19**(223), 353–356 (1695)
8.136 G. Lippman: Relation entre les phenomenes electriques et capillaires, Ann. Chi. Phys. **5**, 494 (1875)
8.137 T. N. Young: An essay on the cohesion of fluids, Phil. Trans. R. Soc. London **95**, 65 (1805)
8.138 B. Berge: Electrocapillarite et mouillage de films isolants par l'eau, C. R. Acad. Sci. Ser. II **317**, 157–163 (1993)
8.139 F. Mugele, J.-C. Baret: Electrowetting from basics to applications, J. Phys. **17**, R705–R774 (2005)
8.140 J. Crassous, C. Gabay, G. Liogier, B. Berge: Liquid lens based on electrowetting: a new adaptive component for imaging applications in consumer electronics, Proc. SPIE **5639**, 143–148 (2004)
8.141 F. Gindele, T. Kolling, F. Gaul: Optical systems based on electrowetting, Proc. SPIE **5455**, 89–100 (2004)
8.142 B. Berge, J. Peseux: Variable focal lens controlled by an external voltage: an application of electrowetting, Eur. Phys. J. E **3**(2), 159–163 (2000)
8.143 B. H. W. Hendriks, S. Kuiper, M. A. J. van As, C. A. Renders, T. W. Tukker: Electrowetting-based variable-focus lens for miniature systems, Opt. Rev. **12**(3), 255–259 (2005)
8.144 C. Gabay, B. Berge, G. Dovillaire, S. Bucourt: Dynamic study of a Varioptic variable focal lens, Proc. SPIE **4767**, 159–165 (2002)
8.145 F. Krogmann, W. Mönch, H. Zappe: A MEMS-based variable micro-lens system, J. Opt. A – Pure Appl. Opt. **8**(7), S330–S336 (2006)
8.146 R. Graham: A variable focus lens and its use, J. Opt. Soc. Am. **30**, 560–563 (1940)
8.147 Bausch & Lomb: Variable Focus Lens, US Patent 2300251 (1942)
8.148 D.-Y. Zhang, N. Justis, Y.-H. Lo: Fluidic adaptive zoom lens with high zoom ratio and widely tunable field of view, Opt. Commun. **249**(1–3), 175–182 (2005)
8.149 A. Werber, H. Zappe: Tunable microfluidic microlenses, Appl. Opt. **44**(16), 3238–3245 (2005)
8.150 J. Chen, W. Wang, J. Fang, K. Varahramyan: Variable-focusing microlens with microfluidic chip, J. Micromech. Microeng. **14**(5), 675–680 (2004)
8.151 D.-Y. Zhang, V. Lien, Y. Berdichevsky, J. Choi, Y.-H. Lo: Fluidic adaptive lens with high focal length tunability, Appl. Phys. Lett. **82**(19), 3171–3172 (2003)
8.152 R. Kuwano, T. Tokunaga, Y. Otani, N. Umeda: Liquid pressure varifocus lens, Opt. Rev. **12**(5), 405–408 (2005)
8.153 A. H. Rawicz, I. Mikhailenko: Modeling a variable-focus liquid-filled optical lens, Appl. Opt. **35**(10), 1587–1589 (1996)
8.154 M. Agarwal, R. A. Gunasekaran, P. Coane, K. Varahramyan: Polymer-based variable focal length microlens system, J. Micromech. Microeng. **14**(12), 1665–1673 (2004)
8.155 D.-Y. Zhang, N. Justis, Y.-H. Lo: Fluidic adaptive lens of transformable lens type, Appl. Phys. Lett. **84**(21), 4194–4196 (2004)
8.156 N. Chronis, G. L. Liu, K.-H. Jeong, L. P. Lee: Tunable liquid-filled microlens array integrated with microfluidic network, Opt. Express **11**(19), 2370–2378 (2003)
8.157 C.-M. Ho, Y.-C. Tai: Annu. Rev. Fluid Mech. **30**, 579–612 (1998)
8.158 F. Schneider, D. Hohlfeld, U. Wallrabe: Miniaturized Electromagnetic Ferrofluid Actuator, ACTUATOR 2006, Proc. 10th Int. Conf. New Actuators, Bremen 2006 (HVG, Bremen 2006) B1.5, pp. 124–127
8.159 L. Dong, A. K. Agarwal, D. J. Beebe, H. Jiang: Adaptive liquid microlenses activated by stimuli-responsive hydrogels, Nature **442**, 551–554 (2006)
8.160 H. Oku, K. Hashimoto, M. Ishikawa: Variable-focus lens with 1-kHz bandwidth, Opt. Express **12**(10), 2138–2149 (2004)
8.161 S. Odenbach (Ed.): *Ferrofluids: Magnetically Controllable Fluids and Their Applications*, Lecture Notes Phys., Vol. 594 (Springer, Berlin, New York 2003)
8.162 F. C. Wippermann, P. Schreiber, A. Bräuer, B. Berge: Mechanically assisted liquid lens zoom system for mobile phone cameras, Proc. SPIE **6289**, 62890T (2006)

8.163 J.A. Armstrong, N. Bloembergen, J. Ducuing, P.S. Pershan: Interactions between light waves in a nonlinear dielectric, Phys. Rev. **127**, 1918–1939 (1962)

8.164 E.J. Lim, M.M. Fejer, R.L. Byer: Second-harmonic generation of green light in periodically poled planar lithium niobate waveguide, Electron. Lett. **25**, 174–175 (1989)

8.165 J. Webjörn, F. Laurell, G. Arvidsson: Blue light generated by frequency doubling of laser diode light ina lithium niobate channel waveguide, IEEE Photon. Technol. Lett. **1**, 316–318 (1989)

8.166 M.M. Fejer, G.A. Magel, D.H. Jundt, R.L. Byer: Quasi phase matched 2nd harmonic generation tuning and tolerances, IEEE J. Quantum Electron. **28**, 2631–2654 (1992)

8.167 L.E. Myers, R.C. Eckardt, M.M. Fejer, R.L. Byer, W.R. Bosenberg, J.W. Pierce: Quasi-phase-matched optical parametric oscillators in bulk periodically poled $LiNbO_3$, J. Opt. Soc. Am. B **12**, 2102–2116 (1995)

8.168 L.E. Myers, G.D. Miller, R.C. Eckardt, M.M. Fejer, R.L. Byer, W.R. Bosenberg: Quasi-phase-matched 1.064-μm-pumped optical parametric oscillator in bulk periodically poled $LiNbO_3$, Opt. Lett. **20**, 52–54 (1995)

8.169 C. Fischer, M.W. Siegrist: Solid-state mid-infrared laser sources, Top. Appl. Phys. **89**, 97–140 (2003)

8.170 D.A. Roberts: Simplified characterization of uniaxial and biaxial nonlinear optical crystals - A plea for standardization of nomenclature and conventions, IEEE J. Quantum Electron. **28**, 2057–2074 (1992)

8.171 P.S. Kuo, K.L. Vodopyanov, M.M. Fejer, D.M. Simanovskii, X. Yu, J.S. Harris, D. Bliss, D. Weyburne: Optical parametric generation of a mid-infrared continuum in orientation-patterned GaAs, Opt. Lett. **31**, 71–73 (2006)

8.172 D.A. Bryan, R. Gerson, H.E. Tomaschke: Increased optical damage resistance in lithium niobate, Appl. Phys. Lett. **44**, 847–849 (1984)

8.173 T.R. Volk, V.I. Pryalkin, N.M. Rubinina: Optical-damage-resistant $LiNbO_3$:Zn crystal, Opt. Lett. **15**, 996–998 (1990)

8.174 Y. Furukawa, K. Kitamura, A. Alexandrovski, R.K. Route, M.M. Fejer, G. Foulon: Green-induced infrared absorption in MgO doped $LiNbO_3$, Appl. Phys. Lett. **78**, 1970–1972 (2001)

8.175 T. Skauli, K.L. Vodopyanov, T.J. Pinguet, A. Schober, O. Levi, L.A. Eyres, M.M. Fejer, J.S. Harris: Measurement of the nonlinear coefficient of orientation-patterned GaAs and demonstration of highly efficient second-harmonic generation, Opt. Lett. **27**, 628–630 (2002)

8.176 G. Rosenman, P. Urenski, A. Agronin, A. Arie, Y. Rosenwaks: Nanodomain engineering in $RbTiOPO_4$ ferroelectric crystals, Appl. Phys. Lett. **82**, 3934–3936 (2003)

8.177 Y. Cho, S. Hashimoto, N. Odagawa, K. Tanaka, Y. Hiranaga: Realization of 10 Tbit/in.^{2} memory density and subnanosecond domain switching time in ferroelectric data storage, Appl. Phys. Lett. **87**, 232907 (2005)

8.178 C. Restoin, S. Massy, C. Darraud-Taupiac, A. Barthelemy: Fabrication of 1D and 2D structures at submicrometer scale on lithium niobate by electron beam bombardment, Opt. Mater. **22**, 193–199 (2003)

8.179 J. Son, Y. Yuen, S.S. Orlov, L. Hesselink: Sub-micron ferroelectric domain engineering in liquid phase epitaxy $LiNbO_3$ by direct-write e-beam techniques, J. Cryst. Growth **281**, 492–500 (2005)

8.180 C.L. Sones, M.C. Wengler, C.E. Valdivia, S. Mailis, R.W. Eason, K. Buse: Light-induced order-of-magnitude decrease in the electric field for domain nucleation in MgO-doped lithium niobate crystals, Appl. Phys. Lett. **86**, 212901 (2005)

8.181 M.C. Wengler, B. Fassbender, E. Soergel, K. Buse: Impact of ultraviolet light on coercive field, poling dynamics and poling quality of various lithium niobate crystals from different sources, J. Appl. Phys. **96**, 2816–2820 (2004)

8.182 E.P. Kokanyan, V.G. Babajanyan, G.G. Demirkhanyan, J.B. Gruber, S. Erdei: Periodically poled structures in doped lithium niobate crystals, J. Appl. Phys. **92**, 1544–1547 (2002)

8.183 I.I. Naumova, N.F. Evlanova, V.A. Dyakov, T.G. Chernevich, O.A. Shustin: Grown PPLN with small period: Selective chemical etching and AFM study, J. Mater. Sci. Mater. Electron. **17**, 267–271 (2006)

8.184 E. Soergel: Visualization of ferroelectric domains in bulk single crystals, Appl. Phys. B **81**, 729–751 (2005)

8.185 C.L. Sones, S. Mailis, W.S. Brocklesby, R.W. Eason, J.R. Owen: Differential etch rates in z-cut $LiNbO_3$ for variable HF/HNO_3 concentrations, J. Mater. Chem. **12**, 295–298 (2002)

8.186 M. Alexe, A. Gruverman (Eds.): *Nanoscale Characterisation of Ferroelectric Materials*, 1 edn. (Springer, Berlin, New York 2004)

8.187 K. Mizuuchi, T. Sugita, K. Yamamoto, T. Kawaguchi, T. Yoshino, M. Imaeda: Efficient 340-nm light generation by a ridge-type waveguide in a first-order periodically poled MgO : $LiNbO_3$, Opt. Lett. **28**, 1344–1346 (2003)

8.188 R. Le Targat, J.-J. Zondy, P. Lemonde: 75%-efficiency blue generation from an intracavity PPKTP frequency doubler, arXiv.org:physics, 0408031 (2004) Unpublished

8.189 M. Bode, I. Freitag, A. Tuennermann, H. Welling: Frequency-tunable 500-mW continuous-wave all-solid-state single-frequency source in the blue spectral region, Opt. Lett. **22**, 1220–1222 (1997)

8.190 C.M. Soukoulis (Ed.): *Photonic Band Gaps and Localization* (Plenum, New York 1993)

8.191 C. M. Bowden, J. P. Dowling, H. O. Everitt (Eds.): Development and applications of materials exhibiting photonic band gaps, J. Opt. Soc. Am. B **10**(2), 279–413 (1993)

8.192 G. Kurizuki, J. W. Haus (Eds.): Principles and applications of photonic band structures, J. Mod. Opt. **41**(2), 171–404 (1994)

8.193 J. D. Joannopoulos, R. D. Meade, J. N. Winn: *Photonic Crystals: Molding the Flow of Light* (Princeton Univ. Press, Princeton 1995)

8.194 C. M. Soukoulis (Ed.): *Photonic Band Gap Materials* (Kluwer, Dordrecht 1996)

8.195 A. Scherer, T. Doll, E. Yablonovitch, E. O. Everitt, J. A. Higgins (Eds.): Special section on electromagnetic cystal structures, design, synthesis and applications, J. Lightwave Technol. (11) **17**, 1928–2207 (1999)

8.196 S. G. Johnson, J. D. Joannopoulos (Eds.): *Photonic Crystals: The Road from Theory to Practice* (Kluwer, Dordrecht 2002)

8.197 K. Sakoda: *Optical Properties of Photonic Crystals* (Springer, Berlin, Heidelberg 2001)

8.198 R. M. De La Rue (Ed.): PECS 2000, Opt. Quantum Electron. **34**(1/3), 1–316 (2002)

8.199 S. Noda, T. Baba: *Roadmap on Photonic Crystals* (Kluwer, Dordrecht 2003)

8.200 K. Inoue, K. Ohtaka (Eds.): *Photonic Crystals: Physics, Fabrication and Applications* (Springer, Berlin, Heidelberg 2004)

8.201 H. Miyazaki, M. Hase, H. T. Miyazaki, Y. Kurosawa, N. Shinya: Photonic material for designing arbitrarily shaped waveguides in two dimensions, Phys. Rev. B **67**, 235109 (2003)

8.202 K. Ohtaka: Energy band of photons and low-energy photon diffraction, Phys. Rev. B **19**, 5057 (1979)

8.203 E. Yablonovitch: Inhibited spontaneous emission in solid-state physics and electronics, Phys. Rev. Lett. **58**, 2059 (1987)

8.204 N. Carlsson, T. Takemori, K. Asakawa, Y. Katayama: Scattering-method calculation of propagation modes in two-dimensional photonic crystals of finite thickness, J. Opt. Soc. Am. B **18**, 1260–1267 (2001)

8.205 K. Ohtaka: Theory I. Basic aspects of photonic bands. In: *Photonic Crystals*, ed. by K. Inoue, K. Ohtaka (Springer, Berlin, Heidelberg 2004) Chap. 3

8.206 Y. Segawa, K. Ohtaka: Other types of photonic crystals. In: *Photonic Crystals*, ed. by K. Inoue, K. Ohtaka (Springer, Berlin, Heidelberg 2004) Chap. 8

8.207 T. E. Ebbessen, H. J. Lezec, H. F. Ghaemi, T. Thio, P. A. Wolff: Extraordinary optical transmission through subwavelength hole arrays, Nature **391**, 667–669 (1998)

8.208 M. Notomi: Theory of light propagation in strongly modulated photonic crystal; Refractionlike behavior in the vicinity of the photonic band gap, Phys. Rev. B **62**, 10696 (2000)

8.209 J. B. Pendry: Negative refraction makes a perfect lens, Phys. Rev. Lett. **85**, 3966 (2000)

8.210 M. Notomi, H. Suzuki, T. Tamamura, K. Edagawa: Lasing action due to the two-dimensional quasiperiodicity of photonic quasicrystals with a Penrose lattice, Phys. Rev. Lett. **92**, 123906 (2004)

8.211 Y. Akahane, T. Asano, B. S. Song, S. Noda: Fine-tuned high-Q photonic-crystal nanocavity, Opt. Express **13**, 1202 (2005)

8.212 T. Yoshie, A. Scherer, G. Khitrova, H. M. Gibbs, G. Rupper, C. Ell, O. B. Shehekin, D. G. Deppe: Rabi splitting with a single quantum dot in a photonic crystal nanocavity, Nature **432**, 200 (2004)

8.213 B. S. Song, S. Noda, T. Asano, Y. Akahane: Ultra-high-Q photonic double-heterostructure nanocavity, Nature Mater. **4**, 207 (2005)

8.214 J. C. Knight, T. A. Birks, P. St. J. Russell, D. M. Atkin: All-silica single-mode optical fiber with photonic crystal cladding, Opt. Lett. **21**, 1547 (1996)

8.215 J. C. Knight, J. Broeng, T. A. Birks, P. St. J. Russell: Photonic band gap guidance optical fiber, Science **282**, 1476 (1998)

8.216 K. Ohtaka: Theory II. Advanced topics of photonic crystals. In: *Photonic Crystals*, ed. by K. Inoue, K. Ohtaka (Springer, Berlin, Heidelberg 2004) Chap. 4

8.217 P. Yeh: *Optical Waves in Layered Media* (Wiley, New York 1988) Chap. 6,7

8.218 K. Ohtaka, T. Ueta, Y. Tanabe: Analog of optics of photonic crystals to that of anisotropic crystals, J. Phys. Soc. Jpn **65**, 3068 (1996)

8.219 E. Cubukcu, K. Aydin, E. Ozbay, S. Foteinopoulou, C. M. Soukoulis: Electromagnetic waves: Negative refraction by photonic crystals, Nature **423**, 604 (2003)

8.220 A. Barrier, M. Mulot, M. Swillo, M. Qui, L. Thylen, A. Talneau, S. Anand: Negative Refraction at Infrared Wavelengths in a Two-Dimensional Photonic Crystal, Phys. Rev. Lett. **93**, 073902 (2004)

8.221 T. Baba: Photonic crystal devices. In: *Photonic Crystals*, ed. by K. Inoue, K. Ohtaka (Springer, Berlin, Heidelberg 2004) Chap. 11

8.222 A. Sugitatsu, T. Asano, S. Noda: Characterization of line-defect-waveguide lasers in two-dimensional photonic-crystal slabs, Appl. Phys. Lett. **84**, 5395 (2004)

8.223 K. Inoue, H. Sasaki, K. Ishida, Y. Sugimoto, N. Ikeda, Y. Tanaka, S. Ohkouchi, Y. Nakamura, K. Asakawa: InAs quantum dot laser utilizing GaAs photonic-crystal line-defect waveguide, Opt. Express **12**, 5502 (2004)

8.224 Y. Sugimoto, Y. Tanaka, N. Ikeda, Y. Nakamura, K. Asakawa, K. Inoue: Low propagation loss of 0.76 dB/mm in GaAs-based single-line-defect two-dimensional photonic crystal slab wave guides up to 1 cm in length, Opt. Express **12**, 1090 (2004)

8.225 E. Kuramochi, M. Notomi, S. Hughes, A. Shinya, T. Watanabe, L. Ramunno: Disoder- induced scattering loss of line-defect waveguides in photonic crystal slabs, Phys. Rev. B **72**, 161318 (2005)

8.226 H. Nakamura, Y. Sugimoto, K. Kanamoto, N. Ikeda, Y. Tanaka, Y. Nakamura, S. Ohkouchi, Y. Watanabe, K. Inoue, H. Ishikawa, K. Asakawa: Ultra-fast photonic crystal/quantum dot all-optical switch for future photonic networks, Opt. Express **12**, 6606 (2004)

8.227 H. Horiuchi, T. Ochiai, J. Inoue, Y. Segawa, Y. Shibata, K. Ishi, Y. Kondo, M. Kanbe, H. Miyazaki, F. Hinode, S. Yamaguchi, K. Ohtaka: Exotic radiation from a photonic crystal excited by an ultra-relativistic electron beam, Phys. Rev. E **74**, 056601 (2006)

8.228 T. Ochiai, K. Ohtaka: Theory of unconventional Smith-Purcell radiation in finite-size photonic crystal, Opt. Express **14**, 7378–7397 (2006)

8.229 D.J.H. Maclean: *Optical Line Systems* (Wiley, Chichester 1996)

8.230 C.K. Kao, G.A. Hockam: Dielectric fiber surface waveguides for optical frequencies, IEE Proc. **133**, 1151 (1966)

8.231 T. Miya, Y. Terunama, T. Hosaka, T. Miyashita: An ultimate low loss single mode fiber at 1.55 μm, Electron. Lett. **15**, 106 (1979)

8.232 T. Moriyama, O. Fukuda, K. Sanada, K. Inada, T. Edahvio, K. Chida: Ultimately low OH content V.A.D. optical fibers, Electron. Lett. **16**, 689 (1980)

8.233 A. Ghatak, K. Thyagarajan: *Optical Electronics* (Cambridge Univ. Press, Cambridge 1989)

8.234 D. Gloge: Weakly guiding fibers, Appl. Opt. **10**, 2252 (1971)

8.235 A. Ghatak, K. Thyagarajan: *Introduction to Fiber Optics* (Cambridge Univ. Press, Cambridge 1998)

8.236 D. Marcuse: Gaussian approximation of the fundamental modes of a graded index fibers, J. Opt. Soc. Am. **68**, 103 (1978)

8.237 D. Marcuse: Interdependence of waveguide and material dispersion, Appl. Opt. **18**, 2930–2932 (1979)

8.238 M.J. Li: Recent progress in fiber dispersion compensators, Proc ECOC, Amsterdam 2001, Tech. Dig. ThM1.1

8.239 R. Ramaswami, K.N. Sivarajan: *Optical Networks: a Practical Perspective* (Morgan Kaufmann, San Francisco 1998)

8.240 Y. Nagasawa, K. Aikawa, N. Shamoto, A. Wada, Y. Sugimasa, I. Suzuki, Y. Kikuchi: High performance dispersion compensating fiber module, Fujikura Rev. **30**, 1 (2001)

8.241 K. Thyagarajan, R.K. Varshney, P. Palai, A. Ghatak, I.C. Goyal: A novel design of a dispersion compensating fiber, Photon. Tech. Lett. **8**, 1510 (1996)

8.242 J.L. Auguste, R. Jindal, J.M. Blondy, Marcou J. Clapeau, B. Dussardier, G. Monnom, D.B. Ostrowsky, B.P. Pal, K. Thyagarajan: 1800 ps/(nm.km) chromatic dispersion at 1.55 μm in dual concentric core fibre, Electron. Lett. **36**, 1689 (2000)

8.243 S. Ramachandran (Ed.): *Fiber Based Dispersion Compensation* (Springer, Berlin, Heidelberg 2007) in press

8.244 R. Kashyap: *Fiber Bragg Gratings* (Academic, San Diego 1999)

8.245 A. Othonos, K. Kalli: *Fiber Bragg Grating: Fundamentals and Applications in Telecommunications and Sensing* (Artech House, Boston 1999)

8.246 P.C. Becker, N.A. Olsson, J.R. Simpson: *Erbium Doped Fiber Amplifiers* (Academic, San Diego 1999)

8.247 W.L. Barnes, R.I. Laming, E.J. Tarbox, P. Morkel: Absorption and emission cross section of Er^{3+} doped silica fibers, IEEE J. Quant. Electron **27**, 1004–1010 (1991)

8.248 E. Desurvire: *Erbium Doped Fiber Amplifiers* (Academic, New York 1994)

8.249 M.A. Arbore, Y. Zhou, H. Thiele, J. Bromage, L. Nelson: S-band Erbium doped fiber amplifiers for WDM transmission between 1488 and 1508 nm, Proc. Optical Fiber Communications Conference, Atlanta 2003, Tech. Dig. WK2

8.250 K. Thyagarajan, C. Kakkar: S-band single stage EDFA with 25 dB gain using distributed ASE suppression, IEEE Photon. Tech. Lett. **16**, 2448–2450 (2004)

8.251 J. Bromage: Raman amplification for fiber communication systems, J. Lightwave Technol. **22**, 79 (2004)

8.252 K. Thyagarajan, C. Kakkar: Fiber design for broadband, gain flattened Raman fiber amplifier, IEEE Photon. Technol. Lett. **15**, 1701–1703 (2003)

8.253 K. Thyagarajan, C. Kakkar: Segmented-clad fiber design for tunable leakage loss, J. Lightwave Technol. **23**(Special issue, Optical Fiber Design), 3444–3453 (2005)

8.254 M.N. Islam: Overview of Raman amplification in telecommunications. In: *Raman Amplifiers for Telecommunications*, Vol. 1, ed. by N. Islam M. (Springer, New York 2004)

8.255 G.P. Agarwal: Fiber optic Raman amplifiers. In: *Guided Wave Optical Components and Devices*, ed. by P. Pal B. (Elsevier, Amsterdam 2006)

8.256 A.R. Chraplyvy: Limitations on lightwave communications imposed by optical-fiber nonlinearities, J. Lightwave Technol. **8**, 1548 (1990)

8.257 R.W. Tkach, A.R. Chraplyvy, F. Forghieri, A.H. Gnauck, R.M. Derosier: Four photon mixing and high speed WDM systems, J. Lightwave Technol. **13**, 841 (1995)

8.258 G.P. Agarwal: *Nonlinear Fiber Optics* (Academic, Boston 1989)

8.259 P. Petropoulos, T.M. Monro, W. Belardi, K. Furusawa, J.H. Lee, D.J. Richardson: 2R-regenerative all-optical switch based on a highly nonlinear holey fiber, Opt. Lett. **26**, 1233 (2001)

8.260 J.K. Ranka, R.S. Windeler, A.J. Stentz: Visible continuum generation in air silica microstructure optical fibers with anomalous dispersion at 800 nm, Opt. Lett. **25**, 25–27 (2000)

8.261 M. Saruwatari: All-optical signal processing for terabit/second optical transmission, IEEE J. Sel. Top. Quantum Electron. **6**, 1363 (2000)

8.262 J. Hansryd, A. Andrekson, A. Westlund, J. Li, P. Hedekvist: Fiber based optical parametric amplifiers and their applications, IEEE Sel. Top. Quantum Electron. **8**, 506 (2002)

8.263 J. K. Ranka, R. S. Windeler: Nonlinear interactions in air-silica microstructure optical fibers, Opt. Photon. News, August 2000, p. 20

8.264 K. Kurokawa, K. Tajima, J. Zhou, K. Nakajima, T. Matsui, I. Sankawa: Penalty free dispersion managed soliton transmission over 100 km low loss PCF, Proc. Optical fiber Communications Conference 2005 Tech. Dig. PDP 21

9. Optical Detectors

Optical detectors are applied in all fields of human activities – from basic research to commercial applications in communication, automotive, medical imaging, homeland security, and other fields. The processes of light interaction with matter described in other chapters of this handbook form the basis for understanding the optical detectors physics and device properties.

This chapter starts with a brief historical sketch of first experiments facilitating development of optical detectors. The overview of photo detector types is followed by the description of the most important figures of merit and different detection regimes.

A detailed description of different types of optical detectors is presented in the following sections. The device structure and physics as well as important materials for fabrication, figures of merit, and brief application notes are given for photoconductors, photodiodes, quantum well photodetectors, semiconductor detectors with intrinsic amplification, charge transfer detectors, photoemissive detectors, and thermal photodetectors. The chapter includes also a brief overview of imaging systems and principles of black and white and color photography.

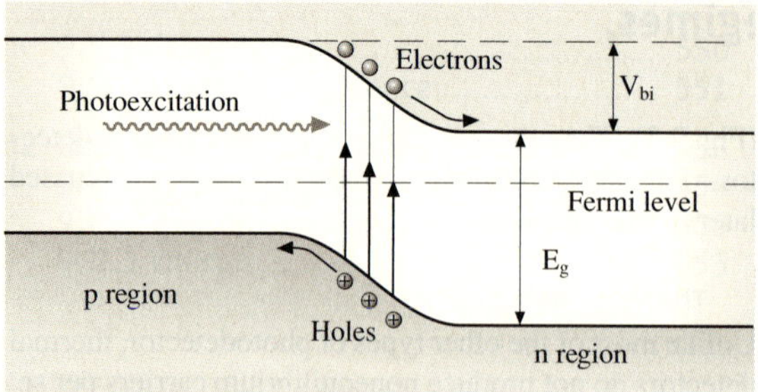

Fig. 9.1 Photovoltaic effect. The contact of p-type and n-type semiconductors creates a barrier with a built-in potential V_{bi}, which facilitates the separation of photogenerated carriers

Photon Noise

Assuming that the photon flux follows Poisson statistics, the noise σ_N for an incident optical photon flux Φ containing $\bar{N}$ photons received over a given period of time Δt can be expressed in units of number of photons per second by

$$\sigma_N = \sqrt{\bar{N}} = \sqrt{\Phi \Delta t}\,. \tag{9.3}$$

We will see later that the relative strengths of different sources of photon noise determine the regimes in which the detector can be operated.

Johnson Noise

Johnson or Nyquist noise is caused by the thermal motion of charge carriers in a resistive element. This is a white source noise. In general, the Johnson noise power transferred through a circuit with an effective frequency bandwidth Δf is described by [9.1]:

$$P_J = kT\Delta f\,, \tag{9.4}$$

where k is the Boltzmann constant and T is the absolute temperature.

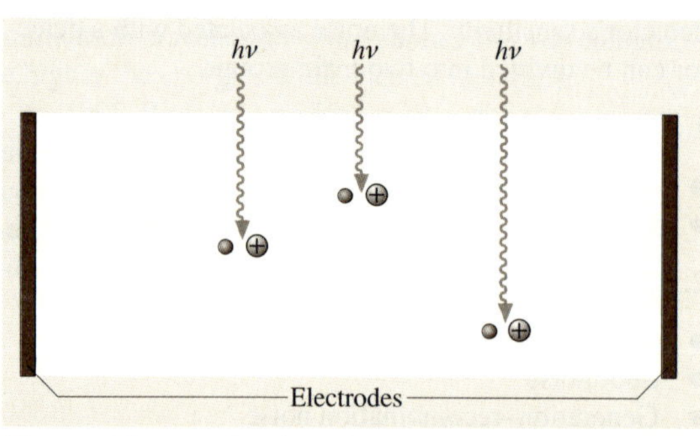

Fig. 9.2 Photoconductive detector. Upon illumination, non-equilibrium carriers are created increasing the conductivity of the detector

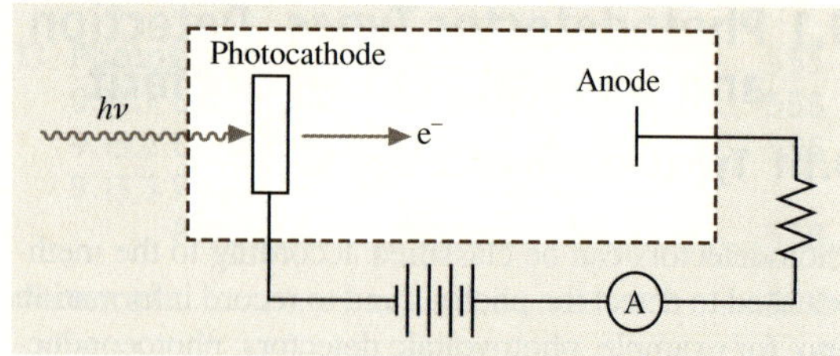

Fig. 9.3 A schematic of a simple photoemissive detector

The Johnson noise voltage V_J (or current i_J) is the voltage (current) that causes the noise power given by (9.4) to be transferred between two equal noise-producing resistors R. Alternatively, the Johnson noise voltage (current) can be calculated from statistical mechanics using the expression for the noise power spectrum of a classical system at equilibrium [9.2]

$$V_J = \sqrt{4kTR\Delta f} \quad \text{and} \quad i_J = \sqrt{\frac{4kT\Delta f}{R}}\,. \tag{9.5}$$

Shot Noise

The shot noise generated in a detector is due to the discrete (stochastic) nature of photoelectron generation. The variance of the shot noise current is

$$i_{sh}^2 = \frac{q^2}{\Delta t^2}\overline{(n-\bar{n})^2} = \frac{q^2\bar{n}}{\Delta t^2} = 2qI\Delta f\,, \tag{9.6}$$

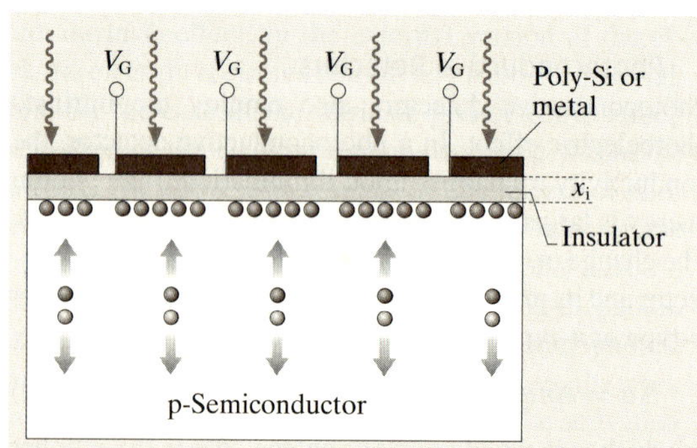

Fig. 9.4 An example of a charge transfer sensor. Each active pixel of the sensor is a capacitor consisting of a electrode (poly-Si or semitransparent metal film) at the top, then an insulating layer of thickness x_i, and a semiconductor layer underneath. Light penetrates the semiconductor bulk through the semitransparent electrode and generates nonequilibrium carriers in the semiconductor bulk. By applying a gate voltage V_G, the nonequilibrium carriers can be separated and stored in the potential well underneath each electrode separately

in which q is the elementary charge, n and $\bar{n}$ are the immediate and average numbers of photoexcited carriers produced in a time period Δt, the photon statistics are assumed to be Poissonian (meaning that $\overline{(n-\bar{n})^2}=\bar{n}$), and the mean photocurrent is $I=\bar{n}q/\Delta t=2\bar{n}q\Delta f$. Finally, the shot noise for the detector is

$$i_{\text{sh}}=\sqrt{2qI\Delta f}\,. \tag{9.7}$$

The shot noise is a white source noise and the expression (9.7) can be integrated over the whole bandwidth of interest Δf to get the total shot noise of the detector. This equation is usually applied only to photon detectors that contain a potential barrier, such as photovoltaic detectors.

Generation–Recombination Noise

The generation–recombination noise is an analog of the shot noise that is applicable to the photoconductor. This type of noise is caused by fluctuations in the generation and recombination of current carriers (due to photons, thermal generation, etc.), and it is expressed by [9.2]

$$i_{\text{G-R}}=2I\left(\frac{\tau\Delta f}{n\left[1+(2\pi f\tau)^2\right]}\right)^{1/2}, \tag{9.8}$$

where the mean current I is due to all sources of current carriers, τ is the carrier lifetime, n is the total number of free carriers, and f is the frequency at which the noise is measured. This type of noise is not a white noise source, since there is a clear frequency dependence. It can be shown that the total generation–recombination shot noise contribution for the photoconductor depends on the photoconductive gain G (which is the number of electrons generated per initial photogenerated electron), and this type of noise can be approximated by

$$i_{\text{G-R}}=2\sqrt{qIG\Delta f} \tag{9.9}$$

for the white noise limit case $\left[(2\pi f\tau)^2\ll 1\right]$. In the opposite case $\Delta f\tau\to\infty$, the generation–recombination shot noise contribution approaches the limit

$$i_{\text{G-R}}=\frac{qIG}{\tau}\,. \tag{9.10}$$

1/*f* Noise

This noise is also known as flicker noise. The power spectrum of this noise falls rapidly with frequency (according to $1/f$). The origin of this noise is not well understood, but is often attributed to imperfect ohmic contacts or to surface-state traps and dislocations [9.1]. The $1/f$ noise component does not present a fundamental limit to sensitivity. The $1/f$ noise can be reduced to negligible levels using current device technology.

Temperature Fluctuations

In thermal detectors, any fluctuations in the temperature of the detection element that are not due to a change in the signal will produce additional noise in the output signal. The temperature fluctuations can be expressed using the analogy with the shot generation–recombination current (9.8), in which the carrier lifetime is substituted by the heat relaxation time constant $\tau_{\text{H}}=\Theta/K$, where K stands for the thermal conductance and Θ is the heat capacity:

$$T_{\text{rms}}=2\left(\frac{Kk\bar{T}^2\Delta f}{K^2+(2\pi f\Theta)^2}\right)^{1/2}, \tag{9.11}$$

In (9.11) we used the notation $\bar{T}$ for the average temperature of the detector.

Microphonic Noise

The microphonic noise is caused by the mechanical displacement of wiring and other components of the detector. Mechanical vibrations in the detector may cause capacitance changes in detection elements, which is the main reason for microphonic noise.

9.1.3 Detection Regimes

Common problems encountered with any type of optical detector include deciding how best to terminate the photodetector with a suitable load resistor and finding the optimum tradeoff between the bandwidth and signal-to-noise ratio.

If we consider the equivalent circuit of a photodetector terminated with a load resistor R_{L}, then the output voltage $V=IR_{\text{L}}$ has a bandwidth (-3 dB high-frequency cutoff) given by

$$\Delta f=\frac{1}{2\pi R_{\text{L}}C}, \tag{9.12}$$

where C is a stray capacitance across the photodetector.

Two main noise contributions are added to the signal. One is the Johnson noise (9.5) of the load resistance $R=R_{\text{L}}$; the other is the shot noise ((9.7) in the case of a photovoltaic detector or (9.8) in the case of a photoconductor). To calculate the shot noise (generation–recombination) current, one should take into account the total photocurrent under illumination, which is the sum of the signal current I_{p} and the dark current I_{dark}:

$$I_{\text{light}} = I_{\text{p}} + I_{\text{dark}} . \quad (9.13)$$

Since the Johnson noise and shot noise are statistically independent, the resulting quadratic mean value of the noise current for the most general case of a photodetector with an intrinsic gain G is

$$i_{\text{n}}^2 = 2q(I_{\text{p}} + I_{\text{dark}})\Delta f G^2 F + \frac{4kT\Delta f}{R_{\text{L}}} . \quad (9.14)$$

For detectors with no intrinsic amplification, the gain $G = 1$ and the excess noise factor $F = 1$.

From (9.12) and (9.14) it can be seen that bandwidth and noise optimization imposes opposite requirements on the value of R_{L}. The largest possible R_{L} is required to minimize the noise current; on the other hand, the smallest possible R_{L} must be used in order to maximize Δf.

It is generally accepted that the best possible detector sensitivity is achieved when the shot noise is greater than the Johnson noise, i. e.,

$$2q(I_{\text{p}} + I_{\text{dark}})\Delta f G^2 F \geq \frac{4kT\Delta f}{R_L} , \quad (9.15)$$

which implies a certain limiting condition on the minimum value of the load resistor. Using (9.14), the signal-to-noise ratio of the detector is calculated as the ratio of the average signal photocurrent $I = I_{\text{p}}G$ to the total noise current i_{n}

$$\begin{aligned} \frac{S}{N} &= \frac{I}{i_{\text{n}}} \\ &= \frac{I_{\text{p}}G}{\left(2q\left(I_{\text{p}} + I_{\text{dark}}\right)\Delta f G^2 F + \frac{4kT\Delta f}{R_{\text{L}}}\right)^{1/2}} . \end{aligned} \quad (9.16)$$

Analyzing (9.16), two detection regimes can be defined, according to whether the signal I_{p} is larger or smaller than the value

$$I_0 = I_{\text{dark}} + \frac{2kT}{qR_{\text{L}}FG^2} . \quad (9.17)$$

Assuming $I_{\text{p}} \gg I_0$ and $F = 1$, we obtain

$$\frac{\text{S}}{\text{N}} = \left(\frac{I_{\text{p}}}{2q\Delta f}\right)^{1/2} , \quad (9.18)$$

where the S/N value is called the *quantum noise limit* of detection. This limit cannot be overcome by any detection system. In fact, (9.18) is a direct consequence of the quantum nature of light and Poissonian photon statistics.

In the small signal regime, $I_{\text{p}} \ll I_0$, the signal-to-noise ratio is

$$\frac{\text{S}}{\text{N}} = \frac{I_{\text{p}}}{(2qI_0\Delta f)^{1/2}} , \quad (9.19)$$

which is the *thermal noise limit* of detection.

9.1.4 Figures of Merit

Certain figures of merit are used to compare the performances of detectors of the same type. One such figure – the signal-to-noise ratio – was introduced in the previous section. Here we will consider other important figures of merit that are generally applicable to almost all photodetectors.

Responsivity

The responsivity $\mathfrak{R}$ is the ratio of the detector output (in amperes or volts) to the intensity of the input optical flux Φ_{p} (in watts). For example, the spectral photocurrent responsivity $\mathfrak{R}_{\text{I}}(\lambda, f)$ of the detector at a given wavelength is

$$\mathfrak{R}_{\text{I}}(\lambda, f) = \frac{I_{\text{p}}}{\Phi_{\text{p}}(\lambda)} , \quad (9.20)$$

where I_{p} is the measured photocurrent. The blackbody responsivity $\mathfrak{R}(T, f)$ is the detector output divided by the incident radiant power from a blackbody source of temperature T modulated at a frequency f that produces the observed output [9.3]

$$\begin{aligned} \mathfrak{R}_{\text{I}}(T, f) &= \frac{I_{\text{p}}}{\int_0^{\alpha} \int_{\lambda} \Phi_{\text{p}}(\lambda)\,\text{d}\lambda} \\ &= \frac{I_{\text{p}}}{A_{\text{source}}\sigma_{\text{SB}}T^4 A_{\text{det}}/\pi L^2} , \end{aligned} \quad (9.21)$$

in which α is a solid angle at which the detector is seen from the source site, A_{source} and A_{det} are the source and detector areas, respectively, L is the distance between the source and detector, and $\sigma_{\text{SB}} = 5.67032 \times 10^{-12}\,\text{W cm}^{-2}\,\text{K}^{-4}$ is the Stefan–Boltzmann constant. Note that the blackbody responsivity is a measure of the response of the detector to the incident radiation, integrated over all wavelengths.

Quantum Efficiency

The quantum efficiency η is the probability that a photoelectron is produced when a photon is incident on the detector. For an incident optical power Φ_{p}, which produces a photocurrent $I_{\text{p}} = \mathfrak{R}_{\text{I}}\Phi_{\text{p}}$, the quantum efficiency is calculated from

$$\eta = \frac{\left(\frac{I_p}{q}\right)}{\left(\frac{\Phi_p}{h\nu}\right)} = \frac{h\nu}{q}\Re_I , \quad (9.22)$$

where q is the elementary charge, h is the Planck constant and ν is the photon frequency.

Noise Equivalent Power

The noise equivalent power (NEP) of the detector is the optical power incident on the detector Φ_p required to produce an output signal equal to the noise output. Sometimes the NEP is specified for a bandwidth of 1 Hz. It can also be defined as being the value of the optical input power needed to produce a signal-to-noise ratio of one. Since the current signal output is $I_p = \Re_I \Phi_p$, the signal-to-noise ratio is

$$\frac{S}{N} = \frac{\Re_I \Phi_p}{i_n} , \quad (9.23)$$

where the rms noise output current i_n is given by (9.14). Placing the left-hand side of (9.23) to unity, we obtain

$$\text{NEP} = \frac{i_n}{\Re_I} . \quad (9.24)$$

In cases when the thermal noise dominates over all other noise sources, the system signal-to-noise ratio is described by (9.19). The noise equivalent power in this case defines the value S_0, known as the device sensitivity

$$S_0 = \text{NEP} = \frac{2}{\Re_I}\sqrt{\frac{kT\Delta f}{R_L}} = \frac{2h\nu}{\eta q}\sqrt{\frac{kT\Delta f}{R_L}} . \quad (9.25)$$

This equation is valid for a photovoltaic device or photoconductor with no intrinsic gain ($G = 1$ and $F = 1$).

In the specific case of photon-limited performance, the system's signal-to-noise ratio is described by (9.18). This mode of operation is typical of photomultiplier tubes, microchannels and some detectors with intrinsic multiplication in the visible and near-infrared spectral ranges. Photon-limited performance is also typical of photovoltaic, photoconductive and some thermal detectors in the infrared spectral range. As mentioned above, the ultimate performance of the detector is reached when the detector noise and amplifier noise are low compared to the photon noise. The photon noise is fundamental, in the sense that it arises not from any imperfection in the detector or its associated electronics, but rather from the detection process itself, as a result of the discrete nature of the electromagnetic field. The radiation falling on the detector is a combination of that from the target (signal) Φ^S (in number of incident photons per second) and that from the background Φ^B (also in number of incident photons per second). An infrared photon detector that achieves background-limited performance is called a "*background-limited infrared photodetector*" (BLIP). The noise equivalent power is then obtained for the BLIP case from (9.24) and (9.18), using the expression (9.22) for the quantum efficiency η, and by noting that the radiant flux $\Phi_p^{B,S} = \Phi^{B,S} h\nu$

$$\text{NEP}_{\text{BLIP}} = h\nu\sqrt{\frac{2\Phi^B \Delta f}{\eta}} = \sqrt{\frac{2h\nu\Phi_p^B \Delta f}{\eta}} . \quad (9.26)$$

Detectivity

The detectivity D of a detector is the reciprocal of the noise equivalent power:

$$D = \frac{1}{\text{NEP}} . \quad (9.27)$$

A more useful figure of merit is the specific detectivity D^* ("dee-star"), which does not depend on the detector area A_{det} and bandwidth Δf

$$D* = D\sqrt{A_{\text{det}}\Delta f} = \frac{\sqrt{A_{\text{det}}\Delta f}}{\text{NEP}} . \quad (9.28)$$

D^* may be used to directly compare detectors of different sizes whose performances have been measured over different bandwidths. The relationship between the spectral D^* and the blackbody D^* is [9.2]:

$$D^*(T, f) = \frac{\int_0^\infty D^*(\lambda, f)\Phi(T, \lambda)\,\mathrm{d}\lambda}{\int_0^\infty \Phi(T, \lambda)\,\mathrm{d}\lambda} . \quad (9.29)$$

For the BLIP case, any shot noise-limited detector will have an D^* given by

$$D^*_{\text{BLIP}}(\lambda, f) = \frac{\sqrt{A_{\text{det}}\Delta f}}{\text{NEP}_{\text{BLIP}}(\lambda, f)} = \frac{1}{h\nu}\sqrt{\frac{\eta A_{\text{det}}}{2\Phi^B}} . \quad (9.30)$$

9.2 Semiconductor Photoconductors

A photoconductor is a light-sensitive resistor that, at its simplest, can consist of a slab of semiconductor with ohmic contacts on opposite sides (Fig. 9.2). The intrinsic photoelectric effect is the dominant mechanism that creates nonequilibrium carriers in the conduction band of an intrinsic photoconductor. In an extrinsic photoconductor, the excitation of shallow impurity levels is involved. The long-wavelength cutoff for this simple case is given by

$$\lambda_c = \frac{hc}{E_g} , \tag{9.31}$$

where E_g is the semiconductor bandgap and c is the speed of light. The conductivity of an unilluminated intrinsic semiconductor device is

$$\sigma_0 = n_0\mu_n q + p_0\mu_p q , \tag{9.32}$$

where n_0, p_0, μ_n, and μ_p are the equilibrium concentrations of free electrons and holes and their mobilities, respectively, and q is the elementary charge. The reciprocal value is called the resistivity and is defined as

$$\rho_0 = \left(n_0\mu_n q + p_0\mu_p q\right)^{-1} . \tag{9.33}$$

Under illumination, nonequilibrium carriers are created at a concentration of

$$\Delta n = \Delta p = \frac{\eta_{ex}\Phi\tau}{A_{det}W} , \tag{9.34}$$

where η_{ex} is the external quantum efficiency, defined above as the probability that a photoelectron is produced when a photon is incident on the detector; Φ is the photon flux in photons per unit time; τ is the carrier lifetime; A_{det} and W are the detector's active area and thickness. Using (9.32) and (9.34), we can obtain the relative change in conductivity of the illuminated photoconductor

$$\frac{d\sigma}{\sigma} = \frac{q\left(\mu_n + \mu_p\right)\eta_{ex}\Phi\tau}{\sigma A_{det}W} . \tag{9.35}$$

The light-driven change in photoconductor resistance R_{det} is described by

$$dR_{det} = -R_{det}\frac{q\left(\mu_n + \mu_p\right)\eta_{ex}\Phi\tau}{\sigma A_{det}W} . \tag{9.36}$$

For an extrinsic semiconductor, the sum $\left(\mu_n + \mu_p\right)$ should be replaced with the mobility of the available carrier.

9.2.1 Photoconductors – Figures of Merit

Responsivity

If the load resistor R_L is connected in series with the photoconductor and the two resistors are biased with a DC voltage V_0, then the light-induced voltage drop across a uniformly illuminated photoconductor can be expressed as

$$dV = \frac{V_0 R_L R_{det}}{\left(R_L + R_{det}\right)^2}\frac{q\lambda\eta_{ex}\tau\left(\mu_n + \mu_p\right)}{\sigma hc A_{det}W}\Phi_p , \tag{9.37}$$

where Φ_p is the input radiant flux (in watts), λ is the wavelength, h is Planck's constant, and c is the speed of light. The voltage and current spectral responsivities are obtained from (9.37)

$$\begin{aligned}\mathfrak{R}_V &= \frac{dV}{\Phi_p} \\ &= \frac{Iq\lambda\eta_{ex}\tau\left(\mu_n + \mu_p\right)}{\sigma hc A_{det}W}\frac{R_L R_{det}}{R_L + R_{det}}\ [\mathrm{V/W}] ,\end{aligned} \tag{9.38}$$

$$\mathfrak{R}_I = \frac{Iq\lambda\eta_{ex}\tau\left(\mu_n + \mu_p\right)}{\sigma hc A_{det}W}\ [\mathrm{A/W}] , \tag{9.39}$$

where $I = \frac{V_0}{R_L + R_{det}}$ is the total DC current through the detector.

Photoconductive Gain

Analyzing the expression (9.39) for the photocurrent responsivity, the photoconductive gain of the intrinsic detector can be defined as

$$G = \frac{I\tau\left(\mu_n + \mu_p\right)}{\sigma A_{det}W} . \tag{9.40}$$

A similar expression for an extrinsic photoconductor containing carriers with a mobility value μ is

$$G = \frac{\tau\mu E}{d} = \frac{\tau}{t} , \tag{9.41}$$

where $E = I/\sigma A_{det}W$ is the electric field across the detector, d is the spacing between the electrodes, and $t = d/\mu E$ is the total carrier transit time between the two electrodes, assuming that the electric field is far from saturation.

Noise

1/f Noise. The $1/f$ dependence holds for the noise power. The noise current/voltage varies as $1/\sqrt{f}$. This noise is always present in photoconductors since they operate at bias current. The empirical equation obtained for the mean square noise current is [9.3]

$$\overline{i_{1/f}^2} = \frac{\text{const} \cdot I^2 \Delta f}{f^\beta} , \quad (9.42)$$

where const is a proportionality constant, Δf and f are the bandwidth and frequency respectively, and β is a constant close to 1.

Johnson Noise. For a semiconductor with an internal resistance of R_{det}, the Johnson noise mean square current is expressed by the right-hand side of (9.5).

Generation–Recombination Noise. The mean square generation–recombination noise current due to photon and thermal excitations of the semiconductor is given by [9.4]

$$\overline{i_{\text{G-R}}^2} = 4q\left(q\eta_{\text{ex}}\Phi A_{\text{det}}G^2 + qG_{\text{th}}G^2\right) , \quad (9.43)$$

where Φ is the incident photon flux, G is the gain, and G_{th} is the thermal generation rate.

Noise Equivalent Power (NEP)

The signal-to-noise ratio for the conductor can be obtained via (9.5) and (9.13) as

$$\frac{\text{S}}{\text{N}} = \frac{\Re_{\text{I}}\Phi_{\text{p}}}{i_{\text{n}}} = \frac{\Re_{\text{I}} h\nu\Phi}{\sqrt{\overline{i_{\text{J}}^2} + \overline{i_{\text{G-R}}^2}}} , \quad (9.44)$$

where $\Re_{\text{I}}$ is the current responsivity given by (9.39), and $\Phi_{\text{p}} = h\nu\Phi$ is the incident radiant power. The related figure of merit – NEP – can be obtained from (9.44) by setting $\frac{\text{S}}{\text{N}} = 1$.

In the case of a photon-limited conductor, the noise equivalent power is determined by the generation–recombination noise current, and it can be expressed using (9.26) and (9.24)

$$\text{NEP} = \frac{\sqrt{\overline{i_{\text{G-R}}^2}}}{\Re_{\text{I}}} = h\nu\sqrt{\frac{2\Phi^{\text{S,B}}\Delta f}{\eta_{\text{ex}}}} , \quad (9.45)$$

where $\Phi_{\text{S,B}}$ is the input signal/background photon flux per unit time on the detector. In the case where the Johnson noise is dominant,

$$\text{NEP} = \frac{\sqrt{\overline{i_{\text{J}}^2}}}{\Re_{\text{I}}} = \frac{h\nu}{q\eta_{\text{ex}}G}\sqrt{2kT\Delta f\left(\frac{1}{R_{\text{det}}} + \frac{1}{R_{\text{L}}}\right)} . \quad (9.46)$$

Detectivity

The specific detectivity for the photon-limited conductor is determined by the expression

$$D^*(\lambda, f) = \frac{1}{h\nu}\sqrt{\frac{\eta_{\text{ex}}A_{\text{det}}}{2\Phi^{\text{S,B}}}} . \quad (9.47)$$

For the Johnson noise limited conductor, the corresponding specific detectivity is

$$D^*(\lambda, f) = \frac{q\eta_{\text{ex}}G}{h\nu}\sqrt{\frac{A_{\text{det}}}{2kT}\left(\frac{R_{\text{L}}R_{\text{det}}}{R_{\text{L}} + R_{\text{det}}}\right)} . \quad (9.48)$$

9.2.2 Photoconductors: Materials and Examples

Lead-Salt-Based Compound Photoconductors

PbS, PbSe, and PbTe cells are the most common examples of semiconductor-based photoconductors. They are sensitive to infrared radiation, have high quantum efficiencies, fast time responses, and can operate at room temperature. Cooled lead salt photoconductors have detectivities that are close to the background limit.

Lead-salt-based photodetectors are manufactured through either chemical deposition or the evaporation of thin ($\approx 1\,\mu\text{m}$) polycrystalline layers of lead salts. The layers are usually sensitized by the controlled introduction of oxygen. It was found that the conductivity of the film changes from n-type to p-type with increasing exposure to oxygen at 300 K.

By controlling parameters such as the dopants used, the deposition temperature, the passivation coating used, and the film thickness, performance characteristics such as resistance, responsivity, time response, spectral response range, noise, and signal-to-noise ratio can be tailored over a wide range. The wavelengths of peak response for different lead salt photodetectors can be tuned within the range of $\approx 2.4\,\mu\text{m}$ to $> 5\,\mu\text{m}$. The spectral response can be extended into infrared by cooling the detector due to the positive temperature coefficient of the energy gap [9.5].

Mercury Cadmium Telluride (MCT) Photoconductors

MCT photoconductors ($Hg_{1-x}Cd_xTe$) that operate over different spectral regions can be created simply by tuning the Cd/Hg ratio. The photoconductive $Hg_{1-x}Cd_xTe$ detector is a good intrinsic detector and can achieve background-limited sensitivities at operating temperatures substantially higher than extrinsic detectors.

$Hg_{1-x}Cd_xTe$ photoconductors are typically fabricated from n-type material with an excess concentration of ionized donors of about $10^{14}\,\mathrm{cm}^{-3}$ [9.1]. This large intrinsic carrier concentration is the result of excess mercury atoms that exceed the stoichiometry and reside interstitially in the lattice. For a typical $Hg_{1-x}Cd_xTe$ detector with a square optically active area and a thickness of 10 μm, the resistance is rather small ($\approx 200\,\Omega$). The disadvantage of such a low-resistance detector is that high currents can flow with relatively low voltage, which produces high dissipation power and limits the detector to applications in the low voltage range.

Selenium Photoconductors

Amorphous selenium (α-Se) is widely used to build the photoconductor arrays used in direct conversion systems for γ-radiation detection.

9.3 Semiconductor Photodiodes

Semiconductor photodiodes are the most popular photodetectors; they are capable of operating in both photovoltaic and photoconductive modes.

9.3.1 Semiconductor Photodiode Principles

p–n Diode in Thermal Equilibrium

Semiconductor photodiodes can be considered to be a variation on regular p–n junction diodes. The p–n junction is obtained by joining together oppositely doped semiconductors. Once in contact, thermodynamic equilibrium is attained by lining up the Fermi levels across the whole structure. Electrons from the n-type semiconductor diffuse into the p region and holes from the p-type semiconductor into the n region, and so a surplus is created of negative electric charge in the p region and a surplus of positive charge in the n region. This creates an electric field that sweeps mobile charges out of the junction region, creating a so-called "space-charge region" depleted of mobile carriers (Fig. 9.5).

Using the band model, the requirement to line up the Fermi levels of n-type and p-type semiconductor species results in a built-in potential V_{bi}, which for the thermal equilibrium case is given by [9.6]

$$V_{bi} = \frac{1}{q}\left(E_i^p - E_i^n\right) = \frac{kT}{q}\ln\frac{N_A N_D}{n_i^2}\,, \tag{9.49}$$

where E_i^p and E_i^n are the intrinsic levels in the neutral p and n regions, N_A and N_D are the acceptor and donor dopant concentrations, and n_i is the intrinsic carrier density, which is obtained using the values for the energy bandgap E_g and the effective densities of states N_V and N_C in the valance band and conduction bands, respectively

$$n_i = \sqrt{N_V N_C}\exp\left(-\frac{E_g}{2kT}\right)\,, \tag{9.50}$$

$$N_V = 2\left(\frac{2\pi m_p kT}{h^2}\right)^{3/2} \quad \text{and}$$

$$N_C = 2\delta\left(\frac{2\pi m_n kT}{h^2}\right)^{3/2}\,. \tag{9.51}$$

The effective masses of the electrons m_n and holes m_p as well as the conduction band degeneracy factor δ (the number of equivalent minima in the conduction band) are specific to each semiconductor.

The space-charge region (depletion region) extends from the junction into the n and p regions. For the simplest case of a two-sided abrupt junction, the width $w = w_n + w_p$ (Fig. 9.5) of the depletion region is found from the requirement that the electric field at the boundary of the space-charge region is zero, and that the potential difference equals the built-in voltage

$$w = w_n + w_p = \sqrt{\frac{2\varepsilon\varepsilon_0\left(N_A + N_D\right)}{qN_A N_D}V_{bi}}\,, \tag{9.52}$$

where ε_0 is the vacuum permittivity, ε is the semiconductor dielectric constant, and w_n and w_p are the space-charge region widths of the n-type and p-type regions of the semiconductor, separated by the junction.

p–n Diode with External Bias

For a forward bias V, the voltage across the junction will decrease from the equilibrium value V_{bi} to $(V_{bi} - V)$, and the width of the space-charge region will shrink correspondingly. For the reverse bias condition, the voltage across the junction will increase to the value $(V_{bi} + V)$, causing an expansion of the space-charge region. The width of the space-charge region (depletion region) for an external bias V is given by

$$w = \sqrt{\frac{2\varepsilon\varepsilon_0\left(N_A + N_D\right)}{qN_A N_D}\left(V_{bi} \mp V\right)}\,, \tag{9.53}$$

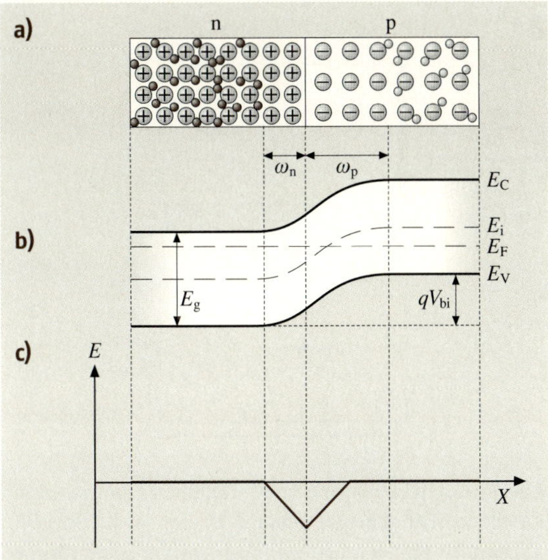

Fig. 9.5a–c An abrupt p–n junction in thermal equilibrium.**(a)** Space-charge distribution: *small circles* show mobile electrons and holes, *large circles* indicate ionized donors and acceptors; **(b)** energy band diagram; **(c)** electric field distribution

in which the $\mp$ signs are for the forward- and reverse-bias conditions, respectively

The total current density J for the ideal photodiode is expressed by the Shockley equation (curve 1 in Fig. 9.6) [9.6]

$$J = J_S \left[\exp\left(\frac{qV}{kT}\right) - 1\right] \equiv q\left(\frac{n_{p0}D_n}{L_n} + \frac{p_{n0}D_p}{L_p}\right)\left[\exp\left(\frac{qV}{T}\right) - 1\right], \quad (9.54)$$

where J_s is the so-called saturation current density, n_{p0} and p_{n0} are the minority carrier concentrations at equilibrium in the p and n regions, respectively, and the carrier diffusion constants D_n and D_p are given by the Einstein relationship

$$D_n = \frac{kT}{q}\mu_n \quad \text{and} \quad D_p = \frac{kT}{q}\mu_p\,, \quad (9.55)$$

in which μ_n and μ_p are the carrier mobilities. The carrier diffusion lengths L_n and L_p are

$$L_n = \sqrt{D_n\tau_n} \quad \text{and} \quad L_p = \sqrt{D_p\tau_p} \quad (9.56)$$

where τ_n and τ_p describe the electron and hole lifetimes. The equilibrium concentration of minority carriers is

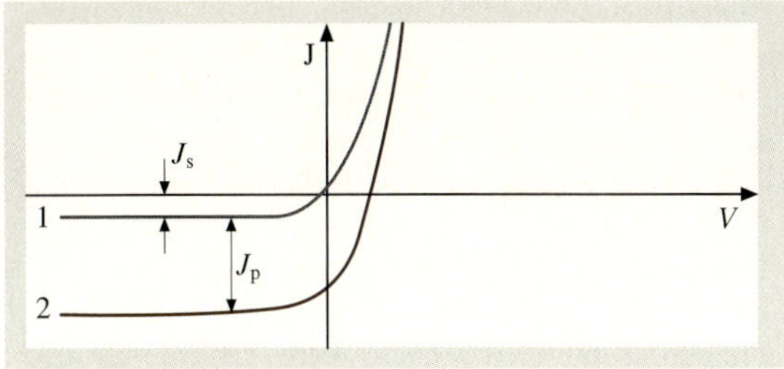

Fig. 9.6 I–V characteristics of an ideal photodiode in the dark (1) and under illumination (2)

determined from $p_{n0} = (n_i^2/n_{n0}) \approx (n_i^2/N_D)$ for the n-type semiconductor and $n_{p0} = (n_i^2/p_{p0}) \approx (n_i^2/N_A)$ for the p-type semiconductor.

Equation (9.54) was obtained for an ideal photodiode with an abrupt junction, using the Boltzmann approximation for carrier concentrations throughout the space-charge layer, assuming that the minority carrier densities are small compared to those of the majority carrier (a low injection assumption), and assuming no generation–recombination current in the space-charge area, which imposes a related requirement on the constant electron and hole currents through the depletion layer.

Using (9.53), one can obtain an expression for the depletion layer capacitance per unit area, which is one of the main parameters that determines the time response of the semiconductor photodiode. For a two-sided abrupt junction,

$$C_j = \frac{\varepsilon\varepsilon_0}{w} = \sqrt{\frac{\varepsilon\varepsilon_0 q N_A N_D}{2\left(N_A + N_D\right)\left(V_{bi} \mp V\right)}}\,. \quad (9.57)$$

p–n Diode Under Illumination

To derive the total photocurrent generated in the bulk of a p–n semiconductor diode, assuming a very shallow p layer on top of the n-type substrate and neglecting the thermally generated current, one must separately consider the photocurrent generated in the space-charge region (of width w) the photocurrent from the neutral region of the bulk of the semiconductor.

Case 1 – Front-Illuminated Photodiode. First consider the situation where light penetrates the photodiode bulk from the p side and the surface p^+ layer is much thinner than the light absorption length $1/\alpha(\delta \ll 1/\alpha$; α is the absorption coefficient) (Fig. 9.7, panel a). This case is typical of so-called front-illuminated photodiodes. Also assume that the crystal thickness is much larger than the light penetration depth, meaning that all light quanta are absorbed during a single passage inside the crystal. We will also neglect the effects of surface carrier recombination to simplify our considerations.

The e–h pairs generated in the space-charge region of the p–n junction by light absorption are separated by the electric field, giving rise to a drift photocurrent J_{dr}. Those carriers that are generated in the neutral, undepleted part of the semiconductor bulk ($x > w$) may also reach the space-charge region via the diffusion mechanism if the diffusion distance is shorter then the carrier diffusion depth L_n or L_p. This current is treated as the diffusion current J_{diff}.

The electron–hole photogeneration rate $G(x)$ in the semiconductor bulk at a distance x from the surface is determined from

$$G(x) = \frac{\partial \Phi(x)}{\partial x} \mathrm{d}x = \Phi_0 \alpha \exp(-\alpha x) , \tag{9.58}$$

in which Φ_0 is the incident optical photon flux per unit area, α is the absorption coefficient, and $\Phi(x) = \Phi_0 \exp(-\alpha x)$ (in accord with the Beer–Lambert law) is the photon flux inside the semiconductor bulk at a distance x from the photodiode surface. The drift current is then

$$J_{dr} = -q \int_0^w G(x) \mathrm{d}x = q\Phi_0 \left[1 - \exp(-\alpha w)\right] . \tag{9.59}$$

For $x > w$, the minority carrier density p_n in the semiconductor bulk is determined by the one-dimensional diffusion equation

$$D_p \frac{\partial^2 p_n}{\partial x^2} - \frac{p_n - p_{n0}}{\tau_p} + G(x) = 0 . \tag{9.60}$$

Since the sample thickness was assumed to be much thicker than $1/\alpha$, the boundary condition at $x = \infty$ is $p_n = p_{n0}$. The boundary condition at $x = w$ is also $p_n = p_{n0}$ because the electric field in the depletion region sweeps away nonequilibrium carriers effectively. Hence, the solution to (9.60) for $x > w$ is given by

$$p_n = p_{n0} - A \exp(-\alpha w) \exp\left\{\frac{w - x}{L_p}\right\} + A \exp(-\alpha x) , \tag{9.61}$$

$$A = \frac{\Phi_0}{D_p} \frac{\alpha L_p^2}{\left(1 - \alpha^2 L_p^2\right)} . \tag{9.62}$$

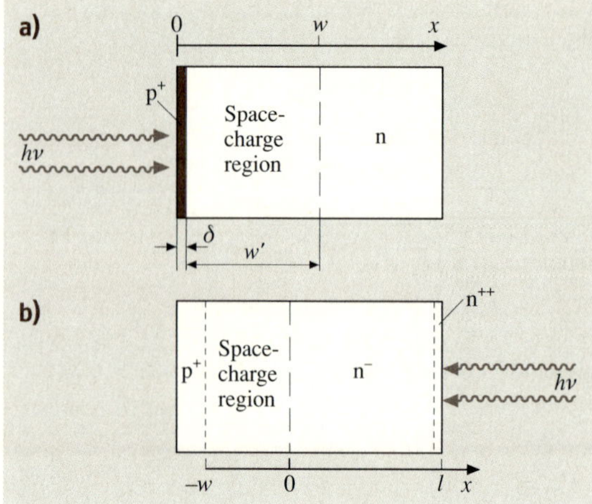

Fig. 9.7 (a) Operation of a front-illuminated p/n junction photodiode. The junction depth δ is very small, so $w \approx w'$. The drift photocurrent originates in the space-charge region and the diffusion current comes from the undepleted n region; **(b)** Back-illuminated photodiode operation. The carrier injection occurs from the back surface of the diode, assuming that the photocurrent only contains the diffusion component

The diffusion current density is then

$$J_{diff} = -qD_p \left(\frac{\partial p_n}{\partial x}\right)_{x=w} = q\Phi_0 \frac{\alpha L_p}{1 + \alpha L_p} \exp(-\alpha w) . \tag{9.63}$$

Combining (9.59) and (9.63) yields the total photocurrent density

$$J_p = J_{dr} + J_{diff} = q\Phi_0 \left(1 - \frac{\exp(-\alpha w)}{1 + \alpha L_p}\right) , \tag{9.64}$$

which shows that the photocurrent J_p is proportional to the incident photon flux Φ_0.

If the thermally generated current (9.54) is not negligible, then the total photodiode current is

$$J_{total} = q \left(\frac{n_{p0} D_n}{L_n} + \frac{p_{n0} D_p}{L_p}\right) \times \left[\exp\left(\frac{qV}{kT}\right) - 1\right] + J_p \tag{9.65}$$

and the corresponding typical current–voltage ($I - V$) characteristic of the photodiode under illumination is shown in Fig. 9.6, curve 2.

If the external leads of the photodetector are shorted, the voltage across the junction is maintained at the built-in potential level and photocurrent will flow. This current is called the *short circuit current* (J_{sc}) and it can be approximated from (9.65) by putting $V = 0$

$$J_{sc} \cong J_p \,. \tag{9.66}$$

On the other hand, if the leads are open, the voltage across the junction will drop to such a value that the flow of photogenerated carriers is compensated for by an increase in the diffusion current given by (9.54). This voltage change appears as an output voltage on the device leads and is called the *open circuit voltage* (V_{oc}), which, ignoring the change of the depletion width under illumination, can be calculated from (9.65) by placing $J_{total} = 0$:

$$V_{oc} = \frac{kT}{q} \ln\left(\frac{\Phi_0\left(\exp(-\alpha w)/(1+\alpha L_p^{-1})\right)}{(n_{p0}D_n)/L_n + (p_{n0}D_p)/L_p}\right) . \tag{9.67}$$

Case 2 – Back-Illuminated Photodiode. In this case, the junction is created on one side of the semiconductor crystal and light penetrates from the other side of the structure. It is interesting to consider the situation when light is absorbed mostly in the neutral region of the semiconductor. To further simplify the description, assume that absorption occurs within a very thin surface layer of the photodiode's backside (absorption coefficient $\alpha \to \infty$). This means that the carrier surface photogeneration rate is described by the δ function

$$G(x) = \Phi_0 \alpha \exp\left[-\alpha\,(l-x)\right] \delta(x-l) \,, \tag{9.68}$$

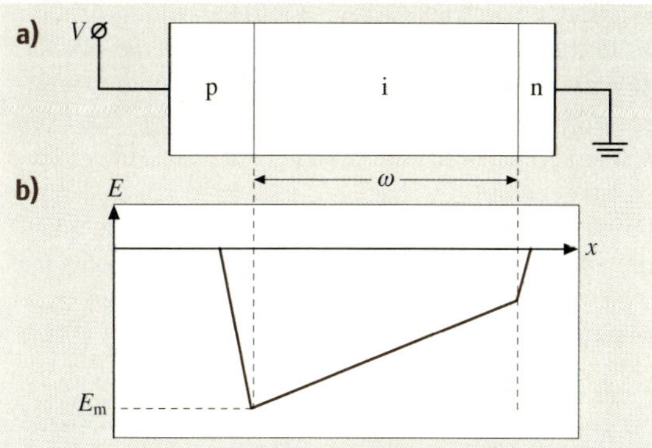

Fig. 9.8 (a) Biased p–i–n photodiode and **(b)** electric field distribution at reverse bias conditions. The depletion width w and the maximum field strength E_m depend on the magnitude of the reverse bias applied. The slope of the electric field distribution in each region depends on the donor/acceptor concentration

where l is the distance between the edge of the depletion region and the surface to which light penetrates (Fig. 9.7, panel b). We will also account for the surface recombination events, which may be important in back-illuminated diodes with extended diffusion regions.

Since the excess carriers are extracted at the edge of the depletion region, the boundary condition at $x = 0$ is $p_n = p_{n0}$. The other boundary condition is obtained from the diffusion current at the surface

$$\begin{aligned} J_{diff}\,|_{x=c} &\equiv qD_p\left(\frac{\partial p_n}{\partial x}\right)_{x=l} \\ &= q\Phi_0 - qS_r\left[p_n(l) - p_{n0}\right] , \end{aligned} \tag{9.69}$$

where S_r is the surface recombination rate in cm/s and $p_n(l)$ is the minority carrier concentration at the surface. For the boundary conditions chosen, the solution to the diffusion equation is

$$\begin{aligned} p_n =& p_{n0} \\ &+ \frac{\Phi_0\left(1 - \frac{S_r}{D_p}\alpha\delta(x-l)L_p^2\right)\sinh(x/L_p)}{(D_p/L_p)\cosh\left(l/L_p\right) + S_r \sinh\left(l/L_p\right)} \\ &+ \frac{\Phi_0\alpha\delta(x-l)L_p^2}{D_p} . \end{aligned} \tag{9.70}$$

The photocurrent in this case contains the diffusion component only and is calculated as

$$\begin{aligned} J_p = J_{diff}\,|_{x=0} &= qD_p\left(\frac{\partial p_n}{\partial x}\right)_{x=0} \\ &= \frac{q\Phi_0}{\cosh(\frac{l}{L_p}) + \frac{S_r L_p}{D_p}\sinh(\frac{l}{L_p})} . \end{aligned} \tag{9.71}$$

p–i–n Photodiode

The efficiency of the transfer of nonequilibrium carriers into the external circuit is much higher in the space-charge region than in the neutral region of the semiconductor. The width w of the space-charge region increases as $(\sqrt{N_{A,D}})^{-1}$ as the donor/acceptor concentrations $N_{A,D}$ decrease; see (9.53). The space-charge layer width will be maximized if an intrinsic layer is inserted between the p-type and n-type layers, resulting in a *p–i–n* structure like that shown in Fig. 9.8. In practice, an intrinsic layer is often substituted for a low-implant concentration layer. The behavior of such a *p–i–n* photodiode is described by the same equations as presented above; however, those equations can be simplified in some cases because the dopant concentrations on different sides of the junction are significantly different.

9.3.2 Photodiodes – Figures of Merit

Quantum Efficiency and Responsivity

The external quantum efficiency of the photodiode is defined as the ratio of the number of carriers producing photocurrent to the number of incident photons Φ_0. Using this definition and (9.64), we obtain for the front-illuminated photodiode

$$\eta_{ex} = \frac{(J_p/q)}{\Phi_0} = 1 - \frac{\exp(-\alpha w)}{1+\alpha L_p} . \tag{9.72}$$

Note that the right-hand side of (9.72) is valid for a relatively large value of αw when light is completely absorbed during the first passage through the semiconductor bulk and nothing is reflected from the backside of the die.

For the case of a back-illuminated photodiode, as described above, the external quantum efficiency is obtained from (9.71):

$$\eta_{ex} = \frac{(J_p/q)}{\Phi_0} = \left[\cosh(\frac{l}{L_p}) + \frac{S_r L_p}{D_p}\sinh(\frac{l}{L_p})\right]^{-1} . \tag{9.73}$$

The internal quantum efficiency is defined as the ratio of the number of electron–hole pairs generated by the light to the number of photons absorbed in the semiconductor bulk:

$$\eta_{in} = \frac{\eta_{ex}}{(1-R_F)} = \frac{(J_p/q)}{\Phi_0(1-R_F)} , \tag{9.74}$$

where R_F is the reflection coefficient of the light incident surface.

The photodiode current responsivity is the ratio of the total output photocurrent I_p (in amperes) to the incident optical power Φ_p(in watts):

$$R_I = \frac{I_p}{\Phi_p} = \frac{\eta_{ex} q}{h\nu} = \frac{\eta_{ex} q\lambda}{hc} \approx \frac{\eta_{ex}\lambda[\mu m]}{1.24}[A/W] , \tag{9.75}$$

where ν is the photon frequency, c is the speed of light in a vacuum, and λ is the wavelength in micrometers.

Carrier Collection Efficiency Function

The internal quantum efficiency $\eta_{in}(\lambda)$ of an ideal photodiode for the case $\alpha w \gg 1$, when all penetrating quanta of light are absorbed by the crystal (of thickness h), and all nonequilibrium carriers participate in the photocurrent, can be calculated using a simple expression

$$\eta_{in}(\lambda) = 1 - \exp[-\alpha(\lambda)h] \equiv \int_0^h dx\alpha(\lambda)\exp[-\alpha(\lambda)x] \tag{9.76}$$

in which x is the photodiode penetration depth starting from the surface. In a real (nonideal) structure, not all of the photogenerated nonequilibrium carriers are collected by the external circuit and participate in the current. This effect is accounted for by the carrier collection efficiency function $P(x)$ [9.7]

$$\eta_{in}(\lambda) = \int_0^h dx P(x)\alpha(\lambda)\exp[-\alpha(\lambda)x] . \tag{9.77}$$

When $\alpha w \gg 1$, the light passes through the photodiode crystal more than once and the expression for the quantum efficiency needs to be modified. When two passages occur, we obtain [9.8]

$$\eta_{in}(\lambda) = \int_0^h dx P(x)\alpha(\lambda)\exp[-\alpha(\lambda)x] + R_B\exp[-\alpha(\lambda)h] \times \int_0^h dx P(h-x)\alpha(\lambda)\exp[-\alpha(\lambda)x] , \tag{9.78}$$

in which the first term on the right-hand side describes the number of carriers collected into the external circuit due to the first passage of radiation inside the crystal, while the second term describes the corresponding number for the radiation reflected back into the crystal from the rear surface of the die (R_B is the die backside reflection coefficient). The backside reflection is considered to be specular reflection here. Assuming that the quantum efficiency $\eta_{in}(\lambda)$ is known from experiment, the carrier collection efficiency function can be retrieved from (9.78) [9.9]. Note that, usually, $P(x) = 1$ inside

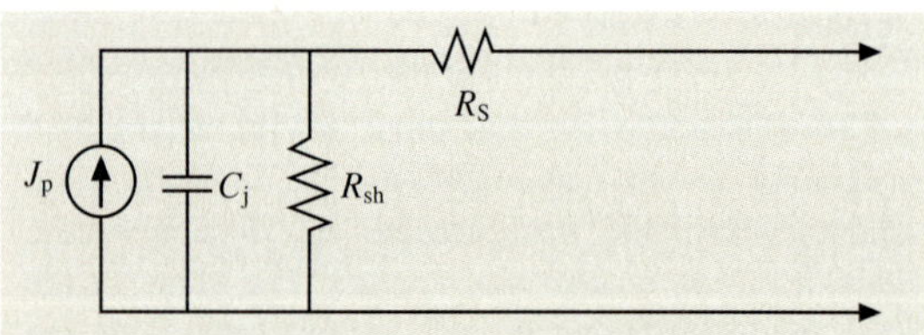

Fig. 9.9 Simplified equivalent circuit of the photodiode

the depletion region, except for the thin surface layer of the photodiode. The expression (9.78) can be easily extended to the case of multiple passages of light inside the semiconductor photodiode bulk.

Shunt Resistance, Series Resistance, and Forward Characteristics

The photodiode equivalent circuit can be thought of as a source of current with a source (or shunt) resistance R_{sh} connected in parallel to it, and a series (or forward) resistance R_s (Fig. 9.9). For the ideal photodiode, $R_{sh} = \infty$ and $R_s = 0$. The photodiode shunt resistance is defined as the dynamic resistance at the origin, and it can be calculated from (9.54)

$$R_{sh} = \frac{\partial V}{\partial I}|_{V=0} = \frac{kT}{qJ_s A_j} \quad (9.79)$$

where A_j is the junction area and J_s is the saturation current density. The shunt resistance is an important characteristic of any photodiode, since it defines the thermal noise current.

The series resistance R_s of a photodiode is defined as the sum of the total resistance R_{cont} of the ohmic contacts and the resistance of the undepleted semiconductor bulk. In the most general case, the series resistance can be expressed as

$$R_s = \frac{1}{A_j}\int_{x_1}^{x_2} \rho(x)\,dx + R_{cont}\,, \quad (9.80)$$

where $\rho(x)$ is the semiconductor bulk resistivity, x_1 is the depletion layer edge and x_2 is the semiconductor bulk boundary. For a planar structure with a shallow abrupt junction like that shown in Fig. 9.5, the series resistance can be calculated as

$$R_s \approx \frac{d-w}{A_j}\rho + R_{cont}\,, \quad (9.81)$$

where d is the substrate thickness, w is the space-charge region width given by (9.53), and ρ is the substrate resistivity, which is $\rho_n = (n\mu_n q)^{-1}$ for the n-type substrate and $\rho_p = (p\mu_p q)^{-1}$ for the p-type substrate.

As it is evident from the Shockley equation (9.54), the junction resistance for the ideal semiconductor diode $R_j = (\partial V/\partial I)$ approaches zero as the forward voltage increases. However, (9.54) does not account for the generation and recombination of carriers in the depletion region, high-injection conditions, imperfect ohmic contacts. These three factors can significantly alter the forward current, even for a relatively small forward bias.

The major recombination–generation process in the depletion region is the capture of the injected carriers, which adds the recombination current J_{rec} to the Shockley equation. It can be shown that this current is proportional to the intrinsic carrier density n_i, the trap density N_T, the carrier capture cross-section σ_{capt}, and the carrier thermal velocity $v_{th} = \sqrt{(3kT/m^*)}$ (with m^* representing for the carrier effective mass) [9.6]

$$J_{rec} \approx \frac{qw}{2}\sigma_{capt} v_{th} N_T n_i \exp\left(\frac{qV}{2kT}\right)\,. \quad (9.82)$$

Combining (9.82) and (9.54), the forward current J_f for the p–n structure can be approximated for $p_{n0} \gg n_{p0}$ and $V_f > (kT/q)$ by

$$J_f \approx q\sqrt{\frac{D_p}{\tau_p}}\frac{n_i^2}{N_D}\exp\left(\frac{qV_f}{kT}\right) + \frac{qw}{2}\sigma_{capt} v_{th} N_T n_i \exp\left(\frac{qV_f}{2kT}\right)\,. \quad (9.83)$$

In general, the experimental forward current can be fitted by the dependence

$$J_f \propto \exp\left(\frac{qV_f}{\gamma kT}\right)\,, \quad (9.84)$$

where γ is an ideality factor that equals 2 when the recombination current dominates and is closer to 1 when the diffusion current becomes dominant. Equation (9.84), which relates the forward current and voltage, is used to estimate how close the photodiode properties are to those of an ideal photodiode.

Breakdown Voltage

As follows from the Shockley equation, the reverse bias current saturates at high voltages if the diffusion of minority carriers into the space-charge region is the only mechanism that builds up the reverse dark current. An additional increase in the reverse dark current occurs due to the generation of carriers in the space-charge region, which may result from the thermal generation of carriers in either the semiconductor substrate itself or within various defects in the semiconductor volume. The latter effect depends on the defects/impurity concentration, while the former can be expressed via the thermal generation rate G_{th} as

$$G_{th} = \beta n_0 p_0 = \beta n_i^2 = \frac{n_i}{\tau_G}\,, \quad (9.85)$$

in which $\beta = (n_i\tau_G)^{-1}$ is a proportionality constant, τ_G is the carrier generation lifetime, and $n_i^2 = n_0 p_0$ is the thermal equilibrium intrinsic carrier concentration, with n_0 and p_0 describing the equilibrium concentrations of electrons and holes. The generation current density within a space-charge region of width w is

$$J_G = qwG_{th} . \quad (9.86)$$

A further increase in the photodiode reverse bias beyond the value at which w extends through the whole semiconductor bulk (the reach-through regime) results in an electric field high enough to cause electrical breakdown, and the reverse bias current increases dramatically. The two mechanisms that are generally considered to be involved are the Zener and avalanche breakdown mechanisms.

Zener Breakdown. This mechanism involves the interaction of the electric field with covalently bound electrons. At a large reverse bias, the conduction band of the n-type region is shifted to well below the top of the valance band of the p-type region of the semiconductor diode. The occupied energy levels in the conduction band of the p-side of the semiconductor line up with the unoccupied energy levels in the conduction band of the n-side, and a strong electric field facilitates the tunneling of the valence electrons through the potential barrier from the p-side towards the n-side of the semiconductor. The barrier height is equal to the semiconductor bandgap E_g, and the tunneling probability ϑ can be found from the expression for tunneling through a triangular barrier [9.10]

$$\vartheta \approx \exp\left(-\frac{8\pi\sqrt{2m}E_g^{3/2}}{3qhE}\right) , \quad (9.87)$$

where m is the mass of an electron, h is Planck's constant, and E is the electric field. The tunneling current density can be estimated as

$$J_{tunn} = qn_Z v_d \vartheta , \quad (9.88)$$

in which n_Z and v_d are the density and the drift velocity of the valance electrons arriving at the Zener barrier, respectively.

Avalanche Breakdown. Avalanche breakdown occurs when the electric field is high enough to trigger multiplication processes during collisions of carriers accelerated by the field with atoms of the substrate. The probability of multiplication depends not only on the value of the electric field but also on the carrier ionization rate and the extent of the avalanche region. The avalanche process can be characterized by the multiplication factor M, which in the case of one type of carrier can be written as [9.11, 12]

$$M_{n,p} = \exp\left(\int_0^L \alpha_{n,p}(x)\,dx\right) , \quad (9.89)$$

where L is the width of the multiplication region, $\alpha_{n,p}(x)$ is the electron/hole ionization rate, which depends on the coordinate x inside the multiplication region. For a simple case of equal ionization rates $\alpha_n = \alpha_p$, the condition for avalanche breakdown can be deduced as $M \to \infty$:

$$\int_0^L \alpha_{n,p}(x)\,dx = 1 . \quad (9.90)$$

Since the ionization rate is a function of the electric field, which varies inside the space-charge region, the breakdown voltage V_B is not easy to calculate precisely. However, the following empirical formula was shown to be valid for different semiconductors [9.13]

$$V_B \approx 60 \left(\frac{E_g(in\ \text{eV})}{1.1}\right)^{3/2} \left(\frac{N_B(in\ \text{cm}^{-3})}{10^{16}}\right)^{-3/4} , \quad (9.91)$$

where N_B is the background dopant concentration on the lightly doped side.

More details on avalanche breakdown will be presented below in the section on avalanche photodiodes.

Noise Current

1/f Noise. The noise current/voltage for this type of noise varies as $1/\sqrt{f}$. The experimentally measured relationship for the mean square noise current is [9.4]

$$\overline{i_{1/f}^2} = \frac{\text{const} I_{DC}^2 \Delta f}{f^\beta} , \quad (9.92)$$

where const is a proportionality constant, Δf and f are the bandwidth and frequency respectively, and β is a constant close to 1. Note that this noise is a function of the direct-current component of the semiconductor photodiode I_{DC} only. In modern designs, this noise is negligibly small.

Johnson Noise. For the semiconductor photodiode, the Johnson noise mean square current is expressed by

$$\overline{i_j^2} = \frac{4kT\Delta f}{R_{eq}} , \quad (9.93)$$

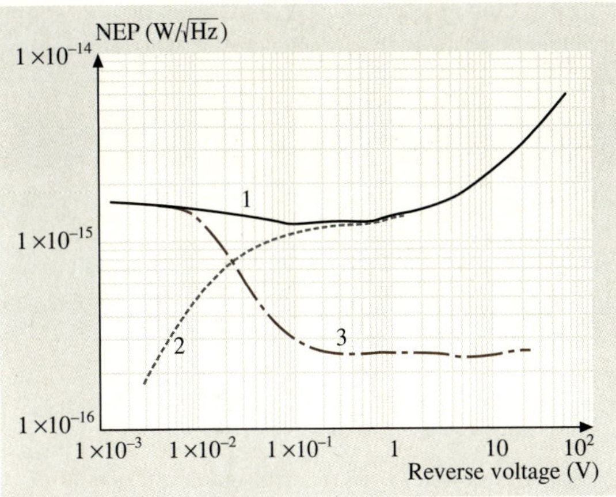

Fig. 9.10 Typical dependence of NEP on the reverse bias for a 1 mm^2 Si photodiode at room temperature (*curve* 1). The Johnson noise component is shown by *curve* 3 and the shot noise by *curve* 2

in which the equivalent resistor is determined as $1/R_{eq} = 1/R_{sh} + 1/R_s$, where R_{sh} and R_s are the shunt and series resistances of the photodiode, respectively. The series resistance is usually much smaller that the shunt resistance and can be neglected.

Shot Noise. The mean square shot noise current of the photodiode consists of two main components: the photon-generated component and the "dark" current component of the semiconductor diode shot noise current:

$$\overline{i_{sh}^2} = \sqrt{2q\,(I_p + I_{dark})\,\Delta f}\,, \tag{9.94}$$

where $I_p = I_s + I_B$ describes the photocurrent produced by the signal and background radiation (I_s and I_B are the signal and background currents, respectively), and I_{dark} accounts for the contribution from all sources of dark current.

Signal-to-Noise Ratio and Noise Equivalent Power (NEP)

The signal-to-noise ratio of a semiconductor photodiode can be found using (9.93) and (9.94)

$$\frac{S}{N} = \frac{\Re_I \Phi_p}{i_n} = \frac{\Re_I h\nu\Phi}{\sqrt{\overline{i_j^2} + \overline{i_{sh}^2}}} \approx \frac{I_p}{\sqrt{2q(I_p + I_{dark})\Delta f + (4kT\Delta f)/(R_{sh})}}\,, \tag{9.95}$$

where $\Re_I$ is the current responsivity given by (9.75), $\Phi_p = h\nu\Phi$ is the incident radiant power, and the approximate equality sign arises from the substitution $R_{eq} \approx R_{sh}$. By placing the signal-to-noise value equal to 1, one can calculate the photodiode NEP as $\text{NEP} = \Phi_p|_{S/N=1} = i_n/\Re_I$. In the case of photon-limited performance, the optical signal is rather strong compared to all sources of noise and $(I_{dark} + 2kT/qR_{sh})/q\Delta f \gg 1$. In this case, the photodiode NEP is determined by the quantum noise associated with the signal photon flux Φ^S (or the background photon flux Φ^B) itself

$$\text{NEP} = h\nu\sqrt{\frac{2\Phi^{S,B}\Delta f}{\eta_{ex}}}\,. \tag{9.96}$$

In the case of a very low optical photon flux, when $I_p \ll (I_{dark} + 2kT/qR_{sh})$, the thermal noise from the shunt resistance and/or the photodiode leakage current noise becomes dominant and the corresponding value of NEP is given by

$$\text{NEP} = \frac{h\nu}{\eta_{ex}}\sqrt{\frac{2\Delta f}{q}\left(I_{dark} + \frac{2kT}{R_{sh}q}\right)}\,. \tag{9.97}$$

A typical plot of NEP versus reverse bias is shown in Fig. 9.10 for a silicon photodiode 1 mm in diameter. At low bias values, the Johnson noise component usually dominates over other noise sources, whereas the shot noise component becomes increasingly large at large reverse bias values, when the dark leakage current increases.

Detectivity

The photodiode detectivity is derived using the general expression for detectivity $D = 1/\text{NEP}$. The specific detectivity for a photon-limited photodiode is determined by the expression

$$D^*(\lambda, f) = D\sqrt{A_{det}\Delta f} = \frac{1}{h\nu}\sqrt{\frac{\eta_{ex}A_{det}}{2\Phi^{S,B}}}\,, \tag{9.98}$$

in which A_{det} is the photodiode active area and the superscripts (S, B) correspond to either signal or background photon flux, whichever is greater. For a Johnson noise limited photodiode with a load resistance R_L, the specific detectivity is

$$D^*(\lambda, f) = \frac{q\eta_{ex}}{h\nu}\sqrt{\frac{A_{det}}{2kT}\left(\frac{R_L R_{sh}}{R_L + R_{sh}}\right)}\,. \tag{9.99}$$

Time Response and Frequency Bandwidth

During the process of light detection by photodiodes, the final step is usually nonequilibrium carrier collection into the external circuitry, which occurs through either drift or diffusion mechanisms, with the time constants τ_{drift} and τ_{diff}, respectively. Since the diffusion and drift processes are independent of each other, the resulting response time is calculated via

$$\tau_r = \sqrt{\tau_{\text{drift}}^2 + \tau_{\text{diff}}^2 + \tau_{\text{RC}}^2}\,, \tag{9.100}$$

in which τ_{RC} is the time constant determined by the terminal capacitance C and the load resistance R_L. The cutoff frequency f_c of the photodetector also defines the 3 dB breakpoint, and is given by

$$f_c = \frac{1}{2\pi\tau_r}\,. \tag{9.101}$$

The upper limit for the carrier diffusion time can be estimated from the following straightforward considerations. The carrier diffusion length is related to the diffusion time and the diffusion constants $D_{n,p}$ for electron and holes via

$$L_{n,p} = \sqrt{D_{n,p}\tau_{\text{diff}}^{n,p}}\,. \tag{9.102}$$

Assuming that the carrier lifetime is long enough to allow nonequilibrium carriers to move without recombination along the carrier concentration gradient, $\tau_{\text{diff}}^{n,p}$

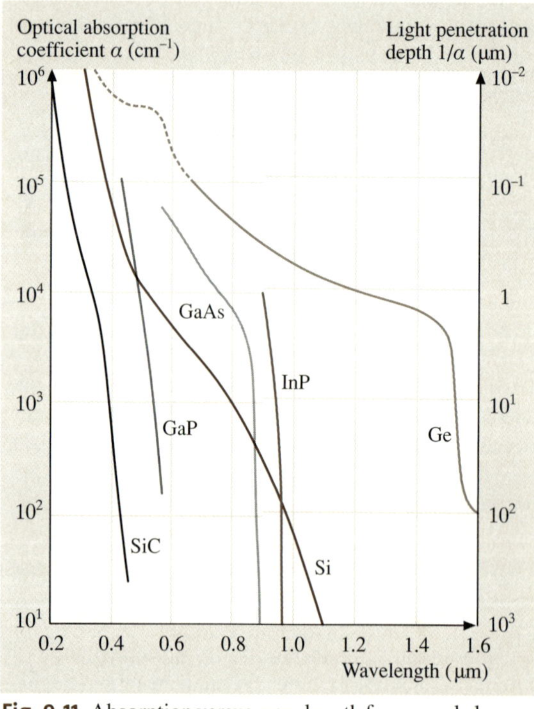

Fig. 9.11 Absorption versus wavelength for several elemental and binary semiconductors

Table 9.1 Lowest energy gaps E_g and carrier mobilities for a number of intrinsic semiconductors at room temperature [9.6, 14]

Material	E_g(eV)	μ_n (cm^2/Vs)	μ_p (cm^2/Vs)
Elementary			
Ge(indirect)	0.67	3900	1900
Si(indirect)	1.12	1500	500
III-V compounds			
InSb	0.23	80000	1000
InAs	0.36	33000	480
GaSb	0.72	5000	1500
InP	1.35	5000	180
GaAs	1.43	8500	400
AlSb(indirect)	1.63	200	500
AlAs(indirect)	2.16	1200	420
GaP(indirect)	2.26	300	150
II-VI compounds			
HgTe(semimetal)	−0.25	25000	350
CdTe	1.48	1000	100
CdSe	1.70	800	
Other			
PbSe	0.26	1500	1500
PbTe(indirect)	0.32	6000	4000
PbS(indirect)	0.41	800	1000

for electrons and holes can be calculated from (9.102) as the time necessary to diffuse through the undepleted region of the semiconductor bulk

$$\tau_{\text{diff}}^{n,p} = \frac{L_{n,p}^2}{D_{n,p}} \equiv \frac{(d-w)^2}{D_{n,p}}\,, \tag{9.103}$$

where d and w are the semiconductor substrate thickness and the depletion region width, respectively. The diffusion constants can be calculated from the known Einstein relation

$$D_{n,p} = \frac{kT\mu_{n,p}}{q}\,, \tag{9.104}$$

where $\mu_{n,p}$ are the carrier mobilities, specific to each semiconductor.

The upper limit for the carrier drift time τ_{drift} across the space-charge region can be estimated using

$$\tau_{\text{drift}}^{n,p} = \frac{w}{v_{\text{drift}}^{n,p}} = \frac{w}{\mu_{n,p}E}\,, \tag{9.105}$$

where the right-hand side states the known phenomenological relationship that the carrier mobility is thc ratio of the average drift velocity to the applied electric filed E

$$v_{\mathrm{drift}}^{\mathrm{n,p}} = \mu_{\mathrm{n,p}} E \,. \tag{9.106}$$

This expression is valid for a relatively low electric field strength, when the effects like electron–phonon scattering can be neglected. At high values of E, the carrier drift velocity saturates and becomes independent of E [9.15].

Ways of reducing the photodiode time response and making the cutoff frequency larger will be discussed below.

9.3.3 Semiconductor Photodiodes – Materials

The choice of semiconductor photodiode material depends primarily on the spectral range of interest. Figure 9.11 shows plots of absorption coefficient versus wavelength for the most common semiconductors. For the region of the spectrum below ≈ 1100 nm, Si is usually used. However, since both Si and the other popular semiconductor material – Ge – are indirect bandgap semiconductors, much of their sensitivity to large regions of the spectrum is due the occurrence of absorption processes involving phonons. Phonon-assisted photon absorption obviously have a much smaller probability of occurrence than the direct valence band to conduction band transitions that occur in direct bandgap semiconductors. This explains their significantly smaller

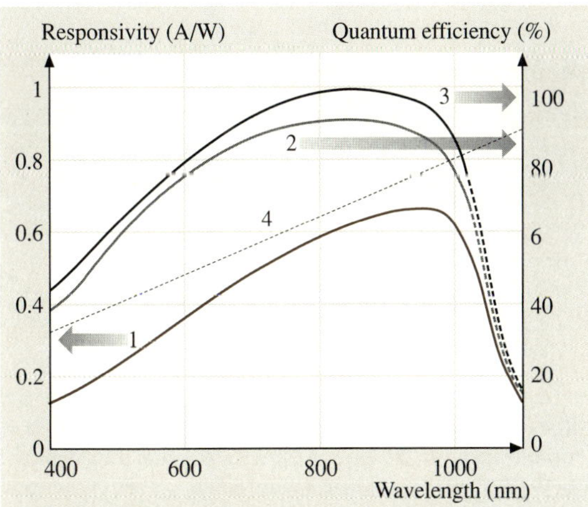

Fig. 9.12 An example of the spectral response of a Si photodiode (*curve* 1). The quantum efficiency of the same photodiode is shown by *curves* 2 (external QE) and *curves* 3 (internal QE). *Curve* 4 shows the dependence of the theoretical responsivity of an imaginary photodiode on $\eta_{\mathrm{ex}} = 100\%$

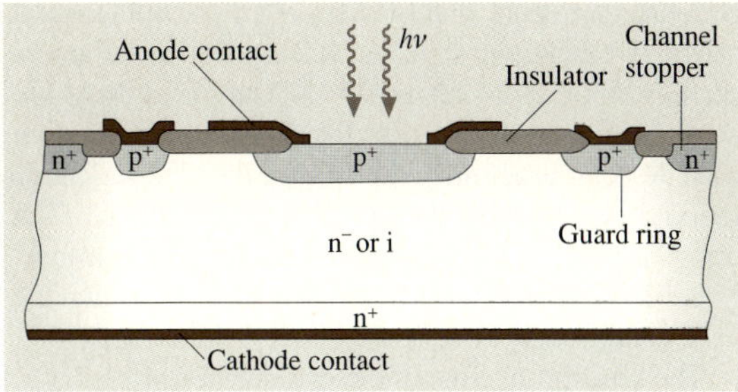

Fig. 9.13 Structure of a p–i–n photodiode (cross-sectional view); p^+ diffusion in the middle of the die forms the active area junction; the p^+ diffusions close to the edges are the guard rings. The spacing from the main (anode) junction used depends on the breakdown and leakage performance required. The n^+ diffusions at the edges are the channel stoppers

absorption coefficients and, correspondingly, the larger absorption lengths observed for Si and Ge than for many compound semiconductors. The band gap energies E_{g} for a number of important semiconductors are summarized in Table 9.1.

The other important photodiode parameter which is directly related to the absorption length is the time response. The time response is normally significantly shorter for direct bandgap semiconductors due to the much thinner absorption volume, which automatically translates into a shorter carrier collection time. However, some specific structural features can greatly improve the response time and we will describe these in the following sections.

The photodiode rise time and time response is a complex function of several parameters, the carrier mobility being among the most important. Table 9.1 shows the carrier mobility values for various semiconductors.

Single-Material Semiconductors

Because Si and Ge are associated with indirect photon absorption, they do not have sharp absorption edges. Si is the semiconductor material most often used for the visible and near-infrared (IR) spectral range. Si photodiodes usually have quantum efficiencies close to 100% at selected intervals within the spectral range from ≈ 400 nm to 1000 nm (Fig. 9.12). A typical spectral responsivity curve for a Si photodiode is also shown in Fig. 9.12. Special front surface treatment allows the high efficiency region of Si photodiodes to be extended to below 400 nm. Among the other advantages of Si photodiodes are their low noise currents (Fig. 9.10). The usual approach used to reduce the dark current in Si-based and other substrate photodiodes is to provide special structures like guard rings and surface leakage stoppers (channel stoppers) – see Fig. 9.13. The guard rings are usually designed to decrease the electric field strength in the depletion region, thus reducing the high-field ionization efficiencies of the host and dopant atoms. The surface leakage stoppers are designed to prevent minority carriers outside the active area from reaching the space-charge region

The absorption edge of Ge extends beyond 1.6 μm, which is advantageous for the purposes of wide spectral range applications but at the same time limits the sensitivity of this material to low photon fluxes because of the high levels of thermally generated dark current. An example of the p–i–n photodiode structure typical of Si and Ge is given in Fig. 9.13.

Binary III–V Semiconductor Photodiodes

III–V binary semiconductors permit the design of photodiodes operating over a wide spectral range from the UV to mid-IR. Direct bandgap III–V group semiconductors have rather small absorption lengths, assuming a relatively high probability of nonequilibrium carrier recombination on surface defects. The latter effect is the main reason for the lower quantum efficiencies of photodiodes based on III–V compounds in comparison with those observed for Si and Ge.

Photodiodes based on III–V compounds usually show increased dark current values, which limit their field of application.

Binary II–VI Semiconductor Photodiodes

II–VI binary semiconductors have also been used in photodiode design, although these materials are not as widely used as III–V compounds. Among those that have been explored are CdTe, which has a bandgap of $\approx 1.48\,\text{eV}$ [9.14] (some sources give a larger value of $\approx 1.6\,\text{eV}$ [9.1]), and HgTe, which is actually a semimetallic compound with a "negative band gap"

Table 9.2 Values of the κ parameter for various semiconductors [9.1]

Material	κ(eV)
$Al_xGa_{1-x}As$	0.27
$GaAs_{1-x}P_x$	0.21
$InAs_{1-x}P_x$	0.27
$InAs_xSb_{1-x}$	0.58
$In_{1-x}Ga_xAs$	0.32
$In_{1-x}Ga_xSb$	0.43

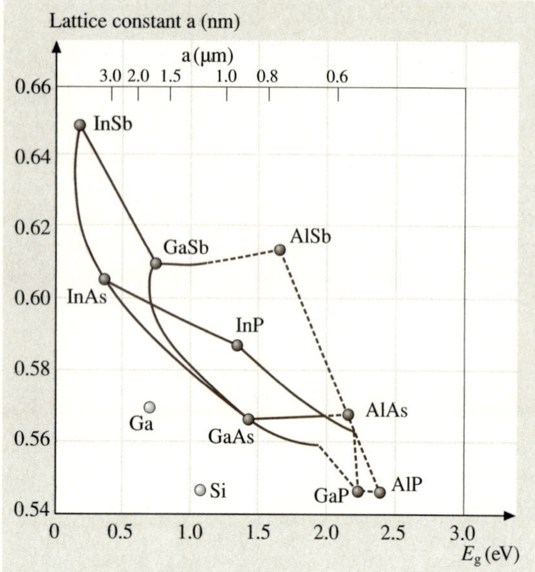

Fig. 9.14 The variation of lattice constant with bandgap for III–V ternary compounds. Direct bandgap material is shown by the *solid line* and indirect bandgap material by the *dashed line*

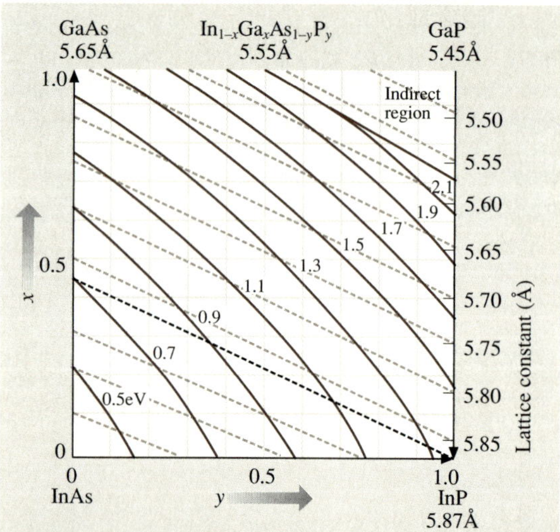

Fig. 9.15 Iso-band gap lines (*solid lines*) and iso-lattice constant lines for the $In_{1-x}Ga_xAs_{1-y}P_y$ quaternary alloy. The *bold dotted line* shows a continuous set of quaternary structures perfectly lattice-matched to InP

of $\leq 0.25\,\text{eV}$. Ternary and quaternary compounds based on II–VI group semiconductors have found much wider application.

Ternary and Quaternary Semiconductor Photodiodes

Ternary $A_{1-x}B_xC$ and quaternary $A_{1-x}B_xC_{1-y}D_y$ compounds are used to build photodiodes with optimized spectral sensitivity ranges. Various formulae have been suggested that may be used to calculate the band gap as a function of the stoichiometric parameters x and y. For example, one of the ideas is based upon the second-order perturbation of the Schrödinger potential, which yields the following expression for the change in the bandgap with x [9.16]

$$E_g(A_{1-x}B_xC) = E_g(AC) + \left[E_g(BC) - E_g(AC) - \kappa\right]x + \kappa x^2 , \quad (9.107)$$

where κ is a positive constant, specific to each compound. Note that in some ternary and quaternary compounds either the direct or indirect bandgap can be the lowest, depending on the values of x and y. Correspondingly, there are two versions of (9.107) for materials that change from direct to indirect gap: AlGaAs is an example. To calculate the direct bandgap value for $Al_{1-x}Ga_xAs$ $(0 < x < 0.44)$, one should always use the direct band gap values of the corresponding binary compounds and vice versa. Table 9.2 shows some calculated κ values for several ternary compounds.

One of the main requirements for building ternary and quaternary structures is the lattice-matching requirement. Figure 9.14 shows the variation in the lattice constant as a function of the band gap for some III V ternary alloys. Note that the successful growth of heterostructures requires that both the band gap and the lattice constant are considered. For example, $Al_{1-x}Ga_xAs$ can be grown on GaAs substrate for any value of x without lattice mismatch-inducing defects. In contrast, for the $Ga_{1-x}As_xP/GaAs$ structure there is a substantial mismatch in lattice constant values between GaAs and GaP, which bars simple high-quality structures from being built over the whole range of x. However, in many cases this problem can be overcome by providing a grading layer between the substrate and the desired compound.

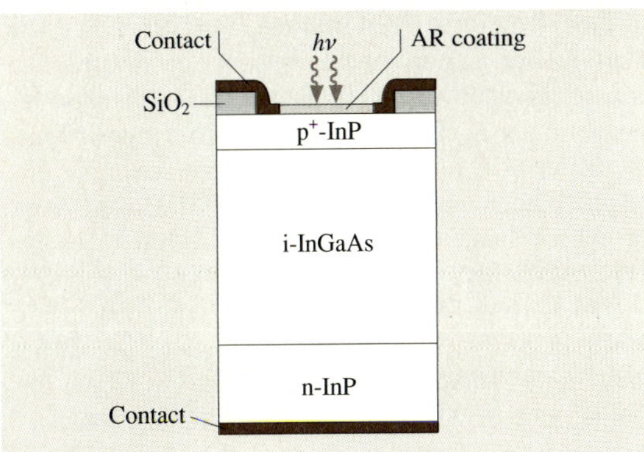

Fig. 9.16 An example of the structure of an InGaAs/InP p–i–n photodiode

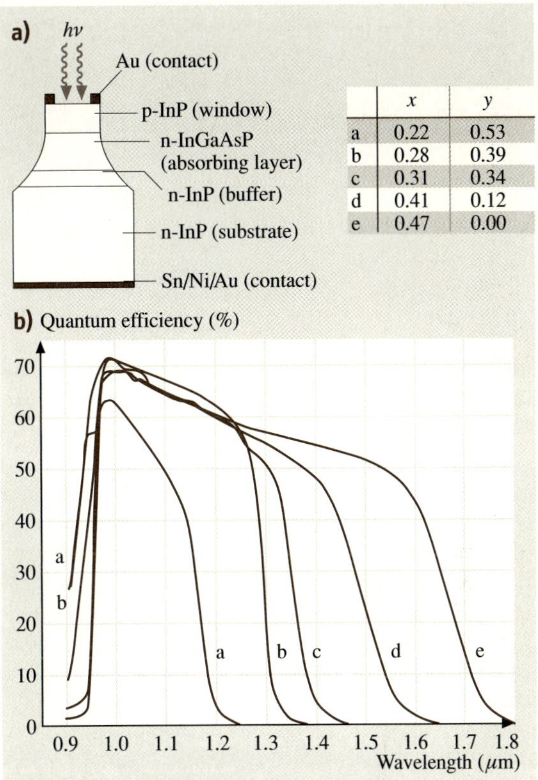

Fig. 9.17 (a) An example of an $In_{1-x}Ga_xAs_{1-y}P_y$ mesa structure photodiode, and (b) external quantum efficiencies for several different quaternary compositions

Instead of using grading layers, quaternary compounds can also be built. Not all quaternary compounds can be ideally matched to a substrate. However, the case of $In_{1-x}Ga_xAs_{1-y}P_y$ provides an exciting exception. The diagram in Fig. 9.15 demonstrates the correlation between the band gap and the lattice constant for this quaternary compound. The solid lines represent iso-bandgap lines. Every bandgap from 0.36 eV (InAs) to 2.2 eV ($In_{0.73}Ga_{0.27}P$) can be created in a direct-gap semiconductor. The dashed lines in Fig. 9.15 represent iso-lattice constant lines. The important feature of this diagram is shown by the bold dotted line, which illus-

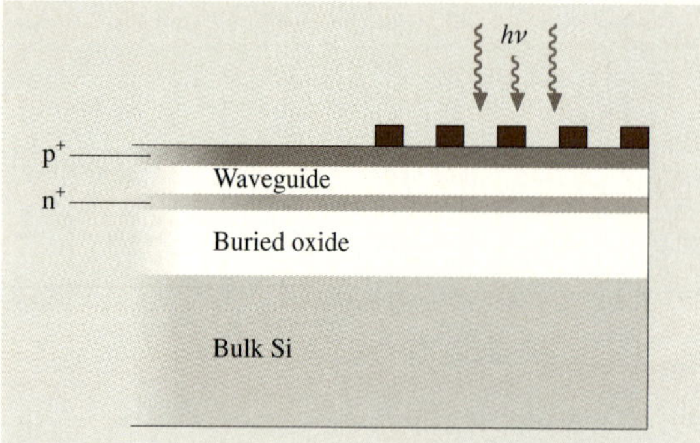

Fig. 9.18 An example of a p–i–n photodiode with a waveguide-grating-coupler at the top: a schematic cross-section

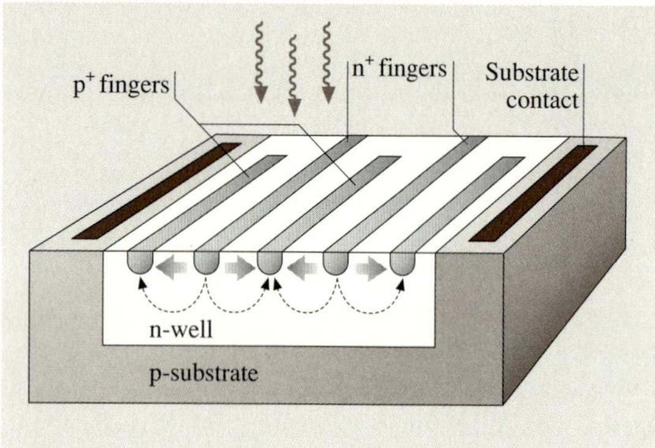

Fig. 9.19 A schematic of an example of a lateral p–i–n photodiode with alternating p^+ and n^+ fingers fabricated using CMOS technology

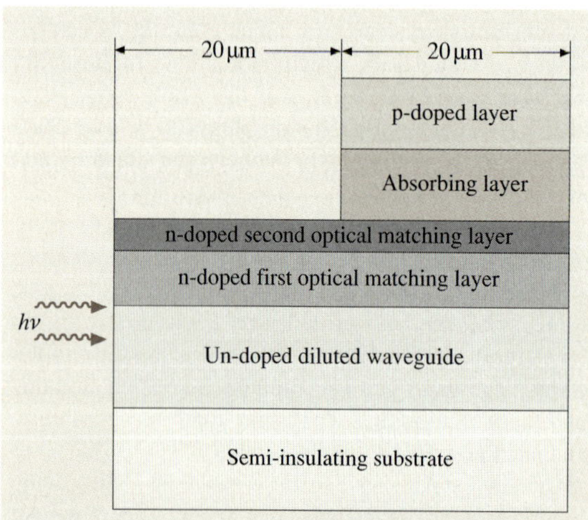

Fig. 9.20 A schematic cross-section of a planar photodiode with a short multimode waveguide coupler

trates a continuous set of quaternary structures with bandgap values from $\approx 1.33\,\text{eV}$ to $\approx 0.5\,\text{eV}$ that are perfectly lattice-matched to InP.

Figure 9.16 shows a typical example of a p–i–n photodiode that employs ternary III–V compounds. The intrinsic layer of $In_{1-x}Ga_xAs$ is grown on top of an n-type InP substrate. The ternary structure is sensitive up to $\approx 1.65\,\mu\text{m}$ for $x = 0.47$. A top layer of p-InP creates the junction and completes the structure. Since InP is transparent to wavelengths above $0.92\,\mu\text{m}$, the device can be used as an efficient p–i–n, back-illuminated photodiode for the wavelength range from 0.92 through to $1.65\,\mu\text{m}$, in which all absorption occurs within the intrinsic layer.

Another example of a quaternary structure p–i–n photodiode is shown in Fig. 9.17, along with typical quantum efficiency curves for a number of stoichiometries.

In terms of II–VI ternary compounds, structures based around the $Hg_{1-x}Cd_xTe$ (mercury–cadmium–telluride, MCT) system had a virtual monopoly over long-wavelength light detection (3–5 μm and 8–14 μm) until recently. Depending on x, this system provides tremendous flexibility when tuning the spectral sensitivity range from visible to far infrared. The bandgap of the alloy $Hg_{1-x}Cd_xTe$ is zero at $x \approx 0.15$ and is only slightly nolinear over the whole range of x up to $E_g = 1.48\,\text{eV}$ at $x = 1$ [9.2]

$$E_g = -0.302 + 1.93x + 5.35\times10^{-4}T(1-2x) - 0.81x^2 + 0.832x^3\,, \tag{9.108}$$

where T is the absolute temperature. Among the drawbacks of MCT photodiodes is a high leakage current, which requires that the device be cooled for noise-free operation.

Fast-Response Photodiodes

Short carrier transit time is one of the determining parameters when designing photodiodes with short response times τ_r. To minimize the carrier transit time, the carrier drift time should be reduced assuming that the photodiode bulk is fully depleted (9.100). However, for indirect bandgap semiconductors with low absorption coefficients, a thick absorption layer is required to achieve high quantum efficiency, and this condition is in conflict with the need to reduce the carrier transit time. There are several ways to overcome this conflict, one of which is to build resonantly enhanced structures; the other is to decouple the photodiode absorption length from the direction of nonequilibrium carrier drift.

The principle of a resonant cavity enhanced photodetector is based on the use of constructive interference in a Fabry–Perot cavity to enhance the optical field inside the photodetector bulk at specific wavelengths [9.17]. Such resonant cavities can be formed using a buried backside reflector and the air/semiconductor top interface. There are various types of backside reflector; the scattering reflector [9.18] and distributed Bragg reflector [9.19] are common examples.

For indirect bandgap semiconductors, it is difficult to achieve both high bandwidth and high efficiency with a vertical design since the length of the absorption region is proportional to the carrier transit time. Decoupling of the absorption length from the motion of nonequilibrium carriers is another efficient method used to improve the speeds and quantum efficiencies of Si- and Ge-based photodiodes. Such decoupling is achieved in lateral surface p–i–n photodiodes on silicon-on-insulator (SOI) substrates, which exploits the idea of using a grating coupler on top of the photodiode [9.20, 21]. The coupler promotes the propagation of the optical beam through a thin, waveguide-type surface layer of Si (Fig. 9.18).

Another way to decouple the absorption length from the carrier motion is to incorporate alternating p+ and n+ fingers in Si substrate [9.22]. In the example shown in Fig. 9.19, the deep carriers are blocked by the junction between the n-well and p-substrate, and the transit time is determined by the fast lateral motion of carriers between the fingers.

In top-illuminated, direct-bandgap semiconductor photodiodes, further reductions in the carrier transit time are limited by surface recombination processes, which decrease the responsivity. Side-illuminated photodiodes, in which the absorption length is decoupled from the carrier transit time, overcome the efficiency–bandwidth tradeoff associated with top-illuminated photodiodes [9.23, 24]. However, they cannot operate at high optical power levels, since carrier generation occurs in a very small volume at the diode input facet.

One of the solutions to this problem is to use evanescently coupled photodiodes that are monolithically integrated with a waveguide [9.25–27]. The evanescent coupling optimizes the distribution of nonequilibrium carriers along the absorption layer. Due to more uniform light absorption, devices that utilize evanescent coupling can achieve saturation currents that are several times higher than achieved in a traditional side-illuminated device. An example of an evanescently coupled photodiode is shown schematically in Fig. 9.20. The structure is usually based on a classical p–i–n photodiode with a very thin ($< 0.5\,\mu m$) undoped InGaAs absorbing layer 20–30 μm in depth and a p^+-doped InP layer on the top. The n^+-doped layer contains InGaAsP structures that may consist of two optical matching layers. The light is delivered from the side and propagates along the undoped diluted waveguide layer, which is a stack of several InP/InGaAsP sandwiches. The bandgaps and the thickness of the optical matching layers are chosen in such a way that they provide a gradual increase in the refractive index from the diluted waveguide towards the thin absorbing layer. Due to the evanescent wave coupling, when the optical wave propagates along the waveguide it is gradually transferred through the optical matching layers to the absorbing layer. A quantum efficiency of higher than 95% can be achieved with

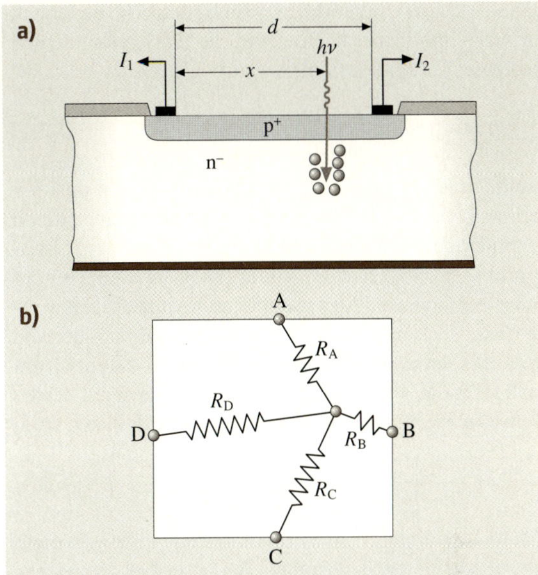

Fig. 9.21 **(a)** Cross-section of the two-terminal position-sensitive detector. **(b)** Unequal resistance paths on the silicon four-terminal sensor

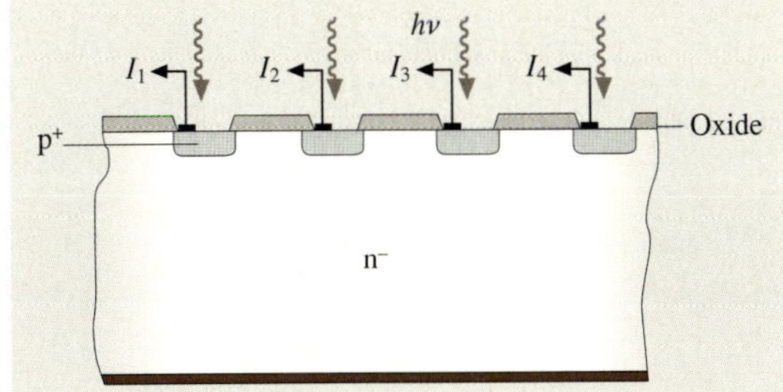

Fig. 9.22 Cross-section of a silicon diode strip detector

a bandwidth of over 50 GHz, which should be beneficial for many applications. However, the fabrication of such structures is rather complex.

Position-Sensitive Photodiodes

Resistive Charge Division Sensor. A silicon resistive charge division sensor, also known as a silicon position sensor, consists of a segment of photodetective silicon (a moderately doped p^+-layer on top of an n-type substrate) with either two or four terminals for signal output and a terminal used for the application of back-bias voltage (Fig. 9.21, panel (a)). Position information is derived from this device by comparing the signal outputs from each terminal. For example, for the two-terminal device shown in Fig. 9.21a, the position can be calculated from

$$x = \frac{I_2}{I_1 + I_2} d\,. \tag{9.109}$$

The focused image causes a photocurrent to flow through the silicon to each terminal. Because the silicon has a given resistance per unit length, more current will flow to the closest terminals and less current to the terminals that are farthest from the focused image (Fig. 9.21, panel (b)).

Position sensors of this type have several advantages over other similar devices. Because there is no gap in the active area, the size of the image is not subject to minimum diameter constraints. A second advantage is that position information is available as long as the image falls somewhere on the active area of the detector. One disadvantage is that the frequency responses of such detectors tend to be lower than those of conventional silicon detectors of the same size because of the series resistance that the photocurrent encounters at the surface of the detector.

Strip Detector. Another way to measure position is to dividing a large-area photodiode into many small, strip-like regions and to read them out separately – see Fig. 9.22. Such strip detectors are produced in the same way as planar photodiodes except that more stringent requirements are placed on mask precision and alignment. The measurement precision depends mainly upon the strip spacing and the readout method.

Quadrant Photodiode. A quadrant photodiode is variation on a strip detector, where the segments are four quadrants, each of which have separate connectors for signal output. Each quadrant functions and behaves like a regular p–i–n or avalanche photodiode. Quadrant photodiodes are widely used in tracking systems. Position information is derived by comparing the signals output from each segment (Fig. 9.23). When the focused image is centered on the quadrant detector, each segment will receive the same amount of optical radiation, and all four signal outputs will be equal (Fig. 9.23, left). As the image moves across the detector surface (corresponding to an angular change in the object being tracked), more radiation falls on one of the segments and less on the opposite segment.

If more optical energy falls above the horizontal line (Fig. 9.23, middle), more photocurrent will flow through the upper segments (A and B) than through the lower segments (C and D). The up–down position of the spot

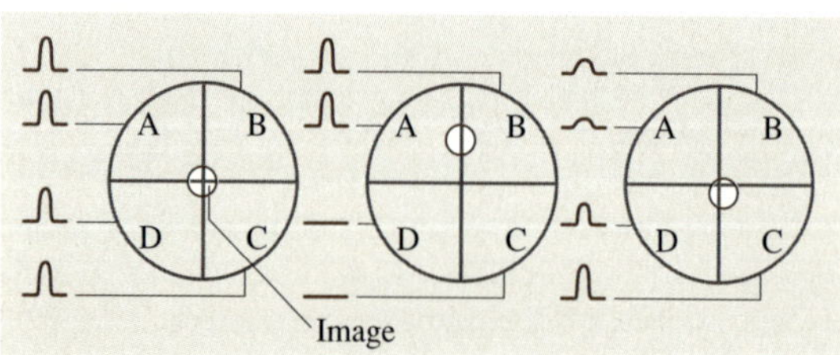

Fig. 9.23 Output from the image at different positions on the quadrant detector

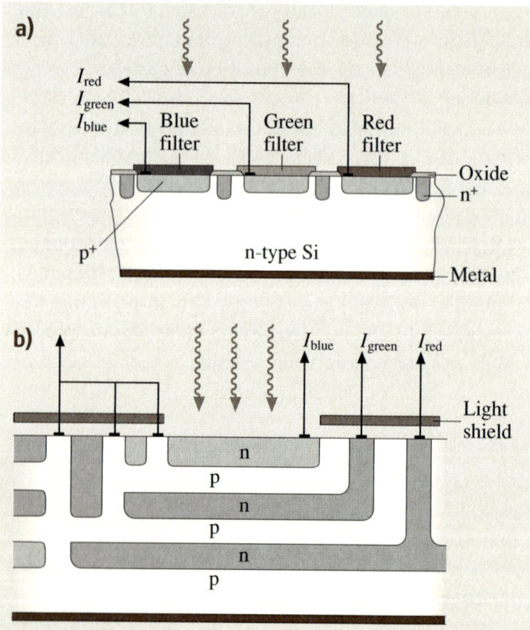

Fig. 9.24 (a) Typical structure of a color-sensitive element with three color filters overlaid over different photodiodes. **(b)** An example of the structure of a vertically arranged color detector. n-Type regions are *hatched*

is characterized by the relative amplitudes of (A + B) − (C + D). Similarly, the term (A + D) − (B + C) provides information on the left–right position.

In most tracking and alignment applications, the parameter of most concern is the ability of the detector to supply a very precise description of the location of the focused image on the face of the detector. In other words, the change in the output signal in relation to the change in the position of the spot is very important. "Sensitivity" is used in this case to describe the ratio

$$\text{Sensitivity} = \frac{\text{Change in output signal}}{\text{Change in position}} . \qquad (9.110)$$

An inherent problem with quadrant photodiodes is the finite width of the separation between segments. If the image is focused to a small spot size, erroneous information will result when the spot falls on the border between segments. This problem is usually avoided by defocusing the image or by using a long-focal-length objective that results in a larger image size and displacement. Additionally, some position information is lost when the entire image falls within one segment. This problem also can be avoided by defocusing or by using long-focal-length optics.

Color-Sensitive Detectors

The conventional way to detect a color image relies on the idea that any color, as perceived by the human eye, is based on the three primary colors: red, green, and blue. Mixing light of these wavelengths in varying proportions can generate other colors. The main principle of operation of any color sensor is therefore based on separating the incident beam into three primary colors and detecting them separately. The methods used to do this exploit two main principles: one uses color overlay filters on top of the active area of the photodetector, while the other performs color selection based on the inherent property of the semiconductor to absorb light of different wavelengths at different depths. A hue diagram is then used to infer the color of the incident light [9.28].

In the first type of color sensor, color detection is performed by using three filters, one for each of the primary colors, and three photodetectors fabricated on the same substrate. A typical structure of this type is shown in Fig. 9.24a. P–n or p–i–n photodiodes are built on an n-type Si wafer with a n^+ guard ring. The filters allow the transmission of one of the primary wavelengths, and the corresponding photodiode measures the intensity of the incident light at that wavelength. The responses of the photodetectors are then used to determine the relative contributions of each of the primary wavelengths to the incident light, and hence to infer its color. Drawbacks to such methods are that it can be complicated to use three sensors with their corresponding filters, and that such a design can take up excessive space when creating a color-sensing array.

Another type of multiple-wavelength sensor employs more than one sensor in a vertically oriented arrangement. Different designs include photodiodes formed in single-crystalline silicon and those formed in amorphous silicon. The operation of these devices is based on the intrinsic filtering property of a semiconductor, which results from the variation in the absorption coefficient with the wavelength of the incident light. An example of such a structure is shown in Fig. 9.24b [9.29]. The layers of the detector have different spectral sensitivities based on their different depths in the semiconductor substrate, doping levels, and biasing conditions. The three n type regions are isolated vertically by p-regions. The blue, green, and red photodiode sensors are formed by the corresponding p/n junctions and are arranged one below the other underneath the sensor surface. The layers of the detector are individually connected to pixel sensor readout circuits.

9.4 QWIP Photodetectors

Quantum well infrared photodetectors (QWIPs) are efficient detectors for the detection of infrared radiation. A QWIP is a photoconductor and a unipolar device. QWIPs are fabricated by alternately growing thin layers of two different wide-bandgap semiconductors. The bandgap discontinuity between two materials creates quantized subbands in the potential wells associated with conduction bands or valence bands. Light absorption initiates electron transitions between the ground and first excited state subbands of multiquantum wells. The structure parameters are designed in such a way that the photoexcited carriers can escape from the potential wells and be collected as photocurrent, which is facilitated by bending the bands using an external electric field. Depending the structure of the QWIP, there can be three different types of electronic transitions: bound-

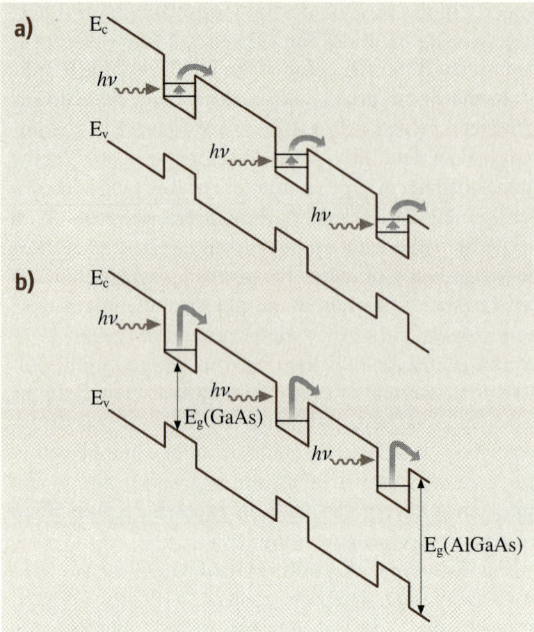

Fig. 9.25a,b Energy band diagram of QWIP under reverse bias showing **(a)** a bound-to-bound transition and **(b)** a bound-to-continuum transition

to-bound, bound-to-continuum, and bound-to-miniband transitions (Fig. 9.25).

9.4.1 Structure and Fabrication of QWIPs

The lattice-matched GaAs/Al$_x$Ga$_{1-x}$As material system is commonly used to create a QWIP structure such

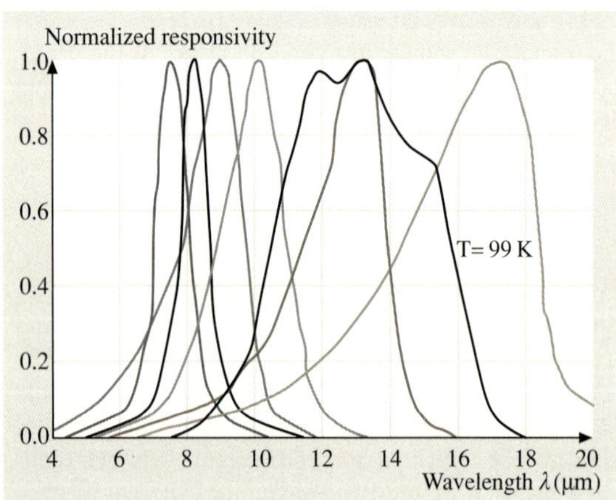

Fig. 9.26 Examples of the responsivity spectra for different GaAs/AlGaAs QWIP structures

as that shown in Fig. 9.25. The quantum well layers of GaAs have a typical thickness of 50–70 Å and are usually n-type doped to $\approx 10^{18}\,\mathrm{cm}^{-3}$. The barrier layers AlGaAs are undoped and have a thickness in the range of 300–500 Å. Typical numbers of periods are 20 to 50. By controlling the quantum well width and the barrier height (which depends on the Al molar ratio of Al$_x$Ga$_{1-x}$As alloy), the intersubband transition energy can be varied over a wide enough range to enable the detection of light at any wavelength in the range 6–20 μm. Fabricated and experimentally investigated QWIPs can already detect wavelengths of up to 35 μm. Other contemporary structures used in QWIPs are InGaAs/InAlAs and InGaAs/InP, which have peak detection wavelengths of 4 μm and 8 μm, respectively.

9.4.2 QWIPs – Properties and Figures of Merit

Spectral Response

The spectral responsivity peak wavelength (λ_p) of a QWIP is determined by the difference in energy between the ground and excited states of the quantum well. Unlike the responsivity spectra of intrinsic infrared detectors, QWIP spectra are much narrower and sharper due to their resonance intersubband absorption. Typically, responsivity spectra of bound and quasi-bound excited state QWIPs are much narrower ($\Delta\lambda/\lambda \approx 10\%$) than those of continuum QWIPs ($\Delta\lambda/\lambda = 24\%$). This is due to the fact that, when the excited state is positioned in the continuum band above the barrier, the range of energies associated with the state becomes wider. The spectral bandwidths of these QWIPs can be further increased by replacing single quantum wells with small superlattice structures (several quantum wells separated by thin barriers) in a multiquantum well structure. Such a scheme creates an excited state miniband due to the overlap of excited state wavefunctions of quantum wells. Energy band calculations based on a two-band model show excited state energy levels spread over greater than 30 meV. Figure 9.26 shows experimentally measured responsivity spectra with FWHM bandwidth $\Delta\lambda > 5$ μm. The broadest responsivity bandwidth demonstrated so far is $\Delta\lambda \approx 6$ μm.

Quantum Efficiency

The QWIP photocurrent is given by the general expression used for photoconductors

$$I_\mathrm{p} = q\Phi\eta_\mathrm{ex}G\,, \tag{9.111}$$

where q is the elementary charge, Φ is the photon flux in reciprocal seconds, η_{ex} is the external quantum efficiency and G is the gain. The quantum efficiency is different from that of a regular photoconductor, since light absorption and carrier generation occurs only in the quantum wells in QWIPs, not homogeneously throughout the structure [9.14]:

$$\eta_{ex} = (1 - R_F)\left[1 - \exp\left(-N_p \alpha N_w L_w\right)\right] E_p P \,, \tag{9.112}$$

where R_F is the front surface reflection, N_p is the number of optical passes, N_w is the number of quantum wells, L_w is the well length, E_p is a term that depends on the bias inside the quantum well, and P is the polarization correction factor.

The photoconductive gain G is similar to that of the standard photoconductor:

$$G = \frac{\tau}{t} = \frac{(\tau \mu V)}{L^2} \,, \tag{9.113}$$

where V is the bias, τ is the minority carrier lifetime and t the transit time across the whole active length L of the QWIP, while μ is the minority carrier mobility. The right-hand side of (9.113) is written assuming that the field strength E is far from saturation, and the carrier drift velocity is given by $v_d = \mu E = \mu (V/L)$.

The photoconductive gain of the QWIP depends on the design of the device. It typically varies from 0.1 to 1. A photoconductive gain G of higher than 1 is possible with QWIPs that have fewer quantum wells.

Noise and Detectivity

The dominant noise observed in QWIP devices is due to the shot noise resulting from the total current in the device. There is no significant thermal noise in QWIPs because the thermally assisted tunneling of carriers is not significant at the operating bias used. To additionally decrease the thermal noise and limit the dark current, the QWIP must be operated at a low temperature. In addition, QWIPs show remarkable noise stability (i. e., very low $1/f$ noise), enabling long integration times.

The QWIP is a type of photodetector that can provide background-limited performance (BLIP), where the detector performance is limited by the noise generated by photons. The signal-to-noise ratio, the detectivity, etc., are independent of the photoconductive gain when the QWIP operates under background-limited conditions.

9.4.3 Applications of QWIPs

There are many ground-based and spaceborne applications that require long-wavelength, large, uniform, reproducible, low cost, low $1/f$ noise, low power dissipation, and radiation-hard infrared focal plane arrays, which often employ QWIPs [9.30]. For example, the absorption lines of many gas molecules, such as ozone, water, carbon monoxide, carbon dioxide, and nitrous oxide occur in the wavelength region from 3 to 18 μm. Thus, infrared imaging systems that operate in the long-wavelength IR (LWIR) region are required in many space applications such as global weather profile monitoring, earth resource mapping, deforestation mapping, and to map the distributions of minor constituents in the atmosphere. In addition, there is great astronomical interest in these very long wavelengths due to the fact that this spectral region is rich in information vital to our understanding of the compositions, structures and energy balances of molecular clouds and star-forming regions of galaxies.

9.5 QDIP Photodetectors

The quantum-dot infrared photodetector (QDIP) is similar to the QWIP except that the potential well is created by a quantum dot instead of a quantum well. In QDIP, the physical dimensions of the active structure are reduced to the de Broglie wavelength, which greatly enhances the degree of carrier confinement. As a result, the spectral response can be controlled even more precisely than for QWIP structures. In addition, the noise characteristics and detectivity of QDIPs can potentially be dramatically improved in comparison with those of QWIP structures.

9.5.1 Structures and Fabrication of QDIPs

There are two basic QDIP structures, one where the photogenerated carriers travel perpendicular to the quantum dot planes, the other where they travel parallel to the planes (Fig. 9.27). The technique used to form quantum dots has improved dramatically following the introduction of an epitaxial synthesis process called the Stranski–Krastanow growth mode [9.31, 32]. Typical structures employ InAs dots on GaAs, InGaAs dots on GaAs, and SiGe dots on Si. As an exam-

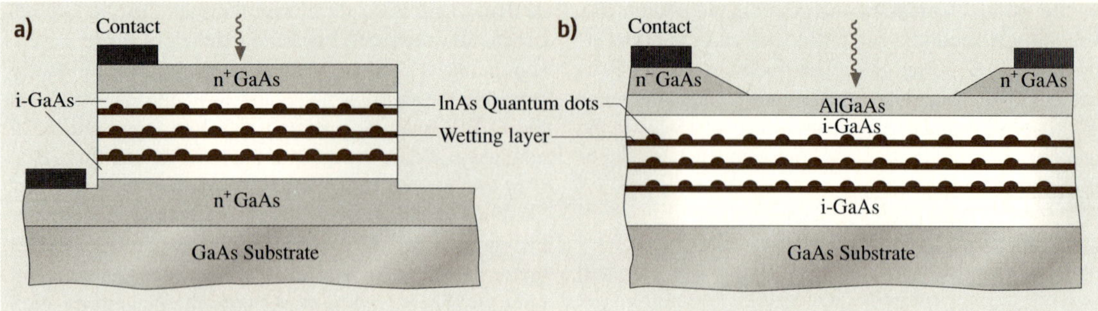

Fig. 9.27a,b Schematic of the cross-sectional structure of a QDIP. The photocurrent flows either **(a)** perpendicular or **(b)** parallel to the quantum dot planes

ple, the QDIP structure can comprise InAs islands on GaAs substrate. At an optimum layer thickness of roughly two monolayers ($\approx 6\,\text{Å}$), the InAs film nucleates into an array of islands (dots) to release the stress. The typical dimensions of the dots are $\geq 150\,\text{Å}$ across the base and $\geq 30\,\text{Å}$ in height. The density of the array is 10^{10}–$10^{11}\,\text{cm}^{-2}$. The ultrathin residual "wetting layer" is usually left underneath the quantum dots.

The doping level in quantum dots is tuned to provide up to five electrons per dot. Alternatively, the dots can be intrinsic but electrons can be introduced by modulation doping the incorporated delta-doped layer, from which electrons are injected and are trapped within the potential well of the quantum dot. The delta-doped layers are fully depleted and do not contribute to the dark current.

To improve light absorption, the quantum dot layers are repeated periodically. The thickness of the intrinsic barrier layer typically ranges from 100 to 500 Å.

The photocurrent from a QDIP is qualitatively described by (9.111), substituting the QDIP photocurrent for the QWIP photocurrent.

9.6 Metal–Semiconductor (Schottky Barrier) and Metal–Semiconductor–Metal Photodiodes

When a metal contacts a semiconductor, a Schottky barrier results. A Schottky-barrier photodiode can operate in two different detection modes. In one of them, electron–hole generation occurs due to the band-to-band excitation of the semiconductor; in the second mode, the carriers are transferred from the metal into the semiconductor across the Schottky barrier. In the front-illuminated version, which is advantageous for the first mode of operation mentioned above, the metal layer that forms the Schottky barrier is very thin (less than $\approx 150\,\text{Å}$), thus allowing more photons to reach the semiconductor bulk. For detectors that use the second mode of operation, backside (through the substrate) illumination is more efficient, since the barrier height $q\varphi_{\text{bn}}$ is always smaller than the energy gap E_{g} and infrared quanta with $q\varphi_{\text{bn}} < h\nu < E_{\text{g}}$ are not absorbed in the semiconductor.

9.6.1 Schottky Barrier Photodiode Properties

Figure 9.28 schematically shows the energy band for a Schottky diode. The metal work function φ_{m} is usually different from the semiconductor work function φ_{s}. The conditions needed to produce a useful barrier to the electrons in an n-type semiconductor–metal contact is $\varphi_{\text{m}} > \varphi_{\text{s}}$. The barrier height is given by

$$q\varphi_{\text{bn}} = q(\varphi_{\text{m}} - \chi)\,, \tag{9.114}$$

where $q\chi$ is the semiconductor electron affinity.

Dark Current and Photocurrent

Metal–semiconductor junctions have properties similar to those of the p–n junction. In particular, the $I-V$ characteristic for dark conditions shows an exponential

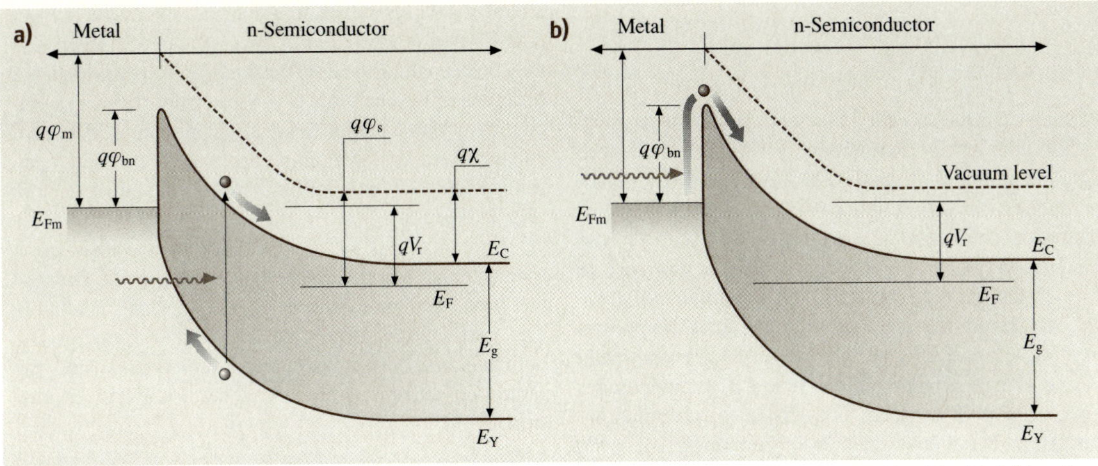

Fig. 9.28a,b Energy band diagram for the Schottky-barrier photodiode operated under reverse bias V_r with **(a)** band-to-band excitation and **(b)** intrinsic photoelectron emission

dependence of I on V for the forward bias, and saturates at a constant value of the reverse bias

$$J_{\text{dark}} = J_s \left[\exp(qV/kT) - 1\right] , \quad (9.115)$$

where the saturation current is

$$J_s = A_R T^2 \exp(-q\varphi_{bn}/kT) \quad (9.116)$$

and A_R is the Richardson constant, which is specific to each semiconductor. The main source of the dark current in Schottky-barrier photodiodes is known to be thermionic emission of majority carriers.

In either mode of operation, the photocurrent from a Schottky-barrier photodiode is given by

$$I_p = (1 - R_\Sigma) q\varphi\eta_{\text{in}} , \quad (9.117)$$

where R_Σ accounts for both the front side reflection of light and absorption by the metal film, η_{in} is the internal quantum efficiency, and Φ is the incident photon flux (in quanta per second).

Quantum Efficiency

For the case of band-to-band excitation, the internal quantum efficiency of an n-type Schottky-barrier photodiode is similar to that of a p–i–n photodiode, and it can be written as

$$\eta_{\text{in}} = 1 - \frac{\exp(-\alpha w)}{1 + \alpha L_p} , \quad (9.118)$$

where α is the absorption coefficient, w is the depletion width, and L_p is the hole diffusion length.

For intrinsic photoemission, the process of photocurrent generation is dependent on the incident quantum energy $h\nu$, and the internal quantum efficiency is given

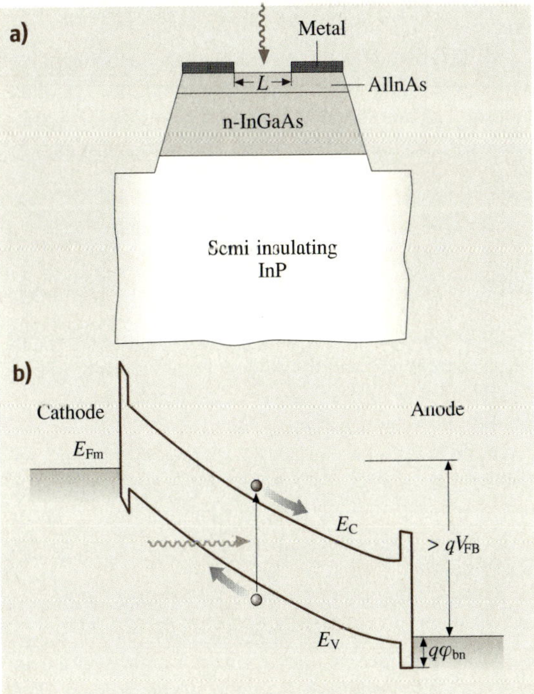

Fig. 9.29 **(a)** Cross-section of the MSM mesa structure photodiode. **(b)** The energy band diagram of the MSM photodiode under a bias beyond flat-band. The increase in the barrier due to the barrier-enhancement layer is shown

by [9.14]

$$\eta_{in} = C_f \frac{(h\nu - q\varphi_{bn})^2}{h\nu} , \qquad (9.119)$$

in which C_F is the so-called Fowler emission coefficient.

Schottky barrier photodiodes do not perform as well as p–n or p–i–n photodiodes; however, they can be advantageous for wavelengths where the semiconductor absorption coefficient is very high ($>10^4\,cm^{-1}$), and most absorption of light occurs within a thin surface layer of the photodiode [9.33]. In a regular p/n structure, the nonequilibrium carrier collection efficiency is very low close to the surface because of pronounced surface charge recombination. If the p semiconductor is replaced with a metal, the surface recombination–generation becomes much less pronounced or even disappears completely, increasing the carrier collection efficiency in the metal–semiconductor depletion layer.

Schottky-barrier photodiodes working in the second mode of operation have also proven to be very useful for infrared light detection, leading to the construction of efficient Schottky-barrier focal plane arrays [9.34–36]. The other advantage of Schottky-barrier photodiodes is their high operational speed.

9.6.2 Metal–Semiconductor–Metal (MSM) Photodiode

MSM photodetectors can be thought of as two Schottky barriers connected back-to-back (Fig. 9.29a). For compound semiconductors, a thin light absorption layer (n-InGaAs in Fig. 9.22a) is deposited on a semi-insulating substrate. A barrier-enhancement layer AlInAs with a thickness of several tens of nanometers may be deposited between the metal and absorption layer to decrease the dark current of the narrow-bandgap semiconductor that constitutes the absorption layer. A barrier enhancement layer fabricated from the wide energy gap semiconductor significantly increases the barrier height (Fig. 9.29b).

Applying bias of any polarity to an MSM photodetector creates one Schottky barrier in the forward direction and the other in the reverse direction. The dark current from the MSM photodetector has both electron and hole current components, and the saturation current is given by the following expression [9.37]

$$\begin{aligned} J_s =& A_{Rn} T^2 \exp(-q\varphi_{bn}/kT) \\ &+ A_{Rp} T^2 \exp\left(\frac{-q\varphi_{bp}}{kT}\right) , \end{aligned} \qquad (9.120)$$

where A_{Rn}, A_{Rp} are the Richardson constants and φ_{bn}, φ_{bp} are the barrier heights at the cathode side and the anode side, respectively (Fig. 9.29b).

The photocurrent first rises with bias and then becomes saturated, indicating that punch-through (complete depletion of the structure) is reached. The saturation voltage corresponds to flat-band conditions, in which the electric field at the anode becomes zero.

MSM photodetectors are advantageous in applications where a very low capacitance is required. The other important advantage of MSM structures is their compatibility with field effect transistor (FET) technology.

9.7 Detectors with Intrinsic Amplification: Avalanche Photodiodes (APDs)

9.7.1 APD: Principles, Basic Properties, and Typical Structures

An APD can be considered to be a family of p–n junction photodiodes working in photoconductive mode and capable of supporting high reverse bias. Upon the generation of an electron–hole (e–h) pair through photon absorption, the electron and/or hole are accelerated in the electric field and gain sufficient kinetic energy to collide with the atoms of the crystal ionizing them. The primary carriers and the products of this impact ionization are also accelerated, creating more nonequilibrium electron–hole pairs. Under favorable conditions, multiple collisions can create an avalanche-type increase in output current. Hence, the avalanche photodiode can be thought of as a type of photoconductor. It can work in either linear or nonlinear modes.

In the avalanche regime, each absorbed photon creates a finite number M of e–h pairs on average. The typical internal gain is tens to hundreds in the linear mode, where the photocurrent is proportional to the incident optical flux.

At high reverse bias, the rate of nonequilibrium carrier extraction on the electrodes falls back the impact ionization rate, which produces a highly nonlinear mode of APD operation – avalanche breakdown. The

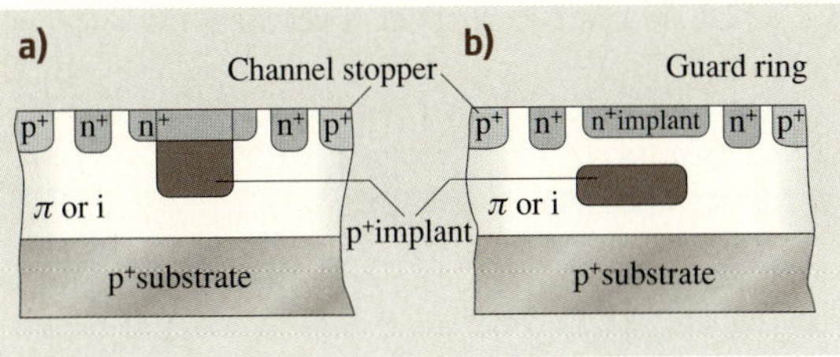

Fig. 9.30 **(a)** Cross-sections of typical reach-through APD structures with separated absorption and multiplication regions. The intrinsic (or π) layer that separates the n^+ implant and the p^+ implant in the structure shown in panel **(b)** greatly improves the photoresponse uniformity and the noise characteristics of the APD

population of nonequilibrium carriers in the high-field region and the associated photocurrent grow exponentially over time. This growth continues until the drop in voltage across the series resistance reduces the voltage drop across the photodiode high-field region, slowing the avalanche and ultimately stopping it. Since the avalanche breakdown is not a destructive effect and is stable against thermal runaway, this nonlinear mode of operation has found many practical applications and has been afforded a specific name – the Geiger mode. High-performance Geiger mode APDs can be used for single photon counting [9.38].

The noise current from an APD has an additional term that goes beyond the simple scaling up of the photocurrent that would flow in a regular photodiode by a factor of M. The origin of this additional noise is the statistic (and stochastic) nature of the multiplication process. Fluctuations in gain produce excess noise, which gets progressively worse as the average gain M of the diode is increased by raising the reverse bias. When the multiplication noise exceeds the noise introduced by downstream circuitry, further increases in gain deteriorate the system's signal-to-noise ratio.

Figure 9.30 shows the APD structures typically employed in practice. Starting material with either n-type or p-type conductivity is chosen based on the maximum value of the minority carrier impact ionization rate. For Si, the electron ionization rate is much higher than that for holes; hence, p-type material is used to build Si APDs.

The diode is fabricated by ion implantation of n-type and p-type dopants. The structure that results, rather than being a simple p–n diode, is either p–π–p–n (Fig. 9.30a) or p–π–p–π–n (Fig. 9.30b). Here π denotes a very lightly – close to intrinsic – p-doped region. A thin layer of lightly π-doped material between the

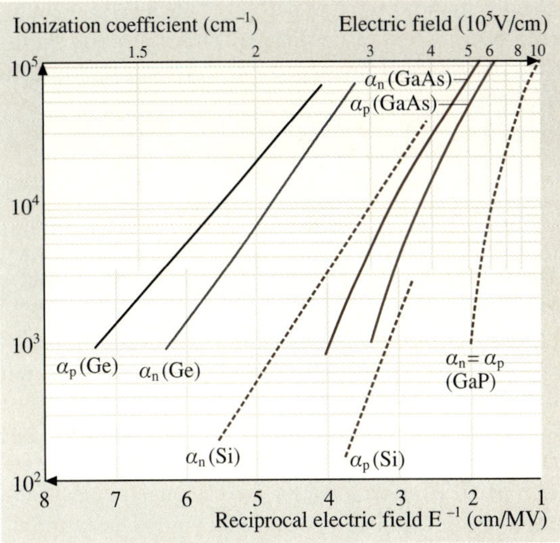

Fig. 9.31 Ionization coefficients for electrons and holes in a number of semiconductors

top n electrode and a highly doped p region in Fig. 9.30b improves the APD noise, response speed and uniformity, as well as the quantum efficiency in comparison with the structure shown in Fig. 9.30a. Both structures in Fig. 9.30 have their absorption regions separated from their multiplication layers. These structures are also called reach-through structures, because the electric field extends all the way from the n/p(π) junction and reaches the p layer. The lower π layer is where the photons are absorbed for either front- or back-illuminated structures. Reverse biasing at the appropriate operating voltage establishes a modest electric field in this photon absorption layer that causes the photoelectron to drift up into the upper p layer. The field in that layer (sometimes called the "p well") is much stronger, sufficient to cause impact ionization, which initiates an avalanche. The photoelectron and the secondary electrons are collected at the top n layer, and the photohole and secondary holes are collected at the substrate.

Note that the p-well exists only in the central portion of the APD. Because of the absence of a p implant, the diode's peripheral portion is a simple p–i–n structure, in which the field is intermediate in strength between the photon-absorption region and the avalanche region. This peripheral diode serves as a "guard ring" that performs two functions. First, it tailors the electric field profile so that avalanche breakdown occurs in the central portion of the diode, not at the periphery. Second, it ensures that electrons generated outside the region directly under the p implant do not drift to the avalanche region, but are collected in the peripheral portion of the n layer without being able to start an avalanche. This collection minimizes the volume from which dark current is multiplied, and therefore minimizes the dark count rate.

9.7.2 APD: Main Characteristics and Figures of Merit

Ionization Rates

The electron and hole ionization rates, α_n and α_p, respectively, are defined as the number of ionizing collisions per unit distance. The ionization rates depend strongly (exponentially) on the threshold electric fields required to overcome different carrier scattering effects. Examples of the experimental ionization rates at 300 K for Si, Ge, GaAs, and GaP versus electric filed are given in Fig. 9.31. The ionization rates can be equal for electrons and holes, as for GaP, or they can significantly differ from each other, as in the cases of Si, Ge, and most compound semiconductors.

Avalanche Gain (Multiplication Factor)

The multiplication factor M is defined as the ratio of the number of carriers leaving the avalanche region to those entering it. Since the ionization rates are strongly dependent on the electric field, which is usually position-dependent, it appears to be almost impossible to explicitly calculate the avalanche gain for any real structure. However, reasonable estimations may be obtained for some practically important cases [9.1, 6, 12].

The low-frequency gain for the electrons/holes is given by

$$M_{n,p} = \left\{ 1 - \int_0^L \alpha_{n,p} \times \exp\left[- \int_x^L (\alpha_{n,p} - \alpha_{p,n}) \, dx' \right] dx \right\}^{-1} , \tag{9.121}$$

where L is the high-field layer width. This expression is simplified for position-independent ionization coefficients, which, for the case of the electron current gain, yields

$$M_n = \frac{(1 - \alpha_p/\alpha_n) \exp\left[\alpha_n L \left(1 - \alpha_p/\alpha_n\right)\right]}{1 - (\alpha_p/\alpha_n) \exp\left[\alpha_n L \left(1 - \alpha_p/\alpha_n\right)\right]} . \tag{9.122}$$

For equal, position-dependent ionization rates [$\alpha_n = \alpha_p = \alpha(x)$]

$$M_p = M_n = \left[1 - \int_0^L \alpha(x) \, dx \right]^{-1} , \tag{9.123}$$

from which we can conclude that the avalanche breakdown ($M_{n,p} \to \infty$) corresponds to the case $\alpha_{n,p} L \to 1$.

The maximum achievable multiplication is limited by the series resistance and the space-charge effect. Combining these two factors into a single equivalent series resistance R, the dependence of the multiplication of photogenerated carriers M_{ph} on the reverse bias V and the breakdown voltage V_B can be empirically described as [9.6]

$$M_{ph} = \frac{I - I_{MD}}{I_p - I_d} = \left[1 - \left(\frac{V - IR}{V_B} \right)^{\gamma} \right]^{-1} , \tag{9.124}$$

where I_p and I are the total primary and multiplied currents, and I_d and I_{MD} are the primary and multiplied dark currents, respectively. The exponent γ depends on the semiconductor material, the doping profile, and wavelength, and it is usually in the range 3–6 [9.14]. For a relatively high light intensity $I_p \gg I_d$ and low parasitic losses $IR \gg V_B$, the maximum value of M_{ph} is given by

$$M_{ph}^{max} \approx \frac{V_B}{\gamma I_p R} . \tag{9.125}$$

For small photocurrents, $I_p \ll I_d$, the maximum avalanche multiplication is:

$$M_{ph}^{max} \approx \frac{V_B}{\gamma I_d R} , \tag{9.126}$$

which means that a high dark current limits the maximum multiplication.

Excess Noise Factor

Due to the statistical nature of the avalanche gain, the mean-square value $\overline{M^2}$ of the gain M is greater than the square M^2 of the mean value. The shot noise of any photodiode is determined by the statistics of the nonequilibrium carriers, electrons and holes. For the p–i–n photodiode, each nonequilibrium carrier generation event creates one electron and one hole only, whereas for each incident photocarrier in the multiplying region of APD three non-equilibrium particles are present: one primary carrier and two newly created secondary carriers. If the spectral density of the noise in the total primary current I_p is $2qI_p$, the spectral density of the

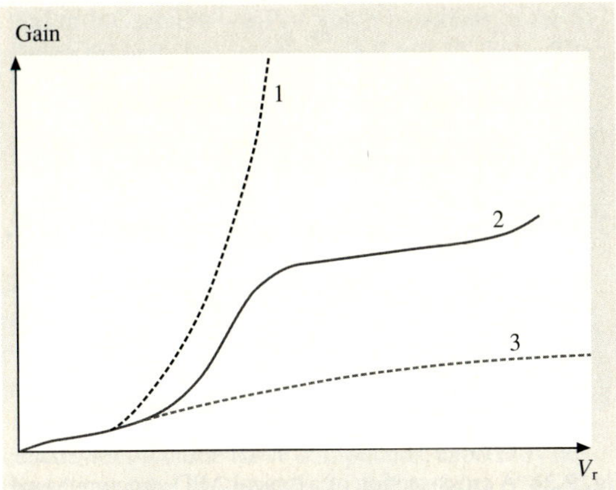

Fig. 9.32 Gain–voltage characteristics of an APD at different stages of processing, as described in the text

noise in the total multiplied current $I = I_p M$ can be written as

$$\overline{i_{sh}^2} = 2qI_p\overline{M^2} = 2qI_pM^2F(M)\,, \tag{9.127}$$

where we put the frequency bandwidth $\Delta f = 1$ and introduce the *excess noise factor* $F(M) = (\overline{M^2}/M^2)$,

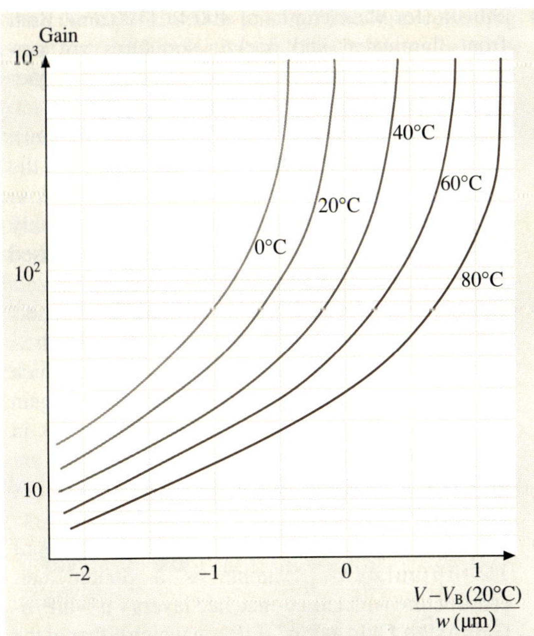

Fig. 9.33 "Universal" gain–voltage characteristics for a reach-through APD. The applied reverse bias V and the breakdown voltage at 20 °C V_B(20 °C) are in volts, while the depletion width w is in microns

which is the ratio of the actual noise to that which would exist if the multiplication process were noiseless.

Since the excess noise depends strongly on the total number of the secondary carriers, its value is large when the electron and hole ionization rates, α_n and α_p, are close to each other. When either $\alpha_n \gg \alpha_p$ or $\alpha_n \ll \alpha_p$ the excess noise decreases. For the ratio (α_n/α_p), which is constant throughout the avalanche region, the excess noise can be approximated by [9.39]

$$F(M) = k_iM + \left(2 - \frac{1}{M}\right)(1 - k_i)\,, \tag{9.128}$$

in which $k_i = (\alpha_p/\alpha_n)$ for the case of electron injection alone, and $k_i = (\alpha_n/\alpha_p)$ for hole injection.

Signal-to-Noise Ratio

The thermal noise in an APD is the same as that for a p–i–n photodiode and it is given by (9.93). The shot noise spectral density of an APD is given by (9.127) above. The signal-to-noise ratio for an avalanche photodiode is then

$$\frac{S}{N} = \frac{I_s}{i_n} \equiv \frac{I - I_{MD}}{i_n} = \frac{\frac{1}{\sqrt{2}}\left(\frac{q\eta_{ex}\Phi_p}{h\nu}\right)M}{\left\{2q\left[I_D' + I_pF(M)M^2\Delta f\right] + \frac{4kT\Delta f}{R}\right\}^{1/2}}\,, \tag{9.129}$$

where I_s and i_n are the average output photocurrent and rms noise current, respectively, I is the total multiplied current, I_{MD} is the multiplied dark current, I_p is the total primary current that is multiplied, I_D' is the component of the dark current (usually the surface leakage current) that is not multiplied, Δf is the noise bandwidth, Φ_p is the incident optical power, and η_{ex} is the external quantum efficiency.

Noise Equivalent Power

Equation (9.129) can be solved for the optical power required to produce a signal-to-noise ratio equal to 1:

$$\mathrm{NEP} = \frac{\sqrt{2}h\nu}{\eta_{ex}}\left(\frac{I_pF(M)\Delta f}{q} + \frac{2I_0\Delta f}{q}\right)^{1/2}\,, \tag{9.130}$$

where

$$I_0 = I_dF(M) + \frac{2kT}{qRM^2} \tag{9.131}$$

is an equivalent current, similar to that introduced by (9.17) of Sect. 9.1.3.

These curves are similar to those of a conventional bipolar transistor with the base current replaced by photocurrent. Note that the photodiode dark current is also amplified by the same amount as the photocurrent.

Since the reverse current, I_{C0}, of the base diode is amplified in the same way as the photocurrent, I_p, the signal-to-noise ratio and the noise equivalent power of the phototransistor are similar to those of a photodiode. For example, the NEP can be written as

$$\mathrm{NEP} = \frac{\sqrt{2}h\nu}{\eta_{ex}}\sqrt{\frac{I_{eq}}{q}} \qquad (9.135)$$

with an equivalent current I_{eq}

$$I_{eq} = I_{CEO}\left(1 + \frac{2h_{fe}^2}{h_{FE}}\right) , \qquad (9.136)$$

where h_{fe} is the common-emitter AC current gain [9.6].

The operating point of the phototransistor may be preset by biasing the base lead; this controls both the speed and gain of the device. The speed of the phototransistor is limited by the base-to-emitter (C_{BE}) capacitance and is specified in terms of the rise time of the device. Since C_{BE} interacts mainly with the load resistor (R_L), the rise time is linearly dependent on R_L. Phototransistors are not as fast as photodiodes. A photodiode is in fact the input section of a phototransistor (collector-to-base section) without an emitter and so it does not suffer from the C_{BE} effect.

9.8.2 Darlington Phototransistor (Photo-Darlington)

A Photo-Darlington is simply a cascaded pair of bipolar transistors (Fig. 9.37). One transistor in the pair serves as a primary photodetector while the other acts as an additional amplifier. The advantage of this design is higher amplification, but the time response is pretty long (several tens of μs).

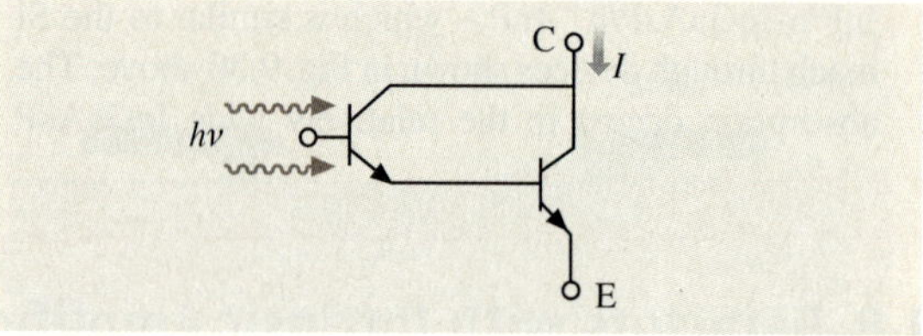

Fig. 9.37 A simplified schematic diagram of a Darlington phototransistor

9.8.3 Field-Effect-Based Phototransistors

Examples of the other type of phototransistors include photosensitive field effect transistors like JFET, MESFET, and MOSFET. These employ various modes of operation, the simplest of which is based on photoconductivity, where photogenerated excess carriers increase the conductivity of the channel. In the mode specific to JFET or MESFET, the gate is not insulated and the gate-to-channel junction can be considered to be a photodiode that has the gate current as the primary photocurrent. For the MOSFET, the gate is insulated and different modes of operation are possible. For example, photogenerated carriers can be trapped at the surface of the semiconductor, which changes the surface potential and induces a change in the channel current.

One interesting design uses the depleted field effect transistor structure (DEPFET) [9.10, 40]. This is based on a combination of the sideward depletion principle and the principle of the field effect transistor. In this structure, a p-channel MOSFET is built on one side of an n-type wafer whereas a p–n diode structure is created on the opposite side. Applying a reverse bias to the diode creates a depletion layer that propagates towards the MOSFET on the wafer's front side. If the MOS gate is biased and an inversion layer is created on the oxide–semiconductor interface, then the minimum potential for the majority carriers (electrons) will occur in the bulk of the wafer, right underneath the MOSFET gate. Upon the photogeneration of electron–hole pairs in the depleted bulk, the minority carriers (holes) will drift in the depleted region towards the diode p/n junction on the back surface of the structure, while the majority carriers will be trapped in the potential well underneath the transistor p-channel. This trapped charge will induce a charge of opposite polarity in the channel, thereby increasing its conductance and the transistor current.

The DEPFET structure is characterized by a smaller overall capacitance, thereby potentially yielding a better noise performance than a photodiode coupled to a transimpedance amplifier. The DEPFET structure allows the amount of charge stored in the MOSFET gate region to be controlled, thus allowing either multiple readouts of stored information or the clearing of that information.

9.9 Charge Transfer Detectors

The detectors described in the previous sections operate primarily as single-element detectors. When designing multi-element arrays, the simplest solution – adding a number of individual detectors together – did not prove to be very fruitful. This became even more evident in applications where high-resolution images with a large number of pixels and a small element size were required. In the early 1970s, a new concept for creating images was introduced, based on the use of charge transfer devices [9.41–44]. This idea appeared to be very promising and triggered extensive research in this area.

The structures of all charge transfer devices – charge-coupled devices (CCD), charge transfer image sensors (CTIS) or charge-coupled image sensors (CCIS), and (more recently) complementary metal–oxide–semiconductor detectors (CMOS) – are based on a metal–insulator–semiconductor (MIS) capacitor. The most important MIS is the metal–oxide–semiconductor (MOS) capacitor made from silicon, which uses silicon dioxide as the insulator. The simplest version of a CCD is an array of closely spaced MOS capacitors. Under the application of a proper sequence of clock voltage pulses, the charges accumulated in the capacitors of the array can be moved in a controlled manner across a semiconductor substrate. This basic mechanism has been applied to build a variety of electronic and optoelectronic devices that perform image sensing, data storage, and signal processing.

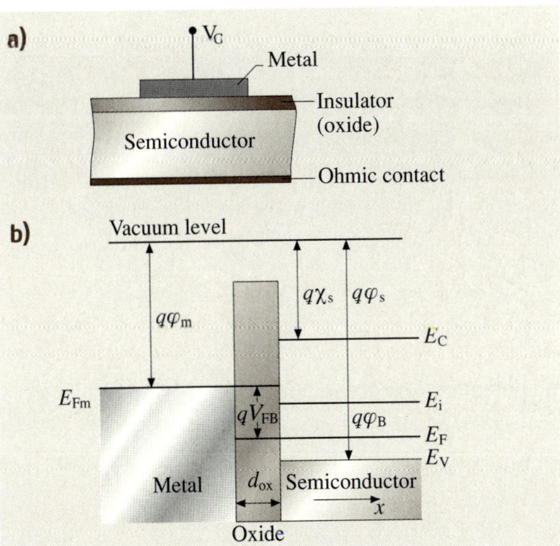

Fig. 9.38 **(a)** Cross-section of a MOS capacitor. **(b)** Simplified energy band diagram of an ideal MOS capacitor biased at $V_G = V_{FB}$

9.9.1 MOS Capacitor

An MOS capacitor is a building block for a charge storage device. A cross-section of such a capacitor is shown in Fig. 9.38a. First consider a MOS capacitor at thermal equilibrium conditions with no fixed or mobile charge in the oxide layer and zero probability that the charge carriers will cross that layer. In this case, the thermal equilibrium can be considered separately in two regions (oxide and semiconductor regions).

Modes of Operation

Flat-Band Conditions. A simplified band diagram of the ideal p-type MOS capacitor at thermal equilibrium and for so-called flat-band conditions is given in Fig. 9.33b. Under these conditions, semiconductor band-bending is absent (the electric field strength throughout the semiconductor is zero) and the electrode voltage applied to the gate V_{FB} corresponds to the difference

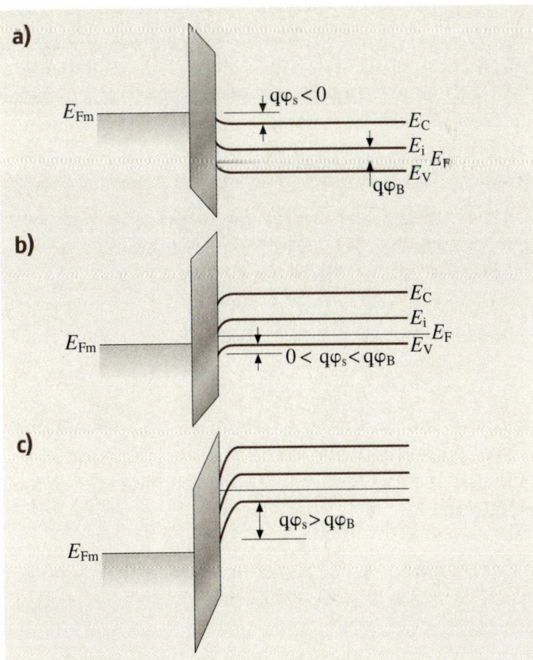

Fig. 9.39a–c Different modes of MOS capacitor operation: **(a)** accumulation, **(b)** depletion, and **(c)** inversion

between the metal and semiconductor work functions

$$V_{FB} = \varphi_m - \varphi_s \,. \tag{9.137}$$

Note that for Si, the barrier height at the Si-SiO$_2$ interface is 3.2 eV for electrons and 4.3 eV for holes. The Fermi level E_F is shifted down from the intrinsic level E_i^p by

$$q\psi_{Bp} = E_i^p - E_F = kT \ln \frac{N_A}{n_i} \,, \tag{9.138}$$

where N_A is the acceptor concentration and n_i is the intrinsic carrier concentration.

Applying a bias to the gate electrode $V_G \neq V_{FB}$ allows the potential ψ_s of the oxide–semiconductor interface to be shifted, which drives the MOS capacitor into different regimes. There can be three distinctly different modes of operation depending on the gate voltage polarity and amplitude: (a) accumulation, (b) depletion, and (c) inversion. These three modes are shown schematically in the band diagrams of Fig. 9.39a–c. The Fermi level remains flat for any operational regime because of zero current flow.

Accumulation Mode. Upon applying a more negative voltage than V_{FB} to the gate of a p-type MOS capacitor, holes will be attracted to the semiconductor–oxide interface, and a thin layer of positive charge (majority carriers) will form in the semiconductor at the boundary. The bands bend upwards near the semiconductor-oxide interface, and the edge of the valence band approaches the Fermi level, thereby filling the potential well at the boundary with free holes. The surface potential energy $q\psi_s$ is negative in this case. Since these free holes are concentrated in a very thin interfacial layer, the surface charge density can be written as

$$Q_{acc} = -\varepsilon_{ox}\varepsilon_0 \frac{V_G - V_{FB}}{d_{ox}} = -C_{ox}(V_G - V_{FB}) \,, \tag{9.139}$$

where ε_{ox} and ε_0 are the oxide dielectric constant and vacuum permittivity, respectively, d_{ox} is the oxide layer thickness, and C_{ox} is the capacitance per unit area of the oxide layer.

Depletion Mode. When a small positive excess voltage $V_G > V_{FB}$ is applied, the majority carriers are repelled from the semiconductor–oxide interface, the bands bend downwards at the interface, and a relatively small positive surface potential $\psi_s (0 < q\psi_s < q\psi_{Bp}$, Fig. 9.39b) is formed.

To find the distribution of potential inside the semiconductor, the depletion approximation is usually used, which assumes that the edge of the depletion region is abrupt and that no mobile majority carriers exist in this depletion region. The potential ψ in the depletion region for the thermal equilibrium case (no photoexcitation, no imperfections in the oxide and interface layers) is described by the Poisson equation

$$\nabla^2\psi = \frac{\partial^2\psi}{\partial x^2} = -\frac{\rho}{\varepsilon_s\varepsilon_0} = \frac{qN_A}{\varepsilon_s\varepsilon_0} \,, \tag{9.140}$$

where ρ is the charge density, N_A is the acceptor concentration in a uniformly doped p-type semiconductor, and ε_s is the dielectric constant of the semiconductor. The electric field $E = -\nabla\psi$ can be found by integrating (9.140) with a constant electric field E_{ox} in the insulator layer assumed as a boundary condition. One more integration provides the potential distribution as a function of depth x inside the semiconductor

$$\psi = \int_0^x E(x)\,dx = \int_0^x \frac{qN_A\,x}{\varepsilon_s\varepsilon_0}\,dx = \frac{qN_A\,x^2}{2\varepsilon_s\varepsilon_0} \,, \tag{9.141}$$

where the potential is assumed to be zero throughout the semiconductor outside the depletion region. The potential varies linearly with distance across the insulator and quadratically in the depletion region. The minimum potential at the interface takes the value

$$\psi_s = \frac{qN_A}{2\varepsilon_s\varepsilon_0} w^2 \,, \tag{9.142}$$

where w is the semiconductor depletion layer width.

The gate voltage V_G is distributed between the voltage drop V_{ox} that charges the oxide capacitance, the work function difference (9.137), and the semiconductor surface potential ψ_s

$$V_G = V_{FB} + V_{ox} + \psi_s \,. \tag{9.143}$$

The potential drop across the oxide is given by

$$\begin{aligned} V_{ox} = E_{ox}d_{ox} &= \frac{qN_A\,w}{\varepsilon_s\varepsilon_0} d_{ox} \equiv \frac{qN_A\,w}{C_{ox}} \\ &= \frac{\sqrt{2qN_A\varepsilon_s\varepsilon_0\psi_s}}{C_{ox}} \,, \end{aligned} \tag{9.144}$$

where on the right-hand side we substituted w for the (9.142). Solving (9.143) and (9.144) for the surface potential at thermal equilibrium yields

$$\psi_s = V_G - V_{FB} + \frac{qN_A\varepsilon_s\varepsilon_0}{C_{ox}^2} - \frac{1}{C_{ox}} \times \sqrt{2qN_A\varepsilon_s\varepsilon_0(V_G - V_{FB}) + \left(\frac{qN_A\varepsilon_s\varepsilon_0}{C_{ox}}\right)^2} . \quad (9.145)$$

The surface layer depletion width w is given by [9.10]

$$w = \sqrt{\frac{\varepsilon_s\varepsilon_0}{qN_A}(V_G - V_{FB}) + \left(\frac{\varepsilon_s}{\varepsilon_{ox}}d_{ox}\right)^2} - \frac{\varepsilon_s}{\varepsilon_{ox}}d_{ox} . \quad (9.146)$$

The capacitance of the MOS structure in the depletion mode can be described as a series capacitance involving the insulator and semiconductor layer capacitance

$$C = \left(\frac{d_{ox0}}{\varepsilon_{ox}\varepsilon_0} + \frac{w}{\varepsilon_s\varepsilon_0}\right)^{-1} . \quad (9.147)$$

Inversion Mode. When a larger positive voltage is applied, the bands bend even more so that the intrinsic level E_i at the semiconductor–oxide interface crosses the Fermi level E_F (Fig. 9.39c). For this mode, the surface potential becomes large ($q\psi_s > q\psi_{Bp}$) and the minority carriers (electrons) at the surface may outnumber the majority carriers (holes). A thin conducting layer of minority carriers forms close to the semiconductor–insulator interface and the surface becomes inverted.

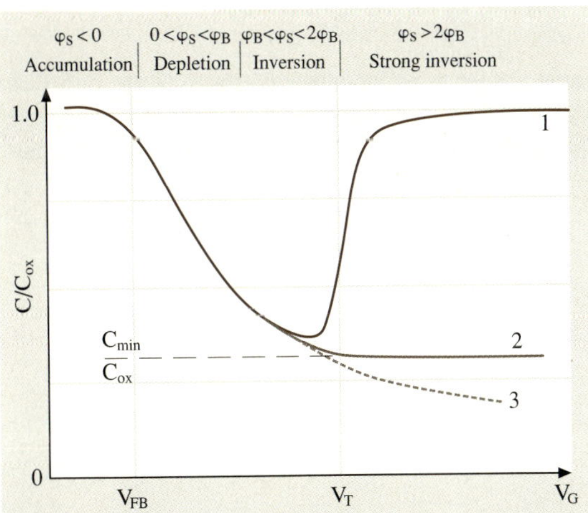

Fig. 9.40 Typical C–V characteristics of a MOS capacitor for a low frequency (*curve* 1) and a high frequency (*curve* 2). *Curve* 3 shows a drop-off in the capacitance in accord with the depletion approximation

If the gate voltage increases further, the minority carrier concentration at the surface surpasses the majority carrier bulk concentration. This situation is called "strong inversion" and it begins at a surface potential of [9.6]

$$\psi_s(\text{inv}) \approx 2\psi_{Bp} = \frac{2kT}{q}\ln\left(\frac{N_A}{n_i}\right) . \quad (9.148)$$

The depletion width reaches its maximum value at

$$w_{max} = \sqrt{\frac{4\varepsilon_s\varepsilon_0\psi_{Bp}}{qN_A}} = \sqrt{\frac{4\varepsilon_s\varepsilon_0 kT\ln(N_A/n_i)}{q^2N_A}} , \quad (9.149)$$

and the gate threshold (turn on) voltage, at which strong inversion occurs is given by

$$V_T = V_{FB} + 2\psi_{Bp} + \frac{d_{ox}}{\varepsilon_{ox}\varepsilon_0}\sqrt{4qN_A\varepsilon_s\varepsilon_0\psi_{Bp}} . \quad (9.150)$$

Further increases above the threshold voltage will only increase the strength of the inversion layer, while the depletion depth remains constant at w_{max}.

MOS Capacitor Characteristics

C–V Characteristics. The dependence of the capacitance on the gate voltage in the MOS structure is shown schematically in Fig. 9.40. Capacitance stays high in the accumulation mode and starts decreasing in the depletion mode due to the increase in the depletion layer width with voltage. When inversion is reached, the capacitance continues to drop until the minimum value

$$C_{min} = \left(\frac{d_{ox}}{\varepsilon_{ox}\varepsilon_0} + \frac{w_{max}}{\varepsilon_s\varepsilon_0}\right)^{-1} \quad (9.151)$$

is reached at the maximum depletion width w_{max}. For strong inversion, the behavior of the capacitance differs between low and high frequencies.

At low frequencies, minority carriers can follow the AC signal to modulate the inversion layer, and the capacitance rises to C_{ox} at $V_G > V_T$ (curve 1 in Fig. 9.40). At high AC frequencies, minority carriers are not generated quickly enough to follow the AC signal. The depletion width either saturates at w_{max} (with a slow ramp of the gate voltage V_G) or it becomes even wider (at fast V_G ramp), so that the overall capacitance of the MOS structure either saturates at C_{min} level (curve 2 in Fig. 9.40) or continues to drop in accord with the depletion layer approximation, which is similar to the abrupt p/n junction approximation (curve 3 in Fig. 9.40).

Part B | 9.9

For the light-induced generation of minority carriers, the equilibrium high-frequency curve, as shown by curves 2 or 3, can be obtained with either a slow or a fast V_G ramp.

MOS Capacitor Under Nonequilibrium Conditions – Concepts of Charge Storage

Under illumination, and for either the depletion or the inversion mode, the photogenerated minority carriers (electrons for the p-type semiconductor) will tend to collect in the potential well ψ_s close to the semiconductor–oxide interface. As the electrons accumulate at the semiconductor interface, the field across the insulator starts to increase, and the surface potential and the depletion width begin to shrink [9.14]

$$\psi_s = V_G - V_{FB} + \frac{qN_A\varepsilon_s\varepsilon_0}{C_{ox}^2} - \frac{Q_{sig}}{C_{ox}} - \frac{1}{C_{ox}} \times \sqrt{2qN_A\varepsilon_s\varepsilon_0\left(V_G - V_{FB} - \frac{Q_{sig}}{C_{ox}}\right) + \left(\frac{qN_A\varepsilon_s\varepsilon_0}{C_{ox}}\right)^2} , \quad (9.152)$$

where Q_{sig} is the signal charge density. The maximum charge density that can be collected in the ideal case is estimated from the condition $\psi_s = 0$

$$Q_{sig}^{max} = C_{ox} V_G , \quad (9.153)$$

which is to be avoided. In practice, the maximum achievable charge density is $\approx 10^{11}\,\mathrm{cm}^{-3}$, which limits the dynamic range of any MOS detector to $\approx 10^4$.

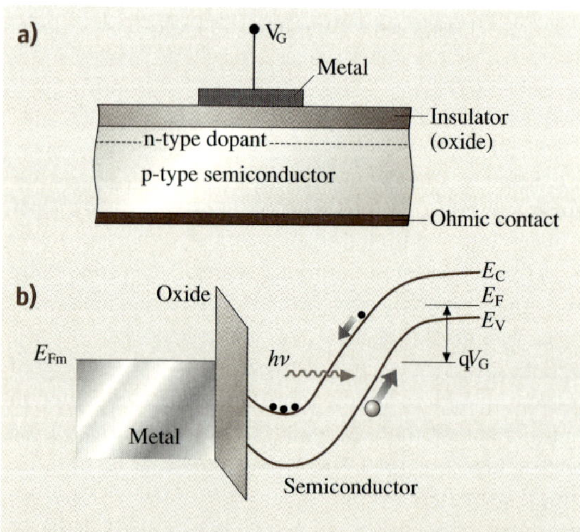

Fig. 9.41 **(a)** Schematic of the structure and **(b)** energy band diagram of a buried-channel MOS capacitor operated under a large positive gate bias V_G

Effects of Dark Currents, Interface Traps and Oxide Charges

Various imperfections deteriorate MOS capacitor performance. For example, mobile oxide charges, oxide trapped charges, and fixed oxide charges near the oxide–semiconductor interface cause a parallel shift in the C–V curve shown in Fig. 9.40 in the V_G direction, although main performance parameters do not suffer significantly from these imperfections. Special attention is usually paid to interface charge traps, which are among the major limitations of MOS capacitor performance. If the charge traps are empty prior to the illumination of the MOS capacitor, they are able to capture the nonequilibrium carriers created by the light and then to release them so slowly that the readout electronics will not detect them correctly.

The effect of interface traps can be substantially reduced by passing a constant background charge to the MOS capacitor. Then, to first order, the interface traps remain permanently filled and interact with the signal charges to a minimal extent. However, the drawback of this method is a reduction in the dynamic range by approximately 10 to 20%.

In addition to light generation, various sources that generate dark current also supply charge to the interface and act as background noise. The dark current density can be expressed as the sum of the currents generated in the depletion region, in the neutral bulk, and at the surface [9.14]

$$J_{dark} = \frac{qn_i w}{2\tau} + \frac{qD_n n_i^2}{L_n N_A} + \frac{qS_r n_i}{2} , \quad (9.154)$$

where τ is the minority carrier lifetime, D_n the diffusion constant, L_n the diffusion length, and S_r is the surface recombination rate. The dark current limits the accumu-

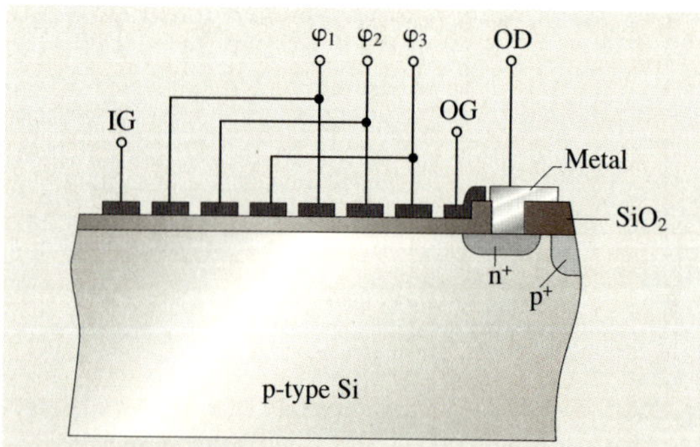

Fig. 9.42 A simplified cross-sectional diagram of an n-channel charge-coupled device

lation time of the MOS detector to a maximum value of

$$t_s = \frac{Q_{sig}^{max}}{J_{dark}} . \quad (9.155)$$

If the gate voltage remains fixed, thermally generated minority carriers will gradually fill the surface potential well and the MOS capacitance will increase toward its low-frequency asymptotic inversion condition C_{ox}.

Buried Channel MOS Capacitor

Previous sections described the surface-channel MOS capacitor. One of the major limitations of this type of MOS structure is the effect of interface traps. To circumvent this problem, the potential well can be buried deeper in the semiconductor bulk, producing a "buried-channel MOS structure" [9.45]. In this structure, the accumulation of nonequilibrium minor carriers is confined to a valley beneath the surface.

Figure 9.41a schematically shows the structure of a buried-channel MOS capacitor built on a p-type semiconductor. The structure has a thin layer ($< 0.5\,\mu m$) of opposite type (n-type) dopant at the semiconductor surface. This layer is fully depleted when the gate pulse is applied. The potential well in this case is formed close the p/n junction and photogenerated carriers are kept away from the interface. The energy band diagram for this capacitor under a high gate pulse is shown in Fig. 9.41b.

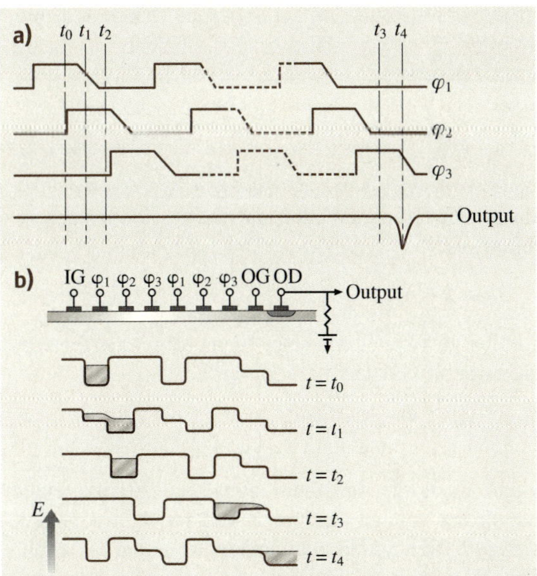

Fig. 9.43 (**a**) Clock waveforms and the output signal from the CCD structure shown in Fig. 9.42; (**b**) Energy levels and charge distribution of the same device with the clock from panel (**a**)

9.9.2 CCDs Employed as Charge-Coupled Image Sensors (CCISs)

A charge-coupled device is an array of MOS capacitors placed close together. The gates are made semitransparent for front-illuminated devices or they can be opaque if the device is back-illuminated. Illumination with light creates nonequilibrium charge carriers via the photoelectric effect under each gate electrode. Applying a proper sequence of gate voltages allows minority carriers to be transferred across the array, thereby delivering an electrical signal proportional to the amount of photons absorbed by each pixel to the output register.

Basic CCD Structure

The CCD photodetection process can be broken down into four steps charge generation, charge collection, charge transfer, and charge measurement. In a CCD array sensor, the charge from each pixel is transferred through a limited number of output nodes (often just one), converted into a voltage, buffered, and sent off-chip as an analog signal. A cross-section of a simple, typical back-illuminated CCD sensor is shown in Fig. 9.42. The device in Fig. 9.42 is a three-phase CCD that consists of a number of n-channel MOS capacitors connected to φ_1, φ_2, and φ_3 clock lines. Those capacitors participate in the charge generation, charge collection, and charge transfer steps. The output gate OG and the output diode OD are the structures needed to detect output charge packets. The input gate IG forms part of a structure that stops charge from flowing outside of the main CCD array. The surface channel stoppers p^+ at the edges of the structure ensure the unidirectional transfer of charge. In order to increase the efficiency of charge generation and collection, CCDs are usually made by using a high resistivity material to deplete up to $10\,\mu m$ of the semiconductor bulk underneath the gate of each MOS capacitor.

Assuming that the charge is generated and collected in the MOS capacitor on the left-hand side only, the charge transfer mechanism can be described using the following scheme. Fig. 9.43a shows the clock waveforms and Fig. 9.43b illustrates the corresponding potential wells and charge distribution for the structure in Fig. 9.42 where the charge is collected by the capacitor on the left-hand side only.

At $t = t_0$, the clock line φ_1 is at high voltage and φ_2 and φ_3 are at low gate voltages. The output diode (OD) is biased with a high positive voltage to prevent the inversion of the surface under the output gate. Thus the surface under the OG is stongly depleted and the OD cannot supply electrons to the main CCD array. If the first MOS capacitor is illuminated, the charge is collected in the well under the first capacitor gate, as shown in Fig. 9.43b for $t = t_0$. At $t = t_1$, the voltage applied to φ_1 returns to the low value while a high voltage is applied to the φ_2 electrodes. The second MOS capacitor becomes depleted and electrons stored under the first gate are then transferred to the well under the second gate. The voltage at φ_1 has a trailing edge because the charge carriers require a finite time to move across the width of the gate. At $t = t_2$, the charge transfer process is completed and the original charge packet is stored under the second gate. This process is repeated until the charge packet reaches the output terminal. At $t = t_4$ the voltages of the φ_3 electrodes return to low values, pushing the electrons towards the output diode OD.

Charge Transfer Mechanisms. Charge is transferred from one gate to another by three mechanisms: (1) self-induced drift (repulsion of like charge); (2) thermal diffusion; and (3) fringe field drift. These mechanisms operate within a free charge transfer model, and the charge transfer process is governed by the continuity equation

$$\frac{\partial n}{\partial t} = \frac{1}{q}\frac{\partial J}{\partial x} , \tag{9.156}$$

where n is the charge carrier concentration, J is the current density, and x is the distance in the direction of transfer. As a rule, self-induced drift accounts for $\approx 95\%$ of the charge transfer and two other mechanisms account for the rest. The expression for the decay of the initial charge packet Q_0 due to the self-induced drift can be expressed via the decay process given by [9.46, 47]

$$Q(t) = Q_0\left(1+\frac{\tau_{\text{si}}}{t}\right) \tag{9.157}$$

where

$$\tau_{\text{si}} = \frac{\pi L^3 HC}{2\mu_{\text{n}} Q_0} \tag{9.158}$$

is the decay time constant, L is the gate length, H is the gate electrode width, C are the total capacitances of the insulator and semiconductor layers, and μ_{n} is the minority carrier mobility.

However, for a small amount of signal charge, thermal diffusion is the dominant transfer mechanism. The total stored charge decreases exponentially with time, and the time constant is [9.6]

$$\tau_{\text{th}} = \frac{4L^2}{\pi^2 D_{\text{n}}} , \tag{9.159}$$

where D_{n} is the minority carrier diffusion constant.

Fringe field drift originates due to the two-dimensional coupling of the electrostatic field from the biased adjacent electrodes. Because the fringe field is present even at very low charge concentrations, the last bit of the signal charge will be transferred by the fringe field.

CCD: Figures of Merit

Spectral Response. The spectral response of a CCIS (CCD) is determined by the same relationships as those that determine the spectral response of a semiconductor photodiode. Since a relatively high-resistivity material is used to manufacture the CCIS, efficient nonequilibrium charge collection occurs from the charge-space (depletion) region, which may be up to 10 μm thick. This is why CCISs have high sensitivity and quantum efficiency over a broad spectral range. For example, Si-based CCISs are typically sensitive to wavelengths from 300 nm to 1000 nm.

Besides the figures of merit of traditional detectors, charge transfer detectors can be characterized by a number of specific parameters discussed below.

Transfer Efficiency. The transfer efficiency η_{tr} is the ratio of the charge Q transferred to the initial charge Q_0 stored

$$\eta_{\text{tr}} = \frac{Q}{Q_0} . \tag{9.160}$$

The related figure of merit is the transfer inefficiency ε_{tr}, defined as

$$\varepsilon_{\text{tr}} \equiv 1 - \eta_{\text{tr}} . \tag{9.161}$$

The charge transfer efficiency for n charge transfer steps can be approximated by [9.3]

$$(\eta_{\text{tr}})^n = \frac{V_{\text{m}} - \Delta V}{V_{\text{m}}} , \tag{9.162}$$

where V_{m} is the maximum signal packet voltage and ΔV is the voltage loss for a lead pulse in a packet. Using (9.162), and by rearranging terms and taking the first two terms of a binomial expansion, one obtains for a large number n of transfer steps

$$\eta_{\text{tr}} \cong 1 - \frac{\Delta V}{n V_{\text{m}}} . \tag{9.163}$$

For the case of incomplete free charge transfer in a clock period, the charge transfer efficiency can be written as [9.6]

$$\eta_{tr} \cong \frac{Q_0 - Q(t)}{Q_0} = 1 - \frac{8}{\pi^2} \exp\left(\frac{-t}{\tau_{th}}\right) . \quad (9.164)$$

Time Response. Charge collection inside the depletion region is a relatively fast process that does not usually limit the operational frequency of the CCIS. Consider other factors that influence the time response and the frequency range of a charge coupled device.

The pixel charge density is given by the summation of various sources of dark current and photocurrent,

$$\frac{dQ}{dt} = J_{dark} + J_{ph} = \frac{qn_i w}{2\tau} + \frac{qD_n n_i^2}{L_n N_A} + \frac{qS_r n_i}{2} + \eta_{ex} q\Phi , \quad (9.165)$$

where the first three terms on the left-hand side are described by (9.154), Φ is the incident optical photon flux for the pixel, and η_{ex} is the external quantum efficiency. Note that the overall detection efficiency of a CCIS can be thought of as a convolution of the quantum efficiency term with the terms that determine the charge transfer efficiency.

The charge transfer efficiency is influenced by different terms in (9.165) for different frequency ranges. Transfer efficiency at high frequencies can be described by the free charge transfer model of (9.156) and it is limited by the clock rates for a given device.

At medium frequencies, the trapping of the signal charge at the interface traps determines the transfer efficiency. Methods used to overcome the problems due to interface traps have been discussed above.

At low clock frequencies, the time response is determined by the dark current (9.165). The low-frequency degradation of the frequency response is due to the buildup of dark current in the charge packets, which distorts the shape and size of the signal charge. To improve the low-frequency response, one must reduce all of the dark current components in (9.165) by minimizing the gate length and choosing a material with a high minority carrier mobility, a long carrier lifetime, a large diffusion length, and a low surface recombination velocity. As mentioned previously, the lower limit frequency response is limited to the inverse of the maximum integration time given by (9.155).

Noise. The most important noise sources in a CCIS are [9.47, 48]:

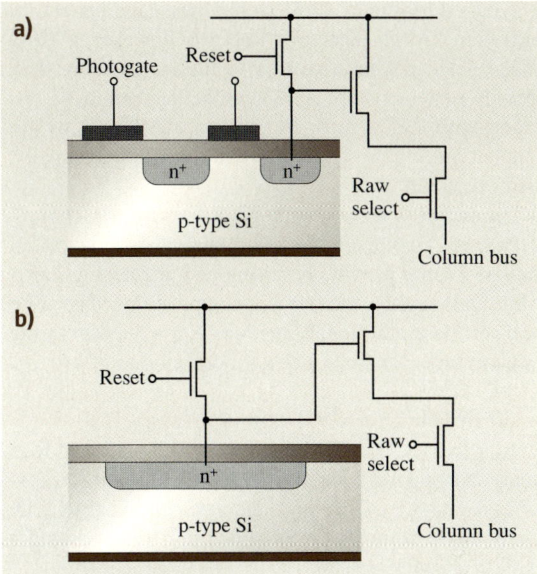

Fig. 9.44a,b Two CMOS sensor pixel designs: **(a)** photogate architecture and **(b)** photodiode architecture

1. The photon shot noise, which provides the basic limit on the sensitivity of the device
$$N_{sh} = \sqrt{2\eta_{ex}\Phi\Delta f} , \quad (9.166)$$
where N_{sh} is given in rms electrons;
2. The dark current shot noise N_{sh_DC}, which can be expressed in rms electrons by
$$N_{sh\_DC} = \sqrt{\frac{2J_{dark}\Delta f}{q}} , \quad (9.167)$$
where J_{dark} is given by (9.154);
3. The charge transfer noise, which is usually negligible due to the use of buried channels;
4. The readout diffusion reset noise, which is usually added after the charge/voltage conversion step.

CCD thermal noise originates in the same way as for photodiodes, and it can be greatly reduced by cooling the detector. Contemporary cooled CCIS can yield photon-limited performance and operate in single photon counting mode.

9.9.3 Complementary Metal Oxide Semiconductor (CMOS) Detectors

Many of the newer imaging devices use a different chip, known as a CMOS chip. CMOS technology is standard computer chip technology that has been around for years. Originally, CCD technology gave better images than CMOS detectors. Recently however, CMOS image detectors have evolved to the stage where they provide images comparable to or better than those obtained with CCD chips. In addition, CMOS chips are simpler and require fewer external devices that support their operation.

In a CMOS sensor, each pixel has its own charge-to-voltage conversion. Such sensors include MOS transistors that provide buffering and addressing capability, and may also include a digitizing circuit. The price paid for these additional functions is a reduction in the area available for front-side light collection.

Basic CMOS Structure and Properties

Similar to a CCD, a CMOS detector also performs four steps: charge generation, charge collection, charge transfer, and charge measurement. However, unlike the CCD, the first three steps are performed within each individual pixel in a CMOS sensor.

In order to convert two-dimensional spatial information into a serial stream of electrical signals, electronic scan circuits read out each pixel in the of CMOS photodetector sequentially. First, the vertical scan circuit selects a row of pixels by setting a high DC voltage on all of the gates of the MOS switches in that row. Next, the horizontal scan circuit selects the pixels in one particular column using the same technique. As a result, only one pixel in the two-dimensional matrix has a high DC voltage at its row and its column switches, which electronically selects it for readout. After the pixel dumps its information into the output stage, it is reset to begin a new integration, and the readout process progresses to the next pixel in the row. This organization, in contrast with CCD architecture, provides random access to each pixel and avoids multiple charge transfers over long distances through semiconductor material.

The most common pixel architectures used for CMOS sensor pixels are photogate conversion and photodiode conversion (Fig. 9.44). The photosensitive element in the photogate design is a MOS capacitor that performs the functions of charge generation and collection. Once the scan circuitry has selected a pixel, a pulse on the transfer gate triggers charge transport from the photosensitive area toward the potential well of a sensing node, created by an n^+ diffusion area at the semiconductor interface (Fig. 9.44a). This node is part of the floating-diffusion amplifier, which performs the charge-to-voltage conversion. The charge-to-voltage conversion in this case is independent of the active area size, which allows a high conversion gain to be achieved.

The photodiode pixel looks very similar to the photogate pixel. However, the additional "transfer gate" is not required since the photodiode capacitance stores the charge and n^+ diffusion is used for the output diffusion of the floating-diffusion amplifier (Fig. 9.44b). For the photodiode pixel, the conversion gain depends on the active area of the photodiode, and it is usually several times smaller than that for the photogate pixel. To allow the detection of low flux levels, most CMOS sensors operate in charge integration mode for both photogate and photodiode designs.

Quantum Efficiency and Spectral Sensitivity

The quantum efficiencies of CMOS detectors are not as high as those of CCD sensors because the CMOS fabrication process requires low-resistivity wafers and relatively low power supply levels, which limits the depletion width of the semiconductor bulk to 1–3 μm. The spectral range across which CMOS sensors are sensitive is usually narrower than the range for CCIS detectors since the absorption length increases towards the red region of the spectrum.

Noise

Noise from a CMOS sensor consists of photon shot noise, dark current shot noise, reset noise, and thermal noise. The noise properties are the same as described above for CCD photodetectors. However, there is an important difference between the noise properties of CCD and CMOS detector arrays. In a CCD, the noise is captured at the highest bandwidth at the output. However, in a CMOS array, due to its column-parallel organization, the noise bandwidth is set to the row readout bandwidth by the column sampling circuitry, which filters the in-pixel noise [9.47].

9.10 Photoemissive Detectors

The operational principle that underpins photoemissive detectors is the external photoelectric effect, which is the emission of electrons from a surface upon exposure to electromagnetic radiation. Detectors with multiple

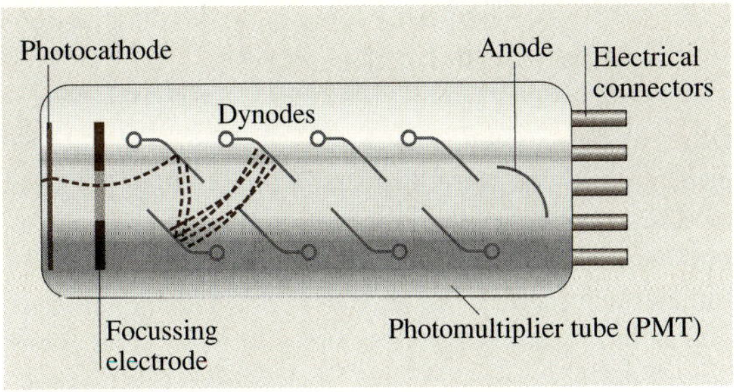

Fig. 9.45 Cross-section of a photomultiplier tube

electrodes have been developed to amplify the resulting signal.

9.10.1 Photoemissive Cell

In their simplest form, photoemissive detectors consist of a metal anode–cathode pair in a vacuum tube. The cathode is made of a material with a very low work function, usually a combination of alkali and other metals. Some common photocathode materials are Ag–CsO–Cs (S-1), Ag–Rb (S-3), and Sb_3Cs (S-4), which are sensitive between 300–1200 nm, 300–1000 nm and 350–600 nm, respectively. Electrons emitted from the cathode following the absorption of a photon are accelerated towards the anode. The photosignal is directly proportional to the intensity of the incident radiation. One method of improving the sensitivity of a photoemissive cell is to add an inert gas, resulting in subsequent ionization.

Figures of Merit

Signal Current. Assuming that the quantum efficiency of a photocathode is η and that the incoming photon flux is Φ, the current signal I_s is equal to

$$I_s = q\eta\Phi \, , \tag{9.168}$$

where q is the elementary charge.

Current Responsivity. The current responsivity of a photoemissive cell is given by the general expression

$$\mathfrak{R}_I = \frac{q\eta\lambda}{hc} \, , \tag{9.169}$$

in which λ is the wavelength, c is the speed of light, and h is Planck's constant.

9.10.2 Photomultiplier

One common way to increase the sensitivity of a photoemissive detector is to use multiple anodes. This kind of a device is called a photomultiplier tube (PMT). Similar to a photoemissive cell, light causes the emission of electrons from a photosensitive cathode. Instead of just one anode, a sequence of electron absorbers that cause enhanced secondary emission, called dynodes, is used in a PMT.

Figure 9.45 shows a schematic of a PMT. The dynodes are connected along a series of resistors that divide up the high voltage applied between the photocathode and the anode. For a typical gain of 200 eV in kinetic energy between adjacent dynodes, each electron is able to set free between four and eight secondary electrons. There are typically ten dynodes in a PMT device, giving a total multiplication factor of about 10^6–10^8.

The dynamic range of a PMT is limited for low signals by the dark current. The biggest contribution to the dark current is the thermal emission of electrons, which can be effectively suppressed by cooling the tube to −40 °C. Other contributions are leakage currents inside the tube and noise generated by incoming ionizing radiation. At the high end, space-charge effects cause the photocurrent to level off with increasing photon flux on the cathode. A cloud of electrons begins to build up behind the last dynode, thus effectively shielding the anode from additional charge carriers.

Photomultipliers are extremely sensitive and can be used for single photon counting purposes [9.49–51]. In this mode the PMT must be operated such that a single photon generates an electron cloud, resulting in the transfer of an anode pulse of several millivolts into a 50 Ω load. Due to the statistical nature of the amplification process, the signal pulse varies in height. A fast discriminator is used to set a threshold that only accepts an analog signal above the noise level and transforms such signals into logic pulses. These pulses are subsequently counted. Setting an appropriate threshold can reduce the thermal noise of a cooled PMT still further, since noise pulses from dynode thermionic emission

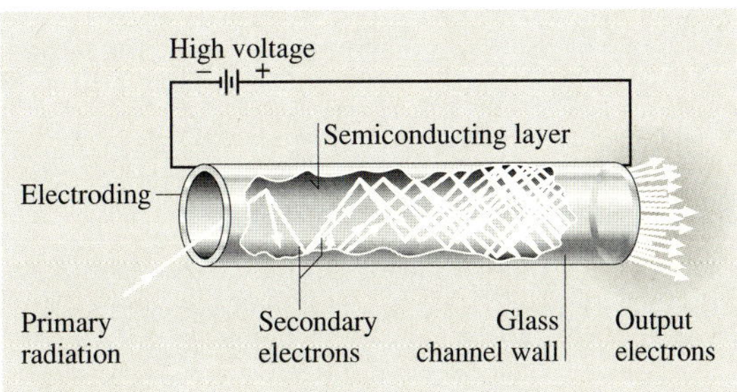

Fig. 9.46 Single-channel electron multipliers (microchannel)

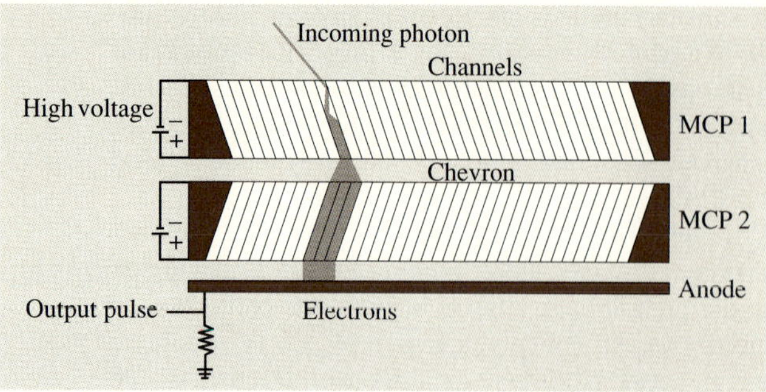

Fig. 9.47 Two microchannel plates in a chevron configuration

have a reduced mean height distribution compared to pulses triggered by an incoming photon.

Figures of Merit

Gain. The PMT gain is

$$G = P_d V^{k \cdot n} , \tag{9.170}$$

where n is the number of dynodes, V is the voltage, P_d is the dynode collection efficiency and k is a constant that depends on the geometry and the dynode material.

Dark Current. Thermionic emission characterizes the PMT dark current I_D

$$I_d = C_1 T^2 e^{-\frac{C_2 \varphi}{k_B T}} , \tag{9.171}$$

where φ is the photocathode work function, k_B is the Boltzmann constant and T is the absolute temperature. C_1 and C_2 are constant factors.

Noise Equivalent Power (NEP). Under dark-current-limited conditions, the noise-equivalent power of a photomultiplier is

$$\text{NEP} = \frac{hc}{\eta\lambda}\sqrt{\frac{2I_d}{q}} , \tag{9.172}$$

Signal-to-Noise Ratio. The signal-to-noise for a photomultiplier is

$$\frac{\text{S}}{\text{N}} \approx \frac{I_c}{\sqrt{2q\Delta f \frac{\delta}{\delta-1}(I_c + 2I_d) + I_A^2}} , \tag{9.173}$$

where I_c is the cathode current, Δf is the bandwidth, I_d is the dark current, I_A is the amplifier noise current, and δ is the secondary emission ratio at each stage.

9.10.3 Single-Channel Electron Multipliers and Microchannel Plates

If the discrete dynode structure described in the previous section on photomultipliers is replaced by a single continuous dynode, the resulting light detector can be much smaller than a PMT but can have similar detection characteristics. These devices are usually referred to as *single-channel electron multipliers*, by their trade name *Channeltrons* [9.52], or, if combined with arrays, as *microchannel plates* (MCPs) [9.53].

A single channel is typically made of a glass tube that has been coated on the inside with first a conductive and then an emissive layer (Fig. 9.46). A high voltage (1000–3000 V) is applied across the length of the tube, which accelerates electrons emitted due to the absorption of photons at the negative end. If the tube is straight, the maximum gain of the device is generally limited to about 10^5. The reason for this is that the high electron density near the anode desorbs ions from the tube walls or ionizes residual gas ions. These positive ions are accelerated back to the cathode, where unwanted secondary electron pulses are created (ion feedback). Therefore, curvature is often introduced into the tube design. Even with this modification some positive ions will still be created in the tube, but due to the short distance they travel and their large masses compared to those of electrons they do not gain enough kinetic energy to set secondary electrons free when they impact the inner tube wall. Depending on the voltage and the resistivity of the conductive material ($\approx 10^{15}\,\Omega$cm), the gain can be tuned to as high as 10^8 for such a device.

Microchannels like those described above can be manufactured to be as small as 10 to 100 μm in diameter. A combination of many of these minute detectors in an array is called a microchannel plate (Fig. 9.47). These plates are very thin, typically around 0.5 mm. Curved channels like those used for the single device are difficult to manufacture. Therefore, a different design is used to suppress the ion feedback. The detector is made of two similar plates that have straight channels at a slight bias angle. The plates are combined in such a way that the charge carriers need to change direction when leaving the first plate before entering the second. This so-called chevron configuration stops positive ions from traveling from the output of the rear plate all the way to the input of the front plate [9.54].

9.11 Thermal Detectors

A wide region of the electromagnetic spectrum, stretching from the UV to the far infrared, can be covered by thermal detectors. In most cases they are a combination of two components. The first one is an absorber in contact with a heat sink, and the second is a temperature-sensitive element (Fig. 9.48). The optical absorbance of the material that converts incoming photons into heat determines their spectral response. By choosing an appropriate material, photons from within a broad wavelength region are absorbed, causing some other physical parameter to change as well. This change is the signal registered by the temperature-sensitive element.

Detector Response

Besides the material of the absorber and the type of temperature sensor used, the sensitivity and the response time of a thermal detector are limited by some general constraints. The temperature change in a thermal detector resulting from a change in the incident light can be calculated as follows. The heat balance equation for a detector with a heat capacity C_{th} that is linked to a heat reservoir by a thermal resistor R_{th} is

$$C_{\text{th}}\frac{\mathrm{d}}{\mathrm{d}t}(\Delta T)+\frac{1}{R_{\text{th}}}\Delta T=\eta_{\text{s}}\Phi_{\text{p}}\,. \tag{9.174}$$

The parameter η_{s} is the surface emissivity of the detector, which is mainly given by the surface finish of the absorber. In general, the change in temperature ΔT and the incident radiant power Φ_{p} are functions of time. For simplicity the latter is assumed to be a sinusoidal excitation

$$\Phi_{\text{p}}=\Phi_{\text{p0}}\,\mathrm{e}^{\mathrm{i}\omega t}\,. \tag{9.175}$$

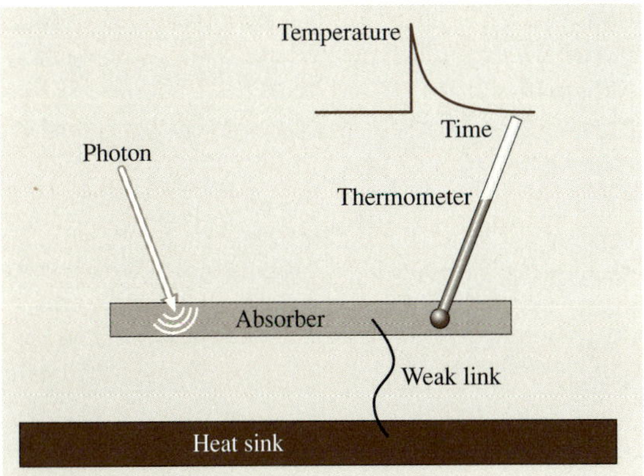

Fig. 9.48 Schematic of a thermal detector

Solving and introducing the thermal time constant τ_{th}

$$\tau_{\text{th}}=R_{\text{th}}C_{\text{th}} \tag{9.176}$$

gives a temperature change of

$$\Delta T=\left|\frac{\eta_{\text{s}}\Phi_{\text{p0}}R_{\text{th}}}{1+\mathrm{i}\omega\tau_{\text{th}}}\,\mathrm{e}^{\mathrm{i}\omega t}\right|=\frac{\eta_{\text{s}}\Phi_{\text{p0}}R_{\text{th}}}{\sqrt{1+\omega^2\tau_{\text{th}}^2}}\,. \tag{9.177}$$

To achieve high sensitivity, the optically induced temperature change can be increased by choosing good thermal insulation. The price to be paid for this improvement is a lower response speed due to a larger τ_{th}. Another option is to decrease the heat capacitance. However, this usually means a smaller detector size, which also reduces the amplitude of the signal.

The temperature-sensitive element that records the change in temperature caused by the irradiation of the absorber can be based on a variety of physical processes, as discussed below. These also contribute to the overall time response of the system.

9.11.1 Mechanical Displacement

Noting the thermal expansion in a solid or liquid is a common way to detect radiation-induced temperature changes. *Bimetal springs* or alcohol-filled *glass thermometers* are examples of devices commonly used to monitor and control the temperature.

The *Golay cell* is an example of a radiation detector that makes use of the expansion of a gas [9.55]. Golay cells are broadband and extremely sensitive, which makes them suitable for detecting wavelengths very far into the infrared part of the spectrum ($>300\,\mu\text{m}$). Another of their advantages is that their operation is mostly independent of the ambient temperature, thus eliminating the need for cooling.

Golay cells consist of a gas reservoir that is heated by the incoming radiation. The heat causes minute pressure changes, which deflect a membrane, which in turn is registered by an optical system. The sensitivity of such a system can reach 10^{-10}–$10^{-11}\,\text{W}/\sqrt{\text{Hz}}$. One variation on this pneumatic detector is the gas-filled *capacitor microphone*.

Advances in microfabrication technology make it possible to create smaller and more sensitive thermal detectors. A good example of this is the *microcantilever*, which is based on a bimetal strip combined with piezoelectric resistance readout, a laser and a position-sensitive photodetector or capacitance meter.

9.11.2 Voltage

If two wires or thin films made of different metals are brought into contact with each other, the difference between the Fermi levels of the materials leads to an electromotive force. This so-called Seebeck effect causes a drop in voltage across the metal/metal junction, the extent of which depends on the temperature. Thermal detectors based on this principle are termed *thermocouples* or *thermoelectric elements*. The measured voltage ΔV is directly proportional to the temperature difference ΔT

$$\Delta V = \alpha_s \Delta T \,, \tag{9.178}$$

where α_s is the Seebeck coefficient (Table 9.3). Using (9.177) and the last equation, it is easy to determine the voltage responsivity of a thermocouple:

$$\Re_v = \frac{\Delta V}{\Phi_{p0}} = \frac{\alpha_s \eta_s R_{th}}{\sqrt{1+\omega^2\tau_{th}^2}} \,. \tag{9.179}$$

For low frequencies ($\omega \gg 1/\tau_{th}$), the voltage response becomes a constant

$$\Re_v = \alpha_s \eta_s R_{th} \,. \tag{9.180}$$

In many applications two metal junctions are combined in series. The first one is thermally linked to the absorber that is exposed to the radiation, whereas the second is kept at a constant reference temperature. To increase the voltage signal an additional junction can be added (a *thermopile*). The voltage response of a thermopile is simply the right-hand side of (9.180) multiplied by the number of junctions.

Using (9.25) and (9.179), the noise equivalent power of a thermocouple can be derived as

$$\mathrm{NEP} = \frac{\sqrt{4kT\Delta f}}{R_v\sqrt{R_{el}}} = \frac{\sqrt{4kT\Delta f}}{\alpha_s\eta_s}\frac{\sqrt{1+\omega^2\tau_{th}^2}}{R_{th}\sqrt{R_{el}}} \,, \tag{9.181}$$

where R_{el} is the ohmic resistance of the detector.

9.11.3 Capacitance

Some dielectric materials, like lithium tantalite or triglycerine sulfate, exhibit spontaneous electrical polarization without the need for an external electrical field. The magnitude of the polarization and therefore the dielectric constant are a function of the temperature. The effect vanishes above a certain threshold (the Curie temperature) completely.

Thermal detectors based on this effect are called *pyroelectric detectors*. They consist of a crystal between two metal electrodes. If the temperature of the device is held constant, leakage currents eventually redistribute charges between the two surfaces facing the electrodes. Thus any internal electric field is canceled out, and under steady state conditions no current or voltage signal is observed between the electrodes. Therefore a pyroelectric detector is AC-coupled. It must be operated at a frequency high enough to prevent leakage current from playing a role. A change in temperature, however, cause charge to move to accommodate for the change in the internal field. An operational amplifier working in either voltage or current mode can pick up this signal.

Responsivity

The magnitude of the electrical readout depends upon the frequency of the input signal. In a crude analysis, one can identify two major contributions that determine this frequency dependence: the thermal response and the electrical response of the system. A more rigorous

Table 9.3 Seebeck coefficients α_s for different materials at 0 °C (in μV/°C)

Material	Seebeck coeff
Aluminum	3.5
Antimony	47
Bismuth	−72
Cadmium	7.5
Carbon	3.0
Constantan	−35
Copper	6.5
Germanium	300
Gold	6.5
Iron	19
Lead	4.0
Mercury	0.6
Nichrome	25
Nickel	−15
Platinum	0
Potassium	−9.0
Rhodium	6.0
Selenium	900
Silicon	440
Silver	6.5
Sodium	−2.0
Tantalum	4.5
Tellurium	500
Tungsten	7.5

treatment of the problem is given, for example, in [9.2, 56].

A temperature change ΔT can generate a pyroelectric charge Q that is proportional to the surface area of the electrodes A

$$Q = \gamma_p A \Delta T \,, \tag{9.182}$$

where γ_p is the pyroelectric constant. The temperature change can be calculated by applying the solution to the heat balance (9.177).

Taking the time derivative and calculating the absolute value yields the change in temperature with time

$$\begin{aligned}\frac{d}{dt}(\Delta T) &= \left| \frac{d}{dt}\left(\frac{\eta_s \Phi_{p0} R_{th}}{1 + i\omega R_{th} C_{th}} e^{i\omega t} \right) \right| \\ &= \frac{\eta_s \Phi_{p0} R_{th} \omega}{\sqrt{1 + \omega^2 \tau_{th}^2}} \,.\end{aligned} \tag{9.183}$$

Combining this result with the derivative of (9.182) gives the electric current induced by a sinusoidal temperature change:

$$I_{el} = \frac{dQ}{dt} = \gamma_p A \frac{d}{dt}(\Delta T) = \frac{\gamma_p \eta_s A \Phi_{p0} R_{th} \omega}{\sqrt{1 + \omega^2 \tau_{th}^2}} \,. \tag{9.184}$$

Using this expression, the current responsivity $\Re_i$ can be written as

$$\Re_I = \left| \frac{I_{el}}{\Phi_p} \right| = \frac{\gamma_p \eta_s A R_{th} \omega}{\sqrt{1 + \omega^2 \tau_{th}^2}} \,. \tag{9.185}$$

In order to get the voltage responsivity $\Re_V$, the electrical capacitance C_{el} and the ohmic resistance R_{el} of the detector must be taken into account (Fig. 9.49). The voltage V_{el} across the output terminals is then

$$V_{el} = \left| \frac{R_{el} I_{el}}{1 + i\omega \tau_{el}} \right| \tag{9.186}$$

where $\tau_{el} = R_{el} C_{el}$ is the electrical time constant. The resulting voltage responsivity is then

$$\begin{aligned}\Re_V &= \left| \frac{V_{el}}{\Phi_p} \right| \\ &= \gamma_p \eta_s A R_{th} R_{el} \frac{\omega}{\sqrt{1 + \omega^2 \tau_{th}^2} \sqrt{1 + \omega^2 \tau_{el}^2}} \,.\end{aligned} \tag{9.187}$$

The frequency response of the system is determined by the values for the thermal and electrical time constants. The thermal response is usually much slower than the electrical response. At low frequencies the current signal generated by the periodic temperature change increases until it levels off at a threshold of about $1/\tau_{th}$. This behavior is also seen for the voltage response. At higher frequencies ($1/\tau_{el}$), however, the electrical capacitance and ohmic resistance of the detector begin to dominate and the voltage starts to drop (Fig. 9.50).

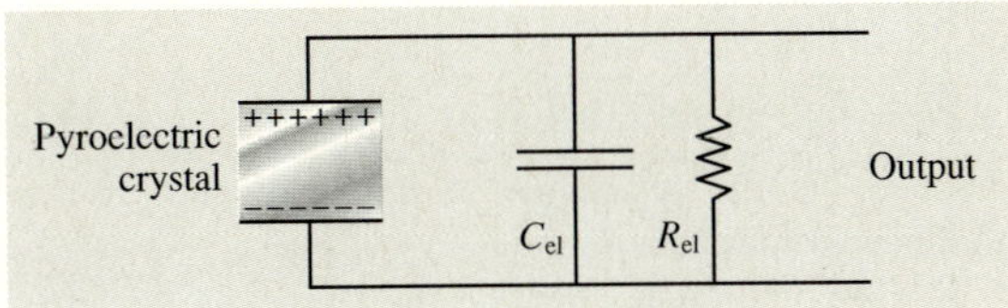

Fig. 9.49 Electrical equivalent circuit for a pyroelectric detector

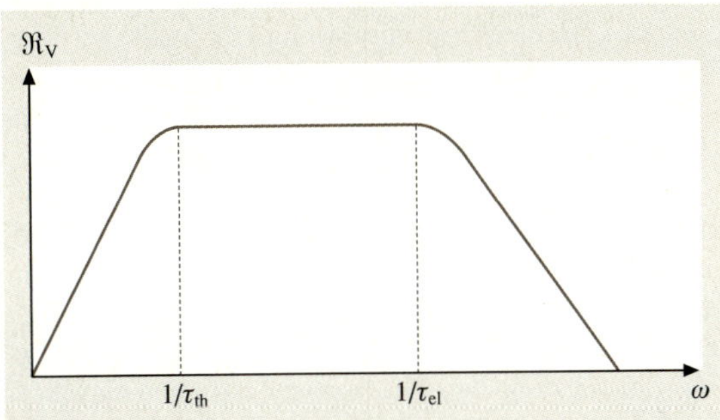

Fig. 9.50 Frequency response of a pyroelectric detector

Specific Detectivity

In the typical case, where the noise performance of a pyroelectric detector is limited by the Johnson noise, the specific detectivity D^* can be obtained using the result for the voltage responsivity (9.187) as well as (9.25) and (9.28):

$$D^* = \frac{\gamma_p \eta_s}{\sqrt{4kT}} A R_{th} \sqrt{A R_{el}} \frac{\omega}{\sqrt{1 + \omega^2 \tau_{th}^2} \sqrt{1 + \omega^2 \tau_{el}^2}} \,. \tag{9.188}$$

It is notable that the performance of the system improves linearly with the product of the detector area and the thermal resistance, as well as with the square root of the product of the area and shunt resistance.

9.11.4 Electrical Resistance

If the heat generated by the optical flux changes the electric resistance of the detector, the device is often called a *bolometer*. To a first-order approximation, the temperature dependence of the resistance R_b of a conductor can be written as

$$R_b = R_0[1 + \alpha_{th}(T - T_0)] , \quad (9.189)$$

where R_0 is the resistance of the conductor at the reference temperature T_0.

In this equation, α_{th} denotes the temperature coefficient of the resistance. This is a material constant and can be either positive (e.g., for metals) or negative (e.g., for semiconductors). In order to detect the change in resistance, an electrical current must be made to flow through the bolometer. Figure 9.51 shows a simple circuit diagram for a bolometer that uses only one load resistor (panel a) and a Wheatstone bridge (panel b).

Since the operating current leads to energy dissipation in the detector and the connecting wires, a detailed calculation of the responsivity must take into account the contribution from Joule heating. Furthermore, depending on the temperature range encountered during operation, heating due to background radiation may also play a role. In the case of semiconductor bolometers, the electrical resistance is not a linear function of the temperature and (9.189) is not valid. The following estimation for the responsivity of a bolometer neglects all of these corrections. For this reason it is only a very crude and general description.

The voltage change across the bolometer is assumed to be proportional to the change in resistance under illumination

$$dV_b = I\,dR_b , \quad (9.190)$$

where R_b is the electrical resistance of the bolometer. From (9.189) and (9.190), one obtains

$$dV_b = I\alpha_{th} R_b\,dT . \quad (9.191)$$

The current through the bolometer and the load resistor R_L is $I = V_b/(R_L + R_b)$, which together with (9.185)

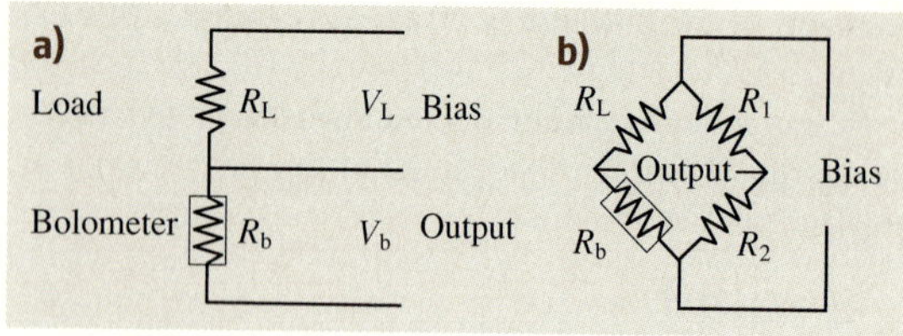

Fig. 9.51a,b Circuit diagram used to bias a bolometer. **(a)** Single load resistor, **(b)** Wheatstone bridge

yields the following equation for the voltage responsivity of a bolometer:

$$\mathfrak{R}_V = \frac{R_b V_b}{R_L + R_b} \frac{\alpha_{th}\eta_s R_{th}}{\sqrt{1 + \omega^2\tau_{th}^2}} . \quad (9.192)$$

Table 9.4 Resistivities and temperature coefficients of resistivity for different materials at room temperature

Material	Resistivity ($10^{-8}\,\Omega$m)	Temperature coefficient (10^{-3}C^{-1})
Silver	1.6	4.1
Copper	1.7	4.3
Aluminum	2.7	4.3
Tungsten	5.4	4.8
Iron	10.1	6.5
Platinum	10.6	3.9
Lead	20.6	4.2
Mercury	95.6	1.0

Metal Strip Bolometer

The bolometer was invented in 1880 by Langley, who measured the change in resistance of platinum strips upon changes in illumination. Other metals were later tested too. A typical value for the temperature coefficient α_{th} is $+0.5\%/^\circ$C (Table 9.4).

Thermistor

For many industrial applications where ruggedness and high performance under challenging environmental conditions are mandatory, thermistors are the detectors of choice. In thermistors, metal oxides serve as radiation-sensitive elements; these oxides are pressed into small beads, disks, wafers, or other shapes, which are sintered at high temperatures. The assembly is coated with epoxy or glass and mounted on an electrically insulating thermally conductive material like sapphire.

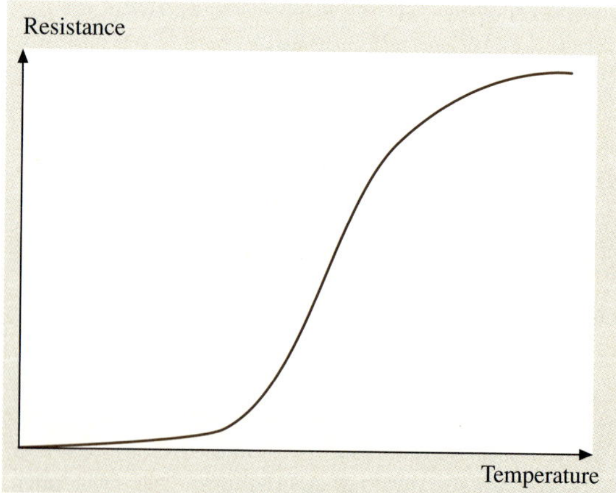

Fig. 9.52 Transition region of a superconductor

The temperature coefficient of this system is typically around 5%/°C.

Johnson noise is the primary contributor to the noise budget of thermistors. Collecting more light on the detector surface can therefore improve the signal-to-noise ratio. Many thermistors incorporate a lens onto the detector surface in order to enhance light collection. These devices have a NEP of $10^{-10}\,\mathrm{W}/\sqrt{\mathrm{Hz}}$ in the spectral range of 1–40 μm for uncooled operation [9.2,57,58].

Semiconductor Bolometer

A variety of low temperature semiconductor bolometers have been developed for applications that demand high-performance detectors [9.59, 60]. At cryogenic temperatures, the phonon modes of the bulk material freeze out and the specific heat becomes very small while the temperature coefficient for the resistance increases. Therefore detectors with very small heat capacitances and high sensitivities are feasible. A good example of such a detector is a germanium bolometer operated at liquid helium temperatures [9.59]. This can be operated near the theoretical limit between 5 and 100 μm. Due to their low heat capacities, Germanium bolometers are very fast thermal detectors with a typical response time of 400 μs. The voltage responsivity of such a device is around 5×10^{-3} V/W.

An even better time response than that obtained with the simple semiconductor bolometer can be realized if germanium is used solely for the temperature sensor and a large-area absorber (typically bismuth) is placed on a sapphire or diamond substrate that is in thermal contact with the semiconductor. The heat capacity of the electrically nonconducting substrate is much (orders of magnitude) smaller than it would be if the whole assembly was made of germanium [9.55].

Superconductive Bolometer

The electrical resistance of a superconductor changes dramatically with temperature, which means that superconducting devices are very sensitive to small changes in temperature the transition from the superconducting to normal conducting state occurs within a narrow transition region at very low temperatures (≈ 4 K). Therefore, liquid helium cooling and millikelvin-resolution temperature control are required for proper operation. Thermometers made of high-temperature superconducting material (e.g., niobium nitride or yttrium barium copper oxide) are much easier to handle. Since they show a more gradual transition at about 77 K, only liquid nitrogen cooling is required. Figure 9.52 shows a typical R–T characteristic for a superconductive bolometer.

As well as the bolometric mode, superconducting detectors can also be operated beyond the transition temperature (nonequilibrium mode) and in the so-called "photon-assisted tunneling mode." In the latter case, changes in the I–V characteristic of the junction are detected.

Part B | 9.12

9.12 Imaging Systems

Imaging arrays usually combine the photoconductors or photodiodes described above with electronic readouts. Imaging systems are very often referred to as "focal plane arrays" (FPAs) – an assembly of individual detector picture elements (pixels) located at the focal plane of the imaging system. In general, the FPA architectures can be classified as either monolithic or hybrid.

In the monolithic approach, some of the multiplexing is done in the detector material itself rather than in an external readout circuit. The basic element of a monolithic array is a MIS (MOS) capacitor. There are a few obvious advantages to the monolithic structure, principally its simplicity and the lower costs associated with a directly integrated structure. Monolithic structures like CCD and CMOS arrays can be built in silicon to form arrays with millions of high-performance pixels for visible and infrared detection (see, e.g., the review by *Rogalsky* and *Bielecki* [9.61]).

Ultraviolet and infrared imagers are most commonly built with a hybrid structure. Visible hybrids have also been built for specific applications (e.g., p–i–n photodiode arrays). Hybrid FPA detectors and multiplexers are fabricated on different substrates and mated with each other by flip-chip bonding or loophole interconnection. In this case, the detector material and the multiplexer can be optimized independently. Indium bump, solder bump, or gold stud bump bonding of readout electronics permit the multiplexing of signals from thousands of pixels onto a few output lines, greatly simplifying the interface between the sensor and the system's electronics. For infrared detection, large-dimension, multiple-element hybrid arrays with several millions of pixels are commonly used. Each element of the IR array performs

Table 9.5 CCD versus CMOS: comparison of features and performance

Feature and performance	CCD	CMOS
Signal out of pixel	Electron packet	Voltage
Signal out of chip	Voltage (analog)	Bits (digital)
Signal out of camera	Bits (digital)	Bits (digital)
Fill factor	High	Moderate
Amplifier mismatch	N/A	Moderate
System noise	Low	Moderate to high
System complexity	High	Low
Camera components	PCB + multiple chips + lens	Chip + lens
Responsivity	Moderate	Slightly better
Dynamic range	High	Moderate
Uniformity	High	Low to moderate
Speed	Moderate to high	Higher
Antiblooming	High to none	High
Biasing and clocking	Multiple, higher voltage	Single, low-voltage

at close to the fundamental photon noise limit within the spectral range between 1 and 40 μm. Materials used to build IR FPAs include PtSi (Schottky-barrier array), InSb, QWIP-based arrays ($GaAs/Al_xGa_{1-x}As$, $In_{0.53}Ga_{0.47}As$ / $In_{0.52}Al_{0.48}As$, $In_{0.53}Ga_{.47}As$ / InP, InGaAsP-based quaternary QWIPs, etc. [9.30]), and HgCdTe.

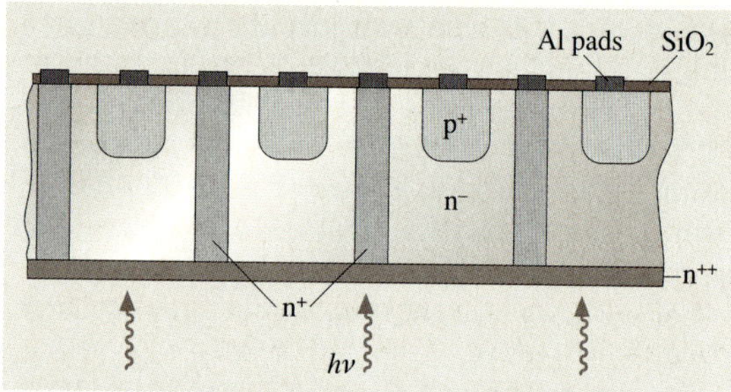

Fig. 9.53 Cross-section of a p–i–n photodiode array manufactured on a thin Si wafer ([9.62])

9.12.1 CCD Arrays and CMOS Arrays

In Sect. 9.9 above, we described the principles of operation and the main properties of charge transfer sensors. A comparison of CCD and CMOS arrays is now presented.

CCD technology has been able to provide top-level detection performance, but at the cost of several drawbacks for the user. The first technology choice requires cheap (relatively) arrays, which CCD technology can provide using thin, back-illuminated devices. On the

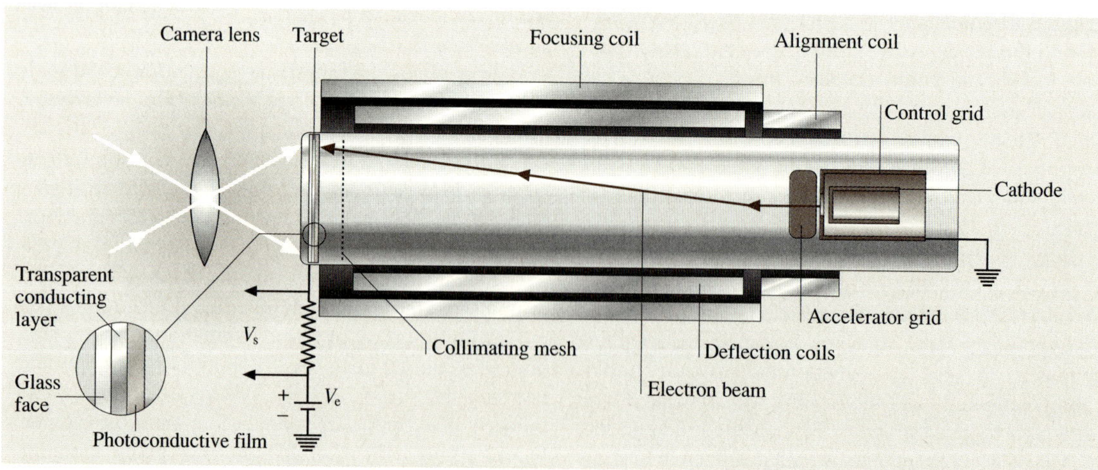

Fig. 9.54 Cross-section of a vidicon

other hand, CMOS detectors have intrinsic advantages (low power consumption, readout rate, noise, radiation hardness, integration capability) that make them well suited to several applications. Thanks to recent efforts to develop optimized detectors compatible with the CMOS core process, CMOS can provide enhanced quantum efficiency (although it remains weaker than CCD in this aspect) [9.47]. Table 9.5 compares some of the important features and properties of CCD and CMOS imaging systems.

9.12.2 p–i–n Photodiode Arrays

Multiple-element 2-D p–i–n and avalanche photodiode arrays have a virtual monopoly in some fields of medical imaging, light detection and ranging (LIDAR), 3-D laser radar (LADAR) systems, laser vibrometers, etc. What makes p–i–n and avalanche photodiode arrays unique is their ability to provide rapid access to any single pixel in the array, a wide dynamic range, high-frequency bandwidth (up to 1 GHz or more), and a high-speed parallel readout regime.

State of the art p–i–n photodiode arrays with very high quantum efficiency over virtually the whole visible and near infrared spectral range were designed on thin (30–100 μm) Si wafers with back-illumination capability [9.62–64]. The features of the structure shown in Fig. 9.53 – narrow neutral region width, isolation of n^+ diffusion between adjacent pixels, and the ability to control the gap width – provide unique properties that make arrays promising for many applications. In particular, rise and fall times of ≈ 1 ns with a small reverse bias of few volts are characteristic of the pixels in such photodiode arrays. The other important feature of the design shown in Fig. 9.53 is very low DC and AC crosstalk between adjacent elements.

9.12.3 Vidicon

A vidicon is a photoconductive detector that was used in the first TV cameras. It consists of a light sensor and an electron beam that provides readout inside a vacuum tube (Fig. 9.54). The photoconductive material, typically antimony sulfide (Sb_2S_3), is suspended in the form of little beads inside a dielectric matrix (mica). The image is focused through an entry window and a transparent and conductive layer (the signal plate) held at positive potential onto the photoconductive material (cathode), which is held at close to ground potential. Electron–hole pairs are generated at the cathode, and the negative charge carriers are pulled by the field towards the anode. An electron beam scans the backside of the cathode, thus depositing electrons onto the illuminated positive areas. This generates a capacitively coupled signal in the signal plate.

9.13 Photography

Photography is undoubtedly one of the oldest methods used to detect light and to create images. The main fields of photography include black-and-white photography, color photography, and holography. This section describes black-and-white and color photographic detectors; optical holography methods are described in detail in a separate chapter.

Photography is a process based on the chemical changes triggered by a photoexcited electron in some types of semiconductors and dye molecules. These changes accumulate during the exposure of the photographic film to light, forming a latent image. Subsequent chemical processing amplifies the image, making it visible.

The compositions used for photographic films and developing chemicals depend on whether a negative or positive image on a transparent film or a positive image on a solid substrate is required. We now consider some basic processes and physics associated with photographic image formation.

9.13.1 Black and White Photography

Basics and Image Creation

The materials most commonly used for black and white photography are silver halide grains (micro crystals of AgBr, AgCl, and AgI), which are included in the active part of the photographic detector. Photographic films consist of multiple layers, each perform a specific function [9.65]. The silver halide grains are suspended in a gelatin binder (creating an emulsion), which is then coated onto a glass or plastic film for mechanical stability. An interface layer is used to improve the bonding of the gelatin to glass, and a protective gelatin layer is usually placed on the top to protect the grains. The backside of the glass or plastic film is covered with an

Table 9.6 Experimental E_g values (in eV) for AgF, AgCl, AgBr, and AgI crystals

Halide	Indirect transition $\Gamma \leftarrow L$	Direct transition at Γ
AgF	2.8±0.3 [9.66]	4.63 [9.66]
AgCl	3.25 [9.67]	5.15 [9.68]
AgBr	2.69 [9.67]	4.29 [9.68]
AgI	2.0±0.3	3.0±0.2

absorptive layer to stop light reflection from the rear surface.

The photographic process is centered on the silver halide grains. Photon absorption in the silver halide results in the creation of a electron–hole pair, with the electron in the conduction band and the hole in the valence band. The process may be described by the following equation (assuming that Br is the halide atom involved):

$$h\nu + Br^- \rightarrow Br + e^- . \qquad (9.193)$$

Here we have taken into account the fact that silver halides are ionic crystals and that the negative charge is associated with Br, and the right-hand side infers that the photoexcited electron can move freely in the conduction band across the AgBr crystal structure. The photoinduced hole at the bromine atom can migrate to the grain surface and react with the gelatin, or it can couple with another hole, forming a bromine molecule. In either case, the bromine atom is excluded from further reactions.

At the same time, photoelectrons can be trapped by the silver atoms,

$$e^- + Ag^+ \Leftrightarrow Ag , \qquad (9.194)$$

which can act as a trap for a second electron,

$$Ag + e^- \rightarrow Ag^- . \qquad (9.195)$$

The resulting negative ion combines with a positive silver ion from the crystal structure, producing a stable two-atom silver molecule

$$Ag^- + Ag^+ \rightarrow Ag_2 . \qquad (9.196)$$

This cluster is still not developable. In further reactions, a silver molecule (9.196) can trap additional positive silver ions from the crystal and photoexcited electrons, thereby forming a cluster with either three or four Ag atoms. A center with a critical size of three or four silver atoms can catalyze the further reduction of silver ions to silver during subsequent chemical processing, thus providing an area of nucleation for the latent image.

Visualization (amplification) of the latent image is accomplished by immersing the exposed film into a chemical reducing agent – known as the "developer." The aggregates of silver atoms act as electrodes and conduct electrons from the reducing agent to the grain, promoting the growth of the silver atom aggregate. The final number of silver atoms in the aggregate exceeds 10^7, which makes the aggregate visible, and remaining silver ions are washed away.

Spectral Response

The primary event associated with photography is the absorption of light by semiconductor microcrystals of silver halides. Silver halides form face-centered cubic crystals with the smallest indirect bandgaps [9.69]. Table 9.6 summarizes the energies of the smallest indirect and direct bandgap transitions for a few silver halide crystals.

Indirect absorption yields a reasonably high photoexcitation rate in the short-wavelength spectral range below 500 nm. The overall efficiency of photographic films is reduced in the UV spectral range due to strong gelatin absorption at $\lambda < 250$ nm. Film response can be extended further into the ultraviolet by using an extremely thin gelatin coating on top of the film.

The extension of the film response into the red is achieved by dye sensitization [9.65]. In sensitized films, photons photoexcite the dye molecule, which is closely attached to the silver halide grain. There are two ways for the photoexcited dye molecule to create conduction electrons in the halide crystal grain. One is through a direct electron transfer process from the dye molecule to the grain [9.70, 71]; the other is through resonant and nonresonant energy transfer processes between the photoexcited dye molecule and the silver halide microcrystal [9.72–74]. In the spectral region to which the dye is sensitized, only the grain surface, as opposed to the entire grain volume, takes part in the initial photon detection.

9.13.2 Color Photography

Additive and Subtractive Processes

Additive and subtractive processes relate the properties of light to color. The additive process is based on the production of color through the combination of two or more primary colors. Thus, combining red, green, and blue in various ways produces all of the colors in the visible spectrum.

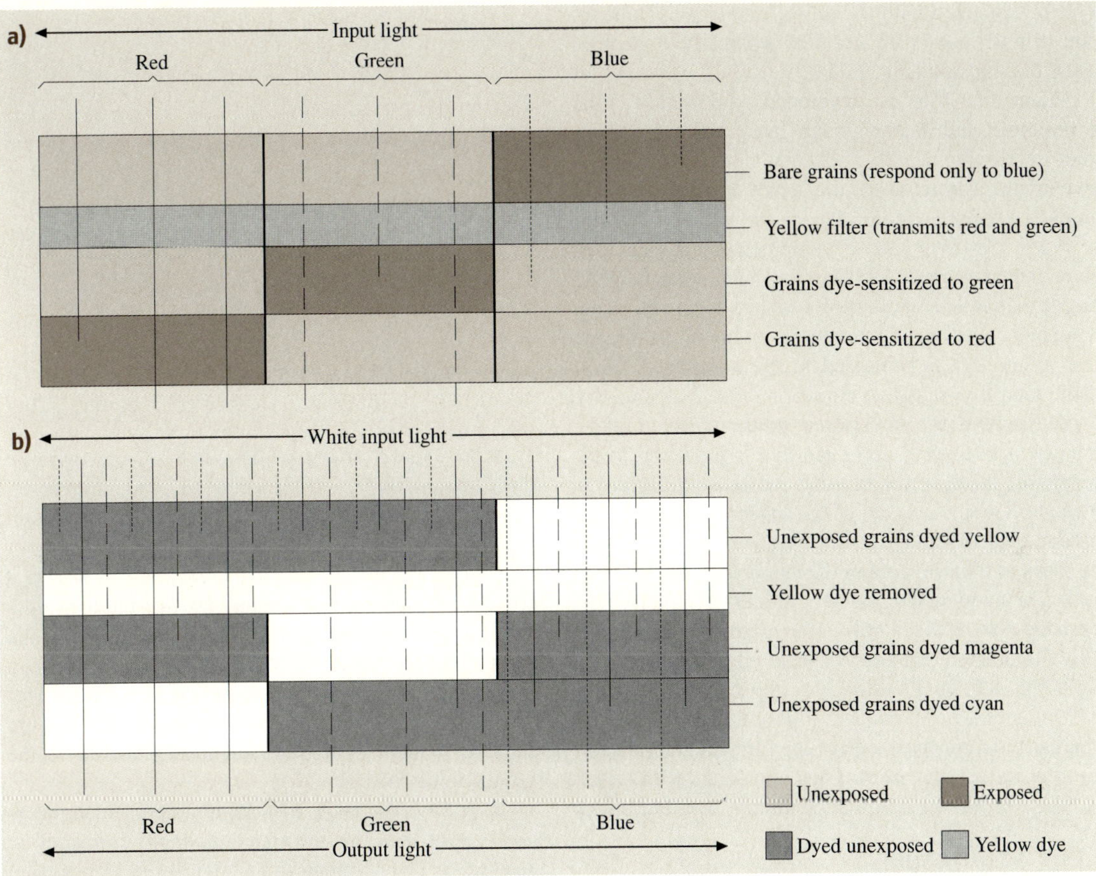

Fig. 9.55a,b Forming a color image using a positive film: **(a)** exposure of the film to the three primary colors; **(b)** development of the positive film

However, the additive process cannot be used in dye-superimposed systems because one layer of an additive primary-color dye removes the whole complementary portion of the light spectrum, and the superposition of two different primary-color dyes will always produce a black color.

The subtractive process is based on the production of color by either the subtraction or the absorption of primary colors. The three primary subtractive colors are yellow, magenta, and cyan, and they are complementary to the additive colors. They each absorb only one primary color. Yellow absorbs blue light, cyan absorbs red light, and magenta absorbs green light. The superposition of two layers with different complementary dyes will always produce a single primary color. Like the majority of photographic film processes, the concept of subtractive color is utilized in the dye-coupled process.

Basics and Image Creation

A variety of methods have been used to produce color photography, but most modern approaches are based on stacking a number of emulsion layers with color couplers on top of each other [9.65, 75]. Figure 9.55 exemplifies the process of color image creation using a positive film. Assume that all layers forming the film are uniform and that the left-hand side of the film is exposed to red light, the center is illuminated with green light, and the right-hand side is illuminated with blue light. The top layer of the structure is an emulsion with bare silver halide grains, which only respond to blue light. A yellow filter removes the blue light, protecting the underlying emulsion layers against their exposure to blue light. This means that those two underlying layers can only be exposed to light containing green and red photons only. One of the underlying layers is sensitized with a dye that responds only to green, whereas the other emulsion layer is sensitized with the red-sensitive dye. Thus, the three primary colors are recorded separately in the three layers of emulsion (Fig. 9.55a).

When the film is developed, the yellow filter is removed and latent image sites are reduced by color-developing agents. The reaction of oxidized color-developing agents with the color couplers in the emulsion produces a color dye via the method described by Fig. 9.55b. First, the development of the exposed silver halide grains amplifies the latent image. After that, development with color couplers forms the complementary dye in the unexposed grains in each layer. The yellow dye is produced in the unexposed grains of the top, blue-sensitive emulsion layer. Magenta dye is produced in the unexposed grains of the middle, green-sensitive layer, and cyan dye is produced in the unexposed areas of the bottom, red-sensitive layer. At the end of the process, all of the silver has been washed out and only the image color dyes remain. As a result, the areas of the film containing stacked yellow and magenta dye layers will transmit or reflect red light (see the left-hand side of Fig. 9.55b) – the same color that originally illuminated this portion of the film; compare with Fig. 9.55a, left. In a similar way, the middle portion of the film in Fig. 9.55b will transmit or reflect green light, whereas the right-hand side of the film will be transparent to or reflect blue light. Thus, the colors with which the film was illuminated will be directly reproduced.

9.13.3 Photography: Properties and Figures of Merit

The properties of photographic films and images are usually described using quantities that have historically proved to be adequate for characterizing photographic images. Below we summarize some of the most important properties and figures of merit.

Detective Quantum Efficiency (DQE)

The latent image formation and the development process usually introduce some degradation into the input signal, resulting in either a loss of information or the addition of noise. The detective quantum efficiency describes this degradation in terms of a ratio of the number of imaginary signal photons N_{out} that would produce, in a perfect detection system, the same information content in the output signal as that received from the real system to the actual number N_{in} of input signal photons [9.50]:

$$\mathrm{DQE} = \frac{N_{out}}{N_{in}} = \frac{(\mathrm{S/N})^2_{out}}{(\mathrm{S/N})^2_{in}} , \quad (9.197)$$

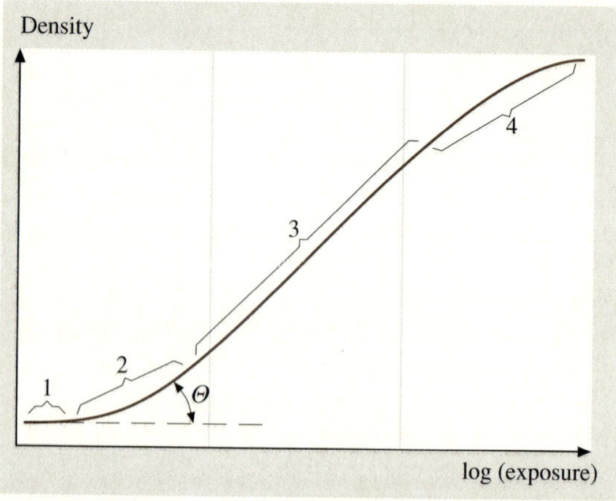

Fig. 9.56 Typical characteristic curve of photographic film. Different portions of the curve describe different exposure levels: (1) gross fog level; (2) underexposure level; (3) normal exposure level; (4) saturation level

where the right-hand side assumes Poisson photon statistics. It is obvious from (9.197) that the DQE is just the external absorptive quantum efficiency η_{ex}, as defined in Sect. 9.1.4 (9.22) as the probability that a photoelectron is produced when the photon is incident on the detector, if there is no subsequent degradation of the signal-to-noise ratio.

In reality, the DQE of a photographic material is much lower than the absorptive quantum efficiency η. On average, up to 20 photons must be absorbed by a grain over a short period of time to provide a grain-developing probability of 50%. This alone reduces the quantum efficiency by a factor of at least 20. Two other effects that reduce the DQE are the nonuniform distribution of grains in the emulsion film and variations in grain properties. These two effects may be expressed quantitatively as the granularity of the film, which refers to the variations in response measured through a tiny fixed aperture for a film that is uniformly exposed to light. When all of these effects are combined, the real DQE for the photographic plate is usually well below 5%.

Some of the methods used to improve the quantum efficiency of photoinduced silver cluster formation in silver halide microcrystals are based on doping the silver halide crystals with dopants that inhibit hole activity, suppressing electron–hole recombination [9.76].

Optical Density and Spectral Sensitivity

The optical density D of the image in the photographic film characterizes the ability of the layers of emulsion to transmit light:

$$D = -\log T \,, \tag{9.198}$$

where T is the transmittance, defined as the amount of light that passes through the developed film divided by the incident amount of light. The density can be measured as either the diffuse density (measured over a hemisphere from an incident collimated beam) or the specular density, which refers to the density measured over a particular solid angle only. The diffuse density is usually used to compare different photographic films.

In photography, the sensitivity can be given as an absolute value or as a relative value (the ratio of the energy required to expose a material compared to some reference). Sensitivity can be measured as the reciprocal of the exposure, expressed in radiant energy flux per unit area Φ_p during an exposure time t, that is required to produce a certain density D_0. The definition of the spectral sensitivity of a photographic emulsion is [9.65]:

$$S(\lambda) = \frac{1}{H(\lambda)} \,, \tag{9.199}$$

in which $H = \Phi_p t$.

Characteristic Curve

The response of a photographic film is usually nonlinear with the intensity of light, as described by the characteristic curve shown in Fig. 9.56. The central portion, 3, of the curve describes the condition at which the net response of the film depends only on the total number of photons received (the reciprocity law). In the two nonlinear regions, the reciprocity law fails.

At low exposure (underexposure, see portion 2 of the characteristic curve), which occurs for either very low exposure times or for very low photon arrival rates, electrons can escape from the trapping centers before a stable development center forms. Note that the characteristic curve usually exhibits a nonzero density even at zero exposure, which is due to both the background radiation from different sources and to chemical development without receiving any light. This level of exposure is called the gross fog level – see portion 1 of the characteristic curve in Fig. 9.56.

High photon rates and long exposure times produce saturation and nonlinearity – see region 4 (overexposure) in Fig. 9.56.

Speed

The speed of a photographic film can be characterized by assuming that the grain must absorb a certain number of photons to become developable [9.50]. This number of photons depends on the grain size. Thus, there is a correlation between the speed and the grain size. For the bare silver halide emulsion (which is sensitive below 500 nm), the whole volume of the grain absorbs and the speed of the film goes as the grain volume times the absorption coefficient. In the case of dye-sensitized grains, only the surface is effective at absorbing photons, so the speed is proportional to the grain surface. For very small grains with diameters close to the wavelength of the light absorbed, diffraction effects reduce the absorption efficiency and hence the emulsion speed.

To improve the speed of the film at low light levels, special efforts must be made to improve the sensitization of the emulsion and the illumination optics.

Contrast and Noise

The slope $\gamma = \tan\theta$ of the linear portion of the characteristic curve in Fig. 9.56 characterizes the contrast of the film. The higher the contrast of the film, the larger the separation between the signal and noise. Emulsions with smaller grains generally achieve higher contrasts, partially because small-grained emulsions are more homogeneous in terms of grain size.

Random variations in the number of grains per resolution element, variations in the properties of the grains themselves, and the stochastic nature of the processes of light absorption, electron transfer, and silver aggregate growth lead to density nonuniformities and detection process noise.

Assuming that the density is defined for each given exposure, the output signal-to-noise ratio can be written as the ratio of the exposure H to the uncertainty in exposure (the noise) σ_H, $(S/N)_{out} = H/\sigma_H$ [9.50]. Not-

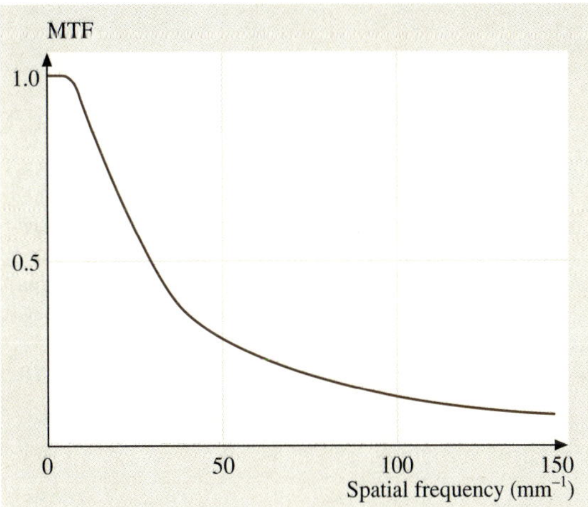

Fig. 9.57 MTF of a typical photographic film

ing that for any small exposure and density variations, the relationship between the uncertainty in exposure σ_H and uncertainty in density σ_D can be expressed as $\sigma_H/\sigma_D \approx dH/dD$, and also noting (from Fig. 9.56) that the contrast $\gamma = dD/d(\log H)$, then the signal-to-noise ratio is

$$\left(\frac{S}{N}\right)_{out} = \frac{H}{\sigma_H} = \frac{H}{\sigma_D}\frac{dD}{dH} = \frac{H}{\sigma_D}\frac{dD}{d(\log H)} = \frac{H}{\sigma_D}\gamma\,, \tag{9.200}$$

where we have taken into account the fact that $d(\log H) = dH/H$.

For Poisson photon statistics, the signal-to-noise ratio for the incident information content is

$$\left(\frac{S}{N}\right)_{in} = \frac{\Phi_p t}{\sqrt{\Phi_p t}} = \sqrt{\Phi_p t} \equiv \sqrt{H}\,, \tag{9.201}$$

where Φ_p is the radiant energy flux per unit area during the exposure time t. Hence, the detective quantum efficiency can be written as

$$\text{DQE} = \frac{H\gamma^2}{\sigma_D^2}\,. \tag{9.202}$$

Resolution and Modulation Transfer Function

Exposing the photographic film to a pattern of alternating black and white lines and then determining the minimum distinguishable line spacing provides a measure of the resolution of the film. This method has its shortcomings, since the film is usually just one part of an imaging system that includes image-forming optics, and it is difficult to separate the effects arising from the different components of the whole system.

A more general concept is to use the modulation transfer function (MTF) to characterize the quality of each component of the imaging system. More specifically, the MTF is the magnitude of the response of the optical system to a variety of spatial frequencies x [9.77]. Performance wise, MTF characterizes the modulation depth of the image $M_{out}(x)$ relative to that of the incident information packet $M_{in}(x)$

$$\text{MTF}(x) = \frac{M_{out}(x)}{M_{in}(x)}\,. \tag{9.203}$$

A typical example of the MTF for a photographic emulsion is presented in Fig. 9.57. It is possible to enhance the sensitivity of the detector to some particular spatial frequencies, resulting in an MTF with a more complex shape than that shown in Fig. 9.57.

The MTF can be calculated from a Fourier transform function (e.g., FFT) of the source object. For example, for a single, ideal, rectangular photodetector pixel, where the one-dimensional spatial response is described by a π-function, the theoretical limit for the MFT is a sinc function, $[\text{sinc}(x) = \sin(x)/x]$. In practice, the closer the FFT of the photodetector spatial response is to the sinc function, the better the imaging properties (resolution, etc.) of the photodetector. Zeros in the FFT function correspond to the harmonics of the fundamental frequency, which is the physical size of the active area in the case of the photodetector.

The MTF is useful because the MTF of an entire linear optical system can be determined by simply multiplying the MTFs of its components. The overall resolution capability of a complex optical system is easily determined in this way.

References

9.1 D. Wood: *Optoelectronic Semiconductor Devices* (Prentice Hall, New York 1994)

9.2 E. L. Dereniak, G. D. Boreman: *Infrared Detectors and Systems* (Wiley, New York 1996)

9.3 E. L. Dereniak, D. G. Crowe: *Optical Radiation Detectors* (Wiley, New York 1984)

9.4 D. L. Smith: Theory of generation-recombination noise in intrinsic photoconductors, J. Appl. Phys. **53**, 7051 (1982)

9.5 T. S. Moss, G. J. Burrell, B. Elis: *Semiconductor Optoelectronics* (Butterworth, London 1973)

9.6 S. M. Sze: *Physics of Semiconductor Devices* (Wiley, New York 2001)

9.7 J. Geist: Quantum efficiency of the p–n junction in silicon as an absolute radiometric standard, Appl. Opt. **18**, 760–762 (1979)

9.8 C. Hicks, M. Kalatsky, R. A. Metzler, A. O. Goushcha: Quantum efficiency of silicon photodiodes in the near infrared spectral range, Appl. Opt. **42**, 4415–4422 (2003)

9.9 A. O. Goushcha, R. A. Metzler, V. N. Kharkyanen, C. Hicks, N. M. Berezetska: Determination of the carrier collection efficiency function of Si photodiode using spectral sensitivity measurements, Proc. SPIE **5353**, 12–19 (2004)

9.10 G. Lutz: *Semiconductor Radiation Detectors Device Physics* (Springer, Berlin, Heidelberg 1999)

9.11 M. Shur: *Physics of Semiconductor Devices* (Prentice Hall, Englewood Cliffs 1990)

9.12 P. P. Webb, R. J. McIntyre, J. Conradi: Properties of avalanche photodiodes, RCA Rev. **35**, 234–277 (1974)

9.13 S. M. Sze, G. Gibbons: Avalanche breakdown voltages of abrupt and linearly graded p-n junctions in Ge, Si, and GaP, Appl.Phys. Lett. **8**, 8111 (1966)

9.14 K. K. Ng: *Complete Guide to Semiconductor Devices* (Wiley, New York 2002)

9.15 A. S. Grove: *Physics and Technology of Semiconductor Devices* (Wiley, New York 1967)

9.16 S. S. Vishnubhatla, B. Eyglunet, J. C. Woolley: Electroreflectance measurements in mixed III-V alloys, Can. J. Phys. **47**, 1661 (1969)

9.17 M. S. Ünlu, S. Strite: Resonant cavity enhanced photonic devices, Appl. Phys. **78**, 607–638 (1995)

9.18 E. Chen, S. Y. Chou: High-efficiency and high-speed silicon metal-semiconductor-metal photodetectors operating in the infrared, Appl. Phys. Lett. **70**, 753–755 (1997)

9.19 M. K. Emsley, O. Dosunmu, M. S. Ünlu: High-speed resonant-cavity-enhanced silicon photodetectors on reflecting silicon-on-insulator substrates, IEEE Photon Technol. Lett. **14**, 519–521 (2002)

9.20 S. M. Csutak, S. Dakshina-Murthy, J. C. Campbell: CMOS-compatible planar silicon waveguide-grating-coupler photodetectors fabricated on silicon-on-insulator (SOI) substrates, IEEE J. Quantum Elect. **38**, 477–480 (2002)

9.21 M. Ghioni, F. Zappa, V. P. Kesan, J. Warnock: A VLSI-compatible high-speed silicon photodetector for optical data links, IEEE Trans. Electron Dev **43**, 1054–1060 (1996)

9.22 J. D. Schaub, S. J. Koester, G. Dehlinger, Q. C. Ouyang, D. Guckenberger, M. Yang, D. Rogers, J. Chu, A. Grill: High speed, lateral pin photodiodes in silicon technologies, Semiconductor Photodetectors 2004, Proc. SPIE **5353**, 1–11 (2004)

9.23 K. Kato, S. Hata, K. Kawano, J. I. Yoshida, A. Kozen: A high efficiency 50 GHz InGaAs multimode waveguide photodetector, IEEE J. Quantum Elect. **28**, 2728–2735 (1992)

9.24 D. Wake, T. P. Spooner, S. D. Perrin, I. D. Henning: 50 GHz InGaAs edge-coupled pin photodetector, Electron. Lett. **27**, 1073–1075 (1991)

9.25 T. Takeuchi, T. Nakata, K. Makita, M. Yamaguchi: High-speed, high-power and high-efficiency photodiodes with evanescently coupled graded-index waveguide, Electron. Lett. **36**, 972–973 (2000)

9.26 S. Demiguel, L. Giraudet, L. Joulaud, J. Decobert, F. Blanche, V. Coupe, F. Jorge, P. Pagnod-Rossiaux, E. Boucherez, M. Achouche, F. Devaux: Evanescently coupled photodiodes integrating a double-stage taper for 40-Gb/s applications – compared performance with side-illuminated photodiodes, J. Lightwave Technol. **20**, 2004–2014 (2002)

9.27 S. Demiguel, N. Li, X. Li, X. Zheng, J. Kim, J. C. Campbell, H. Lu, A. Anselm: Very high-responsivity evanescently coupled photodiodes integrating a short planar multimode waveguide for high-speed applications, IEEE Photonics Technol. Lett **15**, 1761–1763 (2003)

9.28 D. Color Vision. Malacara: *Theory and Applications, SPIE Press Monograph PM105* (Spie, Bellingham 2002)

9.29 R.M. Turner: Vertical color filter detector group and array, US Patent 6864557 (March 8, 2005)

9.30 S. D. Gunapala, S. V. Bandara: Quantum well infrared photodetector (QWIP) focal plane arrays, Semiconduct. Semimet. **62**, 197–282 (2000)

9.31 D. Leonard, M. Krishnamurthy, C. M. Reaves, S. P. Denbaars, P. M. Petroff: Direct formation of quantum-sized dots from uniform coherent islands of InGaAs on GaAs surfaces, Appl. Phys. Lett. **63**, 3203 (1993)

9.32 J. M. Moison, F. Houzay, L. Leprince andF. Barthe, E. Andre, O. Vatel: Self-organized growth of regular nanometer-scale InAs dots on GaAs, Appl. Phys. Lett. **64**, 196 (1994)

9.33 N. Biyikli, I. Kimukin, B. Butun, O. Aytür, E. Ozbay: ITO-Schottky Photodiodes for high-performance detection in the UV–IR spectrum, IEEE J. Select. Top. Quantum Elect. **10**, 759–765 (2004)

9.34 K. C. Hwang, Sheng S. Li, Y. C. Kao: A novel high-speed dual wavelength InAIAs/InGaAs graded superlattice Schottky barrier photodiode for 0.8 and 1.3 micron detection, Proc. SPIE **1371**, 128–137 (1990)

9.35 F. D. Shepherd, A. C. Yang: Silicon Schottky retains for infrared imaging, Tech. Dig. IEEE IEDM **19**, 310–313 (1973)

9.36 W. F. Kosonocky: Review of Schottky-barrier imager technology. In: *Infrared Detectors and Focal Plane Arrays*, Proc. SPIE, ed. by L. Eustace, R. Dereniak, E. Sampson (SPIE, Bellingham 1990) pp. 2–26

9.37 S. M. Sze, D. J. Coleman, A. Loya: Current transport in metal-semiconductor-metal (MSM) structures, Solid State Electron **14**, 1209 (1971)

9.38 B. F. Aull, A. H. Loomis, D. J. Young, R. M. Heinrichs, B. J. Felton, P. J. Daniels, D. J. Landers: Geiger-mode avalanche photodiodes for three-dimensional imaging, Lincoln Lab. J. **13**, 335–350 (2002)

9.39 R. J. McIntyre: Multiplication noise in uniform avalanche diodes, IEEE Trans. Electron Dev. **ED-13**, 164 (1966)

9.40 J. Kemmer, G. Lutz: New semiconductor detector concepts, Nucl. Instrum. Meth **A253**, 356–377 (1987)

9.41 W. S. Boyle, G. E. Smith: Charge coupled semiconductor devices, Bell Syst. Tech. J. **49**, 587 (1970)

9.42 M. F. Tompsett, G. F. Amelio, G. E. Smith: Charge coupled 8-bit shift register, Appl. Phys. Lett. **17**, 111 (1970)

9.43 M. F. Tompsett, G. F. Amelio, W. J. Bertram, R. R. Buckley, W. J. McNamara, J. C. Mikkelsen,

D. A. Sealer: Charge-coupled imaging devices: Experimental results, IEEE Trans. Electron Dev. **18**, 992–996 (1971)

9.44 W. J. Bertram, D. A. Sealer, C. H. Sequin, M. F. Tompsett, R. R. Buckley: Recent advances in charge coupled imaging devices, INTERCON Dig. **292** (1972)

9.45 W. S. Boyle, G. E. Smith: Buried channel charge coupled devices, US Patent 3792322 (1974)

9.46 C. H. Sequin, M. F. Tompsett: *Charge Transfer Devices* (Academic, New York 1975)

9.47 P. Magnan: Detection of visible photons in CCD and CMOS: A comparative view, Nuclear Instrum. Meth. **A504**, 199–212 (2003)

9.48 J. Janesick: *Scientific Charge Coupled Devices* (SPIE, Bellingham, Washington 2001)

9.49 Hamamatsu: *PMT Handbook* (Hamamatsu Corp., Hamamatsu City 2004)

9.50 G. Rieke: *Detection of Light*, Vol. 2nd edn. (Cambridge Univ. Press, Cambridge 1996)

9.51 W. Demtröder: *Laser Spectroscopy*, Vol. 3rd edn. (Springer, Berlin, Heidelberg 2002)

9.52 Burle: *Burle Channeltron Electron Multiplier Handbook for Mass Spectrometry Applications* (Galileo Electro-Optics Corp., Sturbridge 1991)

9.53 J. L. Wiza: Microchannel plate detectors, Nuclear Instruments and methods **162**, 587–601 (1979)

9.54 W. B. Colson, J. McPherson, F. T. King: High-gain imaging electron multiplier, Rev. Sci. Instrum. **44**, 1694–1696 (1973)

9.55 R. De Waard, E. M. Wormser: Description and properties of various thermal detectors, Proc. IRE **47**, 1508–1513 (1959)

9.56 S. G. Porter: A brief guide to pyroelectric detectors, Ferroelectrics **33**, 193–206 (1981)

9.57 E. M. Wormser: Properties of thermistor infrared detectors, J. Opt. Soc. Am. **43**, 15–21 (1953)

9.58 R. W. Astheimer: Thermistor infrared detectors, Proc. SPIE **443**, 95–109 (1984)

9.59 F. J. Low: Low-temperature germanium bolometer, J. Opt. Soc. Am. **51**, 1300–1304 (1961)

9.60 P. L. Richards: Bolometer for infrared and millimeter waves, J. Appl. Phys **76**, 1–24 (1994)

9.61 A. Rogalski, Z. Bielecki: Detection of optical radiation, Bull. Pol. Ac. Tech **52**, 43–66 (2004)

9.62 A.O. Goushcha et al.: Ultra thin back illuminated photodiode array structures and fabrication methods, US Patent 66724735 (2004)

9.63 R. P. Luhta, R. A. Mattson, N. Taneja, P. Bui, R. Vasbo: Back-illuminated photodiodes for multislice CT, Proc. SPIE **5030**, 235–245 (2003)

9.64 S. E. Holland, N. W. Wang, W. W. Moses: Development of low noise, back-side illuminated silicon photodiode arrays, IEEE Trans. Nucl. Sci. **44**, 443–447 (1997)

9.65 W. Thomas: *Handbook of Photographic Science and Engineering* (Wiley, New York 1973)

9.66 A. P. Marchetti, G. L. Bottger: Optical absorption spectrum of AgF, Phys. Rev. B. **3**, 2604–2607 (1971)

9.67 F. C. Brown: *Solid State Chemistry*, ed. by B. Hannay (Plenum, New York 1973)

9.68 B. L. Joesten, F. C. Brown: Indirect optical absorption of AgCl-AgBr alloys, Phys. Rev. **148**, 919–927 (1966)

9.69 S. Glaus, G. Calzaferri: The band structures of the silver halides AgF, AgCl, and AgBr: A comparative study, Photochem. Photobiol. Sci. **2**, 398–401 (2003)

9.70 S. Dahne: The evolution of thinking on the mechanism of spectral sensitization, J. Imaging Sci. Technol. **38**, 101–117 (1994)

9.71 R. W. Gurney, N. F. Mott: The theory of the photolysis of silver bromide and the photographic latent image, Proc. Roy. Soc. **A164**, 151–164 (1938)

9.72 T. Förster: Zwischenmolekulare Energiewanderung und Fluoreszenz, Ann. Phys **2**, 55–75 (1948)

9.73 D. L. Dexter, T. Förster, R. S. Knox: Radiation transfer of energy of electronic excitation between molecules in crystal, Phys. Stat. Sol. **34**, 159 (1969)

9.74 D. L. Andrews, A. A. Demidow (Eds.): *Resonant Energy Transfer* (Wiley, Chichester 1999)

9.75 R. D. Theys, G. Sosnovsky: Chemistry and processes of color photography, Chem. Rev. **97**, 83–132 (1997)

9.76 J. Belloni: Photography: enhancing sensitivity by silver-halide crystal doping, Radiat. Phys. Chem. **67**, 291–296 (2003)

9.77 G. D. Boreman: *Modulation Transfer Function in Optical and Electro-optical systems* (SPIE, Bellingham 2001)

Part C Coherent and Incoherent Light Sources

10. Incoherent Light Sources

Since the invention and industrialization of incandescent lamps at the end of the 19th century electrical lighting has become a commodity in our daily life. Today, incoherent light sources are used for numerous application areas. Major improvements have been achieved over the past decades with respect to lamp efficiency Fig. 10.1, lifetime and color properties.

In the following chapters an overview of various lamp types and their properties is given. They are subdivided by light generation mechanism: thermal emission of radiation close to thermal equilibrium (incandescent lamps), atomic and molecular emission in gas discharge lamps, and emission from solid-state light sources (LEDs).

10.1 Incandescent Lamps

The incandescent lamp is the oldest electrical light source still in widespread use. It can be found in almost any application, especially where comparatively small lumen packages are required and where simplicity and compactness are favored.

10.1.1 Normal Incandescent Lamps

Incandescent lamps produce light by the electrical heating of a metal wire to such a high temperature that radiation in the visible part of the spectrum is emitted [10.1]. The metal wire is mounted in a glass bulb filled with an inert gas (Fig. 10.2).

According to Planck's law the filament must be heated up to at least 2400 K for a white emission color. In view of the efficiency for converting electrical energy into visible light, even higher filament temperatures would be favorable. Unfortunately, increasing the temperature reduces lamp life due to enhanced evaporation rates of the metal. Since tungsten has a high melting point and a low vapor pressure it permits high operating temperature and consequently higher efficiencies can be attained than by any other metal. A possible measure to improve lifetime is the reduction of the tungsten evaporation rate by the addition of rare gases (Kr, Xe). Alternatively, halogens are used in a so-called regenerative cycle

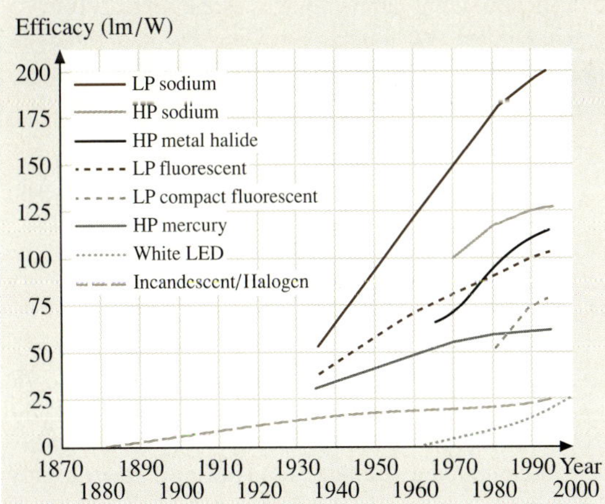

Fig. 10.1 Temporal development of the luminous efficacy of electrical light sources (LP = low pressure, HP = high pressure)

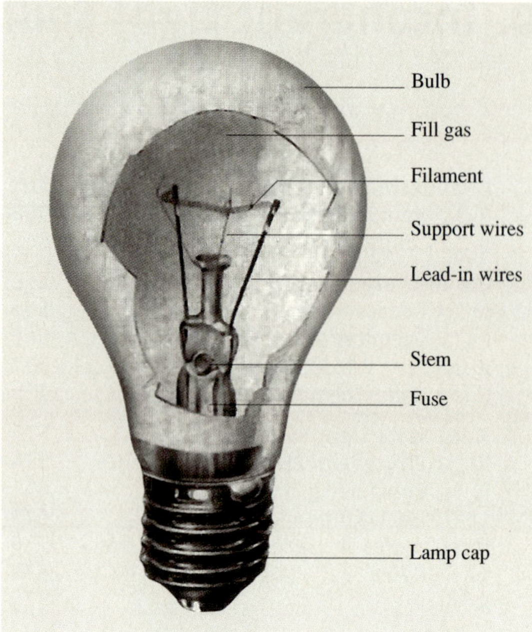

Fig. 10.2 Schematic drawing of an incandescent lamp

effectively to transport tungsten back to the filament. An increase in luminous efficiency is achieved by coiling the tungsten wire. A coiled filament allows higher operation temperatures for a specified lifetime (typically 1000 h). Most common operation parameters yield luminous efficiencies between 8 lm/W and 17 lm/W at filament temperatures of 2400 K and 3100 K, respectively. These values correspond to an energy efficiency of a few percent. Incandescent lamp wattages range up to 2000 W.

10.1.2 Tungsten Halogen Lamps

In a normal incandescent lamp, tungsten evaporates off the filament and condenses on the bulb wall, resulting in so-called blackening. Halogen lamps comprise a halogen, i. e. iodine, bromine, chlorine, added to the normal gas filling. These form volatile tungsten compounds at the glass wall, which are transported back to the hot filament. Here, the tungsten halides are decomposed enabling a so-called chemical transport cycle (Fig. 10.3).

By reduction of the net tungsten evaporation rate, the filament of halogen lamps can be operated at higher temperature compared to standard incandescent lamps. Thus, luminous efficiencies can be increased for reduced lamp size. This allows their application in compact reflectors. Halogen incandescent lamps are available up to 2000 W with luminous efficacies up to 25 lm/W. This value has recently been improved to 35 lm/W by coating the glass bulb with infrared reflective multilayers, and lifetimes up to 2000 h were obtained.

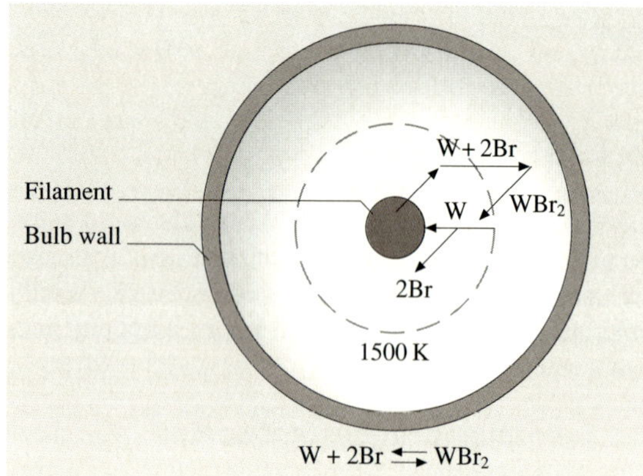

Fig. 10.3 Principle of the chemical transport cycle in tungsten halogen lamps

10.2 Gas Discharge Lamps

10.2.1 General Aspects

A gaseous discharge is obtained by driving an electric current through a gas, typically present between two electrodes. Alternatively, electrodeless microwave-excited discharges and pulsed dielectric-barrier discharges (e.g. used for plasma displays) are known as incoherent light sources.

The actual carriers of the electric current in the gas are electrically charged particles, positive ions and negative electrons. In a neutral nonconductive gas the number of charge carrying particles is extremely small. These particles can be released from the fill gas or the cathode surface by energetic collisions. Many physical factors influence the properties of a gas discharge, the most important ones being the type and pressure of the gas, the electrode material, the operating temperature of the electrodes, the shape and surface structure of the electrodes, the distance between the electrodes, the geometry of the discharge vessel, and the current density. For the purpose of light generation, two main types are distinguished: low-pressure and high-pressure discharge lamps. For lighting applications, both are operated in the arc discharge mode, which is characterized by high current densities $\left(> 1\,\mathrm{A/cm^2}\right)$. To limit the discharge currents, electronic ballasts are used [10.2].

In a low-pressure discharge lamp (gas pressure typically less than 100 Pa) the electrons have a mean free path length larger or of the order of the vessel diameter (e.g. a few cm). Due to low collision rates with the neutral gas atoms they gain high energies (> 1 eV) from the applied electrical field and effectively excite the cold atoms by inelastic collisions. The electrons and atoms are not in thermal equilibrium (Fig. 10.4).

In high-pressure discharge lamps the operating pressure is typically in the range between 10 kPa and 10 MPa. Here, collisions between electrons and atoms or ions are much more frequent, resulting in a thermal equilibrium that is characterized by equal particle temperatures.

In low-pressure discharges atomic line radiation is emitted preferably from resonance transitions of the element with the lowest excitation potential (e.g. Hg: 185 nm and 254 nm, Na: 589 nm). In high-pressure lamps various contributions to the spectrum are obtained: broadened atomic lines (resonance-, van der Waals- and Stark-broadening), molecular radiation bands and quasi-continuous emission due to free–free (Bremsstrahlung) and free–bound (recombination of electrons with ions and atoms) transitions. As a result of these quasi-continuous spectra, the color

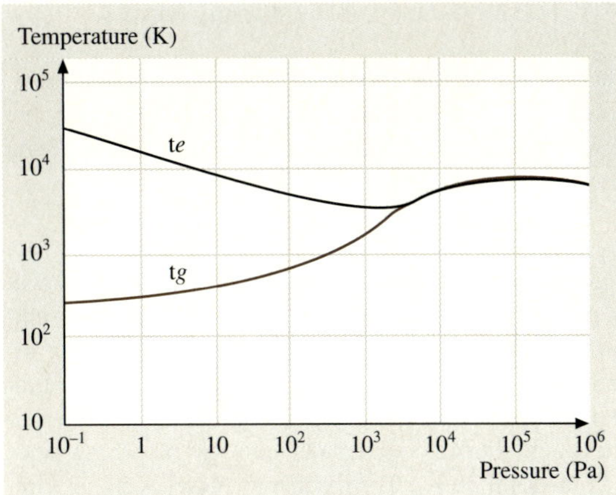

Fig. 10.4 Relationship between electron temperature, gas temperature and gas pressure

rendering properties of high-pressure discharge lamps are fair to excellent, depending on the type of filling [10.3].

10.2.2 Overview of Discharge Lamps

In Fig. 10.5 the most relevant types of discharge lamps are displayed. The light sources are distinguished with respect to their emission spectra and application fields. In the case of low-pressure mercury and xenon excimer lamps luminescent materials are applied for conversion of ultraviolet (UV) radiation into visible light. The temporal improvement in efficiencies is depicted in Fig. 10.1.

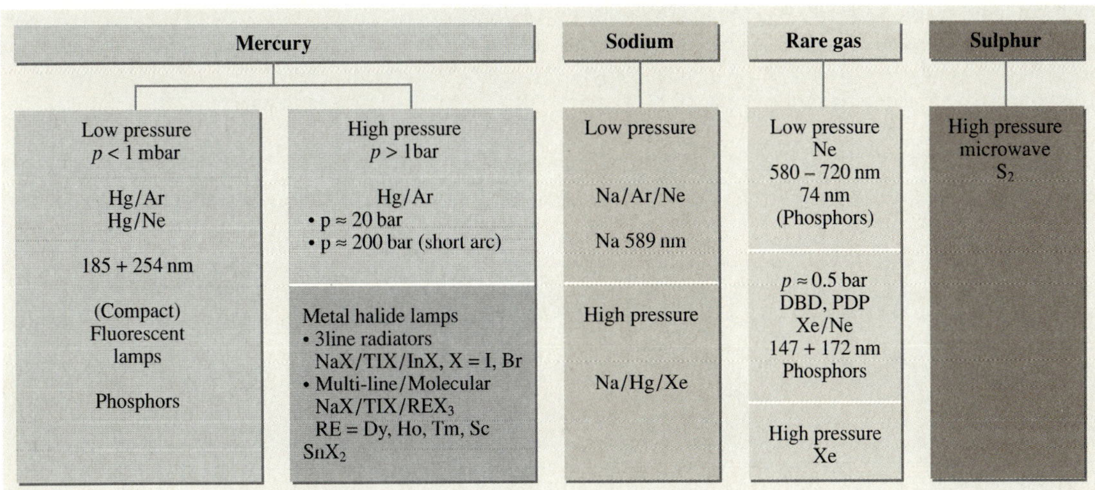

Fig. 10.5 Overview of gas discharge lamps

10.2.3 Low-Pressure Discharge Lamps

The most widely applied radiators in low-pressure discharge lamps are Hg and Na. These elements are the best choice with respect to efficiency for the conversion of electrical input power into radiation.

Low-Pressure Mercury Lamps

The working principle of a low-pressure Hg (also known as a fluorescent lamp) is given in Fig. 10.6. Fluorescent lamps are generally designed in the form of a linear or bent tubular bulb with an electrode sealed in at each end (electrodeless versions are available where the electrical energy is coupled inductively into the discharge vessel via metal coils). The discharge vessel is filled with an inert gas (typically Ar) and a few mg Hg. Since the major part of the emission of Hg atoms (97%) at low pressures (e.g. 5 Pa) is in the ultraviolet, the inner surface of the bulb is coated with a fluorescent powder or phosphor, which converts the UV radiation into visible light. The composition of the phosphor determines the spectral power distribution and the color of the emitted light.

The generation of UV photons is due to transitions of Hg atoms between the excited state levels $^1P_1$, $^3P_1$ and the ground state level $^1S_0$. About 64% of the electrical input power is converted into photons at a wavelength of 185 nm and 254 nm. According to the transitions indicated in Fig. 10.7 only 3% are emitted in the visible part of the spectrum. As a result of the Stokes shift, the overall efficiency is only 28%. This value corresponds to a luminous efficiency of 100 lm/W.

Typical lamp parameters for fluorescent lamps are: electrical input power up to 140 W, luminous efficiency up to 100 lm/W, color temperature between 2700 K and 8000 K. Lamps are available in a large variety of geometries, e.g. cylindrical or circular or U-shaped tubes. The latter are also known as energy-saving or compact fluorescent lamps. Depending on the construction of fluorescent lamps, the lifetime ranges from 5000 to 25 000 h.

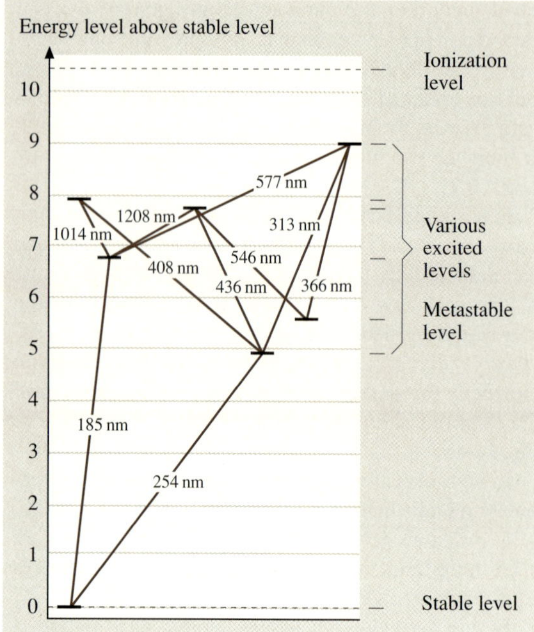

Fig. 10.7 Simplified energy-level scheme and radiative transitions of the Hg atom

Fluorescent Coatings

The most important component of a fluorescent lamp is the fluorescent powder. It is coated onto the inner side of the glass tube. The powder consists of one or several luminescent materials (phosphors), which are in general inorganic compounds doped by transition metals (e.g. Mn^{2+}, Mn^{4+}) or rare-earth ions (e.g. Tb^{3+}, Eu^{3+}). The

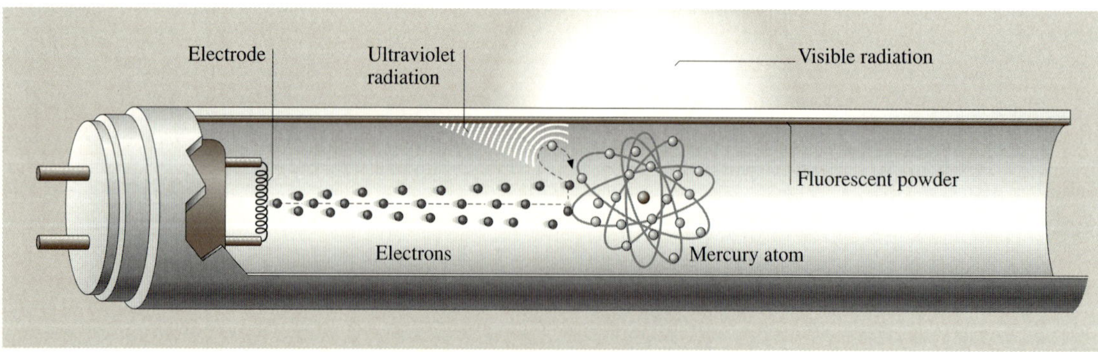

Fig. 10.6 Sketch of a low-pressure mercury discharge (fluorescent) lamp

composition of the fluorescent powder is optimized to convert the absorbing atomic resonance lines of mercury low-pressure discharges (185 nm and 254 nm) into the desired lamp spectrum. In some cases the fluorescent composition also includes so-called sensitizers (e.g. Ce^{3+}) if the first dopant, i. e. the activator, does not sufficiently absorb the mercury lines. Figure 10.8 displays the principle of photon down-conversion in a luminescent material.

The first step of light conversion is the absorption of incident photons either by the activator or the sensitizer. In the case of absorption by the sensitizer the energy is subsequently transferred to the activator. The excited activator ion, for instance Eu^{3+}, decays to the ground state by emitting a photon with a wavelength according to the energy gap between the excited and ground state. The energy difference between the emitted and the absorbed photon (Stokes shift), which is about 50%, is dissipated as heat in the fluorescent powder.

Nowadays, fluorescent compositions with quantum efficiency close to unity are commercially available for many lamp application areas. Table 10.1 gives an overview of most commonly applied luminescent materials. For illumination purposes, halophosphate phosphors, invented in the 1940ies, are still widespread since they yield a white emission spectrum with a reasonable efficiency and color rendering. However, a trichromatic phosphor blend results in higher lamp efficacy and improved color rendition. A typically applied phosphor blend comprises $BaMgAl_{10}O_{17}:Eu$, $LaPO_4:CeTb$, and $Y_2O_3:Eu$. These so-called color 80 lamps (indicated with an [a] in Table 10.1) exhibit a spectral power distribution as depicted in Fig. 10.9.

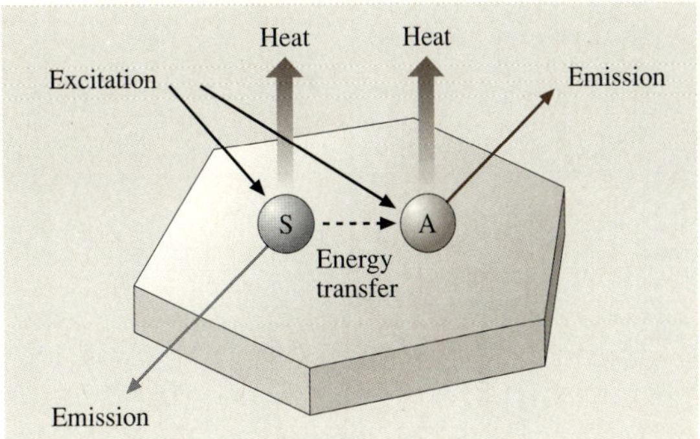

Fig. 10.8 Simplified mechanism of light generation in a fluorescent powder. The abbreviations "S" and "A" indicate sensitizer and activator, respectively

Lamps with an even higher color rendition are obtained by a tetra- or pentachromatic phosphor blend. Color 90 lamps additionally contain phosphors with an emission-band position where the trichromatic phosphor blend does not radiate, viz. in the yellow ($Y_3Al_5O_{12}:Ce$) and deep red ($GdMgB_5O_{10}:Ce, Tb, Mn$) spectral range.

Due to the tunability of the spectra, fluorescent lamps have entered many other application areas be-

Table 10.1 Overview of the most commonly used luminescent materials in fluorescent lamps

Composition of luminescent materials	Emitted peak wavelength (nm)	Color point x, y [a] (after CIE 1931)	Application area
$LaB_3O_6:Bi, Gd$	311	–	Medical lamps
$LaPO_4:Ce$	320	–	Tanning lamps
$BaSi_2O_5:Pb$	350	–	Tanning lamps
$SrB_4O_7:Eu$	368	–	Black light lamps
$Sr_2P_2O_7:Eu$	420	0.167, 0.014	Reprography lamps
$BaMgAl_{10}O_{17}:Eu$	453	0.150, 0.070	Color 80 lamps [b]
$Zn_2SiO_4:Mn$	530	0.256, 0.700	Decoration lamps
$LaPO_4:Ce, Tb$	543	0.343, 0.585	Color 80 lamps [a]
$CeMgAl_{11}O_{19}:Tb$	543	0.350, 0.582	Color 80 lamps [a]
$GdMgB_5O_{10}:Ce, Tb$	543	0.346, 0.531	Color 80 lamps [a]
$Y_3Al_5O_{12}:Ce$	560	0.453, 0.523	Color 90 lamps [a]
$Ca_5(PO_4)_3(F, Cl):Sb, Mn$	575	0.356, 0.377	Halophosphate lamps
$Y_2O_3:Eu$	611	0.643, 0.344	Color 80 lamps [a]
$GdMgB_5O_{10}:Ce, Tb, Mn$	630	0.602, 0.382	Color 90 lamps [a]
$Mg_4GeO_{5.5}F:Mn$	660	0.700, 0.287	Decoration lamps

([a] at 254 nm excitation, [b] see text)

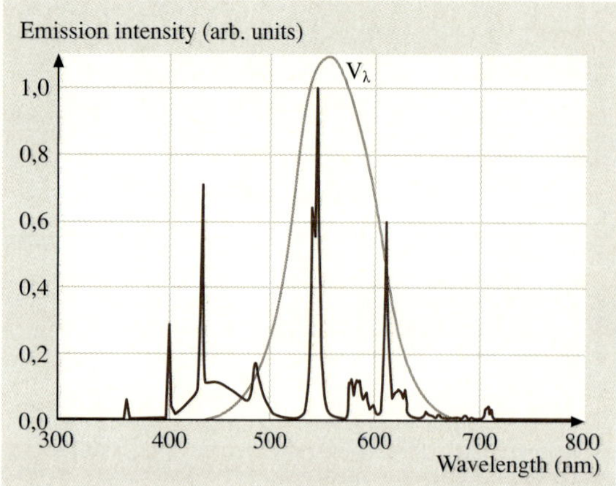

Fig. 10.9 Spectrum of a color 80 fluorescent lamp with a blend of $BaMgAl_{10}O_{17}:Eu$, $LaPO_4:Ce, Tb$, and $Y_2O_3:Eu$ and the eye sensitivity curve V_λ

yond illumination, such as medical and cosmetic skin treatment, decoration, reprography, horticultural lighting. The latter, for instance, comprises a dichromatic phosphor blend consisting of $BaMgAl_{10}O_{17}:Eu$ and $Y_2O_3:Eu$. The spectrum of this mixture is adjusted according to the action spectrum for photosynthesis of green plants.

Other Low-Pressure Discharge Lamps

In low-pressure sodium lamps the discharge has the same origin as in mercury lamps. Sodium emits almost monochromatic yellow light, viz. a doublet consisting of 589.0 nm and 589.6 nm lines (Na D-lines). However, the melting point of Na is higher than Hg. Thus the optimal operating temperature is at about 530 K. At these temperatures the highly reactive Na metal requires a chemically stable wall material such as quartz (SiO_2) or alumina (Al_2O_3). To reduce thermal losses the discharge vessel is mounted inside an outer evacuated bulb Fig. 10.10.

Low-pressure sodium lamps exhibit the highest ever achieved luminous efficiency due to the favorable position of the Na lines with respect to the spectral sensitivity of the human eye, which peaks at a wavelength of 555 nm. At a discharge power of 200 W luminous efficiencies of 200 lm/W (172 lm/W system efficiency) are obtained. A disadvantage is the low rendition of colors due to missing emission at other wavelengths, e.g. in the blue, green and red parts of the spectrum. For this reason their application is limited to street and outdoor lighting where high efficiencies are of importance. In addition, an advantage is their long lifetime, which are of the order of 20 000 h.

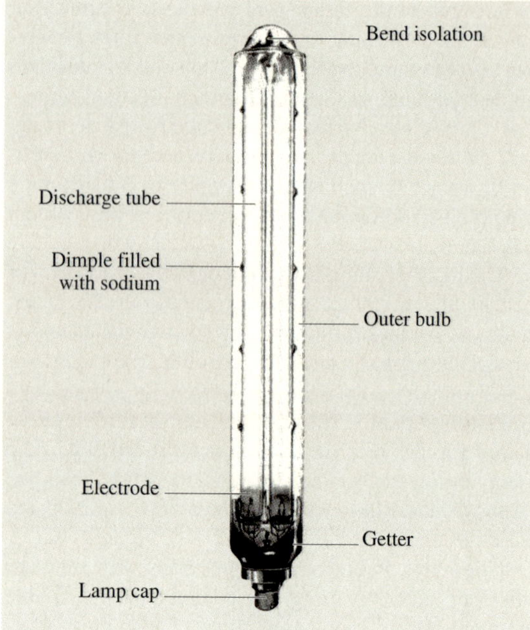

Fig. 10.10 Sketch of a low-pressure sodium lamp

When increasing the Na vapor pressure, as realized in high-pressure Na lamps, the broadened spectral lines improve color rendition significantly (see below).

Besides low-pressure Hg and Na lamps other types of radiators are used for special applications: low-pressure neon discharges and xenon excimer discharges. The former are applied in automotive brake lights and advertisement lighting. Excimer radiation from xenon is used in plasma displays and flat backlights, photocopier lamps and UV purification. The major advantage of these discharges is their fast switching behavior: in contrast to Hg and Na, Xe does not need to be evaporated. Fast switching is achieved with pulsed barrier discharge resulting in the formation of Xe excimers. Atomic and molecular radiation at 147 nm and 172 nm is observed, respectively.

10.2.4 High-Pressure Discharge Lamps

As the pressure is raised in a low-pressure discharge, the rate of elastic collisions increases. Although each collision transfers little energy to heavy particles the plasma is heated effectively due to the high particle densities. To sustain central plasma temperatures in the range 4000–10 000 K a temperature gradient is established. The heat flux along this gradient to the discharge wall is a loss, which limits the radiation efficiency of such arcs to around 60%. Fortunately, selective emission of the plasma allows for efficient radiation in the visible part of the spectrum. High-pressure discharge lamps containing metals like Hg or Na are known.

Besides, other efficient radiators are added to the lamp filling in so-called metal halide lamps. Each combination of metals used generates radiation with a specific color and radiant efficiency. In theory at least, no fewer than 50 different metals can be used to dose metal halide lamps. As a result, lamp manufacturers have introduced various combinations on the market with a wide range of applications.

A technical breakthrough in high-pressure sodium and metal halide lamps has been achieved by the introduction of polycrystalline alumina (sintered Al_2O_3 ceramics) as a wall material. Compared to quartz, a major advantage is its chemical resistance against the hot sodium vapor that is present in high-pressure sodium lamps. In addition ceramic materials offer the possibility to achieve the high wall temperatures needed for effective evaporation of the salt fillings. In quartz envelopes, the maximum temperature is limited by recrystallization of the wall at about 1370 K. This value can be greatly exceeded by using Al_2O_3 instead of SiO_2. On the other hand, vacuum tight sealing of electrodes and the design of the lamps' end construction is more complicated with ceramic materials. This fact is due to their inconvenient thermal expansion behavior and brittleness. In the lighting industry, this is still a current field of research and development of modern light sources.

High-Pressure Mercury Lamps

The construction of a high-pressure mercury lamp is shown in Fig. 10.11. The discharge vessel is filled with mercury, and argon is used as a starting gas [10.1]. Thermal isolation of the discharge vessel is achieved by use of an evacuated outer bulb. The lamp is switched on using a high-voltage ignition pulse, typically of the order of several kV. After ignition, the voltage across the discharge drops down to about 20 V, which is mainly the sum of the cathode and anode fall voltage. At this stage, the lamp is operating as a low-pressure discharge emitting UV radiation mainly at 254 nm. Thermal losses in the arc result in an increase of the wall temperature and thus increasing evaporation of the liquid mercury with time. With a further increase in mercury vapor pressure the radiated energy is concentrated progressively towards the spectral lines of longer wavelengths. When fully run up the lamp establishes a typical mercury pressure 200–1000 kPa and a bluish-white light is produced (Fig. 10.12). At these pressures a part of the emission is still in the UV. By coating the inner surface of the outer

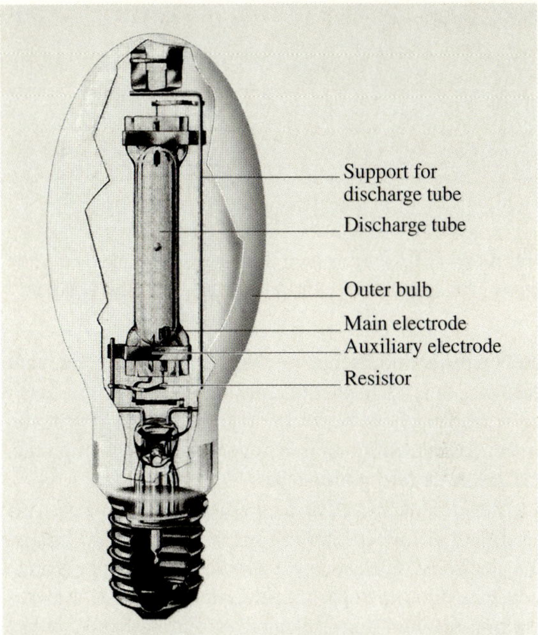

Fig. 10.11 Sketch of a high-pressure mercury lamp. The inner surface of the outer bulb may be coated with phosphor, which additionally converts UV light into visible light

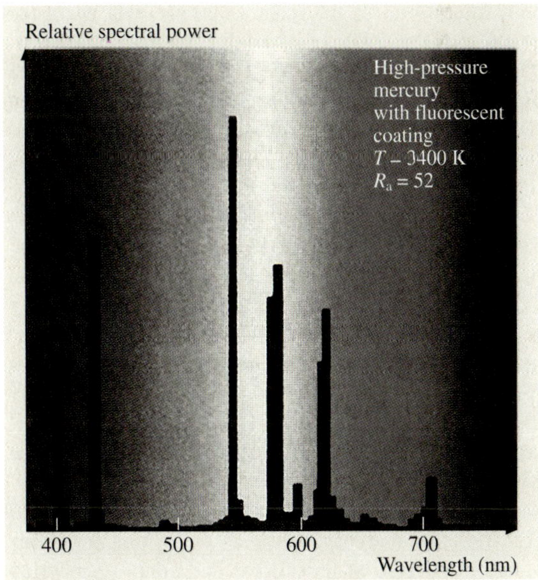

Fig. 10.12 Spectrum of a high-pressure mercury lamp

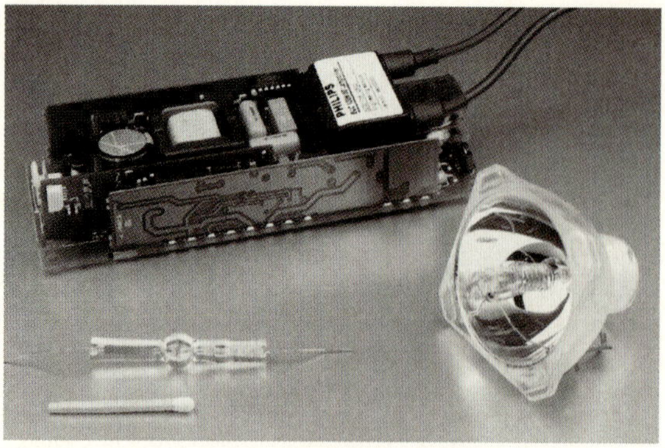

Fig. 10.13 UHP lamp system (lamp, reflector and electronic driver) for application in projection systems and beamers

bulb with a fluorescent powder, e.g. $YVO_4 : Eu$, that converts this UV radiation into visible light, the luminous efficiency can be further increased. In addition, the spectral composition of this light considerably improves the color-rendering properties.

Besides its radiative properties Hg also serves as a buffer gas in today's high-pressure discharge lamps; the electrons are elastically scattered by mercury atoms, which results in a low arc electrical conductivity According to Ohm's law, electrical field strengths of the order of several 10 V/mm are present in the discharge column, which enables an input electrical power at moderate discharge currents. The lamp current must not be too large with respect to electrode stability: electrode temperatures are in the range 2400–3500 K when the electrons are generated thermionically. It should be emphasized that both properties – the volatile character enabling high atom densities and the large cross section for elastic electron scattering – are unique to mercury.

High-pressure mercury lamps are used in various indoor and outdoor applications.

The power range is 50–1000 W with luminous efficiencies around 50–60 lm/W. Color rendition is only moderate to poor. Color temperatures are 3400–6000 K.

A recent development of a high-pressure mercury discharge is the so-called UHP lamp (ultra high performance lamp), which was invented by Philips in the 1990s (Fig. 10.13 [10.4]). This lamp has a very high operating mercury pressure (about 2×10^7 Pa) and a very short arc length of about 1 mm. This point-like light source with its broadened white spectrum is used in projection systems and beamers. In contrast to conventional high-pressure mercury lamps at these high pressures a significant part of the visible radiation is generated not only by pressure-broadened atomic mercury lines but also by Hg_2 molecules. UHP lamps are available on the market in the power range 100–250 W. A key issue for UHP lamps is the maintenance of stable properties of the discharge and electrodes for more than 5000 h. This is achieved with a chemical cycle in which evaporated tungsten is transported back onto the electrodes via compounds including oxygen and bromine. In addition, UHP lamps have to be operated with special electronic gear to control electrode erosion and arc jumping.

Metal Halide Discharge Lamps

Metal halide lamps are similar in construction to high-pressure mercury lamp. The major difference between the two types is that the discharge tube of the former contains metal halides in addition to the mercury, and argon as the starting gas. The metal halides determine the spectral properties of the lamps. As mentioned above mercury serves as a buffer gas that determines the electrical properties, e.g. lamp voltage. The metals are dosed in the form of halides because of their volatile character. High particle densities are needed for efficient radiation. Furthermore, the metal halides are less aggressive with respect to wall corrosion compared to pure metals.

There are three main groups of metal halide lamps: multi-line radiators, molecular radiators and three-band color radiators. In the first type rare-earths and associated elements such as Dy, Ho, Tm or Sc are added to the lamp filling. These species have a large number of transitions at low mean excitation energy. Consequently, such plasmas efficiently emit multi-line radiation in the visible spectrum. Furthermore, NaI and TlI are often added to such lamp fillings to improve luminous efficiency and color properties. The second type is based on molecular radiators such as SnI_2 or $SnCl_2$, which produce a quasi-continuous spectrum. Finally, the third class is based on NaI, TlI and InI_3. Such spectra consist of three color bands in the yellow, green and blue, respectively.

Typical spectra of these three metal halide lamp types are shown in Figs. 10.14a–c. Metal halide lamps are used for a variety of indoor and outdoor lighting applications. The electrical input power ranges from 35 W to more than 1000 W. In general, metal halide lamps have high luminous efficiencies of up to 100 lm/W at good to excellent color rendering. Lifetime is up to 20 000 h, depending on the type of filling used.

A speciality of metal halide lamps is its halide cycle (Fig. 10.15), which is quite similar to the aforementioned tungsten–halogen cycle. When a metal halide lamp first starts, the spectrum is initially that of mercury vapor, since the halides remain unevaporated at the relatively cool wall of the discharge tube (typical wall tempera-

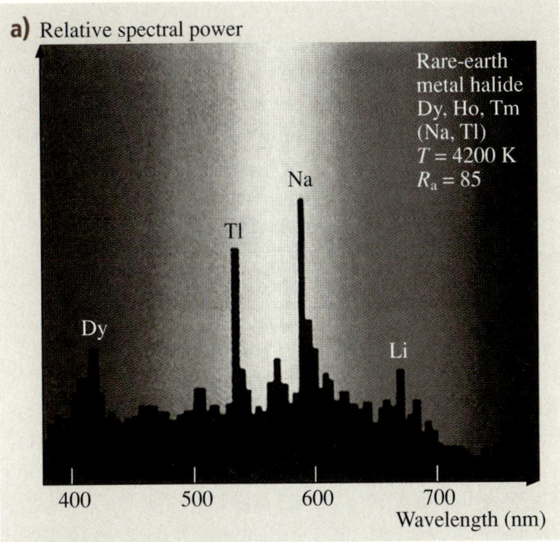

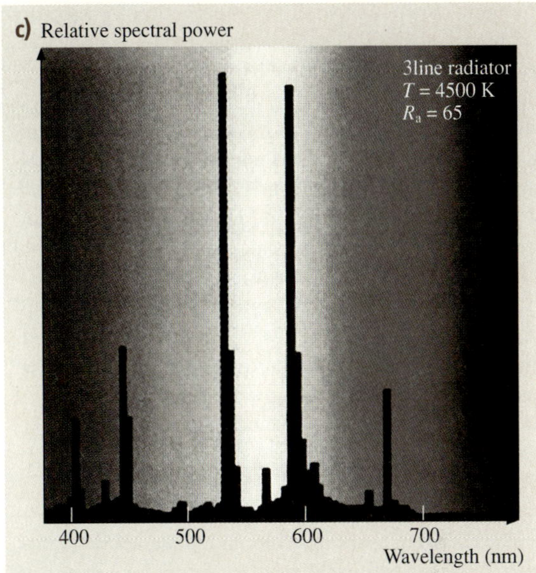

Fig. 10.14 (a) Spectrum of a $NaI/TlI/DyI_3/HoI_3/TmI_3$ lamp for sports and flood lighting. **(b)** Spectrum of a $SnI_2/SnCl_2$ lamp for studio and theater lighting. **(c)** Spectrum of a three-band color (NaI, Tl, InI_3) metal halide lamp. The lithium line is due to the presence of impurities in the wall material (quartz)

ture of 1000 K). As the wall temperature increases, the halides melt and partly vaporize. The vapor is carried into the hot region of the arc by diffusion and convection, where the halide compounds are dissociated into the halogen and metal atoms. As indicated in Fig. 10.14, different halides dissociate at different temperatures. The metal atoms are excited at the hot plasma center, which typically has a temperature of 5000 K. Here the main contribution of atomic radiation takes place. The metals then move nearer the cooler discharge wall where they recombine with the halogen atoms to form once again the halide compound. The whole cycle then repeats itself.

As mentioned above a recent improvement of metal halide lamps was achieved by the introduction of a ceramic wall material (polycrystalline alumina) instead

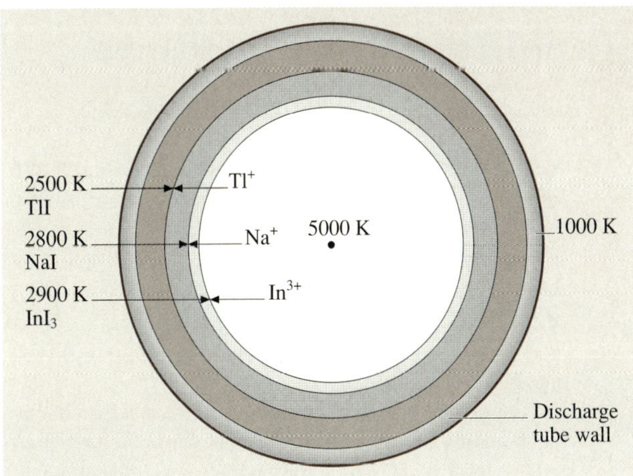

Fig. 10.15 Schematic diagram of the halide cycle of a three-band color metal halide lamp (NaI, TlI, InI_3)

of quartz. Such lamps show improved color stability and long lamp life. However, these lamps are not suitable for optical applications, e.g. in projection systems. The reason is the translucent behavior of polycrystalline Al_2O_3 (PCA) due to multiple scattering of light at grains and pores. Nevertheless, the total transmission for visible light is nearly 95%. The application of PCA to high-power metal halide lamps is also limited due to cracking of the material induced by thermal shock and thermal expansion. In the lighting industry, the applicability of ceramic wall materials is still a current field of research and development of modern light sources. Furthermore, upcoming legislation has initiated several activities on the replacement of mercury in this type of discharge by nontoxic materials [10.3].

Other High-Pressure Discharge Lamps

Besides high-pressure mercury and metal halide lamps, other high-pressure light sources are also known. The so-called blended-light lamp is derived from the conventional high-pressure mercury lamp. The principal difference between the two is that, whereas the latter is dependent on an external ballast to stabilize the lamp current, the blended-light lamp has a built-in ballast in the form of a tungsten filament connected in series with the discharge tube. Consequently, the light from the discharge and the filament combine (or blend). The improved spectral quality is traded off against reduced total efficiency of the system.

In high-pressure sodium lamps, the broadening of the D-lines is used to improve color properties compared to a low-pressure Na discharge. Luminous efficiencies of up to 130 lm/W are obtained. A technical breakthrough has been achieved with the introduction of a ceramic wall material (polycrystalline Al_2O_3). This withstands the hot and highly aggressive Na vapor, resulting in a much longer lamp life compared to quartz (SiO_2).

Another class of high-pressure discharge lamps is represented by the sulphur lamp, invented by Turner et al. at Fusion Systems in 1994. The advantage of such a lamp is their long life due to electrodeless operation. It is operated with a microwave at a wavelength of 2.45 GHz, the electrical input power being of the order of 3 kW. A spherical bulb having a diameter of about 30 mm is filled with sulphur powder. Under continuous operation, sun-like white light is emitted by S_2 molecules, having a partial pressure of about 600 kPa. The high burner efficiency of about 170 lm/W is possible since at these high pressures UV radiation from the sulphur molecules is reabsorbed by the plasma. Thus, emission mainly occurs in the visible part of the spectrum. A technical drawback of microwave lamps is the low energy efficiency of the generator, typically about 65%. Therefore, the overall efficiency of the system is reduced to about 100 lm/W, being of the order of conventional discharge lamps.

Nevertheless, electrodeless discharge lamps are still a hot topic in research and development. A general advantage of these systems is their potentially long life and large luminous flux. The latter, for instance, might be distributed by light pipes for use in industrial and general lighting applications.

10.3 Solid-State Light Sources

10.3.1 Principle of Electroluminescence

In solid-state lamps the electric current is converted into light by recombination of excited states of atoms into the ground state via emission of a photon, i. e. radiative recombination. The nature of the excited state may differ for different types of electroluminescence (EL) principles. Typically the applied voltage has to be at least as large as the emission of the material to fulfill the energy-conservation law.

Besides radiative recombination, many other decay mechanisms are possible in electroluminescence that reduce the overall efficiency, which is typically measured in $w_{opt,out}/w_{el.,in}$, given as a percentage. Weighted by the eye response curve, the use of lumens per Watt (lm/W) gives the optically relevant light output in relation to the invested electrical power. The most important loss factors are non-radiative recombination through phonons or second-order processes converting the electrical power primarily into heat. An important additional loss factor is the outcoupling of the light. Since most EL materials have a higher refractive index than air, part of the light is trapped inside the device by total internal reflection. Due to absorption losses, this light is partly lost, which reduces the overall efficiency. In addition, more specific loss channels will be discussed for the different mechanisms.

General illumination was not feasible using EL devices for a long time, since the total emitted power or lumen package was fairly low and/or the efficiencies were low. This has changed in recent years with the advent of InGaN technology, which turned out to be very stable against high power densities and the resulting heat development. Thus electroluminescence has started to become the third class of illumination devices, creating a new market for lamps, gears and luminaries.

10.3.2 Direct Versus Indirect Electroluminescence

Electroluminescence (EL) in general appears in two distinctly different mechanisms: direct EL and indirect EL. Direct EL describes the direct recombination of electrons and holes in a suitable semiconductor structure. This material can be inorganic, like in most conventional light-emitting diodes (LEDs), which were initially built from GaP and GaAs or more recently and very successful from InGaN, or organic such as, e.g., $\mathrm{Al}q_3$ ($q =$ 8-hydroxychinolinate). In this version, charge carriers are injected from contacts via direct transport with ultimately very little kinetic energy. The carriers meet in the recombination zone, wherein excitons are formed, which successively decay radiatively or non-radiatively; the first process releasing a photon, the latter creating heat. Ideally, the injection barrier from the contacts to the conduction levels of the materials is low and thus the energy necessary to generate recombination and thus light is simply the energy of the emitted light, resulting from the effective band-gap of the material. Thus the theoretically achievable efficiency of direct EL generation is unity.

In contrast, the indirect EL process results from creating excited states through impact ionization. This type of EL is found in classical thin-film (TF) or powder EL devices, also called TFEL or alternating-current thin-film EL (ACTFEL) devices. Here carriers are accelerated by means of an external electrical field. These accelerated electrons collide with ions in the matrix. By impact ionization, these ions or luminescent centers are excited and the radiative recombination of the generated electron–hole pairs releases a photon. Therefore the theoretical efficiency of this process is limited to about 40%. These devices are typically operated using high-voltage high-frequency AC bias and the brightness as well as the spectrum is a function of voltage and frequency. Typically, a few hundred $\mathrm{cd/m^2}$ maximum brightness are achieved in state-of-the-art indirect EL devices. Because efficiency and brightness are limited, we will not consider indirect EL in much detail, since the respective devices have achieved little importance as light sources. They are mainly used in displays, indication or signaling and decorative lighting. It is not expected that dramatic improvements will occur in the near future. The interested reader is referred to detailed descriptions in [10.5].

10.3.3 Inorganic Light-Emitting Diodes (LEDs)

The first practical inorganic semiconductor LEDs were conceived at Texas Instruments in 1961 by Biard and Pitman. These devices emitted in the near infrared because the material used was InGaP with a band-gap of 1.37 eV. One year later the first LED in the visible was built by Holonyiak and co-workers at General Electric. They used a different composition to fabricate a device that emitted red light. From that time on, the use of more refined processing and other materials (such as e.g. AlGaAs) has led to the development of the common LED, mainly used for indicator lights as, for example, shown in Fig. 10.16.

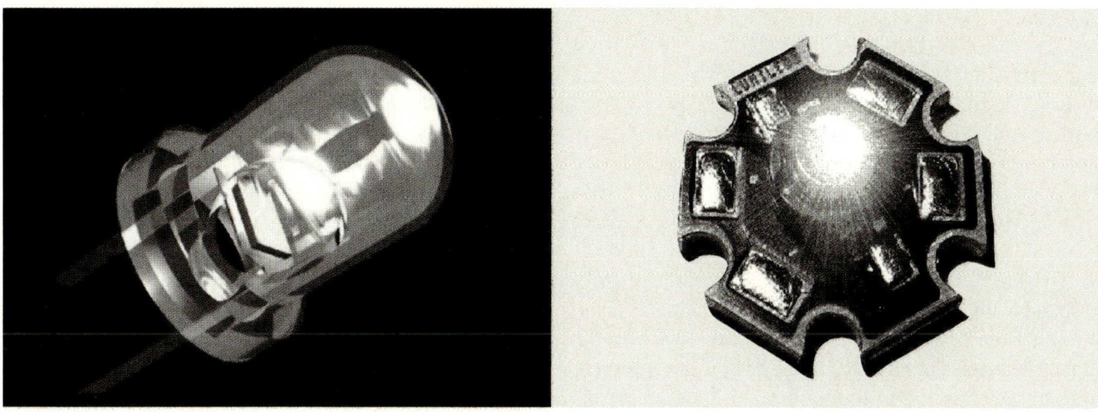

Fig. 10.16 Schematic of a classical indicator and high-power illumination LED

Fig. 10.17 AlInGaP LED shaped as a truncated inverted pyramid (TIP) for efficient light extraction

For several decades, LEDs were not attractive for illumination purposes, mainly for the two following reasons. First, the optical output power (i. e. the lumen package, defining the amount of light emitted from a single LED) was fairly limited, since the packaging and the materials in use did not allow high current densities in the small chips. Since the efficiency was far less than 100%, a lot of heat would have been generated, destroying the semiconductor chip. Secondly, and more important, no real blue LED material was available. For several years II–VI compound semiconductors such as ZnSSe/ZnTe were researched as possible candidates for blue with limited success. However, in 1993 Nakamura et al. achieved the breakthrough for solid-state lighting by proving the successful use of InGaN as a material in LEDs [10.6]. Since the material is chemically very stable, it is on one hand difficult to achieve high-quality epilayers for high-efficiency LED operation, on the other hand, the high stability allows very high current densities without disintegration of the LED material. Therefore the lumen package can be increased to values relevant for lighting applications. More important, however, is the insensitivity of InGaN to defects. Although the defect density in common InGaN LEDs is much higher than what would be acceptable in other material combinations, such as e.g. AlGaAs ($10^9/\mathrm{cm}^3$ vs. $10^6/\mathrm{cm}^3$), the low surface recombination velocity of InGaN and other properties allows efficient LEDs to be built from defect-rich layers. Additionally, with the advent of bright and stable blue LEDs it became possible to generate white light from solid-state devices, either by color conversion of blue LEDs using phosphors or by externally mixing the light from blue, green and red LEDs.

The principal of LEDs is explained in great detail in the semiconductor device literature [10.6–8]. We will limit ourselves here to the use of LEDs for light generation and the respective relevant aspects. Due to the importance of color conversion in generating white light from LEDs, we will discuss in more detail the nature of the phosphors being used.

Pure-Color LEDs

White light is, generally speaking, always a mixture of several pure colors. At least a combination of blue and orange light is needed, which yields a white appearance, however a very low color rendering (CRI, color rendering index), i. e. many colors are poorly reproduced under illumination of such a light source. Increasing the number of colors by combining red, green and blue (RGB) LEDs already enables much higher CRI values and thus a high-quality white light. In the limit of many different visible LEDs, the black-body radiation spectrum of 5500 K (essentially the spectrum of sunlight) is generated, which has, by definition, a CRI of 100. However, due to the finite size of the LED dies and their package, it is rather difficult to achieve homogeneous mixing of the discrete LEDs, at least in a small package. The following paragraph will briefly describe the technology of pure-color LEDs, which are already widely used in all applications, where colored light is necessary, like e.g. signaling. The advantage here is clearly the much higher efficiency, since they replace filtered white light, which has very low efficiency and lifetime.

Two major materials are used in colored LEDs: AlInGaP for red, orange and yellow and InGaN for green and blue (see Fig. 10.18).

The quaternary mixture of AlInGaP is the prime candidate for red to yellow emission colors. It is typically grown on GaAs wafers, since the lattice matching of substrate and epilayer guarantees a low number of defects, which is essential for high efficiency and long lifetime. After deposition of the various layers, the wafer is bonded to a GaP substrate and the original GaAs substrate, which absorbs in the spectral region of the LED emission is removed. This increases the efficiency, since half of the light is emitted towards the substrate and would be lost if the substrate was absorbing. A lot of light is still trapped inside the die due to total internal reflection. Since the refractive index of the material is larger than one, all light emitted under an angle larger than the Brewster angle is reflected back into the crys tal. Due to some residual absorption, e.g. at the metal contacts, multiple internal reflections further diminish the efficiency. By shaping the semiconductor chip in the form of a truncated inverted pyramid, the highest efficiency of a red light source has been obtained by an LED from Lumileds [10.9] (see Fig. 10.17).

This LED is estimated to have close to 100% internal efficiency and an outcoupling efficiency of about 55%, resulting in 102 lm/W wall plug efficiency. This value demonstrates the potential of LED technology as a future lighting source, since this is the most efficient red-light source available. The lumen package of AlInGaP LEDs is, however, limited due to thermal quenching of the emission. For power densities beyond 0.1 W/cm^2, the efficiency and also the lifetime of AlInGaP LEDs drops significantly, since the chip temperature increases.

For the higher energy emission from green to blue and beyond, InGaN has become the material of choice. After several years of intense research, it became possible to create working LEDs in the InGaN material system. The major difficulty to overcome was the problem of p-doping of GaN. This is achieved by doping Mg into the InGaN lattice, which happens at temperatures above 1100 °C in the deposition process. By varying the amount of In in the composition of the LED, the emission can be tuned from UV to green. However, due to segregation effects and subsequent development of lattice defects, the efficiency drops with increasing wavelength, i. e. for high In contents. Comparing the different materials classes, it turns out, that the defect density of InGaN LEDs is much higher than what would be tolerable in more conventional III–V compound semiconductors. This is mainly due to the lack of a lattice-matched substrate material. So far mostly Al_2O_3 (sapphire) with a lattice mismatch of approx. 15% is used. Nevertheless, the devices, with unsurpassed stability and high efficiency, prove that the InGaN system is obviously much less sensitive to defects than other material combinations. The reason for that is not yet entirely clear. However, it has been measured that InGaN has the lowest surface recombination velocity of all III–V composite semiconductors, thus giving excited carriers more chances of radiative recombination. Current efficiencies range from 20% (20 lm/W) for blue to 12% (60 lm/W) at green wavelengths, with continuously increasing tendency in new values literally being published week by week.

As a summary, the efficiencies as a function of wavelength are plotted in Fig. 10.19 together with the eye's sensitivity curve for the two material systems.

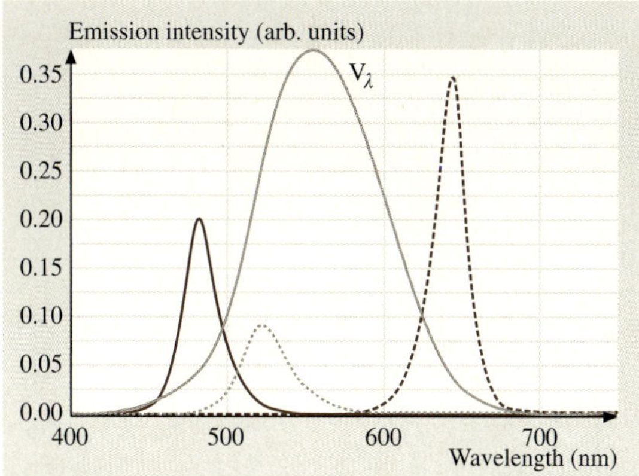

Fig. 10.18 Spectra of blue and green InGaN and red AlInGaP LEDs and the eye sensitivity curve V_λ

White LEDs

As mentioned above, combining the light of several colored LEDs can create white light. Several different concepts have been used, which will be briefly reviewed here (see Fig. 10.20). All of them have in common that a volume of light mixing has to be provided, since the different colors have to be directionally randomized before emission to avoid inhomogeneous illumination colors. This is typically achieved through multiple reflections of the different LEDs in an appropriate housing. A common method is the use of a poly(methyl methacrylate) (PMMA) plate, where the light from different LEDs is coupled. Through total internal reflection, the colors are

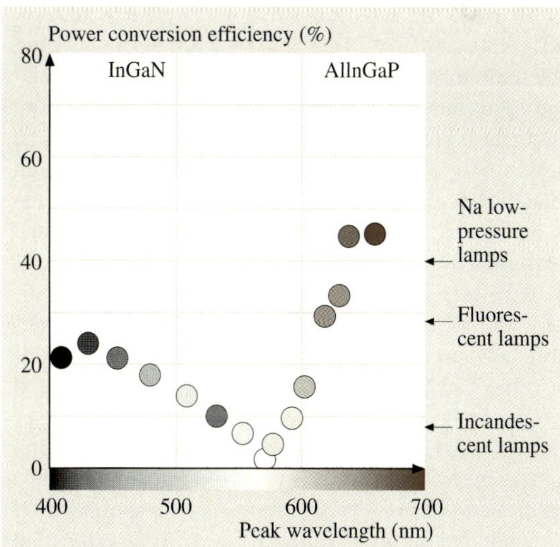

Fig. 10.19 Efficiency of InGaN and AlInGaP LEDs as a function of emission wavelength

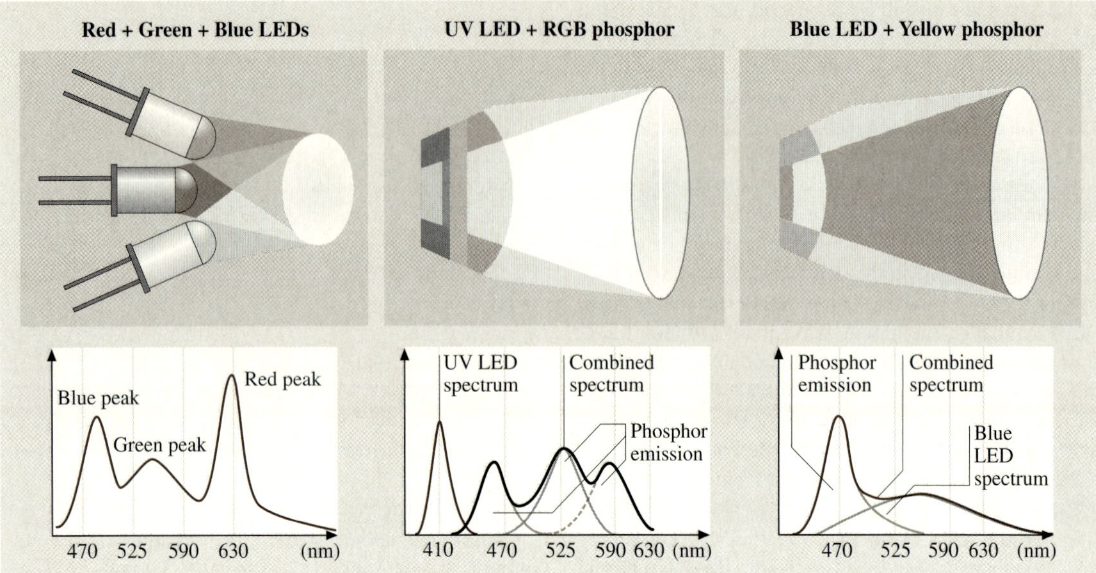

Fig. 10.20 Summary of different methods of generating white light with LEDs

randomized and thus homogeneously mixed to the desired white. A second way to achieve a white color from polychromatic LED modules is to use a diffuser plate in front of the LEDs, again to randomize the spatial distribution of the different colors emitted by the LEDs. Finally tube-like hollow reflectors are being used for the same purpose, however in these systems the de-mixing and thus non-white appearance is often obvious at large angles.

To date, the most common method of achieving white emission from LEDs is by partially converting blue light from the semiconductor to lower-energy emission by means of phosphors, which are deposited onto the LED chip. This implies that the phosphor must show strong absorption of the blue light and a high photochemical stability due to the high excitation density of up to several $100\,\mathrm{W/cm^2}$. The widely used combination of blue InGaN LEDs (430–480 nm) with $(\mathrm{Y, Gd})_3(\mathrm{Al, Ga})_5\mathrm{O}_{12}:\mathrm{Ce}$ (YAG : Ce) thereby dominates the white-LED market today [10.7]. However, more advanced concepts are being realized, including blue LEDs in combination with yellow and red phosphors or red and green phosphors. The former concept aims at the enhancement of red radiation in the entire spectrum to realize illumination-grade warm-white LEDs. The red luminescent materials used in combination with yellow YAG : Ce in so-called red-enhanced LEDs are (Ca, Sr)S : Eu [10.10, 11] and $(\mathrm{Ca, Sr})_2\mathrm{Si}_5\mathrm{N}_8:\mathrm{Eu}$ [10.12]. Red-enhanced LEDs show high CRIs above 90 for low color temperatures between 2500 K and 4000 K.

White LEDs with high color rendering at arbitrary color temperatures between 2500 K and 10 000 K are obtained by applying blue LEDs with a combination of green and red phosphors. *Mueller–Mach* and *Mueller* e.g. found that $\mathrm{SrGa_2S_4}:\mathrm{Eu}$ (535 nm) and SrS : Eu (615 nm) are an appropriate combination for trichromatic LEDs [10.13].

Compared to dichromatic YAG : Ce LEDs, LEDs with a trichromatic spectrum yield higher color rendering while the luminous efficacy is still almost as high.

10.3.4 Organic LEDs

Organic LEDs became an important area of research and development after the publication of direct EL observed from a device incorporating only organic charge-transfer materials and emitters by *Tang* and *van Slyke* from Kodak in 1987 [10.14]. The authors were working with organic photoconductors, which have been used for several decades in copiers and printers. Applying a forward bias to an appropriate device built with these materials, light was generated from directly injected carriers. A second important date was the publication of direct organic EL observed in a single-layer polymer-based device by *Friend* et al. in 1990, making a new class of materials available: semiconducting polymers [10.15].

These marked the beginning of intense R&D in organic EL.

The general build-up of an organic LED (OLED) is represented in Fig. 10.21. The stack typically consists of a substrate (mostly glass, however plastic is considered) with a thin layer (typically 100 nm) of indium–tin oxide (ITO). This is the most prominent representative of transparent conducting materials, which is needed to be able to couple the light out. On top of the ITO layer the organic material is deposited, either as a multilayer stack in small-molecule OLEDs or as a polymer coat in polymeric OLEDs. The top contact, usually the cathode, is built by evaporating reactive materials such as e.g. Ca, Mg or Ba to facilitate injection of electrons into the molecular materials. This is often followed by a less reactive metal, such as Al, to prevent fast oxidation of the top contact. Since many of the organic materials and especially the reactive cathode metal are water- and oxygen-sensitive, a nearly hermetic sealing is required to achieve a practical lifetime of the devices. This is achieved by encapsulating the glass with a second glass or metal can, which often also contains additional getter materials to absorb residual water and oxygen.

In general two types of organic direct EL are distinguished: polymeric and small-molecule organic LEDs. Both systems rely in principle on the same basic mechanisms and, although there are distinctive differences, especially in the case of conjugated polymers, the device set-up is identical for the two versions. Whereas small-molecule OLEDs are deposited using high-vacuum sublimation or gas-phase transport, polymeric OLEDs are typically spin-coated or in the case of displays, ink-jetted.

The conduction mechanism in organic materials is different to inorganic semiconductors. In the molecular solid, after charge injection, the molecule becomes oxidized and reduced, respectively, depending on the nature of the charge injected. If an electron is injected from the molecule to the electrode, a hole is injected into the molecule, which oxidizes the molecule. The charge then moves under the influence of the applied electric field across the molecule and by hopping from the originally excited molecule to the neighboring one. Thus the conductivity of organic layers is rather low, on the order of $10^{-3}\,cm^2/Vs$ in good materials, down to $10^{-8}\,cm^2/Vs$ for nearly insulating ones. The hole mobility is typically larger ($10^{-3}\,cm^2/Vs$) than the electron mobility ($10^{-5}\,cm^2/Vs$). The mobility can be higher in certain materials classes (up to $10^{-1}\,cm^2/Vs$), but those are not typically used in EL devices.

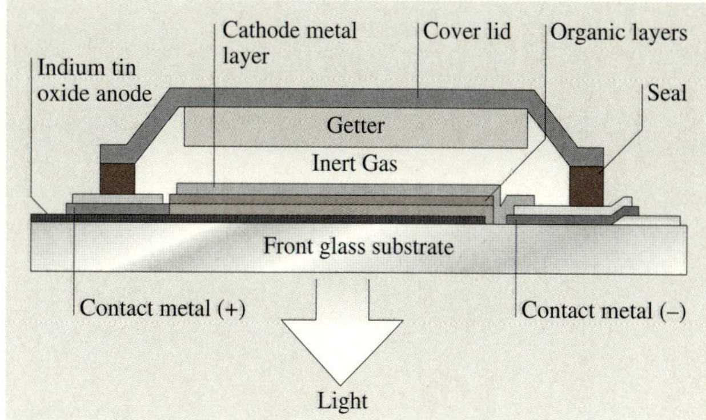

Fig. 10.21 Schematic build-up of an organic LED

The materials fabricated in OLEDs depend on which type of system is being used.

In small-molecule devices the organic layers are usually brought onto the substrate in a layer of multiple materials having defined functions for each material. A typical device layout is shown in Fig. 10.21. On top of the ITO a hole conductor is deposited. This is in most cases a material with high electron affinity to facilitate injection of holes from the anode and high hole mobility to achieve good hole transport. The next layer is than a matrix material with a large band-gap and high triplet energy, doped with an emitter material. These two functions can also be combined into the same material. On top of this layer a hole-blocker is deposited, to prevent the fast-moving holes from flowing directly to the anode. The hole-blocker has to have a very large band-gap and a large electron affinity to efficiently prevent holes from flowing through the structure. Finally an electron conductor/injector is deposited on top to facilitate electron injection into the device.

In polymeric OLEDs all these functions are combined in the material, a conjugated or non-conjugated polymer. Thus the layer sequence is typically reduced to two layers. First, a hole-injection layer, which also serves as a smoothing layer for the roughness of the substrate. This layer is made insoluble and is finally coated with the light-emitting polymer. The total thickness is typically 100–200 nm. Charge transport of electrons and holes as well as emission is provided from the different functional groups of the polymer. A schematic comparison of the different electronic structures of small molecules and polymer OLEDs is shown in Fig. 10.22.

In the recombination zone, electrons and holes meet and create excitons on the excited molecules. These excitons can have either a total spin of 1 or 1/2, thus creating a singlet or a triplet state. The singlet state is

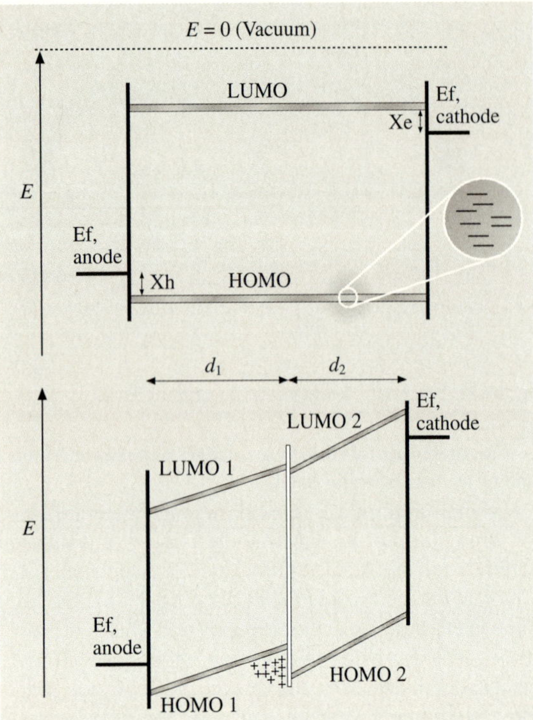

Fig. 10.22 Schematic band diagram of polymeric (single-layer) and small-molecule (double-layer) OLEDs, respectively

optically allowed and therefore able to generate a photon, whereas the triplet state is a forbidden transition, thus no emission of light is possible from these excitons. By statistics, the probability of creating a singlet is only 1/4 for all injected pairs, limiting the principle efficiency of organic EL devices to 25%. However, using other emitters, which have strong spin–orbit coupling it is possible to overcome the selection rules and the probability of forming a radiatively decaying exciton theoretically increases to unity, thus opening the road to very efficient organic LEDs.

Since the delocalization of the electron density across the molecule depends on the steric configuration of the molecule, and the electrical properties depend on the relative position between the molecules in the layer, the film formation is highly important in organic LEDs. Additionally, the coupling of two neighboring molecules can lead to the formation of cooperative electronic states, which exhibit a new excited state, only present in the coupled molecules. This phenomenon is called excimer in the case of identical molecules and exciplex, if two different molecules are involved. The energy of this excited state without ground state is typically lower than the excited states of the single molecules. Therefore, this effect can be used to generate other wavelengths than those available from the constituent molecules only. However, the efficiency of singlet excimers and exciplexes is typically very low.

A crucial problem in organic LEDs (as well as in inorganic ones) is the outcoupling efficiency of the generated light. Due to the fact that many of the materials used have a high refractive index, light generated within the film and emitted at an angle larger than the critical angle of the material suffers from total internal reflection and the light is guided through the structure. Multiple reflections at partially absorbing interfaces reduce the amount of light coupled out. An estimate of the outcoupling ratio, using classical ray optics, gives a value of $1/n^2$, where n is the refractive index of the material. For most materials, this value amounts to 20%, i. e. 4/5 of the light generated does not leave the structure, rendering OLEDs often much less efficient than what could be achieved with improved outcoupling. Intense work is being conducted to identify the most promising methods to improve light outcoupling from thin-film organic LEDs. Recent measurements, using an optically coupled hemispherical lens on top of an OLED substrate indicate that the outcoupling can at least be doubled by appropriate means, improving the efficiency by the same factor of two. This topic is certainly one of the most important areas of OLED research.

Many of the materials used in today's OLEDs are sensitive to moisture and oxygen. Irreversible oxidation occurs at the surface, rendering the material insulating and thus creating a nonemissive (black) spot. These black spots are progressive, i. e. once formed they have a tendency to grow until the OLED is rendered useless. These black spots often start at pinholes in the electrode materials, which in the case of many cathode materials are themselves often the most critical materials with respect to oxidation within the whole OLED stack. To save the OLED from these failure mechanisms, near-perfect encapsulation is necessary, which often includes a getter material put in a sealed package, in close proximity to the OLED cathode to capture any remaining oxygen and water molecules before damage to the OLED occurs. In the near future, it is hoped that thin-film packaging, i. e. the deposition of a nearly impermeable material on top of the OLED will solve this problem.

Although this is one of the most often encountered failure mechanisms of OLEDs, it is not the only one by far. Also, many OLED materials are intrinsically unstable, limiting the useful life of OLEDs and creating the need for more research and development in the field of organic LED materials.

For application as light sources OLEDs have some way to go to achieve high efficiency at high brightness and a long lifetime. However, since all of these properties have been shown in single devices, it is thought possible that OLEDs will be able to deliver. Due to the way in which they generate light, OLEDs will provide a very different light source to existing sources. A diffuse, large area, flat lamp that can eventually become flexible, color-tuneable and transparent is envisioned with OLEDs, creating new potential applications for light sources in the future. To generate white light from OLEDs several possibilities can be used. External mixing of colors (e.g. from pixelated sources), intrinsic broadband (white) emitters, mixed emitters, stacked emitters or phosphor-converted blue or UV-emitting OLEDs (using e.g. the phosphors described above) can be fabricated; the different options having different advantages and disadvantages depending on the applications for the light source. Most estimates establish a time frame of five to ten years before OLEDs will become an important incoherent light source.

10.4 General Light-Source Survey

Table 10.2 gives a summary of the main types of light sources and selected electrical and light technical properties as discussed in this chapter.

Table 10.2 Electrical and technical data of the basic types of light sources (state of affairs 2003)

Light source	Electrical input power (W)	Luminous flux (lm)	Luminous efficacy (lm/W)	Color rendering quality
Incandescent	10–1000	80–15000	8–15	Excellent
Halogen	20–2000	300–60000	15–30	Excellent
Low-pressure Hg discharge	7–150	350–15000	50–100	Good
High-pressure Hg discharge	50–1000	2000–60000	40–60	Good
Metal halide discharge	20–2000	1600–24000	80–120	Good to excellent
Low-pressure Na discharge	20–200	2000–40000	100–200	Poor
High-pressure Na discharge	40–1000	1600–14000	40–140	Moderate to good
Sulphur microwave discharge	up to 5000	up to 450000	80–90 (system)	Good
White dichromatic inorganic LED	1–5	20–150	20–30	Good
White trichromatic inorganic LED	1	20–25	20–25	Excellent
Organic LED (at 1000 cd/m^2)	15 mW (per cm^2)	0.3 lm (per cm^2)	30	Excellent

References

10.1 J. R. Coaton, A. M. Marsden: *Lamps and Lighting* (Wiley, New York 1997)

10.2 W. Elenbaas: *Light Sources* (Crane, Russek, New York 1972)

10.3 M. Born, T. Jüstel: Umweltfreundliche Lichtquellen, Physik J. **2**(2), 43 (2003) In German

10.4 G. Derra, E. Fischer, H. Mönch: UHP-Lampen: Lichtquellen extrem hoher Leuchtdichte, Phys. Bl. **54**(9), 817 (1998)

10.5 Y. A. Ono: *Electroluminescent Devices* (World Scientific, Singapore 1995)

10.6 S. Nakamura, S. Pearton, G. Fasol: *The Blue Laser Diode* (Springer, Berlin, Heidelberg 1997)

10.7 A. Zukauskas, M. S. Shur, R. Caska: *Introduction to Solid-State Lighting* (Wiley, New York 2002) p. 122 and references therein

10.8 S. M. Sze: *Physics of Semiconductor Devices* (Wiley, New York 1981)

10.9 M. R. Krames, M. Ochiai-Holcomb, G. E. Höfler, C. Carter-Coman, E. I. Chen, I.-H. Tan, P. Grillot, N. F. Gardner, H. C. Chui, J.-W. Huang, S. A. Stockman, F. A. Kish, M. G. Craford, T. S. Tan,

C. P. Kocot, M. Hueschen, J. Posselt, B. Loh, G. Sasser, D. Collins: High-power truncated-inverted-pyramid $(Al_xGa_{1-x})_{0.5}In_{0.5}P/GaP$ light emitting diodes exhibiting >50 external quantum efficiency, Appl. Phys. Lett. **75**, 2365 (1999)

10.10 R. Müller-Mach, G. O. Müller, T. Jüstel, P. J. Schmidt: US Patent 2003/0006702

10.11 R. Müller-Mach, G. O. Müller, M. Krames, T. Trottier: High-power phosphor-converted light-emitting diodes based on III-nitrides, IEEE J. Sel. Top. Quantum Electron. **8**, 339 (2002)

10.12 M. Yamada, T. Naitou, K. Izuno, H. Tamaki, Y. Murazaki, M. Kameshima, T. Mukai: Red-enhanced white-light-emitting diode using a new red phosphor, Jpn. J. Appl. Phys. **42**, L20 (2003)

10.13 R. Müller-Mach, G. O. Müller: White light emitting diodes for illumination, SPIE Proc. **3938**, 30 (2000)

10.14 C. W. Tang, S. A. Van Slyke: Organic electroluminescent diodes, Appl. Phys. Lett. **51**, 913 (1987)

10.15 J. H. Borroughes, D. D. C. Bradley, A. R. Brown, R. N. Marks, K. MacKay, R. H. Friend, P. L. Burns, A. B. Holmes: Light-emitting diodes based on conjugated polymers, Nature **347**, 539 (1990)

11. Lasers and Coherent Light Sources

Lasers and Co

This chapter describes lasers and other sources of coherent light that operate in a wide wavelength range. First, the general principles for the generation of coherent continuous-wave and pulsed radiation are treated including the interaction of radiation with matter, the properties of optical resonators and their modes as well as such processes as Q-switching and mode-locking. The general introduction is followed by sections on numerous types of lasers, the emphasis being on today's most important sources of coherent light, in particular on solid-state lasers and several types of gas lasers. An important part of the chapter is devoted to the generation of coherent radiation by nonlinear processes with optical parametric oscillators, difference- and sum-frequency generation, and high-order harmonics. Radiation in the extended ultraviolet (EUV) and X-ray ranges can be generated by free electron lasers (FEL) and advanced X-ray sources. Ultrahigh light intensities up to 10^{21} W/cm^2 open the door to studies of relativistic laser–matter interaction and laser particle acceleration. The chapter closes with a section on laser stabilization.

11.1 Principles of Lasers

11.1.1 General Principles

A laser (an acronym for *light amplification by stimulated emission of radiation*) is a device that produces and amplifies an intense beam of highly coherent, highly directional light. The original proposal of extending the maser (*microwave amplification by stimulated emission of radiation*) idea to the infrared or visible region of the electromagnetic (EM) spectrum, thus giving a laser, was however first made in 1960 by *Maiman* [11.1] using a flash-pumped rod of ruby with polished ends (ruby laser) after a proposal in 1958 by *Townes* and *Schawlow* [11.2]. Nowadays laser devices range in size from semiconductor lasers as small as a grain of salt to solid-state lasers as large as a storage building. Lasers are widely used in industry for, e.g., cutting and welding metals and other materials, in medicine for surgery, in optical communications, in optical metrology, and in scientific research. They are an integral part of such familiar devices as bar-code scanners used in supermarkets, laser printers, compact disc and digital versatile disc (DVD) players. The output wavelength of a laser is determined by the properties of its active medium. Altogether, several thousand lasing lines have been reported ranging from the soft-X-ray down to the far-infrared spectral region and new lines appear frequently in optics and laser journals. Depending upon the type of laser and its operational regime, the corresponding output power may vary from a fraction of a milliwatt to several hundred kilowatts in continuous-wave operation, and from tens of kilowatts to petawatt peak power in pulsed-mode operation.

A laser consists of at least three components (Fig. 11.1):

1. a gain medium that can amplify light by means of the basic process of *stimulated emission*;
2. a pump source, which creates a *population inversion* in the gain medium;
3. two mirrors that form a *resonator* or optical cavity in which light is trapped, traveling back and forth between the mirrors.

The laser beam is usually the fractional part of light trapped in the cavity that escapes from one of the two

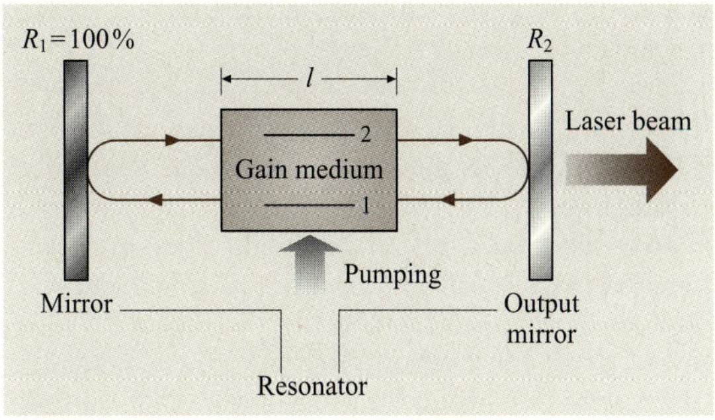

Fig. 11.1 Schematic of a laser oscillator

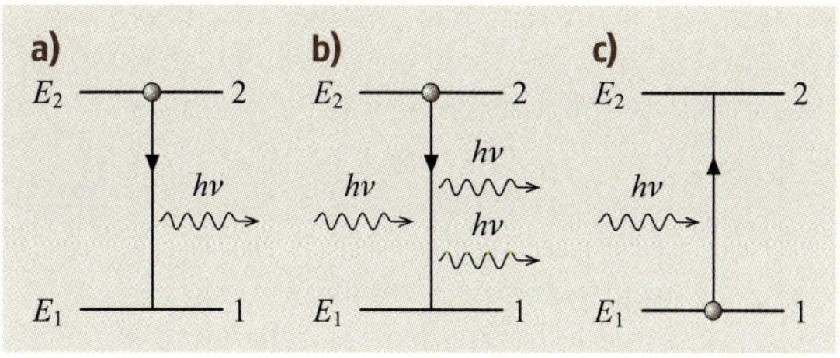

Fig. 11.2a–c Fundamental interaction processes of a quantized atom with an EM wave. **(a)** Spontaneous emission. **(b)** Stimulated emission. **(c)** Absorption

mirrors (the *output coupler*), which has a nonvanishing transmission at the laser oscillation wavelength. The gain medium can be solid (including semiconductors), liquid, or gas, and the pump source can be an electrical discharge, a lamp, or another laser. Other specific components of a laser vary depending on the gain medium and whether the laser is operated continuously or pulsed. Lasers may be in fact divided into two broad groups:

1. continuous wave (CW) or quasi-CW,
2. pulsed.

A CW laser exhibits a steady flow of coherent energy and its output power undergoes little or no change with time. Many gas lasers, such as HeNe and Ar-ion lasers, operate CW; several solid-state lasers, such as Nd^{3+} and Ti^{3+}:Al_2O_3 lasers, are also often operated in CW mode. In pulsed lasers, the output beam power changes with time so as to produce a short optical pulse, usually in a repetitive way and with pulse duration usually ranging from nanoseconds ($1\,\text{ns} = 10^{-9}\,\text{s}$) to femtoseconds ($1\,\text{fs} = 10^{-15}\,\text{s}$). Typical examples of pulsed lasers are many solid-state and liquid lasers, such as Nd:YAG, Ti:Al_2O_3, and dye lasers.

Spontaneous and Stimulated Emission, Absorption

A laser exploits three fundamental phenomena that occur when an electromagnetic wave interacts with a material, namely the processes of spontaneous and stimulated emission, and the process of absorption (Fig. 11.2).

Spontaneous emission and nonradiative decay. Let us consider the energy levels of some given material and indicate by E_1 and E_2 the energies of the ground level, 1, and of an excited level, 2, of the medium. If the atom is initially raised in the excited state 2, it spontaneously tends to decay into the stable ground state 1. If the available energy for the transition, $E_2 - E_1$ is released in the form of an EM wave, the process is referred to as spontaneous, or radiative, emission (Fig. 11.2a). Owing to energy conservation, the frequency ν_0 of the emitted radiation is given by:

$$\nu_0 = \frac{(E_2 - E_1)}{h} , \tag{11.1}$$

where h is the Planck's constant. Since the previous relation can also be written as $E_2 - E_1 = h\nu_0$, we can say that a single photon of energy $h\nu_0$ is emitted during each spontaneous emission process. Note also that this process occurs even if the atom is isolated and no external perturbation is applied. The probability of spontaneous emission can be characterized in the following way. Let us consider an ensemble of atoms and assume that, at time t, there are N_2 atoms per unit volume (*population*) in level 2.

Quantum mechanical calculations show that the rate of decay of these atoms due to spontaneous emission, i. e., $(\mathrm{d}N_2/\mathrm{d}t)_{\text{sp}}$, is proportional to N_2 according to:

$$\left(\frac{\mathrm{d}N_2}{\mathrm{d}t}\right)_{\text{sp}} = -\frac{N_2}{\tau_\text{s}} , \tag{11.2}$$

where τ_s is referred to as the spontaneous emission lifetime and depends on the particular transition involved. The direction, the polarization, and the phase of the emission event are random, so that the overall emitted light by the different atoms of the given population is said to be incoherent.

If the atom is not isolated but interacts, e.g., by collisions with other atoms, the decay from the excited state 2 to the ground state 1 may occur by a release of the internal energy $(E_2 - E_1)$ into some form other than EM radiation (e.g., into kinetic or internal energy of surrounding atoms in a gas or lattice vibrations in a crystal). The phenomenon is referred to as *nonradiative decay*. The corresponding rate of decay from the excited state 2 can usually be expressed in a similar manner as (11.2), namely $(\mathrm{d}N_2/\mathrm{d}t)_{\text{nr}} = -N_2/\tau_{\text{nr}}$, where τ_{nr} is the nonradiative lifetime. When both radiative and nonradiative decay are taken into account, the rate of decay of population N_2 can be written as:

$$\left(\frac{dN_2}{dt}\right)_{decay} = -\frac{N_2}{\tau_s} - \frac{N_2}{\tau_{nr}} \equiv -\frac{N_2}{\tau}, \tag{11.3}$$

where the time constant τ, defined by $1/\tau = 1/\tau_s + 1/\tau_{nr}$, is referred to as the *lifetime* of the excited state 2.

Stimulated emission. If an EM wave of frequency ν close to ν_0 is incident onto the atom while in its excited state 2, the interaction of the wave with the atom may stimulate the atom to decay to level 1 with the simultaneous emission of EM radiation. The process is referred to as *stimulated emission*. In this case, one photon, with the same frequency ν of the incoming radiation and with the same propagation direction, polarization state and phase is emitted (Fig. 11.2b). This is a major distinction to spontaneous emission. It is the fundamental reason why a laser emits coherent light as compared to the incoherent light emitted by other light sources [such as lamps or light-emitting diodes (LEDs)], which exploit spontaneous emission. Quantum mechanical calculations show that the process of stimulated emission can be described by the equation

$$\left(\frac{dN_2}{dt}\right)_{st} = -W_{21}N_2, \tag{11.4}$$

where W_{21} is the rate of stimulated emission. This rate is proportional to the photon flux $F = I/(h\nu)$ of the incoming wave, where I is the wave intensity. We can in fact write $W_{21} = \sigma_{21}F$, where σ_{21} is a quantity having the dimension of an area (the stimulated emission *cross section*), which depends on the characteristics of the given transition and on the frequency difference $\Delta\nu = \nu - \nu_0$, i.e., $\sigma_{21} = \sigma_{21}(\Delta\nu)$. Owing to energy conservation in the process of stimulated emission, the function $\sigma_{21}(\Delta\nu)$ is very narrow at around $\Delta\nu = 0$.

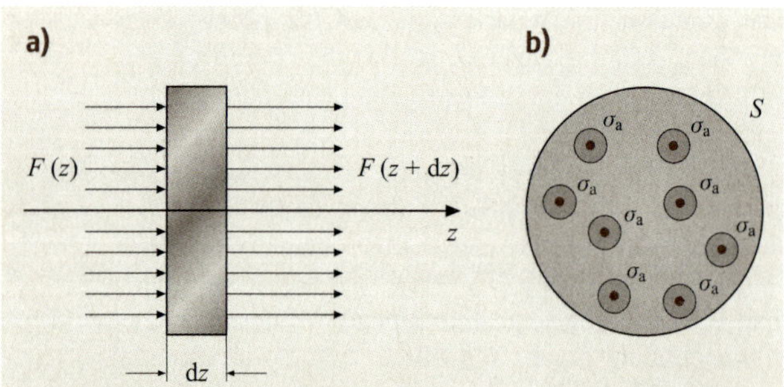

Fig. 11.3a,b Coherent amplification of an EM wave. **(a)** Photon flux-balance diagram in an infinitesimal section dz of the amplifier. **(b)** Physical meaning of the transition cross section

Absorption. Lastly, let us consider the case of an atom initially lying in its ground state 1. In the absence of perturbations, such as collisions with other atoms or with photons, the atom stably remains in this state. However, if an EM radiation of frequency $\nu \simeq \nu_0$ is incident onto the atom, there is a finite probability that the atom will be raised to level 2. The energy difference $E_2 - E_1$ required by the atom to undergo the transition is obtained from the energy of the EM wave, namely one photon of the incoming wave is destroyed. This process is referred to as *absorption* (Fig. 11.2c). As for stimulated emission, the absorption process can be described by

$$\left(\frac{dN_2}{dt}\right)_{a} = -W_{12}N_2, \tag{11.5}$$

where W_{12} is the rate of absorption. Again one can show that $W_{12} = \sigma_{12}F$, where σ_{12} is the absorption cross section. Einstein showed at the beginning of the 20th century that, for nondegenerate levels, $W_{12} = W_{21}$, and thus $\sigma_{12} = \sigma_{21}$. If levels 1 and 2 are g_1-fold and g_2-fold degenerate, respectively, one then has $g_2\sigma_{21} = g_1\sigma_{12}$.

Coherent Amplification of Light

Consider a monochromatic EM plane wave at frequency ν which propagates along the z-direction inside a medium made of a collection of atoms. Let E_1 and E_2 be the energies of two nondegenerate levels 1 and 2 of the atom (this time 1 is not necessarily the ground state). We assume that the resonance frequency of the transition, $\nu_0 = (E_2 - E_1)/h$, is coincident (or very close) to ν. If $F(z)$ is the photon flux of the EM wave at plane z and N_1, N_2 are the populations in levels 1 and 2 (which are assumed to be z-independent), the change dF of the photon flux due to the processes of absorption and stimulated emission along the elemental length dz of the material is given by (Fig. 11.3a)

$$dF = \sigma(N_2 - N_1)F(z)dz, \tag{11.6}$$

where $\sigma \equiv \sigma_{21} = \sigma_{12}$ is the transition cross section. Note that in writing (11.6) we did not consider radiative and nonradiative decays since nonradiative decay does not add new photons whereas spontaneous emission creates photons which are emitted in any direction and thus gives negligible contribution to the incoming photon flux. Most importantly, note that for $N_2 > N_1$, one has $dF/dz > 0$ and the EM wave is amplified during propagation, i.e., the medium acts as a coherent *optical amplifier*. Conversely, for $N_2 < N_1$, one has $dF/dz < 0$

and the medium behaves as an *absorber*. If we let l be the length of the medium, the photon flux $F(l)$ at the output plane is related to the one $F(0)$ at the input plane by the relation $F(l) = F(0)\exp(g)$, where $g = (N_2 - N_1)\sigma l$ is the gain coefficient. The condition for light amplification ($g > 0$) is therefore $N_2 > N_1$, which is often referred to as *population inversion*. We note that in the previous calculation N_1 and N_2 have been assumed independent of the intensity I of propagating wave. This is fine however provided that I is weak enough so as that changes of N_1 and N_2 due to absorption and stimulated emissions can be neglected; however, for strong intensities one needs to account for the phenomenon of *saturation*.

As a final comment, we note that an examination of (11.6) leads to a simple physical interpretation of the transition cross section σ. First, let us suppose that all the atoms of the medium are in the ground state and let us associate with each atom an effective absorption cross section σ_a in the sense that, if a photon enters this cross section, it is absorbed by the atom (Fig. 11.3b). If S is the cross-sectional area of the EM wave, the number of atoms in the element dz of the material is $N_t S\,dz$ (N_t is the total atomic population, i. e., $N_1 = N_t$ and $N_2 = 0$), thus giving a total absorption cross section of $\sigma_a N_t S\,dz$. The fractional decrease of photon flux in the element dz of the material is therefore $(dF/F) = -(\sigma_a N_t S\,dz/S) = -\sigma_a N_t\,dz$. If we compare this expression with (11.6), we conclude that $\sigma_a = \sigma$, so that the meaning we can attribute to σ is that of an effective absorption cross section as just defined.

Population Inversion: The Pumping Process

At thermal equilibrium any material behaves as an absorber. In fact, the distribution of populations N_2^e and N_1^e of levels 1 and 2 at thermal equilibrium is described by the Boltzmann statistics

$$\frac{N_2^e}{N_1^e} = \exp\left(-\frac{E_2 - E_1}{k_B T}\right) , \tag{11.7}$$

where k_B is the Boltzmann constant and T is the absolute temperature. Note that $N_2^e < N_1^e$; in particular N_2^e is negligible as compared to N_1^e if $k_B T \ll E_2 - E_1$; at room temperature ($T = 300\,\mathrm{K}$) one has $k_B T \simeq 208\,\mathrm{cm}^{-1}$. [A reciprocal centimeter is a simple convenient unit for measuring energies in spectroscopy and refers to the inverse of the wavelength of a photon possessing the given energy E. The actual energy E in the SI unit (Joule) can be obtained by multiplying cm^{-1} by hc, with $c = 3\times10^{10}\,\mathrm{cm/s}$ and $h = 6.63\times10^{-34}\,\mathrm{Js}$]. Therefore the condition $N_2^e \ll N_1^e$ is well satisfied for transitions in the near-infrared and visible. To obtain optical amplification instead of optical absorption, we need to create population inversion in the medium by means of a *pumping process* [11.3, 4], which drives the population distribution far from thermal equilibrium. At first sight one might think to achieve population inversion between levels 2 and 1 through the interaction of the material with some strong EM radiation at frequency ν_0, such as that emitted by a flash or arc lamp. Since at thermal equilibrium $N_1^e > N_2^e$, absorption will dominate over stimulated emission, and this will produce an increase of N_2 and a decrease of N_1 from their thermal equilibrium values. However, when N_1 tends to reach the same value of N_2, absorption and stimulated emission processes compensate each other, i. e., the medium tends to become transparent. Such a situation is referred to as *saturation* of the two-level transition. Therefore, owing to saturation, it is impossible to produce population inversion in a two-level system (at least in the steady state). This goal can be achieved, however, when more than two energy levels are considered. Typically three or four energy levels are involved (Fig. 11.4), and correspondingly one speaks of *three-level* or *four-level* lasers. In a three-level laser (Fig. 11.4a), atoms are raised from the ground level 1 to the level 3 through a pumping mechanism. If the material is such that, once an atom excited into level 3 rapidly decays into a lower level 2 (by, e.g., a fast nonradiative decay), then population inversion can be obtained between level 2 and 1. Note that, in a three-level laser scheme, to achieve population inversion it is necessary to raise at least half of the atoms from the ground state 1 to state 3. In a four-level laser (Fig. 11.4b), atoms are again raised from the ground level 0 to an excited state 3, with a rapid decay to the upper laser level 2; however this time laser action takes place from level 2 to an excited lower-lying level 1. Once laser oscillation starts, level 1 is populated by stimulated emission; therefore to maintain popula-

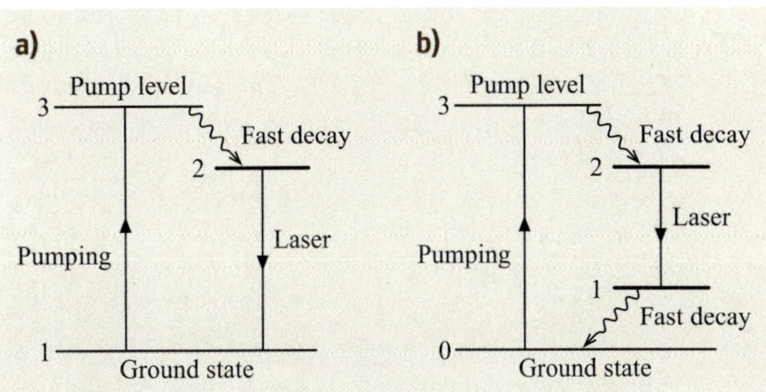

Fig. 11.4a,b Schematic of the energy-level diagram for **(a)** a three-level laser, and **(b)** a four-level laser

tion inversion in stationary conditions (continuous-wave operation), the lower laser level 1 should rapidly depopulate through, e.g., a very fast nonradiative decay into the ground state 0. Compared to a three-level laser, the four-level laser offers the great advantage that population inversion is ideally achieved when just one atom is raised to the pumping level 3. Four-level lasers are thus more used, whenever possible, than three-level lasers. More recently, the so-called *quasi-three-level lasers* have also become a rather important laser category. In these lasers the energy level scheme is similar to that of a four-level laser, however levels 0 and 1 are now nondegenerate sublevels of the ground energy level. The population of the ground state is distributed in all the sublevels according to Boltzmann statistics, and therefore, at room temperature, some population is left in level 1.

The mechanism that allows atom excitation from the ground energy level into the excited pump level 3 is referred to as the *pumping process*. The rate of population of the upper state, 2, due to the pumping, $(\mathrm{d}N_2/\mathrm{d}t)_\mathrm{p}$, can be written as:

$$\left(\frac{\mathrm{d}N_2}{\mathrm{d}t}\right)_\mathrm{p} = W_\mathrm{p} N_\mathrm{g} \,, \tag{11.8}$$

where W_p is the pumping rate and N_g is the population of the ground state. For a four-level laser, one can assume N_g to be constant (and much larger than N_2). In this case one can write from (11.8) $(\mathrm{d}N_2/\mathrm{d}t)_\mathrm{p} = R_\mathrm{p}$, where $R_\mathrm{p} = W_\mathrm{p} N_\mathrm{g}$ is the pump rate per unit volume. The energy required for pumping is generally supplied either optically or electrically. The minimum pump power P_m needed to produce a given pump rate R_p is given by $P_\mathrm{m} = (\mathrm{d}N_2/\mathrm{d}t)_\mathrm{p} V h\nu_\mathrm{mp}$, where $(\mathrm{d}N_2/\mathrm{d}t)_\mathrm{p}$ is the number of atoms per unit volume and time raised to the upper laser level by the pumping process, V is the volume of the active medium, and ν_mp is the minimum pump frequency, given by the difference between the ground level and the upper laser level. For either electrical or optical pumping, the actual pump power P_p turns out to be larger than the minimum value P_m, so that one can define a pump efficiency $\eta_\mathrm{p} = P_\mathrm{m}/P_\mathrm{p}$. Therefore, the relation between the pump rate R_p and the actual pump power P_p is given by:

$$R_\mathrm{p} = \eta_\mathrm{p} \left(\frac{P_\mathrm{p}}{V h\nu_\mathrm{mp}}\right) . \tag{11.9}$$

Optical pumping. In optical pumping by an incoherent source, light from a powerful lamp (usually medium- to high-pressure Xe and Kr flashlamps for pulsed lasers, or high-pressure Kr lamps for CW lasers) is absorbed by the active medium. Solid-state gain media, such as Nd:YAG, and liquid lasers are particularly well suited to optical pumping: absorption lines are in fact very broad and this makes the absorption of the (broadband) light of the lamps efficient. The flashlamps generate significant heat into the material that must be dissipated by water cooling. The pumping source may, however, itself be a laser; in this case one speaks of *laser pumping*. Among the most used laser pump sources, we just mention semiconductor lasers (*diode laser pumping*), argon-ion laser and second-harmonic or third-harmonic generation of Nd lasers. With the advent of powerful and reliable semiconductor lasers, diode-laser pumping is nowadays commonplace for many solid-state and fiber lasers.

Electrical pumping. In gas lasers and semiconductor lasers, the excitation mechanism usually consists of an electrical current flow through the active medium. Gas lasers generally need to be electrically pumped because, due to the small width of their absorption lines, optical pumping would be very inefficient. In gas lasers, electrical pumping is achieved by passing an electric current either CW, at radio frequency or pulsed, directly through the gas itself. During the discharge, ions and free electrons are produced which acquire kinetic energy from the applied field and are able to excite neutral atoms via collisions: $A + e \to A^* + e$, where A^* denotes the atomic species in an excited state. Since electron impact excitation is a nonresonant process, it is a rather efficient pumping method for a gas. In some cases, the gas may contain two species, one of which is first excited by the discharge and then undergoes resonant energy transfer with the other via collision (an example is the HeNe laser). Electrical pumping in semiconductor lasers is achieved by flowing a large current density in a p–n or p–i–n diode. Though optical pumping can be used for semiconductor lasers, electrical pumping proves to be much more convenient.

Pumping processes different from optical or electrical pumping may also be employed in some special lasers; we just mention chemical pumping in chemical lasers, where population inversion is produced directly from an exothermic reaction.

Laser Oscillation

To obtain laser oscillation, the amplifying medium is placed between two mirrors, forming a laser cavity (Fig. 11.1). Light propagating along the cavity axis and passing through the pumped laser medium is reflected back through it, stimulating further emission in the same direction. This means that laser photons undergo multiple reflections within the cavity, being amplified at each pass through the laser medium. One mirror (total reflector) reflects almost all the incident light back through the laser medium while the other (partial reflector or output coupler) transmits a fraction $T_2 = 1 - R_2$, which constitutes the output laser beam. The combination of laser gain medium, pumping source and optical cavity forms a simple laser oscillator: if the amplification is large enough to overcame losses, i. e., when a *threshold condition* is reached, a single photon (which is always present due to quantum noise) can be amplified by several orders of magnitude to produce a huge number of coherently generated photons trapped inside the resonator. In addition, for an open resonator, only the photons which propagate in the paraxial direction of the resonator axis can reach the threshold for oscillation, so that an important property of the output laser beam is that it is *directional*. We can obtain the threshold condition for laser action by a simple argument based on gain/loss balance of light photons in one cavity round-trip. In fact, in one round-trip the photons pass twice through the gain medium and hence the round-trip gain of photons is $\exp(2\sigma Nl)$, where $N = N_2 - N_1$ is the population inversion, and l is the length of the active medium. On the other hand, the round-trip loss for photons can be written as $(1-T_1)(1-T_2)(1-L_i)^2$, where $T_1 = 1 - R_1$ and $T_2 = 1 - R_2$ are the power transmission of the two mirrors whereas L_i accounts for the one-way internal loss in the cavity (due to, e.g., scattering or diffraction loss for open resonators). The population inversion N_c needed to reach threshold (also called *critical inversion*) is simply obtained by equating the round-trip gain and loss. After setting $\gamma_1 = -\ln(1-T_1)$, $\gamma_2 = -\ln(1-T_2)$ and $\gamma_i = -\ln(1-L_i)$, the critical inversion turns out to be given by

$$N_c = \frac{\gamma}{\sigma l}\,, \qquad (11.10)$$

where

$$\gamma = \gamma_i + \frac{\gamma_1 + \gamma_2}{2} \qquad (11.11)$$

is referred to as the single-pass *logarithmic loss* of the cavity. Once the critical inversion is reached, oscillation builds up from spontaneous emission.

To calculate the pump rate R_{cp} needed to reach the threshold condition (the *critical pump rate*), let us consider the four-level laser (Fig. 11.4b). In steady-state conditions and in the absence of lasing, the population accumulated on the upper laser level can simply be calculated from a balance between the number of atoms pumped per unit volume and time in the upper laser level and the number of atoms that decay via radiative and nonradiative ways. Assuming $N_1 \simeq 0$, one obtains $\Delta N \simeq N_2 = R_p\tau$. When $N_2 = N_c$, from the previous equation and with the help of (11.10) the critical pump rate is obtained as $R_{cp} = \gamma/(\sigma\tau l)$. Note that the critical pump rate increases as $\sigma\tau$ decreases. Therefore the product $\sigma\tau$, which depends on the properties of the given transition, can be regarded as a figure of merit for a given laser.

Properties of Laser Beams

Laser radiation shows an extremely high degree of monochromaticity, coherence, directionality and brightness as compared to other noncoherent light sources [11.3].

The *monochromaticity* of laser radiation results from the circumstance that light oscillation sets in at one resonance frequency of the optical cavity, and owing to the balance between gain and loss in CW operation the line width $\Delta\nu_L$ of the oscillating mode is ultimately limited by quantum noise. Conversely, light from incoherent sources which exploit spontaneous emission (including LEDs) has a much lower degree of monochromaticity (as much as 11 orders of magnitude) since the spectral distribution of spontaneously emitted photons is broadened at around the atomic resonance frequency ν_0 owing to various broadening mechanisms. For laser radiation in the visible or near-infrared, line widths $\Delta\nu_L$ as low as a few Hz may, in fact, be achieved in frequency-stabilized laser sources.

The *coherence* of laser radiation refers to either *temporal* or *spatial* coherence. To define *spatial coherence*, let us consider two points P_1 and P_2 that, at a time $t = 0$, belong to the same wavefront, i. e., the phase difference between their electric fields at time $t = 0$ is zero. If the difference $\varphi_2(t) - \varphi_1(t)$ of their phases also remains zero at times $t > 0$, we say that there is perfect spatial coherence between the two points P_1 and P_2. In practice, for any point P_1, in order to have some degree of phase correlation, point P_2 must lie within some finite area around P_1, which is called the *coherence area*. The high degree of spatial coherence of laser radiation stems again from the fact that the spatial field distribution of the beam generated by stimulated emission is a mode of the optical resonator.

To define *temporal coherence* at a given point P, let us consider the phase difference $\varphi(t+\tau) - \varphi(t)$ for the electric field at P at times $t+\tau$ and t. For a given delay τ, if the phase difference is independent of time t, we say that there is temporal coherence over a time τ. If this occurs for any value of τ, the EM wave is said to have perfect temporal coherence. If, conversely, this occurs only for delays τ smaller than a given delay τ_0, the wave is said to have partial temporal coherence, and τ_0 is called the *coherence time* of the EM wave at point P. The concept of temporal coherence is closely related to that of monochromaticity; in fact, for a CW laser one can show that the coherence time τ_0 is related to the laser line width $\Delta\nu_L$ by the simple relation $\tau_0 \approx 1/\Delta\nu_L$. The high degree of temporal coherence of laser radiation is therefore due to its extreme degree of monochromaticity.

The *directionality* of the laser beam is due to the fact that the gain medium is placed inside an *open* optical resonator and, as a consequence, stimulated emission preferentially occurs in the direction orthogonal to the two cavity mirrors (Fig. 11.1), where feedback from the mirrors is most effective and diffraction losses are the smallest. The laser beam emitted from the output coupler shows a divergence angle which, in the ideal case, is limited by diffraction. From diffraction theory, the divergence angle for a monochromatic beam of wavelength λ turns out to be given by

$$\theta_d = \beta\left(\frac{\lambda}{2w}\right), \tag{11.12}$$

where β is a numerical coefficient (of order 1) whose specific value depends on the particular transverse field distribution and $2w$ is the diameter of the beam. For example, considering laser radiation in the visible ($\lambda = 500\,\mathrm{nm}$), the divergence of a laser beam of transverse diameter $2w \approx 1\,\mathrm{cm}$ is solely $\theta_d \approx \lambda/2w \simeq 5\times10^{-5}\,\mathrm{rad}$. This means that, after propagation for a length $L = 1\,\mathrm{km}$, the beam size is increased to solely $w + \theta_d L \approx 6\,\mathrm{cm}$!

The *brightness* of laser radiation is closely related to the directionality and stems from the capability of a laser oscillator to emit a high optical power in a small solid angle of space. For a given emitting source, of area ΔS, if P denotes the optical power delivered in a fractional solid angle $\Delta\Omega$ of space, the brightness of the emitter is defined as $B = P/(\Delta S\Delta\Omega)$. The brightness of a laser source, in which the solid angle of emission $\Delta\Omega$ is determined by its divergence angle θ_d, is given by

$$B = \left(\frac{2}{\beta\pi\lambda}\right)^2 P. \tag{11.13}$$

A laser of moderate power (e.g., a few milliwatts) has a brightness several orders of magnitude greater than that of the brightest conventional sources.

11.1.2 Interaction of Radiation with Atoms

Absorption and Stimulated Emission Rates

Consider a monochromatic EM wave of frequency ν incident on a two-level atom with a transition frequency $\nu_0 = (E_2 - E_1)/h$ close to ν. The calculation of the absorption and stimulated emission rates W_{12} and W_{21} can be done following a semiclassical approach, in which the atom is treated quantum mechanically whereas the EM wave classically [11.3,5–9]. If the atom is initially in its ground state (level 1), the incident wave may induce a transition to level 2 owing to the coupling of the electromagnetic field with the electric and magnetic dipole (and multipole) moments of the atom. The strongest interaction is usually that arising between the electric field $\boldsymbol{E}(t) = \boldsymbol{E}_0\cos(2\pi\nu t)$ of the EM and the electric dipole moment of the transition, which is defined by the matrix dipole element $\mu_{12} = \int u_2^*(\boldsymbol{r})e\boldsymbol{r}u_1(\boldsymbol{r})\mathrm{d}\boldsymbol{r}$, where $\boldsymbol{r}$ is the distance of the electron of the atom, with charge e, from its nucleus, and where $u_1(\boldsymbol{r})$ and $u_2(\boldsymbol{r})$ are the electronic eigenfunctions for the atomic energy levels 1 and 2 respectively. In this electric-dipole approximation, a perturbation theory, which assumes that the interaction between the EM wave and the atom is not disturbed by, e.g., collisions or other phenomena (including spontaneous emission), leads to the following expression for the absorption and stimulated emission rate

$$W_{12} = W_{21} = \frac{2\pi^2}{3n^2\epsilon_0 h^2}|\mu_{12}|^2\rho\delta(\nu - \nu_0), \tag{11.14}$$

where $\rho = n^2\epsilon_0 E_0^2/2$ is the energy density of the EM wave, n is the refractive index of the medium, and δ is the impulse Dirac function. In particular, for a plane wave of intensity $I = c\rho/n$, from (11.14) the following expression for the cross section $\sigma_{12} = W_{12}/F = h\nu W_{12}/I$ can be derived:

$$\sigma_{12} = \frac{2\pi^2}{3n\epsilon_0 ch}|\mu_{12}|^2\nu\delta(\nu - \nu_0). \tag{11.15}$$

Analogously, if initially the electron is on level 2, owing to the interaction with the EM wave there is a probability that the electron decays into state 1 emitting a photon by stimulated emission. The semiclassical perturbation calculation leads for the stimulated emission cross section $\sigma_{21} = h\nu W_{21}/I$ the same expression as that of σ_{12}, i. e., $\sigma_{21} = \sigma_{12}$, provided that the two levels are not degenerate.

The expression for $\sigma_{12} = \sigma_{21}$ as given by (11.15) is unphysical since it implies that the transition probability is zero for $\nu \neq \nu_0$ and ∞ for $\nu = \nu_0$. The inconsistency is removed by observing that the interaction of the monochromatic EM wave with the atom is not perfectly coherent, but it is disturbed by, e.g., collisions with other atoms or with lattice phonons, spontaneous emission and nonradiative decay of the atom. The effect of such disturbing interactions is to broaden the transition line of each atom in the ensemble, in the sense that in (11.15) the Dirac δ function $\delta(\nu-\nu_0)$ should be replaced by a new function $g(\nu-\nu_0)$, symmetric about $\nu=\nu_0$ with $\int g(\nu-\nu_0)d\nu=1$, which is generally given by the Lorentzian function (Fig. 11.5a):

$$g(\nu-\nu_0)=\frac{2}{\pi\Delta\nu_0}\frac{1}{1+[2(\nu-\nu_0)/\Delta\nu_0]^2}\,, \quad (11.16)$$

where $\Delta\nu_0$ depends on the particular broadening mechanism involved. Note that the full width at half-maximum (FWHM) of the Lorentzian function is simply $\Delta\nu_0$. The resonant character of the stimulated emission and absorption processes is maintained since the line broadening $\Delta\nu_0$ is typically several orders of magnitude smaller than ν_0 (e.g., in a low-pressure gas, for a transition in the visible one has $\nu_0\approx 5\times10^{14}$ Hz whereas $\Delta\nu_0\approx 10^6$–10^8 Hz). Since the aforementioned broadening mechanism act in the same way on each atom of the ensemble, it is referred to as a *homogeneous broadening* mechanism.

A somewhat different case occurs when the resonance frequencies ν_0' of the atoms in the ensemble are distributed around a central frequency ν_0 (*inhomogeneous broadening*). This distribution is described by the function $g^*(\nu_0'-\nu_0)$ such that $\int g^*(\nu_0'-\nu_0)\,d\nu_0'=1$ and $g^*(\nu_0'-\nu_0)\,d\nu_0'$ is the fractional part of the atoms in the ensemble whose resonance frequency lies in the interval $(\nu_0', \nu_0'+d\nu_0')$. For the most common mechanisms of inhomogeneous line broadening (such as Doppler broadening in a gas or local-field effects in ionic crystals or glasses) the distribution $g^*(\nu_0'-\nu_0)$ is given by a Gaussian function (Fig. 11.5b):

$$g^*(\nu_0'-\nu_0)=\frac{2}{\Delta\nu_0^*}\left(\frac{\ln 2}{\pi}\right)^{1/2} \times\exp\left(-\frac{4\ln 2(\nu_0'-\nu_0)^2}{\Delta\nu_0^{*2}}\right)\,, \quad (11.17)$$

where $\Delta\nu_0^*$ is the transition line width (FWHM), which depends on the particular broadening mechanism.

Taking into account the simultaneous presence of both homogeneous and inhomogeneous broadening mechanisms, one can show that the cross section $\sigma=\sigma_{12}=\sigma_{21}$ for stimulated emission and absorption assumes the most general form:

$$\sigma=\frac{2\pi^2}{3n\epsilon_0 ch}|\mu_{12}|^2\nu g_t(\nu-\nu_0)\,, \quad (11.18)$$

where the total line shape g_t is given by the convolution:

$$g_t(\nu-\nu_0)=\int g^*(\nu_0'-\nu_0)g(\nu-\nu_0')\,d\nu_0'\,. \quad (11.19)$$

If we consider an ensemble of atoms and indicate by N_1 and N_2 the populations of atoms in the energy levels 1 and 2, as discussed in the previous section a small-signal EM wave of frequency ν which propagates inside the medium experiences amplification for $N_2>N_1$ or absorption for $N_2<N_1$. In the former case, one can introduce the absorption coefficient per unit length $\alpha(\nu)=\sigma(\nu-\nu_0)(N_1-N_2)$, whereas in the latter case one can introduce the gain coefficient per unit length $g(\nu)=\sigma(\nu-\nu_0)(N_2-N_1)$. For a weak-signal wave, so that saturation can be neglected, the intensity of the propagating wave is thus exponentially attenuated or amplified along the propagation direction according to $I(z)=I(0)\exp(-\alpha z)$ or $I(z)=I(0)\exp(gz)$, respectively.

These considerations apply to atomic or molecular transitions which are *electric-dipole* allowed, i.e., for which the dipole matrix element μ_{12} of the transition does not vanish. Transitions between atomic or molecular energy levels with the same parity (e.g., between s states in an atom) are electric-dipole forbidden. This does not mean however that the atom or molecule cannot pass from level 1 to level 2 when interacting with an EM wave; in this case the transition can occur owing to the interaction of the EM wave with, e.g., the magnetic dipole moment (or the electric quadrupole moment) of the transition, though the strength of the cross section describing this process is much smaller than that of an allowed electric dipole transition. For an

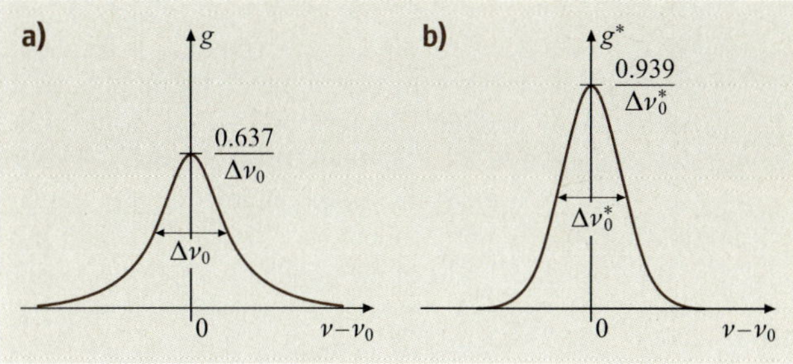

Fig. 11.5 Lorentzian (*left side*) and Gaussian (*right side*) line shapes corresponding to homogeneous and inhomogeneous transition broadening, respectively

electric dipole transition the absorption transition rate W_e is proportional to $\sim |\mu_e E_0|^2$, where μ_e is the electric dipole moment of the transition, E_0 the amplitude of the electric field of the EM wave and $-E_0\mu_e$ is the classical energy of an electric dipole in an external field E_0. Likewise, for a magnetic dipole interaction, the transition rate W_m is proportional to $\sim |\mu_m B_0|^2$, where μ_m is the magnetic dipole of the transition, B_0 is the amplitude of the magnetic field of the EM wave, and $-B_0\mu_m$ is the classical energy of a magnetic dipole in an external magnetic field B_0. By approximating $\mu_e \approx ea$ and $\mu_m \approx \beta$, where $a = 0.529\times10^{-10}$ m is the Bohr radius and $\beta = 9.27\times10^{-24}$ Am^2 is the Bohr magneton, one then obtains:

$$\left(\frac{W_e}{W_m}\right) = \left(\frac{eaE_0}{\beta B_0}\right)^2 = \left(\frac{eac}{\beta}\right)^2 \simeq 10^5 . \qquad (11.20)$$

Spontaneous Emission

Spontaneous emission is the phenomenon by which an atom in an excited energy level tends to decay toward the ground state (even in absence of any external perturbation) by emitting an EM wave, i. e., one photon. A correct explanation of spontaneous emission requires a quantum electrodynamic approach in which both the atom and the EM field are quantized. In practice, due to quantization of the EM field, the mean values of both E^2 and H^2 fields are nonvanishing even in the absence of a (classical) EM wave (zero-point field fluctuations). Such intrinsic fluctuations always perturb the atom in an excited state and trigger its decay toward a lower energy level with the emission of one photon with frequency ν close to the atomic transition frequency ν_0. Atomic decay due to spontaneous emission follows an exponential law with a time constant τ_s, which is referred to as the spontaneous emission lifetime [see (11.2)]. The quantum electrodynamic calculation of τ_s in the electric dipole approximation and for an atom placed in an optical cavity was done by *Weisskopf* and *Wigner* [11.10]. A simple and rigorous calculation of τ_s may be derived using an elegant thermodynamic argument, which was proposed by Einstein well before any development of the quantum electrodynamics [11.11]. Assume that the material is placed inside a black-body cavity whose walls are kept at a constant temperature T. Once the thermodynamic equilibrium is reached, the spectral EM energy density distribution $\rho(\nu)$ inside the cavity is given by the Planck distribution

$$\rho(\nu) = \left(\frac{8\pi\nu^2}{c^3}\right)\frac{h\nu}{\exp(h\nu/kT) - 1} \qquad (11.21)$$

where $\rho(\nu)$ is such that $\rho(\nu)\mathrm{d}\nu$ is the EM energy per unit volume of the cavity associated to modes with frequencies in the interval $(\nu, \nu + \mathrm{d}\nu)$. We note that in (11.21) the factor $(8\pi\nu^2/c^3)$ represents the density of EM modes per unit volume of the cavity, whereas the term $h\nu/[\exp(h\nu/k_BT) - 1]$ is the energy per mode. The populations N_1^e and N_2^e in the atomic levels 1 and 2 at thermal equilibrium are described by a Boltzmann statistics (11.7); on the other hand, at steady state the number of excitation per unit time from level 1 to level 2 due to the absorption of black-body radiation should balance the number of decays per unit time from level 2 to level 1 due to both stimulated emission and spontaneous emission, i. e., $W_{12}N_1^e = W_{21}N_2^e + N_2^e/\tau_s$. One then obtains $1/\tau_s = W_{12}\exp(h\nu_0/kT) - W_{21}$. For a spectrally broad radiation (such as black-body radiation) one can write $W_{12} = W_{21} = \int \mathrm{d}\nu c\sigma(\nu - \nu_0)\rho(\nu)/nh\nu = 2\pi^2|\mu_{12}|^2\rho(\nu_0)/(3n^2\epsilon_0 h^2)$, where we used (11.15). Using this expression for $W_{12} = W_{21}$ and (11.21) for $\rho(\nu_0)$, one finally obtains

$$\tau_s = \frac{3h\epsilon_0 c^3}{16\pi^3\nu_0^3 n|\mu_{12}|^2} . \qquad (11.22)$$

Using a similar thermodynamic argument, one can also show that the spectrum of the photon emitted by spontaneous emission is given by the line shape $g_t(\nu - \nu_0)$ of the transition (11.19), i. e., the probability that the photon emitted by spontaneous emission has a frequency in the range $(\nu, \nu + \mathrm{d}\nu)$ is given by $g_t(\nu - \nu_0)\mathrm{d}\nu$. This property is very interesting since it enables one to measure the transition line shape $g_t(\nu - \nu_0)$ simply in an emission experiment by passing the spontaneously emitted light through a spectrometer of sufficiently high resolution.

To estimate the radiative lifetime τ_s, let us consider for instance an electric-dipole-allowed transition at a frequency corresponding to the middle of the visible range ($\lambda_0 = c/\nu_0 = 500$ nm). Assuming $|\mu_{12}| \simeq ea$ (where $a \simeq 0.1$ nm is the atomic radius), from (11.22) one obtains $\tau_s \simeq 10$ ns. For a magnetic dipole transition $1/\tau_s$ turns out to be 10^5 times smaller, i. e., $\tau_s \simeq 1$ ms. Note that, according to (11.22), τ_s decreases as the cube of the transition frequency, so that the importance of spontaneous emission increases rapidly with frequency. This implies that, as spontaneous emission is often negligible in the middle- to far-infrared, where nonradiative decay dominates, in the X-ray region (e.g., $\lambda_0 < 5$ nm) spontaneous emission is the dominant decay process and τ_s becomes exceedingly short (10–100 fs). Such a short lifetime represents a major challenge for achieving a population inversion in X-ray lasers. It should finally be noted that the rate of spontaneous emission as given by (11.22) can be either enhanced or reduced when the atom radiates in a dielectric cavity whose density of EM modes (i.e., the number of resonant cavity modes per unit frequency and per unit volume) is changed from the value $(8\pi\nu^2/c^3)$ due to, e.g., tight spatial confinement, such as in *microlasers*. For instance, spontaneous decay is fully inhibited for an atom placed inside a photonic crystal whenever the atomic transition frequency ν_0 falls inside a bandgap of the photonic crystal.

Line Broadening Mechanisms

Homogeneous broadening. A line broadening mechanism is said to be homogeneous when it broadens the line of each atom in the same way. In this case the line shape of the single-atom cross section and that of the overall absorption cross section are identical. There are two main homogeneous broadening mechanisms: *collisional broadening* and *natural broadening*.

Collisional broadening occurs in a gas, where the atom may collide with other atoms, ions, free electrons, or with the walls of the container, as well as in a solid, where collisions are due to the interaction of the atom with the lattice phonons. During a process of absorption or stimulated emission of a two-level atom with a monochromatic EM wave, the collisions interrupt the coherent interaction of the EM with the atom; if we write the electronic wave function ψ of the atom during the transition as $\psi = c_1(t)u_1(\boldsymbol{r})\exp[-\mathrm{i}E_1t/\hbar] + c_2(t)u_2(\boldsymbol{r})\exp[-\mathrm{i}E_2t/\hbar]$, assuming that the collision does not induce a decay, it simply introduces a random and rather instantaneous relative phase jump of the coefficients c_1 and c_2, and thus of the oscillating part of the atomic dipole $\mu = \int -e\boldsymbol{r}|\psi(\boldsymbol{r},t)|^2\,\mathrm{d}\boldsymbol{r}$, which is proportional to $c_1c_2^*$. Since in the electric-dipole interaction the radiation–atom coupling is expressed by the energy term $-\mu_e\boldsymbol{E}$, a different but equivalent picture is to assume that it is the phase of the electric field that shows random phase jumps rather than the atomic dipole moment (Fig. 11.6). Therefore we can consider the case of coherent (i.e., not disturbed) interaction of the two-level atom with an EM wave which is not monochromatic but whose spectral content is broadened at around ν owing to the phase jumps. Let $I(\nu') = Ig(\nu'-\nu)$ be the spectral intensity distribution of the EM wave with random phase jumps (Fig. 11.6), where I is the total field intensity and g its spectral shape (with $\int g(\nu'-\nu)\,\mathrm{d}\nu' = 1$). For each fractional spectral component $Ig(\nu'-\nu)\,\mathrm{d}\nu'$ of the EM wave, we may introduce an elemental absorption transition rate $\mathrm{d}W_{12}$ given, according to (11.14), by $\mathrm{d}W_{12} = [2\pi^2|\mu_{12}|^2/(3n\epsilon_0ch^2)]\delta(\nu'-\nu_0)Ig(\nu'-\nu)\,\mathrm{d}\nu'$. The total transition rate, $W_{12} = \int \mathrm{d}W_{12}$, is then given

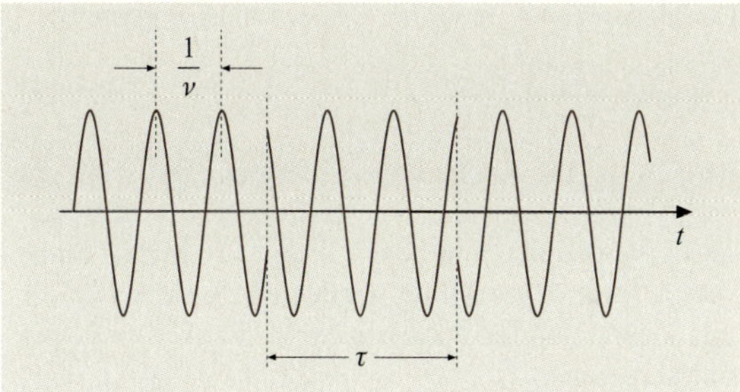

Fig. 11.6 Schematic of the sinusoidal electric field of frequency ν showing random phase jumps at time intervals τ

by $W_{12} = \sigma_{12}(\nu-\nu_0)I/h\nu$, where the cross section σ_{12} is given by (11.15) after replacing the delta function $\delta(\nu-\nu_0)$ with $g(\nu-\nu_0)$. To calculate the line shape g, let us assume that the distribution of the time interval τ between two successive collisions be described by an exponential probability density, $p(\tau) = [\exp(-\tau/\tau_c)]/\tau_c$, where τ_c is the mean value of τ. According to the Wiener–Kintchine theorem, the spectrum g can be calculated as the Fourier transform of the autocorrelation function of the sinusoidal field with phase jumps at time intervals τ (Fig. 11.6). This yields a Lorentzian line shape (11.16) with a FWHM of $\Delta\nu_0$, related to the mean value of collision time τ_c by the equation

$$\Delta\nu_0 = \frac{1}{\pi\tau_c}\,. \tag{11.23}$$

For instance, for an atomic or molecular gas at pressure p and absolute temperature T, from kinetic theory and employing the hard-sphere model of a gas, one has $\tau_c = (2Mk_BT/3)^{1/2}[1/(8\pi pa^2)]$, where M is the atomic mass and a its radius. Note that τ_c is inversely proportional, and hence $\Delta\nu_0$ is directly proportional, to the gas pressure p. As a rough rule of thumb, we can say that, for any atom in a gas, collisions contribute to the line broadening by an amount $(\Delta\nu_{coll}/p) \approx 1\,\mathrm{MHz/Torr}$.

For an atom or an ion in a crystal, collisions occur with lattice phonons. Since the number of phonons in a given lattice vibration is strongly dependent on the lattice temperature T, the corresponding line broadening $\Delta\nu_0$ increases with increasing values of T. For instance, in the Nd:YAG laser Nd^{3+} ions are hosted in the YAG crystal, and collision broadening increases from $\Delta\nu_0 \approx 126\,\mathrm{GHz}$ at room temperature ($T = 300\,\mathrm{K}$) to about 250 GHz at $T = 400\,\mathrm{K}$.

A second homogeneous line broadening mechanism originates from spontaneous emission and is referred to as natural (or intrinsic) line broadening. It can be shown that the natural broadening is again described by a Lorentzian function with a FWHM of $\Delta\nu_0$ given by

$$\Delta\nu_0 = \frac{1}{2\pi\tau_s}\,, \tag{11.24}$$

where τ_s is the spontaneous lifetime. For, e.g., an electric dipole transition at the center of the visible ($\lambda \approx 500\,\mathrm{nm}$), one has $\Delta\nu_{nat} \simeq 16\,\mathrm{MHz}$. Since $\tau_s \approx 1/\nu^3$, the natural line width rapidly increases for transitions at shorter wavelengths (to the UV or X-ray spectral regions).

We finally note that, when the two aforementioned homogeneous line broadening mechanisms act simultaneously, the overall line shape is obtained from the convolution of the two corresponding Lorentzian functions. One then obtains a Lorentzian function with a FWHM given by $\Delta\nu_0 = \Delta\nu_{coll} + \Delta\nu_{nat}$.

Inhomogeneous broadening. A line broadening mechanism is said to be inhomogeneous when it distributes the atomic resonance frequencies over some spectral range. A first case of inhomogeneous broadening is that of ions in ionic crystals or glasses, where the local crystal field induces, via the Stark effect, a local variation of the energy-level separation of the ion. For random local field variations, the corresponding distribution of the transition frequencies $g^*(\nu - \nu_0)$ turns out to be given by a Gaussian function (11.17) with a line width $\Delta\nu_0^*$ that depends on the amount of field inhomogeneity within the crystal or glass.

A second inhomogeneous mechanism, typical of gases, arises from atomic motion and is referred to as *Doppler broadening*. In fact, due to the motion of the atom, the frequency ν' of the EM wave as seen in the rest frame of the atom is shifted as compared to the frequency ν of the wave in the laboratory reference frame according to the relation $\nu' = \nu[1 - (v_z/c)]$, where v_z is the component of the atomic velocity in the propagation z-direction of the EM wave. From the point of view of atom–radiation interaction, this shift is equivalent to a change of the resonance frequency of the atom rather than to a change of the EM frequency. Taking into account the Maxwellian distribution of molecular velocities in a gas, one can show that the distribution of the transition frequencies $g^*(\nu' - \nu_0)$ is given again by a Gaussian function (11.17) with a line width $\Delta\nu_0^*$ given by

$$\Delta\nu_0^* = 2\nu_0 \left(\frac{2k_B T \ln 2}{Mc^2}\right)^{1/2}\,, \tag{11.25}$$

where M is the atomic (or molecular) mass and T the gas temperature. As an example, for the $\lambda = 632.8\,\mathrm{nm}$ line of the HeNe laser, at $T = 300\,\mathrm{K}$ and using the appropriate mass for Ne, one obtains $\Delta\nu_0^* \simeq 1.7\,\mathrm{GHz}$. As a general rule, Doppler broadening in a gas is usually larger than collisional broadening for a gas pressure lower than the atmospheric pressure; collisional broadening, in turn, is usually larger than the natural broadening.

Nonradiative Decay

Besides decaying via radiative emission, an atom in an excited state may decay toward a lower-lying level in a nonradiative way. A first mechanism of nonradiative decay arises from collisions and is called *collisional deactivation*. In this case, for a liquid or a gas, the transition energy is released as excitation and/or kinetic energy of the colliding species, or it is transferred to the container walls; for an ion in a crystal or in a glass the deactivation occurs through the interaction with lattice phonons or glass vibrational modes. When, e.g., the excitation energy of an atomic excited species B^* in a gas is released as a kinetic energy of the species A, the collisional deactivation process occurs via the superelastic collision $B^* + A \to B + A + \Delta E$, where ΔE, the excitation energy to be released, is left as kinetic energy of the colliding partners. When the electronic energy of species B^* is released in the form of the internal energy of species A, we have instead $B^* + A \to B + A^* + \Delta E$, where now ΔE is the difference between the internal energies of the two colliding species. In this latter case, deactivation process is efficient provided that ΔE is appreciably smaller than the thermal energy $k_B T$ of colliding species. A simple description of the nonradiative decay of a given species in an excited state is expressed by a nonradiative lifetime τ_{nr} such that N_2/τ_{nr} is the number of atoms, per unit volume and time, that decay owing to the deactivation process.

Nonradiative decay acts in combination with spontaneous emission and, according to (11.3), the overall lifetime τ of an excited state is given by: $\tau = (1/\tau_{nr} + 1/\tau_s)^{-1}$.

Concluding Remarks

From the preceding discussions, we can say that the most important material parameters of interest for a laser are the transition wavelength λ, the transition cross section at the peak σ_p, the lifetime of the upper laser level τ, and the line width $\Delta\nu_0$ of the transition line shape. These parameters, for the most common gas, liquid and solid-state lasers, are summarized in Table 11.1. We note that, as compared to gas and liquid lasers, the cross sections for solid-state lasers (Nd:YAG, Nd:glass, and Ti^{3+}:Al_2O_3) are relatively small and, correspondingly,

Table 11.1 Main parameters for transitions in most common gas, liquid and solid-state lasers

Transition	λ (nm)	σ_p (cm^2)	τ (μs)	$\Delta\nu_0$
HeNe	632.8	5.8×10^{-13}	30×10^{-3}	1.7 GHz
Ar^+	514.5	2.5×10^{-13}	6×10^{-3}	3.5 GHz
Nd:YAG	1064	2.8×10^{-19}	230	120 GHz
Nd:glass	1054	4×10^{-20}	300	5.4 THz
Rhodamine 6 G	570	3.2×10^{-16}	5.5×10^{-3}	46 THz
Ti^{3+}:Al_2O_3	790	4×10^{-19}	3.9	100 THz

the lifetimes relatively long because in these lasers the transitions are electric-dipole forbidden (or weakly allowed). We note also that the line widths of gas lasers are much smaller than those of solid-state or dye lasers.

11.1.3 Laser Resonators and Modes

As discussed in the introductory section, in a laser oscillator the inverted amplifying medium is placed inside a *laser resonator* or *laser cavity*, which can be viewed as a *trapping box* for light radiation capable of sustaining stationary (i. e., monochromatic) or weakly damped electromagnetic field configurations at some selected optical frequencies [11.3, 5]. Such EM field configurations and the corresponding optical frequencies are called *cavity modes* and *resonance frequencies*, respectively. The most widely used resonators for lasers are open cavities, composed of at least by two plane or spherical mirrors of circular shape (spherical resonators), placed in front each other at some distance L. Except for microlasers, in ordinary lasers the resonator length L is typically much larger than the oscillating wavelength λ, ranging from a few centimeters to a few tens of centimeters, whereas mirror dimensions range from a fraction of centimeter to a few centimeters. The optical cavity is open to reduce drastically the number of modes that can oscillate with low loss. In fact, if the cavity were closed, the number N of resonant modes that might oscillate, i. e., the number of modes whose resonance frequency falls within the gainline of the active medium, is approximately given by $N \approx (8\pi\nu^2/c^3)V\Delta\nu_0$, where $(8\pi\nu^2/c^3)$ is the density of modes, $\Delta\nu_0$ the line width of the gain medium, and V is the cavity volume. Note that V is usually several orders of magnitude larger than λ^3 at optical wavelengths. To estimate N, let us consider an active medium with a narrow line width, such as the $\lambda = 633$ nm transition of the HeNe laser ($\Delta\nu_0^* = 1.7$ GHz). Assume a resonator length $L = 50$ cm closed by a lateral cylinder with diameter $2a = 3$ mm. The cavity volume is then $V = \pi a^2 L$, and the number of cavity modes that fall under the gainline of the HeNe laser is $N \approx 1.2\times10^9$. In the open optical resonant cavities only those modes traveling nearly parallel to the resonator axis experience low losses to allow for laser oscillation. Therefore, the oscillating modes are expected to show a field distribution mostly confined around the optical axis of the resonator and propagating paraxially along it, making the output laser beam highly directional. These cavity modes and corresponding resonance frequencies depend upon three integer numbers n, m and l, which are referred to as *mode indices*. The latter two indices m and l (transverse mode indices) determine the transverse field distribution (i. e., in a plane orthogonal to the paraxial resonator axis) of the mode, whereas the former index n (the longitudinal mode index) determines the longitudinal field distribution (i. e., along the resonator axis) of the mode and gives, in particular, the number of longitudinal nodes of the standing wave. For spherical mirrors with sufficiently wide apertures, transverse modes are expressed by Gauss–Hermite or Gauss–Laguerre functions, depending on rectangular or circular boundary conditions. In particular, the leading-order mode, corresponding to transverse mode indices $m = l = 0$, is a *Gaussian beam* and represents the most common transverse field distribution of any output laser beam. For this reason the study of laser modes is closely related to the propagation properties of Gaussian (or Gauss–Hermite) beams.

Gaussian Beams

The electric field of a monochromatic (and uniformly polarized) light wave propagating at a small angle (i. e., paraxially) along the z-direction of an xyz cartesian system of coordinates can be described as follows:

$$E(x, y, z, t) = E_0 u(x, y, z) e^{i(\omega t - kz)} + \text{c.c.}\,, \quad (11.26)$$

where $\omega = 2\pi\nu$, ν is the *optical frequency*, $k = 2\pi/\lambda$ is the wavenumber, λ is the optical wavelength, $u(x, y, z)$ is the complex field envelope obeying the so-called *paraxial wave equation*, which in case of free-space propagation reads

$$\frac{\partial^2 u}{\partial x^2}+\frac{\partial^2 u}{\partial y^2}-2\mathrm{i}k\frac{\partial u}{\partial z}=0\,. \tag{11.27}$$

Among the solutions of (11.27) that retain their functional form during propagation, the *fundamental Gaussian beam* solution turns out to be particularly suited to describe laser beams both inside and outside the resonator [11.12]. A Gaussian beam is a solution of the paraxial equation (11.27) of the form:

$$u(x,y,z)=\frac{w_0}{w(z)}\exp\left(-\frac{x^2+y^2}{w^2(z)}\right)\times\exp\left(-\mathrm{i}k\frac{x^2+y^2}{2R(z)}\right)\times\exp[\mathrm{i}\varphi(z)]. \tag{11.28}$$

In the preceeding equation, $w(z)$, $R(z)$ and $\varphi(z)$ are given by

$$w(z)=w_0\sqrt{1+\left(\frac{z}{z_\mathrm{R}}\right)^2}\,, \tag{11.29}$$

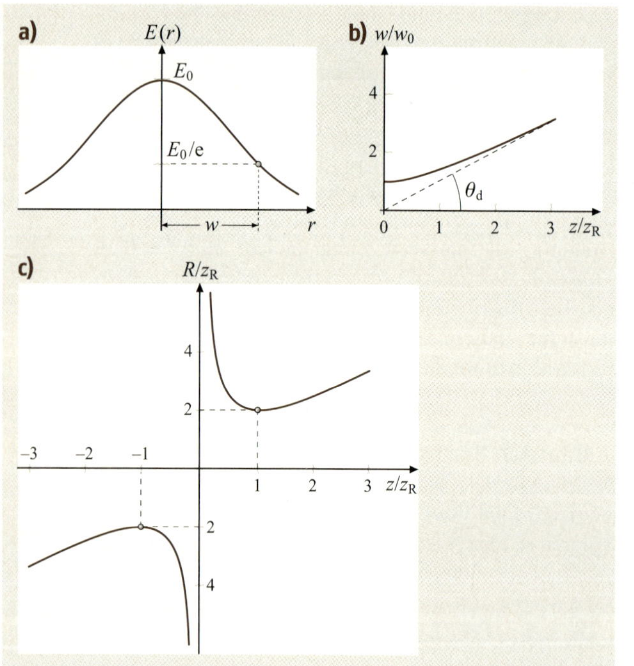

Fig. 11.7a–c The fundamental Gaussian beam in free-space propagation: **(a)** the Gaussian transverse amplitude profile, being $r=(x^2+y^2)^{1/2}$; **(b)** the beam spot size $w(z)$; **(c)** the wavefront radius of curvature $R(z)$

$$R(z)=z\left[1+\left(\frac{z_\mathrm{R}}{z}\right)^2\right], \tag{11.30}$$

$$\varphi(z)=\tan^{-1}\left(\frac{z}{z_\mathrm{R}}\right), \tag{11.31}$$

where $z_\mathrm{R}=\pi w_0^2/\lambda$ is a parameter referred to as the *Rayleigh range*. Note that $u(x,y,z)$ is given by the product of three terms: an amplitude factor with a transverse Gaussian distribution, $(w_0/w)\exp[-(x^2+y^2)/w^2]$ (see Fig. 11.7a); a transverse phase factor, $\exp[-\mathrm{i}k(x^2+y^2)/(2R)]$; and a longitudinal phase factor $\exp(\mathrm{i}\varphi)$. The amplitude factor in (11.28) shows that, while propagating, the beam intensity distribution retains its shape, but its transverse size w, which is called the *beam spot size*, changes along the propagation direction z according to (11.29). Note that $w(z)$ is a symmetric function of z with a minimum spot size $w=w_0$ at the plane $z=0$, which is hence referred to as the *beam waist* (Fig. 11.7b). For $z=z_\mathrm{R}$, one has $w=\sqrt{2}w_0$ so that the Rayleigh range z_R represents the distance from the beam waist at which the beam spot size increases by a factor $\sqrt{2}$. At large distances (i. e., for $z\gg z_\mathrm{R}$), w increases linearly with z, according to $w\approx(w_0/z_\mathrm{R})z$. Hence we can define a beam divergence due to diffraction as $\theta_\mathrm{d}=\lim_{z\to\infty}w(z)/z$, obtaining

$$\theta_\mathrm{d}=\frac{\lambda}{\pi w_0}\,. \tag{11.32}$$

The transverse phase factor in (11.28) has the same form as that for a spherical wave in the paraxial approximation, R playing the role of the radius of curvature of the spherical wavefront. Thus, we can say that a Gaussian beam has an approximately spherical wavefront with a radius of curvature varying along propagation according to (11.30). Note that $R(z)$ is an odd function of z (Fig. 11.7c), showing a minimum $R_\mathrm{min}=2z_\mathrm{R}$ at $z=z_\mathrm{R}$; for $z\gg z_\mathrm{R}$, R increases linearly with z, whereas one has $R\to\infty$ as $z=0$. Thus the wavefront is flat at $z=0$ and, at large distances, its radius increases linearly with z just as for a spherical wave. The longitudinal phase factor φ provides, in addition to the usual phase shift $-kz$ of plane waves, a longitudinal phase shift (sometimes referred to as the *Gouy phase shift*), slowly varying with z from $-(\pi/2)$ to $(\pi/2)$ on going from $z\ll z_\mathrm{R}$ to $z\gg z_\mathrm{R}$.

An important parameter of a Gaussian beam at a given propagation plane z is the so-called *complex-q parameter*, defined by the relation:

$$\frac{1}{q}=\frac{1}{R}-\mathrm{i}\frac{\lambda}{\pi w^2}\,. \tag{11.33}$$

It can be shown that, for a Gaussian beam propagating in free space, the q parameter changes along propagation according to

$$q(z) = z + \mathrm{i}z_\mathrm{R} \,. \tag{11.34}$$

Equations (11.29) and (11.30) can indeed be obtained upon inserting the parameter q from (11.33) into (11.34) and then separating the real and imaginary part of the resulting equation.

The fundamental Gaussian beam described above belongs to a more general set of eigensolutions of (11.27) which can be written as the product of an Hermite polynomial with a Gaussian function. These are known as *Hermite–Gaussian beams*, and assume the following form [11.3, 5]

$$\begin{aligned} u_{l,m}(x,y,z) = & \frac{w_0}{w(z)} H_l\left(\frac{\sqrt{2}x}{w(z)}\right) H_\mathrm{m}\left(\frac{\sqrt{2}y}{w(z)}\right) \\ & \times \exp\left(-\frac{x^2+y^2}{w^2(z)}\right) \\ & \times \exp\left[-\mathrm{i}k\frac{x^2+y^2}{2R(z)}\right] \\ & \times \exp[\mathrm{i}(1+l+m)\varphi(z)] \,, \end{aligned} \tag{11.35}$$

where $w(z)$, $R(z)$ and $\varphi(z)$ are given by (11.29, 30, 31), respectively, and H_l, H_m are Hermite polynomials of order l and m. The lowest-order Hermite–Gaussian beam is obtained by setting $l = m = 0$ in (11.35). These solutions are often referred to as TEM_{lm} beams, where TEM stands for transverse electric magnetic: within the paraxial approximation, both electric and magnetic fields of the EM wave are, in fact, approximately transverse to the z-direction. Note that, for a TEM_{lm} beam, the number of zeros of the field along the x- and y-directions is given by the subscripts l and m, respectively, and therefore the intensity distribution of the TEM_{lm} beam consists of $l+1$ lobes in the horizontal direction and $m+1$ lobes in the vertical direction (Fig. 11.8).

As a final remark, it should be noted that Gauss–Hermite beams maintain their functional shape as they propagate along *an arbitrary* paraxial optical system, described by a paraxial ray matrix $ABCD$. Beam propagation, in this case, is simply ruled by an algebraic relation for the complex q parameter of the Gaussian beam, which is known as the $ABCD$ *law*. In fact, if $z = z_1$ and $z = z_2$ are the input and output planes of the $ABCD$ paraxial optical system, a Gauss–Hermite field distribution $u(x, y, z_1) = H_l(\sqrt{2}x/w_1)H_\mathrm{m}(\sqrt{2}y/w_1)\exp[-\mathrm{i}k(x^2+y^2)/(2q_1)]$ at the input plane $z = z_1$ is transformed, at the output plane

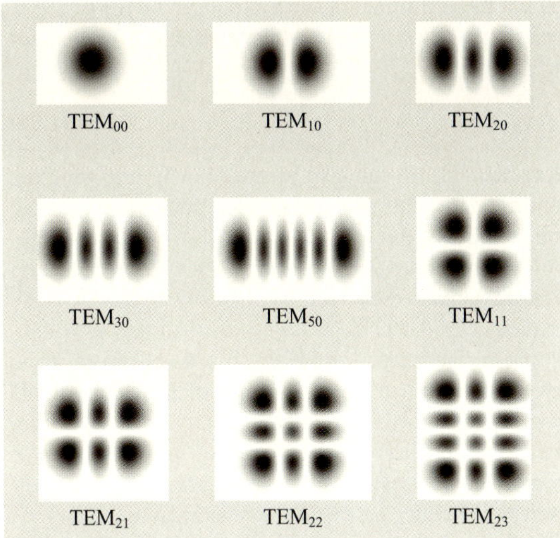

Fig. 11.8 Grey scale intensity patterns of a few low-order Hermite–Gaussian modes

$z = z_2$, into the distribution

$$\begin{aligned} u(x,y,z_2) = & \left(\frac{1}{A+(B/q_1)}\right)^{1+l+m} \\ & \times H_l\left(\frac{\sqrt{2}x}{w_2}\right) H_\mathrm{m}\left(\frac{\sqrt{2}y}{w_2}\right) \\ & \times \exp[-\mathrm{i}k(x^2+y^2)/(2q_2)] \,, \end{aligned} \tag{11.36}$$

where the values q_1 and q_2 of the complex-q parameter at the input ($z = z_1$) and output ($z = z_2$) planes are related by the so-called $ABCD$ *law*

$$q_2 = \frac{Aq_1 + B}{Cq_1 + D} \,. \tag{11.37}$$

As a particular case, note that for free-space propagation from $z_1 = 0$ to $z_2 = z$ one has $A = 1$, $B = z$, $C = 0$, and $D = 1$, so that the $ABCD$ law (11.37) yields (11.34).

Optical Resonators: Introductory Concepts

The simplest optical resonator is the *plane-parallel* or *Fabry–Pérot* resonator, consisting of two plane metallic or dielectric mirrors set parallel to one another [11.13]. At first approximation, the modes of this resonator can be considered as the superposition of two plane EM waves propagating in opposite directions along the cavity axis (Fig. 11.9a). In this approximation, resonance frequencies can be readily obtained by imposing the condition that the cavity length L must be an integer multiple of half-wavelengths, i.e., $L = n(\lambda/2)$, where n is a positive integer. This is a necessary condition for the electric field of the EM standing wave to be zero on the two, e.g. metallic, mirrors. It follows that the resonant frequencies are given by

$$\nu_n = n\left(\frac{c}{2L}\right) . \tag{11.38}$$

Note that (11.38) can also be obtained by imposing the condition that the phase shift of a plane wave due to one round-trip through the cavity must equal an integral number times 2π, i. e., $2kL = 2n\pi$. This condition is readily obtained by a self-consistency argument: if the frequency of the plane wave is equal to that of a cavity mode, the phase shift after one round-trip must be zero (apart from an integer multiple of 2π). Only in this case, in fact, do the amplitudes at any arbitrary point, due to successive reflections, add up in phase to give an appreciable total field. According to (11.38) the frequency difference between two consecutive modes, i. e., differing by 1 in the longitudinal mode index n, is given by

$$\Delta\nu = \frac{c}{2L} . \tag{11.39}$$

This difference is referred to as the *frequency difference* between two consecutive longitudinal (or *axial*) modes. Note that, since the number n indicates the number of half-wavelengths of the mode along the resonator axis, the two consecutive modes have a different longitudinal pattern.

A more general class of laser resonators is the one made of *spherical resonators*, which are formed by two spherical mirrors of radius of curvatures R_1 and R_2, either concave ($R > 0$) or convex ($R < 0$), placed at some arbitrary distance L (Fig. 11.9b). These resonators can be divided into two categories: *stable* resonators and *unstable* resonators. A resonator is said to be unstable when, in bouncing back and forth between the two mirrors, an arbitrary paraxial ray diverges indefinitely away from the resonator axis, either radially or angularly. Conversely, a resonator whose paraxial rays remain bounded is said to be stable. An example of an unstable resonator is shown in Fig. 11.9c. Among stable spherical resonators, symmetric resonators (i. e., having $R_1 = R_2$) are of particular importance; the *confocal* resonator, in which the two spherical mirrors have the focus in common ($R_1 = R_2 = L$), is a noteworthy example of a spherical symmetric resonator. Another important scheme for laser cavities is that employing a *ring* resonator, where the path of the optical rays is arranged to form a closed loop (Fig. 11.9d). Also in this case the resonant frequencies can be obtained by imposing the condition that the total phase shift along the ring path must equal an integer number of 2π. The expression for the resonance frequencies of longitudinal (or axial) modes is thus given by

$$\nu_n = n\left(\frac{c}{L_\mathrm{p}}\right) \tag{11.40}$$

where L_p is the length of the closed loop path. In general, a standing-wave pattern may be formed in ring resonators, because the beam can propagate either clockwise or counterclockwise along the loop. Anyway, unidirectional ring resonators can be realized by means of optical diodes inserted along the beam path.

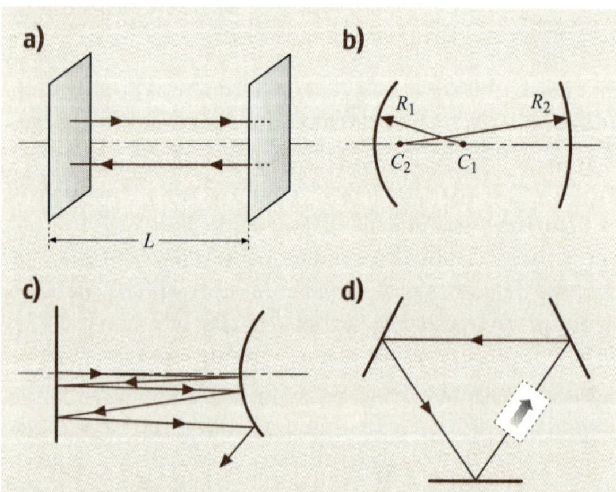

Fig. 11.9 **(a)** Plane-parallel resonator; **(b)** Spherical two-mirror resonator; **(c)** An unstable resonator; **(d)** A unidirectional ring resonator

Stability Condition

In general, a laser cavity can be viewed as made of two spherical mirrors comprising a set of intermediate optical elements like lenses, mirrors, prisms, and so on (Fig. 11.10a). If we define an arbitrary plane β orthogonal to the cavity optical axis, it is intuitive (and can be rigorously demonstrated) that the round-trip propa-

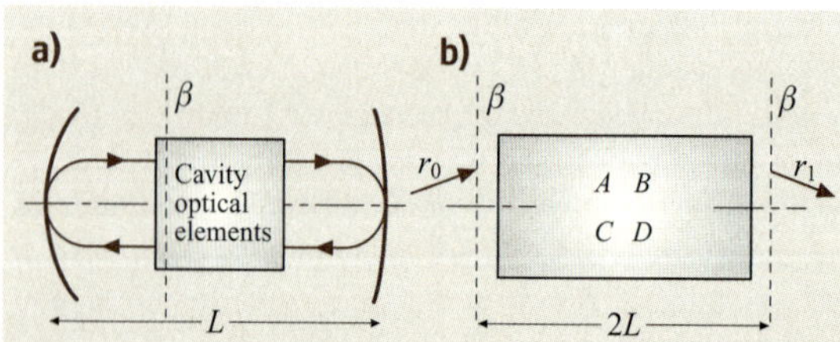

Fig. 11.10 **(a)** A general laser cavity, and **(b)** its equivalent representation given by resonator unfolding with respect to an arbitrary β plane

gation along the cavity is equivalent to the propagation in a suitable optical system S having β as input and output planes, a transformation known as the *unfolding* of the resonator. Paraxial propagation in such a system can be described by means of the $ABCD$ cavity round-trip matrix (Fig. 11.10b). Accordingly, if we let r_0 and r'_0 be, respectively, the transverse coordinate and angle that a ray make with the optical axis when intercepting plane β at time $t=0$, and r_n and r'_n be the coordinate of the same ray after n cavity round-trip, we have

$$\begin{vmatrix} r_n \\ r'_n \end{vmatrix} = \begin{vmatrix} A & B \\ C & D \end{vmatrix}^n \begin{vmatrix} r_0 \\ r'_0 \end{vmatrix} = M^n \begin{vmatrix} r_0 \\ r'_0 \end{vmatrix} . \quad (11.41)$$

Therefore, the optical resonator is stable if and only if, for any initial set of coordinates (r_0, r'_0), the corresponding coordinates (r_n, r'_n) do not diverge as n increases. This condition is met provided that the eigenvalues $\lambda_{1,2}$ of M are, in modulus, not larger than one. Since $\lambda_{1,2} = \exp(\pm i\theta)$, where $\cos(\theta) = (A+D)/2$, the stability condition requires that θ be real, i. e., that

$$\left|\frac{A+D}{2}\right| \leqslant 1 . \quad (11.42)$$

For the particular case of two-mirror resonators, we can go one step further by explicitly calculating the corresponding $ABCD$ matrix. We recall that a given overall matrix can be obtained by the product of matrices of individual optical elements traversed by the beam, with the matrices written in the reverse order compared to ray propagation through the corresponding elements. In our case, the $ABCD$ matrix is then given by the ordered product of the following matrices: reflection from mirror 1, free-space propagation from mirror 1 to 2, reflection from mirror 2, free-space propagation from mirror 2 to 1:

$$\begin{vmatrix} A & B \\ C & D \end{vmatrix} = \begin{vmatrix} 1 & 0 \\ -2/R_1 & 1 \end{vmatrix} \begin{vmatrix} 1 & L \\ 0 & 1 \end{vmatrix} \begin{vmatrix} 1 & 0 \\ -2/R_2 & 1 \end{vmatrix} \begin{vmatrix} 1 & L \\ 0 & 1 \end{vmatrix} .$$

After performing matrix multiplication we obtain

$$\frac{A+D}{2} = \left(1-\frac{L}{R_1}\right)\left(1-\frac{L}{R_2}\right) - 1 . \quad (11.43)$$

It is customary to define dimensionless quantities, referred to as the g_1 and g_2 parameters, defined as $g_1 = 1 - L/R_1$ and $g_2 = 1 - L/R_2$. In terms of these parameters, the stability condition transforms into the very simple relation:

$$0 < g_1 g_2 < 1 . \quad (11.44)$$

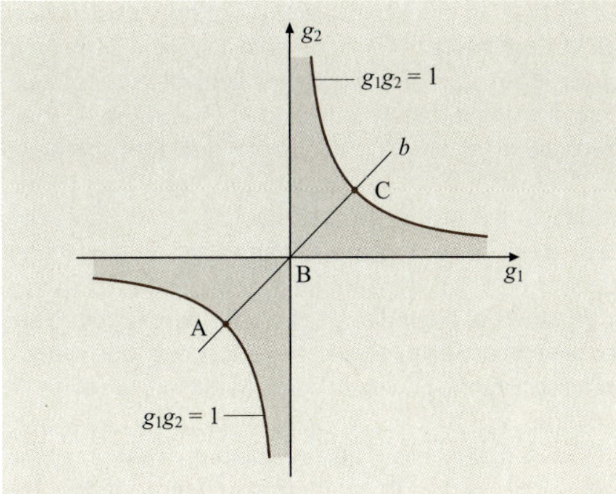

Fig. 11.11 Stability diagram for a general two-mirror spherical resonator. Stable resonators correspond to (g_1, g_2) points lying in the *gray region* of the plane

The stability condition given by (11.44) can be conveniently displayed in the (g_1, g_2)-plane as reported in Fig. 11.11. Stable resonators correspond to those points in the gray region of the plane, excluding those which lie on the boundaries, (i. e., satisfying the conditions $g_1 g_2 = 0$ or $g_1 g_2 = 1$), which are referred to as marginally stable resonators. Note that symmetric resonators (i. e., having mirrors of the same radius of curvature $R_1 = R_2 = R$) lie on the bisector line b. As particular examples of these symmetric resonators, one can see that those corresponding to point A, B and C of the figure are the concentric ($L = 2R$), confocal ($L = R$) and plane ($R = \infty$) resonators, respectively. Since points A, B and C lie on the boundary of the stability region, the corresponding resonators are only marginally stable.

Laser Modes

The modes of an optical resonator are defined as the stationary (i. e., monochromatic) or weakly damped field distributions that can be sustained inside the cavity and that satisfy the boundary conditions imposed by the cavity mirrors. We note that, in open resonators, diffraction losses due to the finite aperture of the mirrors make the modes always leaky. The electric field for a mode in a leaky resonator can then be generally represented as

$$\boldsymbol{E}(x, y, z, t) = \boldsymbol{a}(x, y, z)\cos(\omega t) \times \exp\left[\frac{-t}{2\tau_c}\right] \quad (t > 0) , \quad (11.45)$$

where $\boldsymbol{a}(x, y, z)$ is the mode field distribution, ω is the resonance frequency, and τ_c describes field decay due to cavity losses and is referred to as the *cavity pho-*

ton lifetime. In order to determine the resonator modes, the corresponding resonance frequencies and diffraction losses let us consider the rather general class of resonators with an optical axis z, as shown in Fig. 11.10a. Propagation of an EM wave back and forth between the two end mirrors of the cavity is equivalent to the unidirectional propagation of an EM wave in a periodic sequence of optical elements (e.g., a lens guide) which is obtained by unfolding the resonator, as shown in Fig. 11.10b. Note that, in the unfolding scheme, the end spherical mirrors should be replaced by thin spherical lenses with focal length equal to the mirror radius of curvature.

Let us first consider the propagation of a monochromatic EM field in a periodic lens guide. By writing the electric field along the lens guide as $E(x, y, z, t) = \tilde{E}(x, y, z)\mathrm{e}^{\mathrm{i}\omega t} + \text{c.c.}$, due to the linearity of the Huygens–Fresnel integral the complex field amplitude $\tilde{E}$ after one period of the lens guide can generally be written by an integral transformation, namely (Fig. 11.12)

$$\begin{aligned}&\tilde{E}(x, y, 2L)\\&= \exp(-\mathrm{i}2kL)\\&\quad\times \iint_1 K(x, y; x_1, y_1)\tilde{E}(x_1, y_1, 0)\,\mathrm{d}x_1\,\mathrm{d}y_1\,, \end{aligned} \tag{11.46}$$

where $K(x, y; x_1, y_1)$ is a function of the transverse coordinates of both input and output planes, known as the *propagation kernel*. Note the phase term $(-2kL)$, which represents the phase shift if the wave were a plane wave. The kernel K accounts for all the elements encountered during propagation from input plane 1 ($z = 0$) to output plane 2 ($z = 2L$) and represents, from a physical viewpoint, the field distribution observed at the (x, y)-plane (i. e., at $z = 2L$) when a point-like source at the (x_1', y_1') point is placed at the $z = 0$ input plane. Indeed, if $\tilde{E}(x_1, y_1, 0)$ were a bidimensional Dirac δ-function centered at $x_1 = x_1'$ and $y_1 = y_1'$, i. e., if $\tilde{E}(x_1, y_1, 0) = \delta(x_1 - x_1', y_1 - y_1')$, then from (11.46) one would readily get $\tilde{E}(x, y, 2L) = K(x, y; x_1', y_1')\exp(-2\mathrm{i}kL)$. For a general optical system, the calculation of the kernel K is usually rather complicated. However, within the assumption of infinite aperture of all optical elements, the kernel K is simply expressed in terms of the round-trip resonator-matrix elements $ABCD$ by the relation (Huygens–Fresnel–Kirchhoff kernel)

$$\begin{aligned}&K(x, y; x_1, y_1)\\&= \frac{\mathrm{i}}{\lambda B}\exp\left\{-\frac{\mathrm{i}k}{2B}\right.\\&\quad\left.\times\left[A(x_1^2 + y_1^2) + D(x^2 + y^2) - (2xx_1 + 2yy_1)\right]\right\}\end{aligned} \tag{11.47}$$

and the integral in (11.46) can be extended from $-\infty$ to ∞.

Let us now define a mode of a periodic lens guide as a field distribution which reproduces itself after one guide period except for an overall amplitude reduction, due to lens-guide losses, and a phase shift $\Delta\varphi$ accounting for field propagation. Hence $\tilde{E}(x, y, 2L) = |\tilde{\sigma}|\exp(\mathrm{i}\Delta\varphi)\tilde{E}(x, y, 0)$, where $|\tilde{\sigma}| < 1$. It is now convenient to write the phase shift as $\Delta\varphi = -2kL + \varphi$, where $-2kL$ is the shift of a plane wave and φ is an additional phase term due to the fact that the lens-guide mode is not a plane wave. Hence, for a lens-guide mode we require the condition

$$\tilde{E}(x, y, 2L) = \tilde{\sigma}\exp(-2\mathrm{i}kL)\tilde{E}(x, y, 0)\,, \tag{11.48}$$

where $\tilde{\sigma} = |\tilde{\sigma}|\exp(\mathrm{i}\varphi)$. Substitution of (11.48) into (11.46) yields

$$\begin{aligned}&\iint_1 K(x, y; x_1, y_1)\tilde{E}(x_1, y_1, 0)\,\mathrm{d}x_1\,\mathrm{d}y_1\\&= \tilde{\sigma}\tilde{E}(x, y, 0)\,.\end{aligned} \tag{11.49}$$

Note that the mode distribution $\tilde{E}(x, y, 0)$ is an eigenfunction of a Fredholm homogeneous integral equation of the second kind corresponding to the eigenvalue $\tilde{\sigma}$. As a rather general property, it turns out that the Fredholm integral equation admits a doubly infinite discrete set of confined eigensolutions, which can be distinguished by a pair of integer positive numbers l and m. Accordingly, the corresponding eigenvalues will generally be indicated as $\tilde{\sigma}_{lm} = |\tilde{\sigma}_{lm}|\exp(\mathrm{i}\varphi_{lm})$, with $|\tilde{\sigma}_{lm}| < 1$ owing to

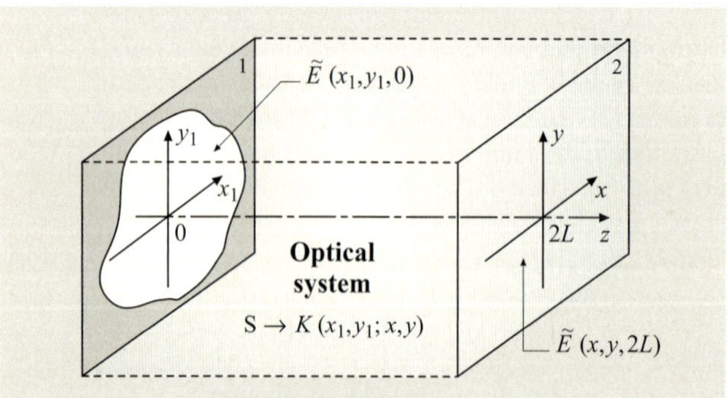

Fig. 11.12 Field calculation at plane 2 ($z = 2L$) when the field $\tilde{E}(x_1, y_1, 0)$ at plane 1 ($z = 0$) is known

the overall losses (namely, diffraction losses, scattering losses from optical elements, and so on) encountered in one period of propagation along the lens guide.

Now, let us return to the problem of cavity mode calculation. In this case, the mode $\tilde{E}$ of the lens guide corresponds to a mode of the resonator of Fig. 11.10a provided that, after one cavity round-trip, the overall phase shift $\Delta\varphi$ accumulated in one cavity round-trip is zero apart from an integer number of 2π, i. e., $\Delta\varphi = -2kL + \varphi_{lm} = -2\pi n$. From this condition and using the relation $k = 2\pi\nu/c$ between the wavenumber k and frequency ν of the mode, one readily obtains the cavity resonance frequencies as

$$\nu_{nlm} = \frac{c}{2L}\left(n + \frac{\varphi_{lm}}{2\pi}\right) . \tag{11.50}$$

Note that we have indicated explicitly that these frequencies depend on the three integer numbers l, m, and n. The integers l and m define the transverse field profiles of the mode (Fig. 11.8), i. e., they represent the number of zeros of the field along the x- and y-coordinates, respectively. The integer n defines the longitudinal field configuration, i. e., the number of zeros of the EM standing wave (nodes) as previously discussed for the plane-parallel resonator (11.38).

For stable resonators with infinite apertures the eigenmodes of the Fredholm equation are given by the Gauss–Hermite functions, and their resonance frequencies can be calculated from (11.50) using the propagation law (11.36). For instance, for the important case of two-mirror spherical resonators, it turns out that $\varphi_{lm} = 2(1+l+m)\cos^{-1}(\pm\sqrt{g_1 g_2})$, where the plus or minus sign depends on whether g_2 (and hence g_1) is positive or negative. The resonance frequencies of the two-mirror spherical resonator are thus given by:

$$\nu_{nlm} = \frac{c}{2L}\left[n + \frac{1+l+m}{\pi}\cos^{-1}(\pm\sqrt{g_1 g_2})\right] . \tag{11.51}$$

For, e.g., a confocal resonator, one has $g_1 = g_2 = 0$, and hence $\nu_{nlm} = [c/(4L)](2n+1+l+m)$. Note that modes with the same value of $(2n+l+m)$ have the same resonance frequency, and they are said to be frequency degenerate. Note also that, for a confocal resonator, the mode spacing is $c/(4L)$.

Photon Lifetime and Cavity Q

The modes of an optical resonator are always leaky and therefore show a finite photon cavity lifetime τ_c. In fact, besides diffraction losses due to finite aperture effects of mirrors or intracavity optical elements, some other loss mechanisms are always present in a laser resonator. For instance, the mirror reflectivity of the output coupler is always smaller than 100%, which means that a fractional part of the photons ϕ stored in the cavity escapes from the resonator at each round-trip. Scattering or absorption losses of intracavity optical elements are also another common cause of photon leakage. To calculate the rate of energy decay in a given cavity mode, let I_0 be the initial intensity corresponding to the field amplitude at a fixed point within the cavity, and let R_1 and R_2 be the (power) reflectivities of the two mirrors and L_i the fractional internal loss per pass, which accounts for scattering, absorption and diffraction losses. The intensity at the same point, after a round-trip time $\tau_R = 2L_e/c$, is $I(\tau_R) = R_1 R_2 (1-L_i)^2 I_0 = I_0 \exp(-2\gamma)$, where L_e is the *cavity optical length* and γ is the logarithmic loss per pass, defined by the relation

$$\gamma = \frac{\gamma_1 + \gamma_2}{2} + \gamma_i \tag{11.52}$$

with $\gamma_1 = -\ln(R_1)$, $\gamma_2 = -\ln(R_2)$, and $\gamma_i = -\ln(1 - L_i)$. In view of the exponential decay law introduced in (11.45), after a round-trip time we must have $I(\tau_R) = I_0 \exp(-\tau_R/\tau_c)$, and thus we conclude that the photon lifetime is given by

$$\tau_c = \frac{\tau_R}{2\gamma} = \frac{L_e}{c\gamma} . \tag{11.53}$$

Having calculated the photon lifetime, the time behavior of the electric field at any point inside the resonator can be written as $E(t) = E_0 \exp(-t/2\tau_c + i\omega t) +$ c.c., where ω is the angular frequency of the mode. The same time behavior then applies for the field of the wave leaving the resonator through the output mirror. Taking the Fourier transform of this field (for $t > 0$) we find that the power spectrum of the emitted light has a Lorentzian line shape with line width (FWHM) given by

$$\Delta\nu_c = \frac{1}{2\pi\tau_c} . \tag{11.54}$$

We can now introduce an important quality factor which is strictly related to the photon lifetime. This is the *cavity Q-factor* which is defined, for any resonant system, as 2π times the ratio between the energy stored in the resonator and the energy lost in one oscillation cycle. Thus a high cavity Q-factor implies low losses in the resonant system. Since in our case the energy stored is $\phi h\nu$ and the energy lost per cycle is $(-\mathrm{d}\phi/\mathrm{d}t)h$ we obtain

$$Q = -\frac{2\pi\nu\phi}{\mathrm{d}\phi/\mathrm{d}t} = \frac{\nu}{\Delta\nu_c} , \tag{11.55}$$

where the exponential decay law $\phi(t) = \phi_0 \exp(-t/\tau_c)$ for the stored cavity photons ϕ has been used and where the expression for $\Delta\nu_c$ given by (11.54) has also been utilized.

As an example, consider a two-mirror spherical resonator with $R_1 = R_2 = R = 0.98$ and assume $L_i \simeq 0$. From (11.53) we obtain $\tau_c = \tau_T/[-\ln(R)] = 49.5\ \tau_T$, where $\tau_T = L/c$ is the transit time of the photon for a single pass in the cavity. Note that the photon lifetime is much longer than the transit time, a typical result of low-loss cavities. If we now assume $L = 90\,\text{cm}$, we get $\tau_T = 3\,\text{ns}$ and $\tau_c \simeq 150\,\text{ns}$. From (11.54) we can then calculate $\Delta\nu_c \simeq 1.1\,\text{MHz}$. Finally, assuming a laser wavelength $\lambda \simeq 630\,\text{nm}$ corresponding to an optical frequency $\nu = 5\times10^{14}\,\text{Hz}$, from (11.55) we have $Q = 4.7\times10^8$. Thus laser resonators have a remarkably high Q-value, which means that a very small fraction of the energy is lost during one oscillation cycle.

11.1.4 Laser Rate Equations and Continuous-Wave Operation

A simple and powerful approach for understanding the basic dynamical behavior of a laser is based on a *rate-equation* model, in which simple balance equations for the total number of atoms undergoing a transition and the total number of photons created or annihilated are written [11.3]. For a more-refined treatment of laser dynamics based on either a semiclassical or a full-quantum electrodynamic approach, which may account for certain phenomena such as dynamical laser instabilities, laser coherence and photon statistics, we refer the reader to more-specialized literature [11.6, 8].

Laser Rate Equations

Let us consider a four-level laser scheme (Fig. 11.4b) and make the following assumptions:

1. the laser transition is homogeneously broadened,
2. the lifetime τ_1 of the lower laser level 1 is short enough that we may neglect the population in level 1,
3. a single longitudinal and transverse mode is oscillating in the cavity,
4. we neglect the precise transverse and longitudinal spatial variation of the cavity mode,
5. we assume uniform pumping of the active medium.

Under these assumptions, we can write the following rate equations for the population inversion $N = N_2 - N_1 \simeq N_2$ in the active medium and the number of photons ϕ of the oscillating mode stored in the cavity:

$$\frac{\mathrm{d}N}{\mathrm{d}t} = R_p - B\phi N - \frac{N}{\tau} \tag{11.56}$$

$$\frac{\mathrm{d}\phi}{\mathrm{d}t} = -\frac{\phi}{\tau_c} + V_a\, B\phi N\,, \tag{11.57}$$

where R_p is the pump rate per unit volume, τ is the lifetime of the upper laser level 2, and τ_c is the photon lifetime for the oscillating mode. In (11.56), the terms R_p, N/τ and $B\phi N = W_{21}N$ account for the pumping process, radiative and nonradiative decay, and stimulated emission, respectively. The constant B, which represents the stimulated transition rate per photon, per mode is related to the transition cross section σ by the simple equation $B = \sigma c/V$, where V is the mode volume in the laser cavity [11.3]. The first term on the right-hand side in (11.57), ϕ/τ_c, represents the number of cavity photons that are lost per unit time due to internal loss, diffraction loss and output coupling through the mirrors. Finally, the second term in (11.57) represents the number of photons (per unit time) that are created in the oscillating mode owing to stimulated emission: since $B\phi N$ represents the number of atoms per unit volume and per unit time that decay creating a photon in the oscillating cavity mode, the total number of photons created per unit time can be expressed as the product of $B\phi N$ with the volume V_a occupied by the cavity mode inside the gain medium. Spontaneous emission in not included in the balance equation (11.57) since only a negligible fraction of spontaneously emitted photons belongs to the oscillating mode. However, spontaneous emission photons are important to allow laser action starting.

The laser output power P_{out} is related to the photon number ϕ by the simple relation

$$P_{out} = \frac{\gamma_2}{2\gamma}(h\nu)\frac{\phi}{\tau_c} = \frac{\gamma_2 c}{2L_e}h\nu\phi\,. \tag{11.58}$$

In fact, $(h\nu)(\phi/\tau_c)$ is the total EM energy lost in the cavity per unit time, and solely a fraction $\gamma_2/(2\gamma)$ of this power is available due to transmission through the output mirror. For a typical CW laser operated in the continuous-wave regime, the number of photons ϕ stored in the cavity may vary from about 10^{10} photons for low-power lasers (such as a HeNe laser delivering $P_{out} = 10\,\text{mW}$ power at $\lambda = 632.8\,\text{nm}$) to 10^{17} photons for high-power lasers (such as a CO_2 laser delivering an output power $P_{out} = 10\,\text{kW}$ at $\lambda = 10.6\,\mu\text{m}$).

Threshold Condition

The population inversion needed to reach the threshold for laser oscillation is simply obtained from (11.57) by imposing $d\phi/dt = 0$. An initially small number of photons turns out to be exponentially damped for $N < N_c$ or exponentially amplified for $N > N_c$, where:

$$N_c \equiv \frac{1}{\tau_c B V_a} = \frac{\gamma}{\sigma l} \quad (11.59)$$

is referred to as the *critical inversion* (or threshold inversion). In this equation, l is the length of the active medium and the threshold condition $\sigma N_c l = \gamma$ is reached when the gain in the inverted medium, $g = \sigma N l$, equals the logarithmic loss γ of the cavity. The pump rate corresponding to the threshold condition is $R_{cp} = N_c/\tau = \gamma/(\sigma l \tau)$; the corresponding pump power at threshold P_{th} is then obtained using (11.9). The perturbation that starts the laser action when the pump rate R_p reaches the critical value R_{cp} is provided by spontaneous emission.

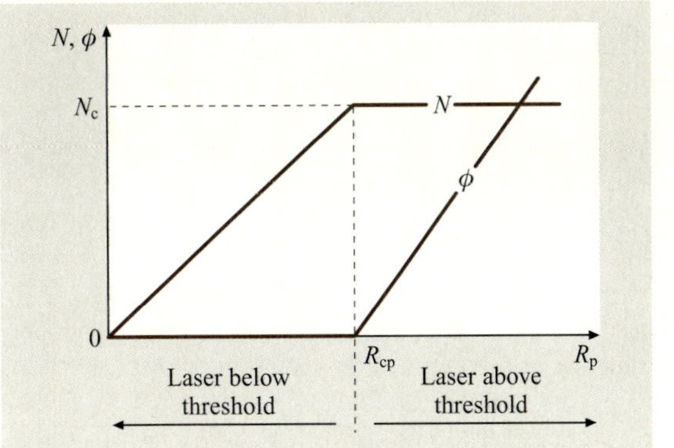

Fig. 11.13 Behavior of population inversion N and photon number ϕ in the oscillating mode versus the pump rate R_p for a four-level laser. R_{cp} is the critical pump rate above which laser action takes place

Output Power and Slope Efficiency

For a pump rate $R_p > R_{cp}$, the rate equations (11.56, 57) admit the solution $N_0 = N_c$ and $\phi_0 = [1/(B\tau)](x - 1)$, corresponding to laser being above threshold (Fig. 11.13). Here $x = R_p/R_{cp} = P_p/P_{th} > 1$ is the above-threshold pump behavior parameter, where P_p is the pump power and P_{th} is its threshold value. The corresponding output laser power can be then calculated from (11.58) and can be cast in the form

$$P_{out} = \eta_s (P_p - P_{th}) , \quad (11.60)$$

where

$$\eta_s = \eta_p \eta_c \eta_q \eta_t . \quad (11.61)$$

Equation (11.60) shows that, within the approximation made, a linear relation is obtained between the output power and the pump power. One can then define the slope efficiency of the laser as $\eta_s = dP_{out}/dP_p$. According to (11.61), η_s is given by the product of four contributions:

1. the *pump efficiency* η_p
2. the *output coupling efficiency* $\eta_c = \gamma_2/(2\gamma)$
3. the laser *quantum efficiency* $\eta_q = (h\nu)/(h\nu_{mp})$
4. the *transverse efficiency* $\eta_t = A_b/A$

where $A_b = V_a/l$ is the transverse mode area in the active medium and A the transverse pumping area. The slope efficiency of a laser may typically vary from less than 1% in low-efficiency lasers (such as in the HeNe laser) to 20–50% or even higher in high-efficiency lasers.

Relaxation Oscillations

One can show that the solution, given above, for lasers above threshold is stable, i.e., that any initial perturbation of the system (e.g., of cavity losses) is damped. When the ratio τ/τ_c between the upper laser level lifetime and cavity photon lifetime is larger than 1 (or much larger, such as in laser transitions which are electric-dipole forbidden), relaxation to the steady state occurs through damped oscillations in both photon number and population inversion. This results in damped oscillations of the output power referred to as *relaxation oscillations*. In solid-state lasers, the frequency of relaxation oscillations typically fall in the 10 kHz–10 MHz region, whereas, in semiconductor lasers, it falls in the GHz region. Relaxation oscillations in slow-gain media (such as in solid-state lasers) triggered by technical noise or by pump power fluctuations are one of the major cause of amplitude noise in the output laser power. Whenever a high degree of intensity stability is required, laser amplitude stabilization may be provided by a suitable active feedback loop.

Laser Tuning

The gain line width of some lasers (e.g., dye lasers or vibronic solid-state lasers) is very wide and, for several applications, one needs to tune the laser output wavelength away from the line center and across the entire available line width. In other cases, different lasing transitions may compete or may potentially be used, and one needs to select one of them. In both circumstances, one can employ a wavelength-selective element inside the laser cavity, which is often referred to as *laser tuner*. For lasers in the middle infrared (such as CO_2 lasers), one generally uses a diffraction grating in the so-called Littrow configuration (Fig. 11.14a) as one of the cavity mirror. Wavelength tuning is simply

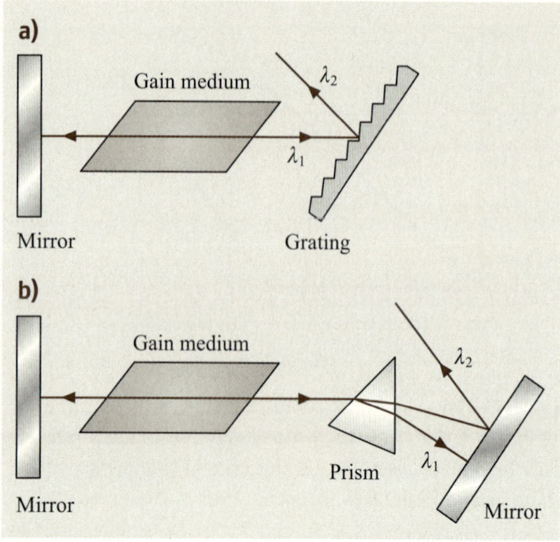

Fig. 11.14 **(a)** Laser tuner based on a diffraction grating in the Littrow configuration. **(b)** Laser tuner based on the use of a dispersive prism

achieved by grating rotation. In the visible or near-infrared spectral regions (such as for the Ar^{3+} laser), a dispersive prism is more commonly used and wavelength tuning is simply achieved by prism, or mirror, rotation (Fig. 11.14b). To reduce insertion losses, the two prisms facets are approximately inclined at Brewster angle. A third wavelength-selective element, which is becoming increasingly popular in the visible or near-infrared, uses a birefringent filter inside the cavity. This device exploits a birefringent plate, inclined at the Brewster angle, which generally changes the polarization state of the intracavity laser beam. In the presence of an intracavity polarizer, or just exploiting the polarizing properties of the Brewster-angle filter, the birefringence filter thus generally produces additional cavity losses. However, at certain wavelengths, λ, the

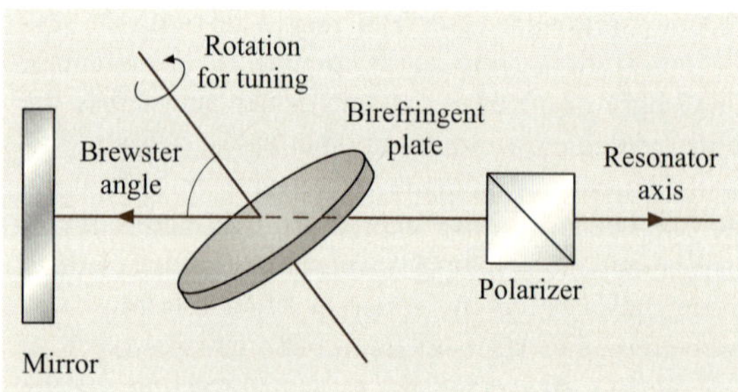

Fig. 11.15 Schematic of a birefringent laser tuner made of an intracavity birefringent plate and polarizer

birefringent plate does not change the polarization state of the beam and thus allows for laser oscillation with low loss. Upon rotation of the plate around the axis orthogonal to its faces (Fig. 11.15) the direction of the birefringence axes of the plates is changed and this correspondingly changes the wavelength at which low loss occurs.

Single Mode Selection and Limit to Laser Monochromaticity

Most often, lasers tend spontaneously to oscillate on several transverse and longitudinal modes, especially when the gainline is relatively broad. The reasons for multimode oscillations are rather involved and their study goes beyond the scope of the present contribution. For many applications, single-mode operation may be required, and therefore one needs to force the laser to oscillate on a single transverse (usually the fundamental TEM_{00} Gaussian mode) and longitudinal mode. For stable resonators, single transverse mode oscillation is easily achieved by placing an aperture inside the cavity of appropriate size in order to increase diffraction loss of higher-order modes with respect to the TEM_{00} mode. In some cases, such as in longitudinally pumped solid-state lasers, limitation on the pump spot size lends itself to TEM_{00} mode selection.

Even when a laser is oscillating on a single transverse mode, it can still oscillate on several longitudinal modes. This usually occurs since the longitudinal-mode separation $\Delta\nu = c/(2L)$ is smaller (or much smaller) than the gain line width $\Delta\nu_0$. For some gas lasers, where the gain line width is relatively small (up to a few GHz), single-longitudinal-mode selection is simply obtained by making the cavity length short enough that the longitudinal-mode separation $\Delta\nu$ becomes larger than $\Delta\nu_0/2$. For, e.g., a HeNe laser ($\Delta\nu_0^* \simeq 1.7\,\mathrm{GHz}$), this condition implies $L \leq 17.5\,\mathrm{cm}$. For solid-state lasers with a bandwidth up to a few hundred GHz, this condition requires cavity lengths in the sub-millimeter region (*microchip lasers*). For lasers with much larger bandwidths (e.g., dye lasers or tunable solid-state lasers), the required length is too small to make the technique of practical relevance and to make the gain in the medium large enough to reach threshold. In these cases, different techniques are employed. For solid-state or dye lasers, the simplest method is to use one (or more) Fabry–Pérot etalons within the cavity, which acts as a narrow frequency spectral-selective element (Fig. 11.16). The etalon thickness and finesse should be designed to ensure single mode selection. This implies that:

1. the half width $\Delta\nu_{FP}/2$ of the transmission peaks of the Fabry–Pérot must be smaller than the longitudinal-mode separation $\Delta\nu = c/(2L)$;
2. the etalon free spectral range $\Delta\nu_{FSR}$ must be larger than the half width of the gain line $\Delta\nu_0/2$ (Fig. 11.16).

For semiconductor lasers, single mode selection is achieved by using a distributed feedback (DFB) structure, in which a longitudinal corrugation of the refractive index in the semiconductor induces frequency mode selection according to Bragg scattering theory.

A special case of single mode selection that deserves mention is that of a laser with a homogeneously broadened line (e.g., Nd:YAG and dye lasers). In this case, multimode oscillation is mainly due to the standing-wave character of the laser mode arising from the interference of counter-propagating waves established between the two cavity mirrors. The use of a ring resonator instead of a linear cavity, in which unidirectional operation is forced by, e.g., an intracavity optical diode, may be enough in this case to achieve single longitudinal-mode operation.

We finally discuss the limit of monochromaticity (and hence of temporal coherence) of a laser. The line width $\Delta\nu_L$ of a laser that oscillates on a single longitudinal mode is ultimately established by spontaneous emission noise. The quantum theory of a laser shows that the spectral shape of emitted light is Lorentzian with a FWHM given by the well-known Schawlow–Townes formula:

$$\Delta\nu_L = \frac{N_2}{N_2 - N_1} \frac{2\pi h\nu_L(\Delta\nu_C)^2}{P} , \tag{11.62}$$

where P is the output power, $\Delta\nu_C = 1/(2\pi\tau_c)$ is the line width of the cold cavity mode, N_2 and N_1 are the steady-state populations in the upper and lower laser levels, respectively, and ν_L is the emission frequency. Typically the line width predicted by the Schawlow–Townes formula is negligibly small compared to that produced by other cavity disturbances (e.g., fluctuations of cavity length or *technical noise*), except for the very important class of semiconductor lasers. For, e.g., a typical HeNe laser oscillating on its red transition ($\lambda = 632.8$ nm), τ_c is of the order of tens of μs, so that $\Delta\nu_L$ is of the order of $\simeq 1$ mHz, which turns out to be much smaller than the line broadening due to technical noise. For instance, a small cavity length change ΔL, due to technical noise, contributes to the frequency broadening $\Delta\nu_L$ by an amount given by $|\Delta\nu_L| = (\Delta L/L)\nu_L$; for $L = 1$ m and $\nu_L = 4.7 \times 10^{14}$ Hz (visible transition), a change of ΔL by just $\approx 10^{-8}$ times the atomic dimension leads to a contribution to $\Delta\nu_L$ comparable to the quantum limit. Conversely, in semiconductor lasers the quantum limit of $\Delta\nu_L$ is considerably larger and typically falls in the MHz range, owing to the much short photon lifetime (τ_c is of the order of few ps). Therefore, the laser line width of a typical semiconductor laser arises from quantum noise.

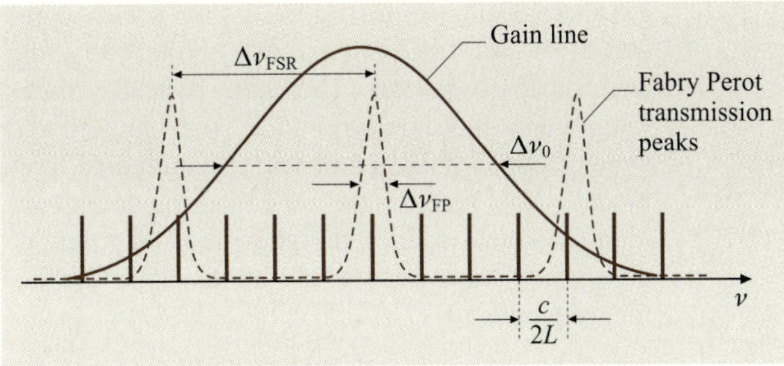

Fig. 11.16 Schematic of single longitudinal mode selection by use of an intracavity Fabry–Pérot etalon

11.1.5 Pulsed Laser Behavior

In lasers operating in the CW or quasi-CW regimes, the maximum achievable optical output power is limited by the maximum available pump power. For high-power CW lasers (such as CO_2 lasers), power levels up to ≈ 100 kW can be reached; however larger power levels, which can be of interest for many applications, are prevented in the CW regime. Transient laser behavior allows one to obtain higher peak powers by concentrating the available energy in a single, short optical pulse or in a periodic sequence of optical pulses [11.3,5]. Additionally, transient laser behavior is a powerful tool for the generation of ultrashort optical pulses, with durations down to ≈ 10 fs in lasers with a broad gainline (notably the Ti^{3+}:Al_2O_3 laser). From a dynamical viewpoint, pulsed laser behavior can be divided into two rather distinct categories:

1. Laser transients occurring on a time scale of the order of the cavity photon lifetime τ_c, i. e., appreciably larger than the cavity round-trip time. This includes the so-called Q-switching and gain-switching regimes, which enable the generation of optical pulses as short as a few nanoseconds with optical peak powers typically in the megawatt range. These are basically single-longitudinal-mode regimes and can be described by means of a rate equation model (11.56, 57).

2. Laser transients occurring on a time scale appreciably shorter (and often much shorter) that the cavity round-trip time. These are basically multi-longitudinal-mode regimes, i. e., they involve the simultaneous oscillation of many longitudinal laser modes, and include the so-called mode-locking regime, which enables the generation of trains of ultrashort laser pulses with durations down to a few femtoseconds.

Laser Q-switching: Dynamical Aspects

Q-switching is a technique that enables the generation of a short optical pulse (of the order of the cavity photon lifetime τ_c) by a sudden switching of the cavity Q factor, i. e., of the cavity loss γ. The change of Q is produced, in principle, upon placing inside the laser cavity an opaque shutter which can be opened or closed. When the shutter is closed (i. e., the cavity Q is low), laser action is prevented and the population inversion N can reach a relatively large value (well above the critical value N_c) due to pumping. When the shutter is opened, the Q factor is suddenly switched to a high value and the laser exhibits a gain $g = \sigma l N$ that greatly exceeds the loss γ. Light emission then occurs via the generation of a short and intense laser pulse. The duration of a Q-switching pulse typically ranges from a few nanoseconds to a few tens of nanoseconds, whereas its peak power is in the megawatt range. To achieve sufficient population inversion when laser action is prevented, a long lifetime τ of the upper laser level is required. Thus Q-switching can be effectively used for electric-dipole-forbidden laser transitions, where τ generally falls in the millisecond range. This is the case of most solid-state lasers (e.g., Nd, Yb, Er in different host materials; Cr-doped materials, such as alexandrite, Cr:LiSAF, and ruby) and some gas lasers (e.g., CO_2 or iodine).

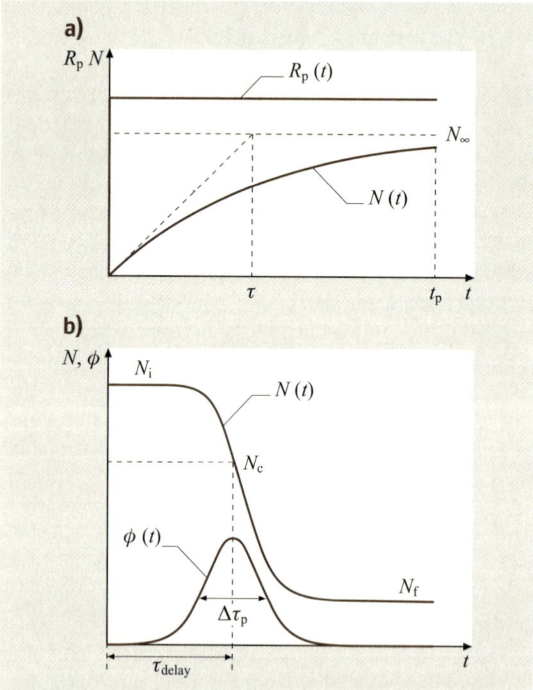

Fig. 11.17a,b Dynamics of fast switching in a four-level laser. **(a)** Temporal behavior of pump rate and population inversion when the cavity Q is low. **(b)** Temporal behavior of population inversion and cavity photons when the cavity Q is suddenly switched to a high value, showing the formation of the Q-switching pulse

To understand the basic dynamics of Q-switching, let us consider a four-level laser and assume that a step pump pulse is applied at time $t = 0$, i. e., $R_p(t) = 0$ for $t < 0$ and $R_p =$ constant for $0 < t < t_p$, where t_p is the duration of the pump pulse; meanwhile the shutter is closed (Fig. 11.17a) and laser action is prevented. From (11.56) with $\phi = 0$, one then obtains that the transient population inversion increases accordingly to the relation $N(t) = N_\infty[1 - \exp(-t/\tau)]$, where the asymptotic value N_∞ is given by $N_\infty = R_p\tau$ (Fig. 11.17a). After a pump time of about 2τ, the population inversion already reaches a value close to its asymptotic limit N_∞, and therefore the pump pulse can be switched off and the shutter opened. In fact, for t_p larger than $\approx 2\tau$ the energy supplied to the medium is not useful for increasing the population inversion any further but is lost as radiative and nonradiative decay. Suppose now that the shutter is opened very rapidly at time t_p (*fast switching*), and take the origin of time at the instant when switching occurs (Fig. 11.17b). From this time on, the evolution of both population inversion and number of photons in the cavity can be numerically computed by solving the rate equations (11.56) and (11.57) with the initial conditions $\phi(0) \simeq 1$ and $N(0) = R_p\tau[1 - \exp(-t_p/\tau)] \equiv N_i$, where the initial condition $\phi(0) \simeq 1$ accounts for the fact that the laser action is started by spontaneous emission (the so-called *extra photon*). The qualitative transient behavior of both N and ϕ is shown in Fig. 11.17b and can be simply understood by observing that, just after the switching time $t = 0$, the gain $g = \sigma N l$ in the gain medium greatly exceeds the single-pass cavity loss γ; therefore the number of photons, which increases nearly exponentially with time starting from the extra photon due to spontaneous emission, typically takes from several hundreds to a few thousand cavity round-trips to reach a sufficient value to saturate the laser transition, thus producing a reduction of the available inversion (Fig. 11.17b). This means that the time delay for the peak of the laser pulse, τ_{delay} typically ranges from several tens to a few hundred of nanoseconds. Note that, on such a time scale, the radiative and nonradiative decays of the population N (which typically occur on the millisecond time scale) are negligible, and therefore population decay just occurs by stimulated emission. The decrease of N as ϕ increases leads, in turn, to a decrease of the gain $g = \sigma Nl$. The peak of the pulse occurs at the delay time, τ_{delay}, such that the population inversion N decreases to its critical value, N_c. In this case, in fact, laser gain g equals cavity loss γ and one has from (11.57) $d\phi/dt = 0$. For $t > \tau_{delay}$, one has $N < N_c$ and $d\phi/dt < 0$. This means that the number of photons now decreases towards zero. Meanwhile, due to stimulated emission, the population inversion also keeps decreasing until the photon pulse has decreased to zero. At this time some population inversion, say N_f, is generally left in the medium after Q-switching (Fig. 11.17b). Note that the quantity $\eta_E = (N_i - N_f)/N_i$ represents the fraction of the energy initially stored in the material that goes into stimulated emitted photons, usually referred to as the inversion, or energy, utilization factor. The duration $\Delta\tau_p$ of the Q-switched pulse, the output energy $E = (1/\tau_c)[\int \phi(t)\,dt](h\nu)(\gamma_2/(2\gamma))$, the energy utilization factor η_E, as well as the time τ_{delay} needed for pulse formation can be derived in a closed form by an analysis of the rate equations (11.56) and (11.57) in which pumping and radiative and nonradiative decay are neglected. One obtains:

$$\Delta\tau_p = \tau_c \frac{(N_i/N_c)\eta_E}{(N_i/N_c) - \ln(N_i/N_c) - 1}\,, \tag{11.63}$$

$$E = \left(\frac{\gamma_2}{2\gamma}\right)(N_i \eta_E V_a)(h\nu)\,, \tag{11.64}$$

$$\tau_{delay} \approx \frac{\tau_c}{(N_i/N_c) - 1} \ln(\phi_p/10)\,, \tag{11.65}$$

where N_i/N_c is the ratio between the initially stored population inversion and critical inversion; ϕ_p is the peak photon number of the Q-switching pulse, given by $\phi_p = V_a(N_i - N_c) - N_c V_a \ln(N_i/N_c)$; and η_E is the energy utilization factor. The value of η_E can be calculated from the implicit equation $(N_i/N_c)\eta_E = -\ln(1-\eta_E)$ from which a plot of (N_i/N_c) vs. η_E can readily be obtained (Fig. 11.18). Note that (11.64) can readily be obtained by a simple energy-balance argument: the stored energy in the medium released as EM wave is in fact equal to $h\nu(N_i - N_f)V_a = h\nu(\eta_E N_i)V_a$, and, out of this energy, only the fraction $[\gamma_2/(2\gamma)]$ goes into the output beam.

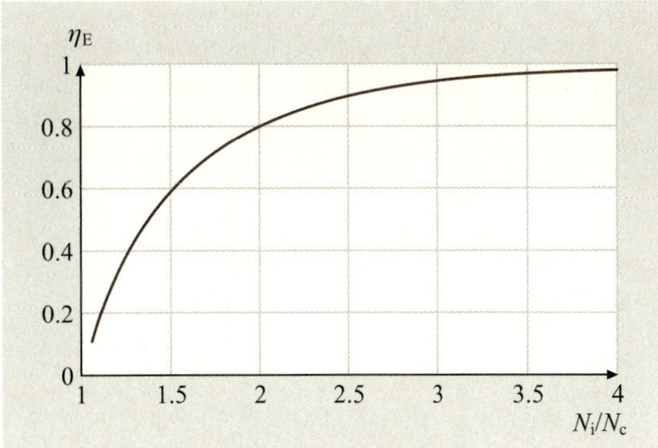

Fig. 11.18 Behavior of the energy utilization factor η_E versus the normalized initial inversion N_i/N_c

Q-switched lasers may generally be operated in two distinct regimes. In so-called pulsed Q-switching, the pump rate $R_p(t)$ generally consists, as explained above, of a pulse with duration comparable to the upper-state lifetime τ. Of course, the pulsed operation can be periodically repeated upon repeating the pulsed pump (generally up to a rate of a few tens of Hz). In so-called CW repetitive Q-switching, the pump rate R_p is held constant and cavity losses are periodically switched from a high to a low value (generally with a rate from a few to a few tens of kHz).

So far we have considered the dynamic behavior corresponding to fast switching, where the switching of the cavity loss is treated as instantaneous (in practice, much shorter than the time τ_{delay}). In the case of *slow switching*, the dynamic behavior is somewhat more complicated and multiple pulses may result, as shown

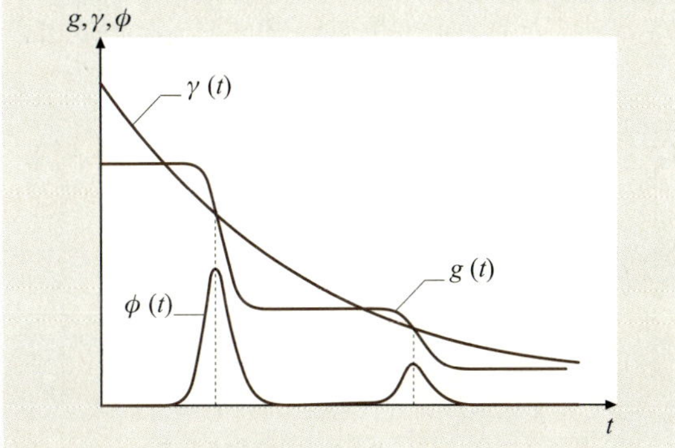

Fig. 11.19 Gain and photon dynamics in a slow-switching laser, showing the formation of multiple pulses

in Fig. 11.19. In the figure, the behavior of the cavity loss γ and laser gain $g = \sigma Nl$ is depicted. The multiple intersections of the slowly decaying curve for the loss $\gamma(t)$ with the gain curve $g(t)$ explain the formation of multiple Q-switching pulses.

Methods of Q-Switching

Several methods have been developed to achieve switching of the cavity Q; the most common are [11.3,4]:

1. Electrooptic Q-switching;
2. Rotating prism;
3. Acoustooptical Q-switching;
4. Saturable absorber Q-switching.

Electrooptic Q-switching. In this case the shutter placed inside the laser cavity is made of a Pockels cell and a polarizer in the configuration shown in Fig. 11.20 (electrooptic shutter). The Pockels cell consists of a suitable nonlinear electrooptic crystal (such as KD*P or lithium niobate for the visible-to-near-infrared region or cadmium telluride for middle-infrared), in which an applied dc voltage induces a change in the crystal refractive indices. The induced birefringence turns out to be proportional to the applied voltage. The transmission axis of the polarizer is set at 45° with respect to the birefringence axes of the crystal. When no dc field is applied to the crystal, no polarization losses are introduced in the cavity by the electrooptic shutter, i. e., cavity losses are low. However, when a dc field is applied such that the phase difference $\Delta\varphi$ between the ordinary and extraordinary waves in the birefringent-induced crystal is equal to $\pi/2$, the Pockels cell operates as a $\lambda/4$ birefringent plate. Therefore, the linearly polarized light, coming from the polarizer, is rotated by 90° after a double pass through the cell and it is then fully reflected out of the cavity by the polarizer. The electrooptic shutter is now closed and cavity Q is zero. The dc voltage to the crystal required to produce the phase shift $\Delta\varphi = \pi/2$ is referred to as the quarter-wave voltage and typically ranges from 1 to 5 kV. To avoid multiple pulses, this voltage must be switched off in a time typically smaller than 20 ns.

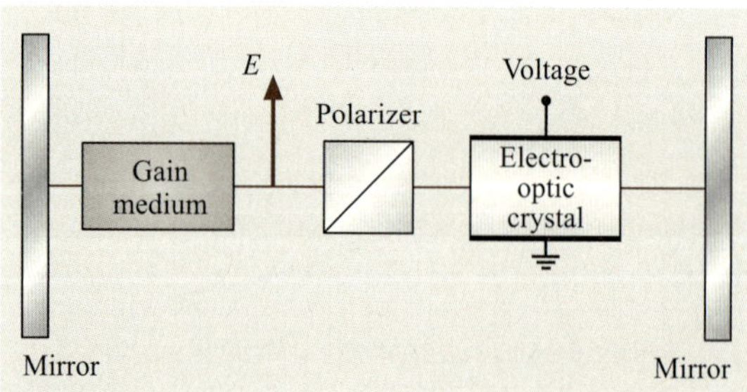

Fig. 11.20 Schematic of a Q-switched laser using a Pockels cell

Rotating prism. In this simple Q-switching technique, one of the cavity mirror is generally made of a rooftop prism and rotation is made through the axis parallel to the other mirror and orthogonal to the prism edge (Fig. 11.21). The high-Q condition is achieved when the prism edge passes through a position parallel to the other cavity mirror. Although rotating prisms are simple and inexpensive devices which can be used at any wavelength, they suffer the limitation arising from the limited rotation speed (≈ 400 Hz). The Q-switching time is then rather long (typically ≈ 400 ns) which often results in the production of multiple pulses (slow switching).

Acoustooptic Q-switching. In this case the shutter consists of an acoustooptic modulator, driven by a radiofrequency (RF) oscillator, which is placed inside the laser cavity. The modulator consists of a transparent block of material (usually fused quartz in the visible to near-infrared or cadmium selenide in the mid- to far-infrared) bonded on one side to a piezoelectric transducer and to an acoustic absorber on the other (Fig. 11.22). When the transducer is on, traveling sound waves are then produced in the material in the direction orthogonal to the plane of the transducer. Due to the photoelastic effect, the resulting strain in the material results in local changes of the material refractive index, i. e., in the generation of an index grating which is traveling along the material itself. Bragg scattering of the laser

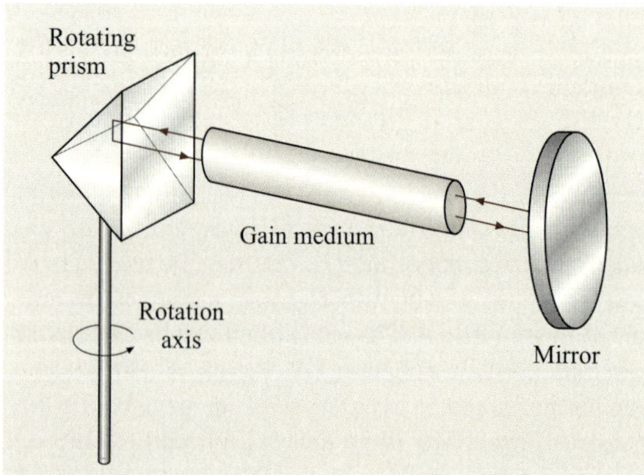

Fig. 11.21 Schematic of a Q-switched laser using a rotating prism

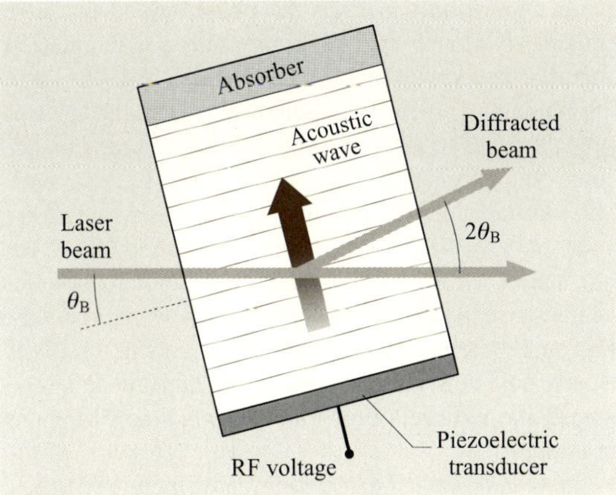

Fig. 11.22 Schematic of an acoustooptic modulator used for laser Q-switching (θ_B is the Bragg angle)

beam propagating across such a grating (Fig. 11.22) thus produces a diffracted beam and hence additional losses in the cavity (low-Q condition). Maximum diffraction efficiency is achieved when the incident light angle θ_B satisfies the Bragg condition $\theta_B = \lambda/(2\lambda_a)$, where λ and λ_a are the optical and acoustical wavelengths, respectively. The high-Q condition is simply obtained by switching off the transducer voltage. Acoustooptic modulators have the advantages of introducing low values of optical insertion losses and can be driven at high repetition rates (several kHz). Therefore they are used mainly for repetitive Q-switching of low-gain CW-pumped lasers (e.g., Nd:YAG or Ar-ion lasers).

Saturable absorber Q-switching. The Q-switching techniques discussed so far use active Q-switches, in the sense that they need an external control source. A notable passive Q-switching technique, in which no external driving control is required, consists of placing a suitable *saturable absorber* inside the laser cavity. This absorber is basically an unpumped two-level medium which absorbs at the laser wavelength with a comparatively low saturation intensity. Thus, due to the phenomenon of saturation, the absorption coefficient of the absorber decreases as the intensity of the intracavity laser beam increases. To model this phenomenon we write the rate equation for the absorber as $(\mathrm{d}N_2/\mathrm{d}t) = \sigma_a(N_1 - N_2)I/(h\nu) - N_2/\tau$, where N_2 and τ are the population and lifetime of the excited level 2, N_1 is the population in the lower (ground) level 1, and $I = I(t)$ is the intracavity laser intensity. In CW operation or if τ is short enough as compared to the changes of $I(t)$, we may assume that $(\mathrm{d}N_2/\mathrm{d}t) \approx 0$ and therefore obtain $N_1 - N_2 \simeq N_t/(1 + I/I_s)$, where $N_t = N_1 + N_2$ is the total absorber population and $I_s \equiv h\nu/(2\sigma\tau)$ is the *saturation intensity* of the absorber. Since now the absorption coefficient is $\alpha = \sigma_a(N_1 - N_2)$, from the last two equations we obtain

$$\alpha = \frac{\alpha_0}{1 + I/I_s}, \tag{11.66}$$

where $\alpha_0 = N_t\sigma_a$ is the unsaturated absorption coefficient of the absorber. Equation (11.66) then shows that, as the intracavity laser intensity I increases, the losses introduced by the saturable absorber decrease and the cavity Q correspondingly increases, i. e., Q-switching is achieved. The detailed dynamics underlying the formation of a Q-switching pulse with the saturable-absorber Q-switch, is however more involved than the one previously discussed. Here we just say that the saturable absorber should have a low value of the saturation intensity so that, when laser action starts, the absorber is bleached earlier (i. e., at lower intensity) than the time when population inversion in the gain medium starts to decrease appreciably owing to stimulated emission.

Typical absorbers used for passive Q-switching consist of dyes in an appropriate solvent; the main drawback of these absorbers is their photochemical degradation, i. e., poor chemical stability, and inadequate thermal properties. Recently, the advent of solid-state absorbers (notably absorbers based on chromium doped into various crystalline hosts) are replacing dye absorbers, thus avoiding the problem of degradation.

Laser Mode-Locking: Dynamical Aspects

Mode-locking is a laser operation regime in which many longitudinal modes of the cavity are simultaneously forced to oscillate with a precise phase relation so that the output laser beam shows a repetitive train of ultrashort optical pulses [11.3, 5]. To achieve mode-locking operation, a suitable device, which is usually referred to as the *mode locker*, must be placed inside the cavity. For a given laser medium, the lower limit to the achievable pulse duration $\Delta\tau_p$ is set by the gain line width ($\Delta\tau_p \gtrsim 1/\Delta\nu_0$), whereas the pulse repetition rate $1/\tau_p$ is usually equal to the difference frequency between two consecutive longitudinal modes $\Delta\nu$ (or an integer multiple of $\Delta\nu$, as for harmonic mode-locking). Therefore, pulse duration, depending upon the gain line width, usually ranges from about 1 ns, in gas lasers, down to 10 fs in broad-bandwidth solid-state lasers. Pulse repetition rate depends, of course, upon the cavity length and usually ranges from about 100 MHz to a few GHz.

Frequency-domain description. The basic principle of laser mode-locking can be explained as follows. Let us consider, for simplicity, a traveling-wave ring laser of optical length L_e, and assume that the separation of the longitudinal cavity modes $\Delta\nu = c/L_e$ be smaller (usually much smaller) than the gain line width $\Delta\nu_0$. In this case, even in the absence of the mode locker, the laser tends spontaneously to oscillate on several longitudinal modes (*free-running* regime). Let $\nu_l = \nu_0 + l\Delta\nu$ be the frequency and $E_l = A_l \exp(\mathrm{i}\varphi_l)$ the complex amplitude of the l-th oscillating longitudinal cavity mode, where l is an integer number and $l = 0$ corresponds to the longitudinal mode closest to the center of the gainline (Fig. 11.23). The electric field $E(z,t)$ inside the laser cavity is then given by the superposition of the oscillating longitudinal modes and can be written in the form $E(z,t) = A(t - z/c)\exp[2\pi\mathrm{i}\nu_0(t - z/c)]$, where z is the longitudinal coordinate measured along the perimetric axis of the ring and the envelope $A(t')$ is given by

$$A(t') = \sum_l A_l \exp(2\pi\mathrm{i}l\Delta\nu t' + \mathrm{i}\varphi_l)\,, \tag{11.67}$$

where $t' = t - z/c$ is a retarded time. Note that, owing to the dependence of the field on $t - z/c$, the field distribution in the cavity is a traveling wave which propagates with the speed of light; therefore we may limit the discussion to the field behavior at a given reference plane $z = 0$, e.g., at the output coupler of the laser. The output laser power, averaged over the rapidly varying optical cycle, will be then proportional to $|A(t)|^2$. Note that, if the mode amplitudes A_l and phases φ_l are constant or slowly varying in time compared to the cavity round-trip time $\tau_R = L_e/c = 1/\Delta\nu$, the signal $A(t)$ is basically periodic in time with a period equal to τ_R. However, the specific form of the signal in one period depends on the precise distribution of the mode amplitudes A_l and, most importantly, on their phases φ_l.

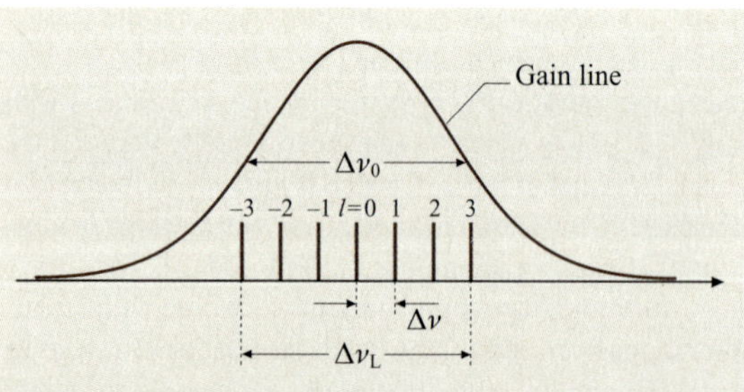

Fig. 11.23 Schematic of a laser oscillating on several longitudinal modes in the frequency domain

In a *free-running laser*, the phases φ_l do not have a precise relation to each other; they may also fluctuate in time. The superposition of N modes with, e.g., the same amplitude $A_l = A_0$ but with randomly distributed phases typically leads to a spiking signal $|A(t)|^2$ made of a periodic sequence of irregular pulses (Fig. 11.24a), each with duration approximately equal to $\Delta\tau_p \approx 1/\Delta\nu_L$, where $\Delta\nu_L = N\Delta\nu$ is the oscillating bandwidth. Note that, since the response time of a conventional photodetector is usually much longer than a few picoseconds, the complex temporal behavior shown in Fig. 11.24a is usually not resolved for free-running multimode lasers, instead its average value – proportional to NA_0^2 – is monitored.

In a *mode-locked laser*, the role of the mode locker is to lock the phases of oscillating modes in a precise manner. The most common and interesting case is that of a mode locker that imposes a *linear phase locking* condition, i. e., $\varphi_l = l\varphi$, where φ is a constant. In this case, one has:

$$A(t'') = \sum_{l=-\infty}^{\infty} A_l \exp(2\pi\mathrm{i}l\Delta\nu t'')\,, \tag{11.68}$$

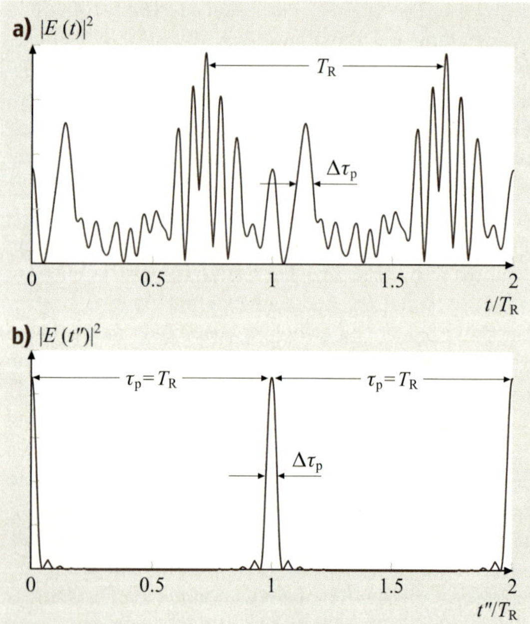

Fig. 11.24a,b Behavior of output power in a laser oscillating on many longitudinal modes with equal amplitude **(a)** Free-running regime, corresponding to random phases ϕ_l. **(b)** Mode-locking regime, corresponding to linear phase locking $\phi_l = l\phi$

where t'' is a translated time given by $t'' = t' + \varphi/(2\pi\Delta\nu)$. Usually, in a mode-locked laser the envelope of the mode amplitudes A_l follows the shape of the gainline, i. e., A_l is maximum at the center of the gainline ($l = 0$) and decreases toward zero as $|l|$ increases; for instance, for active mode-locking in a homogeneously broadened medium, one can show that A_l follows a Gaussian distribution. However, to calculate the series in (11.68) simply, we assume that $A_l = A_0 = \text{const.}$ for $|l| \le N$ and $A_l = 0$ for $|l| > N$, i. e., we assume an odd number $2N+1$ of oscillating modes with the same amplitude. In this case one obtains in (11.68) a geometric progression which can be calculated in a closed form, yielding

$$A(t'') = A_0 \frac{\sin\left[(2N+1)\pi\Delta\nu t''\right]}{\sin(\pi\Delta\nu t'')} . \tag{11.69}$$

A plot of the squared amplitude of the electric field $|A(t'')|^2$ is reported in Fig. 11.24b. Note that a pulse train, at a repetition rate $\tau_p = 1/\Delta\nu$ equal to the cavity round-trip time τ_R, is obtained. The squared amplitude of the electric field at the peak pulse is given by $(2N+1)^2 A_0^2$, whereas the FWHM pulse duration $\Delta\tau_p$ is approximately given by

$$\Delta\tau_p \simeq \frac{1}{(2N+1)\Delta\nu} = \frac{1}{\Delta\nu_L} , \tag{11.70}$$

where $\Delta\nu_L = (2N+1)\Delta\nu$ is again the oscillating bandwidth. Therefore, for broad oscillating bandwidths, phase locking among longitudinal modes leads to the generation of short laser pulses with high peak power. The physical limit to the maximum number of phase-locked modes that can be forced to oscillate is ultimately determined by the gain bandwidth of the active material, i. e., $\Delta\tau_p$ is longer than or equal to $1/\Delta\nu_0$.

In general, when the actual shape (e.g., Gaussian) of the mode amplitudes A_l is taken into account, the overall field amplitude $A(t'')$ can be obtained approximately from (11.68) upon transforming the sum over all modes into an integral, namely $A(t'') \simeq \int_{-\infty}^{+\infty} A(l) \exp(2\pi i l \Delta\nu t'')\, dl$. From this last equation one then sees that the pulse amplitude $A(t'')$ is the Fourier transform of the spectral mode envelope $A(l)$. Therefore, in this case, i. e., under linear phase-locking conditions, the pulse amplitude is said to be transform limited. Note, however, that under phase-locking conditions different from the linear case (e.g., $\varphi_l = \varphi_1 l + \varphi_2 l^2$, as in so-called frequency mode-locking) the mode-locked pulses are no longer transform-limited, i. e., their temporal duration is larger than that predicted by the Fourier limit.

Time-domain description. The previous analysis of mode-locking is often referred to as the *frequency-domain* description since the onset of the periodic pulse train is viewed as due to the coherent superposition of longitudinal modes of the laser cavity. A different description of the mode-locking regime, complementary to the frequency-domain approach, is also possible in the *time domain*. In fact, according to the results shown in Fig. 11.24b and since τ_R is the time for a round-trip, a single pulse of duration $\Delta\tau_p$ is circulating within the laser cavity (Fig. 11.25a). Note that the spatial extension of the pulse $\Delta z = c\Delta\tau_p$, according to (11.70) is given by $\Delta z = L_e/(2N+1)$ where L_e is the length of the ring perimeter. For a sufficiently large number $(2N+1)$ of oscillating modes, Δz is then much smaller than the cavity length L_e. The temporal periodicity of the output laser beam then simply results from the successive transits of the intracavity circulating pulse at the output mirror, which occur at intervals $\tau_p = L_e/c$ equal to the cavity transit time. According to this picture, we readily under-

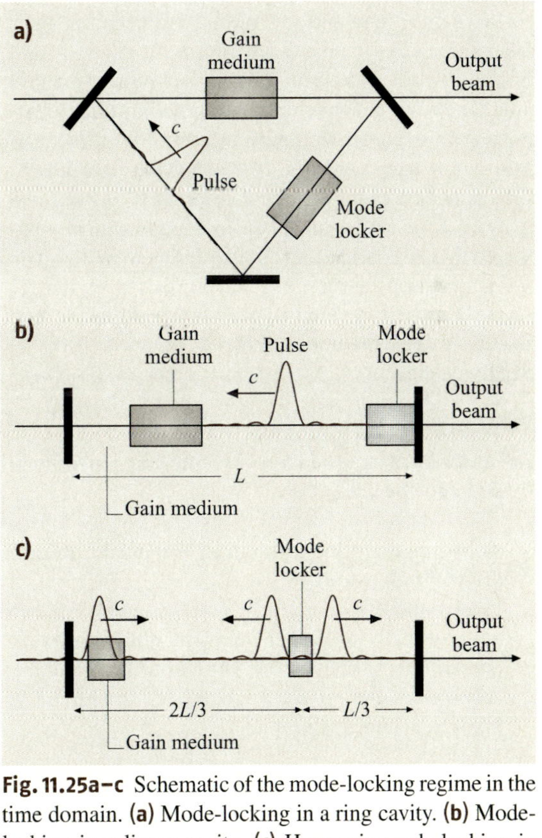

Fig. 11.25a–c Schematic of the mode-locking regime in the time domain. **(a)** Mode-locking in a ring cavity. **(b)** Mode-locking in a linear cavity. **(c)** Harmonic mode-locking in a linear cavity (harmonic order $n = 3$)

stand that the mode-locking regime can be achieved by placing a suitable fast shutter inside the cavity. In fact, if an initially non-mode-locked beam is present within the cavity, its spatial amplitude distribution can be represented as in Fig. 11.24a, with the time t replaced by z/c. By periodically opening the shutter for a short time interval (of the order $\Delta\tau_p$) with period $\tau_p = L_e/c$, possibly at the time when the most intense noisy pulse in Fig. 11.24a reaches the shutter, then only this pulse will survive in the laser cavity, producing the mode-locking condition of Fig. 11.24b. After a transient leading from the temporal pattern of Fig. 11.24a to that of Fig. 11.24b, the mode-locked pulse will consistently reproduce itself after each transit in the cavity.

It should be noted that all considerations made above for a ring laser resonator apply *mutatis mutandis* to a linear (i. e., Fabry–Pérot) laser cavity. In this case, however, self-consistent propagation of the mode-locked pulse in one cavity round-trip requires that the shutter must be placed close to one end mirror of the cavity (Fig. 11.25b). Note that, if the shutter is placed at a distance $L/2$, $L/3$, $\cdots$, L/n from one end mirror and is opened at intervals $\tau_R/2$, $\tau_R/3, \cdots, \tau_R/n$, where L is the cavity length, multiple pulses (precisely $2, 3, \cdots, n$ pulses) may be simultaneously generated and the repetition rate of the pulse train is correspondingly increased by a factor of $2, 3, \cdots, n$ (see, e.g., Fig. 11.25c for the case $n = 3$). Such a mode-locking regime is referred to as *harmonic mode-locking*. Harmonic mode-locking is typically employed in active mode-locked fiber lasers to increase the pulse repetition rate (≈ 1–40 GHz); owing to the relatively long cavity length (≈ 1–10 m), to reach high repetition rates harmonic orders n up to ≈ 1000 are usually needed.

Mode-Locking Methods

The methods to achieve mode-locking can generally be divided into two categories:

1. Active mode locking, in which the mode-locker is driven by an external source.
2. Passive mode locking, in which the mode-locker is not externally driven but exploits some nonlinear optical effect, such as the saturation of a saturable absorber or the nonlinear change of the refractive index in a Kerr medium.

Active mode locking. Active mode locking is usually achieved by placing, inside the laser cavity, either an amplitude modulator, which produces a periodic modulation in time of the cavity loss [amplitude-modulation (AM) *mode locking*], or a phase modulator, which periodically varies the optical length of the resonator [frequency-modulation (FM) *mode locking*]. In lasers with upper-state lifetimes shorter than the cavity round-trip time (e.g., dye lasers), active mode locking can also be achieved by periodic modulation of the laser gain at a repetition rate equal to the longitudinal mode separation $\Delta\nu$ (*synchronous pumping*). We limit here to describe the basic principle of AM mode locking, since it is the most common among the three mentioned techniques. In AM mode locking, the mode locker is usually a Pockels-cell electrooptic modulator for pulsed and high-gain lasers, or an acoustooptic modulator for low-gain lasers. The electrooptic or acoustooptic modulator sinusoidally varies the cavity loss $\gamma(t)$ at a given modulation frequency ν_m (Fig. 11.26). It is thus expected that the mode-locked pulse circulating inside the cavity (Fig. 11.25b) will pass through the modulator at the time t_1 of the cycle where the cavity loss $\gamma(t)$ is minimum. Since the pulse propagating inside the cavity passes again through the modulator at times $t_2 = t_1 + \tau_R$, $t_3 = t_2 + \tau_R$, etc., where $\tau_R = 1/\Delta\nu$ is the cavity round-trip time, a steady mode-locking regime can be reached provided that the synchronization condition $\Delta\nu = \nu_m$ is satisfied. It should be noted that the steady-state pulse duration, $\Delta\tau_p$, is given by the inverse of the oscillation bandwidth, $(2N+1)\Delta\nu$, and, thus, it is ultimately established by the gain bandwidth $\Delta\nu_0$. However, the finite bandwidth of the gain medium influences the steady-state pulse duration in a quite different way for homogeneous or inhomogeneous lines. For an inhomogeneously broadened line, and for a laser well above threshold, the oscillating bandwidth tends to cover the whole gain bandwidth $\Delta\nu_0^*$ even in the absence of AM modulation, and the main role of the mode-locker is just to lock the phases of these oscillating modes. The resulting temporal duration of the mode-locked pulse is therefore given approxi-

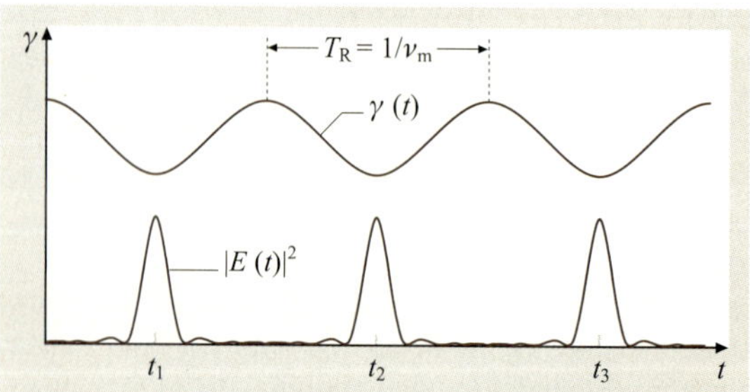

Fig. 11.26 Schematic of AM active mode locking

mately by

$$\Delta\tau_p \simeq \frac{0.44}{\Delta\nu_0^*} . \quad (11.71)$$

Conversely, for a homogeneously broadened gain medium the number of longitudinal modes oscillating in the free-running regime is usually rather modest and the role of the mode-locker is both to enlarge the oscillating bandwidth of the laser (by power transfer from the central to lateral longitudinal modes) and to lock the phases of the oscillating modes. Under steady-state conditions, band enlargement due to the modulator is counteracted by band reduction due to the gain medium and the pulse duration is given by [11.14]

$$\Delta\tau_p \simeq \frac{0.45}{\sqrt{\nu_m \Delta\nu_0}} . \quad (11.72)$$

As an example, we will consider a mode-locked Nd:YAG laser oscillating on its homogeneously broadened line at $\lambda = 1064$ nm. Assuming $\Delta\nu_0 \simeq 126$ GHz ($T = 300$ K), a linear cavity of length $L_e = 1.5$ m and an AM mode-locker placed close to one cavity mirror, the modulator loss must be driven at a frequency $\nu_m = \Delta\nu = c/(2L_e) \simeq$ 100 MHz, and the expected mode-locked pulse duration, according to (11.72), is $\Delta\tau_p \simeq 125$ ps.

It should be noted that in AM mode-locked lasers (as well as in the other active mode locking techniques) even small detunings of the modulation frequency ν_m from the cavity axial mode separation $\Delta\nu$ may result in the destruction of the mode-locking operation. In practice a detuning $|\nu_m - \Delta\nu|/\nu_m$ of the order of $\approx 10^{-4}$ is enough to destroy mode-locking. To obtain stable AM mode-locking, active control of the cavity length is sometimes required, especially when relatively long cavities are employed (such as for AM mode-locked fiber lasers).

Passive mode-locking. There are two main types of passive mode-locking (ML):

1. Fast saturable-absorber ML, which uses the saturation properties of a suitable absorber (e.g., a dye or a semiconductor) with a very short upper-state lifetime;
2. Kerr-lens mode-locking (KLM), which exploits the self-focusing property of a suitable transparent Kerr medium.

Fast saturable-absorber ML [11.15]. Consider a saturable absorber with a low saturation intensity and a relaxation time shorter than the duration of the mode-locked pulses. According to (11.66), the absorption

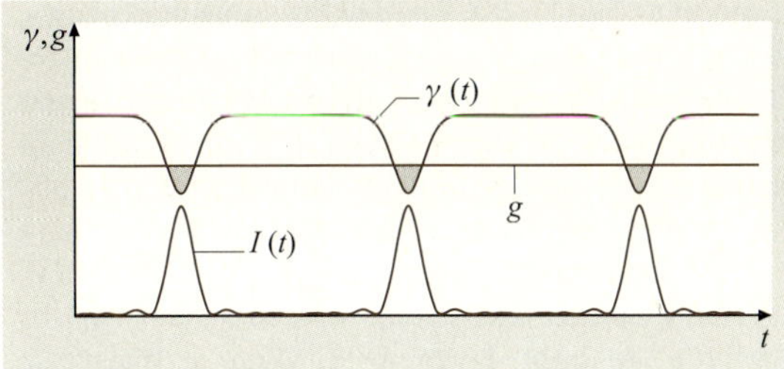

Fig. 11.27 Mode-locking with a fast saturable absorber

loss experienced by a pulse crossing the absorber is dependent on the instantaneous pulse intensity $I(t)$, and decreases as the intensity increases. Thus, starting from the random sequence of light bursts occurring in the unlocked case (Fig. 11.24a), the gain–loss balance will favor the growth and stabilization of the noise pulse with the highest intensity. The steady-state situation, occurring in this case, can be described with the help of Fig. 11.27. Here, gain saturation drives the gain g below the cavity losses $\gamma(t)$ except at those instants where, due to the arrival of the pulse $I(t)$ at the saturable absorber, the losses are reduced due to absorber saturation. During the time intervals corresponding to the hatched regions in Fig. 11.27, the gain is then larger than the instantaneous loss $\gamma(t)$. A so-called window of net gain is thus produced, which tends to increase the peak of the pulse and decrease its wings, i.e., it tends to narrow the pulse. This narrowing is again counteracted by pulse broadening due to the finite amplifier bandwidth until a steady-state pulse duration, whose duration again depends on the inverse of the gain bandwidth, is eventually reached. Good candidates for saturable absorbers must have a short relaxation time τ ($\sim$ a few picoseconds or shorter) and a small saturation intensity, given by $I_0 = h\nu/(2\sigma_a\tau)$. Thus, very large values for the absorption cross section σ_a ($\approx 10^{-16}$ cm^2 or larger) are needed. Ideal absorbers are therefore dye molecules (e.g., cyanine dyes) or, even better, semiconductors. A particularly interesting saturable-absorber geometry consists of integrating a multiple-quantum-well absorber between two mirrors whose spacing is such that the resulting Fabry–Pérot etalon operates in antiresonance. Such a device has been convincingly proven to generate both picosecond and femtosecond laser pulses from several broadband solid-state lasers.

Kerr-lens mode-locking. This technique is based on the use of a nonlinear loss element simply consisting of a nonlinear Kerr medium placed in front of an aperture (Fig. 11.28). The nonlinear medium shows, via the optical Kerr effect, an intensity-dependent refractive index $n = n_0 + n_2 I$, where n_0 is the linear refractive index of the medium, I is the local light intensity and n_2 is a positive coefficient (for a self-focusing medium) which depends on the strength of the nonlinearity (e.g., $n_2 \simeq 4.5 \times 10^{-16}\,\mathrm{cm^2/W}$ for fused quartz and $n_2 \simeq 3.45 \times 10^{-16}\,\mathrm{cm^2/W}$ for sapphire). A light beam with, e.g., a transverse Gaussian intensity distribution $I(r) = I_p \exp[-2(r/w)^2]$ that crosses a thin slice of the Kerr medium of length l thus experiences a transversely varying phase shift $\delta\varphi = 2\pi l n_2 I(r)/\lambda = (2\pi l n_2/\lambda) I_p \exp[-2(r/w)^2]$. Close to the beam center $r = 0$, one can write $\delta\varphi \approx (2\pi l n_2 I_p/\lambda)[1 - 2(r/w)^2]$, i.e., the thin medium introduces a quadratic phase change of the field and thus acts, for $n_2 > 0$, as a positive lens (called a *Kerr lens*) of dioptric power $1/f = 4 n_2 l I_p/(n_0 w^2)$, which increases as the beam intensity I_p increases. If an aperture is placed at some suitable distance from the Kerr medium, a beam with higher intensity will be focused tighter and a higher fraction of the beam will be transmitted through the aperture. Therefore, the Kerr medium with the aperture, like a fast saturable absorber, introduces losses that decrease when the instantaneous pulse intensity is increased, thus leading to mode-locking. Note that, by appropriately controlling the cavity dispersion, the shortest mode-locked pulses ($\approx 6\,\mathrm{fs}$) have been achieved by this technique for Ti^{3+}:Al_2O_3 lasers.

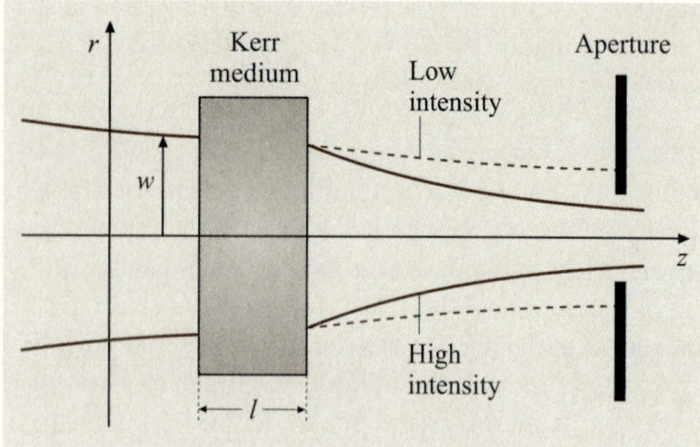

Fig. 11.28 Schematic of a Kerr lens with an aperture used in KLM

11.2 Solid-State Lasers

11.2.1 Basics

Solid-State Lasers Based on Dielectric Insulators

The demonstration of the ruby laser [11.16] (Cr^{3+}:Al_2O_3) in 1960 led to a decade with the realization of a number of crystalline and glass lasers. Especially Nd:YAG [11.17], Nd-doped yttrium aluminum garnet (Nd:$Y_3Al_5O_{12}$), rapidly emerged as one of the most important crystalline dielectric lasers. The first glass laser was a fiber laser [11.18], which decades later led to the development of highly efficient Er-doped fiber amplifiers (EFDA) with important applications in optical communications.

In the last two decades progress in the area of diode-pumped lasers has contributed very much to a renaissance in the field of solid-state lasers. With diode laser pumping it is possible to obtain higher efficiencies and to build rigid all-solid-state devices with simpler and more compact design. Besides Nd^{3+} various efficient diode-pumped rare-earth lasers have been operated with Er^{3+}, Tm^{3+}, Ho^{3+} [11.19–21], and Yb^{3+} [11.22, 23]. In addition, the successful operation of Cr^{3+}-doped [11.24–27] and Ti^{3+}-doped crystals [11.28, 29] as tunable room-temperature lasers has stimulated further research in transition-metal ions. Interesting new results have been obtained with the Cr^{4+} ion [11.30–33] and recently also with the divalent Cr^{2+} ion [11.34].

Nowadays, average and CW output powers for Nd^{3+}- and Yb^{3+}-doped crystals in the kW range are commercially available. Fiber lasers reach output powers in the 100 W range with nearly diffraction-limited beam quality.

Compact solid-state lasers in the visible spectral region (see for instance [11.35]) are of potential interest, especially for display and high-density optical data storage applications. Optical efficiencies of more than 20% with respect to the pump power can be obtained in Nd lasers by internal frequency doubling with nonlinear crystals. An alternative approach enables the generation of visible laser radiation by upconversion schemes, which incorporate energy transfer processes or two-step pump processes as ground-state and excited-state absorption.

In a most general sense lasers based on solids include both dielectric insulators and semiconductors as gain media. However, the development in the past yielded a subdivision into two laser classes based on solids: solid-state lasers (the topic of this chapter) and semiconductor lasers (Sect. 11.3).

Nearly all modern, important solid-state lasers are based on impurity-doped crystals or glasses. Typically, the impurity ions have unfilled electronic shells. So far, only laser ions from the iron, rare-earth and actinide group are known. The most important laser lines of these ions correspond to 4f–4f, 4f–5d, and 3d–3d transitions. In special cases the laser active ions can also be fully substituted into the lattice (stoichiometric laser materials). It should be noted that in the past lattice defects also have been used as laser-active centers (color center lasers); these lasers are however not included in this chapter.

Spectra of Rare-Earth and Transition-Metal Ions in Solids

4f–4f transitions in rare-earth ions. In the free rare-earth ion the electrostatic interaction between the 4f electrons creates a splitting of the energy levels of the $4f^n$ configuration into different LS terms. The resultant wave functions are characterized by the quantum numbers L, S, M_L, and M_S. The electrostatic energy splitting of the ${}^{\{2S+1\}}L$ terms is typically $10^4\,\mathrm{cm}^{-1}$, and each level is $(2L+1)(2S+1)$-fold degenerate with respect to M_L and M_S.

In addition, the energy levels are further split by spin–orbit coupling. If the energetic separation between different LS terms is large compared with the spin–orbit coupling energy there is only a small mixing of the LS terms. If this mixing is very small, the Russel–Saunders approximation holds and the wave functions are characterized by the quantum numbers L, S, J, M_J with a degeneration of $(2J+1)$. The typical splitting of ${}^{\{2S+1\}}L_J$ terms is of the order of $1000\,\mathrm{cm}^{-1}$. Although LS coupling is not strictly valid for rare-earth ions, it is usual to describe the 4f energy states with the Russel–Saunders approximation [11.36].

The interaction of the 4f electrons with the crystal field, i.e., the electrostatic field of the surrounding ligands, results in a Stark splitting of the free ion ${}^{\{2S+1\}}L_J$ terms. This interaction can be treated as a perturbation of the free ion levels. The crystal field splitting and the remaining degeneracy depend on the symmetry of the local crystal field. Lower symmetries increase the number of split levels. However, according to Kramers theorem an odd number of electrons always yields at least twofold degeneracy. The Stark splitting is typically in the energy range of several $100\,\mathrm{cm}^{-1}$. The ground state of a special $4f^n$ configuration (the number of 4f electrons $n=1-14$) can be determined according to Hund's rule. Fig. 11.29 shows the $4f^n$ energy levels of rare-earth ions (Dieke diagram [11.37]).

Electric dipole transitions within the 4f-shell are parity forbidden in the free ion. When doped in a solid, acentric perturbations of the crystal field can however create admixtures of wave functions with opposite parity (for instance $4f^{n-1}5d^1$ states) yielding so-called forced electric dipole transitions. Due to the screening of the outer filled $5s^2$ and $5p^6$ orbitals, the crystal field perturbation is small and electron–phonon coupling is very weak. So, for rare-earth ions at acentric sites one observes electric-dipole zero-phonon transitions with very weak vibronic sidebands. When doping occurs at centric sites, parity remains a good quantum number and all 4f–4f transitions remain electric dipole forbidden. The magnetic-dipole emission cross sections are then very small and are not useful for laser applications.

Further selection rules for electric dipole transitions between $4f^n$ states are:

- $\Delta J \le 6$; $\Delta S = 0$, $\Delta L \le 6$ (Russel–Saunders approximation)
- $J=0 \Leftrightarrow J'=0$ is forbidden.

3d–3d transitions in transition-metal ions. The 3d electrons of transition-metal ions are not shielded and experience a strong perturbation from the crystal field of the surrounding ligand ions. Therefore, the energy-level scheme and also the spectroscopic characteristics of a transition metal ion depend strongly on the strength and the symmetry of the crystalline field originating from the surrounding ions. The energy-level schemes of the transition-metal ions in crystalline hosts are described by so-called Tanabe–Sugano diagrams [11.38]. These diagrams are distinguished by the number of electrons within the 3d electron shell. In these diagrams the energy levels of the transition-metal ion are presented as a function of the crystal field strength (for some examples, see Sect. 11.2.3). For detailed reading appropriate literature can be found in [11.36, 39–42].

Due to the strong interaction with the surrounding ions of the lattice, transition-metal ions mostly exhibit broadband emission because of the electron–phonon coupling between the electronic 3d levels and lattice vibrations. As in the case of 4f–4f transitions, only acentric perturbations can induce electric dipole transitions. Generally the spectra consist of a purely electronic zero-phonon line with vibronic sidebands. In contrast to the situation in rare-earth ions, transition-metal ions at centric sites may also have reasonable transition

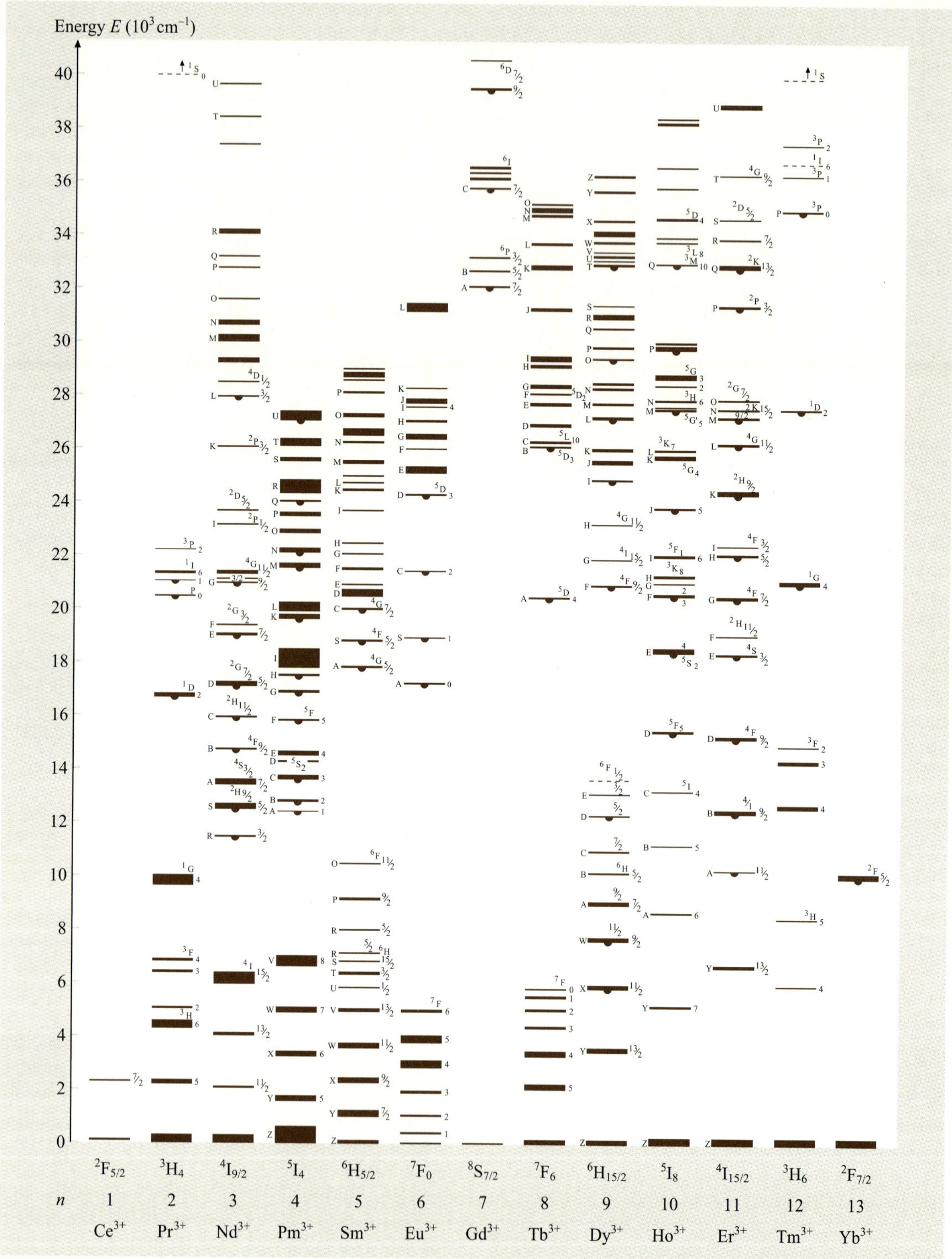

Fig. 11.29 Energy level diagrams of trivalent rare earth ions of RE^{3+} in $LaCl_3$ ("*Dieke-diagram*") [11.37]

probabilities due to dynamical admixtures of wave functions with different parity by acentric phonons. On the other hand, strong electron–phonon coupling can yield higher temperature-dependent nonradiative decay rates compared to 4f–4f transitions. The main interest in transition-metal ions is based on their broad tunability.

Interconfigurational 4f–5d transitions in rare-earth ions. Typically 4f↔5d interconfigurational transitions of rare-earth ions are located in the ultraviolet spectral range. In some cases 4f↔5d transitions are also observed in the visible spectral region. In contrast to 3d↔3d and 4f↔4f transitions, 4f↔5d transitions are electric-dipole allowed, because they obey the parity selection rule. Thus high transition probabilities and hence large absorption and emission cross sections ($\approx 10^{-18}$–$10^{-17}\,\text{cm}^2$) are observed.

Similar to 3d↔3d transitions strong electron–phonon coupling yields broad absorption and emission spectra with spectral half-widths of more than $1000\,\text{cm}^{-1}$. Therefore 4f↔5d interconfigurational transitions are in principle suitable for the generation of tunable laser oscillation.

However, the difficulty in finding suitable pump sources and often also excited-state absorption processes are major drawbacks with respect to the realization of 5d–4f lasers.

Basic Spectroscopic Properties and Laser Parameters

Ground-state absorption. The ground-state absorption of an ion is characterized by the absorption coefficient α and the ground-state absorption cross section σ_{GSA}. These values are derived using the Lambert–Beer law:

$$I(\lambda) = I_0(\lambda)\,e^{-\alpha(\lambda)d} = I_0(\lambda)\,e^{-n_{\text{ion}}\sigma_{\text{GSA}}(\lambda)d}\,, \tag{11.73}$$

where $I(\lambda)$ is the intensity transmitted through the crystal at wavelength λ, $I_0(\lambda)$ is the intensity in front of the crystal, n_{ion} is the ion concentration, $\alpha(\lambda)$ is the absorption coefficient, $\sigma_{\text{GSA}}(\lambda)$ is the absorption cross section, and d is the crystal thickness.

The ground-state absorption spectrum contains information about the energy-level structure of the ion, the cross sections $\sigma_{\text{GSA}}(\lambda)$ and the oscillator strengths of the observed transitions. Note, that the calculation of cross sections needs a second independent measurement of the ion concentration, which is usually done by X-ray microprobe analysis. The relation between the absorption coefficient and the transition-matrix elements can be derived using Fermi's golden rule [11.36]. The energy-level structure of transition-metal ions is theoretically described by the ligand field theory and the angular overlap model (AOM), see for instance [11.43].

Spontaneous emission and emission cross section. The spontaneous emission is characterized by the Einstein coefficient A. In an emission measurement, the photon flux (the number of photons per area per time) is usually measured. The normalization process with respect to the instrument response function can be performed either with respect to photon flux or to the spectral intensity distribution $I_\lambda(\lambda)$ (energy per area per time). The relation between the spectral intensity distribution $I_\lambda(\lambda)$, the Einstein coefficient A ($A = \tau_{\text{rad}}^{-1}$) and the emission cross section $\sigma_{\text{em}}(\lambda)$ is given by the Füchtbauer–Ladenburg equation [11.44]:

$$\sigma_{\text{em}}(\lambda) = \frac{\lambda^5 I_\lambda(\lambda) A}{8\pi n^2 c \int I_\lambda(\lambda)\lambda\,\text{d}\lambda}\,, \tag{11.74}$$

where n is the refractive index and c is the speed of light.

The emission cross sections can also be calculated from the absorption cross sections by the reciprocity method

$$\sigma_{\text{em}}(\lambda) = \sigma_{\text{GSA}}(\lambda)\,\frac{Z_\text{l}}{Z_\text{u}}\exp\left(\frac{E_{\text{zl}} - hc/\lambda}{kT}\right)\,, \tag{11.75}$$

where Z_l and Z_u are the partition functions of the lower and upper energy levels, E_{zl} is the energy of the zero-phonon line of the corresponding transition, k is the Boltzmann constant, and T is the temperature.

For Gaussian band shapes, which are often observed for emission spectra of transition-metal ions, the *McCumber* formula can be used for the determination of the peak emission cross section [11.45, 46]

$$\sigma_{\text{em}} = \sqrt{\frac{\ln 2}{\pi}}\,\frac{A}{4\pi c n^2}\,\frac{\lambda_0^4}{\Delta\lambda}\,, \tag{11.76}$$

where λ_0 is the peak emission wavelength and $\Delta\lambda$ is the full width at half-maximum (FWHM).

Excitation spectra. The excitation spectra allow the determination of those absorption transitions, which yield a specific emission. These measurements are interesting in the case of different absorbing and emitting centers and also for finding energy transfer channels between different optical active centers.

Emission lifetime. The decay time of the metastable level of an ion in a crystal is usually measured after excitation with a short pulse. The measured decay rate (measured as the number of transitions per unit time) is the sum of radiative and nonradiative decay rates. In general, the nonradiative decay of an ion can consist of intra-ionic multiphonon processes and intra-ionic nonradiative transfer processes (for details of energy transfer processes see [11.36]). The equation for the decay is

$$\frac{1}{\tau}=A+\frac{1}{\tau_{nr}}=\frac{1}{\tau_r}+\frac{1}{\tau_{nr}} \quad \text{or}$$
$$W=A+W_{nr}=W_r+W_{nr}\,, \tag{11.77}$$

where τ, τ_r, and τ_{nr} are the total, radiative, and nonradiative decay times, respectively, and W, W_r, and W_{nr} are the total, radiative, and nonradiative decay rates, respectively.

The determination of the Einstein coefficient A is not an easy task, because in a simple decay measurement the combined lifetime is always measured. An additional measurement of the emission quantum efficiency η_{QE} would allow the determination of the radiative and nonradiative decay rate

$$\eta_{QE}=\frac{\tau}{\tau_r}=\frac{W_r}{W}=\frac{W_r}{W_r+W_{nr}}$$
$$\Rightarrow \quad \tau_r=\frac{\tau}{\eta_{QE}} \quad \text{and} \quad \tau_{nr}=\frac{\tau_r\tau}{\tau_r-\tau}\,. \tag{11.78}$$

However, a direct measurement of the quantum efficiency is difficult and usually contains a high error.

Another approach is the indirect determination of the quantum efficiency by analysis of the temperature dependence of the emission lifetime. The two most commonly used models are the simple Mott model of the activation energy [11.47] and the more-sophisticated model of *Struck* and *Fonger* [11.48] using the so-called single-configurational-coordinate model, which describes in a simplified way the interaction between the electronic center and the vibrating crystalline environment.

Excited-state absorption (ESA). Measurements of the excited-state absorption (ESA) spectrum give further insight into the energy-level structure of an ion. In the ground state absorption measurements spin-allowed transitions are mainly observed and often spin-flip transitions are hidden under these absorption bands. Thus the determination of the energy of these spin-flip transitions is often not possible. If the metastable level of the ion has a different spin to the ground state (e.g., Cr^{3+} in strong crystal fields, Mn^{5+}, Fe^{6+}), the ESA spectrum reveals the energetic positions of these states with different spin. By using these data the crystal field parameters can be determined with higher accuracy. The knowledge of ESA processes is also very useful for the determination of their influence on the efficiency of a laser material. ESA may also inhibit gain and hence laser action.

The laser aspect. In this section only the case of steady-state conditions for laser oscillation will be discussed. For more details and for the case of pulsed excitation, see e.g. [11.49–52].

The efficiency of a laser system can be described by the laser threshold P_{thr} and the slope efficiency $\eta=\mathrm{d}P_{out}/\mathrm{d}P_{abs}$, where P_{out} and P_{abs} are the laser output power and the absorbed pump power, respectively. Under the assumptions of only one metastable level (i. e., the upper laser level), ideal overlap between pump beam and resonator mode, homogeneous pump profile, low mirror transmission and low passive losses as well as the absence of excited-state absorption, the following equations hold for continuous-wave operation [11.49, 50]

$$P_{thr}=\frac{h\nu_p}{\eta_p\sigma_{se}(\lambda_l)\,\tau}\times\left[T+L+2d\sigma_{GSA}(\lambda_l)(n_{ion}-n_{thr})\right]\frac{V}{2d}$$
$$\text{(3-level system)}\,, \tag{11.79}$$

$$P_{thr}=\frac{h\nu_p}{\eta_p\sigma_{se}(\lambda_l)\,\tau}(T+L)\frac{V}{2d}$$
$$\text{(4-level system)}\,, \tag{11.80}$$

$$\eta=\eta_p\frac{\lambda_p}{\lambda_l}\frac{T}{T+L}\,, \tag{11.81}$$

where $h\nu_p$ is the energy of pump photon, η_p is the pumping efficiency, i. e., the fraction of absorbed pump photons that are converted into excited ions in the upper laser level, σ_{se} is the stimulated emission cross section, τ is the lifetime of the upper laser level, T is the mirror transmission, L is the passive losses, d is the length of the laser crystal, σ_{gsa} is the ground-state absorption cross section, n is the concentration of laser ions, n_{thr} is the threshold inversion density, λ_p is the pump wavelength, λ_l is the laser wavelength, and V is the pump volume.

In the following, several factors affecting the pump power threshold and the slope efficiency will be discussed.

Influence of the emission quantum efficiency η_{QE}. A quantum efficiency less than unity leads to lifetime shortening and therefore to an increase in the laser threshold, as can be seen from (11.79) and (11.80) after replacing τ by $\eta_{qe}\tau_r$

$$P_{thr} = \frac{h\nu_p}{\eta_p \sigma_{se}(\lambda_l)\eta_{qe}\tau_r} \times \left[T + L + 2d\sigma_{gsa}(\lambda_l)(n_{ion} - n_{thr})\right] \tag{11.82}$$

$$\times \frac{V}{2d} \propto \frac{1}{\eta_{qe}}, \quad \text{(3-level)} \tag{11.83}$$

$$P_{thr} = \frac{h\nu_p}{\eta_p \eta_{qe} \sigma_{se} \tau_r}(T + L)\frac{V}{2d} \propto \frac{1}{\eta_{qe}} \quad \text{(4-level)}. \tag{11.84}$$

The quantum efficiency does not influence the slope efficiency directly (11.81). However in practice a corresponding, significant contribution of nonradiative transitions increases the temperature in the pump volume and usually leads to a lower slope efficiency due to further lifetime reduction and other problems such as thermal lensing.

Influence of passive losses L. Passive losses in a laser system are due to imperfect optical components in the laser resonator. This can be due to, e.g., stray centers and residual absorptions. Passive losses influence both the threshold and the slope efficiency (11.79, 80, 81). In a four-level system, the passive losses can be determined by the Findlay–Clay method [11.53]. For such an analysis, the laser threshold is measured as a function of the output mirror transmission. The following equation holds:

$$P_{thr} = \frac{h\nu_p}{\eta_p \sigma_{se} \tau}\frac{V}{2d}T + \frac{h\nu_p}{\eta_p \sigma_{se} \tau}\frac{V}{2d}L - mT + b\,, \tag{11.85}$$

with slope $m = \frac{h\nu_p}{\eta_p \sigma_{se} \tau}\frac{V}{2d}$ and axis parameter $b = mL$. From a linear fit to $P_{thr} = P_{thr}(T)$ both m and b are determined and thus the value of the passive losses L. An alternative way to determine the passive losses is to rewrite (11.81) for a so-called *Caird plot* [11.54]:

$$\frac{1}{\eta} = \frac{\lambda}{\eta_p \lambda_p}\frac{L}{T} + \frac{\lambda}{\eta_p \lambda_p} = m'\frac{1}{T} + b'\,, \tag{11.86}$$

with slope $m' = \frac{\lambda}{\eta_p \lambda_p}L = b'L$. From a linear fit to $\frac{1}{\eta} = \frac{1}{\eta}\left(\frac{1}{T}\right)$ both m' and b' can be determined and thus the value of the passive losses L.

Influence of excited-state absorption. ESA at the pump wavelength reduces the number of pump photons converted into excited ions in the metastable laser level according to

$$\eta_p = \eta_{p,0}\left(1 - \frac{n_1 \sigma_{ESA}(\lambda_p)}{n_1 \sigma_{ESA}(\lambda_p) + (n_{ion} - n_1)\sigma_{gsa}(\lambda_p)}\right), \tag{11.87}$$

where $\eta_{p,0}$ is the pumping efficiency without ESA, n_{ion} is the concentration of active ions, and n_1 is the population density in the upper laser level. Therefore the pump threshold is increased and the laser slope efficiency is reduced according to (11.79, 80, 81).

ESA on the laser wavelength also affects the laser threshold and the slope efficiency. The stimulated emission cross section $\sigma_{se}(\lambda_l)$ in (11.79) and (11.80) is replaced by $\sigma_{EFF}(\lambda_l) = \sigma_{se}(\lambda_l) - \sigma_{ESA}(\lambda_l)$ and (11.81) has to be extended [11.54]

$$\eta = \eta_p \frac{\lambda_p}{\lambda_l}\frac{\sigma_{se}(\lambda_l) - \sigma_{ESA}(\lambda_l)}{\sigma_{se}(\lambda_l)}\frac{T}{T + L} \propto \frac{\sigma_{EFF}}{\sigma_{se}}\,. \tag{11.88}$$

If $\sigma_{ESA}(\lambda_l) > \sigma_{se}(\lambda_l)$ laser oscillation is not possible.

11.2.2 UV and Visible Rare-Earth Ion Lasers

Lasers based on 5d ↔ 4f transitions of trivalent and divalent rare-earth ions

In this section an overview of lasers oscillating in the ultraviolet and visible spectral range based on transitions of rare-earth ions is given. The first part deals with laser and possible laser systems based on interconfigurational transitions, i. e., $4f^{n-1}5d \rightarrow 4f^n$. In the second part, visible and UV lasers based on intraconfigurational ($4f^n \rightarrow 4f^n$) transitions will be discussed. Both parts deal with crystals as host materials. Finally, in the third part an overview of fiber lasers in the visible spectral range is given.

The 4f↔5d interconfigurational transitions of some divalent (RE^{2+}) and trivalent (RE^{3+}) rare-earth ions are located in the visible and ultraviolet spectral range. They are in principle suitable for the realization of (tunable) laser oscillation. The transitions are electric-dipole allowed and have high transition probabilities and large absorption and emission cross sections of the order of 10^{-17} to $10^{-18}\,cm^2$. Due to their strong electron–phonon coupling, the observed absorption and emission transitions are broad ($> 1000\,cm^{-1}$). Difficulties with respect to laser operation are the nonavailability of efficient and simple excitation sources and the high probability of excited-state absorption and solarization (photoionization), e.g., Ce^{3+}:$LiCaAlF_6$ and Ce^{3+}:$LiSrAlF_6$ depicted in Fig. 11.30.

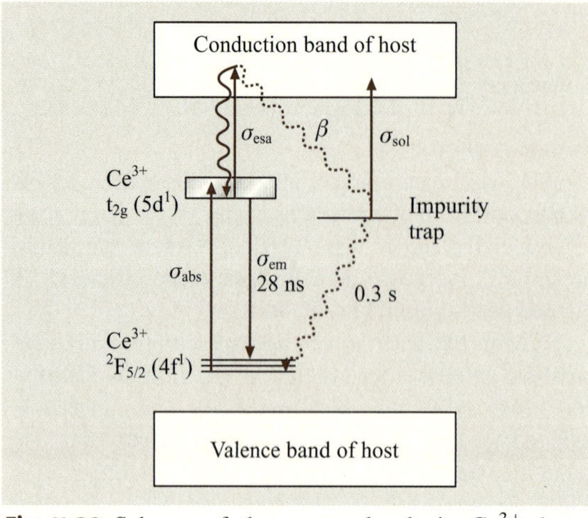

Fig. 11.30 Scheme of the energy levels in Ce^{3+}-doped $LiCaAlF_6$ and $LiSrAlF_6$, where s_{ABS}, s_{EM}, s_{ESA}, and s_{SOL}, are the absorption, emission, ESA, and solarization cross sections, respectively. From [11.55]

In the following the trivalent and divalent rare-earth ions exhibiting short-wavelength emission will be briefly discussed together with a summary of the obtained laser results.

Ce^{3+} *Lasers.* The Ce^{3+} $4f^1$ ground-state configuration splits into the $^2F_{5/2}$ ground state and the $^2F_{7/2}$ excited-state multiplet with an energy separation of $\approx 2000\,cm^{-1}$. The 5d excited configuration consists in an ideal octahedral or cubic symmetry of a threefold orbital degenerated $^2T_2$ state and of a twofold orbital degenerated ^2E level, from which the $^2T_2(^2E)$ state is at lower energy in octahedral (cubic) symmetry. In Fig. 11.31 the room-temperature absorption and emission spectrum of Ce^{3+}:$LiYF_4$ is shown. In general, in the absorption spectrum two to five broad bands are expected, depending on the crystal field symmetry. In emission two bands corresponding to the 5d$\rightarrow$ $^2F_{5/2}$ and 5d$\rightarrow$ $^2F_{7/2}$ transitions with an energetic separation of approximately $2000\,cm^{-1}$ are observed. The high cross sections of the electric-dipole- and spin-allowed $4f^1 \leftrightarrow 5d^1$ transitions correspond to an emission lifetime in the ns range. Ce^{3+}-doped materials were therefore thoroughly investigated for application as tunable solid-state lasers and scintillators, see e.g., the overview articles by Coutts et al. [11.57] and *Dorenbos* [11.58]. Laser oscillation was thus far achieved in $LiYF_4$ [11.56], $LuLiF_4$ [11.59], $LiCaAlF_6$ [11.60], $LiSrAlF_6$ [11.55, 61], LaF_3 [11.62], and BaY_2F_8 [11.63] crystals. The laser data are summarized in Table 11.2. For Ce^{3+}-doped materials the main obstacle with respect to laser oscillation is excited-state absorption assigned to transitions to the conduction band, e.g., as is the case for Ce^{3+}:YAG [11.64, 65]. For Ce^{3+}-doped $LiCaAlF_6$ and $LiSrAlF_6$ the ESA transition leads to a *solarization*, i. e., impurity traps with long lifetimes are populated (Fig. 11.30). With the help of an *anti-solarent* pump, these traps are depleted and higher efficiencies are obtained [11.66, 67].

Pr^{3+} *($4f^2$).* The energy-level scheme of Pr^{3+} is shown schematically in Fig. 11.32. Pr^{3+} has a $4f^2$ ground-state

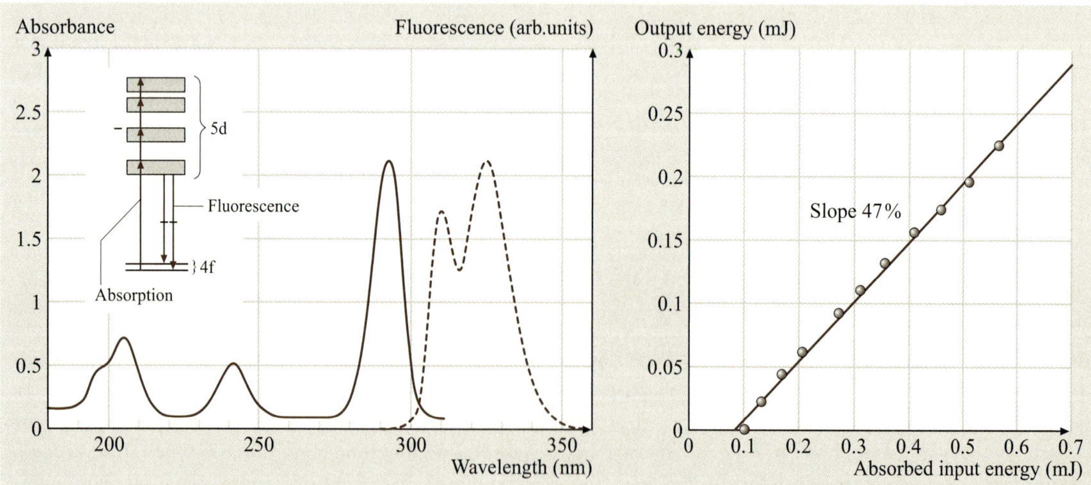

Fig. 11.31 *Left*: Absorption (*solid line*) and emission (*dotted line*) spectra of Ce^{3+}:$LiYF_4$. The inset shows the schematic energy level diagram. From [11.56]. *Right*: Input-output characteristic of a 2% Ce^{3+}, 2% Na:$LiSrAlF_6$

Table 11.2 Overview of lasers based on the 5d→4f transitions of trivalent and divalent rare-earth ions (CVL: copper vapor laser, RS: Raman-shifted)

Laser material	λ_{las} (nm)	E_{out}, P_{out}	η_{sl}	Tuning range (nm)	Pump source	Ref.
Ce^{3+}:$LiYF_4$	325.5				KrF, 249 nm	[11.56]
Ce^{3+}:$LiLuF_4$	309	27 mJ	17%	307.8–311.7 & 323.5–326.5	KrF, 249 nm	[11.75]
		2.1 mJ	55%	307.6–313.5 & 324–328.5	Ce:LiSAF, 290 nm	[11.76,77]
	309	77 μJ		309.5–312.3 & 324.5–327.7	Nd:YAG, 5ω(213 nm)	[11.78]
	309.5	300 mW	38%	305.5–316 & 323–331	CVL, 2ω(289 nm), 7 kHz	[11.79]
	309	67 mW	62%		RS Nd:YAG, 4ω(289 nm), 10 kHz	[11.80]
Ce^{3+}:$LiCaAlF_6$	290		21%		Nd:YAG, 4ω(266 nm)	[11.55]
	289	60 mJ	26%		Nd:YAG, 4ω(266 nm)	[11.81]
	289	30 mJ	39%	284–294	Nd:YAG, 4ω(266 nm)	[11.82]
	289	550 mW	27%	280–311	Nd:YAG, 4ω(266 nm), 1 kHz	[11.83]
	288.5	530 mW	32%	280.5–316	2ω CVL, 271 nm, 7 kHz	[11.79]
	289	0.53 μJ	31%	283–314	Nd:YVO_4, 4ω(266 nm), 1 kHz	[11.84]
	289	230 μJ	49%	280–317	Nd:$YLiF_4$, 4ω(263.3 nm), 0.1–4.3 kHz	[11.85]
Ce^{3+}:$LiSrAlF_6$	290		29%		Nd:YAG, 4ω(266 nm)	[11.55]
	290		47%		Nd:YAG, 4ω(266 nm) + 2ω(532 nm)	[11.66]
				281–315 nm	Nd:YAG, 4ω(266 nm)	[11.86]
Ce^{3+}:LaF_3	286	≈ 5 μJ			KrF, 249 nm	[11.62]
Ce^{3+}:BaY_2F_8	345				XeCl, 308 nm	[11.63]
Nd^{3+}:LaF_3	172				Kr, 146 nm	[11.87,88]
					F_2 Laser, 157 nm	[11.89,90]
Sm^{2+}:CaF_2	708.5 (20 K)			708.5–745 nm (20–210 K)	Xe flash lamp	[11.91,92]

configuration. In this configuration 91 energy levels exist, distributed over 11 manifolds, from which the $^3H_4$ level is the ground state and $^1S_0$ is the highest level at approximately 46 500 cm^{-1}. In the $4f^15d^1$ configuration, 140 energy levels exist. Depending on the crystal field strength and splitting, either the lowest $4f^15d^1$ level or the $^1S_0$ state is lower in energy. Thus after UV excitation, emission occurs either from the $4f^15d^1$ level as a $4f^15d^1 \to 4f^2$ transition or from the $^1S_0$ level as a $4f^2 \to 4f^2$ transition. In any case, the emission spectrum consists of several bands, because several $4f^2$ terminal levels exist. Spectroscopic investigations of the 4f↔5d transitions have so far been carried out for a number of materials, e.g., $Y_3Al_5O_{12}$ [11.68, 69], $YAlO_3$ [11.68, 70, 71], CaF_2, $LiYF_4$ [11.71–73], and $K_5PrLi_2F_{10}$ [11.71]. Thus far laser oscillation based on the $4f^15d^1 \to 4f^2$ or on the $^1S_0 \to 4f^2$ emission of Pr^{3+} has not achieved. The main reason seems to be excited-state absorption into the conduction band, e.g., for $YLiF_4$ [11.73], and into higher-lying $4f^15d^1$ levels, e.g., for $Y_3Al_5O_{12}$ [11.74].

Other RE^{3+} ions. The lowest $4f^{n-1}5d^1$ energy levels for ions from Nd^{3+} to Yb^{3+} are at even higher energies and therefore in the vacuum ultraviolet (VUV) spectral region. Thus the corresponding transitions are not easy to access spectroscopically. A systematic study and detailed analysis of the VUV spectroscopy of trivalent rare-earth ions were performed by Wegh et al. [11.93]. The exploitation of these VUV transitions for laser oscillation still requires much experimental efforts. Laser oscillation on a 5d→4f transition besides for Ce^{3+} has thus far only been realized with Nd^{3+}:LaF_3 at 172 nm with a Kr_2 laser at 146 nm as the pump source [11.87].

Divalent rare-earth ions. Rare-earth ions have a strong tendency to be incorporated into crystals in the trivalent state. In order to obtain the divalent state, the RE^{3+} ion needs to be reduced. Eu^{2+}, Yb^{2+}, Sm^{2+} and, with some limitations, Tm^{2+} can already be obtained in significant amounts during crystal growth by using suitable lattices containing divalent cation sites, appropriate codopants, and a reducing growth atmosphere. The other RE^{2+} ions

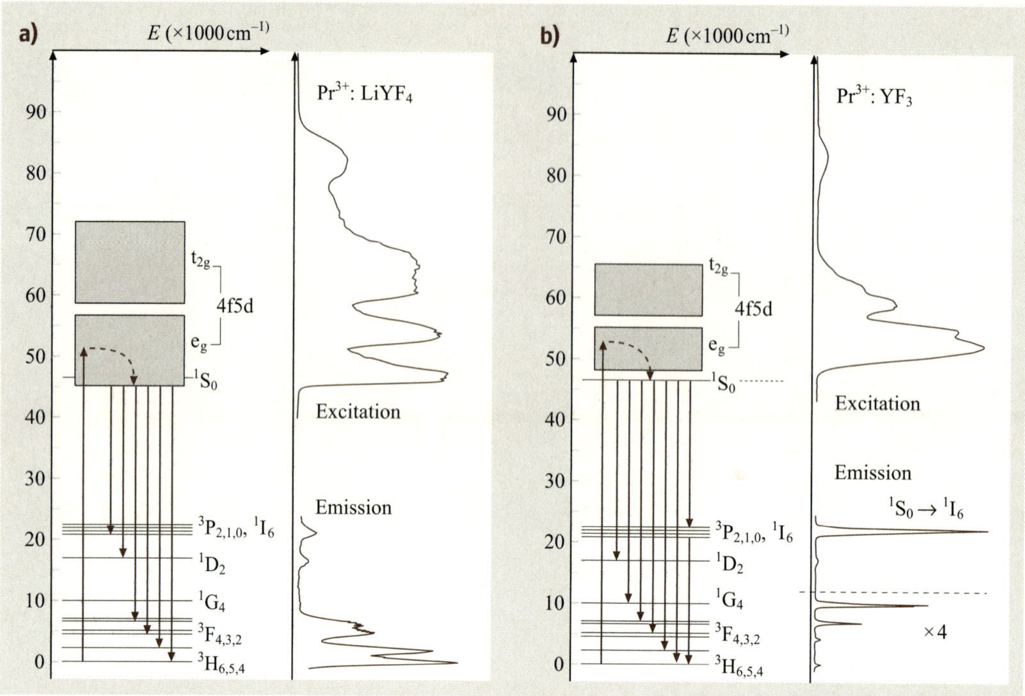

Fig. 11.32a,b Energy-level schemes of Pr^{3+} **(a)** Pr^{3+}:$LiYF_4$, $E(4f^15d^1) < E(^1S_0)$, **(b)** Pr^{3+}:YF_3, $E(4f^15d^1) > E(^1S_0)$. (After [11.94])

are rather difficult to obtain in crystals; usually special growth conditions as well as pre- and post-treatments are necessary, e.g., the use of sealed Ta ampoules during crystal growth, pre-synthesis of REF_2 and $RECl_2$, electron-beam irradiation etc.

5d→4f transitions are expected to occur for systems in which the $4f^{n-1}5d$ level is energetically located within a large energy gap between $4f^n$ multiplets. These ions are Sm^{2+}, Eu^{2+}, Tm^{2+}, and Yb^{2+}.

Sm^{2+} ($4f^6$). The $4f^6$ energy-level scheme of Sm^{2+} is spectroscopically similar to that of Eu^{3+} (Fig. 11.29). For most crystals, Sm^{2+} emission occurs as a $4f^6 \rightarrow 4f^6$ transition from the metastable $^5D_0$ level, which is located at about 15 000 cm^{-1}. In CaF_2, the lowest $4f^55d^1$ level is located just below the $^5D_0$ level and broadband emission occurs. Laser oscillation based on a $4f^55d^1 \rightarrow 4f^6$ transition was realized with Sm^{2+}:CaF_2 [11.91, 92]. The laser wavelength is 708.5 nm at 20 K and increases with temperature to 745 nm at 210 K, which is the highest operation temperature of the Sm^{2+}:CaF_2 laser. This laser is affected by excited-state absorption assigned to transitions to the conduction band and to high-lying $4f^6$ levels [11.95, 96].

Eu^{2+} ($4f^7$). The $4f^7$ energy-level scheme of Eu^{2+} is spectroscopically similar to that of Gd^{3+} (Fig. 11.29). A large energy gap between the $^8S_{7/2}$ ground state and the $^6P_{7/2}$ first excited state exists. Depending on the crystal field strength, the lowest $4f^65d^1$ energy level is located above or below the $^6P_{7/2}$ level. Therefore, either narrow line emission due to the $^6P_{7/2} \rightarrow {}^8S_{7/2}$ transition in the UV region or broadband emission due to the $4f^65d^1 \rightarrow {}^8S_{7/2}$ transition in the blue to yellow spectral range occurs. Laser oscillation based on a $4f^65d^1 \rightarrow 4f^7$ transition of Eu^{2+} has not been realized thus far due to excited-state absorption from the lowest $4f^65d^1$ level to the conduction band and to high-lying $4f^7$ levels [11.96].

Tm^{2+} ($4f^{13}$). The $4f^{13}$ energy-level scheme of Tm^{2+} is spectroscopically similar to that of Yb^{3+} (Fig. 11.29). Laser oscillation of Tm^{2+} was realized in CaF_2 on the $^2F_{5/2} \rightarrow {}^2F_{7/2}$ transition at 1116 nm at temperatures below 27 K [11.97, 98] (Table 11.9). Laser oscillation based on the $4f^{12}5d^1 \rightarrow 4f^{13}$ transition has not been obtained thus far. However, Tm^{2+} is an interesting candidate for laser oscillation on a $4f^{12}5d^1 \rightarrow 4f^{13}$ transition, because there are no high-lying $4f^{13}$ energy levels, which could interfere with the $4f^{12}5d^1$ energy levels and thus act as possible terminal levels for excited-state absorption transitions or quenching processes. The transition between the lowest $4f^{12}5d^1$ level and the ground state is parity allowed, but spin forbidden. Thus the emission cross sections are expected to be about one to two orders of magnitude smaller compared to, e.g., those of the 5d→4f transitions of Ce^{3+}. The $Tm^{2+} \rightarrow Tm^{3+}$ conversion observed under UV/VIS lamp excitation [11.98] indicates the strong tendency of the Tm ion towards the trivalent state and has to be considered as a possible major drawback for laser applications.

Yb^{2+} ($4f^{14}$). The ground-state configuration of the Yb^{2+} ion is $4f^{14}$. This completely filled shell leads to a $^1S_0$ ground state of the free ion and other $4f^{14}$ levels do not exist. In an octahedral crystal field, this state transforms like the $^1A_1$ irreducible representation. The excited $4f^{13}5d^1$ configuration consists in total of 140 energy levels. The $4f^{13}5d$ level splits in a highly symmetric crystal field into a threefold degenerated T_2 and a twofold degenerated E level. The $4f^{13}$ electrons can be considered as a Yb^{3+}-ion configuration with the two manifolds $^2F_{7/2}$ and $^2F_{5/2}$ separated by about $\Delta E_{4f} \approx 10\,000\,\mathrm{cm}^{-1}$. The spin of the 5d electron can either be parallel or antiparallel to that of the $4f^{13}$ core, thus the whole energy-level scheme exists for singlet and for triplet states, from which, according to Hunds' rule, the triplet states are energetically lower. Note furthermore that for the Yb^{2+} free ion the 6s level is energetically lower than the 5d level. Therefore it might be the case that in some materials the $4f^{13}6s$ level is also the lowest excited state. This would lead to a parity- and spin-forbidden transition requiring $\Delta L = 3$ and $\Delta S = 1$. The emission spectra of Yb^{2+}-doped materials consists of a broad band ($\approx 6000\,\mathrm{cm}^{-1}$) with the peak emission wavelength strongly dependent on the host material ($\lambda_{peak} \approx 390–575$ nm) [11.99–103]. Laser oscillation has not been obtained thus far. The main reason is excited-state absorption. In Yb^{2+}:MgF_2 strong ESA transitions in the whole spectral range of absorption and emission are observed, preventing laser oscillation. The ESA cross section is about one order of magnitude larger than the absorption cross section and three orders of magnitude larger than the stimulated emission cross section [11.102, 103].

Lasers based on 4f ↔ 4f transitions of trivalent and divalent rare-earth ions

This section deals with UV and visible lasers based on the 4f↔4f transitions of trivalent and divalent rare-earth ions. Laser oscillation has been realized under flashlamp pumping, direct laser pumping into the upper laser level or higher-lying energy levels and under upconversion pumping.

Pr^{3+} *lasers*. The Pr^{3+} ion is a very interesting and promising ion for obtaining efficient visible laser oscillation. Its energy-level scheme is shown in Fig. 11.33.

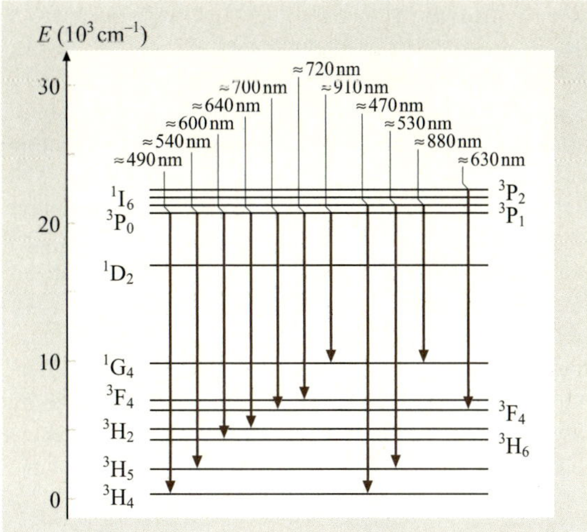

Fig. 11.33 Energy-level scheme of the Pr^{3+} ion. The observed laser transitions in the visible spectral range are indicated by *arrows*

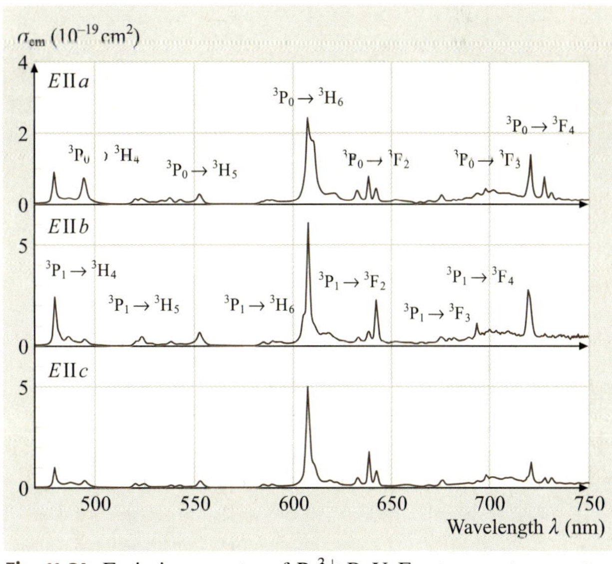

Fig. 11.34 Emission spectra of Pr^{3+}:BaY_2F_8 at room temperature for different polarizations

Laser transitions in the visible spectral range occur from the $^3P_0$ and the thermally populated $^3P_1$ and $^3P_2$ levels. In Fig. 11.34 the emission spectrum of the Pr^{3+}:BaY_2F_8 as an example is shown. The peak cross sections are in the range of ≈ 1 to $5\times10^{-19}\,cm^2$ and thus comparable to the values of the Nd^{3+} $^4F_{3/2} \rightarrow {}^4I_{11/2}$ transition. The highest cross sections are in the orange and red spectral range, therefore the most efficient lasers are expected there. The first Pr^{3+} laser ($\lambda = 1047$ nm) was realized in $CaWO_4$ as early as 1962 by Yariv et al. [11.104]. Since then, laser oscillation has been obtained in more than 20 materials on several transitions and under different pumping conditions, see e.g., the review article by *Kaminskii* [11.105]. The major drawback of Pr^{3+} lasers is their excitation; see the discussion in section on Pr^{3+} laser.

Directly pumped Pr^{3+} lasers. In Fig. 11.35 the absorption spectrum of Pr^{3+}:BaY_2F_8 for the spectral range between 420 nm and 500 nm is shown. Direct pumping of the $^3P_0$ upper laser level is possible for several wavelengths in the blue spectral range corresponding to the $^3H_4 \rightarrow {}^3P_2$, $^3P_1$, $^3P_0$, and $^1I_6$ transitions. In principle the following pumping schemes for direct excitation are possible:

1. Ar^+-ion laser pumping. Ar^+-ion lasers offer the possibility of continuous-wave pumping with a high-quality pump beam. Therefore it allows the characterization of a Pr^{3+}-doped laser material. For Pr^{3+}:$LiYF_4$, laser oscillation on several transitions has been obtained with slope efficiencies up to $\approx 26\%$ and output powers up to ≈ 270 mW [11.106] (Fig. 11.36). However, the wavelength match between the Ar^+ pump and Pr^{3+} absorption is bad (Fig. 11.35). Furthermore Ar^+-ion laser pumping itself is very inefficient, therefore the overall efficiency of a gas-laser-pumped Pr^{3+} laser is very low.
2. Pumping with a frequency-doubled Nd^{3+} laser operating on the $^4F_{3/2} \rightarrow {}^4I_{9/2}$ ground-state transition. These ground-state lasers operate, depending on the host material, between 910 nm and 960 nm, i.e., the frequency doubling yields wavelengths between 455 nm and 480 nm. A Pr^{3+} laser directly pumped by a frequency-doubled Nd^{3+}:YAG ground-state laser operating at 473 nm has been successfully demonstrated by *Heumann* et al. [11.107]. Pr^{3+} laser oscillation occurred at 639.5 nm on the $^3P_0 \rightarrow {}^3F_2$ transition with an output power of nearly 100 mW and a slope efficiency of 12%. Also for frequency-doubled Nd^{3+} ground-state lasers the match between the pump and absorption wavelength is crucial.

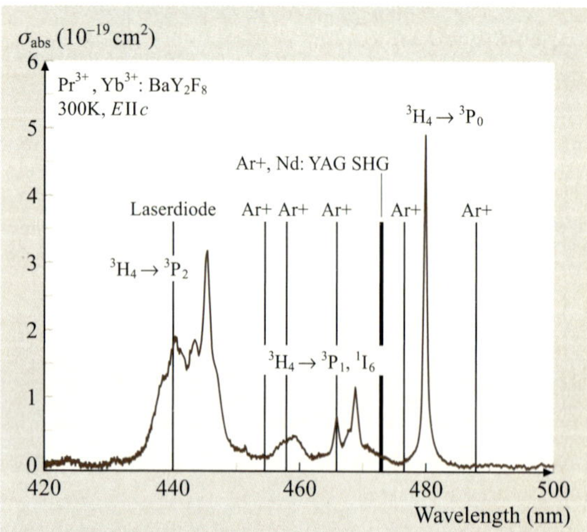

Fig. 11.35 Absorption cross section spectrum of Pr^{3+}, Yb^{3+}:BaY_2F_8 at room temperature. The relevant transitions are assigned. The possible pump laser wavelengths are also shown

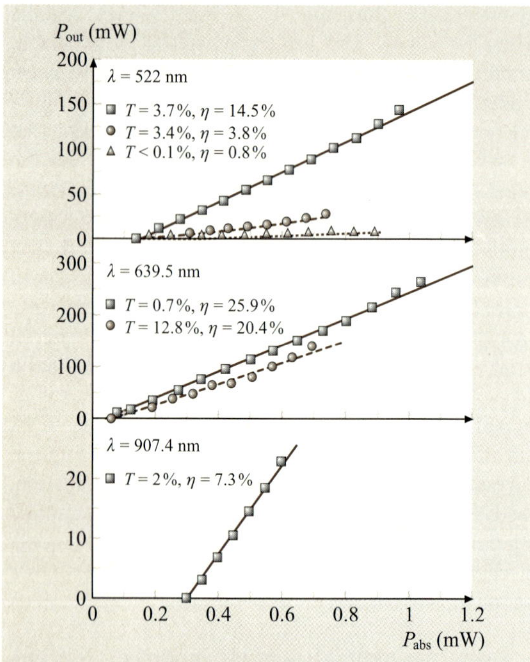

Fig. 11.36 Input–output diagram of the CW Pr^{3+}:$LiYF_4$ laser at different laser wavelengths pumped by an Ar^+-ion laser at 457.9 nm. (After [11.106])

3. Frequency-doubled optically pumped semiconductor lasers (OPS). These lasers are commercially available with powers up to 200 mW [11.108]. In principle, the laser wavelength is adjustable (by choosing the material parameters of the semiconductor) and can be tuned to match the absorption line of the Pr^{3+} ion. Output powers of approximately 35 mW with slope efficiencies of approximately 30% have been obtained [11.109]. Furthermore, *Richter* et al. used an OPS operating at 480 nm as a direct pump source [11.110]. A slope efficiency of 40% and a maximum output power of 75 mW were obtained for $Pr^{3+}:YLiF_4$. For $Pr^{3+}:BaY_2F_8$, the corresponding data are 30% and 51 mW. It is worth noting that intracavity frequency doubling leads to continuous-wave UV radiation of about 19 mW for $Pr^{3+}:YLiF_4$ and $Pr^{3+}:BaY_2F_8$. Due to the multi-wavelength operation of Pr^{3+} lasers, UV generation at 360 nm, 303 nm and even 261 nm seems possible. These pump lasers are still expensive; however power scalability of OPS lasers can be expected making these pump sources very attractive for Pr^{3+}.
4. Blue and UV GaN laser diodes. These diodes operate thus far in the spectral region below 450 nm and with output powers in the mW range. Recently, *Richter* et al. obtained laser oscillation at room temperature with $Pr^{3+}:YLiF_4$ under pumping with a 442 nm GaN laser diode [11.35]. The Pr^{3+} laser emits 1.8 mW at 639.7 nm. The threshold pump power and slope efficiency were 5.5 mW and 24%, respectively. If the development of these diodes towards higher output powers proceeds, laser diodes will certainly be one of the best choices for the direct pumping of Pr^{3+} lasers.
5. Flashlamp pumping [11.111]. Under flashlamp excitation only a small fraction of the emitted radiation of the lamp can be used for the excitation of the Pr^{3+} $^3P_0$ level, because of the narrow line width of the Pr^{3+} absorption lines and of the small spectral range of these lines. Output energies up to 87 mJ and slope efficiencies of 0.3% have been obtained [11.111].
6. Dye-laser pumping. The pump laser wavelength is tunable and thus can be adjusted to match the absorption lines of the Pr^{3+} ion. Therefore, the obtained lasers exhibit a high efficiency. For $Pr^{3+}:LiGdF_4$ lasers, oscillation on several wavelengths and transitions with slope efficiencies up to 37% have been reported [11.112]. However, dye lasers are limited in their practical use.

Table 11.3 lists Pr^{3+} lasers operating in the visible spectral range under direct excitation. In Table 11.4 an overview of the room-temperature laser data for visible Pr^{3+} lasers is given.

Upconversion pumped Pr^{3+} lasers. Because of the difficulties of direct excitation, other pumping schemes using upconversion processes are under investigation for Pr^{3+} lasers. Upconversion describes a process in which the photon energy of the excitation light (pump light) is converted via interaction with the active ions inside the optical material into higher-energy photons [11.113–116]. For Pr^{3+} with its high-lying energy levels upconversion is a suitable way to obtain visible laser oscillation under infrared pumping. Especially the photon-avalanche pumping scheme, which is a combination of different upconversion and energy transfer processes, has been efficiently exploited for the Pr,Yb codoped system. For crystals the $Ti:Al_2O_3$ laser is mainly used as the excitation source, but investigations are being performed on the use of commercially available laser diodes. In fibers, efficient laser operation has already been achieved under laser-diode pumping (see page 633, Visible fiber lasers).

The principle scheme of the photon-avalanche process is shown in Fig. 11.37. A weak ground-state absorption yields the excitation of a few ions into the intermediate (*reservoir*) level. The strong ESA process efficiently brings these ions to the emitting level. An effective feedback mechanism for the reservoir level (in our system a cross-relaxation process) is necessary, which couples the emitting level, the reservoir level and the ground-state level. After these two steps there are two ions in the reservoir level. This cycle repeats and thus the population of the emitting (upper laser) level increases like an avalanche. If the threshold population is reached, the laser oscillation on any transition from the upper laser level may start. The general characteristics are described in detail in [11.113, 114, 117–119]; here only the main points are given for the example of Pr^{3+}, $Yb^{3+}:BaY_2F_8$ [11.120, 121]. The pump-power dependence of the emission intensity exhibits a threshold-like behavior. At this threshold, the slope increases significantly. A second characteristic of the avalanche mechanism is the temporal *S*-shaped evolution of the upconverted emission (Fig. 11.38).

The photon-avalanche process has already been observed in a variety of materials, see e.g. [11.113, 114]. However, laser oscillation has only been obtained in a few materials, see Table 11.5 for Pr^{3+} systems and Table 11.6 for other rare-earth-ion-doped systems. The

Table 11.3 Pr^{3+} lasers in the visible spectral range under direct excitation (bold: laser oscillation on some transitions obtained at room temperature)

Crystal	Transition	Wavelength (nm)	Ref.
$LiYF_4$	$^3P_0 \rightarrow ^3H_4$	479.0	[11.35, 106, 107, 110, 111, 122–126]
	$^3P_1 \rightarrow ^3H_5$	522.0	
	$^3P_0 \rightarrow ^3H_5$	537.8, 545.0	
	$^3P_0 \rightarrow ^3H_6$	604.4, 607.2, 609.2, 613.0	
	$^3P_1 \rightarrow ^3F_2$	615.8, 618, 620.1	
	$^3P_0 \rightarrow ^3F_2$	638.8, 639.5, 644.4	
	$^3P_1 \rightarrow ^3F_3$	670.3	
	$^3P_0 \rightarrow ^3F_3$	695.4, 697.7, 705.5	
	$^3P_1 \rightarrow ^3F_4$	699.4	
	$^1I_6 \rightarrow ^3F_4$	708.2	
	$^3P_0 \rightarrow ^3F_4$	719.5, 720.9, 722.2	
$LiLuF_4$	$^3P_0 \rightarrow ^3H_5$	538	[11.127, 128]
	$^3P_0 \rightarrow ^3H_6$	604.2, 607.1	
	$^3P_0 \rightarrow ^3F_2$	639.9, 640.1	
	$^3P_0 \rightarrow ^3F_3$	695.8, 697.7	
	$^3P_0 \rightarrow ^3F_4$	719.2, 721.5	
$LiGdF_4$	$^3P_1 \rightarrow ^3H_5$	522	[11.106, 112]
	$^3P_0 \rightarrow ^3H_5$	545	
	$^3P_0 \rightarrow ^3H_6$	604.5, 607	
	$^3P_0 \rightarrow ^3F_2$	639	
	$^3P_0 \rightarrow ^3F_3$	697	
	$^3P_0 \rightarrow ^3F_4$	720	
KYF_4	$^3P_0 \rightarrow ^3F_2$	642.5	[11.106, 112]
BaY_2F_8	$^3P_0 \rightarrow ^3H_6$	607.1	[11.110, 125, 129–131]
	$^3P_0 \rightarrow ^3F_2$	638.8	
	$^3P_0 \rightarrow ^3F_3$	693.5–693.8	
	$^3P_0 \rightarrow ^3F_4$	719.1	
LaF_3	$^3P_0 \rightarrow ^3H_6$	598.5, 600.1	[11.132–135]
	$^3P_0 \rightarrow ^3F_4$	719.4, 719.8	
PrF_3	$^3P_0 \rightarrow ^3H_6$	598.4	[11.136]
$LaCl_3$	$^3P_0 \rightarrow ^3H_4$	489.2	[11.137–139]
	$^3P_1 \rightarrow ^3H_5$	529.8	
	$^3P_0 \rightarrow ^3H_6$	616.4, 619.0	
	$^3P_0 \rightarrow ^3F_2$	645.1	
$PrCl_3$	$^3P_0 \rightarrow ^3H_4$	489.2	[11.137, 140]
	$^3P_1 \rightarrow ^3H_5$	529.8, 531	
	$^3P_0 \rightarrow ^3H_6$	617, 620, 622	
	$^3P_0 \rightarrow ^3F_2$	645.2, 647	
$LaBr_3$	$^3P_1 \rightarrow ^3H_5$	532.0	[11.140]
	$^3P_0 \rightarrow ^3H_6$	621.0	
	$^3P_2 \rightarrow ^3F_3$	632.0	
	$^3P_0 \rightarrow ^3F_2$	647.0	
$PrBr_3$	$^3P_0 \rightarrow ^3H_6$	622	[11.137, 140]
	$^3P_0 \rightarrow ^3F_2$	645.1, 649	

Table 11.3 (continued)

Crystal	Transition	Wavelength (nm)	Ref.
$Y_3Al_5O_{12}$	$^3P_0 \rightarrow {}^3H_4$	487.2	[11.141, 142]
	$^3P_0 \rightarrow {}^3H_6$	616	
	$^3P_0 \rightarrow {}^3F_4$	747	
$YAlO_3$	$^3P_0 \rightarrow {}^3H_6$	613.9, 621.3, 621.6	[11.143–147]
	$^3P_0 \rightarrow {}^3F_2$	662	
	$^3P_0 \rightarrow {}^3F_3$	719.5, 719.7, 722	
	$^3P_0 \rightarrow {}^3F_4$	746.9	
	$^1D_2 \rightarrow {}^3F_3$	743.7, 753.7	
$LuAlO_3$	$^3P_0 \rightarrow {}^3H_6$	615.5	[11.144, 145, 147]
	$^3P_0 \rightarrow {}^3F_3$	722.0	
	$^3P_0 \rightarrow {}^3F_4$	749.6	
$SrLaGa_3O_7$	$^3P_0 \rightarrow {}^3H_4$	488	[11.148]
	$^3P_0 \rightarrow {}^3F_2$	645	
$CaWO_4$	$^3P_0 \rightarrow {}^3F_2$	649.7	[11.149]
$Ca(NbO_3)_2$	$^3P_0 \rightarrow {}^3H_6$	610.5	[11.149]
$LiPrP_4O_{14}$	$^3P_0 \rightarrow {}^3H_6$	604.8, 608.5	[11.150]
	$^3P_0 \rightarrow {}^3F_2$	639.6	
	$^3P_0 \rightarrow {}^3F_4$	720.4	
LaP_5O_{14}	$^3P_0 \rightarrow {}^3F_2$	637	[11.151]
	$^3P_0 \rightarrow {}^3F_4$	717	
PrP_5O_{14}	$^3P_0 \rightarrow {}^3F_2$	637.4	[11.152–154]
LaP_5O_{14}	$^3P_0 \rightarrow {}^3F_2$	637.0	[11.151]

Table 11.4 Overview of room temperature laser data of directly pumped Pr^{3+} lasers in the visible spectral range

Host	Transition	λ_{laser} (nm)	Pump	P_{thr}/E_{thr}	P_{out}/E_{out}	η (%)	Ref.
$LiYF_4$	$^3P_1 \rightarrow {}^3H_5$	522.0	457.9 nm	163 mW	144 mW	14.5	[11.106]
	$^3P_0 \rightarrow {}^3H_5$	545.0	CW, Ar^+-ion laser		19 mW		
	$^3P_0 \rightarrow {}^3H_6$	607		110 mW	7 mW	1.2	
	$^3P_0 \rightarrow {}^3F_2$	639.5		8 mW	266 mW	25.9	
	$^3P_0 \rightarrow {}^3F_3$	697		105 mW	71 mW	10.3	
	$^3P_0 \rightarrow {}^3F_4$	720		98 mW	40 mW	7.2	
	$^3P_0 \rightarrow {}^1G_4$	907.4		280 mW	23 mW	7.3	
$LiYF_4$	$^3P_0 \rightarrow {}^3F_2$	639.5	473 nm CW, SHG Nd:YAG	40 mW	≈ 100 mW	12	[11.107]
$LiYF_4$	$^3P_0 \rightarrow {}^3H_6$	613	476 nm CW, Ar^+ – ionlaser		45 mW	≈ 400 fs (ML)	[11.124]
$LiYF_4$	$^3P_0 \rightarrow {}^3F_2$	639.5	442 nm CW, GaN laser diode	5.5 mW	≈ 1.8 mW	24	[11.35]
$LiYF_4$	$^3P_0 \rightarrow {}^3F_2$	639.5	480 nm CW OPS	37 mW	72 mW	40	[11.110]
$LiYF_4$	$^3P_0 \rightarrow {}^3F_2$	639.5	Xe flashlamp (60 μs)	≈ 7 J	87 mJ	≈ 0.3	[11.111]
$LiGdF_4$	$^3P_1 \rightarrow {}^3H_5$	522	468 nm	197 μJ	83 μJ	27	[11.112]
	$^3P_0 \rightarrow {}^3H_5$	545	pulsed dye laser	49 μJ	2 μJ		
	$^3P_0 \rightarrow {}^3H_6$	604.5		144 μJ	32 μJ	37	
	$^3P_0 \rightarrow {}^3H_6$	607		50 μJ	2 μJ		
	$^3P_0 \rightarrow {}^3F_2$	639		4 μJ	98 μJ	32	
	$^3P_0 \rightarrow {}^3F_3$	697		73 μJ	31 μJ	26	
	$^3P_0 \rightarrow {}^3F_4$	720		6 μJ	80 μJ	35	
KYF_4	$^3P_0 \rightarrow {}^3F_2$	642.5	457.9 nm CW, Ar^+-ion laser		15 mW		[11.106]
KYF_4	$^3P_0 \rightarrow {}^3F_2$	642.5	465 nm pulsed dye laser				[11.112]
$YAlO_3$	$^3P_0 \rightarrow {}^3F_4$	746.9	476.5 nm CW, Ar^+-ion laser	25 mW	130 mW	24.6	[11.146, 155]

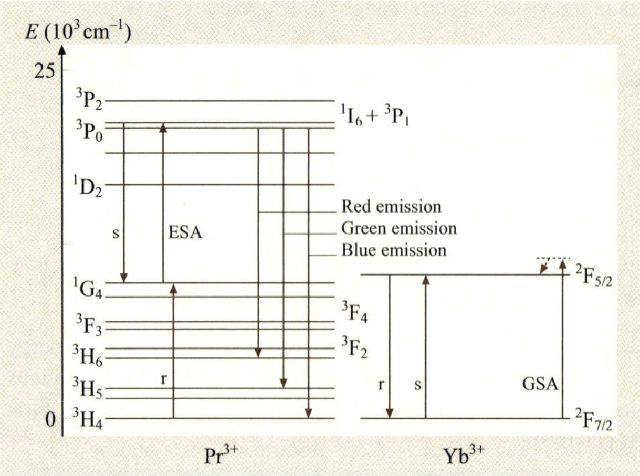

Fig. 11.39 Scheme of the avalanche mechanism in Yb–Pr-doped systems

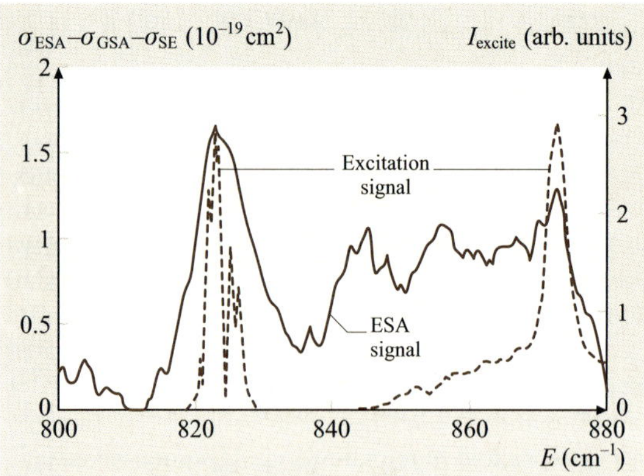

Fig. 11.40 Upconverted emission excitation spectrum and excited-state absorption spectrum of Pr^{3+}, Yb^{3+}:BaY_2F_8. (After [11.120])

Therefore high pump intensities over a long distance are possible, increasing the overall efficiency of the avalanche pumping mechanism, see the section on visible fiber lasers in Sect. 11.2.2.

Er^{3+} *lasers.* The problem for the realization of directly pumped visible Er^{3+} lasers is the lack of efficient pump sources that have a good spectral match to the absorption lines of the Er^{3+} ion. Therefore most visible Er^{3+} laser schemes have been realized by upconversion pumping in the near-infrared spectral range, as shown in Fig. 11.42. Table 11.7 summarizes a few results of direct pumping of Er^{3+} obtained with Xe lamps and a dye laser.

The energy-level schemes of Erbium-doped crystals such as $LiYF_4$ and $LiLuF_4$ [11.127, 128] offer the feasibility of realizing an upconversion laser emitting in the green spectral region ($^4S_{3/2} \rightarrow {}^4I_{15/2}$ transition). As shown in Fig. 11.42 the upconversion excitation of the upper laser level $^4S_{3/2}$ requires sequential two-photon absorption (STPA) either at 810 nm or 970 nm. With a pump wavelength of 810 nm, the ground-state absorption (GSA) $^4I_{15/2} \rightarrow {}^4I_{9/2}$ is followed by nonradiative decay to the $^4I_{11/2}$ level. From there, population is taken to the $^4F_{5/2}$ level by excited-state absorption (ESA). Finally, the $^4S_{3/2}$ level is populated by nonradiative decay. With a pump wavelength around 970 nm, the upconversion excitation scheme of the upper laser level is very similar, however, it involves the GSA process $^4I_{15/2} \rightarrow {}^4I_{11/2}$ and the ESA process $^4I_{11/2} \rightarrow {}^4F_{7/2}$.

Using a Ti:sapphire laser for excitation, various rare-earth-doped fluoride crystals have shown room-temperature CW upconversion lasing in the visible spectral range, e.g., Er^{3+}:$LiYF_4$ [11.181] and Er^{3+}:$LiLuF_4$ [11.182]. Diode pumping of a green erbium-doped upconversion laser in Er^{3+}:$LiLuF_4$ [11.183] and in the mixed fluoride crystal Er^{3+}:$LiKYF_5$ [11.184] has been demonstrated. In the latter case, however, laser operation could only be achieved under chopped excitation with a duty cycle of 20%.

Table 11.8 presents an overview Er-based upconversion lasers in different crystals and at various wavelengths.

Besides $YAlO_3$ most of the interesting candidates are fluorides, due to their relatively small phonon energies and correlated long lifetimes of intermediate states. Rare-earth-doped $LiLuF_4$ exhibits a larger splitting of the manifolds by the crystal field compared to $LiYF_4$ and $LiGdF_4$, which tends to produce more-favorable thermal occupation factors of both the upper and lower laser levels [11.185, 186]. A reasonable overlap between GSA and ESA enables two-step excitation to the $^4S_{3/2}$ level of Er^{3+} using just a single pump wavelength. The emission cross section σ_{em} at 552 nm (π polarization) is $\sigma_{em} = 3.5 \times 10^{-21}\,cm^2$. This value as well as the GSA cross sections around 970 nm are slightly larger than the corresponding cross sections in Er^{3+}(1%):$LiYF_4$. The lifetime τ of the upper laser level $^4S_{3/2}\left[\tau(^4S_{3/2}) = 400\,\mu s\right]$ is slightly longer in $LiLuF_4$ than in $LiYF_4$.

A multipass pumping setup has been used in order to increase the absorbed pump power (Fig. 11.43). Both end faces of the 1.6 mm-long Er^{3+}(1%):$LiLuF_4$ crystal have been prepared with directly coated dielectric mirrors [11.183]. One of these mirrors is highly trans-

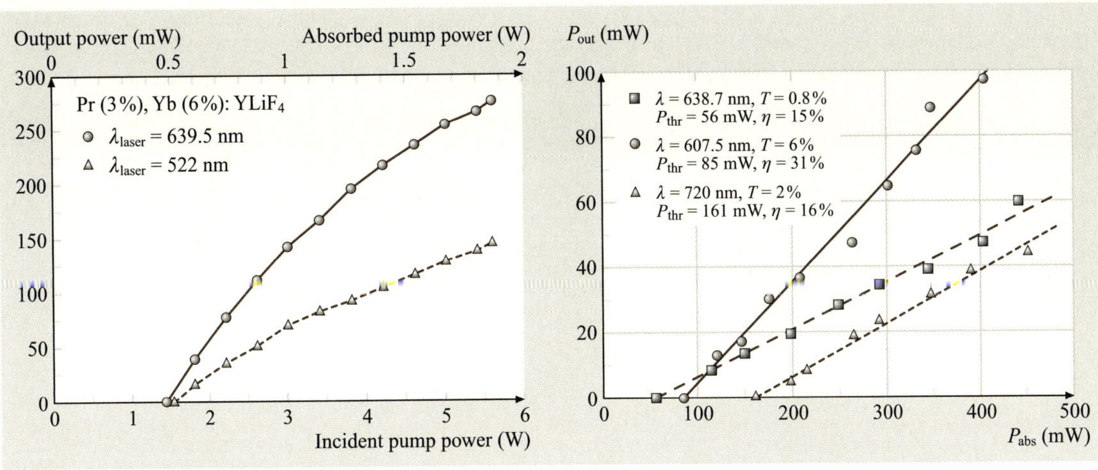

Fig. 11.41 Input–output curves of Pr,Yb:YLF [11.157, 158] (*left*) and Pr,Yb:BYF (*right*) under avalanche pumping with a Ti:sapphire laser

mitting for the pump wavelength and highly reflecting for the laser wavelength, while the other mirror is highly transmitting for the laser wavelength and highly reflecting for the pump wavelength (Fig. 11.43, coatings L and P, respectively). An additional concave mirror with a hole drilled slightly off axis is used to realize up to four passes of the pump radiation through the active volume of the laser crystal. The collimated pump beam is focused through the mirror hole into the crystal. With respect to single-pass pumping significant improvement of the performance of the Er^{3+}:$LiLuF_4$ upconversion laser could be achieved under Ti:sapphire laser excitation by using this setup. The maximum CW output power was 213 mW at an incident pump power of 2.6 W. The slope efficiencies with respect to incident and absorbed pump power were 12% and 35%, respectively. When replacing the Ti:sapphire laser by a 3 W diode, it was possible to realize laser-diode-pumped CW operation of this erbium-doped upconversion laser at room temperature for the first time [11.183]. The maximum output power was 8 mW at an incident pump power of 2.5 W. The absorbed power under four-pass pumping was small and estimated to be between 10% and 12%. The slope efficiency with respect to the absorbed pump power was 14%. Pumped with an optically pumped semiconductor laser operating at 970 nm, continuous-wave laser oscillation at 550 nm with Er^{3+}:$LiLuF_4$ has been achieved with an output power of 500 mW and a slope efficiency of about 30% [11.127, 128, 182, 183].

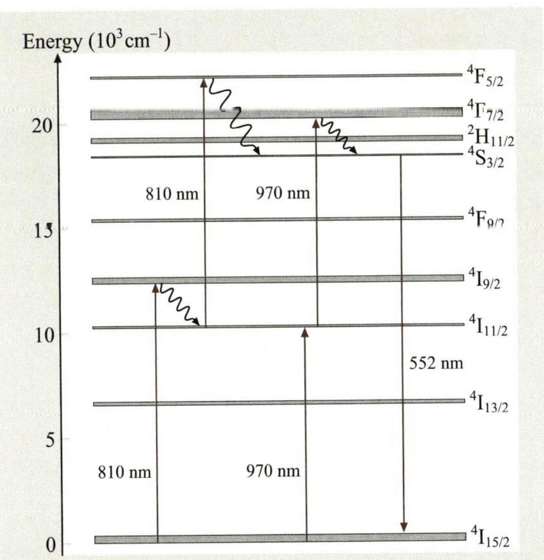

Fig. 11.42 Energy-level scheme of Er^{3+}:$LiLuF_4$ [11.127, 128] with two possible upconversion excitation routes of sequential two-photon absorption (STPA) at 810 nm or 970 nm [11.182, 183] and emission at 552 nm

Other 4f–4f divalent and trivalent rare-earth ion lasers. Besides visible lasers based on 4f–4f transitions of Pr^{3+} and Er^{3+} several other lasers with 4f–4f transitions in the visible spectral region have been realized as shown in Table 11.6. When compared to Pr^{3+} and Er^{3+} lasers, the laser performance of all listed systems in Table 11.6 is worse. However, at least the upconversion pumped Tm^{3+} and Ho^{3+} lasers seem to have some potential for improvement of laser performance in the visible spectral region.

Table 11.8 Visible Er^{3+} lasers (ETU: energy transfer upconversion, STPA: sequential two-photon absorption, CT: cooperative transfer, PA: photon avalanche, CW: continuous wave, p: pulsed, SP: self-pulsed, QS: Q-switched, ML: mode-locked, IR-fl: infrared flashlamp, OPS: optically pumped semiconductor, RT: room temperature)

Host	λ_{laser} (nm)	λ_{pump} (nm)	Pump mechanism	T (K)	Output	η (%)	Ref.
BaY_2F_8	670	IR-fl	ETU	77			[11.169]
$YAlO_3$	550	792 + 840	STPA	≤ 77	0.8 mW, CW	0.2	[11.187]
$YAlO_3$	550	785 + 840	STPA	34	8 mW, CW	1.8	[11.188, 189]
$YAlO_3$	550	807	ETU	7–63	166 mW, CW	13	[11.188]
$YAlO_3$	550	791.3	STPA + looping	7–34	33 mW, CW	3.3	[11.188, 190]
$Y_3Al_5O_{12}$	561	647 + 810	STPA	RT			[11.191]
CaF_2	855	1510	CT	77	64 mW, CW	18	[11.192]
$LiYF_4$	551	797 or 791 (diode)	STPA	≤ 90		0.2	[11.193–196]
$LiYF_4$	551	791 (diode)	ETU / STPA	≤ 90	0.1 mW, SP	0.03	[11.195]
$LiYF_4$	551	802 (diode)	ETU / STPA	≤ 77	2.3 mW, SP		[11.197]
$LiYF_4$	551	797 (diode)	CT	48	100 mW, SP	5.5	[11.198]
$LiYF_4$	702	1500	CT	10	360 μW, CW	0.06	[11.199]
$LiYF_4$	551	810	STPA	RT	40 mW, CW	1.4	[11.181]
$LiYF_4$	551	974	STPA	RT	45 mW, CW	2	[11.200, 201]
$LiYF_4$	551	1500	ETU	80	10 mW, SP	2.9	[11.202]
$LiYF_4$	561	1500	ETU	80	12 mW, SP	3.4	[11.202]
$LiYF_4$	468	1500	ETU	80	0.7 mW, SP	0.2	[11.202]
$LiYF_4$	561, 551, 544	797	ETU	49	467 mW, CW	11	[11.203]
$LiYF_4$	551, 544	1550	CT	9–95	34 mW, CW	8.5	[11.204, 205]
$LiYF_4$	551, 544	1500	CT	≤ 95	0.6 μJ (50 ns), QS	2 mW, ML	[11.206]
$LiYF_4$	551	647 + 810	STPA	RT	0.95 mJ, p	8.5	[11.191]
$LiYF_4$	467	1550	CT	70			[11.116]
$LiYF_4$	469.7	969.3	ETU	≤ 35	2 mW, CW	0.3	[11.207]
$LiYF_4$	469.7	653.2	ETU	≤ 35	6 mW, CW	4.8	[11.207]
$LiYF_4$	560.6	969.3	ETU	≤ 35		2	[11.207]
$LiYF_4$	551	802	ETU	≤ 90	5 mW, SP	2	[11.165]
$LiYF_4$	1230, 850	1530	ETU	110			[11.208]
$LiLuF_4$	552	970 968 (diode)	STPA	RT	213 mW 8 mW	35 14	[11.183]
$LiLuF_4$	552	970 (OPS)	STPA	RT	500 mW (CW) 800 mW (DC: 50)	30	[11.209]
KYF_4	562	647 + 810	STPA	RT	0.95 mJ, p	0.5	[11.191]
$LiKYF_5$	550	488 (Ar^+) 651 (diode) 808 (diode)	STPA	RT	40 mW 50 mW (DC: 20) 150 mW (DC: 20)	18 6 12	[11.184]
$BaYb_2F_8$	670	1540 + 1054 or 1054	ETU	RT			[11.210]
BaY_2F_8	470, 554, 555	792.4	CT	10			[11.211]
BaY_2F_8	552	792.4	CT	40			[11.211]
BaY_2F_8	617, 669	792.4	CT	20			[11.211]
BaY_2F_8	552, 470	≈ 790 or ≈ 970	STPA	10	CW		[11.212]

Blue upconversion laser emission in Tm-doped crystals can be achieved with STPA pumping. The energy levels and upconversion pump mechanisms in Tm^{3+} are illustrated in Fig. 11.44.

Figure 11.44a shows a pure STPA-pumped upconversion laser scheme, whereas in Fig. 11.44b, upconversion pumping also requires an additional cross-relaxation step between two Tm^{3+} ions. This cross-relaxation process populates the intermediate $^3F_4$ state which is the starting level for the second step of the STPA process.

In sensitized upconversion lasers a donor ion D (the sensitizer) absorbs the pump light and transfers its excitation energy to an acceptor ion A. In many cases Yb^{3+} has been used as the donor D for trivalent rare-earth-ion activators A. Through an upconversion energy transfer (ETU), two donor ions transfer their excitation energy successively to an acceptor. Finally the excitation energy of the donor is higher than the energy of an excited Yb^{3+} ion.

Figure 11.45 shows a system with $D = Yb^{3+}$ and $A = Tm^{3+}$. In this case a three-step STPA process is also possible, yielding a variety of laser transitions in the visible region.

Figure 11.46 illustrates the STPA and ETU upconversion mechanisms in Ho^{3+} and in the donor–acceptor system Yb^{3+}–Ho^{3+}. Laser emission can be generated near 750 nm and 550 nm, originating from the metastable $^5S_2$, $^3F_4$ states.

It can be seen in Table 11.6 that, besides STPA and ETU pumping, photon avalanche (PA) pumping has also been used in a few cases (Nd^{3+} and Tm^{3+}).

In Table 11.9 other rare-earth-ion lasers in the visible spectral range based on 4f–4f transitions are listed. The data are taken from [11.213].

Visible fiber lasers

A different and very promising approach to realize efficient upconversion room-temperature laser oscillation is the use of rare-earth-ion-doped fibers. The geometry of the fiber provides waveguiding of both the pump radiation and the stimulated emission, thus long interaction lengths can be realized, yielding high intensities over a long distance, a critical requirement for upconversion lasers. This is one advantage of the fiber concept (for a detailed description of the fiber concept, see the description of Yb fiber lasers in Sect. 11.2.2) compared to bulk materials. Another requirement for visible upconversion lasers is the existence of metastable intermediate levels to act as initial levels for an excited-state absorption or an energy transfer process. Like in crystals, fluoride materials are preferred because of their low phonon energies and generally larger bandgaps. The fiber material of choice is the fluorocirconate glass ZBLAN ($ZrF_4 - BaF_2 - LaF_3 - AlF_3 - NaF$). Another advantage of fibers compared to bulk crystals is that the transitions in glass are broadened and therefore the possibility for resonant transitions or energy transfer is enhanced. An overview of visible fluoride fibers is given in [11.214].

Nearly all rare-earth-ion-doped fluoride fiber lasers can be pumped in the 0.63–1.2 μm region and thus take advantage of the mature semiconductor technology, e.g., AlGaInP (0.63–0.69 μm), GaAlAs (0.78–0.88 μm) and InGaAs (0.90–1.2 μm) and of the highly developed solid-state laser technology, e.g.,

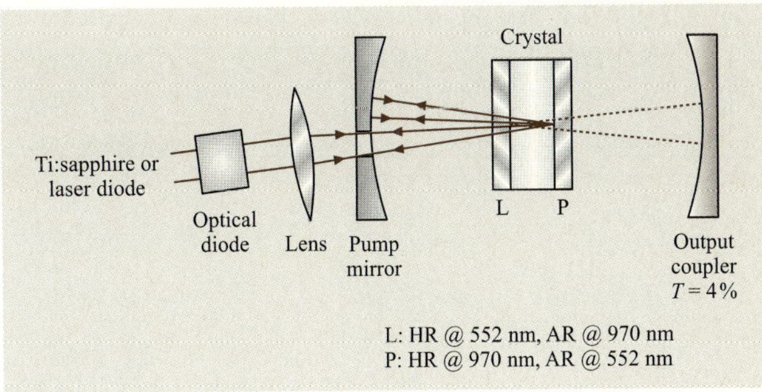

Fig. 11.43 Experimental setup for fourfold-pass pumping [11.183] (HR: high reflection, AR: antireflection)

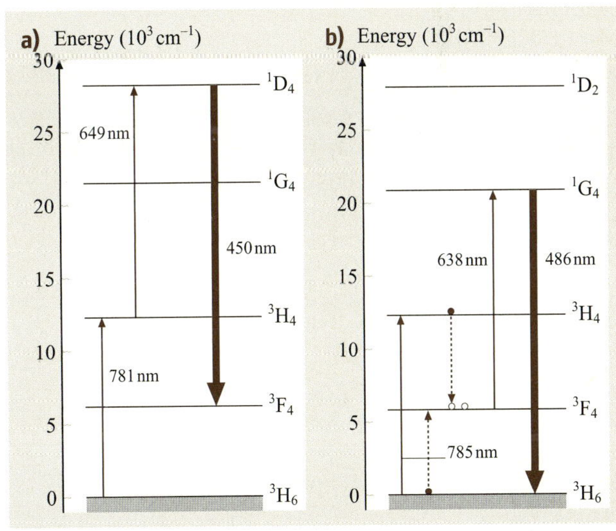

Fig. 11.44a,b Upconversion pump mechanisms for Tm^{3+}. **(a)** Sequential two-photon absorption, **(b)** sequential two-photon absorption with additional cross-relaxation (*dashed line*)

Part C | 11.2

Table 11.9 Visible and UV lasers based on divalent and trivalent rare-earth ions

Dopant ions	Host	Transition	λ_{laser} (nm)	T (K)	Ref.
Sm^{3+}	TbF_3	$^4G_{5/2} \rightarrow {}^6H_{7/2}$	593.2	116	[11.215]
Sm^{2+}	SrF_2	$^5D_0 \rightarrow {}^7F_1$	696.9	4.2	[11.216]
Eu^{3+}	Y_2O_3	$^5D_0 \rightarrow {}^7F_2$	611.3	220	[11.217]
Eu^{3+}	YVO_4	$^5D_0 \rightarrow {}^7F_2$	619.3	90	[11.218]
Gd^{3+}	$Y_3Al_5O_{12}$	$^6P_{7/2} \rightarrow {}^8S_{7/2}$	314.5	300	[11.219]
Tb^{3+}	$LiYF_4$	$^5D_4 \rightarrow {}^7F_5$	544.5	300	[11.220]
Ho^{3+}	CaF_2	$^5S_2 \rightarrow {}^5I_8$	551.2	77	[11.221]
Tm^{2+}	CaF_2	$^2F_{5/2} \rightarrow {}^2F_{7/2}$	1116	< 27	[11.97,98]
Ag^+	KI, RbBr, CsBr		335	5	[11.222]

Nd^{3+} and Yb^{3+} lasers in the near-infrared spectral region.

Visible Pr^{3+} *fiber lasers.* As in crystals, the praseodymium ion is very attractive for visible fiber lasers because of its energy-level scheme (see Fig. 11.33) and the possibility of upconversion pumping by two-step absorption, photon avalanche or energy transfer processes. Laser oscillation at room temperature has been achieved in the red, orange, green and blue spectral ranges. Some of these transitions have even been operated simultaneously [11.223]. Laser oscillation at room temperature originates from the $^3P_0$, $^3P_1$ and $^1I_6$ levels, which are thermally coupled. In Pr^{3+}:ZBLAN, the lifetime of these coupled multiplets is in the range 40–50 μs [11.224,225]. Recently, *Richter* et al. obtained

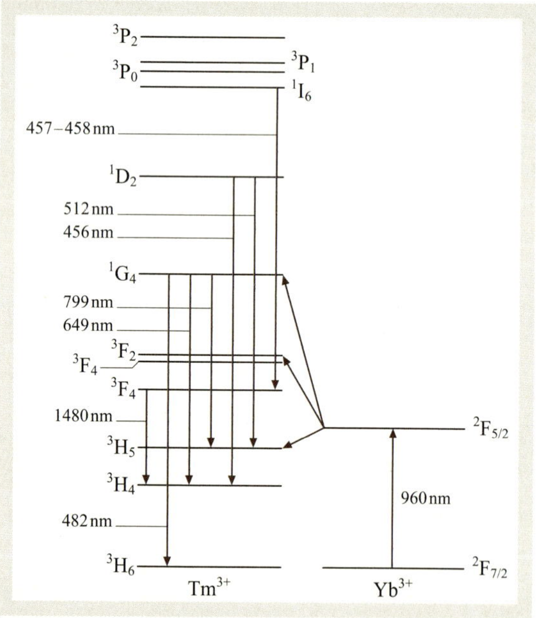

Fig. 11.45 Sensitized upconversion in Yb^{3+},Tm^{3+}-doped BaY_2F_8 after *Trash* and *Johnson* [11.174] with two and three steps in the STPA process

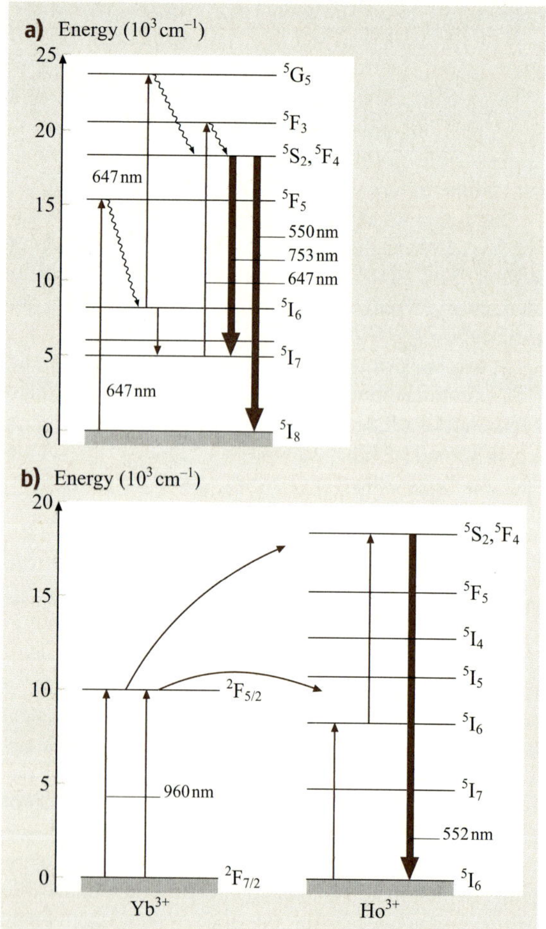

Fig. 11.46a,b STPA (a) and ETU (b) pumping of Ho^{3+} upconversion lasers

Table 11.10 Visible Pr^{3+}-doped ZBLAN fiber lasers (ESA: excited-state absorption (i. e., sequential two-step absorption), ETU: energy transfer upconversion, PA: photon avalanche, RT: room temperature)

Dopant ions	λ_{laser} (nm)	Transition	λ_{pump} (nm)	Pump mechanism	T (K)	Output	η (%)	Ref.
Pr^{3+}	635	$^3P_0 \to ^3F_2$	1010 + 835	ESA	RT	180 mW	10	[11.227]
	605	$^3P_0 \to ^3H_6$	1010 + 835	ESA	RT	30 mW	3.3	
	520	$^3P_1 \to ^3H_5$	1010 + 835	ESA	RT	1 mW		
	491	$^3P_0 \to ^3H_4$	1010 + 835	ESA	RT	1 mW		
Pr^{3+}	635	$^3P_0 \to ^3F_2$	1020 + 840	ESA	RT	54 mW	14	[11.228]
	520	$^3P_1 \to ^3H_5$	1020 + 840	ESA	RT	20 mW	5	
	491	$^3P_0 \to ^3H_4$	1020 + 840	ESA	RT	7 mW	1.5	
Pr^{3+}, Yb^{3+}	635	$^3P_0 \to ^3F_2$	849	ETU	RT	20 mW		[11.229]
Pr^{3+}, Yb^{3+}	635	$^3P_0 \to ^3F_2$	1016 diode + 833 diode	ETU	RT	6.2 mW	3.2	[11.230]
	532	$^3P_1 \to ^3H_5$		ETU	RT	0.7 mW	0.3	
Pr^{3+}, Yb^{3+}	635	$^3P_0 \to ^3F_2$	860 diode	ETU	RT	4 mW	2.2	[11.231]
	602	$^3P_0 \to ^3H_6$	860 diode	ETU		0.2 mW		
Pr^{3+}, Yb^{3+}	520 + 490	$^3P_1 \to ^3H_5$ $+^3P_0 \to ^3H_4$	856 diode		RT	1.4 mW		[11.223]
Pr^{3+}, Yb^{3+}	492	$^3P_0 \to ^3H_4$	1017 diode + 835 diode		RT	1.2 mW	8.5	[11.232]
Pr^{3+}, Yb^{3+}	635–637	$^3P_0 \to ^3F_2$	780–880	PA	RT	300 mW	16.8	[11.161]
	605–622	$^3P_0 \to ^3H_6$	780–880	PA	RT	45 mW	4.6	
	517–540	$^3P_1 \to ^3H_5$	780–880	PA	RT	20 mW	5	
	491–493	$^3P_0 \to ^3H_4$	780–880	PA	RT	4 mW	1.2	
Pr^{3+}, Yb^{3+}	635	$^3P_0 \to ^3F_2$	850	PA	RT	1020 mW	19	[11.162]
Pr^{3+}, Yb^{3+}	635	$^3P_0 \to ^3F_2$	850 diode	PA	RT	440 mW	17	[11.233]
	520	$^3P_1 \to ^3H_5$	850 diode	PA	RT	100 mW	≈ 4	
Pr^{3+}, Yb^{3+}	635	$^3P_0 \to ^3F_2$	850 diode	PA	RT	2 W	45	[11.234, 235]
	520	$^3P_1 \to ^3H_5$	850 diode	PA	RT	0.3 W	17	
Pr^{3+}, Yb^{3+}	491	$^3P_0 \to ^3H_4$	850	PA	RT	165 mW	12.1	[11.236]
	491 + 520	$^3P_0 \to ^3H_4$ $+^3P_1 \to ^3H_5$	850	PA	RT	230 mW	14.3	
	491	$^3P_0 \to ^3H_4$	840 diode	PA	RT	8 mW	≈ 6	
Pr^{3+}, Yb^{3+}	635	$^3P_0 \to ^3F_2$	838 diode	PA	RT	ML: 550 ps (239 MHz)		[11.237]
Pr^{3+}, Yb^{3+}	603 (tunable)	$^3P_0 \to ^3H_6$	840, Ti:Sapphire	PA	RT	55 mW	19	[11.238]
	634	$^3P_0 \to ^3F_2$				100 mW		
Pr^{3+}	635	$^3P_0 \to ^3F_2$	480, diode	Direct	RT	94 mW	41.5	[11.226]

laser oscillation of a Pr^{3+}-doped ZBLAN fiber under direct pumping with a blue semiconductor laser [11.226]. Output powers of 94 mW and a slope efficiency of 41.5% at 635 nm were obtained.

In Table 11.10 the results to date with Pr^{3+}-doped ZBLAN fibers are summarized. The best results obtained thus far, at the most efficient wavelength of 635 nm, are output powers up to 2 W and slope efficiencies up to 45% [11.234]. The efficiencies in the green and blue wavelength regions are about one order of magnitude smaller.

Generally, a double-clad fiber can be used to couple highly divergent pump radiation of near-infrared high-power laser diodes into the large-numerical-aperture inner cladding and propagate it along the fiber axis. On its way down the fiber, the pump radiation is gradually absorbed in the embedded small-numerical-aperture active core. Since the number of transverse radiation

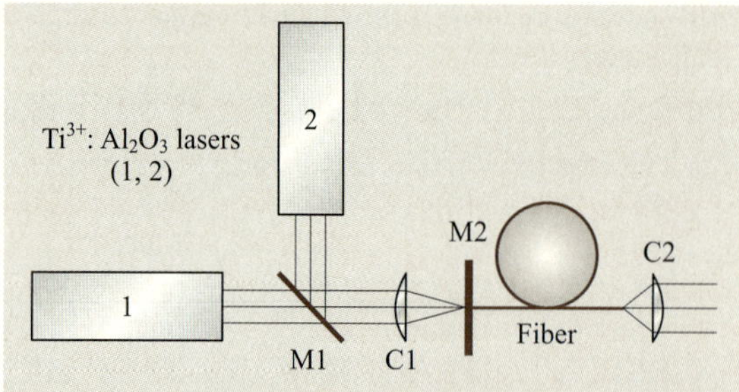

Fig. 11.47 Experimental setup of the Pr^{3+},Yb^{3+}-doped upconversion ZBLAN fiber laser [11.162]. M1 denotes the dichroic mirror, C1 an aspheric lens, M2 the dielectric resonator mirror, and C2 a collimator

modes supported in a cylindrical fiber core depends only on the wavelength of the radiation, the core diameter, and its numerical aperture, the geometry and material of the active fiber core can be chosen to support just one or a few transverse modes of the radiation generated in the active core by near-infrared pumping. Thus, an upconversion double-clad fiber with a suitably chosen rare-earth-ion-doped active core can be used to convert the highly multimode radiation of high-power near-infrared laser-diode arrays into single-mode or few-mode visible laser emission with excellent beam quality. The best results in terms of upconversion laser output power, however, have been achieved when the pump radiation of a high-beam-quality laser source was launched directly into the active core of the described fiber. This is due to the high pump intensities required locally for photon-avalanche upconversion, which can be best provided by diffraction-limited pump radiation focused onto the fiber end face and guided in the rare-earth-ion-doped active core.

For example, Fig. 11.47 shows a demonstration of high-power upconversion laser operation under pumping with two dichroic coupled Ti^{3+}:Al_2O_3 lasers tuned to pumping wavelengths near 850 nm [11.162]. The maximum upconversion-laser output power was 1020 mW at an incident near-infrared total pump power of 5.51 W (Fig. 11.48). The overall slope efficiency with respect to the incident pump power was 19% [11.162].

Also, near-infrared high-power diode-laser bars with beam-shaping optics have been used to pump a Pr^{3+},Yb^{3+}-doped ZBLAN fiber with the common single-clad structure and a large-area multimode active core. In this case, 4.5 W of pump power were launched into the fiber core, generating red output powers in excess of 2 W at 635 nm [11.234]. In the blue spectral range, a Pr^{3+},Yb^{3+}-doped ZBLAN-based upconversion fiber laser has been demonstrated at an emission wavelength of 491 nm ($^3P_0 \rightarrow {}^3H_4$ transition of Pr^{3+}). The pump source was a single-mode diode laser emitting at a wavelength of 840 nm. The maximum blue output power was 8 mW at an incident pump power of 200 mW [11.236].

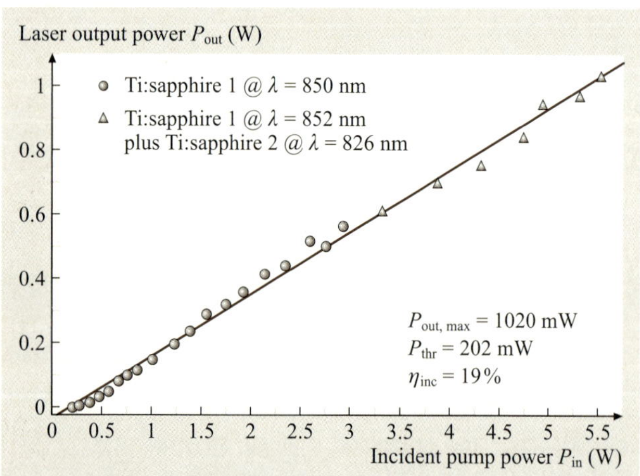

Fig. 11.48 Input–output characteristics of the Pr^{3+},Yb^{3+}-doped upconversion ZBLAN fiber laser [11.162]

Other visible fiber lasers. Laser oscillation in the visible spectral range has also been obtained with other rare-earth ions in ZBLAN. In Table 11.11 an overview of these laser systems is given. A detailed description of these systems and their prospects is given in [11.214].

11.2.3 Near-Infrared Rare Earth Lasers

Nd lasers

most intensively investigated class of solid-state lasers is based on Nd^{3+}-ion-doped materials. The Nd^{3+}-ion offers various groups of laser lines in the near-IR spectral region.

Energy-level scheme. Figure 11.49 shows transitions from the $^4F_{3/2}$ upper laser level into the $^4I_{13/2}$, $^4I_{11/2}$, and $^4I_{9/2}$ manifolds. The specific absorption and emission wavelengths depend on the crystal field, which influences the splitting within a manifold and between different manifolds. As shown in Fig. 11.49 the splitting between the ^4I (and also between ^4F) manifolds is dominated by LS coupling of the 4f electrons and is only affected to second order by covalency effects caused by the crystal field. The energetic splitting between the ^4F

Table 11.11 Overview of room temperature rare earth ion doped upconversion fiber lasers. (in): incident, (l) launched, (abs) absorbed

Active ion	λ_{laser} (nm)	Transition	λ_{pump} (nm)	P_{out} (mW)	η (%)	Ref.
Nd	381	$^4D_{3/2} \rightarrow ^4I_{11/2}$	590	0.076	0.25 (l)	[11.239]
Nd	412	$^2P_{3/2} \rightarrow ^4I_{11/2}$	590	0.470	1.7 (l)	[11.240]
Tm	455	$^1D_2 \rightarrow ^3F_4$	645 + 1064	3	1.5	[11.241]
Tm	480	$^1G_4 \rightarrow ^3H_4$	1130	33	34.6 (abs)	[11.242]
			1123	230	25 (in)	[11.243]
			680 + ≈ 1100	14.8	18.9 (abs)	[11.244–249]
Tm / Yb	480	$^1G_4 \rightarrow ^3H_4$	1070	375		[11.250]
			1065	106	6.6 (in)	[11.251]
			1120, 1140	116	15	[11.252]
Dy	478	$^4F_{9/2} \rightarrow ^6H_{15/2}$	457 (Ar^+)	2.3	0.9	[11.253]
	575	$^4F_{9/2} \rightarrow ^6H_{13/2}$		10	1.5	
Er	540	$^4S_{3/2} \rightarrow ^4I_{15/2}$	801	23	16	[11.254, 255]
Er	540	$^4S_{3/2} \rightarrow ^4I_{15/2}$	970	50	51	[11.256–258]
Ho	544 / 549	$^5F_4 \rightarrow ^5I_8 / ^5S_2 \rightarrow ^5I_8$	≈ 640	40	22.4 (l)	[11.259–263]
Ho	753		647	0.54	3.3	[11.259]
Tm	810		1064	1200	37	[11.264]

and ^4I levels is dominated by Coulomb interaction of the 4f electrons, which is also only effected to second order by covalency effects of the crystal field. Therefore, the transition wavelengths of Nd-doped crystals vary in a certain range near the values given for Nd:$Y_3Al_5O_{12}$ (Nd:YAG) in Fig. 11.49. The strongest and most commonly used laser transition $^4F_{3/2} \rightarrow ^4I_{11/2}$ emits near 1060 nm (Fig. 11.49).

Figures 11.50 and 11.51 show as an example the absorption and emission spectra of Nd:YAG, measured at 300 K. The strongest absorption is located near 808 nm and the strongest laser transition at 1064 nm.

Longitudinal and transversal diode pumping. Nd lasers are usually pumped by lamps or diode lasers. The first generation of Nd-doped solid-state lasers were pumped with CW krypton or pulsed xenon lamps. These lamps have high electrical-to-optical efficiencies of about 70% and are available at reasonable costs. Unfortunately, the overlap of their emission spectra with the narrow 4f–4f absorption spectra of Nd-doped solid-state laser materials is usually poor; thus the electrical to optical pump efficiency is low – typically a few percent.

In modern Nd lasers, diode lasers at 808 nm are employed as pump sources. Laser diodes typically have efficiencies of about 50% and offer several advantages over lamp pumping. Diodes emit spectrally narrow-band radiation (≈ 1–5 nm). Therefore, a much better overlap with the 4f–4f absorption spectra of Nd can be achieved. Because of the good spectral match, the total pump efficiency of diode pumping is much higher than for lamp pumping. In addition the transfer efficiency of pump light into the crystal is much better in the case of diode pumping, so that a total efficiency of 10% to 30% for the conversion of electrical power into laser power is achievable.

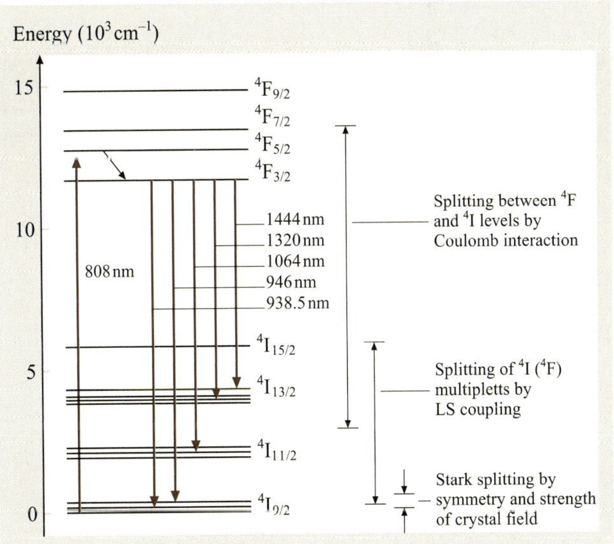

Fig. 11.49 Level scheme, pump and laser transitions of Nd^{3+} in YAG

Part C | 11.2

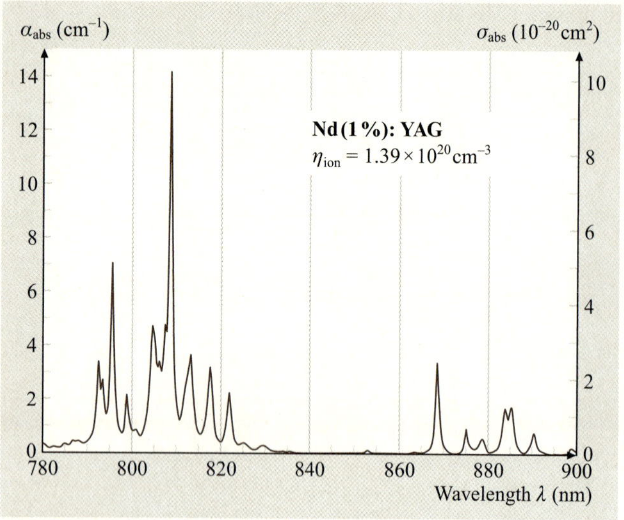

Fig. 11.50 Absorption spectrum of Nd(1%):YAG and absorption cross sections

The selective spectral pumping with diodes also leads to reduced heat deposition in the laser crystal with reduced thermal lensing effects, giving rise to improved beam quality. Diffraction-limited beams can be realized easier.

However, there are also disadvantages of diode lasers. Because of their small active laser cross-sectional area and their relatively large refractive index, laser diode beams have a large divergence. Their beams also possess large beam-quality differences in the transversal and sagittal directions since the active area dimensions are usually $\approx 1\,\mu m \times 100\,\mu m$ and confinement is given only in one dimension. For power scaling, bars or arrays of diodes must be used. This can lead to output powers of up to several tens of watts. Unfortunately, combination of many single diode stripes also broadens the emission spectrum, while the beam quality of the combined radiation is further decreased. So, sophisticated pump optics such as beam shaping must often be employed [11.265].

Several pump geometries can be used with diode laser excitation. The common geometry to pump lasers, which are capable of generating several watts of output power with excellent beam quality, is end pumping (Fig. 11.52). In this geometry, pump radiation is focused into the laser crystal along the laser resonator axis. A very good overlap of pump and laser modes can be provided. Gain media of small volume and with short absorption lengths can be used, and high population inversions at low pump power levels can be achieved. Because most of the pump power is deposited within the volume of the TEM_{00} laser mode, higher-order modes usually cannot oscillate. Excellent beam quality is therefore inherent. However, scaling of end-pumped lasers to output powers above 20 W is difficult to achieve due to

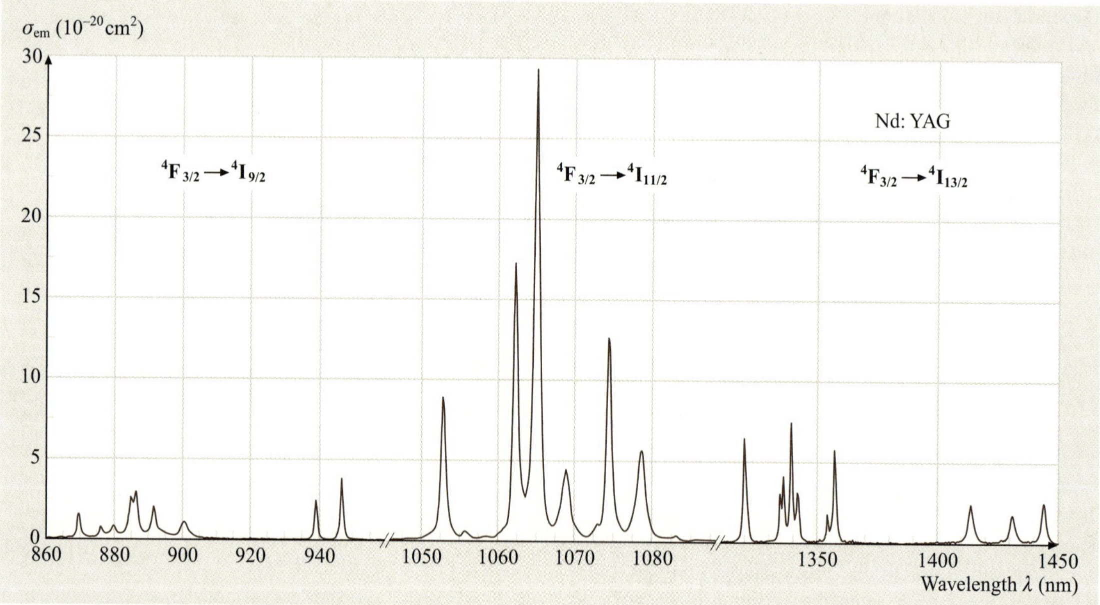

Fig. 11.51 Emission cross section spectra of Nd:YAG

possible thermally induced fracture of the laser crystal within the small volume of pump absorption.

An alternative diode pumping method is side pumping (Fig. 11.53, [11.266]). In this approach, the pumping geometry is principally similar to the arrangement for lamp pumping. One or several linear pump diode bars are positioned around the side surface of the laser crystal. Pump radiation is then usually imaged into the crystal perpendicular to the laser resonator axis. The crystal is relatively homogeneously pumped throughout its total volume, and a smaller excitation density is usually realized than in diode laser end pumped systems. Because of the use of larger gain media, however, more energy can be deposited into the laser crystal and therefore higher output powers can be achieved. Since a larger laser mode is present inside the crystal, higher-order modes also usually oscillate in these lasers. Therefore beam quality is often rather low.

Nd-doped fiber lasers can also be operated in specific configurations, see for example [11.267–278], and are used in many applications. For high-power laser generation, double-clad fibers are usually used. Basic aspects are described below in this chapter (page 650/651).

The most important Nd *lasers.* The most important Nd-doped laser materials feature relatively high emission cross sections, a relatively long upper-state lifetime of Nd^{3+}, a high damage threshold, high mechanical and chemical stability, good thermal conductivity and very good optical quality, see Table 11.12. Many host crystals have been investigated, including yttrium aluminium garnet (YAG), yttrium aluminium perovskite (YAP or YALO), yttrium lithium fluoride (YLF) and yttrium vanadate (YVO). Very efficient and compact diode-pumped lasers with slope efficiencies of over 60% have been demonstrated in many Nd-doped laser materials.

Nd:YAG is still regarded as the most important solid-state laser. Due to their very good optical and mechanical properties, diode-pumped Nd:YAG lasers are robust and reliable. They are in use in many applications. Within the last few years Nd:YAG have also become available as ceramics with high optical quality.

The vanadates Nd:YVO and Nd:GVO emit polarized radiation and exhibit large cross sections and gain.

Nd *laser wavelength and materials.* A variety of Nd lasers have been reported so far. Tables 11.13, 14, 15 list laser materials and their laser wavelengths which have been observed at 300 K on the transitions ${}^4F_{3/2} \rightarrow {}^4I_{9/2}$, ${}^4I_{11/2}$ and ${}^4I_{13/2}$. The tables also list Nd-doped ceramics, which have been developed with very good optical quality and low scattering losses. So, the efficiency of ceramic lasers can be as high as for crystals.

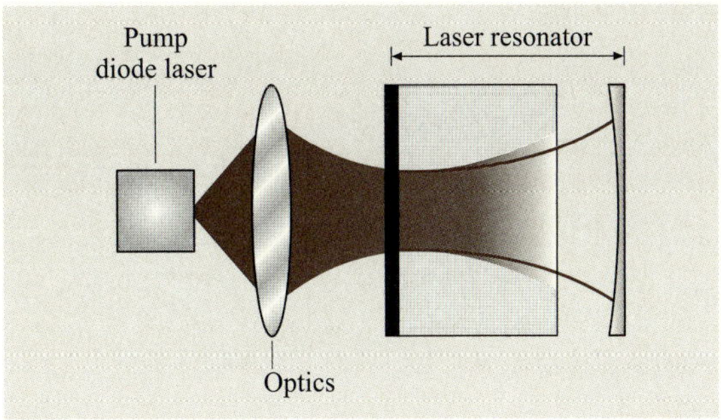

Fig. 11.52 Diode end-pumping scheme

If not specified otherwise, the references for the laser materials can be found in [11.213].

Yb lasers

Coherent oscillation of Yb^{3+} was first observed in YAG at 77 K [11.279]. In 1991 the first realization of a diode-pumped Yb:YAG laser [11.280] at room temperature initialized an intensive renaissance of research on Yb-doped laser materials for laser diode pumping at 300 K. Yb^{3+}-doped solid-state lasers feature several important advantages compared with other rare-earth lasers:

1. Yb^{3+} ions have only two states, the ground state ${}^2F_{7/2}$ and the excited state ${}^2F_{5/2}$, which are separated by an energy of about 10 000 cm^{-1}. Thus, there is no excited-state absorption of the pump and laser radiation (Fig. 11.54).
2. The quantum efficiency of Yb^{3+} is close to unity.

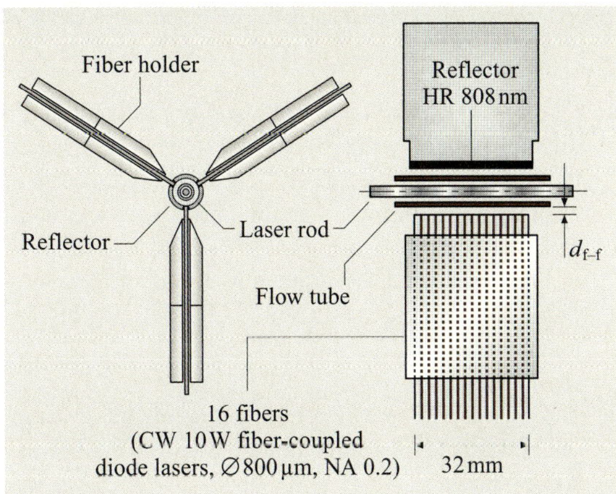

Fig. 11.53 Diode side (transversal) pumping scheme [11.266]

Part C | 11.2

Table 11.12 Important host materials for Nd^{3+} [11.285]

Host	YAG	YAlO (FAP)	YVO	GVO	YLF
Chemical formula	$Y_3Al_5O_{12}$	$YAlO_3$	YVO_4	$GdVO_4$	$YLiF_4$
Lattice symmetry	Cubic	Orthorhombic	Tetragonal	Tetragonal	Tetragonal
Space group	Ia3d	Pnma	$I4_1$/amd	$I4_1$/amd	$I4_1$/a
Lattice constant (Å)	12.00	$a = 5.33$ $b = 7.37$ $c = 5.18$	$a = 7.120$ $c = 6.289$	$a = 7.123$ $b = 6.291$	$a = 5.18$ $c = 10.74$
Density of Nd sites (1% doping)	1.39	1.96	1.255	1.25	1.39
Heat conductivity ($Wm^{-1}K^{-1}$)	11–13	11	~5–12 [a]	~8–12 [a]	6
dn_i/dT ($10^{-6}\,K^{-1}$)	9.9	14.5(a) 9.7(b)	8.5(a) 3.0(c)		−0.9(a) −2.9(c)
dL/dT ($10^{-6}\,K^{-1}$)	8.2	4.4(a) 10.8(b) 9.5(c)	3.1(a) 7.2(c)	1.6(a) 7.3(c)	13(a) 8(c)
Max. phonon energy (cm^{-1})	700	550	850		490
Refractive indices	$n = 1.822$	$n_a = 1.9260$ $n_b = 1.9118$ $n_c = 1.9346$	$n_o = 1.958$ $n_e = 1.2168$	$n_o = 1.972$ $n_e = 1.2192$	$n_o = 1.454$ $n_e = 1.477$
$\tau(^4F_{3/2})$ (μs)	250	160	97	100	500
λ_{abs} (nm)	808	813 ($E \parallel a$)	808	808	792
σ_{abs} ($10^{-20}\,cm^2$)	7.9	7.2 ($E \parallel a$)	60 (π) 12 (σ)	54 (π) 12(σ)	14 (π) 1.2(σ)
$\lambda_{laser}(^4I_{11/2})$ (nm)	1064	1080 ($E \parallel a$)	1064	1063	1047 (π) 1054 (σ)
$\sigma_{em}(^4I_{11/2})$ ($10^{-20}\,cm^2$)	29	25 ($E \parallel a$)	123 (π) 52 (σ)	125 (π) 61 (σ)	1.5 (π) 1.5 (σ)
$\Delta\lambda(^4I_{11/2})$ (nm)	0.8	2.5 ($E \parallel a$)	1.0 (π) 1.5 (σ)	1.2 (π) 1.3 (σ)	1.5 (π) 1.5(σ)
$\lambda_{laser}(^4I_{9/2})$ (nm)	946	930	914 (π) 915 (σ)	912 (π) 912 (σ)	904 (π) 909 (σ)
$\sigma_{em}(^4I_{9/2})$ ($10^{-20}\,cm^2$)	3.9	4.1 ($E \parallel a$)	4.8 (π) 4.3(σ)	6.6 (π) 5.6 (σ)	1.2 (π) 1.3 (σ)
$\Delta\lambda$ (nm)	1.0	2.5 ($E \parallel a$)	2.8 (π) 3.4 (σ)	2.5 (π) 3.3 (σ)	3.0(π) 3.0 (σ)

[a] The published data of heat conductivities of the vanadates YVO and GVO differ considerably [11.281–284]

3. Due to the small Stokes shift and the related small quantum defect (typically $500\,cm^{-1}$) heat generation in the lasing process of Yb^{3+} is small and makes it a suitable ion for high-average-power lasers.
4. The small ionic radius of Yb^{3+} compared to that of other rare-earth ions favors its incorporation into Y-based host crystals such as YAG, allowing higher dopant concentration and thus shorter gain elements such as discs.
5. Yb^{3+} ions exhibit a relatively broad emission band which leads to tunability and the generation of ultra-short pulses.
6. The radiative lifetime of the laser level ranges in different crystals from a few hundred microseconds to a few milliseconds, which implies greater energy storage efficiency, especially for Q-switched operation with diode pumping.

One disadvantage of Yb^{3+}-doped lasers is that they operate in a quasi-three-level scheme with temperature-dependent reabsorption at the laser wavelength. This leads to an increased threshold compared to a four-level scheme, because the pump must bleach the reabsorption losses (see the section on basic spectroscopic properties in Sect. 11.2.1).

Figures 11.55 and 11.56 show as an example the absorption and emission cross section spectra of Yb:YAG [11.286]. When pumped near the absorption

Table 11.13 Laser wavelengths of the $^4F_{3/2} \rightarrow {}^4I_{9/2}$ transition at 300 K [11.213]

Wavelength (nm)	Material
0.8910	$Y_3Al_5O_{12}$
0.8999	$Y_3Al_3O_{12}$
0.901	$Sr_{1-x}La_xMg_xAl_{12-x}O_{19}$ [11.287]
0.9106	$Ba_3LaNb_3O_{12}$
0.910	$LiLuF_4$ [11.285]
0.912	Y_2SiO_5
0.912	YVO_4 [11.288]
0.912	$GdVO_4$ [11.289–291]
0.914	YVO_4 [11.292]
0.916	$LuVO_4$ [11.293, 294]
0.930	$YAlO_3$ [11.213, 295]
0.9312	$YAlO_3$
0.936	$Gd_3Sc_2Ga_3O_{12}$
0.9385	$Y_3Al_5O_{12}$
≈ 0.94	$Y_3Al_5O_{12}$ [11.296]
0.941	$CaY_2Mg_2Ge_3O_{12}$
0.9458–0.9464	$Y_3Al_5O_{12}$ [11.297]
0.946	$Y_3Al_5O_{12}$ [11.213, 298–306]
0.946	$Y_3Al_5O_{12}$ ceramic [11.307]
0.9660	$YAlO_3$
0.966	Sc_2O_3 [11.308]

peaks at 940 nm or 970 nm, the Stokes shift to the lasing wavelength at 1030 nm corresponds to less than 10%, which allows in principle a slope efficiency of more than 90%.

Yb^{3+} ***thin-disc lasers.*** In connection with the power scaling of diode-pumped Yb lasers *Giesen* et al. [11.310] invented the thin-disc laser concept as a pump and resonator design for high-power lasers (Fig. 11.57).

The active medium is a thin circular disc, which is coated for both the pump and laser wavelengths with a highly reflecting (HR) dielectric mirror on the rear side and with an antireflection coating on the front side. The disc is then bonded with the rear HR side to a heat sink. The resonator is formed by the coated crystal and an output coupler. The pump light is provided by fiber-coupled laser diodes which are focused onto the crystal. Since the thin crystal disc only absorbs a small fraction of the pump light during one pass through the crystal, the pump light is reflected back into the disc several times by a parabolic mirror and the folding prism [11.311]. Up to a total of 32 pump light passes may be used.

Due to the geometrical setup a nearly one-dimensional thermal gradient along the laser axis can

Table 11.14 Laser wavelengths of the $^4F_{3/2} \rightarrow {}^4I_{11/2}$ transition at 300 K [11.213]

Wavelength	Material
0.97 (transition $^4F_{5/2}, {}^2H_{9/2} \rightarrow {}^4I_{11/2}$)	$RbPb_2Br_5$ [11.309]
1.0369	CaF_2-SrF_2
1.0370	CaF_2
1.0370–1.0395	SrF_2
1.04065–1.0410	LaF_3
1.0410	CeF_3
1.0412	KYF_4
1.042–1.075	$Na_{0.4}Y_{0.6}F_{2.2}$
1.0445	SrF_2
1.046–1.064	$LiNdP_4O_{12}$
1.0461	CaF_2-YF_3
1.0461–1.0468	CaF_2
1.047	$LiGdF_4$
1.047	$LiNdP_4O_{12}$
1.047	$LiYF_4$
1.047–1.078	NdP_5O_{14}
1.0471	$LiYF_4$
1.0472	$LiLuF_4$
1.0475	$LaBGeO_5$
1.0477	$Li(Nd, La)P_4O_{12}$
1.0477	$Li(Nd, Gd)P_4O_{12}$
1.048	$Li(Bi, Nd)P_4O_{12}$
1.048	$Li(Nd, La)P_4O_{12}$
1.048	$Li(Nd, Gd)P_4O_{12}$
1.048	$K_5(Nd, Ce)Li_2F_{10}$
1.0481	$LiKYF_5$
1.0482	$LaBGeO_5$
1.0482	$NaLa(MoO_4)_2$
1.0486	LaF_3-SrF_2
1.049–1.077	$NaNdP_4O_{12}$
1.0491	$SrAl_{12}O_{19}$
1.0493	$Sr_2Y_5F_{19}$
1.0495	BaY_2F_8
1.0495	GdF_3-CaF_2
1.0497	$SrAl_2O_4$
1.0498	$Ca_2Y_5F_{19}$
1.0498	$SrAl_{12}O_{19}$
1.05	$KNdP_4O_{12}$
1.05	NdP_5O_{14}
1.05	LaP_5O_{14}
1.05	$LiLuF_4$
1.0500	CaF_2-ScF_3
1.0505	$(Nd, La)P_5O_{14}$
1.0505	$5NaF-9YF_3$

Table 11.14 (continued)

Wavelength	Material
1.0506	$5NaF-9YF_3$
1.0507	CdF_2-ScF_3
1.051	$NaNdP_4O_{12}$
1.051	YP_5O_{14}
1.051	$(La, Nd)P_5O_{14}$
1.051	CeP_5O_{14}
1.051	GdP_5O_{14}
1.051	NdP_5O_{14}
1.0511	$(Nd, La)P_5O_{14}$
1.0512	NdP_5O_{14}
1.0512	$(Nd, La)P_5O_{14}$
1.0512	$(Y, Nd)P_5O_{14}$
1.0513	NdP_5O_{14}
1.0515	YP_5O_{14}
1.0515	NdP_5O_{14}
1.052	$(Nd, La)P_5O_{14}$
1.052	$KNdP_4O_{12}$
1.052	$K_5NdLi_2F_{10}$
1.052	$Y_3Al_5O_{12}$
1.0521	BaF_2-YF_3
1.0521	NdP_5O_{14}
1.0521	YF_3
1.0525	YP_5O_{14}
1.0526	BaF_2-GdF_3
1.0528	SrF_2-GdF_3
1.0529	NdP_5O_{14}
1.053	$LiYF_4$
1.053	$(La, Nd)P_5O_{14}$
1.053–1.062	$Ca_3(Nb, Ga)_2Ga_3O_{12}$
1.0530	$LiYF_4$
1.0530	BaY_2F_8
1.0530	CaF_2-LuF_3
1.0530–1.059	$LaMgAl_{11}O_{19}$
1.0531	$LiLuF_4$
1.0532	$LiKYF_5$
1.0534–1.0563	BaF_2-LaF_3
1.0535	$Lu_3Al_5O_{12}$
1.0535–1.0547	$CaF_2-SrF_2-BaF_2-YF_3-LaF_3$
1.0537	BaF_2-CeF_3
1.0539–1.0549	$\alpha-NaCaYF_6$
1.054	$Gd_3Ga_5O_{12}$
1.054	$LaAl_{11}MgO_{19}$
1.054–1.086	$LaAl_{11}MgO_{19}$
1.0540	CaF_2-YF_3
1.0540	BaF_2
1.0540	CaF_2-YF_3

Table 11.14 (continued)

Wavelength	Material
1.0543	BaF_2-CeF_3
1.0543	SrF_2-ScF_3
1.05436	$Ba_2MgGe_2O_7$
1.05437	$Ba_2ZnGe_2O_7$
1.0547	$LaMgAl_{11}O_{19}$
1.05499	CsY_2F_7
1.055	$Na_3Nd(PO_4)_2$
1.055	$K_3(La, Nd)(PO_4)_2$
1.055	$Na_3(La, Nd)(PO_4)_2$
1.0551	$Pb_5(PO_4)_3F$
1.0552	$LaMgAl_{11}O_{19}$
1.0554	$LiNdP_4O_{12}$
1.0554	KY_3F_{10}
1.0555	$CsGd_2F_7$
1.0555	$Ba_5(PO_4)_3F$
1.0556	SrF_2-LuF_3
1.0560	SrF_2-LuF_3
1.0566	$La_2Si_2O_7$
1.0567	SrF_2YF_3
1.0569	$NdGaGe_2O_7$
1.0570	$GdGaGe_2O_7$
1.0572	$LaSr_2Ga_{11}O_{20}$
1.0573	$CaMoO_4$
1.0575	$CsLa(WO_4)_2$
1.0576	$SrAl_4O_7$
1.0576	$SrMoO_4$
1.0576	$La_2Si_2O_7$
1.058	$Gd_3Sc_2Ga_3O_{12}$
1.0582	$Ca_3Ga_4O_9$
1.0582–1.0597	$CaWO_4$
1.0583	$Y_3Sc_2Ga_3O_{12}$
1.0584	$Y_3Sc_2Ga_3O_{12}:Cr$
1.0584	$Y_3Sc_2Ga_3O_{12}$
1.0584	$CaY_2Mg_2Ge_3O_{12}$
1.0585	$YAlO_3$
1.0585	$LiLa(MoO_4)_2$
1.0585	$Sr_5(PO_4)_3F$
1.0585	$CaF_2-SrF_2-BaF_2-YF_3-LaF_3$
1.0585	$KLa(MoO_4)_2$
1.0586	$Sr_5(PO_4)_3F$
1.0586	$PbMoO_4$
1.0587	$KLa(MoO_4)_2$
1.0587	$CaWO_4$
1.0588	$Ca_3(Nb, Ga)_2Ga_3O_{12}$
1.0589	$SrF_2-CeF_3-GdF_3$
1.0589	$Y_3Ga_5O_{12}$

Table 11.14 (continued)

Wavelength	Material
1.05896	$CaMg_2Y_2Ge_3O_{12}$
1.059	(La, Sr)(Al, Ta)O_3
1.059	$BaLaGa_3O_7$
1.059	$NaY(WO_4)_2$
1.059	$Na_{1+x}Mg_xAl_{11-x}O_{17}$
1.059	$Na_2Nd_2Pb_6(PO_4)_6Cl_2$
1.059	$Sr_5(PO_4)_3F$ [11.312]
1.0590	SrF_2-CeF_3
1.0590	$Gd_3Ga_5O_{12}$ waveguide [11.313]
1.0591	$Gd_3Ga_5O_{12}$
1.0591	$Lu_3Sc_2Al_3O_{12}$
1.0591	$LaGaGe_2O_7$
1.0593	$Sr_4Ca(PO_4)_3F$
1.0594	$Lu_3Ga_5O_{12}$
1.0595	$5NaF-9YF_3$
1.0595	$Y_3Sc_2Al_3O_{12}$
1.0595	$NaLa(MoO_4)_2$
1.0595	$BaLaGa_3O_7$
1.0596	$CaAl_4O_7$
1.0596	$Ca_3Ga_2Ge_3O_{12}$
1.0596	SrF_2
1.0597	$Ca_3Ga_2Ge_3O_{12}$
1.0597–1.0583	SrF_2-LaF_3
1.0597–1.0629	$\alpha-NaCaYF_6$
1.0599	$Lu_3Sc_2Al_3O_{12}$
1.0599	$LiGd(MoO_4)_2$
1.05995	$Gd_3Sc_2Al_3O_{12}$
≈ 1.06	$(Gd, Ca)_3(Ga, Mg, Zr)_5O_{12}:Cr$
≈ 1.06	$CaGd_4(SiO_4)_3O$
≈ 1.06	$Ca_4YO(BO_3)_3$ [11.314]
≈ 1.06	$Y_3Al_5O_{12}$ [11.296]
≈ 1.06	$Y_3Sc_{1.0}Al_{4.0}O_{12}$ ceramic [11.315]
≈ 1.06	$Gd_3Sc_2Al_3O_{12}$
≈ 1.06	$GdAl_3(BO_3)_4$ [11.316]
≈ 1.06	$NaLa(MoO_4)_2$
≈ 1.06	$NaGd(WO_4)_2$
≈ 1.06	$NdAl_3(BO_3)_4$
≈ 1.06	$YAl_3(BO_3)_4$ [11.213, 317]
≈ 1.06	$Gd_3Ga_5O_{12}$
≈ 1.06	$GdVO_4$ [11.213, 318–320]
≈ 1.06	YVO_4 [11.319, 320]
≈ 1.06	$La_{0.2}Gd_{0.8}VO_4$ [11.319, 321]
1.060	$Gd_3Ga_5O_{12}:Cr$
1.060	$Ca_4GdO(BO_3)_3$ [11.213, 322–324]
1.060	$Ca_4YO(BO_3)_3$ [11.325]

Table 11.14 (continued)

Wavelength	Material
1.060	LaB_3O_6 [11.326]
1.0600	$Gd_3Ga_5O_{12}$
1.0601	$GdGaGe_2O_7$
1.0603	$Y_3Ga_5O_{12}$
1.0603	$Gd_2(WO_4)_3$
1.0603–1.0632	CaF_2-YF_3
1.0604	$HfO_2-Y_2O_3$
1.0604	$NaLuGeO_4$
1.0605	SrF_2-ScF_3
1.0606	$Gd_2(MoO_4)_3$
1.0606	$Gd_3Ga_5O_{12}$
1.0606	$Gd_3Ga_5O_{12}$ waveguide [11.313]
1.0607	$Sr_3Ca_2(PO_4)_3$
1.0607	CaF_2-ScF_3
1.0608	$ZrO_2-Y_2O_3$
1.0608	$Nd_3Ga_5O_{12}$
1.0608	$NaGaGe_2O_7$
1.0609	$Lu_3Ga_5O_{12}$
1.0609	$NaYGeO_4$
1.061	$Ca_2Al_2SiO_7$
1.061	$BaGd_2(MoO_4)_4$
1.061	$CaMoO_4$
1.061	YVO_4 [11.327]
1.061	$CaLa_4(SiO_4)_3O$
1.0610	$Ca_2Ga_2SiO_7$
1.0610	$7La_2O_3-9SiO_2$
1.0610–1.0627	$Y_3Al_5O_{12}$
1.0612	$Gd_3Sc_2Ga_3O_{12}$
1.0612	$CaLa_4(SiO_4)_3O$
1.0612	$Ca_3(Nb, Ga)_2Ga_3O_{12}$
1.0613	$Ca_4La(PO_4)_3O$
1.0613	$Ba_2NaNb_5O_{15}$
1.0613	$Gd_3Sc_2Ga_3O_{12}$
1.0615	$Ca(NbO_3)_2$
1.0615	$Y_3Al_5O_{12}$
1.0615	$Lu_3Al_5O_{12}$
1.0615	$Y_3Sc_2Ga_3O_{12}$
1.0615	$Ba_{0.25}Mg_{2.75}Y_2Ge_3O_{12}$
1.0615	$NaGdGeO_4$
1.0615	$Y_3Al_5O_{12}$
1.0615–1.0625	$Ca(NbO_3)_2$
1.0618	$Sr_2Ca_3(PO_4)_3F$
1.0618	$SrAl_{12}O_{19}$
1.0618	CaF_2-ScF_3
1.0618	$LaNbO_4$
1.062	$LaSc_3(BO_3)_4$ [11.213, 328]

Table 11.14 (continued)

Wavelength	Material
1.0620	$Gd_3Sc_2Al_3O_{12}$
1.0620	$Lu_3Sc_2Al_3O_{12}$
1.0621	$SrAl_{12}O_{19}$
1.0621	$Gd_3Ga_5O_{12}$
1.0622	$Y_3Sc_2Al_3O_{12}$
1.0623–1.10585	$CaF_2-SrF_2-BaF_2-YF_3-LaF_3$
1.0623	$Lu_3Ga_5O_{12}$
1.0623	CaF_2-LuF_3
1.0623–1.0628	CaF_2
1.0624	$LaNbO_4$
1.0625	$Y_3Ga_5O_{12}$
1.0625	YVO_4
1.0628	$SrWO_4$
1.0629	$Ca_5(PO_4)_3F$
1.0629	$\alpha-NaCaYF_6$
1.0629	$Bi_4Si_3O_{12}$
1.0629–1.0656	CdF_2-YF_3
1.063	$GdVO_4$ [11.329]
1.063	$Gd_3Sc_2Ga_3O_{12}:Cr$
1.063	$SrWO_4$
1.063	$Na_5(Nd,La)(WO_4)_4$
1.063	$NdAl_3(BO_3)_4$
1.063	$(La,Nd)P_5O_{14}$
1.063	$NdAl_3(BO_3)_4$
1.0630	$Ca_5(PO_4)_3F$
1.0632	$CaF_2-YF_3-NdF_3$
1.0632	CaF_2-YF_3
1.0633–1.0653	$\alpha-NaCaCeF_6$
1.06335–1.0638	LaF_3
1.0634	YVO_4
1.0635	LaF_3-SrF_2
1.0635	$NdAl_3(BO_3)_4$
1.0635	$NaLa(WO_4)_2$
1.0635	$(Nd,Gd)Al_3(BO_3)_4$
1.0635	$Bi_4(Si,Ge)_3O_{12}$
1.0635	CaF_2-LuF_3
1.0637–1.0670	$Y_3Al_5O_{12}$
1.06375–1.0672	$Lu_3Al_5O_{12}$
1.0638	CeF_3
1.0638	$CaAl_4O_7$
1.0638	$NaBi(WO_4)_2$
1.0638	NdP_5O_{14}
1.0638	$Ca_3Ga_2Ge_3O_{12}$
1.0638–1.0644	$(Y,Ce)_3Al_5O_{12}$
1.0639	$Ca_3Ga_2Ge_3O_{12}$
1.064	$Y_3Al_5O_{12}$ [11.213, 301, 330–332]
1.064	$Y_3Al_5O_{12}$ ceramic [11.315, 333–346]
1.064	$Y_3Al_5O_{12}:Fe$
1.064	$Y_3Al_5O_{12}:Ti$
1.064	$Y_3Al_5O_{12}:Cr,Ce$
1.064	$Y_3Al_5O_{12}:Ho$
1.064	$Y_3Al_5O_{12}:Er$
1.064	YVO_4 [11.213, 347–350]
1.064	YVO_4 single crystal fiber [11.351]
1.064	LaF_3
1.064	$KGd(WO_4)_2$ [11.352, 353]
1.064	$La_3Ga_{5.5}Ta_{0.5}O_{14}$ [11.354]
1.0640	$La_3Ga_5SiO_{14}$
1.0640–1.0657	CaF_2-CeF_3
1.06405–1.0654	$YAlO_3$
1.0641	$Y_3Al_5O_{12}:Cr$
1.0641	YVO_4
1.0641	$La_3Ga_{5.5}Ta_{0.5}O_{14}$
1.06415	$Y_3Al_5O_{12}$
1.06415	$Y_3Al_5O_{12}$ [11.355]
1.0642	$Ca_3Ga_2Ge_3O_{12}$
1.0642	$NaBi(WO_4)_2$
1.06425	$Lu_3Al_5O_{12}$
1.0643	$SrMO_4$
1.0644	$Bi_4Ge_3O_{12}$
1.0645	CaF_2-LaF_3
1.0645	$La_3Ga_{5.5}Nb_{0.5}O_{14}$
1.0645	$La_3Ga_5SiO_{14}$
1.0645	$YAlO_3$
1.0645	$YAlO:Cr$
1.0646	$Y_3Al_5O_{12}$
1.0646	$KLa(MoO_4)_2$
1.0647	$CeCl_3$
1.0648	YVO_4
1.0649	$CaY_2Mg_2Ge_3O_{12}$
1.065	$GdVO_4$
1.065	$(Nd,Gd)Al_3(BO_3)_4$
1.065	$Sr_5(VO_4)_3Cl$
1.065	$Sr_5(VO_4)_3F$
1.0650	$La_3Ga_5GeO_{14}$
1.0650	$RbNd(WO_4)_2$
1.0652	$CaWO_4$
1.0652	CdF_2-LuF_3
1.0653–1.0633	$\alpha-NaCaCeF_6$
1.0653	$NaLa(MoO_4)_2$
1.0653–1.0665	$NaLa(MoO_4)_2$
1.0654	CaF_2-GdF_3

Table 11.14 (continued)

Wavelength	Material
1.0654	$NdGaGe_2O_7$
1.0656	CdF_2-YF_3
1.0657–1.0640	CaF_2-CeF_3
1.0657	CaF_2
1.0658	$LiLa(MoO_4)_2$
1.0658	$CsNd(MoO_4)_2$
1.0658	$LuVO_4$ [11.356]
1.0659	$GdGaGe_2O_7$
1.066	$Nd(Ga, Cr)_3(BO_3)_4$
1.066	$K_5Nd(MoO_4)_4$
1.066	$K_5Bi(MoO_4)_4$
1.0661	CaF_2
1.0664–1.0672	YVO_4
≈ 1.0665	CdF_2-LaF_3
1.0666	CdF_2
1.0667	CdF_2-GeF_3
1.0667	$NaGd(MoO_4)_2$
1.0668	CdF_2-LaF_3
1.0669	$KY(MoO_4)_2$
1.067	$Ca_3(VO_4)_2$
1.0670	$La_3Ga_5SiO_{14}$
1.0672	$CaY_4(SiO_4)_3O$
1.0672	CdF_2-GdF_3
1.0672	$KGd(WO_4)_2$
1.0672	$La_3Ga_5SiO_{14}$
1.0673	$GaMoO_4$
1.0673	$La_3Ga_5SiO_{14}$
1.0674	$NaY(MoO_4)_2$
1.0675	$LuAlO_3$
1.0675	$Na_2Nd_2Pb_6(PO_4)_6Cl_2$
1.0675	$Nd_3Ga_5SiO_{14}$
1.0675	$Nd_3Ga_5GeO_{14}$
1.068	$Na_2Nd_2Pb_6(PO_4)_6Cl_2$
1.0680	$Nd_3Ga_5GeO_{14}$
1.0682	$Y_3Al_5O_{12}$
1.0687–1.0690	$KY(WO_4)_2$
1.0688	$Ga_3Ga_2SiO_7$
1.0688	$Ca_2Ga_2Ge_4O_{14}$
1.0688	$KY(WO_4)_2$
1.0689	$NdGaGe_2O_7$
1.0690	$GdAlO_3$
1.0690	$Ca_3Ga_2Ge_4O_{14}$
1.0694	$Sr_3Ga_2Ge_4O_{14}$
1.0698	$La_2Be_2O_5$
1.07	KPb_2Br_5 [11.309]
1.07	$RbPb_2Br_5$ [11.309]

Table 11.14 (continued)

Wavelength	Material
1.070	$La_2Be_2O_5$
1.0701	$Gd_2(MoO_4)_3$
1.0701–1.0706	$KLu(WO_4)_2$
1.0706	$KY(WO_4)_2$
1.0706	$LaSr_2Ga_{11}O_{20}$
1.0711	Y_2SiO_5
1.0714	$KLu(WO_4)_2$
1.0714–1.0716	$KLu(WO_4)_2$
1.0715	Y_2SiO_5
1.0716–1.0721	$KLu(WO_4)_2$
1.0720	$CaSc_2O_4$
1.0721	$KLu(WO_4)_2$
1.07255–1.0730	$YAlO_3$
1.0726	$YAlO_3$
1.0729	$YAlO_3$
1.073	$KLu(WO_4)_2$ [11.357]
1.0737	$Y_3Al_5O_{12}$
≈ 1.074	$Y_2O_3-ThO_2-Nd_2O_3$
1.074	$SrAl_{12}O_{19}$
1.0741	Gd_2O_3
1.0741	Y_2SiO_5
1.0742	Y_2SiO_5
1.0746	Y_2O_3
1.0746	Y_2O_3 ceramic [11.358]
1.075	La_2O_2S
1.0757	$Sr_3Ga_2Ge_4O_{14}$
1.0759	$LuAlO_3$
1.0759	Lu_2O_3 ceramic [11.359]
1.0760	$GdAlO_3$
1.0775–1.0845	$CaYAlO_4$
1.0780	$Y_3Al_5O_{12}$
1.0780–1.086	$LaMgAl_{11}O_{19}$
1.0782	Y_2SiO_5
1.0782–1.0815	$YAlO_3$
1.0785	$LuScO_3$
1.0786	$CaAl_4O_7$
1.0786	Y_2O_3 ceramic [11.360]
1.0788	$Ca_2Ga_2SiO_7$
1.0789	Gd_2O_3
1.079	$La_2Be_2O_5$
1.0790	$La_2Be_2O_5$
1.0790	Lu_2SiO_5
1.07925	Lu_2SiO_5
1.0795	$YAlO_3$
1.0795–1.0802	$YAlO_3$
1.0796	$YAlO_3$

Table 11.14 (continued)

Wavelength	Material
1.08	Y_2O_3
1.08	$YAlO_3$ [11.360]
1.080	Lu_2O_3 ceramic [11.359]
1.0804	$LaAlO_3$
1.0806	$CaYAlO_4$
1.0812	Sc_2SiO_5
1.08145	Sc_2SiO_5
1.0817	$LaMgAl_{11}O_{19}$
1.0824	$LaMgAl_{11}O_{19}$
1.082	Sc_2O_3 [11.308]
1.082–1.084	$LaMgAl_{11}O_{19}$
1.0828	$SrAl_4O_7$
1.0829–1.0859	$LiNbO_3$
1.083	$YAlO_3$
1.0832	$LuAlO_3$
1.0832–1.0855	$YAlO_3$
1.0843	$YScO_3$
1.0845	$YAlO_3$
1.0846	$LiNbO_3$
1.085	$LiNbO_3 : Mg$
1.08515	$GdScO_3$
1.0868	$CaSc_2O_4$
≈ 1.0885	$CaF_2 - CeO_2$
1.0885–1.0889	CaF_2
1.0909	$YAlO_3$
1.091	$Ca_4GdO(BO_3)_3$ [11.322]
1.0921	$YAlO_3$
1.0922–1.0933	$LiNbO_3$
1.093	$LiNbO_3$
1.0933	$LiNbO_3$
≈ 1.094	$LiNbO_3 : MgO$
1.0989	$YAlO_3$
1.1054	$Y_3Al_5O_{12}$
1.1119	$Y_3Al_5O_{12}$
1.1158	$Y_3Al_5O_{12}$
1.1225	$Y_3Al_5O_{12}$

be reached within the crystal. This setup minimizes the formation of thermal lenses, and therefore yields better beam quality at high powers compared to a rod laser. Multiple passes of the pump radiation through the crystal increase the absorption efficiency and also the effective pump power density in the crystal. Therefore, the thin disc design is suitable for quasi-three-level systems such as Yb^{3+}. Figure 11.58 shows the input versus output power of a thin-disc Yb:YAG laser [11.311]. CW output

Table 11.15 Laser wavelengths of the $^4F_{3/2} \rightarrow {}^4I_{13/2}$ transition at 300 K [11.213]

Wavelength	Material
1.18 (transition $^4F_{5/2}, ^2H_{9/2} \rightarrow {}^4I_{13/2}$)	$RbPb_2Br_5$ [11.309]
1.3	$KNdP_4O_{12}$
1.3	$NdAl_3(BO_3)_4$
1.302	KYF_4
1.304–1.372	$(La, Nd)P_5O_{14}$
1.3065	$SrAl_{12}O_{19}$
1.307	KYF_4
1.3070	$5NaF - 9YF_3$
1.311–1.334	$NaNdP_4O_{12}$
1.313	$LiYF_4$
1.3133	$LiLuF_4$
1.3150	$Ca_3Ga_2Ge_3O_{12}$
1.316–1.340	$LiNdP_4O_{12}$
1.317	$Li(La, Nd)P_4O_{12}$
1.317	$LiNdP_4O_{12}$
1.3170	CeF_3
1.3175	BaF_2
1.318	$Y_3Al_5O_{12}$
1.318	BaY_2F_8
1.3185	$CaF_2 - GdF_3$
1.3185	$BaF_2 - LaF_3$
1.3185	KY_3F_{10}
1.3187	$Y_3Al_5O_{12}$
1.3188	$Y_3Al_5O_{12}$
1.319	$Y_3Al_5O_{12}$ [11.330, 361]
1.319	$LiNdP_4O_{12}$
1.319	$Y_3Ga_5O_{12}$
1.319	$Y_3Al_5O_{12}$ ceramic [11.362]
1.319–1.325	$(Y, Nd)P_5O_{14}$
1.3190	$Ca_2Y_5F_{19}$
1.3190	$CaF_2 - LaF_3$
1.3190	$CaF_2 - CeF_3$
1.3190	$Sr_2Y_5F_{19}$
1.3190	$\alpha - NaCaCeF_6$
≈ 1.32	$Y_3Al_5O_{12}$ ceramic [11.334]
1.32	$Gd_3Sc_2Ga_3O_{12} : Cr$
1.32	NdP_5O_{14}
1.32	$(La, Nd)P_5O_{14}$
1.32	$K(Nd, Gd)P_4O_{12}$
1.32	$YLiF_4$ [11.363]
1.320	$NaNdP_4O_{12}$
1.3200	$SrF_2 - LuF_3$
1.3200	$BaF_2 - YF_3$
1.3200	$Y_3Al_5O_{12}$

Table 11.15 (continued)

Wavelength	Material
1.3200	YP_5O_{14}
1.3208	$LiLuF_4$
1.3209	$Lu_3Al_5O_{12}$
1.3209	$Ba_5(PO_4)_3F$
1.3212	$LiYF_4$
1.3225	CaF_2
1.323	NdP_5O_{14}
1.323	$(La, Nd)P_5O_{14}$
1.324	$(La, Nd)P_5O_{14}$
1.3245	$CdF_2 - YF_3$
1.3250	$SrF_2 - LaF_3$
1.3250	SrF_2
1.3255	$SrF_2 - CeF_3$
1.3260	$SrF_2 - GdF_3$
1.3270	$CaF_2 - YF_3$
1.3270	BaF_2
1.3270	$Ca_3(Nb, Ga)_2Ga_3O_{12}$
1.328	$Sr_5(PO_4)_3F$ [11.213, 364]
1.3280	$BaF_2 - LaF_3$
1.3285	$\alpha - NaCaYF_6$
1.3285	$SrF_2 - ScF_3$
1.3298	$CsLa(WO_4)_2$
1.3298	$GdGaGe_2O_7$
1.330	$CaF_2 - LuF_3$
1.3300	$Gd_3Ga_5O_{12}$
1.3300	$CdF_2 - ScF_3$
1.3303	$NdGaGe_2O_7$
1.3305	$HfO_2 - Y_2O_3$
1.3305	$Y_3Ga_5O_{12}$
1.3310	LaF_3
1.3310	$Y_3Sc_2Ga_3O_{12}$
1.3310	$NaLuGeO_4$
1.3315	$LaF_3 - SrF_2$
1.3315	$Lu_3Ga_5O_{12}$
1.3315	$Gd_3Ga_5O_{12}$
1.3315	$Ca_3Ga_2Ge_3O_{12}$
1.3317	$Ca_3Ga_2Ge_3O_{12}$
1.3320	CeF_3
1.3320	$ZrO_2 - Y_2O_3$
1.3320	$Ca_3Ga_4O_9$
1.3325	$SrMoO_4$
1.3325	$NaYGeO_4$
1.3326	$Lu_3Al_5O_{12}$
1.3334	$NaGdGeO_4$
1.3338	$Y_3Al_5O_{12}$
1.3340	$CaWO_4$

Table 11.15 (continued)

Wavelength	Material
1.3340	$PbMoO_4$
1.3342	$KLa(MoO_4)_2$
1.3342	$Lu_3Al_5O_{12}$
1.3342	$NaBi(WO_4)_2$
1.3345	$SrAl_4O_7$
1.3347	$Ca_5(PO_4)_3F$
1.3347	$SrWO_4$
1.3350	$KLa(MoO_4)_2$
1.3350	$Y_3Al_5O_{12}$
1.3354	$CaLa_4(SiO_4)_3O$
1.3355	$NaLa(WO_4)_2$
1.3360	$Y_3Sc_2Al_3O_{12}$
1.3360	$Gd_3Sc_2Al_3O_{12}$
1.3360	$Lu_3Sc_2Al_3O_{12}$
13360	$Y_3Sc_2Al_3O_{12}$
1.3360	$CdF_2 - CeF_3$
1.3365	$Ca_2Ga_2SiO_7$
1.3365	$CdF_2 - GaF_3$
1.3365	$CdF_2 - LaF_3$
1.3370	$CaF_2 - YF_3$
1.3370	$CaWO_4$
1.3370	$Gd_3Ga_5O_{12}$
1.3370	$LiLa(MoO_4)_2$
1.3375	$\alpha - NaCaYF_6$
1.338	$Y_3Al_5O_{12}$ [11.365]
1.338	$Y_3Ga_5O_{12}$
1.338	$Sr_5(PO_4)_3F$ [11.366]
1.3380	$Ca(NbO_3)_2$
1.3380	$NaLa(MoO_4)_2$
1.3381	$Y_3Al_5O_{12}$
1.3382	$Y_3Al_5O_{12}$
1.3385	$NaGd(MoO_4)_2$
1.3387	$Lu_3Al_5O_{12}$
1.339	YF_3
1.3390	$CaWO_4$
≈ 1.34	$Y_3Al_5O_{12}$ [11.296, 367]
≈ 1.34	$YAlO_3$ [11.360]
≈ 1.34	$GdVO_4$ [11.213, 319, 320, 368–372]
≈ 1.34	YVO_4 [11.213, 319, 320]
≈ 1.34	$La_{0.2}Gd_{0.8}VO_4$ [11.319, 321]
≈ 1.34	$LuVO_4$ [11.373]
≈ 1.34	$GdAl_3(BO_3)_4$ [11.316]
1.3400	$LiGd(MoO_4)_2$
1.3400	$YAlO_3$
1.3407	$Bi_4Si_3O_{12}$

Table 11.15 (continued)

Wavelength	Material
1.341	$NdAl_3(BO_3)_4$
1.3410	$Y_3Al_5O_{12}$
1.3410	$Lu_3Al_5O_{12}$
1.3410	$YAlO_3$
1.3413	$YAlO_3$
1.3414	$YAlO_3$ [11.213,374,375]
1.3416	$YAlO_3$
1.3418	$Bi_4Ge_3O_{12}$
1.342	YVO_4 [11.347,376–383]
1.3420	$CaAl_4O_7$
1.3425	$Ca(NbO_3)_2$
1.3425	YVO_4
1.3425	$PbMoO_4$
1.3425	CdF_2-YF_3
1.3437	$LuAlO_3$
1.3440	$NaLa(MoO_4)_2$
1.345	$NdAl_3(BO_3)_4$
1.3475	$CaWO_4$
1.3482	$KLu(WO_4)_2$
1.3485	$KY(MoO_4)_2$
1.3493	$Ca_3Ga_2Ge_4O_{14}$
1.35	$KGd(WO_4)_2$ [11.384,385]
1.3500	CdF_2-LuF_3
1.3505	CaF_2-ScF_3
1.351	$La_2Be_2O_5$
1.351	$KGd(WO_4)_2$ [11.353]
1.3510	$KGd(WO_4)_2$
1.3510	$La_2Be_2O_5$
1.3510	$Sr_3Ga_2Ge_4O_{14}$
1.3512	$YAlO_3$
1.3514	$YAlO_3$
1.3520	CdF_2-GdF_3
1.3525	$Ca_2Y_5F_{19}$
1.3525	$KY(WO_4)_2$
1.3525	$Lu_3Al_5O_{12}$
1.3532	$Lu_3Al_5O_{12}$
1.3533	$Y_3Al_5O_{12}$
1.3533	$KLa(MoO_4)_2$
1.3533	$KLu(WO_4)_2$
1.354	$La_2Be_2O_5$
1.3545	$KY(WO_4)_2$
1.3550	$LiNbO_3$
1.3550	$KLu(WO_4)_2$
1.3565	$CaSc_2O_4$
1.3572	$Y_3Al_5O_{12}$
≈ 1.358	Y_2O_3

Table 11.15 (continued)

Wavelength	Material
1.3585	CaF_2-YF_3
1.3585	Y_2SiO_5
1.3585	Lu_2SiO_5
1.3595	LaF_3
1.3600	$\alpha-NaCaYF_6$
1.3628	$LaSr_2Ga_{11}O_{20}$
1.3630	$KLa(MoO_4)_2$
1.3630–2	Sc_2SiO_5
1.365	$La_2Be_2O_5$
1.3657	$KLa(MoO_4)_2$
1.3665	$SrAl_4O_7$
1.3675	LaF_3
1.3680	$SrAl_4O_7$
1.3690	CeF_3
1.3707	$La_3Ga_{5.5}Nb_{0.5}O_{14}$
1.3710	$CaAl_4O_7$
1.3730	$La_3Ga_5GeO_{14}$
1.3730	$La_3Ga_5SiO_{14}$
1.3730	$La_3Ga_{5.5}Ta_{0.5}O_{14}$
1.3745	$LiNbO_3$
1.3760	$LaMgAl_{11}O_{19}$
1.386	YVO_4 [11.377,386]
1.3868	$LaBGeO_5$
1.3870	$LiNbO_3$
1.3885	$CaWO_4$
1.4150	$Y_3Al_5O_{12}$
1.430	$YAlO_3$ [11.387]
1.44	$SrGd_4(SiO_4)_3O$
1.4444	$Y_3Al_5O_{12}$ [11.213,387]
1.486	Sc_2O_3 [11.308]

powers in the kilowatt range from one disc are possible. Further power scaling with several discs in the cavity is also possible [11.311].

Yb *fiber laser.* Fiber lasers present another approach for the generation of high powers with diffraction-limited beam quality. Since the introduction of the double-clad fiber more than two decades ago and with the recent technological advances in the fields of fiber fabrication and beam-shaped high-power diode lasers, the performance of diode-pumped fiber lasers has dramatically improved. Today, fiber lasers can compete with their corresponding bulk crystalline systems in certain applications, especially when fundamental-transverse-mode continuous-wave (CW) laser operation at output powers in the milliwatt to kilowatt range is required.

Basic aspects of fiber lasers. The invention of the double-clad fiber geometry and the holey fiber concept has accelerated the scaling of the output power and hence the success of high-power Er, Nd and Yb fiber lasers. For more-detailed reading a comprehensive introduction to the field of rare-earth-doped fiber lasers can be found in [11.388].

The choice of the fiber material involves a number of considerations: The maximum phonon energy, environmental durability, the drawability and the rare-earth solubility. The maximum phonon energy of the glass sets the overall infrared transparency range of the fiber and the multiphonon relaxation rates that influence the quantum efficiency of radiative electronic transitions by nonradiative decay. The important physical properties of the popular glasses used for optical fibers are shown in Table 11.16.

- Silicate glass. This glass is the most important material used for optical fiber production [11.388,389]. However, the maximum phonon energy is high ($\approx 1100\,cm^{-1}$) and has so far limited the emission wavelength for infrared fiber lasers using this material to approximately 2.2 μm [11.390]. Silica is robust and fibers fabricated from this material involve the very effective modified chemical vapor deposition (MCVD) technique. Reducing the OH^- content in the glass, which has two main absorption peaks in the range 1.3–2.0 μm, reduces the background absorption of fibers [11.391].
- Fluoride glass. These glasses, especially the heavy-metal fluorides [11.392, 393], are used as host materials for mid-infrared fiber lasers. The most widespread fluoride fiber material is ZBLAN [11.394] with a mixture of 53 mol % ZrF_4, 20 mol % BaF_2, 4 mol % LaF_3, 3 mol % AlF_3, and 20 mol % NaF. Since it can be readily drawn into single-mode optical fiber [11.395] it is particularly important to mid-infrared fiber lasers [11.396] and allows for high infrared transparency up to ≈ 6 μm. Nonradiative decay by multiphonon relaxation, however, becomes significant for transitions at wavelengths longer than ≈ 3 μm. In addition to mid-infrared applications, ZBLAN is mostly also used for upconversion fiber lasers which need metastable intermediate pump levels with low multiphonon relaxation rates. An overview of the spectroscopic properties of rare-earth ions doped into ZBLAN is given in [11.397].
- Chalcogenide glasses. Chalcogenides are composed of the chalcogen elements S, Se and Te [11.398, 399]. When rare-earth ions are doped into these glasses [11.400], the radiative transition probabilities, and therefore the absorption and emission cross sections, are high as a result of the high (≈ 2.6) refractive index and the high degree of covalency of the glass. Small phonon energies of 300–450 cm^{-1} produce low rates of multiphonon relaxation for mid-infrared transitions. The low thermal conductivity (Table 11.16) is, however, an important factor

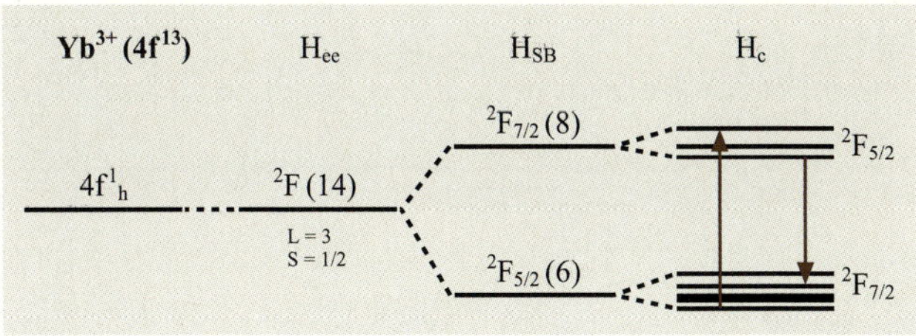

Fig. 11.54 Level scheme of Yb^{3+}: free-ion state (one 4f hole $4f^1_h$), electron–electron interaction (H_{ee}), splitting by spin–orbit interaction H_{SB}, and crystal field H_c. Pump and laser transitions are indicated by *arrows*

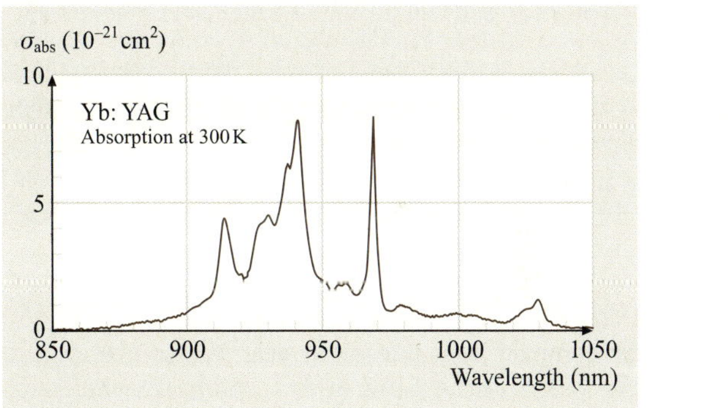

Fig. 11.55 Absorption spectrum of Yb:YAG at 300 K

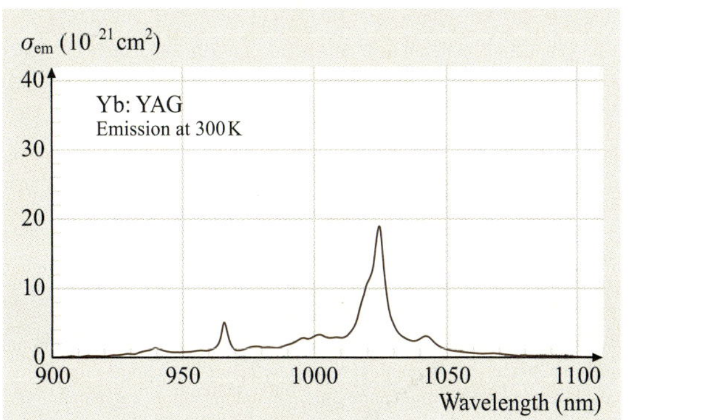

Fig. 11.56 Emission spectrum of Yb:YAG at 300 K

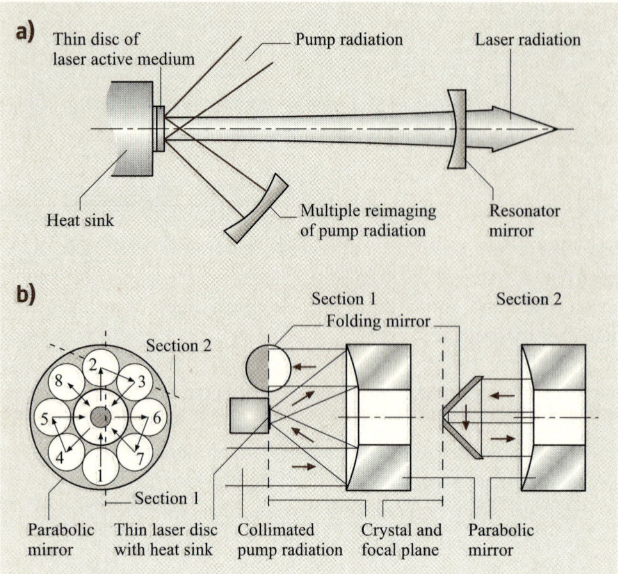

Fig. 11.57 (a) Thin-disc laser setup, (b) multiple-pass pump optics for 16 passes

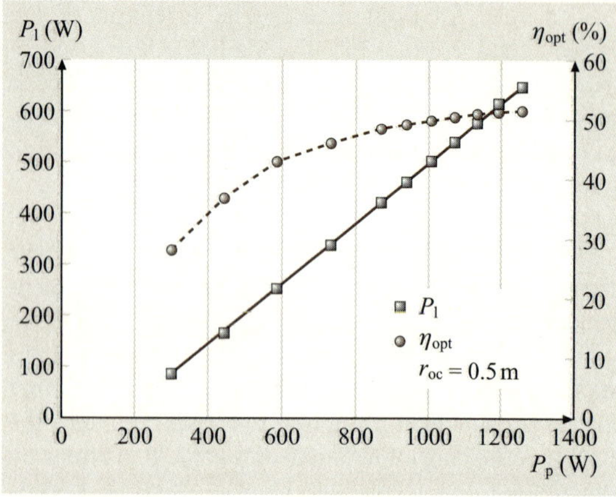

Fig. 11.58 Output power and optical efficiency for a thin-disc Yb:YAG laser using a single disc. (P_p = pump power, P_l = laser output power, η_{opt} = optical-optica efficiency)

to be considered in the design of chalcogenide-based lasers. So far, the most important glasses are the sulfide glasses GaLaS (GLS) [11.401] and GeGaS [11.402] because of the reasonably high rare-earth solubility.

Fiber, pump and resonator geometries. As bulk lasers, fiber lasers can be operated continuous wave, pulsed (including Q-switching) and mode-locked. These operation modes have been investigated intensively for the common laser transitions near 1 μm in Nd^{3+} and Yb^{3+}, and near 1.5 μm in Er^{3+}. However, the small fiber diameter limits the peak power by the damage threshold intensity and, hence, crystalline lasers in bulk geometries are mostly preferred when high-energy short pulses are needed. In an analogous way to the optical excitation of bulk gain media (see the section on longitudinal and transversal pumping in this Sect. 11.2.2), doped optical fibers can be either end pumped (core pumped) or side pumped (cladding pumped). The former method is less scalable since it relies on the use of expensive high-beam-quality pump sources because core areas are usually $< 100\,\mu m^2$. On the other hand, the larger cladding area ($> 10^4\,\mu m^2$) allows for high-power diode-array pumping [11.408–411]. We will here describe the cladding-pumping technique, which is one of the most important developments in high-power fiber-laser technology.

- Fiber designs for cladding pumping. In the design of cladding pumping, the core of the fiber is generally made to guide a single transverse LP_{01} mode. The shape of the multimode pump cladding (Fig. 11.59), however, can be shaped with a number of geometries. The pump cladding, which in turn is surrounded by a low-refractive-index transparent polymer or glass, provides a high numerical aperture (NA) of typically 0.3–0.55. Photonic crystal structures can be also used to improve cladding pumping [11.412] for the operation of fiber lasers at the multi-hundred-watt level. There are three main double-clad-fiber layouts: circular, circular with offset core, and rectangular, as shown schemat-

Table 11.16 Properties of popular fiber materials

Fiber material	Max. phonon energy (cm^{-1})	Infrared transparency (μm)	Propagation losses (λ at minimum) (dB/km)	Thermal conductivity (W/Km)
Silica	1100 [11.403]	< 2.5	0.2 (1.55 μm)	1.38 [11.404]
ZBLAN	550 [11.397]	< 6.0	0.05 (2.55 μm)	0.7–0.8 [11.405]
GaLaS	425 [11.406]	< 8.0	0.5 (3.50 μm)	0.43–0.5 [11.407]

ically in Fig. 11.59. In the case of circular pump cladding [11.53] a portion of the launched pump light is skew to the fiber axis and produces an inner pump beam caustic that never crosses the core. Asymmetric configurations significantly improve the pump beam absorption in the core [11.413,414]. Double-clad pump schemes have been demonstrated with holey or photonic crystal fibers [11.415]. In such fibers single-mode guiding and very large mode areas are possible [11.416].

- Fiber-laser resonators. Typical free-running fiber-laser resonators are shown schematically in Fig. 11.60. In the simplest resonator (Fig. 11.60a), the pump light passes through a dichroic mirror that is highly reflective for the oscillating laser light. Fresnel reflection at the cleaved output end facet of the fiber can provide sufficient feedback for laser oscillation, although with an output-coupler mirror at the output end of the fiber the optical efficiency can be maximized. In an alternative arrangement, the pump light can be launched into the output end of the fiber (Fig. 11.60b). In order to scale the output power, each end of the fiber can be pumped (Fig. 11.60c). Due to its geometry, the fiber provides potentially high pump- and signal-beam intensities without the drawbacks of significant thermal and thermo-optical effects. Its large surface-area-to-volume ratio means that the heat generated in the core is dissipated effectively by radiation and convection from the outer surface of the fiber.

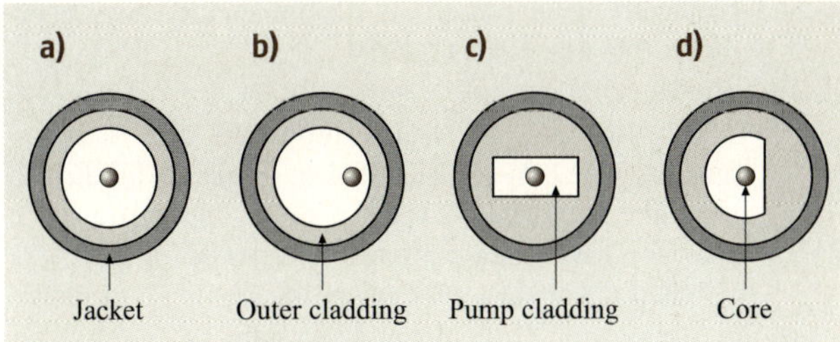

Fig. 11.59a–d Principal double-clad fiber geometries which include **(a)** circular-shaped pump cladding with axially positioned core, **(b)** circular-shaped pump cladding with off-axially positioned core, **(c)** rectangular-shaped pump cladding and **(d)** D-shaped pump cladding

Table 11.17 Material and laser parameters of Yb:YAG

Growth method	Czochralski	[11.417]
Temperature (°*C*)	1930	[11.417]
Crucible	Ir (Re)	[11.417]
Yb distribution coefficient	1.0	[11.417]
Max. doping level (%)	≤ 100	[11.417]
Structure	Cubic	[11.417]
Space group	$Ia3d\text{-}O_h^{10}$	[11.417]
Heat conductivity ($Wm^{-1}K^{-1}$)		
Undoped YAG	11.0	[11.417]
Yb-doped (5%)	6.8	[11.417]
RE-density (10^{21} cm^{-3})	14	[11.417]
λ_{laser} (nm)	1030	[11.417]
	1050	
σ_{em} (cm^2) at 1030 nm	19×10^{-21}	[11.286]
	20×10^{-21}	[11.418]
	21×10^{-21}	[11.419]
σ_{em} (cm^2) at 1050 nm	3×10^{-21}	[11.286]
σ_{abs} (cm^2) at 969 nm	8.3×10^{-21}	[11.286]
	7.7×10^{-21}	[11.419]
σ_{abs} (cm^2) at 941 nm	8.2×10^{-21}	[11.286]
τ_{rad} (μs)	1040	[11.417]
	951	[11.420]

High-power Yb fiber lasers. Yb-doped fiber-lasers have been operated at CW and average output powers in excess of 10 W [11.421–426], in excess of 100 W [11.427–433] and above 1 kW [11.434]. Also, a 30 μm Yb-doped large-mode-area fiber has been used for the amplification of *Q*-switched Nd:YAG pulses at repetition rates between 3 and 50 kHz with average output powers up to 100 W. Pulse energies as high as

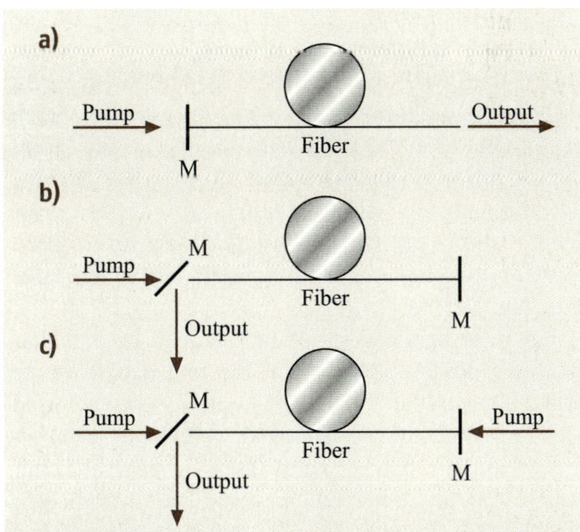

Fig. 11.60a–c Schematic diagram of resonators used for free-running fiber lasers with **(a)** a single-end co-propagating pump, **(b)** a single-end counter-propagating pump and **(c)** dual-end pumps. M represents a mirror

4 mJ with diffraction-limited beam quality have been obtained in this case [11.435].

Yb^{3+} crystalline lasers. The properties and laser parameters of Yb:YAG are listed in Table 11.17. In the meantime, besides Yb:YAG, a lot of Yb-doped laser materials have been explored and tested in CW, *Q*-switched and mode-locked operation. Table 11.18 shows an overview of Yb-doped laser crystals.

Er lasers at 1.5 μm ($^4I_{13/2} \rightarrow {}^4I_{15/2}$)

For many years Er^{3+}-doped materials have been widely investigated for laser applications in the spectral range around 1.6 μm [11.436]. This laser transition ($^4I_{13/2} \rightarrow {}^4I_{15/2}$, see Er^{3+} energy-level scheme depicted in Fig. 11.61) is used for eye-safe lasers for medicine, telecommunication, remote sensing and light detecting and ranging (LIDAR). Suitable erbium-doped materials for the 1.6 μm laser transition should be at first characterized by a high phonon energy, which enables fast depopulation of the pump level $^4I_{11/2}$ via nonradiative decay in order to prevent excited-state absorption (ESA, $^4I_{11/2} \rightarrow {}^2H_{11/2}, {}^4S_{3/2}$) and upconversion $\left[\text{UC1}, ({}^4I_{11/2}, {}^4I_{11/2}) \rightarrow ({}^4I_{15/2}, {}^4F_{7/2})\right]$ from the $^4I_{11/2}$ level and to populate efficiently the $^4I_{13/2}$ upper laser level (Fig. 11.61). The second important condition is that the ESA transition $^4I_{13/2} \rightarrow {}^4I_{9/2}$ should not spectrally overlap with the range of stimulated emission around 1.6 μm and that the upconversion process $\left[\text{UC2}, ({}^4I_{13/2}, {}^4I_{13/2}) \rightarrow ({}^4I_{9/2}, {}^4I_{15/2})\right]$ is weak. Additionally, a significant splitting of the ground-state multiplet is advantageous to achieve a quasi-four-level system. These conditions are best fulfilled by Er^{3+}-doped glasses and fibers, which are to date the most efficient lasers at this transition. However, glasses suffer from poor thermal and mechanical stability, thus Er^{3+}-doped crystalline matrices are still being intensively investigated in order to find suitable crystals for this laser transition.

For most applications of Er^{3+}-doped laser materials, laser diodes operating in the wavelength range around 975 nm (the $^4I_{15/2} \rightarrow {}^4I_{11/2}$ transition) are used as a pump source, thus enabling all-solid-state laser systems. With Ti:sapphire pump lasers in general better laser results are obtained, however, the overall efficiency is low. Another possible pump wavelength lies around 1.5 μm, i. e., directly into the upper laser multiplet. However, for this wavelength region high-power laser diodes are not yet available, therefore usually Er–glass lasers as pump sources are used. In any case, the Er absorption at these wavelengths is rather small, because the absorption cross sections are in the order of 1 to $2\times10^{-20}\,\text{cm}^2$ (Figs. 11.62, 63). Doping levels higher than 1 to 2% are critical, because of increased reabsorption losses and higher rates for energy transfer processes, which depopulates the upper laser level $({}^4I_{13/2}, {}^4I_{13/2}) \rightarrow ({}^4I_{15/2}, {}^4I_{9/2})$. Therefore, to keep the Er^{3+} concentration low but achieve higher absorption, the usual approach is to codope the Er^{3+}-doped laser material with Yb^{3+}, which can be very efficiently pumped around 975–980 nm. Then, the energy transfer process $({}^2F_{5/2}, {}^4I_{15/2}) \rightarrow ({}^2F_{7/2}, {}^4I_{11/2})$ is exploited (Fig. 11.61). The main task for the optimization of Er lasers around 1.55 μm is thus to find the optimum concentration for both dopant ions.

Crystals. The absorption spectra of Er^{3+}:YVO_4 for the $^4I_{15/2} \rightarrow {}^4I_{11/2}$ and $^4I_{15/2} \rightarrow {}^4I_{13/2}$ transitions are shown in Figures 11.62 and 11.63, respectively [11.437]. The peak absorption cross section for the $^4I_{15/2} \rightarrow {}^4I_{11/2}$ transition around 970 nm is up to $2\times10^{-20}\,\text{cm}^2$, for the $^4I_{15/2} \rightarrow {}^4I_{13/2}$ transition around 1500 nm cross sections are higher, approximately up to $4\times10^{-20}\,\text{cm}^2$. The emission spectrum is shown in Fig. 11.63. The peak emission cross sections of the $^4I_{13/2} \rightarrow {}^4I_{15/2}$ emission in Er^{3+}:YVO_4 are up to $2\times10^{-20}\,\text{cm}^2$, in the long-wavelength tail, where the laser oscillation occurs, the peak cross sections are around $0.5\times10^{-20}\,\text{cm}^2$. These values are typical for Er^{3+}-doped crystals, e.g., the peak emission cross section around 1550 nm is $0.31\times10^{-20}\,\text{cm}^2$ for

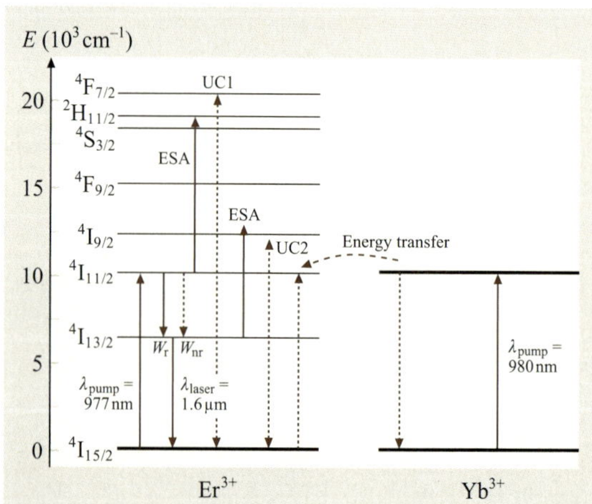

Fig. 11.61 Energy-level diagram of Er^{3+} and Yb^{3+} in a YVO_4 crystal

Table 11.18 Yb^{3+}-doped laser crystals and laser transitions $^2F_{5/2} \rightarrow {}^2F_{7/2}$

Crystal	λ_{laser} (μm)	Pump	T (K)	Output mode	Ref.
$BaCaBO_3F$	1.034	TiS laser	300	p	[11.440]
CaF_2 : Na	≈ 1.046–1.048	Laser diode	300	ML	[11.441]
	1.051	Laser diode	300	Q-switch	[11.441]
CaF_2 : Nd	1.0336	Xe lamp	120	p	[11.442]
	1.030–1.055	TiS laser	300	CW	[11.443]
$Ca_4GdO(BO_3)_3$	-	Laser diode	300	CW	[11.444]
	1.032	TiS laser	300	CW	[11.445]
	1.035–1.088	Laser diode	300	CW	[11.446]
	1.050	Laser diode	300	CW	[11.446, 447]
	1.050	TiS laser	300	CW	[11.447]
	1.082	TiS laser	300	CW	[11.445]
	1.082	Laser diode	300	CW	[11.447]
$Ca_5(PO_4)_3F$	1.043	TiS laser	300	CW	[11.448, 449]
$Ca_3Sr_2(PO_4)_3F$	1.046	TiS laser	300	CW	[11.450]
$Ca_4Sr(PO_4)_3F$	0.985	TiS laser	300	CW	[11.450]
	1.046	TiS laser	300	CW	[11.450]
	1.110	TiS laser	300	CW	[11.450]
$Ca_4YO(BO_3)_3$	1.018–1.087	TiS laser	300	CW	[11.451]
	1.032	Laser diode	300	CW	[11.452]
	1.060	TiS laser	300	CW	[11.453]
	1.084–1.096	TiS laser	300	CW	[11.454]
$Gd_3Ga_5O_{12}$	1.039	Laser diode	300	Q-switch	[11.455]
$Gd_3Ga_5O_{12}$: Nd	1.0232	Xe lamp	77	p	[11.456]
$Gd_3Sc_2Al_3O_{12}$: Nd	1.0299	Xe lamp	77	p	[11.457]
Gd_2SiO_5	1.028–1.093	Laser diode	300	CW	[11.458]
	1.030–1.039	Laser diode	300	CW	[11.459]
	1.045–1.070	Laser diode	300	CW	[11.459]
	1.081–1.097	Laser diode	399	CW	[11.459]
	1.089	Laser diode	300	CW	[11.460, 461]
	1.090	Laser diode	300	CW	[11.462, 463]
	1.091–1.105	Laser diode	300	p	[11.459]
	1.094	Laser diode	300	CW	[11.461]
$GdVO_4$	1.015	Laser diode	300	CW	[11.464]
	1.015	TiS laser	300	CW	[11.465]
	1.015–1.019	Laser diode	300	CW	[11.466]
	1.026–1.031	Laser diode	300	CW	[11.466]
	1.029	Laser diode	300	CW	[11.464]
	1.029	TiS laser	300	CW	[11.465]
	1.040	Laser diode	300	CW	[11.466]
	1.045	Laser diode	300	CW	[11.466]
$(Gd, Y)_2SiO_5$	1.030–1.089	Laser diode	300	CW	[11.458]
$KGd(WO_4)_2$	1.026–1.044	Laser diode	300	CW	[11.467]
	1.030–1.051	Laser diode	300	CW	[11.468]
	1.031–1.0374	Laser diode	300	ML	[11.469]

Table 11.18 (continued)

Crystal	λ_{laser} (μm)	Pump	T (K)	Output mode	Ref.
$KLu(WO_4)_2$	1.030	Laser diode	300	CW	[11.470]
	1.030	TiS laser	300	CW	[11.470]
	1.044	TiS laser	300	CW	[11.471]
	1.0435	Laser diode	300	CW	[11.471]
$KY(WO_4)_2$	1.026	Laser diode	300	CW	[11.417]
	1.026–1.042	Laser diode	300	CW	[11.467]
	1.030	TiS laser	300	CW	[11.472]
	1.048	Laser diode	300	ML	[11.473]
	≈ 1.028	Laser diode	300	ML	[11.474]
	1.030	Laser diode	300	CW	[11.475]
	0.987–1.051	Laser diode	300	CW	[11.476]
$KYb(WO_4)_2$	1.068	TiS laser	300	CW	[11.477]
	1.074	TiS laser	300	qCW	[11.478]
$LaSc_3(BO_3)_4$	1.044	TiS laser	300	CW	[11.417]
	1.045	TiS laser	300	CW	[11.479]
	0.995–1.087	Laser diode	300	CW	[11.480]
$LiGd(MoO_4)_2$	1.027–1.0335	Laser diode	300	CW	[11.481]
$LiNbO_3$-waveguide	1.008	TiS laser	300	CW	[11.482]
	1.030	TiS laser	300	CW	[11.482]
	1.060	TiS laser	300	CW	[11.482]
$LiNbO_3 : MgO$	1.063	Laser diode	300	CW	[11.483]
$Li_6Y(BO_3)_3$	1.040	Laser diode	300	CW	[11.484]
$Lu_3Al_5O_{12}$	1.0297	Xe lamp	77	p	[11.456]
	1.03	Laser diode	175	CW	[11.485]
$Lu_3Al_5O_{12}$: Nd, Cr	1.0294	Xe lamp	77	p	[11.456]
$Lu_3Ga_5O_{12}$: Nd	1.0230	Xe lamp	77	p	[11.456]
Lu_2O_3	≈ 1.029–1.038	TiS laser	300	ML	[11.486]
	1.032	Laser diode	300	CW	[11.417]
ceramics	1.035	Laser diode	300	CW	[11.487]
ceramics	1.079	Laser diode	300	CW	[11.487]
		Laser diode	300	CW	[11.475]
$Lu_3Sc_2Al_3O_{12}$: Nd	1.0299	Xe lamp	77	p	[11.457]
$LuVO_4$	1.0347	Laser diode	300	CW	[11.488]
	1.041	Ti laser	300	CW	[11.488]
	1.0444	Laser diode	300	CW	[11.488]
	1.0527	Laser diode	300	CW	[11.488]
$NaGd(WO_4)_2$	1.016–1.049	TiS laser	300	CW	[11.489]
	1.023	TiS laser	300	CW	[11.490]
	1.033	Laser diode	300	CW	[11.490]
$NaLa(MoO_4)_2$	1.016–1.064	TiS laser	300	CW	[11.491]
	1.017	Laser diode	300	CW	[11.492]
	≈ 1.020	Laser diode	300	Q-switch	[11.492]
	1.023	Laser diode	300	CW	[11.492]
	1.035	Laser diode	300	CW	[11.491]
$NaLa(WO_4)_2$	1.017–1.057	TiS laser	300	CW	[11.493]

Table 11.18 (continued)

Crystal	λ_{laser} (μm)	Pump	*T* (K)	Output mode	Ref.
Sc_2O_3	1.041	TiS laser	300	CW	[11.494]
	1.0416	TiS laser	300	CW	[11.495]
	1.0946	TiS laser	300	CW	[11.495]
ceramic	1.041	Laser diode	300	CW	[11.496]
ceramic	1.094	Laser diode	300	CW	[11.496]
$Sr_5(PO_4)_3F$	0.985	Cr:LiSAF	300	p	[11.497]
	0.985	TiS laser	300	CW	[11.498, 499]
	1.047	TiS laser	300	CW, qCW	[11.500, 501]
		Laser diode	300	p	[11.502]
$Sr_{5-x}Ba_x(PO_4)_3F$	≈ 1.048	TiS laser	300	CW	[11.500]
$(Sr_{0.7}Ca_{0.3})_3Y(BO_3)_3$		Laser diode	300	CW, qCW	[11.503]
$Sr_5(VO_4)_3F$	1.044	TiS laser	300	p	[11.450]
$Sr_3Y(BO_3)_3$		Laser diode	300	CW, qCW	[11.494]
$SrY_4(SiO_4)_3$	1.020–1.095	Laser diode	300	CW	[11.504]
	≈ 1.068	Laser diode	300	ML	[11.504]
$Y_3Al_5O_{12}$	1.016–1.095	TiS laser	300	CW	[11.505]
	1.0293	Xe lamp	77	p	[11.456]
	1.0296	Xe lamp	77	p	[11.279]
	1.023–1.052	Laser diode	300	CW	[11.506]
	1.029	Laser diode	300	CW, qCW	[11.507]
waveguide	1.03	TiS laser	300	CW	[11.508]
	1.030	TiS laser	300	CW	[11.509]
	1.03	Laser diode	300	CW	[11.280, 417, 507, 510]
	1.03	Laser diode	300	*Q*-switch	[11.511, 512]
	1.03	Laser diode	300	ML	[11.513]
	1.03	TiS laser	300	*Q*-switch	[11.514, 515]
waveguide	1.03	Laser diode	300	CW	[11.516]
waveguide	1.03	Laser diode	300	CW	[11.517]
waveguide	1.03	Laser diode	300	*Q*-switch	[11.517]
	1.031	Laser diode	300	p	[11.501]
	1.031	Laser diode	300	CW	[11.518, 519]
	≈ 1.0312	Laser diode	300	ML	[11.519]
	1.0494–1.0504	Laser diode	300	CW	[11.520]
	1.070	Laser diode	300	CW	[11.311]
$Y_3Al_5O_{12}$: Nd	1.0297	Xe lamp	200	p	[11.456]
waveguide	1.03	TiS laser	300	CW	[11.521]
$Y_3Al_5O_{12}$: Nd, Cr	1.0298	Xe lamp	210	P	[11.456]
$YAl_3(BO_3)_4$	≈ 1.040	Laser diode	300	CW	[11.522]
	1.120–1.140	Laser diode	300	CW	[11.523, 524]
$Y_3Ga_5O_{12}$: Nd	1.0233	Xe lamp	77	P	[11.456]
$YLiF_4$	≈ 0.991–1.022	Laser diode	77	P	[11.525]
$YLuSiO_5$	1.014–1.091	Laser diode	300	CW	[11.526]
Y_2O_3 ceramic	≈ 1.076	Laser diode	300	CW	[11.527]
ceramic	≈ 1.076	Laser diode	300	ML	[11.527]
ceramic	1.0767–1.0784	Laser diode	300	CW	[11.520]
ceramic	1.078	Laser diode	300	CW	[11.528]

Table 11.18 (continued)

Crystal	λ_{laser} (μm)	Pump	*T* (K)	Output mode	Ref.
Y_2SiO_5	1.000–1.010	Laser diode	300	CW	[11.529]
	1.082	Laser diode	300	CW	[11.529]
$Y_3Sc_{1.0}Al_{4.1}O_{12}$ ceramic	≈ 1.060	Laser diode	300	CW, ML	[11.530]
YVO_4	1.020–1.027	Laser diode	300	CW	[11.531]
	1.037	TiS laser	300	CW	[11.532]
	1.039	Laser diode	300	CW	[11.532]

Er:$YAlO_3$ [11.436], $0.33\times10^{-20}\,cm^2$ for Er:Y_2SiO_5 [11.438], $0.45\times10^{-20}\,cm^2$ for Er:YAG [11.436], $0.42\times10^{-20}\,cm^2$ for Er:YLF [11.436], and $0.59\times10^{-20}\,cm^2$ for Er:$LaGaO_3$ [11.439].

From the absorption and emission spectra the gain coefficient curves (Fig. 11.64) can be calculated by

$$g = N[P\sigma_{em} - (1-P)\sigma_{abs}]\,,$$

where N is the ion concentration and P is the inversion coefficient, defined as the ratio between the populations in the $^4I_{13/2}$ and the $^4I_{15/2}$ levels. It can be seen that for Er^{3+}:YVO_4 already for an inversion coefficient of $P \approx 0.2$ laser oscillation should be possible within a spectral range from approximately 1530 nm to 1610 nm.

In Fig. 11.65a the input–output characteristics of Ti:sapphire-pumped Er^{3+}:YVO_4 crystals with 0.5% Er and 1% Er doping, operating at 1604 nm, are shown. The slope efficiency with respect to the incident power is about 7–8%. It should be noted that with respect to the absorbed power the 0.5% Er:YVO_4 crystal exhibits a higher slope efficiency. This indicates that for the more highly doped sample (1% Er) the aforementioned loss mechanism of reabsorption and upconversion already take place. The laser threshold is below 200 mW of incident power and below 100 mW of absorbed pump power.

These values for the slope efficiency, the laser threshold and the output power are typical for well-performing Er^{3+}-doped crystals, such as for $Y_3Al_5O_{12}$ [11.533] and $LaSc_3(BO_3)_4$ [11.534].

Recently, high-power and ultra-efficient laser operation of an Er:YAG laser near 1645 nm with approximately 60 W [11.535–538] of output power was demonstrated using a high-power fibre laser at 1532 nm as a pump source (so-called in-band pumping (Fig. 11.65b). This work demonstrates that the small Stokes shift between the pump and laser radiation (1532 nm/1645 nm) yields very efficient Er lasers with slope efficiencies up to 80%.

In Table 11.19 an overview of crystalline room-temperature Er^{3+} lasers is given. Note that lasers on the $^4S_{3/2} \rightarrow {}^4I_{9/2}$ transition are also listed.

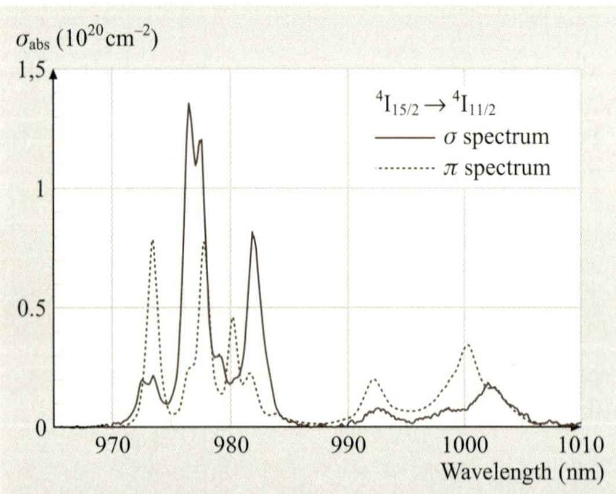

Fig. 11.62 Polarized absorption spectra at room temperature of an Er^{3+}:YVO_4 crystal in the spectral range of the $^4I_{15/2} \rightarrow {}^4I_{11/2}$ transition. (After [11.437])

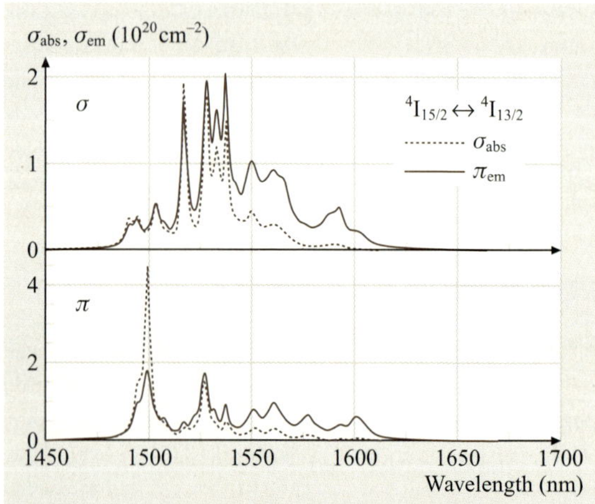

Fig. 11.63 Spectra of absorption cross sections due to the $^4I_{15/2} \rightarrow {}^4I_{13/2}$ transition of Er^{3+} in YVO_4 crystal (*dotted lines*) and emission cross sections ($^4I_{13/2} \rightarrow {}^4I_{15/2}$) (*solid lines*) for σ and π polarization. (After [11.437])

Glass. Laser oscillation in Er^{3+}-doped and Yb^{3+},Er^{3+}-codoped silicate and phosphate glasses was investigated as early as the mid-1960s [11.539–541]. The three-level behavior of the Er^{3+} system and the weak absorption of pump radiation caused by the requirement of low dopant concentration (usually approximately 2 to $5\times10^{19}\,cm^{-3}$) makes it difficult to obtain efficient laser operation in singly Er^{3+}-doped glasses. Therefore, Yb^{3+} codoping is necessary in order to absorb the pump light efficiently at wavelengths around 1 μm. Continuous-wave laser oscillation, *Q*-switched [11.542, 543] and quasi-CW operation with pulse energies up to 35 J and average output powers up to 20 W have been obtained [11.544–548]. The pulse duration of the Xe flashlamps is several milliseconds, thus matching the upper-laser-level lifetime of the Er^{3+} $^4I_{13/2}$ level. Also laser diodes can be efficiently applied as pump sources. *Wu* et al. [11.549] used pulsed laser diodes in a transversal excitation scheme. Using a repetition rate of 50 Hz, a pump pulse duration of 2.5 ms and a peak pump power of ≈ 1.5 kW, an average output power of 8.5 W was obtained, corresponding to an output energy of 170 mJ per pulse. A comparison and discussion of different pumping schemes is given in [11.550]. In all experiments, the main problem of glass materials compared to crystals is the low heat conductivity (for Kigre QE7 0.82 W/mK [11.547] versus 13 W/mK for YAG). Thus, the introduced heating power and therefore also the extractable output power and/or repetition rate is limited and high-power continuous-wave excitation of Er^{3+}-doped glasses is difficult. Despite this, the overall performance of the Kigre QE7 and QX glasses is better than that of Er^{3+}-doped crystals. *Obaton* et al. [11.551] obtained a slope efficiency of 21% in a diode end-pumped setup for a QX glass. *Diening* [11.552] compared Kigre QE7 glass with Er^{3+}:$LaSc_3(BO_3)_4$ and $Y_3Al_5O_{12}$ in the same setup (Fig. 11.66). The achieved output power and slope efficiency for the QE7 glass are about twice as high as for Er^{3+}:$LaSc_3(BO_3)_4$.

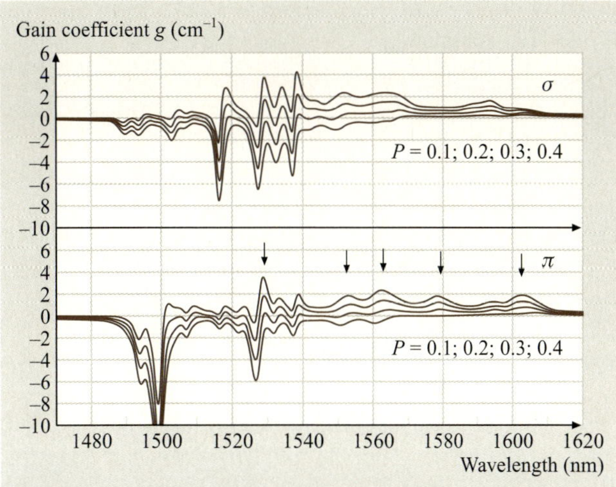

Fig. 11.64 Gain coefficient curves derived for both polarizations for Er^{3+}:YVO_4 in the spectral range of the $^4I_{15/2} \rightarrow {^4I_{11/2}}$ laser transition for four values of the inversion parameter *P* (larger gain values for larger *P*). *Arrows* denote the wavelengths for which laser oscillation was realized. (After [11.437])

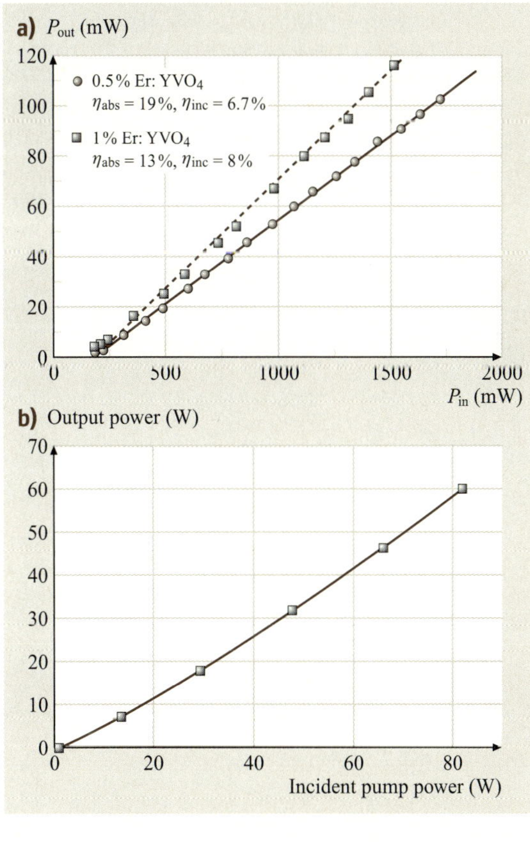

Fig. 11.65 (a) Input–output curve of the CW laser oscillation of Er^{3+} (0.5%):YVO_4 crystal ($\lambda_{laser} = 1604$ nm, output mirror transmission = 1%). The slope efficiencies are given with respect to the absorbed (η_{abs}) and incident (η_{in}) power. (Results from [11.437]). **(b)** In-band pumping of Er:YAG with a fiber laser at 1532 nm and lasing at $\lambda_{laser} = 1645$ nm [11.535]

Table 11.19 Overview of room-temperature Er^{3+}-doped laser operating around 1.6 μm on the $^4I_{13/2} \rightarrow {}^4I_{15/2}$ and on the $^4S_{3/2} \rightarrow {}^4I_{9/2}$ transition

Crystal	λ_{laser} (nm)	Transition	Slope efficiency (%)	Mode of operation	Pump source/Remarks	Ref.
$Ca_2Al_2SiO_7$	1530 nm	$^4I_{13/2} \rightarrow {}^4I_{15/2}$	1.1	CW	Ti:Sapphire 940 nm, 975 nm	[11.553]
	1550 nm	$^4I_{13/2} \rightarrow {}^4I_{15/2}$	1.5	CW	Ti:Sapphire 940 nm, 975 nm	[11.553]
	1555 nm	$^4I_{13/2} \rightarrow {}^4I_{15/2}$	5	CW	Ti:Sapphire 940 nm, 975 nm	[11.553]
$LiNbO_3 : Ti$	1532 nm	$^4I_{13/2} \rightarrow {}^4I_{15/2}$	6	CW	Tl:KCl, 1477 nm	[11.554]
	1563 nm	$^4I_{13/2} \rightarrow {}^4I_{15/2}$	3	CW, pulsed	Tl:KCl 1479 nm, 1484 nm	[11.555]
	1576 nm	$^4I_{13/2} \rightarrow {}^4I_{15/2}$		CW, pulsed	Tl:KCl 1479 nm, 1484 nm	[11.555]
$SrY_4(SiO_4)_3O$	1554 nm	$^4I_{13/2} \rightarrow {}^4I_{15/2}$	0.4	CW	Laser diode 980 nm	[11.556]
$YAlO_3$	1662 nm	$^4S_{3/2} \rightarrow {}^4I_{9/2}$	10.1	CW	Kr^+	[11.557]
	1663 nm	$^4S_{3/2} \rightarrow {}^4I_{9/2}$	0.07	Pulsed	Xe-flash lamp	[11.558]
	1663 nm	$^4S_{3/2} \rightarrow {}^4I_{9/2}$		Pulsed	Xe-flash lamp	[11.559, 560]
	1663.2 nm	$^4S_{3/2} \rightarrow {}^4I_{9/2}$		Pulsed	Xe-flash lamp	[11.561–563]
	1663.2 nm	$^4S_{3/2} \rightarrow {}^4I_{9/2}$		CW	Ar^+, 488 nm	[11.564]
	1667 nm	$^4S_{3/2} \rightarrow {}^4I_{9/2}$	0.02	Pulsed	Xe-flash lamp	[11.565]
	1677.6 nm	$^4S_{3/2} \rightarrow {}^4I_{9/2}$	2.2	CW	Ar^+, 488 nm	[11.564]
	1706 nm	$^4S_{3/2} \rightarrow {}^4I_{9/2}$	0.02	Pulsed	Xe-flash lamp	[11.565]
	1706.1 nm	$^4S_{3/2} \rightarrow {}^4I_{9/2}$		CW	Ar^+, 488 nm	[11.564]
	1729 nm	$^4S_{3/2} \rightarrow {}^4I_{9/2}$	0.02	Pulsed	Xe-flash lamp	[11.565]
	1729.6 nm	$^4S_{3/2} \rightarrow {}^4I_{9/2}$		CW	Ar^+, 488 nm	[11.564]
$LiYF_4$	1620 nm	$^4I_{13/2} \rightarrow {}^4I_{15/2}$		CW	Kr^+, 647 nm	[11.566]
	1640 nm	$^4I_{13/2} \rightarrow {}^4I_{15/2}$		CW	Kr^+, 647 nm	[11.566]
	1664.0 nm	$^4S_{3/2} \rightarrow {}^4I_{9/2}$		Pulsed	Xe flashlamp	[11.567]
	1730 nm	$^4S_{3/2} \rightarrow {}^4I_{9/2}$	0.6	Pulsed	Xe flashlamp	[11.568]
	1732.0 nm	$^4S_{3/2} \rightarrow {}^4I_{9/2}$		Pulsed	Xe flashlamp	[11.569]
$Y_3Al_5O_{12}$	1617 nm	$^4I_{13/2} \rightarrow {}^4I_{15/2}$	10	*Q*-switched (4 kHz)	Er-fiber laser, 1543 nm	[11.570]
	1632 nm	$^4I_{13/2} \rightarrow {}^4I_{15/2}$		Pulsed	Xe flashlamp	[11.571]
	1634 nm	$^4I_{13/2} \rightarrow {}^4I_{15/2}$		Pulsed, intracavity	Er:glass 1549 nm	[11.572]
	1640 nm	$^4I_{13/2} \rightarrow {}^4I_{15/2}$	12.7	CW	Kr^+, 647 nm	[11.533]
	1640 nm	$^4I_{13/2} \rightarrow {}^4I_{15/2}$	0,5	*Q*-switched	Er:glass 1534 nm	[11.573]
	1644 nm	$^4I_{13/2} \rightarrow {}^4I_{15/2}$	7	Pulsed	Er:glass 1535 nm	[11.574]
	1644.9 nm	$^4I_{13/2} \rightarrow {}^4I_{15/2}$		Pulsed	Xe flashlamp	[11.575]
	1645 nm	$^4I_{13/2} \rightarrow {}^4I_{15/2}$	40	Pulsed	Er:glass 1532 nm	[11.576]
	1645 nm	$^4I_{13/2} \rightarrow {}^4I_{15/2}$	40	*Q*-switched	Yb, Er-doped fiber 1530 nm	[11.577]
	1645 nm	$^4I_{13/2} \rightarrow {}^4I_{15/2}$	46	Pulsed	1.5 μm laser diodes	[11.578]
	1645 nm	$^4I_{13/2} \rightarrow {}^4I_{15/2}$	40	CW	Er-fiber laser, 1543 nm	[11.570]
	1645.3 nm	$^4I_{13/2} \rightarrow {}^4I_{15/2}$	81	CW, Q-switched	Yb, Er-doped fiber 1530 nm	[11.535]
	1645.9 nm	$^4I_{13/2} \rightarrow {}^4I_{15/2}$		Pulsed	Xe flashlamp	[11.579]
	1646 nm	$^4I_{13/2} \rightarrow {}^4I_{15/2}$	7	CW	laser diode	[11.580]
	1775.7 nm	$^4S_{3/2} \rightarrow {}^4I_{9/2}$		Pulsed	Xe flashlamp	[11.563, 571, 581]
$Y_3Ga_5O_{12}$	1640 nm	$^4I_{13/2} \rightarrow {}^4I_{15/2}$	0.9	CW	Kr^+, 647 nm	[11.533]
$Y_3Sc_2Ga_3O_{12}$	1643 nm	$^4I_{13/2} \rightarrow {}^4I_{15/2}$	10	Pulsed	Er:glass, 1532 nm	[11.582]
$Lu_3Al_5O_{12}$	1776.2 nm	$^4S_{3/2} \rightarrow {}^4I_{9/2}$		Pulsed	Xe flashlamp	[11.563]

Table 11.19 (continued)

Crystal	λ_{laser} (nm)	Transition	Slope efficiency (%)	Mode of operation	Pump source/Remarks	Ref.
$KGd(WO_4)_2$	1715.5 nm	$^4S_{3/2} \rightarrow ^4I_{9/2}$		Pulsed	Xe flashlamp	[11.583]
	1732.5 nm	$^4S_{3/2} \rightarrow ^4I_{9/2}$		Pulsed	Xe flashlamp	[11.583]
	1733.0 nm	$^4S_{3/2} \rightarrow ^4I_{9/2}$		Pulsed	Xe flashlamp	[11.584]
$KY(WO_4)_2$	1540 nm	$^4I_{13/2} \rightarrow ^4I_{15/2}$	1	CW	Ti:Sapphire	[11.585]
	1737.2 nm	$^4S_{3/2} \rightarrow ^4I_{9/2}$		Pulsed	Xe flashlamp	[11.584]
$KLa(MoO_4)_2$	1730 nm	$^4S_{3/2} \rightarrow ^4I_{9/2}$		Pulsed	Xe flashlamp	[11.586]
$LiLuF_4$	1734.5 nm	$^4S_{3/2} \rightarrow ^4I_{9/2}$		Pulsed	Xe flashlamp	[11.587, 588]
$KLu(WO_4)_2$	1739.0 nm	$^4S_{3/2} \rightarrow ^4I_{9/2}$		Pulsed	Xe flashlamp	[11.584]
$KEr(WO_4)_2$	1737.2 nm	$^4S_{3/2} \rightarrow ^4I_{9/2}$		Pulsed	Xe flashlamp	[11.584]
YVO_4	1604 nm (1531 nm, 1553 nm, 1564 nm, 1580 nm)	$^4I_{13/2} \rightarrow ^4I_{15/2}$	19	CW	Ti:Sapphire	[11.437]
Y_2SiO_5	1617 nm (1545 nm, 1567 nm, 1576 nm)	$^4I_{13/2} \rightarrow ^4I_{15/2}$	5.6	CW	laser diode	[11.580]
Sc_2SiO_5	1558 nm	$^4I_{13/2} \rightarrow ^4I_{15/2}$		CW	Ti:Sapphire 979 nm	[11.589]
	1551 nm	$^4I_{13/2} \rightarrow ^4I_{15/2}$	1.8	CW	Ti:Sapphire 920 nm	[11.589]
	1551 nm	$^4I_{13/2} \rightarrow ^4I_{15/2}$	2.4	CW	laser diode 968 nm	[11.589]
$Sc_2Si_2O_7$	1545 nm	$^4I_{13/2} \rightarrow ^4I_{15/2}$	2.6	CW	Ti:Sapphire 980 nm	[11.589]
	1556 nm	$^4I_{13/2} \rightarrow ^4I_{15/2}$	2.3	CW	Ti:Sapphire 978 nm	[11.589]
$LaSc_3(BO_3)_4$	1563 nm	$^4I_{13/2} \rightarrow ^4I_{15/2}$	6	CW	laser diode 975 nm	[11.534]
$Ca_4YO(BO_3)_3$	1.5–1.6 µm	$^4I_{13/2} \rightarrow ^4I_{15/2}$	26.8	CW	laser diode	[11.590, 591]
$Ca_4GdO(BO_3)_3$	1.54 µm	$^4I_{13/2} \rightarrow ^4I_{15/2}$	15 7	CW Ti:Sapphire 902 nm	laser diode 975 nm	[11.592]

Fibers. Erbium-doped fiber lasers have been extensively studied for their potential use as sources in communication systems operation in the third communication window around 1.55 µm. All these lasers oscillate on the $^4I_{13/2} \rightarrow ^4I_{15/2}$ transition, either in continuous-wave or pulsed mode. In singly Er^{3+}-doped fibers, suitable pump wavelengths for using laser diodes are 810 nm ($^4I_{15/2} \rightarrow ^4I_{9/2}$), 980 nm ($^4I_{15/2} \rightarrow ^4I_{11/2}$) and 1480 nm ($^4I_{15/2} \rightarrow ^4I_{13/2}$). Other possible pump bands are around 660 nm ($^4I_{15/2} \rightarrow ^4F_{9/2}$), 532 nm and 514.5 nm ($^4I_{15/2} \rightarrow ^4H_{11/2}$). The pump wavelengths at 810 nm and 514.5 nm suffer from strong excited-state absorption, yielding a loss of pump photons [11.593]. The gain coefficient of Er-doped fibers is rather high (11 dB/mW [11.594]) due to the fairly high peak emission cross section of $4–7 \times 10^{-21}$ and the long lifetime of the $^4I_{13/2}$ level (8–10 ms) in silica fibers, despite the three-level laser character causing ground-state absorption at this wavelength. Like in Er^{3+}-doped crystals and glasses, concentration quenching occurs. In order to enhance the absorption efficiency without increasing the Er^{3+} concentration and/or fiber length, Yb^{3+} codoping is used, especially when diode pumping between 900 nm and 1000 nm is used. The requirement for efficient operation is – as in the crystals and glasses – efficient energy transfer from the Yb^{3+} $^2F_{5/2}$ level to the Er^{3+} $^4I_{13/2}$ level (Fig. 11.61). In Table 11.20, some of the Er-fiber lasers around 1.55 µm are summarized. For a very detailed discussion of Er^{3+}-doped fiber lasers and amplifier see [11.595]. In summary, Er^{3+}-doped fiber lasers operating around 1.55 mm are extremely efficient. Output powers in the watt range are possible.

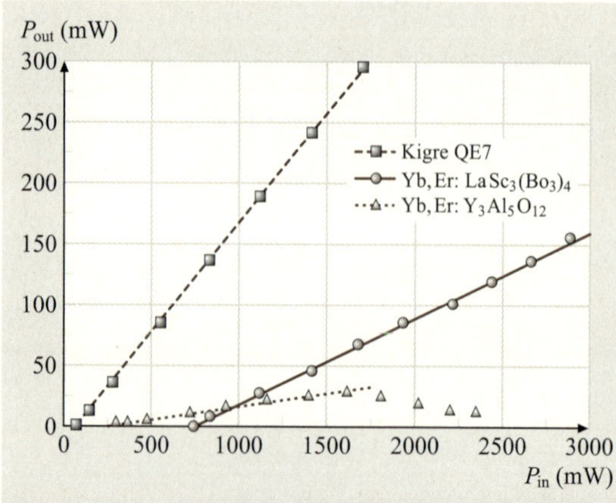

Fig. 11.66 Results of the laser experiments with Kigre QE7 glass, Yb^{3+} (10%), Er^{3+} (0.5%): $LaSc_3(BO_3)_4$ and Yb^{3+}, Er^{3+} (0.5%): $Y_3AL_5O_{12}$. (After [11.552])

Nowadays up to 100 W of output power are commercially available [11.596]. Most efficient pumping occurs at 1480 nm, here slope efficiencies close to the quantum limit of 95% are possible. Pumping at 980 nm is less efficient due to the higher quantum defect, however, at this wavelength highly efficient and reliable high-power laser diodes are available.

Other near-infrared Er^{3+} *lasers.* Room-temperature laser oscillation in Er^{3+}-doped crystals in the near-infrared spectral range has also been observed at other wavelengths. In Yb^{3+}-codoped Er^{3+}:$YLiF_4$ an upconversion pumping scheme was realized for the $^4S_{3/2} \rightarrow {}^4I_{11/2}$ transition at 1234 nm, which allowed continuous-wave room-temperature Ti:sapphire- and diode-pumped laser oscillation [11.597, 598]. Under Ti:sapphire excitation around 966 nm output powers of 160 mW and slope efficiencies of up to 22% were obtained. Note, that without the Yb codoping, the output powers were an order of magnitude lower. Under diode laser pumping at 966 nm, Yb^{3+},Er^{3+}:$YLiF_4$ exhibited output powers of 80 mW and slope efficiencies of 7%.

For this transition Xe flashlamp excitation is also possible allowing room-temperature laser oscillation in $YLiF_4$ [11.599–601], $LuLiF_4$ [11.602] and $YAlO_3$ [11.603]. In $BaYb_2F_8$, the $^4F_{9/2} \rightarrow {}^4I_{13/2}$ transition at 1260 nm exhibited laser oscillation under pulsed Nd laser or Xe flashlamp excitation [11.604]. Laser oscillation on the $^4F_{9/2} \rightarrow {}^4I_{11/2}$ transition was obtained around 1.96 μm in $BaYb_2F_8$ [11.605–607] under pulsed Nd laser and Xe flashlamp excitation.

11.2.4 Mid-Infrared Lasers

Basics

The mid-infrared wavelength range ≈ 1.9–5.0 μm is of interest for a number of applications. Mid-infrared solid-state lasers serve as light sources for spectroscopy, e.g., in remote sensing of the atmosphere, as the frequencies of internal vibrational motion of many molecules can be found in this spectral region. Other applications include medicine, e.g., microsurgery and dentistry in the region of high water absorption around 2.7–3 μm. Laser wavelengths near 2 μm are suitable for tissue welding and lithotripsy.

The first mid-infrared laser was operated in 1960, shortly after the invention of the laser itself. It was operated at 2.6 μm in a calcium fluoride crystal doped with trivalent uranium [11.608]. Pulsed excitation and cooling to low temperatures were typically required for novel laser transitions in the early years. Two years later, a Dy^{2+}-doped CaF_2 laser at 2.36 μm was also demonstrated in continuous-wave operation [11.609, 610]. Among the lasers based on trivalent rare-earth ions, the transitions near 2 μm in Tm^{3+} and Ho^{3+} were operated in $CaWO_4$ in 1962 [11.611,612]. The first observation of coherent emission near 3 μm from erbium ions was reported in 1967 [11.613]. Since those early years, a large number of new host materials has been developed and various new laser transitions in the mid-infrared spectral region have been demonstrated (for a comprehensive overview of ion–host combinations, see [11.614]). Around 1990, Tm^{3+}- and Ho^{3+}-doped solid-state laser systems in $Y_3Al_5O_{12}$ (YAG), $YLiF_4$ (YLF), YVO_4, and $Y_3Sc_2Ga_3O_{12}$ (YSGG) were shown to operate between 1.86 and 2.46 μm [11.615–618], and Er^{3+}-doped lasers in similar host systems cover the wavelength range of 2.66–2.94 μm. Short-pulse lasers at these wavelengths have been demonstrated [11.619, 620].

Nowadays, mid-infrared laser transitions ranging from 1.8 μm up to 7.2 μm [11.621] are known in divalent Dy, trivalent Tm, Ho, Er, Dy, Pr, Tb, and Nd, as well as in trivalent U. The current state of the art in solid-state lasers occurring in rare-earth ions such as Tm^{3+}, Ho^{3+}, Er^{3+} and others and their population mechanisms are discussed in this section on the basis of the spectroscopic properties of these ions. Continuous-wave fundamental-mode power levels ranging from a few mW near 4 μm up to ≈ 100 W near 2 μm have been demonstrated in recent years. Power-scaling methods and their limitations, the possibilities to optimize the population mechanisms and increase the efficiencies of these lasers, as well as the prospects of future mid-infrared lasers in

Table 11.20 Overview over Er^{3+} and Yb^{3+}:Er^{3+}-doped silica fiber lasers. λ_{laser}: laser wavelength, λ_{pump}: pump wavelength, l_{fiber}: fiber length, P_{thr}: laser threshold, η: slope efficiency, P_{out}: output power, P_{pump}: pump power, Ref.: reference, (l): launched, (inc): incident, (abs): absorbed, NA: not available. (After [11.595])

λ_{laser} (nm)	λ_{Pump} (nm)	Er concentration	Yb concentration	l_{fiber} (m)	Remarks	P_{thr} (mW)	η (%)	P_{out} (mW)	max. P_{pump} (mW)	Ref.
1566	514.5	35 ppm Er	-	13	Ar^+ pump	44 (l)	10 (l)	56	600 (l)	[11.622]
1560	532	150 ppm Er_2O_3	-	1	Ring laser	10 (l)	5.1 (l)	1.8	45 (l)	[11.623]
1535	532	100 ppm Er	-	15	Doubled Nd:YAG	NA	28	1000	3600	[11.624]
1560	806	500 ppm Er	-	3.7	Laser diode	10 (l)	16 (l)	8	56 (l)	[11.625]
1620	808	300 ppm Er	-	1.5	Laser diode	3 (abs)	3.3 (abs)	0.13	7 (abs)	[11.626]
1560	980	0.08 wt %	-	0.9	Dye laser	2.5 (abs)	58 (abs)	4.7	11.3 (abs)	[11.627]
1540	980	1100 ppm Er	-	9.5	Ti:Sapphire	> 10 (l)	> 49 (l)	260	540 (l)	[11.628]
1552	1460	1370 ppm Er	-	5	2 laser diodes	37 (l)	14 (l)	8	93 (l)	[11.629]
1552	1470	1370 ppm Er	-	7	laser diode	44 (abs)	6.3 (abs)	1	60 (abs)	[11.630]
1555	1480	45 ppm Er	-	60	laser diode	6.5 (l)	38.8 (l)	3.3	15 (l)	[11.631]
1560	1480	110 ppm Er_2O_3	-	42.6	laser diode	4.8 (abs)	58.6 (abs)	14.2	29 (abs)	[11.632]
1570	810	0.06 wt %	1.3 wt %	1.45	2 laser diodes	12.7 (l)	15.4 (l)	2.3	28 (l)	[11.633]
1560	820	0.08%	1.7%	0.7	Dye laser	3.7 (abs)	7 (abs)	NA	NA	[11.634]
1560	832	0.08%	1.7%	0.7	Dye laser	5 (abs)	8.5 (abs)	NA	NA	[11.635]
1537	962	900 ppm	1.1%	1.6	laser diode	130 (l)	19 (l)	96	620 (l)	[11.636]
1545	980	NA	NA	0.07	laser diode	1 (abs)	25 (inc)	18.6	95 (inc)	[11.637]
1535	1047	0.06%	1.8%	4	Nd:YLF	20 (l)	23 (l)	285	640 (l)	[11.638]
1560	1064	0.08%	1.7%	0.91	Nd:YAG	8 (abs)	4.2 (abs)	1.3 (abs)	80 (abs)	[11.634]
1535	1064	880 ppm	7500 ppm	NA	Nd:YAG	37 (abs)	27 (abs)	NA	NA	[11.639]
1545.6	980/1480	NA	NA	0.07	Laser diode	10 (l)	50 (l)	166	340 (l)	[11.640]

a number of rare-earth ions at transitions in the wavelength range beyond 3 μm and extending to 5 μm are described.

The aspects relevant to rare-earth-ion-doped mid-infrared solid-state lasers, such as the competition between radiative and multiphonon decay and the consequent choice of host materials for these wavelengths will be introduced next. The performance of the most important mid-infrared laser transitions in the wavelength range 2–3 μm will then be discussed in detail: Tm^{3+}-doped lasers at 1.9 μm and 2.3 μm, Ho^{3+}-doped lasers at 2.1 μm and 2.9 μm, Er^{3+}-doped lasers at 2.7–2.9 μm, and Dy^{3+}-doped lasers at 2.9–3.4 μm. At wavelengths beyond 3 μm, it becomes increasingly difficult to find suitable host materials for actively doped laser systems. This statement holds true for glass fibers in the same way as for crystalline materials. The prospects for future mid-infrared solid-state lasers in this wavelength range will be discussed at the end of the section.

Introductions to the fields of mid-infrared solid-state crystalline and fiber lasers can be found in [11.641] and [11.642], respectively.

Decay mechanisms, host materials, and thermal issues

Here a few fundamental aspects of lasers are discussed with emphasis on their impact on mid-infrared solid-state lasers.

Radiative versus multiphonon decay. The choice of host material for mid-infrared solid-state lasers involves a number of considerations. The maximum phonon energy is the most important aspect. The optical transparency range relates to both the size of the bandgap and also the infrared absorption cut-off, hence to the vibrational frequency ν of the anion–cation bonds of the material. For an ordered structure,

$$\nu = \left(\frac{1}{2\pi}\right)\sqrt{\frac{k}{M}}\,, \qquad (11.89)$$

where $M = m_1 m_2/(m_1 + m_2)$ is the reduced mass for two bodies m_1 and m_2 vibrating with an elastic restoring force k. The relative cation–anion bond strength is intimated by the field strength Z/r^2, where Z is the valence state of the cation or anion and r is the ionic radius. Generally, materials composed of large anions and cations with low field strengths display high transparency in the mid-infrared spectral region.

Radiative decay of excited states is in competition with nonradiative multiphonon decay. The maximum phonon energy of the material sets the multiphonon relaxation rates, which influence the quantum efficiency. The rate constant of a multiphonon relaxation process decreases exponentially with the energy gap to the next lower-lying state and with the order of the process, i. e., the number of phonons required to bridge the energy gap [11.643, 646]. As an example, the multiphonon relaxation rates for the common fiber glasses as a function of the energy gap between energy levels are shown in Fig. 11.67.

The influence of multiphonon decay is stronger in oxides than in fluorides because of the smaller atomic mass m_2 of the anion and the larger elastic restoring force k (11.89), due to the stronger covalent bonds in oxides [11.644], both resulting in larger maximum phonon energies in oxides. Typically, nonradiative decay becomes dominant if five or fewer phonons are required to bridge the energy gap [11.645]. Since an energy gap of $\approx 3300\,\mathrm{cm}^{-1}$ corresponds to a transition wavelength of 3 μm, radiative decay prevails for phonon energies below $\approx 600\,\mathrm{cm}^{-1}$, which is roughly the maximum phonon energy of fluorides. Fluorides are, therefore, preferred over oxides as host materials for most mid-infrared laser transitions.

In the example of Fig. 11.68, dominant laser (solid lines) and multiphonon (dotted lines) transitions from the three lowest-energy levels of Er^{3+} are indicated, together with the corresponding lifetimes of the levels in different classes of host materials. In high-phonon-energy oxide host materials, only the 1.5 μm laser transition possesses sufficiently high frequency and large energy gap, resulting in a long $^4I_{13/2}$ lifetime. On the other hand, the $^4I_{11/2}$ lifetime is significantly quenched by multiphonon relaxation and the 2.8 μm laser originating in this level is more easily operated in fluoride host materials. Finally, low-phonon host materials such as chlorides are required to ensure a long $^4I_{9/2}$ lifetime in order to operate the 4.5 μm laser.

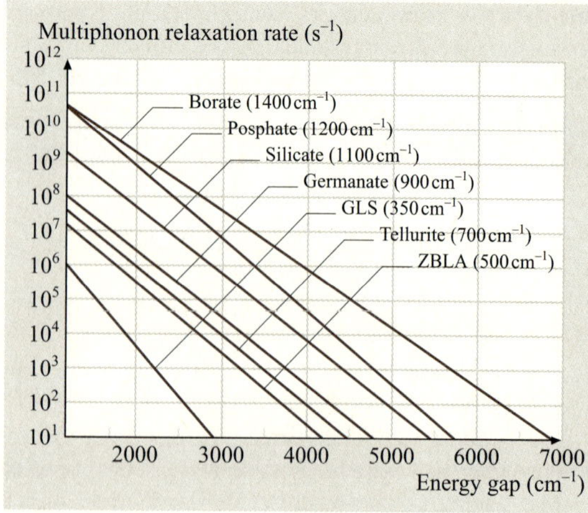

Fig. 11.67 Calculated and measured multiphonon relaxation rates as a function of the energy gap between energy levels for glasses with different maximum phonon energies. (After [11.643, 644])

Host materials for mid-infrared lasers. Crystalline oxide materials have been the laser host materials of choice for several decades, as many oxide crystals are comparatively easy to grow, environmentally stable, and possess high heat conductivities [11.647], fracture limits, and refractive indices, the latter resulting in large absorption and emission cross sections. Most prominent is $Y_3Al_5O_{12}$ (YAG), together with other

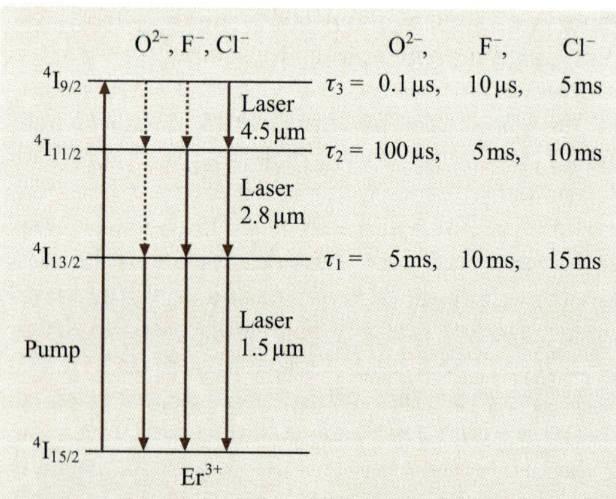

Fig. 11.68 Radiative and laser (*solid lines*) versus nonradiative (*dotted lines*) decay of the lowest three excited states of Er^{3+} in oxide, fluoride, and chloride host materials [11.645]

Part C | 11.2

garnets, but in recent years materials such as mixed garnets, the vanadates YVO_4 and $GdVO_4$, the double tungstates $KY(WO_4)_2$ and $KGd(WO_4)_2$, the sesquioxides Y_2O_3, Sc_2O_3, and Lu_2O_3 and others have shown great promise for efficient lasing when doped with various rare-earth ions. Their performance decreases significantly when lasers are operated at wavelengths above $\approx 2.5-3\,\mu m$ because of their high maximum phonon energies ($\approx 700-900\,cm^{-1}$). Silicate glass is perhaps the most important material used for optical fiber production [11.644, 648], however, in this material the maximum phonon energy of $\approx 1100\,cm^{-1}$ [11.649] is even higher and has so far limited the emission wavelength of mid-infrared fiber lasers using this material to $\approx 2.2\,\mu m$ [11.650]. Silica is robust and involves the very effective modified chemical vapor deposition (MCVD) technique for fiber fabrication. Reducing the OH^- content in the glass, which has two main absorption peaks in the range $1.3-2.0\,\mu m$ [11.651], improves the near-to-mid-infrared utility. Rare-earth ions such as Nd^{3+} and Er^{3+} which have high field strengths have low solubility in silicate glass, which can lead to clustering and microscale phase separation.

The use of fluoride crystals and glasses as host materials for mid-infrared solid-state lasers has found wide acceptance. The heavy-metal fluorides [11.652, 653] are preferred as fiber materials, especially ZBLAN [11.654, 655], a mixture of 53 mol% ZrF_4, 20 mol% BaF_2, 4 mol% LaF_3, 3 mol% AlF_3, and 20 mol% NaF. Since it can be readily drawn into single-mode optical fiber [11.656] it is particularly important to mid-infrared fiber lasers [11.657]. The large atomic weight of the zirconium atom combined with relatively weak bonding provides a maximum phonon energy for ZBLAN of $\approx 550\,cm^{-1}$ [11.658] and allows for high infrared transparency up to $\approx 6\,\mu m$. Multiphonon relaxation, however, becomes significant for transitions at wavelengths longer than $\approx 3-3.5\,\mu m$. Compared to silica, ZBLAN has a lower damage threshold. The crystal field strength is also weaker [11.659]. An overview of the spectroscopic properties of rare-earth ions doped into ZBLAN is given in [11.658]. On the crystalline side, the host materials $LiYF_4$, BaY_2F_8, and their respective isostructural relatives have become the workhorses for many mid-infrared laser transitions [11.660, 661]. Because of their fluoride content, these materials have to be grown under an atmosphere which excludes oxygen.

Among the low-phonon host materials, many compounds naturally possess low heat conductivity and are hygroscopic. This accounts for most of the halides, with increasing hygroscopicity from chloride to iodide. On the other hand, these materials provide phonon energies in the range of $350-150\,cm^{-1}$ [11.662]. Recently, KPb_2Cl_5 and related compounds have emerged as non-hygroscopic, hence promising candidates for mid-infrared lasers [11.663–665]. Chalcogenide glasses are composed of the chalcogen elements S, Se and Te [11.666–668]. They are environmentally durable and have reasonably large glass-forming regions. When the rare-earth ions are doped into these glasses [11.669], the radiative transition probabilities and, therefore, the absorption and emission cross sections are high as a result of the high refractive index (≈ 2.6) of the glass and the high degree of covalency of the rare-earth ion with the surrounding medium. Maximum phonon energies of $300-450\,cm^{-1}$ [11.670] produce low rates of multiphonon relaxation (Fig. 11.67), and therefore high quantum efficiencies. The most important glasses are the sulfide glasses GaLaS (GLS) [11.671] and GeGaS [11.672] because of the reasonably high rare-earth solubility.

Studies into the use of ceramics as host materials for the rare-earths have recently made a lot of progress [11.673]. These ceramics are composed of nanocrystallites of materials such as YAG and can be produced in a simple cost-efficient process at relatively low temperatures. This allows the fabrication of materials with very high melting points [11.674] that are difficult to grow by other techniques such as the Czochralski method. This class of materials is also available in a fiber geometry [11.675]. Ceramic fibers combine the characteristics of crystalline materials such as high absorption and emission cross sections, large thermal conductivity, and even the possibility of doping with transition-metal ions [11.675] with the convenience of guiding the pump and signal light in a fiber. While bulk ceramics have already matured as laser host materials, the losses of ceramic fibers are still comparatively high.

Specific aspects of operating mid-infrared lasers. As higher pump powers become available from laser-diode systems, it is generally recognized that thermal and thermo-optical issues set limitations to the power scalability of end-pumped bulk laser systems. Owing to the unfavorable temperature dependence of thermal and thermo-optical parameters [11.647], the large heat load in the crystal leads, firstly, to a significant temperature increase in the rod, secondly, to strong thermal lensing with pronounced spherical aberrations, and ultimately, to rod fracture in a high-average-power end-pumped system. Thermal management will be required when very high output powers are desired. In particular, for high-power mid-infrared operation, thermal management may be very important because of the decreased quantum efficiency and the consequently higher amount of heat dissipation [11.676].

Due to its geometry, the fiber provides potentially high pump and signal beam intensities without the drawbacks of significant thermal and thermo-optical effects. Its large surface-area-to-volume ratio means that the heat generated from multiphonon relaxation in the core is dissipated effectively by radiation and convection from the outer surface of the fiber. This is especially true for single-clad, core-pumped single-mode fibers where this ratio is highest [11.678]. The larger cladding area ($> 10^4\,\mu m^2$) of double-clad fiber lasers allows for high-power diode-array pumping [11.408, 679–682]. On the other hand, double-clad fibers have a smaller surface-area-to-volume ratio and thermal issues need to be taken into account [11.683–685].

While bulk crystalline mid-infrared lasers matured already during the 1990s, the high costs of fabricating fibers with sufficiently low losses in the mid-infrared region of the spectrum has impeded the necessary research efforts in the field of mid-infrared fiber lasers. However, with the introduction of the double-clad fiber and recent technological advances in the fields of fiber fabrication and beam-shaped high-power diode lasers, the performance of diode-pumped fiber lasers has dramatically improved. Today, mid-infrared fiber lasers can compete with the corresponding bulk crystalline systems in certain applications, especially when fundamental-transverse-mode, CW laser operation at output powers in the milliwatt to the hundred watt range is required.

A large number of techniques for pulsed operation including Q-switching and mode locking of fiber lasers have been explored. These techniques have been investigated intensively for the common laser transitions at 1 μm in Nd^{3+} and Yb^{3+}, and at 1.5 μm in Er^{3+}, and are usually described in combination with these lasers. The small fiber size limits the peak power through the damage-threshold intensity (propagating power per core area) and, hence, crystalline lasers in bulk geometries or optical parametric processes are often preferred when high-energy short pulses are needed. This argument accounts especially for mid-infrared ZBLAN-based fiber lasers, because these fibers possess a lower damage threshold compared to silica fibers. The description of mid-infrared fiber lasers is, therefore, confined to CW operation and specific techniques for pulsed operation of fiber lasers are not discussed in this chapter.

Thulium-doped solid-state lasers at 1.9–2.0 μm and 2.3–2.5 μm

The use of the Tm^{3+} ion for mid-infrared solid-state laser applications has been widespread, partly as a result of the convenient absorption band near 0.79 μm, which allows for direct AlGaAs diode-laser pumping. The primary luminescent transitions of Tm^{3+} relevant to mid-infrared laser emission are the $^3F_4 \rightarrow {}^3H_5$ transition at ≈ 2.3 μm and the $^3H_4 \rightarrow {}^3H_6$ ground-state transition at ≈ 1.9 μm; see the energy-level scheme in Fig. 11.69. The $^3F_4$ level is excited by the 0.79 μm pump wavelength.

Three-level lasers at 1.9–2.0 μm. The first laser emission from Tm^{3+} ions was reported on the phonon-terminated 2 μm transition $^3H_4 \rightarrow {}^3H_6$ in $CaWO_4$:Tm^{3+} in 1962 [11.611]. In 1975, pulsed operation at room temperature was demonstrated in Cr^{3+}-codoped YAG and $YAlO_3$ [11.686]. Cr^{3+} codoping allowed the experimentalist to improve the absorption of flashlamp or ion-laser pump light in the visible spectral range by the active medium and subsequent energy transfer from Cr^{3+} to the Tm^{3+} lasing ions [11.687, 688]. The cross-relaxation process $(^3F_4, {}^3H_6) \rightarrow (^3H_4, {}^3H_4)$ can transform one pump photon absorbed in the $^3F_4$ or a higher-lying energy level into two excitations in the $^3H_4$ upper laser level of the 2 μm transition [11.616, 689] (Fig. 11.69), thereby enhancing the quantum efficiency of this laser by a factor of 2. Laser emission under diode pumping of the $^3F_4$ level at 780–790 nm was achieved in YAG in the late 1980s [11.690]. A single-

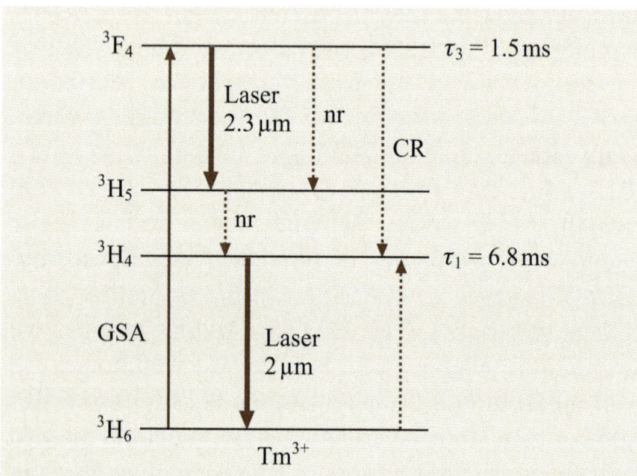

Fig. 11.69 Partial energy-level scheme of Tm^{3+} displaying the measured lifetimes when doped into fluoride glass [11.677], NR and CR represent nonradiative decay and cross-relaxation, respectively

frequency monolithic laser has also been reported in YAG:Tm^{3+} [11.691].

Recently, 14 W, 18 W, and 36 W of output power with high beam quality has been achieved with diode pumping at room temperature in YAG:Tm^{3+} [11.692] and in $LiYF_4$:Tm^{3+} in slab [11.693] and bulk [11.694] geometries, respectively. The latter approach could be scaled to 70 W, however currently with low beam quality [11.695]. Two research groups reported about 115 W and 120 W, respectively, of multi-mode output power from diode-pumped YAG:Tm^{3+} lasers [11.696, 697]. A microchip laser was demonstrated in $GdVO_4$:Tm^{3+} [11.698]. A thin-disc laser has also been demonstrated in YAG [11.699].

The large degree of Stark splitting of the $^3H_6$ ground state, combined with vibronic broadening of the spectrum [11.700], provides the $^3H_4 \rightarrow {}^3H_6$ transition with a very broad emission, spanning ≈ 400 nm in many hosts, which represents one of the broadest luminescent transitions available from any rare-earth ion. Accordingly, its tunability is rather large, ranging from 1.87–2.16 μm in YAG [11.701], 1.84–2.14 μm in YSGG [11.701], 1.93–2.00 μm in $YAlO_3$ [11.702], 1.93–2.09 μm in Y_2O_3 [11.703], 1.93–2.16 μm in Sc_2O_3 [11.703], 1.83–1.97 μm in CaF_2 [11.704], 1.91–2.07 μm in $LiYF_4$ [11.693], 1.85–2.06 μm in BaY_2F_8 [11.705], 1.86–1.99 μm in $GdVO_4$ [11.706], 1.84–1.95 μm in $LuVO_4$ [11.707], 1.79–2.04 μm in $KGd(WO_4)_2$ [11.708], and 1.81–2.03 μm in $NaGd(WO_4)_2$ [11.709]. As for many other transitions, shifts in the center emission wavelength can be achieved by substitution of host ions, e.g., from $Y_3Al_5O_{12}$ to $Lu_3Al_5O_{12}$ [11.710]. The vanadate crystals $GdVO_4$ and YVO_4 as well as double tungstates possess comparatively high absorption coefficients [11.698, 711, 712], allowing pumping also at 805–810 nm, where cheaper and more-reliable pump diodes than at 780–790 nm are available. Exploiting the large gain bandwidth, mode-locked operation of Tm^{3+} 2 μm lasers with pulse durations of 35 ps and 41 ps has been reported in YAG:Tm^{3+} [11.713] and YAG:Cr^{3+},Tm^{3+} [11.714], respectively. Actively [11.715, 716] or passively [11.717] Q-switched laser operation is useful in microsurgery [11.718].

Recent progress in the fields of crystalline epitaxial growth and in-bulk refractive-index modification processes in glasses and crystals has enabled novel solid-state lasers in the waveguide geometry [11.719]. As one of the results, Tm^{3+} waveguide lasers at 2 μm have also been demonstrated in lead germanate glass [11.720], YAG [11.721] with up to 15 W of output power under high-power diode side pumping [11.722], and $KY(WO_4)_2$ [11.723]. The latter laser has, as yet, not shown a performance as good as for the same material in bulk geometry [11.724] or in a waveguide geometry but doped with Yb^{3+} and lasing at 1 μm [11.725]. Epitaxial layers of Tm:$KLu(WO_4)_2$/$KLu(WO_4)_2$ have also been operated with the laser cavity perpendicular to the layer in the 2 μm spectral range [11.726].

The first explorations into fiber lasers utilizing the 1.9 μm ground-state transition related to the dye-laser pumping at 797 nm of a Tm^{3+}-doped silica fiber laser [11.727]. Overlap of the main absorption band with the emission wavelength of AlGaAs diode lasers quickly resulted in diode-laser pumping of these fiber lasers based on either silica [11.728] or fluoride [11.729] glass hosts. The cross-relaxation process $(^3F_4, {}^3H_6) \rightarrow (^3H_4, {}^3H_4)$ and enhancement of excited ions in the $^3H_4$ upper laser level of the 2 μm transition (Fig. 11.69), is highly dependent on the overall concentration of Tm^{3+} ions and competition from multiphonon relaxation from the $^3F_4$ level. Although generally high concentrations of Tm^{3+} in low-phonon-energy glasses enable full exploitation of this beneficial process, it has been shown recently that this cross-relaxation process is resonant in a silica host and hence only moderate (2–3 wt %) Tm^{3+} ion concentrations are required to maximize the benefits of cross-relaxation [11.730].

Also in fibers, the broad emission spectrum allows a large degree of wavelength tunability [11.731]. Recently, tuning ranges of 230 nm from 1.86–2.09 μm [11.732] and 250 nm from 1.72–1.97 μm [11.733] have been demonstrated. Since the Tm^{3+} 1.9 μm transition can be favorably operated in silica fiber (with its higher peak-power damage threshold compared to ZBLAN fiber), pulses in the range of 190–500 fs have been obtained in additive-pulse [11.734] or passive [11.735] mode-locking arrangements using this broad emission spectrum. The smaller emission cross section and the three-level nature of the laser transition resulted in higher pump thresholds compared to standard Nd^{3+}-doped silica fiber lasers. Reabsorption from the ground state of the Tm^{3+} ion has to be overcome because the ground-state multiplet is the lower laser level. Reducing the population of the higher Stark levels of the ground state by cooling the fiber causes emission at shorter wavelengths. Tunability to longer wavelengths can be obtained by variation of the fiber length because of the increased level of reabsorption by the ground state with longer lengths of fiber [11.736].

Early power-scaling experiments involved the use of the convenient 1.064 μm YAG:Nd^{3+} laser which core-pumped the short-wavelength side of the $^3H_5$ level [11.737]. Pumping the long-wavelength side of the $^3H_5$ level with a high-power 1.319 μm YAG:Nd^{3+} laser also yielded efficient output [11.738]. In-band pumping of the transition at 1.57 μm in silica [11.739] and at 1.58–1.60 μm in fluoride glass [11.740, 741] has also been demonstrated. Whilst theoretical modeling of Tm^{3+}-doped silica fiber lasers [11.742] indicates that in-band pumping is the most efficient pump method for silica-based fiber lasers because of the high Stokes efficiency, nevertheless, the wide availability of high-power AlGaAs diode lasers at 790–800 nm and the strong level of cross-relaxation in Tm^{3+}-doped silica means that such diode-cladding-pumped systems in both standing-wave [11.736,743] and ring-resonator [11.744] arrangements are perhaps the most practical ways of producing high output power from this ion (Fig. 11.70). With Yb^{3+} sensitization and pumping at 975 nm, 75 W of output power has been demonstrated [11.745]. Currently, the Tm^{3+}-doped silica fiber laser is the most mature of the mid-infrared fiber-laser systems primarily because of the robustness and convenience offered by the silica glass host. The maximum output power from high-power Tm^{3+}-doped fiber lasers is now ≈ 85 W [11.743], which is comparable to the equivalent diode-pumped Tm^{3+}-doped crystalline laser systems [11.697].

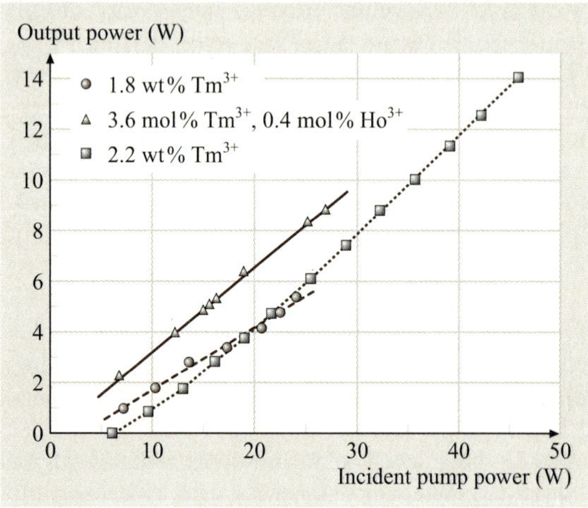

Fig. 11.70 Measured output powers from diode-cladding-pumped fiber lasers using 1.8 wt % Tm^{3+}-doped silica [11.735], 2.2 wt % Tm^{3+}-doped silica [11.746], and 3.6 mol % Tm^{3+}, 0.4 mol % Ho^{3+}-doped fluoride glass [11.747]

Four-level lasers at 2.3–2.5 μm. The mid-infrared four-level CW laser at ≈ 2.3 μm on the transition $^3F_4 \rightarrow {}^3H_5$ has been operated in GSGG:Tm^{3+} and $LiYF_4$:Tm^{3+}, with wavelength tunability ranging from 2.2–2.37 μm [11.748] and 2.2–2.46 μm [11.749], respectively. This laser operates best at low Tm^{3+} concentrations of < 2 at % in order to avoid the aforementioned cross-relaxation, which in this case would depopulate the upper laser level (Fig. 11.69). The lifetime of the lower laser level of the $^3F_4 \rightarrow {}^3H_5$ transition is quite short and leads to a low pump threshold.

Doping Tm^{3+} ions into a ZBLAN fiber offers an increased quantum efficiency of the $^3F_4$ level [11.750–752]. Deliberately designing the fiber to have a relatively low Tm^{3+}-ion concentration reduces cross-relaxation and hence severe lifetime quenching of the $^3F_4$ level. The tunability extends from 2.25 μm to 2.5 μm [11.677]. Simultaneous lasing on the $^3H_4 \rightarrow {}^3H_6$ transition at 1.9 μm produces a two-color fiber laser [11.753]. Applications requiring highly efficient output or multi-mid-infrared-wavelength output will benefit from the use of Tm^{3+}-doped ZBLAN fibers.

Holmium-doped solid-state lasers at 2.1 μm and 2.9 μm

The use of the Ho^{3+} ion as the active dopant for solid-state lasers opens up a number of very useful mid-infrared transitions. In this section, we will concentrate on the $^5I_7 \rightarrow {}^5I_8$ ground-state transition at ≈ 2.1 μm and the $^5I_6 \rightarrow {}^5I_7$ transition at ≈ 2.9 μm; see the energy-level scheme in Fig. 11.71. One of the significant shortcomings of Ho^{3+}, however, is the lack of

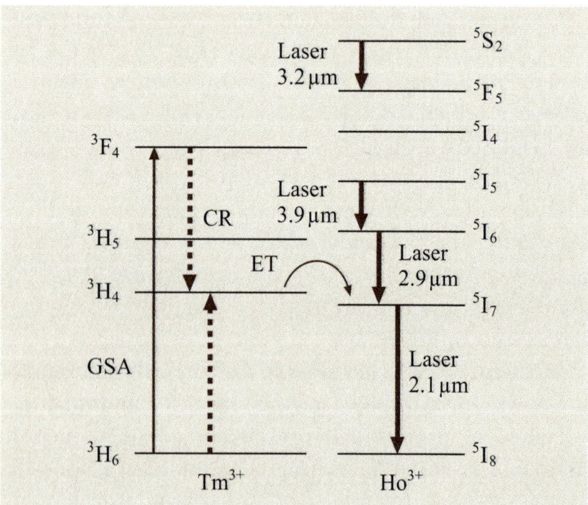

Fig. 11.71 Partial energy-level scheme of Ho^{3+} with a Tm^{3+} sensitizer. ET represents energy transfer

ground-state-absorption (GSA) transitions [11.754] that overlap with convenient high-power pump sources. As a result, many of the early demonstrations of Ho^{3+}-doped room-temperature crystalline CW lasers [11.616] involved sensitizing with Tm^{3+} in order to access the convenient absorption bands and the practical cross-relaxation process Tm^{3+} provides, as we have discussed in Sect. 11.3.1. Energy migration amongst the Tm^{3+} ions and a suitable Tm^{3+}:Ho^{3+} concentration ratio ensures that efficient energy transfer to Ho^{3+} takes place [11.755, 756] (Fig. 11.71).

Three-level lasers at 2.1 μm. The 2 μm transition $^5I_7 \rightarrow {}^5I_8$ in Ho^{3+} ions was first demonstrated as a laser in $CaWO_4$:Ho^{3+} in 1962 [11.612] and in Tm^{3+}-codoped $CaWO_4$:Ho^{3+} in 1963 [11.757]. In 1971, pulsed operation at room temperature on this three-level laser transition was demonstrated in $LiYF_4$ [11.758]. CW laser emission was achieved under Kr^+-laser pumping in Cr^{3+},Tm^{3+}-codoped YSAG:Ho^{3+} and YSGG:Ho^{3+} in 1986 [11.759]. Like in 2 μm Tm^{3+} lasers (Sect. 11.3.1), the Cr^{3+} codopant served as a sensitizer for the absorption of pump light in the visible spectral range and excitation of Tm^{3+} ions by energy transfer. The subsequent excitation of Ho^{3+} ions by energy transfer from Tm^{3+} [11.760, 761] profits from the same Tm^{3+}-Tm^{3+} cross-relaxation as described for 2 μm Tm^{3+} lasers in Sect. 11.3.1 (Fig. 11.71). In the 1980s, laser emission at 2.1 μm in Ho^{3+} under diode pumping of the Tm^{3+} $^3F_4$ level at 780–790 nm with pump thresholds as low as 5 mW was achieved in YAG [11.762–764]. Compact, monolithic, low-threshold laser devices can be achieved in this way [11.765]. Besides many different garnet crystal systems, $LiYF_4$ also regained interest as a host material for CW diode-pumped 2 μm Ho^{3+} lasers around 1990 [11.766, 767]. An Yb^{3+}-codoped, diode-pumped Ho^{3+} laser at 2.1 μm has also been demonstrated [11.768]. Noise suppression [11.769, 770], amplitude and frequency stabilization [11.771–773] of 2 μm Ho^{3+} lasers have been investigated.

Attempts to exploit the rather large gain bandwidth of Ho^{3+} near 2.0–2.1 μm by tuning the emission wavelength were reported in the early 1990s [11.774, 775]. Nowadays, tuning ranges of > 80 nm are achieved in host materials such as mixed YSGG:GSAG [11.776], BaY_2F_8:Ho^{3+} [11.777], and KYF_4 [11.778]. Mode-locking experiments have resulted in 800 ps, 370 ps, and 70 ps pulse durations, obtained in YAG:Cr^{3+},Tm^{3+},Ho^{3+} [11.714], $LiYF_4$:Tm^{3+},Ho^{3+} [11.779], and BaY_2F_8:Tm^{3+},Ho^{3+} [11.780], respectively. In mixed crystals of YSGG:GSAG:Cr^{3+},Tm^{3+},Ho^{3+}, which provide an inhomogeneously broadened, and therefore smoother, gain shape, a pulse duration as short as 25 ps could be achieved [11.776]. *Q*-switched laser operation [11.781–783] has been investigated and applied for microsurgery [11.784].

Besides the cross-relaxation and energy-transfer processes shown in Fig. 11.71, several other energy-transfer processes can occur in Tm^{3+},Ho^{3+}-codoped materials [11.785–789], thereby making the system rather complex and introducing parasitic processes which can deplete the Ho^{3+} $^5I_7$ upper laser level, increase the laser threshold, and diminish the laser efficiency. Rather than codoping the host with Tm^{3+} ions and exciting the Ho^{3+} ions via nonradiative energy transfer from the Tm^{3+} ions, one can directly pump the Ho^{3+} $^5I_7$ upper laser level at 1.9 μm using laser diodes [11.790], the output from a 1.9 μm Tm^{3+} laser [11.791], which provided up to 15 W [11.792] and 19 W [11.694] output power from Ho^{3+}, or a MgF_2:Co laser [11.793]. This approach ensures a low quantum defect and, hence, low heat generation in the laser crystal. This scheme was proven very successful by the use of a high-power Tm^{3+} fiber laser as the pump source, providing 6.4 W of output power and a slope efficiency of 80% versus incident pump power at 1.9 μm [11.794]. Also an efficient 2 μm Ho^{3+} single-frequency ring laser has been demonstrated in this way [11.795].

The first fiber laser configuration making use of this transition employed ZBLANP glass (a variant of ZBLAN) and argon-ion pumping [11.796]. A year later, this was followed by the demonstration of an argon-ion-pumped Ho^{3+}-doped silica fiber laser [11.797]. In both cases, the fiber was singly doped with Ho^{3+}, the output power < 1 mW, and each needed a relatively high pump power to reach laser threshold. Improvements in the output power and efficiency have been made recently with Yb^{3+}-doped silica fiber laser pumping of the $^5I_6$ level [11.798]; however, the output power had only increased to 280 mW because the pump fiber laser was operated at the low-efficiency 1150 nm wavelength. Using the pump wavelength of 1100 nm, which is more optimal with respect to the operation of an Yb^{3+}-doped silica fiber laser, the output power has been increased by approximately an order of magnitude [11.799].

As mentioned above, a practical method of efficiently generating laser emission on the $^5I_7 \rightarrow {}^5I_8$ transition is to codope Ho^{3+} laser ions with Tm^{3+} sensitizer ions. The first demonstration of a fiber laser operating with the Tm^{3+},Ho^{3+} system occurred in 1991 [11.800] when 250 mW was generated at a slope efficiency of 52% from a Ti:sapphire-pumped fluoride fiber laser. A year later [11.801], this work was followed by an increase in the Tm^{3+} concentration to improve cross-relaxation and resulted in a higher slope efficiency being obtained. Demonstration of a Tm^{3+},Ho^{3+}-doped silica fiber laser soon followed [11.802, 803]; however, owing to lower Tm^{3+} concentrations which forces weaker cross-relaxation, significantly lower slope efficiencies were measured, especially when pumped at 1.064 μm [11.804]. When the concentrations are increased and the double-clad pump arrangement used, a significant augmentation of the output power to 5.4 W has been demonstrated [11.805]. To date, the highest output power of 8.8 W from a fiber laser operating on the $^5I_7 \rightarrow {}^5I_8$ transition has been produced by a diode-cladding-pumped Tm^{3+},Ho^{3+}-doped fluoride fiber laser [11.747] (Fig. 11.70). In an analogous way to recent demonstrations in bulk laser systems, tandem-pumping Ho^{3+} with a separate Tm^{3+} laser operating at 1.9 μm may also prove effective in fibers, because it similarly exploits the cross-relaxation process between Tm^{3+} ions but avoids any ETU between Ho^{3+} ions in the $^5I_7$ upper laser level and excited Tm^{3+} ions [11.789]. When sensitizing with Yb^{3+} ions, a Ho^{3+}-doped silica fiber laser at 2.1 μm has been shown [11.806] to operate at moderate efficiency levels despite the fact that the energy transfer from Yb^{3+} to Ho^{3+} is quite nonresonant (Fig. 11.72).

Four-level lasers at 2.9 μm. Laser emission at the transition $^5I_6 \rightarrow {}^5I_7$ near 3 μm in an Ho^{3+}-doped crystal was demonstrated in 1976 [11.808]. Sensitizing Ho^{3+} with Yb^{3+} ions (see the energy-level scheme in Fig. 11.72) in order to exploit the more-favorable absorption features of Yb^{3+} has been used in diode-pumped crystalline lasers for the generation of 2.9 μm output [11.809]. Recent reports on this transition include laser investigations of $YAlO_3$:Ho^{3+} [11.810], diode-pumped Yb^{3+}-codoped YGSS:Ho^{3+} with 10 mJ output energy [11.811], and Cr^{3+},Yb^{3+}-codoped YGSS:Ho^{3+} tunable in the range 2.84–3.05 μm [11.812] and in *Q*-switched operation [11.813]. Since the $^5I_7$ lower laser level is a metastable excited state with a longer lifetime than the $^5I_6$ upper laser level, it is difficult to achieve CW inversion on this transition. Cascade lasing on the $^5I_6 \rightarrow {}^5I_7$ and $^5I_7 \rightarrow {}^5I_8$ transitions at 3 and 2 μm, respectively [11.814, 815], may help to deplete the $^5I_7$ level radiatively, i. e., without significant heat generation. Passive *Q*-switching of this transition has also been demonstrated [11.813].

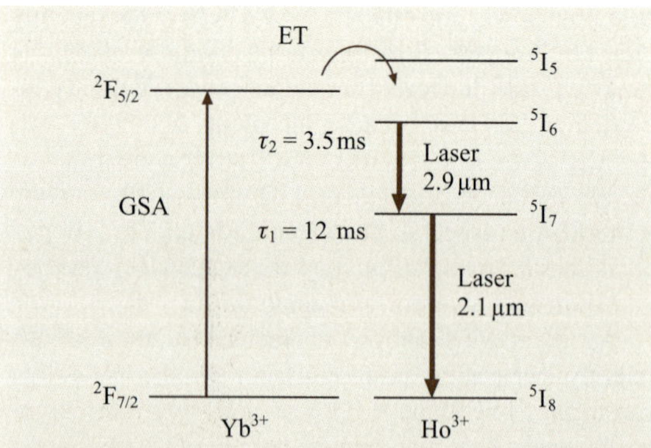

Fig. 11.72 Partial energy-level scheme of Ho^{3+} with Yb^{3+} sensitizer displaying the measured lifetimes of Ho^{3+} when doped into fluoride glass host [11.807]

The combined effect of the infrared absorption cut-off wavelength of ≈ 2.5 μm for pure silica glass and the strong multiphonon relaxation quenching of mid-infrared transitions of rare-earth ions in this host means that four-level fiber lasers operating on the $^5I_6 \rightarrow {}^5I_7$ transition at ≈ 2.9 μm have only involved fluoride glass as the host material. The first demonstration of a fiber laser using this transition [11.816] produced only ≈ 13 mW when pumped at a wavelength of 640 nm. High-power cascade lasing at 2.9 μm and 2.1 μm has been employed to extend the output power to 1.3 W [11.817] by removing bottlenecking at the $^5I_7$ level [11.807] via the second laser transition at 2.1 μm. In an analogous manner to the Er^{3+}-doped fluoride glass system discussed below, the most successful arrangement to date for extracting high power from this transition has involved the use of Pr^{3+} as a desensitizer for the $^5I_7$ energy level: a maximum output power of 2.5 W was produced when the pump wavelength of 1100 nm from a Yb^{3+}-doped silica fiber laser was used [11.818].

An Yb^{3+}-sensitized Ho^{3+}-doped ZBLAN fiber could be directly pumped with diode lasers and may efficiently provide high-power 2.9 μm output without the costly requirement of an intermediate laser system. Initial spectroscopic results look encouraging [11.819], however, the many excited ion interactions that a fluoride host provides may be problematic. A recent demonstration has shown that ion–ion interactions (specifically ETU) in Ho^{3+}-doped fluoride glass are critical to the production of 2.9 μm output from singly Ho^{3+}-doped fluoride glass fiber lasers [11.820].

Erbium-doped solid-state lasers at 2.7–2.9 μm

For a long time, the development of erbium lasers operating on the $^4I_{11/2} \rightarrow {}^4I_{13/2}$ transition near 3 μm was dominated by crystalline systems. The early success of the erbium 3 μm crystal laser has given rise to a significant amount of spectroscopic investigations. This has led to a deep understanding of the rather complex population mechanisms of this laser system and to the development of a large number of suitable host materials.

Crystalline lasers. The first observation of coherent emission near 3 μm from erbium ions was reported in 1967 [11.613]. Yttrium aluminum garnet (YAG) was demonstrated as a host for the erbium 3 μm laser [11.822] in 1975. In 1983, the first CW lasing near 3 μm was obtained in this material [11.823]. At about the same time, it was established [11.824–827] that energy transfer processes [11.828] between neighboring erbium ions in the host lattice play an important role in this laser system. Energy transfer processes can become very efficient at high excitation density [11.829] and govern the population mechanisms of the 3 μm laser at high erbium concentration. In the energy-level scheme of Fig. 11.73a, the important ETU and cross-relaxation processes are introduced. The ETU process $(^4I_{13/2}, {}^4I_{13/2}) \rightarrow (^4I_{15/2}, {}^4I_{9/2})$ leads to a fast depletion of the lower laser level and enables CW operation of a laser transition which, otherwise, could be self-terminating owing to the unfavorable lifetime ratio of the upper compared to the lower laser level. The ETU process from $^4I_{13/2}$ can be so dominant that even under direct pumping of the $^4I_{13/2}$ lower laser level and subsequent excitation of the $^4I_{11/2}$ upper laser level by ETU, 3 μm laser operation was demonstrated in several host materials [11.830].

This ETU process offers another great advantage. Half of the ions that undergo this process are upconverted to the $^4I_{9/2}$ level and, by subsequent multiphonon relaxation, are recycled to the $^4I_{11/2}$ upper laser level from where they can each emit a second laser photon, for a single pump-photon absorption. For a large number of ions participating in this process, a slope efficiency η_{sl} of twice the Stokes efficiency $\eta_{St} = \lambda_{pump}/\lambda_{laser}$ is obtained [11.831], because the quantum efficiency $\eta_{QE} = n_{laser}/n_{pump}$ of pump photons converted to laser photons increases from 1 to 2 (λ and n are the wavelengths and photon numbers of laser and pump transitions, respectively):

$$\eta_{sl} = \eta_{QE}\eta_{St} = 2\eta_{St} . \tag{11.90}$$

This is illustrated in Fig. 11.74.

In a simple rate-equation system which includes the processes shown in Fig. 11.73a the slope efficiency is given by [11.831]

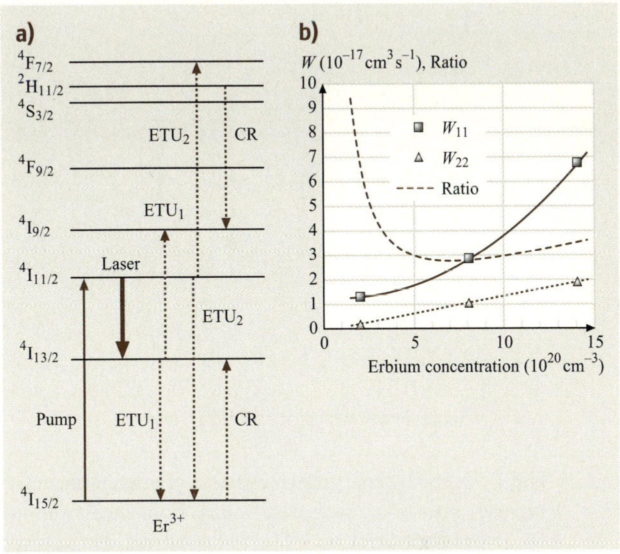

Fig. 11.73 (a) Partial energy-level scheme of erbium indicating the pump and laser transitions, ETU$_1$ from $^4I_{13/2}$, ETU$_2$ from $^4I_{11/2}$, and cross-relaxation (CR) from the thermally coupled $^4S_{3/2}$ and $^2H_{11/2}$ levels. (b) Macroscopic parameters of ETU$_1$ from $^4I_{13/2}(W_{11})$ and ETU$_2$ from $^4I_{11/2}(W_{22})$ and the ratio W_{11}/W_{22} in ZBLAN:Er^{3+} bulk glasses. (After [11.821])

$$\eta_{sl} = \eta_{St}\frac{\ln(1-T)}{\ln[(1-T)(1-L)]}\left(2 - \frac{b_1^2}{b_2^2}\frac{W_{22}}{W_{11}}\right) , \tag{11.91}$$

with T the transmission of the out-coupling mirror, L the internal resonator losses, and b_i and W_{ii}, the Boltzmann factors and ETU parameters of the upper ($i = 2$) and lower ($i = 1$) laser levels, respectively. If ETU occurs only from the lower laser level, i. e., $W_{22} = 0$, we obtain the predicted factor of two increase in slope efficiency from (11.91). The slope efficiency is reduced, however, by the resonator losses, the imperfect mode overlap, and the ETU process from the upper laser level in the case of $W_{22} > 0$. In the investigated host materials, the parameters W_{ii} of both ETU processes increase with increasing erbium concentration because of the influence of energy migration within the erbium $^4I_{11/2}$ and $^4I_{13/2}$ levels on ETU. The slope efficiency of (11.91) is optimum for a maximum ratio W_1/W_2. Spectroscopy of these processes in crystal hosts and laser experiments revealed that the maximum ratio is obtained at dopant concentrations of ≈ 12–15% in BaY_2F_8 [11.832, 833], $\approx 15\%$ in $LiYF_4$ [11.661], $\approx 30\%$ in YSGG [11.834],

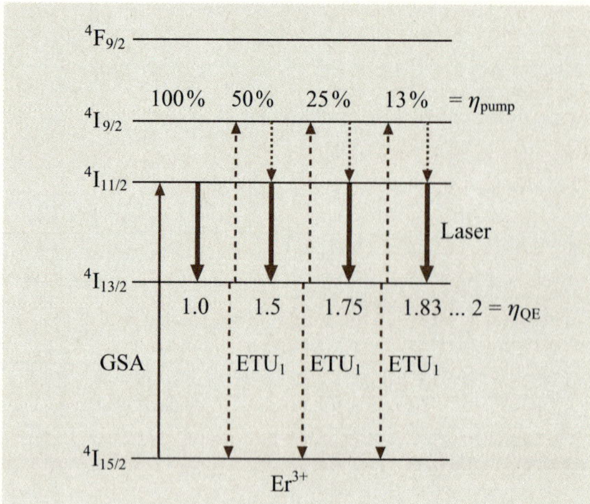

Fig. 11.74 Partial energy-level scheme of erbium illustrating the process of energy recycling from the lower to the upper laser level by ETU. Indicated are the relative pump rate η_p of the upper laser level and the quantum efficiency η_{QE} which increases from 1 to 2 if a large number of ions participate in the process. (After [11.836])

and $\approx 50\%$ in $Y_3Al_5O_{12}$ [11.835]. A trend in this series is the increase of the optimum erbium concentration with phonon energy of the host material.

Energy recycling by ETU is the most efficient way to operate a CW erbium laser near 3 μm. The highest slope efficiency obtained experimentally is currently 50% in $LiYF_4$:15% Er^{3+} [11.837]. The pump wavelength that provides the highest Stokes efficiency of $\eta_{St} = \lambda_{pump}/\lambda_{laser} = 35\%$ is 980 nm, which corresponds to pumping directly into the upper laser level [11.838] (Fig. 11.73a). The highest slope efficiency obtained experimentally [11.837] is currently $\eta_{sl} = 50\%$ in $LiYF_4$:15% Er^{3+}. This result shows that energy recycling is indeed efficient and that slope efficiencies far above the Stokes efficiency can be obtained under CW pumping. Under quasi-CW excitation, the slope efficiency is strongly reduced [11.839], because the lower laser level is much less populated than in the steady-state regime and ETU is less efficient [11.840, 841]. Other operational regimes which deplete the lower laser level without recycling the energy to the upper laser level are less efficient. Consequently, neither co-lasing at the 1.6 μm transition from the lower laser level to the ground state [11.842] nor energy transfer from the erbium lower laser level to a rare-earth codopant [11.660] have reached the efficiency of the recycling regime.

Lifetime quenching of the $^4I_{11/2}$ upper laser level by multiphonon relaxation is stronger in oxide compared to fluoride host materials because of the larger maximum phonon energies in oxides. With an energy gap between the $^4I_{11/2}$ upper and $^4I_{13/2}$ lower laser levels of ≈ 3400–$3500\,\mathrm{cm^{-1}}$, the radiative decay becomes dominant for phonon energies below $\approx 550\,\mathrm{cm^{-1}}$. Since a long lifetime of the $^4I_{11/2}$ upper laser level provides a small pump threshold, fluorides are preferable host materials [11.843] for this laser transition if the pump power is not many times above threshold.

In the 1980s and 1990s, numerous host materials were investigated for CW and pulsed laser operation of Er^{3+} in the 2.7–2.9 μm region. A major role was played by the family of garnet crystals YAG, YSGG, YSAG, YGG, and GGG [11.615, 844, 845]. In the early years, Cr^{3+} codoping was used in order to improve the absorption of broadband flashlamp pump light in the visible spectral range and transfer the absorbed energy from Cr^{3+} to Er^{3+}. Typical output characteristics obtained were 2.7 W average power at a pump energy of 5 J with a repetition rate of 10 Hz [11.845]. Doping levels up to 100% substitution of Y^{3+} by Er^{3+} were tested [11.846]. A new class of host materials, fluoride crystals such as $LiYF_4$, BaY_2F_8, and others, became of importance in the late 1980s [11.847–850]. Laser thresholds as low as 5 mW were obtained under CW excitation by diode lasers [11.849].

In recent years, researchers have obtained CW and quasi-CW diode-pumped output power levels exceeding 1 W at 3 μm from fluoride [11.661] and oxide [11.851–853] crystalline host materials. A significant problem in the energy-recycling regime is increased heat generation due to the multiphonon relaxation $^4I_{9/2} \rightarrow {}^4I_{11/2}$ that follows each ETU process from the lower laser level [11.676]. Glass bulk materials [11.854] suffer from the same thermal and thermo-optical drawbacks as the crystalline bulk materials, with even decreased thermal conductivity in the glass. A possible solution is diode side pumping, which leads to lower excitation densities and correspondingly weaker ETU processes, as well as better heat removal in the slab geometry. The highest output powers of 1.8 W and later 4 W from an erbium 3 μm crystal laser [11.855, 856] have been obtained in this way. A reduced erbium concentration with correspondingly smaller parameters of the ETU processes may aid this approach. However, the efficiency of the energy-recycling regime cannot be reached in this approach.

Other special configurations include the operation of Er^{3+}-doped YAG, GGG, and YSGG lasers in monolithic cavities with output powers up to 0.5 W and tunable single-frequency output [11.857] as well as microlasers in YSGG [11.858]. Pulsed output at 3 μm has been generated from erbium-doped crystalline materials in many configurations and regimes, e.g., under quasi-CW pumping [11.839, 841], active [11.859–868] and passive [11.869–872] Q-switching, and mode-locking [11.619, 869, 873].

Fiber lasers. The erbium-doped fluoride fiber represents a promising alternative for the construction of a compact and efficient all-solid-state laser emitting at the transition at 3 μm. Due to its geometry, the fiber provides large flexibility and potentially high pump and signal beam intensities without the drawbacks of thermal and thermo-optical effects. The first erbium 3 μm fiber laser was demonstrated in 1988 [11.875]. Single-mode [11.876] and diode-pumped [11.877] operation were demonstrated shortly afterwards. Although the lifetime of the $^4I_{13/2}$ lower laser level exceeds that of the $^4I_{11/2}$ upper laser level, CW lasing can be obtained on this four-level-laser transition in ZBLAN (but also in fluoride crystals, see the paragraph on crystalline lasers) without employing special techniques to depopulate the $^4I_{13/2}$ lower laser level, because the lower laser level is not fed significantly by luminescent decay or multiphonon relaxation from the upper laser level [11.878]. In addition, the Stark splitting of the laser levels contributes to population inversion, because the laser transition occurs between a low-lying Stark component of the upper and a high-lying Stark component of the lower laser level [11.836]. During the relaxation oscillations at the onset of lasing, a red-shift of the lasing wavelength is often observed in erbium 3 μm laser systems [11.879–882], because the excitation energy is accumulated in the long-lived $^4I_{13/2}$ lower laser level and the character of the lasing process changes from four-level to three-level lasing [11.836]. For the same reason, the tunability range of a 3 μm CW laser [11.883] is narrowed and red-shifted with increasing pump power.

Pump excited-state absorption (ESA), which is present in Er^{3+} at almost all available GSA wavelengths [11.884], has a major influence on the performance of low-doped, core-pumped erbium 2.7 μm ZBLAN fiber lasers because of the significant amount of ground-state bleaching and excitation of the laser levels under these conditions [11.885]. Pumping at 980 nm directly into the upper laser level provides the highest Stokes efficiency of $\eta_{St} = \lambda_{pump}/\lambda_{laser} = 35\%$ [11.838]. However, ESA at 980 nm from the $^4I_{11/2}$ upper laser level [11.886] is detrimental to lasing. Experimentally, the best pump wavelength [11.885] is near 792 nm, at the peak of ESA from the $^4I_{13/2}$ lower laser level [11.874]; see the measured GSA and ESA cross sections in Fig. 11.75a. Depletion of the lower laser level by ESA favorably results in a redistribution of its population density and overcomes the bottleneck that results from the long lower-level lifetime. However, slope efficiencies obtained in this way were < 15%. Moreover, saturation of the output power at 2.7 μm was observed and the highest reported output powers were in the 20 mW region [11.887, 888]. The excitation of the metastable $^4S_{3/2}$ level (lifetime ≈ 580 μs [11.821]) led to inversion with respect to the $^4I_{13/2}$ level. A second laser transition at 850 nm repopulated the $^4I_{13/2}$ lower laser level of the 2.7 μm transition (Fig. 11.75b), causing the 2.7 μm laser to saturate at low output powers [11.885]. Significant improvement in the performance of this laser system was obtained by deliberately operating a third laser transition $^4S_{3/2} \rightarrow {}^4I_{9/2}$ at 1.7 μm, thereby suppressing the competitive laser at 850 nm and recycling the excitation energy accumulated in the $^4S_{3/2}$ level into the upper laser level; see the energy-level scheme in Fig. 11.75b. The slope efficiency of the 2.7 μm transition increased significantly to 23% [11.889], close to the Stokes efficiency limit of 29% under 800 nm pumping. An output power of 150 mW was demonstrated experimentally [11.889]. Also a three-transition-cascade

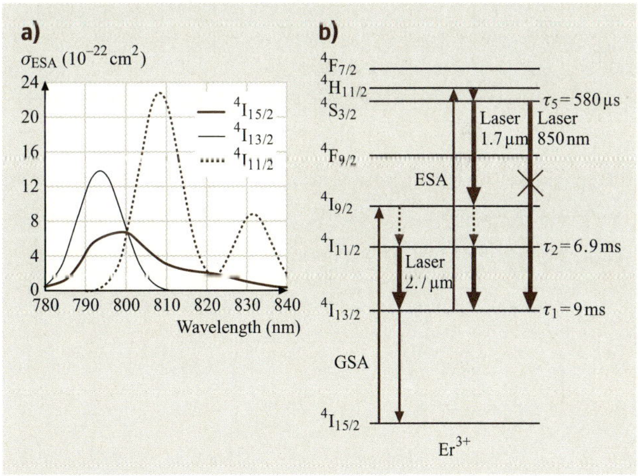

Fig. 11.75 **(a)** Absorption cross sections in ZBLAN:Er^{3+} near 800 nm: GSA $^4I_{15/2} \rightarrow {}^4I_{9/2}$ and ESA $^4I_{13/2} \rightarrow {}^2H_{11/2}$, $^4I_{11/2} \rightarrow {}^4F_{3/2}$, and $^4I_{11/2} \rightarrow {}^4F_{5/2}$ (After [11.874]). **(b)** Partial energy-level scheme of erbium indicating the processes relevant to the ZBLAN:Er^{3+} cascade laser: *lower loop* with GSA to $^4I_{9/2}$, multiphonon relaxation, laser transition at 2.7 μm, luminescent decay, and *upper loop* with ESA to $^2H_{11/2}$, thermal relaxation, laser transition at 1.7 μm, multiphonon relaxation, laser transition at 2.7 μm. Competitive lasing at 850 nm is suppressed in the cascade regime

lasing regime with additional lasing at the transition $^4I_{13/2} \rightarrow {}^4I_{15/2}$ near 1.6 μm was demonstrated [11.890].

In ZBLAN fibers with higher dopant concentrations of typically 1–5 mol % ($\approx 1.6\text{–}8 \times 10^{20}\,\mathrm{cm}^{-3}$) and with the double-clad geometry, ESA is much less important, because the reduced pump intensity with low-brightness diode lasers leads to smaller excitation densities. Currently, the most successful approach towards a high-power erbium 2.7 μm fiber laser is codoping of the fiber with Pr^{3+} [11.891, 892]. This idea was reported already in [11.887, 893, 894] and was proposed for the double-clad fiber laser in [11.895]. In this approach, the Er^{3+} 2.7 μm transition is operated as a simple four-level laser; see the energy-level scheme in Fig. 11.76a. The $^4I_{13/2}$ lower laser level is depopulated by the energy transfer process ET_1 to the Pr^{3+} codopant and fast decay to the ground state by multiphonon relaxation within Pr^{3+}. The energy-transfer process ET_2 from the $^4I_{11/2}$ upper laser level to the Pr^{3+} codopant is weak [11.821]. The strong lifetime quenching of the $^4I_{13/2}$ lower laser level significantly reduces ground-state bleaching and excitation of the laser levels, thus making the influence of ESA negligible, but similarly preventing energy recycling by ETU [11.896]. Each pump photon can at best produce one laser photon in the Er^{3+},Pr^{3+}-codoped system. The theoretical limit of the slope efficiency is given by the Stokes efficiency, which is 29% under 800 nm pumping. Experimentally, a slope efficiency of 17% and an output power of 1.7 W were obtained [11.891] (Fig. 11.76b). Other researchers [11.897] reported output powers of 660 mW. Since ESA from both laser levels is negligible, the system can alternatively be pumped near 980 nm, which provides a Stokes efficiency of 35%. In this way, the experimental slope efficiency could be increased to 25% [11.898]. With improvements in diode-laser technology and an optimized fiber design, an output power of 5.4 W at 2.7 μm with a slope efficiency of 21% could recently be demonstrated from an Er^{3+},Pr^{3+}-codoped ZBLAN fiber laser [11.892, 899].

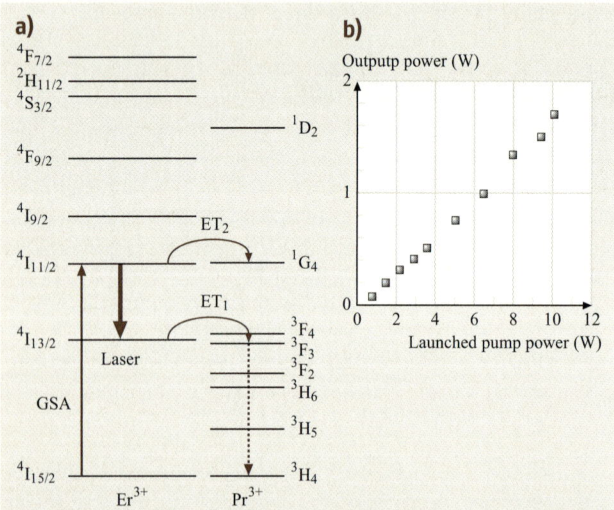

Fig. 11.76 **(a)** Partial energy-level scheme of erbium indicating the processes relevant to the ZBLAN:Er^{3+} lifetime-quenching laser: GSA at 980 nm to the $^4I_{11/2}$ upper laser level (or at 790 nm to the $^4I_{9/2}$ pump level and subsequent multiphonon relaxation to $^4I_{11/2}$), laser transition to the $^4I_{13/2}$ lower laser level, and relaxation to the ground state via energy transfer ET_1 to the Pr^{3+} codopant. The energy transfer ET_2 from the $^4I_{11/2}$ upper laser level to the Pr^{3+} codopant is weak. **(b)** Output power at 2.7 μm under 792 nm pumping. (After [11.891])

The first steps toward pulsed output from erbium 3 μm ZBLAN lasers [11.882, 900, 901] were unsatisfactory in terms of output energies and average powers. There have also been attempts to operate the ZBLAN fiber laser in the energy-recycling regime. The parameters W_{ii} of both ETU processes in ZBLAN bulk glasses [11.821] versus Er^{3+} concentration are shown in Fig. 11.73b. The criterion for optimization of the slope efficiency in (11.90) is maximizing the ratio W_{11}/W_{22}. For Er^{3+} concentrations of > 2–3 mol % at which ETU processes become important, this ratio is ≈ 3, see the *dashed line* in Fig. 11.73b, a more favorable value than reported for $LiYF_4$:Er^{3+} [11.851]. Energy recycling by ETU at high Er^{3+} concentrations [11.902] might lead to output powers at 3 μm on the order of 10 W. In early attempts, two research groups tried to exploit energy recycling [11.903, 904], however the slope efficiencies in these experiments did not exceed the slope efficiencies obtained in Er^{3+},Pr^{3+}-codoped fibers pumped at corresponding pump wavelengths [11.891, 892, 898]. In fibers with relatively large core diameters and therefore transverse multimode operation, singly Er-doped with concentrations smaller than those required for efficient energy recycling, 3 W of output power was obtained [11.905]. With further increase of the core diameter to 90 μm, the highest pulsed output of > 0.5 mJ could be demonstrated by the same researchers. Recently, 8 W of output power with a slope efficiency of 24.4% were demonstrated from a highly Er-doped (60 000 ppm) ZBLAN fiber when dual-end-pumping the fiber with 25 W of launched pump power at 975 nm. This result represents the highest output obtained by the end of 2005 from a fiber laser near 3 μm. Again, the achieved slope efficiency of 24.4% is very similar to that of 25% obtained with the same pump wavelength in a Er^{3+},Pr^{3+}-codoped fiber [11.898]. Therefore, it must be concluded that efficient energy recycling with correspondingly improved slope efficiency similar to the result of 50% in a crystalline host material [11.837] is still lacking.

Dysprosium-doped solid-state lasers at 2.3–2.4 μm and 2.9–3.4 μm

The search for new mid-infrared laser transitions depends entirely on the structure of the energy-level diagram of the rare-earth ions. The Dy^{3+} ion offers a rather dense energy-level scheme in the infrared spectral region, resulting in a range of absorption peaks, as well as a four-level laser transition at 2.3–2.4 μm and a phonon-terminated 3 μm laser transition. The transition at 2.3–2.4 μm was among the first laser transitions reported and the first to be demonstrated in continuous-wave operation [11.609, 610]. Because of the dense energy-level scheme of Dy^{3+}, multiphonon relaxation is an issue of concern in this ion, hence low-phonon host materials may improve its laser performance. Recently, room-temperature laser emission at 2.43 μm was reported in the low-phonon host materials $CaGa_2S_4:Dy^{3+}$ and $KPb_2Cl_5:Dy^{3+}$. At the 3.4 μm transition, room-temperature laser oscillation from Dy^{3+} in $BaYb_2F_8$ has been reported [11.906]. The 1100 nm output from a Yb^{3+}-doped silica fiber laser has been successfully used to pump a Dy^{3+}-doped fluoride fiber laser [11.907]. In this case, a maximum output power of 275 mW was generated with a slope efficiency of only ≈ 5%, however, when the pump wavelength was increased to ≈ 1.3 μm using a YAG:Nd^{3+} laser, the slope efficiency is approximately quadrupled to ≈ 20% [11.908]. Reduced levels of pump ESA are believed to cause this augmentation in the slope efficiency. Future Dy^{3+}-doped fluoride fiber lasers may benefit from further increases in the pump wavelength to 1.7 μm or 2.8 μm.

Before turning our attention to rare-earth-ion-doped solid-state lasers in the wavelength range beyond 3 μm, it should be noted that solid-state lasers based on the actinide ion U^{3+} have also attracted attention in the recent past [11.909, 910].

Solid-state lasers at wavelengths beyond 3 μm

Laser wavelengths longer than typically 3 μm are generally difficult to achieve in solid-state host materials by direct generation from rare-earth or transition-metal ions, because the energy gap between the upper and lower laser level is necessarily small and all the common oxide and fluoride host materials possess maximum phonon energies, which lead to fast multiphonon relaxation of the excitation of the upper laser level. Therefore, many of the laser transitions reported in the literature require cooling of the active device. On the other hand, the attractiveness of this wavelength range for a number of applications has inspired the search for host materials with maximum phonon energies below ≈ 300 cm^{-1}.

In crystalline hosts, a flashlamp-pumped, cooled Er^{3+} laser at 4.75 μm [11.614], a CW diode-pumped, cooled Er^{3+} laser at 3.41 μm [11.911], and a pulsed diode-pumped room-temperature Er^{3+} laser at 4.6 μm [11.912] have been operated. A room-temperature $BaY_2F_8:Ho^{3+}$ laser was operated at 3.9 μm [11.913]. Due to its dense energy-level scheme, Dy^{3+} offers a large range of possible mid-infrared transitions. Recently, room-temperature lasing of Dy^{3+} transitions in the low-phonon host materials $CaGaS_2$ and KPb_2Cl_5 was demonstrated [11.914], in the former material even at 4.31 μm. The longest-wavelength lasers shown to operate in a solid-state material are room-temperature 5 μm and 7 μm lasers in $LaCl_3:Pr^{3+}$ [11.915, 916].

The operation of lasers at wavelengths of 3.22 μm [11.917] and 3.95 μm [11.918] has been obtained from Ho^{3+}-doped ZBLAN fiber and at 3.45 μm [11.919] from Er^{3+}-doped ZBLAN fiber. It was, however, necessary to cool the ZBLAN fiber for the 3.45 μm and 3.95 μm transitions. These two laser transitions span five or six maximum phonon energies in ZBLAN, therefore the lifetime of the upper laser level for each of these transitions is short and engenders an increase in the pump threshold compared to other ZBLAN fiber lasers operating at the shorter mid-infrared wavelengths. In addition, the lower laser levels of these transitions possess quite long lifetimes and some saturation of the output power has been observed [11.920]. This problem (while it can be mitigated with cascaded lasing), combined with the use of inconvenient pump sources has impeded the full utilization of these laser transitions. The 3.95 μm wavelength emitted from the cooled ZBLAN fiber laser is currently the longest laser wavelength that has been generated from a fiber laser.

Generating wavelengths longer than 3 μm from fiber lasers is a task that tests the limits of current glass technology. The need for lower phonon energies has to be balanced with acceptable mechanical, chemical, and thermal properties. Since the highly developed ZBLAN glass is only useful for laser transitions up to 3–3.5 μm, glasses such as the chalcogenides [11.921] will need to fill the gap. It is because these glasses have to be drawn into low-loss fiber that has prevented long-wavelength emission to the extent that is possible in crystalline-based solid-state lasers. Creating efficient, high-power mid-infrared fiber lasers with output wavelengths $> 3\,\mu m$ is at the forefront of current fiber-laser research efforts.

As mentioned above, fiber lasers operating on laser transitions that have wavelengths $> 3\,\mu m$ will need to use glasses, which have very low phonon energies. While rare-earth-ion-doped heavy-metal oxides [11.925] have been studied for 2–3 μm mid-infrared emission; to date, there has been no report of laser action for a fiber laser comprised of such a glass. Heavy-metal oxides do not seem to be suitable for lasers at wavelengths beyond 3 μm, because their maximum phonon energies are comparable to fluoride glasses and are too high for laser transitions beyond 3 μm.

The chalcogenide glasses have been doped with a number of rare-earth ions including Ho^{3+} [11.924], Tm^{3+} [11.923], Tb^{3+} [11.923], Dy^{3+} [11.922], Pr^{3+} [11.926], and Er^{3+} [11.927, 928] for studies into $> 3\,\mu m$ mid-infrared luminescence (Table 11.21). Fiber-laser action has been reported, however, only for an Nd^{3+}-doped GLS glass operating at a wavelength of $\approx 1\,\mu m$ [11.929]. Recent demonstrations of fabricating Bragg gratings [11.930], single-mode fibers [11.931] and holey fibers [11.932] with chalcogenide glass highlight the utility of this glass for fiber-based applications; however, the purity and toxicity of the starting materials and the difficulty of making ultralow-loss fiber currently impede the widespread use of chalcogenide glass for mid-infrared fiber-laser applications. Once these obstacles have been overcome, future $> 3\,\mu m$ fiber lasers will most likely involve the rare-earth ions Pr^{3+}, Nd^{3+}, Dy^{3+}, and Ho^{3+} doped into chalcogenide glass, because most of the important mid-infrared transitions relevant to these ions can be accessed with pump-photon wavenumbers $< 10\,000\,cm^{-1}$. Judicious choice of the overall dopant-ion concentration and the use of particular sensitizer and quenching ions will enable the production of efficient $> 3\,\mu m$ output some time in the future.

Table 11.21 Examples of luminescent transitions investigated as candidates for mid-infrared lasers in sulfide glasses

Ion	λ_{laser} (μm)	Transition	Ref.
Dy^{3+}	3.2	${}^6H_{13/2} \to {}^6H_{15/2}$	[11.922]
Tm^{3+}	3.8	${}^3H_5 \to {}^3H_4$	[11.923]
Ho^{3+}	3.9	${}^5I_5 \to {}^5H_6$	[11.924]
Dy^{3+}	4.3	${}^6H_{11/2} \to {}^6H_{13/2}$	[11.922]
Tb^{3+}	4.8	${}^7F_5 \to {}^7F_6$	[11.923]
Ho^{3+}	4.9	${}^5I_4 \to {}^5I_5$	[11.924]

Conclusions

In the roughly four decades since the demonstration of the first mid-infrared solid-state lasers, thousands of scientific papers have been published which have reported on lasing in novel host materials, replacement of flashlamps by ion-laser and later diode-laser pump sources, ever-growing output powers, higher efficiencies, larger tunability ranges, shorter pulse durations and the like. A general tendency is that, the shorter the wavelength, the better the laser performance. When we approach longer wavelengths in the mid-infrared spectrum, we find that the quality and durability of the required low-phonon-energy host materials decline, Stokes and slope efficiencies decrease, whereas the thermal problems increase. While many crystalline host materials and the corresponding laser techniques have matured during the 1990s, the fast development of high-power, fundamental-mode fiber lasers, which could be witnessed in the 1 μm spectral range, has now also reached the mid-infrared spectral region. However, the assumption that, due to its large surface-to-volume ratio, the fiber geometry might avoid all thermal problems has been questioned by several recent high-power fiber-laser experiments in the near- and mid-infrared spectral region. These phenomena are, in principle, not much different from the situation found in crystalline lasers. Nevertheless, there remain distinct differences between these two host categories. When flexibility of the resonator design, short pulses, and high peak powers are required, crystalline lasers have advantages. On the other hand, fiber lasers are preferred when high beam quality or low pump threshold combined with medium CW output power are desired. The low pump threshold of fiber lasers is an invaluable advantage when cascade-laser operation is required to depopulate the long-lived terminating level of one laser transition by a second laser transition. The comparatively low dopant concentrations that are useful in fiber lasers due to the long interaction lengths can minimize energy dissipation by interionic processes but, equally, limit the exploitation of these processes as a tool to optimize the population mechanisms of a certain laser system, as has been done successfully in several of the mid-infrared crystalline laser systems discussed above. Although still a great challenge with respect to fabrication process and durability, low-phonon crystalline and fiber host materials have the potential to revolutionize CW mid-infrared lasers in the wavelength range between 3–5 μm.

11.2.5 Transition-Metal-Ion Lasers

Basics

In this chapter an overview of transition-metal-ion lasers will be given. The main focus lies on the laser characteristics and results, a more spectroscopically oriented overview is given in [11.933]. Transition metals are elements of the third, fourth and fifth row of elements in the periodic table. Laser oscillation has so far only been obtained with ions of the transition metals of the third row (the Fe row; Ti to Cu). Due to the strong coupling between the electronic levels of the transition-metal ion with the surrounding field established by the crystalline environment, transition-metal-ion lasers are usually tunable over a wide spectral range up to several hundred nanometers. These lasers are of interest for a wide field of applications, e.g., in scientific research, in medicine, for measurement and testing techniques, ultrashort-pulse generation, and communication. They can also be used as coherent light sources for second-harmonic generation, for optical parametric oscillators, and for sum- and difference-frequency generation.

The energy-level scheme and thus also the spectroscopic and laser characteristics of a transition-metal ion in a crystalline field is strongly dependent on the valence state of the ion, the number of ligands (i. e., the coordination number) and the strength and symmetry of the surrounding crystalline field. Therefore it is not possible to draw some kind of Dieke diagram for the transition-metal ions, as it is possible for the trivalent rare-earth ions (see the section on 4f–4f transitions in Sect. 11.2.1). The energy-level schemes of the transition-metal ions in crystalline hosts are in principle described by the so-called Tanabe–Sugano diagrams [11.934–936]. These diagrams are distinguished by the number of electron within the 3d electron shell. In these diagrams the energy of a specific level of the transition-metal ion is depicted as a function of the crystal field strength. We will not describe the quantum-mechanical background to obtain these diagrams, which would be beyond the frame of this handbook. The reader is referred to the appropriate literature [11.937–942].

In comparison to laser systems based on the 4f↔4f transitions of trivalent rare-earth ions, lasers based on 3d↔3d transitions are in general more affected by a higher probability of excited-state absorption, a higher probability of nonradiative decay, and a higher saturation intensity, leading to higher laser thresholds. Often laser oscillation cannot be obtained at all. In the following section, the focus is on specific transition-metal-ion lasers, ordered according to their laser wavelength from visible over the near-infrared to the mid-infrared. This overview includes Ti^{3+} lasers, especially Ti^{3+}:Al_2O_3, Cr^{3+} lasers, Cr^{4+} lasers and finally Cr^{2+} lasers. At the end of this section, other transition-metal-ion lasers are presented, including Co^{2+} and Ni^{2+} lasers. Finally, some general comments about transition-metal-ion lasers will be given in the last section. In Fig. 11.77 an overview of lasers based on 3d↔3d transitions of transition-metal ions is given. It can be seen that almost the whole spectral range between 650 nm and 4500 nm is covered.

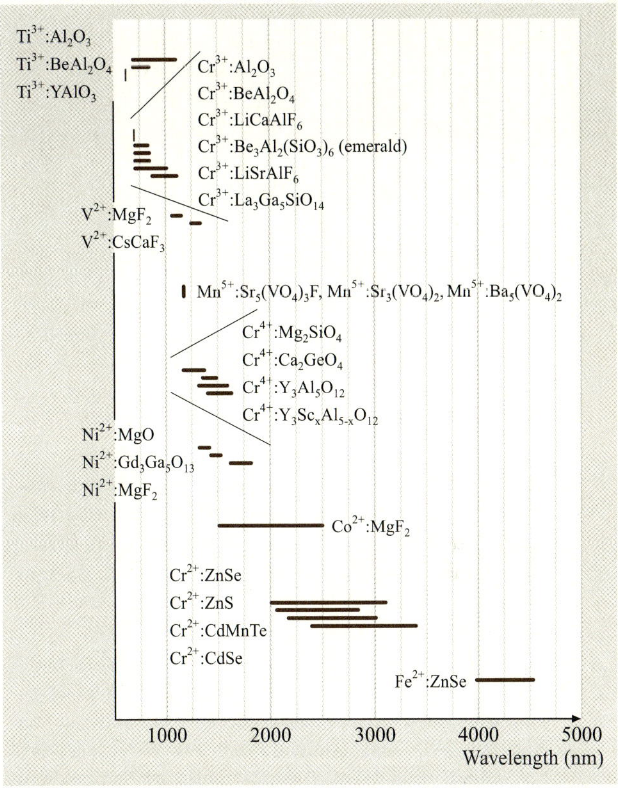

Fig. 11.77 Overview of tunable solid-state lasers based on transition-metal-ion-doped crystals

Overview of Transition-Metal-Ion Lasers

Ti^{3+} lasers. Ti^{3+}-doped Al_2O_3 (Ti:sapphire) has been intensively investigated as a tunable laser material since the first laser operation was reported [11.943, 944]. Efficient laser oscillation was obtained in Al_2O_3 in pulsed [11.943, 945, 946] and continuous-wave [11.945, 947, 948] operation, for further references see [11.213]. The tuning range covers more than 400 nm and spans approximately 670–1100 nm [11.949]. It has a relatively high emission cross section of approximately

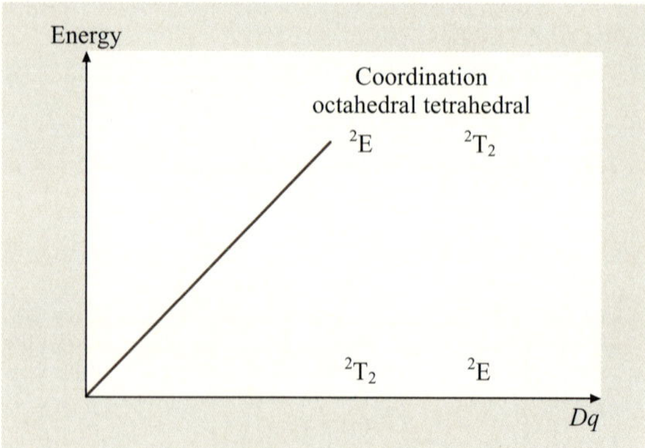

Fig. 11.78 Basic energy-level schemes of $3d^1$ ions in octahedral and tetrahedral coordination

$4.1 \times 10^{-19}\,\mathrm{cm}^2$ [11.950] at the maximum of the emission spectrum. A slope efficiency of 62% has been reported [11.951], which is close to the quantum limit in that experiment of 78% (pump wavelength 589 nm, emission wavelength 750 nm). This indicates the low intrinsic losses of the system.

The Ti^{3+} ion belongs to the $3d^1$ configuration, which is very favorable with respect to laser application because of its simple energy-level scheme (see Figs. 11.78, 11.79). There are only two $3d^1$ energy levels, which diminishes the possibility of excited-state absorption of the laser radiation, a process that limits

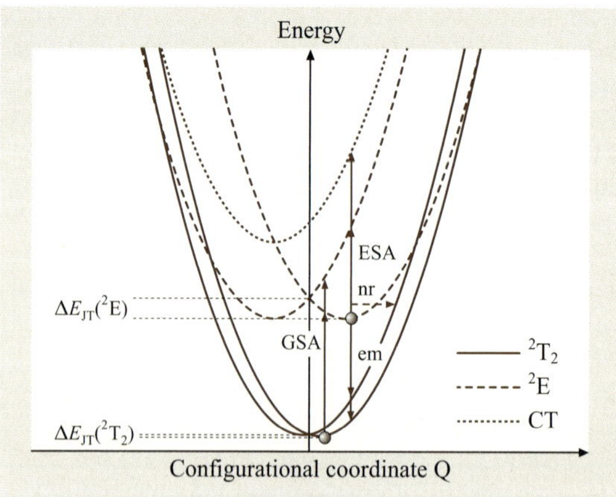

Fig. 11.79 Schematic diagram of the Jahn–Teller effect for octahedrally coordinated d^1-systems. ^2E, $^2T_2$ and CT (charge transfer) are the energy levels, GSA, ESA, and em are ground-state absorption, excited-state absorption, and emission, respectively. nr represents the nonradiative decay via tunneling between the excited state and the ground state. ΔE_{JT} is the Jahn–Teller stabilization energy

the tuning range and efficiency of other transition-metal-ion-doped lasers. The orbital degeneracy of the d levels is removed due to the Jahn–Teller effect, which yields the large absorption and emission bandwidths. The room-temperature absorption and emission spectra are shown in Fig. 11.80. Excited-state absorption transitions occurs to energy levels correlated to charge transfer and conduction-band levels and between the two Jahn–Teller-split components of the upper (^2E) d level. In Ti:sapphire, excited-state absorption was not observed in the spectral range of emission [11.952]. In Table 11.22 the laser-relevant parameters of Ti:sapphire are listed. Besides the favorable spectroscopic and laser characteristic, the Al_2O_3 host material offers a variety of advantageous properties: a high thermal conductivity, as well as mechanical and chemical hardness.

To date, Ti^{3+}:Al_2O_3 is the most common and commercially available tunable solid-state laser. Nowadays, it can be pumped with frequency-doubled Nd lasers at wavelengths around 532 nm, thus efficient all-solid-state laser operation is possible. Formerly, Ar^+-ion laser pumping was applied. In commercial systems, overall efficiencies as high as 30% are obtained.

Besides the broad tuning capability of the Ti:sapphire laser, its ability for ultrashort-pulse generation and amplification is especially exploited. In mode-locked operation, pulses as short as 5 fs [11.953–957] and octave-spanning spectra (e.g., 600 nm to 1200 nm [11.957]) have been obtained.

Laser oscillation with reasonable efficiency has also been reported for Ti^{3+}:$BeAl_2O_4$ [11.958–961]

Table 11.22 Overview of Ti^{3+}:Al_2O_3 laser-relevant parameters [11.948, 950]

Index of refraction	1.76
Absorption cross section	$6.5 \times 10^{-20}\,\mathrm{cm}^2$ ($E \parallel c$)
Fluorescence lifetime	3.2 μs
Fluorescence bandwidth (FWHM)	≈ 200 nm
Peak emission wavelength	790 nm
Peak stimulated emission cross section	$4.1 \times 10^{-19}\,\mathrm{cm}^2$ ($E \parallel c$)
	$2.0 \times 10^{-19}\,\mathrm{cm}^2$ ($E \perp c$)
Quantum efficiency	≈ 0.9–1
Saturation fluence	$0.9\,\mathrm{J/cm}^2$
Dopant concentration	0.1% (weight)
Growth	Czochralski, heat exchange
T_m	2050 °C
Thermal conductivity	28 W/mK
Thermal lens (dn/dT)	$12 \times 10^{-19}\,\mathrm{K}^{-1}$

(Table 11.23). However, the efficiency as well as the tuning range is smaller than Ti^{3+}:Al_2O_3, thus this laser has no commercial application. In $Y_3Al_5O_{12}$ [11.962] and $YAlO_3$ [11.963, 964] crystals doped with Ti^{3+} the observed efficiencies are very low (Table 11.23). The main reason for this is excited-state absorption in the spectral region of emission and absorption [11.952, 965]. In Ti^{3+}-doped systems nonradiative decay processes due to phonon-assisted tunneling between the Jahn–Teller-split excited and ground states also occur, which prevent efficient laser oscillation, e.g., in Ti^{3+}:$Y_3Al_5O_{12}$ [11.966, 967].

Cr^{3+} *lasers.*

Basics. The Cr^{3+} ion is almost always found in octahedral coordination and its energy-level scheme is described by the Tanabe–Sugano diagram shown in Fig. 11.81. In low crystal fields, the first excited state is the $^4T_2$ level, whereas in strong crystal fields the ^2E level is the first excited state. This means, that either broadband emission ($^4T_2 \rightarrow {}^4A_2$) or narrow-line emission ($^2E \rightarrow {}^4A_2$) occurs. The absorption spectra are dominated by quartet–quartet transitions, whereas the excited-state absorption spectra are dominated either by quartet–quartet or doublet–doublet transitions depending on the total spin of the lowest excited state.

Laser characteristics. The first laser was realized in 1960 with ruby, i.e., Cr^{3+}-doped Al_2O_3 [11.968]. In ruby – due to the strong crystal field experienced by the Cr^{3+} ion – the laser oscillation occurs on the $^2E \rightarrow {}^4A_2$ transition. Thus, ruby is a three-level laser and not tunable over a wide range. Ruby is still a commercially available pulsed laser system with peak output powers in the MW range used for applications in measurement and pulsed holography. Ruby has remarkable thermo-mechanical properties which allows high peak power operation, especially in the Q-switched regime. In 1976, *Morris* et al. [11.969] realized with Cr^{3+}-doped $BeAl_2O_4$ (alexandrite) the first tunable laser based on the Cr^{3+} ion. The laser transition is $^4T_2 \rightarrow {}^4A_2$ and terminates in the higher vibronics of the ground state. Therefore, a four-level laser system is realized. To date, the alexandrite laser has found significant commercial applications in industry (e.g., marking, writing and printing), scientific research (e.g., fluorescence dynamics, fluorescence imaging, LIDAR), and

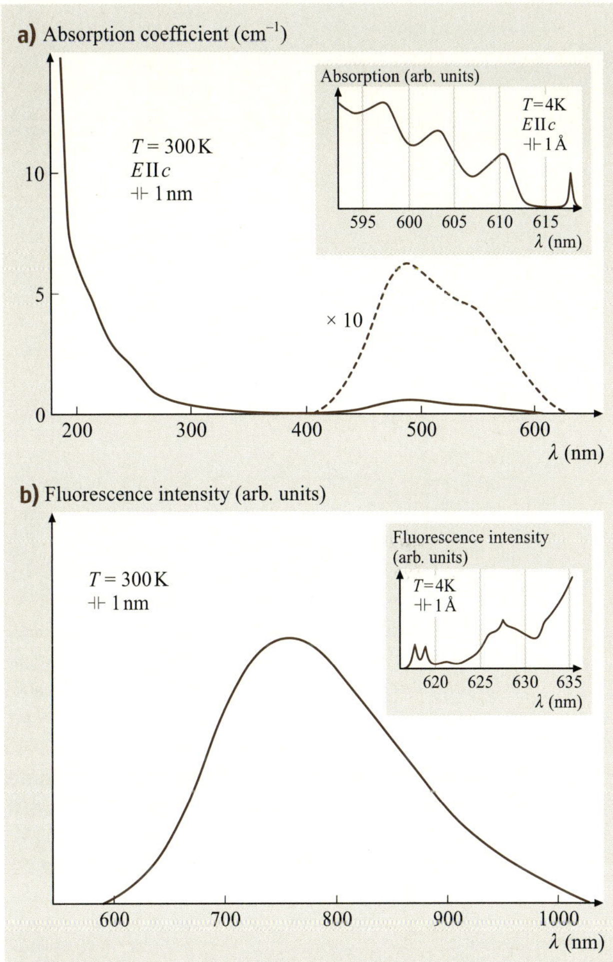

Fig. 11.80a,b Absorption (**a**) and emission (**b**) spectra at room temperature and at 4 K of Ti:sapphire. (After [11.948])

Table 11.23 Overview of other Ti^{3+}-laser materials

Crystal	λ_{laser} (nm)	Tuning range (nm)	Slope efficiency (%)	Mode of operation	Pump source	Ref.
$BeAl_2O_4$	810	730–950	15	Pulsed	SHG Q-switched Nd:YAG (532 nm)	[11.959]
	-	753–949	0.013	Pulsed	flashlamp (10 μs)	[11.961]
$Y_3Al_5O_{12}$	No details given in the reference					[11.962]
$YAlO_3$	615	-	0.3	Pulsed	SHG Q-switched Nd:$YAlO_3$ (540 nm)	[11.963, 964]

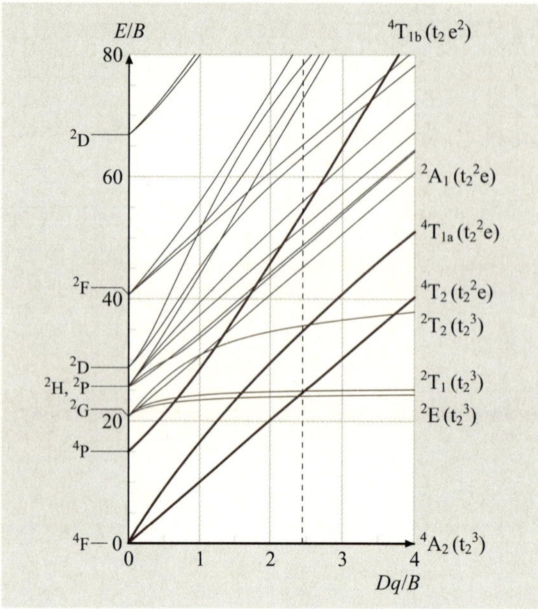

Fig. 11.81 Tanabe–Sugano diagram for octahedrally coordinated $3d^3$ ions, e.g., Cr^{3+} with $C/B = 5.5$. The *dashed vertical line* marks the border between low and high crystal field strength

medicine (e.g., hair and tattoo removal,). Its advantages are its tunability between 700 nm and 820 nm, the high heat conductivity of 23 W/mK, which allows high-power pulsed operation, and its high slope efficiency of up to 51% [11.970, 971]. Even broader laser tuning ranges with similar efficiencies were achieved with the Cr^{3+}-doped colquirite crystals $LiCaAlF_6$, $LiSrAlF_6$, and $LiSrGaF_6$ [11.972] (Table 11.25). However, these crystals suffer from poor thermo-mechanical properties (Table 11.24), which allow only relatively low pump and laser powers. The broad tuning range is exploited for mode-locked operation with pulse lengths down to 9 fs for Cr^{3+}:LiCAF [11.973]. The crystals also offer the possibility of efficient laser-diode pumping with laser diodes at wavelengths around 670–690 nm. Thus compact, mode-locked laser systems in the low-power range are possible [11.974–976]. It is interesting to note that the highest slope efficiency ever obtained with a Cr^{3+}-based laser was achieved with Cr^{3+}:$Be_3Al_2(SiO_3)_6$ (emerald). Also here, however, the thermo-mechanical properties are poor and, furthermore, the emerald crystal is very difficult to grow in laser quality. For the most efficient Cr^{3+} laser materials the relevant laser and materials parameters are listed in Table 11.24. Laser oscillation with Cr^{3+}-doped systems has been realized in more than 30 materials. In Table 11.25 an overview of reported laser systems is given [11.103, 977]. In general the tuning ranges for Cr^{3+} lasers are not as broad as for Ti^{3+}:Al_2O_3 lasers, however, they exhibit the advantage of direct diode-laser pumping around 670 nm, i. e., in the spectral region of the $^4A_2 \rightarrow {}^4T_2$ absorption.

For Cr^{3+} laser systems excited-state absorption (ESA) plays a very important role and is the main reason for the observed large differences in the laser efficiencies. Due to its energy-level scheme (Fig. 11.81), spin-allowed ESA transitions are expected either between the quartet states (for low-crystal-field hosts) or between the doublet states (in strong-field hosts). These transitions cover a wide spectral range, due to the strong electron–phonon coupling. Thus, in general, these ESA transitions overlap with the absorption and emission bands. The influence of the ESA is strongly dependent on the host lattice, i. e., on the crystal field experienced by the Cr^{3+} ion. In general, it is observed that crystals with a medium crystal field for the Cr^{3+} ion are favorable. Furthermore, in some crystals (e.g., the colquirites, alexandrite, emerald etc.) the polarization-dependent selection rules can be exploited. Cubic hosts are generally more strongly affected.

Summary and perspectives for Cr^{3+} lasers. Cr^{3+} lasers are generally interesting for laser applications. High-efficiency diode-pumped and broadly tunable continuous-wave and mode-locked laser oscillation have been obtained in a variety of crystals. Some laser systems are commercially available. However, all of these crystals have specific drawbacks: they have poor thermal mechanical properties, the growing process is difficult or the tuning range is small. Therefore Cr^{3+} lasers are at this time in general not competitive with the Ti^{3+}:Al_2O_3 laser, which offers a broader tuning range, allows shorter pulses in the mode-locked regime and has better thermo-mechanical properties. Ti^{3+}:Al_2O_3 cannot be pumped directly with diode lasers; however, advances in the frequency-doubling technique of neodymium lasers has led to efficient all-solid-state pump lasers which replaced the argon-ion laser as pump. Additionally, pump laser diodes in the spectral region between 630 nm and 700 nm with high beam quality and output power are currently not available in a satisfactory manner. Concerning the research for new Cr^{3+} laser systems one has to take into account that the Cr^{3+} ion has already been investigated in a large number of systems. In principle, new host materials with the perspective of efficient laser operation should have a medium crystal field for the Cr^{3+} ion and polarization-dependent optical prop-

Table 11.24 Materials, spectroscopic and laser parameters of the most important Cr^{3+} lasers

	$LiCaAlF_6$	$LiSrAlF_6$	Alexandrite	Ruby
Structure	Trigonal $P\bar{3}1c$	Trigonal $P\bar{3}1c$	Orthorhombic Pnma	Hexagonal $R\bar{3}c$
Lattice parameters (Å)	5.007 (a) 9.641 (c)	5.084 (a) 10.21 (c)	9.404 (a) 5.476 (b) 4.425 (c)	4.759 (a) 12.989 (c)
Typical Cr^{3+} concentration (cm^{-3})	$\approx 10^{19}-10^{20}$	$\approx 10^{19}-10^{20}$	$\approx 10^{19}-10^{20}$	$\approx 1.6\times10^{19}$
Growth	Czochralski	Czochralski	Czochralski	Czochralski
T_m (°C)	810±10	766±10	1870	2050
Density (g/cm^3)	2.99	3.45	3.69	3.98
Thermal conductivity (W/mK)	4.58 (‖ a) 5.14 (‖ c)	3.0 (‖ a) 3.3 (‖ c)	20	33 (‖ a) 35 (‖ c)
Thermal expansion ($10^{-6}\,C^{-1}$)	22 (‖ a) 3.6 (‖ c)	25 (‖ a) -10 (‖ c)	6 (‖ a) 6 (‖ b) 7 (‖ c)	6.65 (‖ a) 7.15 (‖ c)
n	1.390 (a) 1.389 (c)	1.405 (a) 1.407 (c)	$n_a = 1.7381$ (800 nm) $n_b = 1.7436$ (800 nm) $n_c = 1.7361$ (800 nm)	1.763 (o) 1.755 (e)
$dn/dT(10^{-6}/K)$	-4.2 (‖ a) -4.6 (‖ c)	-2.5 (‖ a) -4 (‖ c)		13.6 (o) 14.7 (e)
σ_{em} ($10^{-20}\,cm^2$)	1.3 (π)	4.8 (π)	0.7	2.5
τ_{em}(300 K) (μs)	170	67	260	3000
$\sigma_{em}\tau(10^{-24}\,cm^{-2}s^{-1})$	2.2	3.2	1.8	75
$\lambda_{peak,em}$ (nm)	763	846	697	694.3
$\Delta\lambda$ (nm)	≈ 120	≈ 200	≈ 75	-
$\Delta\lambda/\lambda_{peak,em}$	≈ 0.16	≈ 0.24	≈ 0.11	-

erties that could help to avoid or at least significantly reduce excited-state absorption in the spectral region of the emission. Furthermore, good thermo-mechanical properties are required.

Cr^{4+} lasers.

Basics. Cr^{4+}-doped crystals have been of interest as tunable, room-temperature laser materials since the late 1980s. In a variety of materials laser oscillation in different operation schemes has been achieved, see Table 11.26 and Table 11.28, where an overview of the obtained laser results is given. The energy-level scheme of the Cr^{4+} ion in crystals can be described with the Tanabe–Sugano diagram shown in Fig. 11.82. The absorption spectra are dominated by the three spin-allowed transitions between the $^3A_2$ ground state and the $^3T_2(^3F)$, $^3T_1(^3F)$, and $^3T_1(^3P)$ excited states. Usually, the energy levels are strongly crystal field dependent and split, thus the absorption spectra of different materials differ significantly. In all materials investigated thus far, broadband emission due to the transition between the $^3T_2$ excited state and the $^3A_2$ ground state is observed. Referring to the Tanabe–Sugano diagram and the indicated area of crystal field values for the Cr^{4+} ion, one would expect – at least in some materials – narrow-line emission. However, due to lattice relaxation and crystal field splitting of the excited state, the $^3T_2$ or one of its crystal field components becomes lower than the ^1E level.

The most efficient laser oscillation of Cr^{4+}-doped materials was realized in Mg_2SiO_4 and $Y_3Al_5O_{12}$ (YAG). The laser data and main spectroscopic data of these materials are listed in Table 11.26 and Table 11.27. A variety of further host materials for the Cr^{4+} ion were also investigated, but with either low efficiency (Table 11.28) or without realization of laser oscillation. A detailed overview of Cr^{4+}-doped systems is given in [11.933].

In all Cr^{4+}-doped materials investigated thus far, there are two main drawbacks for efficient laser oscillation or laser oscillation at all: excited-state absorption and nonradiative decay. Investigation of excited-state

Table 11.25 Free-running laser wavelengths, tuning ranges, laser temperatures, slope efficiencies, operation modes and output power/output energy of Cr^{3+}-laser materials. (CW: continuous wave, p: pulsed, SHG: second harmonic, fl: flash lamp, dc: duty cycle, QS: Q-switched, g-sw: gain-switched, ML: mode locked, * discontinuously)

Host material	Wavelength (nm)	Tuning range (nm)	T (K)	η (%)	Mode	P_{out}/E_{out}	Ref.
$Be_3Al_2(SiO_3)_6$	684.8		300		p (SHG QS Nd:YAG)		[11.988]
	757.4	751–759.2	300		p (fl)	6.8 mJ	[11.989]
	765	728.8–809.0	300	34	CW (Kr^+)	≈ 330 mW	[11.990]
	768	720–840	300	64	q-CW (3% DC)	1.6 W	[11.991, 992]
$LiCaAlF_6$	780	720–840		52.4	CW (5% DC) Kr^+	850 mW	[11.972]
	780	720–840	300	52	CW (5% DC) Kr^+		[11.993]
	780			1.55	p – fl	1.8 J	[11.993]
$LiSrGaF_6$	820	-	300	51	CW, Kr^+	1 W	[11.994]
		800–900	300		QS (10 kHz),	12 μJ	[11.995]
					laser diode		[11.996]
$BeAl_2O_4$	679.9		77				[11.997]
	680.3		77		p (fl)		[11.998, 999]
	680.3	700–800	300		p (fl)		[11.1000]
	750	701–794	300		p (fl)	500 mJ	[11.1000, 1001]
	765	744–788	300	≈ 0.7	q-CW (ac Hg-lamp)	6.5 W	[11.1002]
		745–785	300	2.3	CW (DC Hg-,Xe-lamp)	20 W	[11.1003]
	750	700–820	300		p (fl 125 Hz)	150 W	[11.1004]
	755		300	1.2	CW	60 W	[11.1004]
	750		300		ML (38 ps)		[11.1004]
		752–790	300–583		p (fl), temp. tuning		[11.1005]
		700–820	300–583				[11.970, 971, 997, 1006–1011]
		701–818	300				[11.1000]
		744–788	300				[11.1002]
		701–818	300	2.5	fl	35 W / 5 J	[11.970]
	680.4		300	0.15	fl-QS (20 ns)	500 mJ	[11.970, 1012]
	680.4		300	0.15	fl	400 mJ	[11.970]
	752	726–802	300	51	CW, Kr^+	600 mW	[11.971]
	752	700–820					[11.969]
	753		300	63.8	dye, 645 nm	150 mW	[11.1013]
	753		300		laser diode, ≈ 640 nm	25 mW	[11.1013]
	765		300	28	QS SHG Nd: GVO, 671 nm (80 kHz)	150 mW	[11.1014]
$LiSrAlF_6$	825	780–920	300	36	CW, Kr^+	650 mW	[11.993, 1015]
	865	815–915	300		a-ML (30 ps, 160 fs)	3.5 mW	[11.1016]
	845	780–1010	300	5	p (fl)	2.7 J	[11.1017]
	834		300		CW (laser diode)	20 mW	[11.1018]
	870	858–920	300		laser diode, electronically tuned	4.3 mW	[11.1019]
	849	810–860	300		laser diode	43 mW	[11.1020]
			300		p (fl, 5 Hz)	44 W/ 8.8 J	[11.1021]

Table 11.25 (continued)

Host material	Wavelength (nm)	Tuning range (nm)	T (K)	η (%)	Mode	P_{out}/E_{out}	Ref.
$ScBO_3$	843	787–892	300	25	CW (3%), Kr^+	250 mW	[11.1022]
			300	26	CW (10%), Kr^+	275 mW	[11.972]
$Gd_3Sc_2Ga_3O_{12}$	777	745–805	300	11	CW (1:50 DC)	60 mW	[11.1023]
	785	742–842	300	28	quasi-CW	200 mW	[11.1024–1026]
	790		300	1	p (10 μs, dye)	10 μJ	[11.1027]
		766–820	300	0.06	p – fl	20 mJ	[11.1028]
			300	0.02	p – fl	10 mJ	[11.1029]
				0.57	p – fl	≈ 70 mJ	[11.1030]
$Na_3Ga_2Li_3F_{12}$	791	741–841	300	18.4	CW		[11.54]
$Y_3Sc_2Al_3O_{12}$	769			9	q-CW, Kr^+	50 mW	[11.1031]
$Gd_3Sc_2Al_3O_{12}$	780	750–800	300	0.24	p – fl	200 mJ	[11.1029]
	784	765–801	300	0.12	p – fl	110 mJ	[11.1032]
	784		300	18.5	CW Kr^+	90 mW	[11.1025, 1026, 1033]
	780	750–810	300	0.38	p – fl	260 mJ	[11.1034]
			300		QS (p – fl)	30 mJ	[11.1034]
	784	735–820	300	19	CW, Kr^+	200 mW	[11.1031]
	780	750–803	300	0.24	p – fl	206 mJ	[11.1031]
$SrAlF_5$	921, 935	852–947	300	3.6	q-CW (DC 3%), Kr^+	35 mW	[11.1035]
	910			15	q-CW (DC 2%), Kr^+		[11.1036]
	932	825–1011					[11.1037]
	930	825–1010	300	10	Kr^+		[11.1038]
$KZnF_3$	810, 826		300	1	p – dye (0.5 μs)		[11.1039, 1040]
	790–826	775–825	20–260	0.1	CW, Kr^+		[11.1039, 1040]
		785–865	300	14	CW, Kr^+	85 mW	[11.1041]
		785–865			p – ruby		[11.1042]
		780–845	300	3	CW, Kr^+	55 mW	[11.1043]
	820	766–865	300	14			[11.1044]
$ZnWO_4$		980–1090	77	13	CW	110 mW	[11.1026, 1045]
			300		p – dye		[11.1045]
$La_3Ga_5SiO_{14}$	960	862–1107	300	7.6	CW (3% DC)	80 mW	[11.1046]
		815–1110	300	10	p	10 mJ	[11.1047]
	968						[11.1048]
$Gd_3Ga_5O_{12}$	769	-	300	10	quasi-CW		[11.1023, 1024, 1026]
$La_3Ga_{5.5}Nb_{0.5}O_{14}$	1040	900–1250	300	5	p	10 mJ	[11.1047, 1049]
$Y_3Ga_5O_{12}$	740	-		5	quasi-CW		[11.1023, 1026]
$Y_3Sc_2Ga_3O_{12}$	750	-		5	quasi-CW		[11.1023, 1026]
$La_3Lu_2Ga_3O_{12}$ 830	790–850			3	quasi-CW		[11.1024, 1026, 1048]

Table 11.25 (continued)

Host material	Wavelength (nm)	Tuning range (nm)	T (K)	η (%)	Mode	P_{out}/E_{out}	Ref.
MgO	830	824–878*		2.3	CW, Ar^+	48 mW	[11.1050–1052]
$Al_2(WO_4)_3$	800				q-CW, Kr^+		[11.1053]
$BeAl_6O_{10}$	820	780–920	300	≈ 0.03	p – fl	6 mJ	[11.1054]
	834	795–874	300	-	p – SHG Nd:YAG		[11.1055]
Al_2O_3	692.9 (R2)		300		p – fl		[11.1056]
	693.4		300		p – fl		[11.1057]
	693.4		77		CW, Hg-lamp	4 mW	[11.1058]
			77	12	CW, Ar^+	42 mW	[11.1059]
	694.3 (R1)		300				[11.968, 1060–1065]
	700.9 (N2)		77				[11.1066]
	704.1 (N1)		77				[11.1066]
	767		300				[11.1067]
$Y_3Al_5O_{12}$	687.4	not tunable	≈ 77				[11.1068]
$LiSr_{0.8}Ca_{0.2}AlF_6$	835	750–950	300	1.25	p – fl	1.2 J	[11.1069]
	847	783–945	300		CW, Kr^+	300 mW	[11.1070]
$LiSrCrF_6$	890		300	33	q-CW, TiSa (DC: 2%)	200 mW	[11.1071]
$ScBeAlO_4$	792	740–828	300	31	CW, Kr^+		[11.1072]
$La_3Ga_5GeO_{14}$		880–1220	300	5	p	8 mJ	[11.1047, 1073]
$Sr_3Ga_2Ge_4O_{14}$		895–1150	300	3	p	4 mJ	[11.1047, 1073]
$Ca_3Ga_2Ge_4O_{14}$		870–1210	300	6	p	16 mJ	[11.1047, 1049]
$La_3Ga_{5.5}Ta_{0.5}O_{14}$		925–1240	300	5	p	6 mJ	[11.1047, 1049]
$(Ca, Gd)_3(Ga, Mn, Zr)_5O_{12}$		774–814	300	12	p – ruby	170 mJ	[11.1074]
	≈ 777		300	0.08	p – fl	40 mJ	[11.1074]
$LaSc_3(BO_3)_4$	934		300	0.65	q-CW, Kr^+ (DC: 10%)	3 mW	[11.1075]
$LiInGeO_4$		1150–1480	300		p (g-sw. $Ti:Al_2O_3$)		[11.1076]
$LiScGeO_4$		1220–1380	300		p (g-sw. $Ti:Al_2O_3$)		[11.1076]
$Li:Mg_2SiO_4$	1121	1030–1180	300		p (QS $Cr^{3+}:BeAl_2O_4$)		[11.1077]
	1120, 1130, 1140		285	1.27	CW (Ar^+)	5.5 mW	[11.1077]

absorption has been performed on garnets [11.978–980], forsterite [11.981–984], cunyite [11.985], silicates [11.982] and Wurtzite-type crystals [11.984, 986, 987]. The nonradiative decay via multiphonon relaxation leads to quantum efficiencies far below 100% at room temperature; see the overview given in [11.933]. The preparation of Cr^{4+}-doped crystals requires for most materials special conditions before, during or after the growth process. The Cr^{4+} ion is not as stable in its valence state as Cr^{3+}. Therefore, there is the tendency for incorporation of chromium ions in different valencies in the crystals, especially for materials that do not exhibit an appropriate tetravalent tetrahedrally coordinated lattice site. This is, e.g., the case for $Y_3Al_5O_{12}$, therefore here additional codoping with divalent cations (Mg, Ca) is necessary. But also in Mg_2SiO_4, Cr^{3+} is incorporated into the Mg lattice. No traces of Cr^{3+} were observed in these materials, which do not exhibit an ap-

Table 11.26 Overview of laser results obtained with Cr^{4+}:YAG and Cr^{4+}:Mg_2SiO_4. (CW: continuous wave, lp: long pulse pumped, g-sw: gain switched, DP: diode pumped, ML: mode locked, CF: crystalline fiber)

Crystal	λ_{laser} (nm)	Output	η_{sl} (%)	Tuning range	Mode of operation	Ref.
$Y_3Al_5O_{12}$	1430	7.5 mJ	22	1350–1530	g-sw (55 ns)	[11.1078–1083]
	1450	1900 mW	42	1340–1570	CW ($T = 3\,°C$)	[11.1081, 1083–1085]
	1420	58 mJ	28	1309–1596	lp (200 μs)	[11.1082, 1083, 1086, 1087]
					ML	[11.1088, 1089]
	1440	20 mW	5	1396–1482	CW-ML (26 ps)	[11.1090]
	1520	360 mW	8	1510–1530	CW-ML (120 fs)	[11.1091]
	1510	50 mW	< 1	1490–1580	CW-ML (70 fs)	[11.1092]
	1540	-	-	-	CW-ML (53 fs)	[11.1093]
	1569	30 mW			DP-ML (65 fs)	[11.1094, 1095]
	1450	400 mW			Nd – YVO_4, ML (20 fs)	[11.1096]
	1470	80 mW	5.5	1420–1530	DP	[11.1097]
	1420	150 mW	1.9	-	DP-CF	[11.1098]
	1440				Intracavity Nd:YAG	[11.1099]
Mg_2SiO_4					CW	[11.1100]
	1242		38	-	CW (duty cycle 1:15)	[11.1101, 1102]
		1.1 W	26	-	CW	[11.1103]
					DP	[11.1104, 1105]
					lp	[11.1106]
					flashlamp pumped	[11.1107]
		4.95 mJ	-	1206–1250	flashlamp pumped	[11.1108]
					g-sw	[11.1106, 1109–1113]
	1235			1170–1370	g-sw	[11.1114]
		370 mW	13	1173–1338	g-sw (1.5 kHz/10kHz)	[11.1115]
					ML	[11.1116–1122]
		300 mW	-	-	ML (25 fs)	[11.1122]
	1300				Nd:YAG, ML (14 fs)	[11.1123]
	1260	10 pJ	-	-	DP-ML (1.5 ps)	[11.1124]
	1260	10 mW	5	1236–1300	CW-DP ($T = -10\,°C$)	[11.1104]

propriate lattice site for the Cr^{3+} ion, e.g., Ca_2GeO_4 and Y_2SiO_5. However, the laser results obtained with $Y_3Al_5O_{12}$ and Mg_2SiO_4 indicate that the laser efficiency is not necessarily affected by the presence of Cr^{3+}.

Laser characteristics. In Table 11.26 an overview of the laser results obtained with Cr^{4+}:YAG and Cr^{4+}:Mg_2SiO_4, the materials in which the most efficient laser operation has been obtained, is given. In YAG, the highest slope efficiency obtained thus far in the continuous-wave regime is 42% [11.1085]. The corresponding input–output curve is shown in Fig. 11.83a. Attempts to improve the laser efficiency by changing the crystals composition, i. e., by substituting Lu for Y on the dodecahedral site or Sc for Al on the octahedral site were not successful [11.1086, 1087]. The efficiency decreased significantly. The reason is mainly excited-state absorption, but the lower crystal quality and increased nonradiative rate also contribute [11.1125]. A remarkable characteristic – on first sight – is that the Cr^{4+}:YAG laser oscillates polarized parallel to one of its main crystallographic axes. The laser output was highest when the pump beam of a Nd:YAG laser operating at 1064 nm and propagating along the [001]-axis of the Cr^{4+}:YAG crystal is polarized parallel to one of the crystallographic $< 100 >$-axes of the Cr^{4+}:YAG crystal, and was lowest when its polarization was parallel to one of the $< 110 >$-axes (Fig. 11.83b). The Cr^{4+}:YAG laser output is polarized and maintains its polarization while rotating the pump beam polarization. When the pump beam polarization is parallel to one of the $< 110 >$-axes, the Cr^{4+}:YAG laser polarization

Table 11.27 Parameters of Cr^{4+}-doped $Y_3Al_5O_{12}$ and Mg_2SiO_4

	$Y_3Al_5O_{12}$ (YAG)	Mg_2SiO_4 (Forsterite)
Structure	Ia3d (O_h^{10})	Pbnm (D_{2h}^{16})
Hardness	8.25–8.5	7
Site symmetry	S_4	m
Growth	Czochralski, divalent codopant required	Czochralski
T_m	(1930±20) °C	(1890±20) °C
Cr^{4+} concentration	$\approx 10^{17}-10^{18}\,cm^{-3}$	$\approx 10^{18}-10^{19}\,cm^{-3}$
Thermal conductivity	0.13 W/cmK	0.08 W/cmK
Refractive index (λ_{peak})	1.81	1.669 (a) 1.651 (b) 1.636 (c)
dn/dT (undoped)	$7.7-8.2\times10^{-6}$ /K	9.5×10^{-6} /K
Density	4.56 g/cm^3	3.22 g/cm^3
σ_{abs} (1064 nm)	$\approx 6.5\times10^{-18}\,cm^2$	$\approx 5.0\times10^{-19}\,cm^2$
σ_{em} (λ_{peak})	$\approx 3.3\times10^{-19}\,cm^2$	$\approx 2.0\times10^{-19}\,cm^2$
σ_{ESA} (λ_{peak})	$< 0.3\times10^{-19}\,cm^2$	$< 0.2\times10^{-19}\,cm^2$
τ_{em} (300 K)	4.1 μs	3.0 μs
$\sigma_{em}\tau$	$1.35\times10^{-24}\,cm^{-2}s^{-1}$	$0.6\times10^{-24}\,cm^{-2}s^{-1}$
Quantum efficiency	≈ 0.2	≈ 0.16
$\lambda_{peak,em}$	1380 nm	1140 nm
$\Delta\lambda$	≈ 300 nm	≈ 250 nm
$\Delta\lambda/\lambda_{peak,em}$	≈ 0.22	≈ 0.22

switches [11.1084]. This characteristic can be explained by the crystal structure, the location of the Cr^{4+} ions and the local symmetry they experience. For details, see [11.1126, 1127].

With the Cr^{4+}:Mg_2SiO_4 laser, slope efficiencies of up to 38% [11.1102] and output powers around 1.1 W [11.1103] in continuous-wave operation at room temperature have been realized (Fig. 11.83c. For crys-

Table 11.28 Overview of other Cr^{4+}-doped laser materials. (CW: continuous wave, lp: long pulse pumped, g-sw: gain switched, DP: diode pumped. * laser-active center assigned to Cr^{3+} in [11.1076], Table 11.24)

Crystal	λ_{laser} (nm)	Output	η_{sl} (%)	Tuning range	Mode of operation	Ref.
$Y_3Sc_xAl_{5-x}O_{12}$ (YSAG)	1498 ($x = 0.5$)	23 mJ	10	1394–1628	lp (100 μs)	[11.1087]
	1548 ($x = 1.0$)	4.5 mJ	3	1464–1604	lp (100 μs)	[11.1087]
	1584 ($x = 1.5$)	0.9 mJ	0.5		lp (100 μs)	[11.1087]
$Lu_3Al_5O_{12}$	Not given	50 mW	1.5	-	quasi-CW	[11.1128, 1129]
Ca_2GeO_4						[11.1104, 1130–1132]
	1400	0.4 mJ	6.1	1348–1482	g-sw ($T = 0\,°C$)	[11.1130, 1133]
	1410	20 mW	8.5	1390–1475	CW-DP ($T = -10\,°C$)	[11.1104]
$LiScGeO_4^*$	1300	0.1 mJ	3	1220–1380	g-sw	[11.1134]
Y_2SiO_5						[11.1135, 1136]
	1304	20 mW	0.4	-	quasi-CW (1:8)	[11.1084, 1137]
	1348	0.55 mJ	0.4	-	lp (200 μs)	[11.1084, 1137]
$LiNbGeO_5$	-	-	-	1320–1430	g-sw (110 K)	[11.1138–1140]
$CaGd_4(SiO_4)_3O$	1370	37 μJ	≈ 1		g-sw	[11.1141]
$SrGd_4(SiO_4)_3O$	1440				g-sw	[11.1141]

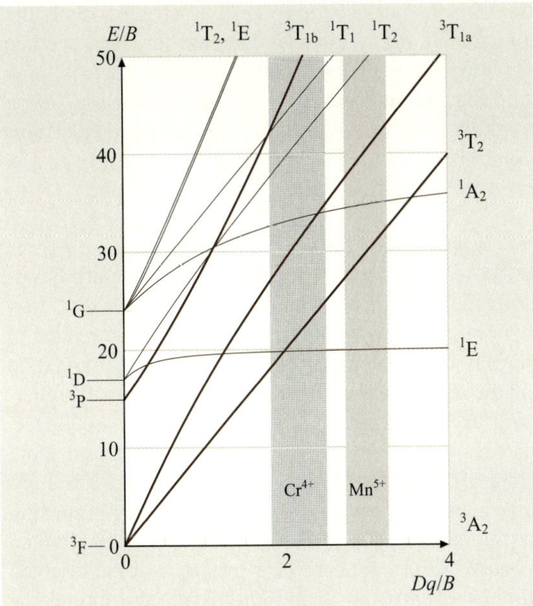

Fig. 11.82 Tanabe–Sugano diagram for tetrahedrally coordinated $3d^2$ ions. The *marked area* correspond roughly to the regions of Dq/B values of Cr^{4+} and Mn^{5+}. For simplicity, C/B was set to 5.6, although it differs for different systems

tals from the forsterite group the attempts to obtain better laser results by substituting the constituent ions of the host lattice were also not successful. A variety of crystals were investigated [11.1142–1144] (see the overview given in [11.933]) but only for Ca_2GeO_4 was laser oscillation obtained. For Ca_2GeO_4 excited-state absorption on the laser wavelength is detrimental [11.985].

Laser oscillation was also obtained for Cr^{4+}-doped Y_2SiO_5, $LiGeNbO_5$, $CaGd_4(SiO_4)_3O$, and $SrGd_4(SiO_4)_3O$. The low efficiencies are probably due to excited-state absorption and the high nonradiative decay rate.

Summary and outlook for Cr^{4+} lasers. Cr^{4+}:YAG and Cr^{4+}:Mg_2SiO_4 are efficient and broadly tunable laser systems for the infrared spectral range including the very interesting region for telecommunication applications around 1.3 μm and 1.55 μm. Mode-locked operation with pulse lengths as short as 20 fs for Cr^{4+}:YAG [11.1096] and 14 fs for Cr^{4+}:Mg_2SiO_4 [11.1123] were obtained. Also direct diode-pumped laser operation was realized, however, with lower efficiencies [11.1094, 1097]. For both systems, power handling is a problem and the ion

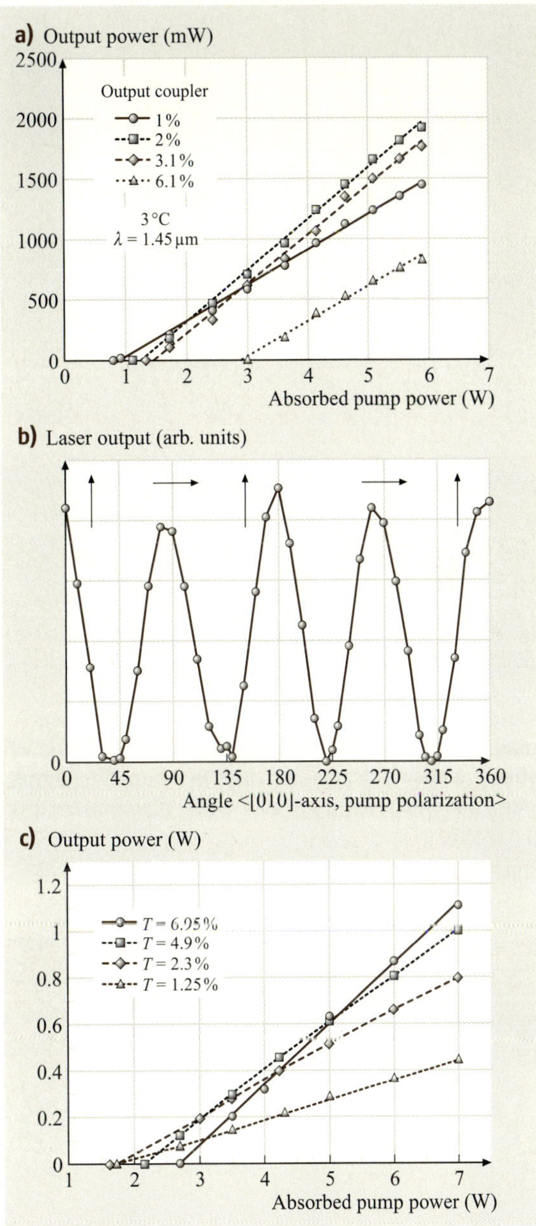

Fig. 11.83 **(a)** Input–output characteristic of a Cr^{4+}:YAG-laser (after [11.1085]), **(b)** Polarization dependence of a Cr^{4+}:YAG laser (after [11.933, 1084]), **(c)** Input–output characteristic of a Cr^{4+}:Mg_2SiO_4 laser (after [11.1103])

concentration in both materials is rather low, yielding a low absorption efficiency of the pump light.

The main obstacle for Cr^{4+} laser materials is excited-state absorption at the emission wavelength, which

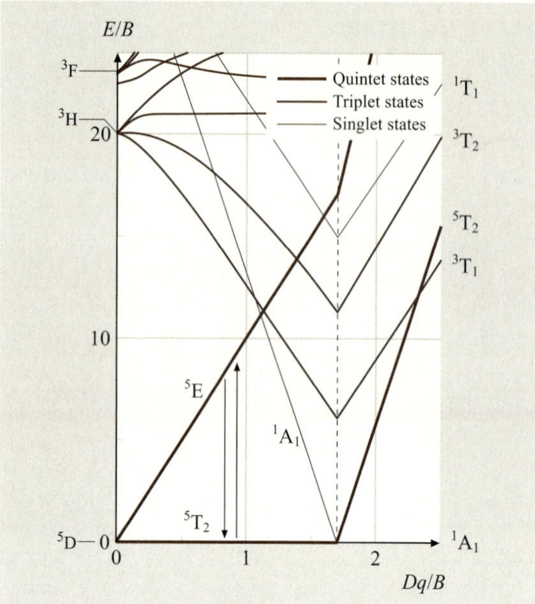

Fig. 11.84 Tanabe–Sugano diagram for tetrahedrally coordinated ions with $3d^4$ configuration, $C/B = 4$

is present in practically all crystals. Its influence is small in systems with an advantageous energy-level scheme and/or a crystal structure that supports strong polarization-dependent selection rules. A second but less important point concerning Cr^{4+} laser materials is the nonradiative decay rate. To decrease this rate, crystals with low-energy phonons and/or small electron–phonon coupling should be used as host materials, e.g., those in which silicon is substituted by germanium or aluminum is substituted by gallium. However, attempts in this direction have not been successful thus far.

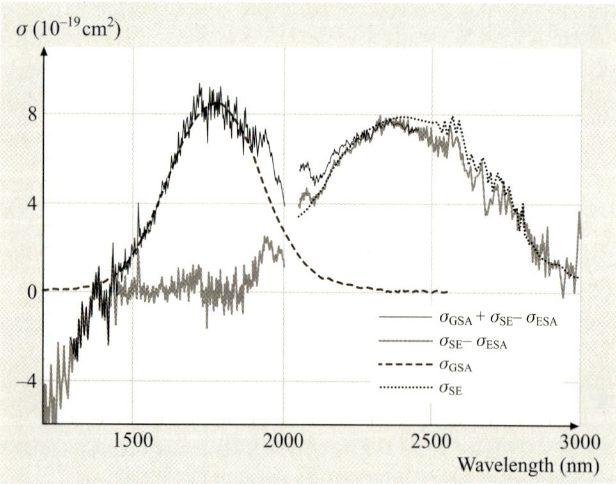

Fig. 11.85 σ_{GSA} (*dashed line*), σ_{SE} (*dotted line*), $(\sigma_{GSA} + \sigma_{SE} - \sigma_{ESA})$ (*thin solid line*), and $(\sigma_{SE} - \sigma_{ESA})$ spectrum (*thick solid line*) of Cr^{2+}:ZnSe at room temperature. The sharp structure beyond 2500 nm is attributed to the water absorption in air and the following normalization process. (After [11.1145])

Cr^{2+} lasers.

Basics. The energy-level scheme of tetrahedrally coordinated Cr^{2+} is shown in Fig. 11.84. The Cr^{2+} ion experiences a low crystal field, e.g., in ZnSe the value of Dq/B is about 0.9. Thus, the $^5T_2$ level is the ground state and the ^5E level is the first excited state, whilst all higher-lying levels are triplet and singlet states. In consequence, the $^5E \to {}^5T_2$ emission is a spin-allowed transition, while all interionic excited-state transitions are spin-forbidden. Such systems are in general very promising for the realization of efficient tunable laser oscillation, because even in the case of a spectral overlap between stimulated emission and excited-state absorption, the transition probabilities for the latter are expected to be about a factor of 10 smaller.

Cr^{2+}-doped chalcogenide crystals have been shown to be efficient and broadband tunable solid-state lasers for the infrared spectral range between 2 μm and 3 μm. Pulsed, continuous-wave, mode-locked and diode-pumped laser operation have been demonstrated in recent years. Possible applications of these mid-infrared lasers include scientific research, remote sensing, trace-gas analysis, medicine, biology, materials processing, and ultrashort-pulse generation.

Materials. The choice of host materials for the Cr^{2+} laser ion is limited due to special conditions. First, the materials have to exhibit a tetrahedrally coordinated lattice site. A divalent lattice site is also preferably, because otherwise a charge-compensation mechanism would have to be established in the lattice. Furthermore, host crystals with low phonon frequencies have to be chosen in order to decrease the possibility of nonradiative decay via multiphonon relaxation. Chalcogenide crystals, with their tetrahedrally coordinated divalent cation lattice sites and with phonon energies lower than $400\,cm^{-1}$, are thus very suitable with respect to the realization of efficient broadband emission in this mid-infrared spectral range. In Table 11.29 some material parameters of the investigated chalcogenide crystals are given in comparison to the data for Al_2O_3. The thermal conductivities of the chalcogenides are rather high and comparable to the values for Al_2O_3. However, the

Table 11.29 Materials parameter of some chalcogenide crystals suitable for Cr^{2+}. The data for $Y_3Al_5O_{12}$ are given for comparison. (*: for CdTe). (After [11.1146–1148]) w: Wurtzite, z: Zincblende

	ZnS	ZnSe	$Cd_{1-x}Mn_xTe$	CdSe	$Y_3Al_5O_{12}$
Structure	Wurtzite: hexagonal Zincblende: cubic	Zincblende: cubic	Zincblende: cubic	Wurtzite: hexagonal	Garnet: cubic
Site symmetry	C_{3v} (hex.) T_d (cubic)	T_d	T_d	C_{3v}	D_2, C_{3i}, S_4
Growth	Vertical Brigdeman	Vertical Bridgeman	Vertical Brigdeman	Vapor transport	Czochralski
T_m (°C)	w: 1700 z: 1020	1525	1070–1092	1250–1350	1930
Hardness (Knoop)	w: 210–240 z: 150–160	130	45*	70	1250
Refractive index	2.29	2.48	2.75	2.57	1.8
Thermal conductivity (W/mK)	w: 17 z: 27	18	≈ 2 (6.2*)	6.2	10
dn/dT (10^{-6}/K)	w: 46	70	107*		9
Transmission range (μm)	w: 0.4–17 z: 0.4–14	0.5–18	1–28*	0.8–18	0.2–10

main disadvantages of the chalcogenide crystals are the growth techniques, which are Bridgman or vapor transport, which usually lead to a lower crystal quality than in the case of growth by the Czochralski method, and the high values for dn/dt, which lead to strong thermal lensing during laser operation, especially in the case of high-power operation.

Spectroscopy. The spectroscopy of Cr^{2+} ions in chalcogenide crystals has been thoroughly investigated in the past [11.1154–1162]. Tetrahedrally coordinated Cr^{2+} ions exhibit a broad band in the absorption spectrum due to the $^5T_2 \rightarrow {}^5E$ transition in the infrared spectral range with a maximum around 1.7–1.9 μm. The emission spectrum due to the $^5E \rightarrow {}^5T_2$ transition is also broad and occurs between 2 μm and 3 μm. The absorption and emission cross sections for the Cr^{2+}-doped chalcogenides are on the order of $\approx 10^{-18}$ cm^2; these are values expected for tetrahedrally coordinated transition-metal ions and are larger than those for Ti^{3+}:Al_2O_3 [11.948]. The room-temperature absorption and emission spectra are shown in Fig. 11.85. The emission lifetimes at room temperature are on the order of several μs and the emission quantum efficiencies are close to unity. The $\sigma_{em}\tau$ product gives an indication of the expected laser threshold, because $P_{thr} \propto (\sigma_{em}\tau)^{-1}$. The values for the Cr^{2+}-doped chalcogenides are higher than in the case of Ti^{3+}:Al_2O_3, so that generally lower laser thresholds

Table 11.30 Overview about the spectroscopic characteristics of Cr^{2+} doped chalcogenide crystals. Data for Ti^{3+}:Al_2O_3 are given for comparison

	ZnS [11.1149–1151]	ZnSe [11.1149–1151]	ZnTe [11.1149–1151]	$Cd_{0.85}Mn_{0.15}Te$ [11.1152]	$Cd_{0.55}Mn_{0.45}Te$ [11.1153]	CdSe [11.1154]	Ti^{3+} : Al_2O_3 [11.948]
σ_{abs} (10^{-20} cm^2)	52	87	123	≈ 270	≈ 170	300	6.5
σ_{em} (10^{-20} cm^2)	75	90	188	270	170	200	45
τ_{em} (300 K) (μs)	8	9	3	1.4	4.8	6	3
$\sigma_{em}\tau$ (10^{-22} cm^{-2}s^{-1})	6.0	8.1	5.6	3.8	8.2	12.0	1.4
η	≈ 0.73	≈ 1	≈ 1	≈ 0.38	≈ 1	≈ 1	≈ 0.9
$\lambda_{peak,em}$ (nm)	2300	2300	2400	2250	2480	2200	800
$\Delta\lambda$ (nm)	≈ 780	1000	≈ 900	≈ 500	770	≈ 550	300
$\Delta\lambda/\lambda_{peak,em}$	≈ 0.34	≈ 0.43	≈ 0.38	≈ 0.22	0.31	≈ 0.25	0.38

there is the possibility to look for different host materials for the tetrahedral Cr^{2+} ion, e.g., $ZnGa_2S_4$, $ZnGa_2Se_4$, $CaGa_2S_4$, and $CaGa_2Se_4$.

Other transition-metal-ion lasers.

V^{2+} lasers. The V^{2+} ion is isoelectronic to the Cr^{3+} ion. Therefore, its energy-level scheme can also be described with the Tanabe–Sugano diagram shown in Fig. 11.81. The absorption spectra of octahedrally coordinated V^{2+} is similar to that of Cr^{3+}, but red-shifted due to the lower valence of the V^{2+} ion. The spectra are dominated by the three broad spin-allowed bands due to transitions between the $^4A_2(^4F)$ ground state and the $^4T_2(^4F)$, $^4T_1(^4F)$, and $^4T_1(^4P)$ excited states. The emission spectra consist of a broad band due to the $^4T_2(^4F) \rightarrow {}^4A_2(^4F)$ transition, also shifted to longer wavelengths compared to the emission spectra of Cr^{3+}. Laser oscillation with V^{2+} on the $^4T_2(^4F) \rightarrow {}^4A_2(^4F)$ transition was realized only in MgF_2 [11.1200–1202] and $CsCaF_3$ [11.1203, 1204]. An overview of the obtained laser results is given in Table 11.31. The laser efficiencies are very low. *Payne* et al. [11.44, 1205] and *Moncorgé* et al. [11.1206] found that ESA is the dominant loss mechanism for laser oscillation. In some materials nonradiative decay also competes with the emission, leading to small emission quantum efficiencies.

Ni^{2+} lasers. The energy-level scheme of octahedrally coordinated Ni^{2+} in crystals can be described with the Tanabe–Sugano diagram shown in Fig. 11.86. In the absorption spectra, three bands according to the spin-allowed transitions from the $^3A_2(^3F)$ ground state to the $^3T_2(^3F)$, $^3T_{1a}(^3F)$, and $^3T_{1b}(^3P)$ excited states are observed. The laser transition of Ni^{2+} lies in the infrared spectral range due to the $^3T_2(^3F) \rightarrow {}^3A_2(^3F)$ transition. Its spectral position is strongly wavelength dependent (Table 11.33). The emission lifetime is typically on the order of ms and the emission quantum efficiency is near unity at room temperature in most materials.

Despite these advantageous spectroscopic data, laser oscillation with Ni^{2+} was obtained only at temperatures below 240 K and only in a few materials (Table 11.33). The absence of laser oscillation at room temperature can be explained by excited-state absorption (ESA), which overlaps with the spectral range of emission. Detailed

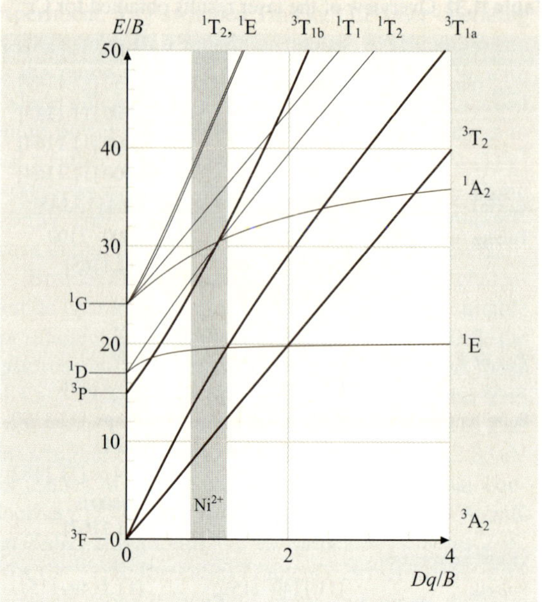

Fig. 11.86 Tanabe–Sugano diagram for the d^8 electron configuration. The *dashed area* corresponds to the Dq/B values of Ni^{2+}

measurements of the ESA for several Ni^{2+}-doped crystals were performed by *Koetke* et al. [11.1207, 1208]. With increasing temperature the excited-state absorption and ground-state absorption bands become broader and thus overlap to a larger extent with the stimulated emission. As a result, the spectral region where $\sigma_{eff} = \sigma_{se} - \sigma_{ESA} > 0$ becomes narrower. Furthermore, the losses due to ground-state absorption increase. However, at low temperatures output powers up to 10 W and slope efficiencies up to 57% were obtained [11.1202].

Mn^{5+} lasers. The Mn^{5+} ion incorporates into crystals mainly at tetrahedrally coordinated lattice sites. Its energy-level scheme can be described with the Tanabe–Sugano diagram shown in Fig. 11.82. Compared to Cr^{4+}, the crystal field is higher due to the higher valence state, thus the Mn^{5+} ion exhibits narrow-line emission due to the $^1E(^1D) \rightarrow {}^3A_2(^3F)$ transition. Optical properties of Mn^{5+} ions in solids have been under investigation for more than 30 years [11.1209–1217].

Table 11.32 Overview of V^{2+}-laser materials

Host material	λ_{laser} (nm)	Tuning range (nm)	T (K)	Mode of operation	P_{out}/E_{out}	η (%)	Ref.
MgF_2	1121	1070–1150	77	pulsed			[11.1200–1202]
$CsCaF_3$	1280	1240–1330	80	CW	≈ 15 μW	0.06	[11.1203, 1204]

Table 11.33 Overview of the results of Ni^{2+} lasers

Host material	λ_{laser} (nm)	Tuning range (nm)	T (K)	Mode of operation	P_{out}/E_{out}	η (%)	Ref.
MgO	1314.4		77	Pulsed			[11.1200]
	1318		80	CW	10 W	57	[11.1202, 1223]
	≈ 1320						[11.1202]
	≈ 1410						[11.1202]
MgF_2		1610–1740	89	CW, CW-Q-qw	1.85 W	28	[11.1202]
		1608–1730	80	CW	185 mW	10	[11.1224]
		1610–1740	80	CW	≈ 100 mW		[11.1225, 1226]
	1670		80	QS (480 ns)	25 mW (1 kHz)		[11.1225]
		1610–1730		ML (23 ps)	≈ 100 mW		[11.1227]
	1623		77	Pulsed			[11.1200, 1228, 1229]
	1630		20–90	CW	1.74 W	37	[11.1230]
		1730–1750	100–200	CW	≈ 0.5 W		[11.1230]
	1636		77–82	Pulsed			[11.1200]
		1674–1676	82–100	Pulsed, CW			[11.1200]
		1731–1756	100–192	Pulsed, CW			[11.1200]
		1785–1797	198–240	Pulsed			[11.1200]
MnF_2	1865		20	Pulsed			[11.1200]
	1915		77	Pulsed			[11.1200]
	1922		77	Pulsed			[11.1200]
	1929		85	CW (exc.)			[11.1200]
	1939		85	CW (exc.)			[11.1200]
$KMgF_3$	1591		77	Pulsed			[11.1231]
$CaY_2Mg_2Ge_3O_{12}$	1460		80	Pulsed	≈ 3.8 mJ	0.7	[11.1202, 1229]
$Gd_3Ga_5O_{12}$		1434–1520	100	Pulsed		6	[11.1207, 1232]

Mn^{5+} laser operation at room temperature was demonstrated by *Merkle* et al. in $Ba_3(VO_4)_2$, $Sr_3(VO_4)_2$, and in $Sr_5(VO_4)_3F$ [11.1218–1220]. The laser transition is realized between the $^1E(^1D)$ excited state and the $^3A_2(^3F)$ ground state, thus these lasers are three-level systems. The efficiency is rather low (laser output energy ~ μJ, $\eta_{sl} \leq 1.6\%$) and laser oscillation from other Mn^{5+} systems has not been reported. The major drawback for these lasers is the excited-state absorption at the stimulated-emission wavelength. Investigations of the excited-state absorption and gain were performed in detail by *Verdún* [11.1217], *Merkle* et al. [11.1218], *Manaa* et al. [11.1221] and *Kück* et al. [11.933, 1222].

To obtain a four-level system, one should search for crystals with low crystal field strengths or with large energy-level splittings. Then the $^3T_2(^3F)$ or one of its crystal field components would be the lowest-energy level. However, thus far Mn^{5+} systems exhibiting broadband emission are not known.

Co^{2+} lasers. The Co^{2+} ion in octahedral coordination exhibits laser oscillation in the mid-infrared spectral region between 1.5 μm and 2.5 μm. Its energy-level scheme can be described with the Tanabe–Sugano diagram for the $3d^7$ electron configuration shown in Fig. 11.87. Three broad and spin-allowed transitions between the $^4T_{1a}$ ground state and the $^4T_2$, $^4A_2$ and $^4T_{1b}$ excited states exist. In MgF_2, these transitions are located around 7000 cm^{-1}, 15 000 cm^{-1}, and 20 000 cm^{-1}, respectively [11.1233]. The emission occurs between 1.5 μm and 2.5 μm, according to the $^4T_2 \rightarrow {}^4T_{1a}$ transition. At low temperatures, the lifetimes are on the order of several ms, while at room temperature the emission is strongly quenched due to nonradiative decay via multiphonon relaxation, yielding very low quantum

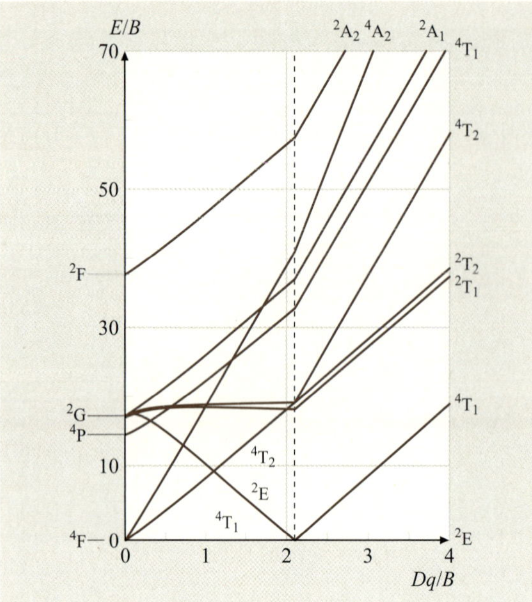

Fig. 11.87 Tanabe–Sugano diagram for the d^7 configuration (after [11.36])

efficiencies [11.1234]. Excited-state absorption overlaps with the $^4T_{1a} \rightarrow {}^4T_2$ absorption band, which is the main pump band for laser operation. In the region of stimulated emission, the excited-state absorption is negligible [11.1233].

Laser oscillation was obtained in the continuous-wave regime only at cryogenic temperatures, while pulsed laser operation was also realized at room temperature for Co^{2+}-doped MgF_2 and $KZnF_3$. An overview of the laser results for Co^{2+} laser systems is given in Table 11.35. The best laser results were obtained for Co^{2+}:MgF_2, which is also a commercial laser system. With Co^{2+}:MgF_2 an overall tuning range of 1.5–2.5 μm was realized with output powers up to 4.2 W, output energies up to 1.6 J and slope efficiencies as high as 65%. As the pump source, Nd:YAG or Nd:glass lasers around 1.3 μm are usually used, however, laser oscillation has also been realized under flashlamp pumping, argon-ion laser excitation and oxygen–iodine laser excitation. Operation regimes are CW, pulsed, Q-switched and mode-locked.

Fe^{2+} lasers. The Fe^{2+} ion has a $3d^6$ electron configuration that is complementary to that of Cr^{2+}. The ^5D free-ion state also splits into a $^5T_2$ and a ^5E state, however, for the Fe^{2+} ion ^5E is the ground state and $^5T_2$ is the first excited state. Consequently, there is just one spin-allowed absorption ($^5E \rightarrow {}^5T_2$) and emission ($^5T_2 \rightarrow {}^5E$) transition. As for the Cr^{2+} ion, all excited-state transitions are spin-forbidden. Thus, one might expect similar laser characteristics for Fe^{2+}-doped lasers as for Cr^{2+} lasers. However, the energy gap between the $^5T_2$ excited state and the ^5E ground state is smaller, thus the emission is at longer wavelengths and the nonradiative decay rate is higher, leading to lifetime shortening and low quantum efficiencies at elevated temperatures. For Fe^{2+}:ZnSe, the emission lifetime first increases from 12 to 120 K from 33 μs to 105 μs and then decreases to about 5 μs at 250 K due to thermally activated multiphonon decay [11.1235, 1236]. At 14 K, the $^5E \rightarrow {}^5T_2$ absorption band is between 2.5 μm and 3.75 μm, and the $^5T_2 \rightarrow {}^5E$ emission band is between 3.7 μm and 4.8 μm.

Laser oscillation of Fe^{2+} was obtained in ZnSe [11.1235–1239] and in n-InP [11.1240]. For Fe^{2+}:ZnSe, room-temperature tunable laser oscillation was recently reported in a gain-switched

Table 11.34 Overview of Mn^{5+}-laser materials. The data are from the literature stated in the text [11.1209, 1222]

	$Ba_3(VO_4)_2$	$Sr_3(VO_4)_2$	$Sr_5(VO_4)_3F$
Structure	Hexagonal, R-3m	Hexagonal, R-3m	Hexagonal, P6_3/m
Site symmetry	C_{3v} (V-site)	C_{3v} (V-site)	C_s (V-site)
Growth	Czochralski/LHPG	Czochralski/LHPG	Czochralski
T_m (°C)	1560		1923
σ_{abs} (10^{-20} cm^2)	≈ 300 (800 nm)	≈ 300 (800 nm)	
σ_{em} (10^{-20} cm^2)	10–20		27 ($E \parallel c$), 13 ($E \perp c$)
σ_{ESA} (10^{-20} cm^2)			14 ($E \parallel c$), 24 ($E \perp c$)
τ_{em} (300 K) (μs)	430–480	525	475–500
$\sigma_{em}\tau$ (10^{-22} cm^{-2}s^{-1})	≈ 0.7		≈ 1.3 ($E \parallel c$), ≈ 0.6 ($E \perp c$)
λ_{laser} (nm)	1181.0	1168.0	1163.7
η_{sl} (%)	0.21	0.08	1.6
P_{out} or E_{out}	≈ 2 μJ	≈ 1 μJ	

Table 11.35 Overview of Co^{2+}-laser materials. (*: versus input power, QS: *Q*-switched operation, ML: mode-locked operation)

Host material	λ_{laser} (nm)	λ_{laser} (nm)	T (K)	Mode of operation	P_{out}/E_{out}	η (%)	Ref.
MgF_2	1860	1500–2000	80	CW	1 W	31	[11.1241]
	1920	1600–2300	80	Pulsed	150 mJ	65	[11.1202, 1242, 1243]
	1920		80	Pulsed	7.3 W (50 Hz)	≈ 23*	[11.1242, 1243]
	1920			QS (220 ns)	25 mJ		[11.1242, 1243]
	1940		225	Pulsed	≈ 11 mJ	14*	[11.1243]
	1920		80		4.2 W	28*	[11.1202, 1243]
	2100	1750–2500	RT	Pulsed	70 mJ	46	[11.1244]
		1650–2010	80	ML (34 ps)			[11.1227]
		1500–2000		QS-ML (200 ps)	400 mW (ave)		[11.1245]
		1600–1900	77	QS (600 ns)	15 mJ		[11.1246]
	1890	1600–2150	77	CW	2 W	32	[11.1247]
	2050		RT	Pulsed	0.2 mJ	29	[11.1248]
	2050		77	Pulsed	1.6 J	44	[11.1249]
	2040	1960–2180	RT	Pulsed	20 mJ	25	[11.1233]
		1630–2080	80	CW	≈ 100 mW	5	[11.1226, 1250]
	1750		77	Pulsed			[11.1200, 1251]
	2060	1800–2450	282	Pulsed	900 mJ	33*	[11.1252]
	1803.5		77	Pulsed			[11.1200, 1251]
	1990		77	Pulsed			[11.1200, 1251]
	2050		77	Pulsed			[11.1200, 1251]
$KMgF_3$		1620–1900	80	CW		20	[11.1253]
	1821		77	Pulsed			[11.1200]
$KZnF_3$		1650–2110	80			30	[11.1253]
	1770	1650–2070	80		120 mW		[11.1254]
	1950	1850–2050	85	CW	≈ 20 mW	≈ 2	[11.1255]
		1700–2150	98	CW	≈ 55 mW	8	[11.1256]
	2024		300	Pulsed	3 mJ	8	[11.1233]
ZnF_2	2165		77	Pulsed			[11.1200, 1251]

mode [11.1237]. The pump source was the second Stokes output of a Nd:YAG laser at 2.92 μm. The output pulse energy was about 1 μJ. The highest output power and slope efficiency of 12 μJ and 8.2%, respectively, were obtained at lower temperatures. In this case, the Fe^{2+}:ZnSe laser was pumped by a pulsed Er^{3+}:YAG laser operating at 2.698 μm [11.1235, 1236]. The laser wavelength is tunable with temperature from 3.98 μm at 15 K to 4.54 μm at 180 K. In n-InP, Fe^{2+} laser oscillation was obtained at 2 K at 3.53 μm, i. e., at the zero-phonon transition [11.1240].

Summary

In this chapter an overview of transition-metal-ion-doped crystals as solid-state laser materials is given. It was shown that they are efficient (usually tunable) laser sources covering a wide spectral range (Fig. 11.77). However, compared to lasers based on $4f^n \rightarrow 4f^n$ transitions of trivalent rare-earth ions, they play only a small role as far as commercialization is concerned. In principle only the Ti^{3+}:Al_2O_3 and – with limitations – the Cr^{3+}:$BeAl_2O_4$ laser can be mentioned. These lasers are mainly used in the field of scientific research. The main reason for this is that their advantages, i. e., their tunability and their capability to generate ultrashort pulses, are not relevant for most industrial applications. They exhibit lower output power than lasers based on $4f^n \rightarrow 4f^n$ transitions of trivalent rare-earth ions (e.g., Nd^{3+}:YAG, Yb^{3+}:YAG) and are more expensive and less efficient than diode lasers. Direct diode-laser pumping is either

Table 11.36 Overview about transition-metal ions sorted according to the corresponding Tanabe–Sugano diagram d^n (TSD-d^n). *Normal*: octahedral coordination; italic: tetrahedral coordination. *Light brown*: laser oscillation in octahedral coordination; *Dark brown*: laser oscillation in tetrahedral coordination. (After [11.933])

Ion	TSD-d^1	TSD-d^2	TSD-d^3	TSD-d^4	TSD-d^5	TSD-d^6	TSD-d^7	TSD-d^8	TSD-d^9
Ti	Ti^{3+}	Ti^{2+}							
V	V^{4+}	V^{3+}	V^{2+}					V^{3+}	V^{4+}
Cr		Cr^{4+}	Cr^{3+}	Cr^{2+}		Cr^{2+}		Cr^{4+}	Cr^{5+}
Mn		Mn^{5+}	Mn^{4+}	Mn^{3+}	Mn^{2+}, Mn^{2+}	Mn^{3+}		Mn^{5+}	Mn^{6+}
Fe				Fe^{2+}	Fe^{3+}, Fe^{3+}	Fe^{2+}		Fe^{6+}	
Co		Co^{3+}	Co^{2+}			Co^{3+}	Co^{2+}		
Ni		Ni^{2+}	Ni^{3+}				Ni^{3+}	Ni^{2+}	
Cu	Cu^{2+}							Cu^{3+}	Cu^{2+}

inefficient or – in the case of Cr^{3+} and Cr^{2+} lasers – requires laser diodes that are not yet available at a satisfactory price and quality on the market.

Transition-metal-ion and especially tunable laser systems exhibit a stronger coupling of the electronic levels to the vibrating lattice of the crystal. This leads to higher possibilities for ESA and nonradiative decay processes compared to the $4f^n \rightarrow 4f^n$ transitions of trivalent rare-earth ions in crystals.

The main problem in realizing efficient laser oscillation is the excited-state absorption from the metastable upper laser level. From (11.87) and (11.88) its effect on laser threshold and slope efficiency becomes clear. It occurs in principle for every electron configuration, either as intra- or interconfigurational transitions or as a transition to charge transfer or conduction-band-related levels. The electron configurations are in general favorable, when intraconfigurational excited-state absorption transitions are not possible, e.g., in the d^1 and d^9 configuration, or are less strong due to selection rules, e.g., in the d^4 and d^6 configurations. Other configurations exhibit more-complex energy-level schemes, therefore intraconfigurational ESA transitions occur with high probability in the spectral regions of emission and excitation. The influence of ESA, however, can even in this case be reduced, e.g., by taking advantage of polarization-dependent transition rules, which result in higher stimulated-emission cross sections than ESA cross sections, as it is the case, e.g., for Cr^{3+}:$LiSrAlF_6$ and Cr^{4+}:$Y_3Al_5O_{12}$.

The role of nonradiative decay for the realization of efficient laser operation is less important, although not negligible. It affects to a first approximation only the laser threshold, which is increased (11.84). However, a nonradiative rate leads to a temperature increase in the pump channel, which also affects the overall laser characteristics. Thus, its general influence depends strongly on the material parameters of the laser system, i. e., mainly of the host material. Materials with a high thermal conductivity and mechanical strength are favored. As an example, Cr^{4+}:$Y_3Al_5O_{12}$ can be given. Here, the quantum efficiency is less than 20%, whereas laser operation with a slope efficiency close to 40% was realized.

Table 11.36 gives an overview of the transition-metal ions in octahedral and tetrahedral coordination investigated to date. They are listed according to their corresponding energy-level diagram (i. e., Tanabe–Sugano diagram). The laser-active transition-metal ions are indicated. Almost all transition-metal ions with different valence states and ligand coordinations have been investigated thus far. Efficient room-temperature laser oscillation was only obtained for Ti^{3+} and Cr^{3+} in octahedral and for Cr^{2+} and Cr^{4+} in tetrahedral coordination. Whether efficient laser operation will also be obtained for other ions depends very much on the host material chosen. For example, the Ti^{3+} ion exhibits efficient laser operation only in Al_2O_3. Therefore, other ions cannot a priori be excluded as efficient laser ions.

11.2.6 Overview of the most Important Laser Ions in Solid-State Lasers

Research on laser materials has created a number of compact, efficient solid-state laser sources for a large variety of applications. Laser materials have been developed for various wavelengths (near-infrared, visible, UV) and power regimes (mW to multi-kW). Special geometries of the active material (microchip, rod, disc, fiber) strongly correlate with the active ion concentrations and cross sections of absorption and gain transitions.

The values of the cross sections depend on the quantum numbers of the final and initial states as well as on the local environment of the active-ion site in the material. The lifetimes of the states are influenced by radiative and nonradiative transitions. So, the microscopic crystal properties play an important role for static and dynamic processes in the laser crystal.

In the near-infrared spectral region high efficiencies have been achieved with diode-pumped oxide- and fluoride-based laser materials doped with the rare-earth ions Nd^{3+}, Tm^{3+}, Ho^{3+}, Er^{3+}, and Yb^{3+}. For high-average-power operation Nd^{3+}- and Yb^{3+}-doped crystals are of greatest interest. Especially the Yb^{3+} ion exhibits very small Stokes losses and minimum heat generation, which reduces thermal lensing and improves beam quality. Transition-metal-doped crystals based on the ions Ti^{3+}, Cr^{2+}, Cr^{3+}, and Cr^{4+} offer broadly tunable radiation within the spectral region 680–3000 nm. In the visible region Er^{3+}-, Tm^{3+}-, and Pr^{3+}-doped laser materials operate at several red, green, and blue transitions with laser diode upconversion and/or direct pumping. So far Ce^{3+} is the only ion with reasonable direct laser performance in the UV.

The following wavelength data provide a rough guide to the spectral range of the various laser ions.

Near-IR rare-earth lasers.

Nd^{3+}	0.9, 1.06, 1.3 μm
Yb^{3+}	1–1.1 μm
Tm^{3+}	2 μm, 1.5 μm (upconversion)
Ho^{3+}	2 μm
Er^{3+}	1.6, 3 μm, 0.85 μm (upconversion)

Visible rare-earth lasers.

Pr^{3+}	0.64 μm (diode-pumped)
Pr^{3+},Yb^{3+}	0.52, 0.63 μm (upconversion)
Er^{3+}	0.55 μm (upconversion)
Tm^{3+}	0.45, 0.48, 0.51, 0.65, 0.79 μm (upconversion)

Frequency doubling of near IR rare-earth lasers (Nd, Yb)

UV rare-earth lasers.

Ce^{3+}	0.3 μm

Frequency doubling of visible rare-earth lasers
Frequency tripling/quadrupling of near-IR lasers

Transition-metal lasers.

Ti^{3+}	0.68–1.1 μm
Cr^{3+}	0.7–1.1 μm
Cr^{4+}	1.2–1.6 μm
Cr^{2+}	2–3 μm

11.3 Semiconductor Lasers

11.3.1 Overview

In crystalline solids the interaction between *atomic energy levels* generates energy bands. A quantum-mechanical treatment provides, in the single-electron approximation, energy bands which overlap which each other or are separated by bandgaps from each other. In semiconductors, we find between the energetically highest band, which is fully occupied by electrons at $T = 0$ K, the valence band (VB), and the energetically lowest band, which is completely empty at $T = 0$ K the conduction band (CB) an energy range, in which we find no allowed energy states, disregarding energy states originating from dopants. This bandgap, E_g, is the energy difference between the lower conduction-band edge E_C and the upper valence-band edge E_V. For laser operation an extreme deviation from thermal equilibrium (nonthermal carrier distribution) is required: carrier inversion. In the CB close to E_C carrier inversion means a much higher density of electrons than holes and in the VB close to E_V a much higher density of holes than electrons. This extremely nonthermal condition is generated by strong electrical or optical pumping of the laser-active material. The electrical carrier injection in most cases is obtained by a p–i–n heterojunction made of III/V semiconductor materials. Choosing InP as a typical example, Fig. 11.88 displays a planarization (projection into 2-D space for simplification) of the real 3-D zinkblende crystal structure with the corresponding band structures below. From the left to the right p-InP, intrinsic i-InP and p-InP are shown.

In the highly p-doped InP bulk regions, at room temperature nearly all acceptors have released their holes into the valence band. Well away from the pn junction the hole field current (drift current) dominates the carrier transport. At the left edge of the device, holes are generated under the p contact. At the right edge of the device, electrons are injected through the n contact.

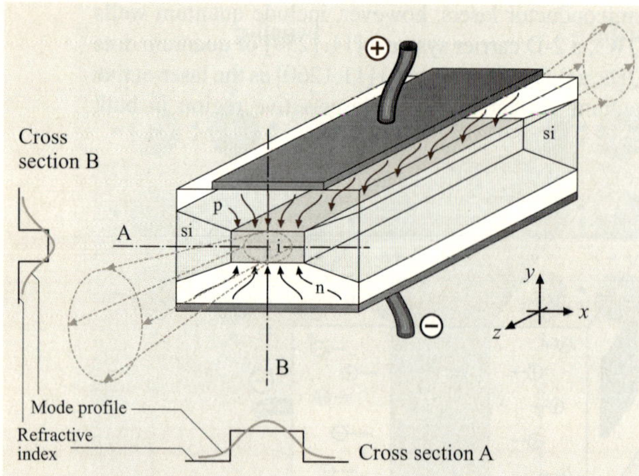

Fig. 11.90 Schematic of a semiconductor laser with 3-D laser-active layer and buried waveguide. Insets: refractive index profiles and mode profiles in the x- and y-directions intersecting the active layer at A and B, respectively

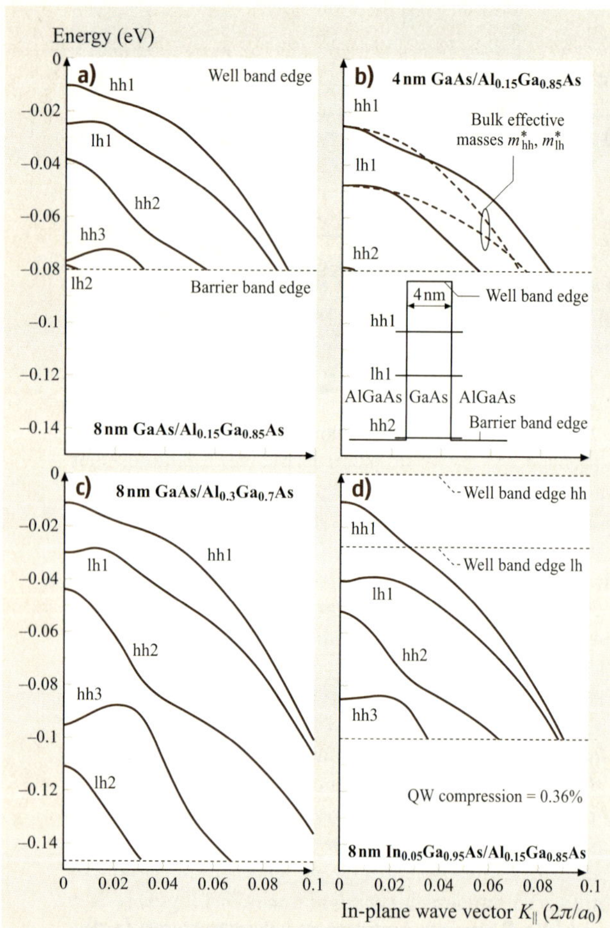

Fig. 11.91 Valence-band structure $\boldsymbol{E}(\boldsymbol{k}_{\|})$ of various Ga(In)As/AlGaAs QWs [11.1261]

lasers (Fig. 11.89e). In Fig. 11.89f two QWs are depicted as an example, providing the emission energy as the difference between the energetically lowest bound states in the wells (note that the ordinate indicates the electron energy).

Finally, electrical and optical confinement in the y-direction (Fig. 11.89c) has to be completed by electrical and optical confinement in the x-direction (Fig. 11.90). Note that this figure is rotated by 90° with respect to Fig. 11.89. In Fig. 11.90 the active material is embedded in the x-direction by semi-insulating (si) material to force the current to move mainly in the active layers (lateral electrical confinement). The si material is chosen to have a higher bandgap, and thus a lower refractive index, than the active material (lateral optical confinement). Also in this case, wave guiding is applied in the x- and z-directions. This is shown in the two insets by the two cross sections A and B, showing the refractive-index profiles and light-intensity profiles. Here, waveguiding in fundamental mode is shown for the case of bulk 3-D active layers.

11.3.2 Resonator Types and Modern Active Layer Materials: Quantum Effects and Strain

Modern optoelectronic semiconductor devices are based on a sequence of materials of different composition. Embedding a film of lower E_g between a material of larger E_g, we obtain quantization effects if the film thickness L_y is on the order of the electron wavelength or below (QW) [11.1259]. Using the Schrödinger equation, we obtain at least one bound state in the CB and VB potential wells, respectively. The electrons and holes are still fully mobile in the x- and z-directions. This means that a wavevector $\boldsymbol{k}_{\|}$ parallel to the heterointerfaces exists. The motion in the y-direction is considerably restricted and statistically described by quantum mechanics. Figure 11.91 displays the VB structure in k-space $[\boldsymbol{E}(\boldsymbol{k}_{\|})]$ obtained by a theoretical model calculation [11.1261] based on the Schrödinger equation for four different Ga(In)As/(Al)GaAs QWs ($\boldsymbol{k}_{\|}$ is oriented parallel to the heterointerfaces).

In most cases in addition the QWs and the barriers in the active layer are additionally strained [11.1262]. Usually, the wells are compressively strained, which means

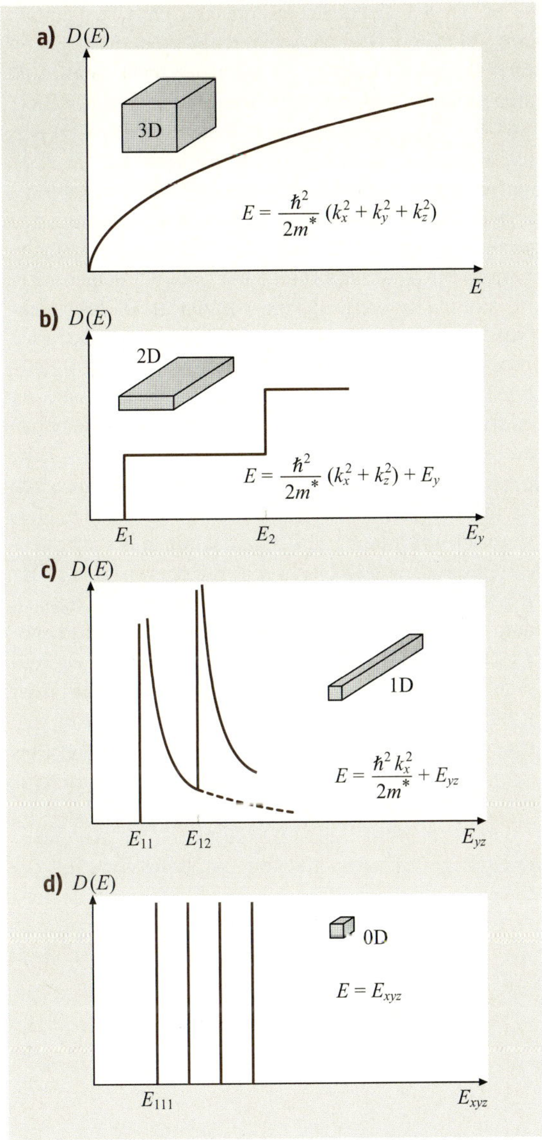

Fig. 11.92a–d Density of states $D(E)$ and corresponding formula of the band structure in parabolic approximation for a **(a)** semiconductor extending in all three spatial directions (3-D), **(b)** quantum well (QW), **(c)** quantum wire, and **(d)** quantum dot (QD)

that the lattice constant of that layer is compressively reduced in the x- and z-directions to match that of the substrate a_o. In many cases, the barriers are in tension, which indicates that the lattice constant of the barriers is increased in the x- and z-directions by tensile strain to be identical to the substrate lattice constant a_o.

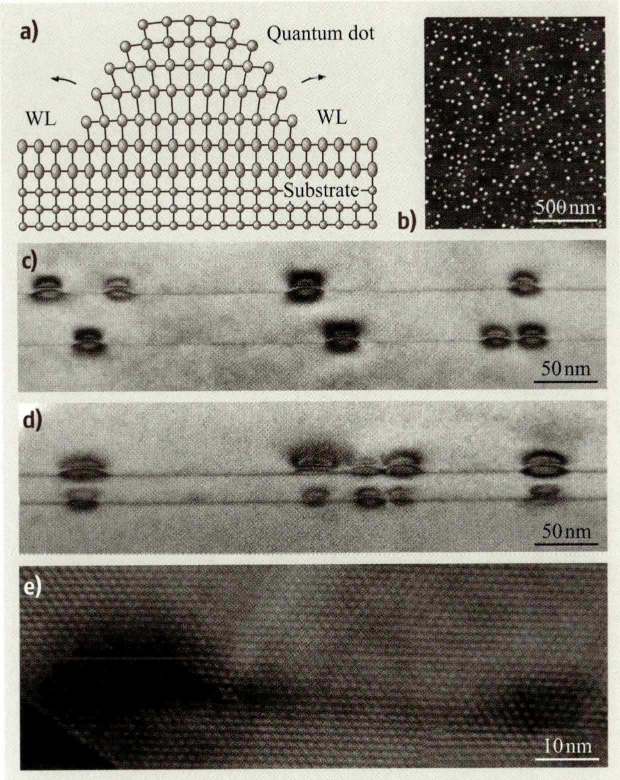

Fig. 11.93 **(a)** From the bottom to the top we see in sequence: the three top monolayers of the GaAs substrate, above two compressively strained monolayers InAs (wetting layer, WL), and above the main part of the QD. **(b)** AFM surface profile of InAs QDs (*bright dots*). **(c,d)** TEM micrograph of a double stack of QDs: **(c)** stack distance 40 nm with uncorrelated dots and **(d)** stack distance 20 nm with vertically aligned dots (**(a–d)** [11.1261]). **(e)** TEM micrograph of a CdSe QD (*dark*) in a ZnSe matrix (*bright*) [11.1263]

Strain and quantization effects in the active layers of semiconductor lasers are used to improve the device properties. For the laser it is beneficial to make the effective masses of electrons and holes as similar as possible, to increase material gain, to reduce threshold, to tailor the density of states and to increase differential gain. This can be obtained by applying strain and/or quantization. In Fig. 11.92, the energy dependence of the density of states is shown for 3-D material to be root-like, and to be constant for 2-D material.

Figure 11.93 depicts examples of InAs [11.1263] and CdSe [11.1264] QDs. Schematically, Fig. 11.93a shows the QD formation in a cross section of the crystal structure. We see the three upper monolayers of the GaAs substrate. During the self-organized QD formation the intentionally integrated strain (detailed explanation below) of the InAs versus GaAs plays an important role. For thermodynamical and elastomechanical reasons, two monolayers of strained InAs are first formed. Continuing the growth, for total-energy reasons, it is of considerable advantage for the semiconductor surface to continue by a localized island-like growth. In this case a possible geometric shape is a pyramid with a base plane directly joined to the top of the InAs monolayers. Experimentally, this phase can be directly studied using an atomic force microscope (AFM) [11.1260]. Figure 11.93b displays in a top view a typical AFM surface profile. In the next process step the QDs are overgrown by GaAs and thereby embedded. Experimentally the final layer sequence can be studied by transmission electron microscopy (TEM) after cleaving. In order to increase the number of QDs in the direction perpendicular to the substrate interface, the whole InAs process is repeated as often as desired after a defined GaAs spacer layer thickness. This provides layered QD arrangements. Figures 11.93c, d show TEM micrographs for two different spacer-layer widths: 40 nm ad 20 nm. We observe a vertical correlation of the dots for the smaller separation. At very high magnification Fig. 11.93e shows QD formations in the II/VI semiconductor system CdSe/ZnSe; note the different scales in the subfigures. This high TEM magnification resolves the individual crystal layers. The CdSe QD is contrasted dark against the surrounding brighter ZnSe.

However, quantization is not limited to a single direction. If we limit the carrier movement in another direction (e.g., in the z- and y-direction as in Fig. 11.92), we obtain a 1-D carrier system and, thus, a quantum wire [11.1265] with hyperbolically shaped density of states branches. Limitation of carrier movement in all three space directions results in quantum dots [11.1260, 1261, 1263–1272], i. e., a zero-dimensional (0-D) carrier system with a δ-like density of states (Fig. 11.92d). For many physical properties of the semiconductor (e.g., carrier mobilities, carrier capture in quantized states and spontaneous or stimulated emission of light), the density of states plays an important role. Using dimensionality and strain efficiently during the design of lasers enables one to enhance desired and suppress undesired properties. Here, modern epitaxy [e.g., metal organic chemical vapour deposition (MOCVD) and molecular beam epitaxy (MBE)] are powerful tools. To date 2-D and 0-D structures have been used successfully for laser devices. 1-D structures have not yet had a breakthrough due to their disadvantageous dynamic properties. QDs are grown today using self-organization in the Stranski–Krastanow growth mode [11.1263, 1265, 1267, 1271]. By growing a large number of QDs with identical quantized energy levels in the laser-active layer, we would benefit from a strongly reduced temperature sensitivity and a very high differential gain and, thus, extremely high bit rates in optical communication. Although many QD lasers have been implemented, the pyramid-like QDs suffer from strong fluctuations in size and thus in their energy levels.

On the basis of these QDs, however, the aforementioned improvements in device properties have not yet be demonstrated. On the other hand, the strong fluctuation causes a strongly spectrally broadened gain profile,

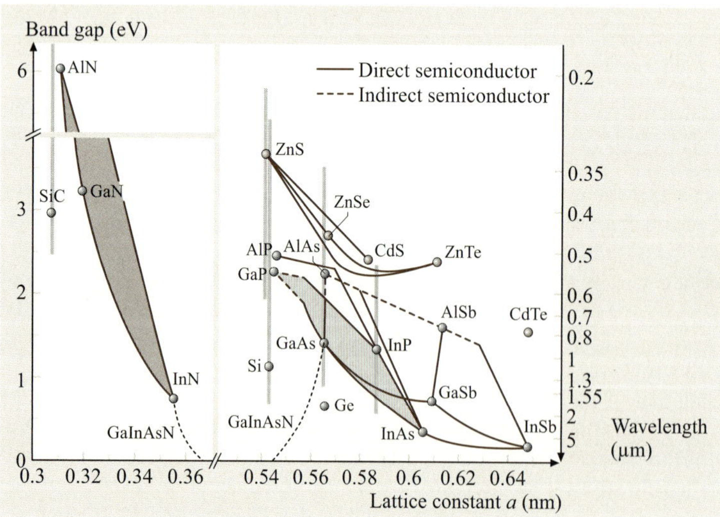

Fig. 11.94 Bandgap energy E_g (*left*) and bandgap wavelength (*right*) as a function of the lattice constant a

which is beneficial for other applications, e.g., short-pulse generation by mode locking or wide spectral tuning of the laser device. However, we should also remember that 10 years were necessary after implementing the first QW lasers to demonstrate that QW lasers were superior to bulk lasers. Similarly, QD lasers will require time before their final breakthrough. However, we believe that QW and QD lasers will share the application area in the future, depending on the specific requirements.

Figure 11.94 shows the dependence of the bandgap on the lattice constant for various III/V and II/VI semiconductors. The brown area corresponds to the quaternary $Al_zGa_{1-x-z}In_xN$. The grey area describes the quaternary $Ga_{1-x}In_xAs_{1-y}P_y$. For high laser efficiencies in interband lasers, active materials with a direct bandgap are required, disregarding for the moment some intraband lasers. Note that simple bulk Si has an indirect band structure and does not provide efficient radiative recombination. Modified Si structures have demonstrated strong luminescence, LED operation or laser oscillation, e.g., from Si/Ge superlattices, QDs in Si and Si-based Raman laser structures [11.1273]. All compound semiconductors, located on one of the grey vertical lines (Fig. 11.94) that indicate the lattice constant of important semiconductor substrate materials, can be lattice-matched to the respective substrate. According to Fig. 11.94 the ternary compounds $Al_{0.48}In_{0.52}As$ ($E_g = 1.43\,eV$) and $Ga_{0.47}In_{0.53}As$ ($E_g = 0.75\,eV$) have the same lattice constant as InP ($E_g = 1.34\,eV$), which is available in wafers today up to a diameter of 150 mm.

QW widths (or QD dimensions), effective masses [more precisely the $E(k)$ functions], and bandgaps of well and barrier materials energetically determine the quantized states and, thus, the spectral gain profile of the laser, i. e., the possible range of laser emission. Figure 11.94 shows that a very large wavelength range can be covered for different applications. Some examples are red and blue lasers which are used for data storage in digital versatile and blue-ray discs, respectively. Emission at about 850 nm is used for short-range optical fibre communication and CD devices, 980 nm lasers for pumping Er-doped fibre amplifiers, while between 1.25 and 1.65 μm the lasers are applied in ultrahigh-bit-rate long-haul optical fibre communication. Visible and infrared wavelengths are very attractive for optical sensing. The range between 0.8 and 1 μm is used for direct laser applications (welding, drilling, cutting and soldering) since it includes lasers with extremely high output power, highest wall-plug efficiencies and lowest price per watt of optical power.

Stimulated emission and optical gain (Fig. 11.95) are essentially determined by the product of the reduced electronic density of states and the Fermi factor $(f_c - f_v)$. This Fermi factor originates from $f_c(1 - f_v) - f_v(1 - f_c)$, namely the *probability of processes photon-*

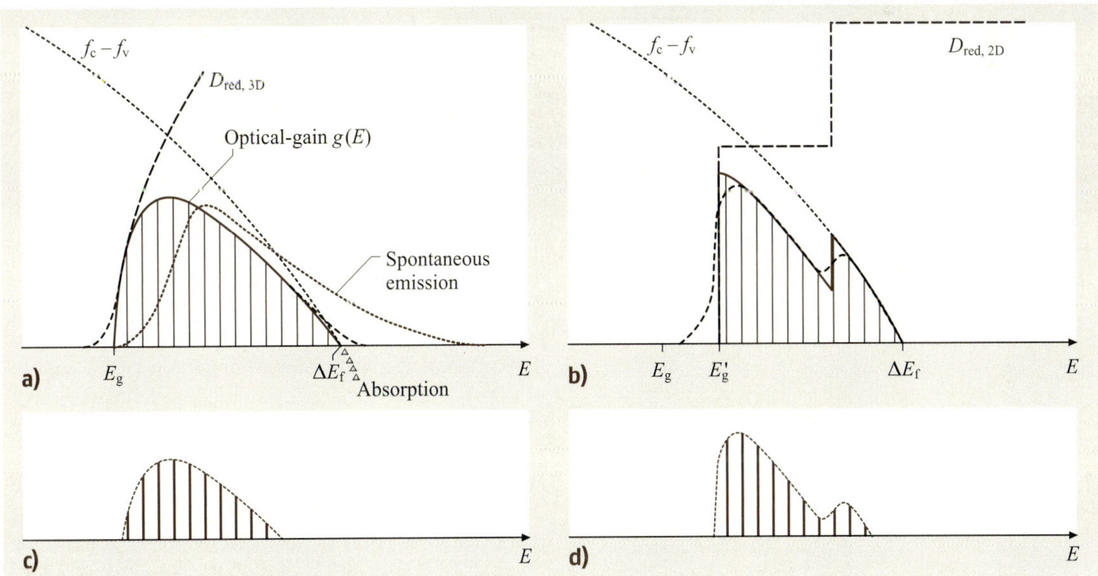

Fig. 11.95a–d Profile of optical material gain, spontaneous emission, Fermi factors and the density of states as a function of energy for **(a)** 3-D und **(b)** 2-D materials. Corresponding net gain profiles in **(c)** and **(d)**

generating photons (emission), i. e., the probability that the CB states are occupied by electrons and VB states are not occupied by electrons $f_c(1-f_v)$ *minus* the *probability of processes destroying photons (reabsorption)*, i. e., the probability of VB states occupied by electrons and CB states not occupied by electrons $f_v(1-f_c)$. The full lines in Fig. 11.95a, b schematically show the spectral material gain profile of 3-D and 2-D semiconductors, respectively. Gain ranges spectrally from the bandgap E_g to the difference between the two quasi-Fermi levels ΔE_F. In reality there are additional effects such as Fermi-level filling, profile broadening [11.1261] (bold broken line) or many-body effects. Material gain minus loss provides net gain (Fig. 11.95c, d). If these laser-active materials are placed in a Fabry–Pérot (FP) laser the net gain can support the modes (vertical lines) as indicated. In Fig. 11.95 the spontaneous emission is shown by the thin broken lines, revealing a different spectral profile, which is important, e.g., for LEDs.

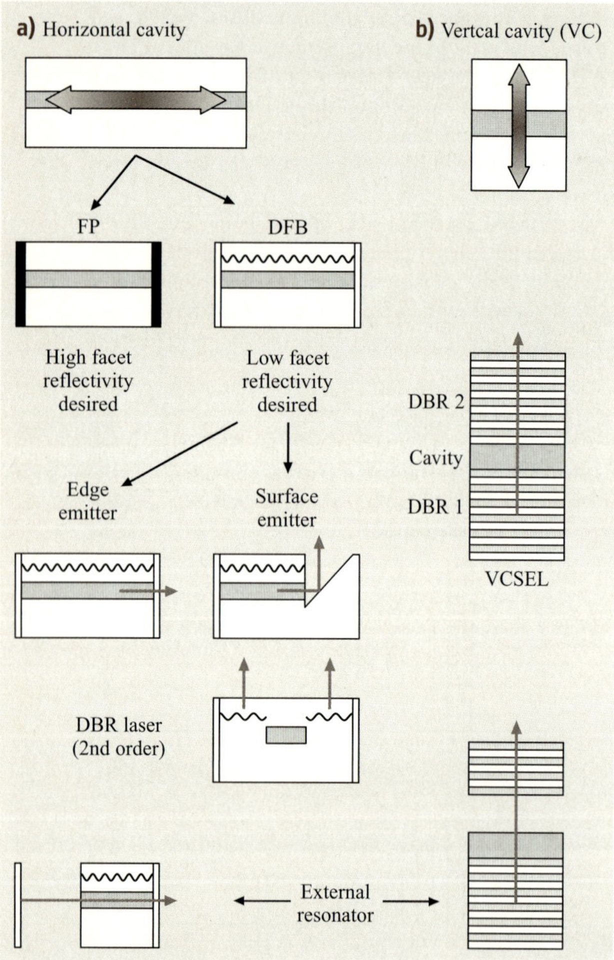

Fig. 11.96 Different geometries of lasers with horizontal (*left*) and vertical resonators (*right*)

Figure 11.96 depicts a classification summary of the most important semiconductor laser geometries. Most of them will be treated in detail below. Generally, the heterointerfaces are located horizontally. We distinguish between horizontal (left) and vertical (right) resonator structures, which is indicated by the orientation of the broad brown double arrow. Thus, we have horizontal cavity lasers (in-plane lasers) and vertical cavity (VC) lasers.

First case: Fabry–Pérot (FP) structures. Here the optical reflection (feedback) is provided by the borders of the resonator (facets). In many cases, the high refractive index difference between the semiconductor and the air is already sufficient, providing an optical reflection coefficient of about 30%. By additional facet coatings this coefficient can be tailored continuously between 0% (antireflection) and 100% (perfect reflection). The resonator modes are given by

$$m_{FP}\left(\frac{\lambda_B}{2n_{eff}}\right)=L\,. \tag{11.92}$$

Intuitively, the resonator length L has to be a positive integer multiple of half of the wavelength of light in the medium, where n_{eff} is the effective refractive index (eigenvalue of the Helmholtz equation, see below) of the waveguide. In the second case, in so-called structures with distributed feedback (DFB), the optical reflection is extended over the whole resonator. A very efficient feedback occurs at the Bragg wavelength λ_B, which is correlated with the DFB grating period Λ via the Bragg condition:

$$m_{DFB}(\lambda_B/n_{eff})=2\Lambda\,. \tag{11.93}$$

m_{DFB} is a positive integer and describes the grating order. An example: for a DFB laser emitting at $\lambda_B=1.55\,\mu m$ having a first-order grating ($m_{DFB}=1$) and an effective refractive index $n_{eff}=3.27$, a grating period of $\Lambda=237$ nm is required. Intuitively, an integer multiple of the wavelength in the medium has to correspond, according to (11.93), to the double grating period. Note that the mathematical structure of (11.92) and (11.93) is identical. In lasers with a typical length of 200 μm we obatin for FP laser modes within the gain profile a very large m_{FP} (on the order of 1000) while for DFB lasers in most cases m_{DFB} is 1 (for a first-order grating).

As expected, light is emitted in horizontal directions for edge-emitting lasers (Fig. 11.96). Surface emitters can be implemented by etching an out-coupling mirror inclined at 45° or by a second-order DFB grating (note the doubled grating period in the figure). In a second-order grating the light field is reflected back horizontally (180°) and coupled out vertically. If the Bragg condition is exactly fulfilled, the emission occurs exactly perpendicular to the surface (90°). However, the larger the deviation in (11.93) in the (>)-direction or (<)-direction, the larger or smaller the angle, respectively. If the grating is interrupted we have a distributed Bragg reflector (DBR) structure. Typically, a central grating-free section is embedded between two DBR sections. In the DFB and DBR structures the horizontally propagating light field alternatively passes two virtual *quasi-layers* of slightly different refractive index. For DBR resonators with *real layers* (see Sect. 11.3.4 or the next section) these operation principles become more evident.

Second case: VCSELs. They are also based on DBR structures, which in contrast to the first case consist of real multiple layers having a very high refractive-index difference. The central cavity has no grating and is embedded between two DBR mirrors, thus also forming an FP-like structure. However, feedback is distributed over the two DBR structures. In most cases the thickness of a single period (equivalent to a pair of neighboring layers) is chosen to equal half the wavelength of light in the medium for the design wavelength. This corresponds to a first-order grating according to (11.93). Since the resonator is oriented perpendicularly and the emission occurs perpendicular to the main chip surface (oriented parallel to the substrate area), this laser is called a vertical-cavity surface-emitting laser (VCSEL). The two mirrors must have a very high reflectivity to reach the laser threshold, since the laser-active layer is relatively thin and has a weak overlap with the light field in the resonator.

Edge-emitting lasers as well as VCSELs will be treated in detail in the following sections. On the basis of both cases (A and B), it is possible to implement lasers with external resonators, as indicated in the last row of Fig. 11.96.

11.3.3 Edge-Emitting Laser Diodes with Horizontal Resonators

As already mentioned, the population inversion between the valence and conduction bands, which is necessary for the coherent amplification of radiation [11.1274], can be obtained by an injection of electrons and holes via a forward-biased pn junction (see Figs. 11.89, 90). The electrons injected into the n semiconductor as well as the holes injected into the p semiconductor diffuse to the pn junction and can recombine there radiatively and generate a photon with energy $\hbar\omega$ (Fig. 11.88). If the external voltage is increased and the carrier density exceeds a critical value in the range of $10^{18}\,\mathrm{cm}^{-3}$, the rate of photon emission becomes higher than the absorption rate, so that an incident wave can be amplified coherently due to stimulated emission. The condition for this population inversion is that the separation between the quasi-Fermi levels E_{f_c} and E_{f_v}, which describe the filling of the conduction and valence bands, is larger than the bandgap E_g (Bernard–Durafourg condition):

$$E_{f_c} - E_{f_v} \geq \hbar\omega \geq E_g\,. \quad (11.94)$$

In this case the semiconductor material is transparent for the generated wave with the wavelength defined by the bandgap. The resulting gain values are very high (in the range of $10^3\,\mathrm{cm}^{-1}$) due to the high carrier density. Lasing occurs if the additional losses of the optical resonator providing the feedback are compensated.

Double-Heterostructure Lasers

The active region, where population inversion is achieved, is very thin in homojunction lasers and the threshold current is very high, since only a small portion of the injected carriers is utilized for the lasing process. Lower threshold currents and continuous operation at room temperature can be achieved with double-heterostructure lasers, where the low-bandgap active layer is sandwiched between n- and p-doped cladding layers with a higher bandgap epitaxially grown on a substrate (Fig. 11.89). The technical realization of these structures is possible if the lattice mismatch between the different layers of the material system does not exceed a critical value. Double heterostructures [11.1257] have three main advantages with respect to laser operation:

- The bandgap difference between the layers is distributed between the valence and conduction bands and creates potential barriers for the injected electrons and holes. For an appropriate choice of doping and applied voltage, a nearly rectangular-shaped potential well can be achieved (Fig. 11.89), which efficiently confines the carriers in the low-bandgap active layer if the potential difference is higher than the thermal activation energy $k_B T$. The width of the active layer d_{act} is defined by the heterostructure geometry. The carriers injected via the pn junction are captured in the potential well and confined to a small volume, thus decreasing the injection current, which is necessary to achieve inversion.

- There is no reabsorption of the radiation emitted from the active layer in the surrounding layers due to the difference in bandgap energies.
- In the material systems used for semiconductor lasers, the low-bandgap active layer has a higher refractive index than the surrounding cladding layers. Therefore, the double heterostructure acts as a dielectric planar slab waveguide, confining the generated optical field to the active region due to the refractive-index difference between the layers. The number of modes supported by this optical waveguide for a given wavelength depends on the thicknesses and the refractive indices of the layers. By a proper choice of the design parameters, a single transverse (perpendicular to the pn junction) mode can be selected, which concentrates the photon density for stimulated emission in the gain region.

Thus, a double heterostructure enables the confinement of carriers and generated photons in the active layer. For bulk active layers, the de Broglie wavelength of the carriers is small compared to the thickness of the active layer, leading to a high degree of carrier confinement. The wavelength of the photons, however, is comparable to the dimension of the structure, so that only a portion of the optical intensity is confined to the active region. The transversal distribution of the light intensity in the double heterostructure is given by the solution of the wave equation of a planar slab waveguide supporting TE and TM modes with an effective index n_{eff} as the respective eigenvalue. Figure 11.89c shows the optical intensity of the fundamental TE mode in a double heterostructure as a function of the transversal coordinate y, revealing the incomplete confinement of the optical field in the active region. The fraction of the mode intensity within the active layer is called the optical confinement or the filling factor Γ_{act}, which is an important design parameter for a semiconductor laser:

$$\Gamma_{act} = \frac{\int_0^{d_{act}} |E(y)|^2\, dy}{\int_{-\infty}^{+\infty} |E(y)|^2\, dy}, \tag{11.95}$$

where $E(y)$ is the electric field. If the light is also confined in the lateral direction of the laser, the definition has to be modified accordingly.

The dependence of the optical confinement factor of the guided modes on the thickness d for a layer structure is shown in Fig. 11.97. For increasing thickness of the waveguide, the confinement factor approaches 1 and additional modes are supported.

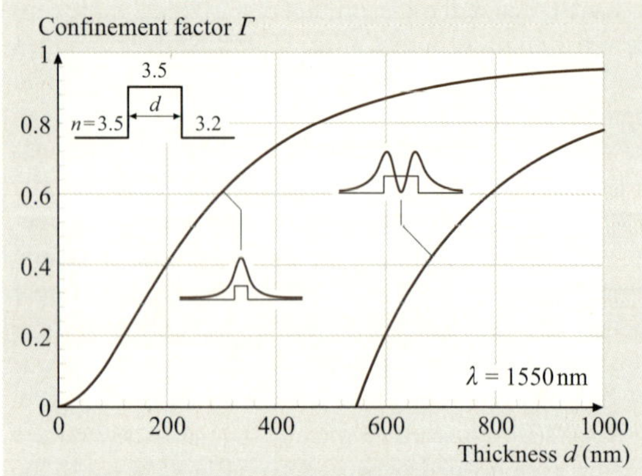

Fig. 11.97 Dependence of the optical confinement factor Γ on the thickness d of an InP/InGaAsP/InP layer structure

An optical mode with an effective refractive index n_{eff} experiences an effective gain g_{eff} in a waveguide with an active layer, which provides the material gain g_{act} and has a refractive index n_{act}

$$g_{eff} = \frac{n_{act}}{n_{eff}} \Gamma_{act} g_{act} . \tag{11.96}$$

The correction factor n_{act}/n_{eff} takes into account the waveguiding effect on the modal gain [11.1275].

Laser Structures

Using modern epitaxial growth methods it is possible to realize semiconductor multilayer structures very precisely so that stable, transverse single-mode operation can be obtained using dielectric waveguiding [11.1276]. For most applications, however, lateral (parallel to the pn junction) patterning of the laser structure is also required to obtain lateral carrier and photon confinement, which is important to obtain stable and efficient laser operation with high spectral purity and good coupling efficiency into a fibre. In addition, lateral confinement of the injection current is necessary to avoid leakage currents bypassing the active region. The confinement of photons, carriers and current has been implemented in many ways, reflecting the specific purpose of the device. Semiconductor lasers can be classified according to the mechanism of lateral waveguiding as gain-guided or index-guided, depending on whether it is the lateral variation of the optical gain or the refractive index that confines the mode. Index-guided lasers can further be subclassified as weakly or strongly index-guided, depending on the magnitude of the lateral refractive-index

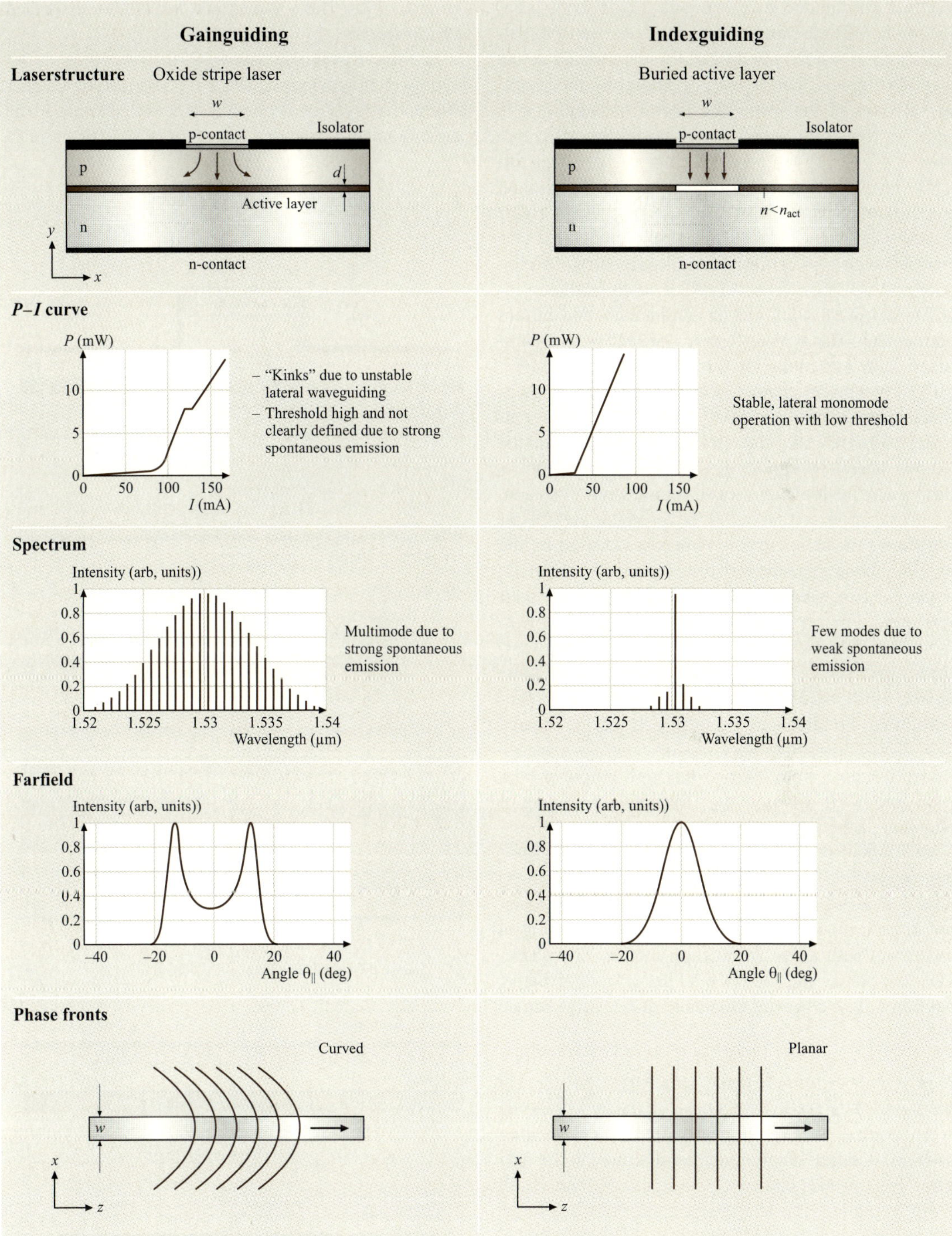

Fig. 11.98 Comparison of gain-guided and index-guided laser structures

step. In a gain-guided laser structure (oxide stripe laser, Fig. 11.98, left) the current is injected via a stripe contact (width $w \approx 5\,\mu m$) in a laterally unpatterned active layer. The optical field is mainly guided by the resulting variation of the gain. The lateral waveguiding is weak, so that even small variations of the refractive index, e.g., due to temperature changes or carrier injection lead to unstable operation. In an index-guided laser structure (buried laser structure, Fig. 11.98, right) the active layer ($w \approx 2\,\mu m$) is laterally embedded into a material with a lower refractive index ($n < n_{act}$), higher bandgap and higher electrical resistivity to facilitate stable lateral waveguiding, carrier confinement and current confinement. The typical features of index- and gain-guided laser structures are summarized in Fig. 11.98. The power–current characteristic (P–I curve) of gain-guided lasers is characterized by a high threshold current (typical 50–100 mA) and kinks originating from the unstable lateral waveguiding. The optical spectrum is multimodal due to the enhanced spontaneous emission. The phase-fronts are curved in the resonator and the far field shows the characteristic twin lobes caused by the laterally inhomogeneous gain distribution. The main advantage of gain-guided lasers is the simplicity of their fabrication. For very large stripe widths (50–100 μm) a so-called broad-area laser results, in which the current is injected laterally and homogeneously via a large contact. Since there is no lateral waveguiding in broad-area lasers, the threshold current is very high (typical a few A) and the multimode emission cannot be coupled efficiently into a fibre. Due to the high output power, broad-area lasers are used, e.g., for optical pumping of solid-state lasers.

Index-guided lasers show stable, lateral single-mode emission with a low threshold current (typical 10 mA) and a P–I curve without kinks. The spontaneous emission in each mode is significantly smaller, resulting in a spectrum with a few dominating modes. The phase-fronts are planar and the far field has a smooth shape, enabling a high coupling efficiency into a single-mode fibre.

Weakly index-guided laser structures. In weakly index-guided lasers the thickness of the waveguiding layer is varied, thus resulting in a lateral waveguiding structure. Lateral single-mode emission can be obtained by a proper choice of the thickness and width of this variation. The lateral index step has to exceed the carrier-induced reduction of the refractive index ($\Delta n \approx 5\times10^{-3}$) so that index-guiding is dominating. Weakly index-guided laser structures can be divided into two categories: ridge waveguide lasers and channeled substrate lasers.

In ridge-waveguide lasers, a rib waveguide is defined by etching a narrow stripe (≈ 3–5 μm) down near to the active layer (at a typical distance of 200 nm). In the case of a metal-clad-ridge waveguide (MCRW) laser in

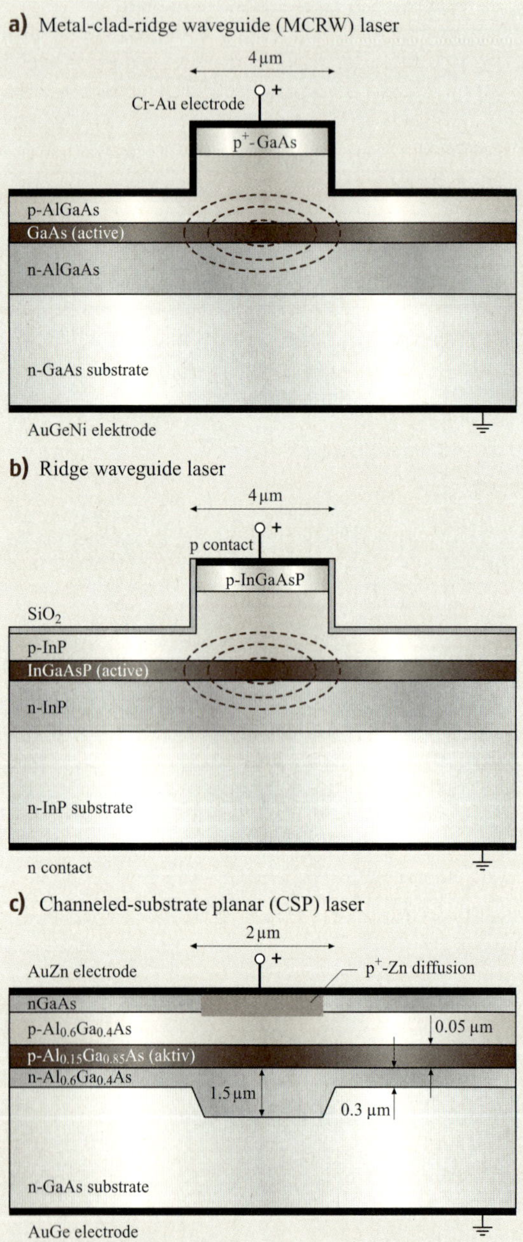

Fig. 11.99 Weakly index-guided laser structures

the material system GaAlAs (Fig. 11.99a) a CrAu electrode is deposited, which gives an electrically conductive contact on the p^+ top layer of the rib and an insulating contact on the p-GaAlAs layer with lower doping along the rib. Thus, the current injection is focused in the region of the active layer below the rib. This current confinement can also be realized by additional isolating layers, e.g., SiO_2 (Fig. 11.99b). The lateral waveguiding is accomplished by the higher refractive index of the semiconductor material compared to the surrounding SiO_2 and air. The etch depth has to be carefully controlled (e.g., by using etch-stop layers) in order to select one lateral mode and minimize current bypass.

The first step in the processing of channeled substrate lasers is the etch of a 2–3 μm-wide and 1 μm-deep channel into the substrate (Fig. 11.99c). In the subsequent epitaxy (e.g., liquid-phase epitaxy) this channel is nearly planarized, thus generating a lateral waveguiding structure, since the active layer has a higher refractive index. Current bypass can be reduced by guiding the injection current using, e.g., localized Zn diffusion.

Weakly index-guided lasers are suitable to obtain low threshold currents of typically 20–40 mA and high output powers with lateral single-mode emission. The weak waveguiding allows for broader active layers compared to buried laser structures, which has a positive effect, e.g., on the series resistance. The low current leakage in ridge waveguide lasers usually results in good linearity of the P–I curve. The emission is more complex than in buried lasers, since index- and gain-guiding are important and even small changes of the refractive index by temperature or current injection can influence the performance. The active layer in weakly index-guided lasers is not affected during processing, so that lateral

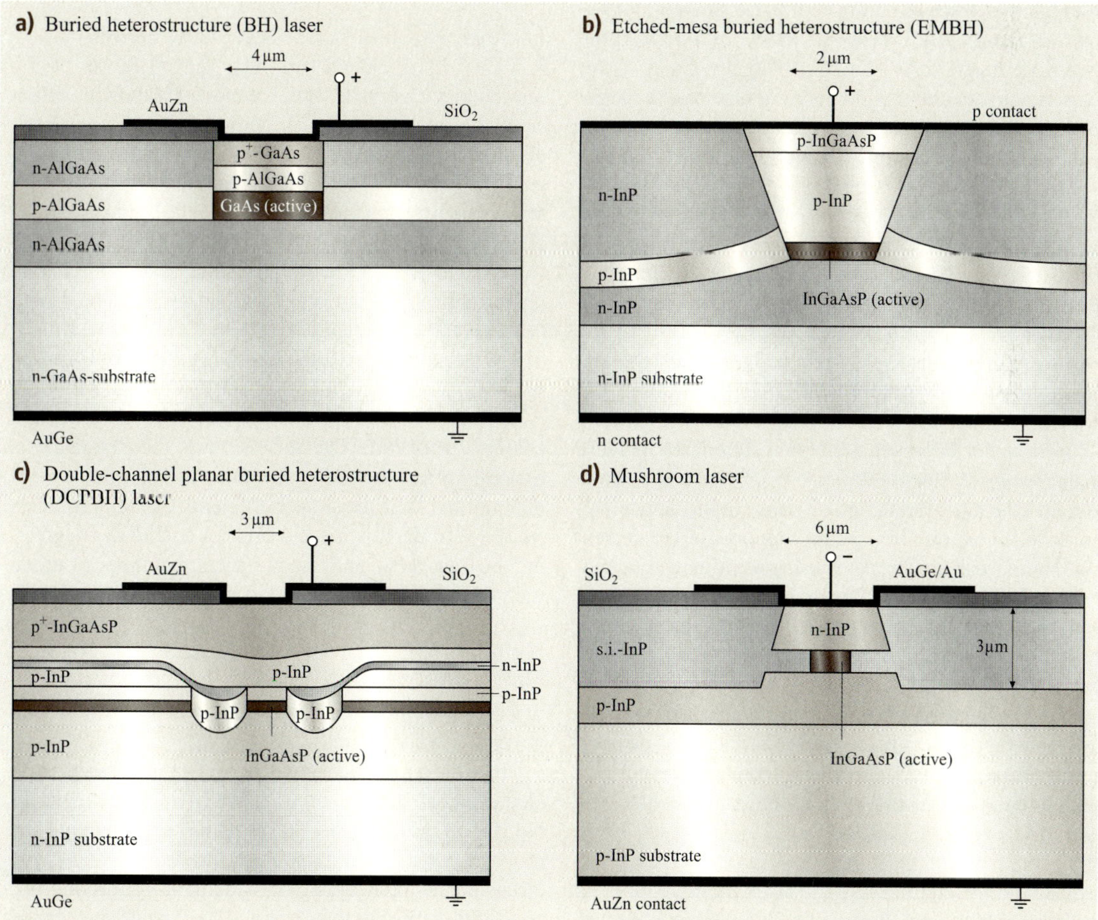

Fig. 11.100 Buried heterostructure lasers

carrier diffusion in the active layer cannot be avoided, but on the other hand, the unpatterned active layer is advantageous for the reliability of the device. Since only one epitaxial growth step is needed, the fabrication of these lasers is significantly easier than buried semiconductor lasers.

Strongly index-guided laser structures. Strong index-guiding can be accomplished by buried laser structures, where a small stripe of the active layer material with a high refractive index is embedded into semiconductor material with lower refractive index and larger bandgap. For this purpose, the active layer has to be patterned and epitaxially regrown afterwards. Lateral single-mode operation is possible, if the resulting stripe width of the active layer does not exceed a critical value defined by the cut-off for the higher lateral modes [11.1277].

Figure 11.100 shows some examples for strongly index-guided laser structures. In the case of a buried-heterostructure (BH) laser structure in the material system GaAs/GaAlAs (Fig. 11.100a) the GaAs active layer is surrounded by AlGaAs, which has a larger bandgap and a lower refractive index. The active layer is grown in the first epitaxial step and structured into 1–2 μm-wide stripes. Using liquid-phase epitaxy (LPE), which planarizes the structure, the regions alongside the stripe are refilled with alternating n- and p-doped AlGaAs layers. This reverse-biased diode structure prevents the current from bypassing the active region. In a similar way, a BH structure can be realized in the long-wavelength material system InGaAsP/InP (Etched-mesa buried heterostructure laser, Fig. 11.100b). The reduction of leakage current using alternating n- and p-doped layers, however, leads to a significant increase of the parasitic electrical capacity of the laser, which degrades the high-frequency response of the device. Alternatively, electrically isolating regions for the current confinement can be fabricated using semi-insulating material (e.g., Fe-doped InP) or by proton implantation. In Double-channel substrate planar buried heterostructure (DCPBH) lasers (Fig. 11.100c) the first epitaxial growth step produces the active InGaAsP-layer and a p-InP cap layer on the n-InP substrate. Then, a double-stripe structure embedding the active region is created by etching. In the subsequent LPE step the generated channels are refilled with p-InP/n-InP and the epitaxial growth of the laser structure is finished up to the InGaAsP top layer. The processing of the mushroom laser (Fig. 11.100d) begins with the etching of 6 μm-wide mesa in the layer structure, which cuts through the active layer grown in the first epitaxy [11.1278]. Using a selective wet chemical etching process the width of the active region is reduced up to 1–2 μm to obtain lateral single-mode operation. The resulting undercut area is epitaxially refilled afterwards, e.g., using vapor phase epitaxy (VPE) with semi-insulating InP [11.1279].

In addition to this strongly index-guided laser structures with a planar active layer, which are useful for integration with a DFB grating, there also exist buried laser structures utilizing nonplanar active layers based on the regrowth of, e.g., V-grooves or mesas.

Using strongly index-guided structures, a very stable and lateral single-mode laser operation with very low threshold currents (< 10 mA) and excellent high-speed characteristics can be obtained, since stable optical waveguiding, carrier confinement and current confinement are combined. The fabrication of these devices, however, is complicated due to the additional epitaxy step.

Edge-emitting Fabry–Pérot laser diodes

In Fabry–Pérot (FP) lasers, the cleaved facets of the semiconductor crystal form the optical resonator, which enables laser operation by providing the optical feedback of the stimulated amplified radiation. This resonator selects the photons generated by stimulated emission with respect to direction and wavelength. The light wave traveling perpendicularly to the facets is amplified if the wavelength matches a longitudinal mode of the resonator [(11.92), Figs. 11.90, 95, 96, 103a]. The lasing process starts if the gain experienced during one round-trip in the cavity equals the losses caused by absorption, scattering and the light output through the facets.

Lasing condition. The length L of the resonator is typically a few hundred micrometers. The intensity reflection and transmission coefficients of the end facets can be estimated using the Fresnel equations, neglecting the transversal and lateral structure of the waveguide with the air ($n = 1$) assumed to be outside the cavity:

$$R = \frac{(n_{\text{eff}} - 1)^2}{(n_{\text{eff}} + 1)^2}\,, \qquad T = \frac{4n_{\text{eff}}}{(n_{\text{eff}} + 1)^2}\,, \tag{11.97}$$

where n_{eff} is the effective refractive index of the waveguide mode considered.

In this one-dimensional model plane waves with electric field amplitude $E(z)$ travel in the longitudinal direction (z) of the FP resonator, experiencing the modal intensity gain g due to stimulated emission. The threshold for laser operation is defined by a round-trip condition requiring that in a stationary state the optical wave remains unchanged after one complete round-trip

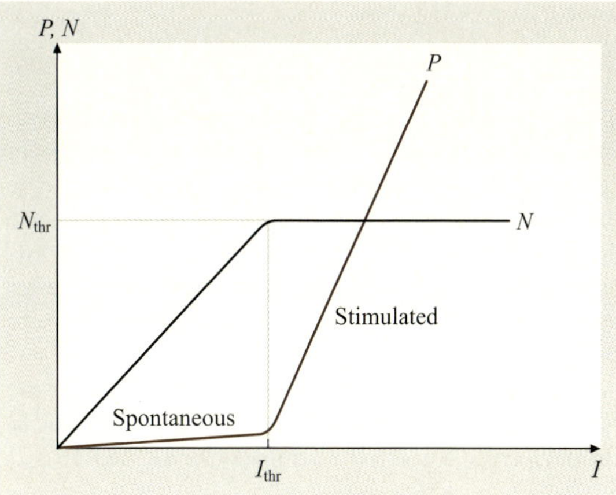

Fig. 11.101 Dependence of the output power (P) and carrier density (N) on the injected current

in the cavity. This round-trip condition gives the mirror losses of the FP resonator

$$g = \alpha_{\mathrm{m}} = \frac{1}{2L} \ln(R_1 R_2) \,, \tag{11.98}$$

where R_1 and R_2 denote the intensity reflection coefficients of the end facets. Since the net gain g in the cavity is composed of the material gain of the semiconductor g_{eff} and the waveguide losses α_s, the lasing condition can be written as

$$g_{\mathrm{thr}} = \frac{n_{\mathrm{act}}}{n_{\mathrm{eff}}} \Gamma_{\mathrm{act}} g_{\mathrm{act}}(N_{\mathrm{thr}}) = \alpha_s + \alpha_{\mathrm{m}} \tag{11.99}$$

if the contribution of spontaneous emission is neglected and N_{thr} denotes the carrier density at threshold. The optical losses of the waveguide are caused by optical scattering from imperfections in the bulk media or at interfaces and free-carrier absorption in the active and cladding layers.

The longitudinal distribution of the photon density $s(z)$ in the active FP resonator is given by the sum of the forward and backward traveling photon densities, which grow exponentially due to the gain g. For a symmetric resonator ($R = R_1 = R_2$) the total photon density $s(z)$ is given by a cosh function (Fig. 11.103a) with a minimum in the middle of the resonator. For a laser diode with as-cleaved facets ($R = 0.28$), the intensity distribution is relatively flat, whereas for antireflection-coated facets a strongly inhomogeneous photon distribution results.

The round-trip condition also gives the optical frequencies of the longitudinal modes of the FP resonator according to $\nu_q = qc/(2Ln_{\mathrm{eff}})$, where $q = 1, 2, 3 \cdots$ and c denotes the vacuum speed of light. The frequency

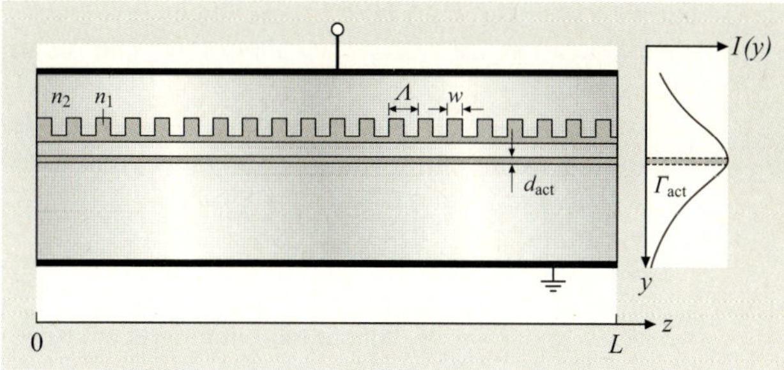

Fig. 11.102 Scheme of a DFB laser structure with an integrated rectangular Bragg grating (duty cycle w/Λ) including the corresponding transversal intensity distribution $I(x)$

separation $\Delta\nu$ of adjacent modes is influenced by the dispersion of the waveguide, which is taken into account by the group refractive index n_{g}

$$\Delta\nu - \frac{c}{2Ln_{\mathrm{g}}} \tag{11.100}$$

with

$$n_{\mathrm{g}} = n_{\mathrm{eff}} + \nu \frac{\mathrm{d}n_{\mathrm{eff}}}{\mathrm{d}\nu} \,. \tag{11.101}$$

For a typical FP laser diode the group index of $n_{\mathrm{g}} \approx 3.5$–$5$ is higher than the effective index $n_{\mathrm{eff}} \approx 3$–$3.5$. The separation of the equidistant mode frequencies is about 150 GHz for a 300 μm-long cavity, which is small compared to the width of the gain curve of

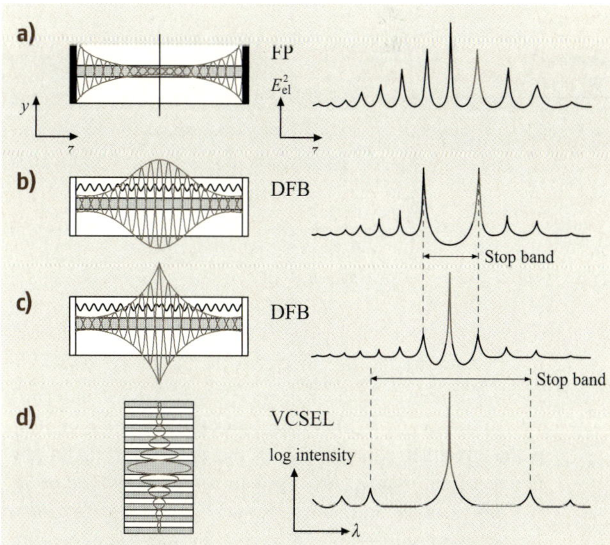

Fig. 11.103 Schematic structural cross sections (*left*) and the corresponding emission spectra (*right*) for various diode lasers

the semiconductor (≈ 5 THz). Thus, longitudinal multimode operation can be expected above threshold for an FP laser diode.

For many applications the degree of single-mode emission is important; it can be quantified by the side-mode suppression ratio (SMSR):

$$\mathrm{SMSR} = 10\log_{10}(P_1/P_2)\,, \tag{11.102}$$

where P_1 and $P_2 \le P_1$ denote the output power of the two strongest modes in the optical spectrum. Typical FP laser diodes achieve a maximum SMSR of approximately 20 dB.

Rate equations. The fundamental static and dynamic properties of semiconductor laser diodes can be modeled using a set of rate equations [11.1280] describing the interaction of electron–hole pairs and photons in the active layer. We consider a strongly index-guiding double heterostructure supporting a single optical mode traveling as a plane wave axially in the cavity. The current I is assumed to be uniformly injected into the active layer with volume V and recombines there, completely neglecting leakage currents. The carrier density within the active layer is treated as homogeneous in the transversal and lateral direction, since the corresponding inhomogeneities of photon density are small and the resulting gradients in carrier density are smoothed out by diffusion. In FP lasers with a sufficiently high mirror reflectivity, the axial variations of the photon density s and the carrier density N can be neglected and the rate equations can be written as:

$$\frac{\mathrm{d}N}{\mathrm{d}t} = \frac{I}{eV} - \frac{N}{\tau_{\mathrm{nr}}} - BN^2 - CN^3 - v_{\mathrm{g}}\frac{n_{\mathrm{act}}}{n_{\mathrm{eff}}} g_{\mathrm{act}}(N,s)s + \frac{F_N(t)}{V}\,, \tag{11.103}$$

$$\frac{\mathrm{d}s}{\mathrm{d}t} = v_{\mathrm{g}}\left[\Gamma_{\mathrm{act}}\frac{n_{\mathrm{act}}}{n_{\mathrm{eff}}} g_{\mathrm{act}}(N,s) - g_{\mathrm{thr}}\right]s + \frac{\Gamma_{\mathrm{act}}\left[R_{\mathrm{sp}} + F_s(t)\right]}{V}\,, \tag{11.104}$$

$$\frac{\mathrm{d}\Phi}{\mathrm{d}t} = \frac{1}{2}\alpha_{\mathrm{H}} v_{\mathrm{g}} \Gamma_{\mathrm{act}}\frac{n_{\mathrm{act}}}{n_{\mathrm{eff}}} g_{\mathrm{act}}(N,s) + F_\Phi(t)\,, \tag{11.105}$$

where $v_{\mathrm{g}} = c/n_{\mathrm{g}}$ is the group velocity of the waveguide, e the electron charge, g_{act} is the material gain of the active layer, Γ_{act} is the optical confinement factor of the active layer and $1/\tau_{\mathrm{nr}}$, B and C are the parameters describing nonradiative, bimolecular and Auger recombination, respectively. Φ is the phase of the complex electric field E, which is connected with the photon number $S = sV/\Gamma_{\mathrm{act}}$ via

$$E(t) = \sqrt{S(t)}\exp[\mathrm{i}\Phi(t)]\,, \tag{11.106}$$

R_{sp} is the time-averaged rate of spontaneous emission into the lasing mode; $F_s(t)$, $F_N(t)$ and $F_\Phi(t)$ represent Langevin noise sources taking into account the statistic nature of the spontaneous emission and the shot-noise character of the carrier recombination and generation. The Langevin forces leading to fluctuations of carrier density and photon density are correlated and have zero mean [11.1280].

The longitudinal excess factor K_z accounts for the enhancement of the spontaneous-emission noise due to the axial dependence of the complex electric field $E(z,t)$

$$K_z(t) = \frac{|\int_0^L |E(z,t)|^2\,\mathrm{d}z|^2}{|\int_0^L E^2(z,t)\,\mathrm{d}z|^2}\,. \tag{11.107}$$

In the case of a transversely single-mode index-guided FP laser the factor K_z is given by [11.1281]

$$K_z^{\mathrm{FP}} = \left(\frac{(\sqrt{R_1}+\sqrt{R_2})(1-\sqrt{R_1R_2})}{\sqrt{R_1R_2}\ln(1/R_1R_2)}\right)^2\,. \tag{11.108}$$

The first rate equation (11.103) can be formally derived from the quantum-mechanical density-matrix formalism. It can be interpreted as a balance of carriers that are injected as a current I and contribute to stimulated emission or are lost for the lasing process via the different recombination processes. The second and third equations (11.104, 105) can be derived from the Maxwell equations with the rotating-wave and slowly varying amplitude approximations.

For semiconductor lasers with a bulk active layer, the dependence of the gain on the carrier density $g_{\mathrm{act}}(N)$ can be approximated as linear [11.1282]

$$g_{\mathrm{act}}(N) = \frac{\mathrm{d}g}{\mathrm{d}N}(N - N_{\mathrm{tr}})\,, \tag{11.109}$$

where $\mathrm{d}g/\mathrm{d}N$ is the differential gain and N_{tr} is the transparency carrier density. In quantum well structures (Figs. 11.91, 92) this dependence is usually described by a logarithmic function [11.1283, 1284]

$$g_{\mathrm{act}}(N) = \frac{\mathrm{d}g}{\mathrm{d}N} N \ln\frac{N}{N_{\mathrm{tr}}} \quad \text{for} \quad g_{\mathrm{act}} \ge 0\,. \tag{11.110}$$

The influence of the photon density on the gain is taken into account by introducing a nonlinear gain coefficient ε according to:

$$g_{\mathrm{act}}(N,s) = \frac{g_{\mathrm{act}}(N)}{1+\varepsilon s}\,. \tag{11.111}$$

This nonlinear gain compression is caused by spectral hole burning and carrier heating, which is significant if the time constant of stimulated emission becomes comparable to the intraband relaxation time.

Refractive index change. In semiconductors the real part of the refractive index depends on the carrier density because of various physical mechanisms. With increasing injection the band-to-band absorption is reduced due to band-filling effects. In addition, the absorption increases due to the reduction of the bandgap (bandgap renormalization resulting from many-body effects) and absorption also increases due to the increasing absorption of free carriers. The resulting total change of the real part of the refractive index, which is related to the gain spectrum via the Kramers–Kronig relation, depends on the wavelength relative to the gain maximum. In the case of 1.5 μm InGaAsP the refractive index decreases with injection. The dependence of the refractive index on the carrier density is theoretically described by the effective line width enhancement or Henry factor α

$$\alpha = \frac{\partial n_{\text{eff}}}{\partial \gamma_{\text{eff}}} , \tag{11.112}$$

where the complex effective refractive index is defined by $n_{\text{eff}} - \mathrm{i}\gamma_{\text{eff}}$. The change of the effective index with carrier density can be written as

$$\delta n_{\text{eff}} \cong \Gamma_{\text{act}} \frac{n_{\text{act}}}{n_{\text{eff}}} \delta n_{\text{act}} = -\Gamma_{\text{act}} \frac{n_{\text{act}}}{n_{\text{eff}}} \frac{\alpha\lambda}{4\pi} \frac{\partial g_{\text{act}}(N)}{\partial N} \delta N . \tag{11.113}$$

The line width enhancement factor is important for the treatment of line width and frequency chirp under modulation [11.1281]. Typically it ranges from 3 to 5, decreasing from the long-wavelength to the short-wavelength side of the gain curve [11.1282].

Steady-state characteristics. The single-mode rate equations can be used to analyze the steady-state behavior of a semiconductor laser. Setting the time derivative to zero in (11.104), we obtain an implicit expression for the photon number S in the case of continuous-wave (CW) operation:

$$S = \frac{R_{\text{sp}}}{v_{\text{g}}[g_{\text{thr}} - \Gamma_{\text{act}} g_{\text{act}}(N, s)]} . \tag{11.114}$$

The number of photons increases as the gain value asymptotically approaches the losses g_{thr}. The small gain difference is compensated by spontaneous emission, which provides the noise input amplified by stimulated emission. Below threshold the photon density is small and (11.103) gives a linear increase of the carrier density according to $N \propto I/eV$. Above threshold the gain is approximately clamped at $g(N_{\text{thr}}) = g_{\text{thr}}$ and the corresponding threshold current, which is defined in the limiting case of vanishing spontaneous emission ($R_{\text{sp}} = 0$), becomes

$$I_{\text{thr}} = \mathrm{e}V \left(\frac{N_{\text{thr}}}{\tau_{\text{nr}}} + BN_{\text{thr}}^2 + CN_{\text{thr}}^3 \right) . \tag{11.115}$$

Using (11.103) the photon number above threshold can then be written as $S = (I - I_{\text{thr}})/ev_{\text{g}}g_{\text{thr}}$. Since the carrier density is clamped at threshold, all injected carriers in excess of the threshold current contribute to stimulated emission and the number of photons increases proportionally to $(I - I_{\text{thr}})$. The total output power $P = v_{\text{g}}\hbar\omega\alpha_{\text{m}}S$ emitted from both facets becomes

$$P = \frac{\hbar\omega}{e} \eta_i \frac{\alpha_{\text{m}}}{g_{\text{thr}}} (I - I_{\text{thr}}) , \tag{11.116}$$

where we assume that only a fraction η_i of the external drive current reaches the active region and the remaining fraction $(1 - \eta_i)$ is lost via leakage current or nonradiative recombination. Thus, the P–I curve of a semiconductor laser diode is a straight line (Fig. 11.101) above threshold with a slope defined by the external quantum efficiency η_{ext}:

$$\eta_{\text{ext}} = \frac{\mathrm{d}P}{\mathrm{d}I} \frac{c}{\hbar\omega} = \eta_i \frac{\alpha_{\text{m}}}{g_{\text{thr}}} , \tag{11.117}$$

which can be interpreted as the ratio of the number of emitted photons to the number of injected electrons per time. The sharpness of the transition from the spontaneous emission below threshold to the stimulated emission above threshold depends on the amount of spontaneous emission into the lasing mode. Leakage currents, thermal effects and spectral hole-burning neglected so far lead to bending of the P–I curve.

Characteristic temperature. The threshold current of a semiconductor laser depends on the temperature T, which can be described phenomenologically by

$$I_{\text{thr}}(T) = I_0 \exp \frac{T}{T_0} , \tag{11.118}$$

where T_0 is the characteristic temperature, which typically ranges between 40 K and 90 K for semiconductor lasers emitting around 1550 nm.

Single-Mode Laser Structures

Transmitters used, e.g., in optical-fibre communication systems should emit light predominantly in a single longitudinal mode since the presence of side modes limits the transmission capacity due to pulse broadening caused by the chromatic dispersion of the fibre. Semiconductor lasers with an FP resonator usually exhibit multimode operation since the gain spectrum is wider than the longitudinal mode spacing and the broadening of the gain profile, which due to spectral hole burning is not perfectly homogeneous, offers several modes with sufficient gain to oscillate. The techniques to achieve reliable longitudinal mode control even under high-bit-rate modulation can be categorized into two main groups:

Short lasers. The discrimination against side modes in FP resonators can be enhanced by reducing the cavity length L. If the mode spacing $\Delta\nu \propto L^{-1}$ becomes comparable to the width of the gain curve, only one mode will oscillate near the gain peak. To obtain stable single-mode operation, however, the lasers must be extremely short. This requires a very good reflectivity of the end facets to overcome the high mirror losses $\alpha_{\mathrm{m}} \propto L^{-1}$, leading to high threshold current densities. The problems of fabricating very short semiconductor devices can be solved using a vertical-cavity surface-emitting laser (VCSEL) structure (Fig. 11.96).

Frequency-selective feedback. The second method to obtain single-mode operation is to incorporate a frequency-selective element in the resonator structure. This can be realized by using coupled cavities, an external grating or a Bragg grating:

Coupled cavities. If one or more additional mirrors are introduced in the FP resonator, the boundary conditions added due to the reflections at each interface severely limit the number of longitudinal modes. To achieve single-mode operation, however, it is often necessary to tune the resonator by changing the drive current or the temperature. Usually the single-mode regime is small so that such structures can only be modulated over a limited current range without mode jumps. In addition, the reproducible fabrication of nearly identical devices turns out to be difficult since the spectral properties strongly depend on the exact lengths of the sections.

External grating. Frequency selection can also be realized by an external grating outside the resonator. The mechanical stability of such lasers, however, is a critical point since the grating is not integrated on the wafer. Consequently, lasers with external gratings are expensive devices (Fig. 11.110a).

Bragg grating. The method most frequently used to achieve single-mode emission is to incorporate a Bragg grating, which creates a periodic variation of the complex refractive index and distributes the feedback throughout the cavity. Dynamic single-mode operation is achieved if the threshold gain for the oscillating mode is significantly smaller than the threshold gain for the other modes. Devices employing Bragg gratings can be classified roughly into three categories: distributed Bragg reflector (DBR), distributed feedback (DFB) (Figs. 11.96, 102, 103, section on Fabry–Pérot structures in Sect. 11.3.2, and Sect. 11.3.6) and gain-coupled (GC) lasers.

In DBR lasers the Bragg grating is etched into passive regions near the cavity ends. The index grating (the variation of the real part of the refractive index) acts as an effective mirror with wavelength-dependent reflectivity and surrounds the central part of the cavity which is active and remains uncorrugated. The longitudinal mode with a wavelength located near the reflectivity maximum of the grating is selected. Since a DBR laser is formed by replacing the mirrors by passive gratings, the properties can be described by an effective mirror model. The transition between the active section and the passive gratings usually complicates the technological realization of in-plane DBR lasers. An important advantage of DBR lasers is that the wavelength can be changed if the grating regions are equipped with separate electrodes that can tune the Bragg frequency via the carrier-induced refractive index change. Thus, DBR gratings are often used in tunable lasers.

In DFB lasers the index grating covers the entire resonator length. At the wavelength corresponding to the corrugation period of the grating, the forward- and backward-traveling waves created by the Bragg scattering are confined in the central part of the cavity so that the mirror losses become a function of the wavelength. The longitudinal mode with the lowest mirror losses corresponding to the most effective concentration of photons in the resonator is selected.

In gain-coupled devices a periodic variation of gain or loss is used to favor a longitudinal mode of the FP resonator. In the ideal case, there is no Bragg scattering at the gain grating, and the longitudinal photon distribution as well as the mirror losses are unchanged compared to the FP cavity. The overlap with the loss or gain grating, however, varies between the different longitudinal modes of the FP resonator. The mode experiencing the largest overlap with the gain grating (or minimum overlap with a loss grating) is selected.

DFB lasers. The spectral properties of DFB lasers [11.1285] are essentially determined by the integrated Bragg grating. The waveguding in such periodic structures can be analyzed by the coupled mode theory [11.1286], which yields approximate analytical solutions describing the light propagation in waveguides with a periodic variation of the complex refractive index by counter-propagating modes exchanging energy by scattering. The strength of the interaction and the amount of feedback in the grating structure are determined by the complex coupling coefficient

$$\kappa = \pi \Delta n / \lambda_0 + \mathrm{i} \Delta g / 4 , \quad (11.119)$$

which is proportional to the variation of the refractive index step Δn, the gain variation Δg and the number of corrugations per length in the grating. The Bragg wavelength λ_B is given by the effective index, the corrugation period Λ and the order m_{DFB} of the grating according to (11.93). For an axially homogeneous first-order grating ($m_{DFB} = 1$) with length L and a perfect antireflection coating on the facets ($R_1 = R_2 = 0$) the theory reveals the following solutions

- In the case of pure index coupling ($\Delta n_{eff} \neq 0$, $\Delta g = 0$) the transmission spectrum turns out to be symmetric with respect to the Bragg wavelength $\lambda_B = 2n_{eff}\Lambda$ where oscillation is forbidden. For a small coupling coefficient, the mode spacing approximately takes the value of an FP resonator $\Delta\lambda = \lambda^2/2n_{eff}L$ but, in contrast to an FP cavity, the threshold gain of the modes is wavelength dependent and increases with growing distance from the Bragg wavelength. A strong coupling produces a transmission stop-band with width $\Delta\lambda \cong \kappa\lambda_B^2/(\pi n_{eff})$ centered at the Bragg wavelength, in which transmission is strongly damped. The two modes with the lowest threshold gain $g \cong 2\pi^2/(\kappa^2 L^3)$ are located at the edges of the stop-band symmetrical to the Bragg wavelength.
- In the case of pure gain coupling ($\Delta g \neq 0$, $\Delta n_{eff} = 0$) the mode degeneracy is removed, which means that the mode with the lowest threshold gain oscillates at the Bragg wavelength symmetrically surrounded by the other modes with higher threshold gain values. The mode spacing is $\Delta\lambda = \lambda^2/2n_{eff}L$ and no stop-band occurs since there is no backscattering at index steps in the grating. The mode selection is due to the different overlap of the standing waves in the FP resonator with the gain grating.

In second-order gratings ($m_{DFB} = 2$) additional scattering occurs in the transversal direction, leading to higher losses. In addition, the coupling coefficient of second-order gratings depends more sensitively on the exact shape of the grating so that it becomes more difficult to control. That is why first-order gratings predominate although the corrugation periods are smaller ($\Lambda \cong 240$ nm for $\lambda = 1.55\,\mu$m). Due to the scattering perpendicular to the optical axis (Fig. 11.96), second-order gratings can be utilized for the vertical emission of light from edge-emitting laser diodes [11.1287, 1288].

Basic properties of index-coupled DFB lasers. The spectrum of an index-coupled DFB laser mainly consists of two degenerate lasing modes at the edges of the stop-band (Fig. 11.103b). This mode degeneracy in index-coupled DFB laser structures is usually removed by incorporating a $\lambda/4$ phase shift in the grating. This can technologically be implemented by inserting an additional section of length $\lambda_0/(4n_{eff}) = \Lambda/2$ in the middle of the grating. The introduction of the $\lambda/4$ phase shift selects the Bragg mode in the middle of the stop-band, revealing the lowest mirror losses so that single-mode operation with an SMSR of > 40 dB can be obtained (Fig. 11.103c).

The axial distribution of the light intensity in the cavity is connected to the mirror losses α_m via the round-trip condition. Thus, a decrease of the mirror losses for the longitudinal modes in a DFB grating is equivalent to increasing longitudinal optical confinement, which means that the photons are concentrated inside the cavity and only a small fraction of the light intensity leaves the resonator through the end facets. The typical longitudinal photon distribution for a DFB laser with and without a $\lambda/4$ phase shift is shown in Fig. 11.103, revealing the pronounced maximum in the middle of the resonator even for moderate coupling coefficients. This strong inhomogeneity of the photon density distribution in DFB lasers leads to an inhomogeneous carrier density distribution above threshold due to the recombination by stimulated emission. With increasing injection the carrier density is depleted in places with a high photon density. This phenomenon is called longitudinal spatial hole burning (LSHB) and has several important consequences for the static and dynamic behavior of DFB lasers above threshold. First, the mode discrimination is influenced since a variation of the carrier density distribution changes the round-trip gain of the various modes having different photon density distributions. Thus, the side-mode suppression can degrade with increasing output power due to LSHB. Second, the mode wavelengths change, even above threshold, since the inhomogeneity of the carrier density caused by LSHB leads to an inhomogeneous axial distribution of the effective refractive index. This effect is utilized in several types of tunable DFB lasers.

Various DFB lasers with a more complex grating structure have been developed in order to obtain a high yield of single-mode devices and a flat axial photon distribution with reduced LSHB. Some examples are:

- $2\times\lambda/8$ phase shifts. The distance between the phase shifts, however, must be optimized to achieve a high yield [11.1289].
- Corrugation pitch modulation (CPM). The DFB grating is divided into three sections. The corrugation period in the central section is slightly higher than in the outer sections so that the phase shift is quasi-continuously distributed along the cavity [11.1290].
- Axial variation of duty cycle by using, e.g., a holographic double-exposure technique [11.1291].
- Axial variation of coupling coefficient by variation of the etch depth [11.1292] or sampled gratings [11.1293].
- Bent waveguides superimposed on homogeneous grating fields can be used to obtain quasi-continuously and arbitrarily chirped gratings with high spatial resolution [11.1294].
- Axially inhomogeneous injection using a three-electrode structure. Spatial hole burning can be compensated for if the injected current density is higher in the central section near the peak of the photon density, thus reducing the gain of the side-modes suffering from the lower current in the outer section [11.1295].

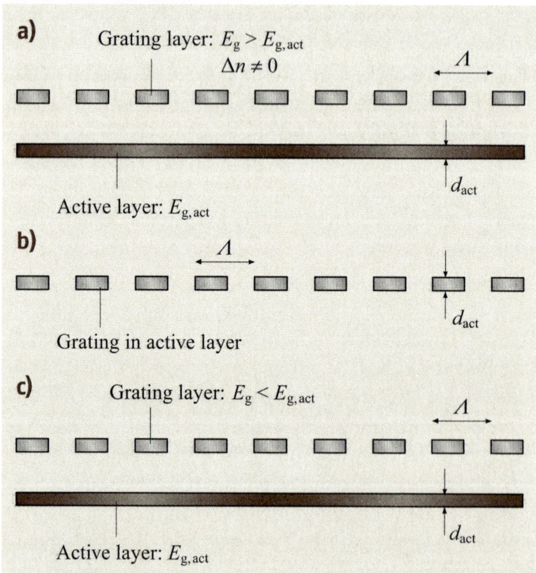

Fig. 11.104a–c Various types of Bragg gratings. **(a)** Index coupling; **(b)** gain coupling; **(c)** loss coupling

The phase relation between the grating and the end facets is difficult to control during the cleaving process of DFB lasers, since the corrugation period for a first-order grating ($\lambda = 1.55\,\mu$m) is typically 240 nm. Experimental and theoretical investigations show that all static and dynamic optical properties of as-cleaved DFB lasers are strongly influenced by this phase relation between the grating and facet. The mirror losses of the various longitudinal modes, the mode discrimination, the intensity distribution, the optical spectra and the dynamic and noise characteristics vary considerably as a function of the end facet phases [11.1296]. Since these end facet phases are distributed randomly after the cleaving process, the yield of good DFB devices is limited. The problem of the uncertain end facet phases can be reduced by appropriate antireflection coatings.

Gain-coupled lasers. In gain-coupled laser structures, a longitudinal mode of the FP resonator is selected by implementing an axial gain or loss grating. In contrast to index-coupled DFB lasers, there is almost no reflection of the light wave in a gain grating. Therefore, the longitudinal intensity distribution and the spectral positions of the various modes are equal to those of an FP resonator. The mode selection is caused by the different overlap of the longitudinal modes with the gain grating. The mode, whose longitudinal field distribution exhibits the largest overlap with the gain grating or the smallest overlap with the loss grating, is selected.

Gain coupling can be realized in different ways:

- The grating structure can be etched directly into the active layer and regrown afterwards with a semiconductor material [11.1297] that is transparent to the laser emission (Fig. 11.104b). The resulting gain coupling in this grating structure is accompanied by a strong index coupling due to the high refractive-index difference between the active layer and the regrown material. With respect to the phase difference between the index and gain grating one distinguishes between in-phase and anti-phase gratings.
- A current blocking pnp layer structure above the active layer is corrugated and regrown, so that a periodic variation of the current density injected into the active layer is achieved [11.1298]. The parasitic index coupling can be kept very small in this type of

gain grating. Due to the Kramers–Kronig relations, however, the gain grating is inevitably accompanied by an index grating.

Loss coupling is obtained if the grating is etched into an additional layer that is separated from the active layer and is absorbing for the emission (Fig. 11.104c). This grating acts as a saturable periodic absorber for the laser emission from the active layer [11.1299]. In contrast to index coupling (Fig. 11.104a) the bandgap E_g of the grating is smaller than the bandgap of the active layer $E_{g,act}$. Usually, the absorbing grating layer has a different refractive index than the material used for the regrowth, so that considerable index coupling occurs. Alternatively, the loss grating can be integrated into a metallization at the surface of the device, so that no epitaxial regrowth is required [11.1300].

There are significant differences between the characteristics of gain- and loss-coupled semiconductor lasers compared to index-coupled DFB lasers:

- In contrast to index-coupled devices, there is almost no reflection of the light wave in an ideal gain grating. Thus, the interference of the reflected wave inside the grating and at the facets is avoided, so that the influence of the end facet phases is strongly reduced compared with index-coupled DFB lasers. Therefore, a high single-mode yield can be obtained without the necessity to use an antireflection coating of the end facets.
- The longitudinal photon distribution is, in the case of a small parasitic index coupling, similar to that of an FP resonator. Thus, the strongly inhomogeneous photon distribution in DFB lasers and the resulting problem of longitudinal spatial hole burning is reduced in gain-coupled devices.

Basics of Lasers with High Modulation Bandwidth

For the highest-bit-rate fibre-optic data communication (1.26–1.68 μm), the ultrafast conversion of the data from the electronic bit sequence into the appropriate optical bit pattern takes place by means of semiconductor lasers. This is done either with a CW laser combined with a subsequent ultrafast optical modulator or via direct modulation of the laser (intensity or frequency modulation). As the simplest example the intensity modulation of the laser is described. The bit pattern available as a sequence of ultrashort current pulses is translated by the laser into the appropriate sequence of ultrashort light pulses, which then propagate e.g. via a fibre towards the receiver. The photodetector retranslates the light pulses back to the electronic bit sequence. Seen in terms of a two-language dictionary, the laser translates ultrafast from electronics into optics and the photodetector ultrafast from optics into electronics.

For this purpose, the electrons injected through the n contact in the semiconductor laser must arrive as fast as possible at the most deeply bound energy states of the conduction-band QWs and the holes injected through the p contact as fast as possible into the ground states of the valence-band QWs (Fig. 11.105). Several retarding physical transport and relaxation processes are involved, whose combined time delay effect can however be minimized [11.1301, 1302]. Owing to the very high doping in the long conduction paths, which start at the contacts, very short dielectric re-

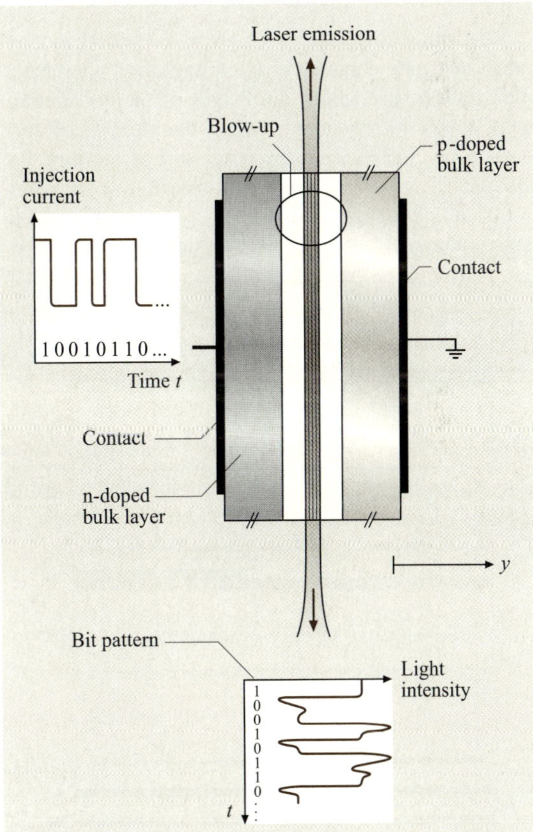

Fig. 11.105 Schematic representation of the direct modulation of a semiconductor laser, translation of an electronic bit pattern (*left*) via a corresponding temporal carrier distribution in the QWs (*center*) into an optical bit pattern (*bottom*)

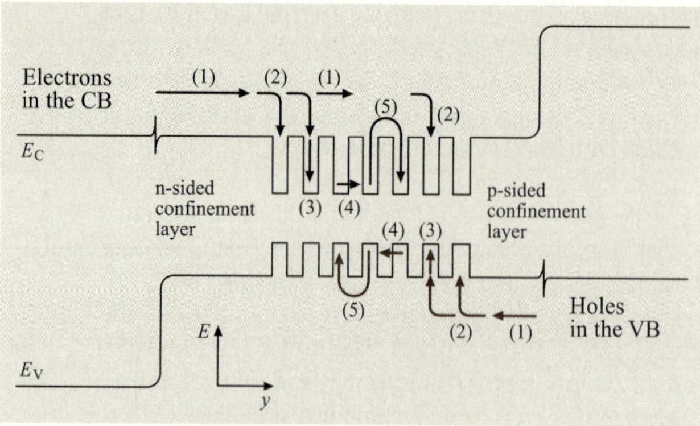

Fig. 11.106 Schematic representation of the charge carrier dynamics close of the active region (here seven QWs) in the band structure in space for an asymmetric laser structure [11.1302] (enlargement of Fig. 11.105 in the area indicated by the ellipse)

laxation times arise there. This can be visualized by means of a pipe completely filled with ping pong balls (high doping throughout the length of the conduction path). A ball injected at one end of the pipe, causes another to be ejected from the other end of the pipe. In this manner, a *current pulse* is transferred to the other end of the pipe almost without delay in heavily doped semiconductors.

In order to reduce optical losses through reabsorption, the confinement layers should be undoped or only slightly doped. Retarding effects arise as a consequence of the charge-carrier transport (number 1 in Fig. 11.106). Using the analogy of the balls and pipe above, a ball in an empty pipe (undoped) must go through the entire length of pipe (the length of a confinement layer) before finally exiting from the other end. Further retarding effects arise from the charge carrier capture (2), relaxation (3) in the respective QW ground states, by the reconciliation of charge carrier inhomogeneities between the individual pots, through tunneling (4) and thermal reemission (5). Since the mobility of the electrons is substantially higher than that of the holes, in asymmetrical laser structures [11.1302] the p-side confinement layers are reduced in thickness in favor of the n side. This favors the transfer of the less-mobile holes. In addition, this asymmetric laser structure design takes into account the different capture probabilities: the capture of holes into the QWs is much more efficient than that of electrons. Thus, a smaller p-sided confinement layer (reservoir of uncaptured electrons giving rise to bit interfering) is beneficial. In today's fastest laser diodes (largest modulation bandwidth) −3 dB modulation frequencies of up to 40 GHz can be obtained.

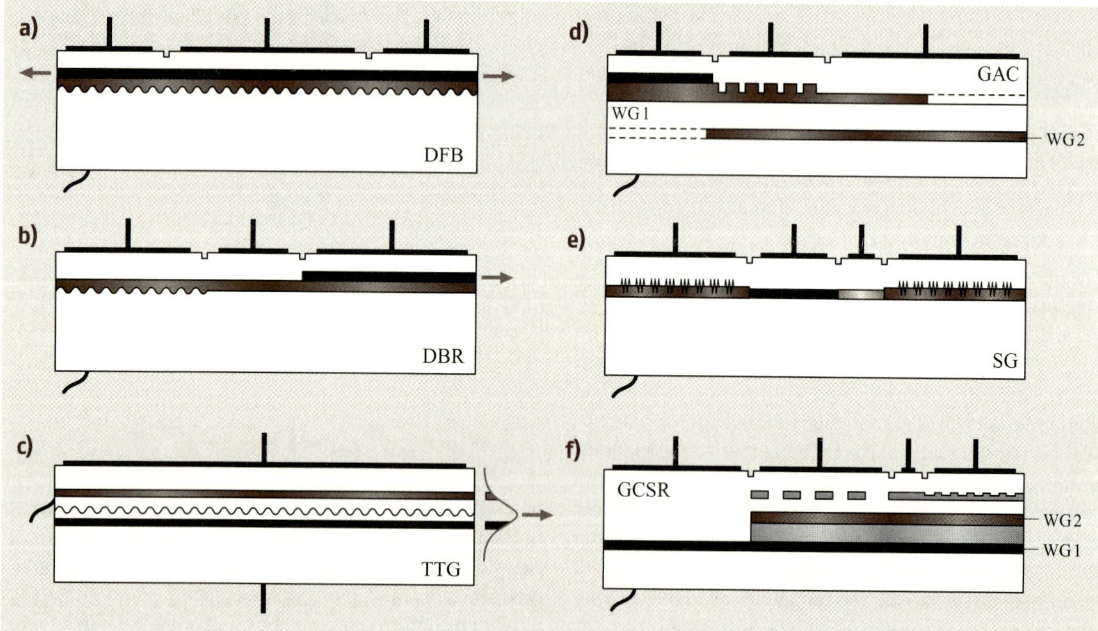

Fig. 11.107a–f Schematic cross-sectional views in the y–z-plane of different laser types: **(a)** three-section DFB, **(b)** three-section DBR, **(c)** TTG, **(d)** GAC, **(e)** SG, which is related to the SSG laser (not displayed) and **(f)** GCSR

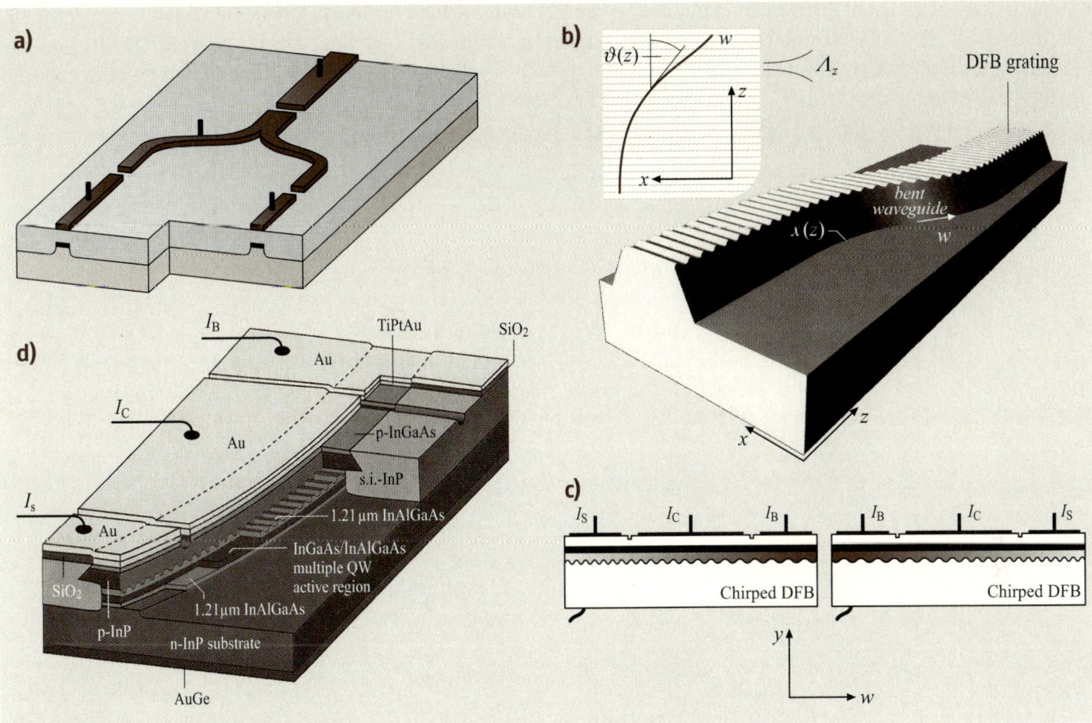

Fig. 11.108a–d Schematic perspective view of **(a)** a Y-laser [11.1303] and **(b–d)** three-section bent-waveguide chirped DFB lasers [11.1304], **(b)** the principle of generating chirped DFB gratings by specifically bent waveguides on homogeneous grating fields, cross section of two independent lasers originating from the bending displayed in **(b)**, **(d)** perspective schematic of a device shown in **(c)**. A scanning electron micrograph (SEM) of such a tunable three-section chirped DFB laser is displayed in Fig. 11.116a

Tunable Lasers

Wavelength tuning of an edge-emitting DFB semiconductor laser can be obtained according to (11.93), in the simplest case, by varying the effective refractive index via the injection currents in different sections, either thermally or by using plasma effects (carrier density variations). In these cases rising injection currents lead

1. to increasing emission wavelengths (red shift) via thermal effects and
2. to a blue shift via the plasma effect.

A superposition of both effects is found in multi-section DFB lasers [11.1305] since gain, tuning and DFB grating exist in all sections (Fig. 11.107a). In order to increase the total tuning range it is necessary to separate gain and tuning [11.1306, 1307]. In the three-section DBR laser (Fig. 11.107b) this is done longitudinally by using a gain section (*right*, the active region is shown in black), a grating-section (*left*) and a phase section (*center*) [11.1308]. The phase section only includes the brown waveguide layer(s) and is used to match the phase of the standing light wave after one round-trip. The gain is controlled, e.g., via the current between the top right contact and the bottom contact. In contrast, a vertical separation is used in the tunable twin guide (TTG) laser [11.1306, 1307, 1309] (Fig. 11.107c). A current from the left-side contact to the top contact only controls the carrier density in the brown waveguiding layer. The gain is controlled via the current between the left side contact and the bottom contact. In the three-section DBR and the TTG laser the tuning is based on the plasma effect, if only forward biasing is used.

Another very important tuning principle is based on two mode combs of slightly different mode spacing. In analogy to a vernier based on two scales [11.1306, 1307], a small detuning of the combs can address individual modes separated widely in the spectrum: only those modes which exactly occur in both combs can oscillate since only this situation provides sufficient reflectivity

Table 11.37 Tuning characteristics of different 1.55 μm laser types, DFB: distributed feedback, DBR: distributed Bragg reflector, TTG: tunable twin guide, SSG: superstructure grating, SG: sampled grating, GAC: grating-assisted coupler, GCSR: grating-assisted co-directional coupler laser with rear sampled grating reflector. The tuning occurs either thermally (t), via the plasma effect (p), via the co-directional coupler (CC) grating or via Vernier effect (V). TR: tuning range

Laser type	Control currents	Tuning (exp.) ($\Delta\lambda$/nm)	Tuning (max. continuous)	Tuning principles	Miscellaneous	Ref.
3 sect. DFB	2	3	3	t, p	Continuous TR	[11.1305]
3 sect. DFB	2	5.5	3	t, p	Bent waveguide, quasi contiunuous TR	[11.1294, 1304, 1312]
3 sect. DBR	3	10	7	p, (t)	Quasi continuous TR	[11.1313]
3 sect. DBR	3	22	7	p, t	Incl. both polarities, quasi continuous TR	[11.1314]
3 sect. DBR	3		7	p	Mode-hop-free, really continuous TR	[11.1315]
TTG	2	7	13	p, (t)	Continuous TR	[11.1306, 1309]
TTG	2	13	13	p, t	Incl. both polarities, continuous TR	[11.1306]
Y	4	51		V, p	No continuous TR	[11.1303]
SSG/SG	4	95		V, p	Quasi contiunuous TR, gaps in the total TR	[11.1310, 1311]
SSG/SG	11	38-50		V, p	Quasi contiunuous TR, *without* gaps in the total TR	[11.1316]
GAC	3	50-70		CC, p	No continuous TR	[11.1262, 1317, 1318]
GCSR	4	100		CC, p	Quasi continuous TR, *without* gaps in the total TR	[11.1319]
GCSR	4	50-114		CC, p	Quasi contiunuous TR, gaps in the total TR	[11.1320]
SSG/SG TTG	3	30-50		V, p	Quasi contiunuous TR	[11.1321]
MG-Y	4-5	46		V, p	Quasi contiunuous TR	[11.1322]

and, thus, sufficient net gain. This principle is applied in the (C^3) laser, Y laser, Mach–Zehnder interferometer laser, superstructure grating (SSG) DBR laser and sampled grating (SG) laser. Two FP cavities of different length and, thus, *two FP mode combs* are used in the Y laser (Fig. 11.108a) and the C^3 laser (two coupled FP lasers of slightly different length, not displayed). Two superstructure mode combs are shifted against each other in the SSG laser [11.1310] and SG DBR laser [11.1311] (Fig. 11.107e). The codirectional coupling in the grating assisted coupler (GAC) laser between two waveguides realizes an additional spectral filter which can be shifted spectrally in a controlled way against an FP mode spectrum. The two waveguides can be laterally or vertically (Fig. 11.107d) coupled. Depending on the size of the grating period, a grating can reflect light (inversion of the propagation wave vector, contra-directional coupling grating, short grating period, e.g., Fig. 11.107a,b,c,e) or change the magnitude of the propagation vector (maintaining propagation direction, codirectional coupling grating, long period, shown in Fig. 11.107d below the center contact). Table 11.37 includes a comparison of different tunable laser types with respect to important device characteristics and properties.

Generally the types discussed up to now cannot provide excellent characteristics in all features such as simple tuning (low number of control parameters), high SMSR, wide tuning, continuous tuning and high efficiency. This motivates the search for promising combinations of the above principles. Using a GAC to select a mode from an SG yields the grating-assisted codirectional coupler laser with rear sampled grating reflector (GCSR) [11.1319, 1320] (Fig. 11.107f). Another laser with 11 sections combines a SSG or a SG with many DBR sections of different grating period each [11.1316]. SSG or the SG laser and the TTG laser has been proposed [11.1321]. Combining the SSG laser and the Y laser yields the modulated grating Y laser [11.1322]. For comprehensive details of tunable lasers we refer to [11.1306, 1307]. Finally, coupled Mach–Zehnder interferometers or arrayed waveguide (AWG) structures can also be used as filter structures in lasers, e.g., in the digitally tunable ring laser using a ladder of ring-resonator filters [11.1323, 1324].

DFB lasers with an axially varying grating period are very attractive to tailor specific device properties. Using bent waveguides on homogeneous DFB gratings ($\Lambda_z = \text{const.}$) enables to generate a local effective pitch length $\Lambda(z)$ that exceeds Λ_z the more, the larger the local tilt angle of the bent waveguide deviates from a vertical intersection of the pitches i. e., the larger the local tilt angle $\vartheta(z)$ is. Figure 11.108b schematically shows a bent waveguide covering a homogeneous grating. Choosing appropriate bending functions $x(z)$ generates chirped DFB gratings with a varying effective pitch length $\Lambda(z) = \Lambda_z / \cos(\vartheta(z))$, as shown in the inset to Fig. 11.108b [11.1294]. The fabrication method is indicated in Fig. 11.108c, which shows two chirped DFB lasers formed after cleaving the structure in Fig. 11.108b in the x-direction in the center of the side of the device oriented in the z-direction. Since the light is strongly guided along the bent w-direction, the light sequentially passes quasi continuously varying pitch lengths $\Lambda(w)$, as displayed in Fig. 11.108c. Since ϑ is largest in the center of Fig. 11.108b, the two chirped DFB lasers have the largest pitch at the center facets. Various applications of chirped DFB gratings have been demonstrated [11.1294, 1304, 1312, 1325], including axially distributed phase shifts, higher SMSR, enhanced single-mode stability and lower line widths. In addition, chirped three-section DFB lasers (Fig. 11.108d) have demonstrated enhanced tuning ranges [11.1294, 1304, 1325] (Table 11.37). The mushroom-type laser structure (Fig. 11.100d) is applied in the chirped three-section DFB laser (Fig. 11.108d), which shows a modulation bandwidth of up to 26 GHz [11.1302].

Further Laser Types with Horizontal Resonators

In the bipolar lasers considered so far the light emission in the active layer takes place through the recombination of an electron in the conduction band with a hole in the valence band (band–band transition or *inter-band* transition). In unipolar lasers the radiative recombination takes place within a single band between two bound states of a potential well (*intra-band* transition, in Fig. 11.109 shown for the conduction band). Through so-called electron recycling, this process is cascaded in stages (explaining the name quantum cascade laser, Fig. 11.109). The band edges are accordingly tilted by the applied voltage (electric field). The multiple-QW structure [11.1326] shown here is based on a uniform QW composition and a different uniform barrier composition. However, there are in each case six different QW widths (the lower sequence of numbers) and barrier widths (the upper sequence of numbers) involved.

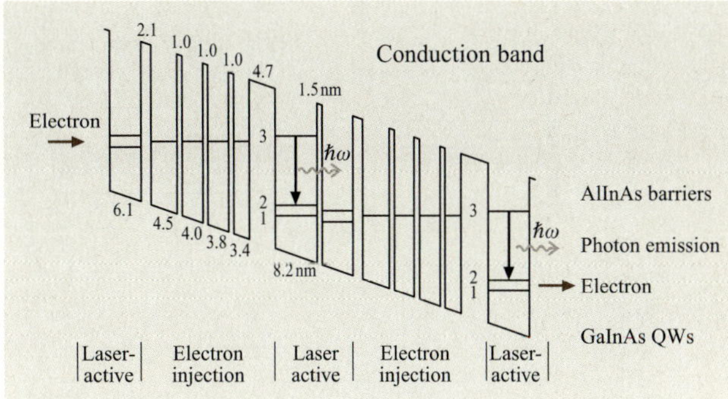

Fig. 11.109 Principle of a quantum cascade laser [11.1326]

Through tailored dimensioning, we ensure that in the case of resonance (for a certain applied voltage) each of the four narrowest QWs exhibits a bound state, and that these four states are energetically identical (i. e., they line up exactly horizontally). Furthermore, the widest QW is dimensioned in such a way that the highest state of the three bound states, marked with a 3 in Fig. 11.109, deviates only insignificantly from the four identical levels mentioned above. Due to the thin barriers and the resulting tunneling processes, an electron injected from the left tunnels horizontally into the excited level 3. After a radiative transition it falls down to level 2. Due to the very thin barriers, the wave functions are generally strongly delocalized. By optimizing the QW geometries the relaxation time from state 2 to state 1 is designed to be very small, so that the maxima of the wave function of state 2 lies on the left and that of state 1 on the right of both of the wider QWs.

This tunneling process proceeds in a stair-like cascade, which is illustrated with just two stages in Fig. 11.109. This example also illustrates very successful recent complex QW structures, which are based on a variety of different quantum cascade structures. Quantum cascade lasers are very successful to produce laser emissions in the mid infrared (3.5–10 μm, in some cases even up to 60 μm), where there are very few suitable semiconductor materials with extremely small bandgaps for the construction of bipolar lasers.

A very elegant way to obtain a longitudinal single-mode oscillation of an FP laser is to use one highly reflecting (HR) and one antireflection (AR) coated facet, with a piece of fibre optically coupled to the latter (Fig. 11.110a) including a DBR grating with a period ((11.93)) such that it filters exactly one single oscillating mode from the spectral amplification profile. This laser design is called a (semiconductor-)fibre laser. In

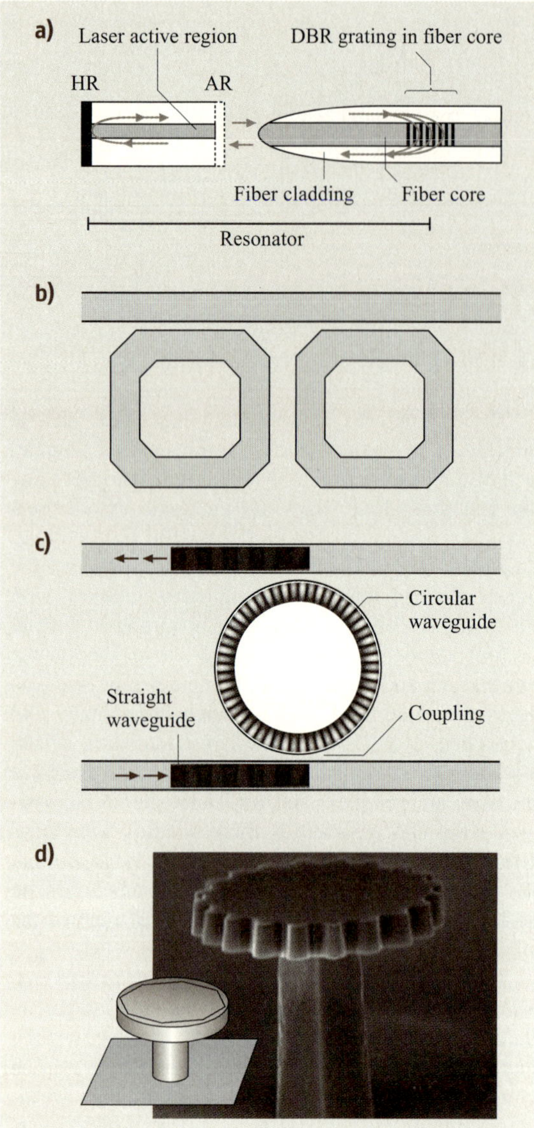

Fig. 11.110a–d Various designs of lasers with horizontal resonator. **(a)** FP laser with an external fibre mirror, **(b)** double ring resonators coupled to a straight waveguide, **(c)** standing wave in a single ring resonator coupled to 2 waveguides, **(d)** DFB microdisc laser [11.1327]; *inset*: FP microdisc laser with *whispering-gallery* mode

contrast, the active medium for optically pumped fibre lasers is a doped fibre.

Laser designs with straight aligned resonators have been considered so far. However, in the ring laser (Fig. 11.110b, c) and the microdisk laser (Fig. 11.110d), the beginning and end of the resonator merge into one another. In the case of waveguiding (Fig. 11.110b) modal solutions result if the length of the circular optical axis is close to an integer multiple of the wavelength in the medium. Similar selection rules apply in microdisc lasers, which can also be illustrated in the ray model with multiple reflections (eight in Fig. 11.110d) at the exterior surfaces. These modes are also known as *whispering-gallery modes* after the acoustic phenomena in the gallery of the St. Paul's cathedral in London and the Gol Gumgaz mausoleum in Bijapur city in India. There, a whispered word is discernible, after multiple reflections from the inner walls of the cupola, back at the ear of the speaker.

11.3.4 Basics of Surface-Emitting Lasers with Vertical Resonators (VCSELs)

The VCSEL presents a large technological challenge in terms of the implementation of DBR mirrors with extremely high reflectivity. In contrast to the edge-emitting DFB or DBR laser, the individual layers of the DBR mirrors in the VCSEL are successively deposited (e.g., by epitaxy). In this way, layer A and the neighboring layer B together make up one period. Since the reflectivity of the DBR mirrors rises with increasing number of periods and refractive index contrast, materials with large refractive index contrast are preferable in order to reduce the total number of periods and thus device cost. The insets in Fig. 11.111 show a DBR mirror (left inset) and the light intensity reflected by the mirror as a function of the wavelength (reflection spectrum, right inset). For $\lambda = 1.55\,\mu\text{m}$, the maximum reflectivity R_{max} in the stop-band is presented in the main picture as a function of the number of periods for different material systems. The data result from theoretical model calculations with consideration of the spectral variations of the refractive indices and absorption coefficients [11.1328, 1329]. Large differences appear in the figure due to different absorption and different refractive index contrast $2(n_A - n_B)/(n_A + n_B)$. For semiconductor epitaxial layers, absorption depends on the doping concentration and is well controllable. Absorption in dielectric layers is more difficult to control and depends strongly on the technological process. Unfortunately, no large refractive index contrasts can be realized in the GaInAsP material system. To achieve a reflectivity of 99.8%, 50 periods, a number that is in practice too high, are needed. In contrast, this is achieved in AlAs/GaAs DBRs with just 20 periods. If the two dielectrics Si_3N_4 and SiO_2 are combined, then even 13 periods are sufficient due to the very high relative refractive index contrast.

This enormously high reflectivity is achieved through constructive interference of the partial waves reflected at all heterointerfaces. Note that with a reflection at the interfaces from the optically thinner to the optically denser medium a phase change of π occurs, but not vice versa [11.1330, 1331]. The condition for constructive interference is fulfilled, e.g., by the combination of $\lambda/4$-thick layers A and B, as is the case in Fig. 11.111. Here, d_A and d_B are the physical thicknesses and $\lambda/4n_A$ and $\lambda/4n_B$ are the optical thicknesses. In general, the following relation applies to achieve high reflectivity using constructive interference

$$n_A d_A + n_B d_A = m\frac{\lambda}{2} \quad (\text{with } m = 1, 3, 5 \dots) . \tag{11.120}$$

On GaAs substrates the almost lattice-matched AlAs/(Al)GaAs combination is ideal for lasers in the wavelength range 800–1300 nm. Although, as with the edge emitters, trials have been made to implement long-wavelength VCSELs on GaAs substrates, 1.55 μm VCSELs are still based on InP. This is due to the fact that the active laser layers (e.g., GaInAsP and AlGaInAs) for the spectral range close to 1.55 μm can be best implemented on InP substrates. There are different possibilities for implementing highly reflecting 1.55 μm DBR mirrors on InP substrates: a kind of pressure bonding technique (wafer fusion) of AlAs/GaAs DBRs [11.1332, 1333] to embed the GaInAsP active region, strongly lattice-mismatched so called pseudomorphic AlAs/GaAs DBRs [11.1334], lattice-matched AlGaInAs, AlAsSb/AlGaAsSb, AlGaAsSb/InP DBRs [11.1335, 1336] or fusing of a low-period GaInAsP/InP DBR with active layers onto a AlAs/GaAs DBR [11.1337].

A resonator results from the combination of two DBR mirrors, where the volume between the DBR mirror-ends towards the center is called the cavity. If the cavity material is passive, then the setup is an optical filter, if it is active then a VCSEL results. Figure 11.112 shows schematically the structure of a VCSEL with two multiple-layer semiconductor DBR mirrors. In the illustration the upper mirror is highly reflective, so the laser light is emitted essentially downwards (brown arrow). The envelope of the vertical intensity distribution discussed below is indicated by the lateral expansion of the brown arrow (Fig. 11.96, depicted in grey-tone variations). The holes are injected from above through the upper p-doped DBR mirror into the active laser zone;

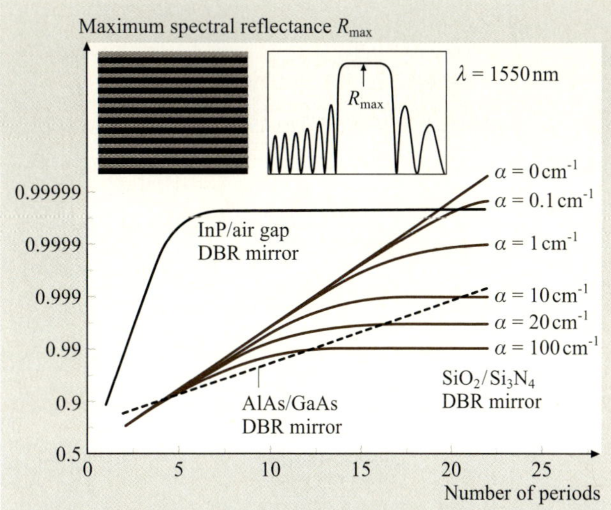

Fig. 11.111 Maximum spectral reflectivity R_{max} in the stop-band for $\lambda = 1.55$ μm as a function of the number of periods for different DBR mirror systems. *Left inset*: cross section of a DBR mirror. *Right inset*: reflectivity as a function of wavelength

the electrons are injected from below through the lower n-doped DBR mirror. The outer curved black arrows indicate the current flow. Here, it is challenging to reduce the electrical resistance of the p-doped DBR mirror by sophisticated doping (impurity concentration profiles) in order to reduce the operating voltages and the buildup of Joule heating. Figure 11.112a schematically depicts the structure of a VCSEL with two dielectric DBR mirrors. The holes and electrons are injected by means of ring contacts, whereby the current flow bypasses the electrically insulating DBR mirrors. Furthermore, of crucial importance is the lateral electrical confinement to enable the production of high electron–hole-pair densities in the central part of the active zone, and thereby very small threshold current densities. Therefore, an electrically insulating ring is built surrounding the lateral laser mode. The realization of this is technologically demanding and is achieved, e.g., by ion implantation or selective oxidation. A sufficiently high thermal conductivity of the DBR mirrors is of further importance in enabling efficient heat dissipation from the active region to the heat sink and, if necessary, of that heat generated inside the DBR mirrors.

For the VCSEL with two dielectric DBR mirrors and a cavity length of $3\lambda/2$ shown in Fig. 11.112a, the electric field of the standing light wave is computed [11.1328, 1329] and presented in Fig. 11.113 together with the entire multilayer structure. In order

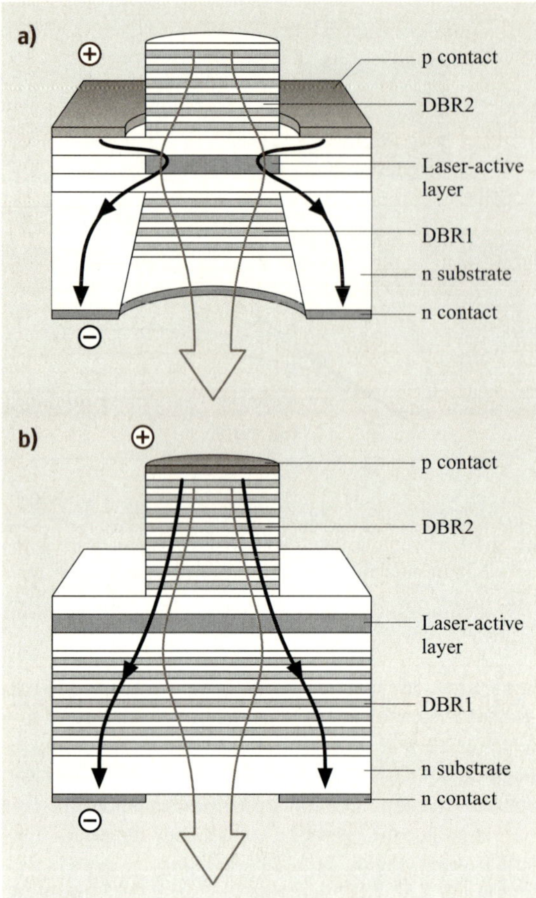

Fig. 11.112a,b Schematic setup of a VCSEL with electrically insulating **(a)** and electrically conducting DBR mirrors **(b)**

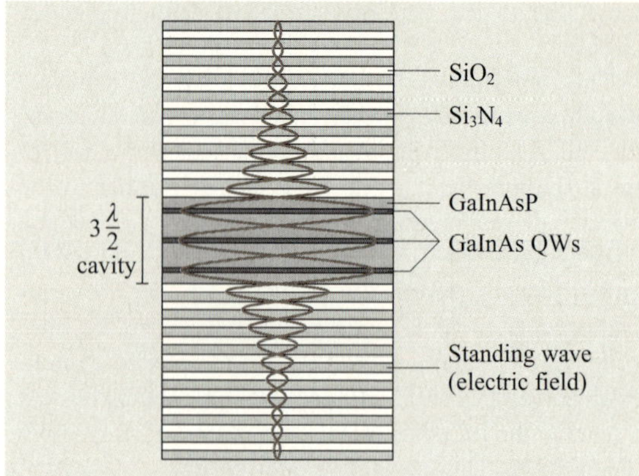

Fig. 11.113 Layered structure of a VCSEL including the standing wave of the electrical field

to achieve a high gain in the laser design, the QWs are positioned at the maxima (antinodes) of the half-waves. Ideally, three QWs, three double QWs or three triple QWs should be placed in the $3\lambda/2$ cavity shown. The envelope of the standing light wave field gradually drops outwards from center due to the distributed reflecting effect of the DBR mirrors. We note again that exactly one quarter of a wavelength is allotted to each mirror layer ($\lambda/4$ layers) and that the nodes for the design wavelength lie exactly at the interfaces.

In the following, the VCSEL spectra are intuitively constructed. To derive this, we consider a semiconductor cavity between two mirrors indicated in the top left of Fig. 11.114. Based on the fixed cavity length, (11.92) leads to the three modes displayed as lines in the spectrum on the right of the figure, which become sharper for higher mirror reflectivities. This means that a cavity of this length is suitable for each of these (laser) wavelengths if the standing wave experiences a sufficiently large optical gain at its peak (antinodes). On the contrary case, for a desired (fixed) laser wavelength λ the cavity length can be selected straight away as $\lambda/2$, $2\lambda/2$ or $3\lambda/2$. If the two mirrors are implement as DBRs (bottom left of the figure), then the line spectrum at the top right is *superposed* with the known reflection spectrum (Fig. 11.114 bottom right) of the DBR mirrors, resulting in the laser spectrum already shown in Fig. 11.103d. The stop-band of the VCSEL is substantially larger because of the higher refractive index contrast in the DBR mirrors compared to that of the DFB laser with the relatively small refractive index contrast between the quasi-layers of the DFB grating. Following from Fig. 11.103c, d the λ/4-phase-shifted DFB laser has many similarities to a VCSEL with a $\lambda/2$ cavity and two very long DBR mirrors with a small refractive index contrast.

Note that the following numerical values are strongly dependent on wavelength and material and that a thoroughly comprehensive treatment would exceed the framework of this volume. For VCSELs, optical power outputs of < 1 mW (lateral single-mode operation) are typical. With a laterally expanded active region, i.e., laterally multimode VCSELs, achieve up to 120 mW. Typical threshold currents lie around 1 mA, with record values of about 0.06 mA (corresponding to 350 A/cm^2). The highest power outputs and the lowest thresholds are achieved in the spectral range 850–1000 nm. Compared to edge emitters the thresholds are amazingly low, but with rather moderate maximum power outputs. In broad-area edge emitters for example, over 6 W [11.1338] and external quantum efficiencies of over 56% [11.1338] have been achieved. While the VCSEL has enormous advantages in the laser–fiber coupling due to the small beam divergence and the symmetric mode profile, the edge emitter incurs substantial overhead costs for light coupling due to the higher divergence and the elliptical mode profile. This is achieved in practice with aspherical lenses or with complicated tapered waveguide structures in the laser or fibre. The VCSEL offers further substantial advantages as it enables simple optical on-wafer testing similar to the testing of integrated electronic circuits. In contrary, edge emitters must be individually isolated for characterization, i. e., at least cleaved into bars, or additionally for surface out-coupling, a 45° mirror (Fig. 11.96) must be implement in combination with etched mirrors.

The modulation bandwidths of VCSELs are typically around 1 GHz (the record being 10 GHz), compared with typical values of 15 GHz (the record being 40 GHz) for edge emitters. If one compares the laser line widths of devices without external resonators, then VCSELs typically attain 200 MHz (the record being 50 MHz) while edge emitters typically attain 1 MHz (the record being 10 kHz). Edge-emitting semiconductor lasers exist in the range between 350 nm and 12 μm, while in the green and mid-IR spectral regions there are as yet either no devices exist or the component lifetimes are not sufficient for practical applications. Electrically pumped VCSELs exist so far within a spectrally much smaller range (420 nm to 2.05 μm) [11.1339–1341]. In VCSELs wavelength tuning is particularly difficult using charge carrier induced (current-induced) or thermal refractive index changes. A typical tuning range of 1 nm is attained in VCSELs compared to 80 nm in single-mode edge emitters. Finally we present a tuning concept for optoelectronic devices with vertical resonators which permits ultrawide wavelength tuning on the basis of only one control parameter.

Micromechanical Tunable Filters and VCSELs

A perpendicular light wave (wavelength λ_D) incident on a highly reflecting DBR mirror is up to 99.9% reflected. The period is normally selected in such a way that λ_D lies at the center of the stop-band (Fig. 11.114 bottom right). When a second identical mirror is positioned parallel at a distance, say, $1.5 \times \lambda_D$ from the first one, as shown in Fig. 11.114, it is experimentally observed that the first mirror does not reflect the wavelength λ_D any more. Although at first sight astonishing, the arrangement of the two DBR mirrors is now almost 100% transparent for λ_D.

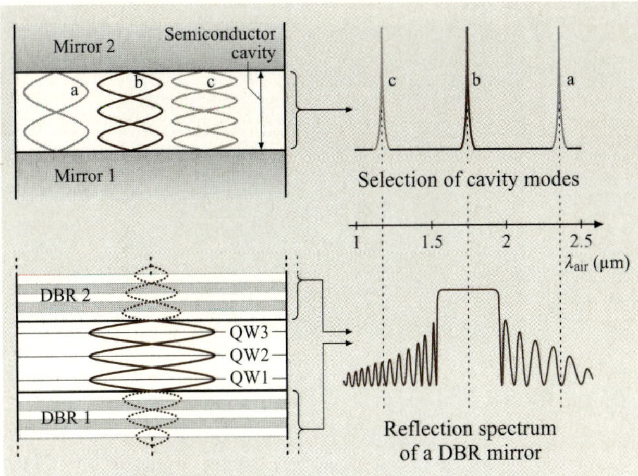

Fig. 11.114 Characteristic spectral components of a VC resonator. Three standing waves in the cavity ((*a–c*) *above left*) cause, as marked by the *bracket*, the build up of the FP modes (*above right*), the DBR mirrors (*bottom left*) are characterized by the reflection spectrum (*bottom right*), as indicated by the two *brackets*. With a very short cavity length only one FP mode is located in the stop band (along the *broken vertical lines*)

For λ_D in a DBR mirror, all reflected partial waves interfere constructively (99.9% reflection) on the incident side of the mirror and destructively on the opposite side. This is the result of complicated multiple reflections and complex zigzag paths (in the simple ray model) portrayed in a simplified manner occur. The light waves penetrate the mirror, but due to the perfect destructive interference, behind the mirrors, they transfer no energy. By the targeted positioning of the second mirror (and adding further boundary surfaces) all partial waves now interfere destructively on the incident side and constructively on the opposite side (behind the second DBR mirror). Furthermore, a standing wave forms in the cavity for λ_D (Fig. 11.114).

Within the stop-band, except within the range of the very sharp filter line (centered at λ_D) all wavelengths are up to 99.9% reflected. In this way extremely high-quality optical filters can be implement, e.g., for fibre-optic telecommunications based on dense wavelength division multiplexing.

Figure 11.111 contains a further material system, so far not discussed, with an extremely high refractive index contrast ($n_{InP} = 3.2$, $n_{air} = 1$). With only four periods, a reflectivity of over 99.8% can be obtained. This unusual structure can be made, e.g., from a semiconductor multilayered structure with alternating InP and $Ga_{0.43}In_{0.57}As$ layers by selective etching of the GaInAs

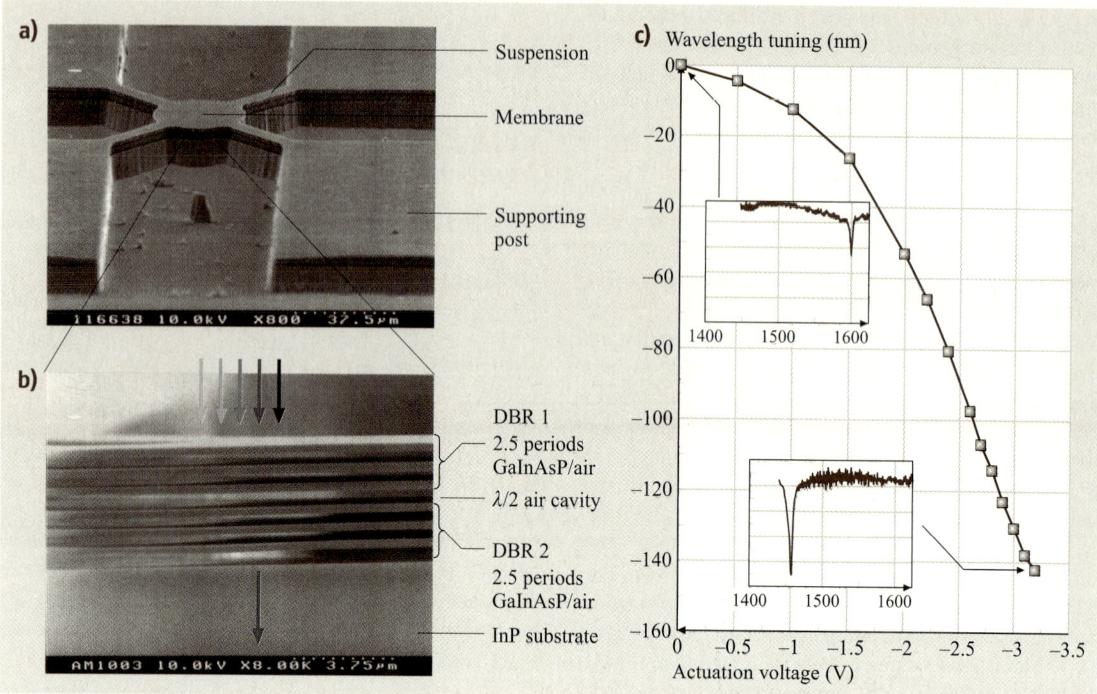

Fig. 11.115a–c Micromachined filter device with vertical resonator based on semiconductor multiple air gaps [11.1342, 1343]. Electron micrograph of the cross section of a filter **(a)**, details in **(b)**, corresponding experimental wavelength tuning as function of the actuation voltage in **(c)**

layers (micromechanical sacrificial-layer technology). Lattice-matching already exists through epitaxy on an InP substrate and, if necessary, an outstanding compatibility with 1.55 μm GaInAsP laser-active layers is available. Figures 11.115a,b show a filter structure with six InP membranes, each of which is fixed to the supporting posts at four points. If a reverse-bias voltage is applied between the upper p-doped and the lower n-doped DBR mirrors the membranes including the sandwiched air cavity can be electrostatically actuated. The cavity length and thus the filter wavelength can be varied in this way with only one control parameter (voltage). With such an optical filter, an enormous continuous wavelength tuning range of 142 nm (Fig. 11.115c) was experimentally achieved with only 3.2 V [11.1328, 1329, 1342, 1343]. In other structures of the same type, a continuous tuning range of as much as 221 nm has been obtained [11.1342]. The same tuning principle occurs in Figures 11.116c,d showing a VCSEL with InP/multiple-air-gap DBR mirrors, GaInAsP laser-active QW layers and an InP substrate.

In conclusion, based on Fig. 11.116 a very interesting parallel between quantum electronics (Fig. 11.116a,b) and quantum photonics (Fig. 11.116c,d) can be drawn. In order to tailor the electronic levels for electrons and holes, semiconductor heterostructures of predetermined composition and thickness are implemented. If the layer thickness of the material with the smaller bandgap is on the order of the magnitude of the electron wavelength, a quantization occurs that leads to defined quantized energy levels in the QWs. These quantized energy levels (eigenvalues) and the corresponding electronic wave functions (eigenfunctions, i. e., modes) are solutions of the Schrödinger equation. The materials, stress and layer thicknesses are set up in this example with the 10 AlGaInAs QWs (Fig. 11.116b) in such a way that the optical emission is at 1.55 μm. These QWs serve as the laser-active medium of an ultrafast semiconductor diode laser (Fig. 11.116a) [11.1302, 1344]. In a very simplified depiction, the multiple QWs define a resonator for electron waves.

The electron microscope micrographs on the right show the exact analogy for photons. In order to select a defined mode in the resonator of a VCSEL, taking into account the refractive indices, layer thicknesses in the order of magnitude of the photon wavelength are chosen. A kind of quantization also arises in this case. The effective refractive indices (eigenvalues) and the corresponding photon wave functions (eigenfunctions, i. e., modes) are solutions of the Helmholtz equation. From the mathematical point of view, the space-dependent parts of the Schrödinger and Helmholtz equations have equivalent behavior. The strong refractive index contrast of InP/air-gap multi-membrane structures is again used in Fig. 11.116d to implement highly reflective DBR mirrors for a 1.55 μm VCSEL (Fig. 11.116c) [11.1342, 1343]. Analogously the periodic refractive index variation defines a resonator for photon waves.

Further analogies in this area are offered by periodic 2-D quantum-dot fields and 2-D photonic crystals, or photonic band structures and electronic band structures.

11.3.5 Edge-Emitting Lasers and VCSELs with Low-Dimensional Active Regions

In commercial semiconductor laser diodes, bulk semiconductor materials (3-D) were exclusively used for the active zones for 25 years. Starting from 1990 edge emitters with QWs (2-D) were implement in research labs and continuously improved. Only after a further five years QW lasers were commercialized and subsequently surpassed the 3-D lasers with lower threshold current densities, higher power outputs, higher characteristic temperatures T_0 and higher bit rates due to more-favorable gain profiles and better electronic confinement. Although quantum wire (1-D) lasers should exhibit even better characteristics, they have so far not been very successful due to technological and geometrical constraints. In particular long carrier transport and carrier-capture times substantially limit the bit rates in quantum wire lasers. Quantum dot QD (0-D) laser structures [11.1263–1272] are a lot more promising. However, a problem at the moment is the strongly varying size of QDs (Fig. 11.93). The theoretically very sharp density-of-state profiles are strongly inhomogeneously spread, so that the predicted high differential gain dg/dn does not arise in practice. Compared with the 3-D and 2-D structures so far no spectrally narrower and higher-gain profile has been obtained. Particularly broad gain profiles can be of advantage however for extremely widely tunable lasers (see the section on tunable lasers in Sect. 11.3.3 and that on micromechanical tunable VCSELs in Sect. 11.3.4). However, as already mentioned, additional research and development effort is expected to provide excellent QD lasers.

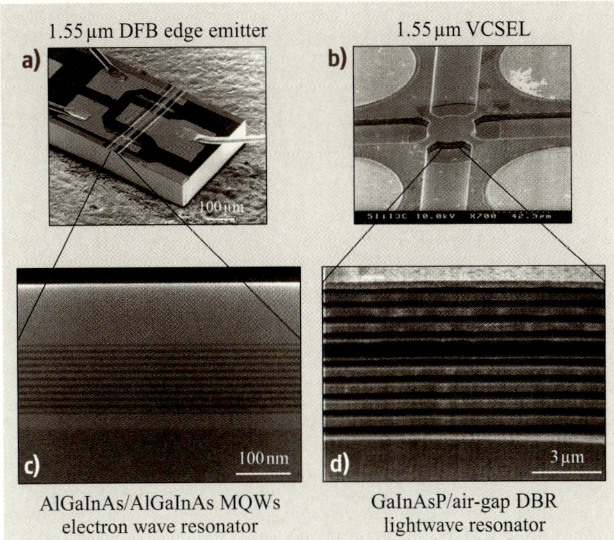

Fig. 11.116 **(a)** Tunable, edge-emitting three-section laser with axially varying DFB grating known as a chirped DFB grating [11.1294], **(b)** strain-compensated multiple QW structure with tensile-strained AlGaInAs barriers and compressively strained AlGaInAs wells[11.1344], **(c)** VCSEL based on multiple InP/air membranes (*centre*), each of which is supported by four suspensions connected to the square supporting posts [11.1343], **(d)** cross section of the vertical resonator, consisting of the laser-active GaInAsP QW region, which is embedded between two InP/air-gap DBR mirrors [11.1342]

11.3.6 Lasers with External Resonators

Both edge emitters and VCSELs can be implemented, as shown in Fig. 11.96, with an external resonator mirror, in contrast to Figs. 11.115 and 11.116c, which involve a relatively large air gap. This can serve several purposes:

1. the extension of the resonator length to achieve narrower line widths,
2. wavelength tuning by inserting a wavelength-selective, rotating element (prism, etalon, grating),
3. mode coupling for the generation of periodic sequences of ultrashort pulses and
4. wavelength conversion by insertion of an optically nonlinear crystal.

Examples of the last of these are lasers in which a frequency-doubling crystal in the air gap of the resonator transforms the IR light, which is emitted from an edge emitter with one side antireflection-coated or a half-cavity VCSEL (active region plus one DBR mirror), into the yellow, green and blue range.

11.4 The CO_2 Laser

The CO_2 laser is one of the most important lasers for industrial, medical and scientific applications. Applications include high-precision material processing, cutting and welding of sheet metal, marking of plastics, cutting of paper and fabrics, surface treatments for steel hardening or medical applications such as tissue coagulation.

CO_2 lasers use a gas mixture of helium, nitrogen and carbon dioxide as the active medium, which is usually excited by an electrical gas discharge. The key characteristics are emission in the mid-infrared at wavelengths around 10 μm and a continuous-wave output power ranging from a few watts for sealed-off miniature laser modules to more than 10 kW for high-power CO_2 lasers with fast gas flow. While most CO_2 lasers operate in a CW mode with quite fast power modulation enabled by electrical excitation, there are also pulsed systems based on Q-switching or the so-called tranversal-excited atmospheric-pressure (TEA) laser. The beam quality is often excellent and in many cases nearly diffraction-limited. This is important for applications such as precision cutting or remote welding. The mid-infrared emission implies that standard glass lenses and glass fiber-optic waveguides cannot be used because of the high infrared absorption of silica-based glasses. Transmitting optics such as focussing lenses or partially reflecting mirror substrates are commonly made of zinc selenide, silicon or germanium. For beam guidance moving-mirror systems mounted on linear and swivel axes are used. Besides the most common emission wavelength of 10.6 μm, special tunable CO_2 lasers can generate laser light at dozens of distinct emission lines between roughly 9.2 μm and 11 μm. Such lasers are used for scientific applications such as molecular spectroscopy and optically pumping of far-infrared gas lasers.

The physical principles and the technical realizations of typical CO_2 lasers are described in the following sections.

11.4.1 Physical Principles

Basic CO_2 Laser Tube Principles

Laser action in a CO_2 gas was first described by *Patel* in 1964 [11.1345, 1346]. The basic physical principles have changed little since then. An excellent and detailed discussion of CO_2 laser fundamentals is given in [11.1347] while additional general laser fundamentals are described in [11.1348] and [11.1349]. Figure 11.117 shows a sketch of a basic longitudinally direct current (DC)-excited CO_2 laser. The laser gas is usually a mixture of CO_2, N_2 and He in a ratio of 1:2:8, for example. The total gas pressure in such a DC-excited laser is a few tens of hPa. The gas is excited by an electrical gas discharge, for example by a steady longitudinal DC glow discharge inside a quartz glass tube. Typical required voltages are 15 kV per meter of discharge length, with a DC current of a few tens of mA. The voltages needed to start the discharge can be considerably higher. To maintain a stable α-type glow discharge [11.1350]

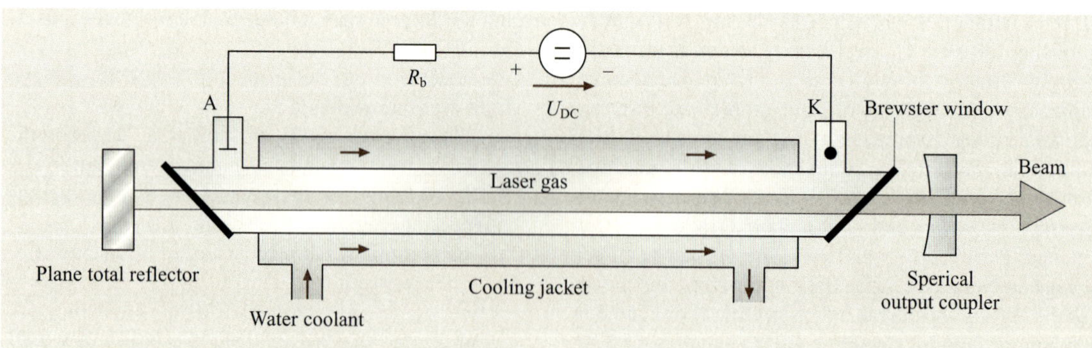

Fig. 11.117 Basic longitudinally DC-excited CO_2 laser

an external ballast resistor R_b is required in series with the high-voltage power supply. This compensates for the differential negative impedance characteristic of the gas discharge and limits the discharge current in order to prevent arcing. Alternatively, power supplies with fast voltage- and current-regulating electronics can be used. Typical lengths of the gas discharge are on the order of one meter, depending on the desired output power.

Such laser tubes can be filled once with the laser gas mixture and are then sealed-off for the lifetime of the laser tube. Other lasers are provided with a gas inlet and outlet to maintain a gas flow through the laser tube. This improves the laser power as fresh laser gas is provided to the laser but also requires additional peripherals such as gas reservoirs and vacuum pumps. Optical windows to seal off the tube can be realized by plane transparent plates at the Brewster angle with respect to the beam axis. One polarization state of the laser light can pass through the Brewster window without unwanted reflections. Thus, the Brewster window also acts as a polarization-selecting element inside the laser resonator.

The CO_2–N_2 molecule mixture can be excited inside the gas discharge by collisions with free electrons [11.1351]. The excited roto-vibrational states provide optical gain at certain wavelengths in the infrared. A simple plano-concave stable optical resonator with two mirrors located at each ends of the discharge tube can be used to provide continuous laser oscillation. The totally reflecting mirror is often made of copper, optionally with additional gold or protective dielectric layers to enhance lifetime and reflectivity. The output coupler can be made of zinc selenide (ZnSe), germanium or silicon, as these materials have good transparency in the mid-infrared, in contrast to common silica-based glasses. The required reflectivity is provided by dielectric coatings on the mirror substrate. As the optical small-signal gain of the active CO_2 gas medium is quite high compared to a helium–neon (HeNe) gas laser and is in the range of $1\,m^{-1}$, reflectivities of output couplers may range from 20% to 90%. Optimum output coupler reflectivities for a maximum laser power are often determined experimentally. Alternatively, they can be calculated by a so-called Rigrod analysis if parameters such as the small-signal gain and the saturation intensity of the medium are known [11.1352, 1353].

Typical efficiencies η_E, defined as the ratio of the extracted optical laser power P_L to the applied electrical power P_E into the discharge, are in the range of 10%. The largest part of the electrical excitation power is dissipated as thermal energy inside the gas volume. To avoid a high thermal population of the lower laser energy levels, the CO_2 laser gas must be kept cool at temperatures of 400–500 K, depending largely on the different CO_2 laser types. A simple method for cooling is a second glass tube, which acts as a cooling jacket coaxial to the laser gas tube with a continuous water flow, as shown in Fig. 11.117.

A rough estimate for these diffusion-cooled DC excited lasers is an output power per discharge length of 80 W/m. This cannot be scaled by using larger cross sections of the laser tube since this would reduce the amount of thermal energy that can be removed from the middle of the gas discharge by diffusion cooling to the walls. The overheating of the gas and the resulting thermal population of the lower energy level of the laser transition would reduce the optical gain. In fact, for optimized cooling the glass tube should be as small as possible without introducing significant aperture losses by truncating the free-space laser beam. Typical diameters are 5–10 mm, depending on the tube length and the optical resonator design [11.1354].

Vibration and Rotation of the CO_2 Molecule

The emission wavelengths of the CO_2 laser are determined by the vibrational and rotational energy levels of the CO_2 molecule, which are discussed in more detail in [11.1355, 1356] or [11.1357]. All data in the following sections are given for the naturally most abundant isotopologue $^{16}O^{12}C^{16}O$. Figure 11.118 shows the three fundamental vibrational modes: the symmetric stretch mode υ_1 along the molecule axis, the bending mode υ_2 with motion of the C atom in a plane perpendicular to the molecule axis, and the asymmetric stretch mode υ_3. The bending mode is twofold degenerate because of the pos-

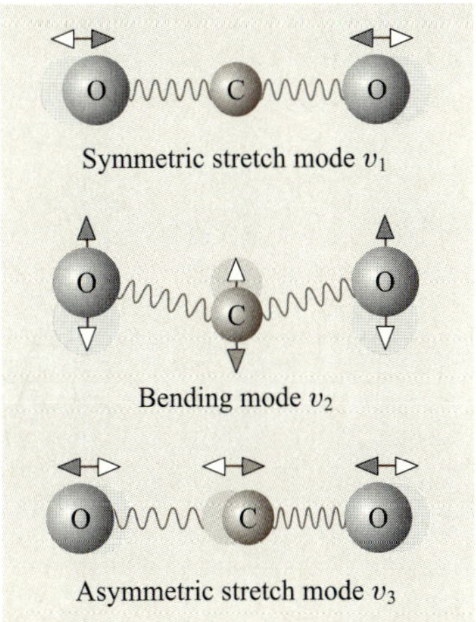

Fig. 11.118 Vibrational modes of the CO_2 molecule

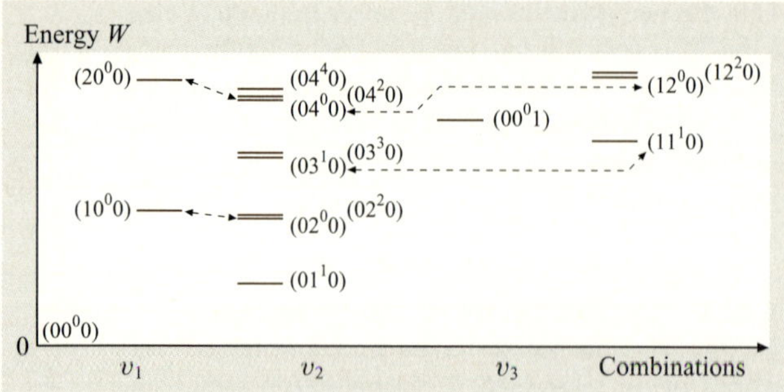

Fig. 11.119 Some of the lowest vibrational energy levels of the CO_2 molecule. The Fermi resonance couples energy levels as indicated by the *dashed lines*

sible bending motions in the two orthogonal coordinates of the plane.

In the classical model of the unperturbed harmonic oscillator each mode has an associated vibrational frequency:

$$f_1 = 40.51\,\text{THz}, \quad f_2 = 20.15\,\text{THz},$$
$$f_3 = 71.84\,\text{THz}\,. \qquad (11.121)$$

In the quantum-mechanical model the energy of any vibrational state of the molecule can only have discrete values called the vibrational energy levels. Using the integer vibrational quantum numbers n_1, n_2 and n_3 that describe the degree of excitation of the vibrational modes, the total vibrational energy W_v of the molecule can be written as a sum over the single vibrations:

$$W_v = hf_1\left(n_1 + \frac{1}{2}\right) + hf_2\,(n_2 + 1) + hf_3\left(n_3 + \frac{1}{2}\right)\,. \qquad (11.122)$$

Even if there is no vibrational excitation present (all $n_i = 0$) and the molecule is in its ground state there is energy stored inside the molecule that cannot be exchanged. Thus, this zero-point energy is neglected in all further energy formulas and diagrams.

To characterize the vibrational state completely another quantum number l is required to define the distribution and the phase of the bending vibration quanta in υ_2 into the two orthogonal coordinates. For each number n_2 there are $n_2 + 1$ values for the angular momentum l:

$$|l| = \begin{cases} n_2,\ n_2-2,\ n_2-4,\ \dots,\ 0\,; & n_2 \text{ even} \\ n_2,\ n_2-2,\ n_2-4,\ \dots,\ 1\,; & n_2 \text{ odd}\,. \end{cases} \qquad (11.123)$$

For example for $n_2 = 2$ the value $l = 0$ describes a linear bending motion, and $l = -2$ or $l = 2$ describes circular motions of the C atom in opposite directions in the plane perpendicular to the molecule axis. A vibrational state of the CO_2 molecule is labeled with the common Herzberg notation $(n_1 n_2^l n_3)$, for example (00^00) for the ground state or (12^20) for a state with simultaneous excitation of several vibrational modes [11.1355].

Figure 11.119 shows the energies of some of the lowest vibrational states. The energy levels observed by emission and absorption of radiation, as shown in Table 11.38, are slightly different than those calculated using the simple-harmonic-oscillator theory (11.122). More-precise calculations using anharmonic corrections and considering the mutual coupling of the three vibrational modes are therefore required.

The so-called Fermi resonance is of particular importance for the laser process in the CO_2 laser. Vibrational states with equal values of the term $2n_1 + n_2$ and $l = 0$ have almost the same energy and are strongly coupled. Due to this Fermi resonance the (10^00) state is shifted up

Table 11.38 Labeling, symmetry/parity classification and energies of some low-lying vibrational states of the CO_2 molecule [11.1358]

State			Frequency and energy		
Herzberg	AFGL	Type	$\bar{\nu}$ (cm^{-1})	f (THz)	W (meV)
(00^00)	00001	Σ_g^+	0	0	0
(01^10)	01101	Π_u	667.380	20.008	82.745
(02^00)	10002	Σ_g^+	1285.408	38.536	159.370
(02^20)	02201	Δ_g	1335.132	40.026	165.535
(10^00)	10001	Σ_g^+	1388.184	41.617	172.113
(03^10)	11102	Π_u	1932.470	57.934	239.596
(03^30)	03301	Φ_u	2003.246	60.056	248.371
(11^10)	11101	Π_u	2076.856	62.263	257.497
(00^01)	00011	Σ_u^+	2349.143	70.426	291.257

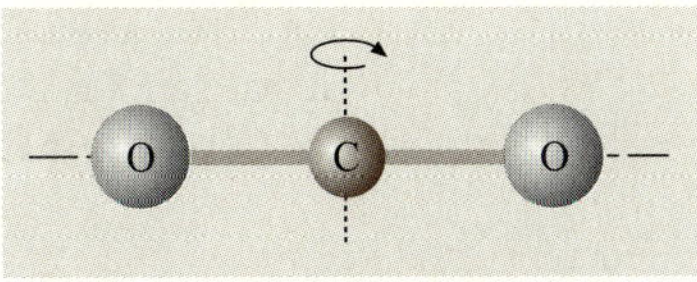

Fig. 11.120 Rotation of the CO_2 molecule

and the (02^00) state is shifted down in energy. The population densities of these states are quickly exchanged by collisions with each other. Also the quantum-mechanical wave functions and the vibrational motion of the two resulting levels are a strong mixture of the unperturbed vibrational states. In the CO_2 laser literature the resulting levels are sometimes labeled (I) for the more (10^00)-like state and (II) for the more (02^00)-like state. Spectroscopic databases, which are an excellent reference for accurate absorption and emission wavelengths of the CO_2 molecule, often use a different labeling scheme than the simple Herzberg scheme, for example the so-called AFGL (Air Force Geophysics Lab, USA) notation, which gives a better treatment of the Fermi resonance [11.1359, 1360].

Superimposed onto the vibration is the rotational motion of the CO_2 molecule, as shown in Fig. 11.120. Only the rotation around an axis perpendicular to the molecule axis has a significant moment of inertia and is considered further. The discrete energy levels in the rigid-rotator model are given by

$$W_r = B \cdot J(J+1)\,, \tag{11.124}$$

where B is the rotational constant and J is the quantum number for the rotational state. The higher the value of J the faster would be the rotation of the molecule in the classical mechanical model. For the (00^01) vibrational state, for example, B has a value of

$$B = 48.0\,\text{meV} = h f_r = h\,11.6\,\text{GHz}\,. \tag{11.125}$$

More precise calculations show that the rotational constant B is slightly dependent on the vibrational state because of the anharmonicity and also dependent on the rotational quantum number J itself because of centrifugal forces.

The total internal energy $W = W_v + W_r$ of the CO_2 molecule is the sum of the vibrational and the rotational energies. The energy stored in the rotational motion is usually much smaller than in the vibration motion. Thus, the rotational energy levels can be thought of as superimposed on each vibrational level.

Emission Lines of the CO_2 Laser

Absorption, spontaneous and stimulated emission of light can occur when the energy of a photon $W_p = h f_p$ equals the energy difference $\Delta W = W_2 - W_1$ between two given energy levels of the respective atom or molecule. Additionally, selection rules based on the conservation of the spin or angular momentum and due to symmetry properties of the wave functions have to be considered. Based on the formalism of vibrational matrix dipole elements only certain transitions are allowed [11.1357]. The strongest allowed CO_2 laser emission lines result from the vibrational transitions (00^01) → (10^00) centered at wavelengths around 10.4 μm (10 μm band) and (00^01) → (02^00) around 9.4 μm (9 μm band). For these regular transitions the following selection rule for the rotational quantum numbers of upper and lower energy levels apply:

$$\Delta J = J_2 - J_1 = \pm 1\,. \tag{11.126}$$

For a given lower rotational state J_1 there are two different possibilities for the emission of photons, as shown in Fig. 11.121. Transitions are named after the rotational quantum number of the lower state. Additionally, the transitions are labeled P (the P branch) for $\Delta J = -1$ and R (the R branch) for $\Delta J = +1$. For

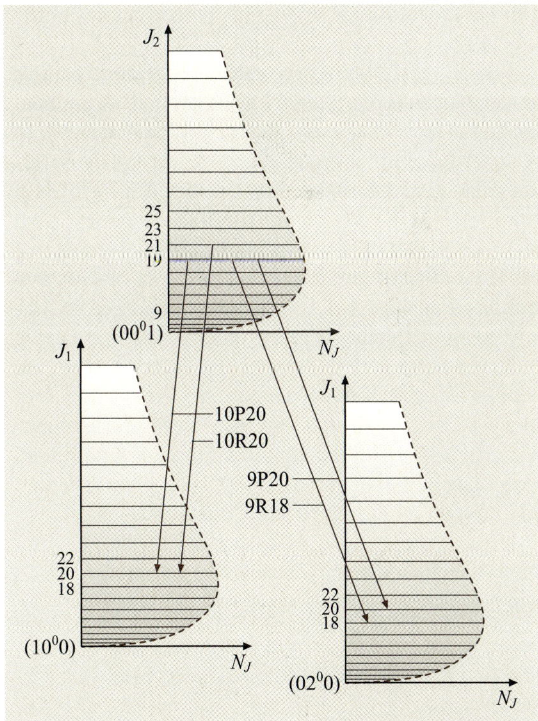

Fig. 11.121 Examples of names for allowed roto-vibrational transitions of the regular bands. The rotational distribution is calculated for $T = 400\,\text{K}$

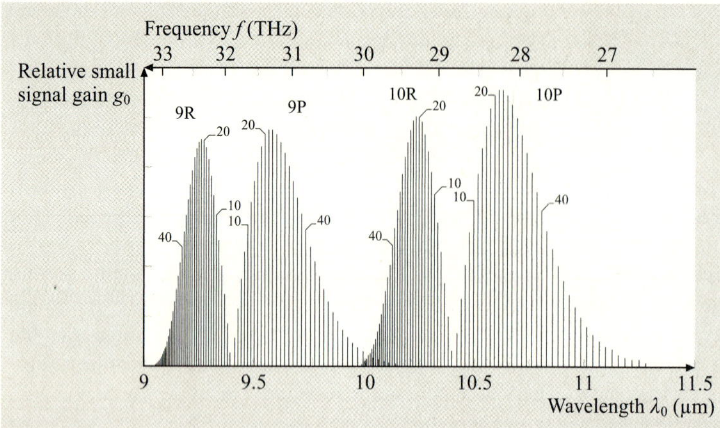

Fig. 11.122 Calculated relative small-signal gain for the regular bands of the CO_2 laser at $T = 500\,K$

example one of the strongest emission lines of the CO_2 laser is labeled 10P(20), indicating a transition from the $10\,\mu m$ band with the lower rotational state $J = 20$ and $\Delta J = -1$. Different transitions have different energy differences ΔW and thus different emission frequencies f or wavelengths λ because of the non-equidistant rotational energies in (11.124). This is the origin of the multitude of possible emission lines of the CO_2 laser.

Figure 11.122 shows the calculated relative gain of the regular bands of a typical CO_2 laser. Both vibrational transitions have characteristic R and P branches due to the superimposed rotational energy levels and the selection rules. Exact line positions are listed in [11.1347] based on [11.1361], or in spectral databases such as [11.1359], which also lists transition dipole moments that are useful for the calculation of the absorption and gain of a transition [11.1362]. Note that in Fig. 11.121 only rotational states of even order J are present for the lower states (10^00) and (02^00). The odd states are missing because of the symmetry properties and parity rules (Σ_g^+) of these states in the $^{16}O^{12}C^{16}O$ isotopologue [11.1356]. Similar considerations are valid for the upper state of the regular band (00^01) (Σ_u^+ state), which only has odd rotational quantum numbers J.

Table 11.39 Possible vibrational lasing transitions

Vibrational band	Comment
$(00^01) \to (10^00)$	Regular band, $10\,\mu m$
$(00^01) \to (02^00)$	Regular band, $9\,\mu m$
$(01^11) \to (11^10)$	Hot band
$(01^11) \to (03^10)$	Hot band
$(00^02) \to (10^01)$	Sequence band
$(00^02) \to (02^01)$	Sequence band

Besides of the regular bands, there are other vibrational transitions with slightly shifted wavelengths that can be operated as lasers: the so-called hot band and sequence band. Table 11.39 shows a summary of vibrational transitions. For laser emission in the sequence or hot bands, lasing in the regular bands must be suppressed, for example by passive absorption cells filled with hot CO_2 gas inside the resonator [11.1347]. Almost all technical lasers operate in the regular bands.

The optical small-signal intensity gain per unit length g_0 of a given transition is proportional to the population density difference of the upper and the lower states, weighted with the respective degeneracy factors:

$$g_0 \sim (N_{n_2 J_2} - \frac{2J_2 - 1}{2J_1 - 1} N_{n_1 J_1})\,. \tag{11.127}$$

The population densities N_{nJ} are the numbers of molecules per unit volume in the roto-vibrational energy level n and J. Within each vibrational state n with a total population of N_n the distribution of the molecules in the different rotational states N_{nJ} is described by a thermal Boltzmann distribution, weighted by the total rotational partition sum and the degeneracy factor for each J:

$$N_{nJ} = N_n \left(\frac{2B}{kT}\right)(2J+1)\exp\left(-\frac{W_r}{kT}\right)\,, \tag{11.128}$$

with the translational gas temperature T and the Boltzmann constant k. The rotational level at the maximum of this distribution at a given temperature can be approximately calculated as

$$J_{max} \approx \sqrt{\frac{kT}{2B}} - \frac{1}{2}\,. \tag{11.129}$$

For typical gas temperatures of $T = 400\,K$ the levels around $J = 19$ have the strongest population. In CO_2

lasers without any further wavelength-selective elements normally only one transition around the 10P(20) line at a wavelength of 10.59 μm will lase, as this one has the strongest gain (Fig. 11.122). Typical values for the small signal gain g_0 are in the range from $0.5\,m^{-1}$ to $1.5\,m^{-1}$.

Even under lasing conditions the rotational distribution will maintain its thermal Boltzmann shape because of fast population exchange between the rotational levels by molecule collisions. Thus, almost all rotational states within the upper vibrational level can contribute to a single lasing transition rather than just the population of the particular upper level, for example the $J = 19$ state of the $(00^0 1)$ vibrational level. This affects mainly the saturation intensity I_s of the optical gain, which is much larger than expected form the population N_{nJ} of a single rotational level J itself. The saturation intensity I_s is in the range from $100\,W\,cm^{-2}$ for sealed-off large-bore DC-excited lasers to over $1000\,W\,cm^{-2}$ for fast-flow systems.

Electrical Excitation and Gas Composition

Without any excitation and in thermal equilibrium the population of the vibrational states is given by a Boltzmann distribution. As the energies of the lower vibrational levels are almost of the same order as the thermal energy $kT = 25.5\,meV$ at 296 K, the thermal population of levels above the ground state cannot be neglected. The population N_n of a given vibrational level n is given by:

$$N_n = N_0 \frac{\exp\left(-\frac{d_n W_n}{kT}\right)}{Q_{vib}} , \tag{11.130}$$

where N_0 is the total density of CO_2 molecules, d_n is the degeneracy factor and W_n the energy of level n, and Q_{vib} is the vibrational total partition sum. Without excitation the population of the upper laser state $(00^0 1)$ is always smaller than the lower laser levels $(10^0 0)$ or $(02^0 0)$. This is the normal case for passive media, which show absorption lines on the respective transition wavelengths.

To achieve an inversion, and thus optical gain, in an active medium an electrical gas discharge is usually used to excite the CO_2 laser. Accelerated electrons inside the discharge collide with the molecules and lose part of their kinetic energy. This energy can be transformed into vibrational excitation energy of the molecules or into kinetic energy by a pure translatory motion. How effective a certain vibrational state of the molecule is excited is described by the excitation cross section, which is dependent on the energy of the electrons. For inversion of the regular CO_2 laser lines, a selective population of the upper laser state $(00^0 1)$ is required. This is barely possible in a pure CO_2 gas discharge as the cross sections for excitation of the different vibrational states are of the same order of magnitude.

Therefore, nitrogen (N_2) is added to the laser gas [11.1351]. As a molecule with two atoms it only has one vibrational mode. As a homonuclear molecule it has no electric dipole moment and hence, no radiative decay of excited vibrational states. Thus, nitrogen can efficiently store energy in its vibrational excited states. More than 50% of all N_2 molecules can be excited to higher vibrational levels inside a gas discharge. The first excited vibrational state, $n = 1$, has nearly the same energy (289 meV) as the upper laser level $(00^0 1)$. Due to resonant collisions between excited nitrogen atoms vibrational energy is readily transferred to the desired CO_2 state

$$N_2(n) + CO_2(00^0 0) \rightleftharpoons N_2(n-1) + CO_2(00^0 1) + \Delta W . \tag{11.131}$$

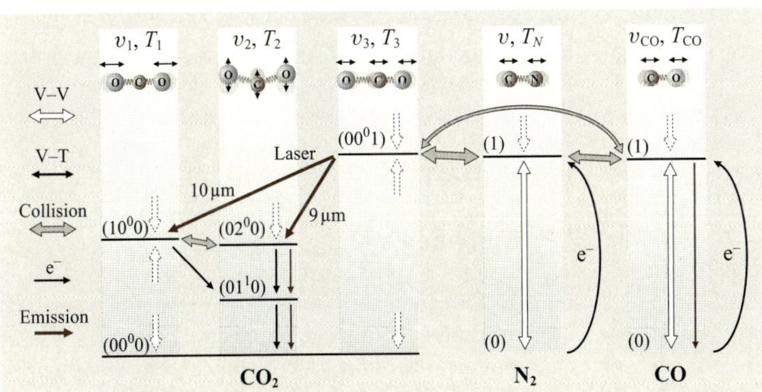

Fig. 11.123 Energy-level diagram of the CO_2 laser process with excitation and decay paths

The energy difference $\Delta W = -2.2\,\text{meV} \ll kT$ is easily available from the thermal kinetic energy of the molecules.

Figure 11.123 shows the energy-level diagram of the CO_2 laser process. Accelerated electrons from the gas discharge excite nitrogen molecules to higher vibrational levels by collisions. These excited nitrogen molecules transfer their energy to the upper laser level (00^01) of the CO_2 molecule and provide a significant population inversion compared to the lower laser levels. From there, two strong lasing transitions by stimulated emission of photons are possible, the $9\,\mu m$ band with (02^00) as the lower laser level and the $10\,\mu m$ band with (10^00).

To maintain a large inversion thermal population of the lower laser levels must be avoided by efficient cooling of the laser gas. Therefore, helium (He) is added to the laser gas. Helium is the gas with the largest thermal conductivity. It also plays an important role for the population decay of the (02^00) and (01^10) levels by vibrational–translational (V–T) relaxation. Thereby, vibrational energy of the CO_2 molecule is transferred to kinetic energy, the translatory motion of the molecules. Radiative decay by spontaneous emission is also possible for the (02^00) and (01^10) levels, but not for the (10^00) level, which shows no change in the dipole moment during vibration. It can reduce its population by V–T relaxation directly to the (01^10) level, but also by the Fermi resonance via collisions and energy exchange with the (02^00) state. In fact, the degree of vibrational excitation of these two vibrational modes is almost the same due to this Fermi coupling. After relaxation to the ground state (00^00) the CO_2 molecule is available for excitation and the laser process again.

Within one vibrational mode, for example v_1 of the CO_2 molecule, there is a strong interaction of the vibrational levels by vibrational–vibrational (V–V) relaxation and excitation processes during collisions:

$$\begin{aligned} &CO_2(n_1') + CO_2(n_1'') \\ &\rightleftharpoons CO_2(n_1'+1) + CO_2(n_1''-1) + \Delta W\,, \end{aligned} \quad (11.132)$$

where the small energy difference ΔW coming from the anharmonic forces can almost be neglected. By this fast thermalization the population distribution of the vibrational levels within one mode v_i can be described by a Boltzmann distribution similar to (11.130) with a specific vibrational temperature T_i for this particular mode. Three vibrational temperatures T_1, T_2, and T_3 are required to describe the three vibrational modes of the CO_2 molecule. Together with the vibrational temperature of the nitrogen T_N and the physical gas temperature T they form the so-called five-temperature model of the CO_2 laser. This model can be used to quantify the degree of vibrational excitation in a given laser system and to calculate the optical gain of different transitions [11.1347, 1363, 1364].

Note that the vibrational temperatures inside a gas discharge can be significantly higher than the usual physical gas temperature T which describes the translatory kinetic energy. For typical laser conditions the temperatures T_3 and T_N can be in excess of 1000 K, even if the physical gas temperature T is only 450 K, for example. This describes the desired high level of excitation of these modes for a strong inversion, especially if there is no laser action to reduce the population in (00^01) by stimulated emission. The vibrational temperatures T_1 and T_2 are almost equal because of their coupling and are slightly larger than the physical temperature, because their population decay to the ground state takes place with a finite rate. Here, as low as possible vibrational temperatures are desired for a low population of the lower laser levels. Of course, no vibrational temperature can be lower than the physical gas temperature T. It is also the temperature T that is valid for the rotational population distribution.

Thus, a gas mixture for a CO_2 laser could be $He : N_2 : CO_2 = 8 : 2 : 1$, with large individual variations between specific laser designs, which are often experimentally defined for maximum laser output. The gas pressure p is typically a few tens of hPa for longitudinally DC-excited lasers, in the range of 100 hPa for lasers with a transversal RF discharge or can equal the atmospheric pressure in so-called TEA (transversaly excited atmospheric pressure) lasers with a pulsed transversal discharge. The gas pressure influences the stability of the gas discharge and the electric field strengths required to start and maintain a stable discharge at the different excitation frequencies and geometries. Together with the applied electric field E inside the discharge the gas pressure p affects the electron energy distribution function (EEDF). For efficient excitation of the desired vibrational nitrogen states electrons with energies of 2–3 eV are required, where the effective cross section for vibrational excitation of N_2 has a maximum [11.1347].

Dissociation and Gas Additives

The complex-shaped electron energy distribution function (EEDF) in such a laser gas discharge always has a tail extending to quite high electron energies. These electrons can start unwanted chemical dissociation processes of the CO_2 molecule inside the gas discharge, for example

$$CO_2 + e^- \rightleftharpoons CO + O^-\,,\ \Delta W = -3.85\,\text{eV}\,, \quad (11.133)$$

$$CO_2 + e^- \rightleftharpoons CO + O + e^-\,,\ \Delta W = -5.5\,\text{eV}\,. \quad (11.134)$$

This is the origin of the complex gas chemistry that leads to a reduced CO_2 concentration and the formation of a significant CO concentration inside the laser gas tube [11.1367]. By the law of mass action a typical equilibrium ratio of 1 : 1 between CO_2 and CO is established if no means to reduce the dissociation are present. Clearly this reduces the overall laser output power and efficiency.

In lasers with a steady gas exchange by a gas flow the laser gas can be replaced by fresh gas from a premix bottle or can be partially regenerated by forced circulation through catalysts to prevent too high a degree of dissociation. Naturally, the gas dissociation is more crucial for sealed-off lasers without any gas exchange. To decrease dissociation processes and maintain a high output power, many different gas additives to the basic mixture have been studied [11.1347]. Table 11.40 shows the composition of some commercial laser gas premixes for high-power CO_2 lasers.

Xenon (Xe) is without doubt the most efficient and most widely used additive for sealed CO_2 lasers [11.1368–1370]. Xe has a first ionizing potential of 12.1 eV, which is some eV less than the other gases. Hence, the same current density in the discharge can be maintained at a lower applied electric field. This in turn shifts the EEDF towards lower energies, which are more efficient for vibrational excitation, and there are fewer fast electrons to cause dissociation processes. On the other hand, Xe with its heavy atomic weight has a very low thermal conductivity. Also it is a very rare and expensive noble gas. Sealed laser mixtures typically contain a few percent of Xe.

Water vapor (H_2O) and molecular hydrogen (H_2) have also been extensively studied [11.1347, 1371]. Both gases react inside the gas discharge, for example with oxygen radicals to form hydroxyl radicals OH. This hydroxyl radical is an efficient catalyst for oxidizing CO back to CO_2. On the other hand, water vapor with its multitude of vibrational states rapidly deexcites the upper laser state (00^01) and quenches the desired population densities of this level. The reported water vapor concentrations for optimum catalysis and yet low vibrational quenching are quite low [11.1347] and were found to be hard to control in real laser systems. In other laser systems no clear proof of an optimum water vapor content was found at all [11.1372, 1373]. As a rule of thump the dew point of the laser gas should be below $-40\,°\text{C}$, equivalent to a water vapor partial pressure of 0.13 hPa. Creeping of water from the cooling system has to be avoided. However, some laser gas premixes may contain small amounts of hydrogen, as deemed appropriate by the manufacturer of the laser system.

Other gas premixes include carbon monoxide (CO) or oxygen (O_2). By the law of mass action both additives can shift the equilibrium of the dissociation reactions (11.133) and (11.134) towards the left. Too high an O_2 concentration has a detrimental influence on the discharge stability and the EEDF, as it is a strongly electronegative gas. CO as a diatomic molecule has similar vibrational energy levels to nitrogen and can also transfer vibrational energy by resonant collisions to the upper laser state of the CO_2 molecule (Fig. 11.123). Another benefit of adding CO and O_2 initially to the gas premix for filling the laser is that the gas is already near its equilibrium composition for the dissociation process, so no overshoot or undershoot of the laser power occurs when the laser is operated for the first time after filling. However, CO is not as effective as nitrogen for vibrational excitation. The energy difference from the CO:($n = 1$) vibrational level to the CO_2:(00^01) level is 25.5 meV and is thus larger than with nitrogen. Additionally, CO has allowed dipole transitions to the ground state by spontaneous emission, which

Table 11.40 Some commercial gas premixes for different industrial CO_2 laser types [11.1365, 1366]

Premix name / (Maker)	Composition He	N_2	CO_2	CO	O_2	Xe	H_2	Laser type, model series
Lasermix® 322 (Linde AG)	65.5	29	5.5					Fast axial-flow, Trumpf TLF Series
Lasermix® 690 (Linde AG)	65	19	4	6	3	3		Diffusion-cooled sealed-off, Rofin Slab DC 0XX
Lasal™ 81 (Air Liquide)	80.8	15	4				0.2	DC excited slow-flow, FEHA SM Series

decreases vibrational excitation compared to nitrogen. Also CO is the only toxic gas of the species discussed here. Care has to be taken when handling such gas mixtures.

Materials for CO_2 Lasers

Materials for the laser tube have to be chosen properly. All common rules for building and preparing vacuum systems apply. Materials with low out-gasing and low vapor pressure must be used. This normally inhibits the application of plastics and polymer-based glues in CO_2 laser tubes. Additionally, the UV radiation of the gas discharge will degrade most organic compounds rapidly. There are only a few epoxy-based adhesives that may be used.

Many low-power sealed-off lasers and even high-power lasers with slow or fast gas circulation use quartz glass tubes to confine the laser gas. Quartz (SiO_2) glass is thermally stable, chemically inert, robust against UV radiation and has low dielectric losses. However, silica-based glasses have high infrared absorption, even for grazing incidence of light to the surface. Thus, the diameter of the quartz glass tube needs to be considerably larger than the laser beam formed by the resonator mirrors. For waveguide lasers (see Sect. 11.4.2) tubes made from other materials must be used. Alumina (polycrystalline Al_2O_3 ceramics) is often used. It is a dense and vacuum-tight ceramic material, which is also dimensionally and thermally stable. Its infrared absorption is lower than quartz glass, making it suitable for hollow dielectric waveguides. Also it has very low dielectric losses at radio frequencies, making it an ideal material for transversal capacitively coupled RF discharge excitation.

Good material choices to confine the gas volume are also inert passivated metals such as stainless steel [11.1374] and passivated aluminium. Steel can easily be welded, both with conventional techniques or with lasers, to built a vacuum-tight laser tube. Welding of aluminum is more difficult. On the other hand, aluminum has a far better thermal and electrical conductivity and a higher infrared reflectivity. This makes aluminum a very good material for waveguide plates and electrodes for transversal RF discharges [11.1375].

Electrolytic copper or oxygen-free high-conductivity (OFHC) copper can also be used because of their excellent thermal and electrical conductivity and infrared reflectivity. On the other hand, copper can be oxidized by oxygen radicals from the dissociation process inside the gas discharge. Copper does not form stable, inert oxides on its surface but rather absorbs and binds the oxygen in the volume. This shifts the equilibrium of the dissociation towards lower CO_2 content in the tube. Therefore, uncoated copper is not well suited as a material in direct contact with the gas discharge. Away from the discharge, however, copper is an excellent material for mirrors. Copper mirrors can be cooled effectively by water channels from the rear side. This is important for resonator and deflecting mirrors inside high-power lasers. For example, with an output coupler reflectivity of 50% the circulating light power inside the resonator is twice the rated laser power of the system. Even if the infrared reflectivity is very high, this can lead to considerable warming and mechanical stress if the mirrors are not cooled. Copper can be electroplated with nickel (Ni) and then coated with gold (Au). This makes the copper surface more stable against oxidation.

Catalysts and Gas Analysis

Much research was carried out to find proper catalysts to reduce the dissociation of CO_2 and to maintain a high output power. In lasers with a gas flow, the laser gas can be circulated through external catalysts. Common catalysts such as hot platinum [11.1376] or platinum on metallic oxides such as tin oxide with a large surface area [11.1377, 1378] can be used. However, fast-flow systems require gas replenishment of a few liters per hour to compensate for residual vacuum leaks or material contamination. As the price of the laser gas mixture is rather low compared to the costs of installing a catalyst, they are usually not applied in fast-flow systems.

For sealed lasers without any gas flow the situation is quite different. The dissociation reactions inside the gas discharge are fast and equilibrium is reached in a tenth of a second within the discharge. On the other hand, diffusion time constants to a nearby gas reservoir are in the range of minutes. Thus, any catalytic surfaces need to be in close contact with the discharge itself. For example, catalytic active platinum cathodes [11.1371] and distributed platinum [11.1379] or sputtered gold coatings on the inside of a longitudinal discharge tube [11.1380] have been reported. For waveguide lasers, gold-coated electrodes were found to be effective in some reports [11.1381, 1382]. The catalytic activity and the advantages for laser output power of these approaches depend on the preparation of the catalytic surfaces, the gas composition and the laser geometry itself, leaving open many questions and an impression of alchemy. Some commercial laser designs embody some sort of these techniques, as deemed best for the given laser design.

To study the effects of catalysts on the gas composition or, more generally, to have more data about the active medium such as rotational and vibrational temperatures various methods can be applied. Mass spectrometers are commonly used for gas analysis. However, short-lived species from the discharge cannot be easily measured. The required pressure reduction from some tens of hPa down to 10×10^{-7} hPa tends to change the original gas compositions. Special techniques in experimental laser systems have been used to measure gas compositions directly from the gas discharge, and have given insight into the complex gas chemistry [11.1367, 1384].

Another method is to study the visible and UV spontaneous emission from electronically excited states inside the gas discharge [11.1385]. This can even be done with the naked eye. Figure 11.124 shows an example spectrum. A gas discharge without much CO appears pink or redish-purple because of vibronic nitrogen emission lines in the red and blue visible spectrum. With a significant carbon monoxide content there are additionally some CO emission lines spread over the visible and a strong emission band in the blue–UV spectral region. Thus, a gas discharge with a higher degree of dissociation looks more blueish-white. The color change can be observed in slow-flow DC-excited gas lasers with glass tubes from the gas inlet to the gas outlet, for example. For quantitative studies fiber-coupled spectrometers with fast readout can be used to gain time-resolved concentration data [11.1383]. With high-resolution spectroscopy, for example of the UV nitrogen emission band, the rotational temperatures of the gas can be measured. Using an imaging system, spatially resolved data for these parameters have even been measured in the discharge gap of an RF-excited high-power slab laser [11.1386].

Absorption spectroscopy with tunable diode lasers is also very effective for measuring species concentrations and roto-vibrational temperatures directly inside the laser tube or gas discharge. To do this, a wavelength-tunable diode laser is adjusted to the various absorption transitions of the relevant molecules (i. e., CO_2, CO, N_2O, NO). By aiming the diode-laser beam through the gas volume and measuring the relative intensities of several rotational and vibrational transitions an accurate diagnostic of the active medium is possible [11.1387–1391]. Recently, the catalytic activity of special gold coatings on electrodes for RF-excited slab lasers was verified using this technique [11.1392].

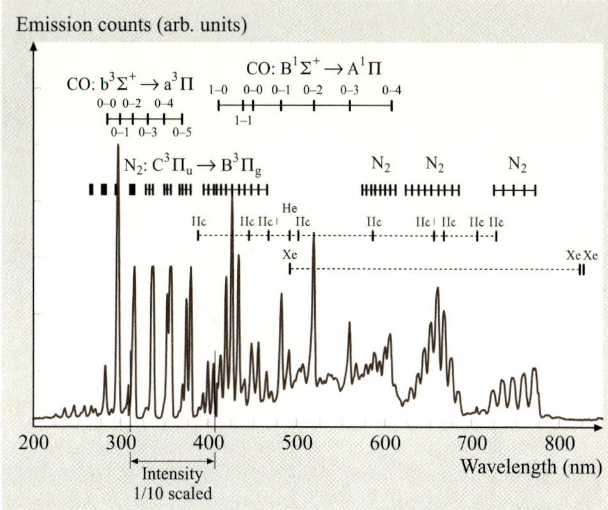

Fig. 11.124 Visible spectrum the sidewards spontaneous emission of a CO_2 laser gas discharge (after [11.1373, 1383])

Output Spectra and Line Broadening

Due to the fast population relaxation between the rotational levels only the rotational transition with the largest gain is usually observed. A similar competition as for the rotational lines is also given for the two regular vibrational transitions. As both have the same upper laser level (00^01), and the population levels of the lower states (10^00) and (02^00) are coupled by the Fermi resonance and have similar decay rates, only the vibrational transition with the largest gain will survive. Typically this is the 10P(20) transition or one of its neighbors for many lasers.

Every roto-vibrational transmission line is broadened by several broadening mechanisms. At gas pressures above 10 hPa collisional or pressure broadening dominates over Doppler broadening. The collision-broadened line shape is described by a Lorentzian function having a line width (FWHM) Δf_L proportional to the total gas pressure p:

$$\Delta f_L = 2p\left(\psi_{CO_2} b_{CO_2} + \psi_{N_2} b_{N_2} + \psi_{He} b_{He}\right) \times \left(\frac{300}{T}\right)^n \tag{11.135}$$

where the ψ_i are the fractions of the respective gases and T is the gas temperature. The temperature exponent n is 0.58 for constant pressure [11.1393], with values ranging from 0.5 to 0.7 also found in the literature. The b_i are the pressure-broadening coefficients resulting from collisions with species i and are given in [11.1393] to

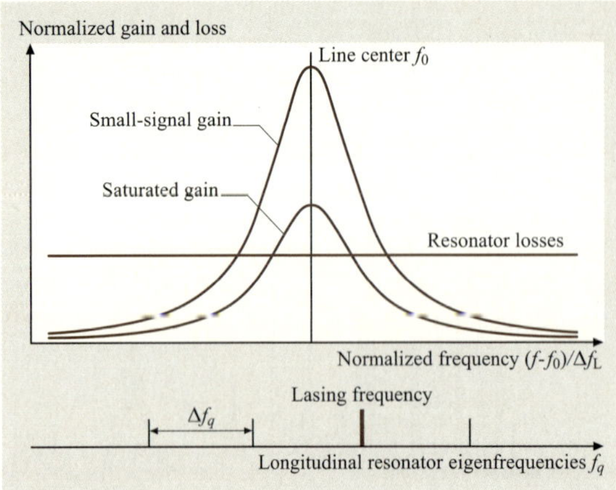

Fig. 11.125 Homogeneous saturation of a Lorentzian gain profile leads to lasing of a single longitudinal resonator eigenfrequency

be:

$$b_{CO_2} = (3.40 - |m| \cdot 0.0272)\,\text{MHz hPa}^{-1} \quad (11.136)$$

$$b_{N_2} = (2.35 - |m| \cdot 0.0127)\,\text{MHz hPa}^{-1} \quad (11.137)$$

$$b_{He} = (1.77 - |m| \cdot 0.00083)\,\text{MHz hPa}^{-1}\,, \quad (11.138)$$

where $m = -J$ for the P branch and $m = J+1$ for the R branch. For example, the most common 10P(20) line at 100 hPa and 500 K in Lasermix 322 from Table 11.40 has a Lorentzian line width (FWHM) of 284 MHz. The Doppler line width at this temperature would be $\Delta f_D =$ 36.2 MHz and can be considered if a Voigt line shape instead of a pure Lorentzian is used [11.1394].

Within this line width one or more longitudinal resonator eigenfrequencies f_q may be located. For an example length for a CO_2 laser resonator of $L_{res} = 2$ m a spacing of the eigenfrequencies $\Delta f_q = c/2L = 75$ MHz results. As the collision-broadened line saturates homogeneously, only the eigenfrequency closest to the center frequency of the transition can lase in the steady state after saturation, as shown in Fig. 11.125. Thus, CO_2 lasers are typically longitudinally single-mode lasers.

Efficiency, Output Power and Cooling

The theoretical limit for the efficiency of a laser is the internal quantum efficiency η_q. For the CO_2 laser this is the ratio of the energy of the laser photons to the energy of the upper laser level (00^01) and is roughly 40%. This is quite high for a gas laser. Real technical lasers have significantly lower efficiencies for several reasons. The electrons inside the gas discharge do not only excite the required vibrational states of the molecules but also other states. Also a significant part of the kinetic electron energy is transferred into pure translatory motion of the molecules and thus transferred into heat. Additional energy is lost in the ionizing processes required to maintain the gas discharge and in chemical dissociation processes. Also not every vibrationally excited molecule contributes to the laser process. Vibrational energy decays by vibrational–translational relaxation, which produces heat again and by radiative decay by spontaneous emission of excited states. Finally, optical energy is dissipated in absorption losses of the mirrors and waveguides or by limited aperture diameters.

Typical electrical efficiencies η_E, defined as the ratio of the extracted laser power P_L to the electrical power P_E applied directly to the discharge tube, are in the range of 10% for slow-flow DC-excited laser, up to 15% for RF-excited waveguide lasers and up to 20% for fast-flowing systems. The overall wall-plug efficiencies η_{tot} as a ratio of laser power P_L to mains supply power P_{mains} are always lower than η_E. This is due to conversion losses in the electronic power supplies to provide the required DC high-voltage or RF power. A considerable amount of energy is also dissipated in the gas circulating system of fast-flow lasers.

The output power of a given CO_2 laser can be roughly estimated by simple thermodynamics. If all aspects of optical design, electrical excitation and gas composition are optimized, electrical efficiencies η_E as discussed above can be assumed. From this point, the laser power is finally limited by the cooling and heat removal capacity of the laser system. The power dissipated by the electric excitation P_{diss} should not heat the laser gas above a certain temperature, for example 450 K.

In diffusion-cooled lasers no forced gas flow contributes to the heat removal. The thermal conductivity of the gas mixture and the diffusion of excited vibrational states to the housing walls determine the thermal transport processes. For a given tolerable temperature rise ΔT of the laser gas in the discharge area, the maximum applicable dissipated thermal power P_{diss} can be calculated by solving the differential heat-transfer equation for the given geometry. For fast-flowing systems the amount of heat removal is given by the specific heat capacity c_P of the gas and the mass flow $\dot{m}$.

With the assumption that almost all electrical excitation power P_E is finally dissipated into heat P_{diss}, the maximum permissable excitation power $P_{E,max}$ for the given temperature rise ΔT is thus known. The maximum possible output power $P_{L,max}$ of this laser system can then be calculated with the assumed electrical efficiency η_E. This exaggerates the importance of the cooling methods for high-power lasers. Different approaches will be discussed in the next section on technical lasers.

11.4.2 Typical Technical Designs

Although CO_2 lasers with a longitudinal DC discharge inside a glass tube as described in Sect. 11.4.1 are still being used and sold because of their simplicity, many other types and realizations of the CO_2 laser have been developed over the last decades. They can by systematically categorized according to various characteristics, as below.

Gas flow inside the laser.

- No gas flow, sealed-off
- Quasi-sealed-off, periodic gas exchange
- Slow gas flow axial to the laser beam
- Fast gas flow axial to the laser beam
- Fast gas flow transversal to the laser beam

Gas cooling.

- Diffusion-cooled, cooled walls of the gas discharge
- Fast gas flow with external heat exchanger

Electrical excitation.

- Longitudinal DC discharge, continuous
- Transversal DC discharge at high gas pressure, pulsed
- Capacitively coupled transversal RF discharge
- Inductively coupled RF discharge
- Microwave-excited gas discharge

Optical resonator.

- Stable optical resonator
- Unstable optical resonator
- Free-space propagation between the mirrors
- Optical waveguide between the mirrors
- Combinations of these in different planes

In principle, almost any of these characteristics could be combined to give specific advantages. The three axes of the optical laser beam, the gas flow and the exciting electrical field can be mutually parallel or orthogonal in general. Some of the most important designs with their specific technical characteristics are discussed further in the next sections. Commercial manufacturers of CO_2 lasers often have quite different approaches, also depending on their own intellectual property and patents.

General design aspects are: as efficient a heat removal as possible, a compact optical resonator for a small footprint even at very high power levels and a rugged and robust design for maintenance-free long-term industrial operation.

More-exotic laser types such as optically pumped CO_2 lasers, black-body radiation-pumped lasers, lasers with electron-beam sustained gas discharges and gas-dynamic CO_2 lasers are not further considered in detail here.

DC-Excited Fast-Axial-Gas-Flow Lasers

Longitudinally DC-excited lasers have the advantage of a rather simple and cost-effective design. Modern DC-excited lasers use sophisticated electronically regulated high-voltage (HV) supply circuits with both voltage and current control to omit the ballast resistor shown in Fig. 11.117. This reduces ohmic losses in the HV circuit. Vacuum-tube-based or more modern semiconductor-based current regulators and switched-mode voltage converters are applied.

For high-power lasers a fast axial gas flow can be used to cool the gas. Lasers with a power in the kW range [11.1396, 1397] were demonstrated in the late 1970s. Figure 11.126 shows an example of a modern DC-excited fast-flow high-power laser. The gas discharge in the gas tube is divided into four independent sections to keep the required high-voltage levels moderate. Two of the discharge and flow tubes shown can be used in parallel, with a U-shaped folded light path and resonator to double the length of the active medium. This is because the output power scales with the length of the gain medium.

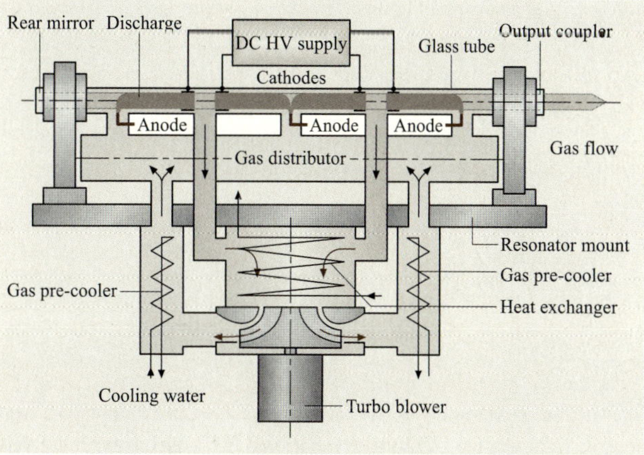

Fig. 11.126 Longitudinally DC-excited high-power CO_2 laser with fast axial gas flow [11.1395]

As discussed in Sect. 11.4.1, the maximum output power of a CO_2 laser is also largely determined by its heat removal capacity. A fast gas circulation can be driven by a turbo blower. From thermodynamic considerations, the maximum laser power $P_{L,max}$ of a fast-flow laser can be calculated with a given electrical efficiency η_E:

$$P_{L,max} = \eta_E P_{E,max} = \left(\frac{\eta_E}{1-\eta_E}\right) P_{th} = \left(\frac{\eta_E}{1-\eta_E}\right) \dot{m}\, c_p\, \Delta T\,, \quad (11.139)$$

where $P_{E,max}$ is the maximum applicable electrical power, P_{th} is the thermal power dissipated in and removed from the discharge, $\dot{m}$ is the mass flow, c_p is the specific heat of the gas mixture, typically $2500\,\mathrm{J\,kg^{-1}\,m^{-3}}$, and ΔT is the maximum gas temperature rise that can be tolerated, typically 250 K. It can be seen clearly that the laser power scales with the mass throughput and thus with the flow velocity, which is limited by the onset of turbulence and the speed of sound. For an assumed efficiency of $P_E = 15\%$, one yields an approximate required mass flow of 0.1 kg/s per kW of laser power. The heat dissipated in the gas is removed by a heat exchanger after the gas has passed the discharge sections. The additional heat resulting from the gas compression by the turbo blower is removed by pre-coolers before the gas enters the discharge tube again.

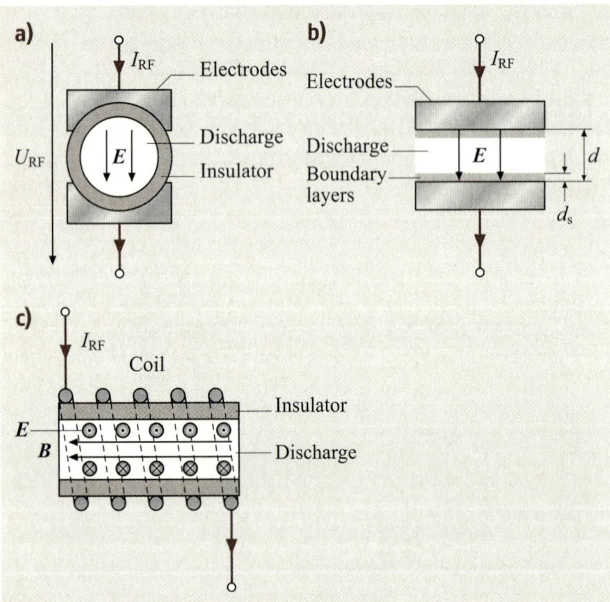

Fig. 11.127a–c Geometries for RF excitation with current I_{RF} and electric field $\boldsymbol{E}$: **(a)** capacitively coupled RF discharge, insulated electrode, **(b)** capacitively coupled RF discharge, non-insulated electrodes, **(c)** inductively coupled RF discharge

Besides the fast gas flow there is also a permanent slow gas exchange with a fresh gas mixture, for example at a rate of 37 l (standard pressure and temperature) per hour in a ratio He : N_2 : CO_2 = 25 : 12 : 2 for a particular commercial system. Longitudinally DC-excited lasers up to several kW laser power are available. Depending on the laser power and the application, such laser systems can emit a mixture of an ideal TEM_{00} mode and a hybrid TEM^*_{10} donut-shaped mode.

RF-Excited Gas Discharges

Radio-frequency (RF) excitation of the gas discharge is widely used in CO_2 lasers. In this context RF means a frequency range of roughly 1–500 MHz. Figure 11.127 shows cross sections of typical geometries found in CO_2 lasers. In capacitively coupled RF (CCRF) discharges the electric field is applied by two metallic electrodes with the gas volume in between. The electric field $\boldsymbol{E}$ in the volume of the gas is sufficiently high to start and maintain a self-sustained glow discharge [11.1350]. The RF voltage U_{RF} and the RF electric field $\boldsymbol{E}$ are typically applied transversally to the laser beam. Thus, much lower voltages U_{RF} are required compared to longitudinally excited discharges. U_{RF} is roughly 100 V per mm discharge gap width d, depending largely on the gas pressure.

The electrodes can be electrically insulated from the discharge by a dielectric material, for example a glass or alumina tube as shown in Fig. 11.127a. No direct contact of the gas with the electrodes is necessary. The current flow is closed by the displacement current, which is $\sim \partial \boldsymbol{E}/\partial t$ according to Maxwell's laws.

Alternatively, the metallic electrodes can be placed in direct contact with the gas volume (Fig. 11.127b). In this case the surface regions on both electrode boundaries have reduced free-electron densities and are called ion sheaths. In these ion sheaths no efficient excitation of vibrational states takes place. Power dissipated in the sheaths does not contribute to the laser power. The width of the sheaths d_s is inversely proportional to the excitation frequency $d_s \sim 1/f$ and is, for example, 0.35 mm at $f = 125$ MHz at a pressure of $p = 90$ hPa [11.1398]. The electrode separation must be significantly larger than this value to ensure sufficient excitation of the gas volume. On the other hand, this reduces the efficiency of diffusion cooling for heat removal from the gas to the electrodes. For small gaps and efficient cooling, excitation frequencies in the 100 MHz range are often chosen.

An optimized value for the electrode separation d is typically 2 mm under these conditions.

With insulated electrodes a so-called dielectric barrier discharge has been reported, which allows efficient excitation at much lower frequencies around 1 MHz while still maintaining a stable discharge with efficient vibrational excitation and good cooling of the gas [11.1400].

The α-type RF discharge is a volume discharge with all ionizing and electron-collision processes taking place throughout the entire gas volume. No ionizing processes at the electrodes are required. Such RF discharges are hence called electrodeless discharges. However, in the case of non-isolated electrodes (Fig. 11.127b) there is also an onset of a γ-type discharge at higher current densities, where surface ionizing processes play an important role [11.1350]. This is usually not the desired discharge type for laser excitation.

RF gas discharges can also be inductively coupled (Fig. 11.127c). The RF current I_{RF} through a coil generates a time-varying magnetic field $\boldsymbol{B}$ which is accompanied by an electric field $\boldsymbol{E}$ according to $\text{curl}\,\boldsymbol{E} = -\partial\boldsymbol{B}/\partial t$. Inductively coupled discharges are not commonly used for CO_2 lasers because the gas cooling and electrical field geometry are not as favorable as for CCRFs.

RF-excited discharges have several advantages. One advantage is that such an electrodeless discharge shows no material sputtering from the cathode or anode and no resulting contamination of the laser gas or degradation by material ablation. Very high energy densities can be applied to the gas volume while still maintaining a stable glow discharge without arcing. Higher gas pressures can be used, typically around 100 hPa. Thus, more active medium and more laser power is available per volume. The required voltages are lower than for DC excitation. Insulation of electric wires and components on the air-pressure side is easily done by sufficient large air gaps or common dielectrics. The impedance of the gas discharge, which can be modeled as a strongly lossy capacitor [11.1401], can be matched to the RF generator impedance with almost lossless reactive components (LC matching circuit). No ohmic ballast resistors are required to stabilize the discharge.

However, there are also some issues to be properly considered in the design of RF-excited lasers. One of those is voltage and discharge homogeneity along the RF electrodes. At a frequency of 100 MHz the vacuum wavelength λ_0 is 3 m. Wave propagation effects and waveguide theory must be applied if structures with one dimension ℓ larger than $\lambda_0/10$ are used. This is often the case for the length of typical RF-excited waveguide lasers. Therefore, regularly spaced inductors parallel to the electrodes are used [11.1369, 1402, 1403]. Figure 11.128 shows an example for an electric circuit model of the electrodes with a gas discharge.

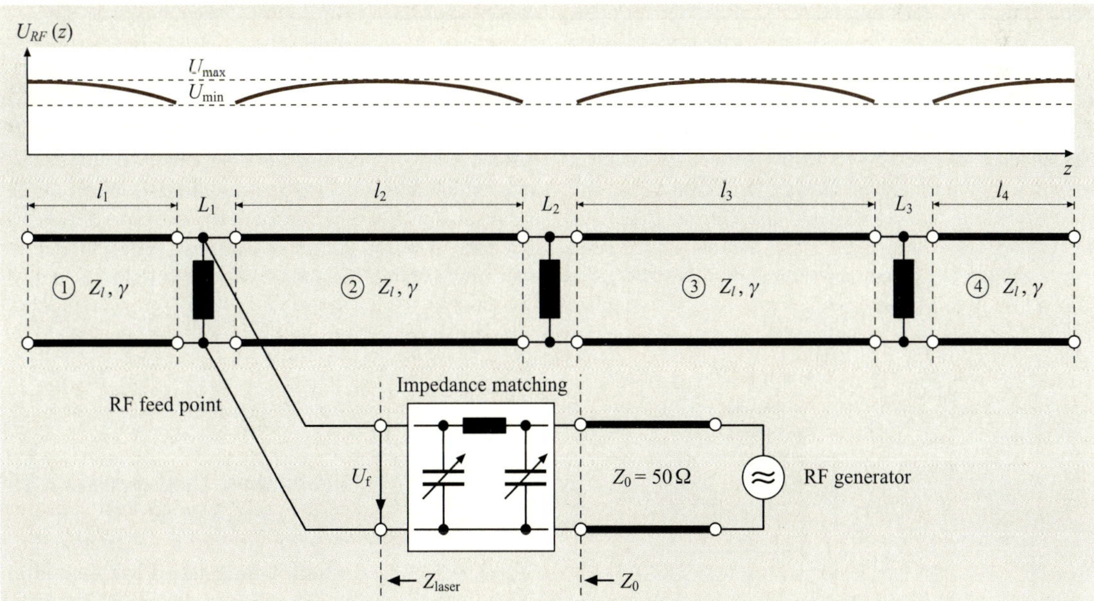

Fig. 11.128 Homogenization of the RF voltage distribution $U_{RF}(z)$ on long electrodes with parallel inductors [11.1399]

The electrical wave propagation along the electrodes is described by a characteristic impedance Z_ℓ and the propagation constant of a lossy transmission line γ. Parallel inductors L_i are used to compensate for the capacity per unit length of the electrodes to homogenize the voltage distribution. The higher the RF frequency, the more inductors in a shorter distance l_i must be used. Also shown is an impedance-matching network to match the complex impedance of the laser Z_{laser} to the system impedance Z_0 of $50\,\Omega$ of common RF generators.

Alternatively, an array of segmented electrodes much smaller than the wavelength can be used. This can be problematic if the electrodes also act as waveguides for the laser light. Mechanical misalignment of the electrode segments may then cause excessive optical losses.

RF power generators are somewhat more complex than DC high-voltage supplies. They must provide roughly ten times the nominal laser power as electrical power at RF frequencies. For low- and medium-power CO_2 lasers up to 500 W, solid-state RF power generators based on semiconductor transistors [mostly metal–oxide–semiconductor field-effect transistors (MOSFETs)] are available at the required frequencies. Such RF generators can be integrated directly into the laser system housing with air or water cooling of the power supply.

For an RF power beyond this level electron-tube-based generators are required. Coaxial power tetrodes, as installed in amplitude-modulation (AM) short-wave radio transmitter stations, are used. This is a mature and reliable technique to generate RF powers of tens of kilowatts in the desired frequency range. Tubes can tolerate large reflected power levels from badly matched loads without permanent damage. Electron tubes are more robust against overvoltage and current transients than transistors. A drawback is that the tubes itself need a high-voltage DC power supply and that they dissipate additional power for cathode heating. The lifetime of such tubes is on the order of 10 000 hours. However, costs for replacement tubes are moderate.

Finally, issues of electromagnetic compatibility (EMC) have to be considered. Therefore, many RF-excited lasers operate in the ISM (Industrial Scientific and Medical) bands of 13.6 MHz, 27 MHz or 40.6 MHz, where less-stringent regulations apply. Systems at different frequencies, especially in the frequency-modulated (FM) radio bands around 100 MHz need to have properly shielded housings, best made of all metal and with contact strips for all service flaps of the laser housing.

Microwave excitation at a frequency of 2.45 GHz has been studied intensively [11.1404–1406]. Powerful magnetrons are very efficient microwave-generating tubes at this frequency and are available at low cost, because of their use in mass-produced products such as kitchen microwave ovens. However, the short wavelength makes it difficult to realize large-area discharges with good uniformity. Additionally, the power density required to maintain a microwave-excited gas discharge is larger than at common RF frequencies and tends to overheat the gas. Thus, microwave-excited CO_2 lasers have been realized mostly in pulsed operation [11.1407] or with a fast gas flow [11.1408]. Diffusion-cooled laser systems have also been demonstrated [11.1409]. In spite of the cost advantages of microwave generation with magnetrons, the mentioned disadvantages and the progress of RF-excited lasers has hindered the commercial success of microwave excitation to date.

Waveguide Lasers

For the laser design in Fig. 11.117 it would be desirable to reduce the diameter of the glass tube to achieve better cooling of the laser gas. This can introduce excessive optical losses by obstructing the laser beam inside the resonator. Losses from such limiting apertures with a radius of a for common Gaussian TEM modes in two-mirror stable resonators of length l are calculated

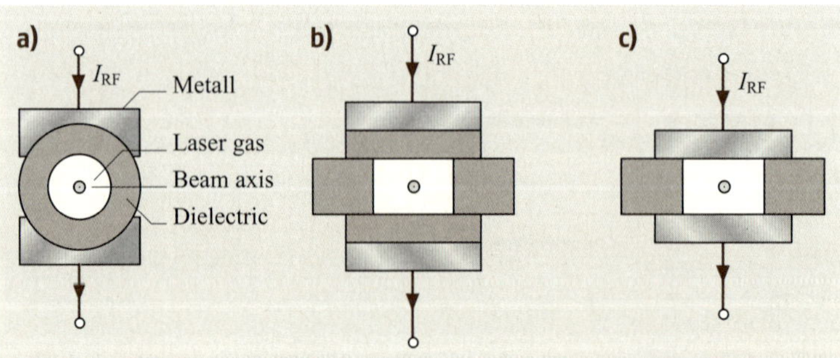

Fig. 11.129a–c Cross sections of some waveguides for CO_2 lasers with RF electrodes: **(a)** circular dielectric waveguide, **(b)** all-dielectric rectangular waveguide, **(c)** hybrid metal–dielectric waveguide

in [11.1410]. These losses rise quickly if the Fresnel number $N = a^2/\lambda l$ gets smaller than 1.

This problem can be overcome if the gas tube has waveguide properties at the optical laser frequency. A comprehensive overview of waveguide laser design is given in [11.1402]. Figure 11.129 shows the principle geometries of waveguides used in CO_2 lasers. The simplest form is a circular hollow dielectric waveguide, which is theoretically described in [11.1411]. Circular metallic waveguides are not an option because they will obviously shorten any electrical field required for the discharge excitation. Rectangular waveguides can have dielectric walls on all four sides or one pair of dielectric and one pair of metallic walls [11.1412, 1413]. Such waveguides support hybrid EH_{mn} modes at optical frequencies with very low losses. These modes can be thought of as a superposition of electromagnetic waves propagating at different angles to the waveguide axis, reflected by the walls at grazing incidence. Therefore, the materials used should have low absorption losses in the infrared. Alumina or beryllium oxide ceramics are a good choice for the dielectrics because of their high mechanical, chemical and thermal stability and their high thermal conductivity for heat removal. Aluminum or gold-plated copper surfaces can be used as metallic waveguide boundaries.

The lowest circular mode EH_{11} is very similar to the Gaussian free-space mode TEM_{00}, with a power overlap of 98%. Thus, waveguide lasers operating only in the lowest possible mode have excellent beam quality.

To build a waveguide laser, such waveguides are filled with laser gas. The excitation of the gas discharge is therefore often realized by a transversal RF discharge. A pair of metallic electrodes are placed on opposite sides of the dielectric waveguide (Figure 11.129a,b) or are part of the waveguide itself (Figure 11.129c). The optical resonator is realized by mirrors at both ends of the waveguide. Design rules for low-loss coupling and good mode discrimination are given in [11.1414]. A good solution is a mirror in a small distance d from the waveguide, which is half the radius of curvature ρ_M of the mirror. Alternatively, plane mirrors close to the waveguide can be used.

The power of a waveguide laser scales almost linearly with the length. Up to 110 W of laser power per meter length have been reported [11.1415].

Many commercial CO_2 lasers use waveguide technology. Figure 11.130 shows the cross section of a particular commercial waveguide laser [11.1416]. The resonator light path is folded in a zigzag shape three times through multiple waveguide channels formed by a ceramic plate and an aluminum profile. This results in a compact laser head design with high power, as the laser power scales with the length of the waveguide. The aluminum profile serves as the RF ground electrode, as one side of the optical waveguide, as a cooling plate and as an RF- and gas-tight housing for sealed operation. The second RF electrode is on top of the ceramic plate. Regularly spaced spiral inductors homogenize the RF voltage distribution along the structure. Air or water cooling can be realized by the cooling channels on the profile. An output power of 100 W is available from a compact package size [11.1417]. Q-switching can optionally be applied to realize pulsed lasers with high peak power, especially for marking applications.

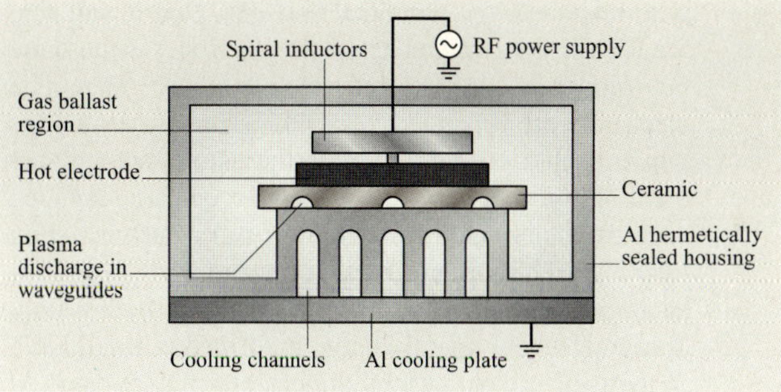

Fig. 11.130 Cross section of a commercial RF excited waveguide laser with a folded resonator light path

Another interesting waveguide laser structure is shown in Fig. 11.131 [11.1418]. The waveguide is formed by two pairs of metallic aluminum electrodes and two ridges from the metallic housing profile. RF power is applied to both electrodes in a push–pull mode

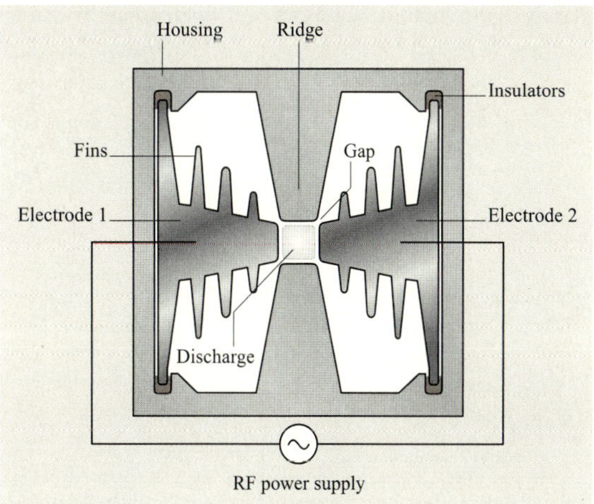

Fig. 11.131 Cross section of an all-metal waveguide laser

with respect to the grounded housing. That means that the RF voltage amplitude on both electrodes is the same but with a 180° phase shift relative to one other. The electrodes are separated from the ridges by small gaps which enable gas circulation in the whole structure for cooling and heat removal. The electrodes are anodized such that the oxide layer provides sufficient insulation to the grounded ridges of the housing. Additional insulators are used to hold the electrodes in the housing. A homogeneous gas discharge is formed in the middle of the structure with lateral dimensions of a few mm. In spite of the small gaps, the structure can act as an optical waveguide in all directions to form a laser beam with good beam quality.

The advantage of this all-metal structure is that no ceramic–metal joints are required. Such joints can potentially cause failure because of the mismatched thermal expansion coefficients. Both the electrodes and the housing can economically be manufactured by extruded aluminum profiles. Heat is easily removed by conduction or passive thermal convection. Without ceramics with their high dielectric constants between the electrodes, the capacitive loading of the electrical waveguide is low. Fewer inductors are thus required for voltage homogenization. The metal housing can be sealed with metal brazing to build long-living sealed-off lasers. A range of medium-power compact lasers up to 240 W is available in this technology [11.1420], partially using folded resonators for power scaling with the waveguide length.

The Slab Laser

Figure 11.132 shows the structure of an RF-excited slab laser. This design has been pioneered by several groups [11.1421–1424], accompanied by the well-known Tulip patent [11.1425]. It is formed by two

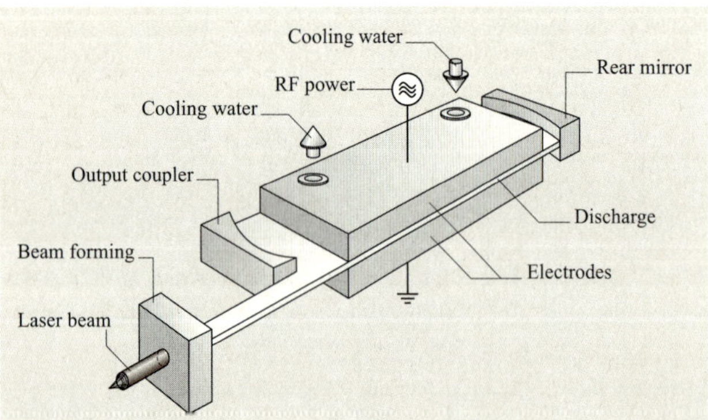

Fig. 11.132 RF-excited slab laser with hybrid waveguide-unstable resonator [11.1419]

large-area metallic electrodes separated by a small gap. The length of the electrodes can be 1 m with a width of some tens of centimeters. A typical gap height is 2 mm. An RF voltage at a frequency in the range of 100 MHz is applied between the electrodes to start a gas discharge in the laser gas mixture between the electrodes. The electrodes are provided with channels for cooling water. Thermal energy dissipated in the gas discharge is effectively removed by conductive or diffusion cooling. Gas discharges with very large areas can be operated stably with this geometry.

In a plane perpendicular to the large-area surfaces the electrode pair acts as a waveguide for optical frequencies. A low-order mode is easily obtained in this waveguide direction. In the plane parallel to the electrode surfaces free-space conditions are valid as there are no limiting boundaries on the sides of the electrodes. With such broad dimensions this would result in a strongly multimode beam profile in this direction, if common stable resonators were applied. Therefore, an unstable resonator is used which has advantages for the extraction of high-quality beams from active laser media with large transversal dimensions [11.1348, 1426]. An unstable resonator is characterized by two mirrors with radii of curvature ρ_1 and ρ_2 and a mirror separation L that do noes fulfill the stability criterion for stable Gaussian beam resonators:

$$0 \leq \left(1 - \frac{L}{\rho_1}\right)\left(1 - \frac{L}{\rho_2}\right) \leq 1 \qquad (11.140)$$

Such a resonator does not reflect a Gaussian beam backwards into itself but rather uses the partial transmission of the beam on one mirror side as the output coupling. It can be shown that the beam profile can have an almost diffraction-limited beam quality in the unstable-resonator plane. In the waveguide plane, the mirrors can be almost flat. Both mirrors can be made of solid copper and can be cooled to withstand very high power densities. By a proper design, the coupling losses between the mirrors and the slab waveguide can be minimized [11.1427].

The laser beam coming directly from this type of resonator has an elliptical shape and has different divergence angles in the waveguide and the free-space plane. This is corrected by beam-shaping optics with at least one cylindrical mirror. Additionally, a spatial mode filter is used to eliminate shadows on the beam wings in the free-space unstable-resonator direction. After that, the beam has a nearly circular shape with an excellent beam quality of $M^2 = 1.1$.

The main advantage of this structure is that the laser power can be scaled by the area A_e of the gas discharge, which extends in two dimensions. This is in contrast to tube-type or waveguide lasers where the power scales only over one dimension, namely the tube length L. From thermodynamic calculations and based on extending results from conventional waveguide lasers, the maximum laser power P_L of a slab with an electrode area A_e and an electrode separation d_e is approximately given by:

$$P_L = C\frac{A_e}{d_e}\,, \quad C \approx 3\frac{\text{W mm}}{\text{cm}^2}\,. \tag{11.141}$$

By this area scaling over A_e, very compact and yet powerful lasers can be made. As with all large RF-excited laser systems, parallel inductors are used for homogeneous RF voltage distribution along the electrodes. The efficient diffusion cooling requires no gas flow, which saves system and operating costs. Medium-power lasers can be operated completely sealed and are available at power levels up to 500 W [11.1429], partially with transistor-based RF generators. High-power slab lasers operate typically with a periodic gas exchange, for example after 72 h of continuous operation, to compensate for residual gas leaks from the large structure. Commercial slab lasers with powers up to 8 kW integrate the slab structure and the electron-tube-based self-oscillating RF generator into one compact housing with an integrated premix gas supply which lasts for approximately 12 months of continuous use [11.1419].

RF Excited Fast-Flow CO_2 Lasers

The basic principle of a fast-axial-gas-flow laser as described in Sect. 11.4.2 can be combined with the advantages of RF excitation. Figure 11.133 shows such a laser system. The gas circulates through glass tubes driven by a turbo radial blower in the center of the structure. The optical resonator is quadratically folded. For an even longer discharge length and more output power, the resonator can be folded in two planes with 16 discharge sections (Fig. 11.134). Optically stable resonator configurations are used. The RF discharge is capacitively coupled with eight pairs of electrodes, two on each side. The electrodes are mutually rotated by an angle of 45° to ensure a homogeneous gain in the cross section, averaged over one resonator cycle. A heat exchanger cools the gas coming from the discharge section and a precooler removes compression heat after the blower before the gas enters the discharge sections again. Typical gas pressures are in the range of 150 hPa. The power scaling with mass flow according to (11.139) is also valid for this

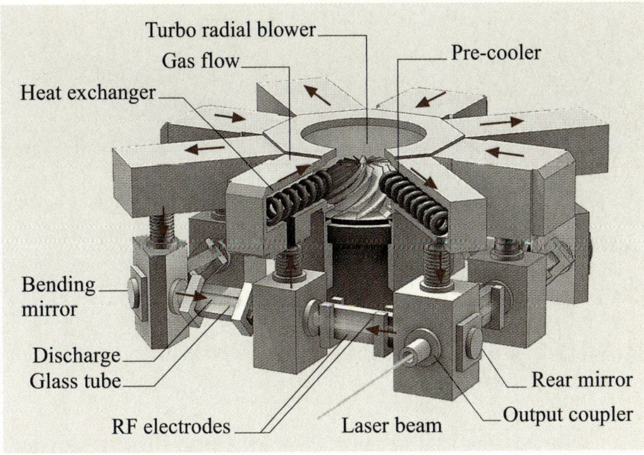

Fig. 11.133 RF-excited fast axial gas flow with a folded resonator [11.1428]

system. Volume flow rates are up to 500 m^3/h per kW extracted laser power with gas velocities of 100 m/s. Lasers with output powers up to 20 kW are commercially available while laser powers above 30 kW have been demonstrated experimentally [11.1430]. The beam quality is near a perfect diffraction-limited TEM_{00} mode with an M^2 of 1.1 for high-quality series up to 4 kW. At the high-power end the 20 kW system still has an M^2 value as low as 5. The wall-plug efficiency of these systems is 10% including the losses in the RF power supply and the gas-circulating system. Electron-tube-based high-power RF oscillators at an industrial ISM standard frequency of 13.6 MHz are used.

Other RF-Excited CO_2 Laser Systems

There are many other commercial laser designs based on RF excitation and with specific resonator or cooling geometrics. Some interesting designs will be discussed briefly here.

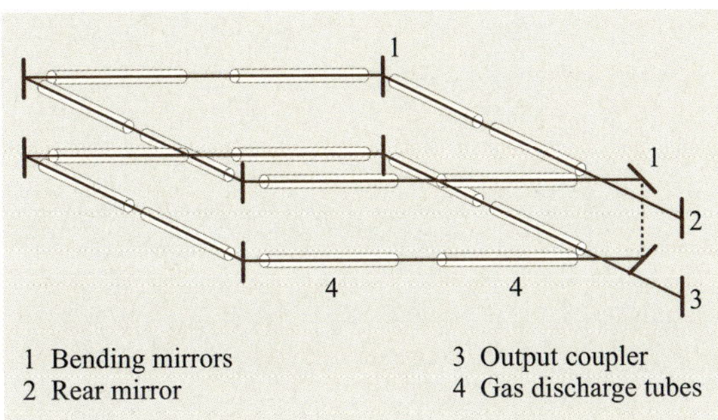

Fig. 11.134 Quadratically folded resonator in two planes [11.1428]

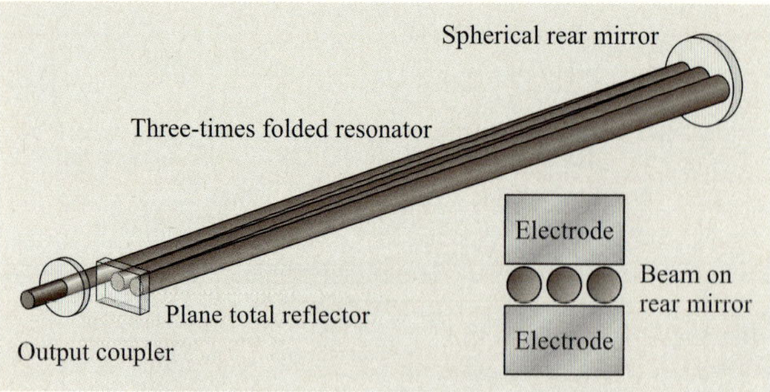

Fig. 11.135 Folded free-space resonator between large-area slab electrodes [11.1431]

Figure 11.135 shows a folded free-space resonator between two large-area RF electrodes. Only three mirrors are required for beam folding. The output coupler and the plane total reflector can even be integrated in one mirror with inhomogeneously reflecting coatings. Compact and robust sealed original equipment manufacturer (OEM) laser modules up to 120 W are available [11.1431].

A combination of large-area electrodes with a hybrid stable–unstable resonator is described in [11.1433] and [11.1434]. In contrast to the slab laser with its plane-parallel electrodes, the electrodes do not act as a waveguide here. Instead, they have a slightly V-shaped surface geometry along the beam propagation direction between the mirrors to select the lowest-order free-space mode in a plane perpendicular to the electrodes, whereas an unstable resonator is used in the free-space plane again. Sealed-off OEM modules in an all-metal design and a maximum power of 400 W can be realized [11.1435].

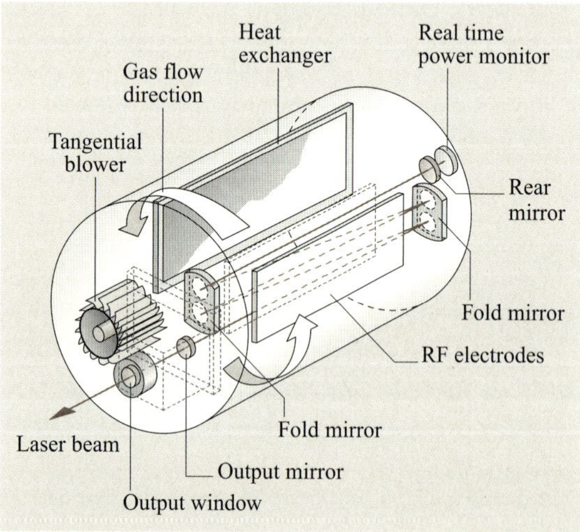

Fig. 11.136 RF-excited laser with fast transversal gas flow [11.1432]

Large-area RF-excited gas discharges can also be combined with a fast gas flow between the electrode plates transversally to the optical resonator and the laser beam (Fig. 11.136). The gas flow is driven by a tangential blower integrated into the large-diameter cylindrical gas vessel. A folded resonator is used to extract the energy from the large active medium. Such lasers can produce 8 kW of laser power from a very compact footprint. The beam quality with $M^2 \approx 5$ is still well suited for welding and surface-treatment applications.

Coaxial laser systems with an annular gas discharge as shown in Fig. 11.137 can be thought of as slab-type lasers with an electrode pair rolled up to form coaxial tubes [11.1436]. RF power is applied to the inner tube, while the outer tube is grounded. This is similar to a coaxial cable, but with an inner-conductor diameter of approximately 10 cm and a discharge gap width of 7 mm, for example. In this design, the electrodes have no optical waveguiding properties, thus such lasers are less susceptible to power and beam-quality variations under mechanical and thermal stress. The resonator is formed by a helical mirror and an axicon mirror, bending the beam to the opposite side of the tube while slightly shifting the azimuthal angle after each pass [11.1437]. In the radial direction the optical resonator configuration is stable, while in the azimuthal direction an unstable resonator is used. The laser is diffusion-cooled by water-cooled electrode tubes. The laser can be operated quasi-sealed-off with a regularly scheduled gas exchange. Compact and robust lasers with 2 kW power can be realized [11.1438]. They are well suited for mounting on moving systems such as robots without bending mirrors for beam steering.

Pulsed TEA Lasers

All the laser designs discussed so far are continuous-wave (CW) lasers by principle. Their power can be modulated by the electrical excitation power with frequencies of a few kHz for DC excitation and up to 100 kHz for some RF-excited lasers. In contrast, the tranversal-excited atmospheric-pressure (TEA) laser is a CO_2 laser that can inherently operate only in pulsed mode, but is capable of generation very high pulse energies and peak powers that are not possible with CW CO_2 lasers. This is achieved by a high gas pressure, at atmospheric pressure or above. A homogeneous DC discharge can only be operated for a short time of 1 μs under these conditions before filamentation and arcing would start. The required voltages at this pressure are in the order of magnitude of 100 kV per meter. Longitudinal discharges would need voltages that are not practical and would not produce stable discharges. Therefore, pulsed transversal discharges are used. Figure 11.138 shows a sketch of a simple TEA laser circuit. A high-voltage power supply (HV) charges a storage capacitor C slowly via the resistors R_{V1} and R_{V2}. When the maximum voltage at the capacitor is reached, a fast high-voltage switch, which is also capable of carrying high currents, is triggered to close the circuit quickly. The capacitor voltage is applied to the electrode pair, where a short but intense discharge transversal to the resonator axis and the laser beam starts. The discharge self-terminates when all the energy stored in the capacitor is dissipated. Afterwards the high-voltage switch resets itself to an open state and charging of the capacitor starts again before the next pulse can be triggered.

With TEA lasers pulse energies in the range of 10 J per liter of active volume and per bar of pressure can be achieved, with peak intensities in the MW range [11.1440, 1441]. Pulse repetition rates are a few 10 Hz for high-power systems and kHz for mini-TEA lasers [11.1442].

Critical design issues are the electrode profile to maintain a stable and homogeneous discharge in the cross section and also over the length of the electrodes. Often pre-ionization techniques by additional electrodes, UV light from corona discharges or even electron beams are therefore applied. The high-voltage switch is also a crucial part. Typically, triggered spark gaps or thyratrons are used, which must be replaced after a given number of pulses. Recently, progress in semiconductor technology has made it possible to replace these vacuum tubes by solid-state switches. A device called a silicon controlled rectifier (SCR) is capable of switching high currents very fast. In combination with a step-up transformer and a saturable inductor/capacitor ladder network the high-voltage level and the fast rise time required for the electrode voltage can be achieved.

Much research was done on TEA lasers in the 1980s, especially in the fields of plasma physics, LIDAR and military applications. Partially due to the success of high-power fast-flow and RF-excited systems, only a few commercial TEA lasers are available. Systems with pulse energies of 2.4 J at a repetition rate of 125 Hz are available, for example. Applications include non-metal processing such as marking and paint or rubber-compound stripping from surfaces in the automotive industry [11.1443].

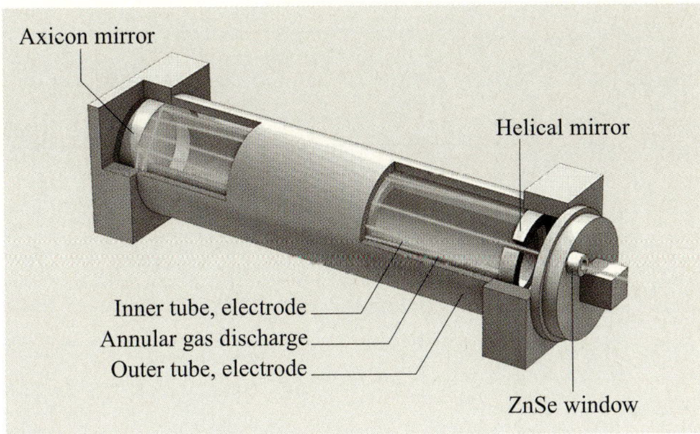

Fig. 11.137 Coaxial laser with an annular gas discharge [11.1439]

Wavelength Selection and Tunable CO_2 Lasers

Without any wavelength-selective elements, a CO_2 laser emits at wavelengths around 10.6 μm in the vicinity of the 10P(20) line. Some commercial lasers are optionally available at the wavelengths of the strongest lines of the other bands, for example at 9.3 μm, 9.6 μm or 10.3 μm (Fig. 11.122). This is done by weakly wavelength-selective elements such as mirror coatings to select one of the four regular emission bands where the strongest line will start to oscillate. For example, this is interesting for plastics machining where materials can show a strong variation of absorption over wavelength.

For precise selection of a particular roto-vibrational transition, a diffraction grating is used. Figure 11.139 shows a DC-excited tunable laser in the Littrow configuration. Depending on the angle α of the incoming beam with respect to the surface normal of the grating, only a light beam at a specific wavelength λ_L is reflected:

$$\lambda_L = 2\Lambda \sin\alpha\,, \qquad (11.142)$$

with Λ being the grating period. For this particular wavelength λ_L the grating acts as a plane reflector on axis to the second resonator mirror. This wavelength is selected

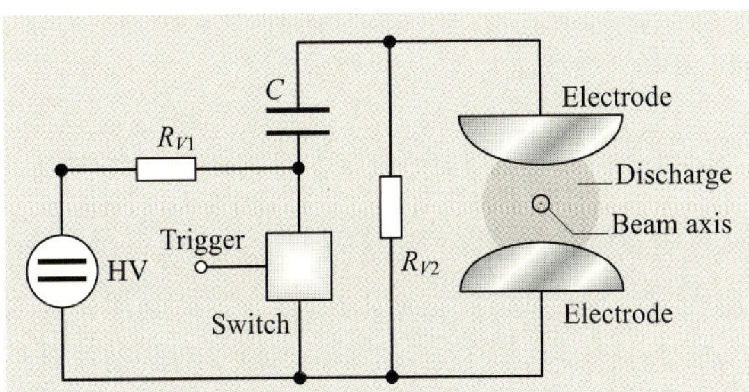

Fig. 11.138 Excitation circuit for a basic TEA laser

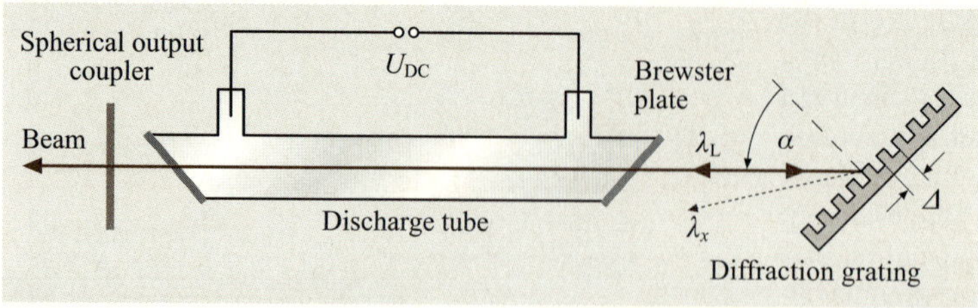

Fig. 11.139 Tunable CO_2 laser in Littrow configuration

Table 11.41 Emission range of a tunable CO_2 laser [11.1444]

Band	Wavelengths (μm) min – center – max	Lines min – center – max
9R	9.158 – 9.271 – 9.367	9R(44) – 9R(20) – 9R(4)
9P	9.443 – 9.552 – 9.836	9P(6) – 9P(20) – 9P(50)
10R	10.095 – 10.247 – 10.365	10R(46) – 10R(20) – 10R(4)
10P	10.441 – 10.591 – 10.936	10P(4) – 10P(20) – 10R(50)

to lase. Other wavelengths λ_x, if present, would be reflected at a different angle not parallel to the resonator axis. They would suffer significant losses from the aperture of the laser tube, which would prevent them from lasing. By rotation of the grating the angle α can be varied to select different roto-vibrational transitions one by one. Such gratings are mostly made of copper or steel with precision-ruled grooves and a gold coating to enhance IR reflectivity. The grating period Λ is chosen to be slightly smaller than the laser wavelength in order to suppress higher diffraction orders. The groove geometry can be adjusted such that the reflectivity in the desired minus-first diffraction order at a given wavelength is maximal (blazed gratings, blazing wavelength).

Table 11.41 shows for example the tuning range of a commercial tunable CO_2 laser. Note that, because of the separation of the rotational lines in the range of 30–90 GHz and the pressure-broadened line width of some 100 MHz, continuous tuning over the entire range is not possible. Within one line width, fine-tuning can be done by changing the resonator length with a piezoelectric element. An output power of 180 W can be achieved for the strongest lines and more than 20 W on the weak lines at the band edges, for example.

Applications include scientific spectroscopy, industrial trace-gas monitoring by absorption spectroscopy or optically pumping of far-infrared molecular gas lasers. By using gas mixtures enriched with special isotopologues, for example $^{13}CO_2$, emission lines at additional wavelengths can be realized. Tables of such wavelengths are given, for example, in [11.1347].

Besides longitudinally DC-excited lasers, RF-excited waveguide lasers are also good candidates for being tuned by a standard Littrow grating. Tunable slab-type lasers with their broad active medium require special grating structures for the extraction of a low-order mode. This can be realized by apodized Littrow gratings where the splitting ratio between the diffraction orders $m = 0$ and $m = -1$ is variable over direction of the grating [11.1445]. Such gratings can act as an output coupler with spatially variably reflectivity for beam forming and as line-selective element for wavelength tuning at the same time. They have been realized on copper mirror substrates with photolithography and a microgalvanic process as well as on silicon substrates with photolithography and anisotropic etching [11.1446].

11.5 Ion Lasers

Ion lasers became a success very soon after their invention in the 1960s. In the following decades, low-power air-cooled argon lasers were made in the tens of thousands for use in printing, color separation, and medical technology. Mid-power ion lasers were a staple of the entertainment industry, used in light shows, special effects, and for holography, as well as in disc mastering. Large-frame ion lasers were ubiquitous in well-equipped chemistry and biology laboratories around the world. Ion lasers were the flagship products of the major commercial laser manufacturers.

But even in their heyday, alternatives to ion lasers were eagerly sought. There were several reasons for this, including early problems with reliability and lifetime, as well as high cost, but the major dissatisfactions resulted from the fundamental low efficiency of the devices. This poor efficiency meant ion lasers were large, of considerable weight, and needed substantial power supplies drawing large utility loads. They also had to be cooled aggressively either by air or water.

With the development of solid-state lasers in the 1990s, in particular diode-pumped lasers with frequency-doubled green output at 532 nm, many expected ion-laser technology to become obsolete. This has not happened. While not nearly at the levels they experienced in the 1980s, sales continue to be steady, ion lasers are still found in their traditional applications, and are a significant contributor to the bottom line of a number of companies, both as new systems and in the replacement market. While there are fewer types of new ion lasers available now, these systems still vary considerably in performance, size, utility requirements, and cost.

An ion-laser system consists of a power supply, control electronics, and the laser head. The laser head itself contains the plasma tube placed within a resonator (or *cavity*) structure. Control of the laser is typically through an interface included in the power supply, which itself can vary considerably in size, complexity, and cooling requirements.

Small ion lasers are air cooled, require only standard wall-plug electrical service, have a laser head roughly the size of a toaster, and are most commonly used to produce only a single blue–green output at 488 nm in the range 5–100 mW. When operated in the low-current range, these lasers can be expected to last for years in a stable environment. Tube replacement when necessary can be performed by the user and is relatively quick and inexpensive.

The greatest variety of product offerings is in the mid-range of performance of 1–5 W of laser output. These systems range from simpler models based on the original ion technology, with air cooling and ceramic plasma tubes, to specialized systems with single-wavelength output intended for specific industrial applications, to only slightly smaller versions of the sophisticated large-frame systems used in advanced laboratories.

A high-performance ion system has a laser head about a meter and half long and can weigh 100 kg. It may require 70 A per phase of three-phase electrical service at 480 V and in addition a cooling-water flow of several liters per minute. But in compensation, these lasers are capable of producing many watts of continuous laser power in a range of selectable wavelengths not available from any other commercial laser system.

Wavelengths available from ion lasers extend from the infrared well into the ultraviolet, with argon, the most common type of ion laser, emitting light primarily in the blue–green to ultraviolet range [11.1447, 1448]. Specific wavelength output is determined either by the choice of the coating for the laser output mirror, or by tuning to the desired wavelength using a Littrow prism placed in the laser resonator. The section below discussing the laser resonator will further describe control of the spectral characteristics of the ion-laser output.

Krypton is not nearly as popular an ion laser as argon, in part because of its lower power, but it remains commercially viable because, like argon, it offers significant levels of continuous output power at wavelengths not easily achieved by other means. For example, more than 1 W CW is available from a single line in the yellow, a color region where laser sources are scarce, and lines in the violet and ultraviolet have important applications as well.

The output power of an ion laser depends on two sets of variables. One set derives from the design of the laser resonator, and includes, for example, the cross-sectional area of the laser beam. This set of variables will be discussed in the section which describes the resonator. The other set of variables that determines laser power results from the physics of the laser process in the plasma created from the noble gas. These variables include the total light amplification available from the plasma.

11.5.1 Ion-Laser Physics

An energy-level diagram is used to show the wavelengths available from a laser material that has been excited into a condition of population inversion. A simplified version illustrating the transitions in singly ionized argon that produce visible laser light is shown in Fig. 11.140 [11.1449]. The transitions shown result from energetic electrons that have decayed to the 4p upper laser level from a still higher level not shown in the figure. From the upper laser level, the electrons are stimulated to fall into the 4s states. The most important individual transitions, which result in the highest output

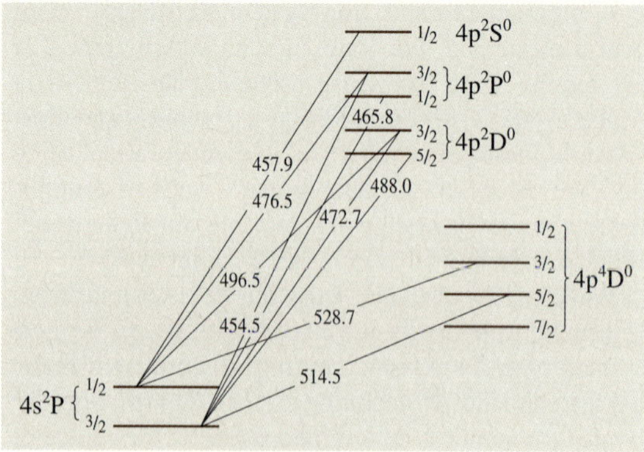

Fig. 11.140 The visible laser transitions of ionized argon (wavelengths in nanometers)

power, are those for 488.0 nm, a blue–green color, and 514.5 nm in the green.

Having many upper laser levels transitioning to only two lower levels raises the possibility of competition for the lower states, which can limit laser output. In the case of argon, however, the lower states depopulate to the ground state very rapidly, avoiding this potential bottleneck. The lifetime of an electron in one of the upper 4p states is about 10 ns, while the lower 4s states make their transition to the ground state in a favorable 0.5 ns. With a proper choice of mirror coating, *all-lines* visible operation can produce laser power from most of these visible lines simultaneously, and the power present in any one of the lines will not be much less than that obtained from single-line operation.

An energy-level diagram can also illustrate problems or requirements in the design of a specific laser system. Figure 11.141 shows the 4p and 4s laser levels of ionized singly argon in relation to the argon ground state. The energy needed to ionize the argon to start the population inversion is about 16 eV, and an additional 20 eV is required to reach the states that decay to the upper laser level. Comparing this total energy input of 36 eV to the energy emitted by the laser transition itself, about 2 eV, it is evident that the efficiency of an argon ion laser will be low. In addition, the temperature of the gas discharge needed to produce electrons with sufficient kinetic energy to achieve the required level of ionization is in the vicinity of 3000 K. This, combined with the low efficiency, indicates that heat management is a primary concern in ion-laser design [11.1450].

Ion lasers use an electric discharge to energize the gas. As can be seen in Fig. 11.141, two collisions with free electrons are required to excite an electron in an

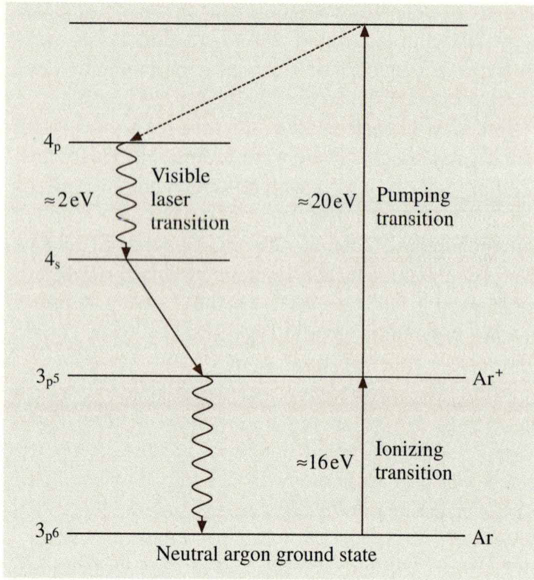

Fig. 11.141 The two-step ion-laser excitation. One electron collision ionizes neutral argon and a second pumps the ion to an excited state from which it decays to the upper laser level

argon ion to the upper level of visible lasing transitions. The gain available from the ionized gas therefore varies as the square of the current density, up to the point where the gain saturates (i. e., other gas-phase processes begin to limit the gain) [11.1451].

The optimum gas pressure that produces the maximum laser output is a balance between competing factors. High pressure provides a greater number of potential ions to stimulate laser emission, while low pressure allows the voltage drop along the plasma more time to accelerate the free electrons to a greater energy between collisions. Different wavelengths have different optimum pressures, and the final choice for a laser designed to operate at multiple wavelengths is a compromise. In general the optimum is in the low-pressure regime. Argon plasma tubes have a fill pressure on the order of 1 Torr [11.1452].

Performance is improved by a magnetic field directed along the length of the plasma discharge, which helps confine the discharge towards the center of the plasma tube [11.1450]. The benefit is greatest for lasers operating at high current. Again, different wavelengths have different optima, and a compromise value of magnetic field strength is employed. Air-cooled lasers typically do not use magnets, as the benefit of a magnetic field for low powers does not justify the added expense, and the addition of magnets would complicate the flow of cooling air.

If the argon is doubly ionized (i. e., stripped of two electrons), a new set of transitions become available that produce ultraviolet laser light [11.1453]. The ground state of Ar^{2+} is about 43 eV above that of the neutral atom, compared to the Ar^{+} ground-state difference of 16 eV mentioned, so operating in the ultraviolet can be expected to be even less efficient. Once above a higher threshold current, UV output rises more rapidly than the visible output does.

The energy diagram for krypton is similar to that for argon, with laser transitions shifted to somewhat lower photon energies in the visible from the upper 5p to the lower 5s bands [11.1454]. Important lines are at 647.1 nm in the red, corresponding to the green 514.5 nm argon line, and at 568.2 nm in the yellow, corresponding to the blue–green line at 488 nm in argon. Unlike argon, however, in krypton these strong lines do experience competition for the same lower state, and so all-lines operation in krypton does not produce the power that might be expected judging from single-line output.

Argon and krypton can be mixed together in the same laser tube to produce a laser offering a dazzling variety of visible wavelengths. A compromise must be reached in the parameters such as fill pressure and magnetic field that allows for the differences in the optimization for each type of laser. Such a mixed-gas laser using broadband mirror coatings has even been used to produce a *white-light* laser, but the argon–krypton mixture becomes unbalanced fairly quickly due to the difference in gas sputtering rates, and the white-light balance is lost as the component wavelengths change in relative strength of output.

11.5.2 Plasma Tube Design

The heart of the ion laser is the plasma tube, which amplifies laser wavelengths available from a low-pressure noble gas by confining a high-current discharge within a narrow bore. The demands placed on the tube structure are extreme, resulting from the need to confine and sustain a discharge as high as 70 A for hours at a time. The bore of the tube must withstand the bombardment of high-energy ions (sputtering) and the heat from the discharge, which can exceed 50 kW under some operating conditions. This heat must be efficiently removed from the tube or tube components will melt or rupture.

Early tube designs relied on cylinders of the ceramic beryllia, or BeO, about 1–2 cm in diameter with a centrally drilled hole of about 1 mm to form the bore. Beryllia cylinders about 10 to 15 cm long are still in use today in lower-power ion lasers, particularly for air-cooled argon. Figure 11.142 shows a cross section of an air-cooled plasma tube. However beryllia proved incapable of sufficiently withstanding the sputtering from the plasmas of high-power or ultraviolet ion lasers, and alternative designs were sought.

Years of intensive research and development have resulted in sophisticated designs for high-current tubes that replace beryllia with a tube structured of composite materials [11.1452, 1455]. These complex (and more expensive) designs can withstand high current densities for an extended period, thus making practical the operation of ion lasers at higher powers or less-efficient

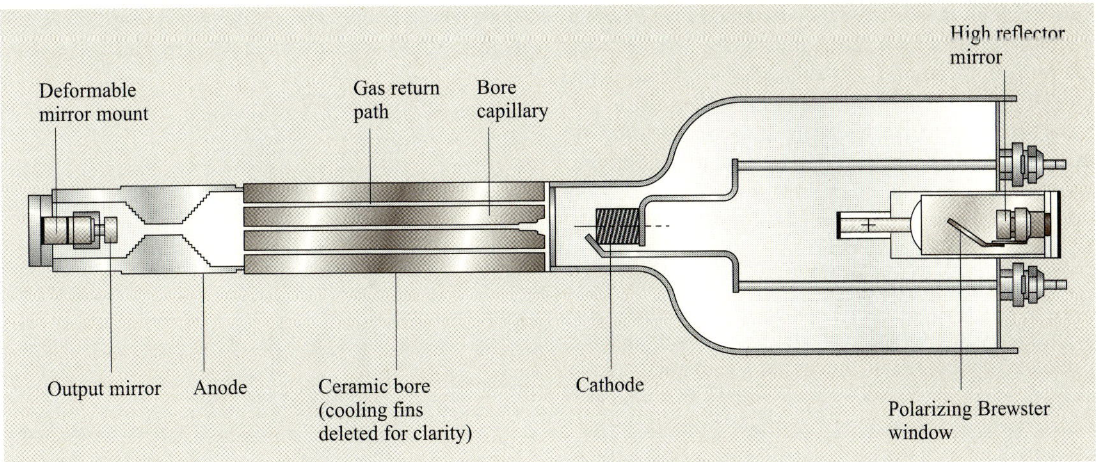

Fig. 11.142 Design of an air-cooled plasma tube

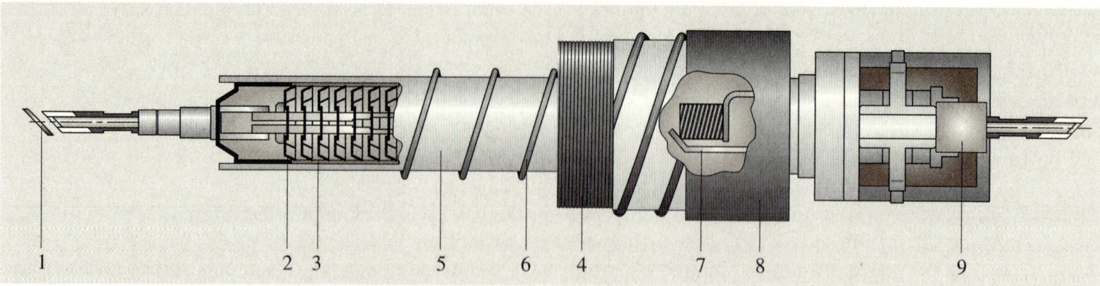

Fig. 11.143 Design of a water-cooled plasma tube

wavelengths. The description that follows is based on the segmented-bore technology used by Spectra-Physics in the manufacture of its water-cooled ion lasers. The overall design is illustrated in Fig. 11.143.

The exterior of the plasma tube is a long, hollow thin-walled ceramic pipe or sleeve a few centimeters in diameter. Inside this sleeve, spaced about one centimeter apart, are many thin discs brazed to the interior of the ceramic. The discs are copper with a central tungsten ring with a hole a few millimeters in diameter. The bore of the plasma tube is defined by the diameter of the hole in the tungsten segments (Fig. 11.144).

The tungsten bore segments are extremely resistant to sputtering action and can withstand the high temperatures of the discharge. The copper discs to which they are bonded conduct the heat of the discharge from the tungsten to the thin ceramic wall of the exterior sleeve, which is bathed outside by a cooling water flow of several liters per minute. The flow of this cooling water is designed to prevent local thermal stresses affecting the tube.

The plasma tube must ensure uniform gas pressure along the length of the bore. In arc discharges in low-pressure gases, the ions lose their momentum in frequent collisions with the tube walls. In contrast, the electrons lose little momentum, so an imbalance is created in the net momentum in the gas. As a result, the neutral gas atoms are driven to the anode end of the plasma tube. To equalize the gas pressure along the length of the tube, holes in the copper discs provide a return path to the cathode end, which also allows the copper discs to help cool the gas in the return flow.

The design of the bore where it opens to the cathode, the so-called *throat* region, is critically important to achieve good laser lifetime. Damage to the bore from sputtering is maximum in the throat. The diameter of the bore segments (or, in lower-power tubes, the conically shaped opening drilled in the beryllia to form the throat) is tapered to match the contour of the electrical field as it enters the bore.

The design of the cathode element itself is also critical. The cathode is helical and is made from a spongy material formed of tungsten powder. It may also include some admixture of calcium, aluminum, barium, and strontium. These additives reduce the work function of the cathode surface so that the temperature needed to produce electron emission is reduced [11.1455].

The efficiency of laser amplification is a function of the gas pressure in the discharge. However ionic sputtering not only erodes the tube materials, but also over time embeds a significant amount of gas in the tube walls and components. Low-power lasers compensate for this effect by widening the body of the plasma tube to form a reservoir, so that the gas lost from sputtering action is a small fraction of the total volume in the tube. Mid- and high-power designs incorporate an active fill system to replenish the gas as it is lost.

To determine when the gas must be replenished, a microprocessor compares the operating voltage of the laser to values stored in a look-up table, taking into account the current, magnetic field and warm-up time. When the voltage indicates that the tube pressure is low, a high-pressure reservoir automatically injects a small quantity of gas into a plenum. A separate valve then

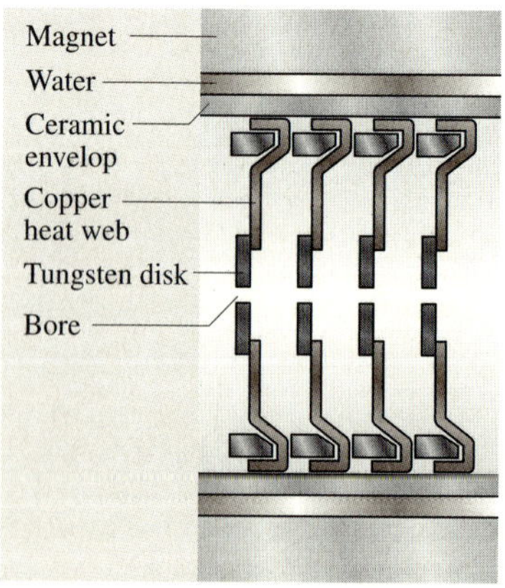

Fig. 11.144 Segmented bore design

opens to allow the plenum to add a precise quantity of gas to the plasma tube. Since this fill process briefly disturbs laser operation, the fill system may be manually disabled for short periods to ensure that critical experiments are not effected.

For air-cooled lasers, which do not employ a magnetic field, copper cooling fins are brazed onto the exterior of the beryllia cylinder that forms the plasma tube. Cooling air is forced over the fins at a rate of a few hundred cubic feet per minute (about $10\,m^3$ per minute). Care must be taken that heat removal is evenly distributed so that the tube does not warp and move the laser beam (the laser output mirror is attached directly to the tube), and the air flow should not induce vibrations in the fins which might translate into motion of the beam.

For larger lasers, the entire plasma tube is encased in a large electromagnet which also forms the confining wall for the water flowing next to the tube. The field chosen for the magnet depends on the diameter of the bore and the details of how the laser is operated, but is usually in the neighborhood of 1000 G. The magnetic field is particularly important for ultraviolet output, which requires stronger magnetic fields for optimum performance.

The ends of the plasma tube are sealed with optics that allow the laser light to exit the tube. These optics are either laser mirrors with dielectric coatings, used in essentially all low-power and some mid-power lasers, or quartz windows oriented at Brewster's angle. Tubes are constructed with Brewster windows to polarize their output, generally oriented vertically. Mirror-sealed tubes will have an internal optic to achieve polarization.

The light emitted from the discharge of high-current tubes poses a challenge for optics. The transitions from the lower laser levels back to the ground state of neutral argon shown in Fig. 11.141 emits a highly energetic photon around 80 nm. This vacuum UV radiation is very deleterious to optical materials. The coatings on the mirrors that seal low-power and some mid-power tubes are made of many alternating layers of dielectric materials, such as silica and alumina, that are capable of withstanding this radiation. Brewster windows used to seal mid-power lasers are typically made of fused silica.

The levels of the vacuum ultraviolet of high-current discharges present a greater problem for the Brewster window seals of these tubes [11.1456, 1457]. Fused silica cannot be used because, when exposed to these radiation levels, it forms color-center defects which then absorb even more radiation in a runaway effect that will degrade laser action. To meet the challenge of sealing high-current plasma tubes, manufacturers use crystalline quartz, carefully cut and oriented on the tube so that the crystalline axis of the two windows on each end of the tube are in alignment.

11.5.3 Ion-Laser Resonators

To complete the ion-laser structure, the plasma tube must be placed between suitable mirrors that provide light feedback for amplification by the discharge. These mirrors are a high reflector at one end of the laser, and a partial transmitter or output coupler that allows a portion of the circulating light to exit the laser. These mirrors, together with the structure that holds them in position, form the laser resonator [11.1458]. For small ion lasers and some mid-frame systems, this resonator structure is very simple: the mirrors are bonded onto the plasma tube itself.

Air-cooled ion lasers sacrifice flexibility for simplicity. Except for the power level, their output characteristics cannot be changed. A mirror position is fixed at manufacture by the plastic deformation of the thin metal tube to which the mirror is bonded. The output wavelength, mode structure of the beam, and so on

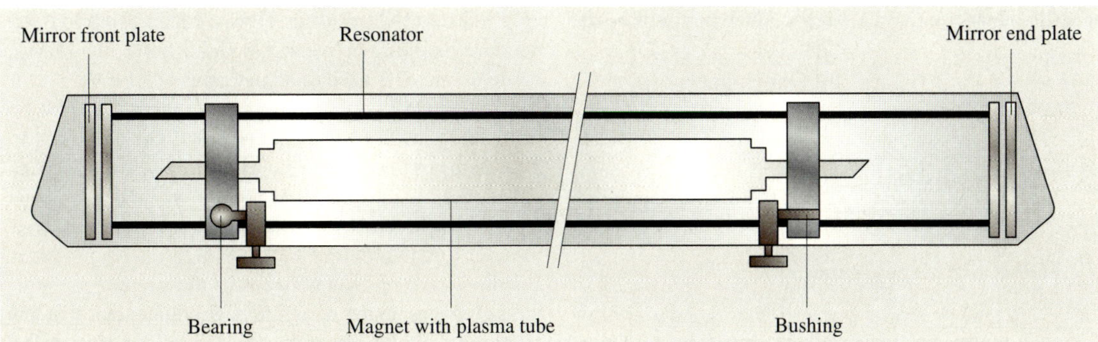

Fig. 11.145 Open resonator design

remain fixed. For many applications this is acceptable, even desirable, but other applications benefit greatly by retaining the flexibility of output that is a hallmark of the ion laser.

The external or open resonator commonly employed with mid- and large-frame systems retains this flexibility. This design requires the use of Brewster-window seals for the plasma tube, as discussed in the preceding section. The resonator mirrors are held in place by a rigid structure that frames the plasma tube (hence the use of *mid-frame* or *large frame* as an indication of laser size). While it outwardly appears simple, the design of the resonator is critical to the performance of the laser.

The resonator frame, as shown in Fig. 11.145, must keep the mirrors aligned to each other and pointing through the laser bore within a very tight tolerance over the entire range of operating and environmental conditions. The choice of materials affects the stability of the resonator. The ideal material has both a low thermal expansion coefficient and a high ability to distribute heat evenly. Graphite and low-expansion compounds such as iron–nickel alloys are typically used for the long rods that form the length of the frame.

Stability also depends on the rigidity of the resonator. Jitter impressed on the laser output due to the microphonic movement of the mirrors can be caused by cooling-water flow, vibration of the resonator structure, and acoustic noise. Isolation of the resonator from both the plasma tube, the magnet, and the cover of the laser head helps reduce jitter.

The mechanical design of the structure is also critical to stability. The most stable configuration is an arrangement of three resonator rods in an equilateral triangle. As a practical matter, this ideal structure does not leave sufficient room for the plasma tube and magnet. As one of the angles of the resonator triangle increases, the resistance to flexure is reduced. The closer the design can be to the ideal equilateral triangle, the better will be the mechanical stability provided for stable power and beam pointing [11.1459].

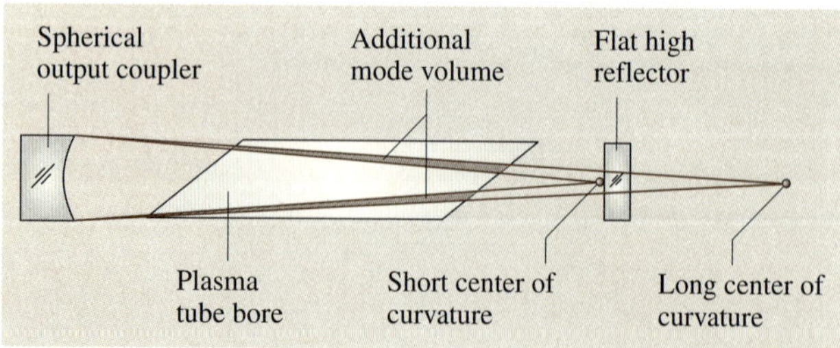

Fig. 11.146 Active mirror positioning allows resonators to extract more power from the discharge

A significant advantage that modern large-frame lasers have over their forerunners is the use of active mirror positioning. The output coupler is mounted on three-point piezoelectric positioner that provides small corrections to the mirror alignment in response to changes in laser output power, most commonly experienced when the laser is warming up. This active resonator also allows for a more aggressive optical design to extract power from the plasma tube.

With active mirror positioning, the cross-sectional area of the beam, and therefore the volume of the plasma discharge used to provide laser amplification, can be increased by enlarging the bore and using a longer-radius output coupler (Fig. 11.146). This is especially beneficial for wavelengths with lower gain. However active control alone cannot provide all of the stability required by this design: the use of long-radius optics requires an extremely rigid resonator structure.

The open resonator allows the user to employ different mirrors to obtain different wavelength output. To obtain *all-lines* argon-ion output in the visible, for example, mirrors are used with coatings that reflect over about 70 nm in the blue and green. A prism on a rotatable mount placed before the high reflector allows the laser to be tuned to single-line output throughout the all-lines range. (The dispersion of the prism directs only one wavelength at a time to the high reflector.) A similar arrangement using different optics enables the same operation for all-lines ultraviolet output.

The open resonator also allows the spatial mode of the beam to be controlled. An adjustable aperture placed between the Brewster window and the output coupler, where the beam is large, provides a variable amount of loss to the edges of the beam while it is inside the resonator. The aperture, when opened to allow the highest power output, will permit a number of transverse modes, or if reduced will result in TEM_{00} mode operation. An aperture diameter between 1.5 and 2 times the TEM_{00} mode diameter is used to achieve this [11.1460].

The spectral content of the laser output may also be changed by insertion of an etalon into the laser resonator. An etalon is a thin optical cavity, such as a plate of glass. When inserted into the resonator, the internal reflections from its surfaces act to narrow the frequency content of the laser output, its *line width*. The resonator produces standing waves of light between its mirrors, called longitudinal modes, many of which have frequencies that fall within the bandwidth of the laser amplifier. The etalon acts as a bandpass filter that introduces variable loss and

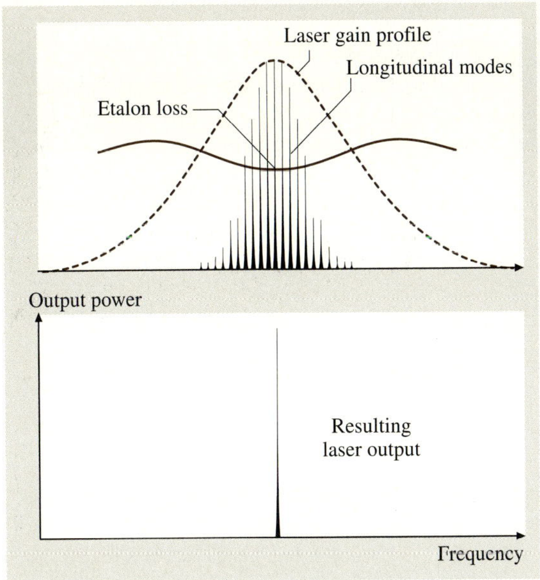

Fig. 11.147 Using an etalon to narrow the spectrum of the laser output

favors oscillation on only one of the longitudinal modes (Fig. 11.147).

Active mirror positioning may also be used to stabilize the line width further, resulting in an even narrower spectral output. The coherence length – the distance over which the output beam maintains a fixed phase relationship – is inversely proportional to the line width. When the laser output is changed from single line to single frequency, the coherence length increases from about 50 mm to 20 m.

A drawback of the open resonator is that the space between the mirrors and the Brewster windows is not vacuum-sealed. Laser performance strongly depends on keeping these optical surfaces clean. In addition, the vacuum UV light from the arc discharge, transmitted by the Brewster windows, converts oxygen to ozone, which is deleterious to the laser optics. This intracavity space must be sealed and its atmosphere controlled to maintain cleanliness and avoid optical damage. Lasers producing short-wavelength UV may employ a nitrogen purge to avoid ozone production. Longer-wavelength lasers can use a simpler technique of placing a catalyst in the enclosed space that converts ozone back into oxygen.

11.5.4 Electronics

A variety of electronic subsystems are required to operate an ion laser, as shown in the block diagram of a large-frame system in Fig. 11.148. Typically packaged in the laser head along with the plasma tube are the start circuit and the light pick-off. Lasers that are not sold as OEM components for a larger system are required to have an interlock switch that will reliably terminate laser operation if the head cover is opened. For water-cooled lasers, the laser head will usually also contain the flow- and temperature-control circuits for the cooling water, and the fill circuitry to monitor and replenish the gas in the plasma tube.

The start circuit provides a high-voltage spark to initiate the discharge. Typically, a pulse of several kV is injected in series with the tube by a pulse transformer or a spark-gap circuit. The light pick-off is the sensor for the feedback loop that measures and regulates the laser output power, and usually consists of a beam splitter with a silicon photodetector. Since an ion laser may be capable of producing a wide spectrum of wavelengths over a considerable range of powers, multiple preamplifier ranges are usually used in the feedback loop, as well as wavelength sensitivity correction color filters or electronic gain modification.

The power supply that drives the plasma tube is almost always packaged as a separate unit. This is because the supply itself is roughly the same volume as the laser head, and separating it from the laser allows much greater flexibility in positioning the laser system. The drive current for the plasma tube is delivered to the laser head through a robust umbilical. Since the power supply also serves as the control interface for the laser, cables

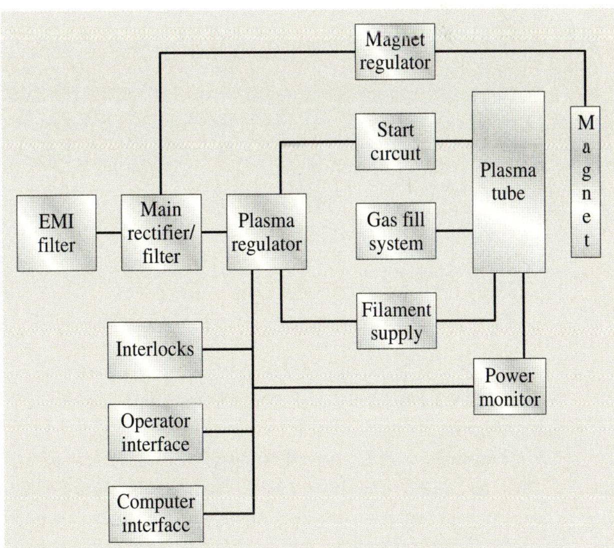

Fig. 11.148 Block diagram of a large-frame ion-laser system

Table 11.42 Ion-laser wavelengths

Powers available in a type of laser (W) (not all wavelengths are available simultaneously)					
Available wavelengths (nm)	Lasing species	Large-frame water cooled (W)	Mid-frame water cooled (W)	Large air-cooled (W)	Small air-cooled (W)
275.4	Argon	0.375			
300.3	Argon	0.5			
302.4	Argon	0.5			
305.5	Argon	0.17			
333.6	Argon	0.4			
334.5	Argon	0.8			
335.8	Argon	0.8			
350.7	Krypton	2.0	0.3		
351.1	Argon	1.5	0.18		
351.4	Argon	0.5	0.06		
356.4	Krypton	0.5	0.12		
363.8	Argon	2.0	0.25		
406.7	Krypton	1.2	0.22		
413.1	Krypton	2.5	0.3		
415.4	Krypton	0.35			
454.5	Argon	0.8	0.18	0.005	0.002
457.9	Argon	1.5	0.55	0.015	0.005
465.8	Argon	0.8	0.24	0.015	0.005
468.0	Krypton	0.5	0.1		
472.7	Argon	1.3	0.32		
476.2	Krypton	0.4	0.1		
476.5	Argon	3.0	1.0	0.025	0.008
482.5	Krypton	0.4	0.05		
488.0	Argon	8.0	2.5	0.1	0.03
496.5	Argon	3.0	1.0	0.03	0.012
501.7	Argon	1.8	0.64	0.015	0.005
514.5	Argon	10.0	3.2	0.1	0.025
528.7	Argon	1.8	0.55	0.01	
530.9	Krypton	1.5	0.25		
568.2	Krypton	0.6	0.25	0.02	
631.2	Krypton	0.2			
647.1	Krypton	3.0	0.8	0.015	
676.4	Krypton	0.9	0.15		
752.5	Krypton	1.2	0.15		
793.1	Krypton	0.1			
799.3	Krypton	0.2	0.03		

that carry monitoring and control signals are typically bundled in the umbilical as well.

Ion-laser power supplies show the same range of variation in size and design as the laser heads. The power supplies of small air-cooled argon units can be as small about 15 cm on a side, while supplies that drive large-frame systems are roughly 0.7 m on a side and weigh almost 100 kg. These larger units may require cooling water. Table 11.43 displays typical service requirements and loads for the three main classes of ion lasers.

The electrical characteristics of the arc discharge are such that it essentially requires a constant-current supply in order to be sustained [11.1458]. The heart of an ion-laser power supply is the plasma current regulator.

Table 11.43 Power supply characteristics of typical ion-laser systems

Ion-laser type	Maximum laser power	Electrical input	Plasma discharge	Discharge power requirement
Air-cooled	20 mW	115 V_{AC} single phase	90 V 8 A	0.7 kW
Small frame	5 W	208 V_{AC} three phase	240 V 55 A	13 kW
Large frame	25 W	480 V_{AC} three phase	550 V 65 A	36 kW

The plasma current needs to be well regulated in order to prevent line ripple and other variations from being impressed on the laser output. The plasma current may be regulated by a linear transistor passbank, or more efficiently by various switching-type power supplies.

In the patented regulator used by Spectra-Physics, a switched resistor regulator consisting of a pulse-width modulator (PWM) controlling a switch transistor is connected to a low-impedance high-power water-cooled resistor. This resistor is connected to the plasma load. The PWM varies the percent of the time that the transistor is turned on and the resistor is conducting current. A parallel capacitor provides a current path to the load when the transistor is off. This regulator looks like a variable resistor that can vary from infinity down to the minimum resistance of the circuit. A shunt regulator is used to provide ripple rejection and high-speed small-signal regulation.

The power supply must power the electromagnet when an axial magnetic field is employed. The electromagnet requires about 10–25% of the power supplied to the arc discharge. Thus a large-frame supply is really two supplies: a 36 kW plasma supply and a regulated 8 kW magnet supply. The power supply must also provide power to heat the tungsten cathode in the plasma tube. Typically a filament transformer provides 15–25 A at about 4 V_{AC} for this purpose.

11.5.5 Ion-Laser Applications

The operator interface to control power and the basic on/off function is through the power supply. In inexpensive models of air-cooled systems, the control interface typically consists of hand-wired knobs and switches on the front panel of the supply. The next step in sophistication is a hand-held system controller with a visual display that attaches to the power supply with a long flexible cable. Upon turn-on, all functions of the laser can be accessed from the system controller. High-end models offer a fully functioning computer interface for automated remote operation and monitoring of the laser.

Ion lasers operate in either current mode or power mode (also called light mode). Current mode holds the plasma discharge current at a fixed value and allows the laser output to vary. Power mode, a much more common method of operating the laser, adjusts the plasma current as necessary to match the output power to a user-requested value. When the current can no longer be increased to achieve the power set-point, it may be time to replace the plasma tube.

Like many modern laser systems, most ion lasers are designed for hands-free operation. Air-cooled argon systems in particular are intended to provide a constant laser source available from a simple flip of a switch. Large-frame systems in laboratory applications in contrast trade on the flexibility they provide the user.

Small air-cooled systems continue to be economical and reliable sources of coherent, polarized. TEM_{00} blue–green laser light. They are usually operated at a power significantly derated from their maximum, which can extend their useful lifetime for years. Biological and medical applications for these lasers include cell sorting. Desoxyribonucleic acid (DNA) sequencing, bacterial analysis, confocal microscopy, and hematology. Many of the dyes used in these applications were originally developed for argon-laser wavelengths.

These lasers are also used in many applications related to producing text and images. The blue beam is valuable for exposing printing plates for high-speed printing, and to provide the color separation required for full-color printing. Similar applications are found in photoprocessing and other photographic sources.

Mid-frame systems are used in many of the same applications as lower-power lasers. In addition, they are used in entertainment, especially for laser light shows, and have laboratory applications in Raman spectroscopy or as a pump sources for tunable laser systems such as a Ti:sapphire. In ophthalmology, the beam can be focused on the retina to repair diabetes – induced retinal detachment, for instance.

The deep-blue and UV outputs of ion lasers have been used in the semiconductor industry for wafer inspection and lithography. The ability of the output to expose photoresist is also used in producing masters for compact discs, which then serve as the molds in the injection-molding manufacture of high-volume CD production.

11.6 The HeNe Laser

The HeNe laser is an electrically pumped continuously emitting gas laser. Its basic principle is a gas discharge in a glass tube filled with a mixture of helium and neon under low pressure. The gas discharge is set up by a cathode and an anode placed at opposite sides of the glass tube. The laser mirrors are usually fixed to the end of the tubes.

The HeNe laser was one of the first lasers to be realized. It was developed in 1960 as the first laser emitting light continuously [11.1461]. However, this was not the well-known bright red line (632.8 nm) that has been discovered, the first laser emitted light at 1.15 μm, which is one of the strongest lines of the HeNe laser. Red laser emission was achieved shortly after, in 1962 [11.1462]. Though other possible transitions delivering light in the visible had been found theoretically, it took a certain time to demonstrate this practically. Improvements in resonator design and the performance of the mirrors had to be achieved. Especially the well-known 543.3 nm line, with its low gain, needed several attempts by researchers until it was demonstrated for the first time in 1970 [11.1463].

Shortly after its first demonstration, the HeNe laser found more and more applications and became the most common laser worldwide with millions of units sold, until laser diodes appeared on the market. More than 40 years after its first appearance the HeNe laser is still of great importance in the worldwide laser market. Around 44 000 units were sold in 2005 [11.1464].

The HeNe laser has been used for adjusting and positioning but also, because of its excellent optical properties, in interferometers, sensors or spectrometers. It was applied in the first scanner tills and even the first CD players were equipped with HeNe lasers. Although these applications have been taken over completely by diode lasers, HeNe lasers still find their use in many fields of analytics, instrumentation, sensor technology, science and education.

Its advantages are excellent beam quality, long lifetime and an unbeatable price–performance ratio. Because of the mature technology and the still large sales numbers manufacturing costs have decreased continuously. Against the trend for shorter product lifetime the HeNe laser has asserted itself for 40 years and is still a competitive product. Many OEM manufactures still choose the HeNe laser for there new products, irrespective of the competition from diode lasers and solid-state lasers. Considering especially the laser diodes we see that, in fact, the diode itself is a rather cheap component. To achieve a beam quality comparable to the HeNe laser, however, requires considerable effort and additional cost, which compensates for the higher costs of a HeNe laser. For example, the strongly elliptical beam of the diode has to be made circular and, in order to avoid wavelength drift, temperature stabilization has to be provided. Another advantage is the longer coherence length of a HeNe laser compared to that of standard diodes.

Further advantages of the HeNe laser are:

- Excellent mode purity, typical $>95\%$ Gaussian TEM_{00}
- A favorable relation between resonator length and resonator width (diameter)
- A nearly diffraction-limited beam
- High beam-pointing stability
- High reproducibility in manufacturing

11.6.1 The Active Medium

Energy-Level Diagram

The most detailed energy-level diagram which we could find in the literature is given in [11.1465]. Nowadays, however, the green and yellow laser lines are of much more importance than the infrared ones, so it was necessary to add these new visible lines to the diagram. It can be seen that neutral helium is excited by electron collisions and transfers its energy by nearly resonant inelastic atomic collisions to the excited states of neon. This means that the lasing atom is the neon, helium is only necessary for energy transfer from the gas discharge to the upper neon levels. The energy transfer from the lower laser levels 3p and 2p ends at the 1s level due to spontaneous emission and the 1s level is depopulated by wall collisions to the ground state of neon. For this reason it is necessary to have a small-diameter discharge tube, in order to secure a quick emptying of the 1s level. It should be mentioned that the rare ^{3}He isotope is used and not ^{4}He. The reason is that the lighter ^{3}He isotope has a higher velocity difference to neon, resulting in better energy transfer to the upper laser levels [11.1466] and consequently a higher gain and output power (about 25% better at 633 nm).

It is known that electron collisions and stimulated absorption of visible light can cause transitions from the 1s to the 2p level, resulting in a higher lifetime of this lower laser level (2p). This leads to a higher inversion for the 3.39 μm laser line as for the other lines, since it has the 3p level as the lower laser level.

Table 11.44 Typical laser transitions

Wavelength (in air) (nm)	Transition	Gain (m^{-1})	Typical power (mW)
543.3	$3s_2$–$2p_{10}$	0.03	0.5–3.0
594.1	$3s_2$–$2p_8$		2.0
611.8	$3s_2$–$2p_6$	0.1	2.0
632.8	$3s_2$–$2p_4$	0.5	0.5–50
640.2	$3s_2$–$2p_2$		
1152.3	$2s_2$–$2p_4$		2.0
1523.1	$2s_2$–$2p_1$	4	1
3391.3	$3s_2$–$3p_4$	100	10

Table 11.44 lists the commercially available laser lines and some typical parameters. The gain [11.1465, 1467] depends strongly on the inversion and the noted values should only give an impression of how the gain changes with wavelength ($g_0 \sim \nu_0^{-3}$). This illustrates that the most difficult laser to build is the green one (543.3 nm).

Another difficulty in making a green laser is that, within each group of energy levels (2s, 3s, 2p, 3p, and so on), the energy distribution of the sublevels (e.g., $2p_1$ to $2p_{10}$) is determined by thermic Ne–Ne collisions, resulting in a Boltzmann distribution, which means that the population rises with falling energy. Thus, the lower laser level for the 543.3 nm line ($2p_{10}$) has a much higher population as the $2p_4$ level for the red line. In order to achieve population inversion for this laser line it is necessary to reduce the number of Ne–Ne collisions by reducing the neon pressure. The consequence is a lower output power at this line and a lower lifetime for green lasers (about 10 000 h) compared to red lasers (> 20 000 h).

Table 11.44 lists only the most important laser lines. A good compilation of laser lines at different elements is given in [11.213], which also contains references to the original literature. Sometimes in an experiment one may have unwanted light from scattering from the discharge tube itself (not laser light). Such discharge lines can be found in [11.1468].

Gas Discharge

The gas discharge is the most important process in the laser, since it has to transform the electrical energy of the power supply into the laser light. This process determines the laser power, power stability, optical noise and lifetime of the laser. Hence, the key know-how of a laser manufacturer is not the resonator design but how to optimize the discharge.

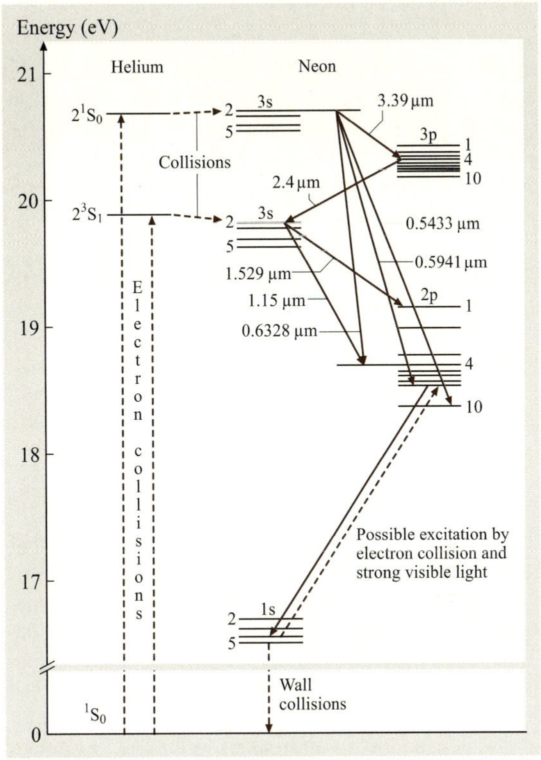

Fig. 11.149 Energy-level diagram of the HeNe laser

As shown in Fig. 11.149 the HeNe laser tube is filled with a helium–neon gas mixture, having a total pressure of about 4–7 mbar and a neon content of about 10%. The operating DC current is in the range of 3.5–11 mA and the corresponding tube voltage is about 1–5 kV. This is a cold-cathode glow discharge, where the active laser medium is formed by the positive column of the discharge, which is located in a capillary of diameters of 0.5–2 mm. Since such a glow discharge has a decreasing voltage–current characteristic curve (see Fig. 11.151),

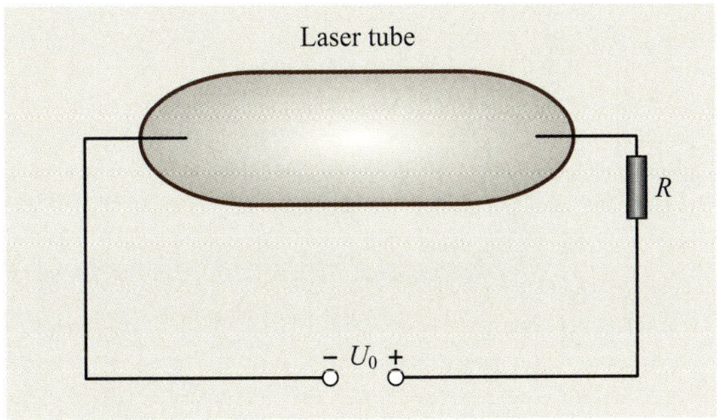

Fig. 11.150 Electrical scheme of the laser

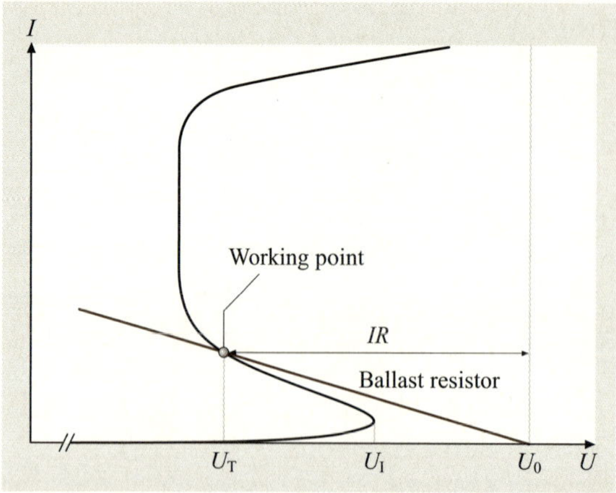

Fig. 11.151 Characteristic voltage–current curve of a HeNe laser

a ballast resistor R of about 60–100 kΩ is used to bring the whole system to an increasing voltage–current curve.

In Fig. 11.151 the black curve shows the electrical behavior of the discharge tube and the brown line the corresponding ballast resistor. After switching on the laser, the power supply has to raise voltage to the ignition point ($U_I \approx 10\,\mathrm{kV}$). After the ignition, the power supply switches to its normal fixed-current operating mode. The corresponding voltage is determined by the crossing point of the discharge curve and the resistor line (voltage at the tube U_T). It is important to choose the resistor in such a way that only one crossing point exists.

Whereas in the tube design a small metal anode is used, the cathode is considerably larger, and is mostly made of aluminium because the cathode has to deliver the electrons for the discharge and to transfer the heat of the cathode fall, the region where the electrons are accelerated from zero to the colliding velocity, outside the laser tube. The capability of the cathode to emit electrons depends on the properties of the cathode surface. Normally the current density at the cathode surface is constant, so the cathode surface must be large enough for the applied operating current, otherwise sputtering occurs and the tube is quickly damaged. The typical cathode size is in the range of some $10\,\mathrm{cm}^2$. In order to avoid sputtering, at the surface a nanometer-scale layer of Al_2O_3 is formed. Normally this oxide layer withstands sputtering for some 10 000 h. If this layer is destroyed, the laser fails within a few hundred hours.

11.6.2 Construction and Design Principles

Construction of a HeNe Laser

Figure 11.152 shows a cross-sectional view of a modern HeNe laser. The principal setup is very similar for all major manufactures of HeNe lasers and differs in details only. The laser tube consists of a glass tube (8) which is melted at both ends to metal end caps (3, 10). To these end caps the laser mirrors (1, 12) are connected using a metal–glass soldering technique. These metal–glass connections provide long-term vacuum sealing of the laser tube. Inside the laser tube the cathode (6) and the gas discharge capillary (7) are located. The glass of the capillary is melted to the glass of the outer tube (8), so the discharge is concentrated inside this capillary. If the mirror (1) is a flat mirror and mirror (12) is concave, the end of the capillary (9) acts as a mode field aperture. If linear polarized laser radiation is required, a Brewster window (2) is placed inside the laser. A favorable position for this window is near the flat mirror, because at this position the beam diameter is the smallest in the tube and there is no angular displacement between the laser beam and the window (the beam is always perpendicular to the mirror surface). The tube is evacuated and filled using the pipe (4), which is sealed after filling the tube by means of a cold soldering process. The spring (13) centers the capillary end. At the mirror holders (3, 10)

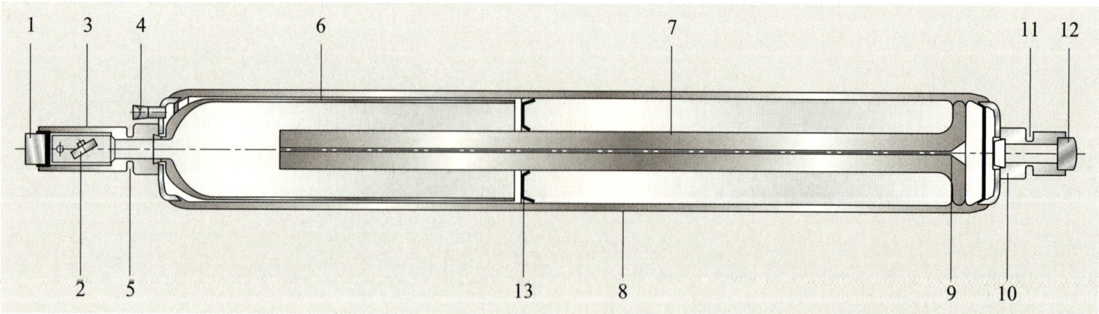

Fig. 11.152 Sectional view of a HeNe laser tube (LASOS Lasertechnik GmbH), see text for explanation

a region of reduced material thickness is formed (5, 11). This is used for the alignment of the mirrors by means of plastic deformation of the metal.

The setup according to Fig. 11.152 represents the basic design of commercial laser tubes up to about 25 mW output power which has been used for about three decades with minor changes only. It provides long-term gas stability, clean optics over the whole lifetime of the laser and a long lifetime of the laser.

In contrast to the present design, early HeNe lasers were not sealed by the mirrors but by a Brewster window at each end. Thus, they used an external resonator, mostly made of mirrors hold on invar rods. This older design is used today only in high-power HeNe lasers (> 35 mW), because such lasers need a very long cavity (> 1 m). Here, the external resonator allows a better stability of the mirror alignment, and the capillary (which suffers from bending due to their own weight) can be aligned too.

Resonator Design

The design goal of a HeNe resonator is mainly to achieve a Gaussian beam profile. The calculation of all beam parameters can be easily done by means of the *ABCD* matrix formalism [11.1469]. In this case, every optical element in the laser is described by a matrix, as described below.

The matrix of a mirror with curvature r

$$M=\begin{pmatrix}1 & 0\\ -2/r & 1\end{pmatrix}. \tag{11.143}$$

The matrix for the free-space propagation of distance s

$$S=\begin{pmatrix}1 & s\\ 0 & 1\end{pmatrix}. \tag{11.144}$$

Starting at the output mirror all single matrices for every element are multiplied for one complete round-trip (RT), resulting in the following calculation:

$$\mathrm{RT}=S\,M_2\,S\,M_1=\begin{pmatrix}A & B\\ C & D\end{pmatrix}. \tag{11.145}$$

In a last step, the $1/\mathrm{e}^2$ beam radius (w) and the wavefront curvature (R) are calculated from this matrix:

The beam radius at the output mirror is

$$w=\sqrt{\frac{\lambda}{\pi}\frac{2|B|}{\sqrt{4-(A+D)^2}}}. \tag{11.146}$$

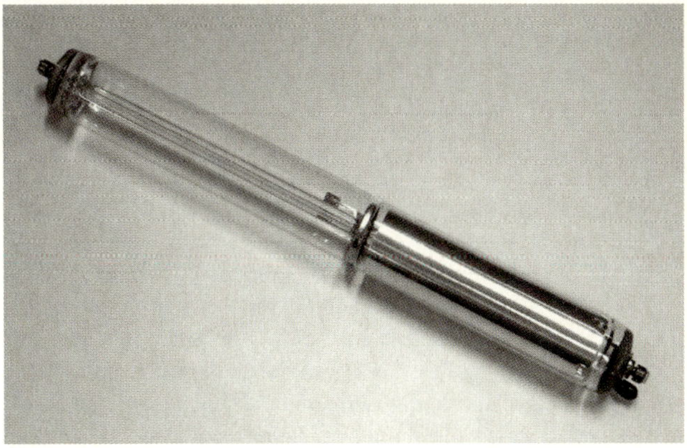

Fig. 11.153 Laser tube with fixed mirrors (LASOS Lasertechnik GmbH)

The wave-front curvature at the output mirror is

$$R=\frac{2B}{D-A}. \tag{11.147}$$

Using the well-known formalism for Gaussian beam propagation, the beam diameter can be calculated at any desired position.

The complex parameter q is defined as:

$$\frac{1}{q}=\frac{1}{R}-\mathrm{i}\frac{\lambda}{\pi w^2}. \tag{11.148}$$

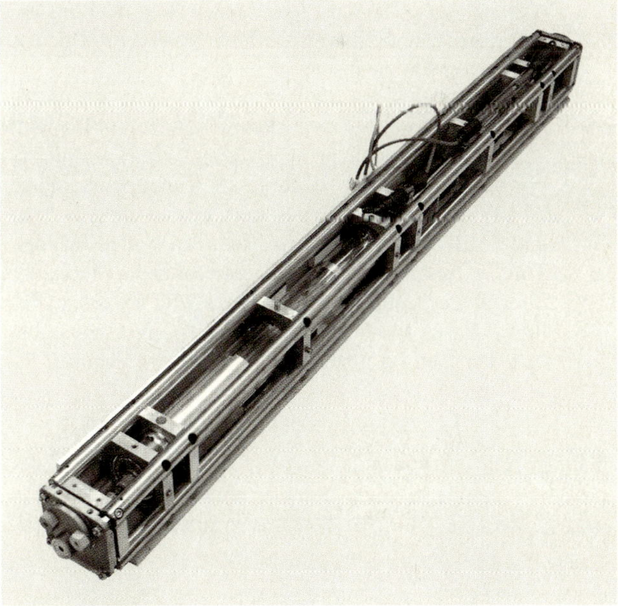

Fig. 11.154 35 mW laser tube with external resonator (LASOS Lasertechnik GmbH)

The propagation of a Gaussian beam through an optical system described by the *ABCD* matrix yields

$$q_{\text{out}} = \frac{Aq_{\text{in}} + B}{Cq_{\text{in}} + D} . \qquad (11.149)$$

The beam diameter at the position of the mode field aperture is important for the laser design (Fig. 11.155). The ratio between this aperture and the beam diameter has to be chosen in such a wa, that the losses for the TEM_{00} mode are low but high enough for the first higher-order transversal modes, e.g., TEM_{01}, in order to suppress the higher modes. This ratio depends on the laser gain and can be calculated theoretically. The practical experience for a common 633 nm laser leads to a ratio between the aperture diameter and beam radius of about 3.5 : 1. For high-power HeNe lasers, this value must be reduced, whereas for the low-gain 543 nm laser it has to be increased.

In practical designs the above formalism is used to balance the resonator design between the beam diameter, required by the application (the beam waist is normally located at the surface of the output mirror), and the laser power, which determines the laser length on one hand, and the available mirror curvatures as well the available aperture diameters and aperture positions on the other.

Scaling Relations

Scaling relations are very useful for the practical design of new lasers. The most important scaling relations are described below. Firstly, the laser power P, which is proportional to the discharge length l (capillary):

$$\frac{P}{l} = \text{const.} \qquad (11.150)$$

For this reason, today HeNe lasers with a power above 25 mW are very rare. These lasers are very long (> 1 m) and need sensitive and expensive external resonators. Thus, in this power region the costs for beam shaping of laser diodes (aspheric lens systems with prisms, cylindrical lenses and/or zoom telescopes) are comparable or lower than for a HeNe laser and the mechanical dimensions of the diode-lasers systems are much smaller (about 15 cm).

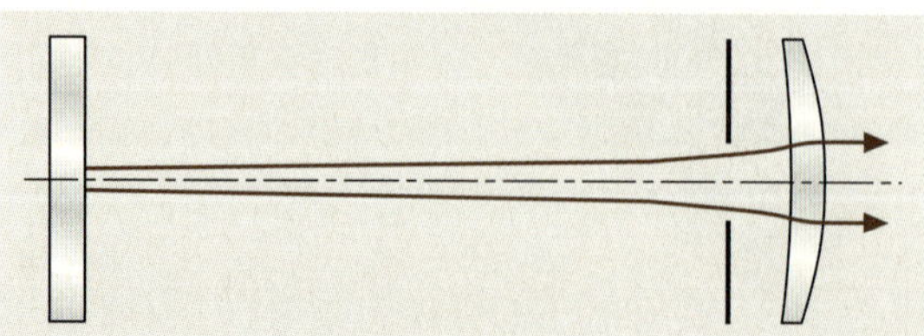

Fig. 11.155 Principle setup of a HeNe resonator with rear and output mirror and a mode-field aperture

The product of the pressure p and capillary diameter d is constant:

$$pd = \text{const.} \qquad (11.151)$$

Following this equation, two systems with the same pd product have comparable discharge properties and the excitation of the upper laser levels is comparable in both cases if the same current is applied. This rule is very useful if a well-running laser must be customized in order to get more power or different beam diameter. If the new capillary diameter has been estimated using the formalism of Sect. 11.6.2 then the corresponding gas pressure for this new design can be calculated by means of (11.151).

The current–density scaling relation is less important because there is only a weak dependence on capillary diameter

$$j\sqrt[4]{d} = \text{const.} \qquad (11.152)$$

The scaling of the small signal gain g_0 with capillary diameter d

$$g_0 d = \text{const.} \qquad (11.153)$$

Equation (11.153) shows that a small capillary diameter d should be achieved in the design process because, with higher gain, losses can be higher too and it is possible to come closer to the theoretical maximum power.

Additionally two electrical relations [11.1470] should be noted: the current density j_K at the cathode

$$\frac{j_K}{p^2} = \text{const.} \qquad (11.154)$$

and the ignition voltage U_I

$$U_I = f\left(\frac{pl}{T}\right) . \qquad (11.155)$$

Equation (11.154) has the consequence that low-pressure laser tubes, like versions for 543 nm, need larger cathodes. Equation (11.155) shows that the voltage necessary to ignite the laser increases with pressure and length and decreases with temperature T.

Laser Line Selection

Today, the HeNe laser has been replaced in the infrared range by laser diodes, but its visible lines, especially those at 632.8 nm, 594.1 nm and 543.3 nm, are widely used in scientific applications such as laser scanning microscopy. Because of this fact, we will only consider the visible range. Except for some educational lasers, all modern HeNe lasers operate at one wavelength only. The reason is that it is only possible by this approach to obtain the maximum power of each laser line and to reduce the optical noise. The operating wavelength is selected from the possible lasing lines by means of the laser mirrors. In the most common design the rear mirror has as high as possible a reflectivity (> 99.9%) for the desired wavelength, and the output coupler will have approx. 1–2% transmission at this line and > 10% at the other lines. So, the unwanted laser lines have losses greater than their gain and will not be amplified. Because of the low gain of the 543.3 nm line, lasers for this wavelength must have output mirrors with transmissions of about 0.05–0.15%.

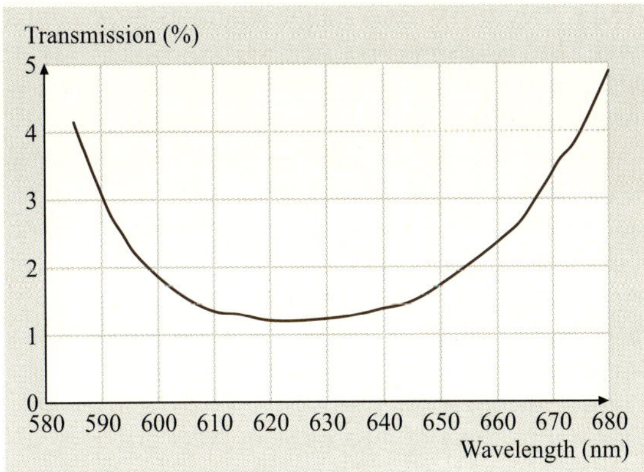

Fig. 11.156 Typical transmission curve of the output mirror for a 632.8 nm laser

3.39 μm Suppression

From Table 11.44 it can be seen that the 3.39 μm line must be suppressed very carefully, because it has a gain that is some orders of magnitude higher than all visible lines. Insufficient suppression of this line causes power loss and power fluctuations at the desired lines. Therefore, both laser mirrors must have a reflectivity for this line below 0.5%. Another approach for the 3.39 μm suppression is to use the Zeeman effect. Applying an inhomogeneous transversal magnetic field to the laser splits the laser levels into the Zeeman sublevels. The gain of the sublevels is the gain of the unsplit level divided by the number of sublevels, so the gain may fall below the losses and this line is suppressed. The frequency difference of the sublevels is the same for all laser lines. However, Doppler broadening of the laser lines depends on their frequencies.

The Doppler broadening of a laser line λ is given by

$$\partial\nu_D = \frac{2}{\lambda}\sqrt{\frac{2kT\ln 2}{m}}\,, \tag{11.156}$$

where k is Boltzmann's constant, T is the gas temperature and m is the mass of the Ne atom.

For visible wavelengths the frequency split caused by the Zeeman effect is lower than the Doppler-broadened line. Thus, the gain is not influenced by the Zeeman effect, whereas for the 3.39 μm line the Zeeman splitting is larger than the Doppler-broadened line and the gain decreases with the number of sublevels. This method is normally used in high-power 633 nm lasers (> 35 mW) or in the 543 nm laser.

Line Width and Coherence Length

The natural line width of a red HeNe laser is about 20 MHz [11.1466] but this line is broadened by two different processes. The first process is collision broadening, leading to a line width of about 500 MHz at typical pressures of 4–6 mbar [11.1466]. This line width is proportional to the gas pressure. The second process is Doppler broadening (11.156)), which results in a width of 1–1.5 GHz for the visible laser lines.

A TEM_{00} laser usually runs at several longitudinal modes, which are spaced by

$$\Delta f = \frac{c}{2L}\,. \tag{11.157}$$

Thus, the longer the resonator (mirror distance L), the more modes can oscillate simultaneously.

As can be seen in Fig. 11.157, the mode spacing is 257 MHz (seven modes) and the gain profile is about 1.5 GHz. A Lamb dip does not occur because the collision broadening is larger than the mode spacing [11.1466]. For interferometric purposes tubes that

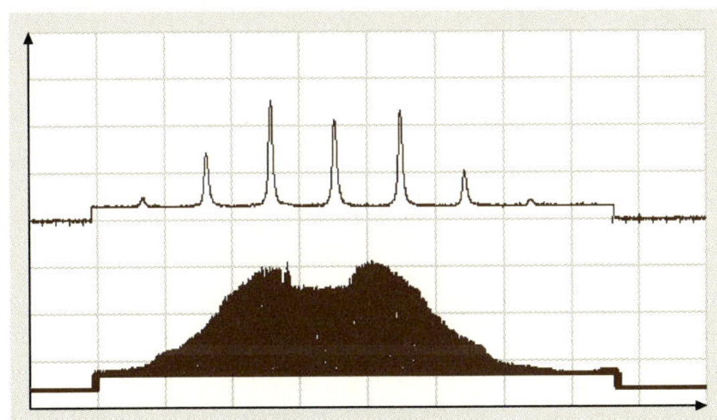

Fig. 11.157 Longitudinal modes (*upper curve*) and corresponding gain profile of a 584 mm-long 20 mW laser at 633 nm (LASOS: LGK 7665 P), measured with a scanning interferometer

exhibit only one or two longitudinal modes are often used. The typical length of such a tube is about 140 mm with a power of 0.5–1.0 mW. Whereas the gain curve is fixed in frequency space, the longitudinal mode pattern shows a movement when the resonator length changes in the dimensions of the wavelength λ, which may result from small ambient-temperature fluctuations. Such a movement causes power fluctuations in the range of about 5% (depending on the laser length). Because in most applications only the sum power of all modes is used, it is possible to stabilize the laser power by increasing the number of modes under the gain profile. This can be done by two methods. The first is to increase the laser length and the second uses a 1 : 1 mixture of the neon isotopes ^{20}Ne and ^{22}Ne instead of natural Ne (90% ^{20}Ne and 10% ^{22}Ne). Because of the heavier nuclear core of the ^{22}Ne isotope the radiated light has a frequency about 800 MHz higher than the light from ^{20}Ne (isotopic shift, [11.1471] p. 333). This means that the difference between these two lines is smaller than the Doppler broadening. Therefore, using a mixture of these two isotopes gives a broader gain curve. The isotopic mixture has no influence on other laser parameters such as the power or noise.

One of the most exciting applications of the HeNe laser is interferometry. For this purpose, the coherence length of the laser is important. The coherence length can by calculated from the line width $\Delta\nu$ by

$$l_c = \frac{c}{\Delta\nu} .$$

The problem of a multimode laser is that it does not run at a single frequency but at several equidistant modes.

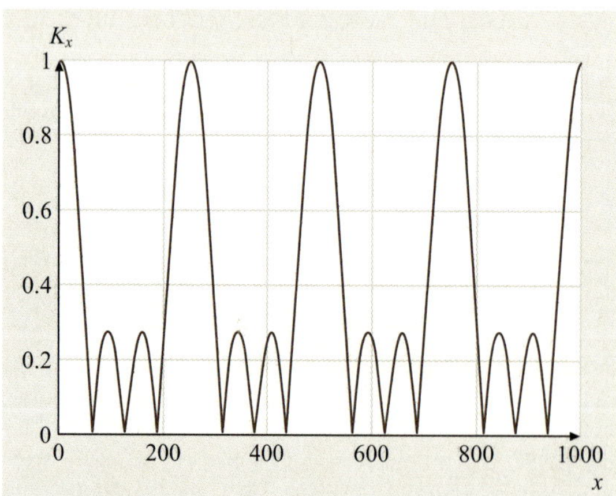

Fig. 11.158 Contrast ratio for the interference of two four-longitudinal-mode laser beams

Figure 11.158 shows the calculation of the measured contrast ratio of a longitudinal multimode laser beam after a Michelson interferometer over the optical path difference. The beam was formed of four laser modes with equal intensity. This picture is overlayed by the decreasing contrast function of the single mode, but the coherence length of the single mode is in the range of kilometers and cannot be seen here. Whereas the interference can be observed over very long distances, the periodic structure makes it necessary to define the coherence length only in the region up to the first minimum. In the case of N longitudinal modes, this first minimum is located at the position

$$l_{\text{min}} = \frac{2L}{N} . \tag{11.158}$$

Thus, for the laser of Fig. 11.157 the coherence length is below 16 cm.

It should be mentioned that in a real laser the different modes have different intensities, but this only changes the amplitudes in Fig. 11.158 and leads to a nonzero intensity at the minima.

11.6.3 Stabilization

In order to obtain a larger coherence length, a common approach is it to use only one of the longitudinal modes. In order to select a single mode, the fact that in an unpolarized laser (without internal Brewster windows) neighboring modes have perpendicular polarization directions can be used. Hence, using a short unpolarized laser with only two modes, it is possible to select one mode by means of an external polarizer.

In some applications the frequency of this mode must be fixed. This can be achieved by controlling the length of the laser tube by means of a heater around the glass tube. One way is to select the two perpendicular modes using a polarizing beam splitter and measure the power

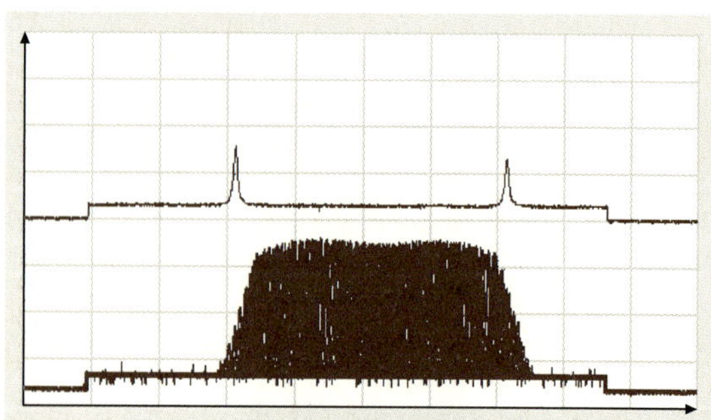

Fig. 11.159 Modes and gain profile of a 140 mm-long laser

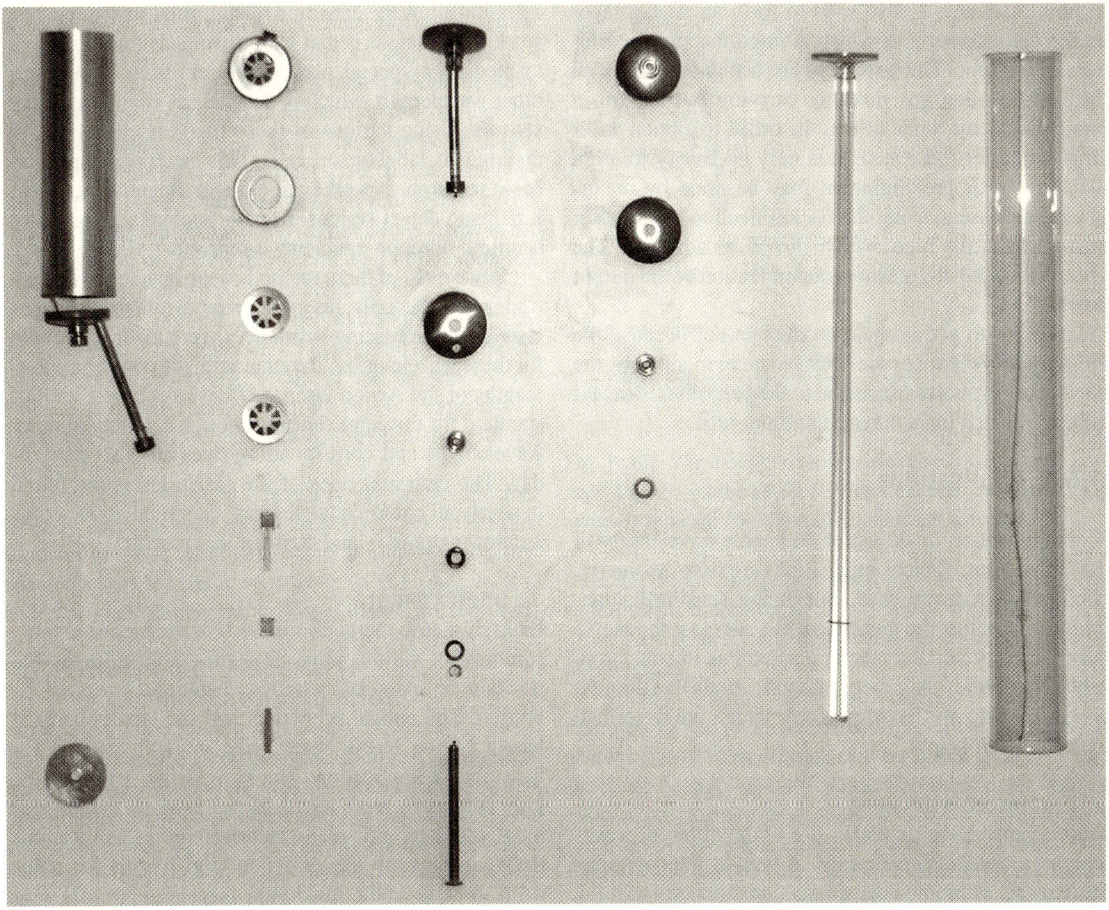

Fig. 11.160 Components of a laser tube

difference between these two modes. By means of the heater, this signal should be held at zero. In this case (Fig. 11.159) both modes are located at the sharp flanks of the gain profile. Here, small frequency changes result in large power changes and exact control of the frequency is possible. This principle is much easier than stabilizing a laser diode, at least with an external resonator with a grating, and the frequency depends only on the atomic properties of the neon. This first approach to frequency stabilization is widely used in the field of mechanical engineering for measurement devices with an accuracy of about 1 nm. A detailed description of frequency stabilization is given in Sect. 11.14.

11.6.4 Manufacturing

Because the first HeNe lasers were built in the 1960s, today the manufacturing technology is a fully developed process, where the processes of cleaning, connecting and the vacuum processes are the key know-how of the manufactures.

After an initial cleaning process of the different components, in a first step the metal parts are soldered together in a high-temperature process (about 1000 °C). In the next step, the metal and glass parts are melted together at temperatures of about 800 °C. During this step, care must be taken that the capillary bore is at the position of the subsequent optical axis. In order to avoid mechanical stress, the different materials must be chosen in such a manner that the thermal expansion coefficients of the metal and glass parts are matched. The third step is soldering the mirrors to the mirror holders (at about 500 °C). After this preparation, the tubes are backed out under vacuum for many hours. Than, by means of an oxygen discharge, the inner surfaces of the tube are cleaned of residual organic traces. The

11.7.2 Technology of Current Excimer Lasers and the N_2 Laser

We start with the fundamentals of excimer lasers and the fluorine laser. Due to its role as a forefather of all gaseous UV lasers, the (molecular) nitrogen laser (N_2 laser) will be briefly dealt with as well. Finally, the technology of discharge-pumped excimer lasers will be briefly outlined.

Excimer Transitions: Unusual Four-Level Laser Systems

Excimer lasers derive their emission from molecules that are generated a priori in an electronically excited state, which decays by emission of (laser) radiation into a repulsive or loosely bound ground state out of which the molecule dissociates. In usual test-tube chemistry the rare gases are noble or inert, i. e., they avoid chemical binding, although special noble-gas compounds do exist also in the electronic ground state. The most important molecules for commercially available excimer lasers are ArF*, KrF*, XeCl*, XeF*, where the asterisk refers to electronic excitation.

The dissociative electronic ground state is responsible for the four-level character of these systems. Generally, four-level laser systems can be operated in continuous manner. However, this is not possible with rare-gas halide excimer lasers because of physical and technical restrictions. A fundamental one is the important role that spontaneous emission plays with respect to the stimulated emission in the ultraviolet spectral region; the ratio of the Einstein coefficient A, which is responsible for spontaneous emission, and that responsible for the stimulated emission in the same transition, B, is proportional to the third power of the transition frequency ν [11.1475]:

$$\frac{A}{B} = \frac{2h\nu^3}{c^2} .$$

Hence stimulated emission can only compete with spontaneous emission if the radiation intensity within the transition $I(\nu)$ is so large that the product $BI(\nu)$ exceeds A, that is, if the system is pumped extremely hard. Electrical power densities of some 100 MW/l must be deposited in the gas volume. In the early days, this excitation density was achieved by using energetic electron beams [11.1472]; nowadays short-pulse transverse electrical discharges are being used. An extension of the pulse duration beyond about 1 μs is hampered by discharge instabilities caused by arcing. The detrimental reaction products and heat must be removed before the next shot so that rapid exchange of the laser gas is required.

After the first experimental demonstration of a bound–free laser system in liquid Xe_2 [11.1476], a large number of other excimer or exciplex molecules were successfully investigated, including homonuclear as well as heteronuclear species [11.1472], but most of these did not gain importance for applications. The door to the systems listed in Fig. 11.161 was opened wide in 1975 [11.1477–1481], with convincing efficiency, wavelengths, and after some technical effort, ease of handling of the laser gases. The gases are primarily the discharge carrier, which is the buffer gas (mostly helium), and, to a much lower extent, the reactants, which are the laser gases.

Molecular Potentials and Reactions Kinetics

The rare-gas halide systems are characterized by two electronic ground states, correlating with the electronic ground states of the rare gas and halogen atoms, which combine to a molecular Σ and a Π state, arising from the p-hole of the halogen atom (Fig. 11.162 [11.1474]). Whereas the Π state is strongly repulsive, the Σ state shows a minimum, mostly with a depth of a few hundred cm^{-1} only, so that thermal energy allows the molecule to dissociate within a few picoseconds. The first electronically excited states correlate with the positive rare gas and negative halogen ions, and hence show a deep minimum. The higher states correlate with the elec-

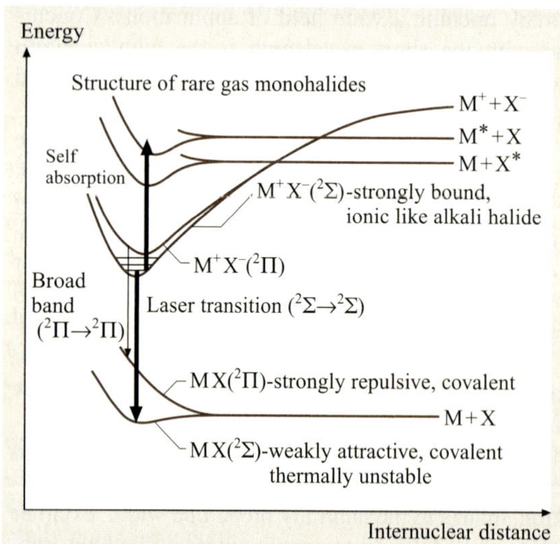

Fig. 11.162 Typical potential energy diagram of a rare-gas halide molecule. M stands for the rare gas, X for the halogen atom [11.1474]

tronic excitations of the neutrals, and their potential is shallower, as is typical for covalent binding. This general scheme holds for the rare-gas halides, except XeF where the potential of the ground state reaches a depth of $1065\,\mathrm{cm}^{-1}$ according to [11.1474].

In the buffer-gas-supported electrical discharge, the excited and ionic species are formed as shown for the KrF* case in Fig. 11.163. The formation of the excited rare-gas halide molecule follows both a *neutral* channel, a chemical exchange reaction, and an *ionic* recombination channel, stabilized by third-body collisions with the buffer-gas atoms. Hence the formation of exciplexes is favored at high pressure, i.e., pressures of a few bars. The buffer gas also provides fast relaxation to the lowest vibrational level within the electronically excited state, followed by a radiative transition to the ground state within a few nanoseconds.

The laser transitions couple with the continuum states above the ground state, showing a homogeneously broadened non-Lorentzian line shape [11.1484]. Hence, excimer lasers can be tuned within a certain bandwidth range if frequency-selective optics are introduced – see the tuning ranges in Fig. 11.161. The homogeneous line broadening turns out to be favorable for the amplification of short pulses.

In more detail, many more excited states that correlate to the excited atomic states exist and are being populated. However, their buffer-gas-induced relaxation feeds the population of the lowest electronically excited state. An overview of the reaction kinetics of both pure rare gases and rare-gas atoms with halogens and the underlying rate constants is given in [11.1474].

The *fluorine laser* is based on bound–bound transitions. With 157 nm it shows the shortest wavelength of all homonuclear halogen lasers [11.1485]. From the two p-holes of the F atoms, three levels belonging to ground electronic configuration are obtained for the F_2 molecule, two of which are bound, so that the stable molecule can be directly electronically excited in an electrical discharge. In addition, excitation transfer from the electronically excited He* and excited F* atoms, and ionic recombination of F^+ and F^- ions, which have a large electron affinity, during interaction with a He atom lead to the population of the upper $^3\Pi$ laser level that decays to the weakly bound lower $^3\Pi$ level. The F_2 laser transitions can be excited in discharges like those in rare-gas halide lasers. Therefore the F_2 laser is frequently mentioned among the commercial excimer lasers – in particular with respect to its short wavelength [11.1486]. Figure 11.164 shows details of the F_2 laser emission together with a *line-selected* spectrum

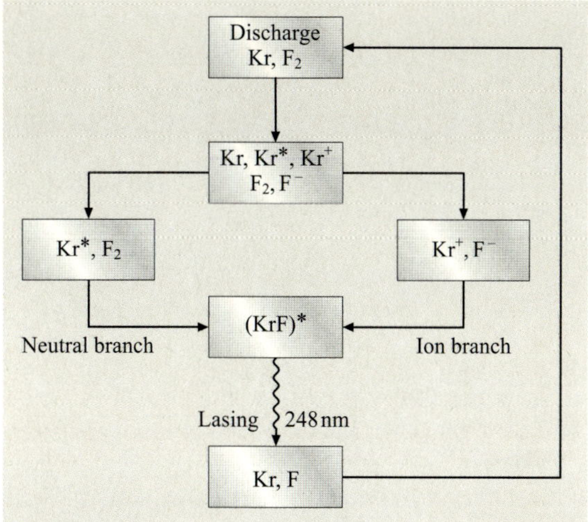

Fig. 11.163 Dominant reaction paths for the generation of KrF* [11.1482]

with a half-width of only 1.034 pm (FWHM, [11.1473, Chap. 6, pp. 97]).

Technical Design Principles for Excimer Lasers

From the physical preconditions described in the two preceding sections it follows that the electrical energy of high density, on the order of $10^{-2}\,\mathrm{J/cm^3}$ [11.1487], must be discharged homogeneously into the laser gas (Fig. 11.165a) on a time scale of a few nanoseconds so that a high-pressure (up to 0.5 MPa) glow discharge can be maintained for as long as possible.

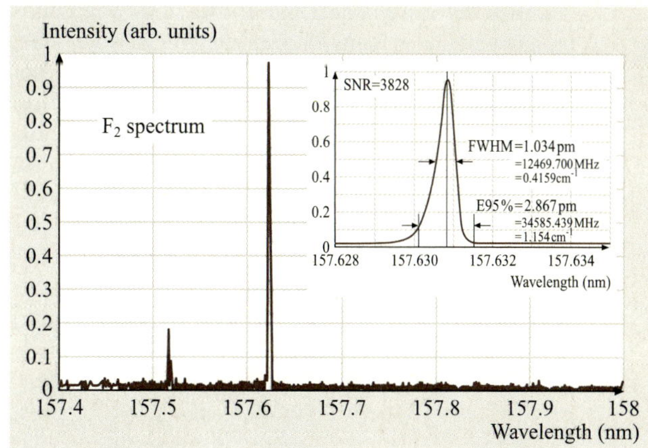

Fig. 11.164 Emission spectrum of the fluorine laser according [11.1483]. *Inset* shows spectral details of the stronger transition after line selection [11.1473]

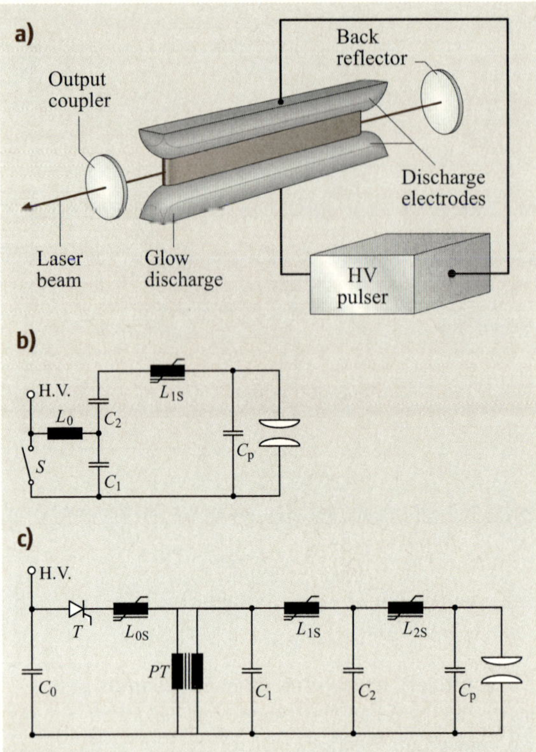

Fig. 11.165 **(a)** Schematic layout of an excimer laser together with gas discharge and laser resonator; **(b,c)** Excitation circuits employing magnetic pulse compression: **(b)** L–C inversion circuit with single-stage pulse compression, **(c)** thyristor-switched circuit using double-stage pulse compression [11.1487]

The electrical charge comes from a storage capacitor or, more effective, a low-impedance pulse-forming line. For a homogeneous ignition, an efficient pre-ionization is required. Pre-ionization by ultraviolet radiation has prevailed over X-rays. Table 11.45 summarizes the different methods of UV pre-ionization. Glow discharges, especially in halogen-containing gases, tend to transform into an arc or sparking discharge due to electron capture by the halogens so that laser emission breaks down. Hence, considerable efforts have been made to stabilize the glow discharge [11.1488, 1489]. In order to supply the charge sufficiently fast, favorable use is made of magnetic pulse compression (Fig. 11.165b,c). The energy switches, formerly thyratrons, are mostly thyristors, gate-turn-off (GTO) thyristors or insulator gate bipolar transistors (IGBT).

To guarantee a long laser-tube lifetime, the choice of electrode material is very important. Considerable material research has taken place and led to different materials for the fluorine and chlorine rare-gas lasers. Even different materials for either the cathode or the anode, or over the profile of one electrode have been used [11.1487]. Rapid transverse gas circulation provides cooling and supply of fresh laser gas for every shot. In order to achieve stable laser output power, halogen gas consumption is compensated by processor-controlled gas injections so that the laser tube can be operated with a single main fill typically for about 1 billion pulses.

Excimer laser resonators (Fig. 11.165a) mostly consist simply of plane windows made of CaF_2, allowing high out-coupling in the range 50–92%. For special applications, unstable resonators or resonators with frequency-selective optics are employed, as well as oscillator-amplifier systems.

The number of technical papers on the technology and construction of excimer lasers and their applications has become numerous. Excimer laser parameters are scalable to a considerable extent. Special excimer lasers have been developed by putting in the foreground either high single-pulse energy (up to 10 J), or high repetition rate (6 kHz), or medium or high average power (up to 1 kW). The efficiencies, calculated from the ratio of laser output energy to stored electrical energy, are typically a few percent ($< 5\%$). Since digital inter-publisher retrieval has become possible [11.1490], nearly every combination of the above can be electronically traced back, in part, over the last three decades. As an example, Table 11.46 shows data specified for an XeCl excimer laser designed for industrial applications.

Beam Characterization of Excimer Lasers

For various applications such as the production of diffractive structures via direct patterning [11.1491, 1492], semiconductor microlithography or eye surgery, a detailed knowledge of the wavefront of the emitted excimer laser radiation is necessary. In addition, all these applications strongly rely on the stability and the precise control of the laser parameters, such as pulse energy, beam width, divergence, pointing stability, uniformity etc. Thus reliable, standardized methods for the evaluation of beam parameters as well as accurate diagnostic tools for UV laser beam characterization are mandatory. While the output energy and the power can be monitored with the standard tools that are used also for other high-power lasers, the recording of spatial beam profiles and directional distributions (wavefronts) requires specific instrumentation adapted to the characteristics of

Table 11.45 Pre-ionization techniques applied in excimer-laser discharges

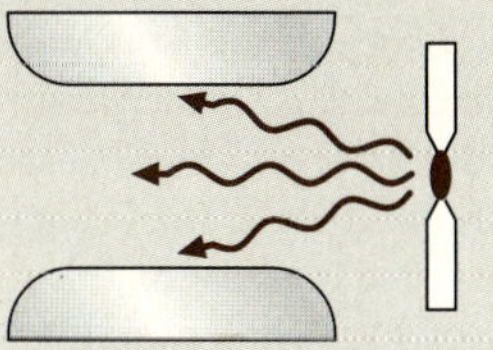	Preionization of the laser gas by UV radiation emitted from spark discharges between pin electrodes mounted in a row parallel to the laser channel. Sparks are being ignited some 10 ns before the main discharge starts so that at least 10^8 electrons/cm^3 are created as a seed for the main discharge.
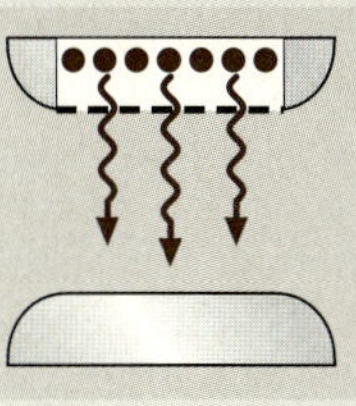	A possible modification of a: spark preionization through one of the electrodes designed as a screen or mesh electrode.
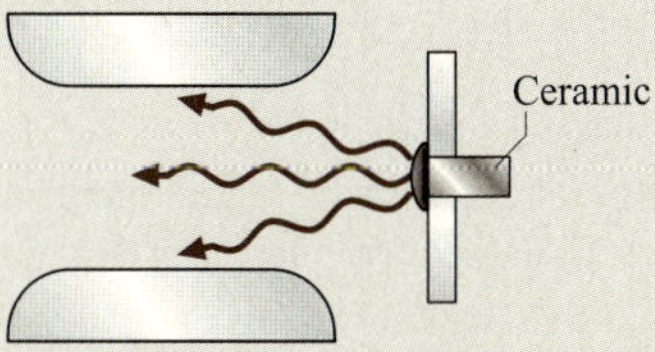	Dielectric(alumina)-surface-guided spark discharge for preionization. Because the discharge spreads, erosion of the pins is reduced considerably, thus the lifetime of the laser gas is increased, and the laser tube can be operated up to 10 billion shots. This design is mostly used in high-average-power excimer lasers.
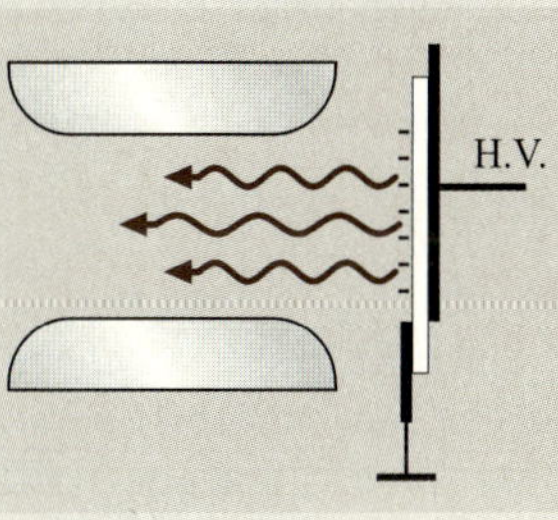	Surface corona discharges on a dielectric sheet emit UV with high spatial uniformity. Consuming a lower amount of energy as compared to sparks, they supply an electron density sufficient for a narrow discharge volume as used in high-repetition-rate excimer lasers, and provide increased discharge electrode lifetime.
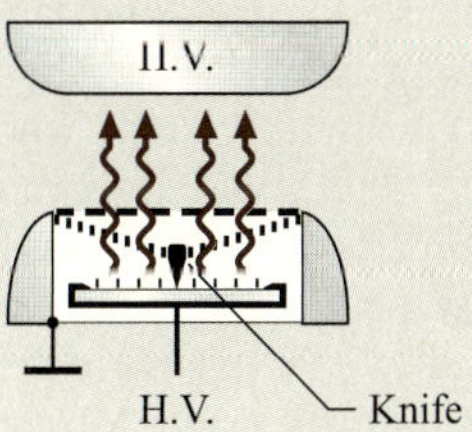	Knife-edge-released discharges creeping on a dielectric surface (similar to the well-known Lichtenberg discharges) cover a large area, thus can be conveniently applied for the preionization of large-aperture excimer lasers. The knife edge is connected to the ground electrode of the main discharge.

Table 11.46 Data from an XeCl excimer laser for industrial use. (Type: Steel 2000, Coherent Lambda Physik 2005)

Maximum stabilized energy (mJ)	Maximum average power (W)	Maximum repetition rate (Hz)	Pulse duration (ns)	FWHM beam dimensions V × H (mm)	Divergence V × H (mrad)	Beam point. stability V × H (mrad)
1050	315	300	29	37 × 13	4.5 × 1.5	0.45 × 0.15

excimer lasers. Besides sensitivity to the various wavelengths in the deep-UV spectral range, such diagnostic systems must also possess detector apertures adapted to the large near-field beam cross section of excimer lasers, and, most important, must guarantee long-term stability of the employed optics and sensors under pulsed high-power UV irradiation. The evaluation of relevant excimer beam propagation parameters is now possible with camera-based profile and wavefront measurements in accordance with current International Organization

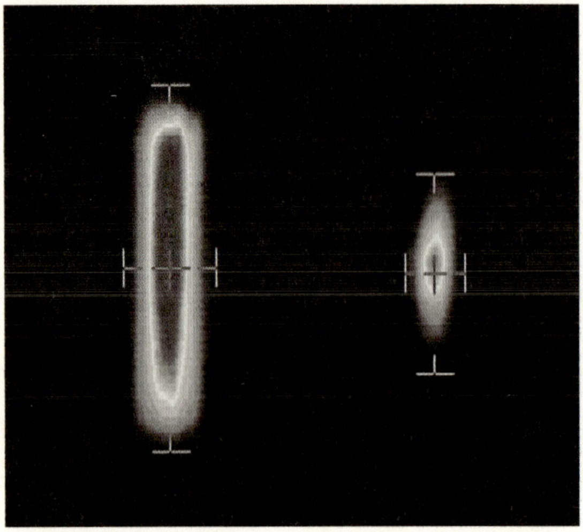

Fig. 11.166 Near-field (*left*) and far-field profile (*right*) of F_2 laser (157 nm, Novaline F1030, Lambda Physik) recorded simultaneously with a camera-based measurement system. The evaluated second-moment beam widths in the horizontal and vertical directions are also indicated

for Standardization (ISO) standards [11.1493, 1494]. Figures 11.166 and 11.167 show results obtained with a Hartmann camera-based measurement system.

Brief Recollection of the N_2 UV laser

The molecular nitrogen laser, realized as early as 1963 with emission in the near infrared [11.1495] and shortly later between 300 and 400 nm with the strongest line at 337 nm [11.1496] (described in detail in 1965 [11.1497, 1498]) quickly became the workhorse for laser spectroscopy. Its predominant use was, and still is, optical pumping of pulsed dye lasers throughout the near-UV, the visible and near-infrared spectrum. In addition, it serves as an excitation source in photochemistry and biological research. The lasing is achieved in high-voltage (10–20 kV) high-speed (few-ns) transverse gas discharges in a broad pressure range from some 10 mbar up to atmospheric pressure (TEA laser), in pure nitrogen and also, with lower efficiency, in air. The 337 nm transition is the most intense of the vibronic transitions (0–0) between the $C^3\Pi_u$ and the $B^3\Pi_g$ states. Since the B state is long-lived (about 10 μs) with respect to the C state (about 10 ns), the laser action is quickly self-terminating.

The pumping is achieved from the X state by direct electronic excitation of the C state, which happens to occur with a larger cross section than that for the B state, so that population inversion can be obtained if the discharge current rises much faster than radiative increase of population of the B state. To emit pulse energies in the mJ range, N_2 lasers must be equipped with low-impedance discharge circuits, such as for example can be realized by embedding the laser channel into a Blümlein transmission line, which enables a rise time of a few nanoseconds. Since the gain is very large, the emission may be stimulated predominantly in one direction when the excitation wave is made to reach the discharge channel synchronous to the light propagation; in this way a 10:1 ratio in favor of one direction can be achieved [11.1499], and fairly short lasers with considerable peak output power (1.2 MW) are possible [11.1500]. Transmission line circuits are thoroughly analyzed in [11.1501]. A pulse energy of 20 mJ has been realized using a fast circuit with magnetic pulse compression [11.1502]. On the other hand, just for demonstration purpose, a nitrogen laser can be easily realized by adjusting two electrodes on a glass sheet and supplying the electrodes with a spark-gap-switched high-voltage source. A home-built nitrogen laser, using standard ignition transformers, spark gaps, refrigerator circulating pumps, and some dielectrics and aluminium foil or copper-layered circuit boards, is precisely described on the web [11.1503].

Historically speaking, N_2 lasers as well as the homonuclear excimer lasers and metastable mercury compounds was proposed as early as 1960 by *Houtermans* [11.1504] – much earlier than the first ruby laser had been realized.

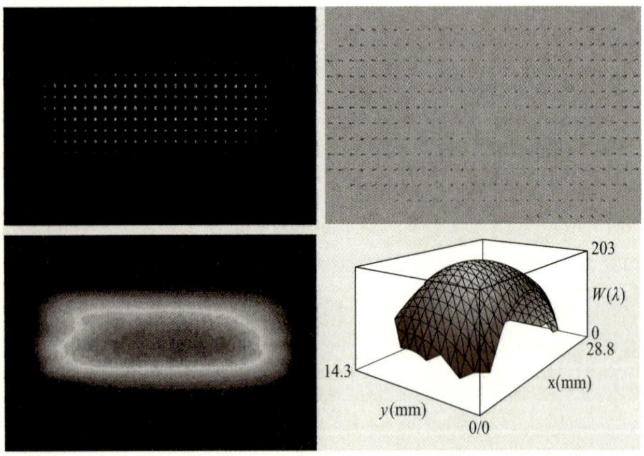

Fig. 11.167 Wavefront measurement of the emission of an F_2 laser (157 nm) using a Hartmann camera (40×30 pinholes, diameter 100 μm, pitch 1 mm). This technique provides comprehensive beam characterization and propagation analysis of excimer lasers from a single measurement

11.7.3 Applications

Physically, the applications of excimer lasers can be grouped according to the benefits from

1. short wavelength, and correspondingly,
2. their high quantum energy,
3. the high pulse energy density obtained by focusing, and
4. the extreme peak power when the pulse energy is compressed into ultrashort pulses.

Property 1 allows exposing and patterning of sub-100 nm structures in photoresists for semiconductor lithography, property 2 allows photochemical near-surface modifications such as color change, as used for marking of plastics, and index-of-refraction change used for the generation of fibre Bragg gratings. Property 3 allows the melting of thin silicon films, used to induce large-grain crystallization in the production line of thin-film transistor (TFT) displays. The properties 1–3 enable microstructuring by material ablation including biological tissue, and property 4 opens the door for the generation of X-rays of extreme brilliance. Besides these technical applications, excimer laser radiation is still widely used for scientific purposes, such as optical pumping of dye lasers. Here, we briefly refer to a few of the technical applications.

Excimer Lasers in the Electronics Industry

Lithography for the generation of semiconductor circuits has been based on excimer laser light sources for more than a decade now, and will continue to do so for a couple of years. Thus this is the economically most important application of these lasers. Because the diffraction-limited minimum (half pitch) that is resolvable in optical projection is given by $k_1\lambda/\text{NA}$, every effort is made to decrease the wavelength λ for the illumination together with the process factor k_1, and to increase the numerical aperture (NA) of the projection system. After extensive

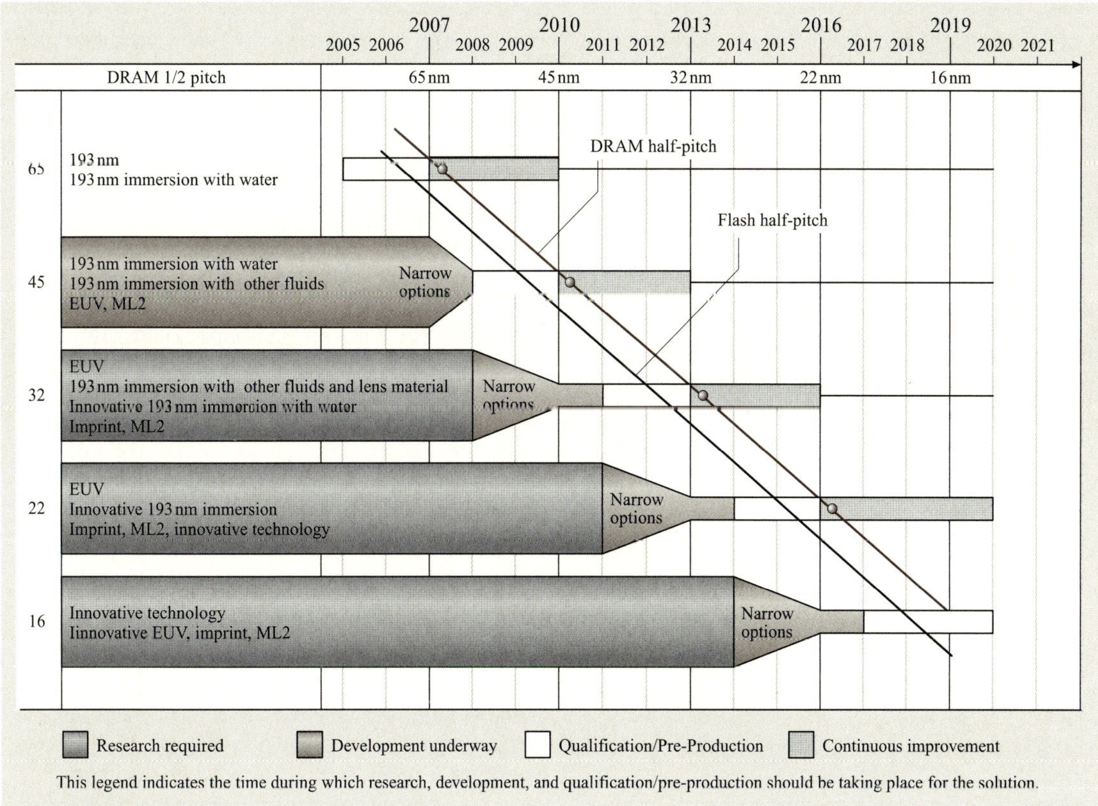

Fig. 11.168 Potential lithography exposure tool solutions according to the *International Technology Roadmap for Semiconductors 2005* [11.1505]. Courtesy ITRS

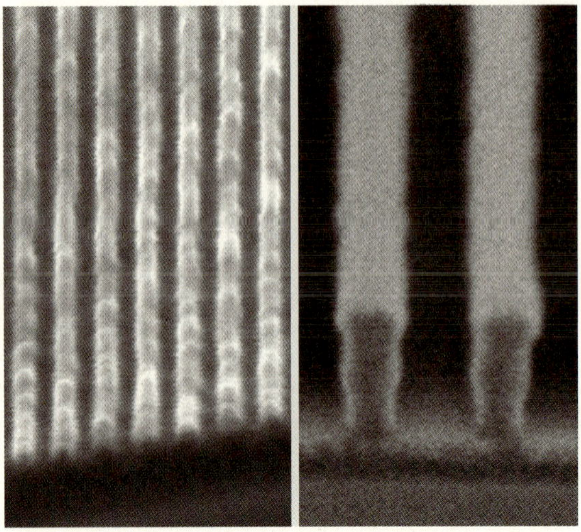

Fig. 11.169 30 nm-wide lines and equally sized spaces obtained at IBM Almaden Research Center (*left*) using 193 nm interferometric high-index immersion lithography, compared to current 90 nm features. Courtesy IBM

use of 248 nm lasers, 193 nm illumination is now state of the art, and increasing the NA by using immersion optics turned out to be a highly successful method to arrive considerably below the diffraction limit. In the International Technology Roadmap for Semiconductors 2005 (Fig. 11.168 [11.1505]) the 45 nm half-pitch for dynamic random-access memories (DRAMs) in production is scheduled for 2010, to be achieved with 193 nm immersion technology. Fluorine lasers (157 nm) are not in discussion any more since immersion fluids with an index of refraction higher than that of pure water are being investigated. These could extend the use of 193 nm down to a 30 nm half-pitch (see Fig. 11.169 from [11.1506]).

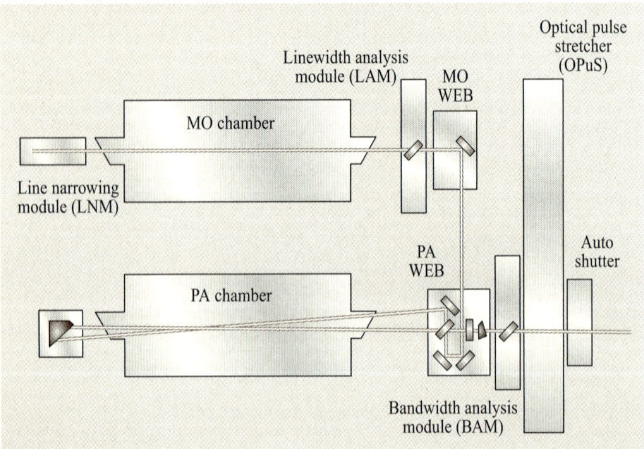

Fig. 11.170 Schematic of a master oscillator power amplifier (MOPA) excimer laser system to be used in semiconductor lithography [11.1507]. Courtesy Cymer Inc.

For the lasers to be used in lithography, there are considerable requirements with respect to line width, wavelength and pulse-to-pulse stability, lifetime of the laser tube and all the optics [11.1508]. Dual-chamber systems distribute the task of narrow band generation in the oscillator and power generation in the amplifier tube; the latter can be designed as a regenerative ring amplifier [11.1508]. Recently, a 60 W system operating at 193 nm was introduced (Fig. 11.170 [11.1507]) with the benefit of improved energy stability of the 6 kHz pulses, so that the exposure dose of the resist can be precisely controlled. This system aims for 45 nm half-pitch production.

Transistors in thin films are the key elements to control the individual pixels in displays. To increase the mobility of the electrons, large silicon crystal grains are produced by applying excimer laser radiation to cause melting and controlled recrystallization, the beam being scanned in lines across the surface [11.1509, 1510]. This process, summarized as excimer laser annealing (ELA), is performed by applying 308 nm radiation that is strongly absorbed by silicon so that thin films on glass can be melted without damage to the substrate [11.1511]. In this way, large-area TFT displays can be manufactured. Even doping imperfections due to incomplete annealing near source/drain junctions can be eliminated by oblique-incidence ELA [11.1512].

Further excimer laser applications that are well established in electronics are printed circuit board (PCB) via drilling [11.1513], wire stripping and wire marking [11.1473]. A rather new technology is the lift-off method to de-bond for example an electronic GaN LED from its sapphire substrate, which is used for crystal growth, by shining the laser through the substrate, thereby ejecting the LED onto its heat-sink electrical interconnect [11.1514]. For future optical coupling of PCBs, attempts to integrate optical mirrors manufactured by ablation should be mentioned [11.1515, 1516].

Processing of Optical, Ceramic, Polymeric, and Biological Materials

Ablation is surely the best-known application of excimer lasers: drilling and microstructuring of glass, quartz, even diamond, ceramics and polymers [11.1517] allows highly precise contouring with high reproducibility. This subject is covered in chapters 11–14 and 16 of [11.1473], which also includes the ophthalmological application including the laser-assisted in situ keratomileusis (LASIK) method ([11.1473, Chap. 19], [11.1518]). Nonablative processing of glass and polymers allows one to generate fiber Bragg gratings in optical waveguides by changing the index of refraction [11.1473, 1519]. Microstructuring is going to play a role also in the upcoming technology of organic light emitting devices (OLEDs) [11.1520].

Femtosecond Excimer Laser Pulses

Another interesting subject of today's excimer technology is their conversion into UV light sources with femtosecond pulse duration. Basically a femtosecond seed pulse at 248 nm is injected into an appropriately modified standard excimer laser resonator to achieve a powerful highly directed short pulse emission. As examples are presented material processing with this radiation and high-intensity studies for the generation of tunable X-rays.

Conversion of a nanosecond excimer-laser system into a fs system is described in [11.1521–1523] and briefly in Fig. 11.171a. The ultrafast UV laser system consists of a Ti:sapphire front laser, a frequency-tripling unit to convert the wavelength of the ultrashort pulses into the UV spectral range, and a specially designed KrF amplifier to boost the energy of the pulses to several mJ. The current laser arrangement, as shown in Fig. 11.171a, uses a commercial Ti:sapphire front-end system (Coherent) delivering pulses of 150 fs duration at a wavelength of 745 nm. After frequency tripling, seed pulses are obtained for the KrF amplifier module, which is the key component of the system. This module is a modified version of a commercial excimer laser. In a three-pass amplification scheme the pulses are amplified up to energies of ≈ 30 mJ at repetition rates exceeding 300 Hz, resulting in an average power of 10 W at 248 nm.

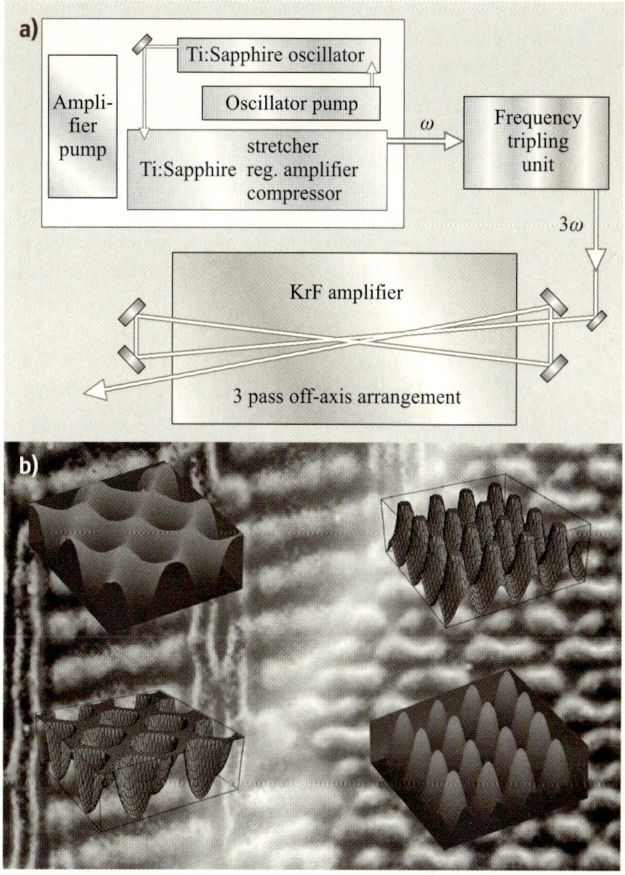

Fig. 11.171 **(a)** Layout of the UV femtosecond hybrid laser system showing Ti:sapphire seed system and three-pass off-axis KrF amplifier. **(b)** Comparison between calculated and measured surface textures for four-beam interference at a phase difference of 0π (*left*) and 0.5π (*right*). The surface contours top left and bottom right were measured via atomic force microscopy (AFM)

UV femtosecond material processing. Nanoscale fabrication of materials is more and more in demand in scientific and industrial applications. The general trend to reduce the size of optomechanical devices and the growing need for assemblies with feature sizes below one micron generates new challenges for laser fabrication techniques.

Short-pulse lasers with picosecond and femtosecond pulse duration offer material processing capabilities with highly decreased damaged area around the irradiated spot, and consequently smaller feature sizes. If the short pulse duration is combined with short wavelengths, unprecedented results can be achieved due to the dependence of the spatial resolution on the wavelength. Thus UV femtosecond laser systems provide superior material processing quality.

Such pulses are ideally suited for sub-micron machining of solid surfaces. Applying various diffractive optical masks in combination with reflective imaging/focusing systems allows the generation of complex 2-D and 3-D structures with feature sizes down to ≈ 200 nm or lower on all materials, including metals, semiconductors and dielectrics. Figure 11.171 shows the calculated and experimentally obtained surface relief structures fabricated with the technique of phase-controlled multiple-beam interference [11.1524].

High-intensity UV femtosecond studies. A leading application of KrF* (248 nm) excimer lasers has been the demonstration of saturated amplification in the multikilovolt X-ray regime that arises from the ability to produce and controllably combine two new highly ordered forms of excited matter:

1. hollow atoms and
2. stable electronically hollow plasma channels.

The realization of this new X-ray source provides a peak brightness that is sufficient for the implementation of a new high-resolution technology for biological microimaging.

Hollow atoms and the cluster concept for X-ray amplification. Previous work [11.1526] on the nitrogen molecule N_2 suggested the possibility of *designing* a new class of molecular materials *optimized* for the efficient production and amplification of X-rays [11.1527]. This idea was immediately tested with Xe clusters [11.1528] with the outcome that the copious production of Xe hollow-atom states [11.1529] emitting both Xe(M) and Xe(L) radiation in the kilovolt spectral region [11.1528] was observed. These data confirmed that the selective excitation of the inner-shell states initially observed in N_2 could be scaled into the multi-kilovolt spectral region. A good example [11.1528, 1529] is given by the characteristic Xe(L) 3d2p hollow-atom emission profile centered at ≈ 2.8 Å shown in Fig. 11.172. *Hollow atoms* are atoms (ions) that intrinsically possess an *inverted* electronic configuration consisting of deeply bound inner-shell vacancies, perhaps multiple, with the simultaneous retention of several electrons in relatively weakly bound outer orbitals. Accordingly, these states are optimally suited for the prompt emission and amplification of X-rays. The demonstration of saturated amplification [11.1525, 1530–1532] followed, together with the ability to form self-trapped plasma channels [11.1533, 1534]] that are well matched to the conditions necessary for Xe(L) excitation in clusters.

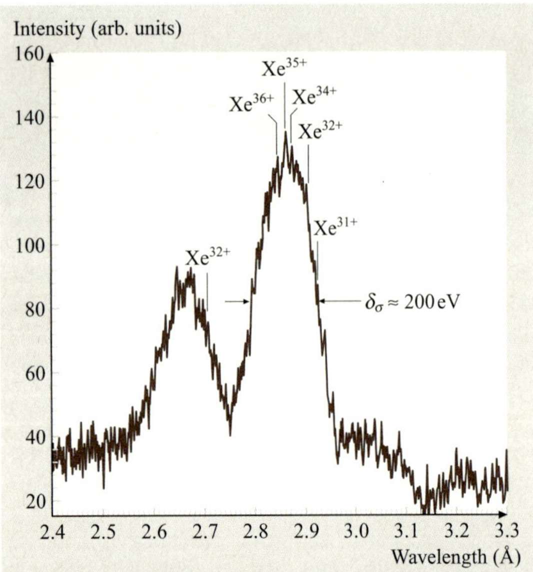

Fig. 11.172 Characteristic spontaneous emission profile of the Xe(L) $3\tilde{d}2p$ hollow-atom spectrum produced from Xe clusters with intense ($\approx 10^{19}$ W/cm^2) femtosecond 248 nm excitation. The splitting between the major and minor lobes arises from the spin–orbit interaction of the 2p vacancy. The full width of the main feature is ≈ 200 eV. The positions of selected charge-state transition arrays (Xe^{31+}, Xe^{32+}, Xe^{34+}, Xe^{35+}, and Xe^{36+}) are indicated. The spectral resolution of these film data is ≈ 4 eV. (After [11.1525])

Multi-kilovolt X-ray amplification with clusters in self-trapped plasma channels. The fundamental power compression and its spatial organization were found in the alliance of the two basic phenomena mentioned above. They are

1. the direct multiphoton excitation of hollow atoms from clusters with ultraviolet radiation and
2. a nonlinear mode of *stable* confined propagation in plasmas resulting from a relativistic/charge-displacement mechanism of self-channeling [11.1533, 1534].

The spectrum typical of the X-ray beam amplified axially in the Xe plasma channel and recorded on film with a von-Hámos-spectrograph is illustrated in Fig. 11.173.

In Fig. 11.174 the axially recorded spectrum presented in Fig. 11.173 is compared to a transversely observed single-pulse spectrum that exhibits deep and broad spectral hole burning [11.1535] that corresponds well with the axially amplified transition arrays. This result demonstrates two key attributes of the Xe(L) system, namely,

1. a high efficiency of energy extraction and
2. a very broad bandwidth (60 eV) for amplification.

In this case, the strongly enhanced lines observed correspond to the Xe^{34+}, Xe^{35+}, and Xe^{36+} charge state arrays of the major lobe shown in Fig. 11.172. These observations, along with additional spectra showing comparable results on several other transition arrays in

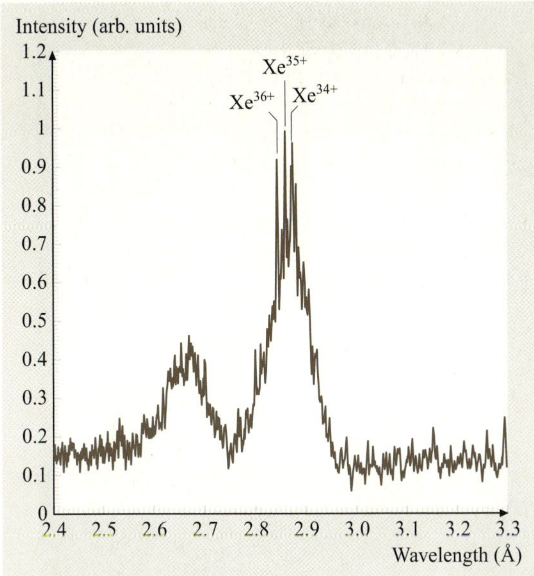

Fig. 11.173 Axially recorded spectrum showing sharp peaks at the positions of the Xe^{34+}, Xe^{35+}, and Xe^{36+} transition arrays. On the basis of geometric considerations concerning the von-Hámos-spectrograph and the ratio of exposures corresponding to the spontaneous signal and the amplified lines, the recorded enhancement of these features is estimated to be minimally $\approx 1.5{-}3.0\times 10^3$ over the strength of the spontaneous emission

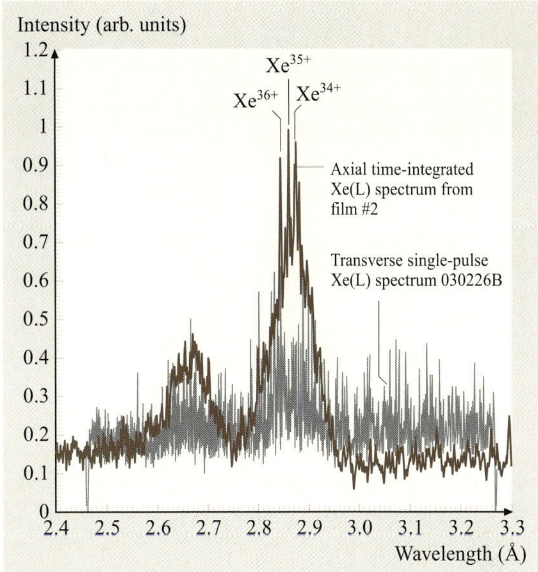

Fig. 11.174 Comparison of the axially detected film #2 shown in Fig. 11.173 with a single pulse spectrum (#030226B) recorded transversely, illustrating deep spectral hole-burning corresponding to the region of the Xe^{34+}, Xe^{35+}, and Xe^{36+} arrays. Since the hole burning descends to the noise level of the detector, efficient energy extraction is evident. The width of the spectral gap is ≈ 60 eV, a value sufficient for amplification of a pulse with a duration of ≈ 30 as

the $\approx 2.71{-}2.93$ Å region, indicate that the hollow-atom states are strongly inverted and that the amplification can be tuned across a substantial fraction of the spectral profile illustrated in Fig. 11.172. An estimate of the peak spectral brightness achieved in the initial experiments [11.1525] gave a value of $\approx 10^{31}-10^{32}$ photons $(s\ mm^2\ mr^2)^{-1}$ and $(0.1\%\ \text{bandwidth})^{-1}$, a range approaching that required for single-molecule imaging in living biological material [11.1536–1538].

In summary, the study of high-intensity interactions with ultraviolet excimer lasers on atoms, molecules, and plasmas over the last 20 years has culminated in the ability to produce new forms of matter that are both highly excited and highly ordered [11.1539]. A consequence is the capacity to achieve saturated X-ray amplification in the multi-kilovolt regime at peak brightness figures sufficient for the implementation of biological microimaging. Basically, the amplification proceeds through the creation of a *highly ordered* excited state [11.1525, 1535] that is comprised of four mutually coupled components: atomic (ionic) matter, plasma electrons, and the two coherent radiative fields, which are the ultraviolet and X-ray waves propagating in the channel. The chief consequence is well-ordered and efficient energy flow conducted through radiation-dominated interactions that are confined to a small phase-space volume. Finally, it is significant to note that *ordered* energy flow was the principal characteristic of excimer systems that made [11.1472] them so abundantly useful over the last three decades. Hence, a key feature of that history appears prominently again in the X-ray range.

11.7.4 Outlook: Radiation in the EUV

The roadmap of the semiconductor industry (Fig. 11.168) shows in which way computer chips with critical dimensions of 32 nm and below are planned to be manufactured: light sources with emission in the extreme ultraviolet (EUV) at a wavelength of 13.5 nm will be used. EUV lithography is considered as the next-generation lithography (NGL) to be established after

Fig. 11.175 EUV microexposure tool from Exitech with an integrated EUV source from XTREME technologies (*left*)

Fig. 11.176 EUV plasma source integrated into an optical system

reaching the physical limits of ultraviolet (immersion) lithography based on ArF excimer lasers.

13.5 nm Technology

The first EUV microexposure tools are already under operation for technology development and feasibility studies (Fig. 11.175).

One of the biggest challenges concerning the introduction of EUV lithography is the development of high-power EUV sources at a reasonable cost. Making the operation of the photolithography manufacturing tools economically viable requires a source with power in the kilowatt range at the 13.5 nm exposure wavelength. In addition, the optical design of the collector optics sets limits on the source size, i. e., an emission volume of a few cubic millimeters. Only the combination of both the high power and small emission volume will enable high optical efficiency and therefore high wafer throughput. The goal is the development and manufacturing of these high-power EUV sources as well as their integration into the optical system of the photolithography tool (Fig. 11.176).

Plasmas are known as efficient emitters of 13.5 nm EUV radiation, if their temperature reaches about 200 000 °C, i. e., approximately 35 times higher than at the surface of the sun. Plasmas can be generated either by an electrical discharge or by means of pulsed laser excitation. Both methods are able to generate small plasma volumes fulfilling the technical requirements of photolithography [11.1540, 1541]. The spatially homogeneously emitted radiation from such plasmas is pulsed with lengths in the nanosecond range (typically). The distribution of the wavelength spectrum is narrow from elements with low atomic numbers and broad from elements with high atomic numbers. In addition, the radiation is almost fully incoherent, as from a thermal Planck emitter.

Today the highest EUV power is achieved in so-called *pinch plasma* sources. However, the heat load on the static electrode configuration can lead to fast erosion and even melting of the surfaces, thereby limiting the power scaling through increasing repetition rates. Alternative technologies with moving electrodes have been investigated and finally a potential solution with rotating disc electrodes (RDE) has been found. A new excitation scheme has been applied to this technology at XTREME technologies, resulting in a world record ($170\,\mathrm{mJ}/2\pi$ sr) of the achieved EUV pulse energy [11.1542]. This effect resulted in a reasonable repetition rate to fulfill the power requirements of high volume manufacturing tools. Further development efforts will be directed to combine the power-scaling capability with the reliability goals for this concept.

11.8 Dye Lasers

11.8.1 Overview

The distinctive feature of dyes as a lasing media is the broad emission band with a typical bandwidth of 50 nm. Laser action from dye molecules was first observed by several research groups in 1966 [11.1543, 1544]. Compared to the gas and solid-state lasers of the 1960s, dye lasers easily excelled both in terms of broad spectral coverage and in versatility of output performance (viz. high per-pulse energy output, long or short pulse operation, flashlamp or laser pumping) Researchers quickly recognized the wavelength tuning capability of dye lasers, which is probably one of the most important operation characteristics in view of applications. In the following years, literally hundreds of organic dyes with emission spectra from the near ultraviolet to the near infrared (300–1200 nm) have been made to lase. Continuous-wave (CW) or ultrashort pulse (femtoseconds) operation of dye laser was also demonstrated. With their broad spectral coverage and narrow linewidth tunability, dye lasers have been the workhorse in scientific laboratories for applications ranging from fundamental physics to clinical medicine for many years. Despite the bad reputation of being prone to various mishaps during operation (e.g., spilling of dye solution being a not infrequent occurrence), there are many loyal fans of liquid dye lasers in the scientific community. This popular support is a tribute to the versatile output performance of these liquid lasers and the simplicity of the technology involved in constructing a liquid dye laser. Almost any research lab can put together a flashlamp with power supply, two mirrors, some common dye (e.g., rhodamine 6 G), and nuts and bolts to build a pulsed liquid dye laser. There are of course numerous user-friendly and powerful liquid dye lasers available. We will first give a short description of the characteristics of liquid dye lasers and then discuss several types that are commonly found in research laboratories for photophysical, photochemical and spectroscopy applications.

11.8.2 General Description

Dyes are organic compounds that contain conjugate double bonds. The presence of the conjugate double bonds renders these compounds optically active. There are over 200 laser dyes that can, in principle, provide spectral coverage from 320 to 1200 nm [11.1545]. The tuning range of each dye is 40–60 nm. When used sequentially, continuous tunable laser action can be obtained from the near ultraviolet to the near infrared. Typically a strongly absorbing and strongly emitting dye is dissolved in a suitable solvent (e.g., a polar solvent such as ethanol or a nonpolar solvent such as chloroform) at concentrations of 10^{-3}–10^{-4} molar to serve as the gain medium. A flashlamp or another laser is used as the pump. Powerful coaxial flashlamp-pumped rhodamine 6 G dye lasers can deliver laser energy up to 400 J per pulse [11.1546]. Average output power of up to 1.2 kW in burst mode were obtained in a flashlamp-pumped dye laser by *Morton* and *Dragoo* [11.1547]. Since the advent of high-power short-wavelength lasers in the 1970s, excimer, nitrogen or frequency-multiplied Nd:YAG lasers are frequently used for pumping liquid dye lasers with output pulse around 10 ns and energy output of tens of millijoules. A saturable absorber (another dye) is used to passively mode-lock the dye laser to generate pulses of about 200 fs without dispersive elements and down to tens of fs with dispersive optical elements [11.1548]. For applications that require single-longitudinal-mode laser output, CW operation of dye lasers is achieved by using a modified flow system to remove the long-lived triplet state of the dye molecule. The linewidth of a free-running jet-stream CW dye laser can be as low as to 2 MHz [11.1549]. In the following sections, we discuss several common dye laser arrangements.

11.8.3 Flashlamp-Pumped Dye Lasers

A flashlamp was the pump source for Maiman's epoch-making ruby laser. Today's flashlamps are still used to pump liquid and solid-state lasers. Linear flash-

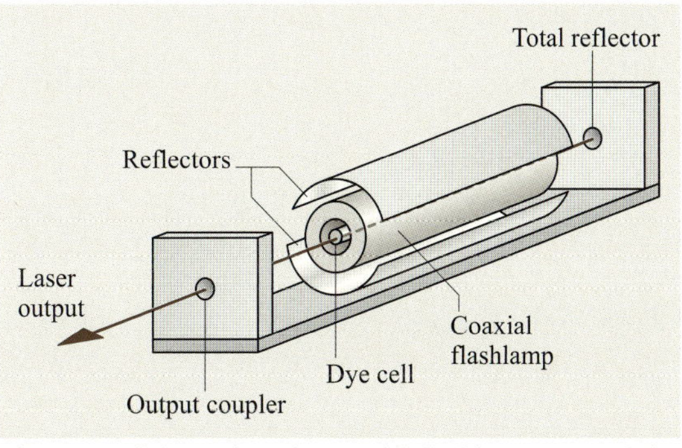

Fig. 11.177 Schematics of a coaxial flashlamp-pumped dye laser

lamps are often the choice for laser pumping and usually consist of a tubular quartz envelope sealed by tungsten electrodes at both ends. The tube is often filled with a heavy rare gas (e.g., xenon or krypton) for high electrical-to-optical conversion efficiency. Still close to 90% of the electrical power ends up as heat, which requires a cooling system for its removal. Often, an elliptical reflector is used to focus the output from the flashlamp into the dye cell. For optimum coupling of the light from the flashlamp into the dye, the dye cell must be placed close to the lamp. Coaxial lamps have emitting surfaces that completely surround the flowing dye. Figure 11.177 shows a coaxial flashlamp-pumped dye laser, a structure commonly employed in many commercial dye as well as solid-state lasers. The dye cell is set at the center of the coaxial, cylindrically shaped xenon flashlamp for optimal pumping. Two broadband mirrors provide the optical feedback. Such a flashlamp-pumped dye laser typically produces a broadband emission (about 10 nm linewidth) centered at the peak of the gain profile with a pulse duration of 1–2 μs. The output energy is about 100 mJ for most commercial flashlamp-pump dye lasers.

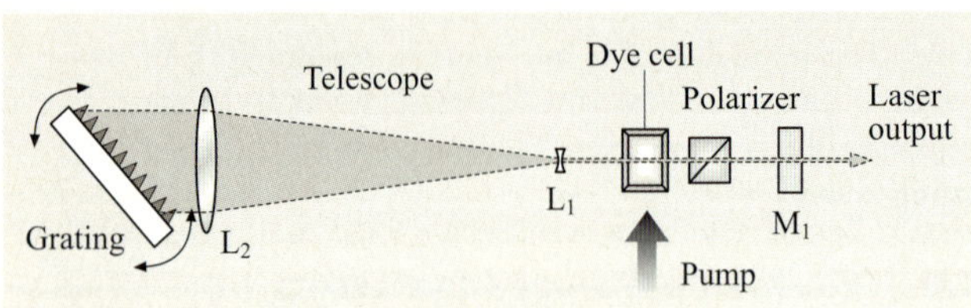

Fig. 11.178 Schematics of a Hänsch-type cavity dye laser. A telescopic lens is used to improve the spectral resolving power of the grating

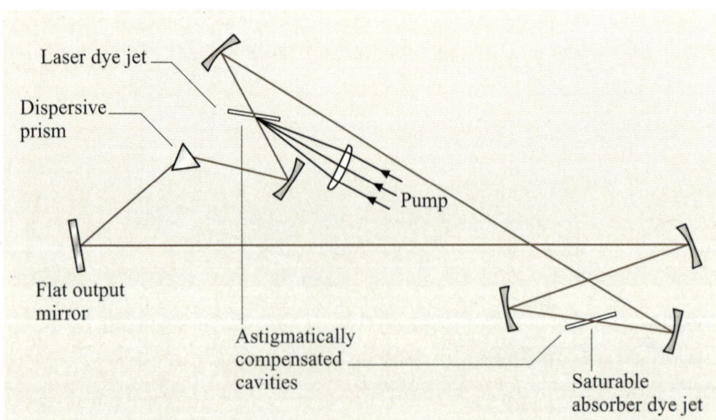

Fig. 11.179 Schematic of a colliding-pulse dye laser for ultrashort-pulse generation

11.8.4 Tunable Dye Lasers Pumped by High-Power Short-Wavelength Lasers

By incorporating wavelength-selective elements inside the resonator cavity, dye laser output can be tuned. Increasingly excimer or frequency-multiplied Nd:YAG lasers are used to pump dye lasers. Excimer, nitrogen and frequency-multiplied Nd:YAG lasers all provide high pump power (peak power of several MW) at short wavelengths, making them highly suitable for pumping dye lasers. Dye lasers pumped by high-power short-wavelength lasers exhibit such high gain that loss introduced by the additional wavelength-selective elements inside the resonator cavity can be overcome. Either prisms or diffraction gratings can be used for wavelength selection, and gratings are generally superior in terms of large dispersion and wavelength-resolving power. Figure 11.178 shows a variant of the telescopic grating resonator originally used by *Hänsch* [11.1550] for the generation of narrow-linewidth laser output. A nitrogen laser provided the pumping. The use of the telescopic lens increases the number of grooves illuminated by the laser light and reduces the light intensity on the grating, preventing damage to the coating on the grating surface. The linewidth of a telescopic-grating cavity dye laser is in the range of 0.1 nm. To further reduce the linewidth to 0.01–0.05 nm, an intracavity etalon (e.g., a coated optical flat or air-spaced Fabry–Pérot cavity) may be introduced. Many variants of the grating-cavity (e.g., the hybrid prism–grating cavity) dye lasers have been developed to achieve narrow-linewidth output [11.1551].

11.8.5 Colliding-Pulse Mode-Locked Dye Lasers

The broad emission band of a dye can be used effectively to generate ultrashort laser pulses, since the theoretical limit of the ultrashort-pulse duration supported by the gain medium is proportional to the inverse of the gain bandwidth. As the emission bandwidth of many laser dyes is 40–50 nm, pulse durations down to tens of fs can be produced. The development of passive mode-locking is essential to the successful demonstration of femtosecond pulse generation in dye lasers. In passive mode-locking, a saturable (nonlinear) absorber (an absorbing dye) with an absorption that matches the emission wavelength of the lasing dye is placed inside the resonator cavity. Ideally, the leading and trailing edges of the optical pulse are removed by the absorber whilst the peak of the pulse, unaffected by the absorber, experiences amplification. The efficiency of the absorber increases considerably if two oppositely traveling pulses interact or collide in the saturable absorber at the same time (colliding-pulse mode locking). This is a result of the two coherent pulses interfering constructively, leading to the reduction of power required for the saturation of the absorber. The colliding-pulse mode-locking configuration can be realized in either a linear or ring cavity. Because of the ease of alignment, the ring cavity is often preferred. Figure 11.179 shows the type of ring laser used originally by *Fork* et al. to generate pulses shorter than 100 fs [11.1552]. Both the laser dye (e.g., rhodamine 6 G) and the absorbing dye (e.g., DODCI) are in the form of a free-flowing stream jet, and a CW argon-ion laser serves as the pump. The radii of curvature of the mirrors must be carefully chosen so that the spot size in the absorber is smaller than that in the gain medium to assure pulse-forming stability.

11.8.6 Tunable Continuous-Wave Dye Lasers

The linewidth of tunable pulsed dye lasers with intracavity dispersive element for narrow-linewidth operation are of the order of 500 MHz. In contrast, a continuous-wave dye laser is capable of delivering laser output with linewidths as narrow as tens of kHz. The high resolving power is ideal for spectroscopy. Such high performance is connected with high cost, complicated design and a high-power pump source (often a CW argon-ion laser with 10–20 W output power). Continuous-wave dye lasers have much lower gain than their pulsed counter part. The loss reduction becomes critical and the removal of the long-lived triplet state is crucial for the stable operation of the laser. A free-flowing jet is used to effect rapid circulation of the dye. Since the gain profile of most dyes may be regarded as homogeneously broadened, one might expect single longitudinal-mode output from a simple grating or prism-cavity CW dye laser. However, single longitudinal-mode operation may not be possible in such linear-cavity dye lasers without intracavity frequency-selecting elements because spatial hole burning can arise gain saturation by the standing wave. Multimode output then results. Additional frequency-selecting elements such as an etalon must be inserted inside the cavity for single-mode operation increasing losses and lowering output power. Figure 11.180 shows a linear (standing-wave) cavity arrangement for a CW dye laser end-pumped by an argon-ion laser for single-mode operation. The dye jet flows perpendicularly to the page. The etalons assure

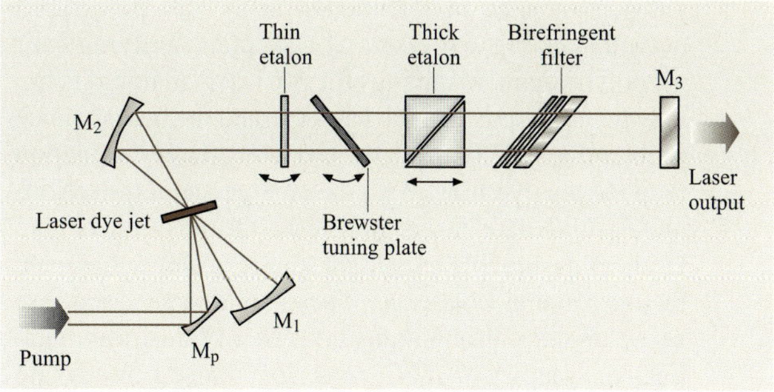

Fig. 11.180 Schematic of a linear-cavity CW dye laser for single-mode operation

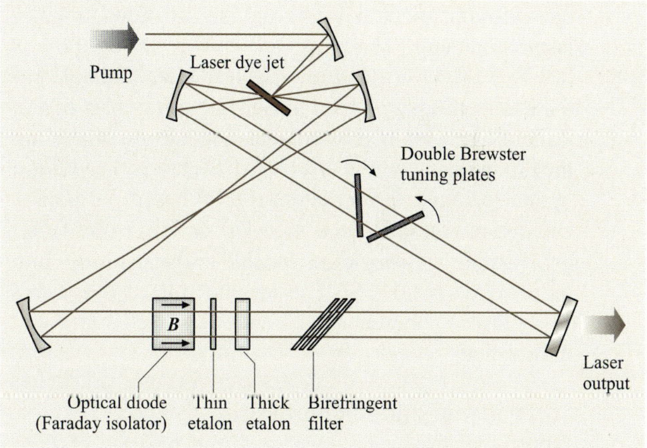

Fig. 11.181 Schematic of a ring-cavity CW dye laser for single-mode operation. Note the use of an optical diode to assure unidirectional propagation of the traveling wave

narrow-linewidth output. The birefringent filters are for wavelength tuning. A ring-cavity supports the propagation of traveling waves and thus is ideal for single-mode operation. Figure 11.181 shows a tunable single-mode ring laser supporting single-mode operation. The optical diode (a Faraday isolator) ensures unidirectional travelling waves thus avoiding spatial hole burning.

11.8.7 Advanced Solid-State Dye Lasers

The contribution of liquid dye lasers to the advance of laser technology, particularly to ultrafast laser technology and the many applications of tunable lasers, cannot be overstated. Interested readers should consult monographs such as *Dye Lasers* by *Schäfer* [11.1553] or *Dye Laser Principles* by *Duarte* and *Hillman* [11.1554] and references therein for more detailed information on the range of applications of dye lasers. The excellent output performance (as mentioned in the last paragraph) notwithstanding, dye lasers earn skepticism among their users. The main misgivings for dye lasers in liquid form, may be the disposal of the used dye and the maintenance problems associated with the physically large circulation loop for the dye flow. With the emergence of transition-metal solid-state lasers in 1980s, the position of dye lasers as the tunable laser light source of choice has been facing a rising challenge. These solid-state lasers are based on the vibronic transitions of 3d transition-metal ions such as Cr^{3+}, Ti^{3+}, and Co^{2+} doped into oxide or fluoride crystals that served as the host [11.1555]. These transitional-metal ion-doped crystals show broad emission bands (typically 200–300 nm) in the visible to near-infrared spectral region and thus are good tunable-laser candidates. The most well known among them is the Ti^{3+}:Al_2O_3 (titanium:sapphire) laser. Tunable radiation in the blue–green and in the ultraviolet can in principle be obtained by frequency doubling or tripling the fundamental output of Ti^{3+}:Al_2O_3 lasers. In addition optical parametric oscillators (OPO) based on nonlinear optical crystals such as ADP or BBO (BaB_2O_4) are capable of providing tunable radiation well into the infrared [11.1556]. The leading role of dye lasers as a tunable coherent-light source is indeed somewhat diminished.

Challenges and Opportunities

Solid-state lasers are often preferred because of their ruggedness and easy maintenance. User-friendly dye lasers must do without the flow loop that circulates the dye solvent. Solid-state dye lasers (SSDL) incorporating the dye molecules in solid matrices appear to be able to combine the cost-effectiveness of liquid dye lasers and the ruggedness of inorganic solid-state lasers. In order for SSDL to be a serious competitor to inorganic solid-state lasers in various fields of applications, the problem of photodegradation of dye molecules must first be dealt with.

Good progress has been made in the synthesis of photostable dyes. The recently synthesized perylene family [11.1557] and the pyrromethene family [11.1558] of laser dyes have been shown to outperform rhodamine-6G in efficiency, tunability and photostability. Equally important to the development of SSDL are the solid-state matrices that serve as host materials where there has also been impressive progress. In the 1980s *Avnir* et al. [11.1559] and *Gromov* et al. [11.1560] demonstrated dye-doped sol-gel materials and dye-doped polymeric materials, respectively, as promising laser media. Both sol-gel materials and polymeric materials were shown to have good chemical stability and wide optical transparency for use as host materials for laser dyes. The pace of research activity has increased considerably since then. In the sections to follow, we will bring the reader up to date with the latest development of solid-state dye lasers. The state of the art of solid-state dye lasers based on pure and hybrid sol-gel materials and on advanced polymers is covered in the next section. The most current development of solid-state dye lasers may be polymeric [11.1561] or sol-gel [11.1562] waveguide lasers using a distributed feedback configuration. These compact lasers produce tunable narrow-linewidth output and appear to be readily integrable into planar optical circuits. This is covered in the penultimate section. Finally the topic of tunable upconverted DFB dye lasers is presented.

Solid-State Dye Lasers Based on a Polymer Host

In 1967, one year after the demonstration of the first liquid dye lasers, *Soffer* and *McFarland* observed lasing in rhodamine-doped poly(methyl methacrylate) (PMMA) [11.1563]. The first polymeric hosts suffered from large thermal coefficients, stress birefringence, optical inhomogeneity and chemical reactivity with laser dyes. The most serious problem of dye-doped polymers is the tendency of aggregation of dye molecules, which effectively quenches fluorescence. As a result, the performance of the initial solid-state dye lasers was less than satisfactory. The properties of PMMA can, however, be improved by purification of the monomer, by introducing alcohol additives [11.1560]

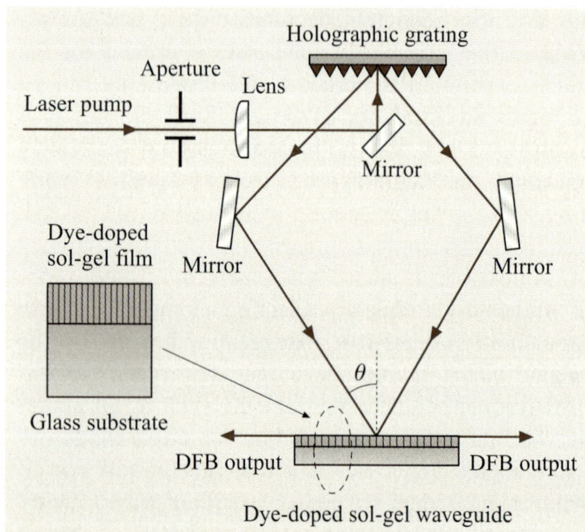

Fig. 11.182 Experimental arrangement of a sol-gel DFB waveguide dye laser

and can be modified through copolymerization with a low molecular-weight polymer [11.1564]. The result is an improved polymer: modified poly(methyl methacrylate) (MPMMA) with the desired optical homogeneity and chemical stability suitable for use as a solid-state dye laser host [11.1565]. Work on the development of new polymeric hosts has been very active. *King* and coworkers showed that the photostability of solid-state dye lasers can be enhanced by deoxygenation of the host [11.1566] and the addition of a triplet-state quencher [11.1567]. Following the idea that photodegradation of dye molecules can be reduced by increasing the rigidity and hence the rate of heat dissipation of the host matrix, *Costela* et al. [11.1568] prepared pyrromethene dye-doped copolymers of methyl methacrylate (MMA) and different methacrylic and acrylic crosslinking polymers. The combined use of a photostable dye and new polymeric hosts results in SSDLs that boost the lifetime exceeding 10^6 shots at a repetition rate of a few Hz when pumped at 532 nm. Narrow-linewidth operation with a linewidth of 1.12 GHz has also been achieved in dye-doped MPMMA using a multiple-prism grating resonator cavity [11.1565]. Further improvement of the polymer host will undoubtedly lead to SSDLs that would gain the acceptance of the common laser users.

Solid-State Dye Lasers Based on a Sol-Gel Host

The sol-gel method is a low-temperature glass-making technique that enables the introduction of organic dyes into inorganic glasses. Porous glass can be obtained via the sol-gel route by hydrolysis and polycondensation of metal alkoxides [11.1568]. Initial studies of sol-gel silica doped with organic dyes indicated that sol-gel materials held good promise as SSDL host materials because of their wide transparency range and the apparently excellent optical and thermal properties of silica [11.1559, 1569]. An added advantage is the high concentration without aggregation and the photostability of a dye when trapped in sol-gel silica, as a result of the isolation of dye molecules in the silica cage. Soon tunable laser action from sulforhodamine-doped sol-gel silica was demonstrated [11.1570]. Lasing and fluorescence properties of a large number of dyes in sol-gel silica that cover the spectral range from the near ultraviolet to the near infrared were fabricated and examined [11.1571]. Several variants of sol-gel materials have been used as solid-state dye-laser host materials with varying degree of success. The first sol-gel material used in dye-laser experiments was in fact a glassy gel obtained by the gelation of a solution and is sometimes called a xerogel [11.1572]. Xerogel derived from inorganic precursors is mechanically fragile and optically lossy due to the presence of numerous pores. The use of organically modified precursors or organic modifiers during the sol-gel process results in organically modified silicates (ORMOSILs) that show improved mechanical strength and much reduced optical loss [11.1565, 1566]. The combined use of the improved sol-gel materials as host and the deoxygenation procedures during sample preparation has lead to sol-gel dye lasers with lifetime exceeding one million shots [11.1573].

One of the attractions of sol-gel materials is the potential that such glassy materials can have properties similar to glasses made via the traditional high-temperature approach. Another attraction is the ability of sol-gel materials to trap both organic and inorganic dopants while showing exceptional chemical stability. Whilst very few polymeric dye lasers work in the blue to near-UV range due to problems related to attenuation and photostability under UV excitation, several UV laser dyes have been doped into sol-gel materials [11.1574, 1575]. Laser action with output wavelengths as short as 340 nm have been observed [11.1575].

From the recent development of polymer or sol-gel dye lasers, one must rely on the role of materials engineering in advancing solid-state dye lasers.

Distributed Feedback Waveguide Dye Lasers

Distributed feedback (DFB) lasers are compact tunable laser sources that produce narrow-linewidth output. The first DFB laser was in fact a solid-state dye laser (dye-doped gelatin film), but it did not operate as a waveguide laser because of the high film thickness [11.1576]. In subsequent developments, the majority of DFB laser work has concentrated on semiconductor lasers that have been of obvious industrial significance. Recently there has been a renewal of interest in organic DFB lasers, particularly those based on waveguide structures. This renewal in interest was caused in part by the application of conjugated polymers as luminescent materials [11.1577, 1578] and the subsequent conjugated polymer laser experiments [11.1579, 1580]. Dye-doped polymeric materials or sol-gel materials can be prepared in a planar waveguide structure by simply using spin-coating or dip-coating. Sol–gel materials have the additional advantage of a larger range of refractive index variation [11.1581, 1582] which allows integrated optics application on a large number of polymer or glass substrates. DFB configuration seems ideally suited for laser output generation in these waveguide structures. Polymeric [11.1583] and sol-gel [11.1562] DFB wave-

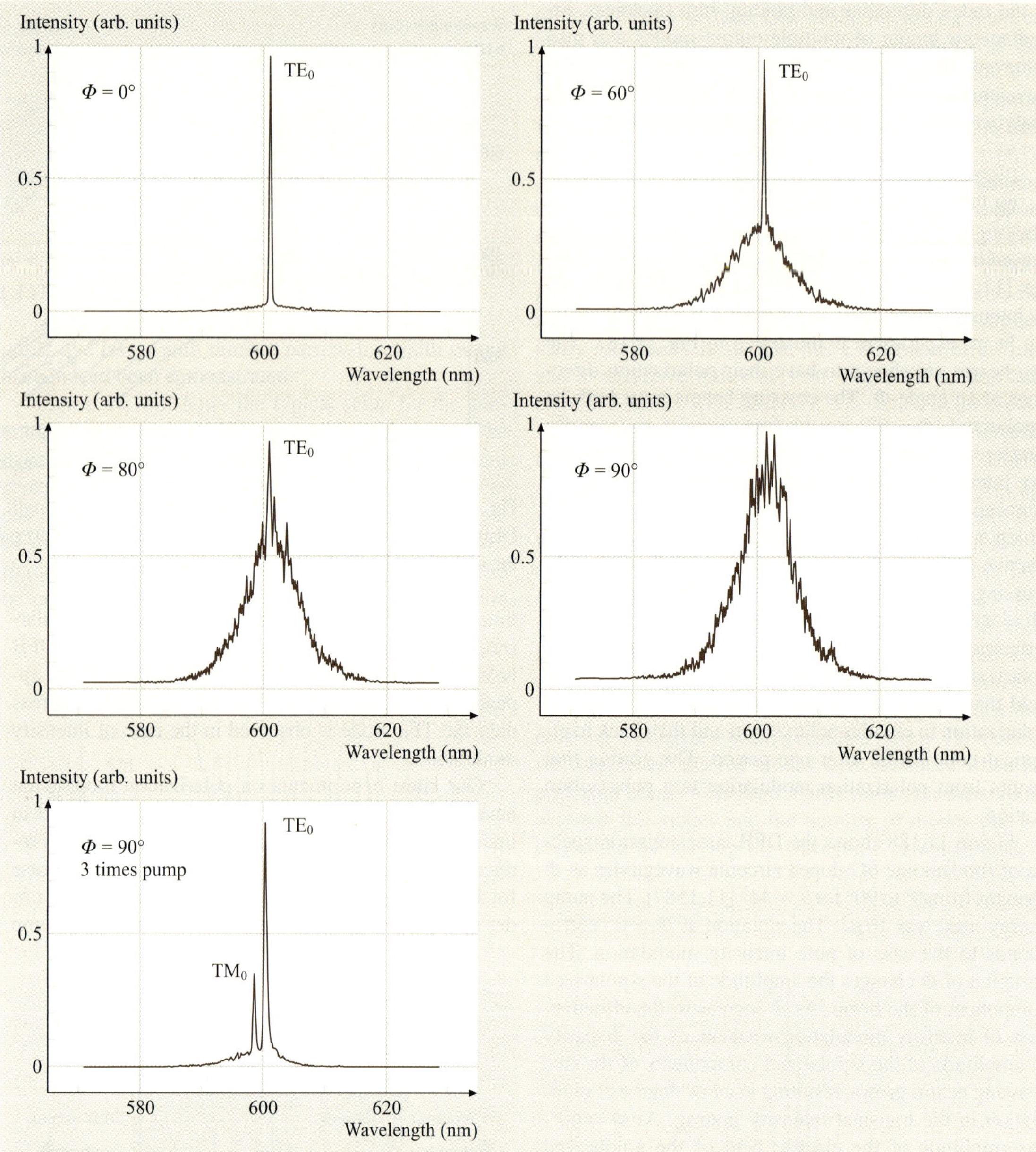

Fig. 11.188 Output emission spectra of sol-gel DFB waveguide dye lasers as Φ varies

energy for DFB lasing by polarization modulation is the same as that by intensity modulation. Further work on DFB lasing by polarization modulation is underway.

Two-Photon-Pumped Solid-State Dye Lasers

Compact visible lasers by direct upconversion of infrared light are versatile light sources that hold promise for many applications in optoelectronics. Many dyes (e.g., R6G, DCM) have been known to emit weakly in the visible when pumped in the near infrared. Indeed, broadband laser emission in the visible was observed in two-photon pumped polymer waveguide and fiber [11.1589, 1590]. A number of dyes with large two-photon pump upconverted absorption cross

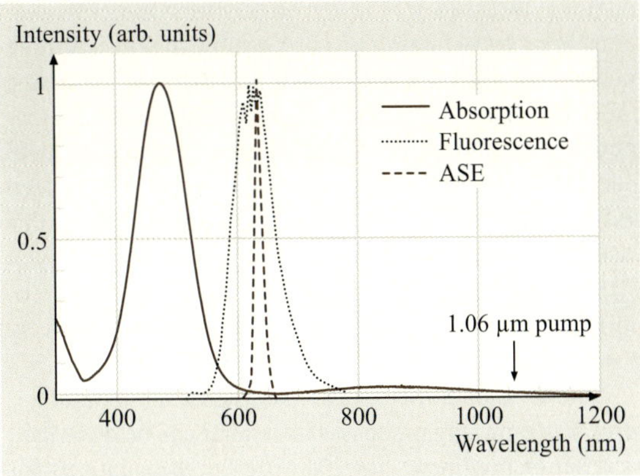

Fig. 11.189 Absorption, two-photon-pumped fluorescence and ASE spectra for an HMASPS-doped zirconia waveguide

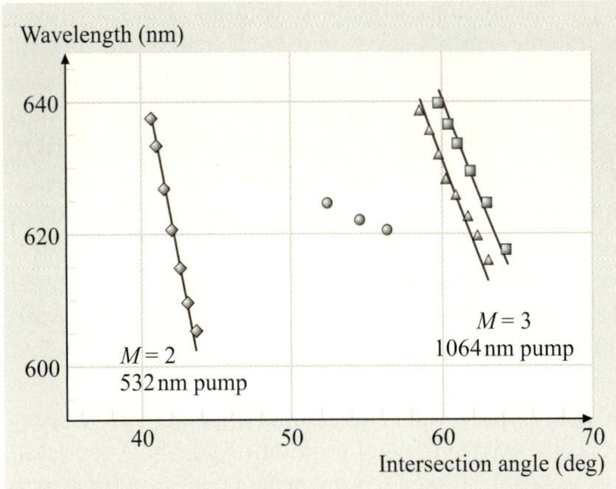

Fig. 11.190 Tuning curves for DFB lasing in HMASPS-doped zirconia waveguides

section have been synthesized recently. In particular, a styrylpyridinium dye: trans-4-[p-(N-hydroxyethyl-N-methylamino) styryl]-N-methylpyridinium p-toluene sulfonate (abbreviated as HMASPS) that shows strong emission in the red [11.1591] was doped in sol-gel zirconia thin films on a glass substrate [11.1592]. Figure 11.189 shows the absorption, two-photon-pumped fluorescence and ASE spectra from the HMASPS thin-film waveguides. A strong emission peak centered at 620 nm is seen. Using two crossing 1.06 μm beams as the pump, DFB lasing was observed in the red with a linewidth as narrow as 45 GHz. Subsequent experiments demonstrated tuning by varying the intersection angle of the two 1.06 μm crossing beams [11.1592]. Figure 11.190 shows the tuning curves for second ($M = 2$) and third ($M = 3$) Bragg orders. A tuning range of 25–30 nm was observed. Extension of the spectral coverage and improved output performance of the two-photon-pumped DFB lasing are expected as more new two-photon dyes are explored for DFB lasing study.

11.9 Optical Parametric Oscillators

Optical parametric oscillators (OPO) are based on nonlinear frequency conversion of laser radiation into two coherent light waves with lower frequencies, called the signal and idler waves. The first optical nonlinear conversion experiment was demonstrated soon after the invention of the laser by the observation of second-harmonic generation of ruby laser radiation in a quartz crystal [11.1593]. This first result initiated intense experimental and theoretical investigations to increase the conversion efficiency and the available wavelength range. In the following several years the physics of the nonlinear frequency conversion was theoretically well understood and all important methods of maximizing the conversion efficiency were elaborated, including quasi-phase matching [11.1594, 1595], birefringence phase matching in crystals [11.1596, 1597] and the requirement for optimum focusing [11.1598]. The first optical parametric oscillator was proposed in 1962 by *Kroll* [11.1599] and realized in 1965 by *Giordmaine* and *Miller* [11.1600], followed by detailed theoretical and experimental investigations [11.1601–1603]. Several review articles display a survey about the state of the art of optical parametric oscillation until the mid 1980s [11.1604–1607].

However, despite this initial efforts, the success of OPOs as a versatile tunable light source fell short of expectations for almost a decade. At that time the reasons were unsuitable birefringence of the crystals to permit phase matching, low transparency and damage thresholds, difficulties in growing large homogenous crystals and the lack of reliable laser sources with good spatial and spectral properties, which are important for efficient frequency conversion.

This situation changed due to significant advances in crystal growth techniques in the mid 1980s for both laser-active and nonlinear crystals [11.1608, 1609], leading to a renaissance of solid-state lasers. The availability of reliable pump sources with high spatial and spectral coherence (in continuous and pulsed operation mode), low-loss optical components and the appearance of new or optimized nonlinear materials (BBO, LiB_3O_5 (LBO), $KTiOPO_4$ (KTP), $KNbO_3$) started a new OPO evolution. Optical parametric oscillators are powerful devices for the efficient generation of widely tunable coherent light. The intense research in OPOs has resulted in reliable OPO systems, which are commercially available. The tunability and the high output power define the optical parametric oscillator as an attractive source in many applications including high-resolution spectroscopy, environmental monitoring, medical research, process control, remote sensing and precision frequency measurements.

Recently a new generation of conversion media have appeared with periodically structured ferroelectric materials for quasi-phase matching. Together with the development of diode-pumped all-solid-state lasers and in particular high-brightness diode-laser systems new perspectives are opened towards modern, compact, efficient, powerful, tunable OPO systems.

11.9.1 Optical Parametric Generation

The theoretical treatment of optical parametric generation has been presented in the early 1960s [11.1610, 1611] and is found in many textbooks and review articles [11.1604–1606,1612–1614]. Therefore, this section will provide only a brief description of the most important equations. The interaction of intense laser fields with a dielectric material induces nonlinear susceptibilities and causes the polarization of the medium to develop new frequency components not present in the incident radiation field.

The lowest-order nonlinear susceptibility $\chi^{(2)}$ is responsible for three-wave mixing processes such as second-harmonic generation (SHG), sum- and difference-frequency mixing (SFG, DFG), optical rectification and the optical parametric generation (OPG). Optical parametric generation with a feedback resonator for at least one of the waves is called optical parametric oscillation (OPO). The relation between the electric field $\boldsymbol{E}$ and the polarization density $\boldsymbol{P}$ in the medium is given for a second-order susceptibility as

$$\boldsymbol{P} = \boldsymbol{\varepsilon}_0(\boldsymbol{\chi}^{(1)}\boldsymbol{E} + \boldsymbol{\chi}^{(2)}\boldsymbol{E}\boldsymbol{E})$$
$$\boldsymbol{P} = \boldsymbol{P}_{\mathrm{l}} + \boldsymbol{P}_{\mathrm{nl}} \tag{11.159}$$

with $\boldsymbol{\varepsilon}_0$ the vacuum permittivity and the susceptibility $\boldsymbol{\chi}^{(k)}$, which are tensors of the rank k for an anisotropic medium. Because $\boldsymbol{\chi}^{(2)}$ is zero in centrosymmetric systems three-wave mixing is only possible in non-isotropic media. Generally, the $\boldsymbol{\chi}^{(2)}$ tensor possesses 27 elements, but many of the components vanish under certain symmetry conditions. The tensor can be reduced even to a scalar if the interacting waves are linearly polarized and monochromatic for a fixed crystal orientation. In this case, the effective nonlinearity is denoted by a coefficient $\chi^{(2)}_{\mathrm{eff}}$. In the literature [11.1614] the second-order susceptibility is often expressed by a coefficient d defined by $d := \chi^{(2)}/2$.

The dynamics of optical parametric generation or three wave-mixing processes in general can be described by a set of nonlinear coupled differential equations for the field amplitudes of the optical waves interacting with the nonlinear medium.

Considering the basic wave equation derived from Maxwell's equations for nonmagnetic dielectric media where $\boldsymbol{P}_{\mathrm{nl}}$ is regarded as source radiation

$$\nabla^2\boldsymbol{E} = \mu_0\varepsilon\frac{\mathrm{d}^2}{\mathrm{d}t^2}\boldsymbol{E} + \mu_0\frac{\mathrm{d}^2}{\mathrm{d}t^2}\boldsymbol{P}_{\mathrm{nl}} \tag{11.160}$$

and assuming

$$\frac{\partial E_1(z)}{\partial z} = \mathrm{i}\kappa_1 E_3(z)E_2^*(z)\mathrm{e}^{\mathrm{i}\Delta kz}\,, \tag{11.161}$$
$$\frac{\partial E_2(z)}{\partial z} = \mathrm{i}\kappa_2 E_3(z)E_1^*(z)\mathrm{e}^{\mathrm{i}\Delta kz}\,, \tag{11.162}$$
$$\frac{\partial E_3(z)}{\partial z} = \mathrm{i}\kappa_3 E_1(z)E_2(z)\mathrm{e}^{-\mathrm{i}\Delta kz}\,, \tag{11.163}$$

with the coefficients $\kappa_{\mathrm{i}} = \omega_{\mathrm{i}}d_{\mathrm{eff}}/n_{\mathrm{i}}c_0 \quad (i = 1, 2, 3)$, d_{eff} the nonlinear coefficient, z the propagation direction and $\Delta k = k_3 - k_2 - k_1$ the phase mismatch.

This set of equations, the *coupled wave equations* represents the interaction of three plane waves in a $\chi^{(2)}$ nonlinear dielectric. They have been solved exactly by *Armstrong* et al. [11.1594] for various input conditions. Depending on the initial conditions for the three complex amplitudes, these equations describe the various types of three wave-mixing processes generated by the single-pass propagation of the electromagnetic field through the crystal, such as second-harmonic generation, sum- and difference-frequency mixing and parametric generation.

The optical parametric generation process where only one wave with frequency ω_3 enters the dielectric medium and generates two waves with the lower frequencies ω_1 and ω_2 is based on the optical parametric fluorescence starting from the quantum noise. This pure quantum mechanical effect was first proposed and studied in 1961 by *Louisell* et al. [11.1615]. A quantum-mechanical model [11.1616] and semiclassical descriptions of the optical parametric fluorescence can be found in the literature [11.1612, 1613, 1617].

The process of parametric amplification starts with a small field amplitude of the signal ($\omega_1 = \omega_s$) or idler wave ($\omega_2 = \omega_i$). The relation between the frequencies is always $\omega_s + \omega_i = \omega_p$ and per the definition $\omega_s \geq \omega_i$.

Introducing the definition for the intensity $I = 1/2\varepsilon_0 nc|E|^2$ the coupled field equations lead to the Manley–Rowe equation [11.1618] which is equivalent to photon conservation.

$$\frac{dI_s}{\omega_s\,dz} = \frac{dI_i}{\omega_i\,dz} = -\frac{dI_p}{\omega_p\,dz} \tag{11.164}$$

The conversion of every pump photon generates a pair of signal and corresponding idler photons. In this sense the relation $\omega_p = \omega_s + \omega_i$ corresponds to energy conservation and $\Delta k = k_p - k_s - k_i = 0$ to the impulse conservation of the parametric process.

Neglecting pump field depletion ($dE_p/dz = 0$) and assuming the presence of only the signal field ($E_i = 0$) in the beginning of parametric process the coupled wave equations can be solved analytically [11.1605, 1606]. The single-pass fractional gain in signal intensity is obtained as

$$G_s(l) = \frac{I_s(z=l)}{I_s(z=0)} - 1 = \Gamma^2 l^2 \frac{\sinh^2 gl}{(gl)^2}\,, \tag{11.165}$$

where l is the length of the nonlinear medium and g the total gain factor

$$g = \sqrt{\Gamma^2 - \left(\frac{\Delta k}{2}\right)^2}\,, \tag{11.166}$$

and Γ the parametric gain factor

$$\Gamma^2 = \frac{2\omega_s\omega_i|d_{eff}|^2 I_p}{n_p n_s n_i \varepsilon_0 c^3}\,. \tag{11.167}$$

The material-dependent part of the gain coefficient $|d_{eff}|^2/(n_p n_s n_i)$ is called the figure of merit (FOM) and classifies the nonlinear quality of the conversion medium.

In the high-gain limit the single-pass gain becomes

$$G_s = \left[1 + \left(\frac{\Delta k}{2g}\right)^2\right] \sinh^2 gl \tag{11.168}$$

which reduces to

$$G_s(l) \cong \frac{1}{4} e^{2\Gamma l} \tag{11.169}$$

when $\Delta k < g$.

In the low-gain limit where $\Gamma^2 l^2 < (\Delta k/2)^2$ the gain of a parametric amplifier is

$$G_s(l) = \Gamma^2 l^2 \frac{\sin^2\left\{\left[\left(\frac{\Delta k}{2}\right)^2 - \Gamma^2\right]^{\frac{1}{2}} l\right\}}{\left[\left(\frac{\Delta k}{2}\right)^2 - \Gamma^2\right] l}\,. \tag{11.170}$$

Considering perfect phase matching $\Delta k = 0$, the single-pass signal gain in the low-gain limit approximates to

$$G_s(l) \cong \Gamma^2 l^2 \propto d_{eff}^2 l^2 \tag{11.171}$$

It can be seen that the magnitude of the parametric gain depends on the intensity of the incoming field, as well as on the material parameters such as nonlinear coefficient, refractive index and the interaction length. However, the main condition for an efficient parametric amplification is the phase-matching condition. As shown in Fig. 11.191 the parametric gain reaches a maximum for $\Delta k = 0$, and decreases symmetrically to zero for $|\Delta kl| = \pi$.

The gain bandwidth is defined by

$$\left|\left(\frac{1}{2}\Delta k^2\right) - \Gamma^2\right|^{1/2} l = \pi\,, \tag{11.172}$$

which reduces for low gain to

$$\left(\frac{1}{2}\right)\Delta kl = \pi\,. \tag{11.173}$$

The gain bandwidth increases with higher gain, but until reaching $\Gamma^2 l^2 \cong \pi$ the bandwidth broadening is small.

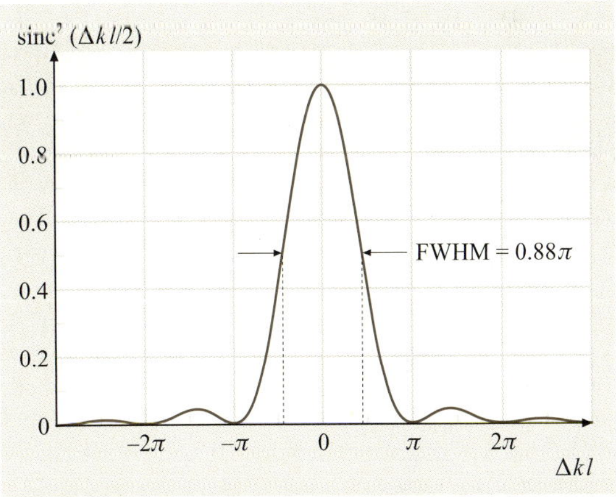

Fig. 11.191 Parametric gain of a nonlinear crystal of length l as a function of the phase mismatch Δkl. The gain is maximum by optimum phase matching $\Delta kl = 0$. The full width at half-maximum (FHWM) of the gain function is 0.88π

11.9.2 Phase Matching

As shown before effective three-wave-mixing in a nonlinear medium is only achieved if the phase-matching condition ($\Delta \boldsymbol{k} = \boldsymbol{k}_p - \boldsymbol{k}_s - \boldsymbol{k}_i = 0$) is fulfilled. Together with the frequency-matching condition ($\omega_p = \omega_s + \omega_i$), this condition implies the spatial phase matching of the three waves which is necessary for constructive interference of the interacting waves within the nonlinear medium.

This conservation of energy and momentum is not possible in materials with isotropic dispersion. Due to the different diffraction indices, the optical fields of the three frequencies differ in phase velocity and the relative phase of the interactive waves varies along the medium. A measure of the phase mismatch is the coherence length $L_c = \pi/|\Delta k|$, which defines the distance over which the relative phase of the interacting waves shifts by π. The propagation beyond the coherence length, which is typically several μm, leads to back-conversion from the generated waves into the pump wave. The oscillating behavior of the generated intensity depending on the interaction length is shown in Fig. 11.192.

Today there are two different techniques to realize a phase-matched conversion process: birefringence phase matching and the quasi-phase matching.

Birefringence Phase Matching

This method uses the dependence of the refractive index on the propagation direction and polarization of electromagnetic waves in birefringence crystals. The waves are distinguished into ordinary (o) and extraordinary (e) waves depending on their polarization direction. In uniaxial crystals the refractive index of the ordinary wave is independent from the propagation direction whereas the refractive index of the extraordinary wave depends on the polar angle between the optical axis and the propagation direction. To achieve phase matching the polarization of the high-frequency pump wave should be selected so that the index of refraction is smallest. The phase-matching condition can be realized by choosing an appropriate angle between propagation direction and optical axis. Angular phase-matched wavelength tuning is then possible by changing the phase-matching angle with regard to the phase-matching condition $n_p\omega_p = n_s\omega_s + n_i\omega_i$. The process is called type I phase matching when both generated waves have the same polarization perpendicular to the polarization of the pump wave, for example (e–oo). For type II phase matching the polarization of the generated waves are oriented perpendicular to each other (e–oe). An analytical solution of the phase matching angle is only possible for type I phase matching in uniaxial crystals, the other processes have to be solved numerically [11.1609]. The analysis in biaxial crystals is more complicate and has only numerical solutions; only in their principle planes can the formalism be reduced to the uniaxial conditions (Table 11.47 [11.1619, 1620]. A consequence of angular birefringence phase matching is the spatial walk-off generated by the birefringence, which reduces the interaction length of the optical waves. Besides the angular phase matching the indices of refraction can be matched by changing the crystal temperature (temperature phase matching). Combined with a propagation direction perpendicular to the principal axis this method is called noncritical phase matching (NCPM) and results in the absence of spatial walk-off and less angular sensitivity, which is important in applications with tight focusing.

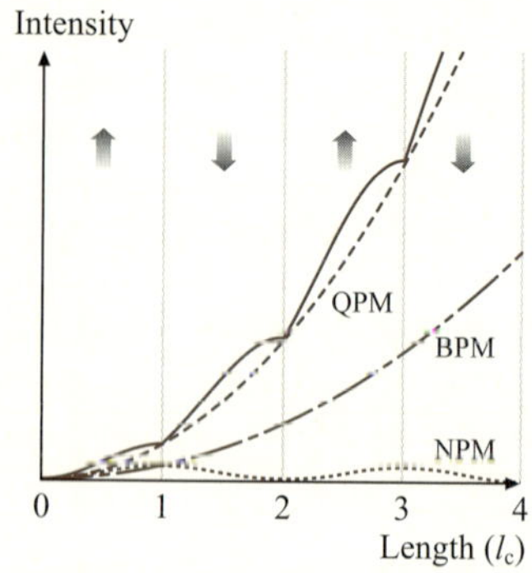

Fig. 11.192 Conversion intensity as function of the propagation length in a nonlinear crystal. NPM: Non-phase-matched process with no significant total conversion intensity. BPM: birefringence phase matching. Intensity increases quadratically with z, with a slope given by $d^2_{\text{eff,BPM}}$. QPM: Quasi-phase-matched process. Intensity increases quadratically with z on average, with a slope given by $d^2_{\text{eff,QPM}} \gg d^2_{\text{eff,BPM}}$

The vector equation of the phase matching condition can be realized in a non-collinear or a collinear way. In contrast to the collinear phase matching is the non-collinear phase matching a vector phase matching. Figure 11.193 shows the non-collinear angle included by the wavevectors. With the choice of a suitable non-collinear angle the angular acceptance is enlarged or the Poynting vector walk-off can be compensated. The conditions for the walk-off compensation are shown in Fig. 11.193a; the Poynting vectors of the ordinary signal and the extraordinary pump wave are paral-

Table 11.47 Refraction indices of ordinary and extraordinary wave for uniaxial and biaxial crystals in the principle planes

Crystal	Extraordinary wave (e)	Ordinary wave (o)
Uniaxial	$n_e(\Theta) = \frac{n_o n_e}{\sqrt{n_e^2 \cos^2 \Theta + n_o^2 \sin^2}}$	$n_o(\Theta) = n_o$
Biaxial, xy-plane	$n_e^{xy}(\Phi) = \frac{n_x n_y}{\sqrt{n_x^2 \cos^2 \Phi + n_y^2 \sin^2 \Phi}}$	$n_o(\Phi) = n_z$
Biaxial, xz-plane	$n_e^{xz}(\Theta) = \frac{n_x n_z}{\sqrt{n_x^2 \cos^2 \Theta + n_z^2 \sin^2 \Theta}}$	$n_o(\Theta) = n_y$
Biaxial, yz-plane	$n_e^{yz}(\Theta) = \frac{n_y n_z}{\sqrt{n_y^2 \cos^2 \Theta + n_z^2 \sin^2 \Theta}}$	$n_o(\Theta) = n_x$

lel. Such walk-off compensation is of advantage for nanosecond OPOs reaching higher conversion efficiencies [11.1621] and necessary for critical-phase-matched OPOs pumped by ultrashort pulses because of the small beam radii [11.1622, 1623]. The diagram of tangential phase matching is shown in Fig. 11.193b. This special case of non-collinear phase matching produces a significant enlarged angular acceptance for the extraordinary pump wave. In addition, group-velocity matching for ultrashort pulses is realized via non-collinear phase matching [11.1624].

Quasi-Phase Matching

An alternative method to achieve phase matching via phase correction is quasi-phase matching (QPM). Here the relative phase between the three waves is corrected using a periodic change in the sign of the nonlinear susceptibility (Fig. 11.192) before the back conversion starts. Although the technique of QPM was proposed even before birefringence phase matching [11.1594], difficulties in fabrication of the periodic structure in the range of the coherence length (typically 1–100 μm) have prevented its realization for a long time. Recent advances in the fabrication of structured ferroelectrics meanwhile have established this technique for efficient frequency-conversion applications.

Quasi-phase matching through periodically poling offers two major advantages over birefringence phase matching. First, the polarization of the interacting waves can be equal. This allows the use of the largest $\chi^{(2)}$ coefficient of the crystal, which reduces the pump threshold and increases the conversion efficiency. The second advantage is that the quasi-phase matching of any combination of pump, signal and idler wavelength can be realized in such materials via periodic poling. In addition, QPM provides design flexibility of nonlinear conversion devices by engineered domain structures. Applications with novel configurations such as chirped gratings for pulse compression [11.1625], fan-out gratings for broad tunability [11.1626], quasi-periodic structures for simultaneous operation of two three-wave-mixing processes [11.1627] and two-dimensional gratings for multiple direction phase-matching processes [11.1628] have been demonstrated.

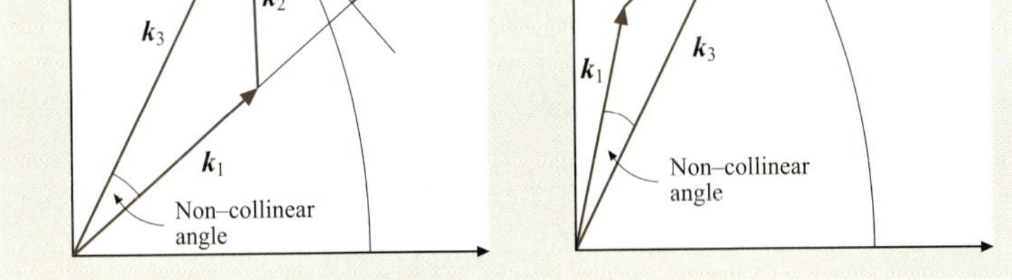

Fig. 11.193a,b Wavevector diagram for different non-collinear phase-matching schemes. **(a)** Walk-off compensation, **(b)** scheme for tangential phase matching

Phase Mismatch and Acceptance Bandwidths

The ideal phase-matched interaction, where the pump wave and idler and signal waves propagate with the same velocity through the nonlinear crystal, $\Delta k = 0$, is affected under real experimental conditions by several parameters such as the divergence and bandwidth of the pump radiation, the bandwidth of the generated radiation and an inhomogeneous temperature distribution. The tolerances of the nonlinear material concerning to these perturbations are expressed by the acceptance bandwidths. They are defined by the FWHM of the parametric gain width (Fig. 11.191) and developed in Taylor series of $\Delta k(0)$:

Angular acceptance

$$\Delta\delta = 2\times 0.886\pi l^{-1}(\delta\Delta k/\delta\delta)^{-1}\ ;$$

Gain band width

$$\Delta\omega_\mathrm{s} = 2\times 0.886\pi l^{-1}(\delta\Delta k/\delta\omega_\mathrm{s})^{-1}\ ;$$

Spectral acceptance

$$\Delta\omega_\mathrm{p} = 2\times 0.886\pi l^{-1}(\delta\Delta k/\delta\omega_\mathrm{p})^{-1}\ ;$$

Temperature acceptance

$$\Delta T = 2\times 0.886\pi l^{-1}(\delta\Delta k/\delta T)^{-1}\ . \qquad (11.174)$$

11.9.3 Optical Parametric Oscillators

An optical parametric oscillator consists of three basic components, namely the pump source, the gain medium, and the feedback resonator. For a qualitative description of the operation principle we consider the most basic OPO setup, as shown schematically in Fig. 11.194.

An intense coherent light field produced by a laser beam propagates through an optically nonlinear crystal with the frequency ω_p. The crystal is placed inside an optical resonator, which resonates for reasons of simplicity with one wave, for example the signal wave. As the nonlinear crystal provides a sufficient nonlinearity, parametric generation takes place and a pump photon is converted into a signal and idler photon fulfilling photon-wise energy conservation $\omega_\mathrm{p} = \omega_\mathrm{s} + \omega_\mathrm{i}$. The resonator feeds the signal wave back into the crystal where it is further amplified through the power transfer from the pump wave, as expressed in (11.164). Optical parametric oscillation starts, when the increased signal intensity surpasses threshold that means the amplified signal wave compensates for the round-trip losses in the resonator (due to mirror transmission, absorption, scattering and diffraction). Reaching this distinct pump intensity, the threshold pump intensity, a significant portion of the pump intensity is converted into signal and idler intensity. This power transfer from the pump to the generated waves reduces the (spatially averaged) pump intensity inside the nonlinear crystal and thus the signal gain. This effect is called gain saturation. It leads to steady-state operation of the OPO, where the signal power generated inside the crystal exactly balances the resonator losses for the signal wave.

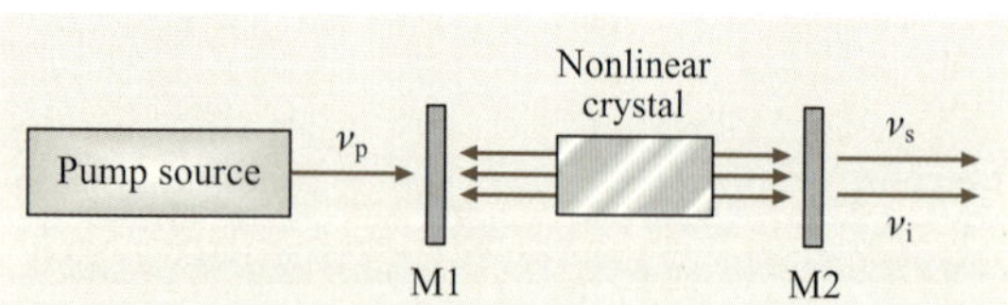

Fig. 11.194 Schematic setup of an OPO. An optical wave with frequency ν_p from the pump source is converted into two output waves with frequencies ν_s and ν_i via three-wave interaction in a nonlinear crystal. M1 and M2 are mirrors of a feedback resonator

As any pair of signal and idler photon can be generated from the initially random vacuum field fluctuations in steady-state operation, the signal and idler frequency pair that has the minimum threshold pump intensity is generated. Thus the generated frequencies, and with them the ratio of the signal and idler frequency, are determined by the frequency dependence of the parametric gain in the crystal and the frequency dependence of the resonator losses.

A detailed description of the behavior of optical parametric oscillators has been published in several reviews [11.1604–1606] and the physics of OPOs is discussed more fundamentally in a number of textbooks about nonlinear optics [11.1612–1614].

In the literature, optical parametric oscillators are mostly classified into three groups: continuous-wave optical parametric oscillators, nanosecond pulsed optical parametric oscillators and synchronously pumped picosecond or femtosecond optical parametric oscillators, depending on the temporal characteristic of their pump laser.

This distinction is not only a practical classification but is important for the theoretical description and the basic performance, such as pump intensity at threshold, conversion efficiency and spectral performance.

11.9.4 Design and Performance of Optical Parametric Oscillators

The performance of an optical parametric system concerning tunability, high output intensity, spectral and spatial coherence is directly related to key elements of the OPO's architecture, which are pump source, nonlinear material and feedback resonator. In general, the physics of the optical parametric process imposes several requirements of potential OPO components.

A simultaneous optimization of all OPO features cannot be fairly done because of material problems and physical conflicts. Therefore, a general description of an optimized OPO design is not possible and the device configuration is always a compromise depending on the requirements of the specific application.

The choice of the pump laser depends together with the selection of the nonlinear material on the required wavelength range and time resolution for a given application. Obviously, the pump laser has to be powerful enough to drive the OPO well above threshold and guarantee stable OPO operation. Since good spatial and spectral beam quality reduces the threshold, preferred pump sources will be high-power lasers of several watts, which also provide an excellent spatial beam quality, and good power and frequency stability. If the available pump power is restricted, low beam divergence is of particular importance to allow focusing into the nonlinear crystal. The deviations in spectral and spatial coherence tolerated by the parametric frequency conversion are determined by the acceptance bandwidths of the nonlinear material. Wavelength tunability of the pump laser offers an additional attractive feature: tuning of the OPO via the pump wavelength.

The selection of the nonlinear material depends on the nonlinear parametric process, the experimental design and the properties of the pump source. Fundamentally, the nonlinear crystal should be distinguished by a broad transparency range, a high optical-damage threshold and large effective nonlinearity. An important parameter is the birefringence of the crystal that has to be high enough to maintain phase matching. However, this effect should not be too large because another consequence of birefringence is spatial walk-off, which reduces the interacting length of the coupled fields. Large spectral and angular acceptance bandwidths are advantageous because real laser systems show finite bandwidths and divergence and in ultrashort optical parametric oscillation, the temporal walk-off and group-velocity dispersion between pump, signal and idler wave should be low. For stable operation of OPO systems with high energies the thermo-mechanical and thermo-optical stability of the crystal should be sufficient. Finally, the crystals should be available in appropriate dimensions and excellent optical homogeneity.

The design of the feedback resonator is strongly connected to the output characteristic of the pump source. If the available pump power is restricted the OPO design may have to be modified for lower threshold, which might however restrict the tuning capabilities of the OPO. Three types of feedback resonators are distinguished, depending on the number of resonating waves. The singly resonant OPO (SRO) has highly reflecting mirrors at the signal or the idler wave. In principle, this configuration is of particular advantage because it should provide continuous wavelength tunability. The disadvantage of these systems are their high pump power at threshold, which exceeds several watts [11.1629, 1630]. Therefore SRO devices are mainly found for pulsed OPO configurations.

Doubly resonant OPO configurations (DRO) where both signal and idler waves are resonated have received much attention, because the pump power at threshold is reduced by one to three orders of magnitude. However, such DRO OPOs do not provide mode-hop-free tunability and often suffer from frequency and power instabilities [11.1631, 1632]. Continuous wavelength tuning requires active stabilization techniques [11.1633, 1634].

The doubly resonant configuration where pump and signal (or idler) waves are resonated is called pump-enhanced SRO (PESRO). Such devices are very attractive because they provide a compromise between the low pump power at threshold of DROs and the wide tunability of SROs when the length of the OPO is carefully locked to a frequency-stabilized pump laser [11.1635]. In addition, a continuous tunability is possible with a pump-enhanced SRO, by using a resonator internal beam splitter to separate the pump spatially from the resonated OPO wave [11.1636]. These waves are then resonated on two separate cavity end mirrors (dual cavity) for independent control of the resonated OPO wave and the length of the pump cavity.

Triply resonant OPOs (TRO) suffer from the combined problems of both types of DROs, consequently they provide less tunability in comparison to SRO and DRO, but they are remarkable because of their low threshold pump power in the range of several mW or below [11.1637, 1638].

Continuous-Wave Optical Parametric Oscillators

Continuous-wave optical parametric oscillators (CW-OPOs) are efficient and highly coherent light sources in the near- and mid-infrared. Due to their tunability, these sources are of considerable interest for high-resolution spectroscopy of molecules and trace-gas detection [11.1639]. Simplified and improved frequency standards are currently being considered, based on the strong frequency and phase correlation of the OPO's signal and idler output waves [11.1640, 1641].

After the first demonstration of CW OPOs in 1968 by *Smith* et al. [11.1601] and *Byer* et al. [11.1602] significant progress is this field was made in the beginning of the 1990s, supported by advances in crystal growth techniques [11.1609], low-loss optical components and CW solid-state lasers.

As discussed before pump power at threshold decreases with increasing cavity finesse for each of the wavelengths. Typical values for the pump power at threshold are listed in Table 11.48 for CW OPOs with the different resonators (TRO, DRO, SRO).

The pump power at threshold of singly resonant OPOs is significantly higher than 1 W, even in a high-finesse cavity with no significant output coupling of the resonant wave. Since most commercially available single-frequency lasers provide maximum pump powers of about 1 W, most CW OPOs demonstrated in the past are doubly or triply resonant devices.

The threshold pump intensity can be calculated from the coupled wave equations in analogy to the parametric gain assuming all three waves (pump, signal and idler) are infinite plane waves, with uniform intensity across the beam, neglecting focusing and double refraction. The gain is considered to be unidirectional and the pump depletion is always small at threshold.

At the steady-state threshold, the gain balances the resonator losses for each resonator round-trip. The threshold condition for an SRO is

$$E_s(0) = (1-\alpha_s)E_s(l_c)\,. \tag{11.175}$$

Where α_s are the round-trip losses of the resonant signal wave and l_c is the crystal length. In the low-gain limit and assuming small losses, the parametric gain coefficient (11.167) can be written [11.1606]

$$\Gamma^2 l_c^2 = 2\alpha_s\,. \tag{11.176}$$

Then we obtain the pump intensity at the SRO threshold

$$I_{p,th} = \alpha_s \frac{n_p n_s n_i}{\omega_s \omega_i d_{eff}^2 l_c^2} \operatorname{sinc}^{-2}\left(\frac{\Delta k l_c}{2}\right)\,. \tag{11.177}$$

Table 11.48 Typical pump power at threshold for CW TRO, DRO, SRO

OPO	Pump threshold (mW)	Ref.
TRO	< 1	[11.1633]
DRO	100	[11.1642]
SRO	2600	[11.1643]

With $\Gamma^2 l_c^2 = \alpha_i \alpha_s$ the pump intensity at the DRO threshold is similarly derived as

$$I_{p,th} = \frac{\alpha_i \alpha_s}{4} \frac{n_p n_s n_i}{\omega_s \omega_i d_{eff}^2 l_c^2} \operatorname{sinc}^{-2}\left(\frac{\Delta k l_c}{2}\right)\,. \tag{11.178}$$

A low threshold is achieved when a long nonlinear crystal with a high figure of merit is used inside a signal resonator with highly reflecting mirrors. In general the pump threshold is lower for higher signal and idler frequencies. However, via the sinc function of the phase mismatch the pump threshold is exposed to an additional and much stronger dependence on the signal and idler wavelength. A minimum threshold is achieved for signal and idler wavelengths with $\Delta k = 0$. As a consequence the OPO always seeks to oscillate as close as possible to the phase-matched wavelengths.

Usually CW OPOs are pumped with focused beams to enhance the pump power, therefore pump, idler and signal waves are expressed as Gaussian beams [11.1644]. In this case the intensity shows an inhomogeneous distribution along the propagation axes z and the intensity and phases are not constant within each of the xy-planes. Only for loose focussing can these inhomogeneities be neglected and is equation (11.177) still a good estimation of the pump intensity at threshold. The pump power at threshold for tight focussing has been derived in an detailed mathematical treatment [11.1614, 1645], which is beyond this scope. For optimized confocal focussing [11.1598], the pump power at threshold is given by

$$P_{p,th} = \frac{4\pi c^2 n_s n_i \alpha_s}{\mu_0 \omega_s \omega_i \omega_p d_{eff}^2 l_c}\,. \tag{11.179}$$

As opposed to the l_c^{-2} dependence of the threshold intensity for plane waves (11.177) for the pump power of Gaussian beams decreases linearly with the crystal length.

Once above threshold a part of the pump power P_p is converted into the appropriate signal and idler waves as long as the effective gain is larger than one. An important parameter is the pump depletion, which occurs until the gain is saturated and the steady state is reached. The analytical description of the internal conversion efficiency η_{int} can be obtained under steady-state conditions assuming small cavity losses and optimum phase matching. The conversion efficiency

$$\eta_{int} = 1 - \frac{P_{out,p}}{P_{in,p}} \tag{11.180}$$

depends on the ratio of the incoming and the transmitted pump power and is a function of the pump ratio

$N_0(N_0 = P_p/P_{p,th})$ [11.1645]. The maximum conversion efficiency for plane waves comes to 100% for a pump ratio of $N_0 = (\pi/2)^2$. The extension of this model to Gaussian intensity distributions with plane wavefronts yields maximal conversion efficiency of 71% supposing a pump ratio of 6.5. The conversion efficiency is reduced by the inhomogeneous intensity distribution of the pump wave as well as distortions of the wavefronts.

Due to parasitical losses in the cavity (Fresnel losses at the surfaces, diffraction and absorption), the measured output power is lower than predicted by the internal efficiency. Therefore, it is useful to determine an external (utilizable) conversion efficiency, which in steady-state conditions is

$$\eta_{ext,s} = \frac{T}{T+L}\frac{\omega_s}{\omega_p}\eta\,, \qquad (11.181)$$

assuming small transmission T of the output mirror and small losses L for the resonant signal wave.

In order to operate the CW OPO continuously with low fluctuations in output power and frequency, the cavity length needs to be stabilized. The main sources of cavity length fluctuations are thermally induced fluctuations of the refractive index, mostly in the crystal, and acoustically and thermally induced perturbations of the position of the cavity mirrors. Fluctuations caused by these sources are usually significant just for low-frequency ranges, i.e., $\nu < 10$ kHz, which can be compensated, for example, by an electronic stabilization device. (For higher-frequency components, i.e., $\nu > 10$ kHz, the noise of the pump laser is significant instead of the cavity length fluctuations). For an OPO with pump enhancement resonance, a change of the cavity length induces a mismatch between the cavity resonance frequency and the pump laser frequency, resulting in an increased OPO threshold and decreased output power. If the OPO is resonant for the signal or the idler wave or for both simultaneously, the change of the cavity length leads to fluctuations of the signal and idler frequencies and to mode hops and cluster hops. One possible method to stabilize the cavity length is to increase the passive stability in a monolithic setup [11.1646]. However, pump-enhanced OPOs and doubly resonant OPOs usually require active control of the cavity length [11.1647]. To stabilize a cavity resonance to a laser frequency several methods have been successfully demonstrated [11.1648–1650]. Long-term stabilization of a doubly resonant parametric oscillator where the frequency instability is below 1 kHz has been demonstrated [11.1651]

One of the most crucial components of CW OPOs is the nonlinear material. For successful CW OPO operation it is of particular importance that the crystals nonlinearity is high and noncritical phase matching is possible. Crystals used so far are noncritical phase-matched $LiNbO_3$, KTP, LBO, $Ba_2NaNb_5O_{15}$ and $KNbO_3$ whereas especially KTP and its arsenate isomorphs $KTiOAsO_4$ (KTA) and $RbTiOAsO_4$ (RTA) turned out to be attractive materials for CW OPOs. Wavelength tuning of NCPM OPOs requires either temperature tuning of the crystal or wavelength tuning of the pump laser. Because of the low-temperature tuning capability of the KTP crystal family wavelength tuning in such OPOs depends on tunable pump sources. Many investigations have been performed with DROs and pump-enhanced OPOs to achieve continuous tunability. A significant increase of the continuous tuning range was realized with the dual-cavity design [11.1634, 1636, 1652]. Low-threshold operation has been demonstrated with a TRO and pump-enhanced SRO KTP OPO pumped with a single-stripe GaAlAs diode laser [11.1652, 1653]. High idler powers of 840 mW in the wavelength range of 2.4–2.9 μm have been achieved with KTP and KTA OPOs designed as an intracavity SRO configuration and pumped by Ti:sapphire lasers [11.1654]. An important step was the realization of a singly resonant CW OPO [11.1629].

In recent years the availability of novel quasi-phase-matched nonlinear material and high-power solid-state lasers enabled significant advances in the development of CW OPOs. The high optical nonlinearity of periodically poled $LiNbO_3$ (PPLN) and long interaction lengths up to 50 mm enables the operation of CW SROs pumped by CW solid-state lasers. Pumping such a 50 mm-long PPLN crystal with the 13.5 W output of a high-power diode-pumped Nd:YAG laser, infrared radiation tunable in the range of 3.25–3.95 μm with an output power of 3.6 W has been reported [11.1655]. The measured pump depletion was as high as 93%. Wavelength tuning was achieved through temperature tuning or by using different poling periods implemented on the crystal. A step toward compact efficient and powerful CW OPOs was performed with the realization of a diode-laser-pumped singly resonant PPLN OPO [11.1656]. The SRO consists of a 38 mm-long PPLN in a four-mirror cavity (Fig. 11.195). Pumping with 25 W of 925 nm laser radiation from an AlGaAs master-oscillator power-amplifier (MOPA) system, 480 mW of single-frequency idler radiation was generated. The tuning range of the idler covers 2.03–2.29 μm. In addition several investigations have been performed with pump-enhanced SROs in

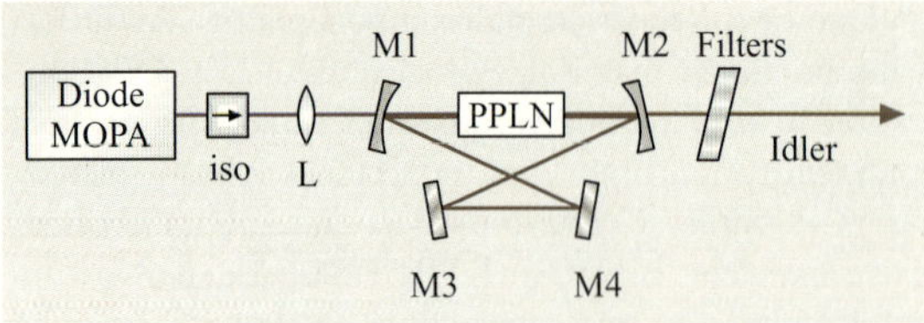

Fig. 11.195 Experimental setup of the diode laser pumped singly resonant OPO (SRO). The pump beam from the MOPA system propagates through a 60 dB isolator (iso) and focuses into the PPLN crystal. All mirrors of the ring cavity are highly reflecting for the signal wave and highly transmitting for the pump and idler wave. The idler output behind M2 is filtered from the residual pump and signal radiation by dielectric filters [11.1656]

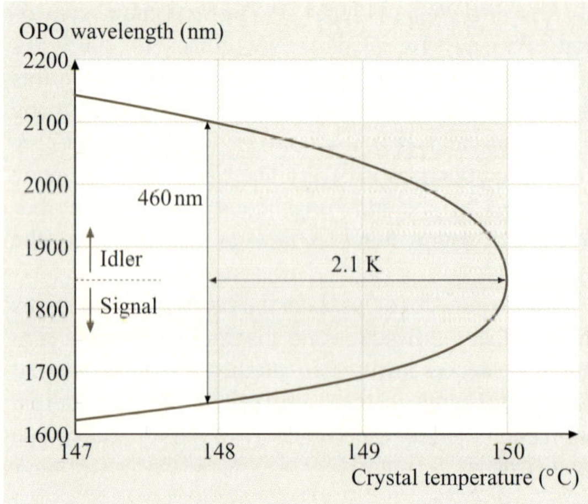

Fig. 11.196 Temperature tuning curve of a near-degenerate PPLT OPO calculated for a pump wave of 925 nm and a grating period of 27.3 μm [11.1660]. *Shaded areas*: parametric gain bandwidth of the 35 mm long PPLT crystal at degeneracy is as large as 460 nm. The Sellmeier coefficients can be found in [11.1661]

PPLN. Tunable radiation was generated in the range of 2.29–2.96 μm with an output power of 140 mW pumped by 800 mW delivered from a miniature diode-pumped single-frequency Nd:YAG laser [11.1657]. An extension of the wavelength into the mid-IR between 4.07–5.26 μm was demonstrated by pumping a PPLN PESRO by CW single-frequency Ti:sapphire radiation [11.1658]. Another pumping design was performed using a high-power fiber laser. The Yb-doped fiber laser generated in a 40 mm-long PPLN crystal mid-IR idler powers in the range 2.98–3.7 μm. Pumping with 8.3 W laser power the OPO generated 1.9 W [11.1659].

Besides the successful demonstration of powerful PPLN OPOs one severe disadvantage of PPLN is the appearance of photorefractive effects, which may damage the crystal. Some effort is made to optimize the growing technique in such a way that these defects can be avoided.

Other QPM materials have also been developed, such as periodically poled KTP (PPKTP) [11.1662–1664], RTA (PPRTA) [11.1665] and Lithium tantalate (PPLT) [11.1660]. Due to the high purity of these materials photorefractive defects do not occur even at high pump levels.

Wide wavelength tunability has been reported in the range 1.5–4 μm in PPLN [11.1666] and 1.55–2.3 μm in PPLT[11.1660] pumping the OPOs near degeneracy, which provides large parametric gain bandwidths for a given poling period, crystal temperature and pump wavelength (Fig. 11.196). In PPLT the wavelength tuning is achieved by changing the crystal temperature within an interval of only 10 K. Operating a CW OPO with subsequent frequency-doubling wavelength tunability from the visible to the mid-IR range between 550–2830 nm has recently been demonstrated. A comparison between the performance of PPKTP and PPLN was performed within these investigations [11.1664].

These cited examples clearly indicate the dynamic in this field. Further research in new ferroelectrics generating radiation beyond 5 μm and the improvement of the existing materials concerning dimensions, homogeneity and structure design will lead to high-efficiency compact integrated optical systems.

Nanosecond Optical Parametric Oscillators

Optical parametric oscillators pumped with nanosecond pulses are the most well-established OPO systems generating powerful tunable radiation in the spectral range from the visible to the near infrared. Since the first demonstrated optical parametric oscillator [11.1600], which was a nanosecond pulsed OPO, reliable nanosecond OPOs were not established before the late 1980s. The appearance of new nonlinear materials with large nonlinearity and high damage thresholds (larger than $1\,J/cm^2$) such as BBO, LBO, and the improvement of the optical quality of KTP or $KNbO_3$ was the start of an intense OPO evolution. These materials together with the optimized spatial beam quality and high peak powers of modern nanosecond Q-switched lasers made it possible to reach the threshold of the single-resonant OPO configuration.

The basic principle of optical parametric oscillation for nanosecond OPOs is the same as for CW oscillators; however, due to the finite length of the pump pulse (typical pulse lengths are 2–30 ns) the OPO does not reach the steady-state regime. In nanosecond OPOs, the cavity length is about a few centimeters, which corresponds to less than 100 round-trips for the pump pulse. The steady-state threshold is no longer valid because of the transient character of pulsed OPOs. The dynamic behavior was considered in a model developed by *Brosnan* and *Byer* [11.1667] using a time-dependent gain analysis. They calculated the buildup to threshold for a signal resonant optical parametric oscillator by assuming a Gaussian temporal profile for the incident pump pulse intensity and a Gaussian spatial distribution for the pump and the signal beams. Additionally their model assumes single pass of the pump wave and considers the effects of mode overlap and spatial walk-off.

They introduced the time-dependent parametric gain coefficient Γ_t

$$\Gamma_t = \sqrt{g_s}\Gamma e^{-\left(\frac{t}{\tau}\right)^2} \quad \text{with} \quad g_s = \frac{\omega_p^2}{\omega_p^2 + \omega_s^2} , \tag{11.182}$$

where τ is the $1/e^2$ intensity half-width of the Gaussian pump pulse. The spatial coupling efficient g_s describe the mode overlap between the resonant signal wave and the pump wave. The influence of the Poynting vector walk-off defines the effective parametric gain length Λ

$$\Lambda = l_\omega \operatorname{erf}\left(\frac{\sqrt{\pi}}{2}\frac{l_c}{l_\omega}\right) \tag{11.183}$$

with the walk-off length, which depends on the birefringence angle ρ,

$$l_\omega = \frac{\sqrt{\pi}}{2}\frac{\omega}{\rho}\sqrt{\frac{\omega_p^2+\omega_0^2}{\omega_p^2+\omega_s^2/2}} . \tag{11.184}$$

The threshold energy fluence for the single-resonant OPO is derived from the coupled wave equations, assuming low pump depletion, as

$$J_{th} = \frac{2.25}{\kappa g_s \Lambda^2}\tau \times \left(\frac{l_r}{2\tau c}\ln\frac{P_{s,th}}{P_{s,0}} + 2\alpha_2 l_c + \ln\frac{1}{\sqrt{R}} + \ln 2\right)^2 \tag{11.185}$$

with $\kappa = \frac{2\omega_s\omega_i d_{eff}^2}{n_p n_s n_i \varepsilon_0 c^3}$, where R is the reflectivity of the mirrors, α_2 are the internal losses of the resonant wave inside the crystal, l_r is the optical cavity length and l_c is the length of the crystal. The threshold identifies the detectable level of the signal energy, which corresponds to an energy of typical 100 μJ, giving a threshold power to noise ratio of $\ln(P_{th}/P_s) = 33$. The time required from the parametric gain of the signal field starting from the noise level P_0 to reach the oscillation threshold P_{th} is called the rise time of the parametric oscillator. The rise time is an important parameter in the design and operation of a parametric oscillator and can be calculated from the time-dependent coupled amplitudes equations. For efficient conversion, the rise time has to be as short as possible. *Pearson* et al. [11.1668] carried out a detailed investigation of the rise time as a function of various experimental parameters, such as pump pulse length, cavity length, cavity losses and the pump ratio N_0, which is the factor between the pump pulse energy and the pump pulse energy at threshold, $N_0 = P_p/P_{p,th}$. Important experimental parameters to minimize the rise time are high pump energy densities and a short cavity length. Beyond the influence of the rise time the threshold energy density can be reduced by using long pump pulse durations τ, optimizing the reflectivity R and avoiding intracavity losses α. Additionally the nonlinearity of the crystal should be large and the length of the crystal should be as long as possible with regard to the limitation through the Poynting vector walk-off.

The spectral bandwidth of a singly resonant optical parametric oscillator depends primarily on the gain bandwidth, which is determined by the dispersion, the birefringence and the length of the nonlinear crystal. In addition, the bandwidth is determined by the spectral properties of the resonator, the correlation between the resonator modes and the characteristics of the pump laser, the wavelength, spectral bandwidth, intensity and divergence. Because of the short pump pulse duration for nanosecond pumped OPOs there is only a weak spectral-mode condensation and the nanosecond OPO does not reach the stationary state, unlike the CW OPO. Usually the nanosecond OPO bandwidth is estimated with the gain bandwidth, as there is no exact analytical description of the spectral bandwidth of pulsed OPOs. Neglecting any saturation *Brosnan* and *Byer* [11.1667] deduced an expression for the spectral condensation depending on the number p of cavity round-trips

$$\Delta\nu(p) = \frac{1}{\sqrt{p}}\Delta\nu . \tag{11.186}$$

A more exact prediction concerning the theoretical bandwidth can be calculated using numerically simulations [11.1669, 1670].

Nanosecond pulsed optical parametric oscillators are powerful devices of widely tunable coherent light. With the nonlinear crystals BBO and LBO such OPOs provide coherent radiation in the entire spectral range from the UV (300 nm) to the near IR (2.5 μm). The advantageous properties of these crystals are high optical damage threshold, high nonlinearities, a wide transparency range and a large birefringence. Since the first successful operation of a BBO OPO reported by *Fan* et al. in 1989 [11.1671] significant progress in OPO technology has been achieved. In the investigations reported so far, BBO OPOs have been pumped with the second, third, and fourth harmonics of a Q-switched Nd:YAG laser at 532 nm [11.1671,1672], 355 nm [11.1672–1677] and 266 nm [11.1672, 1678], and with the fundamental of an XeCl excimer laser at 308 nm [11.1679,1680]. The highest output energies and efficiencies have been obtained with the third harmonic of Q-switched Nd:YAG laser. The generated output energies are in the range 100–200 mJ/pulse and external conversion efficiencies of up to 61% have been reported [11.1672, 1681].

Similar advances have been demonstrated for the LBO OPO that covers almost the same spectral range [11.1669, 1682–1688]. The power densities at threshold for the 355 nm-pumped LBO OPO are typically higher by a factor of two compared to those of equivalent BBO OPOs due to the lower effective nonlinear coefficient. However, LBO possesses a smaller spatial walk-off angle than BBO [$\rho(\text{LBO}) \approx 0.25\rho(\text{BBO})$] allowing the use of longer LBO crystals to compensate the lower nonlinearity. The experimental comparisons of LBO and BBO [11.1669, 1686] demonstrate the advantageous properties of LBO as there are large angular acceptance, low spatial walk-off (which can even be zero for temperature-tuned noncritical phase matching). These properties are valuable if the OPO is to be operated with low-pump-energy laser sources (< 3 mJ) [11.1689], whereas BBO with restricted angular acceptance and larger nonlinearity is more suitable for high-pump-energy laser sources (> 100 mJ).

In the past many investigations were performed primarily for the type I critically phase matched BBO OPO pumped by the third harmonic of a Nd:YAG laser. This well-established standard OPO consists typically of a 12–15 mm-long BBO crystal placed in a linear signal resonant Fabry–Pérot resonator. Changing the phase-matching angle in the range of 10° the OPO covers a tuning range of 400–710 nm for the signal wave and 710–3100 nm for the idler wave. Experimentally the idler tuning is restricted by an increase of absorption in the crystal for wavelengths longer than 2600 nm. The pump energy density at threshold is about 200 mJ/cm^2. Pumped with 2–10 ns-long pulses such OPO devices operate routinely with conversion efficiencies in the range of 20–50% whereby the optical damage threshold of the OPO components limits the pump ratio. The spectral bandwidth of the type I phase-matched 355 nm-pumped BBO OPO varies from 0.2 nm in the blue spectral range ($\lambda = 400$ nm) to more than 10 nm at degeneracy ($\lambda = 710$ nm). The spectral bandwidth of the idler wave equals (expressed in wavenumbers or GHz) that of the corresponding signal frequency.

In the past, many efforts have been made to reduce the spectral bandwidth of OPOs. Significant line narrowing, even to single-mode operation, can be achieved with the insertion of frequency selective elements (such as gratings or etalons) into the OPO resonator or with external injection seeding of the OPO. Although these methods can provide single-mode operation, the disadvantages are obvious. First, the addition of frequency-selective elements increases the resonator length, the cavity losses and thus the oscillation threshold. Therefore, higher pump densities are required and a compromise is needed between the necessary pump power and the damage threshold of the optical components. Second, continuous wavelength tuning may be affected by the complexity of such devices. The spectral line narrowing via injection seeding avoids these disadvantages; the threshold is even reduced, because of a shorter rise time [11.1690]. Nevertheless, successful injection seeding in pulsed OPOs depends on many fundamental parameters of the OPO operation. Besides proper collinear alignment and divergence adaptation between the seed radiation and the seeded OPO wave, the frequency matching between the cavity mode and the seed frequency is essential. Despite these difficulties successful line narrowing was demonstrated with both methods [11.1675, 1681, 1691–1693]. A theoretical description of the spectral characteristic of free-running or injection-seeded OPO systems are given in several numerical models [11.1669, 1670, 1694, 1695].

The investigation and the control of the spatial beam quality of nanosecond OPOs is still a challenge. As a result of large pump-beam diameters, high nonlinear gain in the crystal and a large Fresnel number of the optical resonator the beam quality factor M^2 of the generated radiation are often high > 2–10 and thus far from the diffraction limit ($M^2 = 1$). Most of the numerical models simulating the spatial behavior of nanoseconds OPO are time integrated [11.1696–1699]. Recently time-resolved experimental and numerical investigation of the spectral and spatial dynamics of nanosecond OPO has been reported [11.1700, 1701]. The result of the different investigations is, that the OPO starts on the optical axis with an almost Gaussian beam profile. However, during the buildup of the pulse the fields undergo an inhomogeneous gain caused by the birefringence walk-off in the OPO crystal, pump depletion and back-conversion and as consequence the M^2 value increases with the size of the pump energy, pump depletion and back-conversion.

Some investigations of the spectral and temporal dynamics of the OPO has been performed using a BBO OPO, which consists of a 2–3 mm-long crystal in a 3.5 mm-long cavity [11.1670, 1701]. Due to the wide mode spacing of about 1 cm^{-1} each single mode can be considered. Figure 11.197 shows the statistical mode fluctuation of 10 successive pulses. Thus, the short-cavity OPO proofs the macroscopic manifestation of the zero-point fluctuation of the vacuum.

Based on the numerical and experimental results some improvement on the spectral and spatial performance of nanosecond OPOs were developed based on novel pumping and resonator configurations [11.1621, 1702–1708].

For the generation of wavelengths up to 5 μm the appropriate crystals are KTP, KTA and $KNbO_3$ [11.1709–1712]. Further extension to wavelengths beyond 5 μm can be achieved in $ZnGeP_2$, CdSe and $AgGaSe_2$ [11.1713–1717]. Avoiding two-photon absorption in the materials these crystals have to be operated with pump wavelengths longer than 1.5 μm. Meanwhile nanosecond OPOs are demonstrated with a tuning range up to 14 μm. They provide conversion efficiencies up to 40%, generating output powers of about 5 W or pulse energies of 5 mJ [11.1716, 1717]. The difficulty of finding the appropriate pump lasers for these materials has been overcome with cascaded OPO systems published for $AgGaSe_2$ and CdSe [11.1717–1720].

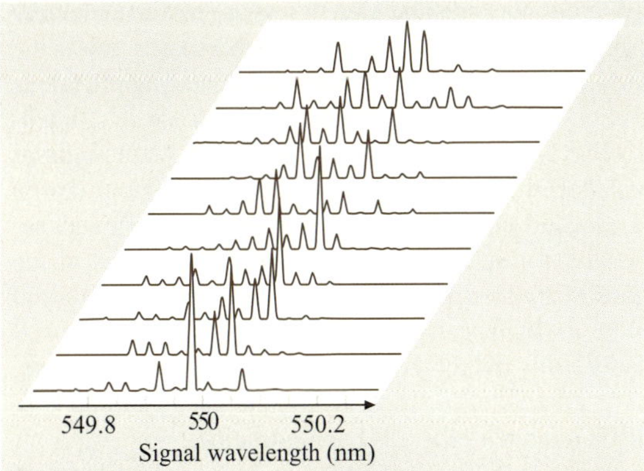

Fig. 11.197 Mode spectra of the signal wave (550 nm) recorded for 10 successive OPO pulses. The OPO is operated at 1.44 times above threshold. The mode spacing is 28.8 GHz (0.96 cm^{-1}) [11.1670]

A new generation of nanosecond OPOs appeared with the availability of periodically poled QPM materials as there are PPLN, PPKTP, PPKTA and PPRTA. These crystals have large effective nonlinearities (8–16 pm/V), up to eight times higher than in a corresponding BBO crystal, and possess long interaction lengths. Typically the aperture of the QPM crystals are 1 mm high. Therefore, these materials are well suited for the generation of mid-IR radiation pumped with high-repetition low-power-pump pulses [11.1721–1723]. With the development of wide large-aperture PPRTA crystals, QPM becomes suitable also for high-power applications [11.1724, 1725]. The large nonlinearity of PPLN enables the operation of efficient single-pass optical parametric oscillators [11.1721, 1726–1728]. More recently tunable single-frequency generation has been demonstrated with injection-seeded PPLN OPOs and PPLN OPGs [11.1729–1731].

Synchronously Pumped Optical Parametric Oscillators

Optical parametric oscillators synchronously pumped by mode-locked solid-state lasers are powerful devices for the generation of tunable ultrashort laser pulses. The first CW synchronously pumped picosecond OPO was demonstrated by *Piskarskas* et al. in 1988 [11.1732] and the first synchronously pumped femtosecond OPO by *Edelstein* et al. in 1989 [11.1733].

Many reports of such OPOs generating pulses in the picosecond or femtosecond regime have been published in recent years [11.1734–1736]. Ultrafast laser systems are attractive pump sources for optical parametric oscillators. They combine high peak pulse intensities providing sufficient nonlinear gain with a moderate energy fluence preventing optical damage of the material. However, remembering the instantaneous character of nonlinear polarization the optical gain is obtained exclusively during the time interval of the pump pulses length. In contrast to nanosecond OPOs, the time interval of ultrashort pulses ($\tau < 100$ ps) is too short to allow a finite number of cavity round-trips. Therefore, the basic operation principle of ultrafast optical parametric oscillators is synchronous pumping to achieve macroscopic amplification of the parametric waves from quantum noise. In this pumping scheme the length of the optical parametric oscillator is matched to that of the pump laser resonator, so that the round-trip transit time in the OPO resonator corresponds to the repetition time of the pump pulses. The consecutive coincidence of parametric and pump wave in the nonlinear crystal amplifies the gain until it overcomes the threshold.

In general, two types of synchronously pumped OPOs are to be distinguished: the pulsed (or quasi-CW) synchronously and the continuous synchronously pumped OPOs. CW synchronously pumped OPOs are pumped with a continuous train of ultrashort pulses as they are emitted by CW mode-locked neodymium lasers or Kerr lens mode-locked Ti:sapphire lasers, while the pump sources of pulsed synchronously pumped OPOs are mode locked Q-switched Nd lasers generating a train of ultrashort pulses contained within a nanosecond or microsecond envelope.

The physical description of CW synchronously pumped OPOs is equivalent to the steady-state formalism for CW OPOs considering the peak pump pulse intensity determining the nonlinear gain. Pulsed synchronously pumped OPOs show the transient behavior of nanosecond OPOs so their analysis has to take into account the rise-time effects caused by the finite time duration of the nanosecond pulsed envelope [11.1667]. An analysis of synchronously pumped OPOs can be found for plane-wave conditions [11.1737] and for Gaussian beams in CW synchronously pumped OPOs [11.1738, 1739].

Most of the early synchronously pumped OPOs were pulsed OPOs [11.1734] because of the significant higher peak pump pulse powers compared to CW mode-locked pump lasers at that time. Picosecond optical parametric oscillation even in the SRO configuration was generated using the nonlinear materials KTP [11.1740], BBO [11.1622] and LBO [11.1741]. However due to the pulsed pump characteristic the output of such OPOs does not consist of a really repetitive pulse train, and the amplitude, intensity and the pulse duration of the output pulse can vary under the pulse envelope. In applications where no high peak powers are needed CW synchronously pumped OPO are advantageous because of their continuous train of output pulses. Meanwhile with the availability of efficient CW mode-locked lasers and more-reliable nonlinear materials the research in this field is focused on CW synchronously pumped OPOs.

In practice, synchronously pumping is a reasonable pumping mechanism only for high repetition rates (> 50 MHz) otherwise the resonator length becomes too long and unpractical. For example, typical repetition rates of Kerr lens mode-locked Ti:sapphire lasers are about 80 MHz, which corresponds to an OPO resonator

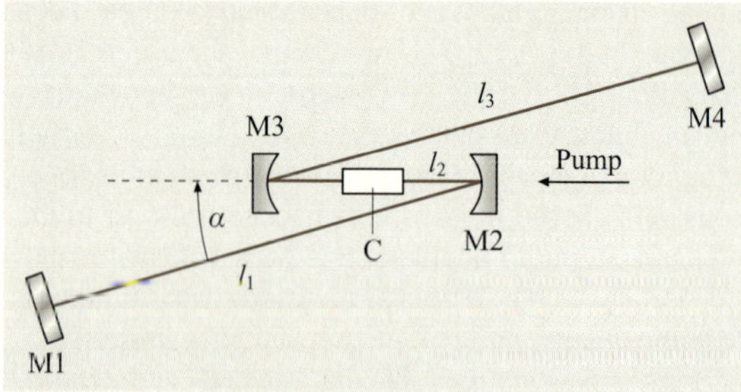

Fig. 11.198 Experimental setup for a linear standing wave resonator consisting of two plane mirrors M1 and M4 and two spherical mirrors M2 and M3. The pump laser beam is focused into the crystal C between the spherical mirrors M2 and M3. The total resonator length depends on the distance between two successively pump pulses

length of 1.85 m. The tolerance of the resonator length detuning can be calculated from $\Delta l_{\text{res}} = \Delta\tau(c/2)$, which is half the geometrical pump pulse length. The theoretical detuning calculated for 1 ps is about 150 μm, however experimentally even shorter detuning tolerances of 30 μm are affordable.

Commonly used resonator designs for synchronously pumped OPOs are ring or standing-wave resonators, where the latter is favorable for compactness reasons (its length is half that of the ring). A typical standing-wave resonator for synchronously pumped OPOs is shown in Fig. 11.198 consisting of two plane and two spherical mirrors.

Picosecond optical parametric oscillators. Picosecond optical parametric oscillators synchronously pumped by CW mode-locked solid-state lasers are powerful devices for the generation of tunable ultrashort laser pulses especially in the IR spectral region. For many applications in high-resolution spectroscopy a high spectral intensity and a narrow bandwidth is required. As the light of such OPOs is almost Fourier limited ($15\,\text{ps} = 1\,\text{cm}^{-1}$) picosecond pulses (with $t > 10\,\text{ps}$) have a sufficient selectivity for spectroscopy, in contrast to pulses in the femtosecond regime. Similar to the nanosecond pulsed OPOs the high peak powers of modern CW mode-locked lasers easily overcome the threshold of SRO resonators. As a consequence, most synchronously pumped CW OPOs operate in the SRO configuration.

In the past, there have been various ways to generate tunable picosecond IR radiation by CW synchronously pumped OPOs. In most synchronously pumped picosecond OPOs, the use of noncritical phase matching prevents a reduction of interaction length due to walk-off between the focused beams inside the nonlinear crystal. Wavelength tuning of noncritical OPOs requires a variation of the pump wavelength or the crystal temperature.

A temperature-tuned picosecond LBO OPO has been reported [11.1742] generating continuously tunable radiation between 650 nm and 2.5 μm with a single crystal. The average output power achieved by this system was below 100 mW. By using tunable pump laser systems like the Kerr-lens mode-locked Ti:sapphire laser pump wavelength tuning offers a convenient tuning mechanism.

A noncritical-phase-matched picosecond KTP OPO [11.1743] generates tunable picosecond pulses in the spectral ranges of 1–1.2 μm for the signal and 2.3–2.9 μm for the idler wave via wavelength tuning of the pump between 720–853 nm. Average output powers of up to 700 mW in 1.2 ps pulses has been achieved with a pump power of 1.6 W at a 82 MHz repetition rate.

The combination of wavelength tuning of the Ti:sapphire laser and temperature tuning in LBO has been performed under the type I noncritical-phase-matching condition, providing continuously tunable picosecond pulses in the spectral range of 1–2.4 μm [11.1744]. The pump power was 1.2 W in 1–2 ps pulses at a 81 MHz repetition rate and generated an average OPO output power of about 325 mW. Using the arsenate isomorphs of KTP, such as KTA and RTA, the wavelength range of the OPO can be extended up to 3.6 μm [11.1745]. Due to their wider transparency range and the lack of grey tracking inside the crystal these materials are advantageous for IR picosecond OPOs.

Further extension of the tuning range of picosecond OPOs into the mid-IR was demonstrated with a critical-phase-matched $AgGaS_2$ OPO pumped by a Q-switched mode-locked Nd:YAG laser [11.1746]. This OPO had an idler tuning range of 3.5–4.5 μm and a maximum output power of 2 W.

Picosecond pulse generation in the visible spectral range has been demonstrated in an LBO OPO pumped with the third harmonic of a CW mode-locked Nd:YLF laser [11.1747]. The generated output power was about 275 mW in the spectral range of 453–472 nm in 15 ps-long pulses at a repetition rate of 75 MHz.

In general, it should be pointed out that the device Ti:sapphire-laser-pumped KTP or KTA OPO is a versatile tool for the generation of tunable picosecond pulses in the entire spectral region from the UV to the near-IR. The frequency-conversion schemes involved, such as second, third, and fourth harmonics of the Ti:sapphire laser, optical parametric oscillation in KTP and frequency doubling of the OPO's signal wave, are summarized in experimental investigations for picosecond and femtosecond radiation [11.1748, 1749].

High-power operation of a singly resonant picosecond OPO has been reported pumping a critical-phase-matched KTP OPO with the fundamental of an mode-locked Nd:YLF laser at 1.053 μm [11.1750]. This device generates 12 ps-long signal pulses at 1.55 μm for 40 ps-long pump pulses at a repetition rate of 76 MHz. A total average output power of 2.8 W was achieved with 800 mW in the idler wave at 3.28 μm.

The new generation of mode-locked lasers are diode-pumped CW mode-locked $Nd:YVO_4$ oscillator–amplifier laser systems. Such an all-solid-state laser system operating at 1.064 μm with 7 ps pulse lengths at repetition rates of about 83 MHz delivers average output powers of 29 W. Pumping a noncritical-phase-matched KTA OPO the OPO output power reaches the multiwatt region providing a signal wavelength of 1.53 μm with an output power of 14.6 W and an idler output power as high as 6.4 W at 3.47 μm in the mid-IR [11.1751]. The total average output power of 21 W corresponds to an external efficiency of 70%. However, both systems operate at a specific signal and idler wavelength without any tunability.

With the availability of QPM nonlinear materials like PPLN, PPKTP, PPRTA new perspectives have arisen in the research of efficient tunable picosecond OPOs. With the advantageous properties of these materials as their large nonlinearities, transparency in the mid IR and the capability to choose an optimized phase-matching condition for the interacting waves, for example noncritical phase matching, they become ideal candidates to realize singly resonant synchronously pumped OPOs with low thresholds, large tunability and efficient output powers. Compact all-solid-state devices have been reported based on PPLN and PPRTA with a pump power threshold as low as 10 mW [11.1752, 1753]. The generated radiation covers the wavelength range from 3.35–5 μm with total output powers of up to 400 mW in 1–5 ps pulses. The power of the idler wave was up to 100 mW. The spectral range of the PPLN OPO has been extended even beyond 5 μm [11.1754] and idler output powers of 0.5 mW are achieved. The operation of a PPLN OPO at GHz repetition rates was demonstrated by pumping the PPLN OPO with an all-solid-state neodymium laser at 10 GHz [11.1755]. Rapid wavelength access was achieved with the use of a mode-locked diode laser master-oscillator power-amplifier (MOPA) system providing 7.8 ps pulses at 2.5 GHz repetition rate. The InGaAs oscillator–amplifier system delivered an output power of 900 mW, which was converted to 78 mW idler power tunable in the spectral range of 2.2–2.8 μm [11.1756].

A high-power picosecond PPLN-OPO was performed by synchronously pumping a PPLN crystal with the fundamental wave of a CW mode-locked Nd:YLF laser. The tuning range of the PPLN OPO is for the signal radiation 1.765–2.06 μm and 2.155–2.61 μm for the idler radiation, respectively. High average output powers for the signal and idler radiation of 2.55 W and 2.4 W have been achieved in 45 ps-long pulses at a 76 MHz repetition rate for an input pump power of 7.4 W, which corresponds to a total external conversion of about 67%. The measured pump depletion was 71% [11.1757].

Much emphasis is placed on further extension of the tuning range of picosecond systems into the IR. For this purpose OPO/OPA and OPG/OPA devices are of great interest [11.1758, 1759] as their output powers and tuning ranges are sufficient to be used for difference-frequency mixing into the mid-IR. The difference frequency mixing in GaSe or CdSe [11.1760–1762] of signal and idler frequencies generated from an PPLN OPO seeded KTP-OPA provides picosecond pulses tunable at 3–24 μm.

Femtosecond optical parametric oscillators. Tunable femtosecond OPOs are attractive light sources for time-resolved spectroscopy of chemical and biological reactions. Femtosecond pulses are characterized by their high peak powers, which easily overcome the threshold of a singly resonant OPO configuration. The short pulse length of about 100 fs is combined with a huge spectral bandwidth of several nanometers ($\approx$ 10 nm) and dispersion effects, such as group velocity dispersion, group velocity mismatch (i. e., temporal walk-off) and spectral acceptance bandwidths can no longer be neglected. The group-velocity dispersion causes a pulse broadening whereas the temporal walk-off lowers the nonlinear gain and/or changes the time characteristic of output pulse. For efficient frequency conversion, the temporal walk-off has to be minimized by an appropriate choice of the crystal length. Typical crystal lengths of femtosecond OPOs are only 1–2 mm, and therefore high nonlinearities are desirable for low operation thresholds. The group velocity dispersion can be minimized via dispersion compensation in the OPO cavity. Additionally to these dispersion effects, other higher-order nonlinear effects come with the high peak pump powers, such as self-phase modulation and cross-phase modulation which result in a chirped OPO output pulse [11.1613].

The start of an extensive development of femtosecond OPOs was initiated with the commercial availability of the Kerr-lens mode-locked Ti:sapphire laser. The first material used in synchronously pumped femtosecond OPOs was KTP [11.1763–1765] in non-collinear critical-phase-matching conditions and collinear noncritical-phase-matching condition. Due to the higher transmission in the mid-IRA OPOs with KTA [11.1766, 1767], RTA [11.1768], $CsTiOAsO_4$ (CTA) [11.1769, 1770], and $KNbO_3$ [11.1771] generate a wider wavelength range of 1–5 μm with pulse durations of less than 100 fs. Typical maximum output powers of such devices are 100–200 mW for the signal and 50–100 mW for the idler wave.

The periodically poled nonlinear materials PPLN, PPRTA, and PPKTP are of particular importance for femtosecond OPOs because noncritical phase matching becomes possible even for the extended wavelength tuning. This feature results in compact, low-threshold, efficient OPO devices. The wavelength tuning of such femtosecond QPM OPOs is performed via grating, pump wavelength or cavity tuning and covers the spectral range from 1.7–5.4 μm in PPLN [11.1772], 1.06–1.22 μm (λ_s) and 2.67–4.5 μm (λ_i) in PPRTA [11.1773], and 1–1.2 μm in PPKTP [11.1774]. Comparing these QPM materials, PPLN and PPRTA have an advantage because of their large transmission range, where PPLN possesses the higher nonlinearity. However PPLN suffers from process-induced photorefractive defects, which may be suppressed by heating the crystal to temperatures higher than 100 °C. Recently MgO-doped PPLN-OPO was investigated [11.1775] allowing reliable OPO operation at room temperature.

Tunable radiation in the range of 4–8 μm have been achieved using a cascaded $AgGaSe_2$ OPO, which has been pumped with the idler of a CTA OPO. This system generates average powers of 35 mW in 300–600 fs-long pulses at a repetition rate of 82 MHz.

Moreover, optical parametric generators and amplifiers are powerful tools for the generation of ultrashort pulses into further infrared regions [11.1776]. The generation of femtosecond pulses in the spectral range 3–12 μm with almost bandwidth limited pulses of 100–200 fs have been reviewed for different parametric processes and a number of relevant materials [11.1777].

Many new perspectives for the generation of intense ultrashort pulses are given by parametric amplification of chirped laser pulses. The basic principle of optical parametric chirped pulse amplification (OPCPA) is that stretched linearly chirped pump pulses are amplified in a non-collinear phase-matched OPA. After the single-pass amplification, the pulses are compressed, providing high energy in short pulse lengths of 10 fs or less. The non-collinear phase-matching scheme is important because of the huge spectral bandwidths of femtosecond pulses. For ideal OPCPA the pump pulse duration corresponds to the length of the chirped pulse. The spatial beam profile should be rectangular (flat-top), and the gain should be constant temporally and spatially. *Dubietis* et al. [11.1776] published the first proposal and demonstration of OPCPA. The first experimental demonstrations of OPCPA were performed in degenerate LBO, KDP and BBO [11.1778–1782]. Amplification of 10^{10} have been achieved in single-pass amplification at degeneracy.

The use of non-collinear phase matching enables the efficient parametrical amplification of nondegenerate wavelengths despite large spectral bandwidth. Recently different sources of supercontinuum have been amplified in BBO and other crystals via frequency-doubled Ti:sapphire lasers with 1 kHz repetition rate [11.1783, 1784]. Ultrashort pulses below 6 fs have been generated, however the achieved pulse energies was lower than 5 μJ.

The generation of higher energies is possible by employing visible picosecond pump pulses and temporally stretched seed pulses. At a center wavelength of 800 nm and a pump wavelength of about 530 nm in BBO and LBO gain bandwidths broader than 2000 cm^{-1} are obtained, which is sufficient to amplify Fourier-limited pulses with 5 fs duration [11.1778]. Due to the large damage thresholds of both materials, high pump intensities can be applied. Recently OPCPA was performed at 800 nm and with a Nd:YAG laser at 532 nm, generating a gain of 4×10^6 in two BBO crystals [11.1785]. The pulse energy was 2 mJ at a 10 Hz repetition rate.

QPM materials also turn out to be advantageous in OPCPA. Broad-bandwidth amplification is not restricted any longer to collinear phase matching at degeneracy or to non-collinear phase matching but can be obtained at practically any center wavelength in the crystals transparency window. Additionally QPM OPA can be increased by using engineered grating profiles, such as chirped gratings [11.1625], fanned gratings [11.1626], and aperiodic gratings [11.1786].

Nondegenerate OPCPA has been demonstrated in PPKTP at a wavelength of 1573 nm pumped by a Nd:YAG laser [11.1787] and close to degeneracy at a wavelength of 1053 nm pumped by a Nd:glass laser [11.1788].

This chapter describes the basic properties and current performance of CW, nanosecond and ultrashort optical parametric oscillators. Today these parametric systems are powerful tunable light sources providing a spectral range from the visible to the mid-IR in all time regimes. The main progress in this field is based on the development of more-sophisticated optical materials and efficient, reliable laser systems. With the realization of QPM materials there was an important breakthrough towards compact widely tunable OPOs. Especially for CW OPOs, QPM materials are the key element for the realization of widely tunable singly resonant OPOs. The high nonlinear gain of QPM materials enables the reduction of the optical parametric oscillation to optical parametric generation, which leads to more-simple devices. The combination of diode-pumped solid-state laser systems and QPM crystals, which already leads to direct diode-pumped OPOs, is the way to future, compact tunable laser systems that may be realized in integrated designs.

Finally it should be mentioned that, besides the importance of the generated OPO radiation for many spectroscopic and technical applications, the OPO itself is still an interesting physical system to be explored in the field of quantum mechanics.

11.10 Generation of Coherent Mid-Infrared Radiation by Difference-Frequency Mixing

The fundamental infrared (IR) or mid-IR region of the electromagnetic spectrum between, say 3–20 μm, is of special interest for many applications, notably molecular spectroscopy [11.1789]. The reason is that most organic and inorganic molecules exhibit strong vibrational–rotational transitions in this wavelength region. This is illustrated in Fig. 11.199, where the absorption ranges of some important functional groups of molecules are plotted for wavelengths of 2–20 μm (top), whereas absorption features of a few selected molecules are depicted in the center. In view of applications in air monitoring the (relative) transmission through the terrestrial atmosphere is plotted at the bottom of Fig. 11.199.

This atmospheric absorption is obviously dominated by the presence of water vapor and CO_2. The atmospheric windows at 3–5 μm and 8–14 μm are clearly visible.

In addition to the ultraviolet–visible and the near-IR range, the mid-IR region has attracted a lot of interest in recent times owing to the numerous applications of

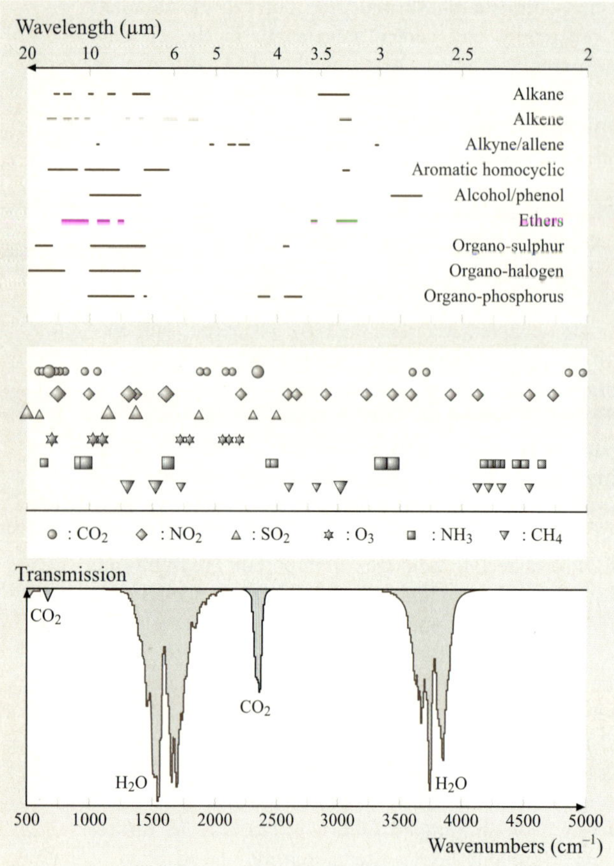

Fig. 11.199 Absorption ranges of important molecular functional groups (*top*) and of selected molecules (*center*). The size of the symbols is related to the absorption strength. Atmospheric transmission is depicted at the *bottom*

trace-gas sensing in rather diverse fields including environmental monitoring and climate research, industrial and process surveillance, workplace safety, agriculture (e.g., ethylene or ammonia detection, surveillance of fermentation), homeland security (e.g., chemical warfare or explosives detection) and medical diagnosis (e.g., human breath monitoring). It has been demonstrated in numerous cases that laser spectroscopy offers the potential for high sensitivity and selectivity, multicomponent capability, and large dynamic range. These are all crucial properties for trace-gas sensing. An important aspect of laser-based sensors is that usually no sample preparation, i. e., neither pretreatment nor pre-concentration, are required, in contrast to many competing techniques. Apart from an appropriate detection scheme, which enables sensitive absorption measurements, a mid-infrared tunable laser source is a prerequisite for successful trace-gas sensing and analysis. In addition to tunability, a narrow line width guarantees sufficient specificity of detection in multicomponent real-world gas mixtures. In the preferred mid-IR range, the choice of coherent sources is, however, rather limited. Table 11.49 lists tunable lasers with wavelengths longer than 3 μm. It essentially includes the conventional well-established line-tunable CO and CO_2 gas lasers, continuously tunable semiconductor lasers (lead-salt diode and quantum cascade lasers), color center lasers, novel crystalline solid-state lasers and nonlinear optical devices [optical parametric oscillators (OPOs) and difference-frequency generation (DFG) sources].

Recent progress both in solid-state lasers and in quantum cascade lasers (QCLs) appears promising. Solid-state laser materials such as Ce^{2+}:ZnSe or Fe^{2+}:ZnSe offer tuning ranges of 2.2–3.1 μm [11.1790] or 4–4.5 μm, respectively (though the latter only when pulsed and with cryogenic cooling) [11.1791]. New developments in QCLs equipped with external cavities yield a continuous tuning range of around 10% of the central wavelength and – at least partly – room-temperature continuous-wave operation [11.1792, 1793]. Despite these attractive prospects the broadest tuning range, best wavelength coverage, room-temperature operation and highest flexibility in terms of wavelength selection is currently still achieved with nonlinear optical devices (DFG and OPOs) [11.1794]. Here we focus on DFG as the scheme that has been widely used in recent times to access mid-IR wavelengths for trace-gas sensing. In fact, these important applications have fostered further developments of DFG systems.

11.10.1 Difference-Frequency Generation (DFG)

Difference-frequency generation (DFG) represents a nonlinear optical effect that is related to the nonlinear susceptibility of second order ($\chi^{(2)}$) of a material. Other related effects are second-harmonic generation (SHG) and sum-frequency generation (SFG). Hence, DFG represents a three-beam interaction process that is mostly used to generate tunable coherent mid-infrared radiation, although it has also been employed for THz generation.

Birefringent, Quasi-Phase Matching and Conversion Efficiency

In difference-frequency generation the light of two laser beams is mixed in a nonlinear crystal and light with a frequency of the difference of the two incident fre-

Table 11.49 Tunable continuous-wave mid-IR laser sources with wavelengths $\geq 3\,\mu m$. RT: room temperature, LN_2: liquid nitrogen (77 K), TE: thermoelectric cooling, SHG: second-harmonic generation, QCL: quantum cascade laser, OPO: optical parametric oscillator, DFG: difference-frequency generation

Laser	Wavelength (μm)	Tuning characteristics	Power	Operation
CO	5–6 (2.7–4, overtone)	only line tunable	50 mW to W	LN_2 cooling, also $\leq 0\,°C$
CO_2	9–11 (4.5–5.5, SHG)	only line tunable	W	RT operation
Lead salt diode	4–30	$\approx cm^{-1}$ mode hop-free	< 0.1 mW	Cryogenic cooling
QCL	4–> 24, THz	cm^{-1} to > 100 cm^{-1} per device	mW	LN_2/TE cooling, also RT
Color center	1–3.3	≈ 0.5 μm for single crystal	100 mW	LN_2 cooling
Solid-state	2.2–3.1	≈ 0.5–1 μm	≤ 1 W	RT operation
OPO	3–16	~ μm for specific setup	≤ 1 W	RT operation
DFG	3–16	~ μm for specific setup	μW to mW	RT operation

quencies is generated. The generated frequency is given by energy conservation:

$$\hbar\omega_p - \hbar\omega_s = \hbar\omega_i \rightarrow \omega_i = \omega_p - \omega_s \,. \tag{11.187}$$

The laser with the highest frequency ω_p is called the pump laser, while the second laser is called the signal laser with a frequency ω_s. The generated idler beam has the lowest frequency ω_i. The phase-matching condition is given by the conservation of momentum:

$$\Delta \boldsymbol{k} = \boldsymbol{k}_p - \boldsymbol{k}_s - \boldsymbol{k}_i = 0 \,, \tag{11.188}$$

where $\Delta \boldsymbol{k}$ is the phase mismatch, and $\boldsymbol{k}_p$, $\boldsymbol{k}_s$, and $\boldsymbol{k}_i$ are the wavevectors of the pump, signal, and idler beam, respectively. In the case of collinear wave propagation the wavevectors can be replaced by $|\boldsymbol{k}| = nc/\lambda$ and (11.188) changes to:

$$\frac{n_p}{\lambda_p} - \frac{n_s}{\lambda_s} - \frac{n_i}{\lambda_i} = 0 \,, \tag{11.189}$$

where n is the refractive index at the corresponding wavelength, λ is the wavelength with the subscripts p, s, and i corresponding to pump, signal and idler, respectively, and c is the speed of light in vacuum. Phase matching in a birefringent crystal can be achieved by

1. angle tuning
2. temperature tuning
3. wavelength tuning (changing the wavelength of the pump and/or signal laser).

The most common way to achieve phase matching with a birefringent crystal is realized by angle tuning, i. e., by rotating the crystal until the phase-matching condition is fulfilled.

For infinite plane waves the idler intensity is given by [11.1795]:

$$I_i = 2\frac{\omega_i^2 d_{eff}^2 L^2 I_s I_p}{c^3 \varepsilon_0 n_p n_s n_i} \mathrm{sinc}^2\left(\frac{\Delta k L}{2}\right) \tag{11.190}$$

where d_{eff} is the effective nonlinear coefficient, L is the crystal length, I is the intensity of the laser beam for pump, signal and idler beams, respectively, ε_0 is the dielectric constant, and Δk is the phase mismatch. Hence the idler intensity scales with the product of the incident intensities $I_s I_p$ and the square of the crystal length L^2. This is valid as long as there is no pump depletion and negligible walk-off between the beams. These effects limit the useful crystal length and they are discussed in more detail in the section on nonlinear crystals. The effective nonlinear coefficient d_{eff} describes the nonlinearity of the crystal seen by the incident light and depends on the crystal structure, the direction of propagation, the polarization of the light, and the nonlinear coefficients d_{ij} (Table 11.51) given by the tensor of the nonlinear susceptibility $\boldsymbol{\chi}^{(2)} = 2\mathbf{d}$.

To achieve phase matching it is taken advantage of the different refractive index for ordinarily and extraordinarily polarized light. Depending on the combination

Table 11.50 Different phase matching types for DFG processes for positive and negative birefringent crystals. PM: phase matching, e: extraordinarily polarized, o: ordinarily polarized, QPM: quasi-phase matching

PM type	Birefringence	Pump beam	Signal beam	Idler beam
I	Positive	o	e	e
I	Negative	e	o	o
II	Positive	o	e	o
II	Negative	e	o	e
III	Positive	o	o	e
III	Negative	e	e	o
QPM		e	e	e

Table 11.51 Nonlinear coefficients d_{ij} measured at wavelength λ_m, transparency range and approximate damage threshold for ns pulses of selected nonlinear optical crystals used in DFG

Crystal	Nonlinear coefficients (pm/V) [11.1796]			λ_m (μm)	Transparency range (μm) [11.1797]	Damage threshold pulsed (MW/cm^2) [11.1797, 1798]
$AgGaS_2$	$d_{14} = 57$	$d_{36} = 20$ $d_{36} = 23.6$		10.6 1.064	0.46–13	25
$AgGaSe_2$	$d_{36} = 33$			10.6	0.7–19	25
$Ba_2NaNb_5O_{15}$	$d_{31} = 12$ [11.1797]	$d_{32} = 12$ [11.1797]	$d_{33} = 16.5$ [11.1797]	1.064	0.37–5	4
$CdGeAs_2$	$d_{36} = 235$			10.6	2.4–18	20–40 [11.1799]
CdSe	$d_{15} = 18$	$d_{31} = -18$ [11.1800]	$d_{33} = 36$ [11.1800]	10.6	0.57–25	60
$CsTiOAsO_4$ (CTA)	$d_{31} = 2.1$	$d_{32} = 3.4$	$d_{33} = 18.1$	1.064	0.35–5.3	500 [11.1801]
GaAs	$d_{14} = 368.7$	$d_{36} = 83$ [11.1800]		10.6	1–17 [11.1799]	60 [11.1799]
GaSe	$d_{22} = 54.4$			10.6	0.62–20	30
$HgGa_2S_4$	$d_{36} = 26$	$d_{31} = 6.7$ [11.1797]		1.064	0.55–13	60
$KNbO_3$	$d_{15} = -17.1$ $d_{32} = -15.8$	$d_{24} = -16.5$ $d_{33} = -27.4$	$d_{31} = -18.3$	1.064	0.4–> 4	180
$KTiOAsO_4$ (KTA)	$d_{31} = 4.2$ $d_{24} = 2.9$ [11.1800]	$d_{32} = 2.8$	$d_{33} = 16.2$	1.064	0.35–5.3	1200
$KTiOPO_4$ (KTP)	$d_{15} = 1.91$ $d_{32} = 4.53$	$d_{24} = 3.64$ $d_{33} = 16.9$	$d_{31} = 2.54$	1.064	0.35–4.5	150
LiB_3O_5 (LBO)	$d_{24} = 0.74$ $d_{15} = 1.03$ $d_{32} = -10$	$d_{31} = 0.8$–1.3 $d_{24} = -0.94$ $d_{33} = -0.94$	$d_{33} = 0$ $d_{31} = 1.09$	1.064 1.079	0.155–3.2	900
$LiIO_3$	$d_{15} = -5.53$	$d_{31} = -7.11$	$d_{33} = -7.02$	1.064	0.28–6	120
$LiInS_2$	$d_{31} = 9.9$	$d_{32} = 8.6$	$d_{33} = 15.8$	10.6	0.35–12.5 [11.1802]	100 [11.1802]
$LiNbO_3$	$d_{31} = -5.95$	$d_{33} = -34.4$ $d_{33} = -27$ [11.1800]		1.064	0.4–5.5 [11.1803]	300
	$d_{31} = -5.77$	$d_{33} = -33.4$		1.150		
	$d_{31} = 3.77$	$d_{33} = -31.8$		1.318		
			$d_{32} = -29.1$	2.120		
	$d_{15} = -5.95$ [11.1803]			–		
	$d_{22} = 3.07$ [11.1803]			–		
$LiTaO_3$	$d_{22} = 2$ [11.1802]	$d_{31} = -1$ [11.1802]	$d_{33} = -21$ [11.1802]	1.064	0.4–5 [11.1802]	–
$RbTiOAsO_4$ (RTA)	$d_{31} = 3.8$	$d_{32} = 2.3$	$d_{33} = 15.8$	1.064	0.35–5.3 [11.1804]	400 [11.1804]
$RTiOPO_4$ (RTP)	$d_{31} = 4.1$	$d_{32} = 3.3$	$d_{33} = 17.1$	1.064	0.35–4.3 [11.1804]	600 [11.1804]
$ZnGeP_2$	$d_{36} = 75$	$d_{14} = 69$ [11.1800]	$d_{25} = 69$ [11.1800]	10.6	0.74–12	3

of polarization of the pump and signal beam and depending on positive or negative birefringence of the crystal, the phase matching is called type I, type II or type III, as defined in Table 11.50. For quasi-phase matching (see below) all three beams have the same polarization.

Equation (11.190) implies a possibility to generate light at the idler frequency without phase matching. I_i has a maximum at $\Delta kL/2 = m\pi/2$, where m is a natural number. This yields the coherence length $l_c = \pi/\Delta k$ after which the newly generated idler light will interfere destructively with the light generated within the previous coherence length. Thus, after twice the coherence length all generated light is destroyed. This can be avoided by changing the polarization of the material by 180° after one or an odd multiple of the coherence length l_c because this will change the sign of the nonlinear coefficient, so the light will interfere constructively and a buildup of the generated light occurs, as illustrated in Fig. 11.200. This means that idler power can be generated without fulfilling the phase-matching condition by using a periodically poled crystal. A quasi-phase matching (QPM) condition can be written by using the grating period $\Lambda = 2l_c$ of the crystal:

$$\frac{n_p}{\lambda_p} - \frac{n_s}{\lambda_s} - \frac{n_i}{\lambda_i} - \frac{1}{\Lambda} = 0\,. \tag{11.191}$$

For the same parameters as for the bulk material less idler power will be generated for quasi-phase matching than for birefringent phase matching by a factor of $(2/\pi)^2$. This factor is often included in d_{eff} by reducing its value by a factor $2/\pi$. Although the efficiency is lower than in the case of birefringent phase matching, more power can often be generated this way because the highest nonlinear coefficient can be used [e.g., for $LiNbO_3$ $d_{33} = -27\,pm/V$ (Table 11.51) instead of $d_{22} = 3.07\,pm/V$, which is relevant for birefringent type II phase matching]. Since all involved polarizations are the same (extraordinarily polarized, Table 11.50), the walk-off angle is zero, therefore longer crystals can be used, resulting in higher idler power. Furthermore, the selection of pump and signal wavelengths is rather flexible. However, the production of periodically poled crystals is difficult and not possible for all crystals. Until now only $LiNbO_3$ (PPLN), $RbTiOAsO_4$ (PPRTA), $KTiOPO_4$ (PPKTP), and $KTiOAsO_4$ (PPKTA) are commercially available periodically poled. These are ferroelectric crystals that are poled by applying a strong electrical field. Another, non-ferroelectric material is GaAs. In the beginning thin plates of GaAs with alternate polarization were stacked manually [11.1805]. Today orientation patterned GaAs, where the periodic poling is implemented during crystal growth [11.1806–1808], is becoming available.

Equation (11.190)) is obtained under the assumption of infinite plane waves and no absorption. For Gaussian beams the relationship between idler power P_i, crystal length L, pump power P_p, signal power P_s, and absorption coefficient α is given by [11.1809–1811]:

$$P_i = P_p P_s \frac{32\pi^2 d_{eff}^2 L}{\varepsilon_0 c n_i \lambda_i^2 (n_s\lambda_p + n_p\lambda_s)} h(\xi, \sigma, \mu, \alpha, L)\,. \tag{11.192}$$

The focusing function $h(\xi, \sigma, \mu, \alpha, L)$ for diffraction-limited Gaussian beams is given by:

$$h(\xi, \sigma, \alpha, L) = \mathrm{Re}\left(\frac{e^{-\frac{\alpha L}{2}}}{4\xi} \times \int_{-\xi}^{\xi} d\tau \int_{-\xi}^{\xi} d\tau' \frac{e^{-i\sigma(\tau-\tau') + \frac{\alpha L}{4\xi}(\tau+\tau')}}{1 + \tau\tau' - i\frac{1+\mu^2}{1-\mu^2}(\tau - \tau')}\right), \tag{11.193}$$

where

$$\xi = \frac{L}{b}, \tag{11.194}$$

$$\mu = \frac{k_s}{k_p} = \frac{n_s\lambda_p}{n_p\lambda_s}, \tag{11.195}$$

$$\sigma = -\pi b\left(\frac{n_p}{\lambda_p} - \frac{n_s}{\lambda_s} - \frac{n_i}{\lambda_i} - \frac{1}{\Lambda}\right). \tag{11.196}$$

Here Λ is the grating period, and b is the confocal parameter of both the pump and signal beam and is given by the minimal beam waist $w: b = k_p w_p^2 = k_s w_s^2$, σ describes the phase mismatch. The focusing function h is discussed further below (see also Figs. 11.204, 205, 206 and 11.208). These equations are valid for both bulk and periodically poled crystals, only d_{eff} changes by a factor $2/\pi$ compared to the case of a bulk crystal. The focusing function describes two competing effects: the efficiency is increased by focusing the beams because of higher intensities, but at the same time reduced because of less collinear wavevectors. A possibility to overcome this problem is to use waveguide periodically poled nonlinear crystals [11.1812]. Here the beams are confined in a waveguide in the crystal leading to collinear wavevectors and high intensities at the same time. With this

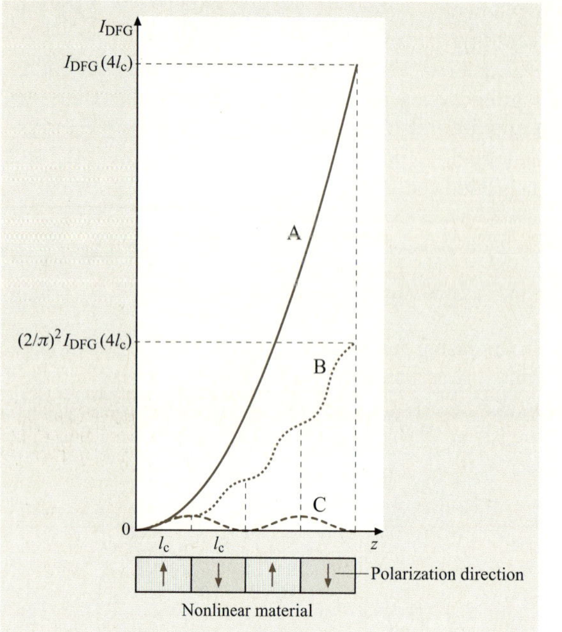

Fig. 11.200 Idler power generation for phase-matching condition $\Delta k = 0$ (A), non-phase-matching condition $\Delta k \neq 0$ (C) and quasi-phase-matching condition in the case of a periodically poled crystal (B). l_c denotes the coherence length [11.1813]

technique continuous-wave (CW) idler powers in the milliwatt range can be achieved whereas with bulk or periodically poled crystals the CW idler power is normally in the microwatt range.

The limit $\xi \to 0$ gives the result for the infinite plane wave because $b \to \infty$ for plane waves. In this case the focusing function reduces to $h \sim \xi$, resulting in an idler power proportional to L^2, as in the case for the plane waves (11.190).

An often used nonlinear crystal for difference-frequency generation in the mid-infrared is $LiNbO_3$ because it can be made periodically poled (known as PPLN) and has a large nonlinear coefficient. Its effective nonlinear coefficient is given by $d_{eff} = 2/\pi d_{33} M_{ij} = -14.4$ pm/V with $d_{33} = -27$ pm/V and $M_{ij} = 0.85$ for $LiNbO_3$ at 3.3 μm [11.1814]. M_{ij} is the Miller factor, which describes the dispersion of the nonlinear coefficient [11.1800, 1815, 1816]. The factor $2/\pi$ is needed when using quasi-phase matching.

To calculate the phase-matching condition the refractive index needs to be known for all wavelengths. It is given by the temperature-dependent Sellmeier equation, e.g., for extraordinary polarized light in $LiNbO_3$ [11.1817]:

$$n_e^2 = 5.35583 + 4.629\times10^{-7}F + \frac{0.100473 + 3.862\times10^{-2}F}{\lambda^2 - (0.20692 - 0.89\times10^{-8}F)^2} + \frac{100 + 2.657\times10^{-5}F}{\lambda^2 - 11.34927^2} - 1.5334\times10^{-2}\lambda^2 . \quad (11.197)$$

Here $F = (T - 24.5)(T + 570.82)$ describes the temperature dependence when T is the temperature in °C and λ is the wavelength in nm. This equation takes also the multiphonon absorption into account, which yields more-accurate data for wavelengths between 4 μm and 5 μm.

When choosing the grating period and the temperature of the crystal, the thermal expansion of the crystal also needs to be taken into account because it also influences the grating period, although much less than the change of the refractive index with temperature. The thermal expansion is (at 25 °C) $a_a = 15\times10^{-6}$ /°C and $a_c = 7.5\times10^{-6}$ /°C [11.1818] where the indices a and c correspond to the crystal axis. The refractive index of $LiNbO_3$ strongly depends on its composition, the Sellmeier equation is for congruent composition, it will be different for stoichiometric composition. Also the refractive index of MgO doped $LiNbO_3$, as often used in experiments, differs from that of ordinary $LiNbO_3$.

Nonlinear Crystals

Numerous crystals show nonlinear optical effects, but only a few are useful for difference-frequency generation. The crystal material needs to be transparent at all wavelengths of pump, signal and idler beams, it should have a high nonlinear coefficient and a high damage threshold. In some cases, e.g., strong dispersion and weak birefringence, phase matching is not possible. These requirements limit the choice of nonlinear crystals. The optical properties of some nonlinear crystals used in DFG systems are listed in Table 11.51. Figure 11.201 shows the absolute nonlinear coefficient and the transparency range of selected crystals for comparison.

For an easier comparison and characterization of crystals and DFG setups reported in the literature the following figure of merit (FoM) is useful:

$$\text{FoM} = \frac{P_i}{P_p P_s L} , \quad (11.198)$$

where P_i, P_p and P_s denote the power of the idler, pump and signal beams, respectively, and L is the crystal

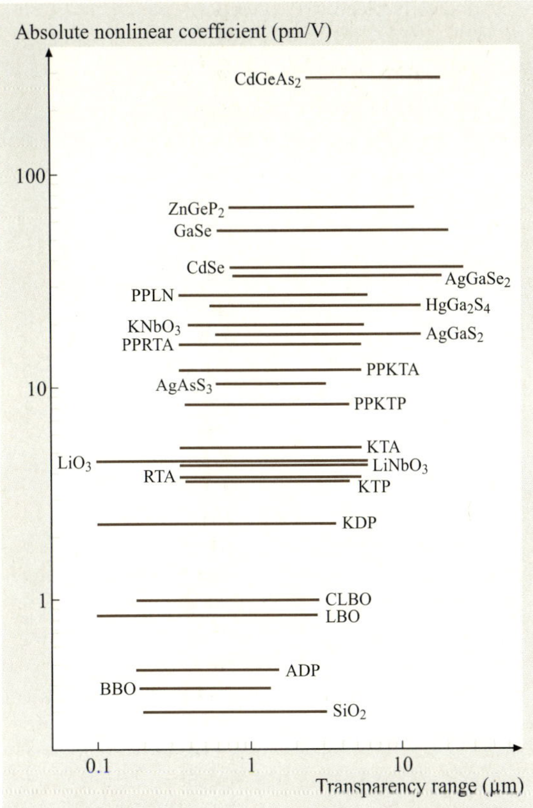

Fig. 11.201 Transparency range of some selected nonlinear crystals used for mid-IR DFG generation as a function of the absolute value of the nonlinear coefficient. For the periodically poled materials (PPLN, PPRTA, PPKTA, PPKTP), the nonlinear coefficient d_{33} is given [11.1813]

length. This way the comparison is independent of the laser powers and crystal lengths chosen. In Table 11.52 this FoM is listed for some representative setups involving various crystals and combinations of pump and signal lasers. Some of the setups generate a CW idler beam by using two CW lasers for pump and signal; other setups have a pulsed idler beam by using two pulsed lasers or one pulsed and one CW laser.

When choosing the pump and signal lasers several issues have to be considered. The laser wavelengths need to be within the transparency range of the crystal, and the wavelengths and the polarizations should be chosen so that phase matching is possible. Another important aspect is the laser power. As the conversion efficiency is rather low, high laser powers are preferred, but too-high powers will damage the crystal, the surface of the crystal or the antireflection (AR) coating on the crystal. The AR coating often has a lower damage threshold than the crystal itself. At high power of the pump laser other effects such as optical parametric generation (OPG) or amplification (OPA) could become even stronger than the difference-frequency generation, resulting in an enlarged line width. If the signal laser has much more power than the pump laser, pump depletion might be a problem. Therefore pump and signal laser and nonlinear crystal have to be carefully matched. To estimate how low the signal power needs to be to avoid pump depletion, the following formula can be used:

$$\eta = \frac{I_\mathrm{i}}{I_\mathrm{p}} = \frac{8\pi^2 d_\mathrm{eff}^2 L^2 I_\mathrm{s}}{\varepsilon_0 n_\mathrm{p} n_\mathrm{s} n_\mathrm{i} c \lambda_\mathrm{i}^2} \ll 1 , \qquad (11.199)$$

where I_i, I_p, and I_s denote the idler, pump, and signal intensity, respectively, d_eff is the effective nonlinear coefficient, L is the crystal length, λ_i is the idler wavelength, ε_0 is the dielectric constant and n is the refractive index with the subscripts p, s, and i referring to pump, signal and idler, respectively. η is the conversion efficiency from pump beam to idler beam, where $\eta = 1$ implies that 100% of the pump beam is converted to the idler beam. This case gives the nonlinear interaction length L_nl, one of several characteristic lengths of a DFG system

$$L_\mathrm{nl} = \sqrt{\frac{\varepsilon_0 n_\mathrm{p} n_\mathrm{s} n_\mathrm{i} c \lambda_\mathrm{i}^2}{8\pi^2 d_\mathrm{eff}^2 I_\mathrm{s}}} . \qquad (11.200)$$

Crystal lengths longer than L_nl will not increase the idler power because of pump depletion. The aperture length L_an is the distance after which the beam is displaced by $2w_0$ because of walk-off effects and is given by:

$$L_\mathrm{an} = \sqrt{\pi} \frac{w_0}{\rho_n} , \qquad (11.201)$$

where w_0 is the minimal beam waist and ρ_n is the walk-off angle. The diffraction length L_{diff} is the length after which the beam diameter has increased by a factor of $\sqrt{2}$. Longer crystals will not increase the idler power. This effect can be calculated more precisely by using the focusing function in (11.192).

$$L_\mathrm{diff} = 4kw_0^2 . \qquad (11.202)$$

For pulsed lasers the interaction length L_qs represents a further important issue

$$L_\mathrm{qs} = \sqrt{\pi}\tau \left(\frac{1}{v_\mathrm{g1}} - \frac{1}{v_\mathrm{g2}} \right)^{-1} , \qquad (11.203)$$

where τ is the pulse duration, v_g1 and v_g2 are the group velocities of the pump and the signal beams, respec-

Table 11.52 Figure of merit (FoM) for different setups and crystals. PP: periodically poled, OP: orientation patterned, ECDL: external-cavity diode laser

Crystal	Pump laser	Signal laser	Idler wavelength (μm)	Idler power (μW)	Figure of merit (μW/(W^2cm))
$LiNbO_3$ bulk [11.1819]	ECDL 795–825 nm 25–30 mW CW	Nd:YAG 1064 nm 1 Watt CW	3.16–3.67	CW 0.030	0.4
PPLN [11.1820]	ECDL 1030–1070 nm, with Yb-fiber amplifier 700 mW CW	Er-fiber laser 1545–1605 nm 5 W CW	2.9–3.5	CW 3500	200
PPLN [11.1821]	Nd:YAG 1064 nm pulsed 6 ns rep. 4–8 kHz 5 kW peak power	ECDL 1500–1600 nm 5 mW CW	3.2–3.7	Pulsed 2000 Average power	4.4×10^{5}[a] 4.4×10^{5}[b]
PPLN [11.1814]	ECDL 808 nm 20 mW CW	Nd:YAG 1064 nm 660 mW CW	3.3	CW 27	410
PPLN [11.1809]	Diode laser master/slave 848–855 nm 78 mW CW	Nd:YAG 1064 nm with Yb-fiber amplifier 5 W CW	4.15–4.35	CW 172	110
PPLN This setup (see Sect. 11.10.3)	ECDL 850–870 nm 125 mW CW	Nd:YAG 1064 nm 2 W CW	4.3–4.7	CW 5–23	4–19
Waveguide PPLN [11.1812]	Diode laser 940 nm 17.5 mW CW	ECDL 1550 nm 20 mW CW	2.30–2.44	CW 400	1.2×10^{5}
$AgGaS_2$ [11.1822]	ECDL 679–683 nm 40 mW CW	Diode laser 786–791 nm 20 mW CW	4.9–5.1	CW 0.1	31
$AgGaSe_2$ [11.1823]	Fabry-Perot diode laser 1290 nm 8 mW CW	ECDL 1504–1589 nm 6 mW CW	7.1–7.3	CW 0.010	52
PPKTP [11.1824]	Nd:YAG 1064 nm 222 mW CW	ECDL 1490–1568 nm with Er-fiber amplifier 34 mW CW	3.2–3.4	CW 0.170	22
PPKTA [11.1825]	Nd:YAG 1064 nm 117.2 mW CW	ECDL 1519 nm 17.4 mW CW	3.45–3.75	CW 0.140	70
PPRTA [11.1826]	Ti:Al_2O_3 laser 710–720 nm 100 mW CW	Ti:Al_2O_3 laser 874–915 nm 200 mW CW	3.4–4.5	CW 10	250
OP GaAs [11.1808]	DFB diode laser 1306–1314 nm 1.5–3.3 mW CW	ECDL 1535–1570 nm with Er-fiber amplifier 1 W CW	7.9	CW 0.038	6
GaSe [11.1827]	Nd:YAG 1064 nm pulsed 20 ps 750 μJ rep. 10 Hz	OPA 1100–4800 nm pulsed 5 ps 35–50 μJ	2.4–28	Pulsed 5 μJ	0.0041[a] 2.1×10^{6}[b]
$LiInS_2$ [11.1828]	Ti:Sapphire 700–810 nm CW	Ti:Sapphire 800–900 nm CW	5.5–11.3	CW -	12.4
$ZnGeP_2$ [11.1829]	OPO signal 1760–1950 nm pulsed 7 ns, rep. 17 Hz 0.95 mJ	OPO idler 2710–2330 nm pulsed 7 ns, rep. 17 Hz 0.95 mJ	5–12	Pulsed 25 μJ	0.2[a] 1.6×10^{6}[b]

[a] FoM calculated with peak powers

[b] FoM calculated with average powers

tively. L_{qs} describes the length after which the pulses from the two beams are separated by τ. Another characteristic length for pulsed laser is the dispersion length L_{ds}, given by

$$L_{ds} = \frac{\tau^2}{g_n} , \qquad (11.204)$$

where g_n is the group-velocity dispersion. This length describes the length after which the pulse duration has doubled. With these characteristic lengths first estimations for crystal lengths and pulse durations can be made.

11.10.2 DFG Laser Sources

The literature on DFG laser sources, new nonlinear crystal materials and applications has become extensive in recent years. By far the largest part of these systems have been, and still are, developed for spectroscopic gas sensing and analysis application. A recent overview of systems is given in reference [11.1813]. In the following we describe a typical DFG setup that has been implemented in our laboratory for measurements on isotopomers of trace gases. Precise isotope ratio measurements of trace gases represent an important contribution to the solution of key questions in various areas, e.g., differentiation between natural and anthropogenic origin of specific compounds. Isotopic compositions are of interest in such diverse fields as ecological CO_2 exchange, volcanic emission, medical diagnostics, extraterrestrial atmospheres, etc.

Detailed Discussion of a DFG Laser Source

Isotopic composition of trace gases such as CO_2, CO or N_2O are of special interest. These molecules with their isotopomers exhibit strong absorption lines between 4.3 to 4.7 μm (Fig. 11.202). Therefore a continuous-wave DFG source, which enables continuous tuning and narrow line width (to differentiate between isotopomers), was implemented. In the following sections the theoretical calculations, the setup and the characterization of this system is described.

Calculations of optimal crystal length and beam parameter. For the wavelengths of interest we chose $LiNbO_3$ as nonlinear optical medium because it has a transmission range from 0.4 μm to 5.0 μm, a large nonlinear coefficient, and it can be produced periodically poled.

However, $LiNbO_3$ has an absorption band at 5 μm so the absorption between 4 μm and 5 μm cannot

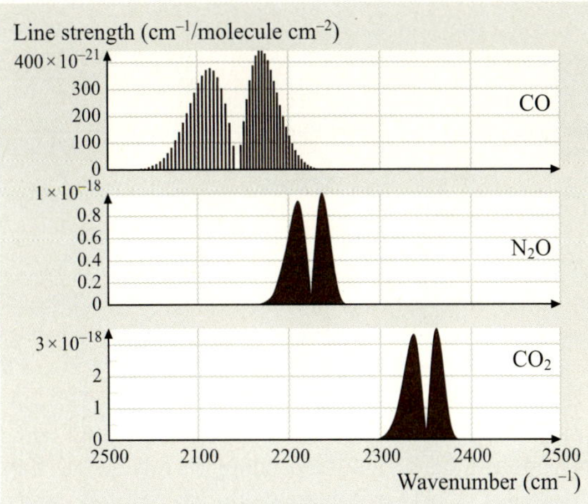

Fig. 11.202 Absorption lines of CO, N_2O and CO_2 as a function of wavenumber in the mid-IR region [11.1830]

be neglected. The absorption coefficient at 4.3 μm is 0.25 cm^{-1}, at 4.6 μm it is 0.55 cm^{-1} and at 4.7 μm 0.75 cm^{-1} [11.1831]. There are other crystals that can be used at this wavelength, e.g., $AgGaS_2$ or $AgGaSe_2$ (Table 11.51), but they are not available periodically poled, so birefringent phase matching has to be used. This means the crystal length needs to be shorter because of walk-off effects and it is not possible to use the maximum nonlinear coefficient. Furthermore alignment is more critical than with a periodically poled crystal. Finally, wavelength tuning is also an important issue. When working with such crystals angle tuning has to be used, which makes wavelength tuning over larger ranges more complicated. Often noncritical phase matching is used, because of its larger acceptance bandwidth, but for a certain idler wavelength the pair of signal and pump wavelength is fixed in this case. This limits the choice of lasers and normally two diode lasers with lower power than other lasers have to be used.

In contrast, when using quasi-phase matching, the phase matching is realized by choosing a grating period, so nearly every combination of lasers can be used (e.g., an external cavity diode laser (ECDL) that can be conveniently wavelength tuned, and a Nd:YAG laser that delivers high laser power). Phase matching is achieved by choosing the correct grating period Λ, which can be adjusted to other wavelengths by changing the temperature of the crystal because of the temperature dependence of the refractive index (11.197). When scanning the wavelength for spectroscopy, the crystal temperature can be changed simultaneously to achieve

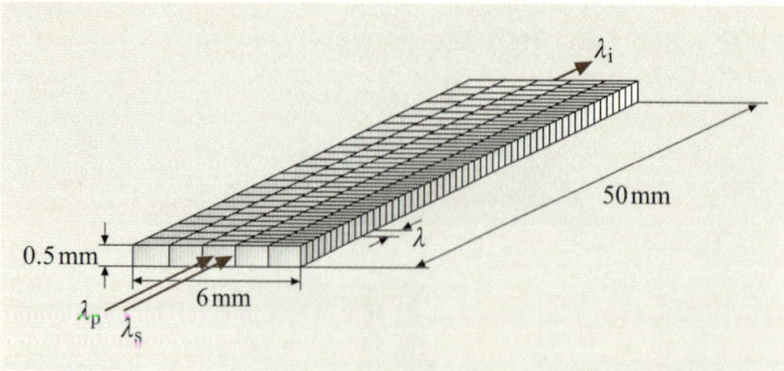

Fig. 11.203 A periodically poled crystal with several gratings for different wavelengths. $\lambda_{p,s,i}$: pump, signal, idler wavelength, respectively; Λ: grating period

mid-IR wavelength ranges of several hundred nm with only one grating period. Depending on wavelength the temperature acceptance bandwidth can be quite large (Fig. 11.209b). Even larger wavelength ranges can be achieved by using a crystal with several gratings, as shown in Fig. 11.203. Typical grating periods for our wavelength range are approximately 23 μm. As a result it appears advantageous to use PPLN for the envisaged 4.3–4.7 μm wavelength range rather than a bulk $AgGaS_2$ or other crystal despite the non-negligible absorption of PPLN.

According to (11.192) the idler power increases with the crystal length but also the absorption increases, therefore there is an optimum crystal length for this wavelength region. The maximum of the focusing function h (11.193) is not at $\sigma = 0$ but, e.g., at $\sigma = 1.3$ for a pump wavelength of 863 nm, a signal wavelength of 1064 nm and a crystal length of 5 cm (Fig. 11.208), so the maximum idler power is not obtained at perfect phase matching, but for a slightly different grating period. However, to simplify the calculations, σ was set to 0. The focusing function and hence the idler power has a clear maximum at about $\xi = 1.3$ (Fig. 11.204). It depends only slightly on wavelength and absorption. Therefore for a chosen crystal length L there is an optimal confocal parameter b.

When the idler power is plotted as a function of crystal length L and of the confocal parameter b, it can be seen that a long crystal combined with a large confocal parameter yields higher idler power (Fig. 11.205). However, in a real experiment the confocal parameter is limited by the thickness of the crystal and crystals longer than 6 cm are not easily commercially available. The problem with longer crystals is that crystal defects have too much influence. The thickness of the crystal is limited by the production process of the periodically poled grating. This is done by applying an electric field to change the orientation in the crystal, which requires very strong field strengths. This limits the thickness to 1 mm. Most crystals have a thickness of 0.5 mm, resulting in a better grating quality than for a thickness of 1 mm.

In Fig. 11.206 it can be seen how the idler power changes for different crystal lengths. The calculations were made with $\sigma = 0$ and keeping $\xi = L/b = 1.3$ at the maximum. The result is that at $\lambda_p = 853$ nm ($\lambda_s = 1064$ nm, $\lambda_i = 4.3$ μm) the optimal length is longer

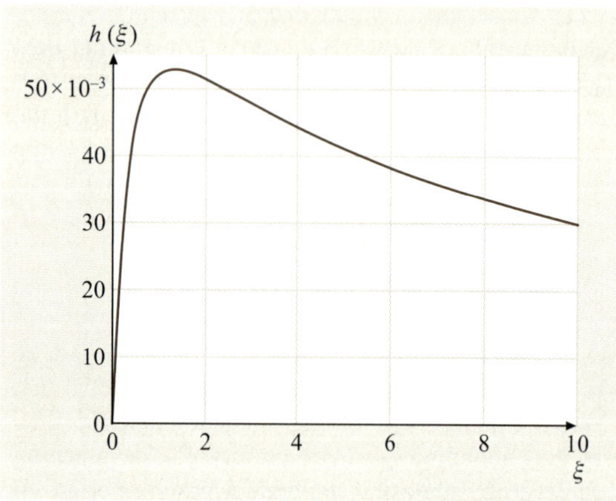

Fig. 11.204 Focusing function $h(\xi)$ versus ξ calculated for $\lambda_p = 868$ nm, $\lambda_s = 1064$ nm, $\alpha = 0.75\,\mathrm{cm}^{-1}$, $\sigma = 0$, and $L = 5$ cm. The maximum is at $\xi = 1.3$ and for small ξ the focusing function h is proportional to ξ. The symbols are explained in (11.192, 193, 194, 195, 196)

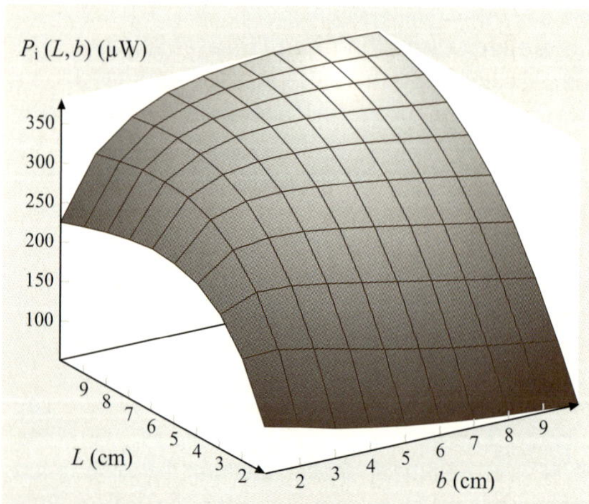

Fig. 11.205 Idler power P_i versus crystal length L and confocal parameter b calculated at a pump wavelength of 853 nm. The idler power increases with increasing crystal length and confocal parameter

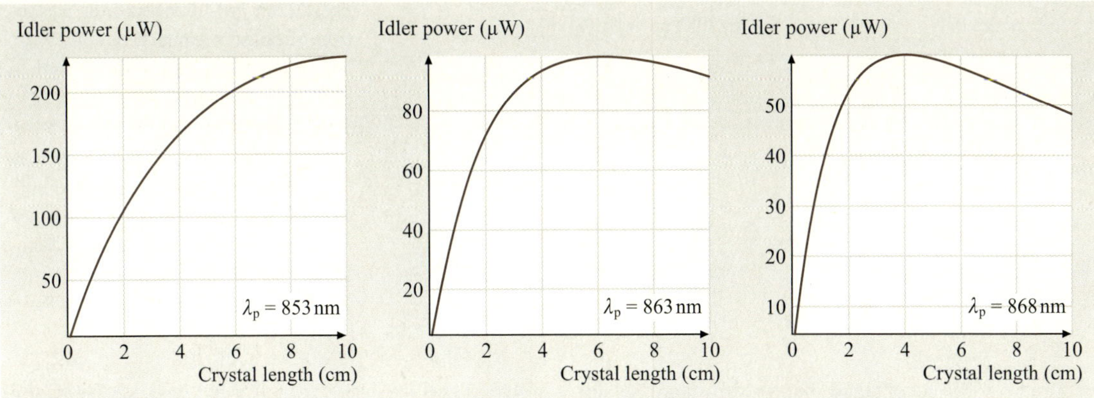

Fig. 11.206 Idler power as a function of crystal length for pump wavelengths λ_p of 853 nm, 863 nm, 868 nm. The signal laser is a Nd:YAG laser fixed at $\lambda_s = 1064$ nm. The calculations were made with $\sigma = 0$ and keeping $\xi = L/b = 1.3$ at the maximum. At 853 nm the ideal crystal length is longer than 10 cm, at 863 nm it is 5.9 cm, at 868 cm it is 4.0 cm

than 10 cm, at $\lambda_p = 863$ nm ($\lambda_i = 4.6\,\mu$m) it is 5.9 cm, and at $\lambda_p = 868$ nm ($\lambda_i = 4.7\,\mu$m) it is 4.0 cm. We chose a crystal length of 5 cm, implying $b = 4$ cm, which gives a minimal beam waist of 0.13 mm. Such a Gaussian beam propagates within the crystal over the whole length.

Based on (11.192), an idler power of 55–170 μW is expected for a pump power of 150 mW, a signal power of 2 W and a crystal length of 5 cm.

Setup and characterization. Our DFG system consists of an external cavity diode laser (EDCL) as pump laser, a continuous-wave Nd:YAG laser as signal laser and a MgO-doped periodically poled $LiNbO_3$ crystal (MgO:PPLN). The ECDL (Sacher TEC-120-850-150) has a power of 150 mW and a wavelength range of 820–875 nm. To cover the idler wavelength range of 4.3–4.7 μm, pump wavelengths of 852–868 nm are needed. The Nd:YAG laser (Innolight Mephisto) has a CW power of 2 W and a wavelength of 1064.5 nm. The MgO:PPLN crystal (HC-Photonics) is 5 cm long and 0.5 mm thick. It has several gratings with periods of 21.45, 22.00, 22.50, 23.10, and 23.65 μm, each with a width of 1.2 mm. For the wavelength of interest in this experiment, only the grating period of 23.1 μm is needed, it is quasi-phase matched for the different wavelengths by changing the temperature from 30 °C to 130 °C. The crystal is antireflection coated for the pump, signal and idler wavelengths.

The laser beams are focused by several lenses (including cylindrical lenses, not shown in Fig. 11.207) so that the minimal beam radius is 0.13 mm within the crystal. The $\lambda/4$ and $\lambda/2$ plates are used to match the beam polarization for quasi-phase matching in the PPLN crystal. A small part of the pump beam is directed to a wavemeter. The recorded pump wavelength then yields the idler wavelength. After the crystal a germanium filter is used to block the near-infrared light. The setup is depicted in Fig. 11.207.

This DFG system was tested and characterized by focusing the mid-IR beam onto a detector (VIGO PDI-2TE-5, TE-cooled) and recording the signal with a lock-in amplifier with a time constant of 100 ms. For modulating the laser power with a frequency of 1.8 kHz, a chopper was placed after the Nd:YAG laser. To find the crystal temperature for phase matching, the temperature was increased in steps of 0.2 °C or 0.5 °C while keeping the pump wavelength constant (Fig. 11.208).

The temperature for phase matching is 6.6–11.2% higher than theoretically calculated (Fig. 11.209a) and

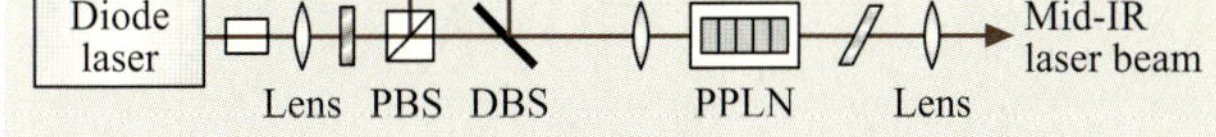

Fig. 11.207 Setup for difference-frequency generation. OI: optical isolator, PBS: polarizing beam splitter, DBS: dichroic beam splitter, PPLN: periodically poled $LiNbO_3$

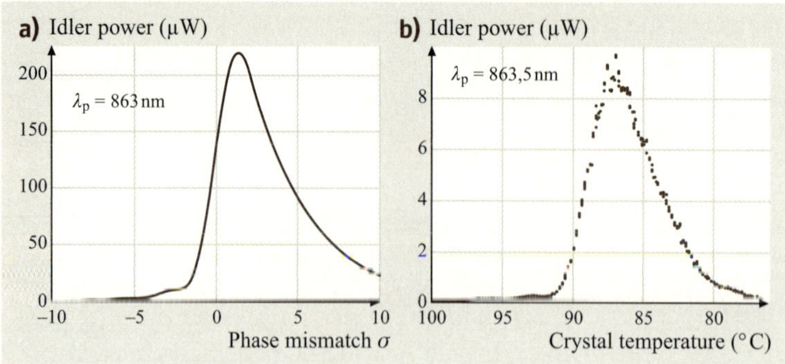

Fig. 11.208 **(a)** Idler power as a function of phase mismatch σ (11.196), calculated for a pump wavelength of 863 nm, a signal wavelength of 1064 nm, and a crystal length of 5 cm. **(b)** Measured idler power as a function of crystal temperature corresponding to phase mismatch, at a wavelength of 863.5 nm and a crystal length of 5 cm. It can be seen that the measured and the calculated curve have a similar shape

the temperature acceptance bandwidth increase for longer wavelengths and is 2.5–3.7 °C (Fig. 11.209b). The power of the CW mid-infrared beam amounts to 23 μW at 4.3 μm and 5 μW at 4.76 μm (Fig. 11.209c). It decreases for increasing wavelengths because of the crystal absorption near 5 μm. The generated idler power is about four to ten times lower than calculated, most probably because of imperfections in the crystal, its grating quality and non-Gaussian beam shape of the ECDL pump laser.

For measurements of the isotopic composition of trace gases, the laser line width should be sufficiently narrow to resolve adjacent molecular lines clearly, also when recorded at reduced gas pressure. The line width of our CW mid-IR source has been determined as approximately 1 MHz and thus fulfills the requirements by far.

Measurement of isotopic composition of trace gases. We measured the isotopic ratios of $^{13}C/^{12}C$ and $^{18}O/^{16}O$ in CO_2 and CO, and of $^{15}N/^{14}N$ in N_2O. N_2O is of special interest because $^{14}N^{15}N^{16}O$ and $^{15}N^{14}N^{16}O$ have the same mass, so they cannot be distinguished by conventional mass spectrometry but laser spectroscopy enables an easy identification. Our measurements were done with the DFG system described above and direct absorption spectroscopy with an astigmatic Herriot cell (New Focus 5611) with an optical path length of 10 m. The setup is shown in Fig. 11.210.

One problem in measuring isotopomers is that the concentration of the main isotopomer is typically a hundred times higher than that of the less-abundant isotopomer. There are two possibilities to overcome this problem, either to measure two lines of similar strength resulting in a strong temperature sensitivity of the measurement, or by choosing lines with similar lower energy levels but with very different line strength. The astigmatic multipass Herriot cell offers the possibility to enter the cell at a different angle than usual, so that the beam leaves the cell after only two passes (Fig. 11.210). This makes it possible to measure two lines of very different line strength by using two different path lengths (the balanced-path-length detection scheme) [11.1832].

As an example, Fig. 11.211 shows the result of a measurement on CO_2 isotopomers in ambient air at a concentration of 350 ppm. The measurements were

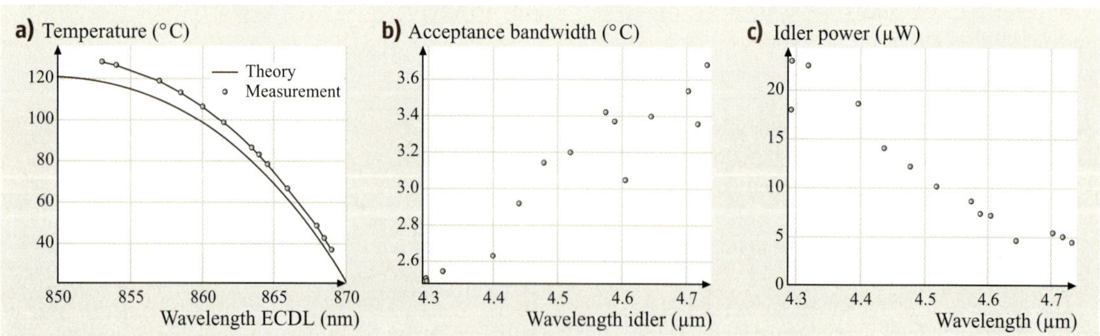

Fig. 11.209 **(a)** Quasi-phase-matching temperature as a function of pump wavelength. **(b)** Acceptance bandwidth as a function of the temperature. **(c)** Idler power as a function of the idler wavelength

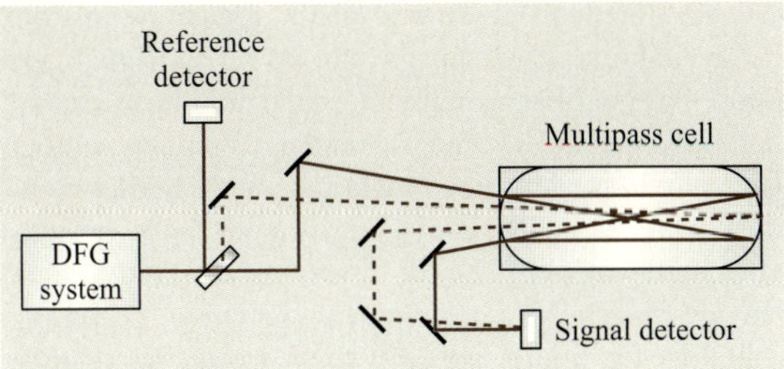

Fig. 11.210 Setup for transmission spectroscopy. In the astigmatic Herriot cell there are two paths, one of 10 m (*solid line*) and one of 40 cm (*dashed line*)

performed at a total pressure of 50 mbar and at room temperature. The experimental data are fitted with Voigt curves to calculate the concentrations of the isotopomers.

The derived isotope ratios of $^{13}C/^{12}C$ for CO_2 are 1.3% ± 0.2% (only long path) and 1.4% ± 0.3% (balanced-path-length detection scheme), which is in good agreement with the natural abundance of 1.1%. The isotope ratio $^{18}O/^{16}O$ of CO_2 is deduced as 0.44% ± 0.06% (only long path) and 0.47% ± 0.11% (balanced-path-length detection scheme), which again is in good agreement with the natural abundance of 0.39%. A more-detailed discussion of these studies, which also includes measurements of CO and N_2O isotopomers, can be found in [11.1833]. A further study on N_2O isotopomers, testing different detection schemes for low concentrations (100 ppm) has been published in [11.1834].

11.10.3 Outlook

Tunable coherent sources in the mid-infrared range play an important role. This is manifested by the ever-growing number of publications. The development is fostered by numerous applications, primarily in gas sensing. The requirements with respect to sensitive and selective monitoring devices with multicomponent capability are manifold: access to a broad wavelength range, broad – preferentially continuous – wavelength tunability, narrow line width (i. e., much narrower than typical molecular absorption line widths), preferentially room-temperature (RT) or near-RT operation, and compact and robust setups for field applications. In this respect, DFG systems represent a very valuable choice. Since their first realization in 1974 [11.1835], DFG-based devices have reached a mature level. Today, their wavelength range can be chosen between, e.g., 2 μm and 19 μm only depending on the available pump and signal sources and nonlinear crystals. The tuning of a DFG source is straightforward and continuous, and the wavelength generated can be accurately determined via the near-infrared input wavelengths. The line width is basically given by the line width of the pump and signal laser. This enables narrow mid-IR line widths as required for high-resolution spectroscopy. In contrast to alternative mid-IR sources, DFG systems are generally room-temperature devices except for the crystal, which may require heating in a small temperature-controlled oven. Line widths of continuous-wave DFG sources are in the MHz range, which makes them attractive

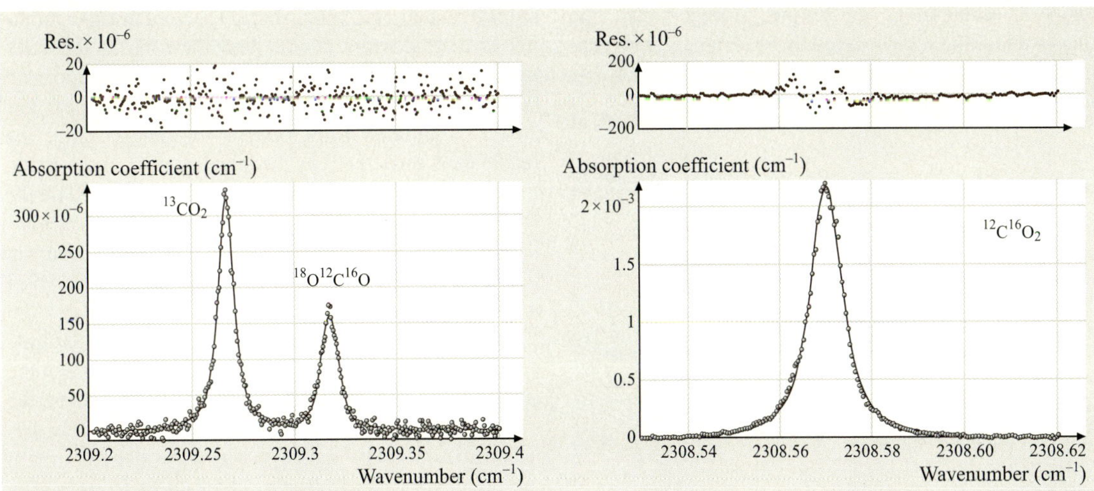

Fig. 11.211 Absorption lines of different isotopomers of 350 ppm CO_2 in ambient air. Experimental data (*dots*) and Voigt fit (*solid line*) for 10 m path length, residuals on top

for studies requiring high spectral resolution such as the briefly discussed isotopomer measurements. DFG sources are rather low-power lasers with CW powers in the μW to mW range. With fiber amplifiers for the near-IR pump and signal sources, higher powers can be achieved if necessary [11.1820]. Recent developments in waveguide technology in PPLN [11.1812] or KTP appear particularly attractive in view of higher conversion efficiencies. Last but not least, a DFG system can be made compact in view of today's diode lasers and diode-pumped solid-state lasers used as pump and signal sources.

In view of the variety of applications a further impetus for DFG systems can be expected. Current developments in several areas appear promising: the availability of compact and powerful signal and pump lasers, fiber amplifiers, new nonlinear crystal materials including organic media to access further wavelength ranges, new materials with higher nonlinear optical coefficients, birefringent bulk crystals with better quality, larger crystal sizes and higher damage thresholds, quasi-phase matching becoming available for more crystals than today, waveguide technology, etc. These developments will contribute to improve the performance of DFG devices, enhance their distribution, lower their costs and hence further increase their role as attractive spectroscopic tools.

11.11 Free-Electron Lasers

The free-electron laser (FEL) is a system consisting of a relativistic electron beam and a radiation field interacting with each other while they propagate through an undulator [11.1836, 1837]. The main components of the free-electron laser are electron accelerator, undulator, and optical resonator (optionally for an FEL oscillator). The undulator is a periodic magnetic structure with a planar, or helical magnetic field that causes periodical transverse deflections of the electron-beam trajectory [11.1838]. Optical resonators are fairly similar to those used in conventional lasers. FEL devices have been realized with nearly all types of accelerators: electrostatic accelerators, RF linear accelerators, induction accelerators, microtrons, storage rings, etc. The wavelength range covered by FELs spans from centimeters down to nanometers [11.1839] (Table 11.53).

Table 11.53 Parameter space of free-electron lasers as of 2006

Radiation	
Wavelength	13 nm–10 mm
Peak power	up to 5 GW
Average power	up to 10 kW
Pulse duration	10 fs to CW
Driving electron beam	
Energy	200 keV–1 GeV
Peak current	1–3000 A
Undulator	
Period	0.5–20 cm
Peak magnetic field	0.1–1 T
Undulator length	0.5–27 m

The scale of the FEL setup is mainly defined by the scale of the driving accelerator. For FELs operating in the millimeter wavelength range this could be a room scale, while unique devices such as VUV and X-ray FELs user facilities have a scale comparable with conventional third-generation synchrotron radiation facilities (Fig. 11.212).

11.11.1 Principle of Operation

The FEL is not actually a laser; it is most closely related to vacuum-tube devices. As with vacuum-tube devices, FEL devices can be divided in two classes: amplifiers and oscillators (Fig. 11.213). An FEL amplifier is a single-pass device, and there is no feedback between the output and input. The FEL oscillator can be considered as an FEL amplifier with feedback. For an FEL oscillator in the optical wavelength range the feedback is carried out by means of an optical resonator. FELs based on the oscillator principle are limited on the short-wavelength side to ultraviolet wavelengths primarily because of mirror limitations. Free-electron lasing at wavelengths shorter than the ultraviolet can be achieved with a single-pass high-gain FEL amplifier.

The field of the electromagnetic wave only has transverse components, so the energy exchange between the electron and the electromagnetic wave is due to the transverse component of the electron velocity. The latter occurs due to periodical wiggling of the electron in the undulator. The driving mechanism of free-electron lasers is the radiative instability of the electron beam, which develops due to the collective interaction of electrons with the electromagnetic field in the undulator. The basic principle of radiation-induced instability can be described within the standard picture for the generation of the synchrotron radiation. Electrons propagate along a sinusoidal path and emit synchrotron radiation in a narrow cone in the forward direction. When an electron beam traverses an undulator, it emits radiation at the resonance wavelength $\lambda = (\lambda_w/2\gamma^2)(1+K^2/2)$. Here, λ_w is the undulator period, $mc^2\gamma$ is the electron beam energy, $K = eH_w\lambda_w/2\pi mc$ is the dimensionless undulator strength parameter, and H_w is the maximum on-axis magnetic field strength of the undulator. Although the electromagnetic wave is always faster than the electrons, a resonant condition occurs such that the radiation slips a distance λ relative to the electrons after one undulator period. The fields produced by the moving charges in one part of the undulator react on moving charges in another part of the undulator. Thus, we deal with some tail–head instability, which leads to a growing concentration of particles wherever a small perturbation started to occur (Fig. 11.214). In the beginning – without microbunching – all N electrons can be treated as individually radiating charges producing the power of the spontaneous emission $\propto N$. With complete microbunching, all electrons radiate almost in phase. This leads to a radiation power $\propto N^2$ and thus to an amplification of many orders of magnitude with respect to the spontaneous emission of the undulator.

The figure of merit for FEL performance is the radiation power gain during one pass of the undulator. The gain primarily depends on the value of the peak beam current. Another quantity of importance for the development of the radiative instability is the electron beam density in the six-dimensional phase space. Low beam quality (large energy spread and emittance) leads to degradation of the FEL gain. Errors in the periodical magnetic structure also degrade FEL performance. Presently accelerator techniques and undulator technology have reached such a level that it is possible to have power gains of the order of 10^7 for the amplification of the radiation of a nanometer wavelength range.

11.11.2 Current Status and Perspective Applications of Free-Electron Lasers

Free-electron lasers hold several potential advantages: continuous tunability of the radiation wavelength, the possibility to obtain high levels of average output power, and the possibility to obtain high conversion efficiencies of the net electrical power to radiation power. An important feature of the FEL radiation is that it has a high degree of transverse coherence. In other words, the FEL radiation can always be focused on to a spot whose size is defined totally by diffraction effects. This feature reveals a wide range of possibilities for FEL applications in the transportation of the radiation over long distances and in obtaining high focused intensities.

Fig. 11.212 Aerial view of the experimental hall for the FLASH user facility in Hamburg (*center*) and the tunnel for the superconducting accelerator and undulator (*covered with grass*). The hall in the upper right corner houses the injector part of the linac. The total length of the FLASH facility is 300 m. The maximum energy of electrons is 1 GeV, and the minimum radiation wavelength is 6 nm. The undulator of the FLASH (photo in the *upper left corner*) is a permanent magnet device (period 2.73 cm, gap 12 mm, peak field 0.47 T). The undulator system is subdivided into six segments, each 4.5 m long

However, FELs are relatively expensive devices, thus their applications are in the fields not covered by conventional radiation sources, in the far-infrared wavelength range and the THz gap (the sub-millimeter wavelength range). Another field for FEL applications is the generation of short-wavelength radiation, from the vacuum ultraviolet down to X-ray range. Technological developments are on the way, aiming at industrial applications of powerful FELs, and the use of FELs for energy transfer in space is also under consideration. Organiza-

Part C | 11.11

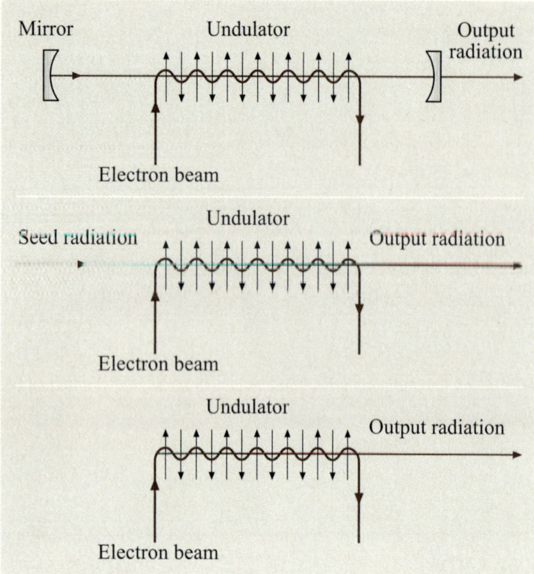

Fig. 11.213 Free-electron laser configurations: oscillator (*top*), seeded amplifier (*middle*), and self-amplified spontaneous emission (SASE) FEL (*bottom*)

tion of user operation at the FEL facilities is similar to that used at synchrotron radiation facilities: the host organization constructs and serves free-electron laser and user beam lines, and users perform their experiments according to the schedule of the facility.

Linac-Based FEL User Facilities

There are several operating linear accelerator (linac)-based FEL user facilities (in China, France, Germany, Japan, Korea, The Netherlands, Russia, and USA) working for scientific applications [11.1840–1853]. The wavelength range covered with these facilities spans from 200 nm to a few hundred micrometers. Typical parameters of the FELs driven by RF linear accelerators are: pulse duration of a few picoseconds, peak power in the MW range, and micropulse repetition rate in the range 10 MHz to 3 GHz. FELs driven by normal conducting RF accelerators operate in pulsed mode with a macropulse repetition rate of 10–100 Hz. The macropulse duration is defined by the length of the RF pulse, typically 1–20 μs. Thus, the average radiation power is in the watt range. FELs driven by superconducting accelerators operate in continuous mode and produce a high level of average output power, up to a few kW. Linac-based FEL facilities are recognized nowadays as a unique tool for scientific applications that require tunable coherent radiation in the infrared wavelength range. The general tendency is that the wavelength region of interest for users at these facilities is moving into the far-infrared and THz gap.

FEL User Facilities at Storage Rings

Several storage rings are equipped with free-electron lasers [11.1854–1857]. Typical parameters of the radiation are: wavelength range of 190–700 nm, pulse duration of some tens of picoseconds, average power of 10–300 mW, and peak power in the kW range. Operation in a locked-mode regime allows the peak power to be increased by an order of magnitude. The general tendency is that users' interest in these facilities is gradually reducing due to the limited features compared with conventional lasers.

High-Average-Power FELs

Recent progress in accelerator technology has paved the way for construction of high-average-power linear accelerators. The quality of electron beams produced by such machines are sufficient to drive free-electron lasers. Application of the energy-recovery technique allows a high level of overall efficiency to be achieved. Pilot facility are already in operation in Japan, Russia, and USA [11.1851–1853]. The demonstrated level of the average radiation power is about 10 kW in the 3–20 μm wavelength range [11.1852], and about 0.4 kW

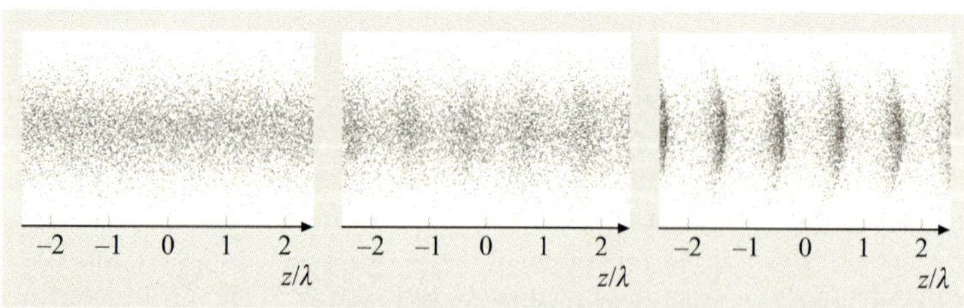

Fig. 11.214 Distribution of the electrons in the electron beam (longitudinal cut). The *left*, *middle*, and *right* snapshots are taken at the beginning, middle, and the end of the undulator. Electron beam density is modulated periodically with the period of the resonance wavelength due to the radiative instability

in the THz gap [11.1851]. Projects of next-generation FELs based on energy-recovery linacs aim at increasing the average power into the tens of kilowatt range, and also to build high-power FELs operating in the UV range [11.1858–1863].

Potential industrial applications of high-average-power FELs involve material processing (for instance, the treatment of polymer surfaces), lithography, isotope separation, and chemical applications. Production of pure isotopes on an industrial scale may have a big impact on future developments. For instance, isotope $_{28}$Si is a radiation-resistant material of great interest for space research and the nuclear energy industry. The thermal conductivity of the pure isotope $_{28}$Si is 50% higher than that in the natural mixture, which seems very attractive for the semiconductor industry. The are also many other isotopes of great practical interest such as $_{13}$C (medical applications) and $_{15}$N (for the study and current control of use of nitrogen fertilizers in agriculture and agrochemistry). The basic process of isotope separation is a selective multiphoton dissociation of molecules. The required resonance wavelength range is in the range 2–50 μm. The required energy in a pulse is no less than 0.1 mJ and monochromaticity of 10^{-2}–10^{-4}, depending on the type of reaction. Industrial production of isotopes requires FELs with average power exceeding 10 kW. Construction of such facilities is on the way, and the first results are expected in the near future.

X-ray Free-Electron Lasers

At the start of this century, we have seen a revolution in synchrotron source intensities. This revolution stemmed from the technique of free-0electron lasers combined with recent progress in accelerator technologies, developed in connection with high-energy linear colliders [11.1866]. X-ray FELS (XFELs) have made a new regime of intensities accessible, thus opening up a fundamentally new physical domain.

A new era of synchrotron radiation research has begun withthe first user experiments on a vacuum UV FEL based on self-amplified spontaneous emission (SASE) [11.1865, 1867]. It is worth mentioning that such an essentially quantum terminology [introduced after amplified spontaneous emission (ASE)] does not reflect the actual physics of the process. The amplification process in the SASE FEL has its origin in density fluctuations in the electron beam. The latter effect is completely classical. The results have been obtained at the TESLA test facility (TTF) at the Deutsches Elektronen-Synchrotron DESY (Hamburg, Germany), using radiation pulses at a wavelength of 98 nm with a 40 fs pulse duration and a peak power of 1.5 GW [11.1864, 1868] (Fig. 11.215). Compared to present-day synchrotron radiation sources its peak brilliance is more than 100 million times higher (Fig. 11.216), the radiation has a high degree of transverse coherence and the pulse duration is reduced from hundreds of picoseconds down to ten-femtosecond time domain. While modern third-generation synchrotron light sources are reaching their fundamental performance limit, recent success in the development of the VUV FEL at DESY has paved the way for the construction of a novel type of light source that will combine most of the positive aspects of both a laser and a synchrotron.

In an X-ray FEL the radiation is produced by the electron beam during a single pass of the undulator [11.1869–1871]. The amplification process starts from shot noise in the electron beam. Any random fluctuations in the beam current correspond to an intensity modulation of the beam current at all frequencies simultaneously – including of course, the frequency to which the undulator is tuned. When the electron beam enters the undulator, the presence of the beam modulation at frequencies close to the resonance frequency initiates the process of radiation. The FEL collective instability in the electron beam produces an exponential growth (along the undulator) of the modulation of the electron

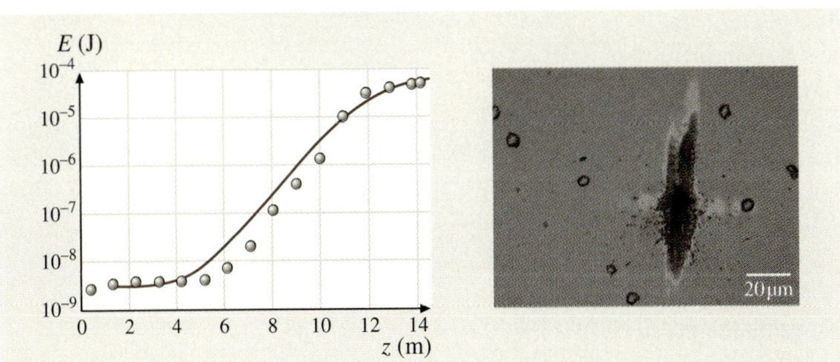

Fig. 11.215 *Left*: Average energy in the radiation pulse versus undulator length for TTF FEL at DESY [11.1864]. *Right*: interaction of powerful VUV radiation with solids [11.1865]. Ablation of a gold target after one pulse of the TTF FEL at DESY. The radiation wavelength is 98 nm, the pulse duration is 40 fs, and the peak power density is about 100 TW/cm^2

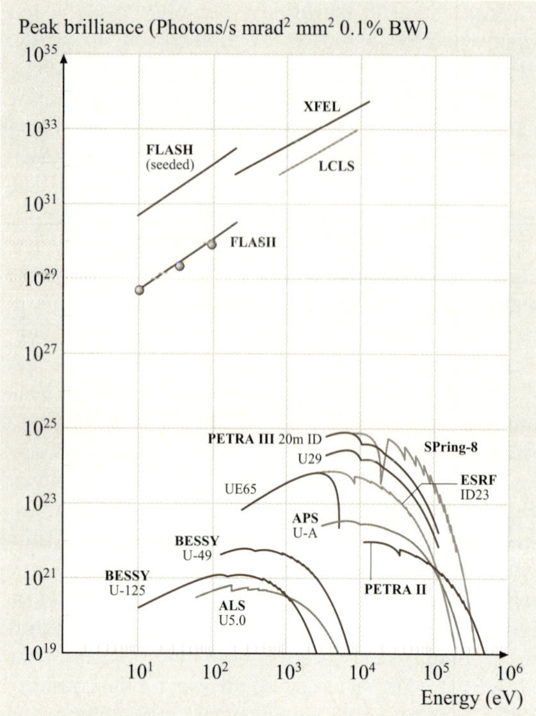

Fig. 11.216 Peak brilliance of X-ray FELs versus third-generation storage rings light sources. Circles show the experimental performance of the FLASH free-electron laser at DESY

density on the scale of undulator radiation wavelength (Figs. 11.214, 215). The fluctuations of current density in the electron beam are uncorrelated, not only in time but also in space. Thus, a large number of transverse radiation modes are excited when the electron beam enters the undulator. These radiation modes have different gains. Obviously, as undulator length progresses, the high-gain modes will predominate increasingly and we can regard the XFEL as a filter, in the sense that it filters from arbitrary radiation field those components corresponding to the high-gain modes. Hence, for a long enough undulator, the emission will emerge in nearly complete transverse coherence. An intensity gain in excess of 10^7 is obtained in the saturation regime. At this level, the shot noise of the electron beam is amplified up to complete microbunching, and all electrons radiate almost in phase, producing powerful, coherent radiation.

The amplification bandwidth of a high-gain FEL amplifier is restricted by the resonance properties of the undulator, that is, by the number of undulator periods N_w within one gain length (the distance over which the power increases by the factor of $\mathrm{e} = 2.718$). The spectrum of transversely coherent fraction of the radiation is concentrated within the narrow band, $\Delta\lambda/\lambda \simeq (2\pi N_\mathrm{w})^{-1}$. The typical amplification bandwidth of the XFEL is of the order of 0.1%. The electron beam in an XFEL transfers enormous peak power. For instance, for typical XFEL parameters (electron energy of 17.5 GeV and peak current 5 kA) it is about 100 TW. The conversion efficiency of kinetic energy of electrons to the light is of the order of the amplification bandwidth, thus the peak power of X-ray radiation is in the multi-GW range (Table 11.54).

Experimental realization of XFELs has developed very rapidly during the last decade. The first demonstration of the SASE FEL mechanism took place in 1997 in the infrared wavelength range [11.1874]. In September 2000, a group at Argonne National Laboratory (ANL) became the first to demonstrate saturation in a visible (390 nm) SASE FEL [11.1875]. In September 2001, a group at DESY (Hamburg, Germany) demonstrated lasing to saturation at 98 nm [11.1864, 1868]. In June 2006 saturation was achieved at 13 nm, the shortest wavelength ever generated by FELs. The experimental results have been achieved at FLASH (*F*ree-Electron-*LAS*er in *H*amburg, Fig. 11.212). Regular user operation of FLASH started in 2005 [11.1872]. Currently FLASH produces GW-level, laser-like VUV radiation pulses with 10–50 fs duration in the wavelength range 13–45 nm. After the energy upgrade of the FLASH linac to 1 GeV planned in 2007, it will be possible to generate wavelengths down to 6 nm.

Recently the German government, encouraged by these results, approved funding of a hard-X-ray SASE

Table 11.54 Main parameters of present and future X-ray FELs [11.1872, 1873]

	2006: FLASH	2013: European XFEL
Radiation		
Wavelength	13–180 nm	down to 0.1 nm
Peak power	up to 5 GW	up to 150 GW
Average power	10 mW	up to 500 W
Pulse duration	10–50 fs	0.2–100 fs
Driving electron beam		
Energy	0.3–0.7 GeV	up to 20 GeV
Peak current	1–3 kA	up to 5 kA
Undulator		
Period	2.73 cm	3.6–8 cm
Peak magnetic field	0.5 T	0.5–1.4 T
Undulator length	27 m	up to 200 m

FEL user facility – the European X-ray free-electron laser [11.1873]. The US department of energy (DOE) has given SLAC the go ahead for the engineering design of the linac coherent light source (LCLS) device to be constructed at SLAC [11.1876]. These devices should produce 100 fs X-ray pulses with over 10 GW of peak power. New X-ray sources will be able to produce intensities of the order of 10^{18} W/cm^2. The main difference between these projects is the linear accelerator: an existing room-temperature linac for LCLS at SLAC, and a future superconducting linac for the European XFEL. The XFEL based on superconducting accelerator technology will enable not only a jump in peak brilliance of ten orders of magnitude, but also an increase by five orders of magnitude in the average brilliance. The LCLS and European XFEL projects are scheduled to start operation in 2009 and 2013, respectively.

11.11.3 Suggested further reading

Books

1. T. C. Marshall: *Free-Electron Lasers* (Macmillan, New York 1985)
2. C. A. Brau: *Free-Electron Lasers* (Academic, Boston 1990)
3. P. Luchini and H. Motz: *Undulators and Free-Electron Lasers* (Clarendon, Oxford 1990)
4. W. B. Colson, C. Pellegrini and A. Renieri (Eds.): *Free Electron Lasers*, Laser Handbook, Vol. 6 (North Holland, Amsterdam 1990)
5. G. Dattoli, A. Renieri, A. Torre: *Lectures on the Free Electron Laser Theory and Related Topics* (World Scientific, Singapore 1993)
6. H. P. Freund, T. M. Antonsen: *Principles of Free Electron Lasers* (Chapman Hall, New York 1996)
7. E. L. Saldin, E. A. Schneidmiller, M. V. Yurkov: *The Physics of Free Electron Lasers* (Springer, Berlin, Heidelberg 1999)

FEL Conference Proceedings

In the period from 1985 to 2002 the FEL Conference Proceedings have been published in *Nuclear Instruments and Methods*, Section A. Starting from 2004 they are collected in electronic form on a WEB page dedicated to accelerator physics http://www.JACoW.org.

11.12 X-ray and EUV Sources

The extension of lasing into the X-ray region has been investigated by a wide variety of methods since the first demonstration of the visible laser in 1960. Although laser operation in the hard-X-ray region, which can be categorized as wavelengths shorter than 0.2 nm, has not been realized, much progress on the development of coherent light sources has been made in the soft-X-ray (0.2–30 nm) and extreme-ultraviolet (EUV, 30–100 nm) regions during these 20 years. There are two major methods to produce coherent short-wavelength radiation in these spectral regions. One is the use of transitions in highly charged ions in high-density plasmas created by laser irradiation of various targets or electric discharge. The other is the generation of very high harmonics of intense laser pulses. Each method has its advantages. X-ray lasers based on high-density plasmas can produce much higher energy per pulse and narrower spectra. High harmonics can be generated with compact, high-repetition rated lasers and produce a wide range of the spectrum from 100 nm to 3 nm. The use of these sources depends on the application and will be complementary.

11.12.1 X-Ray Lasers

The development of an X-ray laser has been one of the elusive dreams of laser physicists. Proposals for excitation schemes for X-ray lasers date back to 1965, when the possibility of achieving soft-X-ray amplification by collisional recombination was first suggested by *Gudzenko* and *Shelepin* [11.1877]. This was followed by proposals for photoionization pumping of X-ray lasers in 1967 [11.1878] and of electron-impact excitation schemes [11.1879]. The latter were inspired in part by the earlier success in the development of visible and ultraviolet ion lasers excited by electron collisions. However, the drastic scaling of the pump power requirements with decreasing wavelength and the low reflectivity of optics at soft-X-ray wavelengths, combined with the short lifetime of the excited levels involved in the lasing process, made the realization of soft-X-ray lasers a very challenging task [11.1880, 1881]. Several experiments carried out during the 1970s and early 1980s yielded the observation of population inversion and gain [11.1882].

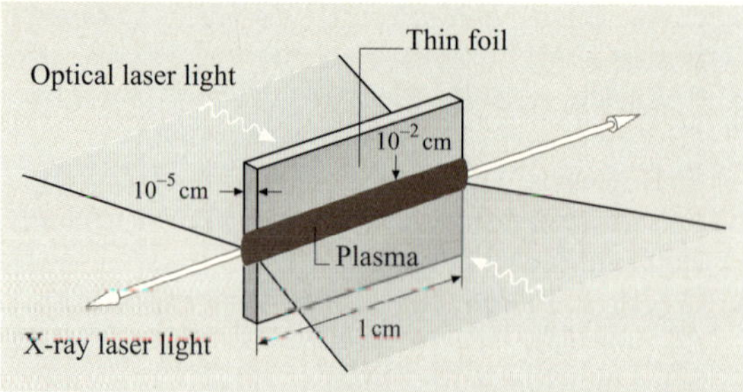

Fig. 11.217 Schematic drawing of an exploding-foil-target X-ray laser

Nevertheless, the experimental demonstration of large amplification at soft-X-ray wavelengths was not realized until 1984. *Matthews* et al. [11.1883] at the Lawrence Livermore National Laboratory reported soft-X-ray amplification at wavelengths of 20.6 and 21.0 nm in Ne-like Se by electron collisional excitation. An exploding-foil-target X-ray laser demonstrated by Matthews et al. and the observed spectra are shown in Figs. 11.217 and 11.218, respectively. In parallel with this, *Suckewer* et al. [11.1884] at Princeton University reported amplification at 18.2 nm in H-like C by recombination excitation. These pioneering works were soon followed by numerous successful soft-X-ray amplification experiments conducted using of the world's most powerful lasers as pump sources [11.1885–1887]. Subsequent experiments achieved soft-X-ray laser operation in the saturated-gain regime [11.1888], and brought several potential applications to fruition [11.1889, 1890]. These applications include microscopy, holography, diagnostics of dense plasmas, and the excitation of nonlinear photoluminescence in crystals. In spite of their pioneering works, these soft-X-ray lasers have not been used as a laboratory tool because both schemes required huge laser energies of several hundred joules and were operated at a low repetition rate. Since saturated amplification at a *water window* wavelength has not been accomplished even with the world largest laser driver [11.1891], a new excitation scheme has been sought for obtaining soft-X-ray lasers operating at shorter wavelengths. Many efforts have been directed in order to reduce the driver laser energy. The reduction of the pump laser energy is essential for the realization of tabletop X-ray lasers for potential applications [11.1892].

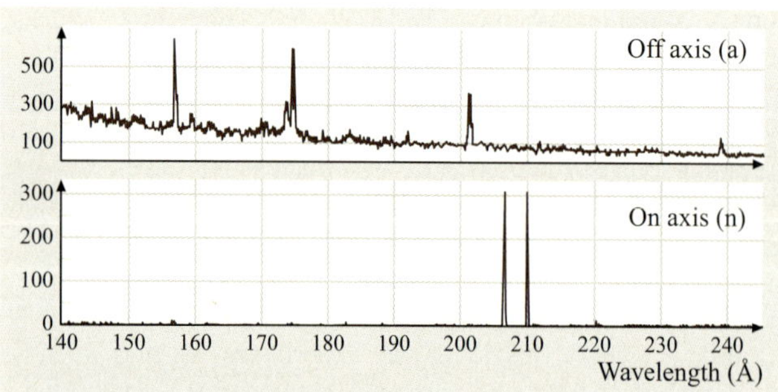

Fig. 11.218a,b Grazing-incidence spectrograph data from **(a)** off axis and **(b)** on axis

Electron Collision Excitation

Ions with specific numbers of electrons have a fully occupied outer-shell structure and survive over a wide range of plasma parameters. This is a significant advantage for plasma X-ray lasers because it results in a high relative abundance of the lasing ions over a wide range of temperature and density. To date, amplification has been observed in those ions. For example, the electronic structure of highly ionized selenium with 24 electrons removed (Se^{24+}) is similar to that of neutral neon (Ne-like), and the transitions are also similar to those of neutral neon [11.1893]. Such a comparison is referred to as isoelectronic scaling of energy levels and transitions.

Figure 11.219 shows a simplified energy-level diagram of a Ne-like ion scheme. The 3p laser upper states are populated mainly by electron collision excitation from the $2p^6$ Ne-like ion ground state. The population inversions are maintained by the rapid radiative decay of the 3s lower states to the ground state. The Ne-like scheme is well studied and is the most robust for electron collisional soft-X-ray lasers. However, it has the disadvantages that a large pump power is required to produce population inversion at a given wavelength. The nickel-like scheme was first proposed by *Maxon* et al. [11.1894] and proved to be useful for shorter-wavelength amplification, below 10 nm [11.1895]. Although the lasing scheme of the Ni-like ions is directly analogous to that of Ne-like ions, its higher quantum efficiency for a given ionization state allows the required pumping energy to be significantly reduced. Recently, gain-saturated amplification in the Ni-like scheme has been obtained at wavelengths as short as 7.3 nm [11.1896].

Transient Collision Excitation

The transient collisional excitation (TCE) scheme first proposed by *Afanasiev* and *Shlyaptsev* [11.1897] is a variation of the electron collisional excitation. TCE has attractive properties such as:

1. the gain coefficients are 1–2 orders of magnitude larger than those obtained for the same transition in the quasi-steady-state regime,
2. the required laser energy for amplification can be reduced greatly, resulting in the realization of tabletop X-ray lasers.

In the TCE scheme, an intense picosecond laser pulse overheats the preformed plasma at a rate faster than other collisional processes. The preformed plasma, containing the lasing ions such as Ne-like and Ni-like ions, is produced by a preceeding low-intensity nanosecond laser pulse. Transient gains in excess of $100\,\mathrm{cm}^{-1}$ have been predicted theoretically. Similar to the quasi-steady-state scheme, the TEC was firstly demonstrated in an Ne-like ion [11.1898] and has been successfully applied to Ni-like ions [11.1899]. Saturated amplification with a gain coefficient as high as $63\,\mathrm{cm}^{-1}$ has been reported from Ni-like Pd at 14.7 nm with a total pump energy as low as 7 J [11.1900]. Figure 11.220 shows the evolution of the 14.7 nm laser line along the plasma column. Recently, this scheme has been extended to lasing at 8.8 nm in Ni-like La [11.1901].

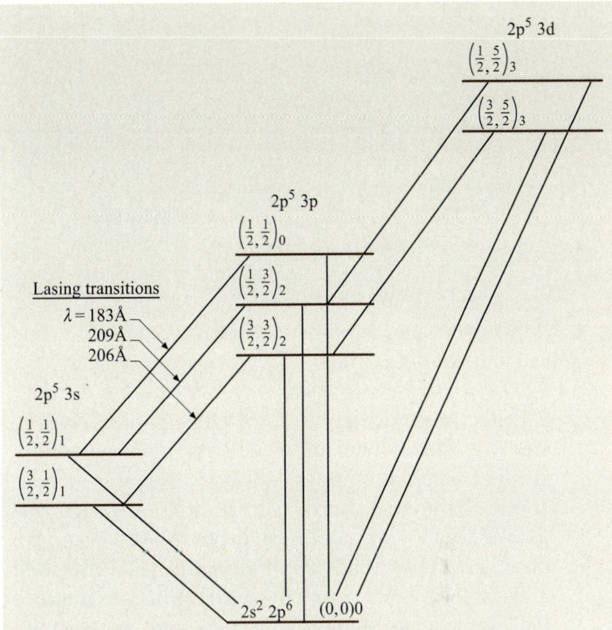

Fig. 11.219 Simplified energy-level diagram for Ne-like Se

Optical-Field Ionization X-Ray Lasers

There are two kinds of plasma production scheme for X-ray laser media by use of high-power lasers. The optical-field ionization (OFI) scheme is completely different from the conventional plasma production scheme. The OFI is direct ionization by a strong optical field, which modifies the Coulomb potential of the atoms or ions to create electrons by tunneling ionization [11.1902]. In contrast to conventional X-ray laser schemes using electron collisional ionization, the OFI scheme requires an intense laser power rather than the large laser energy. Recent advances in the technology of ultrashort-pulse generation and amplification have achieved laser powers high enough to realize OFI X-ray lasers with a tabletop device. The requirement for low pump energy will also allow soft-X-ray lasers to operate at high repetition rates, which is critical for most of the promising applications.

By applying this scheme to low-atomic-number media, a plasma consisting of fully stripped ions and free electrons is produced on a time scale much shorter than their collisional and radiative time. In other words, the temperature of electrons produced by the OFI is not determined by the charge state of ions, but can be controlled separately by ionizing-laser parameters such as the polarization and wavelength [11.1903]. By means of a quasistatic model, *Burnett* and *Corkum* [11.1904] showed that cold dense multiply ionized plasmas could be produced by OFI, which are suitable for recombination X-ray lasers. *Nagata* et al. [11.1905] demonstrated amplification of the Lyman $\alpha(n = 2-1)$ transition at 13.5 nm in H-like Li by the recombination pumping scheme following the OFI. Using a sub-picosecond KrF

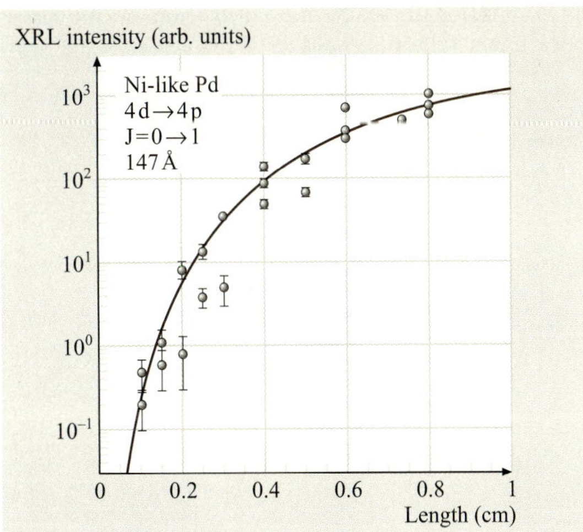

Fig. 11.220 Intensity of 14.7 nm X-ray laser line for 1–8 mm Pd target lengths

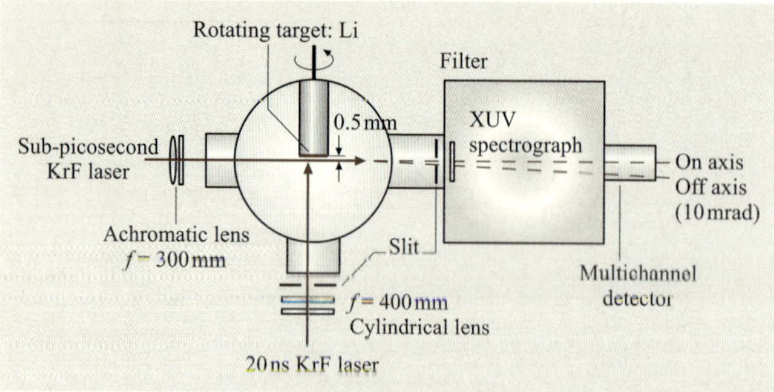

Fig. 11.221 Schematic diagram of the experimental setup of 13.5 nm optical-field ionization X-ray laser

excimer laser focused at $10^{17}\,\mathrm{W/cm^{-2}}$ (Fig. 11.221), singly ionized lithium ions were further ionized to fully stripped states, resulting in population inversion with respect to the ground state of hydrogen ions. A small signal gain of $20\,\mathrm{cm^{-1}}$ and a gain-length product of 4 have been observed. Subsequently, *Korobokin* et al. [11.1906] used a plasma waveguide in a LiF microcapillary to improve the propagation of the pump pulse and successfully increase the gain-length product to 5.5.

On the other hand, *Corkum* et al. also suggested the use of circularly polarized laser pulses for collisional excitation following OFI [11.1907]. Lemoff et al. proposed the specific systems [11.1908] and demonstrated lasing at 41.8 nm in eightfold-ionized Xe [11.1909]. In this experiments, circularly polarized Ti:sapphire laser pulse with an energy of 70 mJ and a duration of 40 fs was focused into a Xe static gas cell to create Pd-like Xe. The intense pump pulse produce the hot electrons which collisionally excite the Pd-like ions to the laser upper level. Recently, *Sebbon* et al. reported the saturated amplification of the 41.8 nm line in Pd-like Xe with 0.6 J, 30 fs Ti:sapphire laser pulses. 5×109 photons per pulse were obtained at gain saturation [11.1910]. They also reported the lasing at 32 nm in nickel-like krypton [11.1911]. The experimental setup is shown in Fig. 11.222.

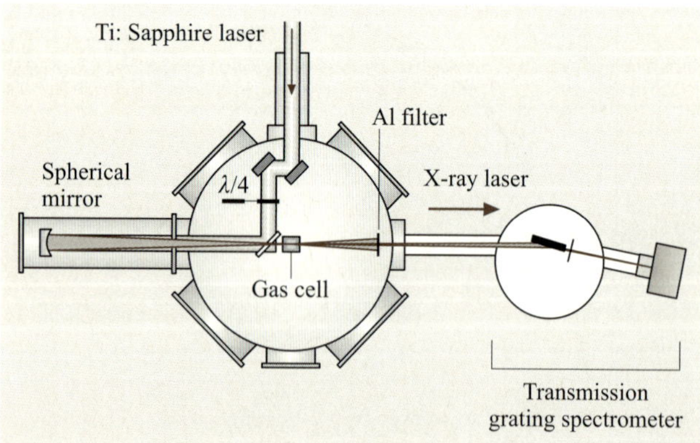

Fig. 11.222 Schematic diagram of the experimental setup of optical-field ionization X-ray laser based on electron collision excitation

Discharge Excitation

Direct excitation of plasma X-ray lasers by an electrical discharge has the potential advantage of high efficiency and compactness over a laser-produced-plasma device. The uniform high-density plasmas required for X-ray amplification can barely by produced by a conventional electric discharge. *Rocca* et al. proposed [11.1912] and demonstrated [11.1913] a capillary discharged soft-X-ray laser operating on the 46.9 nm transition in Ne-like Ar in 1994. A fast electrical discharge in a capillary was used to excite plasma columns up to 20 cm in length with a peak current of 40 kA. Figure 11.223 shows the fast capillary discharge soft-X-ray laser setup. The variation of the spectra as a function of capillary length is shown in Fig. 11.224. A recently developed compact discharge occupies a space of only $0.4\times 1\,\mathrm{m^2}$ in an optical table and produces an average output energy of 0.88 mJ at a repetition rate of 4 Hz. In two-pinhole interference experiments, a high degree of spatial coherence was observed by single-pass amplification in a 36 cm-long capillary. This compact soft-X-ray laser has been used in a variety areas including plasma physics, material characterization, and the characterization of soft-X-ray optics [11.1914].

11.12.2 High-Order Harmonics

Survey

High-order-harmonic generation (HHG) by the interaction of femtosecond high-intensity laser pulses with a gaseous medium has been extensively studied since it has promising potential for use as a coherent extreme ultraviolet and a soft-X-ray (XUV) source as an alternative to soft-X-ray lasers or synchrotron radiation sources. The harmonic spectrum has a very characteristic shape, that is, it falls off for the first few harmonics, then shows a plateau where all the harmonics have the same intensity strength, and finally ends with a sharp cut-off. A typical harmonic spectrum observed by the interaction of ultrashort high-intensity pulses with rare gases is shown in Fig. 11.225. The temporal duration of the harmonics is considered to be shorter than that of the driving laser pulse whose width is typically 100 fs or less. These spectral and short temporal properties make the high-order harmonics an unique coherent source in the XUV region and are opening new application fields.

Approach to Shorter-Wavelength Generation

Since the demonstration of HHG in rare gases using a KrF excimer laser by *McPherson* et al. [11.1915] and a mode-locked Nd:YAG laser by *Ferray* et al. [11.1916] in 1987, much effort has been made to extend the harmonic wavelength to a shorter one with various pumping sources. *Macklin* et al. [11.1917] generated the 109th harmonic (7.4 nm) in neon gas excited with a 806 nm Ti:sapphire laser. *L'Huillier* and *Balcou* [11.1918] observed the 135th harmonic (7.8 nm) using a 1 ps Nd:glass laser. In these experiments, ionization of the neutral rare-gas medium limits the achievable wavelength because of the low effective interaction intensity due to the occurrence of ionization during the interaction. Although ions having larger ionization energy potentially produce higher harmonics, simultaneously existing free electrons cause a large phase mismatch between the pump and high-harmonic waves, resulting in the lowering of cut-off orders. This drawback, which arises from the use of ions as a nonlinear medium, was compensated for to some extent with an ultraviolet pump source [11.1919]. *Nagata* et al. [11.1920] and *Preston* et al. [11.1921] reported the generation of the 37th harmonic (6.7 nm) of a 248 nm KrF excimer laser. In those studies with pumping pulses longer than 100 fs, the shortest wavelength achieved was limited to around 7 nm. More recently, however, progress of ultrashort-pulse laser technology has enabled the use of extremely short pulses with high intensity, which allows the circumvention of the limitation due to ionization [11.1922, 1923]. That is, the effective interaction intensity before the occurrence of ionization can be increase with the extremely short pulses. Consequently, harmonic wavelengths well within the *water window* region was attained, using an ultrashort Ti:sapphire laser operating with a 26 fs pulse duration [11.1924]. With a sub-10 fs Ti:sapphire laser, *Spielmann* et al. [11.1925] obtained coherent continuum emission shorter than 2.5 nm, which corresponds to a photon energy greater than 0.5 KeV. On the other hand, a few approaches from the medium side were reported. Although ions [11.1926, 1927], molecules [11.1928, 1929] and clusters [11.1930, 1931] have been investigated as nonlinear medium in place of rare gases, the output property of the harmonics was hardly improved for practical use.

Theoretical Progress

Theoretical understanding of HHG must be based on two processes:

1. a single-atom response in the driving laser field, and
2. a macroscopic response including propagation effects.

Both responses must be taken into account to make a comparison between the experimental and theoretical results.

The emission property of harmonic radiation from a single atom is determined by the induced atomic polarization, or dipole acceleration, which is calculated from the solution of the time-dependent Schrödinger equation (TDSE). *Krause* et al. [11.1932] showed that the photon energy $E_{\max}$ of the highest harmonic in the plateau is give by $E_{\max} = I_p + 3.17U_p$, where I_p is the ionization energy of the nonlinear medium and $U_p = 9.33 \times 10^{-14} I\lambda^2$ (eV) is the ponderomotive energy of the electron subjected in the pump laser field I (W/cm^2) at a wavelength of λ (μm). In a semiclassical picture [11.1933], the physical origin of this expression is explained in terms of tunneling ionization of the atom, followed by acceleration of the ionized electron in the pump field and recombination with its parent ion. During the acceleration by the optical field, the electron obtains the maximum energy of $3.17U_p$. When this electron recombines with the parent ion, the electron release this energy plus the ionization energy as a harmonic photon. Figure 11.226 shows a semiclassical model of high-order-harmonic generation.

Since calculation based on the TDSE is quite time consuming, it is very difficult to combine the numerical results of TDSE with the propagation equations. The calculation load is considerably relieved by using the model of *Lewenstein* et al. [11.1934, 1935]. This model is based on the strong-field approximation and is valid in the region $U_p > I_p$. *L'Huillier* et al. developed a propagation code coupled to the Lewenstein model and successfully described the harmonics characteristics observed in various experiments [11.1936, 1937].

Phase Matching

In order to increase conversion efficiency and improve the spatial quality of high harmonics, phase matching is essential. It is, however, not easy to satisfy the phase-matching condition along the interaction length because, in contrast with low-order-harmonic generation in the perturbative regime, the dipole phase is dependent on

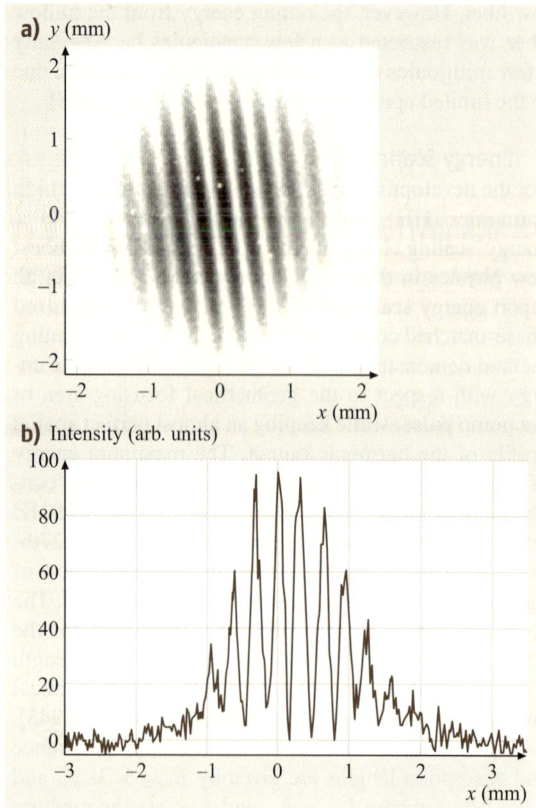

Fig. 11.229a,b Interferogram of a harmonic beam diffracted by double pinholes separated by 100 μm: **(a)** fringe image, **(b)** intensity profile of the interferogram along a horizontal line passing through the center

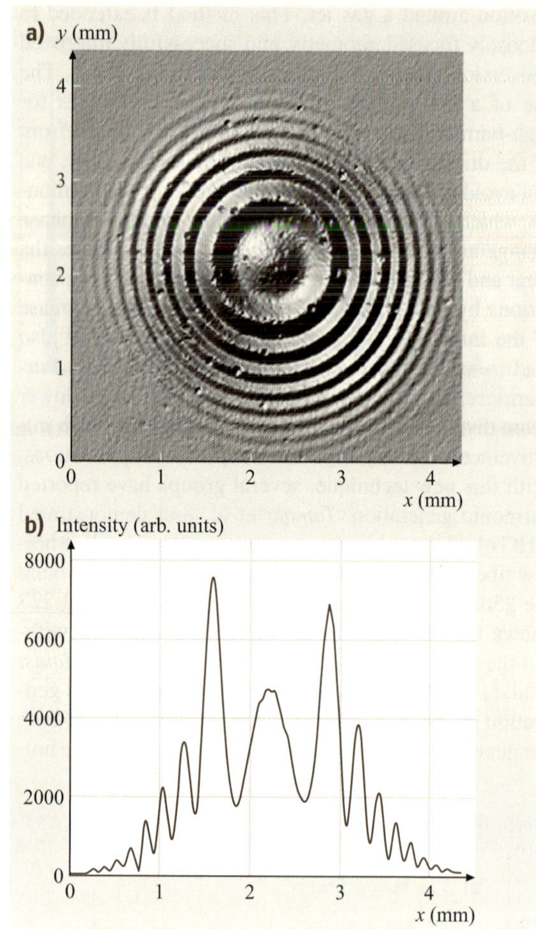

Fig. 11.230 **(a)** Interferogram recorded with a PDI plate with a 10 μm pinhole. **(b)** Intensity profile of interferogram along a horizontal line passing through the center

length, the coherence length, and the absorption length, respectively. The 23rd- and 25th-order harmonics satisfied the optimized condition for this relation. Therefore, those orders were saturated under the experimental conditions. On the other hand, the 27th-order harmonic did not yet satisfy the above conditions, because of low absorption.

Spatial Coherence

Since HHG is based on nonlinear frequency conversion, the spatial and temporal coherences of high-order harmonics are expected inherently to succeed to those of the fundamental laser pulse. However, HHG conducted by use of a tight-focusing geometry in thin gas media did not allow full phase matching to enhance the coherence. Typically multimode components are observed as broadened peaks in the spectrum or the pedestal of the spatial profiles. Recent studies have revealed that macroscopic phase matching can realize highly efficient and spatially well-characterized (nearly Gaussian profile) high-order harmonic light [11.1949, 1950]. Several groups have reported the interferometric measurement of the spatial coherence of high harmonics [11.1951–1954]. When the macroscopic phase matching is achieved in the hollow fiber or loosely focused geometry, the measurements show that the harmonic beams have almost perfect spatial coherence. Figure 11.229 shows the interferogram of the 27th-harmonic beam generated with an Ar-filled hollow fiber pumped with 20 fs, 0.35 mJ Ti:sapphire laser pulses [11.1954]. This interferogram was obtained with two pinholes separated by 100 μm, while the harmonic-beam diameter was measured to be 130 μm. Figure 11.230 also shows the interferogram recorded by point-diffraction interferometry, which indicates that the harmonic beam can be considered as a spherical wave within a phase error of $\lambda/15$ [11.1954].

Applications

In addition to applications in atomic physics [11.1955], high harmonics have been used for solid-state spectroscopy [11.1956] and plasma diagnostics [11.1957]. High intensity of high harmonics is also expected to give rise to nonlinear phenomena in the XUV region. As described above, *Takahashi* et al. [11.1944, 1950] generated a peak power of 130 MW at 62.3 nm in 0.6 torr Xe and 10 MW at 30 nm in 2 torr Ar. When these harmonic pulses are focused with multilayer mirrors, the focused intensity will reach $10^{14}\,\mathrm{W/cm^2}$, which is high enough to cause nonlinear interactions. Theoretical predictions of nonlinear interaction between soft X-rays and matter have been reported by a few groups, such as two-photon ionization of He^+ [11.1958], He double ionization for the autocorrelation of an extreme-ultraviolet (XUV) pulses [11.1959], the advantages of high-intensity short-wavelength radiation for Coulomb explosion imaging [11.1960], and the ionization of cluster targets. These research field can be expected to open a new area in high-intensity physics.

Furthermore, coherent XUV and X-ray pulses are not only useful owing to their short wavelengths, but are also important due to their potential to produce electromagnetic radiation in the range of attoseconds. High-order harmonics have been expected to be the source of attosecond pulses [11.1961, 1962]. Recently, *Hentschel* et al. [11.1963] demonstrated the generation of isolated soft-X-ray attosecond pulses and their temporal characterization by a novel cross-correlation technique using intense few-cycle visible laser pulses.

11.13 Generation of Ultrahigh Light Intensities and Relativistic Laser–Matter Interaction

Modern laser technologies allow the amplification of short laser pulses to energies of some tens of kJ. Additionally, ultrashort pulses containing only a few optical cycles can be generated. By merging these techniques nowadays focused laser beams can reach unprecedented intensities in the range of $10^{21}\,\mathrm{W/cm^2}$ and will reach even higher values in the near future. At these intensities the electric and magnetic field strength is many orders of magnitude higher than those that will ever be possible in a static generation scheme. By applying these fields to a target it becomes possible to gain access to a new interaction regime of light and matter: relativistic optics. This opens a new wide area in experimental science where classical optics meets plasma dynamics, relativistic quantum mechanics, and high-energy physics.

11.13.1 Laser Systems for the Generation of Ultrahigh Intensities

Amplification of Ultrashort Pulses to High Energies

For ultrashort pulses the energy density of light at the surface and in the volume of all the optical elements is limited by the onset of nonlinear effects and laser damage due to the high peak power. Moreover, in a laser amplifier the energy extraction efficiency is a function of the ratio of the energy density and the saturation fluence of the laser material. Hence, a short pulse cannot be amplified efficiently. The chirped pulse amplification (CPA)

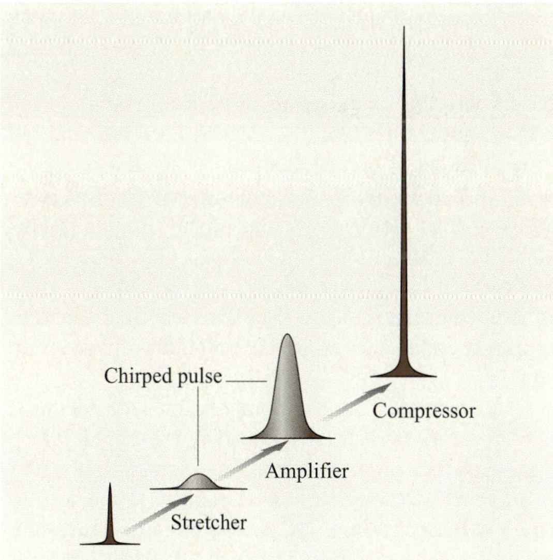

Fig. 11.231 The CPA principle. An ultrashort pulse generated in a mode-locked laser is stretched by adding a spectral phase that group delay different wavelengths of the pulse. After amplification retaining the pulse spectrum and chirp the pulse is recompressed to its original length, resulting in ultrahigh peak power

technique [11.1964] circumvents this difficulty. The principle of CPA is depicted in Fig. 11.231. Ultrashort pulses contain a broad spectral bandwidth according to their Fourier transformation (see Chap. 12). This fact enable the possibility to add a phase shift to different frequencies or wavelengths of the laser pulse just after its generation in a mode-locked laser cavity. The result is a stretched pulse containing a chirp, i. e., the pulse duration is no longer bandwidth limited. The lengthened pulse has accordingly a lower peak power. It can be amplified much more efficiently and compressed to a very short pulse afterwards by adding a spectral phase with an opposite sign to that introduced by the stretcher.

After the invention of the CPA technique tremendous progress in the development of ultrashort pulse lasers towards higher peak power has been witnessed. Nowadays, terawatt (TW) laser pulses can be produced using tabletop laboratory-scale laser systems that operate at repetition rates of 10 Hz or higher. Whereas, even higher power up to the petawatt (PW) can be generated by flashlamp-pumped lasers that have factory-building sizes, like the first PW laser, that was realized in 1999 [11.1965]. The CPA technique is common to all of these devices.

Because pulse stretching and compression are related to each other and are normally based on the same principle, the stretcher and compressor are treated here as a stretcher–compressor pair (SCP). Pulse stretching naturally occurs in the vicinity of dispersion. Hence, the propagation of pulses in dispersive media always leads to pulse stretching. In order to keep the required space and size of optical elements as small as possible highly dispersive optics, i. e., gratings will be used. Nevertheless other elements such as prisms, grating prisms (so called grisms), fibers and chirped fiber Bragg gratings as well as chirped mirrors can be applied. Here we limit ourselves to the simplest case of flat reflective phase gratings, because these offer the largest group delay.

The principle of pulse compression with identical parallel gratings was first described by *Treacy* [11.1966]. A pulse is diffracted by a grating, angularly split into its different wavelengths and propagated to the second parallel-aligned grating. The second grating removes the angular modulation, i. e., redirects all wavevectors of the different wavelets into the same direction. If all the waves are treated as plane waves, the introduced lateral shift can be neglected. For a real, limited beam size the same device of parallel gratings has to be used once more in order to combine the wavelets again spatially. As a side effect this doubles the stretching factor of the arrangement. Section 12.1.3 contains a detailed description of these devices.

For ultrahigh-peak-power lasers a maximum stretching/compression factor, i. e., the ratio of stretched pulse length to bandwidth limited pulse length is required in order to maximize the fluence in the amplifier chain. The stretching factor depends on the line density of the gratings, the center wavelength, and the bandwidth of the pulse. Very long stretching factors of 10 000 or more are required for most broadband rare-earth-doped laser materials in order to allow fluences in the amplifier close to or above the gain-saturation fluence.

If the grating distance is chosen in this way, such that the unclipped spectral bandwidth is twice the spectral full width at half maximum (FWHM) of the pulse, the maximum stretched pulse length τ_{max} is only a function of the grating size L:

$$\tau_{max} = 2\frac{L}{c}\cos(\alpha) , \qquad (11.205)$$

where $\cos\alpha$ is a factor according to the diffraction angle α of the grating and has a maximum of unity in the unrealistic case of a diffracted beam parallel to the grating surface. The factor of 2 refers to the Littrow case, i. e., the input and diffraction angle being equal. The latter normally maximizes the grating diffraction efficiency. For holographic metal-coated diffraction gratings with near-rectangular groove shapes it has been shown that an optimum diffraction efficiency of a metal grating at wavelength λ can be achieved for a grating constant of about $\sqrt{2}\lambda$ [11.1967], which results in a Littrow angle of 45°. This leads to a minimum grating size for an SCP of about 24 cm if a femtosecond pulse has to be stretched to 1 ns. Passing a stretcher or compressor setup again can, in principle, increase the stretching factor, but introduces losses that are most often acceptable for the stretcher but not for the pulse compressor.

Very large grating distances and hence long stretched pulses require a similar accuracy for the compensation of dispersion by a well-matched stretcher and compressor as in the case of ultra-broadband fs pulses, because higher-order dispersion terms increase together with the second-order term. Because most often aberrations of the telescope optics cannot be tolerated, aberration-free all-reflective designs like the Öffner triplet are applied [11.1968]. In special cases aberrations introduced by the stretcher may help to compensate for laser-material dispersion in the amplifier chain [11.1969].

The required grating sizes for the compressor of a high-energy laser system is ruled by the generated group delay in order recompress the pulse and the size

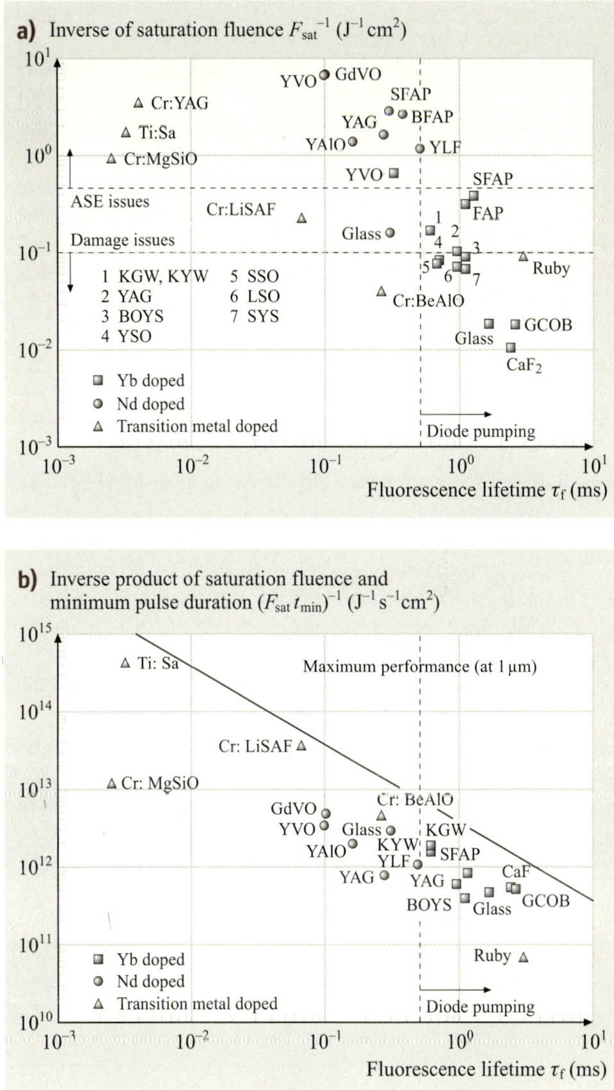

Fig. 11.232 **(a)** Capability of laser materials for energy storage and generation of high energy pulses. The inverse saturation fluence is plotted against the fluorescence lifetime. For a higher gain amplified spontaneous emission (ASE) become an issue whereas for lower gain damage limits efficient energy extraction. Effective pumping will be useful if the fluorescence lifetime exceeds a certain limit. The limits are marked with *dashed* lines. **(b)** Capability of laser materials for energy storage and generation of high peak power. The inverse saturation fluence multiplied by minimum possible pulse width plotted against fluorescence lifetime τ. Differently doped materials are marked with different colors

required to fit the compressor to the beam diameter. Before compression the laser beam has to be expanded to reduce the fluence in the beam to an amount well below the damage threshold of the compressor optics, particularly the gratings. For a metal-coated grating damage thresholds may reach the $0.5\,J/cm^2$ level [11.1967] for half-ps pulses, but will normally be less than $0.25\,J/cm^2$. In the compressor setup these fluences result in intensities that are well above the onset of nonlinear effects in air at normal pressures. In order to prevent the pulse from self-focusing and self-splitting into white-light filaments, pulse compression has to be performed in a vacuum vessel.

In the region of fs to ps pulses the damage threshold for metal coatings is nearly independent of pulse duration, in contrast to dielectrics where the main damage mechanism is linked to nonlinear absorption effects. For pulses longer than 100 fs dielectric gratings may substantially improve overall performance [11.1970]. Damage thresholds two to four times that of gold-coated gratings are reported. These dielectric phase gratings on top of a multilayer mirror can, in principle, show a 100% diffraction efficiency to the minus-first order [11.1971]. Therefore they are favored in high-energy high-peak-power laser systems.

The peak power of a laser system is limited mainly by the possible sizes of the diffraction gratings. A workaround is to add identical smaller gratings to a mosaic grating or tiled grating. The smaller grating tiles have to be coherently phased together in order to make them work like a single one. Using this technique to build a compressor opens the way to further power scaling of high-peak-intensity lasers. Compression of pulses to 650 fs with a tiled grating replacing a meter-sized grating in a high-energy laser system has already been demonstrated [11.1972], as well as pulse compression of a chirped 2 ns pulse down to 150 fs [11.1973] by phasing two 350 mm-wide gold-coated gratings with the help of a six-axis piezo-driven mount.

Laser Materials for High Peak Power

A crucial issue conceiving a high-peak-power laser system is the choice of the gain material. Broad band width and a high cross section for stimulated emission are required. Due to the limited peak power of the pump source a long fluorescence lifetime for energy storage is desired. At higher fluorescence lifetimes either the emission cross section, or the gain bandwidth, or both decrease. For efficient amplification the extracted laser pulse energy density has to be close to the saturation fluence of the gain medium.

Figure 11.232 illustrates the possibility of generating high peak power out of some established laser materials. The inverse of the product of the saturation fluence and the shortest pulse duration indicates the capability of a laser material for high amplification at maximum bandwidth. Assuming Gaussian-shaped gain spectra the correlation between the emission cross section, bandwidth and fluorescence lifetime of a laser material is given by

$$\sigma_{\text{em}} = \frac{c_0^2}{4\pi n^2 \nu^2} \frac{1}{\tau_{\text{f}}} \frac{\sqrt{\ln 2}}{\Delta\nu\sqrt{\pi}} \qquad (11.206)$$

with c_0 the velocity of light in vacuum, n the refractive index, h Planck's constant, ν the center frequency, τ_f the fluorescence lifetime, and $\Delta\nu$ the bandwidth (FWHM). Applying the time–bandwidth product for Gaussian lineshape a criteria for the generation of high peak power at a corresponding fluorescence lifetime depending on the laser wavelength lambda and the refractive index n is obtained

$$\frac{\tau_{\text{f}}}{t_{\text{p}} F_{\text{sat}}} \le 4.26 \times 10^9 \frac{(\lambda[\mu\text{m}])^3}{n^2} \frac{[\text{cm}^2]}{[\text{J}]} . \qquad (11.207)$$

Here the saturation fluence F_{sat} is given by $F_{\text{sat}} = h\nu/\sigma_{\text{em}}$.

In Fig. 11.232 suitable materials for amplification to a high energy level are mapped. Furthermore, there is an optimum region between high gain and low gain, where amplified stimulated emission and damage issues, respectively, are likely to occur.

Recently, it was observed that cryogenic laser materials may show enhanced performance. Either spectroscopic parameters and thermal behavior can be improved by cooling to liquid-nitrogen temperatures [11.1974]. Ti:Sapphire for instance gain, higher efficiency and thermal conductivity. Quasi-three-level systems represented by the Yb-level system show higher emission cross sections and additionally reduced absorption at the laser wavelength, because population of the low laser level is suppressed.

Laser Amplifier Schemes for High Peak Power

Except some examples of high-energy excimer lasers and upcoming free-electron lasers for the generation of high peak power at very short wavelengths typical scientific lasers are based on solid-state lasing materials as presented above, all of which are pumped optically. Because a high flux of pump photons is needed to invert the energy-level population significantly only three options for pumping exist: flashlamps, lasers, and diode lasers. Diode lasers are treated separately because no enhanced additional pumping is required for them, providing a stabilized current is sufficient.

So far the flashlamp is the cheapest generator for a high pump photon flux. In the past flashlamp-pumped solid-state lasers have been scaled to 20 kJ pulse energy in a single beam. Such systems are based on schemes where bundles of meter-size flashlamps pumping several Nd-doped glass discs arranged at Brewster's angle in the laser beam path as sketched in Fig. 11.233. For fusion lasers like the National Ignition Facility in the US (NIF) [11.1975], Laser Megajoule in France (LMJ) [11.1976] or Gekko in Japan [11.1977] several such beam lines are bundled together to form megajoule laser facilities. Such beam lines were the first with the capability to generate petawatt laser pulses [11.1965]. Typical for this kind of lasers are a double-pass main amplifier and a single-pass booster amplifier section.

The drawback of flashlamp pumping is the low efficiency of this scheme. Pump photons are not only generated at the absorption wavelength of the laser material but also at wavelengths not involved in the lasing process. This low efficiency results in substantial heat left inside the laser materials, which are themselves poor heat conductors. The maximum repetition rate of this disc configuration scales with the inverse square of the beam diameter. Amplifiers with beam diameters of some 10 cm can be operated at repetition rates of some shots per day only. With these systems scientific investigations of plasma effects are extremely difficult and expensive. For Nd-doped glasses the gain bandwidth limits the shortest pulses to about 400 fs. To generate 100 TW to PW peak power pulses, energies in the 100 J range are needed, which require large amplifiers. Nevertheless, using these lasers for pumping an OPCPA (see below) opens the door for ultra-broadband amplification of few-

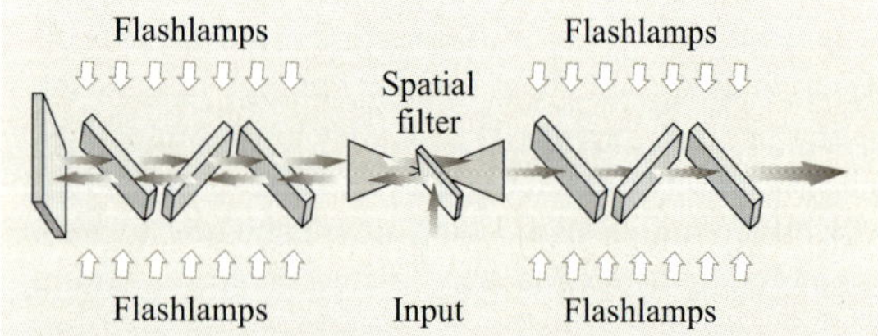

Fig. 11.233 Typical flashlamp pumped slab amplifier configuration. The seed pulse is injected at the spatial filter between a double-pass section and a single-pass booster amplifier

cycle pulses to the 10 J level with reasonable repetition rates [11.1978].

Using laser diodes as the pump source helps to relieve some of these difficulties because their emission spectrum is much narrower and can be fitted to the absorption line of the laser-active medium. Consequently less energy is wasted and less heat is left inside the laser active medium. However, high-power diodes with energy-conversion efficiencies up to 75% and continuous-wave output in the 100 W range are only available in the red and near-infrared spectrum. These diodes are based on double heterostructures of ternary and quaternary semiconductors incorporating GaAs. This spectral range is preferred for pumping rare-earth-doped laser materials, where Nd and Yb are most favored. Broadband gain materials including transition metals such as Cr:LiSAF will attract more interest if suitable diodes providing high brightness are developed. Today diode-pumped infrared frequency-doubled lasers are used to pump Ti:sapphire, the laser material able to amplify the widest spectrum and hence the shortest pulses.

The direct diode-pumped amplification of 150 fs pulses to the joule and 10 J level has already been shown by pumping an Yb-doped fluoride-phosphate glass at 940 nm [11.1979, 1980] with the edge-cooled disc arrangement shown in Fig. 11.234a.

To accomplish experimentalists' demands for higher repetition rates advanced cooling technologies have to be developed even for diode-pumped solid-state lasers (DPSSL). Figure 11.234 shows a variety of DPSSL pumping schemes with different cooling architectures.

In slab lasers, where the various pump schemes depicted in Fig. 11.234b–d are employed, one dimension is used to provide a small distance for heat removal. They differ in the way in which pump light is provided. Expanding this idea to two dimensions results in the cladding pumped fiber amplifier (Fig. 11.234g), which is not or only radiatively cooled like the heat capacity lasers (Fig. 11.234f), which are able to run as long as the temperature does not reach a certain level. These lasers provide bursts of pulses. Thin disc lasers (Fig. 11.234e) allow a very high average power and can produce very good beam profiles but are difficult to scale to larger beam diameters and higher pulse energies. A compromise is the use of a thicker disc at a moderate repetition rate in the same configuration. The thicker disc does not require multiple pump-beam passes for full absorption and provides a larger single-pass amplification than the thin-disc amplifier.

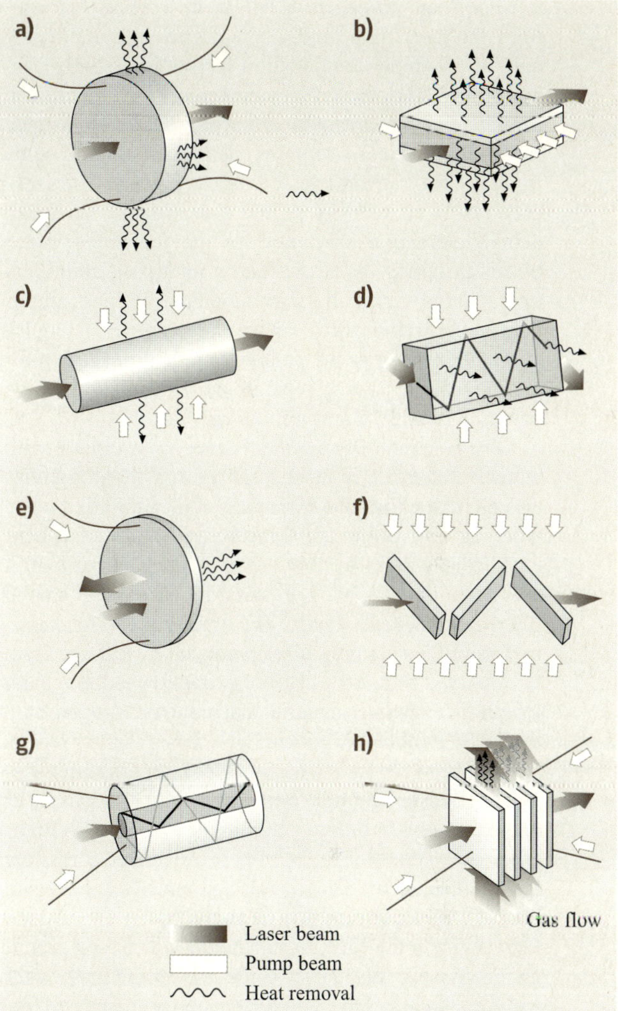

Fig. 11.234a–h Amplifier schemes for diode pumping. Edge cooling: **(a)** discs – conductive cooling **(b)** transverse diode-pumped structured slab and **(c)** transverse diode-pumped slab. Water cooling: **(d)** zigzag slabs and **(e)** thin discs. No cooling: **(f)** heat-capacity lasers and **(g)** fiber amplifiers. Gas cooling: **(h)** thin-disc assemblies

By putting several thin discs together and using the amplifier slab interspace for gas cooling, as in the architecture sketched in Fig. 11.234h, the gain and absorption equals that of the thick disc but efficient cooling is enabled. A diode-pumped laser of this type was developed at the Lawrence Livermore National Laboratory in the US. It is called Mercury [11.1981] and allows an output of 65 J, the highest ns pulse energy from a single DPSSL at a repetition rate of 10 Hz so far reported.

The long propagation length in a fiber that confines the laser light field to the doped and amplifying core allows high amplification factors in a single pass and optical-to-optical efficiencies close to the theoretical limit given by the quantum efficiency. For TEM_{00}-mode radiation the maximum pulse energy is limited to some mJ for CPA. Nevertheless, average powers from several 100 W to the kW region are possible. To reach a high extraction efficiency for the low-energy seed pulses provided by fs oscillators multipass amplifiers or regenerative amplifiers are adopted. Multipass amplifiers can be found with a variety of geometries. All pulse parameters, such as the polarization or propagation direction, can be used to separate pulses from successive passes and finally for out-coupling.

The regenerative amplifier is a seeded oscillator with an active element, in most cases a fast switching Pockels cell, that allows the extraction of the amplified pulse when the gain of the active medium is saturated. It benefits from a stability that stems from gain saturation and beam quality that is ensured by successive spatial filtering in the laser cavity. The drawbacks of regenerative amplifiers are pre-pulses produced by leakage from the cavity in round-trips before extraction and long path lengths in dispersive material that has to be compensated by the SCP of a CPA system.

The last section of a high-power laser amplifier system is a booster amplifier with a low number of passes or one pass only. In order to extract the energy with high-efficiency, saturation fluence has to be reached across the full beam diameter. The result is a so-called top-hat beam profile with a uniform fluence in the center and a steep fluence rise at the edge. In gain media with a long propagation distance this may be achieved automatically after some distance and the length were uniform aperture filling is not guaranteed hardly affects the overall efficiency of the amplifier. For amplifiers with short propagation distances of the pulse in the active medium the seed pulse itself has to be converted into a top-hat profile starting from a Gaussian. A simple method to implement this is by diffraction at a serrated aperture [11.1982].

Because a top-hat beam unlike the Gaussian changes its profile while propagating it has to be relay imaged to successive optical elements of the laser system to avoid hot spots and laser damage. In high-power laser systems relay imaging is often combined with beam expanders and spatial filters [11.1983].

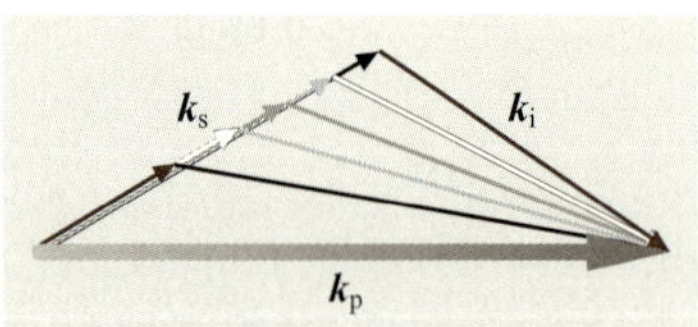

Fig. 11.235 Scheme to achieve broadband optical parametric amplification. In order to achieve broadband signal amplification in a parametric nonlinear optical process using a narrow-band pump different idler wavelengths have different propagating directions in a crystal to fulfill the phase-matching condition while all signal waves propagate collinearly

Broadband Optical Parametric Chirped Pulse Amplification

Alternatively instead of conventional laser amplifiers a parametric amplification process (OPA) can be used for the generation of high-energy broadband pulses. In this second-order nonlinear interaction a pump photon is split into a signal and an idler photon. This process is the reverse of sum-frequency generation. Energy-conservation demands that the sum of the signal and idler frequencies equals the pump frequency. Additionally, fulfilling momentum conservation ensures coherence of the involved waves while propagating through the nonlinear medium. The latter is known as phase matching. The nonlinear medium has to be a crystal in order to show second-order nonlinearity and achieve phase matching. The phase-matching condition also determines the wavelengths of the signal and idler waves.

For a certain direction of pump, signal, and idler waves only one combination of wavelengths can be generated. In order to allow broadband amplification of a signal input wave using a narrow-band pump the different idler wavelengths should have different directions in the crystal. This principle is illustrated in Fig. 11.235 and is called optical chirped pulse amplification (OPCPA) if the signal beam is a chirped stretched pulse [11.1984–1987].

With the OPCPA scheme, energy can be transferred from a narrow-band nanosecond pulse to a stretched broadband signal pulse. The problem of amplification of ultrashort pulses is thereby split into the task of amplification of high-energy laser pulses and the task of implementation of broadband OPA. Comparing the OPCPA with a conventional laser-pumped CPA several advantages can be realized. Firstly no energy is stored in the nonlinear crystal and, except parasitic absorption, no energy is lost in the crystal, which results in no heat being produced in the process. This fact allow easy high-repetition-rate scaling of the amplifiers without phase distortions by thermal effects. Because no resonance to energy levels of a dielectric medium is required, very large band widths at any center wave length can be generated. It was shown that pulses shorter than 10 fs can be amplified. Using crystals with high nonlinear optical coefficients like the most prominent crystal BBO and at sufficient intensities a single pass can provide an amplification of the signal by many orders of magnitude. Regenerative amplifiers are not needed and the problem of pulse leakage from them disappears. Amplification lasts only as long as the pump passes the crystal and therefore pre-pulses and post-pulses as well as back-reflected pulses are not amplified any more and higher pulse contrasts are possible with the OPCPA technique. Lastly the amplified signal beam is an analog of the input signal and unwanted phase distortions of the high-energy pump are carried away with the idler beam like the different $\boldsymbol{k}$ vectors in Fig. 11.236.

In order to use all of these advantages a careful design of OPCPA is required. For instance if an OPCPA stage is driven to its maximum amplification and the process starts to saturate, the impact of intensity variations of the signal to the output is reduced but intensity variations of the pump are transferred to the amplified beam. Additionally, signal and idler waves are generated starting from noise, a problem similar to the amplified spontaneous emission (ASE) in a conventional laser amplifier. To minimize this effect a pump pulse length no longer than the seeding signal pulse and their perfect synchronization is required, which is challenging for ps pulses. Contrariwise ps pulses offer higher intensities at a fixed fluence, which reduces the required crystal length, with the advantage of increasing the amplification bandwidth without changing the crystal orientation. More requirements may arise at different wavelengths involved, like absorption, or second-harmonic generation from the signal or idler beams, amongst others.

Nevertheless, OPCPA is the technique that allows the implementation of an ultrashort pulse option to all coherent high-energy light sources that emit pulses in the nanosecond and picosecond range and transfers them to an ultrahigh-peak-power laser system. A 200 TW 45 fs laser system was recently demonstrated [11.1988] based on OPCPA in a large-aperture KDP crystal. These crystals are used for Pockels cells and frequency converters in fusion laser systems and can be grown with meter-size apertures. This in principle allows multi-petawatt power scaling.

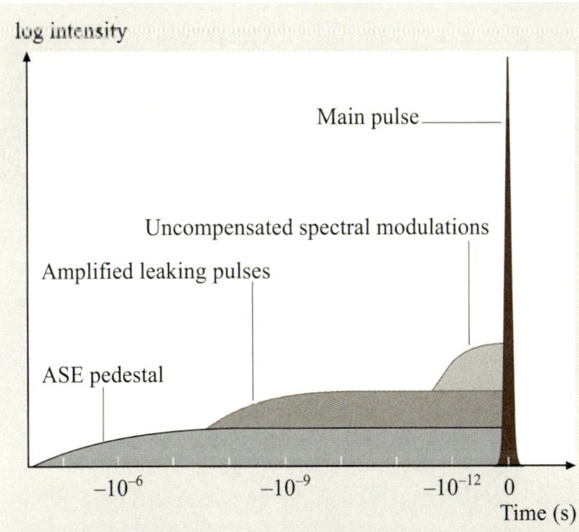

Fig. 11.236 Typical pulse contrast characteristics. The main femtosecond laser pulse is typically preceded by amplified stimulated emission (ASE), amplified pre-pulses and uncompressed chirp

High Demands on Pulse Preparation for Advanced Experiments

The acceleration of electrons and protons by a laser requires light intensities on the order of $10^{20}\,\mathrm{W/cm^2}$, which is many orders of magnitude above the onset of nonlinear effects and atom ionization, which results in Coulomb explosion even in low-density targets such as gases, triggered by lower intensity background radiation of the laser. Experiments are strongly affected by this pre-pulse laser-target interaction [11.1989]. To prevent the target from being destroyed before the main part of the laser pulse arrives, laser output has to be cleaned from its leading pedestal. A typical laser pulse characteristic is shown in Figure 11.236. The long-term pre-pulse pedestal results from ASE of the laser active material, whose duration is closely linked to its fluorescence lifetime. Additionally, regenerative amplifiers that are often used as the first devices in the amplifier chain, always have a certain leakage of pulses from preceding round-trips. The pulses are again amplified in successive multipass and booster amplifiers, generating high-energy pre-pulses on target. Finally, non-bandwidth-limited pulse recompression based on material dispersion, wavelength clipping, aberrations in stretcher and compressor setups and uncompensated higher-order dispersion terms result in pre-pulses and pedestals as shown in Sect. 12.1.2 for frequency-domain-filtered short pulses.

Numerous methods are used to remove this unwanted pre-pulse laser light. Primarily, the total amplification is split into stepwise amplification with additional possibilities for time-domain and spatial pulse filtering. For this purpose fast Pockels cells [11.1990] and spatial filters, which are often combined with beam expanders, are applied. Spatial filters and apertures prevent spontaneously emitted fluorescence light from propagation to the subsequent amplifiers. These measures can often improve pulse contrast ratios considerably [11.1991].

Nonlinear optical effects that favor high intensities can be used as well. Examples of the latter case are saturable absorbers and laser pulse switched gates. The OPCPA inherently embodies such a gate if a rectangular-shaped pump pulse is used. Therefore pre pulse suppression in the OPCPA [11.1992] can be as high as its gain.

Saturable absorbers and self-phase-conjugating mirrors are used for nanosecond laser systems. In the case of a chirped pulse amplification system these techniques require pulse compression before and stretching after filtering. This limits their application to low-energy pulses or prohibits their use if very long stretching factors are needed.

Nonlinear effects can also be used after final pulse recompression. One possibility is frequency conversion by second-harmonic generation. Another pre-pulse suppression method with high efficiency and low loss is the usage of a plasma mirror [11.1993]. Because the laser has to be focused anyway a perfectly transmitting dielectric medium is placed near the focus of the beam. The leading edge of the pulse is transmitted until the intensity reaches the ionization threshold. At this step a plasma is generated with increasing electron density. The plasma occurs on a time scale that does not allow it to expand. If the plasma frequency matches the laser frequency this plasma acts as a perfect mirror and the main pulse is reflected to the target. With a single plasma mirror the contrast ratio for petawatt-class lasers can be improved by factors of about 100 [11.1965].

Experiments with lasers at ultrahigh intensities require the focusing of a high-peak-power laser onto a very small spot. Phase distortions in the beam line can increase the minimum spot size in the focal plane considerably. Adaptive mirrors are used to correct phase fronts [11.1994]. Closed-loop systems allow near-diffraction-limited focusing of ultrashort pulses [11.1995, 1996]. These techniques today allow intensities in the range of $10^{20}-10^{21}\,\mathrm{W/cm^2}$ with terawatt and petawatt laser systems.

11.13.2 Relativistic Optics and Laser Particle Acceleration

The interaction of light and matter at intensities of $10^{20}-10^{21}\,\mathrm{W/cm^2}$ changes considerably when compared to classical, even classical nonlinear optics. At intensities of $10^{13}-10^{15}\,\mathrm{W/cm^2}$ matter becomes ionized and the high intensity laser pulse interacts with a dynamically evolving plasma where the dominant mechanism for the interaction of light and matter is the interaction with free electrons.

In classical optics the interaction of light and matter is described in the following way: the electric field of an electromagnetic wave exerts a force on the bound or free electrons, which then oscillate with the frequency of the wave. The dependence of the oscillating polarization induced in the material by the motion of the electrons on the driving electric field then yields the linear and nonlinear optical constants of the material. Although this picture is inherently classical it also provides the general idea of a quantum-mechanical description of light–matter interaction.

This classical description of light–matter interaction is based on two assumptions, which are valid for sufficiently low light intensities:

1. the force exerted on the electrons by the magnetic field of the electromagnetic wave may be neglected.
2. The speed of the oscillating motion of the electron is small compared to the speed of light.

The breakdown of these two approximations marks the onset of relativistic optics, which is based on a fully relativistic description of the interaction of an electromagnetic wave with matter.

Relativistic Motion of an Electron in an Electromagnetic Wave

We first consider the equation of motion of a free electron in the field of an electromagnetic wave

$$\frac{\mathrm{d}}{\mathrm{d}t}\left(\gamma m\dot{\boldsymbol{r}}\right)=-e\left(\boldsymbol{E}+\dot{\boldsymbol{r}}\times\boldsymbol{B}\right)\,, \tag{11.208}$$

where t is the time, m the electron rest mass, $\dot{\boldsymbol{r}}$ the velocity of the electron, e the elementary charge, $\boldsymbol{E}$ the electric, $\boldsymbol{B}$ the magnetic field of the electromagnetic wave, and

$$\gamma=\left(1-|\dot{\boldsymbol{r}}|^2/c^2\right)^{-\frac{1}{2}}\,,$$

where c is the speed of light in vacuum. For a plane electromagnetic wave of frequency ω and wavenumber $k=\frac{\omega}{c}$ propagating in the z-direction we insert $\boldsymbol{E}=\boldsymbol{E}_0\cos(\omega t-kz)$. Since $\nabla\times\boldsymbol{E}=-\partial\boldsymbol{B}/\partial t$ we obtain: $|\boldsymbol{B}|=|\boldsymbol{E}|/c$. Before solving (11.208) we introduce the normalized quantities: $\hat{t}=\omega t$, $\hat{z}=z\omega/c$, $\boldsymbol{\beta}=\dot{\boldsymbol{r}}/c$ and

$a_0 = \frac{eE_0}{m\omega c}$. Then we rewrite (11.208) in components assuming linear polarization of the electric field in the x-direction:

$$\frac{\mathrm{d}}{\mathrm{d}\hat{t}}(\gamma\beta_x) = a_0(1-\beta_z)\cos\left(\hat{t}-\hat{z}\right) ,$$
$$\frac{\mathrm{d}}{\mathrm{d}\hat{t}}(\gamma\beta_y) = 0 ,$$
$$\frac{\mathrm{d}}{\mathrm{d}\hat{t}}(\gamma\beta_z) = a_0\beta_x\cos\left(\hat{t}-\hat{z}\right) . \qquad (11.209)$$

Although (11.209) may be solved for more-general initial conditions we give here the solution for an electron initially at rest in the origin of our coordinate system, i.e., $\boldsymbol{r}(0)=0$ and $\dot{\boldsymbol{r}}(0)=0$, with the phase parameter

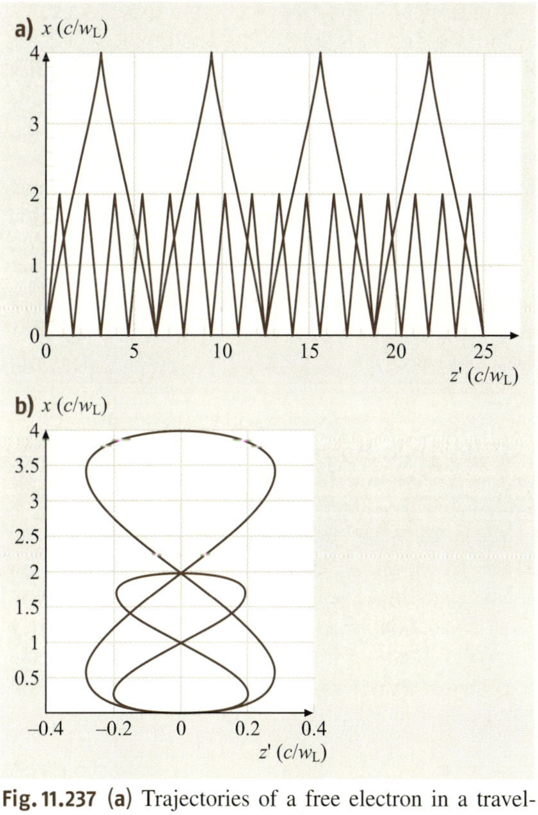

Fig. 11.237 **(a)** Trajectories of a free electron in a traveling electromagnetic wave for different laser field strengths $a_0 = 1$ and $a_0 = 2$ calculated for the initial conditions $\theta_{in} = 0$ and $\beta_{z0} = 0$ in the laboratory frame. **(b)** The trajectories in the moving frame of the electron (indicated by the *primed coordinate*) exhibit the characteristic figure-of-eight movement. Laser field strength and initial conditions corresponding to **(a)** were used

$\Theta = \hat{t} - \hat{z}$:

$$\hat{x} = a_0(1-\cos\Theta) ,$$
$$\hat{y} = 0 ,$$
$$\hat{z} = \frac{a_0^2}{4}\left(\Theta - \frac{1}{2}\sin 2\Theta\right) , \text{ and}$$
$$\gamma^2 = 1 + \frac{a_0^2}{2}\sin^2\Theta . \qquad (11.210)$$

The motion of the electron is displayed in Fig. 11.237 in two different coordinate systems:

1. the laboratory frame and
2. the *average rest frame* of the electron, which is a coordinate system co-moving with the electron in the $\hat{z}'$-direction, where the coordinate $\hat{z}'$ is given by $\hat{z}' = \hat{z} - a_0^2/4\Theta$.

It is immediately apparent that the motion of the electron in the fully relativistic case is quite different from the classical picture, where a solution of (11.208) is obtained for $\gamma = 1$ and $\boldsymbol{B} = 0$.

The electron oscillates, as in the classical case, periodically in the $\hat{x}$-direction, the direction of polarization of the oscillating electric field. However, unlike in the classical case the electron is also accelerated in the $\hat{z}$-direction, the propagation direction of the electromagnetic wave. The magnitude of the electron motion is governed by the parameter a_0, the dimensionless electric field strength [11.1997]. In the classical picture $\gamma = 1$, $\boldsymbol{B} = 0$ which of course becomes incorrect at high velocities; the condition $a_0 = 1$ would correspond to an electron that acquires a maximum oscillation velocity of $\dot{\boldsymbol{r}} = c$. The condition $a_0 \ll 1$, therefore corresponds to the classical case of $|\dot{\boldsymbol{r}}| \ll c$ while $a_0 \gg 1$ describes the extreme relativistic motion of the electron. In practical units a_0 is given by:

$$a_0^2 = \frac{I\lambda^2}{1.37\times 10^{18}\,\mathrm{W\,cm^{-2}\,\mu m^2}} \qquad (11.211)$$

where I is the light intensity and λ its wavelength. For optical wavelength where $\lambda \approx 1\,\mu\mathrm{m}$ intensities above $I \approx 10^{18}\,\mathrm{W/cm^2}$ are therefore called relativistic intensities. In the extreme relativistic case $a_0 \gg 1$ the distance the electron travels in $\hat{z}$-direction during one electric field oscillation is much larger than its excursion in the $\hat{x}$-direction, and a complete reversal of the conditions found in the weakly relativistic case $a_0 < 1$ occurs. The acceleration of electrons in the propagation direction of the electromagnetic wave appears to be in contradiction to the well-known fact derived from energy and momentum conversation that free photons cannot accelerate free electrons. In fact if a light pulse is considered the electron acquires kinetic energy in the rising part of the pulse and looses it again to the electromagnetic wave in the falling part of the pulse. After the pulse has left, the electron experiences a shift in the propagation direction of the pulse but has not acquired a net energy in agreement with the conservation laws for energy and momentum. If, however, the field of the electromagnetic wave is suddenly switched off while the electron moves in the propagation direction the electron cannot give its energy back to the field and maintains its energy acquired from the field. The switching of the field may be achieved, i. e., by shielding the field in a steep plasma gradient or by a guiding channel.

The second interesting feature of the solution of the equation of motion (11.209) is the anharmonic oscillating motion of the electron in the $\hat{x}$- and $\hat{z}$-direction best seen in the co-moving frame (Fig. 11.237b). This means that the electron radiates not only at the frequency of the diving electromagnetic wave but also emits other frequencies. While for $a_0 < 1$ the spectrum contains essentially the even (polarized in the $\hat{z}$-direction) and odd (polarized in the $\hat{x}$-direction) harmonics, the spatial and spectral emission pattern becomes very complex for higher intensities ($a_0 > 1$). This phenomenon is called nonlinear Thomson scattering and has been observed experimentally.

The Ponderomotive Force

The total energy of the electrons driven by the electromagnetic field of a laser is given by

$$E(\boldsymbol{r},t) = \gamma(\boldsymbol{r},t)mc^2 . \tag{11.212}$$

Due to the dependence of γ on the electron velocity $|\dot{\boldsymbol{r}}|$ the energy will vary widely on the length scale of the laser wavelength and the time scale given by the laser frequency. If one is not interested in the fast, mostly oscillatory motion of the electrons one may average over time, yielding:

$$\langle\gamma(r)\rangle = \sqrt{1+\frac{a_0^2(\boldsymbol{r})}{2}} . \tag{11.213}$$

The slow spatial dependence indicated in $a_0^2(\boldsymbol{r})$ may, for example, be due to the intensity variation over the focus of the laser beam, which usually occurs over much larger length scales than a wavelength.

The spatial dependence of the time-averaged energy gives rise to a force called the ponderomotive force

$$\boldsymbol{F}_\text{p} = -\nabla\langle E(\boldsymbol{r})\rangle = -mc^2\nabla\langle\gamma(r)\rangle . \tag{11.214}$$

This force will act on particles that oscillate in spatial regions where the laser intensity is high and push them into regions of lower laser intensity. It is instructive to consider the weakly relativistic limit of the ponderomotive force. With the definition of a_0 we find for $\langle\gamma\rangle - 1 \ll 1$:

$$F_\text{p} = -\frac{e^2}{4m\omega^2}\nabla E_0^2(\boldsymbol{r}) . \tag{11.215}$$

In the classical limit this constitutes the gradient of the time-averaged kinetic energy of the electron oscillating in the laser field [11.1998].

Another way to look at the ponderomotive force is to view regions of high laser intensity as regions of high electromagnetic energy density $W = \varepsilon_0\boldsymbol{E}^2(\boldsymbol{r},t)/2$, which represents a pressure pushing the electrons from regions of high pressure to those of low pressure.

The Optical Properties of a Relativistic Laser Plasma and Relativistic Channeling

When an intense laser pulse at relativistic intensities propagates through a gas consisting of atoms with low atomic number, such as hydrogen or helium, the rising part of the laser pulse fully ionizes the gas and the laser pulse interacts with a fully ionized plasma. For example helium is fully ionized already at an intensity of a few times $10^{16}\,\text{W/cm}^2$ while relativistic intensities usually exceed $10^{18}\,\text{W/cm}^2$. The dielectric constant of a fully ionized plasma is given by

$$\varepsilon_\text{r} = 1 - \frac{\omega_\text{p}^2}{\omega^2} , \tag{11.216}$$

where $\omega_\text{p}^2 = \frac{e^2 n_\text{e}}{\varepsilon_0\gamma m}$ and n_e is the electron density.

Collisions have been neglected here which is a reasonable assumption for laser pulses of durations of less than 1 ps and electron densities of $n_\text{e} \le 10^{22}\,\text{cm}^{-3}$. It should be noted that in a relativistic plasma the plasma frequency ω_p depends on γ and therefore on the laser intensity. The dielectric constant becomes negative for electron densities where the plasma frequency ω_p exceeds the laser frequency ω. In the nonrelativistic case ($\gamma = 1$) an critical electron density n_crit is defined through the condition $\omega_\text{p}^2 = \omega^2$ yielding [11.1998]

$$n_\text{crit} = \frac{\varepsilon_0 m\omega^2}{e^2} . \tag{11.217}$$

For electron densities $n_\text{e} > n_\text{crit}$ the plasma is called dense while it is underdense for $n_\text{e} < n_\text{crit}$. In practical units the critical electron density is given by $n_\text{crit} = 1.1\times10^{21}\,(1\,\mu\text{m}/\lambda)^2\,\text{cm}^{-3}$. With this definition the dielectric constant may be rewritten

$$\varepsilon_\text{r} = 1 - \frac{n_\text{e}}{\gamma n_\text{crit}} \tag{11.218}$$

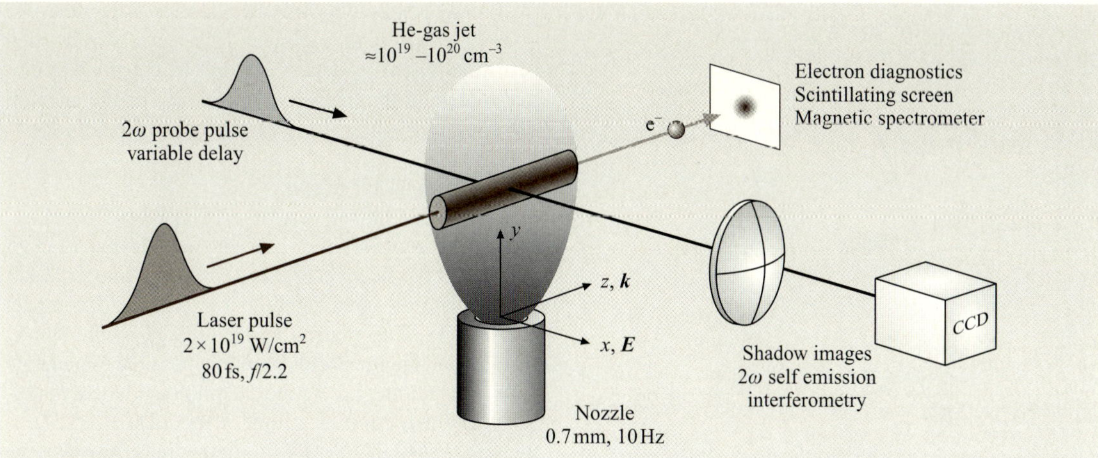

Fig. 11.238 Experimental setup. The main pulse is focused to an intensity of a few 19 W/cm^2 into a pulsed He-gas jet, where a relativistic channel due to relativistic self-focusing is forming. By splitting and frequency doubling part of the main pulse a probe pulse is generated. This probe pulse is used for imaging and observing the laser–plasma interaction on a time scale determined by the probe pulse duration of ≈ 100 fs. The electrons that are accelerated in the relativistic channel are characterized by magnetic spectroscopy and nuclear reactions

Electromagnetic radiation propagates in a medium where $\varepsilon_r > 0$ and is reflected from a medium with $\varepsilon_r < 0$. The condition for light propagation is therefore $n_e/\gamma < n_{crit}$.

The index of refraction of the plasma is given by $n = \sqrt{\varepsilon_r}$. In an underdense plasma n is real and the laser pulse propagates. But n is a complicated function of intensity for two reasons:

1. At the center of the propagating laser pulse the intensity is high. Therefore the ponderomotive force pushes the electrons away from the center and the electron density in the center decreases, increasing the index of refraction.
2. The remaining electrons in the center gain higher energies in the laser field than those sitting at the edges of the laser beam. This increases γ further and thus increases the index of refraction at the center of the laser beam.

This means that a relativistic laser pulse self-modulates the index of refraction in such a way that self-focusing of the laser occurs. As in other self-focusing phenomena known from nonlinear optics self-focusing is countered by diffraction and therefore self-focusing depends on a critical power rather than on intensity. A laser pulse interacting with a fully ionized plasma will therefore undergo self-focusing if its total power P exceeds the critical power P_{crit} [11.1999].

$$P > P_{crit} = \frac{8\pi\varepsilon_0 m^2 c^5}{e^2}\frac{n_{crit}}{n_e} = 17.4\,\text{GW}\frac{n_{crit}}{n_e}\,. \tag{11.219}$$

The dynamic equilibrium of defocusing because of diffraction and self-focusing due to nonlinear effects leads to a guiding effect for the high-intensity laser pulse in a plasma called relativistic channeling. This phenomenon has been observed experimentally. A typical experimental arrangement is shown in Fig. 11.238.

A Ti:sapphire laser ($\lambda = 800$ nm) with a total power of 8 TW is focused to an intensity of about 2×10^{19} W/cm^2 into a helium gas jet from a nozzle with a very well-characterized gas density profile (Fig. 11.239).

Figure 11.239 shows the relativistic channel which extends over a length of about 300 μm, corresponding to about 15 Rayleigh lengths of the laser focusing optics. At the onset of the channel the electron density amounts to 5×10^{19} cm$_0^{-3}$. The critical density is 1.7×10^{21} cm^{-3} at $\lambda = 800$ nm. Therefore the total laser power of 8 TW exceeds the critical power of $P_{crit} \approx 0.6$ TW obtained from (11.219) substantially and relativistic channeling is expected and observed. In addition a probe pulse at $\lambda = 400$ nm was used to measure the electron density in the channel interferometrically (Fig. 11.238). The result shown in Fig. 11.240 gives the electron density profile across the channel at the onset of the channel. It is ap-

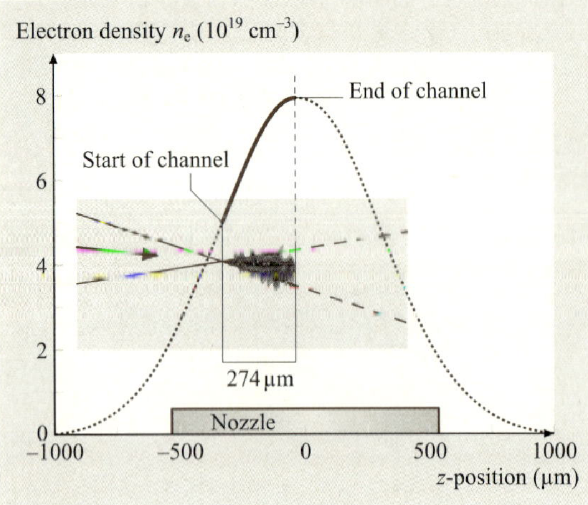

Fig. 11.239 The lateral plasma density profile above the nozzle has a Gaussian shape (*dotted line*). The density along the channel and its position is indicated by the bold line. The inset shows the corresponding observed channel of the same spatial region. The extension of the emission indicates a channel length of 274 µm, about 12 times the Rayleigh length of the laser. The relativistic channel starts near the steepest density gradient and ends close to the maximum density (grad $n_e = 0$)

parent that the electron density at the channel center is depressed, due to the expulsion of electrons because of the ponderomotive force. This result also provides direct evidence for the guiding structure of a relativistic channel.

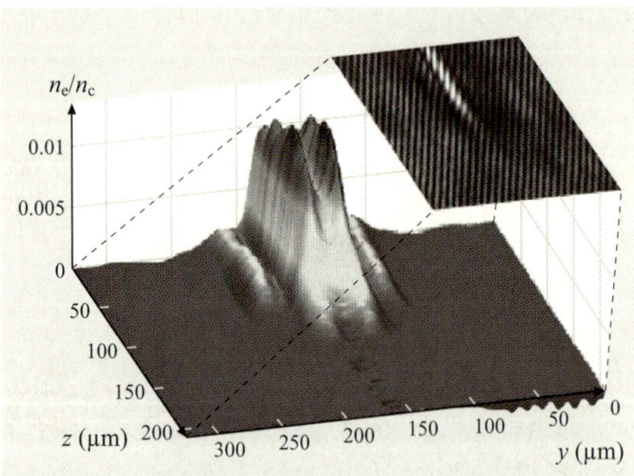

Fig. 11.240 Plasma density at the start of the channel determined by interferometry. The electron density in the walls of the channel rose up to values of $n_e(\text{wall}) = 6 \times 10^{19} cm^{-3}$. The image in the upper right corner shows the interferogram from which the plasma density was obtained

Finally it should be noted that optics in the relativistic regime is always nonlinear. The index of refraction is always intensity dependent due to the ponderomotive force and the intensity dependence of the electron mass.

Electron Acceleration

Since in the relativistic regime of laser–plasma interaction charged particles may be accelerated to high energies, it has been recognized for some time that laser plasmas are an ideal medium for high-field compact accelerators. The plasma as an ionized medium may sustain much higher fields than it is possible to generate with conventional accelerator technology, where material breakdown imposes a limit at less than 100 MV/m. Electric fields in the TV/m range may however be generated in laser plasmas.

In many experiments over the past two decades a variety of acceleration mechanisms have been identified. A selection of these mechanisms will briefly be described in the following with focus on schemes where only a single laser pulse is needed to accelerate initially resting electrons to relativistic energies. All these schemes have in common that a laser pulse is focused into a gas jet, generating an underdense plasma (Fig. 11.238).

Laser wakefield acceleration. When a relativistic laser pulse impinges on a plasma the ponderomotive force expels electrons in both the transverse and longitudinal directions. The electrons expelled from the laser pulse in the backward direction lead to a plasma wave called the laser wakefield (Fig. 11.241) [11.2000].

The plasma wave follows the driving laser pulse with a phase velocity determined by the laser pulse group velocity. The electric fields associated with the plasma wave are now longitudinal. An electron can *ride* on the plasma wave and be accelerated to relativistic energies in the direction of laser propagation. This process is called laser wakefield acceleration (LWFA).

This process is most efficient when the laser pulse length $c\tau$, where τ is the pulse duration, is shorter than the plasma wavelength $\lambda_p = 2\pi c/\omega_p$.

This condition is depicted in Fig. 11.241a. When a large number of electrons acquire a velocity close to the phase velocity of the plasma wave, wave-breaking occurs. The fast electrons are *surfing* on the wake of the plasma wave.

Self-modulated laser wakefield acceleration. If the laser pulse length is longer than the plasma wavelength

$c\tau > \lambda_p$ the laser pulse undergoes a self-modulation instability. The leading edge of the laser pulse drives a plasma wave. The electron density modulation of the plasma wave in turn represents a periodic modulation of the refractive index. It acts on the long laser pulse such that the pulse is self-modulated and breaks up into a train of short pulses (Fig. 11.241b). These shorter pulses now match the conditions for LWFA and can resonantly drive a plasma wave. This self-modulated laser wakefield acceleration (SM-LWFA) is not as efficient as pure LWFA, but still high-energy even quasi-monoenergetic electron beams may be generated.

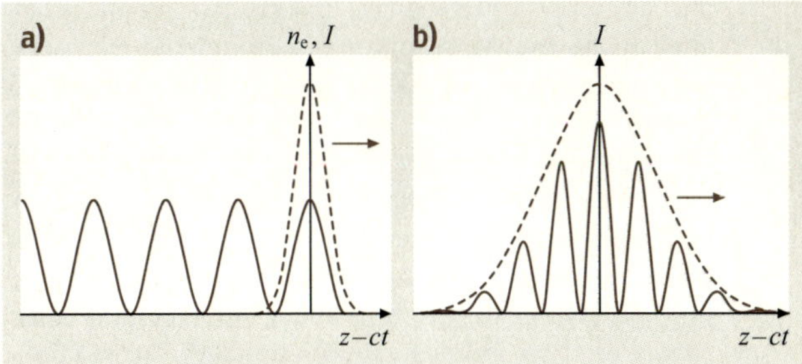

Fig. 11.241 **(a)** LWFA: a short laser pulse ($c\tau \leq p$) shown as a *dashed line* drives a plasma wave. **(b)** SM-LWFA: an initially long laser pulse (*dashed line*) breaks up into a train of shorter pulses that match the condition of LWFA and resonantly drive a plasma wave (after [11.1999])

Direct laser acceleration. Another acceleration process which is quite different in nature to wakefield acceleration is direct laser acceleration (DLA) which is closely related to the formation of a relativistic channel. The ponderomotive force expels electrons from the laser beam axis and generates a radial quasistatic electric field. Electrons that are accelerated along the laser propagation generate an azimuthal magnetic field. The combination of these two fields results in an effective potential well for relativistic electrons. Electrons trapped in this well will oscillate at the frequency $\omega_\beta = \omega_p / \left(2\sqrt{\gamma}\right)$, the betatron frequency. If the trapped electron is moving fast enough along the laser propagation, the laser oscillations may be in phase with the betatron oscillations in the frame of the electron.

In this case, an efficient energy coupling is possible. The energy gained by the electron in this process directly results from the laser field and therefore the name direct laser acceleration is appropriate.

Bubble acceleration. A novel regime of laser wake field acceleration, called *bubble acceleration*, was proposed in 2002 on the basis of particle-in-cell (PIC) simulations [11.2002]. Short ($\tau < 7$ fs) and intense ($a_0 > 1$) laser pulses were predicted to produce quasi-monoenergetic electrons with energies exceeding 100 MeV.

The ponderomotive force also plays an important role for this acceleration mechanism. In a frame moving with an intensive laser pulse propagating through a plasma the electrons are expelled from the center of the pulse in both the longitudinal and transverse directions. It turns out that behind the laser pulse an electron density depression, called a bubble, is generated and electrons the stream around this bubble and enter it again from the back side. In this way a strong electric field in longitudinal direction reaching TV/m is generated, which leads to efficient electron acceleration.

Experiments. Numerous laser acceleration experiments have been carried out, mostly using experimental setups like the one shown in Fig. 11.238. Initially always quasi-exponential electron spectra were obtained with electron temperatures that follow the scaling law

$$kT_e \approx mc^2 \left(\sqrt{1 + \frac{a_0^2}{2}} - 1 \right) , \quad (11.220)$$

which corresponds to the kinetic energy of the electron in the laser field. These exponential spectra

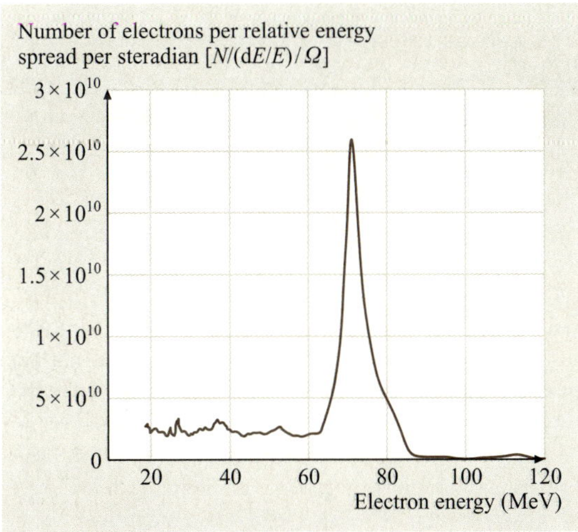

Fig. 11.242 Monoenergetic electron spectrum generated in an experimental setup similar to the one shown in Fig. 11.232 [11.2001]

are not only observed in underdense plasmas but also in the overdense plasmas encountered in laser–solid interactions [11.2003]. For a typical intensity of $I = 10^{20}\,\mathrm{W/cm^2}$ and $\lambda = 1\,\mu\mathrm{m}$, $a_0 = 8.5$ we obtain $kT_e \approx 2.6\,\mathrm{MeV}$ from (11.220).

Only recently, however, has it became possible to generate monoenergetic electron spectra. As predicted by numerical simulations the bubble regime may be reached for high-power, gently focused laser pulses of short pulse duration. Although the requirements of the pure bubble regime have not yet quite been reached experimentally the experiments show that in the transition regime between LWFA and bubble acceleration monoenergetic electron spectra may be generated [11.2001, 2004, 2005]. A typical result is shown in Fig. 11.242.

Ion Acceleration

Relativistic laser–matter interaction also leads to the generation of monoenergetic beam of ions of MeV energies by a mechanism quite different from the mechanism involved in electron acceleration.

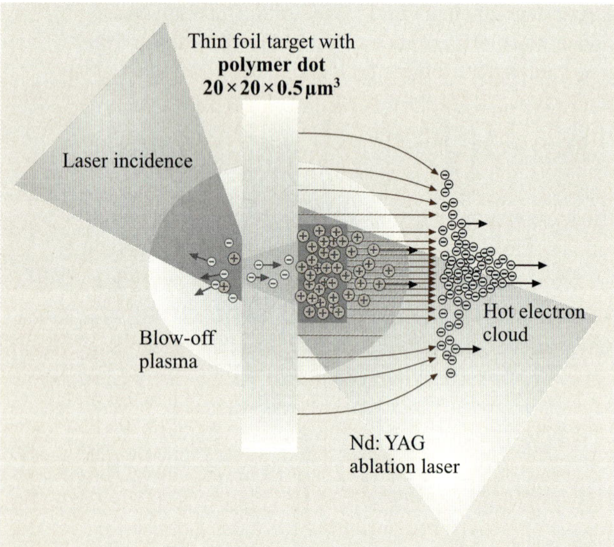

Fig. 11.243 Laser acceleration of protons from the back side of a microstructured target. A terawatt (TW) laser pulse is focused onto the front side of the target foil, where it generates a blow-off plasma and subsequently accelerates electrons. The electrons penetrate the foil, ionize hydrogen and other atoms at the back surface and set up a Debye sheath. The inhomogeneous distribution of the hot electron cloud causes a transversely inhomogeneous accelerating field [target normal sheath acceleration (TNSA)]. Applying a small hydrogen-rich dot on the back surface enhances the proton yield in the central part of the accelerating field, where it is nearly homogenous. These protons constitute the quasi-monoenergetic bunch

As discussed earlier the interaction of an intense light field with matter yields the generation of a hot plasma and the subsequent acceleration of electrons up to relativistic energies. Protons and ions are accelerated by a well-controlled mechanism known as *target normal sheath acceleration* (TNSA) following the initial electron acceleration (Fig. 11.243). Fast electrons are accelerated by an intense laser pulse (intensity $I \geq 10^{19}\,\mathrm{W/cm^2}$) from the surface of a thin metal foil in the forward direction. They penetrate the foil and ionize atoms along their paths. Within about a picosecond, those electrons leaving the target at the rear surface (that is, the back surface with respect to the laser irradiation) build up a quasistatic electric field. The field acts normally to the target surface, has cylindrical symmetry and decreases in the transverse direction. Owing to the ultrashort duration of the electron bunch and its high charge, this field may reach values of several TV/m close to the axis and thus the potential can attain several tens of MeV. Protons and positively charged ions present on the back surface of the foil may be accelerated by this field until they compensate the electron charge. In most cases, the origin of these parasitic protons has been identified to be a hydrocarbon contamination layer on the target surface.

As the duration of the acceleration is ultrashort and the protons (as well as the ions) are at rest before acceleration, comprising a very small phase-space volume, the transverse emittance of the proton beam reaches values as low as a few 10^{-3} mm mrad for 10 MeV protons. However, laser-accelerated ion beams still show an energy spectrum exhibiting a quasi-exponential shape with a distinct cut-off energy. This can be explained by the inhomogeneous distribution of electrons in the sheath, which causes an accelerating field that is inhomogeneous in the transverse direction. For a plane and unstructured target, the transverse dimension of the electric field and hence the source size of the accelerated protons is much larger than the laser's focal spot. Therefore, different parasitic protons experience a range of potentials, resulting in a broad distribution of energies.

Following this understanding of the mechanism of laser acceleration of protons, it was pointed out that the resulting proton energy spectrum has a strong correlation to the spatial distribution of the protons on the target surface. In order to generate high-quality proton beams with monoenergetic features, a bilayered, microstructured target, consisting of a thin high-Z metal foil and a small proton-rich dot on the back surface was proposed. The transverse dimension of such dots is smaller than that of the acceleration sheath, and hence the protons will only be subject to the central – that is, homogeneous – part of the acceleration field. In this configuration, the protons all experience the same electric field and are accelerated in the same potential (Fig. 11.243). The resulting proton beam has a spectrum with a strong monoenergetic peak [11.2006].

Recent experiments with a 10 TW, 600 mJ laser showed that in this way quasi-monoenergetic proton beams containing 10^8 protons with energies of a few MeV and a relative width of $\Delta E/E \approx 10\%$ and a total charge of about 100 pC were generated.

In a different experiment quasi-monoenergetic carbon ions of 4 MeV energy were generated from a very thin carbon layer on a metal target [11.2007].

Applications and Future Developments

Relativistic optics leads to laser particle accelerators that are considerably more compact than classical accelerators. Presently laser accelerators reach maximum energies for monoenergetic electron beams approaching 1 GeV and several MeV for monoenergetic ion beams with a total charge of up to 1 nC per bunch. The scaling for both electron and ion acceleration with laser and target parameters appears to be sufficiently well understood that electron energies exceeding 1 GeV and ion energies of a few hundred MeV in a single acceleration stage appear to be possible in the near future. There are however at least two challenges laser accelerators that have to overcome before widespread applications will become attractive.

In order to reach energies in the TeV range multistaging of both electron and ion accelerators appears necessary. Although energy, beam quality and charge per bunch are already attractive for low-energy applications of laser accelerators their average power presently remains orders of magnitude below the average power of conventional accelerators. This is solely due to limitations in laser technology and may be overcome in the future when high-average-power high-intensity lasers become available.

Laser accelerators have already been used for a number of demonstration experiments. Low-energy nuclear reactions have been induced using high-intensity lasers [11.1999] and nuclear transmutation scenarios employing lasers have been discussed [11.2008]. Laser-generated ion beams carry enough dose for radiation therapy using ion beams, a very promising application because the average power of a laser accelerator is already sufficient for this innovative method of cancer treatment.

With ever-increasing laser intensities other areas of physics such as gravitational physics, elementary particle physics or nonlinear quantum electrodynamics (QED) effects [11.1999] also come into the focus of high-intensity lasers.

11.14 Frequency Stabilization of Lasers

Soon after the development of the first HeNe laser, it was realized that the radiation of a continuous-wave laser would be ideally suited as a measurement tool, provided its frequency was stable and reproducible. Applications of frequency-stabilized lasers include high-resolution laser spectroscopy, quantum optics, optical frequency standards, the determination of fundamental constants, and the detection of gravitational waves. In these various cases, the requirements on frequency-stabilized lasers are quite different. For example, optical frequency standards and optical clocks need the smallest possible uncertainty in the knowledge of the absolute laser frequency whereas gravity-wave detectors need lasers of extremely low frequency noise while the absolute laser frequency is less important. Hence, different stabilization methods have been developed to address the various tasks. The purpose of this chapter is to review the general methods of laser frequency stabilization and to describe some representative examples of frequency-stabilized lasers. The radiations of a variety of stabilized lasers are recommended as optical reference frequencies [11.2009].

Basically, the frequency of any laser is determined within the bandwidth of its amplifying medium by the optical length of the resonator. This optical length in turn is a function not only of its actual geometrical length but also of the refractive index of the gain medium itself, which may depend on several different parameters. The width of the gain profile may vary between a few $10^{-6}\,\nu$ for gas lasers up to approximately 10% for, e.g., dye lasers or laser diodes. Different types of lasers have different noise characteristics. In most cases, the dominant frequency noise is of technical nature and is far above the Schawlow–Townes limit [11.2010]. Since such technical frequency fluctuations are relatively slow they can be controlled by a suitable electronic servo system even if the frequency excursions of the free-running laser are large. Different laser types and the specific requirements on these lasers have led to a variety of stabilization methods. Within the limited space of this article, it is impossible to present a comprehensive and detailed description of all these stabilization schemes. Hence, we concentrate on the description of the basic principles of laser frequency stabilization. For a deeper understanding, the reader is encouraged to study the references given in this article as well as textbooks on laser spectroscopy and optical clocks [11.2011–2014].

We start with a brief discussion of the terms characterizing the frequency behavior of a laser. These are noise, stability, line width, reproducibility, and the uncertainty of the laser frequency. Section 11.14.2 describes the basics of laser frequency stabilization. Examples of stabilized lasers are presented in Sect. 11.14.3. Section 11.14.4 explains a universal method of optical frequency measurements by means of mode-locked femtosecond lasers.

11.14.1 Characterization of Noise, Stability, Line Width, Reproducibility, and Uncertainty of the Laser Frequency

Noise, stability, line width, reproducibility, and the uncertainty of the frequency are important parameters of any frequency-stabilized laser. In general, the laser frequency fluctuates about a mean value which itself may drift and walk randomly. Such variations may be caused, e.g., by changes in the temperature, air pressure, vibrations, acoustics or by fluctuations within the active laser medium itself.

The variations in the laser frequency can be investigated, e.g., by measuring the beat frequency $\nu_B = |\nu_1 - \nu_2|$ between two identical but independently stabilized lasers 1 and 2 (Fig. 11.244). For this purpose, the beams

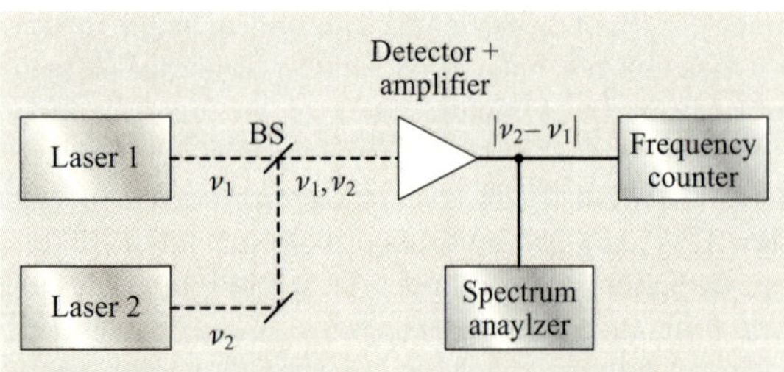

Fig. 11.244 Schematics of a beat-frequency measurement

of two lasers are coaxially combined by a beam splitter and focused onto a fast photodetector. The *beat frequency* ν_B shows up as an oscillation in the power of the combined laser beams. The photodetector transforms it to an oscillation of the photocurrent, provided ν_B is small enough that it can be processed by state-of-the-art electronics.

Frequency fluctuations of a laser can be measured in both the frequency and time domain. Measurements in the frequency domain are often used at higher fluctuation (Fourier) frequencies whereas slower-frequency fluctuations and drifts can be measured conveniently in the time domain (see below).

Spectral Density of Frequency Noise

In the frequency domain, fluctuations $\delta\nu$ at the frequency f can be detected by a frequency discriminator, which converts frequency fluctuations to proportional voltage fluctuations. Since the noise components are not correlated, it is convenient to describe the fluctuations $\delta\nu(f)$ by the mean square $\langle\delta\nu(f)^2\rangle$ which occurs within a bandwidth B at the Fourier frequency f. This value is equivalent to the spectral noise power. The power spectral density is then defined as $S_f = \langle\delta\nu^2\rangle/B$ and the integration of S_f over B results in the total power of the frequency noise within the bandwidth B. For the *relative* power spectral density S_y we get $S_y = \langle(\delta\nu/\nu)^2\rangle/B$.

To a good approximation, the frequency noise of any oscillator can be modeled by a power series of the Fourier frequency

$$S_y = \sum_{\alpha=-2}^{2} c_\alpha f^\alpha . \tag{11.221}$$

Depending on the exponent a the five terms of (11.221) describe

- a random frequency walk for $\alpha = -2$
- flicker $(1/f)$ frequency noise for $\alpha = -1$
- white frequency noise for $\alpha = 0$
- flicker $(1/f)$ phase noise for $\alpha = 1$
- white phase noise for $\alpha = 2$

Superimposed on the noise given in (11.221), we may also find peculiarities in the noise spectrum that are generated by noise sources specific to the individual laser. Such noise may be generated, e.g., by environmental sources such as the excitation of mechanical resonances in the laser resonator, by instabilities of the mechanical setup, or by fluctuations in the refractive index of the laser medium. Furthermore, the laser frequency is sensitive to optical feedback and careful isolation of the laser system from back-scattered light is important for successful operation of stabilized lasers. Usually, the different noise components show up at different Fourier frequencies. Hence, the detection of these noise components helps us to identify the various noise sources and to reduce their influence. In addition, the knowledge of the frequency noise spectrum of a free-running laser is also important for the design and optimization of the frequency control system.

The noise spectra of various types of lasers may be quite different. For example, most of the frequency noise in gas lasers such as HeNe or CO_2 lasers occurs at low Fourier frequencies. It is generated in part by mechanical instabilities of the laser setup. The corresponding frequency excursions $\delta\nu$ may amount to a few MHz at Fourier frequencies up to about 10 kHz. In addition, random-walk and flicker ($1/f$) frequency noise shows up at these low frequencies. Neglecting further possible noise sources originating from plasma resonances of the discharge, the noise spectrum at higher Fourier frequencies may be approximated by white frequency noise. In tunable CW dye lasers, on the other hand, the frequency noise is much larger than in gas lasers. It is generated mostly by variations in the thickness and the refractive index of the dye jet, which is flowing at high speed through the small active volume of the laser. Hence, rather strong frequency fluctuations are observed at Fourier frequencies around 50 kHz. At higher Fourier frequencies the influence of the dye jet decreases and hence the noise also decreases, eventually reaching the photon shot-noise level.

Allan Standard Deviation

Slow variations, in particular drifts and random walk of the laser frequency can be measured conveniently in the time domain by counting the mean frequency ν within a time interval τ. Such measurements are advantageous if changes in the frequency within time intervals between seconds and several hours are of interest. The relative frequency instability in the time domain is characterized by the Allan deviation [11.2015, 2016]

$$\sigma_y(\tau) = \frac{1}{\nu}\left[\frac{1}{2(N-1)}\sum_{n=1}^{N-1}(\nu_{n+1}-\nu_n)^2\right]^{1/2} , \tag{11.222}$$

where the root mean square of the difference between two consecutive frequency measurements $(\nu_{n+1}-\nu_n)$ is used as a measure of the stability. The frequency values ν_{n+1} in (11.222)

$$\nu_{n+1} = \frac{1}{\tau}\int_{i=n\tau}^{(n+1)\tau} \nu_i(t)\,dt \tag{11.223}$$

represent the mean frequency averaged during the n-th interval of duration τ. Since it is difficult to measure the optical frequency directly, it can be down-converted by beating the laser radiation against the radiation of a second laser such that the beat frequency arrives in the radio-frequency range, where it can be counted directly. In this case, the measurement contains the instabilities of both lasers. If the two lasers are identical but independent, we can assume that both lasers contribute equally to the standard deviation and consequently, the measured value corresponds to $\sqrt{2}$ times the Allan deviation $\sigma_y(\tau)$ of a single laser.

The dependence of $\sigma_y(\tau)$ on the averaging time τ contains information about the frequency noise spectrum of the laser. If the relative power spectral density of the noise S_y is known, $\sigma_y(\tau)$ can be calculated by the relation [11.2016]

$$\sigma_y^2(\tau) = \int_0^\infty S_y(f)\frac{(\sin\pi f\tau)^4}{(\pi f\tau)^2}\,df . \tag{11.224}$$

For the different models of the noise process given in (11.221), we can describe the Allan variance $\sigma_y^2(\tau)$ by the relation

$$\sigma_y^2(\tau) = d_\alpha\tau^\beta \tag{11.225}$$

where β is determined by the relation [11.2016]

$$\beta = \begin{cases} -\alpha-1 & \text{for } \alpha \le 1 \\ -2 & \text{for } \alpha > 1 \end{cases} \tag{11.226}$$

and d_α is a constant. The dependence of $\sigma_y^2(\tau)$ on τ is listed for the noise processes of (11.226) in Table 11.55.

Of course, any frequency modulation will also show up in the Allan standard deviation. In the case that the

Table 11.55 Dependence of the Allan deviation on τ for the different noise processes discussed in (11.226)

Noise category	α	β	$\sigma_y^2(2,\tau)$
White frequency noise	0	−1	$\propto 1/\tau$
White phase noise	2	−2	$\propto 1/\tau^2$
$1/f$ (frequency noise)	−1	0	constant
$1/f$ (phase noise)	1	−2	$\propto 1/\tau^2$
Random frequency noise	−2	1	$\propto \tau$

gate time τ is an integer multiple of the modulation period $\tau_m = 1/f_m$, where f_m is the modulation frequency, the influence of the modulation is eliminated. It will be maximum for $\tau_m = (2n+1)/(2f_m)$. With increasing integration times τ, the influence of the modulation washes out and can be neglected, eventually.

As a result of (11.227), we find that the noise process can be estimated from the slope of a log–log $\sigma_y(\tau)$ plot. For example, in a HeNe laser that is frequency stabilized to an absorption line of molecular iodine, $\sigma_y(\tau)$ decreases in a wide range of τ with the square root of τ corresponding to a slope of $-1/2$, (Fig. 11.245) indicating white frequency noise within the corresponding time interval τ. At longer integration times, $\sigma_y(\tau)$ flattens out due to flicker frequency noise and in some cases eventually increases again due to drifts and random frequency walks. For practical applications, the $\sigma_y(\tau)$ plot indicates the minimum integration time τ necessary to achieve a given statistical frequency uncertainty.

Line Width of Laser Radiation

In incoherent light sources such as spectral lamps, where the radiation consists of uncorrelated photons emitted within a certain narrow frequency band, the line width is determined by a superposition of the natural width of the transition and Doppler broadening. It can be measured, for example, by a high-resolution scanning Fabry–Pérot interferometer. In the case of laser radiation, the line broadening is caused by frequency fluctuations of the continuously emitted coherent laser radiation. The frequency fluctuations generate noise sidebands that cause broadening of the laser spectrum. The line width is then determined by the size and spectral extension of these sidebands. In the case of pure harmonic phase modulation, the field E of the laser radiation is given by the relation

$$\begin{aligned} E(t) &= E \sin\left[\Omega_0 t + \delta\phi \sin(\omega t)\right] \\ &= E\Bigg\{ J_0(\delta\phi)\sin(\Omega_0 t) \\ &\quad + \sum_{n=1}^{\infty} J_n(\delta\phi)\exp[\mathrm{i}(\Omega_0 + n\omega)t] \\ &\quad + \sum_{n=1}^{\infty} J_n(-\delta\phi)\exp[\mathrm{i}(\Omega_0 - n\omega)t]\Bigg\}, \end{aligned} \tag{11.227}$$

where $\Omega_0 = 2\pi\nu_0$ and $\omega = 2\pi f$ represent the carrier and modulation frequency, respectively; $\delta\phi = \delta\nu/f$ is the modulation index with $\delta\nu$ the amplitude of the frequency deviation. The amplitude of the n-th sideband given by the n-th order of the Bessel function $J_n(\delta\phi)$ decreases strongly with n for $n > \delta\phi$. Basically, we can distinguish between the two limiting cases of

1. large frequency fluctuations $\delta\nu$ at low Fourier frequencies f and
2. small, fast frequency fluctuations δf in a broad band.

Case 1 is frequently observed with laser resonators, which pick up environmental acoustic or mechanical noise. The mean amplitudes $\langle\delta\nu^2\rangle^{1/2}$ of such fluctuations may be in the range between a few 10 kHz up to several megahertz, whereas they usually occur at Fourier frequencies in a range below 1 kHz. Hence the modulation index $\delta\varphi \gg 1$. In this case, the line width is given by [11.2017]

$$\begin{aligned} \Delta\nu_{\mathrm{FWHM}} &= \left[8\ln(2)\langle\delta\nu^2\rangle\right]^{1/2} \\ &\cong 2.355\langle\delta\nu^2\rangle^{1/2} \end{aligned} \tag{11.228}$$

close to the frequency excursions between the peaks.

If the servo gain of the frequency stabilization is high enough to reduce the frequency fluctuations at low Fourier frequencies, we may observe case 2. Under the assumption that we can neglect frequency drifts and frequency changes at very low Fourier frequencies, we arrive at frequency fluctuations $(\langle\delta\nu\rangle^2)^{1/2} \ll B$ and correspondingly to a small modulation index. In this case, only the amplitudes of the first-order sidebands $J_n(\delta\phi)$ and $J_{-n}(\delta\phi)$ contribute significantly to the line width. If the spectral density S_f is constant, the line profile is Lorentzian and the line width is given by the

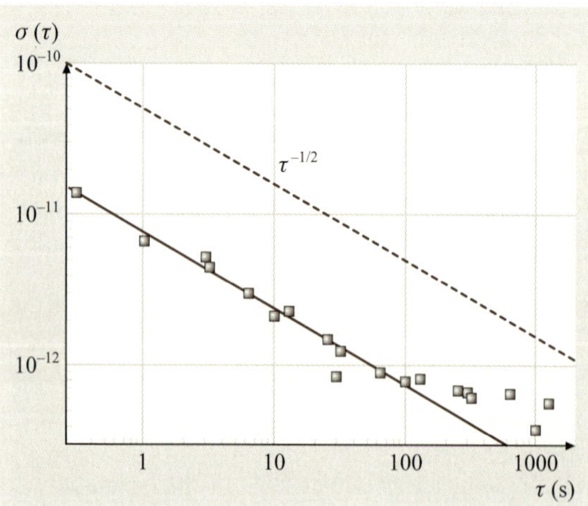

Fig. 11.245 Allan deviation $\sigma_y(\tau)$ versus integration time τ of an iodine-stabilized HeNe laser

relation [11.2017]

$$\Delta\nu_{\text{FWHM}} = \pi S_f . \quad (11.229)$$

Frequency Reproducibility and Uncertainty of Laser Radiation

Whenever the laser frequency is needed as a precise reference, knowledge of the *reproducibility* and *uncertainty* is essential. Since the frequency of the stabilized laser depends on various operational and environmental parameters it is necessary to analyze this dependence and to control these parameters carefully. The *frequency reproducibility* is a measure of the scatter in the frequency values either of an ensemble of differently designed laser systems of the same kind, developed for example in different laboratories. The reproducibility can then be investigated by frequency inter-comparisons between such differently designed lasers. Alternatively, the term *frequency reproducibility* is also used to describe the scatter in the frequency of a single laser caused by the uncertainty in the precise control and optimization of the operation parameters.

The *frequency uncertainty* is a measure of how accurately we can realize the reference frequency and determine its value. Basically, the total uncertainty contains two contributions. The first represents the uncertainty to which the transition frequency of an unperturbed atomic or molecular absorber at rest can be realized. This part contains all values in the uncertainty of frequency shifts caused by environmental and/or operational conditions. Such shifts may be generated, e.g., by a residual Doppler effect, by collisions, and by external fields. The second term in the total uncertainty is linked to the determination of the frequency value. Such frequency measurements have to be referenced to the primary standard of time and frequency, the Cs atomic clock. If the frequency measurement is phase-coherent, the measurement itself does not contribute to the total uncertainty. Under this condition, the uncertainty of the optical frequency is determined only by the uncertainty of the standard itself and that of the Cs reference.

11.14.2 Basics of Laser Frequency Stabilization

In any active frequency control system, the laser frequency is servo-controlled to a reference frequency. This may be provided by an eigenfrequency of a stable optical (Fabry–Pérot) resonator, by the gain profile of the laser itself, or by a suitable atomic or molecular transition. The first case has the advantage that it provides a good signal-to-noise ratio, allowing a fast servo lock. However, since the laser frequency is coupled to the length of an artifact, it may change in time. Stabilization to such artifacts is often applied as an intermediate step to reduce the line width of the laser emission and to allow for high spectral resolution. In a second step, the frequency can then be stabilized to an absolute frequency reference provided by a suitable atomic or molecular absorption line.

The following sections give an introduction to basic methods of laser frequency stabilization. We start with the generation of the error signal and review four prominent examples. The first two methods – the *side-of-the-fringe stabilization* and the *phase-modulation (Pound–Drever–Hall) technique* are used to stabilize the laser frequency to a resonance of a stable optical cavity. Absolute frequency stabilization to an atomic reference is then reviewed and the fundamentals of the electronic servo-control system are described.

Generation of the Error Signal

Basically, the frequency stabilization of a laser converts the frequency deviation $\delta\nu = \nu_l - \nu_0$ between the laser frequency ν_l and the reference frequency ν_0 to an error signal that is proportional to $\delta\nu$. This signal is amplified by a servo amplifier and used to control the laser frequency such that the error signal vanishes. Figure 11.246 shows the basic scheme of a stabilization to an atomic reference frequency.

Side-of-the-fringe stabilization. In commercial stabilized lasers such as tunable dye lasers, it is a well-established practice to stabilize the frequency to a resonance of a stable optical cavity. The frequency

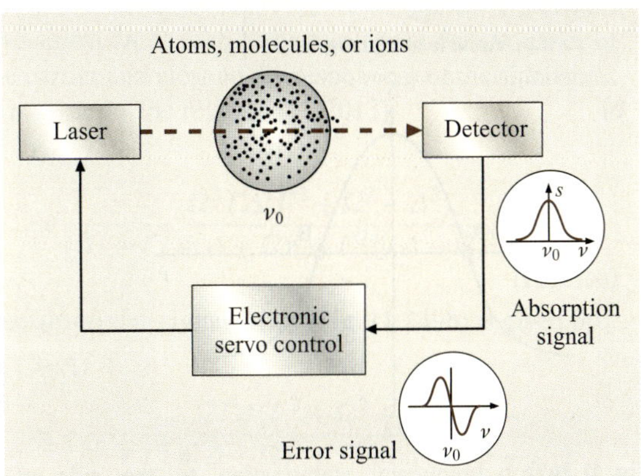

Fig. 11.246 Schematic of a laser frequency stabilization to the center of an absorption line

modulation frequency, and $\Delta/2\pi = (\nu - \nu_0)$ is the frequency detuning.

The central zero crossing of the dispersion signal $D(\Delta)$ (Fig. 11.249a) is used as the error signal for the frequency control. The signal detected in quadrature (Fig. 11.249b) shows a resonance feature when either of the sidebands coincides with the cavity resonance. The different polarities of the two features are caused by the phase difference π between the carrier and the upper sideband and that between the carrier and the lower sideband. Independent of the detection phase, the demodulated signal is always antisymmetric with respect to the central zero crossing. Consequently, a small phase deviation from zero does not change the frequency of the central zero crossing. However, it will slightly change the slope of the frequency discriminator and thereby the gain of the servo loop.

Since the returning beam results from an interference between the frequency components promptly reflected at the entrance mirror and part of the stored radiation leaking out of the cavity, any fast frequency change shows up as a change in the interference signal instantaneously even if the response time of the cavity is much longer. At low Fourier frequencies f, smaller than the half-width $\Gamma/2\pi(2\pi f \ll \Gamma)$, the transient response of the cavity can be neglected and the demodulated signal acts as a *frequency* discriminator. With increasing Fourier frequency, the signal turns to a *phase* discriminator for $2\pi f \gg \Gamma$. Hence, the amplitude $A(f)$ and the phase $\phi(f)$ of the error signal, have a low-pass characteristic which rolls off with $1/f$ for frequencies f above the cut-off frequency $\Gamma/2\pi$ [11.2020]

$$A(f) = A(0)\frac{1}{\sqrt{1+(2\pi f/\Gamma)^2}} ,$$

$$\varphi(f) = -\arctan(2\pi f/\Gamma) . \tag{11.231}$$

If this integrating behavior for $\Gamma > 2\pi f$ is taken into account in the design of the servo amplifier, a servo bandwidth much larger than $\Gamma/2\pi$ can be envisaged. Basically, the servo bandwidth is limited to a few megahertz by delays in the servo system.

If the frequency of the laser is stabilized, the size of the residual error signal provides us with information about the quality of the servo electronics. However, this error signal is by no means a measure of the actually achieved frequency stability or laser line width. There are still frequency fluctuations that are not detected by the electronic servo. In particular, at very high spectral resolution, small variations of the optical length in the cavity can no longer be neglected. The true line width of the laser can be estimated only by methods independent of the actual servo loop, e.g., by beat-frequency comparisons with a second laser or by the use of a second independent frequency discriminator.

To achieve a narrow line width, it is important to isolate the cavity from any distortions, e.g., from vibrations, acoustics, and temperature changes, caused by the environment. For this reason, the resonator is frequently vibration isolated and supported in a temperature-controlled vacuum chamber [11.2020, 2026]. A further cause of possible frequency offsets is a spurious residual amplitude modulation (RAM) of the intrinsic laser beam. Such RAM may be generated, e.g., by an imperfect phase modulator itself. Very often, RAM is also introduced by spurious interferences in the optical setup caused, for example, by light backscattered at the surfaces of optical elements. The servo cannot distinguish between the true error signal and the RAM. Hence, the total of the superimposed signals are servo controlled to zero, leading to a nonzero error signal and consequently to a frequency shift. Usually, this spurious RAM varies with time and we experience frequency fluctuations that are not identified by the error signal of the servo loop. At high spectral resolution, these fluctuations may be orders of magnitude larger than those estimated from the noise in the error signal. Hence, it is important to isolate the cavity from environmental disturbances and to minimize the RAM very carefully.

The narrowest line width achieved with the PDH method in a dye laser was as low as 0.5 Hz [11.2027] and even with diode lasers a line width in the range of approximately 1 Hz could be achieved [11.2021]. The PDH stabilization technique is now used almost exclusively, if a precision stabilization to a cavity resonance is required combined with extremely high spectral resolution.

Stabilization to a Doppler-free atomic reference. The stabilization methods discussed above describe methods to stabilize the laser frequency to a reference provided by the length of an artifact. An *absolute* stabilization requires that the frequency is traced back to a natural constant, e.g., to the center of a suitable atomic or molecular absorption line. Ideally, the frequency of such transition should not depend on environmental or operational parameters such as external fields, atomic collisions, laser power, etc. The atomic reference line should be narrow and provide us with a sufficient signal-to-noise ratio. The atoms may be contained in an absorption cell, in an atomic beam or stored in a trap. At room temperature, the thermal velocity distribution of the atomic absorbers cause a Doppler broadening of the line which is in the range of $\delta\nu/\nu \cong 10^{-6}$ and which is usually much larger than the natural line width of the reference transition. An atom moving with a velocity component v_z in the direction of the laser beam sees a Doppler-shifted frequency

$$\nu_D = \nu_l \left[1 \pm v_z/c + v^2/(2c^2)\right] , \qquad (11.232)$$

where ν_l is the laser frequency and v_z is the velocity component in the direction of the laser beam. In (11.232), the term $\nu_l v_z/c$ represents the first-order Doppler effect. The positive sign is valid for atoms counter-propagating to the laser beam whereas the minus sign belongs to the co-propagating ones. The third term $\nu_l v^2/(2c^2)$ corresponds to the (relativistic) second-order Doppler effect. Whereas the first-order Doppler effect can be strongly suppressed by the application of nonlinear Doppler-free spectroscopic methods such as, e.g., *saturated absorption* [11.2028, 2029] or Doppler-free *two-photon excitation* [11.2030, 2031] the second-order effect can be reduced only by cooling the absorbers.

As a typical example of Doppler-free spectroscopy, we concentrate on the saturated absorption and give a brief description of this method. For simplicity, we neglect the influence of the second order Doppler effect. Let us consider an atomic ensemble which is excited by two counter-propagating laser beams of the same frequency ν_l. If ν_l does not coincide with the line center ν_0 of the transition, the two beams will excite and saturate *two* groups of atoms with velocity components v_z and $-v_z$, of which the Doppler-shifted frequencies ν_D coincide with the laser frequency ν_l. Consequently, two saturation holes are burned into the velocity distribution of the ground state. If we tune the laser frequency closer to the center of the line, $|v_z|$ decreases and the holes approach each other until they overlap for $v_z = 0$ when the laser frequency is tuned to line center. In this singular case, the number of atoms contributing to the absorption decreases and hence less laser power is absorbed, resulting in a narrow Doppler-free absorption dip which is used as a reference for the stabilization. Hence, (first-order) Doppler broadening is suppressed and the width of this dip is determined by the natural width of the transition, by power and collision broadening of the line, and ultimately by the limited interaction time of the atoms moving across the laser beam. To first order, the profile of this dip can be approximated by a Lorentzian line shape.

As a practical example, let us consider an absorbing gas contained in a cell at low pressure. It can be placed inside or outside the laser cavity. If the cell is placed

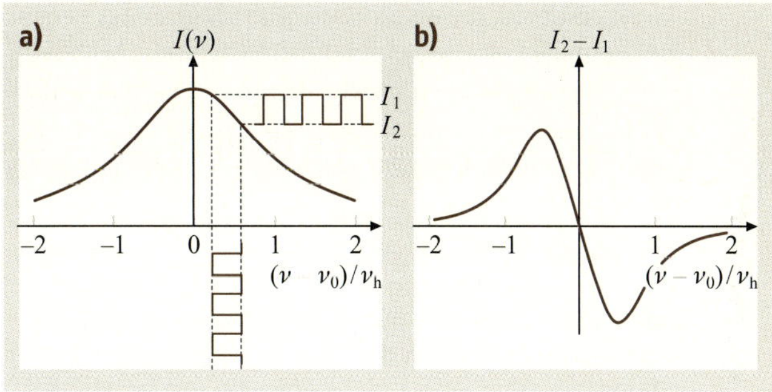

Fig. 11.250 Generation of the error signal by a square-wave frequency modulation

inside, a Doppler-free feature caused by saturated absorption will show up as a small symmetric dip in the tuning curve of the laser at the center of the reference transition. This feature has to be converted to an error signal which is antisymmetric with respect to the detuning $\Delta\nu = \nu_l - \nu_0$. A simple concept to generate the error signal is to switch the laser frequency between two discrete values (Fig. 11.250a). The difference in the corresponding two absorption signals $I_2 - I_1$ leads to an antisymmetric error signal with a zero crossing at the line center (Fig. 11.250b). For small modulation amplitudes, the error signal is proportional to the first derivative of the absorption signal.

Instead of a square-wave modulation, most lasers use a harmonic frequency modulation. Consequently, the laser power will contain harmonics of the modulation frequency. The amplitudes of these harmonics depend on the modulation width and on the detuning. The amplitude of a particular harmonic signal in the laser power can be detected by a photodetector followed by a phase-sensitive detector which is gated by the corresponding harmonic of the modulation frequency. For a symmetric absorption line, the amplitudes of the odd harmonics are antisymmetric and have a zero crossing at $\Delta\nu = 0$. Hence, these odd harmonics are suitable as error signals for frequency stabilization. In many cases, third-harmonic detection is applied in order to suppress a residual slope [11.2032, 2033]. If the modulation width is small compared to the line width, the derived signal represents the corresponding derivative of the saturation signal, e.g., the first derivative for the first harmonic and the third derivative for the third harmonic.

If we approximate the Doppler-free saturation feature by a Lorentzian line profile $I(\nu)$.

$$I(\nu) = \frac{A}{1 + \left[(\nu - \nu_0)/\nu_{1/2}\right]^2} \qquad (11.233)$$

we can estimate the amplitudes of the harmonics versus the detuning for various modulation widths [11.2034]. In (11.233), A and $\nu_{1/2}$ are the height and the half-width (HWHM) of the saturated absorption feature, respectively. The modulated laser frequency $\nu(t)$ can be written as

$$\nu(t) = \nu_0 + \Delta\nu + \delta\nu A \cos(\omega t)\,, \tag{11.234}$$

where $\Delta\nu$ is the detuning of the mean laser frequency from the line center ν_0 and $\delta\nu_A$ is the modulation amplitude. If the modulation frequency $\omega/2\pi$ is small compared to the line width, the detected signal follows the laser frequency $\nu(t)$ closely and we can replace ν in (11.233) by $\nu(t)$ of (11.234), leading to a time-dependent absorption signal

$$I(t) = \frac{A}{1+\left\{[\Delta\nu + \delta\nu_A \cos(\omega t)]/\nu_{1/2}\right\}^2}\,.$$

If we relate $\Delta\nu$ and $\delta\nu_A$ to the half-width (HWHM) $\nu_{1/2}$ we get with the reduced detuning $d_D = \Delta\nu/\nu_{1/2}$ and the reduced modulation amplitude $d_A = \delta\nu_A/\nu_{1/2}$

$$I(t) = \frac{A}{1+[d_D + d_A \cos(\omega t)]^2}\,. \tag{11.235}$$

The signal $I(t)$ is periodic in time and can be expressed by a Fourier series

$$I(t) = \frac{A_0}{2} + \sum_{m=1}^{\infty} A_m \cos(m\omega t)\,, \tag{11.236}$$

where the Fourier coefficient A_m represents the signal amplitude of the m-th harmonic of the modulation frequency $\omega/(2\pi)$. It can be calculated by the relation

$$A_m = \frac{2}{\pi}\int_0^{\pi} I(\tau)\cos(m\tau)\,d\tau\,.$$

For the coefficients A_1 and A_3 we get [11.2034]

$$A_1 = \frac{1}{d_A}\left[(\operatorname{sign} d_D)\,P_- - d_D P_+\right] \tag{11.237a}$$

and

$$A_3 = \frac{1}{d_A^3}\Big\{(\operatorname{sign} d_D)\left[4\left(1-3d_D^2\right)+3d_A^2\right]P_- + \left[4\left(3-d_D^2\right)+3d_A^2\right]d_D P_+ - 16 d_D\Big\} \tag{11.237b}$$

with $P_\pm = \frac{1}{\rho}\sqrt{2(\rho \pm \alpha)}$ and $\rho = \sqrt{\alpha^2 + 4d_D^2}$ and $\alpha = 1 + d_A^2 - d_D^2$.

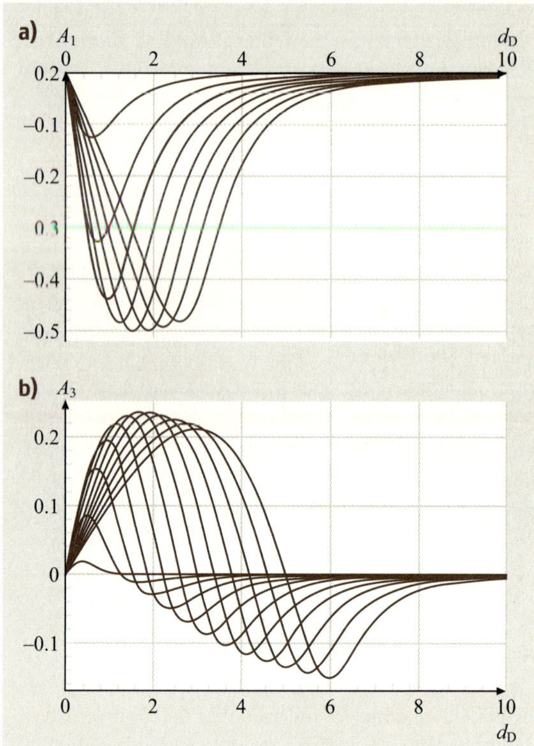

Fig. 11.251 Amplitudes of the first harmonic (A_1) and the third harmonic (A_3) versus the detuning $d_D = \Delta\nu/\nu_{1/2}$. The modulation amplitude was chosen as $d_A = 0.2, 0.6, 1.0, \ldots, 3.0$ for A_1, and $d_A = 0.5, 1.0, 1.5, \ldots, 6.0$ for A_3

A_1 and A_3 are shown versus the detuning d_D for various modulation amplitudes d_A in Fig. 11.251a and Fig. 11.251b, respectively. Both curves have zero crossings at $d_D = 0$. In the central part, their amplitudes depend linearly on the detuning d_D. Hence, both signals A_1 and A_3 can be used as discriminator signals for frequency control. Their slope depends on the modulation width d_A. For small amplitudes $d_A \ll 1$, the slope of the first-harmonic signal depends linearly on d_A and that of the third harmonic increases with the third power of d_A. The largest slope and consequently the highest sensitivity can be obtained with the first harmonic $A_1(d_D)$. In many cases, however, the saturation signal $I(d_D)$ is superimposed on a frequency-dependent (sloping) background. Consequently, a signal is added to the first-harmonic signal $A_1(d_D)$ and the zero crossing is shifted, leading to a frequency offset of the stabilized laser frequency. This shift can be strongly reduced if the third or a higher odd harmonic is applied for the stabilization.

A typical example of a laser utilizing the third-harmonic detection technique is the iodine-stabilized HeNe laser (see Sect. 11.14.3).

Servo Amplifier and Filter

This subsection describes how the error signal is used to control the laser frequency. If the servo control is operating, the initial frequency ν_i of the free-running laser is corrected by the servo loop to ν_S, which is close to the reference frequency ν_0 (Fig. 11.252). The difference $\delta\nu = \nu_S - \nu_0$ represents the residual frequency deviation from the line center ν_0 of the stabilized laser. The deviation $\delta\nu$ is converted to an amplitude $U = C\delta\nu$ by the error signal, where the constant C is proportional to the sensitivity of the frequency discriminator. The signal $C\delta\nu$ is amplified in the following servo amplifier by a factor $g(f)$ and is then transferred to a frequency transducer, which corrects the laser frequency ν_i by $-CDg(f)\delta\nu$. Here D is the sensitivity of the transducer, which transforms the control voltage to a corresponding frequency shift of the laser. The term $g(f)$ characterizes the frequency-dependent gain of the electronic servo amplifier. The servo loop provides a negative feedback to the laser frequency and the corrected value ν_S can be written as

$$\nu_S = \nu_i - CDg(f)\delta\nu \quad (11.238)$$

which can be transformed to

$$\frac{\delta\nu}{\Delta\nu} = \frac{1}{1+CDg(f)}\,, \quad (11.239)$$

where $\Delta\nu = \nu_i - \nu_0$ is the initial frequency offset of the free-running laser.

We see that the servo-control circuit reduces the frequency deviation $\Delta\nu$ of the free-running laser by a factor of $1/[1+CDg(f)]$, where $CDg(f)$ represents the gain of the open servo loop. To achieve a negligibly small value of the residual frequency deviation $\delta\nu$, the servo gain should be as high as possible. In particular, the deviation of the mean frequency averaged over long times should be close to zero. This requires that $g(f \to 0) \to \infty$. In any servo-control system, phase shifts and time delays of the servo loop put limitations on the maximum gain and the bandwidth of the servo-control. A simple technique to provide high servo gain at low frequencies and to reduce the influence of phase shifts and time delays at high frequencies is to reduce the servo gain with increasing Fourier frequency f. Such a characteristic can be obtained by an integrating behavior of the servo gain, i. e., if $CDg(f)$ decreases with $1/f$ for increasing frequencies. In this case, the total transfer function of the control system can be characterized by a unity-gain frequency f_C, the frequency at which the gain of the open servo loop $CDg(f_C) = 1$. At low fluctuation frequencies $f \ll f_C$, the frequency deviations $\Delta\nu(f)$ will be reduced by a factor of f/f_C and an increase of the servo gain corresponds to a proportional increase of the unity-gain frequency f_C. Consequently, in an integrating servo loop a high servo gain requires a large servo bandwidth f_C. In cases where the servo bandwidth f_C is limited and where large frequency fluctuations at low frequencies have to be suppressed, an additional integration can be introduced, leading to a double integrating behavior and consequently to a higher servo gain at low frequencies. The stability of the servo loop requires that this double integration is

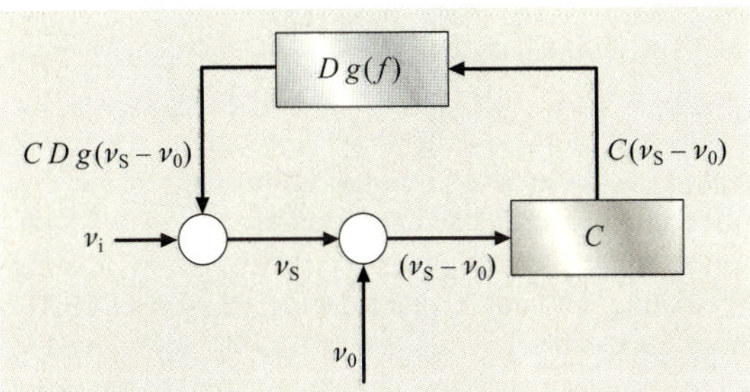

Fig. 11.252 Equivalent circuit of a laser frequency stabilization

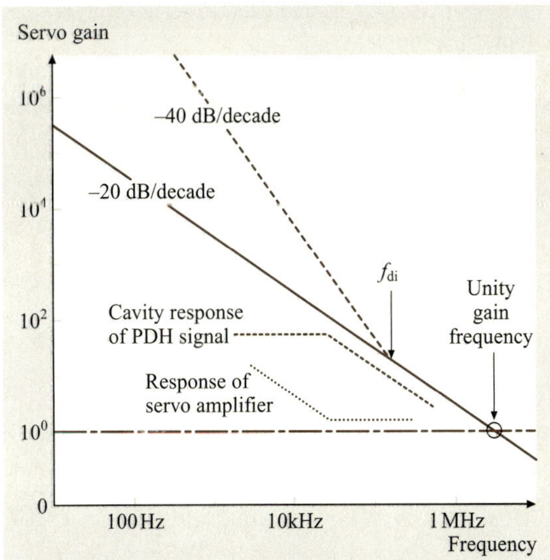

Fig. 11.253 Typical servo gain versus Fourier frequency of a PDH frequency stabilization to a cavity resonance (see text)

effective only at frequencies well below the unity-gain frequency f_C. In practice, the maximum crossover frequency f_{di} between single and double integration should be smaller than $f_C/4$. Figure 11.253 shows a typical dependence of servo gain on the fluctuation frequency of a frequency stabilization to a narrow cavity resonance using the PHD technique.

11.14.3 Examples of Frequency-Stabilized Lasers

This section describes a few examples of stabilized lasers. The radiations of some of them are recommended as reference frequencies for the realization of the length unit *meter* and as references for scientific applications [11.2009]. Frequency-stabilized lasers have been developed covering a wide range of wavelengths from the near UV to the far IR. Of these examples, HeNe lasers are of great practical importance since they are often employed in many applications, e.g., as wavelength standards in interferometers for dimensional metrology. Most of these lasers utilize the frequency of their gain profile itself as the reference for the stabilization. In these cases, the stabilized frequency depends on the discharge parameters, the gas pressure, and gas mixture inside the laser tube and the relative reproducibility is limited to some 10^{-7}. Hence, if the absolute value of the laser frequency is important, it should be compared from time to time with a superior standard, e.g., a HeNe Laser of which the frequency is stabilized to an absorption line of molecular iodine.

With increasing requirements on reproducibility and accuracy, the laser frequency needs to be stabilized to suitable references such as the narrow absorption lines of atoms, molecules, or ions and line broadening caused by the movement of the absorbers needs to be reduced by Doppler-free techniques such as saturated absorption or two-photon excitation. Ultimately any frequency shifts caused by a residual Doppler effect can be strongly reduced if the laser radiation interacts with cold atomic absorbers only.

This section is organized as follows. In the first part, we explain as a typical example the stabilization of a tunable diode laser to the resonance of a stable optical cavity. We then continue with HeNe lasers stabilized to their gain profile. The next section represents laser radiations that are stabilized to suitable absorption lines of atoms, molecules, or ions. In view of its practical importance, we focus on HeNe lasers, which can be stabilized to the absorption lines of molecular iodine. The absorbing particles of these lasers are contained in an absorption cell. In the next step, laser stabilization systems based on atomic or molecular beams are discussed. The last part of this section presents a brief description of lasers stabilized to cold ions and atoms.

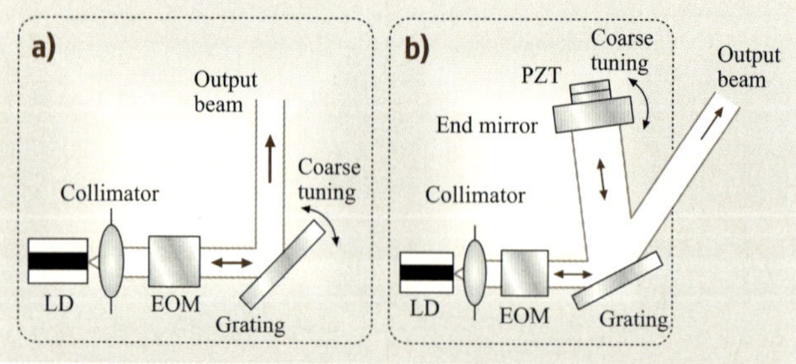

Fig. 11.254a,b Typical setups of extended-cavity diode lasers (ECDLs). **(a)** Littrow configuration, **(b)** Littman configuration (see text)

Frequency Stabilization of Diode Lasers

Tunable lasers are important tools in the field of precision laser spectroscopy. Whereas in the past the visible spectrum was covered mostly by dye lasers, diode lasers are increasingly taking over due to their small size, high efficiency, high power, and reliability. They are available in a wide range of the visible and infrared spectrum. However, if high spectral resolution is required, the line width of most solitary laser diodes, which ranges from 10 MHz to 300 MHz, is orders of magnitude too large. Several methods have been developed to improve the spectral purity of these lasers. Most of them apply optical feedback [11.2035, 2036]. This section discusses a typical example, where the length of the laser resonator is increased by an external reflector [11.2037, 2038]. Compared to solitary laser diodes, the line width of such extended-cavity diode lasers (ECDL) can be reduced to some 100 kHz. Figure 11.254 shows two typical setups of ECDLs in the Littrow and Littman configurations. In both configurations, the beam leaving the antireflection-coated facet of the laser diode (LD) is collimated and directed onto a reflection grating. In the Littrow configuration (Fig. 11.254a), the laser grating acts as the end mirror of the extended laser resonator where the beam is back-reflected into the laser diode in the first diffraction order. The zeroth-order beam is coupled out. Coarse frequency tuning can be performed by rotating the grating whereas a piezoelectric transducer (PZT) and/or an electrooptic modulator (EOM) can be used to fine-tune and control the laser frequency by changing the optical length of the cavity. Frequency tuning can also be achieved by changing the injection current of the laser diode; this method, however, will also change the laser power. The Littrow configuration can be designed to be very compact. Unfortunately, the direction of its output beam changes with any rotation of the grating. This shortcoming is avoided in the Littman configuration [11.2039] where the ECDL contains an additional mirror (Fig. 11.254b). In this configuration the beam deflected in the first diffraction order is directed onto this extra mirror and back-reflected from there into the laser diode. In the Littman configuration, the laser beam passes the grating twice and hence the wavelength selectivity is increased compared to the Littrow configuration. The output beam leaves the grating in the zeroth order. In this configuration, coarse frequency setting is achieved by rotating the end mirror, with the consequence that the direction of the output beam does not change. Fine-tuning of the frequency can be performed in the same way as with the Littrow configuration by means of a PZT or an EOM.

The frequency noise of ECDLs occurs at Fourier frequencies low enough that a further line width reduction below 100 kHz can be achieved by active frequency stabilization to the resonance of an optical cavity [11.2040], as described in Sect. 11.14.2. Precise frequency tuning can be performed, e.g., by changing the length of the optical reference cavity. This method, however, strongly perturbs the stability of the cavity. To achieve the highest possible resolution, it is better to introduce a variable frequency difference between the laser output beam and the cavity resonance. Since the frequency-shifted beam is locked to the cavity resonance, which is not affected, the stabilization circuit will shift the laser frequency, with the consequence that it can be tuned very precisely relative to the stable cavity resonance. Such a shift can be achieved, e.g., by an acoustooptic modulator or an electrooptic modulator. Using the PDH technique and a carefully designed and isolated reference cavity of ultrahigh finesse, a line width as narrow as 1 Hz was recently observed, demonstrating the great potential of diode lasers for applications in precision spectroscopy and optical frequency standards [11.2021].

Frequency Stabilization of HeNe Lasers

HeNe lasers can operate at several different wavelengths within the spectral range between the green ($\lambda = 543$ nm) and the infrared ($\lambda = 3.39\,\mu$m). Regarding the various wavelengths, the emission at $\lambda = 633$ nm is probably most important. HeNe lasers are advantageous since their operation is simple, the intrinsic frequency noise is small, and the output power in the range between 100 μW and some 3 mW is sufficient for many applications in metrology. Provided the neon gas inside the laser tube consists of a single isotope, the width of the Doppler-broadened gain profile is in the range of approximately 1.5 GHz and single- or two-frequency operation can easily be achieved by choosing a sufficiently short length of the laser resonator. Owing to these favored properties and their simplicity, frequency-stabilized HeNe lasers were developed very early [11.2041]. The first stabilization systems – which are still used in many laboratories – apply the gain profile itself as the frequency reference.

Lamb-dip stabilization. The Lamb-dip-stabilized laser was one of the first frequency-stabilized lasers [11.2042]. It makes use of the fact that the output power passes through a local minimum – the *Lamb dip* – when the laser frequency is tuned across the center of the gain curve (Fig. 11.255a). Similar to the Doppler-free minimum of saturated absorption, the Lamb dip is caused by an increased saturation of the gain at the line center when both the forward- and backward-running waves inside the resonator interact with the same velocity group ($v_z = 0$) of atoms. Hence the number of atoms contributing to the laser action is decreased, leading to a reduced laser power at line center. To generate the error signal (Fig. 11.255b), the laser frequency is modulated and the first harmonic in the output power is detected.

Two-mode stabilization. A simple stabilization system can easily be set up, if the HeNe laser is operating at two adjacent longitudinal modes (Fig. 11.256a) and if

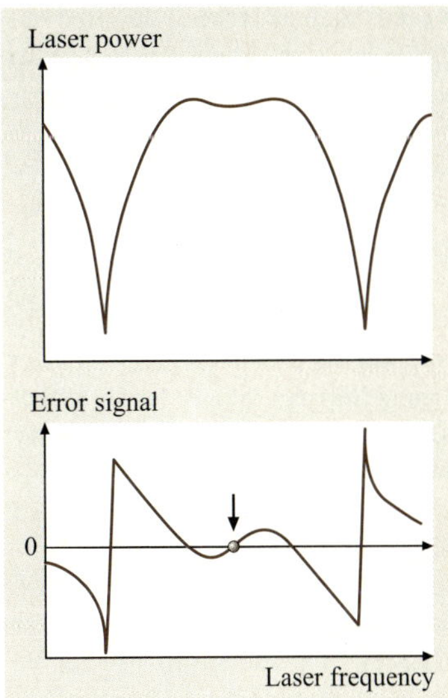

Fig. 11.255a,b Tuning curve **(a)** and error signal **(b)** of a Lamb-dip-stabilized HeNe laser

Part C | 11.14

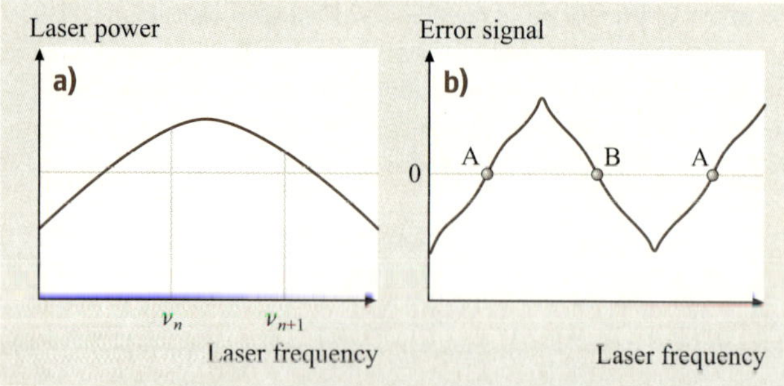

Fig. 11.256a,b Tuning curve (**a**) and error signal (**b**) of a two-mode stabilized HeNe laser

these modes are polarized perpendicular to each other. These conditions are provided, e.g., in commercial HeNe laser tubes of 30 cm length with internal mirrors. Perpendicular polarization is observed since it minimizes the gain competition of two modes. Furthermore, the two planes of polarization are fixed within the tube. These properties allow us to separate the two modes by a polarization divider, e.g., a Wollaston prism, and to monitor the power in each mode by a separate photodetector. The difference between the photocurrents of the two detectors $I_1 - I_2$ depends on the frequencies of the two modes (Fig. 11.256b). It crosses zero when they are distributed symmetrically with respect to the center of the gain curve and the powers in the two modes are identical. Hence, the difference in the photocurrents is used as the error signal for frequency stabilization and no frequency modulation is necessary [11.2043]. Several means can be applied to transfer the control signal to the length of laser tube, such as heating the tube via the discharge current or a separate external heater. With this simple laser stabilization technique, a relative reproducibility of better than 10^{-7} has been demonstrated. During a time interval of five days relative frequency variations as low as 10^{-9} have been demonstrated under favored laboratory conditions [11.2044].

Zeeman stabilization. Another method of frequency stabilization widely applied in commercial HeNe lasers and laser interferometers utilizes the Zeeman effect in a short (≈ 10 cm) single-mode laser. If we apply a longitudinal magnetic field, the Zeeman effect causes a splitting of the laser mode into two submodes ν_+ and ν_- of opposite circular polarization and slightly different frequencies. Depending on the size of the magnetic field, the frequency difference $\nu_+ - \nu_-$ is in the range between 300 kHz and 2 MHz. The two submodes can be separated by a quarter-wave plate followed by a polarizing beam splitter. Similar to two-mode stabilization, the difference of the output powers at ν_+ and ν_- can be utilized as the error signal [11.2045].

Alternatively, the error signal can also be generated by utilizing the frequency difference $\delta\nu = \nu_+ - \nu_-$. It has been shown that $\delta\nu$ changes with the mean laser frequency and that it has a minimum at the line center. This dependence is caused by nonlinear mode-pulling effects in the gain medium. The error signal to find this minimum is generated by modulating the laser frequency and monitoring the beat frequency between the two Zeeman modes by means of a reversible counter [11.2046]. To derive the error signal, the direction of counting is reversed at every half period of the modulation. The accumulated counter content, which corresponds to the integrated error signal, is then fed into a digital-to-analog converter after each period. Its output is used to control the frequency of the laser so that the mean laser frequency coincides with the minimum of the beat frequency. With such lasers a relative frequency reproducibility of 10^{-8} can be achieved over a period of five months under laboratory conditions [11.2046].

Absolute Frequency Stabilization Utilizing Separate Absorption Cells

As indicated above, the frequency of HeNe lasers stabilized to their gain profile may change with time. Such frequency shifts can be reduced by several orders of magnitude if the frequency reference is provided by a suitable narrow absorption line of which the transition frequency is largely independent on the parameters of operation. There are now several different laser systems available in a wide range of the optical spectrum which are stabilized to such frequency references. Some of these have been recommended by the Comité International des Poids et Mesures (CIPM) [11.2009]. Their radiations are listed together with the values of their frequencies, wavelengths, and their relative uncertainties in Table 11.56. Six of the recommended reference frequencies are stabilized to hyperfine-structure components of molecular iodine, which has a rich spectrum of narrow absorption lines in the visible range. They belong to the transition between the electronic B level and the ground state (X level). Since the iodine molecule is heavy, its velocity – and hence the Doppler broadening – is rather low at room temperature. Furthermore, the vapor pressure – causing a pressure-induced shift and broadening – can be controlled conveniently via the temperature of a cooling finger attached to the iodine absorption cell. Four of the iodine references are used to stabilize HeNe lasers at $\lambda \approx 543$ nm, 612 nm, 633 nm, and 640 nm. From this

Table 11.56 Recommended reference wavelengths/frequencies generated by stabilized lasers [11.2009]

	Atom / Molecule	Transition	Wavelength (fm)	Frequency	Rel. std. uncertainty (1 σ)
1	$^{115}In^{+}$	$5s^2\ ^1S_0$–$5s5p\ ^3P_0$	236 540 853.54976	1 267 402 452 899.92 kHz	3.6×10^{-13}
2	^{1}H	1S–2S (Two-photon transition)	243 134 624.62603	1 233 030 706 593.61 kHz	2.0×10^{-13}
3	$^{199}Hg^{+}$	$5d^{10}6s\ ^2S_{1/2}\ (F=0)$–$5d^9 6s^2\ ^2D_{5/2}\ (F=2)$ $\Delta m_F=0$	281 568 867.591969	1 064 721 609 899.143 Hz	1.9×10^{-14}
4	$^{171}Yb^{+}$	$6s\ ^2S_{1/2}\ (F=0)$–$5d\ ^2D_{3/2}\ (F=2)$	435 517 610.73969	688 358 979 309 312 Hz	2.9×10^{-14}
5	$^{171}Yb^{+}$	$^2S_{1/2}\ (F=0, m_F=0)$–$^2F_{7/2}\ (F=3, m_F=0)$	466 878 090.061	642 121 496 772.6 kHz	4.0×10^{-12}
6	$^{127}I_2$	R(56) 32-0, component, a_{10}	532 245 036.104	563 260 223 514 kHz	8.9×10^{-12}
7	$^{127}I_2$	R(127) 11-5 component a_{16} (or f)	632 991 212.58	473 612 353 604 kHz	2.1×10^{-11}
8	^{40}Ca	$^1S_0$–$^3P_1$; $\Delta m_J=0$	657 459 439.29167	455 986 240 494 150 Hz	1.1×10^{-13}
9	$^{88}Sr^{+}$	$5\ ^2S_{1/2}$–$4^2D_{5/2}$	674 025 590.8631	444 779 044 095.5 kHz	7.9×10^{-13}
10	^{85}Rb	$5S_{1/2}(F_g=3)$–$5D_{5/2}(F_e=5)$ two-photon	778 105 421.23	385 285 142 375 kHz	1.3×10^{-11}
11	$^{13}C_2H_2$	$P(16)(\nu_1+\nu_3)$	$\lambda=1\,542\,383\,712$	194 369 569.4 MHz	5.2×10^{-10}
12	CH_4	P(7) ν_3, component of the $F_2^{(2)}$ central hyperfine structure	3 392 231 397.327	88 376 181 600.18 kHz	3×10^{-12}
13	$^{12}C^{16}O_2$	R(10) (00^01)–(10^00) line of the CO_2-laser	10 318 436 884.460	29 054 057 446 579 Hz	1.4 – 13

group, the one operating at 633 nm is of particular importance since it is used in many standard laboratories as a reference for interferometric length measurements, laser wavelength calibrations, and applications in precision laser spectroscopy. It can also act as a reference to calibrate HeNe lasers that are stabilized to their gain profile by beat-frequency comparisons. Hence, the I_2-stabilized HeNe laser at $\lambda = 633$ nm can be regarded as the workhorse in many laboratories.

The iodine molecules are contained at low vapor pressure in a cell. Depending on the laser system, the length of the cell may range from a few centimeters up to several meters. In either case, the content of impurities has to be low in order to prevent any frequency shifts caused by collisions with impurities. For example, in the case of iodine-stabilized lasers, the quality of the iodine filling limits the uncertainty ultimately achievable. Nevertheless, as a result of its rich spectrum, I_2 is often used as the frequency reference for different types of lasers operating within the visible spectrum. Two types of I_2-stabilized lasers, the HeNe laser ($\lambda = 633$ nm) and the frequency-doubled Nd:YAG laser ($\lambda = 532$ nm) will be discussed in the following.

Iodine-stabilized HeNe *laser* ($\lambda = 633$ nm). Figure 11.257 schematically shows a typical setup of an iodine-stabilized HeNe laser. The laser head consists of a HeNe discharge tube and an iodine absorption cell mounted inside the laser resonator. The lengths of the discharge tube and the absorption cell are typically 20 cm and 10 cm, respectively, corresponding to a minimum length of the resonator of approximately 35 cm and a free spectral range of ≈ 430 MHz. Despite of the rather long resonator the laser emits only one single frequency, owing to the loss in the absorption cell. In order to reduce the influence of acoustics and vibrations, the resonator should be made as rugged as possible. The spacers should consist of a material of low thermal expansion. The radii of curvature of the laser mirrors are usually in the range $r = 0.6$–4 m with a radius of $r = 1$ m most commonly used. Both mirrors have a reflectivity of approximately 98%, allowing single-frequency operation over most of the free spectral range. Output powers of up to 300 μW have been achieved, sufficient for most applications in laser spectroscopy and interferometry.

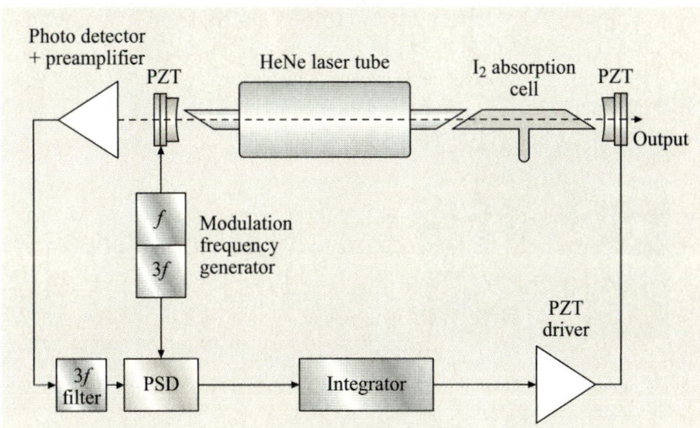

Fig. 11.257 Experimental scheme of an iodine-stabilized HeNe laser

The two mirrors are mounted on PZT actuators, allowing one to change the resonator length and thereby the laser frequency by applying a voltage to the PZTs.

Figure 11.258a shows four absorption features in the tuning curve of the laser. They belong to the hyperfine-structure (HFS) components d, e, f, and g of the transition R(127), $v' = 5$, $v'' = 11$ of $^{127}I_2$. The separation between the components d, e, f, g is approximately 13 MHz and the full width at half-maximum of each line is in the range of 5 MHz, corresponding to a quality factor of $Q = \nu/\delta\nu \approx 10^8$. The height is approximately 0.1% of the laser power. To generate the error signal for the stabilization, the length and thereby the frequency of the laser is modulated harmonically by applying a sine-wave voltage to the left PZT (Fig. 11.257). For the stabilization, third-harmonic detection is used. The third-harmonic signal in the output power is filtered out, demodulated by a phase-sensitive detector (PSD) and passed through a low-pass filter. It then serves as error signal for the stabilization. A typical third-harmonic spectrum of molecular iodine containing seven HFS components within the tuning range of the laser is shown in Fig. 11.258b. Each absorption feature is suitable as a reference for the stabilization.

With the servo loop operating, the length of the laser is stabilized such that the third-harmonic signal of the corresponding HFS component vanishes at its central zero crossing.

The stability and the reproducibility of the I_2-stabilized HeNe laser can be investigated by beat-frequency measurements of two independent laser systems. Figure 11.245 shows the measured Allan standard deviation $\sigma(\tau)$ of an iodine-stabilized HeNe laser versus the integrating time τ. In the range between $\tau = 10$ ms and $\tau = 100$ s, $\sigma(\tau)$ decreases approximately with the square root of τ, corresponding to white frequency noise. The minimum instability of about 2×10^{-13} is observed at an integrating time $\tau \approx 1000$ s.

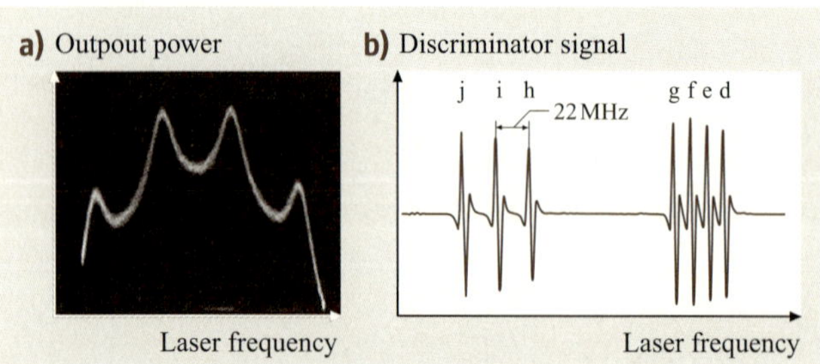

Fig. 11.258a,b Saturation dips (**a**) and error signals (**b**) of an iodine-stabilized HeNe laser ($\lambda = 633$ nm)

The reproducibility of the iodine-stabilized HeNe laser was investigated by measuring the dependence of the laser frequency on the various operation parameters and by frequency inter-comparisons between lasers of different institutes. The laser frequency depends on the modulation width, the iodine vapor pressure, and slightly on the laser power. The coefficients are approximately -6 kHz/Pa and -10 kHz/MHz for the pressure and modulation dependence, respectively. International frequency comparisons between iodine-stabilized lasers have shown that the stabilized frequencies of the majority of iodine-stabilized HeNe lasers coincide within a range of approximately 12 kHz, corresponding to $2.5 \times 10^{-11}\nu$. For specified operation parameters (temperature of the cell: (25 ± 5) °C, temperature of the cold point: (15 ± 0.2) °C, width of frequency modulation between the peaks: (6 ± 0.3) MHz, one-way intracavity beam power: (10 ± 5) mW) a relative standard uncertainty of the recommended frequency value of 2.5×10^{-11} can be achieved [11.2009].

Iodine-stabilized frequency-doubled Nd:YAG lasers ($\lambda = 532$ nm). Frequency-doubled YAG lasers pumped by diode lasers are of particular interest in applications as optical frequency standards for the following reasons. Small lasers providing us with high output powers of 100 mW at 532 nm are commercially available. The intrinsic frequency noise of YAG lasers can be very low. Part of the emission range coincides with strong absorption lines of molecular iodine (Fig. 11.259a) that are suitable as reference frequencies for the stabilization. Figure 11.259b shows the Doppler-free absorption spectrum observed within the continuous tuning range of 5 GHz of a commercial laser. The observed absorption lines represent two sets of hyperfine components belonging to the 32-0, R(57) and the 32-0, P(54) ro-vibrational lines labeled #1104 and #1105, respectively [11.2047]. Any of these transitions is suitable as a frequency reference. Since the frequency difference to the recommended a_{10} component of the 32-0, R(56) transition is known to about 2 kHz an accurate optical frequency can be attributed to each HFS component.

Various methods have been used in different laboratories to generate the error signal. The methods of modulation transfer spectroscopy [11.2048, 2049] and phase-modulation spectroscopy [11.2050,2051] are very powerful to achieve discriminant signal with a high signal-to-noise ratio.

International frequency comparisons show that the frequency reproducibility of these lasers is in the range of 5 kHz [11.2052,2053], mostly limited by potential im-

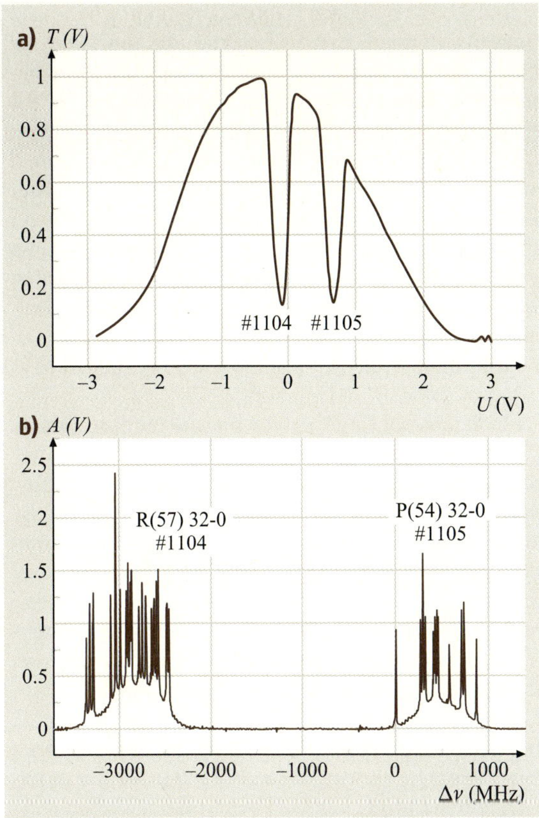

Fig. 11.259a,b Absorption lines of molecular iodine within the emission profile of a commercial frequency-doubled Nd:YAG laser. (**a**) Doppler-limited, (**b**) with Doppler-free excitation, the hyperfine structure is resolved

purities in the absorption cell. Regarding its high power, its compactness, and its high frequency reproducibility, the iodine-stabilized frequency-doubled Nd:YAG laser represents an optical frequency standard with important applications in precision length metrology, interferometry, and spectroscopy.

Laser stabilized to a two-photon transition in rubidium. The radiation of a laser stabilized to the two-photon transition $5S_{1/2}$–$5D_{5/2}$ in rubidium (Fig. 11.260) [11.2054] represents a reference in the near infrared at a wavelength of 778 nm. In this spectral range, easy-to-handle laser diodes of low frequency noise are available, allowing the development of a transportable optical frequency standard of high precision. Furthermore, the laser may provide a precise frequency reference for optical communication systems since twice its wavelength coincides with the transmission band at 1.55 μm. Its optical frequency has been determined [11.2055].

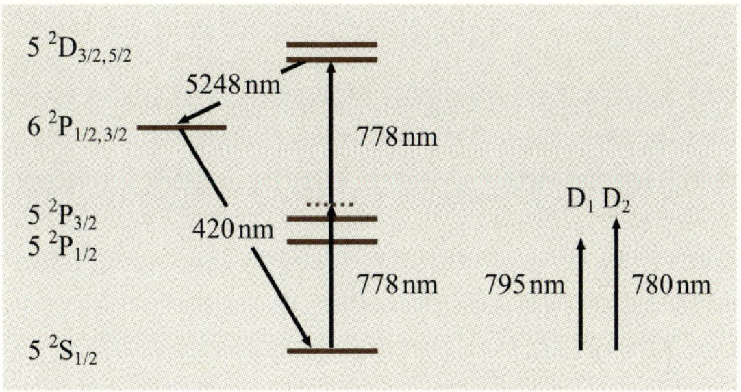

Fig. 11.260 Energy levels of Rb relevant for a two-photon stabilized diode laser

In a simple setup [11.2054], the collimated beam of an extended-cavity diode laser passes through an absorption cell filled with rubidium vapor. The cell is sealed with Brewster windows at its ends and filled with natural rubidium (73% ^{85}Rb and 27% ^{87}Rb). The beam is retro-reflected by a mirror or a cat's eye in order to allow Doppler-free two-photon excitation. Optical feedback into the diode laser is avoided by Faraday isolators. When the laser frequency is scanned through the two-photon resonance, the excitation of the upper level ($5^2D_{5/2}$) is observed via the blue fluorescence (420 nm) of the 6P–5S transition in the cascade of the spontaneous decay 5D → 6P → 5S. With this setup, a frequency stability of $\sigma(\tau) = 3\times10^{-13}\tau^{-1/2}$ has been achieved for integrating times τ up to 2000 s.

It is known that the stabilized frequencies of standards based on two-photon transitions suffer from light shifts. Its magnitude depends linearly on the intensity. Therefore, it is important to control the laser power and to prepare a well-defined laser beam geometry for the excitation. This requirement can be fulfilled if the absorption cell is mounted inside a nondegenerate optical resonator. The use of a resonator also allows a power buildup of the radiation and therefore an increase of the two-photon excitation. Furthermore, it provides exact retro-reflection of the laser beam, which is necessary to suppress residual first-order Doppler shifts. In the experiments [11.2054], the laser frequency was prestabilized in a first step to a resonance of this cavity. It could then be tuned through the two-photon resonance by changing the length of the resonator via a PZT actuator to which one of the mirrors is mounted. Compared to the results of the simple setup described before, the observed frequency stability was approximately the same. However the light shift could be better controlled and consequently the frequency of the transition can be extrapolated more precisely to zero laser power.

Laser-cooled ensembles of Rb atoms have also been used for frequency stabilization to reduce the influence of the second-order Doppler effect [11.2056]. Further evaluations will show if a standard based on cold Rb atoms leads to a significant reduction of the relative frequency uncertainty. Nevertheless, in its present state, the Rb-stabilized laser represents an interesting precision frequency reference.

Stabilized Lasers Based on Atomic or Molecular Beams

Atomic and molecular absorbers are often prepared in beams rather than in absorption cells for various reasons. First, some gases such as hydrogen are not stable as atomic species and need to be prepared just shortly before excitation. Vapors of metals would soon coat the cell windows and strongly decrease their transparency. Second, the use of a collimated beam allows us to utilize an excitation geometry in which the laser beam crosses the atomic beam at a right angle. In the case of single-photon transitions, this may reduce the first-order Doppler effect by several orders of magnitude. Furthermore, for long-lived atomic states, the excitation and detection zones can be separated, leading to an increased signal-to-noise ratio since scattered light from the excitation beam can be better suppressed.

The mean velocity of molecules or atoms at room temperature ranges from approximately 100 m/s to more than 2000 m/s. At this velocity, the interaction time of the particles crossing the laser beam at a right angle, is typically less than 10^{-5} s for beam diameters of a few millimeters. Hence, the corresponding line broadening is more than 0.1 MHz. This broadening can be reduced by applying Ramsey's method of separated field excitation first introduced in the microwave range [11.2057]. For single-photon transitions in the optical regime, where the wavelength of the radiation is typically much smaller than the diameter of the atomic beam, additional steps need to be taken to use separated field excitation. This can be achieved either by blocking well-defined trajectories of the atomic beam [11.2058] or by using three [11.2059] or more [11.2060] spatially separated excitation zones.

An alternative method is a longitudinal excitation where the atoms fly parallel or anti-parallel to the laser beam. This method is suitable in the case of two-photon transitions. For example, one of the recommended radiations (Table 11.56) corresponds to the 1S–2S two-photon transition of atomic hydrogen. In the experiment, a beam of cold atomic hydrogen is formed by guiding hydrogen molecules through a gas discharge and reflecting the atomic particles at a plate which is cooled by liquid helium. The reflector directs the hydrogen atoms on the optical axis of an optical resonator which is tuned close to the 1S–2S two-photon transition. The *longitudinal* excitation inside a resonator allows the strong power buildup necessary for a two-photon excitation, precise mode-matching of the forward- and backward-running waves inside the resonator, and an increased interaction time of the atoms with the laser beam, leading to increased spectral resolution. The relative uncertainty of the transition frequency recommended by the CIPM is as low as 2.0×10^{-13} [11.2009].

In the case of single photon transitions, the transvers excitation by separated laser fields (Fig. 11.262) is frequently applied to achieve high spectral resolution and a good signal-to-noise ratio. It was shown by Bordé that this method leads to an atom interferometer (Ramsey–Bordé interferometer) where the laser beams act as coherent beam splitters for the atomic beam [11.2061–2063]. The prominent advantage of optical Ramsey excitation with separated fields results from the fact that the transit-time broadening and the resolution can be adjusted independently. The former can be increased by choosing short interaction times in each zone, thereby allowing a large fraction of the atoms to contribute to the signal. The resolution, however, is mainly determined by the time of flight between the interaction zones. In the following, we give a brief description of the Ramsey–Bordé interferometer. Utilizing the internal energy structure of the atoms, absorption and induced emission, processes can be used to split or deflect atomic beams. If an atom absorbs a photon from a traveling wave, it also absorbs the photon momentum $\hbar\mathbf{k}$ (Fig. 11.261a). Hence, the excited atom suffers a photon recoil. If the atomic wave was in the excited state

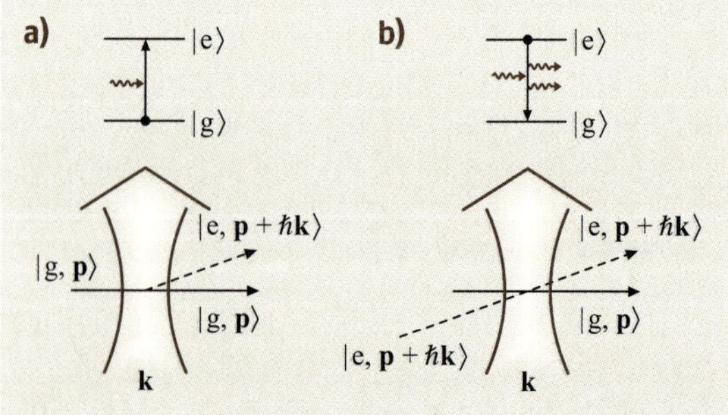

Fig. 11.261 Laser beams as beam splitters for an atomic wave

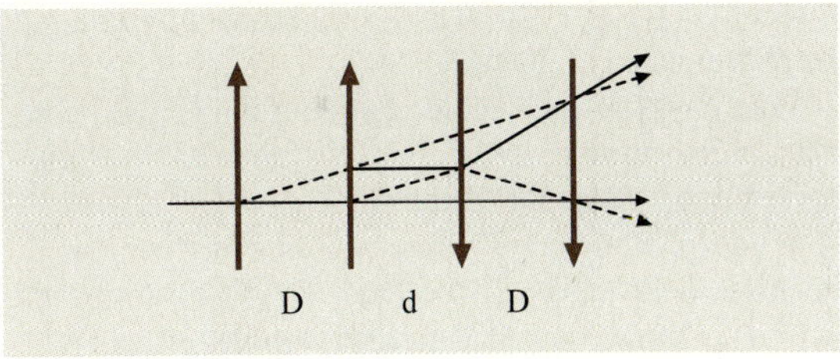

Fig. 11.262 Ramsey–Bordé matter–wave interferometer (the figure shows only those partial beams that lead to an interference structure)

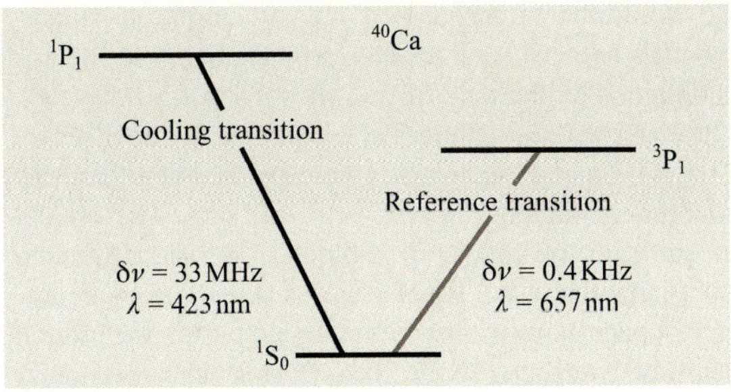

Fig. 11.263 Partial energy-level diagram of ^{40}Ca

before interaction with the laser radiation, the inverse process (induced emission) will occur (Fig. 11.261b). The probability ρ of observing such excitation or emission process depends on the intensity, the frequency of the light field, and the duration of the interaction. In general, the atom will be left in a coherent superposition of the ground and the excited state (Fig. 11.261). Since the atoms have different momenta in both states, it is more appropriate to use the picture of atomic wave packets. The interaction can act, e.g., as a 50% beam splitter ($\rho = 0.5$, $\pi/2$ pulse) or as a mirror ($\rho = 1$, π pulse). In any case, the phase of the light beam is transferred to the deflected partial atomic wave. If the light field is slightly detuned from the atomic resonance, the corresponding energy difference $\hbar\delta\omega$ is also transferred to the deflected partial wave as kinetic energy and consequently the de Broglie wavelength is changed.

The combination of such beam splitters may lead to atom interferometers that are similar to an optical Mach Zehnder interferometer (Fig. 11.262). During the first and each subsequent interaction, the matter wave is coherently split into partial waves with the internal states $|g\rangle$ and $|e\rangle$, which are represented by the solid and the dashed lines in Figs. 11.261 and 11.262. If an absorption or emission is induced, the momentum of the corresponding wave packet is changed by the photon momentum $\hbar\boldsymbol{k}$. At the final interaction, the loop of the trajectories closes and the laser field superimposes the partial waves, resulting in an atom interference. Since the atomic wave packets leave the two outputs of each interferometer in different internal states, the interference structure can be observed by monitoring the population either in the excited or the ground state.

The inter-combination transitions $^1S_0$–$^3P_1$ of the alkaline-earth atoms are well known to represent excellent references for optical frequency standards (see, for example, [11.2065] and references therein). They exhibit narrow natural line widths of about 0.04 kHz (Mg), 0.3 kHz (Ca), and 6 kHz (Sr). In all three cases the frequencies of the $\Delta m_J = 0$ transitions are almost insensitive to electric and magnetic fields.

The inter-combination lines have been investigated in effusive beams of magnesium [11.2066], strontium [11.2067, 2068], and barium atoms [11.2069]. Most work for optical frequency standard applications has probably been performed with Ca atomic beams [11.2070, 2071]. This transition was investigated by various groups [11.2064, 2072–2074]. The radiation of a Ca-stabilized laser is also recommended as a reference frequency by the CIPM [11.2009]. As a typical example let us discuss the layout of a transportable Ca-atomic-beam standard (Fig. 11.264) [11.2064].

A few milliwatts of the available output power of an ECDL – pre-stabilized to a reference frequency of an optical cavity – is sent through a polarization-preserving single-mode fiber to a beam splitter/mirror configuration and is split into two beams 1 and 2 of equal power. From each one of the beams (1 or 2) crossing

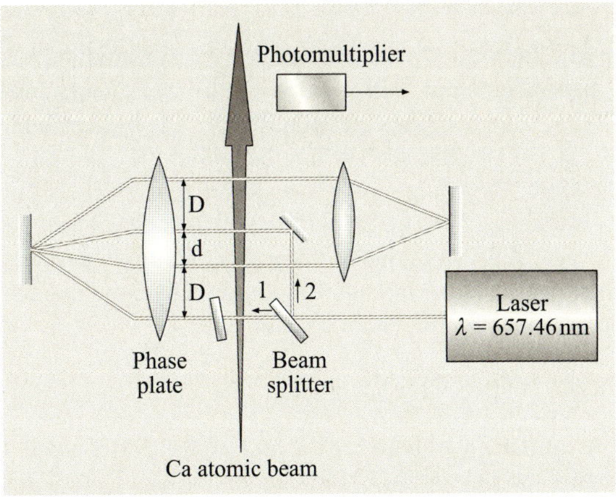

Fig. 11.264 Scheme of a transportable diode-laser frequency stabilized to the $^1S_0$–$^3P_1$ inter-combination transition by means of separated field excitation of a ^{40}Ca beam [11.2064]

the atomic beam perpendicularly an excitation geometry with two pairs of counter-propagating laser beams is obtained by the help of two cat's-eye retro-reflectors. The co-propagating beams are separated by the distance $D = 10$ mm and the distance between the two innermost counter-propagating beams is $d = 13$ mm. The optical layout with the fused silica block is designed in such a way that the part 2 leads to the same excitation geometry as part 1, but its direction with respect to the atomic beam is reversed (Fig. 11.264). During the experiment, the laser beams are propagating either in direction 1 or 2 and the beams propagating in the opposite direction are blocked.

The excited atoms are detected by measuring the fluorescence intensity from the decay of the excited atoms by a photomultiplier. The detected fluorescence intensity of the $\Delta m_j = 0$ transition versus laser frequency detuning $\Delta\omega/2\pi = (\omega - \omega_g)/2\pi$, representing the frequency difference between the laser frequency $\omega/2\pi$ and the frequency $\omega_0/2\pi$ of the Ca inter-combination line, can be described by the expression [11.2060].

$$\begin{aligned}I(\Delta_0) &= \int_0^\infty A(P, v, \Delta_0) f(v) \\ &\times \left\{ \cos\left[2T\left(\Delta_0 + \delta_{rec} + \frac{\omega_0 v^2}{2c^2}\right) + \Phi_L\right]\right. \\ &\left. + \cos\left[2T\left(\Delta_0 - \delta_{rec} + \frac{\omega_0 v^2}{2c^2}\right) + \Phi_L\right]\right\} dv \\ &+ B(P, \Delta_0) . \end{aligned} \quad (11.240)$$

In (11.240), $A(P, v, \Delta_0)$ describes the contribution of a particular atom with velocity v to the signal, and $B(P, v, \Delta_0)$ describes the amplitude of the background of the Doppler-broadened line including the saturation dip, both depending on the laser power P and weakly on the detuning $\delta v = \Delta_0/2\pi$ of the laser frequency. The factor $f(v)$ represents the velocity distribution, where we have neglected the influence of the velocity components perpendicular to the atomic beam.

$$\Phi_L = \Phi_2 - \Phi_1 + \Phi_4 - \Phi_3 \quad (11.241)$$

is a residual phase between the four exciting laser beams having the phases Φ_i in the interaction zone, and $T = D/v$ is the flight time of the atoms between two co-propagating beams. The phase Φ_L – which shifts the frequency of the interference structure – can be detected by reversing the directions of the laser beams (from 1 to 2 or vice versa, Fig. 11.264) and compensated by rotating the phase plate [11.2064].

The phase of each cosine function in (11.240) contains three contributions. Besides Φ_L there are the terms Δ_0 resulting from the detuning, $\delta_{rec} = hk^2/(2m_{Ca}c^2)$ from the photon recoil where $\boldsymbol{k}$ is the wavevector of the laser field and m_{Ca} the mass of a Ca atom, and $\omega_0 v^2/(2c^2)$ from the second-order Doppler effect.

The signal for each velocity group v consists of two cosine functions with a period $1/(2T)$ that is determined by the distance D and the atomic velocity. The two components are separated by the recoil splitting $2\delta_{rec} = 2\pi 23.1$ kHz. For an optimum superposition of the two cosines, the period should be an integer fraction of the recoil splitting. The FWHM line width of the signal is approximately given by $1/(4T)$.

From the measured fluorescence signal versus detuning $\delta v = v - v_o$ (Fig. 11.265) one clearly recognizes in the central part of the saturation dip the two central minima of the two cosine terms separated by the recoil splitting. With increasing detuning the cosine structure is washed out, since all velocity groups v of the atomic beam contribute with a slightly different period. The inset of Fig. 11.265 shows the total detected signal where the Doppler broadening is determined by the degree of collimation and the velocity distribution of the atomic beam, leading to a FWHM of the structure of 7.5 MHz. With an evaluated relative uncertainty of 1.3×10^{-12} combined with a relative stability of $\sigma(\tau) = 9 \times 10^{-13}$ at $\tau = 1$ s, the transportable Ca frequency standard is one order of magnitude superior to

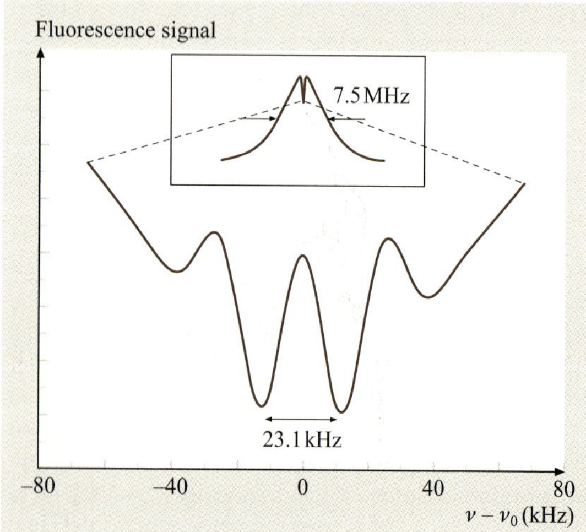

Fig. 11.265 Atom interference structures observed in a thermal beam of Ca atoms [11.2064]

the widely used transportable iodine-stabilized HeNe laser at $\lambda = 633$ nm.

Optical Frequency Standards Based on Laser-Cooled Absorbers

In the case of frequency-stabilized lasers utilizing *thermal* absorbers, the uncertainty in the center of the reference line is ultimately limited by frequency shifts caused by residual phase errors in the optical excitation and by the second-order Doppler shift $\delta\nu/\nu = -v^2/(2c^2)$. Both frequency shifts scale with the velocity of the absorbers and can be reduced by orders of magnitude if the control signal is only generated by slow absorbers. In addition, the increased interaction time of the atoms with the laser beam leads to reduced broadening of the reference line. Hence, the most accurate optical frequency standards are based on cold absorbers. Two typical examples are described in the following subsections. The first describes a standard utilizing a single trapped Yb^+ ion whereas the second part presents a frequency standard based on an ensemble of cold Ca atoms.

Optical frequency standards based on cold, stored ions. The ideal reference for an optical frequency standard would consist of an ensemble of identical but independent atoms at rest in free space. This condition can be approximated in part by a single ion confined in a small volume inside an axially symmetric configuration of electrodes shown schematically in Fig. 11.266 (Paul trap) [11.2075]. Trapping is performed by applying a suitable RF voltage to the electrodes. Such traps allow us to store and cool *single* ions in the field-free region of the trap center.

Ions suitable for precision frequency standards provide a closed transition for cooling and a reference (clock) transition. Cooling of the ion is performed by irradiating it with laser radiation which is tuned slightly below resonance and the cooling occurs by repeated absorption and emission [11.2077, 2078]. When the *clock* transition is probed, the cooling radiation has to be turned off in order to suppress light shifts and strong line broadening. The RF trapping field, however, may stay unchanged, allowing ideally unlimited interrogation times. If several ions are confined in a trap, their repelling forces will lead to an *ion cloud*, extending into the nonvanishing RF field and kinetic energy from the oscillating trapping field will be picked up. Such heating can be avoided if only a single ion is trapped in the center of the trap, where the trapping field vanishes. Since only *one* ion can contribute to the stabilization, it is important to detect the excitation of the reference transition with high efficiency. An efficiency close to 100%

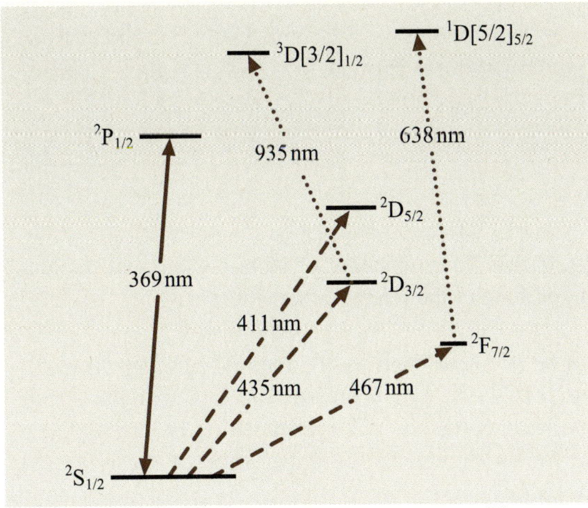

Fig. 11.267 Simplified energy-level diagram of ^{171}Yb

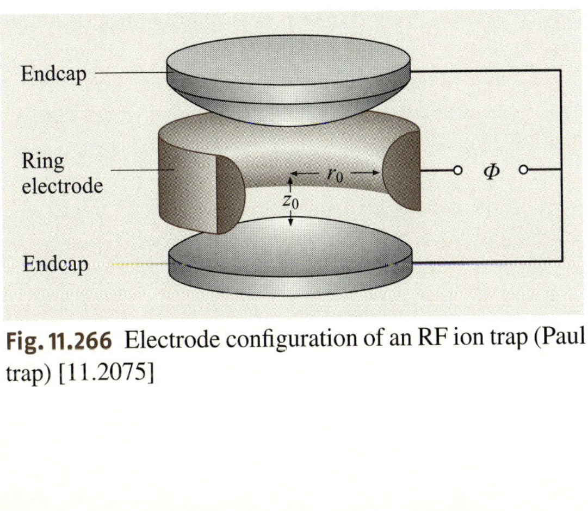

Fig. 11.266 Electrode configuration of an RF ion trap (Paul trap) [11.2075]

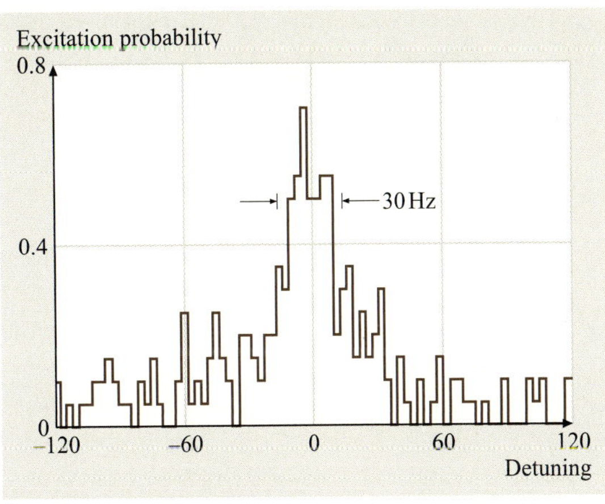

Fig. 11.268 Absorption signal of a single trapped and cooled $^{171}Yb^+$ ion [11.2076]

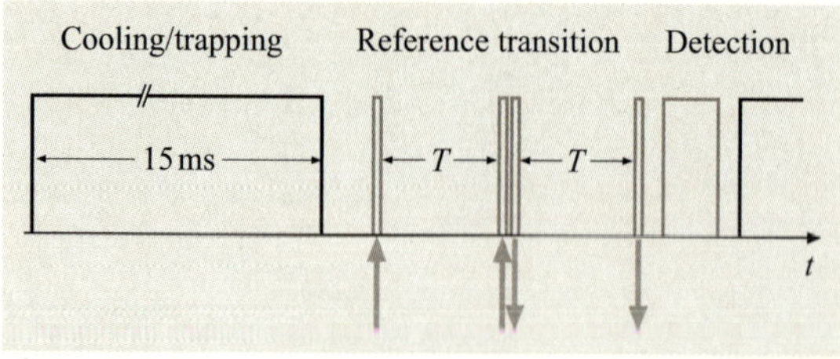

Fig. 11.269 Time sequence of atom cooling and trapping and probing of the reference transition

can be accomplished by the method of *electron shelving* [11.2079], which utilizes the fact that the strong fluorescence on the cooling transition is interrupted if the ion is excited to the upper state of the reference transition.

Several different ions are being investigated for applications in frequency-stabilized lasers. Very promising results have been observed with standards based on the ions of mercury [11.2080], strontium [11.2081, 2082], indium [11.2083], and ytterbium [11.2076, 2084]. As a typical example, we consider an optical standard based on a single ^{171}Yb ion (Fig. 11.267). There are three different clock transitions that can be accessed by the frequency-doubled radiation of diode lasers. The 435 nm clock transition has recently been observed with a line width as narrow as 30 Hz (Fig. 11.268). In the meantime, this linewidth could be further reduced to 10 Hz [11.2085]. Utilizing this reference line a relative uncertainty of 9×10^{-15} has recently been estimated [11.2086], demonstrating the great potential of optical frequency standards based on cold stored ions.

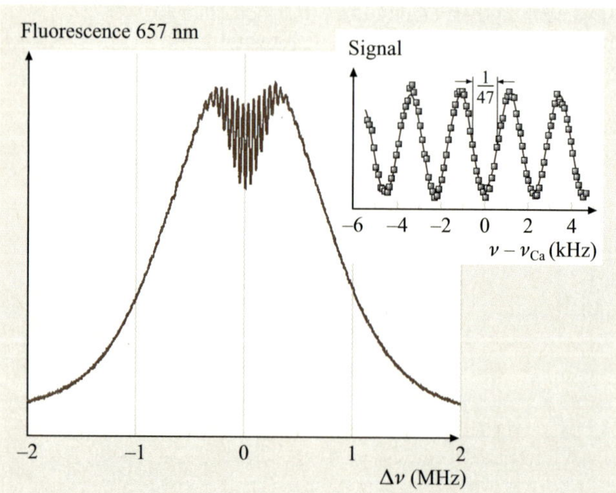

Fig. 11.270 Interference structure of cold, free Ca atoms generated by time-separated excitation

Frequency standards based on cold neutral atoms. Neutral atoms can be cooled and trapped in a magnetooptical trap (MOT) [11.2087] if they have a suitable cooling transition. In contrast to ions, a large number of atoms in the range of 10^7 can be utilized to contribute to the stabilization, leading to an increased signal-to-noise ratio. However, to avoid unwanted frequency shifts, all trapping fields have to switched off when the reference transition is probed. Hence, the interaction time of the free atoms released from the trap is limited by their acceleration in the gravitational field.

Candidates for cold-atom optical frequency standards are silver [11.2088] and the alkali-earth atoms magnesium [11.2089, 2090], strontium [11.2091, 2092], and calcium [11.2065, 2070, 2088]. As a typical example, we describe a frequency-stabilized laser based on cold neutral calcium atoms.

Besides of the narrow reference transition $^1S_0$–$^3P_1$ at $\lambda = 657$ nm (Fig. 11.263), ^{40}Ca has a strong cooling transition $^1S_0$–$^1P_1$ ($\lambda = 423$ nm) that can be used to cool and trap Ca atoms in a MOT [11.2093]. On this transition, the Ca atoms can be cooled to approximately 3 mK. Further cooling to several mikrokelvin can then be achieved in a second step by means of the narrow intercombination transition $^1S_0$–$^3P_1$ utilizing the method of *quench cooling* [11.2094].

In order to achieve high spectral resolution combined with a good signal-to-noise ratio (SNR), the method of separated field excitation in the time domain can be applied, similar to the spatially separated field excitation of an atomic beam. Short pulses of 1 μs duration are used to excite a significant part of the cold ensemble of atoms. The necessary high spectral resolution is then achieved by a sufficiently large time separation T between two consecutive pulses. If the lengths of the pulses are small compared to their separation, the width of the interference fringes $\delta\nu = 1/(4T)$ is inversely proportional to T.

Cooling and trapping of the atoms and probing of the clock transition are performed sequentially (Fig. 11.269). After the atoms have been trapped and cooled for about 15 ms in the first step, the trapping fields (laser beams and magnetic quadrupole field) are switched off, a small homogeneous magnetic quantization field (Helmholtz field) is turned on and the clock transition is probed by two pulsed pairs of counter-propagating laser beams (Fig. 11.269). During this time, the cloud of cold, free Ca atoms expands according to the temperature of about 3 mK at a root-mean-square velocity of $v_{rms} \approx 80$ cm/s. In the third step, the excitation to the $^3P_1$ state is detected by observing the fluorescence of its spontaneous decay to the $^1S_0$ ground state.

If the laser frequency is tuned across the atomic resonance, the fluorescence intensity contains a contribution that oscillates with the cosine of the laser detuning (Fig. 11.270). Similar to spatially separated field excitation, this oscillating behavior can be explained by an atom interference generated by the excitation with time-separated fields [11.2093].

The error signal for the stabilization is generated from the interference signal by modulating the laser frequency and simultaneously measuring the fluorescence intensity. In the most straightforward approach, the frequency is square-wave modulated between two discrete values with the mean frequency tuned close to the center of the central fringe. The difference in the corresponding fluorescence intensities is used as the error signal. This method corresponds to a first-harmonic detection (1f method) of a servo-control system using analog electronics and harmonic modulation. The maximum slope is obtained for a total modulation width of $\delta\nu_{mod} = 1/4T$, i. e., if the frequency alternates between the points of maximum slope of the interference signal. After detection, the error signal is used to step the frequency of the laser spectrometer which corresponds to a digital integrating servo control. The linear drift of the eigenfrequency of the reference resonator can be determined by the servo control and compensated for by adding a corresponding feedforward signal to the signal controlling the laser frequency. Using Ca atoms, cooled on the $^1S_0$–$^1P_1$ transition to approximately 3 mK, a relative uncertainty of $\delta\nu/\nu \cong 2\times10^{-14}$ has been determined [11.2095].

11.14.4 Measurement of Optical Frequencies

In many cases, the use of frequency-stabilized lasers requires precise knowledge of their frequencies. They have to be determined in relation to the primary standard of time and frequency, the Cs atomic clock. Earlier concepts of optical frequency measurements used a harmonic frequency chain of several lasers [11.2096]. For example, measurements of the Ca inter-combination line with such harmonic chain led to a fractional uncertainty of 2.5×10^{-13} [11.2097]. In the meantime alternative powerful methods have been developed to determine optical frequencies. An elegant and successful method is based on a mode-locked femtosecond

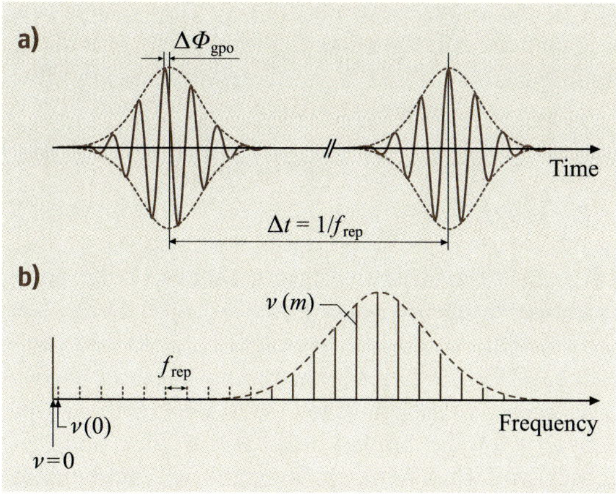

Fig. 11.271a,b Signal of a mode-locked fs laser, shown (a) in the time domain and (b) in the frequency domain

laser [11.2098–2101]. Such lasers emit a continuous train of very short pulses covering a wide range of the optical spectrum of some 100 THz. This range can be further increased to more than one octave utilizing the phase modulation induced by the pulsed laser beam in an optical fiber of low dispersion [11.2102].

In the frequency domain, the continuous train of femtosecond pulses (Fig. 11.271a) corresponds to a comb of frequencies of which the inverse of the pulse separation is equal to the pulse repetition frequency f_{rep} (Fig. 11.271b). The value of an arbitrary comb frequency $\nu(m)$ is determined by

$$\nu(m) = \nu(0) + m f_{rep} , \tag{11.242}$$

where m is an integer representing the respective order number of the comb frequency $\nu(m)$ and f_{rep} is the repetition frequency (Fig. 11.271b). Since the group velocity and the phase velocity of the short light pulses circulating in the laser cavity are slightly different an extrapolation to $m = 0$ leads to the frequency $\nu(0)$ which – in general – does not coincide with the zero frequency (Fig. 11.271b). According to (11.242), the determination of an arbitrary frequency $\nu(m)$ requires the knowledge of f_{rep}, the integer m, and $\nu(0)$. The pulse repetition frequency f_{rep} can be measured phase-coherently against the primary standard of time and frequency. If f_{rep} is larger than approximately 100 MHz it can easily be identified by an interferometric wavelength measurement of moderate relative uncertainty ($\le 10^{-7}$). In order to measure $\nu(0)$, we take advantage of the fact that the comb spans over a frequency range of more than one octave. This allows us to frequency double the low frequency part of the comb spectrum $2\nu(0) + 2m f_{rep}$ and beat it against the corresponding high-frequency part of the comb spectrum $\nu(0) + n f_{rep}$. This *self-referencing* of the frequency comb leads to a beat

$$\begin{aligned} \delta\nu &= 2\nu(0) + 2n f_{rep} - \left[\nu(0) + m\nu_{rep}\right] \\ &= \nu(0) + (2n - m) f_{rep} \end{aligned} \tag{11.243}$$

that contains $\nu(0)$ plus integer multiples of the pulse repetition frequency f_{rep}. Since m, n and f_{rep} are known, $\nu(0)$ can now be measured with very low uncertainty [11.2099, 2100]. Hence, any comb frequency $\nu(m)$ can now be connected phase-coherently to the frequency of the primary standard of time and frequency, the Cs atomic clock. Therefore, we can conclude that the frequency of any other laser emitting within the spectral range of the frequency comb can also be determined phase-coherently by measuring the beat between the laser frequency and a corresponding comb frequency $\nu(m)$. Femtosecond lasers have been applied for the first time to measure the frequency measurements of 1S–2S two-photon transition in hydrogen, leading to the recommended value of the frequency given in Table 11.56 [11.2098].

In the meantime femtosecond lasers are widely used for optical frequency measurements in many laboratories worldwide. The first experiments applied titanium-sapphire femtosecond lasers as frequency comb generators. They are now going to be replaced by femtosecond fibre lasers in view of their high reliability,ease of operation, and accuracy [11.2103]. It has been shown that frequency comb generators based upon fs lasers are well suited for optical frequency measurements of the lowest possible uncertainty. Furthermore, it was demonstrated that frequency ratios could be determined with fs lasers with relative uncertainties as low as 10^{-18} [11.2104]. Basically, in their present state, frequency comb generators based on fs lasers allow the measurement of optical frequencies with uncertainties determined only by the uncertainty of the primary standard of time and frequency and the reproducibility of the stabilized laser whereas the uncertainty of the comb generator itself can still be neglected.

11.14.5 Conclusion and Outlook

The aim of this chapter was to review the foundations and methods of laser frequency stabilization including modern methods of optical frequency measurements. Regarding the wide field of lasers, laser spectroscopy and frequency stabilization and the last development in these fields, this chapter could not be comprehensive within the limited space available. However, we hope that it presents a reasonable overview and gives some introduction to those scientists who are entering the fascinating field of laser frequency stabilization and laser spectroscopy.

The examples of stabilized lasers in Sect. 11.14.3 are discussed to illustrate the variety of methods applied in this field. The development of stabilized lasers is still progressing in various directions. One is the development of efficient, reliable and small lasers, e.g., diode lasers, diode-pumped solid-state lasers, or optical parametric oscillators (OPOs), which may be important to achieve efficient, reliable long-term operation. Another branch is the design of stable optical reference resonators to further decrease the laser line width close to the quantum limit. An important task is to isolate them from seismic influences and environmental disturbances. Here, alternative novel methods are being developed to compensate and reduce the seismic influence. A further reduction of the frequency uncertainty can be envisaged from the refinement of laser cooling. The introduction of optical frequency combs created by mode-locked fs lasers has provided us with a universal tool to determine optical frequencies and optical frequency ratios. This field will certainly be further explored by the development of frequency combs operating continuously over long periods and by increasing the total width of their emission ranges. It is expected that the combination of all these efforts will lead to optical frequency standards and clocks of unprecedented low uncertainty.

References

11.1 T. H. Maiman: Stimulated optical radiation in ruby, Nature **187**, 493 (1960)

11.2 A. L. Schawlow, C. H. Towens: Infrared and optical masers, Phys. Rev. **112**, 1940 (1958)

11.3 O. Svelto: *Principles of Lasers*, 4 edn. (Springer, Berlin, New York 1998)

11.4 W. Koechner: *Solid-State Laser Engineering*, 4 edn. (Springer, Berlin, New York 1996)

11.5 A. E. Siegman: *Lasers* (Univ. Sci. Books, Mill Valley 1986)

11.6 R. Pantell, H. Puthoff: *Fundamentals of Quantum Electronics* (Wiley, New York 1964)

11.7 W. Demtröder: *Laser Spectroscopy*, 2 edn. (Springer, Berlin, Heidelberg 1996)

11.8 M. Sargent, M. O. Scully, W. E. Lamb: *Laser Physics* (Addison-Wesley, London 1974)

11.9 A. Yariv: *Quantum Electronics*, 3 edn. (Wiley, New York 1989)

11.10 V. Weisskopf, E. Wigner: Berechnung der natürlichen Linienbreite auf Grund der Diracschen Lichttheorie, Z. Phys. **63**, 54 (1930)(in German)

11.11 A. Einstein: On the quantum theory of radiation, Z. Phys. **18**, 121 (1917)

11.12 H. Kogelnick, T. Li: Laser beams and resonators, Appl. Opt. **5**, 1550 (1966)

11.13 A. G. Fox, T. Li: Resonant modes in a maser interferometer, Bell Syst. Tech. J. **40**, 453–458 (1961)

11.14 D. J. Kuizenga, A. E. Siegman: FM and AM mode locking of the homogeneous laser Part I: Theory, IEEE J. Quantum Electron. **6**, 694 (1970)

11.15 A. H. Haus: Theory of mode locking with a fast saturable absorber, J. Appl. Phys. **46**, 3049 (1975)

11.16 T. H. Maiman: Stimulated optical radiation in ruby, Nature **187**, 493 (1960)

11.17 J. E. Geusic, H. M. Marcos, L. G. Van Uitert: Laser oscillations in Nd-doped yttrium aluminum, yttrium gallium and gadolinium garnets, Appl. Phys. Lett. **4**, 182 (1964)

11.18 E. Snitzer: Optical maser action of Nd^{+3} in a barium crown glass, Phys. Rev. Lett. **7**, 444 (1961)

11.19 T. Y. Fan, R. L. Byer: Diode-laser pumped solid-state lasers, IEEE J. Quantum Electron. **24**, 895 (1988)

11.20 T. Y. Fan, G. Huber, R. L. Byer, P. Mitzscherlich: Spectroscopy and diode laser-pumped operation of Tm;Ho:YAG, IEEE J. Quantum Electron. **24**, 924 (1988)

11.21 L. Esterowitz: Diode-pumped holmium, thulium, and erbium lasers between 2 and 3 μm operating CW at room temperature, Opt. Eng. **29**, 676 (1990)

11.22 P. Lacovara, H. K. Choi, C. A. Wang, R. L. Aggarwal, T. Y. Fan: Room-temperature diode-pumped Yb:YAG laser, Opt. Commun. **105**, 1089 (1991)

11.23 S. A. Payne, W. F. Krupke, L. K. Smith, L. D. DeLoach, W. L. Kway: Laser properties of Yb in fluoro-apatite and comparison with other Yb-doped gain media, Conf. Lasers Electro-Optics, OSA Tech. Dig. Ser.) **12**, 540 (1992)

11.24 J. C. Walling, O. G. Peterson, H. P. Jenssen, R. C. Morris, E. W. O'Dell: Tunable alexandrite lasers, IEEE J. Quantum Electron. **16**, 1302 (1980)

11.25 B. Struve, G. Huber, V. V. Laptev, I. A. Shcherbakov, E. V. Zharikov: Tunable room-temperature CW-laser action in Cr^{3+}:GdScGa-garnet,, Appl. Phys. B **30**, 117 (1983)

11.26 S. A. Payne, L. L. Chase, L. K. Smith, W. L. Kway, H. W. Newkirk: Laser performance of $LiSrAlF_6 : Cr^{3+}$, J. Appl. Phys. **66**, 1051 (1989)

11.27 R. Scheps: Cr-doped solid-state lasers pumped by visible laser diodes, Opt. Mater. **1**, 1 (1992)

11.28 P. Moulton: Ti-doped sapphire: A tunable solid-state laser, Opt. News **8**, 9 (1982)

11.29 P. Albers, E. Stark, G. Huber: Continuous-wave laser operation and quantum efficiency of titanium-doped sapphire, J. Opt. Soc. Am. B **3**, 134 (1986)

11.30 V. Petričević, S. K. Gayen, R. R. Alfano: Laser action in chromium-activated forsterite for near-infrared excitation: is Cr^{4+} the lasing ion?, Appl. Phys. Lett. **53**, 2590 (1988)

11.31 G. M. Zverev, A. V. Shestakov: Tunable near-infrared oxide crystal lasers, OSA Proc. **5**, 66 (1989)

11.32 W. Jia, B. M. Tissue, K. R. Hoffmann. Lu, W. M. Yen: Near-infrared luminescence in Cr,Ca-doped yttrium aluminium garnet, OSA Proc. Adv. Solid-State Lasers **10**, 87 (1991)

11.33 S. Kück, K. Petermann, G. Huber: Spectroscopic investigation of the Cr^{4+}-center in YAG, OSA Proc. Adv. Solid-State Lasers **10**, 92 (1991)

11.34 R. H. Page, L. D. DeLoach, G. D. Wilke, S. A. Payne, W. F. Krupke: A new class of tunable mid-IR lasers based on Cr^{2+}-doped II-VI compounds, CLEO'95, Baltimore 1995 (Opt. Soc. Am., Washington 1995) CWH5

11.35 A. Richter, E. Heumann, E. Osiac, G. Huber, W. Seelert, A. Diening: Diode pumping of a continuous-wave Pr^{3+}-doped $LiYF_4$ laser, Opt. Lett. **29**, 2638–2640 (2004)

11.36 B. Henderson, G. F. Imbusch: *Optical Spectroscopy of Inorganic Solids* (Clarendon, Oxford 1989)

11.37 G. H. Dieke: *Spectra and Energy Levels of Rare Earth Ions in Crystals*, 1 edn. (Wiley, New York 1968)

11.38 Y. Tanabe, S. Sugano: On the absorption spectra of complex ions, J. Phys. Soc. Jpn. **9**, 766 (1954)

11.39 H. C. Schläfer, G. Gliemann: *Einführung in die Ligandenfeldtheorie* (Akademische Verlagsges., Wiesbaden 1980)

11.40 P. Schuster: *Ligandenfeldtheorie* (Verl. Chemie, Weinheim 1973)

11.41 C. J. Ballhausen: *Introduction to Ligand Field Theory* (McGraw-Hill, New York 1962)

11.42 J. S. Griffith: *Theory of Transition Metal Ions* (Cambridge Univ. Press, Cambridge 1961)

11.43 A. B. P. Lever: *Inorganic electronic spectroscopy* (Elsevier, Amsterdam 1984)

11.44 S. A. Payne, L. L. Chase, G. F. Wilke: Excited-state absorption spectra of V^{2+} in $KMgF_3$ and MgF_2, Phys. Rev. B **37**, 998 (1988)

11.45 D. E. McCumber: Theory of phonon-terminated optical masers, Phys. Rev. **134**, 299 (1964)

11.46 D. E. McCumber: Einstein relations connecting broadband emission and absorption spectra, Phys. Rev. **136**, 954 (1964)

11.47 N. F. Mott: On the absorption of light by crystals, Proc. Soc. A **167**, 384 (1938)

11.48 C. W. Struck, W. H. Fonger: Unified model of the temperature quenching of narrow-line and broad-band emissions, J. Lumin. **10**, 1 (1975)

11.49 O. Svelto: *Principles of Lasers* (Plenum, New York 1989)

11.50 W. Koechner: *Solid-State Laser Engineering* (Springer, Berlin, Heidelberg 1996)

11.51 A. Yariv: *Quantum Electronics* (Wiley, New York 1967)

11.52 A. E. Siegman: *Lasers* (Univ. Sci. Books, Mill Valley 1986)

11.53 D. Findlay, R. A. Clay: The measurement of internal losses in 4-level lasers, Phys. Lett. **20**, 277 (1966)

11.54 J. A. Caird, S. A. Payne, P. R. Staver, A. J. Ramponi, L. L. Chase, W. F. Krupke: Quantum electronic properties of the $Na_3Ga_2Li_3F_{12}:Cr^{3+}$ laser, J. Quant. Electron. **24**, 1077 (1988)

11.55 C. D. Marshall, J. A. Speth, S. A. Payne, W. F. Krupke, G. J. Quarles, V. Castillo, B. H. T. Chai: Ultraviolet laser emission properties of Ce^{3+}-doped $LiSrAlF_6$ and $LiCaAlF_6$, J. Opt. Soc. Am. B **11**, 2054 (1994)

11.56 D. J. Ehrlich, P. F. Moulton, R. M. Osgood: Ultraviolet solid-state Ce:YLF laser at 325 nm, Opt. Lett. **4**, 184 (1979)

11.57 D. W. Coutts, A. J. S. McGonigle: Cerium-doped fluoride lasers, IEEE J. Quantum Elect. **40**, 1430 (2004)

11.58 P. Dorenbos: The 5d level positions of the trivalent lanthanides in inorganic compounds, J. Luminesc. **91**, 155 (2000)

11.59 M. A. Dubinskii, V. V. Semashko, A. K. Naumov, R. Y. Abdulsabirov, S. L. Korableva: A new active medium for a tunable solid-state UV laser with an excimer pump, Laser Phys. **4**, 480 (1994)

11.60 M. A. Dubinskii, V. V. Semashko, A. K. Naumov, R. Y. Abdulsabirov, S. L. Korableva: Ce^{3+}-doped colquiriite. A new concept for a all-solid-state tunable ultraviolet laser, J. Modern Opt. **40**, 1 (1993)

11.61 J. F. Pinto, G. H. Rosenblatt, L. Esterowitz, V. Castillo, G. J. Quarles: Tunable solid-state laser action in $Ce^{3+}:LiSrAlF_6$, Electron. Lett. **30**, 240 (1994)

11.62 D. J. Ehrlich, P. F. Moulton, R. M. Osgood: Optically pumped Ce:LaF3 laser at 286 nm, Opt. Lett. **5**, 339 (1980)

11.63 A. A. Kaminskii, S. A. Kochubei, K. N. Naumochkin, E. V. Pestryakov, V. I. Trunov, T. V. Uvarova: Amplification of the ultraviolet radiation due to the 5d-4f configurational transition of the Ce^{3+} ion in BaY_2F_8, Sov. J. Quantum Electron. **19**, 340 (1989)

11.64 J. F. Owen, P. B. Dorain, T. Kobayasi: Excited-state absorption in $Eu^{+2}:CaF_2$ and Ce^{+3} : YAG single crystals at 298 and 77 K, J. Appl. Phys. **52**, 1216 (1981)

11.65 D. S. Hamilton, S. K. Gayen, G. J. Pogatshnik, R. D. Ghen, W. J. Miniscalco: Optical-absorption and photoionization measurements from the excited states of $Ce^{3+}:Y_3Al_5O_{12}$, Phys. Rev. B **39**, 8807 (1989)

11.66 A. J. Bayramian, C. D. Marshall, J. H. Wu, J. A. Speth, S. A. Payne, G. J. Quarles, V. K. Castillo: Ce : $LiSrAlF_6$ laser performance with antisolarant pump beam, J. Luminesc. **69**, 85 (1996)

11.67 A. J. Bayramian, C. D. Marshall, J. H. Wu, J. A. Speth, S. A. Payne, G. J. Quarles, V. K. Castillo: Ce:LiSrAlF6 laser performance with antisolarant pump beam. In: *OSA Trends Opt. Photon. Adv. Solid-State Lasers*, Vol. 1, ed. by S. A. Payne, C. R. Pollock (Opt. Soc. Am., Washington 1996) pp. 60–65

11.68 E. G. Gumanskaya, M. V. Korzhik, S. A. Smirnova, V. B. Pawlenko, A. A. Fedorov: Spectroscopic characteristics and scintillation efficiency of $YAlO_3$ single crystals activated by cerium, Opt. Spectrosc. **72**, 86 (1992)

11.69 J. Ganem, W. M. Dennis, W. M. Yen: One-color sequential pumping of the 4f5d bands in Pr-doped yttrium aluminum garnet, J. Luminesc. **54**, 79 (1992)

11.70 S. Nicolas, M. Laroche, S. Girard, R. Moncorgé, Y. Guyot, M. F. Joubert, E. Descroix, A. G. Petrosyan: $4f^2$ to 4f5d excited state absorption in $Pr^{3+}:YAlO_3$, J. Phys. Cond. Matt. **11**, 7937 (1999)

11.71 I. Sokólska, S. Kück: Investigation of high-energetic transitions in some Pr^{3+}-doped fluoride and oxide crystals, Proc. SPIE **4412**, 236–241 (2000)

11.72 M. Laroche, A. Braud, S. Girard, J. L. Doualan, R. Moncorgé, M. Thuau, L. D. Merkle: Spectroscopic investigations of the 4f5d energy levels of Pr^{3+} in fluoride crystals by excited-state absorption and two-step excitation measurements, J. Opt. Soc. Am. B **16**, 2269 (1999)

11.73 J. K. Lawson, S. A. Payne: Excited-state absorption of Pr^{3+}-doped fluoride crystals, Opt. Mater. **2**, 225 (1993)

11.74 Y. M. Cheung, S. K. Gayen: Excited-state absorption in $Pr^{3+}:Y_3Al_5O_{12}$, Phys. Rev. B **49**, 14827 (1994)

11.75 T. Kozeki, H. Ohtake, N. Sarukura, Z. Liu, K. Shimamura, K. Nakano, T. Fukuda: Novel design of high-pulse-energy ultraviolet Ce:LiCAF laser oscillator. In: *OSA Trends Opt. Photon.*, Vol. 34, ed. by H. Injeyan, U. Keller, C. Marshall (Opt. Soc. Am., Washington 2000) pp. 400–403

11.76 P. Rambaldi, R. Moncorgé, J. P. Wolf, C. Pédrini, J. Y. Gesland: Efficient and stable pulsed laser operation of Ce : $LiLuF_4$ around 308 nm, Opt. Commun. **146**, 163 (1998)

11.77 P. Rambaldi, R. Moncorgé, S. Girard, J. P. Wolf, C. Pédrini, J. Y. Gesland: Efficient UV laser operation of Ce : $LiLuF_4$ single crystal. In: *OSA Trends Opt. Photon.*, Vol. 19, ed. by W. R. Bosenberg, M. M. Fejer (Opt. Soc. Am., Washington 1998) pp. 10–12

11.78 Z. Liu, H. Ohtake, N. Sarukura, M. A. Dubinskii, R. Y. Abdulsabirov, S. L. Korableva: All-solid-state tunable ultraviolet picosecond $Ce^{3+}:LiLuF_4$ laser with direct pumping by the fifth harmonic of a Nd:YAG laser. In: *OSA Trends Opt. Photon.*, Vol. 19, ed. by W. R. Bosenberg, M. M. Fejer (Opt. Soc. Am., Washington 1998) pp. 13–15

11.79 A. J. S. McGonigle, D. W. Coutts, C. E. Webb: Multi kHz PRF cerium lasers pumped by frequency doubled copper vapour lasers. In: *OSA Trends Opt. Photon.*, Vol. 26, ed. by M. J. Fejer, H. Injeyan, U. Keller (Opt. Soc. Am., Washington 1999) pp. 123–129

11.80 K. S. Johnson, H. M. Pask, M. J. Withford, D. W. Coutts: Efficient all-solid-state Ce:LiLuF laser source at 309 nm, Opt. Commun. **252**, 132–137 (2005)

11.81 Z. Liu, K. Shimamura, K. Nakano, N. Mujilatu, T. Fukuda, T. Kozeki, H. Ohtake, N. Sarukura: Direct generation of 27 mJ, 309 nm pulses from a Ce:LLF oscillator using a large-size Ce:LLF crystal. In: *OSA Trends Opt. Photon.*, Vol. 34, ed. by H. Injeyan, U. Keller, C. Marshall (Opt. Soc. Am., Washington 2000) pp. 396–399

11.82 Z. Liu, S. Izumida, S. Ono, H. Ohtake, N. Sarukura, K. Shimamura, N. Mujilato, S. L. Baldochi, T. Fukuda: Direct generation of 30 mJ, 289 nm pulses from a Ce:LiCAF oscillator using Czochralski-grown large crystal. In: *OSA Trends Opt. Photon.*, Vol. 26, ed. by M. J. Fejer, H. Injeyan, U. Keller (Opt. Soc. Am., Washington 1999) pp. 115–117

11.83 S. V. Govorkov, A. O. Wiessner, T. Schröder, U. Stamm, W. Zschocke, D. Basting: Efficient high average power and narrow spectral linewidth operation of Ce:LICAF laser at 1 kHz repetition rate. In: *OSA Trends Opt. Photon.*, Vol. 19, ed. by W. R. Bosenberg, M. M. Fejer (Opt. Soc. Am., Washington 1998) pp. 2–5

11.84 D. J. Spence, H. Liu, D. W. Coutts: Low-threshold miniature Ce:LiCAF lasers, Opt. Commun. **262**, 238–240 (2006)

11.85 D. Alderighi, G. Toci, M. Vannini, D. Parisi, S. Bigotta, M. Tonelli: High efficiency UV solid state lasers based on $Ce : LiCaAlF_6$ crystals, Appl. Phys. B **83**, 51–54 (2006)

11.86 J. F. Pinto, L. Esterowitz, G. J. Quarles: High performance $Ce^{3+} : LiSrAlF_6/LiCaAlF_6$ UV lasers with extended tunability, Electr. Lett. **31**, 2009 (1995)

11.87 R. W. Waynant, P. H. Klein: Vacuum ultraviolet laser emission from $Nd^{+3} : LaF_3$, Appl. Phys. Lett. **46**, 14 (1985)

11.88 R. W. Waynant: Vacuum ultraviolet laser emission from $Nd^{+3} : LaF_3$, Appl. Phys. B **28**, 205 (1982)

11.89 M. A. Dubinskii, A. C. Cefalas, C. A. Nicolaides: Solid state $LaF_3 : Nd^{3+}$ vuv laser pumped by a pulsed discharge F_2-molecular laser at 157 nm, Opt. Commun. **88**, 122 (1992)

11.90 M. A. Dubinskii, A. C. Cefalas, E. Sarantopoulou, S. M. Spyrou, C. A. Nicolaides, R. Y. Abdulsabirov, S. L. Korableva, V. V. Semashko: Efficient LaF_3:Nd^{3+}-based vacuum-ultraviolet laser at 172 nm, J. Opt. Soc. Am. B **9**, 1148 (1992)

11.91 P. P. Sorokin, M. J. Stevenson: Solid-state optical maser using divalent samarium in calcium fluorid, IBM J. Res. Develop. **5**, 56 (1961)

11.92 Y. S. Vagin, V. M. Marchenko, A. M. Prokhorov: Spectrum of a laser based on electron-vibrational transitions in a $CaF_2 : Sm^{2+}$ crystal, Sov. Phys. JETP **28**, 904 (1969)

11.93 R. T. Wegh, H. Donker, A. Meijerink, R. J. Lamminmäki, J. Hölsä: Vacuum-ultraviolet spectroscopy and quantum cutting for Gd^{3+} in $LiYF_4$, Phys. Rev. B **56**, 13841 (1997)

11.94 I. Sokólska, S. Kück: Observation of photon cascade emission in Pr^{3+}-doped $KMgF_3$, Chem. Phys. **270**, 355 (2001)

11.95 J. K. Lawson, S. A. Payne: Excited-state absorption spectra and gain measurements of $CaF_2 : Sm^{2+}$, J. Opt. Soc. Amer. B **8**, 1404 (1991)

11.96 S. A. Payne, C. D. Marshall, A. J. Bayramian, J. K. Lawson: Conduction band states and the 5d-4f laser transition of rare earth ion dopants, Proc. SPIE **3176**, 68 (1997)

11.97 Z. J. Kiss, R. C. Duncan Jr.: Optical maser action in CaF_2, Proc. IRE **50**, 1532 (1962)

11.98 R. C. Duncan Jr., Z. J. Kiss: Continously operating $CaF_2 : Tm^{2+}$ optical maser, Appl. Phys. Lett. **3**, 23 (1963)

11.99 S. Lizzo: Luminescence of Yb^{2+}, Eu^{2+} and Cu^{+} in solids Ph.D. Thesis (Universiteit Utrecht, Utrecht 1995)

11.100 S. Lizzo, A. Meijerink, G. J. Dirksen, G. Blasse: Luminescence of divalent ytterbium in magnesium fluoride crystals, J. Luminesc. **63**, 223 (1995)

11.101 S. Lizzo, A. Meijerink, G. Blasse: Luminescence of divalent ytterbium in alkaline earth sulphates, J. Luminesc. **59**, 185 (1995)

11.102 S. Kück, M. Henke, K. Rademaker: Crystal growth and spectroscopic investigation of Yb^{2+} doped fluorides, Laser Phys. **11**, 116 (2001)

11.103 S. Kück: Laser-related spectroscopy of ion-doped crystals for tunable solid state-lasers, Appl. Phys. B **72**, 515 (2001)

11.104 A. Yariv, S. P. S. Porto, K. Nassau: Optical maser emission from trivalent praseodymium in calcium tungstate, J. Appl.Phys. **33**, 2519 (1962)

11.105 A. A. Kaminskii: Achievements of modern crystal laser physics, Ann. Phys. (France) **16**, 639 (1991)

11.106 T. Sandrock, T. Danger, E. Heumann, G. Huber, B. H. T. Chai: Efficient continuous wave laser emission of Pr^{3+}-doped fluorides at room temperature, Appl. Phys. B **58**, 149 (1994)

11.107 E. Heumann, C. Czeranowski, T. Kellner, G. Huber: An efficient all-solid-state $Pr^{3+} : LiYF_4$ laser in the visible spectral range. In: *Conf. Lasers Electro-Optics*, OSA Tech. Dig. Ser. (Opt. Soc. Am., Washington 1999) p. 86

11.108 Coherent Inc.: http://www.coherent.com/Lasers/index.cfm?fuseaction=show.page&id=1009&loc=834 (Coherent Inc., Santa Clara 2007)

11.109 E. Osiac, E. Heumann, A. Richter, G. Huber, A. Diening, W. Seelert: Red $Pr^{3+} : YLiF_4$ laser excited by

480 nm optically pumped semiconductor laser. In: *Conf. Lasers Electro-Optics*, OSA Tech. Dig. Ser. (Opt. Soc. Am., Washington 2004)

11.110 A. Richter, N. Pavel, E. Heumann, G. Huber, D. Parisi, A. Toncelli, M. Tonelli, A. Diening, W. Seelert: Continuous-wave ultraviolet generation at 320 nm by intracavity frequency doubling of red-emitting praseodymium lasers, Opt. Express **14**, 3202 (2006)

11.111 A. A. Kaminskii, H. J. Eichler, B. Liu, P. Meindl: $LiYF_4:Pr^{3+}$ laser at 639.5 nm with 30 J flashlamp pumping and 87 mJ output energy, phys. stat. sol. a **138**, K45 (1993)

11.112 T. Danger, T. Sandrock, E. Heumann, G. Huber, B. H. T. Chai: Pulsed laser action of $Pr:GdLiF_4$ at room temperature, Appl. Phys. B **57**, 239 (1993)

11.113 R. Scheps: Upconversion laser processes, Prog. Quantum Electron. **20**, 271 (1996)

11.114 M.-F. Joubert: Photon avalanche upconversion in rare earth laser materials, Opt. Mater. **11**, 181 (1999)

11.115 S. Guy, M.-F. Joubert, B. Jacquier: Photon avalanche and the mean-field approximation, Phys. Rev. B **55**, 8240 (1997)

11.116 M.-F. Joubert, S. Guy, B. Jacquier: Model of the photon-avalanche effect, Phys. Rev. B **48**, 10031 (1993)

11.117 A. Brenier, L. C. Courrol, C. Pedrini, C. Madej, G. Boulon: Excited state absorption and looping mechanism in $Yb^{3+}-Tm^{3+}-Ho^{3+}$-doped $Gd_3Ga_5O_{12}$ garnet, Opt. Mater. **3**, 25 (1994)

11.118 E. Osiac, I. Sokólska, S. Kück: Evaluation of the upconversion mechanisms in Ho^{3+} doped crystals: Experiment and theoretical modelling, Phys. Rev. B **65**, 235119 (2002)

11.119 S. Kück, A. Diening, E. Heumann, E. Mix, T. Sandrock, K. Sebald, G. Huber: Avalanche upconversion processes in Pr, Yb-doped materials, J. Alloy. Comp. **300–301**, 65 (2000)

11.120 E. Osiac, S. Kück, E. Heumann, G. Huber, E. Sani, A. Toncelli, M. Tonelli: Spectroscopic characterisation of the upconversion avalanche mechanism in Pr^{3+}, $Yb^{3+}:BaY_2F_8$, Opt. Mater. **24**, 537 (2003)

11.121 E. Osiac, E. Heumann, S. Kück, G. Huber, E. Sani, A. Toncelli, M. Tonelli: Orange and red upconversion laser pumped by an avalanche mechanism in Pr^{3+}, $Yb^{3+}:BaY_2F_8$, Appl. Phys. Lett. **82**, 3832 (2003)

11.122 L. Esterowitz, R. Allen, M. Kruer, M. Bartoli, L. S. Goldberg, H. P. Jenssen, A. Linz, V. O. Nicolai: Blue light emission by a $Pr:LiYF_4$ – laser operated at room temperature, J. Appl. Phys. **48**, 650 (1977)

11.123 A. A. Kaminskii: Visible lasing of five intermultiplet transitions of the ion Pr^{3+} in $LiYF_4$, Sov. Phys. Dokl. **28**, 668 (1983)

11.124 J. M. Sutherland, P. M. W. French, J. R. Taylor, B. H. T. Chai: Visible continuous-wave laser transitions in Pr^{3+} : YLF and femtosecond pulse generation, Opt. Lett. **21**, 797 (1996)

11.125 D. S. Knowles, Z. Zhang, D. Gabbe, H. P. Jenssen: Laser action of Pr^{3+} in $LiYF_4$ and spectroscopy of Eu^{2+}-sensitized Pr in BaY_2F_8, IEEE J. Quantum Electron. **24**, 1118 (1988)

11.126 A. A. Kaminskii, A. V. Pelevin: Low-threshold lasing of $LiYF_4:Pr^{3+}$ crystals in the 0.72 μm range as a result of flashlamp pumping at 300 K, Sov. J. Quantum Electron. **21**, 819 (1991)

11.127 A. A. Kaminskii: Stimulated emission spectroscopy of Ln^{3+} ions in tetragonal $LiLuF_4$ fluoride, phys. stat. sol. a **97**, K53 (1986)

11.128 A. A. Kaminskii, A. A. Markosyan, A. V. Pelevin, Y. A. Polyakova, S. E. Sarkisov, T. V. Uvarova: Luminescence properties and stimulated emission from Pr^{3+}, Nd^{3+} and Er^{3+} ions in tetragonal lithium-lutecium fluoride, Inorg. Mater. (USSR) **22**, 773 (1986)

11.129 A. A. Kaminskii, B. P. Sobolev, T. V. Uvarova, M. I. Chertanov: Visible stmulated emission of Pr^{3+} ions in BaY_2F_8, Inorg. Mater. (USSR) **20**, 622 (1984)

11.130 A. A. Kaminskii, S. E. Sarkisov: Stimulated-emission spectroscopy of Pr^{3+} ions in monoclinic BaY_2F_8 fluoride, phys. stat. sol. a **97**, K163 (1986)

11.131 A. A. Kaminskii: New room-temperature stimulated-emission channels of Pr^{3+} ions in anisotropic laser crystals, phys. stat. sol. a **125**, K109 (1991)

11.132 A. A. Kaminskii: Stimulated radiation at the transition $^3P_0 \rightarrow {}^3F_4$ and $^3P_0 \rightarrow {}^3H_6$ of Pr^{3+} ions in LaF_3 crystals, Izv. Akad. Nauk. SSSR **17**, 185 (1981)

11.133 A. A. Kaminskii: Some current trends in physics and spectroscopy of laser crystals, Proc. Int. Conf. Lasers, New Orleans 1980, ed. by C. B. Collins (STS, McLean 1981) 328

11.134 R. Solomon, L. Mueller: Stimulated emission at 5985 from Pr^{3+} in LaF_3, Appl. Phys. Lett. **3**, 135 (1963)

11.135 A. A. Kaminskii: Achievements in the fields of physics and spectroscopy of insulating laser crystals. In: *Lasers and Applications, Part I, Proc.*, ed. by I. Ursu, A. M. Prokhorov (CIP, Bucharest 1983) p. 97

11.136 J. Hegarty, W. M. Yen: Laser action in PrF_3, J. Appl. Phys. **51**, 3545 (1980)

11.137 F. Varsanyi: Surface lasers, Appl. Phys. Lett. **19**, 169 (1971)

11.138 K. R. German, A. Kiel, H. J. Guggenheim: Stimulated emission from $PrCl_3$, Appl. Phys. Lett. **22**, 87 (1973)

11.139 Z. Luo, A. Jiang, Y. Huang: Xenon flash lamp pumped self-frequency doubling NYAB pulsed laser, Chin. Phys. Lett. **6**, 440 (1989)

11.140 K. R. German, A. Kiel, H. J. Guggenheim: Radiative and nonradiative transitions of Pr^{3+} in trichloride and tribromide hosts, Phys. Rev. B **11**, 2436 (1975)

11.141 M. Malinowski, M. F. Joubert, B. Jacquier: Simultaneous laser action at blue and orange wavelengths in $YAG:Pr^{3+}$, phys. stat. sol. a **140**, K49 (1993)

11.142 W. Wolinski, R. Wolski, M. Malinowski, Z. Mierczyk: Spectroscopic and laser properties of YAG : Pr^{3+} crystals. In: *Proc. 10th Int. Congr. Laser*, ed. by W. Waidelich (Springer, Berlin, Heidelberg 1992) p. 611

11.143 .A. Kaminskii, A. G. Petrosyan, K. L. Ovanesyan, M. I. Chertanov: Stimulated emission of Pr^{3+} ions in $YAlO_3$ crystals, phys. stat. sol. a **77**, K173 (1983)

11.144 A. A. Kaminskii, A. G. Petrosyan, K. L. Ovanesyan: Stimulated emission spectroscopy of Pr^{3+} ions in $YAlO_3$ and $LuAlO_3$, Sov. Phys. Dokl. **32**, 591 (1987)

11.145 A. A. Kaminskii, K. Kurbanov, K. L. Ovanesyan, A. G. Petrosyan: Stimulated emission spectroscopy of Pr^{3+} ions in orthorhombic $YAlO_3$ single crystals, phys. stat. sol. a **105**, K155 (1988)

11.146 A. Bleckmann, F. Heine, J. P. Meyn, T. Danger, E. Heumann, G. Huber: CW-lasing of Pr:$YAlO_3$ at room temperature. In: *Adv. Solid-State Lasers*, Proc., Vol. 15, ed. by A. A. Pinto, T. Y. Fan (Opt. Soc. Am., Washington 1993) p. 199

11.147 A. A. Kaminskii, A. G. Petrosyan: New laser crystal for the excitation of stimulated radiation in the dark-red part of the spectrum at 300 K, Sov. J. Quantum Electron. **21**, 486 (1991)

11.148 M. Malinowski, I. Pracka, B. Surma, T. Lukasiewicz, W. Wolinski, R. Wolski: Spectroscopic and laser properties of $SrLaGa_3O_7$:Pr^{3+} crystals, Opt. Mater. **6**, 305 (1996)

11.149 A. A. Kaminskii, A. G. Petrosyan, K. L. Ovanesyan: Stimulated emission of Pr^{3+}, Nd^{3+} and Er^{3+} ions in crystals with complex anions, phys. stat. sol. a **83**, K159 (1984)

11.150 C. Szafranski, W. Strek, B. Jezowska-Trzebiatowska: Laser oscillation of a $LiPrP_4O_{12}$ single crystal, Opt. Commun. **47**, 268 (1983)

11.151 M. Szymanski: Simultaneous operation at two different wavelengths of an $PrLaP_5O_{14}$ laser, Appl. Phys. **24**, 13 (1981)

11.152 B. Borkowski, E. Crzesiak, F. Kaczmarek, Z. Kaluski, J. Karolczak, M. Szymanski: Chemical synthesis and crystal growth of laser quality praseodymium pentaphosphate, J. Crystal Growth **44**, 320 (1978)

11.153 M. Szymanski, J. Karolczak, F. Kaczmarek: Laser properties of praseodymium pentaphosphate single crystals, Appl. Phys. **19**, 345 (1979)

11.154 H. Dornauf, J. Heber: Fluorescence of Pr^{3+}-ions in $La_{1-x}Pr_xP_5O_{14}$, J. Luminesc. **20**, 271 (1979)

11.155 T. Danger, A. Bleckmann, G. Huber: Stimulated emission and laser action of Pr_{3+}:doped YAlO, Appl. Phys. B **58**(5), 413 (1994)

11.156 T. Sandrock, E. Heumann, G. Huber, B. H. T. Chai: Continuous-wave Pr,Yb : $LiYF_4$ upconversion laser in the red spectral range at room temperature. In: *OSA Proc. Adv. Solid-State Lasers*, Vol. 1, ed. by S. A. Payne, C. Pollack (Opt. Soc. Am., Washington 1996) p. 550

11.157 E. Heumann, S. Kück, G. Huber: High-power room-temperature Pr^{3+},Yb^{3+}:$LiYF_4$ upconversion laser in the visible spectral range. In: *Conf. Lasers Electro-Optics*, OSA Tech. Dig. Ser. (Opt. Soc. Am., Washington 2000) p. 15

11.158 S. Kück, G. Huber: Diodengepumpte Festkörperlaser, Physikalische Blätter **57**, 43 (2001)

11.159 V. Lupei, E. Osiac, T. Sandrock, E. Heumann, G. Huber: Excited state dynamics in sensitized photon avalanche processes, J. Luminesc. **76**, 441 (1998)

11.160 G. Huber, E. Heumann, T. Sandrock, K. Petermann: Up-conversion processes in laser crystals, J. Luminesc. **72–74**, 1 (1997)

11.161 P. Xie, T. R. Gosnell: Room-temperature upconversion fiber laser tunable in the red, orange, green and blue spectral range, Opt. Lett. **20**, 1014 (1995)

11.162 T. Sandrock, H. Scheife, E. Heumann, G. Huber: High-power continuous-wave upconversion fiber laser at room temperature, Opt. Lett. **22**, 808 (1997)

11.163 H. Scheife, T. Sandrock, E. Heumann, G. Huber: Pr, Yb-doped upconversion fiber laser exceeding 1 W of continuous-wave output in the red spectral range. In: *OSA Trends Opt. Photon.*, Vol. 10, ed. by C. Pollock, W. R. Bosenberg (Opt. Soc. Am., Washington 1997) p. 79

11.164 M. E. Koch, A. W. Kueny, W. E. Case: Photon avalanche laser at 644 nm, Appl. Phys. Lett. **56**, 1083 (1990)

11.165 R. M. Macfarlane, A. J. Silversmith, F. Tong, W. Lenth: CW upconversion laser action in neodymium and erbium doped solids. In: *Proceedings of the Topical Meeting on Laser Materials and Laser Spectroscopy*, ed. by Z. Wang, Z. Zhang (World Scientific, Singapore 1988) p. 24

11.166 R. M. Macfarlane, F. Tong, A. J. Silversmith, W. Lenth: Violet CW neodymium upconversion laser, Appl. Phys. Lett. **52**, 1300 (1988)

11.167 W. Lenth, R. M. Macfarlane: Excitation mechanisms for upconversion lasers, J. Luminesc. **45**, 346 (1990)

11.168 R. J. Thrash, R. H. Jarman, B. H. T. Chai, A. Pham: Upconversion green laser operation of Yb,Ho : KYF_4, Compact Blue Green Lasers Conference, Salt Lake City 1994 (Opt. Soc. Am., Washington 1994) CFA5

11.169 L. F. Johnson, H. J. Guggenheim: Infrared-pumped visible laser, Appl. Phys. Lett. **19**, 44 (1971)

11.170 D. C. Nguyen, G. E. Faulkner, M. Dulick: Blue-green (450-nm) upconversion Tm^{3+} : YLF laser, Appl. Opt. **28**, 3553 (1989)

11.171 R. M. Macfarlane, R. Wannemacher, T. Hebert, W. Lenth: Upconversion laser action at 450.2 and 483.0 nm in Tm : $YLiF_4$. In: *Tech. Dig. Conf. Lasers Electro-Optics* (Opt. Soc. Am., Washington 1990) p. 250

11.172 T. Hebert, R. Wannemacher, R. M. Macfarlane, W. Lenth: Blue continuously pumped upconversion lasing in Tm:$YLiF_4$, Appl. Phys. Lett. **60**, 2592 (1992)

11.173 B. P. Scott, F. Zhao, R. S. F. Chang, N. Djeu: Upconversion-pumped blue laser in Tm:YAG, Opt. Lett. **18**, 113 (1993)

11.174 R.J. Thrash, L.F. Johnson: Upconversion laser emission from Yb^{3+}-sensitized Tm^{3+} in BaY_2F_8, J. Opt. Soc. Am. B **11**, 881 (1994)

11.175 R.J. Thrash, L.F. Johnson: Ultraviolet upconversion laser emission from Yb^{3+} sensitized Tm^{3+} in BaY_2F_8. In: *Adv. Solid State Lasers*, OSA Proc. Ser., Vol. 20, ed. by T. Fan, B. Chai (Opt. Soc. Am., Washington 1994) paper US7

11.176 B.M. Antipenko, S.P. Voronin, T.A. Privalova: Addition of optical frequencies by cooperative processes, Opt. Spectrosc. **63**, 164 (1987)

11.177 F. Heine, V. Ostroumov, E. Heumann, T. Jensen, G. Huber, B.H.T. Chai: CW Yb,Tm:$LiYF_4$ upconversion laser at 650 nm, 800 nm, and 1500 nm. In: *OSA Proc. Adv. Solid-State Lasers*, Vol. 24, ed. by B.H.T. Chai, S.A. Payne (Opt. Soc. Am., Washington 1995) p. 77

11.178 L.F. Johnson, H.J. Guggenheim: New laser lines in the visible from Er^{3+} ions in BaY_2F_8, Appl. Phys. Lett **20**, 474 (1972)

11.179 A.A. Kaminskii, B.P. Sobolev, S.E. Sarkisov, G.A. Senlsenko, V.V. Ryabchenkov, V.A. Federov, T.V. Ovarova: Physiochemical aspects of the preparation, spectroscopy, and stimulated emission of single crystals of $BaLn_2F_8-Ln^{3+}$, Inorg. Mater. (USSR) **18**, 402 (1982)

11.180 R. Brede, T. Danger, E. Heumann, G. Huber: Room temperature green laser emission of Er:$LiYF_4$, Appl. Phys. Lett **63**, 729 (1993)

11.181 F. Heine, E. Heumann, T. Danger, T. Schweizer, G. Huber, B.H.T. Chai: Green upconversion continuous wave Er^{3+}:$LiYF_4$ laser at room temperature, Appl. Phys. Lett. **65**, 383 (1994)

11.182 S. Bär, H. Scheife, E. Heumann, G. Huber: Room-temperature continuous-wave Er^{3+} : $LiLuF_4$ upconversion laser at 552 nm, Conf. Lasers Electro-Optics/Europe 2000, Tech. Dig., Nice 2000 (IEEE, 2000) CTuF3

11.183 E. Heumann, S. Bär, H. Kretschmann, G. Huber: Diode-pumped continuous-wave green upconversion lasing of Er^{3+}:$LiLuF_4$ using multipass pumping, Opt. Lett. **27**, 1699 (2002)

11.184 A. Smith, J.P.D. Martin, M.J. Sellars, N.B. Manson, A.J. Silversmith, B. Henderson: Site selective excitation, upconversion and laser operation in Er^{3+}:$LiKF_5$, Opt. Commun. **188**, 219 (2001)

11.185 M.G. Jani, N.P. Barnes, K.E. Murray, D.W. Hart, G.J. Quarles, V.K. Castillo: Diode-pumped Ho : Tm : $LuLiF_4$ laser at room temperature, IEEE J. Quantum Electronics **33**, 112 (1997)

11.186 E.D. Filer, C.A. Morrison, N.P. Barnes, B.M. Walsh: YLF isomorphs for Ho and Tm laser applications. In: *Advanced Solid State Lasers*, OSA Proc. Ser., Vol. 20, ed. by T. Fan, B. Chai (Opt. Soc. Am., Washington 1994) p. 127

11.187 A.J. Silversmith, W. Lenth, R.M. Macfarlane: Green infrared-pumped erbium upconversion laser, Appl. Phys. Lett. **51**, 1977 (1987)

11.188 R. Scheps: Er^{3+}:$YAlO_3$ upconversion laser, IEEE J. Quantum Electron. **30**, 2914 (1994)

11.189 R. Scheps: Photon avalanche upconversion in Er^{3+}:$YAlO_3$, IEEE J. Quantum Electron. **31**, 309 (1995)

11.190 R. Scheps: Upconversion in Er^{3+}:$YAlO_3$ produced by metastable state absorption, Opt. Mater. **7**, 75 (1997)

11.191 R. Brede, E. Heumann, J. Koetke, T. Danger, G. Huber, B. Chai: Green up-conversion laser emission in Er-doped crystals at room temperature, Appl. Phys. Lett. **63**, 2030 (1993)

11.192 P. Xie, S.C. Rand: Continuous-wave trio upconversion laser, Appl. Phys. Lett. **57**, 1182 (1990)

11.193 W. Lenth, A.J. Silversmith, R.M. Macfarlane: Green infrared-pumped erbium upconversion lasers. In: *Advances in Laser Science III*, AIP Conf. Proc., Vol. 172, ed. by A.C. Tam, J.L. Gole, W.C. Stwalley (AIP, New York 1989) p. 8

11.194 R.A. McFarlane: Dual wavelength visible upconversion laser, Appl. Phys. Lett. **54**, 2301 (1989)

11.195 F. Tong, W.P. Risk, R.M. Macfarlane, W. Lenth: 551 nm diode-laser-pumped upconversion laser, Electron. Lett. **25**, 1389 (1989)

11.196 G.C. Valley, R.A. McFarlane: 1.1-Watt visible upconversion laser modelling and experiment. In: *OSA Proc. Adv. Solid-State Lasers*, Vol. 13, ed. by L.L. Chase, A.A. Pinto (Opt. Soc. Am., Washington 1992) pp. 376–379

11.197 T. Heber, W.P. Risk, R.M. Macfarlane, W. Lenth: Diode-laser-pumped 551 nm upconversion laser in $YLiF_4$: Er^{3+}. In: *OSA Proc. Adv. Solid-State Lasers*, ed. by H.J. Jenssen, G. Dube (Opt. Soc. Am., Washington 1990) p. 379

11.198 R.R. Stephens, R.A. McFarlane: Diode-pumped upconversion laser with 100-mW output power, Opt. Lett. **18**, 34 (1993)

11.199 P. Xie, S.C. Rand: Continuous-wave, fourfold upconversion laser, Appl. Phys. Lett. **63**, 3125 (1993)

11.200 F. Heine, E. Heumann, P. Möbert, G. Huber, B.H.T. Chai: Room temperature cw green upconversion Er^{3+}:$YLiF_4$-laser pumped near 970 nm. In: *Adv. Solid State Lasers*, OSA Proc. Ser., Vol. 24, ed. by B. Chai, S. Payne (Opt. Soc. Am., Washington 1995)

11.201 P.E.-A. Möbert, E. Heumann, G. Huber, B.H.T. Chai: Green Er^{3+}:$YLiF_4$ upconversion laser at 551 nm with Yb^{3+} codoping: a novel pumping scheme, Opt. Lett. **22**, 1412 (1997)

11.202 R.M. Macfarlane, E.A. Whittaker, W. Lenth: Blue, green and yellow upconversion lasing in Er:$YLiF_4$ using 1.5 μm pumping, Electron. Lett. **28**, 2136 (1992)

11.203 R.A. McFarlane: High-power visible upconversion laser, Opt. Lett. **16**, 1397 (1991)

11.204 P. Xie, S.C. Rand: Visible cooperative upconversion laser in Er:$LiYF_4$, Opt. Lett. **17**, 1198 (1992)

11.205 P. Xie, S. C. Rand: Continuous-wave mode-locked visible upconversion laser: erratum, Opt. Lett. **17**, 1882 (1992)

11.206 P. Xie, S. C. Rand: Continuous-wave mode-locked visible upconversion laser, Opt. Lett. **17**, 1116 (1992)

11.207 T. Hebert, R. Wannemacher, W. Lenth, R. M. Macfarlane: Blue and green cw upconversion lasing in Er:$YLiF_4$, Appl. Phys. Lett. **57**, 1727 (1990)

11.208 S. A. Pollack, D. B. Chang, M. Birnbaum: Threefold upconversion laser at 0.85, 1.23, and 1.73 μm in Er:YLF pumped with a 1.53 μm Er glass laser, Appl. Phys. Lett. **54**, 869 (1989)

11.209 E. Heumannn, S. Bär, K. Rademaker, G. Huber, S. Butterworth, A. Diening, W. Seelert: Semiconductor-laser-pumped high-power upconversion laser, Appl. Phys. Lett. **88**, 061108 (2006)

11.210 B. M. Antipenko, S. P. Voronin, T. A. Privalova: Addition of optical frequencies by cooperative processes, Opt. Spectrosc. (USSR) **63**, 768 (1987)

11.211 R. A. McFarlane: Spectroscopic Studies and Upconversion Laser Operation of BaY_2F_8:Er 5%. In: *OSA Proc. Adv. Solid State Lasers*, Vol. 13, ed. by L. L. Chase, A. A. Pinto (Opt. Soc. Am., Washington 1992) pp. 275–279

11.212 R. A. McFarlane: Upconversion laser in BaY_2F_8:Er 5% pumped by ground-state and excited-state absorption, J. Opt. Soc. Am. B **11**, 871 (1994)

11.213 M. J. Weber: *The Handbook of Lasers* (CRC, Boca Raton 1999)

11.214 D. S. Funk, J. G. Eden: Visible fluoride fiber lasers. In: *Rare-Earth-Doped Fiber Lasers and Amplifiers*, 2 edn., ed. by M. J. F. Digonnet (Marcel Dekker, New York 2001)

11.215 B. N. Kazakov, M. S. Orlov, M. V. Petrov, A. L. Stolov, A. M. Tkachuk: Induced emission of Sm^{3+}-ions in the visible region of the spectrum, Opt. Spectrosc. (USSR) **47**, 676 (1979)

11.216 P. P. Sorokin, M. J. Stevenson, J. R. Lankard, G. D. Pettit: Spectroscopy and optical maser action in $SrF_2 : Sm^{2+}$, Phys. Rev. B **127**, 503 (1962)

11.217 . C. Chang: Fluorescence and stimulated emission from trivalent europium in yttrium oxide, J. Appl. Phys. **34**, 3500 (1963)

11.218 J. R. O'Connor: Optical and laser properties of Nd^{3+}- and Eu^{3+}-doped YVO_4, Trans. Metallurg. Soc. AIME **239**, 362 (1967)

11.219 Z. T. Azamatov, P. A. Arsenyev, M. V. Chukichev: Spectra of Gadolinium in YAG single crystals, Opt. Spectrosc. **28**, 156 (1970)

11.220 H. P. Jenssen, D. Castleberry, D. Gabbe, A. Linz: Stimulated emission at 5445 Å in Tb^{3+} : YLF, IEEE J. **QE-9**(6), 665 (1973)

11.221 Y. K. Voronko, A. A. Kaminskii, V. V. Osiko, A. M. Prokhorov: Simulated emission from Ho^{3+} in CaF_2 at 5512 , JETP Lett. **1**, 3 (1965)

11.222 K. Schmitt: Stimulated C'-emission of Ag^+-centers in KI, RbBr, and CsBr, Appl. Phys. A **38**, 61 (1985)

11.223 D. M. Baney, R. Rankin, K. W. Chang: Simultaneous blue and green upconversion lasing in a laser-diode pumped Pr^{3+}/Yb^3 doped fluoride fiber laser, Appl. Phys. Lett. **69**, 1662 (1996)

11.224 A. C. Tropper, J. N. Carter, R. D. T. Lauder, D. C. Hanna, S. T. Davey, D. Szebesta: Analysis of blue and red laser performance of the infrared-pumped praseodymium-doped fluoride fiber laser, J. Opt. Soc. Am. B **11**, 886 (1994)

11.225 Y. Zhao, S. Fleming: Theory of Pr^{3+}-doped fluoride fiber upconversion lasers, IEEE J. Quantum Electron. **33**, 905 (1997)

11.226 A. Richter, H. Scheife, E. Heumann, G. Huber, W. Seelert, A. Diening: Semiconductor laser pumping of contiuous-wave Pr^{3+}-doped ZBLAN fiber laser, Electron. Lett. **41**, 794 (2005)

11.227 G. G. Smart, D. C. Hanna, A. C. Tropper, S. T. Davey, S. F. Carter, D. Szebesta: CW room temperature upconversion lasing at blue, green and red wavelengths in infrared-pumped Pr^{3+}-doped fluoride fiber, Electron. Lett. **27**, 1307 (1991)

11.228 H. M. Pask, A. C. Tropper, D. C. Hanna: A Pr^{3+}-doped ZBLAN fiber upconversion laser pumped by an Yb^{3+}-doped silica fiber laser, Opt. Commun. **134**, 139 (1997)

11.229 J. Y. Allain, M. Monerie, H. Poignant: Red upconversion Yb-sensitised Pr fluoride fiber laser pumped in 0.8 μm region, Electron. Lett. **27**, 1156 (1991)

11.230 D. Piehler, D. Craven, N. Kwong, H. Zarem: Laser-diode-pumped red and green up-conversion fiber lasers, Electron. Lett. **29**, 1857 (1993)

11.231 D. M. Baney, L. Yang, J. Ratcliff, K. W. Chang: Red and orange Pr^{3+}/Yb^{3+} doped ZBLAN fiber upconversion lasers, Electron. Lett. **31**, 1842 (1995)

11.232 Y. Zhao, S. Fleming: All-solid state and all-fiber blue upconversion laser, Electron. Lett. **32**, 1199 (1996)

11.233 H. Zellmer, K. Plamann, G. Huber, H. Scheife, A. Tünnermann: Visible double-cla upconversion fiber laser, Electron. Lett. **34**, 565 (1998)

11.234 H. Zellmer, P. Riedel, A. Tünnermann, M. Kempe: High power multi mode visible upconversion fiber laser in the red spectral range, CLEO/Europe-EQEC Focus Meetings 2001, München 2001, Tech. Dig. (2001) p. 143

11.235 H. Zellmer, P. Riedel, A. Tünnermann, M. Kempe: High-power diode pumped upconversion fiber laser in red and green spectral range, Electron. Lett. **38**, 1250 (2002)

11.236 H. Zellmer, P. Riedel, A. Tünnermann: Visible upconversion lasers in praseodymium-ytterbium-doped fibers, Appl. Phys. B **69**, 417 (1999)

11.237 D. M. Costantini, H. G. Limberger, T. Lasser, C. A. P. Muller, H. Zellmer, P. Riedel, A. Tünner-

mann: Actively mode-locked visible upconversion fiber laser, Opt. Lett. **25**, 1445 (2000)
11.238 M. Zeller, H. G. Limberger, T. Lasser: Tunable Pr^{3+}-Yb^{3+}-doped all-fiber upconversion laser, IEEE Photon. Technol. Lett. **15**, 194 (2003)
11.239 D. S. Funk, J. W. Carlson, J. G. Eden: Ultraviolet (381 nm), room temperature laser in neodymium-doped fluorozirconate fiber, Electron. Lett. **30**, 1859 (1994)
11.240 D. S. Funk, J. W. Carlson, J. G. Eden: Room-temperature fluorozirconate glass fiber laser in the violet, Opt. Lett. **20**, 1474 (1995)
11.241 M. P. LeFlohic, J. Y. Allain, G. M. Stéphan, G. Mazé: Room-temperature continuous-wave upconversion laser 455 nm in a Tm^{3+} fluorozirconate fiber, Opt. Lett. **19**, 1982 (1994)
11.242 I. J. Booth, C. J. Mackechnie, B. F. Ventrudo: Operation of diode laser pumped Tm^{3+} ZBLAN upconversion fiber laser at 482 nm, IEEE J. Quantum Electron. **32**, 118 (1996)
11.243 R. Paschotta, N. Moore, W. A. Clarkson, A. C. Tropper, D. C. Hanna, G. Mazé: 230 mW of blue light from thulium:ZBLAN upconversion fiber laser, Conf. Laser Electro-Optics, Baltimore 1997 (Opt. Soc. Am., Washington 1997) CTuG3
11.244 G. Tohmon, J. Ohya, H. Sato, T. Uno: Increased efficiency and decreased threshold in Tm:ZBLAN blue fiber laser co-pumped by 1.1 µm and 0.68 µm light, IEEE Photon. Technol. Lett. **7**, 742 (1995)
11.245 S. G. Grubb, K. W. Bennett, R. S. Cannon, W. F. Humer: CW room-temperature blue upconversion fiber laser, Electron. Lett. **28**, 1243 (1992)
11.246 S. Sanders, R. G. Waarts, D. G. Mehuys, D. F. Welch: Laser diode pumped 106 mW blue upconversion fiber laser, Appl. Phys. Lett. **25**, 1815 (1995)
11.247 G. Tohmon, H. Sato, J. Ohya, T. Uno: Thulium:ZBLAN blu fiber laser pumped by two wavelengths, Appl. Opt. **36**, 3381 (1997)
11.248 P. Laperle, R. Vallée, A. Chandonnet: Stable blue emission from a 2500 ppm thulium-doped ZBLAN fiber laser, Conf. Laser Electro-Optics, San Francisco 1998 (Opt. Soc. Am., Washington 1998) CTuE1
11.249 P. R. Barber, H. M. Pask, C. J. Mackechnie, D. C. Hanna, A. C. Tropper, J. Massicott, S. T. Daveys, D. Szebesta: Improved laser performance of Tm^{3+} and Pr^{3+}-doped ZBLAN fibers, Conf. Laser Electro-Optics, Anaheim 1994 (Opt. Soc. Am., Washington 1994) CMF3
11.250 H. Zellmer, A. Tünnermann, H. Welling, S. Buteau: All fiber laser system with 0.3 W output power in the blue spectral range, Conf. Laser Electro-Optics, Baltimore 1997 (Opt. Soc. Am., Washington 1997) CTuG3
11.251 H. Zellmer, S. Buteau, A. Tünnermann, H. Welling: All fiber laser system with 0.1 W output power in blue spectral range, Electron. Lett. **33**, 1383 (1997)
11.252 G. Qin, S. Huang, Y. Feng, A. Shirakawa, M. Musha, K.-J. Ueda: Power scaling of Tm^{3+} doped ZBLAN blue upconversion fiber lasers: modeling and experiment, Appl. Phys. B **82**, 6 (2006)
11.253 J. Limpert, H. Zellmer, P. Riedel, G. Mazé, A. Tünnermann: Laser oscillation in yellow and blue spectral range in Dy^{3+} : ZBLAN, Electron. Lett. **36**, 1386 (2000)
11.254 T. J. Whitley, C. A. Millar, R. Wyatt, M. C. Brierley, D. Szebesta: Upconversion pumped green lasing in erbium doped fluorozircnate fiber, Electron. Lett. **27**, 1785 (1991)
11.255 J. F. Massicott, M. C. Brierley, R. Wyatt, S. T. Davey, D. Szebesta: Low threshold, diode-pumped operation of a green, Er^{3+} doped fluoride fiber laser, Electron. Lett. **29**, 2119 (1993)
11.256 D. Piehler, D. Craven, N. Kwong: Green, laser-diode-pumped erbium fiber laser, OSA Topical Metting on Compact Blue/Green Lasers, Salt Lake City 1994 (Opt. Soc. Am., Washington 1994) CFA2
11.257 J. Y. Allain, M. Monerie, H. Poignant: Tunable green upconversion erbium fiber laser, Electron. Lett. **28**, 111 (1992)
11.258 D. Piehler, D. Craven: 11.7 mW green InGaAs-laser-pumped erbium fiber laser, Electron. Lett. **30**, 1759 (1994)
11.259 J. Y. Allain, M. Monerie, H. Poignant: Room temperature CW tunable green upconversion holmium fiber laser, Electron. Lett. **26**, 261 (1990)
11.260 J. Y. Allain, M. Monerie, H. Poignant: Characteristics and dynamics of a room temperature CW tunable green upconversion fiber laser, Proc. 16th European Conference on Optical Communications (Amsterdam 1990) p. 575
11.261 D. S. Funk, S. B. Stevens, S. S. Wu, J. G. Eden: Tuning, temporal, and spectral characteristics of a green ($\lambda \approx 549$ nm) holmium-doped fluorozirconate glass fiber laser, IEEE J. Quantum Electron. **32**, 638 (1996)
11.262 D. S. Funk, J. G. Eden, J. S. Osinski, B. Lu: Green, holmium-doped upconversion fiber laser pumped by a red semiconductor laser, Electron. Lett. **33**, 1958 (1997)
11.263 D. S. Funk: Optical processes and laser dynamics in holmium and neodymium upconversion-pumped visible and ultraviolet fluorozirconate fiber lasers Ph.D. Thesis (University of Illinois, Urbana 1999)
11.264 M. L. Dennis, J. W. Dixon, T. Aggarwal: High power upconversion lasing at 810 nm in Tm:ZBLAN fiber, Electron. Lett. **30**, 136 (1994)
11.265 W. A. Clarkson, D. C. Hanna: Two-mirror beam-shaping technique for high power diode bars, Opt. Lett. **21**(6), 375 (1996)
11.266 D. Golla, M. Bode, S. Knoke, W. Schöne, A. Tünnermann: 62 W CW TEM00 Nd:YAG laser side-pumped by fiber-coupled diode laser, Opt. Lett. **21**(3), 210 (1996)
11.267 P. Glas, D. Fischer, M. Moenster, G. Steinmeyer, R. Iliew, C. Etrich, M. Kreitel, L. E. Nilsson, R. Koppler: Large-mode-area Nd-doped single-

transverse-mode dual-wavelength microstructure fiber laser, Opt. Express **13**(20), 7884 (2005)
11.268 L. B. Fu, M. Ibsen, D. J. Richardson, J. Nilsson, D. N. Payne, A. B. Grudinin: Compact high-power tunable three-level operation of double cladding Nd-doped fiber laser, IEEE Photon. Technol. Lett. **17**(2), 306 (2005)
11.269 H. Jeong, S. Choi, K. Oh: Continuous wave single transverse mode laser oscillation in a Nd-doped large core double clad fiber cavity with concatenated adiabatic tapers, Opt. Commun. **213**(1–3), 33 (2002)
11.270 P. Glas, D. Fischer: Cladding pumped large-mode-area Nd-doped holey fiber laser, Opt. Express **10**(6), 286 (2002)
11.271 E. Rochat, R. Dandliker, K. Haroud, R. H. Czichy, U. Roth, D. Costantini, R. Holzner: Fiber amplifiers for coherent space communication, IEEE J. Sel. Top. Quantum Electron. **7**(1), 64 (2001)
11.272 B. M. Dicks, F. Heine, K. Petermann, G. Huber: Characterization of a radiation-hard single-mode Yb-doped fiber amplifier at 1064 nm, Laser Phys. **11**(1), 134 (2001)
11.273 N. S. Kim, T. Hamada, M. Prabhu, C. Li, J. Song, K. Ueda, A. P. Liu, H. J. Kong: Numerical analysis and experimental results of output performance for Nd-doped double-clad fiber lasers, Opt. Commun. **180**(4–6), 329 (2000)
11.274 I. Zawischa, K. Plamann, C. Fallnich, H. Welling, H. Zellmer, A. Tunnermann: All-solid-state neodymium-based single-frequency master-oscillator fiber power-amplifier system emitting 5.5 W of radiation at 1064 nm, Opt. Lett. **24**(7), 469 (1999)
11.275 E. Rochat, K. Haroud, U. Roth, J. E. Balmer, R. Dandliker, H. P. Weber: High-gain solid-state and fiber amplifier-chain for high-power coherent communication, IEEE Photon. Technol. Lett. **11**(9), 1120 (1999)
11.276 R. Nicolaescu, T. Walther, E. S. Fry, M. Muendel: Ultranarrow-linewidth, efficient amplification of low-power seed sources by a fiber amplifier, Appl. Opt. **38**(9), 1784 (1999)
11.277 T. Miyazaki, K. Inagaki, Y. Karasawa, M. Yoshida: Nd-doped double-clad fiber amplifier at 1.06 μm, J. Lightwave Technol. **16**(4), 562 (1998)
11.278 M. Wegmuller, M. Schurch, W. Hodel, H. P. Weber: Diode-pumped passively mode-locked Nd^{3+}-doped fluoride fiber laser emitting at 1.05 μm: Novel results, IEEE J. Quantum Electron. **34**(1), 14 (1998)
11.279 L. F. Johnson, J. E. Geusic, L. G. Van Uitert: Coherent oscillations from Tm^{3+}, Ho^{3+}, Yb^{3+} and Er^{3+} ions in yttrium aluminum garnet, Appl. Phys. Lett. **7**, 127 (1965)
11.280 P. Lacovara, H. K. Choi, C. A. Wang, R. L. Aggarwal, T. Y. Fan: Room-temperature diode-pumped Yb:YAG laser, Opt. Lett. **16**, 1089 (1991)
11.281 A. Prokohorov: *Spravochnik pr Lazerum, Handbook on Lasers*, Vol. 1 (Sovet-skoe Radio, Moscow 1978)
11.282 L. J. Qin, X. L. Meng, H. Y. Shen, H. Y. Zhu, B. C. Xu, L. X. Huang, H. R. Xia, P. Zhao, G. Zheng: Thermal conductivity and refractive indices of Nd : $GdVO_4$, Cry. Res. Technol. **38**, 793 (2003)
11.283 A. I. Zagummenyi, Y. Zavartsev, P. Studenikin, I. A. Sherbakov, A. Umyskov, P. A. Popov, V. B. Ufimtsev: $GdVO_4$ crystals with Nd^{3+}, Tm^{3+}, Ho^{3+}, and Er^{3+} ions for diode-pumped microchip laser, Proc. SPIE **2698**, 182–192 (1996)
11.284 C. Kränkel, D. Fagundes-Peters, S. T. Fredrich, J. Johannsen, M. Mond, G. Guber, M. Bernhagen, R. Uecker: Continuous wave laser operation of Yb^{3+}:YVO_4, Appl. Phys. B **79**, 543 (2004)
11.285 C. Czeranowsky: Resonatorinterne Frequenzverdopplung von diodengepumpten Neodym-Lasern mit hohen Ausgangsleistungen im blauen Spektralbereich Ph.D. Thesis (University of Hamburg, Hamburg 2002)
11.286 K. Petermann, G. Huber, L. Fornasiero, S. Kuch, E. Mix, V. Peters, S. A. Basun: Rare-earth-doped sesquioxides, J. Luminesc. **87–89**, 973 (2000)
11.287 G. Aka, D. Vivien, V. Lupei: Site-selective 900 nm quasi-three-level laser emission in Nd-doped strontium lanthanum aluminate, Appl. Phys. Lett. **85**(14), 2685 (2004)
11.288 F. Jia, Q. Xue, Q. Zheng, Y. Bu, L. Qian: 5.3 W deep-blue light generation by intra-cavity frequency doubling of Nd : $GdVO_4$, Appl. Phys. B **83**(2), 245 (2006)
11.289 K. Mizuuchi, A. Morikawa, T. Sugita, K. Yamamoto, N. Pavel, T. Taira: Continuous-wave deep blue generation in a periodically poled MgO : $LiNbO_3$ crystal by single-pass frequency doubling of a 912 nm Nd : $GdVO_4$ laser, Jpn. J. Appl. Phys. 2 **43**(10A), L1293 (2004)
11.290 Y. D. Zavartsev, A. I. Zagumennyi, F. Zerrouk, S. A. Kutovoi, V. A. Mikhailov, V. V. Podreshetnikov, A. A. Sirotkin, I. A. Shcherbakov: Diode-pumped quasi-three-level 456 nm Nd . $GdVO_4$ laser, Quantum Electron. **33**(7), 651 (2003)
11.291 C. Czeranowsky, M. Schmidt, E. Heumann, G. Huber, S. Kutovoi, Y. Zavartsev: Continuous wave diode pumped intracavity doubled Nd:$GdVO_4$ laser with 840 mW output power at 456 nm, Opt. Commun. **205**(4–6), 361 (2002)
11.292 Q. H. Xue, Q. Zheng, Y. K. Bu, F. Q. Jia, L. S. Qian: High-power efficient diode-pumped Nd : YVO_4/LiB_3O_5 457 nm blue laser with 4.6 W of output power, Opt. Lett. **31**(8), 1070 (2006)
11.293 L. Zhang, C. Y. Zhang, Z. Y. Wei, C. Zhang, Y. B. Long, Z. G. Zhang, H. J. Zhang, J. Y. Wang: Compact diode-pumped continuous-wave Nd : $LuVO_4$ lasers operated at 916 nm and 458 nm, Chinese Phys. Lett. **23**(5), 1192 (2006)
11.294 C. Y. Zhang, L. Zhang, Z. Y. Wei, C. Zhang, Y. B. Long, Z. G. Zhang, H. J. Zhang, J. Y. Wang: Diode-pumped

continuous-wave Nd : $LuVO_4$ laser operating at 916 nm, Opt. Lett. **31**(10), 1435 (2006)

11.295 J. H. Zarrabi, P. Gavrilovic, S. Singh: Intracavity, frequency-doubled, miniaturized Nd-$YAlO_3$ blue laser at 465 nm, Appl. Phys. Lett. **67**(17), 2439 (1995)

11.296 N. Pavel, V. Lupei, J. Saikawa, T. Taira, H. Kan: Neodymium concentration dependence of 0.94, 1.06 and 1.34 µm laser emission and of heating effects under 809 and 885 nm diode laser pumping of Nd:YAG, Appl. Phys. **82**(4), 599 (2006)

11.297 H. Hara, B. M. Walsh, N. P. Barnes: Tunability of a 946 nm Nd:YAG microchip laser versus output mirror reflectivity and crystal length, Opt. Engineering **43**(12), 3026 (2004)

11.298 J. L. He, H. M. Wang, S. D. Pan, J. Liu H. X. Li, S. N. Zhu: Laser performance of Nd:YAG at 946 nm and frequency doubling with periodically poled $LiTaO_3$, J. Crystal Growth **292**(2), 337 (2006)

11.299 Y. H. Chen, W. Hou, H. B. Peng, A. C. Geng, Y. Zhou, D. F. Cui, Z. Y. Xu: Generation of 2.1 W continuous wave blue light by intracavity doubling of a diode-end-pumped Nd:YAG laser in a 30 mm LBO, Chinese Phys. Lett. **23**(6), 1479 (2006)

11.300 R. Zhou, E. B. Li, H. F. Li, P. Wang, J. Q. Yao: Continuous-wave, 15.2 W diode-end-pumped Nd:YAG laser operating at 946 nm, Opt. Lett. **31**(12), 1869 (2006)

11.301 Y. Lu, B. G. Zhang, E. B. Li, D. G. Xu, R. Zhou, X. Zhao, F. Ji, T. L. Zhang, P. Wang, J. Q. Yao: High-power simultaneous dual-wavelength emission of an end-pumped Nd:YAG laser using the quasi-three-level and the four-level transition, Opt. Commun. **262**(2), 241 (2006)

11.302 Y. Chen, H. Peng, W. Hou, Q. Peng, A. Geng, L. Guo, D. Cui, Z. Xu: 3.8 W of cw blue light generated by intracavity frequency doubling of a 946 nm Nd:YAG laser with LBO, Appl. Phys. B **83**(2), 241 (2006)

11.303 R. Zhou, T. L. Zhang, E. B. Li, X. Ding, Z. Q. Cai, B. G. Zhang, W. Q. Wen, P. Wang, J. Q. Yao: 8.3 W diode-end-pumped continuous-wave Nd:YAG laser operating at 946 nm, Opt. Express **13**(25), 10115 (2005)

11.304 R. Zhou, Z. Q. Cai, W. Q. Wen, X. Ding, P. Wang, J. Q. Yao: High-power continuous-wave Nd:YAG laser at 946 nm and intracavity frequency-doubling with a compact three-element cavity, Opt. Commun. **255**(4–6), 304 (2005)

11.305 C. Czeranowsky, E. Heumann, G. Huber: All-solid-state continuous-wave frequency-doubled Nd:YAG-BiBO laser with 2.8 W output power at 473 nm, Opt. Lett. **28**(6), 432 (2003)

11.306 T. Kellner, F. Heine, G. Huber, S. Kück: Passive *Q*-switching of a diode-pumped 946 nm Nd:YAG laser with 1.6 W average output power, Appl. Opt. **37**(30), 7076 (1998)

11.307 S. G. P. Strohmaier, H. J. Eichler, J. F. Bisson, H. Yagi, K. Takaichi, K. Ueda, T. Yanagitani, A. A. Kaminskii: Ceramic Nd:YAG laser at 946 nm, Laser Phys. Lett. **2**(8), 383 (2005)

11.308 L. Fornasiero, E. Mix, V. Peters, E. Heumann, K. Petermann, G. Huber: Efficient laser operation of Nd : Sc_2O_3 at 966 nm, 1082 nm, and 1486 nm. In: *Adv. Solid-State Lasers*, OSA Trends Opt. Photon., Vol. 26, ed. by M. M. Fejer, H. Injeyan, U. Keller (Opt. Soc. Am., Washington 1999) p. 249

11.309 K. Rademaker, E. Heumann, G. Huber, S. A. Payne, W. F. Krupke, L. I. Isaenko, A. Burger: Laser activity at 1.18, 1.07, and 0.97 µm in the low-phonon-energy hosts KPb_2Br_5 and $RbPb_2Br_5$ doped with Nd^{3+}, Opt. Lett. **30**(7), 729 (2005)

11.310 A. Giesen, H. Huegel, A. Voss, K. Wittig, U. Brauch, H. Opower: Scalable concept for diode-pumped high-power solid-state lasers, Appl. Phys. B **58**(5), 365 (1994)

11.311 C. Stewen, K. Contag, M. Larionov, A. Giesen, H. Huegel: A 1 kW CW thin disc laser, IEEE J. Sel. Top. Quantum Electron. **6**(4), 650 (2000)

11.312 S. Zhao, Q. Wang, X. Zhang, L. Sun, S. Zhang: Laser characteristics of a new crystal Nd : $Sr_5(PO_4)_3F$ at 1.059 µm, Opt. Laser Technol. **28**(6), 477 (1996)

11.313 C. Grivas, T. C. May-Smith, D. P. Shepherd, R. W. Eason: On the growth and lasing characteristics of thick Nd:GGG waveguiding films fabricated by pulsed laser deposition, Appl. Phys. A **79**(4–6), 1203 (2004)

11.314 H. J. Zhang, X. L. Meng, L. Zhu, C. Q. Wang, R. P. Cheng, W. T. Yu, S. J. Zhang, L. K. Sun, Y. T. Chow, W. L. Zhang, H. Wang, K. S. Wong: Growth and laser properties of Nd : $Ca_4YO(BO_3)_3$ crystal, Opt. Commun. **160**(4–6), 273 (1999)

11.315 A. Ikesue, Y. L. Aung: Synthesis and performance of advanced ceramic lasers, J. Am. Ceram. Soc. **89**(6), 1936 (2006)

11.316 M. L. Huang, Y. J. Chen, X. Y. Chen, Y. D. Huang, Z. D. Luo: Study on CW fundamental and self-frequency doubling laser of Nd^{3+} : $GdAl_3(BO_3)_4$ crystal, Opt. Commun. **204**(1–6), 333 (2002)

11.317 P. Dekker, Y. J. Huo, J. M. Dawes, J. A. Piper, P. Wang, B. S. Lu: Continuous wave and *Q*-switched diode-pumped neodymium, lutetium: yttrium aluminium borate lasers, Opt. Commun. **151**(4–6), 406 (1998)

11.318 N. Pavel, T. Taira: Continuous-wave high-power multi-pass pumped thin-disc Nd : $GdVO_4$ laser, Opt. Commun. **260**(1), 271 (2006)

11.319 H. J. Zhang, J. Y. Wang, C. Q. Wang, L. Zhu, X. B. Hu, X. L. Meng, M. H. Jiang: A comparative study of crystal growth and laser properties of Nd:YVO_4, Nd : $GdVO_4$ and Nd : $Gd_xLa_{1-x}VO_4$ ($x = 0.80, 0.60, 0.45$) crystals, Opt. Mater. **23**(1–2), 449 (2003)

11.320 C. Q. Wang, Y. T. Chow, L. Reekie, W. A. Gambling, H. J. Zhang, L. Zhu, X. L. Meng: A comparative study of the laser performance of diode-laser-pumped

$Nd:GdVO_4$ and $Nd:YVO_4$ crystals, Appl. Phys. B **70**(6), 769 (2000)

11.321 C.Q. Wang, H.J. Zhang, Y.T. Chow, J.H. Liu, L. Zhu, J.Y. Wang, X.L. Meng, W.A. Gambling: Spectroscopic and laser properties of $Nd:Gd_{0.8}La_{0.2}VO_4$ crystal, Opt. Laser Technol. **33**(6), 439 (2001)

11.322 G. Lucas-Leclin, F. Auge, S.C. Auzanneau, F. Balembois, P. Georges, A. Brun, E. Mougel, G. Aka, D. Vivien: Diode-pumped self-frequency-doubling $Nd:GdCa_4O(BO_3)_3$ lasers: toward green microchip lasers, J. Opt. Soc. Am. B **17**(9), 1526 (2000)

11.323 D.A. Hammons, M. Richardson, B.H.T. Chai, A.K. Chin, R. Jollay: Scaling of longitudinally diode-pumped self-frequency-doubling Nd:YCOB lasers, IEEE J. Quantum Electron. **36**(8), 991 (2000)

11.324 F. Mougel, F. Auge, G. Aka, A. Kahn-Harari, D. Vivien, F. Balembois, P. Georges, A. Brun: New green self-frequency-doubling diode-pumped $Nd:Ca_4GdO(BO_3)_3$ laser, Appl. Phys. B **67**(5), 533 (1998)

11.325 J.M. Eichenholz, D.A. Hammons, L. Shah, Q. Ye, R.E. Peale, M. Richardson, B.H.T. Chai: Diode-pumped self-frequency doubling in a Nd^{3+}: $YCa_4O(BO_3)_3$ laser, Appl. Phys. Lett. **74**(14), 1954 (1999)

11.326 Y.J. Chen, X.H. Gong, Y.F. Lin, Q.G. Tan, Z.D. Luo, Y.D. Huang: Passively *Q*-switched laser operation of $Nd:LaB_3O_6$ cleavage microchip, J. Appl. Phys. **99**(10), 103101 (2006)

11.327 S.F. Wu, G.F. Wang, J.L. Xiea: Growth of high quality and large-sized $Nd^{3+}:YVO_4$ single crystal, J. Cryst. Growth **266**(4), 496 (2004)

11.328 V.G. Ostroumov, F. Heine, S. Kück, G. Huber, V.A. Mikhailov, I.A. Shcherbakov: Intracavity frequency-doubled diode-pumped $Nd:LaSc_3(BO_3)_4$ lasers, Appl. Phys. B **64**(3), 301 (1997)

11.329 V. Lupei, N. Pavel, Y. Sato, T. Taira: Highly efficient 1063 nm continuous-wave laser emission in $Nd:GdVO_4$, Opt. Lett. **28**(23), 2366 (2003)

11.330 Y. Bo, A.C. Geng, Y.F. Lu, X.D. Yang, Q.J. Peng, Q.J. Cui, D.F. Cui, Z.Y. Xu: A 4.8 W $M^2 = 4.6$ continuous-wave intracavity sum-frequency diode-pumped solid-state yellow laser, Chinese Phys. Lett. **23**(6), 1494 (2006)

11.331 M. Gerber, T. Graf, A. Kudryashov: Generation of custom modes in a Nd:YAG laser with a semipassive bimorph adaptive mirror, Appl. Phys. B **83**(1), 43 (2006)

11.332 G.J. Spühler, T. Südmeyer, R. Paschotta, M. Moser, K.J. Weingarten, U. Keller: Passively mode-locked high-power Nd:YAG lasers with multiple laser heads, Appl. Phys. B **71**(1), 19 (2000)

11.333 J. Lu, H. Yagi, K. Takaichi, T. Uematsu, J.F. Bisson, Y. Feng, A. Shirakawa, K.I. Ueda, T. Yanagitani, A.A. Kaminskii: 110 W ceramic $Nd^{3+}:Y_3Al_5O_{12}$ laser, Appl. Phys. B **79**(1), 25 (2004)

11.334 J.R. Lu, K. Ueda, H. Yagi, T. Yanagitani, Y. Akiyama, A.A. Kaminskii: Neodymium doped yttrium aluminum garnet ($Y_3Al_5O_{12}$) nanocrystalline ceramics – a new generation of solid state laser and optical materials, J. Alloy. Comp. **341**(1–2), 220 (2002)

11.335 C.Y. Wang, J.H. Ji, Y.F. Qi, Q.H. Lou, X.L. Zhu, Y.T. Lu: Kilohertz electro-optic *Q*-switched Nd:YAG ceramic laser, Chinese Phys. Lett. **23**(7), 1797 (2006)

11.336 D. Kracht, D. Freiburg, R. Wilhelm, M. Frede, C. Fallnich: Core-doped ceramic Nd:YAG laser, Opt. Express **14**(7), 2690 (2006)

11.337 Y. Qi, X. Zhu, Q. Lou, J. Ji, J. Dong, Y. Wei: High optical-optical efficiency of 52.5% obtained in high power Nd:YAG ceramic laser, Electron. Lett. **42**(1), 30 (2006)

11.338 Y.F. Qi, X.L. Zhu, Q.H. Lou, J.H. Ji, J.X. Dong, Y.R. Wei: Nd:YAG ceramic laser obtained high slope-efficiency of 62% in high power applications, Opt. Express **13**(22), 8725 (2005)

11.339 L. Guo, W. Hou, H.B. Zhang, Z.P. Sun, D.F. Cui, Z.Y. Xu, Y.G. Wang, X.Y. Ma: Diode-end-pumped passively mode-locked ceramic Nd:YAG Laser with a semiconductor saturable mirror, Opt. Express **13**(11), 4085 (2005)

11.340 J. Lu, H. Yagi, K. Takaichi, T. Uematsu, J.F. Bisson, Y. Feng, A. Shirakawa, K.I. Ueda, T. Yanagitani, A.A. Kaminskii: 110 W ceramic $Nd^{3+}:Y_3Al_5O_{12}$ laser, Appl. Phys. B **79**(1), 25 (2004)

11.341 J.R. Lu, K. Ueda, H. Yagi, T. Yanagitani, Y. Akiyama, A.A. Kaminskii: Neodymium doped yttrium aluminum garnet ($Y_3Al_5O_{12}$) nanocrystalline ceramics – a new generation of solid state laser and optical materials, J. Alloy. Comp. **341**(1–2), 220 (2002)

11.342 J. Lu, M. Prabhu, K. Ueda, H. Yagi, T. Yanagitani, A. Kudryashov, A.A. Kaminskii: Potential of ceramic YAG lasers, Laser Phys. **11**(10), 1053 (2001)

11.343 J.R. Lu, T. Murai, K. Takaichi, T. Uematsu, K. Misawa, M. Prabhu, J. Xu, K. Ueda, H. Yagi, T. Yanagitani, A.A. Kaminskii, A. Kudryashov: 72 W $Nd:Y_3Al_5O_{12}$ ceramic laser, Appl. Phys. Lett. **78**(23), 3586 (2001)

11.344 J.R. Lu, M. Prabhu, J.Q. Xu, K. Ueda, H. Yagi, T. Yanagitani, A.A. Kaminskii: Highly efficient 2% Nd:yttrium aluminum garnet ceramic laser, Appl. Phys. Lett. **77**(23), 3707 (2000)

11.345 I. Shoji, S. Kurimura, Y. Sato, T. Taira, A. Ikesue, K. Yoshida: Optical properties and laser characteristics of highly Nd^{3+}-doped $Y_3Al_5O_{12}$ ceramics, Appl. Phys. Lett. **77**(7), 939 (2000)

11.346 A. Ikesue, K. Yoshida, T. Yamamoto, I. Yamaga: Optical scattering centers in polycrystalline Nd:YAG laser, J. Am. Cer. Soc **80**(6), 1517 (1997)

11.347 Y.K. Bu, Q. Zheng, Q.H. Xue, Y.X. Cheng, L.S. Qian: Diode-pumped 593.5 nm CW yellow laser by type-1 CPM LBO intracavity sum-frequency-mixing, Opt. Laser Technol. **38**(8), 565 (2006)

11.348 T.K. Lake, A.J. Kemp, G.J. Friel, B.D. Sinclair: Compact and efficient single-frequency $Nd:YVO_4$ laser with variable longitudinal-mode discrimination, IEEE Photon. Technol. Lett. **17**(2), 417 (2005)

11.349 J. Liu, J. M. Yang, J. L. He: High repetition rate passively Q-switched diode-pumped Nd:YVO_4 laser, Opt. Laser Technol. **35**(6), 431 (2003)

11.350 J. C. Bermudez, A. V. Kir'yanov, V. J. Pinto-Robledo, M. J. Damzen: The influence of thermally induced effects on operation of a compact diode-side-pumped Nd:YVO_4 laser, Laser Phys. **13**(2), 255 (2003)

11.351 A. S. S. de Camargo, L. A. O. Nunes, D. R. Ardila, J. P. Andreeta: Excited-state absorption and 1064-nm end-pumped laser emission of Nd:YVO_4 single-crystal fiber grown by laser-heated pedestal growth, Opt. Lett. **29**(1), 59 (2004)

11.352 Y. P. Zhang, Y. Zheng, H. Y. Zhang, J. Q. Yao: A laser-diode-pumped 7.36 W continuous-wave Nd:YV04 laser at 1342 nm, Chin. Phys. Lett. **23**(2), 363 (2006)

11.353 Y. M. Wang, M. Lei, J. L. Li, F. M. Zeng, L. J. Zhang, J. H. Liu: Crystal growth and laser characteristics of Nd^{3+} : $KGd(WO_4)_2$, J. Rare Earths **23**(6), 676 (2005)

11.354 H. K. Kong, J. Y. Wang, H. J. Zhang, J. H. Liu, Y. T. Lin, X. F. Cheng, X. B. Hu, X. G. Xu, Z. S. Shao, M. H. Jiang: Growth and laser properties of Nd^{3+} doped $La_3Ga_{5.5}Ta_{0.5}O_{14}$ crystal, J. Cryst. Growth **263**(1–4), 344 (2004)

11.355 A. A. Kaminskii, S. N. Bagayev, K. Ueda, K. Takaichi, H. Yagi, T. Yanagitani: 5.5 J pyrotechnically pumped Nd^{3+} : $Y_3Al_5O_{12}$ cermaic laser, Laser Phys. Lett. **3**(3), 124 (2006)

11.356 J. H. Liu, H. J. Zhang, Z. P. Wang, J. Y. Wang, Z. S. Shao, M. H. Jiang, H. Weber: Continuous-wave and pulsed laser performance of Nd : $LuVO_4$ crystal, Opt. Lett. **29**(2), 168 (2004)

11.357 J. Y. Wang, H. J. Zhang, Z. P. Wang, W. W. Ge, J. X. Zhang, M. H. Jiang: Growth, properties and Raman shift laser in tungstate crystals, J. Crystal Growth **292**(2), 377 (2006)

11.358 G. A. Kumar, J. R. Lu, K. I. Ueda, H. Yagi, T. Yanagitani: Spectroscopic and stimulated emission characteristics of Nd^{3+} in transparent Y_2O_3 ceramics, IEEE J. Quantum Electron. **42**(7–8), 643 (2006)

11.359 J. Lu, K. Takaichi, T. Uematsu, A. Shirakawa, M. Musha, K. Ueda, H. Yagi, T. Yanagitani, A. A. Kaminskii: Promising ceramic laser material: Highly transparent Nd^{3+} : Lu_2O_3 ceramic, Appl. Phys. Lett. **81**(23), 4324 (2002)

11.360 M. Boucher, O. Musset, J. P. Boquillon, E. Georgiou: Multiwatt CW diode end-pumped Nd:YAP laser at 1.08 and 1.34 μm: Influence of Nd doping level, Opt. Commun. **212**(1–3), 139 (2002)

11.361 H. B. Peng, W. Hou, Y. H. Chen, D. F. Cui, Z. Y. Xu, C. Chen, F. D. Fan, Y. Zhu: 28W red light output at 659.5 nm by intracavity frequency doubling of a Nd:YAG laser using LBO, Opt. Express **14**(9), 3961 (2006)

11.362 J. H. Lu, J. R. Lu, T. Murai, K. Takaichi, T. Uematsu, J. Q. Xu, K. Ueda, H. Yagi, T. Yanagitani, A. A. Kaminskii: 36-W diode-pumped continuous-wave 1319 nm Nd:YAG ceramic laser, Opt. Lett. **27**(13), 1120 (2002)

11.363 F. Balembois, D. Boutard, E. Barnasson, M. Baudrier, R. Paries, C. Schwach, S. Forget: Efficient diode-pumped intracavity frequency-doubled CW Nd:YLF laser emitting in the red, Opt. Laser Technol. **38**(8), 626 (2006)

11.364 S. Z. Zhao, Q. P. Wang, X. Y. Zhang, S. T. Wang, L. Zhao, L. K. Sun, S. J. Zhang: Diode-laser-pumped 1.328 μm Nd : $Sr_5(PO_4)_3F$ laser and its intracavity frequency doubling, Appl. Opt. **36**(30), 7756 (1997)

11.365 J. Sulc, H. Jelinkova, K. Nejezchleb, V. Skoda: YAG/V:YAG microchip laser operating at 1338 nm, Laser Phys. Lett. **2**(11), 519 (2005)

11.366 C. Q. Wang, N. Hamelin, Y. T. Chow, X. L. Meng, Z. S. Shao: Low-threshold, high-efficiency, linearly polarized 1.338 μm Nd:S-VAP laser and its frequency doubling, J. Modern Opt. **45**(10), 2139 (1998)

11.367 N. Pavel, V. Lupei, T. Taira: 1.34 μm efficient laser emission in highly-doped Nd:YAG under 885 nm diode pumping, Opt. Express **13**(20), 7948 (2005)

11.368 C. Du, S. Ruan, H. Zhang, Y. Yu, F. Zeng, J. Wang, M. Jiang: A 13.3-W laser-diode-array end-pumped Nd : $GdVO_4$ continuous-wave laser at 1.34 μm, Appl. Phys. B **80**(1), 45 (2005)

11.369 L. J. Qin, X. L. Meng, J. G. Zhang, C. L. Du, L. Zhu, B. C. Xu: Growth and properties of Nd : $GdVO_4$ crystal, Opt. Mater. **23**(1–2), 455 (2003)

11.370 J. Liu, B. Ozygus, J. Erhard, A. Ding, H. Weber, X. Meng: Diode-pumped CW and Q-switched Nd : $GdVO_4$ laser operating at 1.34 μm, Opt. Quantum Electron. **35**(8), 811 (2003)

11.371 H. J. Zhang, C. L. Du, J. Y. Wang, X. B. Hu, X. A. Xu, C. M. Dong, J. H. Liu, H. K. Kong, H. D. Jiang, R. J. Han, Z. S. Shao, M. H. Jiang: Laser performance of Nd : $GdVO_4$ crystal at 1.34 μm and intracavity double red laser, J. Cryst. Growth **249**(3–4), 492 (2003)

11.372 C. L. Du, L. J. Qin, X. L. Meng, G. B. Xu, Z. P. Wang, X. G. Xu, L. Zhu, B. C. Xu, Z. S. Shao: High-power Nd : $GdVO_4$ laser at 1.34 μm end-pumped by laser-diode-array, Opt. Commun. **212**(1–3), 177 (2002)

11.373 H. J. Zhang, J. H. Liu, J. Y. Wang, X. G. Xu, M. H. Jiang: Continuous-wave laser performance of Nd : $LuVO_4$ crystal operating at 1.34 μm, Appl. Opt. **44**(34), 7439–7441 (2005)

11.374 H. Y. Shen, G. Zhang, C. H. Huang, R. R. Zeng, M. Wei: High power 1341.4 nm Nd:$YAlO_3$ CW laser and its performances, Opt. Laser Technol. **35**(2), 69 (2003)

11.375 G. Zhang, H. Y. Shen, R. R. Zeng, C. H. Huang, W. X. Lin, J. H. Huang: The study of 1341.4 nm Nd : $YAlO_3$ laser intracavity frequency doubling by LiB_3O_5, Opt. Commun. **183**(5–6), 461 (2000)

11.376 R. Zhou, X. Ding, W. Q. Wen, X. Q. Cai, P. Wang, J. Q. Yao: High-power continuous-wave diode-end-pumped intracavity frequency dou-

bled Nd : YVO_4 laser at 671 nm with a compact three-element cavity, Chinese Phys. Lett. **23**(4), 849 (2006)
11.377 R. Zhou, E.B. Li, B.G. Zhang, X. Ding, Z.Q. Cai, W.Q. Wen, P. Wang, J.Q. Yao: Simultaneous dual-wavelength CW operation using $^4F_{3/2}-^4I_{13/2}$ transitions in Nd:YVO_4 crystal, Opt. Commun. **260**(2), 641 (2006)
11.378 Y.P. Zhang, Y. Zheng, H.Y. Zhang, J.Q. Yao: A laser-diode-pumped 7.36 W continuous-wave Nd:YVO_4 laser at 1342 nm, Chinese Phys. Lett. **23**(2), 363 (2006)
11.379 H. Ogilvy, M.J. Withford, P. Dekker, J.A. Piper: Efficient diode double-end-pumped Nd:YVO_4 laser operating at 1342 nm, Opt. Express **11**(19), 2411 (2003)
11.380 A. Di Lieto, P. Minguzzi, A. Pirastu, V. Magni: High-power diffraction-limited Nd:YVO_4 continuous-wave lasers at 1.34 µm, IEEE J. Quantum Electron. **39**(7), 903 (2003)
11.381 A. Di Lieto, P. Minguzzi, A. Pirastu, S. Sanguinetti, V. Magni: A 7 W diode-pumped Nd:YVO_4 CW laser at 1.34 µm, Appl. Phys. B **75**(4–5), 463 (2002)
11.382 J.L. He, G.Z. Luo, H.T. Wang, S.N. Zhu, Y.Y. Zhu, Y.B. Chen, N.B. Ming: Generation of 840 mW of red light by frequency doubling a diode-pumped 1342 nm Nd : YVO_4 laser with periodically-poled $LiTaO_3$, Appl. Phys. B **74**(6), 537 (2002)
11.383 A. Sennaroglu: Efficient continuous-wave operation of a diode-pumped Nd:YVO_4 laser at 1342 nm, Opt. Commun. **164**(4–6), 191 (1999)
11.384 S.A. Zolotovskaya, V.G. Savitski, M.S. Gaponenko, A.M. Malyarevich, K.V. Yumashev, M.I. Demchuk, H. Raaben, A.A. Zhilin, K. Nejezchleb: $KGd(WO_4)_2$ laser at 1.35 µm passively Q-switched with V^{3+} : YAG crystal and PbS-doped glass, Opt. Mater. **28**(8–9), 919 (2006)
11.385 A.S. Grabtchikov, A.N. Kuzmin, V.A. Lisinetskii, V.A. Orlovich, A.A. Demidovich, K.V. Yumashev, N.V. Kuleshov, H.J. Eichler, M.B. Danailov: Passively Q-switched 1.35 µm diode pumped Nd:KGW laser with V:YAG saturable absorber, Opt. Mater. **16**(3), 349 (2001)
11.386 R. Zhou, B.G. Zhang, X. Ding, Z.Q. Cai, W.Q. Wen, P. Wang, J.Q. Yao: Continuous-wave operation at 1386 nm in a diode-end-pumped Nd:YVO_4 laser, Opt. Express **13**(15), 5818 (2005)
11.387 H.M. Kretschmann, F. Heine, V.G. Ostroumov, G. Huber: High-power diode-pumped continuous-wave Nd^{3+} lasers at wavelengths near 1.44 µm, Opt. Lett. **22**, 466 (1997)
11.388 M.J.F. Digonnet (Ed.): *Rare-Earth-Doped Fiber Lasers and Amplifiers*, 2 edn. (Marcel Dekker, New York 2001)
11.389 B.J. Ainslie, S.P. Craig, S.T. Davey: The absorption and fluorescence spectra of rare earth ions in silica- based monomode fiber, J. Lightwave Technol. **6**, 287 (1988)
11.390 S.D. Jackson, T.A. King: CW operation of a 1.064 µm pumped Tm-Ho-doped silica fiber laser, IEEE. J. Quantum Electron. **34**, 1578 (1998)
11.391 O. Humbach, H. Fabian, U. Grzesik, U. Haken, W. Heitmann: Analysis of OH absorption bands in synthetic silica, J. Non-Crystal Solids **203**, 19 (1996)
11.392 D.C. Tran, G.H. Sigel Jr., B. Bendow: Heavy metal fluoride glasses and fibers: a review, J. Lightwave Technol. **2**, 566 (1984)
11.393 P.W. France, M.G. Drexhage, J.M. Parker, M.W. Moore, S.F. Carter, J.V. Wright: *Fluoride Glass Optical Fibers* (Blackie, Glasgow, London 1990)
11.394 S.T. Davey, P.W. France: Rare earth doped fluorozirconate glasses for fiber devices, BT J. Technol. **7**, 58 (1989)
11.395 M. Monerie, F. Alard, G. Maze: Fabrication and characterisation of fluoride-glass single-mode fibers, Electron. Lett. **21**, 1179 (1985)
11.396 L. Wetenkamp, G.F. West, H. Többen: Optical properties of rare earth-doped ZBLAN glasses, J. Non-Crystal Solids **140**, 35 (1992)
11.397 L. Wetenkamp: Charakterisierung von laseraktiv dotierten Schwermetallfluorid-Gläsern und Faserlasern Ph.D. Thesis (Technical University of Braunschweig, Braunschweig 1991)
11.398 P.N. Kumta, S.H. Risbud: Rare-earth chalcogenides – an emerging class of optical materials, J. Mater. Sci. **29**, 1135 (1994)
11.399 J.S. Sanghera, J. Heo, J.D. Mackenzie: Chalcohalide glasses, J. Non-Crystal Solids **103**, 155 (1988)
11.400 L.B. Shaw, B. Cole, P.A. Thielen, J.S. Sanghera, I.D. Aggarwal: Mid-wave IR and long-wave IR laser potential of rare-earth doped chalcogenide glass fiber, IEEE. J. Quantum Electron. **48**, 1127 (2001)
11.401 Y.D. West, T. Schweizer, D.J. Brady, D.W. Hewak: Gallium lanthanum sulphide fibers for infrared transmission, Fiber Integrated Opt. **19**, 229 (2000)
11.402 J. Heo, Y.B. Shin: Absorption and mid-infrared emission spectroscopy of Dy^{3+} in Ge-As (or Ga)-S glasses, J. Non-Crystal Solids **196**, 162 (1996)
11.403 R. Reisfeld, M. Eyal: Possible ways of relaxations for excited states of rare earth ions in amorphous media, J. Phys. **46**, C349 (1985)
11.404 D.C. Brown, H.J. Hoffman: Thermal, stress, and thermo-optic effects in high average power double-clad silica fiber lasers, IEEE J. Quantum Electron. **37**, 207 (2001)
11.405 S.M. Lima, T. Catunda, R. Lebullenger, A.C. Hernandes, M.L. Baesso, A.C. Bento, L.C.M. Miranda: Temperature dependence of thermo-optical properties of fluoride glasses determined by thermal lens spectrometry, Phys. Rev. B **60**, 15173 (1999)
11.406 T. Schweizer: Rare-earth-doped gallium lanthanum sulphide glasses for mid-infrared fiber lasers Ph.D. Thesis (University of Hamburg, Hamburg 1998)

11.407 S. M. Lima, T. Catunda, M. L. Baesso, L. D. Vila, Y. Messaddeq, E. B. Stucchi, S. J. L. Ribeiro: Thermal-optical properties of Ga:La:S glasses measured by thermal lens technique, J. Non-Crystal Solids **247**, 222 (1999)

11.408 H. Po, J. D. Cao, B. M. Laliberte, R. A. Minns, R. F. Robinson, B. H. Rockney, R. R. Tricca, Y. H. Zhang: High power neodymium-doped single transverse mode fiber laser, Electron. Lett. **29**, 1500 (1993)

11.409 I. N. Duling III, W. K. Burns, L. Goldberg: High-power superfluorescent fiber source, Opt. Lett. **15**, 33 (1990)

11.410 J. D. Minelly, W. L. Barnes, R. I. Laming, P. R. Morkel, J. E. Townsend, S. G. Grubb, D. N. Payne: Diode-array pumping of Er^{3+}/Yb^{3+} co-doped fiber lasers and amplifiers, IEEE Photon. Technol. Lett. **5**, 301 (1993)

11.411 H. M. Pask, J. L. Archambault, D. C. Hanna, L. Reekie, P. S. J. Russell, J. E. Townsend, A. C. Tropper: Operation of cladding-pumped Yb^{3+}-doped silica fiber lasers in 1 μm region, Electron. Lett. **30**, 863 (1994)

11.412 J. Limpert, T. Schreiber, S. Nolte, H. Zellmer, A. Tünnermann, R. Iliew, F. Lederer, J. Broeng, G. Vienne, A. Petersson, C. Jakobsen: High-power air-clad large-mode-area photonic crystal fiber laser, Opt. Express **11**, 818 (2003)

11.413 M. H. Muendel: Optimal inner cladding shapes for double-clad fiber lasers. In: *Conf. Lasers Electro-Optics*, OSA Tech. Dig. Ser., Vol. 9 (Opt. Soc. Am., Washington 1996) p. 209

11.414 A. Liu, K. Ueda: The absorption characteristics of circular, offset, and rectangular double-clad fibers, Opt. Commun. **132**, 511 (1996)

11.415 K. Furusawa, A. Malinowski, J. H. V. Price, T. M. Monro, J. K. Sahu, J. Nilsson, D. J. Richardson: Cladding pumped ytterbium-doped fiber laser with holey inner and outer cladding, Opt. Express **9**, 714 (2001)

11.416 M. D. Nielsen, J. R. Folkenberg, N. A. Mortensen: Singlemode photonic crystal fiber with effective area of 600 μm^2 and low bending loss, Electron. Lett. **39**, 1802 (2003)

11.417 K. Petermann, D. Fagundes-Peters, J. Johannsen, M. Mond, V. Peeters, J. J. Romero, S. Kutovoi, J. Speiser, A. Giesen: Highly Yb-doped oxides for thin-disc lasers, J. Crystal Growth **275**, 135 (2005)

11.418 L. D. DeLoach, S. A. Payne, L. L. Chase, L. K. Smith, W. L. Kway, W. F. Krupke: Evaluation of absorption and emission properties of Yb^{3+} doped crystals for laser applications, IEEE J. Quantum Electron. **29**, 1179 (1993)

11.419 D. S. Sumida, T. Y. Fan, R. Hutcheson: Spectroscopy and diode-pumped lasing of Yb^{3+}-doped $Lu_3Al_5O_{12}$ (Yb:LuAG). In: *OSA Proc. Adv. Solid-State Lasers*, Vol. 24, ed. by B. H. T. Chai, S. A. Payne (Opt. Soc. Am., Washington 1995) p. 348

11.420 D. S. Sumida, T. Y. Fan: Effect of radiation trapping on fluorescence lifetime and emission cross-section measurements in solid-state laser media, Opt. Lett. **19**, 1343 (1994)

11.421 M. Musha, J. Miura, K. Nakagawa, K. Ueda: Developments of a fiber-MOPA system for the light source of the gravitational wave antenna, Classical and Quantum Gravity **23**(8), S287 (2006)

11.422 I. A. Bufetov, M. M. Bubnov, M. A. Melkumov, V. V. Dudin, A. V. Shubin, S. L. Semenov, K. S. Kravtsov, A. N. Gur'yanov, M. V. Yashkov, E. M. Dianov: Yb-, Er-Yb- and Nd-doped fiber lasers based on multi-element first cladding fibers, Quantum Electron. **35**(4), 328 (2005)

11.423 L. Lombard, A. Brignon, J. P. Huignard, E. Lallier, G. Lucas-Leclin, P. Georges, G. Pauliat, G. Roosen: High power multimode fiber amplifier with wavefront reshaping for high beam quality recovery, Comptes Rendus Physique **7**(2), 233 (2006)

11.424 J. Kim, P. Dupriez, C. Codemard, J. Nilsson, J. K. Sahu: Suppression of stimulated Raman scattering in a high power Yb-doped fiber amplifier using a W-type core with fundamental mode cut-off, Opt. Express **14**(12), 5103 (2006)

11.425 P. Wang, L. J. Cooper, J. K. Sahu, W. A. Clarkson: Efficient single-mode operation of a cladding-pumped ytterbium-doped helical-core fiber laser, Opt. Lett. **31**(2), 226 (2006)

11.426 H. Ohashi, X. Gao, M. Saito, H. Okamoto, M. Takasaka, K. Shinoda: Beam-shaping technique for end-pumping Yb-doped fiber laser with two laser-diode arrays, Jpn. J. Appl. Phys. Part 2 – Letters & Express Letters **44**(16–19), L555 (2005)

11.427 V. Dominic, S. MacCormack, R. Waarts, S. Sanders, S. Bicknese, R. Dohle, E. Wolak, P. S. Yeh, E. Zucker: 110 W fiber laser, Electron. Lett. **35**(14), 1158 (1999)

11.428 J. Limpert, A. Liem, S. Höfer, H. Zellmer, A. Tünnermann, S. Unger, S. Jetschke, H.-R. Müller: 150 W Nd/Yb codoped fiber laser at 1.1 μm. In: *Conf. Lasers Electro-Optics*, OSA Tech. Dig. (Opt. Soc. Am., Washington 2002) pp. 590–591

11.429 C. H. Liu, B. Ehlers, F. Doerfel, S. Heinemann, A. Carter, K. Tankala, J. Farroni, A. Galvanauskas: 810 W continuous-wave and singletransverse-mode fiber laser using 201 μm core Yb-doped double-clad fiber, Electron. Lett. **40**(23), 1471 (2004)

11.430 Y. Jeong, J. K. Sahu, R. B. Williams, D. J. Richardson, K. Furusawa, J. Nilsson: Ytterbium-doped large-core fiber laser with 272 W output power, Electron. Lett. **39**(13), 977 (2003)

11.431 P. Dupriez, A. Piper, A. Malinowski, J. K. Sahu, M. Ibsen, B. C. Thomsen, Y. Jeong, L. M. B. Hickey, M. N. Zervas, J. Nilsson, D. J. Richardson: High average power, high repetition ratepicosecond pulsed fiber master oscillator power amplifier source seeded by a gain-switched laser diode at 1060 nm, IEEE Photon. Technol. Lett. **18**(9–12), 1013 (2006)

11.432 L.J. Cooper, P. Wang, R.B. Williams, J.K. Salm, W.A. Clarkson, A.M. Scott, D. Jones: High-power Yb-doped multicore ribbon fiber laser, Opt. Lett. **30**(21), 2906 (2005)

11.433 A.P. Liu, M.A. Norsen, R.D. Mead: 60 W green output by frequency doubling of a polarized Yb-doped fiber laser, Opt. Lett. **30**(1), 67 (2005)

11.434 Y. Jeong, J.K. Sahu, D.N. Payne, J. Nilsson: Ytterbium-doped large-core fiber laser with 1.36 kW continuous-wave output power, Opt. Express **12**, 6088 (2004)

11.435 J. Limpert, S. Hoefer, A. Liem, H. Zellmer, A. Tuennermann, S. Knoke, H. Voelckel: 100 W average-power, high-energy nanosecond fiber amplifier, Appl. Phys. B **75**, 477–479 (2002)

11.436 S.A. Payne, L.L. Chase, L.-K. Smith, L. Kway, W.F. Krupke: Infrared cross-section measurements for crystals doped with Er^{3+}, Tm^{3+}, and Ho^{3+}, IEEE J. Quantum Electron. **28**, 2619 (1992)and references therein

11.437 I. Sokólska, E. Heumann, S. Kück: Laser oscillation of Er^{3+}:YVO_4 crystals in the spectral range around 1.6 μm, Appl. Phys. B **71**, 893 (2000)

11.438 C. Li, C. Wyon, R. Moncorgé: Spectroscopic properties and fluorescence dynamics of Er^{3+} and Yb^{3+} in Y_2SiO_5, IEEE J. Quantum Electron. **28**, 1209 (1992)

11.439 I. Sokólska: Spectroscopic characterization of $LaGaO_3$:Er^{3+} crystals, Appl. Phys.B **71**, 157 (2000)

11.440 K.I. Schaffers, L.D. DeLoach, S.A. Payne: Crystal growth, frequency doubling, and infrared laser performance of Yb^{3+} : $BaCaBO_3F$, IEEE J. Quantum Electron. **32**, 741 (1996)

11.441 J. Du, X.Y. Liang, Y.G. Wang, L.B. Su, W.W. Feng, E.W. Dai, Z.Z. Xu, J. Xu: 1 ps passively mode-locked laser operation of Na,Yb : CaF_2 crystal, Opt. Express **13**, 7970 (2005)

11.442 M. Robinson, C.K. Asawa: Stimulated emission from Nd^{3+} and Yb^{3+} in noncubic sites of neodymium- and ytterbium-doped CaF_2, J. App. Phys. **38**, 4495 (1967)

11.443 V. Petit, J.L. Camy, R. Moncorge: CW and tunable laser operation of Yb^{3+} in Nd : Yb : CaF_2, Appl. Phys. Lett. **88**(5), 051111 (2006)

11.444 H.J. Zhang, X.L. Meng, P. Wang, L. Zhu, X.S. Liu, X.M. Liu, Y.Y. Yang, R.H. Wang, J. Dawes, J. Piper, S.J. Zhang, L. Sun: Growth of Yb-doped $Ca_4GdO(BO_3)_3$ crystals and their spectra and laser properties, J. Crystal Growth **222**, 209 (2001)

11.445 F. Mougel, K. Dardenne, G. Aka, A. Kahn-Harari, D. Vivien: An efficient infrared laser and self-frequency doubling crystal, J. Opt. Soc. Am. B **16**, 164 (1999)

11.446 F. Auge, F. Balembois, P. Georges, A. Brun, F. Mougel, G. Aka, A. Kahn-Harari, D. Vivien: Efficient and tunable continuous-wave diode-pumped Yb^{3+} : $Ca_4GdO(BO_3)_3$ laser, Appl. Opt. **38**, 976 (1999)

11.447 F. Auge, F. Druon, F. Balembois, P. Georges, A. Brun, F. Mougel, G. Aka, D. Vivien: Theoretical and experimental investigations of a diode-pumped quasi-three-level laser: The Yb^{3+}-doped $Ca_4GdO(BO_3)_3$ (Yb:GdCOB) laser, IEEE J. Quantum Electron. **36**, 598 (2000)

11.448 H. Yang, Z.W. Zhao, J. Zhang, P.Z. Deng, J. Xu, Z.Y. Wei, J. Zhang: Continuous-wave laser oscillation of Yb:FAP crystals at a wavelength of 1043 nm, Chinese Phys. **10**, 1136 (2001)

11.449 S.A. Payne, L.K. Smith, L.D. DeLoach, W.L. Kway, J.B. Tassano, W.F. Krupke: Laser optical and thermomechanical properties of Yb-doped fluoroapatite, IEEE J. Quantum Electron. **30**, 170 (1994)

11.450 S.A. Payne, L.K. Smith, L.D. DeLoach, W.L. Kway, J.B. Tassano, W.F. Krupke, B.H.T. Chai, G. Loutts: Ytterbium-doped apatite-structure crystals: A new class of laser materials, J. Appl. Phys. **75**, 497 (1994)

11.451 L. Shah, Q. Ye, J.M. Eichenholz, D.A. Hammons, M. Richardson, B.H.T. Chai, R.E. Peale: Laser tunability in Yb^{3+} : $YCa_4O(BO_3)_3$ {Yb:YCOB}, Opt. Commun. **167**, 149 (1999)

11.452 H.J. Zhang, X.L. Meng, L. Zhu, P. Wang, X.S. Liu, J. Dawes, P. Dekker, R.P. Cheng, S.J. Zhang, L.K. Sun: Growth, Stark energy level and laser properties of Yb : $Ca_4YO(BO_3)_3$ crystal, Mater. Res. Bull. **35**, 799 (2000)

11.453 A. Aron, G. Aka, B. Viana, A. Kahn-Harari, D. Vivien, F. Druon, F. Balembois, P. Georges, A. Brun, N. Lenain, M. Jacquet: Spectroscopic properties and laser performances of Yb:YCOB and potential of the Yb:LaCOB material, Opt. Mater. **16**, 181 (2001)

11.454 D.A. Hammons, J.M. Eichenholz, Q. Ye, B.H.T. Chai, L. Shah, R.E. Peale, M. Richardson, H. Qiu: Laser action in Yb^{3+} : YCOB (Yb^{3+} : $YCa_4OBO_3)_3$, Opt. Comm. **156**, 327 (1998)

11.455 X.Y. Zhang, A. Brenier, Q.P. Wang, Z.P. Wang, J. Chang, P. Li, S.J. Zhang, S.H. Ding, S.T. Li: Passive *Q*-switching characteristics of Yb^{3+} : $Gd_3Ga_5O_{12}$ crystal, Opt. Express **13**, 7708 (2005)

11.456 G.A. Bogomolova, D.N. Vylegzhanin, A.A. Kaminskii: Spectral and lasing investigations of garnets with Yb^{3+}ions, Sov. Phys. JETP **42**, 440 (1976)

11.457 K.S. Bagdasarov, A.A. Kaminskii, A.M. Kevorkov, A.M. Prokhorov: Rare earth scandium-aluminum garnets with impurity of TR^{3+} ions as active media for solid state lasers, Sov. Phys. Dokl. **19**, 671 (1975)

11.458 J. Du, X.Y. Liang, Y. Xu, R.X. Li, Z.Z. Xu, C.F. Yan, G.J. Zhao, L.B. Su, J. Xu: Tunable and efficient diode-pumped Yb^{3+}:GYSO laser, Opt. Express **14**, 3333 (2006)

11.459 W.X. Li, H.F. Pan, .L. Ding, H.P. Zeng, W. Lu, G.J. Zhao, C.F. Yan, L.B. Su, J. Xu: Efficient diode-pumped Yb : Gd_2SiO_5, Appl. Phys. Lett. **88**, 221117 (2006)

11.460 C.F. Yan, G.J. Zhao, L.H. Zhang, J. Xu, X.Y. Liang, D. Juan, W.X. Li, H.F. Pan, L.G. Ding, H.P. Zeng:

A new Yb-doped oxyorthosilicate laser crystal: $Yb:Gd_2SiO_5$, Solid State Commun. **137**, 451 (2006)

11.461 W.X. Li, H.F. Pan, L.E. Ding, H.P. Zeng, G.J. Zhao, C.F. Yan, L.B. Su, J. Xu: Diode-pumped continuous-wave and passively mode-locked Yb:GSO laser, Opt. Express **14**, 686 (2006)

11.462 Y.H. Xue, C.Y. Wang, Q.W. Liu, Y.F. Li, L. Chai, C.F. Yan, G.J. Zhao, L.B. Su, X.D. Xu, J. Xu: Characterization of diode-pumped laser operation of a novel Yb:GSO crystal, IEEE J. Quantum Electron. **42**, 517 (2006)

11.463 Y.H. Xue, Q.Y. Wang, L. Chai, Q.W. Liu, G.J. Zhao, L.B. Su, X.D. Xu, J. Xu: A novel LD pumped Yb:GSO laser operating at 1090 nm with low threshold, Acta Physica Sinica **55**, 456 (2006)

11.464 J. Petit, B. Viana, P. Goldner, D. Vivien, P. Louiseau, B. Ferrand: Laser oscillation with low quantum defect in $Yb:GdVO_4$, a crystal with high thermal conductivity, Opt. Lett. **29**, 833 (2004)

11.465 J. Petit, B. Viana, P. Goldner, D. Vivien, P. Louiseau, B. Ferrand: Laser osciallation with low quantum defect in $Yb:GdVO_4$, a crystal with high thermal conductivity, Opt. Lett. **29**, 833 (2004)

11.466 J.H. Liu, X. Mateos, H.J. Zhang, J.Y. Wang, M.H. Jiang, U. Griebner, V. Petrov: Characteristics of a continuous-wave $Yb:GdVO_4$ laser end pumped by a high-power diode, Opt. Lett. **31**, 2580 (2006)

11.467 A.A. Lagatsky, N.V. Kuleshov, V.P. Mikhailov: Diode-pumped CW lasing of Yb:KYW and Yb:KGW, Opt. Commun. **165**, 71 (1999)

11.468 J.E. Hellström, S. Bjurshagen, V. Pasiskevicius, J. Liu, V. Petrov, U. Griebner: Efficient Yb:KGW lasers end-pumped by high-power diode bars, Appl. Phys. B **83**, 235 (2006)

11.469 G. Paunescu, J. Hein, R. Sauerbrey: 100 fs diode-pumped Yb:KGW mode-locked laser, Appl. Phys. B **79**, 555 (2004)

11.470 U. Griebner, J.H. Liu, S. Rivier, A. Aznar, R. Grunwald, R.M. Sole, M. Aguilo, F. Diaz, V. Petrov: Laser operation of epitaxially grown $Yb:KLu(WO_4)_2-KLu(WO_4)_2$ composites with monoclinic crystalline structure, IEEE J. Quantum Electron. **41**(3), 408 (2005)

11.471 X. Mateos, R. Solé, J. Gavaldá, M. Aguiló, J. Massons, F. Díaz, V. Petrov, U. Griebner: Crystal growth, spectroscopic studies and laser operation of Yb^{3+}-doped potassium lutetium tungstates, Opt. Mater. **28**, 519–523 (2006)

11.472 A. Aznar, R. Sole, M. Aguilo, F. Diaz, U. Griebner, R. Grunwald, V. Petrov: Growth, optical characterization, and laser operation of epitaxial $Yb:KY(WO_4)_2/KY(WO_4)_2$ composites with monoclinic structure, Appl. Phys. Lett. **85**, 4313 (2004)

11.473 A.A. Lagatsky, E.U. Rafailov, A.R. Sarmani, C.T.A. Brown, W. Sibbett, L.. Smith. Ming: Efficient femtosecond green-light source with a diode-pumped mode-locked $Yb^{3+}:KY(WO_4)_2$ laser, Opt. Lett. **30**, 1144 (2005)

11.474 F. Brunner, T. Südmeyer, E. Innerhofer, F. Morier-Genoud, R. Paschotta: 240 fs pulses with 22 W average power from a mode-locked thin-disk $Yb:KY(WO_4)_2$ laser, Opt. Lett. **27**, 13 (2002)

11.475 M. Larionov, J. Gao, S. Erhard, A. Giesen, K. Contag, V. Peters, E. Mix, L. Fornasiero, K. Petermann, G. Huber, J. Aus der Au, G.J. Spühler, F. Brunner, R. Paschotta, U. Keller, A.A. Lagatsky, A. Abdolvand, N.V. Kuleshov: Thin disk laser operation and spectroscopy characterization of Yb-doped Sesquioxides and Potassium Tungstates. In: *Adv. Solid-State Lasers*, OSA Trends Opt. Photon., Vol. 50, ed. by C. Marshall (Opt. Soc. Am., Washington 2001) p. 625

11.476 M. Hildebrandt, U. Bünting, U. Kosch, D. Haussmann, T. Levy, M. Krause, O. Müller, U. Bartuch, W. Viöl: Diode-pumped Yb:KYW thin-disk laser operation with wavelength tuning to small quantum defects, Opt. Commun. **259**, 796–798 (2006)

11.477 P. Klopp, V. Petrov, U. Griebner, V. Nesterenko, V. Nikolov, M. Marinov, M.A. Bursukova, M. Galan: Continuous-wave lasing of a stoichiometric Yb laser material: $KYb(WO_4)_2$, Opt. Lett. **28**, 322 (2003)

11.478 M.C. Pujol, M.A. Bursukova, F. Güell, X. Mateos, R. Solé, J. Gavaldá, M. Aguiló, J. Massons, F. Díaz: Growth, optical characterization, and laser operation of a stoichiometric crystal $KYb(WO_4)_2$, Phys. Rev. B **65**, 165121 (2002)

11.479 J.J. Romero, J. Johannsen, M. Mond, K. Petermann, G. Huber, E. Heumann: Continuous-wave laser action of Yb^{3+}-doped lanthanum scandium borate, Appl. Phys. B **80**, 159 (2005)

11.480 C. Kränkel, J. Johannsen, M. Mond, K. Petermann, G. Huber: High Power $Yb:LaSc_3(BO_3)_4$ Thin Disk Laser. In: *Adv. Solid State Photon.* (Opt. Soc. Am., Washington 2006) paper WD2

11.481 M. Rico, U. Griebner, V. Petrov, P. Ortega, X.M. Han, C. Cascales, C. Zaldo: Growth, spectroscopy, and tunable laser operation of the disordered crystal $LiGd(MoO_4)_2$ doped with ytterbium, J. Opt. Soc. Am. B **23**, 1083 (2006)

11.482 J.K. Jones, J.P. de Sandra, M. Hempstead, D.P. Shepherd, A.C. Large, A.C. Tropper, J.S. Wilkinson: Channel waveguide laser at 1 μm in Yb-diffused $LiNbO_3$, Opt. Lett. **20**, 1477 (1995)

11.483 M.O. Ramirez, D. Jaque, J.A. Sanz Garcia, L.E. Bausa, J.E. Munoz Santiuste: 74% slope efficiency from a diode-pumped $Yb^{3+}:LiNbO_3:MgO$ laser crystal, Appl. Phys. B **77**, 621 (2003)

11.484 J. Sablayrolles, V. Jubera, J.P. Chaminade, I. Manek-Hönninger, S. Murugan, T. Cardinal, R. Olazcuaga, A. Garcia, F. Salin: Crystal growth, luminescent and lasing properties of the ytterbium doped $Li_6Y(BO_3)_3$ compound, Opt. Mater. **27**, 1681–1685 (2005)

11.485 S.S. Sumida, T.Y. Fan, R. Hutcheson: Spectroscopy and diode-pumped lasing of Yb^{3+}-doped $Lu_3Al_5O_{12}$ (Yb:LuAG). In: *OSA Proc. Adv. Solid State Lasers*, Vol. 24, ed. by B.H.T. Chai, S.A. Payne (Opt. Soc. Am., Washington 1995) p. 348

11.486 U. Griebner, V. Petrov, K. Petermann, V. Peters: Passively mode-locked Yb : Lu_2O_3 laser, Opt. Express **12**, 3125 (2004)

11.487 K. Takaichi, H. Yagi, A. Shirakawa, K. Ueda, S. Hosokawa, T. Yanagitani, A.A. Kaminskii: Lu_2O_3:Yb^{3+} ceramics – a novel gain material for ghigh-power solid-state lasers, phys. stat. sol. (a) **1**, R1 (2005)

11.488 J.H. Liu, X. Mateos, H.J. Zhang, J.Y. Wang, M.H. Jiang, U. Griebner, V. Petrov: Continuous-wave laser operation of Yb : $LuVO_4$, Opt. Lett. **30**, 3162 (2005)

11.489 M. Rico, J. Liu, U. Griebner, V. Petrov: Tunable laser operation of ytterbium in disordered single crystals of Yb : $NaGd(WO_4)_2$, Opt. Express **12**, 22 (2004)

11.490 R. Peters, J. Johannsen, M. Mond, K. Petermann, G. Huber: Yb : $NaGd(WO_4)_2$: Spectroscopic characterisation and laser demonstration, Conference on Lasers and Electro-Optics Europe, Munich 2005 (European Physical Society, Mulhouse 2005) CA9-4-TuE

11.491 M. Rico, J. Liu, J.M. Cano-Torres, A. Garcia-Cortes, C. Cascales, C. Zaldo, U. Griebner, V. Petrov: Continuous wave and tunable laser operation of Yb^{3+} in disordered $NaLa(MoO_4)_2$, Appl. Phys. B **81**, 621 (2005)

11.492 A.V. Mandrik, A.E. Troshin, V.E. Kisel, A.S. Yasukevich, G.N. Klavsut, N.V. Kuleshov, A.A. Pavlyuk: CW and *Q*-switched diode-pumpedlaser operation of Yb^{3+} : $NaLa(MoO_4)_2$, Appl. Phys. B **81**, 1119 (2005)

11.493 J. Liu, J.M. Cano-Torres, E.-B. Cascales, F. Esteban-Betegón, M.D. Serrano, V. Volkov, C. Zaldo, M. Rico, U. Griebner, V. Petrov: Growth and continuous-wave laser operation of disordered crystals of Yb^{3+} : $NaLa(WO_4)_2$ and Yb^{3+} : $NaLa(MoO_4)_2$, Phys. Stat. Sol. (a) **202**(4), R29–R31 (2005)

11.494 V. Peters, E. Mix, L. Fornasiero, K. Petermann, G. Huber, S.A. Basun: Efficient laser operation of Yb^{3+} : Sc_2O_3 and spectroscopic characterization of Pr^{3+} in cubic sesquioxides, Laser Phys. **10**, 417 (2000)

11.495 J. Liu, M. Rico, U. Griebner, V. Petrov, V. Peters, K. Petermann, G. Huber: Efficient room temperature continuous-wave operation of an Yb^{3+} : Sc_2O_3 crystal laser at 1041.6 and 1094.6 nm, Physica Status Solidi A **202**, R19 (2005)

11.496 J. Lu, J.F. Bisson, K. Takaichi, T. Uematsu, A. Shirakawa, M. Musha, K. Ueda, H. Yagi, T. Yanagitani, A.A. Kaminskii: Yb^{3+} : Sc_2O_3 ceramic laser, Appl. Phys. Lett. **83**(6), 1101 (2003)

11.497 A.J. Bayramian, C. Bibeau, R.J. Beach, C.D. Marshall, S.A. Payne, W.F. Krupke: Three-level *Q*-switched laser operation of ytterbium-doped $Sr_5(PO_4)_3F$ at 985 nm, Opt. Lett. **25**, 622 (2000)

11.498 S. Yiou, F. Balembois, P. Georges: Numerical modeling of a continous-wave Yb-doped bulk crystal laser emitting on a three-level laser transition near 980 nm, J. Opt. Soc. Am B **22**, 572 (2005)

11.499 S. Yiou, F. Balembois, K. Schaffers, P. Georges: Efficient laser operation of an Yb:S-FAP crystal at 985 nm, Appl. Opt. **42**, 4883 (2003)

11.500 A.J. Bayramian, C.D. Marshall, K.I. Schaffers, S.A. Payne: Characterization of Yb^{3+}: $Sr_{5-x}Ba_x(PO_4)_3F$ crystals for diode-pumped lasers, IEEE J. Quantum Electron. **35**, 665 (1999)

11.501 L.D. DeLoach, S.A. Payne, L.K. Smith, W.L. Kway, W.F. Krupke: Laser and spectroscopic properties of $Sr_5(PO_4)_3$: Yb, J. Opt. Soc. Am. B **11**, 269 (1994)

11.502 C.D. Marshall, L.K. Smith, R.J. Beach, M.A. Emanuel, K.I. Schaffers, J. Skidmore, S.A. Payne, B.H.T. Chai: Diode-pumped ytterbium-doped $Sr_5(PO_4)_3F$ laser performance, IEEE J. Quantum Electron. **32**, 650 (1996)

11.503 R. Gaume, B. Viana, D. Vivien, J.P. Roger, D. Forunier, J.P. Souron, G. Wallez, S. Chenais, F. Balembois, P. Georges: Mechanical, thermal and laser properties of Yb : $(Sr_{1-x}Ca_x)_3Y(BO_3)_3$ (Yb :CaBOYS) for 1 μm laser applications, Opt. Mater. **24**, 385 (2003)

11.504 F. Druon, S. Chenais, P. Raybaut, F. Balembois, P. Georges, R. Gaume, P.H. Haumesser, B. Viana, D. Vivien, S. Dhellemmes, V. Ortiz, C. Larat: Apatite-structure crystal, Yb^{3+} : $SrY_4(SiO_4)_3O$, for the development of diode-pumped femtosecond lasers, Opt. Lett. **27**, 1914 (2002)

11.505 R. Allen, L. Esterowitz: CW tunable ytterbium YAG laser-pumped by titanium sapphire, Electron. Lett. **31**, 639 (1995)

11.506 T. Taira, J. Saikawa, T. Kobayashi, R.L. Byer: Diode-pumped tunable Yb:YAG miniature lasers at room temperature: Modeling and experiment, IEEE J. Sel. Top. Quantum Electron. **3**(1), 100 (1997)

11.507 H.W. Bruesselbach, D.S. Sumida, R.A. Reeder, R.W. Byren: Low-heat high-power scaling using InGaAs-diode-pumped Yb:YAG lasers, IEEE J. Sel. Top. Quantum Electron. **3**(1), 105 (1997)

11.508 D.C. Hanna, J.K. Jones, A.C. Large, D.P. Shepherd, A.C. Tropper, P.J. Chandler, M.J. Rodman, P.D. Townsend, L. Zhang: Quasi-three level 1.03 μm laser operation of a planar ion-implanted Yb:YAG waveguide, Opt. Commun. **99**, 211 (1993)

11.509 T. Taira, W.M. Tulloch, R.L. Byer: Modeling of quasi-three-level lasers and operation of CW Yb:YAG, Appl. Opt. **36**, 1867 (1997)

11.510 T. Taira, J. Saikawa, T. Kobayashi, R.L. Byer: Diode-pumped tunable Yb:YAG miniature lasers at room temperature: Modeling and experiment, IEEE J. Sel. Top. Quantum Electron. **3**, 100 (1997)

11.511 J. Dong, A. Shirakawa, S. Huang, Y. Feng, K. Takaichi, M. Musha, K. Ueda, A.A. Kaminskii: Stable laser-diode pumped microchip subnanosecond Cr,Yb:YAG self-*Q*-switched laser, Laser Phys. Lett. **2**, 387 (2005)

11.512 O. A. Buryy, S. B. Ubiszkii, S. S. Melnyk, A. O. Matkovskii: The *Q*-switched Nd:YAG and Yb:YAG microchip lasers optimization and comparative analysis, Appl. Phys. B **78**, 291 (2004)

11.513 J. Aus der Au, G. J. Spühler, T. Südmeyer, R. Paschotta, R. Hövel, M. Moser, S. Erhard, M. Karszewski, A. Giesen, U. Keller: 16.2 W average power from a diode-pumped femtosecond Yb:YAG thin disc laser, Opt. Lett. **25**, 859 (2000)

11.514 J. Dong, P. Z. Deng, Y. P. Liu, Y. H. Zhang, G. S. Huang, F. X. Gan: Performance of the self-*Q*-switched Cr,Yb:YAG laser, Chinese Phys. Lett. **19**, 342 (2002)

11.515 J. Dong, P. H. Deng, Y. P. Liu, Y. H. Zhang, J. Xu, W. Chen, X. L. Xie: Passively *Q*-switched Yb:YAG laser with Cr^{4+} : YAG as the saturable absorber, Appl. Opt. **40**, 4303 (2001)

11.516 U. Griebner, H. Schönnagel: Laser operation with nearly diffraction-limited output from a Yb:YAG multimode channel waveguide, Opt. Lett. **24**, 750 (1999)

11.517 C. Bibeau, R. J. Beach, S. C. Mitchell, M. A. Emanuel, J. Skidmore, C. A. Ebbers, S. B. Sutton, K. S. Jancaitis: High-average-power 1 μm performance and frequency conversion of a diode-end-pumped Yb:YAG laser, IEEE J. Quantum Electron. **34**, 2010 (1998)

11.518 P. Burdack, T. Fox, M. Bode, I. Freitag: 1 W of stable single-frequency output at 1.03 μm from a novel, monolithic, non-planar Yb:YAG ring laser operating at room temperature, Opt. Express **14**, 4363 (2006)

11.519 J. Saikawa, T. Taira: Second-harmonic nonlinear mirror CW mode locking in Yb:YAG microchip lasers, Jpn. J. Appl. Phys. **42**, L649 (2003)

11.520 J. Kong, D. Y. Tang, B. Zhao, J. Lu, K. Ueda, H. Yagi, T. Yanagitani: 9.2 W diode-end-pumped Yb : Y_2O_3 ceramic laser, Appl. Phys. Lett. **86**, 161116 (2005)

11.521 N. Sugimoto, Y. Ohishi, Y. Katoh, A. Tate, M. Shimokozno, S. Sudo: A ytterbium- and neodymium-codoped yttrium aluminum garnet-buried channel waveguide laser pumped at 0.81 μm, Appl. Phys. Lett. **67**, 582 (1995)

11.522 P. Wang, J. M. Dawes, P. Dekker, J. A. Piper: Highly efficient diode-pumped ytterbium-doped yttrium aluminum borate laser, Opt. Commun. **174**, 467 (2000)

11.523 J. Li, J. Y. Wang, X. F. Cheng, X. B. Hu, P. A. Burns, J. M. Dawes: Thermal and laser properties of Yb : $YAl_3(BO_3)_4$ crystal, J. Crystal Growth **250**, 458 (2003)

11.524 J. Li, J. Y. Wang, X. B. Hu, Y. G. Liu, J. Piper, P. Dekker: Growth of Yb:YAB crystal and its laser performance, J. Rare Earths **20**, 104 (2002)

11.525 J. Kawanaka, K. Yamakawa, H. Nishioka, K. Ueda: 30 mJ, dioded-pumped, chirped-pulse Yb:YLF regenerative amplifier, Opt. Lett. **28**, 2121 (2003)

11.526 W. X. Li, S. X. Xu, H. F. Pan, L. Ding, H. P. Zeng, W. Guo C. L. Lu, G. J. Zhao, C. F. Yan, . L. Su, J. Xu: Efficient tunable diode-pumped Yb:LYSO laser, Opt. Express **14**, 6681 (2006)

11.527 J. Kong, D. Y. Tang, J. Lu, K. Ueda, H. Yagi, T. Yanagitani: Passively mode-locked Yb : Y_2O_3 ceramic laser with a GaAs-saturable absorber mirror, Opt. Commun. **237**, 165 (2004)

11.528 J. Kong, J. Lu, K. Takaichi, T. Uematsu, K. Ueda, D. Y. Tang, D. Y. Shen, H. Yagi, T. Yanagitani, A. A. Kaminskii: Diode-pumped Yb : Y_2O_3 ceramic laser, Appl. Phys. Lett. **82**(16), 2556 (2003)

11.529 M. Jacquemet, F. Balembois, S. Chenais, F. Druon, P. Georges, R. Gaume, B. Ferrand: First diode-pumped Yb-doped solid-state laser continuously tunable between 1000 and 1010 nm, Appl. Phys. B **78**, 13 (2004)

11.530 A. Ikesue, Y. L. Aung: Synthesis and performance of advanced ceramic lasers, J. Am. Ceramic Soc. **89**(6), 1936 (2006)

11.531 V. E. Kisel, A. E. Troshin, N. A. Tolstik, V. G. Shcherbitsky, N. V. Kuleshov, V. N. Matrosov, T. A. Matrosova, M. I. Kupchenko: Spectroscopy and continuous-wave diode-pumped laser action of $Yb^{3+}:YVO_4$, Opt. Lett. **29**, 2491 (2004)

11.532 C. Kränkel, D. Fagundes-Peters, S. T. Fredrich, J. Johannsen, M. Mond, G. Guber, M. Bernhagen, R. Uecker: Continuous wave laser operation of $Yb^{3+}:YVO_4$, Appl. Phys. B **79**, 543 (2004)

11.533 H. Stange, K. Petermann, G. Huber, E. W. Duczynski: Continuous wave 1.6 μm laser action in Er doped garnets at room temperature, Appl.Phys. B **49**, 269 (1989)

11.534 A. Diening, E. Heumann, G. Huber, O. Kuzmin: High-power diode-pumped Yb,Er:LSB laser at 1.56 μm. In: *Conf. Lasers Electro-Optics 1998*, Tech. Dig. Ser., Vol. 6 (Opt. Soc. Am., Washington 1998) p. 299

11.535 D. Y. Shen, J. K. Sahu, W. A. Clarkson: Highly efficient in-band pumped Er:YAG laser with 60 W of output at 1645 nm, Opt. Lett. **31**(6), 754 (2006)

11.536 D. Garbuzov, I. Kudryashov, M. Dubinskii: 110 W (0.9 J) pulsed power from resonantly diode-laser-pumped 1.6 m Er:YAG laser, Appl. Phys. Lett. **87**, 121101 (2005)

11.537 D. Garbuzov, I. Kudryashov, M. Dubinskii: Resonantly diode laser pumped 1.6 μm-erbium-doped yttrium aluminum garnet solid-state laser, Appl. Phys. Lett. **86**, 131115 (2005)

11.538 S. D. Setzler, M. P. Francis, Y. E. Young, J. R. Konves, E. P. Chicklis: Resonantly pumped eyesafe erbium lasers, IEEE J. Sel. Top. Quant. **11**, 645 (2005)

11.539 E. Snitzer, R. Woodcock: Yb^{3+}-Er^{3+} glass laser, Appl. Phys. Lett. **6**, 45 (1965)

11.540 E. Snitzer: Glass lasers, Proc. IEEE **54**, 1249 (1966)

11.541 E. Snitzer, R.F. Woodcock, J. Segre: Phosphate glass Er^{3+} lasers, IEEE J. Quantum Electron. **4**, 360 (1968)

11.542 S. Hamlin, J. Myers, M. Myers: Eyesafe Lasers: Components, Systems, and Applications, Proc. SPIE **1419**, 100–106 (1991)

11.543 R. Wu, S. Jiang, M. Myers, J. Myers, S. Hamlin: Solid-State Laser and Nonlinear Crystals, Proc. SPIE **2379**, 26 (1995)

11.544 S. Jiang, S. Hamlin, J. Myers, D. Rhonehouse, M. Myers: High-average-power 1.54 μm Er^{3+}:Yb^{3+}-doped phosphate glass laser. In: *Conference on Lasers and Electro-Optics*, OSA Tech. Dig. Ser., Vol. 9 (Opt. Soc. Am., Washington 1996) p. 380

11.545 R. Wu, S.J. Hamlin, J.A. Hutchinson, L.T. Marshall: Laser diode pumped, passively *Q*-switched erbium:glass laser. In: *Adv. Solid-State Lasers*, OSA Trends Opt. Photon., Vol. 10, ed. by C. Pollock, W. Bosenberg (Opt. Soc. Am., Washington 1997) p. 145

11.546 B.I. Denker, A.A. Korchagin, V.V. Osiko, S.E. Sverchkov, T.H. Allik, J.A. Hutchinson: Diode-pumped and FTIR *Q*-switched laser performance of novel Yb-Er Glass. In: *OSA Proc. Adv. Solid-State Lasers*, Vol. 20, ed. by T. Fan, B. Chai (Opt. Soc. Am., Washington 1994) p. 148

11.547 Kigre Inc.: www.kigre.com and references given there (Kigre Inc., Hilton Head 2007)

11.548 J. Taboada, J.M. Taboada, D.J. Stolarski, J.J. Zohner, L.J. Chavey, H.M. Hodnett, G.D. Noojin, R.J. Thomas, S.S. Kumru, C.P. Cain: 100 megawatt power Q-switched Er-glass laser, Proceedings of SPIE, Vol. 6100, 61000B Solid State Lasers XV: Technology and Devices; H.J. Hoffman, R.K.Shori (Eds.), 2006

11.549 R. Wu, J. Myers, M. Myers, T. Wisnewski: 50Hz diode pumped Er:glass eye-safe laser. In: *Adv. Solid-State Lasers*, OSA Trends Opt. Photon., Vol. 26, ed. by M. Fejer, H. Injeyan, U. Keller (Opt. Soc. Am., Washington 1999) p. 336

11.550 B. Majaron, M. Lukač, T. Rupnik: Pumping dynamics in Yb,Er:phospate glasses, Proc. SPIE **1864**, 2 (1993)

11.551 A.F. Obaton, J. Bernard, C. Parent, G.L. Flem, J.M. Fernandez-Navarro, J.L. Adam, M.J. Myers, G. Boulon: New laser material for eye-safe sources: Yb^{3+},Er^{3+}-codoped phospate glasses. In: *Adv. Solid-State Lasers*, OSA Trends Opt. Photon., Vol. 26, ed. by M.J. Fejer, H. Injeyan, U. Keller (Opt. Soc. Am., Washington 1999) p. 655

11.552 A. Diening: Diodengepumpte Festkörperlaser im mittleren Infrarotbereich Ph.D. Thesis (Shaker Verlag, Aachen 1999)

11.553 B. Simondi-Teisseire, B. Viana, A.-M. Lejus, J.-M. Benitez, D. Vivien, C. Borel, R. Templier, C. Wyon: Room temperature CW laser operation at ≈1.55 μm (eye-safe range) of Yb:Er and Yb : Er : Ce : $Ca_2Al_2SiO_7$ crystals, IEEE J. Quantum Electron. **32**, 2004 (1996)

11.554 R. Brinkmann, W. Sohler, H. Suche: Continuous-wave erbium-diffused $LiNbO_3$ waveguide laser, Electron. Lett. **27**, 415 (1991)

11.555 P. Becker, R. Brinkmann, M. Dinand, W. Sohler, H. Suche: Er-diffused Ti : $LiNbO_3$ waveguide laser of 1563 nm and 1576 nm emission wavelengths, Appl. Phys. Lett. **61**, 1257 (1992)

11.556 J. Souriau, R. Romero, C. Borel, C. Wyon: Room-temperature diode-pumped continuous-wave $SrY_4(SiO_4)_3O$:Yb^{3+}, Er^{3+} crystal laser at 1554 nm, Appl. Phys. Lett. **64**, 1189 (1994)

11.557 U. Reimann: Sensibilisierung und Lasereigenschaften von Er^{3+} in Yttrium-Aluminium-Granat und Lanthan-Strontium-Aluminium-Tantalat Diploma Thesis (University of Hamburg, 1991)

11.558 B. Dischler, W. Wettling: Investigation of the laser materials $YAlO_3$:Er and $LiYF_4$:Ho, J. Phys. D **17**, 1115 (1984)

11.559 A.A. Kaminskii, V.A. Fedorov: *Cascade stimulated emission in crystals with several metastable states of* Ln^{3+} *ions*, ed. by A.M. Prokhorov, I. Ursu (Springer, Berlin, New York 1986) p. 69

11.560 A.A. Kaminskii: Cascade laser generation by Er^{3+} ions in $YAlO_3$ crystals by the scheme ${}^4S_{3/2} \rightarrow {}^4I_{9/2} \rightarrow {}^4I_{11/2} \rightarrow {}^4I_{13/2}$, Sov. Phys. Dokl. **27**, 1039 (1982)

11.561 M.J. Weber, M. Bass, G.A. Demars: Laser action and spectroscopic properties of Er^{3+} in $YAlO_3$, J. Appl. Phys. **42**, 301 (1971)

11.562 M.J. Weber, M. Bass, G.A. Demars, K. Andringa, R.R. Monchamp: Stimulated emission at 1.663 μ from Er^{3+} ions in $YalO_3$, IEEE J. Quantum Electron. **6**, 654 (1970)

11.563 A.A. Kaminskii, T.I. Butaeva, A.O. Ivanow, I.V. Mochalov, A.G. Petrosyan, G.I. Rogov, V.A. Fedorov: New data on stimulated emission of crystals containing Er^{3+} and Ho^{3+} ions, Sov. Tech. Phys. Lett. **2**, 308 (1976)

11.564 T. Andreae, D. Meschede, T.W. Hänsch: New CW laser lines in the Er:$YAlO_3$ crystal, Opt. Commun. **79**, 211 (1990)

11.565 M. Dätwyler, W. Lüthy, H.P. Weber: New wavelengths of the $YAlO_3$:Er laser, IEEE J. Quantum Electron. **23**, 158 (1987)

11.566 B. Schmaul, G. Huber, R. Clausen, B.H.T. Chai, P. Li Kam Wa, M. Bass: Er:$LiYF_4$ continuous wave cascade laser operation at 1620 nm and 2810 nm at room temperature, Appl. Phys. Lett. **62**, 541 (1993)

11.567 S.L. Korableva, L.D. Ivanova, M.V. Petrov, A.M. Tkatchuk: Stimulated emission of Er^{3+} ions in $LiYF_4$ crystals, Sov. Phys. Tech. Phys. **26**, 1521 (1981)

11.568 N.P. Barnes, R.E. Allen, L. Esterowitz, E.P. Chicklis, M.G. Knights, H.P. Jenssen: Operation of an Er:YLF laser at 1.73 μm, IEEE J. Quantum Electron. **22**, 337 (1986)

11.569 M. V. Petrov, A. M. Tkatchuk: Optical spectra and multifrequency stimulated emission of $LiYF_4$-Er^{3+} crystals, Opt. Spectrosc. **45**, 81 (1978)

11.570 R. D. Stultz, V. Leyva, K. Spariosu: Short pulse, high-repetition rate, passively *Q*-switched Er:yttrium-aluminum-garnet laser at 1.6 microns, Appl. Phys. Lett. **87**, 241118 (2005)

11.571 G. M. Zverev, V. M. Garmash, A. M. Onischenko, V. A. Pashkov, A. A. Semenov, Y. M. Kolbatskov, A. I. Smirnov: Induced emission by trivalent erbium ions in crystals of yttrium-aluminium garnet, J. Appl. Spectrosc. (USSR) **21**, 1467 (1974)

11.572 K. Spariosu, M. Birnbaum: Intracavity 1.549 μm pumped 1.634 μm Er:YAG lasers at 300 K, IEEE J. Quantum Electron. **30**, 1044 (1994)

11.573 M. B. Camargo, R. D. Stultz, M. Birnbaum: Passive *Q*-switching of the $Er^{3+}:Y_3Al_5O_{12}$ laser at 1.64 μm, Appl. Phys. Lett. **66**, 2940 (1995)

11.574 K. Spariosu, M. Birnbaum: Room-temperature 1.644 micron Er:YAG lasers. In: *Adv. Solid-State Lasers*, Vol. 13, ed. by L. L. Chase, A. A. Pinto (Opt. Soc. Am., Washington 1992) p. 127

11.575 K. O. White, S. A. Schleusener: Coincidence of Er:YAG laser emission with methane absorption at 1645.1 nm, Appl. Phys. Lett. **21**, 419 (1972)

11.576 K. Spariosu, M. Birnbaum, B. Viana: $Er^{3+}:Y_3Al_5O_{12}$ laser dynamics: effects of upconversion, J. Opt. Soc. Am. B **11**, 894 (1994)

11.577 R. C. Stoneman, A. I. R. Malm: High-Power Er:YAG Laser for Coherent Laser Radar, CLEO/IQEC and PhAST Tech. Dig. on CDROM, San Francisco 2004 (Opt. Soc. Am., Washington 2004) CThZ6

11.578 D. Garbuzov, I. Kudryashov, M. Dubinskii: 110 W (0.9 J) pulsed power from resonantly diode-laser-pumped 1.6 μm Er:YAG laser, Appl. Phys. Lett. **87**, 121101 (2005)

11.579 J. R. Thornton, P. M. Rushworth, P. M. Kelly, R. W. McMillan, L. L. Harper: Proceedings 4th Conference Laser Technology (Ann Arbor 1970) 1249

11.580 T. Schweizer, T. Jensen, E. Heumann, G. Huber: Spectroscopic properties and diode pumped 1.6 μm laser performance in Yb-codoped $Er:Y_3Al_5O_{12}$ and $Er:Y_2SiO_5$, Opt. Commun. **118**, 557 (1995)

11.581 A. A. Kaminskii, T. I. Butaeva, A. M. Kevorkov, V. A. Fedorov, A. G. Petrosyan, M. M. Gritsenko: New data on stimulated emission by crystals with high concentration of Ln^{3+} ions, Inorg. Mater. **12**, 1238 (1976)

11.582 K. Spariosu, M. Birnbaum, M. Kokta: Room-temperature 1.643 μm $Er^{3+}:Y_3Sc_2Ga_3O_{12}$ (Er:YSGG) laser, Appl. Opt. **34**, 8272 (1995)

11.583 A. A. Kaminskii, A. A. Pavlyuk, T. I. Butaeva, V. A. Fedorov, I. F. Balashov, V. A. Berenberg, V. V. Lyubchenko: Stimulated emission by subsidiary transitions of Ho^{3+} and Er^{3+}-ions in $KGd(WO_4)_2$ crystals, Inorg. Mater. **13**, 1251 (1977)

11.584 A. A. Kaminskii, A. A. Pavlyuk, A. I. Polyakov, V. V. Lyubchenko: A new lasing channel in a self-activated erbium crystal $KEr(WO_4)_2$, Sov. Phys. Dokl. **28**, 154 (1983)

11.585 N. V. Kuleshov, A. A. Lagatsky, V. G. Shcherbitsky, V. P. Mikhailov, E. Heumann, T. Jensen, A. Diening, G. Huber: CW laser performance of Yb and Er,Yb doped tungstates, Appl. Phys. B **64**, 409 (1997)

11.586 A. A. Kaminskii, I. P. Kozeeva, A. A. Pavlyuk: Stimulated emission of Er^{3+} and Ho^{3+} ions in $KLa(MoO_4)_2$ crystals, phys. stat. sol. a **83**, K65 (1984)

11.587 A. A. Kaminskii: Stimulated emission spectroscopy of Ln^{3+} ions in tetragonal $LiLuF_4$ fluoride, phys. stat. sol. a **97**, K53 (1986)

11.588 A. A. Kaminskii, A. A. Markosyan, A. V. Pelevin, Y. A. Polyakova, S. E. Sarkisov, T. V. Uvarova: Luminescence properties and stimulated emission from Pr^{3+}, Nd^{3+} and Er^{3+} ions in tetragonal lithium-lutecium fluoride, Inorg. Mater. **22**, 773 (1986)

11.589 L. Fornasiero, K. Petermann, E. Heumann, G. Huber: Spectroscopic properties and laser emission of Er^{3+} in scandium silicates near 1.5 μm, Opt. Mater. **10**, 9 (1998)

11.590 P. Burns, J. Dawes, P. Dekker, J. Piper, H. Jiang, J. Wang: CW diode-pumped microlaser operation at 1.5–1.6 μm in Er, Yb:YCOB, IEEE Photon. Technol. Lett. **14**, 1677 (2002)

11.591 P. Burns, J. Dawes, P. Dekker, J. Piper, H. Jiang, J. Wang: 250 mW continuous-wave output from Er,Yb:YCOB laser at 1.5 μm. In: *Advanced Solid State Photonics*, Tech. Dig., Vol. 8 (Opt. Soc. Am., Washington 2003)

11.592 B. Denker, B. Galagan, L. Ivleva, V. Osiko, S. Sverchkov, I. Voronina, J. E. Hellstrom, G. Karlsson, F. Laurell: Luminescent and laser properties of $Yb-Er:GdCa_4O(BO_3)_3$: a new crystal for eye-safe 1.5 μm lasers, Appl. Phys. B **79**, 577 (2004)

11.593 R. I. Laming, S. B. Poole, E. J. Tarbox: Pump excited state absorption in erbium-doped fibers, Opt. Lett. **13**, 1084 (1988)

11.594 M. Shimizu, M. Yamada, M. Horiguchi, T. Takeshita, M. Okayasu: Erbium-doped fiber amplifier with an extremely high gain coefficient of 11 bD/mW, Electron. Lett. **26**, 1641 (1990)

11.595 M. J. F. Digonnet: Continuous-wave silica fiber lasers. In: *Rare-Earth-Doped Fiber Lasers and Amplifiers*, 2 edn., ed. by M. J. F. Digonnet (Marcel Dekker, New York 2001)

11.596 IPG Photonics: product information, http://www.ipgphotonics.com (IPG Photonics, Oxford 2006)

11.597 E. Heumann, P. E. A. Möbert, G. Huber: Room-temperature upconversion-pumped CW Yb,Er:$YLiF_4$ laser at 1.234 μm. In: *OSA Trends Opt. Photon. Adv. Solid-State Lasers*, Vol. 1, ed. by S. A. Payne, C. R. Pollock (Opt. Soc. Am., Washington 1996) p. 288

11.598 E. Heumann, P. E. A. Möbert, G. Huber: Intracavity frequency doubled Yb,Er:$LiYF_4$ upconversion-pumped laser at 617 nm, Exp. Tech. Phys. **42**, 33 (1996)

11.599 S.L. Korableva, L.D. Livanova, M.V. Petrov, A.M. Tkatchuk: Stimulated emission of Er^{3+} ions in $LiYF_4$ crystals, Sov. Phys. Tech. Phys. **26**, 1521 (1981)

11.600 A.M. Tkatchuk, M.V. Petrov, L.D. Livanova, S.L. Korableva: Pulsed-periodic 0.8503 μm YLF:Er^{3+},Pr^{3+} laser, Opt. Spectrosc. (USSR) **54**, 667 (1983)

11.601 M.V. Petrov, A.M. Tkatchuk: Optical spectra and multifrequency stimulated emission of $LiYF_4$-Er^{3+} crystals, Opt. Spectrosc. (USSR) **45**, 81 (1978)

11.602 A.A. Kaminskii, A.A. Markosyan, A.V. Pelevin, Y.A. Polyakova, S.E. Sarkisov, T.V. Uvarova: Luminescence properties and stimulated emission from Pr^{3+}, Nd^{3+} and Er^{3+} ions in tetragonal lithium-lutecium fluoride, Inorg. Mater. **22**, 773 (1986)

11.603 A.A. Kaminskii: Luminescence and multiwave stimulated emission of Ho^{3+} and Er^{3+} ions in orthorhombic $YAlO_3$ crystals, Sov. Phys. Dokl. **31**, 823 (1986)

11.604 B.M. Antipenko, A.A. Mak, B.V. Sinitsyn, O.B. Raba, T.V. Uvarova: New excitation schemes for laser transitions, Sov. Phys. Tech. Phys. **27**, 333 (1982)

11.605 B.M. Antipenko, V.A. Buchenkov, A.A. Nikitichev, B.P. Sobolev, A.I. Stepanov, L.K. Sukhareva, T.V. Uvarova: Optimization of a $BaYb_2F_8$: Er active medium, Sov. J. Quantum Electron. **16**, 759 (1986)

11.606 B.M. Antipenko, A.A. Mak, O.B. Raba, L.K. Sukhareva, T.V. Uvarova: 2 μm-range rare earth laser, Sov. Tech. Phys. Lett. **9**, 227 (1983)

11.607 B.M. Antipenko, A.A. Mak, B.V. Nikolaev, O.B. Raba, K.B. Seiranyan, T.V. Uvarova: Analysis of laser situations in $BaYb_2F_8$:Er^{3+} with stepwise pumping schemes, Opt. Spectrosc. (USSR) **56**, 296 (1984)

11.608 P.P. Sorokin, M.J. Stevenson: Stimulated infrared emission from trivalent uranium, Phys. Rev. Lett. **5**, 557 (1960)

11.609 Z.J. Kiss, R.C. Duncan: Pulsed and continuous optical maser action in CaF_2 : Dy^{2+}, Proc. IRE (Corresp.) **50**, 1531 (1962)

11.610 A. Yariv: Continuous operation of a CaF_2 : Dy^{2+} optical maser, Proc. IRE (Corresp.) **50**, 1699 (1962)

11.611 L.F. Johnson, G.D. Boyd, K. Nassau: Optical maser characteristics of Tm^{3+} in $CaWO_4$, Proc. IRE **50**, 86 (1962)

11.612 L.F. Johnson, G.D. Boyd, K. Nassau: Optical maser characteristics of Ho^{3+} in $CaWO_4$, Proc. IRE **50**, 87 (1962)

11.613 M. Robinson, P.D. Devor: Thermal switching of laser emission of Er^{3+} at 2.69 μ and Tm^{3+} at 1.86 μ in mixed crystals of CaF_2 : ErF_3 : TmF_3^*, Appl. Phys. Lett. **10**, 167 (1967)

11.614 A.A. Kaminskii: *Laser Crystals* (Springer, Berlin, Heidelberg 1990) p. 456

11.615 G. Huber, E.W. Duczynski, K. Petermann: Laser pumping of Ho-, Tm-, Er-doped garnet lasers at room temperature, IEEE J. Quantum Electron. **24**, 920 (1988)

11.616 T.Y. Fan, G. Huber, R.L. Byer, P. Mitzscherlich: Spectroscopy and diode laser-pumped operation of Tm, Ho:YAG, IEEE J. Quantum Electron. **24**, 924 (1988)

11.617 R.C. Stoneman, L. Esterowitz: Efficient, broadly tunable, laser-pumped Tm:YAG and Tm:YSGG CW lasers, Opt. Lett. **15**, 486 (1990)

11.618 J.F. Pinto, L. Esterowitz, G.H. Rosenblatt: Tm^{3+}:YLF laser continuously tunable between 2.20 and 2.46 μm, Opt. Lett. **19**, 883 (1994)

11.619 K.L. Vodopyanov, L.A. Kulevskii, P.P. Pashinin, A.F. Umyskov, I.A. Shcherbakov: Bandwidth-limited picosecond pulses from a YSGG:Cr^{3+} : Er^{3+} laser ($\lambda = 2.79\,\mu m$) with active mode locking, Kvant. Elektron. **14**, 1219 (1987) [English transl.: Sov. J. Quantum Electron. **17**, 776 (1987)]

11.620 E. Sorokin, I.T. Sorokina, A. Unterhuber, E. Wintner, A.I. Zagumennyi, I.A. Shcherbabov, V. Carozza, A. Toncelli, M. Tonelli: A novel CW tunable and mode-locked 2 μm Cr,Tm,Ho:YSGG:GSAG laser. In: *Adv. Solid-State Lasers*, Vol. 19, ed. by W.R. Bosenberg, M.M. Fejer (Opt. Soc. Am., Washington 1998) pp. 197–200

11.621 M.J. Weber: *Handbook of Lasers* (CRC, Boca Raton 2000) p. 1198

11.622 M.S. O'Sullivan, J. Chrostowski, E. Desurvire, J.R. Simpson: High-power narrow-linewidth Er^{3+}-doped fiber laser, Opt. Lett. **14**, 438 (1989)

11.623 P.L. Scrivener, E.J. Tarbox, P.D. Maton: Narrow linewidth tunable operation of Er^{3+}-doped single-mode fiber laser, Electron. Lett. **25**, 549 (1989)

11.624 V.P. Gapontsev, I.E. Samartsev: High-power fiber laser. In: *Proc. Adv. Solid-State Lasers* (Opt. Soc. Am., Washington 1990) p. 258

11.625 R. Wyatt, B.J. Ainslie, S.P. Craig: Efficient operation of array-pumped Er^{3+} doped silica fiber laser at 1.5 μm, Electron. Lett. **24**, 1362 (1988)

11.626 L. Reekie, I.M. Jauncie, S.B. Poole, D.N. Payne: Diode laser pumped operation of an Er^{3+}-doped single mode fiber laser, Electron. Lett. **23**, 1076 (1987)

11.627 W.L. Barnes, P.R. Morkel, L. Reekie, D.N. Payne: High-quantum-efficiency Er^{3+} fiber lasers pumped at 980 nm, Opt. Lett. **14**, 1002 (1989)

11.628 R. Wyatt: High-power broadly tunable erbium-doped silica fiber laser, Electron. Lett. **25**, 1498 (1989)

11.629 K. Susuki, Y. Kimura, M. Nakazawa: An 8 mW CW Er^{3+}-doped fiber laser pumped by 1.46 μm InGaAsP laser diodes, Jpn. J. Appl. Phys. **28**, L1000 (1989)

11.630 Y. Kimura, K. Susuki, M. Nakazawa: Laser-diode-pumped mirror-free Er^{3+}-doped fiber laser, Opt. Lett. **14**, 999 (1989)

11.631 L. Cognolato, A. Gnazzo, B. Sordo, C. Bruschi: Tunable erbium-doped silica fiber ring laser source: design and realization, J. Opt. Commun. **16**, 122 (1995)

11.632 J. L. Wagener, P. F. Wysocki, M. J. F. Digonnet, H. J. Shaw, D. J. Digiovanni: Effects of concentration and clusters in erbium-doped fiber lasers, Opt. Lett. **18**, 2014 (1993)

11.633 W. L. Barnes, S. B. Poole, J. E. Townsend, L. Reekie, D. J. Taylor, D. N. Payne: Er^{3+} Yb^{3+} and Er^{3+} doped fiber lasers, J. Lightwave Technol. **7**, 1462 (1989)

11.634 M. E. Fermann, D. C. Hanna, D. P. Shepherd, P. J. Suni, J. E. Townsend: Efficient operation of an Yb-sensitised Er fiber laser at 1.56 µm, Electron. Lett. **24**, 1135 (1988)

11.635 D. C. Hanna, R. M. Percival, I. R. Perry, R. G. Smart, A. C. Tropper: Efficient operation of an Yb-sensitized Er fiber laser pumped in the 0.8 µm region, Electron. Lett. **24**, 1068 (1988)

11.636 J. D. Minelli, W. L. Barnes, R. I. Laming, P. R. Morkel, J. E. Townsend, S. G. Grubb, D. N. Payne: Diode-array pumping of Er^{3+} / Yb^{3+} co-doped fiber lasers and amplifiers, IEEE Photon. Technol. Lett. **5**, 301 (1993)

11.637 J. T. Kringlebotn, P. R. Morkel, L. Reekie, J.-L. Archambault, D. N. Payne: Efficient diode-pumped single frequency erbium:ytterbium fiber laser, IEEE Photon. Technol. Lett. **5**, 1162 (1993)

11.638 G. G. Vienne, J. E. Caplen, L. Dong, J. D. Minelly, J. Nilsson, D. N. Payne: Fabrication and characterization of Yb^{3+}:Er^{3+} phosphosilicate fibers for lasers, J. Lightwave Technol. **16**, 1990 (1998)

11.639 J. E. Townsend, W. L. Barnes, K. P. Jedrzejewski, S. G. Grubb: Yb^{3+} sensitized Er^{3+} doped silica optical fiber with ultrahigh transfer efficiency and gain, Electron. Lett. **27**, 1958 (1991)

11.640 J. J. Pan, Y. Shi: 166 mW single-frequency outpur power interactive fiber lasers with low noise, IEEE Photon. Technol. Lett. **11**, 36 (1999)

11.641 I. T. Sorokina: Crystalline mid-infrared lasers. In: *Solid-State Mid-Infrared Laser Sources*, Top. Appl. Phys., Vol. 89, ed. by I. T. Sorokina, K. L. Vodopyanov (Springer, Berlin, Heidelberg 2003) pp. 255–349

11.642 M. Pollnau, S. D. Jackson: Mid-infrared fiber lasers. In: *Solid-State Mid-Infrared Laser Sources*, Top. Appl. Phys., Vol. 89, ed. by I. T. Sorokina, K. L. Vodopyanov (Springer, Berlin, Heidelberg 2003) pp. 219–253

11.643 J. M. F. van Dijk, M. F. H. Schuurmans: On the non-radiative and radiative decay rates and a modified exponential energy gap law for 4f-4f transitions in rare-earth ions, J. Chem. Phys. **78**, 5317 (1983)

11.644 M. J. F. Digonnet (Ed.): *Rare-Earth-Doped Fiber Lasers and Amplifiers*, 2 edn. (Marcel Dekker, New York 2001)

11.645 H. U. Güdel, M. Pollnau: Near-infrared to visible photon upconversion processes in lanthanide doped chloride, bromide and iodide lattices, J. Alloy. Comp. **303-304**, 307 (2000)

11.646 L. A. Riseberg, H. W. Moos: Multiphonon orbit-lattice relaxation of excited states of rare-earth ions in crystals, Phys. Rev. **174**, 429 (1968)

11.647 M. Pollnau, P. J. Hardman, M. A. Kern, W. A. Clarkson, D. C. Hanna: Upconversion-induced heat generation and thermal lensing in Nd:YLF and Nd:YAG, Phys. Rev. B **58**, 16076 (1998)

11.648 B. J. Ainslie, S. P. Craig, S. T. Davey: The absorption and fluorescence spectra of rare earth ions in silica-based monomode fiber, J. Lightwave Technol. **6**, 287 (1988)

11.649 R. Reisfeld, M. Eyal: Possible ways of relaxations for excited states of rare earth ions in amorphous media, J. Phys. (Paris) **46**, C349 (1985)

11.650 S. D. Jackson, T. A. King: CW operation of a 1.064 µm pumped Tm-Ho-doped silica fiber laser, IEEE J. Quantum Electron. **34**, 1578 (1998)

11.651 O. Humbach, H. Fabian, U. Grzesik, U. Haken, W. Heitmann: Analysis of OH absorption bands in synthetic silica, J. Non-Crystal Solids **203**, 19 (1996)

11.652 D. C. Tran, G. H. Sigel Jr., B. Bendow: Heavy metal fluoride glasses and fibers: a review, J. Lightwave Technol. **2**, 566 (1984)

11.653 P. W. France, M. G. Drexhage, J. M. Parker, M. W. Moore, S. F. Carter, J. V. Wright: *Fluoride Glass Optical Fibers* (Blackie, Glasgow 1990)

11.654 M. Poulain, M. Poulain, J. Lucas, P. Brun: Verres fluorés au tetrafluorure de zirconium; propriétés optiques d'un verre dopé au Nd^{3+}, Mater. Res. Bull. **10**, 243 (1974)

11.655 S. T. Davey, P. W. France: Rare earth doped fluorozirconate glasses for fiber devices, BT Technol. J. **7**, 58 (1989)

11.656 M. Monerie, F. Alard, G. Maze: Fabrication and characterisation of fluoride-glass single-mode fibers, Electron. Lett. **21**, 1179 (1985)

11.657 L. Wetenkamp, G. F. West, H. Többen: Optical properties of rare earth-doped ZBLAN glasses, J. Non-Crystal Solids **140**, 35 (1992)

11.658 L. Wetenkamp: Charakterisierung von laseraktiv dotierten Schwermetallfluorid-Gläsern und Faserlasern Ph.D. Thesis (Technical University of Braunschweig, Braunschweig 1991)

11.659 Y. D. Huang, M. Mortier, F. Auzel: Stark level analysis for Er^{3+}-doped ZBLAN glass, Opt. Mater. **17**, 501 (2001)

11.660 D. S. Knowles, H. P. Jenssen: Upconversion versus Pr-deactivation for efficient 3 µm laser operation in Er, IEEE J. Quantum Electron. **28**, 1197 (1992)

11.661 T. Jensen, A. Diening, G. Huber, B. H. T. Chai: Investigation of diode-pumped 2.8 µm Er : $LiYF_4$ lasers with various doping levels, Opt. Lett. **21**, 585 (1996)

11.662 M. P. Hehlen, K. Krämer, H. U. Güdel, R. A. McFarlane, R. N. Schwartz: Upconversion in Er^{3+}-dimer systems: Trends within the series $Cs_3Er_2X_9$ (X=Cl,Br,I), Phys. Rev. B **49**, 12475 (1994)

11.663 L. Isaenko, A. Yelisseyev, A. Tkachuk, S. Ivanova, S. Vatnik, A. Merkulov, S. Payne, R. Page, M. Nostrand: New laser crystals based on KPb_2Cl_5 for IR region, Mater. Sci. Eng. B **81**, 188 (2001)
11.664 K. Rademaker, E. Heumann, G. Huber, S. A. Payne, W. F. Krupke, L. I. Isaenko, A. Burger: Laser activity at 1.18, 1.07, and 0.97 μm in the low-phonon-energy hosts KPb_2Br_5 and $RbPb_2Br_5$ doped with Nd^{3+}, Opt. Lett. **30**, 729 (2005)
11.665 K. Rademaker, S. A. Payne, G. Huber, L. I. Isaenko, E. Osiak: Optical pump-probe processes in Nd^{3+}-doped KPb_2Br_5, $RbPb_2Br_5$, KPb_2Cl_5, J. Opt. Soc. Am. B **22**, 2610 (2005)
11.666 P. N. Kumta, S. H. Risbud: Rare-earth chalcogenides – an emerging class of optical materials, J. Mater. Sci. **29**, 1135 (1994)
11.667 J. S. Sanghera, J. Heo, J. D. Mackenzie: Chalcohalide glasses, J. Non-Crystal Solids **103**, 155 (1988)
11.668 L. D. DeLoach, R. H. Page, G. D. Wilke, S. A. Payne, W. F. Krupke: Transition metal-doped zinc chalcogenides: Spectroscopy and laser demonstration of a new class of gain media, IEEE J. Quantum Electron. **32**, 885 (1996)
11.669 L. B. Shaw, B. Cole, P. A. Thielen, J. S. Sanghera, I. D. Aggarwal: Mid-wave IR and long-wave IR laser potential of rare-earth doped chalcogenide glass fiber, IEEE J. Quantum Electron. **48**, 1127 (2001)
11.670 T. Schweizer: Rare-earth-doped gallium lanthanum sulphide glasses for mid-infrared fiber lasers Ph.D. Thesis (University of Hamburg, Hamburg 1998)
11.671 Y. D. West, T. Schweizer, D. J. Brady, D. W. Hewak: Gallium lanthanum sulphide fibers for infrared transmission, Fiber Integrated Opt. **19**, 229 (2000)
11.672 J. Heo, Y. B. Shin: Absorption and mid-infrared emission spectroscopy of Dy^{3+} in Ge-As (or Ga)-S glasses, J. Non-Crystal Solids **196**, 162 (1996)
11.673 J. R. Lu, T. Murai, K. Takaichi, T. Uematsu, K. Misawa, M. Prabhu, J. Xu, K. Ueda, H. Yagi, T. Yanagitani, A. A. Kaminskii, A. Kudryashov: 72 W $Nd:Y_3Al_5O_{12}$ ceramic laser, Appl. Phys. Lett. **78**, 3586 (2001)
11.674 J. Lu, J. Lu, T. Murai, K. Takaichi, T. Uematsu, K. Ueda, H. Yagi, T. Yanagitani, T. Works, A. A. Kaminskii: Optical properties and highly efficient laser oscillation of $Nd:Y_2O_3$ ceramics, Conf. Lasers Electro-Optics Tech. Dig. 2002 (Opt. Soc. Am., Washington 2002) Tech. Dig.
11.675 B. N. Samson, P. A. Tick, N. F. Borrelli: Efficient neodymium-doped glass-ceramic fiber laser and amplifier, Opt. Lett. **26**, 145 (2001)
11.676 M. Pollnau: Analysis of heat generation and thermal lensing in erbium 3 μm lasers, IEEE J. Quantum Electron. **39**, 350 (2003)
11.677 R. M. Percival, S. F. Carter, D. Szebesta, S. T. Davey, W. A. Stallard: Thulium-doped monomode fluoride fiber laser broadly tunable from 2.25 to 2.5 μm, Electron. Lett. **27**, 1912 (1991)
11.678 M. K. Davis, M. J. F. Digonnet, R. H. Pantell: Thermal effects in doped fibers, J. Lightwave Technol **16**, 1013 (1998)
11.679 H. Po, E. Snitzer, R. Tumminelli, L. Zenteno, F. Hakimi, N. M. Cho, T. Haw: Doubly clad high brightness Nd fiber laser pump by GaAlAs phased array. In: *Proceedings of the Optical Fiber Communication Conference* (Opt. Soc. Am., Washington 1989)
11.680 I. N. II. I. Duling, W. K. Burns, L. Goldberg: High-power superfluorescent fiber source, Opt. Lett. **15**, 33 (1990)
11.681 J. D. Minelly, W. L. Barnes, R. I. Laming, P. R. Morkel, J. E. Townsend, S. G. Grubb, D. N. Payne: Diode-array pumping of Er^{3+}/Yb^{3+} codoped fiber lasers and amplifiers, IEEE Photon. Technol. Lett. **5**, 301 (1993)
11.682 H. M. Pask, J. L. Archambault, D. C. Hanna, L. Reekie, P. S. J. Russell, J. E. Townsend, A. C. Tropper: Operation of cladding-pumped Yb^{3+}-doped silica fiber lasers in 1 μm region, Electron. Lett. **30**, 863 (1994)
11.683 L. Zenteno: High-power double-clad fiber lasers, J. Lightwave Technol. **11**, 1435 (1993)
11.684 D. C. Brown, H. J. Hoffman: Thermal, stress, and thermo-optic effects in high average power double-clad silica fiber lasers, IEEE J. Quantum Electron. **37**, 207 (2001)
11.685 N. A. Brilliant, K. Lagonik: Thermal effects in a dual-clad ytterbium fiber laser, Opt. Lett. **26**, 1669 (2001)
11.686 J. A. Caird, L. G. DeShazer, J. Nella: Characteristics of room-temperature 2.3 μm laser emission from Tm^{3+} in YAG and $YAlO_3$, IEEE J. Quantum Electron. **11**, 874 (1975)
11.687 V. A. Smirnov, I. A. Shcherbakov: Rare-earth scandium chromium garnets as active media for solid-state lasers, IEEE J. Quantum Electron. **24**, 949 (1988)
11.688 I. A. Shcherbakov: Optically dense active media for solid-state lasers, IEEE J. Qauntum Electron. **24**, 979 (1988)
11.689 B. M. Antipenko: Cross-relaxation schemes for pumping laser transitions, Zh. Tekh. Phys. **54**, 385 (1984)[English transl. Sov. Phys.-Tech. Phys. **29**, 228 (1984)]
11.690 G. J. Kintz, R. Allen, L. Esterowitz: CW laser emission at 2.02 μm from diode-pumped Tm^{3+}:YAG at room temperature. In: *Conf. Lasers Electro-Optics, Tech. Dig.* (Opt. Soc. Am., Washington 1988)
11.691 T. J. Kane, T. S. Kubo: Diode-pumped single-frequency lasers and *Q*-switched laser using Tm:YAG and Tm,Ho:YAG, OSA Proc. Adv. Solid-State Lasers **6**, 136–139 (1990)
11.692 R. A. Hayward, W. A. Clarkson, D. C. Hanna: High-power diode-pumped room-temperature Tm:YAG and intracavity-pumped Ho:YAG lasers. In: *Adv. Solid-State Lasers*, Vol. 34, ed. by H. Injeyan,

U. Keller, C. Marshall (Opt. Soc. Am., Washington 2000) pp. 90–94

11.693 A. Dergachev, K. Wall, P. F. Moulton: A CW side-pumped Tm:YLF laser. In: *Adv. Solid-State Lasers*, Vol. 68, ed. by M. E. Fermann, L. R. Marshall (Opt. Soc. Am., Washington 2002) pp. 343–350

11.694 P. A. Budni, M. L. Lemons, J. R. Mosto, E. P. Chicklis: High-power/high-brightness diode-pumped 1.9 μm thulium and resonantly pumped 2.1 μm holmium lasers, IEEE J. Select. Top. Quantum Electron. **6**, 629 (2000)

11.695 S. So, J. I. Mackenzie, D. P. Shepherd, W. A. Clarkson, J. G. Betterton, E. K. Gordon: A power-scaling strategy for longitudinally diode-pumped Tm:YLF lasers, Appl. Phys. B **84**, 389 (2006)

11.696 E. C. Honea, R. J. Beach, S. B. Sutton, J. A. Speth, S. C. Mitchell, J. A. Skidmore, M. A. Emanuel, S. A. Payne: 115 W Tm:YAG diode-pumped solid-state laser, IEEE J. Quantum Electron. **33**, 1592 (1997)

11.697 K. S. Lai, P. B. Phua, R. F. Wu, Y. L. Lim, E. Lau, S. W. Toh, B. T. Toh, A. Chng: 120 W continuous-wave diode-pumped Tm:YAG laser, Opt. Lett. **25**, 1591 (2000)

11.698 C. P. Wyss, W. Lüthy, H. P. Weber, V. I. Vlasov, Y. D. Zavartsev, P. A. Studenikin, A. I. Zagumennyi, I. A. Shcherbakov: Emission properties of a $Tm^{3+}:GdVO_4$ microchip laser at 1.9 μm, 120 W continuous-wave diode-pumped Tm:YAG laser, J. Appl. Phys. B **67**, 1–4 (1998)

11.699 N. Berner, A. Diening, E. Heumann, G. Huber, A. Voss, M. Karszewski, A. Giesen: Tm:YAG: A comparison between end pumped laser-rods and the 'thin-disk'-setup. In: *Adv. Solid-State Lasers*, Vol. 26, ed. by M. M. Fejer, H. Injeyan, U. Keller (Opt. Soc. Am., Washington 1999) p. 463

11.700 F. Cornacchia, D. Parisi, C. Bernardini, A. Toncelli, M. Tonelli: Efficient, diode-pumped $Tm^{3+}:BaY_2F_8$ vibronic laser, Opt. Express **12**, 1982 (2004)

11.701 R. C. Stoneman, L. Esterowitz: Efficient, broadly tunable, laser-pumped Tm:YAG and Tm:YSGG CW lasers, Opt. Lett. **15**, 486 (1990)

11.702 R. C. Stoneman, L. Esterowitz: Efficient 1.94 μm Tm:YALO laser, IEEE J. Select. Top. Quantum Electron. **1**, 78–80 (1995)

11.703 L. Fornasiero, N. Berner, B.-M. Dicks, E. Mix, V. Peters, K. Petermann, G. Huber: Broadly tunable laser emission from $Tm:Y_2O_3$ and $Tm:Sc_2O_3$ at 2 μm. In: *Adv. Solid-State Lasers*, Vol. 26, ed. by M. M. Fejer, H. Injeyan, U. Keller (Opt. Soc. Am., Washington 1999) pp. 450–453

11.704 P. Camy, J. L. Doualan, S. Renard, A. Braud, V. Menard, R. Moncorgé: $Tm^{3+}:CaF_2$ for 1.9 μm laser operation, Opt. Commun. **236**, 395 (2004)

11.705 G. Galzerano, F. Cornacchia, D. Parisi, A. Toncelli, M. Tonelli, P. Laporta: Widely tunable 1.94 μm $Tm:BaY_2F_8$ laser, Opt. Lett. **30**, 854 (2005)

11.706 E. Sorokin, A. N. Alpatiev, I. T. Sorokina, A. I. Zagumennyi, I. A. Shcherbakov: Tunable efficient continuous-wave room-temperature $Tm^{3+}:GdVO_4$ laser. In: *Adv. Solid-State Lasers*, Vol. 68, ed. by M. E. Fermann, L. R. Marshall (Opt. Soc. Am., Washington 2002) pp. 347–350

11.707 X. Mateos, J. Liu, H. Zhang, J. Wang, M. Jiang, U. Griebner, V. Petrov: Continuous-wave and tunable laser operation of $Tm:LuVO_4$ near 1.9 μm under Ti:Sapphire and diode laser pumping, phys. stat. sol. (a) **203**, R19 (2006)

11.708 V. Petrov, F. Güell, J. Massons, J. Gavaldà, R. M. Solé, M. Aguilo, F. Díaz, U. Griebner: Efficient tunable laser operation of $Tm:KGd(WO_4)_2$ in the continuous-wave regime at room temperature, IEEE J. Quantum Electron. **40**, 1244 (2004)

11.709 J. M. Cano-Torres, C. Zaldo, M. D. Serrano, V. Petrov, M. Rico, X. Mateos, J. Liu, U. Griebner, F. J. Valle, M. Galan, G. Viera: Broadly tunable operation of Tm^{3+} in locally disordered $NaGd(WO_4)_2$ near 2 microns, Europhoton Conference, Pisa 2006 (European Physical Society, Mulhouse 2006) WeD4

11.710 J. D. Kmetec, T. S. Kubo, T. J. Kane, C. J. Grund: Laser performance of diode-pumped thulium-doped $Y_3Al_5O_{12}$ $(Y,Lu)_3Al_5O_{12}$, and $Lu_3Al_3O_{12}$ crystals, Opt. Lett. **19**, 186 (1994)

11.711 H. Saito, S. Chaddha, R. S. F. Chang, N. Djeu: Efficient 1.94 μm Tm^{3+} laser in YVO_4 host, Opt. Lett. **17**, 189 (1992)

11.712 S. N. Bagaev, S. M. Vatnik, A. P. Maiorov, A. A. Pavlyuk, D. V. Plakushchev: The spectroscopy and lasing of monoclinic $Tm:KY(WO_4)_2$, Quantum Electron. **30**, 310 (2000)[transl. from Kvantovaya Elektronika **30**, 310 (2000)]

11.713 J. F. Pinto, L. Esterowitz, G. H. Rosenblatt: Continuous-wave mode-locked 2 μm Tm:YAG laser, Opt. Lett. **17**, 731 (1992)

11.714 F. Heine, E. Heumann, G. Huber, K. L. Schepler: Mode locking of room-temperature CW thulium and holmium lasers, Appl. Phys. Lett. **60**, 1161 (1992)

11.715 C. Li, J. Song, D. Shen, N. S. Kim, K. Ueda, Y. Huo, S. He, Y. Cao: Diode-pumped high-efficiency Tm:YAG lasers, Opt. Express **4**, 12 (1999)

11.716 C. Li, J. Song, D. Shen, Y. Cao, N. S. Kim, K. Ueda: Flash-lamp-pumped acousto-optic *Q*-switched Cr-Tm:YAG laser, Opt. Rev. **7**, 58 (2000)

11.717 Y. T. Tzong, M. Birnbaum: *Q*-switched 2 μm lasers by use of a $Cr^{2+}:ZnSe$ saturable absorber, Appl. Opt. **40**, 6633 (2001)

11.718 R. Brinkmann, C. Hansen: Beam-profile modulation of thulium laser radiation applied with multimode fibers and effect on the threshold fluence to vaporize water in laser surgery application, Appl. Opt. **39**, 3361 (2000)

11.719 M. Pollnau, Y. E. Romanyuk: Optical waveguides in laser crystals, C. R. Acad. Sci., in press (2007)

11.720 D. P. Shepherd, D. J. B. Brinck, J. Wang, A. C. Tropper, D. C. Hanna, G. Kakarantzas, P. D. Townsend: 1.9 μm operation of a Tm:lead germanate glass waveguide laser, Opt. Lett. **19**, 954 (1994)

11.721 A. Rameix, C. Borel, B. Chambaz, B. Ferrand, D. P. Shepherd, T. J. Warburton, D. C. Hanna, A. C. Tropper: An efficient, diode-pumped, 2 μm Tm:YAG waveguide laser, Opt. Commun. **142**, 239 (1997)

11.722 J. I. Mackenzie, S. C. Mitchell, R. J. Beach, H. E. Meissner, D. P. Shepherd: 15 W diode-side-pumped Tm:YAG waveguide laser at 2 μm, Electron. Lett. **37**, 898 (2001)

11.723 S. Rivier, X. Mateos, V. Petrov, U. Griebner, Y. E. Romanyuk, C. N. Borca, M. Pollnau: $Tm : KY(WO_4)_2$ waveguide laser, Opt. Express, in press (2007)

11.724 X. Mateos, V. Petrov, J. Liu, M. C. Pujol, U. Griebner, M. Aguiló, F. Díaz, M. Galan, G. Viera: Efficient 2 μm continuous-wave laser oscillation of $Tm^{3+} : KLu(WO_4)_2$, IEEE J. Quantum Electron. **42**, in press (2006)

11.725 Y. E. Romanyuk, C. N. Borca, M. Pollnau, U. Griebner, S. Rivier, V. Petrov: Yb-doped $KY(WO_4)_2$ planar waveguide laser, Opt. Lett. **31**, 53 (2006)

11.726 X. Mateos, V. Petrov, U. Griebner, O. Silvestre, M. C. Pujol, M. Aguiló, F. Díaz: Laser operation of a Tm-doped epitaxial tungstate crystal $Tm : KLu(WO_4)_2/KLu(WO_4)_2$ in the 2 μm spectral range, Europhoton Conference, Pisa 2006 (European Physical Society, Mulhouse 2006) WeD1

11.727 D. C. Hanna, I. M. Jauncey, R. M. Percival, I. R. Perry, R. G. Smart, P. J. Suni, J. E. Townsend, A. C. Tropper: Continuous-wave oscillation of a monomode thulium-doped fiber laser, Electron. Lett. **24**, 1222 (1988)

11.728 W. L. Barnes, J. E. Townsend: Highly tunable and efficient diode pumped operation of Tm^{3+} doped fiber lasers, Electron. Lett. **26**, 746 (1990)

11.729 J. N. Carter, R. G. Smart, D. C. Hanna, A. C. Tropper: CW diode-pumped operation of 1.97 μm thulium-doped fluorozirconate fiber laser, Electron. Lett. **26**, 599 (1990)

11.730 S. D. Jackson: Cross relaxation and energy transfer upconversion processes relevant to the functioning of 2 μm Tm^{3+}-doped silica fiber lasers, Opt. Commun. **230**, 197 (2004)

11.731 D. C. Hanna, R. M. Percival, R. G. Smart, A. C. Tropper: Efficient and tunable operation of a Tm-doped fiber laser, Opt. Commun. **75**, 283 (1990)

11.732 W. A. Clarkson, N. P. Barnes, P. W. Turner, J. Nilsson, D. C. Hanna: High-power cladding-pumped Tm-doped silica fiber laser with wavelength tuning from 1860 to 2090 nm, Opt. Lett. **27**, 1989 (2002)

11.733 D. Y. Shen, J. K. Sahu, W. A. Clarkson: High-power widely tunable Tm: fiber lasers pumped by an Er, Yb codoped fiber laser at 1.6 μm, Opt. Express **14**, 6084 (2006)

11.734 L. E. Nelson, E. P. Ippen, H. A. Haus: Broadly tunable sub-500 fs pulses from an additive-pulse mode-locked thulium-doped fiber ring laser, Appl. Phys. Lett. **67**, 19 (1995)

11.735 R. C. Sharp, D. E. Spock, N. Pan, J. Elliot: 190 fs passively mode-locked thulium fiber laser with a low threshold, Opt. Lett. **21**, 881 (1996)

11.736 S. D. Jackson, T. A. King: High-power diode-cladding-pumped Tm-doped silica fiber laser, Opt. Lett. **23**, 1462 (1998)

11.737 D. C. Hanna, I. R. Perry, J. R. Lincoln, J. E. Townsend: A 1 watt thulium-doped CW fiber laser operating at 2 μm, Opt. Commun. **80**, 52 (1990)

11.738 P. S. Golding, S. D. Jackson, P.-K. Tsai, B. C. Dickinson, T. A. King: Efficient high power operation of a Tm-doped silica fiber laser pumped at 1.319 μm, Opt. Commun. **175**, 179 (2000)

11.739 T. Yamamoto, Y. Miyajima, T. Komukai: 1.9 μm Tm-doped silica fiber laser pumped at 1.57 μm, Electron. Lett. **30**, 220 (1994)

11.740 R. M. Percival, D. Szebesta, C. P. Seltzer, S. D. Perrin, S. T. Davey, M. Louka: 1.6 μm semiconductor diode pumped thulium doped fluoride fiber laser and amplifier of very high efficiency, Electron. Lett. **29**, 2110 (1993)

11.741 T. Yamamoto, Y. Miyajima, T. Komukai, T. Sugawa: 1.9 μm Tm-doped fluoride fiber amplifier and laser pumped at 1.58 μm, Electron. Lett. **29**, 986 (1993)

11.742 S. D. Jackson, T. A. King: Theoretical modeling of Tm-doped silica fiber lasers, J. Lightwave Technol. **17**, 948 (1999)

11.743 G. Frith, D. G. Lancaster, S. D. Jackson: 85 W Tm^{3+}-doped silica fiber laser, Electron. Lett. **41**, 687 (2005)

11.744 J. Q. Xu, M. Prabhu, L. Jianren, K. Ueda, D. Xing: Efficient double-clad thulium-doped fiber laser with a ring cavity, Appl. Opt. **40**, 1983 (2001)

11.745 Y. Jeong, P. Dupriez, J. K. Sahu, J. Nilsson, D. Y. Shen, W. A. Clarkson, S. D. Jackson: Power scaling of 2 μm ytterbium-sensitised thulium-doped silica fiber laser diode-pumped at 975 nm, Electron. Lett. **41**, 1734 (2005)

11.746 R. A. Hayward, W. A. Clarkson, P. W. Turner, J. Nilsson, A. B. Grudinin, D. C. Hanna: Efficient cladding-pumped Tm-doped silica fiber laser with high power singlemode output at 2 μm, Electron. Lett. **36**, 711 (2000)

11.747 S. D. Jackson: 8.8 W diode-cladding-pumped Tm^{3+}, Ho^{3+}-doped fluoride fiber laser, Electron. Lett. **37**, 821 (2001)

11.748 G. H. Rosenblatt, J. F. Pinto, R. C. Stoneman, L. Esterowitz: Continuously tunable 2.3 μm Tm:GSGG laser. In: *Proc. Lasers and Electro-Optics Society Annual Meeting* (IEEE, 1993) p. 689

11.749 J. F. Pinto, L. Esterowitz, G. H. Rosenblatt: Tm^{3+} : YLF laser continuously tunable between 2.20 and 2.46 μm, Opt. Lett. **19**, 883 (1994)

11.750 R. Allen, L. Esterowitz: CW diode pumped 2.3 μm fiber laser, Appl. Phys. Lett. **55**, 721 (1989)

11.751 J.Y. Allain, M. Monerie, H. Poignant: Tunable CW lasing around 0.82, 1.48, 1.88 and 2.35 μm in thulium-doped fluorozirconate fiber, Electron. Lett. **25**, 1660 (1989)

11.752 R.G. Smart, J.N. Carter, A.C. Tropper, D.C. Hanna: Continuous-wave oscillation of Tm^{3+}-doped fluorozirconate fiber lasers at around 1.47 μm, 1.9 μm and 2.3 μm when pumped at 790 nm, Opt. Commun. **82**, 563 (1991)

11.753 R.M. Percival, D. Szebesta, S.T. Davey: Highly efficient and tunable operation of 2 color Tm-doped fluoride fiber laser, Electron. Lett. **28**, 671 (1992)

11.754 K. Tanimura, M.D. Shinn, W.A. Sibley, M.G. Drexhage, R.N. Brown: Optical transitions of Ho^{3+} ions in fluorozirconate glass, Phys. Rev. B **30**, 2429 (1984)

11.755 J.K. Tyminski, D.M. Franich, M. Kokta: Gain dynamics of Tm:Ho:YAG pumped in near infrared, J. Appl. Phys. **65**, 3181 (1989)

11.756 V.A. French, R.R. Petrin, R.C. Powell, M. Kokta: Energy-transfer processes in $Y_3Al_5O_{12}$: Tm,Ho, Phys. Rev. B **46**, 8018 (1992)

11.757 L.F. Johnson: Optical maser characteristics of rare-earth ions in crystals, J. Appl. Phys. **34**, 897 (1963)

11.758 E.P. Chicklis, C.S. Naiman, R.C. Folweiler, D.R. Gabbe, H.P. Jenssen, A. Linz: High-efficiency room-temperature 2.06 μm laser using sensitized Ho^{3+} : YLF, Appl. Phys. Lett. **19**, 119 (1971)

11.759 E.W. Duczynski, G. Huber, V.G. Ostroumov, I.A. Shcherbakov: CW double cross pumping of the $^5I_7-^5I_8$ laser transition in Ho^{3+}-doped garnets, Appl. Phys. Lett. **48**, 1562 (1986)

11.760 D.A. Zubenko, M.A. Noginov, V.A. Smirnov, I.A. Shcherbakov: Interaction of excited holmium and thulium ions in yttrium scandium garnet crystals, J. Appl. Spectrosc. **52**, 391 (1990)

11.761 D.A. Zubenko, M.A. Noginov, S.G. Semenkov, V.A. Smirnov, I.A. Shcherbakov: Interionic interactions in YSGG:Cr:Tm and YSGG:Cr:Tm:Ho laser crystals, Sov. J. Quantum Electron. **22**, 133 (1992)

11.762 L. Esterowitz, R. Allen, L. Goldberg, J.F. Weller, M. Storm, I. Abella: Diode-pumped 2 μm holmium laser. In: *Tunable Solid-State Lasers II*, Springer. Ser. Opt. Sci., Vol. 52, ed. by A.B. Budgor, L. Esterowitz, L.G. Deshazer (Springer, Berlin, Heidelberg 1986) pp. 291–292

11.763 G.J. Kintz, L. Esterowitz, R. Allen: CW diode-pumped Tm^{3+}, Ho^{3+} : YAG 2.1 μm room-temperature laser, Electron. Lett. **23**, 616 (1987)

11.764 T.Y. Fan, G. Huber, R.L. Byer, P. Mitzscherlich: Continuous-wave operation at 2.1 μm of a diode-laser-pumped, Tm-sensitized Ho : $Y_3Al_5O_{12}$ laser at 300 K, Opt. Lett. **12**, 678 (1987)

11.765 H. Nakajima, T. Yokozawa, T. Yamamoto, H. Hara, N. Djeu: Power optimization of 2 μm Tm,Ho:YAG laser in monolithic crystal, Jpn. J. Appl. Phys. **33**, 1010 (1994)

11.766 H. Hemmati: 2.07 μm CW diode-laser-pumped Tm,Ho : LiF_4 room-temperature laser, Opt. Lett. **14**, 435 (1989)

11.767 S.A. Payne, L.K. Smith, W.L. Kway, J.B. Tassano, W.F. Krupke: The mechanisms of Tm to Ho energy transfer in $LiYF_4$, J. Phys. **4**, 8525 (1992)

11.768 T. Rothacher, W. Lüthy, H.P. Weber: Diode pumping and laser properties of Yb,Ho:YAG, Opt. Commun. **155**, 68 (1998)

11.769 S. Taccheo, G. Sorbello, P. Laporta, C. Svelto: Suppression of intensity noise in a diode-pumped Tm,Ho:YAG laser, Opt. Lett. **25**, 1642 (2000)

11.770 C. Svelto, S. Taccheo, M. Marano, G. Sorbello, P. Laporta: Optoelectronic feedback loop for relaxation oscillation intensity noise suppression in Tm,Ho:YAG laser, Electron. Lett. **36**, 1623 (2000)

11.771 G.J. Koch, M. Petros, J.R. Yu, U.N. Singh: Precise wavelength control of a single-frequency pulsed Ho,Tm:YLF laser, Appl. Opt. **41**, 1718 (2002)

11.772 P. Laporta, M. Marano, L. Pallaro, S. Taccheo: Amplitude and frequency stabilisation of a Tm,Ho:YAG laser for coherent lidar applications at 2.1 μm, Opt. Lasers Eng. **37**, 447 (2002)

11.773 G. Galzerano, M. Marano, S. Taccheo, P. Laporta: 2.1 μm lasers frequency stabilized against CO_2 lines: comparison between fringe-side and frequency-modulation locking methods, Opt. Lett. **28**, 248 (2003)

11.774 B.T. McGuckin, R.T. Menzies: Efficient CW diode-pumped Tm, Ho:YLF laser with tunability near 2.067 μm, IEEE J. Quantum Electron. **28**, 1025 (1992)

11.775 A. Di Lieto, P. Minguzzi, A. Toncelli, M. Tonelli, H.P. Jenssen: A diode-laser-pumped tunable Ho:YLF laser in the 2 μm region, Appl. Phys. B **57**, 3172 (1993)

11.776 E. Sorokin, I.T. Sorokina, A. Unterhuber, E. Wintner, A.I. Zagumennyi, I.A. Shcherbakov, V. Carozza, A. Toncelli, M. Tonelli: A novel CW tunable and mode-locked 2 μm Cr,Tm,Ho:YSGG:GSAG laser. In: *Adv. Solid-State Lasers*, OSA Trends Opt. Photon., Vol. 19, ed. by W.R. Bosenberg, M.M. Fejer (Opt. Soc. Am., Washington 1998) pp. 197–200

11.777 F. Cornacchia, E. Sani, A. Toncelli, M. Tonelli, M. Marano, S. Taccheo, G. Galzerano, P. Laporta: Optical spectroscopy and diode-pumped laser characteritics of codoped Tm-Ho:YLF and Tm-Ho:BaYF: a comparative analysis, Appl. Phys. B **75**, 817 (2002)

11.778 G. Galzerano, E. Sani, A. Toncelli, G. Della Valle, S. Taccheo, M. Tonelli, P. Laporta: Widely tunable continuous-wave diode-pumped 2 μm Tm,Ho : KYF_4 laser, Opt. Lett. **29**, 715 (2004)

11.779 K.L. Schepler, B.D. Smith, F. Heine, P.A. Budni: Mode-locking of a diode-pumped Tm,Ho:YLF, OSA Proc. Adv. Solid-State Lasers **20**, 257–259 (1994)

11.780 G. Galzerano, M. Marano, S. Longhi, E. Sani, A. Toncelli, M. Tonelli, P. Laporta: Sub-100 ps amplitude-modulation mode-locked Tm,Ho: BaY_2F_8 laser at 2.06 μm, Opt. Lett. **28**, 2085 (2003)

11.781 A. Finch, J. H. Flint: Diode-pumped 6 mJ repetitively-Q-switched Tm,Ho:YLF laser, Conf. Lasers Electro-Optics, Tech. Dig. Ser. **15**, 232 (1995)

11.782 J. R. Yu, U. N. Singh, N. P. Barnes, M. Petros: 125 mJ diode-pumped injection-seeded Ho,Tm:YLF laser, Opt. Lett. **23**, 780 (1998)

11.783 A. M. Malyarevich, P. V. Prokoshin, M. I. Demchuk, K. V. Yumashev, A. A. Lipovskii: Passively Q-switched $Ho^{3+}:Y_3Al_5O_{12}$ laser using a PbSe-doped glass, Appl. Phys. Lett. **78**, 572 (2001)

11.784 M. Frenz, H. Pratisto, F. Konz, E. D. Jansen, A. J. Welch, H. P. Weber: Comparison of the effects of absorption coefficient and pulse duration of 2.12 μm and 2.79 μm radiation on laser ablation of tissue, IEEE J. Quantum Electron. **32**, 2025 (1996)

11.785 M. A. Noginov, S. G. Semenkov, I. A. Shcherbakov, V. A. Smirnov: Energy transfer (Tm to Ho) and upconversion process in $YSGG:Cr^{3+}:Tm^{3+}:Ho^{3+}$ laser crystals, OSA Proc. Adv. Solid State Lasers **10**, 178–182 (1991)

11.786 E. Wintner, F. Krausz, M. A. Noginov, V. A. Smirnov, I. T. Sorokina, C. Spielmann, I. A. Shcherbakov: Interaction of Ho^{3+} and Tm^{3+} ions in $YSGG:Cr^{3+}:Tm^{3+}:Ho^{3+}$ at the strong selective excitation, Laser Phys. **2**, 138 (1992)

11.787 R. R. Petrin, M. G. Jani, R. C. Powell, M. Kokta: Spectral dynamics of laser-pumped $Y_3Al_5O_{12}$:Tm,Ho lasers, Opt. Mater. **1**, 111 (1992)

11.788 L. B. Shaw, R. S. F. Chang, N. Djeu: Measurement of up-conversion energy-transfer probabilities in $Ho:Y_3Al_5O_{12}$ and $Tm:Y_3Al_5O_{12}$, Phys. Rev. B **50**, 6609 (1994)

11.789 G. Rustad, K. Stenersen: Modeling of laser-pumped Tm and Ho lasers accounting for up-conversion and ground-state depletion, IEEE J. Quantum Electron. **32**, 1645 (1996)

11.790 C. D. Nabors, J. Ochoa, T. Y. Fan, A. Sanchez, H. K. Choi, G. W. Turner: Ho:YAG laser pumped by 1.9 μm diode lasers, IEEE J. Quantum Electron. **31**, 1603 (1995)

11.791 C. Bollig, R. A. Hayward, W. A. Clarkson, D. C. Hanna: 2 W Ho:YAG laser intracavity pumped by a diode-pumped Tm:YAG laser, Opt. Lett. **23**, 1757 (1998)

11.792 A. I. R. Malm, R. Hartmann, R. C. Stoneman: High-power eyesafe YAG lasers for coherent laser radar. In: *Adv. Solid-State Lasers*, OSA Trends Opt. Photon., Vol. 94, ed. by G. J. Quarles (Opt. Soc. Am., Washington 2004) pp. 356–361

11.793 D. W. Hart, M. Jani, N. P. Barnes: Room-temperature lasing of end-pumped $Ho:Lu_3Al_5O_{12}$, Opt. Lett. **21**, 728 (1996)

11.794 D. Y. Shen, A. Abdolvand, L. J. Cooper, W. A. Clarkson: Efficient Ho:YAG laser pumped by a cladding-pumped tunable Tm:silica-fiber laser, Appl. Phys. B **79**, 559 (2004)

11.795 D. Y. Shen, W. A. Clarkson, L. J. Cooper, R. B. Williams: Efficient single-axial-mode operation of a Ho:YAG ring laser pumped by a Tm-doped silica fiber laser, Opt. Lett. **29**, 2396 (2004)

11.796 M. C. Brierley, P. W. France, C. A. Millar: Lasing at 2.08 μm and 1.38 μm in a holmium doped fluorozirconate fiber laser, Electron. Lett. **24**, 539 (1988)

11.797 D. C. Hanna, R. M. Percival, R. G. Smart, J. E. Townsend, A. C. Tropper: Continuous-wave oscillation of holmium-doped silica fiber laser, Electron. Lett. **25**, 593 (1989)

11.798 A. S. Kurkov, E. M. Dianov, O. I. Medvedkov, G. A. Ivanov, V. A. Aksenov, V. M. Paramonov, S. A. Vasiliev, E. V. Pershina: Efficient silica-based Ho^{3+} fiber laser for 2 μm spectral region pumped at 1.15 μm, Electron. Lett. **36**, 1015 (2000)

11.799 S. D. Jackson: 2.7 W Ho^{3+}-doped silica fiber laser pumped at 1100 nm and operating at 2.1 μm, Appl. Phys. B **76**, 793 (2003)

11.800 J. Y. Allain, M. Monerie, H. Poignant: High-efficiency CW thulium-sensitized holmium-doped fluoride fiber laser operating at 2.04 μm, Electron. Lett. **27**, 1513 (1991)

11.801 R. M. Percival, D. Szebesta, S. T. Davey, N. A. Swain, T. A. King: Thulium sensitized holmium-doped CW fluoride fiber laser of high-efficiency, Electron. Lett. **28**, 2231 (1992)

11.802 K. Oh, T. F. Morse, A. Kilian, L. Reinhart, P. M. Weber: Continuous-wave oscillation of thulium-sensitized holmium-doped silica fiber laser, Opt. Lett. **19**, 278 (1994)

11.803 C. Ghisler, W. Lüthy, H. P. Weber, J. Morel, A. Woodtli, R. Dändliker, V. Neuman, H. Berthou, G. Kotrotsios: A Tm^{3+} sensitized Ho^{3+} silica fiber laser at 2.04 μm pumped at 809 nm, Opt. Commun. **109**, 279 (1994)

11.804 S. D. Jackson, T. A. King: CW operation of a 1.064 μm pumped Tm-Ho-doped silica fiber laser, IEEE J. Quantum Electron. **34**, 1578 (1998)

11.805 S. D. Jackson, S. Mossman: High power diode-cladding-pumped Tm^{3+}, Ho^{3+}-doped silica fiber laser, Appl. Phys. B **77**, 489 (2003)

11.806 S. D. Jackson, S. Mossman: Diode-cladding-pumped Yb^{3+}, Ho^{3+}-doped silica fiber laser operating at 2.1 μm, Appl. Opt. **42**, 3546 (2003)

11.807 T. Sumiyoshi, H. Sekita: Dual-wavelength continuous-wave cascade oscillation at 3 and 2 μm with a holmium-doped fluoride-glass fiber laser, Opt. Lett. **23**, 1837 (1998)

11.808 A. A. Kaminskii, T. I. Butaeva, A. O. Ivanov, I. V. Mochalov, A. G. Petrosyan, G. I. Rogov, V. A. Fedorov: New data on stimulated emission of crystals containing Er^{3+} and Ho^{3+} ions, Sov. Tech. Phys. Lett. **2**, 308 (1976)

11.809 A. Diening, P. E.-A. Möbert, E. Heumann, G. Huber, B. H. T. Chai: Diode-pumped CW lasing of Yb,Ho : KYF_4 in the 3 μm spectral range in comparison to Er : KYF_4, Laser Phys. **8**, 214 (1998)

11.810 W. S. Rabinovich, S. R. Bowman, B. J. Feldman, M. J. Winings: Tunable laser pumped 3 μm Ho : $YAlO_3$ laser, IEEE J. Quantum Electron. **27**, 895 (1990)

11.811 A. Diening, S. Kück: Spectroscopy and diode-pumped laser oscillation of Yb^{3+},Ho^{3+}-doped yttrium scandium gallium garnet, J. Appl. Phys. **87**, 4063 (2000)

11.812 A. M. Umyskov, Y. D. Zavartsev, A. I. Zagumennyi, V. V. Osiko, P. A. Studenikin: Cr^{3+}:Yb^{3+}:Ho^{3+}:YSGG crystal laser with a continuously tunable emission wavelength in the range 2.84–3.05 μm, Kvant. Elektron. **23**, 579 (1996)[English transl.: Sov. J. Quantum Electron. **26**, 563 (1996)]

11.813 Y. D. Zavartsev, A. I. Zagumennyi, L. A. Kulevskii, A. V. Lukashev, P. A. Studenikin, I. A. Shcherbakov, A. F. Umyskov: *Q*-switching in a Cr^{3+}:Yb^{3+}:Ho^{3+}: YSGG crystal laser based on the $^5I_6-^5I_7$ ($\lambda = 2.92$ μm) transition, Kvant. Elektron. **27**, 13 (1999)[English transl.: J. Quantum Electron. **29**, 295 (1999)]

11.814 Y. D. Zavartsev, V. V. Osiko, S. G. Semenkov, P. A. Studenikin, A. F. Umyskov: Cascade laser oscillation due to Ho^{3+} ions in a (Cr,Yb,Ho):YSGG yttrium-scandium-gallium-garnet crystal, Kvant. Elektron. **20**, 366 (1993)[English transl.: Sov. J. Quantum Electron. **23**, 312 (1993)]

11.815 B. M. Walsh, K. E. Murray, N. P. Barnes: Cr:Er:Tm:Ho: yttrium aluminum garnet laser exhibiting dual wavelength lasing at 2.1 μm and 2.9 μm: Spectroscopy and laser performance, J. Appl. Phys. **91**, 11 (2002)

11.816 L. Wetenkamp: Efficient CW operation of a 2.9 μm Ho^{3+}-doped fluorozirconate fiber laser pumped at 640 nm, Electron. Lett. **26**, 883 (1990)

11.817 T. Sumiyoshi, H. Sekita, T. Arai, S. Sato, M. Ishihara, M. Kikuchi: High-power continuous-wave 3- and 2 μm cascade Ho^{3+} : ZBLAN fiber laser and its medical applications, IEEE J. Select. Top. Quantum Electron. **5**, 936 (1999)

11.818 S. D. Jackson: Single-transverse-mode 2.5 W holmium-doped fluoride fiber laser operating at 2.86 μm, Opt. Lett. **29**, 334 (2004)

11.819 B. Peng, T. Izumitani: Ho^{3+} doped 2.84 μm laser glass for laser knives, sensitised by Yb^{3+}, Rev. Laser Eng. **22**, 9 (1994)

11.820 S. D. Jackson: Singly Ho^{3+}-doped fluoride fiber laser operating at 2.92 μm, Electron. Lett. **40**, 1400 (2004)

11.821 P. S. Golding, S. D. Jackson, T. A. King, M. Pollnau: Energy-transfer processes in Er^{3+}-doped and Er^{3+},Pr^{3+}-codoped ZBLAN glasses, Phys. Rev. B **62**, 856 (2000)

11.822 E. V. Zharikov, V. I. Zhekov, L. A. Kulevskii, T. M. Murina, V. V. Osiko, A. M. Prokhorov, A. D. Savel'ev, V. V. Smirnov, B. P. Starikov, M. I. Timoshechkin: Stimulated emission from Er^{3+} ions in yttrium aluminum garnet crystals at $\lambda = 2.94$ μ, Kvant. Elektron. **1**, 1867 (1974)[English transl.: Sov. J. Quantum Electron. **4**, 1039 (1975)]

11.823 K. S. Bagdasarov, V. I. Zhekov, V. A. Lobachev, T. M. Murina, A. M. Prokhorov: Steady-state emission from a $Y_3Al_5O_{12}$: Er^{3+} laser ($\lambda = 2.94$ μ, $T = 300$ K), Kvant. Elektron. **10**, 452 (1983)[English transl.: Sov. J. Quantum Electron. **13**, 262 (1983)]

11.824 V. I. Zhekov, B. V. Zubov, V. A. Lobachev, T. M. Murina, A. M. Prokhorov, A. F. Shevel: Mechanisms of a population inversion between the $^4I_{11/2}$ and $^4I_{13/2}$ levels of the Er^{3+} ion in $Y_3Al_5O_{12}$ crystals, Kvant. Elektron. **7**, 749 (1980)[English transl.: Sov. J. Quantum Electron. **10**, 428 (1980)]

11.825 V. I. Zhekov, V. A. Lobachev, T. M. Murina, A. M. Prokhorov: Efficient cross-relaxation laser emitting at $\lambda = 2.94$ μ, Kvant. Elektron. **10**, 1871 (1983)[English transl.: Sov. J. Quantum Electron. **13**, 1235 (1983)]

11.826 V. I. Zhekov, V. A. Lobachev, T. M. Murina, A. M. Prokhorov: Cooperative phenomena in yttrium erbium aluminum garnet crystals, Kvant. Elektron. **11**, 189 (1984)[English transl.: Sov. J. Quantum Electron. **14**, 128 (1984)]

11.827 V. I. Zhekov, T. M. Murina, A. M. Prokhorov, M. I. Studenikin, S. Georgescu, V. Lupei, I. Ursu: Coorporative process in $Y_3Al_5O_{12}$: Er^{3+} crystals, Kvant. Electron. **13**, 419 (1986)[English transl.: Sov. J. Quantum Electron. **16**, 274 (1986)]

11.828 D. L. Dexter: A theory of sensitized luminescence in solids, J. Chem. Phys. **21**, 836 (1953)

11.829 M. Pollnau, D. R. Gamelin, S. R. Lüthi, H. U. Güdel, M. P. Hehlen: Power dependence of upconversion luminescence in lanthanide and transition-metal-ion systems, Phys. Rev. B **61**, 3337 (2000)

11.830 S. A. Pollack, D. B. Chang: Ion-pair upconversion pumped laser emission in Er^{3+} ions in YAG, YLF, SrF_2 and CaF_2 crystals, J. Appl. Phys. **64**, 2885 (1988)

11.831 M. Pollnau, R. Spring, C. Ghisler, S. Wittwer, W. Lüthy, H. P. Weber: Efficiency of erbium 3 μm crystal and fiber lasers, IEEE J. Quantum Electron. **32**, 657 (1996)

11.832 M. Pollnau, W. Lüthy, H. P. Weber, T. Jensen, G. Huber, A. Cassanho, H. P. Jenssen, R. A. McFarlane: Investigation of diode-pumped 2.8 μm laser performance in Er : BaY_2F_8, Opt. Lett. **21**, 48 (1996)

11.833 H. J. Eichler, J. Findeisen, B. Liu, A. A. Kaminskii, A. V. Butachin, P. Peuser: Highly efficient diode-pumped 3 μm Er^{3+} : BaY_2F_8 laser, IEEE J. Select. Top. Quantum Electron. **3**, 90 (1997)

11.834 T. Jensen, V. G. Ostroumov, G. Huber: Upconversion processes in Er^{3+} : YSGG and diode pumped laser experiments at 2.8 μm, OSA Proceedings on Adv. Solid-State Lasers **24**, 366–370 (1995)

11.835 R. Groß: Besetzungsdynamik und Wechselwirkungsprozesse in blitzlampengepumpten 3 μm

Er^{3+}-Lasern Ph.D. Thesis (University of Hamburg, Hamburg 1992)

11.836 M. Pollnau, S. D. Jackson: Erbium 3 μm fiber lasers, IEEE J. Select. Top. Quantum Electron. **7**, 30 (2001)Correction, – **8**, 956 (2002)

11.837 C. Wyss, W. Lüthy, H. P. Weber, P. Rogin, J. Hulliger: Emission properties of an optimised 2.8 μm Er^{3+} : YLF laser, Opt. Commun. **139**, 215 (1997)

11.838 R. C. Stoneman, J. G. Lynn, L. Esterowitz: Direct upper-state pumping of the 2.8 μm Er^{3+} : YLF laser, IEEE J. Quantum Electron. **28**, 1041 (1992)

11.839 T. Jensen, G. Huber, K. Petermann: Quasi-CW diode pumped 2.8 μm laser operation of Er^{3+}-doped garnets. In: *Adv. Solid-State Lasers*, Vol. 1, ed. by S. A. Payne, C. R. Pollock (Opt. Soc. Am., Washington 1996) pp. 306–308

11.840 A. M. Prokhorov, V. I. Zhekov, T. M. Murina, N. N. Plantov: Pulsed YAG : Er^{3+} laser efficiency (analysis of model equations), Laser Phys. **3**, 79 (1993)

11.841 M. Pollnau, R. Spring, S. Wittwer, W. Lüthy, H. P. Weber: Investigations on the slope efficiency of a pulsed 2.8 μm Er^{3+} : $LiYF_4$ laser, J. Opt. Soc. Am. B **14**, 974 (1997)

11.842 B. Schmaul, G. Huber, R. Clausen, B. Chai, P. Li Kam Wa, M. Bass: Er^{3+} : $YLiF_4$ continuous wave cascade laser operation at 1620 and 2810 nm at room temperature, Appl. Phys. Lett. **62**, 541 (1993)

11.843 M. Pollnau, T. Graf, J. E. Balmer, W. Lüthy, H. P. Weber: Explanation of the CW operation of the Er^{3+} 3 μm crystal laser, Phys. Rev. A **49**, 3990 (1994)

11.844 E. V. Zharikov, V. V. Osiko, A. M. Prokhorov, I. A. Shcherbakov: Crystals of rare-earth gallium garnets with chromium as active media for solid-state lasers, Inorg. Mater. **48**, 81 (1984)

11.845 P. F. Moulton, J. G. Manni, G. A. Rines: Spectroscopic and laser characteristics of Er,Cr:YSGG, IEEE J. Quantum Electron. **24**, 960 (1988)

11.846 A. M. Prokhorov, A. A. Kaminskii, V. V. Osiko, M. I. Timoshechkin, E. V. Zharikov, T. I. Butaeva, S. E. Sarkisov, A. G. Pretosyan, V. A. Fedorov: Investigations of the 3 μm stimulated emission from Er^{3+} ions in aluminum garnets at room temerature, phys. stat. sol. (a) **40**, K69 (1977)

11.847 G. J. Kintz, R. E. Allen, L. Esterowitz: Diode-pumped 2.8 μm laser emission from Er^{3+} : YLF at room temperature, Appl. Phys. Lett. **50**, 1553 (1987)

11.848 S. A. Pollack, D. Chang, N. L. Moise: Continuous wave and *Q*-switched infrared erbium laser, Appl. Phys. Lett. **49**, 1578 (1986)

11.849 F. Auzel, S. Hubert, D. Meichenin: Multifrequency room-temperature continuous diode and Ar-laser-pumped Er^{3+} laser emission between 2.66 and 2.85 μm, Appl. Phys. Lett. **54**, 681 (1989)

11.850 S. Hubert, D. Meichenin, B. W. Zhou, F. Auzel: Emission properties, oscillator strengths and laser parameters of Er^{3+} in $LiYF_4$ at 2.7 μm, J. Lumin. **50**, 7 (1991)

11.851 T. Jensen: Upconversion-Prozesse und Wirkungsquerschnitte in Er^{3+}-dotierten 3 μm Fluorid- und Granat-Lasern, gepumpt mit CW und quasi-CW Dioden-Arrays Ph.D. Thesis (University of Hamburg, Hamburg 1996)

11.852 R. H. Page, R. A. Bartels, R. J. Beach, S. B. Sutton, L. H. Furu, J. E. LaSala: 1 W composite-slab Er:YAG laser. In: *Adv. Solid-State Lasers*, OSA Trends Opt. Photon., Vol. 10, ed. by C. R. Pollock, W. R. Bosenberg (Opt. Soc. Am., Washington 1997) pp. 214–216

11.853 D.-W. Chen, C. L. Fincher, T. S. Rose, F. L. Vernon, R. A. Fields: Diode-pumped 1 W continuous-wave Er:YAG 3 μm laser, Opt. Lett. **24**, 385 (1999)

11.854 T. Sandrock, A. Diening, G. Huber: Laser emission of erbium-doped fluoride bulk glasses in the spectral range from 2.7 to 2.8 μm, Opt. Lett. **24**, 382 (1999)

11.855 A. Y. Dergachev, J. H. Flint, P. F. Moulton: 1.8 W CW Er:YLF diode-pumped laser, Conf. Lasers Electro-Optics, Tech. Dig. 2000 (Opt. Soc. Am., Washington 2000) 564

11.856 A. Dergachev, P. Moulton: Tunable CW Er:YLF diode-pumped laser. In: *Adv. Solid-State Lasers*, OSA Trends Opt. Photon., Vol. 83, ed. by J. J. Zayhowski (Opt. Soc. Am., Washington 2003) pp. 3–5

11.857 B. J. Dinerman, P. F. Moulton: 3 μm CW laser operations in erbium-doped YSGG, GGG, and YAG, Opt. Lett. **19**, 1143 (1994)

11.858 R. Waarts, D. Nam, S. H. Sanders, J. Harrison, B. J. Dinerman: Two dimensional Er:YSGG microlaser array pumped with a monolithic two-dimensional laser diode array, Opt. Lett. **19**, 1738 (1994)

11.859 S. Schnell, V. G. Ostroumov, J. Breguet, W. A. R. Lüthy, H. P. Weber, I. A. Shcherbakov: Acoustooptic *Q*-switching of erbium lasers, IEEE J. Quantum Electron. **26**, 1111 (1990)

11.860 J. Breguet, A. F. Umyskov, W. A. R. Lüthy, I. A. Shcherbakov, H. P. Weber: Electrooptically *Q*-switched 2.79 μm YSGG:Cr:Er laser with an intracavity polarizer, IEEE J. Quantum Electron. **27**, 274 (1991)

11.861 H. Voss, F. Massmann: Diode pumped *Q* switched erbium lasers with short pulse duration. In: *Adv. Solid-State Lasers*, OSA Trends Opt. Photon., Vol. 10, ed. by C. R. Pollock, W. R. Bosenberg (Opt. Soc. Am., Washington 1997) pp. 217–221

11.862 N. M. Wannop, M. R. Dickinson, A. Charlton, T. A. King: *Q*-switching the erbium-YAG laser, J. Mod. Opt. **41**, 2043 (1994)

11.863 P. Maak, L. Jacob, P. Richter, H. J. Eichler, B. Liu: Efficient acousto-optic *Q*-switching of Er:YSGG lasers at 2.79 μm wavelength, Appl. Opt. **39**, 3053 (2000)

11.864 K. S. Bagdasarov, V. P. Danilov, V. I. Zhekov, T. M. Murina, A. A. Manenkov, M. I. Timoshechkin, A. M. Prokhorov: Pulse-periodic $Y_3Al_5O_{12}$: Er^{3+} laser with high activator concentration, Sov. J. Quantum Electron. **8**, 83 (1978)

11.865 F. Könz, M. Frenz, V. Romano, M. Forrer, H. P. Weber, A. V. Kharkovskiy, S. I. Khomenko: Active and

passive Q-switching of a 2.79 μm Er:Cr:YSGG laser, Opt. Commun. **103**, 398 (1993)
11.866 A. Högele, G. Hörbe, H. Lubatschowski, H. Welling, W. Ertmer: 2.70 μm Cr:Er:YSGG with high output energy and FTIR-Q-switch, Opt. Commun. **125**, 90 (1996)
11.867 H.J. Eichler, B. Liu, M. Kayser, S.I. Khomenko: Er:YAG-laser at 2.94 μm Q-switched by a FTIR-shutter with silicon output coupler and polarizer, Opt. Mater. **5**, 259 (1996)
11.868 M. Ozolinsh, K. Stock, R. Hibst, R. Steiner: Q-switching of Er:YAG (2.9 μm) solid state laser by PLZT electrooptic modulator, IEEE J. Quantum Electron. **33**, 1846 (1997)
11.869 K.L. Vodopyanov, A.V. Lukashev, C.C. Philips, I.T. Ferguson: Passive mode locking and Q-switching of an erbium 3 μm laser using thin InAs epilayers grown by molecular beam epitaxy, Appl. Phys. Lett. **59**, 1658 (1991)
11.870 K.L. Vodopyanov, L.A. Kulevskii, P.P. Pashinin, A.M. Prokhorov: Water and ethanol as bleachable absorbers of radiation in an yttrium-erbium-aluminum garnet laser ($\lambda = 2.94$ μm), Sov. Phys. JETP **55**, 1049 (1982)
11.871 K.L. Vodopyanov, R. Shori, O.M. Stafsudd: Generation of Q-switched Er:YAG laser pulses using evanescent wave absorption in ethanol, Appl. Phys. Lett. **72**, 2211 (1998)
11.872 J. Breguet, W. Lüthy, H.P. Weber: Q-switching of YAG:Er laser with a soap film, Opt. Commun. **82**, 488 (1991)
11.873 B. Pelz, M.K. Schott, M.H. Niemz: Electro-optic mode locking of an Erbium:YAG laser with a RF resonance transformer, Appl. Opt. **33**, 364 (1994)
11.874 M. Pollnau, C. Ghisler, W. Lüthy, H.P. Weber: Cross-sections of excited-state absorption at 800 nm in erbium-doped ZBLAN fiber, Appl. Phys. B **67**, 23 (1998)
11.875 M.C. Brierley, P.W. France: Continuous wave lasing at 2.71 μm in an erbium-doped fluorozirconate fiber, Electron. Lett. **24**, 935 (1988)
11.876 J.Y. Allain, M. Monerie, H. Poignant: Erbium-doped fluorozirconate single-mode fiber lasing at 2.71 μm, Electron. Lett. **25**, 28 (1989)
11.877 R. Allen, L. Esterowitz, R.J. Ginther: Diode-pumped single-mode fluorozirconate fiber laser from the $^4I_{11/2} \rightarrow ^4 I_{13/2}$ transition in erbium, Appl. Phys. Lett. **56**, 1635 (1990)
11.878 R.S. Quimby, W.J. Miniscalco: Continuous-wave lasing on a self-terminating transition, Appl. Opt. **28**, 14 (1989)
11.879 A.A. Kaminskii, A.G. Petrosyan, G.A. Denisenko, T.I. Butaeva, V.A. Fedorov, S.E. Sarkisov: Spectroscopic properties and 3 μm stimulated emission of Er^{3+} ions in the $(Y_{1-x}Er_x)_3Al_5O_{12}$ and $(Lu_{1-x}Er_x)_3Al_5O_{12}$ garnet crystal systems, phys. stat. sol. (a) **71**, 291 (1982)
11.880 V. Lupei, S. Georgescu, V. Florea: On the dynamics of population inversion for 3 μm Er^{3+} lasers, IEEE J. Quantum Electron. **29**, 426 (1993)
11.881 J. Schneider: Kaskaden-Faserlaser im mittleren Infrarot Ph.D. Thesis (Cuvillier Verlag, Göttingen 1996)
11.882 B.C. Dickinson, P.S. Golding, M. Pollnau, T.A. King, S.D. Jackson: Investigation of a 791 nm pulsed-pumped 2.7 μm Er-doped ZBLAN fiber laser, Opt. Commun. **191**, 315 (2001)
11.883 N.J.C. Libatique, J. Tafoja, N.K. Viswanathan, R.K. Jain, A. Cable: 'Field-usable' diode-pumped ≈ 120 nm wavelength-tunable CW mid-IR fiber laser, Electron. Lett. **36**, 791 (2000)
11.884 M. Pollnau, E. Heumann, G. Huber: Time-resolved spectra of excited-state absorption in Er^{3+} doped $YAlO_3$, Appl. Phys. A **54**, 404 (1992)
11.885 S. Bedö, M. Pollnau, W. Lüthy, H.P. Weber: Saturation of the 2.71 μm laser output in erbium doped ZBLAN fibers, Opt. Commun. **116**, 81 (1995)
11.886 R.S. Quimby, W.J. Miniscalco, B. Thompson: Excited state absorption at 980 nm in erbium doped glass, SPIE Vol. **1581**, 72–79 (1991)
11.887 J. Schneider, D. Hauschild, C. Frerichs, L. Wetenkamp: Highly efficient $Er^{3+} : Pr^{3+}$-codoped CW fluorozirconate fiber laser operating at 2.7 μm, Int. J. Infrared Millimeter Waves **15**, 1907 (1994)
11.888 J. Schneider: Mid-infrared fluoride fiber lasers in multiple cascade operation, IEEE Photon. Technol. Lett. **7**, 354 (1995)
11.889 M. Pollnau, C. Ghisler, G. Bunea, M. Bunea, W. Lüthy, H.P. Weber: 150 mW unsaturated output power at 3 μm from a single-mode-fiber erbium cascade laser, Appl. Phys. Lett. **66**, 3564 (1995)
11.890 M. Pollnau, C. Ghisler, W. Lüthy, H.P. Weber, J. Schneider, U.B. Unrau: Three-transition cascade erbium laser at 1.7, 2.7, and 1.6 μm, Opt. Lett. **22**, 612 (1997)
11.891 S.D. Jackson, T.A. King, M. Pollnau: Diode-pumped 1.7 W erbium 3 μm fiber laser, Opt. Lett. **24**, 1133 (1999)
11.892 X.S. Zhu, R. Jain: High power (> 8 Watts CW) diode-pumped mid-infrared fiber lasers, Mid-Infrared Coherent Sources Conference (Barcelona 2005) Tu7
11.893 J.Y. Allain, M. Monerie, H. Poignant: Energy transfer in Er^{3+}/Pr^{3+}-doped fluoride glass fibers and application to lasing at 2.7 μm, Electron. Lett. **27**, 445 (1991)
11.894 L. Wetenkamp, G.F. West, H. Többen: Co-doping effects in $Erbium^{3+}$- and $Holmium^{3+}$-doped ZBLAN glass, J. Non-Crystal Solids **140**, 25 (1992)
11.895 M. Pollnau: The route toward a diode-pumped 1 W erbium 3 μm fiber laser, IEEE J. Quantum Electron. **33**, 1982 (1997)
11.896 S.D. Jackson, T.A. King, M. Pollnau: Modelling of high-power diode-pumped erbium 3 μm fiber lasers, J. Mod. Opt. **47**, 1987 (2000)
11.897 B. Srinivasan, J. Tafoya, R.K. Jain: High-power 'Watt-level' CW operation of diode-pumped 2.7 μm

fiber lasers using efficient cross-relaxation and energy transfer mechanisms, Opt. Express **4**, 490 (1999)
11.898 S. D. Jackson, T. A. King, M. Pollnau: Efficient high power operation of erbium 3 μm fiber laser diode-pumped at 975 nm, Electron. Lett. **36**, 223 (2000)
11.899 X. S. Zhu, R. Jain: Scaling up laser diode pumped mid-infrared fiber laser to 10 Watt level, Conf. Lasers Electro-Optics, Tech. Dig. 2005 (Opt. Soc. Am., Washington 2005) CTuC3
11.900 T. Huber, W. Lüthy, H. P. Weber, D. F. Hochstrasser: *Q*-switching of a diode cladding-pumped erbium fiber laser at 2.7 μm, Opt. Quantum Electron. **31**, 1171 (1999)
11.901 N. J. C. Libatique, J. D. Tafoja, R. K. Jain: A compact diode-pumped passively *Q*-switched mid-IR fiber laser, Conf. Lasers Electro-Optics, Tech. Dig. 2000 (Opt. Soc. Am., Washington 2000) 76
11.902 M. Pollnau, S. D. Jackson: Energy recycling versus lifetime quenching in erbium-doped 3 μm fiber lasers, IEEE J. Quantum Electron. **38**, 162 (2002)
11.903 B. Srinivasan, E. Poppe, J. Tafoya, R. K. Jain: High-power (400 mW) diode-pumped 2.7 μm Er:ZBLAN fiber lasers using enhanced Er-Er cross-relaxation processes, Electron. Lett. **35**, 1338 (1999)
11.904 T. Sandrock, D. Fischer, P. Glas, M. Leitner, M. Wrage, A. Diening: Diode-pumped 1 W Er-doped fluoride glass M-profile fiber laser emitting at 2.8 μm, Opt. Lett. **24**, 1284 (1999)
11.905 T. Segi, K. Shima, T. Sakai, H. Hosoya: 3 μm – band high output erbium-doped fiber lasers, Conf. Lasers Electro-Optics, Tech. Dig. 2004 (Opt. Soc. Am., Washington 2004) CThZ5
11.906 N. Djeu, V. E. Hartwell, A. A. Kaminskii, A. V. Butashin: Room-temperature 3.4 μm Dy : $BaYb_2F_8$ laser, Opt. Lett. **22**, 997 (1997)
11.907 S. D. Jackson: Continuous wave 2.9 μm dysprosium-doped fluoride fiber laser, Appl. Phys. Lett. **83**, 1316 (2003)
11.908 Y. H. Tsang, A. El-Taher, T. A. King, S. D. Jackson: Efficient 2.96 μm dysprosium-doped fluoride fiber laser pumped with a Nd:YAG laser operating at 1.3 μm, Opt. Express **14**, 678 (2006)
11.909 H. P. Jenssen, M. A. Noginov, A. Cassanho: U:YLF, a prospective 2.8 μm laser crystal, OSA Proc. Adv. Solid-State Lasers **15**, 463–467 (1993)
11.910 D. Meichenin, F. Auzel, S. Hubert, E. Simoni, M. Louis, J. Y. Gesland: New room-temperature CW laser at 2.82 μm $U^{3+}/LiYF_4$, Electron. Lett. **30**, 1309 (1994)
11.911 J. F. Pinto, G. F. Rosenblatt, L. Esterowitz: Continuous-wave laser action in Er^{3+} : YLF at 3.41 μm, Electron. Lett. **30**, 1596 (1994)
11.912 S. R. Bowman, S. K. Searles, N. W. Jenkins, S. B. Qadri, E. F. Skelton, J. Ganem: Diode pumped room temperature mid-infrared erbium laser. In: *Adv. Solid-State Lasers*, OSA Trends Opt. Photon., Vol. 50, ed. by C. Marshall (Opt. Soc. Am., Washington 2001) pp. 154–156
11.913 A. M. Tabirian, H. P. Jenssen, A. Cassanho: Efficient, room temperature mid-infrared laser at 3.9 μm in Ho : BaY_2F_8. In: *Adv. Solid-State Lasers*, OSA Trends Opt. Photon., Vol. 50, ed. by C. Marshall (Opt. Soc. Am., Washington 2001) pp. 170–176
11.914 M. C. Nostrand, R. H. Page, S. A. Payne, W. F. Krupke, P. G. Schunemann, L. I. Isaenko: Room-temperature $CaGa_2S_4$: Dy^{3+} laser action at 2.43 and 4.31 μm and KPb_2Cl_5 : Dy^{3+} laser action at 2.43 μm. In: *Adv. Solid-State Lasers*, OSA Trends Opt. Photon., Vol. 26, ed. by M. M. Fejer, H. Injeyan, U. Keller (Opt. Soc. Am., Washington 1999) pp. 441–449
11.915 S. R. Bowman, J. Ganem, B. J. Feldman, A. W. Kueny: Infrared laser characteristics of praseodymium-doped lanthanum trichloride, IEEE J. Quantum Electron. **30**, 2925 (1994)
11.916 S. R. Bowman, L. B. Shaw, B. J. Feldman, J. Ganem: A 7 μm praseodymium-based solid-state laser, IEEE J. Quantum Electron. **32**, 646 (1996)
11.917 C. Carbonnier, H. Többen, U. B. Unrau: Room temperature CW fiber laser at 3.22 μm, Electron. Lett. **34**, 893 (1998)
11.918 J. Schneider: Fluoride fiber laser operating at 3.9 μm, Electron. Lett. **31**, 1250 (1995)
11.919 H. Többen: CW-lasing at 3.45 μm in erbium-doped fluorozirconate fibers, Frequenz **45**, 250 (1991)
11.920 J. Schneider, C. Carbonnier, U. B. Unrau: Characterization of a Ho^{3+}-doped fluoride fiber laser with a 3.9 μm emission wavelength, Appl. Opt. **36**, 8595 (1997)
11.921 R. Reisfeld: Chalcogenide glasses doped by rare earths: structure and optical properties, Ann. Chim. Fr. **7**, 147 (1982)
11.922 T. Schweizer, D. W. Hewak, B. N. Samson, D. N. Payne: Spectroscopic data of the 1.8-, 2.9-, and 4.3 μm transitions in dysprosium-doped gallium lanthanum sulfide glass, Opt. Lett. **21**, 1594 (1996)
11.923 T. Schweizer, B. N. Samson, J. R. Hector, W. S. Brocklesby, D. W. Hewak, D. N. Payne: Infrared emission and ion-ion interactions in thulium- and terbium-doped gallium lanthanum sulfide glass, J. Opt. Soc. Am. B **16**, 308 (1999)
11.924 T. Schweizer, B. N. Samson, J. R. Hector, W. S. Brocklesby, D. W. Hewak, D. N. Payne: Infrared emission from holmium doped gallium lanthanum sulphide glass, Infrared Phys. Technol. **40**, 329 (1999)
11.925 W. H. Dumbaugh, J. C. Lapp: Heavy-metal oxide glasses, J. Am. Ceram. Soc. **75**, 2315 (1992)
11.926 D. W. Hewak, J. A. Medeiros Neto, B. N. Samson, R. S. Brown, K. P. Jedrzejewski, J. Wang, E. Taylor, R. I. Laming, G. Wylangowski, D. N. Payne: Quantum-efficiency of praseodymium doped Ga:La:S glass for 1.3 μm optical fiber amplifiers, IEEE Photon. Technol. Lett. **6**, 609 (1994)

11.927 C. C. Ye, D. W. Hewak, M. Hempstead, B. N. Samson, D. N. Payne: Spectral properties of Er^{3+}-doped gallium lanthanum sulphide glass, J. Non-Crystal Solids **208**, 56 (1996)

11.928 T. Schweizer, D. J. Brady, D. W. Hewak: Fabrication and spectroscopy of erbium doped gallium lanthanum sulphide glass fibers for mid-infrared laser applications, Opt. Express **1**, 102 (1997)

11.929 T. Schweizer, B. N. Samson, R. C. Moore, D. W. Hewak, D. N. Payne: Rare-earth doped chalcogenide glass fiber laser, Electron. Lett. **33**, 414 (1997)

11.930 M. Asobe, T. Ohara, I. Yokohama, T. Kaino: Fabrication of Bragg grating in chalcogenide glass fiber using the transverse holographic method, Electron. Lett. **32**, 1611 (1996)

11.931 R. Mossadegh, J. S. Sanghera, D. Schaafsma, B. J. Cole, V. Q. Nguyen, R. E. Miklos, I. D. Aggarwal: Fabrication of single-mode chalcogenide optical fiber, J. Lightwave Technol. **16**, 214 (1998)

11.932 T. M. Monro, Y. D. West, D. W. Hewak, N. G. R. Broderick, D. J. Richardson: Chalcogenide holey fibers, Electron. Lett. **36**, 1998 (2000)

11.933 S. Kück: Laser-related spectroscopy of ion-doped crystals for tunable solid state-lasers, Appl. Phys. B **72**, 515 (2001)

11.934 Y. Tanabe, S. Sugano: On the absorption spectra of complex ions I, J. Phys. Soc. Jpn. **9**, 753 (1954)

11.935 Y. Tanabe, S. Sugano: On the absorption spectra of complex ions II, J. Phys. Soc. Jpn. **9**, 766 (1954)

11.936 S. Sugano, Y. Tanabe, H. Kamimura: *Multiplets of Transition-Metal Ions in Crystals* (Academic, New York 1970)

11.937 B. Henderson, G. F. Imbusch: *Optical Spectroscopy of Inorganic Solids* (Clarendon, Oxford 1989)

11.938 H. L. Schläfer, G. Gliemann: *Einführung in die Ligandenfeldtheorie* (Akademische Verlagsgesellschaft, Frankfurt am Main 1980)

11.939 P. Schuster: *Ligandenfeldtheorie* (Verlag Chemie, Weinheim 1973)

11.940 A. D. Liehr, C. J. Ballhausen: Inherent configurational instability of octrahedral complexes in E_g electronic states, Ann. Phys. **3**, 304 (1958)

11.941 J. S. Griffith: *The Theory of Transition Metal Ions* (Cambridge Univ. Press, Cambridge 1961)

11.942 A. B. P. Lever: *Inorganic Electronic Spectroscopy* (Elsevier, Amsterdam 1968)

11.943 P. F. Moulton: Ti-doped sapphire: tunable solid-state laser, Opt. News **8**, 9 (1982)

11.944 P. F. Moulton: Pulse-pumped operation of divalent transition-metal lasers, IEEE J. Quantum Electron. **18**, 1185 (1982)

11.945 P. F. Moulton: Spectroscopic and laser characteristics of Ti : Al_2O_3, J. Opt. Soc. Am. B **3**, 125 (1986)

11.946 G. F. Albrecht, J. M. Eggleston, J. J. Ewing: Measurements of Ti^{3+} : Al_2O_3 as a lasing material, Opt. Commun. **52**, 401 (1985)

11.947 P. F. Moulton: Recent advances in transition metal-doped laser. in. In: *Tunable Solid-State Lasers*, Springer Ser. Opt. Sci., Vol. 47, ed. by P. Hammerling, A. B. Budgor, A. A. Pinto (Springer, Berlin, Heidelberg 1985) pp. 4–10

11.948 P. Albers, E. Stark, G. Huber: Continuous-wave laser operation and quantum efficiency of titanium-doped sapphire, J. Opt. Soc. Am. B **3**, 134 (1986)

11.949 R. Rao, G. Vaillancourt, H. S. Kwok, C. P. Khattak: Highly efficient, widely tunable kilohertz repetition rate Ti:sapphire laser pumped by a Nd:YLF laser. In: *Tunable Solid-State Lasers*, OSA Proc. Ser., Vol. 5, ed. by M. L. Shand, H. P. Jenssen (Opt. Soc. Am., Washington 1989) pp. 39–41

11.950 W. Koechner: *Solid-State Laser Engineering* (Springer, Berlin, Heidelberg 1996)

11.951 P. F. Moulton: *Tunable paramagnetic-ion lasers*, Vol. 5, ed. by M. Bass, M. L. Stitch (North Holland, Amsterdam 1985) pp. 203–288

11.952 T. Danger, K. Petermann, G. Huber: Polarized and time-resolved measurements of excited state absorption and stimulated emission in Ti:$YAlO_3$ and Ti : Al_2O_3, Appl. Phys. A **57**, 309 (1993)

11.953 M. S. Pshenichnikov, A. Baltuska, R. Szipöcz, D. A. Wiersma: Sub-5-fs pulses: generation, characterization, and experiments. In: *Ultrafast Phenomena XI*, ed. by W. Zinth, J. Fujimoto, T. Elsasser, D. A. Wiersma (Springer, Berlin, Heidelberg 1998) p. 3

11.954 Z. Cheng, G. Tempea, T. Brabec, K. Ferenc, C. Spielmann, F. Krausz: Generation of intense diffraction-limited white light and 4-fs pulses. In: *Ultrafast Phenomena XI*, ed. by W. Zinth, J. Fujimoto, T. Elsasser, D. A. Wiersma (Springer, Berlin, Heidelberg 1998) p. 8

11.955 A. Baltuska, Z. Wei, M. S. Pshenichnikov, D. A. Wiersma: Optical pulse compression to 5 fs at a 1 MHz repetition rate, Opt. Lett. **22**, 102 (1997)

11.956 M. Nisoli, S. De Silvestri, O. Svelto, R. Szipöcs, K. Ferenc, C. Spielmann, S. Sartania, F. Krausz: Compression of high-energy laser pulses below 5 fs, Opt. Lett. **22**, 522 (1997)

11.957 R. E. Ell, U. Morgner, F. X. Kärtner, J. G. Fujimoto, E. P. Ippen, V. Scheuer, G. Angelow, T. Tschudi, M. J. Lederer, A. Boiko, B. Luther-Davies: Generation of 5 fs pulses and octave spinning spectra directly from a Ti:Sapphire laser, Opt. Lett. **26**, 373 (2001)

11.958 A. I. Alimpiev, G. V. Bukin, V. N. Matrosov, E. V. Pestryakov, V. P. Solntsev, V. I. Trunov, E. G. Tsvetkov, V. P. Chebotaev: Tunable $BeAl_2O_4$:Ti^{3+} laser, Sov. J. Quantum Electron. **16**, 579 (1986)

11.959 E. V. Pestryakov, V. I. Trunov, A. I. Alimpiev: Generation of tunable radiation in a $BeAl_2O_4$: Ti^{3+} laser subjected to pulsed coherent pumping at a high repetion frequency, Sov. J. Quantum Electron. **17**, 585 (1987)

11.960 Y. Segawa, A. Sugimoto, P. H. Kim, S. Namba, K. Yamagishi, Y. Anzai, Y. Yamaguchi: Optical properties and lasing of Ti^{3+} doped $BeAl_2O_4$, Jpn. J. Appl. Phys. **24**, L291 (1987)

11.961 A. Sugimoto, Y. Segawa, Y. Anzai, K. Yamagishi, P. H. Kim, S. Namba: Flash-lamp-pumped tunable Ti : $BeAl_2O_4$ laser, Jpn. J. Appl. Phys. **29**, L1136 (1990)

11.962 T. A. Driscoll, M. Peressini, R. E. Stone, G. Hansen, H. J. Hoffman: Efficient tunable solid-state laser using Ti:sapphire and Ti:YAG. In: *Conf. Lasers Electro-Optics*, Tech. Dig. (Opt. Soc. Am., Washington 1986) p. 106 , paper TUK29

11.963 J. Kvapil, M. Koselja, J. Kvapil, B. Perner, V. Skoda, J. Kubelka, K. Hamel, V. Kubecek: Growth and stimulated emission of YAP:Ti, Czech. J. Phys. B **38**, 237 (1988)

11.964 J. Kvapil, M. Koselja, J. Kvapil, K. Hamel: Growth and stimulated emission of $YAlO_3$. In: *Conf. Lasers Electro-Optics*, Tech. Dig. Ser., Vol. 11 (Opt. Soc. Am., Washington 1989) p. 6 , paper MC2

11.965 T. Wegner, K. Petermann: Excited state absorption of Ti^{3+}:$YAlO_3$, Appl. Phys. B **49**, 275 (1989)

11.966 M. Yamaga, Y. Gao, F. Rasheed, K. P. O'Donnell, B. Henderson, B. Cockayne: Radiative and non-radiative decays from the excited state of Ti^{3+} ions in oxide crystals, Appl. Phys. B **51**, 329–335 (1990)

11.967 F. Bantien, P. Albers, G. Huber: Optical transitions in titanium-doped YAG, J. Luminesc. **36**, 363 (1987)

11.968 T. H. Maiman: Stimulated optical radiation in ruby, Nature **187**, 493 (1960)

11.969 R. C. Morris, C. F. Cline: Chromium-doped beryllium aluminate lasers, US Patent 3997853 (1976)

11.970 J. C. Walling, O. G. Peterson, H. P. Jenssen, R. C. Morris, E. W. O'Dell: Tunable alexandrite lasers, IEEE J. Quantum Electron. **16**, 1302 (1980)

11.971 S. T. Lai, M. L. Shand: High efficiency CW laser-pumped tunable alexandrite laser, J. Appl. Phys. **54**, 5642 (1983)

11.972 S. A. Payne, L. L. Chase, H. W. Newkirk, L. K. Smith, W. F. Krupke: $LiCaAlF_6 : Cr^{3+}$: A promising new solid-state laser material, IEEE J. Quantum Electron. **24**, 2243 (1988)

11.973 P. Wagenblast, U. Morgner, F. Grawert, T. R. Schibli, F. X. Kärtner, V. Scheuer, G. Angelow, M. J. Lederer: Generation of sub-10-fs pulses from a Kerr-lens mode-locked Cr^{3+} : LiCAF laser oscillator by use of third-order dispersion-compensating double-chirped mirrors, Opt. Lett. **19**, 1726 (2002)

11.974 P. M. W. French, R. Mellish, J. R. Taylor, P. J. Delfyett, T. L. Florez: Mode-locked all.solid-state diode-pumped Cr:LiSAF laser, Opt. Lett. **18**, 1934 (1993)

11.975 P. Wagenblast, R. Ell, U. Morgner, F. Grawert, F. X. Kärtner: Diode-pumped 10 fs Cr^{3+} : LiCAF laser, Opt. Lett. **28**, 1713 (2003)

11.976 J.-M. Hopkins, G. J. Valentine, B. Agate, A. J. Kemp, U. Keller, W. Sibbett: Highly compact and efficient femtosecond Cr:LiSAF lasers, IEEE J. Quantum Electron. **38**, 360 (2002)

11.977 L. L. Chase, S. A. Payne: Tunable chromium lasers, SPIE **1062**, 9 (1989)

11.978 S. Kück: Spektroskopie und Lasereigenschaften Cr^{4+}-dotierter oxidischer Kristalle Ph.D. Thesis (University of Hamburg, Hamburg 1994)

11.979 S. Kück, K. L. Schepler, K. Petermann, G. Huber: Excited state absorption and stimulated emission measurements of Cr^{4+}-doped $Y_3Al_5O_{12}$, $Y_3Sc_{0.9}Al_{4.1}O_{12}$, and $CaY_2Mg_2Ge_3O_{12}$. In: *OSA Trends Opt. Photon. Adv. Solid-State Lasers*, Vol. 1, ed. by S. A. Payne, C. R. Pollock (Opt. Soc. Am., Washington 1996) pp. 94–99

11.980 S. Kück, K. L. Schepler, S. Hartung, K. Petermann, G. Huber: Excited state absorption and ist influence on the laser behavior of Cr^{4+}-doped garnets, J. Luminesc. **72-74**, 222 (1997)

11.981 N. V. Kuleshov, V. G. Shcherbitsky, V. P. Mikhailov, S. Hartung, T. Danger, S. Kück, K. Petermann, G. Huber: Excited-state absorption and stimulated emission measurements in Cr^{4+}:forsterite, J. Luminesc. **75**, 319 (1997)

11.982 N. V. Kuleshov, V. G. Shcherbitsky, V. P. Mikhailov, S. Hartung, T. Danger, S. Kück, K. Petermann, G. Huber: Excited-state absorption measurements in Cr^{4+}-doped Mg_2SiO_4 and Y_2SiO_5 laser materials. In: *OSA Trends Opt. Photon. Adv. Solid-State Lasers*, Vol. 1, ed. by S. A. Payne, C. R. Pollock (Opt. Soc. Am., Washington 1996) pp. 85–89

11.983 N. V. Kuleshov, V. G. Shcherbitsky, V. P. Mikhailov, S. Hartung, S. Kück, K. Petermann, G. Huber: Near infrared and visible excited-state absorption in Cr^{4+}:forsterite. In: *OSA Trends Opt. Photon. Adv. Solid-State Lasers*, Vol. 10, ed. by C. R. Pollock, W. R. Bosenberg (Opt. Soc. Am., Washington 1997) pp. 425–430

11.984 S. Hartung: Spektroskopische Untersuchungen breitbandig-emittierender Mn^{3+}- und Cr^{4+}-dotierter oxidischer Kristalle Ph.D. Thesis (University of Hamburg, Hamburg 1997)

11.985 T. C. Brunold, H. U. Güdel, M. F. Hazenkamp, G. Huber, S. Kück: Excited state absorption measurements and laser potential of Cr^{4+} doped Ca_2GeO_4, Appl. Phys. B **64**, 647 (1997)

11.986 S. Hartung, S. Kück, T. Danger, K. Petermann, G. Huber: ESA measurements of Cr^{4+}-doped crystals with Wurtzite-like structure. In: *OSA Trends Opt. Photon.*, Vol. 1, ed. by S. A. Payne, C. R. Pollock (Opt. Soc. Am., Washington 1996) pp. 90–93

11.987 S. Kück, S. Hartung: Comparative study of the spectroscopic properties of Cr^{4+}-doped $LiAlO_2$ and $LiGaO_2$, Chem. Phys. **240**, 387 (1999)

11.988 J. Buchert, A. Katz, R. R. Alfano: Laser action in emerald, IEEE J. Quantum Electron. **19**, 1477 (1983)

11.989 M. L. Shand, J. C. Walling: A tunable emerald laser, IEEE J. Quantum Electron. **18**, 1829 (1982)

11.990 M. L. Shand, S. T. Lai: CW laser pumped emerald laser, IEEE J. Quantum Electron. **20**, 105 (1984)

11.991 S. T. Lai: Highly efficient emerald laser, J. Opt. Soc. Amer. B **4**, 1286 (1987)

11.992 S.T. Lai: Highly efficient emerald laser, AIP conf. proc. **160**, 128 (1987)

11.993 L.L. Chase, S.A. Payne, L.K. Smith, W.L. Kway, H.W. Newkirk, B.H.T. Chai, M. Long: Laser performance and spectroscopy of Cr^{3+} in $LiCaAlF_6$ and $LiSrAlF_6$. In: *Tunable Solid-State Lasers*, Vol. 5, ed. by M.L. Shand, M.P. Jenssen (Opt. Soc. Am., Washington 1989) pp. 71–76

11.994 L.K. Smith, S.A. Payne, W.L. Kway, L.L. Chase, B.H.T. Chai: Investigation of the laser properties of $Cr^{3+}:LiSrGaF_6$, IEEE J. Quantum Electron. **28**, 2612 (1992)

11.995 F. Balembois, F. Druon, F. Falcoz, P. Georges, A. Brun: Performance of $Cr:LiSrAlF_6$ and $LiSrGaF_6$ for continuous-wave diode-pumped *Q*-switched operation, Opt. Lett. **22**, 387 (1997)

11.996 F. Balembois, F. Falcoz, F. Kerboull, F. Druon, P. Georges, A. Brun: Theoretical and experimental investigations of small-signal gain for a diode-pumped *Q*-switched Cr:LiSAF laser, IEEE J. Quantum Electron. **33**, 269 (1997)

11.997 B.K. Sebastianov, Y.I. Remigailo, V.P. Orekhova, V.P. Matrosov, E.G. Tvsetkov, G.V. Bukin: Spectroscopic and lasing properties of alexandrite ($BeAl_2O_4:Cr^{3+}$), Sov. Phys. Dokl. **26**, 62 (1981)

11.998 G.V. Bukin, S.Y. Volkov, V.N. Matrosov, B.K. Sebastianov, M.I. Timoshechkin: Stimulated emission from alexandrite ($BeAl_2O_4:Cr^{3+}$), Sov. J. Quantum Electron. **8**, 671 (1978)

11.999 J.C. Walling, O.G. Peterson: High gain laser performance in alexandrite, IEEE J. Quantum Electron. **16**, 119 (1980)

11.1000 J.C. Walling, H.P. Jenssen, R.C. Morris, E.W. O'Dell, O.G. Peterson: Tunable-laser performance in $BeAl_2O_4:Cr^{3+}$, Opt. Lett. **4**, 182 (1979)

11.1001 J.C. Walling, H.P. Jenssen, R.C. Morris, E.W. O'Dell, O.G. Peterson: Broad band tuning of solid state alexandrite lasers, J. Opt. Soc. Am. B **69**, 373 (1979)

11.1002 J.C. Walling, O.G. Peterson, R.C. Morris: Tunable CW alexandrite laser, IEEE J. Quantum Electron. **16**, 120 (1980)

11.1003 H. Samuelson, J.C. Walling, T. Wernikowski, D.J. Harter: CW arc-lamp-pumped alexandrite lasers, IEEE J. Quantum Electron. **24**, 1141 (1988)

11.1004 J.C. Walling, D.F. Heller, H. Samuelson, D.J. Harter, J.A. Pete, R.C. Morris: Tunable alexandrite lasers: Development and performance, IEEE J. Quantum Electron. **21**, 1568 (1985)

11.1005 S. Guch Jr., C.E. Jones: Alexandrite-laser performance at high temperature, Opt. Lett. **7**, 608 (1982)

11.1006 W.R. Rapaport, H. Samebon: Alexandrite slab laser. In: *Proc. Int. Conf. Lasers 85*, ed. by C.P. Wang (STS Press, McLean 1986) p. 744

11.1007 S. Zhang, K. Zhang: Experiment on laser performance of alexandrite crystals, Chin. Phys. **4**, 667 (1984)

11.1008 M.L. Shand: Progress in alexandrite lasers. In: *Proc. Int. Conf. Lasers 85*, ed. by C.P. Wang (STS Press, McLean 1986) p. 732

11.1009 G. Zhang, X. Ma: Improvement of lasing performance of alexandrite crastals, Chin. Phys. Lasers **13**, 816 (1986)

11.1010 J.E. Jones, J.D. Dobbins, B.D. Butier, R.J. Hinsley: Performance of a 250 Hz, 100 W alexandrite laser system. In: *Proc. Int. Conf. Lasers 85*, ed. by C.P. Wang (STS Press, McLean 1986) p. 738

11.1011 V.N. Lisitsyn, V.N. Matrosov, E.V. Pestryakov, V.I. Trunov: Generation of picosecond pulses in solid-state lasers using new active media, J. Sov. Laser Res. **7**, 364 (1986)

11.1012 J.C. Walling, O.G. Peterson: High gain laser performance in alexandrite, IEEE J. Quantum Electron. **16**, 119 (1980)

11.1013 R. Scheps, J.F. Myers, T.R. Glesne, H.B. Serreze: Monochromatic end-pumped operation of an alexandrite laser, Opt. Commun. **97**, 363 (1993)

11.1014 H. Ogilvy, M.J. Withford, J.A. Piper: Stable, red laser pumped, multi-kilohertz alexandrite laser, Opt. Commun. **260**, 207 (2006)

11.1015 S.A. Payne, L.L. Chase, L.K. Smith, W.L. Kway, H.W. Newkirk: Laser performance of $LiSrAlF_6:Cr^{3+}$, J. Appl. Phys. **66**, 1051 (1989)

11.1016 M.J.P. Dymott, I.M. Botheroyd, G.J. Hall, J.R. Lincoln, A.J. Ferguson: All-solid-state actively mode-locked Cr:LiSAF laser, Opt. Lett. **19**, 634 (1994)

11.1017 M. Stalder, B.H.T. Chai, M. Bass: Flashlamp pumped $Cr:LiSrAlF_6$ laser, Appl. Phys. Lett. **58**, 216 (1991)

11.1018 R. Scheps, J.F. Myers, H.B. Serreze, A. Rosenberg, R.C. Morris, M. Long: Diode-pumped $Cr:LiSrAlF_6$ laser, Opt. Lett. **16**, 820 (1991)

11.1019 Q. Zhang, G.J. Dixon, B.H.T. Chai, P.N. Kean: Electronically tuned diode-laser-pumped $Cr:LiSrAlF_6$ laser, Opt. Lett. **17**, 43 (1992)

11.1020 H.H. Zenzie, A. Finch, P.F. Moulton: Diode-pumped, single-frequency $Cr:LiSrAlF_6$ ring laser, Opt. Lett. **20**, 2207 (1995)

11.1021 D.E. Klimek, A. Mandl: Power scaling of a flashlamp-pumped Cr:LiSAF thin-slab zig-zag laser, IEEE J. Quantum Electron. **38**, 1607 (2002)

11.1022 S.T. Lai, B.H.T. Chai, M. Long, R.C. Morris: $ScBO_3:Cr$ – A room temperature near-infrared tunable laser, IEEE J. Quantum Electron. **22**, 1931 (1986)

11.1023 B. Struve, G. Huber, V.V. Laptev, I.A. Shcherbakov, E.V. Zharikov: Tunable room-temperature CW laser action in Cr^{3+} : GdScGa-garnet, Appl. Phys. B **30**, 117 (1983)

11.1024 B. Struve, G. Huber: Laser performance of Cr^{3+} : Gd(Sc,Ga) garnet, J. Appl. Phys. **57**, 45 (1985)

11.1025 G. Huber, J. Drube, B. Struve: Recent developments in tunable Cr doped garnet lasers. In: *Proc. Int. Conf. Lasers 83*, ed. by R.C. Powell (STS Press, McLean 1983) p. 143

11.1026 G. Huber, K. Petermann: Laser action in Cr-doped garnets and tungstates. In: *Tunable Solid-State Lasers*, ed. by P. Hammerling, A. B. Budgor, A. Pinto (Springer, Berlin, Heidelberg 1985) p. 11

11.1027 B. Struve, G. Huber, V. V. Laptev, I. A. Shcherbakov, E. V. Zharikov: Laser action and broad band fluorescence in Cr^{3+} : GdScGa-garnet, Appl. Phys. B **28**, 235 (1982)

11.1028 E. V. Zharikov, N. N. Ilichev, N. N. Kalitin, V. V. Laptev, A. A. Malyutin, V. V. Osiko, V. G. Ostroumov, P. P. Pashinin, A. M. Prokhorov, V. A. Smirnov, A. F. Umyskov, I. A. Shcherbakov: Tunable laser utilizing an electronic-vibrational transition in chromium in a gadolinium scandium gallium garnet crystal, Sov. J. Quantum Electr. **13**, 1274 (1983)

11.1029 J. Drube, G. Huber, D. Mateika: Flaslamp-pumped Cr^{3+} : GSAG and Cr^{3+} : GSGG: Slope efficiency, resonator design, color centers and tunability. In: *Tunable Solid-State Lasers II*, Springer Ser. Opt. Sci., Vol. 52, ed. by A. B. Budgor, L. Esterowitz, L. G. DeShazer (Springer, Berlin, New York 1986) p. 118

11.1030 M. J. P. Payne, H. W. Evans: Laser action in flashlamp-pumped chromium:GSG-garnet, in. In: *Tunable Solid State Lasers II*, Springer Ser. Opt. Sci., Vol. 52, ed. by A. B. Budgor, L. Esterowitz, L. DeShazer (Springer, Berlin, New York 1986) p. 126

11.1031 J. Drube: Cr:GSAG und Cr:YSAG: Chrom-dotierte Aluminiumgranate als durchstimmbare Festkörperlaser bei Raumtemperatur Ph.D. Thesis (University of Hamburg, Hamburg 1987)

11.1032 J. V. Meier, N. P. Barnes, D. K. Remelius, M. R. Kokta: Flashlamp-pumped Cr^{3+} : GSAG laser, IEEE J. Quantum Electron. **22**, 2058 (1986)

11.1033 J. Drube, B. Struve, G. Huber: Tunable room-temperature CW laser action in Cr^{3+} : GdScAl-garnet, Opt. Commun. **50**, 45 (1984)

11.1034 B. Struve, P. Fuhrberg, W. Luhs, G. Litvin: Thermal lensing and laser operation of flashlamp-pumped Cr:GSAG, Opt. Commun. **65**, 291 (1988)

11.1035 H. P. Jenssen, S. T. Lai: Tunable-laser characteristics and spectroscopic properties of $SrAlF_5$: Cr, J. Opt. Soc. Am. B **3**, 115 (1986)

11.1036 J. A. Caird, P. R. Staver, M. D. Shinn, H. J. Guggenheim, D. Bahnck: Laser-pumped laser measurements of gain and loss in $SrAlF_5$: Cr crystals. In: *Tunable Solid-State Lasers II*, Springer Ser. Opt. Sci., Vol. 52, ed. by A. B. Budgor, L. Esterowitz, L. G. DeShazer (Springer, Berlin, Heidelberg 1986) p. 159

11.1037 J. A. Caird, W. F. Krupke, M. D. Shinn, P. R. Staver, H. J. Guggenheim: Tunable Cr^{4+} : YAG lasers. In: *Advances in Laser Science – I*, ed. by W. C. StWalley, M. Lapp (AIP, New York 1986) pp. 243–244

11.1038 J. A. Caird, W. F. Krupke, M. D. Shinn, P. R. Staver, H. J. Guggenheim: Room temperature $SrAlF_5$: Cr^{3+} laser emission tunable from 825 nm to 1010 nm, Bull. Am. Phys. Soc. 2 **30**, 1857 (1985)

11.1039 U. Brauch, U. Dürr: $KZnF_3$: Cr^{3+} – A tunable solid state NIR-laser, Opt. Commun. **49**, 61 (1984)

11.1040 U. Dürr, U. Brauch, W. Knierim, W. Weigand: Vibronic solid state lasers: transition metal ions in perowskites. In: *Proc. Int. Conf. Lasers 83*, ed. by R. C. Powell (STS Press, McLean 1983) p. 42

11.1041 U. Brauch, U. Dürr: Room-temperature operation of the vibronic $KZnF_3$: Cr^{3+} laser, Opt. Lett. **9**, 441 (1984)

11.1042 M. A. Dubinskii, A. N. Kolerov, M. V. Mityagin, N. I. Silkin, A. P. Shkadarevich: Quasicontinuous operation of a $KZnF_3$: Cr^{3+} laser, Sov. J. Quantum Electron. **16**, 1684 (1986)

11.1043 P. Fuhrberg, W. Luhs, B. Struve, G. Litvin: Single-mode operation of Cr-doped GSGG and $KZnF_3$. In: *Tunable Solid-State Lasers II*, Springer Ser. Opt. Sci., Vol. 52, ed. by A. B. Budgor, L. Esterowitz, L. G. DeShazer (Springer, Berlin, Heidelberg 1986) p. 159

11.1044 R. Y. Abdulsabirov, M. A. Dubinskii, S. L. Korableva, M. V. Mityagin, N. I. Silkin, C. A. Skripto, A. P. Shkadarevich, S. I. Yagudin: Tunable laser based on $KZnF_3$: Cr^{3+} crystal with nonselective pumping, Sov. Phys. Crystallogr. **31**, 353 (1986)

11.1045 W. Kolbe, K. Petermann, G. Huber: Broadband emission and laser action of Cr^{3+} doped zinc tungstate at 1 mm wavelength, IEEE J. Quantum Electron. **21**, 1596 (1985)

11.1046 S. T. Lai, B. H. T. Chai, M. Long, M. D. Shinn: Room temperature near-infrared tunable Cr : $La_3Ga_5SiO_{14}$ laser, IEEE J. Quantum Electron. **24**, 1922 (1988)

11.1047 A. A. Kaminskii, A. V. Butashin, A. A. Demidovich, V. G. Koptev, B. V. Mill, A. P. Shkadarevich: Broadband tunable stimulated emission from octahedral Cr^{3+} ions in new acentric crystals with Ca-gallogermanate structure, phys. stat. sol. (a) **112**, 197 (1989)

11.1048 A. A. Kaminskii, A. P. Shkadarevich, B. V. Mill, V. G. Koptev, A. A. Demidovich: Wide-band tunable stimulated emission from a $La_3Ga_5SiO_{14}$ crystal, Inorg. Mater. **23**, 618 (1987)

11.1049 A. A. Kaminskii, A. P. Shkadarevich, B. V. Mill, V. G. Koptev, A. V. Butashin, A. A. Demidovich: Wideband tunable stimulated emission of Cr^{3+} ions in the trigonal crystal $La_3Ga_{5.5}Nb_{0.5}O_{14}$, Inorg. Mater. **23**, 1700 (1987)

11.1050 S. Kück, E. Heumann, T. Kärner, A. Maaroos: Continuous-wave room-temperature laser oscillation of Cr^{3+} : MgO, Opt. Lett. **24**, 966 (1999)

11.1051 S. Kück, E. Heumann, T. Kärner, A. Maaroos: Continuous wave laser oscillation of Cr^{3+} : MgO. In: *Adv. Solid-State Lasers*, OSA Trends Opt. Photon., Vol. 26, ed. by M. M. Fejer, H. Injeyan, U. Keller (Opt. Soc. Am., Washington 1999) pp. 308–311

11.1052 S. Kück, L. Fornasiero, E. Heumann, E. Mix, G. Huber, T. Kärner, A. Maaroos: Investigation of Cr-doped MgO and Sc_2O_3 as potential laser sources for the near infrared spectral range, Laser Phys. **10**, 411 (2000)

11.1053 K. Petermann, P. Mitzscherlich: Spectroscopic and laser properties of Cr^{3+}-doped $Al_2(WO_4)_3$ and $Sc_2(WO_4)_3$, IEEE J. Quantum Electron. **23**, 1122 (1987)

11.1054 E.V. Pestryakov, V.V. Petrov, V.I. Trunov, A.I. Alimpiev: Generation of tunable radiation on Cr^{3+} ions in a flashlamp-pumped $BeAl_6O_{10}$ crystal, Quantum Electron. **23**, 575 (1993)

11.1055 A.I. Alimpiev, E.V. Pestryakov, V.V. Petrov, V.P. Solntsev, V.I. Trunov, V.N. Matrosov: Tunable lasing due to the $^4T_2$–$^4A_2$ electronic-vibrational transition in Cr^{3+} ions in $BeAl_6O_{10}$, Sov. J. Quantum Electron. **18**, 323 (1988)

11.1056 F.J. McClung, S.E. Schwarz, F.J. Meyers: R_2 line optical maser action in ruby, J. Appl. Phys. **33**, 3139 (1962)

11.1057 R.J. Collins, D.F. Nelson, A.L. Schawlow, W. Bond, C.G.B. Garrett, W. Kaiser: Coherence, narrowing, directionality, and relaxation oscillations in the light emission from ruby, Phys. Rev. Lett. **5**, 303 (1960)

11.1058 D.F. Nelson, W.S. Boyle: A continuously operating ruby optical maser, Appl. Opt. **1**, 181 (1962)

11.1059 M. Birnbaum, A.W. Tucker, C.L. Fincher: CW ruby laser pumped by an argon ion laser, IEEE J. Quantum Electron. **13**, 808 (1977)

11.1060 T.H. Maiman: Optical maser action in ruby, Br. Commun. Electron. **7**, 674 (1960)

11.1061 D. Roess: Analysis of room temperature CW ruby lasers, IEEE J. Quantum Electron. **2**, 208 (1966)

11.1062 V. Evtuhov, J.K. Neeland: Power output and efficiency of continuous ruby lasers, J. Appl. Phys. **38**, 4051 (1967)

11.1063 C.A. Burrus, J. Stone: Room-temperature continuous operation of a ruby fiber laser, J. Appl. Phys. **49**, 3118 (1978)

11.1064 A.N. Kirkin, A.M. Leontovich, A.M. Mozharovskii: Generation of high power ultrashort pulses in a low temperature ruby laser with a small active volume, Sov. J. Quantum Electron. **8**, 1489 (1978)

11.1065 V. Evtuhov, J.K. Neeland: A continuously pumped repetitively *Q*-switched ruby laser and applications to frequency-conversion experiments, IEEE J. Quantum Electron. **5**, 207 (1969)

11.1066 A.L. Schawlow, G.E. Devlin: Simultaneous optical maser action in two ruby satellite lines, Phys. Rev. Lett. **6**, 96 (1961)

11.1067 E.J. Woodbury, W.K. Ng: Ruby laser operation in the near IR, Proc. IRE **50**, 2367 (1962)

11.1068 B.K. Sebastianov, K.S. Bagdarasov, L.B. Pasternak, S.Y. Volkov, V.P. Drekhova: Stimulated emission from Cr^{3+} ions in YAG crystals, JETP Lett. **17**, 47 (1973)

11.1069 B.H.T. Chai, J. Lefaucheur, M. Stalder, M. Bass: Cr : $LiSr_{0.8}Ca_{0.2}AlF_6$ tunable laser, Opt. Lett. **17**, 1584 (1992)

11.1070 H.S. Wang, P. Li Kam Wa, J.L. Lefaucheur, B.H.T. Chai, A. Miller: CW and self-mode-locking performance of a red pumped Cr^{3+} : $LiSr_{0.8}Ca_{0.2}AlF_6$ laser, Opt. Commun. **110**, 679 (1994)

11.1071 L.K. Smith, S.A. Payne, W.F. Krupke, L. DeLoach, R. Morris, E.W. O'Dell, D.J. Nelson: Laser emission from the transition-metal compound $LiSrCrF_6$, Opt. Lett. **18**, 200 (1993)

11.1072 B.H.T. Chai, M.D. Shinn, M.N. Long, S.T. Lai, H.H. Miller, L.K. Smith: Laser and spectroscopic properties of Cr^{3+}-doped $ScAlBeO_4$, Bull. Am. Phys. Soc. **33**, 1631 (1988)

11.1073 A.A. Kaminskii, A.P. Shkadarevich, B.V. Mill, V.G. Koptev, A.V. Butashin, A.A. Demidovich: Tunable stimulated emission of Cr^{3+} ions and generation frequency self-multiplication effect in acentric crystals of Ca-gallogermante structure, Inorg. Mater. **24**, 579 (1988)

11.1074 A.G. Bazylev, A.P. Voitovich, A.A. Demidovich, V.S. Kalinov, M.I. Timoshechkin, A.P. Shkadarevich: Laser performance of Cr^{3+}: $(Gd,Ca)_3(Ga,Mg,Zr)_2Ga_3O_{12}$, Opt. Commun. **94**, 82 (1992)

11.1075 S. Hartung: Cr^{3+}-dotiertes $LaSc_3(BO_3)_4$: Spektroskopie und Lasereigenschaften Diploma Thesis (University of Hamburg, 1994)

11.1076 M. Sharonov, V. Petričević, A. Bykov, R.R. Alfano: Near-infrared laser operation of Cr^{3+} centers in chromium-doped $LiInGeO_4$ and $LiScGeO_4$ crystals, Opt. Lett. **30**, 851–853 (2005)

11.1077 A.V. Gaister, E.V. Zharikov, V.F. Lebedev, A.S. Podstavkin, S.Y. Tenyakov, A.V. Shestakov, I.A. Shcherbakov: Pulsed and CW lasing in a new Cr^{3+} : Li : Mg_2SiO_4 laser crystal, Quantum Electron. **34**, 693–694 (2004)

11.1078 N.B. Angert, N.I. Borodin, V.M. Garmash, V.A. Zhitnyuk, A.G. Okhrimchuk, O.G. Siyuchenko, A.V. Shestakov: Lasing due to impurity color centers in yttrium aluminum garnet crystals at wavelengths in the range 1.35–1.45 μm, Sov. J. Quantum Electron. **18**, 73 (1988)

11.1079 G.M. Zverew, A.V. Shestakov: Tunable near-infrared oxide crystal lasers. In: *OSA Proc. Tunable Solid-State Lasers*, Vol. 5, ed. by M.L. Shand, M.P. Jenssen (Opt. Soc. Am., Washington 1989) pp. 66–70

11.1080 N.I. Borodin, V.A. Zhitnyuk, A.G. Okhrimchuk, A.V. Shestakov: $Y_3Al_5O_{12} - Cr^{4+}$ laser action at 1.34 μm to 1.6 μm, Izv. Akad. Nauk SSSR Ser. Fiz. **54**, 1500 (1990)

11.1081 A.V. Shestakov, N.I. Borodin, V.A. Zhitnyuk, A.G. Okhrimchuk, V.P. Gapontsev: Tunable Cr^{4+}:YAG-lasers. In: *Conf. Lasers Electro-Optics*, OSA Tech. Dig. Ser., Vol. 10 (Opt. Soc. Am., Washington 1991) paper CPDP11

11.1082 W. Jia, H. Eilers, W.M. Dennis, W.M. Yen, A.V. Shestakov: The performance of a Cr^{4+} : YAG laser in the NIR. In: *OSA Proc. Adv. Solid-State*

Lasers, Vol. 13, ed. by L. L. Chase, A. A. Pinto (Opt. Soc. Am., Washington 1992) pp. 28–30

11.1083 H. Eilers, W. M. Dennis, W. M. Yen, S. Kück, K. Petermann, G. Huber, W. Jia: Performance of a Cr:YAG laser, IEEE J. Quantum Electron. **29**, 2508 (1993)

11.1084 S. Kück, J. Koetke, K. Petermann, U. Pohlmann, G. Huber: Spectroscopic and laser studies of Cr^{4+} : YAG and Cr^{4+} : Y_2SiO_5. In: *OSA Proc. Adv. Solid-State Lasers*, Vol. 15, ed. by A. A. Pinto, T. Y. Fan (Opt. Soc. Am., Washington 1993) pp. 334–338

11.1085 A. Sennaroglu, C. R. Pollock, H. Nathel: Efficient continuous-wave chromium-doped YAG laser, J. Opt. Soc. B **12**, 930 (1995)

11.1086 S. Kück, K. Petermann, G. Huber: Near Infrared Cr^{4+}:$Y_3Sc_xAl_{5-x}O_{12}$ lasers. In: *OSA Proc. Adv. Solid-State Lasers*, Vol. 20, ed. by T. Y. Fan, B. H. T. Chai (Opt. Soc. Am., Washington 1994) pp. 180–184

11.1087 S. Kück, K. Petermann, U. Pohlmann, U. Schönhoff, G. Huber: Tunable room-temperature laser action of Cr^{4+}-doped $Y_3Sc_xAl_{5-x}O_{12}$, Appl. Phys. B **58**, 153 (1994)

11.1088 Y. Ishida, K. Naganuma: Compact diode-pumped all-solid-state femtosecond Cr^{4+} : YAG laser, Opt. Lett. **21**, 51 (1996)

11.1089 S. Spälter, M. Böhm, M. Burk, B. Mikulla, R. Fluck, I. D. Jung, G. Zhang, U. Keller, A. Sizmann, G. Leuchs: Self-starting soliton-modelocked femtosecond Cr_{4+} : YAG laser using an antiresonant Fabry-Pérot saturable absorber, Appl. Phys. B **65**, 335 (1997)

11.1090 P. M. W. French, N. H. Rizvi, J. R. Taylor, A. V. Shestakov: Continuous-wave mode-locked Cr^{4+} : YAG laser, Opt. Lett. **18**, 39 (1993)

11.1091 A. Sennaroglu, C. R. Pollock, H. Nathel: Continuous-wave self-mode-locked operation of a femtosecond Cr^{4+} : YAG laser, Opt. Lett. **19**, 390 (1994)

11.1092 P. J. Conlon, Y. P. Tong, P. M. W. French, J. R. Taylor, A. V. Shestakov: Passive mode locking and dispersion measurement of a sub-100-fs Cr^{4+} : YAG laser, Opt. Lett. **19**, 1468 (1994)

11.1093 Y. P. Tong, J. M. Sutherland, P. M. W. French, J. R. Taylor, A. V. Shestakov, B. H. T. Chai: Self-starting Kerr-lens mode-locked femtosecond Cr^{4+} : YAG and picosecond Pr^{3+}:YLF solid state lasers, Opt. Lett. **21**, 644 (1996)

11.1094 S. Naumov, E. Sorokin, I. T. Sorokina: Kerr-lens mode-locked diode-pumped Cr^{4+} : YAG laser. In: *Adv. Solid-State Photon*, Tech. Dig. (Opt. Soc. Am., Sante Fe 2004) paper WE2

11.1095 S. Naumov, E. Sorokin, I. T. Sorokina: Directly diode-pumped Kerr-lens mode-locked Cr^{4+} : YAG laser, Opt. Lett. **29**, 1276 (2004)

11.1096 D. J. Ripin, C. Chudoba, J. T. Gopinath, J. G. Fujimoto, E. P. Ippen, U. Morgner, F. X. Kärtner, V. Scheuer, G. Angelow, T. Tschudi: Generation of 20 fs pulses by a prismless Cr^{4+} : YAG laser, Opt. Lett. **27**, 61 (2002)

11.1097 I. Sorokina, S. Naumov, E. Sorokin, E. Wintner, A. V. Shestakov: Tunable directly diode-pumped continuous wave room-temperature Cr^{4+}:YAG laser. In: *Adv. Solid-State Lasers*, OSA Trends Opt. Photon., ed. by M. M. Fejer, H. Injeyan, U. Keller (Opt. Soc. Am., Washington 1999) pp. 331–335

11.1098 S. Ishibashi, K. Naganuma: Diode-pumped Cr^{4+} : YAG single crystal fiber laser. In: *OSA Trends Opt. Photon. Adv. Solid-State Lasers*, ed. by H. Injeyan, U. Keller, C. Marshall (Opt. Soc. Am., Washington 2000) pp. 426–430

11.1099 K. Spariosu, W. Chen, R. Stultz, M. Birnbaum, A. V. Shestakov: Dual *Q*-switching and laser action at 1.06 µm and 1.44 µm in a Nd^{3+}:YAG–Cr^{4+} : YAG oscillator at 300 K, Opt. Lett. **18**, 814 (1993)

11.1100 V. Petričević, S. K. Gayen, R. R. Alfano: Continuous-wave laser operation of chromium-doped forsterite, Opt. Lett. **14**, 612 (1989)

11.1101 V. Petričević, A. Seas, R. R. Alfano: Forsterite laser tunes in near-IR, Laser Focus World **26**, 109–116 (November 1990)

11.1102 V. Petričević, A. Seas, R. R. Alfano: Slope efficiency measurements of a chromium-doped forsterite laser, Opt. Lett. **16**, 811 (1991)

11.1103 N. Zhavoronkov, A. Avtukh, V. Mikhailov: Chromium-doped forstrite laser with 1.1 W of continuous-wave output power at room temperature, Appl. Opt. **36**, 8601 (1997)

11.1104 J. M. Evans, V. Petričević, A. B. Bykov, A. Delgado, R. R. Alfano: Direct diode-pumped continuous-wave near-infrared tunable laser operation of Cr^4:forsterite and Cr^4 : Ca_2GeO_4, Opt. Lett. **22**, 1171 (1997)

11.1105 L. Qian, X. Liu, F. Wise: Cr:forsterite pumped by broad-area laser diodes, Opt. Lett. **22**, 1707 (1997)

11.1106 V. Petričević, S. K. Gayen, R. R. Alfano: Laser action in chromium-activated forsterite for near-infrared excitation: Is Cr^{4+} the lasing ion?, Appl. Phys. Lett. **53**, 2590 (1988)

11.1107 V. G. Baryshevskii, V. A. Voloshin, S. A. Demidovich, A. E. Kimaev, M. V. Korzhik, M. G. Livshits, M. L. Meilman, B. I. Minkov, A. P. Shkadarevich: Efficient flashlamp-pumped chromium-activated forsterite crystal laser tunable in the infrared, Sov. J. Quantum Electron. **20**, 1297 (1990)

11.1108 A. Sugimoto, Y. Segawa, Y. Yamaguchi, Y. Nobe, K. Yamagishi, P. H. Kim, S. Namba: Flash lamp pumped tunable forsterite laser, Jpn. J. Appl. Phys. **28**, L1833 (1989)

11.1109 V. Petričević, S. K. Gayen, R. R. Alfano, K. Yamagishi, H. Anzai, Y. Yamaguchi: Laser action in chromium-doped forsterite, Appl. Phys. Lett. **52**, 1040 (1988)

11.1110 H. R. Verdún, L. M. Thomas, D. M. Andrauskas, T. McCollum, A. Pinto: Chromium-doped forsterite laser pumped with 1.06 µm radiation, Appl. Phys. Lett. **53**, 2593 (1988)

11.1111 V. Petričević, S. K. Gayen, R. R. Alfano: Chromium-activated forsterite laser, **5**, 77–84 (1989)

11.1112 H. R. Verdún, L. M. Thomas, D. M. Andrauskas, A. Pinto: Laser performance of chromium-aluminum-doped forsterite. In: *OSA Proc. Tunable Solid-State Lasers*, Vol. 5, ed. by M. L. Shand, M. P. Jenssen (Opt. Soc. Am., North Falmouth 1989) pp. 85–92

11.1113 V. Petričević, S. K. Gayen, R. R. Alfano: Near infrared tunable operation of chromium doped forsterite laser, Appl. Opt. **28**, 1609 (1989)

11.1114 V. G. Baryshevsky, M. V. Korzhik, M. G. Livshits, A. A. Tarasov, A. E. Kimaev, I. I. Mishkel, M. L. Meilman, B. J. Minkov, A. P. Shkadarevich: Properties of Forsterite and the Performance of Forsterite Lasers with Lasers and Flashlamp Pumping. In: *OSA Proc. Adv. Solid-State Lasers*, Vol. 10, ed. by G. Dubé, L. Chase (Opt. Soc. Am., Washington 1991) pp. 26–34

11.1115 J. C. Diettrich, I. T. McKinnie, D. M. Washington: Efficient, kHz repetition rate, gain-switched Cr:forsterite laser, Appl. Phys. B **69**, 203 (1999)

11.1116 A. Seas, V. Petričević, R. R. Alfano: CW mode-locked operation of chromium-doped forsterite laser. In: *OSA Proc. Adv. Solid-State Lasers*, Vol. 10, ed. by G. Dubé, L. Chase (Opt. Soc. Am., Washington 1991) pp. 69–71

11.1117 A. Seas, V. Petričević, R. R. Alfano: Continuous-wave mode-locked operation of a chromium-doped forsterite laser, Opt. Lett. **16**, 1668 (1991)

11.1118 A. Seas, V. Petričević, R. R. Alfano: Generation of sub-100-fs pulses from a CW mode-locked chromium-doped forsterite laser, Opt. Lett. **17**, 937 (1992)

11.1119 A. Sennaroglu, C. R. Pollock, H. Nathel: Generation of 48 fs pulses and measurement of crystal dispersion by using a regeneratively initiated self-mode-locked chromium-doped forsterite laser, Opt. Lett. **18**, 826 (1993)

11.1120 A. Seas, V. Petričević, R. R. Alfano: Self-mode-locked chromium-doped forsterite laser generates 50 fs pulses, Opt. Lett. **18**, 891 (1993)

11.1121 Y. Pang, V. Yanovsky, F. Wise, B. I. Minkov: Self-mode-locked Cr:forsterite laser, Opt. Lett. **18**, 1168 (1993)

11.1122 V. Yanovsky, Y. Pang, F. Wise, B. I. Minkov: Generation of 25 fs pulses from a self-mode-locked Cr:forsterite laser with optimized group-delay dispersion, Opt. Lett. **18**, 1541 (1993)

11.1123 C. Chudoba, J. G. Fujimoto, E. P. Ippen, H. A. Haus, U. Morgner, F. X. Kärtner, V. Scheuer, G. Angelow, T. Tschudi: Generation of 20 fs pulses by a prismless Cr^{4+} : YAG laser, Opt. Lett. **26**, 292 (2001)

11.1124 X. Liu, L. Qian, F. Wise, Z. Zhang, T. Itatani, T. Gugaya, T. Nakagawa, K. Torizuka: Diode-pumped Cr:fosterite laser mode locked by a semiconductor saturable absorber, Appl. Opt. **37**, 7080 (1998)

11.1125 S. Kück, K. Petermann, U. Pohlmann, G. Huber: Near-infrared emission of Cr^{4+} doped garnets: Lifetimes, quantum efficiencies, and emission cross sections, Phys. Rev. B **51**, 17323 (1995)

11.1126 N. I. Borodin, A. G. Okhrimchuk, A. V. Shestakov: Polarizing spectroscopy of $Y_3Al_5O_{12}$, $SrAl_2O_4$, $CaAl_2O_4$ crystals containing Cr^{4+}, **13**, 42–46 (1992)

11.1127 S. Kück, K. L. Schepler, S. Hartung, K. Petermann, G. Huber: Excited state absorption and its influence on the laser behavior of Cr^{4+}-doped garnets, J. Luminesc. **72–74**, 222 (1997)

11.1128 J. Zhang, Y. Kalisky, G. H. Atkinson, M. Kokta: Tunable cw laser action of Cr^{4+} : $Lu_3Al_5O_{12}$ at room-temperature. In: *OSA Tech. Dig.* (Opt. Soc. Am., Washington 1995) pp. 182–186

11.1129 Y. Y. Kalisky, J. Zhang, S. R. Rotman, M. R. Kokta: Spectroscopy and laser performance of some rare earth and transition-metal-doped garnets, Proc. SPIE Vol. 2380, p. 24-33, UV and Visible Lasers and Laser Crystal Growth, R. Scheps, M. R. Kokta (Eds.), 1995

11.1130 V. Petričević, A. B. Bykov, R. R. Alfano: Room temperature laser operation of Cr : Ca_2GeO_4, a new near-infrared tunable laser crystal, Adv. Solid-State Lasers Conf., San Francisco 1996 (Opt. Soc. Am., Washington 1996) Post Deadline Paper PDP-2

11.1131 V. Petričević, A. B. Bykov, J. M. Evans, R. R. Alfano: Room-temperature near-infrared tunable laser operation of Cr^{4+} : Ca_2GeO_4, Opt. Lett. **21**, 1750 (1996)

11.1132 V. Petričević, A. B. Bykov, J. M. Evans, A. Seas, R. R. Alfano: Room temperature CW and pulsed near-infrared tunable laser operation of Cr^{4+}: Ca_2GeO_4, Conf. Lasers Electro-Optics 1997, Tech. Dig., Baltimore 1997 (Opt. Soc. Am., Washington 1997) CThT2

11.1133 V. Petričević, A. B. Bykov, J. M. Evans, R. R. Alfano: Room-temperature near-infrared tunable laser operation of Cr^{4+} : Ca_2GeO_4, Opt. Lett. **21**, 1750 (1996)

11.1134 V. Petričević, A. B. Bykov, J. M. Evans, A. Seas, A. Delgado, R. R. Alfano, G. V. Kanunnikov: Pulsed laser operation of Cr^{4+} : $LiScGeO_4$ at 1.3 μm, Conf. Lasers Electro-Optics 1997, Tech. Dig., Baltimore 1997 (Opt. Soc. Am., Washington 1997) CTuE7

11.1135 B. H. T. Chai, Y. Shimony, C. Deka, X. X. Chang, E. Munin, M. Bass: Laser performance of Cr^{4+} : Y_2SiO_5 at liquid nitrogen temperature. In: *OSA Proc. Adv. Solid-State Lasers*, Vol. 13, ed. by L. L. Chase, A. A. Pinto (Opt. Soc. Am., Santa Fé 1992) pp. 28–30

11.1136 C. Deka, B. H. T. Chai, Y. Shimony, X. X. Chang, E. Munin, M. Bass: Laser performance of Cr^{4+} : Y_2SiO_5, Appl. Phys. Lett. **61**, 2141 (1992)

11.1137 J. Koetke, S. Kück, K. Petermann, G. Huber, G. Cerullo, M. Danailov, V. Magni, L. F. Qian, O. Svelto: Quasi-continuous wave laser operation of Cr^{4+}-doped Y_2SiO_5 at room temperature, Opt. Commun. **101**, 195 (1993)

11.1138 A. A. Kaminskii, B. V. Mill, E. L. Belokoneva, A. V. Butashin: Structure refinement and laser properties of orthorhombic chromium containing $LiNbGeO_5$, Inorg. Mater. **27**, 1899 (1991)

11.1139 R. Moncorgé, H. Manaa, A. A. Kaminskii: Spectroscopic investigation of the chromium-doped $LiNbGeO_5$ laser crystal, Chem. Phys. Lett. **200**, 635 (1992)

11.1140 H. Manaa, R. Moncorgé, A. V. Butashin, B. V. Mill, A. A. Kaminskii: Luminescence properties of Cr-doped $LiNbGeO_5$ laser crystal. In: *OSA Proc. Adv. Solid-State Lasers*, Vol. 15, ed. by A. A. Pinto, T. Y. Fan (Opt. Soc. Am., Washington 1993) pp. 343–345

11.1141 R. Moncorgé, H. Manaa, F. Deghoul, C. Borel, C. Wyon: Spectroscopic study and laser operation of Cr^{4+}-doped $(Sr,Ca)Gd_4(SiO_4)_3O$ single crystals, Opt. Commun. **116**, 393 (1995)

11.1142 H. Eilers, U. Hömmerich, S. M. Jacobsen, W. M. Yen: The near-infrared emission of $Cr:Mn_2SiO_4$ and $Cr:MgCaSiO_4$, Chem. Phys. Lett. **212**, 109 (1993)

11.1143 V. Petričević, A. Seas, R. R. Alfano, M. R. Kokta, M. H. Randles: $Cr:Mg_2GeO_4$ and $Cr:CaMgSiO_4$: New potential tunable solid-state laser crystals. In: *Adv. Solid-State Lasers and Compact Blue-Green Lasers Tech. Dig.*, Vol. 2 (Opt. Soc. Am., Washington 1993) pp. 238–240

11.1144 V. Petričević, A. B. Bykov, A. Seas, R. R. Alfano, D. Yao, L. Isaacs, G. V. Kanunnikov: Pulsed laser operation of $Cr^{4+}:LiScGeO_4$ at 1.3 μm, Conf. Lasers and Electro-Optics, Tech. Dig., Baltimore 1997 (Opt. Soc. Am., Washington 1997) CFH1

11.1145 A. V. Podlipensky, V. G. Scherbitsky, N. V. Kuleshov, V. I. Levchenko, V. N. Yakimovich, M. Mond, E. Heumann, G. Huber, H. Kretschmann, S. Kück: Efficient laser operation and continuous wave diode pumping of $Cr^{2+}:ZnSe$ single crystals, Appl. Phys. B **72**, 253 (2001)

11.1146 J. D. Beasley: Thermal conductivities of sdome novel nonlinerar optical materials, Appl. Opt. **33**, 1000 (1994)

11.1147 D. T. F. Marple: Refractive index of ZnSe, ZnTe, and CdTe, J. Appl. Phys. **35**, 539 (1964)

11.1148 W. L. Bond: Measurements of the reefractive indices of several crystals, J. Appl. Phys. **36**, 1674 (1965)

11.1149 L. D. DeLoach, R. H. Page, G. D. Wilke, S. A. Payne, W. F. Krupke: Transition metal-doped zinc chalcogenides: spectroscopy and laser demonstration of a new class of gain media, IEEE J. Quantum Electron. **32**, 885 (1996)

11.1150 L. D. DeLoach, R. H. Page, G. D. Wilke, S. A. Payne, W. F. Krupke: Properties of transition metal-doped zinc chalcogenide crystals for tunable IR laser radiation. In: *OSA Proc. Adv. Solid-State Lasers*, Vol. 24, ed. by B. H. T. Chai, S. A. Payne (Opt. Soc. Am., Washington 1995) pp. 127–131

11.1151 R. H. Page, L. D. DeLoach, K. I. Schaffers, F. D. Patel, R. L. Beach, S. A. Payne, W. F. Krupke: Recent developments in Cr^{2+}-doped II-VI compound lasers. In: *OSA Trends Opt. Photon. Adv. Solid-State Lasers*, Vol. 1, ed. by S. A. Payne, C. R. Pollock (Opt. Soc. Am., Washington 1996) pp. 130–136

11.1152 U. Hömmerich, X. Wu, V. R. Davis, S. B. Trivedi, K. Grasza, R. J. Chen, S. Kutcher: Demonstration of room-temperaturelaser action at 2.5 mm from $Cr_{2+}Cd_{0.85}Mn_{0.15}Te$, Opt. Lett. **22**, 1180 (1997)

11.1153 J. T. Seo, U. Hömmerich, H. Zong, S. B. Trivedi, S. W. Kutcher, C. C. Wang, R. J. Chen: Mid-infrared lasing from a novel optical material: Chromium-doped $Cd_{0.55}Mn_{0.45}Te$, phys. stat. sol. a **175**, R3 (1999)

11.1154 K. L. Schepler, S. Kück, L. Shiozawa: Cr^{2+} emission spectroscopy in CdSe, J. Luminesc. **72–74**, 116 (1997)

11.1155 R. Pappalardo, R. E. Dietz: Absorption spectra of transition ions in CdS crystals, Phys. Rev. **123**, 1188 (1961)

11.1156 C. S. Kelley, F. Williams: Optical absorption spectra of chromium-doped zinc sulfide crystals, Phys. Rev. B **2**, 3 (1970)

11.1157 J. T. Vallin, G. A. Slack, S. Roberts, A. E. Hughes: Infrared absorption in some II-VI compounds doped with Cr, Phys. Rev. B **2**, 4313 (1970)

11.1158 A. Fazzio, M. J. Caldas, A. Zunger: Many-electron multiplet effects in the spectra of 3d impurities in heteropolar semiconductors, Phys. Rev. B **30**, 3430 (1984)

11.1159 H. Nelkowski, G. Grebe: IR-luminescence of ZnS:Cr, J. Luminesc. **1–2**, 88 (1970)

11.1160 G. Grebe, H.-J. Schulz: Luminescence of Cr^{2+} centers and related optical interactions involving crystal-field levels of chromium ions in zinc sulfide, Z. f. Naturforschung **29A**, 1803 (1974)

11.1161 G. Grebe, G. Roussos, H.-J. Schulz: Infrared luminescence of ZnSe:Cr crystals, J. Luminesc. **12–13**, 701 (1976)

11.1162 G. Goetz, A. Krost, H.-J. Schulz: Cr^{2+} (d^4) infrared emission in CdS and CdSe, J. Crystal Growth **101**, 414 (1990)

11.1163 I. T. Sorokina, E. Sorokin, T. J. Carrig: Femtosecond pulse generation from a SESAM mode-locked Cr:ZnSe laser, Conf. Lasers Electro-Optics, Tech. Dig., Long Beach 2006 (Opt. Soc. Am., Washington 2006) CMQ2

11.1164 A. V. Podlipensky, V. G. Scherbitsky, N. V. Kuleshov, V. I. Levchenko, V. N. Yakimovich, A. Diening, M. Mond, S. Kück, G. Huber: 1 W continuous-wave laser generation and excited state absorption measurements in $Cr^{2+}:ZnSe$. In: *Adv. Solid-State Lasers*, OSA Trends Opt. Photon., Vol. 34, ed. by H. Injeyan, U. Keller, C. Marshall (Opt. Soc. Am., Washington 2000) pp. 201–206

11.1165 E. Sorokin, I. T. Sorokina: Tunable diode-pumped continuous-wave $Cr^{2+}:ZnSe$ laser, Appl. Phys. Lett. **80**, 3289 (2002)

11.1166 J. McKay, W. B. Roh, K. L. Schepler: 4.2 W Cr^{2+} : ZnSe face cooled disk laser. In: *Conf. Lasers Electro-Optics*, OSA Tech. Dig. Ser. (Opt. Soc. Am., Washington 2002)

11.1167 A. Giesen, H. Hügel, K. Wittig, U. Brauch, H. Opower: Scalable concept for diode-pumped high-power solid-state lasers, Appl. Phys. B **58**, 365 (1994)

11.1168 T. J. Carrig, G. J. Wagner: Power scaling of Cr^{2+} : ZnSe lasers. In: *Adv. Solid-State Lasers*, OSA Trends Opt. Photon., Vol. 50, ed. by C. Marshall (Opt. Soc. Am., Washington 2001) pp. 406–410

11.1169 W. Alford, G. J. Wagner, J. Keene, T. J. Carrig: High-power and Q-switched Cr:ZnSe lasers. In: *Adv. Solid State Photon.*, OSA Tech. Dig. (Opt. Soc. Am., Washington 2003)

11.1170 I. T. Sorokina, E. Sorokin, S. Mirov, V. Fedorov, V. Badikov, V. Panyutin, K. I. Schaffers: Broadly tunable compact continuous-wave Cr^{2+} : ZnS laser, Opt. Lett. **27**, 1040 (2002)

11.1171 I. T. Sorokina, E. Sorokin, S. Mirov, V. Fedorov, V. Badikov, V. Panyutin, A. Di Lieto, M. Tonelli: Continuous-wave tunable Cr_{2+} : ZnS laser, Appl. Phys. B **74**, 607 (2002)

11.1172 M. Mond, D. Albrecht, E. Heumann, G. Huber, S. Kück, V. I. Levchenko, V. N. Yakimovich, V. G. Shcherbitsky, V. E. Kisel, N. V. Kuleshov, M. Rattunde, J. Schmitz, R. Kiefer, J. Wagner: Laser diode pumping at 1.9 μm and 2 μm of Cr^{2+} : ZnSe and Cr^{2+} : CdMnTe, Opt. Lett. **27**, 1034 (2002)

11.1173 R. H. Page, K. I. Schaffers, L. D. DeLoach, G. D. Wilke, F. D. Patel, J. B. Tassano, S. A. Payne, W. F. Krupke, K.-T. Chen, A. Burger: Cr^{2+}-doped zinc chalcogenides as efficient, widely tunable mid-infrared lasers, IEEE J. Quantum Electron. **33**, 609 (1997)

11.1174 J. McKay, K. L. Schepler, S. Kück: Chromium (II): cadmium selenide laser, Opt. Soc. Am. Annual Meeting, Baltimore 1998 (Opt. Soc. Am., Washington 1998) ThGG1

11.1175 J. T. Seo, U. Hömmerich, S. B. Trivedi, R. J. Chen, S. Kutcher: Slope efficiency and tunability of a Cr^{2+} : $Cd_{0.85}Mn_{0.15}Te$ mid-infrared laser, Opt. Commun. **153**, 267 (1998)

11.1176 A. G. Bluiett, U. Hömmerich, R. T. Shah, S. B. Trivedi, S. W. Kutcher, C. C. Wang: Observation of lasing from Cr^{2+}-CdTe and compositional effects in Cr^{2+}-doped II–VI semiconductors, J. Electron. Mater. **31**, 806 (2002)

11.1177 J. McKay, D. Krause, K. L. Schepler: Optimization of Cr^{2+} : CdSe for efficient laser operation. In: *Adv. Solid-State Lasers*, OSA Trends Opt. Photon., Vol. 34, ed. by H. Injeyan, U. Keller, C. Marshall (Opt. Soc. Am., Washington 2000) pp. 218–224

11.1178 K. Graham, S. B. Mirov, V. V. Fedorov, M. E. Zvanut, A. Avanesov, V. Badikov, B. Ignatev, V. Panutin, G. Shevirdyaeva: Spectroscopic characterization and laser performance of diffusion doped Cr^{2+} : ZnS. In: *Adv. Solid-State Lasers*, OSA Trends Opt. Photon., Vol. 50, ed. by C. Marshall (Opt. Soc. Am., Washington 2001) pp. 561–567

11.1179 J. McKay, W. B. Roh, K. L. Schepler: Extended mid-IR tuning of a Cr^{2+} : CdSelaser. In: *Adv. Solid State Lasers*, OSA Tech. Dig., ed. by M. Fermann, L. Marshall (Opt. Soc. Am., Washington 2002)

11.1180 T. J. Carrig, G. J. Wagner, A. Sennaroglu, J. Y. Jeong, C. R. Pollock: Mode-locked Cr^{2+} : ZnSe laser, Opt. Lett. **25**, 168 (2000)

11.1181 T. J. Carrig, G. J. Wagner, A. Sennaroglu, J. Y. Jeong, C. R. Pollock: Acousto-optic mode-locking of a Cr^{2+} : ZnSe laser. In: *Adv. Solid-State Lasers*, OSA Trends Opt. Photon., Vol. 34, ed. by H. Injeyan, U. Keller, C. Marshall (Opt. Soc. Am., Washington 2000) pp. 182–187

11.1182 I. T. Sorokina, E. Sorokin, A. Di Lieto, M. Tonelli, R. H. Page, K. I. Schaffers: Active and passive mode-locking of Cr^{2+} : ZnSe laser. In: *Adv. Solid-State Lasers*, OSA Trends Opt. Photon., Vol. 50, ed. by C. Marshall (Opt. Soc. Am., Washington 2001) pp. 157–161

11.1183 U. Hömmerich, J. T. Seo, A. Bluiett, M. Turner, D. Temple, S. B. Trivedi, H. Zong, S. W. Kutcher, C. C. Wang, R. J. Chen, B. Schumm: Mid-infrared laser development based on transition metal doped cadmium manganese telluride, J. Luminesc. **87–89**, 1143 (2000)

11.1184 J. McKay, K. L. Schepler, G. Catella: Kilohertz, 2.6 μm Cr^{2+} : CdSe laser. In: *Adv. Solid-State Lasers*, OSA Proc. Trends Opt. Photon., Vol. 26, ed. by M. M. Fejer, H. Injeyan, U. Keller (Opt. Soc. Am., Washington 1999) pp. 420–426

11.1185 G. J. Wagner, T. J. Carrig, R. H. Page, K. I. Schaffers, J.-O. Ndap, X. Ma, A. Burger: Continuous-wave broadly tunable Cr^{2+} ZnSe laser, Opt. Lett. **24**, 19 (1999)

11.1186 G. J. Wagner, T. J. Carrig, R. H. Jarman, R. H. Page, K. I. Schaffers, J.-O. Ndap, X. Ma, A. Burger: High-efficiency, broadly tunable continuous-wave Cr^{2+} : ZnSe laser. In: *Adv. Solid-State Lasers*, OSA Proc. Trends in Optics and Photonics, Vol. 26, ed. by M. M. Fejer, H. Injeyan, U. Keller (Opt. Soc. Am., Washington 1999) pp. 427–434

11.1187 I. T. Sorokina, E. Sorokin, A. Di Lieto, M. Tonelli, R. H. Page, K. I. Schaffers: 0.5 W efficient broadly tunable continuos-wave Cr^{2+} : ZnSe laser. In: *Adv. Solid-State Lasers*, OSA Trends Opt. Photon., Vol. 34, ed. by H. Injeyan, U. Keller, C. Marshall (Opt. Soc. Am., Washington 2000) pp. 188–193

11.1188 I. T. Sorokina, E. Sorokin, A. Di Lieto, M. Tonelli, R. H. Page, K. I. Schaffers: Efficient broadly tunable continuous-wave Cr^{2+} : ZnSe laser, J. Opt. Soc. Am. B **18**, 926 (2001)

11.1189 A. Sennaroglu, U. Demirbas, N. Vermeulen, H. Ottevaere, H. Thienpont: Continuous-wave broadly tunable Cr^{2+}: ZnSe laser pumped by a thulium fiber laser, Opt. Commun. **268**, 115 (2006)

11.1190. V. Podlipensky, V. G. Scherbitsky, N. V. Kuleshov, V. P. Mikhailov, V. I. Levchenko, V. N. Yakimovich, L. I. Postnova, V. I. Konstantinov: Pulsed laser operation of diffusion-doped Cr^{2+} : ZnSe, Opt. Commun. **167**, 129 (1999)

11.1191 R. H. Page, J. A. Skidmore, K. I. Schaffers, R. J. Beach, S. A. Payne, W. F. Krupke: Demonstrations of diode-pumped and grating-tuned ZnSe : Cr^{2+} lasers. In: *Adv. Solid-State Lasers*, OSA Proc. Trends Opt. Photon., Vol. 10, ed. by C. R. Pollock, W. R. Bosenberg (Opt. Soc. Am., Washington 1997) pp. 208–210

11.1192 M. Mond, D. Albrecht, E. Heumann, G. Huber, S. Kück, V. I. Levchenko, V. N. Yakimovich, V. G. Shcherbitsky, V. E. Kisel, N. V. Kuleshov, M. Rattunde, J. Schmitz, R. Kiefer, J. Wagner: Cr^{2+} : ZnSe laser pumped by a 1.57 µm erbium fibre amplifier. In: *Adv. Solid-State Lasers*, OSA Tech. Dig., ed. by M. Fermann, L. Marshall (Opt. Soc. Am., Washington 2002)

11.1193 D. Albrecht, M. Mond, E. Heumann, G. Huber, S. Kück, V. I. Levchenko, V. N-. Yakimovich, V. G. Shcherbitsky, V. E. Kisel, N. V. Kuleshov, M. Rattunde, J. Schmitz, R. Kiefer, J. Wagner: Efficient 100 mW Cr^{2+} : ZnSe laser pumped by a 1.9 µm laser diode. In: *Conf. Lasers Electro-Optics*, OSA Tech. Dig. Ser. (Opt. Soc. Am., Washington 2002) p. CMY5

11.1194 Oral presentations at the *Conference on Lasers and Electro-Optics Europe*, Nice, France, 2000: A. V. Podlipensky, V. G. Scherbitsky, N. V. Kuleshov, V. I. Levchenko, V. N. Yakimovich, A. Diening, M. Mond, S. Kück, G. Huber, paper CWH3 and I. T. Sorokina, E. Sorokin, A. Di Lieto, M. Tonelli, R. H. Page, K. I. Schaffers, paper CWH4

11.1195 M. Mond, E. Heumann, H. Kretschmann, S. Kück, G. Huber, V. I. Levchenko, V. N. Yakimovich, A. V. Podlipensky, V. G. Shcherbitsky, N. V. Kuleshov: Continuous-wave diode pumped Cr^{2+} : ZnSe and high power laser operation. In: *Adv. Solid-State Lasers*, OSA Trends Opt. Photon., Vol. 50, ed. by C. Marshall (Opt. Soc. Am., Washington 2001) pp. 162–165

11.1196 I. T. Sorokina, E. Sorokin, R. H. Page: Room-temperature cw diode-pumped Cr^{2+} : ZnSe laser. In: *Adv. Solid-State Lasers*, OSA Trends Opt. Photon., Vol. 50, ed. by C. Marshall (Opt. Soc. Am., Washington 2001) pp. 101–105

11.1197 M. Mond, D. Albrecht, H. M. Kretschmann, E. Heumann, G. Huber, S. Kück, V. I. Levchenko, V. N. Yakimovich, V. G. Shcherbitsky, V. E. Kisel, N. V. Kuleshov: Erbium fiber amplifier pumped Cr^{2+} : ZnSe laser, phys. stat. sol. (a) **188**, R3 (2001)

11.1198 V. E. Kisel, V. G. Shcherbitsky, N. V. Kuleshov, V. I. Konstantinov, V. I. Levchenko, E. Sorokin, I. Sorokina: Spectral kinetic properties and lasing characteristics of diode-pumped Cr^{2+}: ZnSe single crystals, Opt. Spectrosc. **99**, 663 (2005)

11.1199 I. S. Moskalev, V. V. Fedorov, S. B. Mirov: Multiwavelength mid-IR spatially-dispersive CW laser based on polycrystalline Cr^{2+} : ZnSe, Opt. Express **12**, 4986 (2004)

11.1200 L. F. Johnson, H. J. Guggenheim, R. A. Thomas: Phonon-terminated optical masers, Phys. Rev. **149**, 179 (1966)

11.1201 L. F. Johnson, H. J. Guggenheim: Phonon-terminated coherent emission from V^{2+} ions in MgF_2, J. Appl. Phys. **38**, 4837 (1967)

11.1202 P. F. Moulton: Tunable paramagnetic-ion lasers. In: *Laser Handbook*, Vol. 5, ed. by M. Bass, M. L. Stitch (North Holland, Amsterdam 1985) pp. 203–288

11.1203 U. Brauch, U. Dürr: Vibronic laser action of V^{2+} : $CsCaF_3$, Opt. Commun. **55**, 35 (1985)

11.1204 W. Knierim, A. Honold, U. Brauch, U. Dürr: Optical and lasing properties of V^{2+}-doped halide crystals, J. Opt. Soc. B **3**, 119 (1986)

11.1205 S. A. Payne, L. L. Chase: Excited state absorption of V^{2+} and Cr^{3+} ions in crystal hosts, J. Luminesc. **38**, 187 (1987)

11.1206 R. Moncorgé, T. Benyattou: Excited-state-absorption and laser parameters of V^{2+} in MgF_2 and $KMgF_3$, Phys. Rev. B **37**, 9177 (1988)

11.1207 J. Koetke: Absorption aus angeregten Zuständen in Ni^{2+}- und Er^{3+}-dotierten Kristallen Ph.D. Thesis (Shaker Verlag, Aachen 1994)

11.1208 J. Koetke, K. Petermann, G. Huber: Infrared excited state absorption of Ni^{2+} doped crystals, J. Luminesc. **60–61**, 197 (1993)

11.1209 U. Oetliker, M. Herren, H. U. Güdel, U. Kesper, C. Albrecht, D. Reinen: Luminescence properties of Mn^{5+} in a variety of host lattices: Effects of chemical and structural variation, J. Chem. Phys. **100**, 8656 (1994)

11.1210 J. D. Kingsley, J. S. Prener, B. Segall: Spectroscopy of MnO_4^{3-} in calcium halophosphates, Phys. Rev. **137**, A189 (1965)

11.1211 J. B. Milstein, J. Ackerman, S. L. Holt, B. R. McGarvey: Electronic structures of chromium(V) and manganese(V) in phosphate and vanadate hosts, Inorg. Chem. **11**, 1178 (1972)

11.1212 R. Borromei, L. Oleari, P. Day: Electronic spectrum of the manganate(V) ion in different host lattices, Faraday Trans. II **77**, 1563 (1981)

11.1213 J. Capobianco, G. Cormier, R. Moncorgé, H. Manaa, M. Bettinelli: Gain measurements of Mn^{5+} ($3d^2$) doped $Sr_5(PO_4)_3Cl$ and Ca_2PO_4Cl, Appl. Phys. Lett. **60**, 163 (1992)

11.1214 M. Herren, H. U. Güdel, C. Albrecht, D. Reinen: High-resolution near-infrared luminescence of mangenese(V) in tetrahedral oxo coordination, Chem. Phys. Lett. **183**, 98 (1991)

11.1215 J. Capobianco, G. Cormier, M. Bettinelli, R. Moncorgé, H. Manaa: Near infrared intraconfigurational luminescence spectroscopy of the Mn^{5+} ($3d^2$) ion in Ca_2PO_4Cl, $Sr_5(PO_4)_3Cl$, Ca_2VO_4Cl and Sr_2VO_4Cl, J. Luminesc. **54**, 1 (1992)

11.1216 J. Capobianco, G. Cormier, C. A. Morrison, R. Moncorgé: Crystal-field analyis of Mn^{5+} ($3d^2$) in $Sr_5(PO_4)_3Cl$, Opt. Mater. **1**, 209 (1992)

11.1217 H. R. Verdún: Absorption and emission properties of the new laser-active center in Mn^{5+} in several crystalline hosts. In: *OSA Proc. Adv. Solid-State Lasers*, Vol. 15, ed. by A. A. Pinto, T. Y. Fan (Opt. Soc. Am., Washington 1993) pp. 315–319

11.1218 L. D. Merkle, A. Guyot, B. H. T. Chai: Spectroscopic and laser investigations of $Mn^{5+}:Sr_5(VO_4)_3F$, J. Appl. Phys. **77**, 474 (1995)

11.1219 L. D. Merkle, A. Pinto, H. R. Verdún, B. McIntosh: Laser action from Mn^{5+} in $Ba_3(VO_4)_2$, Appl. Phys. Lett. **61**, 2386 (1992)

11.1220 L. D. Merkle, H. R. Verdún, B. McIntosh: Spectroscopy and laser operation of Mn^{5+}-doped vanadates. In: *OSA Proc. Adv. Solid-State Lasers*, Vol. 15, ed. by A. A. Pinto, T. Y. Fan (Opt. Soc. Am., Washington 1993) pp. 310–314

11.1221 H. Manaa, Y. Guyot, F. Deghoul, R. Moncorgé, L. D. Merkle: Excited state absorption in $Mn^{5+}:Sr_5(VO_4)_3F$, Chem. Phys. Lett. **238**, 333 (1995)

11.1222 S. Kück, K. L. Schepler, B. H. T. Chai: Evaluation of Mn^{5+}-doped $Sr_5(VO_4)_3F$ as a laser material based on excited-state absorption and stimulated-emission measurements, J. Opt. Soc. Am. B **14**, 957 (1997)

11.1223 P. F. Moulton: Lasers and Masers. In: *Handbook of Lasers Science and Technology I*, ed. by M. J. Weber (CRC, Boca Raton 1982) p. 60

11.1224 J. M. Breteau, D. Meichenin, F. Auzel: Etude du laser accordable $MgF_2:Ni^{2+}$, Rev. Phys. Appl. **22**, 1419 (1987)

11.1225 P. F. Moulton, A. Mooradian: Broadly tunable CW operation of $Ni:MgF_2$ and $Co:MgF_2$ lasers, Appl. Phys. Lett. **35**, 838 (1979)

11.1226 A. Mooradian: Transition-metal lasers could provide high power output, Laser Focus **15**, 24 (1979)

11.1227 B. C. Johnson, P. F. Moulton, A. Mooradian: Mode-locked operation of $Co:MgF_2$ and $Ni:MgF_2$ lasers, Opt. Lett. **10**, 116 (1984)

11.1228 L. F. Johnson, R. E. Dietz, H. J. Guggenheim: Optical maser oscillation from Ni^{2+} in MgF_2 involving simultaneous emission of phonons, Phys. Rev. Lett. **11**, 318 (1963)

11.1229 P. F. Moulton: Pulse-pumped operation of divalent transition-metal lasers, J. Quantum Electron. **18**, 1185 (1982)

11.1230 P. F. Moulton, A. Mooradian, T. B. Reed: Efficient CW optically pumped $Ni:MgF_2$ laser, Opt. Lett. **3**, 164 (1978)

11.1231 L. F. Johnson, H. J. Guggenheim, D. Bahnck, A. M. Johnson: Phonon-terminated laser emission from Ni^{2+} ions in $KMgF_3$, Opt. Lett. **8**, 371 (1983)

11.1232 J. Koetke, S. Kück, K. Petermann, G. Huber: Pulsed laser operation of $Ni^{2+}:Gd_3Ga_5O_{12}$, International Quantum Electronics Conference, Anaheim 1994 (Opt. Soc. Am., Washington 1994) QTuE1

11.1233 H. Manaa, Y. Guyot, R. Moncorgé: Spectroscopic and tunable laser properties of Co^{2+}-doped single crystals, Phys. Rev. B **48**, 3633 (1993)

11.1234 M. D. Sturge: The Jahn Teller Effect in Solids. In: *Solid State Physics*, Vol. 20, ed. by F. Seitz, D. Turnbull, H. Ehrenreich (Academic, New York 1967)

11.1235 J. J. Adams, C. Bibeau, R. H. Page, S. A. Payne: Tunable laser action at 4.0 microns from Fe:ZnSe. In: *Adv. Solid-State Lasers*, OSA Proc. Trends Opt. Photon., Vol. 26, ed. by M. M. Fejer, H. Injeyan, U. Keller (Opt. Soc. Am., Washington 1999) pp. 435–440

11.1236 J. J. Adams, C. Bibeau, R. H. Page, D. M. Krol, L. H. Furu, S. A. Payne: 4.0–4.5 μm lasing of Fe : ZnSe below 180 K, a new mid-infrared laser material, Opt.Lett. **24**, 1720 (1999)

11.1237 J. Kernal, V. V. Fedorov, A. Gallian, S. B. Mirov, V. V. Badikov: 3.9 4.8μm gain-switched lasing of Fe : ZnSe at room temperature, Opt. Express **13**, 10608 (2005)

11.1238 V. A. Akimov, V. V. Voronov, V. I. Kozlovskii, Y. V. Korostelin, A. I. Landman, Y. P. Podmarkov, M. P. Frolov: Efficient IR Fe : ZnSe laser continuously tunable in the spectral range from 3.77 to 4.40 μm, Quantum Electron. **34**, 912 (2004)

11.1239 V. V. Voronov, V. I. Kozlovskii, Y. V. Korostelin, A. I. Landman, Y. P. Podmarkov, M. P. Frolov: Laser parameters of a Fe : ZnSe laser crystal in the 85 255K temperature range, Quantum Electron. **35**, 809 (2005)

11.1240 P. B. Klein, J. E. Furneaux, R. L. Henry: Laser oscillation at 3.53 μm from Fe^{2+} in *n*-InP : Fe, Appl. Phys. Lett. **42**, 638 (1983)

11.1241 A. M. Fox, A. C. Maciel, J. F. Ryan: Efficient CW performance of a $Co:MgF_2$ laser operating at 1.5–2.0 μm, Opt. Commun. **59**, 142 (1986)

11.1242 P. F. Moulton: Pulse-pumped operation of divalent transition-metal lasers, IEEE J. Quantum Electron. **18**, 1185 (1982)

11.1243 P. F. Moulton: An investigation of the $Co:MgF_2$ laser system, IEEE J. Quantum Electron. **21**, 1582 (1985)

11.1244 D. Welford, P. F. Moulton: Room-temperature operation of a $Co:MgF_2$ laser, Opt. Lett. **13**, 975 (1988)

11.1245 A. C. Maciel, P. Maly, J. F. Ryan: Simultaneous modelocking and *Q*-switching of a $Co:MgF_2$ laser by loss-modulation frequency detuning, Opt. Commun. **61**, 125 (1987)

11.1246 S. Løvold, P. F. Moulton, D. K. Killinger, N. Menyuk: Frequency tuning characteristics of a *Q*-switched $Co:MgF_2$ laser, IEEE J. Quantum Electron. **21**, 202 (1985)

11.1247 A. di Lieto: Development of a CW $Co:MgF_2$ laser, Opt. Lasers Engineering **39**, 309 (2003)

11.1248 N. P. Vagin, P. G. Kryukov, Y. P. Podmarkov, M. P. Frolov, N. N. Yuryshev: Efficient operation of a Co : MgF_2 crystal laser pumped by radiation from a pulsed oxygen-iodine laser, Quantum Electron. **28**, 288 (1998)
11.1249 R. Y. Abdulsabirov, S. L. Korableva, P. G. Kryukov, A. K. Naumov, Y. P. Podmarkov, V. V. Semashko, M. P. Frolov: Efficient laser pumping of a Co : MgF_2 crystal by radiation with the wavelength 1.3 μm, Quantum Electron. **27**, 589 (1997)
11.1250 P. F. Moulton, A. Mooradian: Broadly tunable CW operation of Ni : MgF_2 and Co : MgF_2 lasers, Appl. Phys. Lett. **35**, 838 (1979)
11.1251 L. F. Johnson, R. E. Dietz, H. J. Guggenheim: Spontaneous and stimulated emission from Co^{2+} ions in MgF_2 and ZnF_2, Appl. Phys. Lett. **5**, 21 (1964)
11.1252 D. M. Rines, P. F. Moulton, D. Welford, G. A. Rines: High-energy operation of a Co : MgF_2 laser, Opt. Lett. **19**, 628 (1994)
11.1253 U. Dürr, U. Brauch, W. Knierim, W. Weigand: Vibronic solid state lasers: Transition metal ions in perowskites, Proc. Int. Conf. Lasers, San Francisco 1983, ed. by R. C. Powell (STS Press, Mc Lean 1983) 142
11.1254 W. Künzel, U. Dürr: Co^{2+}-Doped Perowskites: Laser Materials for the NIR Spectral Region, Appl. Phys. B **28**, 233 (1982)
11.1255 W. Künzel, W. Knierim, U. Dürr: CW infrared laser action of optically pumped Co^{2+} : $KZnF_3$, Opt. Commun. **36**, 383 (1981)
11.1256 K. R. German, U. Dürr, W. Künzel: Tunable single-frequency continuous-wave laser action in Co^{2+} : $KZnF_3$, Opt. Lett. **11**, 12 (1986)
11.1257 H. Kroemer: A proposed class of heterojunction injection lasers, Proc. IEEE **51**, 1782–1784 (1963)
11.1258 Z. I. Alferov: *The double heterostructure concept and its applications in physics, electronics, and technology* (Wiley, Weinheim 2001) Nobel Lecture
11.1259 R. Dingle: Confined carrier quantum states in ultrathin semiconductor heterostructures, Festkörperprobleme, Adv. Solid State Phys. **15**, 21–48 (1975)
11.1260 B. Bushan (Ed.): *Springer Handbook of Nanotechnology* (Springer, Berlin, Heidelberg 2004)
11.1261 J. Piprek: *Semiconductor Optoelectronic Devices*, 1 edn. (Academic, San Diego 2003)
11.1262 I. Kim, R. C. Alferness, U. Koren, L. L. Buhl, B. I. Miller, M. G. Young, M. D. Chien, T. L. Koch, H. M. Presby, G. Raybon, C. A. Burrus: Broadly tunable vertical-coupler filtered tensile-strained InGaAs/InGaAsP multiple quantum well laser, Appl. Phys. Lett. **64**, 2764–2766 (1994)
11.1263 K. Eberl: Quantum-dot lasers, Phys. World **10**, 47–50 (September 1997)
11.1264 A. Pawlis, C. Arens, G. Kirihakidis, K. Lischka, D. Schikora: private communication
11.1265 J. L. Merz, P. M. Petroff: Making quantum wires and boxes for optoelectronic devices, Mater. Sci. Eng. B **9**, 275–284 (1991)
11.1266 N. N. Ledentsov: *Growth Processes and Surface Phase Equilibria in Molecular Beam Epitaxy*, Springer Tracts Mod. Phys., Vol. 156 (Springer, Berlin, Heidelberg 1999)
11.1267 N. N. Ledentsov: Long-wavelength quantum-dot lasers on GaAs substrates: from media to device concepts?, IEEE J. Sel. Top. Quantum Electron. **8**, 1015–1024 (2002)
11.1268 D. Bimberg, M. Grundmann, N. N. Ledentsov: *Quantum Dot Heterostructures* (Wiley-VCH, Weinheim 2001)
11.1269 M. Grundmann (Ed.): *Nano-Optoelectronics* (Springer, Berlin, Heidelberg 2002)
11.1270 M. Grundmann, D. Bimberg: Selbstordnende Quantenpunkte: Vom Festkörper zum Atom, Phys. Bl. **53**, 517 (1997)
11.1271 U. Woggon: *Optical Properties of Semiconductor Quantum Dots*, Springer Tracts Mod. Phys., Vol. 136 (Springer, Berlin, Heidelberg 1997)
11.1272 J. P. Reithmaier, G. Sek, C. Hofmann, S. Kuhn, S. Reitzenstein, L. V. Keldysh, V. D. Kulakovskii, T. L. Reinecke, A. Forchel: Strong coupling in a single quantum dot-semiconductor microcavity system, Nature **432**, 197–200 (2004)
11.1273 O. Boyraz, B. Jalali: Demonstration of a silicon Raman laser, Opt. Express **12**, 5269–5273 (2004)
11.1274 H. C. Casey, M. B. Panish: *Heterostructure Lasers*, Vol. A (Academic, New York 1978)
11.1275 S. Asada: Waveguiding effect on modal gain in optical waveguide devices, IEEE J. Quantum Electron. **27**, 884–885 (April 1991)
11.1276 G. P. Agrawal, N. K. Dutta: *Long-Wavelength Semiconductor Lasers* (Van Nostrand Reinhold, New York 1986)
11.1277 T. Tamir: *Guided-Wave Optoelectronics* (Springer, Berlin, Heidelberg 1988)
11.1278 H. Burkhard, E. Kuphal: InGaAsP/InP mushroom stripe lasers with low CW threshold and high output power, Jpn. J. Appl. Phys. **22**, L721–L723 (November 1983)
11.1279 R. Goebel, H. Janning, H. Burkhard: S.I. InP:Fe hydride-VPE for mushroom type lasers, Proc. 7th Conf. on Semi-insulating Mater. (Ixtapa1992) 125–130
11.1280 K. Petermann: *Laser Diode Modulation and Noise* (Kluwer Academic, Dordrecht 1988)
11.1281 C. H. Henry: Theory of spontaneous emission noise in open resonators and its application to lasers and optical amplifiers, J. Lightwave Technol. **4**, 288–297 (March 1986)
11.1282 L. D. Westbrook, B. Eng: Measurements of dg/dN and dn/dN and their dependence on photon energy in $\lambda = 1.5$ μm InGaAsP laser diodes, IEEE Proc. J. **133**, 135–142 (April 1986)

11.1283 Y. Arakawa, A. Yariv: Quantum well lasers - gain, spectra, dynamics, IEEE J. Quantum Electron. **22**, 1887–1897 (September 1986)

11.1284 P. W. A. Mc Ilroy, A. Kurobe, Y. Uematsu: Analysis and application of theoretical gain curves to the design of multi-quantum-well lasers, IEEE J. Quantum Electron. **21**, 1958–1963 (December 1985)

11.1285 J. Carroll, J. Whiteaway, D. Plumb: *Distributed Feedback Semiconductor Lasers* (IEEE, London 1998)

11.1286 H. Kogelnik, C. V. Shank: Coupled-wave theory of distributed feedback lasers, J. Appl. Phys. **43**, 2327–2335 (May 1972)

11.1287 A. Hardy, R. G. Waarts, D. F. Welch, W. Streifer: Analysis of three-grating coupled surface emitters, IEEE J. Quantum Electron. **26**, 843–849 (May 1990)

11.1288 J. I. Kinoshita: Axial profile of grating coupled radiation from second-order DFB lasers with phase shifts, IEEE J. Quantum Electron. **26**, 407–412 (March 1990)

11.1289 J. E. A. Whiteaway, B. Garrett, G. H. B. Thompson: The static and dynamic characteristics of single and multiple phase-shifted DFB laser structures, IEEE J. Quantum Electron. **28**, 1277–1293 (May 1992)

11.1290 M. Okai, M. Suzuki, T. Taniwatari: Strained multiquantum-well corrugation-pitch-modulated distributed feedback laser with ultranarrow (3.6 kHz) spectral linewidth, Electron. Lett. **29**, 1696–1697 (September 1993)

11.1291 A. Talneau, J. Charil, A. Ougazzaden, J. C. Bouley: High power operation of phase-shifted DFB lasers with amplitude modulated coupling coefficient, Electron. Lett. **28**, 1395–1396 (July 1992)

11.1292 Y. Kotaki, M. Matsuda, T. Fujii, H. Ishikawa: MQW-DFB lasers with nonuniform-depth $\lambda/4$ shifted grating, 17th European Conference on Optical Communications, Paris 1991 (IEEE, Piscataway 1991) 137–140

11.1293 S. Hansmann, H. Hillmer, H. Walter, H. Burkhard, B. Hübner, E. Kuphal: Variation of coupling coefficients by sampled gratings in complex coupled distributed feedback lasers, IEEE J. Sel. Top. Quantum Electron. **1**, 341–345 (June 1995)

11.1294 H. Hillmer, K. Magari, Y. Suzuki: Chirped gratings for DFB laser diodes using bent waveguides, Photon. Technol. Lett. **5**, 10–12 (1993)

11.1295 M. Usami, S. Akiba: Suppression of longitudinal spatial hole-burning effect in $\lambda/4$-shifted DFB lasers by nonuniform current distribution, IEEE J. Quantum Electron. **25**, 1245–1253 (June 1989)

11.1296 T. Matsuoka, H. Nagai, Y. Noguchi, Y. Suzuki, Y. Kawaguchi: Effect of grating phase at the cleaved facet on DFB laser properties, Jpn. J. Appl. Phys. **23**, 138–140 (March 1984)

11.1297 Y. Nakano, Y. Luo, K. Tada: Facet reflection independent, single longitudinal mode oscillation in a GaAlAs/GaAs distributed feedback laser equipped with a gain-coupling mechanism, Appl. Phys. Lett. **55**, 1606–1608 (October 1989)

11.1298 C. Kazmierski, D. Robein, D. Mathoorasing, A. Ougazzaden, M. Filoche: 1.5 μm DFB laser with new current-induced gain gratings, IEEE J. Sel. Top. Quantum Electron. **1**, 371–374 (June 1995)

11.1299 B. Borchert, B. Stegmüller, R. Gessner: Fabrication and characteristics of improved strained quantum-well GaInAlAs gain-coupled DFB lasers, Electron. Lett. **29**, 210–211 (January 1993)

11.1300 T. W. Johannes, A. Rast, W. Harth, J. Rieger: Gain-coupled DFB lasers with a titanium surface Bragg grating, Electron. Lett. **31**, 370–371 (March 1995)

11.1301 H. Hillmer: Optically detected carrier transport in III/V semiconductor QW structures: experiments, model calculations and applications in fast 1.55 μm laser devices, Appl. Phys. B. **66**, 1–17 (1998)invited paper

11.1302 H. Hillmer, A. Greiner, F. Steinhagen, W. Schlapp, E. Binder, T. Kuhn, H. Burkhard: Carrier and photon dynamics in transversally asymmetric AlGaInAs/InP MQW lasers, SPIE Proc. **2693**, 352–68 (1996)invited paper

11.1303 M. Schilling, W. Idler, G. Laube, H. Schweizer, O. Hildebrand: Integrated interferometric injection laser: novel fast and broad-band tunable monolithic light source, IEEE J. Quantum Electron. **27**, 1616–1623 (1991)

11.1304 H. Hillmer, A. Grabmaier, S. Hansmann, H.-L. Zhu, H. Burkhard: Tailored DFB laser properties by individually chirped gratings using bent waveguides, IEEE J. Sel. Top. Quantum Electron. **1**, 356–362 (1995)

11.1305 M. Okai, S. Sakano, N. Chinone: *Wide-range continuous tunable double-sectioned distributed feedback lasers*, 15th Europ. Conf. Opt. Commun. (ECOC, Freiburg 1989) pp. 122–125

11.1306 M.-C. Amann, J. Buus: *Tunable Laser Diodes* (Artech House, Boston 1998)

11.1307 M.-C. Amann, J. Buus, D.J. Blumenthal: *Tunable Laser Diodes and Raleted Optical Sources* (Wiley, Hoboken 2005)

11.1308 Y. Tohmori, S. Suematsu, H. Tsushima, S. Arai: Wavelength tuning of GaInAsP/InP integrated laser with butt-jointed built-in distributed Bragg reflector, Electron. Lett. **19**, 656–657 (1983)

11.1309 M.-C. Amann: Tuning range and threshold current of the tunable twin-guide (TTG) laser, IEEE Photon. Technol. Lett. **1**, 253–254 (1989)

11.1310 H. Ishii, Y. Tohmori, Y. Yamamoto, T. Tamamura, Y. Yoshikuni: Modified multiple-phase-shift superstructure-grating DBR lasers for broad wavelength tuning, Electron. Lett. **30**, 1141–1142 (1994)

11.1311 V. Jayaraman, Z. M. Chuang, L. Coldren: Theory, design, and performance of extended tuning rangesemiconductor lasers with sampled gratings, IEEE J. Quantum Electron. **29**, 1824–1834 (1993)

11.1312 H. Hillmer, B. Klepser: Low-cost edge-emitting DFB laser arrays for DWDM communication systems implemented by bent and tilted waveguides, IEEE J. Quantum Electron. **40**, 1377–1383 (2004)

11.1313 S. Murata, I. Mito, K. Kobayashi: Tuning ranges for 1.5 μm wavelength tunable DBR lasers, Electron. Lett. **24**, 577–579 (1988)

11.1314 M. Öberg, S. Nilsson, T. Klinga, P. Ojala: A three-electrode distributed Bragg reflector laser with 22 nm wavelength tuning range, IEEE Photon. Technol. Lett. **3**, 299–301 (1991)

11.1315 N. Fujiwara, T. Kakitsuka, F. Kano, H. Okamoto, Y. Kawaguchi, Y. Yoshikuni, Y. Tohmori: Mode-hop-free wavelength-tunable distributed Bragg reflector laser, Electron. Lett. **39**, 614–615 (2003)

11.1316 A. J. Ward, D. J. Robbins, G. Busico, E. Barton, L. Ponnampalam, J. P. Duck, N. D. Whitbread, P. J. Williams, D. C. J. Reid, A. C. Carter, M. J. Wale: Widely tunable DS-DBR laser with monolithically integrated SOA: design and performance, IEEE J. Sel. Top. Quantum Electron. **11**, 149–156 (2005)

11.1317 R. C. Alferness, U. Koren, L. L. Buhl, B. I. Miller, M. G. Young, T. L. Koch, G. Raybon, C. A. Burrus: Broadly tunable InGaAsP/InP laser based on a vertical coupler filter with 57-nm tuning range, Appl. Phys. Lett. **60**, 3209–3211 (1992)

11.1318 I. F. Lealman, M. Okai, M. J. Robertson, L. J. Rivers, S. D. Perrin, P. Marshall: Lateral grating vertical coupler filter laser with 58 nm tuningrange, Electron Lett. **32**, 339–340 (1996)

11.1319 R.-J. Rigole, S. Nilsson, T. Klinga, J. Wallin, B. Stalnacke, E. Berglind, B. Stoltz: Access to 20 evenly distributed wavelengths over 100 nm using onlya single current tuning in a four-electrode monolithic semiconductorlaser, IEEE Photon. Technol. Lett. **7**, 1249–1251 (1995)

11.1320 R.-J. Rigole, S. Nilsson, L. Backbom, I. Klinga, J. Wallin, B. Stalnacke, E. Berglind, B. Stoltz: 114-nm wavelength tuning range of a vertical grating assistedcodirectional coupler laser with a super structure grating distributedBragg reflector, IEEE Photon. Technol. Lett. **7**, 697–699 (1995)

11.1321 G. Morthier, B. Moeyersoon, R. Baets: A λ/4-shifted sampled or superstructure grating widelytunable twin-guide laser, IEEE Photon. Technol. Lett. **13**, 1052–1054 (2001)

11.1322 J.-O. Wesström, S. Hammerfeldt, J. Buus, R. Siljan, R. Laroy, H. de Vries: Design of a widely tunable modulated grating Y-branch laser using the additive Vernier effect for improved super-mode selection, 18th IEEE Int. Semiconductor Laser Conf. (Garmisch 2002) TuP16

11.1323 S.-H. Jeong, S. Matsuo, Y. Yoshikuni, T. Segawa, Y. Ohiso, H. Okamoto: Chirped ladder-type interferometric filter for widely tunable laser diode, Electron. Lett. **40**, 990–991 (2004)

11.1324 S. Matsuo, Y. Yoshikuni, T. Segawa, Y. Ohiso, H. Okamoto: A widely tunable optical filter using ladder-type structure, IEEE Photon. Technol. Lett. **15**, 1114–1116 (2003)

11.1325 H. Hillmer, A. Grabmaier, H.-L. Zhu, S. Hansmann, H. Burkhard: Continuously chirped DFB gratings by specially bent waveguides for tunable lasers, IEEE J. Lightwave Technol. **13**, 1905–1912 (1995)

11.1326 J. Faist, A. Tredicucci, F. Capasso, C. Sirtori, D. L. Sivco, J. N. Baillargeon, A. L. Hutchinson, A. Y. Cho: High-power continuous-wave quantum cascade lasers, IEEE J. Quantum Electron. **34**, 336–343 (1998)

11.1327 M. Fujita, T. Baba: Microgear laser, Appl. Phys. Lett. **80**, 2051–2053 (2002)

11.1328 C. Prott, F. Römer, E. Ataro, J. Daleiden, S. Irmer, A. Tarraf, H. Hillmer: Modeling of ultra-widely tunable vertical cavity air-gap filters and VCSELs, IEEE J. Sel. Top. Quantum Electron. **9**, 918–928 (2003)

11.1329 F. Römer, C. Prott, S. Irmer, J. Daleiden, A. Tarraf, H. Hillmer: Tuning efficiency and linewidth of electrostatically actuated multiple air-gap filters, Appl. Phys. Lett. **82**, 176–178 (2003)

11.1330 S. O. Kasap: *Optoelectronics and Photonics – Principles and Practices* (Prentice Hall, London 2001)

11.1331 S. O. Kasap, P. Capper (Eds.): *Springer Handbook of Electronic and Photonic Materials* (Springer, Berlin, New York 2006)

11.1332 D. I. Babic, K. Streubel, R. P. Mirin, N. M. Margalit, J. E. Bowers, E. L. Hu: Transverse-mode and polarisation characteristics of double-fused 1.52 μm vertical-cavity lasers, Electron. Lett. **31**, 653–654 (1995)

11.1333 N. M. Margalit: 64 °C continuous-wave operation of 1.5 μm vertical-cavity laser, IEEE J. Sel. Top. Quantum Electron. **3**, 359–365 (1997)

11.1334 J. Boucart, C. Starck, F. Gaborit, A. Plais, N. Bouche, E. Derouin, J. C. Remy, J. Bonnet-Gamard, L. Goldstein, C. Fortin, D. Carpentier, P. Salet, F. Brillouet, J. Jacquet: Metamorphic DBR and tunnel-junction injection. A CW RT monolithic long-wavelength VCSEL, IEEE J. Sel. Top. Quantum Electron. **5**, 520–529 (1999)

11.1335 M. Ortsiefer, R. Shau, M.-C. Amann: Low-threshold index-guided 1.5 μm long-wavelength vertical-cavity surface-emitting laser with high efficiency, Appl. Phys. Lett. **76**, 2179–2181 (2000)

11.1336 E. Hall, S. Nakagawa, G. Almuneau, J. K. Kim, L. A. Coldren: Room-temperature, CW operation of lattice-matched long-wavelength VCSELs, Electron. Lett. **36**, 1465–1467 (2000)

11.1337 Y. Ohiso, H. Okamoto, R. Iga, K. Kishi, C. Amano: Single transverse mode operation of 1.55 μm buried heterostructure vertical-cavity surface-emitting lasers, IEEE Photon. Technol. Lett. **14**, 738–740 (2002)

11.1338 P. Unger: Introduction to power diode lasers. In: *High-power diode lasers*, Top. Appl. Phys., Vol. 78, ed. by R. Diehl (Springer, Berlin, Heidelberg 2000) pp. 1–54

11.1339 C. J. Chang-Hasnain: Progress and prospects of long wavelength VCSELs, IEEE Opt. Commun. **41**, S30–S34 (February 2003)

11.1340 C. Lauer, M. Ortsiefer, R. Shau, J. Rosskopf, G. Bohm, M. Maute, E. Kohler, M.-C. Amann: Electrically pumped room temperature CW-VCSELs with emission wavelength of 2 μm, Electron. Lett. **39**, 57–58 (2003)

11.1341 J. C. Chang-Hasnain: Tunable VCSEL, IEEE J. Select. Topics Quantum Electron. **6**, 978–986 (2000)

11.1342 A. Hasse, S. Irmer, J. Daleiden, N. Dharmarasu, S. Hansmann, H. Hillmer: Wide continuous tuning range of 221 nm by InP/air-gap vertical-cavity filters, Electron. Lett. **42**, 974–975 (2006)

11.1343 S. Irmer, J. Daleiden, V. Rangelov, C. Prott, F. Römer, M. Strassner, A. Tarraf, H. Hillmer: Ultra low biased widely continuously tunable Fabry-Pérot Filter, Photon. Technol. Lett. **15**, 434–436 (2003)

11.1344 H. Hillmer, R. Lösch, W. Schlapp: Strain-balanced AlGaInAs/InP heterostructures with up to 50 QWs by MBE, J. Crystal Growth **175**, 1120–1125 (1997)

11.1345 C. K. N. Patel, W. L. Faust, R. A. McFarlane: CW laser action on rotational transitions of the $\Sigma_u^+ - \Sigma_g^+$ vibrational band of CO_2, Bull. Am. Phys. Soc. **9**, 500 (1964)

11.1346 C. K. N. Patel: Interpretation of CO_2 optical maser experiments, Phys. Rev. Lett. **12**(21), 588–590 (May 1964)

11.1347 W. J. Witteman: *The CO_2 Laser* (Springer, Berlin, Heidelberg 1987)

11.1348 A. E. Siegman: *Lasers* (Univ. Science Books, Mill Valley 1986)

11.1349 J. Eichler, H. J. Eichler: *Laser*, th4 edn. (Springer, Berlin, Heidelberg 2002)

11.1350 Y. P. Raizer: *Gas Discharge Physics* (Springer, Berlin, Heidelberg 1991)

11.1351 C. K. N. Patel: Selective excitation through vibrational energy transfer and optical maser action in N_2-CO_2, Phys. Rev. Lett. **13**(21), 617–619 (Nov 1964)

11.1352 W. W. Rigrod: Saturation effects in high-gain lasers, J. Appl. Phys. **36**, 2487–2490 (August 1965)

11.1353 G. M. Schindler: Optimum output efficiency of homogeneously broadened lasers with constant loss, IEEE J. Quantum Electron. **16**, 546–549 (May 1980)

11.1354 T. S. Fahlen: CO_2 laser design procedure, Appl. Opt. **12**(10), 2381–2390 (October 1973)

11.1355 G. Herzberg: *Molecular Spectra and Molecular Structure*, Infrared and Raman Spectra of Polyatomic Molecules, Vol. II, nd2 edn. (D. Van Nostrand and Company, Inc., New York 1950)

11.1356 P. F. Bernath: *Spectra of Atoms and Molecules* (Oxford Univ. Press, Oxford 1995)

11.1357 W. Demtröder: *Molecular Physics*, 1 edn. (Wiley-VCH, Weinheim 2005)

11.1358 L. S. Rothman, R. L. Hawkings, R. B. Wattson, R. R. Gamache: Energy levels, intensities, and linewidths of atmospheric carbon dioxide bands, Journal of Quantitative Spectroscopy and Radiation Transfer **48**(5/6), 537–566 (November-December 1992)

11.1359 L. S. Rothman, D. Jacquemart, A. Barbe, D. Chris, Benner, M. Birk, L. R. Brown, M. R. Carleer, C. Chackerian, Jr., K. Chance, L. H. Coudert, V. Dana, V. M. Devi, J.-M. Flaud, R. R. Gamache, A. Goldman, J.-M. Hartmann, K. W. Jucks, A. G. Maki, J.-Y. Mandin, S. T. Massie, J. Orphal, A. Perrin, C. P. Rinsland, M. A. H., Smith, J. Tennyson, R. N. Tolchenov, R. A. Toth, J. Vander Auwera, P. Varanasi, G. Wagner: The HITRAN 2004 molecular spectroscopic database, Journal of Quantitative Spectroscopy and Radiative Transfer **96**(2), 139–204 (December 2005)

11.1360 L. S. Rothman, L. D. G. Young: Infrared energy levels and intensities of carbon dioxide – II, Journal of Quantitative Spectroscopy and Radiation Transfer **25**(6), 505–524 (June 1981)

11.1361 L. Bradley, K. Soohoo, C. Freed: Absolute frequencies of lasing transitions in nine CO_2 isotopic species, IEEE J. Quantum Electron. **22**(2), 234–267 (February 1986)

11.1362 R. T. Gamache, L. S. Rothman: Extension of the HITRAN database to non-LTE applications, Journal of Quantitative Spectroscopy and Radiation Transfer **48**(5/6), 519–525 (November-December 1992)

11.1363 K. Smith, M. Thomson R. *Computer Modelling of Gas Lasers* (Plenum, New York 1978)

11.1364 K. J. Siemsen, J. Reid, C. Dang: New techniques for determining vibrational temperatures, dissociation and gain limitations in CW CO_2 lasers, IEEE J. Quantum Electron. **16**(6), 668–676 (June 1980)

11.1365 Linde AG: Product information, Lasermix® Series (Pullach 2006)

11.1366 Air Liquide GmbH: Product information, LASAL™ Series (Düsseldorf 2006)

11.1367 P. D. Tannen, P. Bletzinger, A. Garscadden: Species composition in the CO_2 discharge laser, IEEE J. Quantum Electron. **10**(1), 6–11 (January 1974)

11.1368 P. O. Clark, J. Y. Wada: The influence of Xenon on seled-off CO_2 lasers, IEEE J. Quantum Electron. **4**(5), 263–266 (May 1968)

11.1369 D. He, D. R. Hall: Influence of xenon on seled-off operation of RF-excited CO_2 waveguide lasers, J. Appl. Phys. **56**(3), 856–857 (August 1984)

11.1370 S. Grudszus, M. März: Influence of gas dissociation and Xenon addition on steady-state microwave-exited CO_2 laser discharges, J. Phys. D **26**(11), 1980–1986 (November 1993)

11.1371 A. L. S. Smith, P. G. Browne: Catalysis in sealed CO_2 lasers, J. Phys. D **7**(12), 1652–1659 (August 1974)

11.1372 U. Berkermann: Massenspektrometrische und spektroskopische Untersuchungen an Plasmen von HF-angeregten CO_2-Lasern Ph.D. Thesis (Ruhr-Universität Bochum, Bochum 1998)

11.1373 J. Schulz: Diffusionsgekühlte, koaxiale CO_2-Laser mit hoher Strahlqualität Ph.D. Thesis (RWTH Aachen, Aachen 2001)

11.1374 A. L. S. Smith, H. Shields, A. E. Webb: Cathode materials for sealed CO_2 waveguide lasers, IEEE J. Quantum Electron. **19**(5), 815–820 (May 1983)

11.1375 W. Haas, T. Kishimoto: Investigations of the gas composition in sealed-off RF-excited CO_2 lasers, Proc. SPIE **1276**, 49–57 (1990)

11.1376 C. Willis: Catalytic control of the gas chemistry of sealed TEA CO_2 lasers, J. Appl. Phys. **50**(4), 2539–2543 (April 1979)

11.1377 D. S. Stark, M. R. Harris: Catalysed recombination of CO and O_2 in sealed CO_2 TEA laser gases at temperatures down to -27 degrees C, J. Phys. E **16**(6), 492–496 (June 1983)

11.1378 D. R. Schryer, G. B. Hoflund (Eds.): *Low-Temperature CO-Oxidation Catalysts for Long-Life CO_2 Lasers*, Vol. 3076 (Langley Research Center, Hampton 1990)

11.1379 P. G. Browne, A. L. S. Smith: Long-lived CO_2 lasers with distributed heterogeneous catalysis, J. Phys. D **7**(18), 2464–2470 (December 1974)

11.1380 J. A. Macken, S. K. Yagnik, M. Samis: CO_2 laser perfomance with a distributed gold catalyst, IEEE J. Quantum Electron. **25**(7), 1695–1703 (July 1989)

11.1381 M. B. Heeman-Ilieva, Y. B. Udalov, K. Hoen, W. J. Witteman: Enhanced gain and output power of a sealed-off RF-excited CO_2 waveguide laser with gold-plated electrodes, Appl. Phys. Lett. **64**(6), 673–675 (February 1994)

11.1382 S. A. Starostin, Y. B. Udalov, P. M. Peters, W. J. Witteman: Catalyst enhanced high power radio frequency excited CO_2 slab laser, Appl. Phys. Lett. **77**(21), 3337–3339 (November 2000)

11.1383 R. Engelbrecht, J. Schulz: Time resolved measurement of gas parameters inside diffusion-cooled CO_2 lasers with NIR diode laser and UV-VIS emission spectroscopy, Proc. SPIE **4184**, 278–281 (2000)

11.1384 Y. Z. Wang, J. S. Liu: Direct mass spectrometric diagnostics for a CO_2 gas laser, J. Appl. Phys. **59**(6), 1834–1838 (March 1986)

11.1385 R. Bleekrode: A Study of the spontaneous emission from $CO_2 - N_2He - H_2$ Laser Discharges $C^3{}_u - B^3{}_g$ emission bands of N_2, IEEE J. Quantum Electron. **5**(2), 57–60 (February 1969)

11.1386 U. Berkermann, A. Liffers, R. Hannemann, G. Jakob, C. Lücking, J. Mentel, G. Schiffner: Investigation of RF excited CO_2 slab laser discharges by measuring N_2 emission bands with high spatial resolution, Proc. SPIE **3092**, 243–246 (1997)

11.1387 C. Dang, J. Reid, B. K. Garside: Dynamics of the CO_2 upper laser level as measured with a tunable diode laser, IEEE J. Quantum Electron. **19**(4), 755–763 (April 1983)

11.1388 D. Toebaert, P. Muys, E. Desoppere: Spatially resolved measurement of the vibrational temperatures of the plasma in a DC-excited fast-axial-flow CO_2 laser, IEEE J. Quantum Electron. **31**(10), 1774–1778 (October 1995)

11.1389 R. Engelbrecht, R. Hocke, H. Brand: Measurement of trace gases in sealed-off CO_2 lasers with tunable diode lasers in the near infrared, ITG-Fachbericht **150**, 399–404 (1998)

11.1390 P. Lorini, P. De Natale, A. Lapucci: Accurate gas diagnostics for sealed-off CO_2-lasers using near-infrared DFB semiconductor lasers, IEEE J. Quantum Electron. **34**(6), 949–954 (June 1998)

11.1391 R. Engelbrecht: Gasanalyse im CO_2-Laser mittels Diodenlaser-Spektroskopie Ph.D. Thesis (Didacta, München 2002)

11.1392 R. Engelbrecht, S. Lau, K. Salffner, H. G. Löhmannsröben: Fasergekoppelte NIR-Diodenlaser-Spektrometer zur simultanen und isotopenaufgelösten Messung von CO und CO_2: Anwendungen in Plasma-Diagnostik und Bodengasanalyse, VDI-Fortschrittsberichte **1959**, 97–114 (2006)

11.1393 R. K. Brimacombe, J. Reid: Accurate measurements of pressure-broadened linewidths in a transversely excited CO_2 discharge, IEEE J. Quantum Electron. **19**(11), 1668–1673 (November 1983)

11.1394 R. J. H. Clark, R. E. Hester (Eds.): *Spectroscopy in Environmental Science*, Advances in Spectroscopy, Vol. 24 (Wiley, New York 1995)

11.1395 Bystronic Laser AG: Product information, Bylaser Series (Niederoenz 2006)

11.1396 P. Wirth: A new fast axial flow CO_2 laser with 1000 watt output power, Proc. SPIE **455**, 21–23 (1984)

11.1397 W. Rath, T. Northemann: Industrial fast axial flow CO_2 laser series from 10 kW to 20 kW, Proc. SPIE **2206**, 185–193 (1994)

11.1398 P. P. Vitruk, H. J. Baker, D. R. Hall: Similarity and scaling in diffusion cooled RF excited carbon dioxide lasers, IEEE J. Quantum Electron. **30**(7), 1623–1634 (July 1994)

11.1399 R. Engelbrecht, R. Schulz, G. Seibert, J. Hagen, P. Schmidt: Gas discharge impedance and transmission line voltages in a RF excited CO_2 slab laser, Frequenz **59**(5-6), 154–157 (May-June 2005)

11.1400 S. Wieneke, C. Uhrlandt, W. Viöl: New additional cooling effect of diffusion cooled sealed-off CO_2 lasers excited by dielectric barrier discharges at about 1 MHz with an all-solid-state generator, Laser Phys. Lett. **1**(5), 241–247 (April 2004)

11.1401 H. J. Baker: Direct measurement of the electrical impedance of narrow gap radio frequency gas discharges in the100 MHz region, Meas. Sci. Technol. **7**, 1630–1635 (August 1996)

11.1402 D. R. Hall, C. A. Hill: Radiofrequency-discharge-excited CO_2 lasers. In: *Handbook of Molecular Lasers*, ed. by P. K. Cheo (Marcel Dekker, New York 1987)

11.1403 A. Lapucci, F. Rosetti, M. Ciofini, G. Orlando: On the longitudinal voltage distribution in radio-frequency discharged CO_2 lasers with large area electrodes, IEEE J. Quantum Electron. **31**, 1537–42 (1995)

11.1404 B. Freisinger, H. Frowein, M. Pauls, G. Pott, J. H. Schafer, J. Uhlenbusch: Excitation of CO_2 lasers by microwave discharges, Proc. SPIE **1276**, 29–40 (1990)

11.1405 J. Nishimae, K. Yoshizawa: Development of CO_2 laser excited by 2.45 GHz microwave discharge, Proc. SPIE **1225**, 340–348 (1990)

11.1406 T. Ikeda, M. Danno, H. Shimazutsu, T. Abe, J. Tanaka: TM_{010}-mode microwave excited high power CO_2 laser using a cylindrical resonant cavity, IEEE J. Quantum Electron. **30**(11), 2657–2662 (November 1994)

11.1407 U. Bielesch, M. Budde, M. Fischbach, B. Freisinger, J. H. Schafer, J. Uhlenbusch, W. Viol: A *Q*-switched multikilowatt CO_2 laser system excited by microwaves, Proc. SPIE **1810**, 57–60 (1993)

11.1408 K. Saito, M. Kimura, T. Uchiyama: Generation of a uniform high-density microwave plasma for CO_2 lasers using orthogonal electric fields, Appl. Phys. B **B82**(4), 621–625 (March 2006)

11.1409 M. März, W. Oestreicher: Microwave excitation of a diffusion cooled CO_2 laser, J. Phys. D **27**(3), 470–474 (March 994)

11.1410 T. Li: Diffraction loss and selection of modes in maser resonators with circular mirrors, Bell Sys. Tech. J. **44**, 917–932 (May 1965)

11.1411 E. A. J. Marcatili, R. A. Schmeltzer: Hollow metallic and dielectric waveguides for long distance optical transmission and lasers, Bell Sys. Tech. J. **43**, 1783–1809 (1964)

11.1412 K. D. Laakmann, W. H. Steier: Waveguides: characteristic modes of hollow rectangular dielectric waveguides, Appl. Opt. **15**(5), 1134–1140 (May 1975)

11.1413 K. D. Laakmann, W. H. Steier: Hollow rectangular dielectric waveguides: errata, Appl. Opt. **15**(9), 2029 (September 1975)

11.1414 J. J. Degnan, D. R. Hall: Finite-aperture waveguide-laser resonators, IEEE J. Quantum Electron. **9**(9), 901–910 (September 1973)

11.1415 B. I. Ilukhin, Y. B. Udalov, I. V. Kochetov, V. N. Ochkin, M. B. Heeman-Ilieva, P. J. M. Peters, W. J. Witteman: Theoretical and experimental investigation of a waveguide CO_2 laser with radio-frequency excitation, Appl. Phys. B **62**(2), 113–127 (February 1996)

11.1416 D. P. Spacht, A. J. DeMaria: The desing of sealed CO_2 lasers continues to improve, Laser Focus World **37**(3), 105–110 (March 2001)

11.1417 Coherent Inc.: Product information GEM CW Series (Santa Clara 2006)

11.1418 P. Laakmann: RF-excited, all-metal gas laser, US Patent 4805182 (1989)

11.1419 Rofin Sinar Laser GmbH: Product information, DC series (Hamburg 2006)

11.1420 Synrad, Inc.: Product information Series48 (Mukilteo 2006)

11.1421 K. M. Abramski, A. D. Colley, H. J. Baker, D. R. Hall: Power scaling of large-area transverse radio frequency discharge CO_2 lasers, Appl. Phys. Lett. **54**(19), 1833–1835 (May 1989)

11.1422 P. E. Jackson, H. J. Baker, D. R. Hall: CO_2 large-area discharge laser using an unstable-waveguide hybrid resonator, Appl. Phys. Lett. **54**(20), 1950–1952 (May 1989)

11.1423 A. D. Colley, H. J. Baker, D. R. Hall: Planar waveguide, 1 kW CW, carbon dioxide laser excited by a single transverse RF discharge, Appl. Phys. Lett. **61**(2), 136–138 (July 1992)

11.1424 R. Nowack, H. Opower, H. Krüger, W. Haas, N. Wenzel: Diffusionsgekühlte CO_2-Hochleistungslaser in Kompaktbauweise, Laser und Optoelektronik **23**(3), 68–81 (1991)(in German)

11.1425 J. Tulip: Carbon Dioxide Slab Laser, US Patent 4719639 (1988)

11.1426 N. Hodgson, H. Weber: *Optische Resonatoren* (Springer, Berlin, Heidelberg 1992)

11.1427 T. Teuma, G. Schiffner, V. Saetchnikov: Fundamental- and high-order-mode losses in slab waveguide resonators for CO_2 lasers, Proc. SPIE **3930**, 144–152 (2000)

11.1428 Trumpf Laser GmbH: Product information, TLF Series (Ditzingen 2006)

11.1429 Coherent Inc.: Product information, Diamond K-Series (Santa Clara 2006)

11.1430 C. Hertzler, R. Wollermann-Windgasse, U. Habich, U. Jarosch, P. Loosen: 30 kW fast axial flow CO_2 laser with RF excitation, Proc. SPIE **2788**, 14–23 (1996)

11.1431 Universal Laser Systems, Inc.: Product information, ULR Series (Scottsdale 2006)

11.1432 Rofin-Sinar Laser GmbH: Product information, HF Series (Hamburg 2006)

11.1433 G. Dunham: Selaed carbon dioxide lasers anter the high-power area, Laser Focus World **35**(3), 105–110 (March 1999)

11.1434 J. A. Broderick, B. K. Jones, J. W. Bethel, E. F. Yelden: Laser system and method for beam enhancement, US Patent 6198759 B1 (2001)

11.1435 Synrad, Inc.: Product information, Firestar series (Mukilteo 2006)

11.1436 D. Ehrlichmann, U. Habich, H. D. Plum: Diffusion-cooled CO_2 laser with coaxial high frequency excitation and internal axicon, J. Phys. D **26**(2), 183–191 (February 1993)

11.1437 D. Ehrlichmann, U. Habich, H. D. Plum, P. Loosen, G. Herziger: Azimuthally unstable resonators for high-power CO_2 lasers with annular gain media, IEEE J. Quantum Electron. **30**(6), 1441–1447 (June 1994)

11.1438 G. Markillie, J. Deile, H. Schlueter: Novel design approach benefits CO_2 laser users, Laser Focus World **39**(10), 75–80 (October 2003)

11.1439 Trumpf Laser GmbH: Product information, TCF Series (Ditzingen 2006)

11.1440 R. T. Brown: CO_2 TEA lasers. In: *Handbook of Molecular Lasers*, ed. by P. K. Cheo (Marcel Dekker, New York 1987)

11.1441 Y. Yanning, W. Chongyi, Y. Lv, T. Rongqing, Z. Jinwen, L. Shiming, Z. Chong: A 3kW average power tunable TEA CO_2 laser, Opt. Laser Technol. **37**(7), 560–562 (October 2005)

11.1442 S. Wieneke, S. Born, W. Viol: Multikilohertz TEA CO_2 laser driven by an all-solid-state exciter, Proc. SPIE **4184**, 291–294 (2001)

11.1443 GSI, Inc.: Product information, Lumonics Impact Series (Rugby 2006)

11.1444 Edinburgh Instruments Ltd.: Product information, PL Series (Livingston 2006)

11.1445 R. Hocke, M. Collischon: Lineselective resonators with variable reflectivity gratings (VRG) for slab-laser geometry, Proc. SPIE **3930**, 52–61 (2000)

11.1446 R. Schulz, M. Collischon, H. Brand, L. P. Schmidt: Lineselective CO_2-lasers with variable line-spacing and variable reflectivity silicon gratings, Proc. SPIE **5777**, 742–745 (2005)

11.1447 W. B. Bridges, A. N. Chester: Visible and UV laser oscillation at 118 wavelengths in ionized neon, argon, krypton, xenon, oxygen, and other gases, Appl. Opt. **4**, 573–580 (1965)

11.1448 A. B. Petersen: Enhanced CW ion laser operation in the range $270 < \lambda < 380$ nm, SPIE Proc. **737**, 106–111 (1987)

11.1449 W. B. Bridges: Laser oscillation in singly ionized argon in the visible spectrum, Appl. Phys. Lett. **4**, 128–130 (1964)

11.1450 E. F. Labuda, E. I. Gordon, R. C. Miller: Continuous-duty argon ion lasers, IEEE J. Quantum Electron. **1**, 273–279 (1965)

11.1451 C. S. Willett: *Introduction to Gas Lasers: Population Inversion Mechanisms* (Pergamon, Oxford 1974)

11.1452 W. B. Bridges, A. S. Halsted: *Gaseous Ion Laser Research*, Tech. Rep. AFAL-TR-67-89; DDC No. AD-814897 (Hughes Research Laboratories, Malibu 1967)

11.1453 J. R. Fendley Jr.: Continuous UV lasers, IEEE J. Quantum Electron. **4**, 627–631 (1968)

11.1454 W. B. Bridges: Ionized gas lasers. In: *Handbook of Laser Science and Technology*, ed. by M. J. Weber (CRC, Boca Raton 1982)

11.1455 A. S. Halsted, W. B. Bridges, G. N. Mercer: *Gaseous Ion Laser Research*, Tech. Rep. AFAL-TR-68-227 (Hughes Research Laboratories, Malibu 1968)

11.1456 G. DeMars, M. Seiden, F. A. Horrigan: Optical degradation of high-power ionized argon gas lasers, IEEE J. Quantum Electron. **4**, 631–637 (1968)

11.1457 M. W. Dowley: Reliability and commercial lasers, Appl. Opt. **10**, 1791–1795 (1982)

11.1458 J. P. Goldsborough: Design of gas lasers. In: *Laser Handbook*, ed. by F. T. Arecchi, E. O. Schulz-DuBois (North-Holland, Amsterdam 1972)

11.1459 S. C. Guggenheimer, D. L. Wright: Controlling the propagation axes of an ion laser, Rev. Sci. Instrum. **62**(10), 2389–2393 (1991)

11.1460 A. Bloom: *Gas Lasers* (Wiley, New York 1968)

11.1461 A. Javan, W. R. Bennet Jr., D. R. Herriot: Population inversion and continuous optical maser oscillation in a gas discharge containing a helium neon mixture, Phys. Rev. Lett. **6**, 106–110 (1961)

11.1462 A. D. White, J. D. Ridgen: Continuous gas maser operation in the visible, Proc. IRE. **50**, 1967 (1962)

11.1463 D. L. Perry: CW laser oscillation at 5433 in neon, IEEE J. Quantum Electron. **7**, 102 (1971)

11.1464 K. Kincade, S. G. Anderson: Laser marketplace 2006, Laser Focus World **42**, 78–93 (January 2006)

11.1465 W. Brunner: *Wissenspeicher Lasertechnik* (VEB Fachbuchverlag, Leipzig 1987) (in German)

11.1466 W. Kleen, R. Müller: *Laser* (Springer, Berlin, Heidelberg 1969) (in German)

11.1467 F. K. Kneubühl, M. W. Sigrist: *Laser* (Teubner, Stuttgart 1991) (in German)

11.1468 G. R. Harrison: *Wavelength Tables with Intensities in Arc, Spark or Discharge Tube* (Wiley, New York 1939)

11.1469 N. Hodgson, H. Weber: *Optische Resonatoren* (Springer, Berlin, Heidelberg 1992) (in German)

11.1470 K. Kupfermüller: *Einführung in die theoretische Elektrotechnik* (Springer, Berlin, Heidelberg 2005) (in German)

11.1471 E. Grimsehl: *Lehrbuch der Physik*, Vol. 4 (Teubner, Leipzig 1975) (in German)

11.1472 C. K. Rhodes (Ed.): *Excimer Lasers*, Top. Appl. Phys., Vol. 30, 2 edn. (Springer, Berlin, Heidelberg 1984)

11.1473 D. Basting, G. Marowsky (Eds.): *Excimer Laser Technology* (Springer, Berlin, Heidelberg 2005)

11.1474 C. A. Brau: Rare gas halogen excimers. In: *Excimer Lasers*, Top. Appl. Phys., Vol. 30, 2 edn., ed. by C. K. Rhodes (Springer, Berlin, Heidelberg 1984) Chap. 4

11.1475 A. Mitchel, M. Zemansky: *Resonance Radiation and Excited Atoms* (Cambridge Uinv. Press, Cambridge 1961) Chap. III

11.1476 N. G. Basov, V. A. Danilychev, Y. M. Popov, D. D. Khodkevich: Laser operating in the vacuum region of the spectrum by excitation of liquid Xenon with an electron beam, Zhourn. Eksperim. Fisika i Tekhnika Pis. Red. **12**, 473 (1970)

11.1477 S. K. Searles, G. A. Hart: Stimulated emission at 281.8 nm from XeBr, Appl. Phys. Lett. **27**, 243 (1975)

11.1478 J. J. Ewing, C. A. Brau: Laser action on the ${}^2\sum^+_{1/2} \rightarrow {}^2\sum^+_{1/2}$ bands of KrF and XeCl and 354 nm laser action on XeF, Appl. Phys. Lett. **27**, 350 and 435 (1975)

11.1479 E. R. Ault, R. S. Bradford Jr., M. L. Bhaumik: High-power xenon fluoride laser, Appl. Phys. Lett. **27**, 413 (1975)

11.1480 J. A. Mangano, J. H. Jacob: Electron-beam-controlled discharge pumping of the KrF laser, Appl. Phys. Lett. **27**, 495 (1975)

11.1481 G. C. Tisone, A. K. Hays, J. M. Hoffman: 100 MW, 248.4 nm, KrF laser excited by an electron beam, Opt. Commun. **15**, 188 (1975)

11.1482 H. von Bergmann, U. Stamm: Principles of excimer lasers. In: *Excimer Laser Technology*, ed. by D. Basting, G. Marowsky (Springer, Berlin, Heidelberg 2005) Chap. 3

11.1483 H. Endert, M. Kauf, R. Pätzel: Excimer laser technology: industry trends and future perspectives, LaserOpto **31**(4), 46 (1999)

11.1484 M. Krauss, F. H. Mies: Electronic structure and radiative transitions of excimer systems. In: *Excimer Lasers*, Top. Appl. Phys., Vol. 30, 2 edn., ed. by C. K. Rhodes (Springer, Berlin, Heidelberg 1984) Chap. 2

11.1485 M. McCuster: The rare gas excimers. In: *Excimer Lasers*, Top. Appl. Phys., Vol. 30, 2 edn., ed. by C. K. Rhodes (Springer, Berlin, Heidelberg 1984) Chap. 3

11.1486 C. Momma, H. Eichmann, A. Tünnermann, P. Simon, G. Marowsky, B. Wellegehausen: Short-pulse amplification in an F_2 gain module, Opt. Lett. **18**, 1180 (1993)

11.1487 H. von Bergmann, U. Rebhan, U. Stamm: Design and technology of excimer lasers. In: *Excimer Laser Technology*, ed. by D. Basting, G. Marowsky (Springer, Berlin, Heidelberg 2005) Chap. 4

11.1488 P. W. Smith: *Pulsed power techniques for discharge pumped visible and ultraviolet excimer lasers*, IEE Colloquium on Pulsed Power Applications, Digest, Vol. 135 (INSPEC, London 1989) pp. 1311–1313

11.1489 D. Mathew, H. M. J. Bastiaens, K.-J. Boller, P. J. M. Peters: Current filamentation in discharge-excited F_2-based excimer laser gas mixtures, Appl. Phys. Lett. **88**, 101502 (2006)

11.1490 http://scitation.aip.org/

11.1491 M. Schulz-Ruhtenberg, J. Ihlemann, G. Marowsky, A. H. Nejadmalayeri, Mi L. Ng, J. Li, P. R. Herman: Fabrication of diffractive phase elements by F-laser ablation of fused silica, Proc. SPIE **5063**, 113 (2003)

11.1492 M. Schulz-Ruhtenberg, J. Ihlemann, J. Heber: Laser patterning of SiO_x-layers for the fabrication of diffractive phase elements for deep UV applications, Appl. Surf. Sci. **248**, 190 (2005)

11.1493 B. Schäfer, K. Mann: Investigation of the propagation characteristics of excimer laser using a Hartmann-Shack sensor, Rev. Sci. Instrum. **71**, 2663 (2000)

11.1494 M. Kunzmann, B. Schäfer, K. Mann: Charakteriserung von Laserstrahlung mittels Hartmann-Shack-Wellenfront-Sensor, Photonik **1**, 52–55 (2005)

11.1495 L. E. S. Mathias, J. T. Parker: Stimulated emission in the band spectrum of nitrogen, Appl. Phys. Lett. **3**(1), 16 (1963)

11.1496 H. G. Heard: Ultra-violet gas laser at room temperature, Nature **4907**, 667 (1963)

11.1497 D. A. Leonard: Saturation of the molecular nitrogen second positive laser transition, Appl. Phys. Lett. **7**(1), 4 (1965)

11.1498 E. T. Gerry: Pulsed-molecular-nitrogen laser theory, Appl. Phys. Lett. **7**(1), 6 (1965)

11.1499 J. Shipman jr.: Travelling wave excitation of high power gas lasers, Appl. Phys. Lett **10**(1), 3 (1967)

11.1500 D. Basting, F. P. Schäfer, B. Steyer: A simple, high power nitrogen laser, Optoelectron. **4**, 43 (1972)

11.1501 J. Schwab, F. Hollinger: Compact high power N_2 laser: circuit theory and design, IEEE J. Quantum Electron. **12**(3), 183 (1976)

11.1502 H. Seki, S. Takemori, T. Sato: Development of a highly efficient nitrogen laser using an ultrafast magnetic pulse compression circuit, IEEE J. Sel. Top. Quantum Electron. **1**(3), 825–829 (1995)

11.1503 M. Csele: http://technology.niagarac.on.ca/people/mcsele/lasers/, Niagara College, Canada; M. Csele: *Fundamentals of Light Sources and Lasers* (Wiley, New York 2004)

11.1504 F. G. Houtermans: Über Maser-Wirkung im optischen Spektralgebiet und die Möglichkeit absoluter negativer Absorption, Helv. Phys. Acta **33**, 933 (1960)(in German)

11.1505 www.itrs.net/Links/2005ITRS/Litho2005.pdf

11.1506 IBM research demonstrates path for extending current chip-making technique, Press release (IBM, 20 February 2006)

11.1507 An Introduction to Ring Technology, White Paper (Cymer Inc., San Diego 2006)

11.1508 R. Pätzel, U. Stamm: Excimer lasers for microlithography. In: *Excimer Laser Technology*, ed. by D. Basting, G. Marowsky (Springer, Berlin, Heidelberg 2005) Chap. 6

11.1509 M. Fiebig: TFT annealing. In: *Excimer Laser Technology*, ed. by D. Basting, G. Marowsky (Springer, Berlin, Heidelberg 2005) Chap. 16.2

11.1510 R. Pätzel, L. Herbst, F. Simon: Laser annealing of LTPS, Proc. SPIE **6106**, 61060A (2006)

11.1511 Coherent Inc.: http://www.coherent.com/Applications/index.cfm?fuseaction=Forms.page&PageID=172 (Coherent Inc., Santa Clara 2007)

11.1512 W.-J. Nam, K.-C. Park, S.-H. Jung, M.-K. Han: OI-ELA poly-Si TFTs for eliminating residual source/drain junction defects, Electrochem. Solid-State Lett. **8**(2), G41–G43 (2005)

11.1513 M. Hessling, J. Ihlemann: Via drilling. In: *Excimer Laser Technology*, ed. by D. Basting, G. Marowsky (Springer, Berlin, Heidelberg 2005) Chap. 11.3

11.1514 J. P. Sercel Associates Inc.:http://www.jpsalaser.com/page.asp?page_id=24 (J. P. Sercel Associates Inc., Hollis 2007)

11.1515 G. van Steenberge, P. Geerinck, S. van Put, P. van Daele: Integration of multimode waveguides and micromirror couplers in printed circuit boards using laser ablation, Proc. SPIE **5454**, 75 (2004)

11.1516 N. Hendrickx, G. van Steenberge, P. Geerinck, J. van Erps, H. Thienpont, P. van Daele: Laser-ablated coupling structures for stacked optical interconnections on printed circuit boards, Proc. SPIE **6185**, 618503 (2006)

11.1517 W. Pfleging, M. Przybylski, H.J. Brückner: Excimer laser material processing: state-of-the-art and new approaches in microsystem technology, Proc. SPIE **6107**, 61070G (2006)

11.1518 D.V. Turnquist: Vision correction with excimer lasers, Engineering in Medicine and Biology Society, Conf. Proc. 26th Annual Int. Conf. IEEE 2004 (IEEE, New York 2004) 5120–6

11.1519 M. Takahashi, A. Sakoh, K. Ichii, Y. Tokuda, T. Yoko, J. Nishii: Photosensitive GeO-SiO films for ultraviolet laser writing of channel waveguides and Bragg gratings with Cr-loaded waveguide structure, Appl. Opt. **42**(22), 4594–4598 (2003)

11.1520 Resonetics: http://www.resonetics.com/PDF/OLED.pdf (Resonetics, Nashua 2007)

11.1521 F.G. Omenetto, K. Boyer, J.W. Longworth, A. McPherson, T. Nelson, P. Noel, W.A. Schroeder, C.K. Rhodes, S. Szatmári, G. Marowsky: High-brightness terawatt KrF* (248 nm) system, Appl. Phys. B **64**(6), 643–646 (1997)

11.1522 J.-H. Klein-Wiele, J. Békési, J. Ihlemann, P. Simon: High speed fabrication of periodic nanostructures, Proc. Third Int. WLT-Conf. Lasers in Manufacturing, Munich 2005, ed. by E. Beyer, F. Dausinger, A. Ostendorf, A. Otto (AT-Fachverlag, Stuttgart 2005) 477

11.1523 S. Szatmári, G. Marowsky, P. Simon: Femtosecond excimer lasers and their applications. In: *Laser Physics and Applications, Laser Systems*, Adv. Mater. Technol., Vol. 1, ed. by G. Herzlger, H. Weber, R. Poprawe (Springer, Berlin, Heidelberg 2007) (in press)

11.1524 J.-H. Klein-Wiele, P. Simon: Fabrication of periodic nanostructures by phase-controlled multiple-beam interference, Appl. Phys. Lett. **83**, 4707–4709 (2003)

11.1525 A.B. Borisov, X. Song, F. Frigeni, Y. Koshman, Y. Dai, K. Boyer, C.K. Rhodes: Ultrabright multikilovolt coherent tunable X-ray source at $\lambda \approx 2.71-2.93$, J. Phys. B **36**, 3433 (2003)

11.1526 G. Gibson, T.S. Luk, A. McPherson, K. Boyer, C.K. Rhodes: Observation of a new inner-orbital molecular transition at 55.8 nm in N_2^{2+} produced by multiphoton coupling, Phys. Rev. A **40**, 2378 (1989)

11.1527 A. McPherson, T.S. Luk, B.D. Thompson, K. Boyer, C.K. Rhodes: Multiphoton induced X-ray emission and amplification from clusters, Appl. Phys. B **57**, 337 (1993)

11.1528 A. McPherson, B.D. Thompson, A.B. Borisov, K. Boyer, C.K. Rhodes: Multiphoton-induced X-ray emission at 4–5 keV from Xe atoms with multiple core vacancies, Nature **370**, 631 (1994)

11.1529 M.W. Clark, D. Schneider, D. DeWitt, J.W. McDonald, R. Bruch, U.I. Safronova, I.Y. Tolstikhina, R. Schuch: Xe L and M X-ray emission following Xe^{44-48+} ion impact on Cu surfaces, Phys. Rev. A **47**, 3983 (1993)

11.1530 A.B. Borisov, X. Song, Y. Koshman, Y. Dai, K. Boyer, C.K. Rodes: Saturated multikilovolt X-ray amplification with Xe clusters: single-pulse observation of Xe(L) spectral hole burning, J. Phys. B **36**, L285 (2003)

11.1531 K. Boyer, A.B. Borisov, X. Song, P. Zhang, J.C. McCorkindale, S. Khan, Y. Dai, P.C. Kepple, J. Davis, C.K. Rhodes: Explosive supersaturated amplification on $3d \rightarrow 2p$ Xe(L) hollow atom transitions at $\lambda \approx 2.7-2.9$, J. Phys. B **38**, 3055 (2005)

11.1532 A.B. Borisov, X. Song, P. Zhang, A. Dasgupta, J. Davis, P.C. Kepple, Y. Dai, K. Boyer: Amplification at $\lambda \approx 2.8$ on Xe(L) $(2\overline{s}2\overline{p})$ double-vacancy states produced by 248 nm excitation of Xe clusters in plasma channels, J. Phys. B **38**, 3935 (2005)

11.1533 A.B. Borisov, A.V. Borovskiy, V.V. Korobkin, A.M. Prokhorov, O.B. Shiryaev, Y.M. Shi, T.S. Luk, A. McPherson, J.C. Solem, K. Boyer, C.K. Rhodes: Observation of relativistic and charge-displacement self-channeling of intense subpicosecond ultraviolet (248 nm) radiation in plasmas, Phys. Rev. Lett. **68**, 2309 (1992)

11.1534 A.B. Borisov, J.W. Longworth, K. Boyer, C.K. Rhodes: Stable relativistic/charge displacement channels in ultrahigh power density ($\approx 10^{21}$ W/cm) plasmas, Proc. Natl. Acad. Sci. USA **95**, 7854 (1998)

11.1535 A.B. Borisov, X. Song, Y. Koshman, Y. Dai, K. Boyer, C.K. Rhodes: Saturated multikilovolt X-ray amplification with Xe clusters: Single-pulse observation of Xe(L) spectral hole burning, J. Phys. B **36**, L285 (2003)

11.1536 J.C. Solem, G.C. Baldwin: Microholography of living organisms, Science **218**, 229 (1982)

11.1537 K. Boyer, J.C. Solem, J.W. Longworth, A.B. Borisov, C.K. Rhodes: Biomedical three dimensional holographic microimaging at visible, ultraviolet and X-ray wavelengths|, Nature Medicine **2**, 939 (1996)

11.1538 R. Netze, R. Wouts, D. van der Spoel, E. Weckert, J. Hajdu: Potential for biomolecular imaging with femtosecond X-ray pulses, Nature **406**, 752 (2000)

11.1539 G. Marowsky, C.K. Rhodes: Hohle Atome – eine neue Form von hochangeregter Materie, Neue Zürcher Zeitung **254**, 42 (1995)(in German)

11.1540 G. Schriever, M. Rahe, W. Neff, K. Bergmann, R. Lebert, H. Lauth, D. Basting: Extreme ultraviolet light generation based on laser-produced plasmas (LPP) and gas-discharge-based pinch plasmas: a comparison of different concepts, Proc. SPIE **3997**, 162–168 (2000)

11.1541 For a survey see: V. Bakshi (Ed.): *EUV sources for lithography*, SPIE Press **PM 149** (2006)

11.1542 U. Stamm, J. Kleinschmidt, D. Bolshukhin, J. Brudermann, G. Hergenhan, V. Korobotchko, B. Nikolaus, M.C. Schürmann, G. Schriever, C. Ziener, V.M. Borisov: Development status of EUV sources

for use in beta-tools and high-volume chip manufacturing tools, Proc. SPIE **6151**, 190–200 (2006)
11.1543 P. P. Sorokin, J. R. Lankard: Stimulated emission observed from an organic dye, chloro-aluminum phthalocyanine, IBM J. Res. Dev. **10**, 162 (1966)
11.1544 F. P. Schäfer, W. Schmidt, J. Voltze: Organic dye solution laser, Appl. Phys. Lett. **9**, 306 (1966)
11.1545 M. Maeda: *Laser Dyes* (Academic, New York 1984)
11.1546 F. N. Baltakov, B. A. Barikhin, L. V. Sukhanov: 400 J pulsed laser using a solution of rhodamine-6G in ethanol, JETP Lett. **19**, 174 (1974)
11.1547 R. G. Morton, V. G. Dragoo: A 200 W average power, narrow bandwidth, tunable waveguide dye laser, IEEE J. Quantum Electron. **17**, 2245 (1981)
11.1548 J.-C. Diels: Femtosecond dye lasers. In: *Dye Laser Principles*, ed. by F. J. Duarte, L. W. Hillman (Academic, New York 1990) pp. 41–132
11.1549 B. Wellegehausen, H. Welling, R. Beigang: High power CW dye lasers, Appl. Phys. **6**, 335 (1974)
11.1550 T. W. Hänsch: Repetitively pulsed tunable dye laser for high resolution spectroscopy, Appl. Opt. **11**, 895 (1972)
11.1551 F. J. Duarte: Technology of pulsed dye lasers. In: *Dye Laser Principles*, ed. by F. J. Duarte, L. W. Hillman (Academic, New York 1990) pp. 239–285
11.1552 R. L. Fork, B. I. Green, C. V. Shank: Generation of optical pulses shorter than 0.1 ps by colliding pulse modelocking, Appl. Phys. Lett. **38**, 671 (1981)
11.1553 F. P. Schäfer (Ed.): *Dye Lasers*, Top. Appl. Phys., Vol. 1 (Springer, Berlin, Heidelberg 1973) pp. 3–540
11.1554 F. J. Duarte, L. W. Williams (Eds.): *Dye Laser Principles* (Academic, New York 1990)
11.1555 J. C. Walling: Tunable paramagnetic-ion solid-state lasers. In: *Tunable Laser*, Top. Appl. Phys., Vol. 59, ed. by L. F. Mollenauer, J. C. White (Springer, Berlin, Heidelberg 1992) pp. 331–398
11.1556 R. W. Boyd: Wave-equation description of nonlinear optical interactions. In: *Nonlinear Optics* (Academic, New York 1992) Chap. 2, pp. 67–94
11.1557 G. Seybold, G. Wagonblast: New perylene and violanthrone dyestuffs for fluorescent collectors, Dyes Pigments **11**, 303 (1989)
11.1558 M. Shah, K. Thangararaj, M. L. Soong, L. T. Wolford, J. H. Boyer, I. R. Politzer, T. G. Pavlopoulos: Pyrromethene-BF2 complexes as laser dyes, Heteroatom. Chem. **1**, 389 (1990)
11.1559 D. Avnir, D. Levy, R. Reisfeld: The nature of silica cage as reflected by spectral changes and enhanced photostability of trapped rhodamine 6G, J. Phys. Chem. **88**, 5956 (1984)
11.1560 D. A. Gromov, K. M. Dyumaev, A. A. Manenkov, A. P. Maslyukov, G. A. Matyushin, V. S. Nechitailo, A. M. Prokhorov: Efficient plastic-host dye lasers, J. Opt. Soc. Am. B **2**, 1028 (1985)
11.1561 Y. Oki, K. Ohno, M. Maeda: Tunable ultrashort pulse generation from a waveguided laser with premixed-dye-doped plastic film, Jpn. J. Appl. Phys. **37**, 6403 (1998)
11.1562 X.-L. Zhu, D. Lo: Sol–gel glass distributed feedback waveguide laser, Appl. Phys. Lett. **80**, 917 (2002)
11.1563 B. H. Soffer, B. B. McFarland: Continuously tunable, narrow-band organic dye lasers, Appl. Phys. Lett. **10**, 266 (1967)
11.1564 R. Sastre, A. Costela: Polymeric solid-state dye lasers, Adv. Mater. **7**, 198 (1995)
11.1565 F. J. Duarte: Solid-state multiple-prism grating dye laser oscillators, Appl. Opt. **33**, 3857 (1994)
11.1566 M. D. Rahn, T. A. King, A. A. Gorman, I. Hamblett: Photostability enhancement of liquid and solid dye lasers, Appl. Opt. **36**, 5862 (1997)
11.1567 M. Ahmad, T. A. King, D.-K. Ko, B. H. Cha, J. Lee: Photostability of lasers based on pyrromethene 567 liquids and solid-state host media, Opt. Commun. **203**, 327–334 (2002)
11.1568 A. Costela, I. Garcia, G. Gomez, O. Garcia, R. Sastre: Laser performance of pyrromethene 567 dye in solid polymeric matrices with different cross-linking degrees, J. Appl. Phys. **90**, 3159–3166 (2001)
11.1569 T. Tani, H. Namikawa, K. Arai, M. Makishima: Photochemical hole-burning study of 1,4-dihydroxyanthraquinone doped in amorphous silica prepared by alcoholate method, J. Appl. Phys. **58**, 3559–3565 (1985)
11.1570 F. Salin, G. Le Saux, P. Georges, A. Brun, C. Bagnall, J. Zarzycki: Efficient tunable solid-state laser near 630 nm using sulforhodamine 640-doped silica gel, Opt. Lett. **14**, 785 (1989)
11.1571 D. Lo, J. E. Parris, J. L. Lawless: Laser and fluorescent properties of dye-doped sol-gel silica from 400 nm to 800 nm, Appl. Phys. B **56**, 385 (1993)
11.1572 M. Canva, P. Georges, J.-F. Perelgritz, A. Brun, F. Chaput, J.-P. Boilot: Perylene-and pyrromethene-doped xerogel for a pulsed laser, Appl. Opt. **34**, 428 (1995)
11.1573 M. Faloss, M. Canva, P. Georges, A. Brun, F. Chaput, J.-P. Boilot: Toward millions of laser pulses with pyrromethene- and perylene-doped xerogels, Appl. Opt. **36**, 6760 (1997)
11.1574 C. Ye, K. S. Lam, K. P. Chik, D. Lo, K. H. Wong: Output performance of a dye-doped sol-gel silica laser in the near UV, Appl. Phys. Lett. **69**, 3800 (1996)
11.1575 K. S. Lam, D. Lo: Lasing behavior of sol-gel silica doped with UV laser dyes, Appl. Phys. B **66**, 427 (1998)
11.1576 H. Kogelnik, C. V. Shank: Stimulated emission in a periodic structure, Appl. Phys. Lett. **18**, 152 (1971)
11.1577 F. Hide, M. A. Diaz-Garcia, B. J. Schwartz, M. R. A. Anderson, Q. Pei, A. J. Heeger: Semiconducting polymers: a new class of solid-state laser materials, Science **273**, 1833 (1997)
11.1578 G. Krazelbinder, G. Leising: Organic solid-state lasers, Rep. Prog. Phys. **63**, 729 (2000)
11.1579 M. D. McGehee, M. A. Diaz-Garcia, F. Hide, R. Gupta, E. K. Miller, D. Moses, A. J. Heeger: Semiconducting polymer distributed feedback lasers, Appl. Phys. Lett. **72**, 1536 (1998)

11.1580 K. P. Kretch, W. J. Blau, V. Dumarcher, L. Rocha, C. Florini, J.-M. Nuzi, S. Pfeiffer, H. Tillman, H. H. Hörhold: Distributed feedback laser action from polymeric waveguides doped with oligo phenylene vinylene model compounds, Appl. Phys. Lett. **76**, 2146 (2000)

11.1581 Y. Sorek, M. Zevin, R. Reisfeld, T. Hurvis, S. Ruschin: Zirconia and zirconia-ormosil planar waveguides prepared at room temperature, Chem. Mater. **9**, 670 (1997)

11.1582 M. Zevin, R. Reisfeld: Preparation and properties of active waveguides based on zirconia glasses, Opt. Mater. **8**, 37 (1997)

11.1583 M. Maeda, Y. Oki, K. Imamura: Ultrashort pulse generation from an integrated single-chip dye laser, IEEE J. Quantum Electron. **33**, 2146 (1997)

11.1584 Y. Oki, T. Yoshiura, Y. Chisaki, M. Maeda: Fabrication of a distributed-feedback dye laser with a grating structure in its plastic waveguide, Appl. Opt. **41**, 5030 (2002)

11.1585 Y. Oki, S. Miyamoto, M. Tanaka, D. Zuo, M. Maeda: Wide-wavelength-range operation of a distributed-feedback dye laser with a plastic waveguide, Opt. Commun. **214**, 277 (2002)

11.1586 J. Wang, G.-X. Zhang, L. Shi, D. Lo, X.-L. Zhu: Tunable multiwavelength distributed-feedback zirconia waveguide lasers, Opt. Lett. **28**, 90–92 (2003)

11.1587 D. Lo, C. Ye, J. Wang: Distributed feedback laser action by polarization modulation, Appl. Phys. B **76**, 649–653 (2003)

11.1588 H. Kogelnik, C. V. Shank: Coupled-wave theory of distributed feedback lasers, J. Appl. Phys. **43**, 2327 (1972)

11.1589 A. Mukherjee: Two-photon pumped upconverted lasing in dye doped polymer waveguides, Appl. Phys. Lett. **62**, 3423 (1993)

11.1590 G. S. He, J. D. Bhawalker, C. F. Zhao, C. K. Park, P. N. Prasad: Upconversion dye-doped polymer fiber laser, Appl. Phys. Lett. **68**, 3549 (1996)

11.1591 D. Wang, G. Y. Zhou, Y. Ren, X. G. Xu, X. F. Cheng, Z. S. Shao, M. H. Jiang: One- and two-photon absorption induced emission in HMASPS doped polymer, Chem. Phys. Lett. **354**, 423–427 (2002)

11.1592 C. Ye, J. Wang, D. Lo: Two-photon-pumped distributed feedback zirconia waveguide lasers, Appl. Phys. B **78**, 539 (2004)

11.1593 P. A. Franken, A. E. Hill, C. W. Peters, G. Weinreich: Generation of optical harmonics, Phys. Rev. Lett **7**, 118–119 (1961)

11.1594 J. A. Armstrong, N. Bloembergen, J. Ducuing, P. S. Pershan: Interactions between light waves in a nonlinear dielectric, Phy. Rev **127**, 1918–1939 (1962)

11.1595 P. A. Franken, J. F. Ward: Optical harmonics and nonlinear phenomena, Rev. Mod. Phys **35**, 23 (1963)

11.1596 J. A. Giordmaine: Mixing of light beams in crystals, Phys. Rev. Lett **8**, 19 (1962)

11.1597 P. D. Maker, R. W. Terhune, M. Nisenoff, C. M. Savage: Effects of dispersion and focusing on production of optical harmonics, Phys. Rev. Lett. **8**, 21 (1962)

11.1598 G. D. Boyd, D. A. Kleinman: Parametric interaction of focused Gaussian light beams, J. Appl. Phys. **39**, 3597 (1968)

11.1599 N. M. Kroll: Parametric amplification in spatially extended media and application to the design of tunable oscillators at optical frequencies, Phys. Rev. **127**, 1207 (1962)

11.1600 J. A. Giordemaine, R. C. Miller: Tunable coherent parameric oscillation in $LiNbO_3$ at optical frequencies, Phys. Rev. Lett. **14**, 973 (1965)

11.1601 R. G. Smith, J. E. Geusic, H. J. Levinstein, J. J. Rubin, S. Singh, L. G. Van Uitert: Continuous optical parametric oscillation in $Ba_2NaNb_2O_{15}$, Appl. Phys. Lett **12**, 308 (1968)

11.1602 R. L. Byer, M. K. Oshman, J. F. Young, S. E. Harris: Visible CW parametric oscillator, Appl. Phys. Lett **13**, 109 (1968)

11.1603 K. Burneika, M. Ignatavichyus, V. Kabelka, A. Piskarskas, A. Stabinis: Parametric light amplification and oscillation in KDP with mode-locked pum, IEEE J. Quantum Electron **8**, 574 (1972)

11.1604 R. G. Smith: Optical parametric oscillators. In: *Lasers*, Vol. 4, ed. by A. K. Levine, A. J. De Maria (Marcel Dekker, New York 1976)

11.1605 S. E. Harris: Tunable optical parametric oscillators, Proc. IEEE **57**, 2096–2113 (1969)

11.1606 R. L. Byer: Optical parametric oscillators. In: *Quantum Electronics*, Nonlinear Optics, Vol. 1, ed. by H. Rabin, C. L. Tang (Academic, New York 1975)

11.1607 T. Y. Fan, R. L. Byer: Progress in optical parametric oscillators, SPIE Proc. **461**, 27 (1984)

11.1608 A. A. Kaminski: *Laser Crystals*, 2nd edn. (Springer, Berlin, Heidelberg 1990)

11.1609 V. G. Dmitriev, G. G. Gurzadyan, D. N.. Nikogosyan: *Handbook of Nonlinear Optical Crystals* (Springer, Berlin, Heidelberg 1991)

11.1610 R. H. Kingston: Parametric amplification and oscillation at optical frequencies, Proc. IRE (Corresp.) **50**, 472 (1962)

11.1611 A. Yariv, W. J. Louisell: Theory of an optical parametric oscillator, IEEE J. Quantum Electron. **2**(9), 418–424 (1966)

11.1612 A. Yariv: *Quantum Electronics* (Wiley, New York 1989)

11.1613 Y. R. Shen: *The Principles of Nonlinear Optics* (Wiley, New York 1984)

11.1614 R. L. Sutherland: *Handbook of Nonlinear Optics* (Marcel Dekker, New York 1996)

11.1615 W. H. Louisell, A. Yariv, A. E. Siegman: Quantum fluctuations and noise in parametric processes, Phys. Rev. **124**, 1646 (1961)

11.1616 D. A. Kleinman: Theory of optical parametric noise, Phys. Rev. **174**, 1027 (1968)
11.1617 L. Carrion, J. P. Giradeau-Montant: Development of a simple model for optical parametric generation, J. Opt. Soc. Am. B **17**, 78 (2000)
11.1618 J. M. Manley, H. E. Rowe: General energy relations in nonlinear reactances, Proc. IRE **47**, 2115 (1959)
11.1619 H. Ito, H. Naito, H. Inaba: Generalized study on angular dependence of induced second-order nonlinear optical polarizations and phase matching in biaxial crystals, J. Appl. Phys **46**, 3992 (1975)
11.1620 J. Yao, W. Sheng, W. Shi: Accurate calculation of the optimum phase-matching parameters in three-wave interactions with biaxial nonlinear-optical crystals, Appl. Opt. **29**, 3927 (1990)
11.1621 L. A. Gloster, Z. X. Jiang, T. A. King: Characterization of an Nd:YAG-pumped $\beta-BaB_2O_4$ optical parametric oscillator in collinear and noncollinear phase-matched configurations, IEEE J. Quantum Electron. **30**, 2961 (1994)
11.1622 S. Burdulis, R. Grigonis, A. Piskarskas, G. Sinkevicius, V. Sirutkaitis, A. Fix, J. Nolting, R. Wallenstein: Visible optical parametric oscillation in synchronously pumped beta-barium borate, Opt. Commun. **74**, 398 (1990)
11.1623 P. E. Powers, R. J. Ellington, W. S. Pelouch, C. T. Tang: Recent advances of the Ti:Sapphire-pumped high repletion rate femtosecond optical parametric oscillator, J. Opt. Soc. Am. B **10**, 2162 (1993)
11.1624 P. Di Trapani, A. Andreoni, C. Solcia, A. Piskarskas: Matching of group velocities in three-wave parametric interaction with femtosecond pulses and applications to travelling wave generators, J. Opt. Soc. Am. B **11**, 2237 (1995)
11.1625 M. A. Arbore, A. Galvanauskas, D. Harter, M. H. Chou, M. M. Fejer: Engineerable compression of ultrashort pulses by use of second-harmonic generation in chirped-period-poled lithium niobate, Opt. Lett. **22**, 1341 (1997)
11.1626 P. A. Powers, T. J. Kulp, S. E. Bisson: Continuous tuning of a continuous-wave periodically poled lithium niobate optical parametric oscillator by use of a fan-out grating design, Opt. Lett **23**, 159 (1998)
11.1627 S.-N. Zhu, N.-B. Ming: Quasi-phase matched third harmonic generation in quasi-periodic optical superlattice, Science **278**, 843 (1997)
11.1628 N. G. R. Broderick, G. W. Ross, H. L. Offerhaus, D. J. Richardson, D. C. Hanna: Hexagonally poled lithium niobate: A two-dimensional nonlinear photonic crystal, Phys. Rev. Lett **84**, 4345 (2000)
11.1629 S. T. Yang, R. C. Eckardt, R. L. Byer: Continuous-wave singly resonant optical parametric oscillator pumped by a single-frequency resonantly doubled Nd:YAG laser, Opt. Lett. **18**, 971 (1993)
11.1630 S. T. Yang, R. C. Eckardt, R. L. Byer: 1.9W-cw ring-cavity KTP singly resonant optical parametric oscillator, Opt. Lett. **19**, 475 (1994)
11.1631 J. A. Giordmaine: Optical parametric oscillation in $LiNbO_3$. In: *Physics of Quantum Electronics*, ed. by P. L. Kelley, B. L. Lax, P. E. Tannenwald (McGraw-Hill, New York 1966)
11.1632 J. Falk: Instabilities in doubly resonant parametric oscillators: A theoretical analysis, IEEE J. Quantum Electron. **7**, 230 (1971)
11.1633 R. C. Eckardt, C. D. Nabors, W. J. Kozlovsky, R. L. Byer: Optical parametric oscillator frequency tuning and control, J. Opt. Soc. Am. B **8**, 646 (1991)
11.1634 D. Lee, N. C. Wong: Stabilization and tuning of a doubly resonant optical parametric oscillator, J. Opt. Soc. Am. B **10**, 1659 (1993)
11.1635 S. Schiller, K. Schneider, J. Mlynek: Theory of an optical parametric oscillator with resonant pump and signal, J. Opt. Soc. Am. B **16**, 1512 (1999)
11.1636 F. G. Colville, M. J. Padgett, M. H. Dunn: Continuos-wave, dual-cavity, doubly-resonant, optical parametric oscillator, Appl. Phys. Lett **64**, 1490 (1994)
11.1637 S. Schiller, R. L. Byer: Quadruply resonant optical parametric oscillation in a monolithic total-internal-reflection resonator, J. Opt. Soc. Am. B **10**, 1696 (1993)
11.1638 C. Richy, K. I. Petas, E. Giacobino, C. Fabre, L. Lugiato: Observation of bistability and delayed bifurcation in a triply resonant OPO, J. Opt. Soc. Am. B **12**, 456 (1995)
11.1639 F. Kühnemann, K. Schneider, A. Hecker, A. A. E. Martis, W. Urban, S. Schiller, J. Mlynek: Photoacoustic trace-gas detection using a CW single-frequency parametric oscillator, Appl. Phys. B **66**, 741–745 (1998)
11.1640 D. Lee, N. C. Wong: Tunable optical frequency division using a phase-locked optical parametric oscillator, Opt. Lett. **17**, 13 (1992)
11.1641 S. Slyusarev, T. Ikegami, S. Oshima: Phase-coherent optical frequency division by 3 of 532 nm laser light a with continuous-wave optical parametric oscillator, Opt. Lett. **24**, 1856 (1999)
11.1642 U. Strössner, A. Peters, J. Mlynek, S. Schiller, J.-P. Meyn, R. Wallenstein: Single-frequency continuous-wave radiation from 0.77 to 1.73 μm generated by a green-pumped optical parametric oscillator with periodically poled $LiTaO_3$, Opt. Lett. **24**, 1602–1604 (1999)
11.1643 W. R. Bosenberg, A. Drobshoff, J. I. Alexander, L. E. Myers, R. L. Byer: Continuous-wave singly resonant optical parametric oscillator based on periodically poled $LiNbO_3$, Opt. Lett. **21**, 713–715 (1996)
11.1644 H. Kogelnik, T. Li: Laser beams and resonators, Appl. Opt **5**, 1567 (1966)
11.1645 J. E. Bjorkholm: Some effects of spatially nonuniform pumping in pulsed optical parametric oscillators, IEEE J. Quantum Electron. **7**, 109 (1971)
11.1646 C. D. Nabors, S. T. Yang, T. Day, R. L. Byer: Coherent properties of a doubly resonant monolithic opti-

cal parametric oscillator, J. Opt. Soc. Am. B **7**, 815 (1990)

11.1647 A. J. Henderson, M. J. Padgett, F. G. Colville, J. Zhang, M. H. Dunn: Doubly resonant OPOs tuning behaviour and stability requirements, Opt. Commun. **119**, 256 (1995)

11.1648 R. L. Barger, M. S. Sorem, J. L. Hall: Frequency stabilization of a CW dye laser, Appl. Phys. Lett **22**, 573 (1973)

11.1649 T. W. Hänsch, B. Couillaud: Laser frequency stabilization by polarization spectroscopy of a reflecting reference cavity, Opt. Commun. **35**, 441 (1973)

11.1650 R. W. P. Drever, J. L. Hall, F. V. Kowalski, J. Hough, G. M. Ford, A. J. Munley, H. Ward: Laser phase and frequency stabilization using an optical resonator, Appl. Phys. B **31**, 97 (1983)

11.1651 R. Al-Tahtamouni, K. Bencheikh, R. Storz, K. Schneider, M. Lang, J. Mlynek, S. Schiller: Long-term stable operation and absolute frequency stabilization of a doubly resonant parametric oscillator, Appl. Phys. B **66**, 733 (1998)

11.1652 M. Scheidt, B. Beier, K.-J. Boller: Frequency-stable operation of a diode-pumped continuous-wave $RbTiOAsO_4$ optical parametric oscillator, Opt. Lett **22**, 1287 (1997)

11.1653 M. Scheidt, B. Beier, R. Knappe, K.-J. Boller, R. Wallenstein: Diode-laser-pumped continuous wave KTP optical parametric oscillator, J. Opt. Soc. Am. B **12**, 2087 (1995)

11.1654 T. J. Edwards, G. A. Turnbull, M. H. Dunn, M. Ebrahimzadeh, F. G. Colville: High-power, continuous-wave, singly resonant intracavity optical parametric oscillator, Appl. Phys. Lett **72**, 1527 (1998)

11.1655 W. R. Bosenberg, A. Drobshoff, J. L. Alexander, L. E. Myers, R. L. Byer: 93% pump depletion, 3.5 W continuous-wave, singly resonant optical parametric oscillator, Opt. Lett **21**, 1336 (1996)

11.1656 M. E. Klein, D. H. Lee, J.-P. Meyn, K.-J. Boller, R. Wallenstein: Singly resonant continuous-wave optical parametric oscillator pumped by a diode laser, Opt. Lett. **24**, 1142 (1999)

11.1657 K. Schneider, P. Kramper, S. Schiller, J. Mlynek: Toward an optical synthesizer: A single frequency parametric oscillator using periodically poled $LiNbO_3$, Opt. Lett. **22**, 1293 (1997)

11.1658 G. A. Turnbull, D. McGloin, I. D. Lindsay, M. Ebrahimzadeh, M. H. Dunn: Extended mode-hop-free tuning using a dual cavity, pump-enhanced optical parametric oscillator, Opt. Lett. **25**, 341 (2000)

11.1659 P. Gross, M. E. Klein, T. Walde, K.-J. Boller, M. Auerbach, P. Wessels, C. Fallnich: Fiber-laser-pumped continuous wave singly-resonant optical parametric oscillator, Opt. Lett. **24**, 1142 (2002)

11.1660 M. E. Klein, D. H. Lee, J.-P. Meyn, B. Beier, K.-J. Boller, R. Wallenstein: Diode pumped continuous wave widely tunable optical parametric oscillator based on periodically poled lithium tantalite, Opt. Lett. **23**, 831 (1998)

11.1661 J.-P. Meyn, M. M. Fejer: Tunable ultraviolet radiation by second harmonic generation in oeriodically poled lithium tantalate, Opt. Lett. **22**, 1214 (1997)

11.1662 T. J. Edwards, G. A. Turnbull, M. H. Dunn, M. Ebrahimzadeh: Continuous-wave, singly resonant optical parametric oscillator based on periodically-poled $KTiOPO_4$, Opt. Express **6**, 58 (2000)

11.1663 A. Garashi, A. Arie, A. Skliar, G. Rosenman: Continuous-wave optical parametric oscillator based on periodically poled $KTiOPO_4$, Opt. Lett. **23**, 1739 (1998)

11.1664 U. Strößner, J.-P. Meyn, R. Wallenstein, A. Arie, P. Irenski, G. Rosenman, J. Mlynek, S. Schiller, A. Peters: Single-frequency CW OPO with ultra-wide tuning range from 550 to 2830 nm, J. Opt. Soc. Am. B **19**, 1419 (2002)

11.1665 T. J. Edwards, G. A. Turnbull, M. H. Dunn, M. Ebrahimzadeh: Continuous-wave, singly resonant optical parametric oscillator based on periodically-poled $RbTiOPO_4$, Opt. Lett. **23**, 837 (1998)

11.1666 M. E. Klein, C. K. Laue, D. H. Lee, K.-J. Boller, R. Wallenstein: Diode pumped singly resonant cw optical parametric oscillator with wide continuous tuning of the near-infrared idler wave, Opt. Lett. **25**, 490 (2000)

11.1667 S. J. Brosnan, R. L. Byer: Optical parametric oscillator threshold and linewidth studies, IEEE J. Quantum Electron. **15**, 415 (1979)

11.1668 J. E. Pearson, U. Ganiel, R. L. Byer: Rise time of pulsed parametric oscillators, IEEE J. Quantum Electron. **8**, 433 (1972)

11.1669 T. Schröder, K.-J. Boller, A. Fix, R. Wallenstein: Spectral properties and numerical modelling of a critically phase-matched LBO optical parametric oscillator, Appl. Phys. B **58**, 425 (1994)

11.1670 A. Fix, R. Wallenstein: Spectral properties of pulsed nanosecond optical parametric oscillators: experimental investigations and numerical analysis, J. Opt. Soc. Am. B **13**, 2484 (1996)

11.1671 Y. X. Fan, R. C. Eckhardt, R. L. Byer, C. Chen, A. D. Jiang: Barium borate optical parametric oscillator, IEEE J. Quantum Electron. **25**, 1196 (1989)

11.1672 A. Fix, T. Schröder, R. Wallenstein: The optical parametric oscillator o beta-barium borate and lithium borate: new sources of powerful tunable laser radiation in the ultraviolet, visible and near infrared, Laser Optoelektron. **23**, 106 (1991)and references therein

11.1673 Y. X. Fan, R. C. Eckhardt, R. L. Byer, J. Nolting, R. Wallenstein: Visible BaB_2O_4 optical parametriv oscillator pumped at 355 nm by a single-axial-mode pulsed source, Appl. Lett. **53**, 2014 (1988)

11.1674 L. K. Cheng, W. R. Bosenberg, C. L. Tang: Broadly tunable optical parametric oscillation in $\beta-BaB_2O_4$, Appl. Lett. **53**, 175 (1988)

11.1675 W. R. Bosenberg, W. S. Pelouch, C. L. Tang: High efficiency and narrow-linewidth operation of a two-crystal $\beta-BaB_2O_4$ optical parametric oscillator, Appl. Phys. Lett. **55**, 1952 (1989)

11.1676 W. R. Bosenberg, C. L. Tang: Type II phase matching in β-barium borate optical parametric oscillator, Appl. Phys. Lett. **56**, 1819 (1990)

11.1677 Y. Wang, Z. Xu, D. Deng, W. Zheng, X. Liu, B. Wu, C. Chen: Highly efficient visible and infrared $\beta-BaB_2O_4$ optical parametric oscillator, Appl. Lett. **58**, 1461 (1991)

11.1678 W. R. Bosenberg, L. K. Cheng, C. L. Tang: Ultraviolet optical parametric oscillation in $\beta-BaB_2O_4$, Appl. Lett. **54**, 13 (1989)

11.1679 H. Komine: Optical parametric oscillation in a beta-barium borate crystal pumped by an XeCl excimer laser, Opt. Lett. **13**, 643 (1988)

11.1680 M. Ebrahimzadeh, A. J. Henderson, M. H. Dunn: An excimer-pumped $\beta-BaB_2O_4$ optical parametric oscillator, IEEE J. Quantum Electron. **26**, 1241 (1990)

11.1681 A. Fix, T. Schröder, R. Wallenstein, J. G. Haub, M. J. Johnson, B. J. Orr: Tunable β-barium borate optical parametric oscillator: operating characteristics with and without injection seeding., J. Opt. Soc. Am. B **10**, 1744 (1993)

11.1682 R. Wallenstein, A. Fix, T. Schröder, J. Nolting: Optical parametric oscillators of Bariumborate and Lithiumborate. In: *Laser Spectroscopy*, Vol. IX, ed. by M. S. Field, J. E. Thomas, A. Mooradian (Academic, New York 1989)

11.1683 Y. Wang, Z. Xu, D. Deng, W. Zheng, B. Wu, C. Chen: Visible optical parametric oscillation in LiB_3O_5, Appl. Phys. Lett. **59**, 531 (1991)

11.1684 K. Kato: Parametric oscillation in LiB_3O_5 pumped at 0.532 μm, IEEE J. Quantum Electron. **26**, 2043 (1990)

11.1685 G. Robertson, A. Henderson, M. H. Dunn: Broadly tunable LiB_3O_5 optical parametric oscillator, Appl. Phys. Lett. **60**, 271 (1992)

11.1686 D. E. Withers, R. Robertson, A. J. Henderson, Y. Tang, Y. Cui, W. Sibbett, B. C. Sinclair, M. H. Dunn: Comparision of lithium triborate and β-barium borate as nonlinear media for optical parametric oscillators, J. Opt. Soc. Am. B **10**, 1737 (1993)

11.1687 F. Hanson, D. Dick: Blue parametric generation from temperature-tuned LiB_3O_5, Opt. Lett. **16**, 205 (1991)

11.1688 M. Ebrahimzadeh, G. Robertson, M. H. Dunn: Efficient ultraviolet LiB_3O_5 optical parametric oscillator, Opt. Lett **16**, 767 (1991)

11.1689 Y. Cui, D. E. Withers, C. F. Rae, C. J. Norrie, Y. Tang, W. Sibbett, B. C. Sinclair, M. H. Dunn: Widely tunable all-solid-state and optical parametric oscillator for the visible and the near infrared, Opt. Lett. **18**, 122 (1993)

11.1690 E. S. Cassedy, M. Jain: A theoretical study of injection tuning of parametric oscillators, IEEE J. Quantum Electron. **15**, 415 (1979)

11.1691 G. Robertson, A. Henderson, M. H. Dunn: Efficient and single-axial mode oscillation of a beta barium borate optical parametric oscillator pumped by an excimer laser, Appl. Phys. Lett. **62**, 123 (1993)

11.1692 W. R. Bosenberg, D. R. Guyer: Broadly tunable and single frequency optical parametric frequency-conversion system, J. Opt. Soc. Am. B **10**, 1716 (1993)

11.1693 J. M. Boon-Engering, L. A. Gloster, W E. van der Veer, I. T. McKinnie, T. A. King, W. Hogervorst: Bandwidth studies of an injection-seeded b-barium borate optical parametric oscillator, Opt. Lett **20**, 2087 (1995)

11.1694 A. V. Smith, W. J. Alford, T. D. Raymond, M. S. Bowers: Comparisionof a numerical model with measured performance of a seeded, nanosecond KTP optical parametric oscillator, J. Opt. Soc. Am. B **12**, 2253 (1995)

11.1695 G. Arisholm, K. Stenerson: Optical parametric oscillators with non-ideal mirrors and single- and multimode pump beams, Opt. Express **4**, 183 (1999)

11.1696 A. Dubois, T. Lepine, P. Georges, A. Brun: OPO radiance optimization using a numerical model, OSA Top. Adv. Solid State Lasers **10**, 394 (1997)

11.1697 A. V. Smith, M. S. Bowers: Image-rotating cavity designs for improved beam quality in nanosecond optical parametric oscillators, J. Opt. Soc. Am. B **18**, 706 (2001)

11.1698 A. V. Smith, D. J. Armstrong: Nanosecond optical parametric oscillatorwith 90° image rotation: design and performance, J. Opt. Soc. Am. B **19**, 1801 (2002)

11.1699 R. Urschel, A. Borsutzky, R. Wallenstein: Numerical analysis of the spatial behaviour of nanosecond optical parametric oscillatorsof beta-barium borate, Appl. Phys. B **70**, 203 (2000)

11.1700 G. Anstett, M. Nittmann, R. Wallenstein: Experimental investigation and numerical simulation of the spatio-temporal dynamics of the light-pulses in nanosecond optical parametric oscillators, Appl. Phys. B **79**, 305 (2004)

11.1701 G. Anstett, R. Wallenstein: Experimental investigation of the spectro-temporal dynamics of the light-pulses of *Q*-switched Nd:YAG lasers and nanosecond optical parametric oscillators, Appl. Phys. B **79**, 827 (2004)

11.1702 R. Urschel, U. Bäder, A. Borsutzky, R. Wallenstein: Spectral properties and conversion efficiency of 355 nm pumped pulsed optical parametric oscillators of beta-barium-borate with noncollinear phasematching, J. Opt. Soc. Am. B **16**, 565 (1999)

11.1703 A. L. Oien, I. T. Mc Kinnie, P. Jain, N. A. Russel, D. M. Warrington, L. A. Gloster: Efficient, low-threshold collinear and noncollinear β-barium borate optical parametric oscillator, Opt. Lett. **22**, 859 (1997)

11.1704 B. C. Johnson, V. J. Newell, J. B. Clark, E. S. McPhee: Narrowbandwidth low-divergence optical parametric oscillator for frequency conversion processes, J. Opt. Soc. Am. B **12**, 2122 (1995)

11.1705 S. Chandra, T. H. Allik, J. A. Hutchinson, M. S. Bowers: Improved OPO brightness with a GRM non-confocal unstable resonator, OSA Top. Adv. Solid-State Lasers **1**, 177 (1996)
11.1706 C. D. Nabors, G. Frangineas: Optical parametric oscillator with Bi-noncollinear Porro prism cavity, OSA Trends Opt. Photon. **10**, 90 (1997)
11.1707 S. Wu, G. A. Blake, Z. Sun: Simple, high-performance type II $\beta-BaB_2O_4$ optical parametric oscillator, Appl. Opt. **36**, 5898 (1997)
11.1708 G. Anstett, G. Göritz, D. Kabs, R. Urschel, R. Wallenstein, A. Borsutzky: Reduction of the spectral width and beam divergence of a BBO-OPO by using collinear type II phasematching and back reflection of the pump beam, Appl. Phys. B **72**, 583 (2001)
11.1709 K. Kato: Parametric oscillation at 3.2 μm in KTP pumped at 1.064 μm, IEEE J. Quantum Electron. **27**, 1137 (1991)
11.1710 C. L. Tang, W. R. Bosenberg, T. Ukachi, R. J. Lane, L. K. Cheng: Optical parametric oscillators, Proc. SPIE **80**, 365 (1992)and references therein
11.1711 W. R. Bosenberg, L. K. Cheng, J. D. Bierlein: Optical parametric frequency conversion properties of $KTiOAsO_4$, Appl. Phys. Lett. **65**, 2765 (1994)
11.1712 R. Urschel, A. Fix, R. Wallenstein, D. Rytz, B. Zysset: Generation of tunable narrow-band midinfrared radiation in a type I potassium niobate optical parametric oscillator, J. Opt. Soc. Am. B **12**, 726 (1995)
11.1713 K. L. Vodopyanov, F. Ganikhanov, J. P. Maffetone, I. Zwieback, W. Ruderman: $ZnGeP_2$ optical parametric oscillator with 3.8 – 12.4 μm tunability, Opt. Lett. **25**, 841 (2000)
11.1714 R. C. Eckardt, Y. X. Fan, R. L. Byer, C. L. Marquardt, M. E. Storm, L. Esterowitz: Broadly tunable infrared parametric oscillator using $AgGaSe_2$, Appl. Phys. Lett. **49**, 608 (1986)
11.1715 R. L. Herbst, R. L. Byer: Single resonant CdSe infrared parametric oscillator, Appl. Phys. Lett. **21**, 189 (1972)
11.1716 N. P. Barnes, K. E. Murray, M. G. Jani, S. R. Harrell: Diode-pumped Ho:Tm:YLF laser pumping an $AgGaSe_2$ parametric oscillator, J. Opt. Soc. Am. B **11**, 2422 (1994)
11.1717 S. Chandra, T. H. Allik, G. Catella, R. Utano, J. A. Hutchinson: Continuously tunable, 6 – 14 μm silver-gallium selenide optical parametric oscillator pumped at 1.57 μm, Appl. Phys. Lett. **71**, 584 (1997)
11.1718 H. Komine, J. M. Fukumoto, W. H. Long, E. A. Stappaerts: Noncritcally phase matched mid-infrared generation in $AgGaSe_2$, IEEE J. Sel. Top. Quantum Electron. **1**, 44 (1995)
11.1719 J. Kirton: A 2.54 μm pumped type II $AgGaSe_2$ mid-IR optical parametric oscillator, Opt. Commun. **115**, 93 (1995)
11.1720 T. H. Allik, S. Chandra: Recent advances in continuously tunable 8 – 12 μm radiation using optical parametric oscillators, Proc. SPIE **3082**, 54 (1997)
11.1721 M. J. Missey, V. Dominic, P. E. Powers, K. L. Schepler: Periodically poled lithium niobate monolithic nanosecond optical parametric oscillators and generators, Opt. Lett. **24**, 1227 (1999)
11.1722 L. E. Myers, W. R. Bosenberg: Periodically poled lithium niobate and quasi-phase-matched optical parametric oscillators, IEEE J. Quantum Electron. **33**, 1663 (1997)
11.1723 U. Bäder, J. Bartschke, I. Klimov, A. Borsutzky, R. Wallenstein: Optical parametric oscillator of quasi-phasematched $LiNbO_3$ pumped by a compact high repetition rate single-frequency passively Q-switched Nd:YAG laser, Opt. Commun. **147**, 95 (1998)
11.1724 H. Karlsson, M. Olson, G. Arvidsson, F. Laurell, U. Bäder, A. Borsutzky, R. Wallenstein, S. Wickström, M. Gustafsson: Nanosecond optical parametric oscillator based on large-aperture periodically poled $RbTiOAsO_4$, Opt. Lett. **24**, 330 (1999)
11.1725 J.-P. Feve, O. Pacaud, B. Boulanger, B. Menaert, J. Hellstrom, V. Pasiskevicius, F. Laurell: Widely and continuously tunable optical parametric oscillator based on a cylindrical periodically poled $KTiOPO_4$ crystal, Opt. Lett. **26**, 1882 (2001)
11.1726 J. J. Zayhowski: Periodically poled lithium niobate optical parametric amplifiers pumped by high.power passively Q-switched microchip lasers, Opt. Lett. **22**, 169 (1997)
11.1727 U. Bäder, J.-P. Meyn, J. Bartschke, T. Weber, A. Borsutzky, R. Wallenstein, R. G. Batchko, M. M. Fejer, R. L. Byer: Nanosecond periodically poled lithium niobate optical parametric generator pumped at 532 nm by a single-frequency passively Q-switched Nd:YAG laser, Opt. Lett. **24**, 1808 (1999)
11.1728 U. Bäder, T. Mattern, T. Bauer, J. Bartschke, M. Rahm, R. Wallenstein, A. Borsutzky: Pulsed nanosecond optical parametric generator based on periodically poled lithium niobate, Opt. Commun. **217**, 375 (2003)
11.1729 M. Rahm, U. Bäder, G. Anstett, J.-P. Meyn, R. Wallenstein, A. Borsutzky: Pulse-to pulse wavelength tuning of an injection seeded nanosecond optical parametric generator with 10 kHz repetition rate, Appl. Phys. B **75**, 47 (2002)
11.1730 S. T. Yang, S. P. Velsko: Frequency-agile kilohertz repetition-rate optical parametric oscillator based on periodically poled lithium niobate, Opt. Lett. **24**, 133 (1999)
11.1731 M. Rahm, G. Anstett, J. Bartschke, T. Bauer, R. Beigang, R. Wallenstein: Widely tunable narrow-linewidth nanosecond optical parametric generator with self-injection seeding, Appl. Phys. **79**, 535 (2004)

11.1732 A. Piskarskas, V. Smil'gyavichyus, A. Umbrasas: Continuous parametric generation of picosecond light pulses, Sov. J. Quantum Electron. **18**, 155 (1988)

11.1733 D. C. Edelstein, E. S. Wachmann, C. L. Tang: Broadly tunable high repetition rate femtosecond optical parametric oscillator, Appl. Phys. Lett. **54**, 1728 (1989)

11.1734 R. L. Byer, A. Piskarskas (Eds.): Optical parametric oscillation and amplificazion, J. Opt. Soc. Am. B **10**, 1656–1791 (1993)

11.1735 W. R. Bosenberg, R. C. Eckhardt (Eds.): Optical parametric devices, Special issue J. Opt. Soc. Am. B **12**, 2083–2322 (1993)

11.1736 M. Ebrahimzadeh, R. C. Eckhardt, M. H. Dunn (Eds.): Optical parametric devices and processes, Special issue J. Opt. Soc. Am. B **16**(9), 1481–1596 (1999)

11.1737 M. F. Becker, D. J. Kuizenga, D. W. Phillion, A. E. Siegmann: Analytical expression for ultrashort pulse generation in mode-locked optical parametric oscillators, J. Appl. Phys. **45**, 3996 (1974)

11.1738 E. C. Cheung, J. M. Liu: Theory of a synchronously pumped optical parametric oscillator in steady state operation, J. Opt. Soc. Am. B **7**, 1385 (1990)

11.1739 E. C. Cheung, J. M. Liu: Efficient generation of ultrashort, wavelength tunable infrared pulses, J. Opt. Soc. Am. B **8**, 1491 (1991)

11.1740 L. J. Bromley, A. Guy, D. C. Hanna: Synchronously pumped optical parametric oscillation in KTP, Opt. Commun. **70**, 350 (1989)

11.1741 M. Ebrahimzadeh, G. J. Hall, A. I. Ferguson: Broadly tunable, all-solid state, visible and infrared picosecond optical parametric oscillator, Opt. Lett. **18**, 278 (1993)

11.1742 G. J. Hall, M. Ebrahimzadeh, A. Robertson, G. P. A. Malcolm, A. I. Ferguson: Synchronously pumped optical parametric oscillator using all solid-state pump lasers, J. Opt. Soc. Am. B **10**, 2168 (1993)

11.1743 A. Nebel, C. Fallnich, R. Beigang, R. Wallenstein: Noncritically phase-matched continuous-wave mode-locked singly resonant optical parametric oscillator synchronously pumped by a Ti:Sapphire laser, J. Opt. Soc. Am. B **10**, 2195 (1993)

11.1744 M. Ebrahimzadeh, S. French, A. Miller: Design and performance of a singly resonant picosecond LiB_3O_5 optical parametric oscillator synchronously pumped by a self-mode locked Ti:Sapphire laser, J. Opt. Soc. Am. B **12**, 2180 (1995)

11.1745 S. French, M. Ebrahimzadeh, A. Miller: High power, high repetition rate picosecond optical parametric oscillator for the near- to mid-infrared, Opt. Lett. **21**, 131 (1996)

11.1746 K. J. McEwan: High-power synchronously pumped $AgGaS_2$ optical parametric oscillator, Opt. Lett. **23**, 667 (1998)

11.1747 D. Wang, C. Grasser, R. Beigang, R. Wallenstein: The generation of blue ps-light-pulses from a CW mode-locked LBO optical parametric oscillator, Opt. Commun. **138**, 87 (1997)

11.1748 R. Beigang, A. Nebel: Frequency conversion of ps pulses: Tunable pulses from the UV to the near infrared. In: *Solid State Lasers: New Developments and Applications*, ed. by M. Inguscio, R. Wallenstein (Plenum, New York 1993) pp. 179–188

11.1749 A. Nebel, H. Frost, R. Beigang, R. Wallenstein: Visible femtosecond pulses by second-harmonic generation of a CW mode-locked KTP optical parametric oscillator, Appl. Phys. B **60**, 453 (1995)

11.1750 C. Grasser, D. Wang, R. Beigang, R. Wallenstein: Singly resonant optical parametric oscillator of $KTiOPO_4$ synchronously pumped by the radiation from a continuous-wave mode-locked Nd:YLF laser., J. Opt. Soc. Am. B **10**, 2218 (1993)

11.1751 B. Ruffing, A. Nebel, R. Wallenstein: All-solid state CW mode-locked picosecond $KTiOAsO_4$ (KTA) optical parametric oscillator, Appl. Phys. B **67**, 537 (1998)

11.1752 G. T. Kennedy, D. T. Reid, A. Miller, M. Ebrahimzadeh, H. Karlsson, G. Arvidsson, F. Laurell: Broadly tunable mid-infrared picosecond optical parametric oscillator based on periodically-poled $RbTiOAsO_4$, Opt. Lett **23**, 503 (1998)

11.1753 P. J. Phillips, S. Das, M. Ebrahimzadeh: High-repetition-rate, all solid state, Ti:Sapphire-pumped optical parametric oscillator for the mid-infrared, Appl. Phys. Lett. **7**, 469 (2000)

11.1754 M. A. Watson, M. V. O'Connor, P. S. Lloyd, D. P. Shepherd, D. C. Hanna, C. B. E. Gawith, L. Ming, P. G. R. Smith, O. Balachninaite: Extended operation of synchronously pumped optical parametric oscillators to longer idler wavelengths, Opt. Lett. **27**, 2106 (2002)

11.1755 S. Lecomte, L. Krainer, R. Paschotta, M. J. P. Dymott, K. J. Weingarten, U. Keller: Optical parametric oscillator with a 10 kHz repetition rate and 100 mW average output power in the spectral region near 1.5 μm, Opt. Lett. **27**, 1714 (2002)

11.1756 A. Robertson, M. E. Klein, M. A. Tremont, K.-J. Boller, R. Wallenstein: 2.5 GHz repetition rate singly-resonant optical parametric oscillator synchronously pumped by a mode-locked diode oscillator amplifier system, Opt. Lett **25**, 657 (2000)

11.1757 K. Finsterbusch, R. Urschel, H. Zacharias: Fourier-transform-limited, high-power picosecond optical parametric oscillator based on periodically poled lithium niobate, Appl. Phys. B **70**, 741 (2000)

11.1758 V. Petrov, E. Noack, R. Stolzenberger: Seeded femtosecond optical parametric amplification in the mid-infrared spectral region above 3 μm, Appl. Opt **36**, 1164 (1997)

11.1759 K. Finsterbusch, R. Urschel, H. Zacharias: Tunable, high-power, narrow-band picosecond IR radiation by optical parametric amplification in KTP, Appl. Phys. B **74**, 319 (2002)

11.1760 K. Finsterbusch, A. Bayer, H. Zacharias: Tunable, narrow-band picosecond radiation in the mid-infrared by difference frequency mixing in GaSe and CdSe, Appl. Phys. B **79**, 457 (2004)

11.1761 A. J. Campillo, R. C. Hyer, S. L. Shapiro: Broadly tunable picosecond infrared source, Opt. Lett. **4**, 325 (1979)

11.1762 A. Dhirani, P. Guyot-Sionnest: Efficient generation of infrared picosecond pulses from 10 to 20 μm, Opt. Lett. **20**, 1104 (1995)

11.1763 W. S. Pelouch, P. E. Powers, C. L. Tang: Ti:Sapphire-pumped, high-repetition-rate femtosecond optical parametric oscillator, Opt. Lett. **17**, 1070 (1992)

11.1764 Q. Fu, G. Mak, H. M. van Driel: High-power, 62 fs infrared optical parametric oscillator synchronously pumped by a 76 MHz Ti:Sapphire laser, Opt. Lett. **17**, 1006 (1992)

11.1765 J. M. Dudley, D. T. Reid, M. Ebrahimzadeh, W. Sibbett: Characteristics of a noncritically phase-matched Ti:Sapphire pumped femtosecond optical parametric oscillator, Opt. Commun. **104**, 419 (1994)

11.1766 P. E. Powers, S. Ramakrishna, L. K. Cheng, C. L. Tang: Optical parametric oscillation with $KTiOAsO_4$, Opt. Lett. **18**, 1171 (1993)

11.1767 D. T. Reid, C. McGowan, M. Ebrahimzadeh, W. Sibbet: Characterization and modeling of a optical parametric oscillator based on KTA and operating beyond 4 μm, IEEE J. Quantum Electron. **33**, 1 (1997)

11.1768 P. E. Powers, C. L. Tang, L. K. Cheng: High repetition rate femtosecond optical parametric oscillator based on $RbTiOAsO_4$, Opt. Lett. **19**, 1439 (1994)

11.1769 P. E. Powers, C. L. Tang: High repetition rate femtosecond optical parametric oscillator based on $CsTiOAsO_4$, Opt. Lett **19**, 37 (1994)

11.1770 G. R. Holtom, R. A. Crowell, L. K. Cheng: Femtosecond mid-infrared optical parametric oscillator based on $CsTiOAsO_4$, Opt. Lett. **20**, 1880 (1995)

11.1771 D. E. Spence, S. Wielandy, C. L. Tang, C. Bosshard, P. Günter: High average power, high-repetition rate femtosecond pulse generation in the 1–5 μm region using an optical parametric oscillator, Appl. Phys. Lett. **68**, 452 (1996)

11.1772 K. C. Burr, C. L. Tang, M. A. Arbore, M. M. Fejer: Broadly tunable mid-infrared femtosecond optical parametric oscillator using all-solid-state-pumped periodically poled lithium niobate, Opt. Lett. **2**, 1458 (1997)

11.1773 D. T. Reid, Z. Penman, M. Ebrahimzadeh, W. Sibbet, H. Karlsson, F. Laurell: Broadly tunable infrared femtosecond optical parametric oscillator based on periodically poled $RbTiOAsO_4$, Opt. Lett. **22**, 1397 (1997)

11.1774 T. Kartaloglu, K. G. Koprulu, O. Aytur, M. Sundheimer, W. P. Risk: Femtosecond optical parametric oscillator based on periodically poled $KTiOAsO_4$, Opt. Lett. **23**, 61 (1998)

11.1775 T. Andres, P. Haag, S. Zelt, J.-P. Meyn, A. Borsutzky, R. Beigang, R. Wallenstein: Synchronuosly pumped femtosecond optical parametric oscillator of congruent and stoichiometric MgO-doped periodically poled lithium niobate, Appl. Phys. B **76**, 241 (2003)

11.1776 A. Dubietis, G. Jonusauskas, A. Piskarskas: Powerful femtosecond pulse generation by chirped and stretched pulse parametric amplification in BBO crystal, Opt. Commun. **144**, 125 (1992)

11.1777 V. Petrov, F. Rotermund, F. Noack: Generation of high-power femtosecond light pulses at 1 kHz in the mid-infrared spectral range between 3 and 12 μm by second-order nonlinear process in optical crystals, J. Opt. A: Pure Appl. Opt. **3**, R1–R19 (2001)

11.1778 I. N. Ross, P. Matousek, M. Towrie, A. J. Langley, J. L. Collier: The prospects for ultrashort pulse duration and ultrahigh intensity using optical parametric chirped pulse amplifiers, Opt. Commun. **144**, 125 (1997)

11.1779 J. L. Collier, C. Hernandez-Gomez, I. N. Ross, P. Matousek, C. N. Danson, J. Walczak: Evaluation of an ultraband high-gain amplification technique for chirped pulse amplification facilities, Appl. Opt. **38**, 7486 (1999)

11.1780 I. N. Ross, J. L. Collier, P. Matousek, C. N. Danson, D. Neely, R. M. Allot, D. A. Pepler, C. Hernandez-Gomez, K. Osvay: Generation of terawatt pulses by use of optical parametric chirped pulse amplification, Appl. Opt. **39**, 2422 (2000)

11.1781 I. Jovanovic, B. J. Comaskey, C. A. Ebbers, R. A. Bonner, D. M. Pennington, E. C. Morse: Optical parametric chirped-pulse amplifier as an alternative to Ti:Sapphire regenerative amplifiers, Appl. Opt. **41**, 2923 (2002)

11.1782 X. Yang, Z. Xu, Y. Leng, H. Lu, L. Lin, Z. Zhang, R. Li, W. Zhang, D. Yin, B. Tang: Multiterawatt laser system based on optical parametric chirped pulse amplification, Opt. Lett. **27**, 1135 (2002)

11.1783 M. R. Armstrong, P. Plachta, E. A. Ponomarev, R. J. D. Miller: Versatile 7 fs optical parametric generation and compression by use of adaptive optics, Opt. Lett. **26**, 1152 (2001)

11.1784 M. Zavelani-Rossi, D. Polli, G. Cerullo, S. de-Silvestri, L. Gallmann, G. Steinmeyer, U. Keller: Few-optical-cycle laser pulses by OPA: broadband chirped mirror compression and SPIDER characterization, Appl. Phys. B **74**, 245 (2002)

11.1785 I. Jovanovic, C. A. Ebbers, C. P. J. Barty: Hybrid chirped-pulse amplification, Opt. Lett. **27**, 1622 (2002)

11.1786 M. A. Arbore, O. Marco, M. M. Fejer: Pulse compression during second-harmonic generation in aperiodic quasi-phase-matched gratings, Opt. Lett. **22**, 865 (1997)

11.1787 F. Rotermund, V. Petrov, F. Noack, V. Pasiskevicius, J. Hellström, F. Laurell, H. Hundertmark, P. Adel, C. Fallnich: Compact all-diode pumped femtosecond laser source based on chirped pulse optical parametric amplification in periodically poled $KtiOPO_4$, Electron. Lett. **38**, 561 (2002)

11.1788 I. Jovanovic, J.R. Schmidt, C.A. Ebbers: Optical parametric chirped-pulse amplification in periodically poled $KTiOPO_4$, Appl. Phys. Lett. **83**, 4125 (2003)
11.1789 P.L. Hanst, S.T. Hanst: Gas measurements in the fundamental infrared region. In: *Air Monitoring by Spectroscopic Techniques*, Chem. Anal. Ser., Vol.127, ed. by M.W. Sigrist (Wiley, New York 1994) pp. 335–470
11.1790 I.T. Sorokina, E. Sorokin, A. Di Lieto, M. Tonelli, R.H. Page, K.I. Schaffers: Efficient broadly tunable continuous-wave Cr^{2+} : ZnSe laser, J. Opt. Soc. Am B **18**, 926–930 (2001)
11.1791 J.J. Adams, C. Bibeau, R.H. Page, D.M. Krol, L.H. Furu, S.A. Payne: 4.0–4.5 μm lasing of Fe:ZnSe below 180 K, a new mid-infrared laser material, Opt. Lett. **24**, 1720–1722 (1999)
11.1792 J. Faist: Continuous-wave, room-temperature quantum cascade lasers, Opt. Photon. News **17**(5), 32–36 (2006)
11.1793 R. Maulini, D.A. Yarekha, J.-M. Bulliard, M. Giovannini, J. Faist, E. Gini: Continuous-wave operation of a broadly tunable thermoelectrically cooled external cavity quantum-cascade laser, Opt. Lett. **30**, 2584–2586 (2005)
11.1794 I.T. Sorokina, K.L. Vodopyanov (Eds.): *Solid-State Mid-Infrared Laser Sources*, Top. Appl. Phys., Vol. 89 (Springer, Berlin, Heidelberg 2003)
11.1795 R.L. Byer, R.L. Herbst: *Parametric oscillation and mixing*, Top. Appl. Phys, Vol.16, ed. by Y.R. Shen (Springer, Berlin, Heidelberg 1977) pp. 81–137
11.1796 R.L. Sutherland: *Handbook of Nonlinear Optics*, 2 edn. (Marcel Dekker, New York 2003)
11.1797 V.G. Dmitriev, G.G. Gurzadyan, D.N. Nikogosyan: *Handbook of Nonlinear Optical Crystals*, Springer Ser. Opt. Sci., Vol. 64, 3 edn. (Springer, Berlin, Heidelberg 1999)
11.1798 Gsänger Optoelektronik GmbH: Datasheet (Gsänger Optoelektronik GmbH, Planegg 1993)
11.1799 Y.R. Shen (Ed.): *Nonlinear Infrared Generation*, Top. Appl. Phys., Vol.16 (Springer, Berlin, Heidelberg 1997)
11.1800 D.A. Roberts: Simplified characterization of uniaxial and biaxial nonlinear crystals: A plea for standardization of nomenclature and conventions, IEEE J. Quantum Electron. **28**, 2057–2074 (1992)
11.1801 V.G. Dmitriev, G.G. Gurzadyan, D.N. Nikogosyan: *Handbook of Nonlinear Optical Crystals* (Springer, Berlin, Heidelberg 1993)
11.1802 Molecular Technology (MolTech) GmbH: Datasheet (MolTech, Berlin 2005), http://www.mt-berlin.com
11.1803 Crystech Inc.: Datasheet (Crystech Inc., Shandong 2005), http://www.crystech.com
11.1804 Raicol Crystals Ltd.: Datasheet (Raicol Crystals Ltd., Yehud 2005), http://www.raicol.com
11.1805 E. Lallier, M. Brevignon, J. Lehoux: Efficient second-harmonic generation of a CO_2 laser with a quasi-phase-matched GaAs crystal, Opt. Lett. **23**, 1511–1513 (1998)
11.1806 D.F. Bliss, C. Lynch, D. Weyburne, K. O'Hearn, J.S. Bailey: Epitaxial growth of thick GaAs on orientation-patterned wafers for nonlinear optical applications, J. Crystal Growth **287**, 673–678 (2006)
11.1807 P.S. Kuo, K.L. Vodopyanov, M.M. Fejer, D.M. Simanovskii, X. Yu, J.S. Harris, D. Bliss, D. Weyburne: Optical parametric generation of a mid-infrared continuum in orientation-patterned GaAs, Opt. Lett. **31**, 71–73 (2006)
11.1808 O. Levi, T.J. Pinguet, T. Skauli, L.A. Eyres, K.R. Parameswaran, J.S. Harris Jr., M.M. Fejer, T.J. Kulp, S.E. Bisson, B. Gerard, E. Lallier, L. Becouarn: Difference frequency generation of 8 μm radiation in orientation-patterned GaAs, Opt. Lett. **27**, 2091–2093 (2002)
11.1809 S. Borri, P. Cancio, P. De Natale, G. Guisfredi, D. Mazzotti, F. Tamassia: Power-boosted difference-frequency source for high-resolution infrared spectroscopy, Appl. Phys. B **76**, 473–477 (2003)
11.1810 G.D. Boyd, D.A. Kleinman: Parametric interaction of focused Gaussian light beams, J. Appl. Phys. **36**, 3597–3639 (1968)
11.1811 T.B. Chu, M. Broyer: Intracavity single resonance optical parametric oscillator, J. Phys. (Paris) **45**, 1599–1616 (1968)
11.1812 T. Yanagawa, O. Tadanaga, Y. Nishida, H. Miyazawa, K. Magari, M. Asobe, H. Suzuki: Simultaneous observation of CO isotopomer absorption by broadband difference-frequency generation using a direct-bonded quasi-phase-matched $LiNbO_3$ waveguide, Opt. Lett. **31**, 960–962 (2006)
11.1813 C. Fischer, M.W. Sigrist: Mid-IR difference frequency generation. In: *Solid-State Mid-Infrared Laser Sources*, Top. Appl. Phys, Vol. 89, ed. by I.T. Sorokina, K.L. Vodopyanov (Springer, Berlin, Heidelberg 2003) pp. 97–140
11.1814 S. Stry, P. Hering, M. Mürtz: Portable difference-frequency laser-based cavity leak-out spectrometer for trace-gas analysis, Appl. Phys. B **75**, 297–303 (2002)
11.1815 M.M. Choy, R.L. Byer: Accurate second-order susceptibility measurements of visible and infrared nonlinear crystals, Phys. Rev. B **14**, 1693–1705 (1972)
11.1816 R.C. Miller: Optical second harmonic generation in piezoelectric crystals, Appl. Phys. Lett. **5**, 17–19 (1964)
11.1817 D.H. Jundt: Temperature-dependent Sellmeier equation for the index of refraction, n_e in congruent lithium niobate, Opt. Lett. **22**, 1553–1555 (1997)
11.1818 R.T. Smith, F.S. Welsh: Temperature dependence of the elastic, piezoelectric, and dielectric constants of lithium tantalate and lithium niobate, J. Appl. Phys. **42**, 2219–2230 (1971)

11.1819 M. Seiter, D. Keller, M.W. Sigrist: Broadly tunable difference-frequency spectrometer for trace gas detection with noncollinear critical phase-matching in $LiNbO_3$, Appl. Phys. B **67**, 351–356 (1998)

11.1820 P. Maddaloni, G. Gagliardi, P. Malara, P. De Natale: A 3.5 mW continuous-wave difference-frequency source around 3 μm for sub-Doppler molecular spectroscopy, Appl. Phys. B **80**, 141–145 (2005)

11.1821 C. Fischer, M.W. Sigrist: Trace-gas sensing in the 3.3 μm region using a diode-based difference-frequency laser photoacoustic system, Appl. Phys. B **75**, 305–310 (2002)

11.1822 A. Khorsandi, U. Willer, P. Geiser, W. Schade: External short-cavity diode-laser for MIR difference-frequency generation, Appl. Phys. B **77**, 509–513 (2003)

11.1823 B. Sumpf, D. Rehle, H.-D. Kronfeldt: A tunable diode-laser spectrometer for the MIR region near 7.2 μm applying difference-frequency generation in $AgGaSe_2$, Appl. Phys. B **67**, 369–373 (1998)

11.1824 K. Fradkin, A. Arie, G. Rosenman: Tunable midinfrared source by difference frequency generation in bulk periodically poled $KTiOPO_4$, Appl. Phys. Lett. **74**, 914–916 (1999)

11.1825 K. Fradkin-Kashi, A. Arie, P. Urenski, G. Rosenman: Mid-infrared difference-frequency generation in periodically poled $KTiOAsO_4$ and application to gas sensing, Opt. Lett. **25**, 743–745 (2000)

11.1826 W. Chen, G. Mourat, D. Boucher, F.K. Tittel: Mid-infrared trace gas detection using continuous-wave difference frequency generation in periodically poled $RbTiOAsO_4$, Appl. Phys. B **72**, 873–876 (2001)

11.1827 Y.K. Hsu, C.W. Chen, J.Y. Huang, C.L. Pan: Erbium doped GaSe crystal for mid-IR applications, Opt. Express **14**, 54845490 (2006)

11.1828 W. Chen, E. Poullet, J. Burie, D. Boucher, M.W. Sigrist, J.J. Zondy, L. Isaenko, A. Yèlisseyev, S. Lobanov: Widely tunable continuous-wave mid-infrared radiation (5.5–11 μm) by difference-frequency generation in $LiInS_2$ crystal, Appl. Opt. **44**, 4123–4129 (2005)

11.1829 S. Haidar, K. Miyamoto, H. Ito: Generation of continuously tunable, 5–12 μm radiation by difference frequency mixing of output waves of a KTP optical parametric oscillator in a $ZnGeP_2$ crystal, J. Phys. D **37**, 3347–3349 (2004)

11.1830 L.S. Rothman, D. Jacquemart, A. Barbe, D.C. Benner, M. Birk, L.R. Brown, M.R. Carleer, C. Chackerian Jr., K. Chance, L.H. Coudert, V. Dana, V.M. Devi, J.-M. Flaud, R.R. Gamache, A. Goldman, J.-M. Hartmann, K.W. Jucks, A.G. Macki, J.-Y. Mandin, S.T. Massie, J. Orphal, A. Perrin, C.P. Rinsland, M.A.H. Smith, R.N. Tolchenov, R.A. Toth, J. Vander Auwera, P. Varanasi, G. Wagner: The HITRAN 2004 molecular spectroscopic database, J. Quantum Spectrosc. Radiat. Transfer **96**, 139–204 (2005)

11.1831 L.E. Myers, R.C. Eckhardt, M.M. Feyer, R.L. Byer, W.R. Bosenberg: Multigrating quasi-phase-matched optical parametric oscillator in periodically poled $LiNbO_3$, Opt. Lett. **21**, 591–593 (1996)

11.1832 J.B. McManus, D.D. Nelson, J.H. Shorter, R. Jimenez, S. Herndon, S. Saleska, M. Zahniser: A high precision pulsed quantum cascade laser spectrometer for measurements of stable isotopes of carbon dioxide, J. Mod. Opt. **52**, 2309–2321 (2005)

11.1833 H. Waechter, M.W. Sigrist: Trace gas analysis with isotopic selectivity using DFG-sources. In: *Mid-Infrared Coherent Sources and Applications*, Nato Science Series II: Math. Phys. Chem., ed. by M. Ebrahimzadeh, I.T. Sorokina (Springer, Berlin, Heidelberg 2006)

11.1834 H. Waechter, M.W. Sigrist: Mid-infrared laser spectroscopic determination of isotope ratios of N_2O at trace levels using wavelength modulation and balanced path length detection, Appl. Phys. B, DOI 10.1007/s00340-007-2576-z (2007)

11.1835 A.S. Pine: Doppler-limited molecular spectroscopy by difference-frequency mixing, J. Opt. Soc. Am. **64**, 1683–1690 (1974)

11.1836 R.H. Pantell, Y. Soncini, H.E. Putthoff: Stimulated photon-electron scattering, IEEE J. Quantum Electron. **11**, 905–907 (1968)

11.1837 J.M.J. Madey: Stimulated emission of bremsstrahlung in a periodic magnetic field, J. Appl. Phys. **42**, 1906–1913 (1971)

11.1838 H. Onuki, P. Elleaume: *Undulators, Wigglers, and Their Applications* (Taylor Francis, New York 2003)

11.1839 W.B. Colson: Short wavelength free electron lasers in 2000, Nucl. Instrum. Meth. A **475**, 397 (2001)

11.1840 G. Ramian: The new UCSB free-electron lasers, Nucl. Instrum. Meth. A **318**, 225–229 (1992)

11.1841 Y.U. Jeong, B.C. Lee, S.K. Kim, S.O. Cho, B.H. Cha, J. Lee, G.M. Kazakevitch, P.D. Vobly, N.G. Gavrilov, V.V. Kubarev, G.N. Kulipanov: First lasing of the KAERI compact far-infrared free-electron laser driven by a magnetron-based microtron, Nucl. Instrum. Meth. A **475**, 47–50 (2001)

11.1842 T. Tomimasu, T. Takii, T. Suzuki, E. Nishimura, S. Ogino, A. Nagai, M. Yasumoto: FEL facilities and application researches at the FELI, Nucl. Instrum. Meth. A **407**, 494–499 (1998)

11.1843 K.W. Berryman, T.I. Smith: First lasing, capabilities, and flexibility of FIREFLY, Nucl. Instrum. Meth. A **375**, 6–9 (1996)

11.1844 J. Xie, J. Zhuang, Y. Huang, Y. Li, S. Lin, C. Mao, R. Ying, Y. Zhong, L. Zhang, G. Wu, Y. Zhang, L. Li, E. Fu, W. Liu: First lasing of the Beijing FEL, Nucl. Instrum. Meth. A **341**, 34–38 (1994)

11.1845 M.J. van der Wiel, P.W. van Amersfoort, FELI. X. Team: FELIX: from laser to user facility, Nucl. Instrum. Meth. A **331**, ABS30–ABS33 (1993)

11.1846 H. A. Schwettman: Challenges at FEL facilities: The Stanford Picosecond FEL Center, Nucl. Instrum. Meth. A **375**, 632–638 (1996)

11.1847 J. M. Ortega, J. M. Berset, R. Chaput, F. Glotin, G. Humbert, D. Jaroszynski, P. Joly, B. Kergosien, J. Lesrel, O. Marcouillé, A. Peremans, R. Prazeres, A. Tadjeddine: Activities of the CLIO infrared facility, Nucl. Instrum. Meth. A **375**, 618–625 (1996)

11.1848 C. A. Brau: The Vanderbilt university free-electron laser center, Nucl. Instrum. Meth. A **318**, 38–41 (1992)

11.1849 M. Yokoyama, F. Oda, K. Nomaru, H. Koike, M. Sobajima, H. Miura, H. Hattori, M. Kawai, H. Kuroda: First lasing of KHI FEL device at the FEL-SUT, Nucl. Instrum. Meth. A **475**, 38–42 (2001)

11.1850 K. Fahmy, G. Furlinski, P. Gippner, E. Grosse, H. Piest, W. Seidel, M. Sczepan, D. Wohlfarth, A. Wolf, R. Wuensch: Properties and planned use of intense THz radiation from ELBE at Dresden-Rossendorf, J. Biol. Phys. **29**, 303–307 (2003)

11.1851 V. P. Bolotin, N. A. Vinokurov, D. A. Kayran, B. A. Knyazev, E. I. Kolobanov, V. V. Kotenkov, V. V. Kubarev, G. N. Kulipanov, A. N. Matveenko, L. E. Medvedev, S. V. Miginsky, L. A. Mironenko, A. D. Oreshkov, V. K. Ovchar, V. M. Popik, T. V. Salikova, S. S. Serednyakov, A. N. Skrinsky, O. A. Shevchenko, M. A. Scheglov: Status of the Novosibirsk Terahertz FEL, Proc. 2004 FEL Conf., Trieste 2004, ed. by R. Bakker, L. Gianessi, M. Marsi, R. Walker (Comitato Conferenze Elettra, Trieste 2004) 226–228

11.1852 S. Benson, D. Douglas, M. Shinn, K. Beard, C. Behre, G. Biallas, J. Boyce, H. F. Dylla, R. Evans, A. Grippo, J. Gubeli, D. Hardy, Ñ, Hernandez-Garcia, K. Jordan, L. Mermmga, G. R. Neil, J. Preble, T. Siggins, R. Walker, G. P. Williams, B. Yunn, S. Zhang, H. Toyokawa: High power lasing in the IR upgrade FEL at Jefferson Lab, Proc. 2004 FEL Conf., Trieste 2004, ed. by R. Bakker, L. Gianessi, M. Marsi, R. Walker (Comitato Conferenze Elettra, Trieste 2004) 229–232

11.1853 N. Nishimori, R. Hajima, R. Nagai, E. J. Minehara: High extraction efficiency observed at the JAERI free-electron laser, Nucl. Instrum. Meth. A **475**, 266–269 (2001)

11.1854 V. N. Litvinenko, S. H. Park, I. V. Pinayev, Y. Wu: Operation of the OK-4/Duke storage ring FEL below 200 nm, Nucl. Instrum. Meth. A **475**, 195–204 (2001)

11.1855 R. P. Walker, J. A. Clarke, M. E. Couprie, G. Dattoli, M. Eriksson, D. Garzella, L. Giannessi, M. Marsi, M. W. Poole, E. Renault, R. Roux, M. Trovo, S. Werin, K. Wille: The European UV/VUV storage ring FEL at ELETTRA: first operation and future prospects, Nucl. Instrum. Meth. A **467–468**, 34–37 (2001)

11.1856 M. Hosaka, M. Katoh, A. Mochihashi, J. Yamazaki, K. Hayashi, Y. Takashima: Upgrade of the UVSOR storage ring FEL, Nucl. Instrum. Meth. A **528**, 291–295 (2004)

11.1857 G. De Ninno, D. Nutarelli, D. Garzella, L. Nahon, M. Hirscha, E. Renault, M. E. Couprie: The super-ACO free electron laser source in the UV and its applications, Radiat. Phys. Chem. **61**, 449–450 (2001)

11.1858 G. N. Kulipanov, A. N. Skrinsky, N. A. Vinokurov: Synchrotron light sources and recent developments of accelerator technology, J. Synchrotron Rad. **5**, 176–178 (1998)

11.1859 G. R. Neil, C. Behre, S. V. Benson, G. Biallas, J. Boyce, L. A. Dillon-Townes, D. Douglas, H. F. Dylla, R. Evans, A. Grippo, D. Gruber, J. Gubeli, C. Hernandez-Garcia, K. Jordan, M. J. Kelley, L. Merminga, J. Mammosser, N. Nishimimori, J. Preble, R. Rimmer, M. Shinn, T. Siggins, R. Walker, G. P. Williams, S. Zhang: Producing ultrashort Terahertz to UV photons at high repetition rates for research into materials, Mater. Res. Soc. Symp. Proc. **850**, MM4.1.1 (2005)

11.1860 E. Minehara: Development and operation of the JAERI superconducting energy recovery linacs, Nucl. Instrum. Meth. A **557**, 16–22 (2006)

11.1861 B. C. Lee, Y. U. Jeong, S. H. Park, Y. G. Lim, S. Miginsky: High-power infrared free electron laser driven by a 352 MHz superconducting accelerator with energy recovery, Nucl. Instrum. Meth. A **528**, 106–109 (2004)

11.1862 Y. Ding, S. Huang, J. Zhuang, Y. Wang, K. Zhao, J. Chen: Design and optimization of IR SASE FEL at Peking University, Nucl. Instrum. Meth. A **528**, 416–420 (2004)

11.1863 M. W. Poole, B. W. J. Mc Neil: FEL options for the proposed UK fourth generation light source (4GLS), Nucl. Instrum. Meth. A **507**, 489–493 (2003)

11.1864 V. Ayvazyan, N. Baboi, I. Bohnet, R. Brinkmann, M. Castellano, P. Castro, L. Catani, S. Choroba, A. Cianchi, M. Dohlus, H. T. Edwards, B. Faatz, A. A. Fateev, J. Feldhaus, K. Floettmann, A. Gamp, T. Garvey, H. Genz, Ch. Gerth, V. Gretchko, B. Grigoryan, U. Hahn, C. Hessler, K. Honkavaara, M. Huening, R. Ischebeck, M. Jablonka, T. Kamps, M. Koerfer, M. Krassilnikov, J. Krzywinski, M. Liepe, A. Liero, T. Limberg, H. Loos, M. Luong, C. Magne, J. Menzel, P. Michelato, M. Minty, U.-C. Mueller, D. Noelle, A. Novokhatski, C. Pagani, F. Peters, J. Pflueger, P. Piot, L. Plucinski, K. Rehlich, I. Reyzl, A. Richter, J. Rossbach, E. L. Saldin, W. Sandner, H. Schlarb, G. Schmidt, P. Schmueser, J. R. Schneider, E. A. Schneidmiller, H.-J. Schreiber, S. Schreiber, D. Sertore, S. Setzer, S. Simrock, R. Sobierajski, B. Sonntag, B. Steeg, F. Stephan, K. P. Sytchev, K. Tiedtke, M. Tonutti, R. Treusch, D. Trines, D. Tuerke, V. Verzilov, R. Wanzenberg, T. Weiland, H. Weise, M. Wendt, I. Will, S. Wolff, K. Wittenburg, M. V. Yurkov, K. Zapfe: Generation of gw radiation pulses from a VUV free-electron laser operating in the femtosecond regime, Phys. Rev. Lett. **88**, 104802 (2002)

11.1865 L. Juha, J. Krasa, A. Cejnarova, D. Chvostova, V.V. Imageek, J. Krzywinski, R. Sobierajski, A. Andrejczuk, M. Jurek, D. Klinger, H. Fiedorowicz, A. Bartnik, M. Pfeifer, P. Kubat, L. Pina, J. Kravarik, P. Kubelmage, Y.L. Bakshaev, V.D. Korolev, A.S. Chernenko, M.I. Ivanov, M. Scholz, L. Ryc, J. Feldhaus, J. Ullschmied, F.P. Boody: Ablation of various materials with intense XUV radiation, Nucl. Instrum. Meth. A **507**, 577–581 (2003)

11.1866 F. Richard, J.R. Schneider, D. Trines, A. Wagner (Eds.): *TESLA: the superconducting electron-positron linear collider with an integrated X-ray laser laboratory*, Technical Design Report, Part I: Executive Summary, Preprint DESY 2001-011 (DESY, Hamburg 2001)

11.1867 H. Wabnitz, L. Bittner, A.R.B. de Castro, R. Doehrmann, P. Guertler, T. Laarmann, W. Laasch, J. Schulz, A. Swiderski, K., von, Haeften, T. Moeller, B. Faatz, A. Fateev, J. Feldhaus, C. Gerth, U. Hahn, E. Saldin, E. Schneidmiller, K. Sytchev, K. Tiedtke, R. Treusch, M. Yurkov: Multiple ionization of atom clusters by intense soft X-rays from a free-electron laser, Nature **420**, 482–485 (2002)

11.1868 V. Ayvazyan, N. Baboi, I. Bohnet, R. Brinkmann, M. Castellano, P. Castro, L. Catani, S. Choroba, A. Cianchi, M. Dohlus, H.T. Edwards, B. Faatz, A.A. Fateev, J. Feldhaus, K. Floettmann, A. Gamp, T. Garvey, H. Genz, Ch. Gerth, V. Gretchko, B. Grigoryan, U. Hahn, C. Hessler, K. Honkavaara, M. Huening, R. Ischebeck, M. Jablonka, T. Kamps, M. Koerfer, M. Krassilnikov, J. Krzywinski, M. Liepe, A. Liero, T. Limberg, H. Loos, M. Luong, C. Magne, J. Menzel, P. Michelato, M. Minty, U.-C. Mueller, D. Noelle, A. Novokhatski, C. Pagani, F. Peters, J. Pflueger, P. Piot, L. Plucinski, K. Rehlich, I. Reyzl, A. Richter, J. Rossbach, E.L. Saldin, W. Sandner, H. Schlarb, G. Schmidt, P. Schmueser, J.R. Schneider, E.A. Schneidmiller, H.-J. Schreiber, S. Schreiber, D. Sertore, S. Setzer, S. Simrock, R. Sobierajski, B. Sonntag, B. Steeg, F. Stephan, K.P. Sytchev, K. Tiedtke, M. Tonutti, R. Treusch, D. Trines, D. Tuerke, V. Verzilov, R. Wanzenberg, T. Weiland, H. Weise, M. Wendt, I. Will, S. Wolff, K. Wittenburg, M.V. Yurkov, K. Zapfe: A new powerful source for coherent VUV radiation: Demonstration of exponential growth and saturation at the TTF free-electron laser, Eur. Phys. J. D **20**, 149–156 (2002)

11.1869 A.M. Kondratenko, E.L. Saldin: Generation of coherent radiation by a relativistic electron beam in an undulator, Part. Accelerators **10**, 207–216 (1980)

11.1870 Y.S. Derbenev, A.M. Kondratenko, E.L. Saldin: On the possibility of using a free electron laser for polarization of electrons in storage rings, Nucl. Instrum. Methods **193**, 415–421 (1982)

11.1871 J.B. Murphy, C. Pellegrini: Free electron lasers for the XUV spectral region, Nucl. Instrum. Methods A **237**, 159–167 (1985)

11.1872 V. Ayvazyan, N. Baboi, J. Bahr, V. Balandin, B. Beutner, A. Brandt, I. Bohnet, A. Bolzmann, R. Brinkmann, O.I. Brovko, J.P. Carneiro, S. Casalbuoni, M. Castellano, P. Castro, L. Catani, E. Chiadroni, S. Choroba, A. Cianchi, H. Delsim-Hashemi, G. Di Pirro, M. Dohlus, S. Duesterer, H.T. Edwards, B. Faatz, A.A. Fateev, J. Feldhaus, K. Floettmann, J. Frisch, L. Froehlich, T. Garvey, U. Gensch, N. Golubeva, H.-J. Grabosch, B. Grigoryan, O. Grimm, U. Hahn, J.H. Han, M.V. Hartrott, K. Honkavaara, M. Huening, R. Ischebeck, E. Jaeschke, M. Jablonka, R. Kammering, V. Katalev, B. Keitel, S. Khodyachykh, Y. Kim, V. Kocharyan, M. Koerfer, M. Kollewe, D. Kostin, D. Kraemer, M. Krassilnikov, G. Kube, L. Lilje, T. Limberg, D. Lipka, F. Loehl, M. Luong, C. Magne, J. Menzel, P. Michelato, V. Miltchev, M. Minty, W.D. Moeller, L. Monaco, W. Mueller, M. Nagl, O. Napoly, P. Nicolosi, D. Noelle, T. Nunez, A. Oppelt, C. Pagani, R. Paparella, B. Petersen, B. Petrosyan, J. Pflueger, P. Piot, E. Ploenjes, L. Poletto, D. Proch, D. Pugachov, K. Rehlich, D. Richter, S. Riemann, M. Ross, J. Rossbach, M. Sachwitz, E.L. Saldin, W. Sandner, H. Schlarb, B. Schmidt, M. Schmitz, P. Schmueser, J.R. Schneider, E.A. Schneidmiller, H.-J. Schreiber, S. Schreiber, A.V. Shabunov, D. Sertore, S. Setzer, S. Simrock, E. Sombrowski, L. Staykov, B. Steffen, F. Stephan, F. Stulle, K.P. Sytchev, H. Thom, K. Tiedtke, M. Tischer, R. Treusch, D. Trines, I. Tsakov, A. Vardanyan, R. Wanzenberg, T. Weiland, H. Weise, M. Wendt, I. Will, A. Winter, K. Wittenburg, M.V. Yurkov, I. Zagorodnov, P. Zambolin, K. Zapfe: First operation of a free-electron laser generating GW power radiation at 32 nm wavelength, Eur. Phys. J. D **37**, 297–303 (2006)

11.1873 M. Altarelli, R. Brinkmann, M. Chergui, W. Decking, B. Dobson, S. Dusterer, G. Gruebel, W. Graeff, H. Graafsma, J. Hajdu, J. Marangos, J. Pflueger, H. Redlin, D. Riley, I. Robinson, J. Rossbach, A. Schwarz, K. Tiedtke, T. Tschentscher, I. Vartaniants, H. Wabnitz, H. Weise, R. Wichmann, K. Witte, A. Wolf, M. Wulff, M. Yurkov (Eds.): *XFEL: The European X-Ray Free-Electron Laser*, Technical Design Report, Preprint DESY 2006-097 (DESY, Hamburg 2006) (see also http://xfel.desy.de)

11.1874 M.J. Hogan, C. Pellegrini, J. Rosenzweig, S. Anderson, P. Frigola, A. Tremaine, C. Fortgang, D.C. Nguyen, R.L. Sheffield, J. Kinross-Wright, A. Varfolomeev, A.A. Varfolomeev, S. Tolmachev, R. Carr: Measurements of gain larger than 10^5 at 12 μm in a self-amplified spontaneous-emission free-electron laser, Phys. Rev. Lett. **81**, 4867–4870 (1998)

11.1875 S.V. Milton, E. Gluskin, N.D. Arnold, C. Benson, W. Berg, S.G. Biedron, M. Borland, Y.-C. Chae, R.J. Dejus, P.K. Den Hartog, B. Deriy, M. Erdmann, Y.I. Eidelman, M.W. Hahne, Z. Huang,

K.-J. Kim, J.W. Lewellen, Y. Li, A.H. Lumpkin, O. Makarov, E.R. Moog, A. Nassiri, V. Sajaev, R. Soliday, B.J. Tieman, E.M. Trakhtenberg, G. Travish, I.B. Vasserman, N.A. Vinokurov, X.J. Wang, G. Wiemerslage, B.X. Yang: Exponential gain and saturation of a self-amplified spontaneous emission free-electron laser, Science **292**, 2037–2041 (2000)

11.1876 J. Arthur, P. Anfmrud, P. Audebert, K. Bane, I. Ben-Zvi, V. Bharadwaj, R. Bionta, P. Bolton, M. Borland, P.H. Bucksbaum, R.C. Cauble, J. Clendenin, M. Cornacchia, G. Decker, P. Den Hartog, S. Dierker, D. Dowell, D. Dungan, P. Emma, I. Evans, G. Faigel, R. Falcone, W.M. Fawley, M. Ferrario, A.S. Fisher, R.R. Freeman, J. Frisch, J. Galayda, J.-C. Gauthier, S. Gierman, E. Gluskin, W. Graves, J. Hajdu, J. Hastings, K. Hodgson, Z. Huang, R. Humphry, P. Ilinski, D. Imre, C. Jacobsen, C.-C. Kao, K.R. Kase, K.-J. Kim, R. Kirby, J. Kirz, L. Klaisner, P. Krejcik, K. Kulander, O.L. Landen, R.W. Lee, C. Lewis, C. Limborg, E.I. Lindau, A. Lumpkin, G. Materlik, S. Mao, J. Miao, S. Mochrie, E. Moog, S. Milton, G. Mulhollan, K. Nelson, W.R. Nelson, R. Neutze, A. Ng, D. Nguyen, H.-D. Nuhn, D.T. Palmer, J.M. Paterson, C. Pellegrini, S. Reiche, M. Renner, D. Riley, C.V. Robinson, S.H. Rokni, S.J. Rose, J. Rosenzweig, R. Ruland, G. Ruocco, D. Saenz, S. Sasaki, D. Sayre, J. Schmerge, D. Schneider, C. Schroeder, L. Serafini, F. Sette, S. Sinha, D. van der Spoel, B. Stephenson, G. Stupakov, M. Sutton, A. Szoke, R. Tatchyn, A. Toor, E. Trakhtenberg, I. Vasserman, N. Vinokurov, D. Waltz, J.S. Wark, E. Weckert, H. Winick, M. Woodley, A. Wootton, M. Wulff, M. Xie, R. Yotam, L. Young, A. Zewail: *Linac Cherent Light Source (LCLS)*, Conceptual Design Report, SLAC-R593 (SLAC, Stanford 2002) (see also http://www-ssrl.slac.stanford.edu/lcls/cdr)

11.1877 G.A. Gudzenko, L.A. Shelepin: Radiation enhancement in a recombining plasma, Sov. Phys. Doklady **10**, 147–149 (1964)

11.1878 M.A. Duguay, P.M. Rentzepis: Some approaches to vacuum UV and X-ray lasers, Appl. Phys. Lett. **10**, 350–352 (1967)

11.1879 R.C. Elton: Extension of 3p to 3s ion lasers into the vacuum ultraviolet region, Appl. Opt. **14**, 97–10197 (1975)

11.1880 R.W. Waynant, R.C. Elton: Review of short wavelength laser research, Proc. IEEE **64**, 1059–1092 (1976)

11.1881 R.C. Elton: *X-Ray Lasers* (Academic, Boston 1990)

11.1882 F.E. Irons, N.J. Peacock: Experimental evidence for population inversion in C^{5+} in an expanding laser-produced plasma, J. Phys. B **7**, 1109–1112 (1974)

11.1883 D.L. Matthews, P.L. Hagelstein, M.D. Rosen, M.J. Eckart, N.M. Ceglio, A.U. Hazi, H. Medecki, B.J. MacGowan, J.E. Trebes, B.L. Whitten, E.M. Campbell, C.W. Hatcher, A.M. Hawryluk, R.L. Kauffman, L.D. Pleasance, G. Rambach, J.H. Scofield, G. Stone, T.A. Weaver: Demonstration of a soft X-ray amplifier, Phys. Rev. Lett. **54**, 110–113 (1985)

11.1884 S. Suckewer, C.H. Skinner, H. Milchberg, C. Keane, D. Voorhees: Amplification of stimulated soft X-ray emission in a confined plasma column, Phys. Rev. Lett. **55**, 1753–1756 (1985)

11.1885 M.D. Rosen, J.E. Trebes, B.J. MacGowan, P.L. Hagelstein, R.A. London, D.L. Matthews, D.G. Nilson, T.W. Phillips, D.A. Whelan: Dynamic of collisional-excitation X-ray lasers, Phys. Rev. Lett. **59**, 2283–2286 (1987)

11.1886 C.J. Keane, N.M. Ceglio, B.J. MacGowan, D.L. Matthews, D.G. Nilson, J.E. Trebes, D.A. Whelan: Soft X-ray laser source development and applications experiments at Lawrence Livermore National Laboratory, J. Phys. B **22**, 3343–3362 (1990)

11.1887 L.B. Da Silva, B.J. MacGowan, S. Mrowka, J.A. Koch, R.A. London, D.L. Matthews, J.H. Underwood: Power measurements of a saturated yttrium X-ray laser, Opt. Lett. **18**, 1174–1176 (1993)

11.1888 J.A. Koch, B.J. MacGowan, L.B. Da Silva, D.L. Matthews, J.H. Underwood, P.J. Baston, S. Mrowka: Observation of gain-narrowing and saturation behavior in Se X-ray laser line profiles, Phys. Rev. Lett. **68**, 3291–3294 (1992)

11.1889 C.H. Skinner: Review of soft X-ray lasers and their applications, Phys. Fluids **3**, 2420–2429 (1991)

11.1890 L.B. Da Silva, T.W. Barbee, R. Cauble, P. Celliers, D. Ciarlo, S. Libby, R.A. London, D. Matthews, S. Mrowka, J.C. Moreno, D. Ress, J.E. Trebes, A.S. Wan, F. Weber: Electron density measurements of high density plasmas using soft X-ray laser interferometry, Phys. Rev. Lett. **74**, 3991–3994 (1995)

11.1891 B.J. MacGowan, S. Maxon, L.B. Da Silva, D.J. Fields, C.J. Keane, D.L. Matthews, A.L. Osterheld, J.H. Scofield, G. Shimkaveg, G.F. Stone: Demonstration of X-ray amplifiers near the carbon K edge, Phys. Rev. Lett. **65**, 420–423 (1990)

11.1892 J.J. Rocca: Table-top soft X-ray lasers, Rev. Sci. Instrum. **70**, 3799–3827 (1999)

11.1893 M.D. Rosen, P.L. Hagelstein, D.L. Matthews, E.M. Campbell, A.U. Hazi, B.L. Whitten, B. MacGowan, R.E. Turner, R.W. Lee: Exploding-foil technique for achieving a soft X-ray laser, Phys. Rev. Lett. **54**, 106–109 (1995)

11.1894 S. Maxon, P. Hagelstein, K. Reed, J. Scofield: A gas puff X-ray laser target design, J. Appl. Phys. **57**, 971–972 (1985)

11.1895 B.J. MacGowan, S. Maxon, P.L. Hagelstein, C.J. Keane, R.A. London, D.L. Matthews, M.D. Rosen, J.H. Scofield, D.A. Whelan: Demonstration of soft X-ray amplification in nickel-like ions, Phys. Rev. Lett. **59**, 2157–2160 (1987)

11.1896 J. Zhang, A.G. MacPhee, J. Lin, E. Wolfrum, R. Smith, C. Danson, M.H. Key, C.L.S. Lewis,

D. Neely, J. Nilsen, G.J. Part, G.J. Tallents, J.S. Wark: A saturated X-ray laser beam at 7 nanometers, Science **276**, 1097–1100 (1997)

11.1897 Y.A. Afanasiev, V.N. Shlyaptsev: Formation of a population inversion of transition in Ne-like ions in steady-state and transient plasmas, Sov. J. Quantum Electron. **19**, 1606–1612 (1989)

11.1898 P.V. Nickles, V.N. Shlyaptsev, M. Kalachnikov, M. Schnürer, I. Will, W. Sandner: Short pulse X-ray laser at 32.6 nm based on transient gain in Ne-like titanium, Phys. Rev. Lett. **78**, 2748–2751 (1997)

11.1899 J. Dunn, A.L. Osterheld, R. Shepherd, W.E. White, V.N. Shlyaptev, R.E. Stewart: Demonstration of X-ray amplification in transient gain nickel-like palladium scheme, Phys. Rev. Lett. **80**, 2835–2838 (1998)

11.1900 J. Dunn, Y. Li, A.L. Osterheld, J. Nilson, J.R. Hunter, V.N. Shlyaptev: Gain saturation regime for laser-driven tabletop, transient Ni-like ion X-ray lasers, Phys. Rev. Lett. **84**, 4834–4837 (2000)

11.1901 T. Kawachi, A. Sasaki, M. Tanaka, M. Kishimoto, N. Hasegawa, K. Nagashima, M. Koike, H. Daido, K. Kato: Observation of strong soft-X-ray amplification at 8.8 nm in the transient collisional-excitation scheme, Phys. Rev. A **69**, 033805 (2004)

11.1902 S. Augst, D. Strickland, D.D. Meyerhofer, S.L. Chin, J.H. Eberly: Tunneling ionization of noble gases in a high-intensity laser field, Phys. Rev. Lett. **63**, 2212–2215 (1989)

11.1903 P.B. Corkum, N.H. Burnett, F. Brunel: Above-threshold ionization in the long-wavelength limit, Phys. Rev. Lett. **62**, 1259–1262 (1989)

11.1904 N.H. Burnett, P.B. Corkum: Cold-plasma production for recombination extreme ultraviolet lasers by optical-field-induced ionization, J. Opt. Soc. Am. B **6**, 1195–1199 (1989)

11.1905 Y. Nagata, K. Midorikawa, S. Kubodera, M. Obara, H. Tashiro, K. Toyoda: Soft-X-ray amplification of the Lyman-a transition by optical-field-induced ionization, Phys. Rev. Lett **71**, 3774–3777 (1993)

11.1906 D.V. Korobkin, C.H. Nam, S. Suckewer, A. Goltsov: Demonstration of soft-X-ray lasing to Groud State in Li III, Phys. Rev. Lett. **77**, 5206–5209 (1996)

11.1907 P.B. Corkum, N.H. Burnett: Multiphoton ionization for the production of X-ray laser plasmas. In: *Short Wavelength Coherent Radiation: Generation and Applications*, OSA Proc., ed. by R.W. Falcone, J. Kirz (Opt. Soc. Am., Washington 1998) pp. 225–229

11.1908 B.E. Lemoff, G.Y. Lin, C.P.J. Barty, S.E. Harris: Femtosecond-pulse-driven, electron-excited XUV lasers in eight-times-ionized noble gases, Opt. Lett. **19**, 569–571 (1994)

11.1909 B.E. Lemoff, G.Y. Lin, C.L. II. I. Gordon, C.P.J. Barty, S.E. Harris: Demonstration of a 10 Hz femtosecocnd-pulse-driven xuv laser at 41.8 in Xe IX, Phys. Rev. Lett **74**, 1574–1577 (1995)

11.1910 S. Sebban, R. Haroutunian, P. Balcou, G. Grillon, A. Rousse, S. Kazamias, T. Marin, J.P. Rousseau, L. Notebaert, M. Pittman, J.P. Chambaret, A. Antonetti, D. Hulin, D. Ros, A. Klisnick, A. Jaegle, A. Carillon, G. Jamelot, J.F. Wyart: Saturated amplification of a collisional pumped optical-field-ionization soft X-ray laser at 41.8 nm, Phys. Rev. Lett **86**, 3004–3007 (2001)

11.1911 S. Sebban, T. Mocek, D. Ros, L. Upcraft, P. Palcou, R. Haroutunian, G. Grillon, B. Rus, A. Klisnick, A. Carillon, G. Jamelot, C. Valentin, A. Rousse, L. Notebaert, M. Pittman, D. Hulin: Demonstration of a Ni-like Kr optical-field-ionization collisional soft X-ray laser at 32.8 nm, Phys. Rev. Lett. **89**, 253901 (2002)

11.1912 J.J. Rocca, O.D. Cortázar, B. Szapiro, K. Floyd, F.G. Tomasel: Fast-discharge excitation of hot capillary plasmas for soft-X-ray amplifiers, Phys. Rev. E **47**, 1299–1304 (1993)

11.1913 J.J. Rocca, V. Shlyaptsev, F.G. Tomasel, O.D. Cortázar, D. Hartshorn, J.L.A. Chilla: Demonstration of a discharge pumped table-top soft-X-ray laser, Phys. Rev. Lett. **73**, 2192–2195 (1994)

11.1914 J. Filevich, K. Kanizay, M.C. Marconi, J.L.A. Chilla, J.J. Rocca: Dense plasma diagnostics with an amplitude-division soft X-ray laser interferometer based on diffraction, Opt. Lett. **25**, 356–358 (2000)

11.1915 A. McPherson, G. Gibson, H. Jara, U. Johann, T.S. Luk, I.A. McIntyre, K. Boyer, C.K. Rhodes: Studies of multiphoton production of vacuum-ultraviolet radiation in the rare gases, J. Opt. Soc. Am. **4**, 595–601 (1987)

11.1916 M. Ferray, A. L'Huillier, X.F. Li, L.A. Lompré, G. Mainfray, C. Manus: Multiple-harmonic radiation of 1064 nm radiation in rare gases, J. Phys. B **21**, L31–L35 (1988)

11.1917 J.J. Macklin, J.D. Kmetec, C.L. II. I. Gordon: High-order harmonic generation using intense femtosecond pulses, Phys. Rev. Lett. **70**, 766–769 (1993)

11.1918 A. L'Huillier, P. Balcou: High-order harmonic generation in rare gases with a 1 ps 1053 nm laser, Phys. Rev. Lett. **70**, 774–777 (1993)

11.1919 N. Sarukura, K. Hata, T. Adachi, R. Nodomi, M. Watanabe, S. Watanabe: Coherent soft-X-ray geneation by the harmonics of an ultrahigh-power KrF laser, Phys. Rev. A **43**, 1669–1672 (1991)

11.1920 Y. Nagata, K. Midorikawa, M. Obara, K. Toyoda: High-order harmonics generation by subpicosecond KrF excimer laser pulses, Opt. Lett. **21**, 15–17 (1996)

11.1921 S.G. Preston, A. Sanpera, M. Zepf, W.J. Blyth, C.G. Smith, J.S. Wark, M.H. Key, K. Burnett, M. Nakai, D. Neely, A.A. Offenberger: High-order harmonics of 248.6 nm KrF laser from helium and neon ions, Phys. Rev. A **53**, R31–R34 (1996)

11.1922 J. Zhou, J. Peatross, M.M. Murname, H.C. Kapteyn, I.P. Christov: Enhanced high-harmonic generation using 25 fs laser pulses, Phys. Rev. Lett. **76**, 752–755 (1996)

11.1923 M. Schnürer, C. Spielman, P. Wobrauschek, C. Streli, N. H. Burnett, C. Kan, K. Frencz, R. Koppitsch, Z. Cheng, T. Brabec, F. Krausz: Coherent 0.5 keV X-ray emission from helium driven by a sub10-fs laser, Phys. Rev. Lett. **80**, 3236–3239 (1998)

11.1924 Z. Chang, A. Rundquist, H. Wang, M. M. Murnane, H. C. Kapteyn: Generation of coherent soft X-ray at 2.7 nm using high harmonics, Phys. Rev. Lett. **79**, 2967–2970 (1997)

11.1925 C. Spielmann, N. H. Burnett, S. Sartania, R. Koppitsch, M. Schnürer, C. Kan, M. Lenzner, P. Wobrauschek, F. Krausz: Generation of coherent X-rays in the water window using 5-femtosecond laser pulses, Science **271**, 661–664 (1997)

11.1926 Y. Akiyama, K. Midorikawa, Y. Matsunawa, Y. Nagta, M. Obara, H. Tashiro, K. Toyoda: Generation of high-power harmonics using laser-produced rare-gas-like ions, Phys. Rev. Lett. **69**, 2176–2179 (1992)

11.1927 S. Kubodera, Y. Nagata, Y. Akiyama, K. Midoriakwa, M. Obara, H. Tashiro, K. Toyoda: High-order harmonic generation in laser-produced ions, Phys. Rev. A **48**, 4576–4582 (1993)

11.1928 Y. Liang, S. Augst, S. L. Chin, Y. Beaudoin, M. Chaker: High harmonic generation in atomic and diatomic molecular gases using intense picosecond laser pulses-a comprison, J. Phys. B **27**, 5119–5130 (1994)

11.1929 H. Sakai, K. Miyazaki: High-order harmonic generation in nitrogen molecules with subpicosecond visible dye-laser pulses, Appl. Phys. **61**, 493–498 (1995)

11.1930 C. Wülker, W. Theobald, D. Ouw, F. P. Schäfer, B. N. Chichkov: Short-pulse laser-produced plasma from C_{60} molecules, Opt. Commun. **112**, 21–28 (1994)

11.1931 T. D. Donnelly, T. Ditmire, K. Neuman, M. D. Perry, R. W. Falcone: High-order harmonic generation in atomic clusters, Phys. Rev. Lett. **76**, 2472–2475 (1996)

11.1932 J. L. Krause, K. J. Schafer, K. C. Kulander: High-order harmonic generation from atoms and ions in the high intensity regime, Phys. Rev. Lett. **68**, 3535–3538 (1992)

11.1933 P. B. Corkum: Plasma perspective on strong field multiphoton ionization, Phys. Rev. Lett. **71**, 1994–1997 (1993)

11.1934 M. L. Lewenstein, P. Balcou, M. Y. Ivanov, A. L'Huillier, P. B. Corkum: Theory of high-harmonic generation by low-frequency laser fields, Phys. Rev. A **49**, 2117–2132 (1994)

11.1935 M. Lewenstein, P. Salieres, A. L'Huillier: Phase of the atomic polarization in high-order harmonic generation, Phys. Rev. A **52**, 4747–4754 (1995)

11.1936 P. Balcou, P. Saliéres, A. L'Huillier, M. Lewenstein: Generalized phase matching conditions for high harmonics: The role of field-gradient forces, Phys. Rev. A **55**, 3204–3210 (1997)

11.1937 M. B. Gaarde, F. Salin, E. Constant, P. Balcou, K. J. Schafer, K. C. Kulander, A. L'Huillier: Spatiotemporal separation of high harmonic radiation into two quantum path components, Phys. Rev. **59**, 1367–1373 (1999)

11.1938 P. Saliéres, A. L'Huillier, M. Lewenstein: Coherent control of high-order harmonics, Phys. Rev. Lett. **74**, 3776–3779 (1995)

11.1939 P. Balcou, A. L'Huillier: Phase-matching effects in strong-field harmonic generation, Phys. Rev. A **47**, 1447–1459 (1993)

11.1940 Y. Tamaki, J. Itatani, M. Obara, K. Midorikawa: Optimization of conversion efficiency and spatial quality of high-order harmonic generation, Phys. Rev. A **62**, 063802 (2000)

11.1941 Y. Tamaki, J. Itatani, Y. Nagata, M. Obara, K. Midorikawa: Highly efficient, phase-matched high-harmonic generation by a self-guided laser beam, Phys. Rev. Lett. **82**, 1422–1425 (1999)

11.1942 A. Rundquist, C. G. Durfee, Z. Chang, C. Herne, S. Backus, M. M. Murnane, H. C. Kapteyn: Phase-matched generation of coherent soft X-rays, Science **280**, 1412–1414 (1998)

11.1943 E. Constant, D. Garzella, P. Breger, E. Mevel, C. Dorrer, C. Le Blanc, F. Salin, P. Agostini: Optimizing high harmonic generation in absorbing gases: model and experiment, Phys. Rev. Lett. **82**, 1668–1671 (1999)

11.1944 E. Takahashi, Y. Nabekawa, T. Otsuka, M. Obara, K. Midorikawa: Generation of highly coherent submicrojoule soft X-rays by high-order harmonics, Phys. Rev. A **66**, 021802 (2002)

11.1945 Y. Tamaki, O. Maya, K. Midorikawa, M. Obara: High-order harmonic generation in a gas-filled hollow fiber. In: *Conf. Lasers Electro-Optics*, OSA Tech. Dig. Ser., Vol. 6 (Opt. Soc. Am., Washington 1998) p. 83

11.1946 Y. Tamaki, Y. Nagata, M. Obara, K. Midoriakwa: Phase-matched high-order-harmonic generation in a gas-filled hollow fiber, Phys. Rev. A **59**, 4041–4044 (1999)

11.1947 C. G. Durfee III., A. R. Rundquist, S. Backus, C. Herne, M. M. Murnane, H. C. Kapteyn: Phase matching of high-order harmonics in hollow waveguides, Phys. Rev. Lett. **83**, 2187–2190 (1999)

11.1948 A. Paul, R. A. Bartels, R. Tobey, H. Green, S. Welman, I. P. Christov, M. M. Murnane, H. C. Kapteyn, S. Backus: Quasi-phase-matched generation of coherrent extreme-ultraviolet light, Nature **421**, 51–54 (2003)

11.1949 E. Takahashi, Y. Nabekawa, M. Nurhuda, K. Midoriakwa: Generation of high-energy high-order harmonics by use of a long interaction medium, J. Opt. Soc. Am. B **20**, 158–165 (2003)

11.1950 E. Takahashi, Y. Nabekawa, K. Midoriakwa: Generation of 10 μJ coherent extreme ultraviolet light by use of high-order harmonics, Opt. Lett. **27**, 1920–1922 (2002)

11.1951 L. Le Deroff, P. Salieres, B. Carre, D. Joyeux, D. Phalippou: Mesurement of the degree of spatial coherence of high-order harmonics using a Fresnel-mirror interferometer, Phys. Rev. A **61**, 043802 (2002)

11.1952 Y. Tamaki, J. Itatani, M. Obara, K. Midorikawa: Highly coherent soft X-ray generation by macroscopic phase matching of high-order harmonics, Jpn. J. Appl. Phys. **40**, L1154–L1156 (2001)

11.1953 R. A. Bartels, A. Paul, H. Green, H. C. Kaypteyn, M. M. Murnane, S. Backus, I. P. Christov, Y. Liu, D. Attowood, C. Jacobsen: Generation of spatially coherent light at extreme ultraviolet wavelength, Science **297**, 376–378 (2002)

11.1954 D. G. Lee, J. J. Park, J. H. Sung, C. H. Nam: Wave-front phase measurements of high-order harmonic beams by use of point-diffraction interferometry, Opt. Lett. **28**, 480–482 (2003)

11.1955 R. Haight, P. F. Seidler: High resolution atomic core level spectroscopy with laser harmonics, Appl. Phys. Lett. **65**, 517–519 (1994)

11.1956 R. Haight, D. R. Peale: Antibonding state on the Ge(111): as surface: spectroscopy and dynamics, Phys. Rev. Lett **70**, 3979–3982 (1993)

11.1957 W. Theobald, C. Wülker, R. Sauerbrey: Temporally resolved measurement of electron densities with ($>10^{23}$ cm^3) high harmonics, Phys. Rev. Lett. **77**, 298–301 (1996)

11.1958 K. Ishikawa, K. Midorikawa: Two-photon ionization of He^+ as a nonlinear optical effect in the soft-X-ray region, Phys. Rev. A **65**, 043405 (2002)

11.1959 T. Nakajima, L. A. A. Nikolopoulos: Use of helium double ionization for autocorrelation of an xuv pulse, Phys. Rev. A **66**, 041402 (2002)

11.1960 A. D. Bandrauk, S. Chelkowski: Dynamic imaging of nuclear wave functions with ultrashort UV laser pulases, Phys. Rev. Lett. **87**, 273004 (2001)

11.1961 S. E. Harris, J. J. Macklin, T. W. Hänsch: Atomic scale temporal structure inherent to high-order harmonic generation, Opt. Commun. **100**, 487–490 (1993)

11.1962 P. B. Corkum, N. H. Burnett, M. Y. Ivanov: Subfemtosecond pulses, Opt. Lett. **19**, 1870–1872 (1994)

11.1963 M. Hentscel, R. Kienberger, C. Spielmann, G. A. Reider, N. Milosevic, T. Brabec, P. Corkun, U. Heinzmann, M. Drescher, F. Krausz: Attosecond metrology, Nature **414**, 509–513 (2001)

11.1964 D. Strickland, G. Mourou: Compression of amplified chirped optical pulses, Opt. Commun. **56**(3), 219–221 (1985)

11.1965 M. D. Perry, D. Pennington, B. C. Stuart, G. Tietbohl, J. A. Britten, C. Brown, S. Herman, B. Golick, M. Kartz, J. Miller, H. T. Powell, M. Vergino, V. Yanovsky: Petawatt laser pulses, Opt. Lett. **24**(3), 160–162 (1999)

11.1966 E. B. Treacy: Optical pulse compression with diffraction gratings, IEEE J. Quantum Electron. **5**(9), 454 (1969)

11.1967 R. D. Boyd, J. A. Britten, D. E. Decker, B. W. Shore, B. C. Stuart, M. D. Perry, L. Li: High-efficiency metallic diffraction gratings for laser applications, Appl. Opt. **34**(10), 1697–1706 (1995)

11.1968 G. Cheriaux, P. Rousseau, F. Salin, J. P. Chambaret, B. Walker, L. F. Dimauro: Aberration-free stretcher design for ultrashort-pulse amplification, Opt. Lett. **21**(6), 414–416 (1996)

11.1969 P. S. Banks, M. D. Perry, V. Yanovsky, S. N. Fochs, B. C. Stuart, J. Zweiback: Novel all-reflective stretcher for chirped-pulse amplification of ultrashort pulses, IEEE J. Quantum Electron. **36**(3), 268–274 (2000)

11.1970 B. W. Shore, M. D. Perry, J. A. Britten, R. D. Boyd, M. D. Feit, H. T. Nguyen, R. Chow, G. E. Loomis, L. F. Li: Design of high-efficiency dielectric reflection gratings, J. Opt. Soc. Am. A **14**(5), 1124–1136 (1997)

11.1971 K. Hehl, J. Bischoff, U. Mohaupt, M. Palme, B. Schnabel, L. Wenke, R. Bödefeld, W. Theobald, E. Welsch, R. Sauerbrey, H. Heyer: High-efficiency dielectric reflection gratings: design, fabrication, and analysis, Appl. Opt. **38**(30), 6257–6271 (1999)

11.1972 T. J. Kessler: Demonstration of coherent addition of multiple gratings for high-energy chirped-pulse-amplified lasers, Opt. Lett. **29**, 635 (2004)

11.1973 J. Hein, M. C. Kaluza, R. Bödefeld, M. Siebold, S. Podleska, R. Sauerbrey: POLARIS an all diode pumped ultrahigh peak power laser for high repetition rate, Lect. Notes Phys. **694**, 47–66 (2006)

11.1974 D. J. Ripin, J. R. Ochoa, R. L. Aggarwal, T. Y. Fan: 165 W cryogenically cooled Yb:YAG laser, Opt. Lett. **29**(18), 2154–2156 (2004)

11.1975 E. I. Moses, R. E. Bonanno, C. A. Haynam, R. L. Kauffman, B. J. MacGowan, R. W. Patterson, R. H. Sawicki, B. M. Van Wonterghem: The national ignition facility: Path to ignition in the laboratory, J. Phys. IV **133**, 57–57 (2006)

11.1976 D. Besnard: The Megajoule laser program – Ignition at hand, J. Phys. IV **133**, 47–47 (2006)

11.1977 K. Mima, K. A. Tanaka, R. Kodama, T. Johzaki, H. Nagatomo, H. Shiraga, Y. Sentoku, N. Miyanaga, H. Azechi, M. Nakai, T. Norimatu, K. Nagai, J. Sunahara, K. Nishihara, T. Taguchi, H. Sakagami: Present status and future prospects of laser fusion research at ILE Osaka university, Plasma Sci. Technol. **6**(1), 2179–2184 (2004)

11.1978 J. Collier, C. Hernandez-Gomez, I. N. Ross, P. Matousek, C. N. Danson, J. Walczak: Evaluation of an ultrabroadband high-gain amplification technique for chirped pulse amplification facilities, Appl. Opt. **38**(36), 7486–7493 (1999)

11.1979 J. Hein, S. Podleska, M. Siebold, M. Hellwing, R. Bödefeld, R. Sauerbrey, D. Ehrt, W. Wintzer:

of an injection laser, Sov. Tech. Phys. Lett. **4**, 438–439 (1978)

11.2038 M. W. Fleming, A. Moradian: Spectral characteristics of external-cavity controlled semiconductor lasers, IEEE J. Quantum Electron. **17**, 44–59 (1981)

11.2039 K. Liu, M. G. Littman: Novel geometry for single-mode scanning of tunable lasers, Opt. Lett. **6**, 117–118 (1981)

11.2040 A. Celikov, F. Riehle, V. L. Velichansky, J. Helmcke: Diode laser spectroscopy in a Ca atomic beam, Opt. Commun. **107**, 54–60 (1994)

11.2041 G. Birnbaum: Frequency stabilization of gas lasers, Proc. IEEE **55**, 1015–1026 (1967)

11.2042 R. A. McFarlane, W. R. Bennett Jr., W. E. Lamb Jr.: Single mode tuning dip in the power output of an He-Ne optical maser, Appl. Phys. Lett. **2**, 189–190 (1963)

11.2043 R. Balhorn, H. Kunzmann, F. Lebowski: Frequency stabilization of internal-mirror helium-neon lasers, Appl. Opt. **11**, 742–744 (1972)

11.2044 D. Ullrich: Frequency stabilization of HeNe lasers by intensity comparison of two longitudinal modes, PTB-Bericht **PTB-F-3**, 1–35 (1988) Braunschweig

11.2045 R. C. Quenelle, L. J. Wuerz: A new microcomputer-controlled laser dimensional measurement and analysis system, Hewlett-Packard J. **34,4**, 3–13 (1983)

11.2046 T. Baer, F. V. Kowalski, J. L. Hall: Frequency stabilization of a 0.633 μm HeNe longitudinal Zeeman laser, Appl. Opt. **19**, 3173–3177 (1980)

11.2047 S. Gerstenkorn, P. Luc: *Atlas du spectre d'absorption de la molecule d'iode (14800 – 20000 cm^{-1})*, Technical Report (Laboratoire Aimé-Cotton, Paris 1978)

11.2048 G. Camy, C. J. Bordé, M. Ducloy: Heterodyne saturation spectroscopy through frequency modulation of the saturating beam, Opt. Commun. **41**, 325–330 (1982)

11.2049 P. A. Jungner, M. Eickhoff, S. D. Swartz, Y. Ye, J. L. Hall, S. Waltman: Stability and absolute frequency of molecular iodine transitions near 532 nm, Proc. SPIE **2378**, 22–34 (1995)

11.2050 G. C. Bjorklund: Frequency-modulation spectroscopy: a new method for measuring weak absorptions and dispersions, Opt. Lett. **5**, 15–17 (1980)

11.2051 J. L. Hall, L. Hollberg, T. Baer, H. G. Robinson: Optical heterodyne saturation spectroscopy, Appl. Phys. Lett. **39**, 680–682 (1981)

11.2052 P. Cordiale, G. Galzerano, H. Schnatz: International comparisons of two iodine stabilized frequency-doubled Nd:YAG lasers at 532 nm, Metrologia **37**, 177–182 (2000)

11.2053 A. Y. Nevsky, R. Holzwarth, M. Zimmermann, T. Udem, T. W. Hänsch, J. von Zanthier, H. Walther, P. V. Pokasiv, M. N. Skvortsov, S. N. Bagayev, H. Schnatz, F. Riehle: Frequency comparison of I_2-stabilized lasers at 532 nm and absolute optical frequency measurement of I_2 absorption lines, Opt. Commun. **192**, 263–272 (2001)

11.2054 Y. Millerioux, D. Touahry, L. Hilico, R. Felder, F. Birabin, B. de Beauvoir: Towards an accurate frequency standard at $\lambda = 778$ nm using a laser diode stabilized on a hyperfine component of the Doppler-free two-photon transitions in rubidium, Opt. Commun. **108**, 91–96 (1994)

11.2055 D. Touahri, O. Acef, A. Clairon, J. J. Zondy, R. Felder, L. Hilico, B. de Beauvoir, F. Birabin, F. Nez: Frequency measurement of the $5S_{1/2}$ (F=3) – $5D_{5/2}$ (F=5) two-photon transition in rubidium, Opt. Commun. **133**, 471–478 (1997)

11.2056 E. Fretel: Spectroscopie à deux photons d'atomes de rubidium dans un piège magnéto-optique Ph.D. Thesis (Conservatoire National des Arts et Métiers, Paris 1997)

11.2057 N. F. Ramsey: A molecular beam resonance method with separated oscillating fields, Phys. Rev. **78**, 695–699 (1950)

11.2058 G. Kramer, C. O. Weiss, B. Lipphardt: *Coherent frequency measurements of the hfs-resolved methane line*, ed. by A. de Marchi (Springer, Berlin, Heidelberg 1989) pp. 181–186

11.2059 Y. V. Baklanov, B. Y. Dubetsky, V. P. Chebotayev: Non-linear Ramsey resonances in the optical region, Appl. Phys. **9**, 171–173 (1976)

11.2060 C. J. Bordé, C. Salomon, S. Avrillier, A. van Lerberge, C. Bréant, D. Bassi, G. Scoles: Optical Ramsey fringes with travelling waves, Phys. Rev. A **30**, 1836–1848 (1984)

11.2061 C. J. Bordé: Atomic interferometry with internal state labeling, Phys. Lett. A **140**, 10–12 (1989)

11.2062 C. J. Bordé: *Matter-wave interferometers: A synthetic approach*, ed. by P. R. Berman (Academic, San Diego 1997) pp. 257–292

11.2063 U. Sterr, K. Sengstock, W. Ertmer, F. Riehle, J. Helmcke: *Atom interferometry based on separated light fields*, ed. by P. R. Berman (Academic, San Diego 1997) pp. 293–362

11.2064 P. Kersten, F. Mensing, U. Sterr, F. Riehle: A transportable optical calcium frequency standard, Appl. Phys. **68**, 27–38 (1999)

11.2065 J. L. Hall, M. Zhu, P. Buch: Prospects for using laser-prepared atomic fountains for optical frequency standards applications, J. Opt. Soc. Am. B **6**, 2194–2205 (1989)

11.2066 U. Sterr, K. Sengstock, J. H. Müller, D. Bettermann, W. Ertmer: The magnesium Ramsey interferometer: Applications and prospects, Appl. Phys. B **54**, 341–346 (1992)

11.2067 A. Celikov, P. Kersten, F. Riehle, G. Zinner, L. D'Evelin, A. Zibrov, V. L. Velichansky, J. Helmcke: External cavity diode laser high resolution spectroscopy of the Ca and Sr intercombination lines for the development of a transportable frequency/length standard. In: *Proc. 49th Annu. IEEE*

Int. Frequency Control Symposium, San Francisco (IEEE, 1995) pp. 153–160

11.2068 G. M. Tino, M. Basanti, M. de Angelis, L. Gianfrani, M. Inguscio: Spectroscopy on the 689 nm intercombination line of strontium using extended – cavity InP/InGaIP diode laser, Appl. Phys. B **55**, 397–400 (1992)

11.2069 A. M. Akulshin, A. Celikov, V. L. Velichansky: Nonlinear Doppler-free spectroscopy of the $6^1S_0 - 6^3P_1$ intercombination transition in barium, Opt. Commun. **93**, 54–58 (1992)

11.2070 R. L. Barger, J. C. Bergquist, T. C. English, D. J. Galze: Resolution of photon-recoil structure of the 6573-A calcium line in an atomic beam with optical Ramsey fringes, Appl. Phys. Lett. **34**, 850–852 (1979)

11.2071 R. L. Barger: Influence of second-order Doppler effect on optical Ramsey fringe profiles, Opt. Lett. **6**, 145–147 (1981)

11.2072 A. Morinaga, F. Riehle, J. Ishikawa, J. Helmcke: A Ca optical frequency standard: Frequency stabilization by means of nonlinear Ramsey resonances, Appl. Phys. B **48**, 165–171 (1989)

11.2073 N. Ito, J. Ishikawa, A. Morinaga: Frequency locking a dye laser to the central Ramsey fringe in a Ca atomic beam and wavelength measurement, J. Opt. Soc. Am. **8**, 1388–1390 (1991)

11.2074 A. S. Zibrov, R. W. Fox, R. Ellingsen, C. S. Weimer, V. L. Velichansky, G. M. Tino, L. Hollberg: High resolution diode laser spectroscopy of calcium, Appl. Phys. **59**, 327–331 (1994)

11.2075 W. Paul: Electromagnetic traps for charged and neutral particles, Rev. Mod. Phys. **62**, 531–540 (1990)

11.2076 C. Tamm, D. Engelke, V. Bühner: Spectroscopy of the electric quadrupole transition $^2S_{1/2}(F-0)-^2D_{3/2}(F=2)$ in trapped $^{171}Yb^+$, Phys. Rev. A **61**, 053405, 1–9 (2000)

11.2077 D. J. Wineland, R. E. Drullinger, F. L. Walls: Radiation pressure cooling of bound resonant absorbers, Phys. Rev. Lett. **40**, 1639–1642 (1978)

11.2078 W. Neuhauser, M. Hohenstatt, P. Toschek, H. Dehmelt: Optical-sideband cooling and visible atom cloud confined in parabolic well, Phys. Rev. Lett. **41**, 233–236 (1978)

11.2079 W. Nagourny, J. Sandberg, H. Dehmelt: Shelved optical electron amplifier observation of quantum jumps, Phys. Rev. Lett. **56**, 2797–2799 (1986)

11.2080 U. Tanaka, S. Bize, C. E. Tanner, R. E. Drullinger, S. A. Diddams, L. Hollberg, W. M. Itano, D. J. Wineland, J. C. Bergquist: The $^{199}Hg^+$ single ion optical clock: recent progress, J. Phys. B **36**, 545–551 (2003)

11.2081 G. Barwood, K. Gao, P. Gill, G. Huang, H. Klein: Development of optical frequency standards based upon the $^2S_{1/2}-^2D_{5/2}$ transition in $^{88}Sr^+$ and $^{87}Sr^+$, IEEE Trans. Instrum. Meas. **50**, 543–547 (2001)

11.2082 L. Marmet, A. A. Madej: Optical Ramsey spectroscopy and coherence measurements of the clock transition in a single trapped Sr ion, Can. J. Phys. **78**, 495–507 (2000)

11.2083 M. Eichenseer, A. Yu Nevsky, J von Zanthier. Ch Schwedes, H. Walther: Towards an indium single-ion optical frequency standard, J. Phys. B **36**, 553–559 (2003)

11.2084 S. A. Webster, P. Taylor, M. Roberts, G. P. Barwood, P. Blythe, P. Gill: The frequency standard using the $^2S_{1/2}-^2F_{7/2}$ octupole transition in $^{171}Yb^+$. In: *Proc. 6th Symp. Frequency Standards and Metrology*, ed. by P. Gill (World Scientific, Singapore 2002) pp. 114–122

11.2085 E. Peik, B. Lipphardt, H. Schnatz, T. Schneider, C. Tamm, S. G. Karschenboim: Limit on the temporal variation of the fine structure constant, Phys. Rev. Lett. **93**, 170801 (2004)

11.2086 J. Stenger, C. Tamm, N. Haverkamp, S. Weyers, H. Telle: Absolute frequency measurement of the 435.5 nm $^{171}Yb^+$-clock transition with a Kerr-lens mode-locked femtosecond laser, Opt. Lett. **26**, 1589–1591 (2001)

11.2087 E. L. Raab, M. Prentiss, A. Cable, S. Chu, D. E. Pritchard: Trapping of neutral sodium atoms with light pressure, Phys. Rev. Lett. **59**, 2631–2634 (1987)

11.2088 W. Ertmer, R. Blatt, J. L. Hall: Some candidate atoms for frequency standards research using radiative cooling techniques. In: *Laser Cooled and Trapped Atoms and Ions*, Nat. Bur. Stand. Spec. Publ., Vol. 653, ed. by W. D. Phillips (NBS, Reading 1983) pp. 154–161

11.2089 N. Beverini, E. Macconi, D. Prereira, F. Strumia, G. Vissani: Production of low velocity Mg and Ca atomic beams by laser light pressure. In: *5th Italian Conf. Quantum Electronics and Plasma Physics*, ed. by G. C. Righin (Italian Physical Society, Bologna 1988) pp. 205–211

11.2090 F. Ruschewitz, J. L. Peng, H. Hinderthür, N. Schaffrath, K. Sengstock, W. Ertmer: Sub-kilohertz optical spectroscopy with time a domain atom interferometer, Phys. Rev. Lett. **80**, 3173–3176 (1998)

11.2091 T. P. Dineen, K. R. Vogel, E. Arimondo, J. L. Hall, A. Gallagher: Cold collisions of Sr^*-Sr in a magneto-optical trap, Phys. Rev. A **59**, 1216–1222 (1999)

11.2092 H. Katori, T. Ido, Y. Isoya, M. Kuwata-Gonokami: Magneto-optical trapping and cooling of strontium atoms down to the photon recoil temperature, Phys. Rev. Lett. **82**, 1116–1119 (1999)

11.2093 T. Kisters, K. Zeiske, F. Riehle, J. Helmcke: High resolution spectroscopy with laser-cooled and trapped calcium atoms, Appl. Phys. B **59**, 89–98 (1994)

11.2094 T. Binnewies, G. Wilpers, U. Sterr, F. Riehle, J. Helmcke, T. E. Mehlstäubler, E. M. Rasel, W. Ertmer: Doppler cooling and trapping on forbidden transitions, Phys. Rev. Lett. **87**, 123002 (2001)

11.2095 G. Wilpers, T. Binnewies, C. Degenhardt, U. Sterr, F. Riehle, J. Helmcke: An optical clock with ultracold neutral atoms, Phys. Rev. Lett. **89**, 230801 (2002)

11.2096 D. A. Jennings, C. R. Pollock, F. R. Peterson, E. Evenson, J. S. Wells, J. L. Hall, H. P. Layer: Direct frequency measurement of the I_2-stabilized HeNe 473 THz (633 nm) laser, Opt. Lett. **8**, 136–138 (1982)

11.2097 H. Schnatz, B. Lipphardt, J. Helmcke, F. Riehle, G. Zinner: First phase-coherent frequency measurement of visible radiation, Phys. Rev. Lett. **76**, 18–21 (1996)

11.2098 N. Niering, R. Holzwarth, J. Reichert, P. Pokasov, T. Udem, M. Weitz, T. W. Hänsch, P. Lemondo, G. Santarelli, M. Abgrall, P. Laurent, C. Salomon, A. Clairon: Measurement of the Hydrogen 1S-2S transition frequency by phase coherent comparison with a microwave cesium fountain clock, Phys. Rev. Lett. **84**, 5496–5499 (2000)

11.2099 D. J. Jones, S. A. Diddams, J. K. Ranka, A. Stentz, R. S. Windeler, J. L. Hall, S. T. Cundiff: Carrier-envelope phase control of femtosecond mode-locked lasers and direct optical frequency synthesis, Science **288**, 635–639 (2000)

11.2100 H. R. Telle, G. Steinmeyer, A. E. Dunlop, D. H. Sutter, U. Keller: Carrier-envelope offset phase control: A novel concept for absolute frequency measurement and ultra-short pulse generation, Appl. Phys. B **69**, 327–332 (1999)

11.2101 J. Stenger, H. R. Telle: Intensity induced mode shift in femtosecond lasers via the nonlinear index of refraction, Opt. Lett. **25**, 1553–1555 (2000)

11.2102 S. A. Diddams, D. J. Jones, J. Ye, S. T. Cundiff, J. L. Hall: Direct link between microwave and optical frequencies with a 300 THz femtosecond laser comb, Phys. Rev. Lett. **84**, 5102–5105 (2000)

11.2103 P. Kubina., P. Adel, F. Adler, G. Grosche, T. W. Hänsch, R. Holzwarth, A. Leitensdorfer, B. Lipphardt, H. Schnatz: Long term comparison of two fiber based frequency comb systems, Opt. Express **13**, 904–909 (2005)

11.2104 J. Stenger, H. Schnatz, C. Tamm, H. R. Telle: Ultraprecise measurement of optical frequency ratios, Phys. Rev. Lett. **88**, 073601 (2002)

12. Femtosecond Laser Pulses: Linear Properties, Manipulation, Generation and Measurement

In this contribution some basic properties of femtosecond laser pulses are summarized. In Sect. 12.1 we start with the linear properties of ultrashort light pulses. Nonlinear optical effects that would alter the frequency spectrum of an ultrashort pulse are not considered. However, due to the large bandwidth, the linear dispersion is responsible for dramatic effects. For example, a 10 fs laser pulse at a center wavelength of 800 nm propagating through 4 mm of BK7 glass will be temporally broadened to 50 fs. In order to describe and manage such dispersion effects a mathematical description of an ultrashort laser pulse is given first before we continue with methods how to change the temporal shape via the frequency domain. The chapter ends with a paragraph on the powerful technique of pulse shaping, which can be used to create complex-shaped ultrashort laser pulses with respect to phase, amplitude and polarization state.

In Sect. 12.2 the generation of femtosecond laser pulses via mode locking is described in simple physical terms. As femtosecond laser pulses can be generated directly from a wide variety of lasers with wavelengths ranging from the ultraviolet to the infrared no attempt is made to cover the different technical approaches.

In Sect. 12.3 we deal with the measurement of ultrashort pulses. Traditionally a short event has been characterized with the aid of an even shorter event. This is not an option for ultrashort light pulses. The characterization of ultrashort pulses with respect to amplitude and phase is therefore based on optical correlation techniques that make use of the short pulse itself. Methods operating in the time–frequency domain are especially useful.

A central building block for generating femtosecond light pulses are lasers. Within only two decades of the invention of the laser the duration of the shortest pulse shrunk by six orders of magnitude from the nanosecond regime to the femtosecond regime. Nowadays femtosecond pulses in the range of 10 fs and below can be generated directly from compact and reliable laser oscillators and the temporal resolution of measurements has outpaced the resolution even of modern sampling oscilloscopes by orders of magnitude. With the help of some simple comparisons the incredibly fast femtosecond time scale can be put into perspective: on a logarithmic time scale one minute is approximately half-way between 10 fs and the age of the universe. Taking the speed of light in vacuum into account, a 10 fs light pulse can be considered as a 3 μm thick slice of light whereas a light pulse of one second spans approximately the distance between earth and moon. It is also useful to realize that the fastest molecular vibrations in nature have an oscillation time of about 10 fs.

It is the unique attributes of these light pulses that open up new frontiers both in basic research and for applications. The *ultrashort pulse duration* for example allows the motion of electrons and molecules to be frozen by making use of so-called pump probe techniques that work similar to strobe light techniques. In chemistry complex reaction dynamics have been measured directly in the time domain and this work was rewarded with the Nobel price in chemistry for A. H. Zewail in 1999. The *broad spectral width* can be used for example in medical diagnostics or – by taking the longitudinal frequency comb mode structure into account – for high-precision optical frequency metrology. The latter is expected to outperform today's state-of-the-art caesium clocks and was rewarded with the 2005 Nobel price in physics for J. L. Hall and T. W. Hänsch. The extreme concentration of a modest energy content in focused femtosecond pulses delivers *high peak intensities* that are used for example in a reversible light matter interaction regime for the development of nonlinear microscopy techniques. The irreversible light matter regime can be for example applied to nonthermal material processing leading to precise microstructures in a whole variety of solid state materials. Finally the *high pulse repetition rate* is exploited, for example, in telecommunication applications.

These topics have been reviewed recently in [12.1]. The biannual international conference series *Ultrafast Phenomena* and *Ultrafast Optics*, including the corresponding conference proceedings, cover a broad range of applications and latest developments.

Besides the specific literature given in the individual chapters some textbooks devoted to ultrafast laser pulses are recommended for a more in-depth discussion of the topics presented here and beyond (see, for example, [12.2–5] and especially for the measurement of ultrashort pulses see [12.6]).

12.1 Linear Properties of Ultrashort Light Pulses

12.1.1 Descriptive Introduction

It is quite easy to construct the electric field of a *Gedanken* optical pulse at a fixed position in space, corresponding to the physical situation of a fixed detector in space. Assuming the light field to be linearly polarized, we may write the real electric field strength $E(t)$ as a scalar quantity whereas a harmonic wave is multiplied with a temporal amplitude or envelope function $A(t)$

$$E(t) = A(t)\cos(\Phi_0 + \omega_0 t) \tag{12.1}$$

with ω_0 being the carrier circular (or angular) frequency. The light frequency is given by $\nu_0 = \frac{\omega_0}{2\pi}$. In the following, angular frequencies and frequencies are only distinguished from each other via their notation. For illustration we will use optical pulses centered at 800 nm, corresponding to a carrier frequency of $\omega_0 = 2.35\,\text{rad/fs}$ (oscillation period $T = 2.67\,\text{fs}$) with a Gaussian envelope function (the numbers refer to pulses that are generated by the widely spread femtosecond laser systems based on Ti:sapphire as the active medium). For simple envelope functions the pulse duration Δt is usually defined by the FWHM (full width at half-maximum) of the temporal intensity function $I(t)$

$$I(t) = \frac{1}{2}\varepsilon_0 c n A(t)^2\,, \tag{12.2}$$

with ε_0 being the vacuum permittivity, c the speed of light and n the refractive index. The factor 1/2 arises from averaging the oscillations. If the temporal intensity is given in W/cm^2 the temporal amplitude $A(t)$ (in V/cm for $n = 1$) is given by

$$A(t) = \sqrt{\frac{2}{\varepsilon_0 c}}\sqrt{I(t)} = 27.4\sqrt{I(t)}\,. \tag{12.3}$$

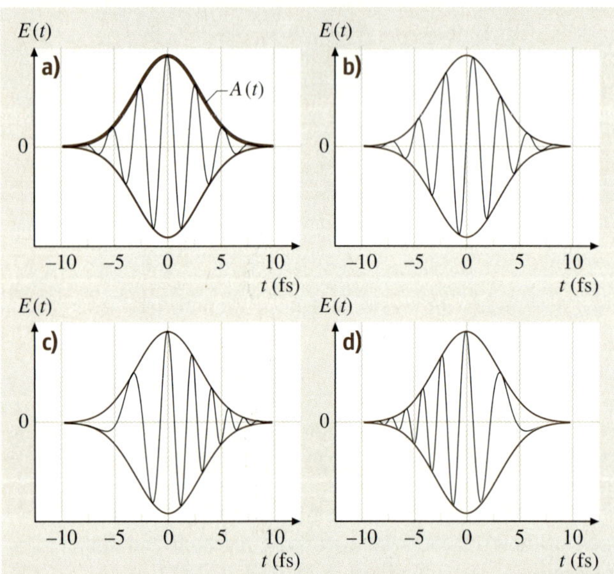

Fig. 12.1a–d Electric field $E(t)$ and temporal amplitude function $A(t)$ for a cosine pulse **(a)**, a sine pulse **(b)**, an up-chirped pulse **(c)** and a down-chirped pulse **(d)**. The pulse duration in all cases is $\Delta t = 5\,\text{fs}$. For **(c)** and **(d)** the parameter a was chosen to be $\pm 0.15/\text{fs}^2$

Figure 12.1a displays $E(t)$ for a Gaussian pulse with $\Delta t = 5\,\text{fs}$ and $\Phi_0 = 0$. At $t = 0$ the electric field strength reaches its maximum value. This situation is called a cosine pulse: for $\Phi_0 = -\pi/2$ we get a sine pulse $E(t) = A(t)\sin(\omega_0 t)$ (Fig. 12.1b) where the maxima of the carrier oscillations do not coincide with the maximum of the envelope $A(t)$ at $t = 0$ and the maximum value of $E(t)$ is therefore smaller than in a cosine pulse. In general Φ_0 is termed the absolute phase or carrier-envelope phase and determines the temporal relation of the pulse envelope with respect to the underlying carrier oscillation. The absolute phase is not important if the pulse envelope $A(t)$ does not significantly vary within one oscillation period T. The longer the temporal duration of the pulses, the more closely this condition is met and the decomposition of the electric field into an envelope function and a harmonic oscillation with carrier frequency ω_0 (12.1) is meaningful. Conventional pulse characterization methods as described in Sect. 12.3 are not able to measure the absolute value of Φ_0. Furthermore the absolute phase does not remain stable in a conventional femtosecond laser system. Progress in controlling and measuring the absolute phase has been made only recently [12.7–10] and experiments depending on the absolute phase are starting to appear [12.11–13]. In the following we will not emphasize the role of Φ_0 any more.

In general, we may add an additional time dependent phase function $\Phi_a(t)$ to the temporal phase term in (12.1)

$$\Phi(t) = \Phi_0 + \omega_0 t + \Phi_a(t) \tag{12.4}$$

and define the momentary or instantaneous light frequency $\omega(t)$ as

$$\omega(t) = \frac{d\Phi(t)}{dt} = \omega_0 + \frac{d\Phi_a(t)}{dt} . \tag{12.5}$$

This additional phase function describes variations of the frequency in time, called a chirp. In Fig. 12.1c,d $\Phi_a(t)$ is set to be at^2. For $a = 0.15/\mathrm{fs}^2$ we see a linear increase of the frequency in time, called a linear up-chirp. For $a = -0.15/\mathrm{fs}^2$ a linear down-chirped pulse is obtained with a linear decrease of the frequency in time. However, a direct manipulation of the temporal phase cannot be achieved by any electronic device. Note that nonlinear optical processes such as, for example, self-phase modulation (SPM) are able to influence the temporal phase and lead to a change in the frequency spectrum of the pulse. In this chapter we will mainly focus on linear optical effects where the spectrum of the pulse is unchanged and changes in the temporal pulse shape are due to manipulations in the frequency domain (Sect. 12.1.3). Before we start, a more mathematical description of an ultrashort light pulse is presented.

12.1.2 Mathematical Description

For the mathematical description we followed the approaches of [12.4,14–19]. In linear optics the superposition principle holds and the real-valued electric field $E(t)$ of an ultrashort optical pulse at a fixed point in space has the Fourier decomposition into monochromatic waves

$$E(t) = \frac{1}{2\pi} \int_{-\infty}^{\infty} \tilde{E}(\omega) e^{i\omega t} \, d\omega . \tag{12.6}$$

The, in general complex-valued, spectrum $\tilde{E}(\omega)$ is obtained by the Fourier inversion theorem

$$\tilde{E}(\omega) = \int_{-\infty}^{\infty} E(t) e^{-i\omega t} \, dt . \tag{12.7}$$

Since $E(t)$ is real-valued $\tilde{E}(\omega)$ is Hermitian, i. e., obeys the condition

$$\tilde{E}(\omega) = \tilde{E}^*(-\omega) , \tag{12.8}$$

where $*$ denotes complex conjugation. Hence knowledge of the spectrum for positive frequencies is sufficient for a full characterization of a light field without dc component we can define the positive part of the spectrum as

$$\begin{aligned} \tilde{E}^+(\omega) = \tilde{E}(\omega) \quad & \text{for } \omega \geq 0 \quad \text{and} \\ 0 \quad & \text{for } \omega < 0 . \end{aligned} \tag{12.9}$$

The negative part of the spectrum $\tilde{E}^-(\omega)$ is defined as

$$\begin{aligned} \tilde{E}^-(\omega) = \tilde{E}(\omega) \quad & \text{for} \quad \omega < 0 \quad \text{and} \\ 0 \quad & \text{for} \quad \omega \geq 0 . \end{aligned} \tag{12.10}$$

Just as the replacement of real-valued sines and cosines by complex exponentials often simplifies Fourier analysis, so too does the use of complex-valued functions in place of the real electric field $E(t)$. For this purpose we separate the Fourier transform integral of $E(t)$ into two parts. The complex-valued temporal function $E^+(t)$ contains only the positive frequency segment of the spectrum. In communication theory and optics $E^+(t)$ is termed the analytic signal (its complex conjugate is $E^-(t)$ and contains the negative frequency part). By definition $E^+(t)$ and $\tilde{E}^+(\omega)$ as well as $E^-(t)$ and $\tilde{E}^-(\omega)$ are Fourier pairs where only the relations for the

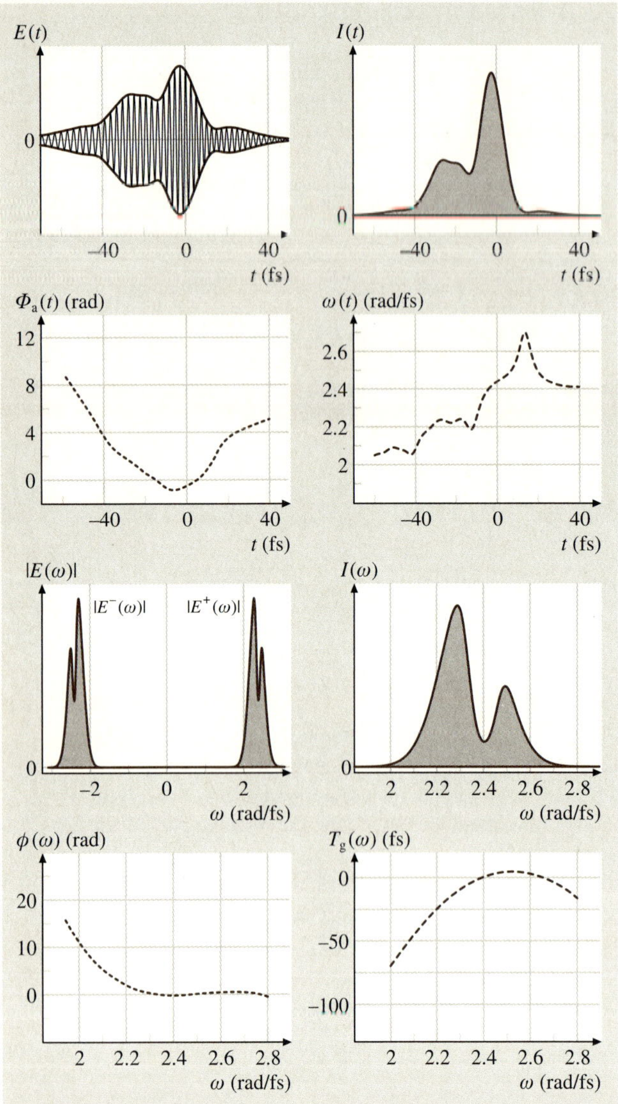

Fig. 12.2 Electric field $E(t)$, temporal intensity $I(t)$, additional temporal phase $\Phi_a(t)$, instantaneous frequency $\omega(t)$, spectrum $|\tilde{E}(\omega)|$, spectral intensity $I(\omega)$, spectral phase $\phi(\omega)$ and group delay $T_g(\omega)$ of a pulse that looks complicated on first glance [having a relatively simple spectral phase $\phi(\omega)$]. When measured with a spectrometer the spectral intensity as a function of wavelength is usually obtained, and the corresponding transformation on the basis of $I(\lambda)\,d\lambda = I(\omega)\,d\omega$ yields $I(\lambda) = -I(\omega)\frac{2\pi c}{\lambda^2}$ where the minus sign indicates the change in the direction of the axis. To avoid phase jumps when the phase exceeds 2π, *phase unwrapping* is employed. That means adding or subtracting 2π to the phase at each discontinuity. When the intensity is close to zero, the phase is meaningless and usually the phase is not plotted in such regions (phase blanking)

positive-frequency part are given as

$$E^+(t) = \frac{1}{2\pi}\int_{-\infty}^{\infty} \tilde{E}^+(\omega)\mathrm{e}^{\mathrm{i}\omega t}\,\mathrm{d}\omega \tag{12.11}$$

$$\tilde{E}^+(\omega) = \int_{-\infty}^{\infty} E^+(t)\mathrm{e}^{-\mathrm{i}\omega t}\,\mathrm{d}t\,. \tag{12.12}$$

These quantities relate to the real electric field

$$\begin{aligned} E(t) &= E^+(t) + E^-(t) \\ &= 2\,\mathrm{Re}\{E^+(t)\} \\ &= 2\,\mathrm{Re}\{E^-(t)\} \end{aligned} \tag{12.13}$$

and its complex Fourier transform

$$\tilde{E}(\omega) = \tilde{E}^+(\omega) + \tilde{E}^-(\omega)\,. \tag{12.14}$$

$E^+(t)$ is complex-valued and can therefore be expressed uniquely in terms of its amplitude and phase

$$\begin{aligned} E^+(t) &= |E^+(t)|\,\mathrm{e}^{\mathrm{i}\Phi(t)} \\ &= |E^+(t)|\,\mathrm{e}^{\mathrm{i}\Phi_0}\,\mathrm{e}^{\mathrm{i}\omega_0 t}\,\mathrm{e}^{\mathrm{i}\Phi_a(t)} \\ &= \sqrt{\frac{I(t)}{2\varepsilon_0 cn}}\,\mathrm{e}^{\mathrm{i}\Phi_0}\,\mathrm{e}^{\mathrm{i}\omega_0 t}\,\mathrm{e}^{\mathrm{i}\Phi_a(t)} \\ &= \frac{1}{2}A(t)\mathrm{e}^{\mathrm{i}\Phi_0}\,\mathrm{e}^{\mathrm{i}\omega_0 t}\,\mathrm{e}^{\mathrm{i}\Phi_a(t)} \\ &= E_c(t)\mathrm{e}^{\mathrm{i}\Phi_0}\,\mathrm{e}^{\mathrm{i}\omega_0 t} \end{aligned} \tag{12.15}$$

where the meaning of $A(t)$, Φ_0, ω_0 and $\Phi_a(t)$ is the same as in Sect. 12.1.1 and $E_c(t)$ is the complex-valued envelope function without the absolute phase and without the rapidly oscillating carrier-frequency phase factor, a quantity often used in ultrafast optics. The envelope function $A(t)$ is given by

$$A(t) = 2|E^+(t)| = 2|E^-(t)| = 2\sqrt{E^+(t)E^-(t)} \tag{12.16}$$

and coincides with the less general expression in (12.1). The complex positive-frequency part $\tilde{E}^+(\omega)$ can be analogously decomposed into amplitude and phase

$$\begin{aligned} \tilde{E}^+(\omega) &= |\tilde{E}^+(\omega)|\,\mathrm{e}^{-\mathrm{i}\phi(\omega)} \\ &= \sqrt{\frac{\pi}{\varepsilon_0 cn}I(\omega)}\,\mathrm{e}^{-\mathrm{i}\phi(\omega)}\,, \end{aligned} \tag{12.17}$$

where $|\tilde{E}^+(\omega)|$ is the spectral amplitude, $\phi(\omega)$ is the spectral phase and $I(\omega)$ is the spectral intensity proportional to the power spectrum density (PSD) – the familiar quantity measured with a spectrometer. From

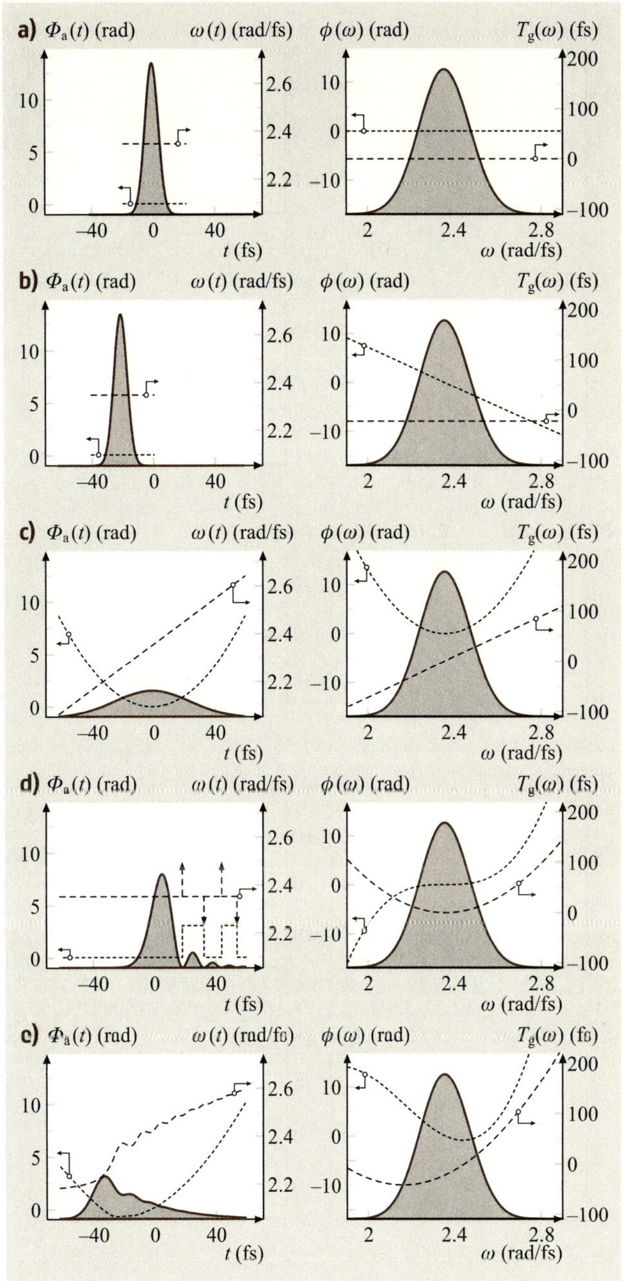

Fig. 12.3a–o Examples for changing the temporal shape of a 800 nm 10 fs pulse via the frequency domain (except **(n)**). *Left*: temporal intensity $I(t)$ (*shaded*), additional temporal phase $\Phi_a(t)$ (*dotted*), instantaneous frequency $\omega(t)$ (*dashed*), *right*: spectral intensity $I(\omega)$ (*shaded*), spectral phase $\phi(\omega)$ (*dotted*) and group delay $T_g(\omega)$ (*dashed*) for a: **(a)** bandwidth-limited Gaussian laser pulse of 10 fs duration; **(b)** bandwidth-limited Gaussian laser pulse of 10 fs duration shifted in time to −20 fs due to a linear phase term in the spectral domain ($\phi' = -20$ fs); **(c)** symmetrical broadened Gaussian laser pulse due to $\phi'' = 200\,\text{fs}^2$; **(d)** third-order spectral phase ($\phi''' = 1000\,\text{fs}^3$) leading to a quadratic group delay. The central frequency of the pulse arrives first, while frequencies on either side arrive later. The corresponding differences in frequencies cause beats in the temporal intensity profile. Pulses with cubic spectral phase distortion have therefore oscillations after (or before) a main pulse depending on the sign of ϕ'''. The higher the side pulses, the less meaningful the FWHM pulse duration; **(e)** combined action of all spectral phase coefficients **(a)**–**(d)**. Phase unwrapping and blanking is employed when appropriate;

(f) π step at the central frequency; **(g)** π step displaced from central frequency; **(h)** sine modulation at central frequency with $\phi(\omega) = 1\sin[20\,\text{fs}(\omega-\omega_0)]$; **(i)** cosine modulation at central frequency with $\phi(\omega) = 1\cos[20\,\text{fs}(\omega-\omega_0)]$; **(j)** sine modulation at central frequency with $\phi(\omega) = 1\sin[30\,\text{fs}(\omega-\omega_0)]$.▶

Amplitude modulation: **(k)** symmetrical clipping of spectrum; **(l)** blocking of central frequency components; **(m)** off center absorption.

Modulation in time domain: **(n)** self-phase modulation. Note the spectral broadening; **(o)** double pulse with pulse to pulse delay of 60 fs ▶▶

(12.8) the relation $-\phi(\omega) = \phi(-\omega)$ is obtained. As will be shown in Sect. 12.1.3 it is precisely the manipulation of this spectral phase $\phi(\omega)$ in the experiment which – by virtue of the Fourier transformation (12.11) – creates changes in the real electric field strength $E(t)$ (12.13) without changing $I(\omega)$. If the spectral intensity $I(\omega)$ is manipulated as well, additional degrees of freedom are accessible for generating temporal pulse shapes at the expense of lower energy.

Note that the distinction between positive- and negative-frequency parts is made for mathematical correctness. In practice only real electric fields and positive frequencies are displayed. Moreover, as usually only the shape and not the absolute magnitude of the envelope functions in addition to the phase function are the quantities of interest, all the prefactors are commonly omitted.

The temporal phase $\Phi(t)$ (12.4) contains frequency-versus-time information, leading to the definition of the instantaneous frequency $\omega(t)$ (12.5). In a similar fashion $\phi(\omega)$ contains time-versus-frequency information and we can define the group delay $T_g(\omega)$, which describes the relative temporal delay of a given spectral component

Table 12.1 Temporal and spectral intensity profiles and time bandwidth products ($\Delta\nu\Delta t \geq K$) of various pulse shapes; $\Delta\nu$ and Δt are FWHM quantities of the corresponding intensity profiles. The ratio $\Delta t_{\text{intAC}}/\Delta t$, where Δt_{intAC} is the FWHM of the intensity autocorrelation with respect to background (Sect. 12.3.2), is also given. In the following formulas employed in the calculations we set $\omega_0 = 0$ for simplicity.

Gaussian: $$E^+(t) = \frac{E_0}{2}\,\mathrm{e}^{-2\ln 2\left(\frac{t}{\Delta t}\right)^2} \qquad \tilde{E}^+(\omega) = \frac{E_0\Delta t}{2}\sqrt{\frac{\pi}{2\ln 2}}\,\mathrm{e}^{-\frac{\Delta t^2}{8\ln 2}\omega^2}\,,$$

Sech: $$E^+(t) = \frac{E_0}{2}\,\mathrm{sech}\left[2\ln\left(1+\sqrt{2}\right)\frac{t}{\Delta t}\right] \qquad \tilde{E}^+(\omega) = E_0\Delta t\frac{\pi}{4\ln\left(1+\sqrt{2}\right)} \times \mathrm{sech}\left(\frac{\pi\Delta t}{4\ln\left(1+\sqrt{2}\right)}\omega\right)\,,$$

Rect: $$E^+(t) = \frac{E_0}{2} \quad t \in \left[-\frac{\Delta t}{2}, \frac{\Delta t}{2}\right],\ 0 \text{ elsewhere} \qquad \tilde{E}^+(\omega) = \frac{E_0\Delta t}{2}\,\mathrm{sinc}\left(\frac{\Delta t}{2}\omega\right)\,,$$

Single sided Exp.: $$E^+(t) = \frac{E_0}{2}\,\mathrm{e}^{-\frac{\ln 2}{2}\frac{t}{\Delta t}} \quad t \in [0, \infty],\ 0 \text{ elsewhere} \qquad \tilde{E}^+(\omega) = \frac{E_0\Delta t}{2\mathrm{i}\Delta t\,\omega + \ln 2}\,,$$

Symmetric Exp.: $$E^+(t) = \frac{E_0}{2}\,\mathrm{e}^{-\ln 2\frac{t}{\Delta t}} \qquad \tilde{E}^+(\omega) = \frac{E_0\Delta t\ln 2}{\Delta t^2\omega^2 + (\ln 2)^2}\,.$$

Shape	$I(t)$	$I(\omega)$	$\Delta\nu\cdot\Delta t$	$\Delta t_{\text{intAC}}/\Delta t$
Gaussian	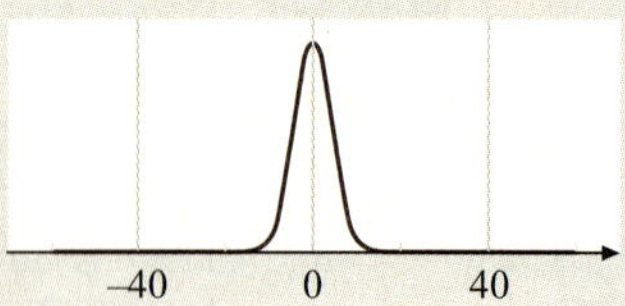	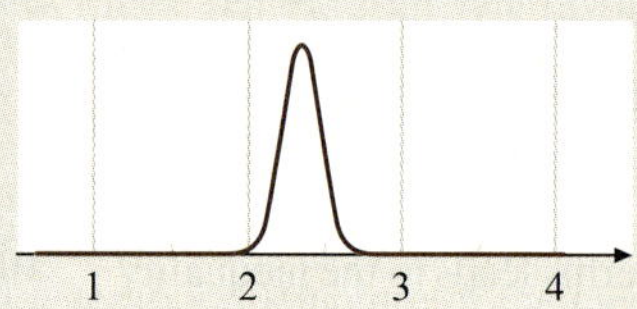	0.441	1.414
Hyperbolic sechant	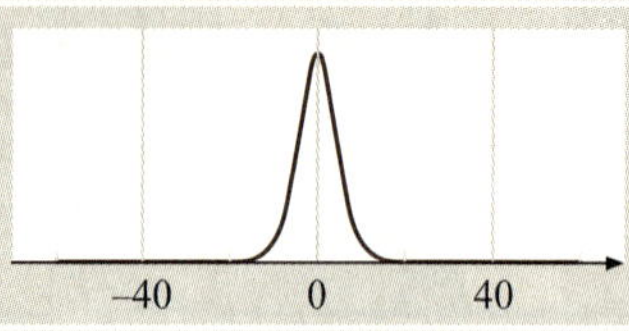	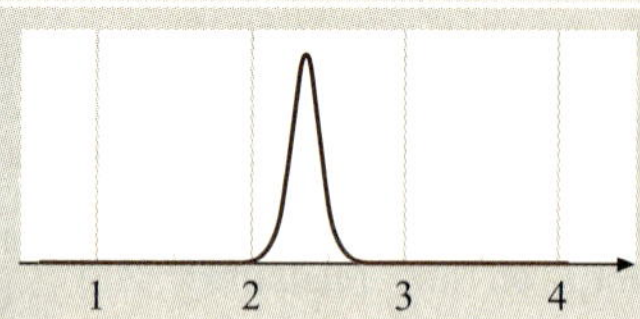	0.315	1.543
Square	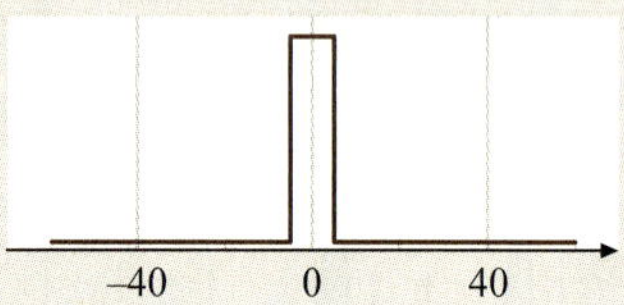	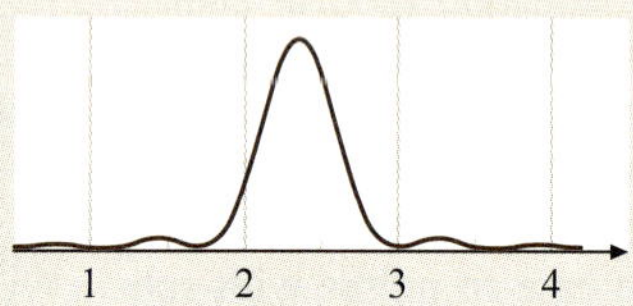	0.886	1.000
Single sided exponential	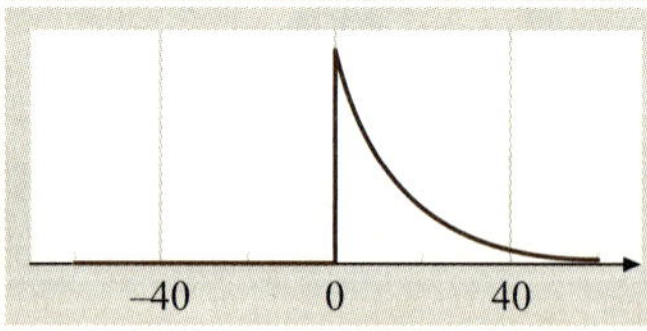	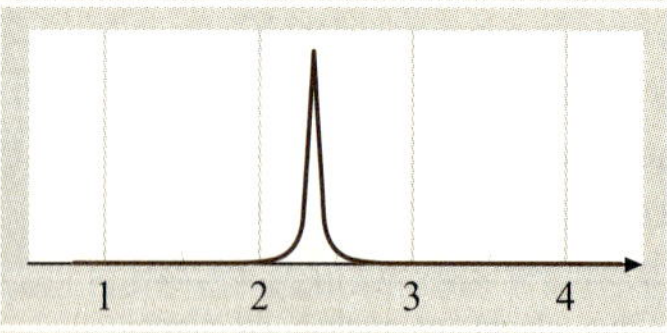	0.110	2.000
Symmetric exponential	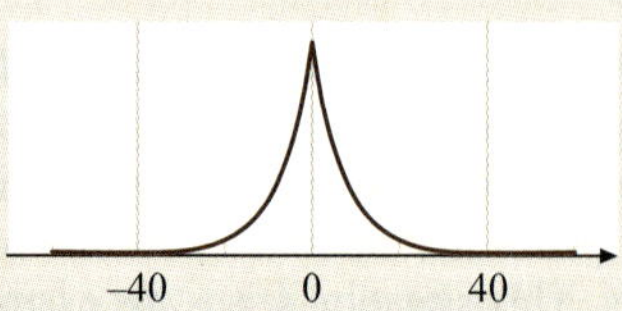	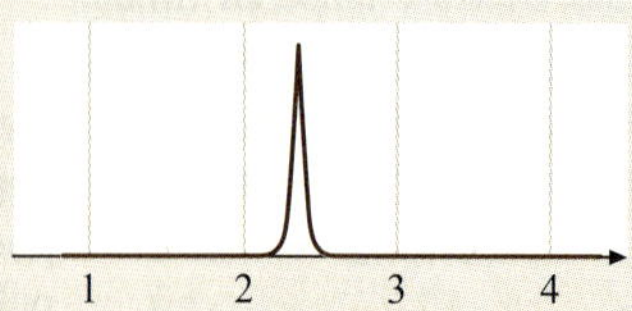	0.142	2.421

Table 12.2 Temporal broadening of a Gaussian laser pulse Δt_{out} in fs for various initial pulse durations Δt and various values of the second-order phase coefficient ϕ'', calculated with the help of (12.26). (The passage of a bandwidth-limited laser pulse at 800 nm through 1 cm of BK7 glass corresponds to $\phi'' = 440\,\text{fs}^2$. For the dispersion parameters of other materials see Table 12.3. Dispersion parameters of further optical elements are given in Sect. 12.1.3.)

	ϕ''						
Δt(fs)	**100 fs^2**	**200 fs^2**	**500 fs^2**	**1000 fs^2**	**2000 fs^2**	**4000 fs^2**	**8000 fs^2**
5	55.7	111.0	277.3	554.5	1109.0	2218.1	4436.1
10	29.5	56.3	139.0	277.4	554.6	1109.1	2218.1
20	24.3	34.2	72.1	140.1	278.0	554.9	1109.2
40	40.6	42.3	52.9	80.0	144.3	280.1	556.0
80	80.1	80.3	81.9	87.2	105.9	160.1	288.6
160	160.0	160.0	160.2	160.9	163.9	174.4	211.7

shown in Sect. 12.1.3. The complex fields for such laser pulses are given by [12.26, 27]

$$\tilde{E}^+_{\text{out}}(\omega) = \frac{E_0 \Delta t}{2} \times \sqrt{\frac{\pi}{2\ln 2}}\, \mathrm{e}^{-\frac{\Delta t^2}{8\ln 2}(\omega-\omega_0)^2}\, \mathrm{e}^{-\mathrm{i}\frac{1}{2}\phi''(\omega_0)(\omega-\omega_0)^2} \tag{12.24}$$

$$E^+_{\text{out}}(t) = \frac{E_0}{2\gamma^{\frac{1}{4}}}\, \mathrm{e}^{-\frac{t^2}{4\beta\gamma}}\, \mathrm{e}^{\mathrm{i}\omega_0 t}\, \mathrm{e}^{\mathrm{i}(at^2-\varepsilon)} \tag{12.25}$$

with

$$\beta = \frac{\Delta t_{\text{in}}^2}{8\ln 2} \quad \gamma = 1 + \frac{\phi''^2}{4\beta^2} \quad a = \frac{\phi''}{8\beta^2\gamma}$$

and

$$\varepsilon = \frac{1}{2}\arctan\left(\frac{\phi''}{2\beta}\right) = -\Phi_0 .$$

For the pulse duration Δt_{out} (FWHM) of the linearly chirped pulse (quadratic temporal phase function at^2) we obtain the convenient formula

$$\Delta t_{\text{out}} = \sqrt{\Delta t^2 + \left(4\ln 2\frac{\phi''}{\Delta t}\right)^2} . \tag{12.26}$$

The statistical definition of the pulse duration derived with the help of the second moment of the intensity distribution uses twice the standard deviation σ to characterize the pulse duration by

$$2\sigma = \frac{\Delta t_{\text{out}}}{\sqrt{2\ln 2}} , \tag{12.27}$$

which is slightly shorter than the FWHM. These values are exact for Gaussian pulses considering only the ϕ'' part and can be used as a first estimate for temporal pulse broadening whenever ϕ'' effects are the dominant contribution (Sect. 12.1.3). Some values of the symmetric pulse broadening due to ϕ'' are given in Table 12.2 and exemplified in Fig. 12.3c.

Spectral phase coefficients of third order, i. e., a contribution to the phase function $\phi(\omega)$ of the form $\frac{1}{6}\phi'''(\omega_0)\cdot(\omega-\omega_0)^3$ are referred to as third-order dispersion (TOD). TOD applied to the spectrum given by (12.21) yields the phase-modulated spectrum

$$\tilde{E}^+_{\text{out}}(\omega) = \frac{E_0 \Delta t}{2} \times \sqrt{\frac{\pi}{2\ln 2}}\, \mathrm{e}^{-\frac{\Delta t^2}{8\ln 2}(\omega-\omega_0)^2}\, \mathrm{e}^{-\mathrm{i}\frac{1}{6}\phi'''(\omega_0)\cdot(\omega-\omega_0)^3} \tag{12.28}$$

and leads to asymmetric temporal pulse shapes [12.27] of the form

$$E^+_{\text{out}}(t) = \frac{E_0}{2}\sqrt{\frac{\pi}{2\ln 2}}\frac{\Delta t}{\tau_0} Ai\left(\frac{\tau - t}{\Delta\tau}\right) \mathrm{e}^{-\frac{\ln 2}{2}\cdot\frac{\frac{2}{3}\tau - t}{\tau_{1/2}}}\, \mathrm{e}^{\mathrm{i}\omega_0 t}$$

$$\text{with } \tau_0 = \sqrt[3]{\frac{|\phi'''|}{2}} \quad \phi^3 = 2(\ln 2)^2\phi'''$$

$$\Delta\tau = \tau_0 \operatorname{sign}(\phi''')$$

$$\tau = \frac{\Delta t^4}{16\phi^3} \quad \text{and} \quad \tau_{1/2} = \frac{\phi^3}{\Delta t^2} \tag{12.29}$$

where Ai describes the Airy function. Equation (12.29) shows that the temporal pulse shape is given by the product of an exponential decay with a half-life period of $\tau_{1/2}$ and the Airy function shifted by τ and stretched by $\Delta\tau$. Figure 12.3d shows an example of a pulse subjected to TOD. The pulse shape is characterized by a strong initial pulse followed by a decaying pulse sequence. Because TOD leads to a quadratic group delay the central frequency of the pulse arrives first, while frequencies on either side arrive later. The corresponding differences in frequencies cause beats in the temporal intensity profile explaining the oscillations after (or before) the main

Table 12.3 Dispersion parameters n, $\frac{dn}{d\lambda}$, $\frac{d^2n}{d\lambda^2}$, $\frac{d^3n}{d\lambda^3}$, T_g, GDD and TOD for common optical materials for $L = 1$ mm. The data were calculated using Sellmeier's equation in the form $n^2(\lambda) - 1 = B_1\lambda^2/(\lambda^2 - C_1) + B_2\lambda^2/(\lambda^2 - C_2) + B_3\lambda^2/(\lambda^2 - C_3)$ and data from various sources (BK7, SF10 from *Schott – Optisches Glas* catalogue; sapphire and quartz from the *Melles Griot* catalogue)

Material	λ (nm)	$n(\lambda)$	$\frac{dn}{d\lambda}\cdot 10^{-2}$ $\left(\frac{1}{\mu m}\right)$	$\frac{d^2n}{d\lambda^2}\cdot 10^{-1}$ $\left(\frac{1}{\mu m^2}\right)$	$\frac{dn^3}{d\lambda^3}$ $\left(\frac{1}{\mu m^3}\right)$	T_g $\left(\frac{fs}{mm}\right)$	GDD $\left(\frac{fs^2}{mm}\right)$	TOD $\left(\frac{fs^3}{mm}\right)$
BK7	400	1.5308	−13.17	10.66	−12.21	5282	120.79	40.57
	500	1.5214	−6.58	3.92	−3.46	5185	86.87	32.34
	600	1.5163	−3.91	1.77	−1.29	5136	67.52	29.70
	800	1.5108	−1.97	0.48	−0.29	5092	43.96	31.90
	1000	1.5075	−1.40	0.15	−0.09	5075	26.93	42.88
	1200	1.5049	−1.23	0.03	−0.04	5069	10.43	66.12
SF10	400	1.7783	−52.02	59.44	−101.56	6626	673.68	548.50
	500	1.7432	−20.89	15.55	−16.81	6163	344.19	219.81
	600	1.7267	−11.00	6.12	−4.98	5980	233.91	140.82
	800	1.7112	−4.55	1.58	−0.91	5830	143.38	97.26
	1000	1.7038	−2.62	0.56	−0.27	5771	99.42	92.79
	1200	1.6992	−1.88	0.22	−0.10	5743	68.59	107.51
Sapphire	400	1.7866	−17.20	13.55	−15.05	6189	153.62	47.03
	500	1.7743	−8.72	5.10	−4.42	6064	112.98	39.98
	600	1.7676	−5.23	2.32	−1.68	6001	88.65	37.97
	800	1.7602	−2.68	0.64	−0.38	5943	58.00	42.19
	1000	1.7557	−1.92	0.20	−0.12	5921	35.33	57.22
	1200	1.7522	−1.70	0.04	−0.05	5913	13.40	87.30
Quartz	300	1.4878	−30.04	34.31	−54.66	5263	164.06	46.49
	400	1.4701	−11.70	9.20	−10.17	5060	104.31	31.49
	500	1.4623	−5.93	3.48	−3.00	4977	77.04	26.88
	600	1.4580	−3.55	1.59	−1.14	4934	60.66	25.59
	800	1.4533	−1.80	0.44	−0.26	4896	40.00	28.43
	1000	1.4504	−1.27	0.14	−0.08	4880	24.71	38.73
	1200	1.4481	−1.12	0.03	−0.03	4875	9.76	60.05

pulse. The beating is also responsible for the phase jumps of π which occur at the zeros of the Airy function. Most of the relevant properties of TOD modulation are determined by the parameter $\Delta\tau$, which is proportional to $\sqrt[3]{|\phi'''|}$. The ratio $\Delta\tau/\Delta t$ determines whether the pulse is significantly modulated. If $|\Delta\tau/\Delta t| \geq 1$, a series of sub-pulses and phase jumps are observed. The sign of ϕ''' controls the time direction of the pulse shape: a positive value of ϕ''' leads to a series of post-pulses as shown in Fig. 12.3d whereas negative values of ϕ''' cause a series of pre-pulses. The time shift of the most intense sub-pulse with respect to the unmodulated pulse and the FWHM of the sub-pulses are on the order of $\Delta\tau$. For these highly asymmetric pulses, the FWHM is not a meaningful quantity to characterize the pulse duration. Instead, the statistical definition of the pulse duration yields a formula similar to (12.26) and (12.27)

$$2\sigma = \sqrt{\frac{\Delta t^2}{2\ln 2} + 8(\ln 2)^2\left(\frac{\phi'''}{\Delta t^2}\right)^2}\,. \tag{12.30}$$

It is a general feature of polynomial phase modulation functions that the statistical pulse duration of a modulated pulse is

$$2\sigma = \sqrt{\tau_1^2 + \tau_2^2}\,, \tag{12.31}$$

where $\tau_1 = \frac{\Delta t}{\sqrt{2\ln 2}}$ is the statistical duration of the unmodulated pulse [cf. (12.21)] and $\tau_2 \propto \frac{\phi^{(n)}}{\Delta t^{n-1}}$ a contribution only dependent on the nth-order spectral phase coefficient. As a consequence, for strongly modulated pulses, when $\tau_2 \gg \tau_1$, the statistical pulse duration increases approximately linearly with $\phi^{(n)}$.

It is not always advantageous to expand the phase function $\phi(\omega)$ into a Taylor series. Periodic phase functions, for example, are generally not well approximated by polynomial functions. For sinusoidal phase functions of the form $\phi(\omega) = A\sin(\omega\Upsilon + \varphi_0)$ analytic solutions for the temporal electric field can be found for any arbitrary unmodulated spectrum $\tilde{E}^+_{\text{in}}(\omega)$. To this end we consider the modulated spectrum (see also Sect. 12.1.3)

$$\tilde{E}^+_{\text{out}}(\omega) = \tilde{E}^+_{\text{in}}(\omega)\,\mathrm{e}^{-\mathrm{i}A\sin(\omega\Upsilon+\varphi_0)}\,, \tag{12.32}$$

where A describes the amplitude of the sinusoidal modulation, Υ the frequency of the modulation function (in units of time) and φ_0 the absolute phase of the sine function. Making use of the Jacobi–Anger identity

$$\mathrm{e}^{-A\sin(\omega\Upsilon+\varphi_0)} = \sum_{n=-\infty}^{\infty} J_n(A)\,\mathrm{e}^{-\mathrm{i}n(\omega\Upsilon+\varphi_0)}\,, \tag{12.33}$$

where $J_n(A)$ describes the Bessel function of the first kind and order n, we rewrite the phase modulation function

$$\tilde{M}(\omega) = \sum_{n=-\infty}^{\infty} J_n(A)\,\mathrm{e}^{-\mathrm{i}n(\omega\Upsilon+\varphi_0)} \tag{12.34}$$

to obtain its Fourier transform

$$M(t) = \sum_{n=-\infty}^{\infty} J_n(A)\,\mathrm{e}^{-\mathrm{i}n\varphi_0}\,\delta(n\Upsilon - t)\,, \tag{12.35}$$

where $\delta(t)$ describes the delta function. Since multiplication in the frequency domain corresponds to convolution in the time domain, the modulated temporal electrical field $E^+_{\text{out}}(t)$ is given by the convolution of the unmodulated field $E^+_{\text{in}}(t)$ with the Fourier transform of the modulation function $M(t)$, i. e., $E^+_{\text{out}}(t) = E^+_{\text{in}}(t) * M(t)$. Making use of (12.35) the modulated field reads

$$E^+_{\text{out}}(t) = \sum_{n=-\infty}^{\infty} J_n(A)\,E^+_{\text{in}}(t - n\Upsilon)\,\mathrm{e}^{-\mathrm{i}n\varphi_0}\,. \tag{12.36}$$

Equation (12.36) shows that sinusoidal phase modulation in the frequency domain produces a sequence of sub-pulses with a temporal separation determined by the parameter Υ and well-defined relative temporal phases controlled by the absolute phase φ_0. Provided the individual sub-pulses are temporally separated, i. e., Υ is chosen to exceed the pulse width, the envelope of each sub-pulse is a (scaled) replica of the unmodulated pulse envelope. The amplitudes of the sub-pulses are given by $J_n(A)$ and can therefore be controlled by the modulation parameter A. Examples of sinusoidal phase modulation are shown in Fig. 12.3h–j. The influence of the absolute phase φ_0 is depicted in Fig. 12.3h and i, whereas Fig. 12.3j shows how separated pulses are obtained by changing the modulation frequency Υ. A detailed description of the effect of sinusoidal phase modulation can be found in [12.28].

12.1.3 Changing the Temporal Shape via the Frequency Domain

For the following discussion it is useful to think of an ultrashort pulse as being composed of groups of quasi-monochromatic waves, that is of a set of much longer wave packets of narrow spectrum all added together coherently. In vacuum the phase velocity $v_{\text{p}} = \omega/k$ and the group velocity $v_{\text{g}} = \mathrm{d}\omega/\mathrm{d}k$ are both constant and equal to the speed of light c, where k denotes the wave number. Therefore an ultrashort pulse – no matter how complicated its temporal electric field is – will maintain its shape upon propagation in vacuum. In the following we will always consider a bandwidth-limited pulse entering an optical system such as, for example, air, lenses, mirrors, prisms, gratings and combinations of these optical elements. Usually these optical systems will introduce dispersion, that is a different group velocity for each group of quasi-monochromatic waves, and consequently the initial short pulse will broaden in time. In this context the group delay $T_{\text{g}}(\omega)$ defined in (12.18) is the transit time for such a group of monochromatic waves through the system. As long as the intensities are kept low, no new frequencies are generated. This is the area of linear optics and the corresponding pulse propagation has been termed linear pulse propagation. It is convenient to describe the passage of an ultrashort pulse through a linear optical system by a complex optical transfer function [12.4, 25, 29]

$$\tilde{M}(\omega) = \tilde{R}(\omega)\,\mathrm{e}^{-\mathrm{i}\phi_{\text{d}}}\,, \tag{12.37}$$

that relates the incident electric field $\tilde{E}^+_{\text{in}}(\omega)$ to the output field

$$\tilde{E}^+_{\text{out}}(\omega) = \tilde{M}(\omega)\tilde{E}^+_{\text{in}}(\omega) = \tilde{R}(\omega)\,\mathrm{e}^{-\mathrm{i}\phi_{\text{d}}}\,\tilde{E}^+_{\text{in}}(\omega)\,, \tag{12.38}$$

where $\tilde{R}(\omega)$ is the real-valued spectral amplitude response describing for example the variable diffraction efficiency of a grating, linear gain or loss or direct amplitude manipulation. The phase $\phi_{\text{d}}(\omega)$ is termed the spectral phase transfer function. This is the phase ac-

Part C | 12.1

cumulated by the spectral component of the pulse at frequency ω upon propagation between the input and output planes that define the optical system. It is this spectral phase transfer function that plays a crucial role in the design of ultrafast optical systems. Note that this approach is more involved when additional spatial coordinates have to be taken into account as, for example, in the case of *spatial chirp* (i.e., each frequency is displaced in the transverse spatial coordinates). Neglecting spatial chirp this approach can be taken as a first-order analysis of ultrafast optical systems. Although inside an optical system this condition might not be met, usually at the input and output all frequencies are assumed to be spatially overlapped for this kind of analysis. Note also that the independence of the different spectral components in this picture does not mean that the phase relations are random – they are uniquely defined with respect to each other. That means that the corresponding pulse in the time domain (by making use of (12.11) and (12.13)) is completely coherent [12.30] no matter how complicated the shape of the femtosecond laser pulse appears. In the first-order autocorrelation function a coherence time of the corresponding bandwidth-limited pulse would be observed. Only in the higher-order autocorrelations the uniquely defined phase relations show up (examples of second-order autocorrelations for phase- and amplitude-shaped laser pulses are given in Fig. 12.28). Figure 12.3f–j exemplifies the temporal intensity, spectral intensity and related phase functions for often employed phase functions. Figure 12.3k–m displays the same quantities for amplitude modulation. Figure 12.3n is an example for self-phase modulation and Fig. 12.3o shows a double pulse with pulse to pulse delay of 60 fs.

In the following discussion we will concentrate mainly on pure phase modulation and therefore set $\tilde{R}(\omega)$ constant for all frequencies and omit it initially. To model the system the most accurate approach is to include the whole spectral phase transfer function. Often however only the first orders of a Taylor expansion around the central frequency ω_0 are needed.

$$\begin{aligned}\phi_\mathrm{d}(\omega) &= \phi_\mathrm{d}(\omega_0)+\phi_\mathrm{d}'(\omega_0)(\omega-\omega_0)\\ &\quad+\frac{1}{2}\phi_\mathrm{d}''(\omega_0)(\omega-\omega_0)^2\\ &\quad+\frac{1}{6}\phi_\mathrm{d}'''(\omega_0)(\omega-\omega_0)^3+\ldots \end{aligned} \tag{12.39}$$

If we describe the incident bandwidth-limited pulse by $\tilde{E}^+_\mathrm{in}(\omega)=|\tilde{E}^+(\omega)|\,\mathrm{e}^{-\mathrm{i}\phi(\omega_0)}\,\mathrm{e}^{-\mathrm{i}\phi'(\omega_0)(\omega-\omega_0)}$ then the overall phase ϕ_op of $\tilde{E}^+_\mathrm{out}(\omega)$ is given by

$$\begin{aligned}\phi_\mathrm{op}(\omega) &= \phi(\omega_0)+\phi'(\omega_0)(\omega-\omega_0)\\ &\quad+\phi_\mathrm{d}(\omega_0)+\phi_\mathrm{d}'(\omega_0)(\omega-\omega_0)\\ &\quad+\frac{1}{2}\phi_\mathrm{d}''(\omega_0)(\omega-\omega_0)^2\\ &\quad+\frac{1}{6}\phi_\mathrm{d}'''(\omega_0)(\omega-\omega_0)^3+\ldots \end{aligned} \tag{12.40}$$

As discussed in the context of (12.19) the constant and linear terms do not lead to a change of the temporal envelope of the pulse. Therefore we will omit in the following these terms and concentrate mainly on the second-order dispersion ϕ'' [also termed the group velocity dispersion (GVD) or group delay dispersion (GDD)] and the third-order dispersion ϕ''' (TOD) whereas we have omitted the subscript d. Strictly they have units of [$\mathrm{fs}^2/\mathrm{rad}$] and [$\mathrm{fs}^3/\mathrm{rad}^2$], respectively, but usually the units are simplified to [fs^2] and [fs^3].

A main topic in the design of ultrafast laser systems is the minimization of these higher dispersion terms with the help of suitably designed optical systems to keep the pulse duration inside a laser cavity or at the place of an experiment as short as possible. In the following we will discuss the elements that are commonly used for the *dispersion management*.

Dispersion due to Transparent Media

A pulse traveling a distance L through a medium with index of refraction $n(\omega)$ accumulates the spectral phase

$$\phi_\mathrm{m}(\omega)=k(\omega)L=\frac{\omega}{c}n(\omega)L\,, \tag{12.41}$$

which is the spectral transfer function due to propagation in the medium as defined above.

The first derivative

$$\frac{\mathrm{d}\phi_\mathrm{m}}{\mathrm{d}\omega}=\phi_\mathrm{m}'=\frac{\mathrm{d}(kL)}{\mathrm{d}\omega}=L\left(\frac{\mathrm{d}\omega}{\mathrm{d}k}\right)^{-1}=\frac{L}{v_\mathrm{g}}=T_\mathrm{g} \tag{12.42}$$

yields the group delay T_g and describes the delay of the peak of the envelope of the incident pulse. Usually the index of refraction $n(\omega)$ is given as a function of wavelength λ, i.e., $n(\lambda)$. Equation (12.42) then reads

$$T_\mathrm{g}=\frac{\mathrm{d}\phi_\mathrm{m}}{\mathrm{d}\omega}=\frac{L}{c}\left(n+\omega\frac{\mathrm{d}n}{\mathrm{d}\omega}\right)=\frac{L}{c}\left(n-\lambda\frac{\mathrm{d}n}{\mathrm{d}\lambda}\right)\,. \tag{12.43}$$

As different groups of the quasi-monochromatic waves move with different group velocities the pulse will be broadened. For second-order dispersion we obtain the group delay dispersion (GDD)

$$\mathrm{GDD} = \phi''_\mathrm{m} = \frac{\mathrm{d}^2\phi_\mathrm{m}}{\mathrm{d}\omega^2} = \frac{L}{c}\left(2\frac{\mathrm{d}n}{\mathrm{d}\omega} + \omega\frac{\mathrm{d}^2 n}{\mathrm{d}\omega^2}\right) = \frac{\lambda^3 L}{2\pi c^2}\frac{\mathrm{d}^2 n}{\mathrm{d}\lambda^2}\,. \tag{12.44}$$

For ordinary optical glasses in the visible range we encounter normal dispersion, i. e., red parts of the laser pulse will travel faster through the medium than blue parts. So the symmetric temporal broadening of the pulse due to ϕ'' will lead to a linearly up-chirped laser pulse as discussed in the context of (12.19) and Fig. 12.3c. In these cases the curvature of $n(\lambda)$ is positive (upward concavity) emphasizing the terminology that positive GDD leads to up-chirped pulses.

For the third-order dispersion (TOD) we obtain

$$\mathrm{TOD} = \phi'''_\mathrm{m} = \frac{\mathrm{d}^3\phi_\mathrm{m}}{\mathrm{d}\omega^3} = \frac{L}{c}\left(3\frac{\mathrm{d}^2 n}{\mathrm{d}\omega^2} + \omega\frac{\mathrm{d}^3 n}{\mathrm{d}\omega^3}\right) = \frac{-\lambda^4 L}{4\pi^2 c^3}\left(3\frac{\mathrm{d}^2 n}{\mathrm{d}\lambda^2} + \lambda\frac{\mathrm{d}^3 n}{\mathrm{d}\lambda^3}\right). \tag{12.45}$$

Empirical formulas for $n(\lambda)$ such as Sellmeier's equations are usually tabulated for common optical materials so that all dispersion quantities in (12.43–12.45) can be calculated. Parameters for some optical materials are given in Table 12.3 for $L = 1$ mm.

Note that in fiber optics a slightly different terminology is used [12.25]. There the second-order dispersion is the dominant contribution to pulse broadening. The β parameter of a fiber is related to the second-order dispersion by

$$\beta = \frac{\left.\frac{\mathrm{d}^2\phi_\mathrm{m}}{\mathrm{d}\omega^2}\right|_{\omega_0}}{L}\left[\frac{\mathrm{ps}^2}{\mathrm{km}}\right], \tag{12.46}$$

where L denotes the length of the fiber. The dispersion parameter D is a measure for the group delay dispersion per unit bandwidth and is given by

$$D = \frac{\omega_0^2}{2\pi c}|\beta|\left[\frac{\mathrm{ps}}{\mathrm{nm\,km}}\right]. \tag{12.47}$$

Angular Dispersion

Transparent media in the optical domain possess positive group delay dispersion leading to up-chirped femtosecond pulses. To compress these pulses, optical systems are needed that deliver negative group delay dispersion, that is systems where the blue spectral components travel faster than the red spectral components. Convenient devices for that purpose are based on angular dispersion delivered by prisms and gratings. We start our discussion again with the spectral transfer function [12.4]

$$\phi(\omega) = \frac{\omega}{c}P_\mathrm{op}(\omega)\,, \tag{12.48}$$

where P_op denotes the optical path length. Equation (12.48) is the generalization of (12.41). The group delay dispersion is given by

$$\frac{\mathrm{d}^2\phi}{\mathrm{d}\omega^2} = \frac{1}{c}\left(2\frac{\mathrm{d}P_\mathrm{op}}{\mathrm{d}\omega} + \omega\frac{\mathrm{d}^2 P_\mathrm{op}}{\mathrm{d}\omega^2}\right) = \frac{\lambda^3}{2\pi c^2}\frac{\mathrm{d}^2 P_\mathrm{op}}{\mathrm{d}\lambda^2} \tag{12.49}$$

and is similar to (12.44). In a dispersive system the optical path from an input reference plane to an output reference plane can be written

$$P_\mathrm{op} = l\cos\alpha\,, \tag{12.50}$$

where $l = l(\omega_0)$ is the distance from the input plane to the output plane for the center frequency ω_0 and α is the angle of rays with frequency ω with respect to the ray corresponding to ω_0. In general, it can be shown [12.4] that the angular dispersion produces negative group delay dispersion

$$\frac{\mathrm{d}^2\phi}{\mathrm{d}\omega^2} \approx -\frac{l\omega_0}{c}\left(\left.\frac{\mathrm{d}\alpha}{\mathrm{d}\omega}\right|_{\omega_0}\right)^2. \tag{12.51}$$

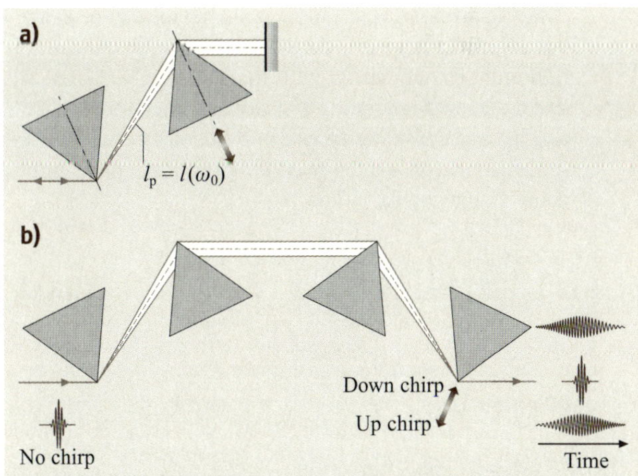

Fig. 12.4a,b Prism sequences for adjustable group delay dispersion. **(a)** Two-prism sequence in double-pass configuration **(b)** four-prism sequence. Note that the spatial distribution of the frequency (spatial chirp) after the second prism can be exploited for phase and/or amplitude manipulations

For pairs of elements (prisms or gratings) the first element provides the angular dispersion and the second element recollimates the spectral components (see, for example, Fig. 12.4). Using two pairs of elements permits the lateral displacement of the spectral components (spatial chirp) to be canceled out and recovers the original beam profile.

Prism Sequences. Prism pairs [12.31] are well suited to introduce adjustable group delay dispersion (Fig. 12.4). Negative group delay dispersion is obtained via the angular dispersion of the first prism where the second prism is recollimating the beam. Recovering the original beam can be accomplished by either using a second pair of prisms or by using a mirror. Inside a laser cavity one can use either the four-prism arrangement or the two-prism arrangement for linear cavities together with a retroreflecting mirror. Outside a laser cavity the two-prism arrangement is often used, where the retroreflecting mirror is slightly off-axis to translate the recovered beam at the entrance of the system to be picked of by an additional mirror. There is also positive group delay dispersion in the system due to the material dispersion of the actual glass path the laser beam takes through the prism sequence. By translating one of the prisms along its axis of symmetry it is possible to change the amount of glass and therefore the amount of positive group delay dispersion. These devices allow a convenient continuous tuning of group delay dispersion from negative to positive values without beam deviation. The negative group delay dispersion via the angular dispersion can be calculated with the help of (12.48) and (12.50). In the case of minimum deviation and with the apex angle chosen so that the Brewster condition is satisfied (minimum reflection losses), the spectra phase introduced by a four-prism sequence $\phi_p(\omega)$ can be used to approximate the group delay dispersion by [12.4]

$$\frac{\mathrm{d}^2\phi_\mathrm{p}}{\mathrm{d}\omega^2} \approx -\frac{4l_\mathrm{p}\lambda^3}{\pi c^2}\left(\frac{\mathrm{d}n}{\mathrm{d}\lambda}\right)^2 \tag{12.52}$$

and the corresponding third-order dispersion yields approximately

$$\frac{\mathrm{d}^3\phi_\mathrm{p}}{\mathrm{d}\omega^3} \approx \frac{6l_\mathrm{p}\lambda^4}{\pi^2 c^3}\frac{\mathrm{d}n}{\mathrm{d}\lambda}\left(\frac{\mathrm{d}n}{\mathrm{d}\lambda}+\lambda\frac{\mathrm{d}^2n}{\mathrm{d}\lambda^2}\right) . \tag{12.53}$$

In order to determine the total GDD and TOD of the four-prism sequence the corresponding contributions of the cumulative mean glass path L (12.44, 12.45) have to be added

$$\begin{aligned}\frac{\mathrm{d}^2\phi_\text{four-prism}}{\mathrm{d}\omega^2} &\approx \frac{\mathrm{d}^2\phi_\mathrm{m}}{\mathrm{d}\omega^2}+\frac{\mathrm{d}^2\phi_\mathrm{p}}{\mathrm{d}\omega^2}\\ &= \frac{\lambda^3 L}{2\pi c^2}\frac{\mathrm{d}^2n}{\mathrm{d}\lambda^2}-\frac{4l_\mathrm{p}\lambda^3}{\pi c^2}\left(\frac{\mathrm{d}n}{\mathrm{d}\lambda}\right)^2 ,\end{aligned} \tag{12.54}$$

$$\begin{aligned}\frac{\mathrm{d}^3\phi_\text{four-prism}}{\mathrm{d}\omega^3} &\approx \frac{\mathrm{d}^3\phi_\mathrm{m}}{\mathrm{d}\omega^3}+\frac{\mathrm{d}^3\phi_\mathrm{p}}{\mathrm{d}\omega^3}\\ &= \frac{-\lambda^4 L}{4\pi^2 c^3}\left(3\frac{\mathrm{d}^2n}{\mathrm{d}\lambda^2}+\lambda\frac{\mathrm{d}^3n}{\mathrm{d}\lambda^3}\right)\\ &\quad+\frac{6l_\mathrm{p}\lambda^4}{\pi^2 c^3}\frac{\mathrm{d}n}{\mathrm{d}\lambda}\left(\frac{\mathrm{d}n}{\mathrm{d}\lambda}+\lambda\frac{\mathrm{d}^2n}{\mathrm{d}\lambda^2}\right) .\end{aligned} \tag{12.55}$$

For a more-detailed discussion and other approaches to the derivation of the total GDD and TOD for a prism sequence see [12.31–35].

In principle one can get any amount of negative group velocity using this method. However a prism distance exceeding 1 m is often impractical. Higher amounts of positive group delay dispersion might be compensated for by the use of highly dispersive SF10 prisms but the higher third-order contribution prevent the generation of ultrashort pulses in the 10 fs regime. Fused quartz is a suitable material for ultrashort pulse generation with minimal higher-order dispersion. For example a four-prism sequence with $l_\mathrm{p} = 50$ cm of fused quartz used at 800 nm yields roughly $\frac{\mathrm{d}^2\phi_\mathrm{p}}{\mathrm{d}\omega^2} \approx -1000\,\mathrm{fs}^2$. Estimating a cumulative glass path of $L = 8$ mm when going through the apexes of the prisms yields $\frac{\mathrm{d}^2\phi_\mathrm{m}}{\mathrm{d}\omega^2} \approx 300\,\mathrm{fs}^2$. In this way a maximum group delay dispersion of +700 fs^2 can be compensated.

Note that in such prism sequences the spatial distribution of the frequency components after the second prism can be exploited. Simple apertures can be used to tune the laser or to restrict the bandwidth. Appropriate phase or amplitude masks might be inserted as well.

Grating Arrangements. Diffraction gratings provide group delay dispersion in a similar manner to prisms. Suitable arrangements can introduce positive as well as negative group delay dispersion (see below). When introducing negative group delay dispersion the corresponding device is termed a *compressor*, while a device introducing positive group delay dispersion is termed a *stretcher*. Grating arrangements have the advantage of being much more dispersive but the disadvantage of introducing higher losses than prism arrangements. As intracavity elements they are used, for example, in high-gain fiber lasers. Outside laser cavities gratings are widely used:

- To compensate for large amounts of dispersion in optical fibers;
- For ultrashort pulse amplification up to the petawatt regime with a technique called chirped pulse amplification (CPA) [12.36]: In order to avoid damage to the optics and to avoid nonlinear distortion of the spatial and temporal profile of the laser beam, ultrashort pulses (10 fs to 1 ps) are typically stretched in time by a factor of 10^3–10^4 prior to injection into the amplifier. After the amplification process the pulses have to be recompressed, compensating also for additional phase accumulated during the amplification process. The topic is reviewed in [12.37];
- For pulse-shaping applications (Sect. 12.1.3);
- In combination with prism compressors to compensate third-order dispersion terms in addition to the group delay dispersion [12.38]. This was the combination employed to establish the long standing world record of 6 fs with dye lasers in 1987 [12.39].

In Fig. 12.5 the reflection of a laser beam from a grating is displayed. The spectrum of an ultrashort laser pulse will be decomposed after reflection into the first order according to the grating equation

$$\sin(\gamma)+\sin(\theta)=\frac{\lambda}{d}\,, \tag{12.56}$$

where γ is the angle of incidence, θ is the angle of the reflected wavelength component and d^{-1} is the grating constant. Blazed diffraction gratings have maximum transmission efficiency when employed in the Littrow

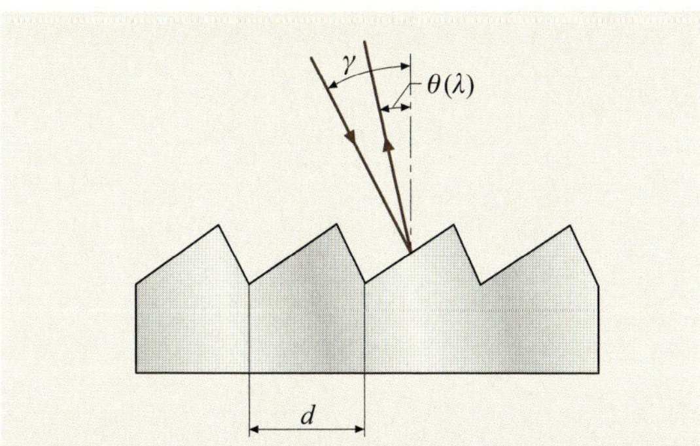

Fig. 12.5 Reflection from a grating: the spectrum of an ultrashort laser pulse will be decomposed after reflection (γ = angle of incidence, $\theta(\lambda)$ = angle of reflection, d^{-1} = grating constant)

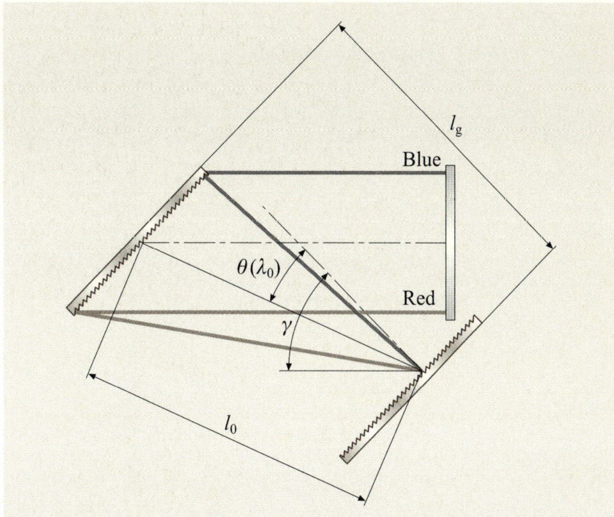

Fig. 12.6 Grating compressor with parallel gratings and a mirror for beam inversion (cf. the corresponding prism setup in Fig. 12.4a). The red spectral components travel a longer optical path than the blue spectral components (l_g denotes the distance between the grating; l_0 denotes the optical path for the center wavelength λ_0 between two gratings; both lengths are used by different authors in deriving the group delay dispersion and the third order dispersion)

configuration, i.e., $\gamma=\theta(\lambda_0)=$ blaze angle. This has the additional advantage that astigmatism is minimized. Blazed gold gratings with an efficiency of 90–95% are commercially available with a damage threshold of $>250\,\mathrm{mJ/cm^2}$ for a 1 ps pulse. For higher efficiency and higher damage threshold dielectric gratings have been developed, for example, dielectric gratings with 98% efficiency at 1053 nm and a damage threshold $>500\,\mathrm{mJ/cm^2}$ for fs pulses are available.

A basic grating compressor (Fig. 12.6) consists of two parallel gratings in a double-pass configuration [12.40]. The first grating decomposes the ultrashort laser pulse into its spectral components. The second grating is recollimating the beam. The original beam is recovered by use of a mirror that inverts the beam. As in such a device the red spectral components experience a longer optical path than the blue spectral components, such an arrangement is suitable to compensate for material dispersion.

In Fig. 12.7 different grating configurations are displayed that produce: (a) zero, (b) positive and (c) negative group delay dispersion. Between the gratings an additional telescope is employed.

In Fig. 12.7a, a so-called zero-dispersion compressor is depicted. The system consists of a telescope that

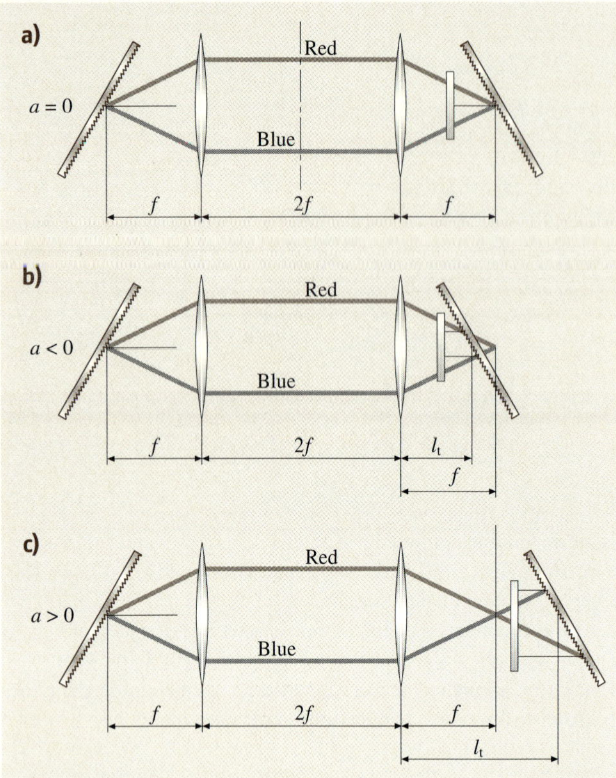

Fig. 12.7a–c Different grating configurations that produce **(a)** zero, **(b)** positive and **(c)** negative group delay dispersion. Arrangement **(a)** corresponds to a zero-dispersion compressor, **(b)** to a stretcher and **(c)** to a compressor. The zero-dispersion compressor is often used in pulse-shaping devices. The *dashed line* in **(a)** indicates the Fourier-transform plane, whereas the stretcher and compressor are key components for chirped pulse amplification

images the laser spot on the first grating onto the second grating. All wavelength components experience the same optical path. In this manner zero net dispersion is obtained. Due to the finite beam size on the grating the components belonging to the same wavelength emerge as a parallel beam and are focused with the lens of focal length f spectrally into the symmetry plane thus providing a Fourier transform plane for pulse shaping, masking or encoding (Sect. 12.1.3 and Fig. 12.11).

Translating one of the gratings out of the focal plane closer to the telescope (Fig. 12.7b) results in an arrangement where the red components travel along a shorter optical path. The device introduces positive group delay dispersion (stretcher).

A compressor is realized by translating the grating away from the focal plane. (Fig. 12.7c).

The dispersion can be further modified by the use of a magnifying telescope. In order to avoid material dispersion in the lenses and to minimize aberration effects, reflective telescopes and especially Öffner telescopes are usually employed [12.41,42].

The phase transfer function ϕ_g for these arrangements can be calculated with the help of a matrix formalism [12.43] and considering the case of finite beam size [12.44].

For a reflective setup (neglecting material dispersion) the group delay dispersion and the third-order dispersion of the three telescope arrangements (magnification = 1) in Fig. 12.7 can be described using a characteristic length L

$$\frac{d^2\phi_g}{d\omega^2} = -\frac{\lambda^3}{\pi c^2 d^2} \frac{1}{\cos[\theta(\lambda)]^2} L\,, \tag{12.57}$$

$$\frac{d^3\phi_g}{d\omega^3} = \frac{d^2\phi_g}{d\omega^2} \frac{3\lambda}{2\pi c} \left\{ 1 + \frac{\lambda}{d} \frac{\tan[\theta(\lambda)]}{\cos[\theta(\lambda)]} \right\}. \tag{12.58}$$

With the help of the grating equation (12.56) $\cos[\theta(\lambda)]$ is given by:

$$\cos[\theta(\lambda)] = \sqrt{1 - \left(\frac{\lambda}{d} - \sin\gamma\right)^2}\,. \tag{12.59}$$

In reflective telescope setups usually only one grating is employed using suitable beam-folding arrangements. This reflects the situation when both gratings in Fig. 12.7 are moved out of the focal plane symmetrically. For the telescope arrangements we therefore obtain as the characteristic length $L = 2fa$. According to Fig. 12.7 the parameter a is determined by the distance of the grating to the lens

$$a = \frac{l_t}{f} - 1 \begin{cases} \text{Compressor: } l_t > f,\ a > 0 \\ \text{Zero-dispersion compressor: : } l_t = f,\ a = 0 \\ \text{Stretcher: } l_t < f,\ a < 0\,. \end{cases} \tag{12.60}$$

For the grating compressor depicted in Fig. 12.6 the characteristic length L is given by

$$L = l_0 = \frac{l_g}{\sqrt{1 - \left(\frac{\lambda}{d} - \sin(\gamma)\right)^2}}\,, \tag{12.61}$$

where l_0 is the optical path length of the center wavelength λ_0 between the gratings and l_g is the distance of the gratings.

For the compressor in Fig. 12.6 we obtain a group delay dispersion of $-1\times10^6\,\text{fs}^2$ [$\lambda = 800\,\text{nm}$, $d^{-1} =$

1200 l/mm, $l_0 = 300$ mm; $\gamma = 28, 6°$ (Littrow)] being orders of magnitude higher than the example given for the prism sequence.

Dispersion due to Interference (Gires–Tournois Interferometers and Chirped Mirrors)

The physics behind dispersion due to interference can be illustrated in the following way [12.25]. Periodic structures transmit or reflect waves of certain frequencies. Strong Bragg-type scattering usually occurs for wavelengths comparable to the periodicity of the structure. In this context the periodicity induces a resonance in the transfer function of the system, which then has dispersion associated with it.

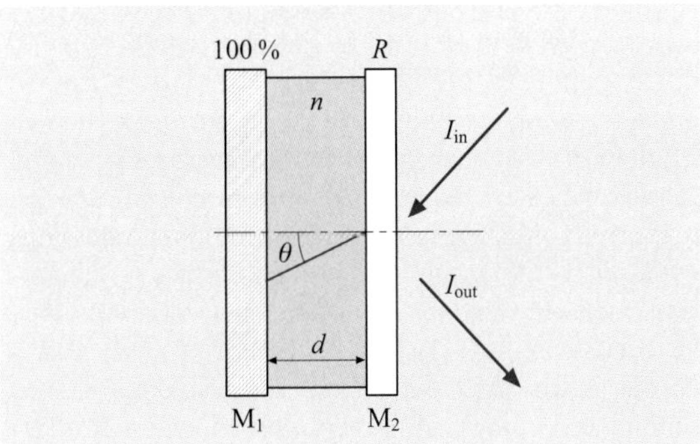

Fig. 12.8 Schematic diagram of a Gires–Tournois interferometer (GTI)

A Gires–Tournois interferometer (GTI) [12.45] is a special case of a Fabry–Pérot interferometer in which one mirror (M1) is a 100% reflector and the top mirror (M2) is a low reflector, typically with a reflectivity of a few percent (Fig. 12.8). The group delay dispersion of such a device is given by (see for example [12.46] or [12.3] and references therein)

$$\frac{d^2\phi_{GTI}}{d\omega^2} = \frac{-2t_0^2(1-R)\sqrt{R}\sin\omega t_0}{(1+R-2\sqrt{R}\cos\omega t_0)^2} . \qquad (12.62)$$

where $t_0 = (2nd\cos\theta)/c$ is the round-trip time of the Fabry-Pérot [12.47], n is the refractive index of the material between the two layers, d is the thickness of the interferometer and θ is the internal angle of the beam between the layers. In this formula material dispersion is neglected and R is the intensity reflectivity of the top reflector. The group delay dispersion can be conveniently tuned either by tilting the device or by changing the interferometer spacing. Increasing t_0 increases the dispersion, but at the same time reduces the frequency range over which the group delay dispersion is constant. These devices are typically used in applications employing pulses larger than 100 fs. For picosecond pulses

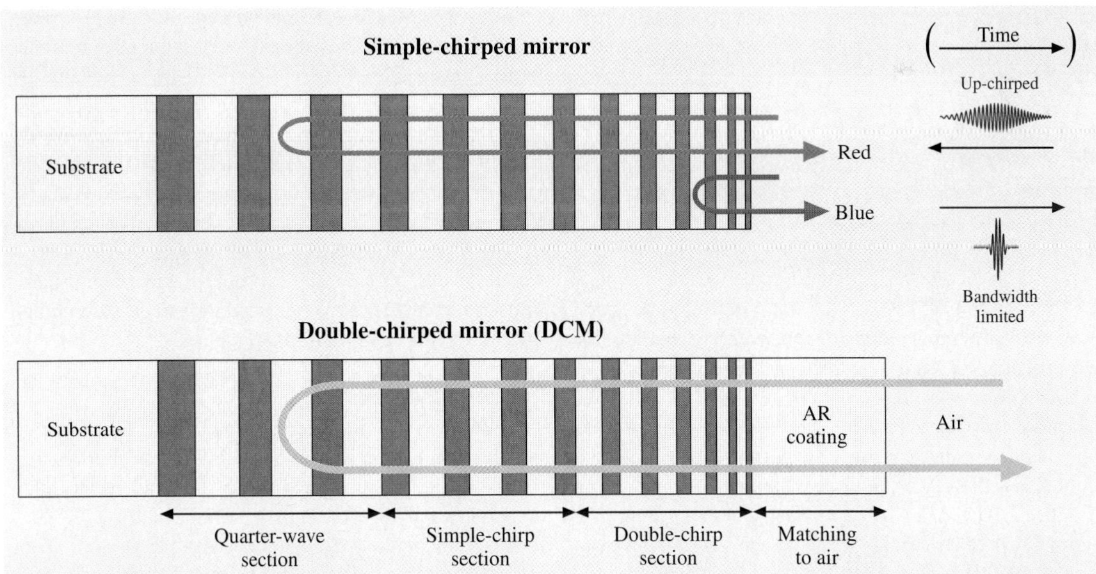

Fig. 12.9a,b Schematic of different types of chirped mirrors: **(a)** simple chirped mirror; the wavelength-dependent penetration depth is depicted. For a proper design, for example, an incoming up-chirped laser pulse can be transformed into a bandwidth-limited pulse after reflection. **(b)** double chirped mirror; impedance matching by an additional antireflection coating on top of the mirror and by a duty-cycle modulation inside the mirror

the mirror spacing is on the order of several mm, for femtosecond lasers the spacing has to be on the order of a few μm. In order to overcome the limitations for femtosecond applications, GTIs were constructed on the basis of dielectric multilayer systems [12.48]. The corresponding spectral transfer functions can be found in [12.4].

Nowadays specially designed dielectric multilayer mirrors offer a powerful alternative for dispersion management. Usually a dielectric mirror consists of alternating transparent pairs of high-index and low-index layers where the optical thickness of all layers is chosen to equal to 1/4 of the Bragg wavelength λ_B. Interference of the reflections at the index discontinuities add up constructively for the Bragg wavelength. If the optical thickness of the layers along the mirror structure is varied, then the Bragg wavelength depends on the penetration depth. Figure 12.9 shows an example where the red wavelength components penetrate deeper into the mirror structure then the blue wavelength components. An up-chirped pulse impinging on the mirror surface can be transformed into a bandwidth-limited pulse after reflection from this mirror. A gradual increase of the Bragg wavelength along the mirror producing a negative group delay dispersion was demonstrated by [12.49] and the corresponding mirror was termed a chirped mirror, allowing for the construction of compact femtosecond oscillators [12.50]. Of course the Bragg wavelength does not have to be varied linearly with the penetration depth. In principle chirp laws $\lambda_B(z)$ can be found for compensation of higher-order dispersion in addition. It was realized, that the desired dispersion characteristics of the chirped mirrors can be spoiled by spurious effects originating from multiple reflections within the coating stack and at the interface to the ambient medium, leading to dispersion oscillations (see the discussion on GTI). An exact coupled-mode analysis [12.51] was used to develop a so-called double-chirp technique in combination with a broadband antireflection coating, in order to avoid the oscillations in the group delay dispersion. Using accurate analytical expressions double chirped mirrors could be designed and fabricated with a smooth and custom-tailored group delay dispersion [12.52] suitable for generating pulses in the two cycle regime directly from a Ti:sapphire laser [12.53]. Double chirping has the following meaning: in conventional chirped mirrors, equal optical lengths of high-index (hi) and low-index (lo) material within one period are employed, i. e., $P_{lo} = P_{hi} = \lambda_B/4$. Double chirping keeps the duty cycle η as an additional degree of freedom under the constraint: $P_{lo} + P_{hi} = (1-\eta)\lambda_B/2 + \eta\lambda_B/2 = \lambda_B/2$. Dispersion oscillations could further be suppressed by a back-side-coated double-mirror design [12.54].

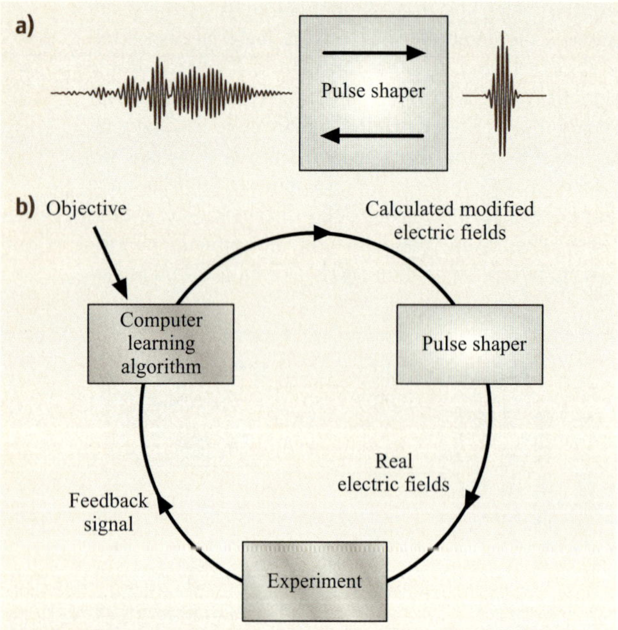

Fig. 12.10 (a) Pulse-shaping issues (schematic): creation of bandwidth-limited pulses from complex-structured pulses (*left to right*). Creation of *tailored* pulse shapes (*right to left*). **(b)** Adaptive femtosecond pulse shaping: a femtosecond laser system (not indicated) and a computer-controlled pulse shaper are used to generate specific electric fields that are sent into an experiment. After deriving a suitable feedback signal from the experiment a learning algorithm calculates a modified electric fields based on the information from the experimental feedback signal and the user-defined control objective. The improved laser pulse shapes are tested and evaluated in the same manner. Cycling through this loop many times results in iteratively optimized laser pulse shapes that finally approach the objective

Pulse Shaping

The methods for dispersion management described so far are well suited to compensate higher-order dispersion terms in linear optical setups such as group delay dispersion and third-order dispersion. Much greater flexibility in dispersion management and the possibility of creating complex-shaped laser pulses with respect to phase, amplitude and polarization state is given with the help of (computer-controlled) pulse-shaping techniques (Fig. 12.10a). The issue was recently reviewed by *Weiner* [12.29, 55].

A new class of experiments emerged in which pulse-shaping techniques were combined with some experimental signal embedded in a feedback learning loop [12.56–59]. In this approach a given pulse shape is evaluated in order to produce an improved pulse shape, which enhances the feedback signal (see Fig. 12.10b). These techniques have an impact on an increasing number of scientists in physics, chemistry, biology and engineering. This is due to the fact that primary light-induced processes can be studied and even actively controlled via adaptive femtosecond pulse shaping. For a small selection of work in various areas see [12.60–71].

Because of their short duration, femtosecond laser pulses cannot be directly shaped in the time domain. Therefore, the idea of pulse shaping is to modulate the incident spectral electric field $\tilde{E}^+_{\text{in}}(\omega)$ by a linear mask, i. e., the optical transfer function, $\tilde{M}(\omega)$ in the frequency domain. According to (12.38) this results in an outgoing shaped spectral electric field $\tilde{E}^+_{\text{out}}(\omega) = \tilde{M}(\omega)\tilde{E}^+_{\text{in}}(\omega) = \tilde{R}(\omega)\text{e}^{-\text{i}\phi_\text{d}}\tilde{E}^+_{\text{in}}(\omega)$. The mask may modulate the spectral amplitude response $\tilde{R}(\omega)$ and the spectral phase transfer function $\phi_\text{d}(\omega)$. Furthermore, polarization shaping has been demonstrated [12.72].

One way to realize a pulse shaper is the Fourier-transform pulse shaper. Its operation principle is based on optical Fourier transformations from the time domain into the frequency domain and vice versa. In Fig. 12.11 a standard design of such a pulse shaper is sketched. The incoming ultrashort laser pulse is dispersed by a grating and the spectral components are focused by a lens of focal length f. In the back focal plane of this lens – the Fourier plane – the spectral components of the original pulse are separated from each other, having minimum beam waists. By this means, the spectral components can be modulated individually by placing a linear mask into the Fourier plane. Afterwards, the laser pulse is reconstructed by performing an inverse Fourier transformation back into the time domain. Optically, this is realized by a mirrored setup consisting of an identical lens and grating. The whole setup – without the linear mask – is called a zero-dispersion compressor since it introduces no dispersion if the $4f$ condition is met (see also Fig. 12.7a). As a part of such a zero-dispersion compressor, the lenses separated by the distance $2f$, form a telescope with unitary magnification. Spectral modulations as stated by (12.38) can be set by inserting the linear mask.

Due to the damage threshold of the linear masks used, cylindrical focusing lenses (or mirrors) are normally used instead of spherical optics. The standard design in Fig. 12.11 has the advantage that all optical

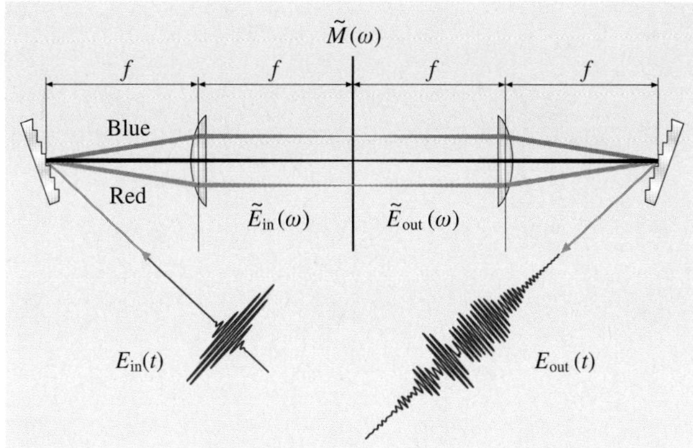

Fig. 12.11 Basic layout for Fourier-transform femtosecond pulse shaping

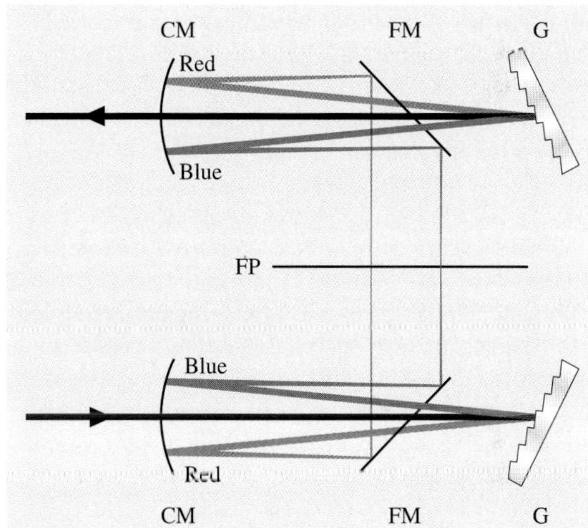

Fig. 12.12 Dispersion-optimized layout for Fourier transform femtosecond pulse shaping. The incoming beam is dispersed by the first grating (G). The spectral components go slightly out of plane and are sagitally focused by a cylindrical mirror (CM) via a plane-folding mirror (FM) in the Fourier plane (FP). Then the original beam is reconstructed by a mirrored setup

components are positioned along an optical line (grating in the Littrow configuration). For ultrashort pulses below 100 fs, however, spatial and temporal reconstruction errors become a problem due to the chromatic abberations introduced by the lenses. Therefore, lenses are often replaced by curved mirrors. In general, optical errors are minimized if the tilting angles of the curved mirrors within the telescope are as small as possible. A folded, compact and dispersion optimized setup is de-

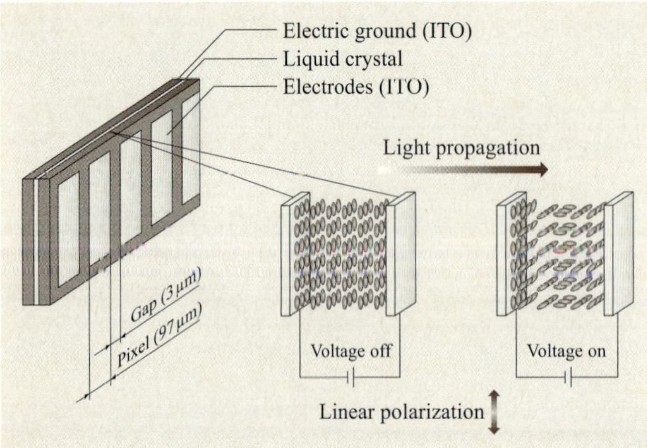

Fig. 12.13 Schematic diagram of an electronically addressed phase-only liquid-crystal spatial light modulator (LC-SLM). By adjusting the voltages of the individual pixels, the liquid-crystal molecules reorient themselves on average partially along the direction of the electric field. This leads to a change in refractive index and therefore to a phase modulation which can be independently controlled for different wavelength components

picted in Fig. 12.12 [12.73]. For ultrashort pulses in the < 10 fs regime prisms have been employed as dispersive elements instead of gratings [12.74].

A very popular linear mask for computer-controlled pulse shaping in such setups is the liquid-crystal spatial light modulator (LC-SLM). A schematic diagram of an electronically addressed phase-only LC-SLM is depicted in Fig. 12.13. In the Fourier plane the individual wavelength components of the laser pulse are spatially dispersed and can be conveniently manipulated by applying voltages at the separate pixels leading to changes of the refractive index . Upon transmission of the laser beam through the LC-SLM a frequency-dependent phase is acquired due to the individual pixel voltage values and consequently the individual wavelength components are retarded with respect to each other. Actual LC-SLMs contain up to 640 pixels [12.75]. In this way, an immensely large number of different spectrally phase modulated pulses can be produced. A phase-only LC-SLM does to a first approximation not change the spectral amplitudes and therefore the integrated pulse energy remains constant for different pulse shapes. By virtue of the Fourier transform properties, spectral phase changes result in phase- and amplitude-modulated laser pulse temporal profiles, as depicted schematically in Fig. 12.14.

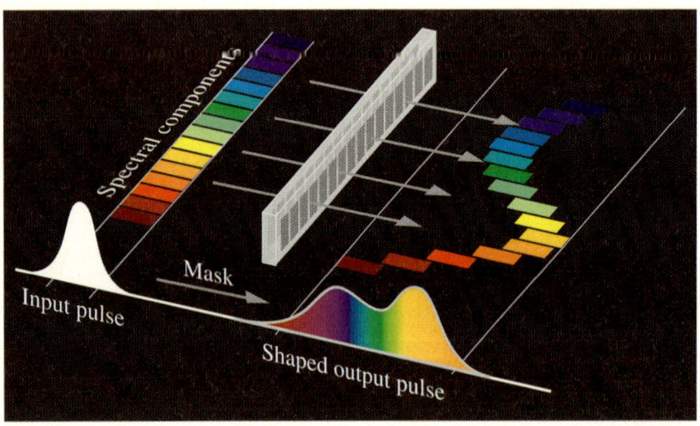

Fig. 12.14 Schematic illustration of shaping the temporal profile of an ultrashort laser pulse by retardation of the spectrally dispersed individual wavelength components in a phase only LC-SLM. The LC-SLM is located in the Fourier plane of the setups displayed in Fig. 12.11 and Fig. 12.12

If such an LC-SLM is oriented at 45° with respect to the linear polarization of the incident light field (either with the help of a wave plate or a suitably designed LC-SLM), polarization is induced in addition to retardance. A single LC-SLM together with a polarizer can be used therefore as an amplitude modulator. However, this also leads to phase modulation depending on the amplitude modulation level. For independent phase and amplitude control dual LC-SLMs are currently used. In such a setup a second LC-SLM is fixed back-to-back at $-45°$ with respect to the linear polarization of light in front of the first LC-SLM and the stack is completed with a polarizer. For an early setup for independent phase and amplitude modulations see [12.76] whereas modern configurations are described in [12.55]. Alternatively, simple amplitude modulation functions $\tilde{R}(\omega)$ can be realized by insertion of absorbing material at specific locations in the Fourier plane, thus eliminating the corresponding spectral components within the pulse spectrum [12.77].

For polarization shaping [12.72] the polarizer is removed and spectral phase modulation can be imposed independently onto two orthogonal polarization directions. The interference of the resulting elliptically polarized spectral components leads to complex evolutions of the polarization state in the time domain. As any element between the LC-SLM stack and the experiment can modify the polarization evolution, dual-channel spectral interferometry and experimentally calibrated Jones matrix analysis have been employed for characterization [12.78]. A representation of a complex polarization shaped pulse is displayed in Fig. 12.15. Such pulses open up an immense range of applications, especially in quantum control, because vectorial proper

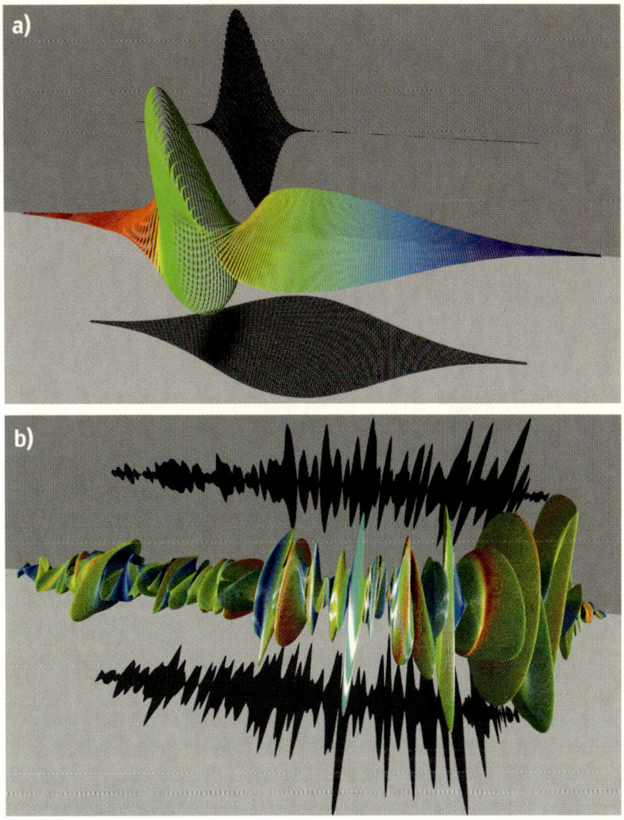

Fig. 12.15a,b Electric field representation for a polarization-modulated laser pulse. Time evolves from left to right, and electric field amplitudes are indicated by the sizes of the corresponding ellipses. The momentary frequency can be indicated by colors, and the shadows represent the amplitude envelopes of the orthogonal electric field components. **(a)** A Gaussian-shaped laser spectrum supporting 80 fs laser pulses is taken for an illustrative theoretical example. **(b)** A complex experimentally realized polarization modulated laser pulse is shown. The width of the temporal window is 7.5 ps (After [12.78])

ties of multiphoton transitions can be addressed [12.79, 80].

Another possibility to realize phase-only pulse shaping is based on deformable mirrors consisting of a small number ($\sim$10) of electrostatically controlled membrane mirrors [12.81]. These devices are placed in the Fourier plane and by a slight out-of-plane tilt upon reflection half of the optics can be saved (see Fig. 12.16 for an illustration). The use of a micro-mirror array with 240 × 200 pixels used in reflection and a waveform update rate larger than 1 kHz has also been demonstrated [12.82].

Acousto-optic modulators (AOMs) can be used for programmable pulse shaping as well. Two different approaches exist.

One approach is depicted in Fig. 12.17 and is reviewed in [12.83, 84]. The AOM crystal is oriented at the Bragg angle to the Fourier plane of a zero-dispersion compressor. In the visible TeO_2 crystals are normally used whereas in the infrared InP crystals are employed. A programmable radio-frequency (RF) signal driving the piezoelectric transducer of the AOM creates an acoustic wave that propagates through the crystal. As light travels at orders of magnitude faster velocity, the acoustic wave can be considered as a fixed modulated grating at the moment the spatially dispersed laser beam hits the crystal. The amplitude and phase of the acoustic wave determine the diffraction efficiency and phase shift at each point in space. The beam is diffracted typically below 1° by the AOM via the photoelastic effect. AOMs can place in the order of thousand independent features onto the spectrum and have a significantly faster update rate than an LC-SLMs. On the other hand the optical throughput of such devices is well below 50% and typical mode-locked laser sources running at 100 MHz repetition rate in general cannot be pulse shaped because the acoustic wave is traveling several tens of microns in 10 ns. This is not a limitation for amplified ultrafast laser systems where the pulse repetition rate is usually slower than the acoustic aperture time, since this allows the acoustic pattern to be synchronized to each amplifier pulse and to be refreshed before the next pulse arrives.

The other AOM approach is based on an acoustooptic programmable dispersive filter (AOPDF) and does not need to be placed in the Fourier plane of a $4f$ device [12.85–87]. A schematic of this device is shown in Fig. 12.18. Again, a programmable signal driving the piezoelectric transducer of the AOM creates an acoustic

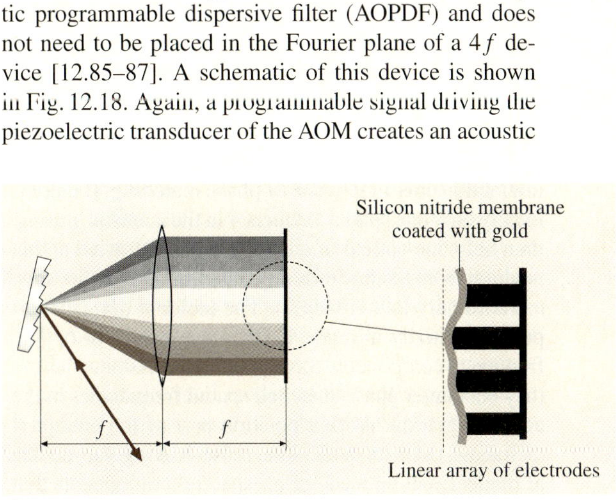

Fig. 12.16 Schematic of a phase-only deformable-mirror pulse shaper

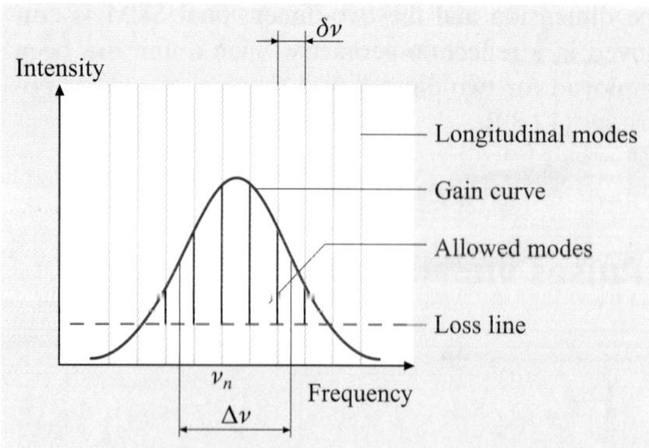

Fig. 12.20 Longitudinal modes in a laser cavity. The spacing $\delta\nu$ of the modes is determined by the cavity length via $\delta\nu = c/2L = 1/T_{RT}$. Only those modes exceeding the loss line will lase. The FWHM of the spectral intensity function $\delta\nu$ is also indicated in addition. In lasers used for pulse generation below 10 fs the number of modes lasing is of the order 10^6

Taking both conditions together the emission spectrum of the laser will consist of those modes which have sufficient gain to lie above the threshold for lasing. The corresponding relationships are depicted in Fig. 12.20. The total electric field $E(t)$ resulting from such a multimode oscillation at a fixed point in space, say at one of the mirrors, is given by

$$E(t) = \sum_{n=0}^{N-1} E_n \sin[2\pi(\nu_0 + n\delta\nu)t + \varphi_n(t)] , \quad (12.65)$$

where N is the number of oscillating modes, $\varphi_n(t)$ is the phase of the nth mode and ν_0 is the lowest-frequency mode above the lasing threshold.

The average laser power output $P(t)$ is proportional to the square of the total electric field. Unless some method of fixing the relative phases $\varphi_n(t)$ of the modes is used they will generally vary randomly in time. This produces a random variation of the average laser power output $P(t)$ as a result of the random interference between modes.

If the phases are fixed with respect to each other $[\varphi_n(t) \to \varphi_n]$, it can be shown that $E(t)$ and accordingly $P(t)$ repeats with the period T_{RT}. In the case that the individual φ_n are randomly fixed, each *noise spike* in the random but periodic laser output power has a duration Δt roughly equal to $1/\Delta\nu$ with $\Delta\nu$ being the FWHM of the spectral intensity function (Fig. 12.23e, f). Within this approach the properties of perfectly mode

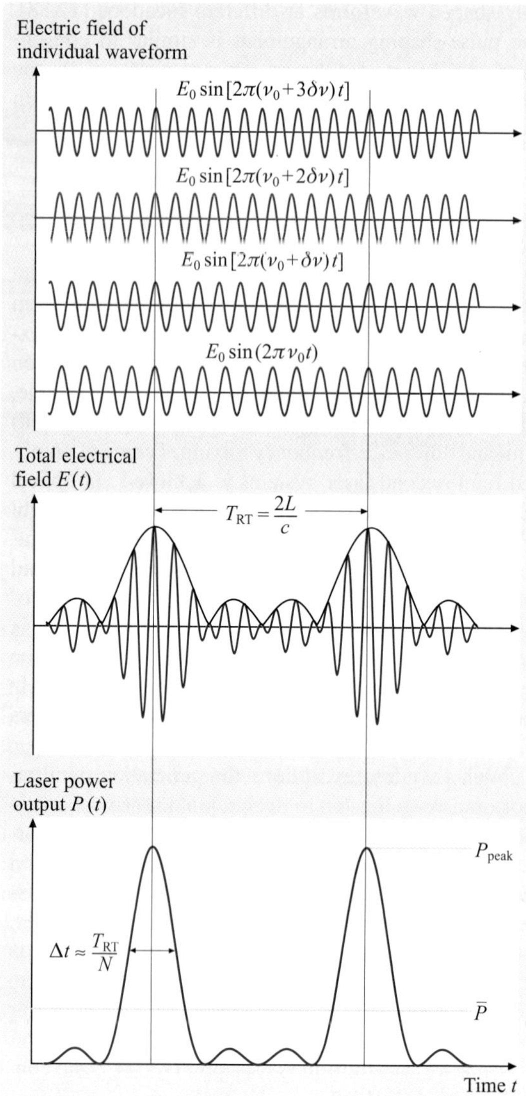

Fig. 12.21 Superposition of four sine waves with equal amplitude E_0, differing in frequency by $\delta\nu$. The electric field of the individual waveforms, the total electric field $E(t)$, its envelope and the output power $P(t)$ as well as the average power $\bar{P}$ are shown

locked lasers are determined by a linear phase relation $\varphi_n = n\alpha$ amongst the modes, that is, a constant phase relation between two adjacent modes. This is the so-called mode-locking condition. To simplify the analysis of this case identical amplitudes $E_n = E_0$ for all modes are assumed corresponding to a square gain profile and for convenience α is set to zero. The summation of (12.65) then yields

$$E(t) = E_0 \sin\left[2\pi\left(\nu_0 + \frac{N-1}{2}\delta\nu\right)t\right]\frac{\sin(N\pi\delta\nu t)}{\sin(\pi\delta\nu t)}\,. \tag{12.66}$$

The resulting electric field consists of a rapid oscillating part at the light central frequency $\nu_c = \nu_0 + \frac{N-1}{2}\delta\nu$ with the envelope $\left|\frac{\sin(N\pi\delta\nu t)}{\sin(\pi\delta\nu t)}\right|$ oscillating with $\delta\nu = 1/T_{RT}$. Averaging the fast oscillation ν_c the output power $P(t)$ is given by

$$P(t) = P_0\left[\frac{\sin(N\pi\delta\nu t)}{\sin(\pi\delta\nu t)}\right]^2 \tag{12.67}$$

where P_0 is the average power of one wave.

A discussion of this equation yields important insight into the properties of laser pulses generated via mode locking.

1. The power is emitted in the form of a train of pulses with a period corresponding to the cavity round-trip time $T_{RT} = \frac{1}{\delta\nu}$.
2. The peak power P_{Peak} increases quadratically with the number N of modes locked together: $P_{Peak} = N^2 P_0$. Mode locking is therefore useful to produce high peak powers and by focusing the laser beam to create high peak intensities; the average power $\bar{P}$ of both a mode-locked and a non-mode-locked laser is given by $\bar{P} = NP_0$.
3. The FWHM pulse duration Δt decreases linearly with the number N of modes locked together or equivalent is approximately the inverse of the gain bandwidth $\Delta\nu$:

$$\Delta t \approx \frac{T_{RT}}{N} = \frac{1}{N\delta\nu} = \frac{1}{\Delta\nu}\,.$$

This is why in the past dye lasers and nowadays solid-state lasers with large gain bandwidths are used to create femtosecond pulses. Ultrafast dye lasers generated pulses as short as 27 fs with around 10 mW of average power [12.99], whereas pulses around 5–6 fs with around 100 mW average power can be produced with Ti:sapphire lasers [12.53, 100]. In general the minimum pulse duration for a given gain profile can be estimated via the bandwidth product introduced in Sect. 12.1.2 and is summarized for various line shapes in Table 12.1.

The basic properties of mode locking are visualized with the help of Figs. 12.21–12.23. Figure 12.21 depicts the Fourier synthesis of a pulse obtained by the superposition of four sine waves with same amplitude and $\varphi_n(t) = 0$ according to (12.65, 12.66, 12.67).

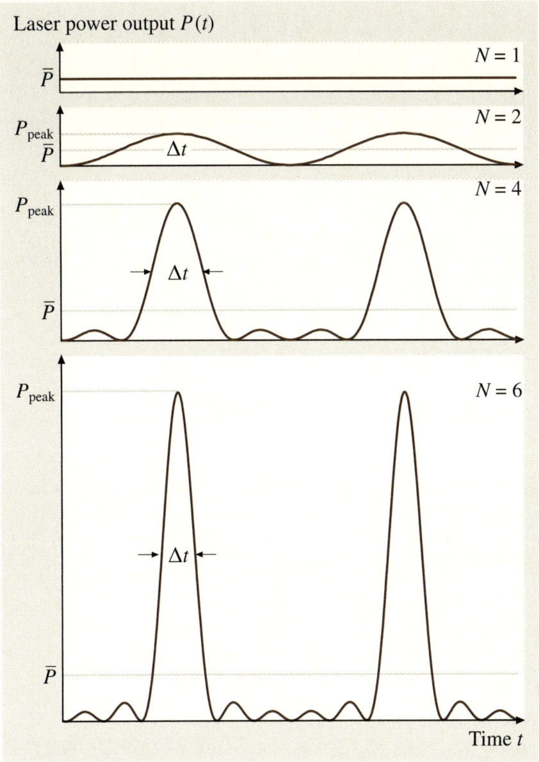

Fig. 12.22 Comparison of the one, two-, four- and six-mode cases. An increase in the number of modes leads to a decrease in pulse duration. The peak power P_{peak} increases quadratically with the number N of modes locked together, whereas the average power $\bar{P}$ of both a mode-locked and a non-mode-locked laser scales linearly with N

In Fig. 12.22 the dependence of the pulse duration and peak power on the number of locked modes is illustrated for this case. Finally in Fig. 12.23 the shape of the average output power is displayed for $N = 10$ equally spaced modes with different relative amplitudes and phase angles according to (12.65).

In the following we will summarize some more-technical related considerations. Mode locking is essentially achieved by applying a periodic loss (or gain) modulation to the intracavity radiation whose period is matched to the cavity round trip time. The mechanisms can be described either in the frequency or time domain.

In the frequency domain one can start the consideration from the lowest-loss longitudinal mode. The periodic modulation at the frequency of the round-trip time leads to sidebands whose frequencies coincide with those of the adjacent longitudinal laser modes. In this

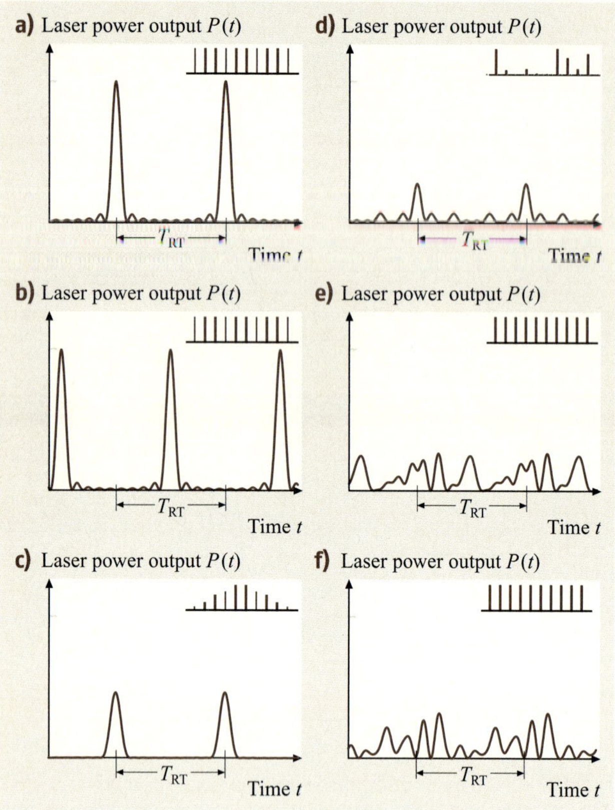

Fig. 12.23a–f Output power for 10 equally spaced modes with different relative amplitudes (as indicated in the *insets*) and phase angles (T_{RT} is the round-trip time): **(a)** linear phase relation $\varphi_n = n\alpha$ amongst the modes (i. e., a constant phase relation between two adjacent modes) with $\alpha = 0$, **(b)** linear phase relation $\varphi_n = n\alpha$ with $\alpha = \pi$, **(c)** Gaussian spectrum with five modes at FWHM and linear phase relation with $\alpha = 0$, **(d)** random spectrum and linear phase relation with $\alpha = 0$, **(e)** constant spectrum and random phase, **(f)** constant spectrum and different random phase

way energy is shifted from one mode to adjacent modes and as a result all longitudinal modes become finally locked in phase. In the time domain, the periodic modulation can be visualized as an intracavity shutter that is open once per round trip time. Such a stationary time window of minimum loss will provide a higher net gain on each round trip for those photons that are concentrated in that time window.

Approaches for providing the periodic modulation are grouped into active and passive schemes, and hybrid schemes that make use of a combination of the two. Active mode locking is obtained with an active element within the laser cavity, for example an acoustooptic modulator, generating a loss modulation. The modulation has to be precisely synchronized with the cavity round trips. Modulating the gain is also possible and can be achieved by synchronous pumping. In this case the amplifying medium of the laser is pumped with the output of another mode-locked laser whereby the cavity round-trip times for both lasers have to be matched. Passive mode locking is obtained by the laser radiation itself that produces the modulation via the interaction with a nonlinear device in the laser cavity. Typical nonlinear devices are some type of saturable absorbers which exhibit an intensity dependent loss as they interact with the laser radiation. This modulation is thus automatically synchronized to the cavity round-trip frequency. Because pulse timing does not have to be externally controlled there is usually no need for synchronization electronics, making passive schemes conceptually simpler compared to active schemes. Originally organic dyes were used as real saturable absorbers, for example to generate picosecond pulses from solid-state lasers and pulses down to 27 fs from dye lasers [12.99]. The shortest pulses nowadays are generated in solid-state laser media, being passively mode-locked using the optical Kerr effect. This approach was originated by [12.101]. Pulses with less than 6 fs are nowadays generated directly from Ti:sapphire lasers with Kerr-lens mode locking [12.53, 100]. At a center wavelength of 800 nm a pulse duration of 5.4 fs contains only two optical cycles at full-width half-maximum of the pulse intensity.

12.3 Measurement Techniques for Femtosecond Laser Pulses

For energy, power, spectrum and spatial beam measurements of ultrashort laser pulses standard laser diagnostic techniques are employed [12.5, 47]. For a measurement of the pulse duration or more interesting of the time-dependent amplitude and phase of an ultrashort laser pulse, dedicated methods have been developed and are described in several textbooks and references therein [12.4–6]. Here the basic ideas and underlying concepts are highlighted.

As the time and frequency domain are related by the Fourier transformation (12.6, 12.7, 12.11, 12.12) it should be sufficient to measure amplitude and phase in only one of the domains. Let us first shortly reflect on the frequency domain. All spectrometers no matter whether diffraction-grating or Fourier-transform devices measure a quantity that is proportional to the spectral intensity (Sect. 12.3.3) and therefore the phase information is lost.

On the other hand direct electronic techniques for temporal pulse-width measurements, consisting of fast photodiodes and high-bandwidth (sampling) oscilloscopes, are limited to the several-picosecond regime. Fast photodiodes are therefore not suited to record the temporal profile of an ultrashort laser pulse. Often they are employed to check on the mode-locked operation of an ultrafast oscillator or to derive synchronization signals for amplification setups or synchronized experiments. The only detector that reaches a time resolution below one picosecond is the streak camera. However, a characterization of ultrashort pulses with respect to amplitude and phase requires optical correlation techniques, especially methods that operate in the time–frequency domain. The latter techniques will be described in more detail.

12.3.1 Streak Camera

The basic principle of a streak camera is depicted in Fig. 12.24. The ultrafast optical signal $I(t)$ to be analyzed is focused on a photocathode, where the signal is converted almost instantaneously into a number of electrons. The electrons then pass through a horizontal pair of accelerating electrodes and hit a phosphor screen after passing an electron multiplier (MCP). The screen is then imaged with the help of a highly sensitive camera (not shown). The temporal resolution relies on the concept of transferring a temporal profile into a spatial profile. This is done by passing the electron pulse between a pair of vertical sweep electrodes. A high voltage is applied to the sweep electrodes at a timing synchronized to the incident light. During this high-speed sweep the electrons arriving at different times are deflected at different angles and consequently hit the MCP at different vertical directions. In this manner the vertical position on the phosphor screen serves as a time axis. The brightness of the signal is proportional to the intensity profile of the incident ultrashort optical signal. The horizontal direction of the image corresponds to the horizontal location of the incident light. For example if the streak camera is used in combination with a polychromator the time variation of the incident light with respect to wavelength can be measured. Time-resolved spectroscopy is therefore one of the application areas of these devices. Commercial devices [12.102, 103] are quoted with a temporal resolution of < 200 fs. Using different photocathode materials a spectral response can be achieved from 115 nm up to 1600 nm. X-ray streak cameras with a temporal resolution of 1.5 ps are quoted as well.

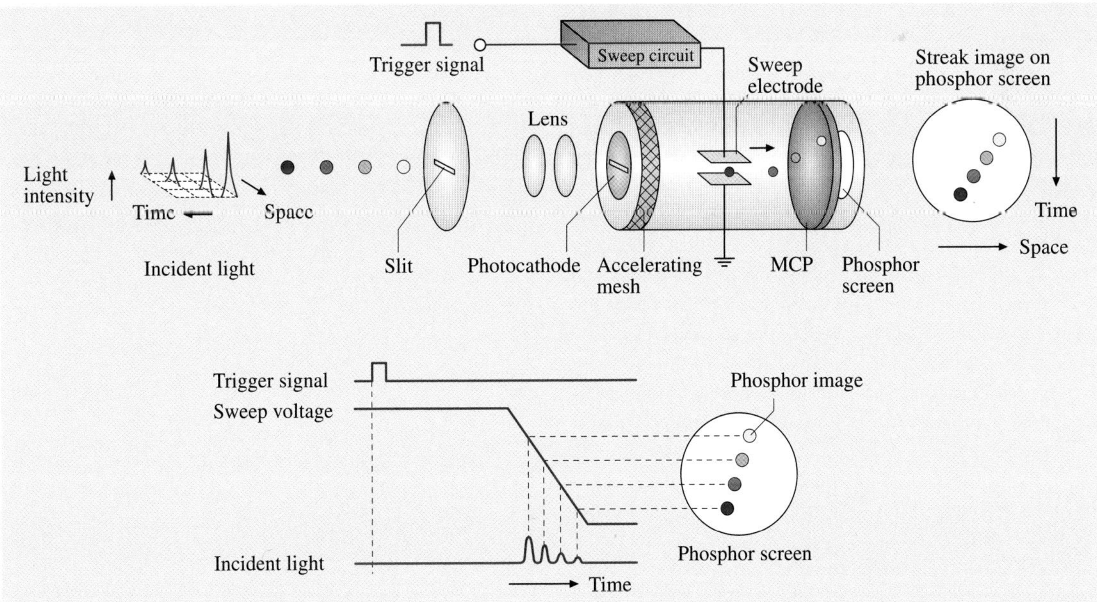

Fig. 12.24 Working principle (*top*) and timing (*bottom*) of a streak camera (After [12.102]). The spatial coordinate might be a wavelength coordinate after having dispersed the ultrashort optical signal with the help of a polychromator

12.3.2 Intensity Autocorrelation and Cross-Correlation

A widely used technique to estimate the pulse duration or to check whether a laser produces pulses rather than statistical intensity fluctuations is to measure the so-called intensity autocorrelation S_{IntAC} [12.104]

$$S_{\text{intAC}}(\tau) = \int_{-\infty}^{\infty} I(t)I(t+\tau)\,dt = \int_{-\infty}^{\infty} I(t)I(t-\tau)\,dt = S_{\text{intAC}}(-\tau)\,. \quad (12.68)$$

This is the time integral of one pulse intensity multiplied by the intensity of a time-shifted replica of the same pulse as a function of the time shift τ. The intensity autocorrelation has its maximum at $\tau = 0$ and is always symmetrical [see (12.68)]. In this fundamental arrangement one pulse serves as a gate to scan the other. It can be realized with any interferometer (for examples see Fig. 12.25) that splits the pulse into two pulses and recombines them with an adjustable time delay between them. Note within that context that, for example a 100 fs pulse duration, corresponds to a spatial extent of 30 μm, a dimension readily measurable with standard translation stages. Measuring the spatial overlap of the two pulses requires a nonlinear process to generate a detection signal proportional to the intensity product of the two pulses. Second-harmonic generation in thin crystals and two-photon absorption in semiconductor photodiodes [12.105, 106] are commonly used (in a *two-photon diode* the photon energy is within the band gap and only simultaneous two-photon absorption can lead to a signal). In the case of frequency-doubling crystals thin crystals have to be used in order to ensure that the ratio of the crystals phase-matching bandwidth to the pulse spectral bandwidth is large. For 100 fs pulses at 800 nm the beta-barium borate (BBO) crystal thickness should

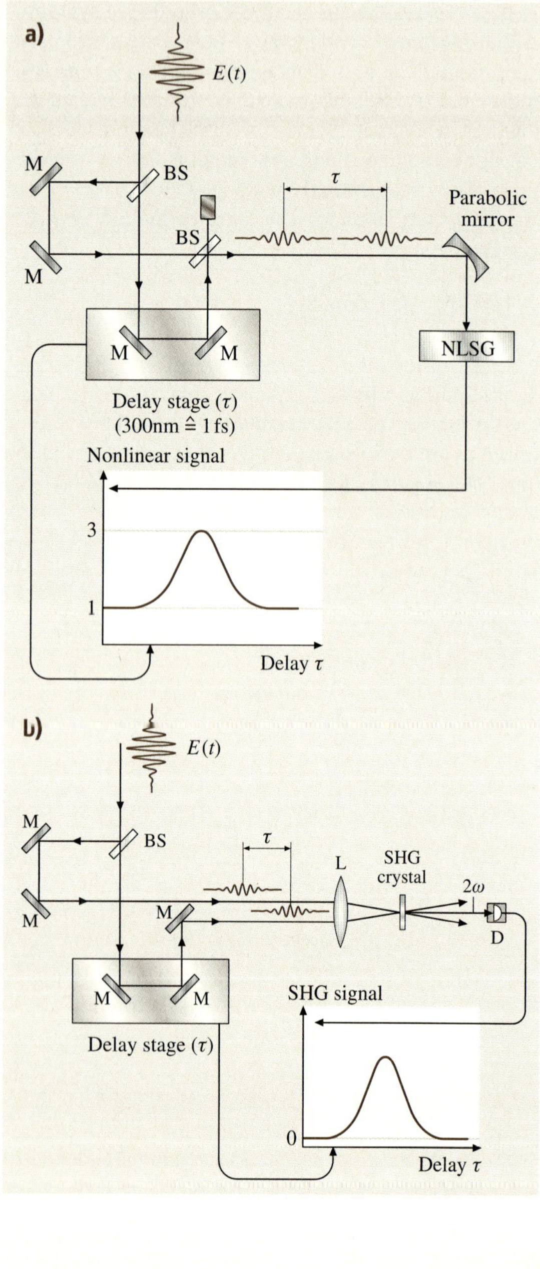

Fig. 12.25a,b Optical layout for autocorrelation setups. **(a)** Collinear autocorrelator (dispersion minimized): the incoming pulse is split into two parts, where one is variably delayed with respect to the other. The pulses are recombined and focused on a nonlinear signal generator (NLSG). Second-harmonic generation in thin crystals and two-photon absorption in semiconductor photodiodes are commonly used for this purpose. Other second-order nonlinear effects can be used as well. The nonlinear signal is measured as a function of delay. If the measurement is performed with interferometric accuracy the interferometric autocorrelation is recorded. If the setup averages the fast oscillations of the light field (see (12.76)) the intensity autocorrelation with background is recorded, having a center-to-offset ratio of 3:1 (see (12.77)).

(b) Non-collinear autocorrelator for recording the background free intensity autocorrelation (M = mirror; BS = beam splitter; SHG = second harmonic generation; D = detector; L = lens) not be thicker than $\approx 100\,\mu\text{m}$ and crystals as thin as $5\,\mu\text{m}$ have been used to measure few-fs pulses [12.6].

The intensity autocorrelation is obtained directly, when the two time-delayed laser pulses are not recombined collinearly but focused at a mutual angle into the thin nonlinear crystal. This leads to the so-called background free intensity autocorrelation. For the collinear setup the intensity autocorrelation is obtained by averaging the fast oscillations of the light field [(12.76) in Sect. 12.3.3].

The collinear intensity autocorrelation has a signal-to-background ratio of 3:1 [(12.77) in Sect. 12.3.3].

The intensity autocorrelation provides only limited information on the pulse shape, because there are infinitely many symmetric and asymmetric pulse shapes that lead to very similar symmetric autocorrelation traces. The procedure to estimate a pulse duration from intensity autocorrelations is to assume a pulse shape and then to calculate the FWHM pulse duration Δt from the known ratio with respect to the FWHM of the intensity autocorrelation Δt_{intAC}. In this approach generally Gaussian shapes or hyperbolic secant shapes are assumed. The ratio $\Delta t_{\text{intAC}}/\Delta t$ for various shapes [12.22] is given in Table 12.1.

If a pulse $I_1(t)$ is characterized, for example, in such a way it can be used to gate a second unknown pulse $I_2(t)$ by measuring the intensity cross-correlation S_{intCC} with a suitable nonlinear second-order signal like, for example, sum- or difference-frequency mixing or two-photon photodiodes

$$S_{\text{intCC}}(\tau) = \int_{-\infty}^{\infty} I_1(t) I_2(t+\tau)\,\mathrm{d}t\,. \tag{12.69}$$

For Gaussian pulse shapes the corresponding FWHM quantities are related by

$$\Delta t_{\text{intCC}}^2 = \Delta t_1^2 + \Delta t_2^2\,. \tag{12.70}$$

In general the second momenta of the individual pulses have to be considered [12.23].

For high-power femtosecond laser systems higher-order cross-correlations $S_{\text{higher order int CC}}$ are a very convenient and powerful tool to determine intensity profiles by making use of nonlinear optical processes of the order $n+1$ and $m+1$

$$S_{\text{higher order intCC}}(\tau) = \int_{-\infty}^{\infty} I_1^n(t) I_2^m(t+\tau)\,\mathrm{d}t\,. \tag{12.71}$$

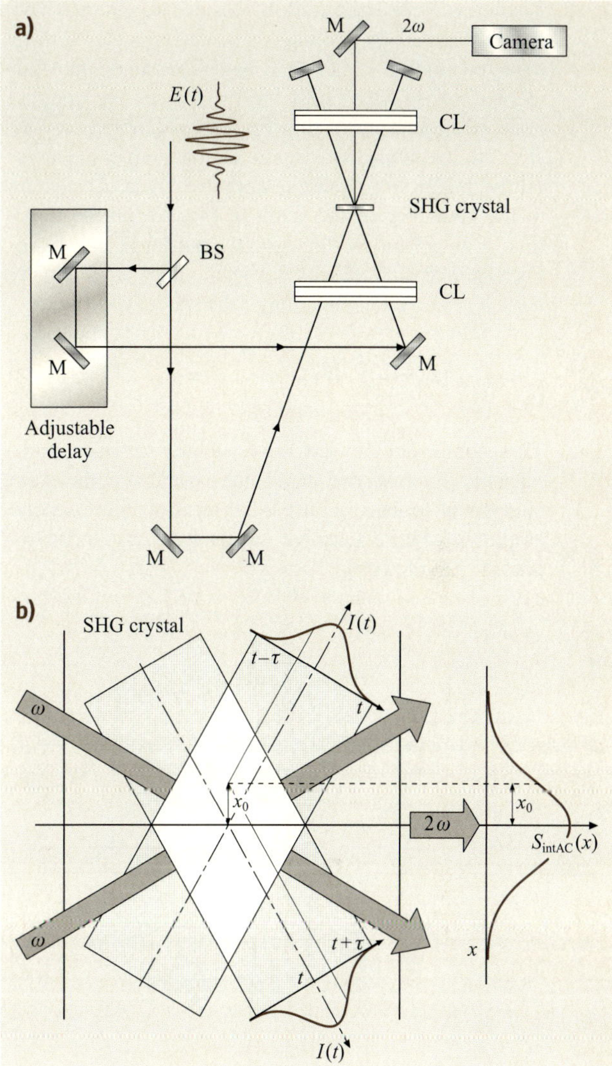

Fig. 12.26 **(a)** Optical layout for a single-shot autocorrelator. The delayed replicas of the incident pulse are focused with the help of a cylindrical lens (CL) onto a second-harmonic generation (SHG) crystal. The spatiotemporal overlap of the two spatially extended pulses is measured via SHG and recorded with a camera (M = mirror; BS = beam splitter). **(b)** Detail of **(a)**. In the region of spatiotemporal overlap, second-harmonic generation is induced via type I phase matching and the autocorrelation in time is transformed into a spatial intensity distribution along the x-axis. (After [12.5])

In this case the corresponding FWHM quantities assuming Gaussian pulse shapes are given by

$$\Delta t_{\text{higher-order intCC}}^2 = \frac{1}{n}\Delta t_1^2 + \frac{1}{m}\Delta t_2^2\,. \tag{12.72}$$

The intensity autocorrelation does not necessarily have to be recorded by moving one interferometer arm as depicted in Fig. 12.25. In a so-called single-shot autocorrelator [12.107, 108] the two pulses are coupled non-collinearly into a thin frequency-doubling crystal (Fig. 12.26). Only in a small region within the crystal the pulses have spatiotemporal overlap. According to the geometry of the setup in Fig. 12.26b the delay time τ is related to the spatial coordinate x_0. Imaging the frequency-doubled signal yields an intensity autocorrelation as a function of the spatial coordinate

$$S_{\text{intAC}}(x_0) = \int_{-\infty}^{\infty} I(x) I(x+x_0)\,\mathrm{d}x\,. \tag{12.73}$$

These single-shot devices are especially suited for high-intensity femtosecond laser pulses and are therefore convenient tools to adjust low-repetition femtosecond amplifiers. Phase-sensitive setups have also been reported [12.108, 109].

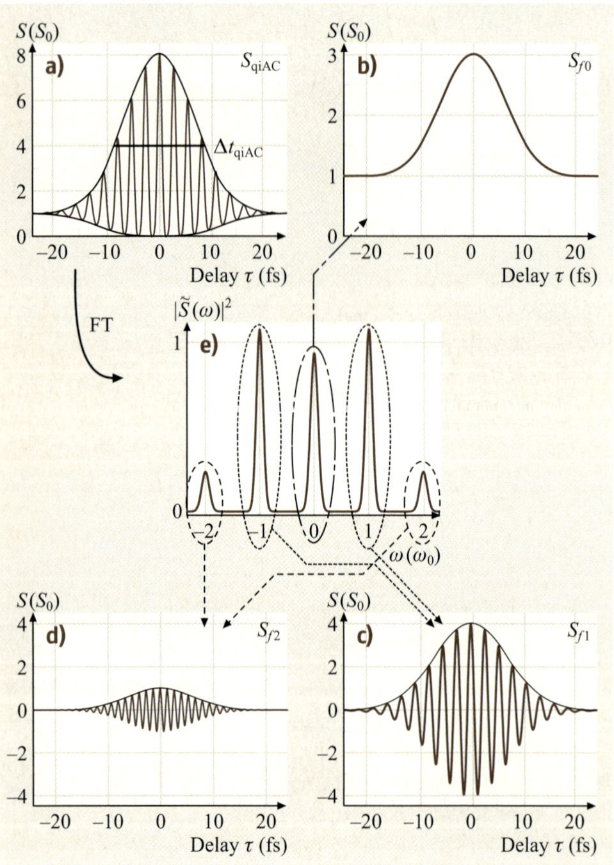

Fig. 12.27a–e Quadratic interferometric autocorrelation (a) and the isolated components S_{f0}, S_{f1} and S_{f2} in the time domain (b)–(d) for a bandwidth-limited Gaussian pulse of 10 fs pulse duration. Note that the offset in (a) introduces an additive value at $\omega = 0$ and (e) is therefore the Fourier transform of the offset-corrected curve. $\Delta t_{\text{quadratic interferometric AC}}$ is indicated in (a) in addition (in the figure qiAC is used as a shorthand notation for quadratic interferometric AC)

12.3.3 Interferometric Autocorrelations

We will now discuss the case of a collinear autocorrelation in more detail. The simplest interferometric signal is that from a linear detector that records the intensity of the recombined pulses. For identical electric fields E of the two pulses, the signal $S_{\text{linear interferometric AC}}$ as a function of their relative delay τ is

$$\begin{aligned} S_{\text{linear interferometric AC}}(\tau) &= \int_{-\infty}^{\infty} [E(t)+E(t+\tau)]^2\,\mathrm{d}t \\ &= 2\int_{-\infty}^{\infty} I(t)\,\mathrm{d}t \\ &\quad + 2\int_{-\infty}^{\infty} E(t)E(t+\tau)\,\mathrm{d}t \end{aligned} \tag{12.74}$$

where we have skipped the prefactors defined in Sects. 12.1.1 and 12.1.2. The signal consists of an offset given by the summed intensity of the two pulses and the interference term that is described by an autocorrelation of the electric field. The Wiener–Khintchine theorem states that the Fourier transformation of the autocorrelation of the electric field yields the spectral density [12.110] – a quantity that is proportional to the spectral intensity $I(\omega)$, which is the basis for Fourier spectroscopy. A linear autocorrelation therefore contains no information beyond the amplitude of the spectrum and the total intensity of the pulse.

A solution to this problem is a nonlinear detector that is sensitive to the squared intensity and yields the signal $S_{\text{quadratic interferometric AC}}$

$$S_{\text{quadratic interferometric AC}}(\tau) = \int_{-\infty}^{\infty} \left\{[E(t)+E(t+\tau)]^2\right\}^2 \mathrm{d}t\,. \tag{12.75}$$

Taking the electric field as $E(t) = \mathrm{Re}[A(t)\mathrm{e}^{\mathrm{i}\Phi_a(t)}\mathrm{e}^{\mathrm{i}\omega_0 t}]$ and defining $S_0 = \int_{-\infty}^{\infty} A^4(t)\,\mathrm{d}t$ in order to normalize one obtains, similar to [12.4, 111],

$$S_{\text{quadratic interferometric AC}}(\tau) = \frac{1}{S_0}\{S_{f0}+S_{f1}+S_{f2}\}$$

with

$$S_{f0} = \int_{-\infty}^{\infty} \left\{A^4(t)+2A^2(t)A^2(t+\tau)\right\} \mathrm{d}t\,,$$

$$S_{f1} = 2\,\mathrm{Re}\left\{ \mathrm{e}^{\mathrm{i}\omega_0\tau} \int_{-\infty}^{\infty} A(t)A(t+\tau) \times \left(A^2(t)+A^2(t+\tau)\right) \mathrm{e}^{\mathrm{i}(\Phi_\mathrm{a}(t+\tau)-\Phi_\mathrm{a}(t))}\,\mathrm{d}t \right\},$$

$$S_{f2} = \mathrm{Re}\left\{ \mathrm{e}^{\mathrm{i}2\omega_0\tau} \int_{-\infty}^{\infty} A^2(t)A^2(t+\tau) \times \mathrm{e}^{\mathrm{i}2(\Phi_\mathrm{a}(t+\tau)-\Phi_\mathrm{a}(t))}\,\mathrm{d}t \right\}, \tag{12.76}$$

where Re denotes the real part. According to (12.48) the signal $S_{\text{quadratic interferometric AC}}$ can be decomposed into three frequency components, S_{f0}, S_{f1} and S_{f2}, at $\omega \approx 0$, $\omega \approx \pm\omega_0$, and $\omega \approx \pm 2\omega_0$, respectively, as illustrated in Fig. 12.27.

S_{f0} corresponds to an intensity correlation with background. It can be obtained either by Fourier filtering or by averaging the fast oscillations in the experiment directly. With (12.76) it follows that this intensity autocorrelation has a center to offset ratio of

$$\frac{S_{f0}(0)}{S_{f0}(\infty)} = \frac{\int_{-\infty}^{\infty} 3A^4(t)\,\mathrm{d}t}{\int_{-\infty}^{\infty} A^4(t)\,\mathrm{d}t} = \frac{3}{1}\,. \tag{12.77}$$

S_{f1} is a sum of two mutual symmetric cross-correlations and depends explicitly on the temporal phase $\Phi_\mathrm{a}(t)$.

S_{f2} represents an autocorrelation of the second harmonic field and is therefore related to the spectral intensity of the second-harmonic spectrum. It also depends on the temporal phase $\Phi_\mathrm{a}(t)$. Note that phase-modulated pulses having the same spectral intensity can have very different spectral intensities after frequency doubling (Fig. 12.28). This has been exploited in recent experiments [12.21, 112, 113]. Making use of a pulse shaper (Sect. 12.1.3) that scans calibrated phase functions and at the same time measuring the second-harmonic spectrum is a noninterferometric method to characterize the spectral phase of ultrashort laser pulses [12.114]. All three components add constructively at $\tau = 0$ and yield a center to background ratio of 8:1. This can be directly seen from (12.75)

$$\frac{S_{\text{quadratic interferometric AC}}(0)}{S_{\text{quadratic interferometric AC}}(\infty)} = \frac{\int_{-\infty}^{\infty} (E+E)^4\,\mathrm{d}t}{\int_{-\infty}^{\infty} E^4\,\mathrm{d}t + \int_{-\infty}^{\infty} E^4\,\mathrm{d}t} = \frac{16\int_{-\infty}^{\infty} E^4\,\mathrm{d}t}{2\int_{-\infty}^{\infty} E^4\,\mathrm{d}t} = \frac{8}{1}\,. \tag{12.78}$$

The center-to-background ratios are used in experiments to check the proper alignment of the interferometer. In order to derive phase information analytical functions, for example Gaussians, can be fitted to the $S_{\text{quadratic interferometric AC}}$ [12.111]. Taking the knowledge of the spectrum into account iterative algorithms that make no assumptions about the underlying pulse shapes have been reported [12.115, 116]. Both approaches deliver meaningful results only for linear chirps and in the case of nearly no noise. The influence of noise on autocorrelation measurements is discussed in [12.4, 117]. This is an important point as most often measurements are performed over an average of pulse trains. Other sources of systematic error in autocorrelation measurements are discussed in [12.6].

The ratio $\Delta t_{\text{intAC}}/\Delta t$ is only valid for the intensity autocorrelation of a bandwidth-limited pulse. For bandwidth-limited Gaussian pulses the FWHM of a quadratic interferometric autocorrelation signal (taken at 4 in a 8:1 plot as displayed in Fig. 12.27a) relates to the pulse duration by

$$\frac{\Delta t_{\text{quadratic interferometric AC}}}{\Delta_t} = 1.6963 \tag{12.79}$$

for bandwidth-limited Gaussian pulses. Figure 12.28 compiles for different pulses the resulting interferometric autocorrelation traces together with the intensity autocorrelation and the spectrum at the second harmonic of the fundamental.

12.3.4 Time–Frequency Methods

As described above, the interferometric autocorrelation, even together with the independently measured spectrum, does not give sufficient information to characterize arbitrary-shaped ultrashort laser pulses with respect to their temporal amplitude $A(t)$ or temporal intensity $I(t)$ and the temporal phase function $\Phi_\mathrm{a}(t)$ or their frequency-domain counterparts (Sect. 12.1.2).

Techniques have emerged that operate not in the time or frequency domain but in the *joint time–frequency domain*, involving both temporal resolution and frequency resolution simultaneously [12.24, 118] and being able to completely determine the pulse shape [12.6].

For illustration purposes we start with an example from music: in order to describe a line of music we use notes. The frequency is indicated by the pitch of the note and the duration of the note indicates how long the frequency has to be held. The sheet of music will

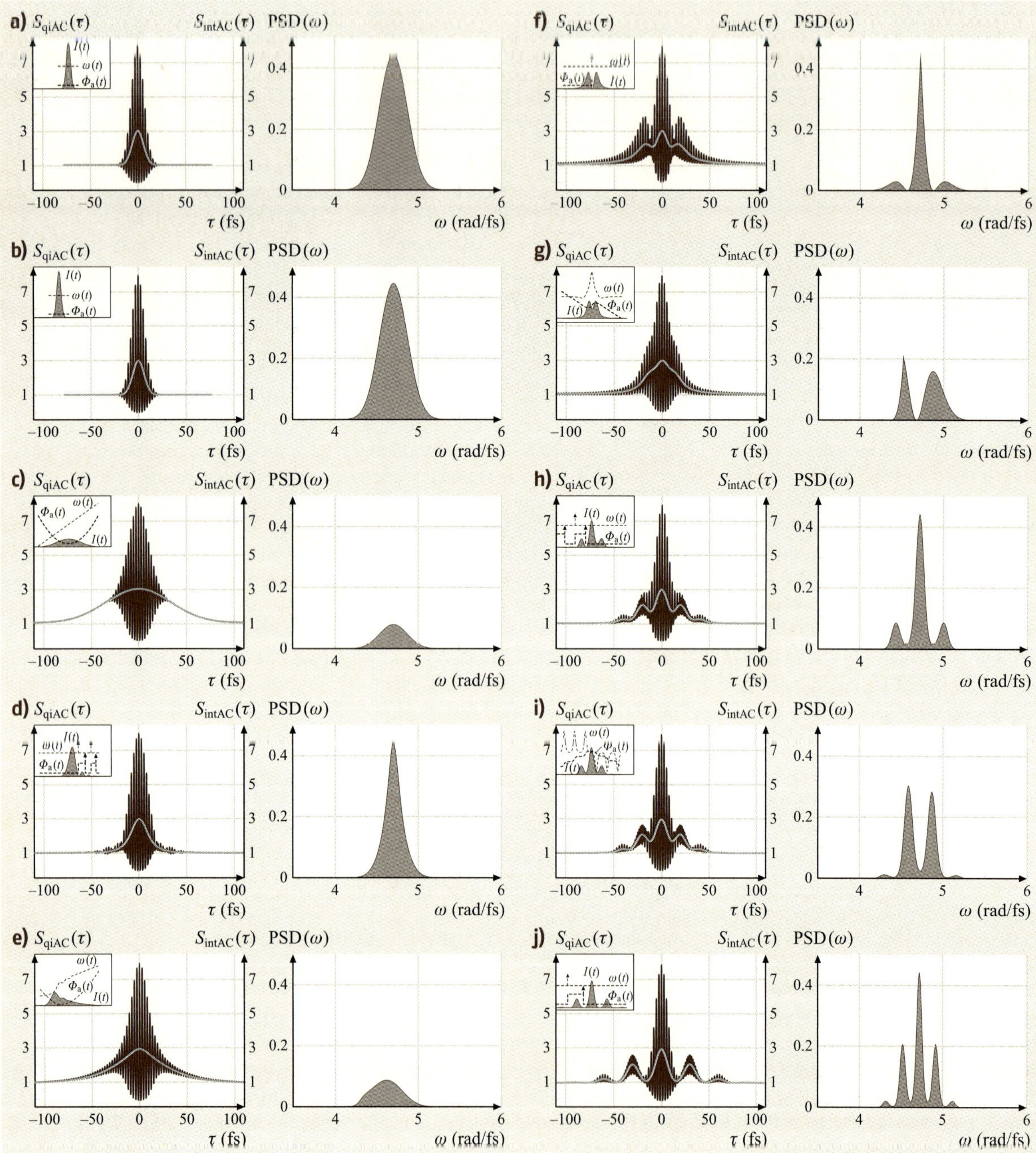

tell us in what order the notes have to be played and additional information like *piano* and *forte* is given to indicate the intensity to be played. The first few notes of Beethoven's fifth symphony are given as an example in Fig. 12.30a. If an orchestra is playing the music and we wish to graphically record the music, a spectrogram $S_{\text{spectrogram}}(\omega, \tau)$ is a useful quantity. A spectrogram of a function $f(t)$ is defined as the energy density spectrum of a short-time Fourier transform $\text{STFT}(\omega, \tau)$

$$S_{\text{spectrogram}}(\omega, \tau) \equiv |\text{STFT}(\omega, \tau)|^2 = \left| \int_{-\infty}^{\infty} f(t) g(t-\tau) \mathrm{e}^{-\mathrm{i}\omega t} \, \mathrm{d}t \right|^2 . \tag{12.80}$$

where $g(t-\tau)$ denotes the gate (or window) function. The concept behind it is simple and powerful. If we want to analyze what is happening at a particular time, we just use a small portion of the signal centered at that time, calculate its spectrum and do it for each instant of

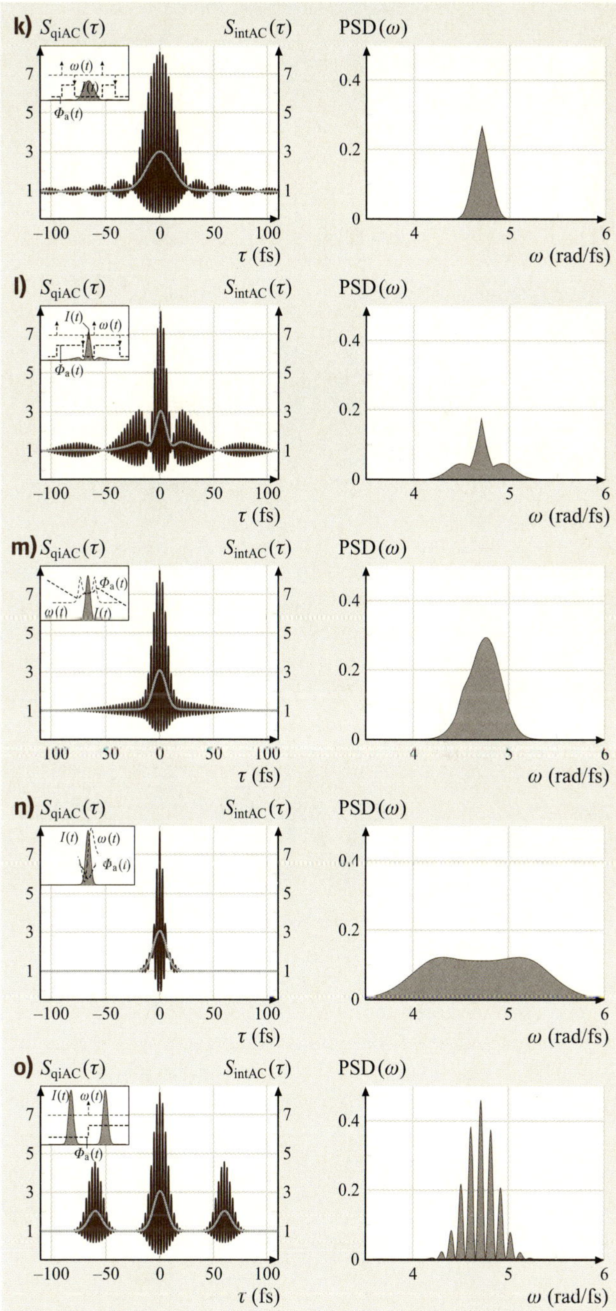

Fig. 12.28a–o *Left*: quadratic interferometric autocorrelation $S_{\text{quadratic interferometric AC}}(\tau)$ (*black*) and intensity autocorrelation $S_{\text{int AC}}(\tau)$ (*grey*) for the pulse shapes of Fig. 12.3 with a central wavelength of 800 nm. The temporal intensity $I(t)$, the additional temporal phase $\Phi_a(t)$ and the instantaneous frequency $\omega(t)$ are shown in the *insets*. *Right*: corresponding power spectrum density PSD(ω) of $S_{\text{qiAC}}(\tau)$ displayed in the region of the second harmonic. Note that for pulses **(a)** to **(j)** the linear spectrum remains unchanged. **(a)** bandwidth-limited Gaussian laser pulse of 10 fs duration, **(b)** bandwidth-limited Gaussian laser pulse of 10 fs duration shifted in time to −20 fs due to a linear phase term in the spectral domain ($\phi' = -20$ fs), **(c)** symmetrical broadened Gaussian laser pulse due to $\phi'' = 200\,\text{fs}^2$, **(d)** third-order spectral phase ($\phi''' = 1000\,\text{fs}^3$) leading to a quadratic group delay, **(e)** combined action of all spectral phase coefficients **(a)**-**(d)**, **(f)** π step at the central frequency, **(g)** π step displaced from the central frequency, **(h)** sine modulation at the central frequency with $\phi(\omega) = 1 \sin[20\,\text{fs}(\omega - \omega_0)]$, **(i)** cosine modulation at the central frequency with $\phi(\omega) = 1 \cos[20\,\text{fs}(\omega - \omega_0)]$, **(j)** sine modulation at the central frequency with $\phi(\omega) = 1 \sin[30\,\text{fs}(\omega - \omega_0)]$, **(k)** symmetrical clipping of spectrum, **(l)** blocking of the central frequency components, **(m)** off-center absorption, **(n)** self-phase modulation. Note the spectral broadening, **(o)** double pulse, with a pulse-to-pulse delay of 60 fs. Note that the second harmonic field $E^2(t) = A^2(t)\mathrm{e}^{2\mathrm{i}\Phi_a(t)}\mathrm{e}^{2\mathrm{i}\omega_0 t}$ gives rise to the autocorrelation function S_{f2} is related to the second harmonic PSD, i. e. squared modulus of the Fourier transform of $E^2(t)$ as displayed in the right column. Hence, the shape of the second harmonic PSD is determined by phase modulation of the linear spectrum and can be used to efficiently control two-photon resonant processes [12.21, 112, 113] ◄►

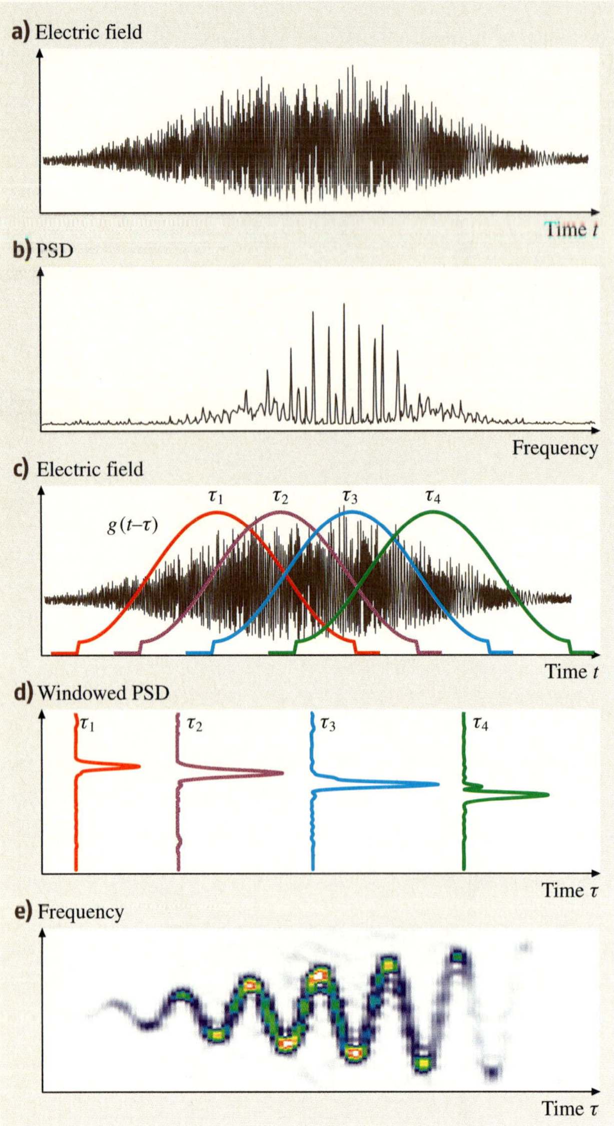

Fig. 12.29a–e Illustration of a short-time Fourier transform of a complicated electric field. **(a)** Electric field as a function of time; **(b)** power spectrum density as a function of frequency; **(c)** gating the electric field; four different time delays are shown; **(d)** power spectrum density for each gate; **(e)** spectrogram, revealing an oscillating instantaneous frequency as a function of time being the origin for the complicated electric field in **(a)**

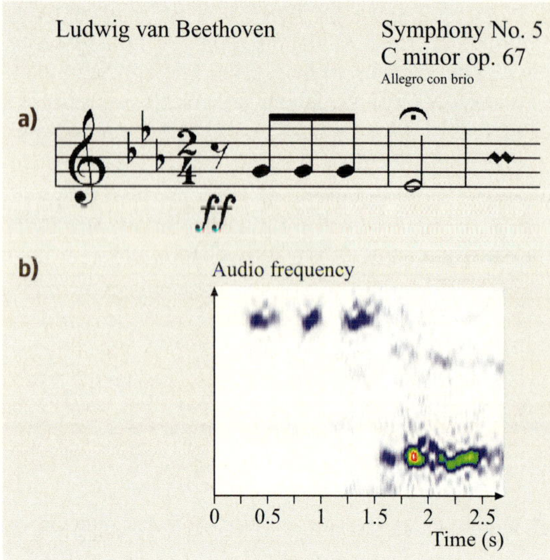

Fig. 12.30a,b Sheet of music **(a)** and corresponding spectrogram **(b)** of Beethoven's fifth symphony

time. A spectrogram corresponding to the beginning of Beethoven's fifth symphony is shown in Fig. 12.30b.

Figure 12.29 shows the concept of STFT on a complicated electric field of a laser. Once an electric field is retrieved in amplitude and phase there are other time – frequency distributions in use for displaying the data like for example the Wigner [12.24, 119–122] and Husimi representations [12.123, 124]. A quantity closely related to the spectrogram is the sonogram $S_{\text{sonogram}}(\omega, \tau)$

$$S_{\text{sonogram}}(\omega, \tau) \equiv \left| \int_{-\infty}^{\infty} \tilde{f}(\omega')\tilde{g}(\omega-\omega')\mathrm{e}^{+\mathrm{i}\omega'\tau}\,\mathrm{d}\omega' \right|^2 , \tag{12.81}$$

where $\tilde{g}(\omega-\omega')$ is a frequency gate in analogy to the time gate $g(t-\tau)$ used in the spectrogram. If $\tilde{g}(\omega)$ is the Fourier transform of $g(t)$ then it can be shown that the sonogram is equivalent to the spectrogram [12.6].

In ultrafast optics, the gate to record the spectrogram or the sonogram is usually the pulse itself.

Spectrogram-Based Methods

Recording a spectrogram is accomplished experimentally by gating the pulse with a variable delayed replica of the pulse in an instantaneous nonlinear optical medium followed by spectrally resolving the gated pulse. The basic optical layout of such a device is almost the same as a non-collinear autocorrelation setup depicted in Fig. 12.25b. Only the detector has to be replaced by a spectrometer and camera system in order to spectrally resolve the gated pulse. The corresponding

Fig. 12.31a–o Calculated FROG traces at corresponding frequencies for various ultrashort pulse shapes with a central wavelength of 800 nm according to the pulses displayed in Fig. 12.3. *Left*: polarization gate (PG) FROG. *Right*: second-harmonic generation (SHG) FROG. The temporal intensity $I(t)$, the additional temporal phase $\Phi_a(t)$ and the instantaneous frequency $\omega(t)$ are shown in the insets as a reminder. **(a)** bandwidth-limited Gaussian laser pulse of 10 fs duration, **(b)** bandwidth-limited Gaussian laser pulse of 10 fs duration shifted in time to -20 fs due to a linear phase term in the spectral domain ($\phi' = -20$ fs), **(c)** symmetrical broadened up chirped Gaussian laser pulse due to $\phi'' = 200\,\text{fs}^2$, **(d)** third-order spectral phase ($\phi''' = 1000\,\text{fs}^3$) leading to a quadratic group delay, **(e)** combined action of all spectral phase coefficients **(a)**–**(d)**, **(f)** π step at the central frequency, **(g)** π step displaced from the central frequency, **(h)** sine modulation at the central frequency with $\phi(\omega) = 1\sin[20\,\text{fs}(\omega - \omega_0)]$, **(i)** cosine modulation at the central frequency with $\phi(\omega) = 1\cos[20\,\text{fs}(\omega - \omega_0)]$, **(j)** sine modulation at the central frequency with $\phi(\omega) = 1\sin[30\,\text{fs}(\omega - \omega_0)]$, **(k)** symmetrical clipping of spectrum, **(l)** blocking of the central frequency components, **(m)** off-center absorption, **(n)** self-phase modulation, (note the spectral broadening), **(o)** double pulse, with a pulse-to-pulse delay of 60 fs ▶▶

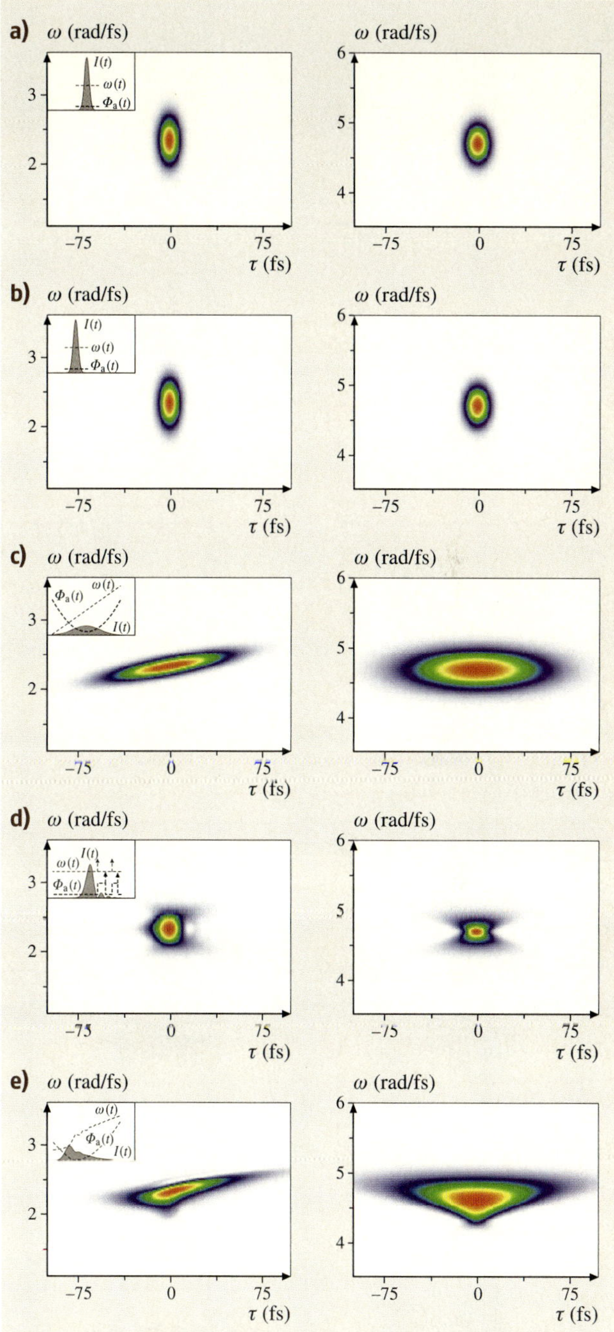

technique has been termed frequency-resolved optical gating (FROG) and is described in great detail in [12.6, 125] and references therein.

Depending on the instantaneous nonlinear optical effect used to gate the pulse in FROG, several different FROG geometries have been investigated [the setup of Fig. 12.25b) would correspond to a second-harmonic-generation (SHG) FROG]. These geometries can also be implemented as single-shot devices, similar to the single-shot autocorrelator depicted in Fig. 12.26. The FROG trace $I_{\text{FROG}}(\omega, \tau)$, that is a plot of frequency (wavelength) versus delay, is a spectrogram of the complex amplitude E_c [see (12.15)]. Neglecting any prefactors, different nonlinear optical effects yield the following expressions according to [12.6, 125].

Polarization-Gate (PG) FROG.

$$I_{\text{FROG}}^{\text{PG}}(\omega, \tau) = \left| \int_{-\infty}^{\infty} E_c(t)\,|E_c(t-\tau)|^2\,\mathrm{e}^{-\mathrm{i}\omega t}\,\mathrm{d}t \right|^2 . \qquad (12.82)$$

In a crossed-polarizer arrangement for the probe pulse this technique makes use of induced birefringence in fused silica in the presence of the gate pulse. The third-order optical nonlinearity is the electronic Kerr effect. The FROG traces obtained by this method are very intuitive (Fig. 12.31).

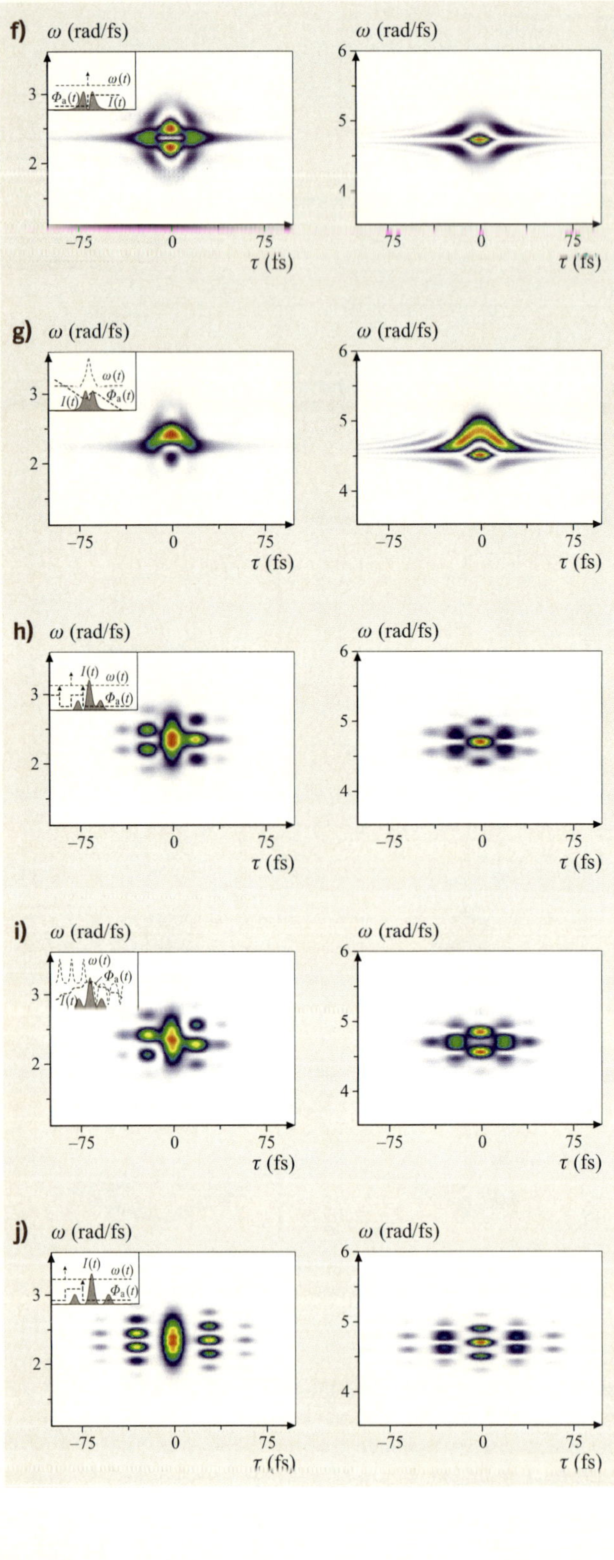

Self-Diffraction (SD) FROG.

$$I_{\text{FROG}}^{\text{SD}}(\omega,\tau)=\left|\int_{\infty}^{\infty} E_{\text{c}}(t)^2 E_{\text{c}}^*(t-\tau)\text{e}^{-\text{i}\omega t}\,\text{d}t\right|^2 . \tag{12.83}$$

In this approach the two beams (with the same polarization) generate a sinusoidal intensity pattern in the nonlinear medium (for example, fused silica) and hence introduce a material grating, which diffracts each beam. One of the diffracted beams is then the signal beam sent to the spectrometer.

Transient-Grating (TG) FROG.

$$I_{\text{FROG}}^{\text{TG}}(\omega,\tau) = \left|\int_{-\infty}^{\infty} E_{\text{c1}}(t)E_{\text{c2}}^*(t)E_{\text{c3}}(t-\tau)\text{e}^{-\text{i}\omega t}\,\text{d}t\right|^2 . \tag{12.84}$$

This is a three-beam setup, where two pulses are overlapped in time and space at the optical Kerr medium (for example fused silica), producing a refractive index grating similar as in SD FROG. In a TG a third pulse is variably delayed and overlapped in the fused silica and is diffracted by the induced grating, producing the signal beam for the spectrometer. The beams in the TG geometry (three input and one output) are kept nearly collinear and form a so-called boxcars arrangement, where the four spots appear in the corners of a rectangle when placing a card into the beams after the nonlinear medium. As TG is a phase-matched process, the sensitivity is higher compared to the SD approach. Depending on which pulse is variably delayed – with the other two coincident in time – the TG FROG trace is mathematically equivalent to the PG FROG (pulse one or three is delayed) or to the SD FROG (pulse two is delayed).

Third-Harmonic-Generation (THG) FROG.

$$I_{\text{FROG}}^{\text{THG}}(\omega,\tau)=\left|\int_{-\infty}^{\infty} E_{\text{c}}(t-\tau)^2 E_{\text{c}}(t)\text{e}^{-\text{i}\omega t}\,\text{d}t\right|^2 . \tag{12.85}$$

This method makes use of third-harmonic generation as the nonlinear process.

Second-Harmonic-Generation (SHG) FROG.

$$I_{\text{FROG}}^{\text{SHG}}(\omega,\tau)=\left|\int_{-\infty}^{\infty} E_{\text{c}}(t)E_{\text{c}}(t-\tau)\mathrm{e}^{-\mathrm{i}\omega t}\,\mathrm{d}t\right|^2 . \tag{12.86}$$

SHG FROG involves spectrally resolving a standard SHG-based non-collinear intensity autocorrelator, which always yields symmetric traces, resulting in a direction of time ambiguity for the SHG FROG. This ambiguity can experimentally be removed, for example, by placing a piece of glass in the beam before the beam splitter to introduce positive chirp or to create satellite pulses via surface reflections. Because of its high sensitivity and as it is based on a standard SHG autocorrelator, this method is widely employed. Examples of SHG FROG traces for various pulse shapes are given in Fig. 12.31.

A comparison of the different approaches is given in Table 12.4. Various calculated traces for common ultrashort pulse distortions for the PG and the SHG FROG geometries are given in Fig. 12.31. Calculated FROG traces for the other beam geometries are given in [12.125]. Measured FROG traces for different geometries are compiled in [12.6].

It is important to note that knowledge of the spectrogram (or sonogram) of the electric field of an ultrashort laser pulse is sufficient to completely determine the electric field in amplitude and phase (besides some ambiguities such as the absolute phase) because a spectrogram is equivalent to the two-dimensional phase-retrieval problem in image science and astronomy [12.126]. In general, phase retrieval is the problem of finding a function knowing only the magnitude (but not the phase) of its Fourier transform. Phase retrieval for a function of one variable is impossible. For example, knowledge of a pulse spectrum does not fully determine the pulse as infinitely many different pulses have the same spectrum (Fig. 12.3a–j). However, phase retrieval for a function of two variables is possible and the FROG trace can be rewritten as the squared magnitude of a two-dimensional Fourier transform [12.125]. Very sophisticated iterative retrieval procedures, which can rapidly retrieve the pulse from the FROG trace with update rates up to several Hz, exist [12.127].

In the following some additional attributes of FROG techniques, which also partially hold for sonogram methods, are summarized:

- As the FROG trace consists of $N \times N$ points, while the intensity and phase on the other hand only have

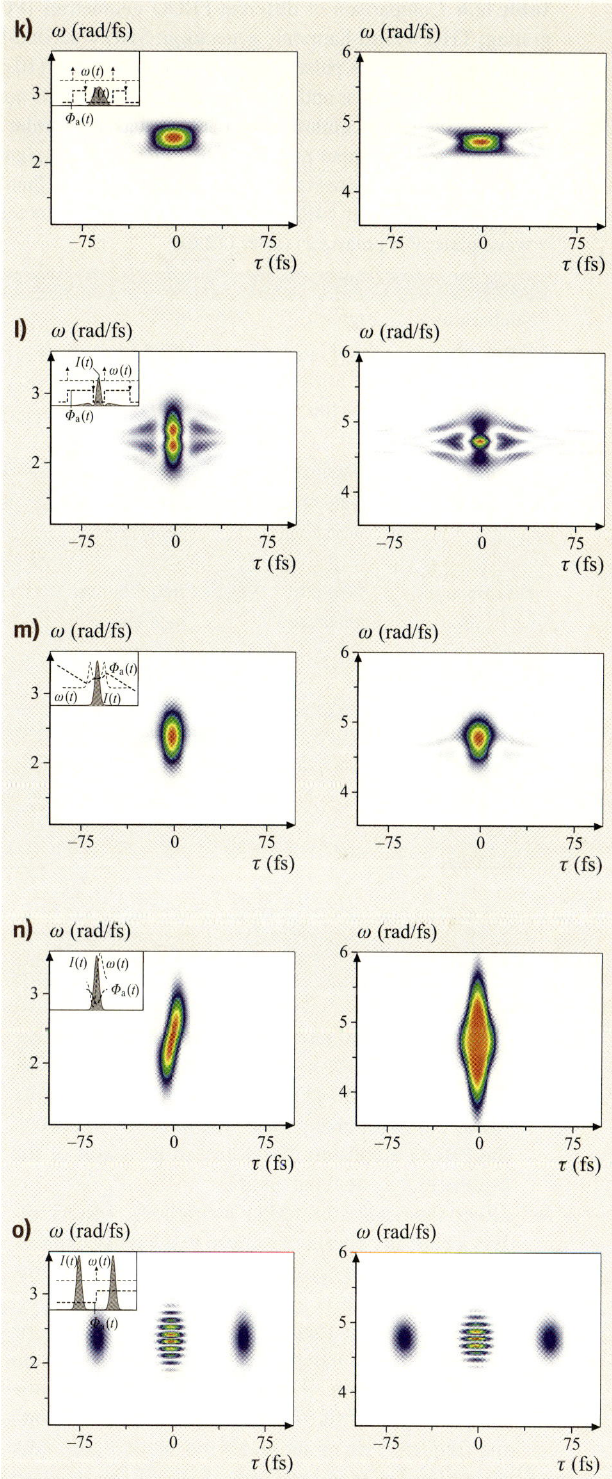

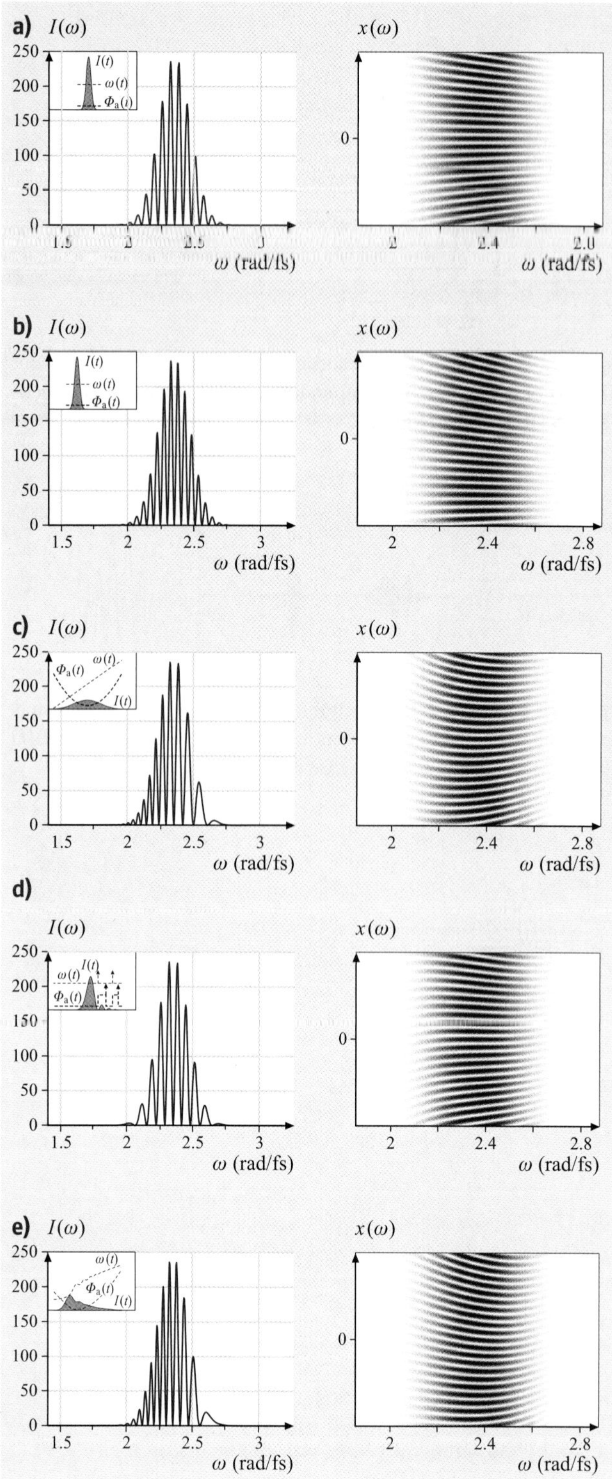

Fig. 12.35a–o Calculated spectral interference (SI) and spatial–spectral interference (SSI) traces for various ultrashort pulse shapes with a central wavelength of 800 nm according to the pulses displayed in Fig. 12.3. *Left* – SI: the time delay between both pulses is 100 fs. *Right*: SSI the angle 2Θ between both beams is 2°. The temporal intensity $I(t)$, the additional temporal phase $\Phi_a(t)$ and the instantaneous frequency $\omega(t)$ are shown in the *insets* as a reminder. **(a)** bandwidth-limited Gaussian laser pulse of 10 fs duration, **(b)** bandwidth-limited Gaussian laser pulse of 10 fs duration shifted in time to -20 fs due to a linear phase term in the spectral domain ($\phi' = -20\,\text{fs}$), **(c)** symmetrical broadened Gaussian laser pulse due to $\phi'' = 200\,\text{fs}^2$, **(d)** third-order spectral phase ($\phi''' = 1000\,\text{fs}^3$) leading to a quadratic group delay, **(e)** combined action of all spectral phase coefficients, **(a)-(d)(f)** π step at the central frequency, **(g)** π step displaced from the central frequency, **(h)** sine modulation at the central frequency with $\phi(\omega) = 1\sin[20\,\text{fs}(\omega-\omega_0)]$, **(i)** cosine modulation at the central frequency with $\phi(\omega) = 1\cos[20\,\text{fs}(\omega-\omega_0)]$, **(j)** sine modulation at the central frequency with $\phi(\omega) = 1\sin[30\,\text{fs}(\omega-\omega_0)]$, **(k)** symmetrical clipping of spectrum, **(l)** blocking of the central frequency components, **(m)** off-center absorption, **(n)** self-phase modulation, (note the spectral broadening) **(o)** double pulse, with a pulse-to-pulse delay of 60 fs ◀, ▶▶

version [12.141]. As the sonogram and the spectrogram are mathematically equivalent FROG retrieval algorithms (being in principle somewhat slower) can also be used in this approach [12.141]. From a practical point of view the method is experimentally more involved than a FROG setup and it is less sensitive, because energy is lost at the filter before the nonlinear medium. In the SHG version, the STRUT apparatus and the FROG apparatus are identical when removing the frequency filter and using a spectrometer as the detector in Fig. 12.32.

12.3.5 Spectral Interferometry

The techniques described so far make use of nonlinear optical processes in order to determine the amplitude and phase of ultrashort laser pulses. Although, with the help of SHG FROG, pulses down to the picojoule regime can be measured in a multishot setup, shaped ultrashort laser pulses might spread out their energy over a time scale of picoseconds and thus prevent characterization with the help of nonlinear processes. However, as these pulses are usually created from an oscillator or an amplifier a well-characterized reference pulse is often available. This allows to make use of highly sensitive linear techniques to determine the amplitude and phase of an ultrashort laser pulse. The technique is named spectral interferometry (SI), frequency-domain interferometry or Fourier-transform spectral interferometry [12.6, 142–145]. The basic SI setup is depicted in Fig. 12.33. A well-characterized reference pulse $E_{\text{ref}}(t)$ and a modified signal pulse $E(t)$ derived from an experiment or a pulse shaper are directed collinearly into a spectrometer. The measured SI spectrum $S_{\text{SI}}(\omega)$ is proportional to the square of the Fourier transform of the sum of the two fields:

$$\begin{aligned} S_{\text{SI}}(\omega) &\propto |\text{Fouriertransform}\,\{E_{\text{ref}}(t)+E(t-\tau)\}|^2 \\ &\propto \left|\tilde{E}_{\text{ref}}(\omega)+\tilde{E}(\omega)\mathrm{e}^{-\mathrm{i}\omega\tau}\right|^2 \\ &\propto \left|\sqrt{I_{\text{ref}}(\omega)}\mathrm{e}^{-\mathrm{i}\phi_{\text{ref}}(\omega)}+\sqrt{I(\omega)}\mathrm{e}^{-\mathrm{i}\phi(\omega)-\mathrm{i}\omega\tau}\right|^2 \\ &= I_{\text{ref}}(\omega)+I(\omega)+\sqrt{I_{\text{ref}}(\omega)}\sqrt{I(\omega)} \\ &\quad\times(\mathrm{e}^{\mathrm{i}\phi_{\text{ref}}(\omega)-\mathrm{i}\phi(\omega)-\mathrm{i}\omega\tau}+c.c.) \\ &= I_{\text{ref}}(\omega)+I(\omega)+2\sqrt{I_{\text{ref}}(\omega)}\sqrt{I(\omega)} \\ &\quad\times\cos(\phi_{\text{ref}}(\omega)-\phi(\omega)-\omega\tau)\,. \end{aligned} \tag{12.87}$$

The phase difference

$$\Delta\phi(\omega)=\phi_{\text{ref}}(\omega)-\phi(\omega) \tag{12.88}$$

can be extracted from the measured $S_{\text{SI}}(\omega)$. Using the arcos function is not recommended because experimental noise can lead to large phase errors [12.145]. Commonly a Fourier-transform technique is used [12.145, 146] where the phase difference is extracted by the Fourier transform of the measured spectrum, ignoring the negative and zero-frequency components and shifting the positive-frequency components to direct current (DC) in order to remove the delay term $\mathrm{e}^{-\mathrm{i}\omega\tau}$. An inverse Fourier transform then yields the phase difference $\Delta\phi(\omega)$. With the help of the known reference phase $\phi_{\text{ref}}(\omega)$, $\phi(\omega)$ is finally obtained.

In the following some attributes of SI are summarized:

- SI requires the spectrum of the reference pulse to contain completely the spectrum of the unknown pulse.
- If the reference pulse and the signal pulse are identical, the phase difference is zero. The remaining oscillations on the spectrum due to the delay term are a convenient tool to adjust, for example, interferometric autocorrelator setups.
- Using the FROG technique for characterization of the reference pulse, the combined technique was termed temporal analysis by dispersing a pair of light E-fields (TADPOLE) [12.6, 147].

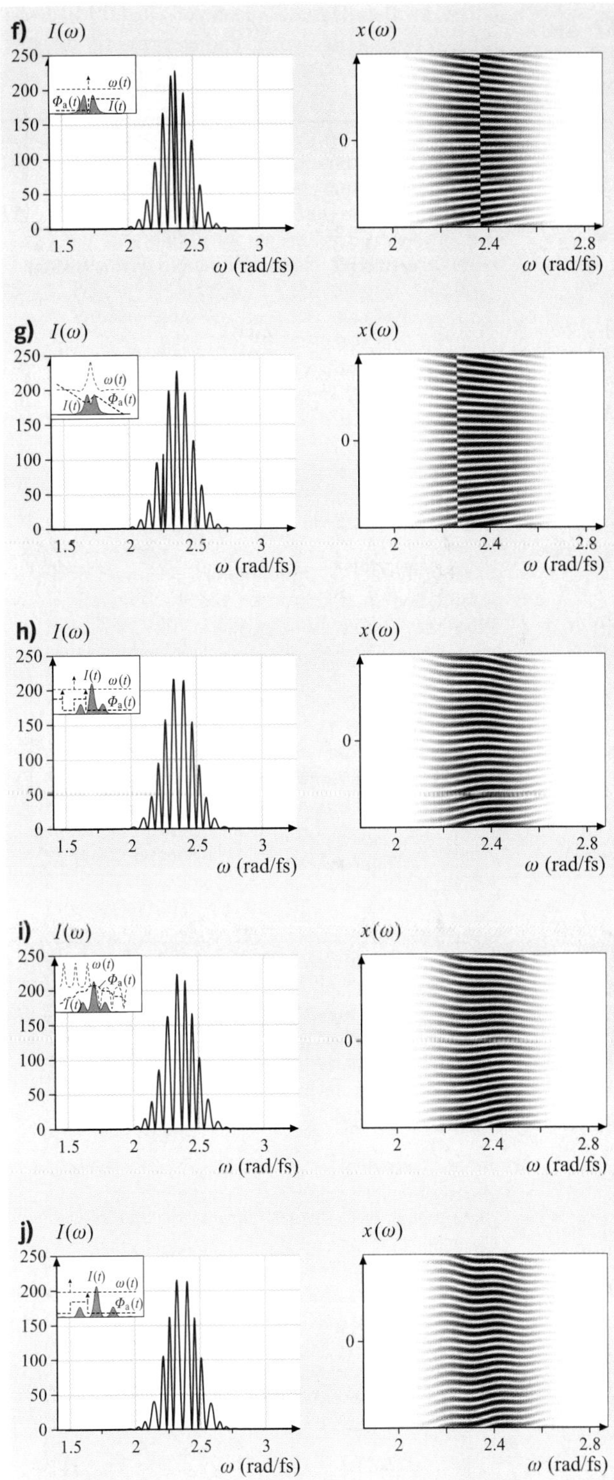

12.32 O. E. Martinez, J. P. Gordon, R. L. Fork: Negative group velocity dispersion using refraction, J. Opt. Soc. Am. A **1**, 1003–1006 (1984)

12.33 F. J. Duarte: Generalized multiple-prism dispersion theory for pulse compression in ultrafast dye lasers, Opt. Quant. Electron. **19**, 223–229 (1987)

12.34 V. Petrov, F. Noack, W. Rudolph, C. Rempel: Intracavity dispersion compensation and extracavity pulse compression using pairs of prisms, Exp. Tech. Phys. **36**, 167–173 (1988)

12.35 C. P. J. Barty, C. L. Gordon, B. E. Lemoff: Multiterawatt 30-fs Ti:Sapphire laser system, Opt. Lett. **19**, 1442–1444 (1994)

12.36 D. Strickland, G. Mourou: Compression of amplified chirped optical pulses, Opt. Commun. **56**, 219–221 (1985)

12.37 S. Backus, C. G. Durfee III, M. M. Murnane, H. C. Kapteyn: High power ultrafast lasers, Rev. Sci. Instrum. **69**, 1207–1223 (1998)

12.38 C. H. B. Cruz, P. C. Becker, R. L. Fork, C. V. Shank: Phase correction of femtosecond optical pulses using a combination of prisms and gratings, Opt. Lett. **13**, 123–125 (1988)

12.39 R. L. Fork, C. H. Brito Cruz, C. H. Becker, C. V. Shank: Compression of optical pulses to six femtoseconds by using cubic phase compensation, Opt. Lett. **12**, 483–485 (1987)

12.40 E. B. Treacy: Optical pulse compression with diffraction gratings, IEEE J. Quant. Electron. **5**, 454–458 (1969)

12.41 A. Suzuki: Complete analysis of a two mirror unit magnification system, Appl. Opt. **22**, 3943 (1983)

12.42 G. Cheriaux, P. Rousseau, F. Salin, J. P. Chambaret, B. Walker, L. F. DiMauro: Aberration-free stretcher design for ultrashort pulse amplification, Opt. Lett. **21**, 414 (1996)

12.43 O. E. Martinez: Matrix formalism for pulse compressors, IEEE J. Quant. Electron. **24**, 2530–2536 (1988)

12.44 O. E. Martinez: Grating and prism compressor in the case of finite beam size, J. Opt. Soc. Am. B **3**, 929–934 (1986)

12.45 F. Gires, P. Tournois: Interferometre utilisable pour la compression d'impulsions lumineuses modulees en frequence, C. R. Acad. Sci. Paris **258**, 6112–6115 (1964)

12.46 P. M. W. French: The generation of ultrashort laser pulses, Rep. Prog. Phys. **58**, 169–267 (1995)

12.47 W. Demtröder: *Laser Spectroscopy* (Springer, Berlin, Heidelberg 1996)

12.48 J. Heppner, J. Kuhl: Intracavity chirp compensation in a colliding pulse mode-locked laser using thin-film interferometers, Appl. Phys. Lett. **47**, 453–455 (1985)

12.49 R. Szipöcs, K. Ferencz, C. Spielmann, F. Krausz: Chirped multilayer coatings for broadband dispersion control in femtosecond lasers, Opt. Lett. **19**, 201–203 (1994)

12.50 A. Stingl, C. Spielmann, F. Krausz, R. Szipocs: Generation of 11 fs pulses from a Ti:sapphire laser without the use of prisms, Opt. Lett. **19**, 204–6 (1994)

12.51 N. Matuschek, F. X. Kärtner, U. Keller: Exact coupled mode theories for multilayer interference coatings with arbitrary strong index modulation, IEEE J. Quant. Electron. **33**, 295–302 (1997)

12.52 N. Matuschek, F. X. Kärtner, U. Keller: Analytical design of double-chirped mirrors with custom-tailored dispersion characteristics, IEEE J. Quant. Electron. **35**, 129–137 (1999)

12.53 D. H. Sutter, G. Steinmeyer, L. Gallmann, N. Matuschek, F. Morier-Genoud, U. Keller, V. Scheuer, G. Angelow, T. Tschudi: Semiconductor saturable-absorber mirror-assisted Kerr-lens mode-locked Ti:sapphire laser producing pulses in the two-cycle regime, Opt. Lett. **24**, 631–633 (1999)

12.54 N. Matuschek, L. Gallmann, D. H. Sutter, G. Steinmeyer, U. Keller: Back-side-coated chirped mirrors with ultra-smooth broadband dispersion characteristics, Appl. Phys. B **71**, 509–522 (2000)

12.55 A. M. Weiner: Femtosecond pulse shaping using spatial light modulators, Rev. Sci. Instrum. **71**, 1929–1960 (2000)

12.56 R. S. Judson, H. Rabitz: Teaching lasers to control molecules, Phys. Rev. Lett. **68**, 1500–1503 (1992)

12.57 T. Baumert, T. Brixner, V. Seyfried, M. Strehle, G. Gerber: Femtosecond pulse shaping by an evolutionary algorithm with feedback, Appl. Phys. B **65**, 779–782 (1997)

12.58 C. J. Bardeen, V. V. Yakolev, K. R. Wilson, S. D. Carpenter, P. M. Weber, W. S. Warren: Feedback quantum control of molecular electronic population transfer, Chem. Phys. Lett. **280**, 151–158 (1997)

12.59 D. Yelin, D. Meshulach, Y. Silberberg: Adaptive femtosecond pulse compression, Opt. Lett. **22**, 1793–1795 (1997)

12.60 A. Assion, T. Baumert, M. Bergt, T. Brixner, B. Kiefer, V. Seyfried, M. Strehle, G. Gerber: Control of chemical reactions by feedback-optimized phase-shaped femtosecond laser pulses, Science **282**, 919–922 (1998)

12.61 T. Brixner, M. Strehle, G. Gerber: Feedback-controlled optimization of amplified femtosecond laser pulses, Appl. Phys. B **68**, 281–284 (1999)

12.62 R. Bartels, S. Backus, E. Zeek, L. Misoguti, G. Vdovin, I. P. Christov, M. M. Murnane, H. C. Kapteyn: Shaped-pulse optimization of coherent emission of high-harmonic soft X-rays, Nature **406**, 164–166 (2000)

12.63 T. Brixner, N. H. Damrauer, P. Niklaus, G. Gerber: Photoselective adaptive femtosecond quantum control in the liquid phase, Nature **414**, 57–60 (2001)

12.64 J. L. Herek, W. Wohlleben, R. Cogdell, D. Zeidler, M. Motzkus: Quantum control of energy flow in light harvesting, Nature **417**, 533–535 (2002)

12.65 J. Kunde, B. Baumann, S. Arlt, F. Morier-Genoud, U. Siegner, U. Keller: Adaptive feedback control of ultrafast semiconductor nonlinearities, Appl. Phys. Lett. **77**, 924–926 (2000)

12.66 T. C. Weinacht, R. Bartels, S. Backus, P. H. Bucksbaum, B. Pearson, J. M. Geremia, H. Rabitz, H. C. Kapteyn, M. M. Murnane: Coherent learning control of vibrational motion in room temperature molecular gases, Chem. Phys. Lett. **344**, 333–338 (2001)

12.67 R. J. Levis, G. M. Menkir, H. Rabitz: Selective bond dissociation and rearrangement with optimally tailored, strong-field laser pulses, Science **292**, 709–713 (2001)

12.68 C. Daniel, J. Full, L. Gonzáles, C. Lupulescu, J. Manz, A. Merli, S. Vajda, L. Wöste: Deciphering the reaction dynamics underlying optimal control laser fields, Science **299**, 536–539 (2003)

12.69 T. Brixner, G. Gerber: Quantum control of gas-phase and liquid-phase femtochemistry, Chem. Phys. Chem **4**, 418–438 (2003)

12.70 M. Wollenhaupt, A. Präkelt, C. Sarpe-Tudoran, D. Liese, T. Baumert: Strong field quantum control by selective population of dressed states, J. Opt. B **7**, S270–S276 (2005)

12.71 C. Horn, M. Wollenhaupt, M. Krug, T. Baumert, R. de Nalda, L. Banares: Adaptive control of molecular alignment, Phys. Rev. A **73**, 031401-1–031401-4 (2006)

12.72 T. Brixner, G. Gerber: Femtosecond polarization pulse shaping, Opt. Lett. **26**, 557–559 (2001)

12.73 A. Präkelt, M. Wollenhaupt, A. Assion, Ch. Horn, C. Sarpe-Tudiran, M. Winter, T. Baumert: A compact, robust and flexible setup for femtosecond pulse shaping, Rev. Sci. Instrum. **74**, 4950–4953 (2003)

12.74 L. Xu, N. Nakagawa, R. Morita, H. Shigekawa, M. Yamashita: Programmable chirp compensation for 6-fs pulse generation with a prism-pair-formed pulse shaper, IEEE J. Quant. Electron. **36**, 893–899 (2000)

12.75 G. Stobrawa, M. Hacker, T. Feurer, D. Zeidler, M. Motzkus, F. Reichel: A new high-resolution femtosecond pulse shaper, Appl. Phys. B **72**, 627–630 (2001)

12.76 M. M. Wefers, K. A. Nelson: Programmable phase and amplitude femtosecond pulse shaping, Opt. Lett. **18**, 2032–2034 (1993)

12.77 A. Präkelt, M. Wollenhaupt, C. Sarpe-Tudoran, A. Assion, T. Baumert: Filling a spectral hole via self-phase modulation, Appl. Phys. Lett. **87**, 121113-1–121113-3 (2005)

12.78 T. Brixner, G. Krampert, P. Niklaus, G. Gerber: Generation and characterization of polarization-shaped femtosecond laser pulses, Appl. Phys. B **74**, 133–144 (2002)

12.79 T. Brixner, G. Krampert, T. Pfeifer, R. Selle, G. Gerber, M. Wollenhaupt, O. Graefe, C. Horn, D. Liese, T. Baumert: Quantum control by ultrafast polarization shaping, Phys. Rev. Lett. **92**, 208301-1–208301-4 (2004)

12.80 M. Wollenhaupt, V. Engel, T. Baumert: Femtosecond laser photoelectron spectroscopy on atoms and small molecules: Prototype studies in quantum control, Ann. Rev. Phys. Chem. **56**, 25–56 (2005)

12.81 E. Zeek, K. Maginnis, S. Backus, U. Russek, M. M. Murnane, G. Mourou, H. C. Kapteyn: Pulse compression by use of deformable mirrors, Opt. Lett. **24**, 493–495 (1999)

12.82 M. Hacker, G. Stobrawa, R. A. Sauerbrey, T. Buckup, M. Motzkus, M. Wildenhain, A. Gehner: Micromirror SLM for femtosecond pulse shaping in the ultraviolet, Appl. Phys. B **76**, 711–714 (2003)

12.83 J. X. Tull, M. A. Dugan, W. S. Warren: High-resolution, ultrafast laser pulse shaping and its applications. In: *Adv. in Magn. and Opt. Resonance*, ed. by W. S. Warren (Academic, New York 1997)

12.84 D. Goswami: Optical pulse shaping approaches to coherent control, Phys. Rep. **374**, 385–481 (2003)

12.85 P. Tournois: Acousto-optic programmable dispersive filter for adaptive compensation of group delay time dispersion in laser systems, Opt. Commun. **140**, 245–249 (1997)

12.86 F. Verluise, V. Laude, Z. Cheng, C. Spielmann, P. Tournois: Amplitude and phase control of ultrashort pulses by use of an acousto-optic programmable dispersive filter: pulse compression and shaping, Opt. Lett. **25**, 575–577 (2000)

12.87 F. Verluise, V. Laude, J. P. Huignard, P. Tournois: Arbitrary dispersion control of ultrashort optical pulses using acoustic waves, J. Opt. Soc. Am. B **17**, 138–145 (2000)

12.88 J. C. Vaughan, T. Feurer, K. A. Nelson: Automated two-dimensional femtosecond pulse shaping, J. Opt. Soc. Am. B **19**, 2489–2495 (2002)

12.89 T. Feurer, J. C. Vaughan, K. A. Nelson: Spatiotemporal coherent control of lattice vibrational waves, Science **299**, 374–377 (2003)

12.90 G. H. C. New: The generation of ultrashort light pulses, Rep. Prog. Phys. **46**, 877–971 (1983)

12.91 J. D. Simon: Ultrashort light pulses, Rev. Sci. Instrum. **60**, 3597–3624 (1989)

12.92 P. M. W. French: Ultrafast solid state lasers, Contemp. Phys. **37**, 283–301 (1996)

12.93 H. A. Haus: Mode-locking of lasers, IEEE J. Quant. Electron. **6**, 1173–1185 (2000)

12.94 G. R. Fleming: *Chemical Applications of Ultrafast Spectroscopy* (Oxford Univ. Pres, New York 1986)

12.95 O. Svelto: *Principles of Lasers* (Plenum, New York 1998)

12.96 A. E. Siegmann: *Lasers* (Univ. Sci. Books, Mill Valley 1986)

12.97 A. Yariv: *Quantum Electronics* (Wiley, New York 1989)

12.98 M. E. Fermann, A. Galvanauskas, G. Sucha: *Ultrafast Lasers – Technology and Applications* (Marcel Dekker, New York 2003)

12.99 J. A. Valdmanis, R. L. Fork: Design considerations for a femtosecond pulse laser balancing self phase modulation, group velocity dispersion, saturable absorption and saturable gain, IEEE J. Quant. Electron. **22**, 112–118 (1986)

12.100 R. Ell, U. Morgner, F. X. Kärtner, J. Fujimoto, E. P. Ippen, V. Scheuer, G. Angelow, T. Tschudi, A. Lederer, A. Boiko, B. Luther-Davies: Generation of 5-fs pulses and octave-spanning spectra directly from a Ti:sapphire laser, Opt. Lett. **26**, 373–375 (2001)

12.101 D. E. Spence, P. N. Kean, W. Sibbett: 60-fs pulse generation from a self-mode-locked Ti:sapphire laser, Opt. Lett. **16**, 42–44 (1991)

12.102 K. K. Hamamatsu Photonics: *Guide to Streak Cameras*, http://www.hamamatsu.com, (1999)

12.103 Y. Tsuchiya: Advances in streak camera instrumentation for the study of biological and physical processes, IEEE J. Quant. Electron. **20**, 1516–1528 (1984)

12.104 E. P. Ippen, C. V. Shank: *Ultrashort Light Pulses*, ed. by S. L. Shapiro (Springer, Berlin, Heidelberg 1977) p. 83

12.105 J. K. Ranka, A. L. Gaeta, A. Baltuska, M. S. Pshenichnikov, D. A. Wiersma: Autocorrelation measurements of 6-fs pulses based on the two-photon-induced photocurrent in a GaAsP photodiode, Opt. Lett. **22**, 1344–1346 (1997)

12.106 W. Rudolph, M. Sheik-Bahae, A. Bernstein, L. F. Lester: Femtosecond autocorrelation measurements based on two-photon photoconductivity in ZnSe, Opt. Lett. **22**, 313–315 (1997)

12.107 F. Salin, P. Georges, A. Brun: Single-shot measurements of a 52-fs pulse **26**, 4528–4531 (1987)

12.108 A. Brun, P. Georges, G. Le Saux, F. Salin: Single-shot characterization of ultrashort light pulses, J. Phys. D **24**, 1225–1233 (1991)

12.109 G. Szabó, Z. Bor, A. Müller: Phase-sensitive single-pulse autocorrelator for ultrashort laser pulses, Opt. Lett. **13**, 746–748 (1988)

12.110 B. W. Shore: *The Theory of Coherent Atomic Excitation*, Vol. 1 (Wiley, New York 1990)

12.111 J. C. Diels, J. J. Fontaine, I. C. McMichael, F. Simoni: Control and measurement of ultrashort pulse shapes (in amplitude and phase) with femtosecond accuracy, Appl. Opt. **24**, 1270–1282 (1985)

12.112 D. Meshulach, Y. Silberberg: Coherent quantum control of two-photon transitions by a femtosecond laser pulse, Nature **396**, 239–242 (1998)

12.113 V. V. Lozovoy, I. Pastirk, A. Walowicz, M. Dantus: Multiphoton intrapulse interference. II. Control of two- and three-photon laser induced fluorescence with shaped pulses, J. Chem. Phys. **118**, 3187–3196 (2003)

12.114 V. V. Lozovoy, I. Pastirk, M. Dantus: Multiphoton intrapulse interference. IV Ultrashort laser pulse spectral phase characterization and compensation, Opt. Lett. **29**, 1–3 (2004)

12.115 J. Peatross, A. Rundquist: Temporal decorrelation of short laser pulses, J. Opt. Soc. Am. B **15**, 216–222 (1998)

12.116 K. Naganuma, K. Mogi, H. Yamada: General method for ultrashort light pulse chirp measurement, IEEE J. Quant. Electron. **25**, 1225–1233 (1989)

12.117 E. W. Van Stryland: The effect of pulse to pulse variation on ultrashort pulsewidth measurements, Opt. Commun. **31**, 93–96 (1979)

12.118 S. Quian, D. Chen: *Joint time-frequency analysis; Methods and Applications* (Prentice Hall, New Jersey 1996)

12.119 M. O. Scully, M. S. Zubairy: *Quantum Optics* (Cambridge Univ. Press, Cambridge 1995)

12.120 W. P. Schleich: *Quantum Optics in Phase Space* (Wiley-VCH, Weinheim 2001)

12.121 H. O. Bartelt, K. H. Brenner, A. W. Lohmann: The Wigner distribution function and its optical production, Opt. Commun. **32**, 32–38 (1980)

12.122 J. Paye, A. Migus: Space time Wigner functions and their application to the analysis of a pulse shaper, J. Opt. Soc. Am. B **12**, 1480–1490 (1995)

12.123 H. W. Lee: Generalized antinormally ordered quantum phase-space distribution functions, Phys. Rev. A **50**, 2746–2749 (1994)

12.124 D. Lalovic, D. M. Davidovic, N. Bijedic: Quantum mechanics in terms of non-negative smoothed Wigner functions, Phys. Rev. A **46**, 1206–1212 (1992)

12.125 R. Trebino, K. W. DeLong, D. N. Fittinghoff, J. N. Sweetser, M. A. Krumbügel, B. A. Richman, D. J. Kane: Measuring ultrashort laser pulses in the time-frequency domain using frequency-resolved optical gating, Rev. Sci. Instrum. **68**, 3277–3295 (1997)

12.126 H. Stark: *Image Recovery: Theory and Application* (Academic, Orlando 1987)

12.127 D. Kane: Recent progress toward real-time measurement of ultrashort laser pulses, IEEE J. Quant. Electron. **35**, 421–431 (1999)

12.128 S. Linden, J. Kuhl, H. Giessen: Amplitude and phasecharacterization of weak blue ultrashort pulses by downconversion, Opt. Lett. **24**, 569–571 (1999)

12.129 J. Zhang, A. P. Shreenath, M. Kimmel, E. Zeek, R. Trebino: Measurement of the intensity and phase of attojoule femtosecond light pulses using optical-parametric-amplification cross-correlation frequency-resolved optical gating, Opt. Express **11**, 601–609 (2003)

12.130 A. Baltuska, M. Pshenichnikov, D. A. Wiersma: Amplitude and phasecharacterization of 4.5-fs pulses by frequency-resolved optical gating, Opt. Lett. **23**, 1474–1476 (1998)

12.131 D. N. Fittinghoff, J. Squier, C. P. J. Barty, J. N. Sweetser, R. Trebino, M. Müller: Collinear type II second-harmonic-generation frequency-resolved optical gating for use with high-numerical aperture objectives, Opt. Lett. **23**, 1046–1048 (1998)

12.132 L. Gallmann, G. Steinmeyer, D. H. Sutter, N. Matuschek, U. Keller: Collinear type II second harmonic-generation frequency-resolved optical gating for the characterization of sub-10-fs optical pulses, Opt. Lett. **25**, 269–271 (2000)

12.133 D. Meshulach, D. Yelin, Y. Silberberg: Real-time spatial-spectral interference measurements of ultrashort optical pulses, J. Opt. Soc. Am. B **14**, 2095–2098 (1997)

12.134 P. O'Shea, M. Kimmel, X. Gu, R. Trebino: Highly simplified device for ultrashort-pulse measurement, Opt. Lett. **26**, 932–934 (2001)

12.135 C. Radzewicz, P. Wasylczyk, J. S. Krasinski: A poor man's FROG, Opt. Commun. **186**, 329–333 (2000)

12.136 S. Akturk, M. Kimmel, P. O'Shea, R. Trebino: Measuring pulse-front tilt in ultrashort pulses using GRENOUILLE, Optics Express **11**, 491–501 (2003)

12.137 S. Akturk, M. Kimmel, P. O'Shea, R. Trebino: Measuring spatial chirp in ulrashort pulses using single-shot frequency-resolved optical gating, Opt. Express **11**, 68–78 (2003)

12.138 M. Winter, M. Wollenhaupt, T. Baumert: Coherent matter waves for ultrafast laser pulse characterization, Opt. Commun. **264**, 285–292 (2006)

12.139 J. L. A. Chilla, O. E. Martinez: Direct determination of the amplitude and the phase of femtosecond light pulses, Opt. Lett. **16**, 39–41 (1991)

12.140 J. P. Foing, J. P. Likforman, M. Joffre, A. Migus: Femtosecond pulse phase measurement by spectrally resolved up-conversion; application to continuum compression, IEEE J. Quant. Electron. **28**, 2285–2290 (1992)

12.141 J. K. Rhee, T. S. Sosnowski, A. C. Tien, T. B. Norris: Real-time dispersion analyzer of femtosecond laser pulses with use of a spectrally and temporally resolved upconversion technique, J. Opt. Soc. Am. B **13**, 1780–1785 (1996)

12.142 Cl. Froehly, A. Lacourt, J. Ch. Vienot: Time impulse response and time frequency response of optical pupils. Experimental confirmation and applications, Nouv. Rev. Opt. **4**, 183–196 (1973)

12.143 J. Piasecki, B. Colombeau, M. Vampouille, Cl. Froehly, J. A. Arnaud: Nouvelle methode de mesure de la reponse impulsionnelle des fibres optiques, Appl. Opt. **19**, 3749–3755 (1980)

12.144 F. Reynaud, F. Salin, A. Barthelemy: Measurement of phase shifts introduced by nonlinear optical phenomena on subpicosecond pulses, Opt. Lett. **14**, 275–277 (1989)

12.145 L. Lepetit, G. Cheriaux, M. Joffre: Linear techniques of phase measurement by femtosecond spectral interferometry for applications in spectroscopy, J. Opt. Soc. Am. B **12**, 2467–2474 (1995)

12.146 M. Takeda, H. Ina, S. Kobayashi: Fourier-transform method of fringe-pattern analysis for computer-based topography and interferometry, J. Opt. Soc. Am. **72**, 156–159 (1981)

12.147 D. N. Fittinghoff, J. L. Bowie, J. N. Sweetser, R. T. Jennings, K. W. DeLong, M. A. Krumbügel, R. Trebino, I. A. Walmsley: Measurement of the intensity and phase of ultraweak, ultrashort laser pulses, Opt. Lett. **21**, 884–886 (1996)

12.148 T. Brixner, A. Oehrlein, M. Strehle, G. Gerber: Feedback-controlled femtosecond pulse shaping, Appl. Phys. B **70**, S119–S124 (2000)

12.149 W. J. Walecki, D. N. Fittinghoff, A. L. Smirl, R. Trebino: Characterization of the polarization state of weak ultrashort coherent signals by dual-channel spectral interferometry, Opt. Lett. **22**, 81–83 (1997)

12.150 C. Iaconis, A. Walmsley: Spectral phase interferometry for direct electric-field reconstruction of ultrashort optical pulses, Opt. Lett. **23**, 792–794 (1998)

12.151 L. Gallmann, D. H. Sutter, N. Matuschek, G. Steinmeyer, U. Keller, C. Iaconis, I. A. Walmsley: Characterization of sub-6-fs optical pulses with spectral phase interferometry for direct electric-field reconstruction, Opt. Lett. **24**, 1314–1316 (1999)

12.152 L. Gallmann, D. H. Sutter, N. Matuschek, G. Steinmeyer, U. Keller: Techniques for the characterization of sub-10-fs optical pulses: a comparison, Appl. Phys. B **70**, 67–75 (2000)

12.153 L. Gallmann, G. Steinmeyer, D. H. Sutter, T. Rupp, C. Iaconis, I. A. Walmsley, U. Keller: Spatially resolved amplitude and phase characterization of femtosecond optical pulses, Opt. Lett. **26**, 96–98 (2001)

12.154 P. Baum, S. Lochbrunner, E. Riedle: Zero-aditional-phase SPIDER: full characterization of visible and sub 20-fs ultraviolet pulses, Opt. Lett. **29**, 210–212 (2004)

Part D Selected Applications and Special Fields

13. Optical and Spectroscopic Techniques

Optical and S

Spectroscopy is the most important method for gaining detailed information on the structure and dynamics of atoms and molecules. The essential criteria of any spectroscopic technique are the attainable spectral resolution and the sensitivity.

Since lasers have been introduced as coherent narrow-band, intense radiation sources, spectroscopy has seen impressive progress. With single-mode lasers the spectral resolution can be greatly increased and finer details, generally hidden within the Doppler width of spectral lines, can be resolved. The available high intensity of lasers allows nonlinear spectroscopic techniques and the possibility of generating ultra-short light pulses has opened access to studies of very fast dynamical processes, such as the breaking of chemical bonds or the time-resolved redistribution of energy pumped into molecules by the absorption of photons.

In this first part we will concentrate on stationary methods of spectroscopy where the spectral resolution, the maximum achievable sensitivity and the development of optimum detectors are the main subjects.

13.1 Stationary Methods

Basic Principles of Spectroscopic Techniques

All spectroscopic techniques can be classified into those suitable for absorption spectroscopy and others adapted to emission spectroscopy. Emission spectroscopy demands the population of excited states that can emit the radiation. This excitation can be achieved by collisions (examples are gas discharges where electron impact excitation is the main excitation mechanism, or hot gases, such as the atmospheres of stars, where collisions between atoms or ions populate excited states) or by photon absorption (e.g. laser-excited fluorescence).

One important criterion for any spectroscopic techniques is its sensitivity, which is defined as the minimum detectable number of photons absorbed or emitted on a transition. This is also a measure for the minimum number of atoms or molecules which can be monitored. In particular for analytical investigations the question of how few atoms or molecules can still be detected in the presence of other species and in a rough environment, is of crucial importance for the selection of the optimum technique. The ultimate sensitivity is reached when single atoms or molecules, for instance as pollutants in a gas, in a liquid or on surfaces can still be measured. One example, where this sensitivity is really needed, is the optical detection of rare isotopes in the presence of other species that are more abundant by many orders of magnitude.

Another essential feature of all spectroscopic techniques is the spectral resolving power R. It is defined as

$$R = |\lambda/\Delta\lambda_{min}| = |\nu/\Delta\nu_{min}| \quad (13.1)$$

where $\Delta\lambda_{min}$ or $\Delta\nu_{min}$ is the minimum spectral interval (measured in wavelengths λ or in frequencies ν) that can still be resolved, i. e. two spectral lines separated by $\Delta\lambda > \Delta\lambda_{min}$ can be recognized as two separated lines. This minimal interval $\Delta\lambda_{min}$ depends on the profile of the spectral lines. A criterium postulated by Lord Rayleigh is the following Fig. 13.1.

Two spectral lines can be regarded as resolved, when the two partly overlapping line profiles have a dip between the two maxima with intensity I_{max}, which does not exceed 0.8 I_{max}.
The spectral resolving power R depends on

1. The spectroscopic instrument used for the separation of different wavelengths of absorbing or emitting transitions. These can be either spectrometers or interferometers.
2. On the line width of the absorption- or emission lines. For gaseous samples at low pressures this is generally the Doppler width, at higher pressures the pressure-broadened line profiles.

Absorption Techniques

Most spectroscopic applications are based on the absorption of radiation. Several techniques have been developed that either use the attenuation of the radiation passing through the absorbing sample, or that monitor various effects caused by the absorption of photons, such as a temperature increase in the sample, the change of electrical conductivity or light emission induced by the absorption [13.1].

We will start with classical absorption spectroscopy, in order to see which factors limit sensitivity.

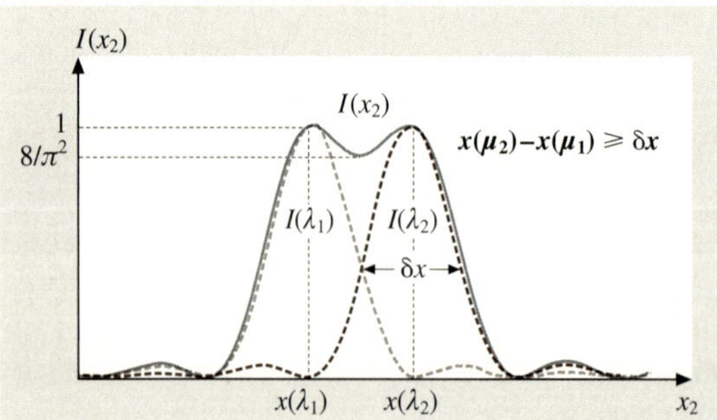

Fig. 13.1 Rayleigh criterium for the resolution of two close spectral line profiles

Classical Absorption Spectroscopy. When a monochromatic electromagnetic wave with frequency ω and intensity I_0 passes through an absorbing sample with absorption path length L, the transmitted intensity is (Fig. 13.2)

$$I_t = I_0 e^{-\alpha L} . \quad (13.2)$$

The spectral absorption coefficient $\alpha(\omega_{ik})$ of a transition $E_k \to E_i$ is determined by the absorption cross section σ_{ik} of the individual molecule on the transition $|i\rangle \to |k\rangle$, the difference $N_k - N_i$ of the populations densities and the ratio g_k/g_i of the statistical weights

$$L(\Omega_{ik}) = [N_k - (g_k/g_i)N_i]\sigma_{ik}(\Omega) , \quad (13.3)$$

where the statistical weights $g = (2J+1)$ give the number of possible orientations of the total angular momentum J of the levels involved.

For small absorptions the exponential function in (13.2) can be expanded into

$$I_t \approx I_0(1-\alpha L) \Rightarrow \Delta I = I_0 - I_t \approx \alpha L I_0 \quad (13.4)$$

and we obtain from (13.3) for an absorption length L

$$\Delta I = [N_k - (g_k/g_i)N_i]\sigma_{ik} L I_0 . \quad (13.5)$$

For non-monochromatic radiation the total absorption depends on the relation between the spectral width of the absorbing transition and the bandwidth of the incident radiation.

For a spectral bandwidth $\Delta\omega_r$ of the radiation and an absorption half-width $\Delta\omega_a$ the relative total absorption is

$$\frac{\Delta I}{I_0} = \frac{L \int \alpha(\omega) I(\omega)\,d\omega}{\int I(\omega)\,d\omega} \approx \bar{\alpha}\frac{\Delta\omega_\alpha}{\Delta\omega_r} , \quad (13.6)$$

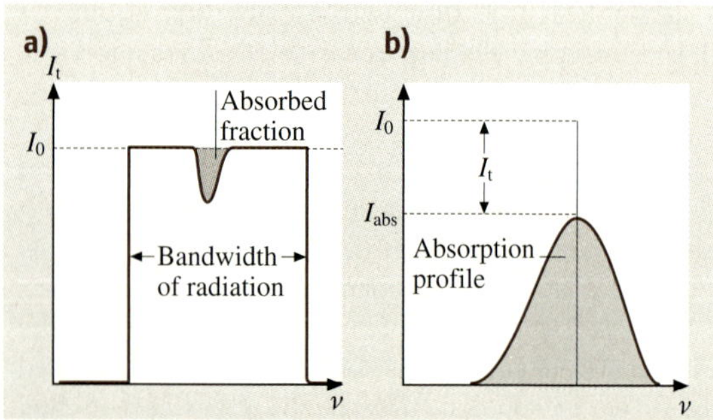

Fig. 13.2a,b Relative absorption by a transition with spectral half-width $\Delta\omega_a$ for different bandwidths $\Delta\omega_r$ of the incident radiation. **(a)** $\Delta\omega_r > \Delta\omega_a$, **(b)** for monochromatic radiation

where the integration extends over all frequencies, although only the interval $\Delta\omega_a$ within the absorption line profile contributes significantly to the absorption. This demonstrates that for $\Delta\omega_r > \Delta\omega_a$ the relative absorption increases with increasing ratio $\Delta\omega_a/\Delta\omega_r$ of absorption line width $\Delta\omega_a$ to radiation bandwidth $\Delta\omega_r$ Fig. 13.2. The use of narrow-band radiation sources is therefore not only advantageous for spectral resolution but also for increasing the sensitivity; this is often overlooked. Spectral profiles of absorption lines can only be measured if the bandwidth $\Delta\omega_r$ of the incident radiation is smaller than the absorption line width $\Delta\omega_a$.

The minimum detectable intensity change ΔI mainly depends on possible fluctuations of the incident intensity I_0 and on other noise sources such as detector noise. We summarize all noise contributions to a total noise ΔI_{noise}. Small absorptions $\Delta I = I_0\alpha L < \Delta I_{\text{noise}}$ need special electronic devices to be detectable. For the intensity change ΔI the detector gives an output signal $S \propto \Delta I$, while the noise contributes the amount $\delta S \propto \Delta I_{\text{noise}}$. A measure of the sensitivity is the achievable signal-to-noise ratio $S/\delta S$. From the condition

$$S \geq \delta S \Rightarrow \Delta I \geq \Delta I_{\text{noise}} , \tag{13.7}$$

we obtain with (13.6) the minimum detectable number density of absorbing molecules as

$$\Delta N = [N_k - (g_k/g_i)N_i] \geq \frac{1}{\sigma_{ik} L(S\delta S)} . \tag{13.8}$$

This shows that for a high sensitivity the absorption path length L should be as long as possible, the signal-to-noise ratio $S/\delta S$ as large as possible, and transitions with a large absorption cross section σ should be selected. We will see later how these requirements can be met experimentally.

Laser Absorption Spectroscopy. The aforementioned demands for high sensitivity and optimum spectral resolution can be best met when narrow-band lasers are used as the radiation sources. Absorption spectroscopy with monochromatic wavelength-tunable lasers resembles in many aspects microwave spectroscopy. The advantages of lasers are their wide tuning range and their possible applications in many spectral regions from the far infrared to the extreme ultraviolet.

A typical experimental set up is shown in Fig. 13.3. The output beam of a tunable laser is split by a 50% beam splitter BS2 into a reference beam and the signal beam that passes through the absorption cell. Another beam splitter BS1 directs a small fraction of the laser beam into a long Fabry–Pérot interferometer with mirror

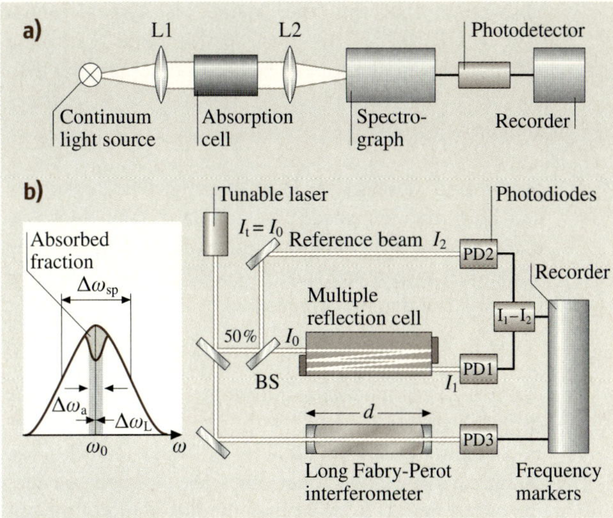

Fig. 13.3a,b Comparison of absorption spectroscopy **(a)** with a broadband incoherent source, **(b)** with a tunable monochromatic laser

separation d which gives equidistant frequency markers separated by $\Delta\nu = c/2d$.

The advantages of this arrangement compared with conventional spectroscopy without lasers can be summarized as follows. (1) There is no need for a spectrometer for wavelength dispersion because the tunable laser is already monochromatic and the absorption lines appear as dips in the transmitted laser intensity I_t when the laser is scanned through the spectrum. (2) The spectral resolution is not limited by any instrument, but solely by the width of the absorption lines, which is generally given by their Doppler width. With Doppler-free techniques Sect. 13.1.4 even this limitation can be overcome. (3) Because of the good collimation of the parallel laser beam a long path length L through the absorption cell can be realized, for instance by multi-path arrangements with spherical mirrors. This enhances the sensitivity (since the absorption $\Delta I = I_0\alpha L$ is proportional to the absorption length L) and allows the detection of weak transitions or of tiny densities of absorbing molecules. (4) The small line width $\Delta\omega_L$ of the laser further increases the sensitivity (13.6). The latter advantage is often overlooked which is, however, not justified, as the following example illustrates:

Example. In conventional spectroscopy with a spectral continuum and a wavelength-selecting spectrometer the grating spectrometer may have a resolution of $\delta\nu = 0.5\,\text{cm}^{-1}$, while the Doppler width of the absorption lines in the visible is typical about $\Delta\nu_a = 0.03\,\text{cm}^{-1}$. This gives for the same absorption αL an intensity change ΔI that is 16 times smaller than for monochromatic radiation, resulting in a 16 times smaller sensitivity.

Modulated Absorption Spectroscopy. The sensitivity can be further enhanced by modulating the laser frequency ω_L at a modulation frequency f, similar to the situation in microwave spectroscopy [13.2]. The incident intensity can then be represented by

$$I(\omega, f) = I_0\{1 + \cos[\omega + a\sin(2\pi ft)t]\}\,. \tag{13.9}$$

This frequency modulation generates sidebands with the frequencies $\omega = \omega_L \pm n\Omega$ with $\Omega = 2\pi f$. The modulation amplitude may be smaller or larger than the line width $\Delta\omega_a$ of the absorption lines. We will first consider the case where it is larger than the line width of the absorbing transitions. In this case the sidebands lie outside the absorption profile if the frequency ω_L is tuned to the center ω_0 of the absorption line.

The experimental set up for modulation spectroscopy can be realized as follows Fig. 13.4. When

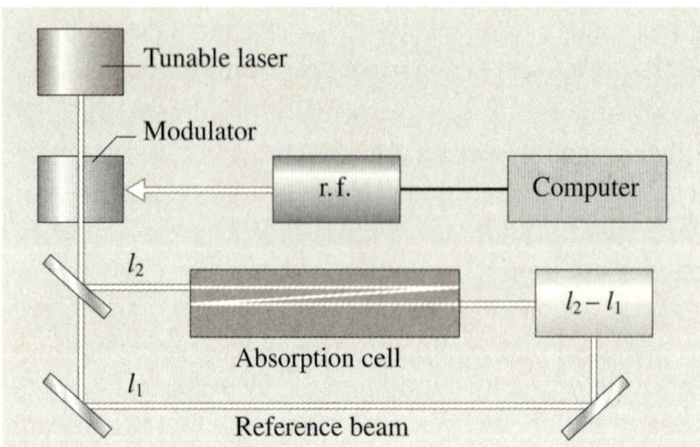

Fig. 13.4 Absorption spectroscopy with a frequency-modulated tunable laser, a reference beam and difference detection

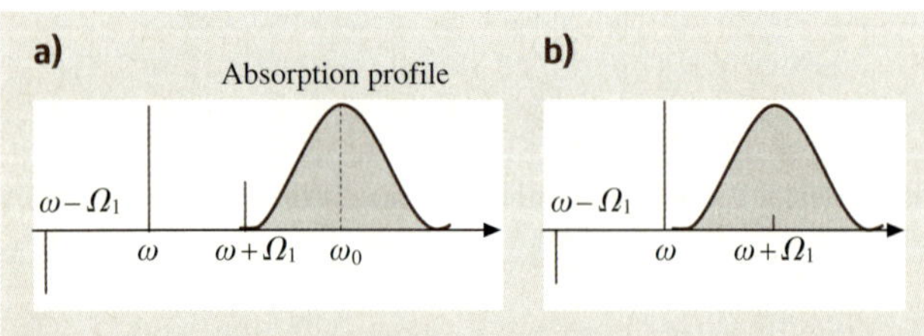

Fig. 13.5 Principle of phase-modulated absorption spectroscopy, where the modulation amplitude is larger than the line width of the absorbing transitions

the laser beam from a tunable laser passes through an electro-optic modulator (a crystal which changes its refractive index proportional to the voltage applied to electrodes on opposite sides of the crystal) the optical path length is periodically changed and with it the phase of the optical wave. This phase modulation results in a frequency modulation because the frequency $\omega = d\phi/dt$ is the derivative of the phase ϕ. In contrast to pure frequency modulation here the phases of the two sidebands are opposite Fig. 13.5. If the transmitted intensity (the sum of the intensities of carrier and sidebands) is detected by a phase-sensitive detector (lock-in) tuned to the modulation frequency Ω, the measured signal is zero if there in no absorption because the lock-in detector measures the two differences $I_c - I_+$ and $I_c - I_-$ of the intensities I_c of the carrier and $I_\pm$ of the two sidebands, which have equal amplitudes but opposite signs and therefore cancel each other.

If, however, one of the sidebands coincides with an absorption line this sideband is attenuated and the balance is perturbed. Fluctuations of the incident intensity appear in the carrier as well as in the sidebands and therefore cancel for the difference measurement.

This eliminates most of the noise and therefore increases the sensitivity. When the carrier frequency coincides with the center frequency of the absorption line, the signal is zero. Since the signal has a zero-crossing at the center frequency of the absorbing transition, this frequency can be determined very accurately.

A further enhancement can be achieved, when the absorption path length L is increased. This can be realized by placing the sample inside an optical multipass configuration, which is shown in Fig. 13.6. It consists of two spherical mirrors with radius of curvature r and a separation of $d \approx r$. The incident laser beam enters through a small hole in one of the mirrors, is reflected by the opposite mirror and is focussed onto the first mirror. After many round trips the beam leaves through the same hole but at another angle, thus avoiding back-reflection into the source. The laser-beam spots on the

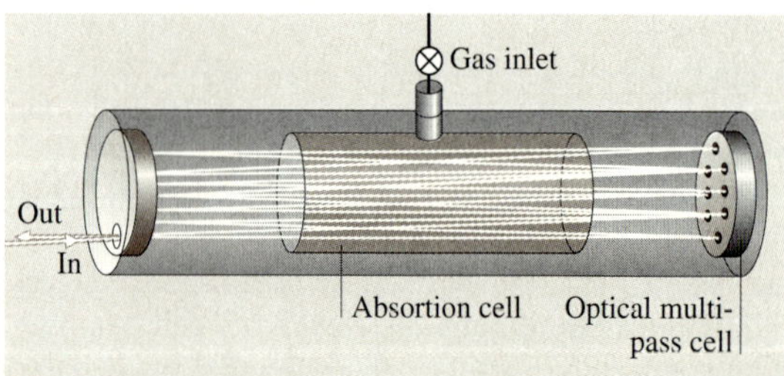

Fig. 13.6 Multipass absorption cell with spherical mirrors

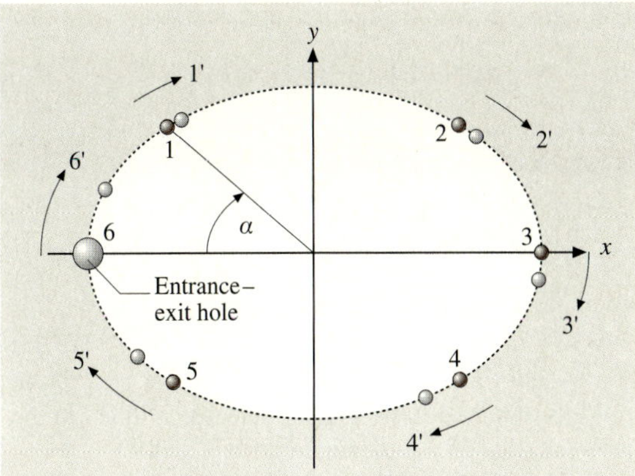

Fig. 13.7 Laser beam spots on the mirror surfaces for successive reflections. *Brown circles* are on mirror M_2, *gray circles* on mirror M_1

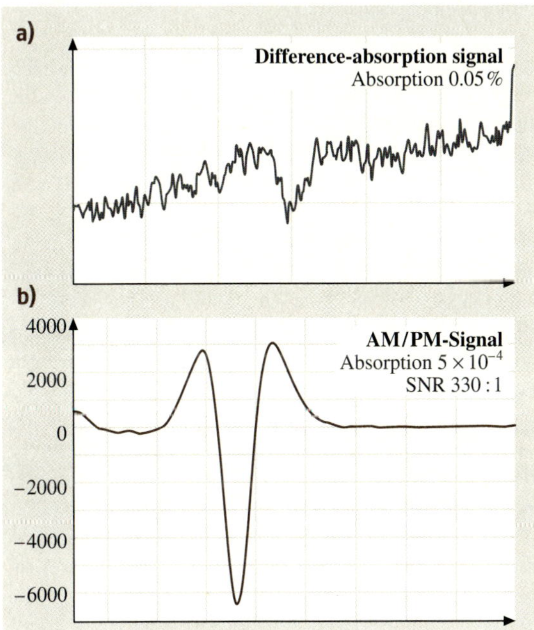

Fig. 13.8a,b Absorption feature of a weak overtone transition in H_2O molecules. **(a)** measured with conventional unmodulated absorption spectroscopy, **(b)** with a phase-modulated (PM) laser and difference detection

mirrors form a circle Fig. 13.7 with an angular distance that can be controlled by changing the mirror separation d. Due to the focusing by the spherical mirrors the spot size remains fairly constant even for many passes between the mirrors. This minimizes the overlap of spots from different reflections. Such an overlap would result in interference that would change with the wavelength λ. When the laser wavelength is tuned over the spectral region of interest, this would result in perturbing features overlapping the spectral lines. The maximum number of round trips is limited by the absorption of the sample and by the reflectivity r of the mirrors.

Example. For example, with a mirror reflectivity r of 98% the transmitted intensity has dropped after N round trips without absorption cell to

$I_t = I_0 r^{-2N} = 0.01765 I_0$ for $N = 100$.

Assuming an absorption coefficient of $\alpha = 10^{-5}\,\mathrm{m}^{-1}$ and an absorption path length of $2NL$ the transmitted intensity becomes for $L = 1$ m is $I_t = I_0 r^{-2N} \mathrm{e}^{-2N\alpha L} = 0.0176\,\mathrm{e}^{-0.002} \approx 0.01765 \times 0.998 \approx 0.01760$.

This illustrates that for weakly absorbing samples the mirror reflectivity rather than the sample absorption is the limiting factor for the number of passes. The enhanced sensitivity is illustrated in Fig. 13.8 where a conventional absorption measurement of a weak overtone absorption line of the H_2O molecule is compared with the signal obtained by the modulation technique. The enhancement of sensitivity amounts to two orders of magnitude.

If the modulation amplitude is smaller than the absorption line width we have the situation depicted in Fig. 13.9. A lock-in detector tuned to the modulation frequency measures the derivative of the absorption profile. From (13.4) we obtain for $\alpha L \ll 1$

$$\frac{\mathrm{d}\alpha}{\mathrm{d}\omega} = -\frac{1}{L I_0}\frac{\mathrm{d}I_t}{\mathrm{d}\omega}\,. \qquad (13.10)$$

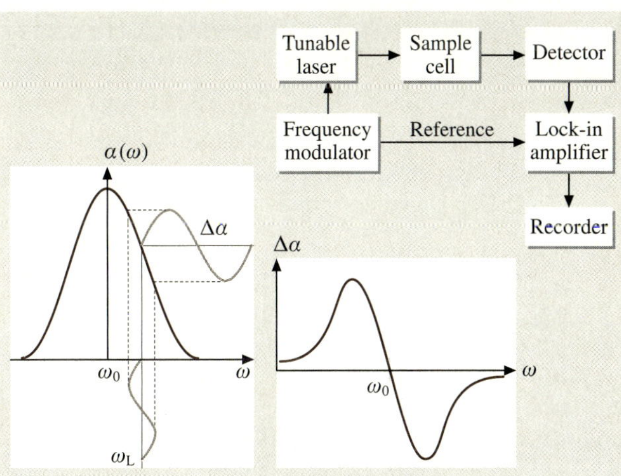

Fig. 13.9 Principle of frequency-modulation spectroscopy with a modulation amplitude smaller than the absorption line width

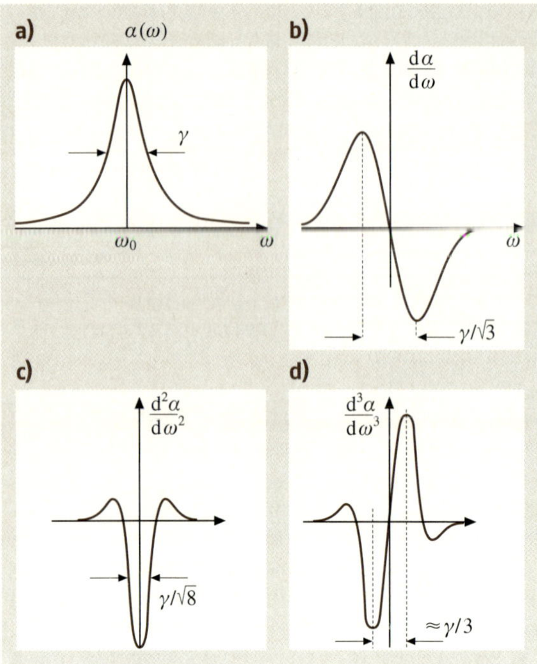

Fig. 13.10 Signal profiles of absorption lines when $\alpha(\omega)$ and its first three derivatives are measured

When inserting the modulated laser frequency

$$\Omega_L = \omega_0(1 + a\ \sin\Omega t) \tag{13.11}$$

this yields the Taylor expansion

$$I_t(\omega) = I_t(\omega_0) + \sum_n \frac{\alpha^n}{n!} \sin^n(\Omega t) \left(\frac{d^n I_t}{d\omega^n}\right)_{\omega_0} . \tag{13.12}$$

According to trigonometric relations the terms $\sin^n \Omega t$ can be replaced by a linear combination of $\sin(n\Omega t)$ and $\cos(n\Omega t)$. Ordering the different terms with respect to their harmonic frequencies $n\Omega$ one obtains

$$\begin{aligned}\frac{\Delta I}{I_0} \approx -aL\Bigg[C + \left(\frac{d\alpha}{d\omega}\right)_{\omega_0} &\sin(\Omega t) \\ - \frac{a}{4}\left(\frac{d^2\alpha}{d\omega^2}\right)_{\omega_0} &\cos(2\Omega t) \\ - \frac{d^2}{24}\left(\frac{d^3\alpha}{d\omega^3}\right)_{\omega_0} &\sin(3\Omega t) + \cdots\Bigg] .\end{aligned} \tag{13.13}$$

This means that the n-th derivative of the absorption profile is obtained if the lock-in detector is tuned to the harmonics $n\Omega$ of the modulation frequency Ω.

In Fig. 13.10 the measured profiles for the first three harmonics are illustrated. Since the technical noise generally decreases with increasing frequency, measurements at higher harmonics often yield a higher signal-to-noise ratio [13.3].

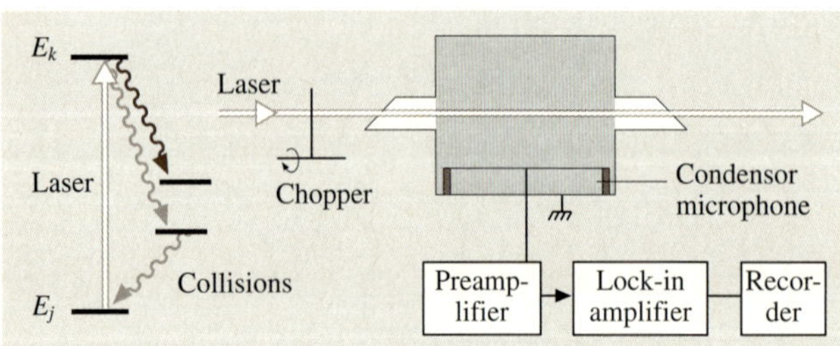

Fig. 13.11 The principle of optoacoustic spectroscopy

Optoacoustic and Optothermal Spectroscopy. With the techniques of absorption spectroscopy discussed so far absorption is detected as the difference between the transmitted and incident intensities. For small absorptions one has to measure a small difference of two large quantities, which has, of course, a larger noise than if the absorbed radiation power could be directly measured. This can be achieved by several techniques, which shall now be discussed. These are based on energy transfer from the absorbed photons to heat, and they monitor various effects caused by the temperature rise of the absorbing sample. We start with optoacoustic spectroscopy [13.4].

An absorption cell with volume V may contain $N = nV$ absorbing molecules, which can be excited into the energy level $E_i = E_k + h\nu$ by absorption of a photon $h\nu$ (Fig. 13.11). If the excited molecule collides with other atoms or molecules in the cell, it can transfer its excitation energy into translational or internal energy of the collision partners. If N_1 molecules are excited and transfer their excitation energy to the N molecules in the cell with f degrees of freedom, the energy $E = (f/2)kT$ of each particle is increased and the temperature of the gas rises by

$$\Delta T = \frac{N_1}{N} h\nu/(fk) . \tag{13.14}$$

For a closed cell the total number N of molecules is constant. A temperature rise therefore increases the gas pressure $p = nkT$ by

$$\Delta p = nk\Delta T = \frac{N_1}{V}\frac{h\nu}{k} . \tag{13.15}$$

The energy of the absorbed photons is converted into kinetic energy, resulting in a pressure rise of Δp.

Until now we have neglected the radiative decay of the excited level. If the emitted fluorescence photons are not absorbed within the cell, their energy is not converted into a pressure increase. With the radiative transition rate R_{rad} and the nonradiative rate R_{nr}, which is caused by collisions, the quantum efficiency is

$$\eta = R_{rad}/(R_{rad} + R_{nr}) = 1/(1 + \tau_{all}/\tau_{nr}) \qquad (13.16)$$

and the transferred energy decreases to

$$\Delta E = N_1(1-\eta)h\nu = N_1 h\nu\tau_{rad}/(\tau_{rad} + \tau_{nr}) \,. \qquad (13.17)$$

The energy conversion increases with the ratio τ_{rad}/τ_{coll} of radiative lifetime to the collisional deactivation time of the excited level.

If the exciting laser beam is chopped at a frequency f that is smaller than the inverse transfer time $(1/\tau_{coll} + 1/\tau_{rad})$ the pressure in the cell is modulated at the frequency f. A sensitive microphone in the side wall of the cell detects these pressure changes which represent acoustic waves. When the chopping frequency is chosen to coincide with one of the acoustical eigenresonances of the cell, standing acoustic waves are generated with amplitudes that are dependent on the acoustic quality factor Q of the cell and which are much higher than for the nonresonant case.

The technique is quite sensitive because the acoustic resonator acts as amplifier and stores the energy converted from the excited molecules into standing acoustic waves for a time $T_r \propto Q$. For illustration Fig. 13.12 shows the optoacoustic spectrum of a very weak overtone band in acetylene, with the different rotational lines. Since the absorbed optical photon energy is converted into acoustic energy the method is named *optoacoustic spectroscopy*.

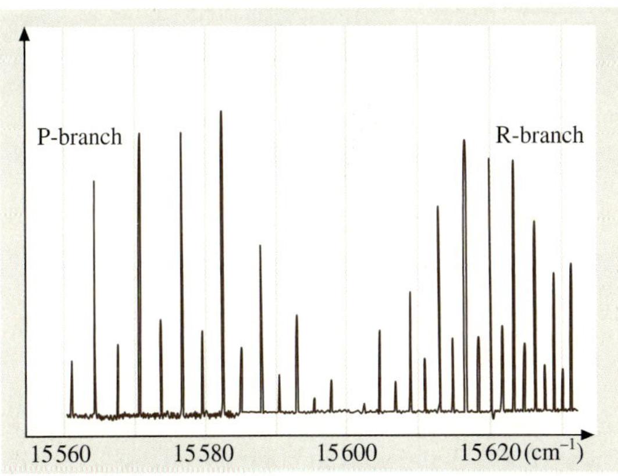

Fig. 13.12 Rotational lines in a vibration overtone transition of acetylene C_2H_2 measured with optoacoustic spectroscopy [13.5]

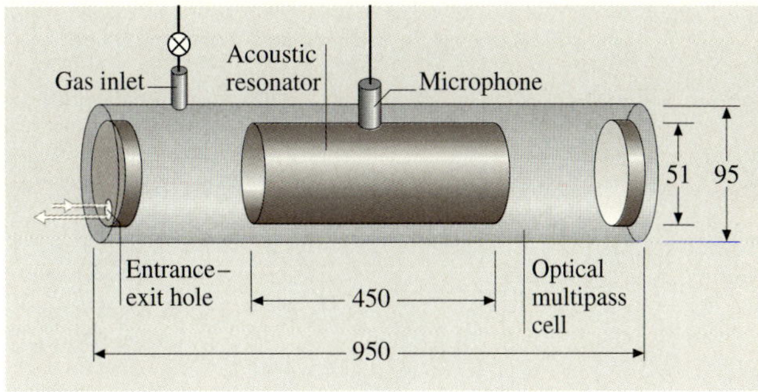

Fig. 13.13 Optoacoustic spectroscopy in an acoustic resonance cell placed inside an optical multipass cell

The sensitivity can be further enhanced when the acoustic cell is placed inside an optical multipass cell Fig. 13.13. The laser beams inside this cell with spherical mirrors form a hyperboloid and excite acoustic standing waves on those locations in the acoustic cell where the standing acoustic waves have maximum amplitude, which optimizes the conversion efficiency for the energy transfer from the excited molecules into the acoustic mode.

The technique can also be applied to the investigation of thin films or of surfaces of solids and liquids. The absorbed incident radiation causes local heating, which results in a thermal deformation. This can be detected with a weak probe laser beam, which is incident on the surface at an angle α against the surface normal. For a flat surface the reflection angle is again α. However, if the surface is deformed by local heating, the local surface normal is inclined and therefore the reflection angle is changing Fig. 13.14 [13.6]. When the surface is heated by a pulsed laser, focussed onto the surface, the change of reflection angle of the continuous-wave (CW)

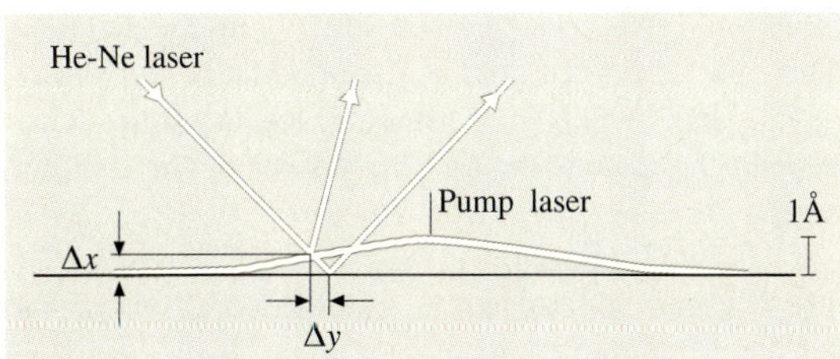

Fig. 13.14 Time-resolved measurements of surface deformations due to local heating induced by a laser pulse [13.6]

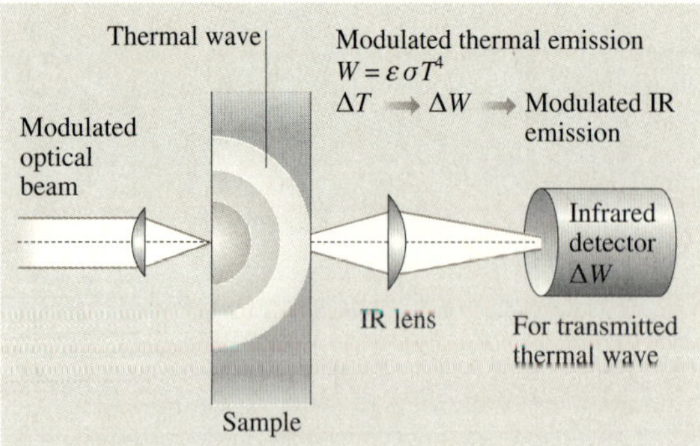

Fig. 13.15 Measurements of thermal waves in a thin metal sheet induced by absorption of a modulated laser beam [13.7]

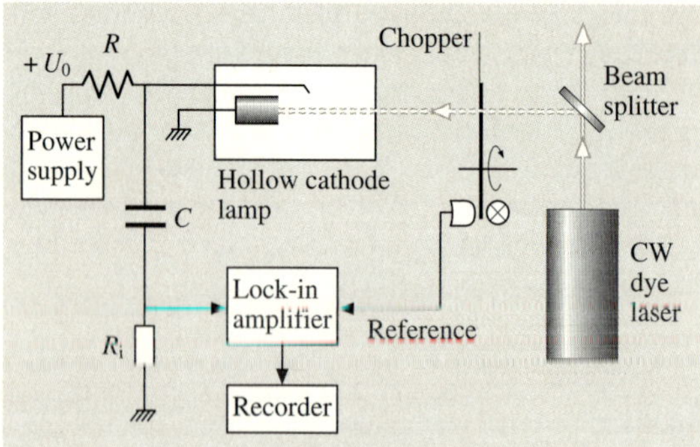

Fig. 13.16 Schematic arrangement for optogalvanic spectroscopy in a hollow cathode discharge

probe laser can be monitored as a function of time. This gives a measure of the time constants of heat transfer from the local spot into the surrounding.

When the heating laser is pulsed or modulated, thermal waves are generated, which transfer the heat away from the local heat source at the focus of the laser. At a distance x from the heat source the temperature reaches its maximum a time

$$\Delta t = (\rho c/2k)x^2 \tag{13.18}$$

after the heating pulse, where ρ is the density, C is the specific heat, and k is the thermal conductivity of the sample material. After the thermal diffusion length

$$x = D = (k/\pi c\rho f)^{1/2} \tag{13.19}$$

the amplitude of the thermal wave generated by a heat source modulated at a frequency f, has decreased to $A_0 \mathrm{e}^{-2\pi} \approx 2\times10^{-3} A_0$.

Measurements of such thermal waves allow the determination of the thermal conductivity of the sample. A possible experimental arrangement is shown in Fig. 13.15. The temperature wave transmitted through the sample can be monitored with an infrared detector. The time constant of this detector limits the time resolution. In the case of a sinusoidally modulated heat source the detected signal is also modulated but shows a phase lag, which is a measure for the time delay between the generation and the arrival of the thermal wave.

Optogalvanic Spectroscopy. For the spectroscopy of high-lying atomic, molecular or ionic levels, gas discharges are convenient because many of these levels are populated by electron impact, in particular metastable levels with long lifetimes. The discharge impedance depends on the number of electrically charged carriers, i. e., electrons and ions, which are created by electron impact ionization from excited levels of the neutral species. If the population of these excited levels is changed, for instance by absorption of photons, the electron density will change and therefore the impedance of the discharge. If the discharge is fed through a ballast resistor R by a power supply with constant voltage, the impedance change causes a corresponding change of the discharge current and of the voltage drop across the resistor R. This forms the basis for optogalvanic spectroscopy [13.8].

The experimental arrangement is shown in Fig. 13.16. The discharge tube (which might be a hollow cathode discharge or a discharge through a capillary, such as that used for the He–Ne laser or the argon-ion laser) is irradiated by the chopped output beam of a tunable laser (dye or semiconductor laser). This causes a modulation Δi of the discharge current at the chopping frequency f and a corresponding voltage modulation $\Delta U = R\Delta i$ across the ballast resistor R, which is coupled through a capacitance C into a lock-in amplifier and recorded by a computer. If the laser wavelength is tuned to a transition $E_i \to E_k$ the population of the two levels changes due to laser-induced transitions by

$$\Delta n_i = n_{i0} - n_{i\mathrm{L}} \; ; \; \Delta n_k = n_{k0} - n_{k\mathrm{L}} \, . \tag{13.20}$$

The corresponding voltage change is then

$$\Delta V = R\Delta i = a[\Delta n_i IP(E_i) - \Delta n_k IP(E_k)] \, , \tag{13.21}$$

where $\mathrm{IP}(E_i)$ is the ionization probability of level E_i. The optogalvanic signals might be positive or negative, depending on the different ionization probabilities of the levels E_i and E_k.

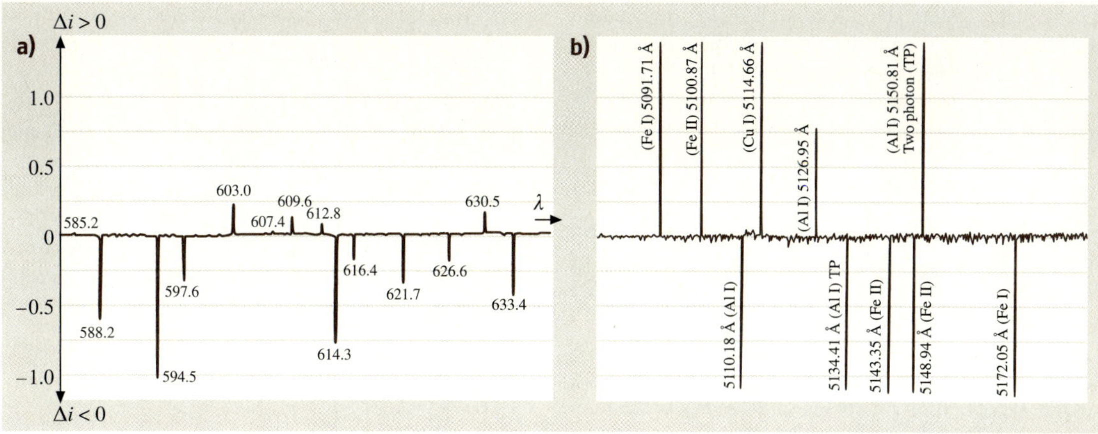

Fig. 13.17 Optogalvanic spectrum of a neon discharge in a hollow cathode covered with a mixture of different metals [13.9]

Most gas discharge tubes are filled with noble gases. However, if the noble gas is mixed with other volatile components their spectra can also be measured by this technique.

Even molecules can be inserted into the discharge. Since some of the molecules might be fragmentized by electron impact the spectra of the mother molecules and their fragments overlap and assignment might be difficult [13.10].

In a hollow cathode discharge the ions that impinge onto the inner walls of the hollow cathode release atoms and ions of the wall material by sputtering processes. With higher discharge currents the spectra of these sputtered materials become increasingly prominent. This is illustrated in Fig. 13.17 where the optogalvanic spectrum of aluminium, copper and iron measured in a hollow cathode discharge is shown.

Investigations of the intensity and time behavior of optogalvanic signals give information on the radiative decay constants of highly excited levels and on collision cross sections for electron impact excitation and ionization. These parameters are important for understanding plasma characteristics.

Ionization Spectroscopy. The most sensitive detection technique is ionization spectroscopy. Here a laser is tuned through the spectrum of interest and excites a selected upper level for each absorbing transition. This excited level is ionized by a second laser Fig. 13.18. If the second laser is sufficiently intense, the ionizing transition can be saturated, which means that every excited molecule is ionized before it can emit a fluorescence photon or is deactivated by collisions. The ions can be collected by an electric field, accelerated and imaged onto the cathode of an open ion multiplier. This is a device that is quite similar to a photomultiplier, except the photocathode is replaced by a metal cathode, where the ions, impinging with energies of a few keV produce electrons by ion bombardment. These electrons are further accelerated and multiplied as in a photomultiplier.

With an optimum design of the ion-collecting system each ion produced through photoionization by the second laser can be imaged onto the ion multiplier and produces a voltage pulse at the multiplier output, which can be counted.

If the first laser with wavelength λ_1 is tuned through the spectral region of interest and the second laser has a fixed wavelength λ_2, the measured ion rate $N_{ion}(\lambda_1)$ essentially gives the absorption spectrum $\alpha(\lambda_1)$ of transitions from the absorbing levels to the levels excited by laser L1. The sensitivity of this method is, however by several orders of magnitude higher than in absorption

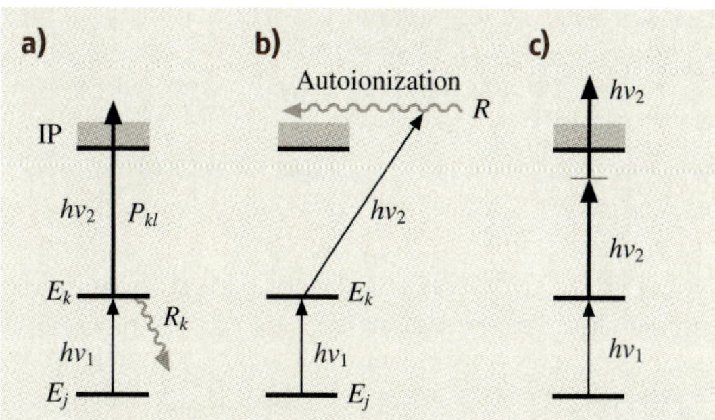

Fig. 13.18a–c Level scheme for multiphoton ionization **(a)** resonant two-photon ionization, **(b)** excitation of auto-ionizing Rydberg states, **(c)** nonresonant two-photon ionization of molecules in a level excited by a third photon

spectroscopy, where the attenuation of the transmitted laser intensity is monitored. In favorable cases single atoms or molecules can be detected [13.11].

With a photon flux N_{L2} of the ionizing laser, the number of ions produced per second is

$$\dot{N}_{\text{ion}} = n_k V \Sigma_{kI} \dot{N}_{L2} , \quad (13.22)$$

where $P_{kI} = \sigma_{kI} \dot{N}_{L2}$ is the ionization probability by laser L_1, and n_k is the stationary population density of the excited level $\langle k|$ in the volume V inside the ionizing laser beam, from which ions can be extracted.

The stationary density n_k of molecules in level $|k\rangle$ is determined by the balance between excitation by laser L_1 and depopulation processes. If $R_k + P_{kI}$ is the total relaxation probability of the excited level $|k\rangle$, we obtain

$$\dot{n}_k = 0 = n_i \dot{N}_{L1} \Sigma_{ik} - n_k(P_{kI} + R_k) \Rightarrow$$
$$n_k = n_i \frac{\dot{N}_{L1} \Sigma_{ik}}{P_{ki} + R_k} . \quad (13.23)$$

With the ion yield efficiency $\eta = P_{kI}/(P_{kI} + R_k)$ this gives for the measured ion rate

$$S = \delta \dot{N}_{\text{ion}} = n_i \dot{N}_{L1} \dot{N}_{L2} V \frac{\Sigma_{ik} \Sigma_{kI}}{P_{kI} + Rk} , \quad (13.24)$$

where δ is the collection efficiency for the ions onto the detector.

Excitation Spectroscopy with Fluorescence Detection. The total fluorescence intensity emitted from an excited level can be measured (without dispersion by a spectrometer) as a function of the wavelength λ of the exciting laser Fig. 13.19. Each absorbed photon creates a fluorescence photon if other deactivation processes of the excited level (for instance collision-induced radiationless transitions) can be neglected. This excitation spectroscopy is a very sensitive version of absorption spectroscopy, as can be seen from the following estimation.

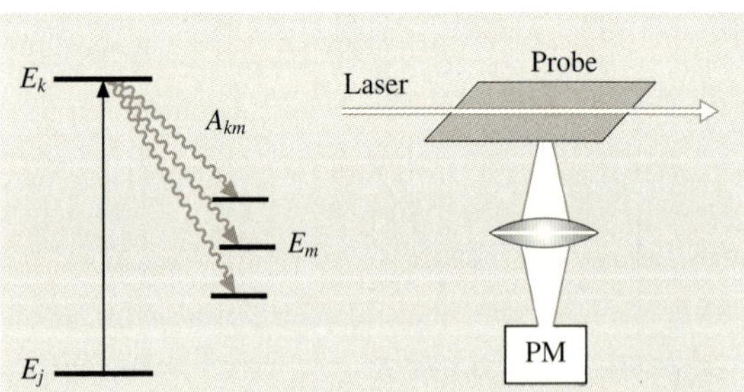

Fig. 13.19 Level scheme and experimental set up for excitation spectroscopy, where the undispersed laser-induced fluorescence is measured as a function of the excitation wavelength (PM = photomultiplier)

The fluorescence photons, imaged onto the cathode of a photomultiplier, create n_{pe} photoelectrons per second. With the quantum efficiency of the photocathode $\eta_{\text{ph}} = N_{\text{pe}}/N_{\text{fl}}$, the rate of absorbed laser photons N_{a}, the fluorescence quantum yield $\eta_{\text{fl}} = N_{\text{fl}}/N_{\text{a}}$, the geometrical collection efficiency δ of the fluorescence photons onto the photocathode, the measured rate of photoelectrons is given by

$$S = N_{\text{pe}} = N_{\text{a}} \eta_{\text{fl}} \eta_{\text{ph}} \delta . \quad (13.25)$$

The absorption rate

$$N_{\text{a}} = n_i \sigma_{ik} N_{\text{L}} A \Delta x \quad (13.26)$$

depends on the number density n_i of molecules in the absorbing level $\langle i|$ within the excitation volume $V = A\Delta x$, the absorption cross section σ_{ik}, the number N_L of laser photons incident per second and cm^2 in a laser beam with cross section A onto the sample with absorption path length Δx.

The collection efficiency δ for fluorescence photons can be enlarged by an elliptical mirror arrangement Fig. 13.20. When the intersection volume V of the laser and molecular beams is placed around one focal point A of the ellipsoid, the fluorescence is imaged at the second focal point B where an optical-fiber bundle transmits

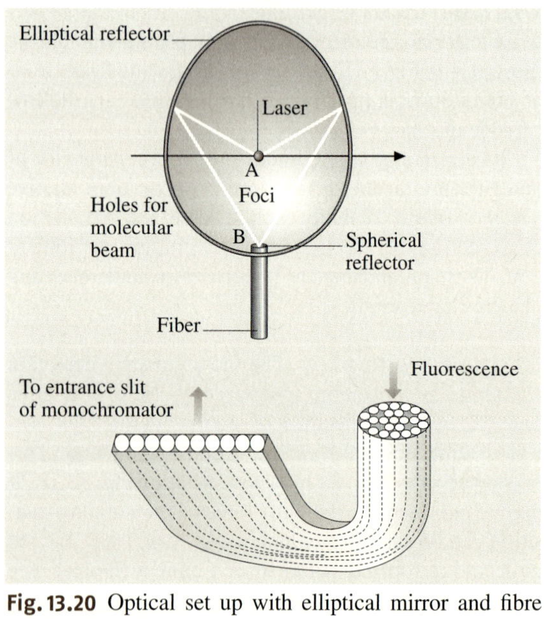

Fig. 13.20 Optical set up with elliptical mirror and fibre bundle for effective collection of fluorescence photons

it either directly to the photomultiplier or to the entrance slit of the spectrometer.

Example. For example with a quantum efficiency $\eta_{ph} = 0.2$ of the multiplier photocathode, a collection efficiency of $\delta = 0.1$, which demands collection optics for the fluorescence, which covers a solid angle of 0.4π, a fluorescence yield $\eta_k = 1$ for the excited level $\langle k|$ and a laser photon flux of $N_L = 3\times10^{18}$ /s, which corresponds to a laser power of 1 W at $\lambda = 500$ nm, an absorption rate $N_a = 10^4$ /s (which means a relative laser absorption of $\Delta I_L/I_L = 3\times10^{-15}$) gives a photoelectron rate $N_{pe} = 200$ /s. If the dark current of the photomultiplier is $N_{pe}(0) = 50$ /s, this gives a signal-to-background ratio of 4, which demonstrates that relative absorptions of less than 10^{-15} can still be detected. This represents a much higher sensitivity than any direct absorption measurement.

Intracavity Absorption. A large enhancement in sensitivity can be achieved when the absorbing sample is placed inside the cavity of the laser [13.12]. This increase of sensitivity has several causes:

1. When the resonator mirrors have transmission $T_1 = 0$ and $T_2 > 0$, the radiation intensity inside the cavity is larger by a factor $q = 1/T_2$ than outside and therefore the number of absorbed photons is multiplied by q. This advantage is quite similar to that of a multipass cell. Observing the laser-induced fluorescence emitted from excited atoms or molecules inside the resonator Fig. 13.21 therefore gives a larger signal proportional to the number of absorbed photons.
2. The output power of the laser depends on the gain of the active medium and the losses inside the laser cavity. Therefore additional losses by absorbing species decrease the output power. This effect is particularly strong if the laser operates close above the lasing threshold Fig. 13.22. Under stationary conditions the population inversion saturates such that the saturated gain

$$g_s = \frac{g_0}{1+I/I_s} = \frac{g_0}{1+P/P_s} = \gamma \tag{13.27}$$

just equals the total losses. Here g_0 is the unsaturated gain and I_s is the saturation intensity Sect. 13.3.1, which decreases the population inversion to one half of its unsaturated value.
This gives for the laser output power

$$P = P_s\frac{g_0-\gamma}{\gamma}\,. \tag{13.28}$$

If additional losses $\Delta\gamma$ are introduced by the absorbing sample the laser power drops to

$$P_a = P - \Delta P = P_s\frac{g_0-\gamma-\Delta\gamma}{\gamma+\Delta\gamma}\,. \tag{13.29}$$

The relative change of the output power is then

$$\begin{aligned}\frac{\Delta P}{P} = \frac{P-P_a}{P} &= \frac{g_0\Delta\gamma}{(g_0-\gamma)(\gamma+\Delta\gamma)}\\ &\approx \frac{g_0}{\gamma}\frac{\Delta\gamma}{g_0-\gamma} fVN\Delta\gamma \ll \gamma\,.\end{aligned} \tag{13.30}$$

Compared with the single-pass absorption $\Delta P = P\alpha L$ for an absorption cell with length L outside the laser resonator the sensitivity enhancement is, with $\Delta\gamma = \alpha L$,

$$Q = \frac{g_0}{\gamma(g_0-\gamma)}\,. \tag{13.31}$$

Just above threshold g_0 is only slightly larger than γ and Q becomes very large. However, the laser output power becomes increasingly unstable as the threshold is approached. Therefore there is an upper limit on the attainable signal-to-noise ratio.

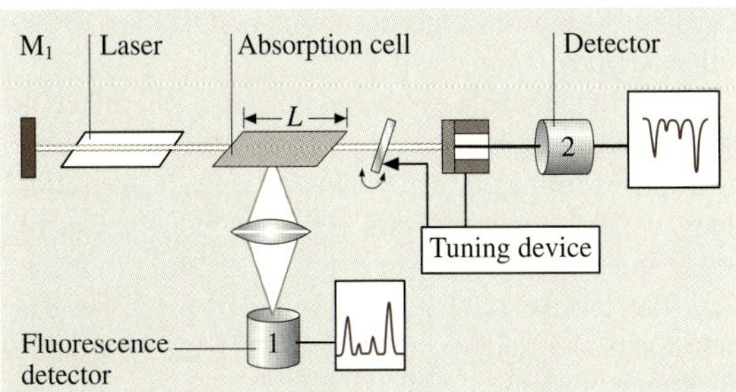

Fig. 13.21 Set up for intracavity absorption, monitored either by the laser-induced fluorescence (13.1) or by the laser output power (13.2)

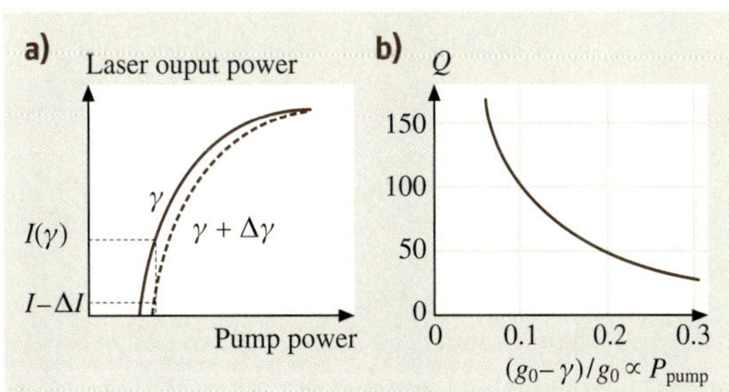

Fig. 13.22 Laser output power as a function of pump power with and without an absorbing sample inside the laser cavity

3. The most sensitive technique for intracavity absorption is based on multimode lasers, where the coupling between different modes is essential for the high sensitivity. This can be understood as follows. Assume that one of the oscillating laser modes coincides with an absorption line of the intracavity sample. The power in this mode is then decreased and this mode causes less saturation of the gain medium. Neighboring laser modes that are closer to the absorbed mode than the saturation line width can take advantage of this decrease in the population inversion and gain intensity. This in turn decreases the gain for the weaker absorbed mode and weakens its power further. This mode coupling can therefore lead to a complete extinction of the absorbing mode. This means that a tiny absorption can cause a drastic change in the intensity of the absorbed mode.

More-detailed calculations show that the relative change of the output power is now

$$\frac{\Delta P}{P} = \frac{g_0 \Delta \gamma}{\gamma (g_0 - \gamma)} (1 + KM) , \qquad (13.32)$$

where M is the number of coupled laser modes and K $(0 \le K \le 1)$ is a measure of the coupling strength. For $K = 0$ we obtain the same result as for a single-mode laser, while for $K = 1$ and a large number M of coupled modes the increase in sensitivity is remarkable.

In real lasers the mode coupling and the mode frequencies fluctuate with time due to thermal instabilities or fluctuations in the refractive index of the liquid gain medium of dye lasers. This introduces additional noise and decreases the sensitivity. It is therefore better to pump the laser with a pump laser whose output power follows a step function that starts at time $t = 0$ and then remains constant. The intracavity absorption is measured at times t with $0 \le t \le t_m$ that are shorter than the mean mode lifetime t_m. The experimental set up is shown in Fig. 13.23 [13.13]. The CW broadband dye laser is pumped by an argon laser, whose output can be switched by the acoustooptical modulator AOM1. The laser output through mirror M_0 is gated by AOM2, dispersed by a spectrograph with a charge-coupled device (CCD) array and monitored by a computer.

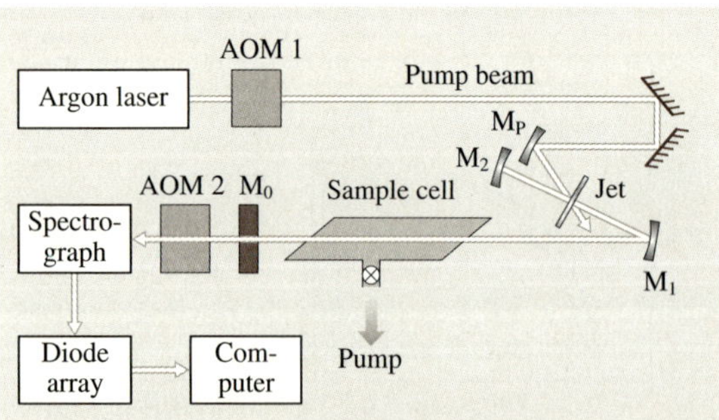

Fig. 13.23 Experimental arrangement for intracavity laser spectroscopy with a step-function pump power and a variable delay for detecting the transmitted dye-laser output

In the course of time the spectral absorption profiles become deeper and narrower until they reach the spectral resolution of the spectrograph.

The maximum effective absorption length achieved with this technique is

$$L_{\text{eff}} = ct_{\max} , \qquad (13.33)$$

where c is the velocity of light and $t_{\max}$ is the maximum observation time.

Example. With a typical delay time $t_{\max} = 100\,\mu\text{s}$ the effective absorption length becomes $L_{\text{eff}} = 3\times 10^8 10^{-4}\,\text{m} = 30\,\text{km}$. If an absorption dip in the dispersed output power of 1% can still be detected, this gives a sensitivity limit for the absorption coefficient of $\alpha_{\min} = 3\times 10^{-9}\,\text{cm}^{-1}$ [13.14].

13.1.1 Absorption and Emission Spectroscopy, Laser-Induced Fluorescence

Until now we have mainly discussed different techniques of absorption spectroscopy. The spectroscopy of radiation emitted from the excited levels of atoms or molecules plays an important role in plasma physics and in particular in astrophysics where the radiation emitted from stars or reflected by planets is the main source of information about extraterrestrial objects.

Emission spectroscopy requires a sufficient population of excited states and needs dispersive instruments to measure the spectrum, i. e. the wavelengths, intensities, line profiles and the degree of polarization of the emission lines.

The excited levels can be populated by thermal collisions if the temperature is sufficiently high (e.g. in the atmospheres of stars), or by electron impact in gas discharges. These mechanisms generally populate many levels and the emission spectrum is accordingly complex. The spectral resolution is limited by the dispersing instrument and reaches about 0.1–0.01 nm for grating spectrometers, while with interferometers the Doppler width of the emission lines can be seen.

The spectrum becomes much simpler if a single upper level can be selectively excited. This is possible in many cases, when a single-mode laser is used for excitation and the fluorescence spectrum emitted by this single level is measured. A typical experimental set up is shown in Fig. 13.24. The fluorescence spectrum emitted by this excited level is dispersed by a spectrometer and the fluorescence lines are recorded by a photomultiplier or a charge-coupled device (CCD) array. If the fluorescence is, for instance, emitted from a single upper rotational–vibrational level (v, J) of a diatomic molecule, the allowed transitions represent a relatively simple vibrational progression with only one or two rotational lines in each vibrational band Fig. 13.23. The relative intensities of these vibrational bands allow the determination of the Franck–Condon factors. Such a laser-induced fluorescence spectrum is indeed much simpler than the emission spectrum from a gas discharge where many upper levels are populated and the total fluorescence is the superposition of all fluorescence series emitted by the different upper levels.

The wave numbers of the fluorescence lines are equal to the differences between the term values of the upper level and the terminating lower levels. If the upper level is the same for all lines the separations of the fluorescence lines just give the energy differences between the levels in the lower state Fig. 13.25. The vibrational spacings yield the vibrational constants of the lower electronic state and the rotational spacings the rotational constants and their dependence on the vibrational level. If the Franck–Condon factors for transitions into high vibrational levels of the electronic ground state are sufficiently large (this depends on the excited upper vibrational level), the vibrational levels in the electronic ground state can be measured up to the dissociation energy. This allows a very accurate determination of the potential curve [13.15].

The selective excitation of a single upper level is only possible if there is no overlap of absorptions lines within their line width. For excitation in a gas cell the spectral width is generally limited by the Doppler width. At higher temperatures many levels in the electronic ground state are thermally populated, which results in a higher density of absorption lines, and the mean distance between different lines is often smaller than the Doppler width. In such cases several upper levels are excited even with a single-mode laser. However, even then the resulting fluorescence spectrum is still simpler than the emission spectrum from a gas discharge.

This problem can be overcome by laser excitation in a cold and collimated molecular beam, where the Doppler width is greatly reduced and the temperature is decreased due to adiabatic cooling during the expansion of the beam from the reservoir through a nozzle into the vacuum chamber.

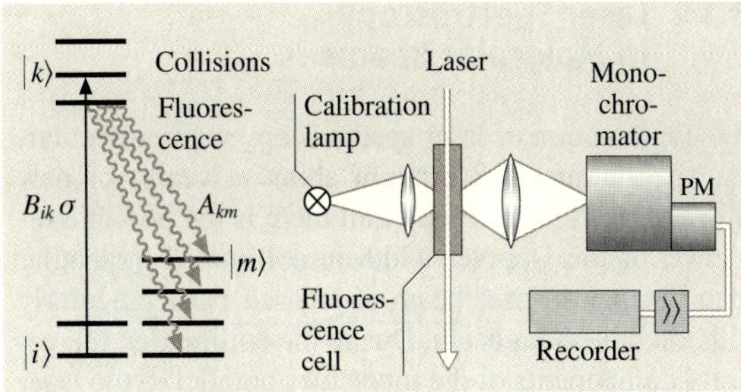

Fig. 13.24 Measurement of dispersed laser-induced fluorescence spectrum with wavelength calibration

The laser-induced fluorescence spectroscopy has proved very useful for the investigation of inelastic collisions in excited states. These collisions transfer the initial population from the selectively excited level $|k\rangle$ into neighboring rotational–vibration levels $|j\rangle$. The fluorescence emitted from these levels is a direct measure of the transfer probability. The ratio $I_{fl}(j)/I_{fl}(k)$ yields the collision cross section for individual transitions if the densities of the collision partners is known. For such measurements the fluorescence must be dispersed by a spectrometer to separate the contributions emitted from the different levels.

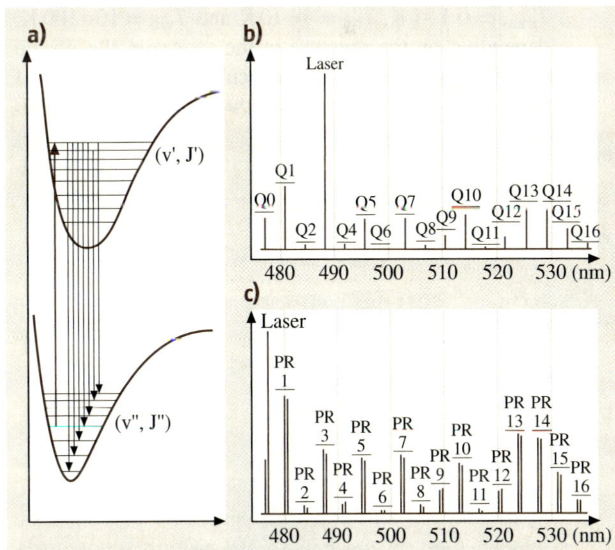

Fig. 13.25a–c Laser-induced fluorescence spectra of the Na_2 molecule for two different selectively excited upper levels: **(b)** $v' = 3$, $J' = 43$,**(c)** $v' = 6$, $J' = 27$ in the $B^1\pi_u$ state

The second crossing point between the molecular beam and two superimposed laser beams is located in the ion source of a mass spectrometer. The first laser excites the molecules into an intermediate level $|i\rangle$ and the second CW argon laser brings them from the state $|i\rangle$ into an energy state just above the ionization threshold. This resonant two-photon ionization is very efficient in producing isotope-selective ions, or in the case of cluster beams to select specific cluster sizes for ionization. This is very helpful for the assignment of the spectra, if many different masses (e.g. isotopes or different cluster sizes in a cluster beam) are present.

For illustration of the attainable spectral resolution Fig. 13.28 shows the hyperfine structure (HFS) of a rotational transition ($J' \leftarrow J''$) in the electronic transition of Na_2 molecules from the $X^1\Sigma_g$ ground state into the mixed $A^1\Sigma_u - a^3\Pi_u$ upper state, where the two states are coupled by spin–orbit coupling [13.18].

Optothermal Spectroscopy in Molecular Beams

The temperature increase caused by the transfer of excitation energy can also be monitored by the change of electrical conductivity with temperature. This is used in the bolometer, which is a detector made of a material with a high temperature coefficient α of electrical conductivity $\sigma = \sigma_0[1+\alpha(T)]$. Generally doped semiconductor materials are used. The coefficient α decreases with decreasing temperature. Therefore a bolometer is generally operated at low temperatures, e.g., at $T = 1.5\,\mathrm{K}$, which can be reached by evaporating ^{4}He in a helium cryostat. Even lower temperatures down into the milliKelvin range can be obtained by a ^{3}He^{4}He dilution refrigerator.

The temperature increase can be caused by the absorption of photons (in this case the bolometer acts as a radiation detector) or by the adsorption of excited

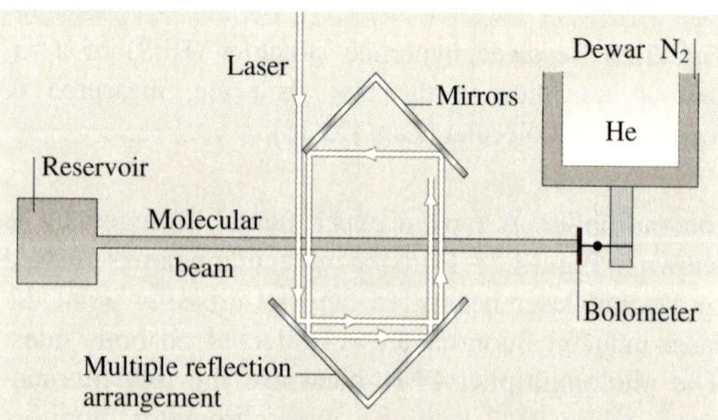

Fig. 13.29 Optothermal spectroscopy in a molecular beam with laser excitation and detection by a cooled bolometer

atoms or molecules at the surface of the bolometer. Here they transfer their excitation energy to the bolometer.

A typical arrangement for molecular spectroscopy in molecular beams is shown in Fig. 13.29 [13.19].

The molecular beam is crossed perpendicular by the beam of a single-mode laser, which excites the molecules into long-lived higher states (for example excited vibrational levels in the electronic ground state or Rydberg states with a long radiative lifetime).

The heat transfer rate from N molecules per sec with excitation energy $\Delta E = h\nu$ is

$$\mathrm{d}Q/\mathrm{d}t = N\Delta E = Nh\nu\,. \qquad (13.41)$$

This results in a temperature increase $\Delta T = T - T_0$ of a bolometer with heat capacity C and heat conduction $G(T-T_0)$ to the surroundings at temperature T_0 without molecules impinging on the bolometer at the temperature T, which can be determined from energy conservation

$$P = Nh\nu = C(\mathrm{d}T/\mathrm{d}t) + G(T-T_0)\,, \qquad (13.42)$$

where the incident power equals the thermal energy increase per sec of the bolometer and the power conducted to the surrounding. Under stationary conditions $(\mathrm{d}T/\mathrm{d}t = 0)$ the temperature rise is obtained from (13.42) as

$$\Delta T = T - T_0 = Nh\nu/G\,. \qquad (13.43)$$

In order to increase the sensitivity of a bolometer, heat conductivity to the surrounding should be as small as possible, i.e. the bolometer has to be thermally well isolated.

In order to subtract the background signal caused by the adsorption of molecules in their ground state, the excitation laser is chopped at a frequency f and the optothermal signal is detected through a lock-in amplifier. Now the energy transferred to the bolometer is time-dependent. The time constant of the bolometer should be shorter than the chopping period.

The time constant depends on the heat capacity C and the heat conductivity G. This can be seen as follows. Assume the transferred power is

$$P = P_0(1 + \mathrm{a}\cos\Omega\mathrm{t})$$
$$\text{with} \quad \mathrm{a} \le 1 \quad \text{and} \quad \Omega = 2\pi f\,. \qquad (13.44)$$

Inserting (13.44) into (13.42) gives

$$T(\Omega) = T_0 + \Delta T[1 + \cos(\Omega t + \phi)]\,. \qquad (13.45)$$

The temperature T is modulated at the same frequency Ω but shows a phase lag ϕ with

$$\tan\phi = \Omega C/G \tag{13.46}$$

against the modulated incident power, which depends on the heat capacity C, the heat conductivity G and the frequency Ω. The amplitude ΔT of the temperature modulation is

$$\Delta T = \frac{\bar{a} P_0 \Sigma}{\sqrt{\Sigma^2 + \Omega^2 P^2}} . \tag{13.47}$$

At the frequency $\Omega_c = G/C$ the amplitude ΔT decreases by a factor of $\sqrt{2}$ against its DC value.

We see that the sensitivity is proportional to $1/G$ but the time constant $\tau \sim G/C$. A sufficiently fast and sensitive detector should have a minimum heat capacity. Since C decreases with temperature, this is again an argument for operating the bolometer at low temperatures [13.20].

In Sect. 13.2 it was shown that the absorption path length should be as large as possible in order to reach a higher sensitivity. Since a single pass through the collimated molecular beam is only about 1 mm, optical arrangements for increasing this absorption length have been designed. In Fig. 13.30 several solutions are illustrated. They consist either of two parallel mirrors, where the incoming laser beam is reflected back and forth between the mirrors, or of a pair of cat-eyes that are shifted against each other by a small amount δ. In the former arrangement the laser beam passes the molecular beam at an angle $90° - \varepsilon$ and the absorption lines are therefore Doppler-shifted by an amount $\Delta\nu = (v/c)\sin\varepsilon$, where v is the velocity of the molecules. The largest advantage can be reached when the crossing point is placed in the beam waist of an external nearly confocal enhancement resonator with highly reflecting mirrors. The incoming laser beam has to be mode-matched to the fundamental mode of the resonator by an appropriate optical design, consisting of two lenses Fig. 13.28c. With such a design an enhancement by a factor 200 of the optical power in the crossing point can be achieved.

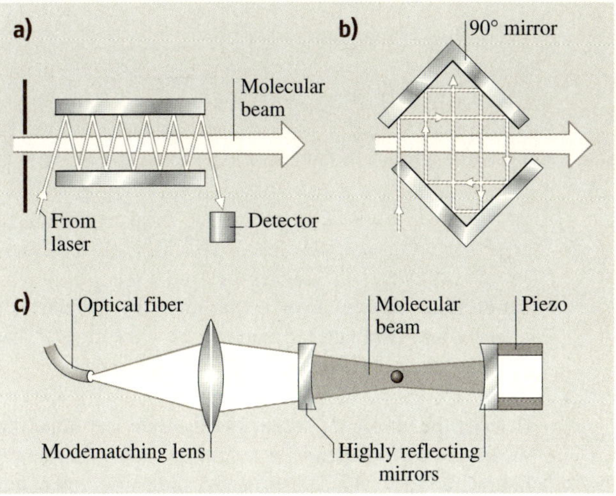

Fig. 13.30a–c Optical arrangements for generating many crossings of the laser beam with the molecular beam. **(a)** two plane-parallel mirrors, **(b)** two cat eyes arrangement, **(c)** crossing point inside a high Q enhancement resonator, where the laser beam is mode-matched to the fundamental mode of the resonator

13.1.3 Nonlinear Laser Spectroscopy

Several spectroscopic techniques that achieve sub-Doppler resolution are based on the nonlinear interaction of molecules with laser radiation. When a plane electromagnetic wave passes in the x-direction through an absorbing gas. We have seen above, that the attenuation $\mathrm{d}I$ of the intensity is related to the absorption coefficient α by

$$\mathrm{d}I = -\alpha I\,\mathrm{d}x . \tag{13.48}$$

The absorption coefficient

$$\alpha = \Delta N \Sigma_{ik}$$
$$\text{with } \Delta N = [N_k - (g_k/g_i)N_i]$$

is determined by the population difference ΔN and the absorption cross section σ. This gives for (13.48) the relation

$$\mathrm{d}I = -[N_k - (g_k/g_i)N_i]\Sigma_{ik} I\,\mathrm{d}x . \tag{13.49}$$

For sufficiently small intensities I the population densities N_i and N_k are not much affected, because relaxation processes can refill the population N_k of the absorbing level Fig. 13.31a. In this case the absorption coefficient α is independent of the intensity I and integration of (13.48) yields *Beer*'s law of linear absorption

$$I = I\,\mathrm{e}^{-\delta N \Sigma x} . \tag{13.50}$$

For higher intensities the absorption rate can exceed the relaxation rates refilling the absorbing level $\langle k|$. This means that the population N_k decreases with increasing intensity I and the absorption coefficient therefore also decreases. Instead of (13.51) we have to write

$$\mathrm{d}I = -\Delta N(I) I \sigma\,\mathrm{d}x . \tag{13.51}$$

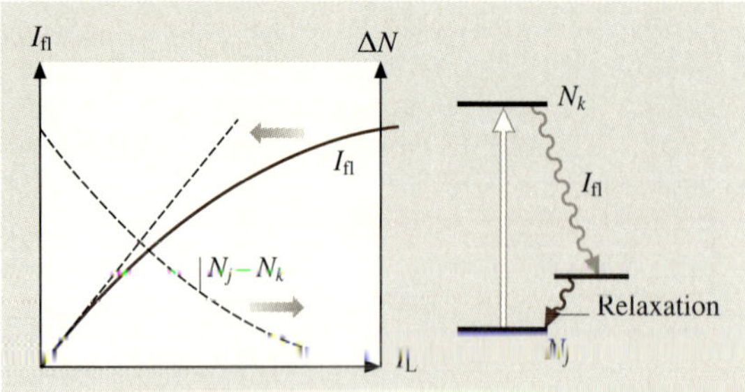

Fig. 13.31a,b Saturation of a transition **(a)** fluorescence intensity and population difference as a function of the excitation intensity, **(b)** level scheme

The change $\mathrm{d}I$ of the intensity and therefore also the absorbed power depends in a nonlinear way on the incident intensity. We can write the intensity-dependent population of the absorbing level $|k\rangle|$ as a power series

$$N_k = N_{k0}(1 - aI - bI^2 - \cdots) \tag{13.52a}$$

with $b \ll a \ll 1$ and a corresponding relation for the upper level $|i\rangle$

$$N_i = N_{i0}(1 + aI - bI^2 \cdots) . \tag{13.52b}$$

For the population difference we obtain Fig. 13.31b:

$$\Delta N = \Delta N_0(1 - 2aI - 2bI^2 - \cdots) . \tag{13.53}$$

Inserting this into (13.51) gives

$$\mathrm{d}I = \Delta N_0 \Sigma\, \mathrm{d}x(I - 2aI^2 - 2bI^3 - \cdots) . \tag{13.54}$$

The first term describes the linear absorption, the second term is quadratically dependent on I and diminishes the absorption because $\mathrm{d}(\Delta N)/\mathrm{d}I < 0$.

The nonlinear absorption can be demonstrated by measuring the intensity of the laser-induced fluorescence $I_{fl}(I_L)$ as a function of the laser intensity Fig. 13.31a. One can see, that I_{fl} first increases linearly with I_L, but for higher laser intensities the increase is sublinear, because the population difference and therefore the absorption coefficient decreases, which causes a smaller relative absorption of the laser intensity $\mathrm{d}I_L/I_L$. For high laser intensities the fluorescence intensity approaches a constant value (saturation) that is limited by the rate of the relaxation processes refilling the absorbing level $|k\rangle$.

This saturation of the absorption can be used for Doppler-free spectroscopy [13.21], as will be outlined in the next section.

Saturation Spectroscopy

We consider a sample of atoms or molecules in the gas phase with a Doppler-broadened absorption profile around the center frequency ω_0. When a monochromatic laser beam with frequency ω passes in the x-direction through the sample, only those molecules can absorb the laser photons that are Doppler-shifted into resonance with the laser frequency. Since the Doppler shift is $\Delta\omega = kv_x$ these molecules must have velocity compo-

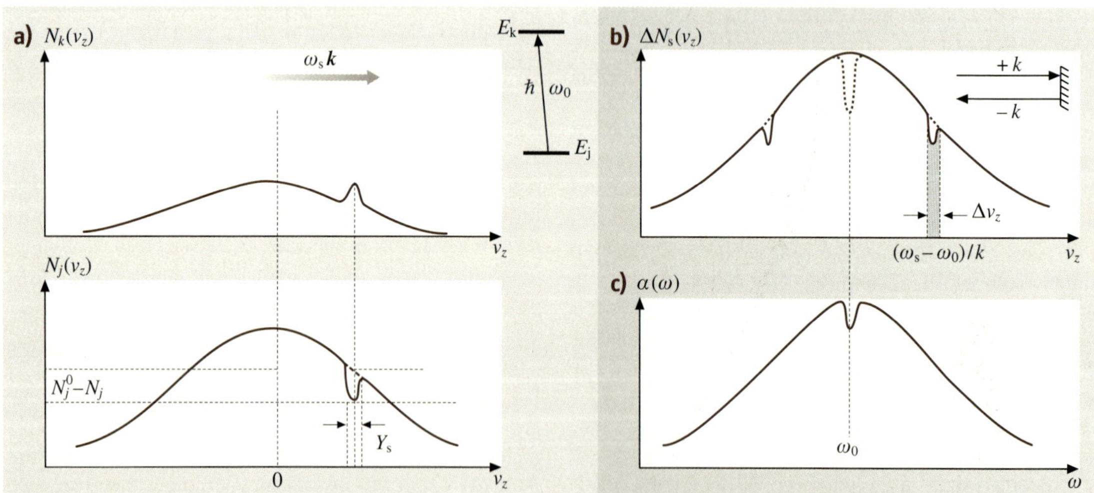

Fig. 13.32a–c Principle of saturation spectroscopy of Doppler-broadened transitions: **(a)** saturation hole and saturation peak in the velocity distribution of lower and upper level population, **(b)** situation for a standing wave, **(c)** Lamb dip in the Doppler-broadened absorption profile

nents $v_{x,}$, that satisfy the relation

$$\omega = \omega_0(1 + kv_x) \pm \gamma \,, \tag{13.55}$$

where γ is the homogeneous line width of the transition. Because of saturation the population N_k in the absorbing level of these molecules decreases and N_i increases accordingly. In the velocity distribution $N_k(v_x)$ of all molecules a narrow dip is burnt with a spectral width γ, while a corresponding peak appears in the distribution $N_i(v_x)$ of molecules in the upper level Fig. 13.32a.

If the laser beam is reflected back into the sample by a mirror, its k vector is reversed and molecules with opposite velocity components interact with the reflected beam Fig. 13.32b. Now two holes, symmetric to $v_x = 0$ are burnt into the velocity distribution of the inversion $\Delta N(v_x)$, i. e., two different velocity classes $\pm(v_x \pm \gamma)$ interact with the two laser beams.

When the laser frequency ω_L is tuned over the Doppler-broadened absorption profile the two holes move towards each other as ω_L approaches the center frequency ω_0 of the molecular transition. For $\omega_L = \omega_0$ only one velocity class in the interval dv_x around $v_x = 0$ interacts with both laser beams. The molecules experience the double intensity and therefore the population inversion ΔN will more decrease for $\omega_L = \omega_0$ than for $\omega_L \neq \omega_0$. The absorption coefficient $\alpha(\omega)$ with a Doppler-broadened profile therefore has a dip around the center frequency ω_0 Fig. 13.32c. This dip, caused by the saturation of the population, is called the Lamb dip, named after Willis Lamb who explained this saturation effect quantitatively.

The width of the Lamb dip equals the natural line width of the molecular transition, broadened by saturation. It is about two orders of magnitude smaller than the Doppler width.

This narrow Lamb dip is used for Doppler-free spectroscopy (called saturation, or Lamb dip spectroscopy). Assume two transitions from a common level $\langle c|$ into two slightly split levels $\langle a|$ and $\langle b|$. If the splitting is smaller than the Doppler width the two Doppler-broadened spectral lines cannot be resolved. However, the two Lamb dips of these transitions are well separated Fig. 13.33.

A possible experimental arrangement is shown in Fig. 13.34. The beam of a wavelength-tunable laser is split by beam splitter (BS) into a strong pump beam and a weak probe beam, which pass in opposite directions through the sample cell. The detector measures the transmitted probe beam intensity as a function of the laser frequency ω_L. Each time the laser frequency coincides with the center frequency of a molecular transition, a Lamb peak appears in the transmitted intensity because the absorption exhibits a dip at this frequency.

The Doppler-broadened background can be eliminated when the pump beam is periodically chopped. Now a lock-in detector measures the difference of the transmitted probe intensity with the pump beam on and

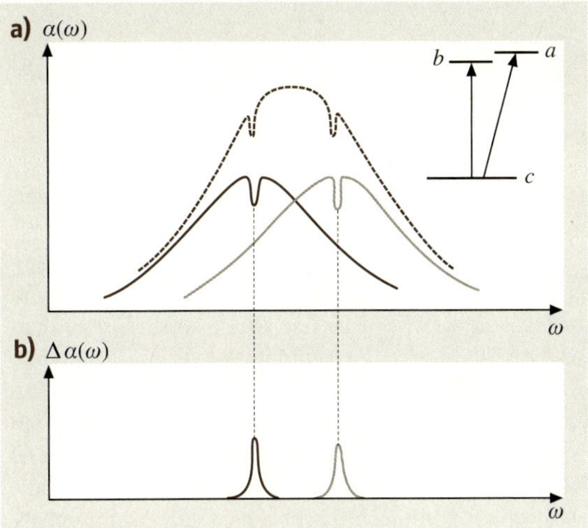

Fig. 13.33 Resolution of the two Lamb dips of two close overlapping Doppler-broadened transitions with a common lower level

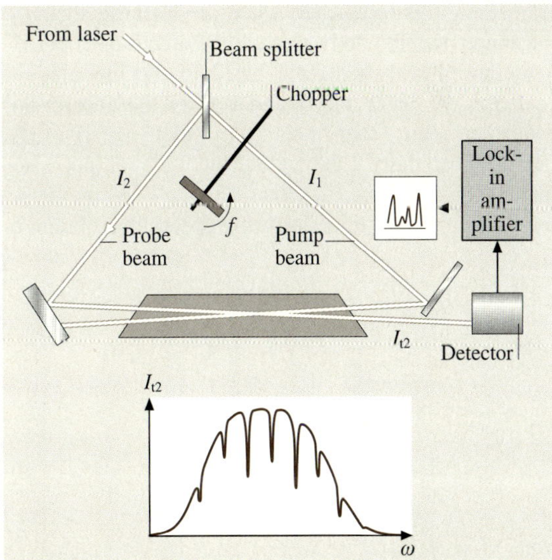

Fig. 13.34 Experimental arrangement for saturation spectroscopy with detection of the transmitted probe beam, resulting in Lamb dips on a Doppler-broadened background

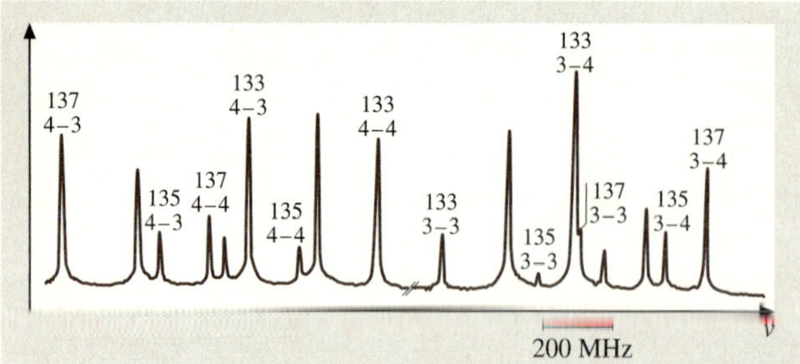

Fig. 13.35 Saturation spectrum of the hyperfine-components of the $6^2S_{1/2} \rightarrow 7^2P$ transition at $\lambda = 459.3$ nm in a mixture of the isotopes ^{133}Cs, ^{135}Cs and ^{137}Cs [13.22]

off. The result is a Doppler-free spectrum Fig. 13.35 with a better signal-to-noise ratio.

Even with low-power lasers saturation spectroscopy can be performed when the sample is placed inside the laser resonator, where the intensity is much higher than outside Fig. 13.36. The standing wave inside the resonator can be composed of a back-and-forth traveling wave. Therefore the conditions for producing a Lamb dip are automatically fulfilled. Either the laser-induced fluorescence from the sample molecules can be monitored, which exhibits the Lamb dips, or the laser intensity transmitted through one of the resonator mirrors is measured, which shows corresponding narrow peaks in the laser output power at the center frequencies of the absorption lines of the sample molecules. If the resonator length is modulated while the laser wavelength is scanned, the first derivative of the Lamb dips or peaks is measured with a lock-in tuned to the modulation frequency f. Since the modulation of the absorption is not harmonic (even if the laser frequency is modulated sinusoidally), the modulated laser output also contains higher harmonics Sect. 13.1.2. When the lock-in is tuned to $3f$ the third derivative spectrum is obtained

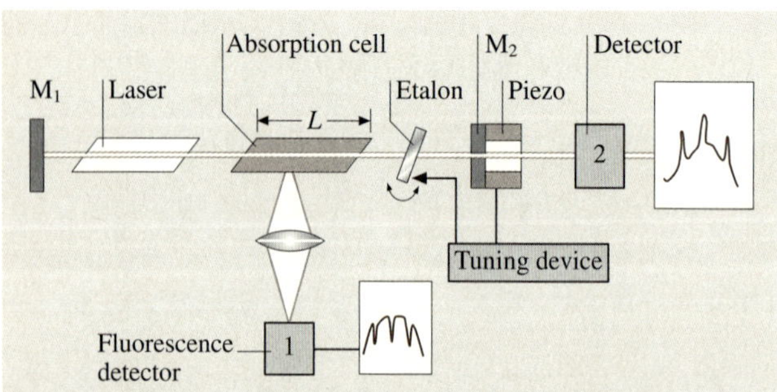

Fig. 13.36 Saturation spectroscopy with the sample inside the laser resonator

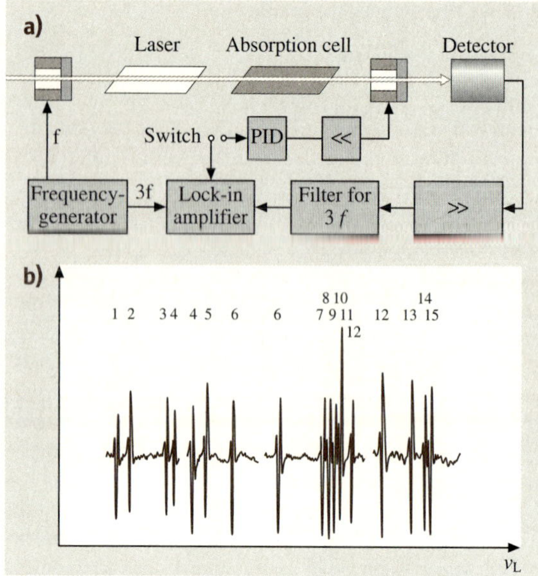

Fig. 13.37a,b Third-derivative technique for saturation spectroscopy inside the laser cavity. **(a)** Experimental arrangement; **(b)** third-derivative saturation spectrum of the hyperfine components of a rotational transition in I_2

Fig. 13.37. Here the Doppler-broadened background is essentially suppressed. Figure 13.36b shows such a third-derivative spectrum for the rotational transition $B^3\Pi_{mathrmu}(v' = 58, J' = 99) \leftarrow X^1\Sigma_g^+(v'' = 1, J'' = 98)$ in the I_2 molecule with its 21 hyperfine components.

Polarization Spectroscopy

While in saturation spectroscopy the selective decrease of absorption by the pump laser was monitored by the counter-propagating probe laser, in polarization spectroscopy it is the birefringence induced by the pump laser that is monitored by the probe laser [13.23]. The principle is shown in Fig. 13.38. The sample is placed between two crossed polarizers P_1 and P_2 that attenuate the linearly polarized probe laser by a factor of $10^5 - 10^7$, depending on the quality of the polarizers. A larger portion of the laser beam is split off by the beam splitter S and acts as a pump beam, which passes in the opposite direction through the sample cell. If the pump laser beam is circularly polarized (e.g., σ^+-polarization), it induces transitions with $\Delta M = +1$, where Mh is the projection of the angular momentum J of the sample molecules onto the direction of the pump beam. Due to saturation some of the M sublevels are partly depopulated (for our example in Fig. 13.38a these are the levels with $M = -2, -1, 0$). This leads to a partial orientation of the molecules because their angular momentum J no longer has an isotropic orientational distribution but points more against the direction of the pump beam.

The linearly polarized probe beam can be composed of a σ^+ and a σ^- component. The two components experience different absorption and a different index of refraction when passing through the sample of orientated molecules. Adding the two components again after they have passed through the sample results in an elliptical polarization of the probe beam with the major axis rotated by a small angle α compared to the initial polarization direction, because of the different phase shifts of the σ^+ and the σ^- components. Now a small component of the electric vector parallel to the transmission axis of the second polarizer exists, which is transmitted to the detector. We will discuss this now more quantitatively.

The linearly polarized probe wave

$$E = E_0 e^{i(\omega t-kz)} \text{with} \quad E_0 = \{E_0\,,0\,,0\} \qquad (13.56)$$

can be composed of a σ^+ and a σ^- component

$$E^+ = E_0^+ e^{i(\omega t-k^+z)}\,, \quad E_0^+ = \frac{1}{2}E_0(\hat{x}+i\hat{y})\,,$$
$$E^- = 0E_0^+ e^{i(\omega t-k^-z)} \quad E_0^- = \frac{1}{2}E_0(\hat{x}-i\hat{y})\,, \qquad (13.57)$$

where $\hat{x}$ and $\hat{y}$ are unit vectors.

The two components experience different absorption coefficients α^+ and α^- and also different refractive indices n^+ and n^-. After a path length L through the sample region that has been orientated by the pump laser the two components are

$$E^+ = E_0^+ e^{i(\omega t-k^+L+i\alpha^+L/2)}\,,$$
$$E^- - E_0^+ e^{i(\omega t-k^-L+i\alpha^-L/2)} \qquad (13.58)$$

Because of the difference $\Delta n = n^+ - n^-$ a phase difference

$$\Delta\Phi = (k^+ - k^-)L = (\omega L/c)(n^+ - n^-) \qquad (13.59)$$

has developed between the two components and because of the difference $\Delta\alpha = \alpha^+ - \alpha^-$ there is also a difference

$$\Delta E_0 = (E_0/2)[e^{-\alpha^+L/2} - e^{-\alpha^-L/2}] \qquad (13.60)$$

in the two amplitudes.

The windows of the absorption cell are slightly bent because of external atmospheric pressure and they therefore show a birefringence which is different for the two circularly polarized components and causes an additional phase shift.

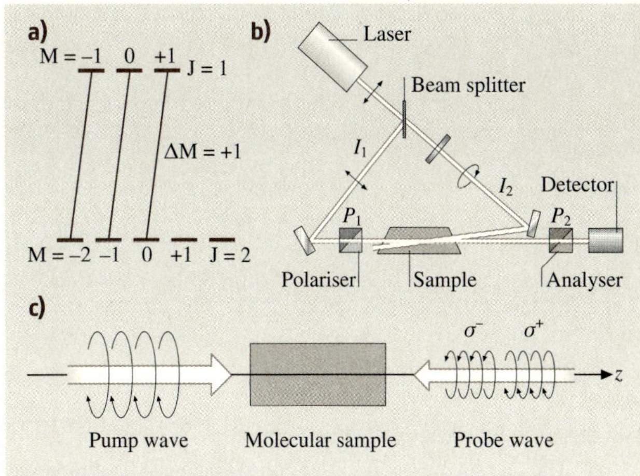

Fig. 13.38 Level scheme and experimental set up for Doppler-free polarization spectroscopy

Behind the exit window the electric field vector of the probe beam is obtained by adding the σ^+ and σ^- components. This gives a slightly elliptical polarized wave with a major axis that is tilted by the small angle θ against the x-axis. This increases the transmittance through the crossed analyzer. The transmitted amplitude becomes (Fig. 13.39)

$$E_t = E_x \sin\theta + E_y \cos\theta\,. \qquad (13.61)$$

The detector signal is proportional to the transmitted intensity. Besides the signal there is also a frequency-independent background due to the residual transmission ξ of the crossed polarizers.

Similar to the case in saturation spectroscopy the line profile of the signals are Doppler-free and are determined by the homogeneous line width γ (natural line width plus collision broadening plus saturation broaden

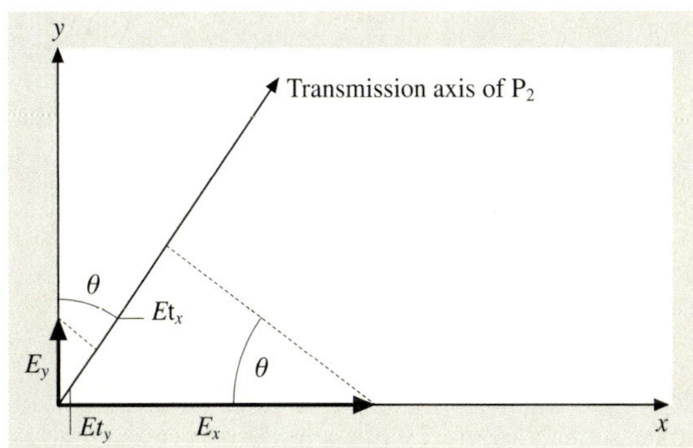

Fig. 13.39 Transmitted polarization components through the analyzer

ing), because only molecules with velocity components $v_x = 0 \pm \gamma k$ can interact with both counter-propagating laser beams.

The final result for the transmitted intensity, measured when the laser frequency is tuned over an absorption line of the sample molecules, is

$$I_t = I_0 e^{-\alpha L} \left(\Sigma^2 + \Theta'^2 + \tfrac{1}{4}\Delta\alpha_\omega^2 + \Theta' \Delta\alpha_0 L \frac{x}{1+x^2} \right) + \left[\tfrac{1}{4}\Delta\alpha_0 \Delta\alpha_\omega L + \left(\frac{\Delta\alpha_0 L}{4}\right)^2 \right] \frac{1}{1+x^2}$$

with

$$x = \frac{\omega_0 - \omega}{\gamma_s/2} . \tag{13.62}$$

The advantages of polarization spectroscopy, compared to saturation spectroscopy are as follows:

1. The technique is essentially background-free. Besides the small residual terms $I_0\xi$ with $\xi = 10^{-5}$–10^{-7} the detector only receives a signal when the pump beam is not blocked. In saturation spectroscopy the pump beam changes the absorption of the probe beam by a small amount and the small difference between two large signals is detected. The sensitivity of polarization spectroscopy is therefore much higher.
2. Uncrossing of the polarizers ($\theta > 0$) increases the dispersion term in (13.62). Such a dispersion signal is ideally suited to stabilize a laser frequency onto the zero-crossing of the signal, which occurs at the center of the Doppler-free absorption line. Choosing $\theta = 0$ yields Lorentzian line profiles, where the magnitude of the signals can be optimized by changing the birefringence of the windows (e.g., by squeezing them gently).
3. It turns out that the signal magnitude and profile depends on the specific molecular transition. For a circularly polarized pump beam the signals of Q-lines ($\Delta J = 0$) are suppressed, while for a linearly polarized pump beam the magnitude of the signals for P- and R-lines ($\Delta J = \pm 1$) decreases rapidly with increasing rotational quantum number J. The assignment of a complex molecular spectrum is therefore facilitated.

Doppler-Free Two-Photon Spectroscopy

For sufficiently strong laser intensities it can happen that an atom or molecule simultaneously absorbs two photons $\hbar\omega_1$ and $\hbar\omega_2$ from two lasers, or two photons $\hbar\omega_1$ from the same laser. Depending on the relative orientations of the two photon spins, transitions with $\Delta L = 0$ or $\Delta L = \pm 2$ are induced. Two-photon transitions are weaker by several orders of magnitude than allowed one-photon transitions. Therefore lasers with sufficient high intensities are demanded for their observation. The probability of two-photon absorption is greatly enhanced if a molecular level E_m is close to the energy $E_k + \hbar\omega_1$ or $E_k + \hbar\omega_2$.

For a two-photon transition from the initial lower level $\langle k|$ to the final level $\langle f|$ in a molecule at rest energy conservation demands

$$E_f - E_k = \hbar(\omega_1 + \omega_2) . \tag{13.63}$$

When the molecule moves with velocity v the frequency ω of the light wave is shifted in the molecular frame to $\omega' = \omega - kv$. The resonance condition (13.63) now becomes

$$E_f - E_k = \hbar(\omega_1 + \omega_2) - hv(k_1 + k_2) . \tag{13.64}$$

If the two photons come from two counter-propagating traveling beams from the same laser, we have the situation that $\omega_1 = \omega_2$ and $k_1 = -k_2$. This has the consequence that the last term in (13.64), which contains the velocity of the molecule, is zero. The two-photon absorption in this case becomes independent of the molecular velocity, which means that all molecules within the velocity distribution contribute to the two-photon absorption, which is now Doppler-free. The experimental arrangement is shown in Fig. 13.40. The sample is placed inside the resonator of a single-mode laser. The two-photon transition is monitored by the laser-induced fluorescence, emitted from the upper level E_f into intermediate levels E_m that are connected to $\langle f|$ by allowed one-photon transitions. A Faraday rotator acts as an optical diode and prevents the reflected beam from passing

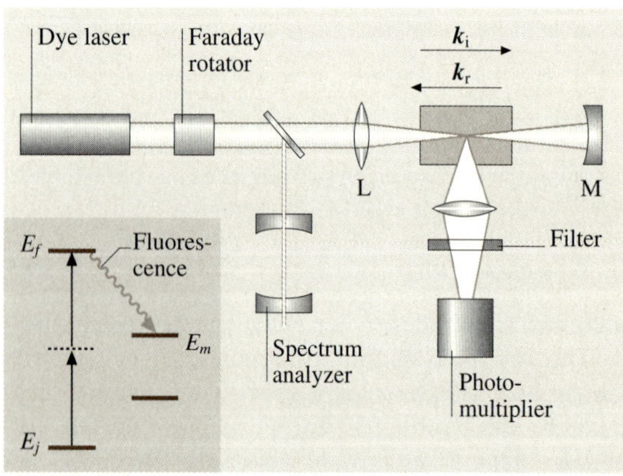

Fig. 13.40 Doppler-free two-photon spectroscopy

back into the laser, because this could cause instabilities in the laser.

When the laser frequency ω is tuned over the two-photon resonance the signal consists of a narrow peak (produced by two photons with opposite k-vector) and a Doppler-broadened background, produced by two photons from the same beam with parallel k-vectors Fig. 13.41. The Doppler width is twice as large as that of a one-photon transition with frequency ω. The probability that the two photons come from opposite beams is twice as large as that for two photons from the same beam. This implies that twice as many molecules contribute to the narrow peak than to the broad background. The narrow peak is therefore $2\Delta\omega_D/\Delta\omega_n$ times as high than the background. Since the Doppler width $\Delta\omega_D$ is about two orders of magnitude larger than the natural line width $\Delta\omega_n$ the background can be neglected.

In Fig. 13.42 the Doppler-free spectrum of lead atoms on the transition $7p^3P_0 \leftarrow 6p^{23}P_0$ at $\lambda = 450.4$ nm illustrates the high spectral resolution allowing the measurement of the isotope shifts [13.24].

More extensive coverage of two-photon spectroscopy can be found in [13.25].

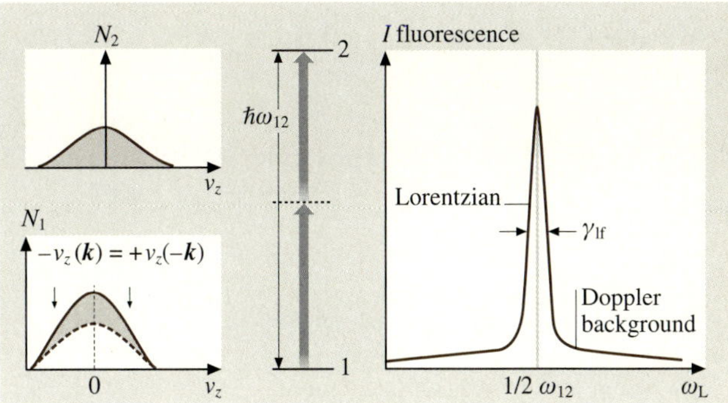

Fig. 13.41 Narrow line profile of Doppler-free two-photon absorption with Doppler-broadened background

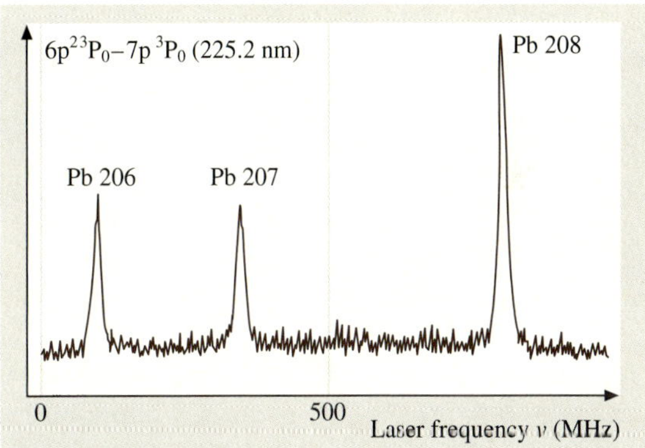

Fig. 13.42 Doppler-free two-photon spectrum of the $6p^23P_0 \rightarrow 7p^3P_0$ transition in the different isotopes of lead [13.24]

13.1.4 Polarimetry and Ellipsometry

Similar to the principle in polarization spectroscopy a technique called *polarimetry* uses the change of the polarization of a light wave which has passed through an anisotropic sample. However, in contrast to polarization spectroscopy the anisotropy is not induced by a polarized pump wave but is inherent to the sample, which consists of so-called *active molecules*. For example, all chiral molecules, such as butanol Fig. 13.43 or blood sugar, turn the polarization vector of a linearly polarized wave by an angle

$$\alpha = (n^+ - n^-)\pi d/\lambda\ , \tag{13.65}$$

where n^+ and n^- are the refractive indices for σ^+ and σ^- light and d is the path length through the sample Fig. 13.44.

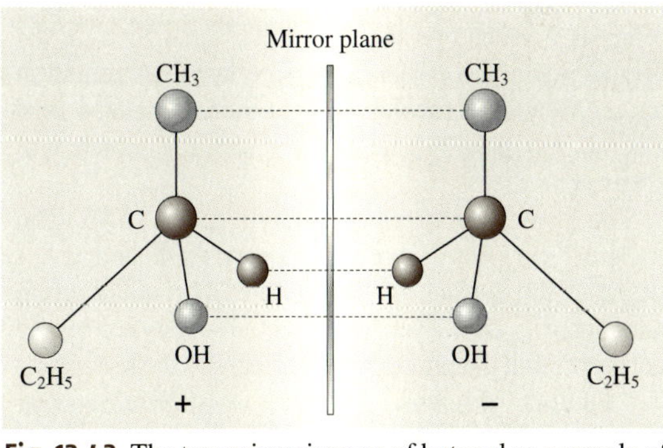

Fig. 13.43 The two mirror images of butanol as example of chiral molecules

In solids optical anisotropy can be induced by internal or external stress. In such cases optical polarimetry can give a two-dimensional picture of the stress distribution. This is used by glass blowers to check whether glass after tempering is stress-free. For mechanical engineering models, Plexiglas can be used to test the stress distribution in frame constructions of buildings and bridges [13.26].

The *spectral ellipsometry* technique uses the polarization characteristics of an optical wave and their changes when the wave is reflected by a surface or transmitted through a sample for sensitive detection of material parameters, such as absorption and birefringence. It is generally applied to solids and liquids, where

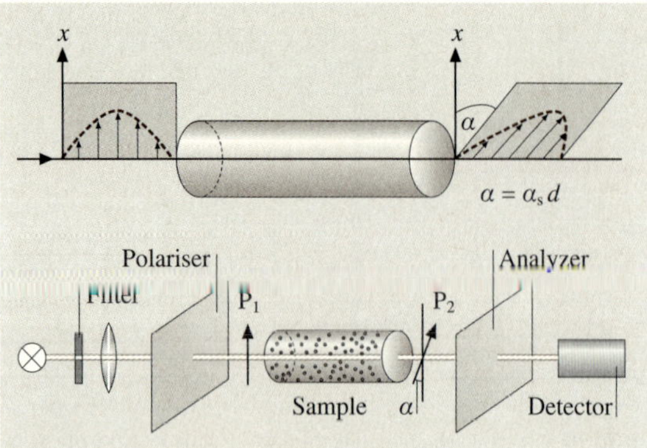

Fig. 13.44 Principle and experimental set up for polarimetry

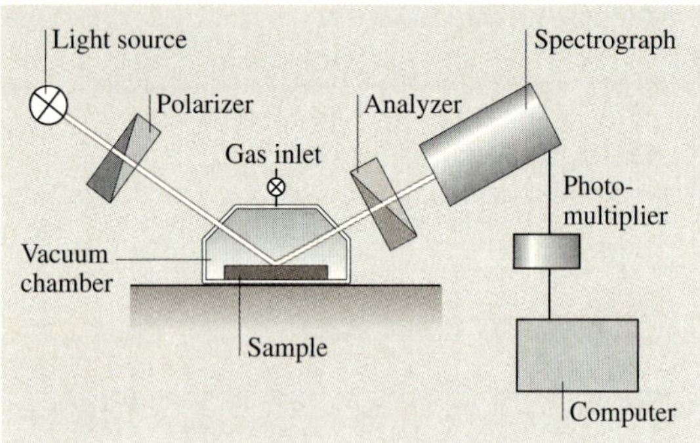

Fig. 13.45 Schematic experimental arrangement for ellipsometry

there are no sharp absorption lines. Therefore the main emphasis is placed on the sensitivity rather than on the spectral resolution. Besides tunable lasers as radiation sources broadband incoherent light sources are also used in combination with a monochromator for wavelength selection.

The basic principle is illustrated in Fig. 13.45. The incident light passes through a polarizer P and is reflected at the surface of the sample. The polarization characteristics of the reflected light are analyzed by the analyzer A and the photomultiplier behind the spectrometer monitors the intensity of a selected spectral interval transmitted through the spectrometer.

Depending on the absorption coefficient $\alpha(\omega)$ light with intensity I_0 penetrates into the sample. After a penetration depth $d = 1/\alpha$ its intensity has decreased to I_0/e and the reflected light with intensity $I_\mathrm{r} = rI_0$ gives information on the complex reflection coefficient r, which also depends on the absorption coefficient α. For linearly polarized incident light the reflected light is gen-

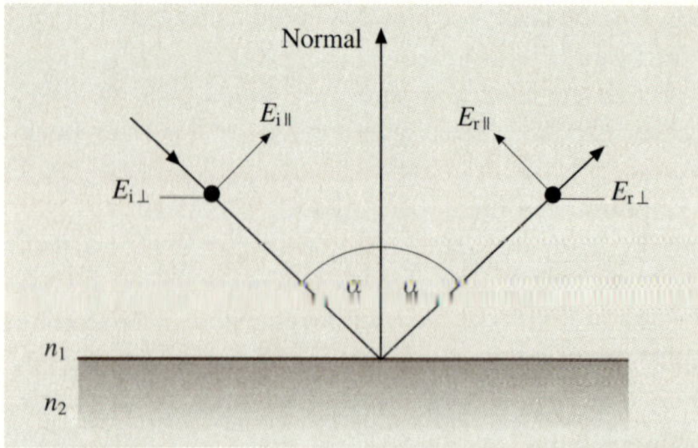

Fig. 13.46 Illustration of the Fresnel formulas for the components of the electric field vectors of the incident and reflected wave

erally elliptically polarized, which gives the name to the method [13.27].

For non-absorbing samples the refractive index n is real and the intensity and the polarization characteristics of the reflected light are governed by the Fresnel formulas. We can compose the electric vector of the incident linearly polarized light of a component $E_\parallel$ parallel and a component $E_\perp$ perpendicular to the plane of incidence Fig. 13.46. The amplitude reflection coefficient for the two components [13.28]

$$P_\parallel = \frac{E_{\mathrm{r}\parallel}}{E_{\mathrm{i}\parallel}} = \frac{n_2 \cos\alpha - n_1 \cos\beta}{n_2 \cos\alpha + n_2 \cos\beta}\,, \tag{13.66a}$$

$$P_\perp = \frac{E_{\mathrm{r}\perp}}{E_{\mathrm{i}\perp}} = \frac{n_1 \cos\alpha - n_2 \cos\beta}{n_1 \cos\alpha + n_2 \cos\beta}\,, \tag{13.66b}$$

is then determined by the refractive indices n_1 and n_2 of the two media and the angle α against the surface normal. Using the Snellius refraction law $\sin\alpha/\sin\beta = n_2/n_1$ we can replace the refractive indices by the refraction angle β and obtain

$$\begin{aligned} P_\parallel &= \frac{\tan(\alpha-\beta)}{\tan(\alpha+\beta)}\,, \\ P_\perp &= \frac{\sin(\alpha-\beta)}{\sin(\alpha+\beta)}\,. \end{aligned} \tag{13.67}$$

The reflected intensity is then

$$\begin{aligned} I_{\mathrm{r}\parallel} &= I_{\mathrm{i}\parallel}\frac{\tan^2(\alpha-\rho)}{\tan^2(\alpha+\rho)}\,, \\ I_{\mathrm{r}\perp} &= \frac{\sin^2(\alpha-\rho)}{\sin^2(\alpha+\rho)}\,. \end{aligned}$$

Since the reflection coefficients for the two components are different, the plane of polarization of the reflected wave is tilted against that of the incident wave. If γ_i is the angle of the electric vector E of the incident wave against the plane of incidence, the angle γ_r of the reflected wave becomes, according to the Fresnel formulas,

$$\tan\gamma_r = \frac{E_{r\perp}}{E_{r\parallel}} = -\frac{\cos(\alpha-\beta)}{\cos(\alpha+\beta)}\tan\gamma_i\,. \tag{13.68}$$

For absorbing materials the index of refraction

$$n = n - \mathrm{i}\kappa \tag{13.69}$$

becomes complex where both the real part n and the imaginary part κ depend on the wavelength. This introduces a phase shift, which is generally different for the two polarization components. The reflected light is elliptically polarized.

13.1.5 Optical Pumping and Double Resonance

We have discussed in Sect. 13.1.1 that emission spectra become much simpler when a single level is selectively populated. Such a selective population or depletion of a level by absorption of photons is called optical pumping, when a noticeable population change ΔN against the thermal population is achieved. The technique of optical pumping is, however, more general than just changing the population of a level. It can also create an alignment of atoms or molecules in the *labeled* level, as we have seen for the example of polarization spectroscopy. This changes the polarization characteristics of a wave transmitted through the sample.

Optical pumping forms the basis for various double-resonance techniques. Assume the pump laser is tuned to a selected transition $|i\rangle \to |k\rangle$ Fig. 13.47 and depletes the lower level $|i\rangle$. If a second radiation field [radiofrequency (RF), microwave, infrared or visible radiation] with tunable frequency is simultaneously applied to the sample, the depleted level will be partly refilled, when the frequency of this second radiation field coincides with a transition $|i\rangle \to |m\rangle$ from the depleted level $|i\rangle$ to other levels $|m\rangle$. This can be monitored either by the increased absorption of a probe laser or by the increase of the fluorescence induced by the pump laser.

If, for example, the pump laser depletes a hyperfine component of an atomic state, the depleted level can be refilled by RF transitions $|ij\rangle \to |in\rangle$ from another HFS component, which is not depleted by the pump Fig. 13.47. When the laser-induced fluorescence is observed as a function of the radiofrequency, a peak is measured at the resonance frequency ω_{inj}, which gives the separation of the two HFS levels. In this case the RF–optical double resonance acts like an amplifier of the RF transitions, because

1. the optical pump increases the population difference ΔN between the two HFS levels and therefore the absorption of the radio frequency (RF).
2. The absorption of one RF photon is monitored by emission of a visible photon, which has an energy about 10^6 times larger than the RF photon and can be detected with a much higher efficiency.

The accuracy of measuring the separation of the levels using radio frequencies is much higher than through the difference of the frequencies of two optical transitions from the two HFS levels into a level in the upper excited state. The RF transition can occur in the lower as well as in the upper state of the optical transition. In the latter case one observes a dip in the fluorescence intensity from level $|k\rangle$ when the RF is tuned through resonance.

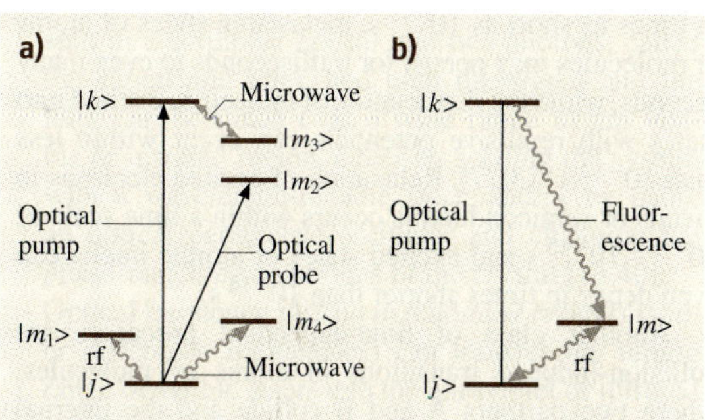

Fig. 13.47 Schematic level scheme for optical pumping and optical–microwave double resonance with fluorescence detection

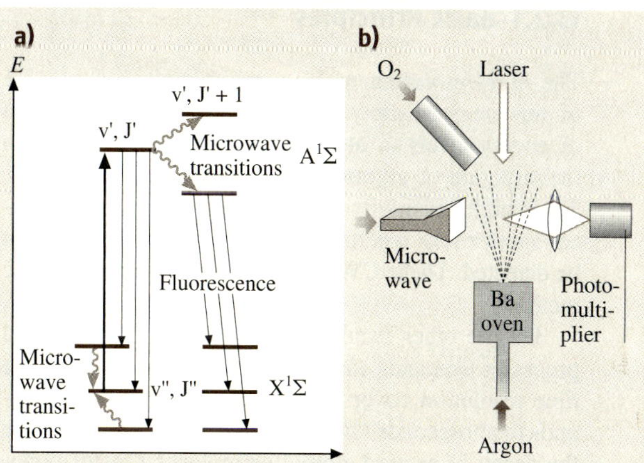

Fig. 13.48 Example of optical–microwave double-resonance spectroscopy in the excited electronic state of BaO [13.29]

Generation of Tunable Femtosecond Pulses with High Peak Powers

The output pulses from a femtosecond oscillator can be amplified in a regenerative amplifier. A schematic diagram of the set up for a high-power femtosecond laser system is depicted in Fig. 13.50 [13.39]. The laser oscillator consists of a Ti:sapphire crystal in a resonator formed by two spherical and a flat mirror. Two prisms serve as dispersion compensators. The crystal is pumped by a mode-locked argon laser. The femtosecond output pulses generally show a frequency chirp (i.e., the optical frequency varies from the leading to the trailing edge of the pulse). These chirped pulses are expanded in time by a grating pair, where the two gratings are not parallel as for pulse compression, but are arranged in such a way that the time profile of the chirped optical pulse becomes broader (pulse stretcher). This pulse stretching is necessary before the pulses are amplified in order to avoid excessively high peak powers in the amplifier, which would destroy the optics. The amplifier, generally a Ti:sapphire crystal pumped by a pulsed solid-state laser, is placed inside a multiple-reflection device that allows the laser pulse to travel many times through the amplifying medium. The energy of the pump pulse must be high enough to restore the inversion of the gain medium between two successive traversals of the amplified pulses (regenerative amplifier). The energy amplification of this multipass device might be as high as 10^6. The repetition rate of the amplified output pulses is limited by that of the pump laser for the regenerative amplifier and is generally not higher than 10 kHz while the mode-locked oscillator before amplification has a repetition frequency of about 100 MHz, which means that only every $10\,000^{th}$ pulse of the mode-locked train is amplified. These amplified pulses are selected by a Pockels cell (PC) and are then further amplified in a second amplifier stage before they are compressed by a parallel grating pair, resulting in ultrashort pulses with peak powers of up to 10 TW [13.43, 44].

Example. The oscillator may deliver output pulses with 50 fs pulse width and 30 nJ energy. This gives a peak power of 600 kW. With an energy amplification of 10^6 the peak power of the amplified and compressed pulse is then 600 GW. If a compression down to 10 fs can be reached the peak power may be as high as 3 TW.

Optical pulse compression can also be realized using nonlinear dispersion in optical transparent materials. By compressing the output pulses of a multistage Ti:sapphire laser amplifier system in a neon-gas-filled hollow fibre, optical pulses with duration of 5 fs and pulse energy of 1 mJ (peak power 0.2 TW) can be generated [13.45].

Optical Parametric Systems

Optical parametric devices, delivering femtosecond pulses with widely tunable wavelengths from the ultraviolet into the near-infrared spectral region, are very useful tools for spectroscopy [13.40, 46–48]. Their basic principle can be described as follows (Fig. 13.51):

A pump photon with energy $\hbar\omega_p$ and wave vector $\boldsymbol{k}_p$ is split by parametric interaction in a nonlinear optical crystal (e.g. β-Barium-Borate BBO) into a signal photon $\hbar\omega_1$ and an idler photon $\hbar\omega_2$, such that energy conservation

$$\omega_p = \omega_1 + \omega_2 \tag{13.70a}$$

and momentum conservation

$$\boldsymbol{k}_p = \boldsymbol{k}_1 + \boldsymbol{k}_2 \tag{13.70b}$$

for the frequencies ω_i and the wave vectors $\boldsymbol{k}_i$ are fulfilled. The phase velocities of the pump, signal and idler waves are matched by proper orientation of the birefringent crystal. Since the extraordinary index of refraction depends on the angle between the beam propagation and the optical axis of the nonlinear crystal, tilting the crystal allows continuous wavelength tuning of the signal and idler waves. If besides the pump wave, a weak signal wave (seed beam) is sent into the nonlinear crystal the signal wave will be enhanced much more than the idler wave. For the generation of high-power ultrashort pulses a non-collinear optical parametric amplifier (NOPA) has proved to be efficient [13.48]. Its basic principle is shown in Fig. 13.52. It consists of three main functional blocks:

1. Continuum generation by focussing a small fraction of the near-infrared (NIR) pump beam onto a sapphire disk

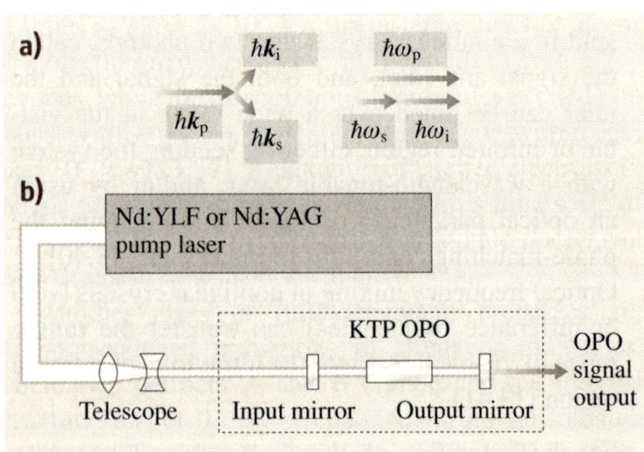

Fig. 13.51 Principle of optical parametric processes

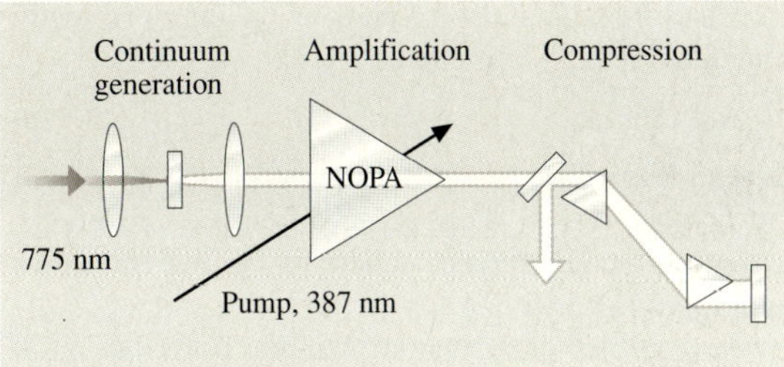

Fig. 13.52 Schematic Illustration of a non-collinear parametric amplifier (NOPA)

2. Parametric amplification of the seed light in a β-barium-borate (BBO) crystal pumped by the frequency-doubled pump light
3. The compression of the broadband output pulses

The difference of this system, compared to amplification by an inverted gain medium is the following: while for conventional amplification the inversion can be stored for a time depending on the relaxation rates of the active medium, in parametric amplification there is only gain during the time of the pump pulse. It is therefore necessary that the seed and pump pulses overlap in time. This can be achieved by using the same laser source for the seed and pump. If for instance, the pump beam is provided by the frequency-doubled output of a Ti:sapphire femtosecond laser and the seed beam comes from the spectral continuum (generated by focussing the output beam of the Ti:sapphire laser at $\lambda = 800$ nm onto a sapphire plate) both beams are generated by the same laser and the time overlap is guaranteed (Fig. 13.53). The phase-matching condition ((13.70a) and (13.70b)) selects from the continuum of difference frequencies

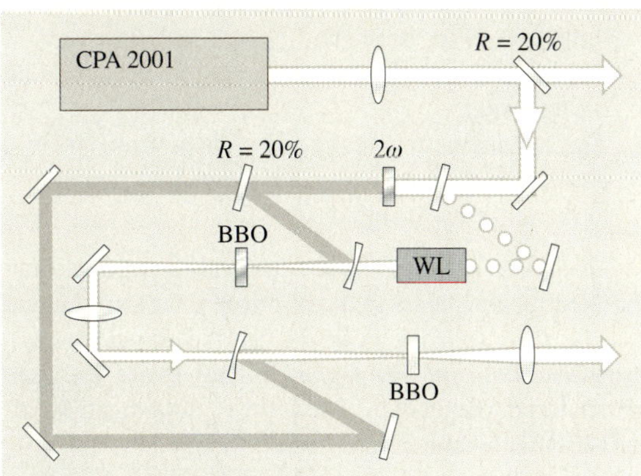

Fig. 13.53 Schematic set up for a widely tunable optical parametric amplifier [13.49]

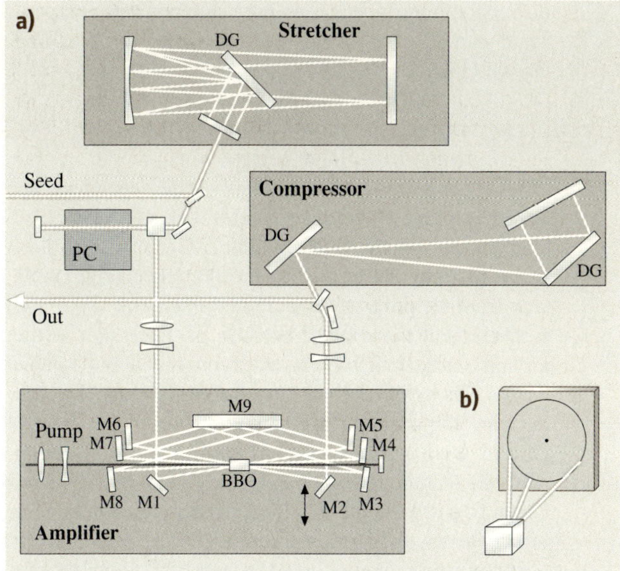

Fig. 13.54 (a) Schematic diagram for multipass non-collinear optical parametric chirped pulse amplifier (DG = diffraction grating, PC = Pockels cell) **(b)** beam alignment in the amplifier. The pump beam (not shown) aims at the center of the circle [13.50]

$\omega = \omega_p - \omega_s$ the desired wavelength for amplification (Fig. 13.52b). If the output of the OPA is mixed with the pump beam in a second BBO crystal, sum or difference frequencies can be generated and this widely tunable intense coherent source proves very useful for spectroscopy across a broad spectral range [13.49].

In Fig. 13.54 a possible experimental set up for multipass non-collinear parametric chirped-pulse amplification is shown [13.50]. The seed source is a Kerr-lens mode-locked Ti:sapphire laser, which delivers a 78 MHz train of pulses with 30 nm spectral width. After passing a pulse stretcher the prolonged pulses enter the multipass parametric amplifier with a BBO crystal, which is pumped by the frequency-doubled output of a Nd:YAG laser with a pulse width of 8 ns. The pump and seed beams cross each other in the BBO crystal at an angle of 2.4°. The amplified pulses are compressed by a grating pair. With such an arrangement terawatt femtosecond pulses can be realized [13.51].

Ultraviolet pulses with duration of 7 fs are efficiently generated by frequency doubling the output of a non-collinear optical parametric amplifier [13.52]. The ultraviolet pulses are tunable between 275 to 335 nm. The acceptance bandwidth of the doubling crystal is increased by a factor of 80 through high-order achromatic phase-matching. The chirp of the visible pulses and the dispersion introduced along the beam path are compensated partially before and partially after the doubling crystal [13.53].

Measuring Spectral and Time Profiles of Ultrashort Optical Pulses

Often ultrashort laser pulses show a time-dependent frequency profile, where the optical frequency of the pulse changes during the pulse width ΔT (chirp). Furthermore, its time and frequency profile may change on its way from the point of generation to the point where the measurement takes place because of dispersion in the air and in any optical elements through which the pulse passes. There are several detection techniques that can measure the time and frequency profile of the pulses. One of them is the frequency-resolved optical gating (FROG) method, which has already been discussed in Chap. 12 of this handbook. Recently, a new technique for direct electric field reconstruction [13.53], which is capable of measuring not only the temporal shape but also the phase of ultrashort pulses directly at the interaction point of a spectroscopic measurement, has been developed. It is called spectral phase interferometry for direct electric field reconstruction (SPIDER) and is a modification of spectral shearing interferometry, which relies on the measurement of interference between two replicas of the wavefront one wants to characterize, displaced by a small amount X along the spatial x-axis (Fig. 13.55). For an electric field

$$E(x)=\sqrt{I(x)}\mathrm{e}^{\mathrm{i}\varphi(x)} \tag{13.71a}$$

which interferes with a spatially shifted replica of itself

$$E(x+X)=\sqrt{I(x+X)}\mathrm{e}^{\mathrm{i}\varphi(x+X)}\,. \tag{13.71b}$$

The interference pattern is recorded by a square-law detector as

$$\begin{aligned}S(x)=&I(x)+I(x+X)\\&+2\sqrt{I(x)}\sqrt{I(x+X)}\cos\left[\varphi(x)-\varphi(x+X)\right].\end{aligned} \tag{13.71c}$$

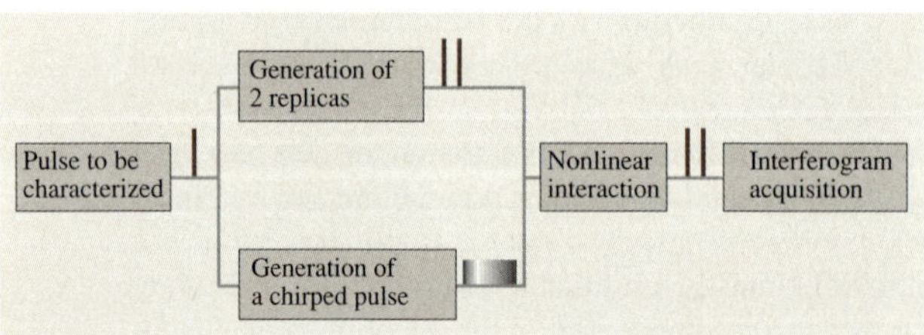

Fig. 13.55 Experimental set up for the SPIDER detection technique [13.54]

The measurement of the intensity at a point x is directly related to the phase difference

$$\Delta\varphi(x)=\varphi(x)-\varphi(x+X) \tag{13.71d}$$

between the phase of the initial wavefront at point x and at point $x+X$. This phase difference can be extracted for the interference pattern by Fourier-transform techniques [13.54].

Now a third pulse that is obtained from the initial pulse by a beam splitter is sent through a dispersive medium where the pulse gets a frequency chirp and becomes much broader. This is superimposed in a nonlinear crystal with the two replicas of the short unknown pulses. During the short pulse the chirped pulse can be regarded as monochromatic but during the delay time $\tau=X/c$ between the two replica pulses the frequency of the chirped pulse changes from ω_c to $\omega_c+\Omega$. Therefore, the sum frequency is accordingly changed. The total signal measured by the square-law detector after sum-frequency generation is given by

$$\begin{aligned}S(\omega_s)=&I(\omega+\omega_c)+I(\omega+\omega_c+\Omega)\\&+2\sqrt{I(\omega+\omega_c)}\sqrt{I(\omega+\omega_c+\Omega)}\\&\times\cos\left[\varphi(\omega+\omega_c)-\varphi(\omega+\omega_c+\Omega)\right].\end{aligned} \tag{13.71e}$$

This signal contains all the information on the phase and spectral profile of the unknown pulse, which can be obtained by a Fourier transform from the measured signal as a function of the delay time τ, which is related to the phase difference $\Delta\varphi=\Omega/\tau$.

The advantages of the SPIDER technique can be summarized as:

1. The experimental set up is simple and does not involve any moving parts,
2. The interferogram is measured with a one-dimensional spectrometer,
3. The acquisition of the experimental trace is done in a single shot,
4. The reconstruction of the pulse shape from the experimental data is quick and shows low sensitivity to the noise background.

One disadvantage is that this technique does not characterize the pulses at the position of the experiment because they split the beam before the actual characterization, thus introducing additional phase through the material dispersion. Therefore a modification of the SPIDER technique has been developed [13.55, 56], called zero-additional-phase SPIDER (ZAP-SPIDER).

Its basic principle is based on interference between a pair of temporally delayed pulses that are spectrally shifted against each other (Fig. 13.56). The unknown pulse to be characterized is sent into a nonlinear crystal where it overlaps with two strongly chirped auxiliary pulses coming from different directions, which are replicas of the same parent pulse. The nonlinear interaction between the three pulses results in two approximately monochromatic pulses at the sum and difference frequencies, which overlap in the BBO crystal with the pulses to be characterized. For visible pulses the crystal is adjusted for sum-frequency mixing and for UV pulses for difference-frequency mixing. Changing the time delay between the unknown pulse and the auxiliary pulses results in mixing of the unknown pulse with different frequencies (because of the strong chirp) of the auxiliary pulses. One thus obtains two spectrally sheared replicas with center frequencies at ω_0 and $\omega_0+\Omega$, which propagate in two different directions because of the phase-matching condition in the nonlinear crystal. The two pulses are recombined with a small delay τ in a spectrograph. The resulting interferogram encodes the phase difference between pairs of sheared frequencies. The complete spectral phase of the unknown pulse is then obtained from a Fourier transform and a special filtering technique [13.56]. A good comparison between different techniques for measuring ultrashort optical pulses can be found in [13.57].

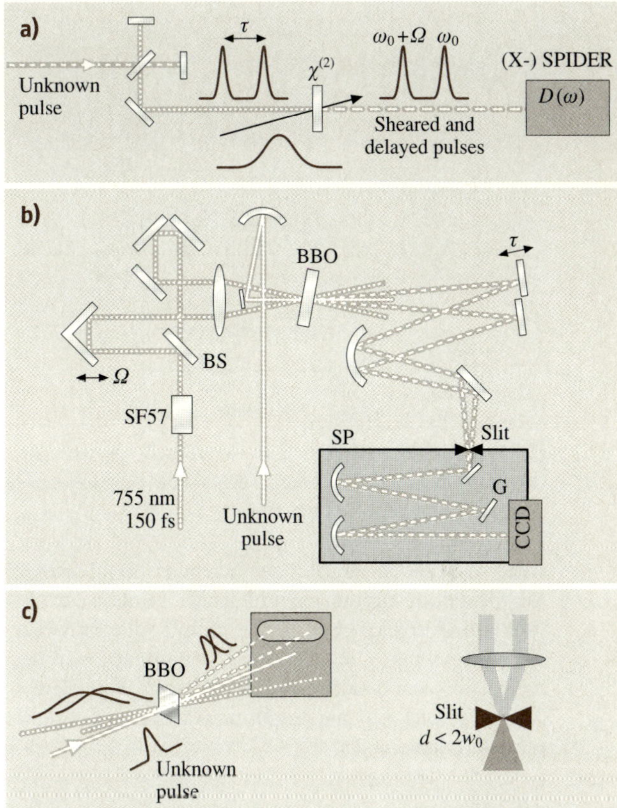

Fig. 13.56 (**a**) Set up for the ZAP-SPIDER method. (**b**) Experimental set up. SF57 = dispersive glass, BS = beam splitter, Ω = delay for adjusting the spectral shear, τ = delay to adjust mean fringe spacing. (**c**) Interference of two spatially separated beams focussed into a thin slit [13.56]

13.2.3 Time-Resolved Spectroscopy

In this section some techniques that allow the study of dynamical problems with a time resolution ranging from microseconds to femtoseconds are described. We will first discuss a sensitive method for absorption spectroscopy transferred to measurements in the time domain. Then an old problem will be presented, namely the determination of the lifetimes of excited states of atoms and molecules. These lifetime measurements yield very accurate values for transition probabilities and give, together with measurements of line intensities, information about the abundance of different elements in the atmospheres of stars.

Cavity-Ring-Down Spectroscopy

During recent years a new very sensitive absorption spectroscopy technique has been developed, named *cavity-ring-down spectroscopy* (CRDS). It measures the absorption of a sample placed inside an optical cavity by monitoring the time decay of the radiation power stored in the cavity [13.58]. Instead of measuring the small difference between the incident and transmitted radiation power in conventional absorption spectroscopy, this technique measures the number of cavity round-trip times after which the power has dropped to $1/\mathrm{e}$ of its initial value. Its basic principle can be understood as follows.

When a laser pulse with input power P_0 is sent into an optical cavity with two highly reflecting mirrors the pulse will be reflected back and forth between the two mirrors (Fig. 13.57a). For each round trip a small fraction will be transmitted through the end mirror M_2 to the detector. With a mirror reflectivity $R_1=R_2=R$, and transmission $T=1-R-A$, where A includes all losses of the cavity (absorption, diffraction, scattering except those losses introduced by the absorbing sample) the transmitted power of the first output pulse is

$$P_1=T^2\mathrm{e}^{-\alpha L}P_0\,, \tag{13.72a}$$

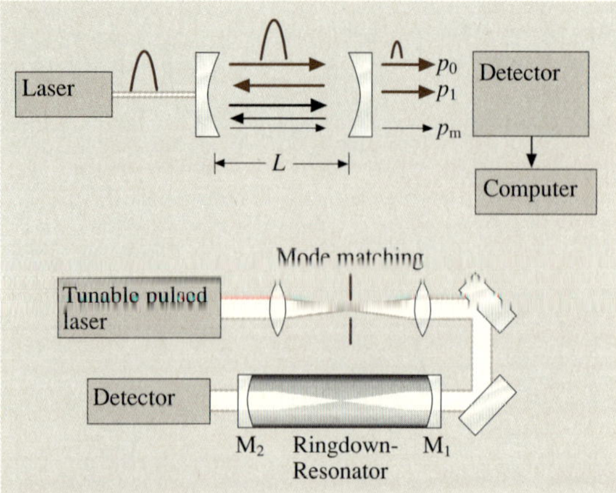

Fig. 13.57 (a) Set up for cavity-ring-down spectroscopy. **(b)** Coupling of incident laser beam into the ring-down cavity with mode-matching optics

where α is the absorption coefficient of the absorbing sample with absorption length L inside the optical cavity with mirror separation d. If the sample fills the whole cavity we obtain $L = d$. For each additional round trip the pulse power decreases by the factor $R^2 \exp(-2\alpha L)$: After n round trips the output power of the $(n+1)$-th pulse has decreased to

$$P_{n+1} = (R\mathrm{e}^{-\alpha L})^{2n} P_1 . \tag{13.72b}$$

With $R = 1-(T+A)$ and $A+T \ll 1 \rightarrow \ln(1-A-T) \approx -(A+T) = 1-R$ this can be written as

$$P_{n+1} = P_1 \,\mathrm{e}^{-2n(1-R+\alpha L)} . \tag{13.72c}$$

The time interval between two successive pulses equals the round-trip time $T = 2d/c$. The envelope of the discrete pulse sequence is the continuously decaying function of the output power from the resonator with the absorbing sample

$$P(t) = P_1 \,\mathrm{e}^{-t/\tau 1} \tag{13.72d}$$

with the time constant (Fig. 13.57b)

$$\tau_1 = (d/c)/(1-R+\alpha L) . \tag{13.73a}$$

The empty cavity ($\alpha = 0$) has the longer decay time

$$\tau_2 = (d/c)/(1-R) . \tag{13.73b}$$

The difference between the reciprocal decay times

$$1/\tau_1 - 1/\tau_2 = (c/d)\alpha L \tag{13.74}$$

directly yields the absorption αL.

Example. $R = 0.999$, $d = L = 1$ m, $\alpha = 10^{-6}\,\mathrm{cm}^{-1} = 10^{-4}\,\mathrm{m}^{-1} \rightarrow \tau_1 = 3.03\,\mu\mathrm{s}$, $\tau_2 = 3.33\,\mu\mathrm{s}$. The difference is small. However, with $R = 0.9999$ we obtain $\tau_1 = 16.5\,\mu\mathrm{s}$ and $\tau_2 = 33\,\mu\mathrm{s}$. This shows how important a high reflectivity of the cavity mirrors is.

A more detailed treatment of CRDS has to take into account the mode structure of the cavity, because the high Q-value of the cavity can only be obtained for those frequencies that match the optical modes of the cavity. For all those wavelengths λ_i contained in a laser pulse, for which $\nu_i = c/\lambda_i \neq m\, c/2d$, the cavity Q, and with it the sensitivity, will be small. Therefore, small absorption lines at those wavelengths may just escape detection. There are several solutions to this problem.

1. The laser pulse is sent into the high-Q cavity without mode-matching optics. It then excites not only longitudinal modes but also many transverse modes. In a non-confocal cavity the frequencies of these transverse modes are closely spaced and fill the frequency distance $\Delta\nu = c/2d$ between the longitudinal modes. Their frequency separation is smaller than the Doppler width of the absorption lines, insuring that all absorption lines are detected. The advantage of this method is the simple experimental set up without mode-matching optics. Its disadvantage is the smaller sensitivity, because the average over all transversal and longitudinal modes yields a smaller Q-value, than for a single longitudinal mode.
2. The phase fronts of the incoming pulses are imaged by a suitable optical system into the cavity in such a way that they coincide exactly with those of the fundamental modes of the cavity (Fig. 13.57b). This gives the highest possible Q-value. However, now the cavity length has to be stabilized onto the laser frequency and has to be synchronously tuned when the laser frequency is tuned through the absorption spectrum, insuring that the cavity mode is always in resonance with the laser frequency. With this mode-matching set up the spectral resolution is limited either by the Doppler width of the absorption lines or by the pulse length ΔT, which gives a Fourier-limited resolution of $\Delta\nu = 1/(2\pi\Delta T)$.

The highest Q-value and therefore the highest sensitivity is achieved with CW lasers where the output beam of the laser is sent through a modulator that forms pulses of the desired length ΔT and a spectral bandwidth that is adapted to the width of the cavity modes. Since this width $\Delta\nu$ is smaller than the Doppler width, one can measure the line profiles and can investigate the

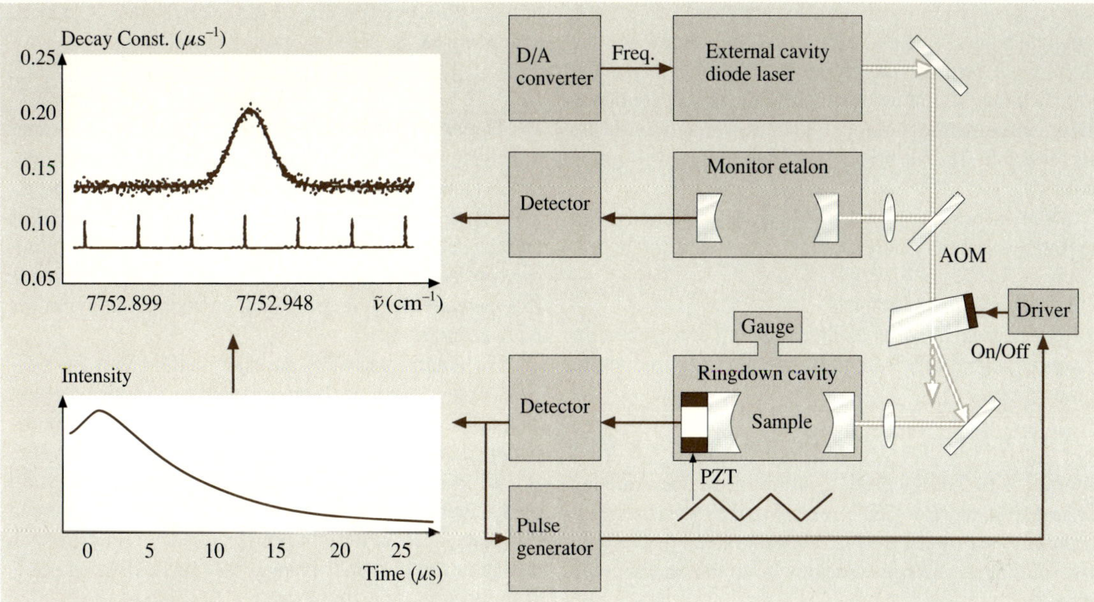

Fig. 13.58 Experimental arrangement for cavity-ring-down spectroscopy with a CW external cavity diode laser [13.58]

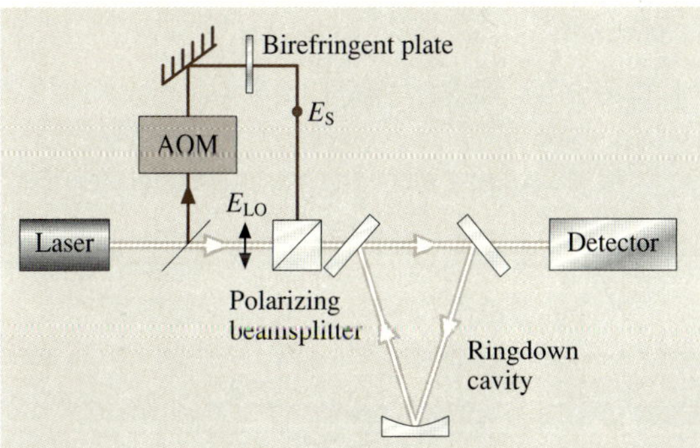

Fig. 13.59 Heterodyne detection of ring-down signals [13.59]

influence of pressure broadening. In Fig. 13.58 a possible set up for cavity-ring-down spectroscopy with an external cavity CW diode laser is shown [13.58].The output beam is pulsed by an acousto-optical modulator and the length of the ring-down cavity is slowly changed by a sawtooth voltage to the piezo, on which one of the mirrors is mounted.

The achievable sensitivity is limited by the accuracy of measuring the decay times τ_1 and τ_2. This accuracy in turn is limited by the noise of the decay curve. The main sources of noise are the intensity fluctuations of the laser and fluctuations and drifts of the cavity length.

A particularly sensitive technique that can eliminate at least part of this noise is based on heterodyne measurements (Fig. 13.59). The output beam of a CW laser is split into two partial beams. One of these is sent directly through mode-matching optics into the cavity, which is stabilized onto the laser frequency. The output signal of the cavity serves as a local oscillator.

The optical frequency of the second beam is shifted by an acoustooptic modulator by exactly the mode separation $\Delta\nu = c/2d$, thus matching the neighboring mode. The amplitude of this *signal beam* is modulated at a frequency of 40 kHz. Both partial beams are again superimposed before they enter the cavity and are in resonance with two adjacent longitudinal cavity modes. The transmitted power

$$P_t \approx |E_s(t) + E_{LO}\,e^{i(2\pi t\Delta\nu+\varphi)}|^2 = |E_s(t)|^2 + |E_{LO}|^2 + 2E_s E_{LO}\cos(2\pi\Delta\nu t+\varphi) \tag{13.75}$$

is measured by the detector. While the second beam consists of pulses with 25 μs pulse width and results in an exponential decay of the transmitted signal, the power of the fist beam is constant and can therefore be used to stabilize the cavity.

The interference term in (13.75) is the product of the large amplitude of the local oscillator (note that the transmitted amplitude of the CW beam $A_t = (1-A)^2 e^{-\alpha L}$ is for $A \ll 1$ and $\alpha L \ll 1$ nearly that of the incident amplitude) and the much smaller amplitude E_s of the modulated signal beam. Measuring the decay of this interference term at the frequency $\Delta\nu = c/2d$ yields a much larger signal as obtained with the conventional cavity ring-down technique, which now decay with the time constant 2τ, but has a much better signal-to-noise ratio.

More information on CRDS can be obtained from the reviews [13.60–62].

Lifetime Measurements

The population N_i of an excited state $|i\rangle$ of a free atom decays exponentially from the value $N_i(0)$ at time $t = 0$, to the value

$$N_i(t) = N_i(0)\mathrm{e}^{-t/\tau} \tag{13.76}$$

at time $t > 0$. The quantity τ is the mean spontaneous lifetime of state $|i\rangle$, which is related to the total transition probability A_i by $A_i = 1/\tau_i$. Typical values of the lifetime of excited valence electrons lie in the nanosecond range.

If the excited atom A collides with other atoms or molecules B, its population can additionally decrease by inelastic collisions. The rate of such deactivating collisions

$$R_i = N_B \sigma_{\text{inel}} \langle v_r \rangle \tag{13.77a}$$

depends on the number density N_B of the collision partners B, the collision cross section σ and the relative velocity v_r. The total transition probability for deactivating the excited state $|i\rangle$ is then increased to

$$A_i^{\text{eff}} = A_i^{\text{spont}} + R_i \tag{13.77b}$$

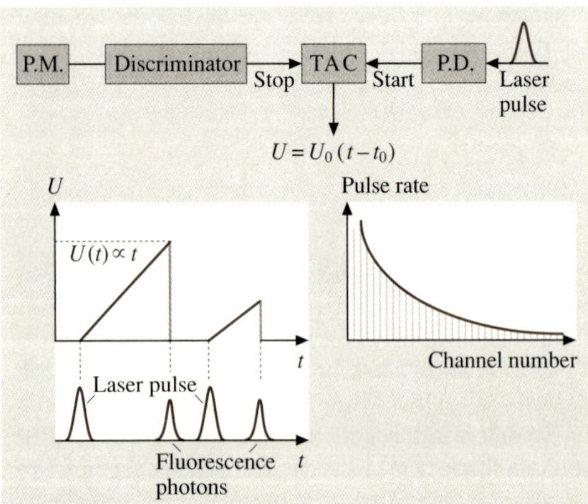

Fig. 13.60 Single-photon delayed coincidence technique

and the effective lifetime of the excited state is reduced according to

$$1/\tau_i^{\text{eff}} = 1/\tau_i^{\text{spont}} + R_i\,. \tag{13.78}$$

The experimental technique for measuring lifetimes depends on the order of magnitude for τ. For lifetimes in the micro- to nanosecond range, the atoms can be excited by a short laser pulse (e.g., a nanosecond pulse) and the decay of the laser-induced fluorescence is monitored with a fast oscilloscope. Digital transient recorders can measure decay curves with a time resolution in the picosecond range.

For small intensities another method is more suitable. It is based on time-correlation spectroscopy of single fluorescence photons [13.63]. Its principle is depicted in Fig. 13.60. The sample atoms or molecules are irradiated by a regular time sequence of short pulses delivered from a mode-locked cavity-dumped CW laser. The cavity-dumping frequency is chosen such that the time interval between two successive pulses is at least three times the expected lifetime of the excited

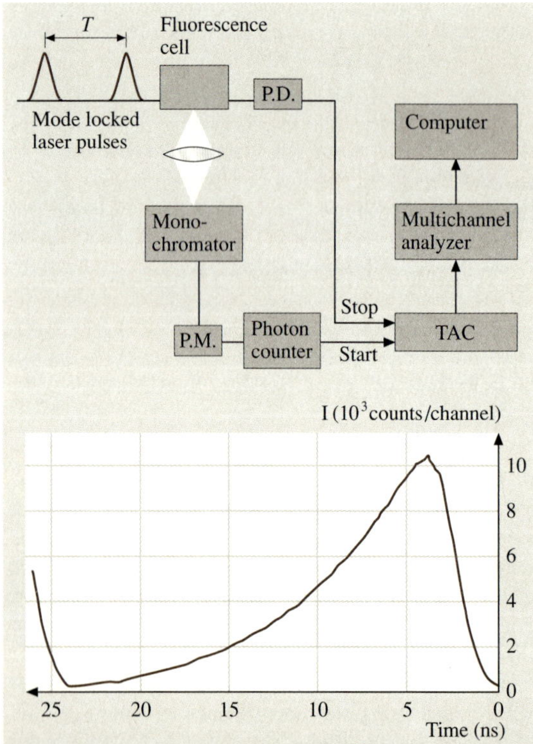

Fig. 13.61 Experimental arrangement for lifetime measurements and decay curve of the excited $B^1\Pi_u$ state of the Na_2 molecule. The time scale runs from *right to left*

molecules. A fraction of each laser pulse is sent to a fast photodiode, which delivers the trigger signal to define the starting time $t = 0$ of a time-to-amplitude converter (TAC). This TAC delivers a ramp voltage $U = at$. The first fluorescence photon detected at time t_1 after the excitation pulse stops the ramp. The output voltage $U = at_1$ is stored in a multichannel analyzer or in a computer. Measuring many events yields the probability distribution of decay times, which gives the exponential decay of (13.76). The detected fluorescence rate should be smaller than the repetition rate $f = 1/T$ of the laser pulses in order to insure that at most one fluorescence photon is detected during the time T between two excitation pulses. With a ramp time of 50 ns and 1000 channels the time resolution is 50 ps. In order to reduce the dead time of the TAC, the time sequence is often inverted: the TAC is started by the fluorescence photon emitted at time t_1 after the excitation pulse and is stopped by the next excitation pulse. Since the repetition rate of the pulses from a mode-locked laser is very stable, the true delay time t_1 can be directly inferred from the measured time difference $T - t_1$. In Fig. 13.61 the whole set up for this method is depicted and a decay curve of an excited level of the Na_2 molecule with the inverted time scale is shown for illustration [13.64].

Pump-and-Probe Techniques

If time resolution in the pico- to femtosecond range is required, direct detection by electronic devices is not fast enough. Here a new technique has been developed, which uses a fast pump pulse that excites or depletes a specific level in atoms or molecules, and a delayed fast probe pulse, that probes the effect of the pump pulse at a controlled time delay (Fig. 13.62). Generally both pulses come from the same laser. The output pulse of the laser is split by a beam splitter into two parts, which

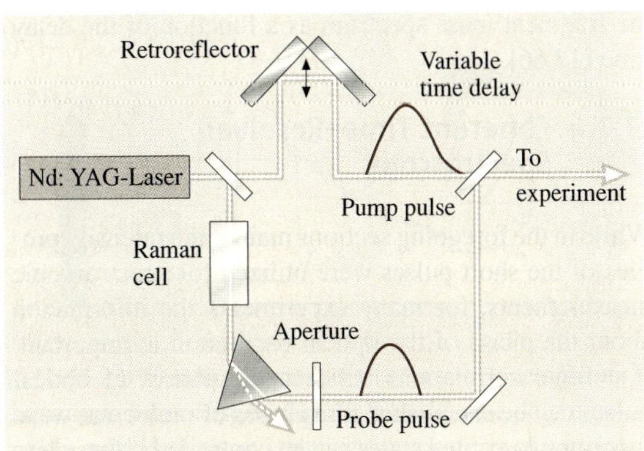

Fig. 13.62 Pump-and-probe technique

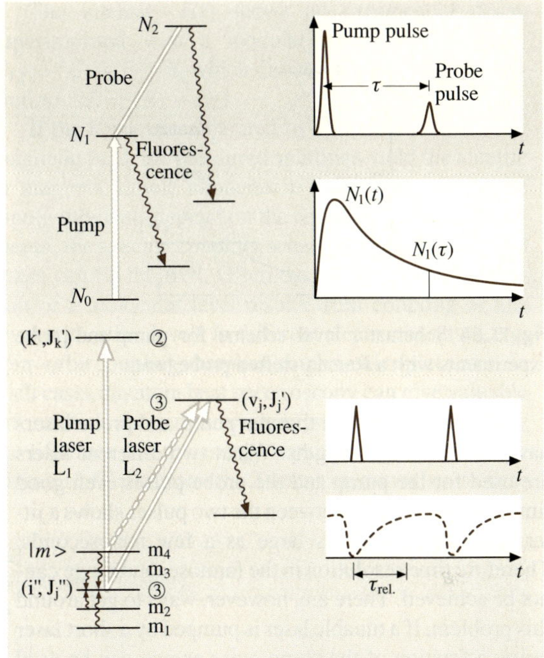

Fig. 13.63 Possible level schemes for the pump-and-probe technique

travel through different path lengths before they are recombined by a second beam splitter. The pump pulse depletes the lower level of the pump transition in a molecule and populates the upper level. The probe pulse can either probe the depletion of the lower level, or can further excite the excited molecules. The intensity of the fluorescence from the excited levels can serve as a monitor (Fig. 13.63a). The intensity of the probe-laser-induced fluorescence from level $|2\rangle$ is measured as a function of the delay time between pump and probe. If the sum of the photon energies is above the ionization limit, the probe laser can even ionize the excited molecules. The measured ion rate $N_{\text{ion}}(t)$ as a function of the delay time t of the probe pulse then gives the time-dependent population of the upper level $|1\rangle$.

The depleted lower level can be refilled by fluorescence from the upper level or by inelastic collisions. The weak probe pulse monitors this refilling process by measuring the fluorescence intensity induced by the probe pulse as a function of the delay time (Fig. 13.63b) because the fluorescence intensity is proportional to the population of the lower level. With this technique a time resolution of a few femtoseconds can be achieved. Note, that the time resolution of the detector can be much worse.

cited levels in the propynal molecule [13.67], where the Fourier transform of the complex measured beat signal yields the energy levels involved.

Molecular Wave-Packet Dynamics

Another example for time-resolved coherent spectroscopy is the investigation of wave packet dynamics of electronically excited molecules. The femtosecond pump laser with optical frequency $\nu = E/h$, time duration Δt and Fourier-limited energy width $\Delta E = \hbar/\Delta t$ coherently excites a group of vibrational levels within the energy range $E \pm \Delta E$. The superposition of their wave functions represents a wave packet that moves with the average vibrational frequency between the inner and outer turning points of the potential. For the example of the Na_2 molecule shown in Fig. 13.68, the wave packet is excited by two-photon absorption from the electronic ground state into the $2^1\Pi_g$ state. The further excitation of the excited molecule by the probe laser depends on the delay between the pump and probe pulses. If the molecule is excited at that time, when the wave packet is at the inner turning point, molecular Na_2^+ ions are formed, while excitation at the outer turning point leads to dissociation into $Na + Na^+$ and produces atomic Na^+

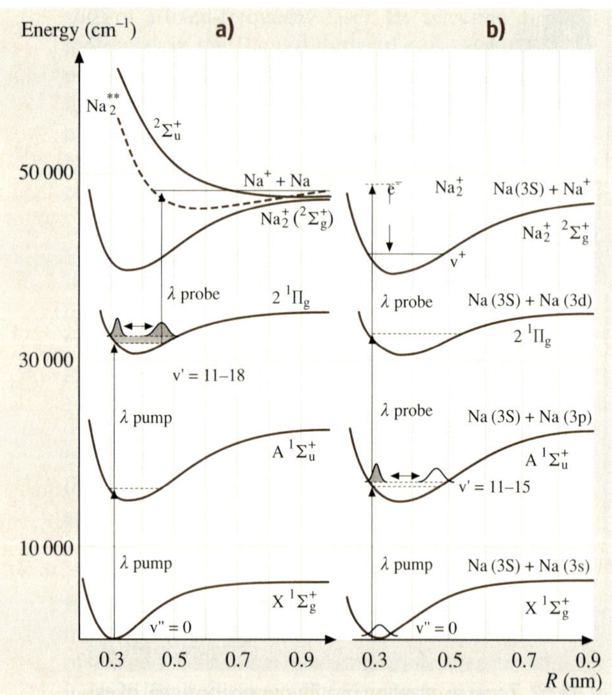

Fig. 13.68 Wave-packet dynamics and coherent control for the dissociation and selective ionization of the Na_2 molecule [13.68]

ions [13.68]. This is an example how a specific product channel can be selected just by choosing the proper time delay between the pump and probe pulses.

Photon Echoes

The time-resolved coherent excitation of molecules can also be used to obtain information not only about the population decay but also about the phase decay. This is realized with the technique of photon echoes which can be understood as follows.

Assume N molecules have been simultaneously excited by a short laser pulse from a lower level $|1\rangle$ into the upper level $|2\rangle$. The total fluorescence power emitted by the transition $|2\rangle \to |1\rangle$ is given by

$$P_{\text{Fl}} = \Sigma\hbar\omega D_{21} , \tag{13.83}$$

where D_{21} is the dipole matrix element for the transition $|2\rangle \to |1\rangle$ and g_1 and g_2 are the statistical weight factors of levels $|1\rangle$ and $|2\rangle$. The sum extends over all N molecules.

With coherent excitation at time $t = 0$ a definite phase relation is established between the wave functions of the N molecules in level $|2\rangle$. We therefore obtain:

$$|\Sigma D_{12}|^2 = |N\, D_{12}|^2 = N^2|D_{12}|^2 , \tag{13.84}$$

whereas for incoherent excitation the phases of the N molecules are random and the result would be $N|D_{12}|^2$. This shows that the total fluorescence power at a time t just after the excitation pulse, where the molecules are still in phase, is N times larger than that for incoherent excitation.

In order to produce photon echoes the molecules are excited at time $t_1 = 0$ by a $\pi/2$ pulse (Fig. 13.69). This is a pulse with a controlled energy that changes the phase of the wave functions of both levels by $\pi/2$ and creates equal populations in levels $|1\rangle$ and $|2\rangle$. Such a system represents an oscillating dipole with oscillation frequency $\omega_{12} = E/\hbar$.

Because of the finite line width $\Delta\omega$ of the absorbing transition $|1\rangle \to |2\rangle$ (for example the Doppler width) the frequencies ω_{12} of the N molecules are not all equal but are distributed within the interval $\Delta\omega$. This causes the phases of the N oscillating dipoles to develop in time at different rates after the end of the $\pi/2$ pulse. At a time $t_2 < T_1$ smaller than the population decay time T_1 but larger than the phase relaxation time T_2 the phases of the N dipoles are randomly distributed.

The molecules are now irradiated at time $t = t_2$ by a second pulse with twice the energy of the first excitation pulse. This pulse inverts the phase of the induced

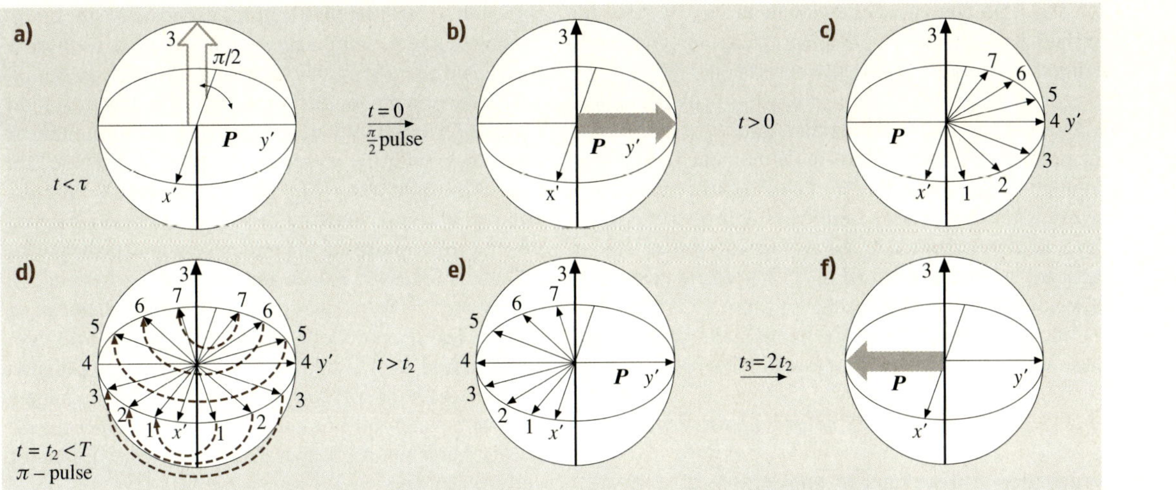

Fig. 13.69 Pulse sequence for the generation of photon echoes and the time evolution of the pseudo-polarization vector, where the photon echo is emitted at $t = 2\tau$

polarization and is therefore called a π pulse. It causes a reversal of the phase development for each dipole. After the time $t_3 = 2t_2$ all molecules are again in phase and they emit a signal which is proportional to N_2^2, where

$$N_2(2t_2) = N_2(0)e^{-2t_2/T_1} \tag{13.85}$$

is the population of molecules in the excited state $|2\rangle$, that have survived the time $2t_2$ after excitation by the first pulse. This signal is called the *photon echo* and is much stronger than the incoherent fluorescence that is emitted at all times $t > 0$ but is only proportional to $N_2(t)$.

The amplitude of the photon echo decreases with increasing time delay due to the population relaxation with the *longitudinal relaxation time* T_1.

A second, generally faster relaxation process is the phase relaxation with the *transverse relaxation time* T_2. This can have several causes:

1. Phase-perturbing collisions change the phase of the oscillating dipoles and therefore prevent them from all being in phase at $t = 2t_2$. Since such collisions give rise to homogeneous line broadening, this contribution to the echo decay is the homogeneous part.
2. The different velocities of the molecules within their Doppler width also cause different phase shifts (the inhomogeneous part).

This second contribution, however, does not prevent the dipoles from restoring their phase at the echo time $t = 2t_2$ and therefore does not diminish the amplitude of the photon echo. The Fourier transform of the decay times of the photon echo therefore gives Doppler-free line widths.

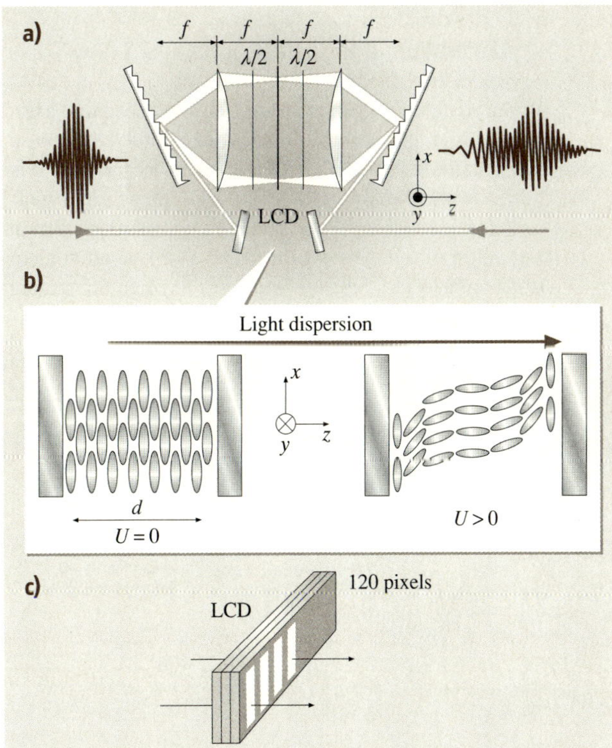

Fig. 13.70 Pulse shaping for coherent control by controlling the refractive index of transparent pixels in a liquid-crystal mask

The time evolution of the system can be best described by the *pseudo-polarization vector*, which is defined as $P = \{P_x, P_y, P_z\}$, where P_x and P_y describe the components of the atomic polarization by the optical wave traveling in the z-direction, and $P_z = D_{12}\Delta N$ is the product of the transition dipole moment D_{12} and the population difference $\Delta N = N_1 - N_2$ between the two levels. The time development of P describes the time-dependent polarization of the molecular system. This is depicted schematically in Fig. 13.70, where relaxation processes have been neglected.

More-detailed information about photon echoes and their applications can be found in [13.70, 71].

13.2.5 Applications of Short Laser Pulses

In this section some selected applications of short laser pulses will be presented, in order to give a feeling for the large variety of possible applications of light pulses in different time domains.

Generation of High Harmonics and X-Ray Spectroscopy

When an intense visible laser pulse is focussed into a noble gas high harmonics of the fundamental wavelength are generated. This can be understood as follows: due to the strong electric field of the light pulse rapid multiphoton ionization of the gas atoms takes place within a fraction of the cycle of the light. The free photoelectrons are accelerated back and forth through the atomic core by the alternating electric field, which changes its direction so fast that the electrons cannot leave the atom. During this acceleration period they emit radiation from the visible down to the X-ray region. Although the period of the electron movement is that of the fundamental light wave, high harmonics are generated due to the nonlinear interaction with the atomic force field and the electric field of the light wave (Fig. 13.71). Coherent X-ray radiation up to photon energies of 700 eV has been observed when a 10 fs pulse of several mJ energy (peak power 10^9 W) is focussed onto a xenon sample, where an intensity of $10^{19}\,\mathrm{W/m^2}$ was reached at a focus diameter of 10 μm. The photon yield was nearly constant over several hundred eV [13.69, 72]. This system represents an intense source of soft X-ray radiation that can compete well with synchrotron radiation in terms of spectral intensity, but is much cheaper than an electron accelerator and storage ring.

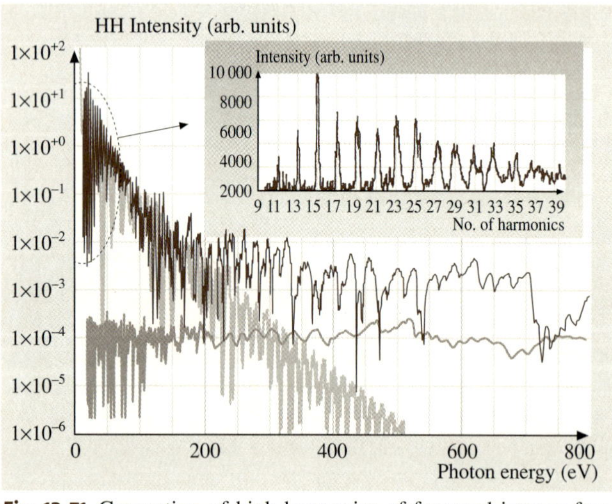

Fig. 13.71 Generation of high harmonics of focussed intense femtosecond pulses [13.69]

Applications in Chemistry and Molecular Dynamics

The dream of chemists is to control chemical reactions by excitation with properly formed laser pulses. This dream has meanwhile become true at least for some favorable cases by a method called coherent control. Its implementation demands very short laser pulses for the following reason: in order to enhance a specific reaction one has to excite specific molecular levels, which decay into the wanted reaction channel. The density of states in electronically excited molecules increases strongly with excitation energy and with the number of atoms in a molecule and the interaction between different states also increases. Even if a single selected level has been excited by photon absorption the excitation energy is spread among many interacting states within a short time, thus washing out the desired selectivity. Therefore the wanted reaction has to start before this energy redistribution takes place. The method of coherent control can overcome these difficulties because it uses femtosecond pulses and prepares a favourable excited state that leads to the optimization of the wanted reaction. Its basic principle is as follows.

The reaction path of an excited polyatomic molecule often depends on the spatial distribution of the molecular wave function in this state, which can be controlled by the shape and frequency distribution of the excitation pulse. When the molecules under investigation are excited by femtosecond pulses the rate of the desired reaction is monitored and serves as the input signal for an optimization procedure where the time profile of the pulse is modified until the optimum reaction rate has been achieved. This optimization can be reached even if the detailed nature of the excited-state wave function is not known. Since the spatial and time distribution of the molecular wave packet is created by the coherent excitation of ultrashort laser pulses with controlled time, frequency and phase profiles, the technique is called *coherent control*. The pulse shaping (Fig. 13.70) can be performed as follows.

The incoming beam is diffracted by an optical grating in the focal plane of a lens. Since the diffraction angle depends on the wavelength, the different spectral components in the pulse are spatially separated and pass through a mask consisting of many liquid-crystal pixels. The refractive index of these pixels can be controlled by an appropriate applied voltage, thus introducing a controlled phase shift of the light passing through them. These phase shifts can be independently changed for the various spectral components. After passing through a second lens a second grating recombines the spatially separated spectral components. Their superposition gives a pulse shape that depends on the amplitudes and the phases of these components and can therefore be changed by the voltages applied to the pixels, which are in turn controlled by the optimization signal. A learning algorithm changes the different voltages in small steps until an overall optimum for the reaction rate has been reached [13.74, 75].

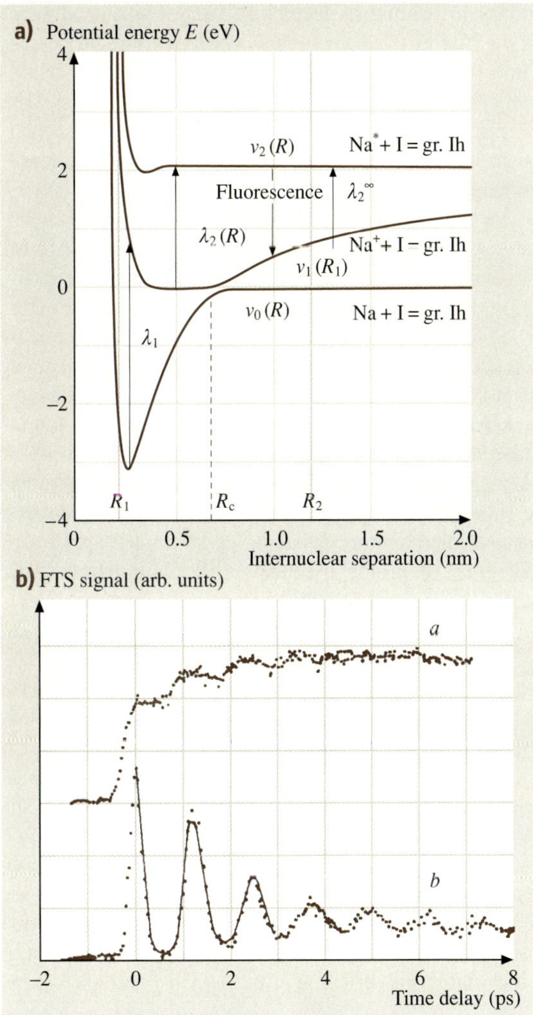

Fig. 13.72a,b Time-resolved spectroscopy of the dissociation of NaI [13.73]. **(a)** Potential diagram with the pump transition at λ_1 and the probe transition at $\lambda_2(R)$. **(b)** Atomic Na fluorescence $I_{Fl}(\Delta t)$ as a function of the delay time Δt with λ_2 tuned to the atomic Na transition (*curve a*) and to $\lambda_2(R)$ with $R < R_c$ (*curve b*)

This technique has been successfully applied to a number of molecules [13.76], even including large biological molecules such as the carotenoid molecules, which play an important role in photosynthetic light-harvesting [13.77].

In many cases excitation of higher vibration levels in the electronic ground state can itself enhance a chemical reaction considerably, in particular if local bonds can be excited. For instance the excitation of a local vibration involving the double bond in ethylene $CH_2{=}CH_2$ will lead to fragmentation into two CH_2 radicals, while excitation of a single bond will split off a hydrogen atom. Here pulse shaping in the infrared spectral region is required in order to excited the desired vibrational mode. Unfortunately there are no liquid masks available for the mid-infrared. It is, however, possible to transfer the pulse shape of a visible pulse by difference-frequency mixing in a nonlinear crystal to a pulse with the difference frequency in the infrared [13.78].

The application of femtosecond spectroscopy to fast molecular processes allows one for the first time to observe in real time the breaking and formation of chemical bonds. We will illustrate this by some examples. The first is the photodissociation of NaI [13.37,73]. The adiabatic potential diagram of NaI (Fig. 13.72) shows an avoided crossing between the repulsive potential of the two interacting neutral atom Na+I and attractive Coulomb potential of the ions Na^++I^-, which is mainly responsible for the strong binding of NaI at small internuclear distances R. If the molecule NaI is excited by a short laser pulse with wavelength λ_1 the excited molecule with reduced mass μ starts to move towards larger distances R with a velocity $v(R) = [(2/\mu)(E - E_{pot}(R)]^{1/2}$. When it reaches the avoided crossing at $R = R_c$ it may either stay on the potential curve $V_1(R)$ and oscillate back and forth between R_1 and R_2, or it may tunnel to the potential curve $V_0(R)$, where it separates into Na+I.

This can be measured with a probe pulse with a wavelength λ_2 tuned into resonance with the transition from $V_1(R)$ into the excited-state potential $V_2(R)$ that dissociates into Na^*+I. Since the dissociation time (a few picoseconds) is very short compared with the lifetime of the excited $Na^*(3p)$ atom,the dissociation NaI* molecule nearly exclusively emits at the atomic resonance fluorescence of $\lambda = 589$ nm. Monitoring the intensity of the atomic fluorescence as a function of the delay time yields the time behavior of the dissociation process. If the probe laser wavelength λ_2 is tuned to to a transition at $R < R_c$ one obtains the lower curve in Fig. 13.72b, showing the oscillation of the system between R_1 and R_2, while for $R > R_c$ the upper curve is obtained, giving the number of molecules in the potential $V_0(R > R_c)$.

The second example is the isomerization of Stilbene, i. e. the photoinduced structural change of excited stilbene form trans- to cis-stilbene (Fig. 13.73). Since the structural change also causes a corresponding change of the rotational constants, a pump-and-probe experiment of the rotating molecule can measure the rotational period and with it the rotational constants [13.79].

Often the molecular structure in excited electronic states is not known and in many cases it might even change with time, due to isomerization processes. Here a very elegant technique has been developed based on time-resolved Laue diagrams of excited molecules. The molecules are excited by a femtosecond pulse. Part of the same laser pulse is focussed onto a target, where a very hot plasma that emits short X-ray pulses in the femto- to attosecond range is produced. These are used to take a Laue diagram of the excited molecule. Varying the time delay between the excitation and the X-ray pulses can follow the structural change of the excited molecule [13.80–82].

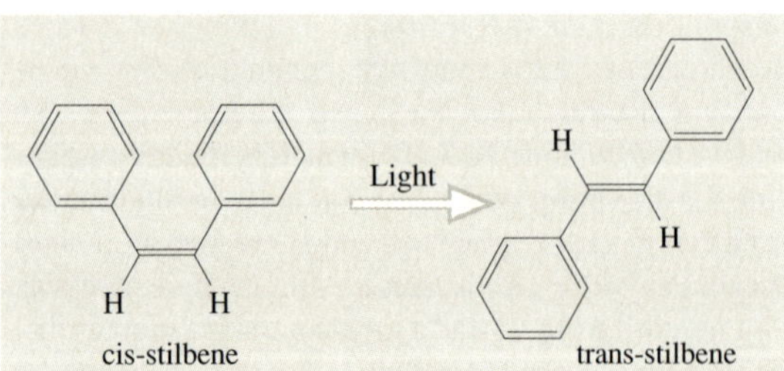

Fig. 13.73 Isomerization of stilbene

Ultrafast Processes in Biology

Many primary processes of biological reactions proceed on a femtosecond time scale. Examples are the visual process in the retina cells of the eye after photon excitation, or the photosynthesis of green plants. The investigation of the different steps in these complex processes demands a combination of spectroscopic techniques with high time resolution and spectral selectivity [13.83].

The first steps in the visual process after the excitation of retinal molecules in the protein rhodopsin in the retina of the eye are isomerization processes where the structure of the molecules in the excited states is changing in such a way that the energy decreases. This can be studied by time-resolved Raman spectroscopy, which gives information on the vibrational frequencies [13.84].

In the photosynthesis of green plants the light-harvesting molecules are carotenoids, which absorb in the blue–green spectral range and transfer their excitation energy to chlorophyll molecules. These transfer processes are in competition with radiationless deactivation. Both kinds of processes proceed on a femtosecond time scale. The final outcome of such transfer processes is proton transfer in chlorophyll, which has been found to be the most important process in the photosynthesis of green plants. It consists of many steps that can be followed by femtosecond spectroscopy. With coherent control, the intermediate processes that decrease the efficiency of proton transfer can be suppressed, thus increasing the yield [13.85].

Applications in Medicine

Although for many medical applications ultrashort lasers do not play an essential role, there are some examples where femtosecond pulses are superior to longer laser pulses [13.86] for the following reasons. Femtosecond lasers allow the application of high peak powers but low energy to human tissues. Because of the low energy, damaging high temperatures are avoided but the resultant nonlinear effects cause absorption in regions of the body which are transparent for linear absorption. By sending the laser pulses through a microscope with a large numerical aperture, small focal spots can be realized and microsurgery is possible. This is, for instance, used for intraocular and intrastromal refractive surgery. The high peak power may result in microplasma formation with subsequent stress waves into the surrounding tissue and chemical reactions. If the limits of unwanted destruction are precisely known, this can be used to create surgical effects. Therefore femtosecond

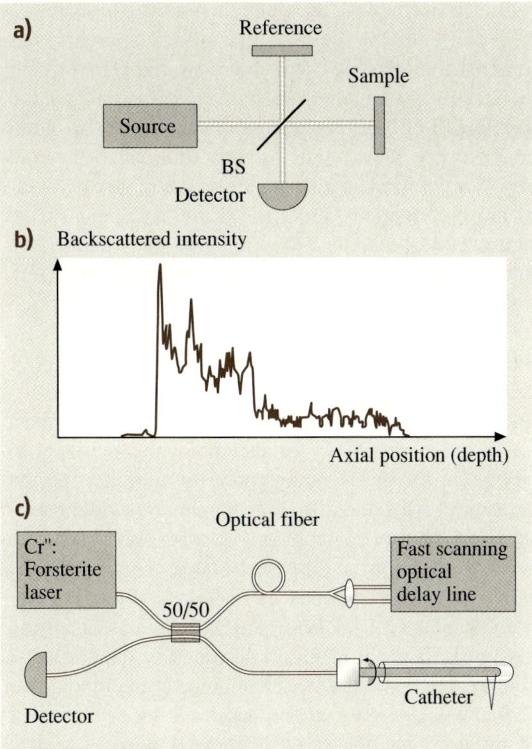

Fig. 13.74 Optical coherence tomography

lasers enable the creation of extremely spatially confined chemical, thermal and mechanical effects in biological media.

Another area where the high time resolution is essential is optical tomography; here the output beam of the laser is split into a reference beam and a part that is sent into selected parts of the body (Fig. 13.74), for instance into the brain. The interference between the reference beam and the radiation backscattered from a thin region of the sample is measured as a function of the delay time between the reference and backscattered signal pulses. Any change in the composition of the tissue changes the interference pattern. The spatial localization becomes better with increasing spectral width of the radiation source. Therefore light-emitting diodes or broadband pulsed lasers are often used. Examples where this technique has been successfully applied are tumors in the brain or in the female breast or plaques in arteries that might cause heart attack [13.87].

A modern device based on an interferometer for optical tomography using optical fibres and a fibre beam splitter is shown in Fig. 13.74b.

Applications in Geodesy and Environmental Research

Short laser pulses are well suited for accurate distance measurements. When a pulse is emitted at time $t = 0$ and is reflected by a target at time $t = t_1$ the backscattered light, collected by a telescope, is received at $t = t_2 = 2t_1$. The distance between the receiver and target is then $d = t_1/c$, where c is the speed of light. With intense picosecond laser pulses the distance between the retro-reflector, placed by astronauts on the moon, and the receiver could be measured within a few centimeters [13.88]. In order to decrease the divergence of the beam caused by diffraction, the laser beam has to be enlarged by a telescope.

For many applications in distance measurements this method has become the standard technique. For shorter distances pulsed diode lasers have sufficient intensity to give a good signal-to-noise ratio and accuracies with uncertainties within the mm–cm range.

For environmental studies of our atmosphere different techniques have been developed (Sect. 13.3). The light detection and ranging (LIDAR) method is based on Mie backscattering by aerosols in the atmosphere. Pollutant molecules in the atmosphere lead to specific absorption of the beam if the laser wavelength is tuned to an absorption line of these molecules, thus decreasing the intensity of the backscattered pulse. For LIDAR measurements the backscattered intensity is alternately measured at an absorption wavelength λ_1 and a wavelength λ_2 where the molecules do not absorb. The ratio of the two backscattered intensities measured at time t

$$I_b(\lambda_1, t) = A\exp(-2\alpha_1 L) \text{ and}$$
$$I_b(\lambda_2, t) = A\exp(-2\alpha_2 L) \qquad (13.86)$$

gives the absorption $L(\alpha_2 - \alpha_1)$. Here $L = t/c$ is the distance to the backscattering target and $\alpha = N\sigma$ is the absorption coefficient, which depends on the density N of the absorbing molecules and the absorption cross section σ. The scattering cross section does not differ much for two close wavelengths λ_1 and λ_2.

If this intensity-ratio signal is measured at two different times t_1 and t_2, the ratio of the two signals just gives the absorption which takes place within the spatial range $\Delta L = L_1 - L_2$ (differential LIDAR) [13.89].

With femtosecond terawatt lasers a long plasma channel can be produced along the laser beam. Due to self-focussing caused by the nonlinear part of the refractive index, in atmospheric air the laser beam shows a series of plasma regions along the beam (like sausages on a string), which emit brilliant white light. These bright plasma spots can be used as spatially well-defined and spectrally broad light source in the atmosphere. If the telescope is directed towards one of these plasma regions the received radiation has passed through the atmospheric layer between the telescope and the light source. The resulting absorption lines of this layer can be spectrally resolved and give for a single laser shot the whole spectrum of absorbing molecules in this layer [13.90]. The advantage of this method is the simultaneous detection of the whole spectrum of all absorbing constituents. Its disadvantage is that it only measures the total absorption integrated over the path length from the radiation source to the detector without giving information about the local variation of the density of absorbing molecules. This disadvantage can be partly compensated by viewing different parts of the plasma channel, which extends over several hundred meters.

Applications of Femtosecond Lasers in Material Science and Processing

The ultimate switching time in semiconductor chips is limited by the relaxation and diffusion times of excited electrons. With the decreasing size of electronic devices the electronic properties of metals are changing with the size of the nanostructures. Here femtosecond spectroscopy can obtain information about the different processes. One example is the investigation of the electron–lattice interaction in single silver nanoparticles [13.91].

Short-pulse lasers are also playing an increasing role in material processing. Although there are many applications of CW lasers and laser pulses with pulse widths in the milli- to nanosecond range for welding or cutting of materials or for drilling holes into specific hard materials, such as diamonds, ceramics, where mechanical drills are hard to use, femtosecond pulses have specific characteristics that make them superior to longer laser pulses in many cases. For example for drilling holes with diameters below 100 μm but a larger depth, femtosecond pulses give cleaner and better-defined holes. This is due to the fact that the material is just evaporated without forming liquid drops that deteriorate the crater edge [13.92, 93].

Generation and Applications of Attosecond Pulses

Although the realization of attosecond pulses is only a few years old, there are already numerous applications. Attosecond pulses are produced by focussing high-intensity femtosecond laser pulses into a sample of noble gas atoms in an atomic beam. The extremely high electric field of the pulse, which might even exceed the Coulomb field of the atom core, generates a highly nonlinear field where the atomic electrons are accelerated and emit high harmonics of the fundamental wavelength of the femtosecond pulse. Since the intensity of the n-th harmonics is proportional to the n-th power of the fundamental intensity, the higher harmonics are preferentially produced during the peak of the femtosecond pulse and their time profile is therefore much narrower than that of the fundamental pulse. From this process ultrashort X-ray pulses are available, which can be used for X-ray diagnostics of ultrafast processes. One example is the production of Laue- iagrams of excited vibrating molecules [13.37, 73]. Since for ordinary Laue diagrams the molecular vibrations wash out the exact geometry of the molecule, pulses with a pulse width short compared to the vibration period can give sharp Laue diagrams for the molecular structure at a specific moment of the vibration. Therefore direct information can be gained about the change of the molecular geometry when the molecule is excited. For biomolecules such a change might be essential for the biological function of the molecule. The experimental technique uses the femtosecond pulse at the fundamental frequency or at its second harmonic for excitation of the molecule and the X-ray attosecond pulse, generated

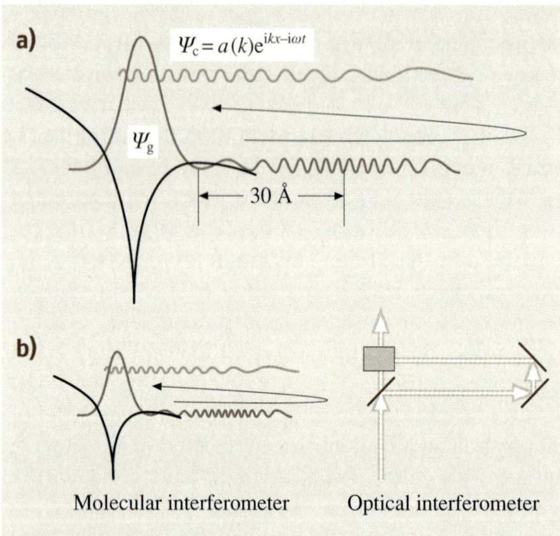

Fig. 13.75 **(a)** One-dimensional cut through an atomic potential in the strong field of a high-power femtosecond laser, showing the paths of the wave packets of the photodetached and returning electron, **(b)** comparison between molecular and optical interferometers [13.87]

as high harmonics by the same femtosecond pulse for diagnostics [13.94].

Another interesting application is the diagnostics of laser-produced high-temperature plasmas and inertially confined fusion. The fast ignition process can occur within attoseconds and the detailed diagnostics of these processes and their time sequence may result in a better understanding and optimization of laser-induced fusion processes [13.95].

The high peak powers attainable with femto- and attosecond pulses can generate electron bunches with high energies reaching beyond 1 MeV. There are serious estimations that even GeV electron pulses can be produced by wake-field acceleration in laser-produced plasmas. This new kind of particle accelerators has found wide interest, since it might also be used for proton acceleration. Although practical realization might still be far away, many laser laboratories and theoretical groups are working on this subject with increasing success [13.96].

A very interesting field is the study of electron dynamics during photo-ionization of atoms or molecules. The ejected electron might be forced by the strong laser field to return to the ion core before it can escape (Fig. 13.75a). The resulting interference between the wave functions of the outgoing and the returning electron give information on the structure of the electronic wavefunction of the orbital from which the electron was released [13.97]. As illustrated in Fig. 13.75b this techniques may be regarded as molecular interferometry, in analogy to optical interferometry. Depending on the phase of the optical field, angular asymmetries of the ejected photoelectrons are observed, which can be utilized to measure the absolute phase of optical fields [13.98].

13.3 LIDAR

13.3.1 Introduction

LIDAR stands for light detection and ranging and is the optical counterpart to radio detection and ranging (RADAR). As the name suggests these are range-resolved mapping techniques, where a microwave transmitter is replaced with a pulsed laser system. Photons backscattered from solid targets or from aerosol particles or molecules in the atmosphere are recorded using an optical telescope and an appropriate detection system. Range information is obtained through the time delay of the backscattered photons. Solid or liquid targets give rise to a sharp echo in the backscattered radiation at a time delay

$$T=\frac{2R}{c}\,, \tag{13.87}$$

where R is the distance and c is the velocity of light. This means that a time delay of 1 μs corresponds to a distance of 150 m. We thus see that for laser pulses of a typical length of 10 ns, resulting in an echo width of 1.5 m (light travels 30 cm in a ns), quite good range resolution can be obtained. This is the basis for normal pulsed laser range-finders. For laser light with a sharp emitted wavelength the elastically scattered echo is detected at the same wavelength. However, weak inelastically scattered Raman components, specific to the molecules irradiated, can also be detected offset from the laser frequency by the molecular vibrational frequency. More prominently, laser-induced fluorescence (LIF) from the target molecules is also observed on the long-wavelength side, in particular if the laser emits at ultraviolet (UV) wavelengths. The LIF signal is frequently spectrally broad, but certain information about the target molecules can still be obtained. In contrast, if a sufficiently powerful laser beam is focused onto a target, the pulses may induce optical breakdown and laser-induced breakdown spectroscopy (LIBS) signals from the free atoms and ions in the hot plasma can be recorded, providing elemental information.

It may be more unexpected to find that a signal from the free atmosphere itself is obtained when the laser pulse traverses it. This is due to elastic Mie and Rayleigh scattering from particles and molecules, respectively. Raman and fluorescence signals can also occur. Fluorescence from atoms or molecules in the atmosphere is normally quenched due to collisional de-excitation, but is prominent if the pressure is low, such as in the Earth's mesosphere. With reference to Fig. 13.76 [13.99], which shows a LIDAR measurement scenario, the intensity of the backscattered radiation, $P(\lambda, R)$, as a function of range R, is given by

$$\begin{aligned}P(\lambda, R) =& CWn_{\mathrm{b}}(R)\sigma_{\mathrm{b}}\frac{\Delta R}{R^2}\\ &\times \exp\left(-2\int_0^R [\sigma(\lambda)N(r)+K_{\mathrm{ext}}(r)]\,\mathrm{d}r\right)\,,\end{aligned} \tag{13.88}$$

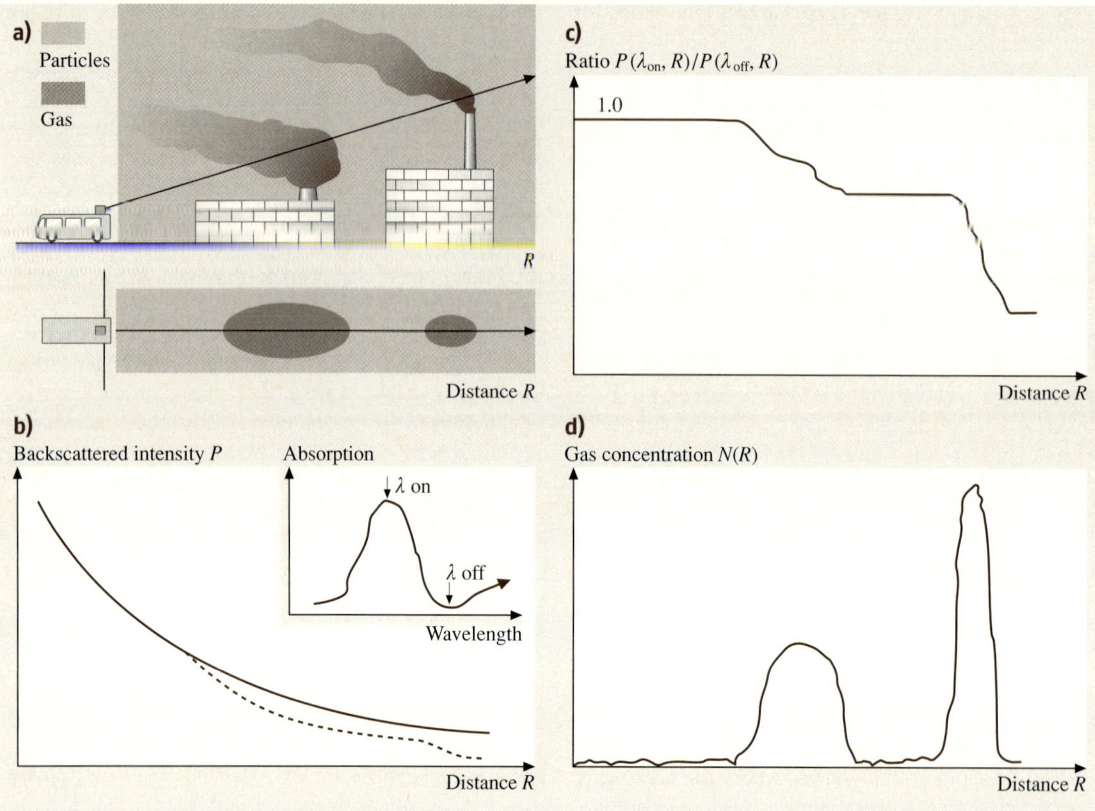

Fig. 13.76a–d The principles of LIDAR **(a)** scenario of LIDAR measurement at an industrial plant, with the atmosphere also seen from above **(b)** LIDAR curves for on- and off-resonant wavelengths **(c)** divided DIAL curve, on/off **(d)** evaluated gas concentrations for the chosen direction (After [13.99])

where C is a system constant, W is the transmitted pulse energy, and $n_b(R)$ is the number density of scattering objects with backscattering coefficient σ_b. Generally, the signal is for realistic atmospheres dominated by the elastic Mie scattering from particles. A main, and expected, observation from (13.88) is that the signal falls off with a $1/R^2$ dependence, basically reflecting the normal illumination law. Further, exponential light attenuation due to absorbing molecules or atoms of a concentration $N(r)$ and with wavelength-dependent absorption cross section $\sigma(\lambda)$ occurs, integrated over the range R, which is traversed twice. There is also an attenuation K_{ext} due to scattering particles with negligible wavelength dependence for small frequency excursions. In the differential absorption LIDAR (DIAL) approach, which is used for the monitoring of minor species such as air pollutants, a tuneable laser is intermittently fired at a wavelength λ_{on} for which a gas of interest absorbs, and at a reference, non-absorbed wavelength λ_{off}, which is located so close that the scattering properties are the same. All on-resonance LIDAR transients are averaged in one computer memory, and all off-resonance (reference) transients are averaged in a second one. By then dividing the two signals, as indicated in (13.89), and in Fig. 13.76, it is possible to eliminate almost all unknown system and atmospheric parameters, leaving a ratio signal that only depends on the differential absorption coefficient for the gas (ideally the off-resonance cross section is zero, but this is not always the case)

$$\frac{P(\lambda_{on}, R)}{P(\lambda_{off}, R)} = \exp\left[-2(\sigma_{on} - \sigma_{off}) \int_0^R N(r)\,dr\right] . \tag{13.89}$$

We note, that the $1/R^2$ dependence is now eliminated, and in particular that the ratio is independent of the range-dependent number of scattering particles. Thus the method works without knowledge of this otherwise very crucial parameter. The range-resolved concentration of gas molecules can be obtained from the slope of the ratio curve, as evaluated in (13.90)) and indicated in Fig. 13.76. Clearly the range of useful measurements is limited by the signal-to-noise ratio, and for very long ranges basically zero is divided by zero. Measurement in different directions allows a three-dimensional picture of the gas distribution to be determined.

$$N(R, R+\Delta R) = \frac{1}{2\Delta R(\sigma_{\text{on}} - \sigma_{\text{off}})} \times \ln \frac{P(\lambda_{\text{off}}, R+\Delta R) P(\lambda_{\text{on}}, R)}{P(\lambda_{\text{on}}, R+\Delta R) P(\lambda_{\text{off}}, R)} . \tag{13.90}$$

LIDAR work was pioneered by *Fiocco* and *Smullin* [13.100]. R. M. Shotland performed initial DIAL work on water vapor [13.101], while early DIAL experiments on an air pollutant (NO_2) were demonstrated by Rothe et al. [13.102] and Grant et al. [13.103]. Our own LIDAR work started in 1975.

LIDAR techniques are extensively discussed in a number of monographs; [13.104–106] and [13.107, 108] represent recent additions. LIDAR achievements are reported in the biennial International Laser Radar Conferences (ILRC), arranged by the International Coordination Group for Laser Atmospheric Studies (ICLAS). The most recent conference was arranged in Nara, Japan, in 2006 [13.109]. This literature provides comprehensive coverage of the field. In the present account, the field will mainly be illustrated by examples from the laboratory of the present author; further information on this work can be found, e.g., in the reviews [13.110–112].

The next section discusses the basic instrumentation used in LIDAR measurements. Then atmospheric applications including meteorology, pollution monitoring and space experiments will be presented. Section 13.3.4 deals with LIDAR experiments on fixed targets, including altimeter, LIF and LIBS LIDAR applications. Unconventional LIDAR-like applications in other fields of science, such as medicine and materials sciences, will then be exemplified. Finally, some conclusions are drawn.

13.3.2 Instrumentation

A LIDAR system consists of a transmitting laser, a receiving telescope, a detector, signal-handling electronics and a computer for capturing the signals and evaluating the data. The system can be placed fixed or be installed in a truck, an airborne platform or a satellite. Means for scanning the laser beam for interrogating a certain area are also frequently provided. The laser transmitter should be pulsed to provide adequate range resolution. For studying particles in the atmosphere, i. e. aerosols, a fixed frequency laser, such as a Q-switched Nd:YAG laser operating at 1064 nm can be used. Flash-lamp pumping is now increasingly being replaced by pumping by powerful diode lasers that provide higher efficiency, lower heat dissipation and longer lifetime. It is also common to use harmonics of the emission at 532, 355 and 266 nm, generated with high efficiency in nonlinear optical crystals. By using a number of such frequencies, information on the size of the scattering particles can be obtained due to the wavelength dependence of the Mie scattering. Clearly, particles both backscatter through σ_b and attenuate through K_{ext}. For evaluation of the particle concentrations and properties a relation between these two quantities must be established, which is a key aspect in inversion algorithms being developed for this purpose. To use the DIAL method a narrow-band tuneable laser must be used, and here dye lasers or titanium-doped sapphire lasers can be employed with the second or third harmonic of the Nd:YAG laser normally used as the pump. Such pump radiation can also be frequency converted over wide wavelength ranges using an optical parametric oscillator (OPO). By nonlinear harmonic generation, or mixing with fixed-frequency radiation, tuneable radiation over a wide range from the deep UV to the mid-IR spectral regions can be obtained. Excimer lasers, crudely tuneable by stimulated Raman scattering in hydrogen or deuterium gas, are sometimes used in the UV region, mostly for ozone monitoring.

The laser radiation is normally sent with a controlled divergence into the atmosphere by using a beam expander. A folding and beam-directing mirror is frequently used, also for directing backscattered radiation into the receiving telescope, which is normally of a Newtonian or Cassegrainian type and with a typical diameter of 10–100 cm. The laser beam typically has a divergence of 1 mrad and the telescope field of view is selected to match this value using an appropriate aperture. Backscattered light is detected by a photomultiplier tube or an avalanche photodiode placed behind a narrow-band filter matching the detection wavelength and blocking out sunlight. Gain-switching is frequently employed to handle the large signal dynamic range otherwise obtained. The signal is digitized with a transient recorder and signal averaging is performed until an adequate signal-to-noise ratio is obtained at the longest ranges of interest. In order to ascertain the validity of (13.90) in DIAL measurements (i. e., that particle scattering did not change between on- and off-resonance measurements) the laser wavelength has to be switched back and forth between adjacent shots.

In the IR region, where photomultiplier tubes having high efficiency and amplification are not available, it can be advantages to use *heterodyne* or *coherent* detection of the LIDAR signal. Then the LIDAR signal is mixed at the detector with a frequency-offset laser frequency and the beat note is detected at an intermediate frequency. This technique is best achieved if the system is equipped with a CW seeding laser which is pulse-amplified in the LIDAR transmitter. The seeding laser is then also a suitable local oscillator for the heterodyne detection. Clearly, Doppler shifts due to moving atmospheric particles occur, and this is a basis for LIDAR detection of wind speed.

When LIF or LIBS signals from solid targets are to be detected the detector is replaced with an optical multichannel analyzer, i. e., a spectrometer followed by a gated and image-intensified CCD. The spectrometer can be fibre-coupled to the focal plane of the telescope, and by properly gating the system, the full LIF or LIBS signal can be captured for each laser shot, with suppression of the ambient light, which can be made negligible during the short gate when the desired signal is present.

An example of a mobile laser radar system is shown in Fig. 13.77 [13.113]. The system, currently used by the Lund University group, has a Nd:YAG-pumped OPO as a transmitter, which covers the spectral range 220–3500 nm by nonlinear optical wavelength extensions. A vertically directed 40 cm-diameter Newtonian telescope is used in conjunction with a rooftop computer-controlled scanning mirror that selects the measurement direction for the aligned transmission and

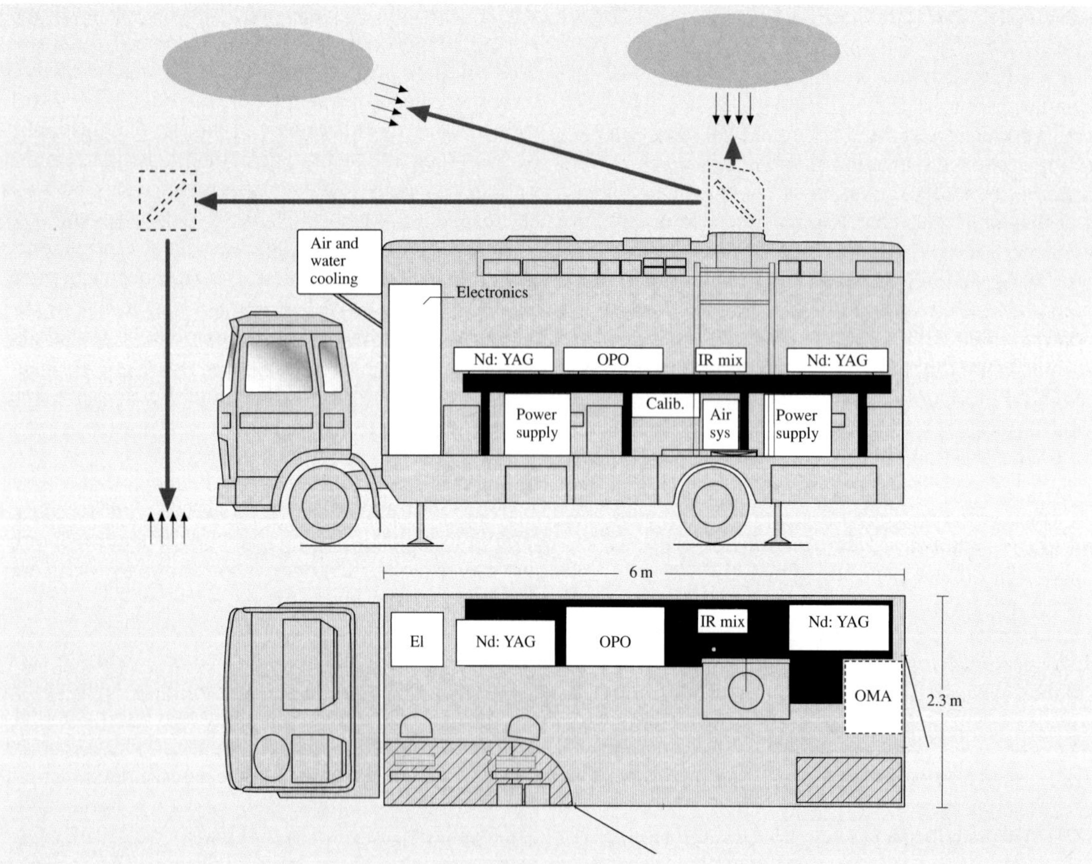

Fig. 13.77 Lund University mobile LIDAR system configured for vertical traverses or free scanning atmospheric measurements, and for water sounding downwards. (After [13.113])

detection lobes. A gated spectrometer is also shown for solid-target applications. Equipped with a motor generator, towed by the truck and providing electricity, a fully autonomous LIDAR system is obtained for land-based and ship-borne applications. The examples of LIDAR experiments shown in the figures of the present paper were all obtained using this system. Clearly, an airborne or satellite-installed LIDAR system must fulfill stringent requirements with regards to weight, efficiency and reliability. We will discuss this further below.

13.3.3 Atmospheric LIDAR Applications

Mesosphere

There are numerous applications of LIDAR techniques for atmospheric sounding. A specialized application is to investigate the low-density layers of atoms, such as sodium, potassium and iron in the mesosphere at heights of about 100 km, where meteorites impinging on the Earth's upper atmosphere are vaporized. The pressure is so low that collisional de-excitation of the laser-excited species (at about 589, 765 and 372 nm, respectively) is negligible and bright atomic fluorescence can be observed. These layers exhibit gravitational-wave patterns and, by probing the line shapes with a narrow-band excitation laser, temperature as well as velocities can be derived from line widths and shifts. A recent utilization of the bright emission is for *laser guide star* applications [13.114], where the localized emission is used for active control of astronomical telescopes to compensate for atmospheric turbulence using *adaptive optics*.

Meteorology

Accurate weather forecasting requires powerful prediction models and equally important, accurate input data. Traditionally, synoptic observations from a global network of observational sites have been used. With the availability of meteorological satellites in geostationary orbits global coverage of many important parameters has been achieved. However, to increase predictive accuracy global wind fields and vertical water-vapor profiles are much needed. Such information can be obtained with LIDAR techniques. Actually, we have already discussed above how Doppler heterodyne techniques can be used to measure range-resolved wind speed. The Doppler shift in the line profile of the scattering molecules and particles can also be observed in direct intensity measurements where a narrow-band transmitter is used to induce the scattering and the so called double-edge technique [13.115, 116] is used to observe wind-induced frequency shifts. By using a Fabry–Pérot interferometer or optically thick atomic or molecular gas cells in the path of the detected light this can be partially blocked out. When the profile is moved in frequency due to the wind, more or less of the light is *scraped off* by the sharp edge filter, and the data can be converted into velocity. After extensive fixed-system and airborne tests the technique is now ready for satellite global coverage within the European Space Agency's (ESA) AEOLUS project, where a frequency-tripled Nd:YAG laser at 355 nm is used.

Water vapor can be measured by DIAL but also more simply using Raman scattering of this high-concentration molecule. By blocking the elastically scattered laser radiation at, e.g., 355 nm and by only monitoring the intensity of the Raman–Stokes shifted water-vapor line at 408 nm through a narrow-band interference filter, vertical profiles for many kilometers are readily obtained.

We have already discussed that, while aerosols and cloud particles yield a strong elastic Mie scattering signal, its interpretation is complex due to the interplay of backscattering and extinction. Accurate measurements of the aerosol influence on the Earth's radiation budget are very important in assessing global change and warming. The uncertainty in the aerosol load is presently the single most important uncertainty in modeling climate forcing [13.117]. By using strong Raman scattering from the uniformly mixed major species nitrogen, in the case for 355 nm excitation centered at 387 nm, it is possible to measure particle extinction independently, since the Raman signal can only be reduced by the particles. In contrast, the elastic signal due to particles is increased by an increasing load of particles but reduced by attenuation. By studying the detected depolarization of primary linearly polarized laser light it is possible to determine the shape of the scattering particles. Spherical particles (such as water droplets) keep the initial polarization, while elongated particles, such as ice needles and snow crystals, exhibit depolarization. A complication is that multiple scattering in spherical particles is also depolarizing. After many years of extensive studies in fixed and airborne LIDAR installations, a dedicated satellite system, (CALIPSO), was launched in 2006, and is now providing global coverage of particle scattering and depolarization using 1064 and 532 nm radiation from a Nd:YAG laser, designed to operate for few years at 20 Hz.

The backscattered signal due to particles and molecules can further be separated by using a Fabry–Pérot interferometer or a gas filter to block the spectrally sharp Mie signal (the particles only move due to the wind), while the wings of the Doppler-broadened Rayleigh scattering is passed (as molecules move at thermal velocities, typically a few hundred meters per second). This technique obviously then also leads to the possibility of measuring range-resolved atmospheric temperature. A further method to measure temperature by LIDAR is to use the rotational Raman signal from nitrogen, which maps out the Boltzmann thermal population distribution on the different rotational levels of the ground-state molecule.

Global warming is a great concern, and is primarily believed to be induced by carbon dioxide emissions due to increasing burning of fossil fuels. Thus, detailed mapping of the sources and sinks of this gas is important, and many projects aimed at DIAL monitoring of CO_2 around 2 μm using Tm,Ho:YLF lasers are now under way.

Minor Species and Pollution

Minor species and pollutant gases in the troposphere can be mapped using the DIAL technique. Raman LIDAR does not have the required sensitivity and fluorescence does not work because of the quenching at atmospheric pressures. As a first illustration we will discuss a very special pollutant, mercury, actually the only one present in the atmosphere as a free atom, rather than as a molecular compound. Disregarding isotopic shifts and the hyperfine structure, mercury atoms have a single transition rather then a multitude of vibrational–rotational transitions over which the molecules are distributed. Thus we automatically gain a sensitivity increase of a factor of about 1000 for mercury, which is needed, since the background levels of this gas is on the ppt ($1:10^{12}$) level. Mercury, atmospherically present in atomic form to about 90%, is emitted from chlor-alkali plants producing sodium hy-

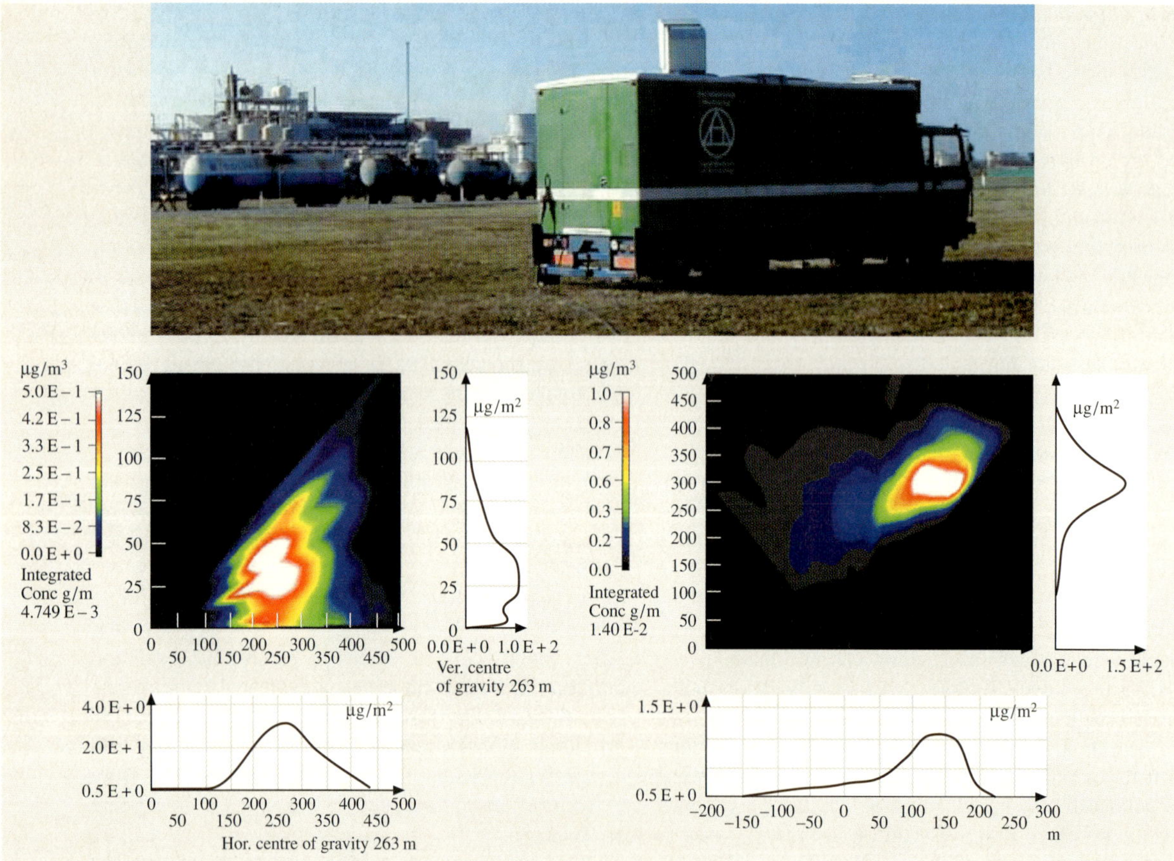

Fig. 13.78 DIAL measurements of atomic mercury. Top, mobile LIDAR system seen at the Rosignano Solvay chlor-alkali plant, Bottom left, vertical scan, and bottom right, horizontal scan. (From [13.118, 119])

droxide (the cathode in the cells for electrolysis of brine is liquid mercury, into which the sodium ions move, forming amalgam, which on washing yields the natron lye). Furthermore, coal-fired power plants and incineration plants including crematories emit mercury. It is also an interesting geophysical tracer gas related to ore deposits, geothermal reservoirs and volcanic activities. DIAL measurements are performed using the 254 nm mercury transition, which can be induced using a frequency-doubled dye laser or OPO. An illustration of atomic mercury DIAL at a chlor-alkali plant is shown in Fig. 13.78 [13.118, 119], where individual LIDAR returns taken fan-like and perpendicularly through the plume downwind from the plant, have been used to generate a plume cross section showing spatially resolved concentration. The area-integrated concentration times the perpendicular wind speed yields the flux, which is the environmentally important parameter. This reasoning can convincingly be supported by a consideration of units:

$$\frac{[g]}{[m^3]}[m][m]\frac{[m]}{[s]} = \frac{[g]}{[s]} \tag{13.91}$$

yielding the proper flux unit g/s. Typical fluxes for plants of this kind can be tens of g/h. Atomic mercury monitoring is considered as a special case study in [13.110] and is also discussed in [13.111] regarding geophysical emissions. We performed LIDAR measurements of mercury at the three largest mercury mines in the world: Almaden (Spain), Abbadia S. Salvatore (Italy) and Idrija (Slovenia).

Another pollutant of even more concern is sulphur dioxide SO_2, which contributes heavily to the acidification of lakes and to atmospheric loading with sulphate particles. Sulphur dioxide is emitted from volcanoes, but the main sources are anthropogenic and are due to the combustion of sulphur-containing fuels. We illustrate SO_2 monitoring with the case of volcanic emission. Figure 13.79 shows data from ship-borne measurements of the Italian volcano Mt. Etna, where the Swedish mobile LIDAR system is placed on the aft deck of the Italian research vessel *Urania*, making a 12 km-long underpass of the volcanic plume, which is encountered at about 3 km height [13.120]. Individual *on* and *off* LIDAR returns are also shown at a particular location to illustrate how the plume cross section is built up. It can be noted that the curves are identical until the height of the plume, where they run apart. Both curves also show extended backscattering due to volcanic aerosols above the smooth background. It is interesting to note that the aerosol plume is slightly below the SO_2 plume due to gravitational separation. This recording yielded a flux of about 50 t/h as the background emission for this active volcano. LIDAR measurements of geophysical gases are discussed in [13.111].

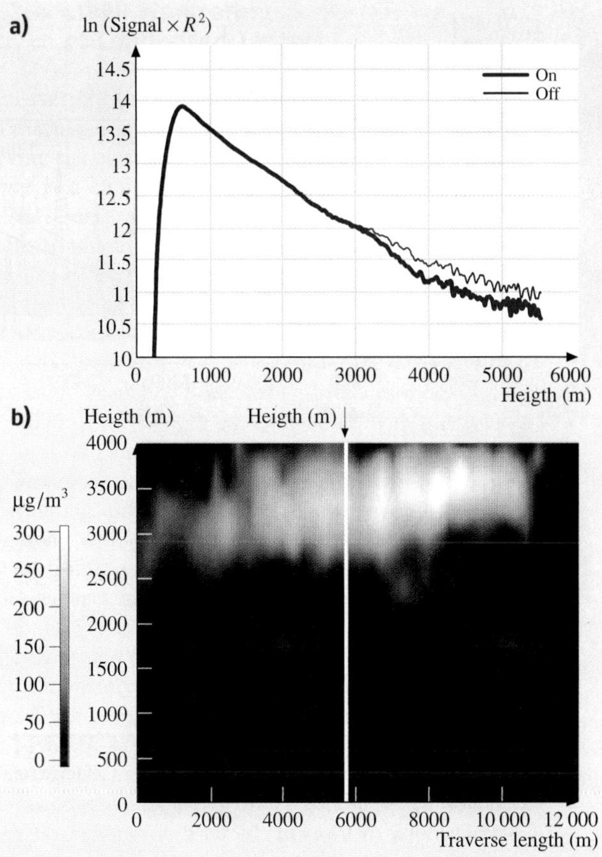

Fig. 13.79 DIAL monitoring of SO_2 from Mt. Etna. The measurements were performed in a 12 km scan with the system firing vertically from the aft deck of the research vessel *Urania*, making a traverse under the volcanic plume off the city of Catania. On- and off-resonance LIDAR curves are shown in the upper part of the figure for one particular location. (After [13.120])

As a third example of LIDAR measurements we will discuss the monitoring of hydrocarbons, which poses special challenges. The primary C–H stretch vibrational transition in hydrocarbons is around 3.4 μm with small variations between different species. The individual rotational structures also lead to heavy spectral overlap between different hydrocarbons. Thus, in the presence of several hydrocarbons it is not possible to use the basic DIAL concept with an *on* and an *off* wavelength,

Fluorescence and Raman LIDAR Studies of Water, Vegetation and Buildings

As discussed in the introduction, the LIDAR echo from a fixed target consists not only of the strong elastic backscattering, but also of laser-induced fluorescence (LIF) and Raman components. These features, red-shifted with regard to the laser pump frequency, can be isolated with a spectrometer, where additional suppression of the elastic return is provided with a colored-glass long-pass filter.

An example of an arrangement for LIDAR studies of condensed targets is given in Fig. 13.81 [13.123]. Frequency-tripled Nd:YAG radiation at 355 nm effectively induces fluorescence, leaving a sufficiently large Stokes-shifted wavelength range to be captured with available efficient detection systems. Two detection arrangements are shown. In the first, the fluorescence light is captured with an optical fibre in the image plane of the telescope and brought to the entrance slit of the spectrometer with a gated and image-intensified CCD detector. Frequently, a multi-fibre arrangement is used, where a cylindrical detection area is transformed into a slit shape to match the input of the spectrometer. In a second imaging arrangement, the laser beam is widened to illuminate a larger area on the target and a special beam-splitting optical arrangement [13.124] is used to capture the same image simultaneously, filtered through four individual bandpass filters, leaving the images detected simultaneously in the four quadrants of the CCD detector. Such an arrangement can be used for small targets and at background illumination conditions allowing an expanded laser beam with correspondingly weak signal generation from any individual target point. More of general purpose is to have a few cm diameter beam with intense radiation (outperforming the sun background in that spot) and then to sweep

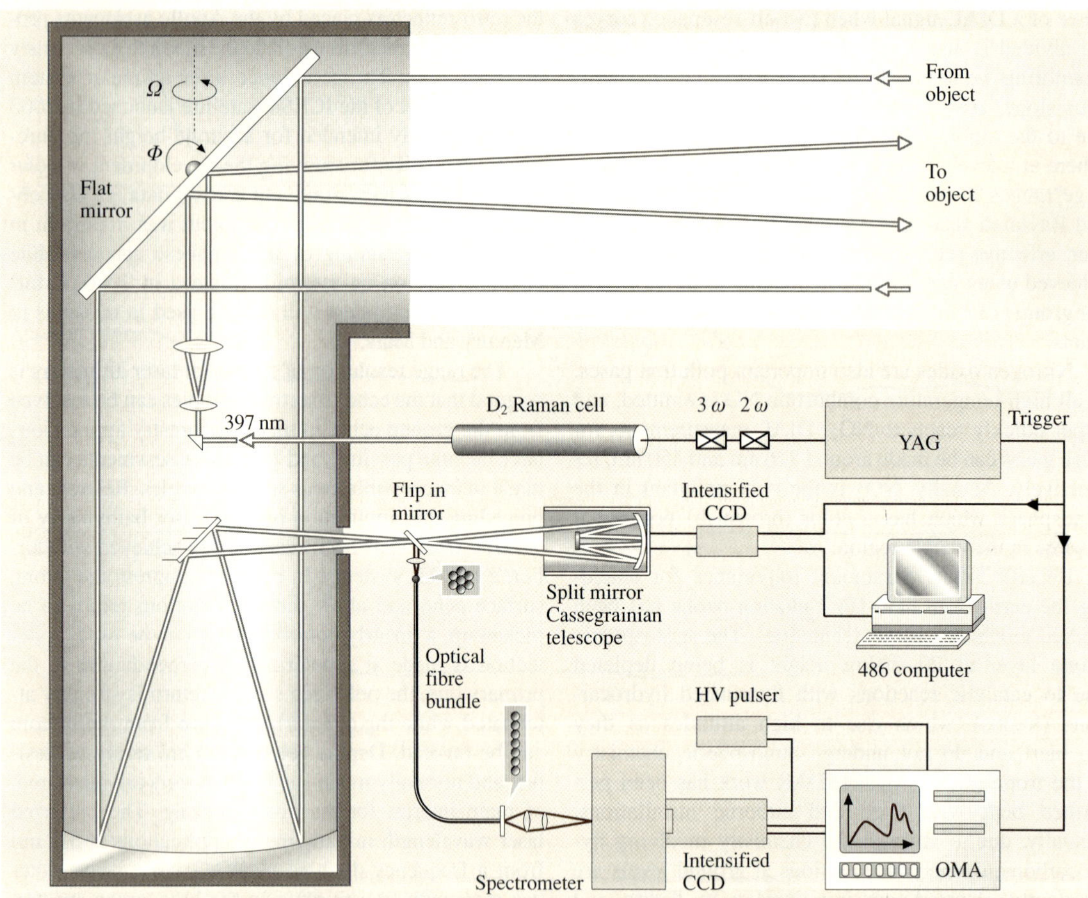

Fig. 13.81 Fluorescence LIDAR system lay out for point monitoring and multi-spectral imaging. (After [13.123])

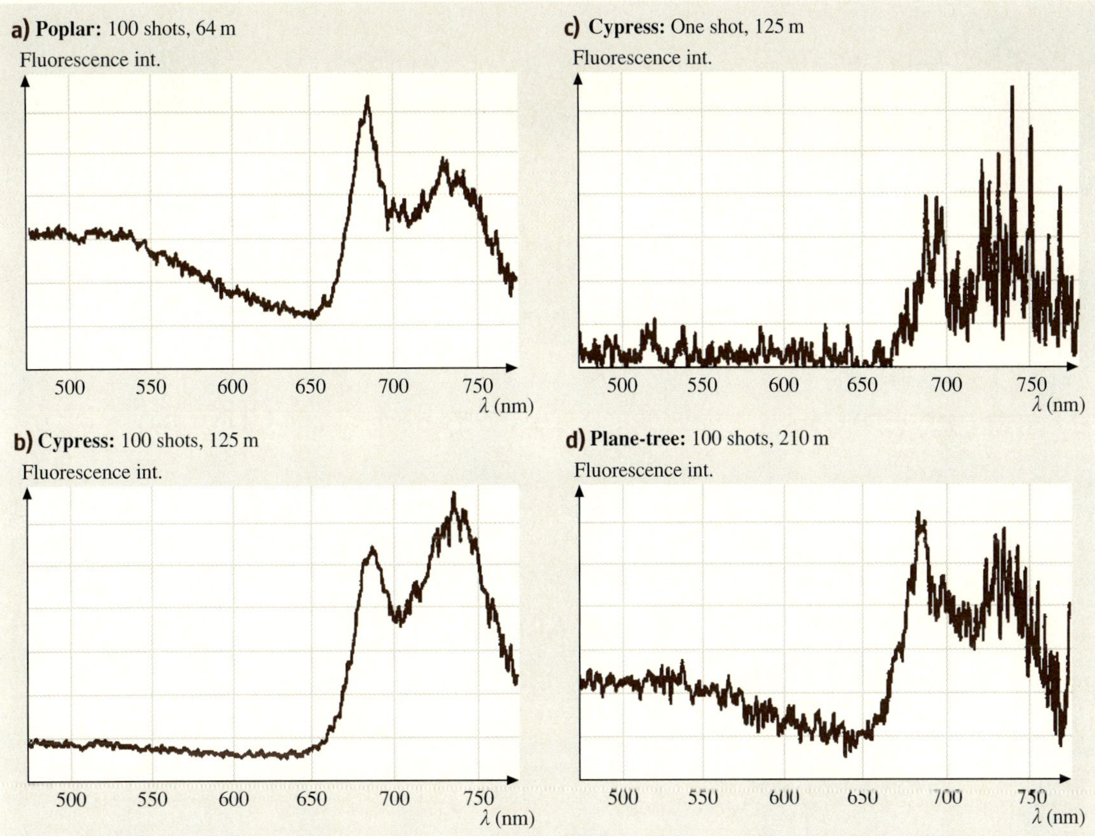

Fig. 13.82 Remote fluorescence LIDAR spectra, recorded at different stand-off ranges. (After [13.125])

this point over the target in a whisk-broom mode. The field is reviewed, e.g., in [13.112].

As an example of individual spectra recorded in such a mode we show in Fig. 13.82 vegetation spectra, captured at distances ranging from 64 to 210 m [13.125]. The dual-peaked chlorophyll signals are clearly seen as well as a blue light distribution related to the leaf wax layer and other leaf structures. The relative intensity of the two red peaks gives information on the chlorophyll concentration, since the first peak is self-absorbed with increasing levels of the pigment. There has been a considerable effort in vegetation monitoring using LIF aiming at the detection of water-related and other environmental stress factors. LIF data may supplement reflectance data now monitored extensively using satellite sensors for land coverage assessment (e.g. LANDSAT, SPOT and ENVISAT sensors).

Cultural heritage studies with spectroscopic methods is an emerging area of laser application. In the same way as specific LIF signatures are obtained from vegetation targets, the surface of stone monuments, such as cathedrals and castles, may also reveal aspects in fluorescence not visible to the naked eye or in reflectance spectroscopy. Following the first point-monitoring applications by *Raimondi* et al. [13.126] we have extended the techniques to multi-spectral imaging with measurement campaigns performed, e.g. at the Lund cathedral (Sweden) [13.123] and the Parma cathedral and baptistery [13.127]. An example from recent measurements at the Övedskloster castle in Southern Sweden is shown in Fig. 13.83. Here the beam was scanned sequentially line by line from a distance of about 40 m [13.128, 129]. Examples of individual spectra recorded are shown in the figure, featuring different amounts of chlorophyll from mostly invisible algal growth. It can be seen that the edges of the tiles are preferred algal locations as are the upper parts of the ornamental urns. The chlorophyll amount is encoded on the colorscale of the pixels having at least a minimum amount of chlorophyll (black).

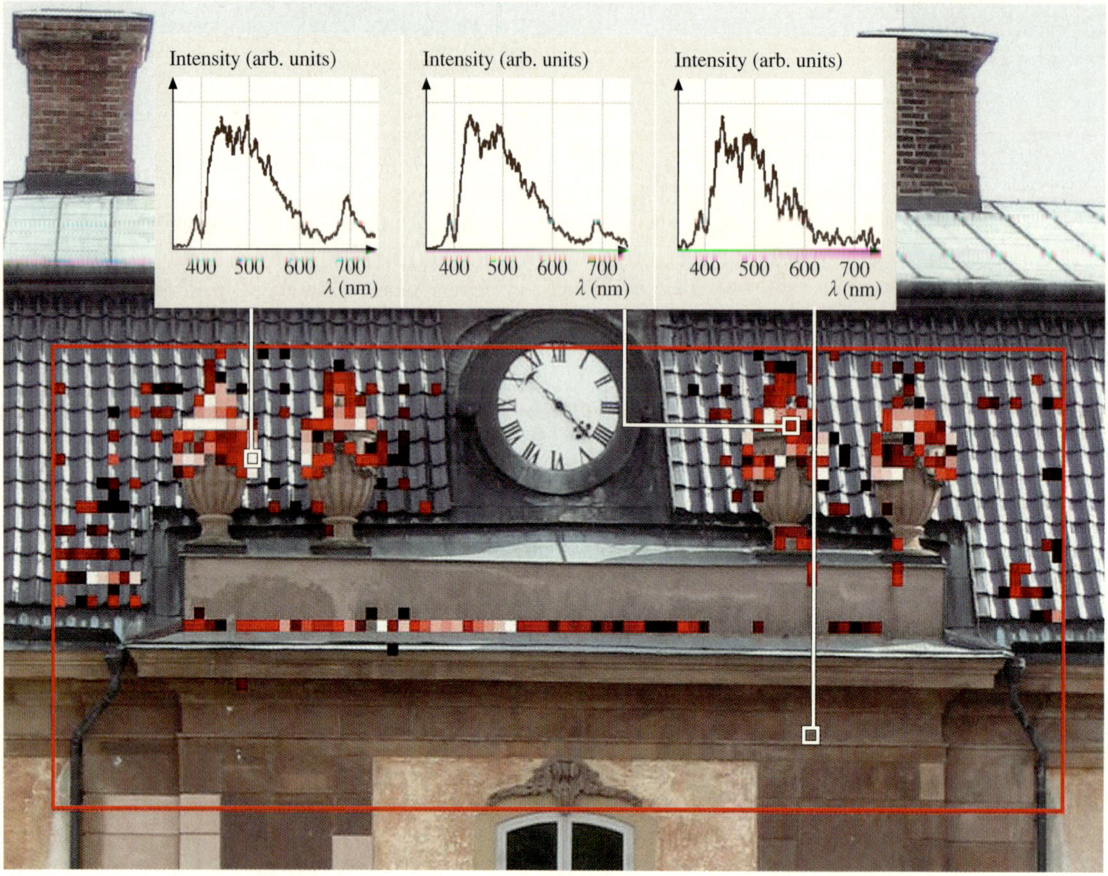

Fig. 13.83 Fluorescence LIDAR imaging of a section of the Övedskloster castle, Sweden. Three spectra illustrate the measurements, and an image with a colorscale indicates the amount of algae invasion. (After [13.128, 129])

A final example is given in Fig. 13.84 with data from a recent measurement campaign [13.130, 131] at the Flavian amphitheater Coliseum, inaugurated in Rome in 80 AD. The monument is constructed out of travertine quarried nearby in Tivoli. Spectra show differences in stone surface even if they appear similar. Chlorophyll invasion is very evident in certain places, and corrosion protection on iron bars applied during historical preservation efforts also has a very characteristic appearance, as shown in the figure, which also includes a photograph from the measuring site.

The first application of fluorescence LIDAR techniques were in the field of water monitoring, as reviewed, e.g., in [13.132, 133]. Land-based, ship- as well as air-borne campaigns have been pursued. An important aspect of water fluorescence monitoring is the assessment or the general level of dissolved organic matter (DOM) in water. DOM gives rise to broad, bluish fluorescence spectrum. By normalizing the DOM signal to the Raman signal due to the water O−H stretch vibrational mode at a Stokes shift of about $3400\,\mathrm{cm}^{-1}$ a convenient intrinsic reference is obtained, which makes the measurement independent of a knowledge of the effective sampling volume. For 355 nm excitation the O−H stretch Raman signal of liquid water occurs at 404 nm. By analyzing the shape of the water Raman peak with a high spectral resolution system it is also possible to measure the depth-resolved water temperature remotely with a bathymetric system.

Other prominent spectral features in LIDAR spectroscopic studies of water are due to chlorophyll, which is present in all pelagic algae, and accessory pigments, e.g. phycocyanine and phycoerytrine, which are present in red algae. The characteristic fluorescence of these pigments peak at 690, 660 and 590 nm, respectively.

Fig. 13.84 Fluorescence LIDAR measurements at the Coliseum, Rome. A photograph with the Swedish LIDAR system in measurement position is shown in the upper part of the figure. Below, individual fluorescence spectra are shown to the *left*, while points with a spectral signature similar to A are shown in the *middle*, and points with a spectral signature similar to D are shown to the *right*. (After [13.130, 131])

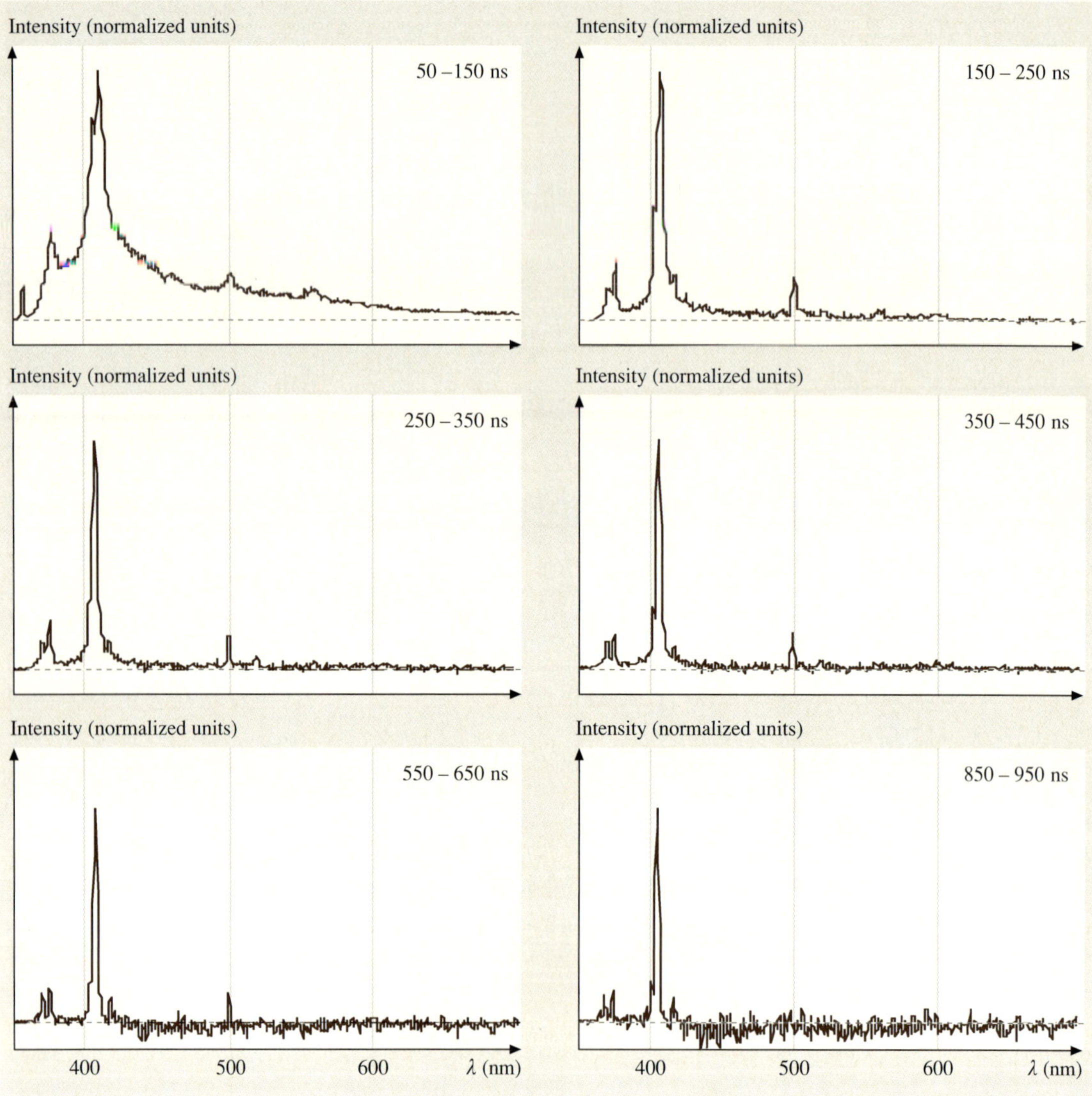

Fig. 13.85 Laser-induced breakdown spectra (LIBS) recorded with a LIDAR system at 60 m stand-off distance for a lead plate for different time delays. The lines become narrower and more well defined as the plasma cools. (After [13.129])

The strongest signals received in aquatic monitoring are from oil spills covering the surface [13.134]. At the same time as a strong oil fluorescence replaces the DOM signal in the blue–green region, the water Raman signal disappears, since even a layer a few micrometers thick becomes optically thick for the exciting UV radiation. Thus oil spills have a very clear spectral signature. Laboratory studies of different types of crude oils and lighter fractions show that oils of specific origins can be identified. Extensive airborne work for oil-spill detection and characterization, complementing data from side-looking airborne radar (SLAR) and synthetic-aperture radar (SAR) has been pursued.

A main limitation of fluorescence LIDAR is its limited range, which makes satellite implementation difficult. However, passive techniques using sunlight-induced fluorescence have been considered for marine oil-slick detection and is now being developed for vegetation monitoring [13.135]. The normal reflectance spectrum of a ground surface carries the dark and narrow Fraunhofer signature lines from the sun with the same contrast as in direct sunlight. However, fluorescence transfers part of the radiation from short to longer wavelengths due to the Stokes shift, and tends to reduce the Fraunhofer line contrast, a phenomenon directly signaling that fluorescence is present.

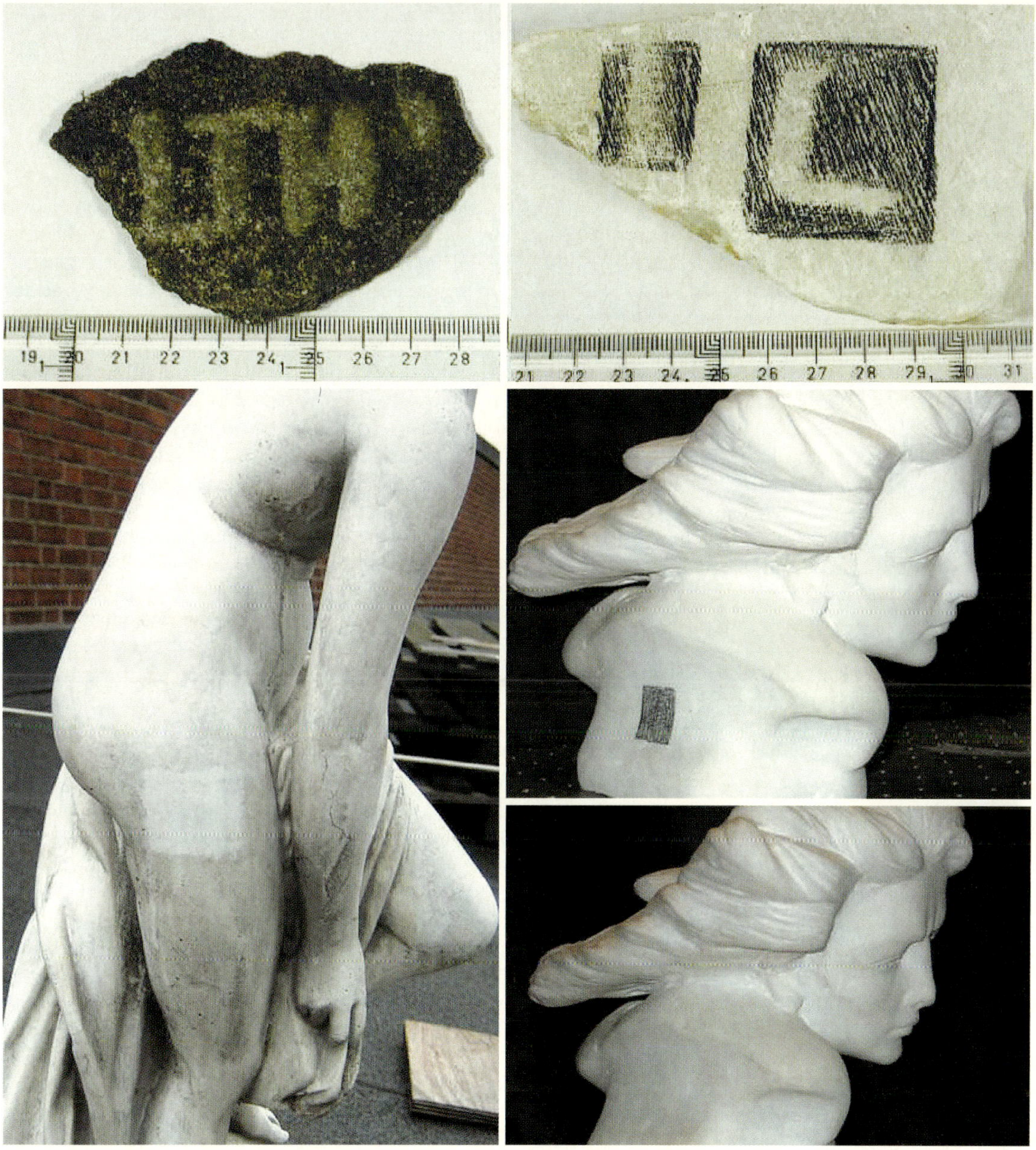

Fig. 13.86 Illustration of remote plasma ablation on different samples placed at 60 m distance from the LIDAR system. *Top*: ablation patterns for concrete and Italian alabaster; *Bottom left*: illustration of the cleaning of an Italian garden statue; *right*: a soiled Carrara marble statue before (*upper figure*) and after (*lower figure*) remote laser cleaning. (After [13.136])

Laser-Induced Breakdown Spectroscopy (LIBS) LIDAR

We will finish this section of LIDAR measurements on condensed targets by a discussion of remote laser-induced breakdown spectroscopy (LIBS). The LIBS technique makes use of focused laser beams from pulsed lasers, where a laser spark is produced at the target surface due to strong heating by the pulse [13.137]. The technique is a valuable analytical analysis method for elemental measurements of the same kind as in flame-emission spectroscopy where vaporization occurs due to the thermal heating. By expanding the LIDAR beam and focusing it is possible to reach a sufficiently small spot size to induce plasma breakdown at a considerable distance if the pulse energy is large enough [13.138]. This technique is also being developed for space exploration, e.g., rock characterization from the Mars landers [13.139]. We have recently performed the first imaging LIDAR LIBS measurements at a distance of 60 m and with pulse energies at 355 nm of the order of 200 mJ [13.136]. The same detection system as for LIF is employed, and by gating a few hundred nanoseconds later it is possible to observe the cooling plasma, in which narrow atomic emission lines are obtained rather than broadened structures observed in the early plasma, where strong fields due to electrons etc. are present. This effect is seen in Fig. 13.85, where a LIBS spectrum from a sheet of lead is observed at 60 m distance for different delays [13.129].

Small amounts of material are ablated in the LIBS plasma. This can be utilized for cleaning of soiled surfaces in the cultural heritage sector. Such remote cleaning is illustrated in Fig. 13.86. Here samples of mortar, alabaster and marble were studied from a distance of 60 m. On white marble samples the cleaning process is self-terminating, since no absorption is obtained once the dirty surface is removed from the original white material. The ablation and cleaning process could in principle also be controlled spectroscopically by simultaneously monitoring the plasma spectrum.

13.3.5 Unconventional LIDAR Applications

LIDAR techniques are conventionally associated with long-range atmospheric measurements. However, a key aspect in these measurements is that they are performed time-resolved and non-intrusively. Using topographic (condensed) targets, the range-resolution aspect is also sometimes suppressed. In the measurement of combustion processes optical techniques are used at typical distances 1 m, but motivated by the remote and non-intrusive nature of optical spectroscopy, avoiding the prohibitive heat and pressures that are characteristic of such systems. Elastic scattering, Raman, fluorescence and coherent anti-Stokes Raman scattering (CARS) techniques are employed [13.140]. Range-resolved measurements in large-scale burners are now being developed making the parallels with LIDAR very close [13.141].

Optical mammography for the detection of breast cancer [13.142] can truly be characterized as short-range LIDAR. Biological tissue is a strongly scattering medium, in which molecular compounds can also be detected using differential absorption. It can be convenient to use white laser light generated by self-phase modulation of femtosecond laser pulses in such measurements, and molecules, such as tumor seeking agents can be detected [13.143]. Such techniques can be extended to the monitoring of pharmaceutical tablets that are strongly scattering [13.144, 145]. Using white-light laser bursts the concentration of the active molecular components in the drug can be evaluated from their imprint in the detected scattered light. These examples constitute multiple-scattering cases much similar to those pertaining to cloud physics.

The LIDAR analogy becomes almost complete in the gas in scattering media absorption spectroscopy (GASMAS) technique [13.146, 147], where free gas enclosures (pores) in highly scattering materials such as wood, foods and construction materials are studied. Normally, CW diode-laser spectroscopy is being applied for path-integrated measurements of the free gas signature (e.g. molecular oxygen and water vapor). Gas concentration determination becomes possible if an independent LIDAR measurement of the time history of the photons inside the sample is being made [13.148]. The GASMAS technique also gives interesting insight into everyday processes such as the drying of materials, e.g. wood [13.149]. The technique has recently been extended to the monitoring of oxygen gas inside the human sinus cavities [13.150], where the laser light is launched through the scattering forehead, passes through the sinus, scatters in the brain, with a fraction returning back through the sinus, passing through the forehead and entering the LIDAR detector. The gas is detected through its sharp differential absorption superimposed on the backscattering spectrum from the bulk molecules having a very slow wavelength dependence (as in atmospheric LIDAR). Some illustrations of unconventional LIDAR work are given in Fig. 13.87.

The many parallels between environmental and medical laser spectroscopy have been pointed out, e.g. in

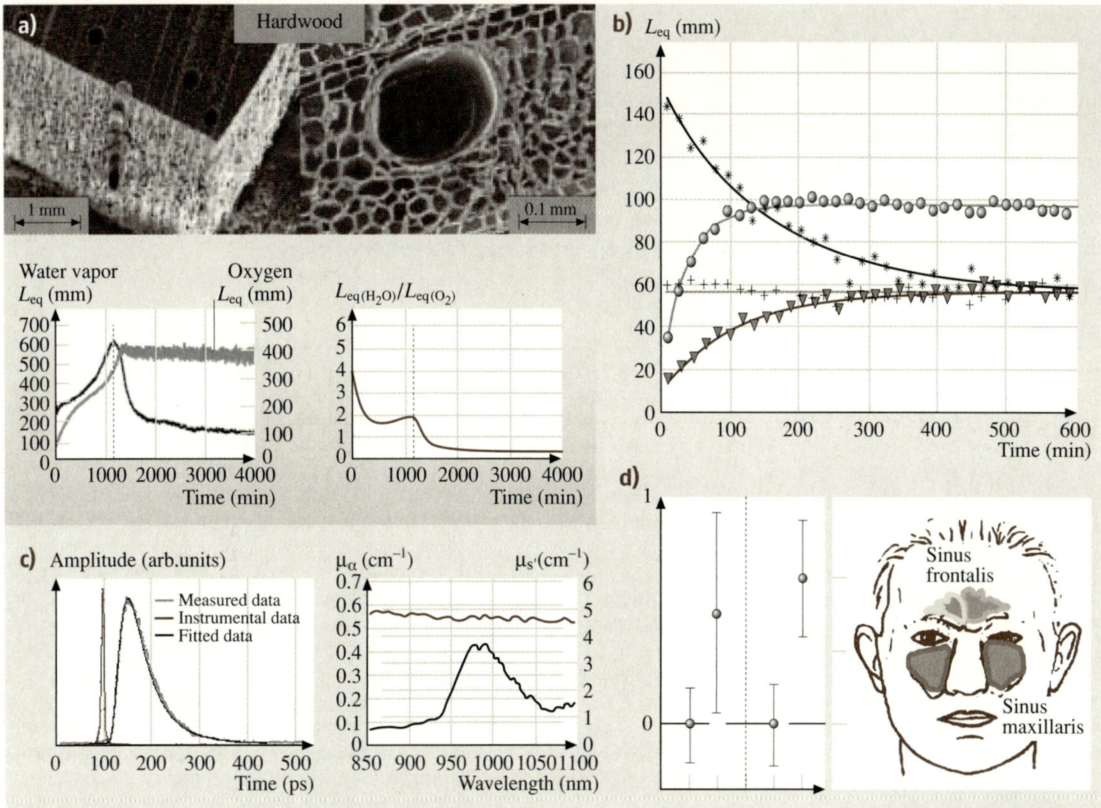

Fig. 13.87a–d Illustration of different kinds of unconventional short-range "LIDAR" measurements **(a)** Monitoring of the drying of wood. Microscope images of wood structure with pores are shown above the temporal development of molecular water and molecular oxygen signals, and their ratio value during a two-day drying process. (After [13.149]) **(b)** Molecular oxygen signals from apple exposed to pure oxygen or nitrogen gas before being taken into ambient air where normal interior gas conditions are restored with a certain time constant. (After [13.147]) **(c)** Time-resolved measurements of light scattered in a white pharmaceutical tablet, and (*right*) observation of the liquid-water absorption spectrum. (After [13.144]) **(d)** Observation of free oxygen gas in the frontal sinus cavities when light is launched into the forehead and backscattered light is detected. Non-zero signals are seen on both sinuses, while zero signal results off the sinuses. (After [13.150])

[13.151], and further aspects of multidisciplinary spectroscopy bridging many disciplines are discussed in [13.152].

13.3.6 Discussion and Outlook

As we have seen, LIDAR techniques provide a wide spectrum of applications regarding the monitoring of the environment. The rapid development of efficient diode-pumped solid-state lasers, sensitive detectors, and fast electronics and data handling have all contributed to make the applications of LIDAR more attractive. With ever more strict considerations regarding the monitoring of the status of the environment, the utilization of LIDARs in surveillance and monitoring systems and networks will be a driving force in the development of the field. Satellite-borne equipment is now planned for many missions, taking advantage of the emerging robustness and reliability of these techniques. A major impact on global atmospheric monitoring and meteorology can be expected.

Recent trends in the LIDAR field include the use of intense laser beams of terawatt peak power and femtosecond duration. Reasonably small, table-top systems have become available after the introduction of the chirped-pulse amplification (CPA) technique [13.153].

Filamentation and channeling due to self-focusing occurs in the atmospheric propagation, keeping the beam together with intense generation of white light through nonlinear effects, most prominently self-phase modulation [13.154]. Free atmosphere measurements, including studies of airborne bio-aerosols, as well as remote LIBS measurements have been demonstrated [13.155, 156]. Clearly, eye-safety considerations are challenging here with laser light present at all wavelengths. LIDAR systems operating outside the human cornea transmission region 400–1400 nm are most convenient for more-general applications due to the greatly relaxed safety aspects.

LIDAR techniques and methods have also stimulated extensions into more unconventional fields, such as biology and medicine. Many LIDAR concepts can be scaled down to the cm level, where new medical diagnostics techniques are now emerging.

References

13.1 W. Demtröder: *Laser Spectroscopy*, 3 edn. (Springer, Berlin, Heidelberg 2003)

13.2 M. Gehrtz, G. C. Bjorklund, E. Whittaker: Quantum-limited laser frequency-modulation spectroscopy, J. Opt. Soc. Am. B **2**, 1510 (1985)

13.3 P. C. D. Hobbs: Ultra-sensitive laser measurements without tears, Appl. Opt. **36**, 903 (1997)

13.4 J. A. Silver: Frequency modulation spectroscopy for trace species detection, Appl. Opt. **31**, 707 (1992)

13.5 G. Talsky: *Derivative Spectrophotometers* (VCH, Weinheim 1994)

13.6 J. Xiu, R. Stroud: *Acousto-Optic Devices: Principles, Design and Applications* (Wiley, New York 1992)

13.7 Ch. Hornberger, W. Demtröder: Photo-acoustic overtone spectroscopy of acethylene in the visible and near infrared, Chem. Phys. Lett. **190**, 171 (1994)

13.8 F. Träger: Surface analysis by laser-induced thermal waves, Laser Optoelektron. **18**, 216 (Sept. 1986)

13.9 St. E. Bialkowski: *Optothermal Spectroscopy. Methods for Chemical Analysis* (Wiley Interscience, New York 1995)

13.10 G. Busse: Nondestructive Evaluation with Thermal Waves. In: *Photoacoustic, Photothermal and Photochemical Processes at Surfaces and in Thin Films*, ed. by P. Hess (Springer, Berlin, Heidelberg 1989)

13.11 B. Barbieri, N. Beverini, A. Sasso: Optogalvanic spectroscopy, Rev. Mod. Phys. **62**, 603 (1990)

13.12 C. R. Webster, C. T. Rettner: Laser optogalvanic spectroscopy of molecules, Laser Focus **19**, 41 (Feb. 1983)

13.13 D. Feldmann: Optogalvanic spectroscopy of some molecules in discharges: NH_2, NO_2, A_2, N_2, Opt. Commun. **29**, 67 (1979)

13.14 K. Narayanan, G. Ullas, S. B. Rai: A two-step optical double resonance study of a Fe-Ne hollow cathode discharge using optogalvanic detection, Opt. Commun. **184**, 102 (1991)

13.15 G. S. Hurst, M. G. Payne, S. P. Kramer, J. P. Young: Resonance ionization spectroscopy and single atom detection, Rev. Mod. Phys. **51**, 767 (1979)

13.16 G. Hurst, M. G. Payne: Principles and applications of resonance ionization spectroscopy. In: *Ultrasensitive Laser Spectroscopy*, ed. by D. S. Kliger (Academic, New York 1983)

13.17 K. Tohama: A simple model for intracavity absorption, Opt. Commun. **15**, 17 (1975)

13.18 V. M. Baev, T. Latz, P. E. Toschek: Laser intracavity absorption spectroscopy, Appl. Phys. B **69**, 172 (1999)

13.19 A. Campargue, F. Stoeckel, M. C. Henevier: High sensitivity intracavity laser spectroscopy: Applications to the study of overtone transitions in the visible range, Spectrochem. Acta Rev. **13**, 69 (1990)

13.20 P. E. Toschek, V. M. Baev: "One is not enough": Intracavity laser spectroscopy with amultimode laser. In: *Laser Spectroscopy and New Ideas*, Springer Ser. Opt. Sci., Vol. 54, ed. by W. M. Yen, M. D. Levenson (Springer, Berlin, Heidelberg 1987)

13.21 M. Raab, M. Weickenmeier, W. Demtröder: The dissociation energy of the cesium dimer, Chem. Phys. Lett. **88**, 377 (1982)

13.22 J. Goodishman: *Interaction Potential Theory I, II* (Academic, New York 1973)

13.23 R. Campargue: *Atomic and Molecular Beams; The State of the Arts 2000* (Springer, Berlin, Heidelberg 2001)

13.24 G. Scoles: *Atomic and Molecular Beam Methods*, Vol. I, II (Oxford Univ. Press, Oxford 1988,1992)

13.25 D. H. Levy, L. Wharton, R. E. Smalley: Laser spectroscopy in supersonic jets. In: *Chemical and Biochemical Applications of Lasers*, Vol. II, ed. by C. B. Moore (Academic, New York 1977)

13.26 J. B. Atkinson, J. Becker, W. Demtröder: Hyperfine structure of the 625 nm band in the $a^3\Pi_u \leftarrow X^1\Sigma_g$ transition for Na_2, Chem. Phys. Lett. **87**, 128 (1982)

13.27 R. E Miller: Infrared laser spectroscopy. In: *Atomic and Molecular Beam Methods*, Vol. II, ed. by G. Scoles (Oxford Univ. Press, Oxford 1992) p. 192

13.28 D. Bassi: Detection Principles. In: *Atomic and Molecular Beam Methods*, Vol. II, ed. by G. Scoles (Oxford Univ. Press, Oxford 1992) p. 153

13.29 S. Mukamel: *Principles of Nonlinear Optical Spectroscopy* (Academic, New York 1999)

13.30 V. S. Letokhov, V. P. Chebotayev: *Nonlinear Laser Spectroscopy* (Springer, Berlin, Heidelberg 1977)

13.31 M. D. Levenson: *Introduction to Nonlinear Spectroscopy* (Academic, New York 1982)

13.32 O. Svelto, S. DeSilvestry, G. Denardo: *Ultrafast Processes in Spectroscopy* (Pleneum, New York 1997) See also the series of Conference Proceedings: *Ultrafast Phenomena I-XIV, Series in Chemical Physics* (Springer, Berlin, Heidelberg 1976–2004)

13.33 C. Rulliere: *Femtosecond Laser Pulses: Principles and Experiments*, 2 edn. (Springer, Berlin, Heidelberg 2004)

13.34 J. C. Diels, W. Rudolph: *Ultrashort Laser Pulses* (Academic, San Diego 1996)

13.35 E. Brabec, F. Krausz: Intense few cycle laser fields, Rev. Mod. Phys. **77**, 545 (2000)

13.36 R. Kienberger, F. Krausz: Attosecond Laser Pulses. In: *Yearbook of Science and Technology* (McGraw-Hill, Maidenhead 2006) p. 19

13.37 A. H. Zewail (Ed.): *Femtochemistry: Ultrafast Dynamics of the Chemical Bond*, Vol. I and II (World Scientific, Singapore 1994)

13.38 See for instance the series on Proc. Int. Conferences on the Physics of Electronic and Atomic Collisions. ICPEAC I–XX (North Holland, Amsterdam)

13.39 A. Rundquist, C. Durfee, Z. Chang, G. Taft, E. Zeek, S. Backus, M. M. Murnane, H.C. Kapteyn, I. Christov, V. Stoev: Ultrafast laser and amplifiere sources, Appl. Phys. B **65**, 161 (1997)

13.40 C. L. Tang, L. K. Cheng: *Fundamentals of Optical Parametric Processes and Oscillators* (CRC, Boca Raton 1998)

13.41 D. Nikogosyan: *Nonlinear Optical Crystals: A Complete Survey* (Springer, Berlin, Heidelberg 2005)

13.42 R. Seiler, Th. Paul, M. Andrist, F. Merkt: Generation of programmable near-Fourier-transform-limited pulses of narrow band laser radiation from the near infrared to the vacuum ultravilolet, Rev. Sci. Instrum. **76**, 103103 (2005)

13.43 G. A. Mourou, C. P. J. Barty, M. D. Pery: Ultra-high intensity lasers: Physics of the extreme on a table top, Phys. Today **51**, 22 (1998)

13.44 R. Clady, G. Coustillier, M. Gastaud, M. Sentis, P. Spiga, V. Tcheremiskine, O. Uteza, L. D. Mikheev, V. Mislavskii, J. P. Chambaret, G. Chériaux: Architecture of a blue high contrast multi-terawatt ultrashort laser, Appl. Phys. B **82**, 347 (2006)

13.45 A. J. Verhoef, J. Seres, K. Schmid, Y. Nomura, G. Tempea, L. Veisz, F. Krausz: Compression of the pulses of a TISapphire laser system to 5 fs at 0.2 terrawatt level, Appl. Phys. B **82**, 513 (2006)

13.46 R. Butkus, R. Danielius, A. Dubietis, A. Piskarskas, A. Stabinis: Progress in chirped pulse optical parametric amplifiers, Appl. Phys.B **79**, 693 (2004)

13.47 E. Riedle, M. Beutler, S. Lochbrunner, J. Piel, S. Schenk, S. Spörlein, W. Zinth: Generation of 10–50 fs pulses tunable through all of the visible and the NIR, Appl. Phys. B **71**, 457 (2000)

13.48 T. Wilhelm, J. Piel, E. Riedle: Sub-20-fs pulses tunable across the visible from a blue-pumped single pass noncollinear parametric converter, Opt. Lett. **22**, 1494 (1997)

13.49 J. Piel, M. Beutler, E. Riedle: 20–50-fs pulses tunable across the near infrared from a blue-pumped noncollinear parametric amplifier, Opt. Lett. **25**, 180 (2000)

13.50 Y. Stepanenko, C. Radzewicz: Multipass noncollinear optical parametric amplifier for femtosecond pulses, Opt. Express **14**, 779 (2006)

13.51 S. Witte, R. Th. Zinkstock, W. Hogervorst, K. S. E. Eikema: Generation of few cycle terawatt light pulses using optical parametric chirped pulse amplification, Opt. Express **13**, 4903 (2005)

13.52 P. Baum, S. Lochbrunner, E. Riedle: Generation of tunable 7 fs ultraviolet pulses, Appl. Phys. B **79**, 1027 (2004)

13.53 I. Z. Kozma, P. Baum, St. Lochbrunner, E. Riedle: Widely tunable sub-30 fs ultraviolet pulses by chirped sum frequency mixing, Opt. Express **23**, 3110 (2003)

13.54 V. Wong, I. A. Walmsley: Characterization of the electric field of ultrashort optical pulses, JOSA B **13**, 2453 (1996)

13.55 C. Iaconis, I. A. Walmsley: Spectral phase interferometry for direct electric field reconstruction of ultrashort optical pulses, Opt. Lett. **23**, 792 (1998)

13.56 P. Baum, St. Lochbrunner, E. Riedle: Zero-additional phase SPIDER: Full characterization of visible and sub-20 fs ultraviolet pulses, Opt. Lett. **29**, 210 (2004)

13.57 G. Stilbenz, C. Roper, Ch. Lienau, Ch. Warmuth, A. S. Wyatt, I. A. Walmsley, G. Steinmeyer: Advanced methods for the characterization of few-cycle light pulses: A comparison, Appl. Phys. B **83**, 511 (2006)

13.58 Y. He, M. Hippler, M. Quack: High-resolution cavity ring-down absorption spectroscopy of nitrous oxide and chloroform using a near infrared cw diode laser, Chem. Phys. Lett. **289**, 527 (1998)

13.59 M. D. Levenson, B. A. Baldus, T. G. Spence, C. C. Harb, J. S. Harris, R. N. Zare: Optical heterodyne detection in cavity ringdown spectroscopy, Chem. Phys. Lett. **290**, 335 (1998)

13.60 G. Berden, R. Peeters, G. Meijer: Cavity ringdown spectroscopy: Experimental schemes and applications, Annu. Rev. Phys. Chem. **19**, 565 (2000)

13.61 K. W. Busch, M. A. Busch (Eds.): *Cavity Ringdown Spectroscopy*, Am. Chem. Soc. Symposium Series 720 (Am. Chem. Soc., Washington 2000)

13.62 G. Berden, R. Peeters, G. Meijer: Cavity ringdown spectroscopy, Rev. Phys. Chem. **19**, 565 (2000)

13.63 D.V. O'Connor, D. Phillips: *Time Correlated Single Photon Counting* (Academic, New York 1984)

13.64 W. Demtröder, W. Stetzenbach, M. Stock, J. Witt: Lifetimes and Franck–Condon factors for the BX system of Na_2, J. Molec. Spectrosc. **61**, 382 (1976)

13.65 See for instance the series: Ultrafast Phenomena, Springer Series in Chemical Physics (Springer, Berlin, Heidelberg 1980–2006)

13.66 W. Fuß, C. Kosmidis, W.A. Schmidt, S.A. Trushin: The lifetime of the perpen-dicular minimum of cis-stilbene observed by dissociative intense laser field ionization, Chem. Phys. Lett. **385**, 423 (2004)

13.67 H. Bitto, J.R. Huber: Molecular quantum beat spectroscopy, Opt. Commun. **80**, 184 (1990)

13.68 T. Baumert, M. Grosser, R. Thalweiler, G. Gerber: Femtosecond time-resolved molecular photoionization: the Na_2-system, Phys. Rev. Lett. **67**, 3753 (1991)

13.69 E. Seres, J. Seres, F. Krausz, Ch. Spielmann: Generation of coherent soft-X-ray radiation extending far beyond the titanium L edge, Phys. Rev. Lett. **92**, 163002–1 (2004)

13.70 I.D. Abella: Echoes at optical frequencies, Progr. Opt. **7**, 140 (1969)

13.71 S. Mukamel: *Principles of Nonlinear Optical Spectroscopy* (Oxford Univ. Press, Oxford 1995)

13.72 J. Seres, E. Seres, A.J. Verhoef, G. Tempea, C. Streli, P. Wobrauschek, V. Yakovlev, A. Scrinzi, C. Spielmann, F. Krausz: Source of coherent kiloelectronvolt X-rays, Nature **433**, 596 (2005)

13.73 L.R. Kundkar, A.H. Zewail: Ultrafast molecular reaction dynamics in real time, Annu. Rev. Phys. Chem. **41**, 15 (1990)

13.74 T. Brixner, N.H. Damrauer, G. Gerber: Femtosecond quantum control, Adv. Atom. Mol. Opt. Phys. **46**, 1–56 (2001)

13.75 D. Zeidler, S. Frey, K.L. Kompa, M. Motzkus: Evolutionary Algorithms and their applications to optimal control studies, Phys. Rev. A **64**, 023420 (2001)

13.76 M. Bergt, T. Brixner, B. Kiefer, M. Strehle, G. Gerber: Controlling the femtochemistry of $Fe(Co)_5$, J. Phys. Chem. **103**, 10381 (1999)

13.77 W. Wohlleben, J. Degert, A. Monmayrant, B. Chatel, M. Motzkus, B. Girard: Coherent control for spectroscopy and manipulation of biological processes, Chem. Phys. Chem. **6**, 850 (2005)

13.78 T. Witte, K.L. Kompa, M. Motzkus: Femtosecond pulse shaping in the mid-infrared by difference-frequency mixing, Appl. Phys. B **76**, 467 (2003)

13.79 P. Gaspard, I. Burghardt (Eds.): *Chemical reactions and their control on the femtosecond timescale*, Adv. Chem. Phys., Vol. 101 (Wiley, New York 1997)

13.80 A.H. Zewail: Diffraction, crystallography and microscopy beyond 3D: Structural dynamics in space and time, Philos. Trans. Soc. A **364**, 315 (2005)

13.81 V.A. Lobastov, R. Srinivasan, C.-Y. Ruan, J.S. Feenstra, A.H. Zewail: Ultrafast diffraction of transient molecular structure in radiationles transitions, J. Phys. Chem. A **105**, 11159 (2001)

13.82 A. Zewail: *Nobel Lecture in Chemistry 1999* (World Scientifique, Singapore 2000)

13.83 V. Sundström: *Femtochemistry and Femtobiology* (World Scientific, Singapore 1997)

13.84 T. Polivka, V. Sundström: Ultrafast dynamics of carotenoid excited states – From solution to natural and artificial systems, Chem. Rev. **104**, 2021 (2004)

13.85 W. Wohlleben, T. Buckup, J.L. Herek, R.J. Cogdel, M. Motzkus: Multichannel carotenoid deactivation in photosynthetic light harvesting as identified by an evolutionary target analysis, Biophys. J. **85**, 4432 (2003)

13.86 A. Vogel, J. Noack, G. Hüttman, G. Paltauf: Mechanisms of femtosecond laser nanosurgery of cells and tissues, Appl. Phys. B **81**, 1015 (2005)

13.87 L. Thrane, M.H. Frosz, D. Levitz, Th.M. Jorgensen, C.B. Andersson, P.H. Hansen, J. Valancuinaite, J. Swartling, St. Andersson-Engels, A. Tycho, H.T. Yura, P.E. Andersen: Extraction of tissue optical properties from optical coherence tomography images for diagnostic purposes, SPIE Proc. **5771**, 139 (2005)

13.88 See for instance: 13th Int. Workshop on Laser Ranging, Washington, Oct. 2002 (Nat. Aeronautics. Space Administration, Goddard Space Flight Center 2003)

13.89 A. Asmann, R. Neuber, P. Rairoux: *Advances in atmospheric remote sensing with LIDAR* (Springer, Berlin, Heidelberg 1997)

13.90 W. Zimmer, M. Rodriguez, L. Wöste: Application perspectives of intense laser pulses in atmospheric diagnostics. In: *Laser in Environmental and Life Sciences*, ed. by P. Hering, J.P. Lay, S. Stry (Springer, Berlin, Heidelberg 2004)

13.91 N. Del Fatti, A. Arbouet, F. Vallee: Femtosecond optical investigation of electron-lattice interactions in an ensemble and a single metal nanoparticle, Appl. Phys. B **84**, 175 (2006)

13.92 G. Rutherford, D. Karnakis, A. Webb, M. Knowles: Optimization of the Laser Drilling Process for Fuel Injection Components, Advanced Laser Application Conference ALAC 2005, Ann Arbor, USA 2005

13.93 N.H. Rizwi: Femtosecond Laser Micromachinig. Current Status and Applications, Riken Rev. **50**, 107 (2003)

13.94 H. Niikura, F. Légaré, R. Hasbani, A.D. Bandrauk, M. Ivanov, D.M. Villeneuve, P.B. Corkum: Sub-laser-cycle electron pulses for probing molecular dynamics, Nature **417**, 917 (2002)

13.95 J.D. Lindl: *Inertial Confinement Fusion* (Springer, Berlin, Heidelberg 1998)

13.96 B.M. Hegelich, B. Albright, P. Audebert, A. Blazevic, E. Brambrink, J. Cobble, T. Cowan, J. Fuchs, J.C. Gauthier, C. Gautier, M. Geissel, D. Habs, R. Johnson, S. Karsch, A. Kemp, S. Letzring, M. Roth, U. Schramm, J. Schreiber, K.J. Witte, J.C. Fernandez: Spectral properties of laser accelerated mid-Z MeV/u ion beams, Phys. Plasmas **12**(5), 056314 (2005)

13.97 P. Corkum: Attosecond imaging: Asking a molecule to paint a self-portrait. In: *Max Born, A Celebration*, ed. by W. Sandner (Max-Born Institut, Berlin Adlershof 2004)

13.98 G. G. Paulus, F. Lindner, H. Walther, A. Baltuska, E. Goulielmakis, M. Lezius, F. Krausz: Measurement of the phase of few-cycle laser pulses, Phys. Rev. Lett. **91**, 253004 (2003)

13.99 H. Edner, K. Fredriksson, A. Sunesson, S. Svanberg, L. Unéus, W. Wendt: Mobile remote sensing system for atmospheric monitoring, Appl. Opt. **26**, 4330 (1987)

13.100 G. Fioccio, L. D. Smullin: Detection of scattering layers in the upper atmosphere (60–140 km) by optical radars, Nature **199**, 1275 (1963)

13.101 R. M. Shotland: Some observation of the vertical profile of water vapour by a laser optical radar, Proc. 4th Symp. Remote Sens. Envir. (Univ. Mich. Ann Arbor 1966) 273

13.102 K. W. Rothe, U. Brinkman, H. Walther: Applications of tuneable dye lasers to air pollution detection: Measurements of atmospheric NO_2 concentrations by differential absorption, Appl. Phys. **3**, 115 (1974)

13.103 W. B. Grant, R. D. Hake Jr., E. M. Liston, R. C. Robbins, E. K. Proctor Jr.: Calibrated remote measurements of NO_2 using differential absorption backscattering technique, App. Phys. Lett. **24**, 550 (1974)

13.104 R. M. Measures: *Laser Remote Sensing: Fundamentals and Applications* (Wiley, New York 1984)

13.105 R. M. Measures (Ed.): *Laser Remote Chemical Analysis* (Wiley Interscience, New York 1988)

13.106 M. Sigrist (Ed.): *Air Pollution Monitoring with Optical Techniques* (Wiley, New York 1993)

13.107 C. Weitkamp (Ed.): *LIDAR: Range-Resolved Optical Remote Sensing of the Atmosphere* (Springer, Berlin, Heidelberg 2005)

13.108 T. Fujii, T. Fukuchi (Eds.): *Laser Remote Sensing* (CRC, Boca Raton 2005)

13.109 C. Nagasawa, N. Sugimoto (Eds.): Proc. 23rd International Laser Radar Conference, Nara, Japan 24–28 July 2006

13.110 S. Svanberg: Differential absorption LIDAR. In: *Air Pollution Monitoring with Optical Techniques*, ed. by M. Sigrist (Wiley, New York 1993) Chap. 3

13.111 S. Svanberg: Geophysical gas monitoring using optical techniques: Volcanoes, geothermal fields and mines, Opt. Lasers Eng. **37**, 245 (2002)

13.112 S. Svanberg: Fluorescence spectroscopy and imaging of LIDAR targets. In: *Laser Remote Sensing*, ed. by T. Fujii, T. Fukuchi (CRC, Boca Raton 2005) Chap. 6

13.113 P. Weibring, H. Edner, S. Svanberg: Versatile mobile LIDAR system for environmental monitoring, Appl. Opt. **42**, 3583 (2003)

13.114 W. Happer, G. MacDonald, C. Max, F. J. Dyson: Atmospheric turbulence compensation by resonant optical backscattering from the sodium layer in the upper atmosphere, J. Opt. Soc. Am. A **11**, 263 (1994)

13.115 C. L. Korb, B. M. Gentry, C. Y. Weng: The edge technique: Theory and application to the LIDAR measurement of atmospheric winds, Appl. Opt. **31**, 4202 (1992)

13.116 C. Flesia, C. Korb: Theory of the double-edged molecular technique for Doppler LIDAR wind measurements, Appl. Opt. **38**, 432 (1999)

13.117 J. T. Houghton (Ed.): *Climate Change 2001 – The Scientific Basis* (Cambridge Univ. Press, Cambridge 2001)

13.118 M. Sjöholm, P. Weibring, H. Edner, S. Svanberg: Atomic mercury flux monitoring using and optical parametric oscillator based LIDAR system, Opt. Express **12**, 551 (2004)

13.119 R. Grönlund, M. Sjöholm, P. Weibring, H. Edner, S. Svanberg: Elemental mercury emissions from chlor-alkali plants measured by LIDAR techniques, Atmos. Environ. **39**, 7474 (2005)

13.120 P. Weibring, J. Swartling, H. Edner, S. Svanberg, T. Caltabiano, D. Condarelli, G. Cecchi, L. Pantani: Optical monitoring of volcanic sulphur dioxide emissions – Comparison between four different remote sensing techniques, Opt. Lasers Eng. **37**, 267 (2002)

13.121 P. Weibring, Ch. Abrahamsson, J. N. Smith, H. Edner, S. Svanberg: Multi-component chemical analysis of gas mixtures using a continuously-tuneable LIDAR system, Appl. Phys. **79**, 525 (2004)

13.122 J. Sandsten, H. Edner, S. Svanberg: Gas visualization of industrial hydrocarbon emissions, Opt. Express **12**, 1443 (2004)

13.123 P. Weibring, Th. Johansson, H. Edner, S. Svanberg, B. Sundnér, V. Raimondi, G. Cecchi, L. Pantani: Fluorescence LIDAR imaging of historical monuments, Appl. Opt. **40**, 6111 (2001)

13.124 H. Edner, J. Johansson, S. Svanberg, E. Wallinder: Fluorescence LIDAR multicolor imaging of vegetation, Appl. Opt. **33**, 2471 (1994)

13.125 M. Andersson, H. Edner, J. Johansson, P. Ragnarson, S. Svanberg, E. Wallinder: Remote Monitoring of vegetation by spectral measurements and multi-colour fluorescence imaging, Proc. ISPRS Symposium (Val d'Isère 1994) 835

13.126 V. Raimondi, G. Cecchi, L. Pantani, R. Chiari: Fluorescence LIDAR monitoring of historical buildings, Appl. Opt. **37**, 1089 (1998)

13.127 D. Lognoli, G. Cecchi, L. Pantani, V. Raimondi, R. Chiari, Th. Johansson, P. Weibring, H. Edner, S. Svanberg: Fluorescence LIDAR imaging of the Parma cathedral and baptistery, Appl. Phys. B **76**, 1 (2003)

13.128 R. Grönlund, J. Hällström, S. Svanberg, K. Barup: Fluorescence LIDAR multispectral imaging of historical monuments – Övedskloster, a Swedish case study, Proceedings of Lacona VI, Vienna, Austria 21-25 September 2005, ed. by J. Nimmricher, W. Kautek, M. Schreiner (Springer, Berlin 2007)

13.129 R. Grönlund, M. Lundqvist, S. Svanberg: Remote imaging laser-induced breakdown spectroscopy and laser-induced fluorescence spectroscopy using nanosecond pulses from a mobile LIDAR system, Appl. Spectrosc. **60**, 853 (2006)

13.130 R. Grönlund, J. Hällström, A. Johansson, L. Palombi, D. Lognoli, V. Raimondi, G. Cecchi, K. Barup, C. Conti, O. Brandt, B. Santillo Frizell, S. Svanberg: Laser-induced fluorescence for assessment of cultural heritage. In: *Laser Remote Sensing*, ed. by T. Fujii, T. Fukuchi (CRC, Boca Raton 2005) p. 723

13.131 J. Hällström, K. Barup, R. Grönlund, A. Johansson, S. Svanberg, G. Cecchi, D. Lognoli, L. Palombi, V. Raimondi, C. Conti: Remote monitoring of conservation status of the Coliseum by laser-induced fluorescence imaging

13.132 F. E. Hoge: Ocean and terrestrial LIDAR measurements. In: *Laser Remote Chemical Analysis*, ed. by R. M. Measures (Wiley Interscience, New York 1988) p. 409

13.133 H. Amann: Laser spectroscopy for monitoring and research in the ocean, Phys. Scr. T **78**, 68 (1998)

13.134 C. E. Brown, M. F. Fingas: Review of the development of laser fluorosensors for oil spill applications, Mar. Pollut. Bull. **47**, 477 (2003)

13.135 I. Moya, L. Camenen, G. Latouche, C. Mauxion, S. Evain, Z. G. Cerovic: An instrument for the measurement of sunlight excited plant fluorescence, Photosynthesis **42**, 65 (1998)

13.136 R. Grönlund, M. Lundqvist, S. Svanberg: Remote imaging laser-induced break-down spectroscopy and remote cultural heritage ablative cleaning, Opt. Lett. **30**, 2882 (2005)

13.137 L. J. Radziemski, T. R. Loree, D. A. Cremers, N. M. Hoffman: Time-resolved laser-induced break-down spectroscopy of aerosols, Anal. Chem. **55**, 1246 (1983)

13.138 S. Palanco, J. M. Baena, J. J. Laserna: Open-path laser-induced plasma spectrometry for remote analytical measurements on solid surfaces, Spectrochim. Acta B **57**, 591 (2002)

13.139 R. C. Wiens, S. Maurice, D. A. Cremers, S. Chevrel: The applicability of laser-induced break down spectroscopy (LIBS) to Mars exploration, Lunar Planet. Sci. **XXXIV**, 1646 (2003)

13.140 K. Kohse-Höinghaus, J. B. Jeffries (Eds.): *Applied Combustion Diagnostics* (Taylor Francis, New York 2002)

13.141 J. Bood, M. Aldén: Development of a picosecond Lidar system for combustion diagnostics, to appear

13.142 R. Berg, O. Jarlman, S. Svanberg: Medical transillumination imaging using short-pulse diode lasers, Appl. Opt. **32**, 574 (1993)

13.143 C. af Klinteberg, A. Pifferi, S. Andersson-Engels, R. Cubeddu, S. Svanberg: In vivo absorption spectroscopy of tumor sensitizers using femtosecond white light, Appl. Opt. **44**, 2213 (2005)

13.144 Ch. Abrahamsson, T. Svensson, S. Svanberg, S. Andersson-Engels, J. Johansson, S. Folestad: Time and wavelength resolved spectroscopy of turbid media using light continuum generated in a crystal fibre, Opt. Express **12**, 4103 (2004)

13.145 C. Abrahamsson, J. Johansson, S. Andersson-Engels, S. Svanberg, S. Folestad: Time-resolved NIR spectroscopy for quantitative analysis of intact pharmaceutical tablets, Anal. Chem. **77**, 1055 (2005)

13.146 M. Sjöholm, G. Somesfalean, J. Alnis, S. Andersson-Engels, S. Svanberg: Analysis of gas dispersed in scattering solids and liquids, Opt. Lett. **26**, 16 (2001)

13.147 L. Persson, H. Gao, M. Sjöholm, S. Svanberg: Diode laser absorption spectroscopy for studies of gas exchange in fruits, Lasers Opt. Eng. **44**, 687 (2006)

13.148 G. Somesfalean, M. Sjöholm, J. Alnis, C. af Klinteberg, S. Andersson-Engels, S. Svanberg: Concentration measurement of gas imbedded in scattering media employing time and spatially resolved techniques, Appl. Optics **41**, 3538 (2002)

13.149 M. Andersson, L. Persson, M. Sjöholm, S. Svanberg: Spectroscopic studies of wood-drying processes, Opt. Express **14**, 3641 (2006)

13.150 L. Persson, K. Svanberg, S. Svanberg: On the potential for human sinus cavity diagnostics using diode laser gas spectroscopy, Appl. Phys. B **82**, 313 (2006)

13.151 S. Svanberg: Environmental and medical applications of photonic interactions, Phys. Scr. **T110**, 39 (2004)

13.152 S. Svanberg: *Atomic and Molecular Spectroscopy – Basic Aspects and Practical Applications*, 4 edn. (Springer, Berlin, Heidelberg 2004)

13.153 D. Strickland, G. Mourou: Compression of amplified chirped optical pulses, Opt. Commun. **56**, 219 (1985)

13.154 A. Braun, G. Korn, X. Liu, D. Du, J. Squier, G. Mourou: Self-channelling of high-peak-power femtosecond laser pulses in air, Opt. Lett. **20**, 73 (1995)

13.155 J. Kasparian, M. Rodrigues, G. Mejean, J. Yu, G. Salmon, H. Wille, R. Bourajou, S. Frey, Y.-B. Andre, A. Mysyrowicz, R. Sauerbrey, J. P. Wolf, L. Wöste: White-light filaments for atmospheric analysis, Science **301**, 61 (2003)

13.156 K. Stelmaszczyk, P. Rohwetter, G. Méjean, J. Yu, E. Salmon, J. Kasparian, R. Ackermann, J.-P. Wolf, L. Wöste: Long-distance remote laser-induced breakdown using filamentation in air, Appl. Phys. Lett. **85**, 3977–3979 (2004)

14. Quantum Optics

Quantum optics is the study of the quantum theory of light at low energies and interactions with bound electronic systems. We discuss physically achievable states of the electromagnetic field, including squeezed states and single photons states, as well as schemes by which they may be generated and measured. Measured systems are necessarily open systems and we discuss how dissipation, noise and decoherence is treated in quantum optics in terms of Markov master equations, quantum trajectories and quantum stochastic differential equations. Quantum optics has recently proved a valuable test-bed to implement new communication protocols such as teleportation and quantum information processing and we discus some of these new schemes including ion traps and linear optics quantum computing.

14.1 Quantum Fields

Quantum optics is the study of the quantum theory of light at low energies. It is a special case of quantum electrodynamics (QED) for the electromagnetic field with frequencies ranging from microwave to ultraviolet and electrons bound in atomic systems. We consider first of all the free electromagnetic field described classically by the vector potential $A(x, t)$. As this obeys the wave equation it may be expanded in terms of plane wave states with two orthogonal transverse polarisations. If we assume the field is bound in a box of volume V, with Dirichlet boundary conditions, we find that

$$A(x,t) = \sum_{n,\nu}\sqrt{\frac{\hbar}{2\epsilon_o\omega_n V}} \times e_{n,\nu}\left(e^{i(k_n.x-\omega_{nt})}\alpha_{n,\nu} + e^{-i(k_n.x-\omega_n t)}\alpha^*_{n,\nu}\right) , \quad (14.1)$$

where $e_{n,\nu}$ are two orthonormal polarisation vectors ($\nu = 1, 2$) which satisfy $k_n \cdot e_{n,\nu} = 0$, as required for a transverse field, and the frequency is given by the dispersion relation $\omega_n = c|k_n|$. The positive- and negative-frequency Fourier amplitudes are, respectively, $\alpha_{n,\nu}$ and $\alpha^*_{m,\nu}$. The corresponding electric field is given by

$$E(x,t) = i\sum_{n,\nu}\sqrt{\frac{\hbar\omega_n}{2\epsilon_o V}} \times e_{n,\nu}\left(e^{i(k_n.x-\omega_n t)}\alpha_{n,\nu} - e^{-i(k_n.x-\omega_n t)}\alpha^*_{n,\nu}\right) . \quad (14.2)$$

Canonical quantisation [14.1] is now carried out by promoting these Fourier amplitudes to the operators $\alpha_{n,\nu} \mapsto a_{n,\nu}$, $\alpha^*_{n,\nu} \mapsto a^\dagger_{n,\nu}$ with bosonic commutation relations

$$[a_{n,\nu}, a^\dagger_{n',\nu'}] = \delta_{\nu\nu'}\delta_{nn'} \quad (14.3)$$

with all other commutations relations zero (we are assuming Coulomb gauge quantisation). The canonical quantisation procedure then gives the Hamiltonian for the free field as

$$H = \sum_k \hbar\omega_k a^\dagger_k a_k , \quad (14.4)$$

where the subscript k stands for both the wave number index, n and polarisation index ν. We have explicitly omitted the constant shift due to the vacuum energy of $\hbar\omega_k/2$ per mode. Typically we now take the continuum limit by

$$\frac{1}{\sqrt{V}}\sum_k \mapsto \frac{1}{(2\pi)^{3/2}}\int \mathrm{d}^3 k a_\nu(\boldsymbol{k}) . \tag{14.5}$$

The canonical commutation relations then take the form

$$\left[a_\nu(\boldsymbol{k}), a^\dagger_{\nu'}(\boldsymbol{k}')\right] = \delta_{\nu\nu'}\delta^3(\boldsymbol{k}-\boldsymbol{k}') . \tag{14.6}$$

In the full theory of QED, the interaction between the electromagnetic field and charged matter is described by coupling between the vector potential and the Dirac spinor field. In quantum optics we only need the low-energy (nonrelativistic) limit of this interaction. This is given by the minimal coupling Hamiltonian [14.2]

$$H = \frac{1}{2m}(\boldsymbol{p}-e\boldsymbol{A})^2 + eV(\boldsymbol{x}) + H_{\text{rad}} , \tag{14.7}$$

where $\boldsymbol{p}$ is the momentum operator for a particle of charge e moving in a Coulomb potential $V(\boldsymbol{x})$. The last term, H_{rad}, is the Hamiltonian of the free radiation field given in (14.4). We now use an occupation-number representation in the antisymmetric sector of the many body Hilbert space for the electronic system based on a set of single-particle states $|\phi_j\rangle$, with position probability amplitudes, $\phi_j(\boldsymbol{x})$, which we take as the bound energy eigenstates of the electronic system without radiation. They could for example be the stationary states of a atom, the quasi-bound states of a single Cooper pair on a mesoscopic superconducting metal island, or the bound exciton states of semiconductor quantum dot. We then define the electronic field operators

$$\hat{\psi}(\boldsymbol{x}) = \sum_j c_j\phi_j(\boldsymbol{x}) , \tag{14.8}$$

where the appropriate commutations relations for the antisymmetric sector are the fermionic forms

$$c_k c_l^\dagger + c_l c_k^\dagger = \delta_{kl} , \tag{14.9}$$

$$c_k c_l + c_l c_k = c_k^\dagger c_l^\dagger + c_l^\dagger c_k^\dagger = 0 . \tag{14.10}$$

In the occupation number representation the Hamiltonian may be written as the sum of three terms, $H = H_{\text{el}} + H_{\text{I}} + H_{\text{rad}}$ where the electronic part is given by

$$\begin{aligned} H_{\text{el}} &= \int \mathrm{d}^3\boldsymbol{x}\hat{\psi}^\dagger(\boldsymbol{x})\left[-\frac{\hbar^2}{2m}\nabla^2 + eV(\boldsymbol{x})\right]\hat{\psi}(\boldsymbol{x}) \\ &= \sum_j E_j c_j^\dagger c_j . \end{aligned} \tag{14.11}$$

The interaction part may be written as the sum of two terms $H_{\text{I}} = H_{\text{I},1} + H_{\text{I},2}$, where

$$\begin{aligned} H_{\text{I},1} = &\int \mathrm{d}^3\boldsymbol{x}\hat{\psi}^\dagger(\boldsymbol{x}) \\ &\times\left\{-\frac{e}{2m}[\boldsymbol{A}(\boldsymbol{x})\cdot\boldsymbol{p} + \boldsymbol{p}\cdot\boldsymbol{A}(\boldsymbol{x})]\right\}\hat{\psi}(\boldsymbol{x}) , \end{aligned} \tag{14.12}$$

$$H_{\text{I},2} = \int \mathrm{d}^3\boldsymbol{x}\hat{\psi}^\dagger(\boldsymbol{x})\left\{\frac{e^2}{2m}[\boldsymbol{A}(\boldsymbol{x})^2]\right\}\hat{\psi}(\boldsymbol{x}) . \tag{14.13}$$

Unless we are dealing with very intense fields for which multi-photon processes are important, the second term $H_{\text{I},2}$ may be neglected.

The dominant interaction energy may then be written as

$$H_{\text{I}} = \hbar\sum_{\boldsymbol{k},n,m} g_{\boldsymbol{k},n,m}(b_{\boldsymbol{k}} + b_{\boldsymbol{k}}^\dagger)c_n^\dagger c_m , \tag{14.14}$$

where the interaction coupling constant is

$$\begin{aligned} g_{\boldsymbol{k},n,m} = &-\frac{e}{m}\left(\frac{1}{2\epsilon_0\hbar\omega_k V}\right)^{1/2} \\ &\times\int \mathrm{d}^3\boldsymbol{x}\phi_n^*(\boldsymbol{x})(\mathrm{e}^{\mathrm{i}\boldsymbol{k}.\boldsymbol{x}}\boldsymbol{p})\phi_m(\boldsymbol{x}) . \end{aligned} \tag{14.15}$$

We now proceed by making the *dipole approximation*. The factor $\mathrm{e}^{\mathrm{i}\boldsymbol{k}x}$ varies on a spatial scale determined by the dominant wavelength scale, λ_0, of the field state. At optical frequencies, $\lambda_0 \approx 10^{-6}$ m. However the atomic wave functions, $\phi_n(\boldsymbol{x})$ vary on a scale determined by the Bohr radius, $a_0 \approx 10^{-11}$ m. Thus we may remove the oscillatory exponential from the integral and evaluate it at the position of the atom $\boldsymbol{x} = \boldsymbol{x}_0$. Using the result

$$[\boldsymbol{p}^2, \boldsymbol{x}] = -\mathrm{i}2\hbar\boldsymbol{p} \tag{14.16}$$

we can write the interaction in terms of the atomic dipole moments

$$\begin{aligned} &\int \mathrm{d}^3\boldsymbol{x}\phi_n^*(\boldsymbol{x})(\mathrm{e}^{\mathrm{i}\boldsymbol{k}.\boldsymbol{x}}\boldsymbol{p})\phi_m(\boldsymbol{x}) \\ &= \mathrm{i}\frac{m}{e}\omega_{nm}\int \mathrm{d}^3\boldsymbol{x}\phi_n^*(\boldsymbol{x})(e\boldsymbol{x})\phi_m(\boldsymbol{x}) , \end{aligned} \tag{14.17}$$

where $\omega_{nm} = (E_n - E_m)/\hbar$.

In the interaction picture the interaction Hamiltonian becomes explicitly time dependent,

$$\begin{aligned} \tilde{H}_{\text{I}}(t) = \hbar &\sum_{\boldsymbol{k},n,m} g_{\boldsymbol{k},n,m}\left(b_{\boldsymbol{k}}\mathrm{e}^{-\mathrm{i}\omega(\boldsymbol{k})t} + b_{\boldsymbol{k}}^\dagger\mathrm{e}^{\mathrm{i}\omega(\boldsymbol{k})t}\right) \\ &\times c_n^\dagger c_m\,\mathrm{e}^{\mathrm{i}\omega_{nm}t} , \end{aligned} \tag{14.18}$$

where the tilde indicates that we are in the interaction picture. If the field is in a state for which the dominant frequency is such that $\omega(\boldsymbol{k}_0) \approx \omega_{nm}$, the field is resonant with a particular atomic transition and we may neglect terms rotating at the very high frequency $\omega(\boldsymbol{k}) + \omega_{nm}$. This is known as the *rotating wave approximation*. This assumes that the field strength is not too large and further that the state of the field does not vary rapidly on a time scale of ω_{nm}^{-1}, i.e. we ignore fields of very fast strong pulses. As a special case we assume that the field is resonant (or near-resonant) with a single pair of levels with $E_2 > E_1$. The interaction picture Hamiltonian in the dipole and rotating wave approximation is then given by

$$\tilde{H}_\mathrm{I} = \hbar \sum_{\boldsymbol{k}} c_1^\dagger c_2 b_{\boldsymbol{k}}^\dagger g_{\boldsymbol{k}} \mathrm{e}^{-\mathrm{i}(\omega(\boldsymbol{k})-\omega_{21})t} + \mathrm{h.c}\,, \tag{14.19}$$

where

$$g_{\boldsymbol{k}} = -\mathrm{i}\,[2\hbar\epsilon_0\omega(\boldsymbol{k})V]^{-1/2}\,\mu_{21}\,\mathrm{e}^{\mathrm{i}\boldsymbol{k}.\boldsymbol{x}_0} \tag{14.20}$$

and

$$\mu_{21} = \omega_a \langle \phi_n | e\boldsymbol{x} | \phi_m \rangle \tag{14.21}$$

with $\omega_a = \omega_2 - \omega_1$.

It is conventional to describe the operator algebra of a two-level system in terms of a peusdo-spin representation by noting that the Pauli operators may be defined as

$$\sigma_z = c_2^\dagger c_2 - c_1^\dagger c_1\,, \tag{14.22}$$

$$\sigma_x = c_2^\dagger c_1 + c_1^\dagger c_2\,, \tag{14.23}$$

$$\sigma_y = -i\left(c_2^\dagger c_1 - c_1^\dagger c_2\right), \tag{14.24}$$

$$\sigma_+ = \sigma_-^\dagger = c_2^\dagger c_1\,. \tag{14.25}$$

The operators $s_\alpha = \sigma_\alpha/2$ (with $\alpha = x, y, z$) then obey the SU(2) algebra for a spin-half system. In terms of these operators we may write the total Hamiltonian for the system of field plus atom in the dipole and rotating wave approximation as

$$H = \sum_{\boldsymbol{k}} \hbar\omega(\boldsymbol{k}) b_{\boldsymbol{k}}^\dagger b_{\boldsymbol{k}} + \frac{\hbar\omega_\mathrm{a}}{2}\sigma_z + \hbar \sum_{\boldsymbol{k}} g_{\boldsymbol{k}} b_{\boldsymbol{k}} \sigma_+ + \mathrm{h.c.} \tag{14.26}$$

If we further restrict the state of the field to include only a single mode, perhaps using a high-Q optical resonator, we arrive at the *Jaynes–Cummings* Hamiltonian,

$$H = \hbar\omega_0 b^\dagger b + \frac{\hbar\omega_a}{2}\sigma_z + \hbar\left(g b \sigma_+ + g^* b^\dagger \sigma_-\right) \tag{14.27}$$

coupling a single harmonic oscillator degree of freedom to a two-level system, which might well be called the standard model of quantum optics [14.3]. The coupling constant g can vary from a few kHz to many MHz. This needs to be compared with the linewidth of the cavity mode, κ, as well as the spontaneous emission rate, γ, of the excited state both of which we have thus far neglected but will discuss in Sect. 14.4. If we can arrange for $g > \kappa, \gamma$ we are in the *strong coupling limit*. This is typically achieved by using very small cavity mode volumes. The resulting physical system is often referred to as *cavity QED*. An example is provided by the experiment of *Aoki* et al. [14.4] in which a cesium atom interacts with the toroidal whispering gallery mode of a micro-resonator as it falls under the action of gravity from a magneto-optical trap. The atomic resonance at is the $6\mathrm{S}_{1/2}$; $F = 4 \rightarrow 6\mathrm{P}_{3/2}$; $F' = 5$ transition in cesium. A coupling constant as large as $g/2\pi = 50$ Mz was achieved. The corresponding cavity and atomic decay rates were $(\kappa, \gamma)/2\pi = (17.9 \pm 2.8, 2.6)$ MHz.

14.2 States of Light

The simplest state of light is the ground state, or vacuum state of the free-field Hamiltonian in (14.4) defined by the tensor product of all states $|0\rangle_k$ such that $a_k|0\rangle_k = 0$. The excited states of the free field are the photon number states defined by the tensor product $|\boldsymbol{n}\rangle = \Pi_k |n_k\rangle_k$ where n_k is the kth component of $\boldsymbol{n}$ and

$$a_k^\dagger a_k |\boldsymbol{n}\rangle = n_k |\boldsymbol{n}\rangle\,. \tag{14.28}$$

These states are defined in terms of the plane-wave expansion of the free field in (14.1) and are thus very delocalised. We can of course expand the field in a completely different set of spatiotemporal modes and define photon-number states accordingly. The physical interpretation of the number states depends on what mode expansion we are using.

The interaction Hamiltonian for the classical electromagnetic field coupled to a classical current four-density $j^\mu(\boldsymbol{x}, t)$ is

$$H_\mathrm{c} = \int \mathrm{d}^3 x e A_\mu(\boldsymbol{x}, t) j^\mu(\boldsymbol{x}, t)\,. \tag{14.29}$$

In quantum optics this may be approximated by treating the current as classical and using the quantised form of the vector potential. The resulting interaction

Hamiltonian takes the form

$$\tilde{H}_{\mathrm{I}}(t) = \mathrm{i}\hbar \sum_{\boldsymbol{k}} \left(E_{\boldsymbol{k}}^*(t) a_{\boldsymbol{k}} \mathrm{e}^{-\mathrm{i}\omega(\boldsymbol{k})t} - E_{\boldsymbol{k}}(t) a_{\boldsymbol{k}}^\dagger \mathrm{e}^{\mathrm{i}\omega(\boldsymbol{k})t} \right) . \tag{14.30}$$

If the source is a monochromatic plane wave at frequency $\omega_0 = c|\boldsymbol{k}|$ we can write this as

$$H_{\mathrm{I}}(t) = \mathrm{i}\hbar(\epsilon a^\dagger - \epsilon^* a) , \tag{14.31}$$

where $a = a_{\boldsymbol{k}_0}$. If the free field starts in the vacuum state at $t = 0$, the state at a later time is a vacuum state for all modes $\boldsymbol{k} \neq \boldsymbol{k}_0$ and for mode $\boldsymbol{k}_0$, the state is

$$|\psi(t)\rangle = \exp[t(\epsilon a^\dagger - \epsilon^* a)]|0\rangle . \tag{14.32}$$

Expanding the exponential and we see that this state may be written in the number state basis as

$$|\psi(t)\rangle = \mathrm{e}^{-|\epsilon t|^2/2} \sum_{n=0}^{\infty} \frac{(\epsilon t)^n}{\sqrt{n!}} |n\rangle_{\boldsymbol{k}_0} . \tag{14.33}$$

We are thus led to define the *coherent states* of a single mode of the field as

$$|\alpha\rangle = D(\alpha)|0\rangle , \tag{14.34}$$

where we have defined the *displacement operator*

$$D(\alpha) = \mathrm{e}^{\alpha a^\dagger - \alpha^* a} \tag{14.35}$$

the name of which is derived from the result

$$D^\dagger(\alpha) a D(\alpha) = a + \alpha . \tag{14.36}$$

Using this last result it is easy to see that $a|\alpha\rangle = \alpha|\alpha\rangle$.

Consider the field operator evaluated for convenience at $\boldsymbol{x} = 0$ and for now ignore the polarisation degree of freedom. We can express this in terms of sine and cosine amplitudes at a carrier frequency ω_{c} as

$$A(t) = A_1(t)\cos(\omega_{\mathrm{c}} t) - A_2(t)\sin(\omega_{\mathrm{c}} 0t) , \tag{14.37}$$

where in the interaction picture

$$A_1(t) = \frac{1}{2} \sum_k \sqrt{\frac{\hbar}{2\epsilon_0 \omega_k V}} \times \left[a_k(t) \mathrm{e}^{-\mathrm{i}(\omega_k - \omega_{\mathrm{c}})t} + a_k^\dagger(t) \mathrm{e}^{\mathrm{i}(\omega_k - \omega_{\mathrm{c}})t} \right] , \tag{14.38}$$

$$A_2(t) = \frac{-\mathrm{i}}{2} \sum_k \sqrt{\frac{\hbar}{2\epsilon_0 \omega_k V}} \times \left[a_k(t) \mathrm{e}^{-\mathrm{i}(\omega_k - \omega_{\mathrm{c}})t} - a_k^\dagger(t) \mathrm{e}^{\mathrm{i}(\omega_k - \omega_{\mathrm{c}})t} \right] . \tag{14.39}$$

We are thus led to define the *quadrature phase operators* for each mode of the field as

$$X = (a + a^\dagger) , \tag{14.40}$$

$$Y = -\mathrm{i}(a - a^\dagger) . \tag{14.41}$$

These operators have the commutation relation $[X, Y] = 2\mathrm{i}$, which is isomorphic to the Heisenberg–Weyl algebra of canonical position and momentum operators and so there is an equivalent uncertainty principle, $\Delta X \Delta Y \geq 1$, where $(\Delta A)^2 \equiv \langle A^2 \rangle - \langle A \rangle^2$. States that satisfy the lower bound in this relation are called minimum-uncertainty states. The vacuum state and coherent state are minimum-uncertainty states with $\Delta X = \Delta Y$. Minimum-uncertainty states for which the quadrature phase operator uncertainties are not equal are called *squeezed states*.

Single-mode squeezed states are generated from the vacuum state by a unitary transformation defined by

$$S^\dagger(\xi)(\tilde{X} + \mathrm{i}\tilde{Y})S(\xi) = \tilde{X}\mathrm{e}^r + \mathrm{i}\tilde{Y}\mathrm{e}^{-r} , \tag{14.42}$$

where $\tilde{X} + \mathrm{i}\tilde{Y} = (X + \mathrm{i}Y)\mathrm{e}^{-\mathrm{i}\phi}$. Thus the canonical pair $(\tilde{X}, \tilde{Y})$ are related by a rotation to the original canonical pair (X, Y). Clearly the uncertainty in the rotated canonical coordinates are $\Delta\tilde{X} = \mathrm{e}^r$, $\Delta\tilde{Y} = \mathrm{e}^{-r}$. The squeezed vacuum state is then defined as $|0, \xi\rangle = S(\xi)|0\rangle$, where

$$S(\xi) = \exp\left[\frac{1}{2}(\xi (a^\dagger)^2 - \xi^* a^2) \right] , \tag{14.43}$$

where $\xi = r\mathrm{e}^{2\mathrm{i}\phi}$. The corresponding transformation of the complex amplitude operators is

$$S^\dagger(\xi) a S(\xi) = a\cosh r + a^\dagger \mathrm{e}^{2\mathrm{i}\phi} \sinh r . \tag{14.44}$$

We may then add a coherent amplitude by displacing the squeezed vacuum state $|\alpha, \xi\rangle = D(\alpha)|0, r\rangle$.

Of more interest are the two-mode squeezed vacuum states which correlate two modes, a and b, of the field through a generalisation of (14.44),

$$S_2^\dagger(\xi) a S_2(\xi) = a\cosh r + b^\dagger \mathrm{e}^{2\mathrm{i}\phi} \sinh r , \tag{14.45}$$

$$S_2^\dagger(\xi) b S_2(\xi) = b\cosh r + a^\dagger \mathrm{e}^{2\mathrm{i}\phi} \sinh r , \tag{14.46}$$

where $S_2(\xi) = \exp[(\xi a^\dagger b^\dagger - \xi^* ab)]$. The two-mode squeezed vacuum state is then defined as

$$|\mathcal{E}\rangle_{ab} = \exp[(\xi a^\dagger b^\dagger - \xi^* ab)]|0\rangle_a \otimes |0\rangle_b . \tag{14.47}$$

The two-mode squeezed state is an entangled state of two modes of an optical field. To see this we note that, for $\xi = r$, a real number, the expansion in number eigenstate basis for each oscillator is

$$|\mathcal{E}\rangle 2_{ab} = \sqrt{(1-\lambda^2)} \sum_{n=0}^{\infty} \lambda^n |n\rangle_a \otimes |n\rangle_b , \tag{14.48}$$

where $\lambda = \tanh r$ and $a^\dagger a|n\rangle_a = n|n\rangle_a$ and $b^\dagger b|n\rangle_b = n|n\rangle_b$.

The entanglement of this state can be viewed in two ways. Firstly as an entanglement between pairs of canonically conjugate physical quantities with a continuous spectrum, the quadrature phases, and secondly as an entanglement between number and phase. We will only discuss the former case here (see [14.5] for the latter case). We can easily show that this state approximates the entangled state first considered by Einstein, Podolsky and Rosen (an EPR state) in the limit $\lambda \to 1$ or $r \to \infty$. The correlation between quadrature phase in each mode is easily seen by calculating the variances in an appropriate set of joint observables. We first define the quadrature phase operators for the two modes

$$\hat{X}_a = a + a^\dagger \,, \tag{14.49}$$

$$\hat{Y}_a = -\mathrm{i}(a - a^\dagger) \,, \tag{14.50}$$

$$\hat{X}_b = b + b^\dagger \,, \tag{14.51}$$

$$\hat{Y}_b = -\mathrm{i}(b - b^\dagger) \,, \tag{14.52}$$

with canonical commutation relations, $[\hat{X}_\nu, \hat{Y}_\mu] = 2\mathrm{i}\delta_{\nu,\mu}$. Then

$$\mathrm{Var}(\hat{X}_a - \hat{X}_b) = 2\mathrm{e}^{-2r} \tag{14.53}$$

$$\mathrm{Var}(\hat{Y}_a + \hat{Y}_b) = 2\mathrm{e}^{-2r} \tag{14.54}$$

where $\mathrm{Var}(A) = \langle A^2\rangle - \langle A\rangle^2$ is the variance. Thus in the limit of $r \to \infty$ the state $|\mathcal{E}\rangle$ approaches a simultaneous eigenstate of $\hat{X}_a - \hat{X}_b$ and $\hat{Y}_a + \hat{Y}_b$. This is the analogue of the EPR state with position replaced by the real quadratures $\hat{X}$ and the momentum replaced by the imaginary quadratures, $\hat{Y}$.

The very strong correlations between the quadrature phase operators for the squeezed state can be seen in the diagonal basis of $\hat{X}_a$, $\hat{X}_b$

$$\begin{aligned}\mathcal{E}(x_a, x_b) &= \langle x_a, x_b|\mathcal{E}\rangle \\ &= (2\pi)^{-1/2} \\ &\quad \exp\left[-\frac{e^{2r}}{8}(x_a - x_b)^2 - \frac{e^{-2r}}{8}(x_a + x_b)^2\right]\end{aligned} \tag{14.55}$$

with $\hat{X}_\alpha|x_\alpha\rangle_\alpha = x_\alpha|x_\alpha\rangle_\alpha$ and $|x, y\rangle = |x\rangle_a \otimes |y\rangle_b$ and where $\alpha \to a, b$.

We now turn to multimode states of the field to describe more-complex states. To this end we assume that the only modes that are excited have the same plane polarisation and are all propagating in the in same direction, which we take to be the x-direction. All other modes are in the vacuum state. The positive-frequency components of the electric field for these modes are then

$$E^{(+)}(x, t) = \mathrm{i}\sum_{n=0}^{\infty}\left(\frac{\hbar\omega_n}{2\epsilon_0 V}\right)^{1/2} a_n\,\mathrm{e}^{-\mathrm{i}\omega_n(t-x/c)} \,. \tag{14.56}$$

In ignoring all the other modes, we are implicitly assuming that all our measurements do not respond to the vacuum state, an assumption which is justified below. Let us further assume that all excited modes of this form have frequencies centred on the *carrier frequency* of $\Omega \gg 1$. Then we can approximate the positive-frequency components by

$$E^{(+)}(x, t) = \mathrm{i}\left(\frac{\hbar\Omega_n}{2\epsilon_0 Ac}\right)^{1/2}\sqrt{\frac{c}{L}}\sum_{n=0}^{\infty} a_n\,\mathrm{e}^{-\mathrm{i}\omega_n(t-x/c)} \,, \tag{14.57}$$

where A is a characteristic transverse area. This operator has dimensions of electric field. In order to simplify the dimensions we now define a field operator that has dimensions of $\mathrm{s}^{-1/2}$. Taking the continuum limit we thus define the positive-frequency operator

$$a(x, t) = e^{-\mathrm{i}\Omega(t-x/c)}\frac{1}{\sqrt{2\pi}}\int_{-\infty}^{\infty}\mathrm{d}\omega' a(\omega')\mathrm{e}^{-\mathrm{i}\omega'(t-x/c)} \,, \tag{14.58}$$

where we have made a change of variable $\omega \mapsto \Omega + \omega'$ and used the fact that $\Omega \gg 1$ to set the lower limit of integration to minus infinity, and

$$\left[a(\omega_1), a^\dagger(\omega_2)\right] = \delta(\omega_1 - \omega_2) \,. \tag{14.59}$$

In this form the moment $n(x, t) = \langle a^\dagger(x, t)a(x, t)\rangle$ has units of s^{-1}. As we show in the next section this moment determines the probability per unit time (the count rate) to count a photon at the space–time point (x, t). The field operators $a(t)$ and $a^\dagger(t)$ can be taken to describe the field emitted from the end of an optical cavity, which selects the directionality.

A multimode coherent state is defined by a multimode displacement operator acting on the vacuum $D|0\rangle$, defined implicitly by

$$D^\dagger a(\omega)D = a(\omega) + \alpha(\omega) \,, \tag{14.60}$$

where consistent with proceeding assumptions, $\alpha(\omega)$ is peaked at $\omega = 0$, which corresponds to a carrier frequency $\Omega \gg 1$. The average field amplitude for this state is

$$\langle a(x,t)\rangle = \mathrm{e}^{-\mathrm{i}\Omega(t-x/c)}\frac{1}{\sqrt{2\pi}}\int_{-\infty}^{\infty}\alpha(\omega)\mathrm{e}^{-\mathrm{i}\omega(t-x/c)}$$
$$\equiv \alpha(x,t)\mathrm{e}^{-\mathrm{i}\Omega(t-x/c)} \qquad (14.61)$$

which implicitly defines the average complex amplitude of the field as the Fourier transform of the frequency-dependent displacements $\alpha(\omega)$. We can also calculate the probability per unit time to detect a photon in this state at space-time point (x, t). This is given by $n(x,t) = |\alpha(x,t)|^2$. Note that in this case the second-order moment $\langle a^\dagger(x,t)a(x,t)\rangle$ factorises, a result characteristic of fields with first-order coherence. A coherent state is closest to our intuitive idea of a classical electromagnetic field.

The multimode single-photon state is defined by

$$|1\rangle = \int_{-\infty}^{\infty} \nu(\omega)a^\dagger(\omega)|0\rangle\,. \qquad (14.62)$$

This state has zero average field amplitude but $n(x,t) = |\nu(t-x/c)|^2$, where $\nu(t)$ is the Fourier transform of $\nu(\omega)$. So while the state has zero average field amplitude there is apparently some sense in which the coherence implicit in the superposition of (14.62) is manifest. We discuss in the next section how this coherence is made explicit through fourth-order interference. For this state the function $\nu(\phi)$ is periodic in the phase $\phi = t - x/c$ and it is not difficult to choose a form with a well-defined pulse sequence. However care should be exercised in interpreting these pulses. They do not represent a sequence of pulses each with one photon, rather they represent a single photon coherently superposed over all pulses. Once a photon is counted in a particular pulse, the field is returned to the vacuum state. A single-photon state is a highly nonclassical state with applications to quantum information processing. A review of current efforts to produce such a state may be found in [14.6].

An important example of a multimode two-photon state is

$$|2\rangle_{ab} = \int_{-\infty}^{\infty}\mathrm{d}\omega_1\int_{-\infty}^{\infty}\mathrm{d}\omega_2\alpha(\omega_1,\alpha_2)a^\dagger(\omega_1)b^\dagger(\omega_2)|0\rangle_{ab}\,, \qquad (14.63)$$

where a and b are two distinguishable spatial or polarisation modes. This state is important in the theory of spontaneous parametric down conversion in which a pump photon at frequency 2Ω is absorbed and simultaneously two photons in each of the modes a and b are produced with frequencies ω_1 and ω_2, respectively, such that $\omega_1 + \omega_2 = 2\Omega$, the condition for energy conservation. A good approximation to the state produced can be written as $|0\rangle + \kappa|2\rangle_{ab}$ where κ is related to the down-conversion efficiency [14.7]. As we have shifted all frequencies by the carrier frequency Ω, we can impose energy conservation as $\omega_1 + \omega_2 = 0$, and choose

$$\alpha(\omega_1,\omega_2) = \alpha(\omega_1)\delta(\omega_1+\omega_2)\,. \qquad (14.64)$$

In other words photons in the symmetric side bands around the frequency of Ω are correlated. Normalisation of the state in (14.63) requires

$$\int_{-\infty}^{\infty}\mathrm{d}\omega|\alpha(\omega)|^2 = 1\,. \qquad (14.65)$$

If we ask for the probability per unit time to count a photon from either mode a or b independently it is easily seen to be a constant in time indicating that we will count a photon in either beam at completely random times. However if we count the coincidence correlation rate $C(t,t') = \langle a^\dagger(t)a(t)b^\dagger(t')b(t')\rangle$ we find

$$C(t,t') = |\alpha(t-t')|^2\,, \qquad (14.66)$$

where $\alpha(t)$ is the Fourier transform of $\alpha(\omega)$. The quantum coherence implicit in the correlations of (14.63) appear as a two-photon correlation with a maximum for coincidence counting. To put it another way, the two photon correlation is a *fourth-order interference* effect as C depends on a fourth-order moment of the field amplitude operators.

14.3 Measurement

At optical frequencies, all measurements ultimately derive from how the field interacts with matter, and it is some property of the material which is ultimately measured. For example if the field is directed to a particular kinds of semiconductor material free carriers can be produced, either as electrons or holes, each refereed to as a photoexcitation event. The resulting charge or current flow may then be amplified to produce a measurable electrical signal which is entirely classical and stochastic. There are two special cases. Firstly if we use light of such low intensity that individual photoexcitation events result in a distinct signal pulse. These pulses can then be counted and different counts recorded per count interval. The statistics of this counting process is ultimately conditioned on the photon-number distribution for the quantum state of the incident light. Secondly we may have light of a higher intensity so that individual counting events cannot be resolved. The classical measurement signal is then a fluctuating current. The noise power spectrum of this current is still conditioned on the quantum state of the input light. This last case occurs if a weak signal is first mixed with a strong coherent light beam by a beam splitter and then the reflected and transmitted beams individually give rise to distinct photocurrents on separate detectors. The strong coherent beam is usually called a *local oscillator*. The interference between the signal and local oscillator enables a phase-sensitive measurement of the field to be made. If it is resonant with the carrier frequency, the photocurrent has a direct current (DC) average signal and we refer to the process as homodyne detection. If the local oscillator is detuned from the signal the photocurrents acquire an alternating current (AC) frequency component equal to the detuning between the local oscillator and the signal carrier frequency. This is called heterodyne detection. We now show how these various classical stochastic signals are conditioned on different field quantities.

14.3.1 Photon Counting

If we assume that the interaction between the detector and signal field is weak, the probability of photoexcitation event per unit time may be calculated from the interaction Hamiltonian and Fermi's golden rule. Typically the interaction Hamiltonian, in the interaction picture takes the form of a dipole interaction in the rotating wave approximation

$$\tilde{H}_\mathrm{I} = \hbar\chi(E^{(+)}(\boldsymbol{x},t)\sigma_+(t) + E^{(-)}(\boldsymbol{x},t)\sigma_-(t)) , \tag{14.67}$$

where g is a typical coupling constant, $E^{(\pm)}(\boldsymbol{x},t)$ are the positive- and negative-frequency components of the field at space time point $(\boldsymbol{x},t)$, while $\sigma_\pm(t)$ are the dipole transition operators. We assume the dipole starts in the ground state $|g\rangle$ defined by $\sigma_-|g\rangle = 0$. The probability that the dipole, located at $\boldsymbol{x}$ is excited per unit time, i.e. the rate of excitation is given by perturbation theory (to second order in χ) [14.8] as

$$p_\mathrm{e}(\boldsymbol{x},t) \propto \langle E^{(-)}(\boldsymbol{x},t)E^{(+)}(\boldsymbol{x},t)\rangle . \tag{14.68}$$

Insofar as an excitation event results in a classical signal, the probability to detect a photon per unit time is then simply,

$$p_1(t) = \gamma\langle E^{(-)}(t)E^{(+)}(t)\rangle , \tag{14.69}$$

where $\gamma \propto \chi^2$ and the other factors that depend on the details of the detector (such as area and spectral sensitivity) and its ability to respond to a photoexcitation event with a detectable signal. The counting process is thus seen to be a classical stochastic random variables $\mathrm{d}N(t)$ which only takes the values 0, 1 in a sufficiently small time interval $\mathrm{d}t$. This is a Poisson point process. The classical average of this process is $\mathcal{E}[\mathrm{d}N(t)] = p_1(t)$. If we now consider a count interval of duration T, the probability to count n photoexcitation events in that time is [14.9]

$$p_n(T) = \left\langle : \frac{[\gamma\bar{I}(T)T]^n}{n!} \exp[-\gamma\bar{I}(T)T] : \right\rangle , \tag{14.70}$$

where the operator $\bar{I}$ is defined by

$$\bar{I}(T) = \frac{1}{T}\int_0^T E^{(-)}(t)E^{(+)}(t)\,\mathrm{d}t , \tag{14.71}$$

where : ... : indicates normal ordering, that is the power-series expansion is done by ordering all the negative-frequency components to the right of the positive-frequency components.

If the count rate is too large to resolve individual counting events, we need to consider a photocurrent. We then define the current as

$$i(t) = e\frac{\mathrm{d}n(t)}{\mathrm{d}t} , \tag{14.72}$$

where $n(t) = \sum_{n=0}^{\infty} n p_n(t)$ is the average number of counts in time t. In terms of the point process $\mathrm{d}N(t,\ n(t)) = \int_0^t \mathrm{d}N(t)$ so that the ensemble average of $i(t)$ is the average current

$$\overline{i(t)} = e\frac{\mathrm{d}\mathcal{E}(n(t))}{\mathrm{d}t} = \gamma\langle E^{(-)}(t)E^{(+)}(t)\rangle . \tag{14.73}$$

The stationary two-time correlation function of the current is then the classical ensemble average,

$$G(\tau) = \lim_{t\to\infty} \mathcal{E}\,[i(t+\tau)i(t)] - \mathcal{E}[i(t+\tau)]\mathcal{E}[i(t)] . \tag{14.74}$$

equation

$$\frac{\mathrm{d}a(t)}{\mathrm{d}t} = -\frac{\mathrm{i}}{\hbar}[H_\mathrm{S}, a(t)] - \frac{\gamma}{2}a(t) + \sqrt{\gamma}a_\mathrm{i}(t)\,. \quad (14.97)$$

In this formulation, the external field input to the cavity plays the role of a quantum noise term. If these equations are linear,

$$\frac{\mathrm{d}\boldsymbol{a}}{\mathrm{d}t} = A\boldsymbol{a} - \frac{\gamma}{2}\boldsymbol{a} + \sqrt{\gamma}\boldsymbol{a}_\mathrm{i}(t)\,, \quad (14.98)$$

where

$$\boldsymbol{a}(t) = \begin{pmatrix} a(t) \\ a^\dagger(t) \end{pmatrix}, \quad (14.99)$$

$$\boldsymbol{a}_{\mathrm{i,o}}(t) = \begin{pmatrix} a_{\mathrm{i,o}}(t) \\ a^\dagger_{\mathrm{i,o}}(t) \end{pmatrix}. \quad (14.100)$$

We can Fourier transform both sides and ignore initial-value terms, as we are primarily interested in the stationary statistics to obtain a linear system of algebraic equations for the frequency components of the various fields. Combining this with (14.96) we find that

$$\boldsymbol{a}_\mathrm{o}(\omega) = -\left[A + \left(\mathrm{i}\omega + \frac{\gamma}{2}\right)I\right] \times \left[A + \left(\mathrm{i}\omega - \frac{\gamma}{2}\right)I\right]^{-1} \boldsymbol{a}_\mathrm{i}(\omega)\,. \quad (14.101)$$

In the case of an empty cavity we get an interaction picture defined with respect to some carrier frequency ω_i for the input external field,

$$a_\mathrm{o}(\omega) = \frac{\frac{\gamma}{2} + \mathrm{i}(\omega - \delta)}{\frac{\gamma}{2} - \mathrm{i}(\omega - \delta)} a_\mathrm{i}(\omega)\,, \quad (14.102)$$

where $\delta = \omega_\mathrm{c} - \omega_\mathrm{i}$. In other words there is simply a frequency-dependent phase shift of each field mode from input to output. Suppose we input a single-photon state defined by

$$|\psi\rangle_\mathrm{i} = \int_{-\infty}^{\infty} \nu(\omega) a_\mathrm{i}^\dagger(\omega)|0\rangle\,. \quad (14.103)$$

After interacting with the cavity this state is transformed to

$$|\psi\rangle_\mathrm{o} = \int_{-\infty}^{\infty} \nu(\omega) a_\mathrm{o}^\dagger(\omega)|0\rangle\,. \quad (14.104)$$

It is then easy to see that the probability per unit time to detect a single photon in the output field is

$$n(t) = \left| \int_{-\infty}^{\infty} \mathrm{d}\omega \left(\frac{\frac{\gamma}{2} + \mathrm{i}(\omega - \delta)}{\frac{\gamma}{2} - \mathrm{i}(\omega - \delta)} \right) \nu(\omega) \mathrm{e}^{-\mathrm{i}\omega t} \right|^2, \quad (14.105)$$

which is a delayed and broadened pulse.

If the quantum Langevin equations are nonlinear the standard procedure is to first find the semiclassical steady states. The semiclassical equations result by ignoring the noise and replacing the creation and annihilation operators by complex numbers, $a \mapsto \alpha$, $a^\dagger \mapsto \alpha^*$. If the steady state, (α_0, α_0^*) is stable fixed point, we write $a = \delta a + \alpha_0$, $a^\dagger = \delta a^\dagger + \alpha_0^*$. The dynamics is then expanded to linear order in δa and $\delta a^\dagger$ and then we proceed as above. An example of this approach is given in [14.15].

Spontaneous emission can also be treated using a master equation. In this case the system is a two-level electronic system, with ground state $|g\rangle$ of energy $\hbar\omega_1$ and excited state $|e\rangle$ with energy $\hbar\omega_2$, representing an electric dipole transition, coupled to the many modes of the radiation field in the dipole and rotating wave approximation. The master equation is

$$\frac{\mathrm{d}\rho}{\mathrm{d}t} = -\frac{\mathrm{i}}{\hbar}[H, \rho] + \gamma(\bar{n}+1)\mathcal{D}[\sigma_-]\rho + \gamma\bar{n}\mathcal{D}[\sigma_+]\rho\,, \quad (14.106)$$

where $\bar{n}$ is the thermal occupation of the radiation field mode at the atomic resonance frequency $\omega_a = \omega_2 - \omega_1$. We have neglected a small term which gives rise to a shift in the atomic transition frequency and which contributes to the Lamb shift. At optical frequencies, $\bar{n} \approx 0$. In the case of a free two-level atom, $H = \frac{\hbar\omega_\mathrm{a}}{2}\sigma_z$ the probability to find the atom in the excited state, $p_\mathrm{e}(t) = \langle e|\rho|e\rangle$ satisfies the equation

$$\frac{\mathrm{d}p_\mathrm{e}}{\mathrm{d}t} = -\gamma p_\mathrm{e}(t) \quad (14.107)$$

with the solution $p_\mathrm{e}(t) = p_\mathrm{e}(0)\mathrm{e}^{-\gamma t}$, which describes spontaneous emission. The dipole polarisation is proportional to the atomic coherence, $\langle e|\rho|g\rangle = \langle\sigma_-\rangle$, which obeys

$$\frac{\mathrm{d}\langle\sigma_-\rangle}{\mathrm{d}t} = -\left(\mathrm{i}\omega_\mathrm{a} + \frac{\gamma}{2}\right)\langle\sigma_-\rangle \quad (14.108)$$

with the solution

$$\langle\sigma_-(t)\rangle = \langle\sigma_-(0)\rangle\,\mathrm{e}^{-(\gamma/2 + \mathrm{i}\omega_\mathrm{a})t}\,. \quad (14.109)$$

The dipole oscillates at the transition frequency and decays, as it radiates.

The radiated field is related to the input field and the local source through an input/output relation in analogy with the case of a cavity discussed above. The positive-frequency components of the field operator take the form

$$E_o^{(+)}(\boldsymbol{x},t) = E_i^{(+)}(\boldsymbol{x},t) - \frac{\omega_a^2}{4\pi\epsilon_0 c^2 r}\left(\boldsymbol{\mu}\times\frac{\boldsymbol{x}}{r}\right) \times \frac{\boldsymbol{x}}{r}\sigma_-(t-x/c)\,, \quad (14.110)$$

where $r=|\boldsymbol{x}|$ is the distance from the course to the point $\boldsymbol{x}$ and $\boldsymbol{\mu}$ is the atomic dipole moment.

If the atom is driven by a classical radiation field, the Hamiltonian becomes [see (14.27) and replace $b\mapsto\beta$]

$$H = \frac{\hbar\omega_a}{2}\sigma_z + \Omega(\sigma_+ e^{-i\omega_L t} + \sigma_- e^{i\omega_L t})\,, \quad (14.111)$$

where Ω is the Rabi frequency and ω_L is the carrier frequency of the driving field. The master equation in an interaction picture at the frequency ω_L is

$$\frac{d\rho}{dt} = -i\frac{\Delta\omega}{2}[\sigma_z,\rho] - i\Omega[\sigma_+ + \sigma_-,\rho] + \gamma\mathcal{D}[\sigma_-]\rho\,, \quad (14.112)$$

where the detuning is $\Delta\omega = \omega_a - \omega_L$. The resulting *Bloch equations* for the atomic moments are linear

$$\frac{d\langle\sigma_-\rangle}{dt} = -\left(\frac{\gamma}{2}+i\Delta\omega\right)\langle\sigma_-\rangle + i\Omega\langle\sigma_z\rangle\,, \quad (14.113)$$

$$\frac{d\langle\sigma_z\rangle}{dt} = -\gamma\left(\langle\sigma_z\rangle+1\right) - 2i\Omega(\langle\sigma_+\rangle - \langle\sigma_-\rangle)\,. \quad (14.114)$$

The solutions for resonance ($\Delta\omega=0$) are

$$\langle\sigma_z(t)\rangle = \frac{8\Omega^2}{\gamma^2+8\Omega^2}\left[1-e^{-3\gamma t/4} \times\left(\cosh\kappa t + \frac{3\gamma}{4\kappa}\sinh\kappa t\right)\right] - 1\,, \quad (14.115)$$

$$\langle\sigma_+(t)\rangle = 2i\Omega\frac{\gamma^2}{\gamma^2+8\Omega^2}\left\{1-e^{-3\gamma t/4} \times\left[\cosh\kappa t + \left(\frac{\kappa}{\gamma}+\frac{3\gamma}{16\kappa}\right)\sinh\kappa t\right]\right\}, \quad (14.116)$$

where

$$\kappa = \sqrt{\frac{\gamma^2}{4} - 16\Omega^2}\,. \quad (14.117)$$

Clearly there is a threshold at $\Omega = \gamma/8$ below which the solutions monotonically approach the steady state and above which they are oscillating. A similar threshold occurs in the solutions for the two-time correlation function $\langle\sigma_+(t)\sigma_-(t+\tau)\rangle_{t\to\infty}$ which determines the spectrum of the scattered light. For $\Omega\gg\gamma$ we find that this spectrum has three peaks at $\omega=\omega_a$ and $\omega=\omega_a\pm\Omega$. This is the *Mollow spectrum* [14.16].

The light scattered by a two-level atom also exhibits *photon anti-bunching*. Consider the conditional probability that given a photon is counted at time t another photon will be counted a time τ later. This is proportional to the second-order correlation function

$$G^{(2)}(t,\tau) = \langle a^\dagger(t)a^\dagger(t+\tau)a(t+\tau)a(t)\rangle\,. \quad (14.118)$$

Usually we are interested in a stationary source so we let $t\to\infty$ and we normalise this by the intensity squared to define

$$g^{(2)}(\tau) = \lim_{t\to\infty}\frac{G^{(2)}(t,\tau)}{\langle a^\dagger(t)a(t)\rangle^2}\,. \quad (14.119)$$

Using the result in (14.110) we can express this directly in terms of correlation functions for the atomic polarisation. As the equations of motion for the atomic variables are linear, the stationary correlation function $\langle\sigma_+(t)\sigma_+(t+\tau)\sigma_-(t+\tau)\sigma_-(t)\rangle_{t\to\infty}$ obeys the same equations [see below (14.141)]. We then find that

$$g^{(2)}(\tau) = 1 - e^{-3\gamma t/4}\left(\cosh\kappa\tau + \frac{3\gamma}{4\kappa}\sinh\kappa\tau\right)\,. \quad (14.120)$$

The result $g^{(2)}(\tau=0)=0$ indicates *photon anti-bunching*, as the probability to count a second photon, immediately after a first one has been counted, vanishes. This is a direct result of the emission process of the source. Photons are emitted when an excited atom relaxes back to the ground state. If a photon is counted, the atom is likely to be in ground state and thus a finite time must elapse before it is re-excited and capable of emitting another one. This prediction, first made by *Carmichael* and *Walls* [14.17], was one of the earliest examples of how quantum optics would differ from a semiclassical description of light. This result has been seen in the resonance fluorescence of a single trapped atom.

14.4.1 Quantum Trajectories

The master equation describes the dynamics of a subsystem by averaging over (tracing out) the properties of the larger *bath* to which it is coupled. Solving the master equation typically results in a mixed state. Any mixed state admits infinitely many decompositions into convex combinations of (nonorthogonal) pure states. In a stochastic unravelling of a master equation we represent the solution at any time as a convex combination of pure states each evolving under a stochastic Schrödinger equation such that if we average over the noise we obtain the solution to the original master equation. This approach leads to a powerful numerical simulation tool as much less memory is required to store a pure quantum state at each time step.

Consider a simple harmonic oscillator coupled to a zero-temperature heat bath. The dynamics, in the interaction picture, is given by the master equation, (14.89) with $\bar{n} = 0$. Solving this equation over a small time interval $\mathrm{d}t$ we can write

$$\rho(t+\mathrm{d}t) = \left\{\rho(t) - \frac{\gamma}{2}[a^\dagger a\rho(t) + \rho(t)a^\dagger a]\,\mathrm{d}t\right\} + \gamma a\rho(t)a^\dagger\,\mathrm{d}t\,. \qquad (14.121)$$

We can think of this as describing photons leaking from a single-mode cavity at Poisson-distributed times. Suppose there were exactly n photons in the cavity at time t so that $\rho(t) = |n\rangle\langle n|$. Then (14.121) would become

$$\rho(t+\mathrm{d}t) = (1-\gamma n\,\mathrm{d}t)|n\rangle\langle n| + \gamma n\,\mathrm{d}t|n-1\rangle\langle n-1|\,. \qquad (14.122)$$

We can think of this as follows. In a small increment of time $\mathrm{d}t$, two events are possible: either a single photon is lost or no photon is lost. If a photon is lost, the state of the field has one less photon so that it changes from $|n\rangle$ to $|n-1\rangle$ and this event will occur with probability $\gamma n\,\mathrm{d}t$. This form results from the last term of (14.121). On the other hand, if no photon is lost the state is unchanged, and this will occur with probability $1-\gamma n\,\mathrm{d}t$, which arises from the first term in (14.121). Thus (14.121) describes a statistical mixture of the two events that can occur in a small time step $\mathrm{d}t$: the first term in square brackets describes the change in the state of the cavity field given that no photon is lost in time interval $\mathrm{d}t$, while the second term describes what happens to the state of the field if one photon is lost in a time interval $\mathrm{d}t$.

If this interpretation is correct it suggests an answer to conditional questions such as: if no photon is lost from time t to $t+\mathrm{d}t$, what is the conditional state of the field? In this case we have no contribution from the last term in (14.121) so the conditional state is the solution to

$$\rho(t+\mathrm{d}t)_\mathrm{c} = \frac{\left\{\rho(t)_\mathrm{c} - \frac{\gamma}{2}[a^\dagger a\rho_\mathrm{c}(t) + \rho(t)_\mathrm{c}a^\dagger a]\,\mathrm{d}t\right\}}{\mathrm{tr}\left\{\rho_\mathrm{c}(t) - \frac{\gamma}{2}[a^\dagger a\rho_\mathrm{c}(t) + \rho_\mathrm{c}(t)a^\dagger a]\,\mathrm{d}t\right\}}$$
$$\approx \rho_\mathrm{c}(t) - \gamma\,\mathrm{d}t\left\{\frac{1}{2}[a^\dagger a\rho_\mathrm{c}(t) + \rho_\mathrm{c}(t)a^\dagger a] - \langle a^\dagger a\rangle_\mathrm{c}(t)\rho_\mathrm{c}(t)\right\}$$

to linear order in $\mathrm{d}t$, where the subscript c is to remind us that we are dealing with a particular conditional state conditioned on a rather special history of null events and $\langle a^\dagger a\rangle_\mathrm{c}(t)$ is the conditional average of the photon number in the state $\rho_\mathrm{c}(t)$.

We can now introduce a *classical* stochastic process, a conditional Poisson process, $\mathrm{d}N(t)$ which is the number of photons lost in time $\mathrm{d}t$. Clearly

$$\mathrm{d}N(t)^2 = \mathrm{d}N(t)\,, \qquad (14.123)$$

$$\mathcal{E}[\mathrm{d}N(t)] = \gamma\langle a^\dagger a\rangle_\mathrm{c}(t)\,, \qquad (14.124)$$

where $\mathcal{E}$ is an average over the classical stochastic variable. In terms of $\mathrm{d}N(t)$ we can now define a *stochastic master equation*

$$\mathrm{d}\rho_\mathrm{c}(t) = \mathrm{d}N(t)\mathcal{G}[a]\rho_\mathrm{c}(t) - \gamma\,\mathrm{d}t\mathcal{H}[a^\dagger a]\rho_\mathrm{c}(t)\,, \qquad (14.125)$$

where we have defined two new super-operators (that map density operators to density operators),

$$\mathcal{G}[A]\rho = \frac{A\rho A^\dagger}{\mathrm{tr}[A\rho c^\dagger]} - \rho\,, \qquad (14.126)$$

$$\mathcal{H}[A]\rho = A\rho + \rho A^\dagger - \mathrm{tr}[A\rho + \rho A^\dagger]\,, \qquad (14.127)$$

for any operator A. Note that, if we take the classical ensemble average over the noise process $\mathrm{d}N(t)$, we recover the original master unconditional master equation in (14.121). The solution to (14.125) is the conditional state at time t conditioned on an entire fine-grained history of jump events (that is to say, the total number of jumps and the time of each jump event). Denote such a history as the sequence of jump times on the interval $[0, t)$ as $h[t] = \{t_1, t_2, \ldots, t_m\}$. The unconditional state is a sum over all such histories

$$\rho(t) = \sum_{h[t]} \mathrm{Pr}(h[t])\rho_\mathrm{c}(h[t])\,, \qquad (14.128)$$

where we have explicitly indicated that the conditional state ρ_c is conditioned on the history of jump events, $h[t]$ in the time interval of interest and $\mathrm{Pr}(h[t])$ is the probability for each history. We have *unravelled* the solution to the master equation in terms of conditional stochastic events. For a point process as considered here the sum over histories has an explicit form in terms of time-ordered integrals [14.18]

$$\rho(t) = \sum_{m=0}^{\infty}\int_0^t \mathrm{d}t_m \int_0^{t_m}\ldots\int_0^{t_1} \mathcal{S}(t-t_m)\mathcal{J}\mathcal{S}(t_m - t_{m-1}) \ldots\mathcal{J}\mathcal{S}(t_1)\rho(0)\,, \qquad (14.129)$$

where the super operators are defined by

$$\mathcal{S}(t)\rho = \mathrm{e}^{-\frac{\gamma}{2}ta^\dagger a}\rho\mathrm{e}^{-\frac{\gamma}{2}ta^\dagger a}, \tag{14.130}$$

$$\mathcal{J} = \gamma a\rho a^\dagger . \tag{14.131}$$

Clearly the probability of a specific jump history is given by

$$\Pr(h[t]) = \mathrm{tr}[\mathcal{S}(t-t_m)\mathcal{J}\mathcal{S}(t_m - t_{m-1}) \ldots \mathcal{J}\mathcal{S}(t_1)\rho(0)] . \tag{14.132}$$

The form of (14.129) indicates that, if we start in a pure state and have access to the entire history of photon loss events $h[t]$, the conditional state $\rho_c(h[t])$ must still be a pure state. This implies that we can write a stochastic Schrödinger for the damped harmonic oscillator

$$\mathrm{d}|\psi_c(t)\rangle = \left[\mathrm{d}N_c(t)\left(\frac{a}{\sqrt{\langle a^\dagger a\rangle_c(t)}} - 1\right) + \gamma\,\mathrm{d}t\left(\frac{\langle a^\dagger a\rangle_c(t)}{2} - \frac{a^\dagger a}{2}\right) - \mathrm{i}H\,\mathrm{d}t\right]|\psi_c(t)\rangle , \tag{14.133}$$

where we have now included the possibility of a Hamiltonian part to the dynamics. We can show the equivalence between this equation and the stochastic master equation by considering the Ito-like expansion

$$\mathrm{d}\left[|\psi_c(t)\rangle\langle\psi_c(t)|\right] = [\mathrm{d}|\psi_c(t)\rangle]\,\langle\psi_c(t)| + |\psi_c(t)\rangle \times [\mathrm{d}\langle\psi_c(t)|] + [\mathrm{d}|\psi_c(t)\rangle|] \times [\mathrm{d}\langle\psi_c(t)|] \tag{14.134}$$

and retaining all terms to first order in $\mathrm{d}t$, noting that $\mathrm{d}N^2 = \mathrm{d}N$.

A point process with a large rate parameter γ can be well approximated on a time scale long compared to γ^{-1} by a white-noise process. This suggests that it must be possible to unravel the master equation in terms of white-noise processes as well as the point process $\mathrm{d}N(t)$. Such master equations give the conditional dynamics conditioned on homodyne and heterodyne measurements on the field leaving the cavity. Here we simply quote the result and show that averaging over the classical noise returns us to the unconditioned master equation.

In the case of the real valued Weiner process $\mathrm{d}W(t)$ we can write the *homodyne* stochastic master equation for a damped simple as

$$\mathrm{d}\rho_c(t) = -\mathrm{i}[H, \rho_c(t)]\,\mathrm{d}t + \mathcal{D}[a]\rho_c(t)\,\mathrm{d}t + \mathrm{d}W(t)\mathcal{H}[a]\rho_c(t) , \tag{14.135}$$

where

$$\mathcal{D}[A]\rho = A\rho A^\dagger + \frac{1}{2}(A^\dagger A\rho + \rho A^\dagger A) \tag{14.136}$$

and $\mathcal{H}[a]$ is given by (14.127).

In terms of a complex-valued white-noise process, $\mathrm{d}W(t) = \mathrm{d}W_1(t) + \mathrm{i}\,\mathrm{d}W_2(t)$, where $\mathrm{d}W_i(t)$ are independent Winer processes, we can write the *heterodyne* stochastic master equation

$$\mathrm{d}\rho_c(t) = -\mathrm{i}[H, \rho_c(t)]\,\mathrm{d}t + \mathcal{D}[a]\rho_c(t)\,\mathrm{d}t + \frac{1}{\sqrt{2}}[\mathrm{d}W_1(t)\mathcal{H}[a]\rho_c(t) + \mathrm{d}W_2(t)\mathcal{H}[-\mathrm{i}a]]\rho_c(t) . \tag{14.137}$$

There is a connection between the quantum jump process and the two-time correlation function discussed in above. We first define a new stochastic process, the rate or the current, as

$$i(t) = \frac{\mathrm{d}N}{\mathrm{d}t} . \tag{14.138}$$

This is a rather singular stochastic process, consisting of a series of delta functions concentrated at the actual jump times. In physical terms this is intended to model the output of an ideal photon counting detector, with infinite response bandwidth, that detects every photon that is lost from the cavity. Define the classical current two-time correlation function

$$G(\tau, t) = \mathcal{E}[i(t+\tau)i(t)] . \tag{14.139}$$

Given the nature of a Poisson jump process, $\mathrm{d}N(t)$ can only take the values 0 or 1, so it is easy to see that we can write the two-time correlation function in terms of the conditional probability to get $\mathrm{d}N(t+\tau) = 1$ given a jump at time t,

$$G(\tau, t)\,\mathrm{d}t^2 = \Pr(\mathrm{d}N(t+\tau) = 1|\,\mathrm{d}N(t) = 1) . \tag{14.140}$$

This conditional probability is given by

$$\Pr[\mathrm{d}N(t+\tau) = 1|\,\mathrm{d}N(t) = 1] = \gamma^2\mathrm{tr}\left[a^\dagger a\mathrm{e}^{\mathcal{L}\tau}a\rho(t)a^\dagger\right]\mathrm{d}t^2 \tag{14.141}$$

where $\mathrm{e}^{\mathcal{L}t}$ is the formal solution to the unconditional master equation evolution written in terms of the abstract generator $\mathcal{L}$ as $\dot{\rho} = \mathcal{L}\rho$. The two-time correlation function is then given by

$$G(\tau, t) = \gamma^2\mathrm{tr}\left[a^\dagger a\mathrm{e}^{\mathcal{L}\tau}a\rho(t)a^\dagger\right] . \tag{14.142}$$

Note that the so-called regression theorem follows directly from the definition of G,

$$\frac{dG(\tau,t)}{d\tau} = \mathcal{L}G(\tau,t) . \qquad (14.143)$$

We usually deal with driven damped harmonic oscillators for which the system settles into a steady state, emitting photons according to the conditional Poisson process derived from the steady-state solution $\rho_\infty = \lim_{t\to\infty}\rho(t)$, so we define the *stationary* two-time correlation function for the current as

$$G(\tau) = \gamma^2 \mathrm{tr}\left[a^\dagger a e^{\mathcal{L}\tau} a\rho_\infty a^\dagger\right] . \qquad (14.144)$$

14.4.2 Simulating Quantum Trajectories

A mixed state for system with a Hilbert space of dimension N requires that we specify N^2 complex matrix elements. On the other hand a pure state requires that we specify only N complex numbers. For this reason numerically solving the master equation is more computationally difficult than solving the Schrödinger equation. We can use the unravelling of a master equation in terms of a stochastic Schrödinger equation to make the numerical solution of master equations more tractable. In this numerical setting, the method of quantum trajectories was independently developed as the *Monte Carlo wavefunction* method [14.19]. We will illustrate the method using the jump process.

Suppose the state at time t is $|\psi(t)\rangle$. Then in a time interval δt, sufficiently short compared to γ^{-1}, the system will evolve to the (unnormalised) state conditioned on no jump having occurred,

$$|\tilde{\psi}(t+\delta t)\rangle = e^{-iH\delta t - \gamma a^\dagger a \delta t/2}|\psi(t)\rangle . \qquad (14.145)$$

To compute this we implement a routine to solve the Schrödinger equation with the effective non-Hermitian Hamiltonian

$$K = H - i\frac{\gamma}{2}a^\dagger a . \qquad (14.146)$$

The norm of this state is the probability that no jump has occurred in the time interval δt,

$$p_0 = \langle\tilde{\psi}(t+\delta t)|\tilde{\psi}(t+\delta t)\rangle \qquad (14.147)$$

$$= 1 - p , \qquad (14.148)$$

where it is easy to see that

$$p = \gamma\delta t\langle\psi(t)|a^\dagger a|\psi(t)\rangle , \qquad (14.149)$$

which we understand to be the probability that a jump takes place in this time interval. We need to ensure that $p \ll 1$.

Let us now chose a random number r uniformly distributed on the unit interval. At the end of the time interval, we compare p and r. If $p < r$ (usually the case) we normalise the state

$$|\psi(t+\delta t)\rangle = \frac{|\tilde{\psi}(t+\delta t)\rangle}{\sqrt{p_0}} \qquad (14.150)$$

and continue the non-Hermitian evolution for a further time step. If however $p > r$, we implement a quantum jump via

$$|\tilde{\psi}(t+\delta t)\rangle \to |\psi(t+\delta t)\rangle = \frac{\sqrt{\gamma}a|\tilde{\psi}(t+\delta t)\rangle}{p/\delta t} . \qquad (14.151)$$

Based on our previous discussions we see that $p/\delta t = \gamma\langle\psi(t)|a^\dagger a|\psi(t)\rangle$, and the jump operation is as described by the first term in (14.133). As the simulation proceeds we accumulate the record of times at which particular jump events occur. That is to say, we have access to a sample fine grained history of the jump process, $h[t]$. However we are primarily interested in solving the master equation. We thus run K trials up to time t, starting from an identical initial state each time, and then form the mixed state

$$\bar{\rho}(t) = K^{-1}\sum_{k=1}^{K}|\psi_k(t)\rangle \qquad (14.152)$$

as a uniform average over the K trials. Strictly speaking the probability of each of the K trials is not uniform, however one can show that, for K sufficiently large, $\bar{\rho}(t) \approx \rho(t)$, with an error that scales as $K^{-1/2}$. In *Mølmer* et al. [14.19] more-general cases are discussed including how to simulate non-zero-temperature master equations or master equations with multiple jump processes.

14.5 Ion Traps

The ability to trap an individual ion and cool it close to the ground state of vibrational motion has enabled precise quantum control to be achieved together with very high-efficiency measurement. The technology originated some three decades ago in the effort to develop ultra-high-precision spectroscopy with particular application to frequency standards. In recent years ion trap technology has led the field in efforts to implement quantum information processing following the seminal proposal of *Cirac* and *Zoller* in 1995 [14.20]. The technology is now entering an even more exciting phase with the development of smaller and more effective traps, ion trap arrays for coherently processing a large number of ions and the integration of traps and optical cavities.

One of the keys to the success of ion traps is the ability to detect the electronic state of a single ion with high efficiency. This is done by laser-induced fluorescence on a dipole-allowed transition, scattering millions of photons per second. The idea goes back to a proposal of *Dehmelt* in 1975 [14.21] and appears in various applications such as electron shelving, cycling transitions and quantum jumps [14.22]. Another key to the success of ion traps is the invention of sub Doppler cooling techniques, particularly resolved side band cooling. This makes it possible to prepare a single ion in a state with very few quanta of vibrational excitation, even so few as to reach the vibrational ground state. A single trapped ion is well approximated by a particle moving in a three-dimensional harmonic potential. It is possible to remove vibrational energy stochastically by laser-induced Raman transitions coupling electronic and vibrational degrees of freedom. In effect, heat is removed from the vibrational motion of of the ion and pumped into the very low temperature heat bath associated with the much higher frequency of the fluorescent radiation.

It is not possible to trap a charged particle in a static potential. This is because the Laplace equation implies there is always one unstable (not trapped) direction in the electrostatic potential. However a time-dependent electric potential can produce an effective harmonic potential for a charge particle.

The quantum description of the centre of mass motion of the ion is given in terms of the eigenstates of the Hamiltonian

$$H = \hbar\nu a_z^\dagger a_z + \hbar\nu_t\left(a_x^\dagger a_x + a_y^\dagger a_y\right) . \tag{14.153}$$

The motion is thus separable into axial (ω_z) and transverse motion and, to be specific, we now concentrate on the axial motion alone, although much of the discussion can be applied to the transverse motion as well. Henceforth, as we neglect the transverse motion, we will drop the subscript on a_z, $a_z^\dagger$.

Many different kinds of ions can be trapped but availability of lasers limits the kinds of ions that can be easily laser cooled. For example, the Wineland group at NIST Colorado uses $^9Be^+$ while the Blatt group in Innsbruck uses $^{40}Ca^+$. When the ion is first trapped it is in a highly excited state of its vibrational motion, corresponding to a temperature of the order of 10^4 K. Cooling typically proceeds in two stages. The first stage is based on Doppler cooling and is very efficient, the second stage is based on resolved sideband cooling.

Ion traps are a remarkably versatile quantum devices for a number of reasons. Firstly, it is possible to coherently couple the vibrational motion and the internal electronic state using an external laser. Secondly, resolved sideband cooling enables the vibrational motion to be prepared in its ground state with probability approaching unity. This is done using an external laser to induce Raman transitions between the ground and excited internal electronic state which absorb one photon and one phonon of vibrational per excitation cycle. Finally, the method of fluorescence shelving enables the internal electronic state of a single trapped ion to measured with efficiency approaching unity. In order to understand these three features we need to begin with a description of how an external laser can couple the vibrational and electronic degrees of freedom.

We will assume that the external laser can couple two internal electronic states, the ground state $|g\rangle$ and the excited state $|e\rangle$. This might involve a direct dipole transition. However for quantum information applications it typically involves a Raman two-photon transition connecting the ground state to an excited metastable state. In either case the Hamiltonian describing the system is

$$H = \hbar\nu a^\dagger a + \frac{\hbar\omega_A}{2}\sigma_z + \frac{\hbar\Omega}{2}\left(\sigma_- e^{i(\omega_L t - k_L\hat{q})} + \sigma_+ e^{-i(\omega_L t - k_L\hat{q})}\right) , \tag{14.154}$$

where $\hat{q}$ is the operator for the displacement of the ion from its equilibrium position in the trap, ν is the trap (secular) frequency, Ω is the Rabi frequency for the two-level transition, ω_A is the atomic transition frequency, and ω_L, k_L are the laser frequency and wave number. There are three frequencies in the problem: ν, ω_A and ω_L. A wide variety of processes can be made dominant by carefully choosing relationships between these three frequencies. Note that the phase of the laser field as seen by the ion is dependent upon the position of the ion. As the ion moves harmonically in the trap the phase is thus modulated at the trap frequency. As we shall see this leads to sidebands in the absorption spectrum for the two-level system.

The ion position operator, in terms of the vibrational raising and lowering operators, is

$$\hat{q} = \left(\frac{\hbar}{2m\nu}\right)^{1/2} (a + a^\dagger) . \qquad (14.155)$$

We now define the Lamb–Dicke parameter, η

$$\eta = k_L \left(\frac{\hbar}{2m\nu}\right)^{1/2} = 2\pi \Delta x_{rms}/\lambda_L \qquad (14.156)$$

where the root-mean-square (r.m.s.) position fluctuations in the oscillator ground state is Δx_{rms}. Then moving to an interaction picture via the unitary transformation

$$U_0(t) = \exp(-\mathrm{i}\nu a^\dagger a\, t - \mathrm{i}\omega_A \sigma_z t) , \qquad (14.157)$$

the interaction Hamiltonian can be written as

$$H_I(t) = \frac{\hbar\Omega}{2}\{\sigma_- \exp[-\mathrm{i}\eta(a\,\mathrm{e}^{-\mathrm{i}\nu t} + a^\dagger\,\mathrm{e}^{\mathrm{i}\nu t})] \times \exp[-\mathrm{i}(\omega_A - \omega_L)t] + \mathrm{h.c}\} \qquad (14.158)$$

The exponential of exponentials make this a complicated Hamiltonian system. However in most ion trap experiments the ion is confined to a spatial region that is significantly smaller than the wavelength of the exciting laser so that we may assume that the Lamb–Dicke parameter is small $\eta < 1$ (typically $\eta \approx 0.01$–0.1). Expanding the interaction to second order in the Lamb–Dicke parameter gives

$$\begin{aligned} H_I(t) = {} & \frac{\hbar\Omega}{2}(1 - \eta^2 a^\dagger a)\left(\sigma_-\,\mathrm{e}^{-\mathrm{i}\delta t} + \sigma_+\,\mathrm{e}^{\mathrm{i}\delta t}\right) \\ & - \mathrm{i}\frac{\hbar\Omega\eta}{2}\left(a\,\mathrm{e}^{-\mathrm{i}\nu t} + a^\dagger\,\mathrm{e}^{\mathrm{i}\nu t}\right)\mathrm{e}^{-\mathrm{i}\delta t}\sigma_- \\ & + \mathrm{i}\frac{\hbar\Omega\eta}{2}\left(a\,\mathrm{e}^{-\mathrm{i}\nu t} + a^\dagger\,\mathrm{e}^{\mathrm{i}\nu t}\right)\mathrm{e}^{\mathrm{i}\delta t}\sigma_+ \\ & - \frac{\hbar\Omega\eta^2}{4}\left(a^2\,\mathrm{e}^{-2\mathrm{i}\nu t} + (a^\dagger)^2\,\mathrm{e}^{2\mathrm{i}\nu t}\right) \\ & \times \left(\mathrm{e}^{-\mathrm{i}\delta t}\sigma_- + \mathrm{e}^{\mathrm{i}\delta t}\sigma_+\right) , \end{aligned}$$

where the detuning of the laser from the atomic frequency is $\delta = \omega - \omega_L$.

By carefully selecting δ to be positive or negative integer multiples of the trap frequency, various resonant terms may be extracted and all time dependent terms neglected. In the first case, known as the carrier excitation, $\delta = 0$, and the resonant terms are

$$H_c = \hbar\Omega(1 - \eta^2 a^\dagger a)\sigma_x \quad \text{(carrier excitation)} , \qquad (14.159)$$

where $\sigma_x = (\sigma_- + \sigma_+)/2$. If we choose $\delta = \nu$ so that the laser frequency is detuned below (to the red of) the carrier frequency by one unit of trap frequency, $\omega_L = \omega_A - \nu$, the resonant terms are

$$H_r = \mathrm{i}\frac{\hbar\eta\Omega}{2}(a\sigma_+ - a^\dagger\sigma_-) \quad \text{(first red sideband excitation)} . \qquad (14.160)$$

This is just the Jaynes–Cummings Hamiltonian except that it involves the absorption of a trap *phonon* as well as one laser photon. On the other hand we can choose $\delta = -\nu$ so that $\omega_L = \omega_A + \nu$ and the laser is detuned one unit of vibrational frequency to the blue of the carrier, the resonant interaction Hamiltonian is

$$H_b = \mathrm{i}\frac{\hbar\eta\Omega}{2}(a^\dagger\sigma_+ - a\sigma_-) \quad \text{(first blue sideband excitation)} . \qquad (14.161)$$

This describes an excitation process that absorbs one photon from the laser and *emits* one trap phonon. Continuing in this way we can define the second red sideband excitation $\delta = 2\nu$ and second blue sideband excitation $\delta = -2\nu$, and so on. Figure 14.2 shows an energy-level diagram that represents the carrier, red and blue sideband excitations.

Once excited to $|e\rangle$ the ion can spontaneously decay to the ground state. For a dipole-allowed transition this can be quite fast, thus enabling another excitation process to occur. In the case of red sideband excitation the net result is to remove one phonon per excitation cycle. This is the basis of sideband cooling. In effect the external laser has coupled the vibrational motion to a very-low-temperature heat bath: the vacuum radiation field at frequency ω_A. Of course it is only possible to address the sidebands if the laser can be accurately tuned to each. As each transition is homogeneously broadened by a width equal to the spontaneous emission rate, γ, we require that $\nu > \gamma$ for resolved sideband cooling. An explicit expression for the spectrum of resonance fluorescence for a single trapped ion follows from the

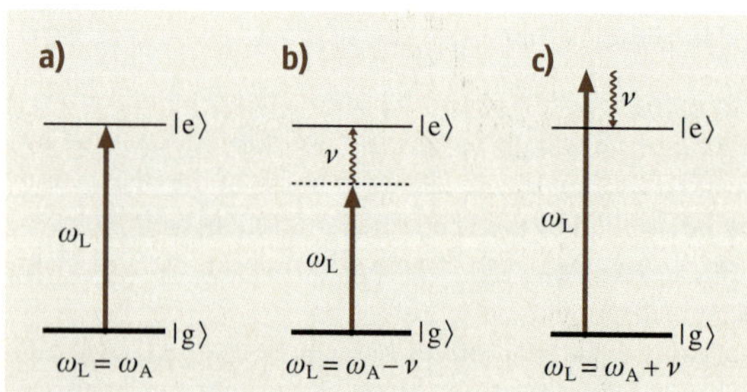

Fig. 14.2a–c Energy-level diagram for **(a)** carrier **(b)** first red sideband and **(c)** first blue sideband excitation

methods given in Sect. 14.4. A detailed calculation in the low intensity limit ($\Omega < \gamma$) for a traveling-wave field was given by *Cirac* et al. [14.23].

Resolved sideband cooling requires that the exciting laser be tuned one unit of trap frequency below the atomic transition frequency. The atom is then excited by the absorption of one laser photon and one trap phonon. In the Lamb–Dicke limit relaxation is dominated by spontaneous emission into the spectral peak at the carrier frequency ($\omega = \omega_A$). Thus on each excitation cycle one unit of vibrational energy is removed on average. A simple rate equation method suffices to understand this phenomenon. We find that the rate of change of the average phonon number is given by

$$\frac{d\bar{n}}{dt} = -\gamma\left(\frac{\eta^2\Omega^2\bar{n}}{2\eta^2\Omega^2\bar{n}+\gamma^2}\right) . \tag{14.162}$$

Note that, as the cooling proceeds, the rate decreases until the vibrational ground state is a steady state. In a more careful treatment we need to consider heating mechanisms, for example off-resonant excitation of the blue sideband [14.24], and the probability of populating the vibrational ground state in the steady state is less than unity. Despite this, resolved sideband cooling care prepare an ion in the vibrational ground state with a probability greater than 99%. Ground state cooling in all three dimensions was first achieved by the NIST group in *Boulder* [14.25]. Other heating mechanisms due to experimental artifacts are often important. For example, fluctuating charge distributions on the trap electrodes lead to a stochastic displacement of the centre of the trap.

In order to read out the state of the ion, an additional auxiliary level can be coupled by a strong laser to one or the other of the ground or excited states, to be specific let us say the ground state Fig. 14.3. If the ion is in the ground state when the probe laser is turned on, fluorescent photons are scattered in all directions and can easily be detected. On the other hand if the ion is in the excited state, it is not resonant with the probe laser and no photons are scattered: the ion remains dark.

If we now apply a weak laser to couple the ground and excited states, incoherent transitions occur $|g\rangle \leftrightarrow |e\rangle$. The net result is that the fluorescent signal due to the probe laser blinks on and off in fashion of a random telegraph process. A typical signal is also shown in Fig. 14.3. In so far as fluorescence indicates that the ion is in the ground state, the random switching of the fluorescence is a direct indicator of *quantum jumps* between the ground and excited states. These jumps were first reported in a number of laboratories [14.22].

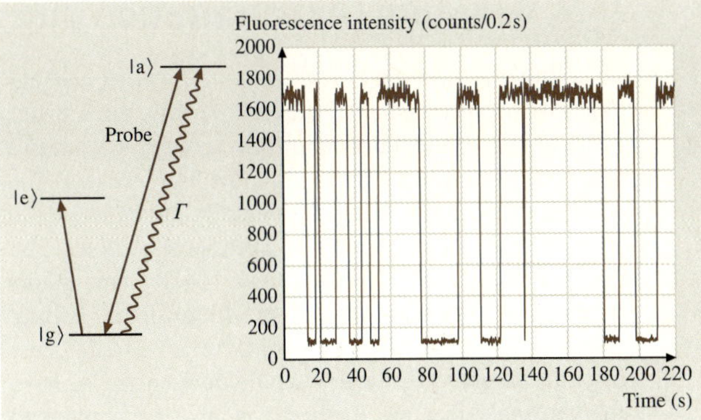

Fig. 14.3 Energy-level diagram showing fluorescence readout of the ground atomic state. A strong probe laser drives a dipole-allowed transition between the ground state $|g\rangle$ and an auxiliary state $|a\rangle$ which decays back to the ground state at a rate Γ scattering many many photons. Also shown (*right*) is the fluorescent signal on the probe transition when a weak laser couples the ground and excited state (after *Leibfried* et al. [14.24] with permission)

The efficiency of the readout is a function of the integration time of the fluorescence signal, that is to say, how long we need to monitor the fluorescence to be sure that we are in period of high intensity. This time must be at least of the order of the average time between photon emission events. In an experiment however there are other sources of error, such as dark counts in the detector and typically the minimum time to distinguish ground and excited states is of the order of 2 ms. We capture the quality of the readout in an overall efficiency is η which is the conditional probability for the ion to be detected in the ground state given that it was prepared in the ground state before the probe laser was turned on.

We now return to the experimental determination of the efficiency of sideband cooling. The objective is to determine the state of vibrational motion by coupling it to the internal state of the ion and then using the fluorescent readout technique described above. At the end of a cooling stage, the electronic state of the ion to is first coupled to its vibrational motion for a time T using the first red and blue sideband transitions. If we write the probability for the atom to found in the excited state after time T as $P_e^R(T)$ and $P_e^B(T)$ for red and blue sideband excitation respectively it can be shown that the mean phonon number $\bar{n}$ is given by $\bar{n}/(1+\bar{n}) = P_e^R(T)/P_e^B(T)$. Thus measurement of the ratio of excitation probability on the first red and blue sideband yields $\bar{n}$ directly.

14.6 Quantum Communication and Computation

Quantum teleportation is a new communication task enabled by quantum entanglement. It permits the transfer of an unknown quantum state from a *client* system (C) provided to a *sender* (A), to a remote *receiver* (B). The sender and receiver share a maximally entangled state, and they can communicate via a classical channel. The original proposal of *Bennett* et al. [14.26] was posed in terms of systems with a two-dimensional Hilbert space (qubits [14.27]). *Furasawa* et al. [14.28], using a proposal of *Braunstein* and *Kimble* [14.29], have demonstrated that the method can also be applied to entangled systems with an infinite-dimensional Hilbert space, specifically for harmonic oscillator states. In that work, a coherent state was teleported using an entanglement resource that consisted of a two mode squeezed vacuum state. The joint measurements required for teleportation are joint quadrature phase on the client system and that part of the entangled resource be shared by the receiver.

Teleportation of continuous variables is possible using a perfect quadrature phase quantum nondemolition (QND) measurement between two optical modes, A and B, to create the entanglement resource. The state that is produced is an optical analogue of the EPR state which had previously been shown by *Vaidman* [14.30] to enable teleportation of continuous observable. The EPR state is not a physical state because quadrature phase eigenstates are infinite-energy states. However we can use arbitrary close approximations to these states in terms of a squeezed vacuum state (14.163). This is essential feature exploited in the scheme of Furasawa et al.

Suppose that at some prior time a two-mode squeezed vacuum state is generated and that one mode is available for local operations and measurements at the sender's location A by observer Alice, while the other mode is open to local operations and measurements in the receiver's location B, by observer Bob. Alice and Bob can communicate via a classical communication channel. Thus Alice and Bob each have access to one of the two entangled subsystems described by

$$|\mathcal{E}\rangle_{\mathrm{AB}} = \sqrt{(1-\lambda^2)}\sum_{n=0}^{\infty}\lambda^n|n\rangle_{\mathrm{A}}\otimes|n\rangle_{\mathrm{B}} \tag{14.163}$$

This state is generated from the vacuum state by the unitary transformation

$$U(r) = \mathrm{e}^{r(a^\dagger b^\dagger - ab)}\,, \tag{14.164}$$

where $\lambda = \tanh r$ and where a, b refer to the mode accessible to Alice and the mode accessible to Bob, respectively.

The sender, Alice, has access to another quantum system, the client (C), in state $|\psi\rangle_{\mathrm{C}}$. Perfect (projective) measurements are made of the joint quadrature phase quantities, $\hat{X}_{\mathrm{C}} - \hat{X}_{\mathrm{A}}$ and $\hat{Y}_{\mathrm{C}} + \hat{Y}_{\mathrm{A}}$ on the client mode and the Alice's part of the entangled mode, A, with the results X and Y, respectively. The conditional state resulting from this joint quadrature measurement is described by the projection onto the state $|X, Y\rangle_{\mathrm{CA}}$ where

$$|X, Y\rangle_{\mathrm{CA}} = \mathrm{e}^{-\frac{\mathrm{i}}{2}\hat{X}_{\mathrm{A}}\hat{Y}_{\mathrm{C}}}|X\rangle_{\mathrm{C}}\otimes|Y\rangle_{\mathrm{A}}\,. \tag{14.165}$$

The (unnormalised) conditional state of total system after the measurement is then seen to be given by

$$|\tilde{\Psi}^{(X,Y)}\rangle_{\mathrm{out}} = {}_{\mathrm{CA}}\langle X, Y|\psi\rangle_{\mathrm{C}}|\mathcal{E}\rangle_{\mathrm{AB}}\otimes|X, Y\rangle_{\mathrm{CA}}\,. \tag{14.166}$$

The state of mode B at the receiver, denoted as Bob, is the pure state

$$|\phi^{(X,Y)}(r)\rangle_{\mathrm{B}} = [P(X, Y)]_{\mathrm{CA}}^{-1/2}\langle X, Y|\psi\rangle_{\mathrm{C}}\otimes|\mathcal{E}\rangle_{\mathrm{AB}} \tag{14.167}$$

with the wave function (in the $\hat{X}_{\mathrm{B}}$ representation),

$$\phi_{\mathrm{B}}^{(X,Y)}(x) = \int_{-\infty}^{\infty}\mathrm{d}x'\,\mathrm{e}^{-\frac{\mathrm{i}}{2}x'Y}\mathcal{E}(x, x')\psi(X+x') \tag{14.168}$$

where $\psi(x) = {}_{\mathrm{C}}\langle x|\psi\rangle_{\mathrm{C}}$ is the wavefunction for the client state we seek to teleport. The kernel is simply the wave function for the two-mode squeezed-state resource.

The state in (14.168) is clearly not the same as the state we sought to teleport. However in the limit of infinite squeezing, $r\to\infty$, we find that $\mathcal{G}(x_1, x_2; r)\to\delta(x_1+x_2)$ and the state of mode B approaches

$$|\phi_{XY}(r)\rangle_{\mathrm{B}} \to \mathrm{e}^{-\frac{\mathrm{i}}{2}Y\hat{X}_{\mathrm{B}}}\mathrm{e}^{\frac{\mathrm{i}}{2}X\hat{Y}_{\mathrm{B}}}|\psi\rangle_{\mathrm{B}} \tag{14.169}$$

which, up to the expected unitary translations in phase space, is the required teleported state.

For finite squeezing the state after Bob's conditional control is not an exact replica of the client state. We can quantify the fidelity of the reproduction by computing the probability that the state received by Bob, after displacement, is the same as the client state. This probability is called the *fidelity* and is given by

$$F = |\langle\psi|\mathrm{e}^{\frac{\mathrm{i}}{2}\mu\hat{X}_{\mathrm{B}}}\mathrm{e}^{-\frac{\mathrm{i}}{2}\nu\hat{Y}_{\mathrm{B}}}|\phi^{(X,Y)}\rangle|^2 \tag{14.170}$$

with $\mu = gY$ and $\nu = gX$, which allows for some flexibility in the choice of displacements in the nonideal case. The quantity g is called the *gain*. In the limit of infinite squeezing we expect $g \to 1$.

In the experimental context, imperfections in the measurements, noise in the classical communication channel, degradation of the entanglement, and imperfections in the local unitary transformations, mean that Bob's state is not precisely the same as the state of the client C. A final step in any teleportation protocol is to check to see to what extent the state is teleported. In other words we need to determine the probability that the teleported state is the one we want, that is we need the fidelity. This of course requires knowledge of the actual state of the mode C. As the fidelity is a probability it must be adequately sampled, so the verification stage requires repeated measurements upon the output state of the teleportation scheme at mode B. The previously mentioned errors act independently from trial to trial which means we must describe the teleported state as a mixed state ρ_{B} in general. In this case the fidelity is given by

$$F = \langle\psi|\rho_{\mathrm{B}}|\psi\rangle\,. \tag{14.171}$$

We can also define an overall measure of performance in terms of the average fidelity $\bar{F}$ obtained by averaging the fidelity over all possible client states $|\psi\rangle$ with some appropriate measure on the set of pure states. We first need to specify the class of client states and the ensemble from which they are drawn. In the case of client states drawn from an ensemble of coherent states we can obtain an explicit result. If A and B share no entanglement $\bar{F} = \frac{1}{2}$. This is the classical boundary for teleportation of a coherent state. A quantum protocol would need to give an average fidelity greater than 0.5.

Quantum teleportation using squeezed states has been demonstrated by the group of *Kimble* at Caltech [14.28]. Related experiments have been performed by the group of di Martini in Rome [14.31] and *Zeilinger* in Innsbruck [14.32]. In order to understand the Caltech experiment we need to explain how some of the formal steps in the preceding analysis are done in the laboratory. We will also need to understand how imperfections, such as photon loss, and noise are included.

In order to effect a joint measurement of the combined quadratures $\hat{X}_C - \hat{X}_A$, $\hat{Y}_C + \hat{Y}_A$, the experiment first combined the client and sender field amplitudes on a 50/50 beam splitter, followed by direct homodyne measurements of the output fields after the beam splitter. After the beam splitter we then make a homodyne measurement of X-quadrature on mode C and the Y-quadrature on mode A. In the case of homodyne detection, the actual measurement records are two photocurrents (I_X, I_P). For unit efficiency detectors, this is an optimal measurement of the corresponding quadratures $\hat{X}_C$, $\hat{Y}_A$. In reality however efficiency is not unity and some noise is added to the measurement results. We shall return to this point later.

The measured photocurrents are a classical stochastic processes and may be sent to the receiver B over a standard communication channel. On receipt of this information the receiver must apply the appropriate unitary operator, a displacement, to complete the protocol. Displacement operators are quite easy to apply in quantum optics. To displace a mode, say B, we first combine it with another mode, prepared in a coherent state with large amplitude, $\alpha \to \infty$, on a beam splitter with very high reflectivity, $R \to 1$, for mode B. If $|\phi\rangle_{\mathrm{B}}$ is the state of B, then after the combination at the beam splitter the state of B is transformed by

$$|\phi\rangle_{\mathrm{B}} \to D(\beta)|\phi\rangle_{\mathrm{B}}\,, \tag{14.172}$$

where $D(\beta) = \exp(\beta b^\dagger - \beta^* b)$ is the unitary displacement operator, and

$$\beta = \lim_{R\to 1}\lim_{\alpha\to\infty} \alpha\sqrt{1-R}\,. \tag{14.173}$$

In terms of the quadrature operators for B the displacement operator can be written

$$D(x, y) = \mathrm{e}^{\mathrm{i}y\hat{X}_{\mathrm{B}} + \mathrm{i}x\hat{Y}_{\mathrm{B}}} \tag{14.174}$$

with $\beta = x + \mathrm{i}y$. A suitable choice of β will produce the required displacements to complete the teleportation protocol. This was achieved by using the measured photocurrents to control the real and imaginary components of the displacement field using electrically controlled modulators. As the measurement records, the photocurrents, are classical stochastic processes they can be scaled by a gain factor, g, to produce the required β.

The experiment included an additional step to verify to what extent the state received by Bob faithfully reproduced the stat of the client field. In this experiment the state of the client was a coherent state. In essence another party, Victor, is verifying the fidelity of the teleportation using homodyne detection to monitor the quadrature variances of the teleported state.

The key feature that indicates success of the teleportation is a drop in the quadrature noise seen by Victor when Bob applies the appropriate unitary operator to his state. This is done by varying the gain g. If Bob simply does nothing to his state ($g = 0$), then Victor simply gets one half of a squeezed state. Such a state has a quadrature noise level well above the vacuum level of the coherent state. As Bob varies his gain, Victor finds the quadrature noise level fall until, at optimal gain, the teleportation is effected and the variance falls to the vacuum level of a coherent state. In reality of course extra sources of noise introduced in the detectors and control circuits limit the extent to which this can be achieved.

In a perfect system the fidelity should be peaked at unit gain. However photon loss in the shared entanglement resource and detector inefficiencies reduce this. In the experiment, the average fidelity at unit gain was found to be $F = 0.58 \pm 0.002$. As discussed previously, this indicates that entanglement is an essential part of the protocol.

14.6.1 Linear Optical Quantum Computing

In 1982 *Feynman* [14.33] suggested there were certain problems that would be difficult to preform on a computer running according to classical mechanics but which would be easy to do on a computer running according to quantum principles. In 1985 *Deutsch* [14.34] showed in more detail what would be required for a quantum computer (QC) and gave examples of problems that might be solved more efficiently on such a machine compared to a classical machine. The promise of quantum computation suggested by Feynman and elaborated by Deutsch was made very apparent in the factoring algorithm of *Shor* in 1994 [14.35]. Shor's algorithm is an efficient factoring algorithm for a quantum computer, while all known algorithms for factoring on a classical computer require an exponentially increasing number of steps as the size of the integer to be factored increases.

How does a quantum computer achieve this enormous increase in efficiency? The answer lies in the quantum superposition principle. Suppose we wish to evaluate a function f on some binary input string x to produce a binary output string $f(x)$. We can code the input and output binary string as the product state of N qubits. The output qubits however are preset to zero. Now we set up a machine so that under unitary quantum evolution the state transforms as

$$|x\rangle|0\rangle \to |x\rangle|f(x)\rangle \,. \tag{14.175}$$

Why do we demand that the transformation be unitary? Consider what happens when we prepare the input qubits in a uniform superposition of all possible input states;

$$\sum_x |x\rangle|0\rangle \to \sum_x |x\rangle|f(x)\rangle \,. \tag{14.176}$$

If the dynamics is unitary the linearity of quantum mechanics ensures that (14.175) implies (14.176). It would appear that in a single run of the machine we have evaluated all possible values of the function.

This is not quite as interesting as it seems. If we measure the output qubits we will get one value at random. That does not seem very useful. To see why it is useful to do this let us ask; when would we ever want to evaluate every value of a particular function? The answer, is when we are not so much interested in a particular value of the function as a *property* of the function. How can we use the superposition state in (14.176) to determine properties of functions? To see this consider a function f that maps the binary numbers $\{0, 1\}$ to $\{0, 1\}$. There must be four such functions, two of which are constant functions with $f(0) = f(1)$, and two have $f(0) \neq f(1)$, so called balanced functions. Suppose now the problem involves determining if a function is balanced or constant. On a classical computer to answer this we need to make two evaluations of the function, $f(0)$ and $f(1)$. We would then need to run the computer twice. However a quantum computer can determine this property in only a single run.

Suppose we have two qubits. One qubit will be used to encode the input data and the other qubit, the output qubit, will contain the value of the function after the machine is run. The output qubit is initially set to 0. The machine might then run according to (14.175). However there is a problem with this expression. If f is a constant function we have two distinct input states unitarily transformed to the same output state. Clearly this is not a reversible transformation and thus cannot be implemented unitarily. The problem is easily fixed however by setting up the machine to evolve the states according to

$$|x\rangle|y\rangle \to |x\rangle|y \oplus f(x)\rangle \,, \tag{14.177}$$

where the addition is defined modulo two and we have allowed all possible settings of both qubits. The unitary transformation which realises this operation has been called the f-controlled NOT gate [14.36]. The input qubit x is the control qubit while the output qubit y is the target. If the value of f on the control qubit is one, the bit on the target is flipped; thus the name. Every unitary transformation on qubits can be realised as suitable networks of simple one- and two-qubit gates using primitive gate operations.

The quantum algorithm that solves this problem is a version of a quantum algorithm first proposed by Deutsch. It proceeds as follows. In the first step we prepare the output qubit in the state $|0\rangle - |1\rangle$ (we ignore normalisation in what follows for simplicity). This can be done using a single-qubit rotation $|1\rangle \to |0\rangle - |1\rangle$. Such a rotation is called a Hadamard transformation. In the second step the input qubit is prepared in the 0 state and is then subjected to a Hadamard gate as well, which immediately produces a superposition of the two possible inputs for the function f. In the third step we couple the input and output qubit via the f-controlled NOT gate. The transformation is

$$(|0\rangle+|1\rangle)(|0\rangle-|1\rangle) \to [(-1)^{f(0)}|0\rangle+(-1)^{f(1)}|1\rangle](|0\rangle-|1\rangle)\,. \tag{14.178}$$

In the last step we apply a Hadamard gate to the input qubit so that

$$[(-1)^{f(0)}|0\rangle+(-1)^{f(1)}|1\rangle](|0\rangle-|1\rangle) \to (-1)^{f(0)}|f(0)\oplus f(1)\rangle(|0\rangle-|1\rangle)\,. \tag{14.179}$$

Thus the input qubit is in state 0 if f is constant and is in state 1 if f is balanced and measurement of the qubit will determine if the function is balanced or constant with certainty in a single run of the machine.

There is a simple quantum optical realisation of this algorithm based on a Mach–Zehnder interferometer, see Fig. 14.4. The interferometer couples two modes of the field, labelled upper (U) and lower (L). A single photon in mode U encodes logical 1 while a single photon in mode L encodes logical 0. At the input a single photon in mode U is transformed by the first beam splitter into a superposition state in which it is in either mode 1 or mode 0. If we encode our qubits so that a $|1\rangle$ corresponds to the photon in mode 1 and a $|0\rangle$ corresponds to a photon in mode 0, the first beam splitter performs a Hadamard transformation. Now we insert into each arm a phase shift ϕ_i which can only be set at 0 or π phase shift. We encode the value of the functions as $\phi_0 = f(0)\pi$ and $\phi_1 = f(1)\pi$. Set the interferometer so that in the absence of the phase shift the photon emerges with certainty at the upper detector, which encodes a 1. The lower detector encodes a zero. It is then clear that if $f(0) = f(1)$ a single photon will emerge at the upper detector, while if $f(0) \neq f(1)$ the photon will be detected at the lower detector, that is the result is a 0.

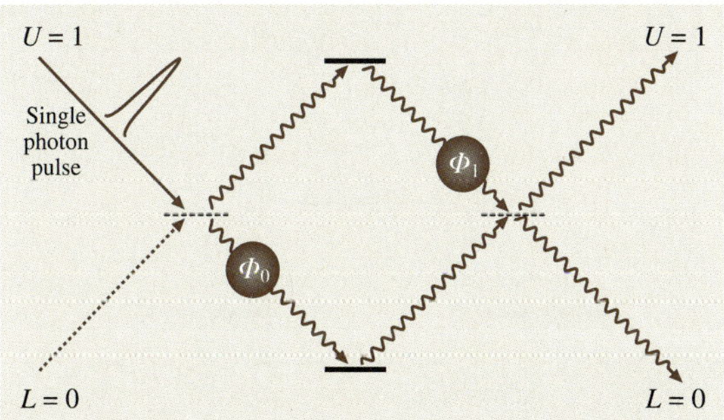

Fig. 14.4 An optical realisation of the Deutsch algorithm in terms of a Mach–Zehnder interferometer. The phase shifts are chosen according to the values of a binary function f as $\phi_0 = f(0)\pi$, $\phi_1 = f(1)\pi$

The previous example illustrates the key features of a quantum algorithm. Firstly it involves unitary transformations of pure quantum states. Secondly we need both single qubit and two qubit interactions to produce entangled states. These were the Hadamard transformation (H-gate) and a controlled NOT transformation (CNOT gate). It turns out that suitable networks of an arbitrary single qubit rotations, together with a controlled NOT gate, can perform any computation involving arbitrarily many qubits.

In the interferometric implementation of Deutsch algorithm we used a simple physical qubit based on a single-photon excitation of one of a pair of spatial modes. This is know as a *dual-rail* logic. The relationship between logical states and the physical photon number state is

$$|0\rangle_{\mathrm{L}} = |1\rangle_1 \otimes |0\rangle_2\,, \tag{14.180}$$

$$|1\rangle_{\mathrm{L}} = |0\rangle_1 \otimes |1\rangle_2\,. \tag{14.181}$$

The modes could be two input modes to a beam splitter distinguished by the different directions of the wave vector, or they could be distinguished by polarisation. In the case of a beam splitter a single qubit gate is easily implemented by the linear transformation

$$a_i(\theta) = U(\theta)^\dagger a_i U(\theta) \tag{14.182}$$

with $U(\theta) = \exp\left[\theta(a_1 a_2^\dagger - a_1^\dagger a_2)\right]$. Thus

$$a_1(\theta) = \cos\theta a_1 - \sin\theta a_2\,, \tag{14.183}$$

$$a_2(\theta) = \cos\theta a_2 + \sin\theta a_1\,. \tag{14.184}$$

The description in the logical basis becomes,

$$|0\rangle_{\mathrm{L}} \to \cos\theta_1|0\rangle_{\mathrm{L}} - \sin\theta_1|1\rangle_{\mathrm{L}}\,, \tag{14.185}$$

$$|1\rangle_{\mathrm{L}} \to \cos\theta_1|1\rangle_{\mathrm{L}} + \sin\theta_1|0\rangle_{\mathrm{L}}\,. \tag{14.186}$$

While single-qubit gates are readily implemented by linear optical devices such as beam splitters, quarter-wave plates, phase shifters etc., two-qubit gates are difficult. In order to implement the controlled phase gate (CSIGN) defined by

$$|x\rangle_{\mathrm{L}}|y\rangle_{\mathrm{L}} \to U_{\mathrm{CP}}|x\rangle_{\mathrm{L}}|y\rangle_{\mathrm{L}} = (-1)^{x.y}|x\rangle_{\mathrm{L}}|y\rangle_{\mathrm{L}}\,. \tag{14.187}$$

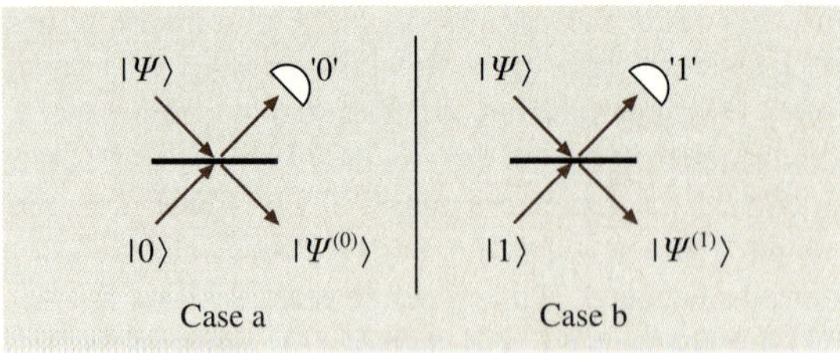

Fig. 14.5 A conditional state transformation conditioned on photon counting measurements

In a dual-rail single-photon code, this can be implemented using a two-mode Kerr nonlinearity. The two-mode generalisation is described by the Hamiltonian

$$H = \hbar\chi a_1^\dagger a_1 a_2^\dagger a_2 . \tag{14.188}$$

At the level of single photons this Hamiltonian produces the transformation, $|x\rangle|y\rangle \to \mathrm{e}^{-\mathrm{i}xy\chi t}|x\rangle|y\rangle$ and it is a simple matter to implement the CSIGN gate in the logical basis for the dual-rail single-photon code.

There are at least two problems in pursuing this approach: (a) the difficulty of realising number states in the laboratory, and (b) the difficulty of producing one-photon phase shifts of the order of π. The second difficulty is very considerable. Third-order optical nonlinearities are very small for a field with such a low intensity as a single photon. However experimental advances may eventually overcome this [14.37, 38].

A quite different approach to achieve large single-photon conditional phase shifts is based on the nonunitary transformation of a state that results when a measurement is made. Consider the situation shown in Fig. 14.5. Two modes of an optical field are coupled via a beam splitter. One mode is assumed to be in the vacuum state (a) or a one-photon state (b), while the other mode is arbitrary. A single-photon counter is placed in the output port of mode 2. What is the conditional state of mode 1 given a count of n photons?

Consider two modes, a_1 and a_2, coupled with a beam-splitter interaction, described by the one-parameter unitary transformation, given in (14.183), (14.184) We now assume that photons are counted on mode a_2 and calculate the conditional state for mode a_1 for two cases: no count and also for a single count at mode a_2. The conditional state of mode a_1 is given by (unnormalised),

$$|\tilde{\psi}^{(i)}\rangle_1 = \hat{\Upsilon}(i)|\psi\rangle_1 , \tag{14.189}$$

where

$$\hat{\Upsilon}(i) =_2 \langle i|U(\theta)|i\rangle_2 , \tag{14.190}$$

with $i = 1, 0$. The probability to observe each event is given by

$$P(i) = \langle\psi|\Upsilon^\dagger(i)\hat{\Upsilon}(i)|\psi\rangle_1 \tag{14.191}$$

which fixes the normalisation of the state,

$$|\psi^{(i)}\rangle_1 = \frac{1}{\sqrt{P(i)}}|\tilde{\psi}^{(i)}\rangle_1 . \tag{14.192}$$

We find that

$$\hat{\Upsilon}(0) = \sum_{n=0}^{\infty} \frac{(\cos\theta - 1)^n}{n!}\left(a_1^\dagger\right)^n a_1^n ,$$
$$\hat{\Upsilon}(1) = \cos\hat{\Upsilon}^{(0)} - \sin^2\theta a_1^\dagger \hat{\Upsilon}^{(0)} a_1 .$$

This can be written more succinctly using normal ordering,

$$\hat{\Upsilon}(0) =: \mathrm{e}^{\ln(\cos\theta)} : . \tag{14.193}$$

In order to see how we can use these kind of transformations to effect a CSIGN gate, consider the situation shown in Fig. 14.6. Three optical modes are mixed on a sequence of three beam splitters with beam splitter parameters θ_i. The *ancilla* modes, a_1 and a_2 are restricted to be in the single-photon states $|1\rangle_2$ and $|0\rangle_3$, respectively. We will assume that the *signal* mode a_0 is restricted to have *at most* two photons, thus

$$|\psi\rangle = \alpha|0\rangle_0 + \beta|1\rangle_0 + \gamma|2\rangle_0 . \tag{14.194}$$

This captures the fact that in the dual-rail encoding a general two-qubit state can have at most two photons. The objective is to chose the beam splitter parameters so that when the two detectors at the output of modes 2 and 3 detect 1 and 0 photons respectively (that is detect no change in their occupation), the signal state is transformed as

$$|\psi\rangle \to |\psi'\rangle = \alpha|0\rangle + \beta|1\rangle - \gamma|2\rangle \tag{14.195}$$

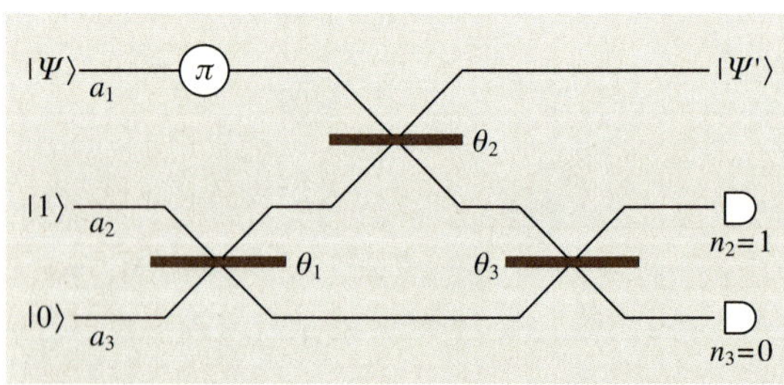

Fig. 14.6 A conditional state transformation on three optical modes, conditioned on photon counting measurements on the ancilla modes a_2, a_3. The signal mode a_1 is subjected to a π phase shift

with a probability that is *independent* of the input state $|\psi\rangle$. This last condition is essential as, in a quantum computation, the input state to a general two-qubit gate is completely unknown. We will call this transformation the NS (for nonlinear sign shift) gate. This can be achieved using: $\theta_1 = -\theta_3 = 22.5°$ and $\theta_2 = 65.53°$. The probability of the conditioning event ($n_2 = 1, n_3 = 0$) is 1/4. Note that we cannot be sure in a given trial if the correct transformation will be implemented. Such a gate is called a *nondeterministic* gate. However the key point is that success is heralded by the results on the photon counters (assuming ideal operation).

We can now proceed to a CSIGN gate in the dual-rail basis. Consider the situation depicted in Fig. 14.7. We first take two dual-rail qubits encoding for $|1\rangle_L|1\rangle_L$. The single-photon components of each qubit are directed towards a 50/50 beam splitter where they overlap perfectly in space and time and produces a state of the form $|0\rangle_2|2\rangle_3 + |2\rangle_2|0\rangle_3$,a effect known as the Hong-Ou-Mandel (HOM) interference [14.39]. We then insert an NS gate into each output arm of the HOM interference. When the conditional gates in each arm work, which occurs with probability 1/16, the state is multiplied by an overall minus sign. Finally we direct these modes towards another HOM interference. The output state is thus seen to be $-|1\rangle_L|1\rangle_L$. One easily checks the three other cases for the input logical states to see that this device implements the CSIGN gate with a probability of 1/16 and successful operation is heralded.

Clearly a sequence of nondeterministic gates is not going to be much use: the probability of success after a few steps will be exponentially small. The key idea in using nondeterministic gates for quantum computation is based on the idea of gate teleportation of *Gottesmann* and *Chuang* [14.40]. In quantum teleportation an unknown quantum state can be transferred from A to B provided A and B first share an entangled state. Gottesmann and Chuang realised that it is possible simultaneously teleport a two-qubit quantum state and implement a two-qubit gate in the process by first applying the gate to the entangled state that A and B share prior to teleportation.

We use a nondeterministic NS gate to prepare the required entangled state, and only complete the teleportation when the this stage is known to work. The teleportation step itself is nondeterministic but, as we see below, by using the appropriate entangled resource the teleportation step can be made near-deterministic. The near deterministic teleportation protocol requires only photon counting and fast feedforward. We do not need to make measurements in a Bell basis.

A nondeterministic teleportation measurement is shown in Fig. 14.8. The client state is a one-photon state in mode 0 $\alpha|0\rangle_0 + \beta|1\rangle_0$ and we prepare the entangled ancilla state

$$|t_1\rangle_{12} = |01\rangle_{12} + |10\rangle_{12} , \qquad (14.196)$$

where mode 1 is held by the sender A and mode-2 is held by the receiver B. For simplicity we omit normalisation constants wherever possible. This an ancilla state is easily generated from $|01\rangle_{12}$ by means of a beam splitter.

If the total count is $n_0 + n_1 = 0$ or $n_0 + n_1 = 2$, an affective measurement has been made on the client state and the teleportation has failed. However if $n_0 + n_1 = 1$, which occurs with probability 0.5, teleportation succeeds with the two possible conditional states being

$$\alpha|0\rangle_2 + \beta|1\rangle_2 \quad \text{if } n_0 = 1, n_1 = 0 , \qquad (14.197)$$

$$\alpha|0\rangle_2 - \beta|1\rangle_2 \quad \text{if } n_0 = 0, n_1 = 1 . \qquad (14.198)$$

This procedure implements a partial Bell measurement and we will refer to it as a nondeterministic teleportation protocol, $T_{1/2}$. Note that teleportation failure is detected and corresponds to a photon-number measurement of the state of the client qubit. Detected number

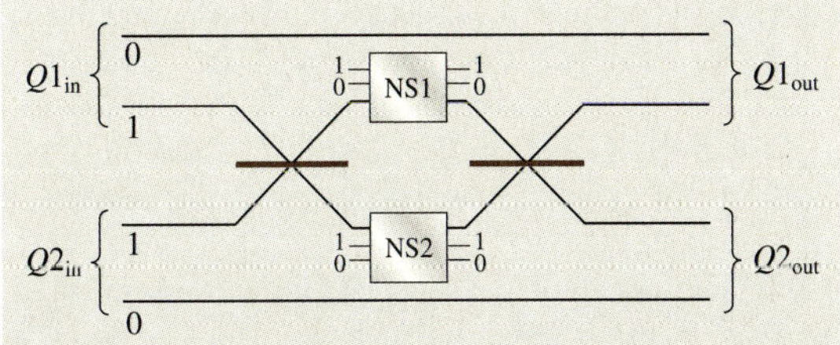

Fig. 14.7 A conditional state transformation conditioned on photon counting measurements. A CSIGN gate that works with probability of 1/16. It uses HOM interference and two NS gates

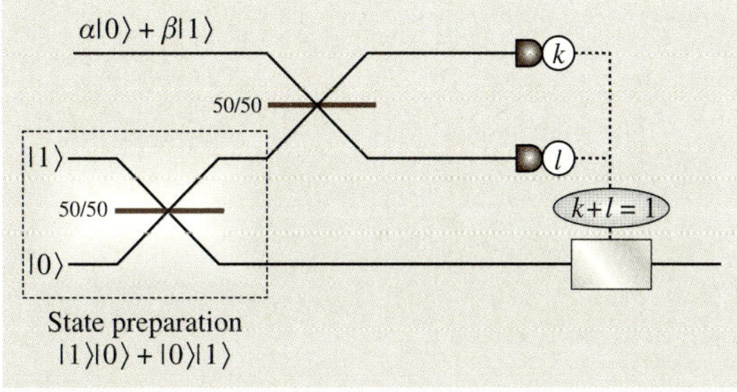

Fig. 14.8 A partial teleportation system for single-photons states using linear optics

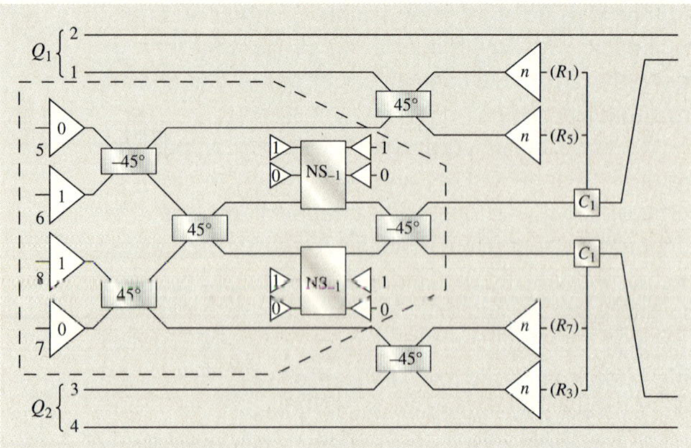

Fig. 14.9 A CSIGN two-qubit gate with teleportation to increase the success probability to 1/4. When using the basic teleportation protocol (T_1), we may need to apply a sign correction. Since this commutes with CSIGN, it is possible to apply CSIGN to the prepared state before performing the measurements, reducing the implementation of CSIGN to a state-preparation (outlined) and two teleportations. The two teleportation measurements each succeed with probability 1/2, giving a net success probability of 1/4. The correction operations C_1 consist of applying the phase shifter when required by the measurement outcomes

measurements are a very special kind of error and can be easily corrected by a suitable error correction protocol. For further details see [14.41]

The next step is to use $T_{1/2}$ to effect a conditional sign flip csign$_{1/4}$ which succeeds with probability 1/4. Note that to implement csign on two bosonic qubits in modes 1, 2 and 3, 4 respectively, we can first teleport the first modes of each qubit to two new modes (labelled 6 and 8) and then apply csign to the new modes. When using $T_{1/2}$, we may need to apply a sign correction. Since this commutes with csign, there is nothing preventing us from applying csign to the prepared state before performing the measurements. The implementation is shown in Fig. 14.9 and now consists of first trying to prepare two copies of $|t_1\rangle$ with csign already applied, and then performing two partial Bell measurements. Given the prepared state, the probability of success is $(1/2)^2$. The state can be prepared using csign$_{1/16}$, which means that the preparation has to be retried an average of 16 times before it is possible to proceed.

To improve the probability of successful teleportation to $1-1/(n+1)$, we generalise the initial entanglement by defining

$$|t_n\rangle_{1\dots 2n} = \sum_{j=0}^{n} |1\rangle^j |0\rangle^{n-j} |0\rangle^j |1\rangle^{n-j} \,. \tag{14.199}$$

The notation $|a\rangle^j$ means $|a\rangle|a\rangle\dots$, j times. The modes are labelled from 1 to $2n$, left to right. Note that the state exists in the space of n bosonic qubits, where the kth qubit is encoded in modes $n+k$ and k (in this order).

To teleport the state $\alpha|0\rangle_0+\alpha|1\rangle_0$ using $|t_n\rangle_{1\dots 2n}$ we first couple the client mode to half of the ancilla modes by applying an $n+1$ point Fourier transform on modes 0 to n. This is defined by the mode transformation

$$a_k \to \frac{1}{\sqrt{n+1}} \sum_{l=0}^{n} \omega^{kl} a_l \,, \tag{14.200}$$

where $\omega = \mathrm{e}^{\mathrm{i}2\pi/(n+1)}$ This transformation does not change the total photon number and is implementable with passive linear optics. After applying the Fourier transform, we measure the number of photons in each of the modes 0 to n. If the measurement detects k bosons altogether, it is possible to show [14.41] that, if $0<k<n+1$, then the teleported state appears in mode $n+k$ and only needs to be corrected by applying a phase shift. The modes $2n-l$ are in state 1 for $0 \le l < (n-k)$ and can be reused in future preparations requiring single bosons. The modes are in state 0 for $n-k<l<n$. If $k=0$ or $k=n+1$ an effective measurement of the client is made, and the teleportation fails. The probability of these two events is $1/(n+1)$, regardless of the input. Note that again failure is detected and corresponds to measurements in the basis $|0\rangle$, $|1\rangle$ with the outcome known. Note that both the necessary correction and the receiving mode are unknown until after the measurement.

The linear optics quantum computing (LOQC) model described above can be drastically simplified by adopting the cluster state method of quantum computation [14.42]. The cluster state model was developed by *Raussendorf* and *Breigel* [14.43] and is quite different from the circuit models that we have been using. In cluster state QC, an array of qubits is initially prepared in a special entangled state. The computation then proceeds by making a sequence of single-qubit measurements. Each measurement is made in a basis that depends on prior measurement outcomes. Nielsen realised that the LOQC mode of [14.41] could be used to efficiently assemble the cluster using the nondeterministic teleportation t_{n}. As we saw the failure mode of this gate constituted an accidental measurement of the qubit in the computational basis. The key point is that such an error does not destroy the entire assembled cluster but merely detaches one qubit from the cluster. This enables a protocol to be devised that produces a cluster that grows on average. The LOQC cluster state method dramatically reduces the number of optical elements required to implement the original LOQC scheme. Of course if large single-photon Kerr nonlinearities were available, the optical cluster state method could be made deterministic [14.44].

A number of LOQC protocols have been implemented in the laboratory. The first was experiment was performed by *Pittmann* and *Franson* [14.45]. This used entangled ancillas that are readily produced as photon pairs in a spontaneous parametric down conversion process. A simplified version of the LOQC model was implemented by *O'Brien* et al. [14.46], based on a proposal of *Ralph* et al. [14.47] for a CNOT gate. The simplification results by firstly setting the beam splitter parameters θ_1 and θ_3 to zero in the NS gate implementation and secondly only detecting photon coincidences at the output. This gate performs all the operations of a CNOT gate but requires only a two-photon input. Detecting only coincidences means that the device must be configured so that correct operation leads to a coincidence detection of both photons at the output. The gate is nondeterministic but gate failures are simply not detected at all. In essence, the control (C) and target (T) qubits act as their own ancilla.

In the University of Queensland (UQ) experiment the two modes of each qubit are distinguished by orthogonal polarisations. This may be converted to spatial mode encoding by using polarising beam splitters and a half-wave plate. The key advantage in using a gate based on two-photon coincidence detection is that spontaneous parametric down-conversion (SPDC) may be used in place of true single-photon sources. An SPDC produces a photon pair in two distinct spatiotemporal modes at random times. There is a small probability of producing more than two photons, but this can be neglected.

A diagnostic of the fidelity of gate operation is provided by state tomography, a reconstruction of the full density matrix of the output state [14.48], when the output is entangled. State tomography requires sampling of the statistics for the measurement outcomes of 16 different two-qubit projections. Given these statistics data inversion can be devised to reconstruct the density matrix for the output state. Given the density matrix, we can then compute its overlap, or fidelity, with respect to the pure ideal entangled state $|\psi^-\rangle$ that the ideal gate would produce. In the case of $|\psi^-\rangle$ the fidelity obtained in the UQ experiment was 0.87 ± 0.08. This is sufficiently high that such a state were it not destroyed in the detection process, would violate a Bell inequality test.

More recent experiments have improved on these early experiments. An NS gate close to the original proposal, was implemented in the Zeilinger group, using a polarization encoding and the four-photon state emitted by spontaneous parametric down-conversion [14.49]. As in the UQ experiment, a coincidence detection configuration was used to signal correct operation of the gate. The experimentally observed conditional phase shift was $1.05 \pm 0.06)\pi$. Future progress on linear optical quantum computing schemes will most likely be based on cluster state implementations. A four-photon cluster state implementation was recently implemented by the Zeilinger group [14.50].

References

14.1 L. H. Ryder: *Quantum Field Theory* (Cambridge Univ. Press, Cambridge 1996)

14.2 R. Loudon: *Quantum Theory of Light* (Oxford Univ. Press, Oxford 1973)

14.3 B.W. Shore, P.L. Knight: The Jaynes–Cummings model, J. Mod. Opt. **40**, 1195–1238 (1993)

14.4 T. Aoki, B. Dayan, E. Wilcut, W.P. Bowen, A.S. Parkins, T.J. Kippenberg, K.J. Vahala, H.J. Kimble: Observation of strong coupling between one atom and a monolithic microresonator, **443**, 671 (2006)

14.5 G.J. Milburn, S.L. Braunstein: Teleportation using squeezed vacuum states, Phys. Rev. A **60**, 937 (1999)

14.6 J. Mod. Optics, Special Issue on Single photon Sources, **51**, Numbers 9–10 (2004)

14.7 A. B. U'Ren, E. Mukamel, K. Banaszek, I.A. Walmsley: Managing photons for quantum information processing, Phil. Trans. R. Soc. A **361**, 1471 (2003)

14.8 R.J. Glauber: The quantum theory of optical coherence, Phys. Rev. **130**, 2529 (1963)

14.9 L. Mandel, E.C.G. Sudarshan, E Wolf: *Theory of photoelectric detection of light fluctuations*, Proc. Phys. Soc. **435** (1964)

14.10 P.L. Kelley, W.H. Kleiner: Theory of electromagnetic field measurement and photoelectron counting, Phys. Rev. A **136**, 316 (1964)

14.11 H.P. Yuen, J.H. Shapiro: Optical communication with two-photon coherent states. III – Quantum measurements realizable with photoemissive detectors, IEEE Trans. Inform. Theory **26**, 78 (1980)

14.12 M.J. Collett, D.F. Walls: Squeezing spectra for nonlinear optical systems, Phys. Rev. A **32**, 2887 (1985)

14.13 C.W. Gardiner: *Handbook of Stochastic Processes* (Springer, Berlin Heidelberg New York 1985)

14.14 H.J. Carmichael: *Statistical Methods in Quantum Optics 1, Master Equations and Fokker Planck Equations* (Springer, Berlin Heidelberg New York 1999)

14.15 D. H. Santamore, Hsi-Sheng Goan, G. J. Milburn, M. L. Roukes: Anharmonic effects on a phonon-number measurement of a quantum-mesoscopic-mechanical oscillator, Phys. Rev. A **70**, 052105 (2004)

14.16 D. F. Walls, G. J. Milburn: *Quantum Optics* (Springer, Berlin Heidelberg New York 1994)

14.17 H. J. Carmichael, D. F. Walls: Proposal for the measurement of the resonant Stark effect by photon correlation techniques, J. Phys. B **9**, L43 (1976)

14.18 M. D. Srinivas, E. B. Davies: Photon counting probabilities in quantum optics, J. Mod. Opt. **28**, 981 (1981)

14.19 K. Mølmer, Y. Castin, J. Dalibard: Monte Carlo wave-function method in quantum optics, J. Opt. Soc. Am. B **10**, 524 (1993)

14.20 J. I. Cirac, P. Zoller: Quantum computations with cold trapped ions, Phys. Rev. Lett. **74**, 4091 (1995)

14.21 H. J. Dehmelt: Proposed dye laser study of 5sÐ4d e2 transition in single Sr+ ion, Bull. Am. Phys. Soc. **20**, 60 (1975)

14.22 J. C. Bergquist, R. G. Hulet, W. M. Itano, D. J. Wineland: Observation of quantum jumps in a single atom, Phys. Rev. Lett. **57**, 1699 (1986)

14.23 J. I. Cirac, R. Blatt, P. Zoller, W. D. Philips: Laser cooling of trapped ions in a standing wave, Phys. Rev. A **46**, 2668 (1992)

14.24 D. Leibfried, R. Blatt, C. Monroe, D. Wineland: Quantum dynamics of single trapped ions, Rev. Mod. Phys. **75**, 281 (2003)

14.25 C. Monroe, D. M. Meekhof, B. E. King, S. R. Jefferts, W. M. Itano, D. J. Wineland, P. L. Gould: Resolved-sideband Raman cooling of a bound atom to the 3D zero-point energy, Phys. Rev. Lett. **75**, 4011 (1995)

14.26 C. H. Bennett, G. Brassard, C. Crepeau, R. Jozsa, A. Peres, W. K. Wooters: Teleporting an unknown quantum state via dual classical and Einstein–Podolsky–Rosen channels, Phys. Rev. Lett. **70**, 1895 (1993)

14.27 B. Schumacher: Quantum coding, Phys. Rev. A **51**, 2783 (1995)

14.28 A. Furasawa, J. L. Sørensen, S. L. Braunstein, C. A. Fuchs, H. J. Kimble, E. S. Polzik: Unconditional quantum teleportation, Science **282**, 706 (1998)

14.29 S. L. Braunstein, H. J. Kimble: Teleportation of continuous quantum variables, Phys. Rev. Lett. **80**, 869 (1998)

14.30 L. Vaidman: Teleportation of quantum states, Phys. Rev. A **49**, 1473 (1994)

14.31 D. Boschi, S. Branca, F. De Martini, L. Hardy, S. Popescu: Experimental realization of teleporting an unknown pure quantum state via dual classical and Einstein–Podolsky–Rosen channels, Phys. Rev. Lett. **80**, 1121 (1998)

14.32 D. Bouwmeester, J.-W. Pan, M. Daniell, H. Weinfurter, M. Zukowski, A. Zeilinger: Experimental quantum teleportation, Nature **390**, 575 (1997)

14.33 R. P. Feynman: Simulating physics with computers, Int. J. Theor. Phys **21**, 467 (1982)

14.34 D. Deutsch: Quantum-theory, the Church-Turing principle and the universal quantum computer, Proc. R. Soc. Lond. A **400**, 97–117 (1985)

14.35 P.Shor: Algorithms for quantum computation: Discrete algorithms and factoring, Proc. 35th Annual Symposium on Foundations of Computer Science (1994). See also LANL preprint quant-ph/9508027

14.36 R. Cleve, A. Ekert, L. Henderson, C. Macchiavello, M. Mosca: On quantum algorithms, LANL quant- ph/ 9903061 17 Mar 1999.

14.37 Q. A. Turchette, C. J. Hood, W. Lange, H. Mabuchi, H. J. Kimble: Measurement of conditional phase shifts for quantum logic, Phys. Rev. Lett. **75**, 4710 (1995)

14.38 W. J. Munro. Kae Nemoto: Nearly deterministic linear optical controlled-NOT gate, Phys. Rev. Lett. **93**, 250502 (2004)

14.39 C. K. Hong, Z. Y. Ou, L. Mandel: Measurement of subpicosecond time intervals between two photons by interference, Phys. Rev. Lett. **59**, 2044 (1987)

14.40 D. Gottesman, I. L. Chuang: Demonstrating the viability of universal quantum computation using teleportation and single-qubit operations, Nature **402**, 390–393 (1999)

14.41 E. Knill, R. Laflamme, G. J. Milburn: Efficient linear optical quantum computation, Nature **409**, 46 (2001)

14.42 M. A. Nielsen: Optical quantum computation using cluster states, Phys. Rev. Lett. **93**, 040503 (2004)

14.43 R. Raussendorf, H. J. Briegel: A one-way quantum computer, Phys. Rev. Lett. **86**, 5188 (2001)

14.44 G. D. Hutchinson, G. J. Milburn: Nonlinear quantum optical computing via measurement, J. Mod. Opt. **51**, 1211–1222 (2004)

14.45 T. B. Pittman, B. C. Jacobs, J. D. Franson: Probabilistic quantum logic operations using polarizing beam splitters, Phys. Rev. A **64**, 062311 (2001)

14.46 J. L. OBrien, G. J. Pryde, A. G. White, T. C. Ralph, D. Branning: Demonstration of an all-optical quantum controlled-NOT gate, Nature **426**, 264 (2003)

14.47 T. C. Ralph, N. K. Langford, T. B. Bell, A. G. White: Linear optical controlled-NOT gate in the coincidence basis, Phys. Rev. A **65**, 062324 (2002)

14.48 D. F. W. James, P. G. Kwiat, W. G. Munro, A. G. White: Measurement of qubits, Phys. Rev. **64**, 052312 (2001)

14.49 K. Sanaka, T. Jennewein, Jian-Wei Pan, K. Resch, A. Zeilinger: Experimental nonlinear sign shift for linear optics quantum computation, Phys. Rev. Lett. **92**, 017902–1 (2004)

14.50 P. Walther, K. J. Resch, T. Rudolph, E. Schenck, H. Weinfurter, V. Vedral, M. Aspelmeyer, A. Zeilinger: Experimental one-way quantum computing, Nature **434**, 169 (2005)

15. Nanooptics

Nanooptics deals with optical near fields, the electromagnetic fields that mediate the interaction between nanometric particles located in close proximity to each other. The projection-operator method is a theoretical description of how a virtual exciton-polariton is exchanged between these particles, corresponding to the nonresonant interaction. The optical near field mediates this interaction, and is represented by a Yukawa function, which means that the optical near-field energy is localized around the nanometric particles like an electron cloud around an atomic nucleus. Its decay length is proportional to the particle size. This chapter is primarily a review of nanophotonics, a leading branch of nanooptics, which is the technology utilizing the optical near field. The true nature of nanophotonics is to realize qualitative innovation in photonic devices, fabrication, and systems by utilizing novel functions and phenomena caused by optical near-field interactions, which are impossible as long as conventional propagating light is used. As evidence of such qualitative innovation, this chapter describes novel nanophotonic devices, nanophotonic fabrication, nanophotonic systems, and extensions related to science.

15.1 Basics

Nanooptics deals with the light generated in or on nanometer-sized particles. It has been applied to open an innovative field of technology, i.e., *nanophotonics*, in order to develop novel photonic devices, fabrications, and systems, as proposed by *Ohtsu* [15.1]. Nanophotonics is a means, for example, to reduce the size of photonic devices, improve the resolution of optical fabrications, and increase the storage density of optical disk memories.

Considering photonic devices, the technologies listed below, which bear some relation to nanophotonics, have recently been developed. Nevertheless, they cannot reduce the size of these devices beyond the diffraction limit of light.

1. **Photonic crystal** [15.2] This is a filter device used to control optical interference and light scattering by installing a subwavelength-sized periodic structure in the device material, as is the case for a diffraction grating. Constructive interference between the scattered light takes place at the center of the material, concentrating the optical energy. In contrast, at the edge of the material, the scattered light interferes destructively. In order to maintain the destructive interference, the rim must be sufficiently larger than the wavelength of light; otherwise the constructive interference is not maintained, and light concentrated at the center leaks to the rim. This means that a photonic crystal involves conventional wave-optical technology and the minimum size of the device is limited by diffraction.
2. **Plasmonics** [15.3] This technology utilizes the resonant enhancement of the light in a metal by exciting free electrons. The optical energy can be concentrated as a plasmon on the metal surface as a result

of the strong interaction with free electrons. However, this technology is based on wave optics in the metal, and so is again limited by diffraction. Related technologies, such as metaoptics [15.4] and optics in materials with a negative refractive index [15.5], are also limited by diffraction.

3. **Silicon photonics** [15.6] This technology involves developing narrow-striped optical waveguides using high refractive-index silicon crystals to confine the light effectively. This is also an application of wave optics in silicon, and is limited by diffraction.
4. **Quantum-dot laser** [15.7] This is a laser device that uses nanometer-sized semiconductor quantum dots as the gain media. Since the quantum dots are much smaller than the wavelength of light, the light cannot be confined in an individual quantum dot efficiently due to scattering and diffraction. To solve this problem, a large number of quantum dots are installed in a conventional laser cavity, which means that the device size is limited by diffraction.

Although the four examples listed above have been popular subjects of study in recent years, they are all based on diffraction-limited wave optics. Even if novel or nanometer-sized materials are used in the future, the size of a photonic device cannot be reduced beyond the diffraction limit as long as propagating light is used for its operation. This also applies to improvements in the resolution of optical fabrication and for increasing the storage density of optical disk memories. To go beyond the diffraction limit, one needs non-propagating nanometer-sized light to induce primary excitation in a nanometer-sized material in such a manner that the spatial phase of the excitation is independent of that of the incident light.

15.2 Nanophotonics Principles

The use of optical near fields has been proposed as a way to transcend the diffraction limit imposed on optical science and technology [15.8]. This proposal holds that an optical near field can be generated on a subwavelength-sized aperture by irradiating the propagating light. It also holds that the size of the spatial distribution of the optical near-field energy depends not on the wavelength of the incident light, but on the aperture size. Although these claims are no more than those in the framework of primitive wave optics, optical near fields have been applied to realize diffraction-free, high-resolution optical microscopy (i. e., optical near-field microscopy), which achieved rapid progress after high-resolution, high-throughput fiber probes were invented and fabricated in a reproducible manner [15.9, 10]. However, in the early stage of these studies, the concept of optical near fields was not clearly discriminated from that of an evanescent wave on a planar material surface (i. e., a two-dimensional topographical material) or that of a guided wave in a subwavelength-sized cross-sectional waveguide (i. e., a one-dimensional topographical material).

In order to distinguish these clearly, note that an evanescent wave is generated by the primary excitations, i. e., electric dipoles, induced near the two-dimensional material surface, which align periodically depending on the spatial phase of the incident light. In contrast, the guided wave in a subwavelength-sized cross-sectional waveguide is generated by the electronic dipoles induced along the one-dimensional waveguide material. They align periodically depending on the spatial phase of the incident light. Examples are the silicon and metallic waveguides used for silicon photonics and plasmonics, respectively. The two-dimensional evanescent wave and one-dimensional guided wave are both light waves, which are generated by periodic alignment of electric dipoles depending on the spatial phase of the incident light.

Unlike these waves, an optical near field is generated by the electronic dipoles induced in a nanometric particle (i. e., a subwavelength-sized zero-dimensional topographical material). Their alignment is independent of the spatial phase of the incident light because the particles are much smaller than the wavelength of the incident light. Instead, it depends on the size, conformation, and structure of the particles. Due to this independence and dependence, optical science and technology beyond the diffraction limit can be realized only by an optical near field, not by an evanescent wave or a guided wave.

Classical electromagnetics explains the mechanisms of optical near-field generation described above (Fig. 15.1): electric dipoles are induced by irradiating a nanometric particle with the incident light. Among the electric lines of forces originating from these electric dipoles, the optical near field is represented by those that originate from the positive charge of the electric dipole and terminate on the negative charge. This does not propagate to the far field. Since the particle is much smaller than the wavelength of the incident light, the alignment of the electric dipoles is determined independently of the spatial phase of the incident light. Therefore, the spatial distribution and decay length of the optical near-field energy depend not on the wavelength of the light, but on the size, conformation, and structure of the particle. Moreover, the scattered light is represented by the closed loop of the electric line of forces, and propagates to the far field.

Methods such as Green's function, numerical calculation using the finite-difference time domain (FDTD) method, have been developed for semiquantitative description of the optical near field based on conventional optical theories (i. e., the theories discussed in other chapters of this handbook) [15.11]. However, conventional optics theories do not provide any physically intuitive pictures of non-propagating nanometric optical near fields because these theories have been developed to describe light waves propagating through macroscopic space or materials. For nanooptics and nanophotonics, a novel theory has been developed based on a framework that is completely different from those of the conventional theories.

This novel theory is based on how one observes an optical near field, i. e., the interaction and energy transfer between nanometric particles via an optical near field. This perspective is essential because the interaction and energy transfer are indispensable for nanophotonic devices and nanophotonic fabrications. That is, in order to observe a non propagating optical near field, a second particle is inserted to generate observable scattered light by disturbing the optical near field (Fig. 15.2). However, the real system is more complicated than that shown in Fig. 15.2 because the nanometric subsystem (the two particles and the optical near field) is buried in the macroscopic subsystem consisting of the macroscopic substrate material and the macroscopic electromagnetic fields of the incident and scattered light (Fig. 15.3).

The premise behind the novel theory is to avoid the complexity of describing all of the behaviors of nanometric and macroscopic subsystems rigorously, since we are interested only in the behavior of the nanometric subsystem. The macroscopic subsystem is expressed as an exciton–polariton, which is a mixed state of material excitation and electromagnetic fields. Since the nanometric subsystem is excited by an electromagnetic interaction with the macroscopic subsystem, the projection-operator method is effective for describing the quantum-mechanical states of these systems [15.12].

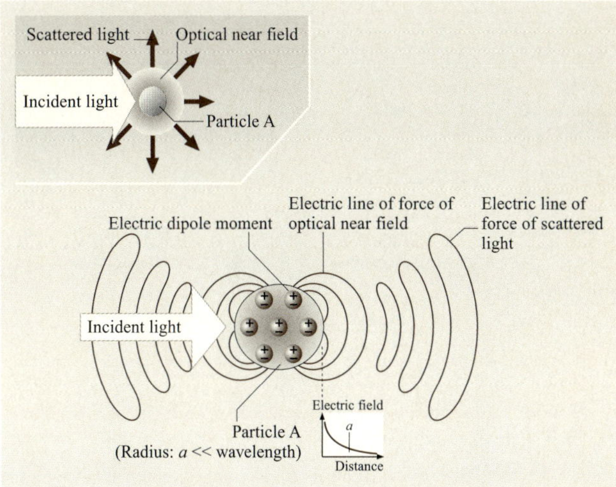

Fig. 15.1 The generation of optical near fields

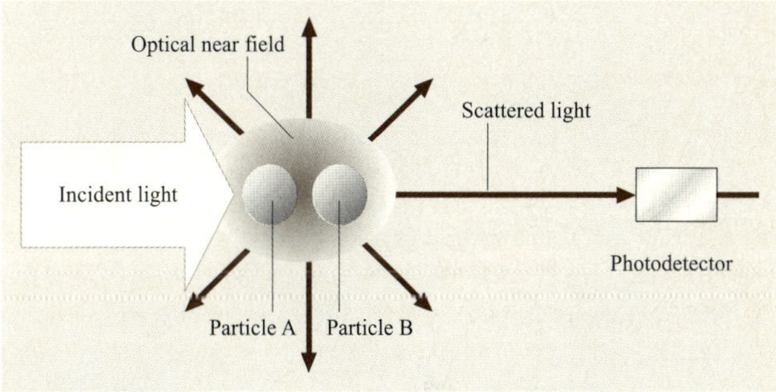

Fig. 15.2 The observation of optical near fields

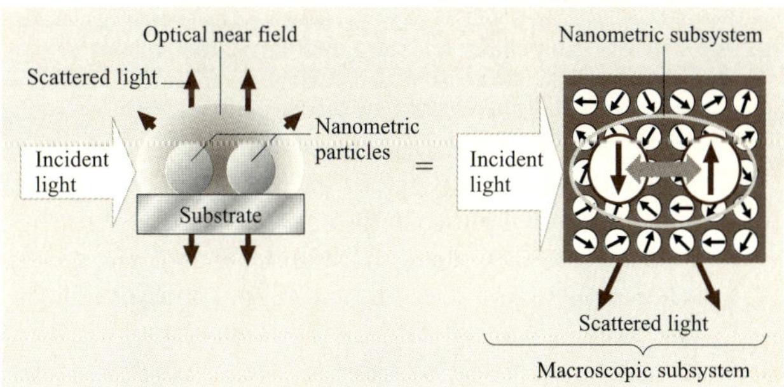

Fig. 15.3 A nanometric subsystem composed of two particles and an optical near field, buried in a macroscopic subsystem

Under this treatment, the nanometric subsystem is regarded as being isolated from the macroscopic subsystem, while the functional form and magnitude of the effective interactions between the elements of the nanometric subsystem are influenced by the macroscopic

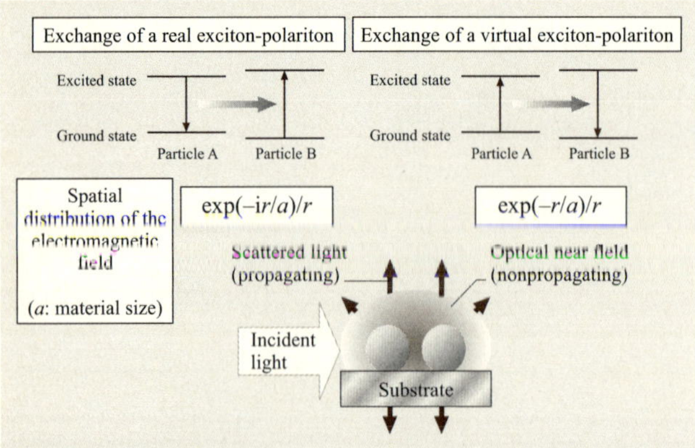

Fig. 15.4 The exchange of a real exciton–polariton and a virtual exciton–polariton

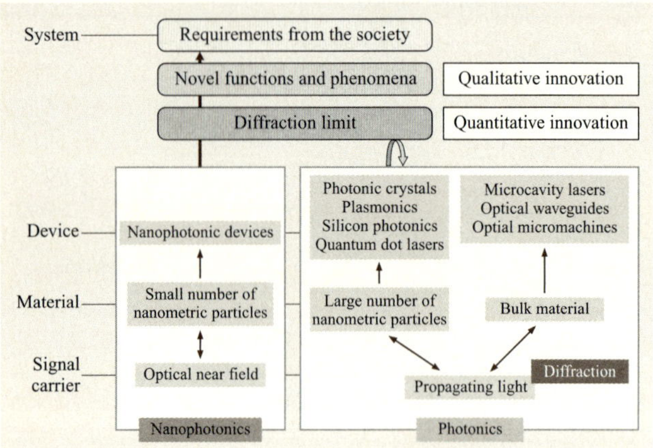

Fig. 15.5 Nanophotonics for realizing qualitative innovation

subsystem. That is, the two nanometric particles can be considered as being isolated from the surrounding macroscopic systems and as interacting by exchanging the exciton–polariton energies.

Since the time required for this local electromagnetic interaction is very short, the uncertainty principle allows the exchange of a virtual exciton–polariton between the two nanometric particles, as well as that of a real exciton–polariton (Fig. 15.4). The former exchange corresponds to the nonresonant interaction between the two particles. The optical near field mediates this interaction, which is represented by a Yukawa function. The Yukawa function represents the localization of the optical near-field energy around the nanometric particles like an electron cloud around an atomic nucleus whose decay length is equivalent to the material size [15.12]. The latter corresponds to the resonant interaction mediated by the conventional propagating scattered light, which is represented by a conventional spherical wave function.

As described, the optical near field is an electromagnetic field that mediates the interaction between nanometric particles located in close proximity to each other. Nanooptics deals with this field. Nanophotonics is the technology that utilizes this field to realize novel device, fabrications, and systems. That is, a photonic device with a novel function can be operated by transferring the optical near-field energy between nanometric particles. In such a device, the optical near field transfers a signal and carries the information. Novel photonic systems become possible by using these novel photonic devices. Furthermore, if the magnitude of the transferred optical near-field energy is sufficiently large, structures or conformations of nanometric particles can be modified, which suggests the feasibility of novel photonic fabrication.

Note that the true nature of nanophotonics is to realize qualitative innovation in photonic devices, fabrications, and systems by utilizing novel functions and phenomena caused by optical near-field interactions, which are impossible as long as conventional propagating light is used (Fig. 15.5). On reading this note, one may understand that the advantage of going beyond the diffraction limit, i. e., quantitative innovation, is no longer essential, and is only a secondary nature of nanophotonics. In this sense, one should also note that optical near-field microscopy, i. e., the methodology used for image acquisition and interpretation in a nondestructive manner, is not an appropriate application of nanooptics and nanophotonics because the magnitude of the optical near-field energy transferred between the probe and sample must be extrapolated to zero.

Although few examples of qualitative innovations in nanooptics exist, nanophotonics has already realized this innovation by correctly utilizing the true nature of the optical near-field interaction, which is reviewed in the following sections.

15.3 Nanophotonic Devices

For the operation of nanophotonic devices, it is essential to transfer the signal and fix the transferred signal magnitude at the output terminal, which can be achieved by transferring and dissipating the optical near-field energy, respectively [15.13]. Several research groups have recently begun similar discussions on this operation (for example, refer to [15.14]). A pair of two closely spaced equivalent quantum dots is used as the input terminal of the nanophotonic device (Fig. 15.6). As a result of the optical near-field interaction between the two quantum dots driven by the input optical signal, the quantized energy levels of the exciton in the quantum dots are split into two. One corresponds to the symmetric state of the exciton, and the other is the antisymmetric state. They represent the respective parallel and antiparallel electric dipole moments induced in the two quantum dots.

The third, larger quantum dot, installed proximal to the input terminal (Fig. 15.7), is used as the output terminal of the device. The higher energy level of the exciton in this quantum dot is tuned to that of the symmetric state of the input terminal, which is made possible by adjusting the size of the third quantum dot. As a result of this tuning, optical near-field energy can be transferred from the input to the output terminal, allowing signal transfer. The excitation transferred to the output terminal is relaxed immediately to the lower energy level in the third quantum dot by coupling with phonons, which fixes the magnitude of the transferred signal.

A device utilizing the symmetric state of the input terminal in the manner shown in Fig. 15.7 is called a *phonon-coupled device*. Examples of phonon-coupled devices include nanophotonic switches, logic gates, such as AND and NOT gates [15.15], content-addressable memory [15.16], and digital-to-analog

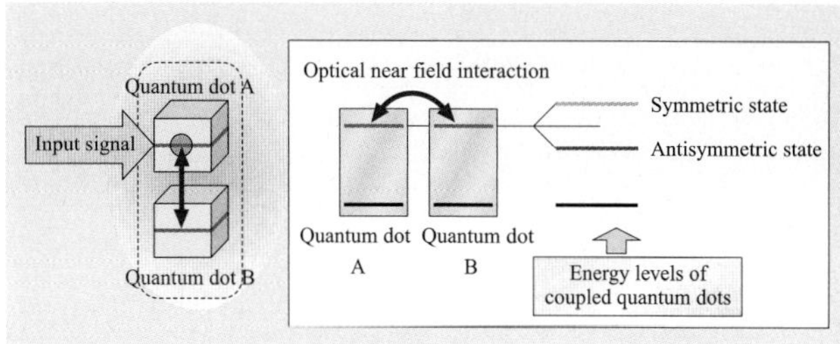

Fig. 15.6 Two closely spaced quantum dots to be used as the input terminal of a nanophotonic device

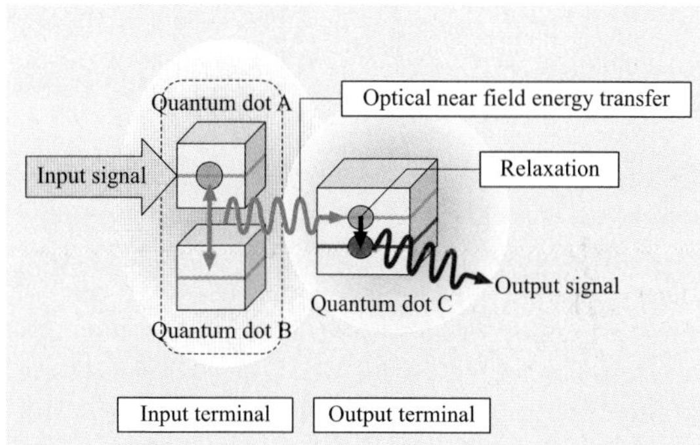

Fig. 15.7 Optical near-field energy transfer from the input to the output terminal and successive fast relaxation

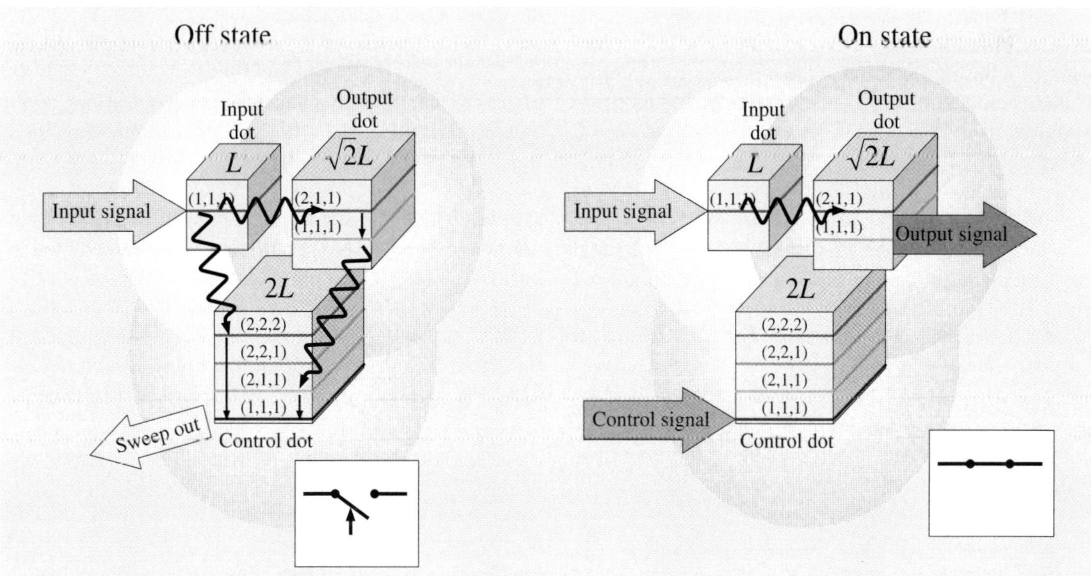

Fig. 15.8 The structure of a nanophotonic switch. L is the size of the quantum dot used as the input terminal

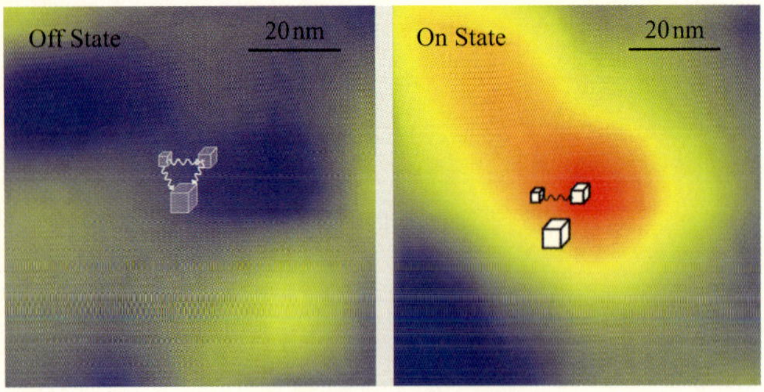

Fig. 15.9 The spatial distribution of the output signal energy at the output terminal of the nanophotonic switch. This switch used three cubic CuCl quantum dots with sizes of 4, 6, and 8 nm, separated by a fixed distance of 3 nm

converters [15.17]. Conversely, a device utilizing the antisymmetric state is called a *propagating light-coupled device*; for example, optical buffer memory and super-radiant-type optical pulse generators [15.18].

As an example of a phonon-coupled device, Fig. 15.8 shows a nanophotonic switch whose operation has been modified slightly from that shown in Fig. 15.7 [15.19]. Three cubic quantum dots with a size ratio of $1:\sqrt{2}:2$ are used as the input, output, and control terminals, respectively. This device utilizes the optical near-field energy transfer between the resonant energy levels of the exciton and successive fast relaxation. Note that the conventional optical transition between two resonant exciton levels in the adjacent quantum dots, e.g., the (1,1,1) and (2,1,1) energy levels of the input and output terminals, is forbidden. That is, no energies are transferred as long as the propagating light mediates the interaction between the quantum dots. In contrast, when using an optical near field, this transition is allowed due to the exchange of the virtual exciton–polariton. This is an example of a qualitatively innovative device operation because it is impossible as long as conventional propagating light is used. Figure 15.9 shows the spatial distributions of the output signal energy on the output terminal for the off and on states of the switching operation. The device is as small as 15 nm. Figure 15.10 represents the dynamic behavior of the output signal, which agrees well with the calculated results based on the virtual exciton–polariton model.

A figure of merit has been used as an important parameter for evaluating device performance. In the case of a switch, it is defined as a combination of device volume, switching time, switching energy, the ratio between on- and off-signal magnitudes, etc. The figure of merit for the nanophotonic switch is 10–100 times larger than those

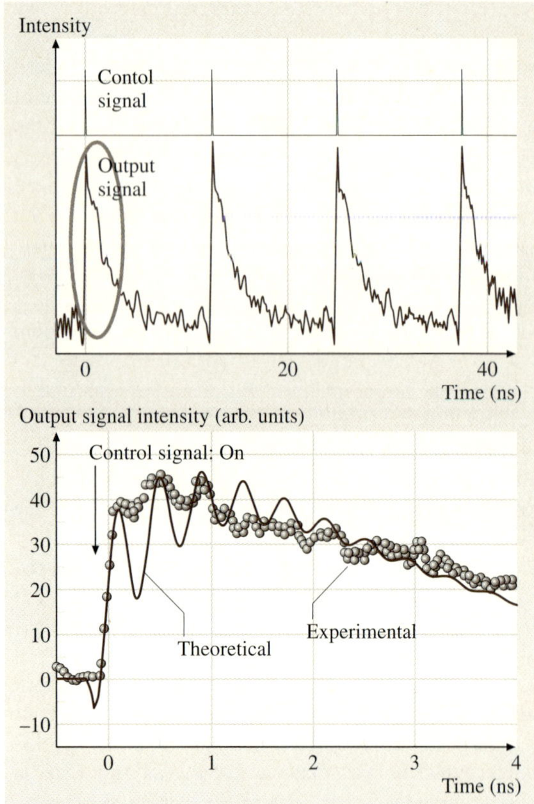

Fig. 15.10 Dynamic behavior of the output signal of a nanophotonic switch. The figure on the *bottom* shows the experimental and theoretical curves of the part of the output signal identified by the ellipse in the figure on the *top*

of conventional photonic switches operated using propagating light (e.g., devices using optical micromirrors, Mach–Zehnder interferometer-type waveguides, optically nonlinear media, and the quantum sublevels in a quantum dot). Furthermore, the magnitude of the heat dissipated from the nanophotonic switch is estimated to be as low as 10^{-12} W in the case of 1 GHz repetitive operation, which is only 10^{-5} times that of a conventional semiconductor transistor. Such a large figure of merit and the ultralow heat dissipation suggest a variety of applications to novel computation and information processing systems.

For applications to advanced systems, nanophotonic devices must be connected to conventional macroscopic photonic devices, such as optical fibers or semiconductor lasers. Figure 15.11 shows a novel device developed for this interconnection that concentrates the propagating light energy to a nanometric region [15.20]. It has

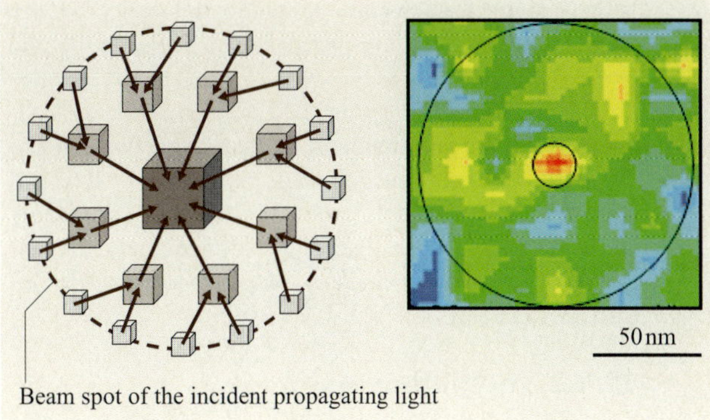

Fig. 15.11 The structure of an optical nanofountain and the spatial distribution of the optical signal energy, which used CuCl quantum cubes of 2–10 nm

smaller quantum dots on the circular rim, medium-sized quantum dots installed inside, and the largest quantum dot fixed at the center. The propagating light energy incident into this circular device is concentrated to the largest quantum dot at the center due to both successive optical near-field energy transfer from the smaller to the larger quantum dots and fast relaxation, as in the case of the nanophotonic switch. This device has been named an *optical nanofountain* and its equivalent numerical aperture is as high as 40.

15.4 Nanophotonic Fabrications

Optical fabrications beyond the diffraction limit are required to produce a variety of conventional devices and to fabricate the nanophotonic devices reviewed in Sect. 15.3. To fabricate nanophotonic devices, several capabilities are required. A variety of materials must be deposited on a substrate and the inaccuracy of their sizes and positions must be as low as 1 nm for efficient, reproducible optical near-field energy transfer. However, conventional fabrication technologies using electron beams, ion beams, and propagating light cannot meet these requirements due to their low resolution, contamination, damage to the substrate, and low throughput.

To meet these requirements, novel technologies must be developed that can be realized by utilizing optical near-field energy transfer. Representative examples of such nanophotonic fabrications are reviewed in this section.

15.4.1 Photochemical Vapor Deposition

Photochemical vapor deposition is a way to deposit materials on a substrate using a photochemical reaction that pre-dissociates metalorganic molecules by irradiating gaseous molecules or molecules adsorbed on the substrate with ultraviolet light. Consequently, the electrons in the molecules are excited to a higher energy level following the Franck–Condon principle. This is an adiabatic process because no molecular vibrations or rotations are excited; i. e., the Born–Oppenheimer approximation is valid. After excitation, the electron transits to the dissociative energy level. As a result of this transition, the molecule is dissociated and the atoms composing the molecule are deposited on the substrate. By using an optical near field as the light source for this deposition scheme, nanometric materials can be fabricated whose size and position are accurately controlled by the spatial distribution of the optical near-field energy. Nanometric metallic particles of Zn and Al and light-emitting semiconductor particles (e.g., ZnO and GaN) have been deposited with size and positional accuracies beyond the diffraction limit [15.21–25]. This is an example of a quantitative innovation.

Although an ultraviolet light with high photon energy must be used for the aforementioned adiabatic process, it has been found that an optical near field with a much lower photon energy (i. e., visible light) can dissociate the molecule. This has been explained by a theoretical model of the virtual exciton–polariton exchange between a metalorganic molecule and the fiber probe tip used to generate the optical near field. In other words, this exchange excites not only the electron, but also molecular vibration. This is a nonadiabatic process, which does not follow the Franck–Condon principle, and therefore the Born–Oppenheimer approximation is no longer valid. Several experimental results have been reported, including the nonadiabatic dissociation of diethylzinc (DEZn) molecules by a visible optical near field and the deposition of nanometric Zn particles [15.26]. The virtual exciton–polariton model explains this nonadiabatic process quantitatively by introducing the contribution of a phonon excited in the fiber probe tip [15.27].

The presented nonadiabatic process suggests that the large, expensive ultraviolet light sources that are indispensable for conventional photochemical vapor deposition are no longer required. It also suggests that the process can even dissociate optically inactive molecules (i. e., inactive to the propagating light), which is advantageous for environment protection because most optically inactive molecules are chemically stable and harmless. For example, optically inactive $Zn(acac)_2$ molecules have been dissociated nonadiabatically using an optical near field to deposit nanometric Zn particles [15.28].

15.4.2 Photolithography

Photolithography is a technology used to carve substrate material. After coating the substrate with a thin film of a photoresist, light is irradiated through a photomask to induce a photochemical reaction in the photoresist. When the aperture on the photomask is smaller than the wavelength of the light, the transmission of the propagating light is sufficiently low while an optical near field is generated at the aperture. Using the photochemical reaction between the optical near field and the photoresist, a nanometric pattern beyond the diffraction limit is formed on the photoresist, and a chemical etching process is subsequently used to carve the substrate. Practical nanophotonic technologies, such as using a two-layered photoresist, have been developed to form deep patterns and to realize a quantitative innovation (Fig. 15.12) [15.29].

As long as an adiabatic process is used for photolithography, ultraviolet light with a high photon energy is required to induce the photochemical reaction via the optical near field. However, photolithography using a nonadiabatic process is possible, as was the case

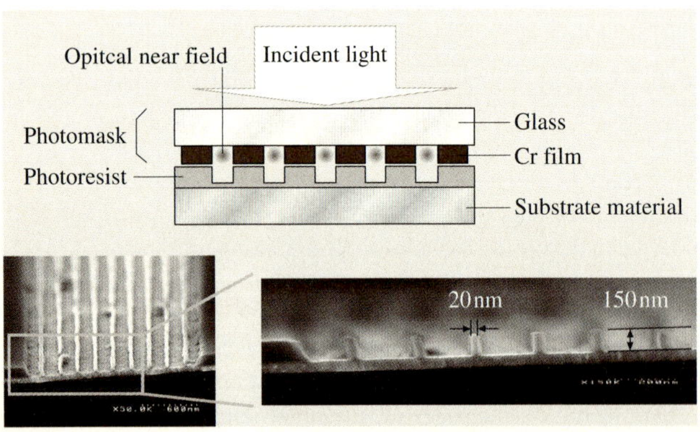

Fig. 15.12 The structure of the photolithography system and the cross-sectional profile of the photoresist after development. The photographs were supplied courtesy of R. Kuroda, Canon Corp

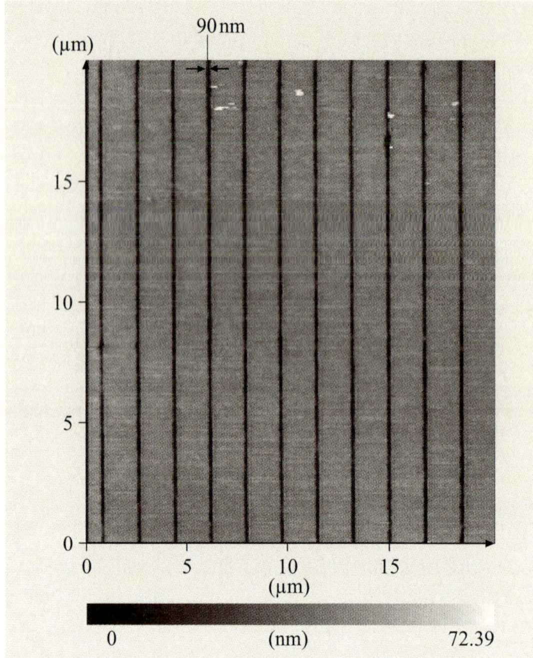

Fig. 15.13 Top view of a photoresist with a corrugated pattern that was fabricated nonadiabatically using optical near fields

in Sect. 15.4.1; i. e., a photochemical reaction can even be induced using visible light with a very low photon energy. Figure 15.13 shows a narrow corrugated pattern fabricated using nonadiabatic photolithography [15.30], which has been analyzed theoretically based on a virtual exciton–polariton model by introducing the contribution of phonons. This result represents a qualitative innovation in photolithography, suggesting that large, expensive ultraviolet light sources are no longer required, and that harmless, chemically stable molecules can be used as the photoresist, even if they are optically inactive. As an example, an optically inactive resist film has been used for electron-beam lithography. Although this film is optically inactive, a photochemical reaction is induced and fine patterns can be fabricated by using a photomask consisting of a two-dimensional array of circular disks (Fig. 15.14) [15.30].

Fabrication using the adiabatic process suffers from the contribution of the small amount of propagating light transmitted through the aperture on the photomask and the plasmon generated on the photomask. This limits the resolution of fabrication and the pattern contrast. In contrast, since the nonadiabatic process is free from these contributions, greater resolution, higher contrast,

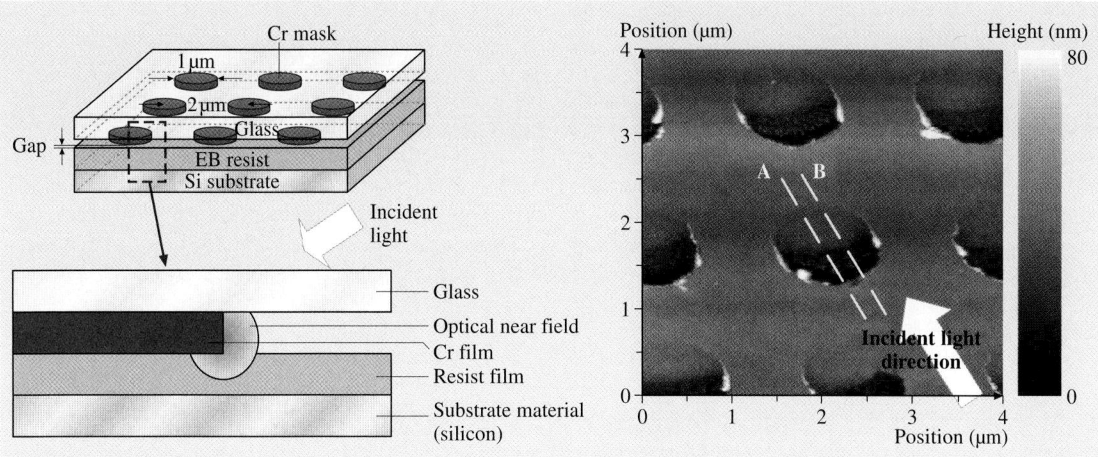

Fig. 15.14 Bird's-eye view of the optically inactive resist film used for electron-beam lithography, fabricated nonadiabatically using optical near fields

and a variety of patterns (e.g., Ts, Ls, and rings) have been realized [15.30].

15.4.3 Self-Organized Deposition and Nanoimprinting

Self-organized deposition is a way to deposit nanometer-sized particles, as discussed in Sect. 15.4.1. In this case, however, a fine structure, e.g., a narrow groove, is first patterned on the substrate instead of using a fiber probe or photomask (Fig. 15.15) [15.31]. While sputtering metallic particles on this patterned substrate, irradiation with propagating light is used to desorb the deposited particles. This irradiation generates an optical near field in the groove, which excites free electron oscillations in the deposited metallic particles. The size of the deposited metallic particles increases with continued sputtering. As a result, the center frequency of the electron oscillation spectrum shifts and its spectral linewidth increases due to the finite path length of the free electrons and resultant damping of the resonant oscillation. The magnitudes of the center frequency shift and linewidth increase are nearly proportional to the particle size if it is sufficiently small [15.32].

When the shifted center frequency is tuned to the optical near-field frequency (i. e., the particle grows to a specific size), the magnitude of the optical near-field energy absorbed by the particle is maximal at this size-dependent optical absorption resonance. Moreover, the desorption rate is maximal under this resonant condition, so a further increase in particle size is prohibited by the balance between the magnitudes of sputtering and desorption [15.33, 34]. Since the size of a deposited particles depends on the photon energy of the optical

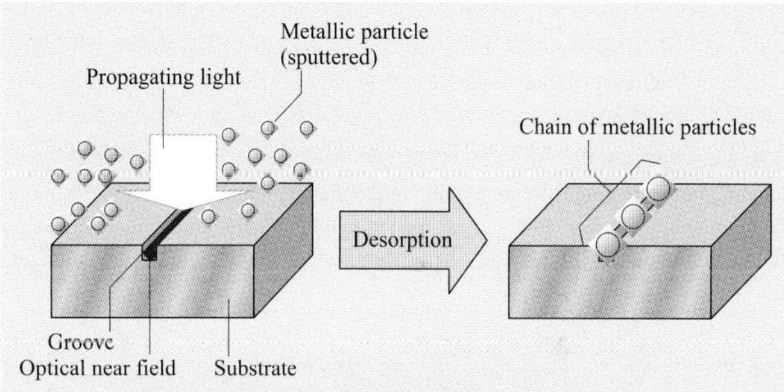

Fig. 15.15 The principle of self-organized deposition

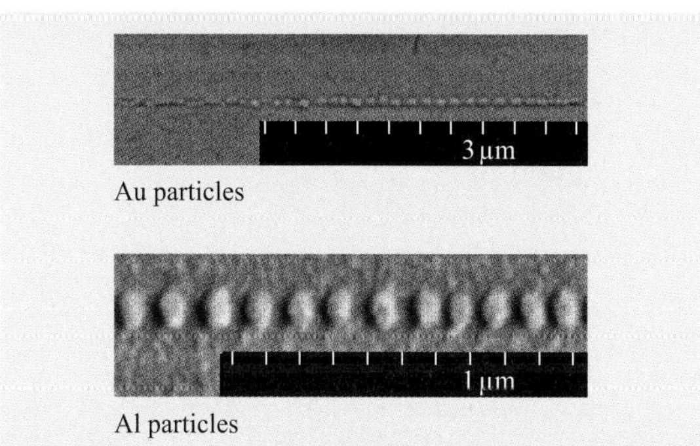

Fig. 15.16 Chains of gold and aluminum particles fabricated via self-organized deposition using red (wavelength = 633 nm) and green (wavelength = 532 nm) light. Their diameters are 100 nm and 70 nm, respectively

near field as the result of this balance, the size can be controlled accurately.

This accurate size-control method forms the first particle, on which a new optical near field is subsequently generated because the propagating light is still being irradiated. As a result of this secondary optical near field, a second particle forms next to the first one, and the size of the second particle is equivalent to that of the first because of size dependent resonant optical absorption [15.33]. Because this formation process occurs successively, a chain of the metallic particles is formed along the groove within the area irradiated by the propagating light (Fig. 15.16). Note that this chain is formed in a self-organized manner, and neither a fiber probe nor a photomask is used. The sizes and separations of these particles are determined accurately by the photon energy of the optical near field.

As a recently developed fabrication technology, light-assisted nanoimprinting has enabled the formation of nanometric structures, without the requirement for a development process [15.35]. By using optical near fields, the adiabatic process forms a pattern that is even smaller than that of the mold [15.36]. Further progress is expected if a nonadiabatic process can also be utilized.

15.5 Extension to Related Science and Technology

Nanophotonics has led nanooptics and realized qualitative innovation. As examples of its extension to related science and technology, this section reviews applications to novel photonic systems and atom photonics.

Nanophotonic systems utilize nanophotonic devices. For example, novel architectures have been proposed for optical signal transmission systems and their performance has been confirmed experimentally. They include interconnections and summation involving an optical nanofountain [15.16], computing using nanophotonic switches and an optical nanofountain [15.17], and broadcasting using multiple nanophotonic switches [15.37]. They have realized quantitative innovations by decreasing the device size and power consumption beyond the diffraction limit. More importantly, qualitative innovation has been realized as a consequence of the novel functions of nanophotonic devices, which are impossible with conventional photonic devices.

An optical/magnetic hybrid disk storage system with a storage density of $1\,\mathrm{Tb/in^2}$ has been developed. Quantitative innovation has already been realized by transcending the diffraction limit of optical storage density [15.38]. Qualitative innovation in optical storage and readout has been also proposed by applying the hierarchy in optical near fields to memory retrieval [15.39].

An example of an extension to related science is atom photonics, which controls the thermal motion of neutral atoms in a vacuum using optical near fields [15.40]. Theoretical studies have examined single-atom manipulation based on the virtual exciton–polariton model [15.41]. Recent experimental studies have examined atom guidance through a hollow optical fiber [15.42], atom-detecting devices [15.43], atom deflectors [15.44], and an atomic funnel [15.45]. Atom photonics will open a new field of science that examines the interaction between virtual exciton–polaritons and a single atom. Furthermore, it can be applied to novel technologies for atomic-level material fabrication.

15.6 Summary

As a result of the development of nanophotonics, which utilizes the local electromagnetic interactions between nanometric particles mediated by optical near fields, nanooptics now exists as a novel field of optics in nanometric space. However, the name nanophotonics is occasionally used for photonic crystals, plasmonics, silicon photonics, and quantum-dot lasers using conventional propagating lights. Here, one should consider a stern warning by C. Shannon on the casual use of the term information theory, which was a trend in its study during the 1950s [15.46]. The term nanophotonics has been used in a similar way, although some work in nanophotonics is not based on optical near-field interactions. For the true development of nanophotonics, one needs deep physical insights into the virtual exciton–polariton and the nanometric subsystem composed of electrons and photons. It is both unethical and misleading to use the name nanophotonics in a casual manner, for example, to obtain a research grant. Since nanooptics should be defined as a field of optical science and technology including nanophotonics, Shannon's warning should be extended to the term nanooptics.

References

15.1 M. Ohtsu: *Progress in Nano-Electro-Optics V*, ed. by M. Ohtsu (Spinger, Berlin, Heidelberg 2006) Preface to Volume V: Based on "nanophotonics" proposed by Ohtsu in 1993, OITDA (Optical Industry Technology Development Association, Japan) organized the nanophotonics technical group in 1994, and discussions on the future direction of nanophotonics were started in collaboration with academia and industry

15.2 E.Yablonovitch: Inhibited spontaneous emission in solid-state physics and electronics, Phys. Rev. Lett. **58**, 2059 (1987)

15.3 V. A. Podolskiy, A. K. Sarychev, V. M. Shalaev: Plasmon modes and negative refraction in metal nanowire composites, Opt. Exp. **11**, 735 (2003)

15.4 R. A. Shelby, D. R. Smith, S. Schultz: Experimental verification of a negative index of refraction, Science **292**, 77 (2001)

15.5 J. B. Pendry: Negative refraction makes a perfect lens, Phys. Rev. Lett. **85**, 3966 (2000)

15.6 H. Rong, A. Liu, R. Nicolaescu, M. Paniccia: Raman gain and nonlinear optical absorption measurements in a low-loss silicon awveguide, Appl. Phys. Lett. **85**, 2196 (2004)

15.7 Y. Arakawa, H. Sakaki: Multidimensional quantum well laser and temperature dependence of its threshold current, Appl. Phys. Lett. **40**, 939 (1982)

15.8 E. A. Synge: A suggested method for extending microscopic resolution into the ultra-microscopic region, Philos. Mag. **6**, 356 (1928)

15.9 M. Ohtsu: *Near-Field Nano/Atom Optics and Technology* (Spinger, Berlin, Heidelberg 1998) pp. 31–69

15.10 M. Ohtsu, H. Hori: *Near-Field Nano-Optics* (Kluwer Academic Plenum, New York 1999) pp. 113–142

15.11 M. Ohtsu, K. Kobayashi: *Optical Near Fields* (Springer, Berlin, Heidelberg 2004) pp. 88–108

15.12 M. Ohtsu, K. Kobayashi: *Optical Near Fields* (Springer, Berlin, Heidelberg 2004) pp. 109–120

15.13 S. Sangu, K. Kobayashi, A. Shojiguchi, M. Ohtsu: Logic and functional operations using a near-field optically coupled quantum-dot system, Phys. Rev. B **69**, 115334 (2004)

15.14 M. Achermann, M. A. Petruska, S. Kos, D. L. Smith, D. D. Koleske, V. I. Klimov: Energy-transfer pumping of semiconductor nanocrystals using an epitaxial quantum well, Nature **429**, 642 (2004)

15.15 T. Kawazoe, K. Kobayashi, K. Akahane, M. Naruse, N. Yamamoto, M. Ohtsu: Demonstration of nanophotonic NOT gate using near-field optically coupled quantum dots, Appl. Phys. B **84**, 243 (2006)

15.16 M. Naruse, T. Miyazaki, F. Kubota, T. Kawazoe, K. Kobayashi, S. Sangu, M. Ohtsu: Nanometric summation architecture based on optical near-field interaction between quantum dots, Opt. Lett. **30**, 201 (2005)

15.17 M. Naruse, T. Miyazaki, T. Kawazoe, K. Kobayashi, S. Sangu, F. Kubota, M. Ohtsu: Nanophotnic computing based on optical near-field interactions between quantum dots, IEICE Trans. Electron. **E88-C**, 1817 (2005)

15.18 A. Shojiguchi, S. Sangu, K. Kitahara, M. Ohtsu: Superradiance and dipole ordering of an n two-level system interacting with optical near fields, J. Phys. Soc. Jpn. **72**, 2984 (2003)

15.19 T. Kawazoe, K. Kobayashi, S. Sangu, M. Ohtsu: Demonstration of a nanophotonic switching operation by optical near-field energy transfer, Appl. Phys. Lett. **82**, 2957 (2003)

15.20 T. Kawazoe, K. Kobayashi, M. Ohtsu: Optical nanofountain: A biomimetic device that concentrates optical energy in a nanometric region, Appl. Phys. Lett. **86**, 103102 (2005)

15.21 V. V. Polonski, Y. Yamamoto, M. Kourogi, H. Fukuda, M. Ohtsu: Nanometric patterning of zinc by optical near-field photochemical vapour deposition, J. Microsc. **194**, 545 (1999)

15.22 Y. Yamamoto, M. Kourogi, M. Ohtsu, V. Polonski, G. H. Lee: Fabrication of nanometric zinc pattern with photodissociated gas-phase diethylzinc by optical near field, Appl. Phys. Lett. **76**, 2173 (2000)

15.23 T. Yatsui, T. Kawazoe, M. Ueda, Y. Yamamoto, M. Kourogi, M. Ohtsu: Fabrication of nanometric single zinc and zinc oxide dots by the selective photodissociation of adsorption-phase diethylzinc using a nonresonant optical field, Appl. Phys. Lett. **81**, 3651 (2002)

15.24 J. Lim, T. Yatsui, M. Ohtsu: Observation of size-dependent resonance of near-field coupling between a deposited Zn dot and the probe apex during near-field optical chemical vapor deposition, IEICE Trans. Electron. **E 88-C**, 1832 (2005)

15.25 Y. Yamamoto, M. Kourogi, M. Ohtsu, G. H. Lee, T. Kawazoe: Lateral integration of Zn and Al dots with nanometer-scale precision by near field optical chemical vapor deposition using a sharpened optical fiber probe, IEICE Trans. Electron. **E85-C**, 2081 (2002)

15.26 T. Kawazoe, Y. Yamamoto, M. Ohtsu: Fabrication of a nanometric Zn dot by nonresonant near-field optical chemical-vapor deposition, Appl. Phys. Lett. **79**, 1184 (2001)

15.27 T. Kawazoe, K. Kobayashi, S. Takubo, M. Ohtsu: Nonadiabatic photodissociation process using an optical near field, J. Chem. Phys. **122**, 024715 (2005)

15.28 T. Kawazoe, K. Kobayashi, M. Ohtsu: Near-field optical chemical vapor deposition using $Zn(acac)_2$ with a non-adiabatic photochemical process, Appl. Phys. B **84**, 247 (2006)

15.29 M. Naya, I. Tsurusawa, T. Tani, A. Mukai, S. Sakaguchi, S. Yasutani: Near-field optical photolithography for high-aspect-ratio patterning using bilayer resist, Appl. Phys. Lett. **86**, 201113 (2005)

15.30 T. Kawazoe, K. Kobayashi, S. Sangu, M. Ohtsu, A. Neogi: Unique Properties of Optical Near Field and their Applications to Nanophotonics. In: *Progress in Nano-Electro-Optics V*, ed. by M. Ohtsu (Springer, Berlin, Heidelberg 2006) pp. 109–162

15.31 T. Yatsui, W. Nomura, M. Ohtsu: Self-assembly of size- and position-controlled ultralong nanodot chains using near-field optical desorption, Nano Lett. **5**, 2548 (2005)

15.32 C. Sönnichsen, T. Franzl, T. Wilk, G. von Plessen, J. Feldmann: Drastic reduction of plasmon damping in gold nanorods, Phys. Rev. Lett. **88**, 077402 (2002)

15.33 K. F. MacDonald, V. A. Fedotov, S. Pochon, K. J. Ross, G. C. Stevens, N. I. Zheludev, W. S. Brocklesby, V. I. Emel'yanov: Optical control of gallium nanoparticle growth, Appl. Phys. Lett. **80**, 1643 (2002)

15.34 T. Yatsui, S. Takubo, J. Lim, W. Nomura, M. Kourogi, M. Ohtsu: Regulating the size and position of deposited Zn nanoparticles by optical near-field desorption using size-dependent resonance, Appl. Phys. Lett. **83**, 1716 (2003)

15.35 M. D. Austin, H. Ge, W. Wu, M. Li, Z. Yu, D. Wasserman, S. A. Lyon, S. Y. Chou: Fabrication of 5 nm linewidth and 14 nm pitch features by nanoimprint lithography, Appl. Phys. Lett. **84**, 5299 (2004)

15.36 T. Yatsui, Y. Nakajima, W. Nomura, M. Ohtsu: High-resolution capability of optical near-field imprint lithography, Appl. Phys. B **84**, 265 (2006)

15.37 M. Naruse, T. Kawazoe, S. Sangu, K. Kobayashi, M. Ohtsu: Optical interconnects based on optical far- and near-field interactions for high-density data broadcasting, Opt. Exp. **14**, 306 (2006)

15.38 T. Matsumoto, Y. Anzai, T. Shintani, K. Nakamura, T. Nishida: Writing 40 nm marks by using a beaked metallic plate near-field optical probe, Opt. Lett. **31**, 259 (2006)

15.39 M. Naruse, T. Yatsui, W. Nomura, N. Hirose, M. Ohtsu: Hierarchy in optical near-fields and its application to memory retrieval, Opt. Exp. **13**, 9265 (2005)

15.40 M. Ohtsu: *Near-Field Nano/Atom Optics and Technology* (Spinger, Berlin, Heidelberg 1998) pp. 218–293

15.41 K. Kobayashi, S. Sangu, H. Ito, M. Ohtsu: Near-field optical potential for a neutral atom, Phys. Rev. A **63**, 013806 (2001)

15.42 H. Ito, T. Nakata, K. Sakaki, M. Ohtsu, K. I. Lee, W. Jhe: Laser Spectroscopy of Atoms Guided by Evanescent Waves in Micron-Sized Hollow Optical Fibers, Phys. Rev. Lett. **76**, 4500 (1996)

15.43 K. Totsuka, H. Ito, T. Kawamura, M. Ohtsu: High spatial resolution atom detector with two-color optical near fields, Jpn. J. Appl. Phys. **41**, 1566 (2002)

15.44 K. Totsuka, H. Ito, K. Suzuki, K. Yamamoto, M. Ohtsu, T. Yatsui: A slit-type atom deflector with near-field light, Appl. Phys. Lett. **82**, 1616 (2003)

15.45 A. Takamizawa, H. Ito, S. Yamada, M. Ohtsu: Observation of cold atom output from an evanescent-light funnel, Appl. Phys. Lett. **85**, 1790 (2004)

15.46 C. E. Shannon: The bandwagon, IRE Trans. Inform. Theory **2**, 3 (1956)

16. Optics far Beyond the Diffraction Limit: Stimulated Emission Depletion Microscopy

In this chapter we show that stimulated emission depletion (STED) microscopy and its derivative concepts are able to radically overcome the diffraction barrier in far-field fluorescence imaging, thus disclosing fluorescent details on the macromolecular scale even with diffracted beams of light.

The optical microscope is an invaluable tool in the life sciences because it is able to noninvasively image structures within cells and tissues. Unfortunately, due to the fact that light propagates as a wave, the smallest possible size of a focal spot in a far-field light microscope is limited by diffraction, putting a lower limit on the size of the structures which can be observed. Concretely, this means that for a lens of semiaperture angle α, the full-width-half-maximum (FWHM) Δr of the main diffraction maximum of the point-spread function (PSF) in the focal plane of the lens is $\Delta r = \lambda/(2n\sin\alpha)$, with λ and n denoting the wavelength of light and the refractive index, respectively [16.1]. If the distance between two objects is smaller than this FWHM, the objects cannot be readily resolved from one another. The diffraction resolution limit is particularly disadvantageous in the life sciences where about 80% of all microscopy applications are carried out with far-field fluorescence systems.

It is sometimes argued that confocal and multiphoton excitation provide subdiffraction resolving power in fluorescence microscopy; in reality neither breaks the diffraction barrier. Confocal microscopy can, in principle, attain a 1.4-fold reduction of the effective FWHM of the focal spot [16.2], because the consecutive point-like illumination and detection in a confocal fluorescence microscope yields an effective PSF that is the product of the excitation and detection PSFs. The PSF multiplication doubles the bandwidth of the optical transfer function of the system. However, this bandwidth increase is rarely realized because the larger spatial frequencies are heavily damped, and therefore swamped by noise. If they are not, the resolution can be improved with the help of a linear image deconvolution by a factor of 2, but no more. Hence, this technique is also conceptually diffraction-limited.

Two- or three-photon excitation fluorescence microscopy benefits from the quadratic or cubic dependence of the fluorescence signal on the focal intensity. While the nonlinearity provided by multiphoton absorption indeed shifts the generation of the signal to the inner part of the excitation spot slightly, a narrowing of the effective spot is usually not obtained. The reason is that the energy gap of the excitation band of the fluorophore to be imaged is finite. Hence it is bridged by the instant absorption of m photons of $1/m$ the energy, which inevitably entails a wavelength of the excitation light m times longer than in the single-photon excitation case. The m-times-longer wavelength renders the focal spot m-times larger, which is not compensated by the nonlinear excitation near the focal center [16.3]. In fact, the resulting spot scales as $\sqrt{m}$. Thus, multiphoton processes do not improve the resolution in absolute terms, and so it is not surprising that multiphoton processes did not open up a new chapter in far-field optical resolution, despite their common use for more than 15 years [16.4]. Moreover, even in the hypothetic case that the requirement for a longer wavelength is not present, a quadratic or cubic intensity dependence would only expand the optical transfer function by a factor of two or three, respectively. Thus, the diffraction barrier would just be shifted but not broken.

The lack of viable concepts for a substantial far-field resolution improvement led to the development of near-field optical microscopy, which abandons focusing altogether [16.5]. Here the spatial resolution is improved by confining the interaction of the sample and the probing light to sub-λ dimensions using a sub-λ-sized scanning aperture or tip. This approach requires tight control of the nanosized stylus, and is limited to imaging surfaces, substantially restricting its range of applicability.

16.1 Principles of STED Microscopy

Stimulated emission depletion (STED) microscopy is the first far-field microscopy concept whose resolution is no longer limited by diffraction [16.6–8]. While a STED microscope still relies on diffracted beams the effective PSF of this microscope can be as sharp as a single molecule and, theoretically, its optical bandwidth can be infinitely large. The key to this surprising feature is a nearly exponential depletion of the fluorescent state by a further beam of light optimized for stimulated emission. Realized at comparatively low intensities, the huge nonlinearity connected with the exponential depletion allows one to squeeze the fluorescent spot to a size corresponding to a fraction of λ. To date, practical STED fluorescence microscopy has displayed a spatial resolution of 33 nm ($\lambda/23$) along the optic axis [16.9]. In the transverse direction in the focal plane, a lateral FWHM of 15–20 nm has been demonstrated with single molecules [16.10] and also in biological samples [16.11]. STED microscopy retains the advantages of far-field optical microscopy, employing regular lenses, and can be used to noninvasively image samples in three dimensions (3-D). In this chapter we briefly explain the physical principles of this emerging technique and illustrate its power by showing recent applications in biological imaging.

In polyatomic organic molecules, stimulated emission is most effective between the vibrationally relaxed sublevel of the first electronic singlet state S_1 and a vibrational sublevel of the ground state S_0^* (Fig. 16.1a) [16.12, 13]. When not subjected to quenching processes, S_1 spontaneously emits a photon within its typical lifetime τ_{fl} of a few nanoseconds, and the S_0^* decays nonradiatively within $\tau_{vib} < 1$ ps. A difference of three orders of magnitude between these lifetimes maintains a population inversion between the two states. Deliberate depletion of the S_1 by stimulated emission [16.6] in these molecules can be optimized by temporally separating the excitation from the stimulated emission. In our fluorescence microscope, the excitation of the fluorescent markers to Franck–Condon vibrational levels in the first singlet state, i.e. $S_0 \rightarrow S_1^*$ (Fig. 16.1a), is performed by pulsed illumination. The molecules are excited by single-photon absorption, followed by ultrafast nonradiative decay to the fluorescent state S_1. For effective depletion of S_1, the molecules are subjected to a synchronized red-shifted STED pulse of duration $\tau_{STED} \approx 300$ ps $< \tau_{fl}$. Since $\tau_{vib} \ll \tau_{STED}$, the vibrational decay in the ground state empties the S_0^* more effectively than the STED pulse could possibly re-excite the molecules to the fluorescent state S_1. Moreover, the S_1 population of molecules residing at a spatial coordinate $\boldsymbol{r}$ decreases exponentially with the fluence $h_{STED}(\boldsymbol{r})$ of the STED pulse, expressed in photons per area per pulse (Fig. 16.1b) [16.7, 9, 14]. Denoting the molecular cross section for stimulated emission by σ, the normalized probability of stimulating an excited molecule to fluoresce at $\boldsymbol{r}$ is given by $F(\boldsymbol{r}) = \exp[-\sigma h_{STED}(\boldsymbol{r})]$.

Here we describe the standard set up of a typical STED microscope (Fig. 16.2). An objective with a high numerical aperture (≈ 1.4) is used to both illuminate and gather light from the sample. Filters and the dichroic beam splitters are optimized for the ideal excitation and stimulating-emission wavelengths of the fluorescent dye used to mark the sample. Excitation is performed with a laser with pulses typically on the order of tens of picoseconds. A quarter-wave plate ensures that the excitation light at the lens is circularly polarized for uniform excitation of the fluorescent markers. The fluorescent light is collected by the same lens and focused onto the core of a fiber guiding the fluorescent light to an avalanche photodiode, where detected photons are counted. Imaged into the focal plane of the objective lens, the core diameter corresponded to ≈ 0.7 times the Airy disk. As a result, the probability of detecting a fluorescent photon is weighted spatially by the detection PSF, $h_{det}(\boldsymbol{r})$, which is the normalized diffraction-limited image of the fiber core. Thus when no STED beam is present, the set up can function as a normal scanning confocal fluorescence microscope of finite pinhole size (given by the core diameter of the fiber, which functions

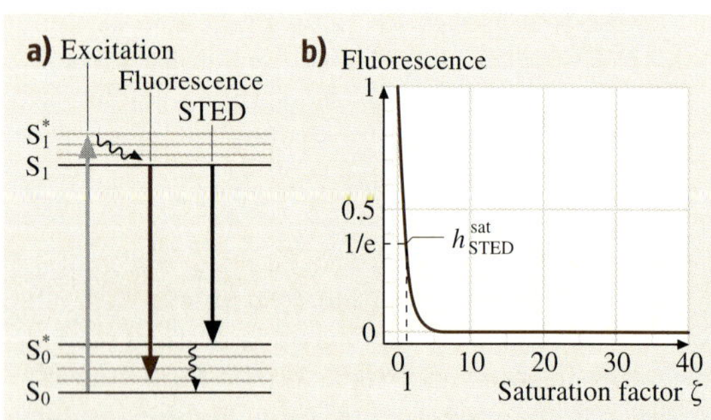

Fig. 16.1a,b Saturated depletion of fluorescent molecule **(a)** Excitation, fluorescence and stimulated emission (STED) in the energy diagram. **(b)** Quenching of the fluorescence depending on the STED intensity, normalized by the saturation intensity

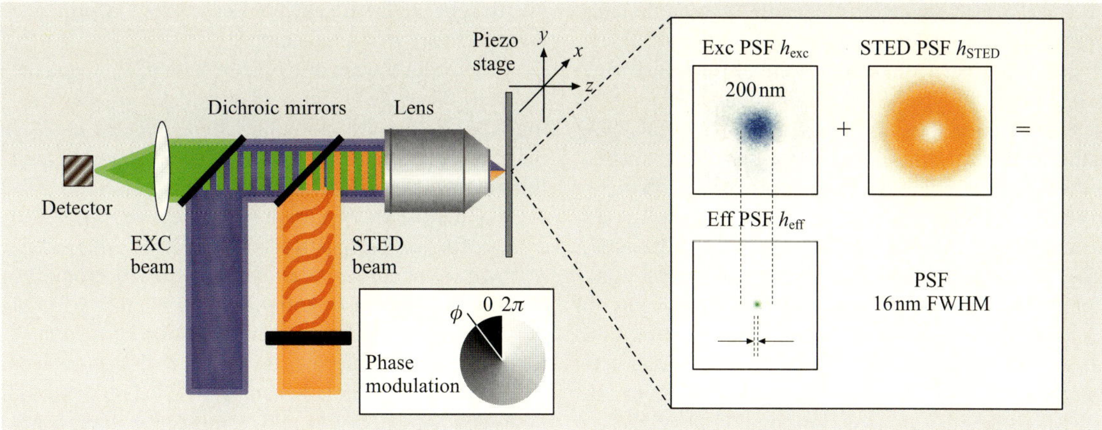

Fig. 16.2 A typical STED microscope. Excitation light forms a regular Airy disk in the lens focal plane (adjacent panel *in blue*) and is overlapped with a *red-shifted* doughnut spot for STED. The doughnut is produced by applying a helical phase delay on the STED beam wavefront $\exp(i\phi)$, with $0 \leq \phi \leq 2\pi$. The linear helical progression of the phase from 0 to 2π ensures that the light field from opposite parts of the entrance pupil of the lens interferes destructively on the optic axis. Overlap of the two spots confines the focal area in which fluorescence is possible to the doughnut centre, yielding the effective fluorescent spot of subdiffraction size displayed in the lower row panel (*in green*). All spots represent measured data and are drawn to scale. The profile of the green effective spot displays a FWHM of 16 nm and a sharp peak

as a pinhole), facilitating the comparison of the normal confocal imaging mode with the STED mode. In the latter, the excitation pulse is immediately followed by a STED pulse of several hundred picoseconds in length. This pulse duration not only avoids the re-excitation from the S_0^* (Fig. 16.1a), but also the two-photon excitation and excessive nonlinear photobleaching of the dye [16.15].

In a regular scanning fluorescence microscope, the fluorescence spot is given by the main maximum of the excitation PSF, i.e., the normalized function $h_{exc}(\boldsymbol{r})$ describing the diffraction-limited intensity spot of the excitation light (Fig. 16.2) [16.2]. The purpose of the STED pulse is to reduce the focal area in which fluorescence is emitted. To attain quenching of only the periphery, the STED pulse is modified to form a zero at the focal center, i.e. a doughnut: $h_{STED}(\boldsymbol{r}=0) \cong 0$ (Fig. 16.2).

To realize the doughnut shape of $h_{STED}(\boldsymbol{r})$, the planar wavefront of the STED beam is subjected to a helical phase delay $P(\phi) = \exp(i\phi)$, with $0 < \phi < 2\pi$ being an angle centered around the optic axis (Fig. 16.2) [16.16]. The phase delay is produced using a spatial light modulator located optically conjugate to the entrance pupil of the lens. The objective transforms this wavefront into a doughnut in the focal plane. Following the exponential dependence on intensity $F(\boldsymbol{r}) = \exp[-\sigma h_{STED}(\boldsymbol{r})]$, the STED pulse suppresses the fluorescence at every point in the focal region, except at the very point $\boldsymbol{r}=0$, where $h_{STED}(\boldsymbol{r})$ equals zero (Fig. 16.2).

The effective PSF $h_{eff}(\boldsymbol{r})$ of this confocalized scanning far-field microscope is given by the product of the excitation and detection PSF, multiplied by the fluorescence probability $F(\boldsymbol{r})$:

$$\begin{aligned} h_{eff}(\boldsymbol{r}) &= h_{exc}(\boldsymbol{r})h_{det}(\boldsymbol{r})F(\boldsymbol{r}) \\ &\equiv h_{conf}(\boldsymbol{r})\exp[-\sigma h_{STED}(\boldsymbol{r})] . \end{aligned} \tag{16.1}$$

By increasing the power of the STED pulse, the fluorescent spot remaining in the center becomes smaller. We define $h_{STED}^{max} \equiv \max[h_{STED}(\boldsymbol{r})]$ as the maximum intensity present in the STED PSF and $h_{STED}^{sat} \equiv 1/\sigma$ as a saturation scaling factor, at which a fraction of $(1-1/e)$ of the fluorophores have been stimulated into the ground state. Subdiffraction resolution prevails for saturation factors $\varsigma \equiv h_{STED}^{max}/h_{STED}^{sat} = \sigma h_{STED}^{max} \gg 1$ [16.17]. Since the function $h_{STED}(\boldsymbol{r})$ is parabolic at $\boldsymbol{r}=0$ to a first approximation, the fluorescent spot size scales with the square root of the STED pulse fluence [16.17, 18]. A Taylor-series expansion of $h_{eff}(\boldsymbol{r})$ yields the FWHM of the effective PSF of the STED microscope as

$$\Delta r \approx \frac{\lambda}{2n \sin\alpha\sqrt{1+\varsigma}} . \tag{16.2}$$

Here λ denotes the wavelength of the depletion pulse, because it is the doughnut that is responsible for the final size of the spot. Under the assumption that the excitation wavelength and λ are not very different, Δr elegantly converges towards the diffraction-limited spot size in the absence of STED, while for $\varsigma \to \infty$ the fluorescence spot size converges to zero. Scanning with a fluorescent focal spot of subdiffraction size Δr delivers images of the same subdiffraction resolution.

It is important to note that, for a significantly narrowed spot ($\varsigma > 5$), the comparatively broad detection PSF $h_{\text{det}}(\boldsymbol{r})$ of the system does not play a role anymore. Therefore, STED microscopy does *not* rely on confocalization, and it is *not* an expansion of confocal microscopy. Conceptually, the confocal pinhole is redundant, because the confinement of the spot to subdiffraction dimensions is due to the STED beam. The confocal pinhole is used to provide other advantages to the system, such as confocal optical sectioning and stray-light suppression. The pinhole does not play any role in breaking the diffraction barrier.

Besides $h_{\text{exc}}(\boldsymbol{r})$ and $h_{\text{STED}}(\boldsymbol{r})$, Fig. 16.2 depicts the measured $F(\boldsymbol{r})$ for the STED intensity employed in this experiment, leading to a very narrow region of remaining fluorescence detection. This figure visualizes that for strong enough fluences $h_{\text{STED}}(\boldsymbol{r})$, the spatial extent of $h_{\text{eff}}(\boldsymbol{r})$ is solely determined by the exponential suppression factor. Neither the excitation nor the detection PSF really matters. In consequence, the same assertion holds for the excitation and the fluorescence wavelength as well as for the presence of the confocal pinhole. Thus, despite the diffraction-limited extent of the local minimum in $h_{\text{STED}}(\boldsymbol{r})$, the exponential factor $F(\boldsymbol{r})$ allows one to squeeze the focal fluorescence spot in principle down to the size of a molecule [16.7].

Challenges toward maximizing both $h_{\text{STED}}^{\max}$ and the detected fluorescent signal in the center of the scanning spot are elevated photobleaching of the fluorescent marker [16.19] and the actual value of the doughnut minimum $h_{\text{STED}}(0)$, which may not be zero. Nonfulfillment of the second condition reduces the desired signal by $1-\exp[-\varsigma h_{\text{STED}}(0)]$. Therefore, $h_{\text{STED}}(0) < 0.02\, h_{\text{STED}}^{\max}$ is imperative to attain high spatial resolution. Photobleaching can be minimized by attending to the population of the molecular triplet state. Elimination of the excitation of molecules trapped in this state has significantly reduced photobleaching in STED microscopy. With this approach, far-field fluorescence microscopy with a focal-plane resolution of 15–20 nm in biological samples has been obtained [16.11]. This corresponds to a 10–12-fold multilateral increase in resolution below the diffraction barrier.

Employing pulse repetition rates <1 MHz usually allows enough time for relaxation of the triplet state between subsequent excitation–depletion cycles. In a number of dyes, this measure enables up to a 30-fold increase in the fluorescent signal compared to earlier 80 MHz STED illumination schemes in which each spot on the sample receives many consecutive excitation and STED pulses. Triplet-state relaxation can also be realized by increasing the scanning speed such that the buildup of the triplet state is prevented. We have termed this illumination scheme triplet-relaxation (T-Rex) STED.

16.2 Nanoscale Imaging with STED

The resolution attainable under T-Rex conditions was first demonstrated on yellow–green beads emitting in the 500–530 nm range (Fig. 16.3). Specified by the manufacturer using electron microscopy, the beads varied by 4 nm around a mean diameter of 24 nm. Whereas confocal recording (Figs. 16.3a,c) yielded undefined blobs, the STED images (Figs. 16.3b,d) discerned virtually every bead in the focal plane [16.11]. Figures 16.3g,h depict line profiles through beads located 42 nm and 39 nm apart, respectively; both are clearly separated.

The function describing the bead image was determined by averaging over 75 beads that were comparatively isolated and of lesser brightness. The result is displayed in Fig. 16.3e, exhibiting a full-width-half maximum (FWHM) of (22.3 ± 2) nm. As given by the convolution of the bead object function with the PSF of the microscope, the bead image function in Fig. 16.3e is just an upper bound for the actual PSF. The actual resolution of the STED microscope is below 20 nm. If the PSF and the bead function are coarsely approximated by a Gaussian, we can assess the actual focal-plane resolution as $22.3/\sqrt{2} = (15.8\pm4)$ nm. Comparison of this value with the resolution limit given by Abbe's criterion in an epifluorescence microscope, $\lambda_{\text{fl}}/2n \sin\alpha = 190$ nm, demonstrates a $\approx$12-fold gain in focal-plane resolution.

We also imaged self-assembling colloidal silica nanospheres on the surface of a glass cover slip in the non-T-Rex mode, precisely at a pulse rate of 80 MHz. The nanospheres were prepared with a fluorescent core

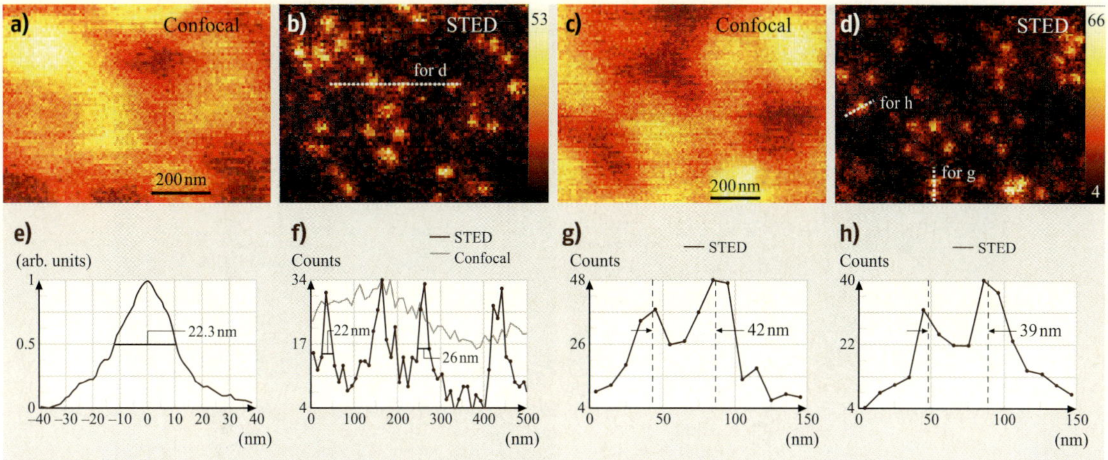

Fig. 16.3a–h Resolution $\leq$20 nm in the focal plane of a STED microscope While the confocal imaging mode ((**a**), (**c**)) fails to resolve the bead agglomeration, the corresponding STED recording ((**b**) and (**d**)) discerns every 24 nm bead. (**e**) Averaged profile of bead images. The 22.3 nm FWHM indicates a lateral resolution of $\approx$ 16 nm in the focal plane (after extraction of the bead size). (**f**) Data as indicated by the dashed line in panel (**b**), both for the confocal and the STED recording. Note the sharp peaks resulting from STED superresolution. ((**g**),(**h**)) Intensity profiles through the data in (**d**) proving the separation capability of the STED microscope

of 68 nm and a non-fluorescent shell of 132 nm diameter, as shown in Fig. 16.4c. The STED beam was adjusted to 14 mW of average power and was tuned to $\lambda = 615$ nm. Whereas the confocal image (Fig. 16.4a) is almost featureless, the STED image in Fig. 16.4b fully disclosed the semicrystalline order of the nanoparticles [16.20]. In particular, grain boundaries, point defects, disclinations, and dislocations became clearly visible.

STED microscopy does not require any mathematical postprocessing. Still, one can take advantage of its optical transfer function (OTF) to strengthen the higher spatial frequencies gained by STED with a linear deconvolution. The OTF is obtained by a Fourier transformation of the effective PSF $h_{\text{eff}}(\boldsymbol{r})$, which unlike in the confocal microscope, depends on the saturation factor ζ and hence on the dye in use. We established $h_{\text{eff}}(\boldsymbol{r})$ for Atto532, and, for the same power, using single labeled antibodies (size < 15 nm). The result is shown in Fig. 16.4d along with the associated OTF. Both the FWHM of 67 nm as well as the enlarged OTF mark a substantial gain in optical bandwidth. The subsequent linear deconvolution of the STED image yields a significantly improved contrast in the image (Fig. 16.4e).

We also deconvolved the confocal image, as shown in the inset of Fig. 16.4a. However, the deconvolved confocal image failed to resolve the nanoparticles, let alone the crystal defects. In fact, the raw STED image provides a higher resolution. Changing the physics of image formation is the more powerful approach.

We also demonstrated the resolution of T-Rex STED microscopy (0.25 MHz pulse rate) with biological immunofluorescence imaging. We labeled secondary antibodies with Atto532, which were then coupled, via primary antibodies, to the proteins of interest. Figure 16.5 displays the protein *heavy subunit of neurofilaments* (NF-H) in the human neuroblastoma cell line SH-SY5Y (retinoic acid-BDNF-differentiated) which establishes crosslinks to organize and stabilize neurofilaments in axons [16.21]. Neurofilaments play an essential role in many neurodegenerative diseases such as Parkinsons'. In contrast to the confocal (Fig. 16.5a) image, the STED (Fig. 16.5d) image identifies neurofilamental substructures of 20–30 nm size [16.11]. This is particularly apparent in Fig. 16.5e, which shows a partial area of (Fig. 16.5d) at a larger magnification, as well as in (Fig. 16.5c) where the details are further enhanced by a linear deconvolution. The line plots of raw data in (Fig. 16.5f) prove that substructures that are only 45 nm apart are separated by a dip of 50%.

The intensity profiles in Figs. 16.3 and 16.5 featuring a FWHM of 22–30 nm demonstrate the fundamental progress in resolution brought about by T-Rex STED microscopy [16.11]. Representing the convolution of the microscope's PSF with the spatial extent of the label-

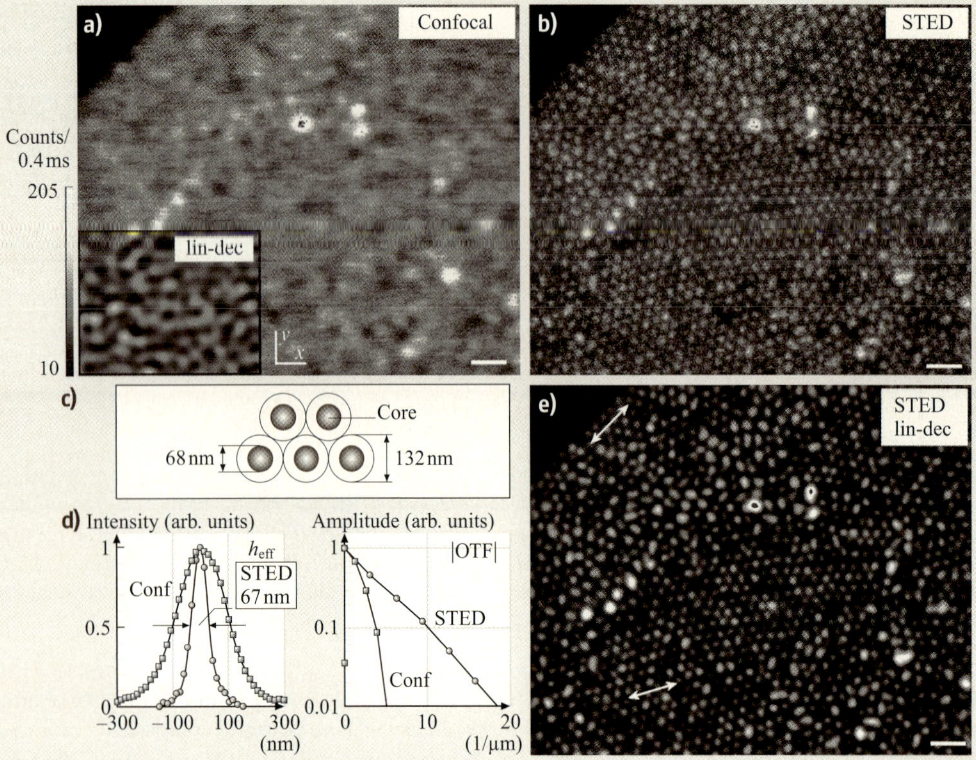

Fig. 16.4a–e Imaging the spatial order of colloidal nanoparticles **(a)** Confocal image and **(b)** corresponding STED image. **(c)** Sketch of silica nanoparticles featuring a fluorescent shell core (dye Atto 532, emission 540–620 nm). **(d)** The effective PSF and OTF, determined from fluorescent point-objects of the same fluorophore as in the core are utilized to enhance the STED-image by linear deconvolution; the result is displayed in panel **(e)** The *double-headed arrows* mark different orientations of the colloidal nanoparticles. For comparison, the *inset* of panel **(a)** shows the linearly deconvolved confocal data. Despite this, only the STED-images (**(b)** and **(e)**) reveal grain boundaries, defects and dislocations in the semicrystalline nanoparticle formation. Scale bars = 500 nm

ing, the FWHM of the profiles are just an upper limit for the actual imaging resolution. In fact, with a complex size of 6–8 nm, the secondary antibodies form an estimated 12–16 nm-diameter volume around the similarly sized primary antibody [16.22]. Thus, the resolution has attained values that are close to the limits set by the labeling itself.

The full subdiffraction imaging potential of STED remains to be realized, and the technique continues to improve after nearly a decade of use. T-Rex STED, for example, can be further improved with relatively simple modifications to increase the image acquisition time and retain its increased resolving power. Though the repetition rate in the T-Rex case (0.25 MHz) is reduced by 320, the effective increase in recording time was only 5- to 10-fold as compared to the 80 MHz case [16.11]. Due to the stronger excitation pumping and triplet-state relaxation, the typical dwell time in our images was 3–8 ms per 10 nm pixel. T-Rex can also be realized by scanning with a speed that exceeds the speed given by the spot size times the pulse repetition rate, e.g. 200 nm × 80 MHz =16 μm/μs, so the molecules do not encounter successive pulses. A compelling advantage of fast scanning is that the total sample illumination of 80 MHz scanning is maintained and the signal gain is still obtained. The recording time can be further reduced by the factor given by the signal gain, i.e., by 150–300 times compared to the scheme used in this work.

Powerful approaches exist to improve the axial resolution as well: the implementation of STED in a 4Pi-microscope [16.7,9,23] or the additional application

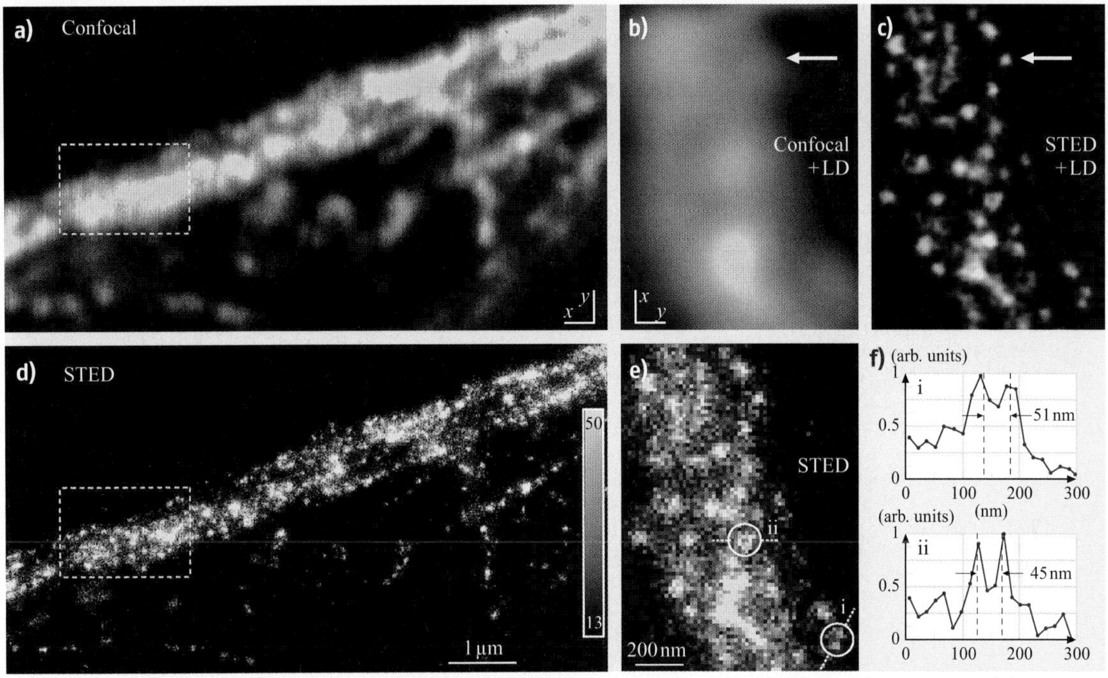

Fig. 16.5a–f Imaging neurofilaments in human neuroblastoma Contrary to the confocal (**a**), the STED (**d**) recording displays details < 30 nm, as highlighted by the comparison of image subregions shown in (**b**) and (**e**) bordered by dashed lines in (**a**) and (**d**). ((**b**) and (**c**)) Subregion after linear deconvolution. Note that the deconvolved confocal image (**b**) does not yield a substantial gain in information. (**f**) Profiles of raw data demonstrate the ability of (undeconvolved) STED data to reveal object structures that are far below the wavelength of light.

of a $h_{\text{STED}}(\boldsymbol{r})$ with strong values above and beneath the focal plane [16.8]. The first alternative involves the addition of the aperture of two opposing lenses, whereas the latter can be realized in a single lens. In fact, $h_{\text{STED}}(\boldsymbol{r})$ functions have been described which improved the axial resolution of a single lens by a factor of about six. As two STED pulses of different structure can be applied simultaneously (within the lifetime of the fluorescent state) the spot can be squeezed both axially and laterally at the same time.

It is important to emphasize that STED is only one possible avenue for attaining resolution enhancement using the concept of the optical switching of fluorescent markers labeling the sample [16.17, 18, 24–26]. The concept of STED microscopy can be generalized to the concept of reversible saturable optical fluorescence transitions (RESOLFT), which can be applied to any switchable fluorescent marker such as the family of fluorescent proteins [16.26]. Moreover, the excitation and reversible-quenching illumination patterns need not be pointwise. As long as the quenching light distribution features locations of zero intensity, the region of fluorescence can be reduced to subdiffraction dimensions according to the square-root law for resolution enhancement derived for STED. Much remains to be done in the field of biological imaging, and developments will no doubt progress quickly in the coming years as the utility of this technique is realized.

References

16.1 M. Born, E. Wolf: *Principles of Optics*, 7 edn. (Cambridge Univ. Press, Cambridge 2002) p. 952

16.2 T. Wilson, C. J. R. Sheppard: *Theory and Practice of Scanning Optical Microscopy* (Academic, New York 1984)

16.3 M. Schrader, K. Bahlmann, S.W. Hell: Three-photon excitation microscopy: Theory, experiment and applications, Optik **104**(3), 116–124 (1997)

16.4 A. Diaspro (Ed): *Confocal and Two-Photon Microscopy* (Wiley–Liss, New York 2002)

16.5 D. W. Pohl, W. Denk, M. Lanz: Optical stethoscopy: Image recording with resolution λ/20, Appl. Phys. Lett. **44**, 651–653 (1984)

16.6 S. W. Hell, J. Wichmann: Breaking the diffraction resolution limit by stimulated emission: Stimulated emission depletion microscopy, Opt. Lett. **19**(11), 780–782 (1994)

16.7 S. W. Hell: Increasing the resolution of far-field fluorescence light microscopy by point spread function engineering. In: *Topics in Fluorescence Spectroscopy*, ed. by R. Lakowicz J. (Plenum, New York 1997) pp. 361–422

16.8 T. A. Klar, S. Jakobs, M. Dyba, A. Egner, S. W. Hell: Fluorescence microscopy with diffraction resolution limit broken by stimulated emission, Proc. Natl. Acad. Sci. USA **97**, 8206–8210 (2000)

16.9 M. Dyba, S. W. Hell: Focal spots of size l/23 open up far-field fluorescence microscopy at 33 nm axial resolution, Phys. Rev. Lett. **88**, 163901 (2002)

16.10 V. Westphal, S. W. Hell: Nanoscale resolution in the focal plane of an optical microscope, Phys. Rev. Lett. **94**, 143903 (2005)

16.11 G. Donnert, J. Keller, R. Medda, M. A. Andrei, S. O. Rizzoli, R. Lührmann, R. Jahn, C. Eggeling, S. W. Hell: Macromolecular-scale resolution in biological fluorescence microscopy, Proc. Natl. Acad. Sci. USA **103**(31), 11440–11445 (2006)

16.12 F. P. Schäfer: *Dye Lasers* (Springer, Berlin, Heidelberg 1973)

16.13 J. R. Lakowicz: *Principles of Fluorescence Spectroscopy* (Plenum, New York 1983)

16.14 L. Kastrup, S. W. Hell: Absolute optical cross section of individual fluorescent molecules, Angew. Chem. Int. Ed. **43**, 6646–6649 (2004)

16.15 M. Dyba, S. W. Hell: Photostability of a fluorescent marker under pulsed excited-state depletion through stimulated emission, Appl. Opt. **42**(25), 5123–5129 (2003)

16.16 K. I. Willig, S. O. Rizzoli, V. Westphal, R. Jahn, S. W. Hell: STED-microscopy reveals that synaptotagmin remains clustered after synaptic vesicle exocytosis, Nature **440**, 935–939 (2006)

16.17 S. W. Hell: Toward fluorescence nanoscopy, Nature Biotechnol. **21**(11), 1347–1355 (2003)

16.18 S. W. Hell: Strategy for far-field optical imaging and writing without diffraction limit, Phys. Lett. A **326**(1–2), 140–145 (2004)

16.19 S. E. Fraser: Crystal gazing in optical microscopy, Nature Biotechnol. **21**(11), 1272–1273 (2003)

16.20 K. Willig, J. Keller, M. Bossi, S. W. Hell: STED microscopy resolves nanoparticle assemblies, New J. Phys. **8**, 106 (2006)

16.21 A. Yuan, R. A. Nixon, M. V. Rao: Deleting the phosphorylated tail domain of the neurofilament heavy subunit does not alter neurofilament transport rate in vivo, Neurosci. Lett. **393**(2-3), 264–268 (2006)

16.22 M. Osborne: Immunofluorescence microscopy of cultured cells. In: *Cell Biology: A Laboratory Handbook*, ed. by J. E. Celis (Academic, San Diego 1998) pp. 462–468

16.23 M. Dyba, S. Jakobs, S. W. Hell: Immunofluorescence stimulated emission depletion microscopy, Nature Biotechnol. **21**(11), 1303–1304 (2003)

16.24 M. G. L. Gustafsson: Nonlinear structured-illumination microscopy: Wide-field fluorescence imaging with theoretically unlimited resolution, Proc. Natl. Acad. Sci. USA **102**(37), 13081–13086 (2005)

16.25 R. Heintzmann, T. M. Jovin, C. Cremer: Saturated patterned excitation microscopy – A concept for optical resolution improvement, J. Opt. Soc. Am. A **19**(8), 1599–1609 (2002)

16.26 M. Hofmann, C. Eggeling, S. Jakobs, S. W. Hell: Breaking the diffraction barrier in fluorescence microscopy at low light intensities by using reversibly photoswitchable proteins, Proc. Natl. Acad. Sci. USA **102**(49), 17565–17569 (2005)

17. Ultrafast THz Photonics and Applications

Ultrafast THz photonics the topic of this chapter describes the union of optical and ultrafast laser capability with electronics to achieve frequency performance and bandwidths extending well into the THz frequency range. With the demonstrated capability to create and to measure subpicosecond electrical signals that are much faster than those produced and measured by any other method, THz photonics will help determine the direction for the development of new materials and ultra high performance technologies of the future.

In the Guided-Wave THz Photonics Section the generation, measurement and applications of these very short electrical pulses are described. They can be used as probes for short electrical pulse studies and the consequent characterization of transmission lines is described, including the measurement of Cherenkov radiation. The performance of transmission lines is compared with metal THz waveguides, whose characterization is also presented. Results for superconducting transmission lines and dielectric THz waveguides are given.

In the Freely Propagating Wave THz Photonics Section recent work demonstrating the generation of freely propagating THz radiation ($1\,\text{THz} = 33.3\,\text{cm}^{-1} = 4.1\,\text{meV}$) via material and electronic excitation by ultrashort laser pulses is presented. The generation of short pulses of THz radiation, by the passage of a short optical pulse through a nonlinear optical material, is also described. A cw photomixer capable of producing tunable radiation by beating together two laser beams is presented. The most developed THz application is THz time-domain spectroscopy (THz-TDS) which is described in detail. The combination of THz-TDS with THz beams will be shown to have some powerful advantages compared to traditional c.w. spectroscopy. The efficacy of THz-TDS is demonstrated by the presented characterizations of water vapor, flames, sapphire, high-resistivity silicon, n and p-type semiconductors, normal and high Tc superconductors, molecular vapors and liquids.

Ultrafast THz photonics describes the union of optical and ultrafast laser capability, for which the elementary excitation is the photon, with electronics to achieve frequency performance and bandwidths extending well into the THz frequency range. Optoelectronics, which is a subfield of photonics, has demonstrated the capability to create and to measure electrical signals that are much faster than those produced and measured by any other method. The subpicosecond electrical pulses produced optoelectronically are more than ten times faster than today's most advanced transistors, which have switching times on the order of 10 ps. The generation of short electrical pulses via optical methods has for some time been accomplished by shorting photoconductive gaps with very short laser pulses [17.1]. The same optical methods can measure the generated electrical pulses by sampling techniques. An alternate measurement approach has been to use the electrooptic effect in a crystal [17.2, 3]. In this case, the field of the electrical pulse is sampled through the rotation of the polarization of the short optical sampling pulse. Both the photoconductive and polarization rotation techniques have been extended into the subpicosecond regime [17.2–11].

Given the capability of generating and measuring these very short electrical pulses, they can be used as probes for short electrical pulse studies. For example, the study of the propagation of such pulses on transmission line structures is an important concern for future ultra-high-performance VLSI (very large scale integration) technologies. In addition to the considerable amount of work that has been done with these short pulses to understand the high-frequency behavior of transmission line structures, they are now being put to use in other areas as well. Because these pulses are much faster than the fastest transistors, they provide a powerful means of studying their switching transients. They can be used to directly measure the impulse response of ultra-high-performance silicon and gallium-arsenide devices [17.7–9, 12–19]. They can be used to evaluate the high-frequency behavior of semiconductor package connectors essential to future generations of computers. In summary, optoelectronic techniques are providing the capability to generate and measure electrical signals in the subpicosecond regime. This optoelectronic capability provides a tool to better understand the properties of existing materials and the components of existing high-performance VSLI technologies. Clearly, this capability will help determine the direction for the development of new materials and ultra-high-performance technologies of the future.

Recently, there has been a great deal of work demonstrating the generation of THz radiation ($1\,\text{THz} = 33.3\,\text{cm}^{-1} = 4.1\,\text{meV}$) via material and electronic excitation by ultrafast laser pulses. Modern integrated circuit techniques have made possible the precise fabrication of micron-sized dipoles, which when photoconductively driven by fs laser pulses, can radiate well into the THz regime [17.20–22]. An alternative and complimentary approach has been to extend radio and microwave techniques into the THz regime through the use of optoelectronic antennas [17.23–32]. The generation of short pulses of THz radiation, by the passage of a short optical pulse through a nonlinear optical material, was demonstrated many years ago [17.33]. A related source is based on the emission of an electromagnetic shock wave due to a volume dipole distribution moving faster than the phase velocity, i. e., electrooptic Cherenkov radiation[17.34, 35]. THz radiation has been generated by photoconductively driving the surface field of semiconductors [17.36, 37] and the intrinsic region of large aperture silicon p-i-n diodes [17.38] with ultrafast laser pulses. A cw photomixer capable of producing tunable radiation has been demonstrated by beating together two laser beams with frequencies above the band gap of a semiconductor [17.39–43].

One type of these sources is based on an optical-type approach, whereby a transient point source of THz radiation is located at the focus of a dielectric collimating lens, followed by an additional paraboloidal focusing and collimating mirror [17.21, 22, 29, 31, 44]. This type of source produces well collimated beams of THz radiation. Matched to an identical receiver, the resulting system has extremely high collection efficiency, a time resolution of less than 150 fs and a frequency range from 0.2 THz to more than 5 THz. This optoelectronic THz system is presently one of the most highly developed and will be the one described in most detail in this chapter, although the alternative systems will also be presented.

Applications of THz optoelectronic systems include THz imaging[17.45–47], THz ranging[17.48, 49] and THz time-domain spectroscopy (THz-TDS) [17.50–52]. Presently, THz-TDS is the most developed and has been shown to be the technique of choice for many research measurements. Consequently, THz-TDS will be the major application described in this article, by means of selected examples of significant measurements. The THz-TDS technique measures two electromagnetic pulse shapes, the input pulse and the propagated pulse, which has changed shape due to its passage through the sample under study. Consequently, via numerical Fourier analyses of the input and propagated pulses, the frequency-dependent absorption and dispersion of the sample can be obtained. The useful frequency range of the method is determined by the initial pulse duration and the time resolution of the detection process. Therefore, with each reduction in the generated electromagnetic pulse width, and/or the time resolution of detection, there is a corresponding increase in the available frequency range.

The combination of THz-TDS with THz beams will be shown to have some powerful advantages compared to traditional continuous wave (cw) spectroscopy. Firstly, the coherent detection of the THz radiation is extremely sensitive. In terms of average power this sensitivity exceeds that of the incoherent liquid-helium-cooled bolometers, by more than 1000 times [17.53]. Secondly, because of the gated and coherent detection, the thermal background, which plagues traditional measurements in this frequency range, is observationally absent [17.53–56].

17.1 Guided-Wave THz Photonics

In this section we will initially present an overview of experimental efforts to obtain a THz ultra-wideband transmission channel. The importance of ultrafast optoelectronics will be shown to be paramount in this effort, whereby the combination of optics, ultrafast laser pulses and electronics has enabled bandwidth and performance increases of as much as 100 times when compared to purely electronic techniques. We will describe in detail how subpicosecond electrical pulses are generated and detected using the photoconductive and electrooptic approaches and go on to discuss how such pulses can be used to study the high-frequency behavior of transmission line structures. The importance of complete characterization of transmission lines and waveguides will be illustrated by select examples, where the characterization was achieved by THz time-domain spectroscopy. For these characterizations it was imperative that single-mode excitation and propagation was achieved. Sliding contact excitation of coplanar transmission lines will be shown to be such a single-mode excitation method. For the THz waveguide investigations, quasioptical coupling techniques were shown to be capable of exciting only a single mode of both metal and dielectric waveguides, even though the many THz bandwidth of the incoming pulse covered as much as 30 waveguide modes. Applications to the study of materials and devices will also be mentioned.

17.1.1 Subpicosecond Electrical Pulses

A typical experimental arrangement for the photoconductive generation and measurement of the subpicosecond electrical pulses is shown pictorially in Figs. 17.1 and 17.2 [17.5, 6]. The pulses are generated on a transmission line structure consisting of two parallel aluminium lines of micron dimensions on a thin silicon layer. As indicated in the diagram, an incoming laser beam consisting of a continuous stream of 0.1 ps (or less) ultrafast light pulses at a repetition rate of typically 100 MHz is divided into two synchronized beams P1 and P2, by a partially transparent mirror. Because the beams follow different paths, the relative timing between P1 and P2 can be precisely adjusted by movement of the prism: 0.15 mm corresponds to 1 ps. The initial (excitation) light pulse P1 strikes the photoconductive substrate (silicon) between the two aluminium lines, shorting them for a fraction of a ps and creating electrical pulses that travel in both directions on the transmission line. The sampling light pulse P2 drives the photoconductive gap at the detection electrode, sampling the electrical pulse as it propagates down the transmission line. The electrical pulse is characterized by measuring the collected electrical charge as a function of the time delay between P1 and P2. Representative electrical pulses measured in this way are shown in Fig. 17.3, where the ps time scale of the phenomena is indicated.

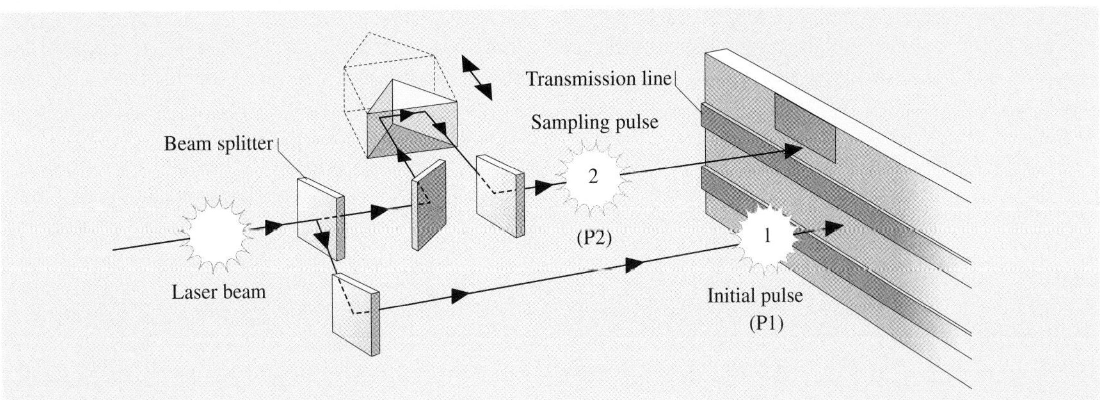

Fig. 17.1 Experimental arrangement for the optoelectronic generation and measurement of subpicosecond electrical pulses

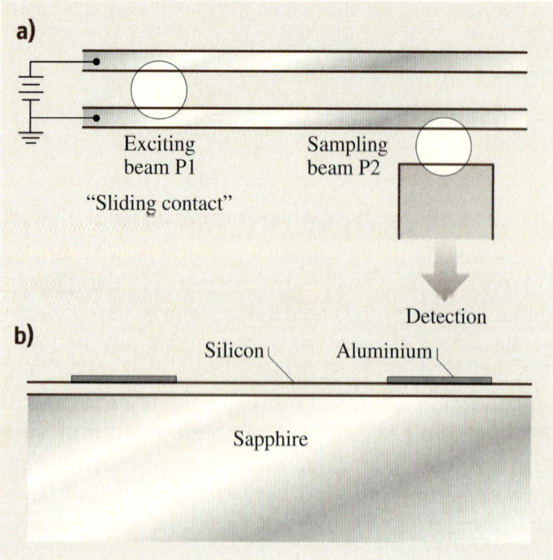

Fig. 17.2 (**a**) Experimental geometry, (**b**) cross section of SOS wafer

17.1.2 Sample Fabrication

The aluminium transmission line structures are fabricated using techniques similar to those used to form the interconnect wiring on semiconductor logic and memory chips. The starting substrate is a commercially available insulating silicon-on-sapphire (SOS) wafer with a 0.5 μm layer of pure undoped silicon epitaxially grown on top, as shown in the cross section in Fig. 17.2b. Typical linewidths are in the range of 0.5–10.0 μm. The metal thickness is in the range of 0.2–1.0 μm, and the length of the line structures varies from 5–25 mm. Numerous contact pads are fabricated along with the line structures to permit attachment of leads. Many different line dimensions and geometries can be fabricated simultaneously on a single 76 mm diameter wafer, since the area occupied by a single line structure is very small compared to the total surface area of the wafer.

At room temperature there are very few free electrons or holes in the silicon layer, and it behaves like an insulator. When an area of the film is struck by an ultrafast 0.1 ps laser pulse, a large number of photocarriers, electrons and holes, are generated almost instantaneously, rendering the area conductive for some length of time thereafter. This time can be made very short by bombarding the film with high-energy oxygen ions to produce crystal damage throughout the 0.5 μm thick silicon layer [17.57].

17.1.3 Generation and Measurement of the Pulses

Once fabrication is complete, chips with one or more transmission line structures are cut from the wafer. The chip to be measured is mounted on a fixture and wire bond connections are made to the contact pads to allow biasing of the transmission line at a voltage V, typically a few volts, and collection of the charge transferred to the detection electrode. The sample is then positioned on an optical table together with the laser, and the beams P1 and P2 are carefully aligned so as to strike the sample as shown in Figs. 17.1 and 17.2. The spot size of the laser is typically 5 μm.

The optoelectronic method of generation of the ultrafast electrical pulses is more precisely illustrated in Fig. 17.2a. This electrical arrangement was driven by 70 fs ultrafast optical laser pulses from a compensated, colliding-pulse, passively mode-locked dye laser producing pulses at a 100 MHz repetition rate. During operation, the P1 pulses short the transmission line by producing carriers in the silicon. These carriers are then trapped (disappear) in less than 600 fs [17.57]. The light pulses essentially turn on and off; they act as a very fast switch. When off, the resistance of the switch is many MΩ, but when on, its resistance is on the order of 1 kΩ. This transient shorting of the line produces an approximately 0.5 ps electrical pulse, which propagates down the transmission line. The generated pulses are measured by the P2 sampling pulses at the detection gap. As the electrical pulse traverses the line adjacent to the detection gap a voltage appears across this gap. If this gap is then photoconductively connected to the line by the P2 pulse, current will flow during the time of this connection. The amount of collected charge will be proportional to the voltage across the gap during the sampling interval. Consequently, the collected charge is measured against the relative time delay between the P1 and P2 pulses, which is mechanically scanned by moving an air-spaced retro-reflector with a computer-controlled stepper motor.

This sampling technique enables measurements down to the subpicosecond regime even though the wires bonded to the chip are low-frequency connections. All of the important time information is contained in the relative time delay between the two laser beams, which can be controlled with great precision. Because the excitation and sampling pulses come at a rate of one every 10 ns, while the relative timing between P1 and P2 is typically swept at a rate of 0.1 ps/s, the consequent slowly varying current reading represents an average of a large number (10^9) of identical events. Although the generated electrical pulses reflect off the end of the transmission line and then echo back and forth on the line, the 10 ns between events allows time for about 100 reflections for complete dissipation due to resistive losses. Thus, the system is electrically quiet prior to being excited with each P1 pulse.

For the generated pulse shown in Fig. 17.3a, the excitation site was very close to the sampling gap so that propagation effects were absent. This measurement illustrates the exceptionally clean pulses and high signal-to-noise ratios that can be obtained with this technique. This short electrical pulse had an amplitude of about 10 mV, and a measured half-width of less than 1 ps. The same pulse is shown on a 10× expanded scale, where the signal-to-noise ratio (S/N) of 500 : 1 (in a single scan) can be seen. In Fig. 17.3b, the

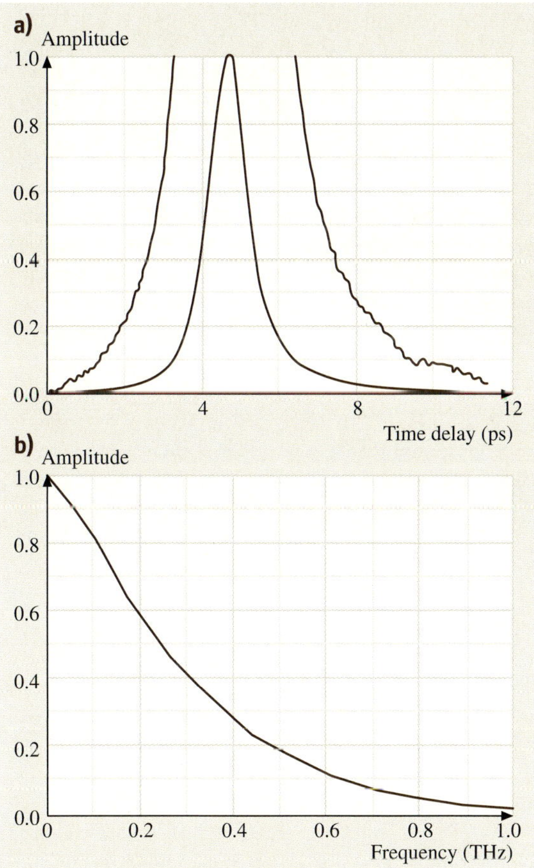

Fig. 17.3 **(a)** Measured ultrafast electrical pulse (*lower trace*), 10× expanded scale (*upper trace*). **(b)** Amplitude spectrum of **(a)**

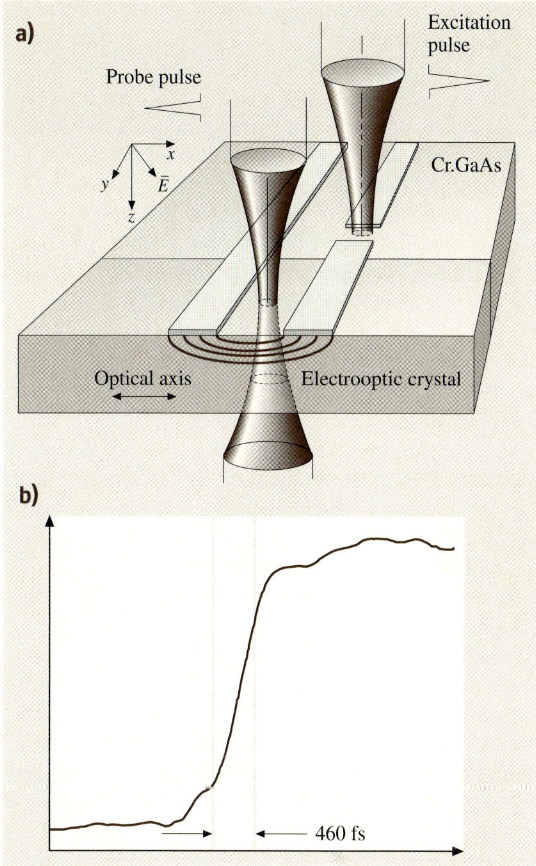

Fig. 17.4 **(a)** Electrooptic sampler arrangement. **(b)** Measured electrical response of the Cr:GaAs photoconductive switch. The 10–90% rise time is 460 fs (Adapted from [17.2, 3])

Fourier transform (amplitude spectrum) of this pulse illustrates the bandwidth available for spectroscopic measurements.

17.1.4 Electrooptic Sampling of Pulses on Transmission Lines

Another measurement technique known as electrooptic sampling[17.2,3,7–12], can measure repetitive electrical transients with subpicosecond resolution and microvolt sensitivity. The technique is based on an electric field coupling between the electrooptic "sampling gate" and the circuit under evaluation and does not require that charge be removed from the circuit. Consequently, electrooptic sampling is a minimally interfering technique that, in certain configurations, can sample electrical signals with micron spatial resolution without making physical contact.

In the example presented in Fig. 17.4, electrooptic sampling was used to measure an electrical transient on a coplanar transmission line generated by photoconductively shorting a 50 μm gap biased at 50 V with the excitation pulse [17.8]. The Cr:GaAs and $LiTaO_3$ crystals were mounted side by side on a glass plate and were ground and polished together to a thickness of 500 μm in order for later fabrication. The transmission line was made by evaporating 0.5 μm of aluminium onto the smooth surface and then using standard photolithographic techniques to define the pattern. The measured 460 fs rise time (10% to 90%) is shown in Fig. 17.4b for which the 50 μm excitation gap was located 200 μm from the sampling point. The probe beam was aligned perpendicularly to the substrate and was focused to a 11 μm diameter between the two 50 μm wide lines separated from each other by 50 μm. The amplitude of the switched signal was 30 mV; the good signal-to-noise ratio indicates the sensitivity of the technique.

Electrooptic sampling reached the impressive combination of a demonstrated rise time of 290 fs and a 40 μm square footprint in the 1 THz bandwidth probe-station developed by *Valdmanis* [17.9]. In other work, a 260 fs rise time electrical transient was measured using electrooptic sampling with a thin lithium tantalate crystal plate overlying a coplanar transmission line [17.10]. An impressive rise time of 150 fs has been demonstrated using a combination of photoconductive excitation and electrooptic sampling together with the total internal reflection probe [17.11]. Using the approach of [17.12] for GaAs circuitry, the chip itself can be used as the nonlinear crystal.

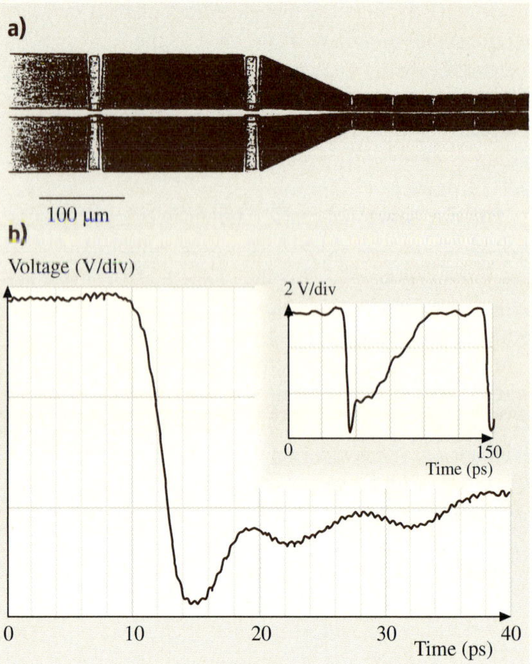

Fig. 17.5 (**a**) Photomicrograph of part of a hyperabrupt-diode nonlinear transmission line showing the taper between the larger-scale and final small-scale sections. (**b**) A 2.5 ps measured fall time shockwave generated on an NLTL with a 25 dBm, 10 GHz sinusoidal input. In the *insert*, the full 100 ps period saw tooth waveform is shown. (Adapted from [17.58])

17.1.5 THz Shockwave Generation on Nonlinear Transmission Lines

The phenomenon of shockwave generation on a fully distributed nonlinear transmission line (NLTL) was first discussed by *Landauer* [17.59] with respect to its use as a parametric amplifier. The shockwave properties were later more fully analyzed [17.60] and the case of periodically loaded [17.61] NLTLs were discussed. Previous studies [17.62] have demonstrated the generation of 3.5 ps shockwaves on a GaAs NLTL having Schottky diodes with uniform doping. In this section the fabrication and results from NLTLs with hyperabrupt varactor diodes will be discussed [17.58]. The increased capacitance variation in hyperabrupt diodes permits using lower loss NLTL designs, that result in a shockwave with higher amplitude and decreased fall time.

The monolithic GaAs NLTL circuit consists of a high-impedance coplanar waveguide (CPW) transmission line with characteristic impedance Z_l that is periodically loaded at spacing τ (in units of time), by Schottky varactor diodes, producing a synthetic transmission line whose propagation velocity is voltage dependent. Several variations of this mixed structure (Fig. 17.5a) having different numbers of large-scale diodes were fabricated with a minimum feature size of 2 μm on GaAs molecular beam epitaxy material with a total active layer thickness of 0.3 μm and a 1.0 μm buried n^+ layer ($6\times10^{18}/cm^3$ doping) for the diode cathode connection.

As a negative-going input voltage transition, with initial voltage V_h, final voltage V_f, and initial fall time t_{in}, propagates along the line, the fall time will decrease with distance due to the differential group delay, $\Delta t = t(V_h) - t(V_f)$, along the wavefront. As the

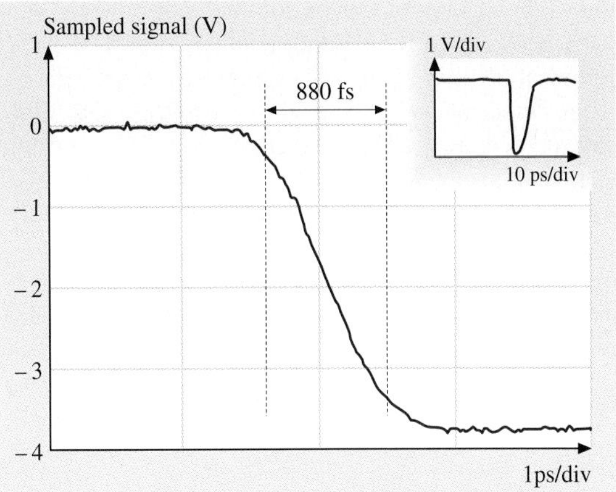

Fig. 17.6 An 880 fs measured-fall-time shockwave generated on a nonlinear transmission line and measured by an on-chip diode sampler at $T = 77$ K. The *inset* is a quarter period of the waveform. (Adapted from [17.63])

fall time decreases, dispersion arising from the varactor cutoff frequency and the periodic-line cutoff frequency competes with the compression arising from the voltage-dependent propagation velocity. An asymptotic final fall time t_{min} is reached, at which point the fall time compression per line section is equal to the fall time broadening per section so that the resulting shockwave then propagates unchanged.

The circuit performance was evaluated by direct electrooptic (EO) sampling[17.12] of the voltage waveforms launched onto the line through microwave wafer probes. Figure 17.5b shows a measured waveform for 25 dBm sine wave excitation of a NLTL consisting of 30 large diodes and 40 small diodes with a measured $\Delta t = 49$ ps and a total length of 7.8 mm. The 10 GHz input had a 10–90% fall time of 30 ps, while the output shock wave had a measured 2.5 ps fall time. The shock amplitude was determined to be about 6 V by calibration of the optical signal to a reference electrical signal.

The first subpicosecond shockwaves generated and measured by entirely electronic means are shown in Fig. 17.6 [17.63]. This measurement was performed with a monolithic NLTL/sampler chip. At room temperature, an approximately 1 V, 1.8 ps fall time was observed with less than 20 fs of jitter for a 25 dBm, 6.56 GHz input. The packaged assembly was then lowered into a dewar flask of liquid nitrogen. Steady improvement in both amplitude and fall time was noted until the packaged circuit was completely immersed, resulting in the minimum fall time of 880 fs with a 3.5 V amplitude.

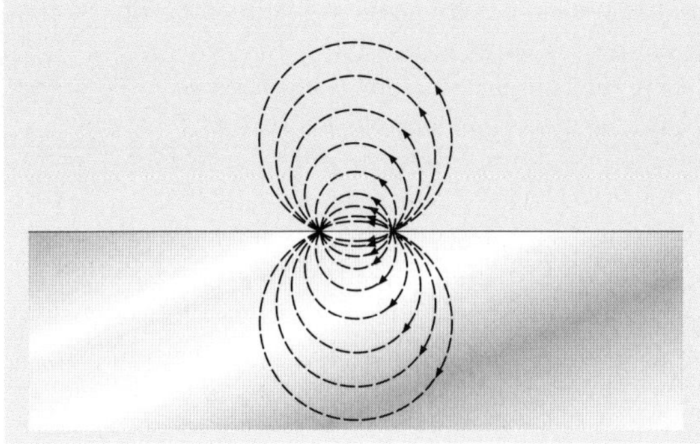

Fig. 17.7 Electrical field lines for the propagating TEM mode

The liquid nitrogen did not significantly load the circuits because of its relatively low permittivity $\varepsilon_r = 1.4$.

17.1.6 Transmission Line Theory

We will now discuss some general aspects of transmission line theory in the quasistatic limit for which the wavelengths involved are large compared to the transverse dimensions of the line [17.64, 65]. For this case the number of transverse electromagnetic (TEM) modes is one less than the number of metal lines making up the transmission line. Consequently, a two line transmission line has a single propagating TEM mode, for which the electric field distribution (shown in Fig. 17.7) is the same as for the static case when the lines are equally and oppositely charged. Any pulse propagating on this line can be described mathematically as a Fourier sum of single frequency components, all with this same TEM modal distribution.

Consideration of the sliding contact excitation site shows that, to first order, charge is simply transferred from one line to the other creating a symmetrical field distribution with respect to the two lines. During the excitation process a current flow is induced between the lines. Localized charge accumulations of opposite sign build up on the segments of the two metal lines under the laser excitation spot, creating a dipolar field distribution similar to that illustrated in Fig. 17.7 and thereby efficiently coupling the sliding contact excitation to the TEM mode.

A most important feature of coplanar transmission lines (of negligible thickness) on an infinite dielectric half-space is that for a constant voltage between the two conducting lines, the electric field lines are the same as for the lines immersed in free space. This result is due to the geometric symmetry with respect to the dielectric boundary and is a consequence of the fact that no electric field lines cross this boundary.

For this simple situation, it has been shown that the phase velocity is given by

$$v_p = c\{2/[\varepsilon(\omega)+1]\}^{1/2} . \quad (17.1)$$

If we define the effective index $n_e = c/v_p$, the expression for the group velocity becomes

$$v_g = c/[n_e + \omega(dn_e/d\omega)] . \quad (17.2)$$

The power absorption coefficient α for the transmission line can be written as the sum of the absorption due to the metal lines α_{ml} and that due to the dielectric α_{dl}:

$$\alpha = \alpha_{ml} + \alpha_{dl} . \quad (17.3)$$

The dielectric loss of the line can be written in the very general form for guided waves

$$\alpha_{dl} = f_f \alpha_d v_{gd}/v_g , \quad (17.4)$$

where the filling factor f_f is defined as the ratio of the energy per unit length in the dielectric to the total energy per unit length on the coplanar line; α_d is the power absorption coefficient in the bulk dielectric and v_{gd} is the group velocity in the dielectric. For the simple case of the coplanar waveguide with the same field pattern in the air and in the dielectric, f_f is given by $f_f = \varepsilon/(1+\varepsilon)$; $v_{gd} = c/\varepsilon^{1/2}$, and α_{dl} is evaluated to be

$$\alpha_{dl} = \alpha_d[\varepsilon/(2+2\varepsilon)]^{1/2} . \quad (17.5)$$

17.1.7 THz-TDS Characterization of Transmission Lines

The powerful technique of THz time-domain spectroscopy (THz-TDS) is especially appropriate for characterizing dielectrics, ultra-wideband transmission lines and waveguides [17.52, 66]. With this technique, two electromagnetic pulse shapes are measured, the input pulse and the propagated pulse, which has changed shape due to its passage through the sample under study. Consequently, via Fourier analyses of the input and propagated pulses, the frequency dependent absorption and dispersion of the sample can be obtained.

The subpicosecond electrical pulses generated on coplanar transmission lines can be used for THz-TDS characterization of the transmission line itself. For the layout shown in Figs. 17.1 and 17.2, the point at which the pulse is generated can be moved with respect to the detection electrode by repositioning the exciting laser beam. This separation can be continuously adjusted and is the equivalent of a "sliding contact." For THz-TDS studies it is important that the pulses propagate as a single mode of excitation on the transmission line. This feature is enabled due to the micron-sized dimensions of the coplanar line and to the sliding contact method of excitation, [17.5], which matches the TEM transverse electromagnetic mode of the transmission line. Thus, as the ultrafast electrical pulse propagates down the transmission line, its pulse shape will only change due to the frequency dependent electrical and magnetic properties of the transmission line, e.g., the metal of the line and the dielectric substrate and radiation processes. Consequently, via THz-TDS, the absorption and dispersion vs. frequency of the line can be obtained [17.67].

The performance of lithographically defined planar transmission lines will be shown to be limited by the resistive loss of the metal lines, Cherenkov radiation loss from the propagating pulse, and absorption by the underlying dielectric. Typically, for THz bandwidth pulses, Cherenkov radiation is the dominant loss process, which can be so severe that after propagating only 1 mm, power at 0.8 THz is reduced to 1/e of its original value. Various techniques have been demonstrated to reduce this radiation. The most straight-forward approach, which is compatible with VLSI processing, is to reduce line dimensions (linewidths and separations) to below 1 μm. However, for such dimensions resistive loss of the lines restricts the bandwidth. Normal superconductors can solve this problem for low frequency applications typically up to a band gap frequency of 0.7 THz, above which the line shows severe loss. As an added benefit, cryogenic temperatures usually lower the THz absorption loss of crystalline substrates. To achieve high-frequency response with submicron linewidths, the high-T_c superconductors with band gaps above 15 THz appear as a possible solution. Early attempts to achieve THz bandwidths with high-T_c transmission lines were limited by the extremely high absorptive loss of the lattice matched YSZ substrate [17.68]. Later investigations used lattice matched lanthanum aluminate substrates with much lower loss and achieved significantly lower propagation loss up to 1 THz at cryogenic temperatures for a high-T_c coplanar transmission line compared to an identical transmission line with gold lines [17.69]. However, a more recent investigation of high-T_c coplanar transmission lines fabricated on a lattice matched MgO substrate gave the disappointing result that at 77 K the attenuation was higher for the high-T_c line than for an identical gold line [17.70]. This situation must be resolved if high-T_c lines are going to realize their promise

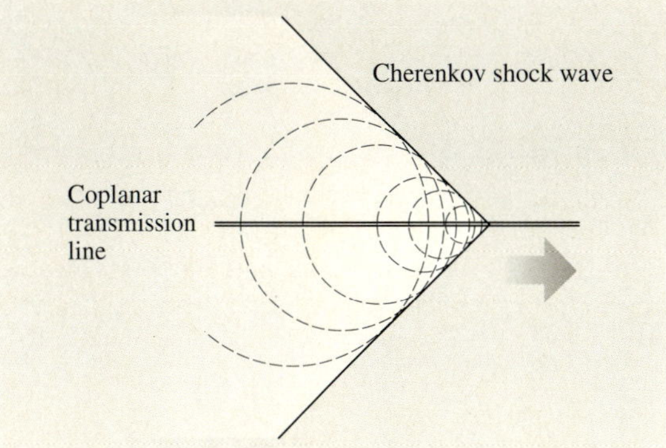

Fig. 17.8 The Cherenkov cone in the dielectric half-space

as the future material for the VLSI interconnect with dimensions below 1 μm and with THz bandwidths.

Cherenkov Radiation

By studying the frequency dependent loss on an Al transmission line on SOS, where the coplanar lines were separated by 15 μm, a powerful radiation process was observed due to a type of Cherenkov radiation [17.71]. When charges move faster than the phase velocity for electromagnetic radiation in a material, Cherenkov radiation is emitted as an electromagnetic shock wave [17.72]. The initial analysis of this effect was for electric monopoles, but the physical picture holds true for higher-order moments as well. The electrical pulses were produced by transferring approximately 2000 electrons between the two lines at the generation site. Consequently, two dipoles with charges of 1000 e and separations of 15 μm move down opposite directions of the line at the measured group velocity of $c/2.45$, significantly faster than the phase velocity of approximately $c/3.3$ for THz radiation in sapphire. Thus, in the dielectric, the situation shown in Fig. 17.8 occurs, which generates radiation of an electromagnetic shock wave in the form of a Cherenkov-type cone.

THz-TDS measurements have characterized this loss process, which is so severe that after propagating only 1 mm the power at 0.8 THz is reduced to $1/e$ of its original value. Previous frequency domain studies have calculated the radiative loss for single-frequency transmission on coplanar metal lines on dielectrics [17.73].

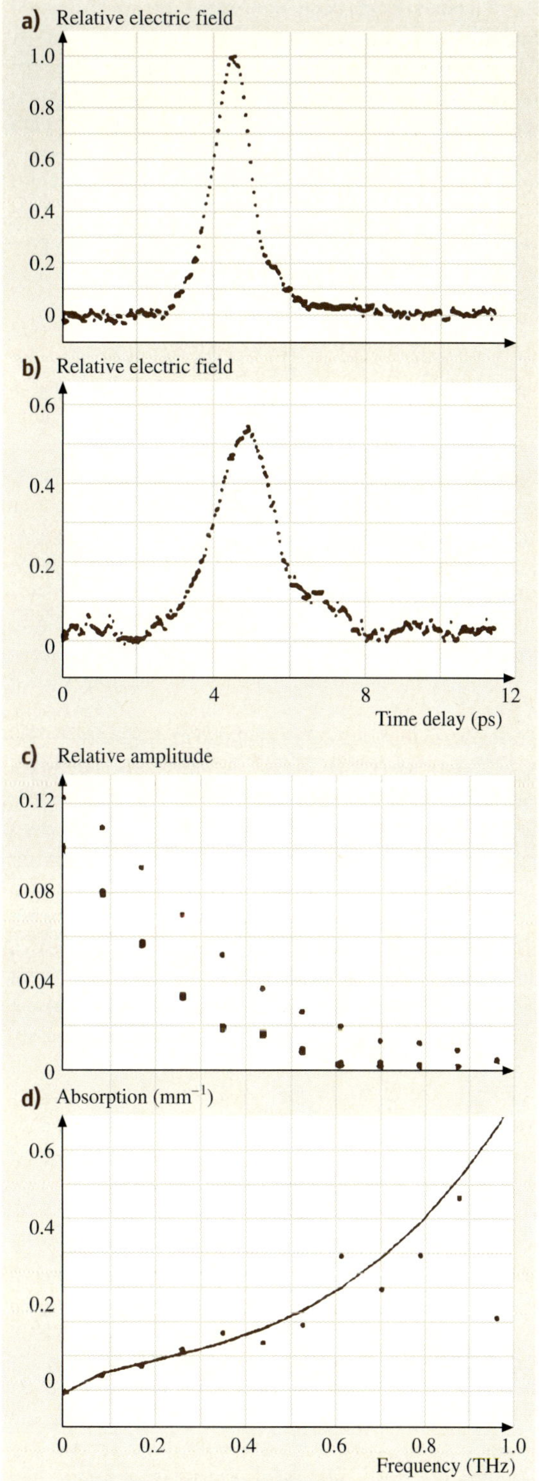

Fig. 17.9 **(a)** Measured initial pulse. **(b)** Measured pulse after 6 mm propagation. **(c)** Amplitude spectrum of pulse (*dots*) compared to propagated pulse (*squares*). **(d)** Experimental amplitude absorption coefficient (*dots*) compared with theory. (Adapted from [17.71])

These results have been confirmed by experimental measurements up to 1 GHz [17.73], where the radiation loss is only 10^{-9} of that presented here. The observed frequency-dependent absorption remains in good agreement with this calculation, even though the THz-TDS measurement extends up to 1 THz. At these high frequencies, the time-domain Cherenkov picture provides a simple intuitive description of the generation of the electromagnetic shock wave and of the radiation process associated with the propagating ultrafast electrical pulse.

In Fig. 17.9a, the measured initial subpicosecond electrical pulse is shown [17.71]. In this case, the spatial separation between the exciting and sampling beams was approximately 50 μm, while the laser spot diameters were 15 μm. When the sliding contact was moved 6 mm from the optical sampling gap the pulse propagation effects shown in Fig. 17.9b were observed. The Fourier analyses of these pulses are shown in Fig. 17.9c, where it is seen that the amplitude spectrum of the initial pulse extends beyond 1 THz and that the transmitted pulse has suffered a significant frequency-dependent loss. From these spectra it is straight-forward to obtain the absorption coefficient as a function of frequency, as shown in Fig. 17.9d.

Evaluating the predicted loss ("absorption") for these conditions [17.71,73,74], the amplitude absorption coefficient is obtained as $\alpha = (0.4\,\text{mm}^{-1})f^3$, where f is the frequency in THz. This result should be compared with the measurements in Fig. 17.9d. The solid line fit to the data in Fig. 17.9d is given by the relationship $\alpha = (0.2\,\text{mm}^{-1})f^{1/2} + (0.65\,\text{mm}^{-1})f^3$. The first term gives the frequency-dependent loss due to the resistive skin effect for the Al lines, while the second term with the f^3 dependence describes the loss due to the radiation. The measured loss due to the radiation is approximately 1.5 times the calculated value. This good agreement, and especially the cubic-frequency dependence, definitely confirms the nature of the effect.

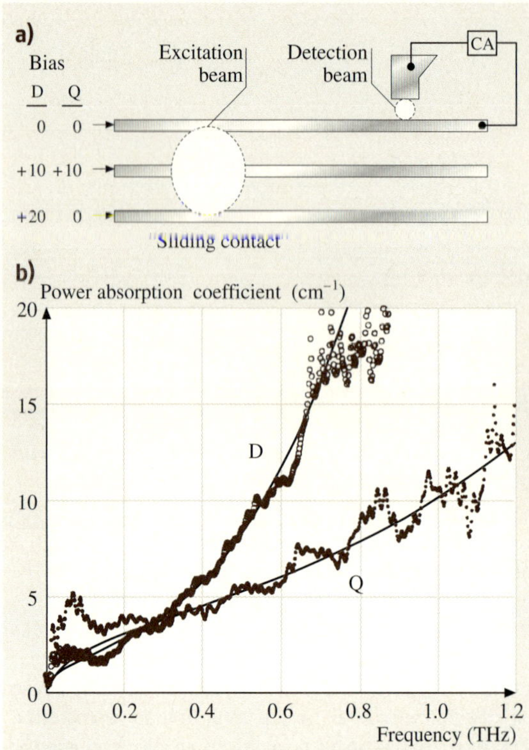

Fig. 17.10 **(a)** Three line coplanar transmission line. The aluminium lines are 4 μm wide separated by 15 μm. Quadrupole electrical pulses are generated with bias Q, dipole pulses with bias D. **(b)** Measured power absorption coefficient compared to theory (Q) quadrupole (odd mode) and (D) dipole even mode pulses. (Adapted from [17.74])

Methods to Eliminate Cherenkov Radiation from Transmission Lines

The Cherenkov radiation loss has been eliminated by the use of silicon-on-insulator materials with a microstrip geometry [17.75] and thin-film microstrip lines [17.76], but the total observed loss owing to the dielectric and the metal was still approximately the same as for the coplanar transmission line. A variety of transmission line structures have been designed to reduce the permittivity mismatch responsible for the Cherenkov radiation [17.77–80]. The ideal transmission line would have no substrate/superstrate mismatch. Along these lines, coplanar air transmission (CAT) lines have been fabricated [17.77]; these coplanar striplines are edge supported by a 200 nm layer of SiO_2 on GaAs. The GaAs below the transmission line is etched away leaving the striplines supported in free space. Rise times as short as 0.8 ps after a propagation length of 2.8 mm were obtained [17.77]. Coplanar striplines have also been fabricated on low permittivity substrates and membranes, on the order of microns thick, demonstrating significantly reduced loss and higher bandwidths [17.78–80].

A new approach to the reduction of radiative loss from transmission lines has been demonstrated [17.74]. Instead of reducing the permittivity mismatch, the mode characteristics of a three line stripline were utilized to generate a quadrupole electrical pulse with significantly reduced loss and dispersion compared to that of the dipole pulse. As shown in Fig. 17.10a, through appropriate bias of the striplines, either the even mode, where the electric fields are parallel between each gap with a dipole field pattern, or the odd mode, which has the antiparallel electric fields in the gaps with a quadrupole field pattern, can be exclusively excited.

Using THz-TDS analysis, the amplitude absorption coefficient $\alpha(f)$ for the quadrupole pulse and dipole pulse are shown in Fig. 17.10b. The absorption for the quadrupole is dramatically reduced over that of the dipole; at 0.8 THz a factor of two difference between the two coefficients is observed. Both curves are fit with the relationship $\alpha(f) = A_{res} f^{1/2} + A_{rad} f^3$, where the resistive A_{res} and the radiative A_{rad} loss coefficients are adjustable parameters. For the dipole pulse A_{res} is $6\,cm^{-1}$, compared to $7\,cm^{-1}$ for the quadrupole. For the dipole pulse, the strong f^3 dependence is seen with a radiative coefficient $A_{rad} = 32\,cm^{-1}$, whereas the quadrupole pulse has the much smaller coefficient $A_{rad} = 3\,cm^{-1}$. This order of magnitude reduction clearly demonstrates that the radiative loss for an electric quadrupole pulse is significantly less than that of a dipole pulse. It is instructive to compare the measured $A_{rad} = 32\,cm^{-1}$ for the $30\,\mu m$-sized dipole with that expected from the measurements of Fig. 17.9, where $A_{rad} = 6.5\,cm^{-1}$ for a $15\,\mu m$ dipole. Because the radiation loss is proportional to the square of this length, Fig. 17.9 would predict an $A_{rad} = 28\,cm^{-1}$ for a $30\,\mu m$-sized dipole in good agreement with observation.

Superconducting Transmission Lines

One of the early applications of THz-TDS was the study of superconducting niobium metal lines [17.81]. Niobium has a superconducting transition temperature of 9.4 K and was cooled in a cryostat to temperatures in the range of 2–8 K. The cryostat had optical windows that allowed for the laser beams to pass through. For the experiment, the same pattern was fabricated in niobium metal on an SOS wafer, which was then ion implanted as before.

With the exciting beam striking very close to the detection electrode, the results are much the same as for aluminium at room temperature. Above the superconducting transition temperature $T_c = 9.4\,K$ for niobium, the propagated pulse was severely broadened by the highly resistive Nb transmission line.

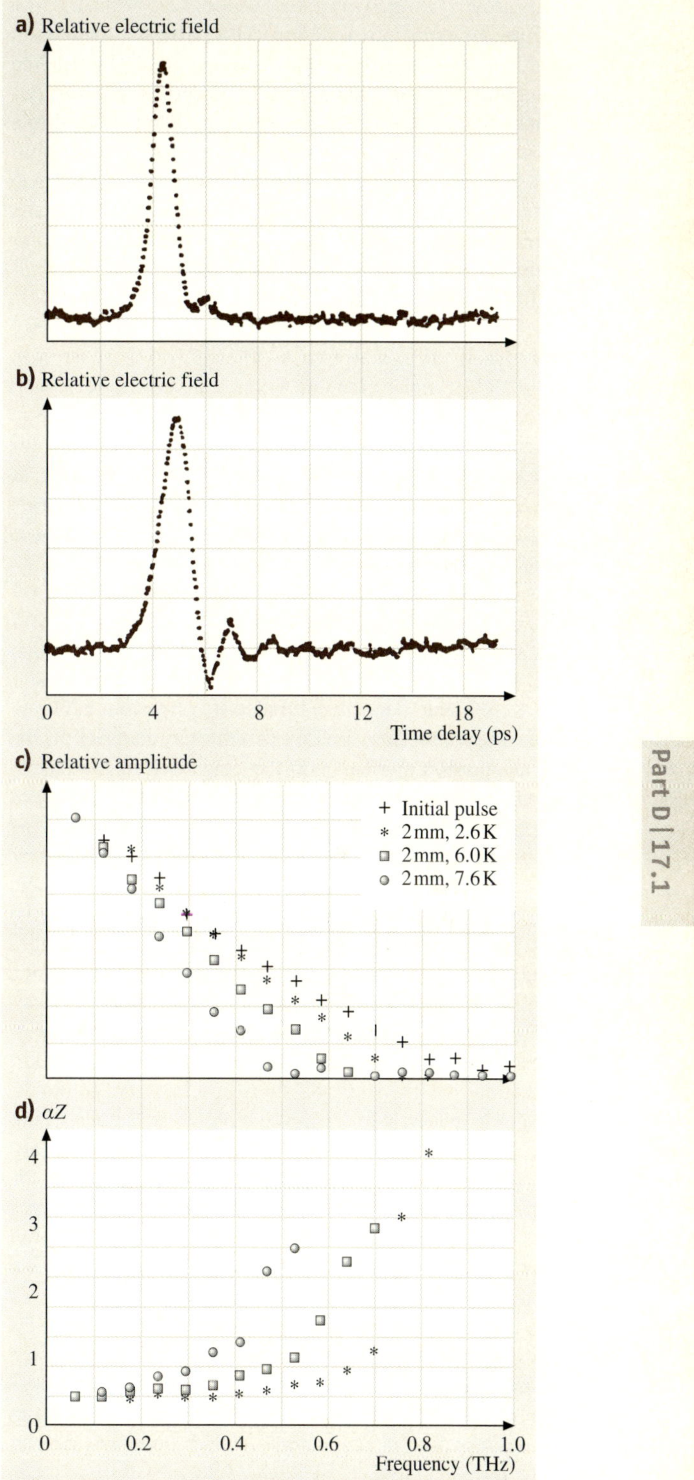

Fig. 17.11 **(a)** Initial electrical pulse. **(b)** Electrical pulse after 2 mm propagation on Nb superconducting transmission line at 2.6 K. **(c)** Fourier transforms of initial and propagated pulses at different temperatures. **(d)** Absorption spectra of transmission line versus temperature

Figure 17.11a shows the pulse shape at 2.5 K and 0.5 mm from the point of excitation. The pulse broadens with propagation and develops a distinct oscillation at the leading edge, as shown in Fig. 17.11b. This oscillatory behavior is characteristic of a superconductor. While a superconductor has no losses at low frequencies, it has significant loss at and above the superconducting gap frequency, corresponding to the energy with which pairs of electrons are bound in the superconducting state. For niobium at 2.6 K this onset is at about 0.7 THz, which is close to the oscillation frequency shown in Fig. 17.11b. When the input and propagated pulses are frequency analyzed, the absorption coefficient has a sharp step at 0.7 THz, corresponding to the superconducting band gap. This result clearly identifies the band gap and illustrates the power of the method. The fact that for frequencies above the band gap, the line behaves as a normal metal has the important technical implication that ordinary superconducting lines cannot be used to propagate 1 ps pulses.

17.1.8 Guided-Wave THz-TDS Characterization of Dielectrics

The frequency-dependent dielectric properties of a material may be measured by placing the material on the transmission line and into the field lines of the propagating pulses. For a typical line composed of 5 μm lines separated by 10 μm, this requires that the material be brought within several microns of the line. For paths several mm long, this is not an easy task. An optoelectronic test that indicates the quality of the optical contact is the measured group velocity v_g of the pulse in the section of the transmission line covered by the dielectric sample. With no sample in place for an SOS chip $v_g = c/2.45$. The expression for the group velocity is easily generalized to the two dielectric case as $v_g = c/[2/(\varepsilon_1 + \varepsilon_2)]^{1/2}$, where ε_1 refers to the substrate and ε_2 refers to the dielectric under study. Consequently, with typical high dielectric constant ($\varepsilon_2 = 10$) samples v_g should be approximately $c/3$. An estimate of the contact with the transmission line is given by a comparison between this calculated value and the measured value. The result can be used to estimate a "filling factor" for the absorption measurement.

An early example of a THz-TDS measurement of a material brought into contact with the transmission line was the observation of some magnetic resonances in erbium iron garnet (ErIG)[17.82]. It is of interest to note here that the interaction is with the magnetic field of the propagating electrical pulse in contrast to the usual

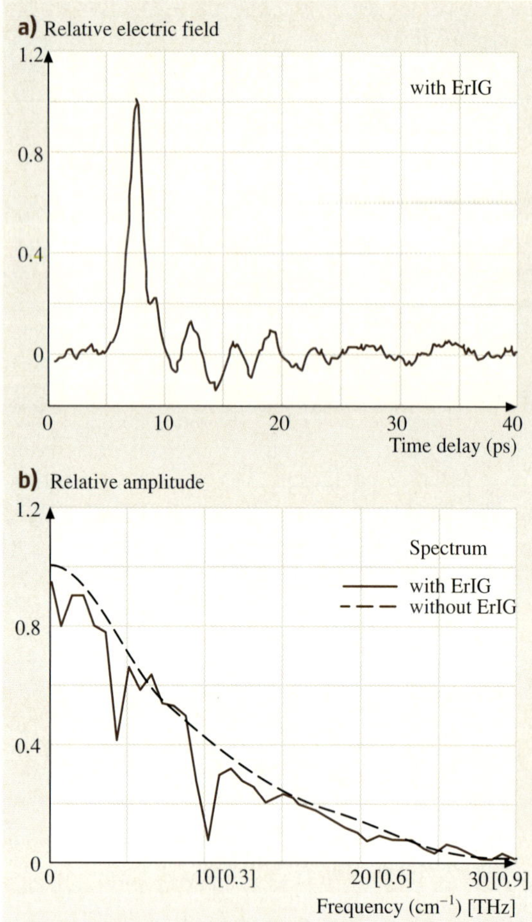

Fig. 17.12 **(a)** Electrical pulses after 4 mm propagation on the line covered with EIG powder. **(b)** Spectrum of propagated pulse with and without ErIG. (Adapted from [17.82])

case where it is the electric field responsible for the observed effects. For this study, an approximately 3 mm section of the line was covered with fine garnet powder. When the ultrafast pulse was propagated through this powder a new dramatic oscillatory time dependence was observed on the output pulse shown in Fig. 17.12a. When this output pulse was frequency analyzed, two well-defined absorption lines appeared as shown in Fig. 17.12b. The line at 10 cm^{-1} had been measured earlier by more traditional spectroscopic techniques and identified as the exchange resonance [17.83], while the line at 4 cm^{-1} had previously not been observed. Earlier measurements using YbIG showed a line at 3 cm^{-1} due to the ferrimagnetic resonance [17.84]. By extrapolating these results to ErIG, the newly observed line appears to be the corresponding ferrimagnetic resonance.

17.1.9 THz Waveguides

THz waveguides have been recently demonstrated to be an alternative to transmission lines [17.85–90]. For bandwidths up to 3.5 THz the attenuation coefficient of metal waveguides has been shown to be less than 1/10 that of lithographically defined transmission lines on dielectric substrates. Within the passband of the waveguide, the measured power coupling into the waveguide was typically 40% of the incoming THz power. Although these waveguides should be quite useful for narrow band or THz-TDS applications, they have very high group velocity dispersion (GVD), which precludes subpicosecond pulse propagation. For the single conductor circular and rectangular metal waveguides the excessive THz pulse broadening is caused by the extreme GVD near the cutoff frequency. This pulse broadening would not occur for the TEM modes of a two-wire coplanar line, a coaxial line, or a parallel-plate metal waveguide, that do not have a cutoff frequency. The group and phase velocities of such TEM modes are determined solely by the surrounding dielectric. Quasioptic coupling techniques are not effective for the complex field patterns of the TEM mode of the two-wire coplanar line or the coaxial line, but are effective for the simple field pattern of the TEM mode of the parallel-plate metal waveguide.

Recent experiments have shown efficient quasioptic coupling of freely propagating subpicosecond pulses of THz radiation into a parallel-plate metal waveguide and the subsequent low-loss, single TEM mode propagation exhibiting negligible GVD [17.89, 90]. Consequently, the ideal THz interconnect [17.66], capable of propagating subpicosecond pulses with minimal loss and no distortion, has been realized.

A promising application of THz metal waveguides appears to be surface-specific frequency-dependent absorption measurements using THz-TDS. For an air-filled waveguide, the observed frequency-dependent absorption coefficient determines the frequency-dependent conductivity, which is difficult to obtain by other techniques. Long-path surface-specific measurements of molecular adsorption to the metal waveguide walls is another area of promise. From experimental measurements made on the absorption of water, we believe that for similar polar molecules, THz-TDS has the sensitivity to characterize the absorption of nanogram quantities of samples of toxic or precious gases contained in a tubular, single conductor THz waveguide. Such enclosed waveguides would also make excellent low-volume sample cells for high-pressure, long-path spectroscopic measurements of toxic or reactive gases and liquids. THz waveguides could enable the observation of nonlinear effects with relatively modest powers, due to the small diameters combined with the long-path, low-loss propagation.

In addition to the above possibilities, THz pulses propagating in the TEM mode open the door to other new research applications. The metal plates could be easily coated with films of other metals and alloys to measure their conductivities. Superconducting plates could be easily characterized. Shorter length waveguides could be constructed of highly doped semiconductors that are typically opaque to THz radiation, thereby enabling the measurement of their frequency-dependent complex conductivity. Simple thin films, more complex Langmuir–Blodgett films, and conducting polymers could coat the waveguide and be characterized by THz-TDS. A unique application for high-power TEM THz pulses would be the study of nonlinear pulse propagation, demonstrating nonlinear coherent effects of samples filling the space between the metal plates. Here, for the first time, a THz pulse can maintain its spatial focus for arbitrarily long paths without temporal broadening, thereby enormously increasing the effects of nonlinear interactions. This situation is similar to the nonlinear enhancement obtained with optical fibers.

Dielectric waveguides do not have the sharp low-frequency cutoff of metal waveguides and thereby extend the low-frequency limit. The initial demonstration [17.87] used single crystal sapphire fibers and achieved single HE_{11} mode propagation with large GVD caused by the extent of the fringing fields increasing with wavelength. The waveguide absorption was that of bulk sapphire. Given a suitable low loss dielectric, such as high-resistivity silicon with a power absorption coefficient of less than $0.05\,cm^{-1}$, dielectric waveguides could have much less absorption than metal waveguides. Such dielectric single-mode THz waveguides would have the promise of an extremely low-loss, flexible interconnect and communications channel, with advantages similar to that of single-mode optical fiber.

Planar dielectric waveguides appear to be especially well adapted for the proposed THz-TDS characterization of surface layers, due to the extensive fringing fields of the propagating THz waves outside the waveguide. For guided wave propagation there is an enormous increase in sensitivity to a thin sample layer, compared to a simple reflection or transmission. Appropriate easily applied samples would include adsorbed molecules, sur-

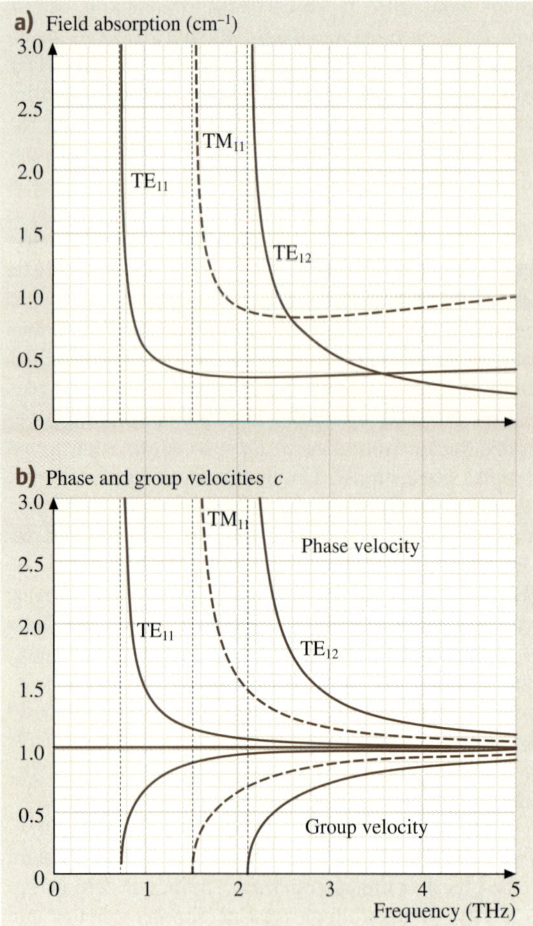

Fig. 17.13 (a) Field absorption and (b) phase and group velocities for the coupled three modes in a 240 μm diameter stainless steel waveguide. (Adapted from [17.86])

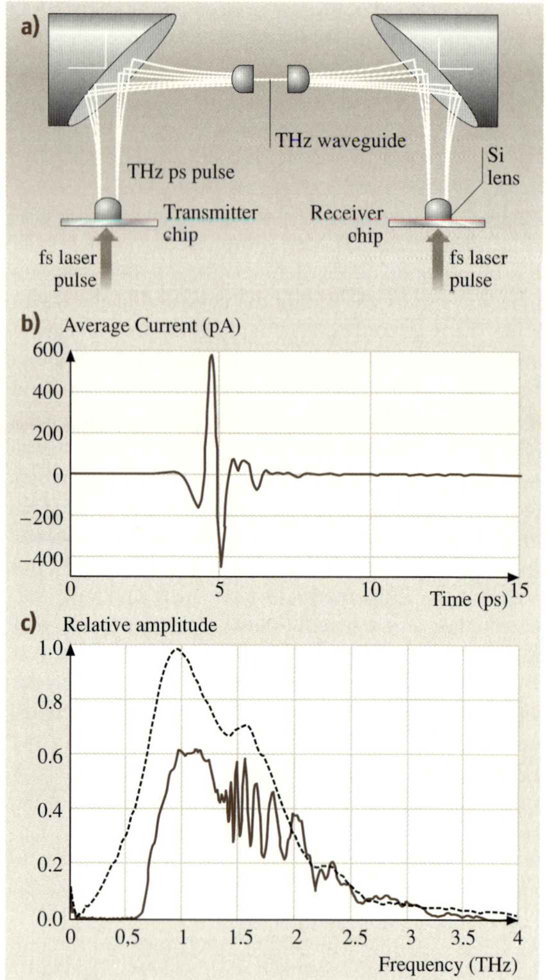

Fig. 17.14 (a) Schematic diagram of the optoelectronic THz-TDS system incorporating quasioptical coupling to the THz waveguide. (b) Measured reference THz pulse. (c) Relative amplitude spectrum of the reference pulse compared to that of the transmitted pulse. (Adapted from [17.85])

face layers, thin films, Langmuir–Blodgett films, films of metals, alloys, and conducting polymers.

Metal Waveguides

Experimental investigations using quasioptical methods to efficiently couple freely propagating subpicosecond pulses of THz radiation into single-conductor, submillimeter metal waveguides and to consequently measure the transmitted pulses from these waveguides have been recently reported [17.85, 86]. Dispersive propagation with low-loss over the frequency band from 0.65 to 3.5 THz was observed with frequency dependent group velocities v_g ranging from $c/4$ to c and phase velocities v_p from $4c$ to c, where $(v_g v_p = c^2)$. Even though the input spectrum overlapped the cutoff frequencies of more than 25 waveguide modes, the linearly polarized incoming THz pulses only significantly coupled into the TE_{11}, TM_{11} and the TE_{12} modes for observations using 24 mm long and 4 mm long stainless steel (SS) tubes with inside diameters of 240 μm and 280 μm, respectively. As shown in Fig. 17.13 [17.91–93], at 1 THz the power absorption coefficient for the dominant TE_{11} mode is $\alpha = 0.7\,\mathrm{cm}^{-1}$ for a 240 μm diameter. The SS waveguide is approximately $1/10$ of the loss compared to lithographically defined coplanar transmission lines [17.66].

The experimental setup for THz waveguides is similar to that of the optoelectronic THz beam system described in the next section, except that at the beam waist centered between the two parabolic reflectors a lens–waveguide–lens system is placed as shown in Fig. 17.14a. For the waveguide system the silicon lens focuses the THz beam to a frequency-independent 1/e-waist diameter of approximately 200 μm, which is coupled into and propagated through the waveguide and is coupled out with the second silicon lens. The reference THz pulse is measured by removing the waveguide and then moving the two silicon lenses to their common focus (the confocal position). Such a reference THz pulse is shown in Fig. 17.14b, where the corresponding amplitude spectrum shown Fig. 17.14c extends from 0.1 THz to 4 THz.

The observed transmitted signal from the 24 mm long waveguide dramatically demonstrated the strong group velocity dispersion of the waveguide [17.85, 86]. The 1 ps input pulse was stretched to about 70 ps, with the high frequencies arriving earlier in time corresponding to negative chirp. The transmitted pulse from a 4 mm long waveguide displays comparatively less cumulative dispersion in Fig. 17.15a (dots). The pulse is spread over 30 ps and shows fewer oscillations. The spectrum of this pulse, Fig. 17.15b (dots), shows the lowest frequency cutoff at about 0.65 THz, which is also calculated to be 0.65 THz for the TE_{11} mode. The comparison of the relative amplitude spectrum of this pulse with that of the reference pulse in Fig. 17.14c, where they both have the same normalization, illustrates the excellent amplitude coupling of the freely propagating THz pulse into the waveguide. Multimode interference was observed to be stronger in the shorter waveguide, starting at about 1.3 THz. The dominant TE_{11} mode has a cutoff frequency of 0.65 THz (77%); the coupled TM_{11} mode has a cutoff at 1.31 THz (20%), and the weakly TE_{12} coupled mode has a cutoff at 1.81 THz (3%). The percent of the total coupled power into each mode is in parentheses.

Using the spectrum of the reference pulse, the output spectrum from the 4 mm waveguide was calculated including the three coupled modes, their respective complex amplitude coupling, as well as their frequency-dependent complex propagation vectors, obtainable as from Fig. 17.13. This calculation overlays as the solid line the measured spectra in Fig. 17.15b; good agreement is obtained for the interference oscillations and the other features of the spectra. The calculated time-domain output pulse (solid line) is given by the inverse Fourier transform of the calculated complex output amplitude spectrum $E_{out}(\omega, z)$ and is compared favorably with the experiment in Fig. 17.15a.

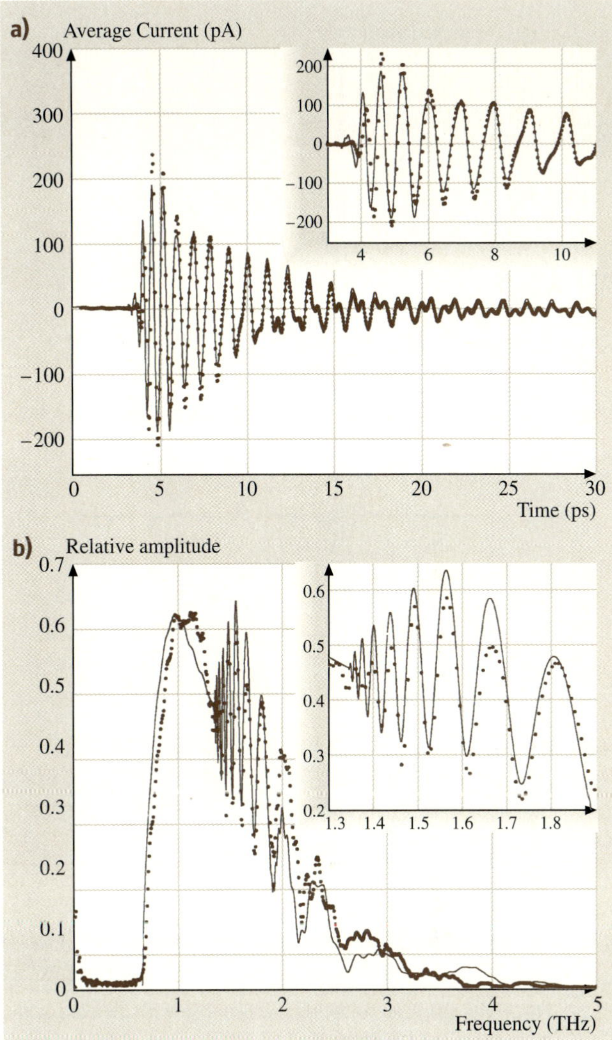

Fig. 17.15 **(a)** Measured THz pulse (*dots*) transmitted through a 4 mm long, 280 μm diameter stainless-steel (SS) waveguide. The *inset* shows the THz pulse on an expanded time scale. **(b)** Amplitude spectrum (*dots*) of the measured transmitted pulse from **(a)**. The *inset* shows the amplitude spectrum on an expanded frequency scale. The *solid curves* are from theory. (Adapted from [17.85])

TEM THz Parallel-Plate Metal Waveguide

In the initial observation [17.89] shown in Fig. 17.16, a freely propagating beam of 0.3 ps FWHM THz pulses was efficiently coupled into a parallel-plate copper waveguide (with a plate separation of 108 μm), and low-

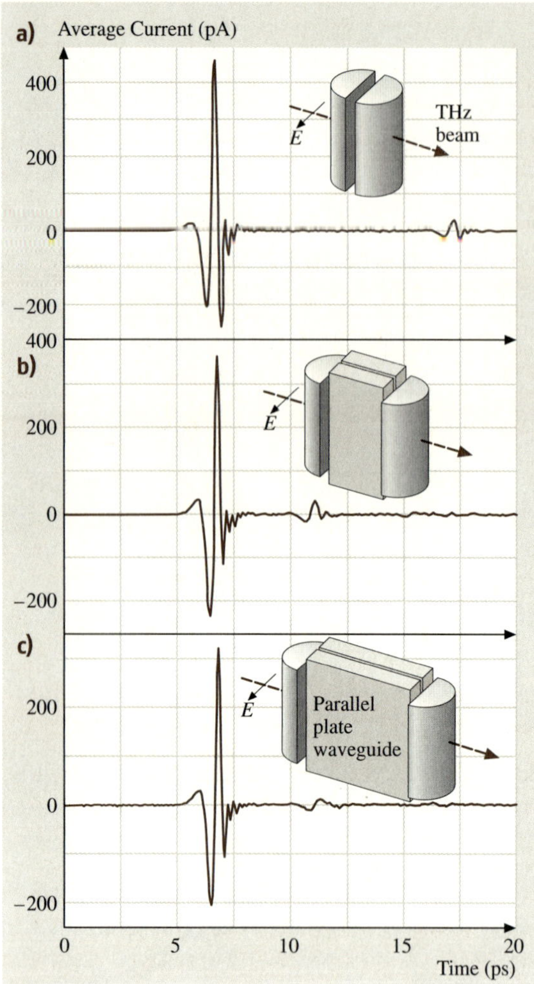

Fig. 17.16 **(a)** Scan of the reference pulse with the confocal cylindrical lens system shown in the *inset*. **(b)**,**(c)** scans of the propagated pulse through the **(b)** 12.6 mm and **(c)** 24.4 mm long waveguides with the lens waveguide lens systems shown in the *insets*. The zero reference time is the same for **(a)**–**(c)**. (Adapted from [17.89])

loss, single TEM mode undistorted pulse propagation was observed within the bandwidth from 0.1–4 THz for a length of 24.4 mm. The experimental setup incorporated planocylindrical lenses to couple the energy into and out of the guide. The lens at the input of the guide

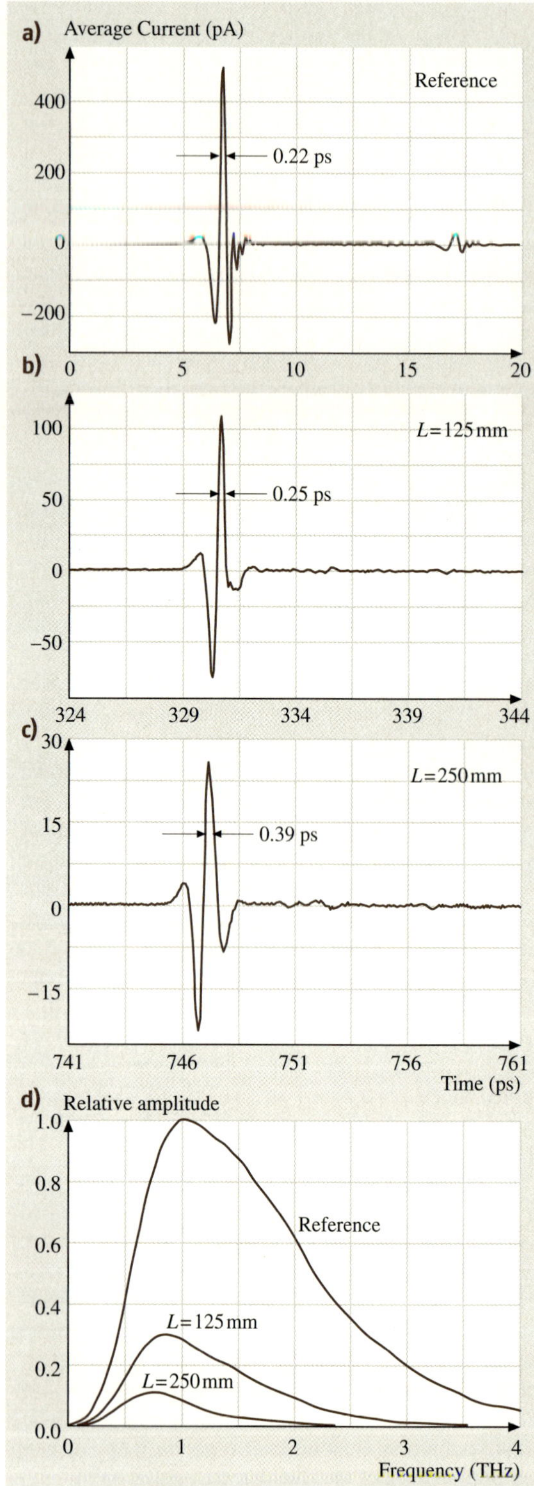

Fig. 17.17 **(a)** Reference pulse and propagated pulse through **(b)** the 125 mm long and **(c)** 250 mm long waveguide. **(d)** Amplitude spectra of the isolated pulses. (Adapted from [17.90])

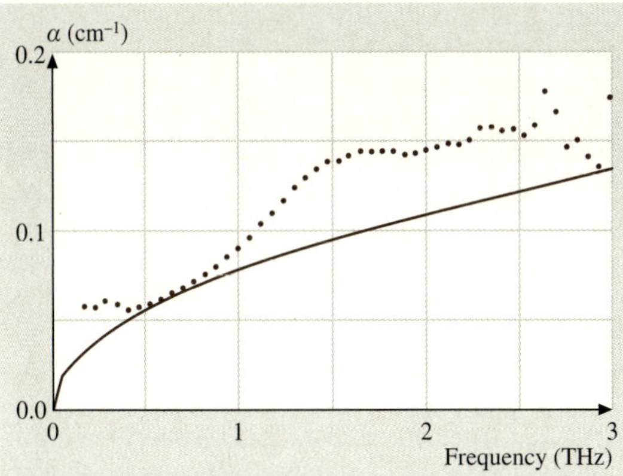

Fig. 17.18 The experimental values (*dots*) compared to theory (*solid line*) for the amplitude attenuation constant α. (Adapted from [17.90])

is used to focus the beam only along one dimension producing an approximately Gaussian beam having an elliptic cross-section with a frequency-independent 1/e-amplitude minor axis of 200 μm at the beam waist, where the entrance face of the guide is located. An identical arrangement is used at the output face.

A more recent demonstration [17.90] is shown in Fig. 17.17, where the FWHM was 0.22 ps for the input reference pulse, 0.25 ps for the output pulse after $L = 125$ mm of TEM propagation, and 0.39 ps for the output pulse after $L = 250$ mm of TEM propagation. Characteristic of TEM propagation, the output pulses exhibit no dispersive pulse broadening [17.89, 90]. The minimal broadening of the output pulses is due to the relative loss in the high-frequency content as seen in the amplitude spectra. As shown in Fig. 17.17d, the FWHM of the amplitude spectra are 1.62 THz, 1.14 THz, and 0.74 THz for the reference, $L = 125$ mm, and $L = 250$ mm pulses, respectively. The smoothness of the output spectra revealing no low-frequency cut-off, confirms single TEM mode propagation [17.85–90]. The associated low loss for this TEM propagation is presented in Fig. 17.18, which gives the measured amplitude attenuation constant compared with theory.

Dielectric Waveguides

An alternative approach to low-loss, guided wave, THz propagation would be to use dielectric waveguides. Such waveguides do not have the sharp low-frequency cut-off of metal waveguides and would thereby extend the low-frequency limit. Dielectric waveguide propagation of subpicosecond THz pulses (in single-crystal sapphire fibers) has now been achieved [17.87]. An incident THz pulse of approximately 0.6 ps duration was observed to undergo considerable reshaping due to the absorptive and dispersive waveguide propagation, resulting in transmitted chirped pulse durations of 10–30 ps. Good agreement between theory and experiment was obtained by analyzing the propagation in terms of the single HE_{11} waveguide mode.

The THz dielectric waveguide work has been extended to planar waveguides based on films, such as high-density polyethylene. This material has low THz absorption and gives very low-loss propagation. Dispersive, low-loss propagation was observed for 2 cm wide by 150 μm thick ribbon waveguides made of high-density polyethylene (HDPE) [17.88]. The large GVD caused extensive THz pulse reshaping and broadening, resulting in positively chirped output pulses. The measured propagated THz pulse through a 10 mm long HDPE waveguide broadened to over 15 ps and showed strong reshaping indicating a frequency chirp. In contrast, the corresponding amplitude spectrum shows much less change. The reference pulse was a half-cycle pulse with a peak to peak pulse width of less than 1 ps; its amplitude spectrum extended from 0.1 to about 3 THz. The amplitude spectrum of the propagated pulses, compared to the spectrum of the reference pulse, clearly exhibits efficient wave-guiding characteristics.

This study demonstrated the feasibility of the plastic ribbon waveguide as a low-loss, single-mode transmission channel for the THz region, capable of utilizing efficient quasioptic coupling. Ultralow loss lines are possible by reducing the thickness of the ribbon such that most of the guided wave energy propagates outside the core region. The performance could be further enhanced by the use of a very low-loss material, such as high-resistivity silicon with a power absorption coefficient of much less than $0.05\,cm^{-1}$ up to 2 THz [17.51]. Consequently, dielectric waveguides could have much less absorption than metal waveguides. Due to the ability to change the group velocity dispersion (GVD) by changing the thickness of the ribbon waveguide, large-frequency regions with essentially no GVD can be obtained.

17.2 Freely Propagating Wave THz Photonics

Recently, there has been a great deal of work demonstrating the generation of THz radiation via material and electronic excitation by ultrafast laser pulses. Modern integrated circuit techniques have made possible the precise fabrication of micron-sized dipoles, which when photoconductively driven by fs laser pulses can radiate well into the THz regime [17.20–22]. An alternative and complimentary approach has been to extend radio and microwave techniques into the THz regime through the use of optoelectronic antennas [17.23–30, 32]. The generation of short pulses of THz radiation by the passage of a short optical pulse through a nonlinear optical material was demonstrated many years ago [17.33]. A related source is based on the emission of an electromagnetic shock wave due to a volume dipole distribution moving faster than the phase velocity, i. e., the electrooptic Cherenkov radiation [17.34, 35], and the electromagnetic shock wave radiated by a surface-dipole distribution propagating faster than the phase velocity [17.71]. THz radiation has been generated by photoconductively driving the surface field of semiconductors [17.36,37] and the intrinsic region of large aperture silicon p-i-n diodes [17.20] with ultrafast laser pulses. A quite efficient source of broadband THz radiation involves the generation of photocarriers in trap-enhanced electric fields with ultrafast laser pulses [17.44, 94]. A cw photomixer capable of producing tunable radiation has been demonstrated by beating together two laser beams with frequencies above the band gap of a semiconductor [17.39–43].

Some of these sources are based on an optical type approach whereby a transient point source of THz radiation is located at the focus of a dielectric collimating lens, followed by an additional paraboloidal focusing and collimating mirror [17.21, 22, 29, 31, 95]. Although this system will be the one described in most detail, alternative approaches will also be presented. One of the most useful versions of the system is based on repetitive, subpicosecond optical excitation of a Hertzian dipole antenna imbedded in a charged coplanar transmission line structure [17.21, 22, 29, 31, 95]. The burst of radiation emitted by the resulting transient dipole is collimated by a THz optical system into a diffraction-limited beam and focused onto a similar receiver structure, where it induces a transient voltage and is detected. The THz optical system gives exceptionally tight coupling between the transmitter and receiver, while the excellent focusing properties preserves the subpicosecond time dependence of the source.

The combination of THz optics with the synchronously-gated, optoelectronic detection process has exceptional sensitivity for repetitively pulsed beams of THz radiation. Via two stages of collimation, a THz beam with a frequency-independent divergence is obtained from the THz transmitter. The THz receiver with identical optical properties collects essentially all of this beam. The resulting tightly coupled system of the THz transmitter and receiver can give strong reception of the transmitted pulses of THz radiation after many meters of propagation. Another reason for the exceptional sensitivity is that the THz receiver is gated. The gating window of approximately 0.6 ps is determined by the laser pulse width and the carrier lifetime in ion-implanted silicon-on-sapphire (SOS). Thus, the noise in the comparatively long time interval (10 ns) between the repetitive THz pulses is not seen by the receiver. A final important feature of the detection method is that it is a coherent process; the electric field of a repetitive pulse of THz radiation is directly measured. Because a repetitive signal is synchronously detected, the total charge (current) from the signal increases linearly with the number of sampling pulses, while the charge (current) from noise increases only as the square root of the number of pulses.

17.2.1 An Optoelectronic THz Beam System

This section describes the optoelectronic generation and detection of freely-propagating femtosecond pulses of THz electromagnetic radiation. Via optoelectronic excitation, a transient point source of THz radiation is generated at the focus of a dielectric collimating lens, followed by an additional paraboloidal focusing and collimating mirror. This source produces well-collimated beams of subpicosecond pulses of THz radiation. Matched to an identical receiver, the resulting system has extremely high collection efficiency. With a demonstrated signal-to-noise ratio of 1000, a time resolution of less than 150 fs, and a frequency range from 0.2 THz to more than 6 THz, the performance of this optoelectronic THz system is limited only by the ballistic acceleration of the photocarriers.

The setup used to generate and detect beams of short pulses of THz radiation is presented in Fig. 17.19. For this example, the transmitting and receiving antennas are identical, each consisting of the antenna imbedded in a coplanar transmission [17.31], as shown in Fig. 17.19a. The antenna is fabricated on an ion-implanted SOS wafer.

The Experimental Setup

The 20 μm wide antenna structure is located in the middle of a 20 mm long coplanar transmission line consisting of two parallel 10 μm wide, 1 μm thick, 5 Ω/mm, aluminium lines separated from each other by 30 μm. In the initial demonstration of this system a colliding-pulse mode-locked (CPM) dye laser, produced 623 nm, 70 fs pulses at a 100 MHz repetition rate in a beam with 10 mW average power. The 5 mW excitation beam was focused onto the 5 μm wide photoconductive silicon gap between the two antenna arms. The 70 fs laser creation of photocarriers causes subpicosecond changes in the conductivity of the antenna gap. When a dc bias voltage of typically 10 V is applied to the transmitting antenna, these changes in conductivity result in pulses of electrical current through the antenna and consequently bursts of electromagnetic radiation are produced. A large fraction of this radiation is emitted into the sapphire substrate in a cone normal to the interface with the radiation pattern presented in [17.95]. The radiation is then collected and collimated by a dielectric lens attached to the backside (sapphire side) of the SOS wafer [17.31]. The dielectric lenses were made of high-resistivity (10 kΩ cm) crystalline silicon with a measured absorption of less than 0.05 cm^{-1} in the presented frequency range [17.51]. The silicon lens is a truncated sphere with a 10 mm diameter and a focal point located at the antenna gap when attached to the back side of the chip. As shown in Fig. 17.19b, after collimation by the silicon lens, the beam diffracts and propagates to a paraboloidal mirror, where the THz radiation is recollimated into a highly directional beam, with beam diameters (10–70 mm) proportional to the wavelength. Thereafter, all of the frequencies propagated with the same 25 mrad divergence. After a typical 50 cm propagation distance this THz beam is detected by the THz receiver, where the paraboloidal mirror focuses the beam

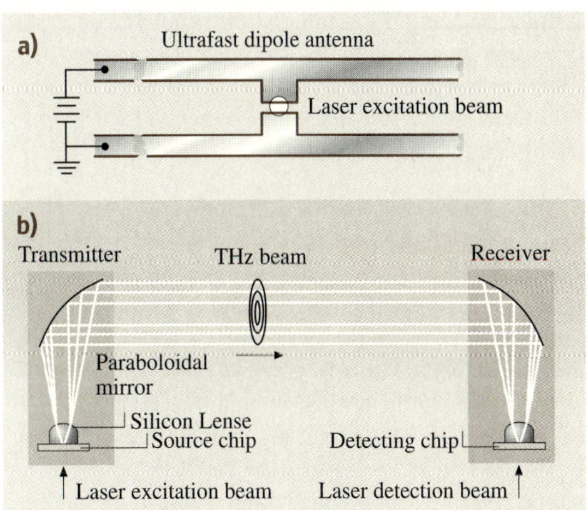

Fig. 17.19 (a) Ultrafast dipolar antenna. (b) THz transmitter and receiver

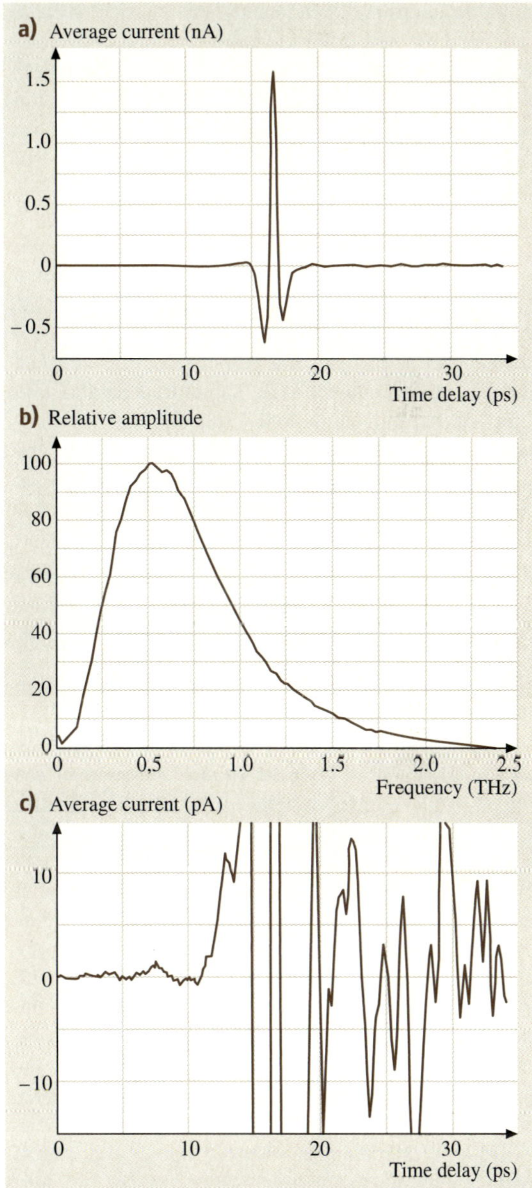

Fig. 17.20 (a) THz pulse measured to 35 ps. (b) Amplitude spectrum to 2.5 THz of the measured pulse shape. (c) THz pulse on a vertical scale expanded by a factor of 100. (Adapted from [17.31])

onto a silicon lens, which focuses it onto a SOS antenna chip, similar to the one used in the emission process. The electric field of the focused incoming THz radiation induces a transient bias voltage across the 5 μm gap between the two arms of this receiving antenna, directly connected to a low-noise current amplifier. The amplitude and time-dependence of this transient voltage is obtained by measuring the collected charge (average current) versus the time delay between the THz pulses and the delayed CPM laser pulses in the 5 mW detection beam. These pulses synchronously gate the receiver, by driving the photoconductive switch defined by the 5 μm antenna gap.

Measurements of Signal-to-Noise Ratio

An early time-resolved measurement [17.31] is shown in Fig. 17.20a. The clean pulse shape is a result of the fast action of the photoconductive switch at the antenna gap, the broadband response of the ultrafast antennas, the broadband THz optical transfer function of the lenses and paraboloidal mirrors, and the very low absorption and dispersion of the silicon lenses. The measured pulse width of 0.54 ps (FWHM) is only an upper limit to the true pulse width, because no deconvolution has been applied to the measurement to take out the response time of the antenna gap. This time-response will be determined in the next section of this article. In Fig. 17.20b, the Fourier transform of the measured signal (Fig. 17.20a) is shown to stretch from about 0.1 to 2.0 THz. This only represents a lower limit to the true extent of the emitted radiation as it contains the frequency response of the receiver. At the low-frequency end, the efficiency of both the emitter and receiver has been shown to be proportional to the length of the antenna, i. e., proportional to the separation between the two lines of the coplanar transmission line. For extremely low frequencies the size of the paraboloidal mirrors will also limit the efficiency. For the high-frequency limit the efficiency of the antenna is strongly reduced when half the wavelength (in the dielectric) of the emitted radiation is no longer small compared to the antenna length. The high-frequency part of the spectrum is also limited by the finite rise time of the current transient and the non-ideal imaging properties of the THz optics.

In Fig. 17.20c the time-resolved signal is shown on a vertical scale expanded by a factor of 100. The structure observable after the main pulse is reproducible and is due to reflections of the electrical pulse on the transmission line, reflections of the THz pulse from the various dielectric interfaces, and absorption and dispersion of residual water vapor. The signal-to-noise ratio in this 4 min scan is more than 10 000 : 1. Another 4 min scan is shown in Fig. 17.21, for which the intensity of the excitation laser beam was reduced to only 15 μW from the 5 mW normally used. The consequent 320-fold reduction in photocurrent indicates a reduction in the power of the THz beam by the factor 1/100 000. However, despite this enormous reduction in power, the peak amplitude is still more than 30 times larger than the rms noise. Based on previous calculations [17.31], the average power in the THz beam during this measurement was about 10^{-13} W. If the power of the THz beam were even further reduced, the detection limit of the THz receiver would be reached at 10^{-16} W, for a signal-to-noise-ratio of unity and a 125 ms integration time. Because the generation and detection of the THz (far-infrared) radiation is coherent, the THz receiver is approximately 1000 times more sensitive than an incoherent helium cooled bolometer [17.53].

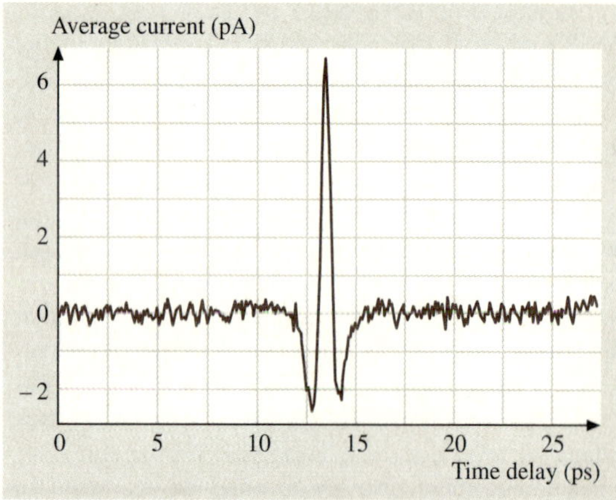

Fig. 17.21 Measured THz pulse to 27.5 ps with a 100 000 times reduction (compared to Fig. 17.20a) of the THz beam power. (Adapted from [17.31])

The Semiconductor Time-Dependent Response Function

In the small antenna limit corresponding to the Hertzian dipole, the generated radiation field is proportional to the time-derivative of the current pulse, which is mainly determined by the intrinsic response of the semiconductor itself. The intrinsic time-domain response function will now be derived for a semiconductor described by the simple Drude formalism. The free carriers are considered as classical point charges subject to random collisions, for which the collision damping is independent of the carrier energy. The frequency-dependent complex conductivity $\sigma(\omega)$ is given by

$$\sigma(\omega)=\sigma_{\text{dc}}\frac{\mathrm{i}\Gamma}{\omega+\mathrm{i}\Gamma}\,, \tag{17.6}$$

where $\Gamma = 1/\tau$ is the damping rate and τ is the average collision time. The dc conductivity is given by $\sigma_{\text{dc}} = e\mu_{\text{dc}}N$, where e is the electron charge, μ_{dc} is the dc mobility and N is the carrier number density. According to Drude, $\mu_{\text{dc}} = e/(m^*\Gamma)$, where m^* is the effective mass. This relationship (17.6) is in good agreement with recent time-domain spectroscopy measurements [17.31] on lightly doped silicon from low frequencies to beyond 2 THz. The following procedure is similar to that of [17.96]. It is helpful to recast the formalism into a frequency-dependent mobility as

$$\mu(\omega)=\mu_{\text{dc}}\frac{\mathrm{i}\Gamma}{\omega+\mathrm{i}\Gamma}\,. \tag{17.7}$$

This relationship shows the fundamental limit to the frequency response of the photoconductive switch. It is important to note that at high frequencies compared to Γ, the imaginary part dominates and falls off with frequency as $1/\omega$. Surprisingly, this allows useful material response to reach up to tens of THz. For example if $\tau = 200$ fs, $\Gamma/2\pi = 0.8$ THz and the response at 8 THz is 0.10 that at low frequencies, similarly, the response at 16 THz is 0.05 that at low frequencies. The dc current density is given by $J_{\text{dc}} = \sigma_{\text{dc}}E$, or equivalently $J_{\text{dc}} = eE\mu_{\text{dc}}N$, where E is a constant electric field. Because of the linearity of the current in N, for a time-dependent carrier density $N(t)$, the time-dependent current density can be written as

$$J(t)=eE\int_{\infty}^{t}\mu(t-t')N(t')\,\mathrm{d}t'\,, \tag{17.8}$$

where $\mu(t-t')$ is the time-domain response function for the mobility. This function is determined by the inverse transform of the frequency-dependent mobility to be the causal function

$$\mu(t-t')=\mu_{\text{dc}}\Gamma\mathrm{e}^{-\Gamma(t-t')} \tag{17.9}$$

which vanishes for negative $(t-t')$.

In order to facilitate the understanding of the photoconductive switch it is useful to rewrite the basic equation (17.8) in the equivalent form,

$$J(t)=eEA\int_{-\infty}^{t}\mu(t-t')\int_{-\infty}^{t'}R_{\text{c}}(t'-t'')I(t'')\,\mathrm{d}t''\,\mathrm{d}t'\,, \tag{17.10}$$

where $I(t'')$ is the normalized intensity envelope function of the laser pulse, A is a constant giving the conversion to absorbed photons/volume and R_{c} is the response function describing the decay of the photogenerated carriers. By defining a new photocurrent response function $j_{\text{pc}}(t-t')$, we can rewrite (17.10) to become

$$J(t)=\int_{-\infty}^{t}j_{\text{pc}}(t-t')I(t')\,\mathrm{d}t'\,, \tag{17.11}$$

where $j_{\text{pc}}(t-t')$ is obtained by evaluating (17.10) with a delta function $\delta(t'')$ laser pulse. Assuming the causal function $R_{\text{c}}(t'-t'') = \exp[-(t'-t'')/\tau_{\text{c}}]$, describing a simple exponential decay of the carriers with the carrier lifetime τ_{c} (significantly longer than the average collision time τ) for positive $(t'-t'')$ and vanishing for negative $(t'-t'')$, and that $\mu(t-t')$ is given by the Drude response of (17.9), the causal response function $j_{\text{pc}}(t^*)$ is then evaluated to be

$$j_{\text{pc}}(t^*)=\frac{\mu_{\text{dc}}eEA\Gamma}{\Gamma-1/\tau_{\text{c}}}\left(\mathrm{e}^{-t^*/\tau_{\text{c}}}-\mathrm{e}^{-t^*/\tau}\right) \tag{17.12}$$

for positive $t^* = (t-t')$ and shown to vanish for negative t^*. In the short-pulse limit of the ultrafast excitation pulses, the time dependence of the photocurrent $J(t)$ is approximately equal to that of the photocurrent response function $j_{\text{pc}}(t^*)$ for positive t^*. For a long carrier lifetime, the time dependence of $j_{\text{pc}}(t^*)$ is described by a simple exponential rise with a rise time on the order of $\tau = 1/\Gamma$, which is equal to 270 fs and 150 fs for the electrons and holes, respectively, in lightly doped silicon as previously measured [17.97, 98]. As these results show,

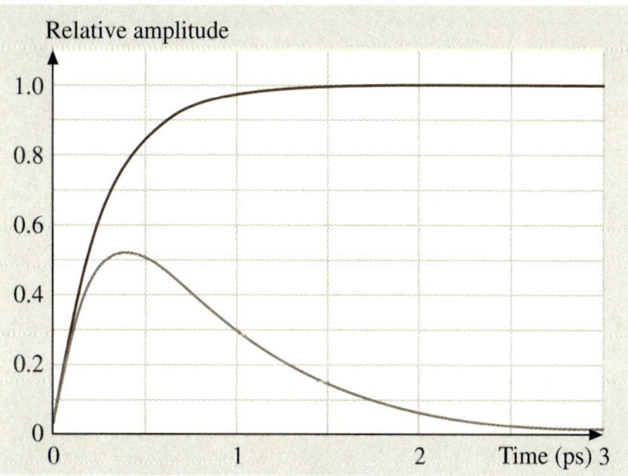

Fig. 17.22 Calculated photoconductive response function to 3 ps with the scattering time $\tau = 270$ fs and infinite carrier lifetime τ_{c} (*upper curve*) and with $\tau_{\text{c}} = 600$ fs (*lower curve*)

the material response can be slow compared to the duration of the ultrafast laser excitation pulses, which can be as short as 10 fs, but is more typically on the order of 60 fs.

The time-dependent response function described by (17.12) is displayed in Fig. 17.22 for the two cases: $\tau = 270$ fs and $\tau_c = \infty$, and $\tau = 270$ fs and $\tau_c = 600$ fs. The result for infinite carrier lifetime has the following intuitive interpretation. After the instantaneous creation of carriers, the initial current and mobility are zero. The carriers then accelerate ballistically, as determined by the applied electric field, their charge and effective mass. This acceleration continues for a time approximately equal to the scattering time τ, after which the velocity and current equilibrate to their steady-state value. This discussion will now be shown to accurately describe the mathematical dependence of (17.12). With $\tau_c = \infty$, (17.12) is equal to

$$j_{pc}(t^*) = \mu_{dc} e E A \left(1 - e^{-t^*/\tau}\right) \tag{17.13}$$

which, for times that are short compared to τ, reduces to

$$j_{pc}(t^*) = \mu_{dc} e E A t^* / \tau \,. \tag{17.14}$$

Remembering that for Drude theory $\mu_{dc} = e/(m^* \Gamma)$, (17.14) is equivalent to

$$j_{pc}(t^*) = A e t^* (e E / m^*) \tag{17.15}$$

which describes the ballistic acceleration eE/m^*.

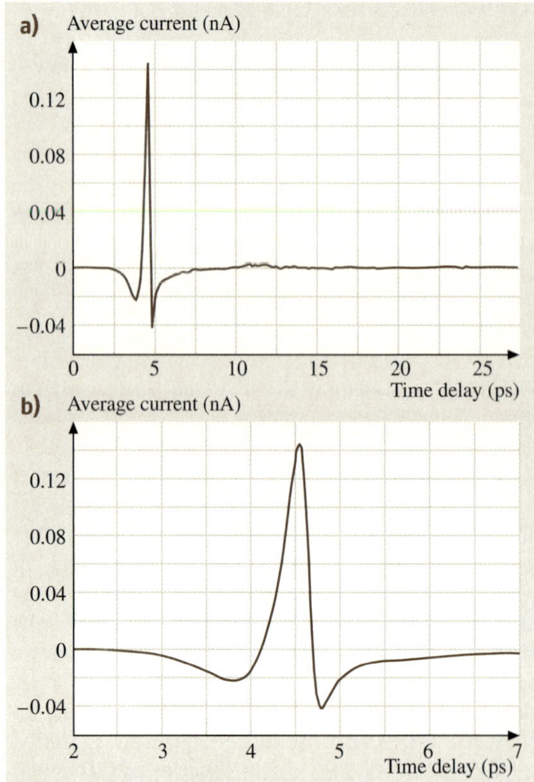

Fig. 17.24 **(a)** THz pulse measured by the ultrafast receiver. **(b)** Measured THz pulse on an expanded time scale (adapted from [17.99])

Measurements of the Time-Dependent Response Function

Here the performance of a new type photoconductive receiver with a measured response exceeding 6 THz is described and analyzed by the above Drude [17.99]. This receiver used a prototype composite optoelectronic chip shown in Fig. 17.23. By using an identical receiver in a high-performance optoelectronic THz beam system as the transmitter, we were able to characterize the receiver with exceptional precision. The antenna structure, used for both the transmitter and receiver, is the same as shown in Fig. 17.19a, but with the reduced size of two parallel 5 μm wide lines separated from each other by 10 μm. A colliding-pulse mode-locked (CPM) dye laser provided the 5 mW focused beams of 623 nm, 80 fs excitation and sampling pulses. Excellent agreement with the Drude theory model was obtained for the photoconductive response.

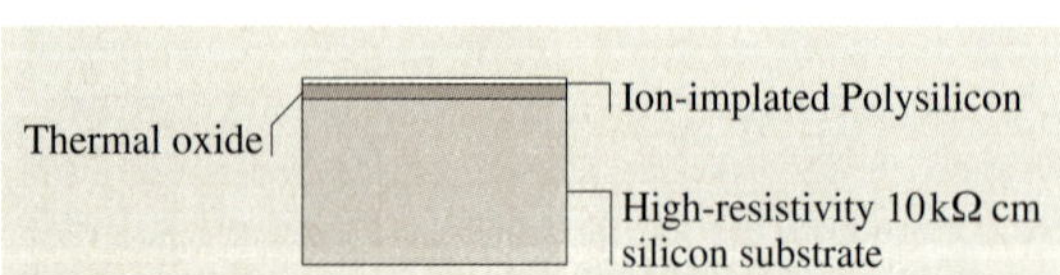

Fig. 17.23 Cross section of the composite chip

Using this receiver, a source spectrum was measured with a FWHM bandwidth almost twice as broad as the initial experimental characterization of a similar system using SOS chips [17.96]. The improved performance was due to the following changes. Firstly, the positions of the paraboloidal mirrors were set to have a unity transfer function for the THz radiation [17.100]. Secondly, exceptional care was taken to match the foci of the silicon lenses to the antenna positions. For on-axis focusing, a series of observations were made using lenses of the same curvature, but with thicknesses varying in steps of 50 μm. In the plane of the chip, the position of the focus was adjusted to ±20 μm. Thirdly, the new composite chip was used to eliminate the absorption of the incoming THz radiation by the sapphire substrate of the

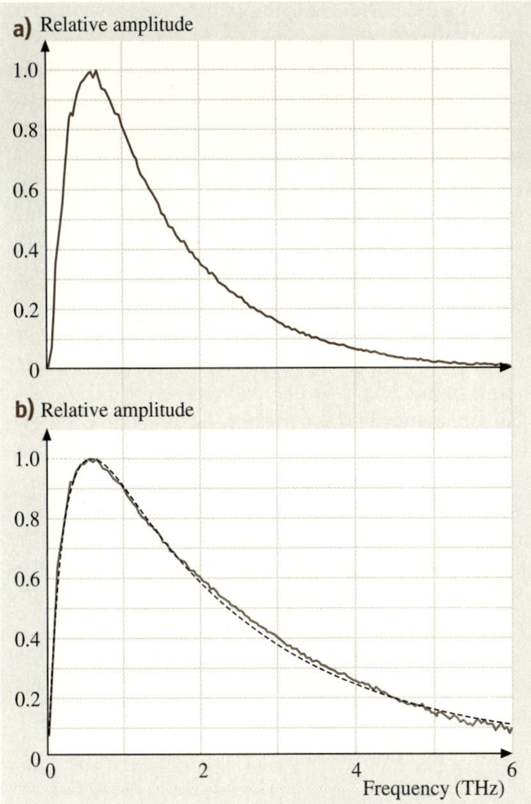

Fig. 17.25 **(a)** The numerical Fourier transform of Fig. 17.24a, equal to the receiver power spectrum. **(b)** Comparison of the calculated receiver amplitude spectrum (*dashed line*) with the measured amplitude spectrum (square root of **(a)**) (adapted from [17.99])

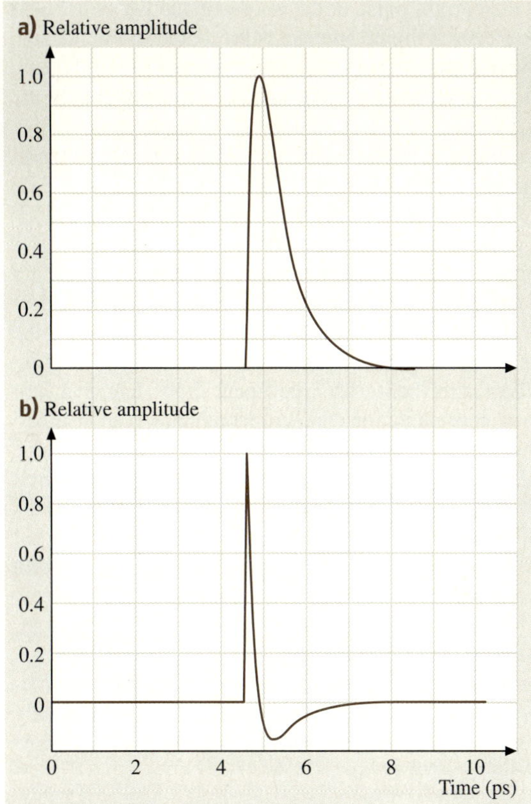

Fig. 17.26 **(a)** Calculated current pulse (11 ps timescale) in semiconductor and antenna. **(b)** Time derivative of the current pulse

previously used SOS detection chip. The 0.5 μm thick active layer of LPCVD polysilicon of the composite chip was annealed at 1000 °C for 1 h in a nitrogen atmosphere and was later implanted with 200 keV and 100 keV oxygen ions both at a dose of $10^{13}/\mathrm{cm}^2$. The underlying 0.7 μm thick thermal oxide layer was grown on a high resistivity silicon substrate.

The THz pulse measured with the ultrafast receiver is shown in Fig. 17.24a. The fall-time (90–10%) from the peak of the pulse to the minimum is only 127 fs, as shown in the expanded view of Fig. 17.24b. This exceptional time resolution demonstrates that a photoconductive receiver can be much faster than is generally realized [17.99], but is consistent with the earlier prediction (for SOS chips) of a 150 fs time resolution [17.96]. The numerical Fourier transform of this pulse peaks at approximately 0.5 THz and extends to 6 THz, as shown in Fig. 17.25a. Because the measured spectrum is the product of the identical transmitter and receiver amplitude spectra, the receiver amplitude spectrum shown in Fig. 17.25b is simply the square root of the measured spectrum of Fig. 17.25a.

The theoretical understanding of the above results will now be presented. Because of the extremely fast electrical response, the current in the ultrafast antenna follows the current $J(t)$ in the semiconductor, where $J(t)$ is determined by the convolution of the photoconductive response function and the 80 fs (FWHM) sech^2-shaped laser driving pulse. Following the proceeding theoretical procedure, the response function is derived from simple Drude theory with a scattering time of 120 fs and a carrier lifetime of 600 fs.

The time-domain response function $j_{\mathrm{pc}}(t^*)$ is assumed to be given by (17.12). This response function is then convolved with the laser pulse, as prescribed by (17.11). Using the above parameters the calculated shape of the current pulse in the photoconductive switch and the Hertzian dipole antenna is presented in Fig. 17.26a. The time derivative of this pulse is given in Fig. 17.26b, where an extremely fast transient, corresponding to the rising edge of the current pulse, is seen. Quantitative evaluation of this time-derivative shows that the Hertzian dipole antenna generates a 150 fs FWHM pulse of THz radiation. Conversely, by the reciprocity principle, 150 fs is also the time resolution of the photoconductive receiver. The numerical Fourier transform of Fig. 17.26b is the predicted amplitude spectrum of the transmitter. In Fig. 17.25b, this spectrum is compared with the measured amplitude spectrum of the receiver. The excellent agreement with experiment shows that, to a very good approximation, the semiconductor response follows Drude theory and that the resulting response time is limited by the ballistic acceleration of the photocarriers. As expected, the calculated amplitude spectrum falls off faster than the intrinsic material response of $\mu(\omega)$ defined by (17.7), due to the 80 fs optical pulsewidth compared to the delta function response.

It has also been shown that the 10 μm long antenna imbedded in the coplanar transmission line has electrical properties much faster than the semiconductor itself. Consequently, the performance is completely determined (and limited) by the intrinsic response time of the semiconductor.

17.2.2 Other THz Transmitters

Because essentially all of the existing and proposed THz applications are constrained by the limiting S/N ratio, any improvement in S/N would provide faster data acquisition rates and would improve the quality of the experimental comparisons. Large increases in S/N would make new applications possible. The most important approach to increase S/N is to increase the transmitter power, while at the same time maintaining or extending the THz bandwidth. This has become particularly appropriate because typical, ultrafast Ti-sapphire mode-locked lasers can produce 500 mW of beam power, compared to the previously used CPM lasers, which produced only on the order of 20 mW of beam power. For example, the GaAs TEF transmitter chip [17.44, 94, 100] (described below) is limited by optical damage to GaAs to a maximum average laser power of only 20 mW in the ultrafast Ti-sapphire driving laser pulse stream. A THz research group at the University of Bath (UK) has attained a 100-fold increase in transmitter power levels from the GaAs TEF chip by focusing a more powerful laser beam (well above the usual operating limits) with a cylindrical lens to produce a highly elliptical focus to minimize saturation and material damage.

A Bell Labs group [17.101], using a field-singularity emitter and a 50 μm dipole receiver antenna fabricated on low-temperature (LT) GaAs, has observed signal strengths of more than 100 times larger than typical results using GaAs or SOS transmitters, and SOS receivers. Although their THz conversion efficiency was essentially the same as for the GaAs TEF transmitter, the increase in THz power and S/N for the Bell Labs work were obtained by driving the LT-GaAs transmitter and the LT-GaAs receiver with optical powers almost an order of magnitude above the allowable operating power levels of the above-described GaAs and SOS transmitter and receiver chips. As shown by this example and earlier work, LT-GaAs has significant potential to extend both the output power and frequency response of photoconductive switches, transmitters and receivers, due to its high resistivity, high carrier mobility and short carrier lifetime [17.41–43, 102–107]. LT-GaAs is much more resistant to optical breakdown than GaAs and is also more resistant to electrical breakdown. In addition, the carrier lifetime can be controlled by annealing temperature and can be made to be lower than 300 fs [17.106, 107]. Currently, the main problem preventing widespread utilization of LT-GaAs is that the commercial LT-GaAs wafer cost is about five times that of either GaAs or SOS.

GaAs Trap Enhanced Field (TEF) THz Source

A different type of high-performance optoelectronic source chip [17.44] is shown in Fig. 17.27d. The simple coplanar transmission line structure consists of two 10 μm wide metal lines separated by 80 μm fabricated on semi-insulating GaAs. Irradiating the metal-semiconductor interface (edge) of the positively biased line with focused ultrafast laser pulses produces synchronous bursts of THz radiation. This occurs because each laser pulse creates a spot of photocarriers in a region of extremely high electric field, called the trap-enhanced field [17.94]. In the initial demonstration of this source, a CPM dye laser provided 60 fs excitation pulses with an average power of 5 mW at 5 μm diameter excitation spot. This GaAs trap-enhanced field (TEF) source chip is completely compatible with the previously described optoelectronic THz beam system. However, this chip is limited only to the transmitter, because of its long carrier lifetime.

The measured THz pulse emitted from the laser excited GaAs TEF source chip with +60 V bias across the transmission line is shown in Fig. 17.27a, and on an ex-

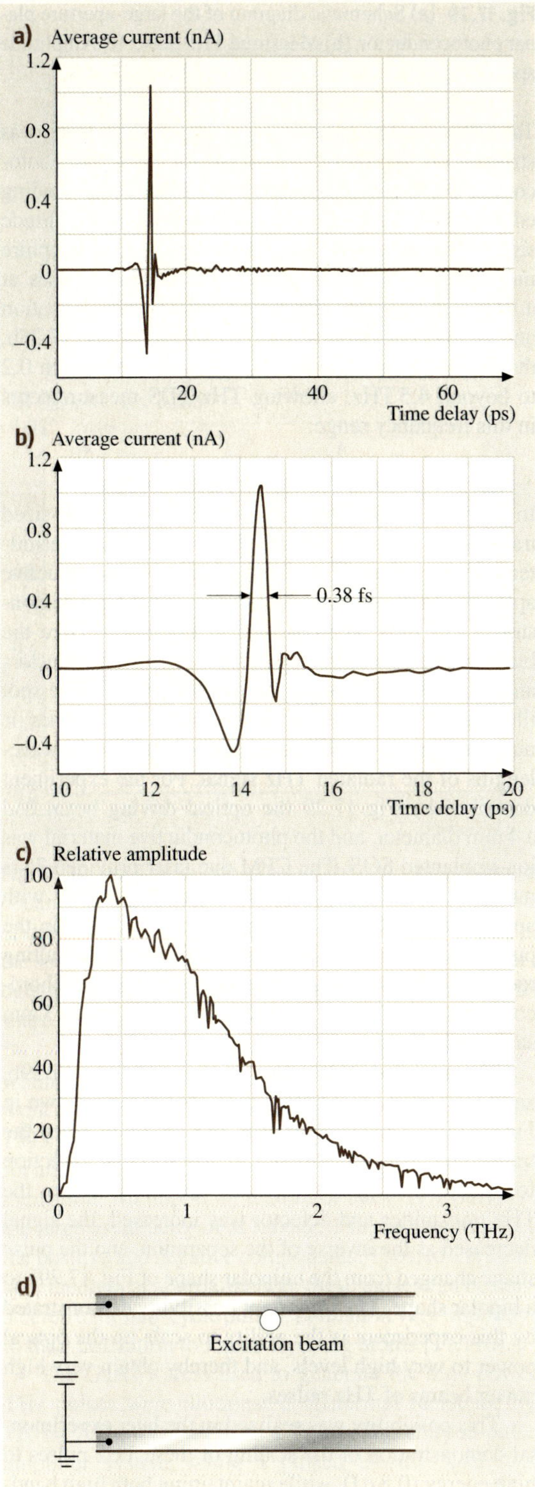

Fig. 17.27 **(a)** Measured THz pulse to 70 ps. **(b)** Measured THz pulse on an expanded 10 ps time scale. **(c)** Amplitude spectrum to 3.5 THz of the THz pulse. **(d)** Source chip configuration used to generate the freely propagating pulses of THz radiation. (Adapted from [17.44])

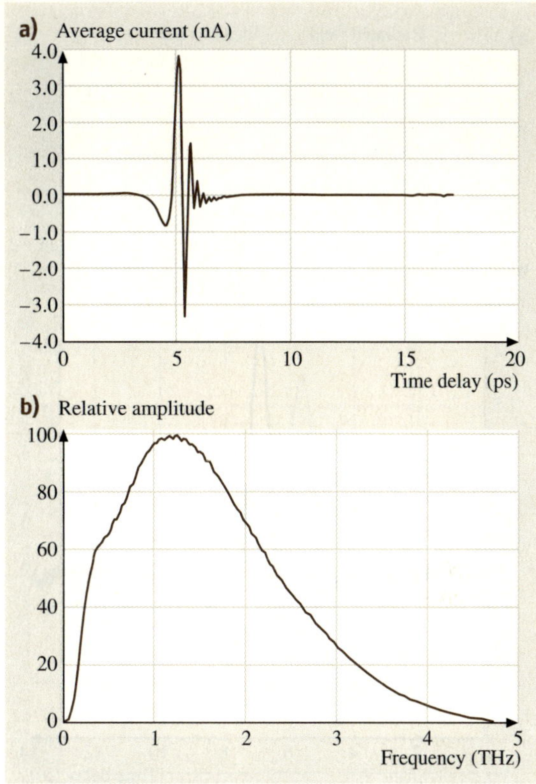

Fig. 17.28 **(a)** Measured THz pulse. **(b)** Amplitude spectrum of the THz pulse

panded time scale in Fig. 17.27b. The measured pulse width with no deconvolution is seen to be 380 fs. At the time these results were obtained [17.44], they were the shortest directly measured THz pulses; the dip on the falling edge was the sharpest feature ever observed with an ion-implanted detector and indicated a response time faster than 190 fs. The amplitude spectrum (numerical Fourier transform) of the pulse extends beyond 3 THz, as presented in Fig. 17.27c. The sharp line structure is due to residual water vapor present in the system.

The GaAs TEF source remains one of the most efficient THz sources today and is the main THz transmitter chip used by the Oklahoma State University Ultrafast THz Research Group. In Fig. 17.28a we show a recent THz pulse measurement using the same type of GaAs

applied across the surface [17.109]. The aluminium electrodes, separated from each other by 1 cm, were connected to the GaAs surface with silver paint. Although THz radiation was observed in both the transmitted and reflected directions, only the transmitted radiation was used. The THz pulse energy was measured both with a helium cooled bolometer and a room-temperature pyroelectric detector.

The power spectrum of the generated THz pulses was obtained by Fourier transform spectroscopy using the interferometer shown in Fig. 17.30a. Typical interferograms measured with the bolometer were 600 fs wide, as shown in Fig. 17.30b, independent of the driving laser intensity, laser polarization, and bias field. The Fourier transform of the interferogram yields the spectral width of 1 THz as shown in the inset of Fig. 17.30b.

High-Surface Field Photoconductors

Another photoconductive THz source that has found significant use and interest requires no external bias voltage and no lithography, because it uses the static internal field at the semiconductor surface, which is perpendicular to the surface [17.36]. Consequently, when such a bare semiconductor wafer is illuminated by femtosecond optical pulses, THz pulses are emitted from the surface in the form of freely propagating electromagnetic beams in the inward and outward directions. The amplitude and phase of the radiated THz pulse depend on the carrier mobility, and the strength, direction and polarity of the internal field.

Figure 17.31a shows the semiconductor illuminated by an unfocused optical beam together with the reflected optical beam and the consequent outward THz radiation and transmitted THz radiation. The CPM dye laser produced 2 nJ, 70 fs, 720 nm output pulses at a 100 MHz repetition rate. The CPM optical beam was split into two beams (30 and 70%), where the stronger excitation beam was modulated by a mechanical chopper at 2 kHz. The weaker sampling beam passed through a variable time delay and was used to gate a dipole antenna fitted with a sapphire lens for detection of the radiated THz pulse. Figure 17.31b shows a THz pulse from semi-insulating InP, detected by the dipole antenna in the outward direction (45° relative to the normal) with a 4 mm diameter laser excitation beam. The measured time response is limited by the dipole detector, which has an optimum frequency of 0.6 THz. The transmitted THz pulse through a 0.4 mm thick semi-insulating InP wafer was measured as a function of the angle of incidence of the excitation beam, by rotation of the wafer. These results are shown in Fig. 17.31c and indicate that the direction of current flow is along the direction of the surface field perpendicular to the wafer surface. Using this surface field technique, THz radiation pulses were generated from InP, GaAs, GaSb, InSb, CdTe, CdSe, and Ge, but not from Si [17.36]. Semi-insulating InP showed the strongest THz radiation; the radiation strength of ⟨100⟩ InP is 2.5 times larger than ⟨111⟩ InP.

It was demonstrated some time ago that the application of an external magnetic field could enhance the THz pulse generation from the surface field [17.111, 112]. Renewed interest in this effect was stimulated by claims that very large enhancements in the average-power of the THz pulse generation result with the application of a large magnetic field [17.113]. Although later investigations showed that these claims were overstated by approximately 100 times, many investigations were performed to characterize all the attributes of the enhancement effect [17.114–116].

Recently, a relatively simple explanation of the B field enhancement has been provided by two groups [17.117, 118]. Firstly, the application of the B field does not increase the transient dipole responsible for the THz radiation. Secondly, the transient dipole and the consequent radiation pattern are reoriented by the B field, with respect to the surface boundaries of the semiconductor crystal. This reorientation of the radiation pattern can dramatically increase the output coupling from the crystal to free space. The coupling problem is quite important because of the high index of refraction of the crystal, which can lead to trapping of the THz radiation in the crystal.

Tunable cw THz Generation by PhotoMixing

A cw photomixer capable of producing tunable radiation can be realized by beating together two laser beams with frequencies above the band gap of a semiconductor [17.39, 40]. Depending upon the generated photocarrier lifetimes and the techniques used to couple out the generated radiation, this method can produce tunable THz radiation at the difference frequency between the two lasers [17.41–43]. The experimental demonstration of cw coherent output over the frequency range from 0.2 to 3.8 THz utilized a low-temperature-grown GaAs (LTG-GaAs, more commonly designated as LT-GaAs) photomixer [17.41–43]. The advantages of LT-GaAs are its short photocarrier lifetime, determined to be 0.27 ps, and the high electrical breakdown field. As shown in the inset of Fig. 17.32a, the LT-GaAs wafer was fabricated with a pattern of interdigitated electrodes covering an area of 20×20 μm and connected to opposite 20×20 μm bonding pads. After electron-beam lithography, the pads

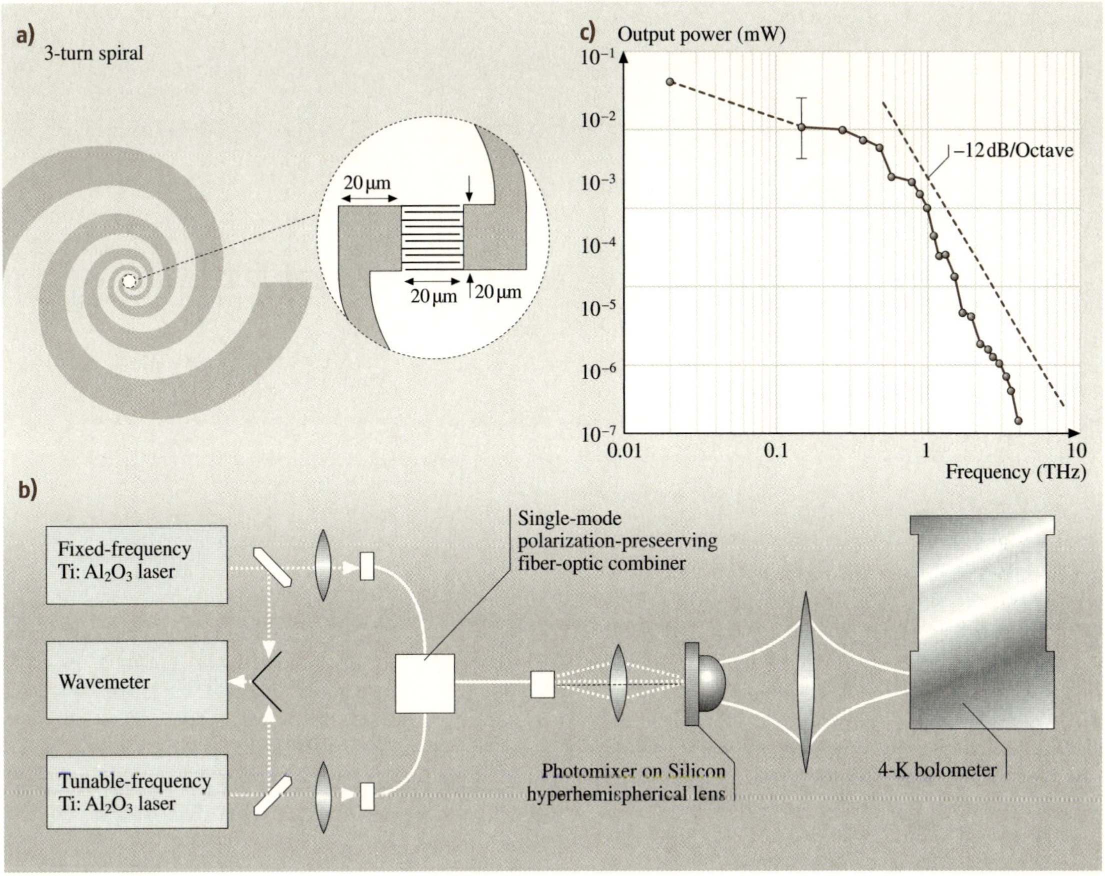

Fig. 17.32 **(a)** Three-turn self-complimentary spiral antenna. The *inset* shows an expanded view of the interdigitated-electrode photomixer structure. **(b)** Experimental setup for the LT-GaAs photomixer. **(c)** Measured output power from the photomixer. (Adapted from [17.43])

were connected to a three-turn self-complimentary spiral antenna shown in Fig. 17.32a, for which the low frequencies radiate from the outer extent of the spiral and the high frequencies from the inner extent. For the inner radius (40 μm) and outer radius (1.8 mm) of the actual antenna, the operating range was estimated to be approximately 60 GHz to 1.0 THz. The photomixer was characterized by the setup shown in Fig. 17.32b, for which the tunable optical pump power was provided by two Ti:Al_2O_3 lasers operating near 780 nm and with 25 mW in each laser beam. The photocurrent generates THz radiation at the tunable difference frequency between the two lasers from the antenna with a bias voltage of 25 V. The emitted radiation propagates predominantly into the GaAs substrate and then into a high-resistivity silicon lens in contact with the backside of the substrate. The THz output beam from the lens is subsequently focused by a plastic lens onto a liquid-helium-cooled Si bolometer. The measured THz output power is shown in Fig. 17.32c.

Electrooptic THz Generation

The generation of pulses of THz (far-infrared radiation) by the passage of a short optical pulse through a nonlinear optical material was demonstrated many years ago [17.33]. For this pioneering experiment 2 ps optical pulses from a mode-locked Nd:glass laser were focused onto and transmitted through millimeter-thick $LiNbO_3$ crystals. The resulting THz radiation was split into two beams. One beam was used for spectral analysis by either a Michelson or a Fabry–Perot interferometer; the other beam was used for normalization. Both beams were separately detected by two n-type InSb (Putley) detectors at 1.5 K in a magnetic field of 5.4 kG. The resultant

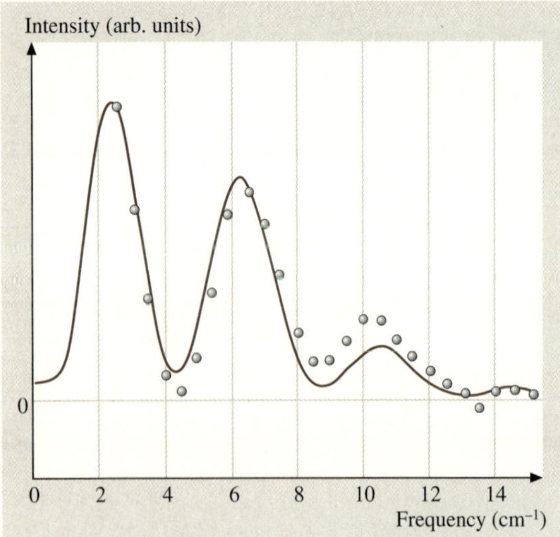

Fig. 17.33 Far-infrared (THz) power spectrum generated by mode-locked pulses in $LiNbO_3$. (Adapted from [17.33])

signals were displayed on an oscilloscope. An example of these early results is shown by the measured far-infrared power spectrum in Fig. 17.33. The experimental points were obtained from the Michelson interferogram and the solid curve is from a theoretical calculation assuming Gaussian laser pulses with a 1.8 ps pulse width. The spectrum can be understood as a product of (1) the spectral content of the mode-locked pulses, (2) a radiation efficiency curve, and (3) a phase-matching curve centered at zero frequency.

The propagation of fs optical pulses in electrooptic materials is also accompanied by the radiation of an extremely fast electromagnetic transient [17.34, 119, 120]. This phenomenon, arising from the inverse electrooptic effect, produces a Cherenkov cone of pulsed radiation having a duration of approximately one cycle and a frequency in the THz range. Although closely resembling classical Cherenkov radiation from relativistic charged particles in dielectric media, it is due to a nonlinear op-

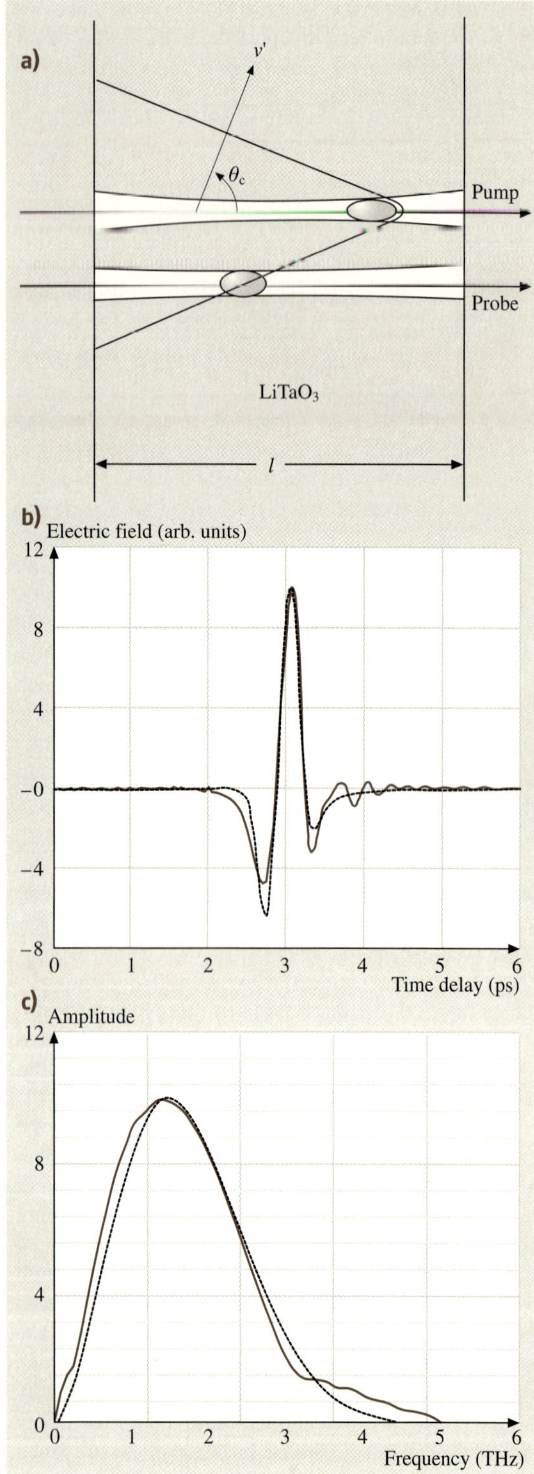

Fig. 17.34 **(a)** Schematic of the experiment used to generate and detect short bursts of Cherenkov THz radiation from fs optical pulses in lithium tantalate. The Cherenkov radiation cone propagates away from the "pump" pulse in a direction θ_c determined by the ratio of the low-frequency phase velocity ($c/6.53$) to the optical group velocity ($c/2.33$). **(b)** Measured electric field (*solid line*) of the THz transient compared with theory (*dashed line*). **(c)** Amplitude spectra of the experimental (*solid line*) and theoretical (*dashed line*) waveforms of **(b)**. (Adapted from [17.119])

tical effect. The charge state of the effective source is neutral, being an extended dipole moment, rather than a point charge; the velocity of the source significantly exceeds the radiation velocity. This situation, indicated in Fig. 17.34a, produces the characteristic Cherenkov conical shock wave. Because the spatial extent of the radiation source is proportional to the intensity envelope of the optical pulse, the radiation field depends on both the time duration and the beam waist of the optical pulse.

As indicated in Fig. 17.34a, two optical pulses were used, one to generate the radiation field, and the other to

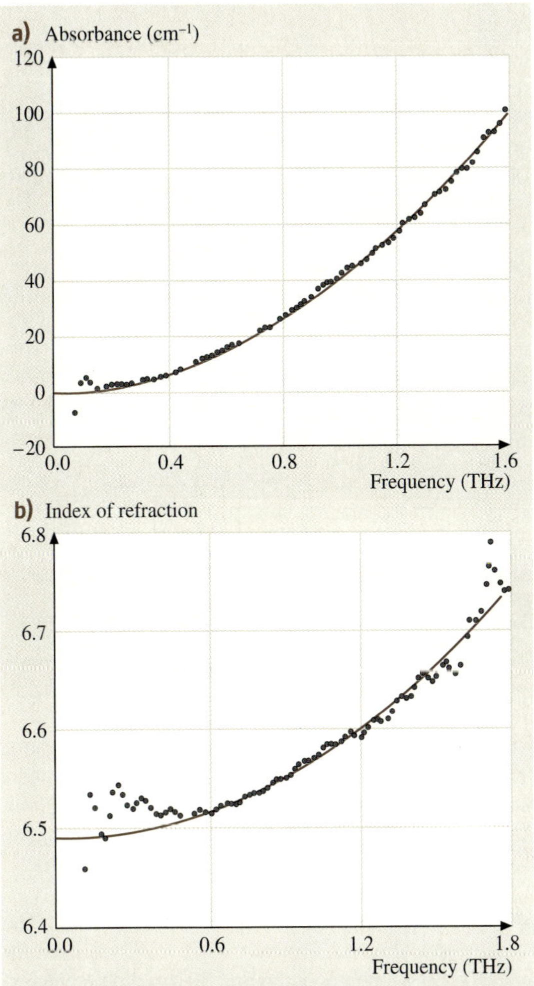

Fig. 17.35 **(a)** Absorbance of lithium tantalate. **(b)** Index of refraction. The *dots* are the measurements and the *solid lines* are the results of fitting the experimental points to a classical single-oscillator model. (Adapted from [17.121])

detect it. The 100 fs, 625 nm optical pulses were obtained from a CPM mode-locked dye laser, with a relatively low pulse energy of 0.1 nJ at a repetition rate of 150 MHz. The configuration of Fig. 17.34a uses electrooptic sampling [17.2, 3, 7–12] to measure the small birefringence produced by the electric field of the radiation pulse. A unique feature of the experiment is the automatic synchronism of the velocities of the radiation field and the probing pulse. As indicated in Fig. 17.34a, the probing pulse "surfs" along the Cherenkov wavefront enabling it to measure the electric field at a stationary point in the waveform by integrating the birefringence along the entire path through the crystal. To obtain the measured waveform, the timing of the probing pulse was slowly swept relative to the generating pulse by a computer controlled variable path length. The measured Cherenkov waveform shown in Fig. 17.34b is extremely fast and approximates a single cycle of frequency of 1.5 THz. The corresponding amplitude spectrum of this waveform is shown in Fig. 17.34c.

The coherent nature of the detection process enables the measurement of the far-infrared properties of materials by observing the change in the waveform due to their dispersion and absorption. Since the time reference is accurately known, phase information is measured as well, and consequently both the real and the imaginary parts of the dielectric function can be obtained [17.121, 122]. Such measurements of the absorption and index of refraction of lithium tantalate are presented below in Fig. 17.35 [17.121].

When a semiconductor is illuminated by an ultrafast laser pulse the sample can emit a THz radiation pulse, as shown in Fig. 17.31. This radiation can be due to the acceleration of the photocarriers in the surface field and for some semiconductors can be also due to optical rectification. It is possible to distinguish between the two processes by directing the incoming light beam at normal incidence to the semiconductor surface. For this case, the radiation from the photocarriers is not coupled out of the semiconductor, and the observed THz pulse radiated in the forward direction is due only to the optical rectification effect. This situation is demonstrated by the results shown in Fig. 17.36 [17.123]. For these results, the mode-locked Ti:sapphire laser illuminating the ⟨111⟩ GaAs sample had output pulse energies greater than 10 nJ, pulse durations of less than 200 fs and a repetition rate of 76 MHz. The laser beam was split into two parts by a beam splitter with a 0.05/0.95 reflection/transmission ratio. The stronger optical beam passed through a variable time delay stage and illuminated the semiconductor sam-

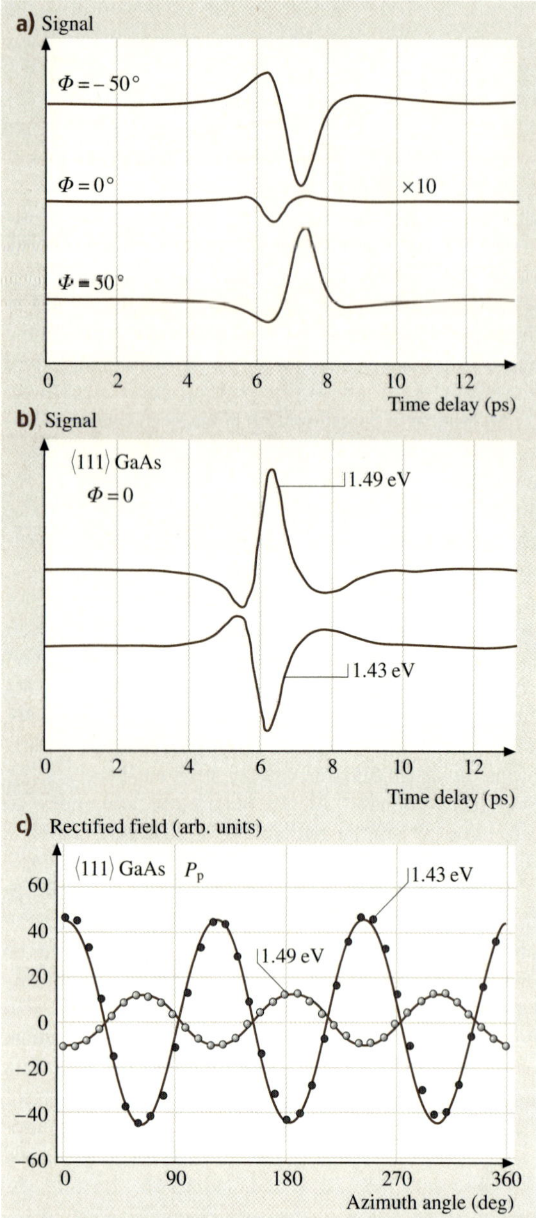

Fig. 17.36 **(a)** THz radiation from ⟨111⟩-oriented GaAs with normal (*center trace*) and ±50° (*upper* and *lower traces*) optical incident angles. **(b)** Emitted THz pulses with incident photon energies of 1.49 eV and 1.43 eV. **(c)** Generated THz pulse amplitude versus crystal azimuth angle. The *open* and *solid circles* are the experimental data at the two photon energies 1.43 eV and 1.53 eV, respectively. The *solid lines* are from theory. (Adapted from [17.123])

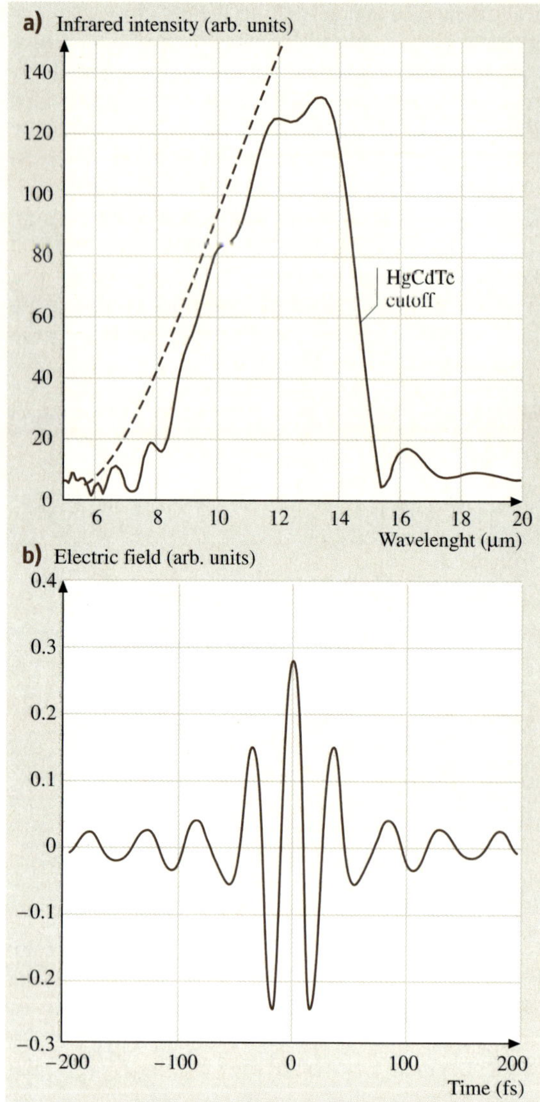

Fig. 17.37 **(a)** Power spectrum of the far-infrared THz emission obtained at normal incidence through the GaAs sample, as measured by the HgCdTe detector. **(b)** THz pulse electric field deduced from the spectrum assuming a negligible dispersion in the thin sample. (Adapted from [17.124])

ple with a spot diameter of approximately 6 mm. The weaker optical beam, typically < 20 mW was used for optical gating a photoconducting 50 μm dipole antenna for detection of the emitted THz pulse.

The optical rectification signal was observed to change dramatically when the laser energy was tuned near the GaAs band gap. A more than 100-fold increase in the signal was measured when the laser energy was tuned from 1.40 to 1.46 eV. The peak signal is reached when the laser energy is just above the band gap. As proof of the nonlinear nature of the effect, the THz signal was measured as a function crystal orientation with respect to the polarization of the laser beam and the expected threefold rotation symmetry was observed as shown in Fig. 17.36c. This work demonstrates the importance of optical rectification as a source of very short THz pulses, whose pulse widths are fundamentally limited only by the driving laser pulse width [17.123].

In another demonstration of optical rectification of ultrafast laser pulses by GaAs, coherent far-infrared THz pulses were generated with a measured spectrum extending from 7 to 15 μm [17.124]. A Ti:sapphire laser, delivering 10–15 fs pulses at a 100 MHz repetition rate, was focused at normal incidence to a spot size of typically 20 μm diameter using a spherical mirror onto a 0.1 mm thick ⟨1, 1, 0⟩ GaAs plate. The resultant THz emission in the forward direction (transmitted through the sample) was directed onto a nitrogen-cooled HgCdTe detector sensitive to wavelengths up to 15 μm. A maximum THz power of 30 nW was measured for an incident laser power of 175 mW. The far-infrared THz spectra were obtained using a modified Michelson interferometer setup, synchronous detection, and numerical Fourier transforms. Figure 17.37a presents the THz power spectrum obtained for the generated THz pulse. The observed strong cutoff at 15 μm originates from the drop off of the spectral sensitivity. Figure 17.37b shows the electric field amplitude constructed from the spectrum of Fig. 17.37a, assuming a constant spectral phase. This work demonstrated that single-cycle THz generation can be extended to the near-infrared [17.124].

Using optical rectification and electrooptic sampling in ⟨110⟩ ZnTe for both the generation and coherent detection of freely propagating pulses of THz radiation, a wideband coherent THz-TDS system has been demonstrated [17.125], as shown in Fig. 17.38a. A mode-locked Ti-sapphire laser producing 130 fs pulses at 800 nm with a 76 MHz repetition rate was used to generate and detect the THz pulses. The 2 mm diameter pump beam, with an average power of 350 mW, was chopped and drove a 0.9 mm thick ⟨110⟩ ZnTe crystal at normal incidence. The pump beam was polarized at an angle of 60° to the ⟨110⟩ crystallographic direction in order to maximize the nonlinear response. The generated THz pulse was incident on an identical ZnTe crystal, after passage through a blocking Teflon filter. The incoming THz field was polarized parallel to the ⟨001⟩ axis of the ⟨110⟩ ZnTe sampling crystal. A 5 μm thick pellicle beam splitter in the THz beam line enabled the co-propagation of the optical probe and the THz beams through the sampling crystal. The probe beam was optic-

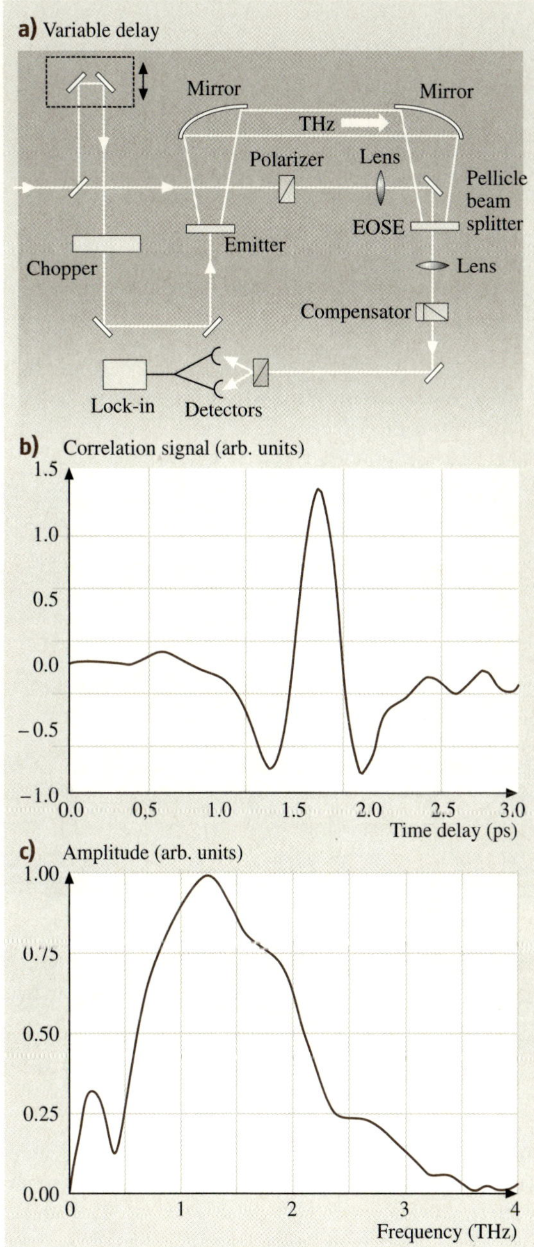

Fig. 17.38 (a) Schematic diagram of the experimental setup. (b) Measured THz pulse. (c) Amplitude spectrum of the measured THz pulse. (Adapted from [17.125])

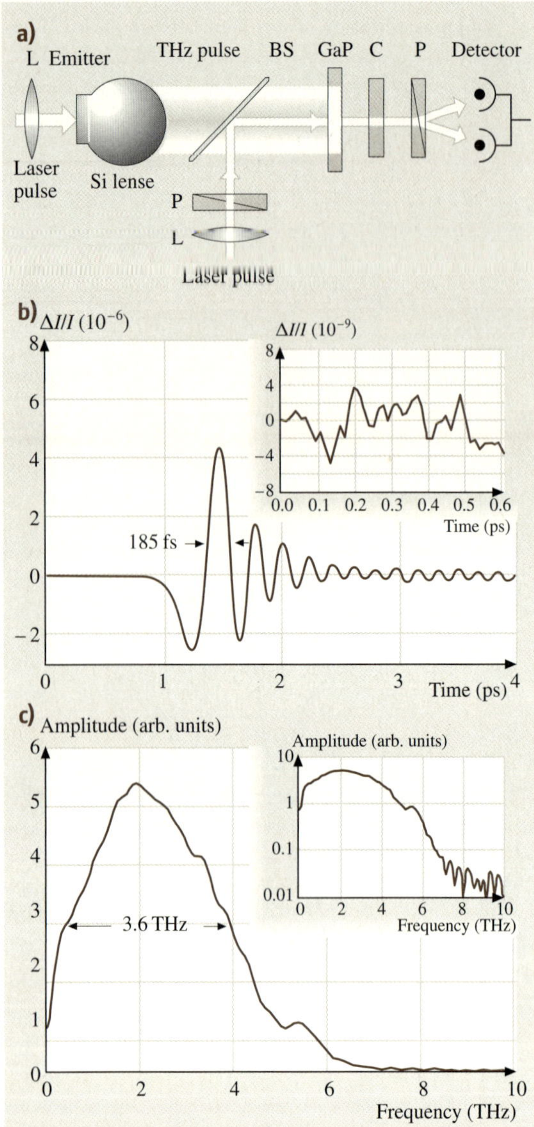

Fig. 17.39 **(a)** Schematic diagram of the experimental setup (C: compensator; P: polarizer; L: lens; BS: pellicle beam splitter). **(b)** Measured THz pulse with a 150 μm thick GaP crystal. *Inset*: noise level at the first 0.6 ps. **(c)** Amplitude spectrum of the measured THz pulse. (Adapted from [17.126])

ally biased at the quarter wave point by a Soleil–Babinet compensator and a crossed polarizer arrangement was used to improve the sensitivity.

The measured 270 fs (FWHM) THz electric field pulse is shown in Fig. 17.38b. The small oscillations after the main peak were due, in part, to the free induction decay of the ambient water vapor excited by passage of the THz beam. In this geometry, both beams underwent multiple internal reflections within the sampling crystal, which were excluded from the data. The amplitude spectrum of the temporal waveform shown in Fig. 17.38c, demonstrates the wideband capability of the system. For this experiment the coherence length decreased rapidly beyond 2 THz due to the THz crystal dispersion.

Another all-optical THz generation and detection system with a bandwidth of 7 THz has been reported [17.126], as shown in Fig. 17.39a. The ultra-broadband THz radiation was generated via optical rectification from ⟨110⟩ oriented GaAs, and coherently detected via free-space electrooptic sampling using a 150 μm thick, ⟨110⟩ GaP crystal. The Ti:sapphire laser source had an output power of up to 2 W and a minimum pulse width of 45 fs. An uncoated 2 μm thick pellicle beam splitter, transparent to the THz beam was used to reflect about 50% of the optical probe beam collinearly along the THz beam into the GaP sensor. An optical chopper modulated the pump laser beam at 1.5 kHz for phase sensitive detection by a lock-in amplifier. Excess common-mode laser noise was effectively canceled by using a pair of identical photodetectors in a balanced detection scheme, so that the detection is primarily shot-noise limited. The laser pump beam with an average power of 1.5 W and 50 fs pulses was focused at normal incidence with a 25 cm focal length lens onto a ⟨110⟩ oriented GaAs crystal directly attached to the flat surface of a 12 mm diameter silicon collimating lens. The peak intensity at the emitter was about 1 GW/cm^2, limited by optical damage to the emitter surface.

Figure 17.39b displays the measured THz pulse. After an average of four consecutive scans, the peak-to-peak noise floor $\Delta I/I$ was below 10^{-8} as shown in the inset. The measured 185 fs THz pulse width (FWHM) was the shortest directly measured pulse duration at that time. The corresponding Fourier transform in Fig. 17.39c shows a 3 dB bandwidth of 3.6 THz with the upper frequency response reaching 7 THz.

17.2.3 Other THz Receivers

The bandwidth of the photoconductive THz receiver is limited most fundamentally by the ballistic acceleration of the generated carriers. In addition, the antenna response can be another limitation. An attractive alternative is based on electrooptic detection, where the receiver response is determined by the very fast electronic nonlin-

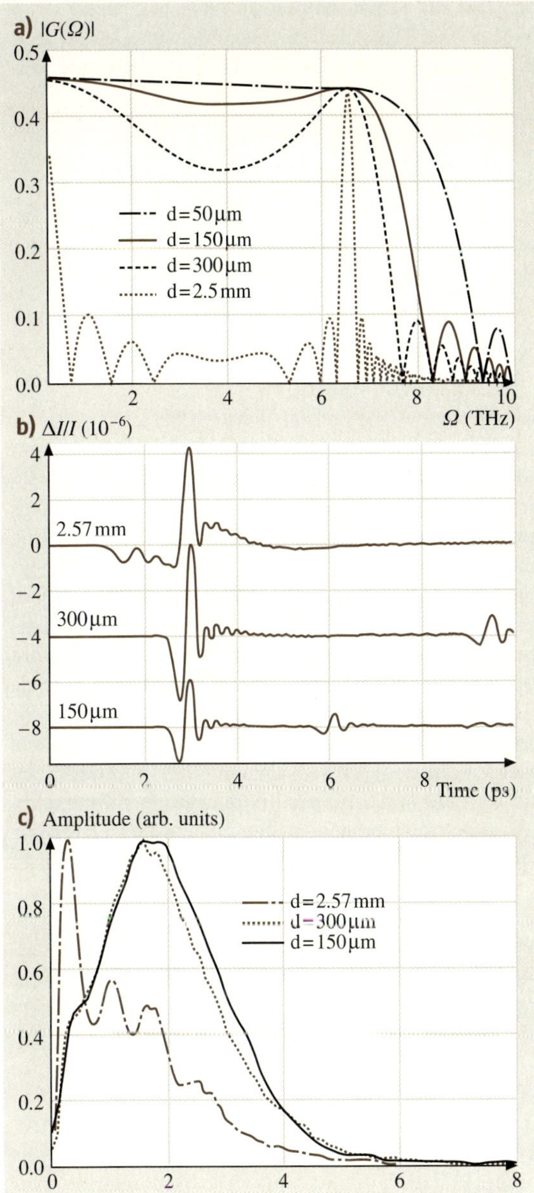

Fig. 17.40 **(a)** Calculated frequency response of GaP sensors with different thickness. **(b)** Measured THz pulses with 2.57 mm, 300 μm, and 150 μm thick ⟨110⟩ GaP detector. **(c)** Amplitude spectra of measured THz pulses. (Adapted from [17.126])

earity. Consequently, the THz receiver bandwidth could in principle be extended to tens of THz.

Electrooptic Detection

The first two demonstrations of electrooptic detection of freely propagating THz radiation were submitted for publication within a few days of each other [17.127, 128]. The Columbia group [17.128] used a 10 μm thick poled polymer film in their electrooptic sampling element and achieved an approximate 1 ps time resolution with a signal-to-noise ratio of 10. The RPI group [17.127] used a 500 μm thick $LiNbO_3$ crystal as their nonlinear element also achieving an approximate 1 ps rise time with a S/N = 10. These initial observations have been markedly improved upon by the same groups, as illustrated by their complete electrooptic THz systems shown in Figs. 17.38 and 17.39. Electrooptic (EO) detection of pulses of THz radiation [17.126, 129] has now demonstrated impressive bandwidths out to 70 THz [17.130]. However, as shown by the recent THz-TDS characterization of the important EO crystal ZnTe [17.131] and another THz-TDS characterization of other EO crystals [17.132], the EO technique is constrained by the THz absorption of the EO crystal, group-velocity matching between the measured THz pulse and the optical probing pulse, and the frequency-dependent nonlinearity [17.130, 131, 133]. This situation requires very thin EO crystals to obtain high-frequency response, concomitant with a reduction in S/N.

An example of these considerations is shown in Fig. 17.40, which describes the THz receiver of the system indicated in Fig. 17.39. Figure 17.40a shows the calculated frequency-dependent efficiency of the electrooptic THz measurement as a function of the GaP crystal thickness. The experimentally measured THz pulses using crystals of different thicknesses are shown in Fig. 17.40b and the corresponding amplitude spectra are shown in Fig. 17.40c.

A very high bandwidth, completely electrooptic system is shown in Fig. 17.41. Here, the measured frequency response extends to 37 THz. For these results the 12 fs Ti:sapphire laser delivered an average power of nearly 500 mW at 800 nm. A 0.45 mm thick ⟨110⟩ oriented GaAs wafer was used as the emitter and a 30 μm thick ⟨110⟩ oriented ZnTe crystal as the electrooptic receiver. A laser power of 350 mW was focused on the GaAs emitter by a gold-coated off-axis parabolic mirror with a 5 cm effective focal length. The resultant THz pulse generated by optical rectification in GaAs was collimated and then focused by the off-axis paraboloidal mirrors. The measured THz pulse is shown in Fig. 17.41b, where the shortest period of oscillation is 31 fs. Figure 17.41c shows the measurement

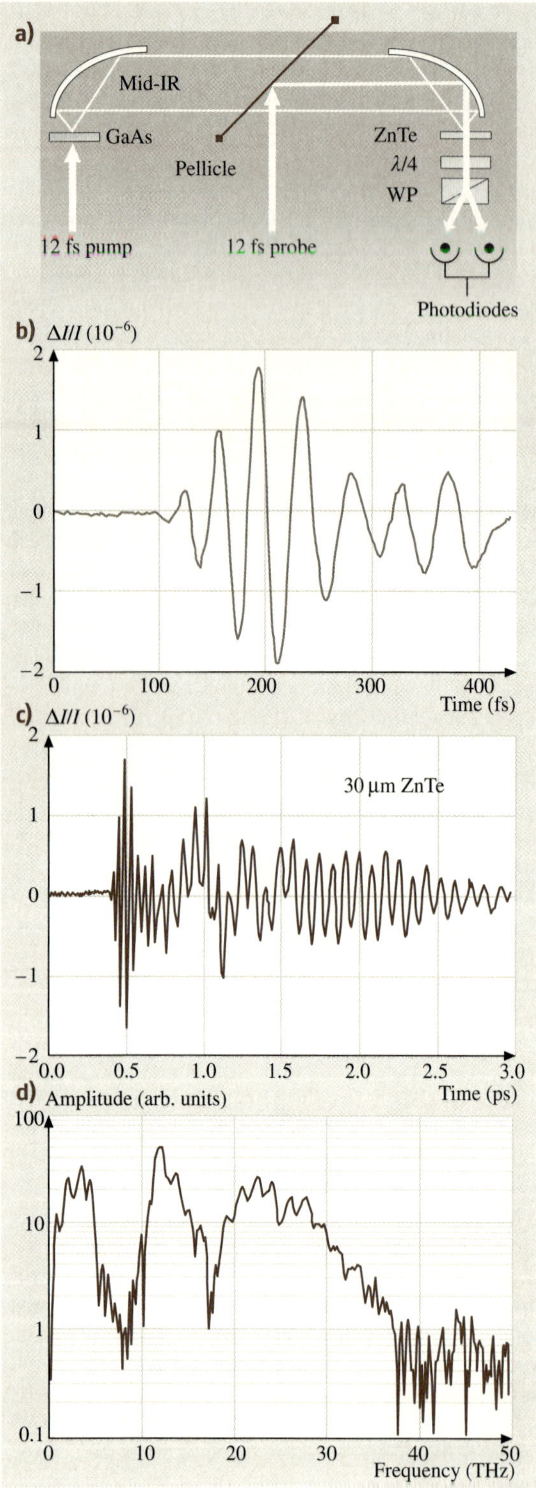

Fig. 17.41 **(a)** Schematic diagram of experimental setup. **(b)** Measured THz pulse. **(c)** THz pulse with a longer scan. **(d)** Amplitude spectrum of the THz pulse. (Adapted from [17.129])

on a longer time scan, and Fig. 17.41d shows the corresponding amplitude spectrum. The highest frequency response reaches to 37 THz. The strong absorption between 5 and 10 THz was caused by the Restrahlen bands of ZnTe and GaAs. The dip around 17 THz was explained in terms of velocity mismatching.

Interferometric Techniques Using Power Detection

An alternative method of source characterization is shown in Fig. 17.30a, which bypasses the problems of receiver bandwidth, and is based on far-infrared interferometric techniques using a power detector. This approach was first demonstrated for THz radiation sources [17.134] by measuring autocorrelation signals with a full-width-at-half-maximum (FWHM) of 230 fs for the THz radiation pulse from laser created carriers accelerated by the surface field of a photoconductive semiconductor. This approach used a single THz radiation source, illuminated by a 10 Hz repetition rate, amplified, 100 fs pulses from a CPM dye laser, together with a Martin–Puplett interferometer and a liquid-helium cooled bolometer. A different interferometric approach, using unamplified, 100 MHz repetition rate, CPM dye laser pulse excitation of a two-source interferometer has since been demonstrated [17.100]. Via this approach, together with fast, scanning-delay-line averaging, orders of magnitude improvements in the signal-to-noise ratio of the measured interferograms were obtained. Using this method, the THz radiation source described in Fig. 17.27d was shown to produce radiation to 6 THz. In addition, a 230 fs FWHM autocorrelation signal and an average power of 30 nW for 4 mW of laser excitation power were measured for this source [17.100]. Other work has used interferometric techniques together with incoherent bolometric detection to study THz emission from cold plasma oscillations [17.135].

17.2.4 THz-TDS with Freely Propagating THz Pulses

The combination of the THz-TDS technique with THz beams has some powerful advantages compared to traditional cw spectroscopy. Firstly, the coherent detection of the far-infrared radiation is extremely sensitive. Although the energy per THz pulse is very low (0.1 fJ), the 100 MHz repetition rate and the coherent detection measures the electric field of the propagated pulse with a S/N ratio of about 10 000 for an integration time of 125 ms [17.31]. In terms of average power this sensitivity exceeds that of liquid-helium-cooled bolometers, by more than 1000 times [17.53]. Secondly, because of the gated and coherent detection, the thermal background, which plagues traditional measurements in this frequency range, is observationally absent. Comparing time-domain spectroscopy with Fourier-transform spectroscopy (FTS), it should be clear that the frequency resolution of the two techniques are similar, as they are both based on a scanning delay line, where to first order the frequency resolution is determined by the reciprocal of the time scan. However, the fact that THz-TDS scans a delay line with a well collimated optical beam does present some experimental advantages. Although for now FTS is superior, being above 4 THz (133 cm^{-1}), the limited power of the radiation sources and the problems

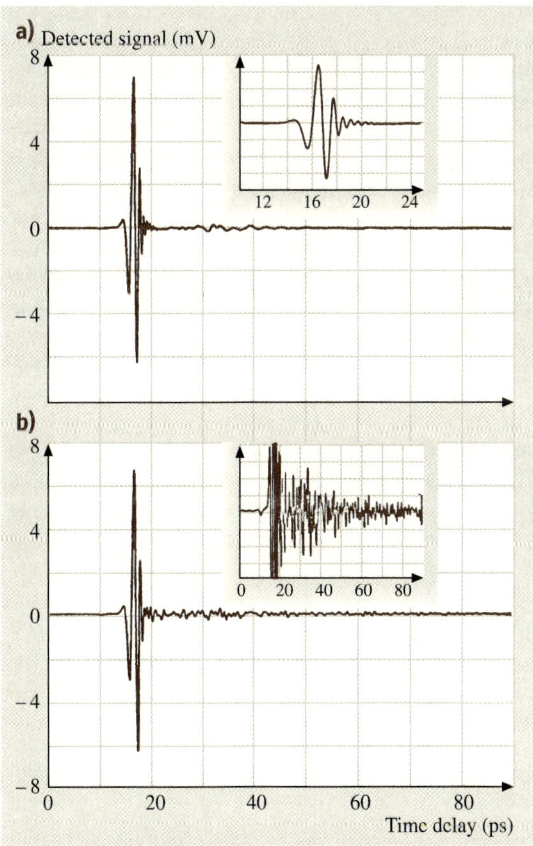

Fig. 17.42 **(a)** Measured freely propagating THz beam in pure nitrogen. *Inset* shows pulse on an expanded time scale. **(b)** Measured THz pulse with 1.5 Torr of water vapor in the enclosure. *Inset* shows pulse on a 20× expanded vertical scale. (Adapted from [17.50])

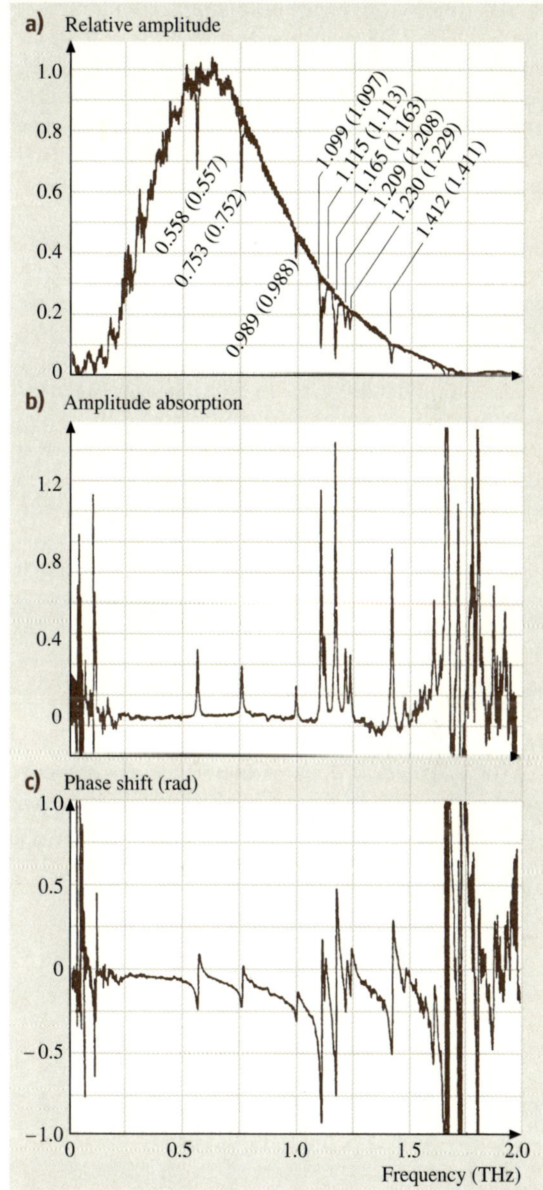

Fig. 17.43 **(a)** Amplitude spectra of Figs. 17.42a,b. **(b)** Amplitude absorption coefficient obtained from **(a)**. **(c)** Relative phase of the spectral components of **(a)**. (Adapted from [17.50])

with the thermal background favor THz-TDS below 4 THz.

We will now describe several different measurement types that illustrate the generality and usefulness of THz-TDS.

THz-TDS Characterization of Water Vapor

The first THz-TDS measurements with the now standard THz optoelectronic system shown in Fig. 17.19 was the characterization of water vapor from 0.25 THz to 1.5 THz, for which the cross-sections of the nine strongest lines were measured with the best accuracy to date [17.50]. For these early measurements 30 μm long source and detector antennas on ion-implanted SOS substrates were used together with MgO [17.68] lenses instead of the currently used, more optimal high-resistivity silicon [17.51]. Figure 17.42a displays the detected THz radiation pulses after propagating through pure nitrogen. This measurement was made in a single 10 min scan of the 200 ps relative time delay between the excitation and detection pulses. When 1.5 Torr of water vapor, corresponding to 8% humidity at 20.5 °C, was added to the enclosure, the transmitted pulse changed to that shown in Fig. 17.42b. The additional fast oscillations are caused by the combined action of the dispersion and absorption of the water vapor lines. The slower and more erratic variations seen in Figs. 17.42a,b result from reflections of the main pulse. They are reproducible and divide out in the data analysis. The inset shows the data on a 20× expanded vertical scale. Here, the oscillations are seen to decay approximately exponentially with an average coherent relaxation time T_2.

The amplitude spectra of Figs. 17.42a,b are compared in Fig. 17.43a, where the strong water absorption lines are clearly observable. The additional structure on both spectra is not noise but results from spurious reflections of the main pulse. At each line the measured frequency with an estimated error of ±0.001 THz and, in parentheses, the literature value are indicated. The corresponding absorption coefficients are displayed in Fig. 17.43b as the negative of the natural logarithm of the ratio of the two amplitude spectra in Fig. 17.43a. Because the electric field is directly measured, the relative phase shift between the two spectra is also obtained as plotted in Fig. 17.43c, without the need for the Kramers–Kronig relations. As expected for a Lorentzian line, the magnitude of the jump in phase, experienced at each resonance, equals the peak absorption.

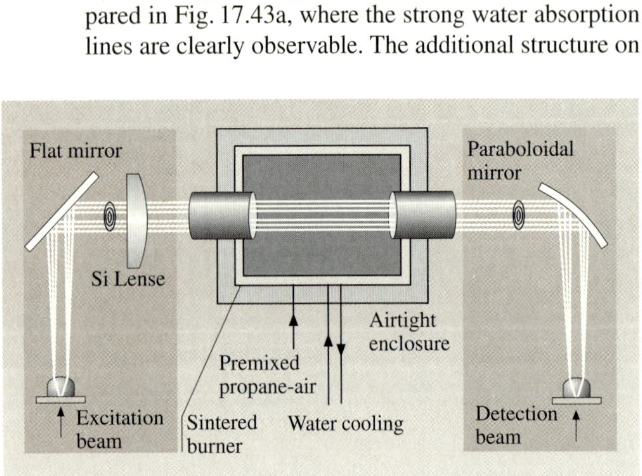

Fig. 17.44 Experimental schematic of the THz-TDS setup and the flat flame burner. (Adapted from [17.55])

THz-TDS Characterization of Flames (Hot Water Vapor)

The first comprehensive THz (far infrared) absorption measurement of flames, covering the region from 0.2 to 2.65 THz (7–88 cm^{-1}) has been accomplished by THz-TDS [17.54]. Such measurements have not been possible with the alternative method of Fourier-transform spectroscopy, because the incoherent bolometric (liquid-helium cooled) detection is overwhelmed by the copious amounts of far-infrared radiation emitted by flames. In contrast, the THz-TDS system is blind this incoherent thermal radiation. For flames, many of the expected constituents, H_2O, CO, OH, NO, and NO_2, have permanent electric-dipole moments and can thereby be simultaneously probed by THz-TDS with approximately 2 GHz spectral resolution. In this THz-TDS observation of a premixed propane–air flame, a large number of absorption lines were initially identified as those of water, CH, and NH_3 [17.54]. The flame temperature of 1300 K was determined by comparing the relative strengths of the water-vapor lines. It was later shown that, except for the one remaining line at 2.865 THz, all of the observed THz spectral, absorption lines in the propane–air flame are from water vapor, and that previously unidentified lines in the flame are due to pure rotational transitions in the ν_2 (010) vibrational state of hot water vapor [17.55].

Later, via THz-TDS and a propane–air flame, measurements of the collisionally broadened pure rotational linewidths of water vapor at high temperatures were performed [17.56]. More than 40 rotational lines were observed in the ground and $\nu_2 = 1$ vibrational levels over the spectral bandwidth from 1 to 2.5 THz (33–83 cm^{-1}). The measured linewidths agree in magnitude with calculated values at the measured temperature of 1490 K; however, contrary to predictions, a pronounced energy

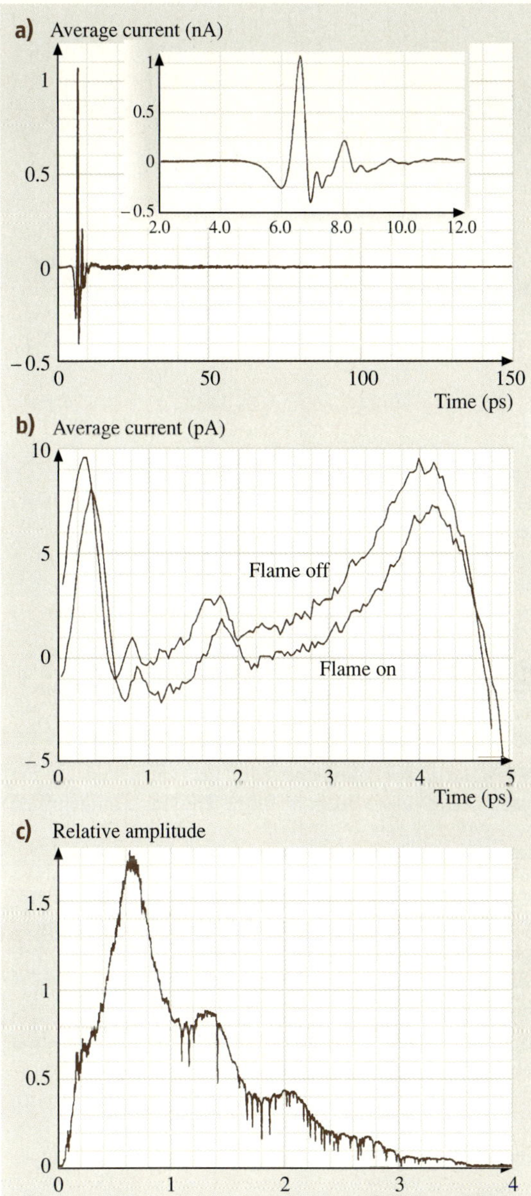

Fig. 17.45 **(a)** Transmitted THz pulse through the 1 : 30 propane–air flame. The *inset* shows the THz pulse in more detail. **(b)** Close-up of the region before the arrival of the THz pulse both with and without the flame. **(c)** Amplitude spectrum of the transmitted THz pulse. (Adapted from [17.55])

dependence was observed. No temperature shift of line centers was observed, and line strengths were in reasonable agreement with tabulated values. The demonstrated ability of THz-TDS to measure linewidths of pure rotational transitions in hot samples can be extended to other samples and gas mixtures. Dopants can be added to the incoming fuel stream to look for important free radicals, transient species and reactants.

The experimental configuration used in this investigation is shown in Fig. 17.44. The path length through the flame was 20.7 cm, and the premixed flame was supported on a 24.8 cm × 17.8 cm sintered bronze water-cooled burner. The burner assembly was enclosed to prevent contamination from room air. The THz transmitter and receiver were enclosed in separate chambers isolated from the burner. The THz beam was coupled into the burner enclosure through 100 μm thick quartz windows fused to 5 cm diameter quartz tubes, permitting the windows to be inserted directly into the flame.

A representative measurement of the flame is shown in Fig. 17.45a; the inset shows the main THz pulse in detail. As illustrated in Fig. 17.45b, there was no increase in background noise with the flame on. Despite the high temperature within the flame, the gated coherent detection and the small acceptance of the THz optics are effective in blocking the incoherent THz radiation that is incident upon the receiver. The normalized amplitude spectrum of the pulse in Fig. 17.45a is shown in Fig. 17.45c. The 0.75 THz oscillatory structure in the spectrum is due to etalon effects of the 100 μm thick quartz windows. The power absorption spectrum, measured by taking the ratio of an average of three scans with the 1 : 30 propane–air flame on to that with the flame off and the burner enclosure purged with dry air, is shown in Fig. 17.46. The measurements fit the calculated line positions to an accuracy of 1 GHz. Twenty-two rotational absorption lines in the ν_2 band were observed; the measured strengths and linewidths agree well with calculations.

THz-TDS Characterization of Sapphire and Silicon

The early THz-TDS measurements [17.51] on single crystal sapphire and silicon were motivated at that time by the need to find the best material for THz lenses in contact with the emitting and detecting chips. The available published data were inadequate for this. Absorption of the THz radiation by the lens material (initially sapphire) imposed an upper limit on the bandwidth of the entire system. The use of silicon lenses, inspired by these measurements of unusually low absorption and dispersion, immediately increased the system bandwidth from

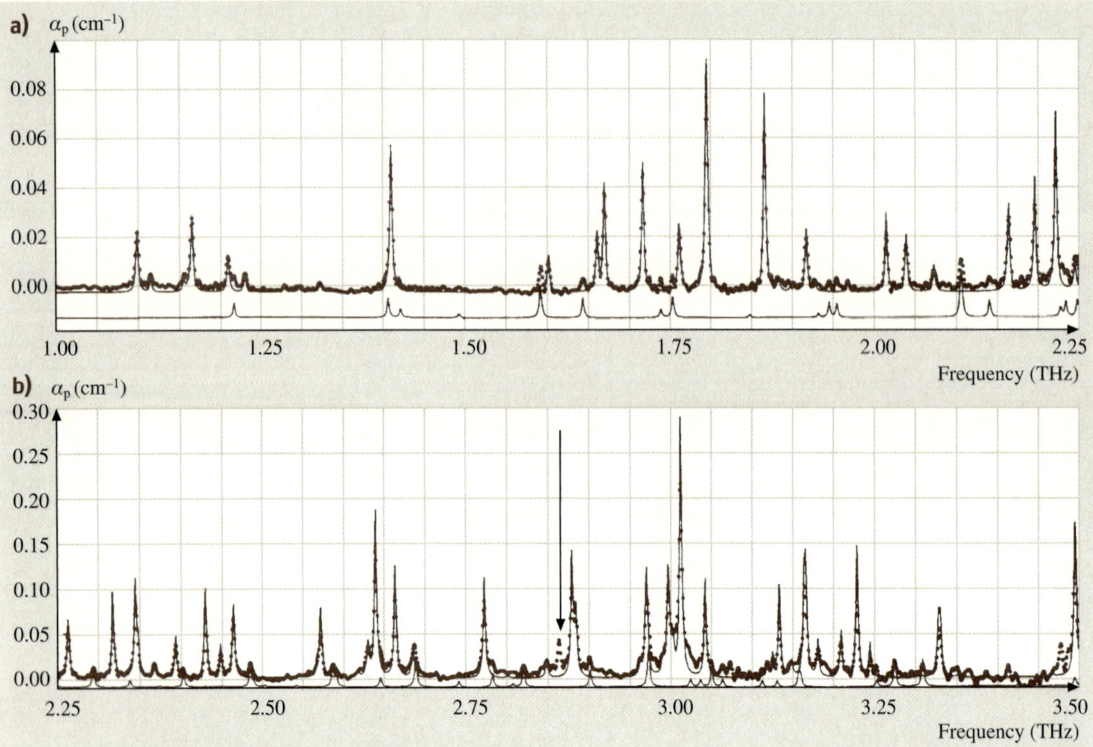

Fig. 17.46 Measured (*dots*) power absorption coefficient of hot water vapor compared with that calculated for the vibrational ground state (*solid curves*) and the first ν_2 vibrational state (*solid curves* offset by 0.01). The single unidentified line is indicated by the *arrow*. (Adapted from [17.55])

2 to 3 THz and gave smoother THz pulses with less ringing structure.

The single crystal sapphire sample was a polished, 57 mm diameter disc, 9.589 mm thick, and with the c-axis in the plane of the disc. A typical THz input pulse incident upon the sample is shown in Fig. 17.47a, and the output pulse (normalized to the input pulse) after propagation through the sample is shown in Fig. 17.47b for which the c-axis of the crystal was perpendicular to the polarization. The reduction in amplitude is due to the reflective loss at both surfaces and to the absorption during passage through the sapphire. The pulse at 73.4 ps delay is the ordinary pulse, while the smaller pulse at 85.1 ps delay is the extraordinary pulse. The ratio of the peak of the ordinary pulse to that of the extraordinary pulse is approximately 25 : 1 and gives the polarization sensitivity of the system. In terms of amplitude, the polarization ratio of the generated THz beam is 5 : 1. The 11.8 ps separation between the two pulses is a measure of the birefringence of sapphire and, neglecting the correction to group velocity due to dispersion, directly gives the difference in the index of refraction between the extraordinary and ordinary ray to be $n_o = 0.37$ compared to the literature value of 0.34 [17.136, 137]. When a full frequency analysis is performed, excellent agreement is obtained with the literature value. The 73.4 ps time delay of the ordinary pulse with respect to the 7.1 ps time delay of the pulse with no sample in place gives the ordinary index of refraction $n_o = 3.07$ in agreement with the literature value [17.136–138]. The absorption coefficient versus frequency determined from these pulses is shown in Fig. 17.48a. Here, we see a monotonic increase in absorption with increasing frequency with the expected quadratic dependence. Due to excessive attenuation caused by the sample being too thick for the weak higher frequency components, the data is considered to be accurate only up to 1.75 THz. Some of the previous work [17.136–138] has been indicated on the curve, and shows rough agreement (within a factor of 2) with the THz-TDS measurement. The relative phases of the Fourier components determine the index of refraction versus frequency as presented in Fig. 17.48b, which

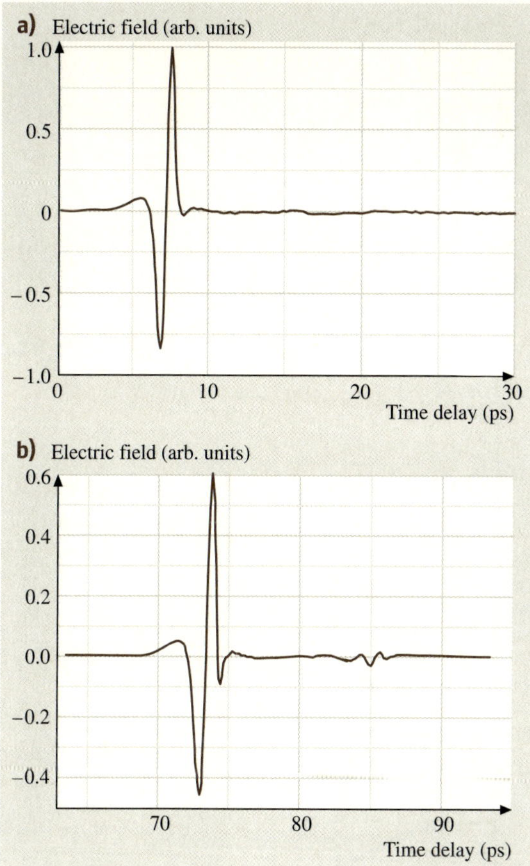

Fig. 17.47 **(a)** Measured input THz pulse. **(b)** Measured THz pulse transmitted through the sapphire crystal. (Adapted from [17.51])

compares reasonably well with the indicated earlier results [17.136–138].

Crystalline silicon is optically isotropic eliminating concern about the polarization of the incident THz beam and crystal orientation. Although there is significant literature concerning the far-infrared properties of silicon below 2 THz there are noteworthy discrepancies among the published data with variations in the measured absorption coefficients of up to 10 times. The main reason

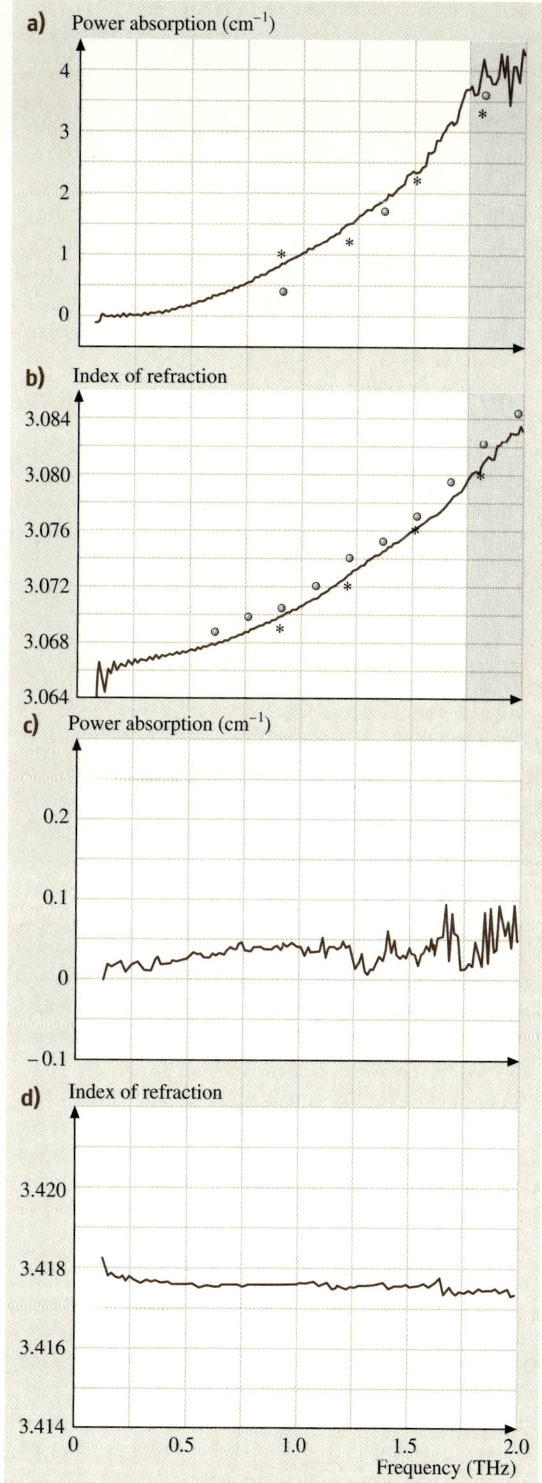

Fig. 17.48a–d THz-TDS measurements of crystalline sapphire and silicon. The *circles* are the data of [17.136, 137]; the *asterisks* are data from [17.138]. **(a)** Power absorption coefficient (ordinary ray) of sapphire. **(b)** Index of refraction (ordinary ray) of sapphire. **(c)** Power absorption coefficient of high-resistivity silicon. **(d)** Index of refraction of high-resistivity silicon. (Adapted from [17.51])

for this confusion is that below 2 THz the results are extremely sensitive to the presence of carriers. THz-TDS measurements [17.97, 98] have shown that for 1 Ω cm, n-type silicon the peak absorption is $100\,cm^{-1}$, and that for 10 Ω cm, n-type, the absorption is $12\,cm^{-1}$. Extrapolating these values to 100 Ω cm, $\alpha = 1\,cm^{-1}$, at 1 kΩ cm, $\alpha = 0.1\,cm^{-1}$, and at 10 kΩ cm, $\alpha = 0.01\,cm^{-1}$. Consequently, unless high-purity, high-resistivity material is used, what is measured is not the properties of the intrinsic semiconductor but that of the carriers due to residual impurities. This problem is most prevalent in the earlier work on silicon with resistivities of 10 Ω cm [17.136] to 100 Ω cm [17.138].

The above TDS measurements were made on a 50 mm diameter, 20.046 mm thick single crystal of high-resistivity (greater than 10 kΩ cm), float-zone silicon. For this material an unprecedented transparency was measured together with a remarkably flat dispersion curve. Throughout the range from low frequencies up to 2 THz the measured absorption coefficient is less than $0.05\,cm^{-1}$, while the index of refraction changes by less than 0.001.

THz-TDS Characterization of n-Type and p-Type Silicon

A THz-TDS measurement of device-grade, doped silicon wafers [17.97, 98] showed that the frequency-dependent absorption and dispersion were completely due to the carriers and not to the host crystal. Consequently, the complex conductance could be characterized over the widest frequency range to date. The samples used were a 283 μm thick wafer of 1.15 Ω cm; n-type and a 258 μm thick wafer of 0.92 Ω cm p-type silicon. The measured absorptions shown in Fig. 17.49a are more than 2000 times greater than that of the host crystal. The clear difference between the n- and p-type material is due to the different dynamic behavior of the electrons and holes. For these measurements the oscillations due to the etalon effects of the sample geometry have been removed numerically. As shown in Fig. 17.49b, the index of refraction is strongly frequency dependent. The agreement with the Drude theory (solid line) is exceptional.

As the THz optical properties of the samples are essentially completely determined by the carrier dynamics,

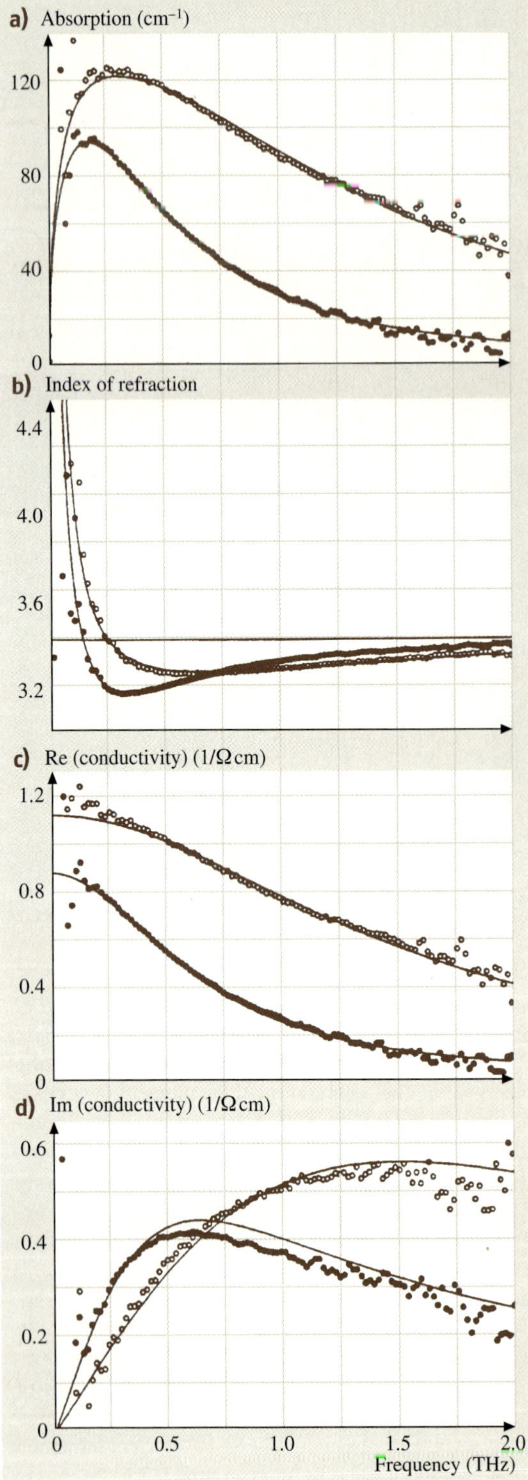

Fig. 17.49a–d THz-TDS results for 1.15 Ω cm, n-type (*dots*) and 0.92 Ω cm, p-type (*circles*) silicon. **(a)** Power absorption. **(b)** Index of refraction. **(c)** Real part of the electric conductivity. **(d)** Imaginary part of the electric conductivity. (Adapted from [17.97, 98])

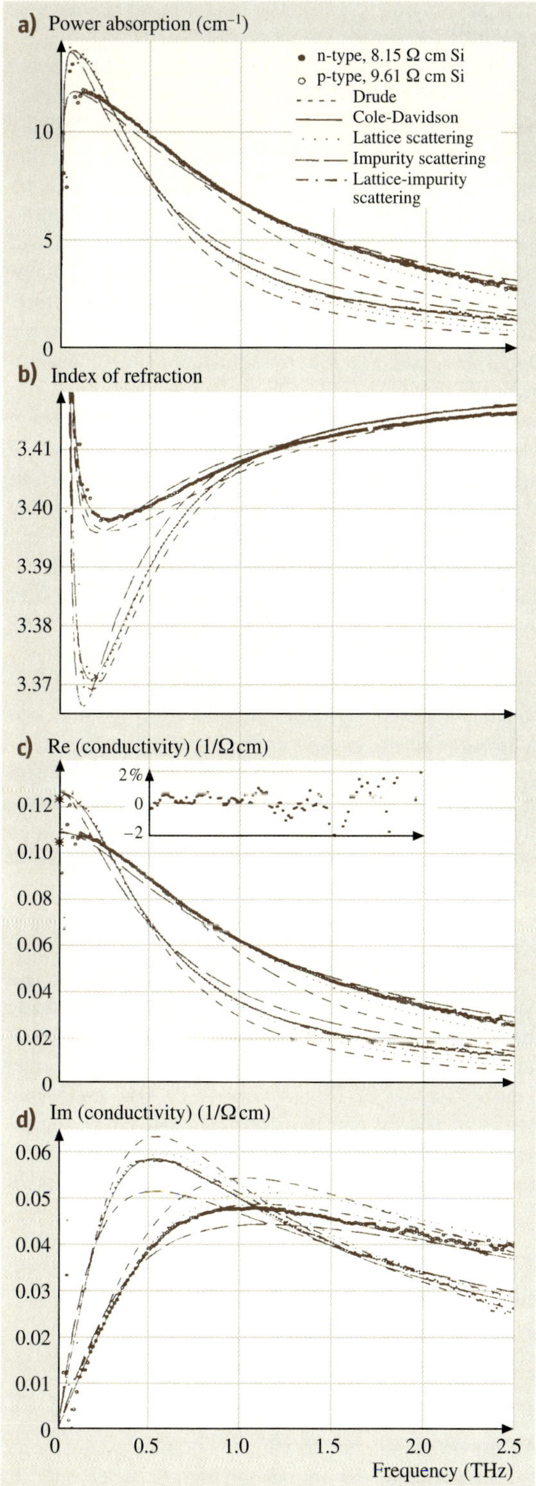

Fig. 17.50a–d THz-TDS measurements for n-type (*dots*) and p-type (*open circles*) silicon compared with five alternative theories. **(a)** Power absorption coefficient. **(b)** Index of refraction. **(c)** Real part of the conductivity. **(d)** Imaginary part of the conductivity. (Adapted from [17.139])

the complex electric conductivity of the doped silicon has also been measured independently of conduction theory and relying on only very general assumptions [17.97, 98]. The electric conductivity is obtained (without any fitting parameters) from the data of Figs. 17.49a and 17.49b; the real part of the conductivity is shown in Fig. 17.49c and the imaginary part in Fig. 17.49d. The extrapolated dc conductivities are 0.89 Ω cm for the p-type material and 1.13 Ω cm for the n-type, compared to the directly measured values of 0.92 Ω cm (p-type) and 1.15 Ω cm (n-type).

The final agreement with Drude theory (solid line) is quite acceptable with the two Drude parameters, the plasma angular frequency ω_p and the damping rate Γ, determined within 5% accuracy from the fits to the data. For 0.92 Ω cm p-type silicon $\omega_p/2\pi = 1.75\,\text{THz}$ and $\Gamma/2\pi = 1.51\,\text{THz}$, while for 1.15 Ω cm n-type silicon $\omega_p/2\pi = 1.01\,\text{THz}$ and $\Gamma/2\pi = 0.64\,\text{THz}$. The measured damping rates and the known effective carrier masses [17.97, 98], determine mobilities of $1680\,\text{cm}^2/\text{V s}$ for the electrons and $500\,\text{cm}^2/\text{V s}$ for the holes. The measured plasma frequencies and effective carrier masses determine the carrier number densities of $1.4\times10^{16}\,\text{cm}^{-3}$ for the p-type and $3.3\times10^{15}\,\text{cm}^{-3}$ for n-type silicon.

More recent definitive THz-TDS measurements of the complex conductivity $\sigma(\omega)$ from dc to 2.5 THz on doped silicon have been recently performed [17.139], as displayed in Fig. 17.50. Compared to the above earlier experimental studies of doped silicon [17.97, 98], these new results have sufficient frequency range and precision to test alternative theories [17.140–143] for the conductivity. However, for both n-type and p-type silicon and over a measured range of more than two orders of magnitude of the carrier density N, agreement was not found with any standard theory, including Drude, lattice-scattering and impurity-scattering theories [17.97, 98, 140–143]. For the first time, a modified $\sigma(\omega)$ of Drude theory to include the Cole–Davidson (C–D) fractional exponent β [17.144], achieved an exceptional fit to all of the measurements. This first application of a C–D type distribution [17.144] to fit $\sigma(\omega)$ of well-ordered crystalline semiconductors extends the demonstrated frequency range of this type distribution by more than three orders of magnitude and demonstrates that fractal conductivity is not just found in disordered material. A later THz-TDS characterization of a very low carrier density sample of n-type neutron-doped float zone silicon obtained an excellent fit with the complex conductivity described by the C–D distribution with $\beta = 0.83$ [17.145].

THz-TDS Characterization of n-Type Gallium Nitride

The THz-TDS characterization of the complex electric conductivity and dielectric response of n-type GaN over the frequency range from 0.1 to 4.0 THz has recently been performed [17.146]. The measured conductivity is well fit by the simple Drude model.

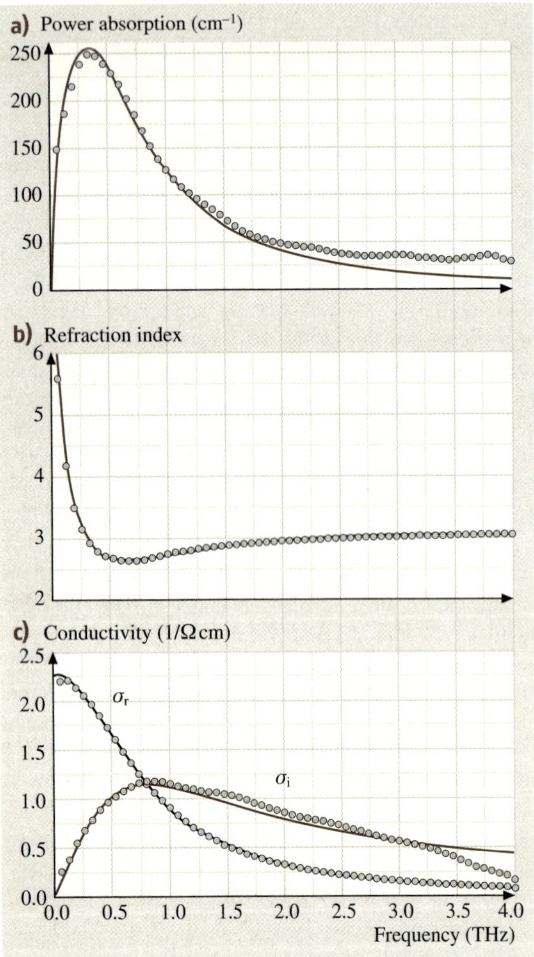

Fig. 17.51a–c THz-TDS measurements of GaN (*open circles*) compared with Drude model (*solid lines*). **(a)** Power absorption coefficient α; **(b)** real part of refractive index n_r; **(c)** Complex electric conductivity σ_r and σ_i. (Adapted from [17.146])

The GaN sample was an unintentionally doped, n-type, 5 mm × 5 mm × 180 μm freestanding crystalline plate. Originally grown on *c*-plane sapphire by the hydride-vapor-phase epitaxy technique (HVPE), the GaN layer was removed from the sapphire substrate by laser liftoff [17.147, 148]. The *c*-axis is perpendicular to the main planes of the sample. The sample was attached to a thin brass plate and centered over a 4 mm diameter hole in the plate, which defines the optical aperture. Another identical clear hole is used to take reference signal designated as the input pulse. For this measurement the standard THz-TDS system was realigned into a $4f$ confocal geometry that enabled excellent beam coupling between the THz transmitter and receiver. In order to compress the THz beam to a diameter comparable to the size of the GaN sample while preserving the low-frequency components, a pair of 25 mm focal length, high-resistivity, silicon lenses, separated by the confocal distance of 50 mm, were placed on the axis of the THz beam between the two paraboloidal mirrors. As a result, midway between the two lenses, a frequency independent 2.8 mm diameter THz beam waist was obtained, on which the 4 mm sample clear aperture was centered.

The resulting measured power absorption coefficient of the sample is shown in Fig. 17.51a, and the measured index of refraction in Fig. 17.51b, as the open circles. The real and imaginary conductivity shown in Fig. 17.51c were obtained from the experimentally determined α and n curves shown in Figs. 17.51a and 17.51b.

The measured absorption, index of refraction and complex conductivity were then theoretically fit using the simple Drude model [17.97, 98]. Three parameters were used to obtain a good fit to the experimental data: $\omega_p/2\pi = 1.82$ THz, $\Gamma/2\pi = 0.81$ THz and the real part of dielectric constant $\mathrm{Re}(\varepsilon_d) = 9.4$. Given the reduced mass $m^* = 0.22\,m_0$ for electrons in GaN [17.149], these parameters correspond to a number density of $N = 0.91 \times 10^{16}/\mathrm{cm}^3$ and a mobility $\mu = 1570\,\mathrm{cm}^2/\mathrm{V\,s}$. In Fig. 17.51a, the material absorption of intrinsic GaN, is responsible for the difference between the Drude fitting curve and the measurement of power absorption at frequencies higher than 2 THz.

THz-TDS Characterization of Normal and High-T_c Superconducting Films

An important subject area is the complex conductivity of normal metal, and normal and high-T_c superconduct ing films. Here, the large dynamic range of the THz systems is the key to transmission measurements with extremely large attenuations from low frequencies up to approximately 6 THz. Consequently, thicker films can be measured and thereby results are obtained which are closer to the desired bulk properties of these materials. With THz-TDS it becomes possible to measure, by non-contact means, the complex conductivities of normal metal and superconducting films. It is to be noted that the THz beams are easily focused to millimeter and even submillimeter spot sizes and thereby can be passed through the windows of optical dewars capable of cooling to cryogenic temperatures.

Direct measurements of the energy gap of normal superconductors have been performed on thin films of niobium [17.150] and lead [17.151]. In the early investigations of high-T_c films, the complex conductivity [17.152] and the real and imaginary parts of the surface impedance [17.153] were measured in both the normal and superconducting states. In another work [17.154], where the real and imaginary parts of the sub-bandgap conductivity σ of $YBa_2Cu_3O_7$ films were directly measured in the frequency range from 0.45 to 2.4 THz (15–80 cm^{-1}), a peak was observed in the real part σ_1 similar to the coherence peak expected in an s-wave BCS superconductor. More recent measurements have extended both the accuracy and frequency range of these measurements [17.155–165].

THz-TDS Characterization of High-T_c Substrates, Nonlinear Crystals and Polymers

The THz transparency of the underlying dielectric substrate is essential to realize the goal of a high-speed coplanar transmission line capable of transmitting a subpicosecond electrical pulse a distance on the order of 10 mm. These considerations become especially important with respect to the realization of a high-bandwidth coplanar transmission line fabricated out of the new high-T_c superconducting materials. The THz-TDS characterization at both room temperature and 85 K of three dielectric substrates used with high-T_c coplanar transmission lines, namely, magnesium oxide, yttria-stabilized zirconia (YSZ), and lanthanum aluminate provided limits on the realizable bandwidths obtainable [17.68]. More recent measurements have characterized another high-T_c substrate [17.166].

Although ZnTe has been the primary EO detection crystal, and even though detailed simulations have been compared with experimental results, the frequency dependent absorption and dispersion of ZnTe had not been measured. Recently, via THz-TDS, extensive measurements of the index of refraction and absorption of ZnTe from low frequencies to 4.5 THz were performed [17.131]. In contrast to what has been previously assumed, it was found that the absorption below 4.5 THz is not dominated by the strong TO-phonon line at 5.3 THz, but instead the measured absorption shows several new phonon lines. In fact, below 3.5 THz no significant absorption from the low frequency wing of the TO-phonon line was observed. However, as previously expected, the behavior of the index of refraction is dominated by the TO-phonon resonance. Another group has reported the THz-TDS characterization of the important electrooptic crystals: $LiNbO_3$, $LiTaO_3$, ZnTe and CdTe [17.132]. This same group has more recently reported the THz-TDS characterization of the nonlinear organic crystal DAST [17.167].

THz-TDS Characterization of Molecular Vapors

Early studies of coherent transients excited by THz pulses involved the use of THz-TDS techniques with N_2O vapor [17.168, 169] and methyl chloride vapor [17.170]. Using a subpicosecond pulse of THz radiation to excite the vapor, the subsequent emission from the vapor of coherent THz pulse trains extending as long as 1 ns was observed. THz-TDS, rotational line-shape investigations have been performed to gain insight into the actual collision process [17.171–173], by measuring the far-wing absorption line profile of ensembles of collision broadened ground state rotational lines of methyl halide vapors out to more than 200 linewidths from resonance, corresponding to frequency offsets as much as 5× the resonant frequency. These observations extend from the regime of the van Vleck–Weisskopf theory to the regime of the Lorentz theory. These measurements have experimentally established the new molecular response theory, which explicitly includes the molecular orientation time during a collision [17.173].

More recent work has studied collision induced tunneling in the methyl halides [17.172] and ammonia [17.174]. In other work the use of THz-TDS for gas sensing has been presented and discussed [17.175, 176]. A gas filter correlation instrument for the THz monitoring has been developed [17.177], and an all electronic THz absorption spectrometer has been demonstrated [17.178].

17.2.5 THz-TDS of Liquids

The efficacy of THz-TDS measurements on liquids has been demonstrated in a series of measurements of the absorption and index of refraction of acetone [17.179], acetonitrile [17.179, 180], liquid ammonia [17.181], benzene [17.182, 183], carbon tetrachloride [17.182, 184], cyclohexane [17.182], ethanol [17.181], methanol [17.179, 181], 1-propanol [17.181], water [17.179, 181, 185–188], and heavy water [17.188]. In addition, nonaqueous mixtures of dipolar liquids have been characterized [17.189], as well as solvated HCl [17.190], and mixtures of water with acetone, acetonitrile, and methanol [17.179]. The obtained spectra typically covered the region from 0.1 to 2 THz, where the dielectric response of the nonpolar liquids is dominated by collision-induced dipole moments. The complex dielectric behavior of liquids in the THz region is different for polar and nonpolar molecules. The absorption in po-

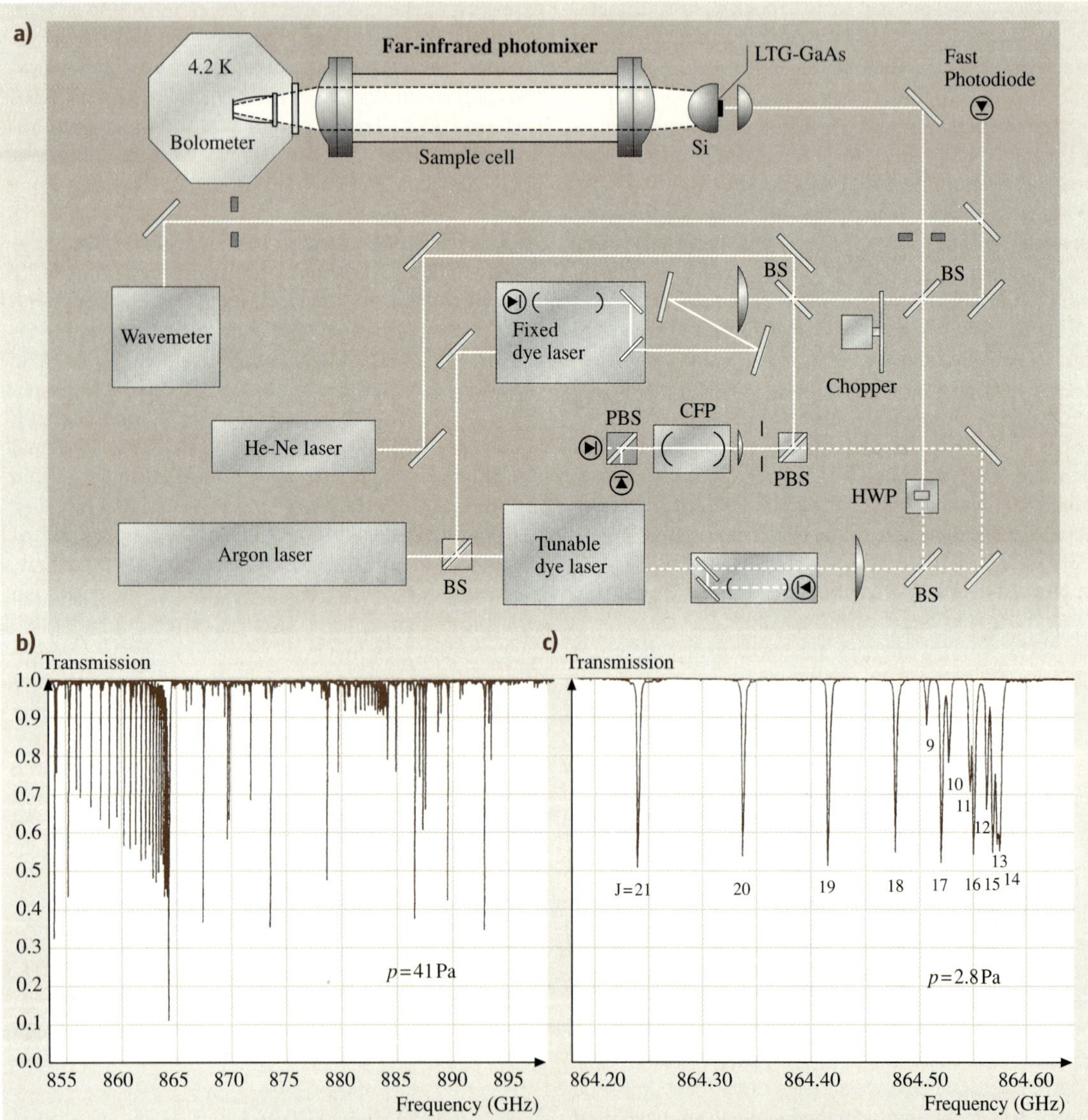

Fig. 17.52 (a) Schematic diagram of the photomixer spectrometer (BS: beam splitter, PBS: polarizer beam splitter, CFP: confocal Fabry–Perot, HWP: half-wave plate). **(b)** Survey scan of SO_2 near the $^rQ_8$ subbranch. **(c)** High-resolution scan of $^rQ_8$ subbranch head. (Adapted from [17.191])

lar molecules is due to the electronic interaction of the incoming field with the molecular permanent dipole moments. The dipoles are aligned by the incoming THz field, while at the same time this alignment is opposed by collision processes. The collision induced relaxation of the polarization has a time constant in the picosecond range, ideal for the THz probing pulses. The much smaller absorption of nonpolar molecules originates in the transient dipole moments induced by collisions in the liquid. The complex dielectric constant can be measured by THz-TDS for both types of liquid samples and useful information thereby obtained.

17.2.6 cw THz Photomixing Spectroscopy

A tunable cw far-infrared (THz) difference-frequency spectrometer has been developed and demonstrated by mixing two single-mode visible dye lasers separated by THz frequencies in an ultra-high-speed photoconductor [17.191]. The photomixer device was fabricated with submicron interdigital electrodes driving a broadband spiral antenna on epitaxial low-temperature-grown GaAs [17.41–43]. The coherent far-infrared radiation was coupled out the backside of the GaAs substrate using an aplanatic Si lens and was detected by a Si-composite bolometer at $T = 4.2$ K. Usable S/N ratios were obtained for radiation from 0.1 to 1.0 THz, using two 584 nm dye lasers. The instrumental linewidth was about 2 MHz, limited by the frequency jitter of the dye lasers. Baseline normalization was necessary to compensate for the laser amplitude fluctuations and for the strong fringes due to standing waves between the source and detector and within the detector dewar. The schematic of the entire system is shown in Fig. 17.52a. The generated THz radiation traverses a 50 cm sample vapor cell using Teflon lenses as windows, from which it is detected by a Si-composite bolometer operated at $T = 4.2$ K. All of the obtained spectra were normalized against laser amplitude fluctuations by dividing the bolometer signal from the THz radiation by the product of the incident laser powers.

The instrument was used to obtain SO_2 self-broadening coefficients for pure rotational transitions in several Q branches. Figure 17.52b shows a 45 GHz scan of the SO_2($P = 41$ Pa) ^rQ_s subbranch region near 865 GHZ normalized to an empty cell trace, each recorded in 5 min with 6000 sample points. Since each point represented 7.5 MHz, it was necessary to pressure broaden the lines slightly to increase the instrumental response because of the 40 ms lock-in time constant. The Doppler width for SO_2 at this frequency is only 1.3 MHz FWHM. Along the zero transmission baseline, the detector noise level in this trace when the lasers are blocked is also shown. The ^rQ_s subbranch head is shown expanded in Fig. 17.52c at lower pressure and with a slower scan rate. A turn around is observed at $J = 14$, whose spacing from the $J = 13$ line of approximately 2.5 MHz is clearly resolved with the instrumental linewidth of 2 MHz.

References

17.1 D. H. Auston: Ultrafast optoelectronics. In: *Ultrashort Laser Pulses Generation and Applications*, Topics Appl. Phys., Vol. 60, ed. by W. Kaiser (Springer, Berlin, Heidelberg 1988) pp. 183–233

17.2 J. A. Valdmanis, G. A. Mourou, C. W. Gabel: Subpicosecond electrical sampling, IEEE J. Quantum Elect. **19**, 664–667 (1983)

17.3 J. A. Valdmanis, G. A. Mourou: Subpicosecond electrooptic sampling: Principles and applications, IEEE J. Quantum Elect. **22**, 69–78 (1986)

17.4 D. E. Cooper: Picosecond optoelectronic measurement of microstrip dispersion, Appl. Phys. Lett. **47**, 33–35 (1985)

17.5 M. B. Ketchen, D. Grischkowsky, T. C. Chen, C.-C. Chi, I. N. Duling III, N. J. Halas, J.-M. Halbout, J. A. Kash, G. P. Li: Generation of sub-picosecond electrical pulses on coplanar transmission lines, Appl. Phys. Lett. **48**, 751–753 (1986)

17.6 D. Grischkowsky, M. B. Ketchen, C.-C. Chi, I. N. Duling III, N. J. Halas, J.-M. Halbout, P. G. May: Capacitance free generation and detection of sub-picosecond electrical pulses on coplanar transmission lines, IEEE J. Quantum Elect. **24**, 221–225 (1988)

17.7 J. M. Wiesenfeld: Electro-optic sampling of high-speed devices and integrated circuits, IBM J. Res. Dev. **34**, 141–161 (1990)

17.8 G. A. Mourou, K. E. Meyer: Subpicosecond electro-optic sampling using coplanar striptransmission lines, Appl. Phys. Lett. **45**, 492–494 (1984)

17.9 J. A. Valdmanis: 1 THz bandwidth prober for high-speed devices and integrated circuits, Electron. Lett. **23**, 1308–1310 (1987)

17.10 D. Krokel, D. Grischkowsky, M. B. Ketchen: Subpicosecond electrical pulse generation using photoconductive switches with long carrier lifetimes, Appl. Phys. Lett. **54**, 1046–1047 (1989)

17.11 U. D. Keil, D. R. Dykaar: Electro-optic sampling and carrier dynamics at zero propagation distance, Appl. Phys. Lett. **61**, 1504–1506 (1992)

17.12 J. L. Freeman, S. K. Diamond, H. Fong, D. M. Bloom: Electro-optic sampling of planar digital GaAs integrated circuits, Appl. Phys. Lett. **47**, 1083–1084 (1985)

17.13 J. M. Wiesenfeld, R. S. Tucker, A. Antreasyan, C. A. Burrus, A. J. Taylor, V. D. Mattera, P. A. Garbinski: Electro-optic sampling measurements of high-speed InP integrated circuits, Appl. Phys. Lett. **50**, 1310–1312 (1987)

17.14 M. Matloubian, H. Fetterman, M. Kim, A. Oki, J. Camou, S. Moss, D. Smith: Picosecond optoelectronic measurement of S-parameters and optical response of an AlGaAs/GaAs HBT, IEEE Trans. Microw. Theory **38**, 683–686 (1990)

17.15 M.Y. Frankel, J. F. Whitaker, G. Mourou: Optoelectronic transient characterization of ultrafast devices, IEEE J. Quantum Elect. **28**, 2313–2324 (1992)

17.16 M.Y. Frankel, J.F. Whitaker, G.A. Mourou, J. A. Valdmanis: Ultrahigh-bandwidth vector network analyzer based on external electrooptic sampling, Solid State Electron. **35**, 325–332 (1993)

17.17 K. Ogawa, J. Allam, N. de B. Baynes, J. R. A. Cleaver, T. Mishima, I. Ohbu: Ultrafast characterization of an in-plane gate transistor integrated with photoconductive switches, Appl. Phys. Lett. **66**, 1228–1230 (1995)

17.18 A. Zeng, M. K. Jackson, M. Van Hove, W. De Raedt: On-wafer characterization of In0.52Al0.48As/In0.53Ga0.47As modulation-doped field-effect transistor with 4.2 ps switching time and 3.2 ps delay, Appl. Phys. Lett. **67**, 262–263 (1995)

17.19 T. Pfeifer, H.-M. Heiliger, T. Loeffler, C. Ohlhoff, C. Meyer, G. Luepke, H. G. Roskos, H. Kurz: Optoelectronic on-chip characterization of ultrafast electric devices: Measurement, techniques and applications, IEEE J. Sel. Top. Quant. **2**, 586–604 (1996)

17.20 D. H. Auston, K. P. Cheung, P. R. Smith: Picosecond photoconducting Hertzian dipoles, Appl. Phys. Lett. **45**, 284–286 (1984)

17.21 Ch. Fattinger, D. Grischkowsky: Point source terahertz optics, Appl. Phys. Lett. **53**, 1480–1482 (1988)

17.22 Ch. Fattinger, D. Grischkowsky: Terahertz beams, Appl. Phys. Lett. **54**, 490–492 (1988)

17.23 G. Mourou, C.V. Stancampiano, A. Antonetti, A. Orszag: Picosecond microwave pulses generated with a subpicosecond laser-driven semiconductor switch, Appl. Phys. Lett. **39**, 295–296 (1981)

17.24 R. Heidemann, Th. Pfeiffer, D. Jager: Optoelectronically pulsed slot-line antennas, Electron. Lett. **19**, 316–317 (1983)

17.25 A. P. DeFonzo, M. Jarwala, C. R. Lutz: Transient response of planar integrated optoelectronic antennas, Appl. Phys. Lett. **50**, 1155–1157 (1987)

17.26 A. P. DeFonzo, M. Jarwala, C. R. Lutz: Optoelectronic transmission and reception of ultrashort electrical pulses, Appl. Phys. Lett. **51**, 212–214 (1987)

17.27 Y. Pastol, G. Arjavalingam, J.-M. Halbout, G.V. Kopcsay: Characterisation of an optoelectronically pulsed broadband microwave antenna, Electron. Lett. **24**, 1318–1319 (1988)

17.28 P. R. Smith, D. H. Auston, M. C. Nuss: Subpicosecond photoconducting dipole antennas, IEEE J. Quantum Elect. **24**, 255–260 (1988)

17.29 M. van Exter, Ch. Fattinger, D. Grischkowsky: High brightness teraHz beams characterized with an ultrafast detector, Appl. Phys. Lett. **55**, 337–339 (1989)

17.30 Y. Pastol, G. Arjavalingam, J.-M. Halbout: Characterization of an optoelectronically pulsed equiangular spiral antenna, Electron. Lett. **26**, 133–134 (1990)

17.31 M. van Exter, D. Grischkowsky: Characterization of an optoelectronic teraHz beam system, IEEE Trans. Microw. Theory **38**, 1684–1691 (1990)

17.32 D. R. Dykaar, B. I. Greene, J. F. Federici, A. F. J. Levi, L. N. Pfeiffer, R. F. Kopf: Log-periodic antennas for pulsed terahertz radiation, Appl. Phys. Lett. **59**, 262–264 (1991)

17.33 K. H. Yang, P. L. Richards, Y. R. Shen: Generation of far-infrared radiation by picosecond light pulses in $LiNbO_3$, Appl. Phys. Lett. **19**, 320–323 (1971)

17.34 D. H. Auston: Subpicosecond electro-optic shockwaves, Appl. Phys. Lett. **43**, 713–715 (1983)

17.35 B. B. Hu, X.-C. Zhang, D. H. Auston: Free-space radiation from electro-optic crystals, Appl. Phys. Lett. **56**, 506–508 (1990)

17.36 X.-C. Zhang, B. B. Hu, J. T. Darrow, D. H. Auston: Generation of femtosecond electromagnetic pulses from semiconductor surfaces, Appl. Phys. Lett. **56**, 1011–1013 (1990)

17.37 X.-C. Zhang, J.T. Darrow, B.B. Hu, D.H. Auston, M.T. Schmidt, P. Tham, E.S. Yang: Optically induced electromagnetic radiation from semiconductor surfaces, Appl. Phys. Lett. **5**, 2228–2230 (1990)

17.38 L. Xu, X.-C. Zhang, D. H. Auston: Terahertz radiation from large aperture Si p-i-n diodes, Appl. Phys. Lett. **59**, 3357–3359 (1991)

17.39 G.J. Simonis, K.G. Purchase: Optical generation, distribution, and control of microwaves using laser heterodyne, IEEE Trans. Microw. Theory **38**, 667–669 (1990)

17.40 D.V. Plant, D.C. Scott, D.C. Ni, H.R. Fetterman: Generation of millimeter-wave radiation by optical mixing in FET's integrated with printed circuit antennas, IEEE Microw. Guided Wave Lett. **1**, 132–134 (1990)

17.41 E. R. Brown, K. A. McIntosh, F. W. Smith, M. J. Manfra, C. L. Dennis: Measurements of optical-heterodyne conversion in low-temperature-grown GaAs, Appl. Phys. Lett. **62**, 1206–1208 (1993)

17.42 E. R. Brown, F. W. Smith, K. A. McIntosh: Coherent millimeter-wave generation by heterodyne conversion in low-temperature-grown GaAs photoconductors, J. Appl. Phys. **73**, 1480–1484 (1993)

17.43 E. R. Brown, K. A. McIntosh, K. B. Nichols, C. L. Dennis: Photomixing up to 3.8 THz in low-temperature-grown GaAs, Appl. Phys. Lett. **66**, 285–287 (1995)

17.44 N. Katzenellenbogen, D. Grischkowsky: Efficient generation of 380 fs pulses of THz radiation by ultrafast laser pulse excitation of a biased metal-semiconductor interface, Appl. Phys. Lett. **58**, 222–224 (1991)

17.45 B. B. Hu, M. C. Nuss: Imaging with terahertz waves, Opt. Lett. **20**, 1716–1719 (1995)

17.46 D. M. Mittleman, R. H. Jacobsen, M. C. Nuss: T-ray imaging, IEEE J. Sel. Top. Quant. **2**, 679–692 (1996)

17.47 D. Mittleman (Ed.): Terahertz imaging. In: *Sensing with Terahertz Radiation* (Springer, Berlin, Heidelberg 2002) pp. 117–154

17.48 R. A. Cheville, D. Grischkowsky: Time domain terahertz impulse ranging studies, Appl. Phys. Lett. **67**, 1960–1962 (1995)

17.49 R. A. Cheville, R. W. McGowan, D. Grischkowsky: Late time target response measured with THz impulse ranging, IEEE Trans. Antenn. Propag. **45**, 1518–1524 (1997)

17.50 M. van Exter, Ch. Fattinger, D. Grischkowsky: Terahertz time-domain spectroscopy of water vapor, Opt. Lett. **14**, 1128–1130 (1989)

17.51 D. Grischkowsky, S. Keiding, M. van Exter, Ch. Fattinger: Far-infrared time-domain spectroscopy with terahertz beams of dielectrics and semiconductors, J. Opt. Soc. Am. B **7**, 2006–2015 (1990) This paper contains a discussion and an extensive reference list describing the development of time-domain-spectroscopy.

17.52 M. C. Nuss, J. Orenstein: Terahertz time-domain spectroscopy. In: *Millimeter and Submillimeter Wave Spectroscopy of Solids*, Topics Appl. Phys., Vol. 74, ed. by G. Gruener (Springer, Berlin, Heidelberg 1998) pp. 7–50

17.53 C. Johnson, F. J. Low, A. W. Davidson: Germanium and germanium-diamond bolometers operated at 4.2 K, 2.0 K, 1.2 K, 0.3 K, and 0.1 K, Opt. Eng. **19**, 255–258 (1980)

17.54 R. A. Cheville, D. Grischkowsky: Far infrared, THz time domain spectroscopy of flames, Opt. Lett. **20**, 1646–1648 (1995)

17.55 R. A. Cheville, D. Grischkowsky: Observation of pure rotational absorption spectra in the ν_2 band of hot H_2O in flames, Opt. Lett. **23**, 531–533 (1998)

17.56 R. A. Cheville, D. Grischkowsky: Foreign and self broadened rotational linewidths of high temperature water vapor, J. Opt. Soc. Am. B **16**, 317–322 (1999)

17.57 F. E. Doany, D. Grischkowsky, C.-C. Chi: Carrier lifetime vs ion-implantation dose in silicon on sapphire, Appl. Phys. Lett. **50**, 460–462 (1987)

17.58 J. W. Rodwell, D. M. Bloom, B. A. Auld: Non-linear transmission line for picosecond pulse compression and broadband pulse modulation, Electron. Lett. **23**, 109–110 (1987)

17.59 C. J. Madden, R. A. Marsland, M. J. W. Rodwell, D. M. Bloom, Y. C. Pao: Hyperabrupt-doped GaAs nonlinear transmission line for picosecond shock-wave generation, Appl. Phys. Lett. **54**, 1019–1021 (1989)

17.60 R. Landauer: Parametric amplification along nonlinear transmission lines, J. Appl. Phys. **31**, 479–484 (1960)

17.61 R. V. Khokhlov: Radiotekhnika i elektronica **6** (1961) 17

17.62 P. Jager, F.-J. Tegude: Non-linear wave propagation along periodic loaded transmission line, Appl. Phys. **15**, 393–397 (1978)

17.63 D. W. Van Der Weide, J. S. Bostak, B. A. Auld, D. M. Bloom: All-electronic generation of 880 fs, 3.5 V shockwaves and their application to a 3 THz free-space signal generation system, Appl. Phys. Lett. **62**, 22–24 (1993)

17.64 E. Collin: *Field Theory of Guided Waves* (McGraw-Hill, New York, Toronto, London 1960)

17.65 K. C. Gupta, R. Garg, I. J. Bahl: *Microstrip Lines and Slotlines* (Artech House, Norwood 1996)

17.66 D. Grischkowsky: Optoelectronic characterization of transmission lines and waveguides by THz time-domain spectroscopy, IEEE J. Sel. Top. Quant. **6**, 1122–1135 (2000) invited paper in the EOS millennium issue

17.67 N. J. Halas, I. N. Duling III, M. B. Ketchen, D. Grischkowsky: Measured dispersion and absorption of a 5 μm coplanar transmission line, Conference on lasers and electro-optics, San Francisco 1986 (Optical Society of America, Washington, D.C. 1986) 328–329

17.68 D. Grischkowsky, S. Keiding: TeraHz time-domain spectroscopy of high Tc substrates, Appl. Phys. Lett. **57**, 1055–1057 (1990)

17.69 M. C. Nuss, P. M. Mankiewich, R. E. Howard, B. L. Straughn, T. E. Harvey, C. D. Brandle, G. W. Berkstresser, K. W. Goossen, P. R. Smith: Propagation of terahertz bandwidth electrical pulses on $Yba_2Cu_3O_{7-\delta}$ transmission lines on lanthanum aluminate, Appl. Phys. Lett. **54**, 2265–2267 (1989)

17.70 C. J. Osbahr, B. H. Larsen, T. Holst, Y. Shen, S. R. Keiding: 2 THz bandwidth electrical pulses on Au and $YBa_2Cu_3O_x$ transmission lines, Appl. Phys. Lett. **74**, 1892–1894 (1999)

17.71 D. Grischkowsky, I. N. Duling III, J. C. Chen, C.-C. Chi: Electromagnetic shock waves from transmission lines, Phys. Rev. Lett. **59**, 1663–1666 (1987)

17.72 J. V. Jelley: *Cerenkov Radiation and its Applications* (Pergamon, New York 1958)

17.73 D. B. Rutledge, D. P. Neikirk, D. P. Kasilingham: Integrated Circuit Antennas. In: *Infrared and Millimeter Waves*, Part II, Vol. 10, ed. by K. J. Button (Academic, New York 1983)

17.74 R. W. McGowan, D. Grischkowsky, J. A. Misewich: Demonstrated low radiative loss of a quadrupole ultrashort electrical pulse propagated on a three strip coplanar transmission line, Appl. Phys. Lett. **71**, 2842–2844 (1997)

17.75 H. Roskos, M. C. Nuss, K. W. Goossen, D. W. Kisker, A. E. White, K. T. Short, D. C. Jacobson, J. M. Poate: Propagation of picosecond electrical pulses on a silicon-based microstrip line with buried cobalt silicide ground plane, Appl. Phys. Lett. **58**, 2604–2606 (1991)

17.76 J.-M. Heiliger, M. Nagel, H. G. Roskos, H. Kurz, F. Schnieder, W. Heinrich, R. Hey, K. Ploog: Low-dispersion thin-film microstrip lines with cyclotene (benzocyclobutene) as dielectric medium, Appl. Phys. Lett. **70**, 2233–2235 (1997)

17.77 D. R. Dykaar, A. F. J. Levi, M. Anzlowar: Ultrafast coplanar air-transmission lines, Appl. Phys. Lett. **57**, 1123–1125 (1990)

17.78 W. H. Knox, J. E. Henry, K. W. Goosen, K. D. Li, B. Tell, D. A. B. Miller, D. S. Chemla, A. C. Gossard, J. English, S. Schmitt-Rink: Femtosecond excitonic optoelectronics, IEEE J. Quantum Elect. **25**, 2586–2595 (1989)

17.79 M. Y. Frankel, R. H. Voelker, J. N. Hilfiker: Coplanar transmission lines on thin substrates for high-speed, low-loss propagation, IEEE Trans. Microw. Theory **42**, 396–402 (1994)

17.80 H.-J. Cheng, J. F. Whitaker, T. M. Weller, L. P. B. Katehi: Terahertz-bandwidth characteristics of coplanar transmission lines on low permittivity substrates, IEEE Trans. Microw. Theory **42**, 2399–2406 (1994)

17.81 W. J. Gallagher, C.-C. Chi, I. N. Duling III, D. Grischkowsky, N. J. Halas, M. B. Ketchen, A. W. Kleinsasser: Subpicosecond optoelectronic study of resistive and superconductive transmission lines, Appl. Phys. Lett. **50**, 350–352 (1987)

17.82 R. Sprik, I. N. Duling III, C.-C. Chi, D. Grischkowsky: Far infrared spectroscopy with subpicosecond electrical pulses on transmission lines, Appl. Phys. Lett. **51**, 548–550 (1987)

17.83 A. J. Sievers III, M. Tinkham: Far infrared spectra of holmium, samarium, and gadolinium iron garnets, J. Appl. Phys. **34**, 1235–1236 (1963)

17.84 L. Richards: Far-infrared magnetic resonance in CoF_2, NiF_2, Kn_iF_3, and YbIG, J. Appl. Phys. **34**, 1237–1238 (1963)

17.85 R. W. McGowan, G. Gallot, D. Grischkowsky: Propagation of ultrawideband short pulses of terahertz radiation through submillimeter-diameter circular waveguides, Opt. Lett. **24**, 1431–1433 (1999)

17.86 G. Gallot, S. Jamison, R. W. McGowan, D. Grischkowsky: THz waveguides, J. Opt. Soc. Am. B **17**, 851–863 (2000)

17.87 S. P. Jamison, R. W. McGowan, D. Grischkowsky: Single-mode waveguide propagation and reshaping of sub-ps terahertz pulses in sapphire fibers, Appl. Phys. Lett. **76**, 1987–1989 (2000)

17.88 R. Mendis, D. Grischkowsky: Plastic ribbon THz waveguides, J. Appl. Phys. **88**, 4449–4451 (2000)

17.89 R. Mendis, D. Grischkowsky: Undistorted guided wave propagation of subpsec THz pulses, Opt. Lett. **26**, 846–848 (2001)

17.90 R. Mendis, D. Grischkowsky: THz interconnect with low loss and low group velocity dispersion, IEEE Microw. Wireless Comp. Lett. **11**, 444–446 (2001)

17.91 S. Ramo, J. R. Whinnery, T. van Duzer: *Fields and Waves in Communication Electronics*, 3rd edn. (Wiley, New York 1994)

17.92 J. C. Slater: Microwave electronics, Rev. Mod. Phys. **18**, 441–512 (1946)

17.93 N. Marcuvitz: *Waveguide Handbook* (Peter Peregrinus, London 1986)

17.94 S. E. Ralph, D. Grischkowsky: Trap-enhanced electric fields in semi-insulators: The role of electrical and optical carrier injection, Appl. Phys. Lett. **59**, 1972–1974 (1991)

17.95 Ch. Fattinger, D. Grischkowsky: Beams of terahertz electromagnetic pulses, Vol. 4, 1989, OSA Proc. on Picosecond electronics and optoelectronics, Salt Lake City, Utah 1989, ed. by T. C. L. G. Sollner, D. M. Bloom (Optical Society of America, Washington, D.C. 1989) 225–231

17.96 D. Grischkowsky, N. Katzenellenbogen: Femtosecond pulses of terahertz radiation: Physics and applications, OSA Proc. 9, Salt Lake City, Utah 1991, ed. by G. Sollner, J. Shah (Optical Society of America, Washington, D.C. 1991) 9–14

17.97 M. van Exter, D. Grischkowsky: Optical and electronic properties of doped silicon from 0.1 to 2 THz, Appl. Phys. Lett. **56**, 1694–1696 (1990)

17.98 M. van Exter, D. Grischkowsky: Carrier dynamics of electrons and holes in moderately-doped silicon, Phys. Rev. B **41**, 12140–12149 (1990)

17.99 N. Katzenellenbogen, H. Chan, D. Grischowsky: New Performance limits of an Ultrafast THz Photoconductive Receiver, Vol. 14 OSA Proc. on Ultrafast Electronics and Optoelectronics, San Francisco (Optical Society of America, Washington, D.C., 1993) pp. 123–125

17.100 S. E. Ralph, D. Grischkowsky: THz spectroscopy and source characterization by optoelectronic interferometry, Appl. Phys. Lett. **60**, 1070–1072 (1992)

17.101 I. Brener, D. Dykaar, A. Frommer, L. N. Pfeiffer, J. Lopata, J. Wynn, K. West, M. C. Nuss: THz emission

from electric field singularities in biased semiconductors, Opt. Lett. **21**, 1924–1926 (1996)
17.102 F. W. Smith, A. R. Calawa, C.-L. Chen, M. J. Manfra, L. J. Mahoney: New MBE buffer used to eliminate backgating in GaAs MESFET'S, IEEE Electron. Device Lett. **9**, 77–80 (1988)
17.103 F. W. Smith, H. Q. Le, V. Diadiuk, M. A. Hollis, A. R. Calawa, S. Gupta, M. Frankel, D. R. Dykaar, G. A. Mourou, T. Y. Hsiang: Picosecond GaAs-based photoconductive optoelectronic detectors, Appl. Phys. Lett. **54**, 890–892 (1989)
17.104 A. C. Warren, J. M. Woodall, J. L. Freeouf, D. Grischkowsky, M. R. Melloch, N. Otsuka: Arsenic precipitates and the semi-insulating properties of GaAs buffer layers grown by low- temperature molecular beam epitaxy, Appl. Phys. Lett. **57**, 1331–1333 (1990)
17.105 A. C. Warren, N. Katzenellenbogen, D. Grischkowsky, J. M. Woodall, M. R. Melloch, N. Otsuka: Subpicosecond, freely propagating electromagnetic pulse generation and detection using GaAs : As epilayers, Appl. Phys. Lett. **58**, 1512–1514 (1991)
17.106 S. Gupta, M. Y. Frankel, J. A. Valdmanis, J. F. Whitaker, G. A. Mourou, F. W. Smith, A. R. Calawa: Subpicosecond carrier lifetime in GaAS grown by molecular beam epitaxy at low temperatures, Appl. Phys. Lett. **59**, 3276–3278 (1991)
17.107 E. S. Harmon, M. R. Melloch, J. M. Woodall, D. D. Nolte, N. Otsuka, C. L. Chang: Carrier lifetime versus anneal in low temperature growth GaAs, Appl. Phys. Lett. **63**, 2248–2250 (1993)
17.108 B. B. Hu, J. T. Darrow, X. C. Zhang, D. H. Auston: Optically steerable photoconducting antennas, Appl. Phys. Lett. **56**, 886–888 (1990)
17.109 D. You, R. R. Jones, P. H. Bucksbaum: Generation of high power sub-single-cycle 500-fs electromagnetic pulses, Opt. Lett. **18**, 290–292 (1993)
17.110 R. R. Jones, D. You, P. H. Bucksbaum: Ionization of Rydberg atoms by subpicosecond half-cycle electromagnetic pulses, Phys. Rev. Lett. **70**, 1236–1239 (1993)
17.111 X. C. Zhang, Y. Jin, T. D. Hewitt, T. Sangsiri, L. E. Kingsley, M. Weiner: Magnetic switching of THz beams, Appl. Phys. Lett. **62**, 2003–2005 (1993)
17.112 X. C. Zhang, Y. Jin, L. E. Kingsley, M. Weiner: Influence of electric and magnetic fields on THz radiation, Appl. Phys. Lett. **62**, 2477–2479 (1993)
17.113 N. Sarukara, H. Ohtake, S. Izumida: High average-power THz radiation from femtosecond laser-irradiated InAs in a magnetic field and its elliptical polarization characteristics, J. Appl. Phys. **84**, 645–656 (1998)
17.114 H. Ohtuke, S. Ono, M. Sakai, Z. Liu, T. Tsukamoto, N. Sarukura: Saturation of THz-radiation power from femtosecond-laser-irradiated InAs in a high magnetic field, Appl. Phys. Lett. **76**, 1398–1400 (2000)
17.115 R. Mclaughlin, A. Corchia, M. B. Johnston, Q. Chen, C. M. Ciesla, D. D. Arnone, G. A. C. Jones, E. H. Linfield, A. G. Davies, M. Pepper: Enhanced coherent THz emission from InAs in the presence of a magnetic field, Appl. Phys. Lett. **76**, 2038–2040 (2000)
17.116 C. Weiss, R. Wallenstein, R. Beigang: Magnetic-field enhanced generation of THz radiation in semiconductor surfaces, Appl. Phys. Lett. **77**, 4160–4162 (2000)
17.117 C. Weiss. Shan, R. Wallenstein, R. Beigang, T. F. Heinz: Origin of magnetic field enhancement in the generation of THz radiation from semiconductor surfaces, Opt. Lett. **26**, 849–851 (2001)
17.118 M. B. Johnston, D. M. Whittaker, A. Corchia, A. G. Davies, E. H. Linfield: Theory of magnetic-field enhancement of surface-field THz emission, J. Appl. Phys. **91**, 2104–2106 (2002)
17.119 D. H. Auston, K. P. Cheung, J. A. Valdmanis, D. A. Kleinman: Cherenkov radiation from femtosecond optical pulses in electro-optic media, Phys. Rev. Lett. **53**, 1555–1558 (1984)
17.120 D. A. Kleinman, D. H. Auston: Theory of electrooptic shock radiation in nonlinear optical media, IEEE J. Quantum Elect. **20**, 964–970 (1984)
17.121 K. P. Cheung, D. H. Auston: A novel technique for measuring far-infrared absorption and dispersion, Infrared Phys. **26**, 23–27 (1986)
17.122 D. H. Auston, K. P. Cheun: Coherent time-domain far-infrared spectroscopy, J. Opt. Soc. Am. B **2**, 606–612 (1985)
17.123 X. C. Zhang, Y. Jin, K. Yang, L. J. Schowalter: Resonant nonlinear susceptibility near the GaAs band gap, Phys. Rev. Lett. **69**, 2303–2306 (1992)
17.124 A. Bonvalet, M. Joffre, J. L. Martin, A. Migus: Generation of ultrabroadband femtosecond pulses in the mid-infrared by optical rectification of 15 fs light pulses at 100 MHz repetition rate, Appl. Phys. Lett. **67**, 2907–2909 (1995)
17.125 A. Nahata, A. S. Weling, T. F. Heinz: A wideband coherent terahertz spectroscopy system using optical rectification and electro-optic sampling, Appl. Phys. Lett. **69**, 2321–2323 (1996)
17.126 Q. Wu, X.-C. Zhang: 7 terahertz broadband GaP electro-optic sensor, Appl. Phys. Lett. **70**, 1784–1786 (1997)
17.127 Q. Wu, X.-C. Zhang: Free-space electro-optic sampling of terahertz beams, Appl. Phys. Lett. **67**, 3523–3525 (1995)
17.128 A. Nahata, D. H. Auston, T. Heinz: Coherent detection of freely propagating terahertz radiation by electro-optic sampling, Appl. Phys. Lett. **68**, 150–152 (1996)
17.129 Q. Wu, X.-C. Zhang: Free-space electro-optics sampling of mid-infrared pulses, Appl. Phys. Lett. **71**, 1285–1286 (1997)
17.130 A. Leitenstorfer, S. Hunsche, J. Shah, M. C. Nuss, W. H. Knox: Detectors and sources for ultrabroad-

band electro-optic sampling: Experiment and theory, Appl. Phys. Lett. **74**, 1516–1518 (1999)
17.131 G. Gallot, R.W. McGowan, Jiangquan Zhang, T.-I. Jeon, D. Grischkowsky: Measurements of the THz absorption and dispersion of ZnTe and their relevance to the electro-optic detection of THz radiation, Appl. Phys. Lett. **74**, 3450–3452 (1999)
17.132 M. Schall, H. Helm, S. R. Keiding: Far infrared properties of electro-optic crystals measured by THz time-domain spectroscopy, Int. J. Infrared Milli. Waves **20**, 595–604 (1999)
17.133 G. Gallot, D. Grischkowsky: Electro-optic detection of THz radiation, J. Opt. Soc. Am. B **16**, 1204–1212 (1999)
17.134 B. I. Greene, J. F. Federici, D. R. Dykaar, R. R. Jones, P. H. Bucksbaum: Interferometric characterization of 160 fs far-infrared light pulses, Appl. Phys. Lett. **59**, 893–895 (1991)
17.135 R. Kersting, K. Unterrainer, G. Strasser, H. F. Kauffmann, E. Gornik: Few-cycle THz emission from cold plasma oscillations, Phys. Rev. Lett. **79**, 3038–3041 (1997)
17.136 D. E. Gray (Ed.): *American Institute of Physics Handbook*, 3rd edn. (McGraw-Hill, New York 1982)
17.137 E. E. Russell, E. E. Bell: Optical constants of sapphire in the far-infrared, J. Opt. Soc. Am. **57**, 543–544 (1967)
17.138 E. V. Loewenstein, D. R. Smith, R. L. Morgan: Optical constants of far-infrared materials, Appl. Opt. **12**, 398–406 (1973)
17.139 T.-I. Jeon, D. Grischkowsky: Nature of conduction in doped silicon, Phys. Rev. Lett. **78**, 1106–1109 (1997)
17.140 R.A. Smith: *Semiconductors* (Cambridge Univ. Press, London 1959)
17.141 J. M. Ziman: *Electrons and Phonons* (Oxford Univ. Press, Oxford 1960)
17.142 D. Long: Scattering of conduction electrons by lattice vibrations in silicon, Phys. Rev. **120**, 2024–2032 (1960)
17.143 M. Vindevoghel, J. Vindevoghel, Y. Leroy: Mean momentum relaxation time and scattering processes from absorption spectra in millimetric and far infrared ranges – Case of n-Si, Infrared Phys. **15**, 161–173 (1975)
17.144 D.W. Davidson, R.H. Cole: Dielectric relaxation in glycerol, propylene glycol, and n-propanol, J. Chem. Phys. **19**, 1484–1490 (1951)
17.145 T.-I. Jeon, D. Grischkowsky: Observation of a Cole–Davidson type complex conductivity in the limit of very low carrier densities in doped silicon, Appl. Phys. Lett. **72**, 2259–2261 (1998)
17.146 W. Zhang, A. K. Azad, D. Grischkowsky: Terahertz studies of carrier dynamics and dielectric response of n-type, freestanding epitaxial GaN, Appl. Phys. Lett. **82**, 2841–2843 (2003)
17.147 D.C. Reynolds, D.C. Look, B. Jogai, A.W. Saxler, S.S. Park, J.Y. Hahn: Identification of the Γ_5 and Γ_6 free excitons in GaN, Appl. Phys. Lett. **77**, 1879 (2000)
17.148 M. K. Kelly, R. P. Vaudo, V. M. Phanse, L. Görgens, O. Ambacher, M. Stutzmann: Large free-standing GaN substrates by hydride vapor phase epitaxy, Jpn. J. Appl. Phys. Part 2 **38**, L217–L219 (1999)
17.149 W.J. Moore, J.A. Freitas Jr., S.K. Lee, S.S. Park, J.Y. Han: Magneto optical studies of free-standing hydride-vapor phase epitaxial GaN, Phys. Rev. B **65**, 081201-1–081201-4 (2002)
17.150 M.C. Nuss, K.W. Goossen, J.P. Gordon, P. M. Mankiewich, M. L. O'Malley, M. Bhushan: Terahertz time-domain measurement of the conductivity and superconducting band gap in niobium, J. Appl. Phys. **70**, 2238–2241 (1991)
17.151 J. F. Federici, B. I. Greene, P. N. Saeta, D. R. Dykaar, F. Sharifi, R.C. Dynes: Cooper pair breaking in lead measured by pulsed terahertz spectroscopy, IEEE Trans. Appl. Supercond. **3**, 1461–1464 (1993)
17.152 J.M. Chwalek, J.F. Whitaker, G.A. Mourou: Submillimetre wave response of superconducting $YBa_2Cu_3O_{7-x}$, using coherent time-domain spectroscopy, Electron. Lett. **27**, 447–448 (1991)
17.153 M.C. Nuss, K.W. Goossen, P.M. Mankiewich, M. L. O'Malley: Terahertz surface impedance of thin $YBa_2Cu_3O_7$ superconducting films, Appl. Phys. Lett. **58**, 2561–2563 (1991)
17.154 M.C. Nuss, P.M. Mankiewich, M.L. O'Malley, E.H. Westerwick, P.B. Littlewood: Dynamic conductivity and coherence peak in $YBa_2Cu_3O_7$ superconductors, Phys. Rev. Lett. **66**, 3305–3308 (1991)
17.155 Ch. Jaekel, Ch. Waschke, H.G. Roskos, H. Kurz, W. Prusseit, H. Kinder: Surface resistance and penetration depth on $YBa_2Cu_3O_{7-\delta}$ thin films on silicon at ultrahigh frequencies, Appl. Phys. Lett. **64**, 3326–3328 (1994)
17.156 S. Spielman, B. Parks, J. Orenstein, D.T. Nemeth, F. Ludwig, J. Clarke, P. Merchant, D.J. Lew: Observation of the quasiparticle hall effect in superconducting $YBa_2Cu_3O_{7-\delta}$, Phys. Rev. Lett. **73**, 1537–1540 (1994)
17.157 R. Buhleier, S. D. Brorson, I. E. Trofimov, J. O. White, H. U. Habermeier, J. Kuhl: Anomalous behavior of the complex conductivity of $Y_{1-x}Pr_xBa_2Cu_3O_7$ observed with THz spectroscopy, Phys. Rev. B **50**, 9672–9675 (1994)
17.158 Y. Liu, J.F. Whitaker, C. Uher, S.Y. Hou, J. M. Phillips: Pulsed terahertz-beam spectroscopy as a probe of the thermal and quantum response of $YBa_2Cu_3O_{7-\delta}$ superfluid, Appl. Phys. Lett. **67**, 3022–3024 (1995)
17.159 B. Parks, S. Spielman, J. Orenstein, D.T. Nemeth, F. Ludwig, J. Clarke, P. Merchant, D.J. Lew: Phase-sensitive measurements of vortex dynamics in the

terahertz domain, Phys. Rev. Lett. **74**, 3265–3268 (1995)
17.160 F. Gao, J.F. Whitaker, Y. Liu, C. Uher, C.E. Platt, M.V. Klein: Terahertz transmission of a $Ba_{1-x}K_xBiO_3$ film probed by coherent time-domain spectroscopy, Phys. Rev. B **52**, 3607–3613 (1995)
17.161 S.D. Brorson, R. Buhleier, L.E. Trofimov, J.O. White, Ch. Ludwig, F.E. Balakirev, H.-U. Habermeier, J. Kuhl: Electrodynamics of high-temperature superconductors investigated with coherent terahertz pulse spectroscopy, J. Opt. Soc. Am. B **13**, 1979–1993 (1996)
17.162 R.D. Averitt, G. Rodriguez, J.L.W. Siders, S.A. Trugman, A.J. Taylor: Artifacts in optical-pump THz probe measurements of $YBa_2Cu_3O_7$, J. Opt. Soc. Am. B **17**, 327–331 (2000)
17.163 I. Wilke, M. Khazan, C.T. Rieck, T. Kaiser, C. Jaekel, H. Kurz: Terahertz surface resistance of high temperature superconducting thin films, J. Appl. Phys. **87**, 2984–2988 (2000)
17.164 I. Wilke, M. Khazan, C.T. Rieck, C. Jaekel, H. Kurtz: Time domain Terahertz spectroscopy as a diagnostic tool for the electrodynamic properties of high temperature superconductors, Physica C **341-348**, 2271–2272 (2000)
17.165 V.K. Thorsmolle, R.D. Averitt, M.P. Maley, L.N. Bulaevskii, C. Helm, A.J. Taylor: C-axis Josephson plasma resonance observed in $Ti_2Ba_2CaCu_2O_8$ superconducting thin films by use of terahertz time-domain spectroscopy, Opt. Lett. **26**, 1292–1294 (2001)
17.166 T. Kiwa, M. Tonouchi: Time domain terahertz spectroscopy of (100) $(LaAlO_3)_{0.3}$–$(Sr_2AlTaO_6)_{0.7}$ substrate, Jpn. J. Appl. Phys. **40**, L38–L40 (2001)
17.167 M. Walther, K. Jensby, S.R. Keiding, H. Takahashi, H. Ito: Far infrared properties of DAST, Opt. Lett. **25**, 911–913 (2000)
17.168 H. Harde, S. Keiding, D. Grischkowsky: THz commensurate echoes: Periodic rephasing of molecular transitions in free-induction decay, Phys. Rev. Lett. **66**, 1834–1837 (1991)
17.169 H. Harde, D. Grischkowsky: Coherent transients excited by subpicosecond pulses of terahertz radiation, J. Opt. Soc. Am. B **8**, 1642–1651 (1991)
17.170 H. Harde, N. Katzenellenbogen, D. Grischkowsky: Terahertz coherent transients from methyl chloride vapor, J. Opt. Soc. Am. B **11**, 1018–1030 (1994)
17.171 H. Harde, N. Katzenellenbogen, D. Grischkowsky: Line-shape transition of collision broadened lines, Phys. Rev. Lett. **74**, 1307–1310 (1995)
17.172 H. Harde, R.A. Cheville, D. Grischkowsky: Collision-induced tunneling in methyl halides, J. Opt. Soc. Am. B **14**, 3282–3293 (1997)
17.173 H. Harde, R.A. Cheville, D. Grischkowsky: Terahertz studies of collision broadened rotational lines, J. Phys. Chem. A **101**, 3646–3660 (1997)
17.174 H. Harde, J. Zhao, M. Wolff, R.A. Cheville, D. Grischkowsky: Time domain spectroscopy on ammonia, J. Phys. Chem. A **105**, 6038–6047 (2001)
17.175 R.H. Jacobsen, D.M. Mittleman, M.C. Nuss: Chemical recognition of gases and gas mixtures with terahertz waves, Opt. Lett. **21**, 2011–2013 (1996)
17.176 D.M. Mittelman, R.H. Jacobsen, R. Neelamani, R.G. Baraniuk, M.C. Nuss: Gas sensing using terahertz time-domain spectroscopy, Appl. Phys. B **67**, 379–390 (1998)
17.177 G. Mouret, W. Chen, D. Boucher, R. Bocquet, P. Mounaix, D. Lippens: Gas filter correlation instrument for air monitoring at submillimeter wavelengths, Opt. Lett. **24**, 351–353 (1999)
17.178 D.W. Van Der Weide, J. Murakowski, F. Keilmann: Gas Absorption spectroscopy with electronic terahertz techniques, IEEE Trans. Microw. Theory **48**, 740–743 (2000)
17.179 D.S. Veneables, Ch.A. Schmuttenmaer: Spectroscopy and dynamics of water with acetone, acetonitrile and methanol, J. Chem. Phys. **113**, 11222–11236 (2000)
17.180 D.S. Venables, C.A. Schmuttenmaer: Far-infrared spectra and associated dynamics in acetonitrile-water mixtures measured with femtosecond THz pulse spectroscopy, J. Chem. Phys. **108**, 4935–4944 (1998)
17.181 J.T. Kindt, C.A. Schmuttenmaer: Far-infrared dielectric properties of polar liquids probed by femtosecond THz pulse spectroscopy, J. Phys. Chem. **100**, 10373–10379 (1996)
17.182 J.E. Pedersen, S.R. Keiding: THz time-domain spectroscopy of nonpolar liquids, IEEE J. Quantum Elect. **28**, 2518–2522 (1992)
17.183 S.R. Keiding: Dipole correlation functions in liquid benzenes measured with THz time domain spectroscopy, J. Phys. Chem. A **101**, 5250–5254 (1997)
17.184 B.N. Flanders, R.A. Cheville, D. Grischkowsky, N.F. Scherer: Pulsed terahertz transmission spectroscopy of liquid $CHCl_3$, CCl_4, and their mixtures, J. Phys. Chem. **100**, 11824–11835 (1996)
17.185 D. Kralj, L. Carin: Wideband dispersion measurements of water in reflection and transmission, IEEE Trans. Microw. Theory **42**, 553–557 (1994)
17.186 L. Thrane, R.H. Jacobsen, P.U. Jepsen, S.R. Keiding: THz reflection spectroscopy of liquid water, Chem. Phys. Lett. **240**, 330–333 (1995)
17.187 C. Ronne, L. Thrane, P.-O. Astrand, A. Wallqvist, K.V. Mikkelsen, S.R. Keiding: Investigation of the temperature dependence of dielectric relaxation in liquid water by THz reflection spectroscopy and molecular dynamics simulation, J. Chem. Phys. **107**, 5319–5331 (1997)
17.188 C. Ronne, P.-O. Astraand, S.R. Keiding: THz spectroscopy of liquid H_2O and D_2O, Phys. Rev. Lett. **82**, 2888–2891 (1999)
17.189 D.S. Venables, A. Chiu, Ch.A. Schmuttenmaer: Structure and dynamics of non aqueous mix-

croscopy. This scheme is recently used in full-field microscopy to obtain magnified high-resolution images of a specimen using Kirkpatrick–Baez mirrors or waveguides.

Currently, all X-ray optics are technology limited in their performance, but significant technological advances have been made, approaching physical limits. Recently, these limits have been addressed theoretically for several X-ray optics, such as waveguides, refractive lenses, and Fresnel zone plates. While beams in the range from 1–10 nm are conceivable with ideal optics, atomic resolution, e.g., at 1 Å seems to be out of reach.

18.1 Interaction of X-Rays with Matter

A hard X-ray photon interacts with an atom mainly through scattering at its electrons or by absorption. Scattering can be both elastic or inelastic. In the cases of inelastic scattering, also called Compton scattering, and photoabsorption the photon is lost for image formation, while elastic scattering is responsible for refractive and diffractive effects.

The refraction of hard X-rays in matter is typically expressed by the index of refraction given in the form

$$n = 1 - \delta + \mathrm{i}\beta\,, \tag{18.1}$$

where δ describes the deviation of the real part of the refractive index from unity and is referred to as the *index of refraction decrement*. As (18.1) suggests (with positive δ), the refractive index of hard X-rays in matter is smaller than unity, i. e., the vacuum is X-ray optically denser than matter. Figure 18.1b illustrates this effect as compared to the refraction of visible light in glass (Fig. 18.1a).
For a given atomic species, δ is given by [18.2]

$$\delta = \frac{N_\mathrm{a}}{2\pi} r_0 \lambda^2 \rho \frac{Z + f'(E)}{A}\,, \tag{18.2}$$

where N_a is Avogadro's constant, r_0 is the classical electron radius, λ and E are the wavelength and energy of the X-rays, respectively, ρ is the mass density of the material, $Z + f'(E)$ is the real part of its atomic form factor in the forward direction, and A is the material's atomic mass. Figure 18.2a shows δ/ρ for different materials as a function of X-ray energy. Away from absorption edges, $f'(E)$ is small and δ is proportional to $\lambda^2 \sim E^{-2}$ and ρ. Since Z/A does not vary much between most elements,

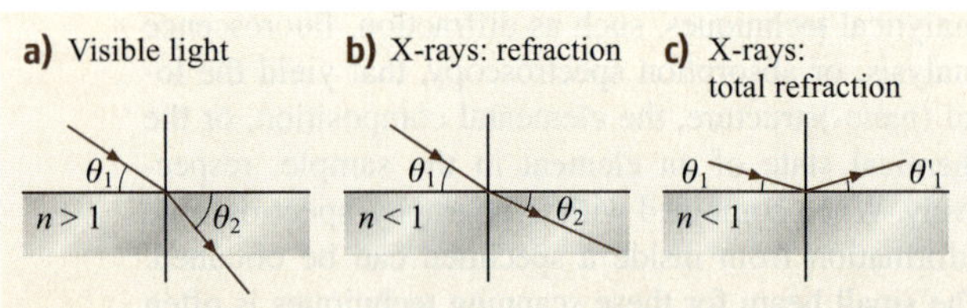

Fig. 18.1 **(a)** For visible light, the refractive index n in matter is larger than one. Therefore, light rays are refracted toward the surface normal when entering matter from the vacuum. **(b)** For hard X-rays, the refractive index n of matter is smaller than one. Therefore, X-rays impinging from vacuum onto a surface are refracted away from the surface normal. **(c)** If the angle of incidence θ_1 falls below the critical angle of total reflection, the X-rays do not propagate deeply into the material but are totally reflected at the surface

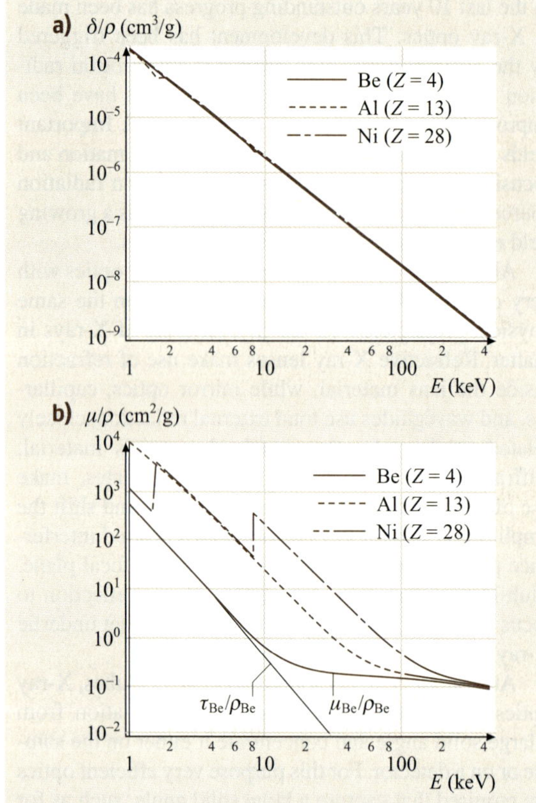

Fig. 18.2 **(a)** δ/ρ (the index of refraction decrement divided by the density of the material) for Be, Al, and Ni as a function of X-ray energy. **(b)** mass attenuation coefficient μ/ρ for the same elements. After [18.1]

δ/ρ varies very little as a function of the atomic species away from absorption edges. As compared to that of visible light in glass, the refraction of hard X-rays is several orders of magnitude weaker. While δ/ρ is of the order of $10^{-6}\,\mathrm{cm^3/g}$ for hard X-rays between 10 and 20 keV (Fig. 18.2a), that of visible light in glass is about $-\delta/\rho \approx 0.2\,\mathrm{cm^3/g}$.

Due to this extremely small refraction, the reflectivity at a surface or interface is extremely low. Therefore, there are no mirrors for hard X-rays reflecting at large angles. However, when an X-ray beam impinges on a plane interface between vacuum (or air) and a medium of index n under a sufficiently small angle θ_1, it is totally reflected (Fig. 18.1c). This total external reflection occurs for θ_1 smaller than the critical angle for total reflection θ_c, i. e.,

$$\theta_1 < \theta_c = \sqrt{2\delta}. \quad (18.3)$$

For hard X-rays, this angle lies below about 0.5 degrees for all materials. In other words, total reflection occurs only at grazing incidence.

The imaginary part β of the refractive index in (18.1) describes the attenuation of X-rays in matter and is related to the linear attenuation coefficient μ by

$$\beta = \frac{\mu\lambda}{4\pi}. \quad (18.4)$$

Lambert-Beer's law

$$I(z) = I_0 \cdot e^{-\mu z} \quad (18.5)$$

describes the transmitted intensity $I(z)$ through a homogeneous piece of material of thickness z, given an incident intensity I_0. The linear attenuation coefficient μ is the inverse of the characteristic length of this exponential decay.

β and μ include photoabsorption as well as the attenuation of the incident beam by elastic (Rayleigh) and inelastic (Compton) scattering. The dependence of μ/ρ on the X-ray energy is shown in Fig. 18.2b. At low energies, photoabsorption τ dominates μ/ρ. As opposed to refraction, photoabsorption is strongly dependent on the atomic number. Between absorption edges it scales approximately like $\tau \sim Z^3/E^3$. With increasing X-ray energy, Compton scattering contributes increasingly to μ/ρ. Compton scattering is only weakly dependent on the atomic species and limits the mass attenuation coefficient from below to about $\mu/\rho > 0.1\,\mathrm{cm^2/g}$ in the energy range from a few keV to several hundred keV. As a consequence, there is no material that is as transparent for hard X-rays as is glass for visible light. For comparison, the attenuation of visible light in glass can be as small as $\mu_{\text{silica}} \approx 10^{-7}\,\mathrm{cm^{-1}}$ for pure silica used for fiber optics [18.3].

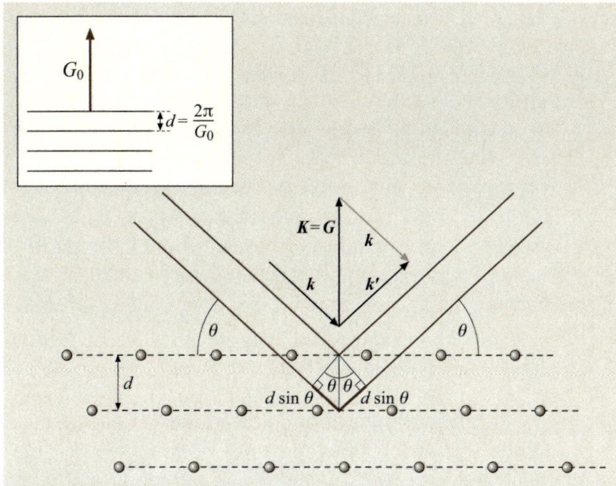

Fig. 18.3 Bragg scattering of hard X-rays from a crystalline material. The lattice planes have a spacing d and the X-rays impinge on the lattice planes under the angle θ

In the general case, elastic scattering contributes little to the attenuation of the transmitted beam. In some cases, however, when for example a Bragg reflection is excited in a crystalline material, elastic scattering can significantly attenuate the transmitted beam. In that case, the X-ray beam can be efficiently diffracted away from the forward and into another direction. Figure 18.3 illustrates this situation, in which the scattered amplitudes from a large number of atoms interfere constructively. This is the case when the path length difference $2d\sin\theta$ between amplitudes reflected from neighboring lattice planes is an integer multiple m of the wavelength λ of the X-rays, i. e.,

$$m\lambda = 2d\sin\theta. \quad (18.6)$$

While elastic scattering preserves the wave number ($k = 2\pi/\lambda$), the wave vector changes its direction

$$\boldsymbol{k}' = \boldsymbol{k} + \boldsymbol{G}, \quad (18.7)$$

where $\boldsymbol{G} = m\boldsymbol{G}_0$ is the reciprocal space vector associated with the lattice planes and the reflection order m (Fig. 18.3). For arbitrary alignment of a monocrystalline sample in a monochromatic beam, this virtually never occurs. However, some X-ray optics, such as crystal or multilayer optics, exploit this effect to monochromatize and focus the X-rays.

18.2 X-Ray Optical Components

A large number of X-ray optical components are described in literature. One reason for this is the difficulty in manipulating X-rays due to their weak interaction with matter. There have been many complementary approaches to this problem, but none has proven significantly superior to the others. Indeed, as technology advances, approaching the physical limits, the performances of different optics become very similar. In this chapter, the most common X-ray optics are described.

18.2.1 Refractive Optics

Rotationally Parabolic Refractive X-ray Lenses

While refractive optics are most common for visible light, the interplay between weak refraction and relatively strong absorption inside the lens material makes the design of refractive lenses for hard X-rays difficult (Sect. 18.1). For instance, at 45°, the angular deviation of the refracted beam is δ, which is typically of the order of 10^{-6} rad, compared to some 17 degrees for visible light at an air–glass interface. From the discovery of X-rays to the 1990s it was therefore believed that refractive lenses for hard X-rays could not be made [18.4–7]. In 1996 it was first demonstrated experimentally that refractive lenses can be fabricated and work despite this weak refraction and relatively strong absorption [18.8]. Since then, a variety of refractive X-ray lenses have been designed [18.9–18].

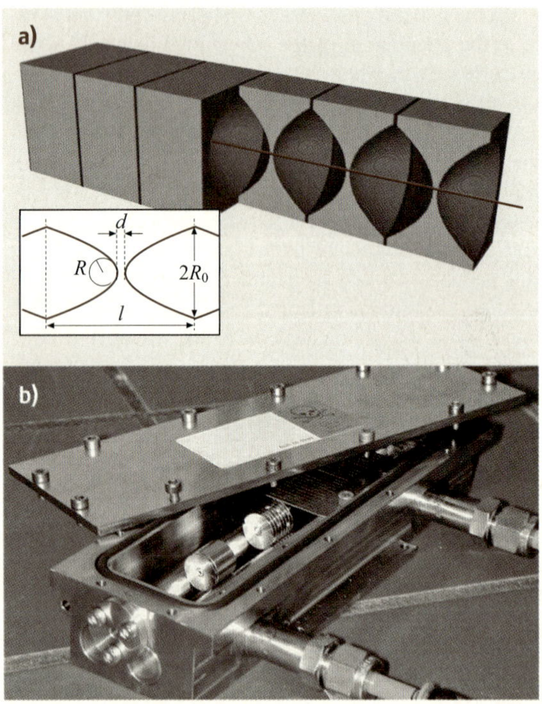

Fig. 18.4 **(a)** Parabolic refractive X-ray lenses. A large number of individual lenses are stacked behind each other to form a refractive X-ray lens. **(b)** Partially assembled parabolic refractive X-ray lens made of beryllium

As the refractive index of hard X-rays is smaller than one ($n < 1$) in any lens material, a focusing lens has to have a concave shape (Fig. 18.4a), and since the refraction of hard X-rays in matter is about six orders of magnitude weaker than that of visible light in glass, the effective lens curvature must be accordingly stronger. For example, a biconvex lens for visible light with a radius of curvature of $R = 1$ m for both surfaces has a focal length of about $f = R/2(n-1) \approx 1$ m. A single X-ray lens with the same focal distance would need to have a radius of curvature R in the micrometer range. It is difficult to fabricate such a lens, in particular with a sufficiently large aperture. This difficulty can be overcome by making the radius of curvature R of an individual lens larger and compensating the lack of refraction by stacking many of these lenses behind each other, as shown in Fig. 18.4a.

As the radius of curvature R has to be as small as possible to limit the number of single lenses, the spherical-lens approximation that is successfully applied in visible light optics does not apply to most X-ray lens designs. It would limit the aperture $2R_0$ of the lens to be much smaller than the radius of curvature R (Fig. 18.4a). To avoid spherical aberration and to allow for apertures that are as large as possible, the lens surfaces have to have a parabolic shape [18.9, 10, 15]. In the paraxial approximation this is the optimal aspherical lens shape. For parabolic lenses the aperture $2R_0$ can be chosen independently of the radius of curvature R Fig. 18.4a.

Attenuation inside the lens is significant for all lens materials (Sect. 18.1), as there is no material that is as transparent in the hard X-ray range as glass is for visible light. Besides a reduction of the transmitted radiation, the attenuation inside the lens material reduces the aperture of the lens, since the thicker material in the outer parts of the lens (Fig. 18.4a) absorbs the X-rays more strongly than the thin material on the optical axis. In order to optimize the transmission and aperture it is thus important to choose the lens material carefully. As μ/ρ increases strongly with increasing atomic number Z ($\mu/\rho \sim Z^3$), atomic species with low Z, such as Li,

Be, B, C, and compounds thereof, are good lens materials [18.8, 10]. In addition, the lenses should be made as thin as possible, reducing the distance d (Fig. 18.4a) between the apices of the parabolas to a minimum.

Ideal for imaging applications, in particular in X-ray microscopy, are rotationally parabolic lenses. At Aachen University, they are made by embossing the lens shape from both sides into the lens material, such as beryllium or aluminium. For the current lens design, the radius of curvature is $R = 200\,\mu\text{m}$ and the aperture $2R_0 \approx 1\,\text{mm}$. Typical focal distances lie in the range 0.3–10 m, depending on the application. Figure 18.4b shows a partly assembled beryllium lens. A single lens is held and centered by a hard metal ring of well-defined diameter. These coin-like single lenses are stacked along two parallel polished shafts to align their optical axes. In this design, the number of lenses can be adjusted starting from a single lens up to 300 lenses in one stack, allowing one to control the optical properties, e.g., focal length, within a wide range in a large energy interval (5 to > 100 keV). The lens is kept in an inert gas atmosphere to avoid corrosion of the beryllium in the intensive X-ray beam.

As the attenuation increases towards the outer parts of the lens, the rays passing the lens in these parts do not contribute as much to the image as those traveling through the lens close to the optical axis. Therefore, the effective aperture D_{eff} describing diffraction at the lens is smaller than the geometric aperture $2R_0$. It typically lies in the range of a few hundred micrometers. Attenuation inside the lens material also reduces the transmission T_p of the optic. Details about these quantities can be found in [18.10, 19].

These optics are ideal for imaging applications with hard X-rays above 5 keV, such as hard X-ray microscopy. They are free of spherical aberration. 100 nm spatial resolution has been achieved with beryllium lenses at 12 keV in an X-ray microscope [18.20]. Potentially, the diffraction limit of these optics can be pushed to below 50 nm. The lenses can be used to generate a small and intensive microbeam, by imaging the X-ray source onto the sample in a strongly reducing geometry. Besides for X-ray microscopy, these optics are commonly used for beam conditioning at third-generation synchrotron radiation sources, since they can withstand the high heat load of modern undulator sources and match their beam size very well. In addition, they find wide application in scanning microscopy, where they are used to generate intensive hard X-ray microbeams.

According to (18.2) the refractive index depends on the X-ray energy, leading to chromatic aberration. For the many experiments carried out with monochromatic synchrotron radiation, chromatic effects are irrelevant. There are, however, also many experiments which require a broad X-ray spectrum or scanning the energy while keeping the focus unchanged. If the energy range is sufficiently small, a slight defocusing of the lenses can be tolerated in some cases [18.21]. In others, one should resort to genuinely achromatic optics, such as total reflection mirrors (Sect. 18.2.2).

Nanofocusing Refractive X-Ray Lenses

To generate beams with a lateral extension well below one micrometer, the geometric image of the source must be made sufficiently small. This is possible by reducing the focal length f of a refractive lens, thus increasing the demagnification ratio. In addition, the diffraction limit decreases with decreasing focal length for refractive lenses. As it is difficult to reduce the focal length of the rotationally parabolic lenses to well below 0.3 m,

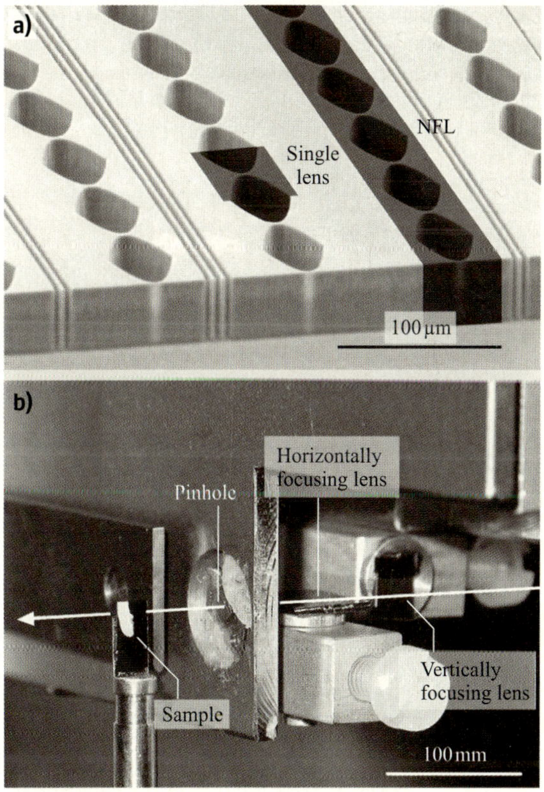

Fig. 18.5 (a) Scanning electron micrograph of nanofocusing lenses. A single lens and a nanofocusing lens are outlined by *dark shaded areas*. (b) Nanoprobe setup: the X-ray beam is focused onto the sample by two crossed nanofocusing lenses

a new type of parabolic lenses with extremely short focal distance was developed [18.22].

In order to achieve focal length in the centimeter range at hard-X-ray energies, the radii of curvature of the individual single lenses (cf. Fig. 18.5a) need to be in the range of a few micrometers. This can for example be achieved with nanofabrication techniques for planar lenses, such as those shown in Fig. 18.5a. They are made of silicon by electron-beam lithography and subsequent deep reactive-ion etching. These planar structures only focus in one dimension. Thus, in order to obtain a point focus, two such lenses with appropriate focal lengths need to be crossed. Figure 18.5b shows a nanoprobe setup based on two crossed nanofocusing lenses (NFLs). With this setup, hard-X-ray beams down to about $50\times 50\,\mathrm{nm}^2$ have been generated at the European Synchrotron Radiation Facility (ESRF) in Grenoble, France [18.23].

With optimized parameters, these optics are expected to generate beams with a lateral size between 10 and 20 nm. This is a physical limit for the given lens design. There are more-complicated lens designs that will in principle allow one to overcome this limit and focus hard X-rays down to below 5 nm [18.24].

18.2.2 Reflective Optics

Mirrors

The reflectivity of a surface or interface is negligible for large angles. Only in grazing incidence below the critical angle of total external reflection θ_c are high reflectivities possible (Sect. 18.1). This makes X-ray mirrors and capillaries very slender, i. e., much longer than wide. According to (18.2) and (18.3) θ_c depends mainly on the X-ray wavelength λ and on the density ρ of the mirror material. The higher the X-ray energy ($E = hc/\lambda$) the smaller the critical angle θ_c. In order to have sufficiently large reflection angles, materials with high density, such as palladium or platinum, are often used.

Two effects reduce the ability of a mirror to reflect X-rays without losses: absorption and surface roughness. X-rays penetrate into the mirror surface to a depth of a few nanometers. Here, they are partially absorbed, so that just below the critical angle the reflectivity is often far below 50%. For heavy elements and just above absorption edges this is particularly strong.

The second detrimental effect to the reflectivity is surface roughness. Due to the short wavelength of hard X-rays roughness brings the reflection amplitudes of a photon in the different areas inside each Fresnel zone out of phase, so that the total amplitude and hence the reflectivity are reduced. For high-quality mirror optics, the surface roughness must be in the 0.1 nm range. This effect and the high requirements on form fidelity make X-ray mirrors very expensive optical components.

For a given grazing incidence geometry with reflection angle θ_1, total-reflection mirrors reflect X-rays up to the energy where $\theta_1 = \theta_c$. They act like a low-pass filter for X-rays and are therefore often used to remove the high-energy content from the beam. In conjunction with a crystal monochromator, this allows one to strongly reduce higher harmonic radiation [$m \geq 2$ in (18.6)] and generate a clean monochromatic beam.

To focus or collimate X-rays, mirrors can be curved. To focus the X-rays, i. e., to image the source from point to point, the mirror surface needs have ellipsoidal shape (Fig. 18.6a). For collimation, i. e., generating a parallel X-ray beam, the mirror surface needs to be parabolic (Fig. 18.6b). To achieve focusing or parallelization in two dimensions with a single reflection, the mirror must have the shape of an ellipsoid or paraboloid of rotation, respectively. Due to the very small reflection angle, this rotational body is very elongated (needle-like) having extremely different curvatures in the sagittal and meridional direction. Since such a mirror is extremely difficult to fabricate with high accuracy, there are only a few realizations. These optics are also referred to as elliptical or parabolic monocapillaries.

To avoid strong sagittal curvature, two reflections from flat mirrors are often used to obtain two-

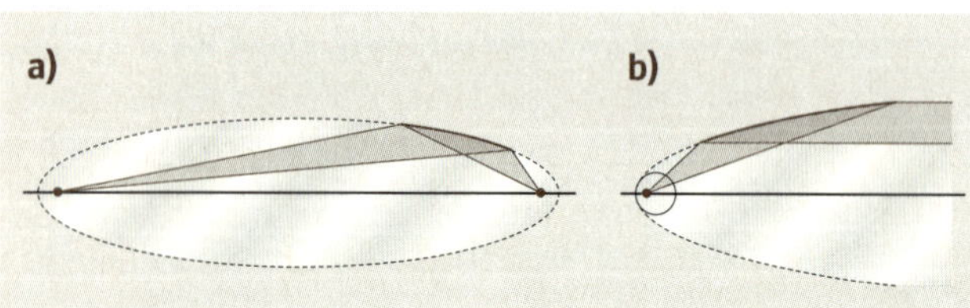

Fig. 18.6 (a) Elliptical mirror for focusing form point to point. (b) Parabolic mirror to collimate the beam from a point source

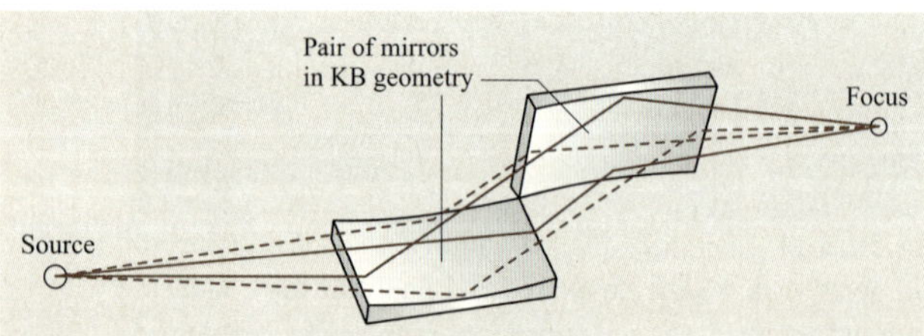

Fig. 18.7 Kirkpatrick–Baez focusing geometry: focusing from point to point using two crossed elliptical mirrors

dimensional focusing. Figure 18.7 illustrates this scheme, which was first described by *Kirkpatrick* and *Baez* in 1948 [18.25]. Today, this technique is widespread as a focusing optic for scanning microscopy at synchrotron radiation sources. While it is relatively difficult to align, it has the great advantage of being achromatic for X-rays below a critical energy given by the maximal angle of incidence. This makes these optics particularly useful for white-beam microdiffraction techniques [18.26] and for absorption spectroscopy studies.

Today, these mirrors can be fabricated with extremely small surface roughness and excellent form fidelity, generating hard X-ray beams with a lateral size as small as 30 nm [18.27]. Since the reflection angles are limited by θ_c, the numerical aperture and thus the diffraction-limited focus size is limited, too. It is expected that beam sizes down to below 20 nm can be reached with total reflection mirrors. One way to overcome this limit is to use multilayer mirrors (Sect. 18.2.3) or use more-complicated multiple-reflection schemes.

Waveguides

Another optical component which is based on total reflection and which has been developed recently are X-rays waveguides. They consist of a substrate made of a material with high density (e.g., nickel), a light-element (e.g., carbon) layer on top of it and a cap layer made of a dense material again (Fig. 18.8). The middle layer is the guiding layer, which is typically a few tens of nm thick. The X-rays are confined inside the waveguide by total external reflection at the side walls defined by the denser materials. The device works in analogy to a waveguide for microwaves.

There are two ways to couple X-rays into a waveguide. A well collimated plane wave can be coupled into the waveguide through the thin cap layer [18.28] or a focused beam can be directly coupled into the waveguide from the side [18.29], as shown in Fig. 18.8a and b, respectively. Only those modes that survive by constructive interference can propagate in the waveguide. Depending on the illumination, different modes can be systematically excited, corresponding to different standing waves in the transverse direction. The first waveguides were planar structures, confining the beam only in one direction. Today, using nanofabrication techniques, two-dimensionally confining wave guides can be made [18.29].

Waveguides are mostly used for two types of applications: to illuminate a sample confined inside the waveguide [18.30] or to generate a very fine beam that is

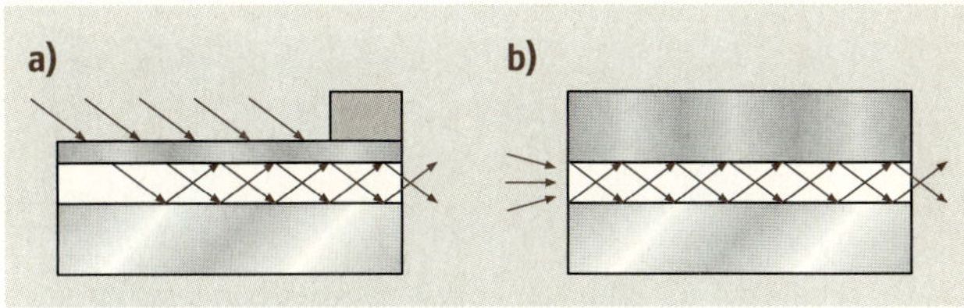

Fig. 18.8 X-ray waveguide. **(a)** The X-rays are coupled into the waveguide through the cap layer. **(b)** Focused X-rays are directly coupled into the waveguide

emerging from its exit. At the exit, the beam size is comparable to the dimensions of the waveguide. Recently, a beam with lateral dimensions of 25 nm × 45 nm was generated [18.29]. This beam can be used for projection microscopy and has a high degree of lateral coherence. If a single mode is excited inside the waveguide, the exiting beam is perfectly coherent. This can be exploited for coherent imaging experiments and holography. The smallest size that X-rays can be confined to by waveguides is also limited by total-reflection effects and is expected to lie slightly below 10 nm [18.31]. Until now, these components are in the stage of development and have not yet found widespread application.

18.2.3 Diffractive Optics

Multilayer Optics

Multilayer optics consist of a series of alternating layers of high- and low-density materials. Figure 18.9a shows a multilayer made of 75 double layers of Ni and B_4C. Figure 18.9b shows the operating principle of these optics: the X-rays impinging at the angle θ are reflected from the multilayer if the Bragg condition (18.6) for the multilayer is fulfilled, i. e., the path-length difference marked in dark brown in Fig. 18.9b is an integer multiple of the wavelength λ. Here, d is the period of the multilayer, which can be chosen freely and typically lies in the range from about 2 to 20 nm, and the resulting Bragg angles lie in the range of degrees for hard X-rays. The reflectivity can be close to one (70 to 90%) for hard X-rays. It depends on the number of layers and the difference in electron density between the two multilayer materials. As for total-reflection mirrors, the reflectivity is also deteriorated by attenuation inside the optic and by roughness of the interfaces.

Similar to crystal optics, multilayers can be used to monochromatize the X-ray beam. The monochromaticity and the angular acceptance depend on the number of periods that contribute to the reflection. This number is either limited by the total number of periods N in the

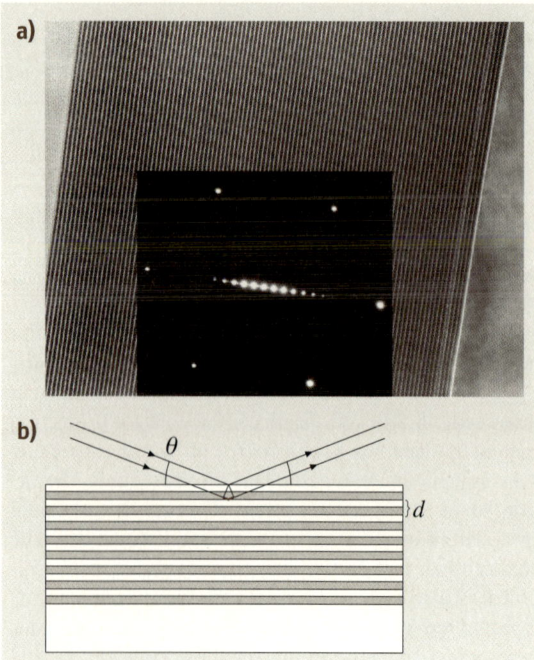

Fig. 18.9 **(a)** High-resolution transmission electron micrograph of a multilayer optic consisting of $N = 100$ double layers of Ni and C. The lattice spacing is $d = 3.2$ nm. The inset shows the electron diffraction pattern from the structure. The clear diffraction peaks along the line through the center of the diffraction pattern and normal to the multilayer stack illustrate the high degree of periodicity. **(b)** Schematic drawing of a multilayer optic. The X-rays are reflected from the multilayer if the Bragg condition (18.6) is fulfilled, i. e. the *dark brown segment* of the lower optical path is an integer multiple of the X-ray wavelength λ

stack or by the extinction length inside the multilayer. A detailed description of the reflectivity of multilayer systems can be found in [18.32].

To use these optics to focus or collimate the X-ray beam in the geometries shown in Fig. 18.10, the spacing d of the multilayer needs to be adapted to the reflection angle that varies along the optic. Such laterally graded multilayers were first introduced by *Schuster* and *Göbel* to capture a large solid angle from an X-ray tube and collimate it for diffraction experiments [18.33, 34]. A variety of optics exist today that either focus or collimate the beam from an X-ray tube. Since the reflection angles are larger than in the total-reflection case (Sect. 18.2.2) there are sagittally curved elliptical and parabolic mirrors in addition to double-bounce systems of Kirkpatrick–Baez (KB) type.

For scanning microscopy, curved multilayer systems in the KB geometry are used with great success, generating very intensive nanobeams down to well below 100 nm [18.35, 36]. These optics have the advantage that they moderately monochromatize the beam at the same time. In this way, an additional crystal monochromator can be avoided. This can be advantageous for experiments that do not require a high degree of monochromatization, since the larger energy band pass leads to a higher flux. Due to the relatively large reflection angles, the numerical aperture of these optics can be relatively large. Ultimately, diffraction-limited foci well below 10 nm are conceivable for this type of optic.

The flexibility in the choice of the lattice parameters and its variation over the mirror open possibilities that cannot be realized with crystal optics. Besides lateral grading of the multilayers, depth grading can also be useful, e.g., to tailor the angular acceptance and energy bandwidth of the device. As there are some applications of these optics in conjunction with laboratory X-ray tubes, there is a larger market for these optics, and they are available from a number of laboratories and commercial companies. Multilayers are also of great importance in the extreme ultraviolet (EUV) range, where they are used in lithographic steppers.

Crystal Optics

The main field of application of crystal optics is monochromatization of X-rays using Bragg reflection (Sect. 18.1). Depending on the crystal, crystal cut, reflection, and the arrangement of different crystals, relative energy bandwidths $\Delta E/E$ from about 10^{-3} to 10^{-7} can be reached, finding wide application at nearly every beam line at synchrotron radiation sources. High energy resolutions are needed in particular in inelastic X-ray scattering and nuclear resonance scattering.

Besides pure monochromatization, crystal optics can also be used to focus X-rays from point to point. As one cannot adjust the d-spacing for varying reflection angles as for multilayer optics (Fig. 18.10) a geometry in which the reflection angle is constant and equal to the Bragg angle θ_B for every point on the optic has to be used. Figure 18.11 shows a bent crystal in the Johann geometry. In the Johann geometry, the Bragg planes are parallel to the crystal's surface. The meridional bending radius R_h is twice the radius of the so-called Rowland circle (Fig. 18.11). To focus also out of plane, the crystal has to be sagittally bent as well. In that direction the bending radius is $R_v = R_h \sin^2 \theta_B$ (Fig. 18.11). Both the source and the focus lie on the Rowland circle, forming an isosceles triangle with the crystal.

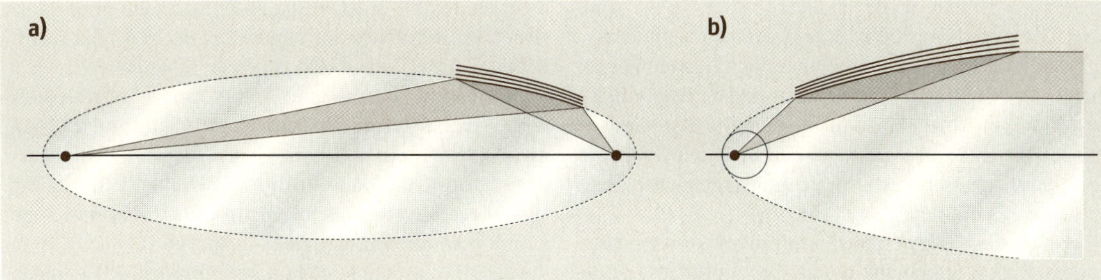

Fig. 18.10 (**a**) Elliptical graded multilayer for focusing. (**b**) Parabolic graded multilayer for focusing and collimation. The period d changes as a function of position on the optic, locally adjusting the Bragg angle to the angle of reflection

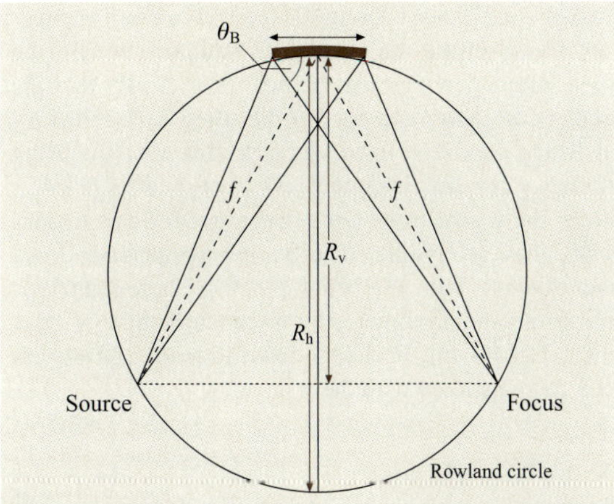

Fig. 18.11 Focusing with a bent crystal optic in the Johann geometry. The source and focus position lie on the Rowland circle

For larger and larger crystals, the surface deviates increasingly from the Rowland circle. To fulfill the Bragg condition strictly in that case, the crystal needs to be cut to follow the Rowland circle when bent. This is called the Johannson geometry, which is very difficult to realize and is therefore not used very often.

The bending of the crystal optic has to be very accurate in order to stay within the width of the Bragg reflection over the whole surface. In addition, the crystal must not be damaged. As this bending procedure is difficult, these optics are not very widespread.

Fresnel Zone Plates

The most important diffractive optical components are Fresnel zone plates. They have found very widespread applications, particularly at energies below a few keV, like for example in the water window. As a genuine imaging optic, they can be used as objective lens in

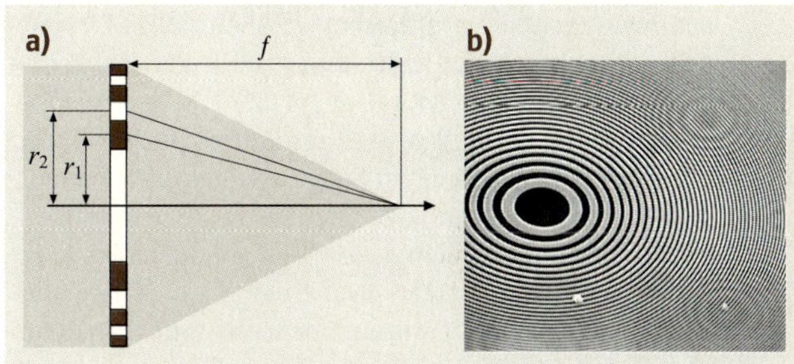

Fig. 18.12 Fresnel zone plate. (**a**) Schematic diagram, (**b**) Scanning electron microscope (SEM) image

full-field microscopy and to generate a small focus for scanning microscopy.

Fresnel zone plates consist in their simplest form of a set of alternating transparent and opaque rings (Fig. 18.12), so-called zones, with an optical path difference of $\lambda/2$ between neighboring zones. The amplitudes of all transparent zones superpose constructively when the radii of all zones are given by

$$r_n = \sqrt{n\lambda f + \frac{n^2\lambda^2}{4}}, \quad n = 0, 1, 2, \ldots, \tag{18.8}$$

where the even zones are transparent and odd zones are opaque. Here, f is the focal length of the optic for a given wavelength λ. In first diffraction order, the zone plate is most efficient, and about 10% of the radiation falling onto its aperture can be focused into this diffraction order. The spatial resolution depends on the numerical aperture $\mathrm{NA} = r_n/f$ and is in this case given by

$$d_t = 1.22\frac{\lambda}{2\mathrm{NA}} = 1.22\Delta r_n, \tag{18.9}$$

which is approximately equal to the width of the outermost zone Δr_n. The factor 1.22 comes from the Rayleigh criterion for the spatial resolution of a circular optic [18.37]. This means that the spatial resolution corresponds to that of the smallest features in the zone plate. For increasing X-ray energy, it becomes increasingly difficult to make a zone plate with opaque zones. Therefore, this type of optic is mostly used in the soft-X-ray regime, in particular in the so-called water window between the absorption edges of carbon and oxygen. As the opaque zones become more transparent, the zone plate loses its efficiency.

For harder X-rays one therefore pursues another scheme. Rather than blanking out the amplitudes of odd zones, one can shift them by π, adding to the constructive interference in the focus. Such phase zone plates are much more efficient than zone plates with opaque zones and have an optimal efficiency of about 40%. Despite this new scheme, the zones still need to have a thickness in the micrometer range for hard X-rays around 10 keV. To reach high spatial resolution efficiently the fabrication of very thin but high zone plate structures is required. While there have been tremendous improvements over the last view years, it is still a challenge to make these structures. However, there are several institutions and commercial sources for these optics. Another phase-shifting optic is a Bragg–Fresnel optic, in which the Fresnel zones are etched into a single crystal surface [18.38]. Tuned to Bragg condition, the X-rays are diffracted from different heights inside the zone plate structure. Refraction introduces the desired phase shift between the even and odd zones.

Currently, in the soft-X-ray range, zone plates with spatial resolution down to 15 nm have been made [18.39]. In the hard X-ray range, the high aspect ratio of the outermost zones currently limits the spatial resolution to about 50 nm [18.40, 41]. For such a high-resolution zone plate, the aperture is typically in the range of a couple of hundred microns and is made of several hundred zones.

As the aspect ratio increases with decreasing outermost zone width, the thin object approximation for the Fresnel zone plates becomes less appropriate. Propagation effects inside the zone plate structure need to be taken into account. It was shown theoretically that the zones of a zone plate need to be tilted to fulfill a local Bragg condition in order not to lose their focusing efficiency for smaller outermost zone widths [18.42]. While these structures are even more difficult to fabricate, they are predicted to be able to generate focal sizes down to 1 nm and below [18.43]. These structures are currently approximated by so-called multilayer Laue lenses [18.44–46]. Focusing down to below 20 nm has been demonstrated with these devices.

References

18.1 C. T. Chantler: Theoretical form factor, attenuation and scattering tabulation for $Z = 1–92$ from $E = 1–10$ eV to $E = 0.4–10.0$ MeV, Journal of Physical and Chemical Reference Data **24**, 71 (1995)

18.2 R. W. James: *The Optical Principles of the Diffraction of X-rays* (Cornell Univ. Press, Ithaca 1967)

18.3 E. Hecht: *Optics* (Addison–Wesley, Reading 1987)

18.4 W. C. Röntgen: *Über eine neue Art von Strahlen*, Sitzungsberichte der phys.-med. Ges. (Physikalisch-medizinische Gesellschaft, Würzburg 1895)

18.5 S. Suehiro, H. Miyaji, H. Hayashi: Refractive lens for X-ray focus, Nature **352**, 385–386 (1991)

18.6 A. Michette: No X-ray lens, Nature **353**, 510 (1991)

18.7 B. Yang: Fresnel and refractive lenses for X-rays, Nucl. Instrum. Methods A **328**, 578–587 (1993)

18.8 A. Snigirev, V. Kohn, I. Snigireva, B. Lengeler: A compound refractive lens for focusing high energy X-rays, Nature **384**, 49 (1996)

18.9 B. Lengeler, C. G. Schroer, M. Richwin, M. Drakopoulos, A. Snigirev, I. Snigireva: A microscope for hard X-rays based on parabolic compound refractive lenses, Appl. Phys. Lett. **74**(26), 3924–3926 (1999)

18.10 B. Lengeler, C. Schroer, B. Benner, M. Richwin, A. Snigirev, I. Snigireva, M. Drakopoulos: Imaging by parabolic refractive lenses in the hard X-ray range, J. Synchrotron Rad. **6**, 1153–1167 (1999)

18.11 V. Aristov, M. Grigoriev, S. Kuznetsov, L. Shabelnikov, V. Yunkin, T. Weitkamp, C. Rau, I. Snigireva, A. Snigirev, M. Hoffmann, E. Voges: X-ray refractive planar lens with minimized absorption, Appl. Phys. Lett. **77**(24), 4058–4060 (2000)

18.12 M. A. Piestrup, J. T. Cremer, H. R. Beguiristain, C. K. Gary, R. H. Pantell: Two-dimensional X-ray focusing from compound lenses made of plastic, Rev. Sci. Instrum. **71**(12), 4375–4379 (2000)

18.13 R. N. Cahn, M. Danielsson, M. Lundqvist, D. R. Nygren: Focusing hard X-rays with old LP's, Nature **404**, 951 (2000)

18.14 E. M. Dufresne, D. A. Arms, R. Clarke, N. R. Pereira, S. B. Dierker, D. Foster: Lithium metal for X-ray refractive optics, Appl. Phys. Lett. **79**(25), 4085–4087 (2001)

18.15 B. Lengeler, C. G. Schroer, B. Benner, A. Gerhardus, M. Kuhlmann, J. Meyer, C. Zimprich: Parabolic refractive X-ray lenses, J. Synchrotron Rad. **9**, 119–124 (2002)

18.16 H. R. Beguiristain, J. T. Cremer, M. A. Piestrup, C. K. Gary, R. H. Pantell: X-ray focusing with compound refractive lenses made from beryllium, Opt. Lett. **27**(9), 778–780 (2002)

18.17 B. Cederström, M. Lundqvist, C. Ribbing: Multi-prism X-ray lens, Appl. Phys. Lett. **81**(8), 1399–1401 (2002)

18.18 B. Nöhammer, J. Hoszowska, A. K. Freund, C. David: Diamond planar refractive lenses for third- and forth-generation X-ray sources, J. Synchrotron Rad. **10**, 168–171 (2003)

18.19 B. Lengeler, C. G. Schroer, M. Kuhlmann, B. Benner, T. F. Günzler, O. Kurapova, F. Zontone, A. Snigirev, I. Snigireva: Refractive X-ray lenses, J. Phys. D **38**, A218–A222 (2005)

18.20 C. G. Schroer, M. Kuhlmann, B. Lengeler, O. Kurapova, B. Benner, C. Rau, A. S. Simionovici, A. Snigirev, I. Snigireva: *Beryllium parabolic refractive X-ray lenses in Design and Microfabrication of Novel X-Ray Optics*, Proc. SPIE, Vol. 4783, ed. by D. C. Mancini (SPIE, Bellingham 2002) pp. 10–18

18.21 C. G. Schroer, M. Kuhlmann, B. Lengeler, M. Richwin, B. Griesebock, R. Frahm, E. Ziegler, A. Mashayekhi, D. Haeffner, J.-D. Grunwaldt, A. Baiker: Mapping the chemical states of an element inside a sample using tomographic X-ray absorption spectroscopy, Appl. Phys. Lett. **82**(19), 3360–3362 (2003)

18.22 C. G. Schroer, M. Kuhlmann, U. T. Hunger, O. Kurapova, S. Feste, F. Frehse, B. Lengeler, M. Drakopoulos, A. Somogyi, A. S. Simionovici, A. Snigirev, I. Snigireva, C. Schug: Nanofocusing parabolic refractive X-ray lenses, Appl. Phys. Lett. **82**(9), 1485–1487 (2003)

18.23 C. G. Schroer, O. Kurapova, J. Patommel, P. Boye, J. Feldkamp, B. Lengeler, M. Burghammer, C. Riekel, L. Vincze, M. Küchler: Hard X-ray nanoprobe based on refractive X-ray lenses, Appl. Phys. Lett. **87**(12), 124103 (2005)

18.24 C. G. Schroer, B. Lengeler: Focusing hard X-rays to nanometer dimensions by adiabatically focusing lenses, Phys. Rev. Lett. **94**, 054802 (2005)

18.25 P. Kirkpatrick, A. Baez: Formation of optical images by X-rays, J. Opt. Soc. Am. **38**(9), 766–774 (1948)

18.26 G. E. Ice, B. C. Larson: Three-dimensional X-ray structural microscopy using polychromatic microbeams, MRS Bull. **29**(3), 170–176 (2004)

18.27 H. Mimura, S. Matsuyama, H. Yumoto, H. Harea, K. Yamamura, Y. Sano, K. Endo, Y. Mori, M. Yabashi, Y. Nishino, K. Tamasaku, T. Ishikawa, K. Yamauchi: *Hard X-ray diffraction-limited nanofocusing with unprecedentedly accurate mirrors in X-Ray Microscopy*, Vol. 7, ed. by Y. Kagoshima (IPAP Conf. Ser., Tokyo 2006) pp. 100–102

18.28 Y. P. Feng, S. K. Sinha, H. W. Deckman, J. B. Hastings, D. S. Siddons: X-ray flux enhancement in thin-film waveguides using resonant beam couplers, Phys. Rev. Lett. **71**(4), 537–540 (1998)

18.29 A. Jarre, C. Fuhse, C. Ollinger, J. Seeger, R. Tucoulou, T. Salditt: Two-dimensional hard X-ray beam compression by combined focusing and waveguide optics, Phys. Rev. Lett. **94**, 074801 (2005)

18.30 M. J. Zwanenburg, H. G. Ficke, H. Neerings: A planar X-ray waveguide with a tunable air gap for the structural investigation of confined fluids, Rev. Sci. Instrum. **71**(4), 1723–1732 (2000)

18.31 C. Bergemann, H. Keymeulen: Focusing X-ray beams to nanometer dimensions, Phys. Rev. Lett. **91**(20), 204801 (2003)

18.32 J. Underwood: Layered synthetic microstructures as bragg diffractors for X-rays and extreme ultraviolet theory and predicted performance, Appl. Opt. **20**(17), 3027–3034 (1981)

18.33 M. Schuster: Parallel-beam coupling into channel-cut monochromators using curved graded multilayers, J. Phys. D: Appl. Phys. **28**, A270–A275 (1995)

18.34 M. Schuster, H. Göbel, L. Brügemann, D. Bahr, F. Burgäzy, C. Michaelsen, M. Störmer, P. Ricardo, R. Dietsch, T. Holz, H. Mai: *Laterallygraded multilayer opticsfor x-ray analysis*, Proc. SPIE, Vol. 3767, ed. by C. A. MacDonald (SPIE, Bellingham 1999) pp. 183–198

18.35 O. Hignette, G. Rostaing, P. Cloetens, A. Rommeveaux, W. Ludwig, A. Freund: *Submicron focusing of hard X-rays with reflecting surfaces at the ESRF, in X-Ray Micro- and Nano-Focusing: Applications and Techniques II*, Proc. SPIE, Vol. 4499, ed. by I. McNulty (SPIE, Bellingham 2001) pp. 105–116

18.36 O. Hignette, P. Cloetens, G. Rostaing, P. Bernard, C. Morawe: Efficient sub 100 nm focusing of hard X-rays, Rev. Sci. Instrum. **76**, 063709 (2005)

18.37 M. Born, E. Wolf: *Principles of Optics* (Cambridge Univ. Press, Cambridge 1999)

18.38 V. Aristov, A. Erko, V. Martynov: Principles of Bragg–Fresnel multilayer optics, Rev. Phys. Appl. **23**, 1623–1630 (1988)

18.39 W. Chao, B. D. Harteneck, J. A. Liddle, E. H. Anderson, D. T. Attwood: Soft X-ray microscopy at a spatial resolution better than 15 nm, Nature **435**(30), 1210–1213 (2005)

18.40 Y. Suzuki, A. Takeuchi, H. Takano, H. Takenaka: Performance test of fresnel zone plate with 50 nm outermost zone width in hard X-ray region, Jap. J. Appl. Phys. **44**, 1994–1998 (2005)

18.41 Xradia, concord; www.xradia.com. (Nov. 15, 2006)

18.42 J. Maser: *Theoretical description of the diffraction properties of zone plates with small outermost zone width in X-Ray Microscopy IV*, ed. by V. V. Aristov, A. I. Erko (Institute of Microelectronics Technology, Russian Academy of Sciences, Institute Microelectronics Technology, Chernogolovka 1994) pp. 523–530

18.43 C. G. Schroer: Focusing hard X-rays to nanometer dimensions using Fresnel zone plates, Phys. Rev. B **74**, 033405 (2006)

18.44 C. Liu, R. Conley, A. T. Macrander, J. Maser, H. C. Kang, M. A. Zurbuchen, G. B. Stephenson:

- Aerosol fluorescence. Excitation of molecules within aerosol particles by radiation can result in (broad band) fluorescence; this process will not be further discussed here.

19.1 Radiation Transport in the Earth's Atmosphere

In the following we introduce the basic quantities pertinent to the propagation of radiation in the atmosphere and discuss the fundamental laws governing radiation transport in absorbing and scattering media, i. e. the atmosphere.

19.1.1 Basic Quantities Related to Radiation Transport

A light source will emit a certain amount W of energy in the form of radiation.

1. The radiant flux Φ is defined as the radiation energy W per unit time (regardless of the direction in which it is emitted):

$$\Phi = \frac{\text{radiated energy}}{\text{time interval}} = \frac{dW}{dt} \quad \left[\frac{\text{W s}}{\text{s}} = \text{W}\right] \tag{19.1}$$

2. The irradiance B is defined as the radiant flux Φ received by an ('illuminated') area A_e:

$$B = \frac{\Phi}{A_e} \quad \left[\frac{\text{W}}{\text{m}^2}\right]; \tag{19.2}$$

3. The radiation intensity is (Ω = solid angle):

$$I = \frac{\Phi}{\Omega} \quad [\text{W/sr}]; \tag{19.3}$$

4. The radiance is (A_s = radiating area):

$$F = \frac{\Phi}{\Omega A_s} \quad \left[\frac{\text{W}}{\text{m}^2\text{sr}}\right]. \tag{19.4}$$

(All areas are assumed to be oriented perpendicular to the direction of propagation of the radiation.)

19.1.2 Absorption Processes

Radiation is absorbed by molecules in the atmosphere (such as ozone, oxygen, or water vapor), aerosol (such as soot) or liquid and solid water particles (cloud droplets (ice crystals)). Absorption of solar ultraviolet (UV) radiation at wavelengths below about 300 nm by atmospheric O_2 and O_3 is important for life on Earth.

19.1.3 Rayleigh Scattering

Elastic scattering (i. e. scattering without change of the photon energy) by air molecules is called Rayleigh scattering. While this is not an absorption process, light scattered out of the probing light beam will normally not reach the detector, thus for narrow beams it is justified to treat Rayleigh scattering as an absorption process. The Rayleigh scattering cross section $\sigma_R(\lambda)$ (in cm^2) is given by [19.1]:

$$\sigma_R(\lambda) = \frac{24\pi^3}{\lambda^4 N_{\text{air}}^2} \frac{\left[n_0(\lambda)^2 - 1\right]^2}{\left[n_0(\lambda)^2 + 2\right]^2} F_K(\lambda) \approx \frac{8\pi^3}{3\lambda^4 N_{\text{air}}^2} \left[n_0(\lambda)^2 - 1\right]^2 F_K(\lambda), \tag{19.5}$$

where: λ denotes the wavelength, $n_0(\lambda)$ is the real part of the wavelength-dependent index of refraction of air, N_{air} is the number density of air (e.g. 2.4×10^{19} molec/cm^3 at 20 °C, 1 atm), $F_K(\lambda) \approx 1.061$ is a correction for anisotropy (polarisability of air molecules).

Note that $n_0(\lambda)^2 - 1 \approx 2[n_0(\lambda) - 1] \propto N_{\text{air}}$, since $n_0 \approx 1$ [in fact n_0 (550 nm) = 1.000293], and $n_0 - 1 \propto N_{\text{air}}$, thus $\sigma_R(\lambda)$ is essentially independent of N_{air}. On that basis a simplified expression for the Rayleigh scattering cross section (in cm^2) is given by *Nicolet* [19.2]

$$\sigma_R(\lambda) \approx \frac{4.02 \times 10^{-28}}{\lambda^{4+x}} \tag{19.6}$$

with

$$x = 0.04 \quad \text{for} \quad \lambda > 0.55\,\mu\text{m},$$
$$x = 0.389\lambda + 0.09426/\lambda - 0.3328 \quad \text{for} \quad 0.2\,\mu\text{m} < \lambda < 0.55\,\mu\text{m}.$$

For these simple estimates the Rayleigh scattering cross section can be written as

$$\sigma_R(\lambda) \approx \sigma_{R_0} \lambda^{-4} \quad \left(\sigma_{R_0} \approx 4.4 \times 10^{-16}\,\text{cm}^2\,\text{nm}^4 \text{ for air}\right). \tag{19.7}$$

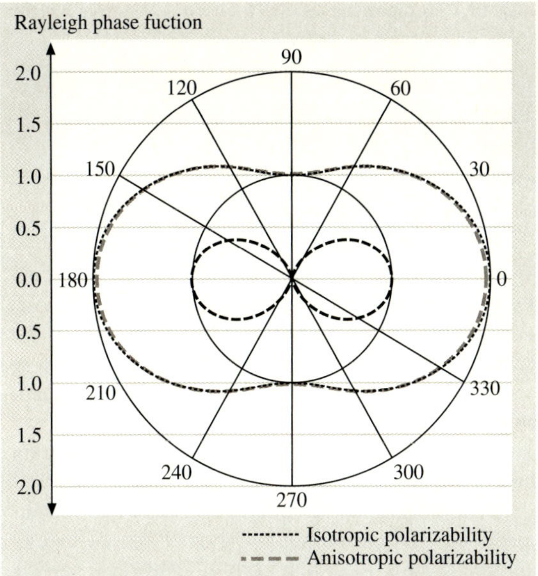

Fig. 19.1 Polar diagram of the Rayleigh scattering phase function $\Phi(\vartheta)$ for unpolarized incident light (*dashed line*, normalized to 2). The contribution of light polarized parallel to the scattering plane shows the $\sin^2\vartheta'$ dependence of a Hertz dipole (*dotted line*, normalized to 1), with $\vartheta' = \pi/2 - \vartheta$ being the angle between dipole axis and the Poynting vector of the incident radiation, while the contribution of light polarized perpendicular to the scattering plane (*drawn line*, normalized to 1) is independent of ϑ. (plot by F. Filsinger)

The extinction coefficient due to Rayleigh scattering $\varepsilon_R(\lambda)$ is then given by

$$\varepsilon_R(\lambda) = \sigma_R(\lambda) N_{air} . \quad (19.8)$$

The Rayleigh scattering phase function (see Fig. 19.1) is given by

$$\Phi(\cos\vartheta) = \frac{3}{4}(1+\cos^2\vartheta) . \quad (19.9)$$

Taking the anisotropy of the polarisability into account the above equations becomes [19.3]:

$$\Phi(\cos\vartheta) = 0.7629(0.9324+\cos^2\vartheta) . \quad (19.10)$$

19.1.4 Raman Scattering

While Raman (and Mie) scattering can be regarded as elastic scattering processes, where no energy is transferred between the scattering particle and the photon, inelastic scattering occurs, if the scattering particle (i. e. molecule) changes its state of excitation during the scattering process. A part of the photons energy is then passed from the photon to the molecule (Stokes lines, $\Delta J = +2$, S-branch) or vice versa (anti-Stokes, $\Delta J = 2$, O-branch). The term rotational Raman scattering (RRS) is used if only the rotational excitation is affected ($\Delta v = 0$). If the vibrational state also changes, the term (rotational-) vibrational Raman scattering (VRS) is used ($\Delta v = \pm 1$). Only discrete amounts of energy given by the difference between the discrete excitation states can be absorbed/emitted. For air (O_2 and N_2) RRS frequency shifts of up to $\pm 200\,\mathrm{cm}^{-1}$ occur, for VRS a vibrational shift of $\pm 2331\,\mathrm{cm}^{-1}$ for nitrogen and $\pm 1555\,\mathrm{cm}^{-1}$ for oxygen has to be added. The VRS is one order of magnitude weaker than the RRS, and RRS is roughly one magnitude weaker than Rayleigh scattering.

In the following we give a quantitative description of rotational- and vibrational-Raman scattering by O_2 and N_2 [19.4–7]. The scattered power density $I_{v,J\to v',J'}$ in $[\mathrm{W/m^2}]$ scattered into the full solid angle 4π involving a transition $(v,\ J \to v',\ J')$ is given by [19.8]:

$$I_{v,J\to v',J'} = I_0 \sigma_{v,J\to v',J'} L N_{air} g_J (2J+1) \frac{1}{Z} \mathrm{e}^{-E(v,J)/kT} , \quad (19.11)$$

where I_0 is the incident power density, N_{air} is the number of molecules in the scattering volume, L is the length of the volume and g_J is the statistical weight factor of the initial rotational state due to the nuclear spin; J and v are the rotational and vibrational quantum numbers, respectively. The factor $(2J+1)$ accounts for the degeneracy due to the magnetic quantum numbers while $\exp[-E(v,\ J)/kT]$ accounts for the population of the initial state of the molecule at temperature T. The state sum Z is given by the product of the rotational state sum Z_{rot} and the vibrational state sum Z_{vib}. The absolute cross section in (19.11) is given by $\sigma_{v,J\to v',J'}$, which can be obtained by integration of the differential cross section $\mathrm{d}\sigma_{v,J\to v',J'}/\mathrm{d}\Omega$ over the entire solid angle Ω. Note that the term differential refers to the solid angle. The energy of the molecule is characterized by the vibrational (v) and rotational (J) quantum numbers and given by

$$E(v,\ J) = E_{vib}(v) + E_{rot}(J) = hc\tilde{\nu}\left(v+\frac{1}{2}\right) + hcBJ(J+1) , \quad (19.12)$$

assuming no coupling between rotation and vibration, B is the rotational constant and $\tilde{\nu}$ is the wave number in

cm^{-1} of the ground state vibration. Allowed transitions are, in that approximation, $\Delta J = 0, \pm 2$ resulting in the Q-, O- and S-branches and $\Delta v = 0, \pm 1$ for vibrational transitions. Due to the temperatures in the Earth's atmosphere only the ground vibrational state is occupied significantly, thus leading only to Stokes transitions of the vibrational states.

Polarization Properties of Vibrational Raman Scattered Light and Line Filling in

The polarization properties of isotropic and anisotropic component of Raman scattered light are described by (19.13), respectively:

$$\frac{I_{\text{parallel}}}{I_{\text{perp}}}(\text{anisotropic}) = \frac{6+\cos^2\Theta}{7} ,$$
$$\frac{I_{\text{parallel}}}{I_{\text{perp}}}(\text{isotropic}) = \cos^2\Theta , \qquad (19.13)$$

where the terms parallel and perpendicular refer to the plane defined by the sun, the scattering point and the observer. Thus only the isotropic part leads to enhanced polarization of scattered light, especially at large scattering angles ($\approx$ solar zenith angle (SZA)). Since the Q-branch of the vibrational band of the Raman-scattered light consists of an isotropic fraction, this will lead to an enhanced degree of polarization in the centre of a Fraunhofer line. But due to the small cross section of vibrational Raman scattering, this enhancement is very small (e.g. $\cong 0.3\%$ for the Ca–I line at 422.7 nm and a resolution of 0.01 nm) and occurs only at large scattering angles. Thus the observed high degrees of polarization at small solar zenith angles cannot be attributed to vibrational Raman scattering [19.9].

Due to the relative small cross section of vibrational–rotational Raman scattering the additional filling in of Fraunhofer or terrestrial absorption lines is around 10% of the filling in due to rotational Raman scattering (see Table 19.1).

Table 19.1 Comparison of the total cross section for the different scattering types for 770 nm, 273 K

Scattering type	Cross section (cm^2)	Ratio (%)
Rayleigh	1.156×10^{-27}	100
O_2 RRS	7.10×10^{-29}	6.1
N_2 RRS	2.94×10^{-29}	2.5
Air RRS	3.82×10^{-29}	3.3
VRS	–	0.1

19.1.5 Mie Scattering

Mie scattering is (after [19.10]) defined as the interaction of light with (particulate) matter of dimensions comparable to the wavelength of the incident radiation. It can be regarded as the radiation resulting from a large number of coherently excited elementary emitters (i. e. molecules) in a particle. Since the linear dimension of the particle is comparable to the wavelength of the radiation, interference effects occur. The most noticeable difference compared to Rayleigh scattering is the usually much weaker wavelength dependence (see below) and a strong dominance of the forward direction in the scattered light. The calculation of the Mie scattering cross section can be very complicated (involving summing over slowly converging series), even for spherical particles, but even more so for particles of arbitrary shape. However, the Mie theory (for spherical particles) is well-developed and a number of numerical models exist to calculate scattering phase functions and extinction coefficients for given aerosol types and particle size distributions see Figs. 19.2, 19.17 [19.11, 12]. The computational effort is substantially reduced by the introduction of an analytical expression for the scattering phase function, which only depends on a few observable parameters. Most common is the Henyey–Greenstein parameterization:

$$\Phi(\cos\vartheta) = \frac{(1-g^2)}{4\pi(1+g^2-2g\cos\theta)^{3/2}} , \qquad (19.14)$$

which only depends on the asymmetry factor g (average cosine of the scattering function):

$$g = \langle\cos\theta\rangle = \frac{1}{2}\int_{-1}^{1} P(\cos\theta)\cos\theta \, \mathrm{d}\cos\theta \qquad (19.15)$$

see e.g. [19.11]. For isotropic scattering [$\Phi(\cos\theta) = $ const] the asymmetry factor $g = 0$ while for a tropospheric aerosol a typical value might be as large as $g \approx 10$.

Tropospheric aerosol is either emitted from the surface (e.g. sea salt, mineral dust, biomass burning) or forms in the gas phase by condensation of chemically formed hygroscopic species (primarily sulphates, nitrates, or oxidized organic material). The aerosol load of the atmosphere, i. e. particle number density and size distribution, depends on the aerosol origin and history. Parameters for typical aerosol scenarios (urban, rural, maritime, background) can be found in the database for the radiative transfer model LOWTRAN [19.13], which includes the extinction coefficients and the asymmetry factors as well as their spectral dependence. Another important aspect is Mie scattering by cloud particles. A radiative transfer model including all cloud effects known to date was, e.g., developed by *Funk* [19.14].

Mie scattering is only partly an absorption process, but by similar arguments as in the case of Rayleigh scattering, for narrow beams it can be treated as an absorption process with an extinction coefficient of

$$\varepsilon_M(\lambda) = \varepsilon_{M0}\lambda^{-\alpha} \tag{19.16}$$

with the Angström exponent α being inversely related to the mean aerosol particle radius. Typically α is found to be in the range 0.5–2.5 with an average value of $\alpha = 1.3$ [19.15, 16]. For the ideal case of an exponential aerosol size distribution:

$$\frac{\Delta N}{\Delta r} = r^{-(\nu+1)} .$$

The Angström exponent is related to the Junge index ν by $\nu = \alpha + 2$ [19.17]. Thus an Angström exponent of 1.3 would correspond to a Junge index of 3.3.

In summary a more comprehensive description of atmospheric extinction (in the presence of a single trace gas species and neglecting Raman scattering) can be expressed as:

$$I(\lambda) = I_0(\lambda)\exp\left\{-L[\sigma(\lambda)c + \varepsilon_R(\lambda) + \varepsilon_M(\lambda)]\right\} \tag{19.17}$$

Typical extinction coefficients due to Rayleigh and Mie scattering at 300 nm are $1.3\times10^{-6}\,\mathrm{cm}^{-1}$ and $1{-}10\times10^{-6}\,\mathrm{cm}^{-1}$, respectively.

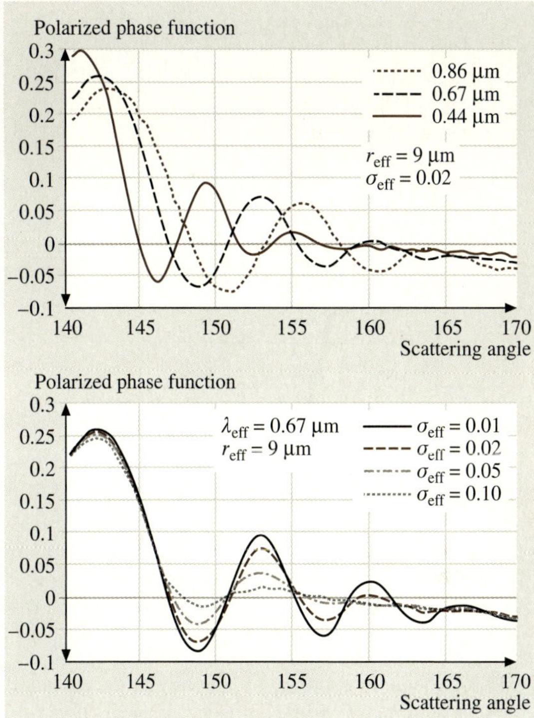

Fig. 19.2 Polarized Mie scattering phase function as a function of scattering angle for cloud droplet having a lognormal particle size distribution with an effective radius $r_{eff} = 9\,\mu$m. *Upper panel*: phase function as a function of wavelength with fixed $\sigma_{eff} = 0.02$ effective size variance; *lower panel*: as a function of effective size variance (courtesy of *Bréon* and *Goloub*, 2003)

19.2 The Radiation Transport Equation

In this section we present the basic equations describing the radiation transport (RT) in absorbing and scattering media as described by the above elementary processes. In physical notation the RT equation is a continuity equation with corresponding source and sink terms (Fig. 19.3); these source and sink terms are assumed to be linear in the radiant flux Φ, intensity I, or radiance F (19.1, 19.3, 19.4).

19.2.1 Sink Terms (Extinction)

First we consider the radiant flux Φ per unit space angle Ω and wavelength λ i.e., the incoming intensity I_λ (19.3) which is attenuated within the distance ds by absorption and scattering. As before we refer to the absorption (a) and scattering (s) coefficients with $\varepsilon_a = N\sigma_a(\lambda)$ and $\varepsilon_s = N\sigma_s(\lambda)$, respectively, with N being the number of absorber or scatterers per volume and $\sigma_i(\lambda)$ being absorption or scattering cross sections. Both processes are commonly referred to as extinction. Accordingly, we obtain the following continuity equation for the incoming intensity traversing the distance ds

$$\begin{aligned} \mathrm{d}I_\lambda &= -[\varepsilon_a(\lambda) + \varepsilon_s(\lambda)]I_\lambda\,\mathrm{d}s \\ &= -[\sigma_a(\lambda) + \sigma_s(\lambda)]nI_\lambda\,\mathrm{d}s , \end{aligned} \tag{19.18}$$

where ε_a is the absorption coefficient, ε_s is the scattering coefficient, $\sigma_a(\lambda)$ is the absorption cross section of the

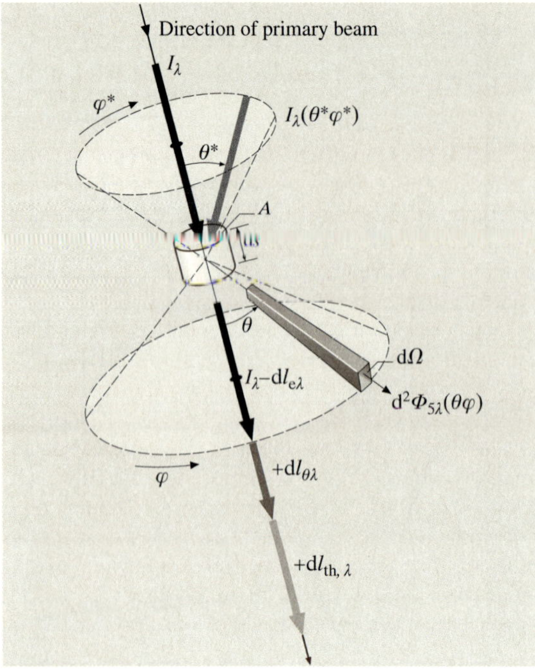

Fig. 19.3 Schematic drawing of the continuity equation for radiative transfer. The continuity assumes linearity of processes in the intensity I_λ. When passing through a distance ds the incoming radiation I_λ is attenuated by extinction (absorption and scattering), $I_{e,\lambda}$. The source term for the outgoing light is the scattered light $dI^*_{s,\lambda}$ from the space angles θ^* and φ^* and thermal radiation $dI^*_{th,\lambda}$ (courtesy of Rödel, 1999)

absorber (molecule), $\sigma_s(\lambda)$ is the scattering cross section of the absorber (molecule), and n is the number of absorbers per unit volume.

19.2.2 Source Terms (Scattering and Thermal Emission)

In the gas phase there are two sources of radiation: thermal emission and scattering, i. e. the radiation removed from the primary beam due to scattering (19.18) reappears as a radiation source.

Scattering

Evidently, the outgoing light receives some intensity $dI^*_{s,\lambda}$, by scattering from all space angles θ^* and φ^*. We introduce a dimensionless scattering function $S(\theta^*,\varphi^*)$

$$S_\lambda(\theta,\varphi)=\frac{4\pi}{\sigma_s}\frac{d\sigma_s(\lambda)}{d\Omega} \tag{19.19}$$

which we integrate over all angles weighted with the incoming intensity $I^*_s,\lambda(\theta^*,\varphi^*)$

$$dI^*_s(\lambda)=\varepsilon_s(\lambda)ds\int_0^{\pi}\int_0^{2\pi}I^*(\lambda,\theta^*,\varphi^*)\times\frac{S(\theta^*,\varphi^*)}{4\pi}d\varphi^*\sin\theta^*d\theta^* \tag{19.20}$$

to obtain the intensity added to the outgoing intensity dI^*_s.

Thermal Emission

Finally the intensity due to thermal emission $dI_{th}(\lambda,T)$ by the volume element ($dV=A\,ds$) is added to the outgoing intensity

$$dI_{th}(\lambda,T)=\varepsilon_a(\lambda)I_p(\lambda,T)ds=\varepsilon_a(\lambda)F_p(\lambda,T)A\,ds \tag{19.21}$$

where, as before, K_a denotes the absorption coefficient and $F_P(\lambda,T)$ denotes the Planck function

$$dF_P(\lambda,T)=\frac{2hc^2}{\lambda^5}\frac{d\lambda}{e^{hc/\lambda kT}-1}\,. \tag{19.22}$$

Combining the above source and sink terms we obtain the radiation transport equation (19.23):

$$\frac{dI(\lambda)}{ds}=-[\varepsilon_a(\lambda)+\varepsilon_s(\lambda)]I(\lambda)+\varepsilon_a(\lambda)I_P(\lambda,T)+\varepsilon_s(\lambda)\times\int_0^{\pi}\int_0^{2\pi}I^*(\lambda,\theta^*,\varphi^*)\frac{S(\theta^*,\varphi^*)}{4\pi}d\varphi^*\sin\theta^*d\theta^* \tag{19.23}$$

19.2.3 Simplification of the Radiation Transport Equation

Frequently simplifications of the radiation transport equation are possible, if only partial systems are of interest.

For instance at short wavelengths (UV, visible) the Planck term can usually be neglected:

$$\frac{dI(\lambda)}{ds}=-[\varepsilon_a(\lambda)+\varepsilon_s(\lambda)]I(\lambda)+\varepsilon_s(\lambda)\times\int_0^{\pi}\int_0^{2\pi}F(\lambda,\theta,\varphi)\frac{S(\theta,\varphi)}{4\pi}d\varphi\sin\theta d\theta\,. \tag{19.24}$$

If thermal radiation (from the atmosphere) is of interest due to its long wavelength, Rayleigh and Mie scattering (by aerosol particles and cloud droplets) can frequently be neglected

$$\frac{dI(\lambda)}{ds} = \varepsilon_a(\lambda)[AF_P(\lambda, T) - I(\lambda)] . \quad (19.25)$$

With the definition of the optical density $d\tau = \varepsilon_a(\lambda)ds$ and after division by A the above equation becomes

$$\frac{dF(\lambda)}{d\tau} = F_P(\lambda, T) - F(\lambda) . \quad (19.26)$$

This latter equation is also known as the Schwarzschild equation.

19.2.4 Light Attenuation in the Atmosphere

When considering the question of the attenuation of radiation in the atmosphere, two (extreme) cases can be distinguished, as illustrated in Fig. 19.4: wide beams (WB, as e.g. the illumination of the Earth's atmosphere by the sun) and narrow beams (NB, as e.g. the light beam emitted by a searchlight-type differential optical absorption spectroscopy (DOAS) light source).

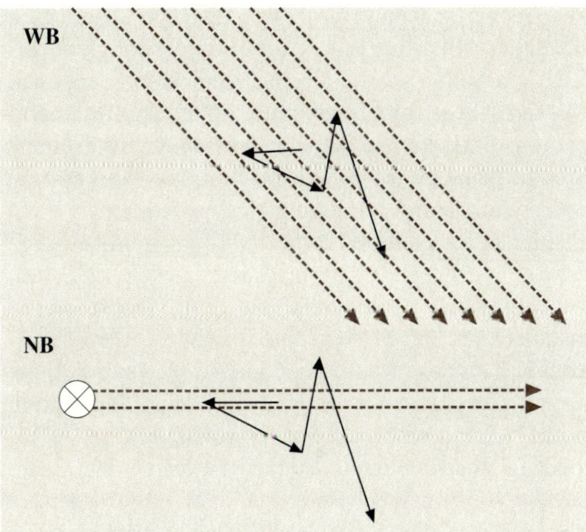

Fig. 19.4 Difference between Wide Beams (WB, *top*) as, e.g., the illumination of the Earth's atmosphere by the sun, and Narrow Beams (NB, *bottom*) as, e.g., the light beam emitted by a searchlight-type light source. In the NB case the probability of a photon being scattered back into the beam after being scattered out of the beam is generally negligible, therefore extinction can be treated like absorption. In the case of a WB the lateral (i.e. perpendicular to the propagation of the incident radiation flux) radiation flux can be neglected. Therefore the scattering has only the effect of reflecting some of the incoming light, as can be calculated by, e.g., a two-stream model (see above)

Wide Beams (WB) in the Atmosphere: The Two-Stream Model

In the following we consider the transport of solar radiation in the atmosphere. Compared to the general form of the radiation transport equation (19.23) the problem is simplified by assuming a flat, horizontally homogeneous and infinite atmosphere. Therefore only the vertical (z) component of the radiance has to be considered.

When radiation enters the atmosphere from above, its vertical component is counted as 'downwelling radiation' $F_\downarrow(\lambda, z)$; upon interaction with atmospheric constituents (gases or aerosol) this radiation is either absorbed or scattered (Fig. 19.5). These extinction processes reduce the amount of downwelling radiation (according to (19.17) where the extension of the light path for non-vertical rays has to be taken into account). While the absorbed radiation causes no further problems, the scattered radiation appears as a source term, which is split into two parts (taking the phase function into account): A downwelling part, which adds to $F_\downarrow(\lambda, z)$ and an upwelling part, which is added to a second, upwelling radiation flux $F_\uparrow(\lambda, z)$. Thermal emission adds to both, $F_\downarrow(\lambda, z)$ and $F_\uparrow(\lambda, z)$. For practical calculation the atmosphere is divided into layers of thickness dz, which change the

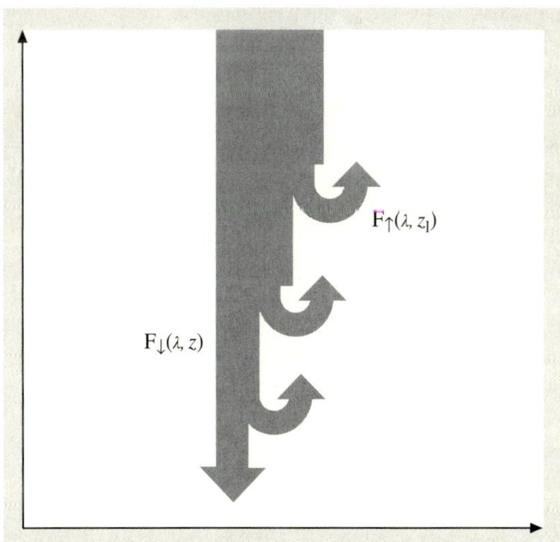

Fig. 19.5 The two-stream model. In a horizontally homogeneous atmosphere radiation transport can be represented by just two 'streams of radiation', the downwelling flux and the upwelling flux

radiance F traversing it by dF. The net radiance is composed of both components (up- and downwelling radiances)

$$F_n(\lambda, z) = F_\uparrow(\lambda, z) - F_\downarrow(\lambda, z) . \tag{19.27}$$

Narrow Beams (NB) in the Atmosphere

In the case of a WB the lateral (i. e. perpendicular to the propagation of the incident radiation flux) radiation flux can be neglected. Therefore the scattering has only the effect of reflecting some of the incoming light, as can be calculated by e.g. a two-stream model (see above).

In contrast, in the case of narrow beams the probability of a photon being scattered back into the beam after being scattered out of the beam is generally negligible (Fig. 19.4), therefore extinction can be treated as absorption.

The attenuation of a light beam by extinction due to atmospheric constituents is in principle described by (19.17). For practical purposes of measurements in the atmosphere, however, (19.17) is oversimplified in that it neglects the presence of other causes of light extinction, including absorption by other molecules present in the atmosphere.

In the natural atmosphere many different molecular species will absorb light. Equation (19.17) must therefore be further extended to

$$\begin{aligned} &I(\lambda) \\ &= I_0(\lambda)\exp\left(-L\{\Sigma[\sigma_i(\lambda)c_i]+\varepsilon_R(\lambda)+\varepsilon_M(\lambda)\}\right) , \end{aligned} \tag{19.28}$$

where $\sigma_i(\lambda)$ and c_i denote the absorption cross section and the concentration of the i-th species, respectively.

19.3 Aerosols and Clouds

Aerosols are a natural constituent of the Earth's atmosphere and their origin is mostly natural, although lately the anthropogenic contribution is increasing. An aerosol is defined as a stable suspension of particles, liquid or solid, in a carrier gas like atmospheric air. Therefore dust, haze, smoke, smog, fog, mist, and clouds can be considered to be specific aerosol types. The total mass of natural aerosols was found be four times larger than due to human activity in 1968, and by the year 2000 the anthropogenic contribution is estimated to have doubled. According to *Junge* and *Manson* [19.18], aerosols can be classified by sizes: particles with diameters below 0.1 mm are named Aitken particles, while those with sizes between 0.1–1 mm are "large" particles, and those larger than 1 mm are called "giant" particles. The aerosols may also be classified according to their origin into marine, continental, rural, remote, background, and urban aerosol, where each category has different characteristics like chemical composition, size distribution, shapes and so on. Aitken nuclei, i. e. the smallest atmospheric aerosol particles, are naturally produced from supersaturated vapors by a process called gas-to-particle conversion (GPC), often in connection with photochemical reactions between gas-phase constituents (Fig. 19.6). Anthropogenic activities such as combustion and industrial processes increasingly lead to the release of aerosol particles, or precursor gases from which aerosol particles are formed, into the atmosphere. Large aerosol particles can be produced by combustion processes and other anthropogenic or natural processes (e.g. the coagulation of smaller Aitken nuclei). Natural sources for giant particles are bulk-to-particle conversion (BPC) processes like sea-salt aerosol production, or desert dust mobilization, while human activity contributes, for example through emission from industries, mining, biomass and wood burning, and agricultural dust generation.

In the atmosphere, typical particle number concentration decreases with increasing particle size since Aitken particles are newly produced; they subsequently coagulate and form fewer larger particles. Accordingly Aitken particles usually dominate the total number of particles in an air sample, but their contribution to the total aerosol volume is small, whilst an increasing contribution is due to the large and giant particles. Conversely, Aitken particles often dominate the surface available for heterogeneous reactions. For particles with sizes ranging from a few nanometers up to tens of micrometers typical number concentrations can be coarsely specified. For example, typical aerosol numbers are: 1) $\approx 1.4\times10^4\,\mathrm{cm^{-3}}$ at the (continental) ground level total concentrations, 2) $\approx 6\times10^2\,\mathrm{cm^{-3}}$ in remote continental regions, 3) $10^5\,\mathrm{cm^{-3}}$ in large cities, 4) $\approx 3\times10^4\,\mathrm{cm^{-3}}$ in small towns, 5) $\approx 5\times10^2\,\mathrm{cm^{-3}}$ over remote oceans, 6) $\approx 50\,\mathrm{cm^{-3}}$ in the clean Arctic, and 7) even lower in the remote Antarctica. The vertical aerosol profiles (in number of particles per cubic centimeter of air) frequently show an exponential decay for the lowest 6 km of the atmosphere. Above

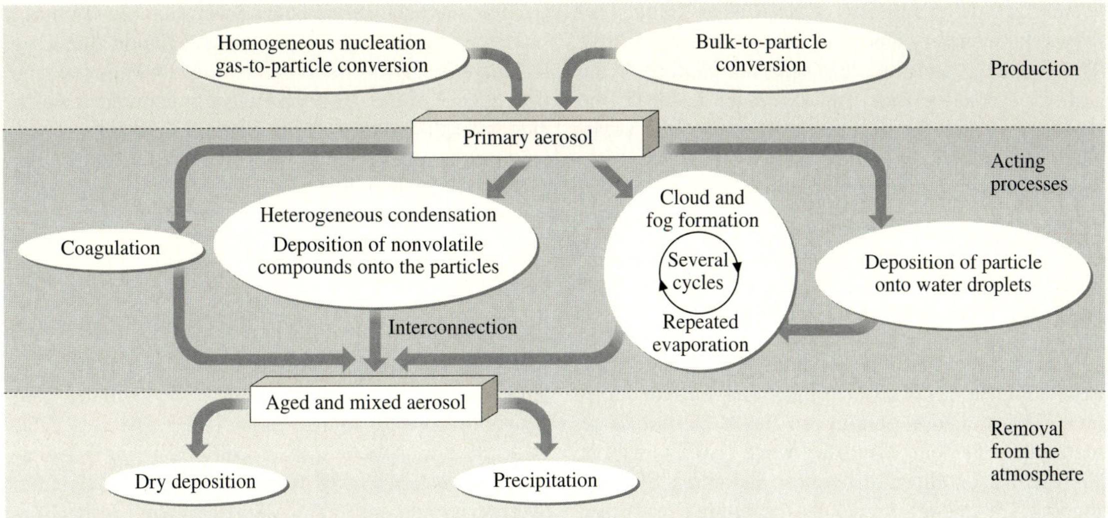

Fig. 19.6 Aerosol-related processes in the atmosphere involving particle formation, processes altering particle characteristics, and removal from the atmosphere

this level, background tropospheric aerosol concentrations (in the Aitken mode) are around 200–300 particles per cm^3. Above the tropopause up to around 30 km, the total number concentration decreases again to a minimum below 1–10 cm^{-3} particles in volcanically quiet periods and 1000 cm^{-3} after major volcanic eruptions.

At any time, about 50% of the globe is cloud covered. Atmospheric clouds are volumes of water-supersaturated air containing hydrometeors, which are microscopic objects (e.g. aerosol particles, cloud droplets, ice and snow crystals) as well as macroscopic objects (snowflakes, graupel grains, raindrops, and hailstones). The individual cloud particles have sizes with radii ranging from 0.001 mm to several hundreds of micrometers, or even centimeters, as in the case of hailstones.

Atmospheric clouds can be classified into four major categories: 1) low-level cloud in the lower troposphere (< 2.5 km), 2) middle-level clouds in the middle troposphere (2.5–8 km), 3) higher clouds in the upper troposphere (8–15 km), and 4) middle (stratospheric) and upper (mesospheric) atmospheric clouds, which themselves form 11 major subgroups (1.1 cumulus, 1.2 stratocumulus, 1.3 stratus, 2.1 altocumulus, 2.2 altostratus, 2.3 nimbostratus, 3.1 cirrus and cirrostratus, 3.2 cirrocumulus, 3.3 cumulonimbus, 4.1 polar stratospheric clouds, and 4.2 noctilucent clouds) of some 100 different cloud types in total.

Both aerosols and clouds have a considerable impact on atmospheric photochemistry and radiation. Their effects in atmospheric photochemistry are most noticeably in: 1) the heterogeneous processing of longer-lived and ozone-friendly halogen species ($ClONO_2$, HCl, $BrONO_2$,...) into ozone-harmful halogen oxides (ClO, BrO) on polar stratospheric cloud particles, which is a major reason for the formation of the Antarctic ozone hole, 2) uptake of acids (e.g., H_2SO_4, HNO_3, HCl, HBr,...) and of N_2O_5 on the stratospheric or tropospheric aerosols and cloud particles, 3) in the heterogeneous conversion of SO_2 into H_2SO_4, or NO_3 and N_2O_5 into HNO_3 and 4) the uptake of species such as NH_3HNO_3, or NH_4HSO_4 into atmospheric aerosols and cloud particles. Eventually when these particles grow large enough (> some μm), the respective gases attached or solved inside are efficiently washed out from the atmosphere.

Aerosols and clouds also have a large effect on the radiative budget of the atmosphere, primarily by increasing the scattering of the incoming solar and outgoing thermal radiation. Clouds affect radiation both through their three-dimensional geometry and the amount, size and nature of the hydrometeors which they contain. In climate models these properties translate into cloud cover at different levels, cloud water content (for liquid water and ice) and cloud droplet (or crystal) equivalent radius. The interaction of clouds and radiation also involves other parameters (asymmetry factor of the Mie diffusion) which depend on cloud composition, and most notably on their phase.

In effect, the occurrence of aerosols and clouds increases the planetary albedo A of the solar short-wave (SW) radiation, and thus lead to a net cooling of the Earth. Conversely clouds, and to a much lesser degree aerosols, increase the infrared albedo B of the atmosphere and thus increase the downwelling atmospheric long-wave (LW) radiation which increases the surface temperature. The subtle balance between cloud impact on the SW and terrestrial long-wave radiation altered by a change in any of the aforementioned parameters, the dependence of the radiation budget on altitude at which clouds occur and the solar inclination (latitude and season) make clouds the most sensitive factors to affect the global climate in both directions (cooling/warming).

19.4 Radiation and Climate

The atmospheric radiation budget largely determines the Earth's climate. Evidently the incoming solar or short-wave (SW) radiation shining on the globe has to be balanced by the outgoing long-wave (LW) radiation. This thermal equilibrium between incoming SW and outgoing LW radiation and can be simply expressed by

$$S_0(1-A)/4 = \varepsilon(1-B)\sigma T_s^4 \tag{19.29}$$

where S_0 is the solar constant (1370 W/m^2), A is the terrestrial SW albedo (0.298), the factor 4 coming from geometric considerations, ε is the LW emissivity (0.96), B is the atmospheric LW albedo (0.34), σ is Boltzmann's constant (5.67×10^{-8} W/m^2/K), and T_s is the average temperature ($T_s = 286$ K) of the Earth's surface. It thus becomes clear that the globe's surface temperature T_s is given by a delicate balance of the parameters S_0, A, and B. In geological and historic

Table 19.2 Compendium of change in recognized periods in climate change and changes in the solar constant S_0

Time scale/period	Cause	Change in solar constant S_0	Observation/problem
4.5 Ga	Secular trend in solar output	Secular 25–30% increase	'Faint Sun problem', i.e. liquid H_2O should not have been there, but was present since 4 Ga before present
450 ka	Change in orbital eccentricity $\Delta\varepsilon = 0.06$	< 0.7 W/m^2 (Milankovitch)	Change in ice volume, $\Delta^{18}O/^{16}O$ in H_2O and $CaCO_3$ in foraminifera
100 ka	Change in orbital eccentricity $\Delta\varepsilon = 0.06$	< 0.7 W/m^2 (Milankovitch)	Change in ice volume, $\Delta^{18}O/^{16}O$ in H_2O and $CaCO_3$ in foraminifera
41 ka	Change in obliquity (axial tilt) due to Earth and Jupiter's orbital plane tilt	June insulation changed by 25% at 80°N, no net change on annual S_0 (Milankovitch)	Change in ice volume, $\Delta^{18}O/^{16}O$ in H_2O and $CaCO_3$ in foraminifera
23 ka		June insulation changed by 100 W/m^2 at 60° N, no net change on annual S_0 (Milankovitch)	Change in ice volume, $\Delta^{18}O/^{16}O$ in H_2O and $CaCO_3$ in foraminifera
1.5 ka	Unclear whether cause is solar, 'eigenmode' of the climate system?		^{14}C and ^{10}Be variation in oceanic sediments, ^{14}C variation in tree rings
210a (Suess) 148a, and 88a (Gleisberg) cycles	Unclear whether cause is solar, 'eigenmode' of the climate system?		^{14}C variation in tree rings, ^{14}C and ^{10}Be variation ice cores,
22a	Sun spot 'overtone', eigenmode in solar magnetic field oscillation	$\Delta S_0 \approx 1.5$ W/m^2, UV radiation changes by 0.37–0.6%	Change in atmospheric circulation?
11a (Schwabe 1843)	Sun spot 'overtone', eigenmode in solar magnetic field oscillation		Change in atmospheric circulation?
4w (27d)	Sun rotation period	Small	?

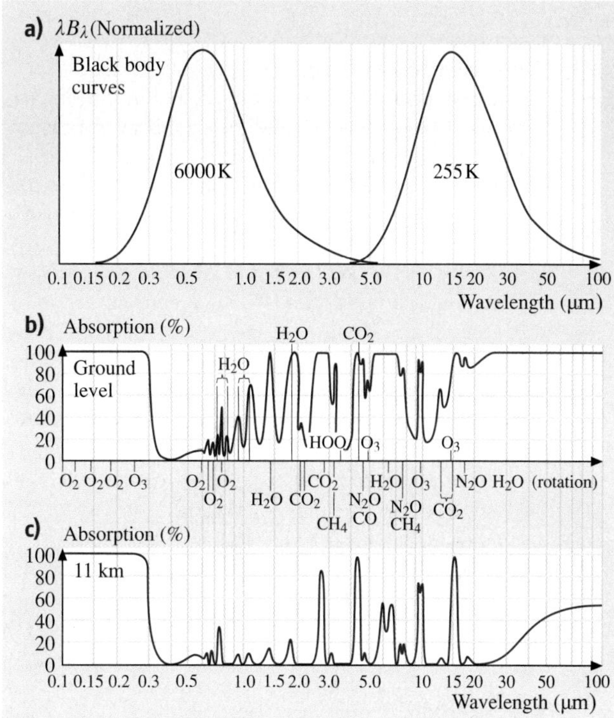

Fig. 19.7a–c Schematic overview of the SW (*left part*) and LW (*right part*) radiative transfer in the atmosphere; *upper panel*: **(a)** incoming solar SW and outgoing LW terrestrial spectrum, **(b)** atmospheric absorption from individual gases at ground level (total atmosphere) and **(c)** at 11 km

times, the solar constant is known to have varied on different timescales, mostly given by changes in the Earth's orbital parameters, as suggested by *Milankovitch* in 1941 [19.19] and solar activity (Table 19.2), causing well recognized and significant effects on the Earth's climate. In this context, most notable is that the size ΔS_0 of any of the known processes listed in Table 19.2 has not been sufficiently large to be held solely responsible for documented ΔT_S changes, which were inferred to be as large as some 10 °C. Therefore, it is well accepted that the global climate system amplifies external disturbances (forcing) through interacting complicated positive and negative feedback cycles (Table 19.2).

Another factor important for T_S is of course the magnitude and variability of the SW albedo A (19.35). This is largely influenced by the amount of cloud and the abundance of atmospheric aerosol; the latter is either due to increasing directly back reflection of SW radiation to space, or by altering the cloud particle formation processes, a process called the indirect aerosol effect [19.20, 21]. In the recent past changes in the global albedo due to increasing atmospheric aerosol abundances, mostly particles containing sulphuric acid has been well recognized and held partly responsible for lower global T_S between the late 1940s to late 1970s [19.20]. Changing global cloud cover is also likely to have occurred in the past. In the tropics, which are a particularly sensitive region for the global energy budget, the cloud cover appears to be highly variable on timescales ranging from days to decades. Accordingly, satellite-based cloud cover records that began in the late

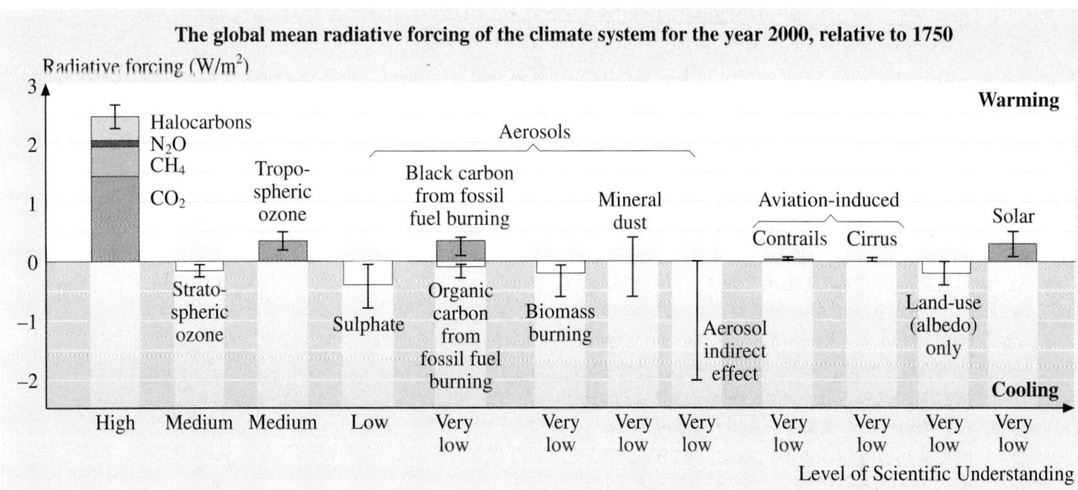

Fig. 19.8 External factors which force climate change (adopted from IPCC, 2001)

1970s are still insufficiently long to establish unambiguously an observation-based relation between a changing A and T_s [19.22].

The atmospheric LW albedo B is responsible for the greenhouse effect of the atmosphere. If B were zero, then according to (19.35), T_s would be 255 K, clearly cooler than the habitable Earth is able to provide ($T_s = 287$ K). The value of B is due to the combined action of the most important atmospheric greenhouse gases H_2O, CO_2, CH_4, O_3, N_2O, and many others, mostly man-made, trace molecules [19.20]. It appears that, even though each greenhouse gas has its own specific spectrum, they combine in a way that only a small fraction of the outgoing LW radiation is not absorbed in the atmosphere, i. e., radiation in the atmospheric window (8–14 μm) (Fig. 19.7).

In 1896, the atmospheric greenhouse effect of CO_2 was already recognized by S. Arrhenius. Accordingly it is undisputed in the scientific community that rising atmospheric CO_2 concentrations affect the atmospheric LW budget and thus T_s. Radiative transfer calculations show that the CO_2 increase from preindustrial 280 ppm to presently 370 ppm alone gives rise to a present radiative forcing of 1.4 W/m^2 and for all greenhouse gases of 2.4 W/m^2 ([19.20], Fig. 19.8). Naively adding the respective radiative forcing in (19.18) leads to a $\Delta T_s = +0.7$ K, precisely the value that is reported for the global average temperature rise in the past 150 y [19.20]. However, due to the complicated interactions in the climate system, it is disputed how the global climate system is and will react in detail to future man-induced LW radiative forcing.

19.5 Applied Radiation Transport: Remote Sensing of Atmospheric Properties

19.5.1 Trace Gases

Measurements of trace gas concentrations (and other quantities such as aerosol distribution or the intensity of the radiation field in the atmosphere) are the experimental prerequisites for the understanding of the physico-chemical processes in the Earth's atmosphere. At the same time the determination of trace gas concentrations in the atmosphere constitutes a challenge for the analytical techniques employed in several respects: It is well known (e.g. *Perner* et al. [19.23]) that species (such as OH radicals) present at mixing rations ranging from as low as 0.1 ppt (one ppt corresponds to a mixing ratio of 10^{-12}, equivalent to about 2.4×10^7 molecules/cm^3) to several ppb (1 ppb corresponds to a mixing ratio of 10^{-9}) can have a significant influence on the chemical processes in the atmosphere. Thus, detection limits from below 0.1 ppt to the ppb range are required, depending on the application. On the other hand, measurement techniques must be specific, i. e. the result of the measurement of a particular species should neither be positively nor negatively influenced by any other trace species simultaneously present in the probed volume of air. Given the large number of different molecules present at the ppt and ppb level, even in clean air, meeting this requirement is also not trivial.

Presently many highly sophisticated techniques for the measurement of atmospheric trace species are in use. Among these, spectroscopic techniques [19.24, 25] offer a series of unique advantages including high (i. e. ppt-level in many cases) sensitivity, very specific detection of a given molecule, inherent calibration, and wall-free operation. These properties, in combination, are difficult to obtain in techniques based on other principles. Here we describe a particular technique, differential optical absorption spectroscopy (DOAS), which has been successfully used in atmospheric measurements for several decades now, while new applications continue to be introduced.

19.5.2 The Fundamentals of DOAS

The DOAS technique [19.26, 27] makes use of the structured absorption of trace gases. Typically spectra are recorded which encompass several hundred spectral channels. Evaluation is done by fitting the spectral structure of the trace gases, thus making use of all the spectral information. DOAS has proven to be particularly useful for the determination of the concentration of unstable species such as free radicals or nitrous acid. In addition, the abundance of aromatic species can be determined at high sensitivity (see below). DOAS, like all spectroscopic techniques, relies on the absorption of electromagnetic radiation by matter. Quantitatively the absorption of radiation is expressed by the Lambert–Beers law:

$$I(\lambda) = I_0(\lambda)\exp[-L\sigma(\lambda)c]\,, \tag{19.30}$$

where $I(\lambda)$ is the intensity after passing through a layer of thickness L, while $I_0(\lambda)$ denotes the initial intensity emitted by the light source. The species to be measured is present at the concentration (number density) N_c. The quantity $\sigma(\lambda)$ denotes the absorption cross section at the wavelength λ; it is a characteristic property of any species. The absorption cross section $\sigma(\lambda)$ can be measured in the laboratory, while the determination of the light path length L is trivial in the case of the arrangement of an artificial light source and detector (Fig. 19.9, arrangement a). Once those quantities are known, the trace gas concentration c can be calculated from the measured ratio $I_0(\lambda)/I(\lambda)$. In contrast to laboratory spectroscopy, the true intensity $I_0(\lambda)$, as would be received from the light source in the absence of any absorption, is usually difficult to determine when measurements are made in the open atmosphere or in smog chambers. The solution lies in measuring the so-called differential absorption. This quantity can be defined as the part of the total absorption of any molecule rapidly varying with wavelength and is readily observable, as will be shown below. Accordingly, the absorption cross section of a given molecule (numbered i) can be split into two portions:

$$\sigma_i(\lambda) = \sigma_{i0}(\lambda) + \sigma_i'(\lambda)\,, \tag{19.31}$$

where $\sigma_{i0}(\lambda)$ varies only slowly with the wavelength λ, for instance describing a general slope, while $\sigma_i(\lambda)$ shows rapid variations with λ, for instance due to an absorption line. The meaning of rapid and slow (or rather smooth and structured) variation of the absorption cross section as a function of wavelength is, of course, a question of the observed wavelength interval and the width of the absorption bands to be detected. Note that extinction due to Rayleigh and Mie scattering can be assumed to be slowly varying with λ. Thus (19.30) becomes

$$\begin{aligned} I(\lambda) = {} & I_0(\lambda)\exp\left\{-L\Sigma\left[\sigma_i'(\lambda)N_{ci}\right]\right\} \\ & \times \exp\left(-L\{\Sigma[\sigma_{i0}(\lambda)c_i] + \varepsilon_R(\lambda) + \varepsilon_M(\lambda)\}\right) \\ & \times A(\lambda)\,, \end{aligned} \tag{19.32}$$

where the first exponential function describes the effect of the structured, differential absorption of trace species, while the second exponential constitutes the slowly varying absorption of atmospheric trace gases as well as the influence of Mie and Rayleigh scattering described by the extinction coefficients $\varepsilon_R(\lambda)$ and $\varepsilon_M(\lambda)$, respectively. The attenuation factor $A(\lambda)$ describes the (slow) wavelength-dependent transmission of the optical system used. Thus we can define a quantity I_0' as the intensity in the absence of differential absorption:

$$\begin{aligned} & I_0'(\lambda) \\ & = I_0(\lambda) \times A(\lambda) \\ & \quad \times \exp\left(-L\{\Sigma[\sigma_{i0}(\lambda)N_{ci}] + \varepsilon_R(\lambda) + \varepsilon_M(\lambda)\}\right). \end{aligned} \tag{19.33}$$

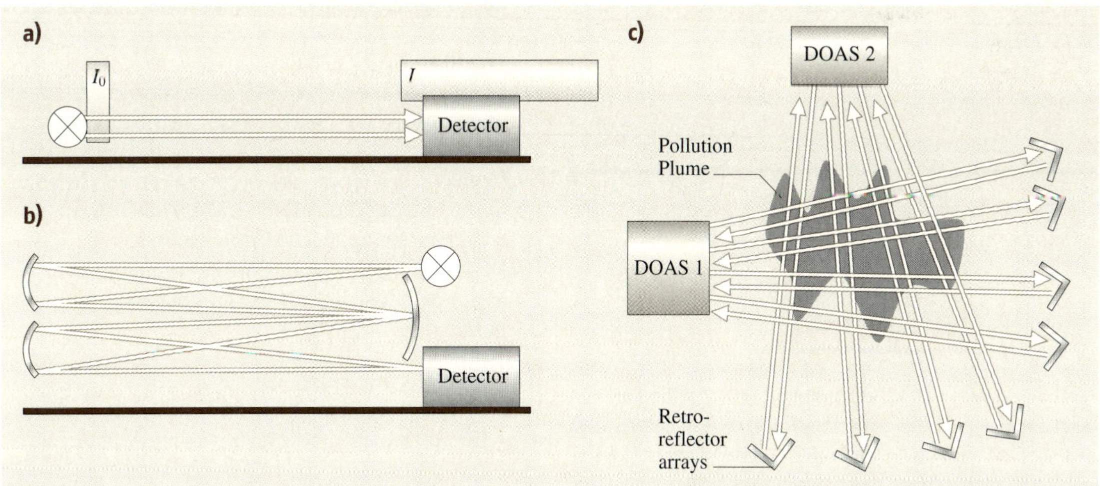

Fig. 19.9a–c The active DOAS principle can be applied in several light-path arrangements and observation modes using artificial light sources such as arc lamps or incandescent lamps in a searchlight-type arrangement, or lasers. The light-path-averaged trace gas concentration is determined in the 'traditional' setup (**a**). In situ concentrations are measured with multiple reflection cells (**b**). The new 'tomographic' arrangement (**c**) employing many DOAS light paths allows mapping of (two- or three-dimensional) trace gas distributions

Fortunately the effect of slowly varying absorbers, i. e. the exponential in (19.33), can be removed by high-pass filtering the spectral data, thus only the first exponential function in (19.32) remains, which is essentially Lambert–Beers law [(19.30) for more than one absorber]. It should be noted that the length of the light path is more difficult to determine – or might actually not have a precise meaning – in applications where direct or scattered sunlight is used (arrangements d, e, and f and Fig. 19.11), in this case the column density S of the trace gas

$$S = \int_0^\infty N_c(l)\,dl \tag{19.34}$$

can still be determined. In the case of a constant trace gas concentration $N_{c0}(l) = c_0$ along a path length L the column density simply becomes $S = N_{c0} L$.

In order to obtain sufficient sensitivity, light-path lengths L of the order of several 100 m to several 1000 m are usually required. For measurements in the free atmosphere, assuming the arrangement a in Fig. 19.9, the trace gas concentration is averaged over the whole length of the light path, which makes the measurement less susceptible to local emissions. On the other hand measurements must frequently be made in a small volume (e.g. in photo-reactors); in these cases multi-reflection systems can be used to fold the light path into a small volume [19.28]. An example of detection limits attainable with multi-reflection systems is given in Table 19.3.

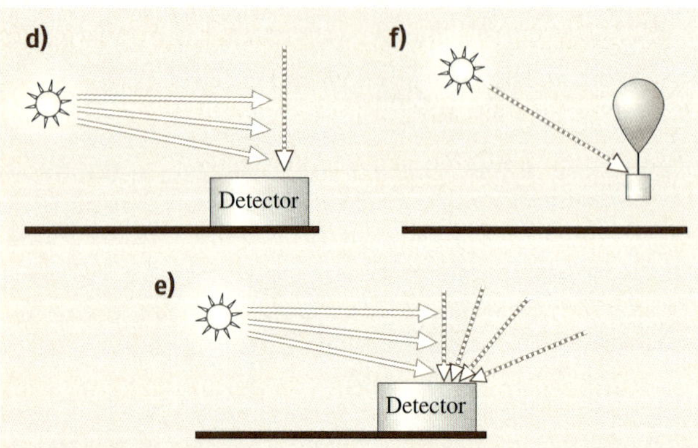

Fig. 19.10d–f The passive DOAS technique can be applied in several light path arrangements and observation modes using natural light sources such as sunlight or moonlight. The zenith scattered light (ZSL-DOAS) set-up **(d)** is most suited for the determination of stratospheric species, while the Multi-AXis (MAX-DOAS) arrangement **(e)** provides highly sensitive trace gas measurements in the atmospheric boundary layer. Direct sunlight observation of trace gases from balloon platforms **(f)** allows the determination of trace gas profiles in the atmosphere

19.5.3 Variations of DOAS

The DOAS technique can be adapted to a large variety of measurement tasks; as a consequence there are many variants of the DOAS technique in use. These can be grouped into active techniques, which employ their own light sources (e.g. Xe-arc lamps or incandescent lamps), and passive techniques, which rely natural light sources (e.g. sunlight, moonshine or starlight).

The most common active DOAS light-path arrangements include (Fig. 19.9):

a) The traditional active long-path systems (usually folded once);
b) Active systems using multi-reflection cells (e.g. white cells);
c) Recently the first tomographic systems allowed the determination of 3-D trace gas distributions by multiple-beam active DOAS.

The most common passive DOAS light path arrangements include (Fig. 19.10):

d) Zenith scattered light (ZSL)-DOAS is well-suited to study stratospheric trace gases. For instance a global network of ZSL-DOAS instruments continuously watches the distribution of stratospheric species (e.g. O_3, NO_2, BrO, OClO).
e) Multi-AXis (MAX)-DOAS allows determination of trace gases and their vertical distribution in e.g. the atmospheric boundary layer or the detection of plumes from cities, stacks, or volcanoes. In addition Airborne Multi-AXis (AMAX)-DOAS is a new technique for application from aircraft.
f) Direct sunlight observation of trace gases from balloon platforms allows the determination of trace gas profiles.

An extremely important, new application is satellite-borne DOAS, pioneered by the GOME and SCIAMACHY instruments [19.29, 30] that allow observation of global stratospheric and tropospheric trace gas distributions (Fig. 19.11):

g) The most basic observation geometry is the nadir view. Here, sunlight backscattered from the Earth and its atmosphere is analyzed by DOAS to yield trace gas slant column densities. Ideally all light recorded by the satellite spectrometer traversed the atmosphere twice.

h) Satellites also allow the observation of scattered sunlight in a limb-viewing geometry. This allows the determination of trace gas profiles (although generally only for the stratosphere).
i) Occultation (sunset, sunrise or starlight) measurements from satellites are an attractive possibility.

In addition it is now possible to determine the photon path-length distribution in clouds by ground-based observation of O_2 and O_4 bands. Here effectively DOAS is reversed: instead of the usual configuration where an unknown concentration is measured while the length of the light path is (at least approximately) known, the photon path-lengths are determined by observing the absorption due to an absorber with known concentration, such as oxygen (O_2) or oxygen dimers (O_4).

19.5.4 Atmospheric Aerosols

The atmospheric aerosol concentration is frequently monitored by light detection and ranging (LIDAR) instruments operated from the ground, aircrafts, or satellites [19.32]. In their most basic configuration they measure the ratio of total extinction due to Mie and

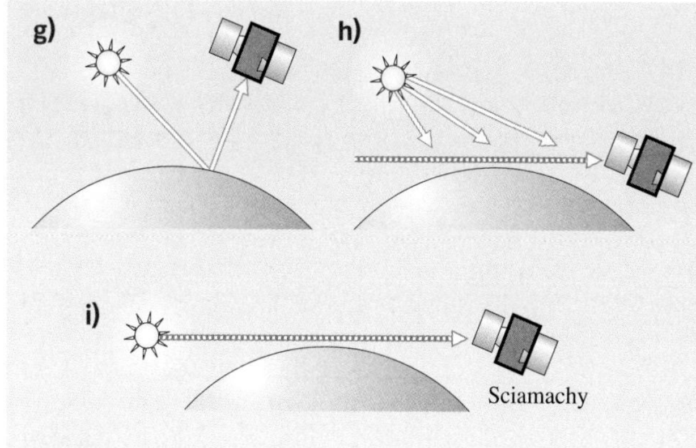

Fig. 19.11g–i Satellite-based DOAS has become an important technique for probing the atmospheric composition on a global scale. The most basic observation geometry is the nadir view (**g**), here sunlight backscattered from the Earth and its atmosphere is analyzed by DOAS to yield trace gas slant-column densities. Ideally all light recorded by the satellite spectrometer traversed the atmosphere twice. Satellites also allow the observation of scattered sunlight in a limb-viewing geometry (**h**). This allows the determination of trace gas profiles (although generally only for the stratosphere). Occultation (sunset, sunrise or starlight) measurements (**i**) from satellite are another attractive possibility to determine trace gas vertical profiles

Table 19.3 Selection of species of atmospheric relevance measurable by DOAS and detection limits. Differential absorption cross-sections and corresponding detection limits (assuming 32 passes in a multi-reflection cell with a 8 m base path) for a series of species to be investigated by DOAS in the Valencia smog chamber. A dielectric mirror coating or DOAS measurements in the open atmosphere would allow longer light paths, resulting in correspondingly improved detection limits

Species	Wavelength of prominent band (nm)	Differential abs. cross section (σ_i') (10^{-20} cm^2/molec.)	Detection limit $L = (40-8) \times 8\,\mathrm{m} = 256\,\mathrm{m}$ (ppt)
ClO	280	350	200
BrO	328	1040	80
IO	427	1700	50
SO_2	300	68	1000
NO_2	430	17	2700
HONO	352	41	2000
O_3	282	10	8000
Glyoxal[b]	ca. 470	≈ 10	≈ 8000
Benzene	253	200[a]	400
Toluene	267	200[a]	400
Phenol	275	3700[a]	20
Xylenes	260–272	≈ 100[a]	≈ 800
Benzaldehyde	285	500[a]	160

[a] data from Trost et al. 1997, R. Volkamer, personal communication 2000 [b] from [19.31]

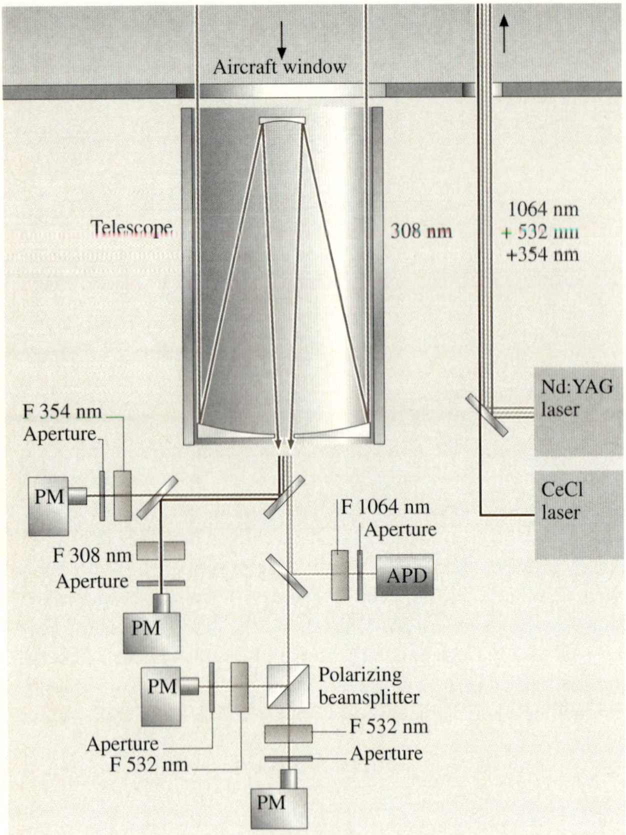

Fig. 19.12 Schematic setup of the DLR-Oberpfaffenhofen aircraft-borne LIDAR system (OLEX). APD: avalanche photo diode; PM: photomultiplier (courtesy G. Ehret/DLR Oberpfaffenhofen)

Rayleigh scattering versus the calculated expectation of pure Rayleigh scattering for a set of absorbing and non-absorbing wavelengths.

The Ozone Lidar EXperiment (OLEX) LIDAR is used as an aerosol and ozone LIDAR (Fig. 19.12). The system consists of two laser transmitters: 1) a Nd:YAG laser operating at a repetition rate of 10 Hz in the fundamental, second harmonic, and third harmonic with output energies of 200 mJ at 1064 nm, 120 mJ at 532 nm, and 180 mJ at 355 nm, and 2) a XeCl laser operating at 10 Hz, emitting 200 mJ at 308 nm. The receiver system is based on a 35 cm Cassegrain telescope with a field of view (FOV) of 1 mrad and a focal length of 500 cm. The system uses five detection channels: one channel for 1064 nm, two channels for 532 nm in two polarization planes, one channel for 355 nm, and one for 308 nm. The 308 and 355 nm channels are specifically dedicated to ozone profiling in the stratosphere, while the other wavelengths are used for aerosol and cloud monitoring. The acquisition electronics can acquire each single signature, with the lasers firing at a 50 Hz repetition rate, and the aerosol distributions measured by the system are available in real time. The total weight of the system is about 270 kg and the total power consumption above 1.6 kW.

In the present example, the OLEX LIDAR was assembled on board the DLR-Falcon aircraft in a down-looking mode that enabled aerosol backscattering measurements below the aircraft cruise altitude. On 24 September 1999, a flight was undertaken leading from Oberpfaffenhofen near Munich to the Adriatic Sea. The

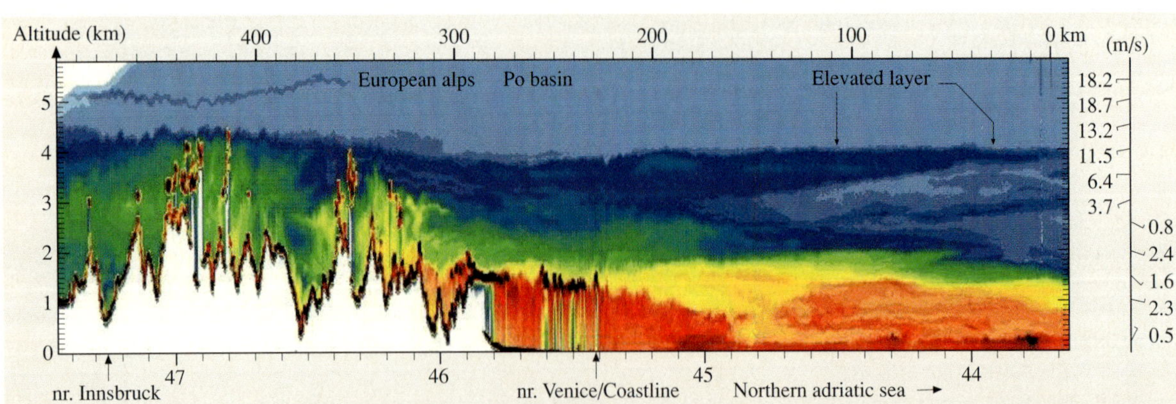

Fig. 19.13 Tropospheric aerosol distribution recorded by OLEX at $\lambda = 1024$ nm during a flight of the Falcon aircraft from Oberpfaffenhofen to the Adriatic Sea on 24 September 1999. The color codes qualitatively indicates the aerosol concentration with the large aerosol concentrations given in yellow and the small in blue color

main scientific objective was to investigate the spatial distribution of made-made aerosols emitted from traffic, industry and house burning into the lower troposphere.

The major detected features were inhomogeneously aerosol-loaded air masses located in the alpine mountain valleys and the fairly uniform planetary boundary layer (PBL) stretching over the whole alpine ridge (Fig. 19.13). Both aerosol layers nicely mark the vertical layering of a vertically layered lower troposphere typical for a high-pressure system at mid-latitudes in fall. Another interesting feature was the structure of the planetary boundary layer (PBL) located over the Adriatic Sea (≈ 2000 m above sea level), which overlaid a very thin marine boundary layer (250–300 m above sea level). This feature resulted from an advection of polluted PBL air masses originating from the North Italian Apennines and the Po Basin by synoptic westerly winds [19.33].

19.5.5 Determination of the Distribution of Solar Photon Path Lengths

A novel application of differential optical absorption spectroscopy (DOAS) is the measurements of path-length distributions of solar photons (photon PDF) transmitted to the ground [19.34–38]. The knowledge of a photon PDFs is of primary interest in the cloudy atmospheric radiative transfer, atmospheric absorption of solar radiation energy and thus for climate. The method relies on the analysis of highly resolved oxygen A-band (762–775 nm) spectra (Fig. 19.14) observed with a telescope with a small field of view (1°) in zenith-scattered sky light. It uses the fact that solar photons are randomly scattered by molecules, aerosols and cloud droplet during their atmospheric transport. Since a fraction of the solar photons become absorbed in the atmosphere, largely depending on the atmospheric opacity for a given wavelength, for each wavelength solar photons travel on different average, long distances in the atmosphere. Spectral intervals with a largely changing atmospheric opacity but known absorber concentration and spectroscopic constants, such as that given by the oxygen A-band, thus contain information on the photon PDFs. Recent studies have been showing that the atmospheric photon PDFs are highly correlating with the total amount of liquid water in the atmosphere, the spatial arrangement of clouds, cloud inhomogeneity and solar illumination [19.34–38]. For example, for thunder-

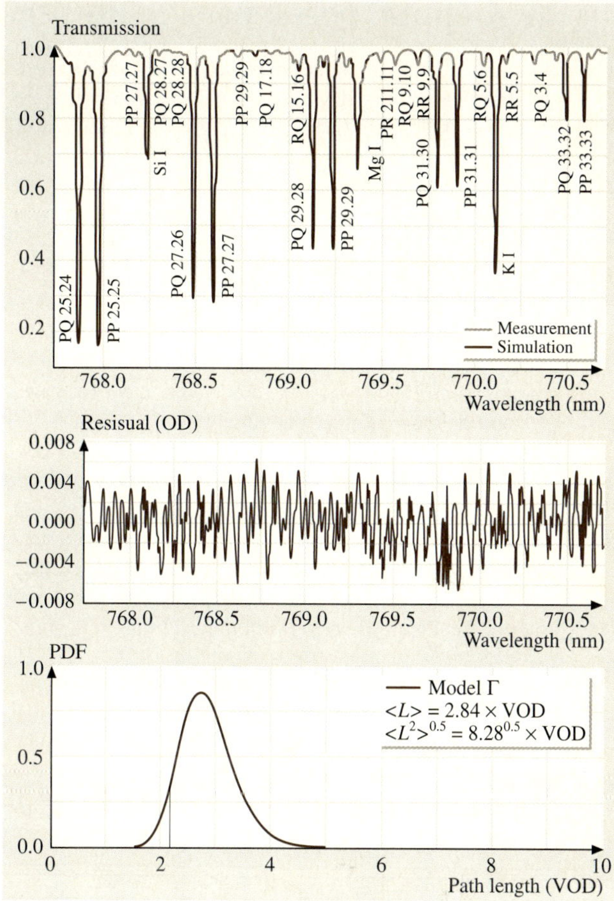

Fig. 19.14 Measured and modeled oxygen A-band spectrum (*upper panel*), inferred residual spectrum (*middle panel*) and inferred photon PDF for the observation between 12:32 and 12:33 UT over Cabauw/NL on 23 September 2001. The *black vertical line* indicates the optical path for the direct sunlight. Photon path lengths are given in units of vertical atmosphere (VOD) or air mass (AM)

storm Cb (cumulus nimbus) clouds, average photon path lengths of up to 100 km have been observed, while for Sc (stratus clouds) typically photon paths are 50–100% larger than for the direct and slant path of the sun's rays. Moreover, it appears that even though the spatial distribution of cloud droplets is inhomogeneous with the moments of the density behaving like multi-fractals, photon PDFs tend to be mono-fractals, mostly due to the so-called radiative smoothing of optically thick clouds [19.39].

19.6 Optical Phenomena in the Atmosphere

Phenomena based on atmospheric optics can be observed nearly every day and everywhere. Examples are rainbows, halos, mirages, coronas, glories, the colors of the sky, sunsets and twilight phenomena, green flashes, noctilucent clouds, polar lights, and many more. All these different phenomena are due to interaction of light with matter present in the atmosphere (Fig. 19.15).

In the following, a brief survey of light scattering by molecules and particles with regard to optical phenomena in the atmosphere will be given. More details on light scattering by particles can be found in a number of monographs [19.40–43]. Several books have focussed on one or more optical phenomena (e.g. [19.44–54]). Some older original publications are available as a collection of articles [19.55] and the proceedings of the regular conferences on atmospheric optics have been published in special issues of the journals of the Optical Society of America [19.56–63], most of them recently put together on a CD-ROM [19.64]. Several videos and movies deal with selected phenomena [19.65–69], some articles focus on connections to the fine arts [19.70–73], a number of books focus on simple experiments as well as observations [19.74,75] and a number of internet sites with a lot of material are also available, Table 19.4.

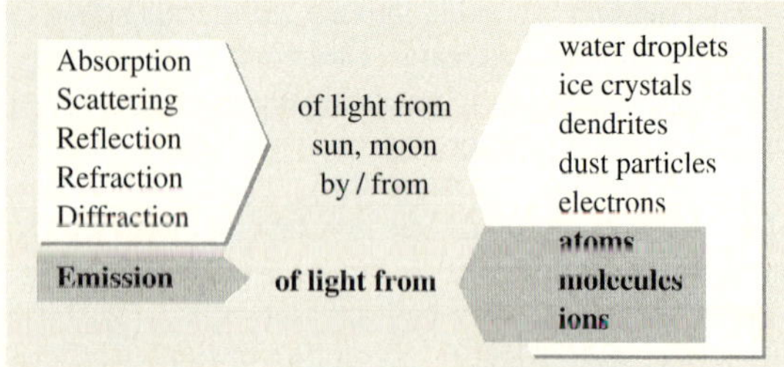

Fig. 19.15 Possible interaction processes of light with constituents of air

19.6.1 Characteristics of Light Scattering by Molecules and Particles

Nearly all optical phenomena in the atmosphere are due to light scattering – mostly sunlight and sometimes moonlight – with the constituents of air. In the following, a brief general survey of light scattering processes in the atmosphere will be given. Consequences for various optical phenomena of the atmosphere will be systematically discussed in the respective sections according to the constituents of the air that are responsible (Table 19.5).

Table 19.4 List of internet sites on atmospheric optics

1.	http://www.polarimage.fi (Images of many phenomena)
2.	http://www.atoptics.co.uk (Images and simulations)
3.	http://www.engl.paraselene.de (Images)
4.	http://www.philiplaven.com (Images and simulations)
5.	http://www.funet.fi/pub/astro/html/eng/obs/meteoptic/links.html (Finnish amateur observer site)
6.	http://www.ursa.fi/english.html (Finnish astronomical association)
7.	http://www.meteoros.de (German halo observer network)
8.	http://mintaka.sdsu.edu/GF/ (Green flashes)
9.	http://thunder.msfc.nasa.gov (Nasa lightning page)
10.	http://www.fma-research.com/ (Lightning, in particular sprites)
11.	http://www.sel.noaa.gov/ (Space weather)
12.	http://www.spaceweather.com/ (Space weather)
13.	http://sunearth.gsfc.nasa.gov/ (Nasa Sun Earth connection, outreach)
14.	http://www.mreclipse.com/ (Eclipses)
15.	http://www.geo.mtu.edu/weather/aurora/ (Auroras)
16.	http://www.exploratorium.edu/auroras/index.html (Auroras)
17.	http://www.amsmeteors.org/ (American meteor society)
18.	http://www.imo.net/ (International meteor organization)
19.	http://liftoff.msfc.nasa.gov/academy/space/solarsystem/meteors/Showers.html (Meteor showers, Nasa)

Table 19.5 Classification of optical phenomena of the atmosphere. Depending on the homogeneity of the air, different light interaction processes correspond to various phenomena, which are indicated by examples

1. Pure homogeneous air

- refraction ⇒ mirages, shape changes of sun/moon at horizon Sect. 19.6.2
- scattering ⇒ blue sky Sect. 19.6.3

2. Inhomogenous atmosphere: Air plus water droplets

- refraction and reflection ⇒ rainbows Sect. 19.6.4
- forward scattering/diffraction ⇒ coronas Sect. 19.6.5
- backward scattering/diffraction ⇒ glories Sect. 19.6.5

3. Inhomogenous atmosphere: Air plus ice crystals

- refraction and reflection ⇒ halos Sect. 19.6.6
- forward scattering/diffraction ⇒ coronas Sect. 19.6.5

4. Inhomogenous atmosphere: Air plus aerosols

- absorption, scattering ⇒ sky colors Sect. 19.6.7
- absorption, scattering ⇒ visibility Sect. 19.6.8
- forward scattering/diffraction ⇒ coronas Sect. 19.6.5

5. Ionized air

- ionization/excitation by solar wind, emission ⇒ auroras Sect. 19.6.9
- ionization/excitation by discharges, emission ⇒ lightning Sect. 19.6.9

Molecular Scattering

For pure air, scatterers are mostly N_2 and O_2 molecules, the greenhouse gases CO_2 and H_2O as well as some traces of other gases such as Ar [19.76]. These scatterers are all much smaller than the wavelength of visible light and one speaks of a molecular atmosphere [19.77]. The vertical distribution of the gaseous components of air follows the barometric formula with a scaling height of about 8 km. The scattering of sunlight from the bound electrons of atoms/molecules is usually called Rayleigh scattering [19.1, 78, 79]. (For historical notes on the problem of the blue sky, see [19.80].)

Scattering from Particles

Normal air usually also contains a variety of different particles, the most predominant being water droplets, ice crystals and snow flakes (dendrites). The size of the water droplets ranges from several micrometers (cloud droplets) to well above 1 mm (raindrops); ice crystals are very often of hexagonal symmetry with sizes on the order of 10–100 μm.

In addition, the atmosphere also contains other liquid and solid particles called aerosols [19.20, 76, 81, 82]. These include many types of particles from dust, oil droplets from forest fires, volcanic ashes as well as anthropogenic droplets/particles with an enormous variety in composition. Typical aerosol particles have size distributions in the range 0.1–10 μm with an average around 1 μm (e.g. [19.83]). The vertical distribution of the particle components of air may also be described by some kind of exponential formula [19.77], however, the scaling height of 1–2 km is much smaller than that for molecules.

The scattering of light from particles in these size ranges differs dramatically from molecular scattering, since the particle size is comparable to or larger than the wavelength of the light. Since atoms/molecules within particles are very close to each other, they are excited coherently, i. e. the total scattering from all molecules within a particle is quite different from the sum of the scattering of all individual molecules within the particle. In 1908, *G. Mie* gave the solution of classical electrodynamics for the simplest case, referring to spherical shapes of the particles [19.10]. Henceforth, this type of light scattering from particles is mostly known as Mie scattering. This name is sometimes also used for extensions to nonspherical particle shapes. For nonabsorbing particles, which are small compared to the wavelength, the Mie scattering gives the same results as Rayleigh scattering. In the general case, however, it shows pronounced differences regarding wavelength dependence, angular dependence, degree of polarization of the scattered radiation as well as scattering cross section per molecule [19.40–42, 77].

Comparison of Rayleigh and Mie Scattering

Wavelength Dependence. Figure 19.16 gives a schematic example of the scattering of light by water molecules as well as water droplets of, say, 10 μm. Whereas molecules or very small particles scatter light mostly according to the inverse fourth power of wavelength, scattering from water droplets is essentially independent of wavelength.

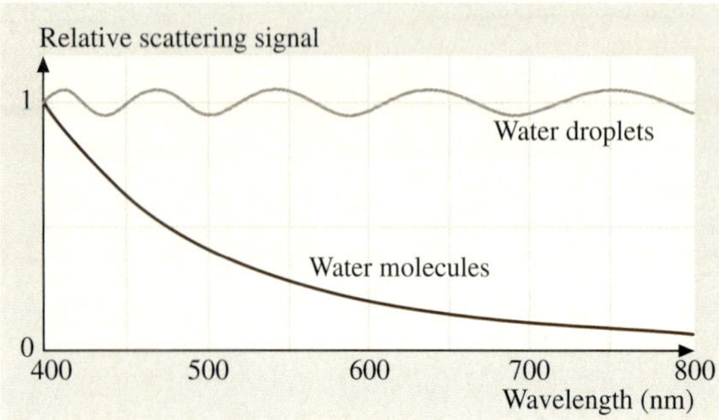

Fig. 19.16 Schematic wavelength dependence for light scattering from water molecules and water droplets of, e.g., 10 μm size

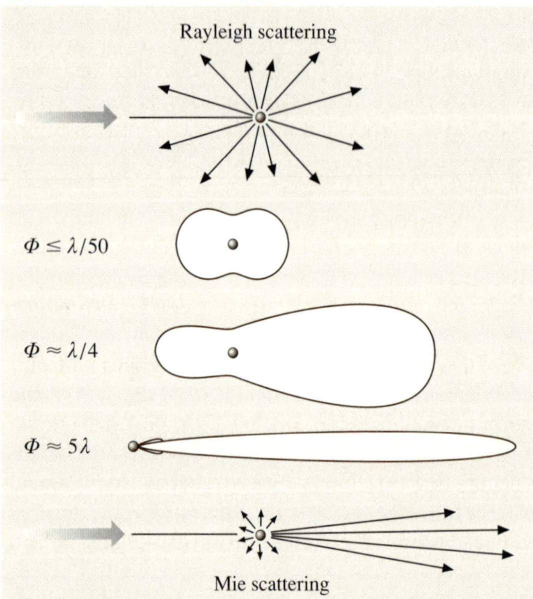

Fig. 19.17 Schematic overview of the angular dependence of light scattering from molecules and very small particles (Rayleigh scattering) to larger particles (Mie scattering)

Angular Dependence. Figure 19.17 gives a schematic overview of how the angular dependence of scattering changes with particle size. Molecules scatter unpolarized light nearly isotropically (see also Fig. 19.1) whereas large particles scatter rather asymmetrically in the forward direction. For a given particle size, the scattering also depends on wavelength.

Degree of Polarization. For Rayleigh scattering, and assuming perfectly spherical air molecules, light scattered at an angle of 90° would be perfectly polarized [19.84]. However, air molecules already show an anisotropy, reducing the polarization to only about 94% [19.77, 79]. For larger particles, the degree of polarization varies considerably as a function of scattering angle for fixed size. These patterns, however, strongly depend on size [19.40, 41, 77]. In addition, the degree of polarization at a fixed angle varies as a function of wavelength.

Scattering Cross Section per Molecule:. The enormous differences between incoherent and coherent scattering become apparent when comparing light scattering from isolated molecules to those of the same molecules within a particle, e.g. water molecules within a water droplet or a raindrop [19.77]. For example, the scattering per molecule that belongs to a cloud droplet of 1 μm is about 10^9 times larger than scattering by an isolated water molecule and still a factor of about 100 larger than a molecules in a 1 mm raindrop.

Single and Multiple Scattering, Optical Mean Free Path and Air Mass

Light from the sun (or moon) has to traverse the atmosphere before reaching the eye of an earthbound observer. It is attenuated by scattering and absorption, depending on the number of scatterers present in the line of sight. If absorption can be neglected, one may use the scattering coefficient

$$\beta_S = N\sigma \,, \tag{19.35}$$

where N is the number of scatterers per unit volume and σ is the extinction cross section, $1/\beta_S$ defines the scattering mean free path, i. e. the average distance a photon has to travel before being scattered. (If absorption is included, an additional absorption coefficient β_A is introduced and β_S will be replaced by $\beta = \beta_S + \beta_A$.) Using β_S, one may define the optical thickness τ between two points A and B of the atmosphere as

$$\tau = \int_A^B \beta_S(x)\,dx\,. \tag{19.36}$$

The optical thickness is a measure of the physical thickness in units of the scattering mean free path. For light along a radial path, i. e. coming from the zenith, the amount of air to be traversed is least and one refers to the normal optical thickness, which is also called the optical depth. For a pure molecular atmosphere, this is depicted in Fig. 19.18. In the visible spectral range, its values are small compared with unity, hence, a photon is unlikely to be scattered more than once, i. e. single scattering predominates.

The amount of air to be traversed by light, the so called air mass, depends on the sun's elevation. Air mass (AM) is defined to be unity for the light source in the zenith. It increases for smaller sun elevation angles. As a matter of fact, it is the optical thickness, relative to its zenith value. The exact value depends on the surface pressure. At 5° elevation (85° zenith angle) its value is about 10 and it ultimately reaches a value of about 40 at grazing incidence [19.51, 77, 85, 86], i. e. for the case of sunrise or sunset. In this case, the light must pass about 40 times the amount of air compared to the case of the zenith.

Combining the air mass with the normal optical thickness gives the optical thickness for arbitrary sun elevation. Since AM reaches large values, optical thicknesses may become large compared with unity, i. e. multiple scattering events become likely for low sun elevations.

19.6.2 Mirages

Mirages are due to the propagation of light waves in air that has gradients in the index of refraction [19.44–48, 51]. In general, light which is propagating in clear air (no particles) is partly scattered according to Rayleigh scattering, while the rest propagates undisturbed in the forward direction. The scattered light consists of sideways scattering and forward scattering, the latter being in phase with the incident light. As a consequence, light in clear air is slightly weakened by sideways scattering, the forward-propagating light being the superposition of the incident light and the forward-scattered light. The weakening due to sideways scattering will be discussed below (blue sky). Here, the phenomena associated with the forward-propagating light will be

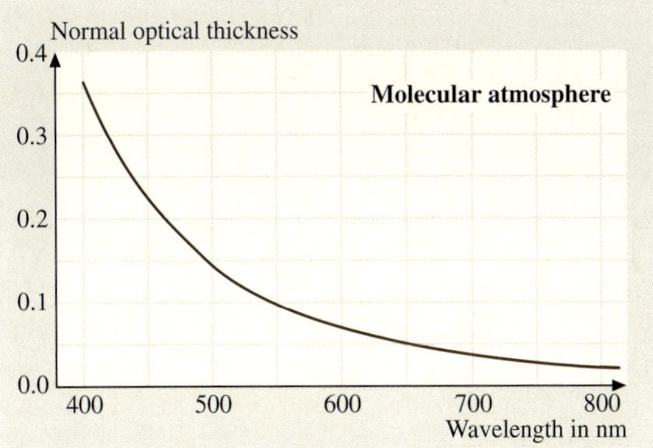

Fig. 19.18 Normal optical thickness of a purely molecular atmosphere (data from [19.3])

discussed. Scattering effects by particles are neglected here.

In homogeneous media, the forward propagation of light is expressed in terms of the refractive index n (for tables of n [19.3]). n depends on temperature T (in K), pressure p (in mbar) and humidity. Equation (19.37) [19.87] gives an approximation for n of dry air

$$n = 1 + \frac{77.6p}{T} \times 10^{-6}\,. \tag{19.37}$$

For example, $T = 300\,\mathrm{K}$ and $p = 1030\,\mathrm{mbar}$ yields $n \approx 1.000266$. Increasing the air temperature by 40 K (e.g., air directly above the ground which is heated by the sun) gives $n \approx 1.000235$ (for humid air, the numbers only change slightly). These small changes of $\Delta n \approx 3\times 10^{-5}$ are responsible for all kinds of mirages.

The most simple mirage effect is known from astronomy: if an observer looks at an object in the sky at a certain altitude, the actual position of the object is lower in the sky. This is due to the refraction of light upon propagation in the atmosphere with a gradient in n giving rise to curved light paths.

At the zenith, the deviation is zero, at 45° zenith distance it amounts to 1′ (1 minute of an arc) increasing to more than 38′ at an angle of 90°. This has to be compared with the angular size of the sun or moon, which is about 30′, i. e. half a degree. Since the refraction is only about 28′ at a zenith distance of 89.5°, the sun or the moon in an otherwise undisturbed atmosphere (no inversion layers) will thus be flattened [19.47], see also [19.88–92].

Mirages are images of objects which may be seen in addition to the real objects. They can be seen whenever there are temperature differences in the atmosphere,

Fig. 19.19 Inferior mirage: a palm tree scatters sunlight in all directions. The atmosphere is heated close to the ground. Therefore the index of refraction is smaller there, giving rise to curved light paths. Direct and refracted light can enter the eye of an observer, which is interpreted as an object on top of an inverted image (after [19.48])

which give rise to unusual gradients in the index of refraction. Similarly to astronomical refraction, gradients in n give rise to curved light paths. If warm air is above the ground, n is smaller close to the ground than above it. This leads to inferior mirages like the well-known wet streets on hot summer days. The formation of inferior mirages is illustrated in Fig. 19.19. A prominent feature of a mirage is the so-called vanishing line of the object [19.48]. Light below certain object points has no chance whatsoever to reach the eye of the observer, i. e. parts of the object cannot be seen, irrespective of the direction into which the light is scattered.

In the case of an inversion layer in the atmosphere, the air at a given height is warmer than below, which may give rise to superior mirages. Similarly to inferior mirages, the propagation of light with curved paths in this case can give rise to multiple images on top of the object. In nature, combinations of inferior and superior mirages may occur. Also, heated vertical walls may result in lateral mirages.

The angular size of mirages is usually about 0.5–1.0°. For observations, binoculars and cameras with telephoto lenses are used. The mirages are often distorted and flicker due to local density fluctuations in the air.

Theoretical progress in understanding mirages came with the computer. The propagation of light rays in a given atmosphere, defined by the vertically varying index of refraction $n(z)$, is computed theoretically using ray-tracing methods. The various methods [19.93–96] differ by their choice of geometry and $n(z)$. Such simulations prove very helpful in understanding the general characteristics of mirages, like e.g. so-called mock mirages [19.97, 98].

Recent research has explained many unusual mirage phenomena quantitatively, including long-range mirages like the Novaya–Zemlya effect from atmospheric ducting [19.99–102] and small-scale movements seen within superior mirages when atmospheric gravity (buoyancy) waves are present [19.103, 104].

Superior mirages seen over large distances (70–100 km) can result from fairly complex atmospheric temperature profiles [19.95]. Also, the origin of double inferior mirages as well as unusual horizontal stripes in inferior mirages has been analyzed theoretically [19.96]. Very recently, focus had been on very precise measurements and comparison to modeled astronomical refraction of the setting sun [19.105, 106] and the occurrence of very bright superior mirages [19.107].

In addition to analysis of observations and theoretical models, inferior mirages and superior mirages with multiple images may also be studied quantitatively or as demonstration experiments in the laboratory [19.65, 108–112].

19.6.3 Clear Sky: Blue Color and Polarization

Light scattering in pure gases is due to the electrons in atoms and molecules. As a result, the scattered light intensity varies approximately as $1/\lambda^4$ (Fig. 19.16), i. e., red light is scattered much less than blue light [19.1, 78, 79]. When looking at the daylight sky illuminated by the sun one mostly sees scattered light and, hence, a blue sky (In Sect. 19.6, the qualitative term intensity which is often not properly defined (see [19.113]), does usually mean either radiance, given in W/m^2 sr, or spectrally resolved radiance in W/m^2 sr nm).

The scattered light is strongly polarized at a scattering angle of 90° [19.84]. Concerning the whole sky, the polarization varies and also has neutral points, named after Arago and Babinet [19.44, 46, 51]. In reality, even if very clean air is used, the maximum degree of polarization is about 94%, see Sect. 19.6.1. The atmosphere usually also contains an appreciable number of particles, which show Mie scattering with different angular polarization dependencies. In addition, there may be contributions of multiply scattered light as well as backscattering from the surface of the Earth. Overall, the degree of polarization typically reaches values around 80%. Obviously, this depends on the concentration of particles and is thus related to the transmission of the atmosphere [19.114]. More details on scattering effects and recent research will be given in Sect. 19.6.7.

19.6.4 Rainbows

Rainbows are due to scattering of sunlight from raindrops in the atmosphere [19.43–48,51]. The drops are in the size range between 10 μm (white fogbow/cloud bow) and several millimeters. One usually assumes spherical droplets, however, drop shapes may vary depending on size [19.115, 116] or they may even oscillate [19.117]. Drop formation and size distributions have been extensively studied [19.118]. The effect of nonsphericity on rainbow phenomena is discussed, e.g., in [19.119, 120].

The observed bows are mostly due to single scattering events. Hence, the complex problem of light scattering from millions of raindrops in a rain shower can be reduced to the scattering of sunlight from a single raindrop. Since raindrops are large compared to the wavelength of light, geometrical optics already gives a rough description. Figure 19.20 depicts a single raindrop and parallel light rays from the sun.

The rays are characterized by their impact parameter b. Each ray is reflected and refracted upon hitting an interface between air and water. For clarity, only the light paths that give rise to the primary rainbow are sketched. They are defined by refraction into the droplet, internal reflection and refraction back into the air. Similarly higher-order rainbows may be constructed by allowing for multiple internal reflections. The central ray (1) is not deflected by the refractions and exits at a scattering angle of 180° with respect to the incident ray. Ray number (2) suffers a scattering angle of about 170° and so on. For increasing impact parameter the scattering angle δ decreases until a minimum is reached at an impact parameter of about $0.86\,R$ (ray 7). The flatness of this shallow minimum (Fig. 19.20) at an angle of about 138° is responsible for the fact that much more light is scattered into this direction – geometrically along the surface of a cone – than in others, since many incoming light rays of slightly differing impact parameters will emerge in the same output direction. As a consequence, an observer who is looking at a cloud of raindrops that is illuminated from the back by sunlight, will observe a bow of angular size 42° (Fig. 19.21).

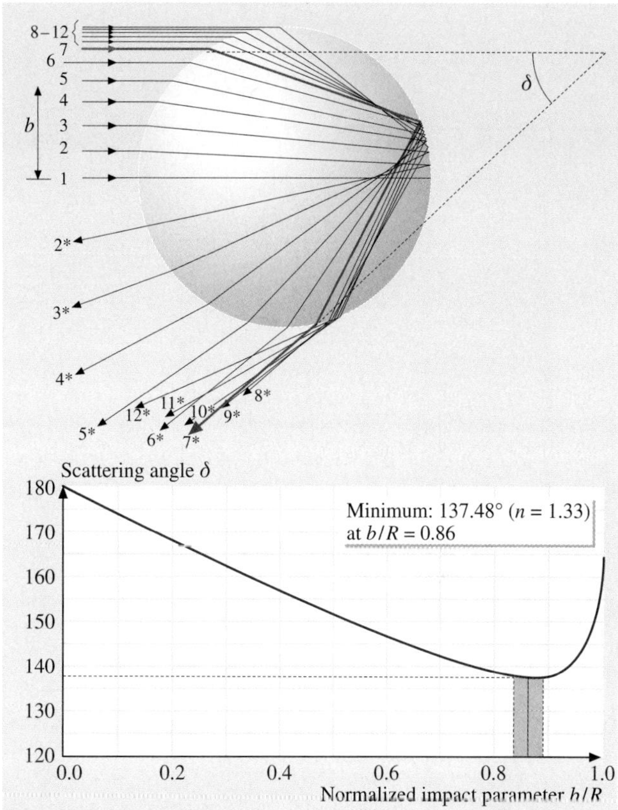

Fig. 19.20 Simple explanation of a rainbow from a single raindrop in terms of geometrical optics. Parallel light rays, characterized by the impact parameter b, are scattered due to refraction, internal reflection and refraction (for clarity other light paths are omitted). The scattering angle δ as a function of b exhibits a shallow minimum, i. e. many incoming light rays with different values of b will be scattered in the same direction

The rainbow angle, i. e. the minimum of δ in the plot of Fig. 19.20, can be easily calculated (e.g. [19.47, 121]) with geometrical optics by finding the minimum of the scattering angle $\delta(\alpha_{\text{inc}})$ with respect to the angle of incidence α_{inc} of the light rays onto the droplet. Assuming $n_{\text{air}} \approx 1$ one finds for the primary bow

$$\delta(\alpha_{\text{inc}}) = 2\alpha_{\text{inc}} - 4\alpha_{\text{refr}} + \pi\,, \quad \text{with}$$
$$\cos\alpha_{\text{inc}} = \sqrt{\frac{n^2-1}{3}}\,. \tag{19.38}$$

Part D | 19.6

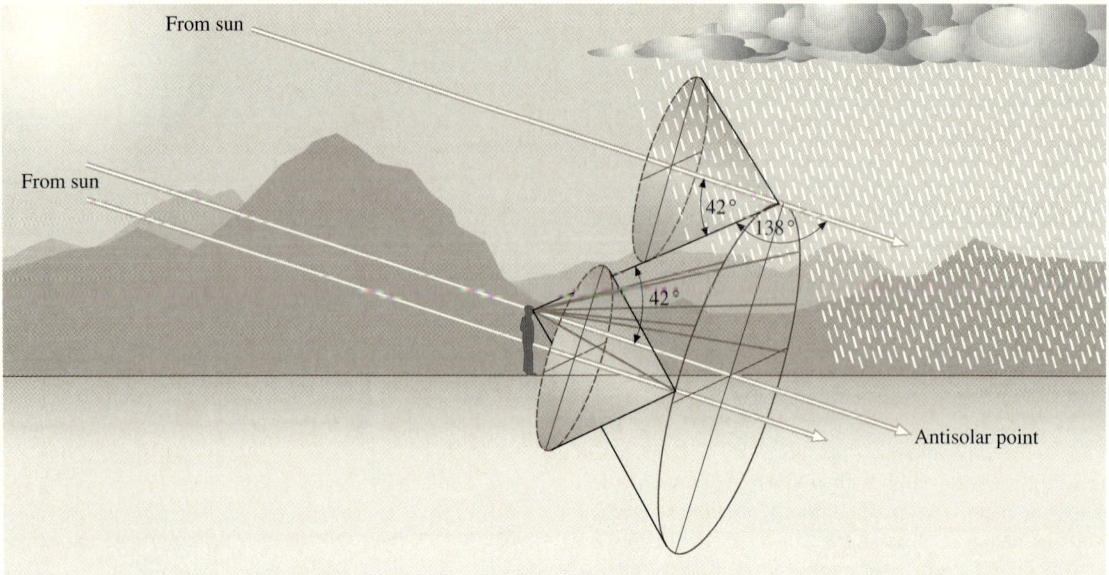

Fig. 19.21 Observation of a rainbow. Each raindrop scatters light preferably in a cone of the rainbow angle. Hence, an observer will see more light when looking in the direction of the surface of a cone that is centered in a droplet. The whole rainbow is due to millions of drops, the effect of which is an observable bright feature, geometrically on the surface of another cone centered in the eye of the observer and making an angle of 42° with regard to the antisolar point (after [19.48]). Its surface touches the surfaces of the scattered light cones of each individual droplet

Here α_{refr} denotes the angle of the refracted ray within the raindrop and n is the index of refraction of water. For $n = 1.33$, Snell's law gives a scattering angle of $\delta(\alpha_{inc}) = 137.5°$. Due to dispersion ($n_{650\,nm} = 1.331$, $n_{400\,nm} = 1.343$) the rainbow angles depend on wavelength. Taking into account the finite size of the sun of about 0.5° one finds an angular width of about 2.2° for the primary bow. The analysis is easily extended to higher-order rainbows [19.47], most of which may, however, only be observed in the laboratory [19.110, 122–125].

Several rainbow features can only be explained by wave optics. Light from the rainbow is appreciably polarized perpendicular to the plane of incidence, since the angle of incidence for the internal reflection is close to the Brewster angle [19.126]. Also, careful measurements showed that the light intensity shows an interference pattern, named after Airy [19.50, 127, 128], which differs markedly from the prediction of geometrical optics (Fig. 19.22).

Until the beginning of the 20th century, wave optics features were computed using Airy's theory, which gives reasonable results for droplet sizes down to 10 μm; for smaller sizes, it fails. Improvements came with Mie theory [19.10, 40, 41], however, its numerical solutions in terms of scattering efficiencies as a function of angle for size distributions of the droplets and finite angular size of the solar disk had to wait for the development of computers. Similarly, Debye developed

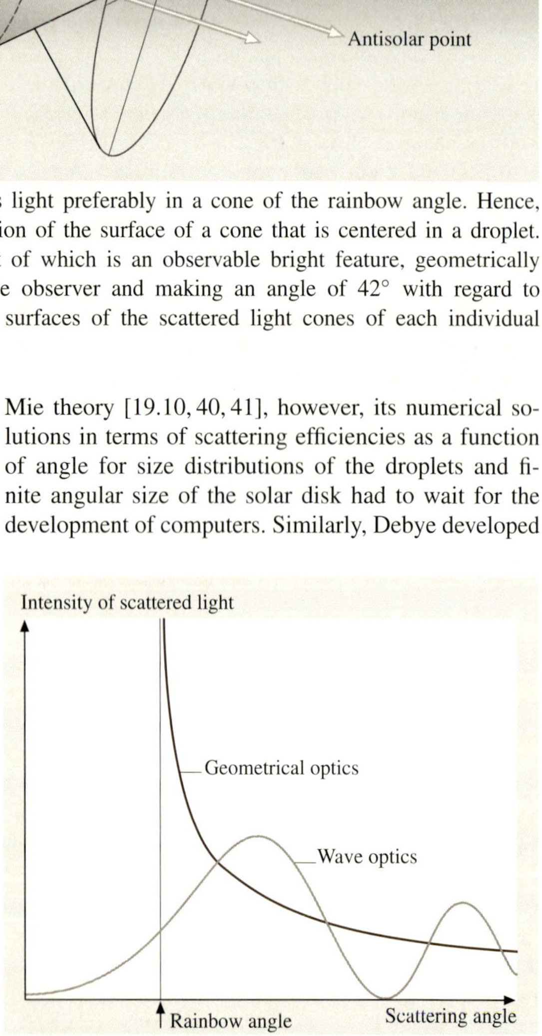

Fig. 19.22 Differences between the theories of Descartes and Airy for the scattered light intensity of raindrops

a tool of how to decompose the partial wave amplitudes for scattering of a plane wave into individual contributions of certain types. This later led to the complex angular-momentum theory [19.129]. Application of the mathematics of catastrophe theory to light scattering showed that the rainbow is a manifestation of the so-called fold caustic [19.130].

Here, results of the full electrodynamic treatment, i. e. Mie theory, are presented [19.40,41,131–133]. Figure 19.23 depicts results for monochromatic blue light of $\lambda = 450\,\mathrm{nm}$, scattered by a lognormal distribution of water droplets with mean radii ranging from $2\,\mu\mathrm{m}$ to $200\,\mu\mathrm{m}$. The full width at half maximum of the size distribution is 10% of the mean radius.

Obviously, large droplets show Airy rings at the rainbow angles. For other wavelengths, these curves would be slightly shifted due to dispersion. The overlapping of the interference rings for all wavelengths leads to so-called supernumerary arcs (e.g. [19.134, 135]). Decreasing the droplet size leads to a broadening of the rainbows due to diffraction. In this case, the overlapping of all wavelengths leads to a white rainbow, also called a fog bow. If the size becomes too small, no rainbows are observable. The other features in Fig. 19.23 will be discussed in Sect. 19.6.5.

There are many other interesting features of rainbows. For example the first-order rainbow often visually appears brighter at the base of the bow than at the top of the bow. This is due to the fact that, in a rain shower, raindrops are distorted in shape to oblate spheroids. Light reaching an observer from the base of a rainbow traverses a water droplet in the horizontal plane where the droplet cross section is circular. In contrast, light from the top of the rainbow traverses the water droplet in the vertical plane where the droplet cross section is elliptical. Since the scattered light intensity due to the internal reflection is less in the elliptical cross section than that in the circular cross section the base of the rainbow is brighter than the top of the bow (the longer optical path from the top may also lead to increased extinction).

Similar explanations [19.136] are possible for other features such as, e.g., that the supernumeraries of the first-order rainbow often appear most vividly at the top of the bow or that supernumeraries of the second-order rainbow are almost never seen. Also, the effects of the electric fields [19.137] and acoustic forces accompanying thunderstorms can have observable consequences on rainbows [19.117]. Visibility and general brightness of rainbows depend on multiply scattered and absorbed

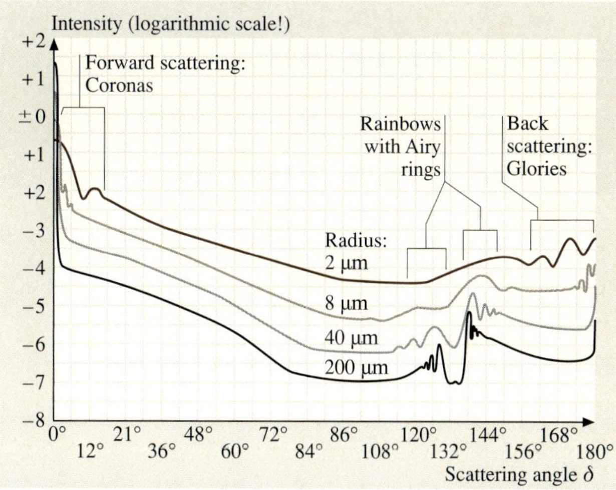

Fig. 19.23 Angular distribution of blue light scattered by water droplets with mean radii ranging from $2\,\mu\mathrm{m}$ and $8\,\mu\mathrm{m}$ (fog and cloud droplets) to $200\,\mu\mathrm{m}$ (small raindrops) (computation by E. Tränkle)

light [19.138, 139]. Even infrared rainbows have been detected [19.140].

In addition to analysis of observations and theoretical models, rainbow phenomena may also be studied in the laboratory either for quantitative comparison to theory or just as demonstration experiments using either single droplets or cylinders of water or other materials [19.109, 110, 122–124, 136, 141, 142].

Present investigations of rainbow phenomena deal, e.g., with experiments on higher-order rainbows from acoustically levitated water drops [19.143] or cylinders [19.144, 145] and total internal reflection rainbows [19.146] as well as new simulations of rainbows and fogbows [19.147, 148].

19.6.5 Coronas, Iridescence and Glories

The forward and backward scattering of small water droplets can also give rise to very distinct features [19.43, 44, 47, 48, 51]. First, colored concentric rings, called coronas, around the sun or the moon with angular distances up to, e.g., 15° may be observed, when looking through thin clouds. The middle whitish part around the light source is often called an aureole. If, second, colorful clouds are observed at larger distances from the sun (up to 45°), one usually speaks of cloud iridescence. Third, backscattering that leads to colorful concentric rings at angles close to 180° with respect to the light source are called glories. Figure 19.24 gives an

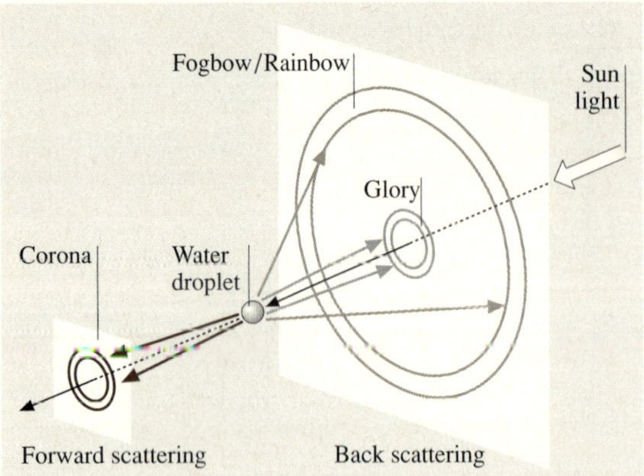

Fig. 19.24 Overview of geometries of light-scattering phenomena from a water droplet

overview of the light-scattering geometries of rainbows, coronas and glories due to water droplets.

Coronas

In its most spectacular form, coronas can be observed as a concentric series of three or four brilliantly colored rings around the sun or moon. In its simplest form, the corona is just represented by the aureole, which is a white disk near the sun or moon bordered by a bluish ring and terminated by a reddish-brown band.

Coronas are easily explained using Mie theory [19.149, 150]. Coronas occur as strong forward-scattering phenomena (Fig. 19.23) for very small nearly monodisperse sizes of typically about 10 μm. More simply, coronas are treated as diffraction phenomena by water droplets. The first minimum for light of wavelength λ diffracted by a circular aperture of diameter $2R = D$ is given by

$$D \sin\phi = 1.22\lambda \,. \tag{19.39}$$

For droplets of nearly uniform size, the diffraction angles ϕ obviously depend on wavelength, giving rise to rather pure perceived colors of the coronas. In this case, the average droplet size may be estimated from the angular size of the rings from (19.39). To obtain the approximate angle for the red band of the solar corona, it is customary to use $\lambda = 570\,\text{nm}$ in (19.39) since the positions of the red bands are believed to correspond to the minima for green light [19.151]. (Alternatively to 570 nm, 490 nm has been proposed [19.152].) For example, $2R = 10\,\mu\text{m}$ gives $\phi \approx 4°$. (Note that there will be deviations between the correct Mie theory and diffraction theory for droplets with sizes of 10 μm and below.) A much better way to judge droplet sizes is to fit the colored rings of computer simulations to those of photos of natural observations [19.150].

An aureole may be produced by relatively large particles that tightly compress the rings, or by the presence of a broad particle size distribution that causes the colored rings to overlap.

Simple diffraction theory has problems for drops with sizes only slightly greater than the incident wavelength. In this case, Mie theory must be applied, since the amplitudes of the diffracted and transmitted light are comparable in the forward-scattering direction, leading to interference between the two components that can significantly alter the scattering pattern.

The nature of the cloud particles that cause coronas has been the subject of controversy [19.152, 153]. Observations of some coronas due to very high clouds suggested that not only supercooled spherical cloud droplets, but also ice crystals may be responsible for coronas. However, ice crystals can assume a variety of shapes and orientations that should not give very pure coronal colors. The issue was settled by experiments which simultaneously used polarization LIDAR data as well as photographic evidence of coronas to demonstrate that cirrus clouds composed of hexagonal ice crystals with particle sizes ranging from 12 to 30 μm (which is unusually small for most cirrus clouds) can generate multiple-ringed colored coronas [19.152]. Later the study also included aircraft probes of the particles [19.153]. In one case a weak corona and a halo display (see below) of the hexagonal crystals were observed simultaneously within the same cloud. This is possible, since laboratory experiments indicate a lower limit of about 20 μm for the sizes of ice crystals that can produce halo effects.

Some while ago, it was also noted that coronas need not be spherical. Elliptical coronas have been reported at special times in early summer. They were explained as diffraction phenomena from nonspherical birch and pine pollen grains [19.154–156]. Also, split coronas due to local variations of droplet size or changes between water and ice have been observed [19.157]. Recently, new simulations have been reported [19.158]. Special coronas, called Bishop's rings, can be produced by volcanic ashes, as has been reported after the explosion of Krakatau in 1883.

Iridescence

Cloud iridescence is a brilliant display, which occurs quite often when appropriate thin clouds are close to the sun or moon or if edges of thick clouds happen to have rather monodisperse droplets. Iridescence is also observed, though more rarely, at much larger angular separation from the sun of up to 45° and it may be easily observed and analyzed within contrails [19.151]. If interpreted using diffraction, like coronas, the drop sizes $2R$ are in the range 2–4 μm. These sizes are indeed quite rare, though possible near the edges of growing or evaporating clouds.

Whereas, coronas are generally observed at smaller angles for mostly monodisperse droplet populations with sizes of about 10 μm, iridescence usually occurs at larger angles and requires cloud droplet sizes on the order of a few microns in diameter with broad size distributions. These conditions are often found in the vicinity of the visible cloud margin where there is also a sharp gradient of the drop sizes. Thus, coronas and iridescence appear to require quite different cloud microphysical conditions to be present, and should also be most frequently observed in different angular regions [19.151]. Recent studies deal with coronas and iridescence in mountain wave clouds [19.157] as well as rare iridescence cases in cirrus clouds [19.159, 160].

Glories

The colored rings of glories, sometimes also called spectre of the Brocken are observed in a backscattering geometry, if sunlight illuminates fog or clouds with very small water droplets. Nowadays, observations can be performed easily when flying in airplanes and observing the shadow of the plane on nearby clouds.

Early interpretations could not use Mie theory (since computers were not available), but tried to understand the phenomenon in simpler terms. The assumption of diffraction of reflected light by the droplets (diffraction from circular apertures) did not work, since intensities and angular distances of the pattern are different. However, diffraction from circular rings could explain the phenomenon [19.40, 161]. For this to happen, glories must be due to light entering the raindrops at grazing incidence (this interpretation was later supported by Mie theory [19.162]). Then, the scattered light is more or less exiting at the opposite side of a droplet, i. e. at scattering angles of 180°. In terms of geometrical optics, this is not possible, however, wave theory allows for surface waves at the boundary. In the backward direction, all contributions interfere constructively to produce the glory. These model assumptions were supported by experiments with microscopic water droplets supported by submicroscopic spider webs [19.163]. Later on, this model was further developed very successfully [19.129]. Nowadays, glories are explained by using Mie theory. Glories are present in Fig. 19.23 as oscillatory structures in the angular range from about 160 to 180° for droplets of 4 and 16 μm size. The more uniform the droplet size, the more rings may be observed.

Knowing from coronas that colorful rings may also be produced by fairly monodisperse assemblies of ice crystals and even tree pollen, one may ask whether the glories that are often observed from airplane windows may also be due to ice crystals in high-altitude cirrus clouds.

From the above interpretation of Mie theory, the glory is due to circumferential, i. e., surface-wave ray paths which are unique to spheres. Glories are, hence, produced by spherical scatterers. Based on photographic evidence, it was therefore investigated [19.164] whether glories may be also due to spherical or near-spherical ice crystals? In conclusion it was found that ice grows in amorphous and presumably spherical shapes only at unlikely frigid temperatures. Recent studies have, however, suggested that small near-spherical ice particles (diameters ranging from 9 to 15 μm) can sometimes, though very rarely, occur near the tops or along the margins of some ice clouds, and these may give rise to observable glories. Very recently, new simulations for glories were reported [19.147, 165, 166].

19.6.6 Halos

Halos are caused by the refraction and/or reflection of sunlight by ice crystals in the atmosphere [19.43–48, 51, 52, 66, 167]. They can already be understood in terms of geometrical optics. Ice crystals that cause good halo displays are usually hexagonal plate or column crystals with sizes in the range of 20–100 μm [19.168]. Such crystals form very often, depending on the temperature and the amount of water vapor in the air [19.169]. Good documentation with micrographs can be found in the literature [19.52, 170].

A great number of different halos may be observed in nature and in the laboratory since there is an enormous variety of possible types of ray paths through the crystals and crystals may have different orientations in the air.

The most simple halos are parhelia, also called sun dogs, which are often observed in cirrus clouds (and sometimes in contrails [19.171]). They are due to oriented plate crystals, i. e. crystals falling with their principal axes vertical. This orientation mode is often possible and is, e.g., similar to movements of leaves (or sky divers), which maximize their air resistance as they fall [19.118]. If light rays from the sun enter a prism face

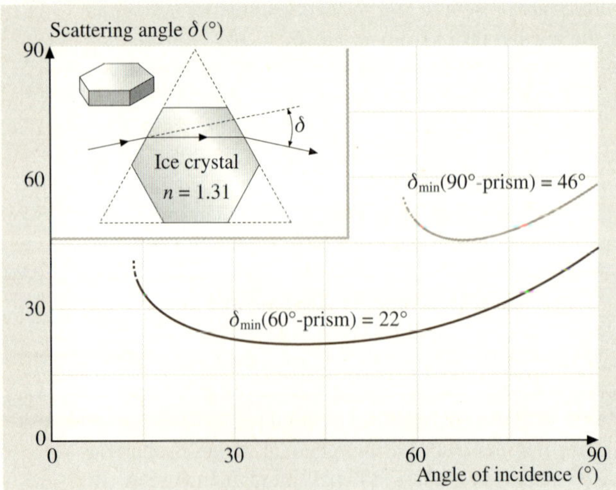

Fig. 19.25 Ray paths within hexagonal ice crystals ($n = 1.31$) show an angle of minimum deviation of 22°. For clarity, other light paths are omitted

of a crystal and exit an alternate prism face, the situation corresponds to refraction from a 60° prism. The scattering angle as a function of angle of incidence shows a flat minimum (Fig. 19.25) that is at a scattering angle of 22° for ice crystals with $n = 1.31$. Similarly to the explanation of the rainbow, more light is deflected in this direction. Therefore an observer looking at the sun or the moon through a thin cirrus cloud will see more light at an angular distance of 22° from the light source. For randomly oriented crystals, the resulting display will be

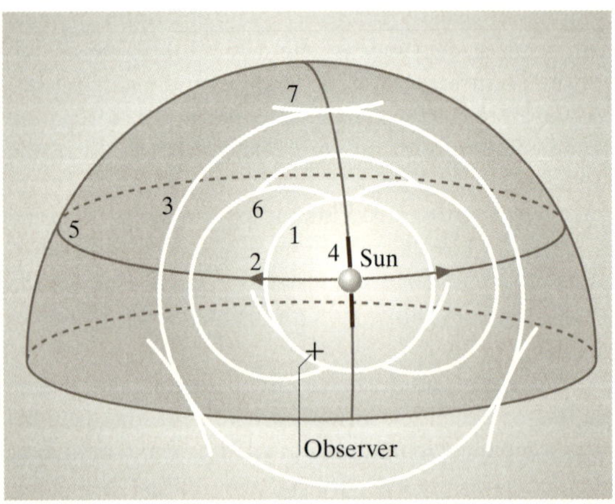

Fig. 19.26 A schematic survey of some halos which may be observed (1: 22° halo, 2: parhelia, 3: 46° halo, 4: sun pillar, 5: parhelic circle, 6: circumscribed halo, 7: circumzenithal arc). These displays all depend on sun elevation

a circular halo, centered at the light source, for oriented crystals, light will be deflected to both sides of the light source, giving rise to parhelia. The color of the parhelia, reddish towards the sun and bluish tails at the outside, is due to dispersion, i. e. different minimum deviation angles for the various wavelengths.

Similarly, other halos, often referred to as arcs, are due to other ray paths. A very brief survey of some common halos (today, more than 40 different types are known) is shown in Fig. 19.26. The circumzenith arc (No. 7 in Fig. 19.26) is due to rays which enter the top basal face and exit a prism face of crystals with vertical symmetry axis. For random orientation, ray paths which refer to such a 90° prism lead to circular halos of 46° (No. 3 in Fig. 19.26). Pure reflection halos – not colored, since refraction does not contribute – are e.g. sun pillars (No. 4 in Fig. 19.26). They arise when sunlight is reflected at grazing incidence from the horizontal faces of plate crystals oriented with their axes vertical. Similarly oriented crystals also give rise to the parhelic circle (No. 5 in Fig. 19.26), which is due to reflected sunlight from crystal faces lying in the vertical plane.

Singly oriented column crystals – column crystals with axes horizontal but otherwise unconstrained – give rise to the upper and lower tangent arcs. Much more rarely, column crystals orient with two prism faces horizontal. Crystals having these Parry orientations give rise to Parry arcs. Still other crystal orientation modes are possible. Furthermore, crystals occasionally have pyramidal faces that produce halos at other angular distances, referred to as odd-radius halos.

The modern theory of halos started in the 17th century [19.44, 172]. In recent decades, enormous progress has been made. This is due to computer techniques, the application of more traditional, conceptual mathematics [19.173], experiments which collected atmospheric ice crystals during observation of halo displays in cold climates, and to dedicated observers who have documented many new halos.

Probably the biggest step was made by the introduction of computer simulations which allow the computation of complex halos from many different types of crystal shapes, sizes and orientation. They gave a lot of new information, especially regarding intensity variations within a given halo [19.48, 174, 175]. Various arcs could be directly related to the crystals and the ray paths that were producing them. The theoretical displays could be easily compared with photographs or drawings of real halo displays.

Many excellent displays were documented by Finnish or German observer networks ([19.176], see also Table 19.4) and scientists in Antarctica [19.52]. Computer simulations were also used to explain old documented displays dating back to the 18th century [19.177]. Analysis of halos also include polarization effects, started by systematic studies in the seventies [19.53, 178]. Halo polarimetric studies should allow the detection and identification of birefringent crystals in extraterrestrial atmospheres [19.179–181]. Very recently, the focus had been on comparison of observations with the simulation of halo polarization profiles, the size and shape of the crystals in the model being based on sampled ice crystals [19.182].

Halo phenomena have also been studied in the laboratory for quantitative comparison to theory and as demonstration experiments using prisms or hexagons made of glass, lucite or other materials [19.43, 66, 109, 110, 112].

19.6.7 The Color of the Sun and Sky

The Color of the Sun

Light scattering in the atmosphere is responsible for the color of the sun or moon, as perceived by earthbound observers. Outside the atmosphere, solar irradiation has a very broad spectrum, ranging from UV to the deep infrared [19.183]. Neglecting processes in the sun's atmosphere, it more or less resembles the emission of a black body of about 5900 K. This radiation is interpreted as white by human eyes (the exact color may be computed from spectra according to color metric formulae, see e.g. [19.76]). In the Earth's atmosphere, the radiation is absorbed and scattered. Compared to the outside spectrum, the irradiation curve is lowered and shows characteristic absorption features, mostly from H_2O, CO_2 and oxygen (see Fig. 19.7).

Color changes occur depending on air mass, i.e. sun elevation. Starting with a purely nonabsorbing molecular atmosphere, atmospheric transmission has been computed for air mass AM = 1 [19.3]. Therefrom the transmission at arbitrary zenith angles φ and starting at height h of an observer may be computed according to [19.86, 114]

$$T(\varphi, h) = T(0, h)^{\mathrm{AM}(\varphi,h)} \quad (19.40)$$

Results for the transmission of red ($\lambda = 630$ nm), green ($\lambda = 530$ nm) and blue ($\lambda = 430$ nm) light for a molecular atmosphere and AM = 1, ($\varphi = 0°$, zenith), AM = 10 ($\varphi \approx 85°$) and AM = 38 ($\varphi = 90°$, horizon) are given in Table 19.6.

Table 19.6 Transmission of a molecular atmosphere ($h = 0$) for sun in zenith (AM = 1: $\varphi = 0°$) or close to the horizon (AM = 10: $\varphi = 85°$, AM = 38: $\varphi = 90°$), $T(\varphi = 0°)$ after [19.3]

λ(nm)	T(AM = 1)	T(AM = 10)	T(AM = 38)
630	0.945	0.625	0.117
530	0.892	0.320	0.013
430	0.764	0.068	0.000036

Whereas the zenith sun is still regarded as white, the sun's spectrum near the horizon is drastically changed in favor of red and green light. Above the horizon, yellow colors may be observed and close to the horizon, red dominates. Since, simultaneously, the radiance of the sun decreases drastically, the brightness also decreases with φ. Naked-eye observations are, however, still not possible in very clear air even if $h = 0$ (sea level). In general, however, haze due to water vapor and aerosols leads to additional weakening.

Aerosols are very important for optical variations within the atmosphere [19.15, 16]. The difference of observations and theoretical spectra, including selective absorption of molecules within the air, is attributed to aerosols [19.114].

Extraordinary color changes of the sun can sometimes occur, if clouds of nearly uniformly sized particles are present in the line of sight. For example, the forest fires in Canada in 1950 ejected large numbers of oil droplets in the air. These were transported over long distances and led to blue sun observations even in Edinburgh [19.74, 184]. Similar events may be due to volcanic eruptions.

Sky Colors

Obviously, sky colors are due to many factors [19.43–49, 51]. A simple qualitative explanation of typical sky colors at sunset can be given according to Fig. 19.27.

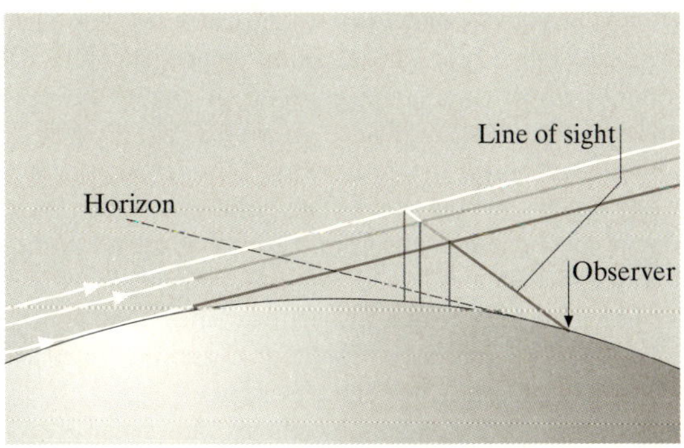

Fig. 19.27 Geometry for qualitative explanation of sky colors after sunset (see text for details)

The sun has just set, i. e. it is below the horizon. Sunlight still illuminates portions of the atmosphere. The color that an observer perceives in a given direction is due to contributions of scattered and attenuated light along a line of sight in the respective direction. In simple terms, the amount of radiation entering the observers eye depends on, first, the light path through the atmosphere before scattering, second, the height of the scattering event, which determines the density of scatterers, third, the scattering angle, and fourth, the optical thickness along the line of sight, i. e. the attenuation due to scattering and/or absorption. The subtle interplay of these four parameters determines the spectrum of light reaching the eye of an observer and hence the perceived color.

Different locations along the line of sight will contribute to the scattered light with varying intensity as well as spectral composition. The light beam most distant from the Earth will reach the line of sight, nearly unattenuated, and a broad spectrum is available for scattering. Due to the large height, only a small amount of radiation is scattered towards the observer, with blue dominating. This light is further attenuated before reaching the observer. The final spectral composition of the light depends on the optical thickness that must be traversed. However, blue is most strongly attenuated, i. e. overall this light will not contribute much and will not dominate the color. The closer the light passes to the ground, the more its spectrum is shifted towards the red. Although less light will intersect the line of sight compared to the large height rays, it will contribute more, since it interacts at lower heights, i. e. with a much larger density of scatterers. Taking into account the further shift of the scattered light spectrum towards the red due to scattering along the line of sight shows that the sky color will be dominated by yellowish and red colors.

The brightness and color of the clear sky has also been investigated for various zenith angles φ [19.185]. For high sun elevations, the zenith sky of a pure Rayleigh atmosphere is blue (as observed for clear air), however, it would turn yellowish after sunset. This is in contrast to the observed blue color. The solution of this puzzle are the Chappuis absorption bands of ozone in the wavelength range between 500 and 700 nm. For high sun elevations, the ozone absorption contribution to sky light is small compared to Rayleigh scattering whereas it dominates for very low sun elevations.

Aerosols within the atmosphere are usually not distributed homogeneously, but in the form of accumulations, carried by the wind, thus undergoing constant qualitative changes, including condensation processes [19.114]. As a consequence, the extinction coefficient for various atmospheric compositions, e.g., for different types of clouds, and low- or high-altitude haze varies considerably [19.76]. For simplicity, let us now assume just a single aerosol layer. Its influence on sky colors is simply to intensify the red and yellow colors in the forward direction. This is due to the fact that the scattering angles are rather small and that aerosol particles show very strong forward scattering, compared to Rayleigh scattering (Fig. 19.17).

Similar qualitative arguments may also be applied to the phenomenon of the so-called blue mountains. During the day, the sky is illuminated by the sun. If the line of sight to a distant mountain is close to the horizon, the radiation entering the eye of an observer is due to two contributions. First, it consists of the light originally scattered by the mountain towards the observer. This light is attenuated due to scattering along the line of sight. Second, the observer will see sunlight that is scattered by the air along the line of sight. This air light will be blue in a Rayleigh atmosphere. Observing a scenery with mountains at various distances will change the amount of the two contributions: the more distant, the smaller the first will be and the more important the second contribution (see the discussion on visibility in Sect. 19.6.8). As a consequence, distant objects will appear to have a faint blue color [19.45, 51]. This scheme can be used to get a measure for the thickness of the atmosphere [19.186].

Very often, one observes that sky light close to the horizon is whitish. This can be understood in a similar way. Since no objects except the sky are observed, the first contribution from the above argument vanishes and only air light has to be considered. Sunlight which is scattered in a length interval close to the observer contributes a large amount of blue light due to Rayleigh scattering at low heights above the ground. Light which is scattered at larger distances from the observer is scattered at larger heights, i. e. contributes smaller amounts of predominantly blue light. In addition this light is attenuated and spectrally modified by additional scattering along the line of sight. As a result, light which originates far from the observer contributes small amounts of multiply scattered mostly red light. All contributions along the line of sight add up to give an overall white color. Since light from distant portions will cease to contribute if scattered out of the line of sight, the overall brightness has a limit [19.45, 51, 77].

Observations from Mauna Loa of the polarization and color ratio during twilight often show deviations from the clear-sky average. These were attributed to cloud or haze shadow effects and the importance of stratospheric dust layers was emphasized [19.187]. Stratospheric dust clouds from volcanic eruptions do strongly affect the degree of polarization of sky light and also shift the positions of neutral points by more than 15° [19.188]. Such investigations strongly profit from the development of LIDAR systems and the respective theories of backscattering from small particles [19.189–192].

Modern experiments simultaneously measure the sky light radiance distribution at three wavelengths, the degree of polarization, and the plane of polarization [19.193]. Recently, the color coordinates of sky light have been investigated, showing that sky light colors have a wide range of chromaticity curves, depending strongly on location. Sometimes difficulties arise in separating luminance changes from chromaticity changes [19.194]. Astonishingly, clear daytime skies show a local maximum of radiance near the astronomical horizon. Its angular width and elevation vary with the solar elevation, azimuth relative to the sun, and aerosol optical depth. This is understood in terms of second-order scattering processes [19.195].

When the sun is seen through thin clouds, it usually has a sharp edge, but occasionally is fuzzy. This is due to certain clouds with ranges of cloud optical thicknesses depending on the cloud particle sizes [19.196].

Present investigations of sky light deal, e.g., with digital imaging of clear-sky polarization [19.197], twilight modeling [19.198] as well as color and luminance asymmetries in clear [19.199] and overcast [19.200] skies. Sky light measurements have also been reported for Mars, based on data from the Viking and Pathfinder missions. The Martian atmosphere has a high dust-particle content resulting in very high sky brightness and the sky colors vary strongly with the angular position of the sun [19.201].

Theoretical modeling of atmospheric extinction greatly profited from computer technology. Nowadays, the transmission of the atmosphere is modeled by extensive databases. The most prominent of these are the HITRAN and LOWTRAN databases [19.76, 202]. In particular, the LOWTRAN program calculates the transmission spectrum for molecular absorption, Rayleigh scattering and aerosol extinction at a moderate resolution of $20\,\mathrm{cm}^{-1}$.

19.6.8 Clouds and Visibility

Clouds

Clouds are visible aggregates of water droplets or ice crystals, suspended in the air and grown around condensation nuclei [19.118]. Fog or clouds have droplet sizes around or above $10\,\mu\mathrm{m}$, with densities ranging from 10 to $1000\,\mathrm{droplets/cm}^3$. On average we will assume $300\,\mathrm{droplets/cm}^3$ [19.203]. Droplets of $5-10\,\mu\mathrm{m}$ radius will then correspond to volume fractions f of about $10^{-7}-10^{-6}$.

Clouds exhibit several optical phenomena, the most obvious being their white color, but certain types of clouds show particular phenomena. Cirrus and cirrustratus clouds may give rise to coronas, iridescence and halos, contrails of aircraft engines are cirrus-like trails of condensed vapor with characteristics similar to cirrus clouds, and cumulonimbus clouds relate to thunderstorms, i. e. lightning. Here, the color of clouds, their absorption and extinction will be briefly discussed. The effect of photon path length distributions within clouds is discussed in the contribution of *U. Platt* and *K. Pfeilsticker* (Sect. 19.5.5).

If illuminated by white light, clouds appear white (lower brightness is often interpreted as grey or dark). Since water droplets are mostly nonabsorbing in the visible, the color must be due to scattering. Scattering of light from monodisperse water droplets of about $10\,\mu\mathrm{m}$ shows Ripple structures (Fig. 19.16) [19.40,41], for size distributions this is washed out and scattering is mostly independent of wavelength in the visible spectral range. Although this is a sufficient condition for clouds to be white, it is not necessary [19.204]. As a matter of fact, a suspension of Rayleigh scatterers (such as fat droplets in milk), i. e. selective scatterers, also give a white color due to multiple scattering.

The criterion for multiple scattering to take place is that the optical thickness τ is large compared to unity. A molecular atmosphere had $\tau < 1$, therefore the explanation of the blue sky needs only single scattering events. For clouds, τ for a distance x can be estimated by the following argument: within clouds, the extinction coefficient is more or less constant, hence $\tau = \beta x$. The extinction coefficient β can be estimated from Mie theory calculations. Droplets of about $10\,\mu\mathrm{m}$ in diameter have extinction cross sections of about twice the geometrical cross section (this is the extinction paradox, [19.41]), i. e. of the order of $10^{-6}\,\mathrm{cm}^2$. For concentrations of $300/\mathrm{cm}^3$, this leads to extinction coefficients β of the order of $0.05\,\mathrm{m}^{-1}$. A very simple estimate [19.77] can also be made by starting from $\beta = N\sigma$ and substituting the number density N by the volume fraction $f \approx N\varnothing^3$ and the cross section by approximately the geometrical cross section $\sigma \approx \varnothing^2$ to find

$$\beta \approx f/\varnothing \qquad (19.41)$$

for $f = 10^{-6}$ and 10-μm particles, this also gives the order of 0.1/m.

Hence, a cloud of several meters thickness corresponds to the optical thickness of the whole normal atmosphere. The transmission is given by $T = \mathrm{e}^{-\beta x}$ and for $\beta = 0.05\,\mathrm{m}^{-1}$ a distance of 20 m leads to $\tau = 1$ ($T = 1/\mathrm{e}$), $x = 100\,\mathrm{m}$ to $\tau = 5$ ($T \approx 7\times10^{-3}$), and $x = 200\,\mathrm{m}$ to $\tau = 10$ ($T \approx 5\times10^{-5}$), which is sufficient to obscure the disk of the sun [19.77]. Low transmission automatically leads to high reflection, i.e. thick clouds have a high albedo which explains the bright white color observed from airplanes above clouds. In conclusion, multiple scattering dominates for any cloud that is optically thick and composed of nearly nonabsorbing particles; upon illumination by white light it will look white. The color of clouds may change during sunrise and sunset if spectrally filtered light illuminates the clouds, giving rise to spectacular sceneries.

The very small transmission of sunlight through thick clouds also accounts for the fact that the bases of such clouds are usually very dark. If droplets near the base grow to form raindrops, the extinction coefficient changes. The volume fraction stays about the same, but raindrops have a size of, say, 1 mm, hence β in a rain shower is about a factor 100 smaller than within the clouds. This explains, why it is easily possible to look through very heavy rainfalls.

Visibility

Visibility is a measure of how far one can see through the air. Even in the clearest air – assuming that no aerosol particles are present and that one is not restricted by the Earth's curvature – the horizontal visibility is limited to a few hundred kilometers. This is due to the scattering and absorption of the components of the air (vertically, the atmosphere is optically much thinner, hence we may see the stars at night). In meteorology, visibility refers to the observation of distant dark objects against a bright background, usually the sky. The basis for a theoretical derivation [19.77, 205–209] is the contrast ratio C, giving the normalized difference between the background intensity of the object and the intensity of the air light from the line of sight. One finds for the resulting maximum distance D

$$D = -\ln(C)/\beta\,, \qquad (19.42)$$

where β is the scattering coefficient. The sensitivity of the human eye has a maximum at $\lambda = 555\,\mathrm{nm}$, therefore one usually evaluates (19.42) for green light. D then just depends on the threshold contrast C, i.e. the lowest visual brightness contrast, a person can see. It is common to assume $C = 0.02$ although it varies from individual to individual. This gives $D = 3.912/\beta$. Values of D for various β are shown in Table 19.7. Obviously, there is an enormous range of distances from several meters in dense fog to several hundred kilometers in very clear air. Hence, the visibility is an indicator of the aerosol and/or water droplet content.

19.6.9 Miscellaneous

There is a great number of other optical phenomena associated with the atmosphere. In the following, these will be very briefly described and references will be given.

Green flashes or green suns are phenomena when either the last rays at sunset or the first rays at sunrise look green or when, at low sun elevations, the upper rim of the sun looks green. In the first case, just a very small portion of the sun is above the horizon, in the second case, the sun's disk is usually highly distorted due to mirage effects. The explanations of these phenomena require refraction and the scattering of the sunlight as well as mirage effects [19.103, 210, 211].

Noctilucent clouds are blue, white or silvery looking clouds within the stratosphere at heights of about 80 km

Table 19.7 Typical values for visibility in molecular atmosphere and examples if aerosols (natural/anthropogenic) and water droplets are present

	β	D	
Molecular atmosphere	$0.013\,\mathrm{km}^{-1}$	300 km	Extremely excellent visibility
Some particles	$0.050\,\mathrm{km}^{-1}$	80 km	Excellent visibility
Many particles	$0.500\,\mathrm{km}^{-1}$	8 km	Poor visibility
Thick fog	$0.050\,\mathrm{m}^{-1}$	80 m	Extremely poor visibility

which can still be illuminated by sunlight 1–2 h after sunset [19.212–214]. They often show wavelike features with wavelengths in the range from 1–100 km.

Interesting colorful phenomena may be observed during eclipses. For example, the colors and brightness of lunar eclipses are directly related to the scattering of sunlight within the Earth's atmosphere [19.215, 216]. Other examples include polarization of sky light [19.217,218], twilight phenomena [19.219], and illuminance measurements [19.220] during solar eclipses.

Shadows are regions in space that are shielded from light. They give rise to very prominent daily-life phenomena in atmospheric optics such as the crepuscular rays, sometimes called sun beams or shadow beams [19.45,48,51,221] or the fact that mountain shadows are always triangular, irrespective of the contour of the mountain [19.222, 223].

A very personal experience, based on our visual system, is the so-called moon illusion, i. e. the fact that the sun or the moon looks much bigger near the horizon than when it is higher in the sky [19.224, 225]. This phenomenon is not reproduced on photos.

In addition to phenomena with external light sources such as the sun, moon or stars, a number of optical phenomena in the atmosphere produce light. The most well known are auroras, multicolored, diffuse and slowly moving lights, usually seen only in high altitudes against the clear dark night sky. They are due to air molecules and atoms, which are excited by collisions with electrons from the solar wind [19.51, 226, 227]. At times of increased solar activity, they may also be observed at low latitudes. These events are accurately predictable, since the solar wind needs time to reach the Earth (see the websites on space weather in Table 19.4).

Lightning, one of the most common and spectacular natural phenomena, is due to electric charging effects in thunderstorm clouds. The exact microphysical processes that are responsible for the charging are still under dispute [19.228–230]. Recent research has revealed a whole new class of phenomena: lightning-related transient luminous events (TLE), including sprites, blue jets, elves, trolls etc. [19.231]. These TLE extend from the top of thunderclouds at heights of up to 100 km and raise serious questions about possible hazards for aerospace operations. In addition, promising efforts are under way concerning lightning protection by initiation of the discharge using intense femtosecond lasers [19.232].

Meteors are sometimes faint, sometimes brilliant streaks of light with luminous trails in the night sky. Meteor incorporates all phenomena when a cosmic particle enters the atmosphere. Outside the atmosphere, they are called meteoroids and, if they reach ground, meteorites. The visual effects are due to heating of particles in the atmosphere due to frictional forces. Small particles heat up until they evaporate. Thereby, they collide with air molecules and ionize a channel of air. Light is then emitted upon recombination [19.233, 234].

References

19.1 Lord Rayleigh, J.W. Strutt: On the transmission of light through an atmosphere containing small particles in suspension and on the origin of the blue of the sky, Philos. Mag. **XLVII**, 375–384 (1899)

19.2 M. Nicolet: On the molecular scattering in the terrestrial atmosphere: An empirical formula for its calculation in the homosphere, Planet. Space Sci. **32**, 1467–1468 (1984)

19.3 R. Penndorf: Tables of the refractive index for standard air and the Rayleigh scattering coefficient for the spectral region between 0.2 and 20 μm and their application to atmospheric optics, J. Opt. Soc. Am. **47**, 176–182 (1957)

19.4 M. Bussemer: Der Ring-Effekt: Ursachen und Einfluß auf die spektroskopische Messung stratosphärischer Spurenstoffe Ph.D. Thesis (Univ. Heidelberg, Heidelberg 1993)

19.5 J. Burrows, M. Vountas, H. Haug, K. Chance, L. Marquard, K. Muirhead, U. Platt, A. Richter, V. Rozanov: *Study of the Ring Effect*, Final Report European Space Agency (ESA), Noordvijk, The Netherlands, Contract 109996/94/NL/CN (1995)

19.6 H. Haug, K. Pfeilsticker, U. Platt: Vibrational Raman scattering in the atmosphere, (1996) unpublished

19.7 C. E. Sioris, W. F. J. Evans: Filling in of Fraunhofer and gas-absorption lines in sky spectra as caused by rotational Raman scattering, Appl. Opt. **38**, 2706–2713 (1999)

19.8 H.W. Schrötter, H.W. Klöckner: Raman scattering cross-sections in gases. In: *Raman Spectroscopy of Gases and Liquids*, ed. by A. Weber (Springer, Berlin, Heidelberg 1979)

19.9 D. Clarke, H.M. Basurah: Polarisation measurements of the ring effect in the daytime sky, Planet. Space. Sci. **37**, 627–630 (1989)

19.10 G. Mie: Beiträge zur Optik trüber Medien, speziell kolloidaler Metalllösungen, Ann. Phys. **25**, 377–445 (1908) in German

19.11 H.C. Van de Hulst: *Multiple Light Scattering, Tables, Formulas and Applications*, Vol. 1,2 (Academic, London 1980)

19.12 W.J. Wiscombe: Improved Mie scattering algorithms, Appl. Opt. **19**, 1505–1509 (1980)

19.13 R.G. Isaacs, W.-C. Wang, R.D. Worsham, S. Goldberg: Multiple scattering lowtran and fascode models, Appl. Opt. **26**, 1272–1281 (1987)

19.14 O. Funk: Photon path length distributions for cloudy skies; Oxygen a-band measurements and radiative transfer calculations Ph.D. Thesis (Univ. Heidelberg, Heidelberg 2000)

19.15 A. Angström: On the atmospheric transmission of sun radiation and on dust in the air, Geogr. Annal. Stockholm **11**, 156–166 (1929)

19.16 A. Angström: On the atmospheric transmission of sun radiation and on dust in the air, Geografiska Annal. **12**, 130 (1930)

19.17 C.E. Junge: *Air Chemistry and Radioactivity*, Vol. 4 (Academic, New York 1963)

19.18 C.E. Junge, J.E. Manson: Stratospheric aerosols studies, J. Geophys. Res. **66**, 2163–2182 (1961)

19.19 M. Milankovitch: *Canon of Insolation and the Ice-Age Problem*, Royal Serbian Acad. Sp. Pub., Vol. 132 (Royal Serbian Acad., Belgrade, Yugoslavia 1941)

19.20 J.T. Houghton, Y. Ding, D.J. Griggs, M. Noguer, P.J. van der Linden, X. Dai, K. Maskell, C.A. Johnson (Eds.): *Climate Change 2001: The Scientific Basis, published for Intergovernmental Panel on Climate Change (IPCC)* (Cambridge Univ. Press., Cambridge 2001)

19.21 S. Twomey: Pollution and the planetary albedo, Atmos. Environ. **8**, 1251–1256 (1974)

19.22 B.A. Wielicki, T. Wong, R.P. Allan, A. Slingo, J.T. Kiehl, B.J. Soden, C.T. Gordon, A. Levin, J. Miller, S.-K. Yang, D.A. Randall, F. Robertson, J. Susskind, H. Jacobowitz: Evidence for Large Decadal Variability in the Tropical Mean Radiative Energy Budget, Science **295**, 841–844 (2002) Issue 5556

19.23 D. Perner, U. Platt, M. Trainer, G. Huebler, J.W. Drummond, W. Junkermann, J. Rudolph, B. Schubert, A. Volz, D.H. Ehhalt, K.J. Rumpel, G. Rumpel, G. Helas: Tropospheric OH concentrations: A comparison of field data with model predictions, J. Atmos. Chem. **5**, 185–216 (1987)

19.24 M.W. Sigrist (Ed.): *Air Monitoring by Spectroscopic Techniques*, Chemical Analysis Series, Vol. 127 (Wiley, New York 1984)

19.25 U. Platt: Modern methods of the measurement of atmospheric trace gases, J. Phys. Chem. Chem. Phys. 'PCCP' **1**, 5409–5415 (1999)

19.26 U. Platt, D. Perner: *Measurements of atmospheric trace gases by long path differential UV/visible absorption spectroscopy: Optical and Laser Remote Sensing*, ed. by D.K. Killinger, A. Mooradien, Springer Ser. Opt. Sci. **39**, 95–105 (1983)

19.27 U. Platt: *Differential optical absorption spectroscopy (DOAS), Air Monitoring by Spectroscopic Techniques*, Chemical Analysis Series, Vol. 127, ed. by M.W. Sigrist (Wiley, New York 1994) pp. 27–84

19.28 R. Volkamer, T. Etzkorn, A. Geyer, U. Platt: Correction of the oxygen interference with UV spectroscopic (DOAS) measurements of monocyclic aromatic hydrocarbons in the atmosphere, Atmos. Environ. **32**, 3731–3747 (1998)

19.29 J.P. Burrows, E. Hölzle, A.P.H. Goede, H. Visser, W. Fricke: SCIAMACHY – Scanning imaging absorption spectrometer for atmospheric chartography, Acta Astronautica **35**(7), 445–451 (1995)

19.30 J.P. Burrows, M. Weber, M. Buchwitz, V.V.. Rozanov, A. Ladstätter-Weissenmayer, A. Richter, R. De Beek, R. Hoogen, K. Bramstedt, K.-U. Eichmann, M. Eisinger: The global ozone monitoring experiment (GOME): Mission concept and first scientific results, J. Atmos. Sci. **56**, 151–175 (1999)

19.31 R. Volkamer, P. Spietz, J.P. Burrows, U. Platt: High-resolution absorption cross-section of Glyoxal in the UV/vis and IR spectral ranges, J. Photochem. and Photobiol. A: Chemistry **172**, 35–46 (2005) DOI: 10.1016/j.jphotochem.2004.11.011

19.32 C. Weitkamp (ed.): *Lidar – Range-Resolved Optical Remote Sensing of the Atmosphere* (Springer, Berlin, Heidelberg, New York 2005)

19.33 S. Nyeki, K. Eleftheriadis, U. Baltensperger, I. Colbeck, M. Fiebig, A. Fix, C. Kiemle, M. Lazaridis, A. Petzold: Airborne Lidar and in-situ aerosol observations of an elevated layer, Leeward of the European Alps and Apennines, Geophys. Res. Lett. **29**, 1852 (2002) doi:10.1029/2002 GL 014897

19.34 K. Pfeilsticker, F. Erle, O. Funk, H. Veitel, U. Platt: First geometrical path lengths probability density function derivation of the skylight from spectroscopically highly resolving oxygen A-band observations, 1. Measurement technique, atmospheric observations, and model calculations, J. Geophys. Res. **103**, 11483–11504 (1998)

19.35 K. Pfeilsticker: First geometrical path lengths probability density function derivation of the skylight from spectroscopically highly resolving oxygen A-band observations. 2. Derivation of the Lévy-index for the skylight transmitted by midlatitude clouds, J. Geophys. Res. **104**, 4101–4116 (1999)

19.36 V. Veitel, O. Funk, C. Kurz, U. Platt, K. Pfeilsticker: Geometrical path length probability density function of the skylight transmitted by mid-latitude cloudy skies; Some case studies, Geophys. Res. Lett. **25**, 3355–3358 (1998)

19.37 Q.-L. Min, L.C. Harrison, E. Clothiaux: Joint statistics of photon path length and cloud optical depth: Case studies, J. Geophys. Res. **106**, 7375–7386 (2001)

19.38 O. Funk, K. Pfeilsticker: Photon path lengths distributions for cloudy skies: Oxygen A-band measurements and model calculations, Annal. Geophysicae **21**, 615–626 (2003)

19.39 C. Savigny, O. Funk, U. Platt: Radiative Smoothing in zenith-scattered skylight transmitted to the ground, Geophys. Res. Lett. **26**, 2949–2952 (1999)

19.40 H. C. van de Hulst: *Light Scattering by Small Particles* (Dover, New York 1981)
19.41 C. F. Bohren, D. R. Huffman: *Absorption and Scattering of Light by Small Particles* (Wiley, New York 1983)
19.42 U. Kreibig, M. Vollmer: *Optical Properties of Metal Clusters*, Springer Ser. Mat. Sci., Vol. 25 (Springer, Berlin, Heidelberg 1995)
19.43 M. Vollmer: *Lichtspiele in der Luft–Atmosphärische Optik für Einsteiger* (Spektrum-Elsevier, Heidelberg 2006)
19.44 J. M. Pernter, F. M. Exner: *Meteorologische Optik*, 2nd edn. (Braumüller, Wien 1922)
19.45 M. G. J. Minnaert: *Light and Color in the Outdoors*, 1 edn. (Springer, Berlin, Heidelberg 1993) in Dutch 1937
19.46 W. J. Humphreys: *Physics of the Air* (Dover, New York 1963) p. 1929 reprint
19.47 R. A. R. Tricker: *Introduction to Meteorological Optics* (Elsevier, New York 1970)
19.48 R. Greenler: *Rainbows, Halos, and Glories* (Cambridge Univ. Press, Cambridge 1980)
19.49 A. Meinel, M. Meinel: *Sunsets, Twilights, and Evening Skies* (Cambridge Univ. Press, Cambridge 1983)
19.50 C. F. Boyer: *The Rainbow: From Myth to Mathematics* (Princeton Univ. Press, Princeton 1987) revised printing
19.51 D. K. Lynch, W. Livingston: *Color and Light in Nature*, 2 edn. (Cambridge Univ. Press, Cambridge 2001)
19.52 W. Tape: *Atmospheric Halos* (Am. Geophys. Soc., Washington 1994)
19.53 G. P. Können: *Polarized Light in Nature* (Cambridge Univ. Press, Cambridge 1985)
19.54 R. Lee, A. B. Fraser: *The Rainbow Bridge* (Penn. State Press, University Park 2001)
19.55 C.F. Bohren (Ed.): *Selected papers on Scattering in the Atmosphere*, SPIE Milestone Series MS7 (SPIE, Bellingham 1989)
19.56 OSA (ed): *Proceedings of 1st conference on atmospheric optics*, J. Opt. Soc. Am., Vol. 68 (Optical Society of America, Washington, D.C. 1979)
19.57 OSA (ed): *Proceedings of 2nd conference on atmospheric optics*, J. Opt. Soc. Am., Vol. 73 (Optical Society of America, Washington, D.C. 1983)
19.58 OSA (ed): *Proceedings of 3rd conference on atmospheric optics*, Vol. 4 (Optical Society of America, Washington, D.C. 1987)
19.59 OSA (ed): *Proceedings of 4th conference on atmospheric optics*, Appl. Opt., Vol. 30 (Optical Society of America, Washington, D.C. 1991)
19.60 OSA (ed): *Proceedings of 5th conference on atmospheric optics*, Appl. Opt., Vol. 33 (Optical Society of America, Washington, D.C. 1994)
19.61 OSA (ed): *Proceedings of 6th conference on atmospheric optics*, Appl. Opt., Vol. 37 (Optical Society of America, Washington, D.C. 1998)
19.62 OSA (ed): *Proceedings of 7th conference on atmospheric optics*, Appl. Opt., Vol. 42 (Optical Society of America, Washington, D.C. 2003)
19.63 OSA (ed): *Proceedings of 8th conference on atmospheric optics*, Appl. Opt., Vol. 44 (Optical Society of America, Washington, D.C. 2005)
19.64 C. L. Adler (Ed.): *On Minnaert's Shoulders: Twenty Years of the Light and Color Conferences*, Vol. 1 (Opt. Soc. Am., Washington 1999) Classic Preprints on CD-ROM
19.65 R. Greenler: *The Mirage, the Discovery of Greenland and the Green Flash*, video available from www.blueskyassociates.com
19.66 R. Greenler: *Sunlight and Ice Crystals in the Skies of Antarctica*, video available from www.blueskyassociates.com
19.67 R. Greenler: *Red Sunsets, Black Clouds and the Blue Moon: Light Scattering in the Atmosphere*, video available from www.blueskyassociates.com
19.68 M. Engler: *Fata Morganen–Zauberspiegel am Horizont*, German TV film first broadcast 1996 by French–German TV program ARTE
19.69 M. Engler: *Fata Morgana–Naturwunder und Zauberspuk*, German TV film first broadcast 2001 by French–German TV program ARTE
19.70 S. Gedzelman: Atmospheric optics in art, Appl. Opt. **30**, 3514 (1991)
19.71 S. Rother: *Der Regenbogen, eine malereigeschichtliche Studie* (Böhlau, Köln 1992) in German
19.72 K. Sassen: Rainbows in the Indian rock art of desert western America, Appl. Opt. **30**, 3523 (1991)
19.73 K. Sassen: Possible halo depictions on the prehistoric rock art of Utah, Appl. Opt. **33**, 4756 (1994)
19.74 C. F. Bohren: *Clouds in a Glass of Beer* (Wiley, New York 1987)
19.75 E. A. Wood: *Science from your Airplane Window* (Dover, New York 1975) 1st edn. 1968
19.76 M. Bass: *Handbook of Optics*, Vol. 1, ed. by E. van Stryland, D. Williams, W. Wolfe (McGraw Hill, New York 1995)
19.77 C. F. Bohren: Atmospheric optics, Encyclop. Appl. Phys. **12**, 405 (1995)
19.78 Lord Rayleigh, J. W. Strutt: On the light from the sky, its polarization and colour, Philos. Mag. **XLI**, 107, 274 (1871)
19.79 A. T. Young: Rayleigh scattering, Phys. Today **35**, 42–48 (1982)
19.80 G. Hoeppe: *Blau – Die Farbe des Himmels* (Spektrum Akad. Verlag, Heidelberg 1999) in German
19.81 K. Bullrich: Scattered radiation in the atmosphere and the natural aerosol, Adv. Geophys. **10**, 99 (1964)
19.82 S. Twomey: *Atmospheric Aerosols* (Elsevier, Amsterdam 1977)
19.83 T. E. Graedel, P. J. Crutzen: *Atmospheric Change: An Earth System Perspective* (Freeman, New York 1993)
19.84 E. Hecht: *Optics*, 3 edn. (Addison Wesley, San Francisco 1998)

19.176 M. Pekkola: Finnish halo observing network: Search for rare halo phenomena, Appl. Opt. **30**, 3542 (1991)

19.177 R. Greenler: Sunlight, ice crystals, and the sky archaeology, Proc. Roy. Inst. Great Britain **65**, 47 (1994)

19.178 G. P. Können: Polarization and intensity distributions of refraction halos, J. Opt. Soc. Am. **73**, 1629 (1983)

19.179 G. P. Können, J. Tinbergen: Polarimetry of a 22° halo, Appl. Opt. **30**, 3382 (1991)

19.180 G. P. Können, A. A. Schoenmaker, J. Tinbergen: A polarimetric search for ice crystals in the upper atmosphere of venus, Icarus **102**, 62 (1993)

19.181 G. P. Können: Symmetry in halo displays and symmetry in halo-making crystals, Appl. Opt. **42**, 318 (2003)

19.182 G. P. Können, H. R. A. Wessels, J. Tinbergen: Halo polarization profiles and sampled ice crystals: Observations and interpretation, Appl. Opt. **42**, 309 (2003)

19.183 E. Boeker, R. van Grondelle: *Environmental Science* (Wiley, New York 2001)

19.184 R. Wilson: The blue sun of 1950 September, Mon. Not. R. Astron. Soc. **111**, 478 (1951)

19.185 E. O. Hulburt: Explanation of the brightness and color of the sky, particularly the twilight sky, J. Opt. Soc. Am. **43**, 113 (1953)

19.186 M. Vollmer: Estimating the Thickness of the Atmosphere by Rayleigh Scattering, Am. J. Phys. **71**, 979–983 (2003)

19.187 F. E. Volz: Zenith polarization and color ratio during twilight, Appl. Opt. **20**, 4172 (1981)

19.188 K. L. Coulson: Effects of the El Chichon volcanic cloud in the stratosphere on the polarization of light from the sky, Appl. Opt. **22**, 1036 (1983)

19.189 F. G. Fernald, B. M. Herman, J. A. Reagan: Determination of aerosol height distributions by lidar, J. Appl. Meteor. **11**, 482 (1972)

19.190 F. G. Fernald: Analysis of atmospheric LIDAR observations: Some comments, Appl. Opt. **23**, 652 (1984)

19.191 J. D. Klett: Lidar inversion with variable backscatter/extinction ratios, Appl. Opt. **24**, 1638 (1985)

19.192 J. Bosenberg, D. Brassington, P. C. Simon (Eds.): *Instrument Development for Atmospheric Research and Monitoring: Lidar Profiling, DOAS and Tunable Diode Laser Spectroscopy* (Springer, Berlin, Heidelberg 1997)

19.193 Y. Liu, K. Voss: Polarized radiance distribution measurements of skylight II: Experiment and data, App. Opt. **36**, 8753 (1997)

19.194 R. L. Lee: Twilight and daytime colors of the clear sky, Appl. Opt. **33**, 4629 (1994)

19.195 R. L. Lee: Horizon brightness revisited: Measurements and a model of clear sky radiances, Appl. Opt. **33**, 4620 (1994)

19.196 J. R. Linskens, C. F. Bohren: Appearance of the sun and moon seen through clouds, Appl. Opt. **33**, 4733 (1994)

19.197 R. L. Lee: Digital imaging of clear sky polarization, Appl. Opt. **37**, 1465 (1998)

19.198 R. L. Lee, J. Hernández-Andrés: Measuring and modeling twilight's purple light, Appl. Opt. **42**, 445 (2003)

19.199 J. Hernández-Andrés, R. L. Lee, J. Romero: Color and luminance asymmetries in the clear sky, Appl. Opt. **42**, 458 (2003)

19.200 R. L. Lee, J. Hernández-Andrés: Colors of the daytime overcast sky, Appl. Opt. **44**, 5712 (2005)

19.201 N. Thomas, W. J. Markiewicz, R. M. Sablotny, M. W. Wuttke, H. U. Keller, J. R. Johnson, R. J. Reid, P. H. Smith: The color of the martian sky and its influence on the illumination of the martian surface, J. Geophys. Res. **104**, 48795 (1999)

19.202 http://cfa-www.harvard.edu/hitran//, http://www.hitran.com

19.203 C. S. D. Ahrens: *Meteorology Today*, Vol. 4 (West Publ. Comp., St. Paul 1991)

19.204 C. F. Bohren: Multiple scattering of light and some of its observable consequences, Am. J. Phys. **55**, 524 (1987)

19.205 H. Koschmieder: Theorie der horizontalen Sichtweite, Beitr. Phys. freien Atmosphäre **XII**, 33 (1924) in German

19.206 H. Koschmieder: Theorie der horizontalen Sichtweite II: Kontrast und Sichtweite, Beitr. Phys. freien Atmosphäre **XII**, 171 (1925) in German

19.207 S. Q. Duntley: The reduction of apparent contrast by the atmosphere, J. Opt. Soc. Am. **38**, 179 (1948)

19.208 N. Mason, P. Hughes: *Introduction to Environmental Physics* (Taylor and Francis, London 2001)

19.209 M. Z. Jacobson: *Atmospheric Pollution* (Cambridge Univ. Press., Cambridge 2002)

19.210 G. E. Shaw: Observations and theoretical reconstruction of the green flash, Pure Appl. Geophys. **102**, 223 (1973)

19.211 A. T. Young: Green flashes and mirages, Opt. Photon. News **10**, 31 (1999)

19.212 F. H. Ludlam: Noctilucent clouds, Tellus **IX**, 341 (1957)

19.213 M. Gadsden, W. Schröder: *Noctilucent Clouds* (Springer, Berlin, Heidelberg 1989)

19.214 G. E. Thomas: Noctilucent clouds, Rev. Geophys. **29**, 553 (1991)

19.215 M. Littmann, K. Willcox: *Totality, Eclipses of the Sun* (Univ. Hawaii Press, Honolulu 1991)

19.216 P. S. Harrington: *Eclipse* (Wiley, New York 1997)

19.217 G. E. Shaw: Sky brightness and polarization during the 1973 African eclipse, Appl. Opt. **14**, 388 (1975)

19.218 G. P. Können: Skylight polarization during a total solar eclipse: A quantitative model, J. Opt. Soc. Am. A **4**, 601 (1987)

19.219 E. H. Geyer, M. Hoffmann, H. Volland: Influence of a solar eclipse on twilight, Appl. Opt. **33**, 4614 (1994)

19.220 K.-P. Möllmann, M. Vollmer: Measurements and predictions of the illuminance during a solar eclipse, Eur. J. Phys. **27**, 1299–1314 (2006)
19.221 D. K. Lynch: Optics of sun beams, J. Opt. Soc. Am. A **4**, 609 (1987)
19.222 W. Livingston, D. Lynch: Mountain shadow phenomena, Appl. Opt. **18**, 265 (1979)
19.223 D. K. Lynch: Mountain shadow phenomena 2: The spike seen by an off-summit observer, Appl. Opt. **19**, 1585 (1980)
19.224 H. E. Ross, C. Plug: *Thy Mystery of the Moon Illusion* (Oxford Univ. Press, Oxford 2002)
19.225 G. R. Lockhead, Myron L. Wolbarsht: Toying with the moon illusion, Appl. Opt. **30**, 3504 (1991)
19.226 S.-I. Akasofu: The dynamic aurora, Sci. Am. **261**, 54 (May 1989)
19.227 J. A. Whalen, R. R. O'Neil, R. H. Pritchard: The aurora. In: *Handbook of Geophysics and the space environment*, ed. by A. S. Jursa (Air Force Geophysics Laboratory, US Air Force 2001) Chap. 12
19.228 R. Feynman: *The Feynman Lectures on Physics*, Vol. II (Addison Wesley, San Francisco 1963) Chap. 9
19.229 M. A. Uman: *Lightning* (Dover, New York 1984) reprint of 1969 ed. by Mc Graw Hill
19.230 E. R. Williams: The electrification of thunderstorms, Sci. Am. **260**, 48 (Nov. 1988)
19.231 W. A. Lyons, R. A. Armstrong, E. A. Bering III, E. R. Williams: The hundred year hunt for the sprite, EOS Trans. Am. Geophys. Union **81**, 373 (2000)
19.232 M. Rodriguez: Triggering and guiding megavolt discharges by use of laser-induced ionized filaments, Opt. Lett. **27**, 772–774 (2002)
19.233 P. Moore: *The Data Book of Astronomy* (IOP, Bristol 2000)
19.234 J. B. Kaler: *Astronomy* (Harper Collins, New York 1994)

20. Holography and Optical Storage

The term holography is composed of the Greek words *holos* (= whole) and *graphein* (= to record, to write), and thus summarizes the key aspects of its underlying principle: recording the complete wavefront of an object, i. e., its intensity as well as its phase. Interference and diffraction phenomena are employed to record and retrieve the full information, a technique pioneered by Dennis Gabor in 1948. He was honored with the Nobel prize in Physics in 1971, reflecting the general impact of holography on modern physics.

Holography plays an essential role in today's science and industry. Relevant applications making use of its principle have been developed, including three-dimensional (3-D) displays and holographic cameras, interferometers for nondestructive material analysis, archival data storage systems, diffractive optical systems, and embossed display holograms for security features. The success of holography was made possible in particular by the availability of coherent laser-light sources. In the meantime holography has even been performed using microwaves, neutrons, electrons, X-rays, and acoustic waves.

The first part of this chapter is devoted to holography itself. It provides an introduction to the historical development and reviews the principle of wavefront reconstruction. This section also includes an overview of hologram classification, recording/read-out geometries, holographic techniques and recording materials. Special emphasis is given to explaining the principles of some of the most important holographic applications, finishing with a brief insight into a few of the latest discoveries making use of Gabor's principle, such as holographic scattering and neutron diffractive optics.

The second part of this chapter addresses trends in optical storage, focussing on holographic data storage. It highlights different approaches to achieving increased optical storage density. This section also discusses the historical development of optical storage, the need for increased storage densities (and hence storage capacities) and the role of optical storage systems in today's life. Various approaches to increasing the areal density of optical storage systems are introduced. Next, the advantages of and approaches to volume optical recording that are currently under consideration for future generations of optical storage systems are presented. The state of the art as well as physical and technical attempts to realize holographic data storage are discussed in detail.

20.1 Introduction and History

In May 1948 *Dennis Gabor* (Fig. 20.1) introduced the principle of wavefront reconstruction in the scientific journal Nature [20.1]. In the frame of his one-page contribution "A new microscopic principle" he discussed the high-resolution visualization of microscopic objects and emphasized the impact of his idea for the improvement of electron microscopes. The word *holography*, as his principle was designated shortly thereafter, is a compound of the Greek words ὅλος (*holos* = whole) and γράφειν (*graphein* = to record, to write) and summarizes the key aspects of Gabor's principle: the recording of the complete/overall wavefront from an object, i. e., recording of the object within all three dimensions. Holography is therefore a serious extension of photography, which is limited to the recording of the projected image of three-dimensional objects. In standard photography, the intensities/amplitudes of the wavefront of the object are recorded without the appertaining phase information. Photography does not include the depth information, i. e., a look around the object is not possible. In holography, viewing the reconstructed image is identical to the observation of the real wavefront of the object. Within ge-

Fig. 20.1 *Dennis Gabor* (1900–1979, born in Budapest, Hungary), Nobel Prize Laureate 1971 (from [20.10])

Table 20.1 Important milestones during the first decades of holography (After [20.7])

1948	Dennis Gabor	*A new microscopic principle*	[20.1]
1962	Emmett Leith and Juris Upatnieks	Invention of the off-axis technique	[20.2]
1963	Yuri N. Denisyuk	White-light reflection hologram	[20.3]
1965	N. L. Hartmann	Off-axis recording geometry for reflection holograms	[20.4]
1965	Robert Powell and Karl Stetson	First paper on holographic interferometry	[20.5]
1967	Shankoff and Pennington	Dichromatic gelatin for hologram recording	[20.6]
1967	World Book Encyclopedia Science Yearbook	First mass-distributed hologram	[20.7]
1967	Larry Siebert	First hologram of a person by pulsed laser light	[20.8]
1968	Stephen A. Benton	White-light transmission holography	[20.9]
1974	Michael Foster	Embossing technique for mass production of holograms	[20.7]
1968	Cranbrook Academy of Art in Michigan	First holographic art exhibition	
1971	Lloyd Cross	San Francisco School of Holography	
1971	Dennis Gabor	Nobel prize	[20.10]
1972	Lloyd Cross	Invention of the integral hologram	
1975	Rich Rallison	First production of glass sandwich dichromate holograms	
1976	Victor Komar	Prototype for a projected holographic movie	[20.11]
1976	New York	Foundation of the Museum of Holography	
1983	MasterCard International, Inc.	First use of hologram technology in bank card security	

ometrical limits the object can be observed from various perspectives and it is possible to look behind the object.

Holography uses the interference phenomenon for recording both the amplitude and the phase information. The first holograms were realized by Gabor with the light of a mercury-arc lamp, that hence contained distortions and an extraneous twin image. The further history of holography is punctuated by the development of coherent laser-light sources in the 1960s, the real success of holography being initiated in 1962: *Emmett Leith* and *Juris Upatnieks* read Gabor's paper and "simply out of curiosity" they adapted Gabor's principle to an off-axis geometry that they borrowed from their work on side-reading radar. The result was the first laser transmission hologram of three-dimensional objects (a toy train and bird) [20.2, 12, 13]. At the same time *Yuri N. Denisyuk* combined holography with the findings of *Gabriel Lippmann* on natural color photography [20.14] and produced a white-light reflection hologram [20.3]. For the first time, a hologram could be reconstructed using the light from an ordinary incandescent light bulb. In 1965 the patent for the off-axis recording geometry for reflection holograms was issued to *N. Hartmann* [20.4] – a technique that emerged as a standard in holographic recording. The invention was due to at least three teams of co-workers: E. Leith, J. Upatnieks, A. Kozma, J. Marks, and N. Massey (University of Michigan); G. Stroke, A. Labeyrie (University of Michigan) with K. Pennington and L. Lin (Bell Labs); C. Schwartz and N. Hartmann (Batelle Memorial Institute). Table 20.1 gives an overview of the impressive milestones within the first decades of the history of holography. A recent review of the history of holography is given in [20.15]. Nowadays, holography touches many fields of our everyday life including arts, archeology, health and medicine, biotechnology, advertising and communication. Gabor's principle continues to be a topic of research and potentially will open up novel perspectives, e.g., in optical data storage (cf. Sect. 20.5). The impact of this invention was honored with the Nobel prize in 1971 [20.10]. *Dennis Gabor* died on February 8, 1979.

20.2 Principles of Holography

20.2.1 Recording of Holograms and Wavefront Reconstruction

The holographic principle makes use of the interference phenomenon, as interference allows the transfer of the phase information of a wavefront into an intensity pattern. Recording both the amplitude and phase of the wavefront of an object, i.e. recording a hologram, is performed by interference of a reference wave with complex amplitude R and a coherent signal wave S using

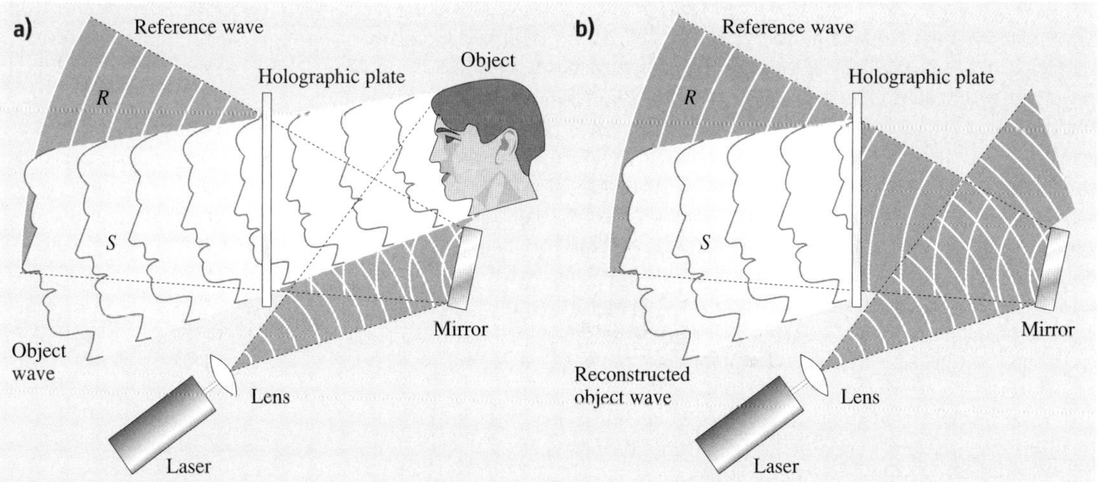

Fig. 20.2 **(a)** Recording a hologram and **(b)** wavefront reconstruction. The intensity pattern from the interfering reference R and signal S waves is recorded by the holographic plate thus representing a hologram. A complete reconstruction of the signal wave S appears by light diffraction if the hologram is exposed to the reference wave only

a holographic plate Fig. 20.2. For the sake of simplicity we do not consider vectorial properties of the waves at the moment. The wavefront of the object is reconstructed by light diffraction if the hologram is exposed only to the reference wave.

The simplest hologram is formed by the interference of two plane waves with non-collinear propagation directions. In this case a pattern with sinusoidally modulated amplitude is recorded on the holographic plate, i. e., an *elementary holographic grating*. Exposure of this grating to the reference wave leads to diffraction along the former direction of the signal wave. Thus the wavefront of the plane signal wave is reconstructed.

More complex wavefronts are reconstructed in an analogous way. Let us consider the reference and signal waves with complex amplitudes R and S. Then the intensity I of the interference pattern can be expressed by

$$I \propto |R+S|^2 = |R|^2 + |S|^2 + R^*S + RS^* , \quad (20.1)$$

where * denotes the complex conjugate. Exposing a photochromic film (such as a holographic plate) to the interference pattern, the absorption coefficient and hence the transmission T will become spatially modulated. In the case of a linear response of the film to intensity, i. e., $T = T_0 - \Delta T I$, the illumination of the holographic plate with the reference wave results in the reconstructed wave

$$RT = R\left[T_0 - \Delta T(|R|^2 + |S|^2 + R^*S + RS^*)\right] . \quad (20.2)$$

Here, T_0 is the transmission prior to illumination and ΔT is a parameter that characterizes the photosensitivity of the film. The important term in (20.2) is $\Delta T R R^* S = \Delta T |R|^2 S$. The factor $\Delta T |R|^2$ is constant and phase independent, but the signal wave S is completely revealed with *amplitude and phase*. The other terms in (20.2) describe the transmitted reference wave and further diffracted waves. Separation of these particular waves from each other can be established by choosing a proper recording geometry (spatial separation via off-axis geometry, cf. Sect. 20.2.3).

The *amplitude* and *phase* information of the recorded object is stored in each singular point of the holographic plate, representing the particular viewpoint onto the object. Thus the lateral dimension of the holographic plate is the limit for the visual angle onto the recorded object. Reconstruction is even possible by exposure of only a small part of the hologram with the reference wave.

In summary, the holographic principle only requires a source of coherent waves, an object, and a photosensitive medium with sufficient spatial resolution. It is worth noting that there is no restriction on the nature of the radiation or its wavelength and thus any (partially) coherent wave can be applied, such as electrons, neutrons, X-rays, microwaves, photons and acoustic waves. All of these have been utilized for holographic recording within the last decades. As for the variety of waves, the spectrum of photosensitive media for hologram recording is rather broad and ranges from photochromic films, ^{3}He detectors (neutrons) to germanium detectors or avalanche photodiodes (X-rays).

20.2.2 Classification Scheme

The fundamental principle of recording and reconstructing an object wave can be realized in a great deal of variants, predetermined by the available recording materials, the optical setups (geometries) or techniques for recording and read out, etc. Therefore one could classify holograms, e.g., with respect to the recording scheme, the read-out configuration or the hologram properties [20.16]. The classification scheme suggested in this section primarily considers the reconstruction properties of the particular holograms. We note that such properties are closely associated with the geometrical schemes for recording (Sect. 20.2.3) and/or the type of holographic recording medium (Sect. 20.2.5), introduced later in this chapter.

The first step in hologram classification is to distinguish between *phase* or *amplitude*, *thin* or *thick*, *transmission* or *reflection* and *static* or *dynamic holograms* (see Fig. 20.3). Note, that the combination of these particular types yields a precise designation as well as sufficient information about the hologram. Our previous example from Sect. 20.2.1 is a static thick amplitude transmission hologram.

Furthermore, the particular recording and reconstruction properties are considered, which expands the variety of hologram types. Among these are: *rainbow holograms*, *polarization holograms*, *holographic*

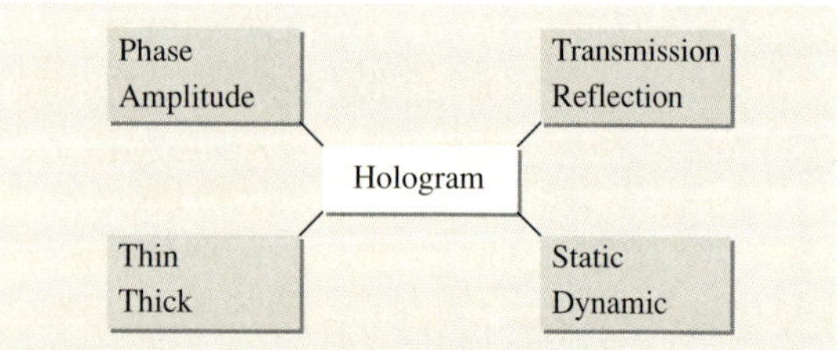

Fig. 20.3 Holograms are primarily classified as phase or amplitude, thin or thick, transmission or reflection and static or dynamic holograms

stereograms, *composite holograms*, *multiplexed holograms* (Sect. 20.2.4), *Fresnel holograms*, *Fraunhofer holograms*, *Fourier holograms*, *image-plane holograms*, and *embossed holograms*. The corresponding hologram properties will be presented in the following sections.

Phase and Amplitude Holograms

In Sect. 20.2.1 we assumed that the recording process results in a spatial modulation of the transmission of the holographic plate. The diffracted reference wave is then affected in its amplitude. Thus the hologram commonly is called an *amplitude* or *absorption hologram*. Figure 20.4 shows the microscopic structure of an *amplitude hologram*. In other photosensitive media the refractive index or the thickness is changed upon illumination with light. In this case the phase of the reference wave is affected during diffraction, i. e., reconstruction of the object wave. Therefore this type of hologram is called a *phase hologram*. Usually the latter is more favorable, because the intensity of the reconstructing reference wave is not damped by material absorption, so that holograms with considerable thickness can be recorded.

Thin and Thick Holograms

Taking into account the effective hologram thickness and the amplitude of the absorption or phase modulation, holograms can be classified as *thin* or *thick*. Note, that effective thickness d_{eff} of the hologram has to be distinguished from the thickness of the medium d in which the hologram is recorded. They are related by $d_{eff} \leq d$. The case $d_{eff} = d$ is valid for non-absorbing media and diameters of reference and signal waves much larger than d. A rough rule of thumb defines holograms with a thickness of about the light wavelength or less as thin holograms. However, a sharp boundary does not exist and only the extremal cases are unambiguously defined [20.18]. For thick holograms it is mandatory to fulfill the Bragg condition for reconstruction:

$$sK = k_S - k_R \quad \text{or} \tag{20.3}$$

$$s\lambda = 2\Lambda \sin\theta_B \,. \tag{20.4}$$

The Bragg condition is a result of energy and momentum conservation. Here, $\boldsymbol{K}$ is the grating vector of the elementary grating, $\boldsymbol{k}_{S,R}$ are the wave vectors of the signal and reference wave, respectively, and s is the diffraction order. Equation (20.4) is a scalar version of (20.3) for the particular case that the grating vector is perpendicular to the surface normal vector $\hat{\boldsymbol{N}}$. Equation (20.4) imposes a strict relation on the wavelength λ, the angle of incidence (Bragg angle θ_B) and the grating spacing $\Lambda = 2\pi/|\boldsymbol{K}|$ to reconstruct the hologram efficiently. At the same time, higher diffraction orders ($s > 1$) are depleted; they are a characteristic property only for thin holograms. The Bragg condition is illustrated by a wave-vector diagram in (Fig. 20.5a). The sharpness of the Bragg condition is inversely proportional to the dimensions of the grating. Assuming a symmetric geometry, i. e., $\boldsymbol{K} \perp \hat{\boldsymbol{N}}$, and a grating with thickness d yields an uncertainty of the Bragg condition of $2\pi/d$. This region is indicated by the shaded area in Fig. 20.5. As long as the Bragg mismatch vector $\delta\boldsymbol{k}$ resides within this area, Bragg diffraction can be observed (Fig. 20.5b). The modulus of the mismatch is proportional to the angular and wavelength mismatch ($\Delta\theta$ and $\Delta\lambda$). By this means, an angular accuracy to reconstruct a thick hologram can be in the region of less than 0.001°, a fact that is utilized for angular and wavelength multiplexing (Sect. 20.2.4).

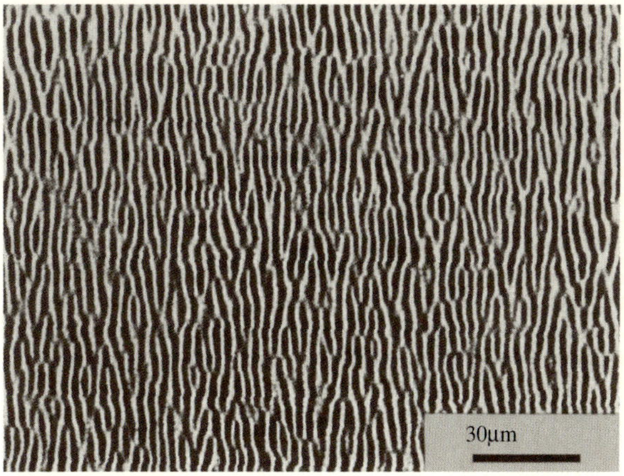

Fig. 20.4 Hologram microstructure of a diffuse object (After [20.17])

Transmission and Reflection Holograms

The classification into *transmission* and *reflection holograms* takes the direction of the wave vectors of the reference and reconstructed signal wave with respect to the holographic medium into account. This is schematically illustrated in Fig. 20.6 for two extremal cases. A transmission hologram is characterized by the fact that both reference and signal wave are transmitted through the hologram, i. e., the signal wave is reconstructed by forward diffraction processes. For reflection holograms only the reference wave is transmitted. Thus, the latter is realized if the sign of the dot product between the surface normal vector $\hat{\boldsymbol{N}}$ and the wave vectors $\boldsymbol{k}_R$ and $\boldsymbol{k}_S$, respectively, differ. Reflection holograms require recording media with thicknesses much larger than the grating

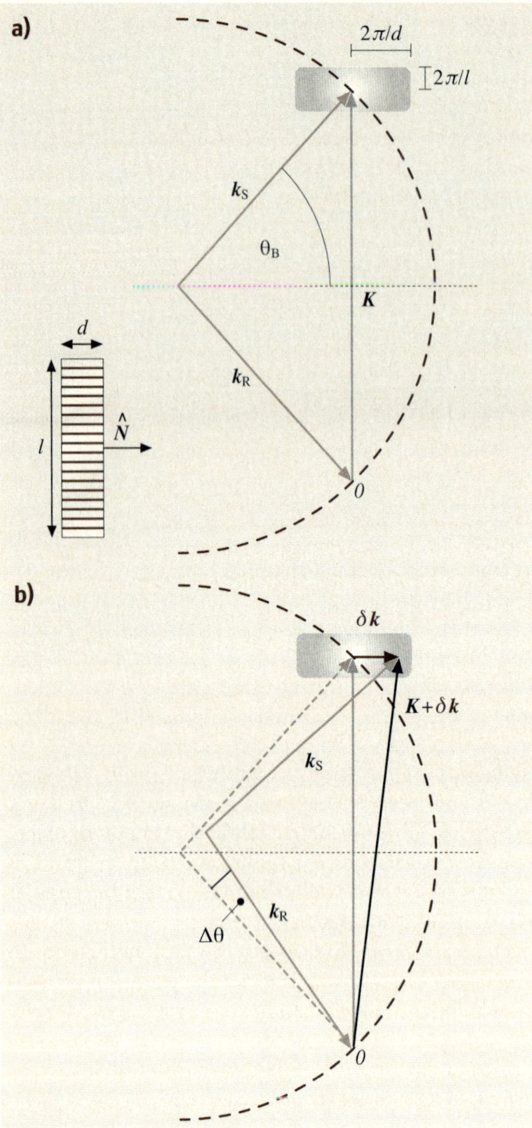

Fig. 20.5 Wave vector diagram: energy and momentum conservation in reciprocal space. **(a)** Exact Bragg matching. **(b)** Slight angular mismatch, $|\delta \boldsymbol{k}| < 2\pi/d$. Reconstruction with lower efficiency is still possible as long as the Bragg mismatch vector $\delta \boldsymbol{k}$ ends within the grey area

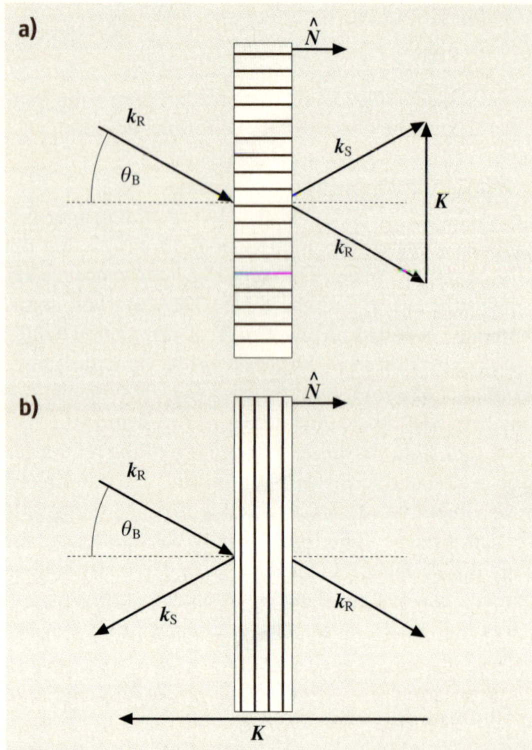

Fig. 20.6 Directions of the wave vectors of the reference and signal waves with respect to the holographic recording medium for **(a)** a transmission hologram, and **(b)** a reflection hologram

spacing. Hence they usually represent *thick reflection holograms*. A benefit of a thick reflection hologram or *white-light reflection hologram* is its high wavelength selectivity, which acts as a narrow-band wavelength filter for the reference light and allows reconstruction with white-light sources.

Static and Dynamic Holograms

Holograms that do not affect the reference and signal waves during recording, e.g., in holographic media that require further (chemical) processing, are called *static holograms*. In contrast an interaction between the recorded hologram and the recording beams leads to *dynamic holograms*: the recording beams themselves coherently diffract from the hologram and thus influence their own light modulation. Obviously, this feedback subsequently leads to changes of the hologram. This nonlinear dynamic interaction can result in an amplification of weak signal beams. As a consequence the contrast ratio of the interference pattern and thus of the hologram is a function of the material thickness. The ratio might be enhanced, e.g., as a function of the propagation coordinate in transmission holograms. Further coupling between the recorded hologram and the incoming interference pattern is shown to result in light amplification [20.19], which has been successfully applied for image [20.20] or signal amplification [20.21].

Rainbow Holograms

The invention of *rainbow holograms* stems from *Benton* [20.9, 22], allowing the reconstruction of bright, sharp monochromatic images from transmission holograms with white light. For this purpose the hologram information is reduced by losing the vertical parallax, which is an unimportant feature of holography, as three-dimensional viewing depends essentially on the horizontal parallax. Rainbow holograms are produced by copying a transmission hologram (Sect. 20.2.4) in which the area illuminated by the reference wave is strongly reduced with a horizontal slit. As a result, reconstruction of a rainbow hologram with white light splits the image in the vertical plane to form a continuous (rainbow) spectrum. Thus the reconstructed image can be observed at any part of the spectrum in the corresponding color.

Polarization Holograms

The holographic principle demands reconstruction of the complete object wavefront, usually of its amplitude and phase distribution. However, the object may additionally influence the polarization field of the wavefront, e.g., the electric field vector is influenced by reflection processes on specular surfaces. *Polarization holograms* allow the reconstruction of the polarization field of the signal wave as well and endow the holographic medium with specific recording and reconstruction properties. According to Table 20.2 we distinguish between the materials by their property to record and/or to reconstruct the polarization field. The type I and III media are obviously not suited for recording polarization holograms, although type I is applied to record and reconstruct standard holograms, i. e., phase and amplitude of wavefronts. Type III holograms are of interest, since the reconstructed polarization field is a fingerprint of the material properties and can thus be used to analyze the recording medium.

A sensitivity to light polarization is known in various media, e.g., media with photoinduced dichroism or photoinduced birefringence (type I, type III) [20.23, 24].

Table 20.2 Type I media allow only recording of the light polarization. Type II media allow both recording and reconstructing the polarization field. Type III media superimpose a material-specific polarization field to the reconstructed wavefront

	Record	Reconstruct
Type I	✓	×
Type II	✓	✓
Type III	×	✓

Recording and reconstruction of polarization holograms was further shown in photorefractive electro-optic crystals such as iron-doped $LiNbO_3$ [20.25], where the photovoltaic tensor provides the reconstruction of the polarization field (type I, type II and type III). In recent decades further photosensitive materials have been discovered that allow efficient type II recording of the polarization field, among them crystals with a center of inversion [20.26] (Sect. 20.2.5).

Holographic Stereograms and Composite Holograms

Holographic stereograms are recorded by photographing the object from various perspectives in the first step and subsequent holographic recording of the particular photographs at corresponding positions of the stereogram in the second step [20.27, 28]. The advantage of holographic stereograms is that coherent laser light is not required for the primary recording process, and thus it is even possible to record large objects or living subjects. Each hologram reconstructs a two-dimensional photographic image of the object with a particular perspective defined by the spatial position on the stereogram. Obviously, the resolution of the vertical and horizontal parallax depends on the dimension and shape of the particular hologram (slides or frames [20.29]). Modified recording techniques even allow reconstruction with white light [20.30].

Composite holograms are made up from a multitude of single holograms, which – in contrast to holographic stereograms – reconstruct a three-dimensional image of the object from different perspectives [20.31]. This allows one to reconstruct a 360° view of the object at the expense of losing vertical parallax and strong object distortions when the image is viewed with both eyes.

Fresnel Holograms, Fraunhofer Holograms

If the object is placed a short distance from the hologram, the wavefront is described by Fresnel's approximation to the Fresnel–Kirchhoff diffraction integral [20.32]. This type of near-field holograms is called a *Fresnel hologram*. Each particular point of the image is associated with an emerging spherical wave and the hologram resembles the complex wave interference pattern. Thus each particular point of the hologram can be addressed by a reference wave to reconstruct the complete image but is limited to the corresponding perspective. *Fraunhofer holograms* are generated if the Fraunhofer diffraction pattern is recorded, which appears with small objects that are placed in the far-field. In this case the recorded wavefront is proportional to the Fourier transform of the spatial amplitude distribution of the object. Fraunhofer holograms can be applied to inline holography (Sect. 20.2.3), circumventing the usual problem associated with the presence of the conjugate image. As early as 1963 Fraunhofer holography was applied to visualize distributions of aerosol particles [20.33].

Fourier Holograms, Image-Plane Holograms

Recording the Fourier transform of the complex amplitude of the object and reference waves [20.34] yields *Fourier holograms*. Recording can be performed, for example, by using a lens in the optical path. Fourier holograms offer the useful property that the reconstructed image does not move when the hologram is translated in its own plane, owing to a general property of Fourier transforms [20.35]. In *image-plane holograms* the real image of the object instead of the object itself is recorded [20.36–38]. This can be done, for example, by imaging the object onto the holographic medium with a lens. Thus each point of the object directly relates to a single point of the hologram, i. e., the object information is not distributed over the entire hologram. Image-plane holograms are highly sensitive to mechanical defects.

Embossed Holograms

Holograms can be embossed onto various substrate materials by a stamp that carries the inverse hologram as a surface relief. Embossed holograms are usually generated in four steps:

- A hologram is copied to a photoresist (Sect. 20.2.5), so that the fringe pattern appears as a spatial relief structure.
- A metallic stamping master with the inverted relief structure, commonly called the shim, is made by electrolytic processing of the photoresist.
- This metal stamp is embossed onto a substrate, e.g., a thermoplast material, under heat and pressure.
- The embossed hologram results by coating the substrate relief with a thin silver layer and subsequent hardening by electron beams.

The required minimum spatial resolution of 200 lines/mm ensures sufficient quality in reconstructing the signal wave. Nearly any condensed material can be applied as a substrate for the production of embossed holograms, which are profitable in high volumes. Embossed holograms are therefore of interest in commercial industries and as security features (Sect. 20.3.6).

20.2.3 Recording Geometries

Transmission, 90° and Reflection Geometry

Recording geometries are classified according to the directions of the reference and signal beam, and thus of the grating vector, with respect to the surface normal of the medium. Three principal geometries are distinguished: transmission geometry, reflection geometry and 90° geometry, as schematically shown in Fig. 20.7. In the transmission geometry (Fig. 20.7a) the reference and signal waves enter the holographic recording material through the same entrance face. The resulting interference pattern is characterized by the grating vector $\boldsymbol{K}$, which is perpendicular to this surface normal in the symmetric case ($\boldsymbol{K} \perp \hat{N}$). In order to obtain high diffraction efficiency combined with sharp angular selectivity it is favorable to utilize thick storage media and large Bragg angles.

In reflection geometry (Fig. 20.7b) the beams enter the holographic material from opposite faces of the medium, so that in the symmetric case the grating vector is parallel to the surface normal ($\boldsymbol{K} \parallel \hat{N}$). Note, that the Bragg condition in the form of (20.4) is only valid for the particular case of transmission geometry in the symmetric case ($\boldsymbol{K} \perp \hat{N}$) [20.39]. The reflection geometry provides optimal Bragg selectivity for wavelength multiplexing (Sect. 20.2.4), especially for $\theta_{\mathrm{B}} = 0°$. The thickness of the holographic recording medium enhances the selectivity. The angular range that can be used in the different geometries is reduced by the refractive indices of the storage medium. This reduction can be important for materials with high refractive indices.

In the 90° geometry (Fig. 20.7c) the two beams enter the holographic recording medium via two mutually orthogonal faces. Usually $\theta_{\mathrm{B}} = 0$ is chosen, so that the angle between the grating vector and the surface normal is 45°. In this geometry the storage medium has a cubic shape. The highest angular selectivity is realized, and in addition the medium is thick enough to obtain efficient gratings with a high selectivity due to the beam diameters.

Inline Off-Axis Geometries

A recording geometry is designated as *inline* if the wave vectors of the reference and signal wave are collinear (Fig. 20.8). The inline transmission geometry (or *Gabor geometry*) was introduced by Gabor when recording his first holograms [20.1]. The requirements on the coherence length of the recording waves and the spatial resolution of the holographic medium are low. These obvious advantages are accompanied by the

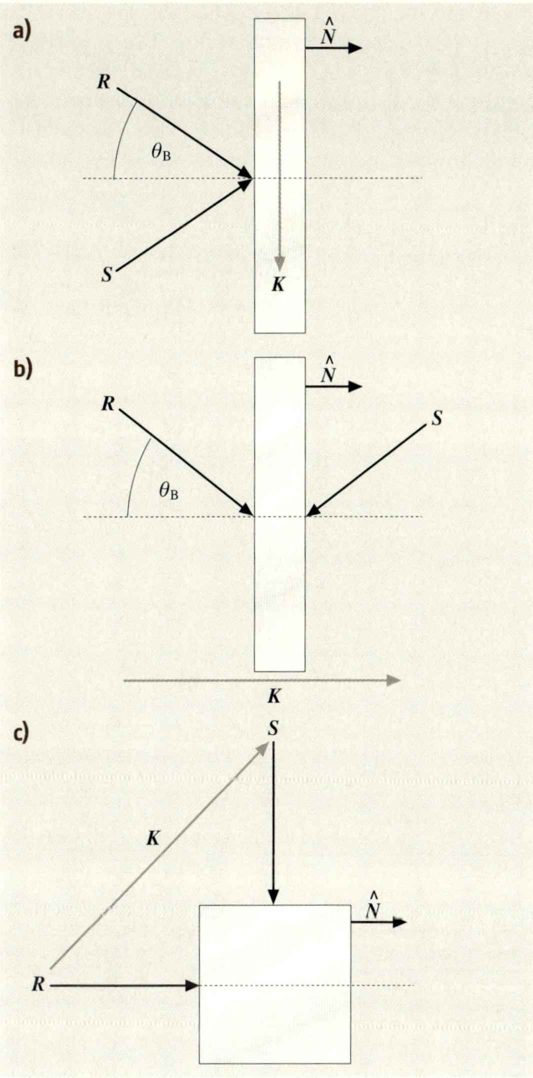

Fig. 20.7 Principal geometries for hologram recording. **(a)** Transmission geometry: reference and signal beams penetrate the holographic recording medium through the same entrance face. In the symmetric case, the grating vector is perpendicular to the surface normal $\hat{N}$ of the medium. **(b)** Reflection geometry: reference and signal beams enter the holographic recording medium through opposite entrance faces. In the symmetric case, the grating vector is parallel to the surface normal. **(c)** 90° geometry: reference and signal beams enter the holographic recording medium through two mutually orthogonal entrance faces, usually with a crossing angle of 90° and $\theta_B = 0$. The grating vector encloses an angle of 45° with the surfaces of the medium

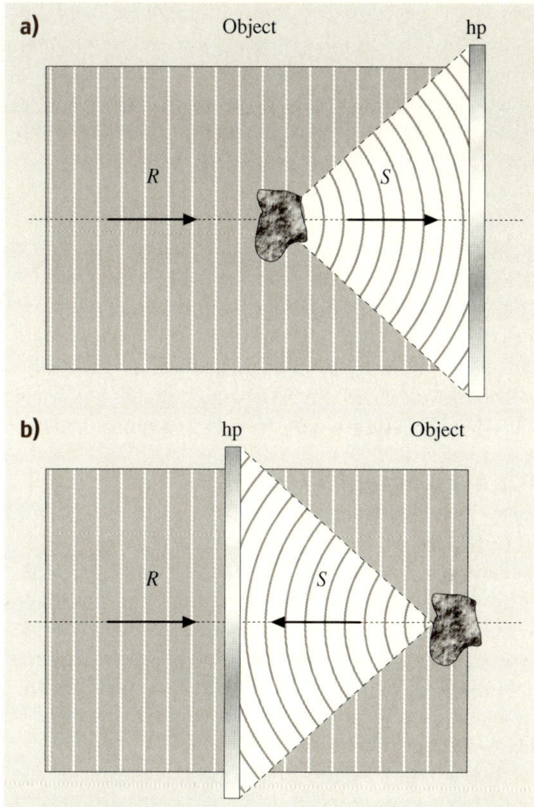

Fig. 20.8 **(a)** Inline transmission geometry (or Gabor geometry), **(b)** Inline reflection geometry (or Lippmann–Bragg geometry or Denisyuk geometry). hp: holographic plate

disadvantage that the image and reference waves are reconstructed in the same propagation direction, leading to an overlap of various images. The inline geometry is still popular for recording of white-light reflection holograms (*Lippmann–Bragg geometry* or *Denisyuk geometry*), because only a few optical components – at least a laser and a lens for beam expansion – are required in addition to a transparent thick holographic medium.

The *off-axis geometry* introduced by Leith and Upatnieks was enabled after coherent laser systems became available. It spatially separates reference and signal beam, both in recording and reconstruction. An off-axis transmission geometry is sketched in Fig. 20.2 and can be applied to the recording of reflection holograms as well.

360° Geometry, Edge-Lit Geometry

The recording of a hologram in a 360° geometry was first demonstrated in 1965 [20.40] and allows one to reconstruct the complete object information from any perspective [20.41] (also called *full-view holograms*). For recording and reconstruction, a cylindrical holographic film that surrounds the object is used. The expanded reference wave propagates along the symmetry axis, thus illuminating both the film and the object. *Edge-lit holograms* [20.42] are recorded with an expanded reference wave entering at one edge of the holographic medium, thus illuminating the entire photosensitive area. An advantage of this geometry is the opportunity to reconstruct the image using a point source, e.g., a laser diode, integrated at the hologram edge. Thus a space-consuming expansion of the reconstruction wave is superfluous and edge-lit holograms are well-suited for compact, portable holographic displays.

20.2.4 Holography Techniques

Hologram Multiplexing

The Bragg condition is of utmost importance in holography as it allows the superimposition of a multitude of holograms in the same volume element of the holographic medium. The number (#) of holograms superimposed in the same volume element of the holographic recording material is limited by the decrease of the efficiency of the particular hologram, $\eta \sim 1/\#^2$ [20.43]. Figure 20.9a shows schematically the concept of *hologram multiplexing* using the example of *angular multiplexing* [20.44, 45]. Each recorded grating vector occupies a certain volume in the reciprocal space that is inversely proportional to the corresponding grating dimensions (see Sect. 20.2.2). This is equivalent to saying that Bragg diffraction will occur as long as the mismatch is confined within this volume. Three elementary holographic gratings characterized by their grating vectors $\boldsymbol{K}_{1,2,3}$ are recorded (multiplexed) within the same volume element. The recording waves embrace the same angle, i. e., the grating spacing is constant, but vary in their Bragg angle $\theta_{1,2,3}$. The definition of Bragg angle is adopted from [20.39]. The holograms are reconstructed, completely unaffected from each other, if the angular mismatch $\Delta\theta_B$ for reconstruction of the particular hologram exceeds a certain value $|\delta \boldsymbol{k}|$, which is represented by the shaded area.

In an analogous way the Bragg condition allows *wavelength multiplexing* [20.45–47]. Here holograms are recorded with different writing wavelengths (cf. Fig. 20.9b). Other techniques tune the phases of the reference wave separately for each hologram (*phase-coded multiplexing* [20.48]) or shift the spatial position of the interfering beams slightly with respect to the previously recorded hologram (*shift multiplexing* [20.49]), i. e., phase and spatial shift are further Bragg selective measures. As early as 1962 *Denisyuk* [20.3, 50] and *van Heerden* [20.51] discussed the possibility of using the high selectivity of the Bragg condition for volume data storage in order to obtain extremely high data storage

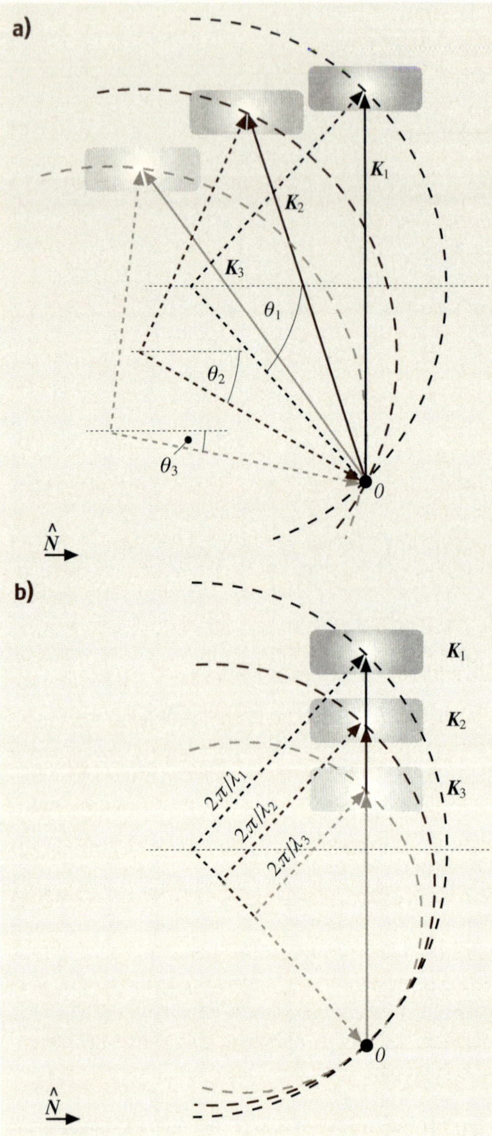

Fig. 20.9 The *shaded area*, proportional to $(dl)^{-1}$ (see Sect. 20.2.2), represents the region in reciprocal space, that is occupied by a particular recorded grating. Thus, grating vectors with overlapping areas would produce cross-talk. **(a)** Angular multiplexing. **(b)** Wavelength multiplexing

densities. The high Bragg selectivity further provides a high diffraction efficiency [20.39, 52], the suppression of the conjugated image [20.53], the reconstruction of holograms with white light [20.37, 54] and the recording of multicolor holograms (*color holography*) [20.55, 56].

A second class of multiplexing methods takes the recording geometry and the properties of the recording material into account and is not based on the Bragg selectivity. Such *fractal methods* [20.57] can therefore be applied to thin holograms as well. As an example holograms can be recorded within different volume elements of the holographic medium without affecting each other (*spatial multiplexing* [20.58, 59]). Other methods include *out-of-plane angular multiplexing* [20.60–62], *peristrophic multiplexing* [20.63] and *out-of-plane shift multiplexing* [20.64]. In addition, the combination of different multiplexing techniques greatly enhances the number of superimposed holograms and has successfully been shown to have a high impact on holographic data storage [20.62, 65–71].

Copying Holograms

White-light reflection holograms are usually produced by a direct copy from a transmission hologram, called the transmission *master hologram*. In the copying process the real image is generated from the master and can thus be used as a signal wave to record a copy in reflection geometry. As a result the initial virtual image of the master is reconstructed from the copy. This process offers some unique advantages: (a) a multitude of identical white-light reflection holograms can be fabricated without the presence of the original object, (b) the reconstructed object of the master hologram can be placed before, in, and behind the plane of the hologram copy, (c) the copying process does not require the high mechanical stability of the optical setup in a *contact copy*, i. e. the master and hologram copy are next to each other, and (d) the contrast ratio between the signal and reference wave can be adjusted and thus optimized. However, interference due to reflected waves between the master and copy, mechanical deformations of the master or copy and the presence of higher diffraction orders by reconstructing the master may result in unwanted image distortions, such as *Talbot* fringes.

Digital Holography

An interference pattern of wavefronts can be recorded digitally by using computer-addressable photosensitive detectors with sufficient spatial resolution, such as charge-coupled devices (CCDs). This opens up the field of *digital holography*, sometimes also called *television (TV) holography* or *electro-holography* [20.72]. The recording geometry for digital holography is limited by the low spatial resolution of the photosensitive digital array. Recording in the off-axis geometry with a standard array resolution of about 100 lines/mm limits the maximum angle between the reference and signal waves to a few degrees. Thus, only small objects at large distances can be recorded [20.73] encouraging the development of concepts to overcome this limitations, such as *phase-shifting digital holography* [20.74].

Digital holography further allows the reconstruction of digitally recorded fringe patterns by optoelectronic devices, i. e., by sending the information of the fringe pattern to a computer-addressable array such as a spatial light modulator (SLM). The SLM pixels manipulate the phase or amplitude of transmitted (or reflected) waves and hence it is possible to reconstruct the spatial distribution of the phase or amplitude of the signal wave. Such digital holograms offer several advantages, as the fringe pattern sent to the SLM can be manipulated digitally and actively. This allows, e.g., the suppression of the annoying twin and conjugate images by digital filtering or switching between different fringe patterns. The latter is useful for active optics or real-time holographic displays. Digital holograms are finally limited by the spatial resolution and the particular modulation depth of the phase or amplitude of the computer-addressable array. The field of applications for digital holograms is of increasing importance for *computer-generated* or *synthetic fringe patterns* and will be presented in more detail in Sect. 20.3.5.

20.2.5 Holographic Recording Materials

Holographic recording with high contrast of the reconstructed waves requires a photosensitive recording material with an optical resolution down to a light wavelength of $\approx 0.2\,\mu m$ (= 5000 lines/mm), i. e., more than a factor of 5–10 better than photographic films. Besides the optical resolution a linear photosensitive response (linear transfer characteristic), low noise from background scattering, the possibilities to reuse the holographic recording material as well as to record phase and/or amplitude holograms are key characteristics of a recording material. The properties of holographic recording materials have been of interest from the beginning of holography and a first review appeared as early as 1977 [20.75, 76]. At present the variety of holographic recording materials basically can be classified into reversible and irreversible media, see Table 20.3.

Table 20.3 Some of the common reversible and irreversible holographic recording materials. Still, photographic emulsions and hardened dichromated gelatin [20.6] are the most convenient and widely used materials for display holograms, whereas photoresists are ideal for the production of nickel masters for embossed holograms (cf. Sect. 20.2.2). Photochromic materials keep their importance for holographic interferometry (cf. Sect. 20.3.3) and high storage densities in holographic data storage is based on thick volume phase holograms recorded in photorefractive crystals (Sect. 5.9.3)

Reversible media	Irreversible media	
	(+) chemical processing	(−) chemical processing
Electro-optic crystals[a]	Photographic emulsions[a,b]	Photopolymers[a]
Photochromic crystals[b]	Dichromated gelatin[a]	Chalkogenide glasses[a,b]
Unconventional photorefractive media[a]	Photoresists, Photolacquer[a]	Polymer-dispersed liquid crystals[a]
Photo-thermoplastics[a]		
Photodichroism[b]		

[a] Recording of phase holograms
[b] Recording of amplitude holograms

In recent years the variety of holographic recording media has been extended to new classes of materials. The search for alternatives was in particular triggered by the development of applications such as holographic data storage (Sect. 20.5). At present there is pronounced scientific interest in reversible high-efficient materials with fast response or long-term properties. In the course of this research a new class of materials became apparent, characterized by their unique photosensitive behavior, despite the fact that the physical origin for their photosensitivity is completely different [20.77–80]. These are centrosymmetric crystals, so that the linear electro-optic effect is forbidden by symmetry. Photorefraction and photochromism were discovered, e.g., in centrosymmetric molecular crystals [20.81] where the photosensitivity originates from photoinduced generation of long-lived metastable electronic states. Here, the unique properties are: highly efficient recording of polarization holograms, a photoreversible recording process, and the possibility to tune the spectral range of photosensitivity and the storage time over a wide range simply by material design. Further promising candidates from this new group of holographic recording materials are photorefractive garnets, e.g. $Gd_3Ga_5O_{12}$ [20.82] (Fig. 20.10). Another novel, fascinating type of holographic materials is a composite of liquid crystals and polymers, so-called polymer-dispersed liquid crystals. The initial stage is a homogeneous solution, that is a mixture of mesogenic molecules and photosensitive monomers. Upon holographic exposure, a photopolymerization reaction takes place more rapidly in the bright regions of the optical interference pattern and consequently the monomers diffuse to these regions while the liquid-crystalline molecules congregate in the dark regions [20.83]. The

Na_2(NP) CdF_2: Ga, Y GGG: Ca TGG: Ce PDLC

Fig. 20.10 Representatives of novel photorefractive materials: $Na_2[Fe(CN)_5NO] \times 2H_2O(Na_2[NP])$, (courtesy of Theo Woike, Institute of Mineralogy and Geochemistry, University of Cologne, Germany), CdF_2 : Ga, Y (courtesy of Institute of Crystallography of Russian Academy of Sciences, Moscow, Russia), $Gd_3Ga_5O_{12}$: Ca (GGG:Ca, courtesy of László Kovács, Research Institute for Solid State Physics and Optics, Hungarian Academy of Sciences, Budapest, Hungary), $Tb_3Ga_5O_{12}$: Ce (TGG:Ce, courtesy of Lothar Ackermann, FEE, Idar-Oberstein, Germany), and a polymer-dispersed liquid crystal (PDLC) between glass substrates advantages over other holographic media are the large amplitudes of the light-induced change of the refractive index, and the ability to tune this amplitude using external electric fields [20.84, 85].

20.3 Applications of Holography

From the beginning, Gabor's principle of wavefront reconstruction primarily was applied in the field of wave optics. Gabor himself emphasized the possibility to enhance the optical resolution in electron microscopy by holography, which was finally demonstrated, but not until 1990 [20.86]. It is therefore not surprising that the first dominant applications of holography were in the field of *high-resolution three-dimensional displays*. This initiated the entry of holography into the fields of arts and commerce, accompanied by great interest from the film industry and the military. The first museum for holography opened its doors already in 1976 in New York City.

Holographic interferometry developed as a further key application for holography, as was already recognized in 1965 [20.5]. The technique allows the visualization of object deformations and movements with a resolution down to the order of the light wavelength – even in real time. Holographic interferometry has had a great impact in *object recognition*, which has only recently attracted interest since it was adapted to work with digital holography [20.87].

Wavefront reconstruction with computer-generated holograms has driven the important field of *diffractive optics* in applied holography. Today it is possible to generate any phase and intensity wavefront distribution using synthetic holograms. In this way, beams can be deflected, focused or shaped with high precision and velocity, which is a basic property for a lot of other common optical applications, such as optical switches and filters in optical telecommunication.

During recent decades holography has been applied with microwaves, neutrons, electrons [20.88–91], X-rays [20.92–94], and acoustic waves. The latter, e.g., finds its applications in nondestructive material testing, in medicine for diagnostic purposes, in ultrasound microscopy, in seismology or in underwater positioning. In contrast, microwave holography has had its impact in understanding the important microwave techniques such as radar, doppler radar, antennae with synthetic apertures, etc. [20.95].

At present the field of applications is still growing and new discoveries in the field of holography are currently being recognized, such as second-harmonic holography [20.96–99], holographic creation of photonic crystals [20.84, 100, 101] or self-adaptive intra-cavity holography in lasers [20.102, 103], which mean that holography remains an exciting field in optics. In the following we present some selected impressive applications – beyond the field of science – to give an insight into this broad spectrum of applied holography: holographic data storage, holography in archaeology, holographic interferometry, holography in medicine and biology, applications of computer-generated holograms, holograms as security features, holographic scattering for material analysis, atomic-resolution holography and neutron optics.

20.3.1 Holographic Data Storage

The application of holography for optical data storage was introduced by *van Heerden* in 1963 [20.51]. He discussed the recording of multiple holograms within the same volume element, which was successfully demonstrated by *Leith* using a photographic plate as the storage medium three years later [20.45]. High-density recording using the angular multiplexing technique was not applied to holographic data storage until 1973 [20.104].

Today, after 40 years, a holographic data storage system is still not commercially available. Most of the technological components that are necessary for such a system, such as, e.g., the laser, camera, and two-dimensional liquid-crystal display (SLM), were available in the past, with breakthroughs in specification and cost being achieved in recent years. However, the search for a holographic storage material fulfilling all the application requirements is still in progress, since holographic storage demands unique specifications such as high light sensitivity and nearly no light absorption. Today we find a multitude of development platforms that have been built in the last four decades. They prove the principles and demonstrate the different ways to store a large amount of data at a sufficient high rate [20.105–108]. An insight into holographic data storage is given in [20.109, 110] and Sect. 20.5.

20.3.2 Holography in Archaeology

Holography is ideal for archival recording not only of digital data but also of any valuable or fragile, analogue three-dimensional objects, such as museum artifacts. The technique is of increasing significance in the field of archaeology, as it allows the production of true replicas of the original for distribution throughout the world [20.111, 112]. In particular color holography is applied to achieve optimum results, as shown in Fig. 20.11 [20.113, 114]. Besides presentation or display purposes, objects can seldom be studied in places remote from the museum and this makes it possible to compare and even to assembly pieces excavated from different places in the world.

20.3.3 Holographic Interferometry

The principle of *holographic interferometry* with its broad variety of applications is already a key topic of several books [20.115–118]. The technique was first discovered in 1965 [20.5, 119–123] and is also referred to as *hologram interferometry* or *holometry*. In the following we introduce three of the basic inventions of *holographic interferometry*: double-exposure, real-time, and time-average interferometry.

Double-Exposure Interferometry

The technique of *double-exposure interferometry* is applied to visualize and/or to determine object deformations or dilatations down to the scale of the light wavelength [20.124–127]. The object is holographically recorded twice, before and after the object deformation or dilatation. Thus the hologram reconstructs two signal waves at the same time and a fringe pattern results by signal wave interference. The fringe pattern is correlated to the phase differences of the object wavefronts between the two exposures. Holographic interferometry is advantageous when compared with non-holographic techniques, since it allows the visualization of changes in the object's wavefront arising at two different times. The *heterodyne technique* was the first automated analysis in holographic interferometry with sufficiently high precision [20.128–130]. Two reference waves of slightly different frequencies are used for hologram recording. In this way, each pixel of the reconstructed image is time-dependent and can be monitored by photodiodes. As a result the resolution in the object's wavefront distortion is less than 1 nm. However, the analysis of the fringe pattern is ambiguous when the scale of object deformation exceeds the light wavelength because of the phase-shift periodicity of modulo 2π in the fringe pattern. This problem is solved in part by the *phase-shift technique*. Here, two reference waves of the same wavelength but at different recording angles are used for recording and reconstruction [20.131–133]. Thus the intensity of the fringe pattern is a function of both the object's wavefront distortion and the phase difference between the two reference waves. Impressing an external phase shift on one of the reference beams during reconstruction, the intensity of any object point oscillates modulo 2π as a function of the external phase shift. The phase distortion of the object wavefront is then equal to the phase difference between the reference and reconstruction waves, both of which are automatically determined with high precision.

Fig. 20.11 Hologram of a plate of clay with cuneiform writings exhibited in a museum in Nicosia (Cyprus). As the hologram reconstructs the three-dimensional object information it is possible to decipher the cuneiform writings, which contain important information in their relief structure. (Photo courtesy of G. von Bally, Laboratory of Biophysics, University of Münster, Germany)

Real-Time Interferometry

The optical principle of *real-time interferometry* [20.134] is equivalent to double-exposure interferometry, but allows the determination of the kinetics of phase distortion of the object's wavefront. The hologram is recorded only once in a holographic recording material and is reconstructed within the same optical setup. Thus, the reconstructed signal wave interferes with the object wave itself and any distortion of the object's wavefront is immediately visualized by a corresponding fringe pattern in the reconstructed image. In this way, deformations or dilatations of the object can be investigated with respect to the holographically recorded object wave in real time, i. e., the reconstructed signal wave of the hologram acts as the initial measure.

Time-Average Interferometry

Linear or harmonic oscillations are visualized with *time-average interferometry* [20.136, 137], as shown for a violin at various frequencies in Fig. 20.12. The key idea is that the contrast of the hologram is reduced by moving objects, i. e., the recorded amplitude of the phase or amplitude modulation decreases due to movement of the interference pattern during the recording process. This feature is a disadvantage in holography as it requires extremely high mechanical stability of the optical setup (less than $\lambda/10$) during holographic recording. However, in the case of linear or harmonic oscillations, the intensity of the reconstructed image is shown to follow a Bessel function, so that it is a function of the phase amplitude of the periodic object deformation. Thus it is possible to visualize amplitudes of oscillating membranes with high precision. In order to overcome the problem of the mechanical stability of the optical setup due to mechanical coupling with the oscillating objects, *real-time time-average interferometry* is commonly applied, including a phase-shift technique for the reference beam [20.138, 139]. A hologram of the unexcited object is recorded and the method of real-time interferometry is used to visualize the oscillation amplitudes. Stroboscopic reconstruction [20.140–142] with the same frequency as the object oscillation can also be applied to visualize object deformations restricted to single frequencies.

20.3.4 Holography in Medicine and Biology

The fields of medicine and biology have impressively demonstrated how different holographic inventions, such as holographic interferometry and digital holography, can be applied together to give powerful results, and how the life sciences already make extensive use of holography.

Endoscopic holography was the first important medical application of holography. This method combines the advantages of endoscopic intra-cavity observation with holography, and thus offers the physician new in situ diagnostic capabilities. Recording of a hologram with the use of a single fiber for both the signal and the reference beam was demonstrated as early as 1976 [20.143]. Further improvements used fiber bundles [20.144] and gradient-index optical imaging systems [20.145] or even combined with an articulated arm [20.146]. The problem

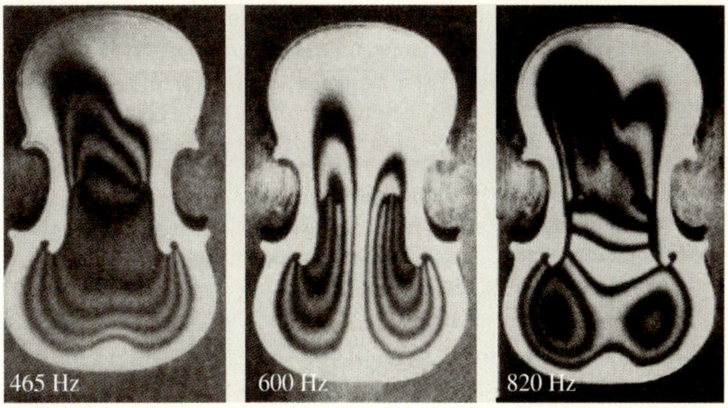

Fig. 20.12 Reconstructed double-exposed hologram of resonant eigenmodes of a violin at **(a)** 465 Hz, **(b)** 600 Hz, and **(c)** 820 Hz. The *brightest areas* indicate oscillation nodes. (After [20.135])

of the cumbersome equipment of conventional holography was overcome by migrating from analog to digital recording [20.147] using numerical reconstruction, enabling morphology studies of cavities and tissues even at the cellular level.

Holography is applied as a nondestructive optical technique to image the properties of biological objects and tissues and to detect abnormalities such as tumors. In medicine, holographic interferometry in its different variants is the method of choice to perform in situ and even in vivo detection of tissue. Phase-shifting holographic interferometry was demonstrated to work for breast-cancer detection [20.148] by using mild pressure stressing of the breast and pulsed laser light. Pulsed digital holography has also been employed as a noninvasive technique to measure the elasticity of biological tissues, e.g., a human hand, pig tissue or human tumors in gel, by exciting them by short-shock mechanical pulses prior to recording phase-difference maps [20.149, 150].

Imaging an object located below the flesh of a human being, i. e., recording a hologram through a living person, has been performed by using a low-intensity continuous-wave laser [20.151] and the principle of first-arriving light [20.152]. Digital holography has been applied as a gating method, which allows a rapid succession of holograms to be made, thus increasing the signal-to-noise ratio.

True three-dimensional representation (*holographic display*) is another facet of the use of holography in cardiology, where the third dimension might be an electric potential, time or any other variable of interest [20.153]. Standard medical imaging techniques such as magnetic resonance imaging or computerized tomography provide two-dimensional transect images of the organ or object to be examined. Employing digital holography a three-dimensional hologram can be obtained, which considerably facilitates the interpretation of the data [20.154, 155]. Utilizing a photorefractive film in conjunction with digital Fourier holography, it has been demonstrated that abnormalities detected in mammography, which usually vary in size, can be picked out according to their size and then visualized with high resolution down to about 10 μm [20.156].

Digital holographic microscopy is a useful noninvasive method for the visualization of transparent objects, e.g., living cells or tumors with high accuracy. In contrast to phase-contrast microscopy it provides a full-field, high-resolution amplitude measurement and even detects the quantitative distribution of the optical path length created by the specimen. Therefore information on the morphology and the refractive-index pattern can be gained, which allows sub-wavelength axial accuracy [20.157, 158].

20.3.5 Diffractive Optics with Computer-Generated Holograms

Since the early days of holography the generation of holograms by means of simply computing the interference pattern has attracted attention [20.159–162] and is still one of the fastest developing topics in holography. The main advantage is that the object only has to be known in mathematical terms and does not need to exist physically. Starting with one-bit holograms drawn by plotters, which were photographically reduced and recorded on a film, computer-generated holograms (CGHs) have developed greatly with increasing computing power and the availability of high-resolution eight-bit spatial light modulators (SLM) with refresh rates of up to 1 kHz [20.163, 164].

Various applications making use of CGHs have been realized, in particular as device for beam steering and manipulation, i. e., as computer-generated holographic optical elements. CGHs are employed as so-called *laser tweezers*, i. e., tools for micromanipulation of particles in both physics [20.165, 166] and life sciences [20.167], using the principle of optical trapping [20.168]. Another fascinating application is the realization of non-diffracting beams by means of CGH [20.169]. Contrary to standard Gaussian beams, a fundamental Bessel beam does not undergo diffraction, i. e., the transverse optical intensity distribution does not change during propagation [20.170]. Optical phase singularities can be created with CGHs, calculated by the interference between a plane wave and superposition of the TEM_{10} and TEM_{01} Gauss–Hermite optical-resonator modes (hybrid doughnut mode) [20.171]. Moreover, CGHs are utilized for quality control of optical systems, e.g., for the characterization of aberrations. In particular, the assembly and adjustment of multicomponent optical systems can be monitored [20.172].

20.3.6 Security Aspects of Holography

A valuable, practical and widespread application of holography is its use for authentication or security devices. Holograms on credit cards, banknotes and identity documents are state of the art today, see e.g. [20.173, 174]. However, the impact of holography as a security feature becomes much more impressive, e.g., in optical data transmission. Data must not only be transferred in a fast but also a secure way, which is realized using *holographic encryption*. Benefits of this technique over others are that optical processing provides the maximum number of degrees of freedom and that optical encryption is ideal for information stored in holographic media. One promising approach is to process data using a double-random phase encryption technique [20.175] and record them as a digital hologram [20.176]. Sophisticated geometries and elaborate techniques such as phase-shifting interferometry [20.177] allow fast and secure encryption/decryption using electronic, optoelectronic or even all-optical approaches. The latter is inexpensive in time, whereas pure electronic decryption requires a huge computer load. Further possibilities are orthogonal phase-coded multiplexing with a random phase mask in the reference-beam path [20.178], fabrication of CGHs using incoherent multiplexing (synthetic double-exposure) by electron-beam lithography [20.179], or encrypted optical storage in a photorefractive crystal using angular multiplexing [20.180]. Besides its other advantages over standard methods holography provides the unique possibility of watermarking three-dimensional objects [20.181], again employing phase-shifting interferometry and digital holography.

20.3.7 Holographic Scattering for Material Analysis

Light scattering in holographic recording materials is a disadvantage for wavefront reconstruction as it generates a serious scattering background in the reconstructed image and thus strongly affects the signal-to-noise ratio. Scattering in holographic recording materials with nonlinear response to intensity shows up as a bright intensity

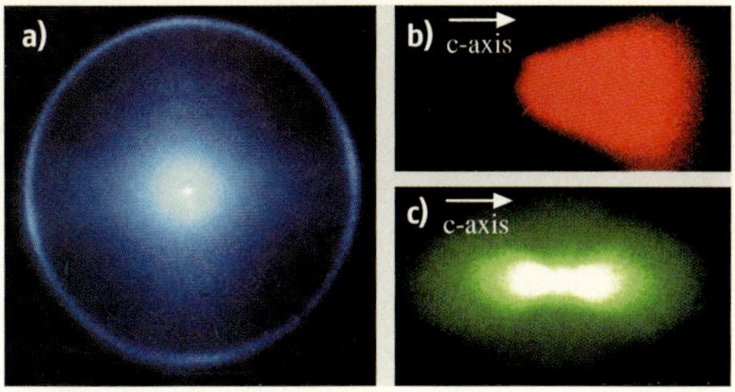

Fig. 20.13 Far-field nonlinear holographic light scattering as observed in various recording materials in the steady state. **(a)** Holographic scattering ring in $Na_2[Fe(CN)_5NO] \times 2H_2O$: the scattered light is orthogonally polarized to the incident pump beam polarization. The diameter of the ring depends on the birefringence [20.189], **(b)** photoinduced light scattering in $Sr_{0.61}Ba_{0.39}Nb_2O_6 : Ce$: the scattering lobe is directed antiparallel to the orientation of the polar c-axis [20.183], **(c)** beam fanning in $LiTaO_3 : Fe$: the two scattering lobes are directed parallel to the polar c-axis [20.190]. One lobe shows higher intensity which indicates the orientation of the spontaneous polarization

distribution with a characteristic shape. The scattering appears over a wide apex angle and is a function of the exposure. At present such scattering phenomena have gained considerable interest in the scientific field of optical nonlinearities [20.182]: besides investigations of their physical origin, *holographic scattering* phenomena are useful for material analysis [20.183]. It has been demonstrated that holographic scattering can be applied as a nondestructive contact-free optical method, e.g., to investigate phase transitions [20.184, 185] and ferroelectric properties [20.186, 187]. Figure 20.13 shows the example of a far-field steady-state scattering pattern observed in three different holographic recording materials. The diameter of the ring in Fig. 20.13a allows the determination of the birefringence [20.188]. Photoinduced light scattering in $Sr_{0.61}Ba_{0.39}Nb_2O_6 : Ce$ (SBN:Ce, Fig. 20.13b) and $LiTaO_3$ (Fig. 20.13c) can be used to determine the direction of the polar c-axis. In SBN:Ce, the scattering lobe is oriented in a direction antiparallel to the polar axis.

20.3.8 Atomic-Resolution Holography

The key technique for the determination of atomic and molecular structures, in particular for investigations of crystallographic structures, has been diffraction for a long time. However, this method is not a true physical solution, as it only yields information about the intensities and not about the amplitudes of the diffracted waves. Holography, on the other hand, is known to overcome this obstacle by recording the interference pattern of the reference and object waves. However, its spatial resolution is limited by the wavelength of radiation and also by the resolution of the recording medium. The first limitation is lifted simply by using electrons, high-energy electromagnetic waves (X-rays, γ-rays) or neutrons with wavelengths in the range of Å for holographic recording. To overcome the second limitation *Szöke* proposed a scheme to study the atomic order of solids by employing atomic-resolution holography [20.191], an idea that was first experimentally demonstrated by using electrons [20.192]. In this experiment a 3-D reconstruction of the relative positions of atoms was performed by evaluating the angular distributions of the scattering intensities. The intensity patterns are formed by interference between electrons that are emitted from a source atom and reach the detector either directly (reference wave) or after scattering from neighboring atoms (object wave). This is called the inside-source concept, where a point-like source is produced inside the investigated material. The inverse scheme exchanges the role of the source and detector (inside detector), i. e., the interference pattern of the incident and scattered waves is investigated using a point-like detector in the material (see [20.193, 194] and Fig. 1 in [20.195]).

Several experimental approaches for atomic-resolution holography have been demonstrated, employing electrons, positrons [20.196, 197], X-rays, γ-rays, and neutrons, each of them characterized by its specific interaction with matter. Therefore, electrons and positrons are mostly used for investigating the structure of surfaces because of their low penetration depth. X-rays are sensitive to the electronic properties, whereas with neutrons the nuclear structure is probed.

Various mechanisms of electron emission from a localized atomic source have been utilized, e.g., Kikuchi scattering [20.192], the photoelectric effect [20.198–200], low-energy electron diffraction [20.196, 201, 202] or Auger-electron diffraction [20.203]. Atomic-resolution X-ray holography was first performed by using fluorescent X-radiation [20.204], later by employing a multi-energy inverse mode [20.195] or a combination of these [20.205, 206]; γ-ray holography was experimentally demonstrated using the example of an ^{57}Fe film in the inverse scheme [20.207]. Only recently an advanced γ-ray holography technique has been developed that allows the distinction between atoms of the same kind that occupy inequivalent crystallographic sites, a task where X-rays and neutrons failed. Here, coherent effects in the Mößbauer nuclear absorption are utilized, which provide local spectroscopic information [20.208]. The latest development is atomic-resolution neutron holography, suggested by *Cser* et al. [20.194] and subsequently realized in both experimental schemes, the *inside source* [20.209, 210] and *inside detector* arrangement [20.211].

Still there is a steady improvement of the evaluation techniques, e.g., complex γ- or X-ray holography to remove the disturbing twin image without making measurements too time consuming [20.212, 213].

20.3.9 Neutron Diffractive Optics

In 1990 it was demonstrated that a holographically recorded refractive-index pattern could be reconstructed not only with light but also with cold neutrons [20.214]. Obviously, the illumination induced refractive-index changes for both light and neutrons. In analogy to light optics this phenomenon was called the *photo-neutron-refractive effect* [20.215]. The efficient diffraction of cold neutrons from light-induced gratings was a huge step in the development of neutron diffractive optics by means of optical holography, as such methods were not previously possible. For thermal neutrons (with wavelengths in the sub-nanometer range) single crystals are usually used as diffractive optics, for ultra-cold neutrons (with wavelengths of about 5 nm and more) mechanically ruled gratings are used. The method of preparing gratings by means of light optical holography closed the gap between these two approaches. Nowadays, mirrors and beam splitters can be produced, culminating in the successful setup of an interferometer for cold neutrons [20.216,217]. In Fig. 20.14a the angular dependence of the diffraction efficiency for a holographically prepared volume grating that acts almost like a mirror is shown. Figure 20.14b depicts the results of an interferometric measurement utilizing three holographically prepared thick diffraction gratings as beam splitters and a mirror. The interferometer may be regarded as one of the most useful neutron optical devices as it provides information about the wave function, i. e., *amplitude and phase*, in contrast to standard scattering or diffraction techniques where only the intensity is measured. The

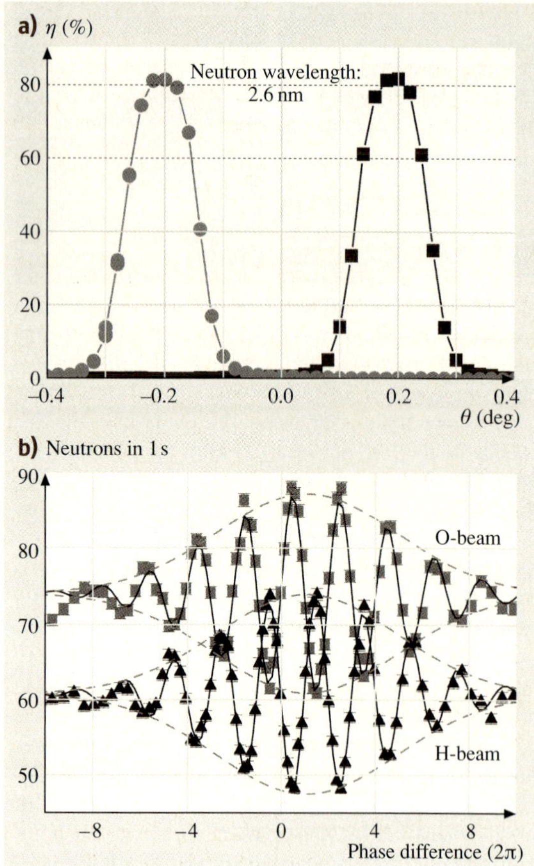

Fig. 20.14 (**a**) Diffraction efficiency for neutrons with 2.6 nm wavelength. (**b**) Interference fringes obtained by introducing a phase difference between two partially coherent neutron beams. *Squares* and *triangles* show the measured intensity of the output beams as a function of the phase shift. Wavelength 2 nm

principal future application of such an instrument could be the investigation of mesoscopic structures and their kinetics in the fields of condensed-matter physics and engineering, chemistry, and biology. A recent review on neutron optics with gratings employing the photo-neutron-refractive effect and light optical holography is given in [20.218].

20.4 Summary and Outlook

There is no doubt that holography is an important field in optics and one of the key inventions of the 20th century. It is stressed that the idea of Dennis Gabor could not have its full impact until the development of light sources with sufficient coherence – nearly 20 years after discovery. Obviously, Gabor was quite ahead of his time when inventing holography.

Holography offers a multitude of applications in various fields of our life. Now, new developments are making use of the wavefront reconstruction principle. As it will probably have an increasing influence on our life, the history of holography will continue to be an exciting journey.

20.5 Optical Data Storage

After the correction of vision with glasses, contact lenses or reshaping of the cornea, and after the recording of still and moving images on film or electronic cameras, the most pervasive application of optics in our daily life is optical data storage: the storage and retrieval of data using optics close to the diffraction limit and a miniature solid-state laser, all realized at very affordable prices.

The distribution of audio information on a removable disk – sound stored as modulations in a flat surface – has had a long and successful track record. At first, analog signals were imprinted on shellac and then later vinyl platters. However, recently sound has been almost exclusively distributed as and replayed from digitally encoded data, molded into the surface of compact polymer disks. This migration was driven by consumer demand for higher fidelity, i. e., higher bandwidth in recording, replication and playback, which in turn required dramatic increases in the amount of stored information. At first the diameter of the disk was increased, but this soon reached its practical limit. With a mechanical stylus, higher areal storage density increased the susceptibility to wear and tear, which forced a transition from mechanical detection to a noncontact optical scheme.

In the form of the read-only compact disk (CD), optical data storage became the dominant distribution vehicle for music (CD) and later for computer software with the CD read-only memory (CD-ROM) [20.225, 226]. With the ferocious appetite of consumers for ever more information at ever higher data rates, the CD and the CD-ROM have undergone a metamorphosis into the digital versatile disk (DVD) [20.227]. The DVD standard offers higher areal density per layer, and with as many as four layers of prerecorded information, can provide sufficient read-out bandwidth and capacity for the distribution of several hours worth of high-quality compressed video. Since its introduction into the market, DVD devices have reportedly experienced the fastest growth of quarterly sales and market penetration of any consumer electronics technology ever, as of the last quarter of 2002. But with the advent of high-definition television (HDTV), even higher data rates and capacity will be needed, requiring yet another increase in areal density. Such increases have motivated several proposed next-generation formats, including the BluRay® optical disk format. Table 20.4 compares the pertinent characteristics of the CD, DVD and the proposed BluRay formats, highlighting the significant technical progress that has been made.

In addition to this distribution of prerecorded content on removable and interchangeable media, the desire of consumers to record information by and for themselves has been an important economic and technical driver. The wax-coated drum of Edison, with analog information engraved with a vibrating needle, has given way to the CD and DVD *burner*, with carefully modulated and interleaved digital data imprinted with a flashing laser beam. Recordable optical disks rely on using high

Table 20.4 Comparison between CD, DVD, and the proposed BluRay format [20.219–224]

	CD	DVD	BluRay
Capacity (single layer)	0.65 GB	4.7 GB	23.3, 25, or 27 GB
Laser wavelength	780 nm	650 nm	405 nm
Numerical aperture	0.45	0.6	0.85
Track density	16 000 /in.	34 000 /in.	79 000 /in.
Minimum length pit ($\sim$ 2–3 bits)	0.833–0.972 μm	0.4–0.44 μm	0.138–0.16 μm
Areal density (user)	0.39 Gb/sq.in.	2.77 Gb/sq.in.	13.7, 14.7, or 15.9 Gb/sq.in.
Reference velocity	1.2 m/s	3.49–3.84 m/s	4.55–5.28 m/s
Data rate	1.47 Mb/s	11.08 Mb/s	36 Mb/s

laser power to write marks that low laser power can still read. The high laser power must modify the local optical properties of the recording medium to provide a significant change of signal; conversely, the low read-out power must *not* affect the properties of the media so as to protect any recorded data. Processes such as ablation of a dye layer are irreversible and thus are well suited for *write-once, read-many times* (WORM) applications. The prevalent implementation of this technique is the CD R: WORM media with a form factor and functionality nearly identical to a CD or CD-ROM once it has been recorded. Other recording processes – foremost the change between amorphous and crystalline structure in a thin metal, semiconductor or alloy film – can be reversible and allow erasable or rewritable (R/W) media. For these types of media, the thermal characteristics are as important as the optical parameters. One has to ensure that, after short transient heating with the tightly focussed laser, each very small spot will cool fast enough to quench the film rapidly into an amorphous phase. On the other hand, the same R/W medium must cool slowly after a large area is uniformly illuminated, allowing recrystallization of the film and erasure of previously recorded data. The issues of optical, thermal and device engineering for WORM and R/W recording are extremely complex and go beyond the scope of this brief summary. Details can be found, for example, in [20.226].

In addition to the pervasive 120 mm-diameter CD/DVD disk, many other disk diameters have been used. One early niche application was the laser disk, used for the distribution of high-quality prerecorded movies before DVD-level areal densities and data compression were available. For high-end data storage applications, diameters of 5.25″, 12″ and up to 14″ have been commonly used to increase the capacity per platter up to 200 GB. Smaller disks for small-form-factor, portable consumer electronics devices are becoming increasingly popular. Examples are MP3 players, digital still and digital video cameras using optical disk media with diameters as small as 1″.

The facts that the optical recording media are removable, have high storage capacity and a relatively low price also make them a prime candidate for massive data warehouses. Jukeboxes for home audio and video applications allow a home user to have all the multimedia content that they own at their fingertips. In a similar fashion, commercial data libraries and warehouses use data silos with robotics and industrial-strength CD-R recorders and CD-ROM players to provide massive amounts of near-line storage at low cost.

Optical read out of cheaply replicated, injection-molded CDs, CD-ROMs, and the rapidly emerging DVD disks has clearly come to dominate multimedia content and software distribution. On the other hand, general-purpose read–write information storage is still the exclusive domain of magnetic recording. Magnetic recording in the form of hard disk drives delivers higher areal densities, higher data rates and faster access times than optical storage, while supporting millions of write/erase cycles and many years of retention time at very competitive cost. To obtain these high performance characteristics, the removability offered by the once ubiquitous floppy disk had to be sacrificed. Thus the key advantages of optical recording are clearly the removability and interchangeability of the media, as well as the parallel replication of prerecorded disks in seconds via injection molding. Optical recording on WORM media, typically in the form of burning a CD-R, plays an important role as inexpensive backup for data stored on magnetic disks, and is trying to compete with magnetic tapes for archival data storage in large libraries. Attempts to bring a commercially viable optical tape technology to market have not succeeded to date [20.228].

In 1997, 2000 and again in 2003, the National Storage Industry Consortium (NSIC) convened meetings of 50 leading industry experts to develop three comprehensive optical disk storage roadmaps [20.219–221]. These roadmaps summarized the anticipated component, system and market development. While the overall trends were predicted correctly, technical and economic challenges considerably slowed down the actual pace of development.

Key drivers for future developments were expected to be the rapid expansion of the Internet and the emergence of HDTV, with both calling for higher bandwidth and storage capacity. The Internet has expanded, but high-speed broadband service to the home is not yet pervasive. Standards issues, broadcaster resistance and concern about copyright management have delayed the advent of HDTV. At the same time, the pervasive use of computers and storage in consumer electronics devices (and eventually common household appliances) is requiring more storage in smaller form factors. Thus all of these requirements are still driving optical data storage to reach ever higher storage densities [20.219–221].

The remainder of this article highlights different approaches towards achieving this goal of increased optical storage density, either by increasing the density at the surface of an optical recording material, or by utilizing the unique capability of optics to access the volume of suitable media.

20.6 Approaches to Increased Areal Density

For many years, the demands for better data storage have been met by evolutionary advances: increases in the areal density and other performance specifications of magnetic and optical recording devices. The areal density of hard disks has been growing continuously through more than seven orders of magnitude over the last 50 years, but during the last five years the growth accelerated to a compounded annual growth rate of slightly more than 100%. Optical storage, with a much shorter track record, has increased in storage density by a factor of five from the original CD standard to the DVD format, and another factor of five to the proposed next-generation BluRay format (Table 20.4). The reason for this dramatic difference between the increase of aerial densities for magnetic and optical recording is that optical data storage has specialized on applications that require removability of the optical recording media. Since high-performance magnetic recording does not support removability, hard drives need not work with the media of previous generations. Thus the only standards to satisfy are those on data input/output and form factor, which has encouraged strong technical competition and resulted in tremendous progress. Optical storage, however, is dominated by removable, interchangeable media with backward compatibility. This compatibility facilitates the introduction of each new generation of technology in the market, due to the large existing base of recorded material, but forces a time-consuming standards process for each higher-density generation.

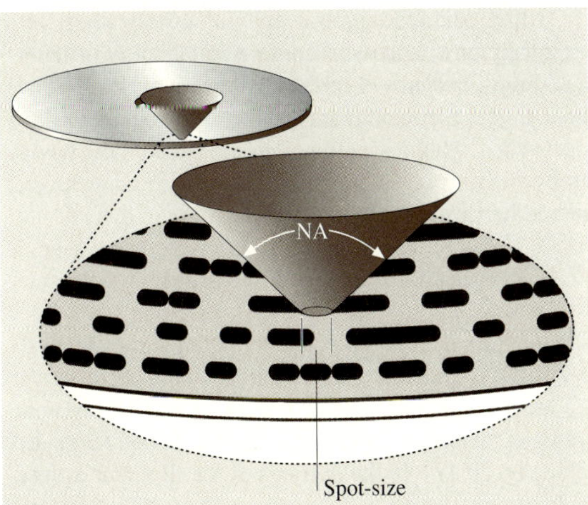

Fig. 20.15 Conventional optical storage uses a tightly focussed laser beam to access individual bits in a single layer

Still, optical storage has substantial potential as a storage technology. While the limits of magnetic recording are still being debated [20.229], the limits of conventional optical recording are well understood [20.219–221]. Current optical storage technology is already working close to the optical diffraction limit. However, significant future increases in density are possible by taking advantage of the wavelength and/or numerical-aperture dependence of the diffraction limit, or by going beyond it with near-field techniques. In addition, if the signal/noise ratio is sufficient then gray-scale techniques allow the storage of more than one bit per location [20.230].

The diameter of the diffraction-limited spot is directly proportional to the laser wavelength λ and inversely proportional to the numerical aperture (NA) of the imaging lens (Fig. 20.15). The area A of the spot is then proportional to the square of these parameters [20.231]

$$A \sim \left(\frac{\lambda}{\mathrm{NA}}\right)^2 , \tag{20.5}$$

and the resulting maximum areal density is simply the inverse of this area times b, the number of bits per spot,

$$\mathcal{D} \sim b \left(\frac{\mathrm{NA}}{\lambda}\right)^2 . \tag{20.6}$$

As described in Table 20.4, the differences in capacity and data rate between the CD, DVD and BluRay formats is a clear consequence of reducing the diffraction-limited spot size of the focus at the medium. Other factors were also involved in the density improvement found in the newer standards – such as stronger modulation coding, signal processing, error correction, and more aggressive tolerancing – but it is unclear how much more improvement could be extracted from these areas in future optical disk standards.

20.6.1 Short-Wavelength Lasers

Early optical storage products used infrared wavelengths between 830 nm and 780 nm, simply because these were the only diode lasers available with the reliability, optical power, quantities and cost that the industry required for a consumer product. As the development from CD to DVD to BluRay highlights, the obvious path to further increases in storage density has been the development of suitable short-wavelength lasers. While blue lasers have been used in laboratories around the world for high-density recording studies, they have yet to show up in consumers products. Instead of leapfrogging from 800 nm to 400 nm lasers (using GaN) and thus immediately quadrupling the storage density, the development has been gradual. Magneto-optical (MO) drives employ visible wavelengths of 650–685 nm, while DVDs rely on 635–650 nm lasers [20.219]. Continued advances to even shorter wavelength have been made with novel diode laser materials and concepts. Progress with prototypes of green ZnSe lasers at 490 nm has been substantial and blue GaN lasers with wavelength as short as 375 nm have been explored [20.219]. The NSIC roadmaps predicted the general availability of optical storage products with these short-wavelength lasers around 2002 [20.219, 220]. That this has not occurred reflects both technical difficulties in fabricating cheap, long-lifetime blue lasers as well as economic reality: HDTV and broadband Internet, the applications that were to drive the need for this technology, are penetrating the market much more slowly than expected. However, other promising evolutionary approaches might provide significant improvements in areal storage density while the performance, lifetime, and cost of blue laser diodes slowly improve.

20.6.2 Increased Numerical Aperture

Instead of reducing the laser wavelength, a corresponding increase in numerical aperture can achieve the same increase in areal density. The increase in density when making the transition from CD to DVD was partially due to an increase in NA. There are, however, practical limits to the increase in NA, such as manufacturing tolerances for diffraction-limited optics. The key issue here is depth of focus, δ, which is directly proportional to the wavelength, but inversely proportional to the square of the numerical aperture [20.231]:

$$\delta \sim \frac{\lambda}{\mathrm{NA}^2} . \tag{20.7}$$

To maximize density, the data layer of the media should be as close as possible to the focus of a diffraction-limited read-out beam. Inevitably, manufacturing tolerances on the drive and media make a fast focusing servo essential. For ultimate performance, even dual-stage servos with multiple lenses are being considered. Obviously, these issues become more difficult as the optical depth of focus decreases. Smaller depth of focus requires that the thickness of the cover layer over the recorded data (which takes up some of the working distance of the optics and forces correction for spherical aberration) decreases, and indeed the BluRay format calls for a cover layer only 100 μm thick [20.224].

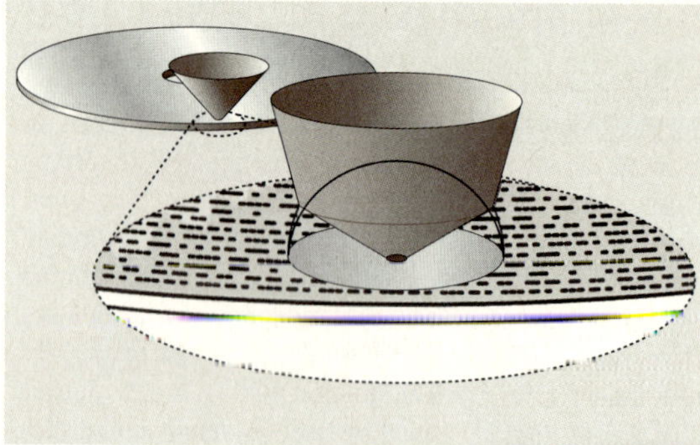

Fig. 20.16 A solid-immersion lens (SIL) can increase the effective NA beyond 1.0, further increasing density but requiring evanescent coupling between the SIL and disk

Although the maximum NA in air is 1.0, microscopists have long improved their resolution beyond this by using oil-immersion objective lenses. Rather than coating optical disks with oil, however, numerical apertures greater than 1 can similarly be attained by bringing a solid-immersion lens into close proximity of the disk media [20.232], as shown in Fig. 20.16. If the gap between the bottom of the lens and the media is smaller than the light wavelength, then evanescent coupling of light across this gap illuminates a subwavelength-diameter spot on the media. Similar to the read–write head in magnetic recording, it is possible to fly a solid immersion lens on an air bearing over the media [20.232]. This approach to high storage density, while pursued aggressively for several years, has yet to deliver a successful commercial storage product. In addition, because of the susceptibility of this tiny air gap to external contaminants on the disk, such near-field approaches would seemingly have to sacrifice media removability: one of the very advantages that optical storage has historically had to offer.

20.6.3 Magnetic Super-resolution

In magnetooptic recording, another interesting effect can be utilized to increase the resolution. In this technology, bits are stored by heating a magnetic film locally with a laser in the presence of an externally applied magnetic field (Fig. 20.17). In the portions of the film that exceed the Curie temperature, the local magnetization will orient with the applied field. This magnetization change is then sensed optically by detecting the slight rotation of the polarization due to the Faraday effect [20.231]. Here,

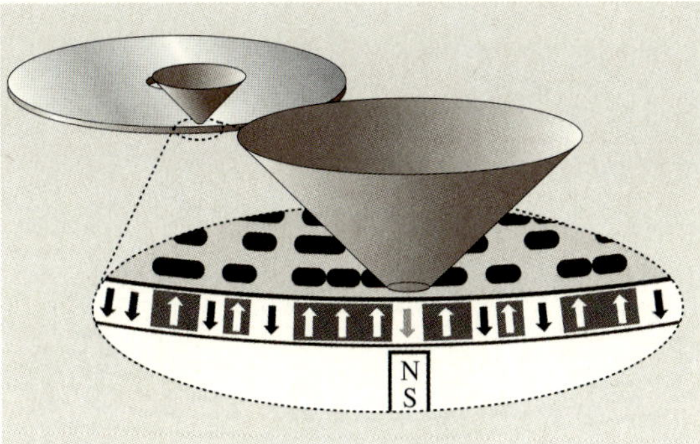

Fig. 20.17 Magneto-optical storage uses a tightly focussed laser beam to record bits, using local heating of a magnetic film above its Curie temperature in the presence of an external magnetic field

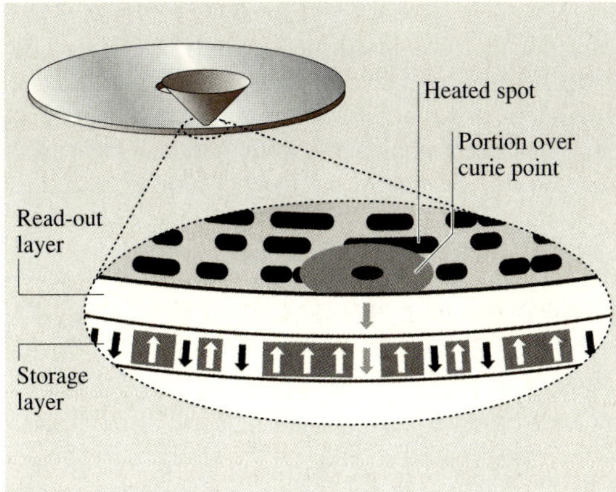

Fig. 20.18 Magnetic super-resolution uses two layers, with the heat from the laser spot transferring just the bits of interest from the bottom storage layer into the top read-out layer

thermal diffusion processes rather than optics can dominate the resolution. A short exposure with a laser beam of nonuniform spatial profile, such as a simple Gaussian beam, can produce a temperature distribution that exceeds the Curie temperature only for a brief time, and only in the very center of the beam. Subsequent heat diffusion away from the spot quickly brings the temperature below the Curie point, resulting in magnetic reversal in an area of sub-optical-wavelength dimensions, including marks as small as 100 nm [20.233].

However, such marks must still be optically detected. With dozens of such marks within the diameter of each focussed spot, deconvolving each mark individually from the aggregate optical signal would be quite challenging. An elegant technique called magnetic super-resolution (MSR) can overcome this read-out problem using the highly nonlinear behavior of thermal diffusion [20.233]. The MSR technique adds a magneto-optic read-out layer above the magnetic storage layer. For read out of a particular sub-wavelength spot, this top layer is briefly heated with a laser just as in the recording process described above: the power density in the center of the beam barely exceeds the Curie temperature. As above, the magnetization of the hot spot orients with the applied field, which in this case is the field due to the recorded bit in the lower layer. The surrounding area of the upper layer does not reach the Curie point and is not affected. After the magnetization at that location has been *revealed*, the read-out layer contains only the sub-resolution spot with reversed magnetization surrounded by a magnetically homogeneous background (Fig. 20.18). This sub-resolution spot can then be optically detected with sufficient signal/noise ratio, despite the much larger diameter of the focussed laser beam. Magnetic super resolution is, therefore, a near-field technique, where the aperture is created in close proximity to the recording layer by illuminating a cover layer with a suitable laser.

Taking advantage of similar magnetic multilayer techniques that have been recently introduced to increase the areal density of conventional magnetic recording, further increases in optical storage density are possible by introducing additional layers for magnetic amplification (magnetically amplified magneto-optic system or MAMMOS [20.234]).

20.7 Volumetric Optical Recording

Both magnetic and conventional optical data storage technologies, where individual bits are stored as distinct magnetic or optical changes on the surface of a recording medium, are approaching physical limits beyond which individual bits may be too small or too difficult to store and retrieve. Storing information throughout the volume of a medium – not just on its surface – offers an intriguing high-capacity alternative.

Three-dimensional optical storage literally opens up another dimension to increase the capacity of storage media. In principle, a volume element with the dimensions of the wavelength should suffice to store one bit. The volumetric density that can be achieved in the diffraction limit would then scale with $1/\lambda^3$. Trade-offs between density and data rate make it possible, at least in principle, to forgo some of this density to obtain blistering data rates. Some of the techniques that have been proposed do not require mechanical motion, enabling access times in the range of tens of microseconds, albeit to a small storage volume with moderate capacity.

20.7.1 Volumetric Addressing Techniques

Several approaches for volumetric optical data storage have been explored. These can be distinguished by the method used to address the stored data. Some of the techniques simply extend the multilayer nature of conventional optical storage already begun with the two-layer DVD standard, while others take advantage of the 3-D character of optics. The techniques described here are:

- Adjusting focus to access data on a particular layer;
- Using an interferometer (sensing differences in path length) to address a layer;
- Addressing a particular point, line, or plane in a medium by intersecting two laser beams;
- Using material with extremely narrow spectral sensitivities to address data; and
- Addressing data by the spacing and direction of interference fringes, known as K-vector or holographic addressing

20.7.2 Addressing by Depth of Focus

Depth of focus, a phenomenon familiar to anyone that has used a microscope, is already an integral part of the current DVD standard. To increase the capacity per disk over the single-layer limit of 4.7 GB per disk [20.235], the standard can double this capacity with two data layers, or quadruple it with a two-sided, two-layer-per-side option (the disk must be flipped to access data with a single objective, or a second optical pick-up is required to address both sides of such a data carrier simultaneously).

Using depth of focus has the clear advantage of being similar to existing CD and DVD technology, and it could possibly be compatible with the large installed base of optical storage devices. As shown in Fig. 20.19, the key disadvantage is that the individual layers must be widely separated to avoid crosstalk errors on read out, or tracking problems with the focus servo. The required diffraction-limited performance is difficult to accomplish over a sizable depth range, limiting the maximum number of layers.

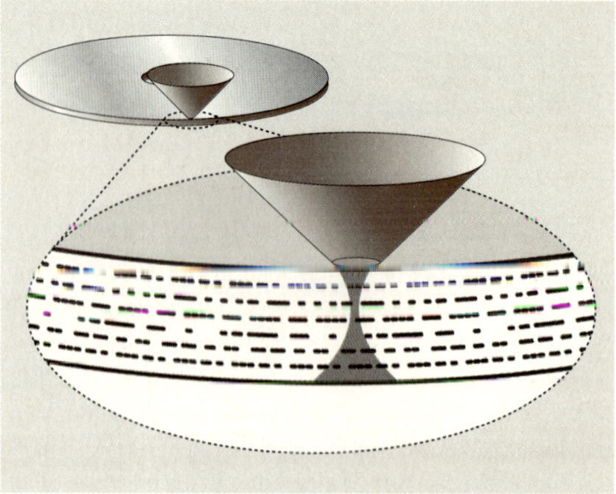

Fig. 20.19 The effective optical areal density can be increased by using depth of focus to access multiple bit layers within a disk

One way to improve this is to use fluorescence for read out: since the light from a mark is now incoherent, the effects of interference and scattered light can be reduced and a higher signal-to-noise ratio (SNR) obtained. These reduced noise effects trade off with the omnidirectional output of the fluorescence from the illuminated voxel (volumetric pixel).

Multilayer fluorescent disks (MFD) employ media with many (10 to possibly as many as 100) layers [20.236]. Each layer contains pits or grooves filled with fluorescent material, which absorbs the incident laser light and re-emits incoherent red-shifted fluorescent light. Portions of the disk manufacturing process are similar to CD/DVD fabrication: a mastering process (here for each individual layer) followed by replication via hot embossing or stamping. The pits are filled with fluorescent dyes – which require appropriate absorption/emission spectra, high conversion efficiency, short time response, and high saturation levels – and then the layer substrates are bonded together. Such ROM media fluorescent dyes are under development, as well as WORM media based on thermal bleaching and multiphoton processes for red and green light [20.236]. Reversible photochromic material have been demonstrated in the lab.

MFD drives are similar to CD/DVD drives with differences in the optical head: dichroic mirrors and filters to separate the fluorescent signal light from the excitation light, optics for spherical aberration correction, and a more sensitive detector. Tracking and focusing techniques are similar to other optical disks, with additional electronics and possibly data encoding needed to find and maintain focus on the different layers. A ROM system with more than 20 layers has been demonstrated at CD density (650 nm-wavelength semiconductor laser, 1 mW in the reading spot, three-spot tracking, astigmatic focusing) [20.236]. The SNR for the experimental disk was better than 30 dB (1.5 MHz bandwidth) [20.236].

In addition, optical cards with 10-layer media and 1 micron-sized marks have been demonstrated. CCD read out was used, image processing was applied to read data. This could lead to MFD-ROM systems with 10–20 layers at DVD density (approximately 50–100 GB per disk), and optical ROM cards with 1 GB capacity, followed later by WORM versions of these devices [20.236]. As with CDs and DVDs, the advent of blue lasers would be expected to support an increase in capacity by a factor of four, although this would require the dye chemistry to be re-engineered to these wavelengths.

20.7.3 Two-Photon Absorption for Addressing of a Bit Cell

Another means of limiting the interaction volume of the addressing beam in the medium is by using two-photon absorption, which scales with the square of the beam intensity. This effectively shortens and narrows the focus volume where the intensity is sufficient for writing. These approaches often require femtosecond lasers with high peak intensities, but could enable media with hundreds of layers allowing capacities approaching 1 terabyte per disk. Here the media is initially homogeneous (not layered), with two-photon effects used to either initiate or quench fluorescence at each voxel [20.238, 239]. Confocal microscopy has been used to detect extremely small marks in initial studies, but it is unclear whether such an approach would work on a rapidly spinning disk.

Further gains in depth resolution can be realized by crossing two laser beams of the same or different wavelength, using two-photon absorption to initiate a change of the media's optical properties only in the commonly illuminated volume [20.237, 240], as shown in Fig. 20.20. Here just one volume element within a larger block of recording media can be selected for recording or read out without affecting adjacent volume elements, thus avoiding inter-track and inter-layer interference. Extensions of this technique use columns or sheets of light to address a large number of volume elements in parallel, although these extensions make diffraction-limited performance for sub-micron-sized marks more difficult to obtain.

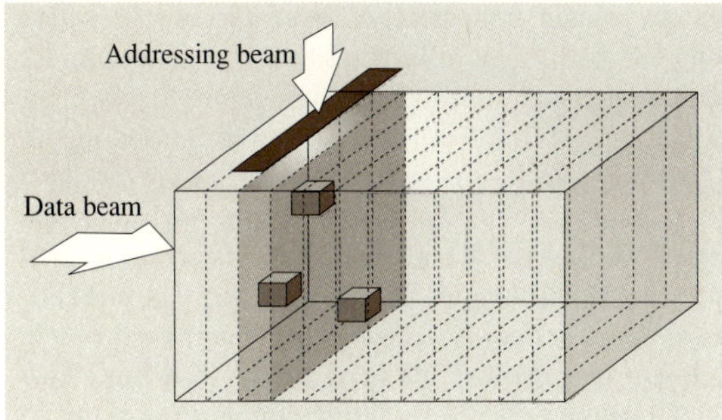

Fig. 20.20 Orthogonal beams can write and read data in parallel into a three-dimensional volume, using two-photon fluorescence at the intersection of the beams (After [20.237])

Even so, effective areal densities greater than $100\,\text{Gb/in}^2$ would appear achievable by using many layers with relatively large marks (i.e., greater than 1 μm). Once written, the marks (or columns or pages) may be nondestructively read by the two-photon fluorescence. Demonstrated capabilities include the recording and reading of media with more than 100 data layers, recording tracks of $2\times2\,\mu\text{m}^2$ data marks, and the construction of several proof-of-principle portable read-out systems [20.237, 240, 241].

This two-photon 3-D optical storage technology seeks to provide disk drive systems with high capacity (100–500 GB/disk) and high data rates (1–10 Gb/s) using inexpensive, easily manufactured, and long-lived plastic media. The drive technologies are potentially backward compatible with conventional DVD media. The media will be removable and offers the possibility of wide wavelength and mechanical tolerances in drives for near-line/online servers.

Efficient two-photon absorption recording requires moderate average laser power (50 mW) but high peak intensity at the focused spot (on the order of GW/cm^2), necessitating the use of a pulsed laser source for recording. Suitable diode-pumped solid-state pulsed lasers are available, albeit at a significant price and size disadvantage as compared to semiconductor laser diodes. These laser requirements remain the largest risk to this approach.

20.7.4 Interferometry

Taking advantage of a relatively short coherence length and variable path difference between two laser beams, one can obtain constructive interference somewhere along the path of a collimated laser beam to address volume elements throughout the depth of the media [20.242]. While similar to the conventional depth of focus approach, it is different enough to be not easily compatible with CD or DVD. Interferometric optical storage would have the advantage of diffraction-limited performance throughout thick media, and that it is similar to optical coherence domain reflectometry, widely explored as a method for imaging through thick biological samples [20.243]. While simple in concept, it does require interferometric stability and has not been explored in detail for data storage.

20.7.5 Persistent Spectral Hole Burning (PSHB)

Spectral hole burning uses media that can support dopants with extremely selective spectral response (Fig. 20.21). Accessing this narrow spectral response requires a tunable narrow-band laser, which tends to be a complex and expensive light source. Thus high storage densities and storage volumes are essential to amortize the cost of this light source over a large storage capacity. Unfortunately, these very high densities are currently only possible at cryogenic temperatures, adding additional complexity and cost. However, the latency and the data transfer rates that could theoretically be achieved by this technique are superior, allowing spectral hole burning to play in the memory as well as in the storage arena.

Several groups have been pursuing variations of spectral hole burning for data storage and memory applications [20.244, 245]. PSHB can alternatively be addressed using time-domain interference techniques, in which holograms laid down in frequency spectrum allow time sequences to be replayed as *photon echoes* [20.246, 247]. Both time- and spectral-access alternatives support spatial multiplexing across numerous locations, and would support the prospect of combining this technique with the spatial holographic techniques to be described below. Crosstalk and distortions between spectral channels, and photon flux are typically the important limiting factors. Time-domain access relieves the requirement for laser tunability, although the laser source does still need to be narrow band and extremely stable. It can also enable content addressability (accessing data by its correlation or similarity with search data). However, efficient time-domain addressing schemes are yet to be developed and random access to data is difficult to provide, with *first-in first-out* (FIFO) being the standard.

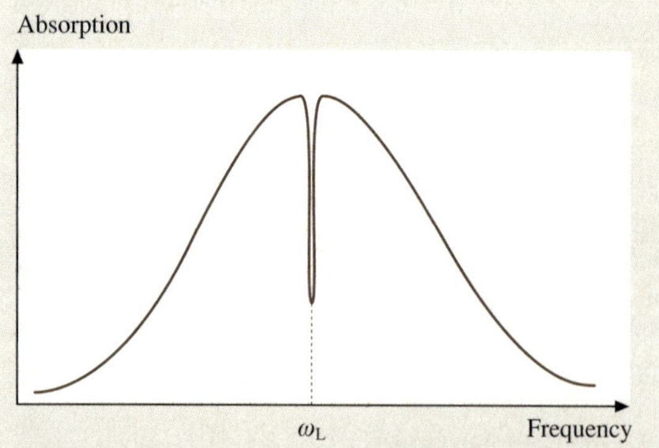

Fig. 20.21 After illumination with a narrow-band laser at ω_L, a persistent and very narrow spectral hole has been *burned* into the absorption spectrum

Tunable solid-state narrow-band lasers are currently being developed both for these PSHB applications as well as for telecommunications applications. Depending on the choice of PSHB medium, the required frequency tuning range spans either 10 GHz (0.25 Å at about 750 nm), 200 GHz (15 Å at 1500 nm) or 2 THz (30 Å at 650 nm). None of these tuning ranges is extremely difficult – although mode hopping can be problematic – but opto-mechanical approaches would sacrifice one of the key features of spectrally sensitive storage: rapid random access.

20.7.6 Holographic Storage

Holographic data storage is a volumetric approach which, although conceived decades ago, has made recent progress toward practicality with the appearance of lower-cost enabling technologies, significant results from long-standing research efforts, and progress in holographic recording materials [20.248, 249].

A hologram preserves both the phase and amplitude of an optical wavefront of interest – called the object beam – by recording the optical interference pattern between it and a second coherent optical beam (the reference beam). Figure 20.22a shows this process. The reference beam is designed to be simple to reproduce at a later stage, such as a simple collimated plane wave. These interference fringes are recorded if the two beams have been overlapped within a suitable photosensitive media. The bright and dark variations of the interference pattern create chemical and/or physical changes in the media, preserving a replica of the interference pattern as a change in absorption, refractive index or thickness.

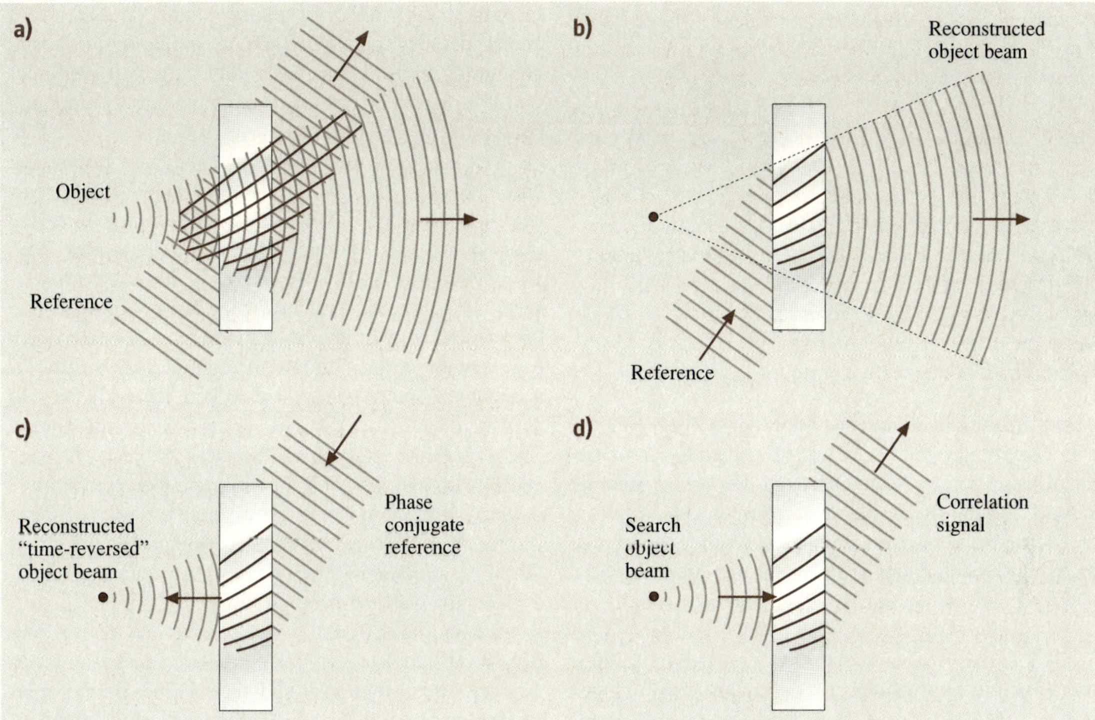

Fig. 20.22 How to record and read data using holograms. **(a)** Holographic storage of a single data bit. The spherical wave from a single pixel interferes with a coherent plane wave in the reference beam. The resulting interference pattern changes the refractive properties of the photosensitive medium. **(b)** The hologram is read out using the original reference beam, which is diffracted by the stored interference pattern to reconstruct the original spherical wavefront. An image of this beam can be formed on a single detector pixel, resulting in the retrieval of a single bit. **(c)** The hologram can also be read out by illuminating it with a counter-propagating (or *phase-conjugate*) reference beam, which reconstructs a phase conjugate copy of the original object beam. This beam returns to its original point of origin, where the bit value can be read without requiring a high-quality imaging system [20.251–255]. **(d)** A third way to retrieve data involves illuminating it with a diverging object beam, which reconstructs the original plane-wave reference beam. This beam can be focused onto a detector and provides an optical measurement of the correlation between the stored data and the illuminating object beam [20.250, 256]. This technique can allow one to search the stored data according to its content, rather than according to its address [20.257–259]. (After [20.260])

When the recording is illuminated by a read-out beam that is similar to the original reference beam, some of the light is diffracted to reconstruct a weak copy of the object beam (Fig. 20.22b). If the object beam originally came from a 3-D object, then the reconstructed hologram makes a virtual image of the 3-D object reappear [20.250]. Interestingly, a backward-propagating or phase-conjugate reference wave, illuminating the stored grating from the back side, reconstructs an object wave that also propagates backward toward its original source (Fig. 20.22c), creating a real but pseudoscopic image.

Although holography was conceived in the late 1940s, it was not considered a potential storage technology until the development of the laser in the 1960s. The resulting rapid development of holography for displaying 3-D images led researchers to realize that holograms could also store data at a volumetric density of as much as $1/\lambda^3$ [20.261]. In contrast to conventional data storage, where each bit of data is assigned to a particular location within the storage volume or upon the storage surface, holographic storage distributes data throughout a volume in a delocalized way. Data are transferred to and from the storage material as 2-D images composed of thousands of pixels, with each pixel representing a single bit of information. No one location in the crystal is responsible for storing that one bit; each is distributed

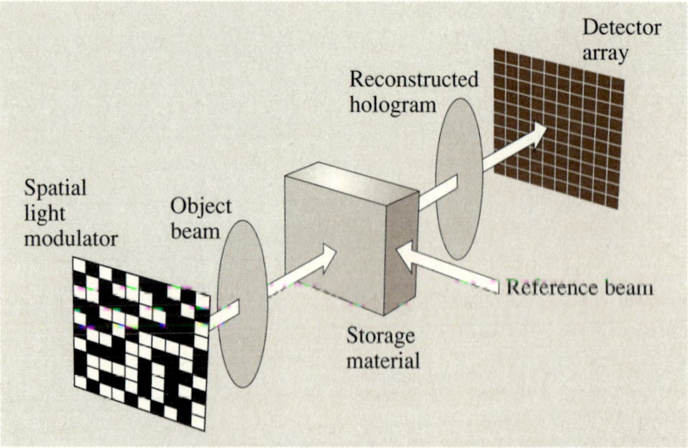

Fig. 20.23 Data are imprinted onto the object beam by shining the light through a pixelated input device called a spatial light modulator. A pair of lenses image the data through the storage material onto a pixelated detector array, such as a charge-coupled device (CCD). A reference beam intersects the object beam in the storage material, allowing the holograms to be stored and retrieved later. (After [20.260])

throughout the associated recorded interference fringes. The direction and spacing of those particular fringes ensures that light arrives at a particular photodetector (within the large detector array) only when a particular read-out beam is incident.

The $1/\lambda^3$ theoretical density limit can thus be intuitively understood as a crosstalk limit forced by diffraction. Given that each reconstructed object beam must pass through an aperture A to reach a detector array, then two sets of fringes that differ in direction by less than $\lambda/\sqrt{A}$ will be indistinguishable due to diffraction. Given that the contributions of the fringe spacings will be integrated over a thickness of L, two sets of fringes that differ in spacing by less than λ/L cannot be individually reconstructed. So roughly L/λ data pages of A/λ^2 pixels each can be holographically stored within a volume $V = AL$. While this simple argument ignores the role of the bulk index of refraction (as well as real-world media and noise issues), $1/\lambda^3$ would be impressive density performance, corresponding (for green light) to the storage of 1 terabyte of data in a cubic centimeter.

Since each data page is retrieved by an array of photodetectors in parallel (Fig. 20.23), rather than bit-by-bit, the holographic scheme promises fast read-out rates as well as high density [20.249, 260, 262–265], because laser beams can potentially be moved rapidly without inertia, unlike the actuators in disk drives. With the inherent parallelism of its page-wise storage and retrieval, a very large compound data rate could be reached by having a large number of relatively slow (and therefore cheap) parallel channels. If a thousand holograms, each containing a million pixels, could be retrieved every second, for instance, then the output data rate would reach 1 Gb/s.

Despite this attractive potential and fairly impressive early progress [20.266, 267], holographic storage has been – for nearly four decades – an intriguing but elusive alternative to conventional data storage techniques. With the recent availability of relatively low-cost components, such as liquid-crystal displays as input devices and camera chips from electronic cameras and camcorders for detector arrays, interest in commercializing holographic storage efforts was rekindled. Recently, largely as part of an effort that was sponsored by the United States Defense Advanced Research Projects Agency (DARPA), members of the holographic data storage systems (HDSS) consortium were able to demonstrate impressive performance with experimental platforms: at the IBM Almaden Research Center a data density of $250\,\text{Gb/in}^2$ was achieved [20.268]; at Stanford University a data rate of 10 Gb/s was reached for the read out of holographic data [20.269], and a team at the Rockwell Science Center demonstrated access times on the order of 10 microseconds [20.270]. Each of these demonstrations were the product of extensive studies of recording physics, systems trade-offs, signal processing, and coding techniques. These experiments also showed that it will be difficult, but not impossible, to obtain a combination of all three desirable performance characteristics in one hardware platform.

A rather unique feature of holographic data storage is associative retrieval, or content-addressable data storage [20.258]. If a partial or search data pattern is imprinted on the object beam, illuminating the stored holograms reconstructs all of the reference beams, each weighted by the similarity between the search pattern and the content of their particular data page. By determining, for example, which reference beam has the highest intensity, the closest match to the search pattern can be found without initially knowing its address.

20.7.7 Holographic Multiplexing

If a hologram is recorded in a thin material – as on many credit cards – the read-out beam can differ in angle or wavelength from the reference beam used for recording the image; the scene will still appear. However, if the hologram is recorded in a thick material, the reconstructed object beam will only appear when the read-out beam is nearly identical to the original reference beam.

For any read-out beam, some of the incident optical power is diffracted by the recorded hologram to create a diffracted wavefront. In a thick hologram, this diffracted wavefront accumulates energy from throughout the thickness of the storage material. The Bragg condition applies when the diffracted wavefront is momentum matched to the read-out beam and grating. For holographic media that record an exact copy of the interference fringes, this occurs when the read-out beam is identical in wavelength and incidence angle to the original recording beam. Away from this condition, the discrepancy between the wavefront that would be momentum matched (to the read-out light and the grating) and the wavefront that can actually propagate (the closest solution of the wave equation) represents a phase error. Thus the wavefront diffracted by the front portion of the hologram, after propagating through most of the thick material, finds itself out-of-phase with the wavefront diffracted by the rear portions of the hologram. The integration of this phase error over the thickness of the volume hologram creates Bragg selectivity: the hologram disappears as the angle or wavelength is tuned away from the Bragg condition.

As the material becomes thicker, accessing a stored volume hologram requires tight tolerances on the stability and repeatability of the wavelength and the angle provided by the laser and read-out optics. However, Bragg selectivity also opens up a tremendous opportunity: a small storage volume can now store multiple superimposed holograms, each distributed throughout the entire volume yet selectively accessible with its original reference beam. Any particular data page can then be read out independently by illuminating the stored gratings with the reference wave that was used to store that particular page. The theoretical limits for the storage density of this technique are around tens of terabits per cubic centimeter [20.248, 249].

Several different techniques [20.271] have been developed to define a set of suitable reference beams. It has already been implied that tuning the incidence angle [20.272, 273] or wavelength [20.274] will multiplex holograms. The former has been used much more than the latter, simply because rotating a mirror through large angles has been easier to implement than rapidly – and widely – tunable lasers.

Instead of recording one hologram per incidence angle, it is possible to use all the incidence angles for each hologram, imposing a unique phase on each of these beamlets individually. The number of holograms that can be superimposed depends on the number of orthogonal phase codes. Such phase-code multiplexing [20.275, 276] is widely investigated because spatial light modulators could be used to apply these phases rapidly without mechanical motion. One issue is the need for low phase error (both random and deterministic [20.277, 278]) in these devices; another is the impact of grating degeneracy on any phase codes arranged in a 2-D pattern, a consideration that is unfortunately often overlooked.

Another improvement upon angle multiplexing actually exploits this grating degeneracy. Although holographic data pages do not disappear when changing the incidence angle out of plane, they are displaced on the detector array as the diffracted beam moves to remain in the plane formed by the incident beam and the grating vector. Once the data page slides completely off the detector array, the same Bragg angle can be used to store a second hologram. This is referred to as fractal multiplexing, because one backs off from $(A/\lambda)^2$ pixels per page to some fractal dimension (say, $(A/\lambda)^{2-x}$), allowing an increase in the number of stored pages to $(L/\lambda)\times(A/\lambda)^x$ [20.279]. Using a combination of angle and fractal multiplexing, as many as 10 000 holograms have been stored in a 1 cm^3 volume [20.280, 281]. Techniques analogous to angle and fractal multiplexing, called shift and peristrophic multiplexing, have been developed for multiplexing holograms in thin disks and are described later.

20.7.8 Media

Media for holographic storage has long been one of the primary focus points for researchers in this field. There are two major classes of holographic storage media: write-once media, typically to be used as thin (0.2–2 mm) disks and accessed through disk rotation or translation; and read–write media, typically kept stationary and accessed by beam steering.

20.7.9 Write-Once Read-Many

A material that permanently stores volume holograms must generally support some irreversible photochemical reaction or photophysical process, triggered by the bright regions of the optical interference pattern, that then leads to changes in index of refraction or absorption. For example, a photopolymer material (as its name would suggest) polymerizes in response to optical illumination: material diffuses from darker to brighter regions so that short monomer chains can bind together to form long molecular chains [20.282–286]. Because this diffusion process can be phototriggered, sensitivities can be made high enough to support holographic recording with single short pulses [20.287]. While high sensitivity is desirable in a commercial application, so that holograms can be rapidly recorded with a cheap, low-power laser, high sensitivity can also cause problems. For instance, some of the media volume may be inadvertently affected by partial exposure as nearby spots are recorded, forcing either a reduction in sensitivity or in areal density (e.g. guard bands). In contrast to photopolymers, the illuminated molecules in a so-called direct-write or photochromic material undergo a local change in their absorption or index of refraction, driven by photochemistry or photoinduced molecular reconfiguration. Examples include photoaddressable polymers [20.288], and binding of absorbers to polymer hosts (such as phenanthraquinone (PQ) to polymethylmethacrylate (PMMA) [20.289]).

Both types of materials are inexpensive to make in bulk, but both can have problems reproducing the object beam faithfully. Photopolymers tend to shrink during recording, distorting the reconstructed pixelated images [20.290]. Direct-write media respond both to the rapid variations of the interference pattern encoded with data and to long-range brightness variations across the illuminated spot. Such effects also distort the reconstructed data pages. These problems can be minimized by careful system design, such as signal-processing techniques that can compensate for shifted and distorted data pages [20.291, 292], and optical illumination systems that deliver beams with extremely uniform brightness [20.293].

Although problems with shrinkage, scattering and dynamic range remain, recent developments in these write-once materials have overcome previous difficulties with poor optical quality and excessive absorption and led to fairly thick samples (0.5–1 mm). Together with recently developed multiplexing techniques that use peristrophic beam rotation [20.294], spherical [20.295] or randomly speckled [20.296] reference beams to increase the number of holograms that can be superimposed in thin media, these developments have brought write-once read-many holographic storage systems to the stage where several commercialization efforts are underway.

20.7.10 Read–Write

In contrast to the organic WORM media, most erasable holographic materials tend to be inorganic photorefractive crystals doped with transition metals or rare-earth ions [20.297–300]. These crystals are often available in centimeter-thick samples and include lithium niobate, strontium barium niobate, and barium titanate, doped with iron, cerium, manganese or other dopants.

For details on these materials please see Sect. 20.5 on photorefractive crystals by Buse and Krätzig. Before discussing the important issue of volatility in photorefractive crystals, we describe some alternative read–write holographic storage media. These include photorefractive polymers [20.301–304], bacteriorhodopsin [20.305–307], and the DX center in semiconductor materials [20.308]. These materials are difficult to obtain in the thicknesses that would be required for competitive capacities, and also have their own idiosyncrasies. While photorefractive polymers can achieve large index changes very rapidly and provide many avenues for tuning through constituent substitutions, they require large voltages to create the orientational analogue of the electro-optic effect and tend to have fairly short dark lifetimes (seconds to minutes) [20.301–304]. Bacteriorhodopsin can be tuned by genetic and chemical modifications [20.305, 307] and does not require an external electric field. However, volatility and dynamic range are serious issues [20.306, 307], and the required operating wavelengths tend to be tightly tied to the protein's innate photocycle [20.305–307]. Essentially, bacteriorhodopsin acts much like a write-once direct-write saturable material, where readout fills in the holograms. By completing the photocycle with a second wavelength, holograms are erased and new ones can be written in the photosensitive molecules reset to their initial state. Finally, at low temperatures ($< 150\,K$), the persistent photoconductivity exhibited by the DX center in semiconductors offers an opportunity for writing strong phase holograms [20.308]. Photoexcited electrons persist in the conduction band without decaying, leading to large index changes. This DX-center material also acts as a saturable material. Here, raising the temperature erases the holograms, since the photoelectrons now have enough thermal energy to make it back to the original ground state [20.308].

20.7.11 Nonvolatile Read–Write Storage

Unfortunately, the ability of a photorefractive material to erase through charge re-excitation also results in the undesired erasure of stored holograms during any subsequent optical exposure. This means that holograms erase while other holograms are being recorded, and worse still, while any of the holograms superimposed within that same volume are being read out. In addition, there is gradual erasure in the dark through thermal excitation [20.309, 310].

These erasure effects can be counteracted during recording by carefully scheduled the sequence of exposures, to ensure that the final diffraction efficiencies will be equal [20.256, 311]. The first holograms are recorded to somewhat higher diffraction efficiencies with longer exposures, so that as subsequent holograms are recorded and these first ones decay, all final diffraction efficiencies are equal [20.311]. As a result of the recording schedule, the effect of erasure is mitigated at some small cost in effective recording rate. (Recording schedules are also used in write-once media, to compensate for changes in sensitivity as a function of recording exposure [20.312]. This can include any pre-exposure required before the material begins to change its index of refraction.)

The more crucial issue is erasure of the stored holograms during read out. With some photorefractive crystals, stored holograms can be fixed – made semi-permanent and resistant to erasure during read out – by separate thermal [20.313–317] or electronic [20.318, 319] processes. Unfortunately, these fixing processes affect all the stored holograms within a volume simultaneously, only preserve a fraction of the original grating strength, and tend to be slow and cumbersome. For instance, it is not clear how to thermally fix holograms in one sub-volume while not affecting holograms in any of the immediately neighboring sub-volumes.

Another proposed solution to the volatility problem is periodic copying and refreshing of the pixelated data pages [20.320]. Here the data pages are read out periodically and rewritten into the memory. Most of the experimentally demonstrated schemes have involved methods to reinforce the original hologram, but copying the data page into a separate storage location should also work. The problems with all of these is their complexity, the possibility of loss of data fidelity in all the copying and re-copying, and the reduced performance from having the system inaccessible to user-generated read and write commands during the refresh operations.

A third approach to nonvolatile holographic storage is to read the hologram with a wavelength different to that used for recording [20.321, 322]. The idea is that if the absorption at this new read-out wavelength is much lower, the erasure will take place much more slowly. The systems problem with this is that not all of the spatial frequencies in the hologram will be Bragg-matched simultaneously, and there will be a trade-off between how much of the page will be visible, whether the pixels in the page will land where they are supposed to, and how hard one can push towards the theoretical density limit of $1/\lambda^3$ [20.321, 322]. These systems issues grow worse as the ratio between the two wavelengths grows larger; but given the broad absorption spectra in these crystals, significant changes in absorption require large ($\sim 50\%$) changes in wavelength.

A fourth method for achieving nonvolatile storage in photorefractive materials is by recording at a wavelength of light which is only absorbed by the crystal in the presence of an additional gating beam of different wavelength [20.323–327]. This additional beam is present only during recording and is switched off while the information is read out, allowing the data to be retrieved without erasure. The recorded interference fringes thus remain Bragg-matched to the read-out wavelength, and both read-out and diffracted wavefronts experience low absorption losses.

Conventional photorefractive materials can be optimized for this gated, two-color recording process by changing the way in which they are fabricated or by adding multiple dopants. For instance, the two-color response of lithium niobate can be enhanced by changing the ratio of lithium to niobium in the compound [20.324], or by doping it with both manganese and iron atoms [20.325]. Gated, two-color photorefractive materials have received much attention in the past few years, leading to improvements in both the sensitivity and dynamic range of the materials (increasing both the speed with which data can be written and the capacity) [20.324, 325, 327]. Further improvements, however, will be needed before prototypes can be built.

20.7.12 Phase-Conjugate Read Out for Read-Write Systems

Recent experimental demonstrations of holographic data storage have concentrated on demonstrating high density and fast read out. High areal density can be achieved in holographic data storage by carefully balancing interpixel crosstalk (introduced by the small aperture through which each data page is focused) against the loss of signal associated with recording multiple holograms. For instance, an equivalent areal density of 394 bits/μm^2 (80 × larger than single-layer DVD) was recently demonstrated [20.268]. (The equivalent volumetric density was 1.1% of $1/\lambda^3$). Conversely, fast read-out rates can be attained by reading out large data pages in rapid succession. An optical read-out rate of 10 Gbits/s at moderate density ($\sim$ 10 bits/μm^2) was recently demonstrated, and a full system read-out rate of 1 Gbit shown (including the camera and decoding hardware) [20.328]. Both of the demonstrations reached these specifications by combining large *megapixel* data pages of 1024 × 1024 pixels with the short focal length optics needed for high density.

However, extending read–write holographic storage to high capacity without sacrificing fast access means that this same high density must be achieved at many storage locations without moving the storage media. The demands this would impose upon optical imaging performance are quite severe – thus the further improvements in capacity that could be achieved by simply designing better lenses would be limited to commercially uninteresting values. However, several researchers have long proposed bypassing these imaging constraints with phase-conjugate read out [20.251–254], as shown in Fig. 20.24. Once a hologram is recorded, the wavefront reconstructed by a phase-conjugate read-out beam will retrace the path of the incoming object beam in reverse, canceling out any accumulated phase errors from lens aberrations or material imperfections. This allows data pages to be retrieved with high fidelity using image confinement in fiber-type media [20.251, 252], an inexpensive lens, or even without imaging lenses for an extremely compact system [20.253, 254]. However, many pairs of phase-conjugate reference beams are needed to read the many different holograms recorded within the same volume – and maintaining these beams over long periods of time would be impossible from a practical point of view.

One solution to this problem is to separate the phase-conjugation and hologram storage processes into two successive steps with a *buffer* hologram [20.255]. Holograms can then easily be multiplexed at a large number of separate storage locations using only one SLM and one detector array. With gated, two-color media, the long-term storage material does not absorb the information-bearing beam until the gating light is present [20.255]. With the phase-conjugate read-out, total internal reflection could be used to confine the image-bearing beam within a small cross section without sacrificing the ability to retrieve this image at the detector array [20.251, 252, 255]. Such a system only requires a single pair of phase-conjugate beams, generated either by carefully alignment or with a self-pumped phase-conjugate mirror [20.255].

A second proposed solution to phase-conjugate read out is to attain high capacity from multiple compact modules, each created by attaching an SLM, a detector array, and the storage media directly to a pair of beam splitters [20.253]. The phase-conjugate readout allows the whole system to remain extremely compact, and density can be further augmented by increasing the page size. It has been shown that phase-conjugate read out can retrieve pages with pixel pitches as small as 1 μm [20.329]. This approach has the advantage that the object beam need not be confined with total internal reflection because it is never allowed to propagate far from the SLM. But it does require inexpensive components, since the capacity per set of components is relatively low. Both the compact and buffer-enable phase-conjugate systems still require a low-power and convenient method for supplying thousands of unique reference beams to hundreds or thousands of spatial locations, using either micromechanical mirrors, liquid-crystal beam steerers, individually addressable lasers, or wavelength-tunable lasers.

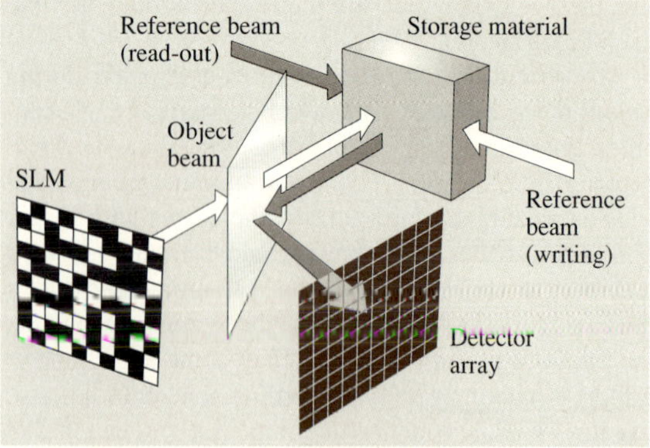

Fig. 20.24 By adding a counter-propagating readout beam, a phase-conjugate object beam can be reconstructed which retraces the original path of the object beam in reverse. Proper placement of a polarizing beam splitter and detector array can allow high-fidelity data retrieval without expensive imaging optics

The successful use of phase-conjugation in holographic storage should enable compact and affordable high-capacity systems, with only a moderate increase in the overall system complexity. Obviously, such systems still await a recording material that supports both read–write access and nonvolatile storage [20.325, 326, 330]. Even so, there remain other serious issues that must be addressed before commercialization. Thermal stability must be good lest the interference patterns change spacing and orientation as the media expands or contracts with temperature. Good mechanical and laser stability are also required (the media and interference fringes must not move during exposures, and motion afterwards can cause the reconstructed optical signals to veer away from their assigned detector pixels). Fortunately, the stability during recording will become easier to attain as exposure times decrease through improvements in material sensitivity and increases in available laser power.

Despite all of the techniques for removing and suppressing volatility, it is unlikely that any read–write holographic storage material will be truly nonvolatile: most likely the data will slowly fade over several months or years due to thermal effects (slow excitation from electron traps, diffusion of compensating ions) or through residual absorption. So while blocks of read–write media may be removable from the read–write head (which enables something like a petabyte jukebox), the media will probably need to remain within the jukebox so that data can be periodically refreshed.

20.7.13 Write-Once Systems Using Spinning Disks

In contrast to read–write holographic systems, progress in write-once materials research (especially photopolymers [20.285,286,331]) has brought write-once systems to the stage where prototypes for the first holographic storage products are being developed.

Beyond the problems of perfecting the media (in characteristics such as dynamic range, scatter, sensitivity, shelf-life before and after recording, and thermal expansion properties) are the systems engineering issues of building robust holographic data storage devices around a spinning disk format. What makes this even more challenging is that the obvious application areas (low-cost data archiving, possible next-generation distribution format for data and multimedia) call for inexpensive and robust disk readers (as well as cheap media). The first systems problem is the interplay between high rotation speed (needed for low latency) and the need for a high-power, compact pulsed laser to read and write with single pulses. And then there are the difficulties of getting the pulse to the right spot (tracking, focusing, synchronizing pulses to disk rotation), and getting the reconstructed data page to the detector array (compensating for tracking, tilt, disk jitter). *Zhou* et al. have demonstrated tracking for low-density holographic disks [20.332]. They showed both tracking and tilt compensation: the former by measuring the data page rotation to synchronize the beam shutter (on a continuous wave (CW) laser), and the latter by tuning the reference beam angle so that data pages landed squarely on the pixelated detector array [20.332].

To reach high density, the reference beam must cover a wide spread of incidence angles, so good antireflection coatings are essential to keep power from being lost in Fresnel reflections (and this increases media cost). To get the best density while suppressing aberrations in the imaging system, the object beam should enter the holographic disk media at normal incidence. As this must be done with a short-working-distance optical system, the delivery of writing and reading beams around these imaging optics (without increasing the scatter into them) is further complicated. Although a read-only transmission-geometry head can avoid passing the reference beam past input optics, transmission geometry implies that the read head is split into two parts on either side of the rotating disk (and both sides must be carefully aligned).

Several novel multiplexing methods have been developed to allow holograms to be superimposed very densely, even in thin disks. High density can be reached with peristrophic multiplexing, at the cost of a fairly complicated read head that rotates the reference beam around the normal to the disk surface [20.294]. In contrast, by using either a spherical [20.295] or a randomly speckled [20.296] reference beam, the motion of the spinning disk can allow the reference beam to reconstruct stored holograms selectively with an extremely simple read heads. If this shift multiplexing is done with a spherical reference beam, then holograms can be packed densely along one line (i. e., along the track), but only sparsely along the orthogonal direction (tracks must be widely spaced) [20.295]. Speckle-shift, or correlation, multiplexing using a random phase plate or diffuser [20.296] can allow dense packing in both radial and along-track dimensions, but this advantage does not come for free. Essentially, the size of the random speckles determines the disk motion needed to make each hologram disappear through destructive interference [20.296]. This should be small to maximize density, but not as small as the innate disk wobble and jitter of an inexpensive disk and spindle. On the other hand, the destructive interference depends on the number of random speckles that are spatially integrated as the reconstructed hologram transits the thickness of the disk. So while smaller speckle lead to better inter-page crosstalk SNR, they also make the read-out conditions so selective that holograms might not be reliably found with inexpensive components. Another consideration is any noise from gratings and index changes recorded into the highly sensitive WORM recording media by the speckle pattern itself.

These systems difficulties do not prevent one from building systems that can write and read holograms on spinning disks – several working demonstrations have been shown [20.332, 333]. For instance, Orlov et al., working at Stanford on the final systems demonstrator for the DARPA-sponsored holographic data storage systems (HDSS) program, built a system capable of 10 Gbit/s optical read out, and 1 Gbit/s end-to-end electronic read out, at greater than DVD areal densities on a disk spinning at > 300 rpm [20.328]. The spindle was so accurate that holograms could be incrementally recorded over several rotations (i. e., the accuracy and repeatability were at interferometric levels). However, any commercial product will need to use much smaller and cheaper components, without sacrificing the high density, the fast read-out rate, and the ability to robustly write and read holograms on the fly.

An alternate approach to wavelength multiplexing is to use micro-holograms [20.335,336]. Here each micro-hologram occupies a few square microns of surface area, and can either extend throughout the thickness of the disk or exist in one of several thick layers. Multiple bits of data are written at each micro-hologram by means of reflection gratings, which can be read out by active wavelength multiplexing (laser light scanned across the appropriate spectral band) [20.335], wavelength multiplexing (white light in, colored light out containing data) [20.336] or by angle multiplexing [20.335]. The beams are confined either by the focussed beam itself within a thin film [20.335], or by a micro-fiber within the material [20.336].

20.7.14 Content-Addressable Storage

With a conventional memory or data storage device, a user must supply an address at which the desired data is located. In volume holographic data storage, this implies that the data – which were once imprinted on an object beam and stored within the volume – can be read out later by illuminating the volume with the correct addressing reference beam (Figs. 20.15a and 20.15b). However, this hologram can also be illuminated by the object beam (Fig. 20.15d), reconstructing all of the angle-multiplexed reference beams that were used to record data pages into the volume (Fig. 20.25). The amount of power diffracted into each output beam is proportional to the similarity between the input data page being displayed on the spatial light modulator and the stored data page. The set of output beams can be focused onto a detector array, so that each beam forms its own correlation peak. The stored pages that match the input page can be identified by setting a threshold on the detected optical signal. If the patterns that make up these pages correspond to the various data fields of a database, and if each stored page represents a data record, then this optical correlation process can be used to compare an entire digital database against the search argument simultaneously.

This search parallelism gives content-addressable holographic data storage an inherent speed advantage over a conventional serial search. This is particularly true for searches on complex queries through large databases, where an index for every possible search query becomes untenable. For example, it would take a conventional software-based search ≈ 40 s to go through one million records, each containing 1 kB of data, if this gigabyte of data has to be pulled off a hard drive for each search. Connecting a gigabyte of dynamic random-access memory (DRAM) to a 1 GHz microprocessor could reduce this search time to ≈ 1 s. In comparison, an appropriately designed holographic system could search the same records in about 1 millisecond. Spatial light mod-

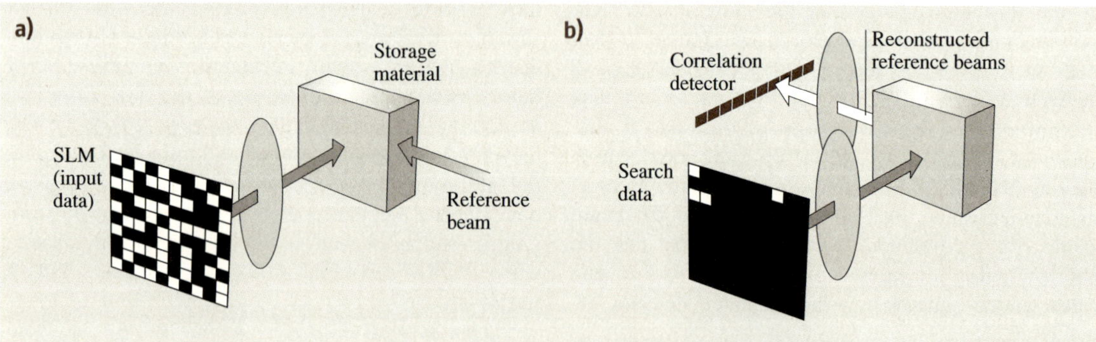

Fig. 20.25 (**a**) After multiple holograms are stored, each with a structured data page modulated onto the object beam and the reference beam introduced at a unique angle, then (**b**) illuminating the stored holograms with a new data-bearing object beam (the search data page) will reconstruct all of those reference beams in parallel. The optical power in each is proportional to the 2-D cross-correlation between the search page and the associated stored page, making it possible to search an entire database with a single optical exposure. (After [20.258, 334])

ulators capable of 10 kHz frame rates (e.g., one search every 100 μs) are also becoming commercially available, although the potential performance speed of such a holographic content-addressable memory depends on having sufficient hologram output power (diffraction efficiency times read-out power) as well as on the requisite modulator speed.

The key to the massive parallelism is to arrange multiple storage volumes, which can each store around 1000 holograms, along the path of the data-bearing object beam [20.334]. As the optically encoded search beam passes through each sub-volume, a small amount of power diffracts from the holograms that are significantly similar to the search information [20.258,334]. By using one photodetector per hologram, millions of analog similarity metrics could be measured simultaneously. Exact matches to a query [20.257–259], or records that are just similar to a query [20.258,334,337] could be identified.

However, since the detected analog result of each correlation is subject to random noise, it is possible that database records which sufficiently match the search argument may be overlooked in favor of records that *almost* sufficiently match it. Fortunately, a hybrid system can combine the speed advantage of the holographic content-addressable memory with the digital precision of serial electronics [20.258, 334]. By passing both matching and near-matching records from the holographic front-end to a subsequent electronic processor, the probability of overlooking even one matching record can be made arbitrarily low ($< 10^{-12}$) while retaining much of the speed advantage [20.258]. Given sufficient signal-to-noise ratio and a spatial light modulator of merely 1 kHz frame rate, a database of 1 million one-kilobyte records could be searched in less than 0.5% of the time required by a 1 GHz microprocessor connected to 10 GB of DRAM.

If the ability to search thousands or millions of holograms in parallel can be demonstrated and a suitable nonvolatile holographic recording media developed, volume holographic content-addressable data storage could be an attractive method for rapidly searching vast databases with complex queries.

20.8 Conclusion

The evolutionary approaches that are based on current technologies and have the potential to be forward compatible – such as the use of depth of focus – have the lowest risks in terms of media, storage device and cost. However, the benefits for potential density gains are also limited, especially if the key features of optical data storage – removability and interchangeability – are to be retained. And the window of opportunity for becoming a viable storage technology is closing.

On the other hand, the revolutionary approaches promise much higher densities at substantially higher risk. Some of this risk arises because laboratory development tends to demonstrate features individually, avoiding effects which only show up when performance is pushed across-the-board. For instance, as density and data rates increase, maintaining an acceptable signal-to-noise ratio and accessing each desired record without crosstalk become severe concerns that cannot be considered independently. However, some of the more exotic techniques have additional attributes that may make them advantageous for niche applications.

These revolutionary approaches are several years away from product prototypes. Issues here are mainly the need for better media, and the relatively high complexity and cost of the drive. Applications that take advantage of some of the unique properties will in all likelihood be targeted by first product designs. Whether revolutionary optical storage technologies will be able to become mainstream, or whether conventional optical disk storage will continue to evolve, will depend on both technical feasibility and commercial viability. The demand is there, but the jury is still out on the best approach to meet the ferocious appetite of the information age.

References

20.1 D. Gabor: A new microscopic principle, Nature **161**, 777 (1948)

20.2 E. N. Leith, J. Upatnieks: Reconstructed wavefronts and communication theory, J. Opt. Soc. Am. **52**, 1123 (1962)

20.3 Y. N. Denisyuk: On the reproduction of the optical properties of an object by the wave field of its scattered radiation, Opt. Spectrosc. (USSR) **15**, 279 (1963)

20.4 N. L. Hartmann: Wavefront reconstruction with incoherent light, US Patent 3532406 (1965)

20.5 R. L. Powell, J. H. Hemmye: Holography and hologram interferometry using photochromic recording materials, J. Opt. Soc. Am. **56**, 1540 (1966)

20.6 T. A. Shankoff: Phase holograms in dichromated gelatin, Appl. Opt. **7**, 2101 (1968)

20.7 http://www.holophile.com

20.8 L. Siebert: Front-lighted pulse laser holography, Appl. Phys. Lett. **11**, 326 (1967)

20.9 S. Benton: Hologram reconstructions with extended incoherent sources, J. Opt. Soc. Am. **59**, 1545 (1969)

20.10 W. Odelberg (ed.): *From Les Prix Nobel en 1971* (Nobel Foundation, Stockholm 1972)

20.11 V. G. Komar, V. I. Mandrosov, G. Sobolev, D. A. Tsyrulnikov: Image projection onto a holographic screen, Kvantovaya Elektron. **2**, 193 (1975)

20.12 E. N. Leith, J. Upatnieks: Wavefront reconstruction with continuous-tone objects, J. Opt. Soc. Am. **53**, 1377 (1963)

20.13 E. N. Leith, J. Upatnieks: Wavefront reconstruction with diffused illumination and three-dimensional objects, J. Opt. Soc. Am. **54**, 1295 (1964)

20.14 Nobel Foundation: *Nobel Lectures: Physics 1901–1921* (Elsevier, Amsterdam 1967)

20.15 S. F. Johnston: Reconstructing the history of holography, SPIE Proc. **5005**, 455 (2003)

20.16 P. Hariharan: *Optical holography*, 2 edn. (Cambridge Univ. Press, Cambridge 1996)

20.17 G. Groh: *Holographie* (Berliner Union, Stuttgart 1973)

20.18 T. K. Gaylord, M. G. Moharam: Thin and thick gratings: Terminology clarification, Appl. Opt. **20**, 3271 (1981)

20.19 P. Yeh: *Introduction to Photorefractive Nonlinear Optics* (Wiley, New York 1993)

20.20 F. Laeri, T. Tschudi, J. Albers: Coherent CW image amplifier and oscillator using two-wave interaction in a $BaTiO_3$-crystal, Opt. Commun. **47**, 387 (1983)

20.21 M. Kaczmarek, R. W. Eason: Very-high-gain single-pass two-beam coupling in blue Rh : $BaTiO_3$, Opt. Lett. **20**, 1850 (1995)

20.22 S. A. Benton: White light transmission/reflection holographic imaging. In: *Applications of Holography and Optical Data Processing*, ed. by E. Marom, A. A. Friesem, E. Wiener-Avnaer (Pergamon, Oxford 1977) p. 401

20.23 S. D. Kakichashvili: Method of recording phase polarization holograms, Kvantovaya Elektron **1**, 1435 (1974)

20.24 T. Todorov, L. Nikolova, N. Tomova: Polarization holography. 1: A new high-efficiency organic material with reversible photoinduced birefringence, Appl. Opt. **23**, 4309 (1984)

20.25 S. G. Odulov: Spatially oscillating photo-voltaic current in iron-doped lithium-niobate crystals, Sov. Phys. JETP Lett. **35**, 10 (1982)

20.26 M. Imlau, T. Woike, R. Schieder, R. A. Rupp: Holographic recording with orthogonally polarized waves in centrosymmetric $Na_2[Fe(CN)_5NO] \cdot 2H_2O$, Europhys. Lett. **53**, 471 (2001)

20.27 J. T. McCrickerd, N. George: Holographic stereogram from sequential component photographs, Appl. Phys. Lett. **12**, 10 (1968)

20.28 D. J. DeBitetto: Holographic panoramic stereograms synthesized from white light recordings, Appl. Opt. **8**, 1740 (1969)

20.29 J. D. Redman, W. P. Wolton, E. Shuttleworth: Use of holography to make truly 3-dimensional X-ray images, Nature **220**, 58 (1968)

20.30 N. D. Haig: 3-dimensional holograms by rotational multiplexing of 2-dimensional films, Appl. Opt. **12**, 419 (1973)

20.31 M. C. King: Multiple exposure hologram recording of a 3-d image with a 360 degree view, Appl. Opt. **7**, 1641 (1968)

20.32 M. Born, E. Wolf: *Principles of Optics*, 7 edn. (Cambridge Univ. Press, Cambridge 2002)

20.33 B. J. Thompson: Fraunhofer diffraction patterns of opaque objects with coherent background, J. Opt. Soc. Am. **53**, 1350 (1963)

20.34 A. Vanderlugt: Signal-detection by complex spatial-filtering, IEEE T. Inform. Theory **10**, 139 (1964)

20.35 J. W. Goodman: *Introduction to Fourier Optics* (McGraw-Hill, San Francisco 1968)

20.36 L. Rosen: Focused-image holography with extended sources, Appl. Phys. Lett. **9**, 337 (1966)

20.37 G. W. Stroke: White-light reconstruction of holographic images using transmission holograms recorded with conventionally-focused images and in-line background, Phys. Lett. **23**, 325 (1966)

20.38 G. B. Brandt: Image plane holography, Appl. Opt. **8**, 1421 (1969)

20.39 H. Kogelnik: Coupled wave theory for thick hologram gratings, AT&T Tech. J. **48**, 2909–2947 (1969)

20.40 R. Hioki, T. Suzuki: Reconstruction of wavefronts in all directions, Jpn. J. Appl. Phys. **4**, 816 (1965)

20.41 T. H. Jeong, P. Rudolf, A. Luckett: 360° holography, J. Opt. Soc. Am. **56**, 1263 (1966)

20.42 L. H. Lin: Edge-illuminated hologram, J. Opt. Soc. Am. **60**, 714 (1970)

20.43 R. J. Collier, C. B. Burckhardt, L. H. Lin: *Optical Holography* (Academic, Orlando 1971)

20.44 D. L. Staebler, J. J. Amodei, W. Phillips: Multiple storage of thick holograms in $LiNbO_3$, IEEE J. Quantum Elect. **8**, 611 (1972)

20.45 E. N. Leith, A. Kozma, J. Upatnieks, J. Marks, N. Massey: Holographic data storage in three-dimensional media, Appl. Opt. **5**, 1303 (1966)

20.46 G. A. Rakuljic, V. Leyva, A. Yariv: Optical data storage using orthogonal wavelength multiplexed volume holograms, Opt. Lett. **17**, 1471 (1992)

20.47 S. Yin, H. Zhou, F. Zhao, M. Wen, Z. Yang, J. Zhang, F. T. S. Yu: Wavelength multiplexed holographic storage in a sensitive photorefractive crystal using visible-light tunable diode laser, Opt. Commun. **101**, 317 (1993)

20.48 C. Denz, G. Pauliat, G. Roosen, T. Tschudi: Volume hologram multiplexing using a deterministic phase encoding method, Opt. Commun. **85**, 171 (1991)

20.49 G. Barbastathis, M. Levene, D. Psaltis: Shift multiplexing with spherical reference waves, Appl. Opt. **35**, 2403 (1996)

20.50 Y. N. Denisyuk: Photographic reconstruction of the optical properties of an object in its own scattered radiation field, Sov. Phys. Doklady **7**, 543 (1962)

20.51 P. J. Van Heerden: Theory of optical information storage in solids, Appl. Opt. **2**, 393 (1963)

20.52 L. H. Lin: Hologram formation in hardened dichromated gelatin films, Appl. Opt. **8**, 963 (1969)

20.53 A. A. Friesem: Holograms in thick emulsions, Appl. Phys. Lett. **7**, 102 (1965)

20.54 G. W. Stroke, A. E. Labeyrie: White-light reconstruction of holographic images using Lippmann-Bragg diffraction effect, Phys. Lett. **20**, 368 (1966)

20.55 K. S. Pennington, L. H. Lin: Multicolor wavefront reconstruction, Appl. Phys. Lett. **7**, 56 (1965)

20.56 A. A. Friesem, R. J. Fedorowicz: Recent advances in multicolor wavefront reconstruction, Appl. Opt. **5**, 1085 (1966)

20.57 G. Barbastathis, D. Psaltis: *Volume Holographic Multiplexing Methods*, Springer Ser. Opt. Sci., Vol. 76 (Springer, Berlin, Heidelberg 2000) pp. 21–62

20.58 H.-Y. S. Li, D. Psaltis: Three-dimensional holographic disks, Appl. Opt. **33**, 3764 (1994)

20.59 D. Psaltis: Parallel optical memories, Byte **17**, 179 (1992)

20.60 H. Lee, X.-G. Gu, D. Psaltis: Volume holographic interconnections with maximal capacity and minimal cross talk, J. Appl. Phys. **65**, 2191 (1989)

20.61 F. H. Mok, G. W. Burr, D. Psaltis: Angle and space multiplexed holographic random access memory (HRAM), Opt. Memory Neural Networks **3**, 119 (1994)

20.62 G. Burr, F. Mok, D. Psaltis: Angle and space multiplexed holographic storage using the 90-degrees geometry, Opt. Commun. **117**, 49 (1995)

20.63 K. Curtis, A. Pu, D. Psaltis: Method for holographic storage using peristrophic multiplexing, Opt. Lett. **19**, 993 (1994)

20.64 G. Barbastathis, A. Pu, M. Levene, D. Psaltis: Holographic 3D Disks Using Shift Multiplexing, SPIE Proc. **2514**, 355 (1995)

20.65 F. H. Mok, M. C. Tackitt, H. M. Stoll: Storage of 500 high-resolution holograms in a $LiNbO_3$ crystal, Opt. Lett. **16**, 605 (1991)

20.66 A. Pu, K. Curtis, D. Psaltis: A new method for holographic data storage in polymer films. In: *Nonlinear Optics: Materials, Fundamentals and Applications Meeting* (IEEE, New York 1994) p. 433

20.67 S. Campbell, X. Yi, P. Yeh: Hybrid sparse-wavelength angle-multiplexed optical data storage system, Opt. Lett. **19**, 2161 (1994)

20.68 D. Psaltis, F. Mok: Holographic memories, Sci. Am. **273**, 70 (1995)

20.69 A. Pu, D. Psaltis: High-density recording in photopolymer-based holographic three-dimensional disks, Appl. Opt. **35**, 2389 (1996)

20.70 A. Pu, D. Psaltis: *Holographic 3D disks using shift multiplexing*, CLEO96, Vol. 9 (OSA, Washington 1996) p. 165

20.71 A. Pu, D. Psaltis: *Holographic data storage with 100 bits/μm^2 density. Optical Data Storage Topical Meeting* (IEEE, New York 1997) p. 48

20.72 S. Matthews: A light touch, Laser Focus World **40**, 137 (2004)

20.73 U. Schnars, W. Jüptner: Direct recording of holograms by a CCD target and numerical reconstruction, Appl. Opt. **33**, 179 (1994)

20.74 I. Yamaguchi, T. Zhang: Phase-shifting digital holography, Opt. Lett. **22**, 1268 (1997)

20.75 H. M. Smith: *Holographic recording Materials*, Top. Appl. Phys., Vol. 20 (Springer, Berlin, Heidelberg 1977)

20.76 P. Hariharan: Holographic recording materials – recent developments, Opt. Eng. **19**, 636 (1980)

20.77 R. A. Linke, T. Thio, J. D. Chadi, G. E. Devlin: Diffraction from optically written persistent plasma gratings in doped compound semiconductors, Appl. Phys. Lett. **65**, 16 (1994)

20.78 A. I. Ryskin, A. S. Shcheulin, B. Koziarska, J. M. Langer, A. Suchocki, I. I. Buczinskaya, P. P. Fedorov, B. P. Sobolev: CdF_2 : In: A novel material for optically written storage of information, Appl. Phys. Lett. **67**, 31 (1995)

20.79 B. Sugg, H. Nürge, B. Faust, R. Niehüser, H.-J. Reyher, R. A. Rupp, L. Ackermann: The photorefractive effect in terbium gallium garnet, Opt. Mater. **4**, 343 (1995)

20.80 T. Woike, S. Haussühl, B. Sugg, R. A. Rupp, J. Beckers, M. Imlau, R. Schieder: Phase gratings in the visible and near-infrared spectral range realized by metastable electronic states in $Na_2[Fe(CN)_5NO] \cdot 2H_2O$, Appl. Phys. B **63**, 243–248 (1996)

20.81 M. Imlau, S. Haussühl, T. Woike, R. Schieder, V. Angelov, R. A. Rupp, K. Schwarz: Holographic recording by excitation of metastable electronic states in $Na_2[Fe(CN)_5NO] \cdot 2H_2O$, A new photorefractive effect, Appl. Phys. B **68**, 877 (1999)

20.82 M. A. Ellabban, M. Fally, R. A. Rupp, L. Kovács: Light-induced phase and amplitude gratings in centrosymmetric Gadolinium Gallium garnet doped with Calcium, Opt. Express **14**, 593 (2006)

20.83 C. C. Bowley, G. P. Crawford: Diffusion kinetics of formation of holographic polymer-dispersed liquid

crystal display materials, Appl. Phys. Lett. **76**, 2235 (2000)
20.84 M. J. Escuti, J. Qi, G. P. Crawford: Tunable face-centered-cubic photonic crystal formed in holographic polymer dispersed liquid crystals, Opt. Lett. **28**, 522 (2003)
20.85 G. P. Crawford: Electrically switchable Bragg gratings, Opt. Photon. News **14**, 54 (2003)
20.86 E. Völkl, H. Lichte: Electron holograms for subangstrom point resolution, Ultramicrosc. **32**, 177 (1990)
20.87 B. Javidi, E. Tajahuerce: Three-dimensional object recognition by use of digital holography, Opt. Lett. **25**, 610 (2000)
20.88 W. D. Rau, P. Schwander, F. H. Baumann, W. Höppner, A. Ourmazd: Two-dimensional mapping of the electrostatic potential in transistors by electron holography, Phys. Rev. Lett. **82**, 2614 (1999)
20.89 M. R. McCartney, M. A. Gribelyuk, J. Li, P. Ronsheim, J. S. McMurray, D. J. Smith: Quantitative analysis of one-dimensional dopant profile by electron holography, Appl. Phys. Lett. **80**, 3213 (2002)
20.90 E. Völkl, L. F. Allard, D. Joy: *Introduction to Electron Holography* (Kluwer Academic, Dordrecht 1999)
20.91 A. Tonomura: *Electron Holography*, Springer Ser. Opt. Sci., Vol. 70, 2 edn. (Springer, Berlin, Heidelberg 1999)
20.92 R. A. London, M. D. Rosen, J. E. Trebes: Wavelength choice for soft X-ray laser holography of biological samples, Appl. Opt. **28**, 3397 (1989)
20.93 M. Howells, C. Jacobsen, J. Kirz: X-ray holograms at improved resolution: a study of zymogen granules, Science **238**, 514 (1987)
20.94 J. E. Trebes, S. B. Brown, E. M. Campbell, D. L. Matthews, D. G. Nilson, G. F. Stone, D. A. Whelan: Demonstration of X-ray holography with an X-ray laser, Science **238**, 517 (1987)
20.95 E. N. Leith: Quasi-holographic techniques in the microwave region, Proc. IEEE **59**, 1305 (1971)
20.96 A. Andreoni, M. Bondani, M. A. C. Potenza, Y. N. Denisyuk: Holographic properties of the second-harmonic cross correlation of object and reference optical wave fields, J. Opt. Soc. Am. B **17**, 966 (2000)
20.97 Y. N. Denisyuk, A. Andreoni, M. Bondani, M. A. C. Potenza: Real-time holograms generated by second-harmonic cross correlation of object and reference optical wave fields, Opt. Lett. **25**, 890 (2000)
20.98 M. Bondani, A. Andreoni: Holographic nature of three-wave mixing, Phys. Rev. A **66**, 33805 (2002)
20.99 M. Bondani, A. Allevi, A. Brega, E. Puddu, A. Andreoni: Difference-frequency-generated holograms of two-dimensional objects, J. Opt. Soc. Am. B **21**, 280 (2004)
20.100 M. Campbell, D. N. Sharp, M. T. Harrison, R. G. Denning, A. J. Turberfield: Fabrication of photonic crystals for the visible spectrum by holographic lithography, Nature **404**, 53 (2000)
20.101 Y. V. Miklyaev, D. C. Meisel, A. Blanco, G. von Freymann, K. Busch, W. Koch, C. Enkrich, M. Deubel, M. Wegener: Three-dimensional face-centered-cubic photonic crystal templates by laser holography: fabrication, optical characterization, and band-structure calculations, Appl. Phys. Lett. **82**, 1284 (2003)
20.102 N. Huot, J. M. Jonathan, G. Pauliat, P. Georges, A. Brun, G. Roosen: Laser mode manipulation by intracavity dynamic holography: application to mode selection, Appl. Phys. B **69**, 155 (1999)
20.103 S. Y. Lam, M. Damzen: Self-adaptive holographic solid-state dye laser, Opt. Commun. **218**, 365 (2003)
20.104 L. D'Auria, J. P. Huignard, E. Spitz: Holographic read-write memory and capacity enhancement by 3D storage, IEEE Trans. Magn. **9**, 83 (1973)
20.105 J. Heanue, M. Bashaw, L. Hesselink: Volume holographic storage and retrieval of digital data, Science **265**, 749 (1994)
20.106 M. P. Bernal, H. Coufal, R. K. Grygier, J. A. Hoffnagle, C. M. Jefferson, R. M. MacFarlane, R. M. Shelby, G. T. Sincerbox, P. Wimmer, G. Wittmann: A precision tester for studies of holographic optical storage materials and recording physics, Appl. Opt. **35**, 2360 (1996)
20.107 I. McMichael, W. Christian, D. Pletcher, T. Y. Chang, J. H. Hong: Compact holographic storage demonstrator with rapid access, Appl. Opt. **35**, 2375 (1996)
20.108 G. W. Burr, J. Ashley, H. Coufal, R. K. Grygier, J. A. Hoffnagle, C. M. Jefferson, B. Marcus: Modulation coding for pixel-matched holographic data storage, Opt. Lett. **22**, 639 (1997)
20.109 M. Imlau, T. Bieringer, S. G. Odoulov, T. Woike: Holographic data storage. In: *Nanoelectronics and Information Technology. Advanced Electronic Materials and Novel Devices*, ed. by R. Waser (Wiley-VCH, Weinheim 2003) Chap. 27, pp. 661–686
20.110 H. J. Coufal, D. Psaltis, G. T. Sincerbox (eds.): *Holographic Data Storage*, Springer Ser. Opt. Sci., Vol. 76 (Springer, Berlin, Heidelberg 2000)
20.111 F. Dreesen, G. von Bally: High resolution colorholography for archaeological and medical applications,optics within life sciences. In: *Optics within Life Science*, ed. by C. Fotakis, T. G. Papazoglou, C. Kapouzos (Springer, Berlin, Heidelberg 2000) p. 349
20.112 F. Dreesen, G. von Bally: Color rendering in reflection holography. In: *Optical Technologies in the Humanities*, Ser. Opt. Within Life Sci., Vol. 4, ed. by D. Dirksen, G. von Bally (Elsevier, Amsterdam 1996) p. 79
20.113 G. von Bally, F. Dreesen, V. B. Markov, A. Roskhop, E. V. de Haller: Recording of color holograms on PFG-03Ts, Tech. Phys. Lett. **21**, 76 (1995)
20.114 G. von Bally, D. Dreesen, A. Roshop, E. de Haller, G. Wernicke, N. Demoli, U. Dahms, H. Gruber,

W. Sommerfeld: Holographic methods in cultural heritage preservation and evaluation. In: *Optical Methods in Biomedical and Environmental Sciences*, Ser. Opt. Within Life Sci., Vol. 3, ed. by H. Ohzu, S. Komatsu (Elsevier, Amsterdam 1994) p. 297

20.115 Y. I. Ostrovsky, M. Butusov, G. V. Ostrovskaya: *Interferometry by Holography*, Springer Ser. Opt. Sci., Vol. 20 (Springer, Berlin, Heidelberg 1980)

20.116 C. M. Vest: *Holographic Interferometry* (Wiley Interscience, New York 1979)

20.117 G. Wernicke, W. Osten: *Holografische Interferometrie* (Physik-Verlag, Weinheim 1982)

20.118 T. Kreis: *Holographic interferometry*, Akad. Ser. Opt. Metrol., Vol. 1 (Akademie, Berlin 1996)

20.119 K. A. Stetson, R. L. Powell: Interferometric hologram evaluation and real-time vibration analysis of diffuse objects, J. Opt. Soc. Am. **55**, 1694 (1965)

20.120 R. J. Collier, E. T. Doherty, K. S. Pennington: Application of Moiré techniques to holography, Appl. Phys. Lett. **7**, 223 (1965)

20.121 R. E. Brooks, L. O. Heflinger, R. F. Wuerker: Interferometry with a holographically reconstructed comparsion beam, Appl. Phys. Lett. **7**, 248 (1965)

20.122 M. H. Horman: An application of wavefront reconstruction to interferometry, Appl. Opt. **4**, 333 (1965)

20.123 B. P. Hildebrand, K. A. Haines: Interferometric measurements using the wavefront reconstruction technique, Appl. Opt. **5**, 172 (1966)

20.124 K. A. Haines, B. P. Hildebrand: Surface-deformation measurement using the wavefront reconstruction technique, Appl. Opt. **5**, 595 (1966)

20.125 L. O. Heflinger, R. F. Wuerker, R. E. Brooks: Holographic interferometry, J. Appl. Phys. **37**, 642 (1966)

20.126 N. Abramson: The holo-diagram: A practical device for making and evaluating holograms, Appl. Opt. **8**, 1235 (1969)

20.127 J. E. Sollid: Holographic interferometry applied to measurements of small static displacements of diffusely reflecting surfaces, Appl. Opt. **8**, 1587 (1969)

20.128 K. A. Stetson: Method of vibration measurements in heterodyne interferometry, Opt. Lett. **7**, 233 (1982)

20.129 R. J. Pryputniewicz: Pulsed laser holography in studied of bone motions and deformations, Opt. Eng. **24**, 832 (1985)

20.130 R. Thalmann: Heterodyne and quasi-heterodyne holographic-interferometry, Opt. Eng. **24**, 824 (1985)

20.131 T. Tsuruta, N. Shiotake, Y. Itoh: Hologram interferometry using 2 reference beams, Jpn. J. Appl. Phys. **7**, 1092 (1968)

20.132 G. S. Ballard: Double-exposure holographic interferometry, J. Appl. Phys. **39**, 4846 (1968)

20.133 E. Marom, F. M. Mottier: 2-reference-beam holographic interferometry, J. Opt. Soc. Am. **66**, 23 (1976)

20.134 G. M. Brown, R. M. Grant, G. W. Stroke: Theory of holographic interferometry, J. Acoust. Soc. Am. **45**, 1166 (1969)

20.135 E. Jansson, N. E. Molin, H. Sundin: Resonances of a violin body studied by hologram interferometry and acoustical methods, Phys. Scr. **2**, 243 (1970)

20.136 A. D. Wilson, D. H. Strope: Time-average holographic interferometry of a circular plate vibrating simultaneously in 2 rationally related modes, J. Opt. Soc. Am. **60**, 1162 (1970)

20.137 R. Tonin, D. A. Bies: Time-averaged holography for study of 3-dimensional vibrations, J. Sound Vibrat. **52**, 315 (1977)

20.138 C. C. Aleksoff: Time average holography extended, Appl. Phys. Lett. **14**, 23 (1969)

20.139 F. M. Mottier: Time-averaged holography with triangular phase modulation of the reference wave, Appl. Phys. Lett. **15**, 285 (1969)

20.140 E. Archbold, A. E. Ennos: Observation of surface vibration modes by stroboscopic hologram interferometry, Nature **217**, 942 (1968)

20.141 B. M. Watrasiewicz, P. Spicer: Vibration analysis by stroboscopic holography, Nature **217**, 1142 (1968)

20.142 P. Shajenko, C. D. Johnson: Stroboscopic holographic interferometry, Appl. Phys. Lett. **13**, 44 (1968)

20.143 D. Hadbawnik: Holographische Endoskopie, Optik **45**, 21 (1976)

20.144 M. Yonemura, T. Nishisaka, H. Machida: Endoscopic hologram interferometry using fiber optics, Appl. Opt. **20**, 1664 (1981)

20.145 G. von Bally, W. Schmidthaus, H. Sakowski, W. Mette: Gradient-index optical systems in holographic endoscopy, Appl. Opt. **23**, 1725 (1984)

20.146 G. von Bally, E. Brune, W. Mette: Holographic endoscopy with gradient-index optical imaging system and optical fibers, Appl. Opt. **25**, 3425 (1986)

20.147 O. Coquoz, R. Conde, F. Taleblou, C. Depeursinge: Performances of endoscopic holography with a multicore optical fiber, Appl. Opt. **34**, 7186 (1995)

20.148 D. B. Sheffer, W. Loughry, K. Somasundaram, S. K. Chawla, P. J. Wesolowski: Phase-shifting holographic interferometry for breast cancer detection, Appl. Opt. **33**, 5011 (1994)

20.149 S. Schedin, G. Pedrini, H. J. Tiziani, A. K. Aggarwal: Comparative study of various endoscopes for pulsed digital holographic interferometry, Appl. Opt. **40**, 2692 (2001)

20.150 M. d. S. Hernández-Montes, C. Pérez-López, F. Mendoza Santoyo, L. M. Muñoz Guevara: Detection of biological tissue in gels using pulsed digital holography, Opt. Express **12**, 853 (2004)

20.151 H. Chen, M. Shih, E. Arons, E. Leith, J. Lopez, D. Dilworth, P. C. Sun: Electronic holographic imaging through living human tissue, Appl. Opt. **33**, 3630 (1994)

20.152 N. H. Abramson, K. G. Spears: Single pulse light-in-flight recording by holography, Appl. Opt. **28**, 1834 (1989)

20.153 I. Bukosza: Three-dimensional representation of ventriculography using contour-line holography, Appl. Opt. **31**, 2485 (1992)

20.154 C. Liu, C. Yan, S. Gao: Digital holographic method for tomography-image reconstruction, Appl. Phys. Lett. **84**, 1010 (2004)

20.155 M.-K. Kim: Tomographic three-dimensional imaging of a biological specimen using wavelength-scanning digital interference holography, Opt. Express **7**, 305 (2000)

20.156 S.-R. Kothapalli, P. Wu, C.S. Yelleswarapu, D.V.G.L.N. Rao: Medical image processing using transient Fourier holography in bacteriorhodopsin films, Appl. Phys. Lett. **85**, 5836 (2004)

20.157 D. Carl, B. Kemper, G. Wernicke, G. von Bally: Parameter-optimized digital holographic microscope for high-resolution living-cell analysis, Appl. Opt. **43**, 6536 (2004)

20.158 P. Marquet, B. Rappaz, P.J. Magistretti, E. Cuche, Y. Emery, T. Colomb, C. Depeursinge: Digital holographic microscopy: a noninvasive contrast imaging technique allowing quantitative visualization of living cells with subwavelength axial accuracy, Opt. Lett. **30**, 468 (2005)

20.159 A. Kozma, D. L. Kelly: Spatial filtering for detection of signals submerged in noise, Appl. Opt. **4**, 387 (1965)

20.160 A.W. Lohmann, D.P. Paris: Binary Fraunhofer holograms, generated by computer, Appl. Opt. **6**, 1739 (1967)

20.161 W.H. Lee: Sampled Fourier transform hologram generated by computer, Appl. Opt. **9**, 639 (1970)

20.162 W.-H. Lee: Binary synthetic holograms, Appl. Opt. **13**, 1677 (1974)

20.163 H. Melville, G. F. Milne, G. C. Spalding, W. Sibbett, K. Dholakia, D. McGloin: Optical trapping of three-dimensional structures using dynamic holograms, Opt. Express **11**, 3562 (2003)

20.164 W. J. Hossack, E. Theofanidou, J. Crain: High-speed holographic optical tweezers using a ferroelectric liquid crystal microdisplay, Opt. Express **11**, 2053 (2003)

20.165 E.R. Dufresne, G.C. Spalding, M.T. Dearing, S.A. Sheets, D.G. Grier: Computer-generated holographic optical tweezer arrays, Rev. Sci. Instrum. **72**, 1810 (2001)

20.166 D.G. Grier, A.A. Sawchuk: Dynamic holographic optical tweezers: transforming mesoscopic matter with light, Trends Opt. Photon. (OSA) **90**, 84 (2003)

20.167 A. Jesacher, S. Fürhapter, S. Bernet, M. Ritsch-Marte: Diffractive optical tweezers in the Fresnel regime, Opt. Express **12**, 2243 (2004)

20.168 A. Ashkin, J.M. Dziedzic, J.E. Bjorkholm, S. Chu: Observation of a single-beam gradient force optical trap for dielectric particles, Opt. Lett. **11**, 288 (1986)

20.169 A. Vasara, J. Turunen, A.T. Friberg: Realization of general nondiffracting beams with computer-generated holograms, J. Opt. Soc. Am. A **6**, 1748 (1989)

20.170 J. Durnin: Exact solutions for nondiffracting beams. I. the scalar theory, J. Opt. Soc. Am. A **4**, 651 (1987)

20.171 N. R. Heckenberg, R. McDuff, C. P. Smith, A. G. White: Generation of optical phase singularities by computer-generated holograms, Opt. Lett. **17**, 221 (1992)

20.172 A. R. Agachev, N. P. Larionov, A. V. Lukin, T. A. Mironova, A. A. Nyushkin, D. V. Protasevich, R. A. Rafikov: Computer-generated holographic optics, J. Opt. Technol. **69**, 871 (2002)

20.173 I.M. Lancaster: Holograms and authentication: meeting future demands. In: *Practical Holography XVIII*, Mater. Appl., Vol. 5290, ed. by T.H. Jeong, H.I. Bjelkhagen (SPIE, Bellingham 2003) p. 318

20.174 D. Weber, J. Trolinger: Novel implementation of nonlinear joint transform correlators in optical security and validation, Opt. Eng. **38**, 62 (1999)

20.175 P. Réfrégier, B. Javidi: Optical image encryption based on input plane and Fourier plane random encoding, Opt. Lett. **20**, 767 (1995)

20.176 B. Javidi, T. Nomura: Securing information by use of digital holography, Opt. Lett. **25**, 28 (2000)

20.177 E. Tajahuerce, O. Matoba, S.C. Verrall, B. Javidi: Optoelectronic information encryption with phase-shifting interferometry, Appl. Opt. **39**, 2313 (2000)

20.178 J. F. Heanue, M. C. Bashaw, L. Hesselink: Encrypted holographic data storage based on orthogonal-phase-code multiplexing, Appl. Opt. **34**, 6012 (1995)

20.179 N. Yoshikawa, M. Itoh, T. Yatagai: Binary computer-generated holograms for security applications from a synthetic double-exposure method by electron-beam lithography, Opt. Lett. **23**, 1483 (1998)

20.180 O. Matoba, B. Javidi: Encrypted optical storage with angular multiplexing, Appl. Opt. **38**, 7288 (1999)

20.181 S. Kishk, B. Javidi: Watermarking of three-dimensional objects by digital holography, Opt. Lett. **28**, 167 (2003)

20.182 M.A. Ellabban, M. Fally, R.A. Rupp, T. Woike, M. Imlau: Holographic Scattering and its Applications. In: *Recent Developments in Applied Physics*, Vol. 4, ed. by S.G. Pandalay (Transworld Publishing, Trivandrum 2001) pp. 241–275

20.183 M. Imlau, M. Goulkov, M. Fally, T. Woike: Characterization of polar oxides by photo-induced light scattering. In: *Polar Oxides: Properties, Characterization and Imaging*, ed. by U. Böttger, S. Tiedke, R. Waser (Wiley, New York 2005) Chap. 9, pp. 163–188

20.184 M. Goulkov, M. Imlau, R. Pankrath, T. Granzow, U. Dörfler, T. Woike: Temperature study of photoinduced wide-angle scattering in cerium-doped strontium barium niobate, J. Opt. Soc. Am. B **20**, 307–313 (2003)

20.185 M. Goulkov, T. Granzow, U. Dörfler, T. Woike, M. Imlau, R. Pankrath, W. Kleemann: Temperature dependent determination of the linear electrooptic coefficient r_{33} in $Sr_{0.61}Ba_{0.39}Nb_2O_6$ single crystals by means of light-induced scattering, Opt. Commun. **218**, 173–182 (2003)

20.186 M.Y. Goulkov, T. Granzow, U. Dörfler, T. Woike, M. Imlau, R. Pankrath: Study of beam-fanning hysteresis in photo- refractive SBN:Ce: light-induced and primary scattering as functions of polar structure, Appl. Phys. B **76**, 407–416 (2003)

20.187 M. Goulkov, M. Imlau, T. Granzow, T. Woike: Beam fanning reversal in the ferroelectric relaxor $Sr_{0.61}Ba_{0.39}Nb_2O_6$ at high external electric fields, J. Appl. Phys. **94**, 4763 (2003)

20.188 S. Hausfeld, M. Imlau, T. Weisemöller, M. Fally, T. Woike: Parametric scattering upon light-induced generation of metastable molecular states. In: *Trends in Optics and Photonics*, Vol. 99, ed. by G. Zhang, D. Kip, D. Nolte, J. Xu (OSA, Washington 2005) p. 405

20.189 M. Imlau, R. Schieder, R.A. Rupp, T. Woike: Anisotropic holographic scattering in centrosymmetric sodium nitroprusside, Appl. Phys. Lett. **75**, 16 (1999)

20.190 M. Goulkov, S. Odoulov, T. Woike, J. Imbrock, M. Imlau, H. Hesse: Holographic light scattering in photorefractive crystals with local response, Phys. Rev. B **65**, 195111 (2002)

20.191 A. Szöke: X-ray and electron holography using a local reference beam. In: *Short Wavelength coherent radiation: Generation and Applications*, Vol. 147, ed. by D.T. Attwood, J. Boker (AIP, New York 1986) pp. 361–367

20.192 G.R. Harp, D.K. Saldin, B.P. Tonner: Atomic resolution electron holography in solids with localized sources, Phys. Rev. Lett. **65**, 1012 (1990)

20.193 G. Faigel, M. Tegze: X-ray holography, Rep. Prog. Phys. **62**, 355 (1999)

20.194 L. Cser, G. Krexner, G. Török: Atomic-resolution neutron holography, Europhys. Lett. **54**, 747 (2001)

20.195 T. Gog, P.M. Len, G. Materlik, D. Bahr, C.S. Fadley, C. Sanchez-Hanke: Multiple-energy X-ray holography: Atomic images of hematite (Fe_2O_3), Phys. Rev. Lett. **76**, 3132 (1996)

20.196 S.Y. Tong, H. Huang, X.Q. Guo: Low-energy electron and low-energy positron holography, Phys. Rev. Lett. **69**, 3654 (1992)

20.197 A. Hamza, P. Asoka-Kumar, W. Stoeffl, R. Howell, D. Miller, A. Denison: Development of positron diffraction and holography at LLNL, Radiat. Phys. Chem. **68**, 635 (2003)

20.198 J.J. Barton: Photoelectron holography, Phys. Rev. Lett. **61**, 1356 (1988)

20.199 P.M. Len, J.D. Denlinger, E. Rotenberg, S.D. Kevan, B.P. Tonner, Y. Chen, M.A. van Hove, C.S. Fadely: Holographic atomic images from surface and bulk W(110) photoelectron diffraction data, Phys. Rev. B **59**, 5857 (1999)

20.200 S. Omori, Y. Nihei, E. Rotenberg, J.D. Denlinger, S. Marchesini, S.D. Kevan, B.P. Tonner, M.A. van Hove, C.S. Fadley: Differential photoelectron holography: A new approach for three-dimensional atomic imaging, Phys. Rev. Lett. **88**, 055504 (2002)

20.201 D.K. Saldin, P.L. de Andres: Holographic LEED, Phys. Rev. Lett. **64**, 1270 (1990)

20.202 H. Wu, S. Xu, S. Ma, W.P. Lau, M.H. Xie, S.Y. Tong: Surface atomic arrangement visualization via reference-atom-specific holography, Phys. Rev. Lett. **89**, 216101 (2002)

20.203 H. Li, B.P. Tonner: Real-space interpretation of X-ray-excited Auger-electron diffraction from Cu(001), Phys. Rev. B **37**, 3959 (1988)

20.204 M. Tegze, G. Faigel: X-ray holography with atomic resolution, Nature **380**, 49 (1996)

20.205 G. Tegze, M. Faigel, S. Marchesini, M. Belakhovsky, A.I. Chumakov: Three dimensional imaging of atoms with isotropic 0.5 Å resolution, Phys. Rev. Lett. **82**, 4847 (1999)

20.206 M. Tegze, G. Faigel, S. Marchesini, M. Belakhovsky, O. Ulrich: Imaging light atoms by X-ray holography, Nature **407**, 38 (2000)

20.207 P. Korecki, J. Korecki, T. Ślęzak: Atomic resolution γ-ray holography using the Mössbauer effect, Phys. Rev. Lett. **79**, 3518 (1997)

20.208 P. Korecki, M. Szymnoński, J. Korecki, T. Ślęzak: Site-selective holographic imaging of iron arrangements in magnetite, Phys. Rev. Lett. **92**, 205501 (2004)

20.209 B. Sur, R.B. Rogge, R.P. Hammond, V.N.P. Anghel, J. Katsaras: Atomic structure holography using thermal neutrons, Nature **414**, 525 (2001)

20.210 L. Cser, G. Török, G. Krexner, M. Prem, I. Sharkov: Neutron holographic study of palladium hydride, Appl. Phys. Lett. **85**, 1149 (2004)

20.211 L. Cser, G. Török, G. Krexner, I. Sharkov, B. Faragó: Holographic imaging of atoms using thermal neutrons, Phys. Rev. Lett. **89**, 175504 (2002)

20.212 P. Korecki, G. Materlik, P. Korecki: Complex γ-ray hologram: Solution to twin images problem in atomic resolution imaging, Phys. Rev. Lett. **86**, 1534 (2001)

20.213 Y. Takahashi, K. Hayashi, E. Matsubara: Complex X-ray holography, Phys. Rev. B **68**, 052103 (2003)

20.214 R.A. Rupp, J. Hehmann, R. Matull, K. Ibel: Neutron diffraction from photoinduced gratings in a PMMA matrix, Phys. Rev. Lett. **64**, 301 (1990)

20.215 M. Fally: The photo-neutronrefractive effect, Appl. Phys. B **75**, 405–426 (2002)

20.216 U. Schellhorn, R.A. Rupp, S. Breer, R.P. May: The first neutron interferometer built of holographic gratings, Physica B **234-236**, 1068–1070 (1997)

20.217 C. Pruner, M. Fally, R.A. Rupp, R.P. May, J. Vollbrandt: Interferometer for cold neutrons, Nucl. Instrum. Meth. A **560**, 598 (2006)

20.218 M. Fally, C. Pruner, R.A. Rupp, G. Krexner: *Neutron physics with photorefractive materials*, Springer Ser. Opt. Sci., Vol. 115 (Springer, Berlin, New York 2007) pp. 317–349

20.219 NSIC: *NSIC-OIDA Optical Disk Storage Roadmap* (National Storage Industry Consortium and Optoelectronics Industry Development Association, San Diego 1997)

20.220 NSIC: *NSIC Optical Disk Storage Roadmap* (National Storage Industry Consortium, San Diego 2000)

20.221 NSIC: *NSIC Optical Disk Storage Roadmap* (National Storage Industry Consortium, San Diego 2003)

20.222 http://www.mpeg.org/MPEG/DVD/

20.223 http://www.ee.washington.edu/conselec/CE/kuhn/doit96/dhome.htm

20.224 http://www.sony.net/SonyInfo/News/Press/200202/02-0219E/

20.225 A.B. Marchant: *Optical Recording* (Addison-Wesley, Boston 1990)

20.226 M. Mansuripur, G. Sincerbox: Principles and Techniques of Optical Data Storage, Proc IEEE **85**(11), 1780–1796 (1997)

20.227 ECMA-267 Standard: 120 mm DVD – Read-only disk, http://www.ecma.ch, Dec. 1999

20.228 W.S. Oakley: A novel digital optical tape recorder, Proc. SPIE **2604**, 265 (1995)

20.229 D.A. Thompson, J.S. Best: The future of magnetic data storage technology, IBM J. R. Devel. **44**(3), 311–322 (2000)

20.230 T.L. Wong, M.P. O'Neill: Multilevel optical recording, J. Magn. Soc. Jpn. **25**(3), 433–436 (2001)

20.231 M. Mansuripur: *The Physical Principles of Magneto-optical Recording* (Cambridge Univ. Press, Cambridge 1995)

20.232 B.D. Terris, H.J. Mamin, D. Rugar: Near-field optical data storage, Appl. Phys. Lett. **68**(2), 141–143 (1996)

20.233 M. Kaneko, K. Aratani, M. Ohta: Multilayered Magnetooptical Disks for Magnetically Induced Superresolution, Jpn. J. Appl. Phys. **31**(2B), 568 (1992)

20.234 H. Awano, S. Ohnuki, H. Shirai, N. Ohta: Magnetic amplifying magneto-optical system. In: *Optical Data Storage*, Proc. SPIE, Vol. 3109, ed. by H. Birecki, J. Kwiecien (SPIE, Bellingham 1997) p. 83

20.235 Y.V. Martynov, H.A. Wierenga: Migration path of optical storage drives and media, J. Inf. Storage Proces. Syst. **2**(1), 93–100 (2000)

20.236 C3D, White paper, Constellation 3D, (Jun 2000), http://www.c-3d.net/whitepaper.html

20.237 S. Hunter, F. Kiamilev, S. Esener, D.A. Parthenopoulos, P.M. Rentzepis: Potentials of two-photon based 3D optical memories for high performance computing, Appl. Opt. **29**(14), 2058–2066 (1990)

20.238 S. Kawata, Y. Kawata: Three-dimensional optical data storage using photochromic materials, Chem. Rev. **100**(5), 1777–1788 (2000)

20.239 K. Yamasaki, S. Juodkazis, M. Watanabe, H.B. Sun, S. Matsuo, H. Misawa: Recording by microexplosion and two-photon reading of three-dimensional optical memory in polymethylmethacrylate films, Appl. Phys. Lett. **76**(8), 1000–1002 (2000)

20.240 F.B. McCormick, H. Zhang, A. Dvomikov, E. Walker, C. Chapman, N. Kim, J. Costa, S. Esener, P. Rentzepis: Parallel access 3D multilayer optical storage using 2-photon recording. In: *Advanced Optical Data Storage: Materials, Systems, and Interfaces to Computers*, Proc. SPIE, Vol. 3802, ed. by P.A. Mitkas, Z.U. Hasan, H.J. Coufal, G.T. Sincerbox (SPIE, Bellingham 1999) pp. 173–182

20.241 I. Cokgor, F.B. McCormick, A.S. Dvornikov, M.M. Wang, N. Kim, K. Coblentz, S.C. Esener, P.M. Rentzepis: Multilayer disk recording using 2-photon absorption and the numerical simulation of the recording process. In: *Optical Data Storage*, Proc. SPIE, Vol. 3109, ed. by H. Birecki, J. Kwiecien (SPIE, Bellingham 1997) pp. 54–55

20.242 S.R. Chinn, E.A. Swanson: Multilayer optical storage by low-coherence reflectometry, Opt. Lett. **21**(12), 899–901 (1996)

20.243 D. Huang, E.A. Swanson, C.P. Lin, J.S. Schuman, W.G. Stinson, W. Chang, M.R. Hee, T. Flotte, K. Gregory, C.A. Puliafito, J.G. Fujimoto: Optical Coherence Tomography, Science **254**(5035), 1178–1181 (1991)

20.244 W.E. Moerner: Molecular electronics for frequency domain optical storage: persistent spectral hole-burning a review, J. Molec. Electr. **1**(1), 55–71 (1985)

20.245 W.E. Moerner (ed.): *Persistent Spectral Hole Burning: Science and Applications* (Springer, New York 1988)

20.246 T.W. Mossberg: Time-domain frequency-selective optical data storage, Opt. Lett. **7**(2), 77–79 (1982)

20.247 E.S. Maniloff, A.E. Johnson, T.W. Mossberg: Spectral data storage using rare-earth-doped crystals, MRS Bulletin **24**(9), 46–50 (1999)

20.248 G. Sincerbox (ed.): *Selected papers on holographic data storage*, SPIE Milestone Series, Vol. MS95 (SPIE, Bellingham 1994)

20.249 H.J. Coufal, D. Psaltis, G. Sincerbox (eds.): *Holographic Data Storage* (Springer, Berlin 2000)

20.250 J.W. Goodman: *Introduction to Fourier Optics*, 2nd edn. (McGraw-Hill, New York 1996)

20.251 F. Ito, K. Kitayama, H. Oguri: Holographic Image Storage in $LiNbO_3$ Fibers with Compensation for Intrasignal Photorefractive Coupling, J. Opt. Soc. Am. B **9**(8), 1432–1439 (1992)

20.252 A. Aharoni, M.C. Bashaw, L. Hesselink: Distortion-Free Multiplexed Holography in Striated Photorefractive Media, Appl. Opt. **32**(11), 1973–1982 (1993)

20.253 J. J. P. Drolet, E. Chuang, G. Barbastathis, D. Psaltis: Compact, integrated dynamic holographic memory with refreshed holograms, Opt. Lett. **22**(8), 552–554 (1997)

20.254 F. Zhao, K. Sayano: High density phase-conjugate holographic memory with phase-only image compressors, Opt. Mem. Neur. Net. **6**(4), 261–264 (1997)

20.255 G. W. Burr, I. Leyva: Multiplexed phase-conjugate holographic data storage with a buffer hologram, Opt. Lett. **25**(7), 499–501 (2000)

20.256 D. Psaltis, D. Brady, K. Wagner: Adaptive optical networks using photorefractive crystals, Appl. Opt. **27**(9), 1752–1759 (1988)

20.257 B. J. Goertzen, P. A. Mitkas: Volume holographic storage for large relational databases, Opt. Eng. **35**(7), 1847–1853 (1995)

20.258 G. W. Burr, S. Kobras, H. Hanssen, H. Coufal: Content-addressable data storage by use of volume holograms, Appl. Opt. **38**(32), 6779–6784 (1999)

20.259 P. A. Mitkas, G. W. Burr: Volume holographic optical correlators. In: *Holographic Data Storage*, ed. by H. J. Coufal, D. Psaltis, G. T. Sincerbox (Springer, Berlin 2000) pp. 429–445

20.260 J. Ashley, M.-P. Bernal, G. W. Burr, H. Coufal, H. Guenther, J. A. Hoffnagle, C. M. Jefferson, B. Marcus, R. M. Macfarlane, R. M. Shelby, G. T. Sincerbox: Holographic data storage, IBM J. Res. Dev. **44**(3), 341–368 (May 2000)

20.261 P. J. van Heerden: Theory of optical information storage in solids, Appl. Opt. **2**(4), 393–401 (1963)

20.262 D. Psaltis, F. Mok: Holographic Memories, Scient. Amer **273**(5), 70–76 (1995)

20.263 J. F. Heanue, M. C. Bashaw, L. Hesselink: Volume holographic storage and retrieval of digital data, Science **265**(5173), 749–752 (1994)

20.264 J. H. Hong, I. McMichael, T. Y. Chang, W. Christian, E. G. Paek: Volume holographic memory systems: techniques and architectures, Opt. Eng. **34**(8), 2193–2203 (1995)

20.265 D. Psaltis, G. W. Burr: Holographic data storage, Computer **31**(2), 52 (1998)

20.266 W. C. Stewart, R. S. Mezrich, L. S. Cosentin, E. M. Nagle, F. S. Wendt, R. D. Lohman: Experimental Read-Write Holographic Memory, RCA Review **34**(1), 3–44 (1973)

20.267 L. D'Auria, J. P. Huignard, C. Slezak, E. Spitz: Experimental holographic read-write memory using 3D storage, Appl. Opt. **13**(4), 808–818 (1974)

20.268 G. W. Burr, C. M. Jefferson, H. Coufal, M. Jurich, J. A. Hoffnagle, R. M. Macfarlane, R. M. Shelby: Volume holographic data storage at areal density of 250 gigapixels/in^2, Opt. Lett. **26**(7), 444–446 (2001)

20.269 S. S. Orlov, W. Phillips, E. Bjornson, Y. Takashima, P. Sundaram, L. Hesselink, R. Okas, D. Kwan, R. Snyder: High-transfer-rate high-capacity holographic disk data-storage system, Appl. Opt. **43**(25), 4902–4914 (2004)

20.270 J. A. Ma, T. Chang, S. Choi, J. Hong: Ruggedized digital holographic data storage with fast access, Opt. Quant. Electr. **32**(3), 383–392 (2000)

20.271 G. Barbastathis, D. Psaltis: Volume holographic multiplexing methods. In: *Holographic Data Storage*, ed. by H. J. Coufal, D. Psaltis, G. T. Sincerbox (Springer, Berlin 2000) pp. 21–62

20.272 D. L. Staebler, J. J. Amodei, W. Phillips: Multiple storage of thick holograms in $LiNbO_3$, IEEE J. Quantum Elect. **8**(6), 611 (1972)

20.273 F. H. Mok, M. C. Tackitt, H. M. Stoll: Storage of 500 high-resolution holograms in a $LiNbO_3$ crystal, Opt. Lett. **16**(8), 605–607 (1991)

20.274 G. A. Rakuljic, V. Leyva, A. Yariv: Optical data storage by using orthogonal wavelength-multiplexed volume holograms, Opt. Lett. **17**(20), 1471–1473 (1992)

20.275 J. E. Ford, Y. Fainman, S. H. Lee: Array interconnection by phase-coded optical correlation, Opt. Lett. **15**(19), 1088–1090 (1990)

20.276 C. Denz, G. Pauliat, G. Roosen, T. Tschudi: Volume hologram multiplexing using a deterministic phase encoding method, Opt. Commun. **85**(2–3), 171–176 (1991)

20.277 Z. Q. Wen, Y. Tao: Orthogonal codes and cross-talk in phase-code multiplexed volume holographic data storage, Opt. Commun. **148**(1–3), 11–17 (1998)

20.278 K. T. Kim, B. C. Cho, E. S. Kim, S. K. Gil: Performance analysis of phase-code multiplexed holographic memory, Appl. Opt. **39**(23), 4160–4167 (2000)

20.279 D. Psaltis, X. Gu, D. Brady: Fractal sampling grids for holographic interconnections, Proc. SPIE **963**, 468–474 (1988)

20.280 G. W. Burr: *Volume holographic storage using the 90° geometry*, PhD thesis (California Institute of Technology, Pasadena, Calif. 1996)

20.281 X. An, D. Psaltis, G. W. Burr: Thermal fixing of 10000 holograms in $LiNbO_3$: Fe, Appl. Opt. **38**(2), 386–393 (1999)

20.282 W. S. Colburn, K. A. Haines: Volume hologram formation in photopolymer materials, Appl. Opt. **10**(7), 1636–1641 (1971)

20.283 R. T. Ingwall, M. Troll: Mechanism of Hologram Formation in DMP-128 Photopolymer, Opt. Eng. **28**(6), 586–591 (1989)

20.284 L. Dhar, M. G. Schnoes, T. L. Wysocki, H. Bair, M. Schilling, C. Boyd: Temperature-induced changes in photopolymer volume holograms, Appl. Phys. Lett. **73**(10), 1337–1339 (1998)

20.285 R. T. Ingwall, D. Waldman: Photopolymer systems. In: *Holographic Data Storage*, ed. by H. J. Coufal, D. Psaltis, G. T. Sincerbox (Springer, Berlin 2000) pp. 171–198

20.286 L. Dhar, M.,G. Schnoes, H. E. Katz, A. Hale, M. L. Schilling, A. L. Harris: Photopolymers for digital holographic data storage. In: *Holographic Data Storage*, ed. by H. J. Coufal, D. Psaltis, G. T. Sincerbox (Springer, Berlin 2000) pp. 199–208

21. Laser Safety

After a short introduction on current applications of lasers at the beginning of this chapter, Sect. 21.1 gives some historical remarks on laser safety.

In Sect. 21.2 biological interactions and the effects of laser radiation on human tissue are described. Absorption, penetration and transmission of optical radiation are dealt with in detail. The wavelength-dependent transmission through the eye from the cornea to the retina and the role of the pigment epithelium as a selectively absorbing layer are illustrated quantitatively. The optical gain as a result of the focusing capability of the eye is demonstrated together with the increase of irradiance on the retina relative to the power density of the laser beam in front of the eye. State-of-the-art descriptions of photochemical, photothermal and photoionizing effects on biological tissue in general are given and the effects of laser radiation, especially on the various parts of the eye, are described. Retinal and non-retinal injuries of the eye are analyzed and illustrated. In addition, injuries to the skin are shown.

The topic of Sect. 21.3 is maximum permissible exposure. Its relevance to the prevention of short- and long-term effects is described and some remarks on safety or reduction factors are made. The relationship between radiant exposure and irradiance is derived and its general perspective to maximum power values and exposure duration is demonstrated. For the eye, the influence of physiological factors such as eye movements is discussed in more detail for the case of a continuous wave exposure from a point-source. In addition, the concept of an angular subtense to describe the irradiation from an extended optical source is analyzed. Furthermore, recently achieved results are described which show that the strong belief in aversion responses and especially in the blink reflex as a reliable physiological reaction is no longer valid, since no more than about 20% of people perform a blink reflex and even less avert the head.

International standards and regulations on laser safety worldwide are summarized in Sect. 21.4.

Laser classes are described in Sect. 21.5 of this chapter according to the classification scheme of the international standard IEC 60825-1. The meanings of accessible emission limits and time base are explained. A comprehensive description of the various laser classes is given and valuable hints about dealing with each laser class are also described. It is shown how the nominal ocular hazard distance might be calculated from given data.

In Sect. 21.6 protective measures are described. As is usually the case technical and engineering measures rank more highly than administrative and personal protective measures. The protective measures are assigned to the respective laser classes. In addition the role of a laser safety officer is described.

In Sect. 21.7 special recommendations for the most hazardous situations are given and common unsafe procedures are listed. Finally a special topic deals with laser pointers.

Lasers are used in various applications in industry, science, and medicine, for materials processing, distance measurement, alignment, industrial and medical heat treatment etc. In addition, there is increasing public use in the entertainment field and for simple laser pointers.

The laser is often hidden in a protective enclosure and the laser beam is not visible like in an optical fiber communication system, compact disc (CD) player, laser printer, to name but a few. In other applications, such as diagnostic devices, laser endoscopes, or devices for cutting, welding and soldering, the laser beam irradiates a target either to measure levels of some substance or to treat a material. The objective is to deliver enough energy to heat or even vaporize the respective material, while the construction of the laser product and the mode of operation ensure that no harmful laser radiation can escape from the laser device or laser product. In some cases so much optical radiation, i. e. laser radiation and/or secondary and collateral optical radiation, can be scattered or produced during material processing procedures, that adverse health effects cannot be completely excluded.

Normally, lasers used in industrial applications are not a source of special health concern under normal operating conditions, but the situation may be completely different during maintenance or service, where the laser beam is freely accessible and a potentially hazardous situation might arise if the necessary safety measures are not in place.

Special care has to be taken in research and development laboratories, where the open beam is accessible and intrabeam viewing (direct viewing into the beam) may involuntarily happen.

Since there are a large number of devices where lasers are incorporated and since the wavelengths cover the whole optical spectrum, and the power ranges from few micro- or milliwatts to petawatts, various hazardous situations have to be considered in laser safety. In addition not only continuous wave (CW) lasers but also pulsed ones are in use, even in the femtosecond range, i. e. a few mW/cm^2 as well as power densities such as 10^{21} W/cm^2 are achieved in special cases.

21.1 Historical Remarks

Due to the fact that the first laser, which had originally been named an optical maser by Th. Maiman in 1960, emitted a red beam at 694.3 nm in a pulsed mode [21.1] and the first He–Ne lasers emitted in the near-infrared [21.2] or at 632.8 nm, experience was limited to a few special laser types and their corresponding wavelengths in the beginning of the laser era.

Initially, only common-sense safety guidelines could be developed based on a qualitative understanding of the interactions. There was no quantitative understanding at that time.

A dramatic change was observed when ion lasers, such as the argon-ion laser, became commercially available in 1966 and output powers of a few watts became available in science, medicine and entertainment in the subsequent years.

The recommended corneal exposure levels, mainly given for the ruby-laser wavelength varied at that time between 6.8×10^{-4} mW/cm^2 to 1 mW/cm^2, i. e. a factor of about 1 500 [21.3]. Less-conservative laser safety factors were used as more biological data became available; this tendency toward an increase in the recommended safe exposure levels was quite different to that for other physical or chemical agents.

The increasing number of laser users, the spread of laser applications into new fields, and the lack of adequate measures for protection resulted in an increasing number of accidents in these early years. Damage to the eyes was noted relatively often, whereas injuries to the skin, such as burning or even charring, occurred very seldom, since powerful infrared lasers such as the Nd:YAG or CO_2 laser at 1.064 μm and 10.6 μm, respectively, were not as widespread at that time.

The increasing number of laser accidents prompted the establishment of a conference dealing with the topic of laser accidents and how to prevent them; laser safety has been regarded as an integral part of laser application development since then.

The fact that there has never been a single organization that had the competence or legislative power to establish an applicable system of safety procedures, rules and protective measures worldwide, to be respected by all laser users irrespective of their geographic location or national affiliation, resulted in many different systems of regulations and recommendations.

One of the first supranational recommendations was published in 1982 by the World Health Organization (WHO) as the *Environmental Health Criteria* [21.4].

Since optical radiation is normally defined in the wavelength range from 100 nm to 1 mm the same is true for laser radiation, i. e. a very large range of laser wavelengths has to be taken into account. In addition, lasers might be used as a continuous-wave emitter or in a pulsed mode with pulse durations from milliseconds to a few femtoseconds. Simultaneously there exists a huge range of power and/or energy densities that can be achieved with the various laser types. This fact clearly shows that it is not possible to allocate a laser to either a safe or to an unsafe group of products.

Therefore more than two so-called laser classes have been used from the very beginning in international, regional and national standards. Today laser products are categorized into as many as seven different laser classes in order to take account of the various parameters of emitted radiation and the associated hazard level.

In order to protect against adverse health effects of exposure to laser radiation, so-called exposure limits (ELs) or maximum permissible exposure values (MPEs) have been introduced, based on the best available scientific and experimental knowledge and have been adapted whenever new experience justified a change of the former values.

21.2 Biological Interactions and Effects

21.2.1 Fundamental Interactions

Laser radiation might hit every site on the surface of a human being. Due to their anatomical and optical characteristics, the eyes are most vulnerable to laser radiation from about 400 nm to about 1400 nm, i. e. covering the visible and the near-infrared regions of the optical spectrum, where optical radiation is transmitted from the cornea to the retina Fig. 21.1.

The eye is divided into two parts: the front or anterior chamber, which is bounded by the cornea, the iris, and the lens; and the back or posterior eye ball, which is bounded by the retina and contains the gel-like vitreous humour (vitreous body). If we look in the direction of the laser beam shown in Fig. 21.1 into the eye with an ophthalmoscope we will see an image of the ocular fundus. It looks reddish, but the prominent retinal vessels can be clearly seen. Other distinctive features are the whitish optic disc (blind spot), and the fovea. The fovea is a small depression in the retinal surface that may be more highly pigmented than the surrounding retina and is the area of most acute vision. The fovea is the center of the macula. The macula is responsible for detailed and color vision.

The retina consists of several layers of nerve cells (horizontal, bipolar, amacrine and ganglion cells) that cover the photosensitive rods and cones; i. e. light falling on the retinal surface has to pass through the different layers of cells before it reaches the photoreceptors. Underneath the layer of rods and cones is a layer called the retinal pigment epithelium (RPE), which contains a brownish black pigment called retinal melanin, and beneath this is a layer of fine blood vessels, the choriocapillaris. The final absorbing layer is the choroid, which contains both pigmented cells and blood vessels. The outside of the eye ball is called the sclera Fig. 21.1.

The eyes and the skin are the organs that are most susceptible to potential damage from laser radiation. The degree of interaction between laser radiation and biological tissue is determined mainly by its optical characteristics. Only the laser radiation that is transmitted through the respective surface and not reflected can interact with the tissue structures. Since the tissue consists of various substructures and components such as vessels, cells, membranes, and proteins, the inhomogeneous materials mainly demonstrate Mie scattering, since the components and areas are larger than the respective laser wavelength.

Some of the photons are scattered hundreds or even thousands of times per centimeter, but the most important fundamental interaction is absorption. Only in this case is the energy carried by the photons transformed into other kinds of energy, namely the excitation of

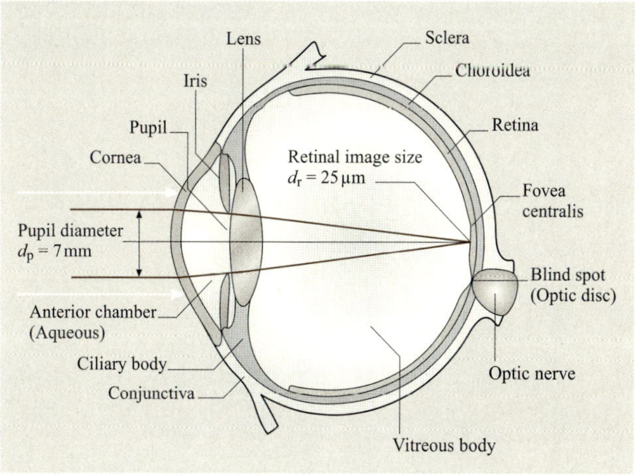

Fig. 21.1 Schematic drawing of the human eye with the standardized pupil diameter

atoms and the vibration of molecules. This increases the heat content of the respective tissue volume.

Normally the interaction of laser radiation with biological tissue can be described as a function of the irradiance, given in units of W/m^2 or radiant exposure in J/m^2.

Irradiance and radiant exposure correspond to the power and energy density and describe the power of a laser beam at a given position in space at a given time; the exposure duration is one of the most relevant quantities describing the extent of a biological effect.

Comparing the interaction of laser radiation and organic material with inorganic material, the main difference is given by the fact that specific processes may happen in a living tissue as a result of the time course of the interaction where destructive and repair processes have to be taken into account, i. e. the result of laser irradiation is not only dependent on the physical parameters of the laser beam but on biological capabilities and characteristics as well.

The reflection from a light-colored surface is normally much larger than from a dark-pigmented one, especially in the visible part of the optical spectrum. Between 400 nm and 1400 nm a lot of photons are backscattered from the biological tissue, i. e. generalized reflection coefficients as large as about 50% are possible. In the case that only absorption takes place

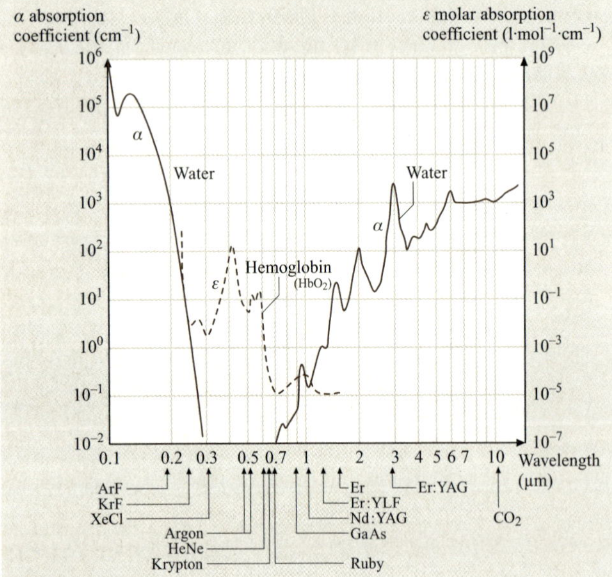

Fig. 21.2 Absorption of water and hemoglobin as a function of wavelength

the power density or irradiance $E(z)$ decreases exponentially according to the Lambert–Beer law

$$E(z) = E_0 \exp(-\alpha z) , \tag{21.1}$$

where α is the absorption coefficient of the respective medium given in cm^{-1}, E_0 is the irradiance just below the tissue surface (incident irradiance), and z denotes the optical axis (direction of the laser beam).

From (21.1) the depth of penetration z_0 can be derived as

$$z_0 = \frac{1}{\alpha} , \tag{21.2}$$

i. e. the laser beam penetrates a less-absorptive tissue more than a translucent or opaque material. In human tissue the absorption is determined by the components of the corresponding structures.

Absorption and Penetration Depth of Water

Since many parts of the human body contain a lot of water, the wavelength dependence of absorption in water, which is shown in Fig. 21.2, determines the absorption characteristics of tissue with a high water content.

In this graph the increase of absorption in the ultraviolet due to the excitation of the constituent atoms can be clearly seen. In addition the infrared spectrum is a fingerprint of the fundamental valence vibrations at about 6 μm and 3 μm, and of the respective overtones (second harmonics and above) at shorter wavelength. The decreased strength of absorption of the harmonics, which normally decreases by a factor of about 10 between orders of harmonics, can be seen in Fig. 21.2.

In addition to the absorption coefficient of water, that of hemoglobin is shown in Fig. 21.2. The absorption of the red chromophore of blood fills the gap in the water absorption to a certain extent, i. e. in human tissue both components normally have to be considered in calculations.

With respect to (21.2) the penetration depth in water, where the irradiance decreases to $1/e$ of the amount at the surface, might be calculated from Fig. 21.2 and is shown in Fig. 21.3.

Transmission and Absorption of the Human Eye

Since the human eye has a thickness of about 25 mm between the surface of the cornea and the retina, and since the ocular media, especially the anterior chamber and the vitreous body (cf. Fig. 21.1) might be considered from optical data to be like water, it can be seen from Fig. 21.3 that optical radiation between UV-A and about 1500 nm reaches the retina. Due to the fact that the cornea and the lens show some absorption in the ultraviolet the range of ocular transmission is given, as shown in Fig. 21.4, as 400–1400 nm.

In addition to the ocular transmission, Fig. 21.4 shows the wavelength-dependent absorption of the retinal pigment epithelium, which is anatomically located between the retina and the choroid. To estimate the most hazardous wavelength region both curves for transmission T and absorption A in Fig. 21.4 have to be multiplied to give the actual percentage of optical radiation that is absorbed in the most sensitive layer of the human eye [21.6]. The product TA may be thought of as the action spectrum for retinal injury.

Since the eyes are the doors to the brain, function as the gateway to the optical environment, and possess the capability to focus optical radiation onto a relatively small spot on the retina, which in correct vision is located in the fovea Fig. 21.1, the irradiance is increased by a factor of about 500,000. This optical gain is equivalent to the quotient of the area of the beam at the front of the eye to the small spot on the retina. Since the pupil limits the extension of the beam Fig. 21.1 the ratio is given by

$$\left(\frac{\text{largest diameter of the pupil}}{\text{smallest spot on the retina}}\right)^2 \approx \left(\frac{7\ \text{mm}}{10\ \mu\text{m}}\right)^2 \approx 500\,000 \qquad (21.3)$$

Due to the special characteristics of a laser beam, namely that it is:

- monochromatic,
- coherent (capable of interference), and
- directional (laser generally radiate in a narrow beam),

a 60 W laser with a 4 mm-diameter beam produces about 480 000 mW/cm^2 on a surface 1.0 m away, whereas a conventional 60 W light bulb (with a theoretical efficiency of 100%) produces only about 0.5 mW/cm^2 at the same distance. For comparison the irradiance of the sun is about 100 mW/cm^2.

Laser radiation is distinguished from most other known kinds of radiation by its beam collimation. Considering the capability of the human eye to focus the beam onto the retina, a laser beam with a power density of about 1 mW/cm^2 in the plane of the pupil or at the cornea might be transformed into an irradiance of up to 500 W/cm^2 due to the optical gain in irradiance from the cornea to the retina. This characteristic has to be taken into account in the wavelength region of 400–1400 nm.

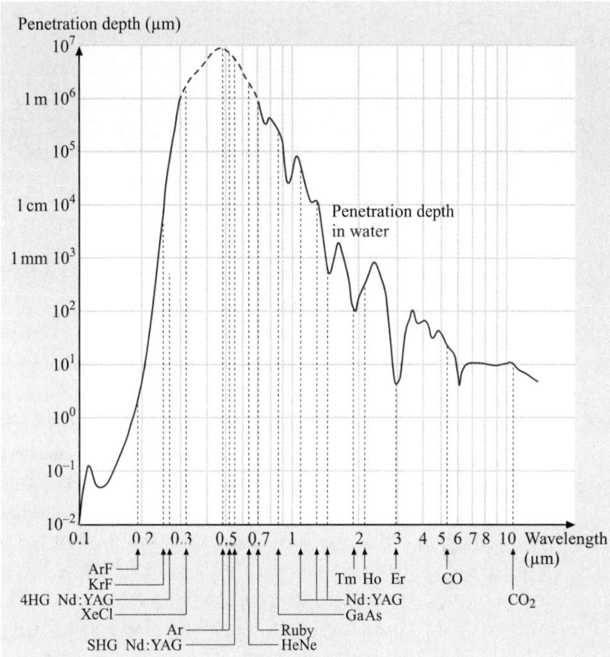

Fig. 21.3 Depth of penetration in water as function of wavelength (cf. some often used laser lines)

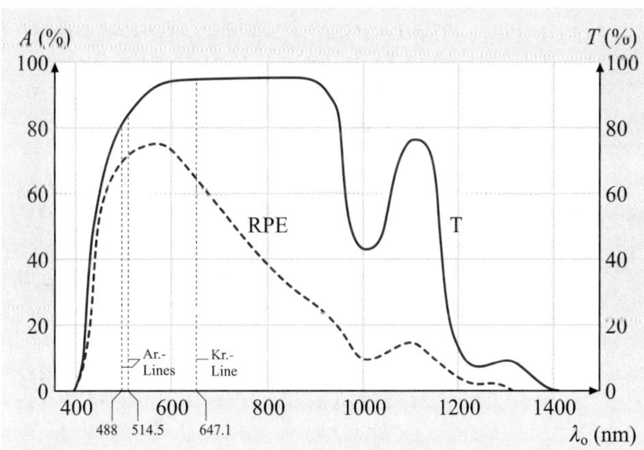

Fig. 21.4 Transmission T of the eye from the cornea to the retina and absorption A of the retinal pigment epithelium (RPE) as a function of wavelength λ_0 [21.5]

Due to the fact that the transmission is nonzero in this wavelength region, it is often called the *retinal hazard region*, but wavelengths outside this band should not be described as *eye-safe*, since other parts of the eye are vulnerable to shorter or longer wavelengths and adverse health effects may happen there.

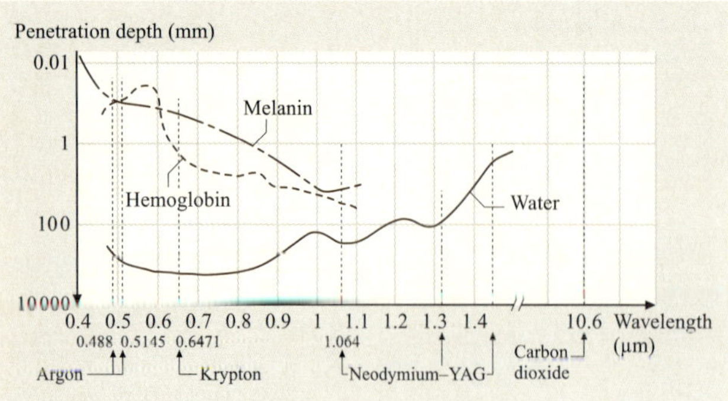

Fig. 21.5 Penetration depth of water, hemoglobin and melanin as a function of wavelength

The most likely effect of intercepting a laser beam with the eye is a thermal burn that destroys the retinal tissue. Since retinal tissue does not regenerate, the damage is permanent.

In addition to the absorption of water, other chromophores have important characteristics for eye and skin. Figure 21.5 compares the penetration depth of water, hemoglobin and melanin.

Photochemical, Photothermal and Photoionizing Effects

In general the irradiance or radiant exposure can result in a large spectrum of biological effects, which can be divided into photochemical, photothermal and photoionizing effects.

The irradiance or power density is the radiant power incident per unit area upon a surface expressed in watts per square meter (W/m^2) and the radiant exposure or energy density is the time integral of the irradiance, expressed in joules per square meter (J/m^2). In some cases the radiance is used to describe the respective bioeffect. The radiance is the radiant flux or power output per unit solid angle per unit area, expressed in watts per square meter per steradian ($Wm^{-2}sr^{-1}$).

The irradiance E is given as

$$E = \frac{\mathrm{d}P}{\mathrm{d}A} \text{ in W/m}^2 \tag{21.4}$$

and the radiant exposure H

$$H = \int_0^t E(t) \cdot \mathrm{d}t \text{ in J/m}^2 \tag{21.5}$$

where $\mathrm{d}P$ is the power expressed in watts [W], $\mathrm{d}A$ is the surface in square meters [m^2], and t is the time or duration of the exposure, expressed in seconds [s].

Values of $E(t)$, E and/or H come from measurements, or may be calculated from data provided by the manufacturer of the laser product.

Irradiances up to $1\,W/cm^2$ normally result in reversible processes such as photostimulation or photochemical reactions in the skin, especially for exposure durations longer than 1 s. However, there exist photochemical reactions at higher irradiances that may also result in permanent damage. Photochemical reactions generally follow the law of Bunsen and Roscoe, and for exposure durations of the order of 1–3 h or less (where repair mechanisms come into play), the threshold expressed as a radiant exposure is constant over a wide range of exposure durations.

At higher irradiances between about $10\,W/cm^2$ and $10^5\,W/cm^2$ in the millisecond regime up to a few seconds the biological tissue becomes heated and denaturation, i. e. coagulation of proteins, results in cell necrosis (cell death). If a temperature level of 100 °C is achieved, vaporization of cell substances such as water happens.

At even higher irradiances up to about $10^{10}\,W/cm^2$ and exposure durations as short as microseconds and somewhat shorter, as in the case of Q-switched pulses, so-called nonthermal effects are initiated. These result in the ablation of superficial layers of the tissue caused by mechanical disruption, where only a minimal temperature rise can be measured in the circumference of the irradiated area and the respective necrotic zones extend only some micrometers, i. e. are comparable to the cell dimensions.

In the case of very short exposure durations at power densities/irradiances up to $10^{12}\,W/cm^2$ so-called optical breakdown happens. The electrical field amplitudes might have values of 1–100 MV/cm, which is comparable to the atomic or intramolecular Coulomb field, and therefore a microplasma is created, resulting in an ionizing effect. The rapid expansion of this plasma results in a shockwave and has the capability to destroy even the hardest substances in the human body.

If a photoacoustic effect occurs in the eye a shockwave in the retinal tissue is induced that causes a rupture of the tissue. This damage is permanent, as with a retinal burn. Acoustic damage is actually more destructive than a thermal burn. Acoustic damage usually affects a greater area of the retina, and the threshold energy for this effect is substantially lower. Table 21.1 summarizes these various effects.

All these effects are used in various medical treatments for the benefit of patients. In this case the application of laser radiation is under medical super-

Table 21.1 Effects of laser radiation on biological tissue

Effect		
Photochemical	**Photothermal**	**Photoionizing**
Biostimulation	Hyperthermia	Ablation
	Coagulation	Disruption
	Carbonization	Fragmentation
	Vaporization	
$t = 10–1000\,s$	$t = 1\,ms–1\,s$	$t = 10\,ps–10\,\mu s$
$\leq$ 100 mW	1–100 W	
$\leq$ 50 mW/cm^2	1–100 W/cm^2	

vision and control, whereas involuntary exposure can result in a hazardous situation and/or damage of the respective tissue. This can be very dramatic if the eyes are involved and therefore maximum permissible exposure limits have been established, based on the best available knowledge of these effects as a function of irradiance, radiant exposure, exposure duration, wavelength, size of the irradiated area etc.

21.2.2 Effects of Laser Radiation on the Eye and Skin

Since laser radiation might be generated at nearly every wavelength in the range between 100 nm and 1 mm, various effects have to be considered as a result of acute or chronic exposure. Table 21.2 gives an overview of the various processes and effects that might predominantly happen in these wavelength regions in the eyes or skin, since these organs are the most susceptible to damage by laser radiation.

The anatomical and physiological characteristics of the eye result in the most dangerous wavelength region of 400–1400 nm, where transmission of laser radiation from the cornea to the retina occurs, as can be seen in Fig. 21.4.

In this case most of the optical energy is absorbed by the photopigments (chromophores) of the retinal pigment epithelium, which has a layer thickness of about 10 μm, whereas not more than about 15% of the incident

Table 21.2 Effects of laser radiation as a function of wavelength region in the eye and skin

Wavelength region (nm)	Eye	Skin
UV-C: 100 nm to 280 nm	Photokeratitis (corneal inflammation; cornea surface absorbs all) Photoconjunctivitis	Erythema (reddening of the skin) Pre-carcinoma
UV-B: 280 nm to 315 nm	Photokeratitis Photoconjunctivitis Cataractogenesis (cataract formation)	Increased pigmentation Enhanced skinphotoaging (accelerated skin aging) Erythema Oedema Pre-carcinoma Carcinoma (skin cancer)
UV-A: 315 nm to 400 nm	Cataract formation (lens is principal absorber)	Acute pigmentation Enhanced skinphotoaging Skin burns Carcinoma (skin cancer)
VIS: 400 nm to 700 nm	Photochemical and photothermal damage to the retina	Photosensitive reactions (photosensitization) Skin burns (reddening of skin, blisters, charring)
IR-A: 700 nm to 1400 nm	Cataract formation from heating proteins in the lens Photothermal retinal damage	Skin burns (reddening of skin, blisters, charring, damage to underlying organs)
IR-B: 1400 nm to 3000 nm	Cataract formation (corneal or lenticular opacities) Photothermal damage (burns) to the cornea	Oedema of the skin Skin burns
IR-C: 3000 nm to 1 mm	Photothermal damage of the cornea	Skin burns

laser radiation is absorbed in the chromophores located in the cones and/or rods. As a result thermal damage is initiated in the RPE, but extends to the receptor cells in the retina and may, in addition, destroy the neuroretina, which contains horizontal, bipolar and amacrine cells and finally projects onto the ganglion cells. Hazardous irradiation in the human eye can extend as far as the choroid.

Retinal Injuries

Photothermal damage might appear as a minimal lesion, where only a few receptor cells are destroyed, but at higher optical energy values much larger, permanent destruction may happen. A mild burn on the retina appears yellow or grey, and a white patch is produced at greater power. A lesion which is only visible as a slight discoloration is a threshold lesion. If a victim is looking directly into the beam (intrabeam viewing), he might receive a foveal burn, i. e. his center of vision is destroyed. If this region is damaged, the effect may appear initially as a blurred white spot obscuring the central area of vision; however, within two or more weeks, it may change to a black spot. Ultimately, the victim may cease to be aware of this blind spot (scotoma) during normal vision. However, it can be revealed immediately on looking at an empty visual scene such as a blank sheet of white paper. Figure 21.6 shows two examples of retinal injuries from a visible laser beam.

A burn in the peripheral retina may only cause a small scotoma, which may not be detected subjectively or even during a systematic eye examination.

Photoacoustic retinal damage may be associated with an audible pop at the time of exposure. Visual disorientation due to retinal damage may not be apparent to the victim until considerable thermal damage has occurred. Hole burning and bleeding might also happen with sufficiently high power levels, i. e. in addition to field defects, hemorrhage and debris projected into the vitreous humor can permanently impair vision.

This kind of injury will be clearly noticed by the victim, whereas smaller retinal lesions might not even be recognized and can barely be found in a routine ophthalmoscopic inspection.

Exposure to a visible laser beam can be detected by the bright color flash of the emitted wavelength and an afterimage of its complementary color, e.g., a green 532 nm laser light would normally produce a green flash followed by a red afterimage.

Since laser radiation at wavelengths above about 700 nm is invisible to the human eye, near-infrared radiation has to be regarded as particularly hazardous and might result in serious injuries. Exposure to the Q-switched Nd:YAG laser beam (1064 nm) is especially hazardous and may initially go undetected because the beam is invisible and the retina lacks pain sensory nerves. At these longer wavelengths photothermal effects still dominate for exposure durations in excess of 10 s, which is important when using near-infrared lasers in the CW mode.

The real extension of a retinal injury is very often not apparent directly after the laser irradiation, but several hours later. This is related to time-dependent processes that are incurred with thermal damage such that inflammation and repair processes have to be taken into account.

At lower irradiance values chronic injuries may result from laser radiation in the wavelength region of 400–600 nm, when the respective exposure duration is long enough, i. e. several seconds or more [21.7]. The explanation of such effects is given by a photochemically initiated process, which is not connected to a measurable temperature increase of the tissue. However a small temperature rise in the retina appears to be synergistic with the photochemical effect. Photoretinitis is one of these special radiation-induced effects, which may even result in blindness, i. e. due to the photochemical processes the respective photoreceptors are no longer able to fulfill their normal functions.

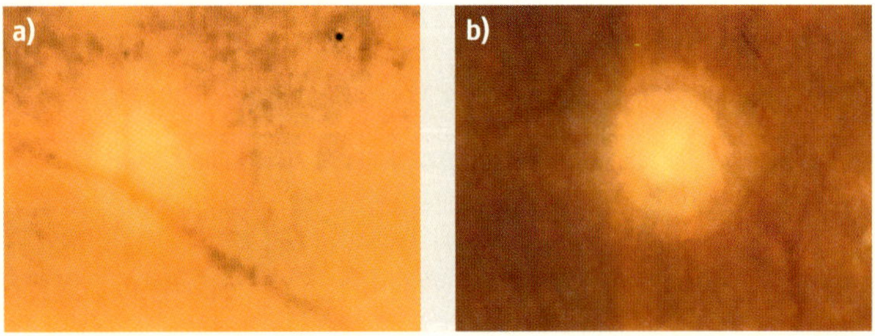

Fig. 21.6a,b Photothermal retinal lesions from a visible laser beam; **(a)** small lesion of about 150 μm, **(b)** lesion diameter about 500 μm (different magnifications)

Damaging effects can also be the direct result of specific molecular absorption of a given light. Rather than releasing the energy, however, the species undergoes a chemical reaction that is unique to its excited state. This is what is known as the photochemical effect.

Non-retinal Injuries

In addition to retinal injuries, hazards are also possible in the ultraviolet and the mid- and far-infrared spectrum to the anterior parts of the eye, i. e. the cornea, conjunctiva and the lens.

Similar to the photochemically induced, so-called athermal retinal injuries, photokeratitis and photoconjunctivitis, which are inflammations of the cornea and the conjunctiva, respectively, are both wavelength dependent and can be described by their respective action spectra. This is a characteristic of all photochemical effects, which show a relatively strong wavelength dependence and a maximum sensitivity.

In the UV-A part of the spectrum, the intensities or exposure durations required to produce photokeratitis or erythema are much greater than for so-called actinic UV radiation (UV-B and UV-C). The cornea and the skin are nearly as sensitive in the actinic UV region, but corneal damage is more painful.

In addition to these photochemical effects, lenticular cataracts may be formed by ultraviolet [21.8] and infrared radiation.

Exposure to the invisible carbon-dioxide laser beam (10 600 nm) can be detected by a burning pain at the site of exposure on the cornea or conjunctiva. A corneal burn may be manifested as a surface irregularity or a white opacity Fig. 21.8. If the lesion is minor, i. e. restricted to the outer cellular layer, then it will heal and disappear within about 48 hours. Deeper burns may cause permanent damage.

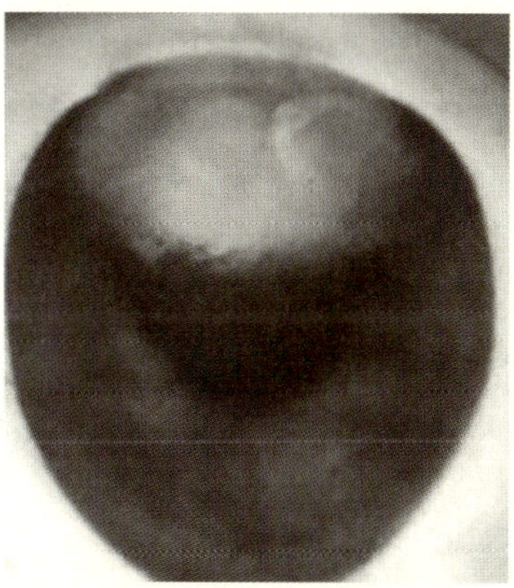

Fig. 21.7 Damage of the eye with a Nd:YAG laser (deep coagulation of the anterior parts; illumination with and scattering of an aimed He–Ne laser beam)

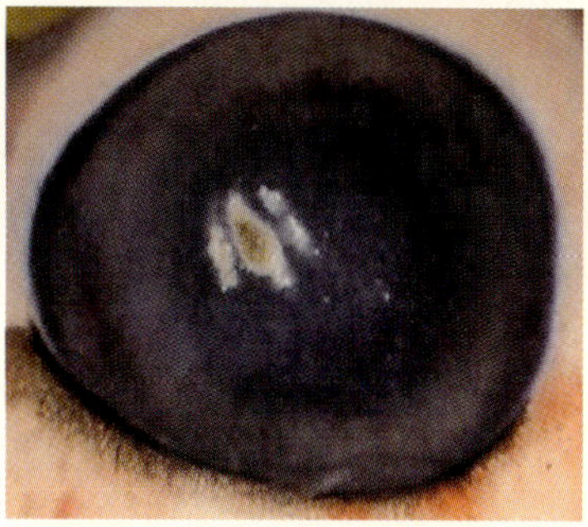

Fig. 21.8 Damage of the eye with a CO_2 laser with the mode pattern visible on the cornea

Figure 21.7 shows an example of a 1 s exposure with a 100 W Nd:YAG laser (1.064 μm) of an animal eye in a laboratory trial, where it can be seen that even the anterior parts of the eye are destroyed by coagulation. Retinal damage was not the goal of this special investigation, but has certainly happened due to the high transmission at this wavelength and the focusing capabilities of the eye. In Fig. 21.8 the damaging capability of a short exposure (about 0.5 s) of a 60 W CO_2 laser of the cornea is illustrated, where even the mode pattern of the laser beam can be seen.

Injuries to the Skin

In contrast to eye injuries any damage to the skin is of minor importance as far as skin burns are concerned, although even third-degree burns might not be completely excluded at higher laser power levels. Injury to the skin normally results from temperatures exceeding 45–60 °C, but this value is also time dependent.

Special care has to be taken if ultraviolet radiation is present, since erythema (skin reddening resulting from vasodilatation), oedema (swelling) and even carcinoma might result from acute (short-term) or chronic (long-term) exposure, respectively.

Ultraviolet radiation may not only arise from UV lasers, such as e.g. excimer lasers, but from so-called secondary optical radiation, which comes from various material processing applications like cutting and welding with high-power Nd:YAG or CO_2 lasers. Concerning its biological effect, it is not important whether it is a coherent or incoherent radiation which interacts with biological tissue, but this characteristic clearly determines the smallest spot on the retina.

Exposure of the skin to high-power laser beams (1 W or more) can cause burns. At the sub-five-watt level, the heat from the laser beam will cause a flinch reaction before any serious damage occurs. The sensation is similar to touching any hot object, you tend to pull your hand away or drop it before any major damage occurs. With

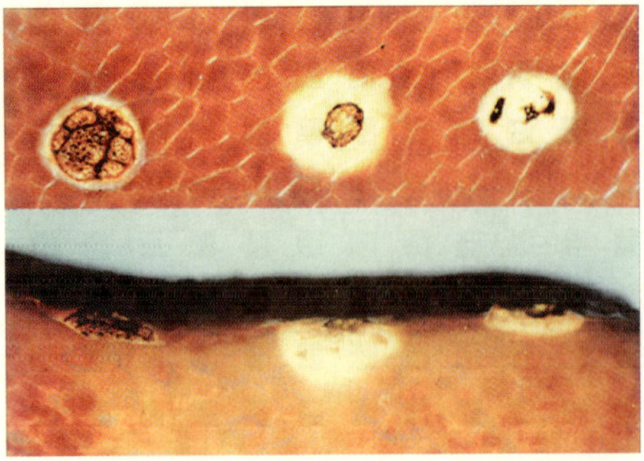

Fig. 21.9 Surface reactions (*top*) and penetration depth (*below*) of various laser beams (*left* to *right*: CO_2, Nd:YAG, and argon-ion laser), beam diameter: about 4 mm, tissue: liver

higher-power lasers, a burn can occur even though the flinch reaction may rapidly pull the affected skin out of the beam. These burns can be quite painful as the affected skin can be cooked, and forms a hard lesion that takes considerable time to heal.

Figure 21.9 shows examples of tissue reactions and penetration depth achieved with an argon-ion laser, a CO_2 and a Nd:YAG laser. As can clearly be seen the penetration depth is largest for near-infrared wavelengths, as might be explained from Figs. 21.4 and 21.5, whereas due to the high absorption in water or hemoglobin the penetration depth is low with the CO_2 (10.6 μm) and argon-ion (450–515 nm) lasers.

In addition to incoherent optical radiation other secondary effects have to be taken into account when lasers are applied. To name but a few, the respective hazards might arise from electrical currents and high voltages, electromagnetic high-frequency fields, optical pump radiation, X-rays, explosive atmospheres, inflammable materials, toxic and infectious or even carcinogenic liquids or solid materials.

The various so-called secondary physical and chemical agents associated with laser products and applications must be regarded in the risk assessment or analysis and shall be restricted to a given limit of exposure or dose.

21.3 Maximum Permissible Exposure

21.3.1 Threshold values and ED-50

In order to protect human beings from any hazardous laser radiation, maximum permissible exposure (MPE) values have been established. The MPEs represent that level of laser radiation to which, under normal circumstances, persons may be exposed without suffering adverse effects. The MPE levels represent the maximum level to which the eye or skin can be exposed without consequential injury immediately or after a long time and are related to the wavelength of the radiation, the pulse duration or exposure duration, the tissue at risk and, for visible and near-infrared radiation at 400–1400 nm, the size of the retinal image.

Whereas these MPEs were very conservative at the beginning many values were increased in later years due to improved investigation methods and more-sophisticated models taking into account the various effects as a function of wavelength and exposure duration.

Since in many cases there is no distinct threshold value for humans the MPEs have been derived from experimental results on different animal species. Therefore a certain degree of uncertainty exists as far as extrapolations from animal experiments to effects in humans have to be made. However all currently available scientific knowledge indicates that the respective exposure limits (ELs) provide a safe level of protection against known biological laser radiation effects.

By agreement, the MPEs have mainly been established at 1/10 of the so-called ED-50 value (effective dose 50%), which is the value of irradiance or radiant exposure where an ophthalmoscopically visible lesion exists in 50% of the irradiated cases. The ED-50 is also often referred as the *threshold*. Since the slope of the probit curve is relatively steep it might be expected that at 10% of the ED-50 no harmful effects can be found and therefore the MPE values include a safety or reduction factor.

Not only individual susceptibility but biological and biophysical variables affect the data and therefore it is not possible to define a single safety or reduction factor between the threshold of an injury (observed effect) and the exposure limit.

As has been shown in different laboratory studies microscopically visible injuries usually appeared at between 25% and 50% of ED-50, but never at values less than 10% [21.9]. A safety or reduction factor of about 2 in the UV and 5–10 in the visible part of the spectrum is used for most corneal and retinal effects [21.9].

MPE values should be used as guides in the control of exposures, and should not be regarded as precisely defined dividing lines between safe and dangerous levels. In any case, exposure to laser radiation should be as low as possible.

The extent of injury varies considerably with wavelength and therefore it is necessary to adjust the exposure limits and the MPEs for different wavelengths, but in a simple manner, i. e. by correction factors given as straight lines (potential functions or constants) in a logarithmic illustration. For example, this factor increases from 1 to 5 as the wavelength increases from 700 nm to 1050 nm. As a consequence of the increased absorption of laser radiation in the ocular media (Fig. 21.4) another factor has been introduced in the MPEs, which increases from 1 to 8 in the wavelength region between 1150 nm and 1200 nm.

At exposure durations exceeding about 10 s the difference between the ocular exposure limits for shorter wavelengths, i. e. below about 500 nm, and longer ones, increases with greater exposure durations since short-wavelength visible laser radiation causes predominantly photochemical retinal injuries, i. e. the MPEs are much lower for longer exposure durations at shorter wavelengths.

Since exposure limits are based on experimental results and on understanding of the possible photochemical, photothermal, photomechanical and photoionizing interactions with biological tissues that should not cause adverse effects, it can easily be understood that there was and still is a lack of accurate data for some wavelengths, exposure durations, beam parameters and so on. Therefore worst-case exposure conditions have been assumed.

In addition, new and improved data have shown that some of the earlier exposure limits had over-conservative safety factors and revised exposure limits have been recommended.

One could certainly follow the contours of the experimentally achieved bioeffect threshold curves, as far as they exist, resulting in more gradual exposure limit curves, but up to now step functions and simple mathematical functions have been used for the MPE values.

MPE values were derived to preclude not only acute effects but delayed adverse effects too. Since to date only limited data are available for chronic, long-term, exposures, delayed effects might not be completely excluded. Until now no MPE values have been given for extremely long exposures, i. e. more than eight hours. On the other hand intrabeam exposure situations seldom last more than 100 s. However, situations where long-term scattered or diffusely reflected laser radiation is present should be included in a risk analysis, especially in the case of occupational exposure.

21.3.2 MPE Values for the Eye

MPE values are given for the cornea and for the skin. The values for the cornea have been derived for worst-case situations, i. e. a pupil diameter of 7 mm and a stable image of the respective laser beam on the retina. Since these conditions might happen very seldom, e.g. with an anesthetized eye, the MPE values for the eye implicitly include an additional safety factor.

The pupillary aperture averages the radiant exposure and the irradiance. For many years a 7 mm aperture, corresponding to the dark-adapted, maximal, dilated pupil, has been standard. This is equivalent to a worst-case situation. The risk does not decrease proportionally with decreasing pupillary area when a point source is viewed [21.10]. In the visible and near-infrared region, since laser radiation is focused on the retina, it does not matter whether the laser beam is 1 mm or 2 mm in diameter when it enters the pupil. Ultimately all the optical power/energy is concentrated into approximately the same spot on the retina.

Based on diffraction theory a retinal image diameter of about 3–6 μm would be expected, but the human eye is not such a good optical device and therefore a minimal image size of about 25 μm has been chosen for safety considerations, although somewhat smaller spots have been reported.

Point Sources and Eye Movements

In the case of CW exposure, physiological factors have to be taken into account whenever exposure situations are analyzed. Pupillary activity (pupillary reflex), eye movements (saccades, micro-saccades, drift, tremor), breathing, heart beat, blood flow and other bodily movements are the physiological factors that influence the exposure duration, being greatest for point sources and least for large, extended sources. On the other hand the person's visual task is also governed by visual fixations, i. e. there are periods when the retina fixes an object only on the fovea. The fixation time is normally about 200–400 ms, but depending on the level of processing or cognition, longer times might not be excluded. The eye movement studies of *Ness* et al. [21.10] clearly showed that the retinal image can remain remarkably still for a few seconds. It turned out in investigations that correction factors for eye movements were only important for viewing conditions longer than 10 s [21.9].

It has been shown that the expected image diameter of about 25 μm, which is equivalent to an angular subtense (visual angle) α of 1.5 mrad for an object of 150 μm at a distance of 100 mm, which is taken as a very conservative minimal distance of accommodation, and is given by the focal length of 17 mm, is blurred due to eye movements to a diameter of about 50 μm within an exposure duration of 250 ms for unrestrained head viewing, to 75 μm for 10 s, and that it is rare to achieve an illuminated zone as small as 135 μm for 100 s [21.10].

As is stated in a revision of the laser guidelines given by the International Commission on Non-ionizing Radiation Protection (ICNIRP) [21.11] the earlier recommendations of the exposure limits have not been changed as far as exposure durations between 1 ns and 10 s are regarded, but due to the fact that the image is blurred by physiological reactions like various eye and body movements, the safety factor is further increased. On the other hand the MPEs for longer exposure duration, i. e. longer than 10 s, have been increased, especially from 2001.

The accumulated radiant exposure in the retinal image area directly predicts the extent and position of photochemical retinal injury such as photoretinitis, i. e. fixation and eye movement studies have permitted the determination of the so-called *blue-light hazard* exposure limits. It is important to note that thermal retinal injury thresholds decrease for increasing retinal image or spot size due to radial heat flow during and after exposure.

Due to the influence of eye movements, the photothermal exposure limit is given as a constant irradiance of 1 mW/cm^2 for point sources (i. e. $\alpha \leq 1.5$ mrad) in the visible part of the spectrum for exposure durations longer than 10 s. For longer exposure durations the photochemical exposure limit becomes the important part, especially below 500 nm.

It has been shown by modeling the exposure situation for longer exposure durations, taking into account experimentally achieved data relating to eye movements, that thermal damage of the retina is not possible with a laser beam power of 0.4 mW at the 632.8 nm wavelength of a He–Ne laser, which is equivalent to an irradiance of 1 mW/cm^2 through a 7 mm pupil [21.12].

With the same model and a higher absorption in the green–blue part of the spectrum (cf. Fig. 21.4) it can be shown that a beam power of 5 mW can be sufficient to increase the retinal temperature by about 18 K, and can therefore result in protein denaturation, i. e. coagulation of the area on the retina. It should be taken into account that most of the thermal denaturation of proteins (coagulation) occurs after the pulse, which means that the geometry of the retinal exposure influences the lesion size. Radial heat flow is responsible for a relatively strong dependence of retinal injury on retinal image or spot size [21.13].

In addition long-term irradiation with wavelengths shorter than about 500 nm can result in predominant photochemical damage of the retina. This is the main reason that exposure limits are lower for shorter wavelengths for longer exposures.

Extended Sources

In order to take into account that there is a difference concerning the image size and the dynamics of the thermal processes, an additional factor has been included in the MPE values. This is important for extended optical sources. Such cases might happen when diffuse surface reflections, transmission through a diffusing translucent material or e.g. laser-diode arrays have to be treated in safety considerations. Due to the optical imaging relationships for extended sources in the 400–1400 nm wavelength region the correction factor increases from 1 for point sources and collimated beams up to 100/1.5 for a maximum angular subtense of 100 mrad, equivalent to a retinal image diameter of 1.7 mm. If the extension of the respective source is larger than 10 mm no further increase of the MPE values occurs.

For collimated beams (focus or "focal point" at infinity) and point sources the angular subtense, i. e. the visual angle subtended by a circular source at a distance of e.g. 100 mm at the eye is denoted by α. If the value of α is greater than $\alpha_{\text{min}} = 1.5$ mrad, so-called extended-source conditions apply. The angle α should not be confused with the beam divergence, which is a characteristic of the laser. The beam divergence is the far-field plane angle of the cone defined by the beam diameter, and the beam diameter at a point in space is the diameter of the smallest circle that contains a certain amount of the total laser power (or energy). For the purpose of standards dealing with laser safety the percentage is taken to be 63%. In the case of a Gaussian beam, 63% corresponds to a diameter where the irradiance (radiant exposure) falls to 1/e of its central peak value. It should be borne in mind that the beam diameter is very often taken in textbooks at the point where the intensity, i. e. the power density, falls to 1/e^2. In this case the amount of power/energy passing through an aperture with such a diameter would be about 86%. The relationship between both diameters is simply a factor of $\sqrt{2}$ in the case of a Gaussian beam (TEM_{00} or fundamental mode), i. e. the diameter corresponding to a decrease to $1/e^2$ of the maximum value is larger by this factor compared to the diameter used in safety standards.

The consideration of image size in the exposure limits is done by applying another correction factor, which is related to the angular subtense. This simplifies the MPE values, since the point-source exposure limits simply have to be multiplied by this factor to yield the linearly increased extended-source exposure limits, i. e. limits applicable to extended sources are based on the intrabeam viewing exposure limits modified by a correction factor α/α_{min}.

For biological effects, the irradiance or radiant exposure on the retina is important. These two quantities are directly proportional to the radiance or integrated radiance of the optical source, which are inherent characteristics of the source. This means that the energy entering the eye increases as the square of the diameter of the retinal image when an extended source is viewed [21.5]. Because of heat dissipation by the choroidal blood flow this image-size dependence fails above about 100 mrad, which is equivalent to an image size of 1.7 mm.

For visual angles subtending the source that exceed 100 mrad the exposure limits increase as the square of the angle or can be expressed as a constant radiance.

Based on many years of experience in laser safety it is important to recognize that almost all hazardous ocular exposure conditions occur for intrabeam point-source or collimated-beam conditions. Extended sources very seldom pose a realistic hazard, but nevertheless should be treated with care, in order not to exceed the respective MPEs.

21.3.3 MPEs given as Radiant Exposure and Irradiance

As an example the MPE values for the radiant exposure H_{MPE} and the irradiance E_{MPE} are given for the wavelength region 400–700 nm as [21.11]

$$H_{MPE} = 18t^{0.75} C_E \text{ in J/m}^2 \tag{21.6}$$

and

$$E_{MPE} = \frac{H_{MPE}}{t} = 18t^{-0.25} C_E \text{ in W/m}^2 , \tag{21.7}$$

where $C_E = \alpha/\alpha_{min}$ for $\alpha_{min} = 1.5\,\text{mrad} \le \alpha \le \alpha_{max} = 100\,\text{mrad}$ and t is the exposure duration.

Because the thermal injury threshold, expressed as the radiant power entering the eye, decreases as the exposure duration raised to the -0.25 power, moderate increases in the exposed retinal area will compensate for the increased risk for longer viewing duration. A reduction of e.g. only 44% results in a tenfold increase in the exposure duration. On the other hand doubling the optical power and therefore the irradiance requires that the exposure duration is decreased by a factor of 16, i. e. only 6.25% of the former value are allowed, in order to stay below the MPE-value.

From (21.6) and (21.7) the maximum optical power P_{MPE} can be calculated as

$$P_{MPE} = E_{MPE} A = E_{MPE} \frac{\pi}{4} d_p^2 \approx 0.7t^{-0.25} \text{ in mW} , \tag{21.8}$$

where the diameter d_p of the pupil has to be taken as 7 mm according to IEC 60825-1 [21.14] and t is measured in seconds.

Table 21.3 shows some examples of maximum power and the corresponding maximum exposure duration $t \approx 0.23/P^4$ derived from (21.8).

A pupil size of 7 mm is certainly unrealistically large for longer exposure durations since the pupillary reflex, which decreases the pupil diameter with a velocity of some millimeters per second and starts after a latency period of 0.3–1 s, limits the incident optical power in the case of beams with a somewhat larger diameter. In addition, it is conservatively assumed that the retinal thermal injury threshold varies inversely with the retinal spot size.

For photochemically induced retinal injury such a relationship does not exist, but the threshold values are wavelength- and dose (i. e. radiant exposure)-dependent, i. e. the photochemical thresholds expressed as irradiances decrease inversely with lengthening exposure duration.

Outside the retinal hazard region, i. e. below 400 nm and above 1400 nm, the source size is not important

Table 21.3 Maximum power P_{MPE} and exposure duration t (examples given for point sources and collimated beams, i. e. $C_E = 1$)

P_{MPE}/mW	0.39	0.5	0.6	0.7	1.0	2.0	4.0	5.0
t/ms	10 000	3 700	1 800	1 000	250	15	1	0.37

Table 21.4 Laser classes according to IEC 60825-1 and additional advice for safety

Class 1
The accessible laser radiation is non-hazardous under all reasonable foreseeable conditions. These conditions are kept during the intended use. Class 1 lasers do not emit harmful levels of radiation. Although neither photochemical nor photothermal damage is possible it might not be excluded that glare, dazzling and reduction in color vision due to afterimages might happen, especially when the emitted power is close to the upper limit of this laser class.
Class 1M
The accessible laser radiation is restricted to the wavelength region between 302.5 nm and 4000 nm, since this is the preferred range, where optical instruments are used. The accessible laser radiation does not pose any hazard for the eyes as far as the beam diameter is not decreased by optical instruments like eye-loupes, lenses, or telescopes. As long as no optical instrument that decreases the beam diameter is used the hazard is similar to the one in class 1. However, the use of optically collecting instruments might result in hazards comparable to those for classes 3R or 3B. The optical power is limited to class 3B.
Class 2
The accessible laser radiation is restricted to the visible part of the optical spectrum, i. e. 400 nm to 700 nm. The laser radiation is non-hazardous for short-term exposure (up to 0.25 s), even for the eyes. Additional radiation outside the visible spectrum has to fulfill the conditions of class 1. Since the eyes are not in danger as far as accidental and short-term exposure is regarded, laser products of class 2 may be applied without any additional protective measures if it is ensured that neither a deliberate intrabeam viewing of more than 0.25 s nor a repeated intrabeam viewing into a specular reflected laser beam could happen. However, it should be taken into account that neither the blink reflex nor aversion responses limit the exposure duration to 0.25 s, as is stated in IEC 60825-1, since these physiological reactions do not occur to a high enough degree. It has been shown in a recent study that the frequency of the blink reflex is less than about 20% and aversion responses occur even less frequent [21.15–18]. These findings do not state that class 2 laser are no longer safe, but that users of such lasers should be instructed to perform active protective reactions, i. e. close the eyes actively and avert the head in the case of intrabeam viewing. These measures can increase the safety of laser class 2 sufficiently and prevent a violation of the MPE values. It is believed that it is possible to stare long enough into a class 2 laser to cause damage to the eye. In the case of continuous-wave (CW) operation, i. e. for emission durations longer than 0.25 s, the optical power is limited to 1 mW, as far as collimated beams or point sources are considered. For extended sources the power might be increased by a correction factor depending on the angular subtense up to a maximum of about 66.7 mW.
Class 2M
The accessible laser radiation is restricted to the visible optical spectrum (400–700 nm) and does not pose any hazard to the eyes as long as the beam diameter is not decreased by optical instruments such as eye-loupes, lenses, or telescopes. Additional radiation outside the visible spectrum has to fulfill the conditions of class 1M. As long as no optical instruments that decrease the beam diameter are used the hazard is similar to that in class 2. Therefore the same recommendations are valid for this class. In addition the use of optical collecting instruments might result in hazards comparable to those for classes 3R or 3B. The optical power is limited to class 3B.

Table 21.4 (cont.)

Class 3R
The accessible laser radiation is between 302.5 nm and 10^6 nm (1 mm) and might be hazardous for the eyes. Laser products belonging to this class are potentially dangerous, but the risk of eye damage is less than for class 3B due to the fact that the AELs are restricted to a much lower level. The optical power or energy is a maximum of five times the accessible emission limit (AEL) of class 2 between 400 nm and 700 nm and five times the AEL of class 1 for all other wavelengths. In the case of CW operation the optical power is limited to 5 mW for collimated beams and point sources between 400 nm and 700 nm. Due to the fact that the maximum power is not large compared to class 3B minor protective measures and safety requirements have to be fulfilled. However, anyhow laser belonging to this class should be treated with care since a fivefold exceeding of the MPE could result in minimal but nevertheless persistent defects on the retina. This is especially important to consider since aversion responses including the blink reflex might not limit the exposure duration as required in the visible part of the optical spectrum.
Class 3A
In the former edition of the international standard IEC 60825-1 this class was used for laser products that did not pose a hazard either for short-term exposure up to 0.25 s in the visible part of the spectrum or for long-term exposure at all other wavelengths, as long as no optical instruments capable of reducing the beam diameter were used. This laser class is not hazardous when viewed momentarily with the naked eye, but poses severe eye hazards when viewed through optical instruments (e.g., microscopes and binoculars). This special class has a double limit; first a power ceiling of 5 mW in the visible part of the optical spectrum and simultaneously a 25 W/m^2 limit, to restrict the power entering a 7 mm pupil to 1 mW. This particular class has resulted in quite a lot of trouble, mainly due to the fact that some standards never incorporated these double limits. In addition instead of class 3A these laser products are labeled class IIIA, IIIa or 3a, which resulted in many mistakes. Moreover, due to national regulations many imported laser products have been classified incorrectly, e.g. a class IIIA laser product according to the ANSI standard has to be classified as a 3B product wherever the EN 60825-1 standard is applied and the manufacturer requirements according to this standard have to be fulfilled. In addition the user has to be familiar with the different demands concerning class 3B instead of class IIIA, which should not be confused with class 3A. In some cases there was a difference in labeling for class IIIA, i. e. if the power density did not exceed 2.5 mW/cm^2 a *caution* label was used and if $E > 2.5$ mW/cm^2 a *danger* label was used. The misunderstanding arising from this situation has been overcome with the introduction of the new class 3R, where reduced requirements have to be fulfilled, but a certain risk, although not very high, does exist, especially for worst-case conditions, such as the adjustment and alignment of the beam in intrabeam viewing and accommodation of the eyes. Voluntarily staring into the beam would increase the risk of retinal eye damage; therefore care has to be taken when the MPE values are exceeded.
Class 3B
The accessible laser radiation is hazardous for the eyes and often even for the skin. Intrabeam viewing is especially dangerous. The classification of laser products into this class is based on the conditions that a reflection of a beam from a perfect diffuse reflector might be safely viewed from a distance of 13 cm for a 10 s period. However, diffuse reflection seldom exists and instead partially direct reflections should be considered in a risk assessment, especially when the maximum power of 500 mW in this laser class is used. Even active aversion responses including active closure of the eyes are not sufficient to protect against laser radiation from such a laser. In addition to the fact that the MPE values of the eye are exceeded the same is true in general for the skin too.

Table 21.4 (cont.)

Class 4
The accessible laser radiation is very dangerous for the eyes and dangerous for the skin. Even diffuse scattered laser radiation might be hazardous.
In addition this laser radiation can result in fire and explosion.
Lasers belonging to this class are regarded as high power lasers, where the power or energy exceeds the AELs of class 3B.
The laser radiation of class 4 laser products is so intense that damage can be expected for all kinds of exposure of the eyes or the skin.
Special care has to be taken concerning fire and explosion hazards.

and therefore the exposure limits are given without a correction factor based on the angular subtense.

In addition to alterations in the exposure limits for exposure duration above 10 s, exposure limits for pulse durations between 1 ns and 100 fs are recommended in the revised exposure limit guidelines [21.9]. The main reasons for this are that the corresponding data have not appeared to be consistent until recently and that the underlying damage mechanisms for sub-nanosecond laser-induced injury have not been well understood.

The development of exposure limits for very short laser pulses has been very difficult because of the different mechanisms involved. It was found that the damage produced by nonlinear effects does not follow the same relationship with wavelength, pulse duration, and retinal image size as thresholds from thermochemical (photothermal) and thermoacoustic injury or damage mechanisms [21.19, 20]. In addition self-focusing of the ocular media further concentrates incident laser energy and increases the retinal radiant exposure.

For many years the ICNIRP has published guidelines on limits of exposure to laser radiation [21.9, 11], which have become part of various laser-safety standards. In the ICNIRP guidelines recommendations are given that have been accepted in various standards, concerning the treatment of single-pulse laser exposure and exposure of the eye to any continuous-wave laser [21.11].

Aversion Responses Including the Blink Reflex; Active Protective Reactions

For exposure of the eye to a CW laser an aversion response time of 0.25 s is recommended as an adequate hazard criterion, if purposeful staring into a visible laser beam is neither intended nor anticipated. A maximum exposure duration of 10 s is recommended for hazard analysis for either unintended or purposeful viewing conditions for ocular exposures of the eyes in the near-infrared. These two recommendations are based on expected natural behavior.

There was a strong belief in aversion responses and especially in the blink reflex as a reliable physiological reaction if a bright light is viewed. A laser source certainly represents a very bright light. For example a 1 mW laser has a luminance (brightness) that exceeds even the luminance of the sun at noon. However, as has been shown in recent investigations, a class 2 laser is not sufficient to stimulate more than about 20% of people to perform a blink reflex [21.21, 22]. Aversion responses such as head and body movements are even less frequent and therefore active protective measures such as those given in the adapted recommendations for class 2 and 2M (Table 21.4) are strongly recommended to prevent any hazard that might result from omission of the expected natural physiological behavior and prolonged exposure.

Non-Visible Laser Radiation and Repetitively Pulsed Lasers

A duration of 10 s is regarded as a sufficient for near-infrared exposure situations, since in this case eye movements will provide a natural physiological exposure limitation and thus eliminate the need to consider exposure durations greater than 10 s. Unusual situations, where even longer exposure durations have to be taken into account, should be treated in a special hazard analysis.

In the case of repetitive laser exposures UV and non-UV lasers are treated differently. For UV lasers the exposure dose is additive regardless of the repetition rate, whereas for visible and infrared laser the exposure limit per pulse for repetitively pulsed intrabeam viewing must be reduced by a correction factor of $N^{-0.25}$, where N is the number of pulses, i. e. the product of the pulse repetition frequency (PRF) and the total exposure duration (e.g. 0.25 s or 10 s).

The thermal repetitive-pulse exposure obeys what is referred to as the additivity rule where the net effect is modified by the described correction factor [21.11]. This reduces the exposure limit for a single pulse in a train of pulses. However, if multiple pulses exist within a shorter duration than a wavelength-dependent minimal time, e.g. 18 μs for 400–1050 nm, the pulses are treated like a single pulse.

Although there is some lack of knowledge in special exposure situations, current knowledge of bioeffects from the last 40 years seems to be sufficient to establish guidelines for exposure limits. There is no special need for separate MPEs for the general population compared to occupational exposure as is the case for electromagnetic fields with frequencies below 300 GHz. Exposure limit values or maximum exposure limits are limits on exposure to laser radiation that are based directly on established health effects and biological considerations. Compliance with these limits will ensure that people exposed to artificial sources of laser radiation are protected against all known adverse health effects.

Exposure levels below the MPE should not result in adverse health effects, since the exposure limits incorporate the knowledge of scientific research and practical experience, based upon the best available information.

21.4 International Standards and Regulations

In the USA, the manufacturer of a laser product has to fulfill the legal requirements given in the Food and Drug Administration/Center for Devices and Radiological Health (FDA/CDRH) regulation 21CFR1040.10 [21.23], i. e. by the federal government, while the establishment of standards for laser users is left to the individual states.

The federal government of the USA has established requirements covering the manufacture in or import into the USA of any product containing lasers. These regulations provide for product classification and require certain product features and labels in accordance with the product class. The categories utilized are classes I, IIa, II, IIIa, IIIb, and IV, corresponding to increasing levels of hazard.

Standards exist on an international, regional and national basis. As far as laser safety is regarded the International Electrotechnical Commission (IEC) published a first standard in 1984. This standard IEC 825 [21.24] has been updated or amended several times and has been valid in its current edition since 2001 as IEC 60825-1 [21.14]. A second edition of this international laser safety standard will be published in 2007.

Since many national standard organizations are members of the IEC this standard is accepted as a horizontal product safety standard by many countries around the world. Due to various agreements this international standard is transferred into regional ones, such as the European standard EN 60825-1, which is published by the Commission Européenne Normalisation Electrotechnique (CENELEC) for its member states. All national standards based on regional ones such as the European Normalization (EN) standards must only be translated into the language of the respective country, i. e. alterations or modifications are not allowed. This results in national editions of the fundamental laser safety standard such as DIN EN 60825-1 in Germany, BS EN 60825-1 in the UK, and AF EN 60825-1 in France.

One important disparity between the CDRH requirements and the IEC standard reflects the fact that the IEC is not an enforcement body but only publishes standards written by expert committees. Implementation and enforcement depend on the laws and regulations of the individual nations.

The European standard EN 60825-1, which is equivalent today to IEC 60825-1, applies throughout the European Union (EU) and the European Free Trade Association (EFTA) countries and became increasingly important as it is referred to as a harmonized standard in the official journal of the EU under the *Low-Voltage Directive*.

However, in addition to the international standard IEC 60825 other standardizing organizations have developed their own standards dealing with laser products and laser safety. For a long time the standard ANSI Z136.1 [21.25, 26], which is edited by the American National Standards Institute (ANSI), has been a competitor on a standardization market place. The ANSI standard provides guidance for the safe use of lasers and laser systems for four laser classes, i. e. 1, 2, 3 and 4, where the medium-power lasers are divided into two subclasses, 3a and 3b. For the purpose of the ANSI standard, laser products which have been classified earlier than 2000 as class IIa under the federal laser product performance standard should be treated the same as class 1. Due to a *laser notice* [21.27] published by the CDRH/FDA there is now a relatively wide-reaching agreement between IEC 60825-1 and ANSI Z136.1. In addition ANSI Z136.1 includes an appendix where harmonization with the IEC 60825-1 is described in more detail and some differences are stressed.

Since standards are being developed as a market-regulating or supporting element in many cases, national and/or regional divergence is sometimes economically explainable, but for laser safety in general, a consistent and widely agreed level should exist. This is currently the case.

In addition to the general product safety standard IEC 60825-1 other safety requirements are specified in other relevant standards, e.g. for medical lasers [21.28], information technology or machinery [21.29].

For the safety of optical-fiber communication systems the standard IEC 60825-2 [21.30] provides guidance concerning products that enable laser radiation including light-emitting diode (LED) radiation for communications applications. Another part of IEC 60825 describes laser guards [21.31]. IEC 60825-12 is one of the youngest standards of the IEC 60825 series and describes the safety with free-space optical communication systems [21.32]. In total, 15 parts of IEC 60825 are currently published or exist as drafts.

21.5 Laser Hazard Categories and Laser Classes

Three aspects of the use of lasers need to be taken into account in the evaluation of the possible hazards and in the application of control measures:

- the capability of the laser or laser system to injure people. This includes any consideration of human access to the main exit port or any subsidiary port,
- the environment in which the laser is used, and
- the level of training of the personnel who operate the laser or who may be exposed to its radiation.

The practical means for evaluation and control of laser radiation hazards is to classify laser products according to their relative hazard potential, and then to specify appropriate controls for each class. The use of the classification system will, in most cases, preclude any requirement for radiometric measurements by the user. In many cases it is therefore unnecessary to make use of the MPEs directly. The experience from laser use in industrial, scientific and medical applications has permitted the development of a system of laser hazard categories.

Laser products are assigned, therefore, to different laser classes. This should aid the user in hazard evaluation of the laser and provide the necessary information concerning protective and control measures. Thus the classification of laser products is based on the increasing degree of hazard or risk posed by the laser product. Currently there exist seven classes, namely 1, 1M, 2, 2M, 3R, 3B, and 4 in IEC 60825-1. In addition, the laser class 3A exists, but new laser products are no longer classified into this special category.

One of the main objects of the international standard IEC 60825-1 is to protect persons from laser radiation in the wavelength range 180 nm to 1 mm by indicating safe working levels of laser radiation and by introducing a system of classification of lasers and laser products according to their degree of hazard. In addition, requirements for both user and manufacturer to establish procedures and supply information are laid down so that proper precautions can be adopted. To ensure adequate warning to individuals of hazards associated with accessible radiation from laser products, signs, labels and instructions have to be used.

The increasing classification number applies not only to an increasing hazard, i. e. the capability to cause an injury, but to more protective measures as well, which form the requirements to be specified by the manufacturer and used by the user. The manufacturer is required to perform the classification and to label the laser product accordingly.

21.5.1 Accessible Emission Limits

The classification of laser products is based on so-called accessible emission limits (AEL), i. e. maximal power or energy values. The respective values are given in tables for the different laser classes in the product standard IEC 60825-1.

The determination and derivation of the AELs, although generally based upon the MPEs, necessitated a risk analysis and determination of reasonably foreseeable exposure conditions.

A main characteristic of the AELs is that the respective MPEs, which are valid for the time base of 30 000 s or 100 s, are not exceeded for laser class 1. The

AELs of class 1 are simply derived by multiplication of the respective MPEs by the area of the averaging aperture as specified, e.g. 7 mm in the case of visible laser radiation. The time base of 30 000 s is used when intentional or deliberate and long-term exposure is intended, whereas 100 s is used when unintentional exposure is expected. However, in the case of ultraviolet radiation 30 000 s always has to be used, since this shorter optical radiation shows dose-related effects.

The classification of a laser product has to be done by the manufacturer for normal use and maintenance under reasonable, foreseeable so-called single-fault conditions.

Maintenance implies tasks which are necessary to ensure routine performance of the laser, i. e. maintenance might be done by the laser user. This includes tasks such as cleaning and replenishing of expendables. Beam access is usually not required. This should not be confused with service, which is performed less frequently and usually requires beam access, e.g. to replace resonator mirrors or to repair components of the laser product.

The AEL for class 2 is derived by assuming an exposure duration of 0.25 s, i. e. the time base for class 2, and multiplying the MPE for 0.25 s for visible laser radiation by the area of a 7 mm aperture.

21.5.2 Description of the Laser Classes

The classes can be described as in Table 21.4. Besides the definitions in IEC 60825-1 some additional advice is given in order to improve the safety of working with lasers considering the latest findings, particularly for low-power lasers.

The maximum power of a laser product belonging to class 1M or 2M is limited to the AEL of class 3B. When the exposure with optical instruments is unlikely, the use of class 1M and 2M lasers does represent a small risk, but if it is reasonably foreseeable that optical instruments might be used, controls and restrictions are required.

One of the main reasons to alter the former classification system has been that LED products have been included in the scope of the international standard IEC 60825-1 whenever the word laser is used. The newly defined measurement conditions for classes 1M and 2M had the effect that fewer LED products were classified at unrealistically high levels, e.g. 3B. In the meantime is was decided to remove LEDs from the scope of IEC 60825-1 in the second edition, which will be published in 2007 and to treat most of them as lamps according to the Commission International de l'Eclairage (CIE) standard S009 [21.33]. An international dual-logo standard IEC 62471/CIE S009:2002 currently exists as a first edition (2006-07) on photobiological safety of lamps and lamp systems [21.34].

Risk Analysis

In order to perform a risk analysis and to classify a laser product the characteristics of the laser itself and its applications must be known. This means that the following data in particular should be available:

- wavelength,
- mode of operation [continuous wave (CW), single pulsed, Q-switched, repetitively pulsed (scanning), mode locked, very short pulses (femtosecond laser)],
- optical power (W) or optical energy (J or J/pulse),
- pulse duration,
- pulse repetition frequency (PRF),
- beam diameter,
- beam divergence,
- mode profile.

It is worth mentioning that for safety calculations the beam diameter and the beam divergence are taken at the $1/e$ decrease of the irradiance and not at $1/e^2$, which means that the corresponding values according to the standard are a factor of $\sqrt{2}$ less.

Beam divergence and Nominal Ocular Hazard Distance

The beam produced by a laser does spread according to its divergence, which is the rate at which the beam spreads. The laser radiation covers a continually larger area as it travels away from the laser product. This angle is measured in radians or milliradians and is usually very small compared to the beam from other light sources. A typical divergence value for a laser is 1 milliradian (1 mrad). This means that the beam spreads such that its spot increases in diameter at a rate of 1 mm for every 1 m (1000 mm) from the laser.

Because the beam spreads at a fixed and measurable angle, the diameter of the beam can be determined at any desired distance from the laser. To approximate the beam size (diameter D) for long distances z one can use a linear equation:

$$D = (\text{divergence angle}) \cdot (\text{range}) + a = \Theta z + a \,, \tag{21.9}$$

where z is the distance (range), θ is the divergence angle, and a is the diameter of the beam at the aperture of the laser product.

When estimating the situation near the laser aperture, i. e. in the so-called near field the following equation has to be used:

$$D = [(\Theta z)^2 + (a)^2]^{1/2} . \tag{21.10}$$

From (21.9) the nominal ocular hazard distance (NOHD) can be calculated as

$$z = \text{NOHD} \geq \frac{1}{\Theta}\left[\sqrt{D^2} - a\right] \tag{21.11}$$

and together with (21.8)

$$\text{NOHD} \geq \frac{1}{\Theta}\left[\sqrt{\frac{4P}{\pi E_{\text{MPE}}}} - a\right] \tag{21.12}$$

is obtained. This equation might be used to estimate the safe distance from a given laser source with a given optical power and beam divergence for the respective MPE value.

21.6 Protective Measures

The majority of injuries involve the eye and, to a lesser extent, the skin. The injuries occur mainly during alignment procedures, or because protective eyewear is either inappropriate or not used.

In order to reduce the possibility of injury by minimizing unnecessary accessible radiation and to give improved control of the laser radiation hazards through protective features and provide safe usage of laser products by specifying user control measures IEC 60825-1 has been published.

Laser products require certain built-in safety features, depending on the class to which they have been assigned by the manufacturer.

Usually technical and engineering protective measures shall be applied to protect against hazardous laser radiation. The trivial and most effective solution is total enclosure of the laser (encapsulation) and of all accessible beam paths. If this is not possible partial beam enclosure or restricted access to beam paths, and administrative controls have to be applied whenever possible instead of using personal protective equipment (PPE) such as laser eye protectors (laser safety goggles).

The safest laser products rely upon engineered safety features. Technical measures might be divided into such measures that are part of the laser product itself or that have to be installed as a consequence of the degree of hazard of the respective laser product. For example safety interlocks can isolate the laser if anyone enters an area of high risk.

21.6.1 Technical and Engineering Measures

Table 21.5 gives an overview of technical and engineering measures as a function of the laser class and degree of hazard, whereas Table 21.6 gives some information concerning administrative measures. Protective housings with interlocks, service access panel with interlocks, master switch, key control, viewing portals/display screens, safe collecting optics, and enclosed or encapsulated open beam paths are all types of engineering controls. In addition a remote interlock connector, beam-stop attenuator, activation warning systems, a controlled area where the nominal hazard zone (NHZ) has to be determined are required. Equipment labels and area posting are required or recommended for class 3B and class 4 lasers.

For more-specific and detailed information on manufacturing requirements the standard is obligatory. The warning labels for class 2, 3R (in the wavelength range 400–1400 nm) or 4 read as follows according to IEC 60825-1:

LASER RADIATION DO NOT STARE INTO BEAM CLASS 2 LASER PRODUCT

LASER RADIATION AVOID DIRECT EYE EXPOSURE CLASS 3R LASER PRODUCT

or

LASER RADIATION AVOID EYE OR SKIN EXPOSURE TO DIRECT OR SCATTERED RADIATION CLASS 4 LASER PRODUCT

The laser hazard symbol, the safety alert symbol and information carried by the sign are different in IEC 60825-1 and ANSI Z136.1.

21.6.2 Administrative Measures

Administrative measures are an important backup for technical measures and controls. Administrative controls might be given by written standard operating procedures (SOP), maintenance and service procedures,

Table 21.5 Technical and engineering measures

Measure	1	1M	2	2M	3R	3B	4
Walls					Matt, light, diffuse reflecting		
Shielding					High absorption for the laser wavelength and effective against secondary radiation		
Nominal ocular hazard distance (NOHD)					Labeling of the boundaries/limits		
Beam warning; emission warning					Optical or acoustical at the entrance; on top of the laser product; only for non-visible radiation	Optical or acoustical at the entrance, on top of the laser product	
Emergency switch	Dependent on the product specific risk analysis						
Protective housing	Class 1 should be the goal						
Safety interlock					Designed to prevent removal of the panel; reliable implementation		
Remote control	Necessary if class 3B or 4 are embedded					At the door contact or connection of an emergency stop	
Key-switch, key control						Limited subgroup of people; remove key when laser is not used	
Viewing optics			Install laser filter; AEL of class 1/1M must not be exceeded				
Scanning	Scan failure shall not cause laser product to exceed its classification						
Class label	Required wording		Warning label/hazard symbol and required wording on explanatory label				
Aperture label					Specified wording required		
User information	Operation manuals must specify product classification and contain instructions for safe use						

Table 21.6 Administrative measures for the different laser classes. These are examples according to accident insurance at the workplace in Germany only; different regulations may be valid at other countries, institutions etc.

Measure	1	1M	2	2M	3R	3B	4
Advice	Required when a laser class 3R, 3B or 4 is embedded				Registration of the laser product		
LSO (laser safety officer)	In general not required				Required*; written appointment		
NOHD NHZ (nominal hazard zone)	In general not required, labeling necessary if the beam belongs to class 2 or 2M and passes the working place				Determination of the boundaries/limits; eventually for a limited duration; movable shielding during maintenance; the area must be posted with appropriate warning signs		
Laser protective filters	Not required as long as intrabeam viewing is not mandatory				Required** if engineering and administrative procedures are not practicable; ambient luminance should eventually be increased		
Special training	Not required	Required; written confirmation; at least annually					
Beam paths		Block the beam at the end of its path; avoid specular reflections					

* According to IEC 60825-1 an LSO is not required for class 3R laser products for visible emission

**According to IEC 60825-1 eye protection is not required for class 3R; ANSI Z136.1 requires an LSO for embedded class 3a too if service is performed output emission limitations, education and training, authorized personnel, controlled access and a limit on the number of spectators to name but a few. Lights and notices should warn people when they approach a laser hazard zone.

Laser Safety Officer

According to IEC 60825-1 the presence of a laser safety officer (LSO) is not required when working with lasers classified less than class 3B. But since class 3R has been part of the former class 3B up to 2001 the requirement is still valid according to the regulations applicable in Germany as far as occupational laser safety is concerned [21.35, 36]. This is especially justified since the MPEs might be exceeded working with class 3R lasers and the LSO is a person who is aware of the safe working practices with these laser products. The special responsibility of the LSO may be regulated according to national regulations. In some countries the LSO is an adviser or consultant only, in others it is a certified person, i. e. one who is knowledgeable in the evaluation and control of laser hazards and has responsibility for oversight of the control of laser hazards.

Training and Education

Training and education is certainly one of the most important things which has to be done in advance of using certain lasers. Operation of class 1M and class 2M laser products where there is a possibility to use optical instruments, class 3R, class 3B and class 4 laser products, can represent a hazard not only to the user but also to other people over a considerable distance. Because of this hazard potential, only persons who have received training to an appropriate level should be placed in control of such systems. The training, which may be given by the manufacturer or supplier of the laser product, the laser safety officer, or by an approved external organization, should include, but is not limited to:

- Familiarization with system operating procedures
- the proper use of hazard control procedures, warning signs, etc.
- the need for personnel protection
- accident reporting procedures
- bioeffects of the laser beam upon the eye and the skin

21.6.3 Personal Protective Equipment

Last but not least, protective equipment may be used to prevent hazardous situations or even damage to the eyes. This PPE should be in compliance with the respective standards. In Europe PPE have to meet the essential safety requirements stated in a directive of the European Economic Community (EEC)/EU and must be labeled with a conformité Européenne (CE) mark.

Concerning the adequate PPE, i. e. laser safety goggles, one has to determine wavelength and maximum viewing duration. The maximum irradiance or radiant exposure is calculated and from this data an adequate level of protection is chosen.

In general the following should be considered when specifying suitable protective eyewear:

- Wavelength of operation
- radiant exposure or irradiance
- MPE value(s)
- level of protection/optical density of eyewear at laser output wavelength
- visible-light transmission requirements
- resistance or radiant exposure or irradiance, at which damage to eyewear occurs
- need for prescription glasses
- comfort and ventilation
- degradation or modification of absorbing media, even if temporary or transient,
- strength of materials (resistance to shock),
- peripheral vision requirement
- any relevant national regulations.

Protective eyewear should be comfortable to wear, provide as wide a field of view as possible, maintain a close fit while still providing adequate ventilation to avoid problems of misting up and provide adequate visual transmittance. Care should be taken to avoid, as far as possible, the use of flat reflecting surfaces which might cause hazardous specular reflections. It is important that the frame and any side-pieces should give equivalent protection to that afforded by the lens(es)/filters. In addition all laser eye-protectors shall be clearly labeled with information adequate to ensure the proper choice of eyewear with particular lasers.

21.6.4 Beyond Optical Hazards

In some laser applications, mainly with class 4 laser products, in addition to optical hazards, control measures can be necessary for electrical and fire hazards, X-rays, noise, and airborne contaminants.

Most lasers utilize high voltages that can be lethal. Some lasers require hazardous or toxic substances to operate (i. e., chemical dye, excimer lasers). The solvents used in dye lasers are flammable. High-voltage pulses or flash lamps may cause ignition. Flammable materials may be ignited by direct beams or specular reflections from high-power continuous wave (CW) infrared lasers.

21.6.5 Future Regulations

The current edition of the international standard IEC 60825-1 includes a detailed user's guide to administrative regulations and controls. However, guidelines for users in standards are only informative and not normative, since it is the right of the various countries to set local regulations according to the given laws, decrees and regulations etc.

In the European Union a *Directive of the European Parliament and of the Council on the minimum health and safety requirements regarding the exposure of workers to the risks arising from physical agents (artificial optical radiation)* has been published as the 19th directive within the remit of article 16(1) of directive 89/391/EEC [21.37]. Member states of the European Union shall bring into force the laws, regulations and administrative provisions necessary to comply with the Directive 2006/25/EC by 27 April 2010.

It is one of the main goals under the treaty of the council of the EU, by means of directives, to adopt minimum requirements for encouraging improvements, especially in the working environment, to guarantee a better level of protection of the health and safety of workers. This will be done by the introduction of minimum health and safety requirements regarding the exposure of workers to the risks caused by physical agents; in this case by artificial optical radiation, which includes laser radiation.

The directive lays down minimum requirements, thus giving member states of the EU the option of maintaining or adopting more-stringent provisions for the protection of workers, in particular the fixing of lower exposure limit values. Adherence to the exposure limit values should provide a high level of protection as regards the health effects that may result from exposure to optical radiation.

Since the exposure limits in the European directive are in agreement with the given values in the ICNIRP guidelines it might be expected that there is no strong influence as far as current regional laser safety regulations in Europe are affected. In order to facilitate implementation of the Directive 2006/25/EC the Commision shall draw up a practical guide to the articles dealing with the determination of exposure and assessment of risks as well as provisions aimed at avoiding or reducing risks.

Laser safety is treated in several textbooks, handbooks and publications of various institutions. In addition to the international, regional and national standards, more-detailed information on laser safety in general can be found in a recent publication [21.38], but nevertheless even more historical texts should be regarded as a source of information, especially if background knowledge is needed [21.39,40]. Laser safety is dealt with on many Internet web-sites, but many of these need a thorough update.

21.7 Special Recommendations

Up to now the number of accidents is low relative to the number of applications of lasers worldwide. However, nevertheless there is a risk, especially during situations such as beam alignment and working with high-power lasers, where even reflected beams might be hazardous.

The most dangerous situations and causes of ocular beam accidents besides exposure during alignment are misaligned optics, lack of training, improper restoration, intentional intrabeam viewing, use of multiple wavelengths, and last but not least when PPE are not utilized.

Laser types such as ruby, argon ion, krypton ion, dye, He:Cd, SHG–Nd:YAG (second-harmonic generated, so-called green-lasers at 532 nm), He:Ne and many diode lasers (semiconductor lasers) all belong to lasers whose beam will be focused on the retina due to the fact that the ocular media are transparent. The same is true for Nd:YAG too, but in this case the beam is nonvisible, which enhances the risk. This is especially true for Q-switched Nd:YAG lasers, which have caused a lot of ocular incidents. In nearly all cases eye protection normally available was not worn in ocular incidents.

However, even with laser safety goggles one cannot be sure of being adequately protected, since there exist a lot of eye protectors that do not guarantee the resistance against a laser beam stated in the European standard EN 207 [21.41], rather they are rated according to optical density only. This is an important characteristic of safety goggles, but might not be sufficient for high-power laser radiation. The European standard ensures not only a specified safety level for the laser eye-protector but resistance against damaging laser radiation of 10 s.

In addition to laser eye-protectors laser adjustment eye-protectors are used, especially during adjustments. Laser adjustment eye-protectors according to the relevant European standard [21.42] reduce the incident laser radiation to a value according to class 2 conditions, i. e. the safety is based on short-term exposure, and the blink reflex has been chosen as an argument. Due to the new findings laser adjustment eye-protectors should be used with care in order not to trust in this expected physiological reaction, but to perform active protective reactions (e.g. close the eyes and turn the head away) in case of intrabeam viewing with this kind of goggles.

Laser Pointers

It is not possible to close this chapter without the following remarks on the special topic of laser pointers.

On the subject of laser pointers ICNIRP states that "based upon current medical knowledge a 5 mW laser is incapable of producing a permanent retinal thermal injury for a momentary (0.25 s) exposure under ideal optical conditions of exposure" [21.43]. In addition it is said: "Although an unintentional momentary intrabeam viewing will not produce a permanent retinal injury, it is theoretically possible to produce retinal photocoagulation in an eye with perfectly clear ocular media by staring at a collimated 5 mW laser beam for more than 10 s" [21.43, 44]. According to the ICNIRP one is assured that "in adults, the pupillary, blink, and aversion responses terminate accidental laser pointer exposures in less than 0.25 s". This clearly says that there is no realistic risk for permanent eye damage from a laser pointer, but at least on that score the statement is wrong as has been shown in lab and field trials with more than 1500 volunteers [21.22], since the frequency of aversion responses including the blink reflex is less than 20%. It cannot be denied that physiological eye movements exist as have been described above, but nonetheless one should avoid shining a laser pointer into anyone's eye and it should be common sense not to stare into the beam of a laser pointer. Any person engaging in a trial to find out whether there is a risk or not could end up with a small but permanent eye injury. This is especially worthy of note since laser pointers are mass-produced devices where quality control is not a particular feature, which is supported by the fact that specimens with a power of more than 10 mW have been found. It is therefore wise that some countries have restricted the sale or use of laser pointers above class 2 in order to increase safety against retinal injury. In addition even green class 1 laser pointers could do the job, since the photopic brightness is much higher compared to red laser-diode pointers.

Hazardous Situations

The following list contains some reasons for hazardous situations and common unsafe practices that are causes of preventable laser accidents:

- High retinal irradiance results when the beam is focused by the lens of the eye to create an extremely high retinal hazard risk for visible and near-infrared lasers
- low beam divergence resulting in a large intrabeam hazard distance
- not wearing protective eyewear during alignment procedures
- not wearing protective eyewear in the laser control area
- available eye protection not used
- wearing the wrong eyewear
- misaligned optics and upwardly directed beams
- equipment malfunction
- improper methods of handling high voltage
- intentional exposure of unprotected personnel
- bypassing of interlocks, door and laser housing
- insertion of reflective materials into beam paths
- lack of pre-planning
- turning on power supply accidentally
- operating unfamiliar equipment
- lack of protection from non-beam hazards
- high irradiance at the focal point of an external lens creating greater skin hazard risk

In order to prevent laser accidents it is highly recommended to use the lowest possible or practical power, especially during alignments. It is recommended that the plane of the laser beam be above or below the level of a seated or standing person.

Perhaps the most important recommendation is never to aim a laser beam at a person, especially not at a person's eye. As quickly as it takes to blink – spontaneously or voluntarily – a laser beam can severely damage the eye. Though these injuries are relatively rare, they are permanent and can be avoided.

References

21.1 T. Maiman: Stimulated optical radiation in ruby masers, Nature **187**, 493–494 (1960)

21.2 A. Javan, W. R. Bennett, D. R. Herriott: Population inversion and continuous optical maser oscillation in a gas discharge containing a He-Ne mixture, Phys. Rev. Lett. **6**, 106–110 (1961)

21.3 D. H. Sliney: The development of laser safety criteria biological considerations. In: *Laser applications in medicine and biology*, Vol. 1, ed. by M. L. Wolbarsht (Plenum, New York 1971) pp. 163–238

21.4 World Health Organization (WHO): *Lasers and Optical Radiation, Environmental Health criteria, no. 23* (WHO, Geneva 1982)

21.5 H.-D. Reidenbach: *Hochfrequenz- und Lasertechnik in der Medizin* (Springer, Berlin, Heidelberg 1983)

21.6 W. J. Geeraets, E. R. Berry: Ocular spectral characteristics as related to hazards from lasers and other light sources, Am. J. Ophthalmol. **66**, 15–20 (1968)

21.7 W. T. Ham Jr., H. A. Mueller, D. H. Sliney: Retinal sensitivity to damage from short wavelength light, Nature **260**, 153–156 (1976)

21.8 W. T. Ham Jr., H. A. Mueller, J. J. Ruffolo Jr., D. Guerry III, R. K. Guerry: Action spectrum for retinal injury from near ultraviolet radiation in the aphakic monkey, Am. J. Ophthalmol. **93**, 299–302 (1982)

21.9 International Commission on Non-Ionizing Radiation Protection (ICNIRP). Guidelines on limits of exposure to laser radiation of wavelengths between 180 nm and 1000 µm, Health Phys. **71**, 804–819 (1996)

21.10 J. W. Ness, H. Zwick, B. Stuck, D. J. Lund, J. M. Molchany, D. H. Sliney: Eye movements during fixation: implications to laser safety for long duration viewing, Health Phys. **78**, 1–9 (2000)

21.11 International Commission on Non-Ionizing Radiation Protection (ICNIRP): Revision of guidelines on limits of exposure to laser radiation of wavelengths between 400 nm and 1.4 µm, Health Phys. **79**, 431–440 (2000)

21.12 B. J. Lund: Computer model to investigate the effect of eye movements on retinal heating during long-duration fixation on a laser source, J. Biomed. Opt. **9**, 1093–1102 (2004)

21.13 D. Courant, L. Court, B. Abadie, B. Brouillet: Retinal damage threshold from single-pulse laser exposures in the visible spectrum, Health Phys. **56**, 637–642 (1989)

21.14 International Electrotechnical Commission: *IEC 60825-1: 1993 + A1:1997 + A2:2001: Safety of laser products – Part 1: Equipment classification, requirements and user's guide* (IEC, Geneva 2001)

21.15 H.-D. Reidenbach: Aversion responses including the Blink reflex: Psychophysical behaviour and active protection reactions as an additional safety concept for the application of low power lasers in the visible spectrum, ILSC, Los Angeles 2005, ed. by Laser Institute of America (Laser Institute of America, Orlando, FL 2005) 67

21.16 H.-D. Reidenbach, A. Wagner: Ein Beitrag zum Lidschlussreflex bei inkohärenter optischer Strahlung (Abstract in English), 31. Jahrestagung des Fachverbandes für Strahlenschutz NIR 99, Bd. II, Köln 1999, ed. by N. Krause, M. Fischer, H.-P. Steimel (TÜV-Verlag, Köln 1999) 935–946

21.17 H.-D. Reidenbach: First surveys regarding the blink reflex with low power lasers, Laser Bioeffects Meeting, Paris 2002, ed. by Commissariat a l'Energie Atomique Direction des Sciences du Vivant, Austrian research Centers Health Physics, International Commission on Non-Ionizing Radiation (Commissariat a l'Energie Atomique Direction des Sciences du Vivant, Austrian research Centers Health Physics, International Commission on Non-Ionizing Radiation, Paris 2002) 11-1–11-17

21.18 H.-D. Reidenbach, J. Hofmann, K. Dollinger, M. Seckler: A critical consideration of the blink reflex as a means for laser safety regulations, paper 8c5, IRPA 11, Madrid 2004, ed. by Spanish Radiation Protection Society (Spanish Radiation Protection Society, Madrid, Spain 2004) 805–811

21.19 C. P. Cain, G. D. Noojin, D. X. Hammer, R. J. Thomas, B. A. Rockwell: Artificial eye for in vitro experiments of laser light interaction with aqueous media, J. Biomed. Opt. **2**, 88–94 (1997)

21.20 W. P. Roach, T. E. Johnson, B. A. Rockwell: Proposed maximum permissible exposure limits for ultrashort laser pulses, Health Phys. **76**, 349–354 (1999)

21.21 H.-D. Reidenbach, H. Warmbold, J. Hofmann, K. Dollinger: First experimental results on eye protection by the blink reflex for laser class 2, Biomed. Technik/Biomed. Engin. **46 suppl. 1**, 428–429 (2001)

21.22 H.-D. Reidenbach, K. Dollinger, J. Hofmann: Results from two research projects concerning aversion responses including the blink reflex, SPIE Proc. **B5688**, 429–439 (2005)

21.23 U. S. Food and Drug Administration - Center of Devices and Radiological Health (FDA/CDRH): *Code of Federal Regulations. Title 21 – Food, Drugs, sec. 1040.10 Laser products (21CFR1040.10)* (U.S. Food, Drug Administration, Rockville 2001)

21.24 International Electrotechnical Commission: *IEC 825-1 Safety of laser products – Part 1: Equipment classification, requirements and user's guide.* (IEC, Geneva 1984)

21.25 American National Standards Institute: *ANSI Z136.1-2000: American National Standard for the Safe Use of Lasers.* (The Laser Institute of America, Orlando 2000)

21.26 American National Standards Institute: *ANSIZ136.1-1973: American National Standard for the Safe Use of Lasers.* (The Laser Institute of America, Orlando 1973)

21.27 U.S. Food and Drug Administration - Center of Devices and Radiological Health (CDRH/FDA): Laser products – conformance with IEC60825-1, Am.2, IEC60601-2- 22; Final guidance for industry, FDA, Laser Notice **50** (2001) (Doc. issued on July 26, 2001)

21.28 International Electrotechnical Commission: *IEC 60601-2-22:1995: Medical electrical equipment – Part 2: Particular requirements for the safety of diagnostic and therapeutic laser equipment.* (IEC, Geneva 1995)

21.29 International Electrotechnical Commission/International Standards Organization: *IEC/ISO 11553-1: Safety of machinery – Laser processing machines – Part 1: General safety requirements.* (IEC/ISO, Geneva 2005)

21.30 International Electrotechnical Commission: *IEC 60825-2:2004-06Ed.3: Safety of laser products – Part 2: Safety of optical fibre communication systems.* (IEC, Geneva 2005)

21.31 International Electrotechnical Commission: *IEC 60825-4 Ed.2.0b:2006: Safety of laser products – Part 4: Laser guards.* (IEC, Geneva 2006)

21.32 International Electrotechnical Commission: *IEC 60825-12:2004: Safety of laser products – Part 12: Safety of free space optical communication systems used for transmission of information* (IEC, Geneva 2004)

21.33 Commission Internationale de L'Eclairage (CIE): *CIE S 009/E:2002: Photobiological Safety of Lamps, Lamp Systems* (CIE Central Bureau, Vienna 2002)

21.34 International Electrotechnical Commission/Commission Internationale de l'Eclairage: *IEC 62471/CIE S009:2002, First edition 2006-07, Photobiological safety of lamps and lamp systems* (IEC, Geneva 2006)

21.35 Berufsgenossenschaft der Feinmechanik und Elektrotechnik: *Berufsgenossenschaftliche Information BGI832 Betrieb von Lasereinrichtungen – Anwendung der Unfallverhütungsvorschrift "Laserstrahlung" BGV B2 auf neue Laserklassen und MZB-Werte nach DIN EN 60825-1 (VDE 0837-1) (in German)* (Heymann, Köln 2003)

21.36 Berufsgenossenschaft der Feinmechanik und Elektrotechnik: *Unfallverhütungsvorschrift BGV B2 Laserstrahlung vom 1. April 1988 in der Fassung vom 1. Januar 1997 mit Durchführungsanweisungen vom Oktober 1995, (bisherige VBG 93), (in German)* (Heymann, Köln 1997)

21.37 European Parliament and Council of the European Union: Directive 2006/25/EC of the European Parliament and of the Council of 5 April 2006 on the minimum health and safety requirements regarding the exposure of workers to risks arising from physical agents (artificial optical radiation)(19th individual Directive within the meaning of Article 16(1) of Directive 89/391/EEC, Official Journal of the European Union **114**, 38–59 (2006)

21.38 R. Henderson, K. Schulmeister: *Laser Safety* (Institute of Physics, Bristol 2003)

21.39 D. H. Sliney, M. L. Wolbarsht: *Safety with lasers and other optical sources* (Plenum, New York 1980)

21.40 D. H. Sliney (Ed.): Selected Papers on Laser Safety, SPIE Press, Bellingham **MS117** (1995)

21.41 European Committee for Electrotechnical Standardization: *EN207:2002-12: Personal eye-protection – Filters and eye-protection against laser radiation (laser eye-protectors) (includes amendment A)* (CENELEC, Bruxelles 2002)

21.42 European Committee for Electrotechnical Standardization: *EN208:2002-12: Personal eye-protection – Eye-protection for adjustment work on lasers and lasersystems (laser adjustment eye-protectors) (includes amendment A)* (CENELEC, Bruxelles 2002)

21.43 International Commission on Non-Ionizing Radiation Protection (ICNIRP): ICNIRP Statement on Laser Pointers, Health Phys. **77**, 218–220 (1999)

21.44 W.T. Ham Jr., W.J. Geeraets, H.A. Mueller, R.C. Williams, A.M. Clarke, S.F. Cleary: Retinal burn thresholds for the helium-neon laser in the rhesus monkey, Arch. Ophthalmol. **84**, 797 (1970)

Acknowledgements

A.1 The Properties of Light

by Richard Haglund

The author thanks Prof. Emil Wolf for helpful discussions, and gratefully acknowledges the financial support of a Senior Scientist Award from the Alexander von Humboldt Foundation and of the Medical Free-Electron Laser program of the Department of Defense (Contract F49620-01-1-0429) during the preparation of this chapter.

A.4 Nonlinear Optics

by Aleksei Zheltikov, Anne L'Huillier, Ferenc Krausz

We acknowledge the support of the European Community's Human Potential Programme under contract HPRN-CT-2000-00133 (ATTO) and the Swedish Science Council.

A.5 Optical Materials and Their Properties

by Klaus Bonrad

The author of Sect. 5.9.2 is grateful to Dr. Thomas Däubler, Dr. Dirk Hertel, and Dr. Frank Voges for fruitful and stimulating discussions.

C.11 Lasers and Coherent Light Sources

by Gerd Marowsky, Uwe Brinkmann

As well as the acknowledgements mentioned in the footnotes within the text, we would like to acknowledge very helpful discussions with A. Borisov, A. Görtler, J. Ihlemann, R. Pätzel and C. Peth. The patient assistance of T. Eggers and J. Jethwa is greatly appreciated as their support has been central to the edited version of the text. The part *Beam Characterization of Excimer Lasers* is based upon a contribution from K. Mann and coworkers, Laserlaboratorium Göttingen. The part *UV Femtosecond Material Processing* is based on a contribution from P. Simon and coworkers, Laserlaboratorium Göttingen. The part *High-Intensity UV Femtosecond Studies* is based on a contribution from C. K. Rhodes and coworkers from the University of Chicago, Illinois. The part 13.5 nm *Technology* is based on a contribution from U. Stamm and coworkers, XTREME technologies, Göttingen and Jena, Germany.

by Dennis Lo[†]

This work was supported in part by RGC Earmarked Research Grant of the Hong Kong SAR Government CUHK 4366/99E, and CUHK 4233/03E.

by Helen Wächter, Markus W. Sigrist

The authors thank a number of coworkers for their valuable input, notably R. Bartlome, Dr. C. Fischer, D. Marinov, Dr. J. Rey, M. Stahel, and Dr. D. Vogler. The financial support by the Swiss National Science Foundation and ETH Zurich for the isotopomer studies is gratefully acknowledged.

by Jürgen Helmcke

In writing the chapter on frequency-stabilized lasers, the author has greatly benefited from fruitful cooperation and helpful discussions with his colleagues at PTB, in particular with Drs. Fritz Riehle, Harald Schnatz, Uwe Sterr, and Harald Telle. Special thanks belong to Dr. Fritz Riehle for his careful and critical reading of the manuscript. Part of the work discussed in this chapter was supported by the Deutsche Forschungsgemeinschaft (DFG) under SFB 407.

C.12 Femtosecond Laser Pulses: Linear Properties, Manipulation, Generation and Measurement

by Matthias Wollenhaupt, Andreas Assion, Thomas Baumert

Finally the authors would like to acknowledge Marc Winter for help in preparing various figures and like to thank Andrea Klumpp as well as Marc Winter for carefully proofreading the manuscript.

D.13 Optical and Spectroscopic Techniques

by Sune Svanberg

The author gratefully acknowledges a most stimulating collaboration with a large number of colleagues and graduate students in the field of laser-based remote sensing. This work was supported by the Swedish Space Board, the Swedish Natural Sciences Research Council, the FORMAS Research Council, the Knut and Alice Wallenberg Foundation and the European Community within the EUREKA-LASFLEUR (EU380), and the Access to Large Scale Facility/Research Infrastructure programmes.

D.16 Optics far Beyond the Diffraction Limit: Stimulated Emission Depletion Microscopy

by Stefan W. Hell

The results summarized in this chapter have been accomplished by collaborative research within the department.

We thank M. Bossi and R. Medda for preparation of samples, and A. Schönle, C. Eggeling and V. Westphal for valuable discussions.

D.20 Holography and Optical Storage

by Mirco Imlau, Martin Fally

The authors acknowledge financial support by the Austrian Science Fund (P-15642) and the Deutsche Forschungsgemeinschaft. Mirco Imlau is indebted for a visiting professorship with the University of Vienna that considerably facilitated this work. We thank Romano Rupp for valuable comments and remarks.

About the Authors

Andreas Assion

Chapter C.12

Femtolasers Produktions GmbH
Vienna, Austria
andreas.assion@femtolasers.com

Andreas Assion joined Femtolasers in January of 2005. Prior to joining Femtolasers, he worked with ultrafast lasers on the observation and control of quantum optical phenomena in atoms and molecules. He earned his diploma and doctorate studying molecular dynamical effects, including coherent control of complex molecules. After a post-doctoral position with the German Space Agency, he completed his Habilitation 2004.

Thomas E. Bauer

Chapter A.5, Sect. 5.7

JENOPTIK Polymer Systems GmbH
Coating Department
Triptis, Germany
thomas.bauer@jenoptik-ps.de

Thomas Bauer is a physicist working at Jenoptik Polymer Systems GmbH (former WAHL optoparts) as head of coating department. His main areas of interest are plastic optics in general and coating of plastic optics in particular.

Thomas Baumert

Chapter C.12

Universität Kassel
Institut für Physik
Kassel, Germany
baumert@physik.uni-kassel.de

Professor Baumert received is Ph.D. with Prof. Gerber, University of Freiburg, Germany in 1992. Further positions in his carreer were: 1992–1993, post doc with Prof. Zewail, Caltech, Pasadena; 1993–1997, "Habilitation", University of Würzburg, Germany; 1998–1999, head of LIDAR group, DLR Oberpfaffenhofen, Germany; 1999 Full Professor of Experimental Physics at University of Kassel, Germany. His research area: Femtosecond spectroscopy and ultrafast laser control of matter. Awards: Gödecke thesis award (1992), Heisenberg-Scholarship of DFG (1997–1998), Philip-Morris-Award (2000).

Dietrich Bertram

Chapter C.10

Philips Lighting
Aachen, Germany
dietrich.bertram@philips.com

Prior to joining the CTO office at Philips Lighting as technical officer solid-state lighting, Dietrich Bertram headed a project on LED light sources at Philips Research. His background education is physics, where he obtained a masters degree at Marburg University in epitaxy of III-V materials and a Ph.D. from the Max-Planck Institute of Solid State Research, Stuttgart, Germany.

Klaus Bonrad

Chapter A.5, Sects. 5.9.2, 5.10

Schott Spezialglas AG
Division Luminescence Technology
Mainz, Germany
klaus.bonrad@schott.com

Klaus Bonrad studied chemistry in Darmstadt and received his Ph.D. in Mainz at the Max-Planck-Institute of Polymer Research for synthesis and characterisation of electrooptical macrocycles. After a post-doc position at Virginia Polytechnic Institute and State University in Blacksburg/USA he worked for IBM and SCHOTT Spezialglas AG in the field of organic light emitting diodes developing large area displays in Mainz.

Matthias Born

Chapter C.10

Philips Research Laboratories Aachen
Aachen, Germany
matthias.born@philips.com

Matthias Born is a physicist and joined Philips Research Aachen, Germany, in 1992. He is leading several projects about plasma physics and diagnostics of gas discharges with a major topic on mercury-free lamps for general and automotive lighting applications. He is also working as a professor for physics at the Heinrich-Heine-University of Düsseldorf.

Annette Borsutzky

Chapter C.11, Sect. 11.9

Technische Universität Kaiserslautern
Fachbereich Physik
Kaiserslautern, Germany
borsu@uni-kl.de

Annette Borsutzky studied physics in Bielefeld and Hannover, Germany, where she received in 1992 her Dr. rer. nat. working on nonlinear frequency mixing in crystals and gases. Joining the university of Kaiserslautern studies of optical parametric oscillators, diode-pumped solid state lasers as well as the characterization of new nonlinear and laser active materials are at the center of her work.

Hans Brand

Chapter C.11, Sect. 11.4

Friedrich-Alexander-University of Erlangen-Nürnberg LHFT
Department of Electrical, Electronic and Communication Engineering
Erlangen, Germany
hans@lhft.eei.uni-erlangen.de

Hans Brand received the degrees Dipl.-Ing. in 1956, Dr.-Ing in 1962 and Dr.-Ing. habil. in 1962 at the RWTH Aachen, Germany. In 1969 he became professor at the Chair for Microwave Engineering at the FAU Erlangen, Germany. His main fields of research are microwaves, millimeter wave and terahertz components and systems as well as gas laser and infrared laser technology. In 1996 he became Fellow of the IEEE. He is emeritus since 1998.

Robert P. Breault

Chapter B.7

Breault Research Organization, Inc.
Tucson, AZ, USA
bbreault@breault.com

Robert P. Breault is the Chairman and founder of the Breault Research Organization. He works on stray light analysis and suppression. He is the author of the APART stray light analysis program, used to analyze the Hubble telescope and many others. He received the B.S. in mathematics from Yale University, and his M.S. and Ph.D. in optical sciences from the University of Arizona. He is a fellow of SPIE and founder and Co-chairman of the Arizona Optics Industry Association.

Matthias Brinkmann

Chapter A.5, Sects. 5.1, 5.1.9, 5.10

University of Applied Sciences Darmstadt
Mathematics and Natural Sciences
Darmstadt, Germany
brinkmann@fh-darmstadt.de

Dr. Brinkmann is a professor of optical engineering at the University of Applied Sciences Darmstadt, Germany. He obtained his Ph.D. degree in Physics from the Ruhr-Universität Bochum,Germany in 1997 for his work on high-temperature superconductors. Prior to joining the University of Applied Sciences Darmstadt, he worked as a staff scientist and research manager at Schott Glas corporate research in Mainz, Germany. His main scientific focus covered thermal material properties of glass and microstructured optical glass for photonic applications. His current research activities include integrated waveguide optics and diffractive microoptics for various Photonic applications.

Uwe Brinkmann

Chapter C.11, Sect. 11.7

Bovenden, Germany
ubri@arcor.de

Uwe Brinkmann obtained his education in physics at the universities Munich, Heidelberg, Hannover, and worked in laser research at the Universität Cologne before joining Lambda Physik, Göttingen, as Head of Research and Development. Since 1988 self-employed, he edited the German periodical Laser und Optoelektronik over 20 years and is a contributing editor to Laser Focus World since 1987.

Robert Brunner

Chapter B.8, Sect. 8.1

Carl Zeiss AG
Central Research and Technology
Jena, Germany
r.brunner@zeiss.de

Robert Brunner graduated from the University of Ulm in 1994 and received his Ph.D. degree in the field of near-field optical microscopy. Since 1998 he works at the Research Center of Carl Zeiss, where he is the responsible Lab Manager for microstructured optics. His current research interests are hybrid diffractive/refractive optics, subwavelength structures, refractive microoptics, and high resolution optics.

Geoffrey W. Burr

IBM Almaden Research Center
San Jose, CA
burr@almaden.ibm.com

Chapter D.20

Geoffrey W. Burr received his B.S. in Electrical Engineering (EE) and B.A. in Greek Classics from the State University of New York at Buffalo in 1991. That year Eta Kappa Nu selected him as the Alton B. Zerby Outstanding EE Senior in the U.S. He received his M.S. and Ph.D. in Electrical Engineering from the California Institute of Technology in 1993 and 1996, respectively, under the supervision of Professor Demetri Psaltis. Since that time, Dr. Burr has worked at the IBM Almaden Research Center in San Jose, California, where he is currently a Research Staff Member. He has worked extensively in holographic data storage, volume holography, signal processing and systems tradeoffs in data storage, and optical information processing. Dr. Burr's current research interests also include nanophotonics, numerical modeling for design optimization, and phase-change nonvolatile memory. He is a member of SPIE, OSA, IEEE, Eta Kappa Nu, and Tau Beta Pi.

Karsten Buse

University of Bonn
Institute of Physics
Bonn, Germany
kbuse@uni-bonn.de

Chapter A.5, Sects. 5.9.3

Karsten Buse received his Ph.D. from the University of Osnabrück, Germany. Since 2000 he is holding the Heinrich Hertz professorship for physics at the University of Bonn. His research focus is on nonlinear-optical and photosensitive dielectric materials like nonlinear and photorefractive crystals. He authored and co-authored more than 150 publications and more than 20 patents in this field.

Carol Click

Schott North America
Regional Research and Development
Duryea, PA, USA
carol.click@us.schott.com

Chapter A.5, Sects. 5.1.2, 5.1.3, 5.10

Dr. Click received her Ph.D. from the University of Missouri – Rolla in ceramic engineering focused in contamination in phosphate laser glasses. She is now involved in the research and development necessary to commercialize Schott's inorganic low temperature reactive bonding technology for producing light-weighted Zerodur optics and precision optical components.

Mark J. Davis

Schott North America
Regional Research and Development
Duryea, PA, USA
mark.davis@us.schott.com

Chapter A.5, Sects. 5.5, 5.10

Mark Davis earned his Ph.D. in Geology from Yale University in 1996, focusing on the kinetics of nucleation in glass-forming melts and related topics. Since then, his research has centred on the development of new glass-ceramic materials for a range of applications, in addition to continued efforts towards a more fundamental understanding of crystallization processes.

Wolfgang Demtröder

TU Kaiserslautern
Department of Physics
Kaiserslautern, Germany
demtroed@physik.uni-kl.de

Chapter D.13, Sects. 13.1, 13.2

Wolfgang Demtröder studied Physics, Mathematics and Science of Music at the Universities of Münster, Tübingen and Bonn. He received his Ph.D. in 1961 with Prof. Paul in Bonn. He was research assistant in Freiburg, visiting fellow at JILA in Boulder, Colorado and since 1970 he is Professor of Physics at the University of Kaiserslautern, and visiting Professor at the Universities at Stanford (USA), Kobe (Japan), New South Wales in Sydney (Australia) and at the Technical University in Lausanne (Switzerland). His fields of research are high resolution laser spectroscopy of molecules and metal-clusters, time resolved spectroscopy, spectroscopy of collision processes. He received the Max Born Prize in 1994 and the Heisenberg Medal 2001.

Henrik Ehlers

Laser Zentrum Hannover e.V.
Department of Thin Film Technology
Hannover, Germany
h.ehlers@lzh.de

Chapter A.6

Henrik Ehlers is working in the field of optical thin films. He studied physics in Hannover, Germany, and is currently head of the Process Development Group in the department of Thin Film Technology at the Laser Zentrum Hannover. The focus of the group is on R&D in modern deposition processes, in situ process monitoring, and advanced process automation.

Rainer Engelbrecht

Friedrich-Alexander-University of Erlangen-Nürnberg
Department of Electrical, Electronic and Communication Engineering
Erlangen, Germany
rainer@lhft.eei.uni-erlangen.de

Chapter C.11, Sect. 11.4

Rainer Engelbrecht studied electrical engineering at the University of Erlangen-Nürnberg and received his diploma and Dr. degree in 1995 and 2001, respectively. The doctoral thesis was on gas analysis in CO_2 lasers by diode laser spectroscopy. His current research fields are nonlinear fiber optics, Raman fiber lasers and low-noise photo receivers.

Martin Fally

University of Vienna
Faculty of Physics, Department for Experimental Physics
Vienna, Austria
martin.fally@univie.ac.at

Chapter D.20, Sect. 20.1

Martin Fally earned both his Ph.D. in physics (1996) and his habilitation in solid-state physics (2003) from the Vienna University, Austria. Since then he is Associate Professor at the Department for Experimental Physics. In 2003–2004 he held a Mercator Visiting- Professorship at the University of Osnabrück, Germany. He authored or co-authored more than 40 publications in the fields of structural phase transitions (experimental and theoretical), quasi one-dimensional systems (theoretical), neutron-scattering, neutron-diffraction, photorefractive materials, holographic scattering (experimental and theoretical). In 2001 he was awarded the Prize of the City of Vienna for research in natural sciences.

Yun-Hsing Fan

University of Central Florida
College of Optics and Photonics
Orlando, USA
yfan@mail.ucf.edu

Chapter A.5, Sects. 5.3, 5.10

Ms. Yun-Hsing Fan is currently a Ph.D. candidate at the School of Optics/CREOL, University of Central Florida. Her current research is to develop novel electronic liquid crystal (LC) lenses and fast-response infrared phase modulators for optical communications. Her future work will focus on polarization-independent LC lens and fast switching polymer-network LC modulators for far-infrared and visible regions.

Enrico Geißler

Carl Zeiss AG
Central Research and Technology
Jena, Germany
e.geissler@zeiss.de

Chapter B.8, Sect. 8.2

Enrico Geißler completed his studies of electrical engineering at the University of Applied Sciences Jena, Germany, in 1998. Since graduation he has been at the Research Center of Carl Zeiss, where he is currently Senior Scientist for Digital Visualization Systems. His current research interests are spatial light modulators and MEMS.

Ajoy Ghatak

Indian Institute of Technology Delhi
Physics Department
New Delhi, India
ajoykghatak@yahoo.com

Chapter B.8, Sect. 8.7

Ajoy Ghatak has published more than 170 research papers in international journals and is the co-author (with Professor Thyagarajan) of six books. He is a Fellow of the Optical Society of America (OSA) and is the recipient of the 2003 OSA Esther Hoffman Beller award, the International Commission for Optics Galileo Galilei award and the CSIR S.S. Bhatnagar award. His areas of interests are fiber optics and quantum mechanics.

Alexander Goushcha

SEMICOA
Costa Mesa, CA, USA
ogoushcha@semicoa.com

Chapter B.9

Dr. Alexander Goushcha (aka Gushcha) is a Chief Scientist and CTO at SEMICOA, a California-based manufacturer of high-reliability silicon transistors and optoelectronics. He obtained his Ph.D. degree in Physics from the Institute for Physics, Ukrainian Academy of Sciences in Kyiv (Ukraine). He has been working in the fields of semiconductor physics and technology, biophysics and molecular electronics, and nonlinear optics at the Institute for Physics, Kyiv, Ukraine, MPI Strahlenchemie, Mülheim a.d. Ruhr, Germany, and UC Riverside, CA. Dr. Goushcha is the author of about 100 technical papers in referred journals and holds 10 patents and patent applications.

Daniel Grischkowsky

Oklahoma State University
Electrical and Computer Engineering
Stillwater, OK, USA
grischd@ceat.okstate.edu

Chapter D.17

Daniel R. Grischkowsky is a Regents Professor and the Bellmon Professor of Optoelectronics at Oklahoma State University. He received his B.S. from Oregon State University in 1962 and his Ph.D. degree in physics from Columbia University in 1968. In 1969 he joined the IBM Watson Research Center, Yorktown Heights, New York, where he developed and experimentally verified the adiabatic following model in 1972. In 1982 his research group developed the optical-fiber pulse compressor, and later in 1989 developed the technique of THz time-domain spectroscopy (THz-TDS). In 1993 he relocated to Oklahoma State University to pursue THz-TDS applications. He is a fellow of The American Physical Society (APS), The Institute of Electrical and Electronics Engineers (IEEE) and The Optical Society of America (OSA). He was awarded the Boris Pregel Award (1985) by the New York Academy of Sciences for the development of the optical fiber pulse compressor, the R.W. Wood Prize (1989) from OSA for distinguished contributions to the field of optical pulse compression, particularly for pioneering work on the use of optical fibers for generating ultrashort pulses of light, and the William F. Meggers Award (2003) from OSA for seminal contributions to the development and application of THz time-domain spectroscopy.

Richard Haglund

Vanderbilt University
Department of Physics and Astronomy
Nashville, TN, USA
richard.haglund@vanderbilt.edu

Chapter 1

Professor Halund earned his Ph.D. in experimental nuclear physics from the University of North Carolina, Chapel Hill. He was staff member, Los Alamos National Laboratory from 1975 to 1984. Since 1984 he is Professor of Physics at Vanderbilt University. He was Alexander von Humboldt awardee in 2003. His current research activities are in nonlinear optics in metal and metal-oxide nanoparticles; size and dimensional effects in metal-insulator transitions; and ultrafast mid-infrared laser processing of polymers and organic materials.

Stefan Hansmann

Al Technologies GmbH
Darmstadt, Germany
shansmann@al-technologies.de

Chapter C.11, Sect. 11.3

Stefan Hansmann received his Ph.D. from the Technical University of Darmstadt for his work on simulation and realization of DFB laser diodes. He worked 10 years in the field of optoelectronics at the research center of Deutsche Telekom and become the head of a research group focusing on the application of photonic technologies in telecommunication. Thereafter he served as a technical manager in several companies of III/V semiconductor industry and is now the chief technical officer of AL Technologies GmbH in Darmstadt, commercializing high speed InP based semiconductor laser technology.

Ferenc Krausz

Max-Planck-Institut für Quantenoptik
Garching, Germany
ferenc.krausz@mpq.mpg.de

Chapter A.4, Sect. 4.2

Ferenc Krausz was awarded his M.S. in Electrical Engineering at Budapest University of Technology in 1985, his Ph.D. in Quantum Electronics at Vienna University of Technology in 1991, and his "Habilitation" degree in the same field at the same university in 1993. He joined the Department of Electrical Engineering as Associate Professor in 1998 and became Full Professor in the same department in 1999. In 2003 he was appointed as Director of Max Planck Institute of Quantum Optics in Garching, Germany, and since October 2004 he has also been Professor of Physics and Chair of Experimental Physics at Ludwig Maximilian's University of Munich. His research has included nonlinear light-matter interactions, ultrashort light pulse generation from the infrared to the X-ray spectral range, and studies of ultrafast microscopic processes. By using chirped multilayer mirrors, his group made intense light pulses comprising merely a few wave cycles available for a wide range of applications and utilized them for pushing the frontiers of ultrafast science into the attosecond regime. His most recent research focuses on attosecond physics: the control and real-time observation of the atomic-scale motion of electrons. He co-founded Femtolasers GmbH, a Vienna-based company specializing in cutting-edge femtosecond laser sources.

Eckhard Krätzig

University of Osnabrück
Physics Department
Osnabrück, Germany
eckhard.kraetzig@uos.de

Chapter A.5, Sects. 5.9.3

Eckhard Krätzig received his Ph. D. degree in physics from the Johann Wolfgang Goethe University of Frankfurt/Main, Germany in 1968. Then he joined the Philips Research Laboratories Hamburg, where he headed the Solid State Physics Group. Since 1980 he has been a Professor of Applied Physics at the University of Osnabrück. During the last years his research interests were focused on photorefractive effects and light-induced charge-transport phenomena.

Stefan Kück

Physikalisch-Technische Bundesanstalt
Optics Division
Braunschweig, Germany
stefan.kueck@ptb.de

Chapter C.11, Sect. 11.2

Dr. Kück is working group leader for laser radiometry at the Physikalisch-Technische Bundesanstalt, the German National Metrology Institute. He obtained his Ph.D. in 1994 and habilitated in 2001 in the field of solid-state lasers. His main research topic is the development of new methods, procedures and standards for the high-precision measurement of laser power and laser pulse energy.

Anne L'Huillier

University of Lund
Department of Physics
Lund, Sweden
anne.lhuillier@fysik.lth.se

Chapter A.4, Sect. 4.2

Anne L'Huillier defended her Ph.D. thesis in Paris in 1986. She worked at the Commissariat à l'Energie Atomique in Saclay, France, until 1995 and then moved to Lund University, Sweden, where she became professor in 1997. Her current research is on the generation of high-order harmonics of laser light in gases and its application to attosecond science. In 2003, she got the Julius Springer prize for Applied Physics together with F. Krausz. She became member of the Royal Swedish Academy of Sciences in 2004.

Bruno Lengeler

Aachen University (RWTH)
II. Physikalisches Institut
Aachen, Germany
lengeler@physik.rwth-aachen.de

Chapter D.18

Bruno Lengeler is emeritus professor of physics and former head of a physics institute at Aachen university. He is a solid-state physicist who has worked for many years on spectroscopy and imaging with synchrotron radiation, in particular on the development of parabolic refractive X-ray lenses. He was Director of Research at the European Synchrotron Radiation Facility.

Martin Letz

Schott Glas
Materials Science, Central Research
Mainz, Germany
martin.letz@schott.com

Chapter A.5, Sects. 5.1.1, 5.1.4, 5.1.6, 5.1.7, 5.10

Dr. Martin Letz studied physics at the universities of Braunschweig, Stuttgart (Germany) and Tartu (Estonia) and finished his Ph.D. in 1995 at Stuttgart University in the field of theoretical solid state physics with a work on magnetic polarons. In the following he worked as a post-doctoral fellow at Queen's University, Kingston (Canada) and at the University of Mainz (Germany). During this time he performed investigations on statistical physics of strongly correlated quantum-mechanical systems of strongly correlated classical systems, light scattering and on the dynamics of the glass transition in molecular fluids. In 2001 Martin Letz joined the central research of Schott Glass.

Gerd Leuchs

University of Erlangen-Nuremberg
Institute of Optics, Information and Photonics
Erlangen, Germany
leuchs@physik.uni-erlangen.de

Chapters A.2, A.3

Gerd Leuchs studied physics and mathematics at Cologne and received his Ph,D. in 1978. After research years in USA, he headed the German gravitational wave detection group, then became technical director at Nanomach AG, Switzerland. Since 1994 he has been Professor of Physics at the University of Erlangen, since 2003 also director of the Max Planck research group of optics, information and photonics.

Norbert Lindlein

Friedrich-Alexander University of Erlangen-Nürnberg
Max-Planck Research Group
Institute of Optics Information and Photonics
Erlangen, Germany
norbert.lindlein@optik.physik.uni-erlangen.de

Chapters A.2, A.3

Norbert Lindlein received in 1996 his Ph.D. from the Friedrich-Alexander University Erlangen-Nürnberg (Germany). In 2002 he finished his habilitation in physics and is a member of the Physics Faculty of the University of Erlangen-Nürnberg since. His research interests include the simulation and design of optical systems, diffractive optics, microoptics and optical measurement techniques using interferometry or Shack-Hartmann wavefront sensors.

Stefano Longhi

University of Politecnico di Milano
Department of Physics
Milano, Italy
longhi@fisi.polimi.it

Chapter C.11, Sect. 11.1

Stefano Longhi is Associate Professor of Physics of Matter at the Polytechnic Institute of Milan. He has authored more than 100 papers in the fields of laser physics, photonics, nonlinear and quantum optics. Professor Longhi is Fellow of the Institute of Physics and member of the J. Physics-B editorial board. In 2003 he was awarded with the Fresnel Prize of the European Physical Society.

Ralf Malz

LASOS Lasertechnik GmbH
Research and Development
Jena, Germany
ralf.malz@lasos.com

Chapter C.11, Sect. 11.6

Ralf Malz studied physics at the Friedrich-Schiller-Universität Jena 1984-1989 and received his Ph.D. in1993 at the same university. Then he worked at Carl-Zeiss-Jena and later LASOS in the field of CO_2 waveguide and slab lasers. His current work is on argon-ion, HeNe lasers, diode-laser modules and fiber coupling.

Wolfgang Mannstadt

Schott AG
Research and Technology Development
Mainz, Germany
wolfgang.mannstadt@schott.com

Chapter A.5, Sects. 5.1.4, 5.8, 5.10

Dr. Wolfgang Mannstadt studied physics and received his Ph.D. from the Philipps University in Marburg. His main research field is the materials simulation with ab initio DFT methods. He received a Feodor-Lynen fellowship from the Humboldt Foundation and worked as a research assistent in the group of Prof. A.J. Freeman at the Northwestern University in Evanston. His current work in the R&D department at Schott is focuses on the simulation of materials properties with DFT and nanostructured optical materials.

Gerd Marowsky

Laser-Laboratorium Göttingen e.V.
Göttingen, Germany
gmarows@gwdg.de

Chapter C.11, Sect. 11.7

Dr. Gerd Marowsky graduated in 1969 from the University of Göttingen, Germany in the fields of experimental and theoretical physics and mineralogy, and is currently director of Laser-Laboratorium Göttingen. He has made numerous scientific contributions to the general field of lasers and high-field interactions. Current research interests include quantum electronics in general, lasers, laser applications in environmental research, nonlinear optics, nonlinear inorganic and organic materials, and applications of short-duration ultraviolet laser pulses. Dr. Marowsky is well know in Canada as the Sector-Coordinator in Materials–Physical Technologies, a Science and Technology Agreement that supports numerous collaborations between Canadian and German scientists. Gerd Marowsky is also professor at the University of Göttingen, Germany and holds adjunct professor positions in Electrical and Computer Engineering at both Rice University, Houston and the University of Toronto.

Dietrich Martin

Carl Zeiss AG
Corporate Research and Technology
Microstructured Optics Research
Jena, Germany
d.martin@zeiss.de

Chapter B.8, Sect. 8.4

Dietrich Martin received the B.Sc. (University of Technology, Dresden) in 1992, the M.Sc. (University of Oldenburg) in 1997. He earned his Ph. D. from the University of Kassel in 2002 working on optical properties of alkali metal clusters. He joined the Corporate R&D Department of Carl Zeiss and is engaged in research on microstructured and variable optical components.

Bernhard Messerschmidt

GRINTECH GmbH
Research and Development, Management
Jena, Germany
messerschmidt@grintech.de

Chapter B.8, Sect. 8.3

Bernhard Messerschmidt studied physics at the Jena University, Germany and graduated with a Ph.D. in 1998 from the Fraunhofer Institute of Applied Optics in Jena on modelling and optimization of ion exchange processes in glass for the generation of GRIN lenses. From 1994 to 1995, he was a research fellow at the Institute of Optics, University of Rochester, New York. In 1999, he established the company Grintech as a co-founder and is currently one of the principal managers of Grintech and responsible for research and development.

Katsumi Midorikawa

RIKEN
Laser Technology Laboratory
Saitama, Japan
kmidori@riken.jp

Chapter C.11, Sect. 11.12

Dr. Katsumi Midorikawa is a director and chief scientist of the Laser Technology Laboratory at RIKEN. He received his Ph. D. degree from Keio University in Electrical Engineering. His research interests currently focus on ultrashort high-intensity laser – matter interaction and its application, including the generation of coherent X-ray and femotosecond laser processing.

Gerard J. Milburn

The University of Queensland
Center for Quantum Computer Technology
School of Physical Sciences
St. Lucia, QLD, Australia
milburn@physics.uq.edu.au

Chapter D.14

Milburn's research is largely in the fields of quantum optics, quantum measurement and control, and quantum computing and has published more than 200 papers and three books. He is a Fellow of the Australian Academy of Science and The American Physical Society. He is currently an Australian Research Council Federation Fellow at The University of Queensland.

Kazuo Ohtaka

Chiba University
Center for Frontier Science
Photonic Crystals
Chiba, Japan
ohtaka@cfs.chiba-u.ac.jp

Chapter B.8, Sect. 8.6

In 1965 Dr. Kazuo Ohtaka graduated from Department of Applied Physics, University of Tokyo where he was assistant professor since 1967. Since 1998 he was Professor of Department of Applied Physics of Chiba University and since 2000 he is Professor of the Center for Frontier Science, Chiba University. In 2000 he obtained the Best Papers Award of the Physical Society of Japan. He is currently working in the field of photonic crystals, their fundamentals and applications.

Motoichi Ohtsu

Department of Electronics Engineering
The University of Tokyo
Tokyo, Japan
ohtsu@ee.t.u-tokyo.ac.jp

Chapter D.15

Dr. Ohtsu is a Professor of the University of Tokyo. As a founder of nanophotonics, he is a director of several national projects for nanophotonic devices, storage and fabrication. He has authored more than 400 technical papers and 50 books. He holds 100 patents. He is a fellow of the Optical Society of America and has been awarded more than ten prizes from academic institutions including the I. Koga gold metal from URSI. He was also awarded the Medal with Purple Ribbon from the Japanese Government.

Roger A. Paquin

Advanced Materials Consultant
Oro Valley, AZ, USA
materials.man@att.net

Chapter A.5, Sects. 5.9.4, 5.10

Roger Paquin is an independent materials consultant specializing in materials and processes for dimensionally stable components for optical and precision instrument systems, with emphasis on Be, SiC and composites for mirrors and structures. He has published over 50 papers and book chapters and teaches short courses on the subject. Mr. Paquin is a Fellow of SPIE, The International Society for Optical Engineering.

Klaus Pfeilsticker

Universität Heidelberg
Institut für Umweltphysik
Fakultät für Physik
und Astronomie
Heidelberg, Germany
klaus.pfeilsticker@iup.uni-heidelberg.de

Chapter D.19, Sect. 19.1

Dr. Klaus Pfeilsticker is a professor of physics at the University of Heidelberg since 2004. Before he worked at the Max-Planck-Institut für Kernphysik, Heidelberg, the Alfred Wegner Institut, Bremerhaven, the Research Center, Jülich and at the National Oceanic and Atmospheric Administration (NOAA), Boulder. His main research interests are the photochemistry and the radiative transfer of the atmosphere. His recent research focuses on the photochemistry, budget and trend of reactive halogen species in the upper troposphere and stratosphere, the spectral solar irradiance and its variability, and photon path length distribution of solar photons in clear and cloudy skies.

Ulrich Platt

Universität Heidelberg
Institut für Umweltphysik
Fakultät für Physik
und Astronomie
Heidelberg, Germany
ulrich.platt@iup.uni-heidelberg.de

Chapter D.19, Sect. 19.1

Dr. Ulrich Platt is a professor of physics at the University of Heidelberg since 1989, before he worked at the research Center Jülich and at the University of California, Riverside. His main interests are atmospheric chemistry of free radicals and spectroscopic measurements of atmospheric constituents. He is the co-inventor of the DOAS technique. His current research centres on reactive halogen species in the troposphere and their role in the tropospheric chemistry as well as on remote sensing of trace gas distributions in the atmosphere.

Markus Pollnau

University of Twente
MESA+ Institute for Nanotechnology
Enschede, The Netherlands
m.pollnau@ewi.utwente.nl

Chapter C.11, Sect. 11.2.4

Markus Pollnau obtained his Diploma and Ph.D. in physics from the Universities of Hamburg, Germany, and Bern, Switzerland, respectively. Following research at the University of Southampton, UK, and the Swiss Federal Institute of Technlogy, Lausanne, Switzerland, he was appointed full professor at the University of Twente, The Netherlands, in 2004. Currently, he works on light generation in integrated dielectric structures. He has co-authored over 200 international publications.

Steffen Reichel

SCHOTT Glas
Service Division Research
and Technology Development
Mainz, Germany
steffen.reichel@schott.com

Chapter A.5, Sects. 5.1.1, 5.1.6, 5.1.7, 5.1.8, 5.10

Dr. Reichel is an electrical engineer with extensive experience in optics and electromagnetics. He got his Ph.D. in Er-doped fiber amplifiers and worked on several topics in electromagnetics, wave and laser optics, optical fibers/waveguides, and geometrical optics. He is Senior Member of the IEEE and worked for Lucent Technologies and is now manager of the Physical Science Group at Schott.

Hans-Dieter Reidenbach

University of Applied Sciences Cologne
Institute of Communications Engineering
Institute of Applied Optics and Electronics
Cologne, Germany
hans.reidenbach@fh-koeln.de

Chapter D.21

Dr. Reidenbach is Professor at the University of Applied Sciences Cologne and head of the research laboratory Medical Technology. He obtained the Dr.-Ing. degree from the University of Erlangen. His scientific work resulted in new applications of laser beams, incoherent optical radiation and high frequency currents in operative endoscopy, transanal surgery and interstitial thermotherapy. Currently his research is on optical irradiation and psychophysical behaviour.

Hongwen Ren

University of Central Florida
College of Optics and Photonics
Orlando, FL, USA
hren@mail.ucf.edu

Chapter A.5, Sects. 5.9, 5.10

Dr. Hongwen Ren received his Ph.D. degree from Changchun Institute of Optics, Fine Mechanics and Physics, Chinese Academy of Sciences in 1998. After that, he was faculty member of the North Liquid Crystal R&D Center, Chinese Academy of Sciences as an assistant professor. In August 2001, he joined the College of Optics & Photonics, University of Central Florida (UCF) as a research Scholar. Dr. Ren's current research interests and projects are liquid crystal/polymer dispersions, nanoliquid crystal device, and adaptive e-lens.

Detlev Ristau

Laser Zentrum Hannover e.V.
Department of Thin Film Technology
Hannover, Germany
d.ristau@lzh.de

Chapter A.6

Detlev Ristau is a physicist with an extensive research background in optical thin film technology. He received his Ph.D. from the University of Hannover in 1988 and authored more than 200 technical papers. Current research activities include the development and precise control of ion processes as well as the measurement of the power handling capability and losses of optical components.

Simone Ritter

Schott AG
Division Research and Technology
Development
Material Development
Mainz, Germany
simone.ritter@schott.com

Chapter A.5, Sects. 5.3, 5.10

Dr. Simone Ritter studied Chemistry in Leipzig and Tübingen and received her Ph.D. 1994 for syntheses, characterization, structures and reactivity of complexes with rhenium-nitrogen-multi-bonds. For the last 7 years, she worked as scientific referent for coloured and optical glasses at Schott. Her research involves development and characterization of glasses with new optical properties.

Evgeny Saldin

Deutsches Elektronen Synchrotron (DESY)
Hamburg, Germany
evgueni.saldin@desy.de

Chapter C.11, Sect. 11.11

Dr. Evgeny Saldin is an expert in the field of physics of charged particle beams, accelerators, and free electron lasers. He has authored a book on free electron lasers and more than a hundred papers in peer-reviewed.

Roland Sauerbrey

Chapter C.11, Sect. 11.13

Forschungszentrum
Dresden-Rossendorf e.V.
Dresden, Germany
r.sauerbrey@fzd.de

Roland Sauerbrey is the Scientific Director of the Forschungszentrum Dresden-Rossendorf and a professor at the Technical University of Dresden. After receiving his Ph.D. in physics from the University of Würzburg in 1981 he worked as a professor at Rice University in Houston, Texas. In 1994 he moved to the Friedrich-Schiller-University in Jena where he stayed until 2006 as a professor of physics. During the last 20 years he has been actively involved in the emerging field of relativistic light – matter interaction and the development of ultrashort pulse, ultrahigh-intensity lasers. He is a fellow of the Optical Society of America and the Institute of Physics.

Evgeny Schneidmiller

Chapter C.11, Sect. 11.11

Deutsches Elektronen Synchrotron (DESY)
Hamburg, Germany
evgeny.schneidmiller@desy.de

Evgeny Schneidmiller is an expert in the field of physics of charged particle beams, accelerators, and free electron lasers. He has authored a book on free electron lasers and more than a hundred papers in peer-reviewed journals.

Bianca Schreder

Chapter A.5, Sects. 5.1.5, 5.6, 5.10

Schott Glas
Division Research and Technology Development
Material Development
Mainz, Germany
bianca.schreder@schott.com

Dr. Bianca Schreder studied Chemistry in Würzburg and received her Ph.D. from the Department of Physical Chemistry for her work on Laser Spectroscopy on II–VI-Semiconductor Nanostructures. She is now working in the Research and Development Department at Schott Glas, Mainz. Her work involves investigation and development of glass systems with special optical properties.

Christian G. Schroer

Chapter D.18

Dresden University of Technology
Institute of Structural Physics
Dresden, Germany
schroer@physik.tu-dresden.de

Christian G. Schroer made his doctoral studies in mathematical physics at the Research Centre Jülich (doctoral degree University of Cologne in 1995). After a visit as postdoctoral fellow to the University of Maryland, he worked as a research and teaching assiociate at Aachen University in the field of X-ray optics and microscopy. After his Habilitation in 2004, he was a staff scientist at DESY in Hamburg until he responded to a call to the chair of Structural Physics of Condensed Matter at Dresden University of Technology in early 2006.

Markus W. Sigrist

Chapter C.11, Sect. 11.10

ETH Zurich, Institute of Quantum Electronics
Department of Physics
Zurich, Switzerland
sigrist@iqe.phys.ethz.ch

Markus W. Sigrist is Professor of Physics at ETH Zurich (Switzerland) and Adjunct Professor at Rice University in Houston (USA). His current research involves development and implementation of tunable mid-infrared laser sources and sensitive detection schemes for spectroscopic trace gas analyses in environmental, industrial and medical applications. He published 2 books and over 150 papers. He is Fellow of OSA and Topical Editor of Applied Optics.

Glenn T. Sincerbox

University of Arizona
Optical Sciences
Tucson, AZ, USA
sinbox@cox.net

Chapter D.20, Sect. 20.2

Glenn Sincerbox is currently a Professor Emeritus of the College of Optical Sciences of the University of Arizona where he was a Professor of Optical Sciences and the Director of the Optical Data Storage Center; a Center that performed leading-edge research on advanced optical storage materials, systems and techniques. Prior to that, Mr. Sincerbox was with IBM Research for 34 years holding numerous technical and management positions.He has published over 50 technical papers and presented over 60 papers. He holds 40 US Patents and has 70 patent publications. His research was primarily in the field of optical storage with emphasis on holographic storage. He is a fellow of the OSA and has been active for over 15 years in the International Commission for Optics holding positions as vice president and treasurer.

Elisabeth Soergel

University of Bonn
Institute of Physics
Bonn, Germany
soergel@uni-bonn.de

Chapter B.8, Sect. 8.5

Elisabeth Soergel studied physics at the Ludwig-Maximilian-University in Munich. She received her Diploma and Ph.D. from the Max-Planck Institute for Quantum Optics in Garching in the field of scanning probe microscopy. After a postdoc stay at the IBM research laboratory in Rüschlikon, Switzerland, she joined the University of Bonn in 2000 with the main research field on visualization of ferroelectric domains by scanning probe techniques.

Steffen Steinberg

LASOS Lasertechnik GmbH
Jena, Germany
steffen.steinberg@lasos.com

Chapter C.11

Steffen Steinberg studied physics and received his doctoral degree at Friedrich-Schiller-University Jena, with a work on manipulation of laser light with integrated optical devices. Later his work concentrated on different applications of gas and solid-state laser technology especially laser display technology and fiber optical devices. Currently he is working as Sales Manager at LASOS in Jena, Germany.

Sune Svanberg

Lund University
Division of Atomic Physics
Lund, Sweden
sune.svanberg@fysik.lth.se

Chapter D.13, Sect. 13.3

Sune Svanberg made his Ph.D. in Physics at Göteborg University in 1972. He became professor and head of the Atomic Physics Division, Lund University, in 1980. He is also Director of the Lund Laser Centre, a European Large Scale Infrastructure. He is fellow of the American Physical Society and the Optical Society of America, member of 5 academies and three-fold honorary doctor. His present research fields include basic atomic laser spectroscopy and applications of laser spectroscopy to environmental and medical research.

Orazio Svelto

Politecnico di Milano
Department of Physics
Milan, Italy
orazio.svelto@fisi.polimi.it

Chapter C.11, Sect. 11.1

Orazio Svelto is Professor of Quantum Electronics at the Politecnico di Milano and the Director of the Milan Section at the Institute of Photonics and Nanotechnologies (IFN), Milan, Italy, belonging to the Italian National Research Council (CNR). His current research activities include ultrashort-pulse generation and applications, physics of laser resonators and techniques of mode selection, laser applications in biology and biomedicine, and physics of solid-state lasers, also including the generation of femtosecond laser pulses, down to the record value of 3.8 fs recently established by the group, and to the applications of these ultrashort pulses. He is the author of Principles of Lasers (1998). Professor Svelto is Fellow of the Optical Society of America and the Institute of Electrical and Electronics Engineers, and a member of several Italian academies including the "Accademia dei Lincei". He was the recipient of the Italgas Prize for research and technology innovation, the Quantum Electronics Prize of the European Physical Society and the Charles H. Townes Award of the Optical Society of America.

Bernd Tabbert

Semicoa
Engineering Department
Costa Mesa, CA, USA
btabbert@semicoa.com

Chapter B.9

Bernd Tabbert received his Ph.D. from the University of Heidelberg, Germany in 1994. For several more years his research focused on optical studies of impurity atoms and bubbles in cryogenic liquids including a project at the University of California, Los Angeles. He is now working as an engineering manager for a semiconductor manufacturer developing sensors for medical X-ray applications.

K. Thyagarajan

Indian Institute of Technology Delhi
Physics Department
New Delhi, India
ktrajan@physics.iitd.ernet.in

Chapter B.8, Sect. 8.7

Professor Thyagarajan has published more than 125 research publications and is the co-author (with Professor Ajoy Ghatak) of six books. He has held visiting positions at Thomson-CSF, France and the University of Florida. He is a Fellow of the Optical Soceity of America and was awarded the title of "Officier dans l'ordre des Palmes Académiques" by the French Government in 2003. His current research interests are optical fiber amplifiers and guided wave nonlinear optics.

Mary G. Turner

Engineering Synthesis Design, Inc.
Tucson, AZ, USA
mary.t@engsynthesis.com

Chapter B.7

Dr. Turner has been involved with the development of optical software and conducted training on optical design and optical software for the previous 12 years. She wrote a chapter on "Optical Aberrations" in The Optics Encyclopedia and on "Reflecting and Catadioptric Objectives" in the Optical Engineer's Desk Reference. She is a fellow of SPIE and a member of SID and OSA.

Giuseppe Della Valle

Polytechnic Institute of Milan
Department of Physics
Milan, Italy
giuseppe.dellavalle@polimi.it

Chapter C.11, Sect. 11.1

Dr. Giuseppe Della Valle received his Ph.D. in Physics from the Polytechnic Institute of Milan, Italy, in 2005. He is currently Assistant Professor of Physics at the Physics Department of the same Institute.. His research activity is mainly devoted to solid-state lasers and optical devices based on doped glass substrates operating in the near infrared for telecom and metrological applications.

Michael Vollmer

University of Applied Sciences
Brandenburg
Department of Physics
Brandenburg, Germany
vollmer@fh-brandenburg.de

Chapter D.19, Sect. 19.2

Michael Vollmer studied physics in Heidelberg where he received his Ph.D. and habilitation, working on optical spectroscopy of metal clusters. Presently he is Professor of Experimental Physics working in the fields of infrared thermal imaging, spectroscopy, atmospheric optics, and didactics of physics. He is author of two books and member of the editorial boards of several journals.

Silke Wolff

SCHOTT Spezialglas AG
Department of Research & Technology
Development, Material Development
Optical Glasses
Mainz, Germany
silke.wolff@schott.com

Chapter A.5, Sects. 5.2, 5.10

Dr. Silke Wolff, laureate of R&D award, is an expert in glass development with an origin in analytical chemistry. Her main areas of research are innovative optical glasses, including market and customer-related material development, application, optimization of production processes and patent protection. Actual topics are exceptional optical properties, suitability for reheat hot forming and/or precision moulding and environmental compatibility.

Matthias Wollenhaupt

Universität Kassel
Institut für Physik
Kassel, Germany
wollenhaupt@physik.uni-kassel.de

Chapter C.12

Matthias Wollenhaupt did research in high-resolution laser spectroscopy with applications to aerospace and atmospheric chemistry. His current research in femtosecond laser spectroscopy is focused on pulse-shaping techniques for the design of tailored femtosecond laser pulses as a tool to control ultrafast light-induced processes. His scientific interests range from basic research, such as quantum control of molecular dynamics to applications of non-linear optics and materials processing.

Shin-Tson Wu

University of Central Florida
College of Optics and Photonics
Orlando, FL, USA
swu@mail.ucf.edu

Chapter A.5, Sects. 5.9, 5.9.1, 5.10

Dr. Wu is a Fellow of the IEEE, SID and OSA. He is a recipient of the IEEE Outstanding Engineer Award, SID Special Recognition Award, ERSO (Taiwan) Special Achievement Award, Hughes Team Achievement Award, and Hughes Annual Outstanding Paper Award. Professor Wu has co-authored 2 books: "Reflective Liquid Crystal Displays" (2001), and "Optics and Nonlinear Optics of Liquid Crystals" (1993), 4 book chapters, over 200 journal papers, and 18 issued patents.

Helen Wächter

ETH Zurich, Institute of Quantum Electronics
Department of Physics
Zurich, Switzerland
waechter@phys.ethz.ch

Chapter C.11, Sect. 11.10

Helen Wächter is a Ph.D. student at the Institute of Quantum Electronics of ETH Zurich after receiving her physics diploma from ETH in 2003. Her main research interests are in the field of IR laser spectroscopy and trace-gas monitoring. This includes the development of new coherent light sources by difference frequency generation (DFG) and new sensitive detection schemes. Her emphasis is on isotope-selective trace-gas analysis.

Mikhail Yurkov

Deutsches Elektronen Synchrotron (DESY)
Hamburg, Germany
mikhail.yurkov@desy.de

Chapter C.11, Sect. 11.11

Dr. Mikhail Yurkov is an expert in the field of physics of charged particle beams, accelerators, and free electron lasers. He has authored a book on free electron lasers and more than a hundred papers in peer-reviewed journals.

Aleksei Zheltikov

M.V. Lomonosov Moscow State University
Physics Department
Moscow, Russia
zheltikov@phys.msu.ru

Chapter A.4, Sect. 4.1

Aleksei Zheltikov's current research is related to nonlinear-optical processes in photonic-crystal fibers and nanostructures.

Detailed Contents

Part A Basic Principles and Materials

Detailed Cont.

Detailed Cont.

Part C Coherent and Incoherent Light Sources

Detailed Cont.

Part D Selected Applications and Special Fields

Subject Index

B

C

D

Subject Index

Subject Index

F

G

H

I

J

K

L

S

T

Subject Index

U

V

Periodic Table of the Elements

18 VIIIA	IUPAC Notation / CAS Notation
2 He	Atomic Number / Element Symbol
	Unstable Nuclei

Main Groups

1 IA	2 IIA	13 IIIA	14 IVA	15 VA	16 VIA	17 VIIA	18 VIIIA	Shells
1 H							2 He	K
3 Li	4 Be	5 B	6 C	7 N	8 O	9 F	10 Ne	K–L
11 Na	12 Mg	13 Al	14 Si	15 P	16 S	17 Cl	18 Ar	K–L–M
19 K	20 Ca	31 Ga	32 Ge	33 As	34 Se	35 Br	36 Kr	–L–M–N
37 Rb	38 Sr	49 In	50 Sn	51 Sb	52 Te	53 I	54 Xe	–M–N–O
55 Cs	56 Ba	81 Tl	82 Pb	83 Bi	84 Po	85 At	86 Rn	–N–O–P
87 Fr	88 Ra							–O–P–Q

Subgroups

3 IIIB	4 IVB	5 VB	6 VIB	7 VIIB	8 VIII (1)	9 VIII (2)	10 VIII (3)	11 IB	12 IIB	Shells
21 Sc	22 Ti	23 V	24 Cr	25 Mn	26 Fe	27 Co	28 Ni	29 Cu	30 Zn	L M N
39 Y	40 Zr	41 Nb	42 Mo	43 Tc	44 Ru	45 Rh	46 Pd	47 Ag	48 Cd	–M–N–O
57 La	72 Hf	73 Ta	74 W	75 Re	76 Os	77 Ir	78 Pt	79 Au	80 Hg	–N–O–P
89 Ac	104 Rf	105 Db	106 Sg	107 Bh	108 Hs	109 Mt	110 Ds	111 Rg	112	–O–P–Q

Lanthanides (Shells –N–O–P)

58 Ce	59 Pr	60 Nd	61 Pm	62 Sm	63 Eu	64 Gd	65 Tb	66 Dy	67 Ho	68 Er	69 Tm	70 Yb	71 Lu

Actinides (Shells –O–P–Q)

90 Th	91 Pa	92 U	93 Np	94 Pu	95 Am	96 Cm	97 Bk	98 Cf	99 Es	100 Fm	101 Md	102 No	103 Lr